Introductory
Semiconductor Electronics

Nigel P. Cook

PRENTICE HALL
Englewood Cliffs, New Jersey   Columbus, Ohio

To my loving wife, Dawn,
and beautiful children,
Candice and Jonathan,
whose love inspires me.

This book is dedicated to
all of the innovative, energetic,
and motivational instructors
who go out of their way to
prove that mathematics, science,
and technology are within
everyone's grasp and can be
made easy, interesting, and fun.

Library of Congress Cataloging-in-Publication Data
Cook, Nigel P.
Introductory semiconductor electronics / Nigel P. Cook.
p. cm.
Includes index.
ISBN 0-13-249855-3
1. Solid state electronics. I. Title.
TK7871.85.C6187 1995
621.3815—dc20 95-14481

Cover Photo: Michael A. Smith
Editor: Dave Garza
Production Editor: Sheryl Glicker Langner
Project Management: Elm Street Publishing Services, Inc.
Cover Designer: Tom Mack
Production Manager: Pamela D. Bennett
Marketing Manager: Debbie Yarnell
Supplements Editor: Judith Casillo
Illustrations: Rolin Graphics Inc.

This book was set in Times Ten by The Clarinda Company and was printed and bound by R. R. Donnelley & Sons Company. The cover was printed by Phoenix Color Corp.

Printed in the United States of America
10 9 8 7 6 5 4 3 2 1

ISBN: 0-13-249855-3

Prentice-Hall International (UK) Limited, *London*
Prentice-Hall of Australia Pty. Limited, *Sydney*
Prentice-Hall of Canada, Inc., *Toronto*
Prentice-Hall Hispanoamericana, S. A., *Mexico*
Prentice-Hall of India Private Limited, *New Delhi*
Prentice-Hall of Japan, Inc., *Tokyo*
Simon & Schuster Asia Pte. Ltd., *Singapore*
Editora Prentice-Hall do Brasil, Ltda., *Rio de Janeiro*

Preface

TO THE STUDENT

The early pioneers in electronics were intrigued by the mystery and wonder of a newly discovered science, whereas people today are attracted by its ability to lend its hand to any application and accomplish almost anything imaginable. If you analyze exactly how you feel at this stage, you will probably discover that you have mixed emotions about the journey ahead. On one hand, imagination, curiosity, and excitement are driving you on, while apprehension and reservations may be slowing you down. Your enthusiasm will overcome any indecision you have once you become actively involved in electronics and realize that it is as exciting as you ever expected it to be.

ORGANIZATION OF THE TEXTBOOK

This textbook has been divided into five basic parts. Chapters 1 and 2 introduce semiconductor electronics and review dc/ac electronics. Chapters 3 through 7 cover diodes and diode circuits. Chapters 8 through 14 cover transistors and transistor circuits, and Chapters 15 through 18 cover integrated circuits. Finally, in Part V, Chapter 19 covers thyristors and transducers.

PART I Introduction, Review, and Semiconductor Principles
- Chapter 1 Electricity and Electronics—Analog and Digital
- Chapter 2 Semiconductor Principles

PART II Diodes and Diode Circuits
- Chapter 3 Junction Diodes
- Chapter 4 Zener Diodes and Light-Emitting Diodes
- Chapter 5 Diode Power Supply Circuits
- Chapter 6 Diode Multiplier, Clipper, and Clamper Circuits
- Chapter 7 Special Application Diodes

PART III Transistors and Transistor Circuits
- Chapter 8 Bipolar Junction Transistors (BJTs)
- Chapter 9 Bipolar Transistor Digital Circuits
- Chapter 10 Bipolar Transistor Low-Frequency Amplifier Circuits
- Chapter 11 Bipolar Transistor High-Frequency Amplifier Circuits
- Chapter 12 Bipolar Transistor Oscillator Circuits
- Chapter 13 Junction Field Effect Transistors (JFETs)
- Chapter 14 Metal Oxide Semiconductor Field Effect Transistors (MOSFETs)

PART IV Integrated Circuits
- Chapter 15 Integrated Circuit Construction
- Chapter 16 Operational Amplifiers

The material covered in this book has been logically divided and sequenced to provide a gradual progression from the known to the unknown and from the simple to the complex. Since the topics covered in *Introductory DC/AC Circuits* and *Introductory Semiconductor Electronics* form the foundation on which your understanding of electronics will rest, a great deal of effort has been made to ensure that the style, format, approach, and content of this book are compatible with you, the student.

DEVELOPMENT, CLASS TESTING, AND REVIEWING

The first phase of development was conducted in the classroom with students and instructors as critics. Each topic was class-tested by videotaping each lesson, and the results were then evaluated and implemented. This feedback was invaluable and enabled me to fine-tune my presentation of topics and instill understanding and confidence into the students.

The second phase of development was to forward a copy of the revised manuscript to several instructors at schools throughout the country. These technical and topical critiques helped to mold the text into a more accurate form.

The third and final phase was to class-test the final revised manuscript and then commission the last technical review in the final stages of production.

ILLUSTRATED TOUR OF TEXTBOOK FEATURES

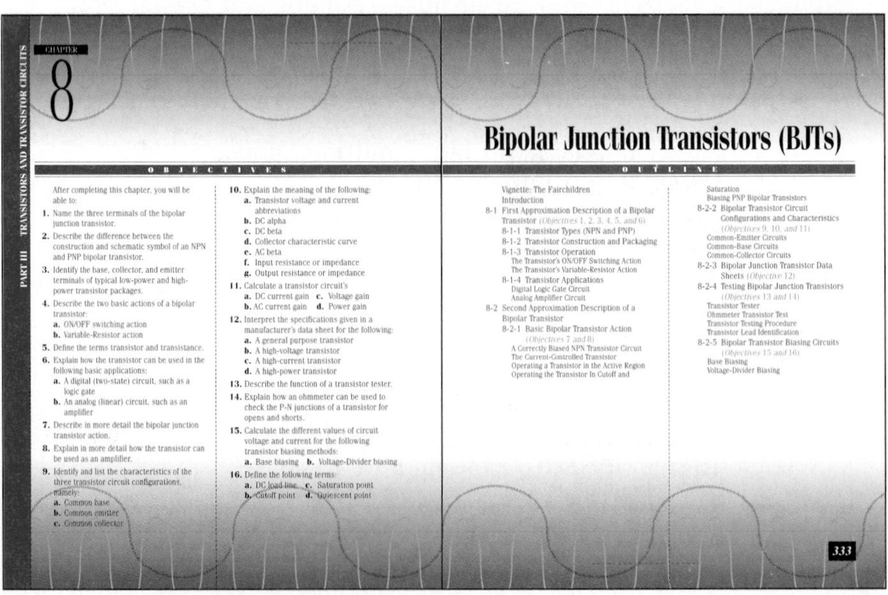
PART III TRANSISTORS AND TRANSISTOR CIRCUITS

CHAPTER 8

Bipolar Junction Transistors (BJTs)

OBJECTIVES

After completing this chapter, you will be able to:

1. Name the three terminals of the bipolar junction transistor.
2. Describe the difference between the construction and schematic symbol of an NPN and PNP bipolar transistor.
3. Identify the base, collector, and emitter terminals of typical low-power and high-power transistor packages.
4. Describe the two basic actions of a bipolar transistor:
   a. ON/OFF switching action
   b. Variable-Resistor action
5. Define the terms transistor and transistance.
6. Explain how the transistor can be used in the following basic applications:
   a. A digital (two-state) circuit, such as a logic gate
   b. An analog (linear) circuit, such as an amplifier
7. Describe in more detail the bipolar junction transistor action.
8. Explain in more detail how the transistor can be used as an amplifier.
9. Identify and list the characteristics of the three transistor circuit configurations, namely:
   a. Common base
   b. Common emitter
   c. Common collector
10. Explain the meaning of the following:
    a. Transistor voltage and current abbreviations
    b. DC alpha
    c. DC beta
    d. Collector characteristic curve
    e. AC beta
    f. Input resistance or impedance
    g. Output resistance or impedance
11. Calculate a transistor circuit's
    a. DC current gain c. Voltage gain
    b. AC current gain d. Power gain
12. Interpret the specifications given in a manufacturer's data sheet for the following:
    a. A general purpose transistor
    b. A high-voltage transistor
    c. A high-current transistor
    d. A high-power transistor
13. Describe the function of a transistor tester.
14. Explain how an ohmmeter can be used to check the P-N junctions of a transistor for opens and shorts.
15. Calculate the different values of circuit voltage and current for the following transistor biasing methods:
    a. Base biasing b. Voltage-Divider biasing
16. Define the following terms:
    a. DC load line c. Saturation point
    b. Cutoff point d. Quiescent point

OUTLINE

Vignette: The Fairchildren
Introduction
8-1 First Approximation Description of a Bipolar Transistor (*Objectives* 1, 2, 3, 4, 5, *and* 6)
8-1-1 Transistor Types (NPN and PNP)
8-1-2 Transistor Construction and Packaging
8-1-3 Transistor Operation
The Transistor's ON/OFF Switching Action
The Transistor's Variable-Resistor Action
8-1-4 Transistor Applications
Digital Logic Gate Circuit
Analog Amplifier Circuit
8-2 Second Approximation Description of a Bipolar Transistor
8-2-1 Basic Bipolar Transistor Action (*Objectives* 7 *and* 8)
A Correctly Biased NPN Transistor Circuit
The Current-Controlled Transistor
Operating a Transistor in the Active Region
Operating the Transistor In Cutoff and Saturation
Biasing PNP Bipolar Transistors
8-2-2 Bipolar Transistor Circuit Configurations and Characteristics (*Objectives* 9, 10, *and* 11)
Common-Emitter Circuits
Common-Base Circuits
Common-Collector Circuits
8-2-3 Bipolar Junction Transistor Data Sheets (*Objective* 12)
8-2-4 Testing Bipolar Junction Transistors (*Objectives* 13 *and* 14)
Transistor Tester
Ohmmeter Transistor Test
Transistor Testing Procedure
Transistor Lead Identification
8-2-5 Bipolar Transistor Biasing Circuits (*Objectives* 15 *and* 16)
Base Biasing
Voltage-Divider Biasing

333

Each chapter opening lists performance-based objectives with each objective directly correlated to a chapter section on the facing page.

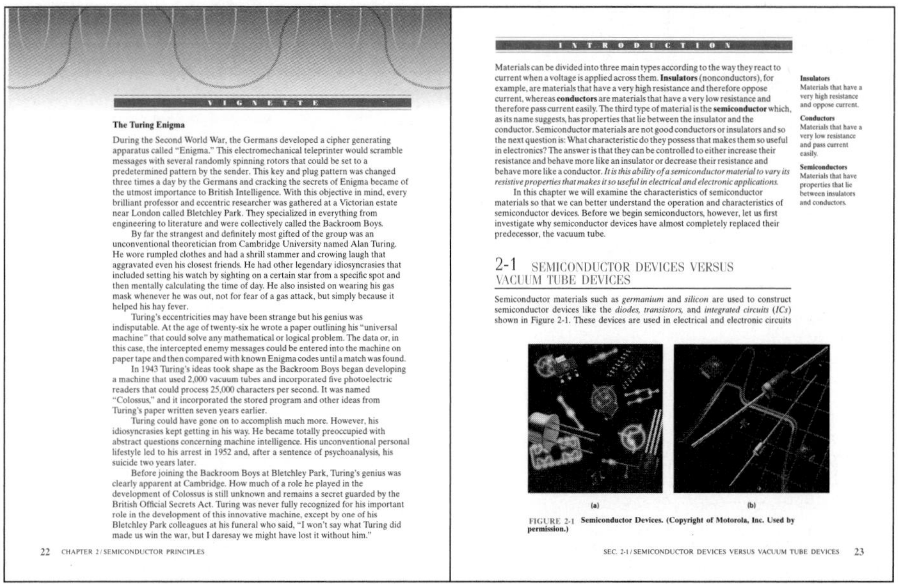

VIGNETTE

The Turing Enigma

During the Second World War, the Germans developed a cipher generating apparatus called "Enigma." This electromechanical teleprinter would scramble messages with several randomly spinning rotors that could be set to a predetermined pattern by the sender. This key and plug pattern was changed three times a day by the Germans and cracking the secrets of Enigma became of the utmost importance to British Intelligence. With this objective in mind, every brilliant professor and eccentric researcher was gathered at a Victorian estate near London called Bletchley Park. They specialized in everything from engineering to literature and were collectively called the Backroom Boys.

By far the strangest and definitely most gifted of the group was an unconventional theoretician from Cambridge University named Alan Turing. He wore rumpled clothes and had a shrill stammer and crowing laugh that aggravated even his closest friends. He had other legendary idiosyncrasies that included setting his watch by sighting on a certain star from a specific spot and then mentally calculating the time of day. He also insisted on wearing his gas mask whenever he was out, not for fear of a gas attack, but simply because it helped his hay fever.

Turing's eccentricities may have been strange but his genius was indisputable. At the age of twenty-six he wrote a paper outlining his "universal machine" that could solve any mathematical or logical problem. The data or, in this case, the intercepted enemy messages could be entered into the machine on paper tape and then compared with known Enigma codes until a match was found.

In 1943 Turing's ideas took shape as the Backroom Boys began developing a machine that used 2,000 vacuum tubes and incorporated five photoelectric readers that could process 25,000 characters per second. It was named "Colossus," and it incorporated the stored program and other ideas from Turing's paper written seven years earlier.

Turing could have gone on to accomplish much more. However, his idiosyncrasies kept getting in his way. He became totally preoccupied with abstract questions concerning machine intelligence. His unconventional personal lifestyle led to his arrest in 1952 and, after a sentence of psychoanalysis, his suicide two years later.

Before joining the Backroom Boys at Bletchley Park, Turing's genius was clearly apparent at Cambridge. How much of a role he played in the development of Colossus is still unknown and remains a secret guarded by the British Official Secrets Act. Turing was never fully recognized for his important role in the development of this innovative machine, except by one of his Bletchley Park colleagues at his funeral who said, "I won't say what Turing did made us win the war, but I daresay we might have lost it without him."

22 CHAPTER 2 / SEMICONDUCTOR PRINCIPLES

INTRODUCTION

Materials can be divided into three main types according to the way they react to current when a voltage is applied across them. **Insulators** (nonconductors), for example, are materials that have a very high resistance and therefore oppose current, whereas **conductors** are materials that have a very low resistance and therefore pass current easily. The third type of material is the **semiconductor** which, as its name suggests, has properties that lie between the insulator and the conductor. Semiconductor materials are not good conductors or insulators and so the next question is: What characteristic do they possess that makes them so useful in electronics? The answer is that they can be controlled to either increase their resistance and behave more like an insulator or decrease their resistance and behave more like a conductor. *It is this ability of a semiconductor material to vary its resistive properties that makes it so useful in electrical and electronic applications.*

In this chapter we will examine the characteristics of semiconductor materials so that we can better understand the operation and characteristics of semiconductor devices. Before we begin semiconductors, however, let us first investigate why semiconductor devices have almost completely replaced their predecessor, the vacuum tube.

Insulators Materials that have a very high resistance and oppose current.

Conductors Materials that have a very low resistance and pass current easily.

Semiconductors Materials that have properties that lie between insulators and conductors.

2-1 SEMICONDUCTOR DEVICES VERSUS VACUUM TUBE DEVICES

Semiconductor materials such as *germanium* and *silicon* are used to construct semiconductor devices like the *diodes, transistors,* and *integrated circuits (ICs)* shown in Figure 2-1. These devices are used in electrical and electronic circuits

(a) (b)

FIGURE 2-1 **Semiconductor Devices. (Copyright of Motorola, Inc. Used by permission.)**

SEC. 2-1 / SEMICONDUCTOR DEVICES VERSUS VACUUM TUBE DEVICES 23

Motivational stories detail the people behind the electronics industry, and a conversational introduction reviews what has been previously covered and what is about to be covered.

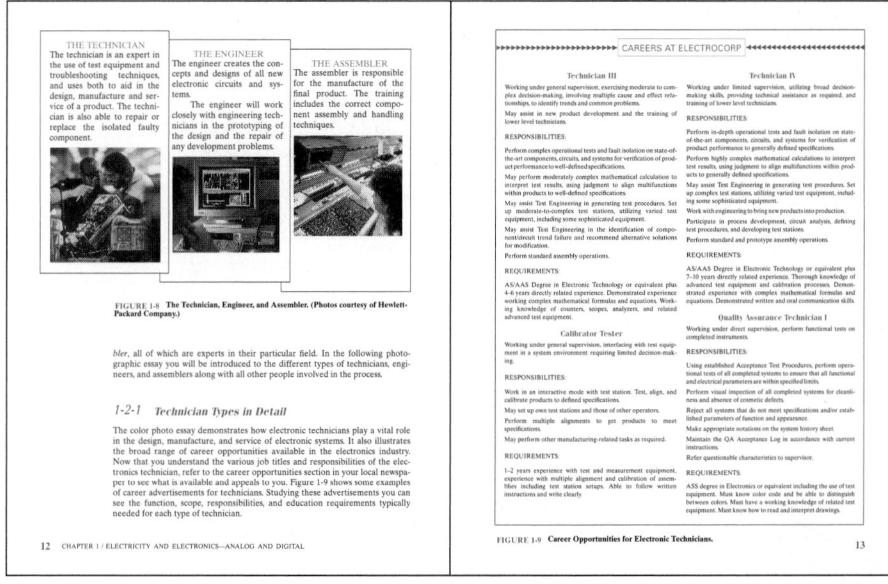

THE TECHNICIAN
The technician is an expert in the use of test equipment and troubleshooting techniques, and uses both to aid in the design, manufacture and service of a product. The technician is also able to repair or replace the isolated faulty component.

THE ENGINEER
The engineer creates the concepts and designs of all new electronic circuits and systems.

The engineer will work closely with engineering technicians in the prototyping of the design and the repair of any development problems.

THE ASSEMBLER
The assembler is responsible for the manufacture of the final product. The training includes the correct component assembly and handling techniques.

FIGURE 1-8 **The Technician, Engineer, and Assembler. (Photos courtesy of Hewlett-Packard Company.)**

bler, all of which are experts in their particular field. In the following photographic essay you will be introduced to the different types of technicians, engineers, and assemblers along with all other people involved in the process.

1-2-1 Technician Types in Detail

The color photo essay demonstrates how electronic technicians play a vital role in the design, manufacture, and service of electronic systems. It also illustrates the broad range of career opportunities available in the electronics industry. Now that you understand the various job titles and responsibilities of the electronics technician, refer to the career opportunities section in your local newspaper to see what is available and appeals to you. Figure 1-9 shows some examples of career advertisements for technicians. Studying these advertisements you can see the function, scope, responsibilities, and education requirements typically needed for each type of technician.

12 CHAPTER 1 / ELECTRICITY AND ELECTRONICS—ANALOG AND DIGITAL

CAREERS AT ELECTROCORP

Technician III

Working under general supervision, exercising moderate to complex decision-making, involving multiple cause and effect relationships, to identify trends and common problems.

May assist in new product development and the training of lower level technicians.

RESPONSIBILITIES:

Perform complex operational tests and fault isolation on state-of-the-art components, circuits, and systems for verification of product performance to well-defined specifications.

May perform moderately complex mathematical calculation to interpret test results, using judgment to align multifunctions within products to well-defined specifications.

May assist Test Engineering in generating test procedures. Set up moderate-to-complex test stations, utilizing varied test equipment, including some sophisticated equipment.

May assist Test Engineering in the identification of component/circuit trend failure and recommend alternative solutions for modification.

Perform standard assembly operations.

REQUIREMENTS:

AS/AAS Degree in Electronic Technology or equivalent plus 4–6 years directly related experience. Demonstrated experience working complex mathematical formulas and equations. Working knowledge of counters, scopes, analyzers, and related advanced test equipment.

Calibrator Tester

Working under general supervision, interfacing with test equipment in a system environment requiring limited decision-making.

RESPONSIBILITIES:

Work in an interactive mode with test station. Test, align, and calibrate products to defined specifications.

May set up own test stations and those of other operators.

Perform multiple alignments to get products to meet specifications.

May perform other manufacturing-related tasks as required.

REQUIREMENTS:

1–2 years experience with test and measurement equipment, experience with multiple alignment and calibration of assemblies including test station setups. Able to follow written instructions and write clearly.

Technician IV

Working under limited supervision, utilizing broad decision-making skills, providing technical assistance as required, and training of lower level technicians.

RESPONSIBILITIES:

Perform in-depth operational tests and fault isolation on state-of-the-art components, circuits, and systems for verification of product performance to generally defined specifications.

Perform highly complex mathematical calculations to interpret test results, using judgment to align multifunctions within products to generally defined specifications.

May assist Test Engineering in generating test procedures. Set up complex test stations, utilizing varied test equipment, including some sophisticated equipment.

Work with engineering to bring new products into production.

Participate in process development, circuit analysis, defining test procedures, and developing test stations.

Perform standard and prototype assembly operations.

REQUIREMENTS:

AS/AAS Degree in Electronic Technology or equivalent plus 7–10 years directly related experience. Thorough knowledge of advanced test equipment and calibration processes. Demonstrated experience with complex mathematical formulas and equations. Demonstrated written and oral communication skills.

Quality Assurance Technician I

Working under direct supervision, perform functional tests on completed instruments.

RESPONSIBILITIES:

Using established Acceptance Test Procedures, perform operational tests of all completed systems to ensure that all functional and electrical parameters are within specified limits.

Perform visual inspection of all completed systems for cleanliness and absence of cosmetic defects.

Reject all systems that do not meet specifications and/or established parameters of function and appearance.

Make appropriate notations on the system history sheet.

Maintain the QA Acceptance Log in accordance with current instructions.

Refer questionable characteristics to supervisor.

REQUIREMENTS:

ASS degree in Electronics or equivalent including the use of test equipment. Must know color code and be able to distinguish between colors. Must have a working knowledge of related test equipment. Must know how to read and interpret drawings.

FIGURE 1-9 **Career Opportunities for Electronic Technicians.**

13

Chapter 1 includes extensive information about career opportunities for the electronics technician providing interest in and motivation for the material which follows.

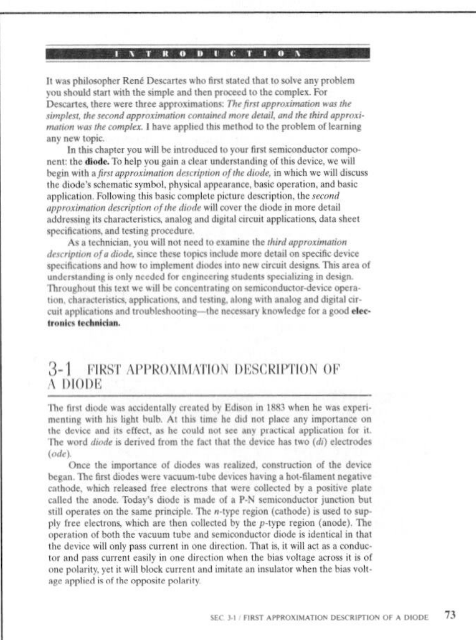

INTRODUCTION

It was philosopher René Descartes who first stated that to solve any problem you should start with the simple and then proceed to the complex. For Descartes, there were three approximations: *The first approximation was the simplest, the second approximation contained more detail, and the third approximation was the complex.* I have applied this method to the problem of learning any new topic.

In this chapter you will be introduced to your first semiconductor component: the **diode.** To help you gain a clear understanding of this device, we will begin with a *first approximation description of the diode,* in which we will discuss the diode's schematic symbol, physical appearance, basic operation, and basic application. Following this basic complete picture description, the *second approximation description of the diode* will cover the diode in more detail addressing its characteristics, analog and digital circuit applications, data sheet specifications, and testing procedure.

As a technician, you will not need to examine the *third approximation description of a diode,* since these topics include more detail on specific device specifications and how to implement diodes into new circuit designs. This area of understanding is only needed for engineering students specializing in design. Throughout this text we will be concentrating on semiconductor-device operation, characteristics, applications, and testing, along with analog and digital circuit applications and troubleshooting—the necessary knowledge for a good **electronics technician.**

3-1 FIRST APPROXIMATION DESCRIPTION OF A DIODE

The first diode was accidentally created by Edison in 1883 when he was experimenting with his light bulb. At this time he did not place any importance on the device and its effect, as he could not see any practical application for it. The word *diode* is derived from the fact that the device has two (*di*) electrodes (*ode*).

Once the importance of diodes was realized, construction of the device began. The first diodes were vacuum-tube devices having a hot-filament negative cathode, which released free electrons that were collected by a positive plate called the anode. Today's diode is made of a P-N semiconductor junction but still operates on the same principle. The *n*-type region (cathode) is used to supply free electrons, which are then collected by the *p*-type region (anode). The operation of both the vacuum tube and semiconductor diode is identical in that the device will only pass current in one direction. That is, it will act as a conductor and pass current easily in one direction when the bias voltage across it is of one polarity, yet it will block current and imitate an insulator when the bias voltage applied is of the opposite polarity.

SEC 3-1 / FIRST APPROXIMATION DESCRIPTION OF A DIODE 73

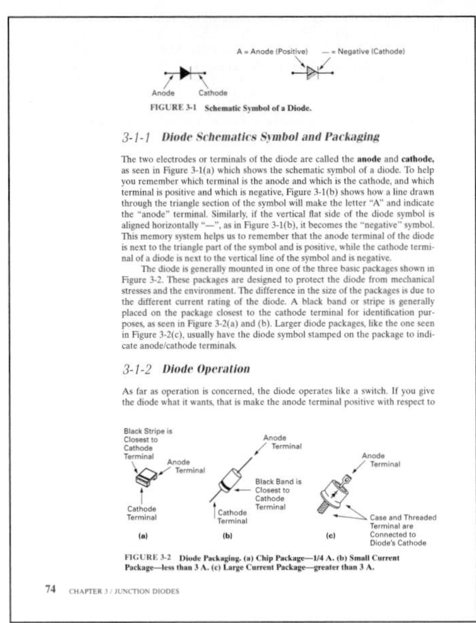

A = Anode (Positive) — = Negative (Cathode)
Anode Cathode

FIGURE 3-1 Schematic Symbol of a Diode.

3-1-1 ***Diode Schematics Symbol and Packaging***

The two electrodes or terminals of the diode are called the **anode** and **cathode,** as seen in Figure 3-1(a) which shows the schematic symbol of a diode. To help you remember which terminal is the anode and which is the cathode, and which terminal is positive and which is negative, Figure 3-1(b) shows how a line drawn through the triangle section of the symbol will make the letter "A" and indicate the "anode" terminal. Similarly, if the vertical flat side of the diode symbol is aligned horizontally "—", as in Figure 3-1(b), it becomes the "negative" symbol. This memory system helps us to remember that the anode terminal of the diode is next to the triangle part of the symbol and is positive, while the cathode terminal of a diode is next to the vertical line of the symbol and is negative.

The diode is generally mounted in one of the three basic packages shown in Figure 3-2. These packages are designed to protect the diode from mechanical stresses and the environment. The difference in the size of the packages is due to the different current rating of the diode. A black band or stripe is generally placed on the package closest to the cathode terminal for identification purposes, as seen in Figure 3-2(a) and (b). Larger diode packages, like the one seen in Figure 3-2(c), usually have the diode symbol stamped on the package to indicate anode/cathode terminals.

3-1-2 ***Diode Operation***

As far as operation is concerned, the diode operates like a switch. If you give the diode what it wants, that is make the anode terminal positive with respect to

Black Stripe is Closest to Cathode Terminal
Anode Terminal
Cathode Terminal
(a)
Anode Terminal
Black Band is Closest to Cathode Terminal
Cathode Terminal
(b)
Anode Terminal
Case and Threaded Terminal are Connected to Diode's Cathode
(c)

FIGURE 3-2 Diode Packaging. (a) Chip Package—1/4 A. (b) Small Current Package—less than 3 A. (c) Large Current Package—greater than 3 A.

74 CHAPTER 3 / JUNCTION DIODES

The approximations method of teaching and presenting concepts leads to a dramatic improvement in student comprehension. In the "first approximation description," a general overview of the device's symbol, packaging, operation, and typical application is covered. The "second approximation description" examines the device's characteristics, testing, and circuit applications in more detail.

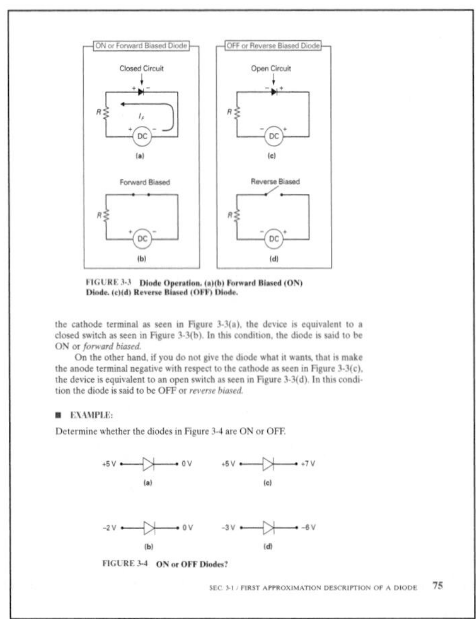

ON or Forward Biased Diode
Closed Circuit
R
I_F
DC
(a)
Forward Biased
R
DC
(b)
OFF or Reverse Biased Diode
Open Circuit
R
DC
(c)
Reverse Biased
R
DC
(d)

FIGURE 3-3 Diode Operation. (a)(b) Forward Biased (ON) Diode. (c)(d) Reverse Biased (OFF) Diode.

the cathode terminal as seen in Figure 3-3(a), the device is equivalent to a closed switch as seen in Figure 3-3(b). In this condition, the diode is said to be ON or *forward biased.*

On the other hand, if you do not give the diode what it wants, that is make the anode terminal negative with respect to the cathode as seen in Figure 3-3(c), the device is equivalent to an open switch as seen in Figure 3-3(d). In this condition the diode is said to be OFF or *reverse biased.*

■ EXAMPLE:

Determine whether the diodes in Figure 3-4 are ON or OFF.

+5 V — 0 V (a)
+5 V — +7 V (c)
−2 V — 0 V (b)
−3 V — −6 V (d)

FIGURE 3-4 ON or OFF Diodes?

SEC 3-1 / FIRST APPROXIMATION DESCRIPTION OF A DIODE 75

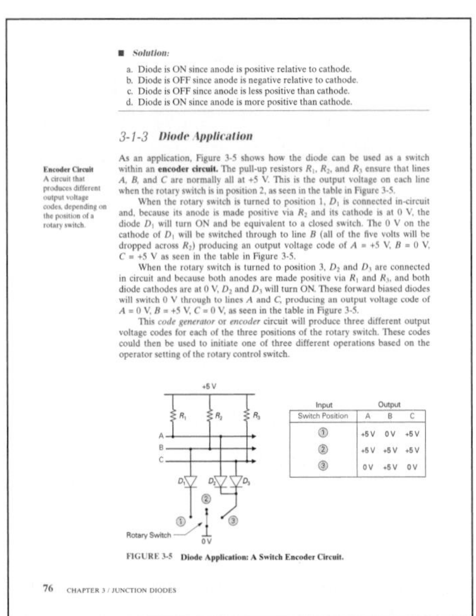

■ *Solution:*

a. Diode is ON since anode is positive relative to cathode.
b. Diode is OFF since anode is negative relative to cathode.
c. Diode is OFF since anode is less positive than cathode.
d. Diode is ON since anode is more positive than cathode.

3-1-3 ***Diode Application***

Encoder Circuit A circuit that produces different output voltage codes, depending on the position of a rotary switch.

As an application, Figure 3-5 shows how the diode can be used as a switch within an **encoder circuit.** The pull-up resistors R_1, R_2, and R_3 ensure that lines A, B, and C are normally all at +5 V. This is the output voltage on each line when the rotary switch is in position 2, as seen in the table in Figure 3-5.

When the rotary switch is turned to position 1, D_1 is connected in-circuit and, because its anode is made positive via R_2 and its cathode is at 0 V, the diode D_1 will turn ON and be equivalent to a closed switch. The 0 V on the cathode of D_1 will be switched through to line B (all of the five volts will be dropped across R_2) producing an output voltage code of $A = +5$ V, $B = 0$ V, $C = +5$ V as seen in the table in Figure 3-5.

When the rotary switch is turned to position 3, D_2 and D_3 are connected in circuit and because both anodes are made positive via R_1 and R_3, and both diode cathodes are at 0 V, D_2 and D_3 will turn ON. These forward biased diodes will switch 0 V through to lines A and C, producing an output voltage code of $A = 0$ V, $B = +5$ V, $C = 0$ V, as seen in the table in Figure 3-5.

This *code generator* or *encoder* circuit will produce three different output voltage codes for each of the three positions of the rotary switch. These codes could then be used to initiate one of three different operations based on the operator setting of the rotary control switch.

+5 V
R_1 R_2 R_3
A B C
D_1 D_2 D_3
Rotary Switch
0 V

Input	Output		
Switch Position	A	B	C
①	+5 V	0 V	+5 V
②	+5 V	+5 V	+5 V
③	0 V	+5 V	0 V

FIGURE 3-5 Diode Application: A Switch Encoder Circuit.

76 CHAPTER 3 / JUNCTION DIODES

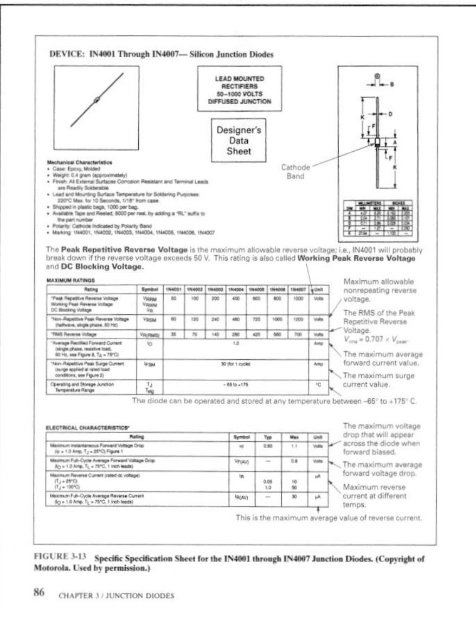

DEVICE: IN4001 Through IN4007— Silicon Junction Diodes

Mechanical Characteristics

- Case: Epoxy, Molded
- Weight: 0.4 gram (approximately)
- Finish: All External Surfaces Corrosion Resistant and Terminal Leads are Readily Solderable
- Lead and Mounting Surface Temperature for Soldering Purposes: 220°C Max. for 10 Seconds, 1/16" from case
- Shipped in plastic bags, 1000 per bag.
- Available Tape and Reeled, 5000 per reel, by adding a "RL" suffix to the part number
- Polarity: Cathode Indicated by Polarity Band
- Marking: 1N4001, 1N4002, 1N4003, 1N4004, 1N4005, 1N4006, 1N4007

DIM	MILLIMETERS MIN	MILLIMETERS MAX	INCHES MIN	INCHES MAX
A	4.07	5.20	0.160	0.205
B	2.04	2.71	0.080	0.107
D	0.71	0.86	0.028	0.034
F	—	1.27	—	0.050
K	27.94	—	1.100	—

The **Peak Repetitive Reverse Voltage** is the maximum allowable reverse voltage; i.e., IN4001 will probably break down if the reverse voltage exceeds 50 V. This rating is also called **Working Peak Reverse Voltage** and **DC Blocking Voltage.**

MAXIMUM RATINGS

Rating	Symbol	1N4001	1N4002	1N4003	1N4004	1N4005	1N4006	1N4007	Unit
*Peak Repetitive Reverse Voltage Working Peak Reverse Voltage DC Blocking Voltage	V_{RRM} V_{RWM} V_R	50	100	200	400	600	800	1000	Volts
*Non-Repetitive Peak Reverse Voltage (halfwave, single phase, 60 Hz)	V_{RSM}	60	120	240	480	720	1000	1200	Volts
*RMS Reverse Voltage	$V_{R(RMS)}$	35	70	140	280	420	560	700	Volts
*Average Rectified Forward Current (single phase, resistive load, 60 Hz, see Figure 8, T_A = 75°C)	I_O	1.0							Amp
*Non-Repetitive Peak Surge Current (surge applied at rated load conditions, see Figure 2)	I_{FSM}	30 (for 1 cycle)							Amp
Operating and Storage Junction Temperature Range	T_J T_{stg}	−65 to +175							°C

ELECTRICAL CHARACTERISTICS*

Rating	Symbol	Typ	Max	Unit
Maximum Instantaneous Forward Voltage Drop (i_F = 1.0 Amp, T_J = 25°C) Figure 1	v_F	0.93	1.1	Volts
Maximum Full-Cycle Average Forward Voltage Drop (I_O = 1.0 Amp, T_L = 75°C, 1 inch leads)	$V_{F(AV)}$	—	0.8	Volts
Maximum Reverse Current (rated dc voltage) (T_J = 25°C) (T_J = 100°C)	I_R	0.05 1.0	10 50	µA
Maximum Full-Cycle Average Reverse Current (I_O = 1.0 Amp, T_L = 75°C, 1 inch leads)	$I_{R(AV)}$	—	30	µA

FIGURE 3-13 **Specific Specification Sheet for the IN4001 through IN4007 Junction Diodes. (Copyright of Motorola. Used by permission.)**

86 CHAPTER 3 / JUNCTION DIODES

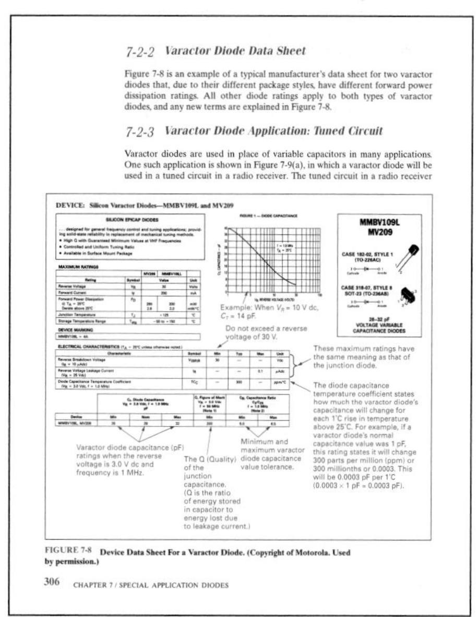

7-2-2 ***Varactor Diode Data Sheet***

Figure 7-8 is an example of a typical manufacturer's data sheet for two varactor diodes that, due to their different package styles, have different forward power dissipation ratings. All other diode ratings apply to both types of varactor diodes, and any new terms are explained in Figure 7-8.

7-2-3 ***Varactor Diode Application: Tuned Circuit***

Varactor diodes are used in place of variable capacitors in many applications. One such application is shown in Figure 7-9(a), in which a varactor diode will be used in a tuned circuit in a radio receiver. The tuned circuit in a radio receiver

DEVICE: Silicon Varactor Diodes—MMBV109L and MV209

SILICON EPICAP DIODES

. . . designed for general frequency control and tuning applications; providing solid-state reliability in replacement of mechanical tuning methods.

- High Q with Guaranteed Minimum Values at VHF Frequencies
- Controlled and Uniform Tuning Ratio
- Available in Surface Mount Package

FIGURE 7-8 **Device Data Sheet For a Varactor Diode. (Copyright of Motorola. Used by permission.)**

306 CHAPTER 7 / SPECIAL APPLICATION DIODES

Data sheet annotations are included to explain the meaning of the manufacturer's key characteristics.

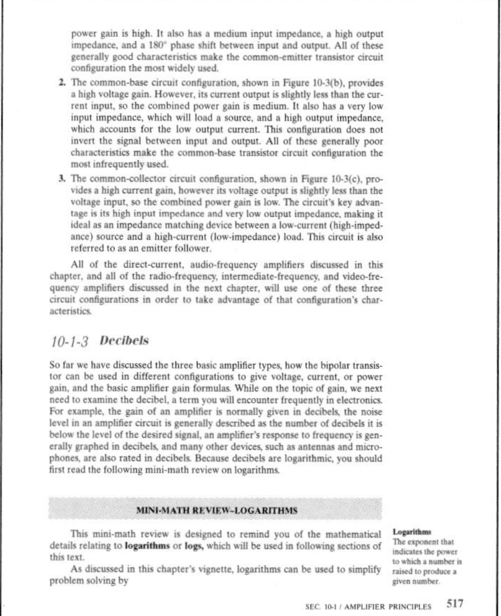

power gain is high. It also has a medium input impedance, a high output impedance, and a 180° phase shift between input and output. All of these generally good characteristics make the common-emitter transistor circuit configuration the most widely used.

2. The common-base circuit configuration, shown in Figure 10-3(b), provides a high voltage gain. However, its current output is slightly less than the current input, so the combined power gain is medium. It also has a very low input impedance, which will load a source, and a high output impedance, which accounts for the low output current. This configuration does not invert the signal between input and output. All of these generally poor characteristics make the common-base transistor circuit configuration the most infrequently used.
3. The common-collector circuit configuration, shown in Figure 10-3(c), provides a high current gain, however its voltage output is slightly less than the voltage input, so the combined power gain is low. The circuit's key advantage is its high input impedance and very low output impedance, making it ideal as an impedance matching device between a low-current (high-impedance) source and a high-current (low-impedance) load. This circuit is also referred to as an emitter follower.

All of the direct-current, audio-frequency amplifiers discussed in this chapter, and all of the radio-frequency, intermediate-frequency, and video-frequency amplifiers discussed in the next chapter, will use one of these three circuit configurations in order to take advantage of that configuration's characteristics.

10-1-3 ***Decibels***

So far we have discussed the three basic amplifier types, how the bipolar transistor can be used in different configurations to give voltage, current, or power gain, and the basic amplifier gain formulas. While on the topic of gain, we next need to examine the decibel, a term you will encounter frequently in electronics. For example, the gain of an amplifier is normally given in decibels, the noise level in an amplifier circuit is generally described as the number of decibels it is below the level of the desired signal, an amplifier's response to frequency is generally graphed in decibels, and many other devices, such as antennas and microphones, are also rated in decibels. Because decibels are logarithmic, you should first read the following mini-math review on logarithms.

MINI-MATH REVIEW–LOGARITHMS

This mini-math review is designed to remind you of the mathematical details relating to **logarithms** or **logs,** which will be used in following sections of this text.

As discussed in this chapter's vignette, logarithms can be used to simplify problem solving by

Logarithms The exponent that indicates the power to which a number is raised to produce a given number.

SEC. 10-1 / AMPLIFIER PRINCIPLES 517

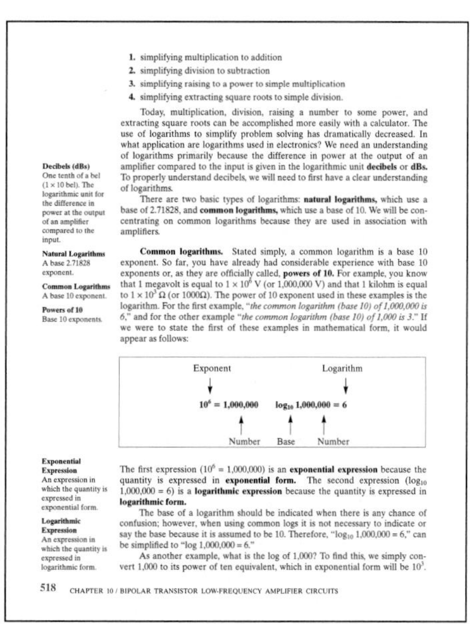

1. simplifying multiplication to addition
2. simplifying division to subtraction
3. simplifying raising to a power to simple multiplication
4. simplifying extracting square roots to simple division.

Today, multiplication, division, raising a number to some power, and extracting square roots can be accomplished more easily with a calculator. The use of logarithms to simplify problem solving has dramatically decreased. In what application are logarithms used in electronics? We need an understanding of logarithms primarily because the difference in power at the output of an amplifier compared to the input is given in the logarithmic unit **decibels** or **dBs.** To properly understand decibels, we will need to first have a clear understanding of logarithms.

Decibels (dBs) One tenth of a bel (1 × 10 bel). The logarithmic unit for the difference in power at the output of an amplifier compared to the input.

There are two basic types of logarithms: **natural logarithms,** which use a base of 2.71828, and **common logarithms,** which use a base of 10. We will be concentrating on common logarithms because they are used in association with amplifiers.

Natural Logarithms A base 2.71828 exponent.

Common Logarithms A base 10 exponent.

Powers of 10 Base 10 exponents.

Common logarithms. Stated simply, a common logarithm is a base 10 exponent. So far, you have already had considerable experience with base 10 exponents or, as they are officially called, **powers of 10.** For example, you know that 1 megavolt is equal to 1×10^6 V (or 1,000,000 V) and that 1 kilohm is equal to $1 \times 10^3\ \Omega$ (or 1000Ω). The power of 10 exponent used in these examples is the logarithm. For the first example, "*the common logarithm (base 10) of 1,000,000 is 6,*" and for the other example "*the common logarithm (base 10) of 1,000 is 3.*" If we were to state the first of these examples in mathematical form, it would appear as follows:

The first expression ($10^6 = 1{,}000{,}000$) is an **exponential expression** because the quantity is expressed in **exponential form.** The second expression ($\log_{10}$ 1,000,000 = 6) is a **logarithmic expression** because the quantity is expressed in **logarithmic form.**

Exponential Expression An expression in which the quantity is expressed in exponential form.

The base of a logarithm should be indicated when there is any chance of confusion; however, when using common logs it is not necessary to indicate or say the base because it is assumed to be 10. Therefore, "$\log_{10}$ 1,000,000 = 6," can be simplified to "log 1,000,000 = 6."

Logarithmic Expression An expression in which the quantity is expressed in logarithmic form.

As another example, what is the log of 1,000? To find this, we simply convert 1,000 to its power of ten equivalent, which in exponential form will be 10^3.

518 CHAPTER 10 / BIPOLAR TRANSISTOR LOW-FREQUENCY AMPLIFIER CIRCUITS

Mini-Math Reviews are included in appropriate places to review the necessary mathematics needed for the discussion at hand.

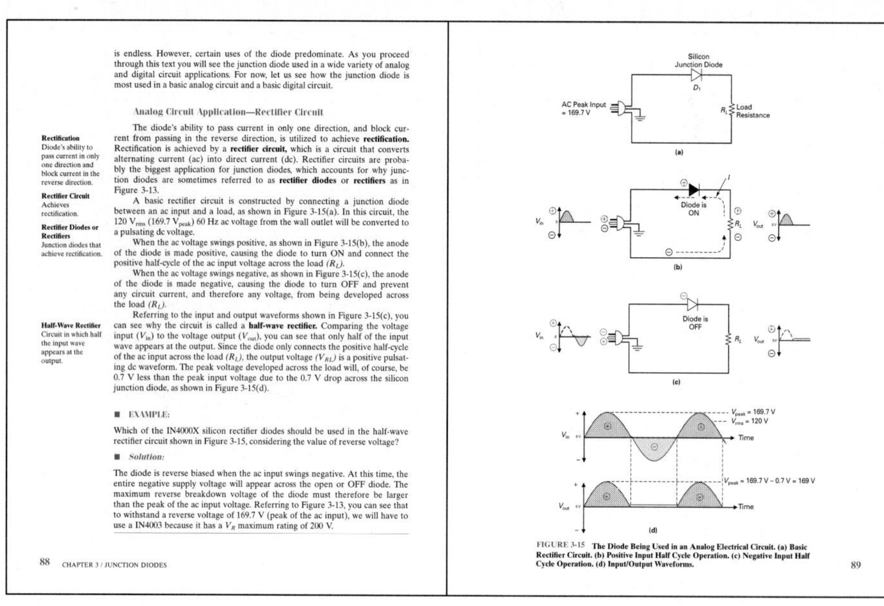

is endless. However, certain uses of the diode predominate. As you proceed through this text you will see the junction diode used in a wide variety of analog and digital circuit applications. For now, let us see how the junction diode is most used in a basic analog circuit and a basic digital circuit.

Analog Circuit Application—Rectifier Circuit

Rectification Diode's ability to pass current in only one direction and block current in the reverse direction.

Rectifier Circuit Achieves rectification.

Rectifier Diodes or Rectifiers Junction diodes that achieve rectification.

The diode's ability to pass current in only one direction, and block current from passing in the reverse direction, is utilized to achieve **rectification.** Rectification is achieved by a **rectifier circuit,** which is a circuit that converts alternating current (ac) into direct current (dc). Rectifier circuits are probably the biggest application for junction diodes, which accounts for why junction diodes are sometimes referred to as **rectifier diodes** or **rectifiers** as in Figure 3-13.

A basic rectifier circuit is constructed by connecting a junction diode between an ac input and a load, as shown in Figure 3-15(a). In this circuit, the 120 V_{rms} (169.7 V_{peak}) 60 Hz ac voltage from the wall outlet will be converted to a pulsating dc voltage.

When the ac voltage swings positive, as shown in Figure 3-15(b), the anode of the diode is made positive, causing the diode to turn ON and connect the positive half-cycle of the ac input voltage across the load (R_L).

When the ac voltage swings negative, as shown in Figure 3-15(c), the anode of the diode is made negative, causing the diode to turn OFF and prevent any circuit current, and therefore any voltage, from being developed across the load (R_L).

Half-Wave Rectifier Circuit in which half the input wave appears at the output.

Referring to the input and output waveforms shown in Figure 3-15(c), you can see why the circuit is called a **half-wave rectifier.** Comparing the voltage input (V_{in}) to the voltage output (V_{out}), you can see that only half of the input wave appears at the output. Since the diode only connects the positive half-cycle of the ac input across the load (R_L), the output voltage (V_{RL}) is a positive pulsating dc waveform. The peak voltage developed across the load will, of course, be 0.7 V less than the peak input voltage due to the 0.7 V drop across the silicon junction diode, as shown in Figure 3-15(d).

■ EXAMPLE:

Which of the IN4000X silicon rectifier diodes should be used in the half-wave rectifier circuit shown in Figure 3-15, considering the value of reverse voltage?

■ *Solution:*

The diode is reverse biased when the ac input swings negative. At this time, the entire negative supply voltage will appear across the open or OFF diode. The maximum reverse breakdown voltage of the diode must therefore be larger than the peak of the ac input voltage. Referring to Figure 3-13, you can see that to withstand a reverse voltage of 169.7 V (peak of the ac input), we will have to use a IN4003 because it has a V_R maximum rating of 200 V.

88 CHAPTER 3 / JUNCTION DIODES

FIGURE 3-15 **The Diode Being Used in an Analog Electrical Circuit. (a) Basic Rectifier Circuit. (b) Positive Input Half Cycle Operation. (c) Negative Input Half Cycle Operation. (d) Input/Output Waveforms.**

89

Revolutionary blending of analog and digital electronics prepares the student for the increasing digital environment. A running glossary defines all new terms in the margin of the text.

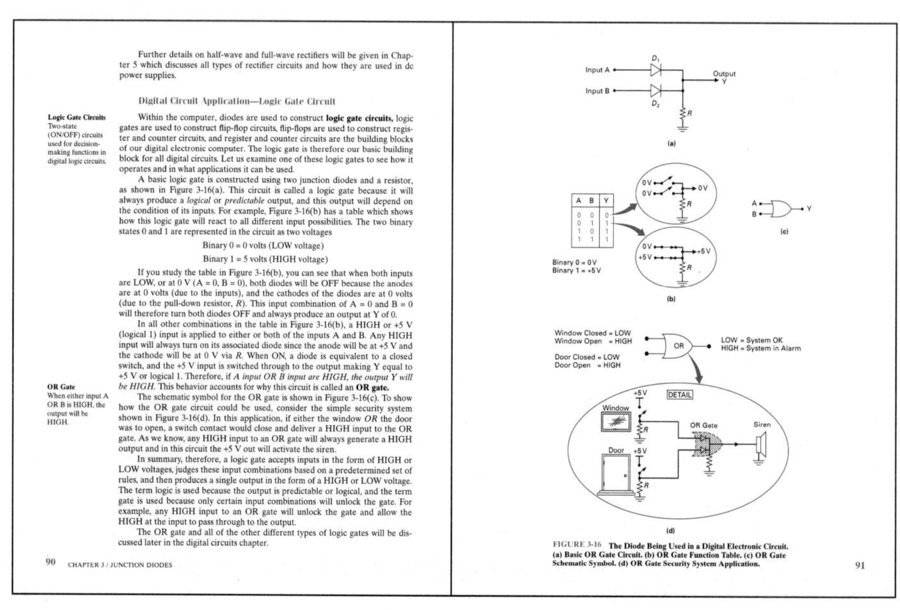

Further details on half-wave and full-wave rectifiers will be given in Chapter 5 which discusses all types of rectifier circuits and how they are used in dc power supplies.

Digital Circuit Application—Logic Gate Circuit

Logic Gate Circuits Two-state (ON/OFF) circuits used for decision-making functions in digital logic circuits.

Within the computer, diodes are used to construct **logic gate circuits,** logic gates are used to construct flip-flop circuits, flip-flops are used to construct register and counter circuits, and register and counter circuits are the building blocks of our digital electronic computer. The logic gate is therefore our basic building block for all digital circuits. Let us examine one of these logic gates to see how it operates and in what applications it can be used.

A basic logic gate is constructed using two junction diodes and a resistor, as shown in Figure 3-16(a). This circuit is called a logic gate because it will always produce a *logical* or *predictable* output, and this output will depend on the condition of its inputs. For example, Figure 3-16(b) has a table which shows how this logic gate will react to all different input possibilities. The two binary states 0 and 1 are represented in the circuit as two voltages

Binary 0 = 0 volts (LOW voltage)

Binary 1 = 5 volts (HIGH voltage)

If you study the table in Figure 3-16(b), you can see that when both inputs are LOW, or at 0 V (A = 0, B = 0), both diodes will be OFF because the anodes are at 0 volts (due to the inputs), and the cathodes of the diodes are at 0 volts (due to the pull-down resistor, *R*). This input combination of A = 0 and B = 0 will therefore turn both diodes OFF and always produce an output at Y of 0.

In all other combinations in the table in Figure 3-16(b), a HIGH or +5 V (logical 1) input is applied to either or both of the inputs A and B. Any HIGH input will always turn on its associated diode since the anode will be at +5 V and the cathode will be at 0 V via *R*. When ON, a diode is equivalent to a closed switch, and the +5 V input is switched through to the output making Y equal to +5 V or logical 1. Therefore, if *A input OR B input are HIGH, the output Y will be HIGH.* This behavior accounts for why this circuit is called an **OR gate.**

OR Gate When either input A OR B is HIGH, the output will be HIGH.

The schematic symbol for the OR gate is shown in Figure 3-16(c). To show how the OR gate circuit could be used, consider the simple security system shown in Figure 3-16(d). In this application, if either the window *OR* the door was to open, a switch contact would close and deliver a HIGH input to the OR gate. As we know, any HIGH input to an OR gate will always generate a HIGH output and in this circuit the +5 V out will activate the siren.

In summary, therefore, a logic gate accepts inputs in the form of HIGH or LOW voltages, judges these input combinations based on a predetermined set of rules, and then produces a single output in the form of a HIGH or LOW voltage. The term logic is used because the output is predictable or logical, and the term gate is used because only certain input combinations will unlock the gate. For example, any HIGH input to an OR gate will unlock the gate and allow the HIGH at the input to pass through to the output.

The OR gate and all of the other different types of logic gates will be discussed later in the digital circuits chapter.

90 CHAPTER 3 / JUNCTION DIODES

FIGURE 3-16 **The Diode Being Used in a Digital Electronic Circuit. (a) Basic OR Gate Circuit. (b) OR Gate Function Table. (c) OR Gate Schematic Symbol. (d) OR Gate Security System Application.**

91

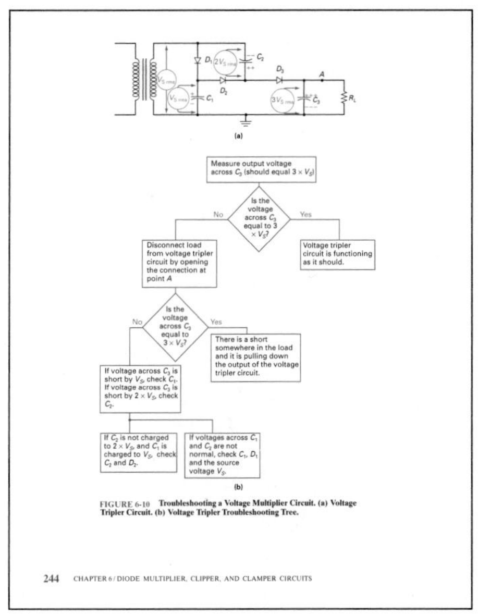

FIGURE 6-10 **Troubleshooting a Voltage Multiplier Circuit. (a) Voltage Tripler Circuit. (b) Voltage Tripler Troubleshooting Tree.**

244 CHAPTER 6 / DIODE MULTIPLIER, CLIPPER, AND CLAMPER CIRCUITS

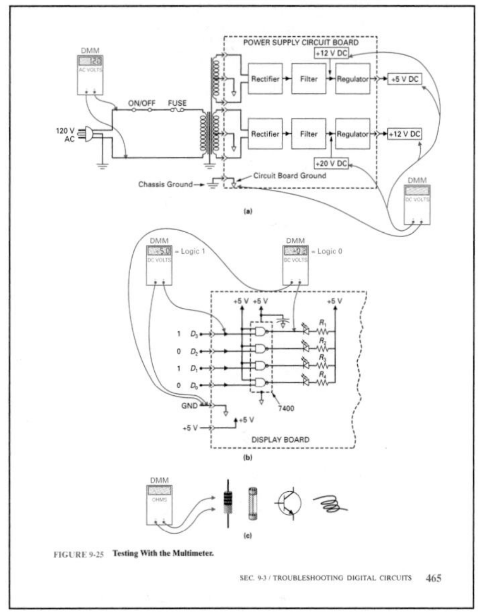

FIGURE 9-25 **Testing With the Multimeter.**

SEC. 9-3 / TROUBLESHOOTING DIGITAL CIRCUITS 465

A strong troubleshooting and testing emphasis prepares the student for the working world.

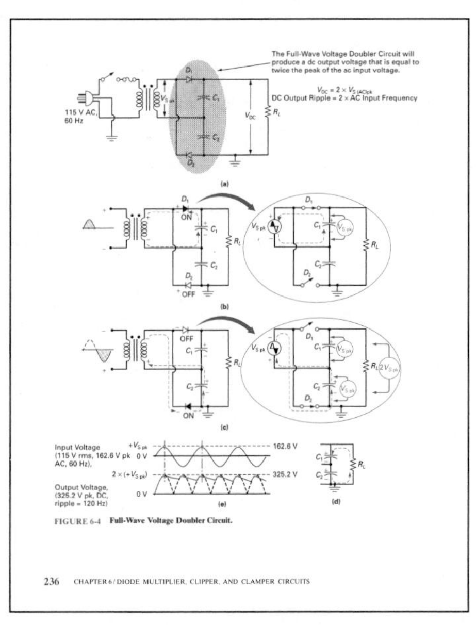

FIGURE 6-4 **Full-Wave Voltage Doubler Circuit.**

236 CHAPTER 6 / DIODE MULTIPLIER, CLIPPER, AND CLAMPER CIRCUITS

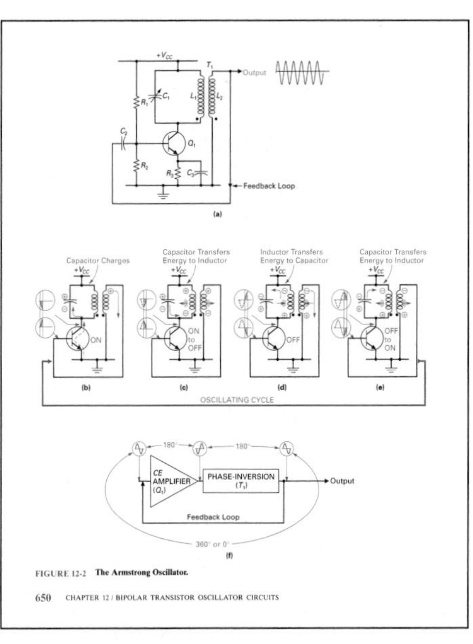

FIGURE 12-2 **The Armstrong Oscillator.**

650 CHAPTER 12 / BIPOLAR TRANSISTOR OSCILLATOR CIRCUITS

Unique action diagrams bring the operation of a circuit to life by showing its various operating states.

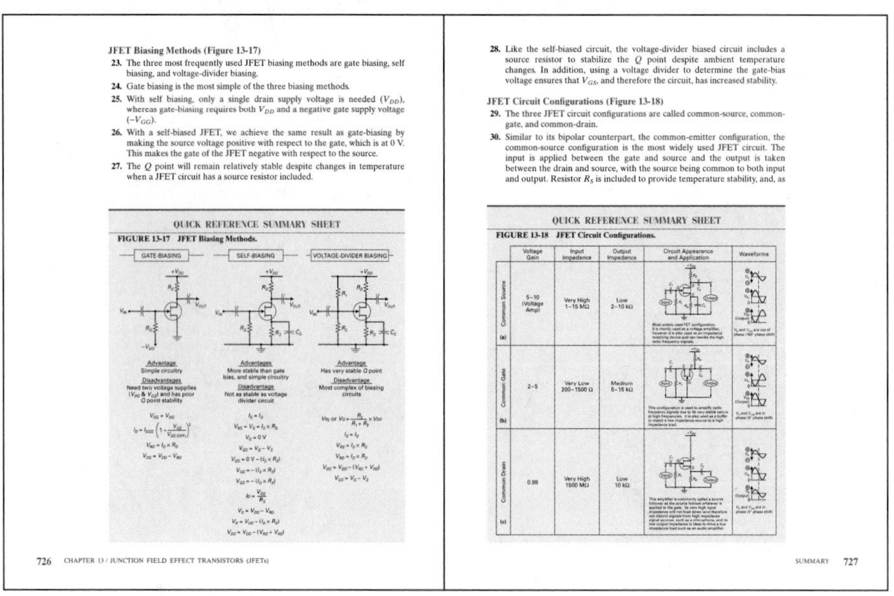

JFET Biasing Methods (Figure 13-17)

23. The three most frequently used JFET biasing methods are gate biasing, self biasing, and voltage-divider biasing.
24. Gate biasing is the most simple of the three biasing methods.
25. With self biasing, only a single drain supply voltage is needed (V_{DD}), whereas gate-biasing requires both V_{DD} and a negative gate supply voltage ($-V_{GG}$).
26. With a self-biased JFET, we achieve the same result as gate-biasing by making the source voltage positive with respect to the gate, which is at 0 V. This makes the gate of the JFET negative with respect to the source.
27. The Q point will remain relatively stable despite changes in temperature when a JFET circuit has a source resistor included.

QUICK REFERENCE SUMMARY SHEET

FIGURE 13-17 JFET Biasing Methods.

GATE-BIASING
SELF-BIASING
VOLTAGE-DIVIDER BIASING
Advantage
Simple circuitry
Disadvantages
Need two voltage supplies (V_{DD} & V_{GG}) and has poor Q point stability
Advantages
More stable than gate bias, and simple circuitry
Disadvantage
Not as stable as voltage divider circuit
Advantage
Has very stable Q point
Disadvantage
Most complex of biasing circuits

726 CHAPTER 13 / JUNCTION FIELD EFFECT TRANSISTORS (JFETs)

28. Like the self-biased circuit, the voltage-divider biased circuit includes a source resistor to stabilize the Q point despite ambient temperature changes. In addition, using a voltage divider to determine the gate-bias voltage ensures that V_{GS}, and therefore the circuit, has increased stability.

JFET Circuit Configurations (Figure 13-18)

29. The three JFET circuit configurations are called common-source, common-gate, and common-drain.
30. Similar to its bipolar counterpart, the common-emitter configuration, the common-source configuration is the most widely used JFET circuit. The input is applied between the gate and source and the output is taken between the drain and source, with the source being common to both input and output. Resistor R_S is included to provide temperature stability, and, as

QUICK REFERENCE SUMMARY SHEET

FIGURE 13-18 JFET Circuit Configurations.

	Voltage Gain	Input Impedance	Output Impedance	Circuit Appearance and Application	Waveforms
Common Source (a)	5–10 (Voltage Amp)	Very High 1–15 MΩ	Low 2–10 kΩ		
Common Gate (b)	2–5	Very Low 200–1500 Ω	Medium 5–15 kΩ		
Common Drain (c)	0.98	Very High 1500 MΩ	Low 10 kΩ		

SUMMARY 727

Quick Reference Summary Sheets within the end-of-chapter summary provide a visual reference of all key facts, symbols, circuits, and formulas.

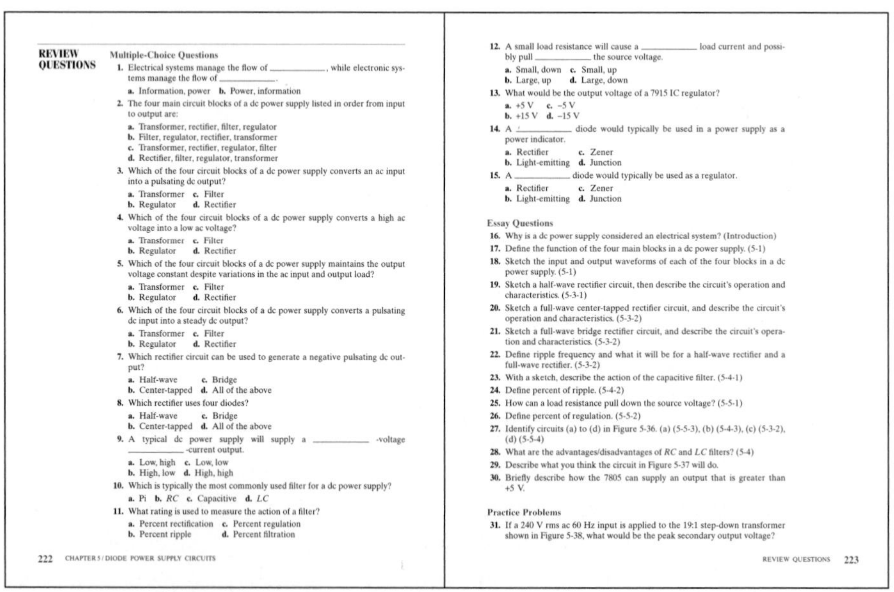

REVIEW QUESTIONS

Multiple-Choice Questions

1. Electrical systems manage the flow of ________, while electronic systems manage the flow of ________.
   a. Information, power b. Power, information
2. The four main circuit blocks of a dc power supply listed in order from input to output are:
   a. Transformer, rectifier, filter, regulator
   b. Filter, regulator, rectifier, transformer
   c. Transformer, rectifier, regulator, filter
   d. Rectifier, filter, regulator, transformer
3. Which of the four circuit blocks of a dc power supply converts an ac input into a pulsating dc output?
   a. Transformer c. Filter
   b. Regulator d. Rectifier
4. Which of the four circuit blocks of a dc power supply converts a high ac voltage into a low ac voltage?
   a. Transformer c. Filter
   b. Regulator d. Rectifier
5. Which of the four circuit blocks of a dc power supply maintains the output voltage constant despite variations in the ac input and output load?
   a. Transformer c. Filter
   b. Regulator d. Rectifier
6. Which of the four circuit blocks of a dc power supply converts a pulsating dc input into a steady dc output?
   a. Transformer c. Filter
   b. Regulator d. Rectifier
7. Which rectifier circuit can be used to generate a negative pulsating dc output?
   a. Half-wave c. Bridge
   b. Center-tapped d. All of the above
8. Which rectifier uses four diodes?
   a. Half-wave c. Bridge
   b. Center-tapped d. All of the above
9. A typical dc power supply will supply a ________-voltage ________-current output.
   a. Low, high c. Low, low
   b. High, low d. High, high
10. Which is typically the most commonly used filter for a dc power supply?
    a. Pi b. RC c. Capacitive d. LC
11. What rating is used to measure the action of a filter?
    a. Percent rectification c. Percent regulation
    b. Percent ripple d. Percent filtration

222 CHAPTER 5 / DIODE POWER SUPPLY CIRCUITS

12. A small load resistance will cause a ________ load current and possibly pull ________ the source voltage.
    a. Small, down c. Small, up
    b. Large, up d. Large, down
13. What would be the output voltage of a 7915 IC regulator?
    a. +5 V c. −5 V
    b. +15 V d. −15 V
14. A ________ diode would typically be used in a power supply as a power indicator.
    a. Rectifier c. Zener
    b. Light-emitting d. Junction
15. A ________ diode would typically be used as a regulator.
    a. Rectifier c. Zener
    b. Light-emitting d. Junction

Essay Questions

16. Why is a dc power supply considered an electrical system? (Introduction)
17. Define the function of the four main blocks in a dc power supply. (5-1)
18. Sketch the input and output waveforms of each of the four blocks in a dc power supply. (5-1)
19. Sketch a half-wave rectifier circuit, then describe the circuit's operation and characteristics. (5-3-1)
20. Sketch a full-wave center-tapped rectifier circuit, and describe the circuit's operation and characteristics. (5-3-2)
21. Sketch a full-wave bridge rectifier circuit, and describe the circuit's operation and characteristics. (5-3-2)
22. Define ripple frequency and what it will be for a half-wave rectifier and a full-wave rectifier. (5-3-2)
23. With a sketch, describe the action of the capacitive filter. (5-4-1)
24. Define percent of ripple. (5-4-2)
25. How can a load resistance pull down the source voltage? (5-5-1)
26. Define percent of regulation. (5-5-2)
27. Identify circuits (a) to (d) in Figure 5-36. (a) (5-5-3), (b) (5-4-3), (c) (5-3-2), (d) (5-5-4)
28. What are the advantages/disadvantages of RC and LC filters? (5-4)
29. Describe what you think the circuit in Figure 5-37 will do.
30. Briefly describe how the 7805 can supply an output that is greater than +5 V.

Practice Problems

31. If a 240 V rms ac 60 Hz input is applied to the 19:1 step-down transformer shown in Figure 5-38, what would be the peak secondary output voltage?

REVIEW QUESTIONS 223

An extensive end-of-chapter test bank, as illustrated here and on the following page, tests the student's understanding with multiple-choice, essay, and practice problems and troubleshooting questions.

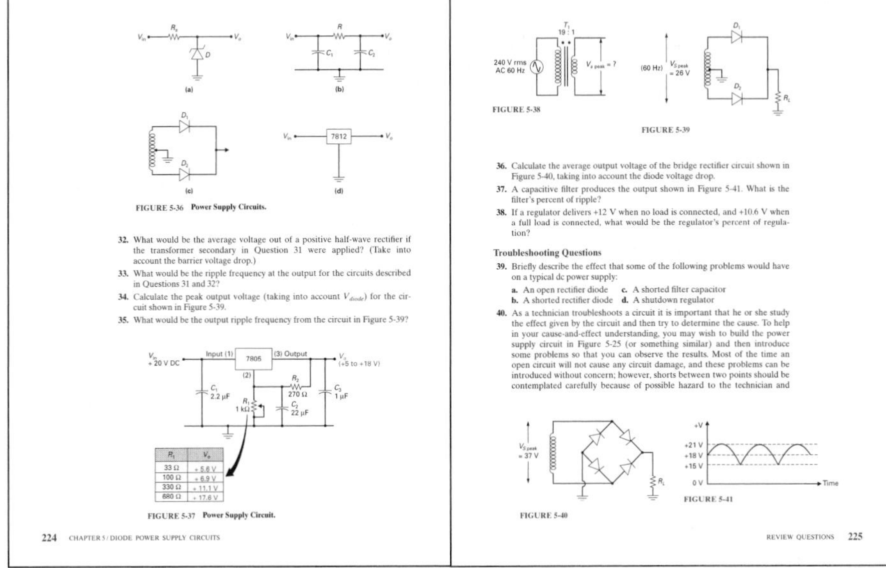

FIGURE 5-36 **Power Supply Circuits.**

32. What would be the average voltage out of a positive half-wave rectifier if the transformer secondary in Question 31 were applied? (Take into account the barrier voltage drop.)

33. What would be the ripple frequency at the output for the circuits described in Questions 31 and 32?

34. Calculate the peak output voltage (taking into account V_{diode}) for the circuit shown in Figure 5-39.

35. What would be the output ripple frequency from the circuit in Figure 5-39?

R_1	V_o
33 Ω	+5.6 V
100 Ω	+6.9 V
330 Ω	+11.1 V
680 Ω	+17.6 V

FIGURE 5-37 **Power Supply Circuit.**

224 CHAPTER 5 / DIODE POWER SUPPLY CIRCUITS

FIGURE 5-38

FIGURE 5-39

36. Calculate the average output voltage of the bridge rectifier circuit shown in Figure 5-40, taking into account the diode voltage drop.

37. A capacitive filter produces the output shown in Figure 5-41. What is the filter's percent of ripple?

38. If a regulator delivers +12 V when no load is connected, and +10.6 V when a full load is connected, what would be the regulator's percent of regulation?

Troubleshooting Questions

39. Briefly describe the effect that some of the following problems would have on a typical dc power supply:

a. An open rectifier diode **c.** A shorted filter capacitor
b. A shorted rectifier diode **d.** A shutdown regulator

40. As a technician troubleshoots a circuit it is important that he or she study the effect given by the circuit and then try to determine the cause. To help in your cause-and-effect understanding, you may wish to build the power supply circuit in Figure 5-25 (or something similar) and then introduce some problems so that you can observe the results. Most of the time an open circuit will not cause any circuit damage, and these problems can be introduced without concern; however, shorts between two points should be contemplated carefully because of possible hazard to the technician and

FIGURE 5-40

FIGURE 5-41

REVIEW QUESTIONS 225

TO THE INSTRUCTOR

Many people have asked me if there is one single factor involved with student retention. My answer is that whether a student enjoys electronics or hates it, and therefore stays in the program or leaves, is directly dependent on the instructor. I'm sure we have all seen the power we possess over a class of students. If this is used in a positive way, we can retain students despite any obstacles.

The responsibility of being the single most important factor in a student's education can be intimidating, unless we feel that we are fully supported. This support comes in the form of:

- a good textbook and set of teaching aids
- a good lab manual and set of test equipment
- a good course curriculum, and
- some teaching tricks of the trade

TEXTBOOK, LAB MANUAL, AND TEACHING AIDS

In all introductory texts, readability and relativity should be prime concerns to provide a complete understanding of the topics presented. However, in the pursuit of simplicity, boredom is often the price to be paid. Students are more involved in the learning process when they understand how the subject relates to their personal needs, career opportunities, and interests. Consequently, logical, current, and dynamic avenues of approach are needed to generate student enthusiasm, interest, and persistence.

This text has been specifically designed around the above philosophies, with a **complete picture introduction, real-life vignettes** at the beginning of each chapter, the use of **concept analogies** to take the student from the known to the unknown, a constant reference to **actual applications,** and a color insert that describes the many **career opportunities** available in the area of electronics.

The objective of this textbook, and its associated lab manual, is to provide a student with all of the theoretical and practical knowledge needed in the area of semiconductor electronics to become an electronics technician. By definition, **an electronics technician must be able to diagnose, isolate, and repair electrical and electronic circuit and system malfunctions.** In order to achieve this goal, he or she must have a thorough knowledge of electronics, test equipment, and troubleshooting techniques. This text has been designed specifically with this objective in mind and therefore contains **all aspects of semiconductor electronics** along with a strong emphasis on **practical applications, test equipment,** and **testing and troubleshooting techniques.**

Throughout this text I have also tried to provide to the student the basis for gaining *intelligence* along with an education. All too often education is mistaken for intelligence. Education is simply the acquisition of information. For example, just because I have acquired a first-rate set of woodworking tools does not make me a good carpenter. To be a good carpenter, I must first develop carpentry skills. There are many elements in that nebulous, indefinable quality we call "intelligence"; however, generally it falls into our understanding of **analogies,** which measure our ability to see relationships, and **mathematics, reasoning,** and **logic,** which measure our ability to think logically and make use of the facts we know. Therefore, instead of simply stating a formula and requiring a student to memorize it, which would be simply supplying an education, I have introduced the student to the quantities involved, compared them to something they know, and then studied the relationships to build up a formula in the same way as the pioneers. These analogies will develop a student's ability to see relationships while the study of relationships will develop good reasoning and logic skills, and therefore instill "intelligence." Although this takes a little more time in the early stages, the rewards and ease of understanding in the later stages will compensate for the initial effort.

With the lab manual, my good friend Gary Lancaster and I have tried to create **practical** experiments that inspire and challenge students. Our experiments were purposefully organized to work in conjunction with the textbook and translate all of its theoretical material into practical experience.

Along with the textbook and lab manual you will find a good set of teaching aids that will support your teaching effort.

ELECTRONIC TECHNICIAN TRAINING: INDUSTRY FEEDBACK

To be competitive, "electronic manufacturing companies" and "electronic servicing companies" must make a choice: either they compete with foreign nations on their terms with *lower wages, increased hours,* and *demoralized workers,* or they restructure the work organization and upgrade worker skills for a *high-skill, high-performance organization.* The high school graduate and experienced worker are therefore faced with the same simple choice: *high skills or low wages*. Since "high

TABLE 1 Areas of Training Most Useful for Job Performance

ELECTRONIC MANUFACTURERS	ELECTRONIC SERVICING
1. Troubleshooting Techniques	1. Customer Relations
2. Use of Test Equipment	2. Troubleshooting Techniques
3. Circuit Recognition and Analysis	3. Use of Test Equipment
4. Adaptability and Teamwork Skills	4. Adaptability and Teamwork Skills
5. Customer Relations	5. Circuit Recognition and Analysis

skills" would be the most obvious choice, the next question is, what skills should the electronics technician have to be competitive in today's and tomorrow's workforce? Posing this question to advisory board members and electronic panels from more than 40 electronic companies, the following feedback was obtained.

1. *Training must provide a core of knowledge for an electronic technician.* Training must produce an electronic technician. By definition, an electronic technician is a person having a complete understanding of the operation, application, and testing of electronic devices, circuits, and systems. Table 1 indicates in priority which training skills industry felt were most useful.

   Table 1 lists test equipment and troubleshooting equipment rate as high priorities. Since test equipment is a technician's tool of the trade, Table 2 lists, in order, the test equipment that industry felt should have the most emphasis.

2. *Technician must be trained to adapt to new technologies.* Training should supply "intelligence along with an education." By definition, education is the acquisition of information and an indication of the ability to make such acquisitions, and intelligence is our ability to think logically and make use of facts we know. This is dependent on our mathematical, logic, and reasoning skills, and our ability to continually learn and see relationships. Concentrating on math skills, logic and reasoning skills, learning skills, and relationship skills develops intelligence, which in turn develops better adaptability and organizational and problem solving skills. Success in these areas will ensure career advancement. To address each of these skills in a little more detail:

   a. *Math Skills:* As a tool, math should be integrated into appropriate position and, once taught, should be instantly applied so as to demonstrate need.

   b. *Logic and Reasoning Skills:* To give only an education, a student would be told to "memorize a formula." To give an education along with intel-

TABLE 2 Test Equipment Most Useful for Job Performance

1. Digital Multimeter	5. Power Supply
2. Oscilloscope	6. Logic Analyzer
3. Logic Probe and Pulser	7. Signal Generators
4. Automatic Test Equipment	

ligence, training should explain why certain quantities are included in a formula and why they are either proportional or inversely proportional. In short, work first on concept comprehension.

   c. *Learning Skills:* The ability to adapt to new technology is directly dependent on an employee's ability to study and learn independently. Reading comprehension is therefore a necessity. To encourage this skill, training must be conversational and interactive, not dry and cold.

   d. *Relationship Skills:* Use analogies to go from the known to the unknown. Learning, retention, and comprehension are enhanced when parallels are explored, and this will in turn develop organizational and relationship skills.

3. *Training must develop better communication, teamwork, and customer service skills.* Industry's drive for high quality and reduced cost has led to the integration of electronic system design, manufacture, and service. There is a need, therefore, for increased teamwork at all levels of personnel, and technicians must have good communication and interpersonal skills. From the standpoint of customer service skills, training should address how to work under pressure with customers on-site, dress professionally, and explain technical topics to nontechnical people. This can be achieved by having students constantly practice communicating their science (written and spoken).

4. *Training should educate students so that they have a better understanding of business practices, including manufacturing and production processes, and technical and business data.* Training should include tours of electronic companies and details on development procedures, job titles and responsibilities (technical and nontechnical), work ethics, environments, and so on.

5. *The electronic technician's role is evolving due to the complexity of electronic systems, the increase in computer-driven systems, and the economics of unit replacement versus component replacement.* To address these points:

   a. Since computer work stations are becoming an increasingly important tool, training must provide a complete understanding of computer applications in industry.

   b. As equipment becomes more complex, analytical and troubleshooting skills are critical for technicians. However, since yesterday's system is today's component, troubleshooting is moving away from a focus on discrete components to a subsystem and unit (functional) level of maintenance.

   c. Since nearly all design, manufacturing, and service equipment are programmable, the ratio of a technician's involvement in hardware versus software has changed:

      Late 1980s: 80 percent Hardware, 20 percent Software
      Present: 60 percent Hardware, 40 percent Software

   Training should therefore include an introduction to computers and their applications at an early stage in the program. Technicians must be "application software proficient," and, therefore, the following should be covered: DOS, Windows, UNIX, Word Processing, Databases, Spreadsheets, Circuit Simulation Software, Basic Programming (such as C), and Networks.

Most employers agreed that the electronic technician's role is continually evolving to keep pace with the ever-changing electronics industry. As educators, we are also having to continually adapt to the needs of industry so that our graduates have the skills for both career and advancement.

TEACHING TECHNIQUES

Over the past 15 years I have taught thousands of students all aspects of electronics. In this time I have acquired a considerable amount of feedback regarding students' needs and responses to their introduction to the science of electricity and electronics. In this section I would like to share some of my teaching techniques that have enabled me to obtain a high level of student retention.

1. Using concept analogies wherever possible to take the student from the known to the unknown.
2. Keeping the objective "to produce an electrical/electronics technician" in mind at all times, and therefore concentrating on delivering a complete knowledge of electricity, electronics, test equipment, and troubleshooting techniques.
3. Explaining all topics with analogies, reasoning, and logic first, and reinforcing the topic with mathematics second. This provides for giving the students intelligence along with an education.
4. Assuming no prior knowledge of electricity, electronics, or technical mathematics. Starting with the complete picture or world of electronics, and including mini-math reviews where they are needed.
5. Encouraging student enthusiasm, interest, and persistence by including real-life vignettes of entrepreneurs, actual applications, and a career opportunity tour.

In addition, there are a few other teaching methods that have yielded a very good response in my classes, and so I will mention them to you for your consideration.

6. Application Presentations: These 20-minute presentations, generally on the first day of each week, describe and demonstrate systems such as a Jacob's Ladder, laser, computer system, a variety of application software, robots, test equipment, and so on. These presentations seem to generate a lot of student enthusiasm since they demonstrate the exciting end objective of their studies.

   I have also turned the tables and had students give presentations on any electrical or electronic equipment. Although the topics seem to be limited to video games, remote controlled toys, and car stereo systems, the presentations have helped develop necessary student communication skills.

7. Schematic Descriptions and Troubleshooting: Once a certain knowledge of electricity, electronics, and test equipment has been mastered, I begin to

apply this knowledge to actual system schematics. To cover a circuit fully, we follow four steps.

a. First we study the purpose of a circuit, its inputs and outputs, and its schematic diagram and circuit description.
b. Second, we compare the schematic to the actual printed circuit board so that we can identify the components.
c. Next we study the troubleshooting guides or charts and analyze the process of diagnosing and isolating a problem.
d. The final step is for me to introduce a problem and then have a team of normally two students diagnose, isolate, and repair the problem.

8. First Approximations of a Subject: It was philosopher René Descartes who first stated that to solve any problem you should start with the simple and then proceed to the complex. For Descartes there were three approximations; the **first approximation** was the simplest, the **second approximation** contained more detail, and the **third approximation** was the complex. I have applied this method to my teaching of all topics of electricity and electronics and have noticed a dramatic improvement in student comprehension. To explain this in a little more detail, a first approximation is a general description of a subject in which the key points and purposes are outlined. Following this complete picture description, we step through the details of the chapter, which is a second approximation of the subject. In keeping with the main objective, only the first and second approximations of a subject are needed for a technician since they cover purpose, construction, symbol, operation, characteristics, applications, testing, and troubleshooting. The third approximation of a subject covers design and therefore is only covered by engineering students.

These first approximations give the student a view of the complete picture, instead of having to wait until we finish all of the pieces in the chapter and then trying to connect them all together.

Like me, you may wish to develop first approximation presentations for all topics, especially the more difficult to grasp subjects, such as series-parallel circuits, capacitance, inductance, transistors, and so on.

If you should have any other ideas that you have found in your experience would assist other instructors, please send them to me in care of Prentice Hall.

INSTRUCTOR MOTIVATION: "THE NUTTY PROFESSOR"

I have been teaching electronics for nearly 15 years. In that time, teaching has given me many memorable and rewarding experiences. In contrast, it has been the source of many frustrating moments which have nearly brought me to the brink of madness.

As you are probably already aware, our chosen profession will most often work us day and night, and most likely will not make us a fortune. The key question is, therefore, why do we do it? The answer is simple. It comes from the personal interactions between the instructor and his or her students. It comes in the form of many intangibles, such as the delight of bringing a drink of knowledge to those who thirst for it, the challenge of the unknown and searching for answers, the pride in turning confusion and bewilderment into understanding and comprehension, and from turning an unskilled person into a working professional. There are mutual benefits to be reaped just from being involved with the learning process. The exchange of knowledge develops for both of us a better understanding of the subject and strong communicative skills, producing a student who is effective at work and in society as a whole.

Teaching is extremely demanding and often very frustrating, but the rewards well compensate the effort. It is a profession seldom understood except by those who really enjoy being involved with life. This vocation may be a "Mad, Mad, Mad World," but it's *my* world.

ACKNOWLEDGMENTS

A special thank-you goes to the professional, creative, and friendly people at Prentice Hall, namely: my editor, David Garza; my marketing manager, Debbie Yarnell; my sales manager, Todd Rossell; my production editors, Sheryl Langner and Martha Beyerlein; and Judy Casillo, Maria Klimek, and Sylvia Huning.

A special thank-you goes to the husband and wife team of Steve and Marilyn Howe, who worked tirelessly and in a timely fashion to ensure a high level of quality and accuracy.

My appreciation and thanks are extended also to the following instructors who have reviewed and contributed greatly to the development of this textbook: Thomas Armstrong, Texas A & M University; C. R. Collins; John Hamilton, Spartan School of Aeronautics; Richard E. Hoover, Owens Technical College; Stephen A. Howe, Wentworth Technical School; Raymond Klein; Oleh M. Kuritza, College of DuPage; Gary Lancaster, National Education Centers; Leo R. Majewski, Dunwoody Institute; and Marcus Rasco, DeVry Institute of Technology—Irving, Texas.

Nigel P. Cook

Contents

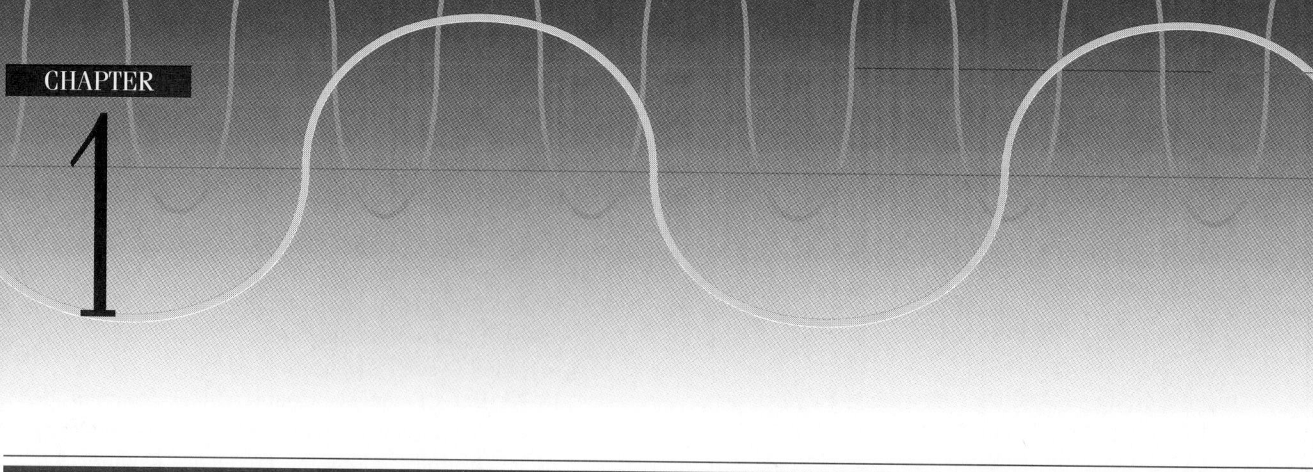

CHAPTER 1

OBJECTIVES

After completing this chapter, you will be able to:

1. Define the following terms:
   a. Electronics
   b. Electricity
   c. Electrical components, circuits, and systems
   d. Electronic components, circuits, and systems
   e. Analog electronic circuits
   f. Analog electrical circuits
   g. Digital electronic circuits
   h. Digital electrical circuits
   i. Linear circuits
   j. Two-state system
   k. Binary number system
2. Define the responsibilities of the following:
   a. The electronics technician
   b. The electronics engineer
   c. The electronics assembler
3. List several different types of electronic technicians and explain what part of the electronic product development process they are involved in.
4. Describe the steps involved in the development of an electronic system.
5. Describe the basic broad function, scope, responsibilities, and education requirements for each technician job classification.

Electricity and Electronics—Analog and Digital

OUTLINE

VIGNETTE

Communication Skills

Josiah Williard Gibbs was born in New Haven, Connecticut, in 1839. He graduated from Yale in 1858 and was appointed professor of mathematical physics at Yale in 1869, a position he held until his death in 1903. In his lifetime Gibbs wrote many important papers on the equilibrium of heterogeneous substances, the elements of vector analysis, and the electromagnetic theory of light.

Many historians rank Gibbs along with Newton and Einstein; however, he remains generally unknown to the public. This fact is due largely to his inability to communicate clearly and effectively. Strangely, for all of his technical genius, he just could not explain himself, a frustration that plagued him throughout his life. It took scientists years to comprehend what he was trying to explain, and as one scientist joked, "It was easier to rediscover Gibbs than to read him."

Note: This vignette carries a strong message. No matter how great a technical genius a person might become, if he or she cannot communicate this technical capability, they will never be understood or appreciated. In a recent survey of industry's needs for the technician of today and tomorrow, this point was addressed directly in the following way: "Training must develop better communication skills. Industry's drive for high quality and reduced cost has led to the *integration of design, manufacture, and service.* There is a need for increased teamwork at all levels of personnel." Technicians must have good communication and interpersonal skills. This can be achieved by constantly practicing communicating (in written and spoken form) your science.

INTRODUCTION

At this point you have reached a milestone in your education in electricity and electronics. You have completed your course in "Introductory DC/AC Electronics," and you are about to begin the next phase entitled "Introductory Semiconductor Devices and Circuits."

In this chapter, we will start by reviewing the complete picture of electricity and electronics, and then introduce what will be covered in the next phase. This chapter will also address careers in the electronics industry.

1-1 REVIEW OF ELECTRICITY AND ELECTRONICS

By definition, **electronics** is the branch of technology or science that deals with the use of *components* to control the flow of **electricity** in a vacuum, gas, liquid, semiconductor, conductor, or superconductor. Both electrical and electronic components, circuits, and systems control electron flow; however, their applications are distinctly different.

Electronics
Science related to the behavior of electrons in devices.

Electricity
Science that states that certain particles possess a force field, which with electrons is negative and with protons is positive.

To properly manage power, electrical devices must perform such functions as *generating, distributing,* and *converting* electrical power.

To properly manage information, electronic devices must perform such functions as *generating, sensing, storing, retrieving, amplifying, transmitting, receiving,* and *displaying* information.

Some systems are designed specifically to manage the flow of power and therefore are only electrical, while other systems are designed to manage both power and information. For example, a television contains both electrical components and circuits that manage the flow of electrical power from the wall outlet and also electronic components and circuits that manage the flow of information or TV signals from the antenna or cable. The electrical circuits are needed because they supply power to the electronic circuits which in turn manage the flow of audio (sound) and video (picture) information signals.

Electrical Components
Components, circuits, and systems that manage the flow of *power.*

Electronic Components
Components, circuits, and systems that manage the flow of *information.*

1-1-1 The Tree of Electricity and Electronics

Figure 1-1(a) illustrates the tree of electricity and electronics. Working from the bottom up, you can see that everything rests on the four basic electrical roots: voltage, current, resistance, and power. Electrical and electronic components were developed to generate and control these four basic electrical phenomena. When a group of components are interconnected, they form a circuit. Just as components are the building blocks for circuits, circuits are in turn the building blocks for electronic equipment or systems.

Figure 1-1(b) breaks up your electronics course into four basic steps, which correlate to the basic blocks shown in the tree. In your previous *Introductory DC/AC Electronics* course, you completed, in detail, all of Step 1 (Basics of Electricity) and Step 2 (DC/AC Devices). In this *Introductory Semiconductor Devices and Circuits* course, you will be covering the second half of Step 2 (Semiconductor Devices), and be introduced to Step 3 (Electrical and Electronic Circuits). These topics are covered in this text and are shown shaded in Figure 1-1(b).

After completing Steps 1, 2 and 3, you will have obtained a good knowledge of electrical and electronic devices and circuits and be ready to apply this knowledge to communication, data processing, consumer, industrial, test and measurement equipment, and biomedical system applications. These six different branches or classifications of electrical and electronic equipment are shown at the top of the tree in Figure 1-1(a) and listed under Step 4 in Figure 1-1(b).

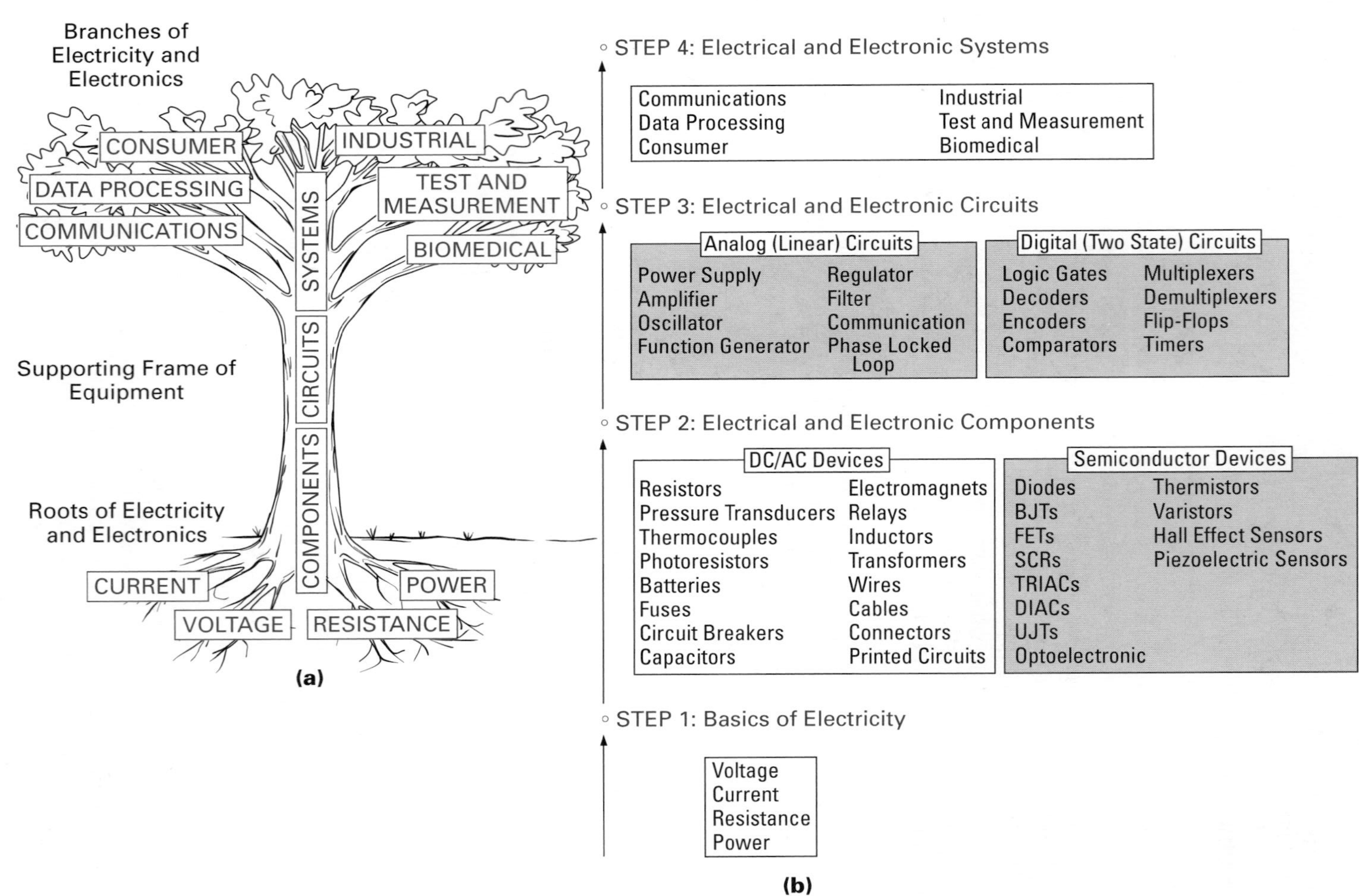

Branches of Electricity and Electronics
CONSUMER
DATA PROCESSING
COMMUNICATIONS
INDUSTRIAL
TEST AND MEASUREMENT
BIOMEDICAL
SYSTEMS
CIRCUITS
COMPONENTS
Supporting Frame of Equipment
Roots of Electricity and Electronics
CURRENT
VOLTAGE
RESISTANCE
POWER
(a)
STEP 4: Electrical and Electronic Systems
Communications
Data Processing
Consumer
Industrial
Test and Measurement
Biomedical
STEP 3: Electrical and Electronic Circuits
Analog (Linear) Circuits
Power Supply
Amplifier
Oscillator
Function Generator
Regulator
Filter
Communication
Phase Locked Loop
Digital (Two State) Circuits
Logic Gates
Decoders
Encoders
Comparators
Multiplexers
Demultiplexers
Flip-Flops
Timers
STEP 2: Electrical and Electronic Components
DC/AC Devices
Resistors
Pressure Transducers
Thermocouples
Photoresistors
Batteries
Fuses
Circuit Breakers
Capacitors
Electromagnets
Relays
Inductors
Transformers
Wires
Cables
Connectors
Printed Circuits
Semiconductor Devices
Diodes
BJTs
FETs
SCRs
TRIACs
DIACs
UJTs
Optoelectronic
Thermistors
Varistors
Hall Effect Sensors
Piezoelectric Sensors
STEP 1: Basics of Electricity
Voltage
Current
Resistance
Power
(b)

FIGURE 1-1 The Tree of Electricity and Electronics.

1-1-2 Analog and Digital

Analog and **digital** are two other terms that need to be further explained, since all of the electrical and electronic circuits listed in Step 3 of Figure 1-1(b) were classified in one of these two categories. To begin, let us examine **analog electronic circuits** and **analog electrical circuits.**

Analog
Relating to devices or circuits in which the output varies in direct proportion to the input.

Digital
Relating to devices or circuits that have outputs of only two distinct levels or steps, for example, on-off, 0-1, open-closed, and so on.

Analog Electronic Circuits
Electronic circuits that represent information in a continuously varying form.

Analog Electrical Circuits
A power circuit designed to control power in a continuously varying form.

Analog Signals, Components, and Circuits

Figure 1-2(a) shows an electronic circuit designed to amplify speech information detected by a microphone. One of the best ways to represent information electrically is to have a voltage change in direct proportion to the information it is representing. In the example in Figure 1-2(a), the *pitch and loudness* of the sound waves applied to the microphone should control the *frequency and amplitude* of the voltage signal from the microphone. Therefore, this output voltage signal is said to be an analog of the input speech signal. The word "analog" means "similar to," and in Figure 1-2(a) the electronic signal produced by the microphone is an analog (or similar) to the speech signal since a change in speech loudness or pitch will cause a corresponding change in voltage amplitude or frequency.

In Figure 1-2(b), a light detector or solar cell converts light energy into an electronic signal. This electronic signal represents the amount of light present because changes in voltage amplitude result in a change in light level intensity. Once again, the output electronic signal is an analog (or similar) to the light signal.

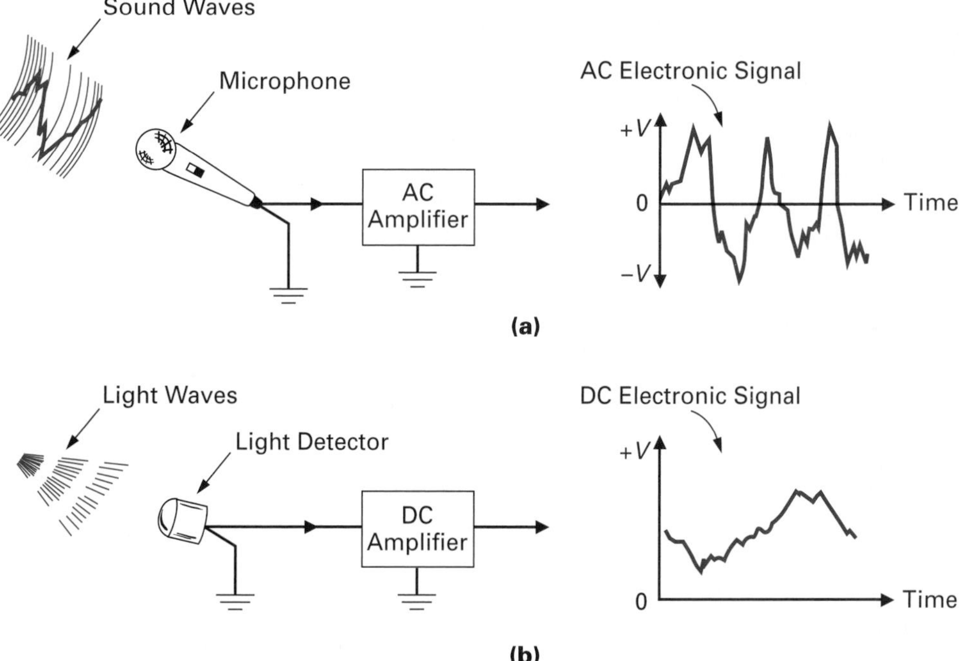

FIGURE 1-2 Analog Electronic Signals, Components, and Circuits. (a) AC Information Signal. (b) DC Information Signal.

Therefore, Figure 1-2 indicates two analog electronic (information) circuits. The microphone in Figure 1-2(a) generates an *AC analog signal* that is then amplified by an *AC amplifier circuit.* The microphone is considered an *analog device or component* and the amplifier is an *analog circuit.* The light detector would also be an analog component; however, in this example, it generates a *DC analog signal* that is then amplified by a *DC analog amplifier circuit.* Both of the electronic or information signals in Figure 1-2 vary smoothly and continuously, in accordance with the natural quantities they represent (sound and light).

Figure 1-3 shows a simple example of an *analog electrical (power) circuit.* The amount of current through the light bulb, and therefore the lamp's brightness, can be varied by the dimmer. The brightness of the bulb is an analog (or similar) to the angular position of the dimmer control. When used in this way, the light bulb and dimmer are called analog components, and this analog electrical circuit can be used to produce an infinite number of brightness levels.

Linear Circuits
A circuit in which the output varies in direct proportion to the input.

Analog circuits are often called **linear circuits** since a linear output, by definition, varies in direct proportion to the input. This linear circuit response is evident in both the analog electronic and electrical circuits shown in Figure 1-2 and 1-3. For example, refer to the graph in Figure 1-3 and you can see that the bulb's brightness (output) varies in direct proportion to the angular position of the dimmer (input).

Digital Signals, Components, and Circuits

Digital Electronic (Information) Circuits
Electronic circuits that encode information into a group of pulses consisting of HIGH or LOW voltages.

With **digital electronic (information) circuits,** information is first *encoded* into a group of pulses, such as the example seen in Figure 1-4(a). This code consists of a series of HIGH and LOW voltages, in which the HIGH voltages are called "1s" (ones) and the LOW voltages are called "0s" (zeroes). As an example, Figure 1-4(b) lists the "American Standard Code for Information Interchange" which is abbreviated ASCII (pronounced "askey"). Referring to Figure 1-4(a), you will notice that the "1101001" information or data stream code corresponds to the lower case "i" in the ASCII table shown highlighted in Figure 1-4(b). This means that whenever the lower-case "i" key on a computer keyboard is pressed, a "1101001" code is generated, thereby encoding the information "i" into a group of pulses.

The early digital systems made use of ten levels or voltages, and each of these voltages corresponded to one of the ten digits in the decimal number system

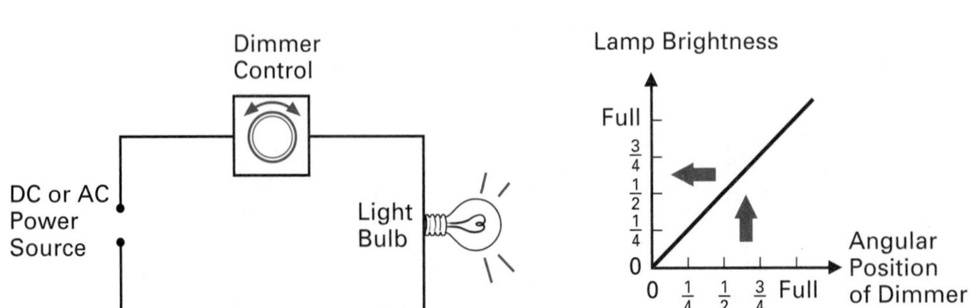

FIGURE 1-3 **Analog Electrical Components and Circuit.**

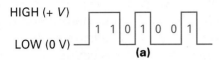

HIGH (+ V)

1 1 0 1 0 0 1

LOW (0 V)

(a)

The American Standard Code for Information Interchange (ASCII)

Character	Code	Character	Code
•	0 1 0 0 0 0 0	P	1 0 1 0 0 0 0
!	0 1 0 0 0 0 1	Q	1 0 1 0 0 0 1
“	0 1 0 0 0 1 0	R	1 0 1 0 0 1 0
#	0 1 0 0 0 1 1	S	1 0 1 0 0 1 1
$	0 1 0 0 1 0 0	T	1 0 1 0 1 0 0
%	0 1 0 0 1 0 1	U	1 0 1 0 1 0 1
&	0 1 0 0 1 1 0	V	1 0 1 0 1 1 0
‘	0 1 0 0 1 1 1	W	1 0 1 0 1 1 1
(	0 1 0 1 0 0 0	X	1 0 1 1 0 0 0
)	0 1 0 1 0 0 1	Y	1 0 1 1 0 0 1
*	0 1 0 1 0 1 0	Z	1 0 1 1 0 1 0
+	0 1 0 1 0 1 1	[	1 0 1 1 0 1 1
,	0 1 0 1 1 0 0	/	1 0 1 1 1 0 0
–	0 1 0 1 1 0 1	]	1 0 1 1 1 0 1
.	0 1 0 1 1 1 0	^	1 0 1 1 1 1 0
/	0 1 0 1 1 1 1	_	1 0 1 1 1 1 1
0	0 1 1 0 0 0 0	'	1 1 0 0 0 0 0
1	0 1 1 0 0 0 1	a	1 1 0 0 0 0 1
2	0 1 1 0 0 1 0	b	1 1 0 0 0 1 0
3	0 1 1 0 0 1 1	c	1 1 0 0 0 1 1
4	0 1 1 0 1 0 0	d	1 1 0 0 1 0 0
5	0 1 1 0 1 0 1	e	1 1 0 0 1 0 1
6	0 1 1 0 1 1 0	f	1 1 0 0 1 1 0
7	0 1 1 0 1 1 1	g	1 1 0 0 1 1 1
8	0 1 1 1 0 0 0	h	1 1 0 1 0 0 0
9	0 1 1 1 0 0 1	i	1 1 0 1 0 0 1
:	0 1 1 1 0 1 0	j	1 1 0 1 0 1 0
;	0 1 1 1 0 1 1	k	1 1 0 1 0 1 1
<	0 1 1 1 1 0 0	l	1 1 0 1 1 0 0
=	0 1 1 1 1 0 1	m	1 1 0 1 1 0 1
>	0 1 1 1 1 1 0	n	1 1 0 1 1 1 0
?	0 1 1 1 1 1 1	o	1 1 0 1 1 1 1
@	1 0 0 0 0 0 0	p	1 1 1 0 0 0 0
A	1 0 0 0 0 0 1	q	1 1 1 0 0 0 1
B	1 0 0 0 0 1 0	r	1 1 1 0 0 1 0
C	1 0 0 0 0 1 1	s	1 1 1 0 0 1 1
D	1 0 0 0 1 0 0	t	1 1 1 0 1 0 0
E	1 0 0 0 1 0 1	u	1 1 1 0 1 0 1
F	1 0 0 0 1 1 0	v	1 1 1 0 1 1 0
G	1 0 0 0 1 1 1	w	1 1 1 0 1 1 1
H	1 0 0 1 0 0 0	x	1 1 1 1 0 0 0
I	1 0 0 1 0 0 1	y	1 1 1 1 0 0 1
J	1 0 0 1 0 1 0	z	1 1 1 1 0 1 0
K	1 0 0 1 0 1 1	(	1 1 1 1 0 1 1
L	1 0 0 1 1 0 0	¦	1 1 1 1 1 0 0
M	1 0 0 1 1 0 1	)	1 1 1 1 1 0 1
N	1 0 0 1 1 1 0	~	1 1 1 1 1 1 0
O	1 0 0 1 1 1 1	DEL	1 1 1 1 1 1 1

(b)

FIGURE 1-4 Two-State Information. (a) Information or Data Code. (b) ASCII Codes.

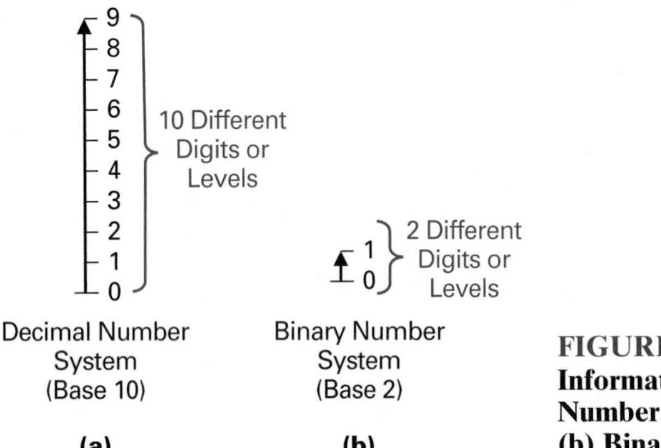

FIGURE 1-5 Representing Information Using the Digits of a Number System. (a) Decimal. (b) Binary.

(0 = 0 V, 1 = 1 V, 2 = 2 V, 3 = 3 V, up to 9 = 9 V). These circuits were very complex because they had to generate one of ten levels and sense the difference between all ten levels. This complexity led to inaccuracy because some circuits would periodically confuse one level for a different level. *The solution to the problem of circuit complexity and inaccuracy was solved by adopting a two-state system instead of a ten-state system.* Using a two-state system, you can generate codes for any number, letter, or symbol, as seen in Figure 1-4(b). Just as the ten-state system was based on the decimal (base 10) number system, the two-state system is based on the base 2 number system which is called **binary.** Figure 1-5(a) shows how the familiar decimal system has ten digits or levels labeled 0 through 9. The base 2 binary number system, shown in Figure 1-5(b), only makes use of two digits in the number scale, and therefore only the digits "0" and "1" exist in binary. These two states are typically represented as two different values of voltage (0 = LOW voltage, 1 = HIGH voltage). Using combinations of *binary digits* (abbreviated "bits"), we can represent information as a binary code or digital signal.

Binary
Number system having only two levels and used in digital electronics.

As an example of a digital electronic (information) circuit, Figure 1-6 illustrates how a switch and resistor could be used to generate the binary digits of an ASCII code. In this example, the two states of binary are represented as an electronic signal in which switch closed = 0 V (binary "0") and switch open = +5 V (binary "1").

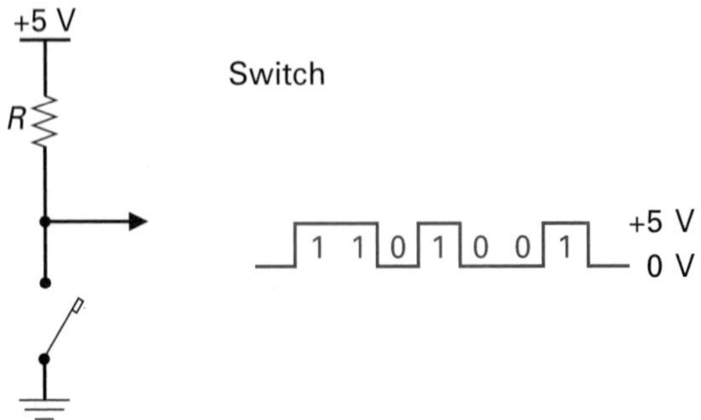

FIGURE 1-6 Digital Electronic (Information) Circuit.

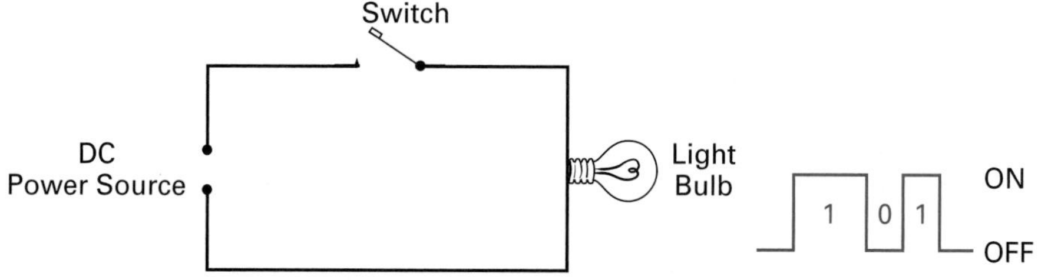

FIGURE 1-7 **Digital Electrical (Power) Circuit.**

In contrast, Figure 1-7 shows a simple **digital electrical (power) circuit.** In this circuit, the switch will turn the light bulb ON or OFF and generate one of two states: switch open = light bulb OFF (binary 0) or switch closed = light bulb ON (binary 1).

Digital Electrical (Power) Circuit
An electrical circuit designed to turn power either ON or OFF (two-state).

SELF-TEST REVIEW QUESTIONS FOR SECTION 1-1

1. ____________ is the branch of technology or science that deals with the use of components to control the flow of ____________ in a vacuum, gas, liquid, semiconductor, conductor, or superconductor.
2. If a temperature sensor generates a dc voltage that can be anywhere between 0 V and +15 V depending on the ambient heat, this signal would be considered a/an:
   a. Analog electronic signal **c.** Analog electrical signal
   b. Digital electronic signal **d.** Digital electrical signal
3. A house thermostat basically contains two temperature-controlled switches that either turn on HEAT or COOL depending on the temperature settings of both and the ambient temperature. The ON/OFF information from these temperature controlled switches is considered a/an:
   a. Analog electronic signal **b.** Digital electronic signal
4. Why are analog circuits also referred to as linear circuits?

1-2 CAREERS IN ELECTRONICS

Most electronic system manufacturers are normally quite happy to organize a tour of their facility to students studying electronics. As a motivational tool this exercise is invaluable; however, since most departments are toured on an "as you come across them" basis, confusion arises as to how all of the departments work together to design, manufacture, market, sell, and service a line of electronic products. In the photographic tour featured in this chapter's color insert, you will be introduced to the inner workings of a typical electronics company.

Before we begin the photo essay however let us first introduce the people that are directly involved with electronics. Figure 1-8 briefly defines the responsibilities of the *electronics technician, electronics engineer,* and *electronics assem-*

THE TECHNICIAN

The technician is an expert in the use of test equipment and troubleshooting techniques, and uses both to aid in the design, manufacture and service of a product. The technician is also able to repair or replace the isolated faulty component.

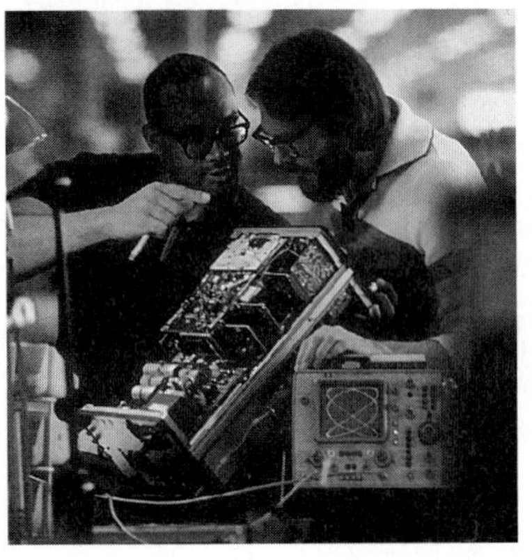

THE ENGINEER

The engineer creates the concepts and designs of all new electronic circuits and systems.

The engineer will work closely with engineering technicians in the prototyping of the design and the repair of any development problems.

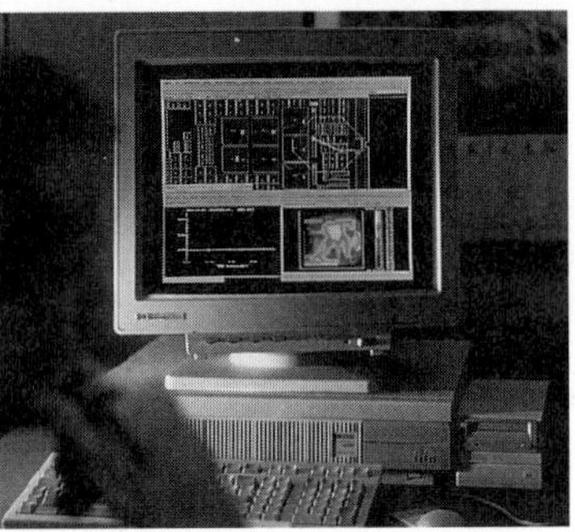

THE ASSEMBLER

The assembler is responsible for the manufacture of the final product. The training includes the correct component assembly and handling techniques.

FIGURE 1-8 The Technician, Engineer, and Assembler. (Photos courtesy of Hewlett-Packard Company.)

bler, all of which are experts in their particular field. In the following photographic essay you will be introduced to the different types of technicians, engineers, and assemblers along with all other people involved in the process.

1-2-1 Technician Types in Detail

The color photo essay demonstrates how electronic technicians play a vital role in the design, manufacture, and service of electronic systems. It also illustrates the broad range of career opportunities available in the electronics industry. Now that you understand the various job titles and responsibilities of the electronics technician, refer to the career opportunities section in your local newspaper to see what is available and appeals to you. Figure 1-9 shows some examples of career advertisements for technicians. Studying these advertisements you can see the function, scope, responsibilities, and education requirements typically needed for each type of technician.

Electronic Technicians in the Product Development Process

In the following 16 pages of this color insert you will be introduced to the many different types of "Electronic Technician." By definition, "An electronics technician must be able to diagnose, isolate, and repair electronic circuit and system malfunctions." In order to achieve this goal an electronic technician must have a thorough knowledge of electronics, test equipment and troubleshooting techniques.

In the basic block diagram below you can see the steps followed by a typical electronics company in the development of an electronic system. As you can see, these companies employ a variety of electronic technicians, all of which play a very important role in the development of an electronic system. In the following pages of this color insert we will examine these electronic technicians in more detail.

Electronic Technicians in the System Development Process

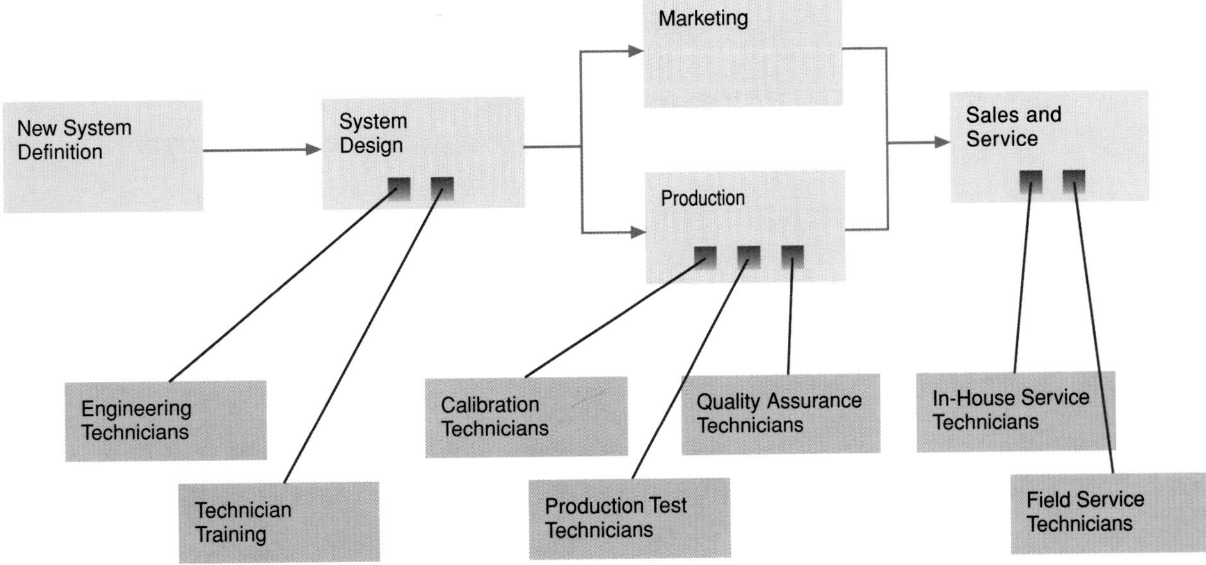

Development of an Electronic Product

The flow chart below expands on the basic block diagram on the previous page, showing the order from top to bottom, in which an electronic product is developed from conception to shipping. This photographic essay also shows the responsibilities of each person and how he or she fits into this process, giving you a clearer picture of the many career opportunities available.

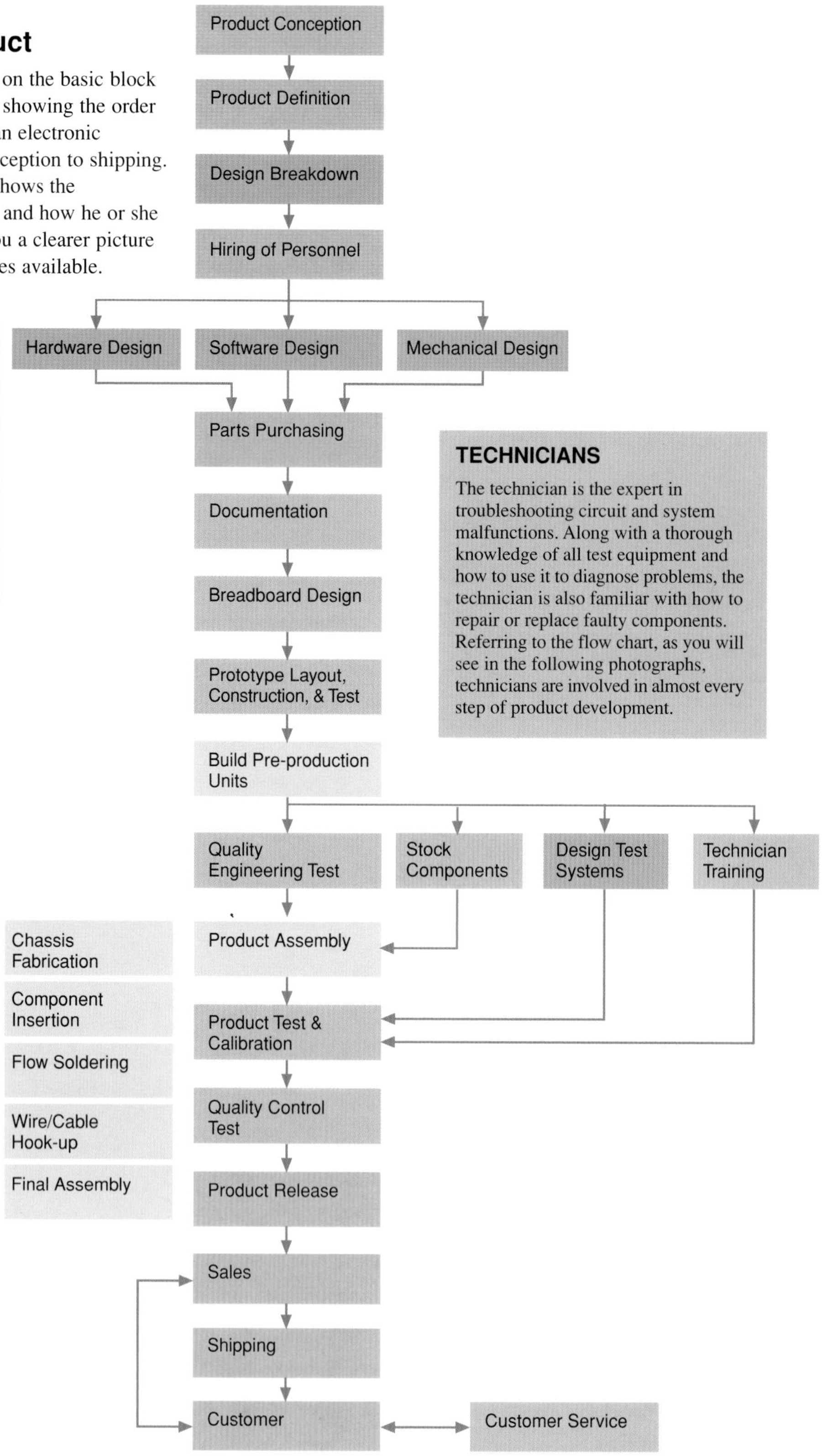

ENGINEERS

The engineer is responsible for designing the electronic system and supervising its development. During this process, the engineer relies heavily on the engineering technician to assist in the prototyping of the design and the diagnosis and repair of problems.

TECHNICIANS

The technician is the expert in troubleshooting circuit and system malfunctions. Along with a thorough knowledge of all test equipment and how to use it to diagnose problems, the technician is also familiar with how to repair or replace faulty components. Referring to the flow chart, as you will see in the following photographs, technicians are involved in almost every step of product development.

ASSEMBLERS

The assembler is responsible for putting the electronic product together and is trained in the correct component handling and assembly techniques.

The first step in the process can be seen in this photograph. The key company managers meet to define the new product. This new product will have to meet the needs of the ever-expanding electronics industry. During this meeting, the new product's budget, target dates, and key features will be outlined.

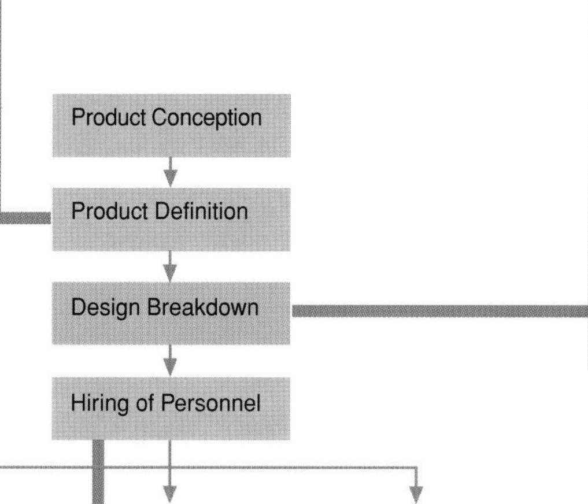

Once the product has been defined, it is the job of engineering to specify and design all major components of the new equipment. In this photograph, the design of the new product is being broken up into smaller tasks by an engineering manager and these task projects are being assigned to different engineers and engineering technicians.

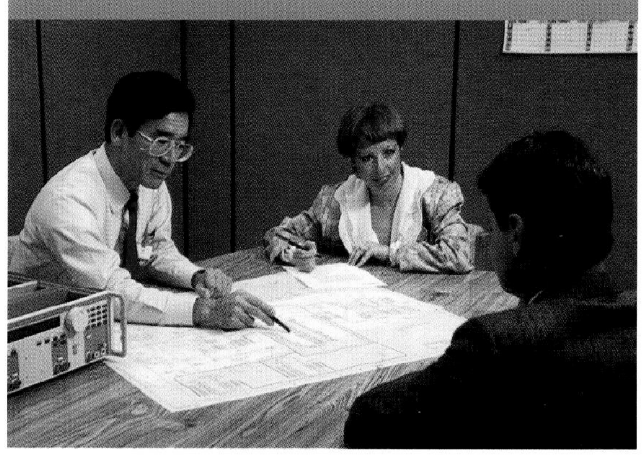

New personnel may be needed by the company to develop and produce this new product. Here, a prospective employee is receiving a technical and personnel interview for a technician position.

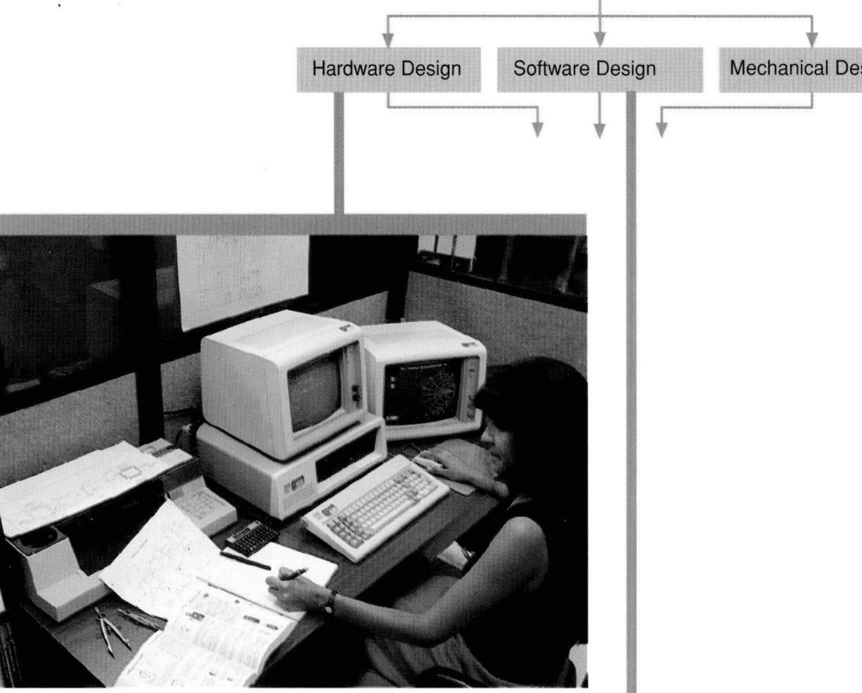

The hardware design engineer shown is entering the newly designed circuit, which consists of many interconnected electronic components, into a computer. The equipments hardware is the physical cables, components, and circuits you can see inside every piece of electronic equipment.

At the heart of many electronic products produced today is a computer controlling operation. The software design engineer is entering a list of instructions into the equipments computer memory, and it is this program of instructions that will control how the equipment operates.

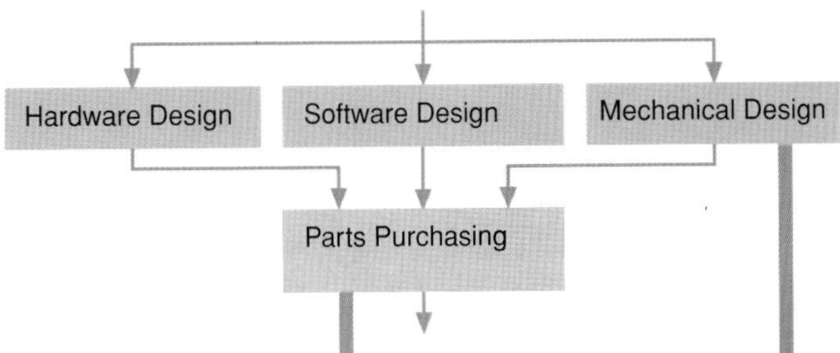

The software program resides within the equipment hardware, and both need to be encased inside a chassis or enclosure. In this photograph, a mechanical design engineer is designing the equipment housing and front panel.

Once the design is complete, the equipment parts have to be ordered so the product can be constructed. In this photograph, the engineering technician on the left is examining the variety of components available from the supplier, while the purchasing agent on the right is comparing costs.

Documentation

From the moment of conception to the final product, the unit is documented with mechanical drawings, schematic diagrams, and written instructions.

All documentation generated on a product is used to produce an operator's and maintenance manual that will accompany the finished unit when shipped. In this photograph, you can see a technical writer overseeing the drafting of some line art by an illustrator.

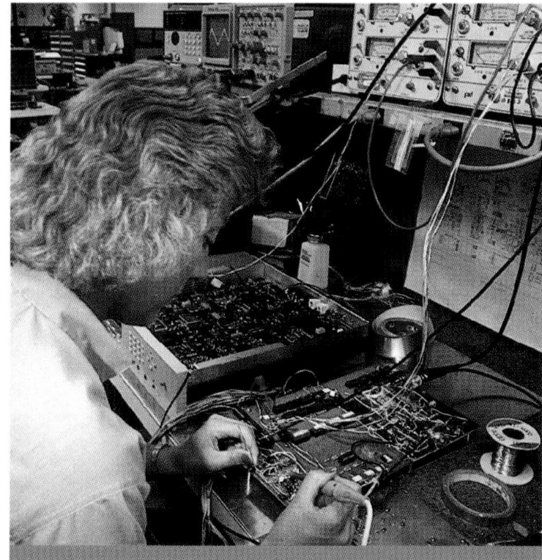

From sketches supplied by the engineers, a breadboard model of the design is constructed. The breadboard model is an experimental arrangement of a circuit in which the components are temporarily attached to a flat board. In this arrangement, the components can be tested to prove the feasibility of the circuit. A breadboard facilitates making easy changes when they are necessary. Here you can see an engineering technician breadboarding the design.

Breadboard Design

Prototype Layout, Construction, & Test

The engineer does not consider the final location of the components in constructing the breadboard model. At the next stage, however, a prototype working model completely representative of the final, mass-produced product is hand-assembled. The breadboard is replaced by a printed circuit board (PCB). In this scene, an engineering technician is producing a PCB layout from the design schematic diagrams.

The newly constructed prototype seen here is undergoing a complete evaluation of its mechanical and electrical form, design and performance.

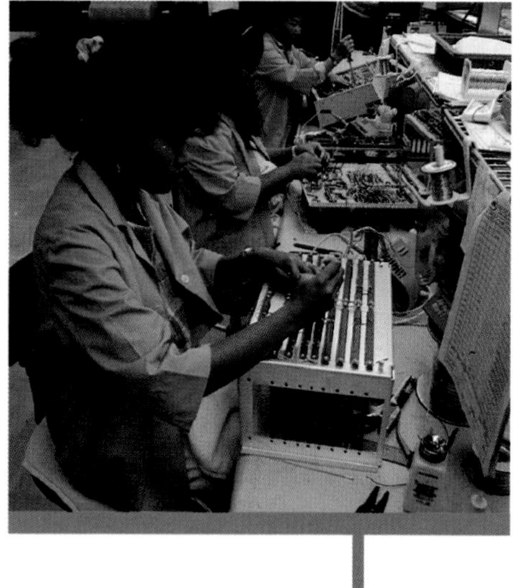

An approved prototype will initiate the production of several pre-production units, which can be seen in this photograph. The pathways these units take through construction will be mapped and used for mass assembly of the new product.

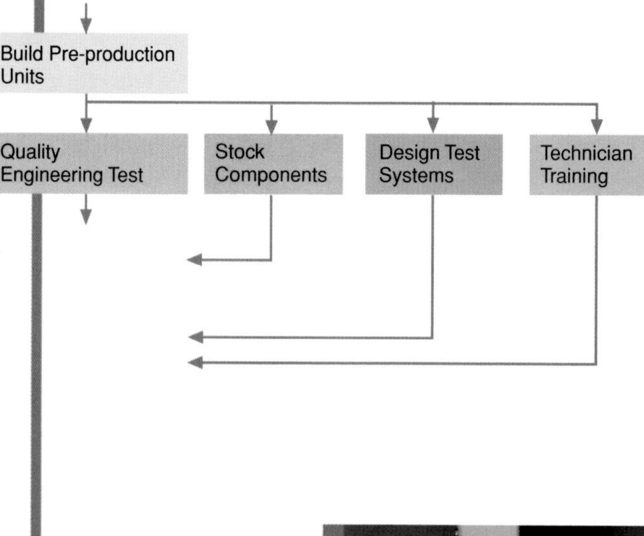

One of the pre-production units is taken through an extensive series of tests to determine whether it meets the standards listed. In this photograph, a quality assurance (QA) technician is evaluating the new product as it is put through an extensive series of tests.

The stockroom holds all of the raw materials needed to build products. In the foreground of this photograph, incoming components are being inspected and values verified, while in the background the parts kits needed for the many phases of assembly are being requisitioned.

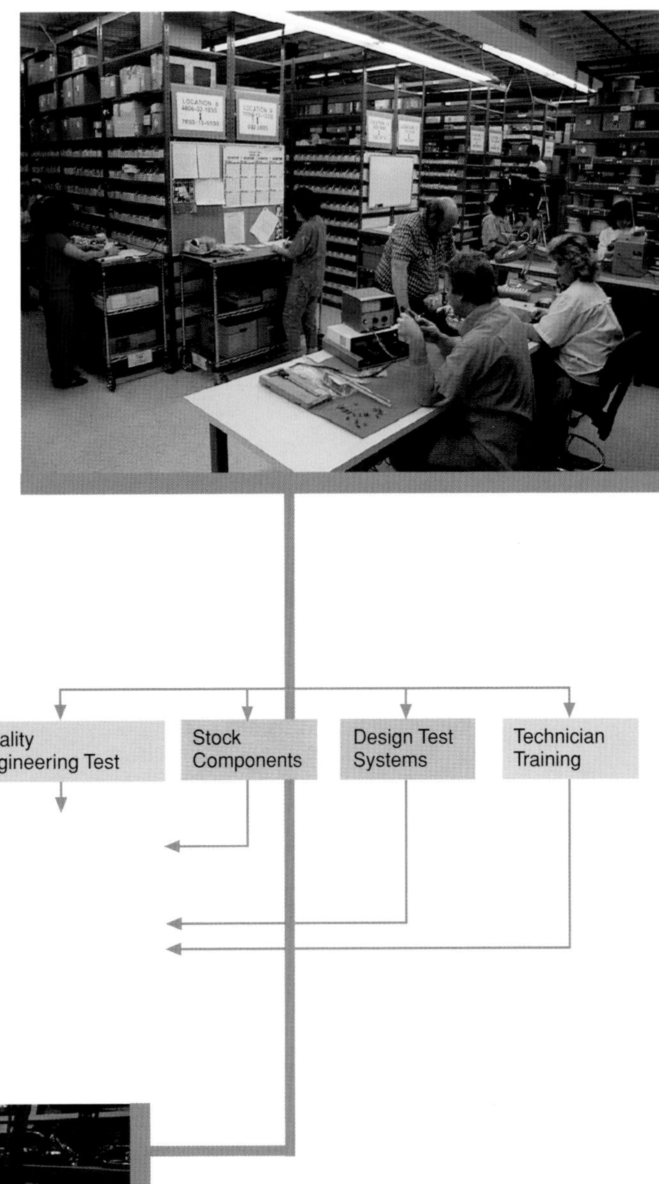

In this photograph, prepping of the many electronic components is taking place. In this process, wires, cables, and components are pre-cut to save time during the product assembly.

Once the assembled units come out of production, they will need to be tested. In this phase of the process, a test engineer is designing an automatic test system for the new product.

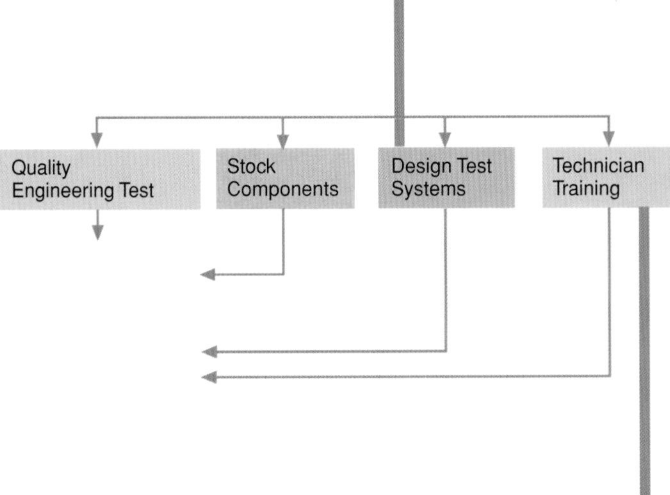

Training is a very important function. In this photograph, engineering technicians, production test technicians, quality control technicians, and customer service technicians are being taught the operation and component level maintenance of the new equipment.

The components which make up the unit are grouped into several areas where complex precision assembly takes place. This photograph of the machine shop shows the chassis or housing for the equipment being fabricated.

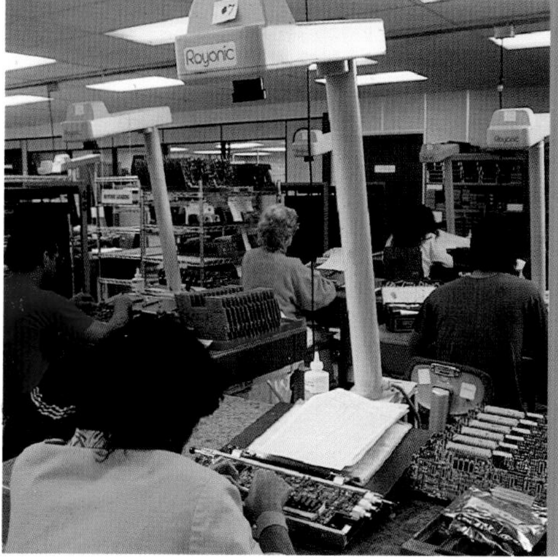

Product Assembly

- Chassis Fabrication
- Component Insertion
- Flow Soldering
- Wire/Cable Hook-up
- Final Assembly

In this photograph, all of the electronic components are being inserted into their respective positions within the printed circuit boards.

Once the boards have been filled with components, they are run through a flow- or wave-soldering machine. This machine solders the components to the board by moving the printed circuit board over a flowing wave of molten solder in a solder bath.

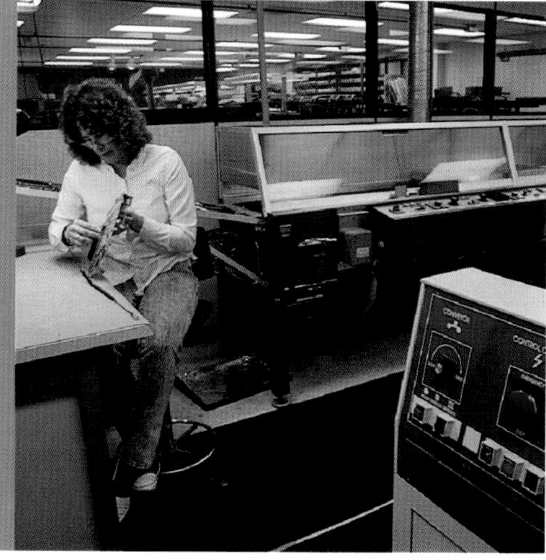

The wires and cables that interconnect all of the separate boards and units within the equipment are added at this stage.

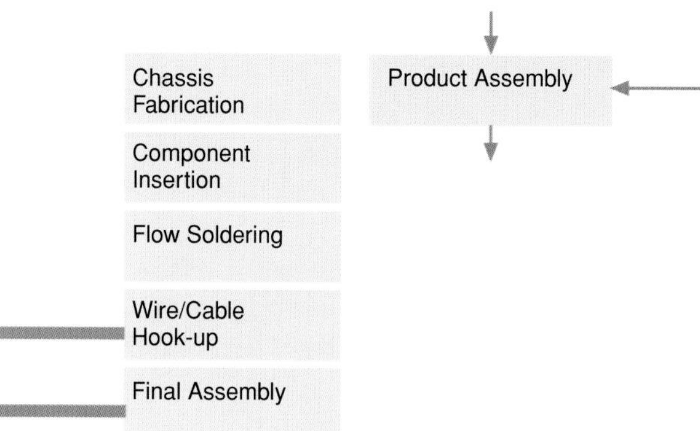

Final assembly of the product is taking place in this photograph. The equipment will have the remaining units inserted, and its front and rear panels will be mounted and connected.

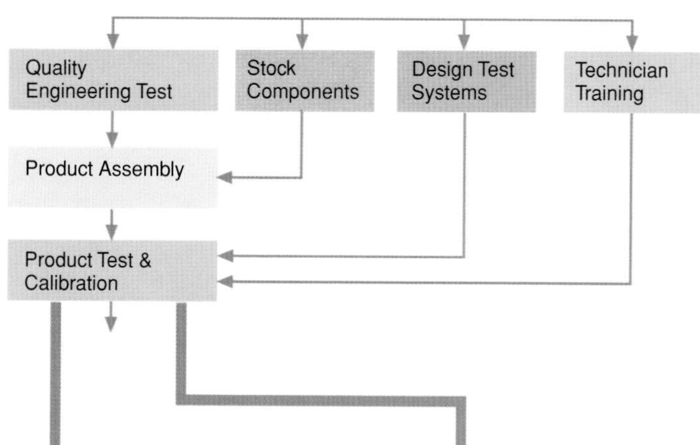

The more complex problems are handled by the production test technicians seen in this photograph. Once the system is fully operational, it is calibrated by a calibration technician.

After leaving assembly, the unit is hooked up to a test system and subjected to various testing procedures. The automatic test equipment (ATE) found at these stations perform many tests that would be too time-consuming for a technician to do manually.

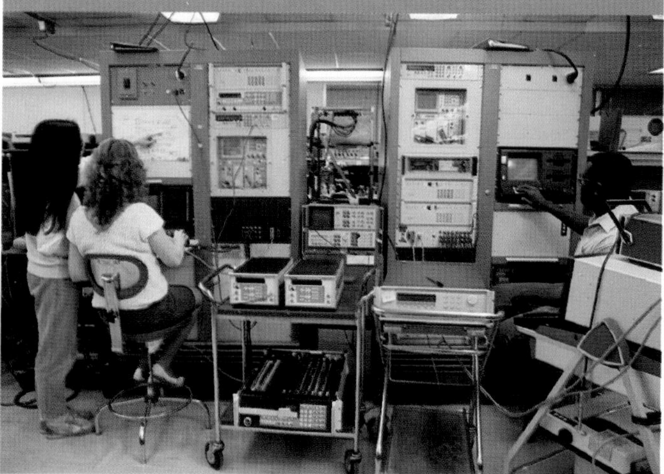

The quality control technicians seen in this photograph are performing the extensive series of electrical and mechanical final inspection tests. These will ensure that the performance standards listed in the unit's specifications are being met.

Quality Control Test

Product Release

From the time of product definition, the marketing personnel have been planning the advertising brochures and sales approach in preparation for the release of the product.

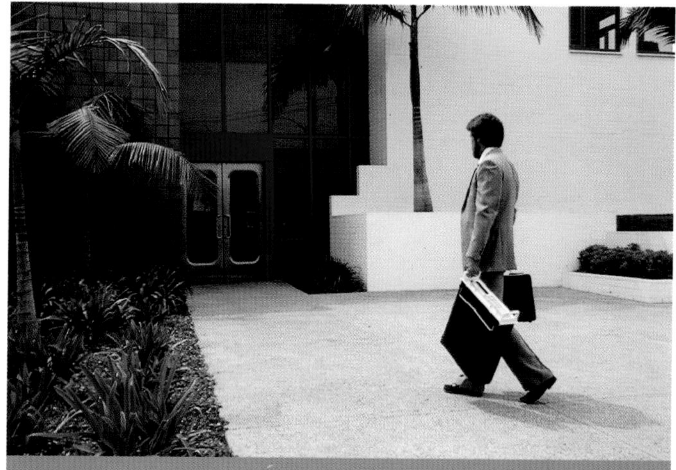

In this photograph, an applications engineer is making a sales call and will demonstrate to the customer the new product with all of its features and possible applications.

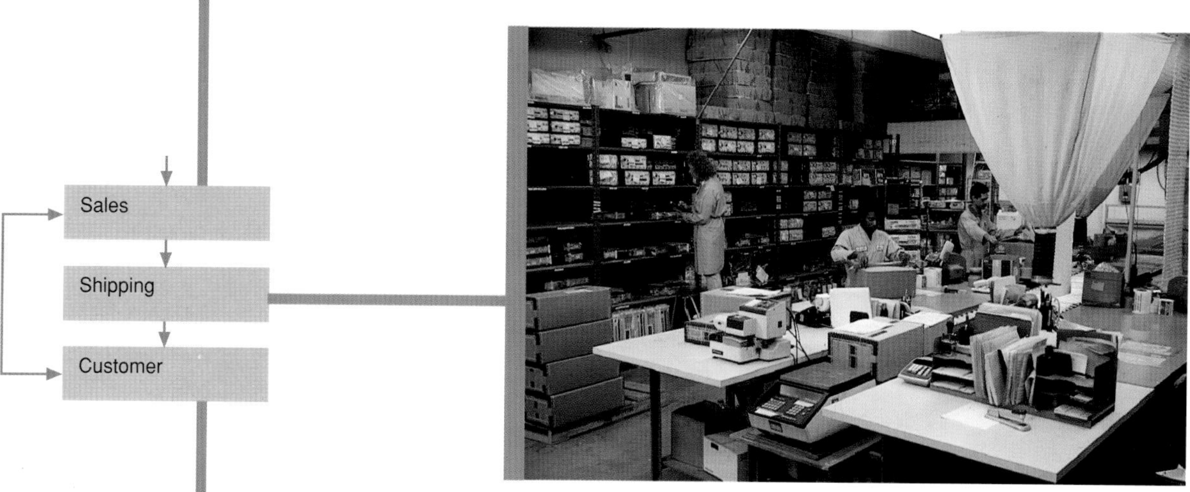

In the background of this photograph, you can see all of the equipment ready to be shipped. In the foreground, a delicate electronic unit is being carefully packaged to prevent damage during transit.

In this photograph, a customer is receiving instruction on the operation of the purchased unit.

Once the customer has received the electronic equipment, customer service provides assistance in maintenance and repair of the unit through direct in-house service or at service centers throughout the world. This photograph shows some in-house service technicians troubleshooting problems on returned units.

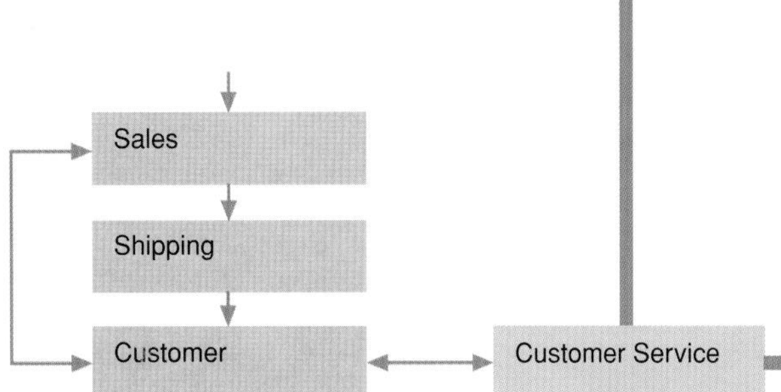

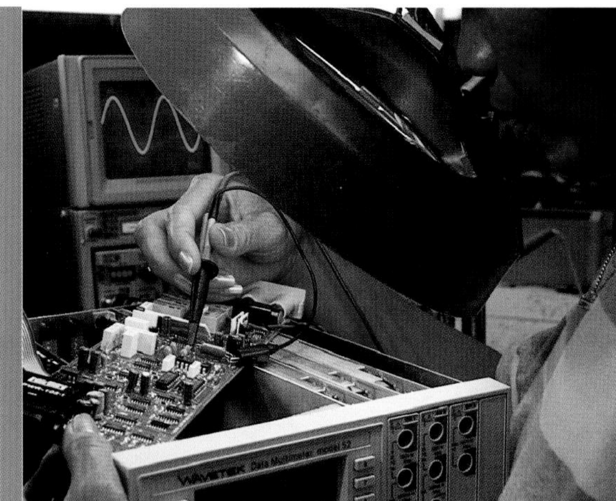

In this photograph, another in-house service technician is performing a test on a returned faulty unit containing the latest surface mount devices (SMD).

The field service technician seen here has been requested by the customer to make a service call on a malfunctioning unit that currently is under test.

▶▶▶▶▶▶▶▶▶▶▶▶▶▶▶▶▶▶▶▶▶▶▶ CAREERS AT ELECTROCORP 

Technician III

Working under general supervision, exercising moderate to complex decision-making, involving multiple cause and effect relationships, to identify trends and common problems.

May assist in new product development and the training of lower level technicians.

RESPONSIBILITIES:

Perform complex operational tests and fault isolation on state-of-the-art components, circuits, and systems for verification of product performance to well-defined specifications.

May perform moderately complex mathematical calculation to interpret test results, using judgment to align multifunctions within products to well-defined specifications.

May assist Test Engineering in generating test procedures. Set up moderate-to-complex test stations, utilizing varied test equipment, including some sophisticated equipment.

May assist Test Engineering in the identification of component/circuit trend failure and recommend alternative solutions for modification.

Perform standard assembly operations.

REQUIREMENTS:

AS/AAS Degree in Electronic Technology or equivalent plus 4–6 years directly related experience. Demonstrated experience working complex mathematical formulas and equations. Working knowledge of counters, scopes, analyzers, and related advanced test equipment.

Calibrator Tester

Working under general supervision, interfacing with test equipment in a system environment requiring limited decision-making.

RESPONSIBILITIES:

Work in an interactive mode with test station. Test, align, and calibrate products to defined specifications.

May set up own test stations and those of other operators.

Perform multiple alignments to get products to meet specifications.

May perform other manufacturing-related tasks as required.

REQUIREMENTS:

1–2 years experience with test and measurement equipment, experience with multiple alignment and calibration of assemblies including test station setups. Able to follow written instructions and write clearly.

Technician IV

Working under limited supervision, utilizing broad decision-making skills, providing technical assistance as required, and training of lower level technicians.

RESPONSIBILITIES:

Perform in-depth operational tests and fault isolation on state-of-the-art components, circuits, and systems for verification of product performance to generally defined specifications.

Perform highly complex mathematical calculations to interpret test results, using judgment to align multifunctions within products to generally defined specifications.

May assist Test Engineering in generating test procedures. Set up complex test stations, utilizing varied test equipment, including some sophisticated equipment.

Work with engineering to bring new products into production.

Participate in process development, circuit analysis, defining test procedures, and developing test stations.

Perform standard and prototype assembly operations.

REQUIREMENTS:

AS/AAS Degree in Electronic Technology or equivalent plus 7–10 years directly related experience. Thorough knowledge of advanced test equipment and calibration processes. Demonstrated experience with complex mathematical formulas and equations. Demonstrated written and oral communication skills.

Quality Assurance Technician I

Working under direct supervision, perform functional tests on completed instruments.

RESPONSIBILITIES:

Using established Acceptance Test Procedures, perform operational tests of all completed systems to ensure that all functional and electrical parameters are within specified limits.

Perform visual inspection of all completed systems for cleanliness and absence of cosmetic defects.

Reject all systems that do not meet specifications and/or established parameters of function and appearance.

Make appropriate notations on the system history sheet.

Maintain the QA Acceptance Log in accordance with current instructions.

Refer questionable characteristics to supervisor.

REQUIREMENTS:

AS/AAS degree in Electronics or equivalent including the use of test equipment. Must know color code and be able to distinguish between colors. Must have a working knowledge of related test equipment. Must know how to read and interpret drawings.

FIGURE 1-9 Career Opportunities for Electronic Technicians.

IC ELECTRONICS

Customer Service Technician III

Working under minimal supervision, perform all work assignments given by direct supervisor. Interface between the customer, marketing and/or engineering departments to solve technical problems.

RESPONSIBILITIES:

Utilizing all appropriate tools and test equipment to perform timely, quality repair and calibration of all systems returned by customers.

When appropriate, instruct customers in the proper method of calibration and repair of systems. Instruction may take place at the factory or at the customer's facility.

Aid marketing in solving customer application and technical problems via the telephone or at the customer's facility at marketing's discretion.

Evaluate manuals and other customer documents for errors or omissions.

Assist with training of Customer Service technicians.

Work with QA, Production, and Engineering in technically improving systems performance, serviceability, and reliability.

When appropriate, instruct customers and service technicians in proper soldering techniques meeting company standards.

REQUIREMENTS:

AS/AAS Degree in Electronics or equivalent plus 5–6 years experience troubleshooting analog and digital systems at least 1–3 years of which should be in a service environment. Must be able to read and understand flow charts, block diagrams, schematics, and truth tables. Must be able to operate and utilize test equipment such as oscilloscopes, counters, voltmeters, and analyzers. Must be able to use basic engineering principles to evaluate problem areas.

A demonstrated ability to effectively communicate and work with customers, and suggest alternative applications for product utilization is also required.

Engineering Tech III

Working under minimal supervision from Project Engineer, will be assigned to specific, complex project development of a major project.

RESPONSIBILITIES:

Test and document complete system performance.

Modify and improve circuit and total system performance with direction from Project Engineer.

Coordinate all necessary documents to complete the transfer of new product to manufacturing including calibration and troubleshooting procedures.

Work with the Project Engineer to coordinate and maintain time scheduling, documentation accuracy, and work assignments of Engineering Techs I and II.

Train and assist Calibration Technicians as required.

REQUIREMENTS:

AS/AAS Degree in Electronics or equivalent plus 5 years Industrial Technician. Ability and skill to complete prototypes with minimum supervision. Ability to train manufacturing personnel for instrument calibration.

Customer Service Technician II

Working under moderate supervision, perform all work assignments given by lead tech or direct supervisor. Work necessary overtime as assigned by supervisors.

RESPONSIBILITIES:

Utilizing all appropriate tools, troubleshoot and repair customer systems in a timely, quality manner to the general component level.

Working with basic test equipment, perform timely quality calibration of customer systems to specifications.

When appropriate, instruct others in proper soldering techniques meeting company standards.

Timely repair of QA rejects.

Aid marketing in solving customer problems via the telephone or at the customers' facility at marketing's discretion.

When appropriate, aid marketing with sales applications.

Help QA, Production, and Engineering in solving field problems.

Evaluate manuals and other customer documents for errors or omissions.

REQUIREMENTS:

AS/AAS Degree in Electronics or equivalent plus 4–5 years experience troubleshooting analog and/or digital systems, at least 6–12 months of which should be in a service environment. Must be able to read and understand flow charts, block diagrams, schematics, and truth tables.

FIGURE 1-9 (Continued)

QA Technician II

Working under general supervision, perform inspection and tests as delegated by departmental management.

RESPONSIBILITIES:

Using established Acceptance Test Procedures, perform operational tests of all completed systems to ensure that all functional and electrical parameters are within specified limits.

Perform visual inspection of all completed systems for cleanliness and absence of cosmetic defects.

Reject all systems that do not meet specifications and/or established parameters of function and appearance.

Make appropriate notations on the Instrument History Sheet.

Maintain the QA Acceptance Log in accordance with current instructions.

Interface with contract customer quality reps for verification of customer requirements.

REQUIREMENTS:

AS/AAS Degree in Electronics (or equivalent), plus 1–3 years experience in testing electronic equipment. Must know color code and be able to distinguish between colors. Must have a working knowledge of related test equipment. Must know how to read and interpret blueprints and drawings. Must be capable of making responsible judgmental decisions on minor abnormalities.

Engineering Tech II

Working under limited supervision from senior engineers, will be assigned to a specific, moderately simple, product development project.

RESPONSIBILITIES:

Build breadboard of electronic circuits designed by engineer(s).

Test and document each circuit performance.

Completion and calibration of simple prototype units.

Test and document system performance.

Aid Engineer(s) in a checkout of P.C. layout, assembly drawings, parts lists, calibration, and QA procedures to prepare QA audit package of the product.

Train and assist Calibration Technician(s) in pre-production phase.

REQUIREMENTS:

AS/AAS Degree in Electronics or equivalent plus 3–4 years experience, including two years industrial experience and 1–2 years experience as an Engineering Technician. Ability and skill to complete prototype and pre-ERN units with minimum supervision. Working knowledge of common electronic components in the specific areas required by the project.

Engineering Tech I

Working under close supervision, perform all work assignments as given by all levels of engineers.

RESPONSIBILITIES:

Breadboard electronic circuits from schematics.

Test, evaluate, and document circuits and system performance under the engineer's direction.

Check out, evaluate, and take data for the engineering prototypes including mechanical assembly of prototype circuits.

Help to generate and maintain preliminary engineering documentations. Assist engineers and senior Engineering Technicians in ERN documentation.

Maintain the working station equipment and tools in orderly fashion.

Support the engineers in all aspects of the development of new products.

REQUIREMENTS:

AS/AAS Degree in Electronics or equivalent plus 1–2 years technician experience. Ability to read color code, to solder properly, and to bond wires where the skill is required to complete breadboards and prototypes. Ability to use common machinery required to build prototype circuits. Working knowledge of common electronic components: TTL logic circuits, op-amps, capacitors, resistors, inductors, semiconductor devices.

Customer Service Tech I

Working under moderate supervision, perform all work assignments given by lead tech or direct supervisor. Work necessary overtime as assigned by supervisors.

RESPONSIBILITIES:

Utilizing all appropriate tools, troubleshoot and repair customer systems in a timely, quality manner, to the general component level.

Working with basic test equipment, perform timely quality calibration of customer systems to specifications.

Timely repair of QA rejects.

Solder and desolder components where appropriate, meeting company standards.

Aid marketing in solving customer problems via the telephone.

When appropriate, instruct customers in the proper methods of calibration and repair of products.

REQUIREMENTS:

AS/AAS Degree in Electronics or equivalent plus 2–3 years experience troubleshooting analog and/or digital systems at least 6–12 months of which should be in a service environment. Must be able to read and understand flow charts, block diagrams, schematics, and truth tables. Must be able to operate and utilize test equipment such as oscilloscopes, counters, voltmeters, and analyzers.

Ability to effectively communicate and work with customers.

FIGURE 1-9 (Continued)

DIGITAL DYNAMICS CORP.

Technician I

Working under close supervision, performing all work assignments, and exercising limited decision-making.

RESPONSIBILITIES:

Perform routine, simple operational tests and fault isolation on simple components, circuits, and systems for verification of product performance to well-defined specifications.

May perform standard assembly operations and simple alignment of electronic components and assemblies.

May set up simple test equipment to test performance of products to specifications.

REQUIREMENTS:

AS/AAS Degree in Electronic Technology or equivalent work experience.

Technician II

Working under moderate supervision, exercising general decision-making involving simple cause and effect relationships to identify trends and common problems.

RESPONSIBILITIES:

Perform moderately complex operational tests and fault isolation on components, circuits, and systems for verification of product performance to well-defined specifications.

Perform simple mathematical calculations to verify test measurements and product performance to well-defined specifications.

Set up general test stations, utilizing varied test equipment, including some sophisticated equipment.

Perform standard assembly operations.

REQUIREMENTS:

AS/AAS Degree in Electronic Technology or equivalent plus 2–4 years directly related experience. Demonstrated experience working mathematical formulas and equations. Working knowledge of counters, scopes, spectrum analyzers, and related industry standard test equipment.

Q.A. Technician III

Working under limited supervision, perform all duties and functions and Final Inspections. Leads are usually chosen from this level.

RESPONSIBILITIES:

Using established Acceptance Test Procedures, perform operational tests of all completed systems to ensure that all functional and electrical parameters are within specified limits.

Perform independent analysis of first articles to customer/company specifications.

Perform key product audits for design integrity by functional and physical configuration audit.

Perform visual inspection of all completed systems for cleanliness and absence of cosmetic defects.

Reject all systems that do not meet specifications and/or established parameters of function and appearance.

Make appropriate notations on the system history sheet.

Maintain the Final Inspection Log in accordance with current instructions.

Interface with contract customer quality reps for verification of customer requirements.

REQUIREMENTS:

ASS Degree in Electronics (or equivalent), plus 3–5 years experience in testing electronic equipment. Must have thorough knowledge of procedures, specifications, and test equipment, and the ability to identify and develop appropriate configuration of test equipment. Must be able to understand and utilize software programs for test purposes, and have thorough knowledge of mechanical parameters necessary for the assembly of manufactured products.

FIGURE 1-9 (Continued)

SELF-TEST REVIEW QUESTIONS FOR SECTION 1-2

1. Which one of these six different branches of electronics do you consider the most interesting?
2. Briefly describe in what capacity you would like to be involved in electronics.

SUMMARY

1. Electronics is the branch of technology or science that deals with the use of components to control the flow of electricity in a vacuum, gas, liquid, semiconductor, conductor, or superconductor.
2. Electrical components, circuits, and systems manage the flow of power. To properly manage power, these electrical devices must perform such functions as generating, distributing, and converting electrical power.
3. Electronic components, circuits, and systems manage the flow of information. To properly manage information, these electronic devices must perform such functions as generating, sensing, storing, retrieving, amplifying, transmitting, receiving, and displaying information.
4. Electronic circuits manage information, while electrical circuits manage power.
5. Semiconductor devices or components are generally the controlling element in electrical and electronic circuits.
6. The word "analog" means "similar to" and, in the case of analog electronic signals, information is represented as a voltage or current that changes in accordance with the quantities it represents.
7. Analog electronic signals and analog electrical power vary smoothly and continuously.
8. Analog circuits are often called linear circuits because their analog signal outputs generally vary in direct proportion to the input.
9. A digital electronic signal consists of a group of binary digits (1s and 0s) which make up a code. This two-state or binary code signifies a quantity or value with the two binary digits being represented as a LOW and HIGH voltage.
10. The technician is an expert in the diagnosis and repair of problems or malfunctions within electrical and electronic systems.
11. The engineer creates the concepts and designs of all new electrical and electronic systems.
12. The assembler is responsible for the fabrication of electronic systems.
13. Some of the many different technician classifications are:
    - **a.** Engineering Technician
    - **b.** Calibration Technician
    - **c.** Production-Test Technician
    - **d.** Quality Assurance (QA) Technician
    - **e.** Service Technician

NEW TERMS

American Standard Code for Information Interchange (ASCII)
Analog
Analog Circuits
Analog Electrical Circuits
Analog Electronic Circuits
Binary
Calibration Technician
Digital
Digital Circuits
Digital Electrical Circuits
Digital Electronic Circuits

Electrical Components
Electricity
Electronic Components
Electronics
Electronics Assembler
Electronics Engineer
Electronics Technician
Engineer
Engineering Technician
Field-Service Technician
In-House Service Technician
Linear Circuits
Product Test Technician
Quality Assurance Technician
Semiconductor Devices
Service Technician
Technician
Two-State Circuits

REVIEW QUESTIONS

Multiple-Choice Questions

1. __________ components, circuits, and systems manage the flow of *power.*
   a. Electronic **b.** Electrical **c.** Digital **d.** Analog
2. __________ components, circuits, and systems manage the flow of *information.*
   a. Electronic **b.** Electrical **c.** Digital **d.** Analog
3. __________ devices perform such functions as generating, sensing, storing, retrieving, amplifying, transmitting, receiving, and displaying information.
   a. Electronic **b.** Electrical **c.** Digital **d.** Analog
4. __________ devices perform such functions as generating, distributing, and converting electrical power.
   a. Electronic **b.** Electrical **c.** Digital **d.** Analog
5. Which of the following is an example of a semiconductor component?
   a. Fuse **c.** Bipolar Junction Transistor (BJT)
   b. Transformer **d.** All of the above
6. Which of the following is an example of an analog electrical circuit?
   a. Amplifier **c.** Power supply
   b. Logic gate **d.** ON/OFF lighting circuit
7. Which of the following is an example of an analog electronic circuit?
   a. Amplifier **c.** Power supply
   b. Logic gate **d.** ON/OFF lighting circuit
8. Which of the following is an example of a digital electrical circuit?
   a. Amplifier **c.** Power supply
   b. Logic gate **d.** ON/OFF Lighting circuit
9. Which of the following is an example of a digital electronic circuit?
   a. Amplifier **c.** Power supply
   b. Logic gate **d.** ON/OFF lighting circuit
10. Since digital circuits are based on the binary number system, they are often referred to as __________ circuits. On the other hand, the output of ana-

log circuits vary in direct proportion to the input which is why analog circuits are often called ____________ circuits.

a. ten-state, linear **c.** digital, ten-state
b. linear, two-state **d.** two-state, linear

Essay Questions

11. Define "electronics." (1-1)

12. What is the difference between electrical components, circuits, and systems, and electronic components, circuits, and systems? (1-2)

13. Define the following: (1-1)

a. Analog electronic circuit **c.** Digital electronic circuit
b. Analog electrical circuit **d.** Digital electrical circuit

14. Why are analog circuits also called linear circuits? (1-1-2)

15. Why do digital systems have information coded into binary codes instead of decimal codes? (1-1-2)

16. State the responsibilities of the following people: (1-2)

a. The technician **b.** The engineer **c.** The assembler

17. List some of the different technician types. (1-2)

18. Briefly describe the steps involved in the development of an electronics product. (1-2)

19. What responsibilities do the following people have? (1-2)

a. Applications Engineer **c.** Parts-Purchasing Agent
b. Marketing Person **d.** Pre-Production Unit Assembler

20. What level of education is generally needed for most technician positions? (1-2)

CHAPTER

2

OBJECTIVES

After completing this chapter, you will be able to:

1. Compare semiconductor devices to their predecessor the vacuum tube.

2. Describe why semiconductor devices have almost completely replaced vacuum tubes in most applications.

3. Explain the atom's subatomic particles.

4. Describe the terms:
   - **a.** Neutral atom
   - **b.** Negative ion
   - **c.** Positive ion
   - **d.** Valence shell

5. State the laws of attraction and repulsion.

6. Explain the difference between an atom, an element, a molecule, and a compound.

7. State the relationship between the number of valence electrons in an atom and its conductivity.

8. List the three semiconductor elements that are used to construct components for electrical and electronic circuit applications.

9. Describe why semiconductor atoms will form a crystal lattice structure due to covalent bonding, and define the term "intrinsic."

10. Explain an atom's energy gaps and energy levels along with the terms:
   - **a.** Energy gap
   - **b.** Conduction band
   - **c.** Excited state
   - **d.** Hole
   - **e.** Electron-Hole pair
   - **f.** Recombination
   - **g.** Lifetime

11. Explain why semiconductor materials, and therefore semiconductor devices, have a negative temperature coefficient of resistance.

12. Define the term "hole flow," and describe why the total current in a semiconductor is equal to the sum of the electron-flow and hole-flow currents.

13. Explain why pure semiconductor materials are doped and why they remain electrically neutral.

14. Explain the similarities and differences between *n*-type and *p*-type semiconductor materials, and define the terms:
   - **a.** Extrinsic semiconductor
   - **b.** Majority carriers
   - **c.** Minority carriers

15. Describe the following in relation to the P-N junction:
   - **a.** The junction
   - **b.** The depletion region
   - **c.** The barrier voltage

16. Explain how P-N junctions can be:
   - **a.** Forward biased
   - **b.** Reverse biased

17. Calculate and define the terms:
   - **a.** Forward voltage (V_F) drop
   - **b.** Forward current (I_F)
   - **c.** Reverse voltage (V_R) drop

18. Define the following terms:
   - **a.** Diffusion current
   - **b.** Leakage or reverse current (I_R)

Semiconductor Principles

OUTLINE

VIGNETTE

The Turing Enigma

During the Second World War, the Germans developed a cipher generating apparatus called "Enigma." This electromechanical teleprinter would scramble messages with several randomly spinning rotors that could be set to a predetermined pattern by the sender. This key and plug pattern was changed three times a day by the Germans and cracking the secrets of Enigma became of the utmost importance to British Intelligence. With this objective in mind, every brilliant professor and eccentric researcher was gathered at a Victorian estate near London called Bletchley Park. They specialized in everything from engineering to literature and were collectively called the Backroom Boys.

By far the strangest and definitely most gifted of the group was an unconventional theoretician from Cambridge University named Alan Turing. He wore rumpled clothes and had a shrill stammer and crowing laugh that aggravated even his closest friends. He had other legendary idiosyncrasies that included setting his watch by sighting on a certain star from a specific spot and then mentally calculating the time of day. He also insisted on wearing his gas mask whenever he was out, not for fear of a gas attack, but simply because it helped his hay fever.

Turing's eccentricities may have been strange but his genius was indisputable. At the age of twenty-six he wrote a paper outlining his "universal machine" that could solve any mathematical or logical problem. The data or, in this case, the intercepted enemy messages could be entered into the machine on paper tape and then compared with known Enigma codes until a match was found.

In 1943 Turing's ideas took shape as the Backroom Boys began developing a machine that used 2,000 vacuum tubes and incorporated five photoelectric readers that could process 25,000 characters per second. It was named "Colossus," and it incorporated the stored program and other ideas from Turing's paper written seven years earlier.

Turing could have gone on to accomplish much more. However, his idiosyncrasies kept getting in his way. He became totally preoccupied with abstract questions concerning machine intelligence. His unconventional personal lifestyle led to his arrest in 1952 and, after a sentence of psychoanalysis, his suicide two years later.

Before joining the Backroom Boys at Bletchley Park, Turing's genius was clearly apparent at Cambridge. How much of a role he played in the development of Colossus is still unknown and remains a secret guarded by the British Official Secrets Act. Turing was never fully recognized for his important role in the development of this innovative machine, except by one of his Bletchley Park colleagues at his funeral who said, "I won't say what Turing did made us win the war, but I daresay we might have lost it without him."

INTRODUCTION

Materials can be divided into three main types according to the way they react to current when a voltage is applied across them. **Insulators** (nonconductors), for example, are materials that have a very high resistance and therefore oppose current, whereas **conductors** are materials that have a very low resistance and therefore pass current easily. The third type of material is the **semiconductor** which, as its name suggests, has properties that lie between the insulator and the conductor. Semiconductor materials are not good conductors or insulators and so the next question is: What characteristic do they possess that makes them so useful in electronics? The answer is that they can be controlled to either increase their resistance and behave more like an insulator or decrease their resistance and behave more like a conductor. *It is this ability of a semiconductor material to vary its resistive properties that makes it so useful in electrical and electronic applications.*

Insulators
Materials that have a very high resistance and oppose current.

Conductors
Materials that have a very low resistance and pass current easily.

Semiconductors
Materials that have properties that lie between insulators and conductors.

In this chapter we will examine the characteristics of semiconductor materials so that we can better understand the operation and characteristics of semiconductor devices. Before we begin semiconductors, however, let us first investigate why semiconductor devices have almost completely replaced their predecessor, the vacuum tube.

2-1 SEMICONDUCTOR DEVICES VERSUS VACUUM TUBE DEVICES

Semiconductor materials such as *germanium* and *silicon* are used to construct semiconductor devices like the *diodes, transistors,* and *integrated circuits* (*ICs*) shown in Figure 2-1. These devices are used in electrical and electronic circuits

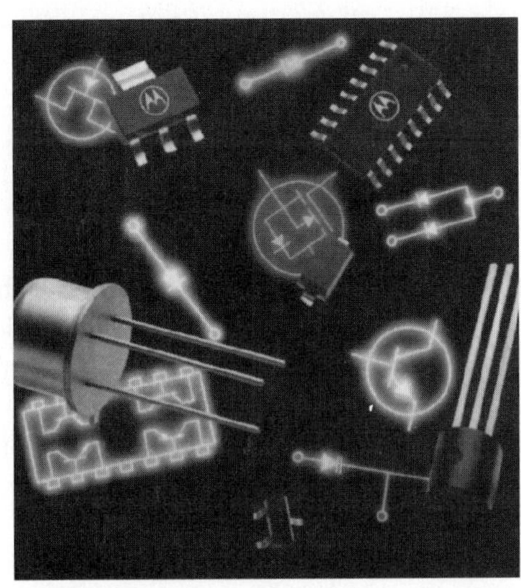

(a)

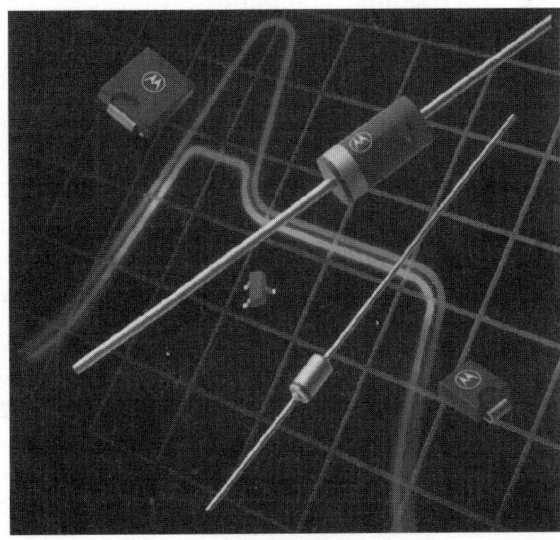

(b)

FIGURE 2-1 Semiconductor Devices. (Copyright of Motorola, Inc. Used by permission.)

to control current and voltage, so as to produce a desired result. For example, a diode could be used as the controlling element in a rectifier circuit that would convert ac to pulsating dc. A transistor, on the other hand, could be made to act like a variable resistance so it could amplify a radio signal. Conversely, an integrated circuit could be used to generate an oscillating signal or be made to perform arithmetic operations.

These semiconductor devices first became available in 1960. From about 1920 to 1960 the controlling element in all electrical and electronic circuits was the vacuum tube. Historians say that the only way to fully understand the present and the near future is to be familiar with the past. Let us now consider the specific reasons for this transition from vacuum tubes to semiconductor devices.

Vacuum Tube or Thermionic Valve
An electron tube evacuated to such a degree that its electrical characteristics are essentially unaffected by the presence of residual gas or vapor. Eventually replaced by the transistor for amplification and rectification.

Triode
A three-electrode vacuum tube that has an anode, cathode, and control grid.

Cathode
A negative electrode or terminal.

Anode
A positive electrode or terminal.

Tetrode
A four-electrode electron tube that has an anode, cathode, control grid, and an additional electrode.

Pentode
A five-electrode electron tube that has an anode, cathode, control grid, and two additional electrodes.

2-1-1 *Vacuum Tube Devices*

The term **vacuum tube** or **thermionic valve** is used to describe a variety of special devices that made possible radio, television, telecommunications, radar, sonar, computers, and many more systems between 1920 and 1960. These vacuum tubes were all enclosed in a sealed glass container that had been pumped free of air (hence the name vacuum) and had electric connections to its internal parts through the base of the container. Two of the most frequently used vacuum tubes were the **diode** and **triode** which are shown in Figure 2-2.

All vacuum tubes contain a *thermionic electron emitter,* which is a specially treated metal electrode that will emit electrons when heated by a heater. This effect was first demonstrated in 1877 by British physicist Joseph Thomson. It was not until 1904 that the first vacuum tube was invented by British engineer John Fleming. He called it the diode since it contained two (*di*) electrodes (*ode*). These two electrodes were the thermionic electron emitter that released the electrons, the **cathode,** and a plate electrode that collected the electrons, the **anode.** The main elements of the diode valve, or *Fleming diode,* as it was named at that time, are shown in Figure 2-2(a). This diode was used as a rectifier in power supply circuits (where it converted ac to dc), and as a *detector* in receiver circuits (where it extracted the radio signal from the carrier wave). The diode's main disadvantage was that it could not amplify or increase the amplitude of small signals. This disadvantage was overcome by U.S. scientist Lee De Forest in 1907 when he introduced a third electrode called the *control grid* between the cathode and anode. This control grid was used to increase or decrease the number of electrons reaching the anode. By controlling the grid, the triode valve, named because it contained three (*tri*) electrodes (*ode*), could be made to amplify (increase in amplitude) small radio signals. The triode valve was used in the first commercial radio receivers in the 1920s, and would later make possible television and other communication systems. The main elements of the triode valve are shown in Figure 2-2(b).

In the late 1920s other vacuum tubes emerged such as the **tetrode** and **pentode,** and improvements were introduced to the existing types. These electronic devices dominated the electronics industry until 1960 when the transistor took over.

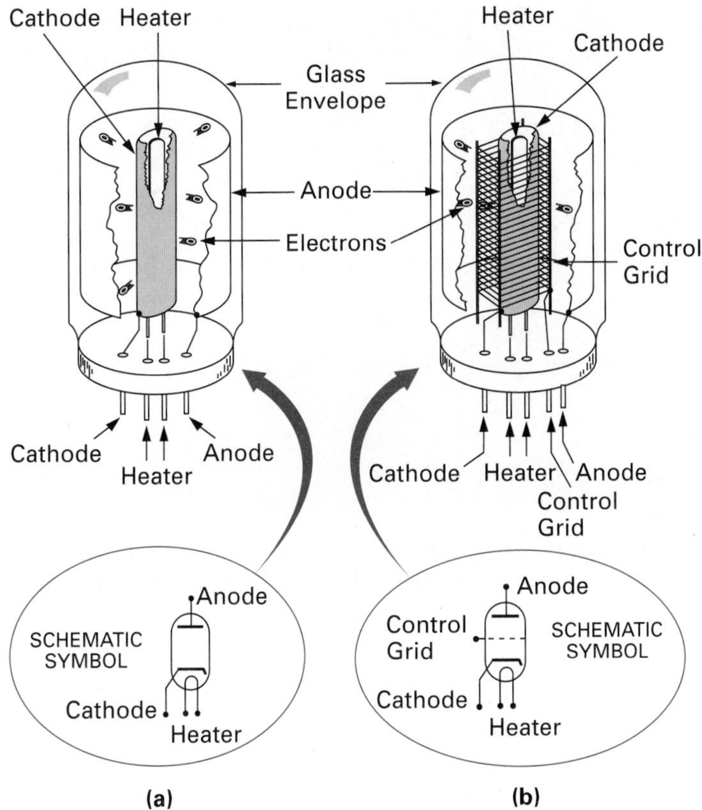

FIGURE 2-2 Vacuum Tube Devices. (a) Diode Vacuum Tube. (b) Triode Vacuum Tube.

Vacuum Tube Limitations

The vacuum tube was replaced by the transistor in so many applications because it suffered from a number of physical drawbacks. The first is that vacuum tubes are fragile and can be easily damaged by shock or vibration. The second is that the heater required a great deal of power, and the third is their relatively large size. The second and third disadvantages made the vacuum tube unsuitable for battery operated portable equipment, and, in addition, the tube heater had only a certain life span which meant that the tubes needed to be replaced regularly. To illustrate these disadvantages, Figure 2-3 compares the *electronic numeric integrator and computer (ENIAC),* which was unveiled in 1946, to today's programmable pocket calculator. The ENIAC weighed 38 tons, measured 18 feet wide and 88 feet long and used 17,486 vacuum tubes. These vacuum tubes produced a great deal of heat and developed frequent faults requiring constant maintenance. Today's programmable pocket calculator on the other hand, makes use of semiconductor integrated circuits and is far more powerful, portable, and reliable than the ENIAC. Another advantage is cost: in 1946 the ENIAC cost $400,000 to produce whereas a scientific calculator can be purchased today for less than $40.

FIGURE 2-3 Solid State Systems versus Vacuum Tube Systems. (a) The Electronic Numerical Integrator and Computer (ENIAC). (b) Today's Pocket Calculator.

Present Day Vacuum Tube Applications

Despite all of the vacuum tube's limitations, it is still used today in some areas of science and technology. In industry, 100 kW triode vacuum tubes are made mechanically and electrically very rugged and are used to generate Radio Frequency (RF) power at frequencies from 100 kHz to around 30 MHz. In communications, vacuum tubes are used to generate the high-frequency and high-power outputs needed for radio and television transmitters. In science, vacuum tubes are used in fusion research and linear accelerators where experimental tubes are being operated at 100 kV. **Cathode ray tubes (CRTs)** are still used extensively in televisions and computer monitors; however, the semiconductor color-active matrix display is beginning to take over.

Cathode Ray Tube
A vacuum tube in which electrons emitted by a hot cathode are focused into a narrow beam by an electron gun and then applied to a fluorescent screen. The beam can be varied in intensity and position to produce a pattern or picture on the screen.

2-1-2 *Solid State Devices*

The most significant development in electronics since World War II has been a small semiconductor device called the transistor. It was first introduced in 1948 by its inventors William Schockley, Walter Bratten, and John Bardeen in the Bell Telephone Laboratories and was described as a **solid state device.** This term was used because the transistor contained a solid semiconductor material between its input and output pins, unlike its predecessor the vacuum tube which had a vacuum between its input and output pins.

Solid State Device
Uses a solid semiconductor material, such as silicon, between the input and output whereas a vacuum tube has vacuum between input and output.

The first *point-contact transistor* unveiled in 1948 was extremely unreliable, and it took its inventors another twelve years to develop the superior *bipolar junction transistor* (*BJT*) and make it available in commercial quantities.

In 1960, many electronic system manufacturers began to use the bipolar junction transistor instead of the vacuum tube in low-power and low-frequency applications. Research and development into semiconductor or solid state devices mushroomed and a variety of semiconductor devices began to appear. A different type of transistor emerged called the *field effect transistor* (*FET*), which had characteristics similar to those of the vacuum tube. Once it was discovered that semiconductor materials could also generate and sense light, a new line of *optoelectronic devices* became available. Later it was discovered that semiconductor materials could sense magnetism, temperature, and pressure and, as a result, a variety of sensor devices or transducers (energy converters) appeared on the market. Along with all these different types of semiconductor devices, a wide variety of semiconductor diodes emerged that could rectify, regulate, and oscillate at high frequencies. Even to this day it is clear that we have not yet seen all the potential value of semiconductors. Figure 2-4 illustrates many of these semiconductor or solid state devices.

Discrete Components
Separate active and passive devices that were manufactured before being used in a circuit.

Although semiconductor diodes and transistors are still widely used as individual or **discrete components,** in 1959 Robert Noyce discovered that more than one transistor could be constructed on a single piece of semiconductor material. Soon other components such as resistors, capacitors, and diodes were added with transistors and then interconnected to form a complete circuit on a single chip or piece of semiconductor material. This integrating of various components on a single chip of semiconductor was called an *integrated circuit* (*IC*) or *IC chip.* Today the IC is used extensively in every branch of electronics with hundreds of thousands of transistors and other components being placed on a chip of semiconductor no bigger than this ■. Figure 2-5 illustrates some of the different types of integrated circuits.

Like an evolving species, semiconductors have come to dominate the products of which they used to be only a part. For example, there used to be 400 components in a typical cellular telephone. Now there are 40, and soon only 3 or 4 IC chips will make up the entire phone circuitry. Today the semiconductor

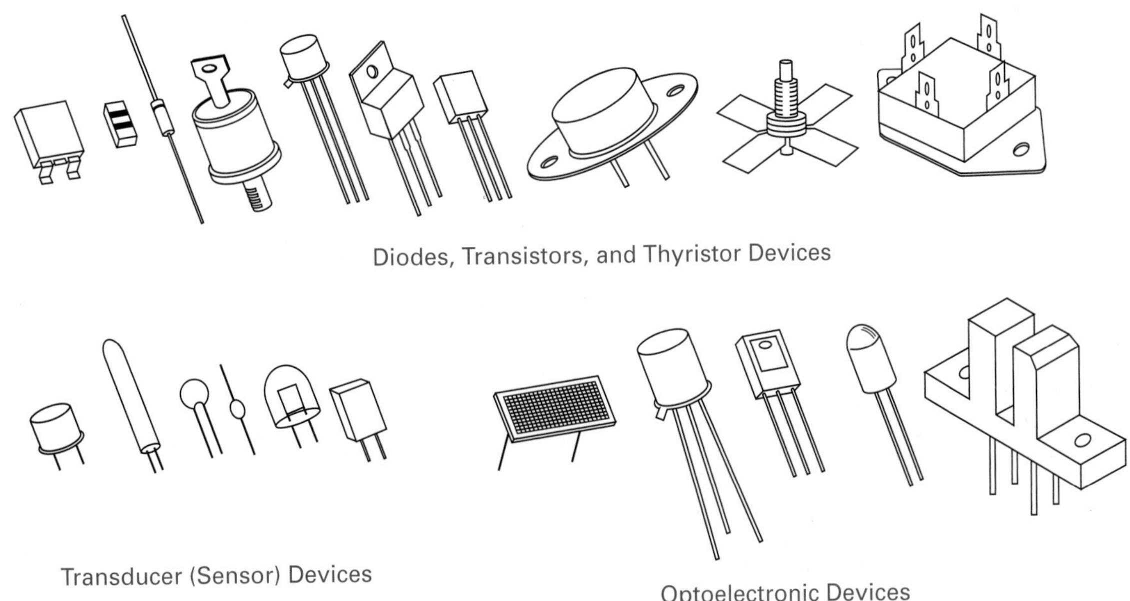

FIGURE 2-4 Discrete Semiconductor (Solid State) Devices.

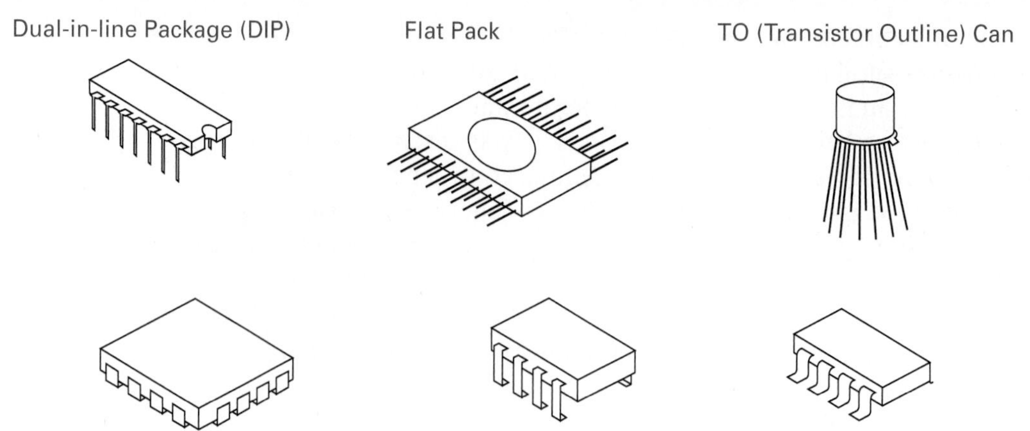

FIGURE 2-5 Semiconductor Integrated Circuits (ICs).

business—once regarded as a technical sideshow—occupies center stage and is key to the development of new products for all industries.

SELF-TEST REVIEW QUESTIONS FOR SECTION 2-1

1. Name the three most frequently used semiconductor devices in electrical and electronic equipment.
2. What four main advantages do semiconductor devices have over vacuum tube devices?
3. The main function of a semiconductor device is to control the ____________ or ____________ in an electrical or electronic circuit.
4. Devices made from semiconductor materials are often called:
   a. vacuum tube devices
   b. heated cathode devices
   c. solid state devices
   d. ENIAC devices

Element
There are 107 different natural chemical substances, or elements, that exist on earth. These can be categorized as gas, solid, or liquid.

Atom
The smallest particle of an element.

2-2 THE STRUCTURE OF MATTER

All of the matter on the earth and in the air surrounding the earth can be classified as being either a solid, liquid, or gas. A total of approximately 107 different elements are known to exist in, on, and around the earth. An **element,** by definition, is a substance consisting of only one type of atom; in other words, every element has its own distinctive atom, which makes it different from all the other elements. This **atom** is the smallest particle into which an element can be divided without losing its identity, and a group of identical atoms is called an element, shown in Figure 2-6.

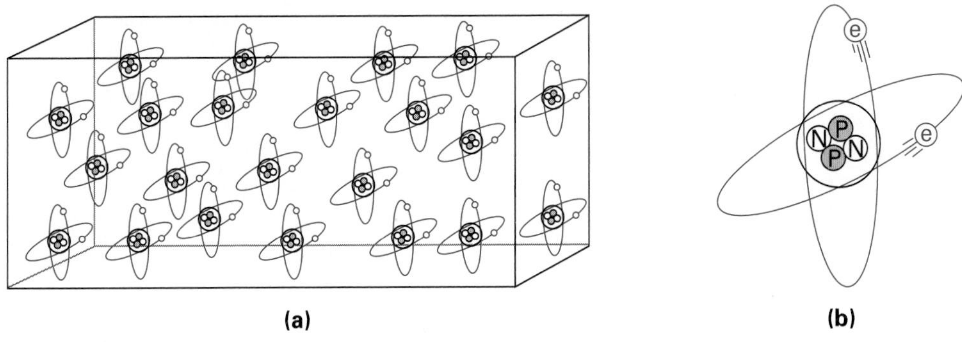

FIGURE 2-6 **Elements and Atoms. (a) Element: Many Similar Atoms. (b) Atom: Smallest Unit.**

For the sake of discussion, let us take a small amount of either a solid, a liquid, or a gas and divide it into two pieces. Then we divide a resulting piece into two pieces, and keep repeating the process until we finally end up with a tiny remaining part. Viewing the part under the microscope, as shown in Figure 2-7, the substance can still be identified as the original element as it is still made up of many of the original solid, liquid, or gas atoms. A small amount of gold, for example, the size of a pinpoint, will still contain several billion atoms. If the element subdivision is continued, however, a point will be reached at which a single atom will remain. Let us now analyze the atom in more detail.

2-2-1 The Atom

The word atom is a Greek word meaning a particle that is too small to be subdivided. At present, we cannot clearly see the atom; however, physicists and

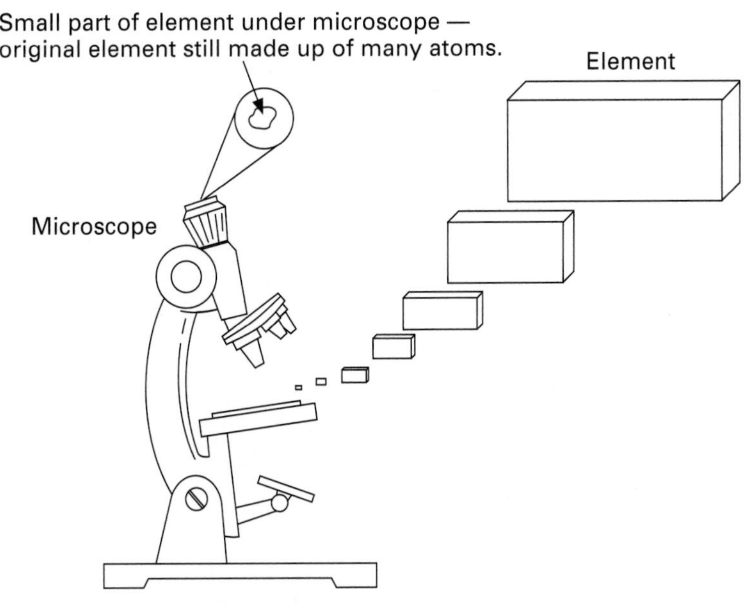

FIGURE 2-7 **An Element under the Microscope.**

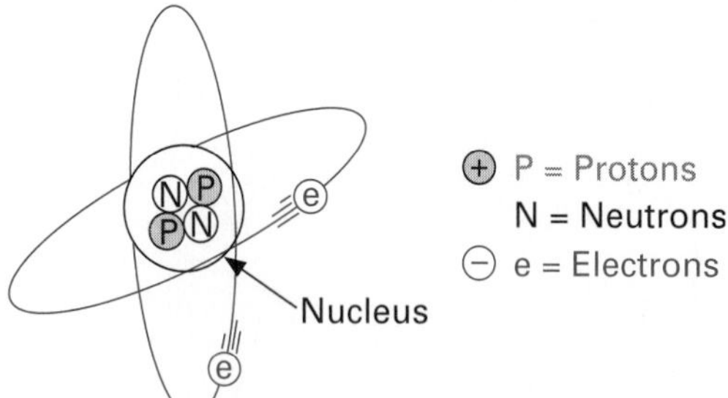

FIGURE 2-8 The Atom.

researchers do have the ability to record a picture as small as 12 billionths of an inch (about the diameter of one atom), and this image displays the atom as a white fuzzy ball.

Subatomic
Particles such as electrons, protons, and neutrons that are smaller than atoms.

In 1913, a Danish physicist, Neils Bohr, put forward a theory about the atom, and his basic model outlining the **subatomic** particles that make up the atom is still in use today and is illustrated in Figure 2-8. Bohr actually combined the ideas of Lord Rutherford's (1871–1937) nuclear atom with Max Planck's (1858–1947) and Albert Einstein's (1879–1955) quantum theory of radiation.

The three important particles of the atom are the *proton,* which has a positive charge, the *neutron,* which is neutral or has no charge, and the *electron,* which has a negative charge. Referring to Figure 2-8, you can see that the atom consists of a positively charged central mass called the *nucleus,* which is made up of protons and neutrons surrounded by a quantity of negatively charged orbiting electrons.

Atomic Number
Number of positive charges, or protons, in the nucleus of an atom.

Table 2-1 lists the periodic table of the elements, in order of their atomic number. The **atomic number** of an atom describes the number of protons that exist within the nucleus.

The proton and the neutron are almost 2000 times heavier than the very small electron, so if we ignore the weight of the electron, we can use the fourth column in Table 2-1 (weight of an atom) to give us a clearer picture of the protons and neutrons within the atom's nucleus. For example, a hydrogen atom, shown in Figure 2-9(a), is the smallest of all atoms and has an atomic number of 1, which means that hydrogen has a one-proton nucleus. Helium, however [Figure 2-9(b)], is second on the table and has an atomic number of 2, indicating that two protons are within the nucleus. The **atomic weight** of helium, however, is 4, meaning that two protons and two neutrons make up the atom's nucleus.

Atomic Weight
The relative weight of a neutral atom of an element, based on a neutral carbon atom having an atomic weight of 12.

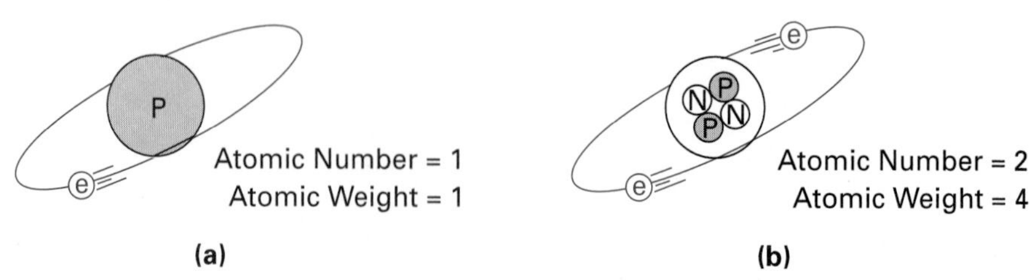

FIGURE 2-9 (a) Hydrogen Atom. (b) Helium Atom.

The number of neutrons within an atom's nucleus can subsequently be calculated by subtracting the atomic number (protons) from the atomic weight (protons and neutrons). For example, Figure 2-10 illustrates a beryllium atom:

Beryllium
Atomic number: 4 (protons)
Atomic weight: 9 (protons and neutrons)

If the number of protons is 4, the number of neutrons is 5 (9 – 4 = 5).

A **neutral atom** or *balanced atom* is one that has an equal number of protons and orbiting electrons, so the net positive proton charge is equal but opposite to the net negative electron charge, resulting in a balanced or neutral state. For example, Figure 2-11 (p. 34) illustrates a copper atom, which is the most commonly used metal in the field of electronics. It has an atomic number of 29, meaning that 29 protons and 29 electrons exist within the atom when it is in its neutral state.

Neutral Atom
An atom in which the number of positive charges in the nucleus (protons) is equal to the number of negative charges (electrons) that surround the nucleus.

Orbiting electrons travel around the nucleus at varying distances from the nucleus, and these orbital paths are known as **shells or bands.** The orbital shell nearest the nucleus is referred to as the first or K shell. The second is known as the L, the third is M, the fourth is N, the fifth is O, the sixth is P, and the seventh is referred to as the Q shell. There are seven shells available for electrons (K, L, M, N, O, P, and Q) around the nucleus, and each of these seven shells can only hold a certain number of electrons, as shown in Figure 2-12 (p. 34). The outermost electron-occupied shell is referred to as the **valence shell or ring,** and these electrons are termed *valence electrons.* In the case of the copper atom in Figure 2-10, a single valence electron exists in the valence N shell.

Shells or Bands
An orbital path containing a group of electrons that have a common energy level.

Valence Shell or Ring
The outermost shell formed by electrons.

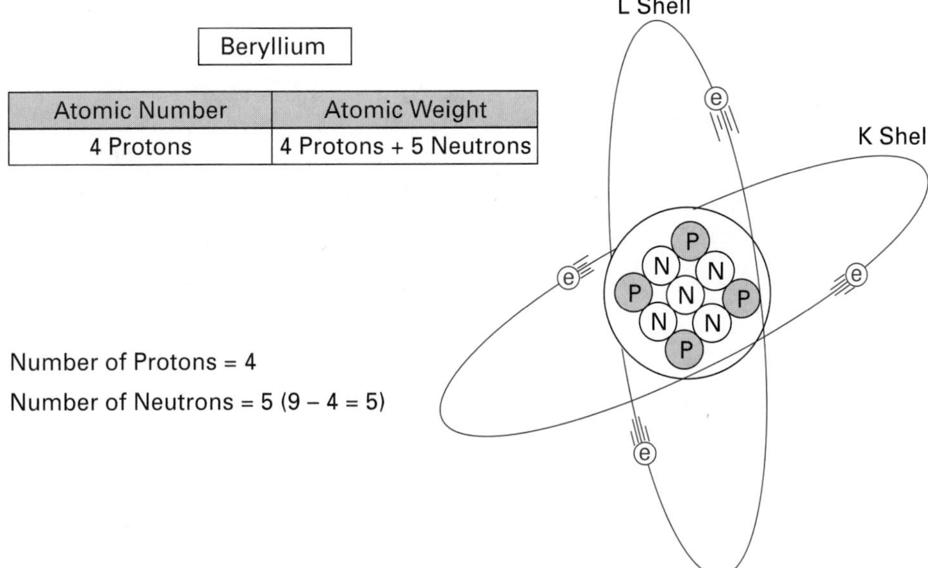

FIGURE 2-10 Beryllium Atom.

TABLE 2-1 Periodic Table of the Elements

ATOMIC NUMBER	ELEMENT NAME	SYMBOL	ATOMIC WEIGHT	ELECTRONS/SHELL K	L	M	N	O	P	Q	DISCOVERED	COMMENT
1	Hydrogen	H	1.007	1							1766	Active gas
2	Helium	He	4.002	2							1895	Inert gas
3	Lithium	Li	6.941	2	1						1817	Solid
4	Beryllium	Be	9.01218	2	2						1798	Solid
5	Boron	B	10.81	2	3						1808	Solid
6	Carbon	C	12.011	2	4						Ancient	Semiconductor
7	Nitrogen	N	14.0067	2	5						1772	Gas
8	Oxygen	O	15.9994	2	6						1774	Gas
9	Fluorine	F	18.998403	2	7						1771	Active gas
10	Neon	Ne	20.179	2	8						1898	Inert gas
11	Sodium	Na	22.98977	2	8	1					1807	Solid
12	Magnesium	Mg	24.305	2	8	2					1755	Solid
13	Aluminum	Al	26.98154	2	8	3					1825	Metal conductor
14	Silicon	Si	28.0855	2	8	4					1823	Semiconductor
15	Phosphorus	P	30.97376	2	8	5					1669	Solid
16	Sulfur	S	32.06	2	8	6					Ancient	Solid
17	Chlorine	Cl	35.453	2	8	7					1774	Active gas
18	Argon	Ar	39.948	2	8	8					1894	Inert gas
19	Potassium	K	39.0983	2	8	8	1				1807	Solid
20	Calcium	Ca	40.08	2	8	8	2				1808	Solid
21	Scandium	Sc	44.9559	2	8	9	2				1879	Solid
22	Titanium	Ti	47.90	2	8	10	2				1791	Solid
23	Vanadium	V	50.9415	2	8	11	2				1831	Solid
24	Chromium	Cr	51.996	2	8	13	1				1798	Solid
25	Manganese	Mn	54.9380	2	8	13	2				1774	Solid
26	Iron	Fe	55.847	2	8	14	2				Ancient	Solid (magnetic)
27	Cobalt	Co	58.9332	2	8	16	2				1735	Solid
28	Nickel	Ni	58.70	2	8	16	2				1751	Solid
29	Copper	Cu	63.546	2	8	18	1				Ancient	Metal conductor
30	Zinc	Zn	65.38	2	8	18	3				1746	Solid
31	Gallium	Ga	69.72	2	8	18	4				1875	Liquid
32	Germanium	Ge	72.59	2	8	18	4				1886	Semiconductor
33	Arsenic	As	74.9216	2	8	18	5				1649	Solid
34	Selenium	Se	78.96	2	8	18	6				1818	Photosensitive
35	Bromine	Br	79.904	2	8	18	8				1898	Liquid
36	Krypton	Kr	83.80	2	8	18	8				1898	Inert gas
37	Rubidium	Rb	85.4678	2	8	18	8	1			1861	Solid
38	Strontium	Sr	87.62	2	8	18	8	2			1790	Solid

All matter exists in one of three basic states: solids, liquids, and gases. The atoms of a solid are fixed in relation to one another but vibrate in a back-and-forth motion, unlike liquid atoms, which can flow over each other. The atoms of a gas move rapidly in all directions and collide with one another. The far-right column of Table 2-1 indicates whether the element is a gas, a solid, or a liquid at room temperature and normal pressure.

TABLE 2-1 (*continued*)

ATOMIC NUMBER[a]	ELEMENT NAME	SYMBOL	ATOMIC WEIGHT	K	L	M	N	O	P	Q	DISCOVERED	COMMENT
				ELECTRONS/SHELL								
39	Yttrium	Y	88.9059	2	8	18	9	2			1843	Solid
40	Zirconium	Zr	91.22	2	8	18	10	2			1789	Solid
41	Niobium	Nb	92.9064	2	8	18	12	1			1801	Solid
42	Molybdenum	Mo	95.94	2	8	18	13	1			1781	Solid
43	Technetium	Tc	98.0	2	8	18	14	1			1937	Solid
44	Ruthenium	Ru	101.07	2	8	18	15	1			1844	Solid
45	Rhodium	Rh	102.9055	2	8	18	16	1			1803	Solid
46	Palladium	Pd	106.4	2	8	18	18	0			1803	Solid
47	Silver	Ag	107.868	2	8	18	18	1			Ancient	Metal conductor
48	Cadmium	Cd	112.41	2	8	18	18	2			1803	Solid
49	Indium	In	114.82	2	8	18	18	3			1863	Solid
50	Tin	Sn	118.69	2	8	18	18	4			Ancient	Solid
51	Antimony	Sb	121.75	2	8	18	18	5			Ancient	Solid
52	Tellurium	Te	127.60	2	8	18	18	6			1783	Solid
53	Iodine	I	126.9045	2	8	18	18	7			1811	Solid
54	Xenon	Xe	131.30	2	8	18	18	8			1898	Inert gas
55	Cesium	Cs	132.9054	2	8	18	18	8	1		1803	Liquid
56	Barium	Ba	137.33	2	8	18	18	8	2		1808	Solid
57	Lanthanum	La	138.9055	2	8	18	18	9	2		1839	Solid
72	Hafnium	Hf	178.49	2	8	18	32	10	2		1923	Solid
73	Tantalum	Ta	180.9479	2	8	18	32	11	2		1802	Solid
74	Tungsten	W	183.85	2	8	18	32	12	2		1783	Solid
75	Rhenium	Re	186.207	2	8	18	32	13	2		1925	Solid
76	Osmium	Os	190.2	2	8	18	32	14	2		1804	Solid
77	Iridium	Ir	192.22	2	8	18	32	15	2		1804	Solid
78	Platinum	Pt	195.09	2	8	18	32	16	2		1735	Solid
79	Gold	Au	196.9665	2	8	18	32	18	1		Ancient	Solid
80	Mercury	Hg	200.59	2	8	18	32	18	2		Ancient	Liquid
81	Thallium	Tl	204.37	2	8	18	32	18	3		1861	Solid
82	Lead	Pb	207.2	2	8	18	32	18	4		Ancient	Solid
83	Bismuth	Bi	208.9804	2	8	18	32	18	5		1753	Solid
84	Polonium	Po	209.0	2	8	18	32	18	6		1898	Solid
85	Astatine	At	210.0	2	8	18	32	18	7		1945	Solid
86	Radon	Rn	222.0	2	8	18	32	18	8		1900	Inert gas
87	Francium	Fr	223.0	2	8	18	32	18	8	1	1945	Liquid
88	Radium	Ra	226.0254	2	8	18	32	18	8	2	1898	Solid
89	Actinium	Ac	227.0278	2	8	18	32	18	9	2	1899	Solid

[a]Rare earth series 58–71 and 90–107 have been omitted

2-2-2 ***Laws of Attraction and Repulsion***

For the sake of discussion and understanding, let us theoretically imagine that we are able to separate some positive and negative subatomic particles. Using these separated protons and electrons, let us carry out a few experiments. Studying Figure 2-13, you will notice that:

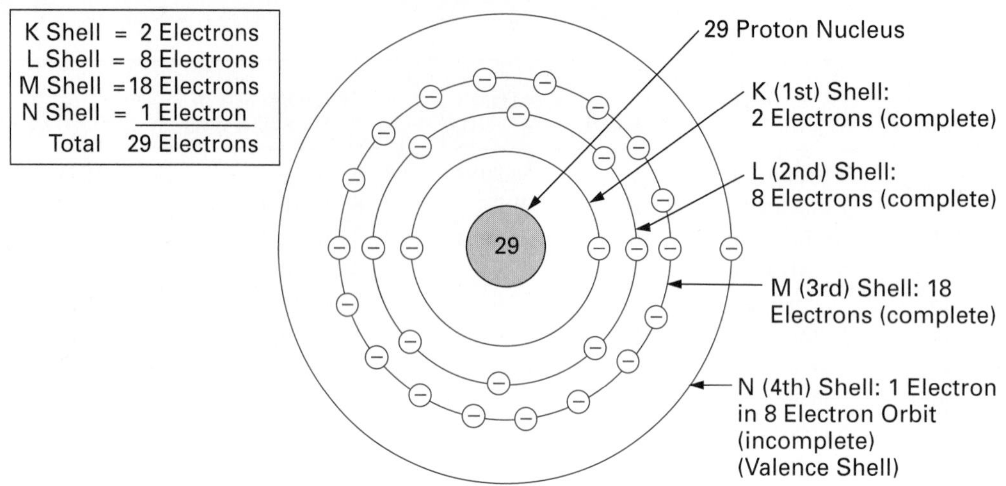

FIGURE 2-11 Copper Atom.

1. *Like charges* (positive and positive or negative and negative) repel one another. (Figures 2-13 (a) and (b))
2. *Unlike charges* (positive and negative or negative and positive) attract one another. (Figure 2-13(c))

Orbiting negative electrons are therefore attracted toward the positive nucleus, which leads us to the question of why the electrons do not fly into the atom's nucleus. The answer is that the orbiting electrons remain in their stable orbit due to two equal but opposite forces. The centrifugal outward force exerted on the electrons due to the orbit counteracts the attractive inward force (centripetal) trying to pull the electrons toward the nucleus due to the unlike charges.

Due to their distance from the nucleus, valence electrons are described as being loosely bound to the atom. The electrons can, therefore, easily be dislodged from their outer orbital shell by any external force to become a free electron.

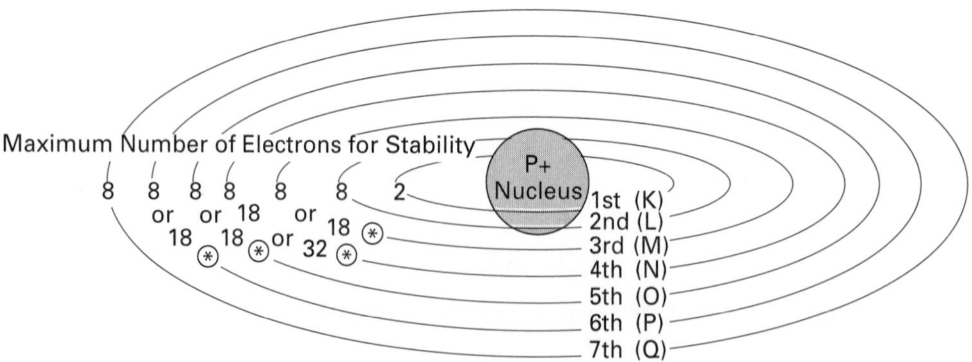

*The maximum number of electrons in these shells is dependent on the element's place in the periodic table.

FIGURE 2-12 Electrons and Shells.

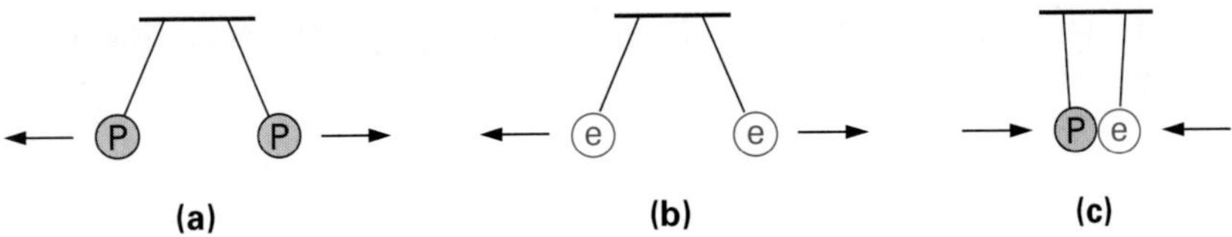

P = Proton (positive)
e = Electron (negative)

FIGURE 2-13 Attraction and Repulsion. (a) Positive Repels Positive. (b) Negative Repels Negative. (c) Unlike Charges Attract.

2-2-3 *The Molecule*

An atom is the smallest unit of a natural element, and an element is a substance consisting of a large number of the same atom. Combinations of elements are known as **compounds,** and the smallest unit of a compound is called a **molecule,** just as the smallest unit of an element is an atom. Figure 2-14 summarizes how elements are made up of atoms, and compounds are made up of molecules.

Compound
A material composed of united combinations of elements.

Molecule
The smallest particle of a compound that still retains its chemical characteristics.

Water is an example of a liquid compound in which the molecule (H_2O) is a combination of an explosive gas (hydrogen) and a very vital gas (oxygen). Table salt is another example of a compound; here the molecule is made up of a highly poisonous gas atom (chlorine) and a potentially explosive solid atom (sodium). These examples of compounds each contain atoms that, when alone,

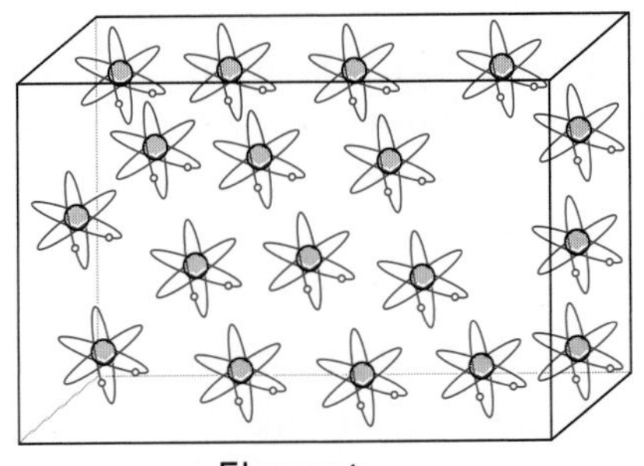

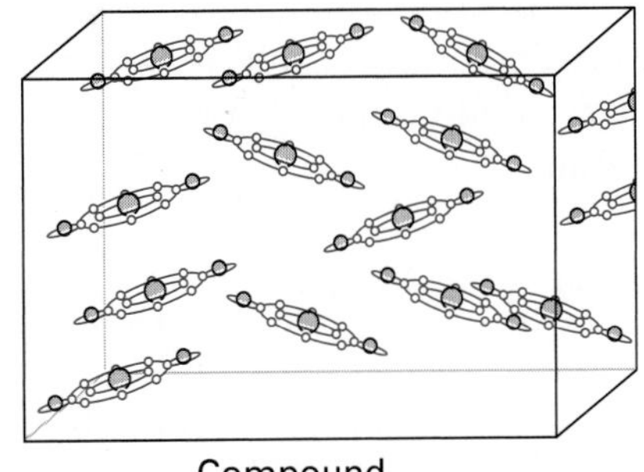

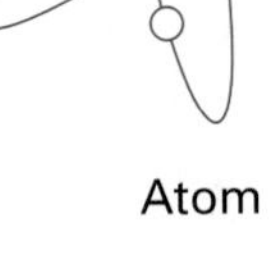

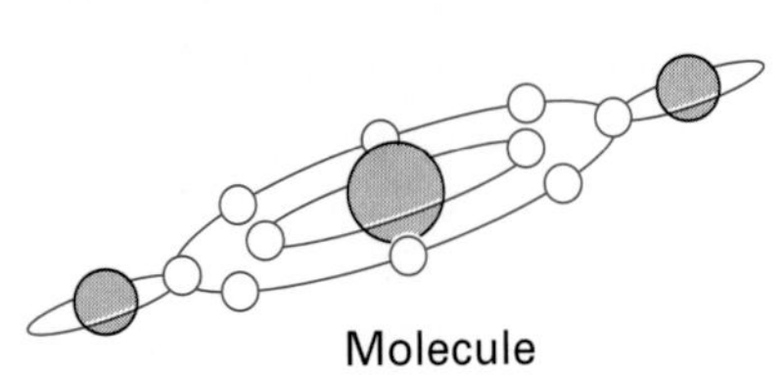

FIGURE 2-14 (a) An Element Is Made Up of Many Atoms. (b) A Compound Is Made Up of Many Molecules.

are both poisonous and explosive, yet when combined the resulting substance is as ordinary and basic as water and salt.

SELF-TEST REVIEW QUESTIONS FOR SECTION 2-2

1. Define the difference between an element and a compound.
2. Name the three subatomic particles that make up an atom.
3. What is the most commonly used metal in the field of electronics?
4. State the laws of attraction and repulsion.

2-3 SEMICONDUCTOR MATERIALS

A semiconductor material is one that is neither a conductor nor a nonconductor (insulator). This means simply that it will not conduct current as well as a conductor or block current as well as an insulator. Some semiconductor materials are pure or natural elements such as carbon (C), germanium (Ge), and silicon (Si), while other semiconductor materials are compounds.

Silicon and germanium are used most frequently in the construction of semiconductor devices for electrical and electronic applications. Germanium is a brittle grayish-white element that may be recovered from the ash of certain types of coals. Silicon, the most popular semiconductor material due to its superior temperature stability, is a white element normally derived from sand. Let us now examine the silicon, germanium, and carbon semiconductor atoms in more detail.

2-3-1 *Semiconductor Atoms*

Figure 2-15 illustrates the silicon, germanium, and carbon atoms. The silicon atom has 14 protons in its nucleus and 14 electrons in three orbital paths distributed as 2, 8, and then 4 electrons in its valence shell. The germanium atom has 32 protons within its nucleus and 32 electrons in four orbital paths distributed as 2, 8, 18, and finally 4 electrons in the valence band or shell. The carbon atom has 6 protons in its nucleus and 6 orbiting electrons in two orbital paths distributed as 2 and 4 electrons in the valence shell. The question is: What do all these atoms have in common? The answer is that all semiconductor atoms have *four valence electrons.*

The valence shell of an atom can contain up to 8 electrons, and it is the number of electrons in this valence shell that determines the conductivity of the atom. For example, an atom with only 1 valence electron would be classed as a good conductor whereas an atom having 8 valence electrons, and therefore a complete valence shell, would be classed as an insulator.

To summarize, Figure 2-15 shows three semiconductor atoms, all of which contain four valence electrons. Since the number of valence electrons determines

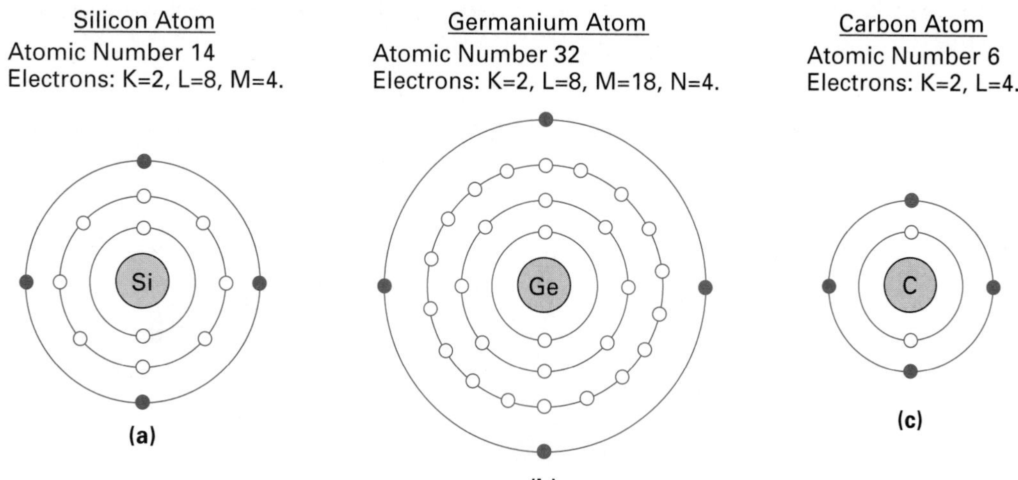

FIGURE 2-15 Semiconductor Atoms with Their 4 Valence Shell Electrons.

the conductivity of the element, semiconductor atoms are midway between conductors (which have 1 valence electron) and insulators (which have 8 valence electrons). Silicon and germanium are used to manufacture semiconductor devices, whereas carbon is combined with other elements to construct resistors.

2-3-2 Crystals and Covalent Bonding

So far we have discussed only isolated atoms. When two or more similar semiconductor atoms are combined to form a solid element, they automatically arrange themselves into an orderly lattice-like structure or pattern known as a **crystal,** as shown in Figure 2-16(a). This pattern is formed because each atom shares its four valence electrons with its four neighboring atoms. Since each atom shares one electron with a neighboring atom, two atoms will share two, or a pair, of electrons between the two cores. These two atom cores are pulling the two electrons with equal but opposite force and it is this pulling action that holds the atoms together in this solid crystal-lattice structure. The joining together of two semiconductor atoms is called an **electron-pair bond** or **covalent bond.** When many atoms combine, or bond, in this way the result is a crystal (smooth, glassy, solid) lattice structure. To illustrate this bonding process, each atom in Figure 2-16(a) has been drawn as a square and each valence shell has been drawn as an octagon (eight-sided figure) so that we can easily see which electrons belong to which atom. As you can see, the atom in the center of the diagram has 4 valence electrons (shown at the corners of the Si square), and shares one electron from each of its four neighbors.

Crystal
A solid element with an orderly lattice-like structure.

Electron-Pair Bond or Covalent Bond
A pair of electrons shared by two neighboring atoms.

Figure 2-16(b) shows a larger view of a silicon crystal structure. All of the atoms in this structure are electrically stable because all of their valence shells are complete (they all contain eight electrons). These completed valence shells cause the pure semiconductor crystal structure to act as an insulator since it will not easily give up or accept electrons. Pure semiconductor materials, which are

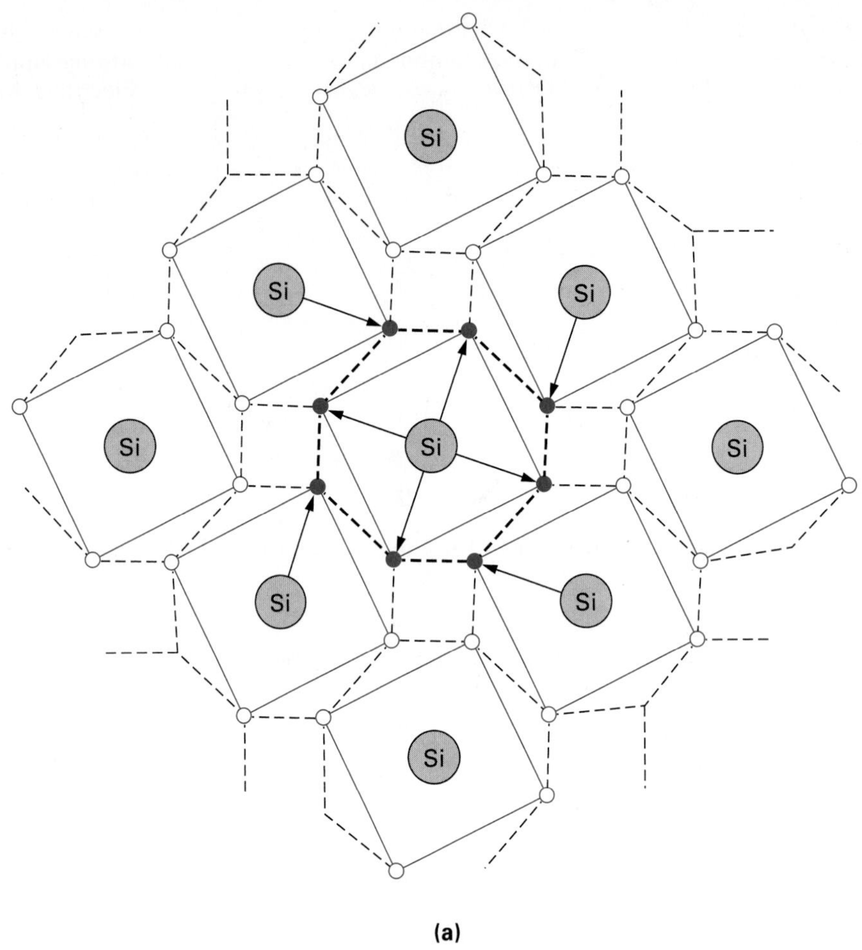

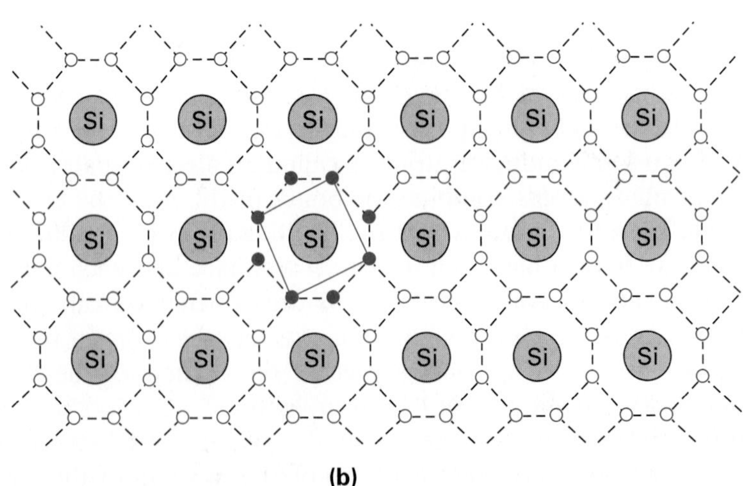

FIGURE 2-16 Covalent Bonding. (a) Silicon Atoms Sharing Valence Electrons. (b) Silicon Crystal Lattice Structure.

often called **intrinsic** materials, are therefore very poor conductors. Once this pure material is available, it must then be modified by a *doping* process to give it the qualities necessary to construct semiconductor devices. Silicon is most frequently used to construct solid state or semiconductor devices such as diodes and transistors because germanium has poor temperature stability and carbon crystals (diamonds) are too expensive to use.

Intrinsic Semiconductor Materials
Pure semiconductor materials.

2-3-3 *Energy Gaps and Energy Levels*

Let us now examine the relationships between electrons and orbital shells in a little more detail so that we can better understand charge and conduction within a semiconductor material.

As mentioned previously, there are seven shells available for electrons (K, L, M, N, O, P, and Q) around the nucleus. Electrons must travel or orbit in one of these orbital paths because they cannot exist in any of the spaces between orbital shells. Each orbital shell has its own specific energy level. Therefore, electrons traveling in a specific orbital shell will contain the shell's energy level. Figure 2-17 shows an example of an atom's orbital shell energy levels. The energy levels for each shell increase as you move away from the nucleus of the atom. The valence shell and the valence electrons will always have the highest energy level for a given atom. The space between any two orbital shells is called the **energy gap.** Electrons can jump from one shell to another if they absorb enough energy to make up the difference between their initial energy level and the energy level of the shell that they are jumping to. For example, in Figure 2-17 the valence shell has an energy level of 1.0 **electron-volts (eV).** Because this atom has three orbital shells, the valence shell will be energy level 3 (e3). The second energy level or orbital shell (e2) has an energy level of 0.6 eV. Therefore, for an electron to jump from energy level 2 (shell 2) to energy level 3 (e3 or valence shell), it will have to absorb a value of energy equal to the difference between e2 and e3. This will equal:

$$1.0 \text{ eV} - 0.6 \text{ eV} = 0.4 \text{ eV}$$

In this example, when either heat, light, or electrical energy was applied, one of the electrons in shell 2 (e2) absorbed 0.4 electron volts of energy and jumped to valence shell (e3).

If a valence (e3) electron absorbs enough energy it can jump from the valence shell into the **conduction band.** The conduction band is an energy band in which electrons can move freely or wander within a solid. When an electron jumps from the valence shell into the conduction band, it is released from the atom and no longer travels in one of its orbital paths. The electron is now free to move within the semiconductor material and is said to be in the **excited state.** An excited electron in the conduction band will eventually give up the energy it absorbed in the form of light or heat and return to its original energy level in the atom's valence shell.

When an electron jumps from the valence shell or band to the conduction band, it leaves a gap in the covalent bond called a **hole.** This action is shown in Figure 2-18(a). A hole is created every time an electron enters the conduction band. This action creates an **electron-hole pair.**

Energy Gap
The space between two orbital shells.

Electron-Volt (eV)
A unit of energy equal to the energy acquired by an electron when it passes through a potential difference of 1 V in a vacuum.

Conduction Band
An energy band in which electrons can move freely within a solid.

Excited State
An energy level in which a nucleus may exist if given sufficient energy to reach this state from a lower state.

Hole
The gap in the covalent bond left when an electron jumps from the valence shell or band to the conduction band.

Electron-Hole Pair
When an electron jumps from the valence shell or band to the conduction band, it leaves a gap in the covalent bond called a hole. This action creates an electron-hole pair.

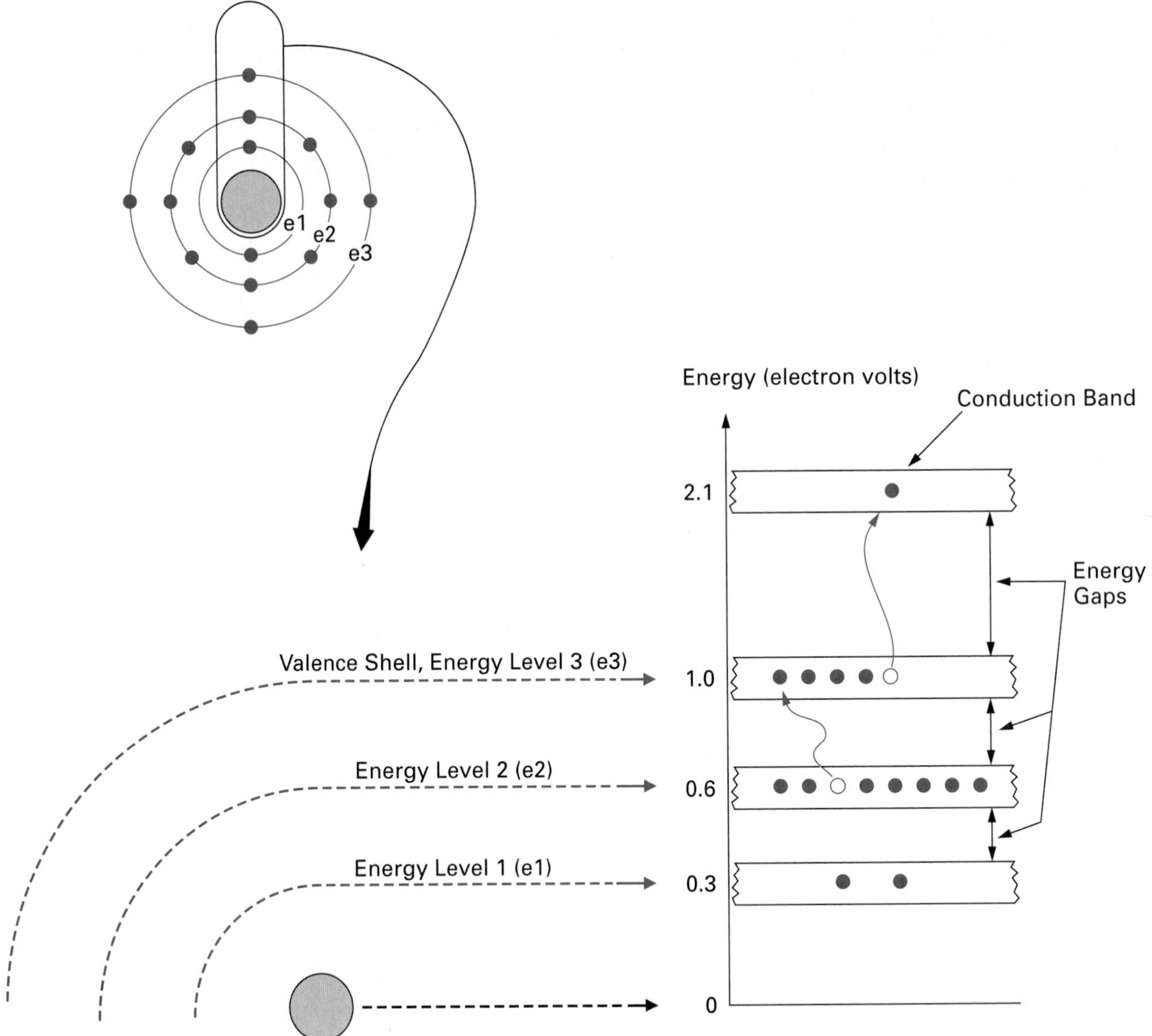

FIGURE 2-17 **An Atom's Orbital Shell Energy Levels.**

Recombination
Within microseconds, an electron in the conduction band will give up its energy and fall into one of the valence shell holes in the covalent bond. This action is called recombination.

It only takes a few microseconds before a free electron in the conduction band will give up its energy and fall into one of the valence shell holes in the covalent bond. This action is called **recombination** and is shown in Figure 2-18(b). The time difference between an electron jumping into the conduction band (becoming a free electron) and then falling back into a hole (recombination) is called the **lifetime** of the electron-hole pair.

Lifetime
The time difference between an electron jumping into the conduction band and then falling back into a hole.

2-3-4 Temperature Effects on Semiconductor Materials

At extremely low temperatures the valence electrons are tightly bound to their parent atoms, preventing valence electrons from drifting between atoms. There-

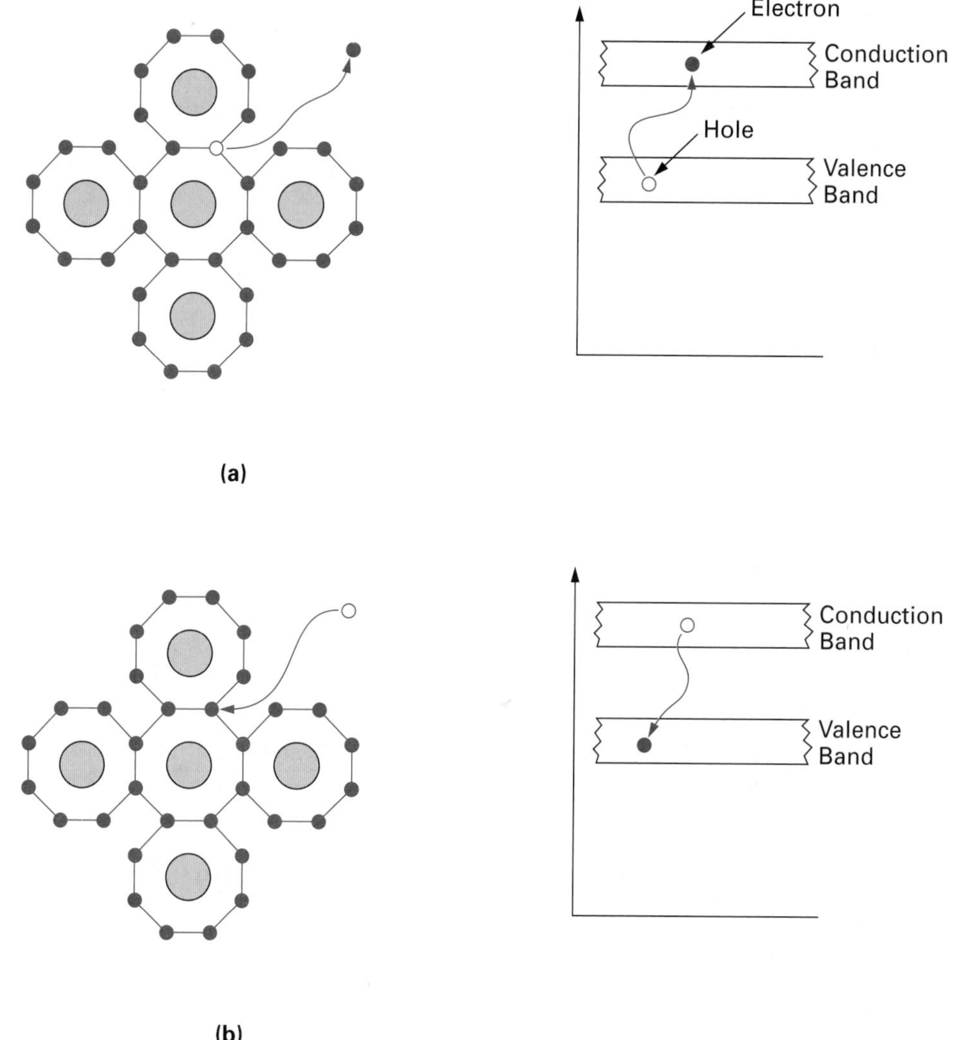

FIGURE 2-18 Valence Band and Conduction Band Actions. (a) Generating an Electron-Hole Pair. (b) Recombination.

fore, pure or intrinsic semiconductor materials function as insulators at temperatures close to absolute zero (−273.16°C or −459.69°F).

At room temperature, however, the valence electrons absorb enough heat energy to break free of their covalent bonds creating electron-hole pairs, as shown in Figure 2-19. Therefore, *the conductivity of a semiconductor material is directly proportional to temperature, in that an increase in temperature will cause an increase in the semiconductor material's conductance.* This means; *an increase in temperature (T↑) will cause an increase in a semiconductor's conductivity (C↑) and current (I↑).* This is why all circuits containing a semiconductor device tend to consume more current once they have warmed up.

Stated another way, *semiconductor materials, and therefore semiconductor devices, have a negative temperature coefficient of resistance which means as temperature increases (T ↑), their resistance decreases (R ↓).*

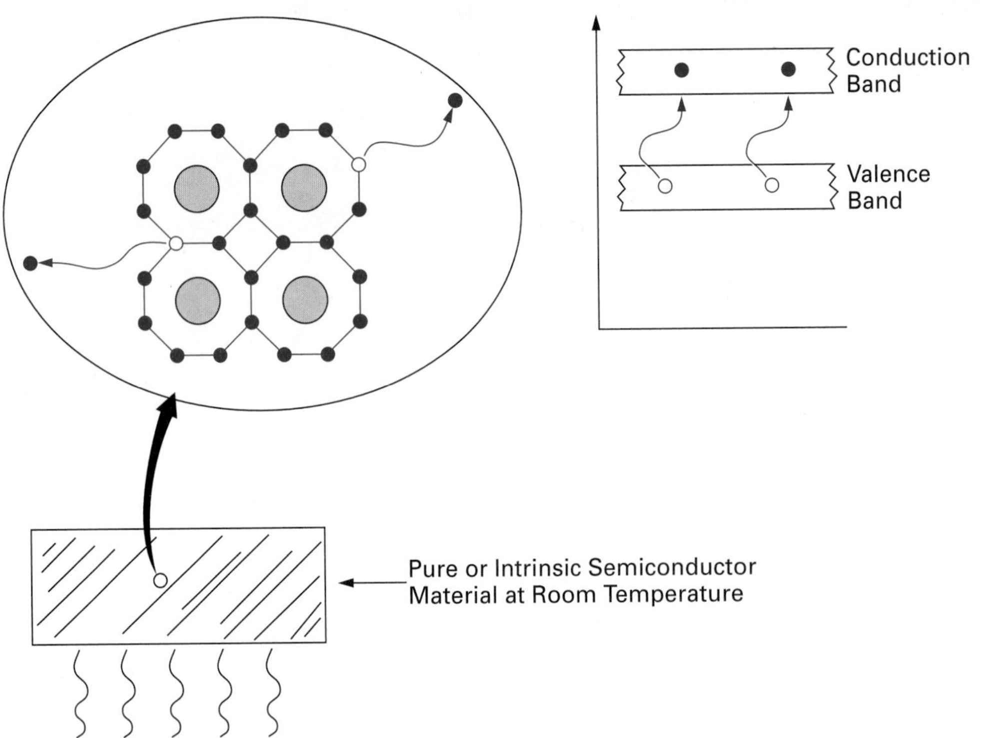

FIGURE 2-19 Temperature Effects on Semiconductor Materials.

2-3-5 *Applying a Voltage Across a Semiconductor*

Free Electrons
An electron that is able to move freely when an external force is applied.

If a voltage was applied across a room-temperature section of intrinsic semiconductor material, **free electrons** in the conduction band would make up a small electrical current as shown in Figure 2-20. In this illustration you can see how the negatively charged free electrons are attracted to the positive terminal of the voltage source. For every free electron that leaves the semiconductor material on the right side and travels to the positive terminal of the source, another electron is generated at the negative terminal of the voltage source and is injected into the left side of the semiconductor material. These injected electrons are captured by holes in the semiconductor material (recombination). As you can see from this illustration, current in a semiconductor material is made up of both electrons and holes. The holes act like positively charged particles while the electrons act like negatively charged particles. As electrons jump between atoms in a migration to the positive terminal of the source voltage, they leave behind them holes which are then filled by other advancing electrons. These advancing electrons leave behind them other holes, making it appear as though these holes are traveling towards the negative terminal of the source voltage. This **hole flow** is a new phenomenon to us, and it is one of the key differences between a semiconductor and a conductor. With conductors we were only interested in free-electron flow, but with semiconductors we must consider the movement of free electrons (negative charge carriers) and the apparent movement of holes (positive charge carriers).

Hole Flow
Conduction in a semiconductor when electrons move into holes when a voltage is applied.

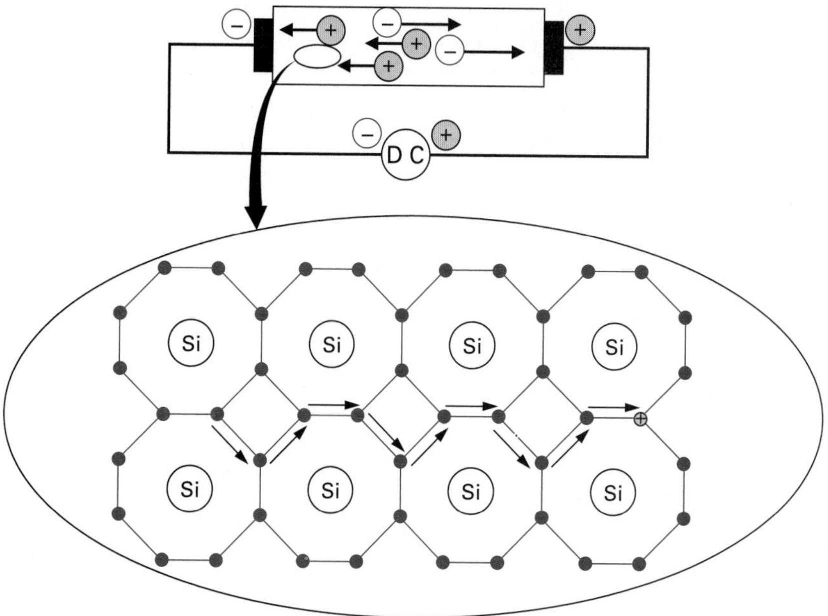

FIGURE 2-20 Electron Flow and Hole Flow in an Intrinsic Semiconductor.

In summary, therefore, *when a potential difference is applied across a semiconductor, the electrons move towards the positive potential and the holes travel towards the negative potential. The total current flow is equal to the sum of the electron flow and the hole flow currents.*

SELF-TEST REVIEW QUESTIONS FOR SECTION 2-3

1. What do all semiconductor atoms have in common?
2. The electrical conductivity of an element is determined by the number of electrons in the valence shell. Semiconductor atoms are midway between conductors, which have ____________ valence electron(s), and insulators, which have ____________ valence electron(s).
3. Each of the semiconductor atoms in a crystal lattice shares its electrons with four neighboring atoms. This joining of atoms is called a ____________ ____________.
4. Semiconductor materials have a ____________ temperature coefficient of resistance, which means as temperature increases, resistance ____________.
5. The number of electron-hole pairs within a semiconductor will increase as temperature ____________.
6. When a pure or ____________ semiconductor is connected across a voltage, free electrons travel towards the ____________ terminal of the applied voltage, whereas holes appear to travel towards the ____________ terminal of the applied voltage.

2-4 DOPING SEMICONDUCTOR MATERIALS

Doping
The process wherein impurities are added to the intrinsic semiconductor material either to increase the number of free electrons or to increase the number of holes.

Extrinsic Semiconductor
A semiconductor whose electrical properties are dependent on impurities added to the semiconductor crystal.

At room temperature pure or intrinsic semiconductors will not permit a large enough value of current. Therefore, some modification has to be applied in order to increase the semiconductor's current carrying capability or conductivity. **Doping** is a process wherein impurities are added to the intrinsic semiconductor material either to increase the number of free electrons (negative doping) or to increase the number of holes (positive doping).

Basically, there are two types of impurities that can be added to semiconductor crystals. One type of impurity is called a *pentavalent material* because its atom has five (*penta*) valence electrons. The second type of impurity is called a *trivalent material* because its atoms have three (*tri*) valence electrons. A doped semiconductor material is referred to as an **extrinsic semiconductor** material because it is no longer pure.

2-4-1 n-Type Semiconductor

Figure 2-21(a) shows how a semiconductor material's atoms will appear after pentavalent atom impurities have been added. The pentavalent atoms, which are listed in Figure 2-21(b), can be added to molten silicon to create, when cooled, a crystalline structure that has an extra electron due to the pentavalent (5 valence-electron impurity) atoms. The fifth pentavalent electron is not part of the covalent bonding and requires little energy to break free and enter the conduction band, as shown in Figure 2-21(c). Because millions of pentavalent atoms are added to the pure semiconductor, there will be millions of free electrons available for flow through the material.

n-Type Semiconductor
A material that has more valence-band electrons than valence-band holes.

Majority Carriers
The type of carrier that makes up more than half the total number of carriers in a semiconductor device.

Minority Carriers
The type of carrier that makes up less than half of the total number of carriers in a semiconductor device.

Even though the doped semiconductor material has millions of free electrons, the material is still electrically neutral. This is because each arsenic atom has the same number of protons as electrons and so do the silicon atoms. Therefore, the overall number of protons and electrons in the semiconductor is still equal and the result is a net charge of zero. However, because we now have more electrons than valence-band holes, the material is called an ***n*-type semiconductor.** *n*-Type semiconductors have more conduction-band electrons than valence-band holes. The electrons are therefore called the **majority carriers** and the valence-band holes are called the **minority carriers.** In Figure 2-21(c) you can see the abundance of conduction-band electrons. The holes in the valence band are few and are generated by thermal energy because the semiconductor is at room temperature.

When a voltage is applied across an *n*-type semiconductor, as shown in Figure 2-21(d), the additional free conduction-band electrons travel toward the positive terminal of the dc source. The applied voltage will cause extra electrons to break away from their covalent bonds to create holes, resulting in an increase in current and conductivity. Although the total current flow in this *n*-type semiconductor is the sum of the electron and hole currents, the conduction band electrons make up the majority of the flow.

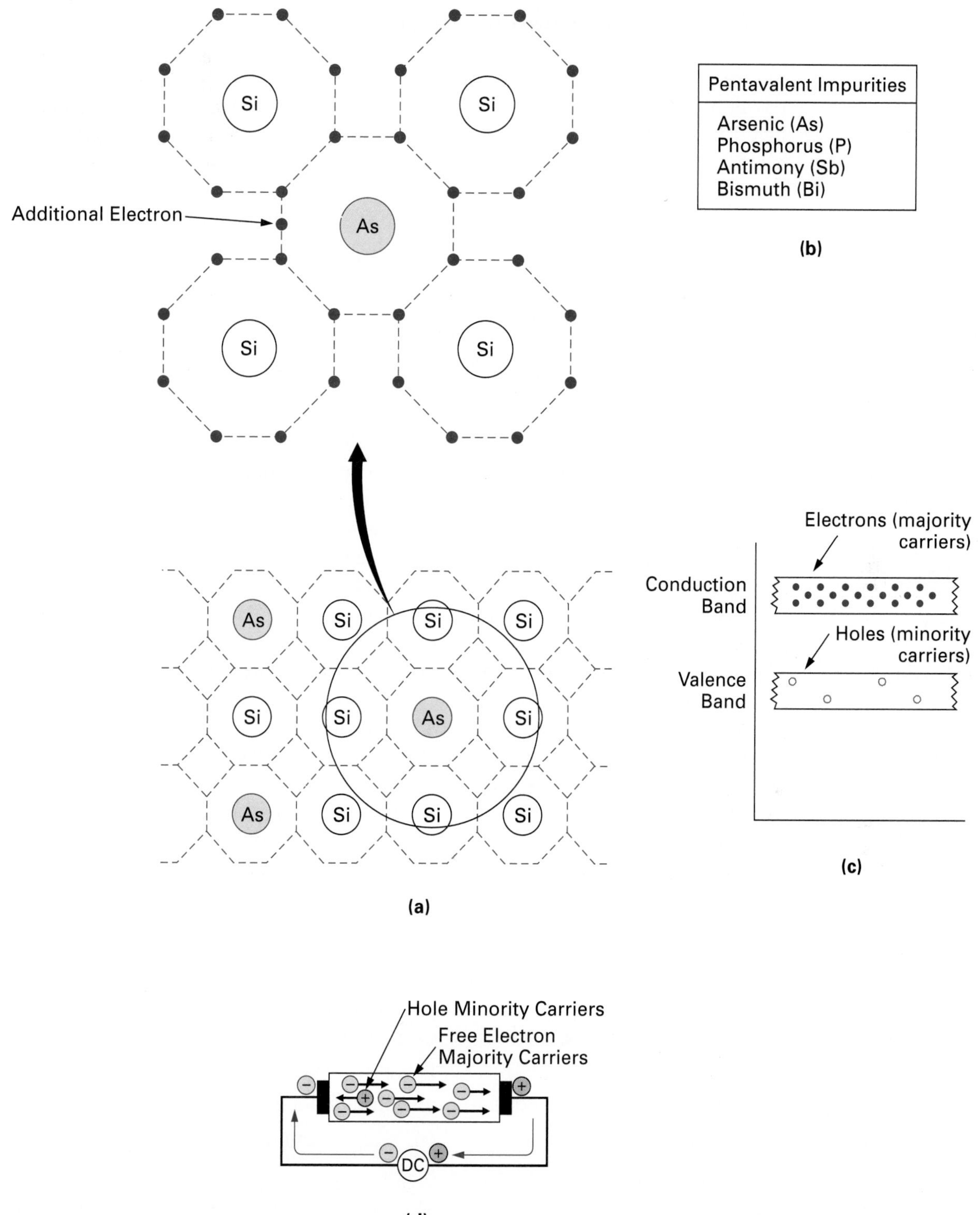

FIGURE 2-21 Adding Pentavalent Impurities to Create an *n*-Type Semiconductor Material.

2-4-2 p-Type Semiconductor

Figure 2-22(a) shows how a semiconductor material's atoms will appear after trivalent atom impurities have been added. The trivalent atoms, which are listed in Figure 2-22(b), can be added to molten silicon to create, when cooled, a crystalline structure that has a hole in the valence band of every trivalent (3 valence-electron impurity) atom. Instead of an excess of electrons, we now have an excess of holes. Because millions of trivalent atoms are added to the pure semiconductor, there will be millions of holes available for flow through the material.

Even though the doped semiconductor material has millions of holes, the material is still electrically neutral. This is because each aluminum atom has the same number of protons as electrons and so do the silicon atoms. Therefore the overall number of protons and electrons in the semiconductor is still equal and the result is a net charge of zero. However, because we now have more valence band holes than electrons the material is called a ***p*-Type semiconductor.** *p*-Type semiconductors have more valence-band holes than conduction-band electrons. The holes are called the **majority carriers** and the electrons are called the **minority carriers.** In Figure 2-22(c) you can see the abundance of valence-band holes. The few electrons in the conduction band are generated by thermal energy because the semiconductor is at room temperature.

***p*-Type Semiconductor**
A material that has more valence-band holes than valence-band electrons.

When a voltage is applied across a *p*-type semiconductor, as illustrated in Figure 2-22(d), the large number of holes within the material will attract electrons from the negative terminal of the dc source into the *p*-type semiconductor. These holes appear to move because each time an electron moves into a hole it creates a hole behind it, and the holes appear to move in the opposite direction to the electrons (towards the negative terminal of the dc source). The applied voltage will cause some electrons to break away from the covalent bond resulting in an increased current and conductivity. Although the total current flow in this *p*-type semiconductor is the sum of the hole and electron currents, the valence-band holes make up the majority of the flow.

SELF-TEST REVIEW QUESTIONS FOR SECTION 2-4

1. Why are impurities added to pure semiconductor materials?
2. Pentavalent atoms add ____________ to semiconductor crystals, to create ____________ type semiconductors.
3. Trivalent atoms add ____________ to semiconductor crystals, to create ____________ type semiconductors.
4. In an *n*-type semiconductor the majority carriers are ____________, whereas in a *p*-type semiconductor the majority carriers are ____________.

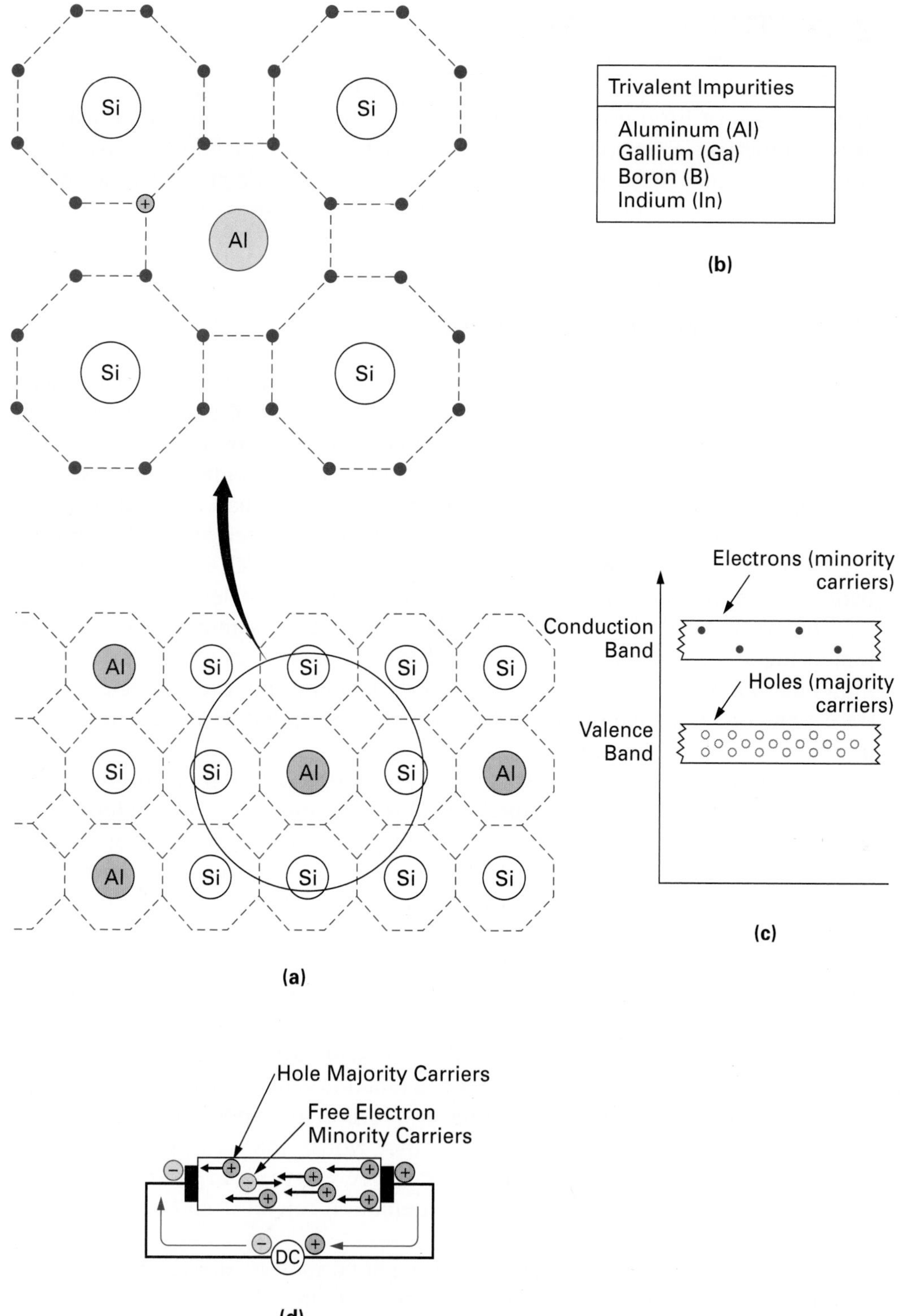

FIGURE 2-22 Adding Trivalent Impurities to Create a *p*-Type Semiconductor Material.

2-5 THE P-N JUNCTION

P-N Junction
The point at which two opposite doped materials come in contact with one another.

On their own, *n*-type semiconductor materials and *p*-type semiconductor materials are of little use. Together, however, these two form a **P-N semiconductor junction.** Semiconductor devices such as diodes and transistors are constructed using these P-N junctions, which give specific current flow characteristics. In this section we will examine the characteristics of the P-N junction in detail.

2-5-1 *The Depletion Region*

Figure 2-23(a) shows the individual *n*-type and *p*-type materials. The *n*-type material is represented as a block containing an excess of electrons (solid circles), while the *p*-type material is represented as a block containing an excess of holes (open circles). The energy diagrams below the two semiconductor sections show the differences between the two materials. Because different impurity atoms were added to the pure semiconductor material, the atomic make-up of the *n*-type and *p*-type materials is slightly different, which is why the valence bands and conduction bands are at slightly different energy levels.

Figure 2-23(b) shows the two *n*-type and *p*-type semiconductor sections joined together. A manufacturer of semiconductor devices would not join two individual pieces in this way to create a P-N junction. Instead, a single piece of pure semiconductor material would have each of its halves doped to create a *p*-type and *n*-type section.

The point at which the two oppositely doped materials come in contact with one another is called the **junction.** This junction of the two materials now permits the free electrons in the *n*-type material to combine with the holes in the *p*-type material as shown in Figure 2-23(c). As free electrons in the *n* material cross the junction and combine with holes in the *p* material, they create negative ions (atoms with more electrons than protons) in the *p* material, and leave behind positive ions (atoms with less electrons than protons) in the *n* material, as shown in Figure 2-23(d). An area or region on either side of the junction becomes emptied or depleted of free electrons and holes. This small layer containing positive and negative ions is called the **depletion region.**

Depletion Region
A small layer on either side of the junction that becomes empty, or depleted, of free electrons or holes.

As the ion layer on either side of the junction builds up, it has the effect of diminishing and eventually preventing any further recombination of free electrons and holes across the junction. In other words, the negative ions in the *p* region near the junction repel and prevent free electrons in the *n* region from recombination. This action prevents the depletion region from becoming larger and larger.

Barrier Potential or Barrier Voltage
The potential difference, or voltage, that exists across the junction.

These positive ions or charges and negative ions or charges accumulate a certain potential. Since these charges are opposite in polarity, a potential difference or voltage called the **barrier potential** or **barrier voltage** exists across the junction as shown in Figure 2-23(e). At room temperature, the barrier voltage of a silicon P-N junction is approximately 0.7 V, and a germanium P-N junction is approximately 0.3 V.

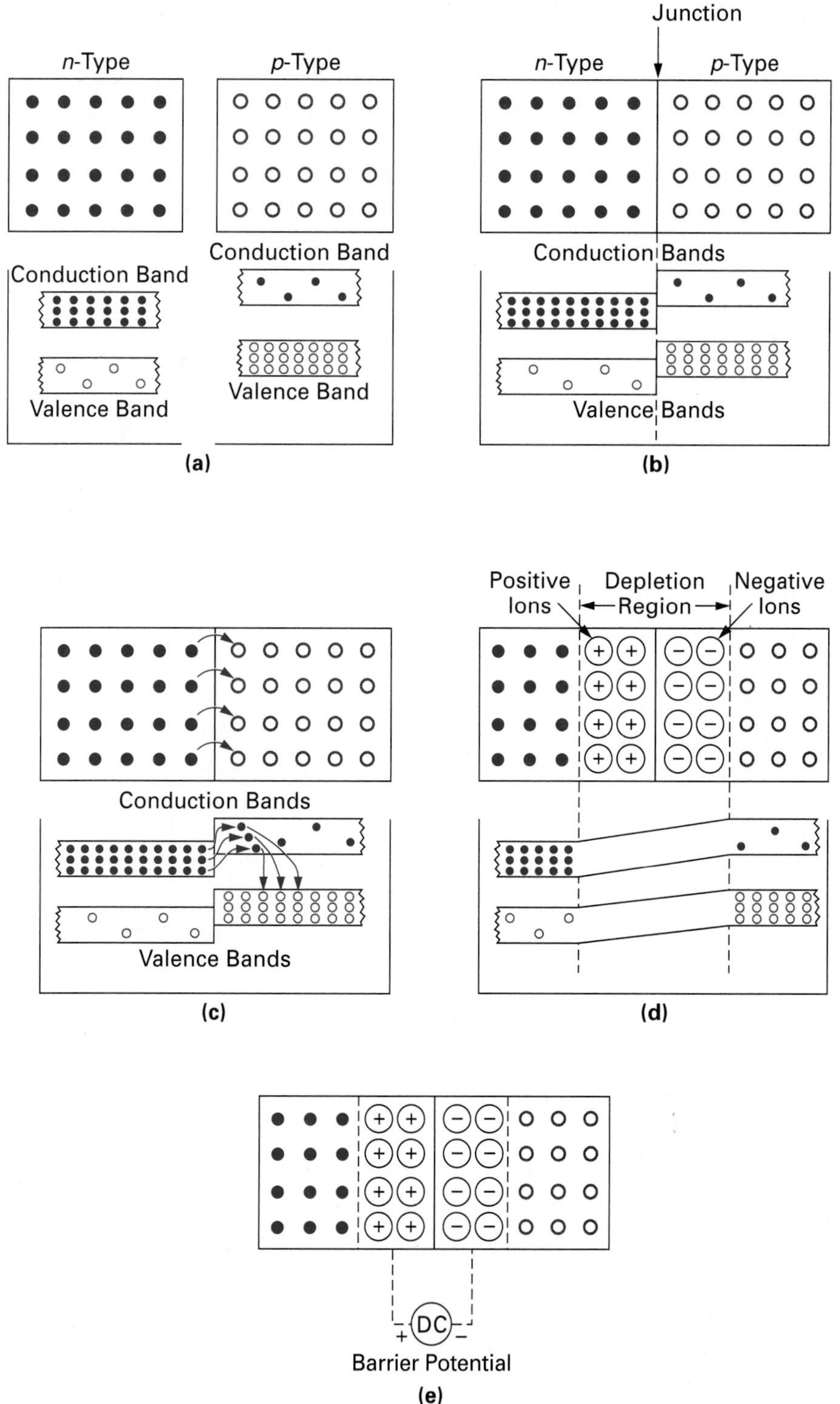

FIGURE 2-23 The Depletion Region.

2-5-2 Biasing a P-N Junction

Bias Voltages
The DC voltages applied to control a device's operation.

Semiconductor devices are constructed using P-N junctions. These P-N junctions need voltages of a certain amplitude and polarity to control their operation. These voltages, which incline or cause the device to operate in a certain manner, are known as **bias voltages.** Bias voltages control the width of the depletion region, which in turn controls the resistance of the P-N junction and, therefore, the amount of current that can pass through the P-N junction or semiconductor device.

To be specific, a small depletion region ($dr\downarrow$) will offer a small P-N junction resistance ($R\downarrow$) and therefore permit a large P-N junction current ($I\uparrow$). In this instance, the P-N semiconductor junction is said to be **forward biased** and acts like a conductor.

Forward Biased
A small depletion region at the junction will offer a small resistance and permit a large current. Such a junction is forward biased.

On the other hand, a large depletion region ($dr\uparrow$) will offer a large P-N junction resistance ($R\uparrow$) and therefore only permit a small P-N junction current ($I\downarrow$). In this instance the P-N semiconductor junction is said to be **reverse biased** and acts like an insulator.

Reverse Biased
A large depletion region at the junction will offer a large resistance and permit only a small current. Such a junction is reverse biased.

Forward Biasing a P-N Junction

Figure 2-24 shows in detail why a forward biased P-N junction will pass current with almost no opposition (act like a conductor). To begin, Figure 2-24(a) shows a P-N junction with wires attached. A resistor has been included to limit the amount of current passing through the P-N junction to a safe level. Energy diagrams have also been included on the right side of each part of Figure 2-24 to show the relationship between the conduction band and valence band of the *p* and *n* regions.

Let us now connect a dc voltage across the P-N junction to see how it reacts. This is shown in Figure 2-24(b). The negative potential of the dc source has been applied to the *n* region and the positive potential of the dc source has been applied to the *p* region. Referring to the energy diagram in Figure 2-24(b), you can see that the conduction band electrons in the *n* region are repelled by the negative voltage source towards the junction. On the opposite side, valence band holes in the *p* region are repelled by the positive voltage source towards the junction. A forward-conducting current will begin to flow if the external source voltage is large enough to overcome the internal barrier voltage of the P-N junction. In this example we will assume that the dc source voltage is 10 volts and therefore this will be more than enough to overcome the silicon P-N junctions barrier potential of 0.7 volts.

Conduction through the P-N junction is shown in Figure 2-24(c). When forward biased, a P-N junction will act as a conductor and have a low but finite resistance value that will cause a corresponding voltage drop across its terminals. This **forward voltage (V_F) drop** is approximately equal to the P-N junction's barrier voltage:

Forward Voltage Drop (V_F)
The forward voltage drop is equal to the junction's barrier voltage.

$$\text{Forward Voltage } (V_F) \text{ for Silicon} = 0.7 \text{ V}$$

$$\text{Forward Voltage } (V_F) \text{ for Germanium} = 0.3 \text{ V}$$

Figure 2-24(c) shows how a voltmeter can be used to measure the forward voltage drop of 0.7 V across a silicon P-N junction when it is forward biased.

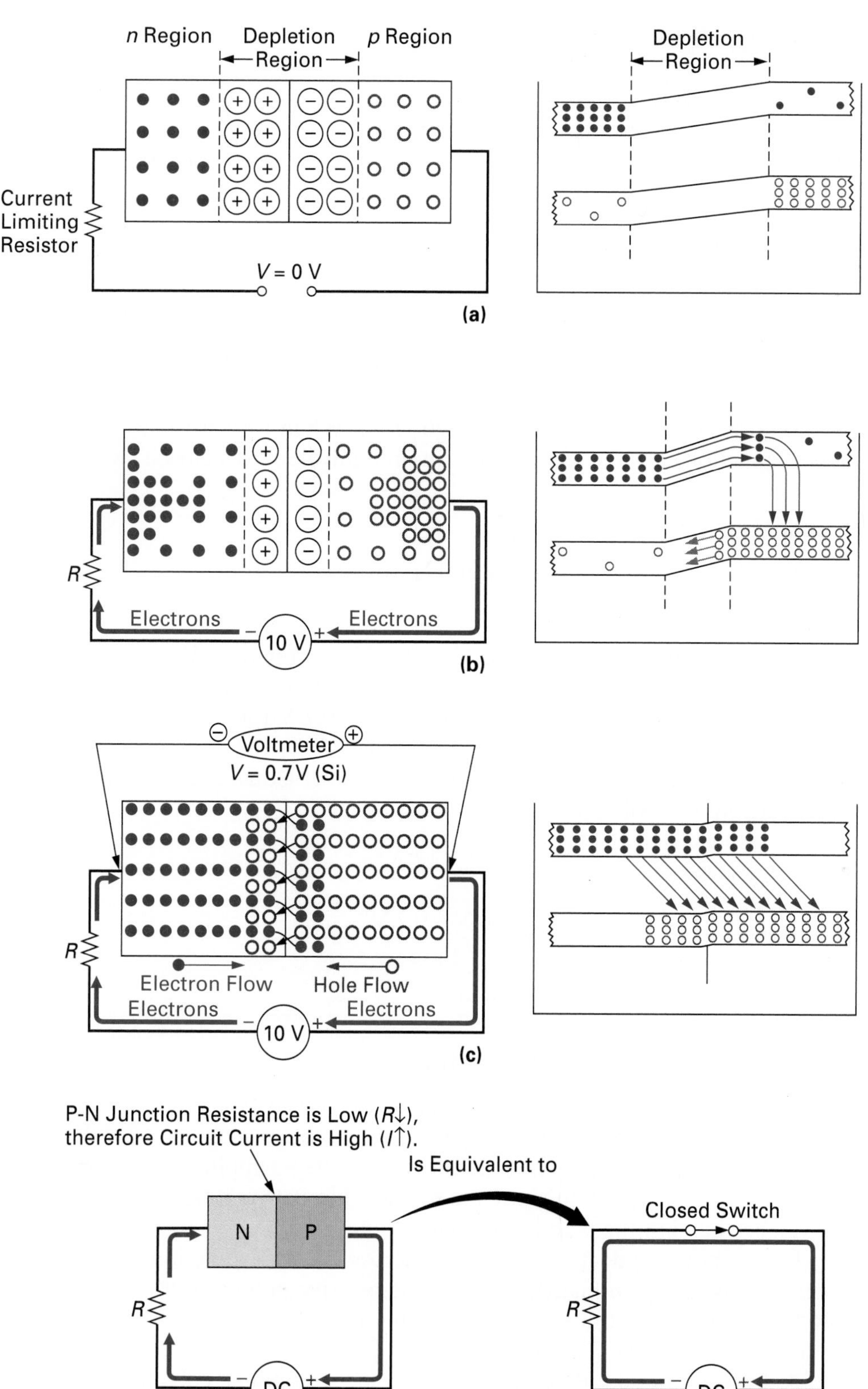

FIGURE 2-24 Forward Biasing a P-N Semiconductor Junction.

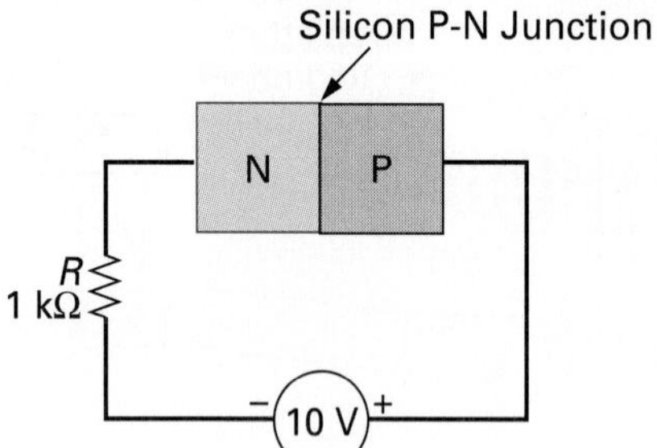

FIGURE 2-25 A P-N Junction Circuit.

To summarize, Figure 2-24(d) shows that when a P-N junction is forward biased ($+V \rightarrow p$ region, $-V \rightarrow n$ region) the P-N junction resistance is low ($R\downarrow$), and therefore the circuit current is high ($I \uparrow$). When forward biased therefore, the P-N junction acts like a conductor and is equivalent to a closed switch.

■ **EXAMPLE:**

Calculate the current for the circuit in Figure 2-25.

■ ***Solution:***

The silicon P-N junction is forward biased ($+V \rightarrow p$ region, $-V \rightarrow n$ region). The applied voltage of 10 V will be more than enough to overcome the silicon P-N junction forward voltage drop of 0.7 volts ($V_F = 0.7$ V for silicon). Since 10 volts is applied, and the P-N junction is dropping 0.7 V, the remaining voltage of 9.3 V is being dropped across the 1 kΩ resistor. Consequently, the forward-biased current (I_F) will equal:

$$I_F = \frac{V_S - V_{P\text{-}N}}{R}$$

$$= \frac{10\text{ V} - 0.7\text{ V}}{1\text{ k}\Omega}$$

$$= 9.3\text{ mA}$$

Reverse Biasing a P-N Junction

Figure 2-26 shows in detail why a reverse biased P-N junction will reduce current to almost zero (act like an insulator). To begin, Figure 2-26(a) shows a P-N junction with wires attached and no voltage being applied. Energy diagrams have again been included to show the relationship between the conduction band and valence band of the p and n regions.

Let us now connect a dc voltage across the P-N junction to see how it reacts. This is shown in Figure 2-26(b). The positive potential of the dc source is now being applied to the n region, and the negative potential of the dc source has been applied to the p region.

A forward biased P-N junction is able to conduct current because the external bias voltage forces the majority carriers in the n and p regions to combine at the junction. In this instance, however, the dc bias voltage polarity has been reversed, causing free electrons in the n region to travel to the positive ter-

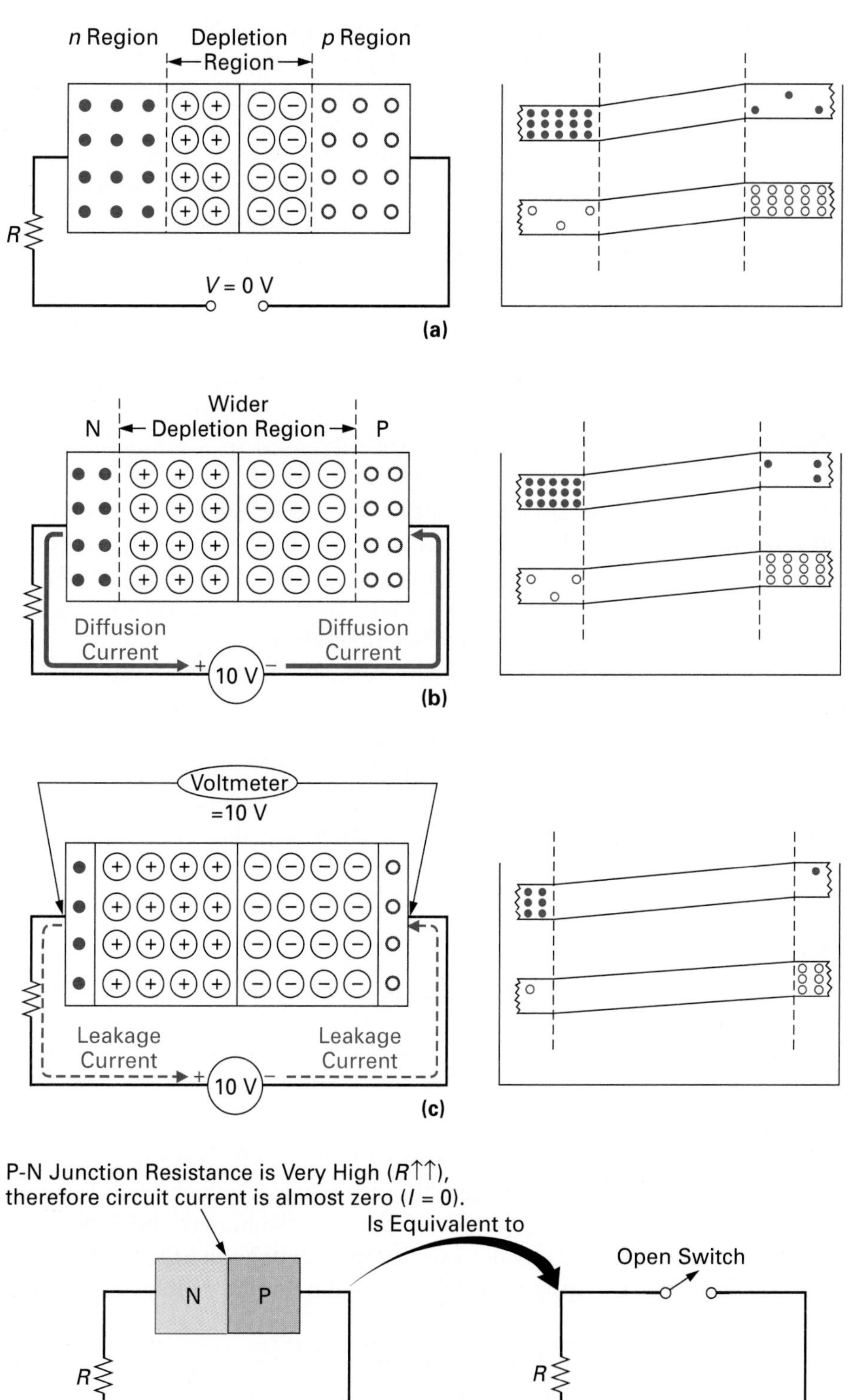

FIGURE 2-26 Reverse Biasing a P-N Semiconductor Junction.

minal of the voltage source leaving behind a large number of positive ions at the junction. This increases the width of the depletion region. At the same time, electrons from the negative terminal of the source are attracted to the holes in the *p* region of the P-N junction. These electrons fill the holes in the *p* region near the junction creating a large number of negative ions. This further increases the width of the depletion region. The current that is present at the time the depletion layer is expanding is called the **diffusion current.** Referring to Figure 2-26(b), you can see that the depletion region is now wider than the unbiased P-N junction shown in Figure 2-26(a).

Diffusion Current
The current that is present when the depletion layer is expanding.

The ions on either side of the junction build up until the P-N junction's internal-barrier voltage is equal to the external-source voltage, as shown by the voltmeter in Figure 2-26(c). When reverse biased, therefore, the **reverse voltage (V_R) drop** across a P-N junction is equal to the source or applied voltage. At this time the resistance of the junction has been increased to a point that current drops to zero.

Reverse Voltage Drop (V_R)
The reverse voltage drop is equal to the source voltage (applied voltage).

Actually, an extremely small current called the **leakage current** or **reverse current (I_R)** will pass through the P-N junction, as shown in Figure 2-26(c). It is present because the minority carriers (holes in the *n* region, electrons in the *p* region) are forced towards the junction where they combine, producing a constant small current. The current in the P-N junction is still considered to be at zero because the leakage or reverse current is so small (nanoamps in silicon diodes).

Leakage Current or Reverse Current (I_R)
The extremely small current present at the junction.

To summarize, Figure 2-26(d) shows that when a P-N junction is reverse biased ($+V \rightarrow n$ region, $-V \rightarrow p$ region), the P-N junction resistance is extremely high ($R \uparrow \uparrow$), and the circuit current is effectively zero ($I = 0$ amps). When reverse biased, therefore, the P-N junction acts like an insulator and is equivalent to an open switch.

■ **EXAMPLE:**

Referring to Figure 2-27(a) and (b), calculate each circuit's:

a. current value
b. P-N junction voltage drop

■ ***Solution:***

The P-N junction in Figure 2-27(a) is reverse biased ($+V \rightarrow n$ region, $-V \rightarrow p$ region); therefore, the P-N junction resistance is extremely high ($R \uparrow \uparrow$), and the circuit current is effectively zero ($I = 0$). When reverse biased, the P-N junction acts like an insulator and is equivalent to an open switch, and the voltage developed across the open P-N junction will equal the source voltage applied. For Figure 2-27(a);

$$\text{Circuit Current} = 0$$
$$\text{P-N Junction Voltage Drop} = V_S = 6\text{ V}$$

The P-N junction in Figure 2-27(b) is forward biased ($+V \rightarrow p$ region, $-V \rightarrow n$ region). Therefore, the P-N junction resistance is low ($R\downarrow$), and the circuit current is high ($I \uparrow$). When forward biased, the P-N junction acts like a conductor and is equivalent to a closed switch, and the P-N junction's voltage drop

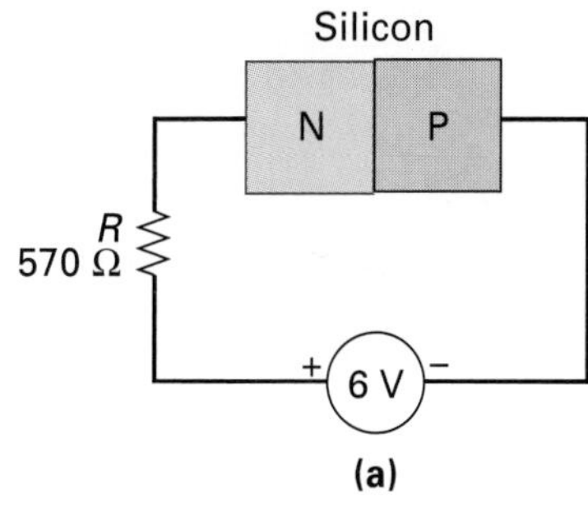

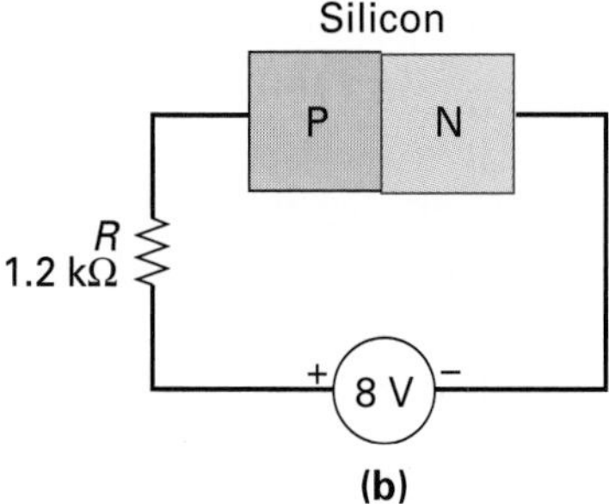

FIGURE 2-27 P-N Junction Circuit Examples.

will equal the forward voltage (V_F) for a silicon P-N junction. For Figure 2-27(b);

$$\text{Circuit Current} = I_F = \frac{V_S - V_{P\text{-}N}}{R}$$
$$= \frac{8\text{ V} - 0.7\text{ V}}{1.2\text{ k}\Omega}$$
$$= 6.08\text{ mA}$$
$$\text{P-N Junction Voltage Drop} = V_F = 0.7\text{ V}$$

SELF-TEST REVIEW QUESTIONS FOR SECTION 2-5

1. When a P-N junction is formed, a ______________ region is created on either side of the junction.
2. The barrier voltage within a silicon diode is:
   a. 700 mV **b.** 7.0 V **c.** 0.3 V **d.** None of the above
3. A P-N junction is forward biased when its P terminal is made positive relative to its N terminal. (True/False)
4. A reverse biased P-N junction acts like a/an ______________ switch, whereas a forward biased P-N junction acts like a/an ______________ switch.

SUMMARY

1. Materials can be divided into three main types according to the way they react to current when a voltage is applied across them. Insulators (nonconductors) are materials that have a very high resistance and therefore oppose current, whereas conductors are materials that have a very low resistance and therefore pass current easily. The third type of material is the semiconductor which, as its name suggests, has properties that lie between the insulator and conductor.
2. Semiconductor materials are neither a good conductor nor insulator. Their advantage is that they can be controlled to either increase their resistance and behave more like an insulator, or decrease their resistance and behave more like a conductor. It is this ability of a semiconductor material to vary its resistive properties that makes it so useful in electrical and electronic applications.
3. Semiconductor materials such as germanium and silicon are used to construct semiconductor devices such as diodes, transistors, and integrated circuits (ICs). These devices are used in electrical and electronic circuits to control current and voltage, so as to produce a desired result.
4. Semiconductor devices first became available in 1960. From about 1920 to 1960, the controlling element in all electrical and electronic circuits was the vacuum tube.
5. The term vacuum tube or thermionic valve is used to describe a variety of special devices that made possible radio, television, telecommunications, radar, sonar, computer, and many more systems between 1920 to 1960. These vacuum tubes were all enclosed in a sealed glass container, which had been pumped free of air (hence the name vacuum), and had electric connections to its internal parts through the base of the container. Two of the most frequently used vacuum tubes were the diode and triode.
6. The vacuum tube was replaced by the transistor in so many applications because it suffered from a number of physical drawbacks. The first is that vacuum tubes are fragile and can be easily damaged by shock or vibration. The second is that the heater required a great deal of power, and the third is their relatively large size. The second and third disadvantages made the vacuum tube unsuitable for battery-operated portable equipment, and, in addition, the tube heater had only a certain life span, which meant that the tubes needed to be replaced regularly.
7. Despite all of the vacuum tube limitations, it is still used today in many areas of science and technology. In industry, 100 kW triode vacuum tubes are made mechanically and electrically very rugged and are used to generate radio frequency (RF) power at frequencies from 100 kHZ to around 30 MHZ. In communications, vacuum tubes are used to generate the high-frequency and high-power outputs needed for radio and television transmitters. In science, vacuum tubes are used in fusion research and linear accelerators where experimental tubes are being operated at 100 kV. Cathode Ray Tubes (CRTs) are still used extensively in televisions and computer monitors.

8. The transistor was first introduced in 1948 by its inventors William Schockley, Walter Bratten, and John Bardeen and was described as a solid state device. This term was used because the transistor contained a solid semiconductor material between its input and output pins, unlike its predecessor the vacuum tube, which had a vacuum between its input and output pins.
9. The first point-contact transistor unveiled in 1948 was extremely unreliable, and it took its inventors another twelve years to develop the superior Bipolar Junction Transistor (BJT) and make it available in commercial quantities.
10. Research and development into semiconductor or solid state devices has resulted in a different type of transistor called the Field Effect Transistor (FET) which has characteristics similar to those of the vacuum tube. Once it was discovered that semiconductor materials could also generate and sense light, a new line of optoelectronic devices became available. Later it was discovered that semiconductor materials could sense magnetism, temperature, and pressure and, as a result, a variety of sensor devices or transducers (energy converters) appeared on the market. Along with all these different types of semiconductor devices, a wide variety of semiconductor diodes emerged that could rectify, regulate, and oscillate at high frequencies.
11. Although semiconductor diodes and transistors are still widely used as individual or discrete components, in 1959 Robert Noyce discovered that more than one transistor could be constructed on a single piece of semiconductor material. Soon other components such as resistors, capacitors, and diodes were added with transistors and then interconnected to form a complete circuit on a single chip or piece of semiconductor material. This integrating of various components on a single chip of semiconductor was called an Integrated Circuit (IC) or IC chip. The IC is used extensively in every branch of electronics with hundreds of thousands of transistors and other components being placed on a chip of semiconductor.
12. All matter on, in, and around the earth can be classified as being either a solid, a liquid, or a gas.
13. An element is a material consisting of only one type of atom.
14. Protons, neutrons, and electrons are subatomic particles that make up the atom.
15. The atomic number of an atom describes the number of protons within the nucleus, whereas the atomic weight of an atom can be used to describe the number of protons and neutrons within the atom's nucleus.
16. Elliptically orbiting electrons travel in paths or shells that are labeled K, L, M, N, O, P, and Q and extend out from the nucleus.
17. Like charges repel one another, while unlike charges attract.
18. An atom is the smallest particle of an element, whereas a molecule (which is the combination of two or more atoms) is the smallest part of a compound.

19. A semiconductor material is one that is neither a conductor nor a nonconductor (insulator). This means that it will not conduct current as well as a conductor or block current as well as an insulator.

20. Some semiconductor materials are pure or natural elements such as carbon (C), germanium (Ge), and silicon (Si), while other semiconductor materials are compounds.

Semiconductor Atoms (Figure 2-28)

21. All semiconductor atoms have four valence electrons. The valence shell of an atom can contain up to 8 electrons, and it is the number of electrons in this valence shell that determines the conductivity of the atom. An atom with only 1 valence electron would be classed as a good conductor whereas an atom having 8 valence electrons, and therefore a complete valence shell, would be classed as an insulator.

22. When two or more similar semiconductor atoms are combined to form a solid element, they automatically arrange themselves into an orderly lattice-like structure or pattern known as a crystal. Since each atom shares one electron with a neighboring atom, two atoms will share two, or a pair, of electrons between the two cores. These two-atom cores are pulling the two electrons with equal but opposite force and it is this pulling action that holds the atoms together in this solid crystal lattice structure. The joining together of two semiconductor atoms is called an electron-pair bond or covalent bond.

23. All of the atoms in pure semiconductors are electrically stable because all of their valence shells are complete (they all contain eight electrons). Pure semiconductor materials, called intrinsic materials, are therefore very poor conductors.

QUICK REFERENCE SUMMARY SHEET

FIGURE 2-28 Semiconductor Atoms.

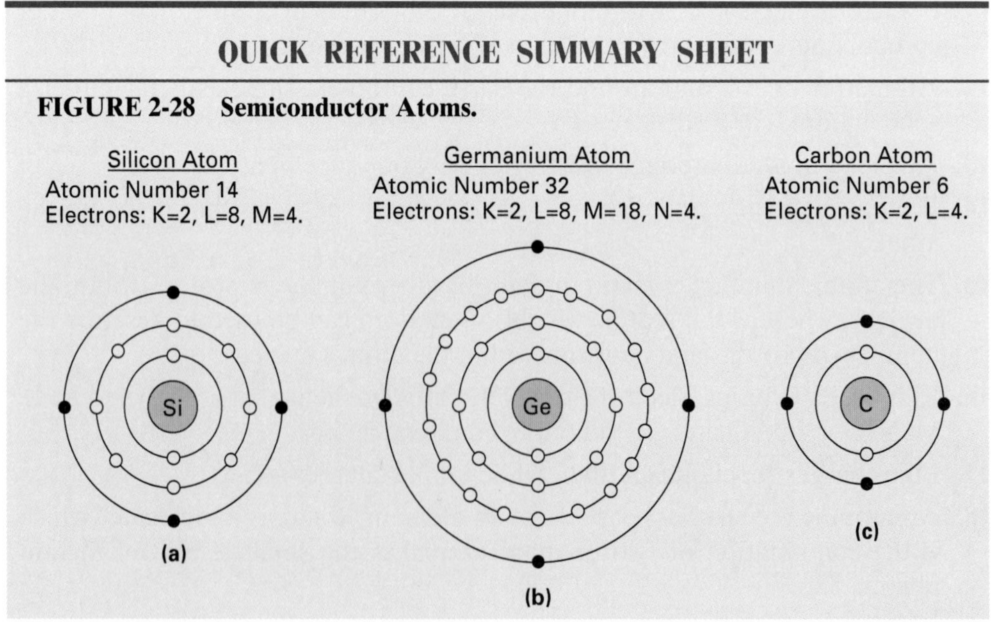

24. A pure semiconductor must be modified by a doping process to give it the qualities necessary to construct semiconductor devices.
25. Silicon is most frequently used to construct solid state or semiconductor devices such as diodes and transistors because germanium has poor temperature stability and carbon crystals (diamonds) are too expensive to use.
26. Electrons must travel, or orbit, in one of the orbital paths since they cannot exist in any of the spaces between orbital shells. Each orbital shell has its own specific energy level. The energy levels for each shell increase as you move away from the nucleus of the atom. The valence shell, and therefore the valence electrons, will always have the highest energy level for a given atom. The space between any two orbital shells is called the energy gap.
27. Electrons can jump from one shell to another if they absorb enough energy to make up the difference between their initial energy level and the energy level of the shell that they are jumping to.
28. If a valence electron absorbs enough energy it can jump from the valence shell into the conduction band. The conduction band is an energy band in which electrons can move freely or wander within a solid. When an electron jumps from the valence shell into the conduction band, it is released from the atom and no longer travels in one of its orbital paths. The electron is now free to move within the semiconductor material and is said to be in the excited state.
29. When an electron jumps from the valence shell, or band, to the conduction band, it leaves a gap in the covalent bond called a hole.
30. The conductivity of a semiconductor material is directly proportional to temperature, in that an increase in temperature will cause an increase in the semiconductor material's conductance. Stated another way, semiconductor materials, and therefore devices, have a negative temperature coefficient of resistance that means as temperature increases ($T\uparrow$), their resistance decreases ($R\downarrow$).
31. When a potential difference is applied across a semiconductor, the electrons move towards the positive potential and the holes travel towards the negative potential, and the total current flow is equal to the sum of the electron flow and the hole flow currents. The holes act like positively charged particles while the electrons act like negatively charged particles.
32. At room temperature, pure or intrinsic semiconductors will not permit a large enough value of current. Doping is a process whereby impurities are added to the intrinsic semiconductor material either to increase the number of free electrons (negative doping) or to increase the number of holes (positive doping).

Creating an *n*-Type and *p*-Type Semiconductor Material (Figures 2-29 and 2-30)

33. There are two types of impurities that can be added to semiconductor crystals. One type of impurity is called a pentavalent material because its atom has five (*penta*) valence electrons. The second type of impurity is called a

FIGURE 2-29 Creating an *n*-Type Semiconductor Material.

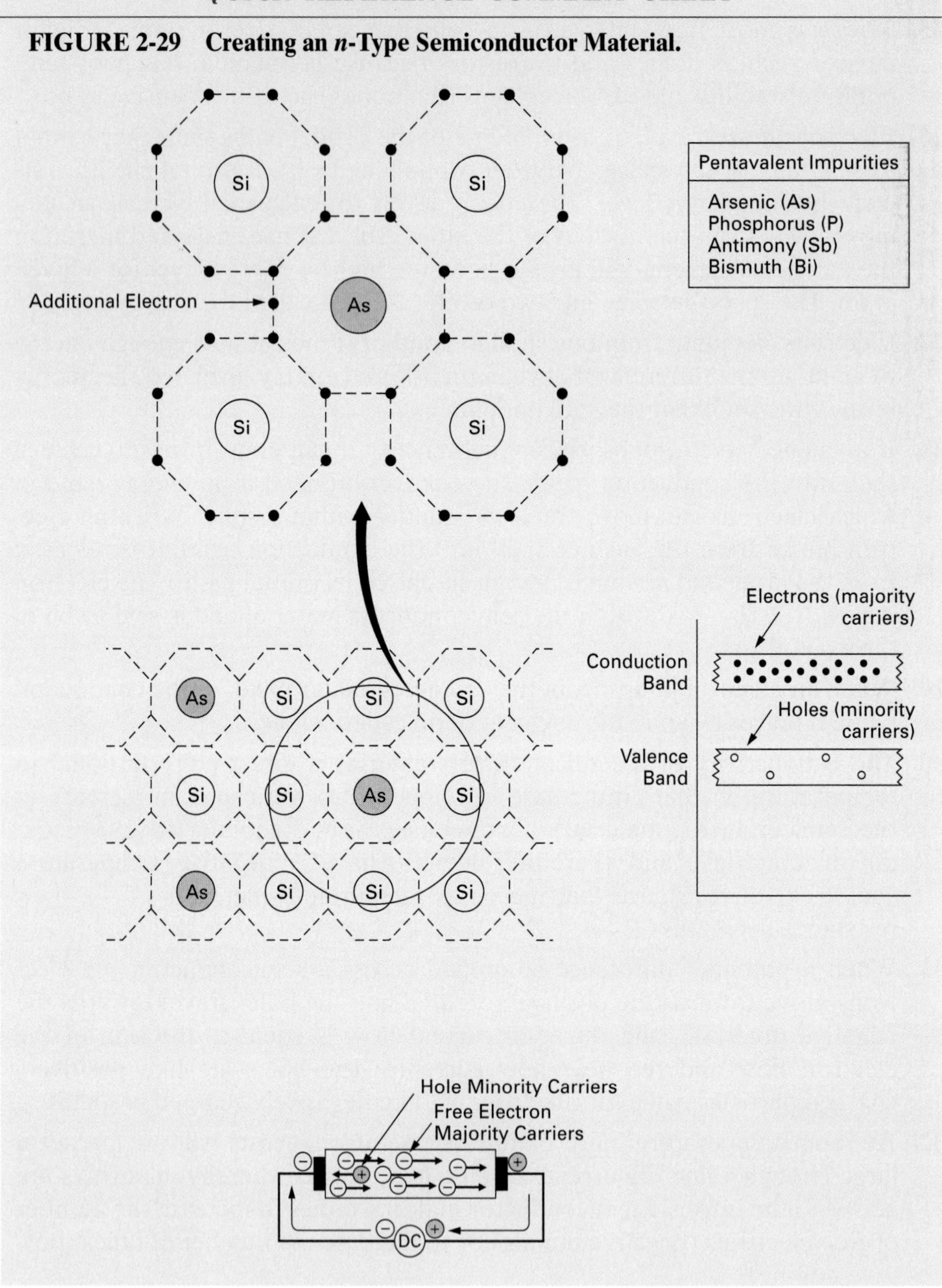

QUICK REFERENCE SUMMARY SHEET

FIGURE 2-30 Creating a *p*-Type Semiconductor Material.

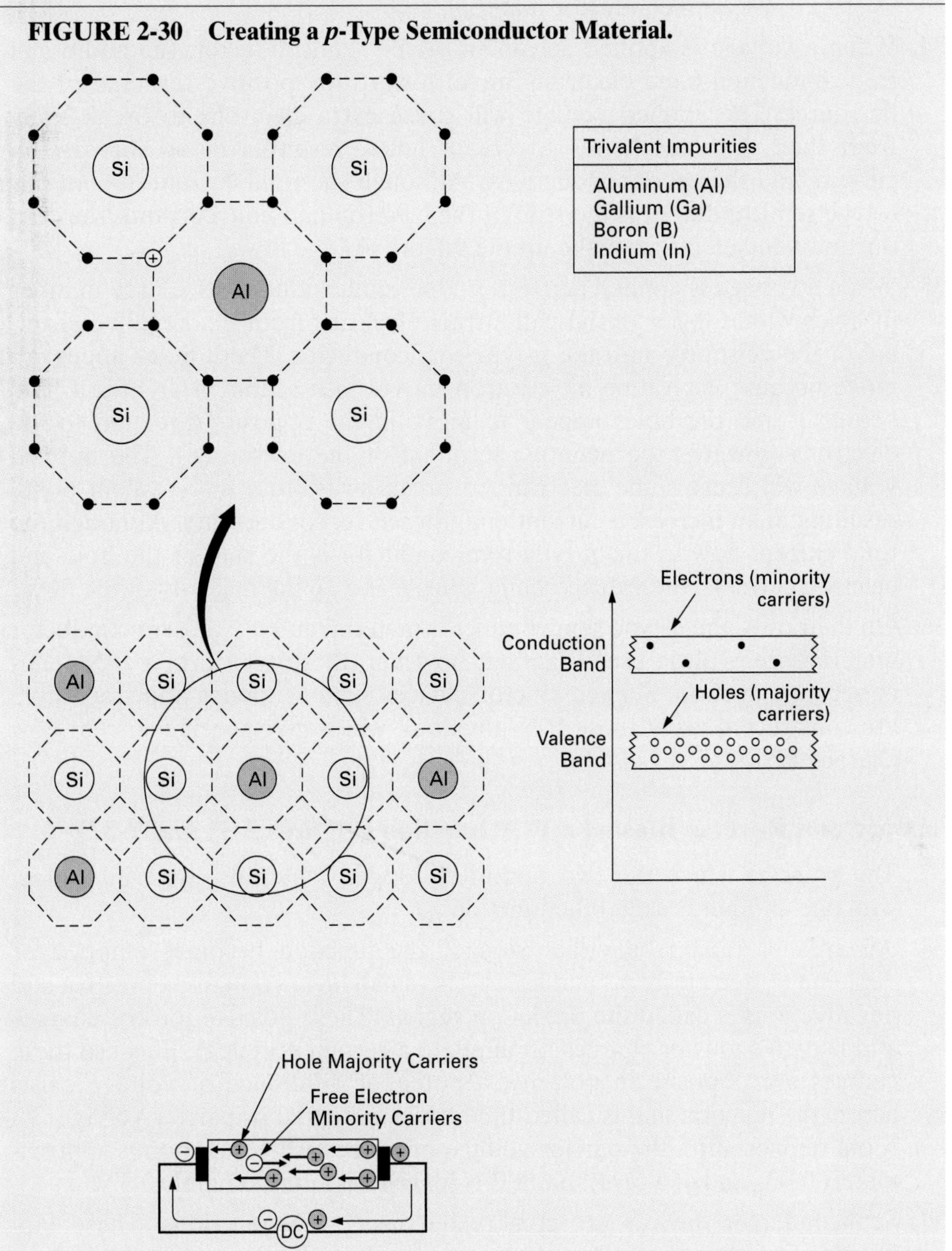

trivalent material because its atoms have three (*tri*) valence electrons. Because a doped semiconductor material is no longer pure it is referred to as an extrinsic semiconductor material.

34. When a voltage is applied across an *n*-type semiconductor, the additional free conduction-band electrons travel toward the positive terminal of the dc source. The applied voltage will cause extra electrons to break away from their covalent bonds to create holes, resulting in an increase in current and therefore conductivity. Although the total current flow in this *n*-type semiconductor is the sum of the electron and hole currents, the conduction-band electrons make up the majority of the flow.

35. When a voltage is applied across a *p*-type semiconductor, the large number of holes within the material will attract electrons from the negative terminal of the dc source into the *p*-type semiconductor. These holes appear to move because each time an electron moves into a hole, it creates a hole behind it and the holes appear to move in the opposite direction to the electrons (towards the negative terminal of the dc source). The applied voltage will cause some electrons to break away from the covalent bond, resulting in an increased current and, therefore, conductivity. Although the total current flow in this *p*-type semiconductor is the sum of the hole and electron currents, the valence-band holes make up the majority of the flow.

36. On their own, the *n*-type semiconductor material and *p*-type semiconductor material are of little use. Together, however, these two form a P-N semiconductor junction. Semiconductor devices such as diodes and transistors are constructed using these P-N junctions, which give specific current flow characteristics.

Forward and Reverse Biasing a P-N Junction (Figures 2-31 and 2-32)

37. The point at which the two oppositely doped materials come in contact with one another is called the junction.

38. An area or region on either side of the junction becomes emptied or depleted of free electrons and holes. This small layer containing positive and negative ions is called the depletion region. These positive ions or charges and negative ions or charges accumulate a certain potential. Because these charges are opposite in polarity, a potential difference or voltage exists across the junction and is called the barrier potential or barrier voltage. At room temperature, the barrier voltage of a silicon P-N junction is approximately 0.7 V, and of a germanium P-N junction is approximately 0.3 V.

39. Semiconductor devices are constructed using P-N junctions. These P-N junctions need voltages of a certain amplitude and polarity to control their operation. These voltages, which incline or cause the device to operate in a certain manner, are known as bias voltages. Bias voltages control the width of the depletion region, which in turn controls the resistance of the P-N junction and therefore the amount of current that can pass through the P-N junction or semiconductor device.

40. To be specific, a small depletion region ($dr\downarrow$) will offer a small P-N junction resistance ($R\downarrow$) and therefore permit a large P-N junction current ($I\uparrow$). In

QUICK REFERENCE SUMMARY SHEET

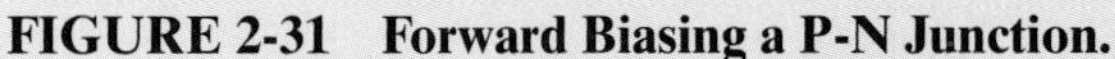

FIGURE 2-31 Forward Biasing a P-N Junction.

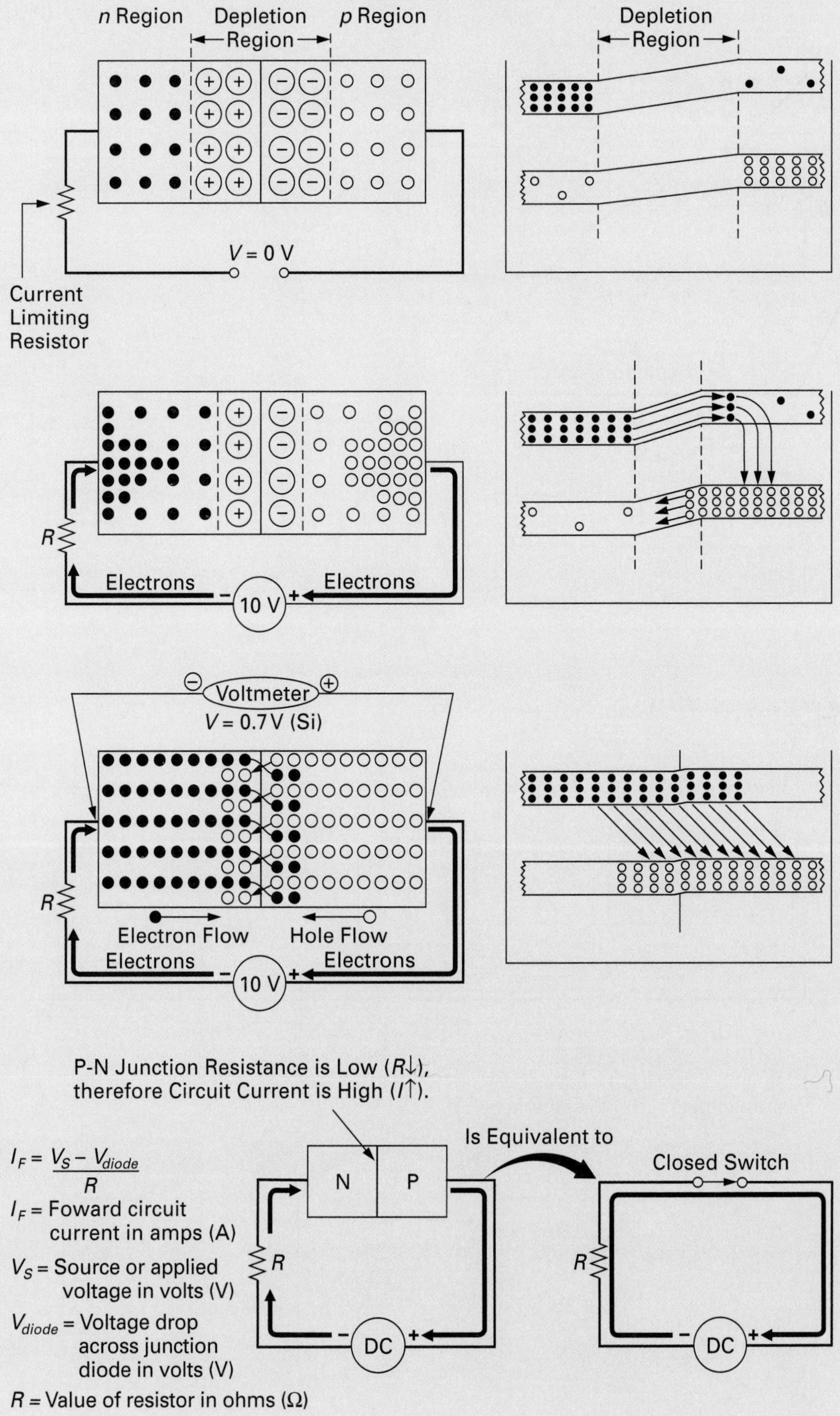

$$I_F = \frac{V_S - V_{diode}}{R}$$

I_F = Foward circuit current in amps (A)

V_S = Source or applied voltage in volts (V)

V_{diode} = Voltage drop across junction diode in volts (V)

R = Value of resistor in ohms (Ω)

FIGURE 2-32 Reverse Biasing a P-N Junction.

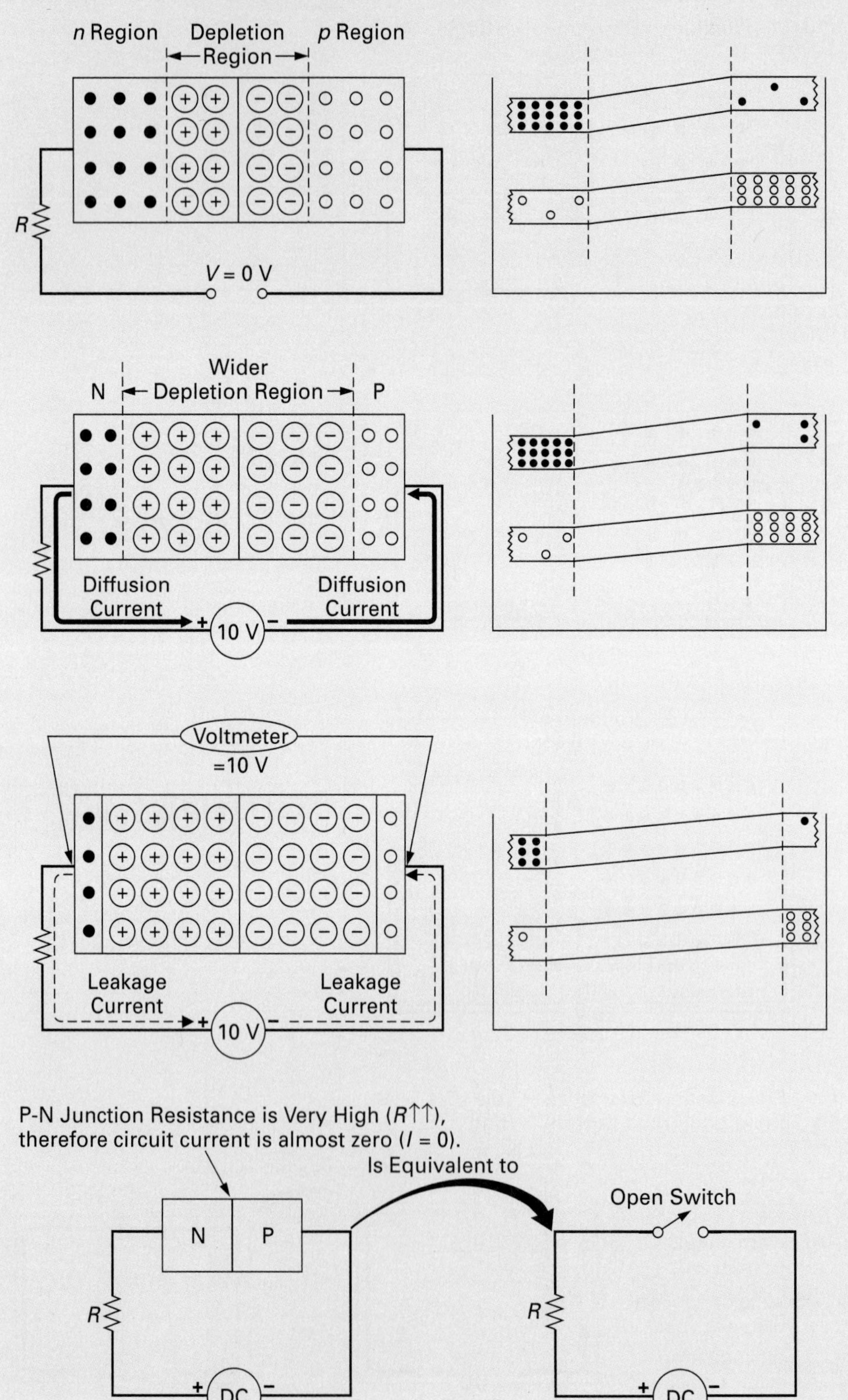

this instance the P-N semiconductor junction is said to be forward biased and acts like a conductor.

41. On the other hand, a large depletion region ($dr\uparrow$) will offer a large P-N junction resistance ($R\uparrow$) and therefore only permit a small P-N junction current ($I\downarrow$). In this instance the P-N semiconductor junction is said to be reverse biased and acts like an insulator.

42. When a P-N junction is forward biased ($+V \rightarrow p$ region, $-V \rightarrow n$ region), the P-N junction resistance is low ($R\downarrow$); therefore, the circuit current is high ($I\uparrow$). When forward biased, the P-N junction acts like a conductor and is equivalent to a closed switch.

43. When a P-N junction is reverse biased ($+V \rightarrow n$ region, $-V \rightarrow p$ region), the P-N junction resistance is extremely high ($R\uparrow\uparrow$); therefore, the circuit current is effectively zero ($I = 0$ A). When reverse biased, the P-N junction acts like an insulator and is equivalent to an open switch.

NEW TERMS

Anode
Atom
Atomic Number
Atomic Weight
Bands
Barrier Potential
Barrier Voltage
Bias Voltages
Biasing
Bipolar Junction Transistor (BJT)
Carbon
Cathode
Compound
Conduction Band
Conductors
Control Grid
Covalent Bond
Crystal
Depletion Region
Detector
Diffusion Current
Diodes
Discrete Components
Doping
Electron
Electron Flow
Electron-Pair Bond
Electron-Hole Pair
Electronic Numeric Integrator and Computer (ENIAC)
Element
Energy Gap
Energy Level
Excited State
Field Effect Transistor (FET)
Fleming Diode
Forward Biased
Forward Voltage Drop
Free Electron
Germanium
Hole
Hole Flow
IC Chip
Impurity Atom
Insulators
Integrated Circuit (IC)
Intrinsic
Junction
Leakage Current
Lifetime
Like Charges
Majority Carrier
Minority Carrier
Molecule
Negative Ion
Negative Temperature Coefficient of Resistance
Neutral Atom
Neutron
n-Type Semiconductor
Optoelectronic Devices
Pentavalent Material
Pentode
P-N Semiconductor Junction
Point-Contact Transistor
Positive Ion
Proton
p-Type Semiconductor
Recombination
Reverse Biased
Reverse Current
Reverse Voltage Drop
Semiconductor
Shells

Silicon	Thermionic Valve	Vacuum Tube
Solid State Device	Transistors	Valence Electrons
Subatomic	Triode	Valence Ring
Tetrode	Trivalent Material	Valence Shell
Thermionic Electron Emitter	Unlike Charges	

REVIEW QUESTIONS

Multiple-Choice Questions

1. What is the atomic number of silicon?

a. 14 **b.** 16 **c.** 10 **d.** 32

2. How many valence electrons are normally present in the valence shell of a semiconductor material?

a. 2 **b.** 4 **c.** 6 **d.** 8

3. Adding trivalent impurities to an intrinsic semiconductor will produce a/an ____________ material.

a. Extrinsic **b.** *n*-type **c.** *p*-type **d.** Both (a) and (b) are true

4. What is the majority carrier in an *n*-type material?

a. Holes **b.** Electrons **c.** Neutrons **d.** Protons

5. Adding pentavalent impurities to an intrinsic semiconductor will produce a/an ____________ material.

a. Extrinsic **b.** *n*-type **c.** *p*-type **d.** Both (a) and (b) are true

6. What are the majority carriers in a *p*-type semiconductor?

a. Holes **b.** Electrons **c.** Neutrons **d.** Protons

7. A semiconductor material has a ____________ temperature coefficient of resistance, which means that as temperature increases its resistance ____________.

a. Positive, increases **c.** Negative, increases
b. Positive, decreases **d.** Negative, decreases

8. A hole is considered to be ____________.

a. Negative **b.** Positive **c.** Neutral **d.** Both (b) and (c) are true

9. Intrinsic semiconductors are doped to increase their ____________.

a. Resistance **b.** Conductance **c.** Inductance **d.** Reactance

10. As temperature increases, a semiconductor acts more like a/an ____________.

a. Conductor **b.** Insulator

11. A negative ion has more:

a. Protons than electrons **c.** Neutrons than protons
b. Electrons than protons **d.** Neutrons than electrons

12. A positive ion has:

a. Lost some of its electrons **c.** Lost neutrons
b. Gained extra protons **d.** Gained more electrons

13. The resistance of a semiconductor material is more than the resistance of:

    a. Glass **b.** Copper **c.** Ceramic **d.** Both (a) and (c) are true

14. The basic function of a semiconductor device in an electrical or electronic circuit is to:

    a. Control current **c.** Increase the price of the equipment
    b. Control voltage **d.** Both (a) and (b) are true

15. For a silicon P-N junction, $V_F = ?$

    a. The value of the applied voltage **b.** 300 mV **c.** 0.7 V **d.** 10 V

Essay Questions

16. What key advantages do solid state devices have over vacuum tube devices? (2-1)
17. What two semiconductor materials are most frequently used in the manufacture of semiconductor devices? (2-3)
18. What do all semiconductor atoms have in common? (2-3-1)
19. Define the following terms:

    a. Covalent Bond (2-3-2)
    b. Intrinsic Semiconductor (2-3-2)
    c. Conduction Band (2-3-3)
    d. Hole Flow (2-3-5)
    e. Doping (2-4)
    f. P-N Semiconductor Junction (2-5)
    g. Depletion Region (2-5-1)
    h. P-N Junction (2-5-1)
    i. Majority Carriers (2-4)
    j. Minority Carriers (2-4)

20. Why do semiconductor materials, and therefore devices, have a negative temperature coefficient of resistance? (2-3-4)
21. Semiconductor devices are made from intrinsic or pure semiconductor materials. True or False? (2-4)
22. What is the relationship between the number of valence electrons in an atom and the conductivity of an element? (2-3-1)
23. Describe the differences between:

    a. An *n*-type semiconductor material (2-4-1)
    b. A *p*-type semiconductor material (2-4-2)

24. Why are doped semiconductor materials still electrically neutral? (2-4-1)
25. How is a P-N junction formed? (2-5)
26. What is a depletion region, and how is it formed? (2-5-1)
27. What is the barrier voltage of:

    a. A silicon P-N junction (2-5-1) **b.** A germanium P-N junction (2-5-1)

28. Describe what occurs when a P-N junction is:

    a. Forward biased (2-5-2) **b.** Reverse biased (2-5-2)

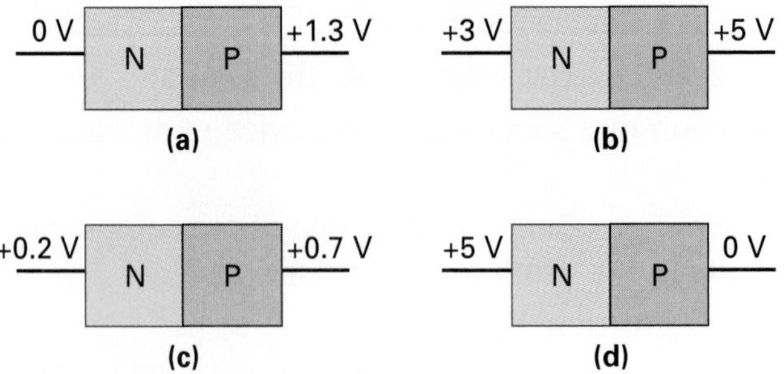

FIGURE 2-33 **Biased P-N Junctions.**

29. Define the following:

a. Forward Voltage Drop (2-5-2) **b.** Leakage Current (2-5-2)

30. Describe why a P-N junction acts like a switch. (2-5-2)

Practice Problems

31. Which of the silicon P-N junctions in Figure 2-33 are forward biased, and which are reverse biased?

32. Determine the current for the circuits shown in Figure 2-34.

33. What would be the voltage drop (V_F) across each of the P-N junctions shown in Figure 2-34?

34. What would be the voltage drop across each of the resistors in Figure 2-34?

35. Which of the P-N junctions in Figure 2-34 are equivalent to open switches, and which are equivalent to closed switches?

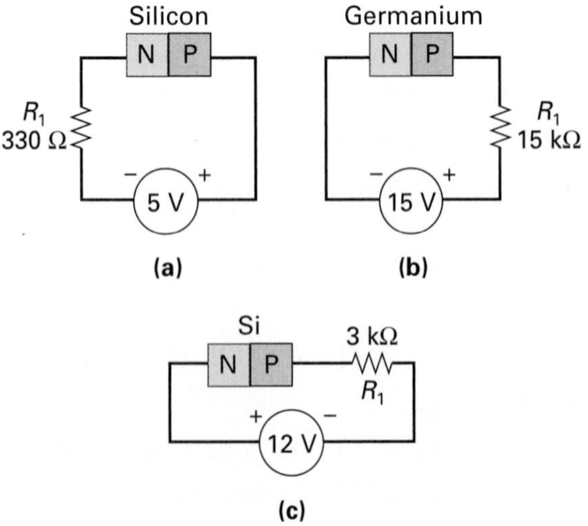

FIGURE 2-34 **P-N Junction Circuit.**

CHAPTER 3

OBJECTIVES

After completing this chapter, you will be able to:

1. Name and identify the terminals of a junction diode package and its schematic symbol.
2. Determine whether a diode is forward or reverse biased by observing the applied bias voltage's polarity.
3. Show how the diode can be used in a basic circuit application.
4. Explain how a junction diode contains a P-N junction.
5. Describe how the *p*-type and *n*-type material of a diode react to forward and reverse bias voltages.
6. Explain the forward and reverse characteristics of junction diodes.
7. Interpret the graphically plotted voltage-current *(V-I)* and temperature characteristic curves for a typical junction diode.
8. Describe how the junction diode is used in the following circuit applications:
   a. A half-wave rectifier circuit
   b. A logic OR gate
9. Determine whether a junction diode is functioning normally by testing it with an ohmmeter.

Junction Diodes

OUTLINE

VIGNETTE

A Problem with Early Mornings

René Descartes was born in Brittany, France, in 1596. At the age of eight he had surpassed most of his teachers at school and was sent on to the Jesuit College in La Flèche, one of the best in Europe. It was here that his genius in mathematics became apparent; however, due to his extremely delicate health, his professors allowed him to study in bed until midday.

In 1616 he had an urge to see the world and so he joined the army which made use of Descartes' mathematical genius in military engineering. While traveling, Descartes met Dutch philosopher Isaac Beekman who convinced him to leave the army and, in his words, "turn his mind back to science and more worthier occupations."

After leaving the army Descartes traveled looking for some purpose, and then on November 10th, 1619, he found it. Descartes was in Neuberg, Germany, where he had shut himself in a well-heated room for the winter. It was on the eve of St. Martin's that a freezing blizzard forced Descartes to retire early. That night he described having an extremely vivid dream that clarified his purpose and showed him that physics and all sciences could be reduced to geometry and were therefore all interconnected like a chain.

In his time, and to this day, he is heralded as an analytical genius. In fact, Descartes' procedure can still be used as a guide to solving any problem.

Descartes' four-step procedure for solving a problem:

1. Never accept anything as true unless it is clear and distinct enough to exclude doubt from your mind.
2. Divide the problem in as many parts as necessary to reach a solution.
3. Start with the simplest things and proceed step by step towards the complex.
4. Review the solution so completely and generally that you are sure nothing was omitted.

For me, this four-step procedure has been especially helpful as a troubleshooting guide for system and circuit malfunctions.

Descartes' fame was so renowned that he was asked in 1649 to tutor Queen Christina of Sweden. The Queen demanded that her lessons begin at 5 o'clock in the morning, which conflicted with Descartes' lifetime practice of remaining in bed until midday. After several unsuccessful attempts to change her majesty's mind, and with pressure being applied by the French ambassador, Descartes agreed to the early morning lessons. A short time later on his way to the palace one cold winter morning, Descartes caught a severe chill and died within two weeks.

INTRODUCTION

It was philosopher René Descartes who first stated that to solve any problem you should start with the simple and then proceed to the complex. For Descartes, there were three approximations: *The first approximation was the simplest, the second approximation contained more detail, and the third approximation was the complex.* I have applied this method to the problem of learning any new topic.

In this chapter you will be introduced to your first semiconductor component: the **diode.** To help you gain a clear understanding of this device, we will begin with a *first approximation description of the diode,* in which we will discuss the diode's schematic symbol, physical appearance, basic operation, and basic application. Following this basic complete picture description, the *second approximation description of the diode* will cover the diode in more detail addressing its characteristics, analog and digital circuit applications, data sheet specifications, and testing procedure.

As a technician, you will not need to examine the *third approximation description of a diode,* since these topics include more detail on specific device specifications and how to implement diodes into new circuit designs. This area of understanding is only needed for engineering students specializing in design. Throughout this text we will be concentrating on semiconductor-device operation, characteristics, applications, and testing, along with analog and digital circuit applications and troubleshooting—the necessary knowledge for a good **electronics technician.**

3-1 FIRST APPROXIMATION DESCRIPTION OF A DIODE

The first diode was accidentally created by Edison in 1883 when he was experimenting with his light bulb. At this time he did not place any importance on the device and its effect, as he could not see any practical application for it. The word *diode* is derived from the fact that the device has two (*di*) electrodes (*ode*).

Once the importance of diodes was realized, construction of the device began. The first diodes were vacuum-tube devices having a hot-filament negative cathode, which released free electrons that were collected by a positive plate called the anode. Today's diode is made of a P-N semiconductor junction but still operates on the same principle. The *n*-type region (cathode) is used to supply free electrons, which are then collected by the *p*-type region (anode). The operation of both the vacuum tube and semiconductor diode is identical in that the device will only pass current in one direction. That is, it will act as a conductor and pass current easily in one direction when the bias voltage across it is of one polarity, yet it will block current and imitate an insulator when the bias voltage applied is of the opposite polarity.

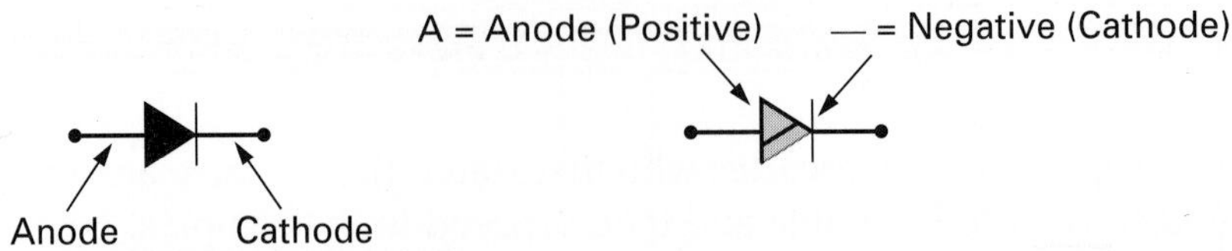

FIGURE 3-1 Schematic Symbol of a Diode.

3-1-1 *Diode Schematics Symbol and Packaging*

The two electrodes or terminals of the diode are called the **anode** and **cathode,** as seen in Figure 3-1(a) which shows the schematic symbol of a diode. To help you remember which terminal is the anode and which is the cathode, and which terminal is positive and which is negative, Figure 3-1(b) shows how a line drawn through the triangle section of the symbol will make the letter "A" and indicate the "anode" terminal. Similarly, if the vertical flat side of the diode symbol is aligned horizontally "—", as in Figure 3-1(b), it becomes the "negative" symbol. This memory system helps us to remember that the anode terminal of the diode is next to the triangle part of the symbol and is positive, while the cathode terminal of a diode is next to the vertical line of the symbol and is negative.

The diode is generally mounted in one of the three basic packages shown in Figure 3-2. These packages are designed to protect the diode from mechanical stresses and the environment. The difference in the size of the packages is due to the different current rating of the diode. A black band or stripe is generally placed on the package closest to the cathode terminal for identification purposes, as seen in Figure 3-2(a) and (b). Larger diode packages, like the one seen in Figure 3-2(c), usually have the diode symbol stamped on the package to indicate anode/cathode terminals.

3-1-2 *Diode Operation*

As far as operation is concerned, the diode operates like a switch. If you give the diode what it wants, that is make the anode terminal positive with respect to

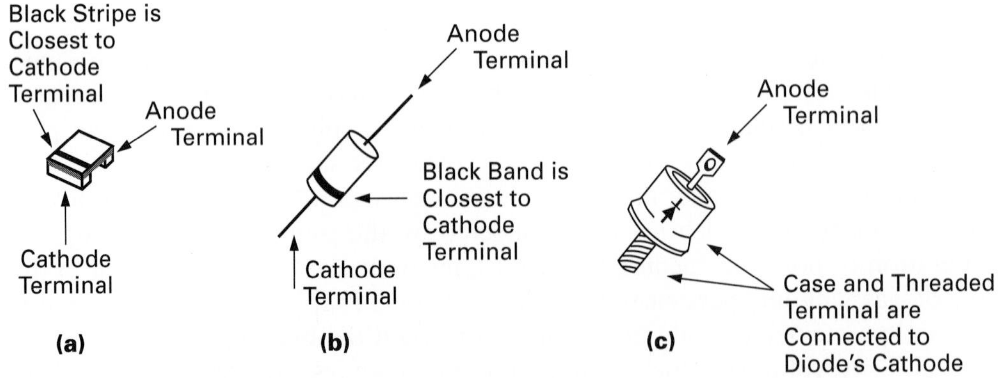

FIGURE 3-2 Diode Packaging. (a) Chip Package—1/4 A. (b) Small Current Package—less than 3 A. (c) Large Current Package—greater than 3 A.

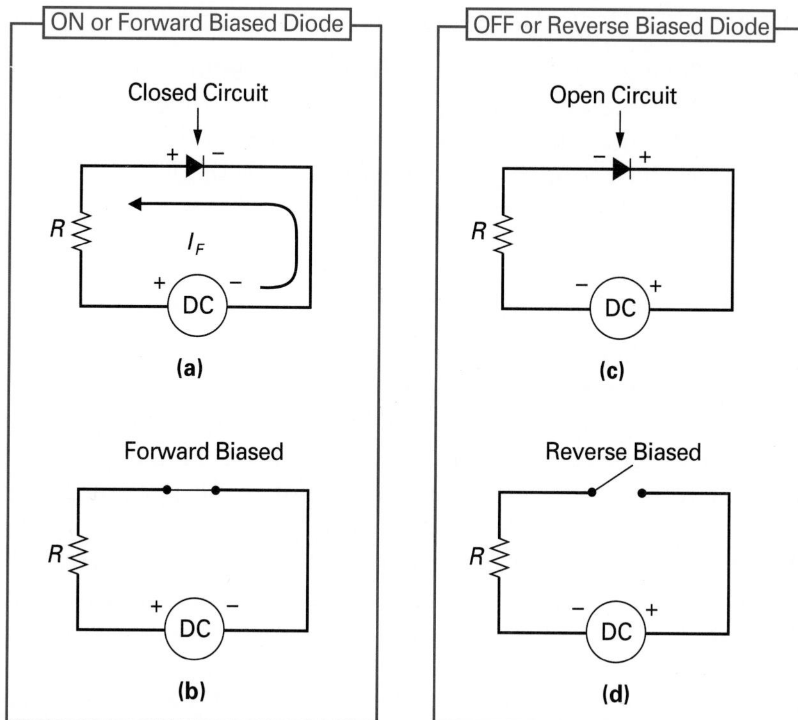

FIGURE 3-3 Diode Operation. (a)(b) Forward Biased (ON) Diode. (c)(d) Reverse Biased (OFF) Diode.

the cathode terminal as seen in Figure 3-3(a), the device is equivalent to a closed switch as seen in Figure 3-3(b). In this condition, the diode is said to be ON or *forward biased.*

On the other hand, if you do not give the diode what it wants, that is make the anode terminal negative with respect to the cathode as seen in Figure 3-3(c), the device is equivalent to an open switch as seen in Figure 3-3(d). In this condition the diode is said to be OFF or *reverse biased.*

■ **EXAMPLE:**

Determine whether the diodes in Figure 3-4 are ON or OFF.

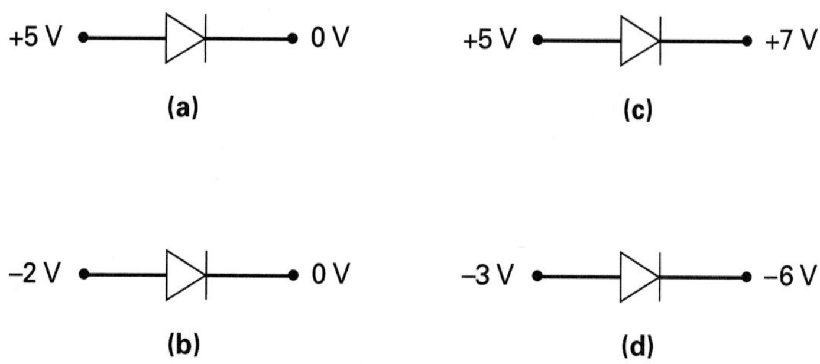

FIGURE 3-4 ON or OFF Diodes?

■ ***Solution:***

a. Diode is ON since anode is positive relative to cathode.
b. Diode is OFF since anode is negative relative to cathode.
c. Diode is OFF since anode is less positive than cathode.
d. Diode is ON since anode is more positive than cathode.

3-1-3 Diode Application

Encoder Circuit
A circuit that produces different output voltage codes, depending on the position of a rotary switch.

As an application, Figure 3-5 shows how the diode can be used as a switch within an **encoder circuit.** The pull-up resistors R_1, R_2, and R_3 ensure that lines *A, B,* and *C* are normally all at +5 V. This is the output voltage on each line when the rotary switch is in position 2, as seen in the table in Figure 3-5.

When the rotary switch is turned to position 1, D_1 is connected in-circuit and, because its anode is made positive via R_2 and its cathode is at 0 V, the diode D_1 will turn ON and be equivalent to a closed switch. The 0 V on the cathode of D_1 will be switched through to line *B* (all of the five volts will be dropped across R_2) producing an output voltage code of A = +5 V, B = 0 V, C = +5 V as seen in the table in Figure 3-5.

When the rotary switch is turned to position 3, D_2 and D_3 are connected in circuit and because both anodes are made positive via R_1 and R_3, and both diode cathodes are at 0 V, D_2 and D_3 will turn ON. These forward biased diodes will switch 0 V through to lines *A* and *C,* producing an output voltage code of A = 0 V, B = +5 V, C = 0 V, as seen in the table in Figure 3-5.

This *code generator* or *encoder* circuit will produce three different output voltage codes for each of the three positions of the rotary switch. These codes could then be used to initiate one of three different operations based on the operator setting of the rotary control switch.

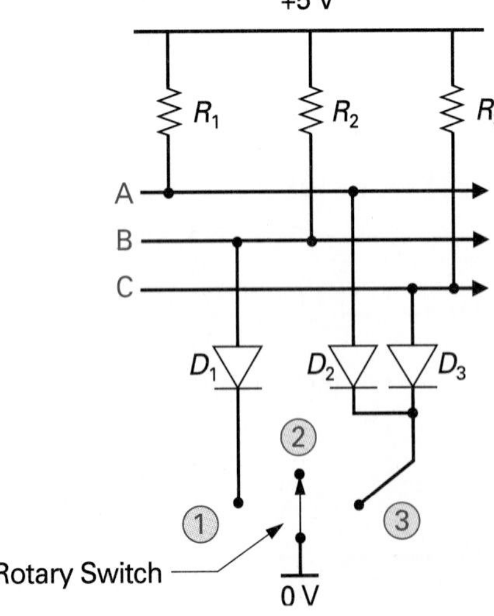

Input	Output		
Switch Position	A	B	C
1	+5 V	0 V	+5 V
2	+5 V	+5 V	+5 V
3	0 V	+5 V	0 V

FIGURE 3-5 Diode Application: A Switch Encoder Circuit.

SELF-TEST REVIEW QUESTIONS FOR SECTION 3-1

1. Name the two terminals of a diode.
2. The diode operates like a ____________.
3. An ON diode is said to be ____________ biased, while an OFF diode is said to be ____________ biased.
4. Would a diode be ON or OFF if its anode had +6 V applied and its cathode had +9 V applied?

3-2 SECOND APPROXIMATION DESCRIPTION OF A JUNCTION DIODE

Now that we understand the basic operation and application of the diode, let us examine its characteristics in more detail.

3-2-1 *The P-N Junction Within a Junction Diode*

In the previous chapter, we discussed how a pure or intrinsic semiconductor material could be doped with a pentavalent element or trivalent element to obtain the two basic semiconductor types. The first type is called an *n*-type semiconductor because its majority carriers are electrons, while the second type is called *p*-type semiconductor because its majority carriers are holes. On their own, the *n*-type semiconductor and *p*-type semiconductor are of little use. Together, however, these two form a P-N semiconductor junction. A manufacturer of semiconductor devices would not join two individual pieces to create a P-N junction. Instead, a single piece of pure semiconductor material would have each of its halves doped to create a *p*-type and *n*-type section or region.

Semiconductor devices such as diodes and transistors are constructed using these P-N junctions. A diode, for example, has only one P-N junction and is created by doping a single piece of pure semiconductor to produce an *n*-type and *p*-type region. A bipolar junction transistor, on the other hand, has two P-N junctions and is created by doping a single piece of pure semiconductor with three alternate regions (NPN or PNP). As mentioned in Chapter 2, the point at which these two opposite-doped materials come in contact with each other is called a junction, which is why these devices are called **junction diodes** and *bipolar junction transistors.*

Junction Diode
A semiconductor diode whose ON/OFF characteristics occur at a junction between the *n*-type and *p*-type semiconductor materials.

Figure 3-6 illustrates the schematic symbol for the *junction diode,* and the inset shows how it contains one P-N junction. The *n*-type region is called the *cathode,* while the *p*-type region is called the *anode.*

Since a junction diode is basically a P-N semiconductor junction, the diode will operate in exactly the same way as the P-N junction described in Chapter 2.

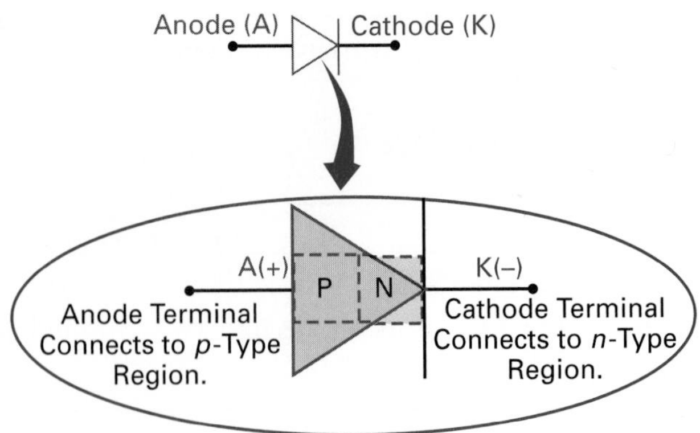

FIGURE 3-6 The P-N Junction Within a Junction Diode.

3-2-2 *Biasing a Junction Diode*

Bias Voltage
Voltage that inclines or causes the diode to operate in a certain manner.

Semiconductor diodes are constructed using P-N junctions. These P-N junctions need voltages of a certain amplitude and polarity to control their operation. These voltages, which incline or cause the diode to operate in a certain manner, are known as **bias voltages.** Bias voltages control the width of the depletion region, which in turn controls the resistance of the junction and, therefore, the amount of current that can pass through the P-N junction diode.

Forward Biasing a Diode

Figure 3-7(a) shows how a junction diode can be forward biased. Like the P-N junction, the junction diode's operation is determined by the polarity of the applied voltage. In this figure the negative terminal of the applied voltage is connected to the *n* region of the diode, and the positive terminal of the applied voltage is connected to the *p* region of the diode ($+V \rightarrow p$ region, $-V \rightarrow n$ region). Free electrons are repelled from the *n* region by the negative source and attracted to the positive terminal of the voltage source. This forward-conducting electron flow will only occur if the external source voltage is large enough to overcome the internal barrier voltage of the junction diode. For a silicon diode, the external source voltage must be equal to or greater than 0.7 V, whereas for a germanium diode, the applied voltage must be equal to or greater than 0.3 V. The resistor is added in series with the diode to limit the forward current to a safe level because an excessive current will generate more heat than the diode can dissipate, causing the diode to burn out.

A forward biased diode will conduct current as long as the external bias voltage is of the correct polarity and amplitude. Figure 3-7(a) shows the direction of forward current (I_F). This electron-flow current is from the negative terminal of the applied voltage to the *n* region of the diode, through the diode to the *p* region, and then to the positive terminal of the applied voltage source. This means that forward (electron flow) current passes through the diode symbol from the bar to the triangle. In other words, forward electron flow is actually

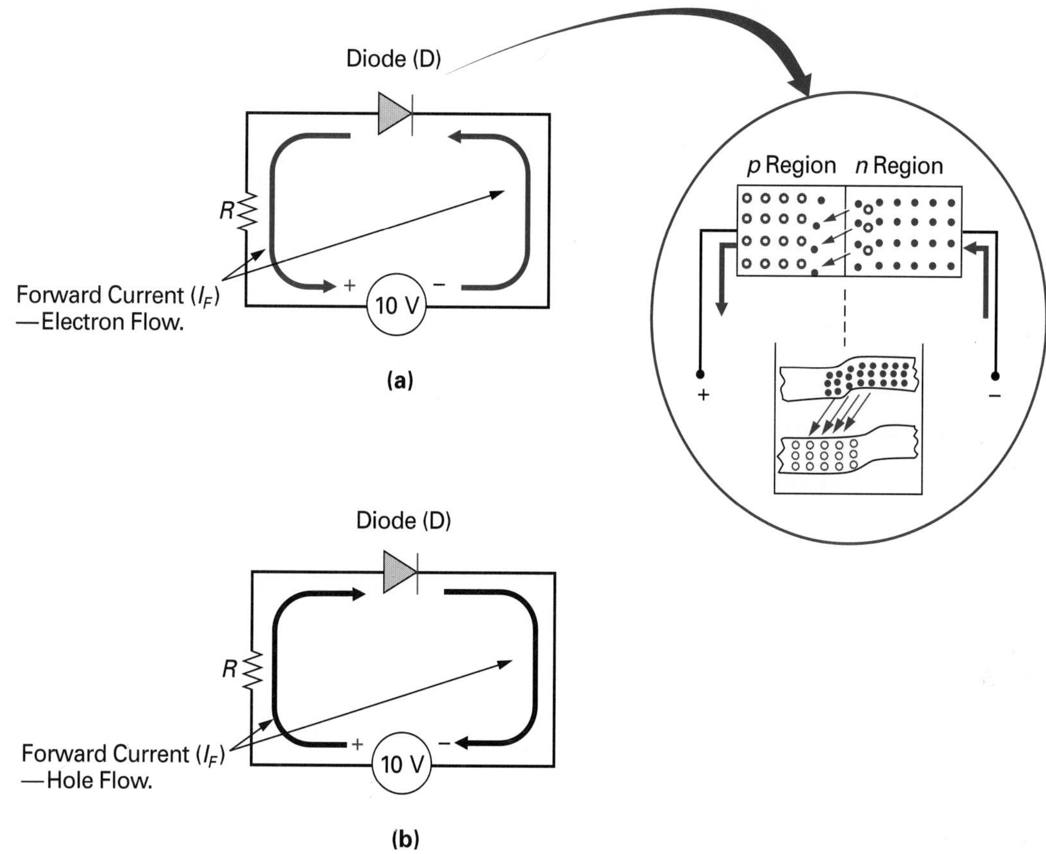

FIGURE 3-7 A Forward Biased P-N Junction Diode. (a) Electron Flow. (b) Hole (Conventional) Flow.

traveling against the arrow formed by the diode's symbol. To further explain this, let us compare Figure 3-7(a), which shows forward electron flow, to Figure 3-7(b), which shows forward hole flow. As previously discussed in Chapter 2, apparent hole flow is in the opposite direction to electron flow. That is, electron flow is from negative to positive, while hole flow, or conventional current flow, is from positive to negative. The diode symbol actually points in the direction of forward hole flow, and therefore the symbol reflects conventional flow. Remember, when a P-N junction or diode is forward biased, the electrons move towards the positive terminal of the applied voltage, and the holes travel towards the negative terminal of the applied voltage. The total current flow is equal to the sum of the electron flow and the hole flow currents.

Like the P-N junction, a diode, when forward biased, has a low but finite resistance value that will cause a corresponding voltage drop across its terminals. This voltage drop is approximately equal to the barrier voltage of the diode (Si = 0.7 V, Ge = 0.3 V). Knowing the diode's forward voltage drop, the value of applied voltage, and the value of circuit resistance, we can calculate the value of forward current. You will recognize this formula because it is identical to the one used in Chapter 2 for the P-N junction.

$$I_F = \frac{V_S - V_{diode}}{r}$$

I_F = Forward circuit current in amps (A)
V_S = Source or applied voltage in volts (V)
V_{diode} = Voltage drop across junction diode in volts (V)
R = Value of resistor in ohms (Ω)

■ **EXAMPLE:**

Calculate the value of current for the circuit shown in Figure 3-8.

■ ***Solution:***

The diode is forward biased because the applied voltage is connected so that its positive terminal is applied to the *p* region (anode) and the negative terminal is applied to the *n* region (cathode). Because a silicon diode is being used, the forward voltage drop will be 0.7 V. With an applied voltage of 8.5 V and a circuit resistance of 1.2 kΩ, the circuit current will equal

$$I_F = \frac{V_S - V_{diode}}{R}$$

$$I_F = \frac{8.5\text{ V} - 0.7\text{ V}}{1.2\text{ k}\Omega}$$

$$I_F = \frac{7.8\text{ V}}{1.2\text{ k}\Omega}$$

$$I_F = 6.5\text{ mA}$$

Reverse Biasing a Diode

Figure 3-9 shows how a junction diode can be reverse biased. Once again, the junction diode's operation is determined by the polarity of the applied voltage. In this figure the positive terminal of the applied voltage is connected to the

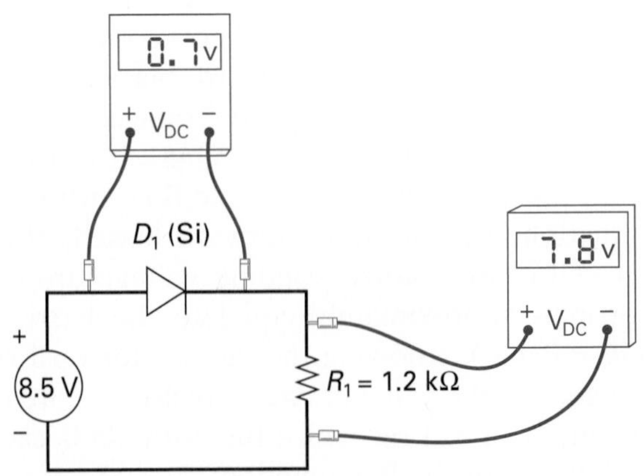

FIGURE 3-8 A P-N Junction Diode Circuit.

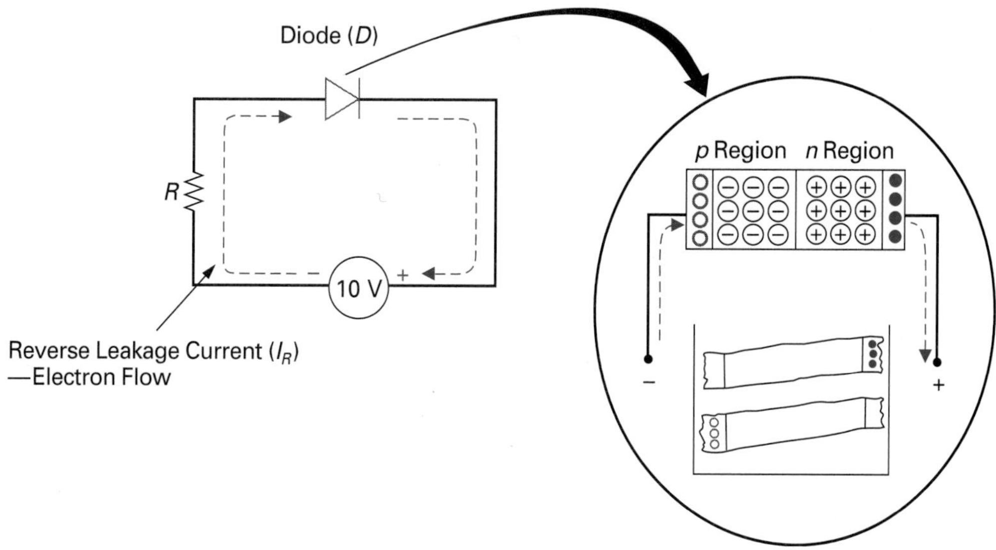

FIGURE 3-9 A Reverse Biased P-N Junction Diode.

n region of the diode, and the negative terminal of the applied voltage is connected to the *p* region of the diode ($+V \rightarrow n$ region, $-V \rightarrow p$ region). This applied voltage polarity will reverse bias the junction diode since free electrons in the *n* region are attracted and therefore travel to the positive terminal of the applied source voltage. At the same time, holes feel the attraction of the negative terminal of the applied voltage. As a result, the depletion region will increase in width until its internal voltage is equal, but opposite to, the external applied voltage. At this time the diode will stop conducting, cutting off all current flow.

In actual fact, a very small current leaks through the diode when it is reverse biased. This current is extremely small (microamps to nanoamps in value) and in most cases can be ignored. It is caused by minority carriers, which are holes in the *n* region and electrons in the *p* region, that are forced towards the junction by the applied source voltage.

■ EXAMPLE:

Calculate the voltage drop across the diodes and the resistors in Figure 3-10.

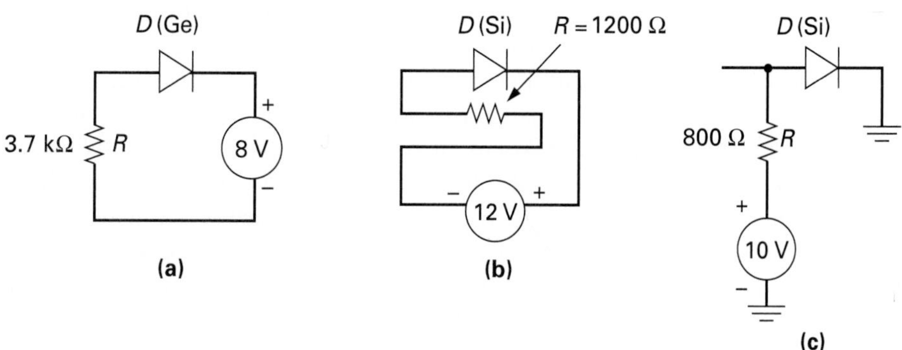

FIGURE 3-10 P-N Junction Diode Biasing Examples.

■ ***Solution:***

a. The germanium diode in Figure 3-10(a) is reverse biased ($+V \rightarrow n$ region, $-V \rightarrow p$ region), and is therefore equivalent to an open switch. Since all of the applied voltage will always appear across an open in a series circuit

$$V_{diode} = V_S = 8 \text{ V}$$
$$V_R = 0 \text{ V}$$

b. The silicon diode in Figure 3-10(b) is also reverse biased ($+V \rightarrow n$ region, $-V \rightarrow p$ region), and is therefore equivalent to an open switch. Since all of the applied voltage will always appear across an open in a series circuit:

$$V_{diode} = V_S = 12 \text{ V}$$
$$V_R = 0 \text{ V}$$

c. The silicon diode in Figure 3-10(c) is forward biased ($+V \rightarrow p$ region, $-V$ or ground $\rightarrow n$ region), and is therefore equivalent to a closed switch. The voltage drop across a forward biased silicon diode is approximately equal to the barrier voltage of the diode, which is 0.7 V. The voltage drop across the resistor will therefore be equal to the difference between the applied source voltage (V_S) and the voltage drop across the diode (V_{diode}).

$$V_{diode} = 0.7 \text{ V}$$
$$V_R = V_S - V_{diode} = 10 \text{ V} - 0.7 \text{ V} = 9.3 \text{ V}$$

3-2-3 *The Junction Diode's Characteristic Curve*

Now that we know how the junction diode operates, it is time to examine the diode's characteristics in a little more detail. To help us analyze the P-N junction diode's voltage, current, and temperature characteristics at various values, we will plot the diode's characteristics on a graph, since a picture is generally worth a thousand words.

Graph Origin
Center of the graph where the horizontal axis and vertical axis cross.

The graph in Figure 3-11 shows how much current will pass through a typical junction diode when it is both forward biased or reverse biased. The center of the graph, where the horizontal axis and vertical axis cross, is called the **graph origin.** This origin is a zero point for both voltage and current. For instance, voltage is plotted on the horizontal axis, with forward bias voltages (V_F) increasing positively to the right of the origin and reverse bias voltages (V_R) increasing negatively to the left of the origin or zero voltage point. Conversely, current is plotted on the vertical axis of this graph, with forward current (I_F) increasing positively above the origin and reverse current (I_R) increasing negatively below the origin or zero current point. Manufacturers of diodes create a graph like the one seen in Figure 3-11 by applying various values of forward and reverse voltages. The result is a continuous curve called a voltage-current or *V-I* characteristic curve. Let us now examine the forward diode characteristics (upper right quadrant) and the reverse diode characteristics (lower left quadrant) in more detail.

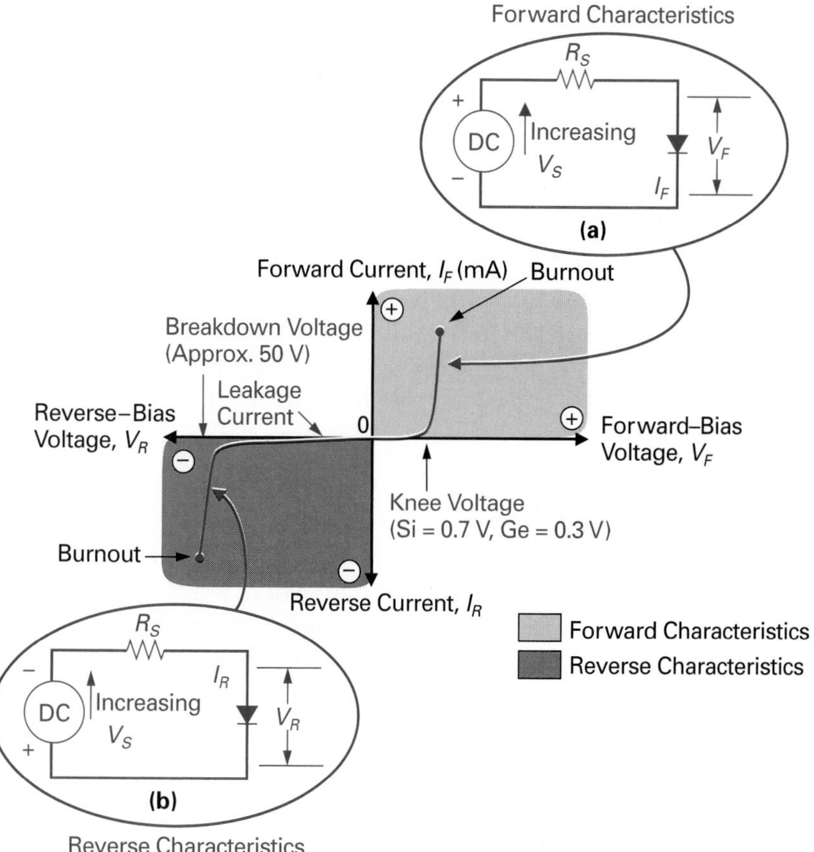

FIGURE 3-11 The Junction Diode Voltage/Current Characteristic Curve.

Forward Characteristics

The upper right quadrant of the four sections in Figure 3-11 shows what forward current will pass through the diode when a forward bias voltage is applied. As you can see from the inset, the diode is forward biased by applying a positive potential to its anode and a negative potential to its cathode. To review, when the forward bias voltage exceeds the diode's internal barrier voltage, its resistance drops to almost zero, resulting in a rapid increase in forward current. In this instance the diode is said to be ON, and is equivalent to a closed switch.

These characteristics can be seen in the forward curve in Figure 3-11. Beginning at the graph origin and following the curve into the forward quadrant, you can see that the forward current through a diode is extremely small until the forward bias voltage exceeds the diode's internal barrier voltage, which for silicon is 0.7 V and for germanium is 0.3 V. Once the forward bias voltage exceeds the diode's internal barrier voltage, the forward current through the diode increases rapidly at a linear rate. The point on the forward voltage scale at which the curve suddenly rises resembles the shape of a human knee, which is why this point is called the **knee voltage.** This knee voltage is just another name for the diode's internal barrier voltage, which for a silicon diode is about 0.7 V.

Knee Voltage
The point in the forward voltage scale of the *V-I* characteristic curve at which the curve suddenly rises and resembles the shape of a human knee.

Referring back to the linearly increasing current portion of the forward curve in Figure 3-11, you will notice that although there is a large change in forward current, the forward voltage drop across the diode remains almost constant between 0.7 V to 0.75 V.

The amount of heat produced in the diode is proportional to the value of current through the diode ($P\uparrow = I^2\uparrow \times R$). For example, an IN4001 diode, which is a commonly used low-power silicon diode, has a manufacturer's maximum forward (I_F max.) rating of 1 A. If this value of current is exceeded, the diode will begin generating more heat than it can dissipate and burn out. A series current limiting resistor *(R_S)* is generally always included to limit the forward current, as shown in the inset in Figure 3-11. Although the series resistor will limit forward current, it cannot prevent a damaging forward current if enough pressure or forward voltage is applied ($V\uparrow = I\uparrow \times$ R).

Reverse Characteristics

The lower left quadrant of the four sections in Figure 3-11 shows what reverse current will pass through the diode when a reverse bias voltage is applied. As you can see from the inset, a diode is reverse biased by applying a negative potential to its anode and a positive potential to its cathode. To review, when reverse biased, the internal barrier voltage of a diode will increase until it is equal to the external voltage. In this instance, current is effectively reduced to zero and the diode is said to be OFF and equivalent to an open switch.

These characteristics can be seen in the reverse curve in Figure 3-11. Beginning at the graph origin and following the curve into the reverse quadrant, you can see that the reverse current through the diode increases only slightly (approximately 100 μA). Throughout this part of the curve the diode is said to be blocking current because the leakage current is generally so small and is ignored for most practical applications. If the reverse voltage *(V_R)* is further increased, a point will be reached where the diode will break down, resulting in a sudden increase in current. This excessive current is due to the large external reverse bias voltage that is now strong enough to pull valence electrons away from their parent atoms, resulting in a large increase in minority carriers. The point on the reverse voltage scale at which the diode breaks down and there is a sudden increase in reverse current is called the **breakdown voltage.** Referring to the reverse curve in Figure 3-11, you can see that most silicon diodes break down as the reverse bias voltage approaches 50 V. For example, the IN4001 low-power silicon diode has a reverse breakdown voltage (which is sometimes referred to as the **Peak Inverse Voltage** or **PIV**) of 50 V listed on its manufacturer's data sheet. If this reverse bias voltage is exceeded, an avalanche of continuously rising current will eventually generate more heat than can be dissipated, resulting in the destruction of the diode.

Breakdown Voltage or Peak Inverse Voltage (PIV)
The point on the reverse voltage scale at which the diode breaks down and there is a sudden increase in the reverse current.

Temperature Characteristics

As discussed previously in Chapter 2, semiconductor materials, and therefore diodes, have a negative temperature coefficient of resistance. This means as temperature increases ($T\uparrow$), their resistance decreases ($R\downarrow$). Let us now consider these temperature effects on diodes since they may be critical in some applications.

Figure 3-12 shows how the forward and reverse currents through a diode can be affected by changes in temperature. The forward voltage drop across a conducting diode is inversely proportional to temperature, which means that the voltage drop across a diode will be less at higher temperatures. This change in forward voltage drop, however, is very small and does not greatly affect the diode's operation in most applications.

On the other hand, the diode's reverse current is greatly affected by changes in temperature. To review, the leakage current in a diode is caused by the minority carriers in the *n* and *p* regions. At low temperatures, the reverse leakage current is almost zero. At room temperature, the reverse leakage current has increased; however, it is still too small to have any adverse effects on circuit applications. At higher temperatures, however, the reverse leakage current increases to a point that the reverse biased diode is no longer equivalent to an open switch. This means that if a circuit malfunction generated excessive heat, or if a system's cooling fan failed, the system's diodes might not switch OFF when reverse biased, causing additional system problems. As a general rule, the reverse leakage current tends to double for every temperature increase of 10° C.

3-2-4 A Junction Diode's Specification Sheet

A manufacturer's **specification sheet** or **data sheet** details the characteristics and maximum and minimum values of operation for a given device. Generally, engineers will study these details to determine whether a specific device can be incorporated into a circuit design. As a technician, you should be familiar with some of the basic operating limits of a device so that you can isolate component malfunctions within a circuit by determining whether the device is operating to specifications. For example, if a specific diode is placed in a circuit in which its maximum rating is exceeded in some instances, the fault is not with the compo-

Specification Sheet or Data Sheet
Details the characteristics and maximum and minimum values of operation of a device.

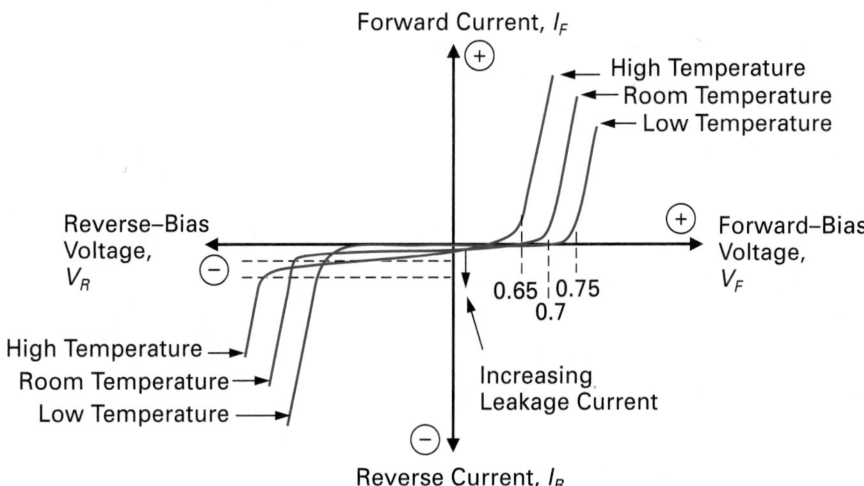

FIGURE 3-12 Temperature Effects on a Junction Diode's *V-I* Characteristic Curve.

DEVICE: IN4001 Through IN4007— Silicon Junction Diodes

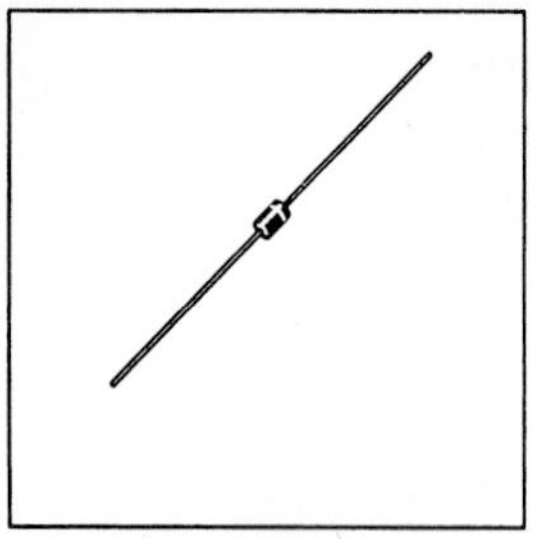

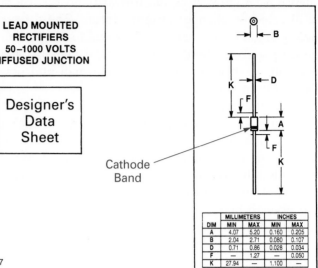

LEAD MOUNTED RECTIFIERS 50–1000 VOLTS DIFFUSED JUNCTION

Designer's Data Sheet

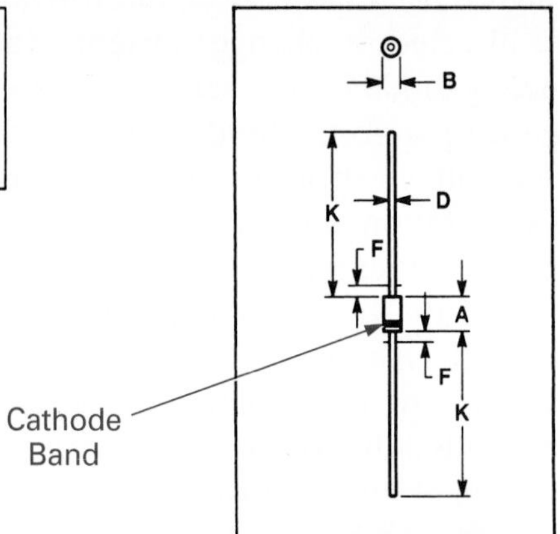

Cathode Band

DIM	MILLIMETERS MIN	MILLIMETERS MAX	INCHES MIN	INCHES MAX
A	4.07	5.20	0.160	0.205
B	2.04	2.71	0.080	0.107
D	0.71	0.86	0.028	0.034
F	—	1.27	—	0.050
K	27.94	—	1.100	—

Mechanical Characteristics

- Case: Epoxy, Molded
- Weight: 0.4 gram (approximately)
- Finish: All External Surfaces Corrosion Resistant and Terminal Leads are Readily Solderable
- Lead and Mounting Surface Temperature for Soldering Purposes: 220°C Max. for 10 Seconds, 1/16″ from case
- Shipped in plastic bags, 1000 per bag.
- Available Tape and Reeled, 5000 per reel, by adding a "RL" suffix to the part number
- Polarity: Cathode Indicated by Polarity Band
- Marking: 1N4001, 1N4002, 1N4003, 1N4004, 1N4005, 1N4006, 1N4007

The **Peak Repetitive Reverse Voltage** is the maximum allowable reverse voltage; i.e., IN4001 will probably break down if the reverse voltage exceeds 50 V. This rating is also called **Working Peak Reverse Voltage** and **DC Blocking Voltage.**

MAXIMUM RATINGS

Rating	Symbol	1N4001	1N4002	1N4003	1N4004	1N4005	1N4006	1N4007	Unit
*Peak Repetitive Reverse Voltage Working Peak Reverse Voltage DC Blocking Voltage	V_{RRM} V_{RWM} V_R	50	100	200	400	600	800	1000	Volts
*Non–Repetitive Peak Reverse Voltage (halfwave, single phase, 60 Hz)	V_{RSM}	60	120	240	480	720	1000	1200	Volts
*RMS Reverse Voltage	$V_{R(RMS)}$	35	70	140	280	420	560	700	Volts
*Average Rectified Forward Current (single phase, resistive load, 60 Hz, see Figure 8, $T_A = 75°C$)	I_O	1.0							Amp
*Non–Repetitive Peak Surge Current (surge applied at rated load conditions, see Figure 2)	I_{FSM}	30 (for 1 cycle)							Amp
Operating and Storage Junction Temperature Range	T_J T_{stg}	– 65 to +175							°C

Maximum allowable nonrepeating reverse voltage.

The RMS of the Peak Repetitive Reverse Voltage. $V_{rms} = 0.707 \times V_{peak}$.

The maximum average forward current value.

The maximum surge current value.

The diode can be operated and stored at any temperature between –65° to +175° C.

ELECTRICAL CHARACTERISTICS*

Rating	Symbol	Typ	Max	Unit
Maximum Instantaneous Forward Voltage Drop ($i_F = 1.0$ Amp, $T_J = 25°C$) Figure 1	v_F	0.93	1.1	Volts
Maximum Full–Cycle Average Forward Voltage Drop ($I_O = 1.0$ Amp, $T_L = 75°C$, 1 inch leads)	$V_{F(AV)}$	—	0.8	Volts
Maximum Reverse Current (rated dc voltage) ($T_J = 25°C$) ($T_J = 100°C$)	I_R	 0.05 1.0	 10 50	μA
Maximum Full–Cycle Average Reverse Current ($I_O = 1.0$ Amp, $T_L = 75°C$, 1 inch leads)	$I_{R(AV)}$	—	30	μA

The maximum voltage drop that will appear across the diode when forward biased.

The maximum average forward voltage drop.

Maximum reverse current at different temps.

This is the maximum average value of reverse current.

FIGURE 3-13 Specific Specification Sheet for the IN4001 through IN4007 Junction Diodes. (Copyright of Motorola. Used by permission.)

nent but with the circuit design. In this situation, simply replacing the component will not solve the problem. The device will have to be replaced with a diode which has a greater maximum rating.

Figure 3-13 shows the specific specification sheet for the IN4001 through IN4007 series of silicon junction diodes. This data sheet serves as a good example for showing the amount of details that are normally supplied to design engineers by device manufacturers. As a technician, you would mainly be interested in the diode's maximum reverse voltage, maximum forward current, and average voltage drop. For the IN4001, these values are:

Maximum reverse voltage (V_R) = 50 V

Maximum forward current (I_O) = 1 A

Average voltage drop ($V_{F(av)}$) = 0.8 V

■ **EXAMPLE:**

Would any of the maximum ratings of the IN4001 diode in Figure 3-14 be exceeded?

■ ***Solution:***

The maximum reverse voltage for an IN4001 is 50 V, and since this reverse voltage is not being exceeded by the applied voltage, the diode is being operated within this specification.

The maximum forward current will be:

$$I = \frac{V_S - V_{diode}}{R}$$

$$I = \frac{12\text{ V} - 0.8\text{ V}}{10\ \Omega} = 1.12\text{ A}$$

Since this is in excess of the 1 A maximum listed in the specification sheet, the diode will more than likely burn out due to excessive current and, therefore, heat.

3-2-5 *Junction Diode Applications*

It would be safe to say that the junction diode is used in almost every electronic and electrical system. Like resistors and capacitors, the list of diode applications

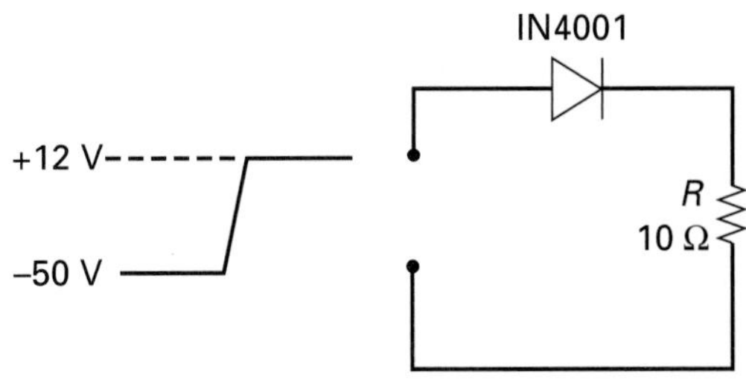

FIGURE 3-14 An IN4001 Diode Example Circuit.

is endless. However, certain uses of the diode predominate. As you proceed through this text you will see the junction diode used in a wide variety of analog and digital circuit applications. For now, let us see how the junction diode is most used in a basic analog circuit and a basic digital circuit.

Analog Circuit Application—Rectifier Circuit

Rectification Diode's ability to pass current in only one direction and block current in the reverse direction.

Rectifier Circuit Achieves rectification.

Rectifier Diodes or Rectifiers Junction diodes that achieve rectification.

The diode's ability to pass current in only one direction, and block current from passing in the reverse direction, is utilized to achieve **rectification.** Rectification is achieved by a **rectifier circuit,** which is a circuit that converts alternating current (ac) into direct current (dc). Rectifier circuits are probably the biggest application for junction diodes, which accounts for why junction diodes are sometimes referred to as **rectifier diodes** or **rectifiers** as in Figure 3-13.

A basic rectifier circuit is constructed by connecting a junction diode between an ac input and a load, as shown in Figure 3-15(a). In this circuit, the 120 V_{rms} (169.7 V_{peak}) 60 Hz ac voltage from the wall outlet will be converted to a pulsating dc voltage.

When the ac voltage swings positive, as shown in Figure 3-15(b), the anode of the diode is made positive, causing the diode to turn ON and connect the positive half-cycle of the ac input voltage across the load (R_L).

When the ac voltage swings negative, as shown in Figure 3-15(c), the anode of the diode is made negative, causing the diode to turn OFF and prevent any circuit current, and therefore any voltage, from being developed across the load (R_L).

Half-Wave Rectifier Circuit in which half the input wave appears at the output.

Referring to the input and output waveforms shown in Figure 3-15(c), you can see why the circuit is called a **half-wave rectifier.** Comparing the voltage input (V_{in}) to the voltage output (V_{out}), you can see that only half of the input wave appears at the output. Since the diode only connects the positive half-cycle of the ac input across the load (R_L), the output voltage (V_{RL}) is a positive pulsating dc waveform. The peak voltage developed across the load will, of course, be 0.7 V less than the peak input voltage due to the 0.7 V drop across the silicon junction diode, as shown in Figure 3-15(d).

■ **EXAMPLE:**

Which of the IN4000X silicon rectifier diodes should be used in the half-wave rectifier circuit shown in Figure 3-15, considering the value of reverse voltage?

■ ***Solution:***

The diode is reverse biased when the ac input swings negative. At this time, the entire negative supply voltage will appear across the open or OFF diode. The maximum reverse breakdown voltage of the diode must therefore be larger than the peak of the ac input voltage. Referring to Figure 3-13, you can see that to withstand a reverse voltage of 169.7 V (peak of the ac input), we will have to use a IN4003 because it has a V_R maximum rating of 200 V.

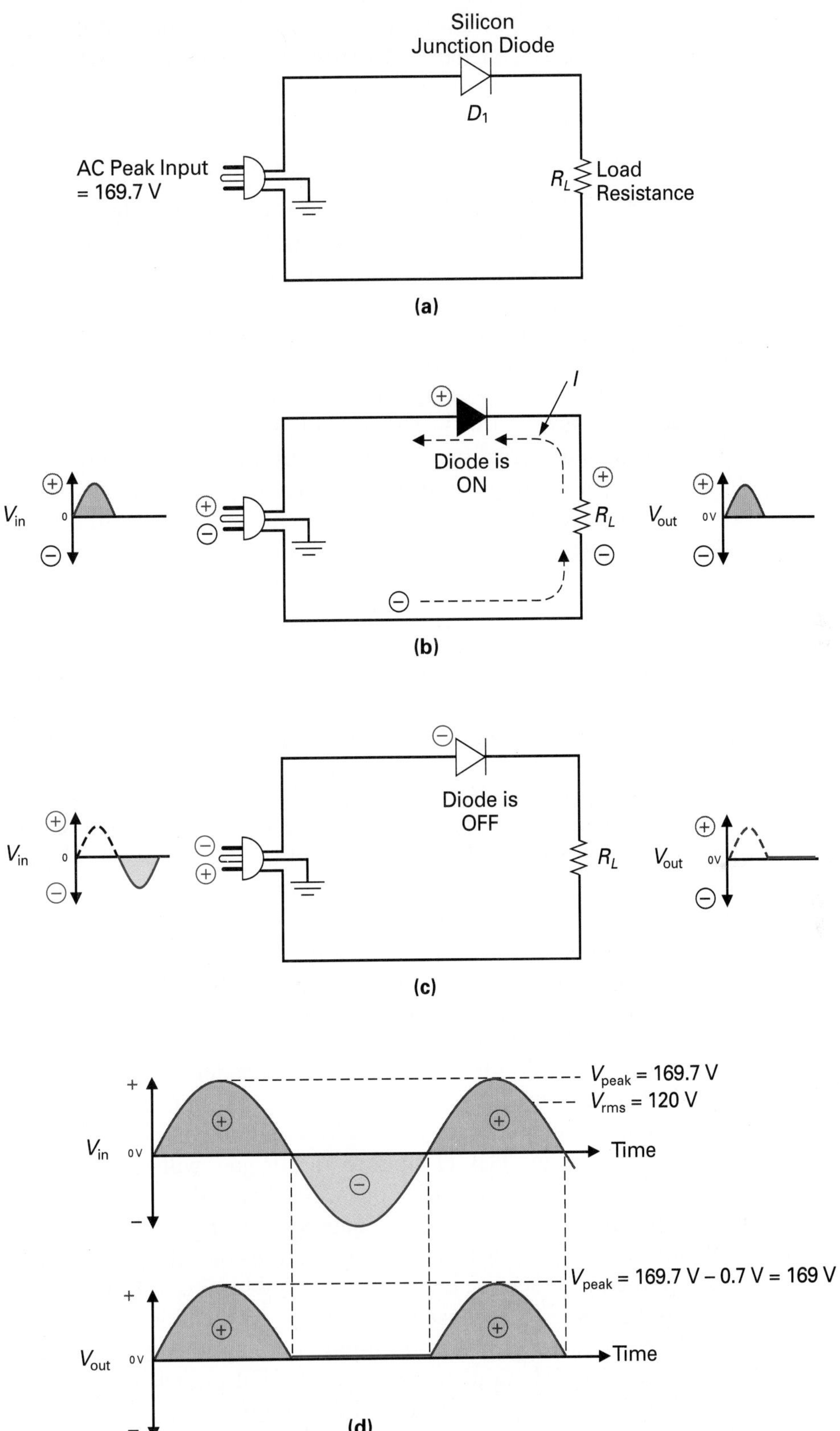

FIGURE 3-15 The Diode Being Used in an Analog Electrical Circuit. (a) Basic Rectifier Circuit. (b) Positive Input Half Cycle Operation. (c) Negative Input Half Cycle Operation. (d) Input/Output Waveforms.

Further details on half-wave and full-wave rectifiers will be given in Chapter 5 which discusses all types of rectifier circuits and how they are used in dc power supplies.

Digital Circuit Application—Logic Gate Circuit

Logic Gate Circuits Two-state (ON/OFF) circuits used for decision-making functions in digital logic circuits.

Within the computer, diodes are used to construct **logic gate circuits,** logic gates are used to construct flip-flop circuits, flip-flops are used to construct register and counter circuits, and register and counter circuits are the building blocks of our digital electronic computer. The logic gate is therefore our basic building block for all digital circuits. Let us examine one of these logic gates to see how it operates and in what applications it can be used.

A basic logic gate is constructed using two junction diodes and a resistor, as shown in Figure 3-16(a). This circuit is called a logic gate because it will always produce a *logical* or *predictable* output, and this output will depend on the condition of its inputs. For example, Figure 3-16(b) has a table which shows how this logic gate will react to all different input possibilities. The two binary states 0 and 1 are represented in the circuit as two voltages

Binary 0 = 0 volts (LOW voltage)

Binary 1 = 5 volts (HIGH voltage)

If you study the table in Figure 3-16(b), you can see that when both inputs are LOW, or at 0 V (A = 0, B = 0), both diodes will be OFF because the anodes are at 0 volts (due to the inputs), and the cathodes of the diodes are at 0 volts (due to the pull-down resistor, *R*). This input combination of A = 0 and B = 0 will therefore turn both diodes OFF and always produce an output at Y of 0.

In all other combinations in the table in Figure 3-16(b), a HIGH or +5 V (logical 1) input is applied to either or both of the inputs A and B. Any HIGH input will always turn on its associated diode since the anode will be at +5 V and the cathode will be at 0 V via *R*. When ON, a diode is equivalent to a closed switch, and the +5 V input is switched through to the output making Y equal to +5 V or logical 1. Therefore, if *A input OR B input are HIGH, the output Y will be HIGH.* This behavior accounts for why this circuit is called an **OR gate.**

OR Gate When either input A OR B is HIGH, the output will be HIGH.

The schematic symbol for the OR gate is shown in Figure 3-16(c). To show how the OR gate circuit could be used, consider the simple security system shown in Figure 3-16(d). In this application, if either the window *OR* the door was to open, a switch contact would close and deliver a HIGH input to the OR gate. As we know, any HIGH input to an OR gate will always generate a HIGH output and in this circuit the +5 V out will activate the siren.

In summary, therefore, a logic gate accepts inputs in the form of HIGH or LOW voltages, judges these input combinations based on a predetermined set of rules, and then produces a single output in the form of a HIGH or LOW voltage. The term logic is used because the output is predictable or logical, and the term gate is used because only certain input combinations will unlock the gate. For example, any HIGH input to an OR gate will unlock the gate and allow the HIGH at the input to pass through to the output.

The OR gate and all of the other different types of logic gates will be discussed later in the digital circuits chapter.

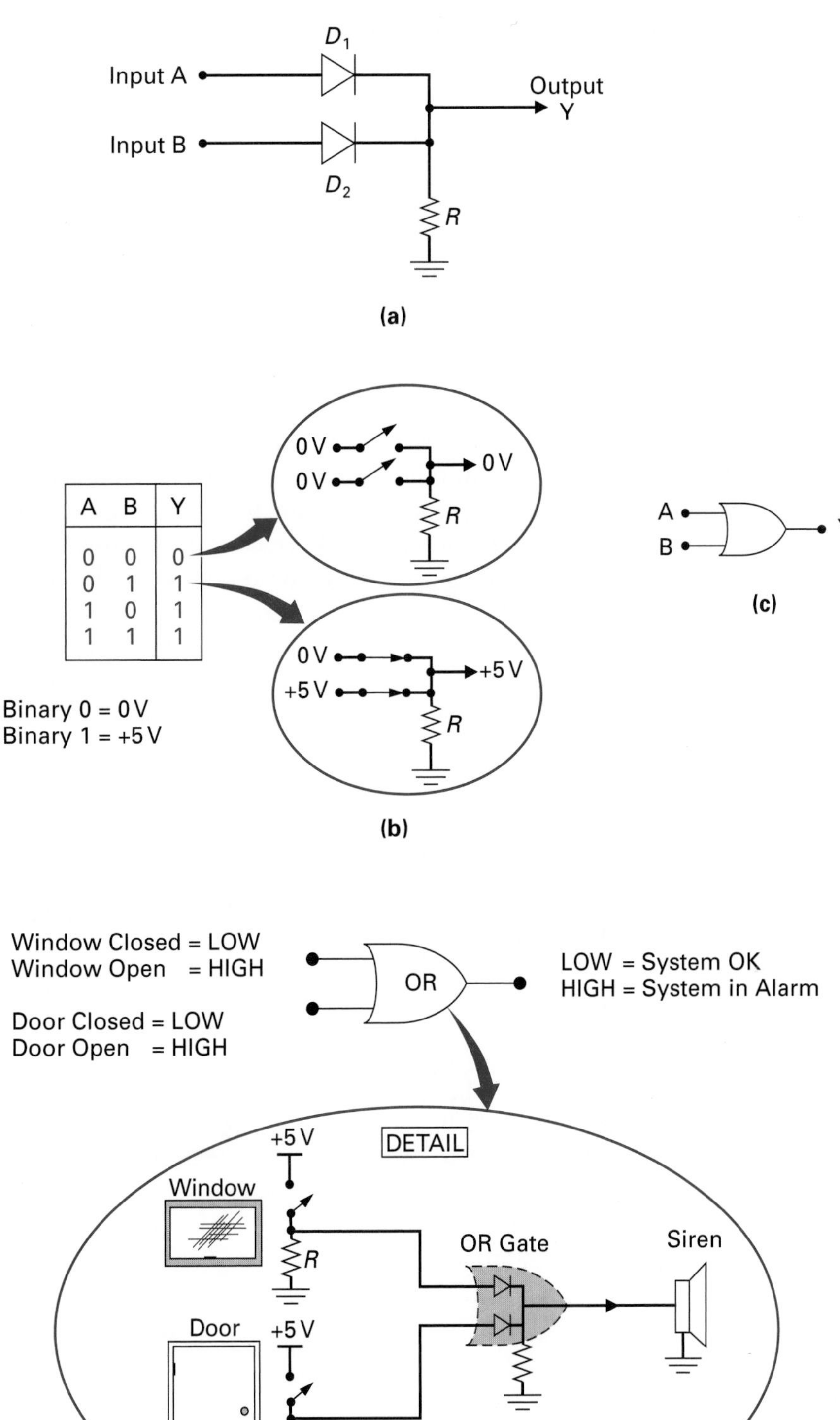

FIGURE 3-16 The Diode Being Used in a Digital Electronic Circuit. (a) Basic OR Gate Circuit. (b) OR Gate Function Table. (c) OR Gate Schematic Symbol. (d) OR Gate Security System Application.

3-2-6 *Testing Junction Diodes*

Diodes that are operated beyond their maximum forward and reverse ratings will, more than likely, malfunction. These malfunctions can result in one of two types of failure. The diode may burn out and then act as a permanently open switch (less common), or effectively melt the semiconductor material and act as a permanently closed switch (more common). A defective diode will not be able to be switched ON or OFF. Instead, it will either remain permanently OFF or permanently ON.

Figure 3-17 shows how an ohmmeter can be used to check whether a diode has malfunctioned or is operating correctly. A good diode should display a very low resistance when it is biased ON, and a very high resistance when it is biased OFF. Figure 3-17(a) shows how a diode can be forward biased by an ohmmeter's internal battery (+ lead to anode, – lead to cathode), and if good, should display a low value of resistance (typically less than 10 Ω). Figure 3-17(b) shows how the diode is then flipped over and reverse biased by the ohmmeter's internal battery (+ lead to cathode, – lead to anode). If the diode is good, the ohmmeter should display a very high resistance (typically greater than 1000 MΩ). Since this value is generally off the ohmmeter's scale, you will probably have the display showing OL or OR. This is what you should expect since it means that the reverse biased diode's resistance is so high that it is over the range, or off the scale, selected.

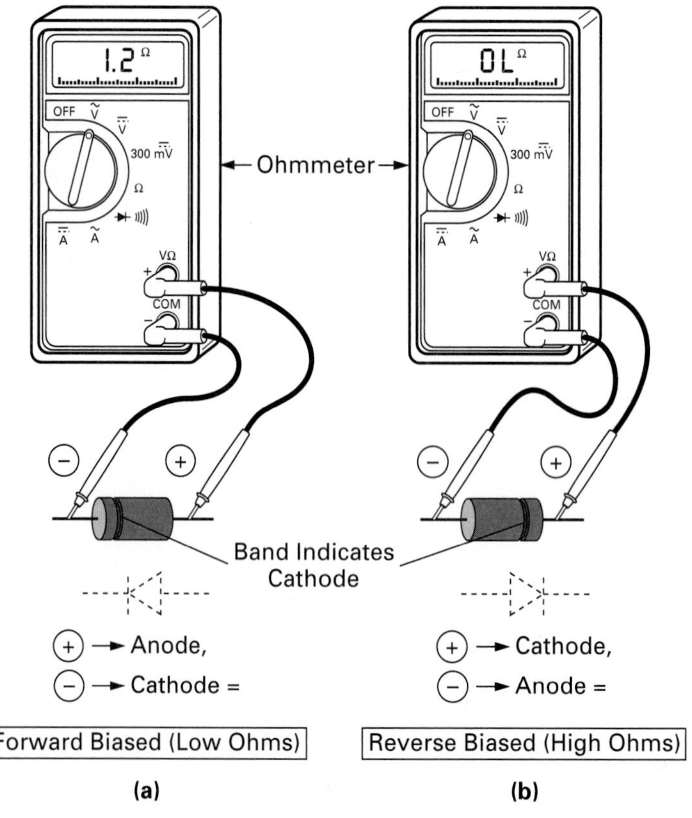

FIGURE 3-17 Testing Diodes with an Ohmmeter.

A diode that has been damaged and is permanently shorted between its anode and cathode will display a LOW resistance on the ohmmeter when it is both forward and reverse biased by the ohmmeter. Finding the shorted diode in a circuit may not be the end of your troubles, since it may be one of the neighboring components that initially caused the diode's ratings to be exceeded. Furthermore, the shorted diode may have allowed a damaging current to pass through to another part of the circuit causing additional damage.

A diode that has been damaged and is permanently opened between its anode and cathode will display a HIGH resistance on the ohmmeter when it is both forward and reverse biased by the ohmmeter. Once again, the destruction of the diode generally occurs when the maximum voltage or current ratings of the diode have been exceeded. Remember that the diode may not have just malfunctioned on its own; the problem may have been caused by one of the neighboring components. In this instance, however, an open diode rarely damages other associated components because current is blocked by the fuse-like action of the diode's destruction.

As you proceed through this text, you will see diodes used in a variety of circuit applications, and you will see what symptoms a malfunctioning diode will display in these circuits.

■ **EXAMPLE:**

Figure 3-18 shows the results from testing four diodes with an ohmmeter. Which of the diodes tested good, which tested bad, and for what reason?

■ ***Solution:***

a. The diode in Figure 3-18(a) has a low forward resistance and a high reverse resistance and is switching ON and OFF correctly.

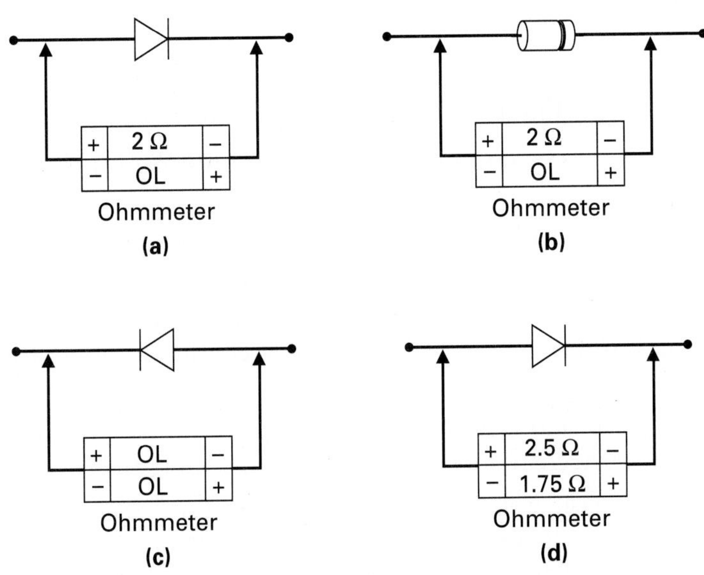

FIGURE 3-18 Testing Diode Examples.

b. The diode in Figure 3-18(b) has a low forward resistance and a high reverse resistance and is switching ON and OFF correctly.
c. The diode in Figure 3-18(c) has a high forward resistance and a high reverse resistance and is not switching ON and OFF. It seems to be remaining permanently OFF, indicating an open between its anode and cathode terminals.
d. The diode in Figure 3-18(d) has a low forward resistance and a low reverse resistance and is not switching ON and OFF. It seems to be remaining permanently ON, indicating a short between its anode and cathode terminals.

SELF-TEST REVIEW QUESTIONS FOR SECTION 3-2

1. What is the typical forward voltage drop across a silicon diode?
2. What barrier voltage has to be overcome in order to forward bias a silicon diode?
3. How many P-N junctions are within a junction diode?
4. What is PIV, and what would it be for a typical silicon diode?
5. As the temperature of a diode increases, the forward voltage drop ____________. (increases or decreases)

SUMMARY

1. The first diode was accidentally created by Edison in 1883 when he was experimenting with his light bulb.
2. The word *diode* is derived from the fact that the device has two (*di*) electrodes (*ode*).
3. A diode will act as a conductor and pass current easily in one direction when the bias voltage across it is of one polarity, yet it will block current and imitate an insulator when the bias voltage applied is of the opposite polarity.

The Junction Diode (Figure 3-19)

4. The two electrodes or terminals of the diode are called the anode and cathode.
5. A black band or stripe is generally placed on the package closest to the cathode terminal for identification purposes.
6. If you make the anode terminal of a diode positive with respect to the cathode terminal, the device is equivalent to a closed switch and is said to be ON, or forward biased.
7. If you make the anode terminal of a diode negative with respect to the cathode, the device is equivalent to an open switch and is said to be OFF, or reverse biased.
8. Semiconductor devices such as diodes and transistors are constructed using P-N junctions. A diode for example, has only one P-N junction and is created by doping a single piece of pure semiconductor to produce an *n*-type

QUICK REFERENCE SUMMARY SHEET

FIGURE 3-19 The Junction Diode.

Schematic Symbol **Memory Aid for Terminal Identification**

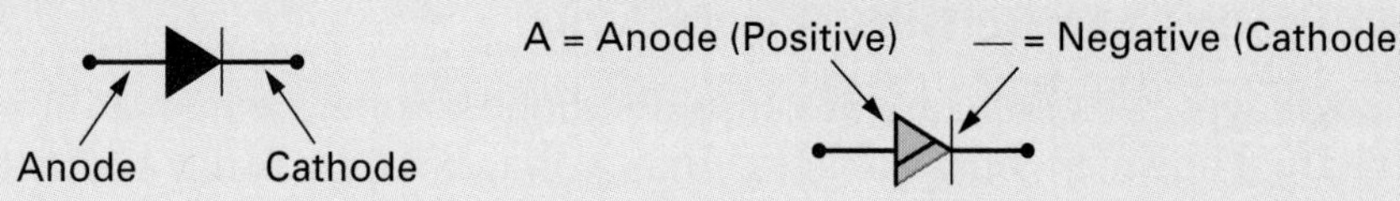

Package Types

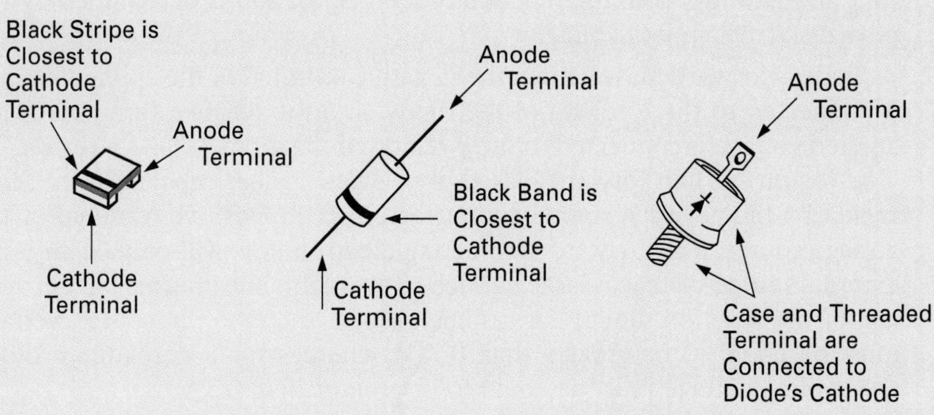

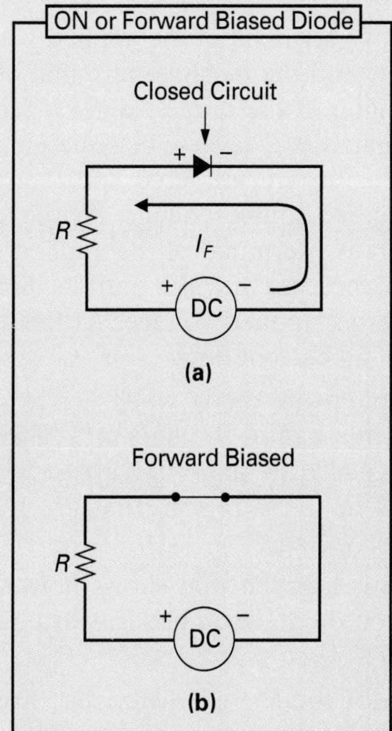

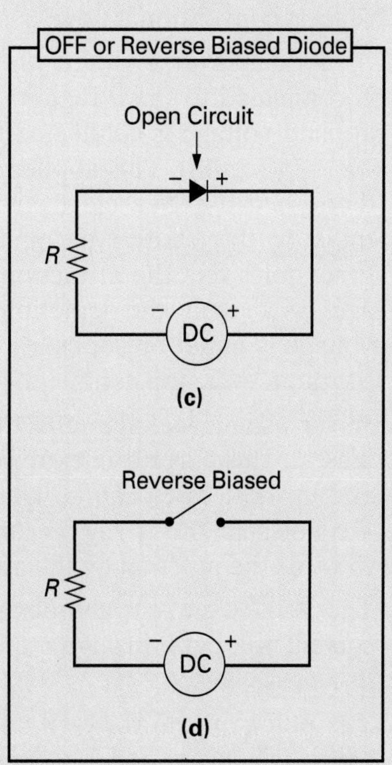

and *p*-type region. A bipolar junction transistor, on the other hand, has two P-N junctions and is created by doping a single piece of pure semiconductor with three alternate regions (NPN or PNP).

9. The *n*-type region is called the cathode, while the *p*-type region is called the anode.

Junction Diode Characteristics (Figure 3-20)

10. The voltages, which incline or cause the diode to operate in a certain manner, are known as bias voltages. Bias voltages control the width of the depletion region, which in turn controls the resistance of the junction and therefore the amount of current that can pass through the P-N junction diode.

11. Like the P-N junction, the junction diode's operation is determined by the polarity of the applied voltage.

12. A diode is forward biased when the negative terminal of the applied voltage is connected to the *n* region of the diode, and the positive terminal of the applied voltage is connected to the *p* region of the diode ($+V \rightarrow p$ region, $-V \rightarrow n$ region). When forward biased, free electrons are repelled from the *n* region by the negative source and attracted to the positive terminal of the voltage source. This forward-conducting electron flow will only occur if the external source voltage is large enough to overcome the internal barrier voltage of the junction diode. For a silicon diode, the external source voltage must be equal to or greater than 0.7 V, whereas for a germanium diode the applied voltage must be equal to or greater than 0.3 V.

13. The diode symbol actually points in the direction of forward hole flow and reflects conventional flow.

14. A diode is reverse biased when the positive terminal of the applied voltage is connected to the *n* region of the diode, and the negative terminal of the applied voltage is connected to the *p* region of the diode ($+V \rightarrow n$ region, $-V \rightarrow p$ region). This applied voltage polarity will reverse bias the junction diode because free electrons in the *n* region are attracted and therefore travel to the positive terminal of the applied source voltage. At the same time, holes feel the attraction of the negative terminal of the applied voltage. As a result, the depletion region will increase in width until its internal voltage is equal but opposite to the external applied voltage. At this time, the diode will stop conducting, cutting off all current flow.

15. A very small leakage current leaks through the diode when it is reverse biased. This current is extremely small (microamps to nanoamps in value) and in most cases can be ignored. It is caused by minority carriers, which are holes in the *n* region and electrons in the *p* region, that are forced towards the junction by the applied source voltage.

16. The voltage-current characteristic curve is a graph that shows how much current will pass through a typical junction diode when it is forward biased or reverse biased.

17. The point on the forward voltage scale of the *V-I* characteristic curve at which the curve suddenly rises and resembles the shape of a human knee is

QUICK REFERENCE SUMMARY SHEET

FIGURE 3-20 Junction Diode Characteristics.

Voltage/Current Characteristics

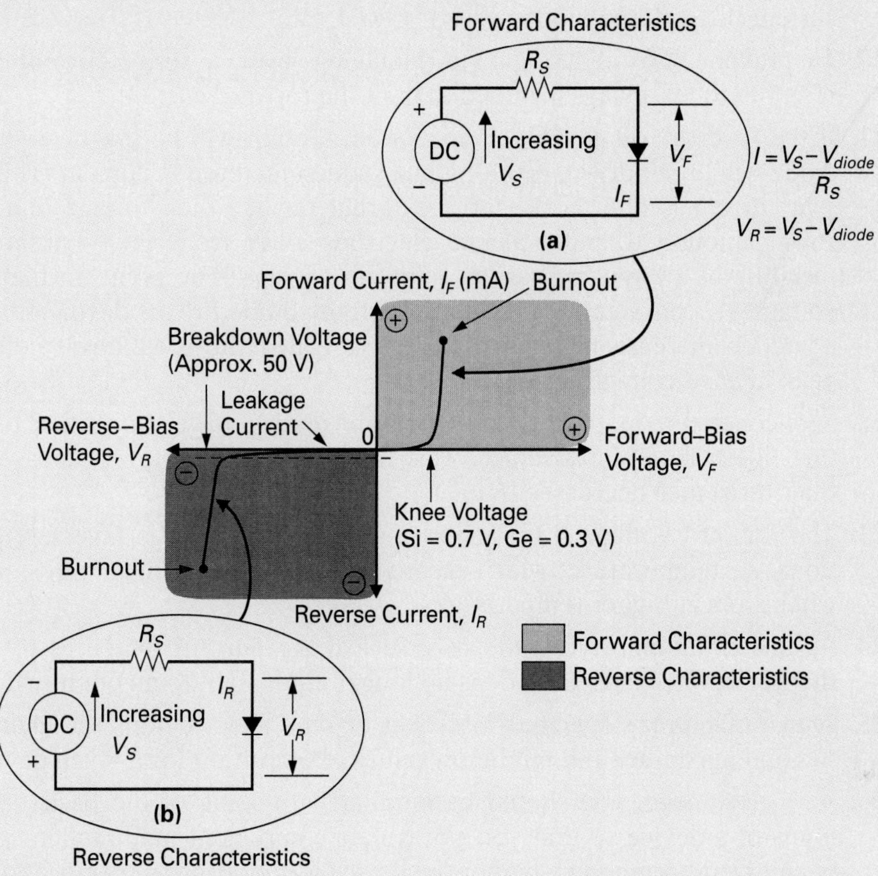

Temperature Characteristics

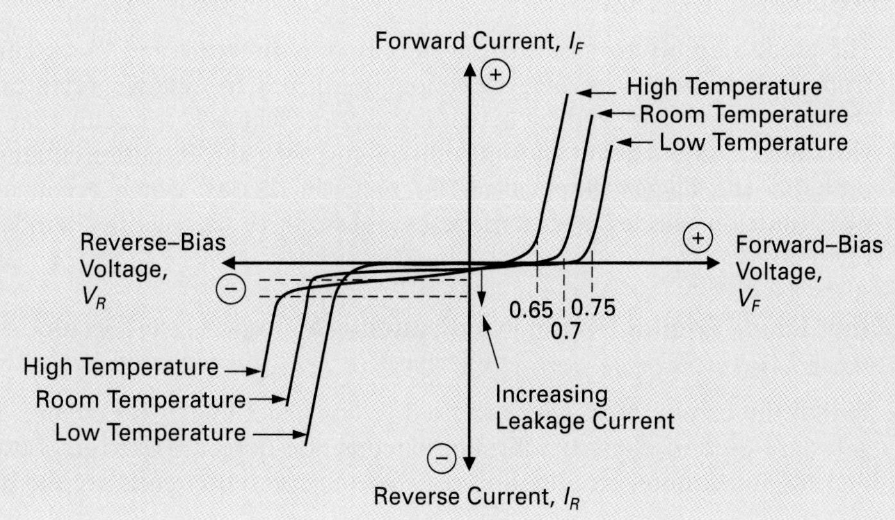

called the knee voltage. This knee voltage is just another name for the diode's internal barrier voltage, which for a silicon diode is about 0.7 V.

18. The forward voltage drop across the diode remains almost constant between 0.7 V to 0.75 V.

19. The amount of heat produced in the diode is proportional to the value of current through the diode ($P\uparrow = I^2\uparrow \times R$).

20. To prevent current rising to a damaging level, a series current limiting resistor (R_S) is always included to limit the forward current.

21. If the reverse voltage (V_R) is increased, a point will be reached where the diode will break down, resulting in a sudden increase in current. This excessive current is due to the large external reverse bias voltage that is now strong enough to pull valence electrons away from their parent atoms, resulting in a large increase in minority carriers. The point on the reverse voltage *V-I* characteristic scale at which the diode breaks down and there is a sudden increase in reverse current is called the breakdown voltage, or peak inverse voltage (PIV).

22. Semiconductor materials, and therefore diodes, have a negative temperature coefficient of resistance. This means as temperature increases ($T\uparrow$), their resistance decreases ($R\downarrow$).

23. The forward voltage drop across a conducting diode is inversely proportional to temperature, which means that the voltage drop across a diode will be less at higher temperatures.

24. At higher temperatures, the reverse leakage current increases to a point that the reverse biased diode is no longer equivalent to an open switch.

25. A manufacturer's specification sheet or data sheet details the characteristics and maximum and minimum values of operation for a given device.

26. As a technician, you should be familiar with some of the basic operating limits of a device so that you can isolate component malfunctions within a circuit by determining whether the device is operating to specifications.

Junction Diode Analog Circuit Application—A Rectifier Circuit (Figure 3-21)

27. The diode's ability to pass current in only one direction and block current from passing in the reverse direction is utilized to achieve rectification. Rectification is achieved by a rectifier circuit, which is a circuit that converts alternating current (ac) into direct current (dc). Rectifier circuits are probably the biggest application for junction diodes, which accounts for why junction diodes are sometimes referred to as rectifier diodes or rectifiers.

Junction Diode Digital Circuit Application—A Logic Gate Circuit (Figure 3-22)

28. Within the computer, diodes are used to construct logic gate circuits, logic gates are used to construct flip-flop circuits, flip-flops are used to construct register and counter circuits, and register and counter circuits are the build-

QUICK REFERENCE SUMMARY SHEET

FIGURE 3-21 Junction Diode Analog Circuit Application—A Rectifier Circuit.

Basic Circuit

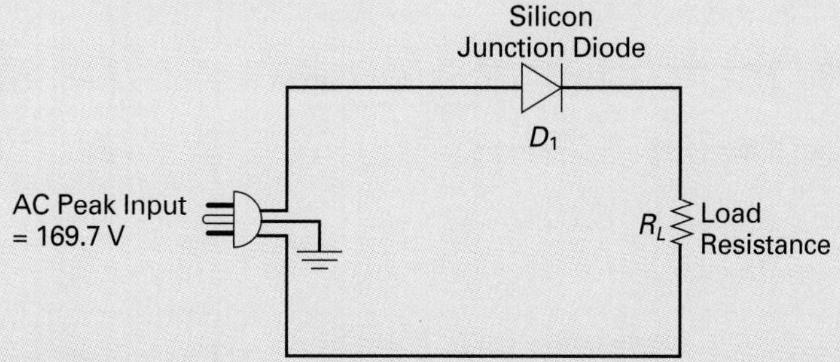

Operation of Circuit During Positive Alternation

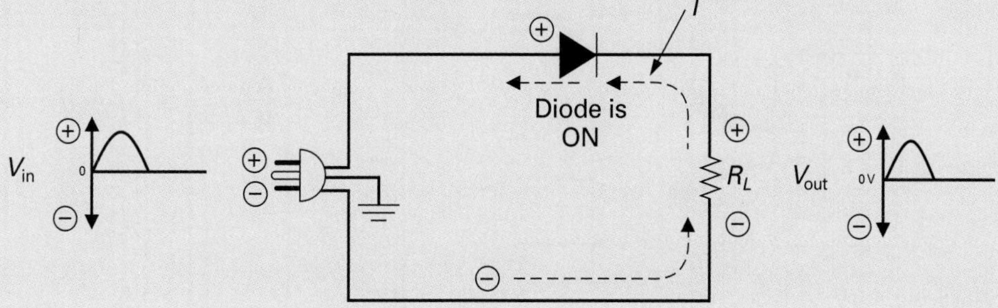

Operation of Circuit During Negative Alternation

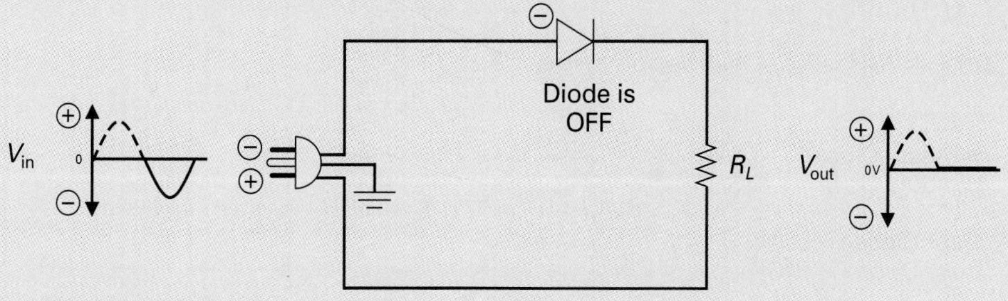

Input/Output Waveforms

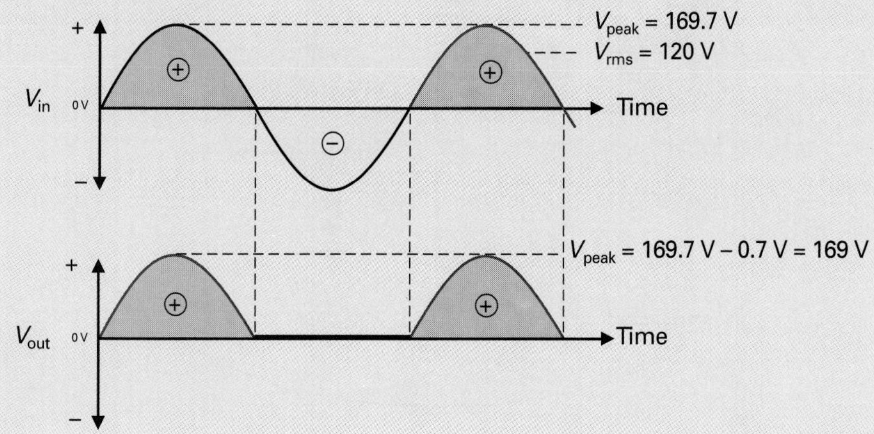

FIGURE 3-22 Junction Diode Digital Circuit Application—A Logic Gate Circuit.

Basic OR Gate Circuit

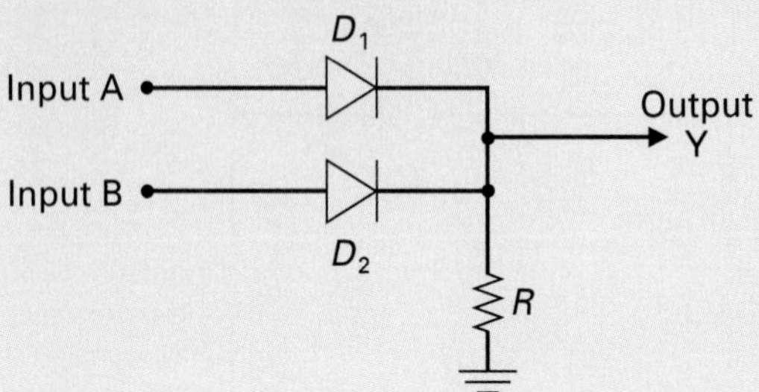

OR Gate Function Table

OR Gate Schematic Symbol

A	B	Y
0	0	0
0	1	1
1	0	1
1	1	1

Binary 0 = 0 V
Binary 1 = +5 V

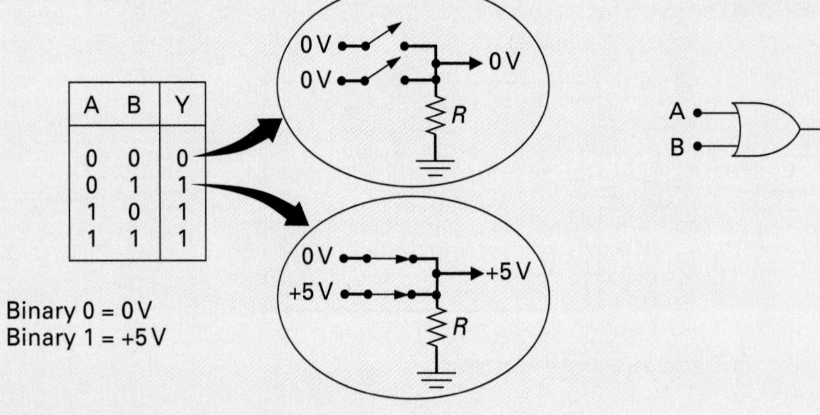

OR Gate Application — Basic Security System

Window Closed = LOW
Window Open = HIGH

Door Closed = LOW
Door Open = HIGH

LOW = System OK
HIGH = System in Alarm

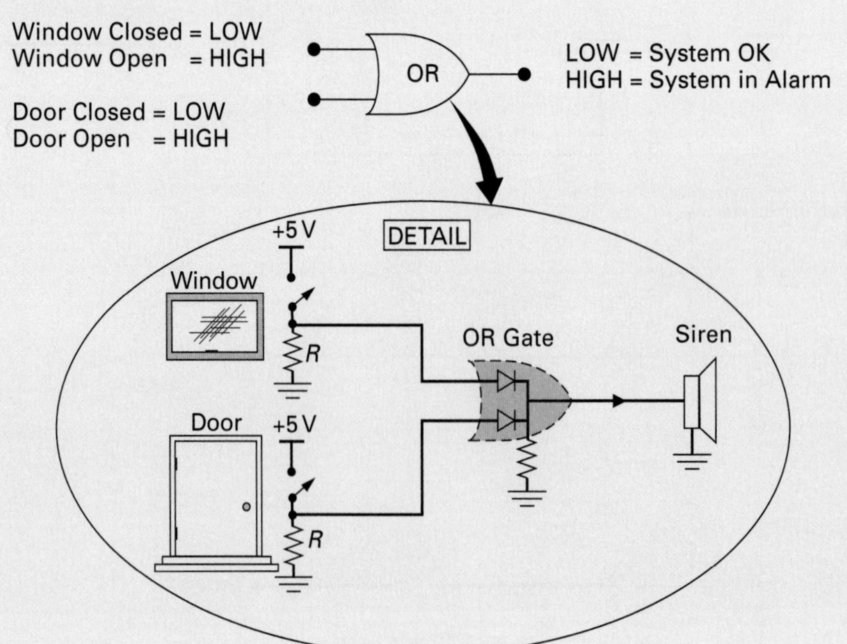

ing blocks of our digital electronic computer. The logic gate is, therefore, our basic building block for all digital circuits.

29. A logic gate accepts inputs in the form of HIGH or LOW voltages, judges these input combinations based on a predetermined set of rules, and then produces a single output in the form of a HIGH or LOW voltage. The term logic is used because the output is predictable or logical, and the term gate is used because only certain input combinations will unlock the gate.

Junction Diode Testing (Figure 3-23)

30. Diodes that are operated beyond their maximum forward and reverse ratings will, more than likely, malfunction. These malfunctions can result in one of two types of failure. The diode may burn out and then act as a permanently open switch (less common), or effectively melt the semiconductor material and act as a permanently closed switch (more common). A defective diode will no longer be able to be switched ON or OFF. Instead, it will either remain permanently OFF or permanently ON.

31. A good diode should display a very low resistance when it is biased ON and a very high resistance when it is biased OFF. If good, when a diode is forward biased by an ohmmeter it should display a low value of resistance

QUICK REFERENCE SUMMARY SHEET

FIGURE 3-23 Junction Diode Testing.

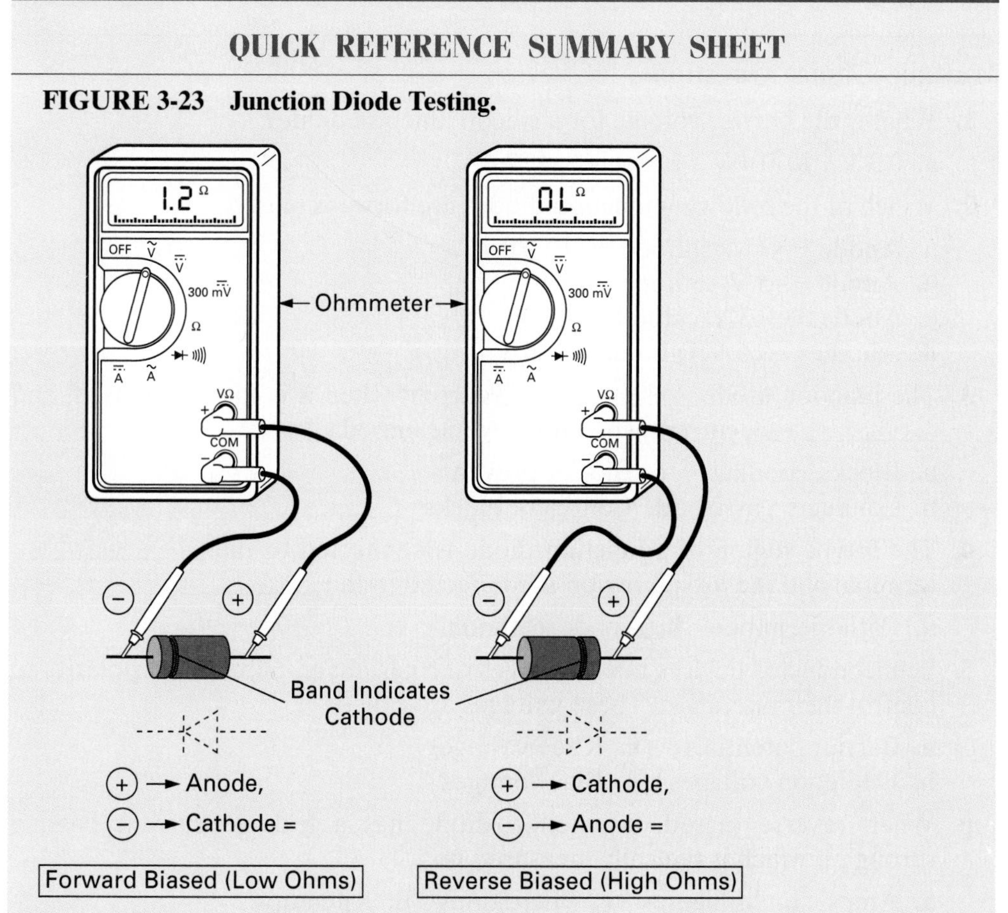

(typically less than 10 Ω). When a diode is then flipped over and reverse biased by the ohmmeter's internal battery (+ lead to cathode, – lead to anode), the ohmmeter should display a very high resistance (typically greater than 1000 MΩ). Since this value is generally off the ohmmeter's scale, you will probably have the display showing OL or OR.

32. A diode that has been damaged and is permanently shorted between its anode and cathode will display a LOW resistance on the ohmmeter when it is both forward and reverse biased by the ohmmeter.

33. A diode that has been damaged and is permanently opened between its anode and cathode will display a HIGH resistance on the ohmmeter when it is both forward and reverse biased by the ohmmeter.

NEW TERMS

Anode
Bias Voltage
Breakdown Voltage
Cathode
Characteristic Curve
Data Sheet
Diode
Encoder Circuit
Graph Origin
Half-Wave Rectifier
Junction Diode
Knee Voltage
Leakage Current
Logic Gate
OR Gate
Peak Inverse Voltage
Rectification
Rectifier
Rectifier Circuit
Specification Sheet

REVIEW QUESTIONS

Multiple-Choice Questions

1. What is the barrier voltage for a silicon junction diode?

a. 0.3 V **b.** 0.4 V **c.** 0.7 V **d.** 2.0 V

2. Which of the following junction diodes are forward biased?

a. Anode = +7 V, cathode = +10 V
b. Anode = +5 V, cathode = +3 V
c. Anode = +0.3 V, cathode = +5 V
d. Anode = –9.6 V, cathode = –10 V

3. The junction diode ______________ current when it is forward biased, and ______________ current when it is reverse biased.

a. Blocks, conducts **c.** Blocks, prevents
b. Conducts, passes **d.** Conducts, blocks

4. The *n*-type region of a junction diode is connected to the ______________ terminal and the *p*-type region is connected to the ______________.

a. Cathode, anode **b.** Anode, cathode

5. Semiconductor devices need voltages of a certain amplitude and polarity to control their operation. These voltages are called:

a. Barrier potentials **c.** Knee voltages
b. Depletion voltages **d.** Bias voltages

6. When reverse biased, a junction diode has a leakage current passing through it which is typically measured in:

a. Amps **b.** Milliamps **c.** Microamps **d.** Kiloamps

7. What happens to the forward voltage drop across the diode (V_F) if temperature increases? V_F will:

a. Decrease **c.** Remains the same
b. Increase **d.** Be unpredictable

8. When forward biased, a junction diode is equivalent to a/an ____________ switch, whereas when it is reverse biased it is equivalent to a/an ____________ switch.

a. Open, closed **c.** Open, open
b. Closed, closed **d.** Closed, open

9. The black band on a diode's package is always closest to the ____________.

a. Anode **c.** *p*-type material
b. Cathode **d.** Both (a) and (c) are true

10. When current dramatically increases, the voltage point on the diode's forward *V-I* characteristic curve is called the:

a. Breakdown voltage **c.** Barrier voltage
b. Knee voltage **d.** Both (b) and (c) are true

11. When current dramatically increases, the voltage point on the diode's reverse *V-I* characteristic curve is called the:

a. Breakdown voltage **c.** Barrier voltage
b. Knee voltage **d.** Both (b) and (c) are true

12. A logic gate is:

a. A circuit that converts ac to dc
b. An analog circuit
c. A two-state decision making circuit
d. A circuit that converts dc to ac

13. A rectifier is:

a. A circuit that converts ac to dc
b. An analog circuit
c. A two-state decision making circuit
d. A circuit that converts dc to ac

14. What resistance should a good diode have when it is reverse biased?

a. Less than 10 Ω **c.** Between 120 Ω and 1.2 kΩ
b. More than 1000 MΩ **d.** Both (b) and (c) are true

15. What resistance should a good diode have when it is forward biased?

a. Less than 10 Ω
b. More than 1000 MΩ
c. Between 120 Ω and 1.2 kΩ
d. Both (b) and (c) are true

Essay Questions

16. What is a junction diode? (3-1)

17. Sketch the schematic symbol for a diode and label its terminals. (3-1-1)

18. Describe the basic operation of a diode. (3-1-2)
19. What is the relationship between a P-N junction and a diode? (3-2-1)
20. Explain in detail the differences between a forward biased diode and a reverse biased diode. (3-2-2)
21. Referring to the junction diode's *V-I* characteristic curve, describe the: (3-2-3)

    a. Forward bias curve **b.** Reverse bias curve
22. Describe the temperature characteristics of a junction diode. (3-2-3)
23. List the following specifications for the IN4004 junction diode: (3-2-4)

    a. V_R **b.** $I_{F(av)}$ **c.** $I_{R(av)}$
24. What is a rectifier circuit? (3-2-5)
25. Sketch a half-wave rectifier circuit and its input and output wave shapes. (3-2-5)
26. A diode is a ____________ within a protective package. (3-2-1)
27. What is a logic gate? (3-2-5)
28. Sketch an OR gate circuit and a table showing what outputs will be produced for all combinations of digital inputs. (3-2-5)
29. What type of diode malfunction is most common? (3-2-6)
30. How can an ohmmeter be used to test for an open or short within a diode? (3-2-6)

Practice Problems

31. Which of the silicon diodes in Figure 3-24 are forward biased and which are reverse biased?
32. Calculate I_F for the circuits in Figure 3-25.
33. What would be the voltage drop across each of the diodes in Figure 3-25?
34. What would be the voltage drop across each of the resistors in Figure 3-25?

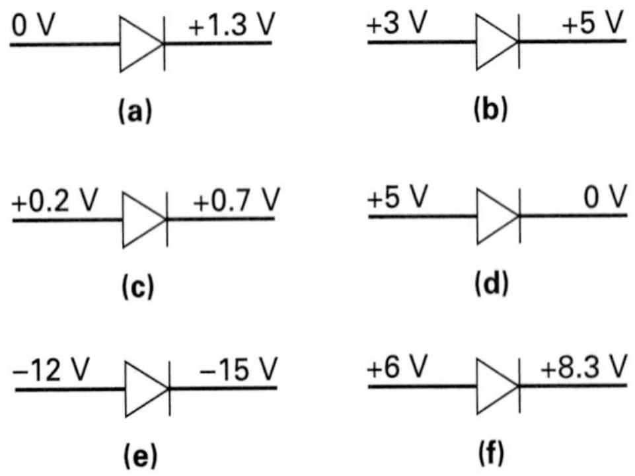

0 V +1.3 V
(a)
+3 V +5 V
(b)
+0.2 V +0.7 V
(c)
+5 V 0 V
(d)
−12 V −15 V
(e)
+6 V +8.3 V
(f)

FIGURE 3-24 Biased Junction Diodes.

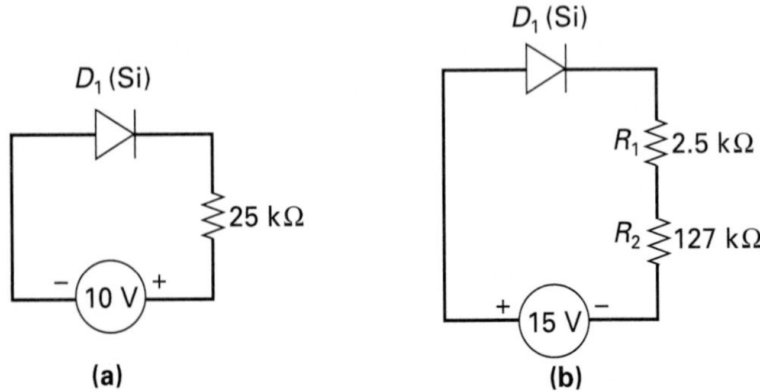

FIGURE 3-25 Forward Current Examples.

Troubleshooting Questions

35. Referring to the switch encoder circuit in Figure 3-26, first determine what digital codes will be generated at A, B, and C for each of the switch positions 1, 2 and 3. Next, determine what problems would occur for each of the following circuit malfunctions or conditions:

a. What would happen if D_1 were to open permanently?
b. Would a 200 mA fuse protect the +5 V supply voltage?

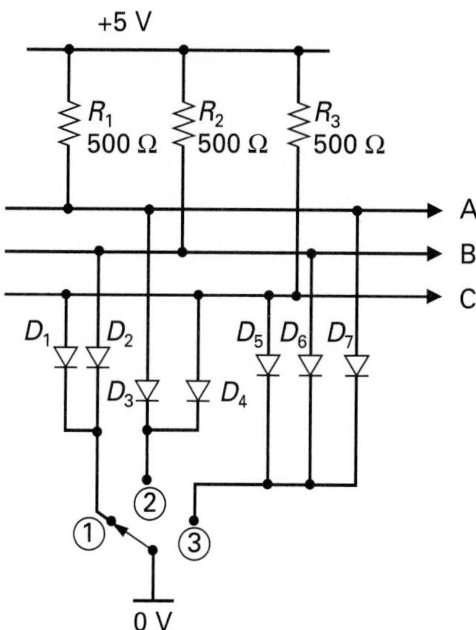

Input	Output		
Switch Position	A	B	C
1			
2			
3			

FIGURE 3-26 **A Switch Encoder Circuit.**

CHAPTER

4

OBJECTIVES

After completing this chapter, you will be able to:

1. Sketch the schematic symbol and describe the different package types for the zener diode.
2. Describe the forward and reverse voltage-current characteristics of a typical zener diode and the key parameters that govern its operation.
3. Explain the basic operating principles of the zener diode.
4. Interpret a typical manufacturer's data sheet to determine whether a zoner is suitable as a substitute for a faulty zener diode.
5. Describe how the zener diode can be used in the following applications:
   a. A voltage regulator circuit
   b. As a voltage reference in a comparator circuit
6. Discuss how zener diodes are tested.
7. Sketch the schematic symbol, and describe the different package types for the light emitting diode or LED.
8. Describe the forward and reverse voltage-current characteristics of a typical LED and the key parameters that govern its operation.
9. Explain the basic operating principles of the LED.
10. Interpret a typical manufacturer's data sheet to determine whether it is suitable as a substitute for a faulty LED.
11. Describe how the LED can be used in the following applications:
    a. As a power and signal indicator
    b. As a two-color, two-state indicator
    c. In multisegment displays
    d. In security and fiber optic systems
12. Discuss how LEDs and multisegment LED displays are tested.
13. Describe the three-step troubleshooting process, and apply it to a typical circuit containing P-N junction diodes, zener diodes, and light-emitting diodes.

Zener Diodes and Light Emitting Diodes

OUTLINE

VIGNETTE

Wireless

Lee De Forest was born in Council Bluffs, Iowa on August 26, 1873. He obtained his PhD from Yale Sheffield Scientific School in 1899 with a thesis that is recognized as the first paper on radio communications in the United States.

De Forest was an astounding practical inventor; however, he wasn't a good businessman and many of his inventions were stolen and exploited by business partners.

In 1902, he invented an electrolytic radio detector and an alternating-current transmitter while working for the Western Electric Company. He also developed an optical sound track for movies which was rejected by film producers. Later another system was used that was based on De Forest's principles.

In 1907, De Forest patented his *audion* detector or triode (three-electrode) vacuum tube. This tube was a more versatile vacuum tube than the then available diode (two-electrode) vacuum tube, or *Fleming Valve,* invented by English electrical engineer, Sir John Fleming. By adding a third electrode, called a control grid, De Forest made amplification possible because the tube could now be controlled. The triode's ability to amplify as well as rectify led to the development of radio communications, or wireless as it was called, and later to television. De Forest also discovered that by cascading (connecting end-to-end) amplifiers, a higher gain could be attained. This ability led to long distance communications because weak long-range input signals could be increased in magnitude to a usable level.

In 1905, he began experimenting with speech and music broadcasts. In 1910 De Forest transmitted the singing voice of Enrico Caruso and, in so doing, became one of the pioneers of radio broadcasting.

Slowly the usefulness of the triode as a generator, amplifier, and detector of radio waves became apparent. However, it was not until World War I that the audion became an invaluable electronic device and was manufactured in large quantities.

As well as working on the technical development of radio, De Forest gave many practical public demonstrations of wireless communications. He became internationally renowned and was often referred to as "the father of radio." Before his fame and fortune, however—just before everyone became aware of the profound effect that his audion tube would have on the world—an incident occurred. While trying to sell stock in his company, De Forest was arrested on charges of fraud and his device, which was to launch a new era in electronics and make possible radio and television, was at this time called "a strange device like an incandescent lamp . . . which has proved worthless."

INTRODUCTION

In the previous chapter we studied the *basic P-N junction diode.* In 1960, many electronic system manufacturers began to use the semiconductor P-N junction diode instead of the vacuum tube diode in low-power and low-frequency applications. Like all inventions, once the basic P-N junction diode appeared on the market and was in wide use, it inspired others to investigate all of its possibilities. Research and development into all semiconductor, or solid state, devices mushroomed at this time resulting in a variety of new semiconductor devices, including several different types of diodes. Today there are many different types of semiconductor diodes available, all of which have different characteristics and are suited for different applications. After the basic P-N junction diode, the two most widely used diodes are the **zener diode** and the **light emitting diode.**

In this chapter, we will be examining the operation and characteristics of the zener diode and light emitting diode in preparation for their many different circuit applications. For example, in Chapter 5 we will see how the basic P-N junction, zener, and light emitting diode can be used to construct dc power supply circuits. In later chapters we will see how these and other diodes are used in a variety of circuit applications such as rectifying, regulating, detecting, clipping, clamping, multiplying, switching, displaying, and oscillating.

4-1 THE ZENER DIODE

In the previous chapter we discussed the forward and reverse characteristics of a basic P-N junction diode. When forward biased, a basic junction diode will turn ON and be equivalent to a closed switch. When reverse biased, a basic junction diode will turn OFF and be equivalent to an open switch. In all basic P-N junction diode circuit applications, the external bias voltage was always kept below the maximum forward current rating and reverse breakdown voltage rating because this would result in a damaging value of current.

It was American inventor Clarence Zener who first began investigating in detail the reverse breakdown of a diode. He discovered that the basic P-N junction diode would conduct a high reverse current when a sufficiently high reverse bias voltage was applied across its terminals. This large external reverse voltage would cause an increase in the number of minority carriers in the *n* and *p* regions as valence electrons are pulled away from their parent atoms. Clarence Zener was fascinated by the reverse breakdown characteristics of a basic P-N junction diode and, after many years of study, developed a special diode that could handle high values of reverse current and still safely dissipate any heat that was generated. These special diodes—constructed to operate at voltages that are equal or greater than the reverse breakdown voltage rating—were named **zener diodes** after their inventor.

Zener Diode
Diodes constructed to operate at voltages that are equal or greater than the reverse breakdown voltage rating.

Let us now examine the zener diode's schematic symbol and package types and then the operation and characteristics of the zener diode.

4-1-1 Zener Diode Symbol and Packages

Figure 4-1(a) shows the two schematic symbols used to represent the zener diode. As you can see, the zener diode symbol resembles the basic P-N junction diode symbol in appearance; however, the zener diode symbol has a zig-zag bar instead of the straight bar. This zig-zag bar at the cathode terminal is included as a memory aid since it is "Z" shaped and will always remind us of zener.

Figure 4-1(b) shows two typical low-power zener diode packages, and one high-power zener diode package. The surface mount low-power zener package has two metal pads for direct mounting to the surface of a circuit board, while the axial lead low-power zener package has the zener mounted in a glass or epoxy case. The high power zener package is generally stud mounted and contained in a metal case. These packages are identical to the basic P-N junction diode low-power and high-power packages. Once again, a band or stripe is used to identify the cathode end of the zener diode in the low-power packages, whereas the threaded terminal of a high-power package is generally always the cathode.

4-1-2 Zener Diode Voltage-Current (V-I) Characteristics

Figure 4-2 shows the *V-I* (voltage-current) characteristic curve of a typical zener diode. This characteristic curve is almost identical to the basic P-N junction diode's characteristic curve. For example, when forward biased at or beyond

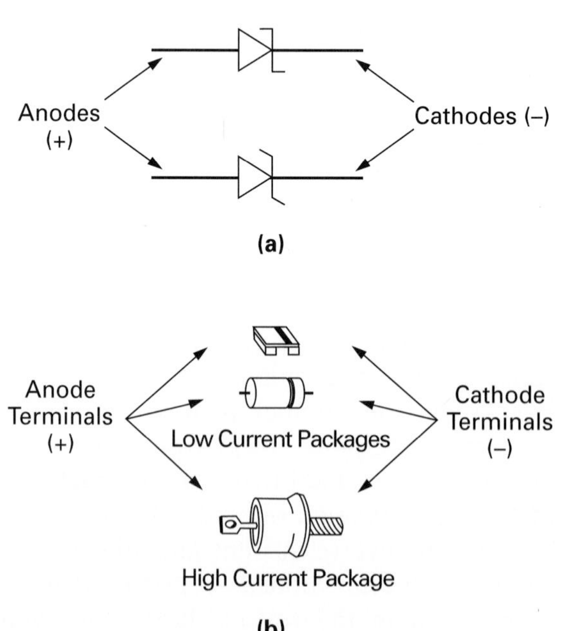

FIGURE 4-1 The Zener Diode. (a) Zener Diode Schematic Symbols. (b) Packages.

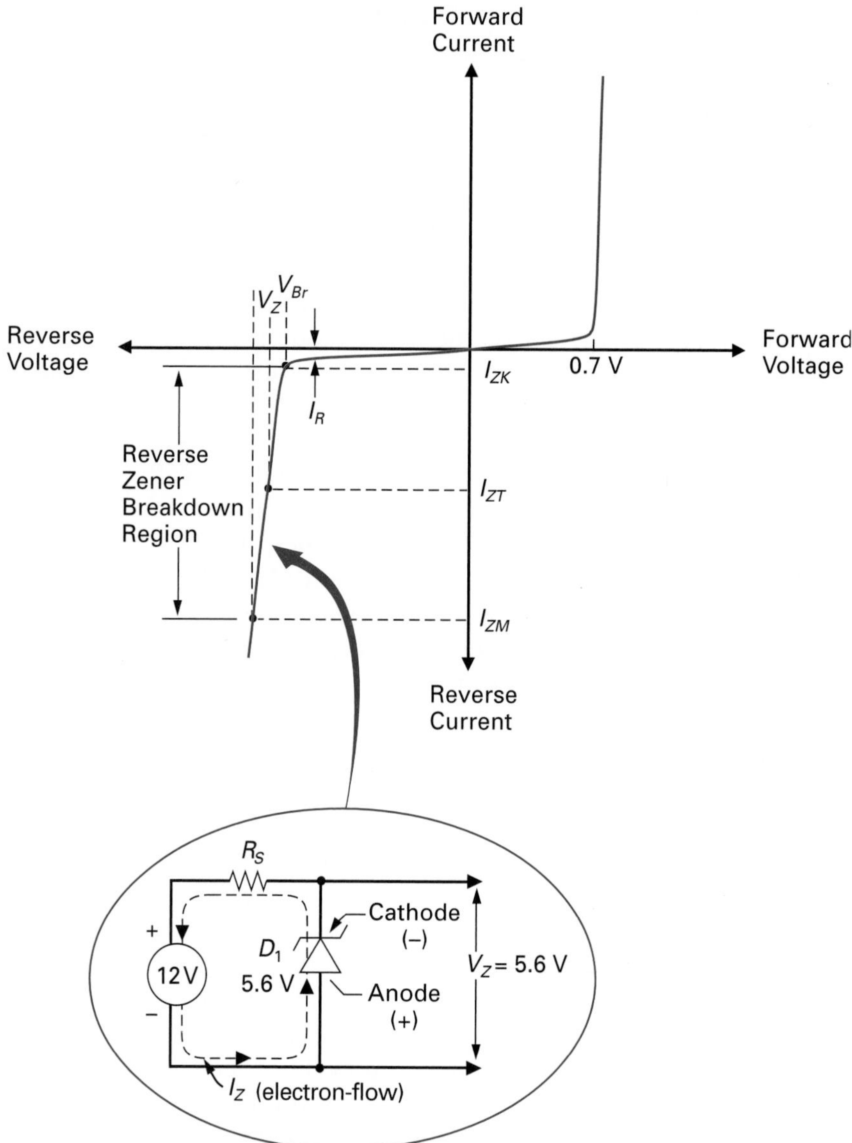

FIGURE 4-2 The Zener Diode Voltage-Current Characteristic Curve.

0.7 V, the zener diode will turn ON and be equivalent to a closed switch; whereas, when reverse biased, the zener diode will turn OFF and be equivalent to an open switch. The main difference, however, is that the zener diode has been specifically designed to operate in the reverse breakdown region of the curve. This is achieved, as can be seen in the inset in Figure 4-2, by making sure that the external bias voltage applied to a zener diode will not only reverse bias the zener diode (+→ cathode, – → anode) but also be large enough to drive the zener diode into its reverse breakdown region. Because the zener diode will generally always be operated in the reverse breakdown region, let us now discuss this reverse breakdown region of the *V-I* curve in a little more detail.

Zener Reverse Breakdown Region

Referring to the zener reverse breakdown region of the curve in Figure 4-2, you can see that when the external reverse-bias voltage is high enough, the zener diode will break down and conduct a high reverse current. This reverse current through the diode will increase at a rapid rate as the external reverse bias voltage is increased. The rapid increase in current through the diode is caused by the zener diode's sudden drop in resistance because the external reverse-bias voltage is now large enough to tear valence electrons away from their parent atoms.

Reverse Leakage Current (I_R)
The undesirable flow of current through a device in the reverse direction.

Zener Breakdown Voltage (V_{Br})
The point on the reverse bias voltage scale where the current through the zener suddenly increases.

Zener Knee Current (I_{ZK})
The knee-shaped point on the *V-I* curve at which a minimum value of reverse current will pass through the diode.

Zener Current (I_Z)
The current through the zener diode.

Zener Voltage (V_Z)
The voltage drop across the zener when it is being operated in the reverse zener breakdown region.

Zener Test Current (I_{ZT})
The specified value of current that is passing through the zener diode when the reverse voltage rating is measured.

As the reverse voltage across the zener diode is increased from the graph origin (which represents 0 volts), the value of **reverse leakage current (I_R)** begins to increase. The reverse zener breakdown region of the curve begins when the current through the zener suddenly increases. This point on the reverse bias voltage scale is called the **zener breakdown voltage (V_{Br}).** At this knee-shaped point, a minimum value of reverse current, known as the **zener knee current (I_{ZK}),** will pass through the zener. If the applied voltage is increased beyond V_{Br}, the value of **zener current (I_Z)** will increase. The voltage drop developed across the zener when it is being operated in the reverse zener breakdown region is called the **zener voltage (V_Z).** Comparing the voltage developed across the zener (V_Z) to the value of current through the zener (I_Z), you may have noticed that *the voltage drop across a zener diode (V_Z) remains almost constant when it is operated in the reverse zener breakdown region, even though current through the zener (I_Z) can vary considerably. This ability of the zener diode to maintain a relatively constant voltage regardless of variations in zener current is the key characteristic of the zener diode.*

Zener Voltage Rating

The breakdown voltage of a zener diode is controlled by adjusting the doping to control the width of the zener diode's depletion layer. By controlling the depletion layer's width, a manufacturer can create a zener that will break down at a specific voltage. Generally, manufacturers rate zener diodes based on their zener voltage (V_Z) rather than their breakdown voltage (V_{Br}). A wide variety of zener diode voltage ratings are available ranging from 1.8 V to several hundred volts. For example, many of the frequently used low-voltage zener diodes have ratings of 3.3 V, 4.7 V, 5.1 V, 5.6 V, 6.2 V, and 9.1 V. All of these zener voltage ratings are nominal values that represent the reverse voltage that will be developed across the zener diode when a specified value of current, called the **zener test current (I_{ZT})**, is passing through the zener diode. As can be seen in Figure 4-2, the zener test current is typically at a midpoint in the reverse zener breakdown region, midway between the low zener knee current (I_{ZK}), and the *maximum zener current (I_{ZM})*.

Like other components such as resistors and capacitors, a zener diode's voltage rating (V_Z) has a specified value and a tolerance. The following example shows how to calculate the tolerance for a typical zener diode.

■ **EXAMPLE:**

What will be the zener voltage range for a 5.6 V ± 10% zener diode?

■ ***Solution:***

$$10\% \text{ of } 5.6\text{ V} = 0.56\text{ V}$$
$$5.6\text{ V} - 0.56\text{ V} = 5.04\text{ V}$$
$$5.6\text{ V} + 0.56\text{ V} = 6.16\text{ V}$$
$$5.04 \text{ volts to } 6.16 \text{ volts}$$

Figure 4-3 shows the small *change in zener voltage (ΔV_Z)*, and the corresponding larger *change in zener current (ΔI_Z)*.

Zener Power Dissipation

Zener diodes have a **maximum power dissipation (P_D)** rating that indicates the maximum power that the zener diode can safely dissipate. A variety of zener diodes are available with maximum power ratings from several hundred milli-

Maximum Power Dissipation (P_D)
Maximum power that the zener diode can safely dissipate.

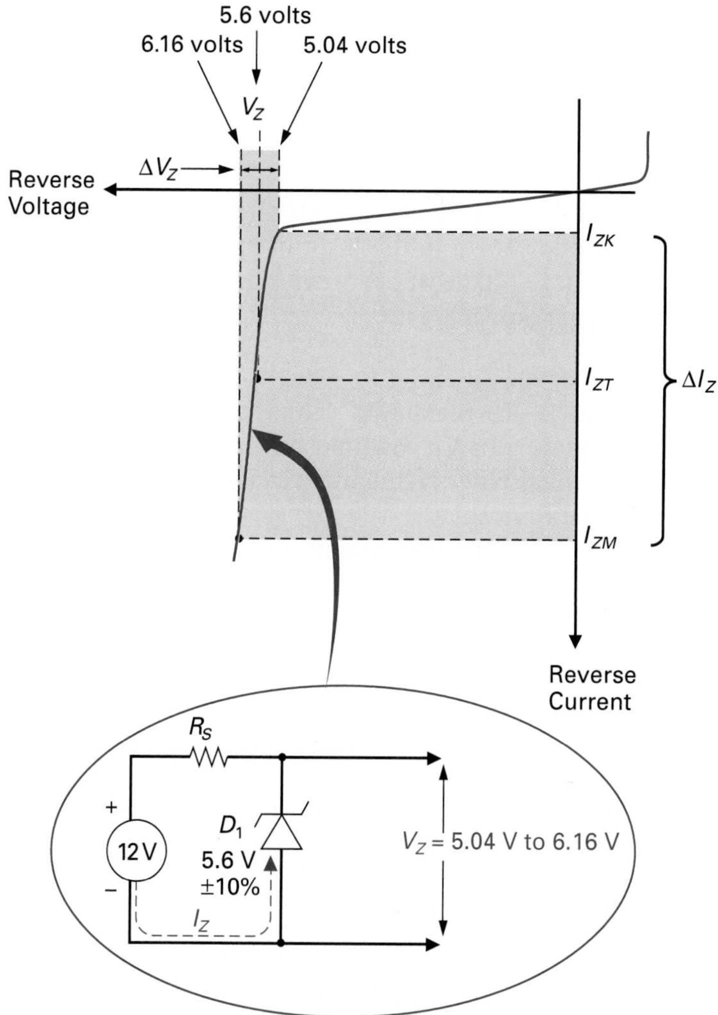

FIGURE 4-3 Small Zener Voltage Change.

watts up to 100 watts. However, most of the more frequently used zener diodes have maximum power ratings of 500 mW and 1 W. This power rating is generally given for a specific operating temperature of 50°C. Like most semiconductors, zener diodes have a negative temperature coefficient of resistance, which means as temperature increases their resistance will decrease resulting in an increase in current and therefore heat. This means that the amount of power that a zener diode can safely dissipate will decrease as the operating temperature increases.

Zener Maximum Reverse Current

If the maximum zener current (I_{ZM}) of a zener diode is exceeded, the diode will more than likely be destroyed because this value of current will generate more heat than the zener can safely dissipate (P_D). Manufacturers of zener diodes generally specify this value of maximum zener current (I_{ZM}) in the device data sheet. If a value of maximum zener current (I_{ZM}) is not specified, it can be calculated by using the zener diode's power dissipation rating (P_D), the zener voltage rating (V_R), and the power formula,

$$I_{ZM} = \frac{P_D}{V_Z}$$

■ **EXAMPLE:**

Suppose the zener diode shown in the inset in Figure 4-4 has a power rating of 500 mWatts, and a zener voltage of 5.6 V with a ±10% tolerance. Calculate the value of maximum zener current.

■ ***Solution:***

As previously calculated, the maximum value of zener voltage would be 5.6 V plus 10% of 5.6 V equals 6.16 V. Now that we have the maximum voltage limit and maximum power dissipation value, we can use the power formula to calculate the maximum zener current

$$I_{ZM} = \frac{P_D}{V_Z}$$

$$I_{ZM} = \frac{500 \text{ mW}}{6.16 \text{ V}}$$

$$I_{ZM} = 81.17 \text{ mA}$$

Referring to the reverse zener breakdown region in Figure 4-4, you can see I_{ZM} shown at the lower end of the reverse characteristic curve.

Zener Circuit Voltage and Current

In the example in the inset in Figure 4-4, a 5.6 V zener diode is connected across a 12 V source. The 12 V input voltage polarity is connected so that it reverse biases the zener diode. In order for this circuit to operate properly, the input voltage (V_{in}) must be higher than the breakdown voltage of the zener

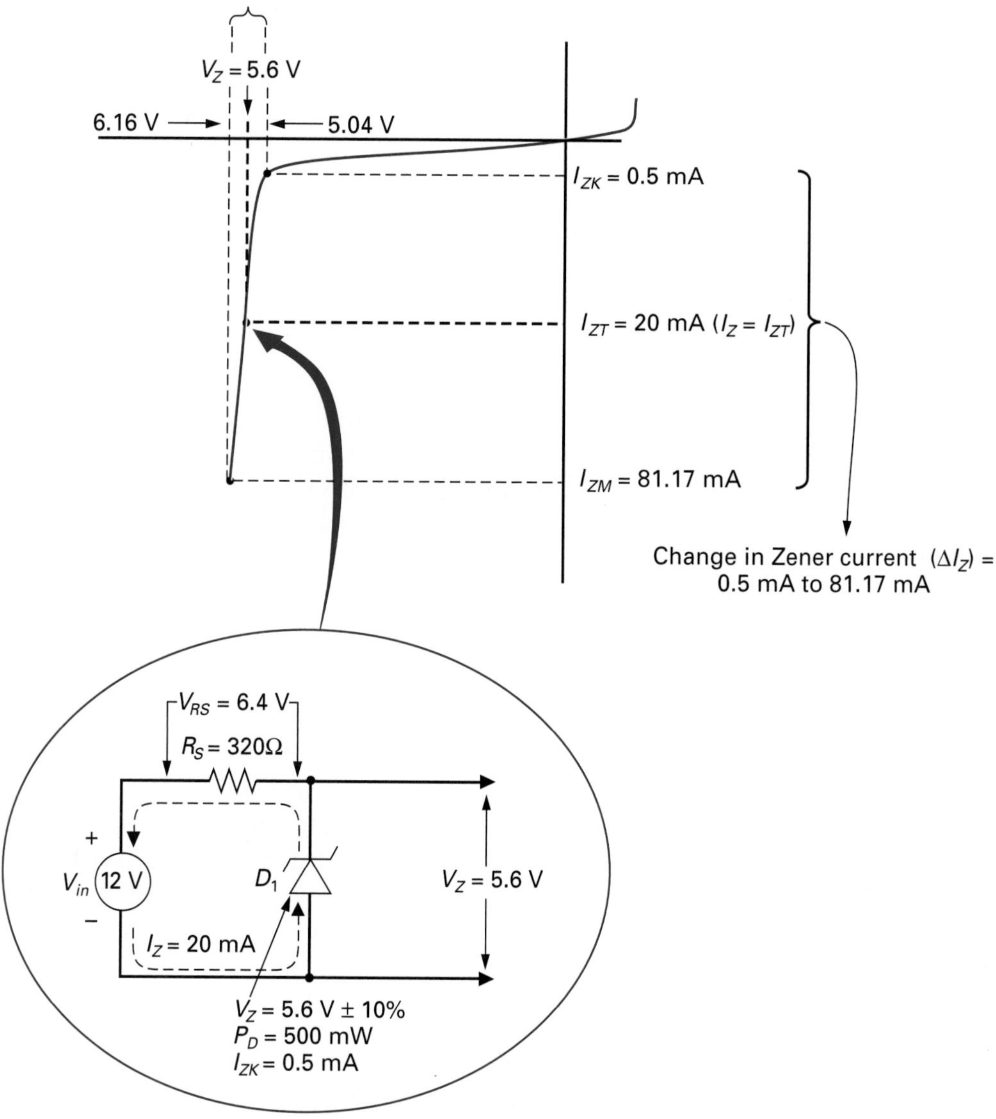

FIGURE 4-4 Zener Circuit Voltage and Current Calculations.

diode (V_Z). In this instance, the input voltage of 12 V is large enough to send the 5.6 V zener into its reverse breakdown region. As far as the voltage drops across each of the components in the inset in Figure 4-4, the voltage across the zener diode will be equal to the zener diode's voltage rating (5.6 V), and the voltage developed across the series resistor (V_{RS}) will be equal to the difference between the diode's zener voltage and the input voltage, therefore,

$$V_{RS} = V_{in} - V_Z$$

$$= 12\text{ V} - 5.6\text{ V}$$
$$= 6.4\text{ V}$$

As we know, once the series resistor's voltage drop (V_{RS}) and resistance (R_S) are known, we can calculate current.

$$I = \frac{V_{RS}}{R_S}$$

$$I = \frac{6.4 \text{ V}}{320 \ \Omega}$$

$$I = 20 \text{ mA}$$

Since the series resistor (R_S) and the zener diode (D_1) are in series with one another, the current through the zener diode (I_Z) will be the same as the current through the series resistor ($I_{RS} = I_Z = 20$ mA).

Zener Impedance

Zener Impedance (Z_Z)
Opposition offered by a zener diode to current.

Zener Impedance (Z_Z) is the opposition offered by a zener diode to current. A manufacturer calculates the value of zener impedance by varying the zener current above and below the zener test current value and then monitoring the change in zener voltage. Once the change in zener current (ΔI_Z) and corresponding change in zener voltage (ΔV_Z) is known, Ohm's Law can be used to calculate zener impedance with the following formula:

$$Z_Z = \frac{\Delta V_Z}{\Delta I_Z}$$

Applying this formula to the example in the inset in Figure 4-2, we can calculate the zener diode's impedance. Previously, we determined that the change in zener voltage (ΔV_Z) is from 5.04 volts to 6.16 volts. If the change in zener current (ΔI_Z) is from 0.5 mA (zener knee current, I_{ZK}) to 81.17 mA (maximum zener current, I_{ZM}), the zener diode's impedance will be

$$Z_Z = \frac{\Delta V_Z}{\Delta I_Z}$$

$$Z_Z = \frac{6.16 \text{ V} - 5.04 \text{ V}}{81.17 \text{ mA} - 0.5 \text{ mA}}$$

$$Z_Z = \frac{1.12 \text{ V}}{80.67 \text{ mA}}$$

$$Z_Z = 13.9 \ \Omega$$

4-1-3 *Zener Diode Data Sheets*

Like the basic P-N junction diode's data sheet, a zener diode data sheet details the characteristics and maximum and minimum values of operation for a given device. Figure 4-5 contains the data sheet for a series of silicon zener diodes. Studying this specification sheet along with the associated notes, you will be able to find the characteristics and maximum and minimum ratings previously discussed. Remember that, as a technician, you should be familiar with some of the basic operating limits of a device so that you can isolate component malfunc-

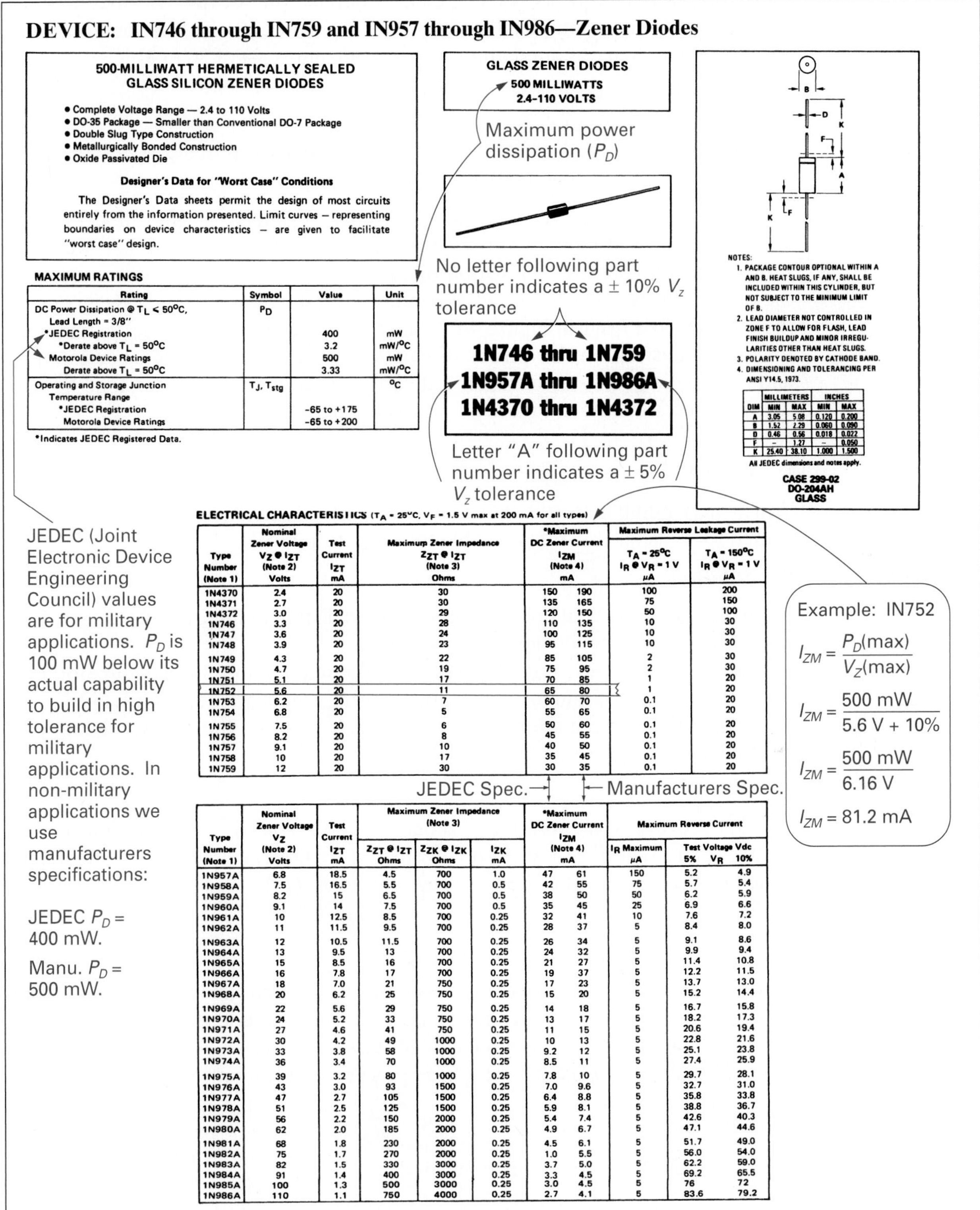

DEVICE: IN746 through IN759 and IN957 through IN986—Zener Diodes

500-MILLIWATT HERMETICALLY SEALED GLASS SILICON ZENER DIODES

- Complete Voltage Range — 2.4 to 110 Volts
- DO-35 Package — Smaller than Conventional DO-7 Package
- Double Slug Type Construction
- Metallurgically Bonded Construction
- Oxide Passivated Die

Designer's Data for "Worst Case" Conditions

The Designer's Data sheets permit the design of most circuits entirely from the information presented. Limit curves – representing boundaries on device characteristics – are given to facilitate "worst case" design.

GLASS ZENER DIODES
500 MILLIWATTS
2.4–110 VOLTS

1N746 thru 1N759
1N957A thru 1N986A
1N4370 thru 1N4372

MAXIMUM RATINGS

Rating	Symbol	Value	Unit
DC Power Dissipation @ $T_L \leq 50^oC$, Lead Length = 3/8"	P_D		
*JEDEC Registration		400	mW
*Derate above $T_L = 50^oC$		3.2	mW/°C
Motorola Device Ratings		500	mW
Derate above $T_L = 50^oC$		3.33	mW/°C
Operating and Storage Junction Temperature Range	T_J, T_{stg}		°C
*JEDEC Registration		−65 to +175	
Motorola Device Ratings		−65 to +200	

*Indicates JEDEC Registered Data.

NOTES:
1. PACKAGE CONTOUR OPTIONAL WITHIN A AND B. HEAT SLUGS, IF ANY, SHALL BE INCLUDED WITHIN THIS CYLINDER, BUT NOT SUBJECT TO THE MINIMUM LIMIT OF B.
2. LEAD DIAMETER NOT CONTROLLED IN ZONE F TO ALLOW FOR FLASH, LEAD FINISH BUILDUP AND MINOR IRREGULARITIES OTHER THAN HEAT SLUGS.
3. POLARITY DENOTED BY CATHODE BAND.
4. DIMENSIONING AND TOLERANCING PER ANSI Y14.5, 1973.

DIM	MILLIMETERS MIN	MILLIMETERS MAX	INCHES MIN	INCHES MAX
A	3.05	5.08	0.120	0.200
B	1.52	2.29	0.060	0.090
D	0.46	0.56	0.018	0.022
F	–	1.27	–	0.050
K	25.40	38.10	1.000	1.500

All JEDEC dimensions and notes apply.

CASE 299-02
DO-204AH
GLASS

ELECTRICAL CHARACTERISTICS (T_A = 25°C, V_F = 1.5 V max at 200 mA for all types)

Type Number (Note 1)	Nominal Zener Voltage V_Z @ I_{ZT} (Note 2) Volts	Test Current I_{ZT} mA	Maximum Zener Impedance Z_{ZT} @ I_{ZT} (Note 3) Ohms	*Maximum DC Zener Current I_{ZM} (Note 4) mA		Maximum Reverse Leakage Current T_A = 25°C I_R @ V_R = 1 V µA	Maximum Reverse Leakage Current T_A = 150°C I_R @ V_R = 1 V µA
1N4370	2.4	20	30	150	190	100	200
1N4371	2.7	20	30	135	165	75	150
1N4372	3.0	20	29	120	150	50	100
1N746	3.3	20	28	110	135	10	30
1N747	3.6	20	24	100	125	10	30
1N748	3.9	20	23	95	115	10	30
1N749	4.3	20	22	85	105	2	30
1N750	4.7	20	19	75	95	2	30
1N751	5.1	20	17	70	85	1	20
1N752	5.6	20	11	65	80	1	20
1N753	6.2	20	7	60	70	0.1	20
1N754	6.8	20	5	55	65	0.1	20
1N755	7.5	20	6	50	60	0.1	20
1N756	8.2	20	8	45	55	0.1	20
1N757	9.1	20	10	40	50	0.1	20
1N758	10	20	17	35	45	0.1	20
1N759	12	20	30	30	35	0.1	20

Type Number (Note 1)	Nominal Zener Voltage V_Z (Note 2) Volts	Test Current I_{ZT} mA	Maximum Zener Impedance (Note 3) Z_{ZT} @ I_{ZT} Ohms	Z_{ZK} @ I_{ZK} Ohms	I_{ZK} mA	*Maximum DC Zener Current I_{ZM} (Note 4) mA		Maximum Reverse Current I_R Maximum µA	Test Voltage Vdc V_R 5%	Test Voltage Vdc V_R 10%
1N957A	6.8	18.5	4.5	700	1.0	47	61	150	5.2	4.9
1N958A	7.5	16.5	5.5	700	0.5	42	55	75	5.7	5.4
1N959A	8.2	15	6.5	700	0.5	38	50	50	6.2	5.9
1N960A	9.1	14	7.5	700	0.5	35	45	25	6.9	6.6
1N961A	10	12.5	8.5	700	0.25	32	41	10	7.6	7.2
1N962A	11	11.5	9.5	700	0.25	28	37	5	8.4	8.0
1N963A	12	10.5	11.5	700	0.25	26	34	5	9.1	8.6
1N964A	13	9.5	13	700	0.25	24	32	5	9.9	9.4
1N965A	15	8.5	16	700	0.25	21	27	5	11.4	10.8
1N966A	16	7.8	17	700	0.25	19	37	5	12.2	11.5
1N967A	18	7.0	21	750	0.25	17	23	5	13.7	13.0
1N968A	20	6.2	25	750	0.25	15	20	5	15.2	14.4
1N969A	22	5.6	29	750	0.25	14	18	5	16.7	15.8
1N970A	24	5.2	33	750	0.25	13	17	5	18.2	17.3
1N971A	27	4.6	41	750	0.25	11	15	5	20.6	19.4
1N972A	30	4.2	49	1000	0.25	10	13	5	22.8	21.6
1N973A	33	3.8	58	1000	0.25	9.2	12	5	25.1	23.8
1N974A	36	3.4	70	1000	0.25	8.5	11	5	27.4	25.9
1N975A	39	3.2	80	1000	0.25	7.8	10	5	29.7	28.1
1N976A	43	3.0	93	1500	0.25	7.0	9.6	5	32.7	31.0
1N977A	47	2.7	105	1500	0.25	6.4	8.8	5	35.8	33.8
1N978A	51	2.5	125	1500	0.25	5.9	8.1	5	38.8	36.7
1N979A	56	2.2	150	2000	0.25	5.4	7.4	5	42.6	40.3
1N980A	62	2.0	185	2000	0.25	4.9	6.7	5	47.1	44.6
1N981A	68	1.8	230	2000	0.25	4.5	6.1	5	51.7	49.0
1N982A	75	1.7	270	2000	0.25	1.0	5.5	5	56.0	54.0
1N983A	82	1.5	330	3000	0.25	3.7	5.0	5	62.2	59.0
1N984A	91	1.4	400	3000	0.25	3.3	4.5	5	69.2	65.5
1N985A	100	1.3	500	3000	0.25	3.0	4.5	5	76	72
1N986A	110	1.1	750	4000	0.25	2.7	4.1	5	83.6	79.2

FIGURE 4-5 Device Data Sheet for a Series of Zener Diodes. (Copyright of Motorola. Used by permission.)

tions within a circuit by determining whether the device is operating to specifications. For example, the 5.6 V IN752 zener diode has the following specifications when operated at a zener test current (I_{ZT}) of 20 mA:

Zener Voltage (V_Z) = 5.6 V ± 10%
Zener Impedance (Z_Z) = 11 Ω
Manufacturer's recommended Maximum Zener Current (I_{ZM}) = 80 mA
Manufacturer's recommended Power Dissipation (P_D) = 500 mW

In some instances, you may need to replace a defective zener diode. Figure 4-6 shows a typical manufacturer's zener diode selection sheet. When we were substituting one basic P-N junction diode for another, we were concerned with the device's forward current and maximum reverse voltage ratings. Referring to the columns and rows in Figure 4-6, you can see that the two critical parameters for zener substitution are power dissipation (P_D), which is listed horizontally, and zener voltage (V_Z), which is listed vertically. To help you understand how to use the selection table in Figure 4-6, let us try an example.

■ **EXAMPLE:**

Which part number would you use to order a replacement for a faulty axial leaded 5.6 V zener that must be able to dissipate 1.25 W?

■ ***Solution:***

Referring to the vertical nominal zener voltage listing in Figure 4-6, we must first locate the row that corresponds to a V_Z of 5.6 V.

Secondly, we must refer to the horizontal maximum power dissipation listing in the same table and locate a wattage rating that is greater than 1.25 W.

Finding the intersection between the 5.6 V row and the 1.5 W column, you should locate the part number IN5919B.

As previously mentioned, a zener's reverse voltage remains almost constant V_Z, and therefore the power dissipated by the zener diode (P_D) is determined by the amount of current through the zener (I_Z).

$$P_Z\updownarrow = V_Z \times I_Z\updownarrow$$

Manufacturer's data sheets, like the example shown in Figure 4-6, list the maximum power that the zener can dissipate (P_D), which can typically range from 0.25 W to 100 W. Some data sheets, however, may include the maximum zener current rating (I_{ZM}). If this is not included, as was the case in Figure 4-6, it can easily be calculated by using the transposed power formula mentioned earlier.

$$I_{ZM} = \frac{P_D}{V_Z}$$

Let us now use this formula in an example.

■ **EXAMPLE:**

Referring to the table in Figure 4-6, what would be the maximum current that can be handled by an IN5919A (5.6 V, 1.5 W) zener diode?

DEVICE: Silicon Zener Diodes—1.8 V to 400 V, 225 mW to 5 W

Horizontal listing indicates maximum power dissipation (P_D)

Vertical listing indicates nominal zener voltage (V_Z).

Nominal Zener Breakdown Voltage	225 mW Surface Mount SOT-23		500 mW Surface Mount SOD-123	500 mW Low Level Surface Mount SOD-123	500 mW Surface Mount SOD-123	3 Watt Surface Mount SMB
Volts	Anode, Cathode, No Connection. Plastic Case 318-07 TO-236AB		Plastic Case 425, Style 1			Plastic Case 403A Cathode = Notch
1.8				MMSZ4678T1		
2.0				MMSZ4679T1		
2.2				***MMSZ4680T1***		
2.4	BZX84C2V4LT1	MMBZ5221BLT1	MMSZ2V4T1	MMSZ4681T1	***MMSZ5221BT1***	
2.5		MMBZ5222BLT1			***MMSZ5222BT1***	
2.7	BZX84C2V7LT1	MMBZ5223BLT1	MMSZ2V7T1	MMSZ4682T1	MMSZ5223BT1	
2.8		MMBZ5224BLT1			MMSZ5224BT1	
3.0	BZX84C3V0LT1	MMBZ5225BLT1	MMSZ3V0T1	MMSZ4683T1	***MMSZ5225BT1***	
3.3	BZX84C3V3LT1	MMBZ5226BLT1	MMSZ3V3T1	MMSZ4684T1	MMSZ5226BT1	1SMB5913BT3
3.6	BZX84C3V6LT1	MMBZ5227BLT1	MMSZ3V6T1	***MMSZ4685T1***	MMSZ5227BT1	1SMB5914BT3
3.9	BZX84C3V9LT1	***MMBZ5228BLT1***	MMSZ3V9T1	MMSZ4686T1	MMSZ5228BT1	1SMB5915BT3
4.3	BZX84C4V3LT1	MMBZ5229BLT1	MMSZ4V3T1	MMSZ4687T1	***MMSZ5229BT1***	1SMB5916BT3
4.7	***BZX84C4V7LT1***	MMBZ5230BLT1	MMSZ4V7T1	***MMSZ4688T1***	***MMSZ5230BT1***	1SMB5917BT3
5.1	***BZX84C5V1LT1***	***MMBZ5231BLT1***	***MMSZ5V1T1***	***MMSZ4689T1***	***MMSZ5231BT1***	1SMB5918BT3
5.6	***BZX84C5V6LT1***	***MMBZ5232BLT1***	***MMSZ5V6T1***	***MMSZ4690T1***	***MMSZ5232BT1***	1SMB5919BT3
6.0		MMBZ5233BLT1			MMSZ5233BT1	
6.2	***BZX84C6V2LT1***	***MMBZ5234BLT1***	***MMSZ6V2T1***	MMSZ4691T1	***MMSZ5234BT1***	1SMB5920BT3
6.8	BZX84C6V8LT1	***MMBZ5235BLT1***	MMSZ6V8T1	MMSZ4692T1	***MMSZ5235BT1***	1SMB5921BT3
7.5	BZX84C7V5LT1	MMBZ5236BLT1	MMSZ7V5T1	MMSZ4693T1	MMSZ5236BT1	1SMB5922BT3
8.2	BZX84C8V2LT1	MMBZ5237BLT1	MMSZ8V2T1	MMSZ4694T1	MMSZ5237BT1	1SMB5923BT3
8.7		MMBZ5238BLT1		MMSZ4695T1	MMSZ5238BT1	
9.1	BZX84C9V1LT1	MMBZ5239BLT1	MMSZ9V1T1	MMSZ4696T1	MMSZ5239BT1	1SMB5924BT3
10	BZX84C10LT1	***MMBZ5240BLT1***	MMSZ10T1	MMSZ4697T1	***MMSZ5240BT1***	1SMB5925BT3
11	BZX84C11LT1	MMBZ5241BLT1	MMSZ11T1	MMSZ4698T1	MMSZ5241BT1	1SMB5926BT3
12	***BZX84C12LT1***	***MMBZ5242BL***T1	MMSZ12T1	MMSZ4699T1	***MMSZ5242BT1***	1SMB5927BT3
13	BZX84C13LT1	MMBZ5243BLT1	MMSZ13T1	MMSZ4700T1	MMSZ5243BT1	1SMB5928BT3
14		MMBZ5244BLT1		MMSZ4701T1	MMSZ5244BT1	
15	BZX84C15LT1	***MMBZ5245BLT1***	MMSZ15T1	MMSZ4702T1	***MMSZ5245BT1***	1SMB5929BT3
16	BZX84C16LT1	MMBZ5246BLT1	MMSZ16T1	MMSZ4703T1	***MMSZ5246BT1***	1SMB5930BT3
17		MMBZ5247BLT1		MMSZ4704T1	MMSZ5247BT1	
18	***BZX84C18LT1***	***MMBZ5248BLT1***	***MMSZ18T1***	MMSZ4705T1	***MMSZ5248BT1***	1SMB5931BT3
19		MMBZ5249BLT1		MMSZ4706T1	MMSZ5249BT1	
20	BZX84C20LT1	***MMBZ5250BLT1***	MMSZ20T1	MMSZ4707T1	***MMSZ5250BT1***	1SMB5932BT3
22	BZX84C22LT1	MMBZ5251BLT1	MMSZ22T1	MMSZ4708T1	MMSZ5251BT1	1SMB5933BT3
24	BZX84C24LT1	MMBZ5252BLT1	MMSZ24T1	MMSZ4709T1	***MMSZ5252BT1***	1SMB5934BT3
25		MMBZ5253BLT1		MMSZ4710T1	MMSZ5253BT1	
27	BZX84C27LT1	***MMBZ5254BLT1***	MMSZ27T1	MMSZ4711T1	MMSZ5254BT1	1SMB5935BT3
28		MMBZ5255BLT1		MMSZ4712T1	MMSZ5255BT1	
30	BZX84C30LT1	MMBZ5256BLT1	MMSZ30T1	MMSZ4713T1	***MMSZ5256BT1***	1SMB5936BT3
33	BZX84C33LT1	***MMBZ5257BLT1***	MMSZ33T1	MMSZ4714T1	MMSZ5257BT1	1SMB5937BT3
36	BZX84C36LT1	MMBZ5258BLT1	MMSZ36T1	MMSZ4715T1	MMSZ5258BT1	1SMB5938BT3
39	BZX84C39LT1	MMBZ5259BLT1	***MMSZ39T1***	MMSZ4716T1	MMSZ5259BT1	1SMB5939BT3
43	BZX84C43LT1	MMBZ5260BLT1	MMSZ43T1	MMSZ4717T1	MMSZ5260BT1	1SMB5940BT3
47	BZX84C47LT1	MMBZ5261BLT1	MMSZ47T1		***MMSZ5261BT1***	1SMB5941BT3
51	BZX84C51LT1	MMBZ5262BLT1	MMSZ51T1		MMSZ5262BT1	1SMB5942BT3
56	BZX84C56LT1	MMBZ5263BLT1	MMSZ56T1		MMSZ5263BT1	1SMB5943BT3
60		MMBZ5264BLT1			MMSZ5264BT1	
62	BZX84C62LT1	MMBZ5265BLT1	MMSZ62T1		MMSZ5265BT1	1SMB5944BT3
68	BZX84C68LT1	MMBZ5266BLT1	MMSZ68T1		MMSZ5266BT1	1SMB5945BT3
75	BZX84C75LT1	MMBZ5267BLT1	MMSZ75T1		MMSZ5267BT1	1SMB5946BT3
82		MMBZ5268BLT1			MMSZ5268BT1	1SMB5947BT3
87		MMBZ5269BLT1			MMSZ5269BT1	
91		MMBZ5270BLT1			MMSZ5270BT1	1SMB5948BT3
100					MMSZ5271BT1	1SMB5949BT3
110						1SMB5950BT3
120						1SMB5951BT3
130						1SMB5952BT3
150						1SMB5953BT3
160						1SMB5954BT3
170						
180						1SMB5955BT3
200						1SMB5956BT3

Nominal Zener Breakdown Voltage	1 Watt Cathode = Polarity Band		1.3 Watt Cathode = Polarity Band			1.5 Watt Cathode = Polarity Band	3 Watt Cathode = Polarity Band	5 Watt Cathode = Polarity Band
Volts	Glass Case 59-03 (DO-41)	Plastic Surmetic 30 Case 59-03 (DO-41)	Glass Case 59-03 (DO-41)		Plastic Surmetic 30 Case 59-03 (DO-41)			Plastic Surmetic 40 Case 17
3.3	1N4728A	MZP4728A	BZX85C3V3RL			1N5913B		1N5333B
3.6	1N4729A	MZP4729A	***BZX85C3V6RL***			1N5914B		1N5334B
3.9	1N4730A	MZP4730A	BZX85C3V9RL	MZPY3.9RL	MZD3.9RL	1N5915B	3EZ3.9D5	1N5335B
4.3	1N4731A	MZP4731A	BZX85C4V3RL	MZPY4.3RL	MZD4.3RL	1N5916B	***3EZ4.3D5***	1N5336B
4.7	1N4732A	MZP4732A	BZX85C4V7RL	MZPY4.7RL	MZD4.7RL	1N5917B	3EZ4.7D5	1N5337B
5.1	***1N4733A***	MZP4733A	BZX85C5V1RL	MZPY5.1RL	MZD5.1RL	***1N5918B***	3EZ5.1D5	1N5338B
5.6	***1N4734A***	MZP4734A	BZX85C5V6RL	MZPY5.6RL	MZD5.6RL	1N5919B	3EZ5.6D5	1N5339B
6.0								1N5340B
6.2	***1N4735A***	MZP4735A	BZX85C6V2RL	MZPY6.2RL	MZD6.2RL	***1N5920B***	3EZ6.2D5	1N5341B
6.8	***1N4736A***	MZP4736A	BZX85C6V8RL	MZPY6.8RL	MZD6.8RL	1N5921B	3EZ6.8D5	1N5342B
7.5	1N4737A	MZP4737A	BZX85C7V5RL	MZPY7.5RL	MZD7.5RL	1N5922B	3EZ7.5D5	1N5343B
8.2	***1N4738A***	MZP4738A	BZX85C8V2RL	MZPY8.2RL	MZD8.2RL	1N5923B	3EZ8.2D5	1N5344B
8.7								1N5345B
9.1	1N4739A	MZP4739A	BZX85C9V1RL	MZPY9.1RL	MZD9.1RL	1N5924B	3EZ9.1D5	1N5346B
10	***1N4740A***	MZP4740A	BZX85C10RL	MZPY10RL	MZD10RL	1N5925B	3EZ10D5	1N5347B
11	***1N4741A***	MZP4741A	BZX85C11RL	MZPY11RL	MZD11RL	1N5926B	3EZ11D5	1N5348B
12	***1N4742A***	MZP4742A	BZX85C12RL	MZPY12RL	MZD12RL	1N5927B	3EZ12D5	1N5349B
13	1N4743A	MZP4743A	BZX85C13RL	MZPY13RL	MZD13RL	1N5928B	3EZ13D5	1N5350B
14							3EZ14D5	1N5351B
15	***1N4744A***	MZP4744A	BZX85C15RL	MZPY15RL	MZD15RL	***1N5929B***	3EZ15D5	1N5352B
16	***1N4745A***	MZP4745A	BZX85C16RL	MZPY16RL	MZD16RL	1N5930B	3EZ16D5	1N5353B
17							3EZ17D5	1N5354B
18	***1N4746A***	***MZP4746A***	BZX85C18RL	MZPY18RL	***MZD18RL***	1N5931B	3EZ18D5	1N5355B
19							3EZ19D5	1N5356B
20	***1N4747A***	MZP4747A	BZX85C20RL	MZPY20RL	MZD20RL	1N5932B	3EZ20D5	1N5357B
22	***1N4748A***	MZP4748A	***BZX85C22RL***	MZPY22RL	MZD22RL	1N5933B	3EZ22D5	1N5358B
24	***1N4749A***	***MZP4749A***	***BZX85C24RL***	MZPY24RL	MZD24RL	1N5934B	3EZ24D5	1N5359B
25								1N5360B
27	***1N4750A***	MZP4750A	BZX85C27RL	MZPY27RL	MZD27RL	1N5935B	3EZ27D5	1N5361B
28							3EZ28D5	1N5362B
30	***1N4751A***	***MZP4751A***	BZX85C30RL	MZPY30RL	MZD30RL	***1N5936B***	3EZ30D5	1N5363B
33	***1N4752A***	MZP4752A	BZX85C33RL	MZPY33RL	MZD33RL	1N5937B	3EZ33D5	1N5364B
36	1N4753A	MZP4753A	BZX85C36RL	MZPY36RL	MZD36RL	1N5938B	3EZ36D5	1N5365B
39	1N4754A	MZP4754A	BZX85C39RL	MZPY39RL	MZD39RL	1N5939B	3EZ39D5	1N5366B
43	1N4755A	MZP4755A	BZX85C43RL	MZPY43RL	MZD43RL	1N5940B	3EZ43D5	1N5367B
47	1N4756A	MZP4756A	BZX85C47RL	MZPY47RL	MZD47RL	1N5941B	3EZ47D5	1N5368B
51	1N4757A	MZP4757A	BZX85C51RL	MZPY51RL	MZD51	1N5942B	3EZ51D5	1N5369B
56	1N4758A	MZP4758A	BZX85C56RL	MZPY56RL	MZD56	1N5943B	3EZ56D5	1N5370B
60								1N5371B
62	1N4759A	MZP4759A	BZX85C62RL	MZPY62RL	MZD62	1N5944B	3EZ62D5	1N5372B
68	1N4760A	MZP4760A	BZX85C68RL	MZPY68RL	MZD68	1N5945B	3EZ68D5	1N5373B
75	1N4761A	MZP4761A	BZX85C75RL	MZPY75RL	MZD75	1N5946B	3EZ75D5	1N5374B
82	1N4762A	MZP4762A	BZX85C82RL	MZPY82RL	MZD82	1N5947B	3EZ82D5	1N5375B
87								1N5376B
91	1N4763A	MZP4763A	BZX85C91RL	MZPY91RL	MZD91	1N5948B	3EZ91D5	1N5377B
100	1N4764A	MZP4764A	BZX85C100RL	MZPY100RL	MZD100	1N5949B	3EZ100D5	1N5378B
110		1M110ZS5			MZD110	1N5950B	3EZ110D5	1N5379B
120		1M120ZS5			MZD120	1N5951B	3EZ120D5	1N5380B
130		1M130ZS5			MZD130	1N5952B	3EZ130D5	1N5381B
140							3EZ140D5	1N5382B
150		1M150ZS5			MZD150	1N5953B	3EZ150D5	1N5383B
160		1M160ZS5			MZD160	1N5954B	3EZ160D5	1N5384B
170							3EZ170D5	1N5385B
180		1M180ZS5			MZD180	***1N5955B***	3EZ180D5	1N5386B
190							3EZ190D5	1N5387B
200		1M200ZS5			MZD200	1N5956B	3EZ200D5	1N5388B
220							3EZ220D5	
240							3EZ240D5	
270							3EZ270D5	
300							3EZ300D5	
330							3EZ330D5	
360							3EZ360D5	
400							3EZ400D5	

FIGURE 4-6 Manufacturer's Zener Diodes Selection Sheet. (Copyright of Motorola. Used by permission.)

■ ***Solution:***

$$I_{ZM} = \frac{P_D}{V_Z}$$

$$I_{ZM} = \frac{1.5\text{ W}}{5.6\text{ V}} = 268\text{ mA}$$

4-1-4 Zener Diode Applications

Like the basic P-N junction diode, the zener diode is used in almost every electronic and electrical system. Although the zener diode can be forward and reverse biased, it is most widely used in applications where it is continually reverse biased and operating in the reverse zener breakdown region. As you proceed through this text you will see the zener diode used in a wide variety of analog and digital circuit applications. For now, let us see how the junction diode is used in a basic analog circuit and in a basic digital circuit.

Analog Circuit Application—A Voltage Regulator

All electronic systems that use the 120 V ac power from the wall outlet as a power source have an internal dc power supply that converts the ac input into the needed dc supply voltages for all of the circuits within the system. The ac from the electric company fluctuates during the day as the demand by consumers changes. For example, in the early morning hours when demand is low the ac input could be as high as 125 V ac, whereas at midday when the demand is at its peak the ac is being heavily loaded and the voltage could be pulled down to as low as 105 V ac. The circuits within electronic systems require dc power supplies that have a very tight tolerance. For example, many digital electronic circuits use a 5 V dc power supply that must only vary plus or minus a quarter of a volt (4.75 V to 5.25 V). If the dc supply voltage exceeds this tolerance, the circuit may not operate properly or be damaged.

Voltage Regulator Circuit
Maintains the output voltage of a voltage source constant despite variations in the input voltage and load resistance.

A circuit is therefore needed that will stabilize or regulate the input voltage to provide an unvarying output voltage. This is achieved with a **voltage regulator circuit,** which maintains the output voltage of a voltage source constant despite variations in the input voltage and the load resistance. Figure 4-7(a) shows how a zener diode can be used to regulate a dc source voltage, which varies between 20 V and 30 V, and produce an unvarying 12 V output. To begin with, the zener diode must be connected so that the polarity of the source voltage reverse biases the diode. Secondly, the source voltage must exceed the zener diode's reverse-breakdown voltage (V_Z) in order to send the zener into its reverse-breakdown region. A series resistor (R_S) is included to allow enough current to pass through the diode so that it operates within its reverse zener breakdown region, and yet limit current so that it does not exceed the maximum zener current. Since R_S and the zener diode are in series with one another, the current through both devices can be calculated with the formula:

$$I_S = \frac{V_{in} - V_Z}{R_S}$$

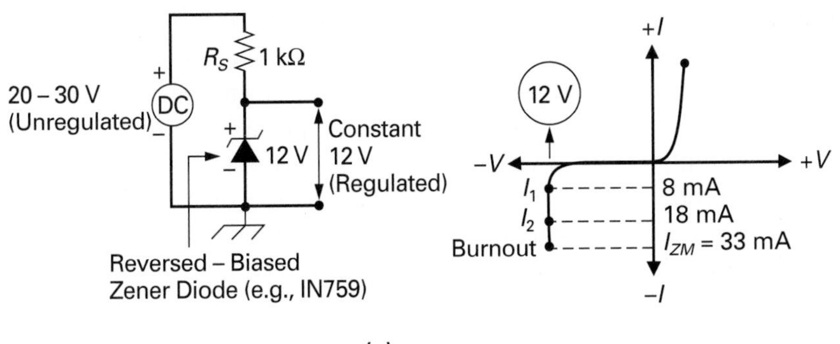

(a)

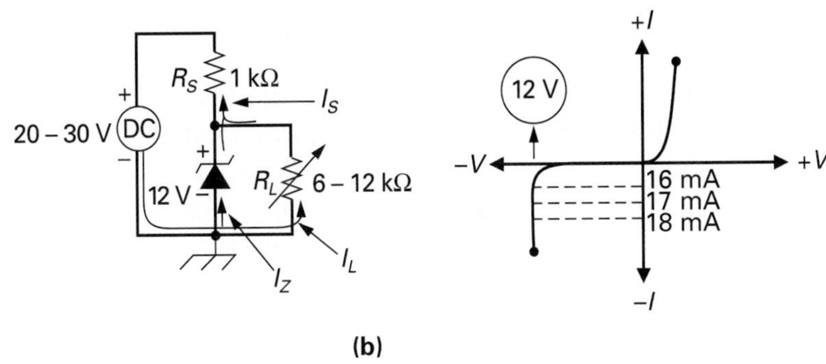

(b)

FIGURE 4-7 The Zener Diode in a Voltage Regulator Circuit. (a) Regulating the 12 V Output Despite Variations in the Source Voltage. (b) Regulating the 12 V Output Despite Variations in the Load Resistance.

I_S = Series Current
V_{in} = Input Source Voltage
V_Z = Zener Voltage
R_S = Series Resistor Value

In the example circuit in Figure 4-7(a), the input source voltage can vary anywhere from 20 V to 30 V. Let us calculate the value of series current for the two extreme low- and high-input voltages:

When V_{in} = **20 V**

$$I_S = \frac{V_{in} - V_Z}{R_S}$$

$$I_S = \frac{20\text{ V} - 12\text{ V}}{1\text{ k}\Omega} = 8\text{ mA}$$

When V_{in} = **30 V**

$$I_S = \frac{V_{in} - V_Z}{R_S}$$

$$I_S = \frac{30\text{ V} - 12\text{ V}}{1\text{ k}\Omega} = 18\text{ mA}$$

As the dc input voltage increases and decreases, it causes an increase and decrease in the dc current through the zener diode and the series resistor. Referring to the *V-I* curve in Figure 4-7(a), you can see that because the zener is operating within its reverse-breakdown region, a large swing in the dc input voltage (20 V to 30 V) will only result in a change in zener current (8 mA to 18 mA) and not in zener voltage. The zener voltage, which is applied to the output and is therefore the regulated output voltage, remains constant at 12 volts.

The zener diode is able to maintain the voltage drop across its terminals constant by continually changing its resistance or impedance (Z_Z) in response to a change in input voltage. For example, an input voltage increase ($V_{in}\uparrow$) will result in a circuit current increase and zener current increase ($I_Z\uparrow$), which will be countered by an equal but opposite zener diode impedance decrease ($Z_Z\downarrow$). An input voltage decrease ($V_{in}\downarrow$) will result in a circuit current decrease and zener current decrease ($I\downarrow$), which will be countered by a zener diode impedance increase ($Z_Z\uparrow$). Therefore, a change in zener diode current (I_Z) is always accompanied by an equal but opposite change in zener diode impedance (Z_Z). It is this equal but opposite change in I_Z and Z_Z that causes a cancellation, resulting in a constant output zener voltage. Since the zener diode's voltage remains constant due to its changing impedance, all of the input voltage change will appear across the series resistor which has a fixed or constant resistance. In summary:

When $\boldsymbol{V_{in}\downarrow}$ **to 20 V:**
$I_Z\downarrow$ to 8 mA, causes $Z_Z\uparrow$ to 1500 Ω ($Z_Z = V_Z/I_Z = 12\text{ V}/8\text{ mA} = 1500\ \Omega$) and therefore V_Z remains constant.

$$V_Z = I_Z \times Z_Z = 8\text{ mA} \times 1500\ \Omega = 12\text{ V} \qquad (V_Z = I_Z\downarrow \times Z_Z\uparrow)$$
$$V_{RS} = V_{in} - V_Z = 20\text{ V} - 12\text{ V} = 8\text{ V}$$

When $\boldsymbol{V_{in}\uparrow}$ **to 30 V:**
$I_Z\uparrow$ to 18 mA, causes $Z_Z\downarrow$ to 666.7 Ω ($Z_Z = V_Z/I_Z = 12\text{ V}/18\text{ mA} = 666.7\ \Omega$) and therefore V_Z remains constant.

$$V_Z = I_Z \times Z_Z = 18\text{ mA} \times 666.7\ \Omega = 12\text{ V} \qquad (V_Z = I_Z\uparrow \times Z_Z\downarrow)$$
$$V_{RS} = V_{in} - V_Z = 30\text{ V} - 12\text{ V} = 18\text{ V}$$

As mentioned in the beginning of this section, *a voltage regulator circuit must maintain the output voltage of a voltage source constant despite variations in the input voltage and the load resistance.* Up to now, we have only seen voltage regulation from one side of the coin. We have seen how the voltage regulator circuit maintains the output voltage constant despite variations in the input voltage. However, as the voltage regulator circuit definition states, it must also maintain the output voltage constant despite changes in load resistance.

In Figure 4-7(b), a load resistance has been connected to the output of the voltage regulator circuit. This load resistance represents the overall resistance of a system, circuit, or device when it is connected to the output of the regulator circuit. The load, which could be a television for example, is represented in Figure 4-7(b) as a variable resistor. This is because most loads do change their resistance when they are in operation. For example, a television draws more current, and therefore has a lower resistance, when its volume or brightness controls are turned up. Since the voltage drop across a load is determined by the resistance of the load, a continual change in load resistance would normally cause a continual change in voltage drop and therefore supply voltage. If the load was connected directly across the input voltage without the regulator circuit, the voltage supplied would change continuously with every change in load resistance. Our zener regulator circuit now has two variables to regulate—a continuous change in input voltage (between 20 V to 30 V) and a continuous change in output load resistance (between 6k Ω to 12 kΩ).

To operate properly, the load, which in this example we will think of as a pocket television, must be supplied with a constant 12 V power supply. Before beginning our calculations, let us study the basics of the circuit in Figure 4-7(b). Since the zener diode and load resistance are in parallel with one another, the current through the zener diode (I_Z) will combine with the current through the load (I_L) to form the total series current (I_S) which will all pass through the series resistor (R_S). Therefore

$$I_Z + I_L = I_S$$

Let us now calculate all the possible combinations of maximum and minimum values of input voltage and load resistance:

When $\boldsymbol{V_{in}}$ **= 20 V:** $V_{RS} = V_{in} - V_Z = 20\text{ V} - 12\text{ V} = 8\text{ V}$, and therefore

$$I_{RS} = \frac{V_{RS}}{R_S} = \frac{8\text{ V}}{1\text{ k}\Omega} = 8\text{ mA}$$

$$R_L = 6\text{ k}\Omega$$

$$I_L = \frac{V_L}{R_L} = \frac{12\text{ V}}{6\text{ k}\Omega} = 2\text{ mA}$$

$$I_Z = I_S - I_L = 8\text{ mA} - 2\text{ mA} = 6\text{ mA}$$

$$R_L = 12\text{ k}\Omega$$

$$I_L = \frac{V_L}{R_L} = \frac{12\text{ V}}{12\text{ k}\Omega} = 1\text{ mA}$$

$$I_Z = I_S - I_L = 8\text{ mA} - 1\text{ mA} = 7\text{ mA}$$

When $\boldsymbol{V_{in}}$ **= 30 V:** $V_{RS} = V_{in} - V_Z = 30\text{ V} - 12\text{ V} = 18\text{ V}$, and therefore

$$I_{RS} = \frac{V_{RS}}{R_S} = \frac{18\text{ V}}{1\text{ k}\Omega} = 18\text{ mA}$$

$$R_L = 6\text{ k}\Omega$$

$$I_L = \frac{V_L}{R_L} = \frac{12\text{ V}}{6\text{ k}\Omega} = 2\text{ mA}$$

$$I_Z = I_S - I_L = 18\text{ mA} - 2\text{ mA} = 16\text{ mA}$$

$$R_L = 12\text{ k}\Omega$$

$$I_L = \frac{V_L}{R_L} = \frac{12\text{ V}}{12\text{ k}\Omega} = 1\text{ mA}$$

$$I_Z = I_S - I_L = 18\text{ mA} - 1\text{ mA} = 17\text{ mA}$$

As you can see from these calculations, the zener voltage, and therefore the output load voltage because the zener and load are in parallel, is kept at a constant 12 V despite changes in the input voltage and the output load resistance. Voltage regulator circuits will be discussed in more detail in the following dc power supply circuits chapter and in later circuits chapters.

Digital Circuit Application—Voltage Reference for a Comparator

Figure 4-8 shows a low-voltage battery sense circuit. The two series connected six-volt batteries make up a 12 V dc supply voltage for several circuits that are not shown in this illustration. The IN752 zener diode and the LM311N comparator make up the low-voltage battery sense circuit that also operates from the battery voltage. Whenever the battery voltage falls below 10 V, the comparator sends a low-voltage output signal (logic 1).

Reference Voltage The voltage developed across a zener that is applied to the positive input of a comparator.

The comparator, which is discussed in a later chapter in more detail, has two inputs labeled "negative(–)" and "positive(+)". The 5.6 V zener diode develops a **reference voltage** that is applied to the positive input of the comparator. The battery voltage is applied across the voltage divider made up of R_1 and R_2, with the voltage developed across R_2 being applied to the negative input of the comparator. If the batteries are fully charged (12 V), the voltage across R_2, and therefore at the comparator's negative input, will be greater than the voltage reference of 5.6 V at the positive input.

$$V_{R2} = \frac{R_2}{R_1 + R_2} \times V_S = \frac{56\ \text{k}\Omega}{44\ \text{k}\Omega + 56\ \text{k}\Omega} \times 12\ \text{V} = 6.72\ \text{V}$$

The inputs to the comparator are therefore "not true" because the positive input is less than the negative input. In this instance, the output of the comparator will be LOW (logic 0 – 0 V).

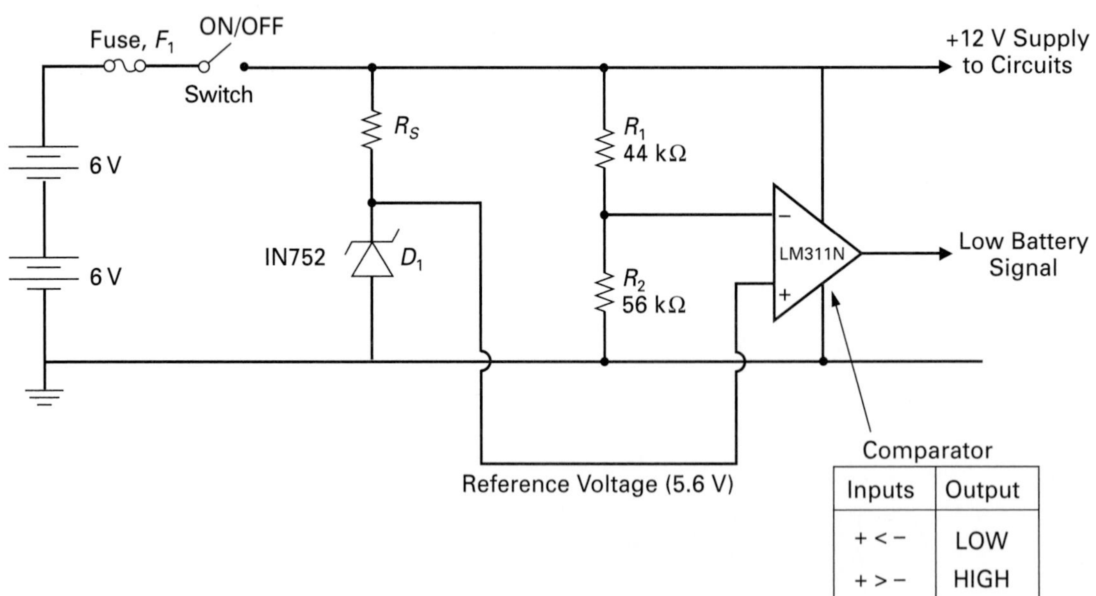

Inputs	Output
+ < –	LOW
+ > –	HIGH

FIGURE 4-8 A Low-Voltage Battery Sense Circuit.

If the battery voltage falls below 10 V (for example, 9.9 V), the voltage appearing at the negative input to the comparator will fall below 5.6 V.

$$V_{R2} = \frac{R_2}{R_1 + R_2} \times V_S = \frac{56\ \text{k}\Omega}{44\ \text{k}\Omega + 56\ \text{k}\Omega} \times 9.9\ \text{V} = 5.54\ \text{V}$$

When this happens, the inputs to the comparator will be "true" since the positive input of the comparator will be more positive than the negative input. In this instance, the output of the comparator goes HIGH (logic 1, approximately 10 V) signaling a low-battery condition.

In this circuit you can see how a zener diode can be used to provide a reference voltage to a two-state digital comparator, which is used to signal either a "battery good" or a "battery low" condition.

4-1-5 *Testing Zener Diodes*

Because a zener diode is designed to conduct in both directions, we cannot test it with the ohmmeter as we did the basic P-N junction diode. The best way to test a zener diode is to connect the volt meter across the zener while it is in circuit and power is applied, as seen in Figure 4-9. If the voltage across the zener is at its specified voltage, then the zener is functioning properly. If the voltage across the zener is not at the nominal value, then the following checks should be made:

1. Check the source input voltage. If this voltage (V_{in}) does not exceed the zener voltage (V_Z), the zener diode will not be at fault because the source voltage is not large enough to send the zener into its reverse breakdown region.
2. Check the series resistor (R_S) to determine that it has not opened or shorted. An open series resistor will have all of the input voltage developed across it and there will be no voltage across the zener. A shorted series resistor will not provide any current limiting capability and the zener could possibly burn out.

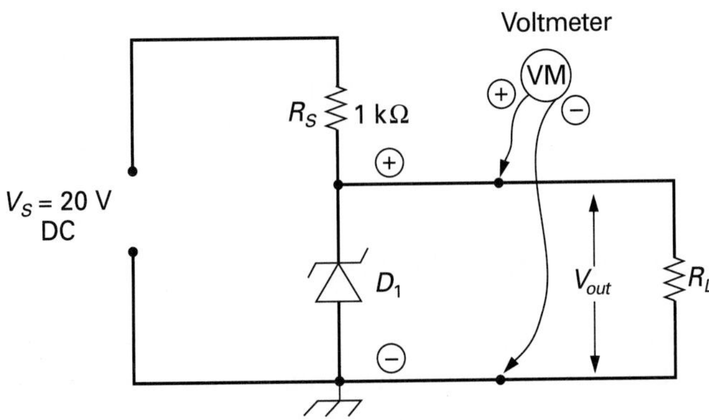

FIGURE 4-9 Zener Diode Testing.

3. Check that there is not a short across the load because this would show up as 0 V across the zener and make the zener look faulty. To isolate this problem, disconnect the load and see if the zener functions normally.

   If these three tests check out okay, the zener diode is probably at fault and should be replaced.

SELF-TEST REVIEW QUESTIONS FOR SECTION 4-1

1. The zener diode is designed specifically to operate at voltages exceeding breakdown. True or false?
2. In most applications, a zener diode is ____________ biased.
3. What is the difference between a zener diode's schematic symbol and a basic P-N junction's symbol?
4. Will the voltage across the series resistor in a zener voltage regular circuit remain constant as the input voltage changes?
5. A voltage regulator circuit should deliver a constant output voltage despite variations in the ____________ and the ____________.

4-2 THE LIGHT-EMITTING DIODE

Light-Emitting Diode (LED)
A semiconductor device that produces light when an electrical current or voltage is applied to its terminals.

Optoelectronic
Combines optics and electronics.

The **light-emitting diode (LED)** is a semiconductor device that produces light when an electrical current or voltage is applied to its terminals. In other words, it converts electrical energy into light energy. It is classified as an **optoelectronic** device since it combines optics and electronics. The LED is probably the most widely used light source in electronic equipment, having replaced the incandescent lamp due to its longer life expectancy and lower operating power.

The LED is basically a P-N junction diode and, like all semiconductor diodes, it can be either forward biased or reverse biased. When forward biased, it will emit energy in response to a forward current. This emission of energy may be in the form of heat energy, light energy, or both heat and light energy depending on the type of semiconductor material used. It is the type of material used to construct the LED that determines the color, and therefore frequency, of the light emitted. For example, different compounds are available that will cause the LED to emit red, yellow, green, blue, white, orange, or infrared light when it is forward biased.

4-2-1 LED Symbol and Packages

Figure 4-10(a) shows the two schematic symbols used most frequently to represent an LED. The two arrows leaving the diode symbol represent light.

A typical LED package is shown in Figure 4-10(b). The package contains the two terminals for connection to the anode and cathode and a semi-clear case which contains the light-emitting diode and a lens. Looking at this illustration,

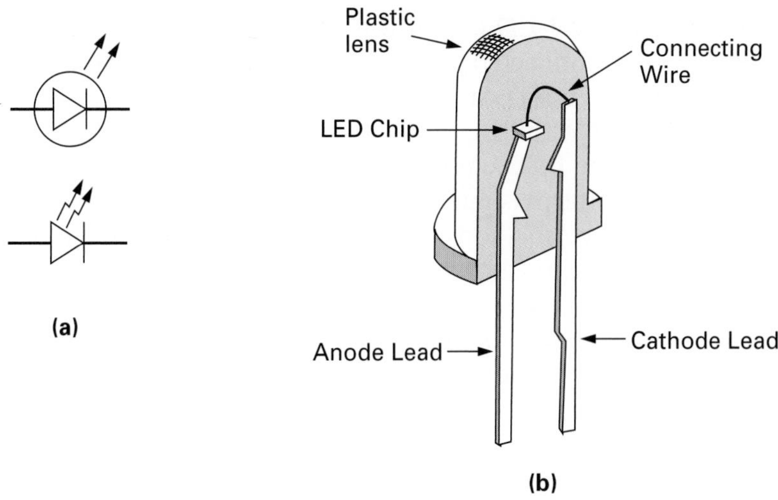

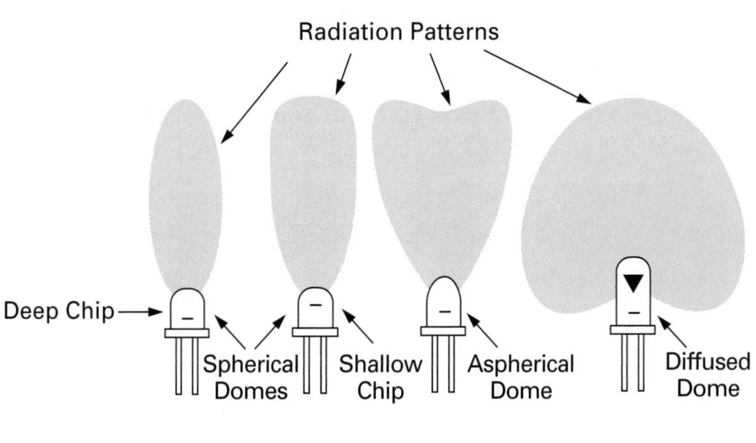

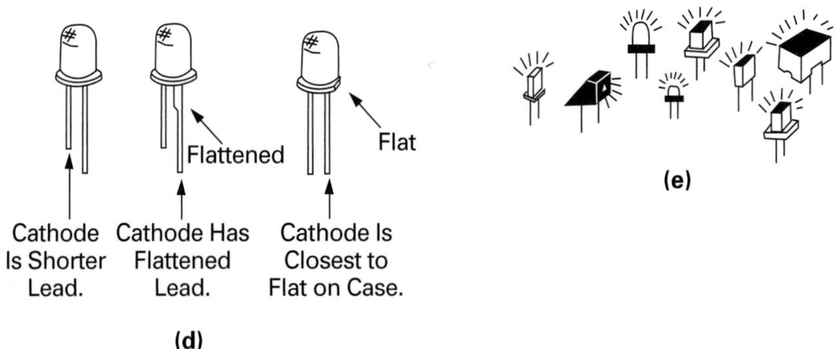

FIGURE 4-10 The Light-Emitting Diode (LED). (a) Schematic Symbol. (b) Construction. (c) Radiation Patterns. (d) Lead Identification. (e) Package Types.

you can see that the LED chip is directly connected to the cathode lead, while the anode lead is connected to the LED chip by a thin wire. The dome-shaped top of the plastic (epoxy) case serves as the lens and acts as a magnifier to conduct light away from the LED chip. By adjusting the lens material, lens shape, and the distance the LED chip is from the lens, manufacturers can obtain a variety of radiation patterns, some of which are shown in Figure 4-10(c).

There are three methods used to identify the anode and cathode leads of the LED, and these are shown in Figure 4-10(d). In all three cases, the cathode lead is distinguished from the anode lead by either having its lead shorter, its lead flattened, or nearest to the flat side of the case.

To give you some idea of the variety of LEDs available, Figure 4-10(e) shows many of the more frequently used package styles.

4-2-2 LED Characteristics

Light-emitting diodes have *V-I* characteristic curves that are almost identical to the basic P-N junction diode as shown in Figure 4-11.

Studying the forward characteristics, you can see that LEDs have a high forward voltage rating (V_F is normally between +1 V to +3 V) and a low maxi-

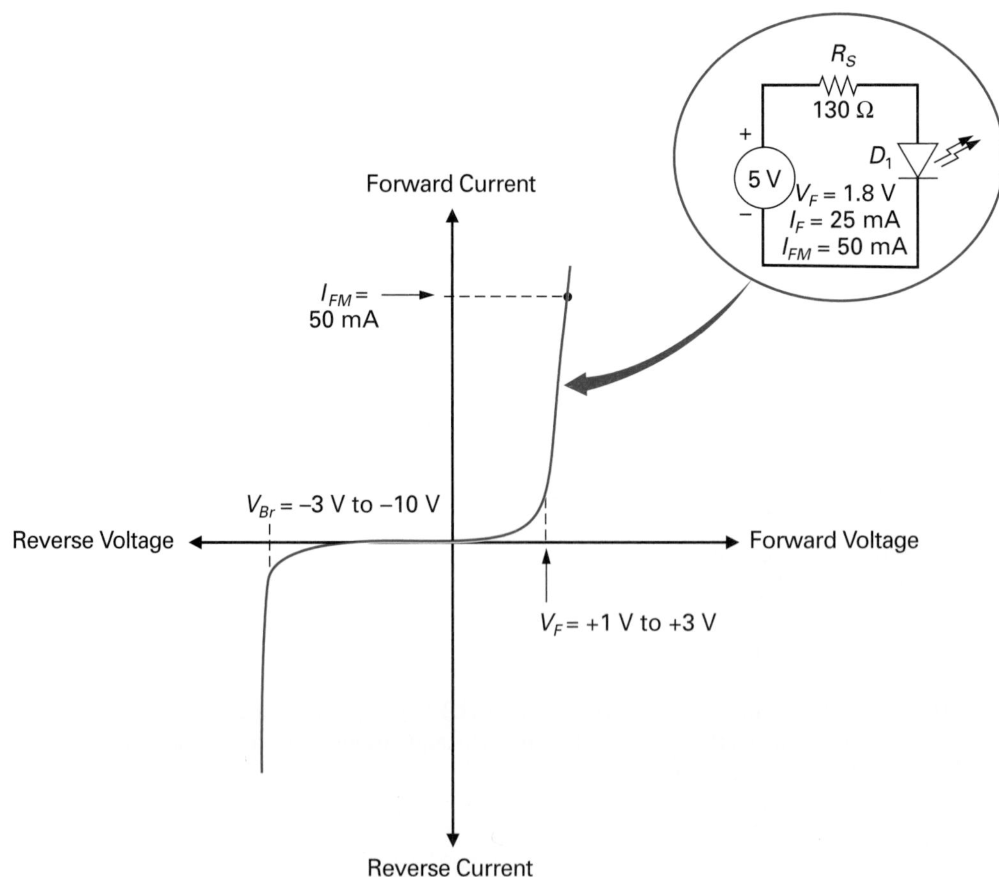

FIGURE 4-11 The LED Voltage-Current Characteristic Curve.

mum forward current rating (I_F is normally between 20 mA to 50 mA). In most circuit applications, the LED will have a forward voltage drop of 2 V, and a forward current of 20 mA. The LED will usually always need to be protected from excessive forward current damage by a series current-limiting resistor, which will limit the forward current so that it does not exceed the LED's *maximum forward current* (I_{FM} or I_{FMAX}) rating, as seen in the inset in Figure 4-11. The circuit in the inset shows how to forward bias an LED. In this example circuit, the source voltage of 5 V is being applied to a 130 Ω series resistor (R_S) and an LED with an I_{FM} rating of 50 mA and a V_F rating of 1.8 V. To calculate the value of current in this circuit, we would first deduct the 1.8 V developed across the LED (V_{LED}) from the source voltage (V_S) to determine the voltage drop across the resistor (V_{RS}). Then using Ohm's law, divide V_{RS} by R_S to determine current.

$$I_S = \frac{V_S - V_{LED}}{R_S}$$

$$= \frac{5\ \text{V} - 1.8\ \text{V}}{130\ \Omega} = \frac{3.2\ \text{V}}{130\ \Omega} = 24.6\ \text{mA}$$

Studying the reverse characteristics, you can see that LEDs have lower reverse-breakdown voltage values than junction diodes (V_{Br} is typically –3 V to –10 V), which means that even a low reverse voltage will cause the LED to break down and become damaged.

4-2-3 LED Operation

Figure 4-12 shows the basic operation of a light-emitting diode. When forward biased, the negative terminal of the dc source injects electrons into the *n*-type region, or cathode, of the LED. These electrons travel towards the P-N junction.

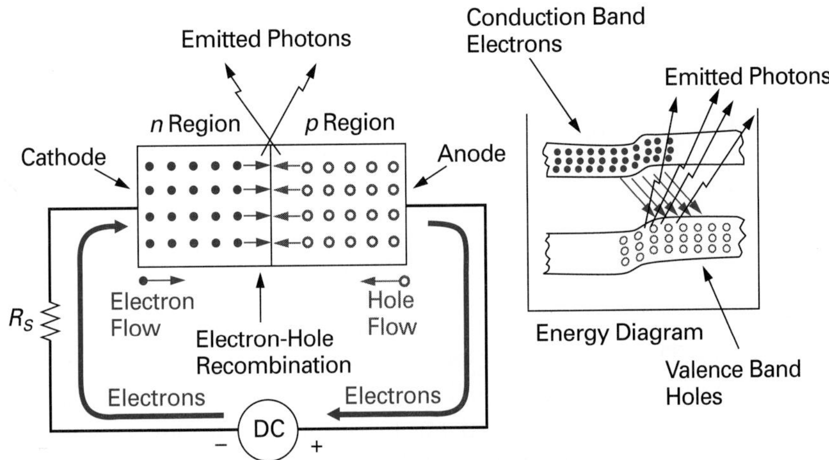

FIGURE 4-12 Basic Operation of an LED.

At the same time, electrons are attracted out of the p-type region, or anode, and travel towards the positive terminal of the source voltage making it appear as though holes are moving towards the P-N junction. The electrons from the n-type region and the holes from the p-type region combine at the P-N junction. Since the electrons are at a higher energy level (conduction band), they release energy as they drop into a valence hole. The energy is a small packet of light called a **photon.** Most forward biased LEDs release billions of photons per second as billions of electrons combine with holes at the P-N junction. The semi-transparent semiconductor material used to construct LEDs allows the light energy to escape where it is captured and focused by the dome-shaped lens of the LED package.

Photon
A small packet of light.

The basic P-N junction diode also generates photons when holes and electrons recombine at the P-N junction. However, because silicon is opaque (not transparent), the light simply cannot escape.

4-2-4 *LED Data Sheet*

Figure 4-13 shows a typical LED data sheet. Studying this specification sheet along with the associated notes, you will be able to find the LED's key characteristics. Remember that, as a technician, you should be familiar with some of the basic operating limits of a device so that you can isolate component malfunctions within a circuit by determining whether the device is operating to specifications. For example, the MLMP = 3000 series of light emitting diodes has the following key specifications:

Maximum Forward Current (I_{FM} or Average Forward Current) = 50 mA
Forward Voltage Drop (V_F) = 1.4 V to 2.0 V
Reverse Breakdown Voltage (V_R) = 5 V

■ **EXAMPLE:**

What value of resistor would be needed to limit the current to its maximum forward rating if an MLMP = 3000 LED was connected to a 10 V source?

■ ***Solution:***

Since the MLMP = 3000 LED will drop a maximum of 2 V when forward biased, the series current limiting resistor will drop the remaining 8 V ($V_S - V_F = V_{RS}$, 10 V − 2 V = 8 V). With V and I known, we can use Ohm's law to calculate the value of R_S.

$$R_S = \frac{V_S - V_F}{I_F} = \frac{10\text{ V} - 2\text{ V}}{50\text{ mA}} = 160\ \Omega$$

The brightness of an LED is determined by the value of current that passes through the LED. Therefore, more current will mean more light, and less current will mean less light. Doubling the value of R_S will halve the current and halve the LED's brightness, giving it poor visibility. Generally, a value of R_S is chosen that will keep the forward current just below its maximum current rating. In this example, a good choice would be 168 Ω, because this is the next standard-

DEVICE: HLMP- 3000 Series—Red LED Lamps

T-1¾ (5 mm) Red Solid State Lamps

Technical Data

HLMP-3000
HLMP-3001
HLMP-3002
HLMP-3003
HLMP-3050

Features

- **Low Cost, Broad Applications**
- **Long Life, Solid State Reliability**
- **Low Power Requirements: 20 mA @ 1.6 V**
- **High Light Output:** 2.0 mcd Typical for HLMP-3000; 4.0 mcd Typical for HLMP-3001
- **Wide and Narrow Viewing Angle Types**
- **Red Diffused and Non-diffused Versions**

Description

The HLMP-3000 series lamps are Gallium Arsenide Phosphide light emitting diodes intended for High Volume/Low Cost applications such as indicators for appliances, smoke detectors, automobile instrument panels, and many other commercial uses.

The HLMP-3000/-3001/-3002/-3003 have red diffused lenses whereas the HLMP-3050 has a red non-diffused lens. These lamps can be panel mounted using mounting clip

HLMP-0103. The HLMP-3000/-3001 lamps have 0.025" leads and the HLMP-3002/-3003/-3050 have 0.018" leads.

Package Dimensions

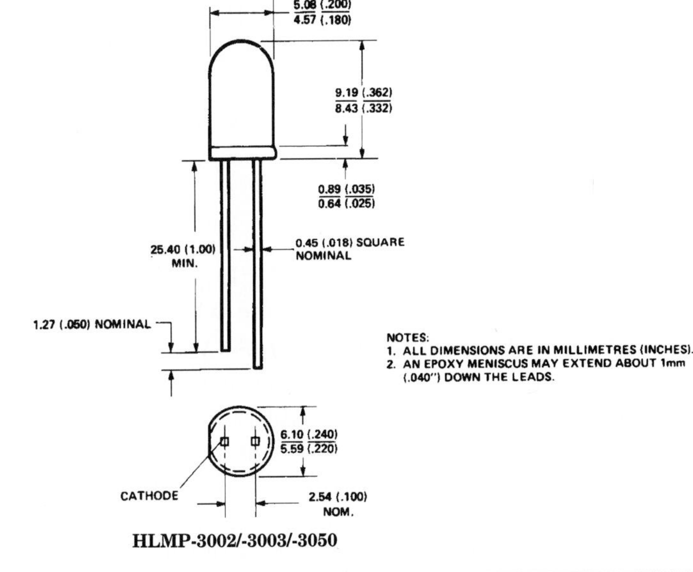

HLMP-3002/-3003/-3050

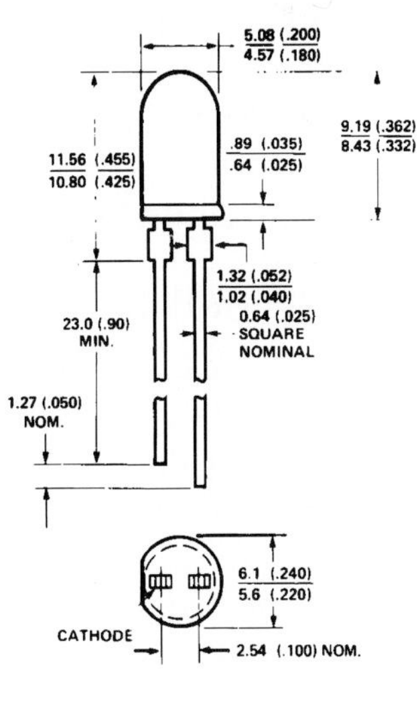

HLMP-3000/-3001

FIGURE 4-13 Device Data Sheet for a Series of Light Emitting Diodes. (Courtesy of National Semiconductor Corporation.)

value resistor (see standard resistor values in Appendix). Its value would reduce forward current to just below its maximum current rating and ensure a good level of brightness.

$$I_S = \frac{V_S - V_{LED}}{R_S} = \frac{10\text{ V} - 2\text{ V}}{168\ \Omega} = \frac{8\text{ V}}{168\ \Omega} = 47.6\text{ mA}$$

4-2-5 *LED Applications*

Like the basic P-N junction diode, the light emitting diode is used in almost every electronic and electrical system. As you proceed through this text you will see the LED used in a wide variety of circuit applications. For now, let us see some examples of LED applications.

LED Indicator Circuits

The individual LED is most commonly used as a *power and signal level indicator,* both applications of which can be seen in Figure 4-14.

Figure 4-14(a) shows how an LED can be used as a power indicator on the front panel of a system, such as a CD player. Because the LED is powered by the unit's internal power supply, as shown in the inset, the LED will only turn ON and indicate power is present when the unit is supplying power to its internal circuits.

Figure 4-14(b) shows how three LEDs can be used as signal level indicators to display the code generated by an encoder circuit. For example, when the rotary switch is in position 2, the junction diode D_3 will turn ON and pull line B LOW. This LOW (0 V) signal will not be able to forward bias the LED D_7, and it will not light up. Lines A and C, however, will be pulled HIGH by the resistors R_1 and R_3, and these HIGH outputs will forward bias LEDs D_6 and D_8 and they will light up. Therefore, the code generated by the encoder circuit when it is in position 2 is "HIGH-LOW-HIGH," and this will be displayed on the LEDs as "ON-OFF-ON."

Bi-Color LEDs

Two-color LEDs were developed so that a single LED package could be used, on an aircraft control panel for example, to indicate whether something was functioning properly (green) or was a problem (red). As you can see in the schematic symbol in Figure 4-15(a), these multicolor LEDs actually contain two *reverse-parallel connected* LEDs. This means that two different color LEDs are connected in parallel but in the reverse direction inside one package, as seen in the device's data in the inset in Figure 4-15(a). Figure 4-15(b) shows how this device will operate when a 1 Hz (1 cycle per second) 5 V peak square wave is applied to the LED and its current limiting resistor. When the input voltage is positive, the right or red LED will be forward biased and will turn ON. When the input voltage reverses and is negative, the left or green LED is forward biased and will turn ON. Because a 1 Hz input will have a half-second positive

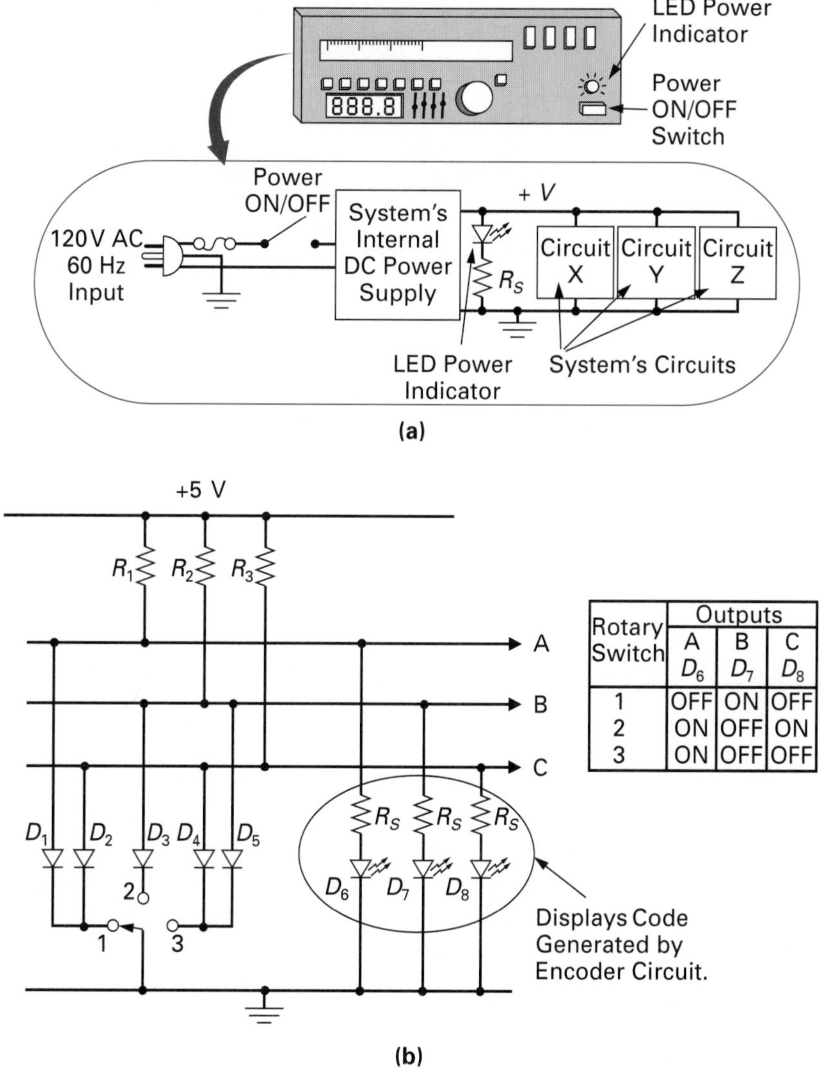

Rotary Switch	Outputs		
	A D_6	B D_7	C D_8
1	OFF	ON	OFF
2	ON	OFF	ON
3	ON	OFF	OFF

FIGURE 4-14 Single LED Applications. (a) Power Indicator. (b) Signal Indicator.

alternation and a half-second negative alternation, the LED package will flash red for half a second, then green for half a second, and then repeat the cycle continuously. You may wish to try this experiment in lab. If you do, try increasing the frequency of the applied voltage until the switching is so rapid that the two colors blend and appear as yellow.

Seven-Segment Display
Seven LEDs mounted in seven segments. By turning ON a combination of segments, the numbers 0 through 9 can be displayed.

Multisegment LED Displays

One of the biggest applications of LEDs is in the multisegment display. The **seven-segment display,** which is shown in Figure 4-16(a), is probably the most widely used multisegment LED display. Its seven segments are labeled in a clockwise direction—*a, b, c, d, e, f, and g*—as shown in the inset in Fig-

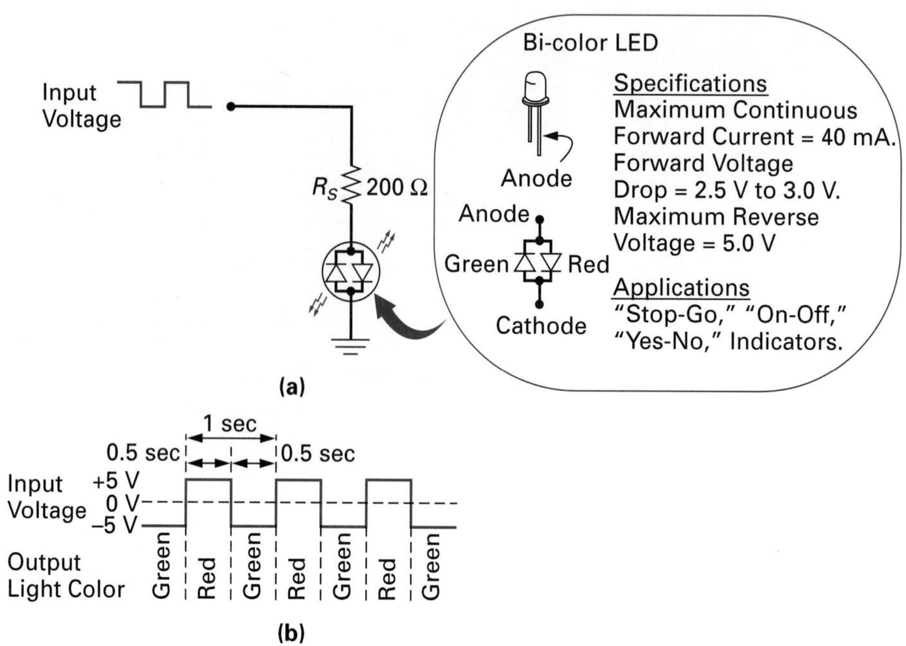

FIGURE 4-15 Bi-Color LED Applications.

ure 4-16(a). As can be seen in Figure 4-16(b), the display has seven LEDs mounted in seven segments. Some package types include an eighth LED in a hole which is used to indicate a decimal point (dp). As an example, Figure 4-16(c) shows the pin number assignment for a TIL312 seven-segment and two-decimal point display. The pin numbers are shown inside parentheses. By turning ON a combination of segments, the numbers 0 through 9 can be displayed, as shown in Figure 4-16(d). For example, by turning ON all of the segments, the display will show the number 8. If only segments b and c are turned ON, the display will show the number 1.

Digital Displays
Indicate a quantity using decimal digits.

Analog Displays
Indicate a quantity with the amount of pointer movement.

Generally, in all display applications such as automotive, instrumentation, aircraft, audio, and appliance, more than one LED display is used, as shown in Figure 4-16(e). These **digital displays** indicate a quantity using decimal digits. The older **analog displays** used a moving pointer and scale to indicate a quantity, as shown in Figure 4-16(f), with the amount of pointer movement being an analog, or similar, to the magnitude of the quantity.

Common-Anode Connection
Common connection between the anodes of LEDs.

Common-Cathode Connection
Common connection between the cathodes of LEDs.

These segmented displays are available in either a **common-anode** or a **common-cathode** configuration, as shown in Figure 4-16(g) and (h).

The configuration in Figure 4-16(g) is called a common-anode connection because there is a common connection between all the anodes of the LEDs. As you can see in Figure 4-16(g), a common-anode connection will have its "common" connected to a positive supply voltage. To light up an LED, therefore, the cathodes of the LEDs in this configuration will have to be switched LOW. The circuit connections in Figure 4-16(b), and the TIL312 in Figure 4-16(c), are both "common-anode" seven-segment display types.

The configuration in Figure 4-16(h) is called a common-cathode connection because there is a common connection between all the cathodes of the LEDs.

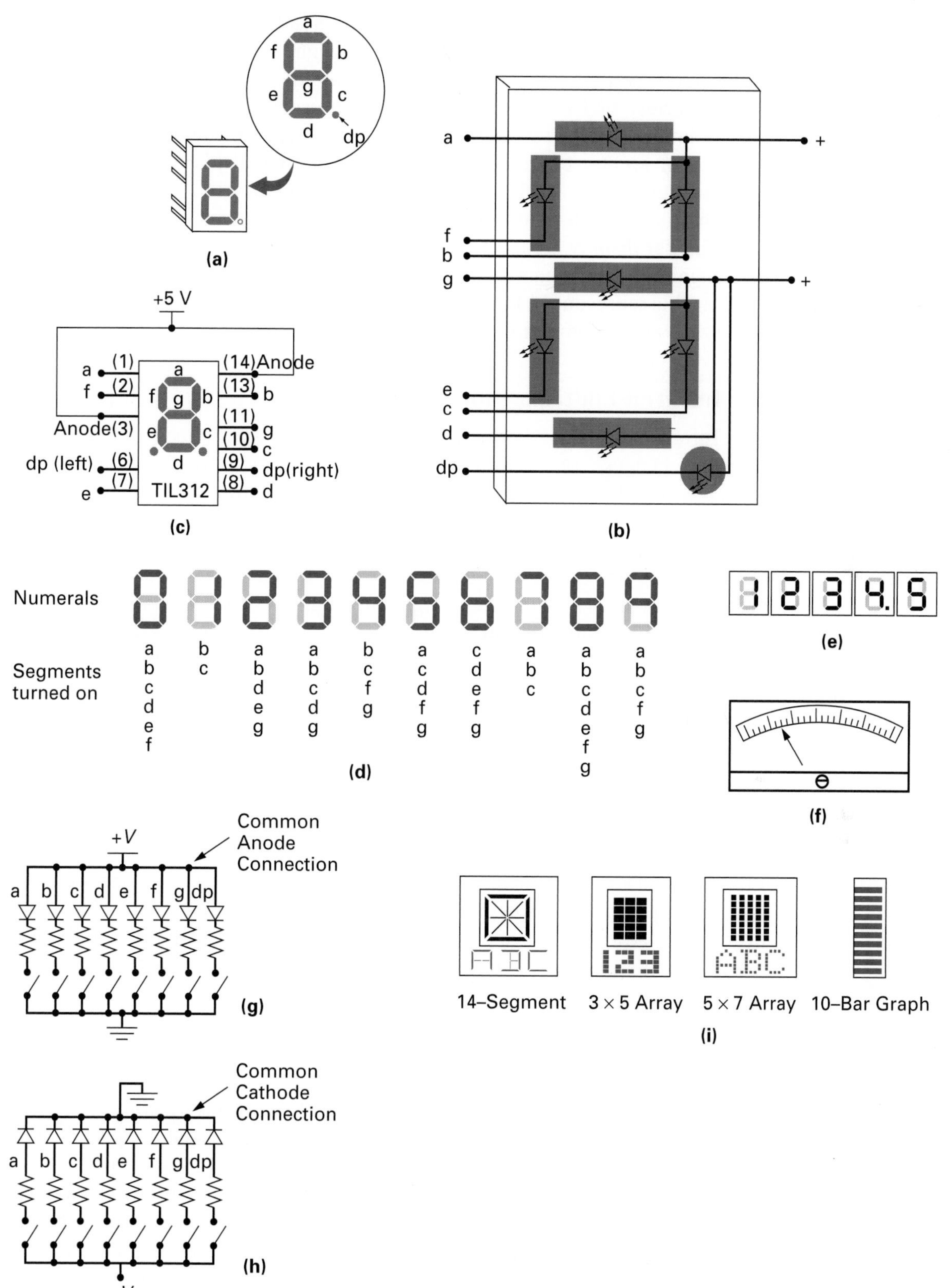

FIGURE 4-16 **Multisegmented LED Displays.**

As you can see in Figure 4-16(h), a common-cathode connection will have its "common" connected to ground or 0 V. To light up an LED, the anodes of the LEDs in this configuration will have to be switched HIGH.

A variety of LED display types are shown in Figure 4-16(i). The need to display letters resulted in a display with more segments such as the 14-segment display. For greater character definition, **dot matrix displays** are used. The most popular type consists of 35 small LEDs arranged in a grid of five vertical columns and seven horizontal rows to form a 5 × 7 matrix.

Dot Matrix Displays
Greater character definition with dot matrix displays.

Bar Display
Displays the magnitude of a quantity by the number of activated segments.

The **bar display** is rapidly replacing analog meter movements for displaying the magnitude of a quantity. For example, the magnitude of a quantity such as fuel can be represented by the number of activated segments with no bars lit being zero and all bars lit being maximum.

Infrared-Emitting Diodes (IREDs)

Infrared-emitting diodes are used in a few specialized applications. For example, because the infrared light cannot be seen by the human eye, the IRED is ideally suited for security applications—an intruder would inadvertently break the invisible beam setting off the alarm. The infrared-emitting diode is also used as a light transmitter in fiber optic communications. Fiber optic IREDs are similar to the conventional LEDs, however, they are more precisely designed to emit more light in a tighter beam. Fiber optics is a technology in which information, such as voice telephone signals, are converted to light, by an IRED for instance, and then transmitted along the inside of a thin, flexible glass or plastic fiber. Both of these applications use specially designed IREDs that will be discussed later when you are studying security and communication systems.

4-2-6 *Testing LEDs*

Light-emitting diodes have a very long life expectancy and are much more rugged than their predecessor, the small incandescent light bulb. They do, however, break down and have to be replaced with either exactly the same device or a similar device with the same characteristics. These key characteristics are:

Maximum Forward Current

Forward Voltage Drop

Reverse Breakdown Voltage

To understand how to test an LED, let us first see how a simple LED circuit should operate and then examine what would happen if the circuit developed a malfunction. An example circuit is shown in Figure 4-17(a).

In this circuit, a comparator compares a zero volts reference to a changing input voltage. The LED in this circuit is functioning as a level indicator, and will turn ON whenever the input voltage goes positive. The comparator, which is discussed in a later chapter in more detail, has two inputs labeled "negative (–)" and "positive (+)." Whenever the input voltage to the positive input of the comparator is below 0 V, the comparator's inputs are "not true" (+ is less than –),

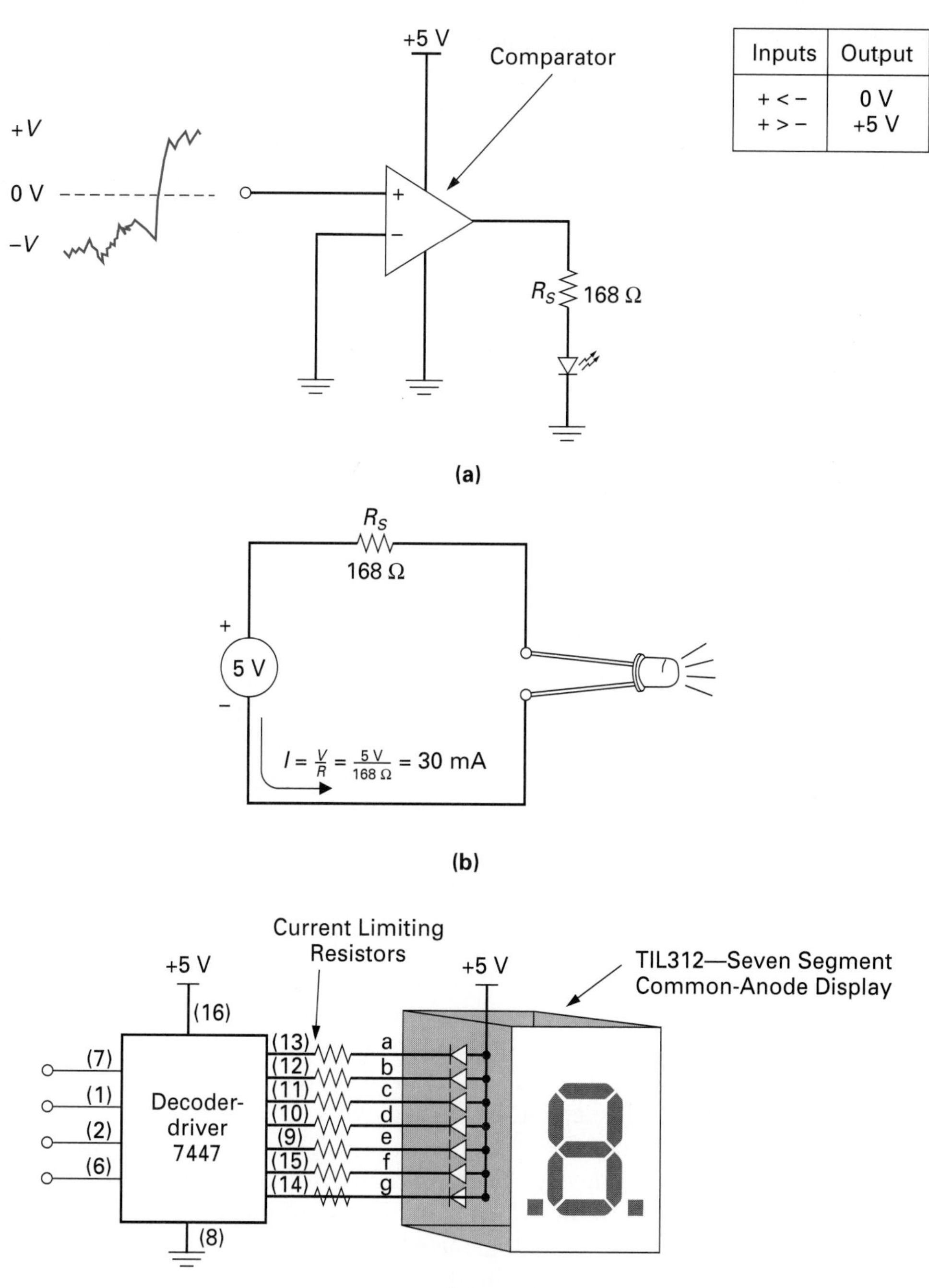

Inputs	Output
+ < −	0 V
+ > −	+5 V

FIGURE 4-17 Testing LEDs. (a) Troubleshooting a Single LED Circuit. (b) LED Test Circuit. (c) Multisegment LED Display Circuit Troubleshooting.

and therefore the output is LOW. In this instance, a LOW or 0 V output will not activate the LED. Whenever the input voltage to the positive input of the comparator is above 0 V, the comparator's inputs are "true" (+ is greater than −), and therefore the output is HIGH. In this instance, a HIGH or +5 V output will turn ON the LED.

Circuits containing LEDs give us a visual indication that there is a problem because the LEDs do not turn ON when the circuit condition is right. Like all circuit malfunctions however, the problem is not always obvious. For example, if the LED in Figure 4-17(a) did not turn ON at any time, the following checks should be made:

1. Check that the input voltage is going positive.
2. Check that the comparator's output is +5 V when the input is positive. If the comparator's output does not switch from 0 V to +5 V when the input goes positive, you should check the +5 V and 0 V supply voltages to the comparator and the ground at the negative input. After these checks, you should disconnect the comparator's output to see if it functions normally because a shorted output will be pulled LOW constantly.
3. Check the series current-limiting resistor. An open resistor will prevent power from reaching the LED, and like all series opens, will develop all of the source voltage across its terminals. While a shorted resistor will not limit the forward current, it will probably burn out the LED. Replacing the burnt-out LED will not cure the problem because the shorted resistor will simply burn out the new LED.

As you can see, there are many elements in any circuit that can contribute to a false diagnosis of the problem. However, by systematically checking all of the logical possibilities you can isolate the circuit fault.

Figure 4-17(b) shows how to construct a simple LED test circuit using a dc power supply and a series current-limiting resistor. If a circuit problem is isolated to the LED, the new LED should be tested with a similar circuit to ensure it is operating correctly before it is inserted into the circuit.

Multisegment LED displays, such as the one seen in Figure 4-17(c), are driven by digital integrated circuits called decoder-drivers which will be discussed in the digital chapter. Like all troubleshooting, it is impossible to isolate a circuit problem unless you fully understand how the circuit should normally operate. At this time, without the full understanding of the decoder-driver circuit, it will be difficult to isolate any faults in this circuit. We can, however, discuss how to check the multisegment LED display.

To begin with, if none of the segments are lighting up, check the supply voltage and ground to all devices in the circuit. To isolate the most common multisegment LED display fault, which is the failure of one or more segments to light up, you must apply the same logic used for the single LED circuit. That is, first determine whether the LED in the multisegment display has failed or whether the *decoder-driver circuit* is not delivering the correct output. For example, a LOW at pin 13 of the decoder-driver should turn ON the "a" segment in the seven-segment display. By following the same steps as the ones we applied to the single LED circuit (the decoder-driver's output voltage, the series current-limiting resistor, and so on) you will be able to diagnose the circuit problem. It is not possible to repair a single LED in a multisegment display. If the fault is isolated to one of the segments, the entire display will have to be replaced. Remember to always check replacements to see that they are compatible (common-anode or cathode, I_F rating, and so on).

SELF-TEST REVIEW QUESTIONS FOR SECTION 4-2

1. Light emitting diodes emit radiation as a result of the recombination of ___________ and ___________ at the P-N junction.
2. What determines the color of light produced by the LED?
3. What would be the typical voltage drop across a forward biased LED?
   - **a.** 0.7 V
   - **b.** 2 V
   - **c.** 0.3 V
   - **d.** 5.6 V
4. A ___________ is normally always included to limit I_F to just below its maximum.
5. The amount of light produced by an LED is directly proportional to the amount of ___________.

4-3 TROUBLESHOOTING DIODE CIRCUITS

In this section we will discuss how to troubleshoot a typical circuit containing both the zener diode and light-emitting diode discussed in this chapter and the P-N junction diode discussed in the previous chapter.

The procedure for fixing a failure can be broken down into three basic steps:

Step 1: DIAGNOSE

The first step is to determine whether a problem really exists. To carry out this step, a technician must collect as much information as possible about the system, circuit, and components used and then diagnose the problem.

Step 2: ISOLATE

The second step is to apply a logical and sequential reasoning process to isolate the problem. In this step, a technician will operate, observe, test, and apply troubleshooting techniques in order to isolate the malfunction.

Step 3: REPAIR

The third and final step is to make the actual repair and then test the circuit.

Steps 1 and 2 can be collectively called **troubleshooting,** which by definition is *the process of diagnosing and locating breakdowns in equipment by means of systematic checking and analysis.* To assist you in troubleshooting, you have troubleshooting tools (test equipment), documentation (technical or service manuals), and experience. The experience can only be acquired with practice, and because very few of your lab experiments will work perfectly the first time you construct them, you will gain a wealth of experience as you experiment in lab. Try to remember that, although it seems frustrating at the time, the more problems you have to resolve with your lab experiments, the better.

Troubleshooting
The process of diagnosing and locating breakdowns in equipment by means of systematic checking and analysis.

Like cars, we expect electronic systems to operate indefinitely without ever malfunctioning. If they ever do malfunction, it always seems to be at the worst possible time and generally results in a loss of time, money, and productivity. When a technician is called to troubleshoot a system failure, his or her main objective is to diagnose, isolate, and repair the system as soon as possible. The time factor involved in this process is generally always critical, because a system that is not operating properly is not productive and is not being used for its main objective.

To be an effective *electronic technician* or *troubleshooter,* you must have a thorough knowledge of electronics, test equipment, troubleshooting techniques, and equipment repair. Since "practice makes perfect," let us examine the three-step troubleshooting process in more detail and apply it to a typical circuit containing junction diodes, a zener diode, and light emitting diodes.

4-3-1 Typical Diode Circuit

As an example, Figure 4-18(a) shows the schematic diagram for a seven-segment encoder and display circuit. A 120 V ac input from the wall outlet is switched and fused before being applied to a dc power supply, which generates a +12 V dc output. This +12 V output is applied to a 5.1 V zener regulator circuit, which supplies power to a seven-segment encoder and display circuit.

Figure 4-18(b) shows how the LOW and HIGH codes generated by the encoder circuit will turn ON and OFF the necessary LEDs in the seven-segment display to produce the digits 0 through 9. For example, when the rotary switch is in position 0, a single P-N junction diode makes line *G* LOW. This will turn segment "g" of the seven segment display OFF, while the pull up resistors R_1 through R_6 will make lines *A* through *F* HIGH and therefore turn ON segments "a" through "f." This condition is shown in the first row of the table in Figure 4-18(b).

The remaining rows in Figure 4-18(b) show what HIGH and LOW codes are generated and applied to the seven-segment display by the junction diodes in the encoder circuit for each position of the rotary switch.

Now that we have an understanding of the circuit's operation, let us apply our three step troubleshooting procedure to it.

4-3-2 Step 1: Diagnose

It is extremely important that you first understand how a system, circuit, and all of its components are supposed to work so that you can determine whether or not a problem exists. If you were preparing to troubleshoot the seven-segment encoder and display circuit in Figure 4-18, your first step should be to read through the circuit description and review the operation and function of each device used in the circuit until you feel completely comfortable with the correct operation of the circuit.

For other circuits and systems, technicians usually refer to service or technical manuals, which generally contain circuit descriptions (sometimes called "theory of operation") and troubleshooting guides. As far as each of the devices are

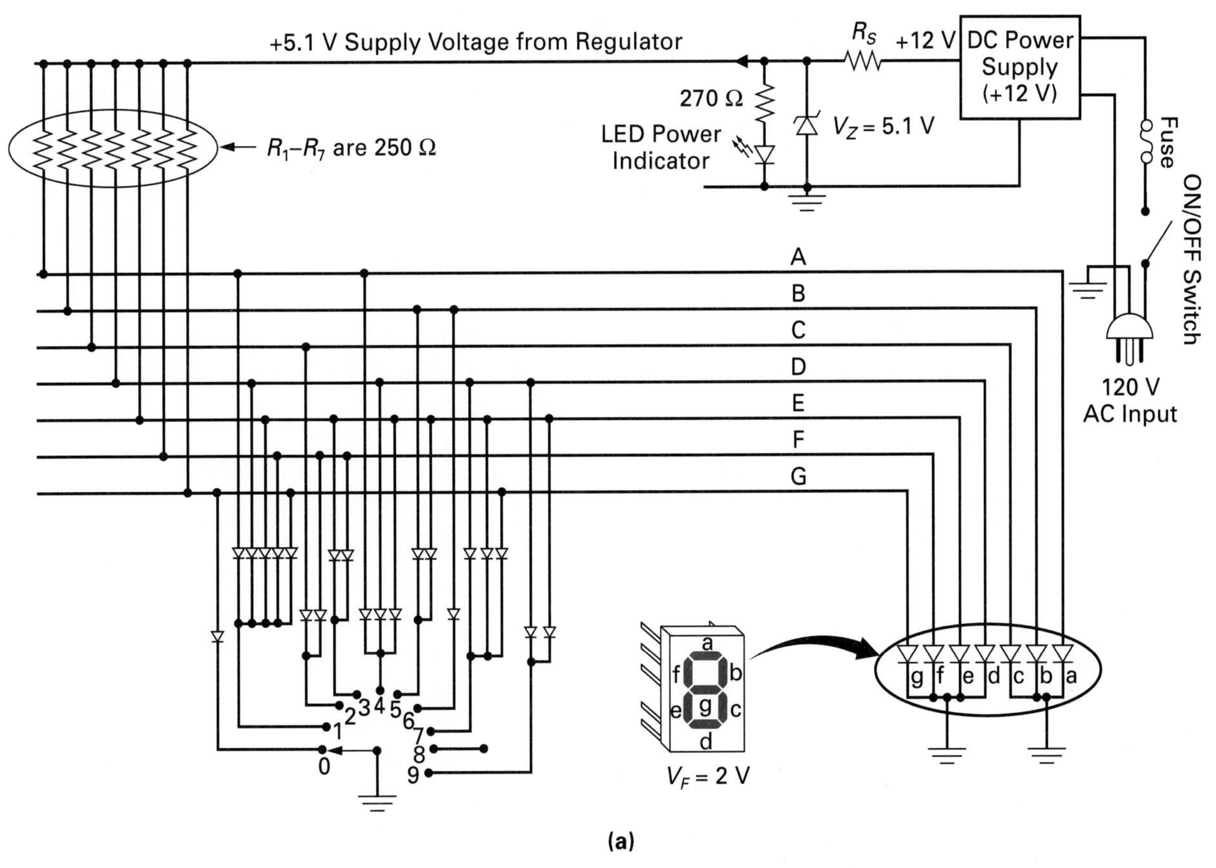

(a)

	a	b	c	d	e	f	g
0	HI	HI	HI	HI	HI	HI	LO
1	LO	HI	HI	LO	LO	LO	LO
2	HI	HI	LO	HI	HI	LO	HI
3	HI	HI	HI	HI	LO	LO	HI
4	LO	HI	HI	LO	LO	HI	HI
5	HI	LO	HI	HI	LO	HI	HI
6	HI	LO	HI	HI	HI	HI	HI
7	HI	HI	HI	LO	LO	HI	LO
8	HI	HI	HI	HI	HI	HI	HI
9	HI	HI	HI	LO	LO	HI	HI

(b)

FIGURE 4-18 A Seven-Segment Encoder and Display Circuit.

concerned, technicians refer to manufacturer's data books, which contain a full description of the device.

Referring to all of this documentation before you begin troubleshooting will speed up and simplify the isolation process. Once you are fully familiar with the operation of the circuit, you will be ready to *diagnose the problem* as either an *operator error* or as a *circuit malfunction.*

Many technicians bypass this data collection step and proceed straight to the isolating phase of the troubleshooting procedure. If you are fully familiar with the circuit's operation and the components used within the circuit, this shortcut would not hurt your performance. However, to use an old expression which should be kept in mind throughout any troubleshooting procedure as a sort of brake to stop you from racing past the problem, remember, "less haste, more speed." Take the time to familiarize yourself with the circuit's physical layout, circuit diagram, and the specification of all the components used before you go to Step 2.

Here are some examples of operator errors that can be mistaken for circuit malfunctions if a technician is not fully familiar with the operation of the seven-segment encoder and display circuit in Figure 4-18.

SYMPTOMS	DIAGNOSIS
1. The rotary switch is in position 8, but nothing is being shown on the seven-segment display.	Check first that the power supply is plugged in and turned ON because either of these operator errors will give the appearance of a circuit malfunction.
2. All of the segments are turned ON.	If the rotary switch is in position 8, the circuit is operating normally. However, if the rotary switch is in any other position, there is a circuit malfunction.
3. The display does not count through the digits 0 to 9 and then repeat the sequence.	This circuit does not cycle automatically through the digits 0 to 9. It displays the digit that is selected by the rotary switch.

Once you have determined that the problem is not an operator error but, in fact, a circuit malfunction, proceed to Step 2 and isolate the circuit problem.

4-3-3 ***Step 2: Isolate***

A technician spends most of the time in this phase of the process isolating a circuit problem. Steps 1 (diagnose) and 3 (repair) may only take a few minutes to complete compared to Step 2, which could take up to a few hours to complete. However, with practice and a good logical and sequential reasoning process, you can quickly isolate even the most obscure of problems.

The directions you take as you troubleshoot a circuit malfunction may be different for each problem. However, certain troubleshooting techniques apply to all problems. Let us now review some of these techniques and apply them to our seven-segment encoder and display circuit in Figure 4-18.

Cause and Effect Troubleshooting Studying the effects from the faulty circuit and reasoning out the cause.

1. Use a **cause and effect troubleshooting process,** which means study the effects you are getting from the faulty circuit and then try to reason out what could be the cause. For example:

EFFECT	CAUSE
1. No display of any type	Power problem to entire circuit. a. No 5.1 V b. No ground connection to rotary switch, seven-segment display c. Junction diodes, zener diode, or seven segment display incorrectly oriented
2. Display shows incorrect digit for every switch position	Problem must be common to all. Check: a. Rotary switch b. Seven-segment display c. Wiring
3. Display has only one segment failing	Problem must be confined to that segment's control or display. Check: a. Control line's pull-up resistor b. Control line connection to encoder c. Control line connection to seven-segment display d. Display segment's connecting pin

2. Check first for *obvious errors:*

a. Power supply fuse

b. Wiring errors if circuit is newly constructed

c. Devices are incorrectly oriented. For example, if the zener diode, the junction diodes, or the seven-segment display were placed in the circuit in the opposite direction to what is shown in Figure 4-18, the device would not operate as it should and neither would the circuit.

3. Use your *senses* to check for broken wires, loose connections, overheating or smoking components, leads or pins not making contact, and so on.

4. Apply the **half-split method** of troubleshooting first to the entire system, then to a circuit within the system, and then to a section within the circuit to help speed up the isolation process. You have probably been using this half-split troubleshooting method for years without really knowing you were doing it. The process involves choosing a point roughly in the middle of the problem area. By making a test at this midpoint you can determine whether the problem exists either before this point or after this point. For example, if a midpoint is tested and the voltage at that point is good, you should proceed onto the second half of the circuit, whereas if the voltage at the midpoint is incorrect you should proceed to test the first half of the circuit. Once you have determined which half of the circuit is malfunctioning, you should again choose a midpoint and repeat the process to narrow down the fault area until it is localized.

Half-Split Method The process of choosing a point roughly in the middle of the problem area, testing this midpoint, then proceeding on to the second half of the circuit, and so on.

5. Always test a circuit using the voltmeter for *power* and *ground*. If a circuit does not function in any way, this would be an obvious first test. However, many circuits can develop strange symptoms due to a lack of power and therefore this is always a good check to make initially.

Substitution Substituting a suspect device with a known working device.

Troubleshooting Guides A graphical means to show the sequence of tests to be performed on a system or circuit that is malfunctioning.

6. **Substitution** can be used to help speed up your troubleshooting process. Once the problem is localized to an area, you may wish to substitute a suspect device with a known working device to see if the problem can be quickly remedied.
7. Make use of any circuit descriptions or **troubleshooting guides** in technical manuals. These troubleshooting charts are a graphical means to show the sequence of tests to be performed on a system or circuit that is malfunctioning. By completing these tests and following a path based on the results obtained, a technician can localize a problem. Following is an example of a basic troubleshooting tree for the seven-segment encoder and display circuit in Figure 4-18.

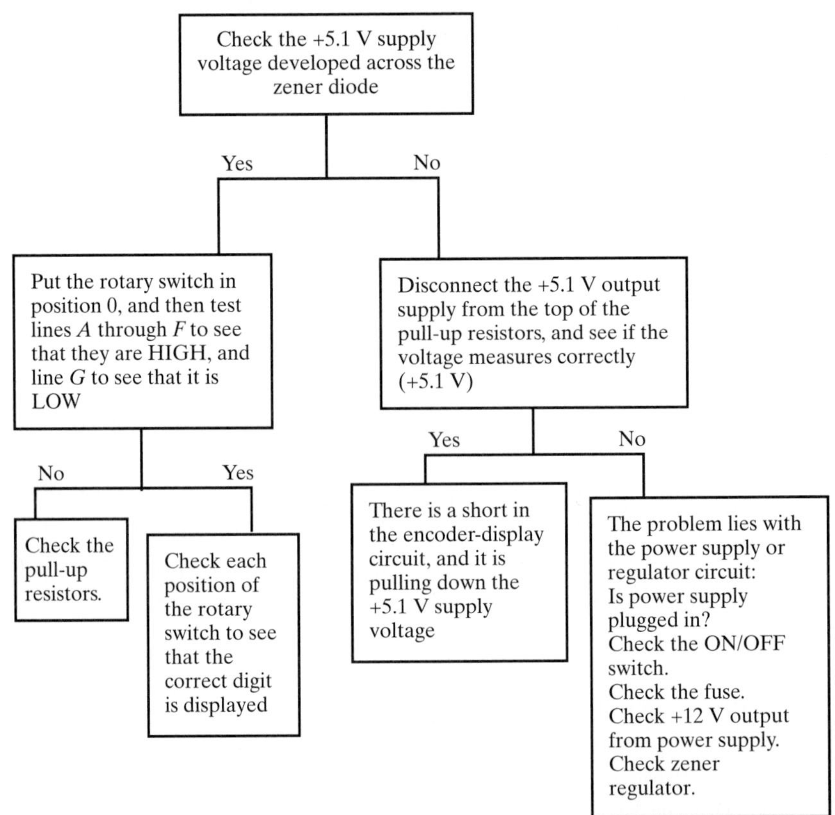

Let us now try a few more examples of problems that could occur with the seven-segment encoder and display circuit shown in Figure 4-18.

■ **EXAMPLE:**

What would happen to the circuit in Figure 4-18 if the zener diode were to short?

■ ***Solution:***

If the zener diode were to short, there would be no voltage developed across the zener and no voltage supplied to the encoder-display circuit.

In most cases, a zener regulator diode short would provide such a low resistance path across the output dc power supply (because the 12 V is now only connected across R_S) that the large increase in current would probably blow the fuse protecting the circuit.

■ **EXAMPLE:**

What would happen to the circuit in Figure 4-18 if one of the encoder P-N junction diodes were to open?

■ ***Solution:***

If one of the encoder diodes opens, it will be unable to ground its respective control line (*A* through *G*) when it is selected by the rotary switch. The output line will not be switched LOW by this open diode and the segment it controls will not be switched OFF.

■ **EXAMPLE:**

What would happen to the circuit in Figure 4-18 if one of the current-limiting resistors were to short?

■ ***Solution:***

If one of the resistors R_1 through R_7 were to short, the lack of current limiting would cause an excessive current through the connected components in the encoder and display circuit. These devices would probably be damaged.

■ **EXAMPLE:**

What would happen to the circuit in Figure 4-18 if one of the LEDs in the seven-segment display were to open?

■ ***Solution:***

An open seven-segment LED will have no effect on the rest of the circuit; however, that segment will never turn ON.

■ **EXAMPLE:**

What would happen to the circuit in Figure 4-18 if one of the current limiting resistors were to open?

■ ***Solution:***

If one of the resistors R_1 through R_7 were to open, power would be blocked from reaching the connected encoder control line and its respective LED segment. The result would be that the affected LED will never turn ON.

As we complete each chapter in this text, we will continue to apply these troubleshooting methods to a variety of circuits. You will also be introduced to many more new testing techniques which will not only help you in lab but also help prepare you for the future.

4-3-4 Step 3: Repair

The final step is to repair the circuit, which could involve simply removing a wire clipping or some excess solder, resoldering a broken connection, reconnecting a connector, or some other easy repair. In most instances, however, the repair will involve the replacement of a faulty component. For a circuit that has been constructed on a breadboard or prototyping board, the removal and replacement of the component is simple. However, when a printed-circuit board is involved, you should make a note of the component's orientation (in the case of diodes, seven-segment displays, and other similar devices), and observe good desoldering and soldering practices.

When the circuit has been repaired, always perform a final test to see that the circuit is fully operational.

SELF-TEST REVIEW QUESTIONS FOR SECTION 4-3

1. What are the three basic troubleshooting steps?
2. To be an effective electronic technician or troubleshooter, you must have a thorough knowledge of ____________, ____________, ____________, and ____________.
3. What is a troubleshooting tree?
4. What would happen to the circuit in Figure 4-18 if one of the control lines (*A* through *G*) were to open between the encoder and the display?

SUMMARY

1. After the basic P-N junction diode, the two most widely used diodes are the zener diode and the light-emitting diode.
2. The basic P-N junction diode, zener diode, and light-emitting diode can be used in a variety of circuit applications such as rectifying, regulating, detecting, clipping, clamping, multiplying, switching, displaying, and oscillating.

The Zenor Diode (Figure 4-19)

3. It was American inventor Clarence Zener who first began investigating in detail the reverse breakdown of a diode. He discovered that the basic P-N junction diode would conduct a high reverse current when a sufficiently high reverse bias voltage was applied across its terminals. This large external reverse voltage would cause an increase in the number of minority carriers in the *n* and *p* regions as valence electrons are pulled away from their parent atoms. Clarence Zener was fascinated by the reverse breakdown characteristics of a basic P-N junction diode and, after many years of study, developed a special diode that could handle high values of

QUICK REFERENCE SUMMARY SHEET

FIGURE 4-19 The Zenor Diode.

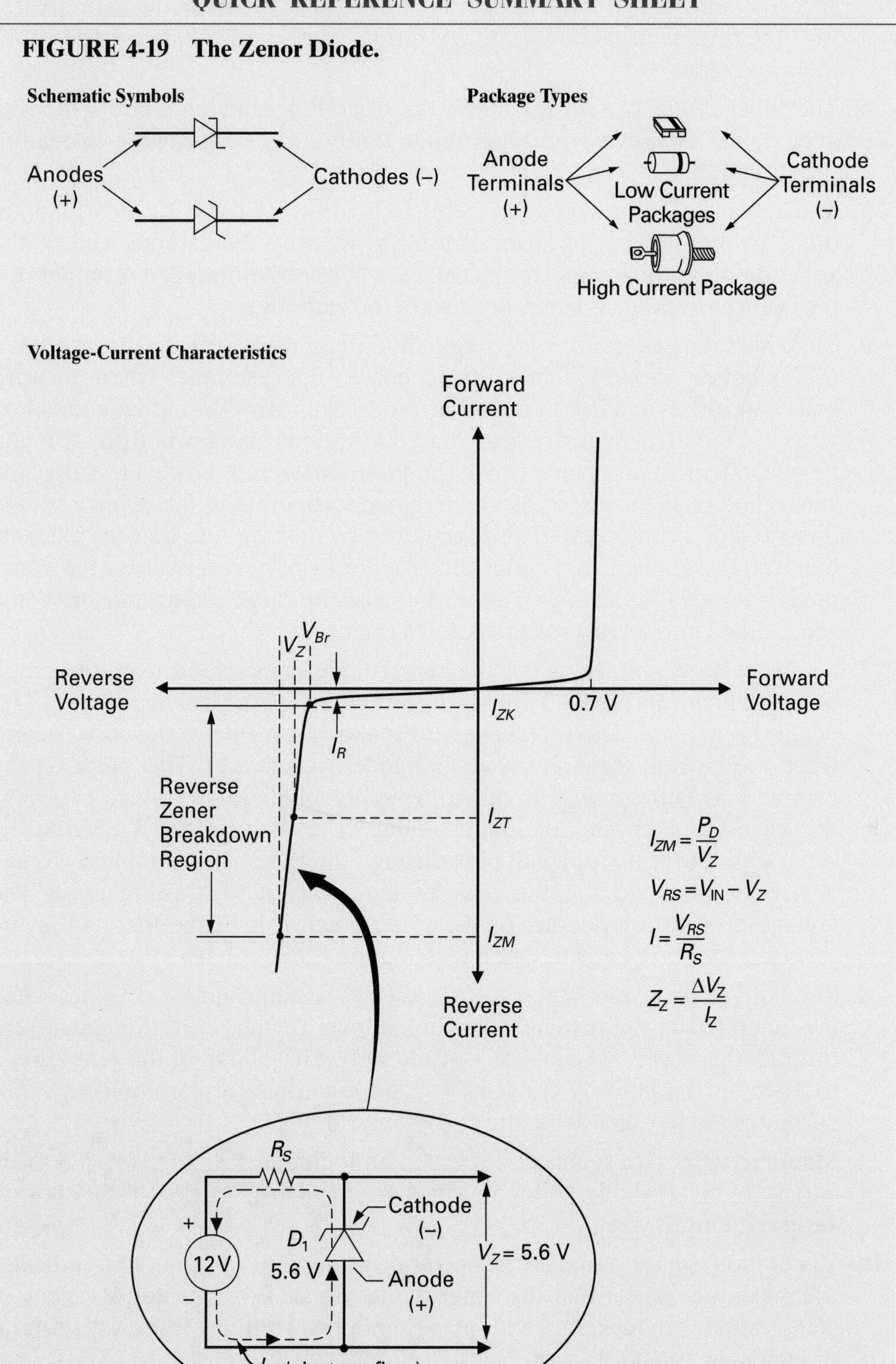

reverse current and still safely dissipate any heat that was generated. These special diodes, constructed to operate at voltages that are equal to or greater than the reverse breakdown voltage rating, were named zener diodes after their inventor.

4. The zener diode symbol resembles the basic P-N junction diode symbol in appearance. However, the zener diode symbol has a zig-zag bar instead of the straight bar.
5. The zener diode packages are identical to those of the basic P-N junction diode in that a band or stripe is used to identify the cathode end of the zener diode in the low-power packages, whereas the threaded terminal of a high-power package is generally always the cathode.
6. The characteristic curve of a zener diode is almost identical to the basic P-N junction diode's characteristic curve. For example, when forward biased at or beyond 0.7 V, the zener diode will turn ON and be equivalent to a closed switch. When reverse biased, the zener diode will turn OFF and be equivalent to an open switch. The main difference, however, is that the zener diode has been specifically designed to operate in the reverse breakdown region of the curve. This is achieved by making sure that the external bias voltage applied to a zener diode will not only reverse bias the zener diode (+ → cathode, − → anode) but also be large enough to drive the zener diode into its reverse breakdown region.
7. As the reverse voltage across the zener diode is increased from the graph origin (which represents 0 volts), the value of reverse leakage current (I_R) begins to increase. The reverse zener breakdown region of the curve begins when the current through the zener suddenly increases. This point on the reverse bias voltage scale is called the zener breakdown voltage (V_{Br}). At this knee-shaped point, a minimum value of reverse current, known as the zener knee current (I_{ZK}), will pass through the zener. If the applied voltage is increased beyond V_{Br}, the value of zener current (I_Z) will increase. The voltage drop developed across the zener when it is being operated in the reverse zener breakdown region is called the zener voltage (V_Z).
8. The voltage drop across a zener diode (V_Z) remains almost constant when it is operated in the reverse zener breakdown region, even though current through the zener (I_Z) can vary considerably. This ability of the zener diode to maintain a relatively constant voltage regardless of variations in zener current is the key characteristic of the zener diode.
9. Manufacturers rate zener diodes based on their zener voltage (V_Z). A wide variety of zener diode voltage ratings are available ranging from 1.8 V to several hundred volts.
10. Zener diodes have a maximum power dissipation (P_D) rating that indicates the maximum power that the zener diode can safely dissipate. A variety of zener diodes are available with maximum power ratings from several hundred mW up to 100 W.
11. If the maximum zener current (I_{ZM}) of a zener diode is exceeded, the diode will more than likely be destroyed because this value of current will generate more heat than the zener can safely dissipate (P_D).

12. The voltage developed across a zener diode will be equal to the zener diode's voltage rating, and the voltage developed across the circuit's series resistor will be equal to the difference between the diode's zener voltage and the input voltage.
13. Zener Impedance (Z_Z) is the opposition offered by a zener diode to current.
14. When substituting one basic P-N junction diode for another, we were concerned with the device's forward current and maximum reverse voltage ratings. For zener diodes the two important characteristics are power dissipation (P_D) and zener voltage (V_Z).
15. Although the zener diode can be forward and reverse biased, it is most widely used in applications where it is continually reverse biased and operating in the reverse zener breakdown region.

Zenor Diode Applications (Figure 4-20)

16. A voltage regulator circuit maintains the output voltage of a voltage source constant despite variations in the input voltage and the load resistance.
17. A zener diode is able to regulate its voltage because any change in zener diode current (I_Z) is always accompanied by an equal but opposite change in zener diode impedance (Z_Z). It is this equal but opposite change in I_Z and Z_Z that causes a cancellation, resulting in a constant output zener voltage.

Zenor Diode Testing (Figure 4-21)

18. The best way to test a zener diode is to connect the volt meter across the zener while it is in circuit and power is applied. If the voltage across the zener is at its specified voltage, then the zener is functioning properly. If the voltage across the zener is not at the nominal value, you should check the input voltage, the series resistor, the load, and then the zener.

The Light Emitting Diode (Figure 4-22)

19. The light-emitting diode (LED) is a semiconductor device that produces light when an electrical current or voltage is applied to its terminals. In other words, it converts electrical energy into light energy. It is classified as an optoelectronic device because it combines optics and electronics.
20. The LED is probably the most widely used light source in electronic equipment. It has replaced the incandescent lamp due to its longer life expectancy and lower operating power.
21. When forward biased, the LED will emit energy in response to a forward current. This emission of energy may be in the form of heat energy, light energy, or both heat and light energy depending on the type of semiconductor material used. It is the type of material used to construct the LED that determines the color, and therefore frequency, of the light emitted. For example, different compounds are available that will cause the LED to emit red, yellow, green, blue, white, orange, or infrared light when it is forward biased.

QUICK REFERENCE SUMMARY SHEET

FIGURE 4-20 Zenor Diode Applications.

Voltage Regulator Circuit

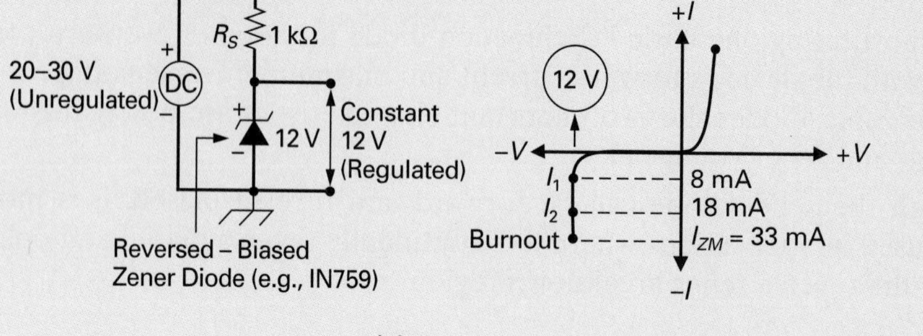

$$I_S = \frac{V_{IN} - V_Z}{R_S}$$

$$I_{RS} = \frac{V_{RS}}{R_S}$$

$$I_Z + I_L = I_S$$

(a)

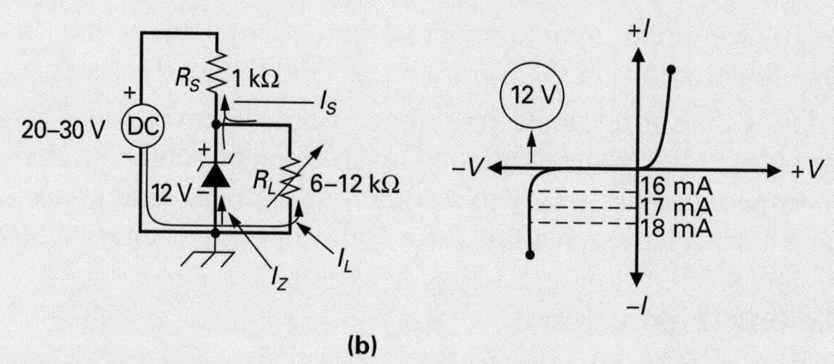

(b)

Voltage Reference for a Low-Voltage Battery Sense Circuit

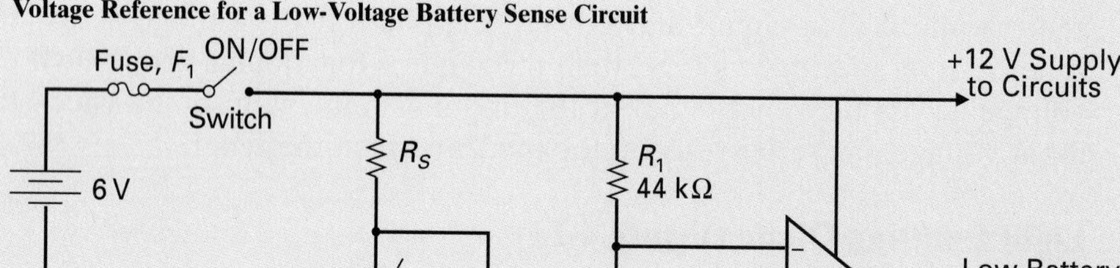
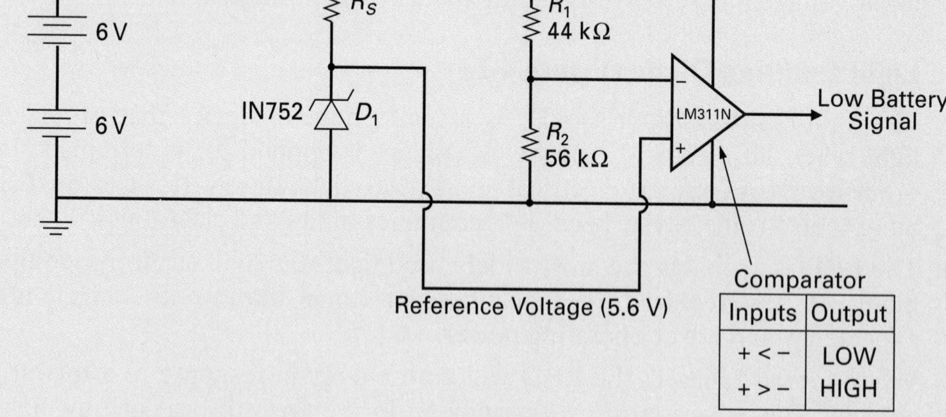

Inputs	Output
+ < −	LOW
+ > −	HIGH

22. The LED schematic symbol has two arrows leaving the standard diode symbol that represent light.

23. The package contains the two terminals for connection to the anode and cathode, and a semi-clear case which contains the light emitting diode and a lens. The dome-shaped top of the plastic (epoxy) case serves as the lens and

QUICK REFERENCE SUMMARY SHEET

FIGURE 4-21 Zenor Diode Testing.

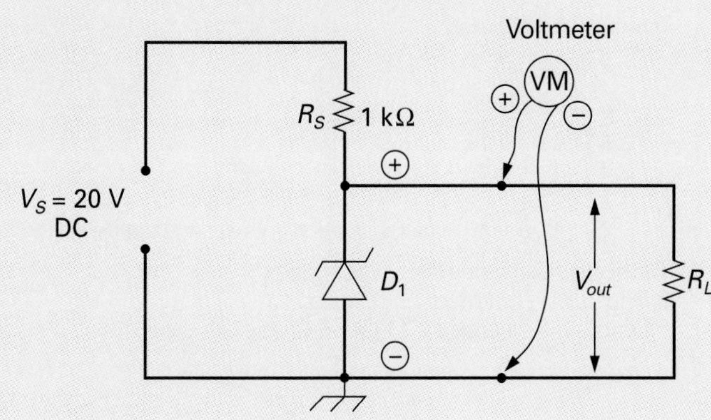

acts as a magnifier to conduct light away from the LED chip. By adjusting the lens material, lens shape, and the distance of the LED chip from the lens, manufacturers can obtain a variety of radiation patterns.

24. There are three methods used to identify the anode and cathode leads of the LED. In all three cases the cathode lead is distinguished from the anode lead by either having its lead shorter, its lead flattened, or the side of the case nearest to the cathode lead.

25. LEDs have a high forward voltage rating (V_F is typically +1 V to +3 V) and a low maximum forward current rating (I_F is typically 20 mA to 100 mA). These ratings mean that the LED will develop more voltage across its terminals when it is forward biased. The low forward current ratings means that the LED will usually always need to be protected from excessive current damage by a series current-limiting resistor.

26. LEDs have low reverse breakdown voltage values (V_{Br} is typically –3 V to –10 V), which means that only a low reverse voltage will cause the LED to break down and become damaged.

27. When forward biased, the negative terminal of the dc source injects electrons into the *n*-type region, or cathode, of the LED. These electrons travel towards the P-N junction. At the same time, electrons are attracted out of the *p*-type region, or anode, and travel towards the positive terminal of the source voltage making it appear as though holes are moving towards the P-N junction. The electrons from the *n*-type region and the holes from the *p*-type region combine at the P-N junction. Because the electrons are at a higher energy level (conduction band), they release energy as they drop into a valence hole. The energy is a small packet of light called a photon. Most forward biased LEDs release billions of photons per second as billions of electrons combine with holes at the P-N junction. The semi-transparent semiconductor material used to construct LEDs allows the light energy to escape where it is captured and focused by the dome shaped lens of the LED package.

QUICK REFERENCE SUMMARY SHEET

FIGURE 4-22 The Light Emitting Diode.

Schematic Symbol

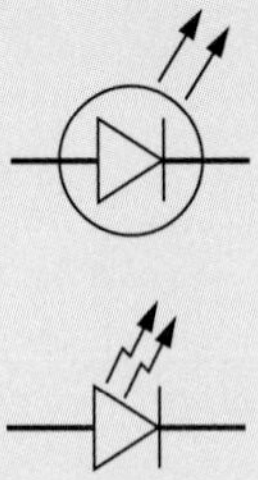

Terminal Identification

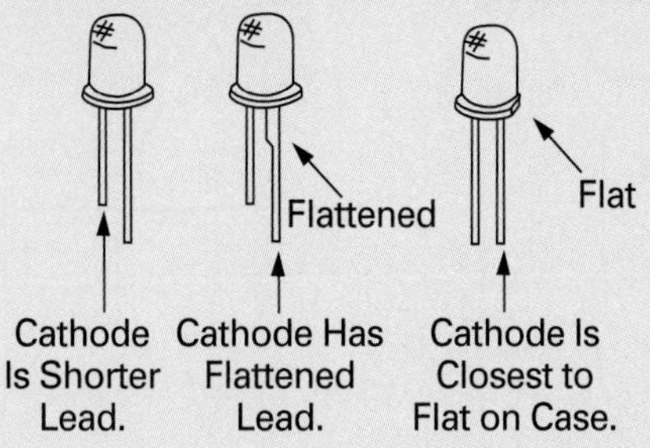

Flattened

Flat

Cathode Is Shorter Lead.

Cathode Has Flattened Lead.

Cathode Is Closest to Flat on Case.

LED Characteristics

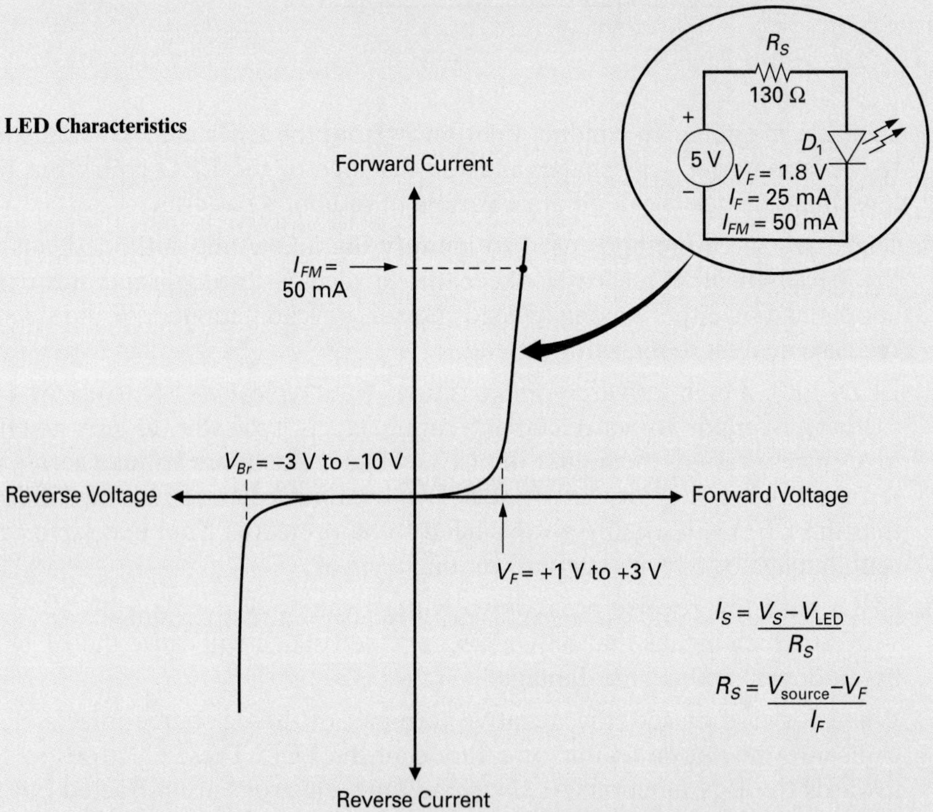

R_S

130 Ω

+

5 V

−

D_1

V_F = 1.8 V

I_F = 25 mA

I_{FM} = 50 mA

Forward Current

I_{FM} = 50 mA

V_{Br} = −3 V to −10 V

Reverse Voltage

Forward Voltage

V_F = +1 V to +3 V

Reverse Current

$$I_S = \frac{V_S - V_{LED}}{R_S}$$

$$R_S = \frac{V_{source} - V_F}{I_F}$$

LED Operation

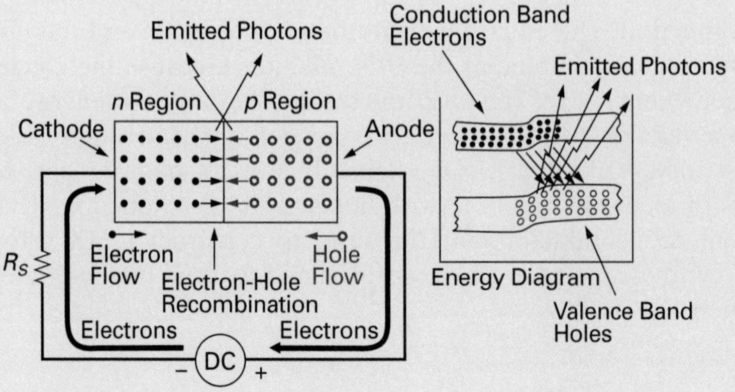

Emitted Photons

n Region

p Region

Cathode

Anode

R_S

Electron Flow

Electron-Hole Recombination

Hole Flow

Electrons

Electrons

− DC +

Conduction Band Electrons

Emitted Photons

Energy Diagram

Valence Band Holes

28. The light-emitting diode's key specifications are maximum forward current, forward voltage drop, and maximum reverse voltage.

29. The brightness of an LED is determined by the value of current that passes through the LED. Therefore, more current will mean more light, and less current will mean less light.

LED Applications

30. The individual LED is most commonly used as a power indicator and signal level indicator.

31. Two-color LEDs were developed so that a single LED package could be used to indicate one of two states such as stop-go, on-off, yes-no, and so on.

32. One of the biggest applications of LEDs is in the multisegment display.

Multisegment LED Displays (Figure 4-23)

33. The seven-segment display is probably the most widely used multisegment LED display. Its seven segments are labeled in a clockwise direction—*a, b, c, d, e, f, and g.* By turning ON a combination of segments, the numbers 0 through 9 can be displayed.

34. Segmented LED displays are available in either a common-anode or a common-cathode configuration. In the common-anode configuration, there is a common connection between all the anodes of the LEDs. The common-anode connection has its "common" connected to a positive supply voltage. To light up an LED, the cathodes of the LEDs in this configuration will have to be made LOW. In the common-cathode configuration, there is a common connection between all the cathodes of the LEDs. The common-cathode connection has its "common" connected to zero volts. The anodes of the LEDs in this configuration will have to be made HIGH to light up an LED.

Testing LED's

35. Infrared-emitting diodes are used in security and fiber optic communication applications.

36. Light-emitting diodes have a very long life expectancy and are much more rugged than their predecessor, the small incandescent light bulb.

37. A problem with an LED circuit is normally easily noticed because the LED fails to light up. As with all circuit troubleshooting, you can isolate the circuit fault by systematically checking all of the logical possibilities.

Troubleshooting Diode Circuits

38. The procedure for fixing a failure can be broken down into three basic steps:

Step 1: DIAGNOSE

The first step is to determine whether a problem really exists. To carry out this step, a technician must collect as much information as possible about the system, circuit, and components used, and then diagnose the problem.

QUICK REFERENCE SUMMARY SHEET

FIGURE 4-23 Multisegment LED Display.

7 Segment Package
a
f
b
e
g
c
d
dp
7 Segment Package Construction
a
f
b
g
e
c
d
dp
+
+
Common-Anode Device
+5 V
a (1)
f (2)
Anode(3)
dp (left) (6)
e (7)
(14)Anode
(13) b
(11) g
(10) c
(9) dp(right)
(8) d
a
f g b
e c
d
TIL312
Activated Segments for Digits 0 Through 9
Numerals
Segments turned on
a b c d ef
b c
a b d e g
a b c d g
b cf g
a c df g
c d ef g
a b c
a b c d ef g
a b cf g
Common–Anode Connection
+V
Common Anode Connection
a b c d e f g dp
Common–Cathode Connection
Common Cathode Connection
a b c d e f g dp
+V
Other Multisegment LED Types
14–Segment
3 ∞ 5 Array
5 ∞ 7 Array
10–Bar Graph

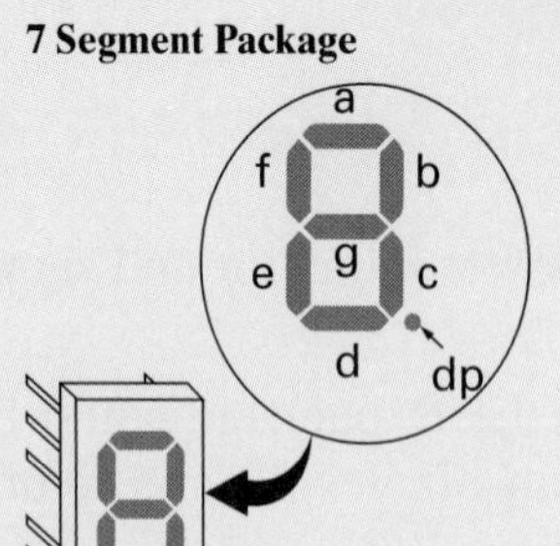

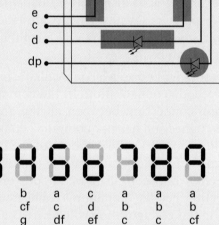

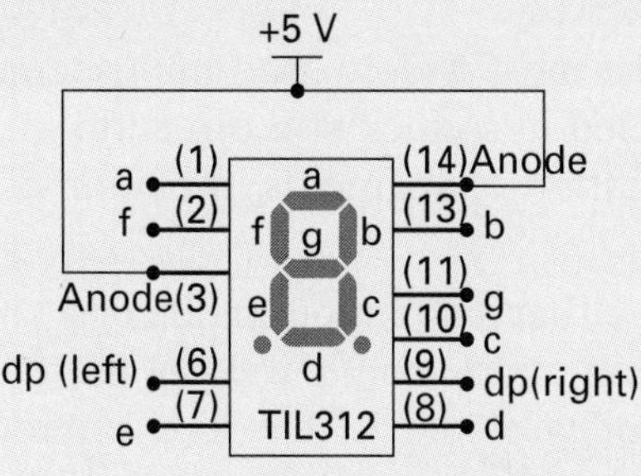

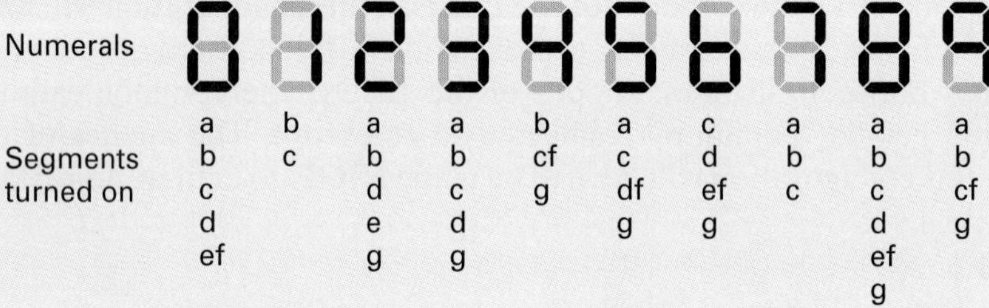

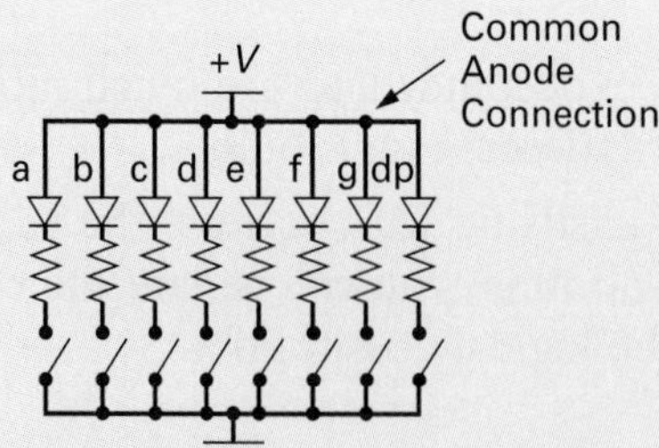

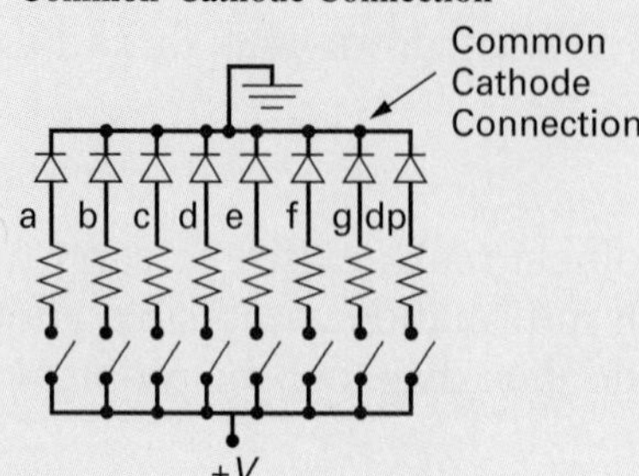

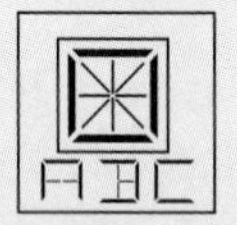

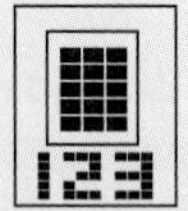

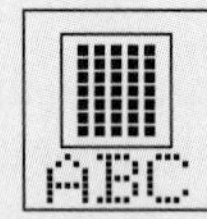

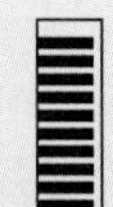

Step 2: ISOLATE

The second step is to apply a logical and sequential reasoning process to isolate the problem. In this step, a technician will operate, observe, test and apply troubleshooting techniques in order to isolate the malfunction.

Step 3: REPAIR

The third and final step is to make the actual repair and then test the circuit.

39. Troubleshooting is the process of diagnosing and locating breakdowns in equipment by means of systematic checking and analysis.

40. To assist you in troubleshooting, you have troubleshooting tools (test equipment), documentation (technical or service manuals), and experience.

41. To be an effective electronic technician or troubleshooter you must have a thorough knowledge of electronics, test equipment, troubleshooting techniques, and equipment repair.

42. Referring to all of the documentation before you begin troubleshooting will generally speed up and simplify the isolation process. Once you are fully familiar with the operation of the circuit, you will be ready to diagnose the problem as either an operator error or as a circuit malfunction.

43. The directions you take as you troubleshoot a circuit malfunction may be different for each problem. However, certain troubleshooting techniques apply to all problems.

1. Use a cause and effect troubleshooting process.
2. Check first for obvious errors.
3. Use your senses.
4. Apply the half-split method.
5. Always first test a circuit using the voltmeter for power and ground.
6. Substitution can be used to help speed up your troubleshooting process.
7. Make use of any troubleshooting guides or circuit descriptions.

44. When the circuit has been repaired, always perform a final test to see that the circuit is now fully operational.

NEW TERMS

Analog Displays
Bar Display
Bi-Color LED
Cause and Effect Troubleshooting
Circuit Malfunction
Common-Anode Display
Common-Cathode Display
Comparator
Decoder-Driver Circuit
Diagnose
Digital Displays
Dot Matrix Display
Fiber Optic Communications
Half-Split Method
Infrared-Emitting Diodes
IRED
Isolate
LED
Light-Emitting Diode
Maximum Power Dissipation
Maximum Zener Current
Multisegment LED Display

Operator Error
Optoelectronic
Photo
Power Indicator
Reference Voltage
Regulator
Repair
Reverse Leakage Current
Reverse-Parallel Connected
Seven Segment Display
Signal Level Indicator
Substitution
Troubleshooting
Troubleshooting Guide
Voltage Regulator Circuit
Zener Breakdown Voltage
Zener Current
Zener Diode
Zener Impedance
Zener Knee Current
Zener Reverse Breakdown Region
Zener Test Current
Zener Voltage

REVIEW QUESTIONS

Multiple-Choice Questions

1. The ____________ diode is designed to withstand high reverse currents that result when the diode is operated in the reverse breakdown region.

   a. Basic P-N junction **c.** Zener
   b. Light emitting **d.** Both (a) and (c) are true

2. When a zener diode's breakdown voltage is exceeded, the reverse current through the diode increases from a small leakage value to a high reverse current value.

   a. True **b.** False

3. When operating in the reverse breakdown region, the ____________ the zener will vary over a wide range, while the ____________ the zener will vary by only a small amount.

   a. Voltage drop across, forward current through
   b. Forward current through, voltage drop across
   c. Reverse current through, forward current through
   d. Reverse current through, forward drop across

4. The zener diode's symbol is different from all other diode symbols due to its:

   a. Z shaped cathode bar **c.** Straight bar cathode
   b. Two exiting arrows **d.** None of the above

5. The ability of a zener diode to maintain a relatively constant ____________ regardless of variations in zener ____________ is the key characteristic of a zener diode.

   a. Current, voltage **c.** Current, impedance
   b. Impedance, voltage **d.** Voltage, current

6. A 12 V ± 5% zener diode will have a voltage drop in the ____________ to ____________ range.

   a. 10.8 V to 13.2 V **c.** 5.04 V to 18.96 V
   b. 11.4 V to 12.6 V **d.** 11.88 V to 12.12 V

7. A 9.1 V zener diode has a power rating of 10 W. What is the diode's value of maximum zener current?

   a. 1.1 mA **b.** 91 mA **c.** 1.1 A **d.** None of the above

8. A/an ____________ circuit maintains the output voltage of a voltage source constant despite variations in the input voltage and the load resistance.

   a. Encoder **b.** Logic gate **c.** Comparator **d.** Voltage regulator

9. The zener diode is able to maintain the voltage drop across its terminals constant by continually changing its ____________ in response to a change in input voltage.

   a. Impedance **b.** Voltage **c.** Power rating **d.** Both (a) and (c) are true

10. In a voltage regulator circuit, the voltage developed across the zener diode remains constant and therefore any changes in the input voltage must appear across the:

    a. Load **b.** Series resistor **c.** Source terminals **d.** Zener diode

11. The light emitting diode is a semiconductor device that converts ____________ energy into ____________ energy.

    a. Chemical, electrical **c.** Electrical, light
    b. Light, electrical **d.** Heat, electrical

12. The ____________ lead of an LED is distinguished from the other terminal by its longer lead, flattened lead, or its close proximity to the flat side of the case.

    a. Anode **b.** Cathode

13. The forward voltage drop across a typical LED is usually:

    a. 0.7 V **b.** 0.3 V **c.** 5 V **d.** 2.0 V

14. Since the forward voltage rating of an LED remains almost constant, the output power of an LED is directly proportional to its:

    a. Impedance **c.** Reverse voltage
    b. Forward current **d.** Both (a) and (c) are true

15. If all seven cathodes in a seven-segment display are connected to 0 V, the display is referred to as a common ____________ configuration and requires a ____________ voltage input to turn on an LED segment.

    a. Anode, LOW **c.** Anode, HIGH
    b. Cathode, LOW **d.** Cathode, HIGH

Essay Questions

16. Who invented the zener diode? (4-1)
17. Sketch the zener diode schematic symbol. (4-1-1)
18. What is the zener diode's key characteristic? (4-1-2)
19. Define the following terms: (4-1-2)

    a. Reverse leakage current
    b. Zener breakdown voltage
    c. Zener knee current
    d. Zener current

e. Zener voltage
f. Zener impedance
g. Maximum zener power dissipation
h. Zener test current
i. Maximum zener current

20. How do manufacturers change the zener's voltage rating? (4-1-2)
21. Describe the relationship between a zener's voltage rating, power dissipation rating, and maximum zener current rating. (4-1-2)
22. What are the key characteristics given in a zener diodes's data sheet? (4-1-3)
23. Sketch an application circuit for the zener diode, and then briefly describe the circuit's purpose and operation. (4-1-4)
24. What is a voltage regulator circuit? (4-1-4)
25. How does a zener diode maintain a constant output voltage when the input voltage continually changes? (4-1-4)
26. Describe how a voltage regulator circuit using a zener diode will maintain a constant voltage to the load despite changes in the load resistance. (4-1-4)
27. LED is an abbreviation for ____________. (4-2)
28. Sketch the schematic symbol used to represent an LED. (4-2-1)
29. What elements are contained in a typical LED package? (4-2-1)
30. What method(s) is used to distinguish an LED's cathode from its anode? (4-2-1)
31. What are the key characteristics given in an LED's data sheet? (4-2-4)
32. Describe the relationship between forward current and an LED's brightness. (4-2-4)
33. In what application(s) is the individual LED most commonly used? (4-2-5)
34. In what application is the bi-color LED used? (4-2-5)
35. What is a multisegment LED display? (4-2-5)
36. What is the difference between an analog display and a digital display? (4-2-5)
37. What is the difference between a common-cathode LED display and a common-anode LED display? (4-2-5)
38. What is a seven-segment LED display? Sketch the typical package for this device. (4-2-5)
39. In what applications are bar displays used, and would a bar display be classified as an analog or a digital display? (4-2-5)
40. In what application(s) is the infrared emitting diode (IRED) used? (4-2-5)

Practice Problems

41. In reference to the polarity of the applied voltage, which of the circuits in Figure 4-24 are correctly biased for normal zener operation?
42. In reference to the magnitude of the applied voltage, which of the circuits in Figure 4-24 are correctly biased for normal zener operation?

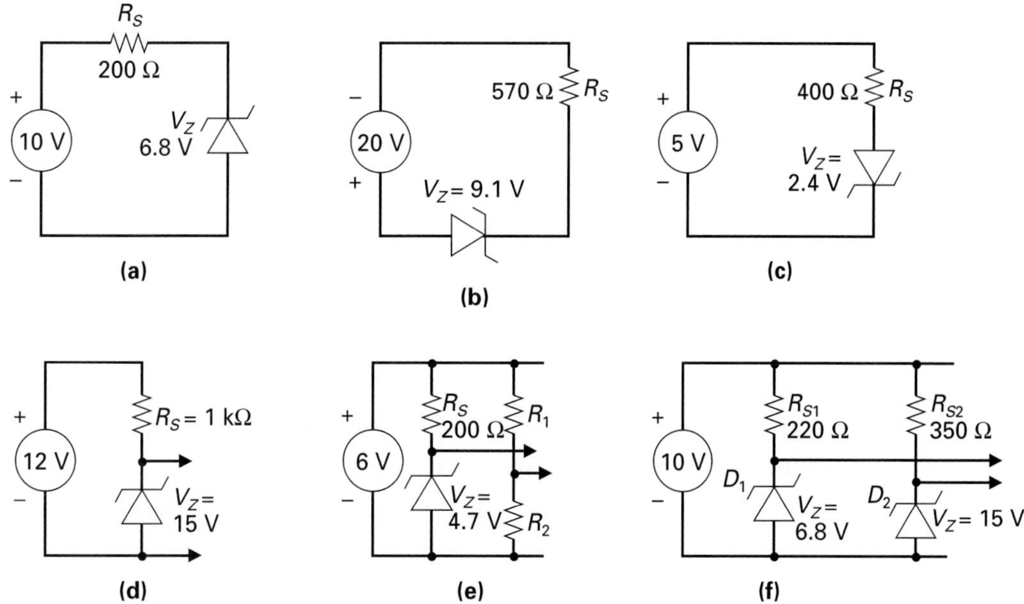

FIGURE 4-24 **Biasing Voltage Polarity and Magnitude.**

43. Calculate the value of circuit current for each of the zener diode circuits in Figure 4-24.

44. Would a 1 watt zener diode have a suitable power dissipation rating for the circuit in Figure 4-24(a)?

45. What wattage or maximum power dissipation rating would you choose for the zener diode in Figure 4-24(e)?

46. What would be the regulated output voltage and polarity at points *X* and *Y* for the circuits shown in Figure 4-25 (a) and (b)?

47. Knowing that the zener diode must always be reverse biased, how could we change the circuit connection in Figure 4-25(a) and (b) to obtain opposite polarity supply voltages?

48. Calculate the value of circuit current for both the circuits shown in Figure 4-25.

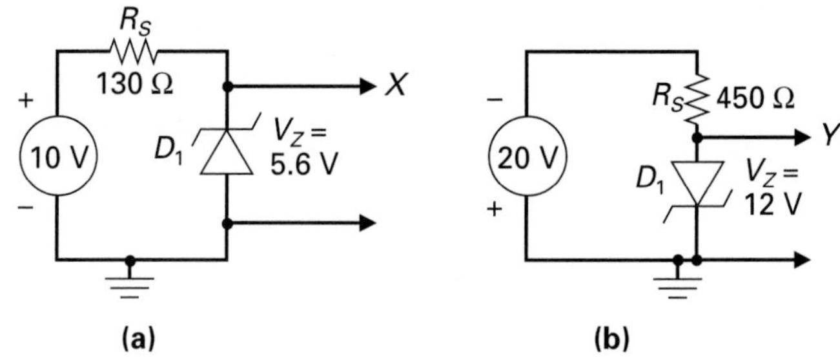

FIGURE 4-25 **Voltage Regulator Circuits (Unloaded).**

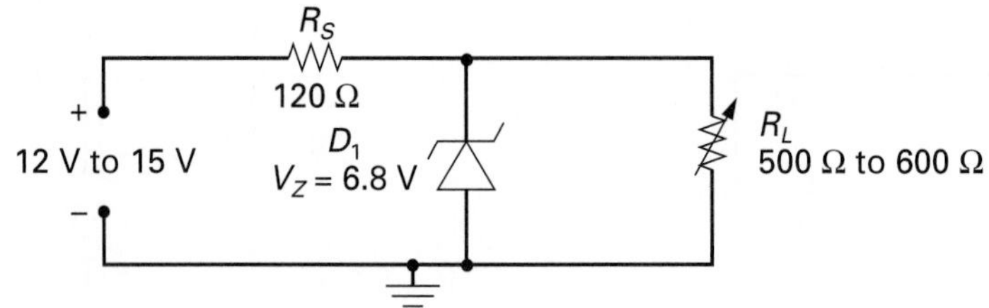

FIGURE 4-26 A Voltage Regulator Circuit (Loaded).

49. Calculate the value of I_{RS}, I_Z, I_{RL}, and V_{RL} for the two extreme low and high voltages shown in Figure 4-26. Assume R_L remains constant at 500 Ω.
50. Calculate the values of I_{RS}, I_{RL}, and I_Z for all values of maximum and minimum input voltage and load resistance for the circuit in Figure 4-26.
51. Which of the light-emitting diodes in Figure 4-27 are biased correctly?
52. Calculate the value of circuit current for each of the LEDs in Figure 4-27.
53. Would a maximum forward current rating of 18 mA be adequate for the LED in Figure 4-27(a)?
54. Would a maximum reverse voltage rating of 5 V be adequate for the LEDs in Figure 4-27(c)?
55. Figure 4-28 shows how two basic P-N junction diodes and a resistor can be used to construct an OR gate circuit. In this circuit, an LED has been connected to the output so that any HIGH or positive 5 V output will turn ON the LED. Considering the *A* and *B* input combinations shown in the table, indicate whether the LED will be ON or OFF for each of these input conditions.
56. Assuming a 0.7 V voltage drop across the P-N junction diode and a 2 V voltage drop across the LED, what would be the value of current through the LED if one of the inputs in Figure 4-28 was HIGH (+5 V)?
57. Which of the bi-color LEDs will be ON in Figure 4-29 when the input voltage is +10 V, and which will be ON when the input voltage is –10 V?
58. In Figure 4-29, a basic P-N junction diode (D_1) has been included across R_1 so that when the input voltage goes negative this diode will bypass the additional current-limiting resistor R_1. When the input voltage is positive,

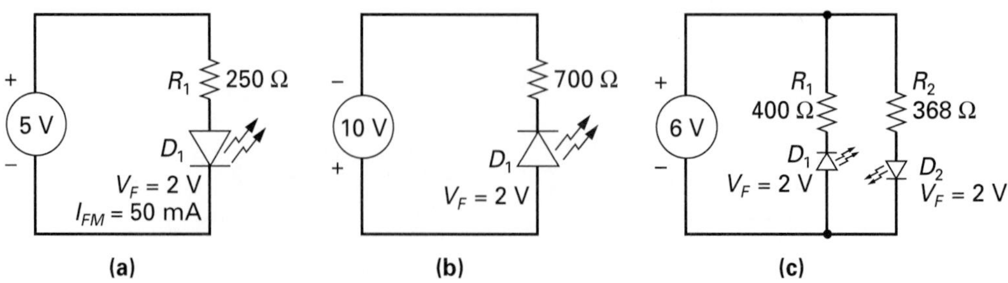

FIGURE 4-27 Biasing Light Emitting Diodes.

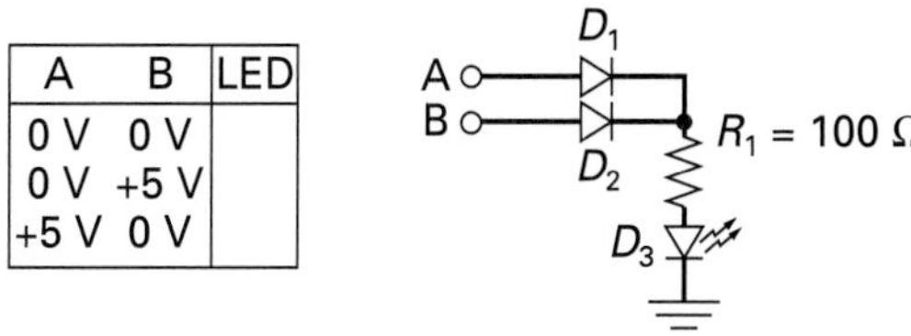

FIGURE 4-28 An OR Gate Circuit with an LED Output.

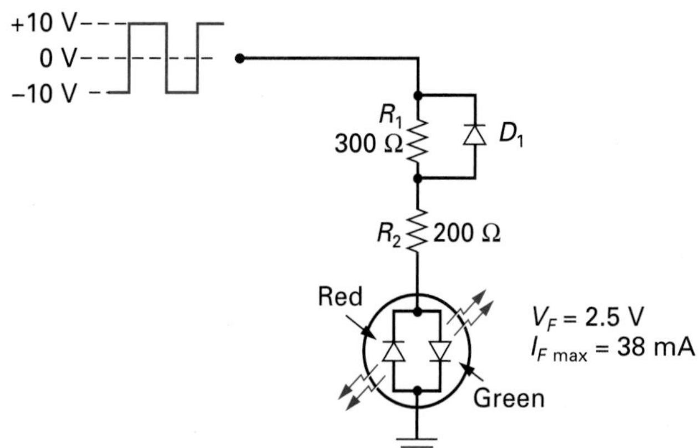

FIGURE 4-29 A Bi-Color Output Display Circuit.

D_1 is reversed biased and therefore both R_1 and R_2 will limit the value of series current. Which of the colors in the bi-color LED will be brighter?

59. Calculate the value of green and red LED current for the circuit in Figure 4-29.

60. In Figure 4-30 a comparator is used to compare 0 V (at the negative input) to a sine wave which varies between +5 V and –5 V (at the positive input). When the comparator's inputs are "true" (positive input is positive with respect to negative input), the comparator's output will be switched HIGH (+12 V). When the comparator's inputs are "not true" (positive input is negative with respect to the negative input), the comparator's output will be switched LOW (–12 V). What value of series current-limiting resistor should be used to ensure that the LEDs are bright, but do not burn out, for the +12 V and –12 V source voltages? Aim for a forward current that is about 90% of the maximum rating.

61. Would the 4-LED display in Figure 4-31 (D_{11}-D_{14}) be classified as a common-anode or common-cathode display?

62. In regard to Figure 4-31, construct a table to show the HIGH and LOW encoder outputs on *A, B, C,* and *D* for each of the rotary switch positions.

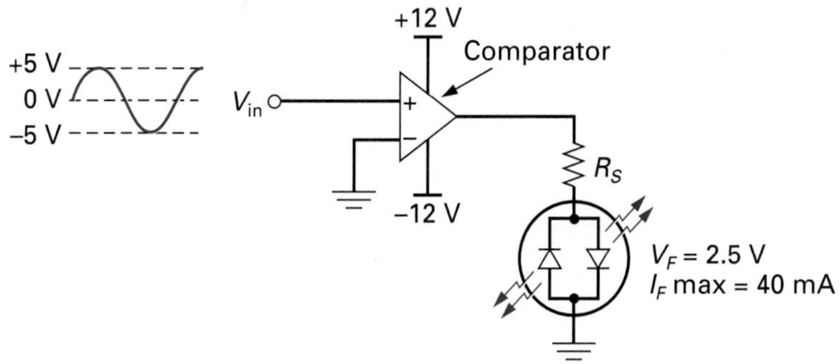

FIGURE 4-30 A Comparator Circuit with a Bi-Color Output Display.

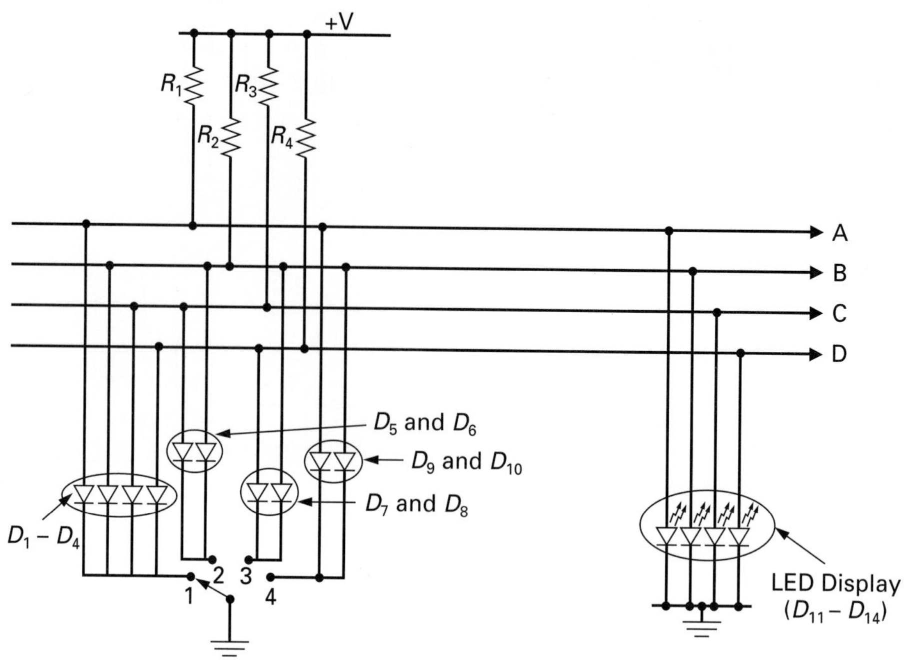

FIGURE 4-31 A Switch Encoder Circuit with an LED Output Display.

Also show in the table which of the LEDs are ON or OFF for each of these codes.

63. In Figure 4-32, seven switches are used to turn ON and OFF the seven LEDs in a seven-segment display. Construct a table to show which switches will be CLOSED and which will be OPEN to display the digits 0 through 9 on the seven-segment display.

64. Are the seven-segment displays in Figure 4-32 and Figure 4-33 common-anode or common-cathode types?

65. Figure 4-33(a) shows how a 5.1 V zener voltage regulator circuit can be used to supply power to a seven-segment encoder and display circuit. Figure 4-33(b) shows how the LOW and HIGH codes generated by the encoder circuit will turn ON and OFF the necessary LEDs in the seven-segment display to produce the digits 0 through 9. For example, when the rotary switch is in position 0, a single P-N junction diode makes line G LOW. This will turn segment "g" of the seven-segment display OFF, while the pull up resistors R_1 through R_6 will make lines A through F HIGH and therefore turn ON segments "a" through "f." This condition is shown in the first row of the table in Figure 4-33(b).

Calculate the following encoder circuit details:

a. The amount of current that will be drawn by each of the basic P-N junction diodes in the encoder circuit.

b. What rotary switch position will cause the maximum value of current, and what will this current value be?

Calculate the following display circuit details:

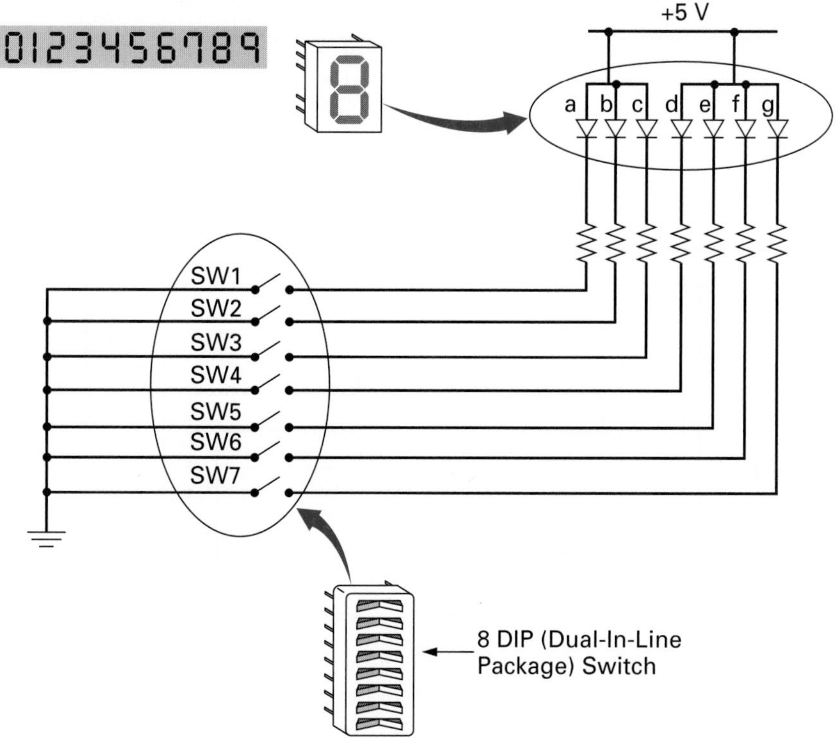

FIGURE 4-32 A Seven-Segment Display Circuit.

c. The amount of current that will be drawn by each of the LEDs in the seven-segment display.
d. What digit on the seven-segment display will cause the maximum value of current, and what will this current value be?

Calculate the following encoder-display circuit details:

e. Which path draws more current: an encoder diode or a display diode?
f. When the digit 8 is being displayed, a total of 99.2 mA of current is being drawn from the 5.1 V source. When the digit 1 is being displayed, a total of 112.8 mA is being drawn from the 5.1 V source. Because the source only sees the encoder-display circuit as a single load resistance, what will this resistance be equal to?

Troubleshooting Questions

66. Describe briefly how you would test a zener diode.

67. Describe briefly how you would test a light-emitting diode.

68. Define troubleshooting.

69. Describe the three steps used to locate and repair a circuit malfunction.

70. In regard to Figure 4-33, describe what symptoms would occur from the following circuit malfunctions:

a. The zener diode shorts
b. The zener diode opens

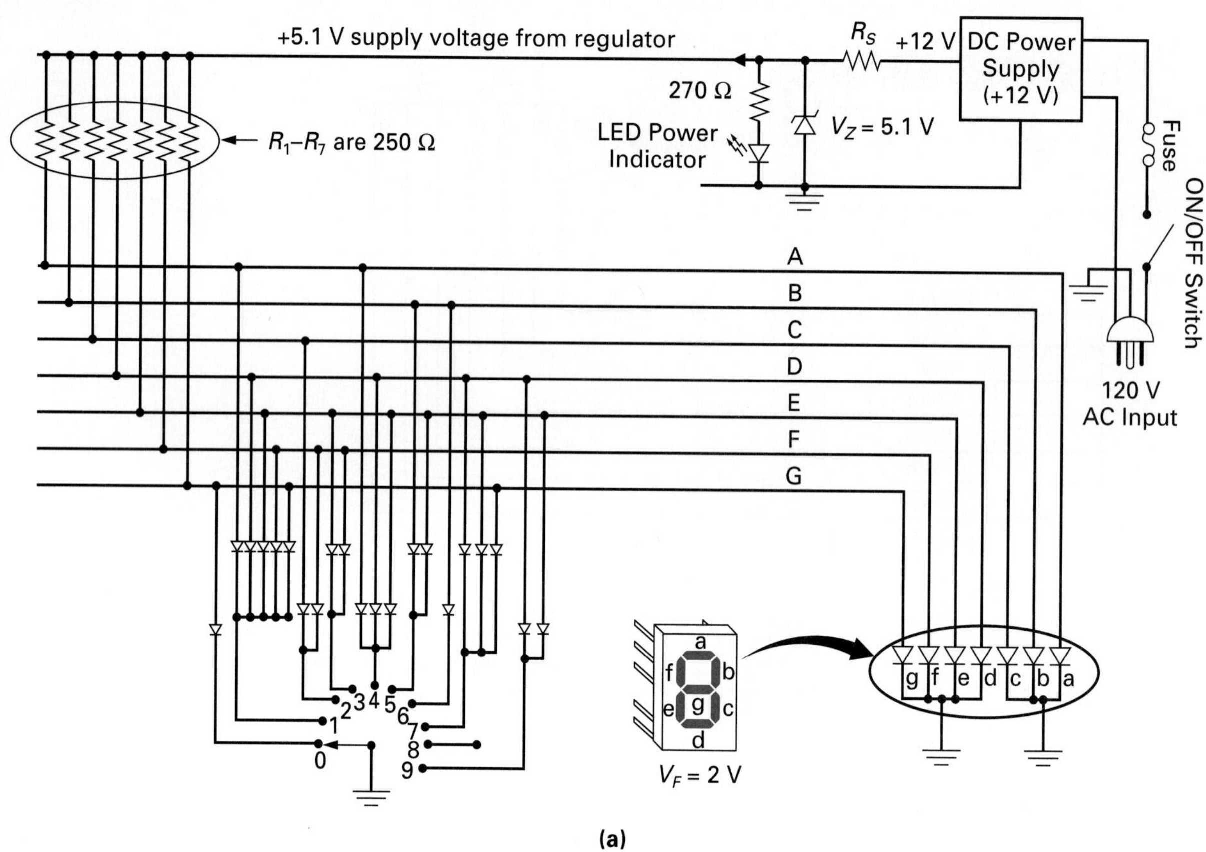

(a)

	a	b	c	d	e	f	g
0	HI	HI	HI	HI	HI	HI	LO
1	LO	HI	HI	LO	LO	LO	LO
2	HI	HI	LO	HI	HI	LO	HI
3	HI	HI	HI	HI	LO	LO	HI
4	LO	HI	HI	LO	LO	HI	HI
5	HI	LO	HI	HI	LO	HI	HI
6	HI	LO	HI	HI	HI	HI	HI
7	HI	HI	HI	LO	LO	HI	LO
8	HI	HI	HI	HI	HI	HI	HI
9	HI	HI	HI	LO	LO	HI	HI

(b)

FIGURE 4-33 A Seven-Segment Encoder and Display Circuit.

c. R_S opens
d. R_S shorts
e. The dc supply voltage decreases to 10 V
f. The dc supply voltage increases to 15 V
g. One of the encoder diodes opens
h. One of the encoder diodes shorts

i. There is an open in one of the control lines A through G between the encoder and the seven-segment display
j. One of the seven-segment LEDs opens
k. One of the seven-segment LEDs shorts
l. One of the resistors R_1 through R_7 shorts
m. One of the resistors R_1 through R_7 opens

CHAPTER

5

OBJECTIVES

After completing this chapter, you will be able to:

1. Define the function of the transformer, rectifier, filter, and regulator in a dc power supply.
2. Describe the characteristics of and differences between the half-wave rectifier, the full-wave center-tapped rectifier, and the full-wave bridge rectifier.
3. Interpret the key characteristics of a typical power-rectifier diode and bridge rectifier module from a manufacturer's data sheet.
4. Calculate the average output voltage and ripple frequency of all rectifier types.
5. Describe the characteristics of the capacitive filter, *RC* filter, and *LC* filter.
6. Define and calculate percent of ripple.
7. Describe the relationship between source and load.
8. Define and calculate percent of regulation.
9. Explain how a zener diode can be connected to act as a power supply regulator.
10. Describe the operation and characteristics of an IC regulator.
11. Interpret the key characteristics of a typical IC regulator from a manufacturer's data sheet.
12. Explain the function of each device in a typical 120 V/240 V ac input +5 V, +12 V, and −12 V dc output power supply.
13. Describe the three-step troubleshooting process and apply it to a typical dc power supply circuit.
14. Examine some of the more common dc power supply circuit malfunctions such as transformer, rectifier, filter, and regulator problems.

Diode Power Supply Circuits

OUTLINE

VIGNETTE

Conducting Achievement

Werner Von Siemens (1816–1892) was born in Lenthe, Germany. He became familiar with the recently developed electric telegraph during military service. Later in 1847, together with skilled mechanic J. G. Halske, he founded the electrical firm of Siemens and Halske. This firm, under Siemens's guidance, became one of the most important electrical undertakings in the world. Siemens invented cable insulation, an armature for large generators, and the dynamo, or electric generator, which converts mechanical energy to electric energy.

Karl Wilhelm Siemens (1823–1883), Werner's brother, is also well known for his work in the fields of electricity and heat. At the age of nineteen he visited England expressly to patent his electroplating invention and never left, making England his home. From 1848 onward, Wilhelm represented his brother's firm in London and became an acknowledged authority. The company installed much of the overland telegraph cable then in existence, as well as an underwater telegraph cable, using submarines. One month before he died, Wilhelm was knighted as Sir William Siemens in acknowledgment of his achievements.

In the family tradition, Alexander Siemens (1847–1928), a nephew of William, went to England in 1867 and worked his way up, beginning in the Siemens's workshops. In 1878 he became manager of the electric department and was responsible for the installation of electric light at Godalming, Surrey, the first English town to be lit with electricity. Like many other members of the family, he patented several inventions and, after the death of Sir William, he became company director.

In honor of the Siemens family's achievements, conductance (G) is measured in the unit siemens (S), with a resistance or impedance of 1 Ω being equal to a conductance or admittance of 1 S.

INTRODUCTION

As mentioned previously *electronic systems are designed to manage the flow of information.* In order for them to achieve this function, all of the electronic circuits within the system require certain constant dc supply voltages. If only a small amount of power is needed, batteries can be used to deliver a dc supply, as is the case with portable equipment such as calculators, watches, cassette radios, multimeters, and so on. With larger electronic systems such as computers, television sets, music systems, and video systems, the dc supply voltages will be obtained from a dc power supply, which is generally a subsystem within the main system. The dc power supply converts a 120 V ac 60 Hz input into the desired dc supply voltages, and it is therefore an electrical system because *electrical systems are designed to manage the flow of power.*

In Chapter 3 we discussed the P-N junction diode, while in Chapter 4 we discussed the zener and light-emitting diodes. In this chapter we will see how we can use all of these diodes to construct a dc power supply. This chapter is also of great importance to our study of troubleshooting, since greater than 90% of all electronic system malfunctions are power related.

5-1 BLOCK DIAGRAM OF A DC POWER SUPPLY

Figure 5-1(a) shows the four basic blocks of a dc power supply. Almost every piece of electronic equipment that makes use of 120 V ac as a source of power will have a built-in dc power supply, as shown in Figure 5-1(b). Stand-alone dc power supplies, such as the one shown in Figure 5-1(c), are also available for use in laboratory experimentation.

Let us now refer to the block diagram of the dc power supply in Figure 5-1(a) and describe the function of each block.

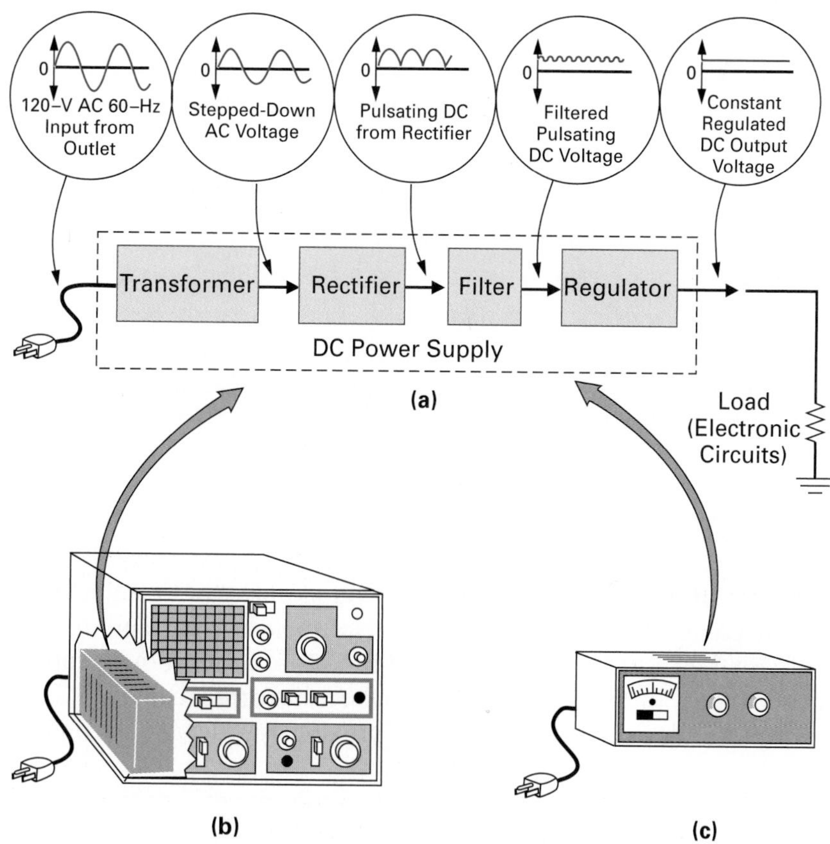

FIGURE 5-1 DC Power Supply. (a) Block Diagram. (b) Built-in Subsystem. (c) Stand-Alone Unit.

Since the final voltage desired is generally not 120 V, a transformer is usually included to step the ac line voltage up or down to a desired value. Electronic circuits generally require low-voltage supply values such as 12 and 5 V dc, and a step-down transformer would be used. For example, a step-down transformer with a turns ratio of 10:1 would reduce the 120 V ac input to a 12 V ac output. The output current capability of this transformer will be 1:10, which will be ideal because most electronic circuits require low supply voltages with high current capacity.

As shown by the wave forms in Figure 5-1(a), the rectifier converts the stepped-down ac input from the transformer to a pulsating dc output. This pulsating dc output could not be used to power an electronic circuit because of the continuous changes between zero volts and a peak voltage. The filter smoothes out the pulsating dc ripples into an almost constant dc level, as seen in the waveform after the filter.

The final block is called a regulator, and although there appears to be no difference between the regulator's input and output waveforms, it provides a very important function. The regulator maintains the dc output voltage from power supply constant, or stable, despite variations in the ac input voltage or variations in the output load resistance.

Many dc power supplies have several rectifier, filter, and regulator stages, depending on how many dc output voltages are desired for the electronic system. In the following sections we will examine these electrical circuits in more detail and then combine all of them in a working dc power supply circuit.

SELF-TEST REVIEW QUESTIONS FOR SECTION 5-1

1. List the four basic circuit blocks that make up a dc power supply.
2. Which circuit block converts ac to dc?
3. Which circuit block converts the high ac input voltage into a low ac output voltage?
4. The ____________ maintains the dc output voltage constant despite variations in the ac input voltage and the output load resistance.

5-2 TRANSFORMERS

The turns ratio of a transformer in a dc power supply can be selected to either increase or decrease the 120 V ac input. With most electronic equipment, a supply voltage of less than 120 V is required, and therefore a step-down transformer is used. The secondary output voltage V_S from the transformer can be calculated with the following formula, which was introduced previously in dc/ac electronic theory:

$$V_S = \frac{N_S}{N_p} \times V_p$$

To apply this formula to an example, refer to the power supply transformer shown in Figure 5-2. This transformer can be connected so that it delivers the same secondary voltage for either a 120 V or a 240 V rms ac input. For example, when the two primary windings are connected in parallel, as shown in Figure 5-2(a), the transformer turns ratio is 6:1 step-down, and therefore the secondary voltage will be:

$$V_S = \frac{N_S}{N_p} \times V_p$$

$$= \frac{1}{6} \times 120\ \text{V}_{rms} = 20\ \text{V}_{rms}$$

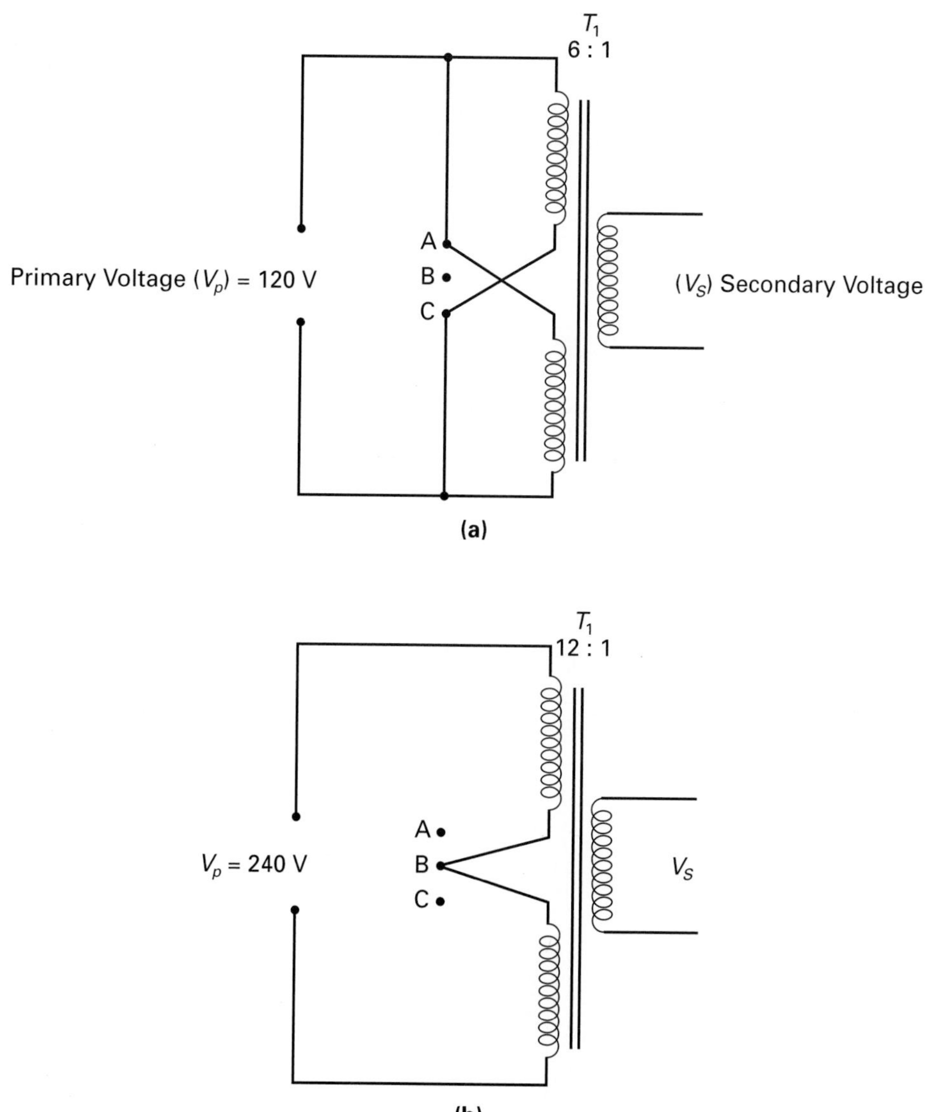

FIGURE 5-2 A 120V/240V Power Supply Transformer. (a) 120V Connection. (b) 240V Connection.

When the two primary windings are connected in series, as shown in Figure 5-2(b), the transformer turns ratio is 12:1 step-down, and therefore the secondary voltage will be

$$V_S = \frac{N_S}{N_p} \times V_p$$

$$= \frac{1}{12} \times 240 \text{ V}_{rms} = 20 \text{ V}_{rms}$$

By doubling the number of primary turns, we can accept twice the input voltage and deliver the same output voltage.

■ **EXAMPLE:**

Calculate the peak secondary output voltage of the transformer shown in Figure 5-3.

■ ***Solution:***

If the rms input is 120 V, the peak input will be

$$V_p = V_{rms} \times 1.414$$

$$= 120 \text{ V} \times 1.414 = 169.7 \text{ V}$$

Now that V_p peak is known, V_S peak can be calculated.

$$V_S = \frac{N_S}{N_p} \times V_p$$

$$= \frac{1}{14} \times 169.7 \text{ V} = 12.12 \text{ V}$$

SELF-TEST REVIEW QUESTIONS FOR SECTION 5-2

1. Most dc power supplies employ a step ______________ transformer.
2. If a 120 V rms ac input was applied to a 5:1 step-down transformer, what would be the secondary voltage?

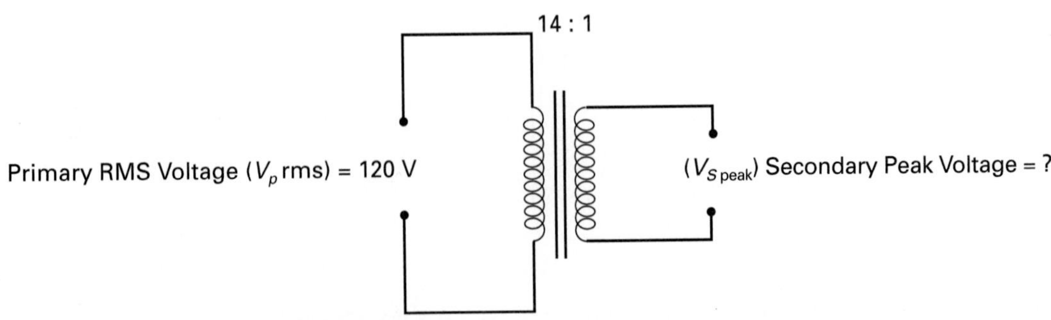

FIGURE 5-3 Transformer Coupling.

5-3 RECTIFIERS

The P-N junction diode's ability to switch current in only one direction makes it ideal for converting two-direction alternating current into one-direction direct current. In this section we will discuss the three basic diode rectifier circuits: the half-wave rectifier, the full-wave center-tapped rectifier, and the full-wave bridge rectifier.

5-3-1 *Half-Wave Rectifiers*

The **half-wave rectifier circuit** is constructed simply by connecting a diode between the power supply transformer and the load, as shown in Figure 5-4(a). When the secondary ac voltage swings positive, as shown in Figure 5-4(b), the anode of the diode is made positive, causing the diode to turn ON and connect the positive half-cycle of the secondary ac voltage across the load (R_L). When the secondary ac voltage swings negative, as shown in Figure 5-4(c), the anode of the diode is made negative, and therefore the diode will turn OFF. This will prevent any circuit current, and no voltage will be developed across the load (R_L).

Half-Wave Rectifier Circuit
A circuit that converts ac to dc by only allowing current to flow during one-half of the ac input cycle.

Output Voltage

Figure 5-4(d) illustrates the input and output waveforms for the half-wave rectifier circuit. The 120 V ac (rms) input, or 169.7 V ac peak input, is applied to the 17:1 step-down transformer, which produces an output of:

$$V_S = \frac{N_S}{N_p} \times V_p$$

$$= \frac{1}{17} \times 169.7 \text{ V (peak)} = 10 \text{ V (peak)}$$

Because the diode will only connect the positive half-cycle of this ac input across the load (R_L), the output voltage (V_{RL}) is a positive pulsating dc waveform of 10 V_{peak}. In this final waveform in Figure 5-4(d), you can see that the circuit is called a half-wave rectifier because only half of the input wave is connected across the output.

The average value of two half-cycles is equal to 0.637 V_{peak}. Therefore, the average value of one half-cycle is equal to 0.318 V_{peak} (0.637/2 = 0.318):

$$V_{av} = 0.318 \times V_{S\,peak}$$

In the example in Figure 5-4, the average voltage of the half-wave output will be:

$$V_{av} = 0.318 \times V_{S\,peak}$$

$$V_{av} = 0.318 \times 10 \text{ V}$$

$$= 3.18 \text{ V}$$

To be more accurate, there will, of course, be a small voltage drop across the diode due to its barrier voltage of 0.7 V for silicon and 0.3 V for germa-

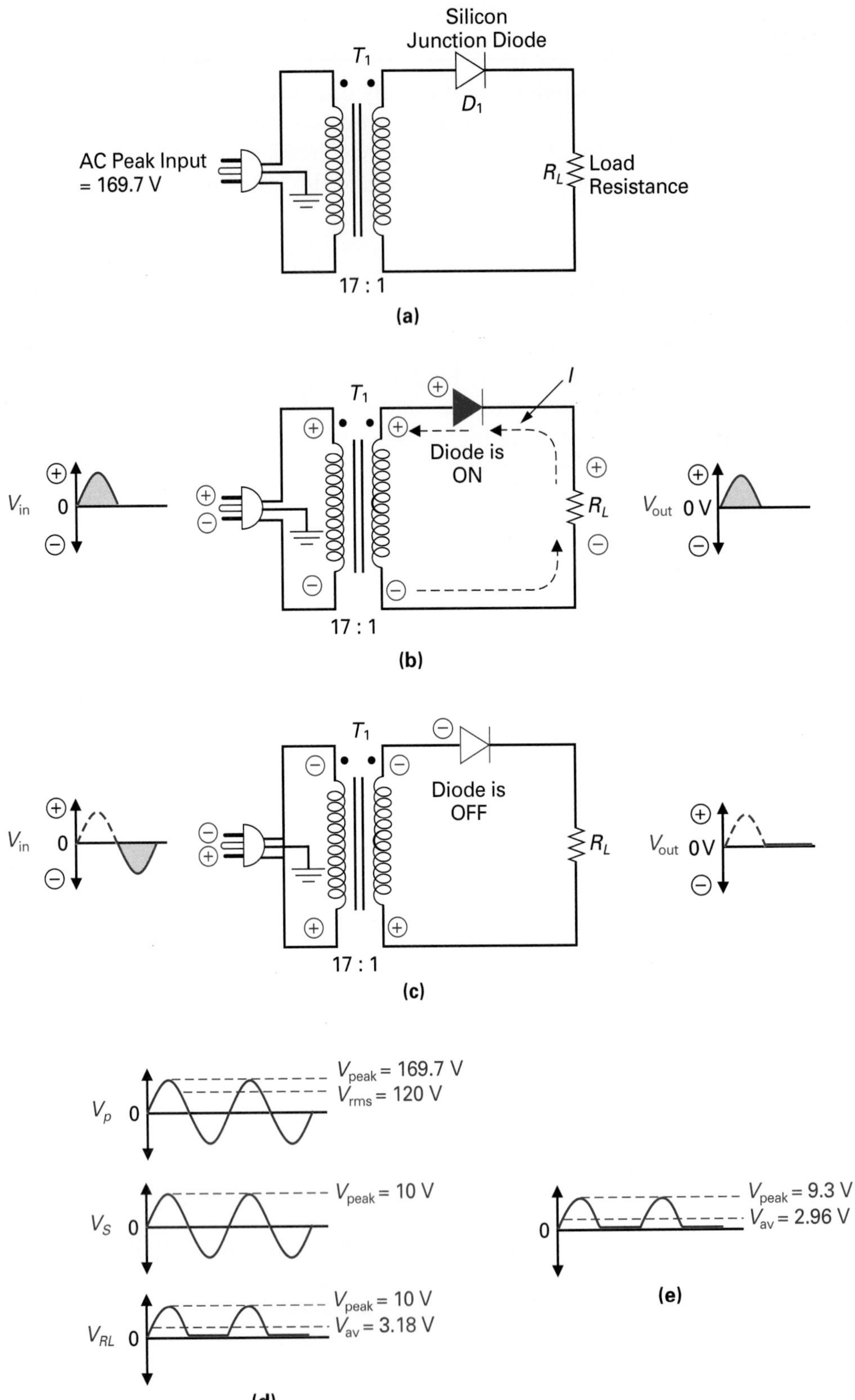

FIGURE 5-4 Half-Wave Rectifier. (a) Basic Circuit. (b) Positive Input Half-Cycle Operation. (c) Negative Input Half-Cycle Operation. (d) Input/Output Waveforms. (e) Half-Wave Output Minus Diode Barrier Voltage.

nium. The output from the circuit in Figure 5-4 would actually have a peak of 9.3 V (10 V – 0.7 V), and therefore an average of 2.96 V (0.318 × 9.3 V), as shown in Figure 5-4(e).

$$V_{out} = V_S - V_{diode}$$

Another point to consider is the reverse breakdown voltage of the junction or rectifier diode. When the input swings negative, as illustrated in Figure 5-4(c), the entire negative supply voltage will appear across the open or OFF diode. The maximum reverse breakdown voltage, or peak inverse voltage (PIV) rating of the rectifier diode, must therefore be larger than the peak of the ac voltage at the diode's input.

Output Polarity

The half-wave rectifier circuit can be arranged to produce either a positive pulsating dc output, as shown in Figure 5-5(a), or a negative pulsating dc output, as shown in Figure 5-5(b). Studying the difference between these circuits you can see that in Figure 5-5(a) the rectifier diode is connected to conduct the positive half-cycles of the ac input, while in Figure 5-5(b), the rectifier diode is reversed so that it will conduct the negative half-cycles of the ac input. By changing the direction of the diode in this manner, the rectifier can be made to produce either a positive or a negative dc output.

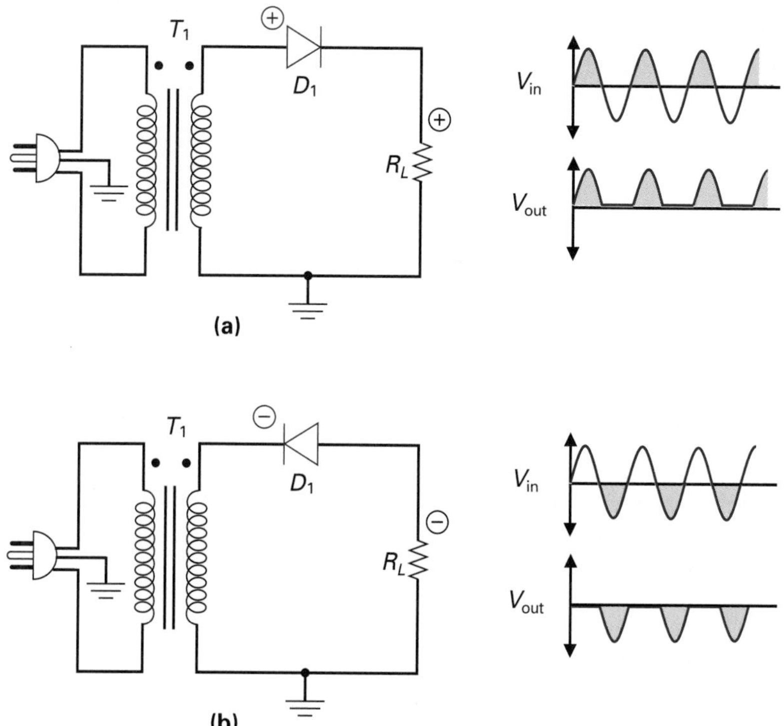

FIGURE 5-5 Changing the Output Polarity of a Half-Wave Rectifier. (a) Positive Pulsating DC. (b) Negative Pulsating DC.

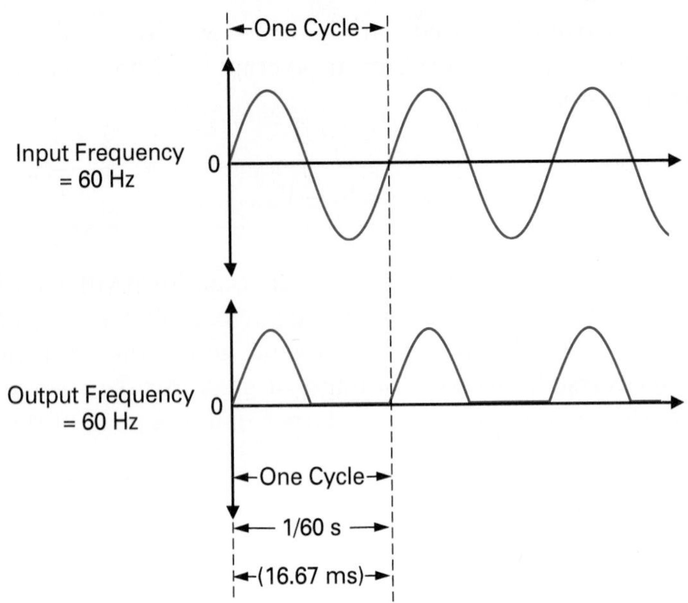

FIGURE 5-6 Ripple Frequency of a Half-Wave Rectifier.

Ripple Frequency

Referring to the input/output waveforms of the half-wave rectifier circuit shown in Figure 5-6, you can see that one output ripple is produced for every complete cycle of the ac input. Consequently, if a half-wave rectifier is driven by the 120 V ac 60 Hz line voltage, each complete cycle of the ac input and each complete cycle of the output will last for one-sixtieth or 16.67 ms (1/60 = 1 ÷ 60 = 16.67 ms). The frequency of the pulsating dc output from a rectifier is called the ripple frequency, and for half-wave rectifier circuits, the

Output pulsating dc ripple frequency = input ac frequency

■ **EXAMPLE:**

Calculate the following for the rectifier circuit shown in Figure 5-7:

a. Output polarity
b. Peak and average output voltage, taking into account the diode's barrier potential
c. Output ripple frequency

■ ***Solution:***

a. Negative pulsating dc

b.

$$V_{in\ peak} = 169.7 \text{ V}$$
$$V_S = \frac{N_S}{N_p} \times V_p$$

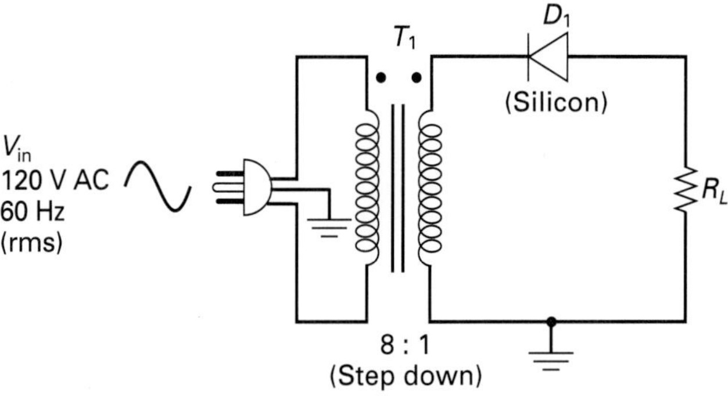

FIGURE 5-7 Half-Wave Rectifier.

$$= \frac{1}{8} \times 169.7\text{ V (peak)} = -21.2\text{ V (peak)}$$

$$V_{out} = V_S - V_{diode}$$

$$= 21.2\text{ V} - 0.7\text{ V} = -20.5\text{ V}$$

$$V_{av} = 0.318 \times V_{out}$$

$$= 0.318 \times 20.5\text{ V} = -6.5\text{ V}$$

c. Output ripple frequency = input frequency = 60 Hz

5-3-2 Full-Wave Rectifiers

The half-wave rectifier's output is difficult to filter to a smooth dc level because the output voltage and current are applied to the load for only half of each input cycle. In this section we will examine two full-wave rectifier circuits which, as their name implies, switch both half-cycles (or the full ac input wave) of the input through the load in only one direction.

Center-Tapped Rectifier

The basic **full-wave center-tapped rectifier** circuit, which is shown in Figure 5-8(a), contains a center-tapped transformer and two diodes. The center tap of the transformer secondary is grounded (0 V) to create a 180° phase difference between the top and bottom of the secondary winding.

Full-Wave Center-Tapped Rectifier
A rectifier circuit that make use of a center tapped transformer to cause an output current to flow in the same direction during both half-cycles of the ac input.

When the ac input (V_{in}) swings positive, the circuit operates in the manner illustrated in Figure 5-8(b). A positive voltage is developed on the top of T_1 secondary turning ON D_1, while a negative voltage is developed on the bottom of T_1 secondary turning OFF D_2. This will permit a flow of electrons up through the load as indicated by the dashed current (I) line, developing a positive output half-cycle across the load (V_{out}).

When the ac input (V_{in}) swings negative, the circuit operates in the manner illustrated in Figure 5-8(c). A negative voltage is developed on the top of T_1 secondary turning OFF D_1, while a positive voltage is developed on the bottom of T_1 secondary turning ON D_2. This will permit a flow of electrons up through the load as indicated by the dashed current (I) line, again developing a positive output half-cycle across the load (V_{out}).

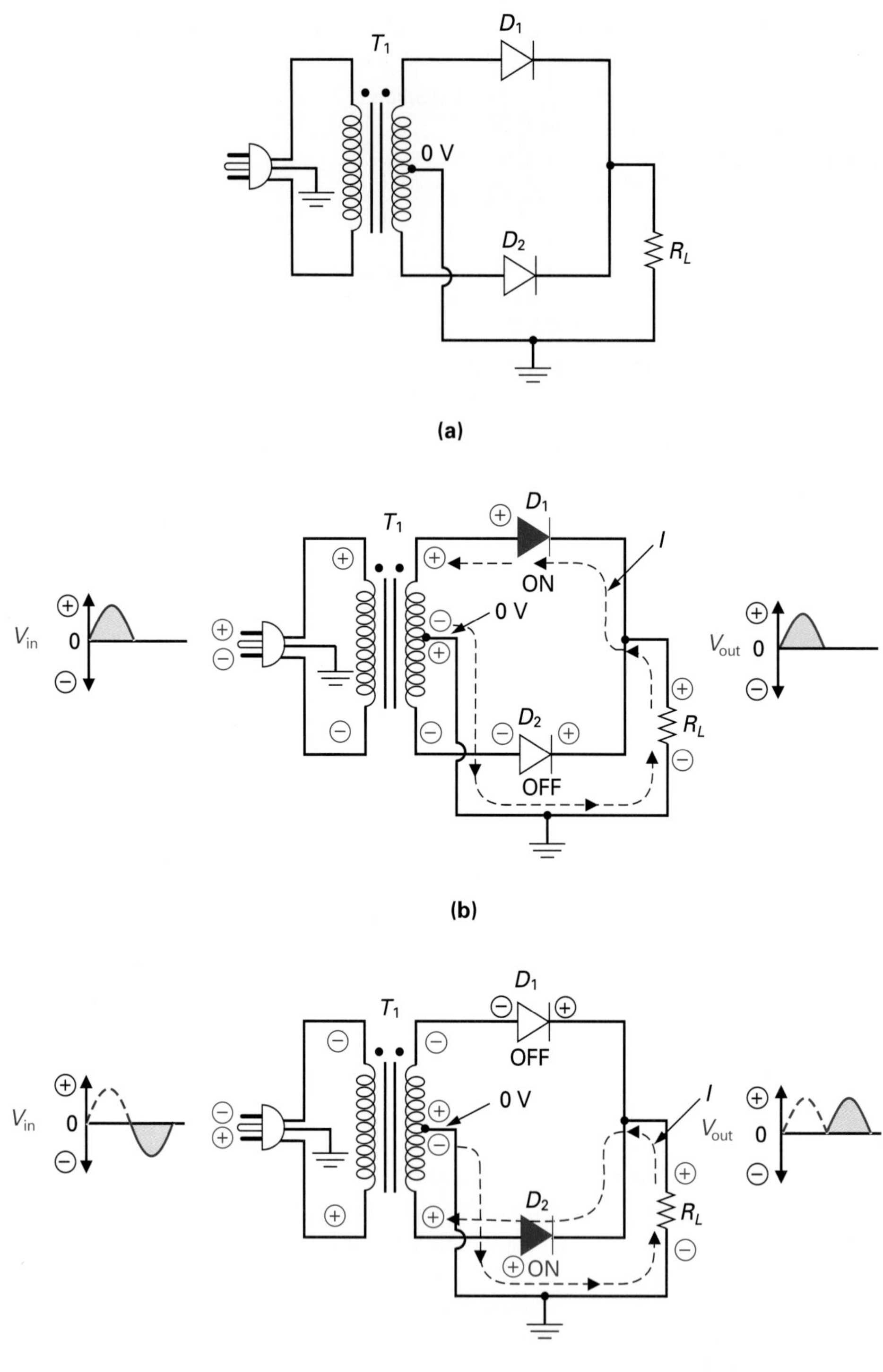

FIGURE 5-8 Full-Wave Center-Tapped Rectifier. (a) Basic Circuit. (b) Positive-Input Half-Cycle Operation. (c) Negative-Input Half-Cycle Operation.

The two diodes and center-tapped transformer in this circuit switch the two-direction ac current developed across the secondary through the load in only one direction and develop an output voltage across the load of the same polarity. The input/output voltage details of a full-wave center-tapped rectifier are shown in Figure 5-9. A 1:1 turns ratio has been selected in this example to emphasize how the center tap in the secondary of the transformer causes the secondary voltage (V_S) to be divided into two equal halves (V_{S1} and V_{S2}). Referring to the waveforms, you can see that the ac line voltage of 120 V rms, or 169.7 V_{peak} (V_{in}), is applied to the primary of the transformer T_1 (V_p). A 1:1 turns ratio ensures that this same voltage will be developed across the secondary winding ($V_S = V_p = 169.7\ V_{peak}$). This secondary voltage will be split in two as indicated by the waveforms V_{S1} and V_{S2}, which will each have a peak voltage of 1/2 V_S ($V_{S1\,peak} = V_{S2\,peak} = 1/2\ V_{S\,peak} = 1/2 \times 169.7\ V = 84.9\ V_{peak}$). Because V_{S1} will be connected across the output during one half-cycle, and V_{S2} will be connected across the output during the other half-cycle, the output voltage (V_{out}) will also have a peak voltage equal to 1/2 V_S or 84.9 V. Compared to a half-wave rectifier, therefore, whose peak output voltage is equal to the peak secondary voltage, the full-wave center-tapped rectifier seems to do worse, delivering a peak output of 1/2 V_S. However, the full-wave center-tapped rectifier compensates for the halving of the peak output voltage by doubling the number of half-cycles at the output, compared to a half-wave rectifier. The average output voltage (V_{av}) can be calculated for a full-wave rectifier with the following formula:

$$V_{av} = 0.636 \times \frac{1}{2} V_{S\,peak}$$

■ **EXAMPLE:**

Calculate the average output voltage from a half-wave and full-wave center-tapped rectifier if $V_{S\,peak} = 169.7$ V.

■ ***Solution:***

Half-wave rectifier:

$$V_{av} = 0.318 \times V_{S\,peak}$$
$$= 0.318 \times 169.7\ V = 54\ V$$

Full-wave center-tapped rectifier:

If $V_{S\,peak} = 169.7$ V, then $V_{S1\,peak}$ and $V_{S2\,peak} = 84.9$ V

$$V_{av} = 0.636 \times \frac{1}{2} V_{S\,peak}$$
$$= 0.636 \times 84.9\ V = 54\ V$$

As you can see from this example, even though the peak output of the full-wave center-tapped rectifier was half that of the half-wave rectifier, the average output was the same because the full-wave center-tapped rectifier doubles the number of half-cycles at the output, compared to a half-wave rectifier.

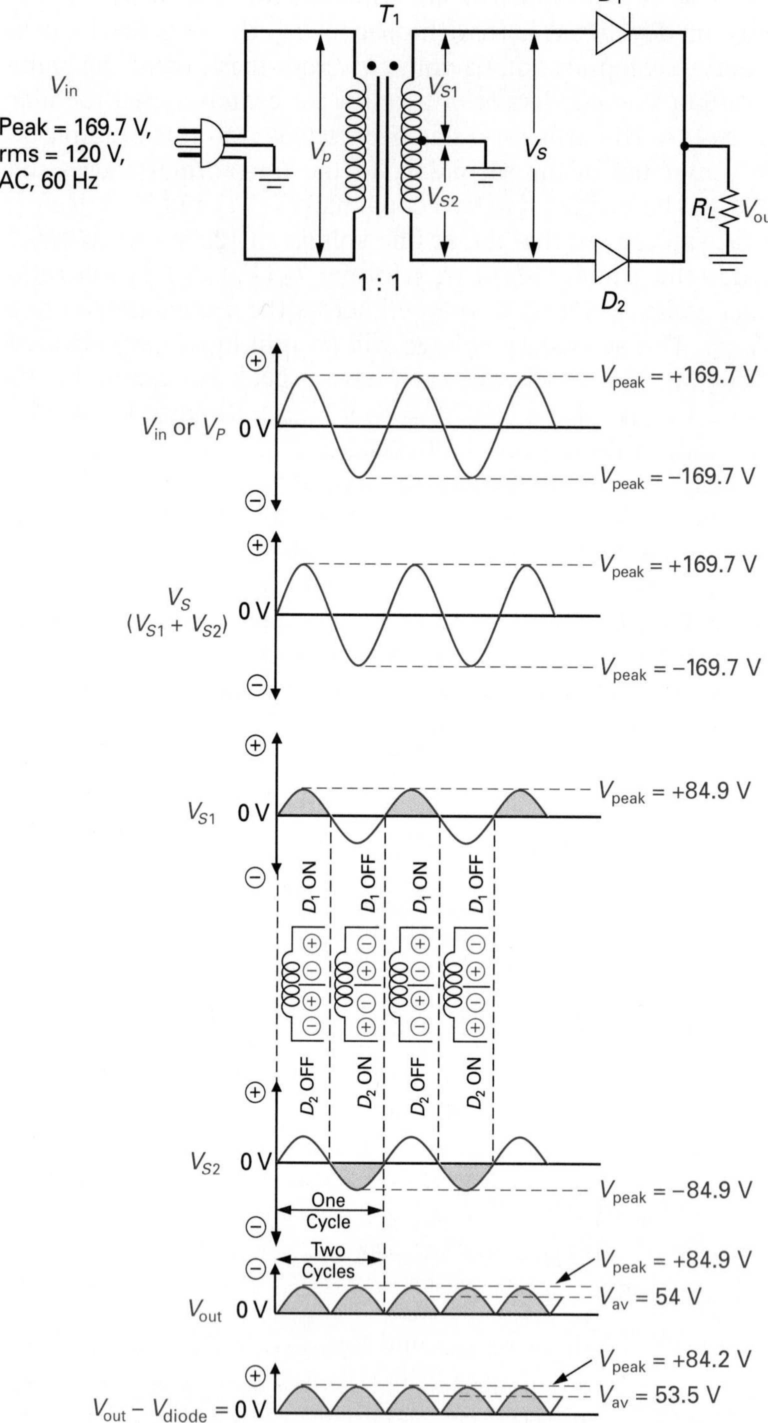

FIGURE 5-9 Input/Output Waveforms of a Full-Wave Center-Tapped Rectifier.

Because two half-cycles appear at the output for every one cycle at the input, as shown in Figure 5-9, the ripple frequency will be twice that of the input frequency, and this higher frequency will be easier to filter or smooth of fluctuations.

Output pulsating dc ripple frequency = 2 × input ac frequency

To be completely accurate, we should take into account the barrier voltage drop across the diodes. Since only one diode is ON for each half-cycle, the peak output voltage will only be less 0.7 V (silicon diode) as shown by the last waveform in Figure 5-9. Therefore

$$V_{out} = \frac{1}{2} V_{S\,peak} - 0.7\text{ V}$$

With regard to the peak inverse voltage, if you refer back to Figure 5-8(b), you will see that the full secondary peak voltage ($V_{S\,peak}$) appears across the OFF diode D_2 for one half cycle of the input. Similarly, the full $V_{S\,peak}$ voltage appears across the OFF D_1 during the other half cycle of the input, as shown in Figure 5-8(c). Both diodes must therefore have a peak inverse voltage (PIV) rating that is larger than the peak secondary voltage ($V_{S\,peak}$). For the circuit in Figure 5-9, the diode's maximum reverse voltage rating (PIV) must be greater than 169.7 V.

Bridge Rectifier

By center tapping the secondary of the transformer and having two diodes instead of one, we were able to double the ripple frequency and ease the filtering process. However, that seems to be the only advantage the full-wave center-tapped rectifier has over the half-wave rectifier, and the price we pay is having a more expensive center-tapped transformer and an extra diode.

With the **bridge rectifier circuit** shown in Figure 5-10(a), we can have the peak secondary output voltage of the half-wave circuit and the full-wave ripple frequency of the center-tapped circuit. This circuit was originally called a "bridge" rectifier because its shape resembled the framework of a suspension bridge. Figures 5-10(b) and (c) illustrate how the bridge rectifier circuit will behave when an ac input cycle is applied.

Bridge Rectifier Circuit
A full-wave rectifier circuit using four diodes that will convert an alternating voltage input into a direct voltage output.

When the ac input (V_{in}) swings positive, the circuit operates in the manner illustrated in Figure 5-10(b). A positive potential is applied to the top of the bridge, causing D_2 to turn ON, while a negative potential is applied to the bottom of the bridge, causing D_3 to turn ON. With D_2 and D_3 ON, and D_1 and D_4 OFF, electrons will flow up through the load as indicated by the dashed current line (I), developing a positive output half-cycle across the load (V_{out}).

When the ac input (V_{in}) swings negative, the circuit operates in the manner illustrated in Figure 5-10(c). A negative potential is applied to the top of the bridge, causing D_1 to turn ON, while a positive potential is applied to the bottom

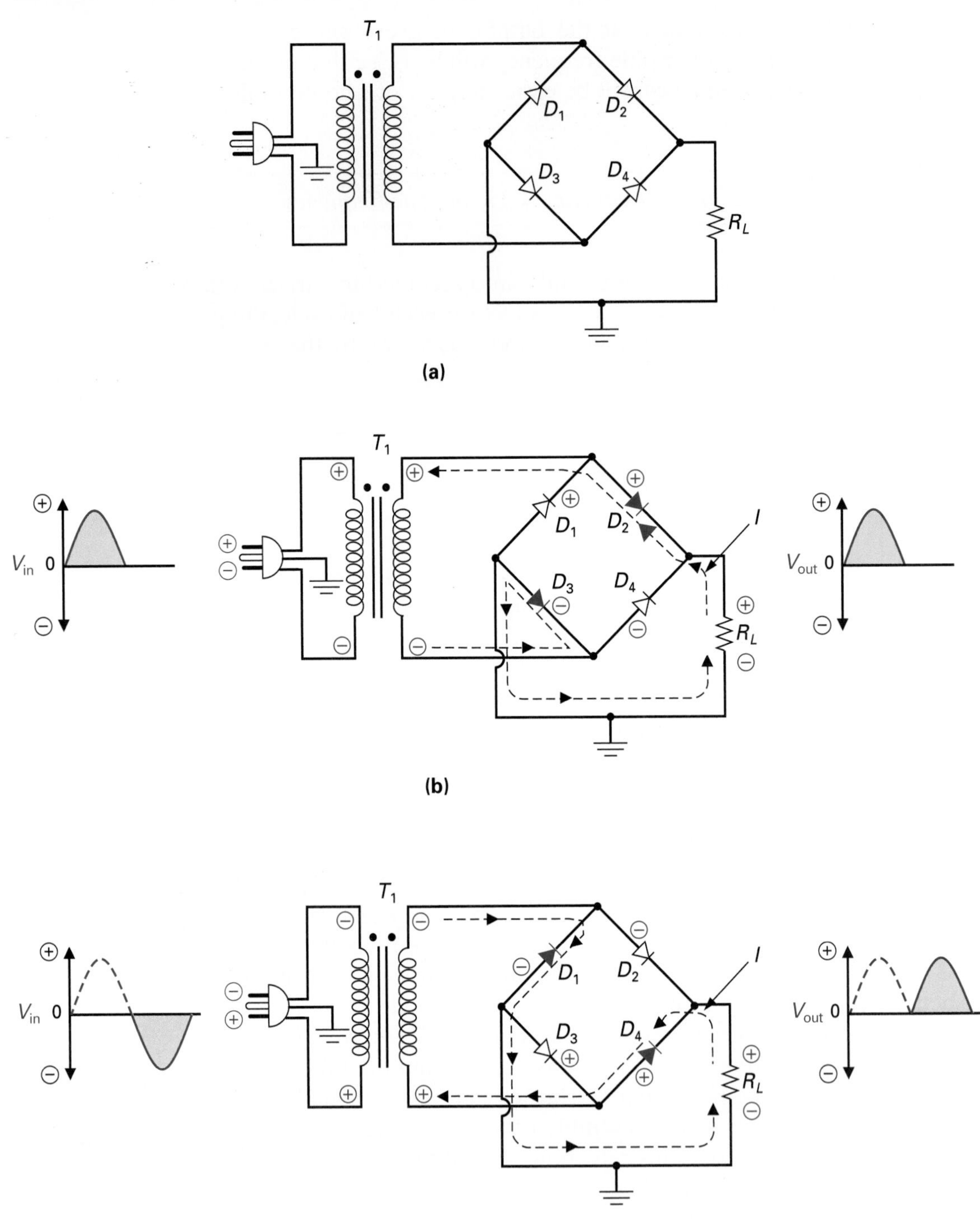

FIGURE 5-10 **Full-Wave Bridge Rectifier. (a) Basic Circuit. (b) Positive Half-Cycle Operation. (c) Negative Half-Cycle Operation.**

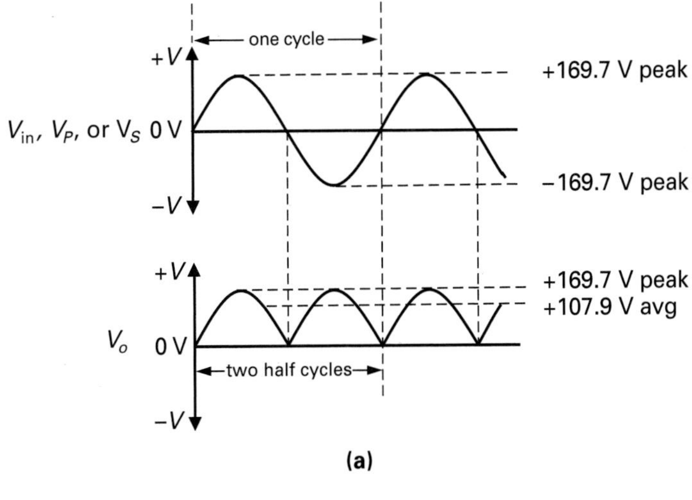

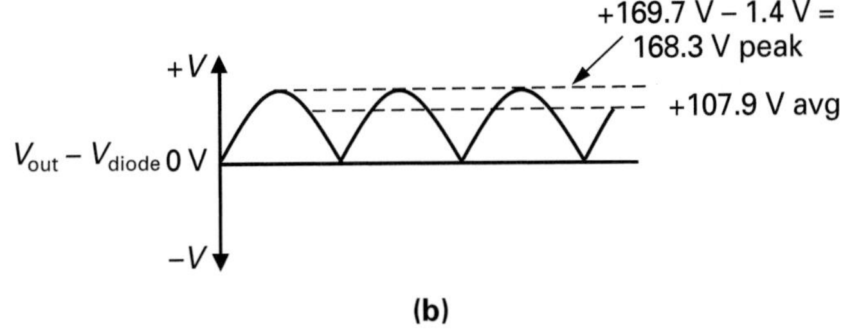

FIGURE 5-11 Input/Output Waveforms of a Full-Wave Bridge Rectifier.

of the bridge, causing D_4 to turn ON. With D_1 and D_4 ON and D_2 and D_3 OFF, electrons will flow up through the load as indicated by the dashed current line (I), again developing a positive output half-cycle across the load (V_{out}).

Like the center-tapped rectifier, the bridge rectifier switches two half-cycles of the same polarity through to the load. However, unlike the center-tapped rectifier, the bridge rectifier connects the total peak secondary voltage across the load. A secondary peak voltage of 169.7 V ac would produce a pulsating dc peak across the load of 169.7 V, or the same voltage, as shown in Figure 5-11(a). The average voltage in this case would be larger than that of a center-tapped rectifier, and equal to

$$V_{av} = 0.636 \times V_{S\,peak}$$

$$= 0.636 \times 169.7 \text{ V} = 107.9 \text{ V}$$

■ **EXAMPLE:**

Calculate the average output voltage from a center-tapped and bridge rectifier if $V_{S\,peak} = 169.7$ V.

■ ***Solution:***

Center-tapped:

$$V_{av} = 0.636 \times 1/2\ V_S$$
$$= 0.636 \times 84.9\ \text{V}$$
$$= 54\ \text{V}$$

Bridge:

$$V_{av} = 0.636 \times V_S$$
$$= 0.636 \times 169.7\ \text{V}$$
$$= 107.9\ \text{V}$$

As you can see from this example, unlike the center-tapped rectifier that only connects half of the peak secondary voltage across the load, the bridge rectifier connects the total peak secondary voltage across the load.

Because two half-cycles appear at the output for every one cycle at the input, as shown in Figure 5-11(a), the ripple frequency will be twice that of the input frequency, and this higher frequency will be easier to filter or smooth of fluctuations.

Output pulsating dc ripple frequency = 2 × input ac frequency

To be completely accurate, we should take into account the barrier voltage drop across the diodes. Because two diodes are ON for each half-cycle and both diodes are connected in series with the load, a 1.4 V (2 × 0.7 V) drop will occur between the peak secondary voltage (V_S) and the peak output voltage, as shown in Figure 5-11(b).

$$V_{out} = V_{S\,peak} - 1.4\ \text{V}$$

With regard to the peak inverse, if you refer to Figure 5-12, you can see that the full $V_{S\,peak}$ voltage appears across the two OFF diodes. Therefore, each

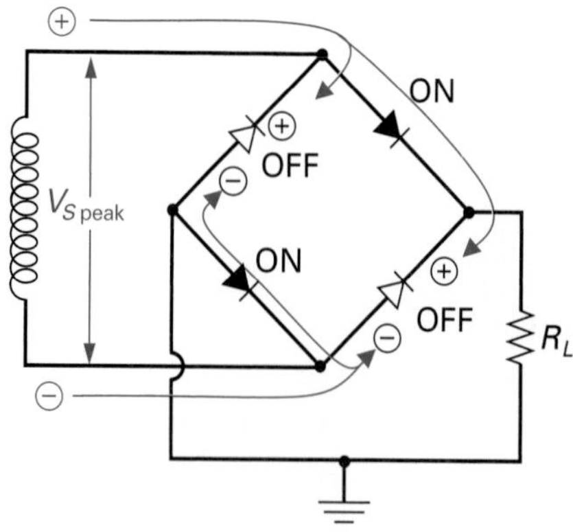

FIGURE 5-12 Peak Inverse Voltage.

diode must have a peak inverse voltage (PIV) rating that is greater than the peak secondary voltage (V_S). For the circuit and values given in Figures 5-10 and 5-11, the diode's maximum reverse voltage rating (PIV) must be greater than 169.7 V.

5-3-3 *Power Rectifier Device Data Sheets*

As previously mentioned, a P-N junction diode is often referred to as a power-rectifier diode because it is probably most often used as a rectifier in a dc power supply. These power-rectifier junction diodes must have higher ratings than the small-signal junction diodes discussed in Chapter 3 because they are used in power. For example, a typical small-signal diode will have the following ratings:

A forward voltage drop (V_F) of 0.7 V

A maximum forward current ($I_{F\,max}$) of 1 A

A maximum reverse voltage ($V_{R\,max}$) of 50 V

On the other hand, a power-rectifier diode, such as the series shown in Figure 5-13, is designed specifically for power supply circuit applications and will have the following much higher ratings:

A forward voltage drop (V_F) of about 1 V

A maximum forward current ($I_{F\,max}$) of 50 A

A maximum reverse voltage ($V_{R\,max}$) of up to 150 V

Figure 5-14 shows how semiconductor manufacturers have taken a single piece of semiconductor material and constructed an entire circuit. This integrated circuit (IC) module contains four power-rectifier diodes interconnected to form a full-wave bridge rectifier. Figure 5-14 shows how to connect the four terminals of the module. The two ac terminals are connected to the transformer's secondary, while the two dc terminals are connected across the load. The integration of components in this way, making "yesterday's circuit today's component," has the advantages of lower assembly costs (only one device to connect instead of four) and reduced equipment size. Since the module still contains four power-rectifier diodes, the formulas for a bridge rectifier still apply in exactly the same way.

SELF-TEST REVIEW QUESTIONS FOR SECTION 5-3

1. A rectifier converts an ____________ input into a ____________ output.
2. A ____________ -wave rectifier uses only one diode.
3. What would be the ripple frequency out of a half-wave rectifier if a 220 V 50 Hz ac input were applied?
4. Which full-wave rectifier circuit does not need to have a transformer?

DEVICE: Power Rectifier Diodes

DESCRIPTION

Designed to meet the efficiency demand of switching type power supplies, these devices are useful in many switching applications.

The low thermal resistance and forward voltage drop of this series allows the user to replace DO-5 size devices in many applications.

FEATURES

- Low Forward Voltage
- Very Fast Switching
- Low Thermal Resistance
- High Surge Capability
- Mechanically Rugged
- Both Polarities Available

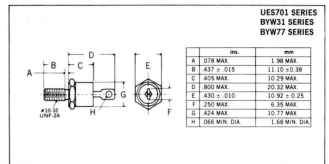

MECHANICAL SPECIFICATIONS

UES701 SERIES
BYW31 SERIES
BYW77 SERIES

	ins.	mm
A	.078 MAX.	1.98 MAX.
B	.437 ± .015	11.10 ±0.38
C	.405 MAX.	10.29 MAX.
D	.800 MAX.	20.32 MAX.
E	.430 ± .010	10.92 ± 0.25
F	.250 MAX.	6.35 MAX.
G	.424 MAX.	10.77 MAX.
H	.066 MIN. DIA.	1.68 MIN. DIA.

#10-32 UNF-2A

RECTIFIERS

High Efficiency, 25 A

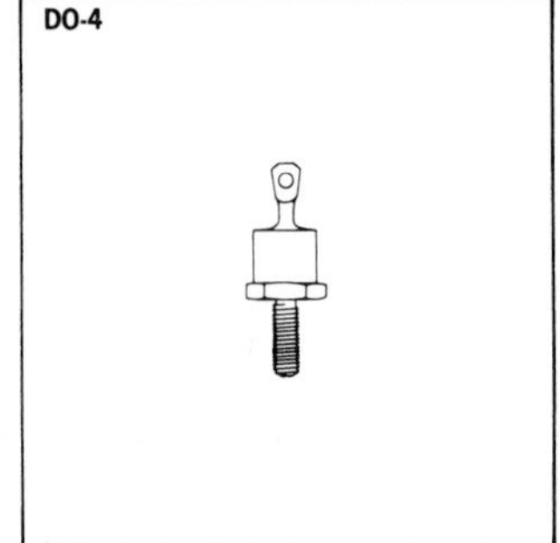

ABSOLUTE MAXIMUM RATINGS

	UES701	UES702	UES703
Peak Inverse Voltage, V_R	50V	100V	150V
Repetitive Peak Inverse Voltag, V_{RRM}	50V	100V	150V
Non-Repetitive Peak Inverse Voltage, V_{RSM}	50V	100V	150V
Maximum Average D.C. Output Current I_O @ T_C	25A @ 100°C		
RMS Forward Current, $I_{F(RMS)}$	40A		
Non-Repetitive Sinusoidal Surge Current (8.3mS), I_{FSM}	400A		
Thermal Resistance, Junction to Case, $R_{\theta JC}$	1.5°C/W		
Storage Temperature Range, T_{STG}	−55°C to +175°C		
Maximum Operating Junction Temperature, $T_{J\ MAX}$	+175°C		

ABSOLUTE MAXIMUM RATINGS

	BYW31-50	BYW31-100	BYW31-150	BYW77-50	BYW77-100	BYW77-150
Peak Inverse Voltage, V_R	50V	100V	150V	50V	100V	150V
Repetitive Peak Inverse Voltage, V_{RRM}	50V	100V	150V	50V	100V	150V
Non-Repetitive Peak Inverse Voltage, V_{RSM}	50V	100V	150V	50V	100V	150V
Maximum Average D.C. Output Current, I_o @ T_C = 100°C	25A @ 100°C			30A @ 107°C		
RMS Forward Current, $I_{F(RMS)}$	40A			50A		
Non-Repetitive Sinusoidal Surge Current (8.3mS), I_{FSM}	320A			500A		
Thermal Resistance, Junction to Case, $R_{\theta JC}$	1.5°C/W			1.5°C/W		
Storage Temperature Range, T_{STG}	−55°C to +150°C			−55°C to +150°C		
Maximum Operating Junction Temperature, $T_{J\ MAX}$	+150°C			+150°C		

ELECTRICAL SPECIFICATIONS

Type	Maximum Reverse Voltage V_R	Maximum Forward Voltage V_F		Maximum Reverse Current I_R		Maximum Reverse Recovery Time t_{RR}
		T_C = 25°C	T_C = 125°C	T_C = 25°C	T_C = 125°C	
UES701 UES702 UES703	50V 100V 150V	0.95V @ I_F = 25A	0.825V @ I_F = 25A	20μA @ Rated V_R	4mA @ Rated V_R	35ns(1)
		T_C = 25°C	T_C = 100°C	T_C = 25°C	T_C = 100°C	
BYW31-50 BYW31-100 BYW31-150	50V 100V 150V	1.3V @ I_F = 100A	0.85V @ I_F = 20A	20μA @ Rated V_R	2.5mA @ Rated V_R	50ns(2)
		T_C = 25°C	T_C = 100°C	T_C = 25°C	T_C = 100°C	
BYW77-50 BYW77-100 BYW77-150	50V 100V 150V	1.1V @ I_F = 63A	V_F / I_F: 0.75V / 10A 0.85V / 20A 1.2V / 100A	25μA @ Rated V_R	2.5mA @ Rated V_R	50ns(2)

FIGURE 5-13 Device Data Sheet for a Series of Power Rectifier Diodes. (Used with permission from Microsemi Corp.–Watertown. © 1982 Microsemi Corp.–Watertown, Watertown, MA, U.S.A.

DEVICE: GBPC 12, 15, 25 and 35 Series—Full-Wave Bridge Rectifier Modules

GBPC - W Wire leads

GBPC - Standard

FEATURES

- ♦ This series is UL recognized under component index, file number E54214
- ♦ The plastic package has Underwriters Laboratory flammability recognition 94V-0
- ♦ Integrally molded heatsink provide very low thermal resistance for maximum heat dissipation
- ♦ Universal 3-way terminals; snap-on, wire wrap-around, or P.C.B. mounting
- ♦ High forward surge current capabilities
- ♦ Chip junctions are glass passivated
- ♦ Typical I_R less than 0.3 μ A
- ♦ High temperature soldering guaranteed: 260°C /10 seconds at 5lbs., (2.3 kg) tension

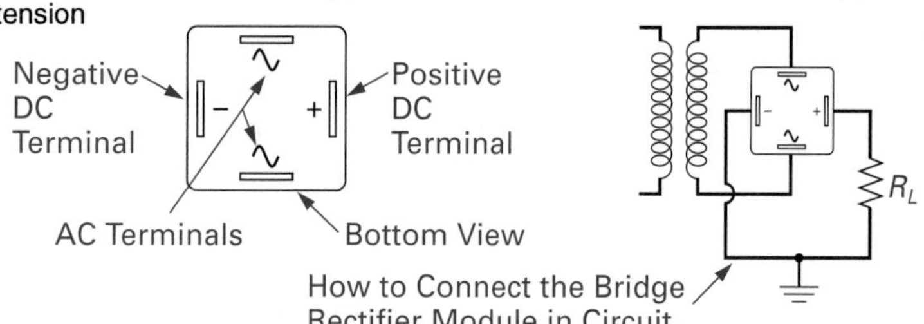

MAXIMUM RATINGS AND ELECTRICAL CHARACTERISTICS

Ratings at 25°C ambient temperature unless otherwise specified.

	SYMBOLS	GBPC12,15,25,35 005	01	02	04	06	08	10	UNITS
Maximum repetitive peak reverse voltage	V_{RRM}	50	100	200	400	600	800	1000	Volts
Maximum RMS voltage	V_{RMS}	35	70	140	280	420	560	700	Volts
Maximum DC blocking voltage	V_{DC}	50	100	200	400	600	800	1000	Volts
Maximum average forward rectified output current (SEE FIG.1) GBPC12 / GBPC15 / GBPC25 / GBPC35	$I_{(AV)}$	12.0 / 15.0 / 25.0 / 35.0							Amps
Peak forward surge current single sine-wave superimposed on rated load (JEDEC Method) GBPC12 / GBPC15 / GBPC25 / GBPC35	I_{FSM}	200.0 / 300.0 / 300.0 / 400.0							Amps
Rating (non-repetitive, for t greater than 1 ms and less than 8.3 ms) for fusing GBPC12 / GBPC15 / GBPC25 / GBPC35	I^2t	160.0 / 375.0 / 375.0 / 660.0							A^2sec
Maximum instantaneous forward voltage drop per leg at GBPC12 I_F=6.0A / GBPC15 I_F=7.5A / GBPC25 I_F=12.5A / GBPC35 I_F=17.5A	V_F	1.1							Volts
Maximum reverse DC current at rated DC blocking voltage per leg T_A=25°C / T_A=125°C	I_R	5.0 / 500.0							mA
RMS isolation voltage from case to leads	V_{ISO}	2500.0							Volts
Typical junction capacitance per leg (NOTE 1)	C_J	300.0							pF
Typical thermal resistance per leg (NOTE 2) GBPC12-25 / GBPC35	$R_{\Theta JC}$	1.9 / 1.4							°C/W
Operating junction storage temperature range	T_J, T_{STG}	-55 to +150							°C

FIGURE 5-14 Device Data Sheet for a Series of Full-Wave Bridge Rectifier Modules. (Courtesy of General Instrument Power Semiconductor Division.)

5-4 FILTERS

The filter in a dc power supply converts the pulsating dc output from the half-wave or full-wave rectifier into an unvarying dc voltage, as was shown in the waveforms in the basic block diagram in Figure 5-1. In this section, we will examine the three basic types of filters: the **capacitive filter,** the ***RC* filter,** and the ***LC* filter.**

Capacitive Filter
A capacitor used in a power supply filter system to suppress ripple currents while not affecting direct currents.

***RC* Filter**
A selective circuit which makes use of a resistance-capacitance network.

***LC* Filter**
A selective circuit which makes use of an inductance-capacitance network.

5-4-1 *The Capacitive Filter*

Figure 5-15(a) shows a positive-output half-wave rectifier and capacitive filter circuit. The waveforms in Figure 5-15(b), (c), and (d) show how the capacitive filter will respond to the half-wave pulsating dc output from the rectifier. Let us examine how the filter operates by referring to the dashed lines in the circuit in Figure 5-15(a), the output waveform from the rectifier shown in Figure 5-15(b), and the small-value capacitive filter waveform shown in Figure 5-15(c). When the ac input swings positive, the diode is turned ON and the capacitor charges as indicated by the black dashed charge current line in Figure 5-15(a). The charge time-constant will be small because no resistance exists in the charge path except for that of the resistance of the connecting wires (charge $\tau\downarrow = R\downarrow \times C$). This charge time is indicated by the gray shaded section between 0 and 90° in Figure 5-15(c). When the ac input begins to fall from its positive peak (at 90°), the diode is turned OFF by the large positive potential on the diode's cathode being supplied by the charged capacitor, and the decreasing positive potential on the diode's anode being supplied by the input. With the diode OFF, the capacitor begins to discharge as indicated by the colored dashed discharge current line in Figure 5-15(a). The discharge time constant is a lot longer than the charge time because of the load resistance (discharge $\tau\uparrow = R\uparrow \times C$). The decreasing slope in Figure 5-15(c) illustrates the decreasing voltage across the capacitor, and therefore across the load at the output (V_{out}), as the capacitor discharges. As the ac input and the rectifier's pulsating dc cycle repeat, the output (V_{out}) is, as shown in Figure 5-15(c), an almost constant dc output with a slight variation or ripple above and below the average value.

Figure 5-15(d) shows how a larger-value capacitor will make the charge and discharge time constants longer ($\tau\uparrow = R \times C\uparrow$), therefore decreasing the amount of ripple and increasing the average output voltage.

Figure 5-16(a) shows a positive output voltage full-wave bridge rectifier and capacitive filter circuit. Figure 5-16(b), (c), and (d) show the output waveforms from the rectifier and the filtered output from a small-value and a large-value capacitor filter. The ripple frequency from a full-wave rectifier is twice that of a half-wave rectifier, and therefore the capacitor does not have too much time to discharge before another positive half-cycle reoccurs. This results in a higher average voltage than in a half-wave circuit, and if a larger-value capacitor is used, the ripple will be even less because of the greater time constant ($\tau\uparrow = R \times C\uparrow$), causing an increased average voltage, as shown in Figure 5-16(d).

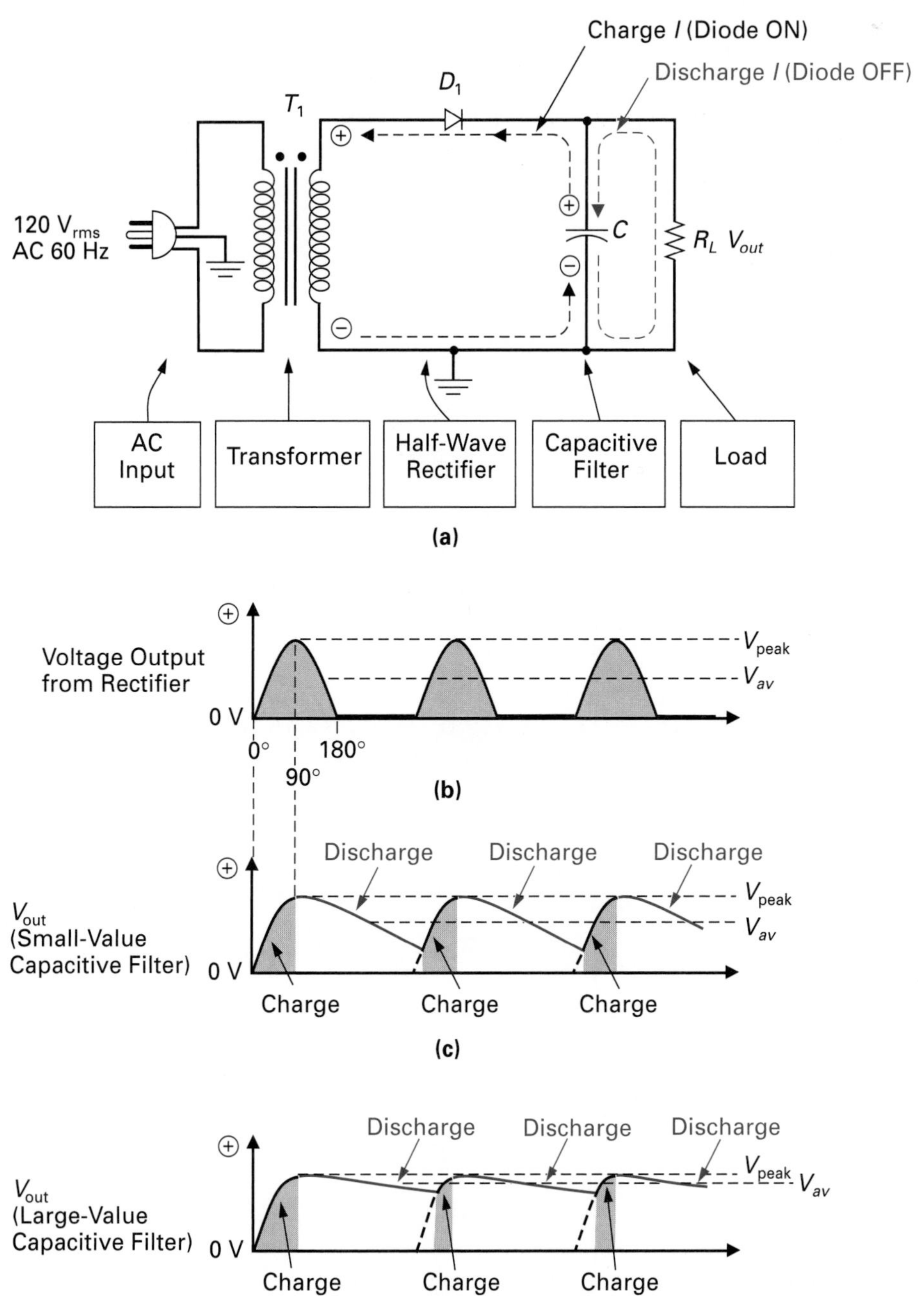

FIGURE 5-15 Capacitive Filtering of a Half-Wave Rectifier Output.

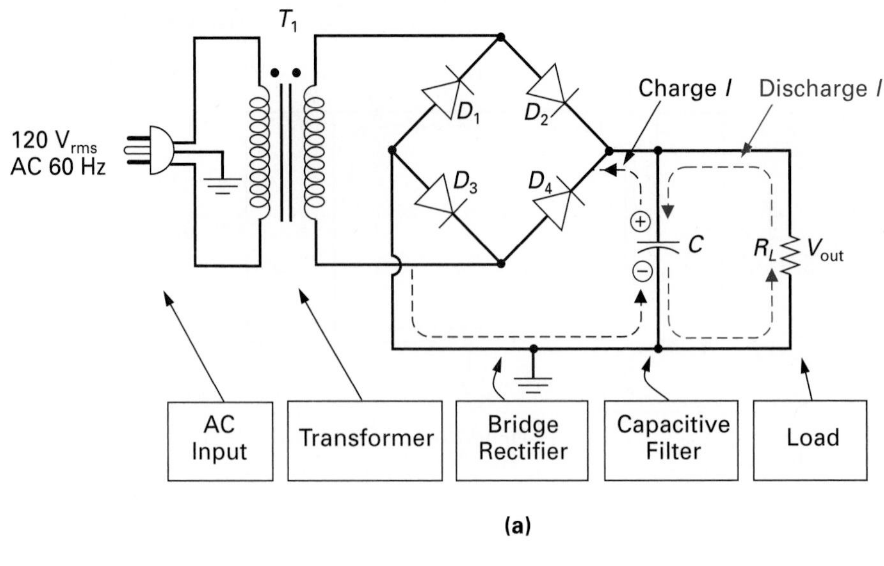

(a)

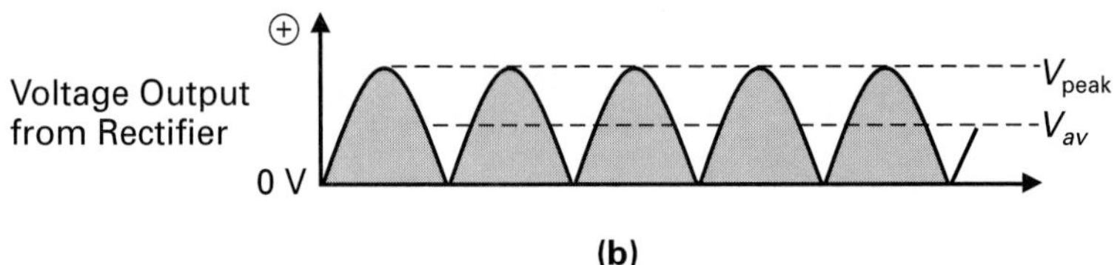

(b)

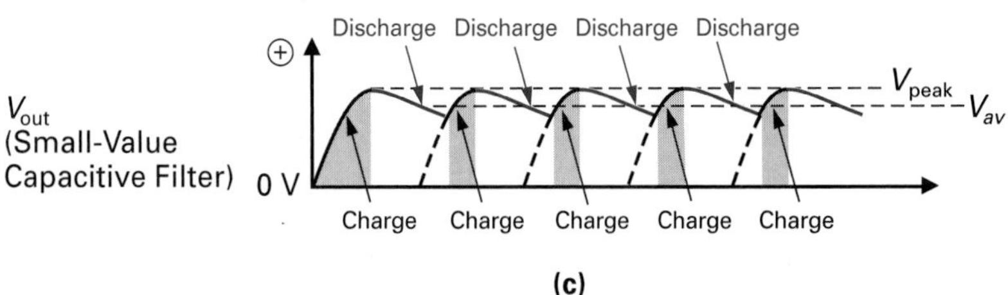

(c)

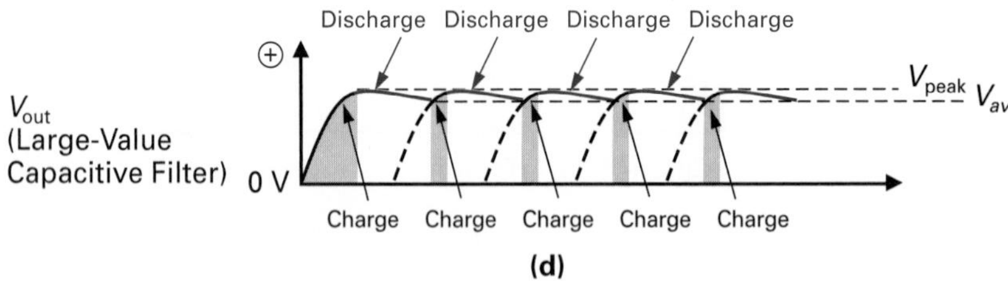

(d)

FIGURE 5-16 Capacitive Filtering of a Full-Wave Rectifier Output.

5-4-2 *Percent of Ripple*

In the previous two capacitive filter circuits, you could see that even though the output from a filter should be a constant dc level, there is a slight fluctuation or ripple. This fluctuation is called the percent ripple and its value is used to rate the action of the filter. It can be calculated with the following formula:

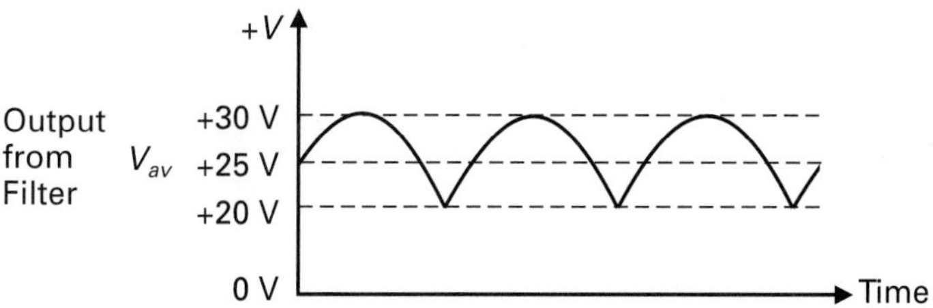

FIGURE 5-17 Percent of Ripple Out of a Filter.

$$\text{Percent ripple} = \frac{V_{rms} \text{ of ripple}}{V_{av} \text{ of ripple}} \times 100$$

To help show how this is calculated, let us apply this formula to the example in Figure 5-17. In this example, the filter's output is fluctuating between +20 and +30 V, and therefore the ripple's peak-to-peak value is

$$\text{Peak-to-peak of ripple} = 20 \text{ to } 30 \text{ V}$$
$$= 10 \text{ V}_{pk\text{-}pk}$$

A peak-to-peak value of 10 V means that the ripple has a peak value of

$$\text{Peak of ripple} = 1/2 \text{ of pk-pk value}$$
$$= 1/2 \text{ of } 10 \text{ V}$$
$$= 5 \text{ V}_{pk}$$

Once the peak of the ripple is known, the ripple rms value can be calculated:

$$\text{RMS of ripple} = 0.707 \text{ of peak}$$
$$= 0.707 \text{ of } 5 \text{ V}$$
$$= 3.54 \text{ V}$$

The average of the ripple is approximately midway between peaks, which in the example in Figure 5-17 is +25 V. Inserting these values in the formula, we can calculate the percent of ripple:

$$\% \text{ ripple} = \frac{V_{rms} \text{ of ripple}}{V_{av} \text{ of ripple}} \times 100$$
$$= \frac{3.54 \text{ V}}{25 \text{ V}} \times 100$$
$$= 14\%$$

This means that the average voltage out of the rectifier (+25 V) will fluctuate 14%.

5-4-3 *RC Filters*

The low-pass pi (π) filter shown in Figure 5-18 could be used to further reduce ripple, due to its ability to pass dc but block any fluctuations. It is called a π filter because it contains two vertical parts and one horizontal part

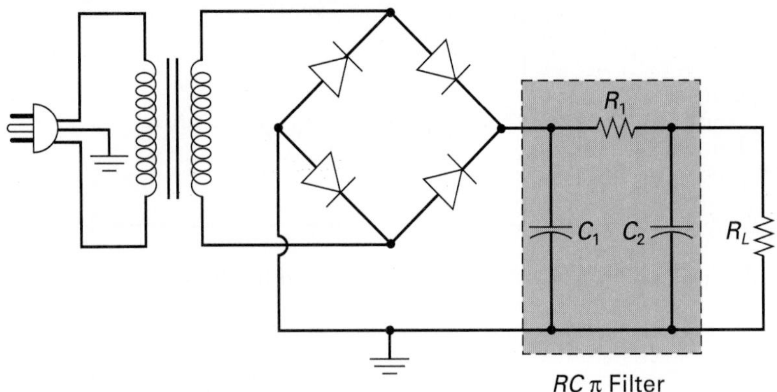

FIGURE 5-18 ***RC* π Filter.**

and therefore resembles π. The first capacitor, C_1, acts in the same manner as the capacitor in the capacitive filter. The components R_1 and C_2, however, form a voltage divider. Because capacitive reactance increases as frequency decreases ($X_C\uparrow$ is inversely proportional to $f\downarrow$), most of the dc voltage will appear across C_2 and be applied to the load, while little of the ac ripple will appear across C_2 and be applied to the load. In summary, the capacitor blocks the dc component and directs it to the load while shunting the ac component away from the load.

5-4-4 *LC Filters*

Replacing R_1 in the *RC* π filter with an inductor L_1, as shown in Figure 5-19, will increase the efficiency of the filter because the inductor will oppose current without generating heat. The inductor, L_1, will offer a very low opposition or reactance to dc ($X_L\downarrow \propto f\downarrow$) and a very high reactance to the ac ripple ($X_L\uparrow \propto f\uparrow$). Although less power will be wasted, the inductor will be more expensive than the resistor.

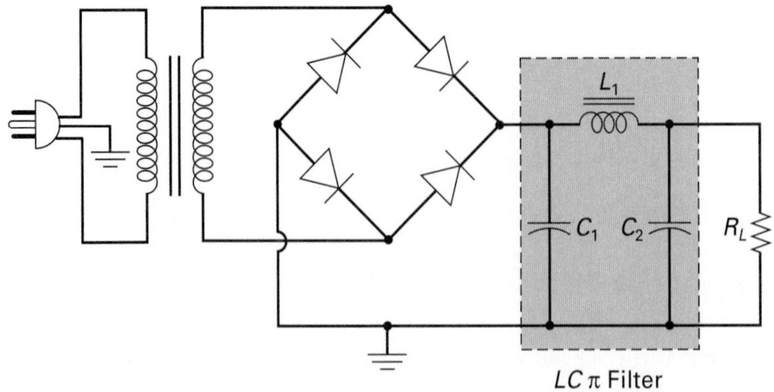

FIGURE 5-19 ***LC* π Filter.**

SELF-TEST REVIEW QUESTIONS FOR SECTION 5-4

1. Would the filter in a dc power supply be considered:
   a. High pass
   b. Low pass
   c. Bandpass
   d. Band stop
2. The percent ripple of a filter is proportional/inversely proportional to the value of the capacitor.
3. Which component would be connected in shunt, and which component would be connected in series to form a low-pass *RC* filter?
4. The most basic type of power supply filter is a capacitor in ________ with the load. (series/parallel)

5-5 REGULATORS

The **voltage regulator** in a dc power supply maintains the dc output voltage constant despite variations in the ac input voltage and the output load resistance. To help explain why these variations occur, let us first examine the relationship between a source and its load.

5-5-1 *Source and Load*

Ideally, a dc power supply should convert all of its *ac electrical energy input* into a *dc electrical energy output.* However, like all devices, circuits, and systems, a dc power supply is not 100% efficient and, along with the electrical energy output, the dc power supply generates wasted heat. This is why a dc power supply can be represented as a voltage source with an internal resistance (R_{int}), as shown in Figure 5-20(a). The internal resistance, which is normally very small (in this example, 1 Ω), represents the heat energy loss or inefficiency of the dc power supply. The load resistance (R_L) represents the resistance of the electronic circuits in the electronic system. Since R_{int} and R_L are connected in series with one another, the 10 V supply in this example will be developed proportionally across these resistors, resulting in

$$V_{R_{int}} = 0.1 \text{ V}$$
$$V_{R_L} = \frac{9.9 \text{ V}}{10 \text{ V total}}$$

If the load resistance (R_L) were to decrease dramatically to 1 Ω, as shown in Figure 5-20(b), the decrease in load resistance ($R_L\downarrow$) would cause an increase in load current ($I_L\uparrow$). This current increase would cause an increase in the heat generated by the power supply ($P_{R_{int}}\uparrow = I^2\uparrow \times R$). This loss is seen at the output of the dc power supply, which is now delivering only a 5 V output to the load because the 10 V source voltage is being divided equally across R_{int} and R_L.

$$V_{R_{int}} = 5 \text{ V}$$

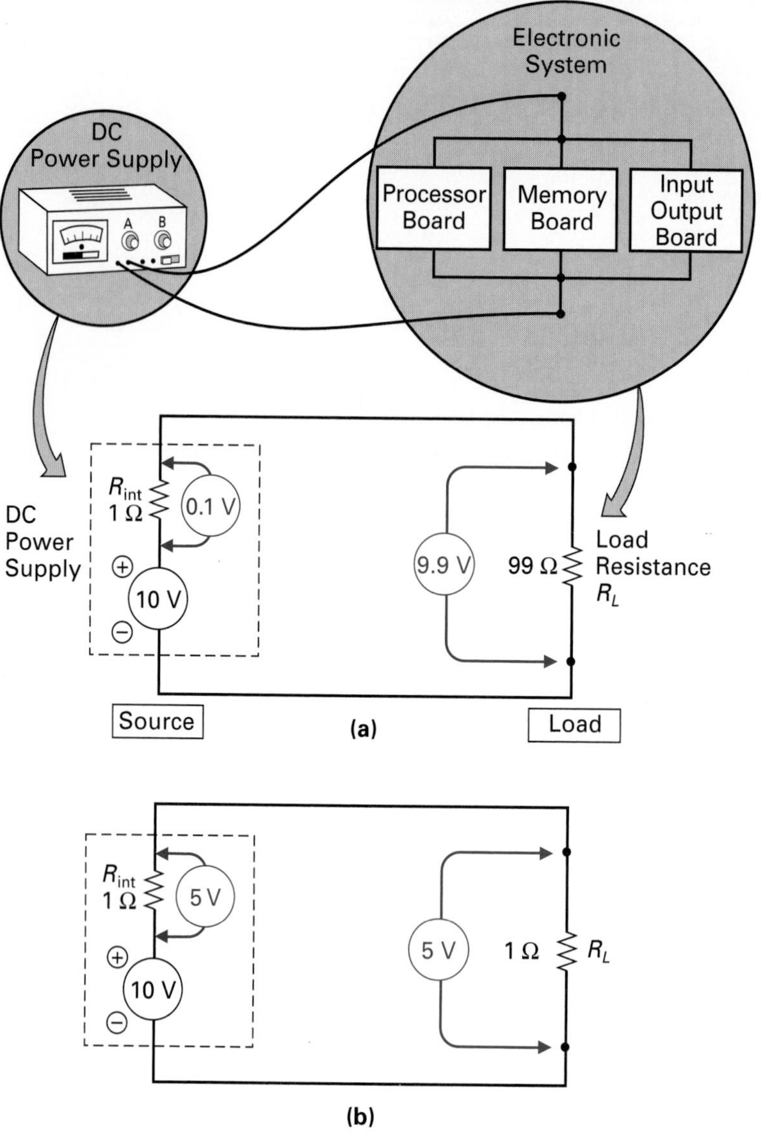

FIGURE 5-20 Source and Load.

$$V_{R_L} = \frac{5\ \text{V}}{10\ \text{V total}}$$

If the load resistance decreases, the load current will increase, and if too much current is drawn from the source, it will pull down the source voltage. Although a load resistance will generally not change as dramatically as shown in Figure 5-20(b), it will change slightly as different control settings are selected, and this will *load the source*. For example, the load resistance of the electronic circuits in your music system will change as the volume, bass, treble, and other controls are adjusted. This load resistance change would affect the output of a dc power supply if a regulator were not included to maintain the output voltage constant despite variations in load resistance.

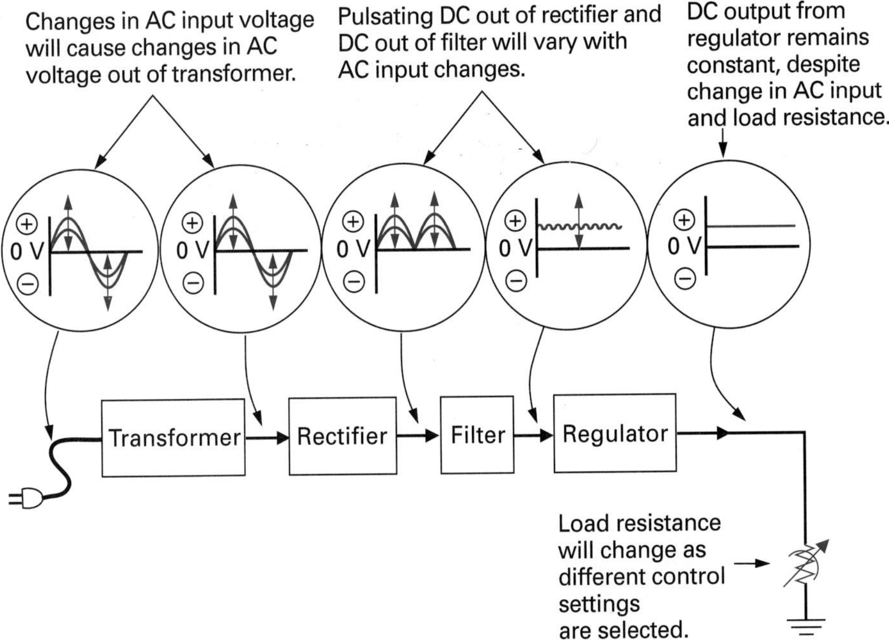
Changes in AC input voltage will cause changes in AC voltage out of transformer.
Pulsating DC out of rectifier and DC out of filter will vary with AC input changes.
DC output from regulator remains constant, despite change in AC input and load resistance.
0 V
0 V
0 V
0 V
0 V
Transformer
Rectifier
Filter
Regulator
Load resistance will change as different control settings are selected.

FIGURE 5-21 Function of a Regulator.

To the electric company, each user appliance is a small part of its load, and it is the source. The ac voltage from the electric company can be anywhere between 105 V and 125 V ac, rms (148 to 177 V ac, peak), depending on consumer use, which depends on the time of day. When many appliances are in use, the load current will be high and the overall load resistance low, causing the source voltage to be pulled down. If a dc power supply did not have a regulator, these different input ac voltages from the electric company would produce different dc output voltages, when we really want the dc supply voltage to the electronic circuits to always be the same value no matter what ac input is present. This is the other reason that a regulator is included in a dc power supply; it maintains the dc output voltage constant despite variations in the ac input voltage.

In summary, a regulator is included in a dc power supply to maintain the dc output voltage constant despite variations in the output load resistance and the ac input voltage, as shown in Figure 5-21.

5-5-2 *Percent of Regulation*

The percent of regulation is a measure of the regulator's ability to regulate—or maintain constant—the output dc voltage. It is calculated with the formula

$$\text{Percent regulation} = \frac{V_{nl} - V_{fl}}{V_{nl}} \times 100$$

V_{nl} = no-load voltage

V_{fl} = full-load voltage

If we apply this formula to the example in Figure 5-20, we get

$$\% \text{ regulation} = \frac{V_{nl} - V_{fl}}{V_{nl}} \times 100$$

$$= \frac{10\text{ V} - 5\text{ V}}{10\text{ V}} \times 100 = 50\%$$

In the ideal situation, a regulator would be included and would maintain the output voltage constant between no-load and full-load, resulting in a percent regulation figure of

$$\% \text{ regulation} = \frac{V_{nl} - V_{fl}}{V_{nl}} \times 100$$

$$= \frac{10\text{ V} - 10\text{ V}}{10\text{ V}} \times 100$$

$$= 0\%$$

Most regulators achieve a percent regulation figure that is not perfect (0%) but is generally in single digits (2% to 8%).

5-5-3 *Zener Regulator*

In Chapter 4, it was shown how the zener diode could be used as a voltage regulator. Figure 5-22 shows how a zener diode (D_5) and a series resistor (R_1) would be connected in a dc power supply circuit to provide regulation. An increase in the ac input voltage would cause an increase in the dc output from the filter, which would cause an increase in the current through the series resis-

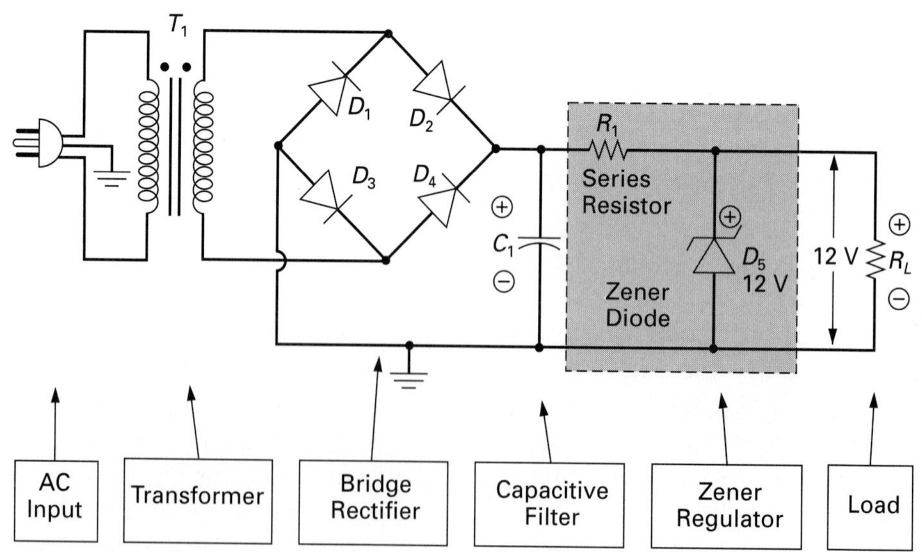

FIGURE 5-22 Zener Diode Regulator.

tor (I_S), the zener diode (I_Z), and the load (I_L). The reverse-biased zener diode will, however, maintain a constant voltage across its anode and cathode (in this example, 12 V) despite these input voltage and current variations, with the additional voltage being dropped across the series resistor. For example, if the output from the filter could be between +15 and +20 V, the zener diode would always drop 12 V, while the series resistor would drop between 3 V (when input is 15 V) and 8 V (when input is 20 V). Consequently, the zener regulator will maintain a constant output voltage despite variation in the ac input voltage.

From the other standpoint, the zener regulator will also maintain a constant output voltage despite variations in load resistance. The zener diode achieves this by increasing and decreasing its current (I_Z) in response to load resistance changes. Despite these changes in zener and load current however, the zener voltage (V_Z), and therefore the output voltage (V_{out} or V_{R_L}), always remains constant. The disadvantage with this regulator is that the series connected resistor will limit load current and, in addition, generate unwanted heat.

5-5-4 *The IC Regulator*

Most dc power supply circuits today make use of integrated circuit (IC) regulators, such as the one shown in Figure 5-23. These IC regulators contain about 50 individual or discrete components all integrated on one silicon semiconductor chip and then encapsulated in a three-pin package. The inset in Figure 5-23 shows how all of the IC regulator's internal components form circuits which are shown as blocks. For example, a short-circuit protection and thermal shutdown circuit will protect the IC regulator by turning it OFF if the load current drawn exceeds the regulator's rated current, or if the heat sink is too small and the IC regulator is generating heat faster than it can dissipate it.

The three terminals of the IC regulator are labeled "input" (pin 1), "output" (pin 3), and "ground" (pin 2). The package is generally given the identification code of either 78XX or 79XX. The 78XX (seventy-eight hundred) series of regulators are used to supply a positive output voltage, with the last two digits specifying the output voltage. For example, 7805 = +5 V, 7812 = +12 V, and so on. The 79XX (seventy-nine hundred) series of regulators are used to supply a negative output voltage, with the last two digits specifying the output voltage. These regulators will deliver a constant regulated output voltage as long as the input voltage is greater than the regulator's rated output voltage and can deliver a maximum output current of up to 1.5 A if properly heat sunk.

Figure 5-24 shows the manufacturer's data sheet for the 7800 series of voltage regulators. Referring to the device's internal circuitry, you can see the large number of components which are incorporated within this single three-terminal package, including a zener diode which is being used to provide an internal voltage reference. Additional voltage regulator ICs will be discussed in Chapter 18.

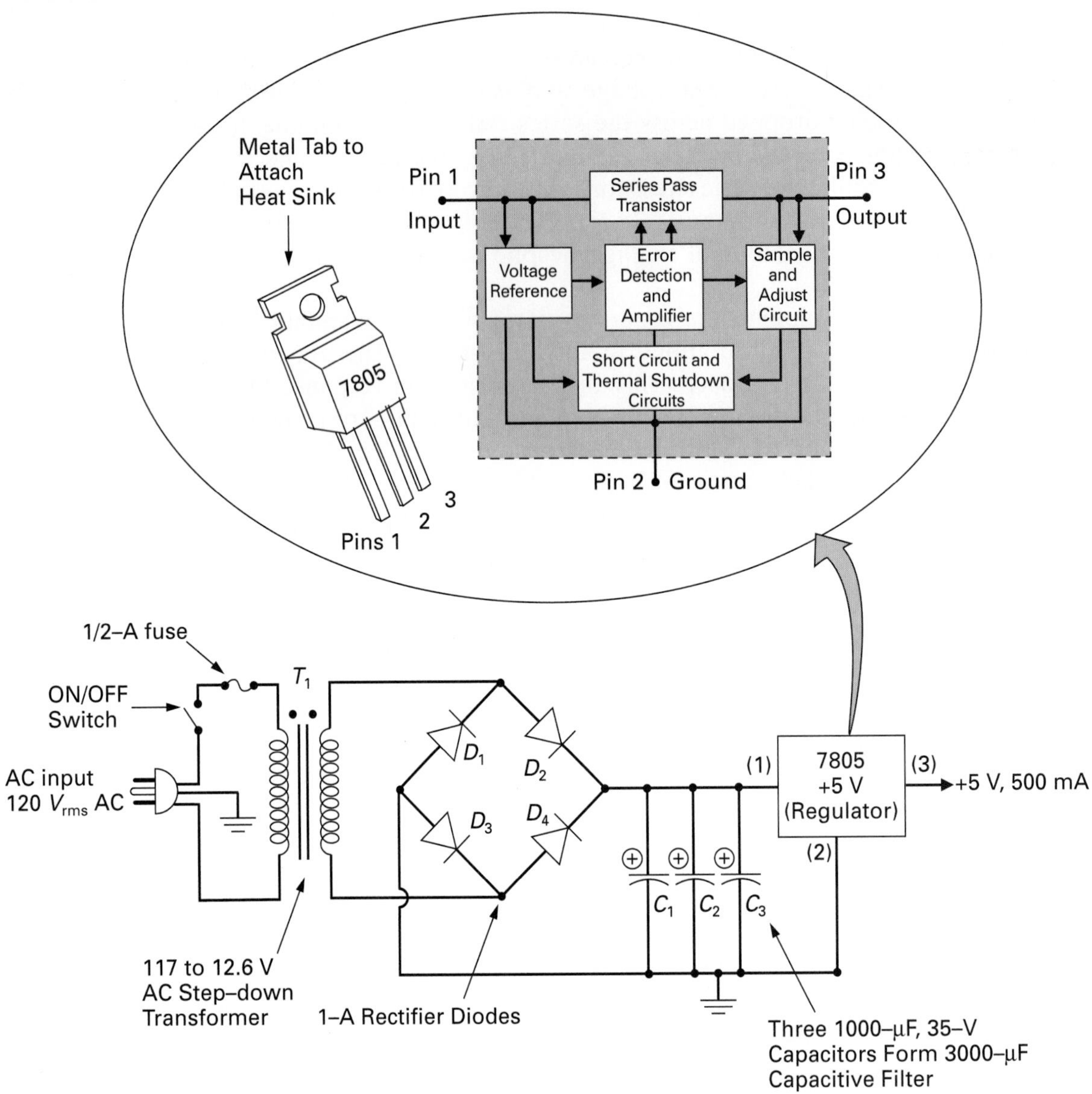

FIGURE 5-23 Integrated Circuit Regulator within a Power Supply Circuit.

SELF-TEST REVIEW QUESTIONS FOR SECTION 5-5

1. Why is a voltage regulator needed in a dc power supply?
2. Normally, would a large or small load resistance pull down the source voltage?
3. Is the zener diode in a zener regulator circuit connected in series or in parallel with the load?
4. Indicate the output voltages of the following IC regulators:
   a. 7812 **b.** 7905 **c.** 7815 **d.** 7912

DEVICE: LM78XX Series—IC Voltage Regulators

General Description

The LM78XX series of three terminal regulators is available with several fixed output voltages making them useful in a wide range of applications. One of these is local on card regulation, eliminating the distribution problems associated with single point regulation. The voltages available allow these regulators to be used in logic systems, instrumentation, HiFi, and other solid state electronic equipment. Although designed primarily as fixed voltage regulators these devices can be used with external components to obtain adjustable voltages and currents.

The LM78XX series is available in an aluminum TO-3 package which will allow over 1.0A load current if adequate heat sinking is provided. Current limiting is included to limit the peak output current to a safe value. Safe area protection for the output transistor is provided to limit internal power dissipation. If internal power dissipation becomes too high for the heat sinking provided, the thermal shutdown circuit takes over preventing the IC from overheating.

Considerable effort was expanded to make the LM78XX series of regulators easy to use and mininize the number of external components. It is not necessary to bypass the output, although this does improve transient response. Input bypassing is needed only if the regulator is located far from the filter capacitor of the power supply.

For output voltage other than 5V, 12V and 15V the LM117 series provides an output voltage range from 1.2V to 57V.

Features

- Output current in excess of 1A
- Internal thermal overload protection
- No external components required
- Output transistor safe area protection
- Internal short circuit current limit
- Available in the aluminum TO-3 package

Voltage Range

LM7805C	5V
LM7812C	12V
LM7815C	15V

Absolute Maximum Ratings

Input Voltage (V_O = 5V, 12V and 15V)	35V
Internal Power Dissipation (Note 1)	Internally Limited
Operating Temperature Range (T_A)	0°C to + 70°C
Maximum Junction Temperature	
(K Package)	150°C
(T Package)	150°C
Storage Temperature Range	−65°C to + 150°C
Lead Temperature (Soldering, 10 sec.)	
TO-3 Package K	300°C
TO-220 Package T	230°C

Metal Can Package
TO-3 (K)
Aluminum

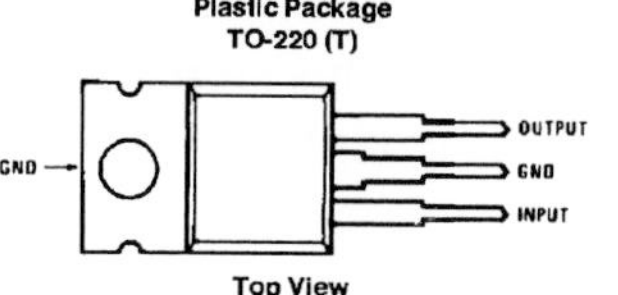

Bottom View

Plastic Package
TO-220 (T)

GND — OUTPUT — GND — INPUT

Top View

Schematic and Connection Diagrams

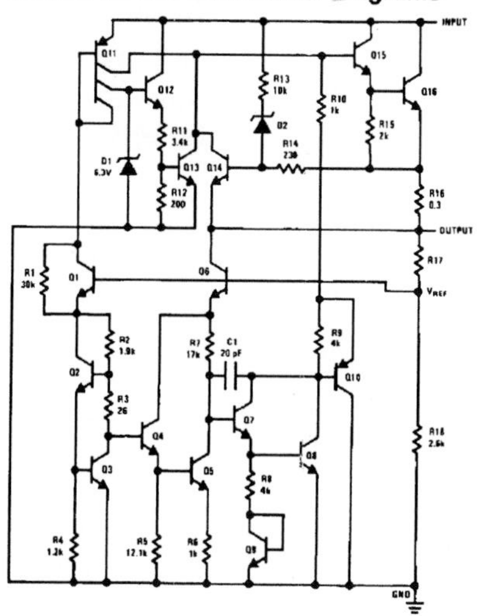

Electrical Characteristics LM78XXC (Note 2) 0°C ≤ Tj ≤ 125°C unless otherwise noted.

Output Voltage				5V			12V			15V			Units
Input Voltage (unless otherwise noted)				10V			19V			23V			
Symbol	Parameter	Conditions		Min	Typ	Max	Min	Typ	Max	Min	Typ	Max	
V_O	Output Voltage	Tj = 25°C, 5 mA ≤ I_O ≤ 1A		4.8	5	5.2	11.5	12	12.5	14.4	15	15.6	V
		P_D ≤ 15W, 5 mA ≤ I_O ≤ 1A		4.75		5.25	11.4		12.6	14.25		15.75	V
		V_{MIN} ≤ V_{IN} ≤ V_{MAX}			(7.5 ≤ V_{IN} ≤ 20)			(14.5 ≤ V_{IN} ≤ 27)			(17.5 ≤ V_{IN} ≤ 30)		V
ΔV_O	Line Regulation	I_O = 500 mA	Tj = 25°C		3	50		4	120		4	150	mV
			ΔV_{IN}		(7 ≤ V_{IN} ≤ 25)			14.5 ≤ V_{IN} ≤ 30)			(17.5 ≤ V_{IN} ≤ 30)		V
			0°C ≤ Tj ≤ +125°C			50			120			150	mV
			ΔV_{IN}		(8 ≤ V_{IN} ≤ 20)			(15 ≤ V_{IN} ≤ 27)			(16.5 ≤ V_{IN} ≤ 30)		V
		I_O ≤ 1A	Tj = 25°C			50			120			150	mV
			ΔV_{IN}		(7.5 ≤ V_{IN} ≤ 20)			(14.6 ≤ V_{IN} ≤ 27)			(17.7 ≤ V_{IN} ≤ 30)		V
			0°C ≤ Tj ≤ +125°C			25			60			75	mV
			ΔV_{IN}		(8 ≤ V_{IN} ≤ 12)			(16 ≤ V_{IN} ≤ 22)			(20 ≤ V_{IN} ≤ 26)		V
ΔV_O	Load Regulation	Tj = 25°C	5 mA ≤ I_O ≤ 1.5A		10	50		12	120		12	150	mV
			250 mA ≤ I_O ≤ 750 mA			25			60			75	mV
		5 mA ≤ I_O ≤ 1A, 0°C ≤ Tj ≤ +125°C				50			120			150	mV
I_Q	Quiescent Current	I_O ≤ 1A	Tj = 25°C			8			8			8	mA
			0°C ≤ Tj ≤ +125°C			8.5			8.5			8.5	mA
ΔI_Q	Quiescent Current Change	5 mA ≤ I_O ≤ 1A				0.5			0.5			0.5	mA
		Tj = 25°C, I_O ≤ 1A				1.0			1.0			1.0	mA
		V_{MIN} ≤ V_{IN} ≤ V_{MAX}			(7.5 ≤ V_{IN} ≤ 20)			(14.8 ≤ V_{IN} ≤ 27)			(17.9 ≤ V_{IN} ≤ 30)		V
		I_O ≤ 500 mA, 0°C ≤ Tj ≤ +125°C				1.0			1.0			1.0	mA
		V_{MIN} ≤ V_{IN} ≤ V_{MAX}			(7 ≤ V_{IN} ≤ 25)			(14.5 ≤ V_{IN} ≤ 30)			(17.5 ≤ V_{IN} ≤ 30)		V
V_N	Output Noise Voltage	T_A = 25°C, 10 Hz ≤ f ≤ 100 kHz			40			75			90		μV
$\frac{\Delta V_{IN}}{\Delta V_{OUT}}$	Ripple Rejection	f = 120 Hz	I_O ≤ 1A, Tj = 25°C or	62	80		55	72		54	70		dB
			I_O ≤ 500 mA, 0°C ≤ Tj ≤ +125°C	62			55			54			dB
		V_{MIN} ≤ V_{IN} ≤ V_{MAX}			(8 ≤ V_{IN} ≤ 18)			(15 ≤ V_{IN} ≤ 25)			(18.5 ≤ V_{IN} ≤ 28.5)		V
R_O	Dropout Voltage	Tj = 25°C, I_{OUT} = 1A			2.0			2.0			2.0		V
	Output Resistance	f = 1 kHz			8			18			19		mΩ
	Short-Circuit Current	Tj = 25°C			2.1			1.5			1.2		A
	Peak Output Current	Tj = 25°C			2.4			2.4			2.4		A
	Average TC of V_{OUT}	0°C ≤ Tj ≤ +125°C, I_O = 5 mA			0.6			1.5			1.8		mV/°C
V_{IN}	Input Voltage Required to Maintain Line Regulation	Tj = 25°C, I_O ≤ 1A			7.5		14.6			17.7			V

For example, the output voltage of a 7805 could be between 4.8 V to 5.2 V for a 10 V input. The output voltage therefore has a 4% tolerance.

For example, the output of a 7805 will change a maximum of 50 mV as the load current changes from 5 mA to 1.5 A, and will change a maximum of 25 mV as load current changes from 250 mA to 750 mA.

FIGURE 5-24 Device Data Sheet for the LM78XX Series of Voltage Regulators. (Courtesy of National Semiconductor Corporation.)

5-6 TROUBLESHOOTING A DC POWER SUPPLY

In this section we will discuss how to troubleshoot a typical dc power supply circuit. To begin with, however, let us review the three-step troubleshooting procedure introduced in the previous chapter. The procedure for fixing a failure can be broken down into three basic steps:

Step 1: DIAGNOSE

The first step is to determine whether a problem really exists. To carry out this step, a technician must collect as much information as possible about the system, circuit, and components used, and then diagnose the problem.

Step 2: ISOLATE

The second step is to apply a logical and sequential reasoning process to isolate the problem. In this step, a technician will operate, observe, test, and apply troubleshooting techniques in order to isolate the malfunction.

Step 3: REPAIR

The third and final step is to make the actual repair and then test the circuit.

Let us examine this three-step troubleshooting process in more detail, and apply it to a typical dc power supply circuit.

5-6-1 *Typical Power Supply Circuit*

As an example, Figure 5-25 shows the schematic diagram for a dc power supply circuit that can generate a +5 V, +12 V, and –12 V dc output from a 120 V or 240 V ac input. You should recognize most of the pieces of this picture because this circuit contains nearly all of the devices discussed in the previous sections of this chapter.

The power supply ON/OFF switch (S_1) switches a 120 V ac input to the 120 V primary-winding connection, or a 240 V ac input to the 240 V primary-winding connection of the transformer (T_1). By doubling the number of primary turns, we can accept twice the input voltage and deliver the same output voltage. The upper secondary winding of T_1 supplies a bridge rectifier module (BR_1), that generates a +12 V peak full-wave pulsating dc output, which is filtered by C_1 and regulated by the 7805 (U_1) to produce a +5 V dc 750 mA output. The +5 V output will turn on a "power ON" LED (D_3), which is mounted on the power supply printed circuit board (PCB) along with the filter and regulator. R_1 is a current-limiting resistor for D_3.

The lower secondary winding of T_1 supplies the two half-wave rectifier diodes D_1 and D_2, which generate –50 V and +50 V peak outputs, respectively. These inputs are filtered by C_2 and C_3 and then regulated by a 7912 (U_2), which generates a –12 V dc 500 mA output, and by a 7812 (U_3), which generates a +12 V dc 500 mA output. The devices C_2, C_3, U_2, and U_3 are also all mounted on the power supply printed circuit board (PCB).

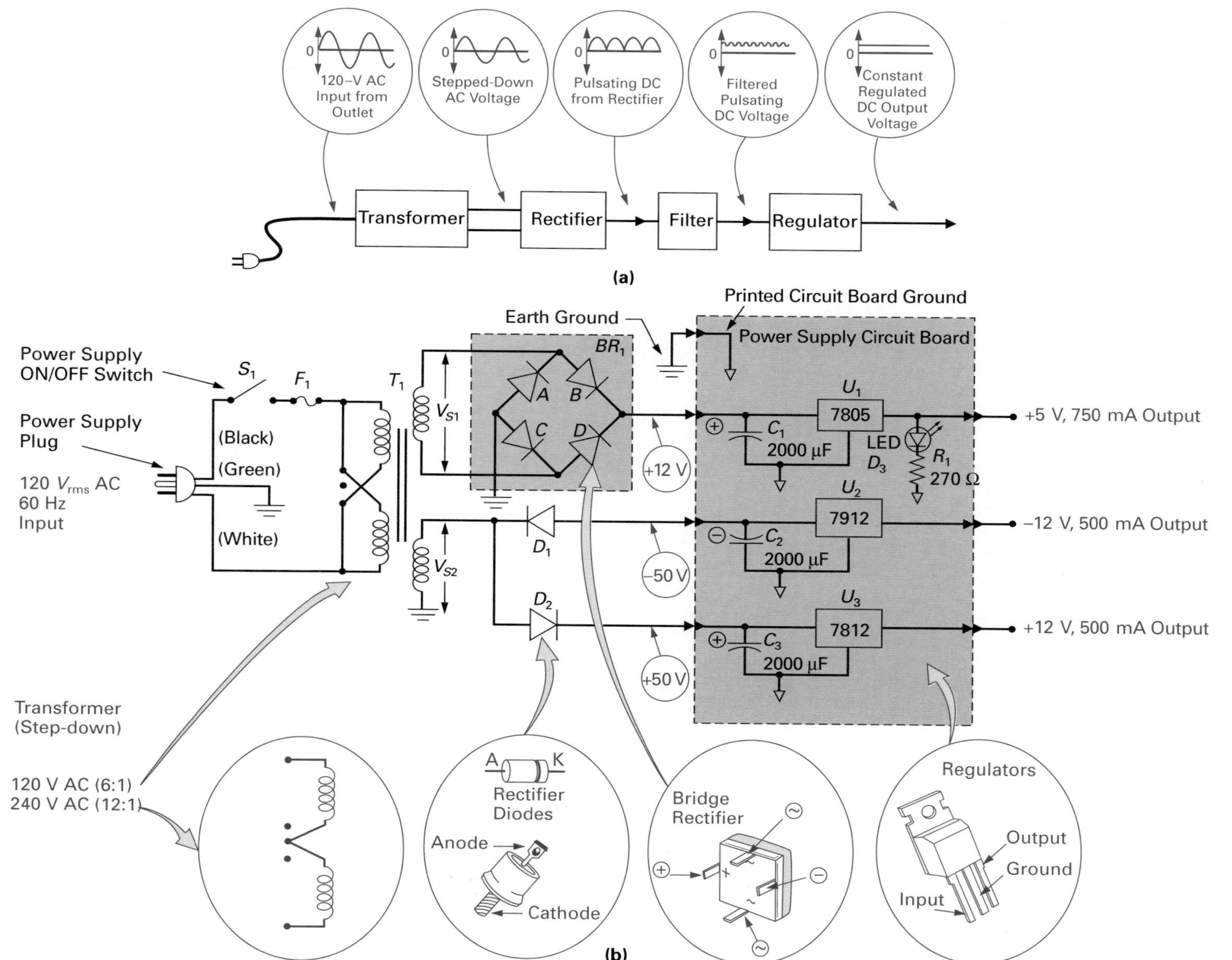

FIGURE 5-25 DC Power Supply Circuit. (a) Block Diagram. (b) Circuit Diagram.

5-6-2 *Step 1: Diagnose*

It is extremely important that you first understand how a system, circuit, and all of its components are supposed to work so that you can determine whether or not a problem exists. If you were preparing to troubleshoot the dc power supply circuit in Figure 5-25, your first step should be to read through the circuit description and review the operation of each device used in the circuit until you feel completely confident with the correct operation of the circuit.

For other circuits and systems, technicians usually refer to service or technical manuals which generally contain circuit descriptions and troubleshooting guides. As far as each of the devices are concerned, technicians refer to manufacturer data books that contain a full description of the device.

Referring to all of this documentation before you begin troubleshooting will generally speed up and simplify the isolation process. Once you are fully familiar with the operation of the circuit, you will be ready to *diagnose the problem* as either an *operator error* or as a *circuit malfunction*.

Here are some examples of operator errors that can be mistaken for circuit malfunctions if a technician is not fully familiar with the operation of the dc power supply circuit in Figure 5-25.

SYMPTOMS	DIAGNOSIS
1. No +5 V, -12 V or +12 V output.	Check first that the power supply is plugged in and turned ON because either of these operator errors will give the appearance of a circuit malfunction.
2. The +5 V LED is ON, but the −12 V and +12 V LEDs are not lit.	There is no LED indicator for the −12 V and +12 V outputs, therefore this is a normal condition.

Once you have determined that the problem is not an operator error but, in fact, a circuit malfunction, proceed to Step 2 and isolate the circuit problem.

5-6-3 *Step 2: Isolate*

A technician will generally spend most of the time in this phase of the process isolating a circuit problem. Steps 1 (diagnose) and 3 (repair) may only take a few minutes to complete compared to Step 2 which could take a few hours to complete. However, with practice and a good logical and sequential reasoning process you can quickly isolate even the most obscure of problems.

The directions you take as you troubleshoot a circuit malfunction may be different for each problem. However, certain troubleshooting techniques apply to all problems. Let us now review some of these techniques and apply them to our dc power supply circuit in Figure 5-25.

1. Check first for **obvious errors:**
   a. Power fuse blown
   b. Wiring errors if circuit is newly constructed

c. Devices are incorrectly oriented. For example, if the rectifier diodes, bridge rectifier module, IC regulators, or the LED were placed in the circuit in the opposite direction to what is shown in Figure 5-25, that device would not operate as it should and therefore neither would the circuit.

2. Use your **senses** to check for broken wires, loose connections, overheating or smoking components, leads or pins not making contact, and so on.

3. and **4.** Use a **cause and effect troubleshooting process** which involves studying the effects you are getting from the faulty circuit and then trying to reason out what could be the cause. Also, apply the **half-split method** of troubleshooting first to the entire system, then to a circuit within the system, and then to a section within the circuit to help speed up the isolation process. To explain how to use these two methods with the example circuit in Figure 5-25, let us examine a few examples.

■ **EXAMPLE:**

The dc power supply in Figure 5-25 was being used to supply power to several circuits in an electronic system. If a power problem were to occur, how would we apply the half-split method?

■ ***Solution:***

As you know, the half-split process involves first choosing a point roughly in the middle of the problem area. By making a test at this midpoint you can determine whether the problem exists either before or after this point. For example, Figure 5-26 shows how this half-split method can be applied to a power supply problem. By testing the output of the dc power supply (+5 V, −12 V, +12 V) you can determine whether the power problem is in the power supply or within the circuits that are being supplied power.

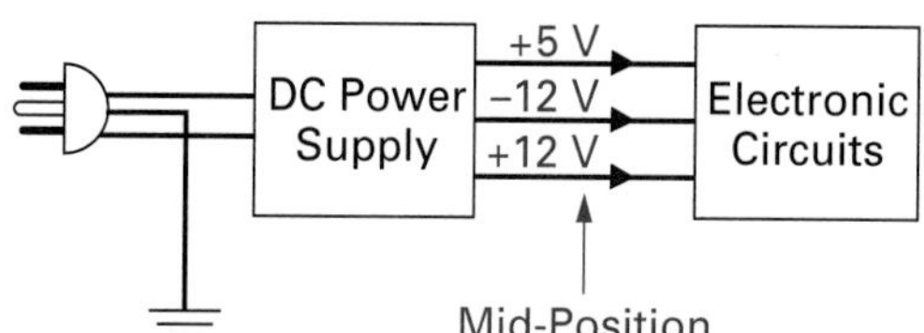

FIGURE 5-26 Using the Half-Split Method to Isolate a System Problem.

a. If the dc supply voltages are present and of the correct value, then the dc power supply is functioning normally.
b. If these dc supply voltages are not present or are not of the correct value, then we will need to isolate the problem.

In some instances you will have to be careful not to assume the obvious when applying the half-split method. For example, let us imagine that we have tested the output of the dc power supply and its output voltage is incorrect. Your first instinct would be to assume that the problem lies within the dc power supply. However, as we discovered in this chapter, *a short in the load can pull down a source, making it appear to be supplying a faulty output.* Therefore, most problems need to be further isolated because the problem could be in the source or it could be in the load. Let us look at another example in which the source appears to be the problem when, in fact, the fault is in the load.

■ **EXAMPLE:**

The +5 V output from a dc power supply in Figure 5-27 measures +2 V on the multimeter. How would we proceed to troubleshoot this problem?

■ ***Solution:***

Figure 5-27 shows how a +5 V output from a dc power supply could be pulled down to +2 V due to a short or low resistance path in one of the electronic circuits that are being supplied power. These problems can cause the power supply input fuse to blow or the IC regulator to disconnect its output due to the excessive current being drawn by the short. Because the output of a power supply generally goes to several boards, and within each board it is connected across many devices, it becomes very difficult to isolate a short. If the fuse does not blow and the regulator IC does not switch OFF, a short will definitely load or pull down the dc supply voltage. At first, it appears that the power supply itself is malfunctioning because of the decreased output voltage, but once you disconnect the output of the dc power supply (at point A) from the external circuit boards, the +5 V should go back up to its correct voltage indicating that the problem lies in the load and not in the source. To isolate the faulty circuit board, you will have to reconnect the supply voltage (at point A) and then disconnect each circuit board one at a time in the following way to isolate the faulty board:

a. First disconnect board 1 at point B and then check the +5 V output from the dc power supply to see if it has returned to its normal voltage. If the dc supply voltage at point A remains LOW, then the short is not within board 1 and you should reconnect this board.

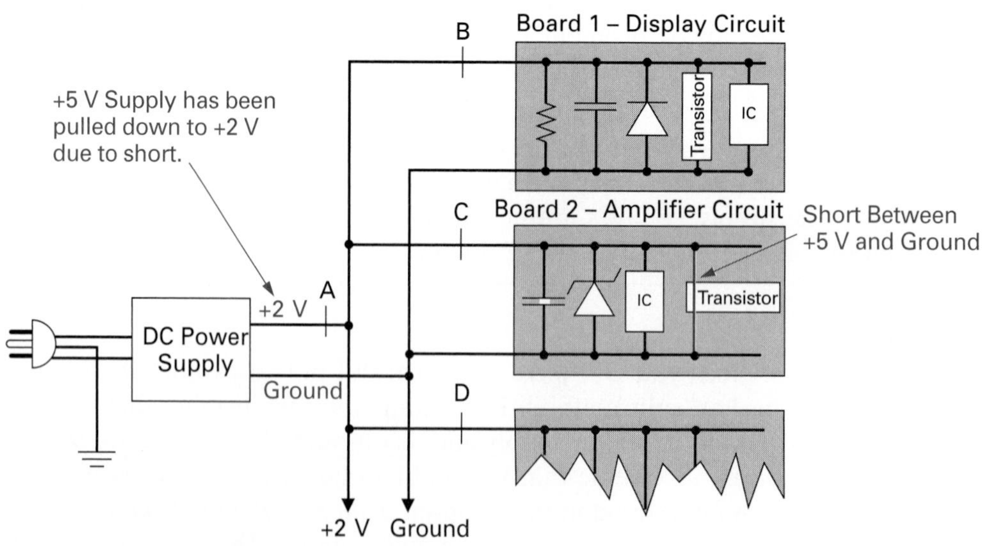

FIGURE 5-27 Locating Power Line Shorts.

b. Disconnect board 2 at point *C* and again check the +5 V output from the dc power supply at point *A* to see if it has returned to its normal voltage. If its output voltage remains LOW, then the short is not within board 2, and you should reconnect this board. In the example in Figure 5-27, however, the short does exist within board 2 and, once this board is disconnected at point *C*, the output from the dc power supply at point *A* will return to +5 V.

c. If the fault was not in board 2, you would have continued to disconnect each board until you isolated the short.

Once the faulty board is located, you will have to troubleshoot that circuit until the short is isolated. For example, if this board were a seven-segment encoder and display circuit, such as the one discussed in the previous chapter, we would follow the troubleshooting procedure discussed previously for that board. A dramatic approach to localizing a difficult-to-find short on a complex printed circuit board is to freeze spray the entire board using an aerosol freeze can. Once the board is reconnected to the power supply, the short will become quickly visible since its path of excess current will generate heat and therefore defrost faster than the rest of the board.

What would happen if an open developed in one of the lines connecting power to a circuit? Let us answer this question by again looking at another example.

■ **EXAMPLE:**

How would you isolate the open shown in Figure 5-28?

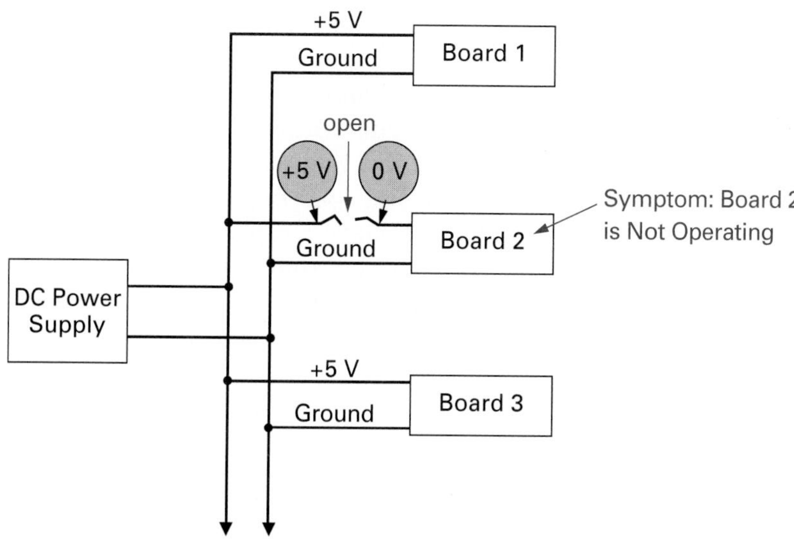

FIGURE 5-28 Locating Power Line Opens.

■ ***Solution:***

An open in any of the dc supply lines connecting power to a circuit will cause that circuit not to operate. These opens can be easily traced with a voltmeter to determine the point at which voltage is no longer present.

In summary, once you have disconnected the circuit load from a dc power supply's output, you can isolate the problem as being either a source fault or load fault, because, *if the power supply's output voltage goes back to its rated voltage after it has been disconnected from the circuit load, the problem is with the load. Whereas, if the output voltage remains LOW or at zero after the source has been disconnected from the load, the problem is with the source.*

Assuming that we have isolated the problem to the power supply, let us now examine several types of common failures. Once again, we will apply these problems to the example circuit in Figure 5-25.

■ **EXAMPLE:**

What would happen to the circuit in Figure 5-25 if the power supply's fuse were to blow?

■ ***Solution:***

A short in the power supply or the load will cause an excessive current to be drawn from the secondary of the transformer. The primary of the transformer will try to respond and deliver the power needed by increasing its current. Since the power supply's fuse (F_1) is connected in series with the transformer's primary, as shown in Figure 5-25, the increase in primary current will blow or open the fuse. When a fuse needs to be replaced, always make sure that the replacement has the same voltage and current rating. In most cases, power supplies use a **slow-blow fuse** which allows a surge of current to pass through the power supply when it is first turned on. This surge current is momentary, usually lasting for less than a second, while all of the system circuits are warming up. If the surge or high current persists for longer than a second, however, the fuse will blow and protect the power supply. Under no circumstances should you ever use a fuse that has a higher current rating because this will permit a higher value of current to pass through to the power supply and load circuits. Never be tempted to "defeat" or bypass the fuse if you do not have a replacement by placing a piece of wire across the fuse terminals. This could introduce other faults, start a fire, or possibly electrocute somebody, since there is now nothing to prevent a damaging value of current.

Slow-Blow Fuse
Allows a surge of current to pass through the power supply when it is first turned on.

■ **EXAMPLE:**

What would happen to the circuit in Figure 5-25 if the power supply transformer (T_1) developed the following problems:

a. An open primary or secondary winding
b. A shorted primary or secondary winding

c. A short between the primary winding and its grounded case, or between the secondary and its grounded case

■ ***Solution:***

a. An open in the primary or secondary winding will cause the power supply's output voltage to drop to zero. Making a few checks with the voltmeter, you will find that you have primary ac voltage but no secondary ac voltage. If you unplug the power supply and disconnect the transformer and check the resistance of its windings, you will find the open winding using the ohmmeter (it will have an infinite resistance).
b. A shorted primary or secondary will provide a low resistance path, resulting in an excessive primary current and a blown fuse. If the fuse does not blow because the current is not quite high enough, the shorted transformer will produce almost no output voltage and be generating a lot of heat. Once again, by unplugging and disconnecting the transformer and then checking its primary and secondary winding resistance, you should find the short (the transformer in Figure 5-25 should typically have a primary resistance of 50 Ω, and a secondary resistance of about 10 Ω).
c. If either the primary winding or secondary winding were to short to the transformer's casing (which is generally grounded), the result will be a low-resistance path to ground, resulting in an excessive primary current and a blown fuse. Once again, by unplugging and disconnecting the transformer and then checking the winding-to-case resistance for both the primary and secondary, you should be able to find out if a winding has shorted to the case. This winding-to-case resistance should normally be infinite ohms.

Once you have determined that the transformer has developed an internal problem, you will have to get a replacement because these problems cannot be repaired. Be sure that the replacement has the same transformer rating.

■ **EXAMPLE:**

What would happen to the circuit in Figure 5-25 if one of the rectifier diodes were to open in the

a. Half-wave rectifiers
b. Full-wave rectifier

■ ***Solution:***

a. An open rectifier diode is quite a common power supply problem. If a half-wave rectifier diode such as D_1 or D_2 in Figure 5-25 were to open, the symptoms would be easy to diagnose because no pulsating dc would be applied to the filters and regulators, and there would not be a −12 or +12 V output from the power supply. Like all opens, all of the supply voltage (V_{S2}) will appear across the defective device's terminals when measured with the multimeter.
b. If one of the diodes in the bridge rectifier in Figure 5-25 were to open, it would prevent current flow for one-half of the input cycle. This would result in a half-wave output that would be difficult to filter, and if the aver-

age voltage into U_1 is less than +5 V, the regulator would not function, producing no +5 V output.

■ **EXAMPLE:**

What would happen to the circuit in Figure 5-25 if one of the rectifier diodes were to short in the

a. Half-wave rectifiers
b. Full-wave rectifier

■ ***Solution:***

a. A shorted half-wave rectifier diode, such as D_1 or D_2 in Figure 5-25, would connect ac to the filter capacitor, causing the filter capacitor to charge and discharge. Because a constant dc voltage is not being applied to the regulator, there would be no dc output voltage.
b. If a diode, such as B, in the bridge rectifier module BR_1 in Figure 5-25 were to short, it would act like a forward-biased diode when diodes B and C are turned ON. However, when diodes A and D are forward biased, the ON D and shorted B will short the secondary winding, causing an excessive current. It is likely that this high current will burn open the shorted diode B and the ON diode D.

■ **EXAMPLE:**

What would happen to the circuit in Figure 5-25 if a filter capacitor were leaky or were to short?

■ ***Solution:***

A shorted or leaky filter capacitor will generally cause an excessive current that will usually burn open the rectifier diodes, producing no dc output. If the leakage is not too large, the problem will be an increase in the ripple voltage into the regulator.

■ **EXAMPLE:**

What would happen to the circuit in Figure 5-25 if a regulator's output current rating was exceeded?

■ ***Solution:***

Referring to Figure 5-25, you can see that each output of the regulator has a rated voltage and maximum load current value. If this current is exceeded due to a short in the load, the regulator will shut down to protect itself. These built-in short-circuit protection and thermal shutdown circuits normally prevent the regulator from damage. In most cases, if the regulator's input is fine but there is no output voltage, the regulator has probably shut itself down, due to a short in the load or an insufficient heat sink. By disconnecting the output of the regulator from the load, you should be able to determine if the problem is the source or the load.

As we complete each chapter in this text, we will continue to apply these troubleshooting methods to a variety of circuits. You will also be introduced to many more new testing techniques which will not only help you in lab but also help prepare you for the future.

5-6-4 *Step 3: Repair*

The final step is to repair the circuit, which could involve simply removing a wire clipping or some excess solder, resoldering a broken connection, reconnecting a connector, or some other easy repair. In most instances, however, the repair will involve the replacement of a faulty component. For a circuit that has been constructed on a breadboard or prototyping board, the removal and replacement of the component is simple. However, when a printed circuit board is involved you should make a note of the component's orientation (in the case of rectifier diodes, rectifier modules, electrolytic capacitors, IC regulators and other similar devices), and observe good desoldering and soldering practices.

When the circuit has been repaired, always perform a final test to see that the circuit is now fully operational.

SELF-TEST REVIEW QUESTIONS FOR SECTION 5-6

Refer to Figure 5-25.

1. What is the average voltage of the +12 V peak out of BR_1?

2. Could we generate +15 V dc output by replacing the 7812 with a 7815?

SUMMARY

1. If only a small amount of power is needed, batteries can be used to deliver a dc supply, as is the case with portable equipment such as calculators, watches, cassette-radios, multimeters, and so on.
2. With larger electronic systems such as computers, television sets, music systems, and video systems, the dc supply voltages will be obtained from a dc power supply, which is generally a subsystem within the main system.
3. The dc power supply converts a 120 V ac 60 Hz input into the desired dc supply voltages.
4. Since the final voltage desired is generally not 120 V, a transformer is usually included to step the ac line voltage up or down to a desired value. Electronic circuits generally require low-voltage supply values such as 12 and 5 V dc, and therefore a step-down transformer would be used.
5. The rectifier converts the stepped-down ac input from the transformer to a pulsating dc output.
6. The filter smoothes out the pulsating dc ripples into a constant dc level, as seen in the waveform after the filter.

7. The regulator maintains the dc output voltage from power supply constant or stable, despite variations in the ac input voltage or variations in the output load resistance.

Power Supply Transformers (Figure 5-29)

8. The turns ratio of a transformer in a dc power supply can be selected to either increase or decrease the 120 V ac input. With most electronic equipment, a supply voltage of less than 120 V is required, and therefore a step-down transformer is used.

QUICK REFERENCE SUMMARY SHEET

FIGURE 5-29 Power Supply Transformers.

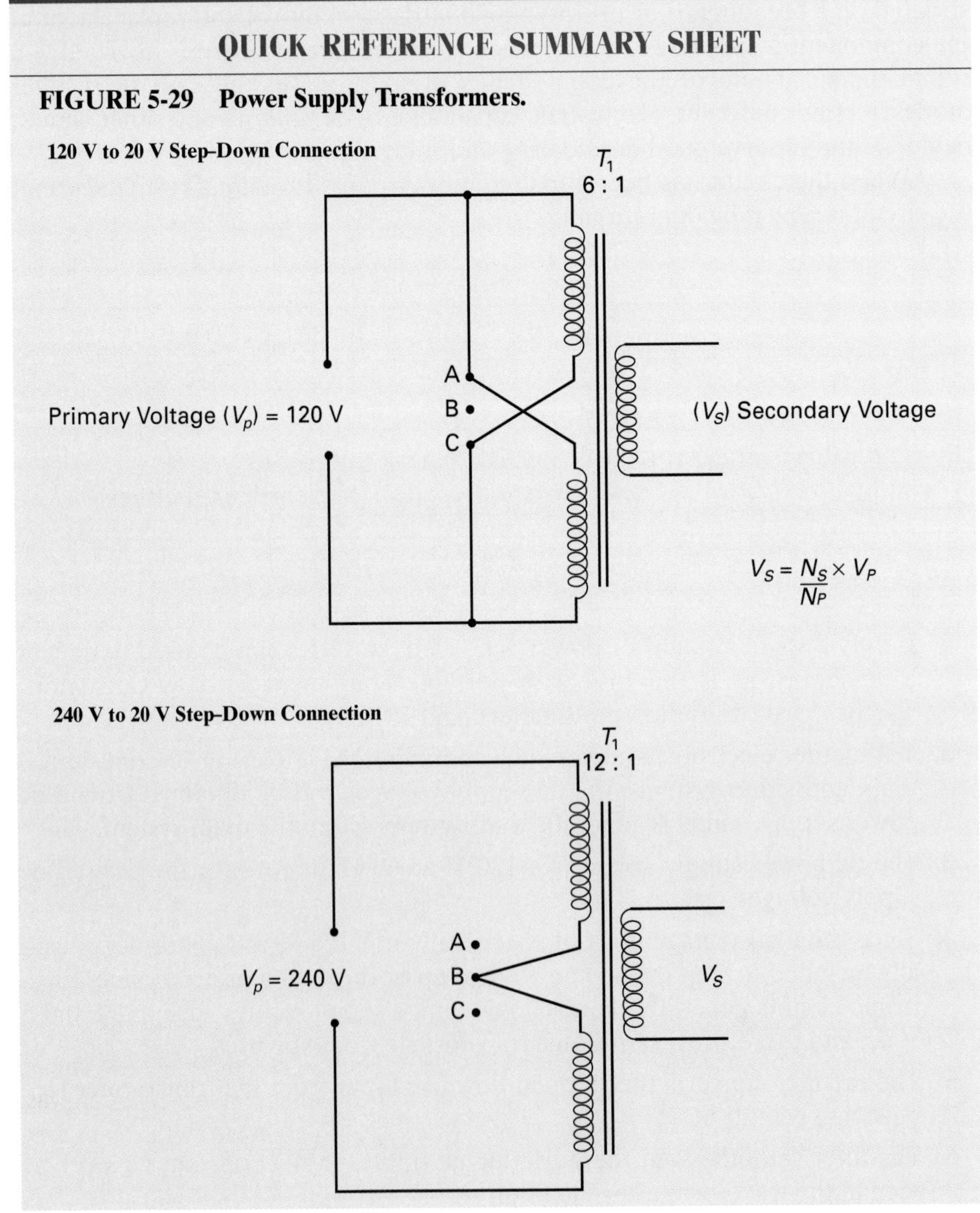

Half-Wave Rectifier Circuits (Figure 5-30)

9. The P-N junction diode's ability to switch current in only one direction makes it ideal for converting two-direction alternating current into one-direction direct current.
10. The half-wave rectifier circuit is constructed simply by connecting a diode between the power supply transformer and the load.
11. When the secondary ac voltage swings positive, the anode of the diode is made positive, causing the diode to turn ON and connect the positive half-cycle of the secondary ac voltage across the load (R_L). When the secondary ac voltage swings negative, the anode of the diode is made negative, and therefore the diode will turn OFF. This will prevent any circuit current, and therefore no voltage will be developed across the load (R_L).
12. The circuit is called a half-wave rectifier because only half of the input wave is connected across the output.
13. To be more accurate, there will, of course, be a small voltage drop across the diode due to its barrier voltage of 0.7 V for silicon and 0.3 V for germanium.
14. When the input swings negative, the entire negative supply voltage will appear across the open or OFF diode. The maximum reverse breakdown voltage, or peak inverse voltage (PIV) rating of the rectifier diode, must therefore be larger than the peak of the ac voltage at the diode's input.
15. By changing the direction of the rectifier diode, the rectifier can be made to produce either a positive or a negative dc output.
16. The frequency of the pulsating dc output from a rectifier is called the ripple frequency, and for half-wave rectifier circuits, the output pulsating dc ripple frequency = input ac frequency.

Full-Wave Center-Tapped Rectifier Circuits (Figure 5-31)

17. The half-wave rectifier's output is difficult to filter to a smooth dc level because the output voltage and current are applied to the load for only half of each input cycle.
18. The full-wave rectifier circuit switches both half-cycles (or the full ac input wave) of the input through the load in only one direction.
19. The basic full-wave center-tapped rectifier circuit contains a center-tapped transformer and two diodes. The center tap of the transformer secondary is grounded (0 V) to create a 180° phase difference between the top and bottom of the secondary winding.
20. The two diodes and center-tapped transformer in this circuit switch the two-direction ac current developed across the secondary through the load in only one direction and therefore develop an output voltage across the load of the same polarity.
21. Even though the peak output of the full-wave center-tapped rectifier was half that of the half-wave rectifier, the average output was the same because the full-wave center-tapped rectifier doubles the number of half-cycles at the output, compared to a half-wave rectifier.

QUICK REFERENCE SUMMARY SHEET

FIGURE 5-30 Half-Wave Rectifier Circuits.

Basic Circuit

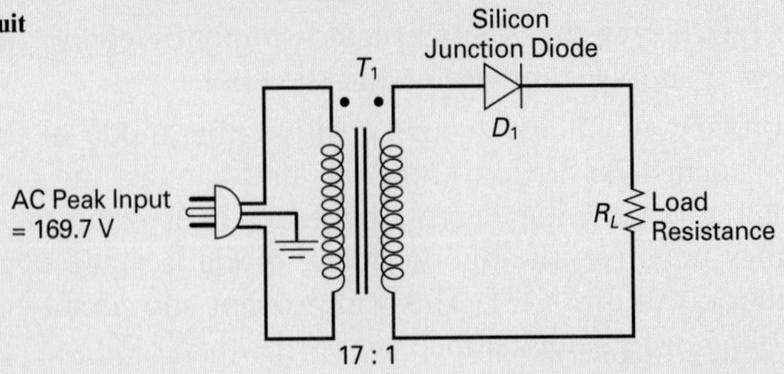

Positive Input Half-Cycle Operation

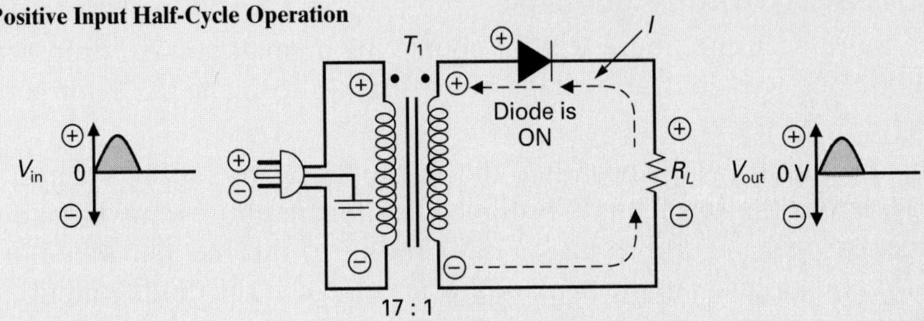

Negative Input Half-Cycle Operation

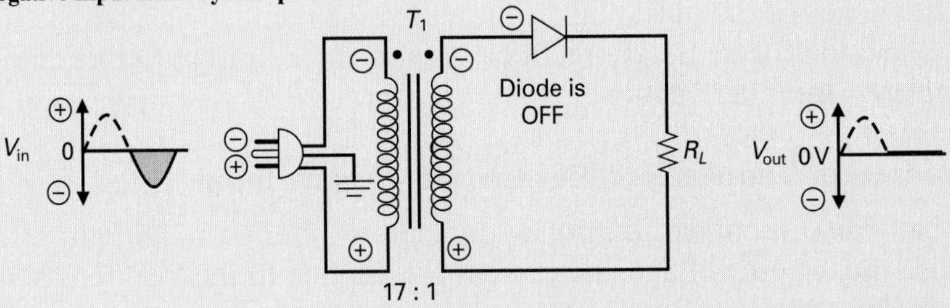

Input/Output Waveforms

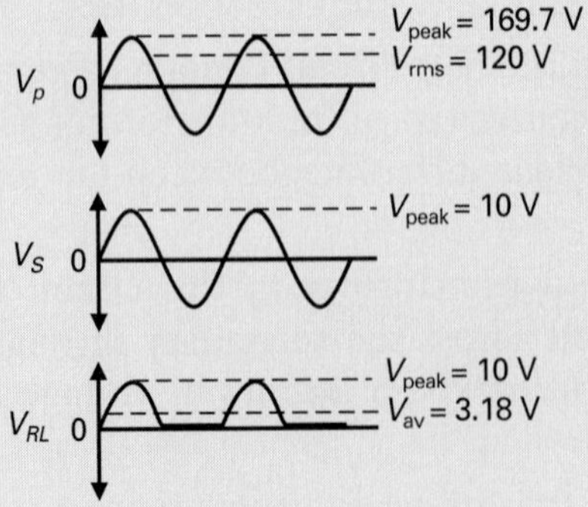

Half-Wave Output Minus Diode Barrier Voltage

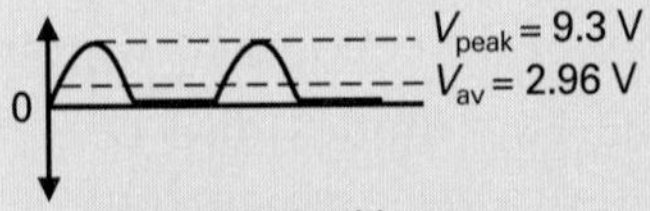

$V_{av} = 0.318 \times V_{S\,(peak)}$

$V_{out\,(peak)} = V_{S\,(peak)} - V_{diode}$

Ripple freq. = Input freq.

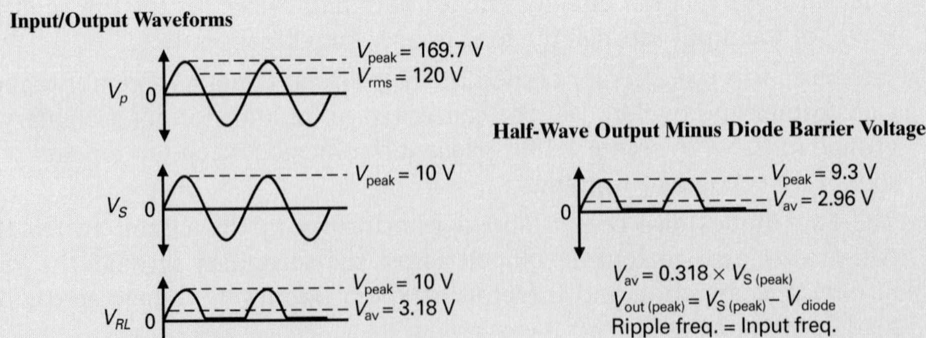

QUICK REFERENCE SUMMARY SHEET

FIGURE 5-31 Full-Wave Center-Tapped Rectifier Circuits.

Basic Circuit

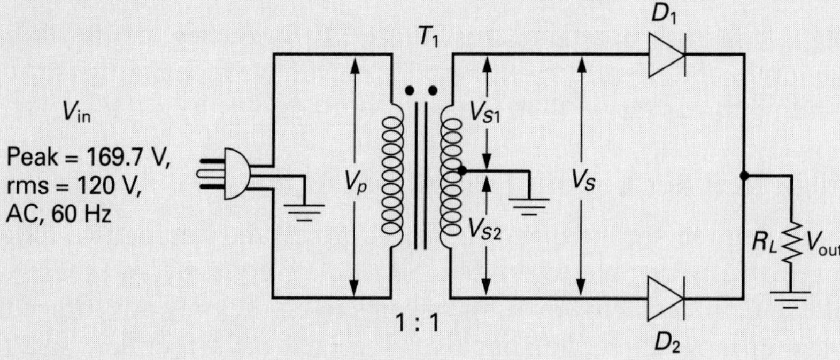

Input/Output Waveforms

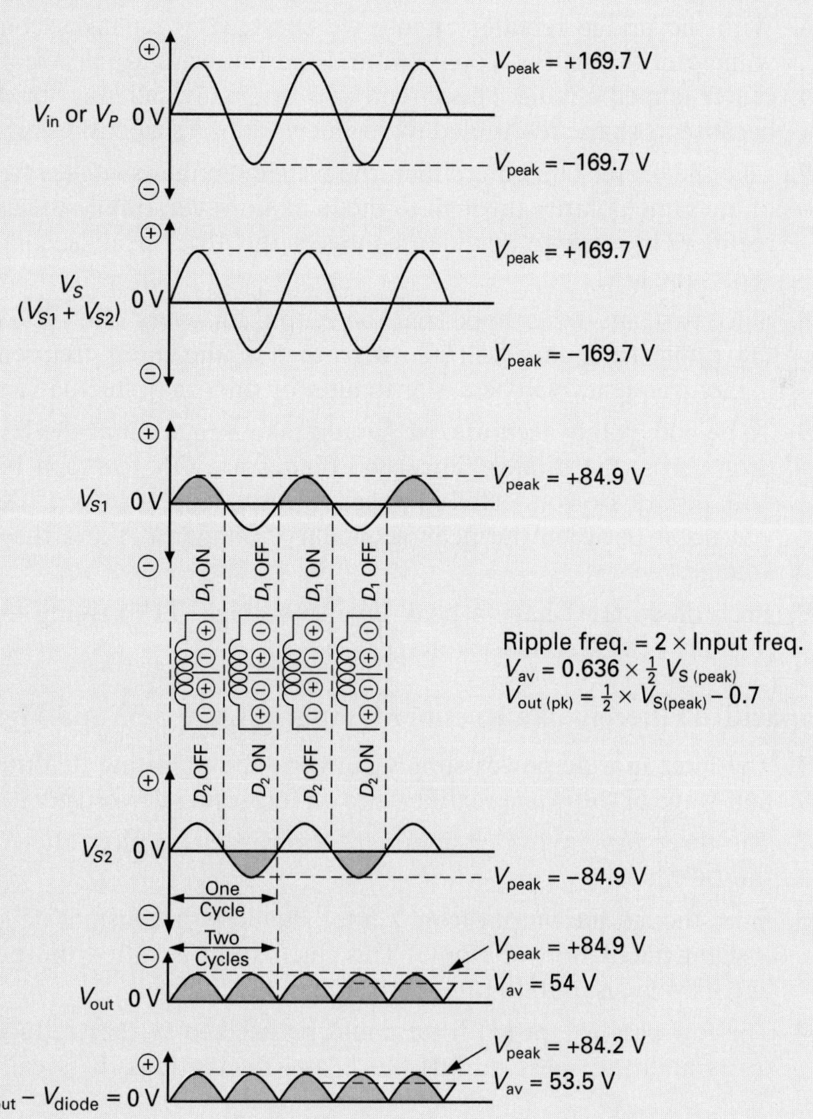

22. Since two half-cycles appear at the output for every one cycle at the input, the ripple frequency will be twice that of the input frequency, and this higher frequency will be easier to filter or smooth of fluctuations.

23. Since only one diode is on for each half-cycle, the peak output voltage will only be less 0.7 V (silicon diode).

24. The full $V_{S\ \text{peak}}$ voltage appears across the OFF D_1 during the other half cycle of the input, and therefore both diodes must have a peak inverse voltage (PIV) rating that is larger than the peak secondary voltage ($V_{S\ \text{peak}}$).

Full-Wave Bridge Rectifier Circuits (Figure 5-32)

25. By center tapping the secondary of the transformer and having two diodes instead of one, we were able to double the ripple frequency and therefore ease the filtering process. However, that seems to be the only advantage the full-wave center-tapped rectifier has over the half-wave rectifier, and the price we pay is having to have a more expensive center-tapped transformer and an extra diode.

26. With the bridge rectifier circuit, we can have the peak secondary output voltage of the half-wave circuit and the full-wave ripple frequency of the center-tapped circuit. This circuit was originally called a "bridge" rectifier because its shape resembled the framework of a suspension bridge.

27. Like the center-tapped rectifier, the bridge rectifier switches two half-cycles of the same polarity through to the load; however, unlike the center-tapped rectifier, the bridge rectifier connects the total peak secondary voltage across the load.

28. Since two half-cycles appear at the output for every one cycle at the input, the ripple frequency will be twice that of the input frequency, and this higher frequency will be easier to filter or smooth of fluctuations.

29. To be completely accurate, we should take into account the barrier voltage drop across the diodes. Since two diodes are ON for each half-cycle and both diodes are connected in series with the load, a 1.4 V (2 X 0.7 V) drop will occur between the peak secondary voltage (V_S) and the peak output voltage.

30. Each diode must have a peak inverse voltage (PIV) rating that is greater than the peak secondary voltage(V_S).

Capacitive Filtering of a Rectifier Output (Figure 5-33 and Figure 5-34)

31. The filter in a dc power supply converts the pulsating dc output from the half-wave or full-wave rectifier into an unvarying dc voltage.

32. The three basic types of filters are the capacitive filter, the *RC* filter, and the *LC* filter.

33. Even though the output from a filter should be a constant dc level, there is a slight fluctuation or ripple. This fluctuation is called the percent ripple and its value is used to rate the action of the filter.

34. The low-pass *RC* pi (π) filter could be used to further reduce ripple, due to its ability to pass dc but block any fluctuations. It is called a π filter

QUICK REFERENCE SUMMARY SHEET

FIGURE 5-32 Full-Wave Bridge Rectifier Circuits.

Positive Input Half-Cycle Operation

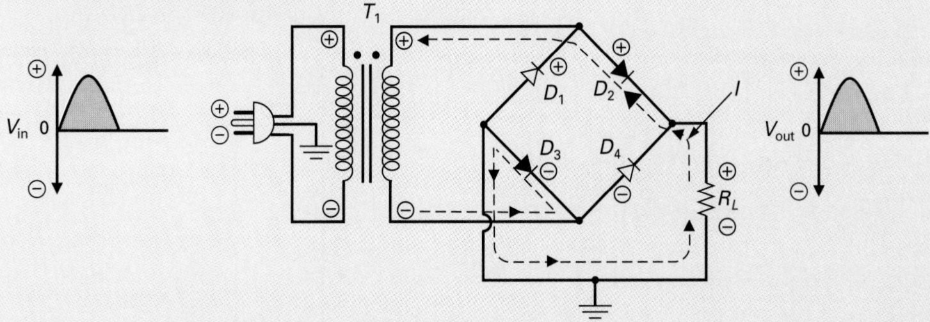

Negative Input Half-Cycle Operation

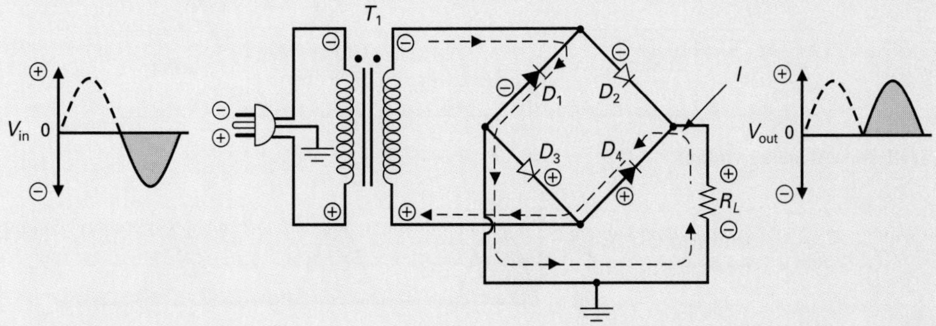

Input/Output Waveforms

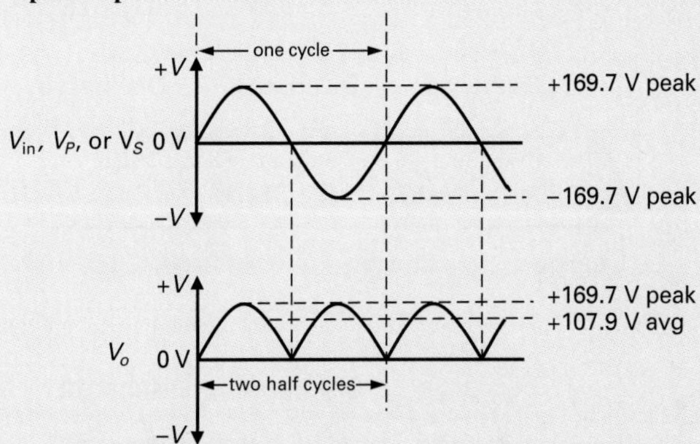

Full-Wave Output Minus Diode Barrier Voltage

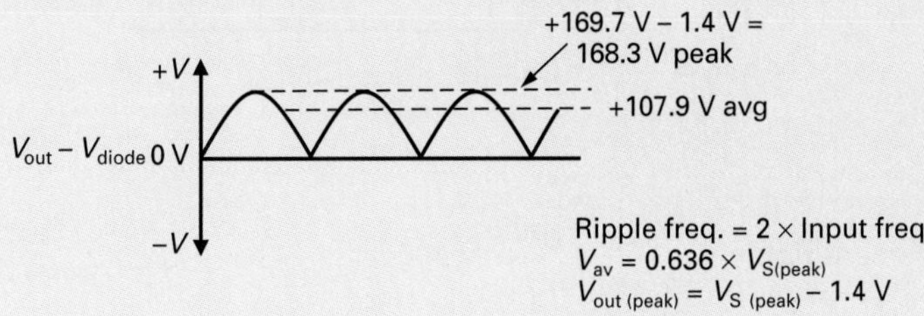

Ripple freq. = 2 × Input freq.

$V_{av} = 0.636 \times V_{S(peak)}$

$V_{out\ (peak)} = V_{S\ (peak)} - 1.4\ V$

QUICK REFERENCE SUMMARY SHEET

FIGURE 5-33 Capacitive Filtering of Half-Wave Rectifier Output.

Basic Circuit

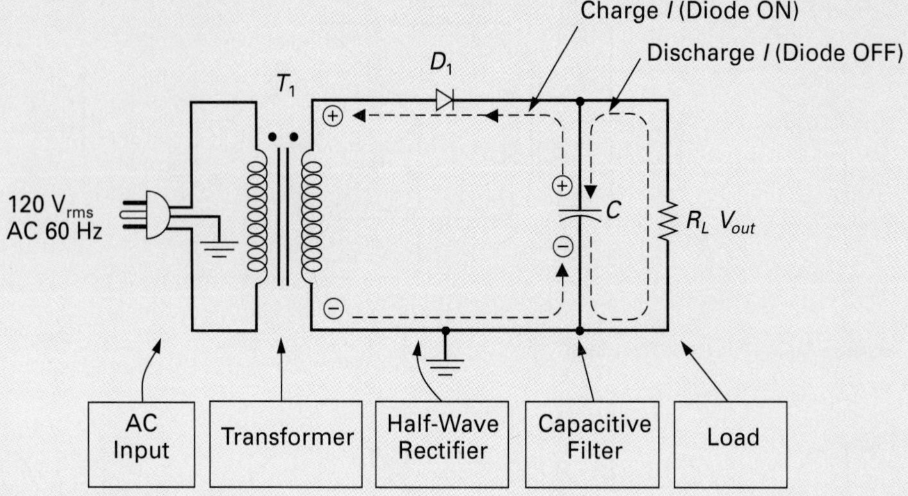

Half-Wave Rectifier Output

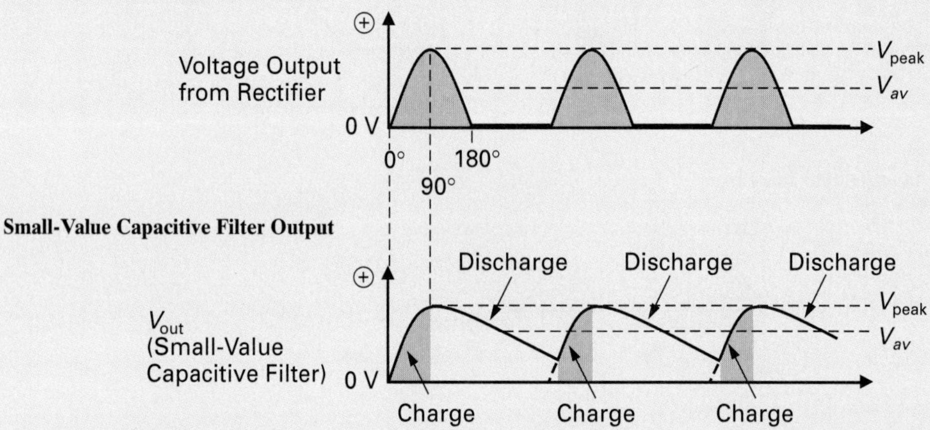

Small-Value Capacitive Filter Output

Large-Value Capacitive Filter Output

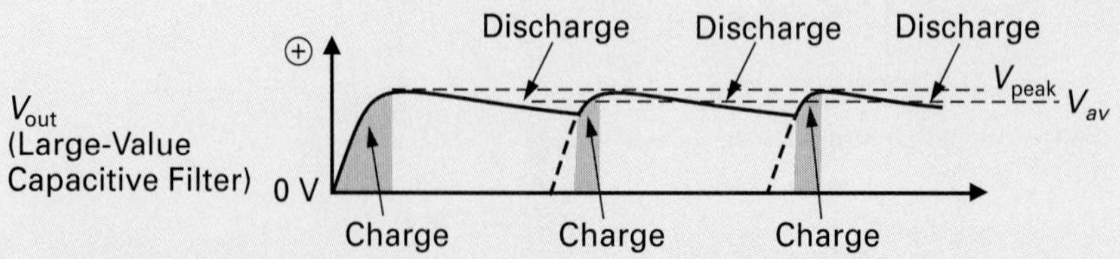

$$\text{Percent ripple} = \frac{V_{rms} \text{ of Ripple}}{V_{av} \text{ of Ripple}} \times 100$$

QUICK REFERENCE SUMMARY SHEET

FIGURE 5-34 Capacitive Filtering of Full-Wave Rectifier Output.

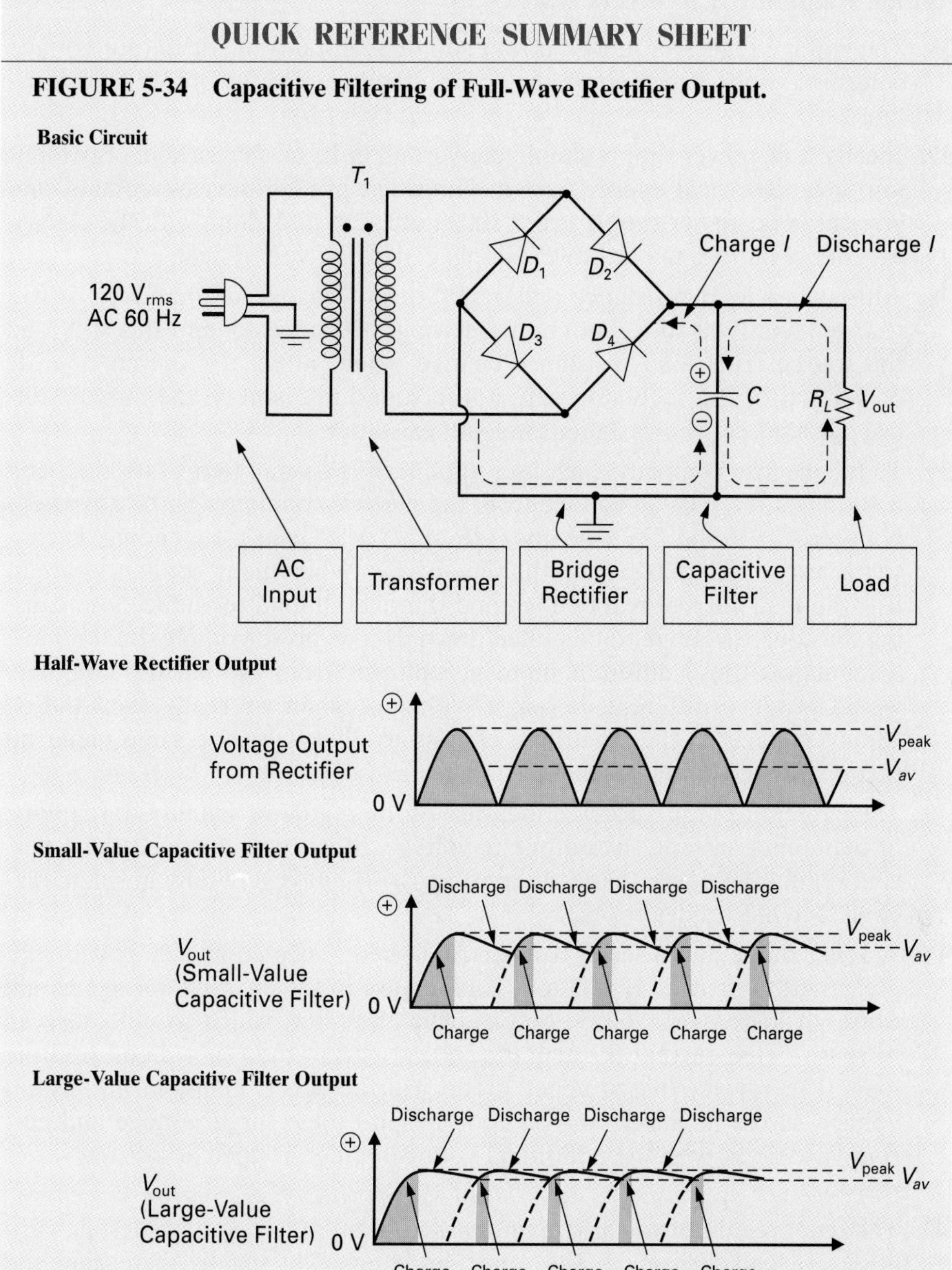

because it contains two vertical parts and one horizontal part and therefore resembles π.

35. The *LC* filter replaces the series resistor in the *RC* π filter with an inductor L_1 to increase the efficiency of the filter since the inductor will oppose current without generating heat.

Voltage Regulator Circuits (Figure 5-35)

36. The voltage regulator in a dc power supply maintains the dc output voltage constant despite variations in the ac input voltage and the output load resistance.

37. Ideally, a dc power supply should convert all of its ac electrical energy input into a dc electrical energy output. However, like all devices, circuits, and systems, a dc power supply is not 100% efficient and along with the electrical energy output, the dc power supply generates wasted heat.

38. Although a load resistance will generate not change dramatically, it will change slightly as different control settings are selected, and this will load the source. This load resistance change would affect the output of a dc power supply if a regulator were not included to maintain the output voltage constant despite variations in load resistance.

39. To the electric company, each user appliance is a small part of its load, and it is the source. The ac voltage from the electric company can be anywhere between 105 V and 125 V ac rms (148 to 177 V ac peak), depending on consumer use, which depends on the time of day. When many appliances are in use, the load current will be high and the overall load resistance low, causing the source voltage to be pulled down. If a dc power supply did not have a regulator, these different input ac voltages from the electric company would produce different dc output voltages, when we really want the dc supply voltage to the electronic circuits to always be the same value no matter what ac input is present.

40. The percent of regulation is a measure of the regulator's ability to regulate, or maintain constant, the output dc voltage. Most regulators achieve a percent regulation figure that is not perfect (0%) but is generally in single digits (2% to 8%, typically).

41. A zener diode and a series resistor could be connected in a dc power supply circuit to provide regulation. An increase in the ac input voltage would cause an increase in the dc output from the filter, which would cause an increase in the current through the series resistor, the zener diode, and the load. The reverse-biased zener diode will, however, maintain a constant voltage across its anode and cathode despite these input voltage and current variations, with the additional voltage being dropped across the series resistor.

42. The zener regulator will also maintain a constant output voltage despite variations in load resistance. The zener diode achieves this by increasing and decreasing its current (I_Z) in response to load resistance changes. Despite these changes in zener and load current however, the zener voltage (V_Z), and therefore the output voltage (V_{out} or V_{RL}), always remains constant.

43. The zener regulator's disadvantage is that the series connected resistor will limit load current, and, in addition, generate unwanted heat.

44. Most dc power supply circuits today make use of integrated circuit (IC) regulators. These IC regulators contain about fifty individual or discrete components all integrated together on one silicon semiconductor chip and then encapsulated in a three-pin package.

QUICK REFERENCE SUMMARY SHEET

FIGURE 5-35 Voltage Regulator Circuits.

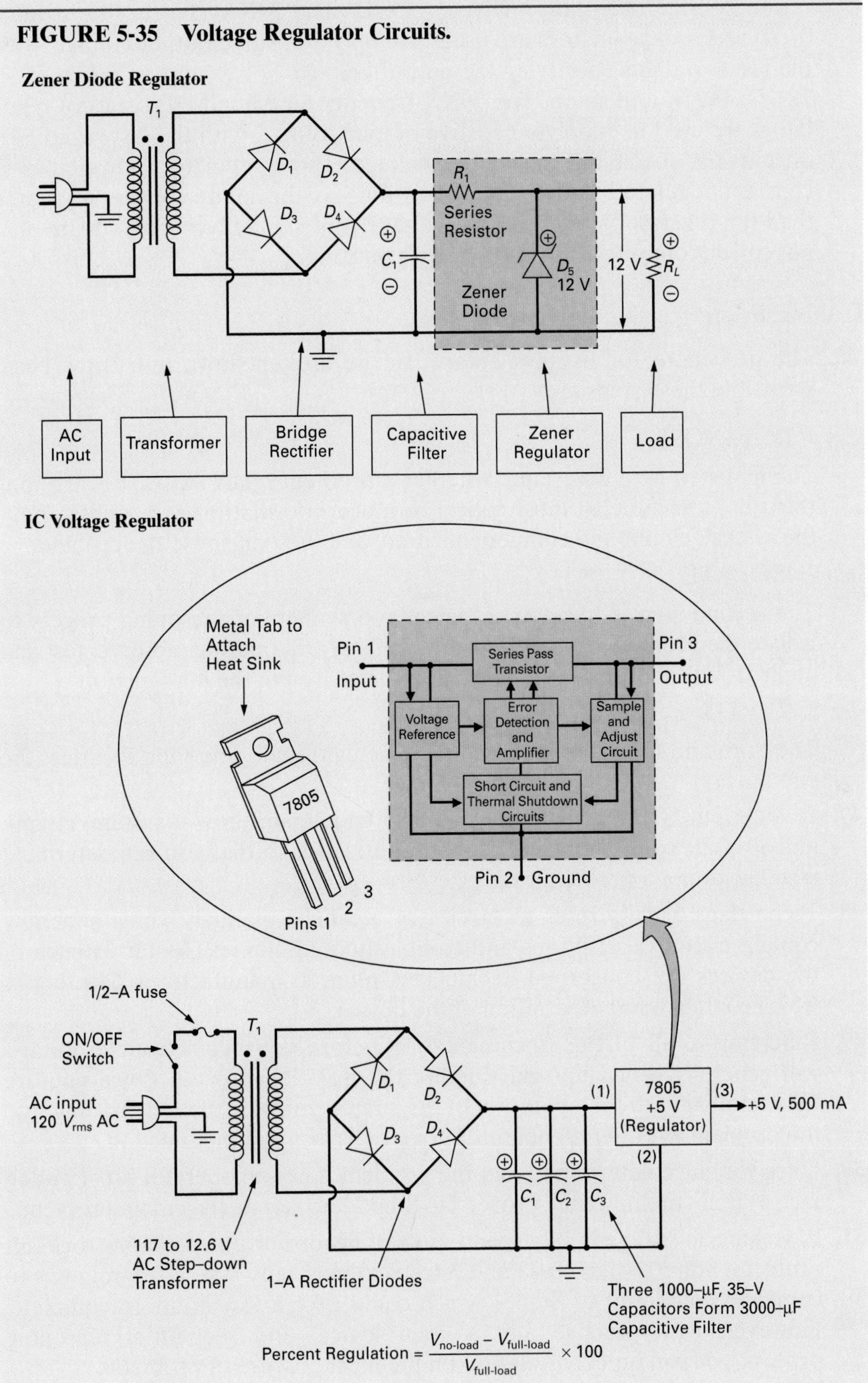

45. The three terminals of the IC regulator are labeled "input" (pin 1), "output" (pin 2), and "ground" (pin 3). The package is generally given the identification code of either 78XX or 79XX. The 78XX (seventy-eight hundred) series of regulators are used to supply a positive output voltage, with the last two digits specifying the output voltage. For example, 7805 = +5 V, 7812 = +12 V, and so on. The 79XX (seventy-nine hundred) series of regulators are used to supply a negative output voltage, with the last two digits, once again, specifying the output voltage. These regulators will deliver a constant regulated output voltage as long as the input voltage is greater than the regulator's rated output voltage, and can deliver a maximum output current of up to 1.5 A if properly heat sunk.

Troubleshooting a DC Power Supply

46. The procedure for fixing a failure can be broken down into three basic steps, and these steps are:

Step 1: DIAGNOSE

The first step is to determine whether a problem really exists. To carry out this step, a technician must collect as much information as possible about the system, circuit and components used, and then diagnose the problem.

Step 2: ISOLATE

The second step is to apply a logical and sequential reasoning process to isolate the problem. In this step, a technician will operate, observe, test and apply troubleshooting techniques in order to isolate the malfunction.

Step 3: REPAIR

The third and final step is to make the actual repair, and then final test the circuit.

47. It is extremely important that you first understand how a system, circuit, and all of its components, are supposed to work so that you can determine whether or not a problem exists.

48. Technicians usually refer to service or technical manuals which generally contain circuit descriptions and troubleshooting guides. As far as each of the devices are concerned, technicians refer to manufacturer data books which contain a full description of the device.

49. Referring to all of this documentation before you begin troubleshooting will generally speed up and simplify the isolation process. Once you are fully familiar with the operation of the circuit, you will be ready to diagnose the problem as either an operator error or as a circuit malfunction.

50. Once you have determined that the problem is not an operator error but, in fact, a circuit malfunction, proceed to Step 2 and isolate the circuit problem.

51. A technician will generally spend most of his or her time isolating a circuit problem. Steps 1 (diagnose) and 3 (repair) may only take a few minutes to complete compared to Step 2 which could take a few hours to complete. However, with practice and a good logical and sequential reasoning process, you can quickly isolate even the most obscure of problems.

52. The directions you take as you troubleshoot a circuit malfunction may be different for each problem. However, certain troubleshooting techniques apply to all problems.

   a. Check first for obvious errors.

      1. Power fuse
      2. Wiring errors if circuit is newly constructed
      3. Devices are correctly oriented. For example, if the rectifier diodes, bridge rectifier module, IC regulators or the LED were placed in the circuit in the opposite direction to what is shown in Figure 5-25, that device would not operate as it should and therefore neither would the circuit.

   b. Use your senses to check for broken wires, loose connections, overheating or smoking components, leads or pins not making contact, and so on.

   c. Use a cause and effect troubleshooting process, which means study the effects you are getting from the faulty circuit and then try to reason out what could be the cause.

   d. Apply the half-split method of troubleshooting first to the entire system, then to a circuit within the system, and then to a section within the circuit to help speed up the isolation process. Once you have disconnected the circuit load from a dc power supply's output, you can isolate the problem as being either a source fault or load fault. If the power supply's output voltage goes back to its rated voltage after it has been disconnected from the circuit load, the problem is with the load. Whereas if the output voltage remains LOW or at zero after the source has been disconnected from the load, the problem is with the source.

53. The final step is to repair the circuit, which could involve simply removing a wire clipping or some excess solder, resoldering a broken connection, reconnecting a connector, or some other easy repair. In most instances, however, the repair will involve the replacement of a faulty component. For a circuit that has been constructed on a breadboard or prototyping board, the removal and replacement of the component is simple. However, when a printed circuit board is involved, you should make a note of the component's orientation, and observe good desoldering and soldering practices.

54. When the circuit has been repaired, always perform a final test to see that the circuit is now fully operational.

NEW TERMS

Bridge rectifier
Capacitive filter
Center-tapped rectifier
Full-wave rectifier
Full-wave voltage doubler circuit
Half-wave rectifier
LC filter
Percent of regulation
Percent of ripple
RC filter
Regulator
Ripple frequency
Voltage-Double Circuit
Voltage-Tripler Circuit

REVIEW QUESTIONS

Multiple-Choice Questions

1. Electrical systems manage the flow of ____________, while electronic systems manage the flow of ____________.

   a. Information, power **b.** Power, information

2. The four main circuit blocks of a dc power supply listed in order from input to output are:

   a. Transformer, rectifier, filter, regulator
   b. Filter, regulator, rectifier, transformer
   c. Transformer, rectifier, regulator, filter
   d. Rectifier, filter, regulator, transformer

3. Which of the four circuit blocks of a dc power supply converts an ac input into a pulsating dc output?

   a. Transformer **c.** Filter
   b. Regulator **d.** Rectifier

4. Which of the four circuit blocks of a dc power supply converts a high ac voltage into a low ac voltage?

   a. Transformer **c.** Filter
   b. Regulator **d.** Rectifier

5. Which of the four circuit blocks of a dc power supply maintains the output voltage constant despite variations in the ac input and output load?

   a. Transformer **c.** Filter
   b. Regulator **d.** Rectifier

6. Which of the four circuit blocks of a dc power supply converts a pulsating dc input into a steady dc output?

   a. Transformer **c.** Filter
   b. Regulator **d.** Rectifier

7. Which rectifier circuit can be used to generate a negative pulsating dc output?

   a. Half-wave **c.** Bridge
   b. Center-tapped **d.** All of the above

8. Which rectifier uses four diodes?

   a. Half-wave **c.** Bridge
   b. Center-tapped **d.** All of the above

9. A typical dc power supply will supply a ____________ -voltage ____________ -current output.

   a. Low, high **c.** Low, low
   b. High, low **d.** High, high

10. Which is typically the most commonly used filter for a dc power supply?

    a. Pi **b.** *RC* **c.** Capacitive **d.** *LC*

11. What rating is used to measure the action of a filter?

    a. Percent rectification **c.** Percent regulation
    b. Percent ripple **d.** Percent filtration

12. A small load resistance will cause a ____________ load current and possibly pull ____________ the source voltage.

 a. Small, down c. Small, up
 b. Large, up d. Large, down

13. What would be the output voltage of a 7915 IC regulator?

 a. +5 V c. –5 V
 b. +15 V d. –15 V

14. A ____________ diode would typically be used in a power supply as a power indicator.

 a. Rectifier c. Zener
 b. Light-emitting d. Junction

15. A ____________ diode would typically be used as a regulator.

 a. Rectifier c. Zener
 b. Light-emitting d. Junction

Essay Questions

16. Why is a dc power supply considered an electrical system? (Introduction)
17. Define the function of the four main blocks in a dc power supply. (5-1)
18. Sketch the input and output waveforms of each of the four blocks in a dc power supply. (5-1)
19. Sketch a half-wave rectifier circuit, then describe the circuit's operation and characteristics. (5-3-1)
20. Sketch a full-wave center-tapped rectifier circuit, and describe the circuit's operation and characteristics. (5-3-2)
21. Sketch a full-wave bridge rectifier circuit, and describe the circuit's operation and characteristics. (5-3-2)
22. Define ripple frequency and what it will be for a half-wave rectifier and a full-wave rectifier. (5-3-2)
23. With a sketch, describe the action of the capacitive filter. (5-4-1)
24. Define percent of ripple. (5-4-2)
25. How can a load resistance pull down the source voltage? (5-5-1)
26. Define percent of regulation. (5-5-2)
27. Identify circuits (a) to (d) in Figure 5-36. (a) (5-5-3), (b) (5-4-3), (c) (5-3-2), (d) (5-5-4)
28. What are the advantages/disadvantages of *RC* and *LC* filters? (5-4)
29. Describe what you think the circuit in Figure 5-37 will do.
30. Briefly describe how the 7805 can supply an output that is greater than +5 V.

Practice Problems

31. If a 240 V rms ac 60 Hz input is applied to the 19:1 step-down transformer shown in Figure 5-38, what would be the peak secondary output voltage?

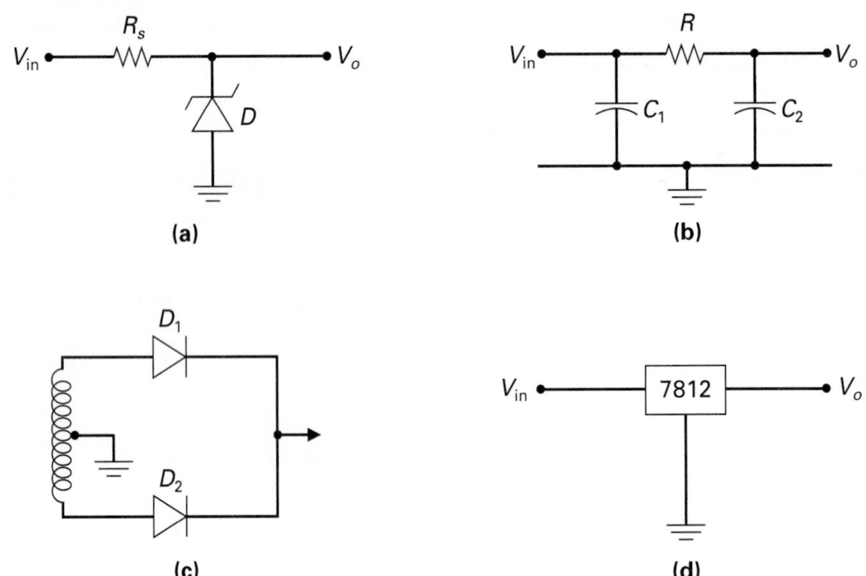

FIGURE 5-36 Power Supply Circuits.

32. What would be the average voltage out of a positive half-wave rectifier if the transformer secondary in Question 31 were applied? (Take into account the barrier voltage drop.)

33. What would be the ripple frequency at the output for the circuits described in Questions 31 and 32?

34. Calculate the peak output voltage (taking into account V_{diode}) for the circuit shown in Figure 5-39.

35. What would be the output ripple frequency from the circuit in Figure 5-39?

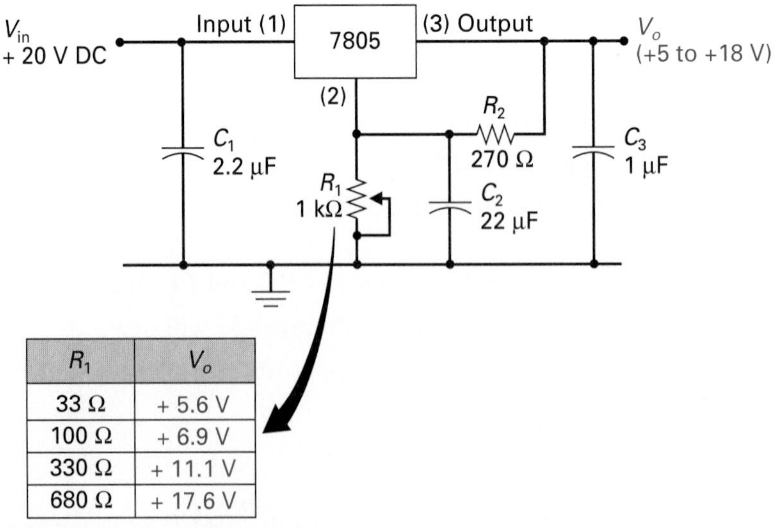

R_1	V_o
33 Ω	+ 5.6 V
100 Ω	+ 6.9 V
330 Ω	+ 11.1 V
680 Ω	+ 17.6 V

FIGURE 5-37 Power Supply Circuit.

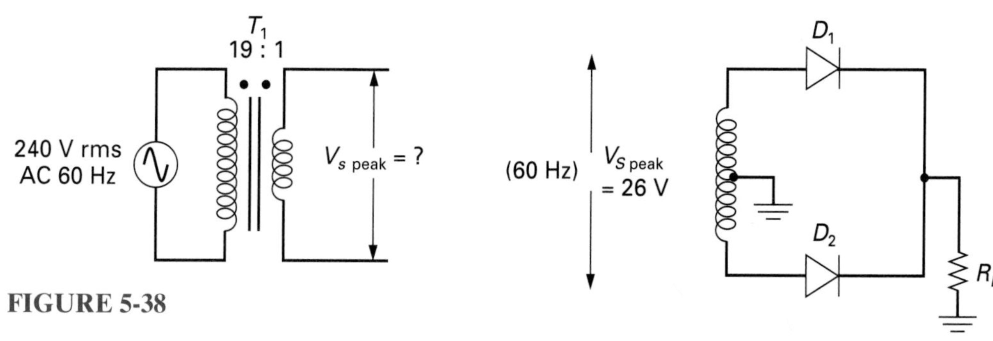

FIGURE 5-38

FIGURE 5-39

36. Calculate the average output voltage of the bridge rectifier circuit shown in Figure 5-40, taking into account the diode voltage drop.

37. A capacitive filter produces the output shown in Figure 5-41. What is the filter's percent of ripple?

38. If a regulator delivers +12 V when no load is connected, and +10.6 V when a full load is connected, what would be the regulator's percent of regulation?

Troubleshooting Questions

39. Briefly describe the effect that some of the following problems would have on a typical dc power supply:

a. An open rectifier diode
b. A shorted rectifier diode
c. A shorted filter capacitor
d. A shutdown regulator

40. As a technician troubleshoots a circuit it is important that he or she study the effect given by the circuit and then try to determine the cause. To help in your cause-and-effect understanding, you may wish to build the power supply circuit in Figure 5-25 (or something similar) and then introduce some problems so that you can observe the results. Most of the time an open circuit will not cause any circuit damage, and these problems can be introduced without concern; however, shorts between two points should be contemplated carefully because of possible hazard to the technician and

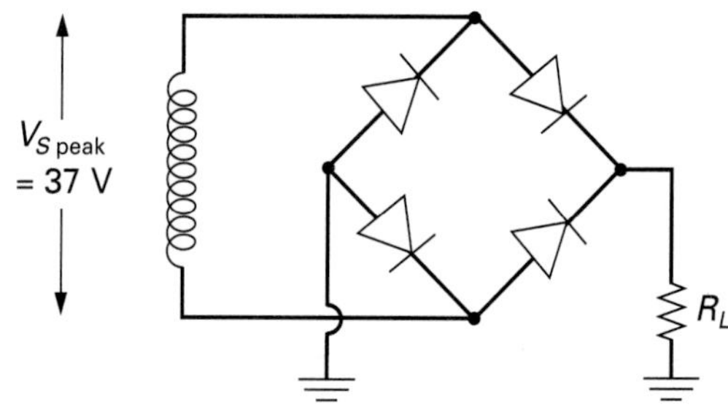

FIGURE 5-40

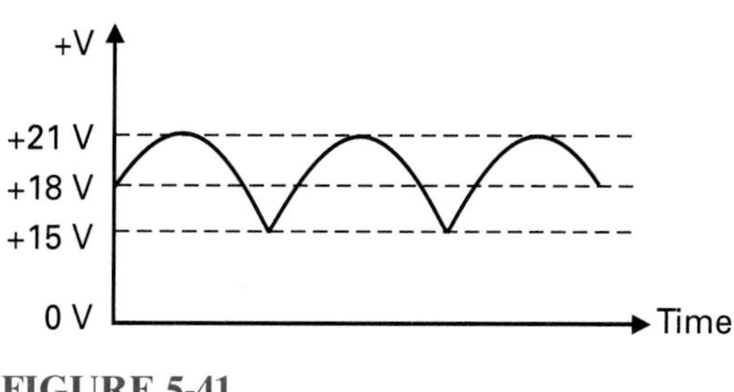

FIGURE 5-41

damage to the circuit. Here are a few examples of circuit problems for Figure 5-25. Describe what circuit problems they will produce:

a. Remove F_1 (Caution: Like all fuses, the 120 V ac will be present across the fuse terminals when it is removed.)
b. Reverse the orientation of D_2 or D_1
c. Open one of the inputs to the bridge rectifier
d. Disconnect one of the filtering capacitors
e. Disconnect the ground input to the 7805
f. Short the 7912 output to ground
g. Reverse the orientation of the LED

CHAPTER

6

OBJECTIVES

After completing this chapter, you will be able to:

1. Define the function performed by a voltage-multiplier circuit, clipper circuit, and a clamper circuit.
2. Describe the operation and characteristics of the following voltage multiplier circuits:
   a. The half-wave voltage-doubler circuit
   b. The full-wave voltage-doubler circuit
   c. The voltage-tripler circuit
   d. The voltage-quadrupler circuit
3. Explain how a voltage-multiplier circuit can be used in a circuit application such as an oscilloscope's dc power supply.
4. Describe the troubleshooting approach and typical circuit malfunctions of a voltage-multiplier circuit.
5. Discuss the operation and characteristics of the following series clipper circuits:
   a. The basic negative and positive series clipper circuits
   b. The biased negative and positive series clipper circuits
6. Explain the basics of electronic communications and how a series clipper circuit can be used in an amplitude modulation (AM) modulator and demodulator circuit.
7. Discuss the operation and characteristics of the following shunt clipper circuits:
   a. The basic negative and positive shunt clipper circuits
   b. The biased negative and positive shunt clipper circuits
   c. The variable shunt clipper circuit
   d. The zener shunt clipper circuit
   e. The symmetrical zener shunt clipper circuit
8. Explain how a shunt clipper circuit can be used in a variety of transient protection circuit applications.
9. Describe the troubleshooting approach and typical circuit malfunctions of series and shunt clipper circuits.
10. Discuss the operation and characteristics of the positive and negative clamper circuits.
11. Explain the basics of television communications and how a clamper circuit can be used in a television receiver circuit.
12. Describe the troubleshooting approach and typical circuit malfunctions of a clamper circuit.

Diode Multiplier, Clipper, and Clamper Circuits

OUTLINE

VIGNETTE

Business Was Dead

Business in 1892 was not going at all well for Almon Strowger, a funeral home director in the small town of La Porte, Indiana. Strangely enough, the telephone, which promoted business for most shopkeepers, was Strowger's downfall. This was because the wife of the owner of a competing funeral parlor served as the town's telephone operator. When people called the operator and asked to be put through to the funeral parlor, they were naturally connected to the operator's husband.

As the old expression states, "necessity is the mother of invention," and for Strowger it became a case of do or die. Making up his mind to cure the problem, he began to study the details of telephone communications. He designed a mechanism that made telephone connections automatic and thus bypassed the town operator. The switch he invented consisted of two parts—a ten-by-ten array of terminals (called the bank) arranged in a cylindrical arc and a moveable switch (called the brush). The brush stepped along and around a cylinder and could be put in one of one hundred positions to connect to one of one hundred terminals. The position of the brush was driven by an electromagnet which responded to pulses produced by the telephone dial.

The "Strowger switch" brought about the first automatic telephone exchange system in the world, with improved versions still used extensively in offices up to the 1960s.

INTRODUCTION

While rectification is by far the most common use of the diode, there are many other electrical and electronic circuits that make use of the diode's one-way, or unilateral, switching characteristics. In the previous chapters, we have seen the junction diode used in rectifying, encoding, and decision making (logic gate) circuit applications. We have also seen the zener diode used in voltage regulating and voltage referencing applications and the light-emitting diode used in indicator and displaying applications. These are only a few of the many applications of these diodes. In this chapter, we will examine several other diode-circuit applications, including voltage multiplying, signal detecting, waveform shaping, circuit protecting, and dc restoring.

First, we will continue our discussion from the previous chapter on dc power supply circuits by examining the *voltage-multiplier circuit.* This circuit uses a few diodes to generate a large dc supply voltage that is some multiple of the ac peak input voltage. For example, an oscilloscope's or television's dc power supply would typically use a voltage multiplier circuit to convert the ac input of 115 V to a 1000 volt dc supply for an electrode of the cathode-ray tube.

The second type of circuit is used to change the shape of a waveform and is called a *clipper circuit.* This circuit uses a diode to clip or eliminate an unwanted portion of the ac input signal and converts one type of signal to another. For example, a clipper circuit could be used to clip off the top section and bottom section of a sine wave and change the waveform's shape from sine to almost square.

The third type of circuit is used to change or restore the dc reference of an alternating waveform and is called a *clamper circuit.* This circuit uses a diode to clamp or lock the top or bottom of a waveform to a dc reference voltage. For example, a dc restorer could be used to change a 10 V peak square wave that varies above and below 0 V to a 10 V peak square wave that varies above and below a dc reference voltage of +5 V.

6-1 VOLTAGE MULTIPLIER CIRCUITS

As mentioned in the previous chapter, because most electronic circuits require a dc supply voltage that is lower than 115 V, a step-down transformer is normally always included. This step-down transformer will reduce the 115 V ac input voltage from the wall outlet to a lower ac voltage and then a rectifier, filter, and a regulator will produce a low-value dc supply voltage, as shown in Figure 6-1(a).

In some instances, a circuit or device may require a dc supply voltage that is larger than 115 V and a step-up transformer will have to be used, as shown in Figure 6-1(b). In this example circuit, you can see that the transformer steps the 115 V ac input up to 200 V so that the rectifier, filter, and regulator can produce two dc supply voltages of 178 V and 153 V. Step-up transformers are expensive, and if a wide range of dc supply voltages are needed, it becomes much more economical to use a voltage-multiplier circuit to obtain the higher value dc supply voltages needed.

Voltage-multiplier circuits can replace rectifier and filter circuits because they employ diodes and capacitors in almost exactly the same way to convert an ac input to a dc output. The big difference is that, as their name implies, *voltage-multiplier circuits produce a dc output voltage that is some multiple of the peak ac input voltage.* For example, a **voltage-doubler circuit** will produce a dc output voltage that is twice the peak of ac input voltage, whereas a **voltage tripler** will produce a dc output voltage that is three times the peak of the ac input voltage.

Voltage-Doubler Circuit
Produces a dc output voltage that is twice the peak of the ac input voltage.

Voltage-Tripler Circuit
Produces a dc output voltage that is three times the peak of the ac input voltage.

Obtaining a greater output voltage from a multiplier circuit makes it seem as though we get more out than we put in. This is not the case because voltage multipliers do not internally generate power. If the output voltage of a multiplier is increased by some multiple, the output current of the circuit will be decreased by the same multiple. For example, a voltage doubler will deliver twice the output voltage ($V \times 2$), but only half the output current ($I/2$); and therefore the output power from a voltage multiplier circuit will never be greater than the input power ($P = V\uparrow I\downarrow$). In fact, since the diodes in a voltage multiplier circuit will dissipate some value of power, there will be slightly less power at the output compared to the input.

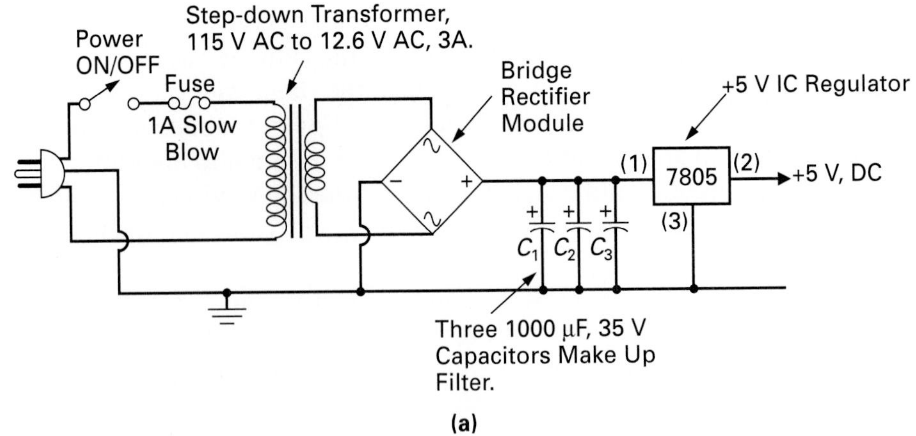

(a)

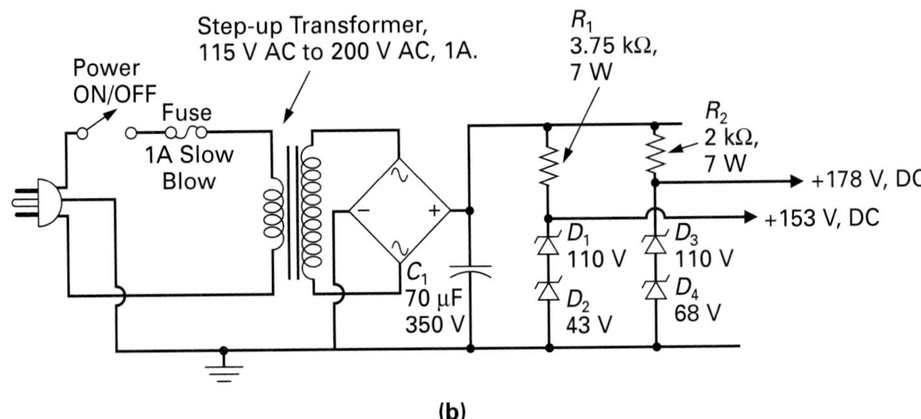

(b)

FIGURE 6-1 Using a Transformer To Obtain Higher DC Supply Voltages.

6-1-1 *Voltage-Doubler Circuits*

Like the rectifier, there are two types of voltage doubler circuits: the half-wave voltage doubler and the full-wave voltage doubler. To begin with, let us examine the simplest of the voltage multiplier circuits, the half-wave voltage doubler.

Half-Wave Voltage-Doubler Circuit

The half-wave voltage-doubler circuit is made up of two diodes and two capacitors and is shown in Figure 6-2(a). This circuit will produce a dc output voltage that is approximately twice the peak value of the ac sine wave input from the secondary of the transformer. In this example, we will assume that the transformer has a 1:1 turns ratio and the 115 V rms, 60 Hz primary voltage will be present at the secondary and applied to the half-wave voltage-doubler circuit. This 115 V rms, secondary voltage will have a peak value of:

$$115 \text{ V} \times 1.414 = 162.6 \text{ V}$$

Figure 6-2(b) shows how the circuit will respond to the negative half cycle of the input sine wave. When the input voltage swings negative, the top of the transformer secondary is made negative and this potential will forward bias D_1

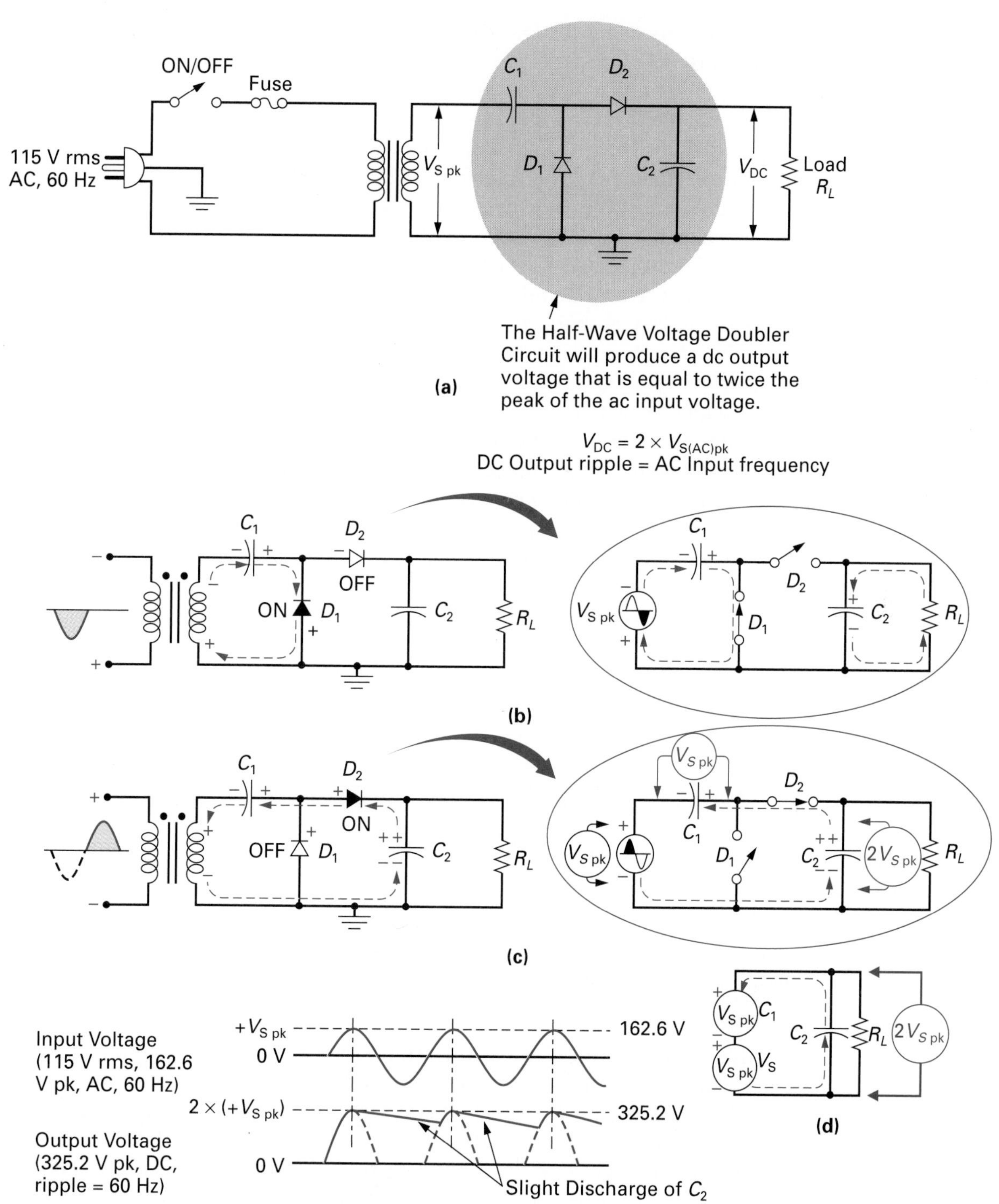

FIGURE 6-2 Half-Wave Voltage Doubler Circuit.

and reverse bias D_2. With D_1 ON, capacitor C_1 will charge to the polarity shown until its plate-to-plate voltage is equal to the peak value of the input sine wave (162.6 V). To simplify the operation of the circuit during this negative half-cycle of the input, the inset in Figure 6-2(b) represents D_1 as a closed switch and D_2 as an open switch. In this equivalent circuit it can be seen how C_1 is connected across the secondary of the transformer by D_1 and will charge to the peak value of the input sine wave. Because there is almost no resistance in the charge path of C_1, the capacitor will charge very quickly to $V_{S\,pk}$.

Figure 6-2(c) shows how the circuit will respond to the positive half-cycle of the input sine wave. When the input voltage swings positive, the top of the transformer secondary is made positive and this potential will forward bias D_2 and reverse bias D_1. Referring to the equivalent circuit in the inset in Figure 6-2(c), you can see what effect diodes D_1 and D_2 will have when they are represented as an open and closed switch. Diode D_2 has now connected the transformer secondary voltage and the charge voltage of C_1 to the output capacitor C_2. Because the transformer's secondary peak voltage (V_{Spk}) and the capacitor C_1's peak charge voltage (V_{Spk}) are now *series-aiding voltage sources* (+/– → +/–), capacitor C_2 will charge to the sum of these two series peak voltages ($C_2 = 2 \times V_{Spk}$). Because C_2 is connected across the load R_L, the dc output voltage will equal twice that of the peak secondary ac voltage, as shown in the simplified diagram in Figure 6-2(d).

$$V_{out} = 2 \times V_{S\,pk}$$

$$= 2 \times 162.6 \text{ V}$$
$$= 325.2 \text{ V}$$

When V_S again switches polarity and returns to the condition shown in Figure 6-2(b), the cycle will repeat with C_1 charging to the peak of the transformer's secondary voltage. During this half-cycle, you may have noticed that the output capacitor C_2 is able to discharge through the load, as shown in the inset in Figure 6-2(b). The time constant formed by the output capacitor and the load resistance is normally very high, and therefore during this negative half-cycle of the input, C_2 will only discharge slightly. This slight discharge of C_2 during the negative alternation of the input voltage can be seen in the output voltage waveform in Figure 6-2(e). Looking closely at this output voltage waveform, you can see that the circuit is called a half-wave voltage doubler because the output capacitor C_2 is charged for only half of the ac input cycle, producing an output that closely resembles that of the filtered half-wave rectifier circuit. Comparing the input voltage waveform to the output voltage waveform in Figure 6-2(e), you can see that for every one cycle of the input (positive-to-positive peak) there is one cycle of the output (positive-to-positive peak). The output dc ripple frequency of a half-wave voltage doubler will therefore equal the input ac frequency:

$$f_{out\ ripple} = f_{in\ ac}$$

Output dc ripple frequency = the input ac frequency

In the circuit in Figure 6-2, the output ripple frequency will equal 60 Hz.

■ **EXAMPLE:**

By reversing the directions of the diodes and capacitors in Figure 6-2, you can change the circuit from a *positive half-wave voltage doubler* to a *negative half-wave doubler.* Sketch a negative half-wave doubler circuit and show the C_1 and C_2 charge paths for each alternation of the input. In addition, what would be the peak of the dc output voltage from this circuit, and what would be its ripple frequency?

■ ***Solution:***

Figure 6-3 shows the component orientation and connection for a negative half-wave doubler circuit. During the positive alternation of the ac input, D_1 will turn ON and allow C_1 to charge, with the polarity shown, to the peak of the ac input. During the negative alternation of the ac input, D_2 will turn ON and connect the two series-aiding peak voltages of V_S and the charge across C_1, across the output capacitor C_2. The result will be that C_2 will charge to twice the peak of the ac input, and this dc negative voltage will be applied across the load.

$$115 \text{ V ac, rms} \times 1.414 = 162.6 \text{ V ac peak}$$

$$\begin{aligned} V_{out} &= 2 \times V_{S\,pk} \\ &= 2 \times 162.6 \text{ V} \\ &= -325.2 \text{ V} \end{aligned}$$

Output dc ripple frequency = the input ac frequency = 60 Hz

Full-Wave Voltage-Doubler Circuit

The half-wave voltage doubler has two disadvantages. The first is that the ripple frequency of 60 Hz is difficult to filter, and the second is that an expensive output capacitor is required because it must have a voltage rating that is more than twice the peak of the ac input.

The **full-wave voltage-doubler circuit** is shown in Figure 6-4(a). As you can see, this circuit uses the same number of components as the half-wave voltage doubler (two diodes, two capacitors), and yet it overcomes the two disadvantages of difficult filtering and the need for a high output capacitor voltage rating.

Full-Wave Voltage-Doubler Circuit
A rectifier circuit that doubles the output voltage by charging capacitors during both alternations of the ac input—making use of the full ac input wave.

Figure 6-4(b) shows how the circuit will respond to the positive half-cycle of the input sine wave. When the input voltage swings positive, the top of the transformer secondary is made positive and this potential will forward bias D_1

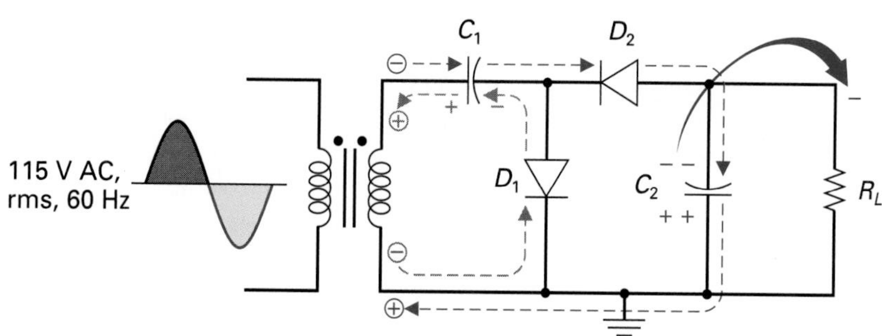

FIGURE 6-3 Negative Half-Wave Voltage Doubler.

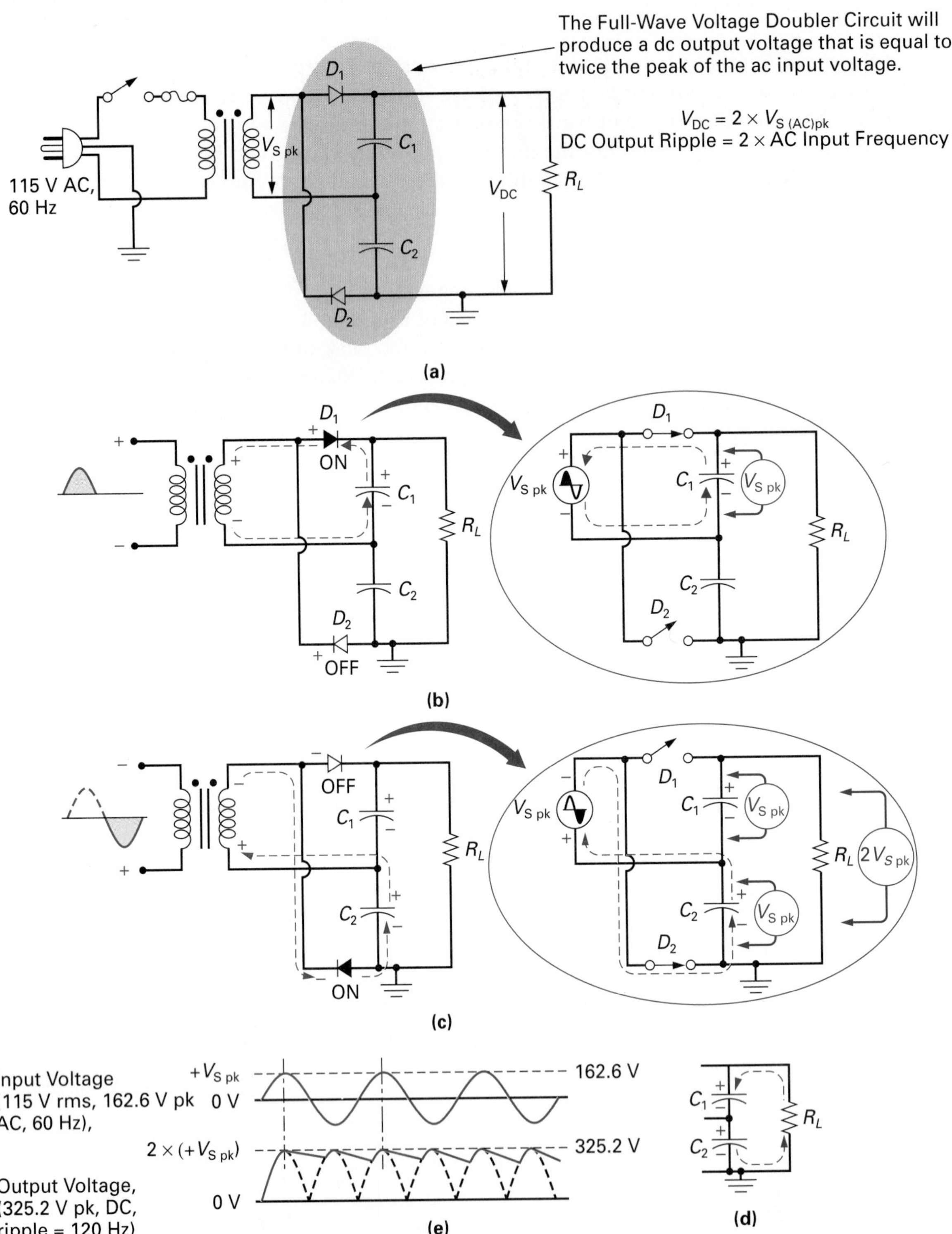

FIGURE 6-4 Full-Wave Voltage Doubler Circuit.

and reverse bias D_2. With D_1 ON, capacitor C_1 will charge to the polarity shown until its plate-to-plate voltage is equal to the peak value of the input sine wave (162.6 V). To simplify the operation of the circuit during this positive half-cycle of the input, the inset in Figure 6-4(b) represents D_1 as a closed switch and D_2 as an open switch. In this equivalent circuit it can be seen how C_1 is connected across the secondary of the transformer by D_1, and therefore it will charge to the peak value of the input sine wave. Because there is almost no resistance in the charge path of C_1, the capacitor will charge very quickly to $V_{S\,pk}$.

Figure 6-4(c) shows how the circuit will respond to the negative half-cycle of the input sine wave. When the input voltage swings negative, the top of the transformer secondary is made negative and this potential will forward bias D_2 and reverse bias D_1. Referring to the equivalent circuit in the inset in Figure 6-4(c), you can see what effect diodes D_1 and D_2 will have when they are represented as an open and closed switch. In this equivalent circuit it can be seen how C_2 is connected across the secondary of the transformer by D_2, and will charge to the peak value of the input sine wave. As there is almost no resistance in the charge path of C_2, the capacitor will charge very quickly to $V_{S\,pk}$. Because capacitors C_1 and C_2 are now *series-aiding voltage sources* (+/– → +/–), the sum of these two series peak voltages will be applied across the load.

$$V_{out} = 2 \times V_{S\,pk}$$

$$= 2 \times 162.6\ \text{V}$$

$$= 325.2\ \text{V}$$

Once the capacitors C_1 and C_2 are charged, the diodes D_1 and D_2 will only conduct during the peaks of the ac input. Between these peaks, C_1 and C_2 will discharge through the load as shown in Figure 6-4(d). The time constant formed by the output capacitors and the load resistance is normally very high, and therefore C_1 and C_2 will only discharge slightly. This slight discharge of C_1 and C_2 can be seen in the output voltage waveform in Figure 6-4(e). Looking closely at this output voltage waveform, you can see that the circuit is called a full-wave voltage doubler because the output capacitors C_1 and C_2 are charged for both half-cycles of the ac input, producing an output that closely resembles that of the filtered full-wave rectifier circuit. Comparing the input voltage waveform to the output voltage waveform in Figure 6-4(e), you can see that for every one cycle of the input (positive-to-positive peak) there are two cycles of the output (positive-to-positive-to-positive peak). The output dc ripple frequency of a full-wave voltage doubler will therefore equal twice that of the input ac frequency:

$$f_{out\ ripple} = 2 \times f_{in\ ac}$$

Output dc ripple frequency = 2 × *the input ac frequency*

In the circuit in Figure 6-4, the output ripple frequency will equal 120 Hz. The full-wave voltage doubler overcomes both disadvantages of the half-wave circuit by having a higher ripple frequency which is easier to filter. Because the output

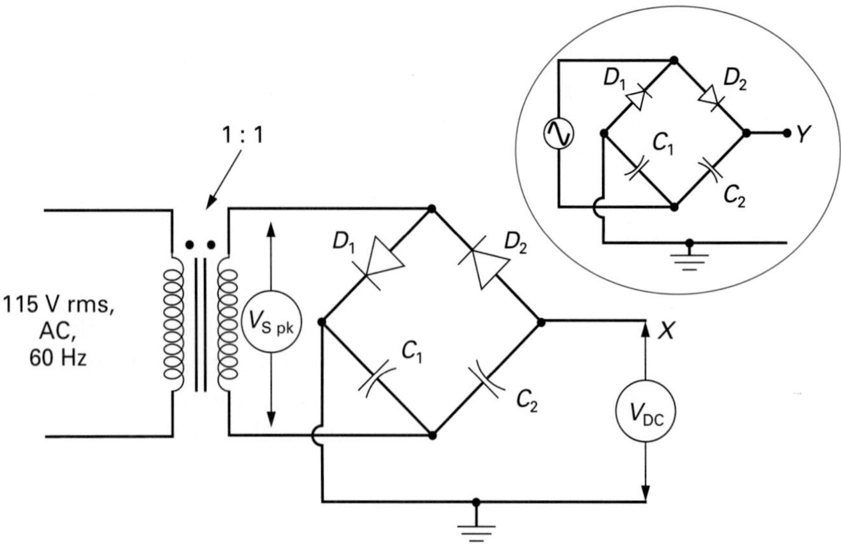

FIGURE 6-5 A DC Power Supply Circuit.

voltage is split across C_1 and C_2, each capacitor need only have a voltage rating that is slightly more than the peak of the ac input voltage.

■ **EXAMPLE:**

Identify the circuit in Figure 6-5 and then answer the following questions:

a. What would be the polarity and value of the output voltage at point *X,* and from the inset circuit at point *Y?*
b. What would be the output ripple frequency from both circuits?

■ ***Solution:***

The circuit in Figure 6-5 is a negative full-wave voltage doubler circuit, and the circuit in the inset in Figure 6-5 is a positive full-wave voltage doubler.

a. 115 V ac, rms × 1.414 = 162.6 V ac, peak

Point *X:*

$$V_{out} = 2 \times V_{S\,pk} = 2 \times (-162.6\text{ V}) = -325.2\text{ V}$$

Point *Y:*

$$V_{out} = 2 \times V_{S\,pk} = 2 \times (+162.6\text{ V}) = +325.2\text{ V}$$

b. Output dc ripple frequency = 2 × input ac frequency = 120 Hz

6-1-2 Voltage-Tripler Circuit

A voltage-tripler circuit will produce a dc output voltage that is three times the peak of the ac input voltage. Figure 6-6 shows how three diodes and three capacitors can be connected to form a voltage-tripler circuit. To explain how

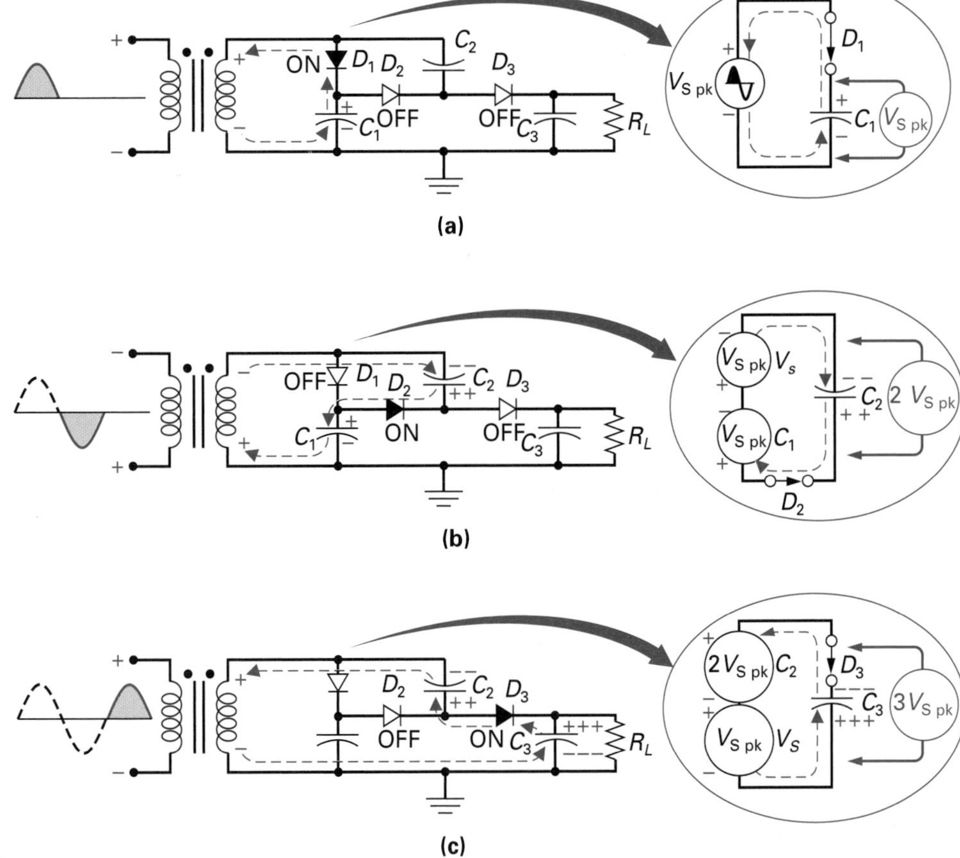

FIGURE 6-6 Voltage-Tripler Circuit.

this circuit operates, we will examine how it responds to three half cycles of the ac input sine wave. These three circuit states are shown in Figure 6-6(a), (b) and (c).

As can be seen in Figure 6-6(a), the first positive half cycle will turn D_1 ON and allow C_1 to charge to the peak of the ac secondary voltage. This condition is shown in the simplified diagram in the inset in Figure 6-6(a).

Figure 6-6(b) shows what will occur when the secondary voltage swings negative. Diode D_2 will be forward biased because its anode has been made positive by the charge on C_1, and its cathode has been made negative by the potential at the top of the transformer's secondary. Diode D_2 has now connected the transformer's secondary voltage, and the charge voltage of C_1 across the capacitor C_2, as shown in the inset in Figure 6-6(b). Because the transformer's secondary peak voltage ($V_{S\,pk}$), and the capacitor C_1's peak charge voltage ($V_{S\,pk}$) are now series-aiding voltage sources (+/– → +/–), capacitor C_2 will charge to the sum of these two series peak voltages ($C_2 = 2 \times V_{S\,pk}$).

Figure 6-6(c) shows what will happen at the peak of the next positive half cycle of the input. Diode D_3 will be forward biased because its anode has been made positive by the charge on C_2, and its cathode has been made negative by the potential at the bottom of the transformer's secondary. Diode D_3 has now connected the transformer's secondary voltage, and the charge voltage of C_2 across the capacitor C_3 as shown in the inset in Figure 6-6(c). Because the transformer's secondary peak voltage ($V_{S\ pk}$) and C_2's double peak charge voltage ($2 \times V_{S\ pk}$) are now series-aiding voltage sources (+/– → +/–), capacitor C_3 will charge to the sum of these two series peak voltages, which is three times the secondary peak voltage:

$$C_3 = V_{S\,pk} + (2 \times V_{S\,pk}) = 3 \times V_{S\,pk}$$

■ **EXAMPLE:**

What would be the peak output voltage across the load from the voltage-tripler circuit in Figure 6-6, if the ac input was 115 V rms?

■ ***Solution:***

$$115 \text{ V ac, rms} \times 1.414 = 162.6 \text{ V ac, peak}$$

$$\begin{aligned} V_{out} &= 3 \times V_{S\,pk} \\ &= 3 \times 162.6 \text{ V} \\ &= 487.8 \text{ V} \end{aligned}$$

6-1-3 *Voltage-Quadrupler Circuit*

A voltage-quadrupler circuit will produce a dc output voltage that is four times the peak of the ac input voltage. Figure 6-7 shows how four diodes and four capacitors can be connected to form a voltage-quadrupler circuit. Studying this circuit, you may have noticed that it contains two half-wave voltage-doubler circuits. The components D_1, C_1, D_2, C_2 form the upper half-wave voltage doubler, while D_3, C_3, D_4, C_4 make up the lower half-wave voltage doubler. Because these two circuits are connected in parallel, they will operate in unison to charge both C_2 and C_4 each to twice the peak of the ac input voltage. The series combination of the voltages developed across C_2 and C_4 will be four times the peak of

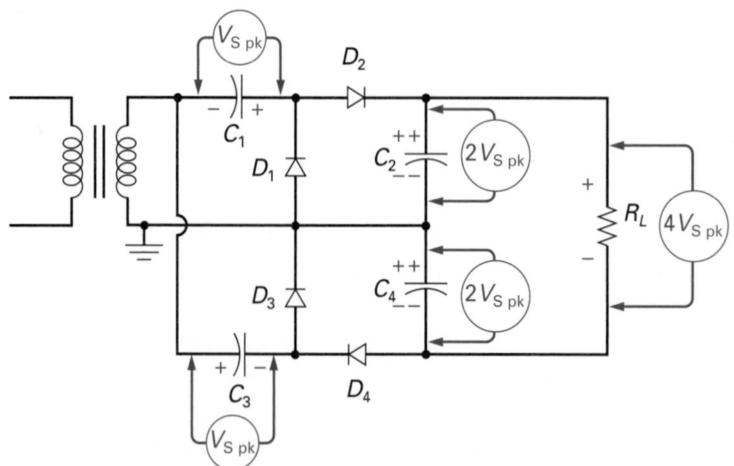

FIGURE 6-7 Voltage-Quadrupler Circuit.

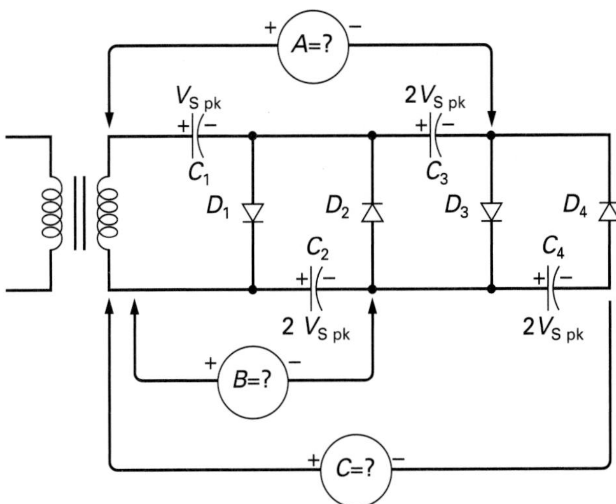

FIGURE 6-8 Voltage Multiplier Output Voltages.

the ac input voltage ($4 \times V_{S\,pk}$), and it is this voltage that will be applied across the load.

As mentioned previously, if the output voltage of a multiplier is increased by some multiple, the output current of the circuit will be decreased by the same multiple. Voltage quadruplers and higher order multiplier circuits generally have poor regulation because the amount of current they can supply is so limited. These circuits are only used in low current applications where variations in the voltage supplied can be tolerated.

■ **EXAMPLE:**

Referring to the voltage multiplier circuit in Figure 6-8, what multiple of the secondary peak voltage will be obtained at voltage points *A*, *B* and *C?*

■ ***Solution:***

$$A = 3 \times V_{S\,pk}$$
$$B = 2 \times V_{S\,pk}$$
$$C = 4 \times V_{S\,pk}$$

6-1-4 Voltage Multiplier Application— An Oscilloscope's DC Power Supply

One application for the voltage multiplier can be seen in the oscilloscope dc power supply circuit shown in Figure 6-9.

As you can see from the variety of output voltages from this dc power supply, an oscilloscope's electronic devices and circuits require several different dc supply voltages ranging from +5 V to –1200 V. From previous circuits you will recognize the 115 V/240 V transformer primary connections and the rectifier,

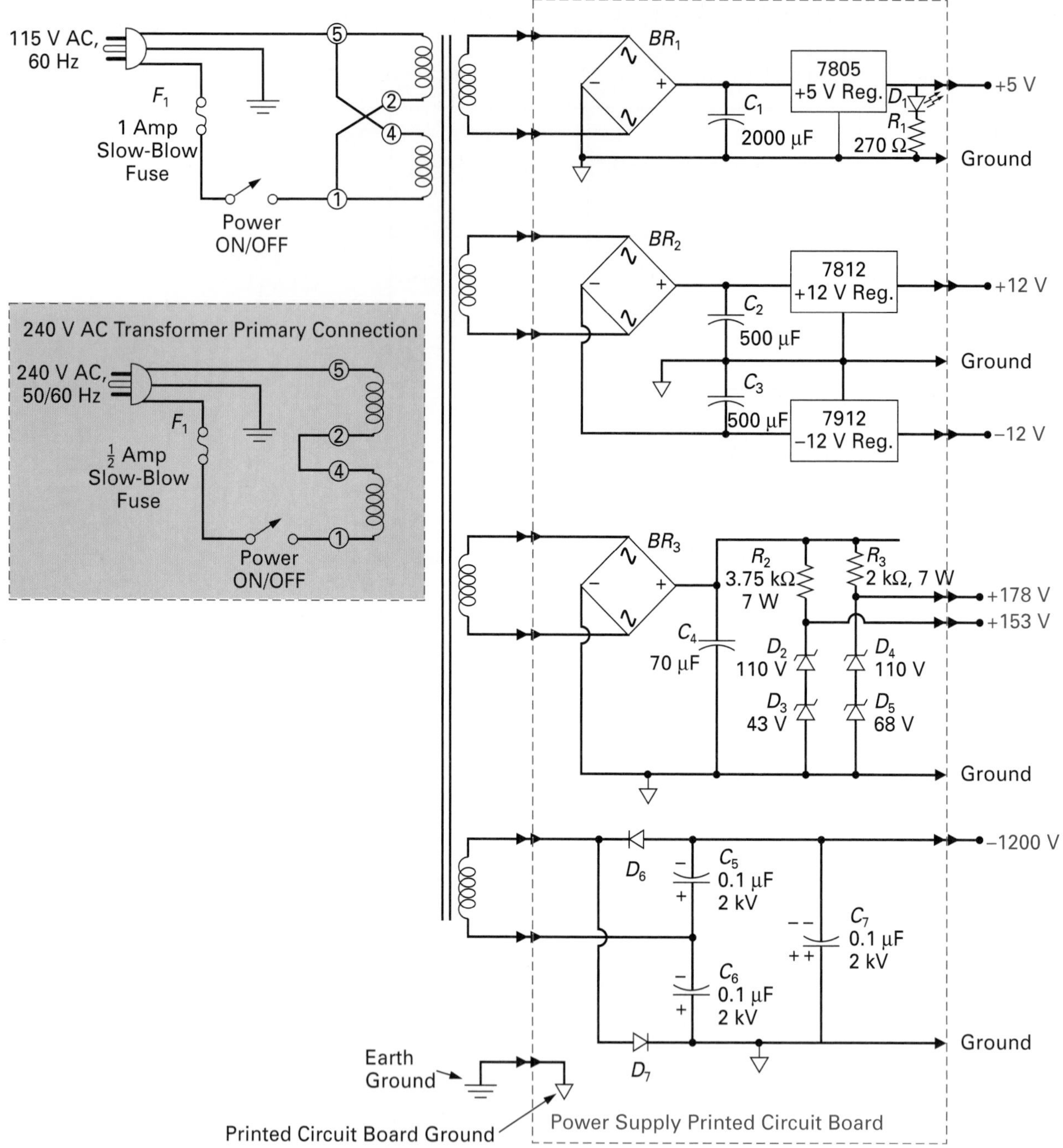

FIGURE 6-9 An Oscilloscope's DC Power Supply.

filter, and regulator circuits for the +5 V, +12 V, −12 V, +178 V, and +153 V dc supply voltages. The lower secondary of the transformer is connected to a full-wave voltage-doubler circuit that generates a −1200 V supply voltage for the oscilloscope's deflection plates. The components D_6, C_5, D_7, and C_6 form the basic negative full-wave voltage-doubler circuit, with the additional filtering capacitor C_7 being included to reduce the ripple output from this circuit.

6-1-5 Troubleshooting Voltage-Multiplier Circuits

In this section, we will discuss how to troubleshoot a typical voltage-multiplier circuit. As before, always try to apply the three-step troubleshooting procedure: first diagnose whether a problem really exists, then proceed to isolate the problem by applying a logical and sequential reasoning process, and finally repair the circuit and perform a final test.

As an example, let us apply our troubleshooting procedure to the voltage tripler circuit shown in Figure 6-10(a). As we discovered in previous sections, the final capacitor output voltage is dependent on the preceding capacitors charging properly. Because their charging is dependent on the correct operation of the switching diodes, any of the multiplier's circuit components can affect the final output voltage. This interdependency means that a failure in any of the components in a voltage multiplier circuit will affect the final output voltage. As a troubleshooter, your difficulty is to try and isolate which component in the circuit is causing the resulting problem.

To explain the logic behind testing a typical voltage multiplier, we will examine the troubleshooting procedure for the voltage tripler circuit shown in Figure 6-10(a). To help clarify the logical procedure for troubleshooting this circuit, let us follow the steps shown in the troubleshooting chart given in Figure 6-10(b). Looking at the top block, you can see that the best approach is to first test the output voltage across C_3. This voltage should be three times the secondary peak voltage if the circuit is operating correctly. In actual fact, the diodes in the voltage-tripler circuit will drop some voltage and, due to the ripple caused by the capacitor's charge and discharge cycle, the voltage measured will not exactly be equal to three times the secondary peak voltage. Keep this in mind when measuring all of the voltages developed across the capacitors in any multiplier circuit. Remember that your voltmeter is calibrated to display the rms value of the ripple which, based on only a slight fluctuation, should be close to the peak value. Returning to the troubleshooting tree in Figure 6-10(b), you can see that once we have measured the output voltage across C_3, we can follow one of two paths based on whether the voltage reading was correct (the circuit is functioning normally) or incorrect. If the output voltage from the voltage multiplier circuit is not correct (in most cases the voltage will be lower than expected), your next step should be to disconnect the load from the voltage tripler circuit by opening the connection at point *A*. If the output from the voltage tripler returns to its normal value, there is a short somewhere in the load and it is pulling down the output voltage of the multiplier circuit. If, however, the output voltage across C_3 remains at its low inaccurate value, the problem exists somewhere in the circuit before point *A*. You may recognize the previous troubleshooting technique as the "half-split method." Now that we have determined that the problem is somewhere in the voltage multiplier circuit, we can use our understanding of the circuit's operation to further isolate the problem. For example, if the voltage across C_3 is short by V_S, then we should check the voltage across C_1, whereas if the voltage across C_3 is short by two times V_S, we should check the voltage across C_2. If C_2 is not charged to twice the secondary voltage, but C_1 is charged to V_S, then the problem has been isolated to C_2 and D_2. If, however, the voltages across C_1 and C_2 are not normal, then we should check C_1, D_1, and the ac source voltage.

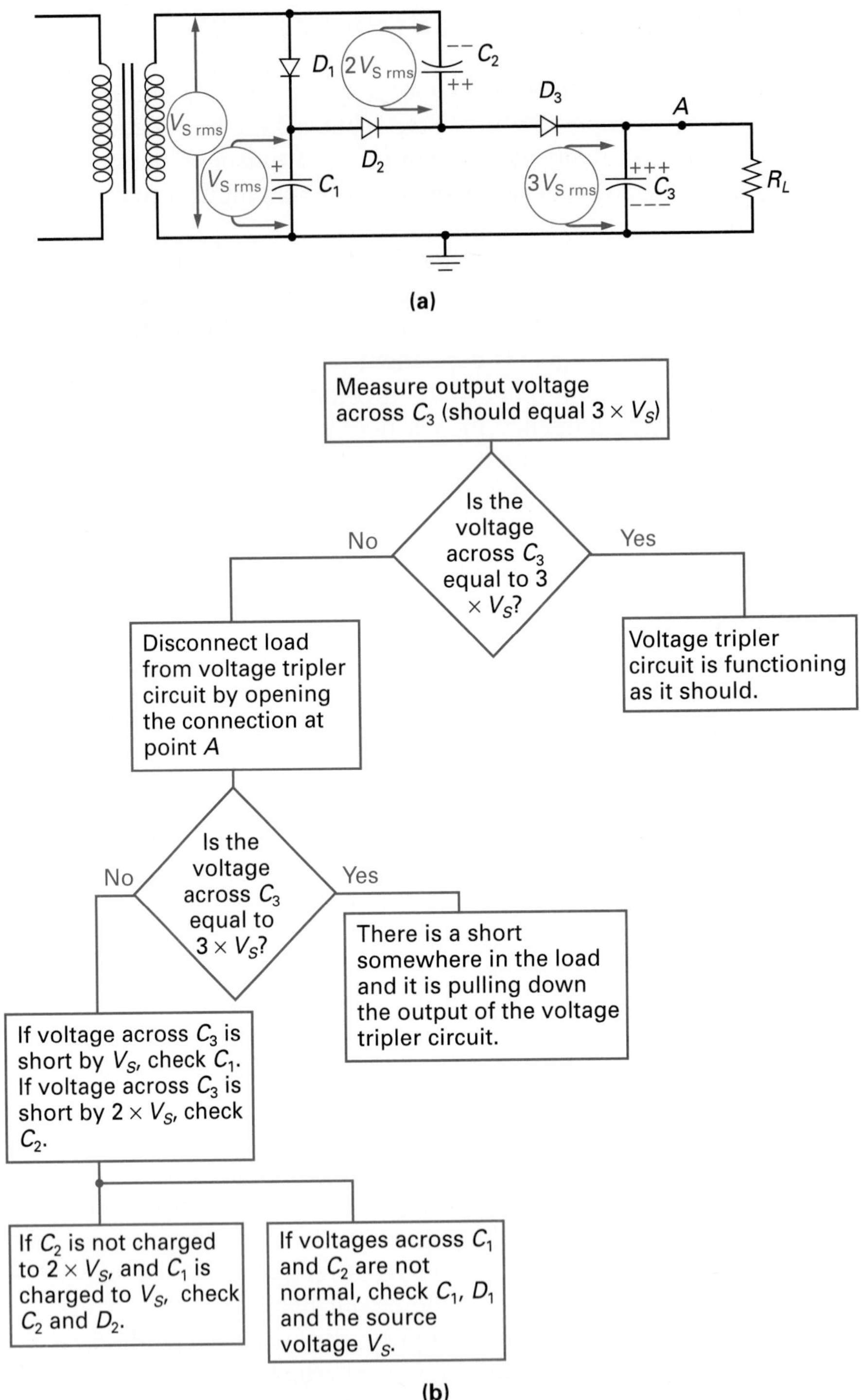

FIGURE 6-10 Troubleshooting a Voltage Multiplier Circuit. (a) Voltage Tripler Circuit. (b) Voltage Tripler Troubleshooting Tree.

SELF-TEST REVIEW QUESTIONS FOR SECTION 6-1

1. A half-wave voltage doubler whose input is 115 V rms will produce an output that is about ____________ volts, peak.
2. What are the two disadvantages of the half-wave voltage doubler compared to the full-wave voltage doubler?
3. What would be the peak output from a voltage tripler and voltage quadrupler circuit for a 115 V rms, ac input?
4. Assuming a 115 V rms, 60 Hz input, what would be the output ripple frequency from a half-wave voltage doubler circuit and full-wave voltage doubler circuit?
5. Which is easier to filter—a high or low ripple frequency?

6-2 DIODE CLIPPER CIRCUITS

A **clipper circuit** is used to cut off or eliminate an unwanted section of a waveform. It is used in basically one of two applications:

a. In many applications, *it is used to remove a natural part of the waveform.* For example, the half-wave rectifier is a basic clipper circuit since it eliminates either the positive or negative alternation of the ac input signal.

b. In other applications, *it is used to prevent a voltage from exceeding a certain value.* In these instances, the clipper circuit is often called a **limiter** because it will limit a high amplitude voltage pulse or other signal.

6-2-1 *Series Clipper Circuits*

The diode is an ideal clipper since it can be connected in one of two ways and biased at a certain reference level to clip off a certain part of the input waveform. There are two types of clipper circuits: the **series clipper,** which contains a diode that is in series with the load, and the **shunt clipper,** which has a diode that is in shunt, or in parallel, with the load. In this section we will examine the different series clipper circuits and their applications.

Basic Series Clipper Circuits

Figure 6-11 shows the two basic series clipper circuits along with their input and output waveforms. Comparing these circuits to the previously discussed half-wave rectifier circuits, you will notice that they are identical.

The circuit shown in Figure 6-11(a) is called a **negative series clipper** because this circuit has a diode connected in *series* with the load or output, and its orientation is such that it will *clip* off the *negative* alternation of the ac input. Referring to the waveforms in Figure 6-11(b), you can see that during the positive alternation of the ac input, the diode is forward biased and equivalent to a closed switch, as shown in the inset. The positive alternation is connected to the

Clipper Circuit or Limiter
Used to eliminate an unwanted section of waveform.

Series Clipper
A circuit that will clip off part of the input signal. Also known as a limiter since the circuit will limit the ac input. A series clipper circuit has a clipping or limiting device in series with the load.

Shunt Clipper
A circuit that will clip off part of the input signal. Also known as a limiter since the circuit will limit the ac input. A shunt clipper circuit has a clipping or limiting device in shunt with the load.

Negative Series Clipper
A circuit that has a diode connected in series with the load or output, and its orientation is such that it will clip off the negative alternation of the ac input.

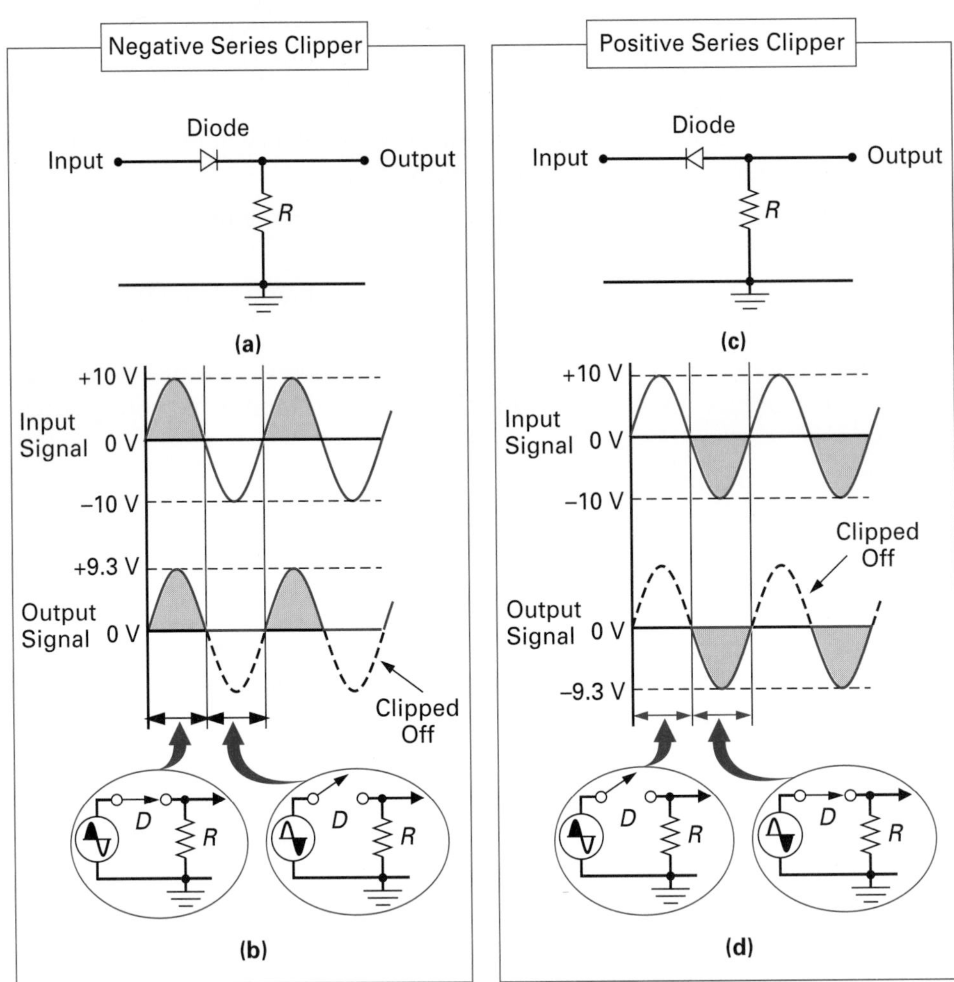

FIGURE 6-11 Basic Series Clipper Circuits.

output; however, there will be a 0.7 V loss in voltage between input and output due to the series connected diode.

$$V_{out} = V_{in} - 0.7 \text{ V}$$

$$V_{out} = 10 \text{ V} - 0.7 \text{ V} = 9.3 \text{ V}$$

During the negative alternation of the ac input, the diode is reverse biased and equivalent to an open switch as shown in the inset in Figure 6-11(b). The negative alternation will be blocked or clipped by the series diode, and the output voltage will be 0 V.

Positive Series Clipper
A circuit that has a diode connected in series with the load or output, and its orientation is such that it will clip off the positive alternation of the ac input.

The circuit shown in Figure 6-11(c) is called a **positive series clipper** since this circuit has a diode connected in *series* with the load or output, and its orientation is such that it will *clip* off the *positive* alternation of the ac input. Comparing this circuit to the one in Figure 6-11(a), you can see that the only difference is the direction of the diode. Here, the positive alternation of the ac input will be clipped off, and the negative alternation of the ac input will be passed to the output. Referring to the waveforms in Figure 6-11(d), you can clearly see this result. Once again the amplitude of the output voltage takes into account the 0.7 V drop due to the series connected diode.

As mentioned earlier, the construction of these circuits is no different than that of the half-wave rectifier circuits discussed in detail in the previous chapter. The difference between the two is their application: a half-wave rectifier circuit is used in a dc power supply to convert ac to dc, whereas a series clipper circuit is used to clip off or eliminate a certain part of an ac signal. To draw a parallel, this "same construction/different application" situation could be compared to the electromagnet and inductor. They are also constructed in exactly the same way but used in two completely different applications.

■ **EXAMPLE:**

Referring to the circuit in Figure 6-12(a), sketch the waveforms that would be obtained for each of the test points 1, 2 and 3. Label each waveform with their voltage peak values and time periods.

■ ***Solution:***

Figure 6-12(b) shows the waveforms that will be obtained at test points 1, 2 and 3 for the circuit in Figure 6-12(a). The differentiator circuit (made up of C_1 and R_1) is included to convert the square wave input to the positive and negative spike signal shown at test point 2. Like all differentiator circuits, the values of R and C are chosen so that the RC time constant is short compared to the half-cycle period of the input square wave. In this example, the time constant is equal to 12 ms ($\tau = R \times C = 1\ \text{k}\Omega \times 12\ \mu\text{F} = 12$ ms). Even taking into account 5 time constants (which is the time it takes for a capacitor to fully charge or discharge), the time of 60 ms is still short compared to the half-cycle period of the input square wave which is 100 ms. The negative series clipper circuit will clip or block the negative spike from the differentiator circuit, and therefore the output signal at TP_3 will be a positive 5 V spike.

A circuit such as the one shown in Figure 6-12 is often used in applications in which a positive pulse is needed from a square wave. This narrow pulse could be used to activate or trigger a circuit into operation only for the duration of the positive spike. This kind of operation is called **positive-edge triggering** because the circuit is only being triggered into operation on the positive edge of the square wave input. If a positive series clipper circuit was used in place of the negative series clipper in Figure 6-12(a), the positive spike would be clipped and the negative spike output could provide **negative-edge triggering.**

Positive-Edge Triggering
Circuit only triggered into operation on the positive edge of the square wave input.

Negative-Edge Triggering
Circuit only triggered into operation on the negative edge of the square wave input.

Biased Series Clipper Circuits

With the basic series clippers discussed in the previous section, the clipping level was at 0 V with the direction of the diode determining whether everything above or below 0 V was eliminated. With **biased series clipper circuits,** the clipping level can be changed by applying a bias voltage to the series connected diode. The two basic types of biased series clipper circuits are shown in Figure 6-13.

In Figure 6-13(a), the cathode of the diode is set at +5 V by a dc source. This +5 V dc supply will set the clipping level because the diode cannot conduct

Biased Series Clipper Circuit
A circuit in which the clipping level can be altered by applying a bias voltage to the series connected diode.

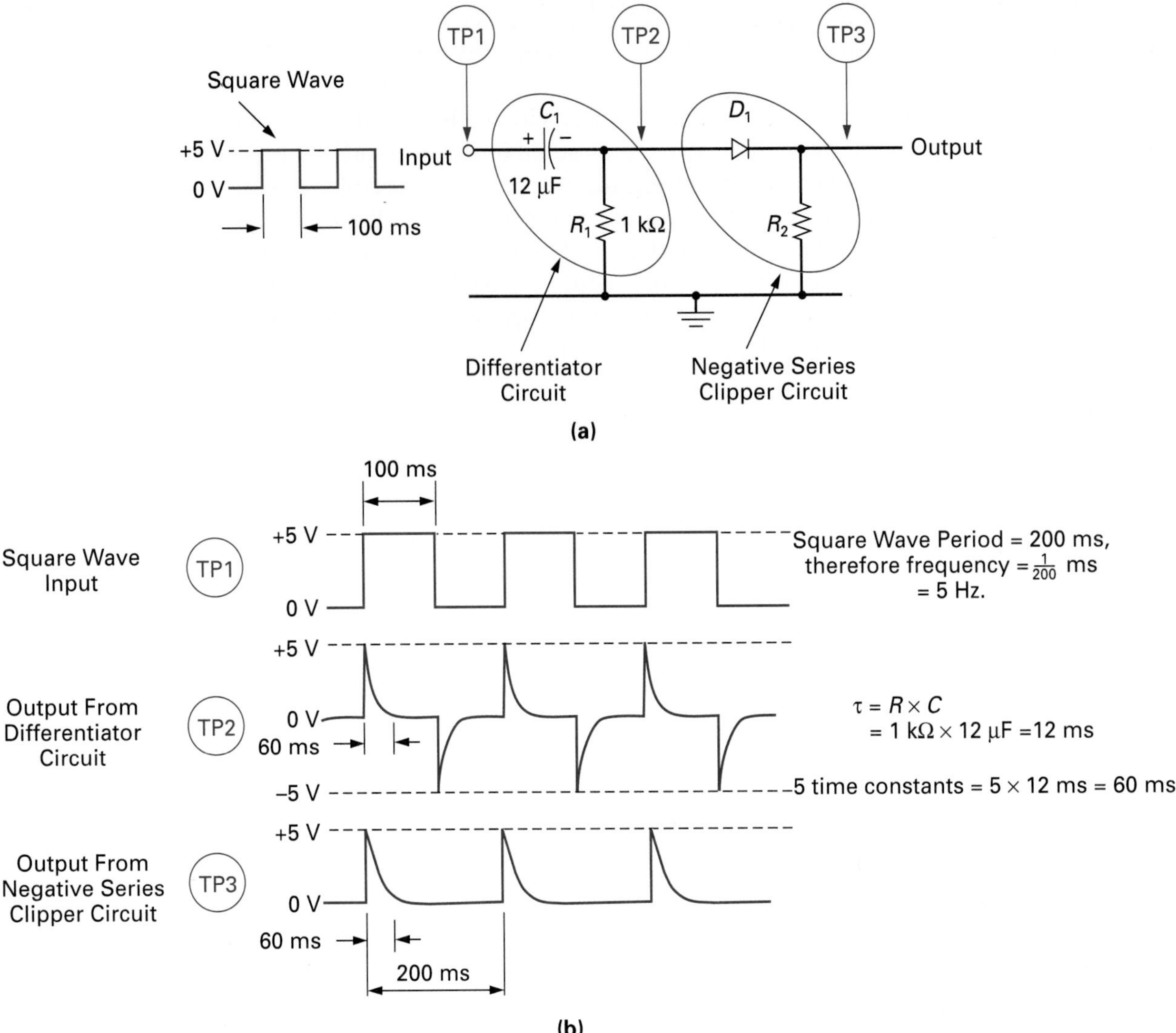

FIGURE 6-12 Positive Pulse Generator Circuit.

until the input signal overcomes the +5 V supply and the diode's 0.7 V barrier voltage. Referring to the input/output waveforms in Figure 6-13(b), you can see that the output remains at +5 V except when the positive ac input exceeds this voltage level. At this positive peak of the ac input waveform, the diode will conduct and switch this portion of the input signal through to the output. The peak output will, of course, be 0.7 V less than the input due to the voltage drop across the series connected diode. As you can see in the output waveform in Figure 6-13(b), part of the positive alternation and all of the negative alternation will be clipped off.

Figure 6-13(c) shows how the circuit can be converted to a *biased positive series clipper* by reversing the direction of the diode and the polarity of the bias voltage. The –5 V dc supply will set the clipping level because the diode cannot conduct until the input signal overcomes the –5 V supply and the diode's 0.7 V

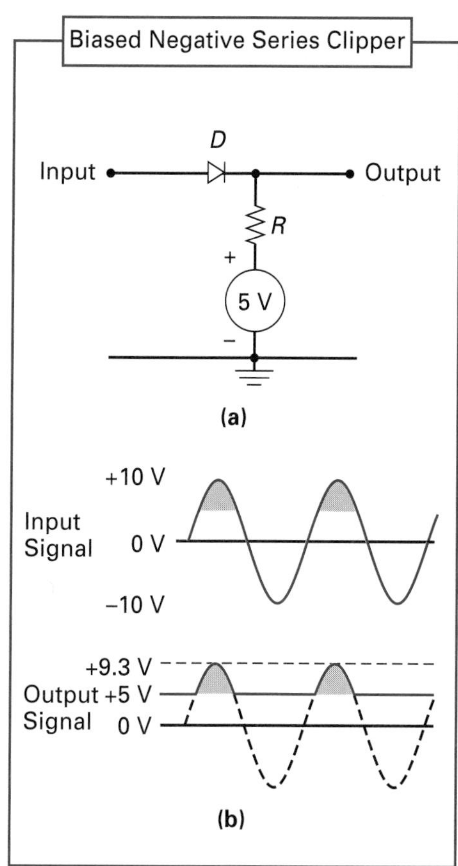

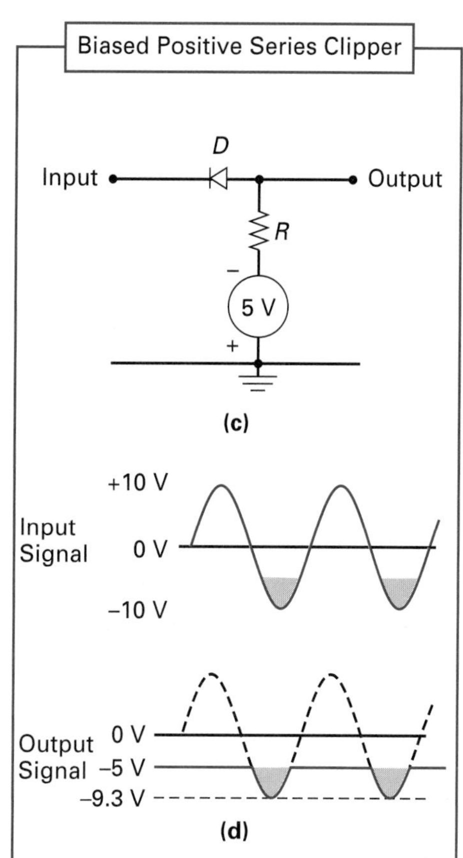

FIGURE 6-13 Biased Series Clipper Circuits.

barrier voltage. Referring to the input/output waveforms in Figure 6-13(d), you can see that the output remains at –5 V except when the negative ac input exceeds this voltage level. At this negative peak of the ac input waveform, the diode will conduct and switch this portion of the input signal through to the output. The peak output will, of course, be 0.7 V less than the input due to the voltage drop across the series connected diode. As you can see in the output waveform in Figure 6-13(d), part of the negative alternation and all of the positive alternation will be clipped off.

Series Clipper Application—Amplitude Modulation (AM) Circuits

Electronic communication circuits will be covered in detail in a later chapter. For now, however, let us introduce the topic and discuss two important communication circuits, both of which can be constructed to make use of a series clipper circuit.

Communication is defined as *"a process by which information is exchanged."* The two basic methods of transferring information are the *spoken word* and the *written word.* Communication of the spoken word began with face-to-face contact and then evolved into telephone and radio communications. Communication of the written word began with hand-carried letters and evolved into newspapers, the mail system, the telegraph, and, now, electronic mail.

The distance we could communicate became important, and systems of long distance communication were developed that began with smoke signals and beating drums. It was, however, radio that first enabled us to significantly increase our communication range.

As an example, let us examine a simple radio communication system. Figure 6-14(a) shows the block diagram of a basic voice or *audio* radio communicator. With radio communications, the *sound waves* are first converted to *electrical waves* by the microphone and then these electrical signals are converted to *electromagnetic or radio waves* by the transmitting antenna. If the voice, or audio (which is a Greek word meaning "I hear"), signal was transmitted at its original frequency (20 Hz to 20 kHz), three basic problems would occur:

a. First, the transmitting antenna would have to be nearly 25 kilometers long.

b. Second, the systems transmitting the low-frequency audio signals would be very inefficient.

c. Third, and most important, if every station were to transmit the same audio frequencies, they would all interfere with one another.

Carrier Frequency
The transmitting frequency assigned to radio stations.

Modulation
The process of converting a low-frequency information signal into a high-frequency carrier signal.

Amplitude Modulation
The amplitude of the high-frequency carrier is changed or varied to follow the amplitude of the low-frequency information signal.

These three problems can be solved by assigning each station that transmits radio signals its own transmitting frequency, called a **carrier frequency.** For example, your local AM radio stations each have their own carrier frequency in the AM radio band, and this high frequency acts as a carrier for the stations' speech and music information. With every station transmitting their information at a different frequency, there will be no interference between stations. Each station, however, will have to convert their lower frequency audio signals to their assigned higher carrier frequency before transmitting. This process of converting a low-frequency information signal to a high-frequency carrier signal is called **modulation.**

Figure 6-14(b) shows how a series clipper circuit can be used to achieve **amplitude modulation (AM).** With AM, *the amplitude of the high-frequency carrier is changed or varied to follow the amplitude of the low-frequency information signal.* This is best seen by looking at the change in the waveforms between the two input waveforms at test points 1 and 2 (TP1 and 2), and the output waveform at test point 3 (TP3). Studying the output waveform (TP5), you can see that the amplitude of the carrier varies at the same rate as the audio signal. In fact, the envelope of the carrier (which is shown as a dotted line) is an exact replica of the information signal input.

The circuit in Figure 6-14(a) achieves amplitude modulation of the carrier wave in the following way. The input information signal (at TP1) and the carrier (at TP2) are applied to R_1 and R_2 and combine to produce the waveform shown at TP3. As you can see, the carrier at TP3 is now varying or riding on the varying level of the information signal. With switch 1 open, the negative series clipper circuit, formed by D_1 and R_3, will produce a series of positive pulses that will vary in amplitude in the same way as the audio signal, as shown in the waveform at TP4. If we were to now close SW_1, the positive pulses of current will flow through to the tank circuit made up of C_1 and L_1. This tank circuit is tuned to the carrier frequency, and each positive pulse from D_1 will cause the tank to resonate or ring. The flywheel action of this tank will produce a negative half-cycle that has the same amplitude as the positive half-cycle. In other words, a larger positive pulse

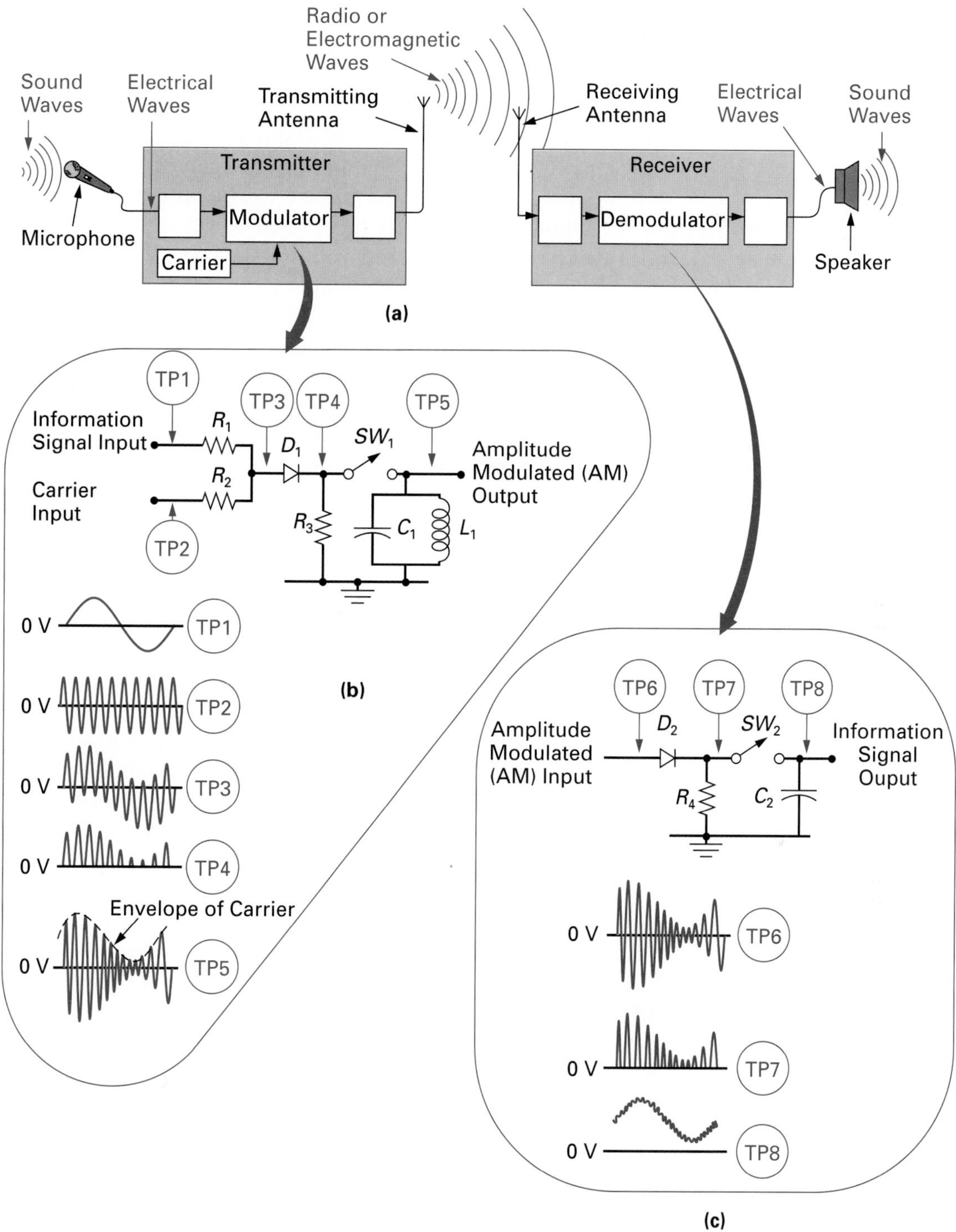

FIGURE 6-14 **AM Communication Circuits. (a) Block Diagram of Basic AM Transmitter and Receiver. (b) Diode AM Modulator Circuit. (c) Diode AM Demodulator or Detector Circuit.**

from the clipper circuit will produce a corresponding larger negative pulse from the tank. Similarly, a low amplitude positive pulse will produce a low-amplitude negative pulse, as long as the tank circuit's Q is not too high. As you can see in the final amplitude modulated wave at TP5, the amplitude of the high frequency carrier is now varying in accordance with the low frequency information signal.

The switch (SW_1), which would not normally be included in this circuit, is there to help in the circuit explanation so that we can see the progression of the waveforms between input and output. If this switch was not present, the waveform at TP5 would also be seen at TP4, and we would not be able to see the difference between the output of the series clipper and the final output of the AM modulator.

As can be seen in the block diagram in Figure 6-14(a), a circuit has to be included at the receiving end to recover the information signal from the radio-frequency (RF) carrier because the human ear cannot hear high radio frequency signals. The circuit that extracts the audio information from the high-frequency carrier is called a **demodulator** or **detector circuit.**

Demodulator or Detector Circuit The circuit that extracts the audio information from the high-frequency carrier.

Figure 6-14(c) shows a simple diode AM demodulator or detector circuit. The input to this circuit is the amplitude modulated wave that has been picked up by the receiving antenna and is shown at TP6. With SW_2 open, diode D_2 and R_4 act as a negative series clipper circuit, clipping off the negative alternations of the input and producing a set of amplitude varying positive pulses across R_4, as shown at TP7. With SW_2 closed, C_2 will charge quickly to the peak of each positive pulse. Between positive pulses, C_2 will try to discharge through R_4. However, the RC time constant of R_4 and C_2 is chosen because a long time constant ensures C_2 will discharge only slightly. The final output developed across C_2, and shown at TP8, will follow the envelope of the receiver's AM input wave. This final output information signal will have a slight ripple; however, the ripple will not be noticed because the carrier frequency is so much higher than the frequency of the information signal. Comparing the final output from the receiver's demodulator circuit (at TP8), you can see that it varies in exactly the same way as the original information signal input to the transmitter modulator circuit (TP1).

Once again, switch SW_2 was included in this circuit to help in the circuit explanation. If this switch was not present, the waveform at TP8 would also be seen at TP7, and we would not be able to see the difference between the output of the series clipper and the final output of the AM demodulator.

6-2-2 Shunt Clipper Circuits

As previously mentioned, there are two types of clipper circuits: the **series clipper,** which contains a diode that is in series with the load, and the **shunt clipper,** which has a diode that is in shunt, or in parallel, with the load. In this section, we will examine the different shunt clipper circuits and their applications.

Basic Shunt Clipper Circuits

Figure 6-15 shows the two basic shunt clipper circuits along with their input and output waveforms. These shunt clipper circuits achieve exactly the same results as the series clipper circuits, but operate in almost an opposite way. To

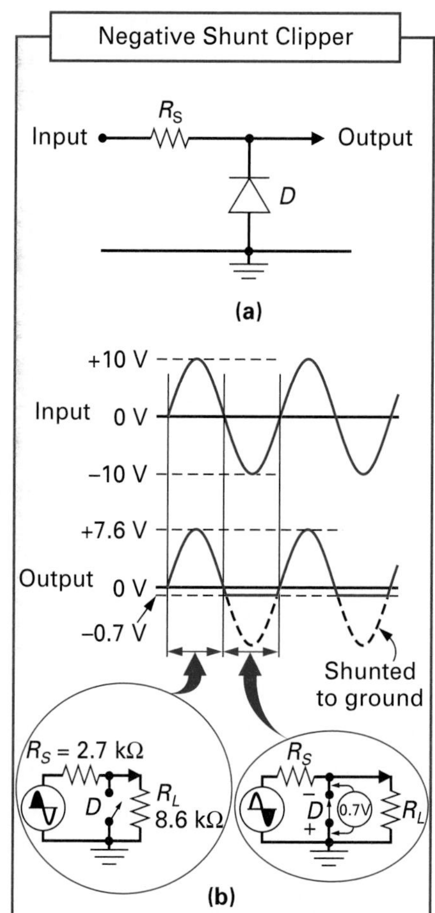

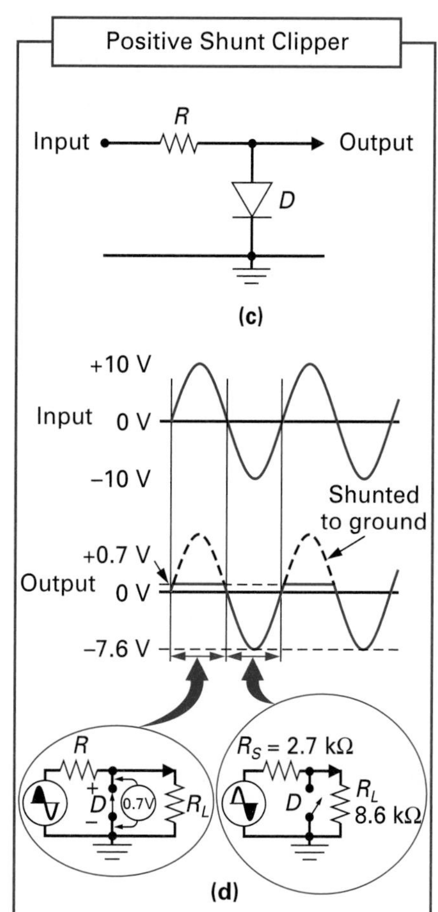

FIGURE 6-15 Basic Shunt Clipper Circuits.

explain this, the series clipper produced an output when the diode was forward biased and no output when the diode was reverse biased. As you will see in the operation of the shunt clipper, it will short the input signal to ground when it is forward biased and produce an output when the diode is reverse biased. Let us now examine the two basic shunt clipper circuits in more detail.

The circuit shown in Figure 6-15(a) is called a **negative shunt clipper** since this circuit has a diode connected in *shunt* with the load or output, and its orientation is such that it will *clip* off the *negative* alternation of the ac input by shunting it to ground. Referring to the waveforms in Figure 6-15(b), you can see that during the positive alternation of the ac input, the diode is reverse biased and equivalent to an open switch, as seen in the inset. With the diode effectively removed from the circuit, the output voltage developed across the load can be calculated using the standard voltage-divider formula.

Negative Shunt Clipper
A circuit that has a diode connected in shunt with the load or output, and its orientation is such that it will clip off the negative alternation of the ac input.

$$V_{out} = \frac{R_L}{R_L + R_S} \times V_{in}$$

Applying this formula to the example values given in the inset for the positive alternation in Figure 6-15(b), the peak output voltage will be as follows:

$$V_{out} = \frac{R_L}{R_L + R_S} \times V_{in}$$

$$V_{out} = \frac{8.6\ \text{k}\Omega}{8.6\ \text{k}\Omega + 2.7\ \text{k}\Omega} \times 10\ \text{V} = 7.6\ \text{V}$$

As you can see from this example, although the output signal will resemble the positive alternation of the input in shape, the peak output will be less than the peak input voltage due to the voltage division of R_S and R_L.

During the negative alternation of the ac input signal, the shunt clipper diode is forward biased, as seen in the inset in Figure 6-15(b). The forward biased diode will develop its usual drop of 0.7 V and because the diode is connected in parallel with the load, the load voltage will equal the forward voltage drop across the diode. To state this with a formula, the output voltage will equal

$$V_{out} = -0.7\ \text{V}$$

During the negative alternation, the output voltage will equal –0.7 V, with the remaining voltage being dropped across the series resistor, R_S.

$$V_{R_S} = -10\ \text{V}_{peak} - (-0.7\ \text{V}) = -9.3\ \text{V}_{peak}$$

During this negative alternation of the ac input, the reason the series resistor R_S is included is clear. When the shunt diode is forward biased, R_S acts as a current-limiting resistor. Without R_S in the circuit, the diode would short the ac input voltage to ground, which would probably cause the diode's maximum forward current rating to be exceeded. In most applications, the value of R_S is chosen to be a low value so that when the diode is reverse biased the voltage drop across R_S will be small, and most of the input signal voltage will be developed across the load.

To clip off, or shunt to ground, the positive alternation of the ac input, we simply change the diode's direction as shown in Figure 6-15(c). Referring to the waveforms for the **positive shunt clipper**, shown in Figure 6-15(d), you can see that during the positive alternation of the ac input the diode is forward biased and therefore the output is +0.7 V. During the negative alternation of the ac input, the diode is reverse biased and, if we assume the same values of R_S and R_L, the negative peak output will be –7.6 V due to the voltage divider action of R_S and R_L.

Positive Shunt Clipper
A circuit that has a diode connected in shunt with the load or output, and its orientation is such that it will clip off the positive alternation of the ac input.

■ **EXAMPLE:**

Referring to Figure 6-16, calculate the peak of the square wave's output voltage.

■ ***Solution:***

When the input signal goes negative, the shunt clipper diode will conduct and shunt the signal to ground. When the input signal is positive, however, the diode is reverse biased and the series resistor and the load form a voltage divider, with the following voltage being developed across the output or load:

$$V_{out} = \frac{R_L}{R_L + R_S} \times V_{in}$$

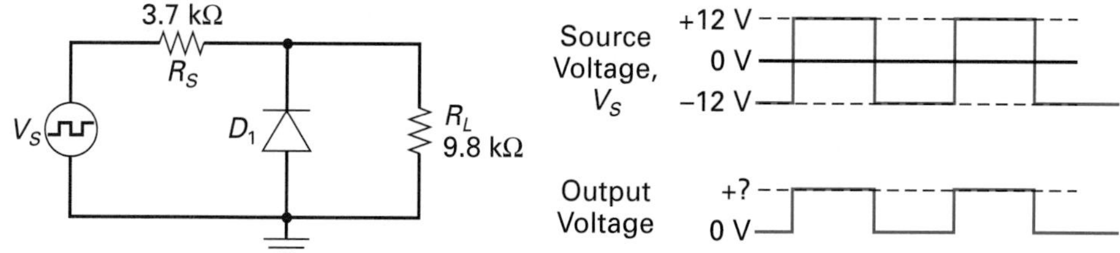

FIGURE 6-16 Positive Square Wave Generator Circuit.

$$V_{out} = \frac{9.8\text{ k}\Omega}{9.8\text{ k}\Omega + 3.7\text{ k}\Omega} \times 12\text{ V} = \frac{9.8\text{ k}\Omega}{13.5\text{ k}\Omega} \times 12\text{ V} = 8.7\text{ V}$$

Biased Shunt Clipper Circuits

Like the series clipper circuit, shunt clipper circuits can have their clipping level adjusted by introducing a bias voltage. Figure 6-17 illustrates how a bias voltage can be applied to the shunt connected diodes to adjust the point at which a certain section of the input signal is eliminated.

Figure 6-17(a) shows how a shunt connected diode and a bias voltage can be connected to clip off or shunt to ground a section of the negative alternation. Referring to the waveforms in Figure 6-17(b), you can see that during the positive alternation of the ac input the diode is reverse biased and therefore equivalent to an open switch, as shown in the inset. Assuming the values given in the inset for the positive alternation in Figure 6-17(b), the output voltage developed across the load will be:

$$V_{out} = \frac{R_L}{R_L + R_S} \times V_{in}$$

$$V_{out} = \frac{9.3\text{ k}\Omega}{9.3\text{ k}\Omega + 1.3\text{ k}\Omega} \times 10\text{ V} = \frac{9.3\text{ k}\Omega}{10.6\text{ k}\Omega} \times 10\text{ V} = 8.8\text{ V}$$

Referring back to Figure 6-17(a), you can see that the anode of the diode is connected to –5 V. This means that the diode can only conduct when the input voltage exceeds the bias voltage of –5 V and the diode's barrier voltage of 0.7 V, a total of –5.7 V. Referring to the waveforms in Figure 6-17(b), you can see that when the negative alternation of the ac input exceeds –5.7 V, the diode is forward biased and the input signal is clipped at this voltage.

$$V_{out} = V_{bias} + V_F$$

$$= -5\text{ V} + (-0.7\text{ V}) = -5.7\text{ V}$$

Figure 6-17(c) shows how the positive peak of the ac input signal can be clipped by reversing the direction of the diode and the polarity of the bias voltage from Figure 6-17(a). Referring to the waveforms for this biased positive shunt clipper circuit in Figure 6-17(d), you can see that when the input exceeds +5.7 V, the diode is forward biased and the input signal will be clipped at this voltage.

$$V_{out} = V_{bias} + V_F = +5\text{ V} + (+0.7\text{ V}) = +5.7\text{ V}$$

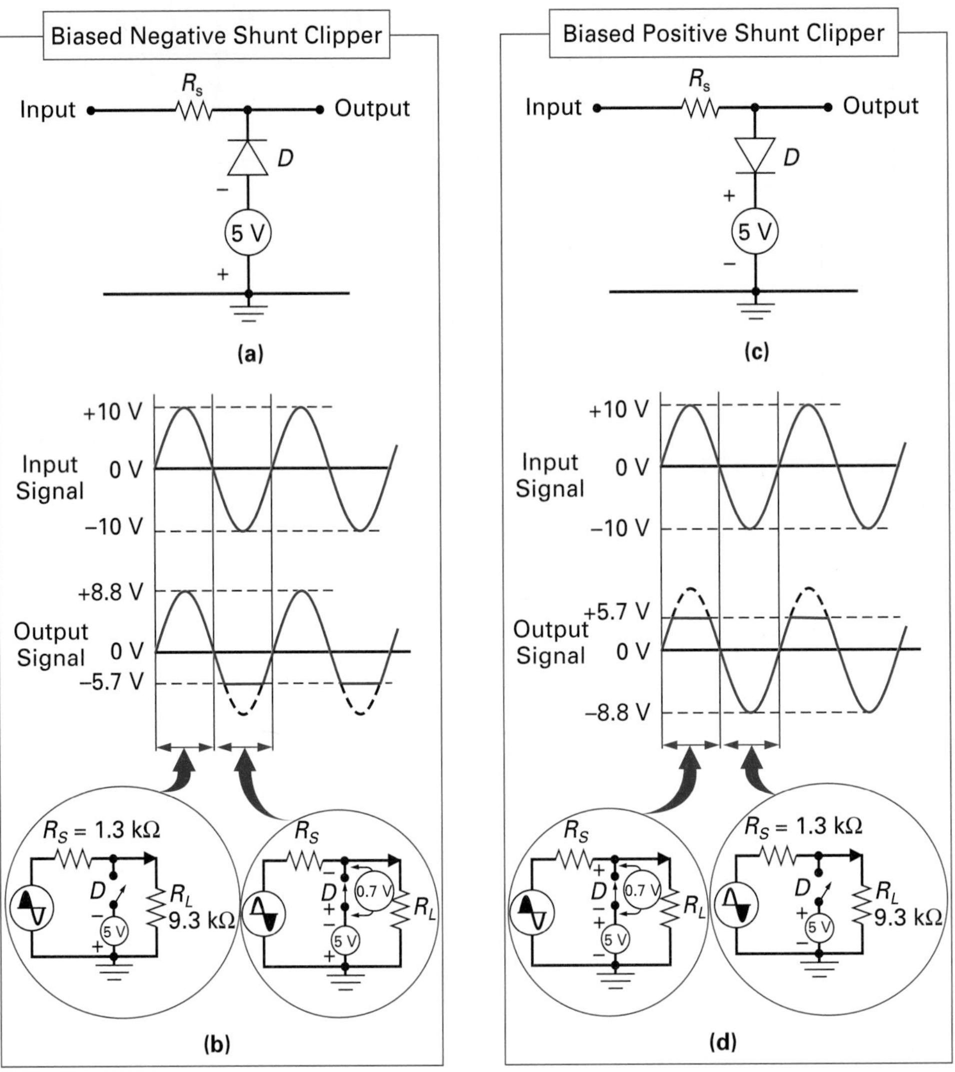

FIGURE 6-17 Biased Shunt Clipper Circuits.

During the negative alternation of the ac input signal, the diode is reverse biased and therefore equivalent to an open switch, as shown in the inset in Figure 6-17(d). Assuming the values given in this inset, the output voltage developed across the load will be

$$V_{out} = \frac{R_L}{R_L + R_S} \times V_{in}$$

$$V_{out} = \frac{9.3 \text{ k}\Omega}{9.3 \text{ k}\Omega + 1.3 \text{ k}\Omega} \times (-10 \text{ V}) = -8.8 \text{ V}$$

Variable Shunt Clipper Circuit

Figure 6-18 shows how a potentiometer can be used to adjust the positive shunt clipping level. In this circuit, R_1 will vary the value of positive biasing voltage applied to the cathode of the shunt clipping diode. When the positive alter-

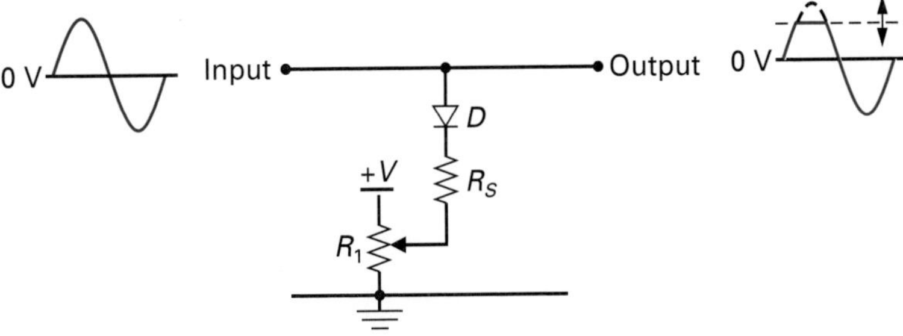

FIGURE 6-18 Variable Shunt Clipper Circuit.

nation of the ac input signal exceeds the positive voltage set at the diode's cathode and the diode 0.7 V barrier voltage, the diode will turn ON and limit the positive output to this voltage. By reversing the diode's direction and the polarity of the bias voltage, the circuit could be made to act as a variable negative shunt clipper.

Zener Shunt Clipper Circuit

A zener shunt clipper circuit makes use of the zener diode's forward biased switching action and its reverse bias zener action to clip both the positive and negative alternation of the input signal. Figure 6-19(a) shows how differently a clipper circuit will react if a zener diode is used instead of a P-N junction diode. When the input signal goes positive, the zener diode will remain OFF until the input signal reaches the diode's zener voltage (V_Z). At this time, the zener diode will conduct and hold the output voltage at this level. When the input signal goes

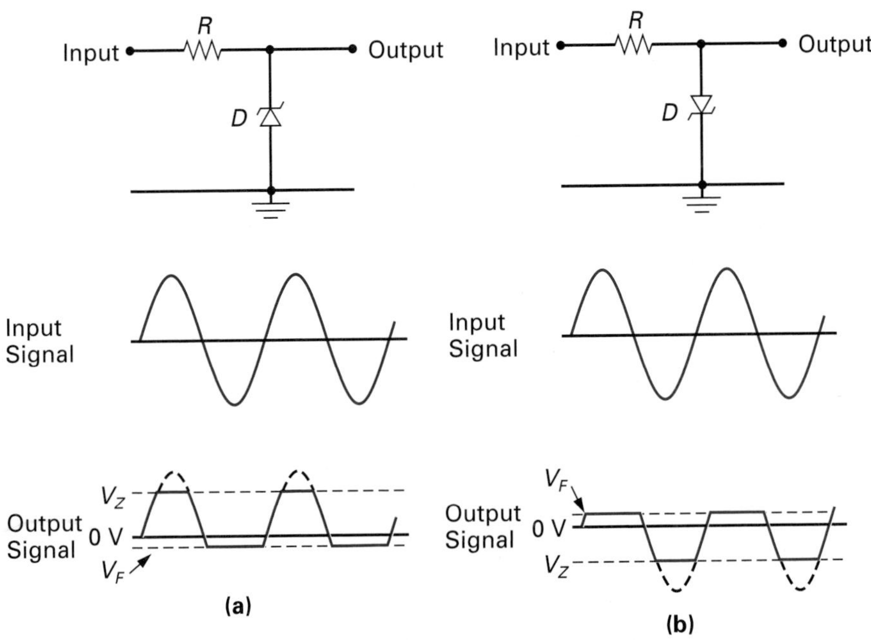

FIGURE 6-19 Zener Shunt Clipper Circuits.

negative, the zener diode will be forward biased and clip the input signal's negative alternation at –0.7 V (V_F).

If we were to reverse the direction of the zener diode, the output signal clipping would be reversed, as shown in Figure 6-19(b). In this instance, the positive alternation of the input signal would be limited to +0.7 V (V_F), and the negative alternation of the input would be clipped at the zener voltage (V_Z).

Symmetrical Zener Shunt Clipper Circuit

Figure 6-20(a) show how two back-to-back zener diodes can be used to limit both input signal peaks to produce a clipped symmetrical output. Referring to the input/output waveforms in Figure 6-20(b), you can see that both the positive and negative alternation have a simplified equivalent circuit shown in the insets.

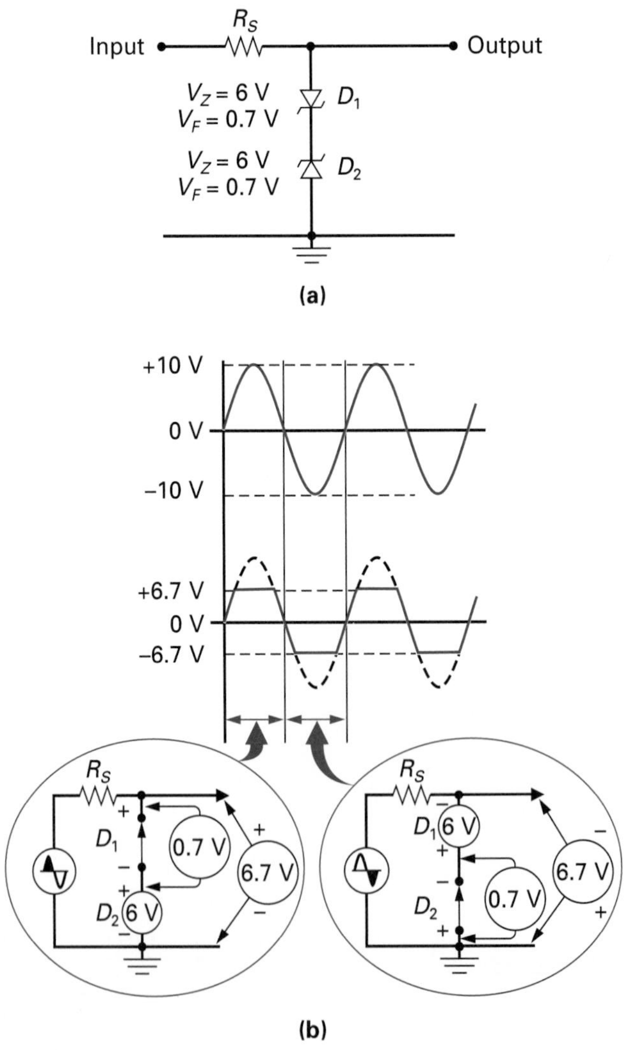

FIGURE 6-20 Symmetrical Zener Shunt Clipper Circuit.

When the input signal swings positive and exceeds +6.7 V, both zener diodes will conduct. Diode D_1 will conduct because it is forward biased, and D_2 will conduct because it has been sent into its reverse breakdown region, and its zener voltage rating is 6 V. The output will be limited or clipped in this case to

$$V_{out} = V_Z + V_F$$

$$= 6\text{ V} + 0.7\text{ V} = +6.7\text{ V}$$

When the negative alternation of the input exceeds –6.7 V, diode D_1 will drop its zener voltage of 6 V because it is operating in its reverse breakdown region, and D_2 will drop 0.7 V because it is forward biased. The output will therefore be clipped to

$$V_{out} = -V_Z + (-0.7\text{ V})$$

$$= -6\text{ V} + (-0.7\text{ V}) = -6.7\text{ V}$$

In summary, if the input signal exceeds these two extremes of +6.7 V and –6.7 V, the diodes will conduct and shunt the peaks to ground. Between these two voltages, however, neither of the diodes can conduct, and the input signal is passed through to the output.

Shunt Clipper Circuit Applications

In most applications, shunt clipper circuits are used to prevent a voltage from exceeding a certain value. In these instances, the shunt clipper circuit is often called a limiter, because it will limit an unwanted high amplitude voltage or current pulse or spike. These voltage spikes or current surges are often referred to as **transient voltages or currents.** By definition, these are pulses, or sudden momentary cycles, that occur in a circuit because of a sudden change in voltage or load. Figure 6-21 shows three examples of how a shunt clipper or limiter circuit can be used in a circuit to eliminate transient voltages or currents.

Transient Voltages or Currents
Pulses, or sudden momentary cycles, that occur in a circuit because of a sudden change in voltage or load.

In Figure 6-21(a), a symmetrical zener shunt circuit is being used to protect a dc power supply from the transients that can occur in the ac voltage from the wall outlet, as seen in the waveform in the inset. These zeners, which would typically have a V_Z rating of 200 V, will limit these transients to 200 V and prevent these large spikes from being coupled through the transformer and dc power supply to the electronic equipment. Generally, sensitive equipment, such as computers, will plug into a *surge protection power strip* instead of directly into the wall outlet. This surge protection power strip will include a symmetrical zener shunt clipper circuit, or a special *transient suppressor diode,* which will be discussed in detail in the following chapter.

In Figure 6-21(b), the shunt connected clipper diodes D_3 and D_4 are being used to prevent the input voltage to circuit B from going below 0 V and above +5 V. Many low voltage circuits, such as digital circuits, will be damaged if their input voltages exceed a voltage range. Electromagnetic Interference (EMI) and changes in voltage or load can inject voltage spikes onto input lines. As can be seen in the waveforms in the inset in Figure 6-21(b), without shunt clipper cir-

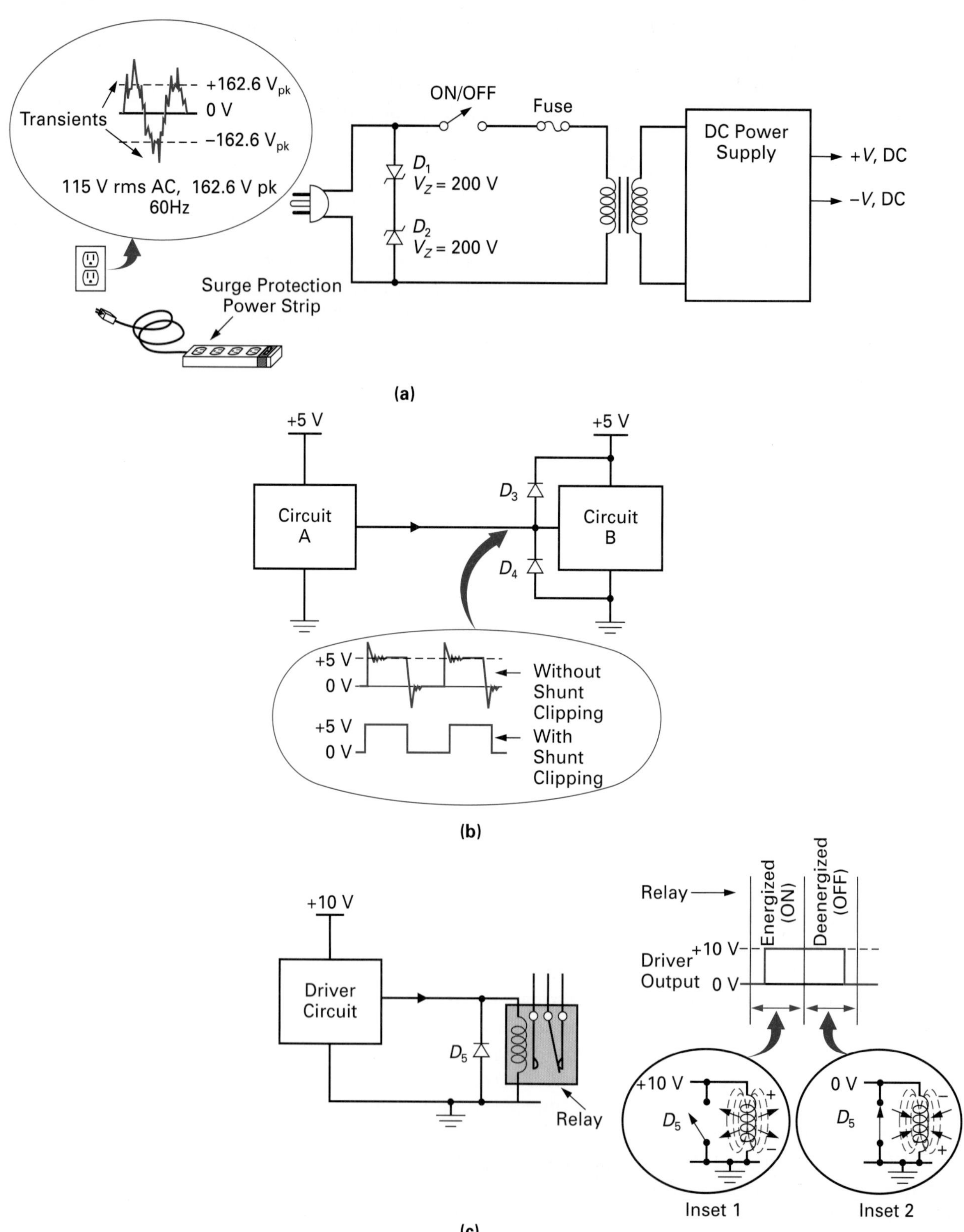

FIGURE 6-21 Transient Protection Circuits.

cuits these transients will be coupled directly into circuit B and possibly cause damage. By including these shunt connected diodes, we can limit the input voltage to a high of 5.7 V with D_3 and a low of –0.7 V with D_4.

In Figure 6-21(c), a shunt connected clipper diode is being used to protect a driver circuit from a counter emf transient that is always generated by a coil. In this example, the coil is the electromagnet of a relay. Referring to the associated waveform in Figure 6-21(c), you can see that when the driver circuit's output goes to +10 V, the relay coil is energized. Because D_5 is reverse biased, it will have no effect on the circuit's operation at this time, as shown in inset 1. However, when the driver's output drops from +10 V to 0 V, the relay coil is deenergized and the collapsing magnetic field will induce an emf in the coil that is of the same amplitude but of opposite polarity to that of the energizing voltage, as shown in inset 2. If this –10 V counter emf were allowed to go back to the driver circuit, it could damage the components in the driver circuit because they are now caught between a –10 V counter emf at the output and a +10 V supply voltage (20 volts of pressure). By including the clipper diode D_5 in shunt or parallel with the coil, the –10 V counter emf will forward bias the diode and be shunted to ground. You will often find these shunt clipper diodes connected in parallel with devices that have coils, such as speakers, motors, and so on.

6-2-3 *Troubleshooting Clipper Circuits*

In this section we will discuss how to troubleshoot a series and shunt clipper circuit. As before, always try to apply the three-step troubleshooting procedure: first diagnose whether a problem really exists, then proceed to isolate the problem by applying a logical and sequential reasoning process, and, finally, repair the circuit and perform a final test.

Figure 6-22(a) shows a negative series clipper circuit, its usual input/output waveforms, and a table listing the more common faults. Before we begin with these circuit faults, however, let us determine that the problem actually exists within the clipper circuit by using the half-split method. For example, if there is a no output signal (or a low amplitude signal) from the series clipper circuit at point A, your first step should be to disconnect the load from the clipper circuit by opening the connection at point A. If the output from the clipper circuit returns to normal, there is a short somewhere in the load and it is pulling down the output signal of the clipper circuit. If, however, there is still no output voltage signal, the problem exists somewhere in the clipper circuit or source before point A. Now that we have determined that the problem is somewhere in the clipper circuit, we can use our understanding of the circuit's operation to further isolate the problem, as listed below.

CAUSE	EFFECT
D_1 Open	There will be no output. All of the source voltage will appear across the series open.
D_1 Short	There will be no clipping action. Output waveform will be the same as input.
R_1 Short	There will be no output, due to shunting action of R_1. Source fuse may blow due to short when input is positive.

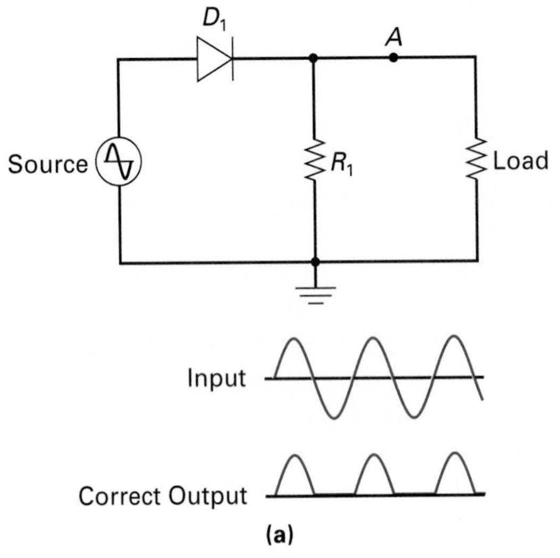

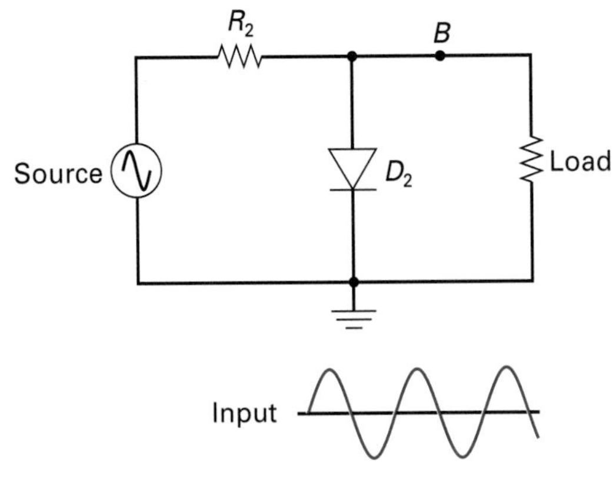

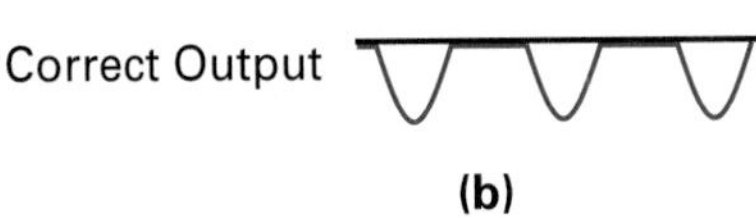

FIGURE 6-22 Troubleshooting a Clipper Circuit. (a) Series Clipper. (b) Shunt Clipper.

Figure 6-22(b) shows a positive shunt clipper circuit, its usual input/output waveforms, and a table listing the more common faults. Like the series clipper, you should first isolate whether there is a clipper circuit fault or a problem in the load. If the problem exists in the shunt clipper circuit, use your understanding of the circuit's operation to further isolate the problem, as listed below.

CAUSE	EFFECT
D_2 Open	There will be no clipping action. Output waveform will be the same as input.
D_2 Short	There will be no output, since D_2 will shunt both the positive and negative alternation to ground.
R_2 Open	There will be no output. All of the source voltage will appear across the series open.

SELF-TEST REVIEW QUESTIONS FOR SECTION 6-2

1. What are the two basic types of clipper circuits?
2. Which type of clipper circuit has a diode connected in parallel with the load?
3. Which of the following clipper circuits could be used to remove the positive spike of a differentiator circuit's output?
   a. Basic negative series clipper
   b. Basic positive series clipper
   c. Basic positive shunt clipper
   d. Both (b) and (c)
4. What other name is used to describe a clipper circuit?

6-3 DIODE CLAMPER CIRCUITS

Unlike the clipper circuit, the **clamper circuit** does not change the shape of the waveform. The clamper circuit is included to shift a waveform's dc reference so that it alternates above and below a dc reference voltage instead of alternating above and below 0 V. The two basic clamper circuits are detailed in Figure 6-23. These circuits are the **positive clamper** and the **negative clamper.**

Clamper Circuit
Shifts the waveform's dc reference so that it alternates above or below a dc reference voltage instead of alternating above and below 0 V.

Positive Clamper
Shifts the reference or start point of the input waveform to a positive dc voltage.

Negative Clamper
Shifts the reference or start point of the input waveform to a negative dc voltage.

As you can see in Figure 6-23(a), a negative clamper will shift the reference or start point of the input waveform to a negative dc voltage. In this example, the circuit will "clamp" or lock the top of the sine wave to 0 V without changing the shape or amplitude of the input waveform. As a result, our 20 V peak-to-peak sine wave input signal will now alternate 10 V above and 10 V below a –10 V reference voltage, as seen in the output signal.

A positive clamper will shift the reference or start point of the input waveform to a positive dc voltage, as shown in Figure 6-23(b). In this example, the circuit will "clamp" or lock the bottom of the sine wave to 0 V without changing the shape or amplitude of the input waveform. As a result, our 20 V peak-to-peak sine wave input signal will now alternate 10 V above and 10 V below a +10 V reference voltage, as seen in the output signal.

To explain how these circuits achieve this change in dc reference, the illustrations in Figure 6-23(c) and (d) show how these clamper circuits operate. Clamper circuits will perform this operation on any waveshape. However, in these circuits we will consider a square wave input because its two states of HIGH and LOW make it easier for us to gain a clear understanding of the circuit's operation.

6-3-1 *Negative and Positive Clamper Circuit Operation*

In Figure 6-23(c), the negative clamper circuit made up of C_1 and D_1 is being used to clamp or lock the top of a 20 V, pk-pk, square-wave input to 0 V. Referring to the output waveform, you can see that the resulting output signal is of exactly the same amplitude and shape; however, it now alternates above and below a negative dc reference voltage of –10 V. The circuit achieves this in the

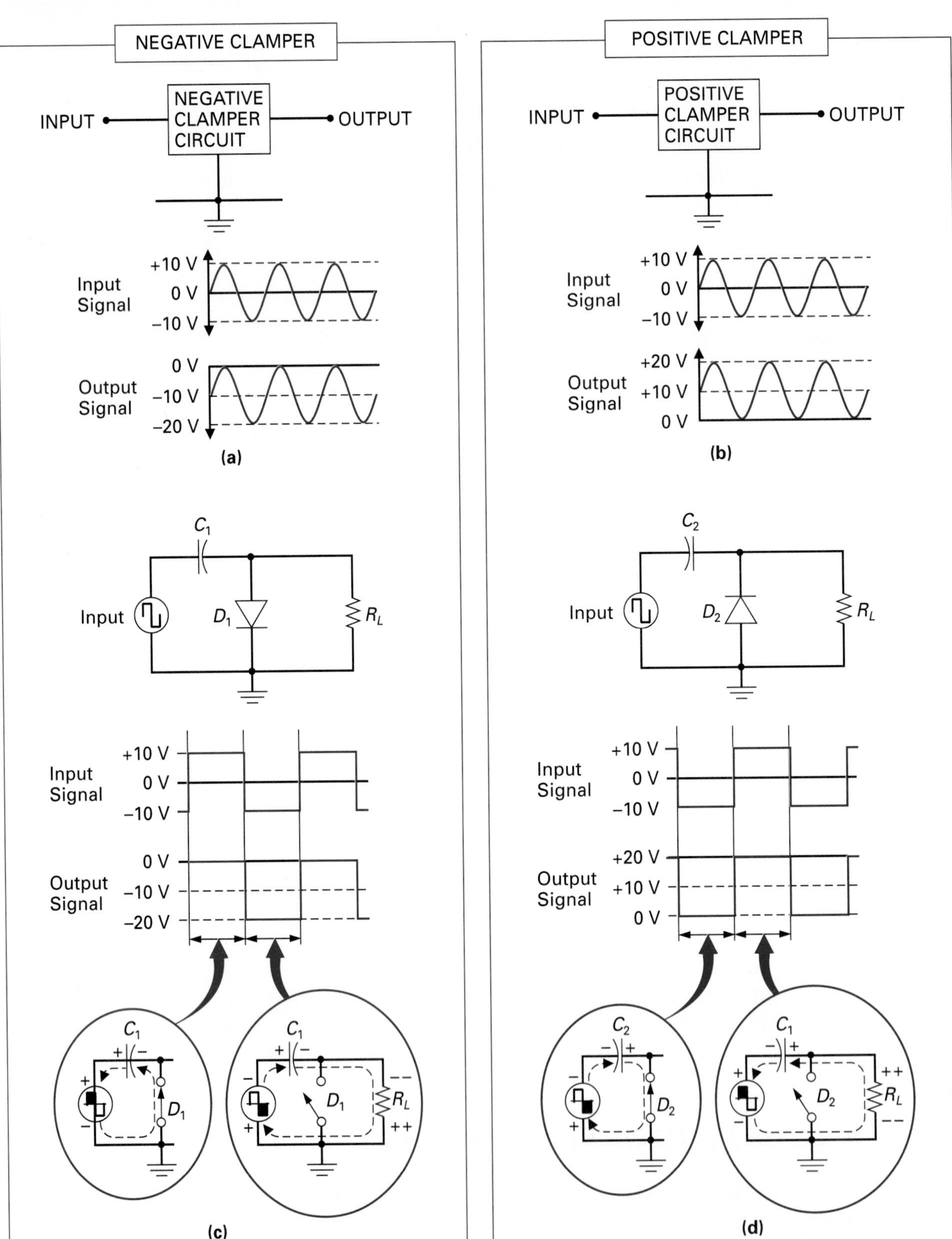

FIGURE 6-23 Diode Clamper or DC Restorer Circuits.

following way. When the input swings positive, D_1 will be forward biased and C_1 will charge quickly to 10 V, as shown in the inset for the positive alternation in Figure 6-23(c). Because the input voltage and the capacitor voltage are series-opposing, the output voltage at this time will be

$$V_{in} + V_c = V_{out}$$

$$(+10\text{ V}) + (-10\text{ V}) = 0\text{ V}$$

When the input swings negative, however, D_1 will be reverse biased and C_1 will discharge through the load, as shown in the inset for the negative alternation in Figure 6-23(c). Because the input voltage and the capacitor voltage are series-aiding, the output voltage at this time will be:

$$V_{in} + V_c = V_{out}$$
$$(-10\text{ V}) + (-10\text{ V}) = -20\text{ V}$$

Comparing the input waveform to the output waveform in Figure 6-23(c), you can see that the shape and amplitude of the waveform has been maintained. However, we now have a new dc reference of –10 V instead of a dc reference of 0 V, due to the top of the waveform being clamped to 0 V.

To be a little more accurate, when the diode D_1 is forward biased during the positive alternation of the square wave input, it will develop a +0.7 V drop across its terminals and this will be applied to the output. Also, the time constant for charging C_1 is small because its only resistance is that of D_1 ($\tau = R_{D1} \times C_1$). However, it may take a few cycles of the input signal before C_1 is fully charged to the peak of the positive input voltage (+10 V) if the input signal's period is not sufficient. Because a total of five time constants will be needed to charge C_1, this time can be calculated with the following formula;

$$\tau_{charge} = 5\,(R_{D1} \times C_1)$$

By contrast, the discharge time constant of C_1, which is dependent on the value of C_1 and the load's resistance, is arranged to be large so that C_1 will only discharge slightly between positive cycles of the input. The time it would take C_1 to fully discharge can be calculated with the formula

$$\tau_{discharge} = 5\,(R_L \times C_1)$$

■ **EXAMPLE:**

Referring to the negative clamper circuit shown in Figure 6-23(c), calculate the following for these values:

$$C_1 = 2\ \mu\text{F}$$
$$D_1 \text{ forward resistance} = 6\ \Omega$$
$$\text{Square wave input freq.} = 13.33\text{ kHz, +10 V to }-10\text{ V, 0 V ref.}$$
$$R_L = 20\text{ k}\Omega$$

a. The time it takes C_1 to fully charge.
b. If C_1 will have time to fully charge during the first positive alternation of the input.
c. The output voltage change and dc reference voltage ignoring the 0.7 V drop across D_1.
d. The time it takes C_1 to fully discharge.

■ ***Solution:***

a.

$$\begin{aligned}\tau_{charge} &= 5\,(R_{D1} \times C_1) \\ &= 5\,(6\ \Omega \times 2\ \mu F) \\ &= 5 \times (0.000012) \\ &= 60\ \mu s\end{aligned}$$

b. If the input frequency equals 13.33 kHz, the signal's period will be equal to 75 μs (1/13.33 kHz). Because the positive alternation will only last for 37.5 μs (75 μs/2), it will take C_1 two positive alternations of the input to fully charge and the output waveform to settle to its "steady state" negative clamped output.

c. The output voltage will be a 20 V, pk-pk, square-wave with a dc reference of −10 V.

d.

$$\begin{aligned}\tau_{discharge} &= 5\,(R_L \times C_1) \\ &= 5\,(20\ k\Omega \times 2\ \mu F) \\ &= 5 \times 0.04 \\ &= 200\ \mu s\end{aligned}$$

Since this discharge time is large compared to the input signal's negative alternation period (37.5 μs), the capacitor will only slightly discharge during negative cycles.

In Figure 6-23(d), the positive clamper circuit made up of C_2 and D_2 is being used to clamp or lock the bottom of a 20 V, pk-pk, square-wave input to 0 V. Referring to the output waveform, you can see that the resulting output signal is of exactly the same amplitude and shape; however, it now alternates above and below a positive dc reference voltage of +10 V. The circuit achieves this in the following way. When the input swings negative, D_2 will be forward biased and C_2 will charge quickly to 10 V, as shown in the inset for the negative alternation in Figure 6-23(d). Since the input voltage and the capacitor voltage are series-opposing, the output voltage at this time will be

$$V_{in} + V_c = V_{out}$$

$$(-10\ V) + (+10\ V) = 0\ V$$

When the input swings positive, however, D_2 will be reverse biased and C_2 will discharge through the load, as shown in the inset for the positive alternation in Figure 6-23(d). Because the input voltage and the capacitor voltage are series-aiding, the output voltage at this time will be

$$V_{in} + V_c = V_{out}$$

$$(+10\ V) + (+10\ V) = +20\ V$$

Comparing the input waveform to the output waveform in Figure 6-23(d), you can see that the shape and amplitude of the waveform has been maintained. However, we now have a new dc reference of +10 V instead of a dc reference of 0 V, due to the bottom of the waveform being clamped to 0 V.

6-3-2 ***Clamper Circuit Application—Television Receiver Circuit***

Electronic communication circuits will be covered in detail in a later chapter. For now, let us introduce the basics of television reception because a TV receiver is one application that makes use of a clamper circuit.

Figure 6-24(a) shows the basic elements of a TV broadcasting system. A television system converts a succession of visual images into corresponding electric signals and then transmits these signals by radio or over wires (cable) to distant receivers (TV sets) which will respond to these signals and reproduce the original images. Every television broadcast consists of two separate signals: an audio (sound) signal generated by a microphone and a video (picture) signal generated by a camera tube. To explain the video in more detail, one TV picture is made up of about 150,000 small dots or picture elements called "pixels." The video camera contains a special camera tube that generates a video signal by scanning the screen horizontally line by line, covering 525 horizontal lines in succession from the top to the bottom of the picture in 1/30th of a second. These 525 lines make up one "frame" of a TV picture.

Figure 6-24(b) shows the block diagram of a typical TV receiver. This system contains many circuits that process the incoming television signal. All of these circuits are powered by voltages from the television's internal dc power supply. To isolate the bias voltages of one circuit from the next, "coupling capacitors" are connected between circuits to couple or pass the alternating television signal, but block or isolate each circuit's dc bias voltages.

Figure 6-24(c) shows a typical television video signal. This signal will reproduce the original image by creating a pattern of light and dark. For example, at 75% of maximum the screen is black, while at 12.5% of maximum the screen is white. Between these two extremes are various levels of grey. As this signal is processed and amplified by the various circuits between the receiving antenna and the cathode ray tube, it alternates above and below 0 V since none of the coupling capacitors allow any dc component to be passed. The receiver's cathode ray tube must produce an electron beam which is synchronized with the transmitter's electron beam from the television camera tube. Variations in the flow of electrons will create the right pattern of light and dark to reproduce the original picture. A positive dc clamper circuit is included in the final stage of a television receiver to restore the dc component of the video signal after ac amplification. The resulting dc voltage serves as the bias voltage for the grid of the CRT. Referring to the waveforms in Figure 6-24(d), you can see how the dc clamper circuit has restored the dc reference voltage of the TV video signal. This positive voltage signal will now control the grid of the cathode ray tube and either increase (more positive) or decrease (less positive) the number of electrons that travel to the screen of the CRT and generate one horizontal line of

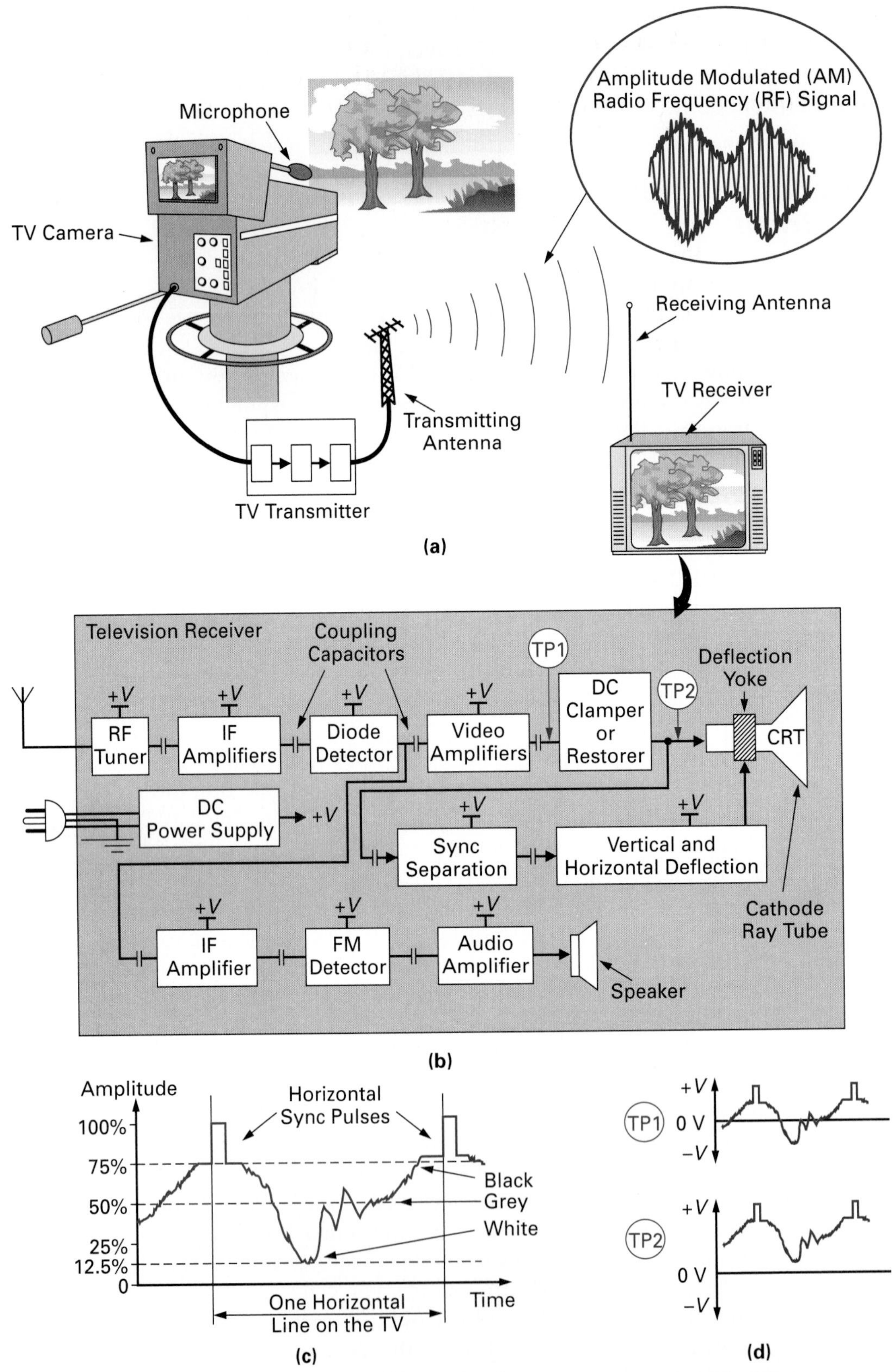

FIGURE 6-24 Television Communication Circuit. (a) Basic TV System. (b) Block Diagram of TV Receiver. (c) Complex Video Waveform. (d) Input/Output Waveform of DC Restorer Circuit.

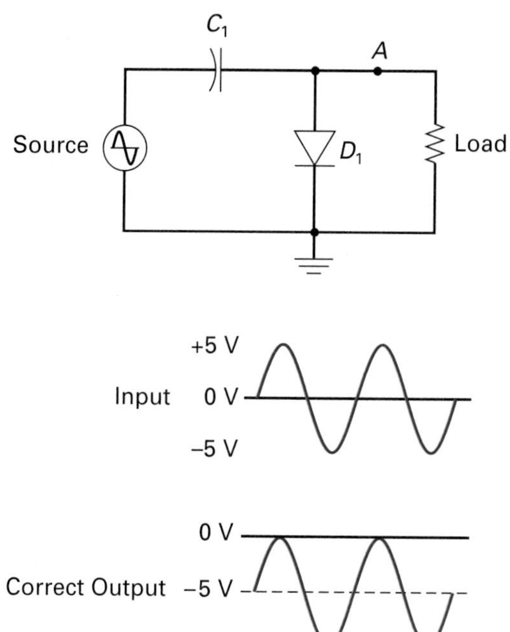

FIGURE 6-25 Troubleshooting a Clamper Circuit.

video. This dc restoring action of the circuit accounts for why dc clamper circuits are often called **dc restorer circuits.**

DC Restorer Circuit DC clamper circuit.

6-3-3 Troubleshooting Clamper Circuits

In this section, we will discuss how to troubleshoot a clamper circuit. As before, always try to apply the three-step troubleshooting procedure: first, diagnose whether a problem really exists, then proceed to isolate the problem by applying a logical and sequential reasoning process, and, finally, repair the circuit and perform a final test.

Figure 6-25 shows a negative clamper circuit, its usual input/output waveforms, and a table listing the more common faults. Before we begin with these circuit faults, however, let us use the half-split method to determine that the problem actually exists within the clamper circuit. For example, if there is a no output signal (or a low amplitude signal) from the clamper circuit at point *A*, your first step should be to disconnect the load from the clamper circuit by opening the connection at point *A*. If the output from the clamper circuit returns to normal, there is a short somewhere in the load and it is pulling down the output signal of the clamper circuit. If, however, there is still no output voltage signal, the problem exists somewhere in the clamper circuit or source before point *A*. Now that we have determined that the problem is somewhere in the clamper circuit, we can use our understanding of the circuit's operation to further isolate the problem, as listed below.

CAUSE	EFFECT
D_1 Open	There will be no clamping action. AC Input will be coupled to output by C_1, and output will be centered at O V.

D_1 Short	Output will be O V, and source fuse may blow due to heavy current dosage and discharge of C_1 as input alternates.
C_1 Open	There will be no output signal. All of the source voltage will appear across the series open.
C_1 Short	Positive alternation of input will be applied to load, while negative alternation of input will be shunted to ground, causing source fuse to probably blow.
C_1 Leaky	Uneven charge and discharge of C_1 will cause output signal shape to be distorted, and inability of C_1 to hold charge will cause dc reference of output signal to continuously vary.

SELF-TEST REVIEW QUESTIONS FOR SECTION 6-3

1. What other name is used to describe a clamper circuit?
2. What are the two basic types of clamper circuits?
3. A positive sawtooth control signal is being used to sweep the trace of an oscilloscope from left to right on the display. Due to several stages of capacitor coupled amplifiers, the ramp is centered at 0 V. What type of clamper circuit would be used to convert the ramp to a positive signal that starts at 0 V and increases positive?

SUMMARY

1. The voltage multiplier circuit is used to generate a large dc supply voltage that is some multiple of the ac peak input voltage.
2. The clipper circuit is used to clip or eliminate an unwanted portion of the ac input signal and convert one type of signal to another.
3. The clamper circuit is used to clamp or lock the top or bottom of a waveform to a dc reference voltage.

Voltage Multiplier Circuits (Figure 6-26)

4. Most electronic circuits require a dc supply voltage that is lower than 115 V and so a step-down transformer is normally always included. In some instances, a circuit or device may require a dc supply voltage that is larger than 115 V and a step-up transformer will have to be used. Step-up transformers are expensive, and if a wide range of dc supply voltages are needed, it becomes much more economical to use a voltage multiplier circuit to obtain the higher value dc supply voltages needed.
5. Voltage multiplier circuits can replace rectifier and filter circuits because they employ diodes and capacitors in almost exactly the same way to convert an ac input to a dc output. However, voltage multiplier circuits produce a dc output voltage that is some multiple of the peak ac input voltage.
6. A voltage-doubler circuit will produce a dc output voltage that is twice the peak of ac input voltage.
7. If the output voltage of a multiplier is increased by some multiple, the output current of the circuit will be decreased by the same multiple.

QUICK REFERENCE SUMMARY SHEET

FIGURE 6-26 Voltage Multiplier Circuits.

Half-Wave Voltage Doubler (Positive DC Output Voltage)

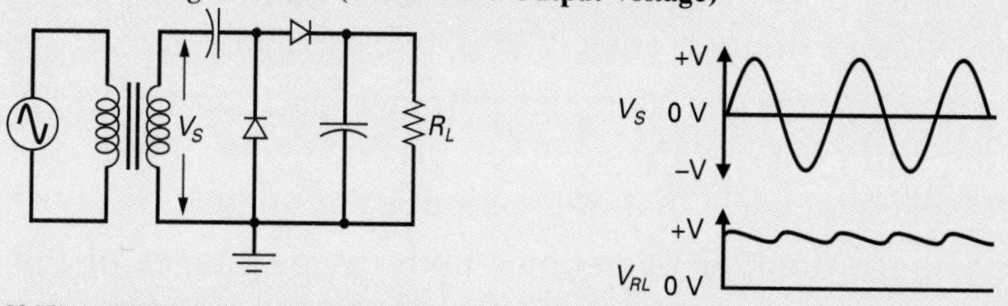

New Formulas

$V_{RL} = 2 \times V_{S\,pk}$

$f_{out\,ripple} = f_{in\,AC}$

Half-Wave Voltage Doubler (Negative DC Output Voltage)

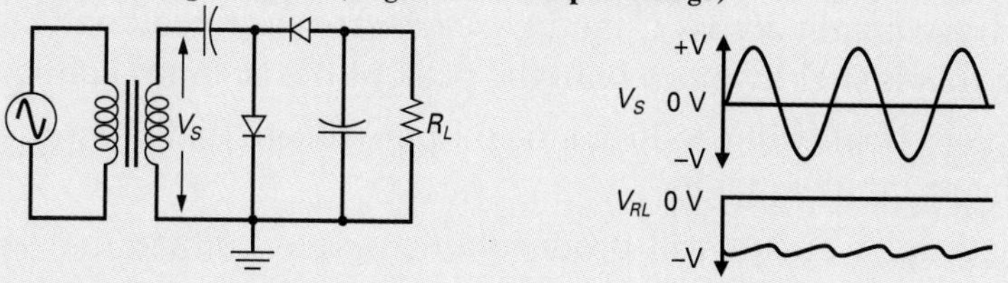

$V_{RL} = 2 \times V_{S\,pk}$

$f_{out\,ripple} = f_{in\,AC}$

Full-Wave Voltage Doubler (Positive DC Output)

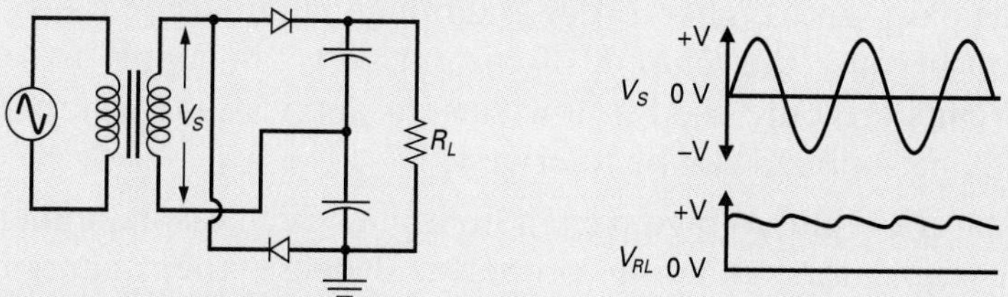

$V_{RL} = 2 \times V_{S\,pk}$

$f_{out\,ripple} = 2 \times f_{in\,AC}$

Voltage Tripler Circuit (Positive DC Output)

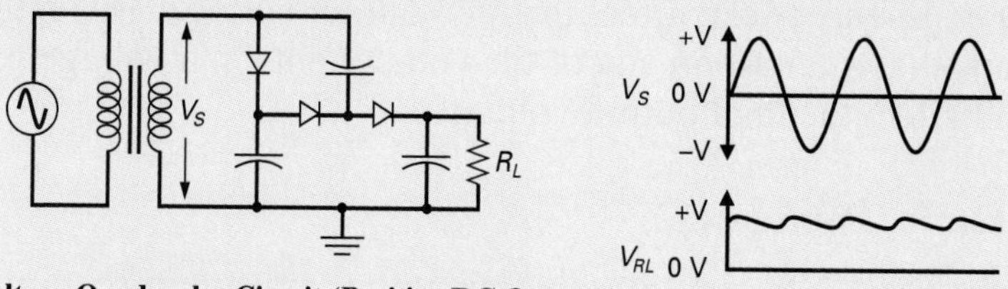

$V_{RL} = 3 \times V_{S\,pk}$

$f_{out\,ripple} = 2 \times f_{in\,AC}$

Voltage Quadrupler Circuit (Positive DC Output)

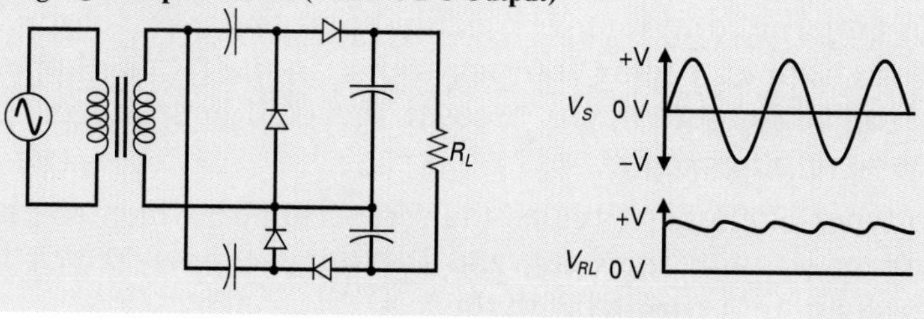

$V_{RL} = 4 \times V_{S\,pk}$

$f_{out\,ripple} = 2 \times f_{in\,AC}$

8. There are two types of voltage-doubler circuits: the half-wave voltage doubler and the full-wave voltage doubler.
9. The output dc ripple frequency of a half-wave voltage doubler will equal the input ac frequency.
10. By reversing the directions of the diodes and capacitors, you can change a positive half-wave voltage doubler to a negative half-wave doubler.

11. The half-wave voltage doubler has two disadvantages. The first is that the ripple frequency of 60 Hz is difficult to filter, and the second is that an expensive output capacitor is required because it must have a voltage rating that is more than twice the peak of the ac input.
12. The full-wave voltage doubler circuit uses the same number of components as the half-wave voltage doubler (two diodes, two capacitors).
13. With the full-wave voltage doubler, the output capacitors C_1 and C_2 are charged for both half cycles of the ac input, producing an output dc ripple frequency that will equal twice that of the input ac frequency.
14. The full-wave voltage doubler overcomes both disadvantages of the half-wave circuit by having a higher ripple frequency that is easier to filter. Because the output voltage is split across C_1 and C_2, each capacitor need only have a voltage rating that is slightly more than the peak of the ac input voltage.
15. A voltage-tripler circuit will produce a dc output voltage that is three times the peak of the ac input voltage.
16. A voltage-quadrupler circuit will produce a dc output voltage that is four times the peak of the ac input voltage.
17. Voltage quadruplers and higher order multiplier circuits generally have poor regulation because the amount of current they can supply is so limited. These circuits are only used in low-current applications where variations in the voltage supplied can be tolerated.
18. The final capacitor output voltage from a multiplier circuit is dependent on the preceding capacitors charging properly. Because their charging is dependent on the correct operation of the switching diodes, any of the multiplier's circuit components can affect the final output voltage. This interdependency means that a failure in any of the components in a voltage multiplier circuit will affect the final output voltage.

Series Clipper Circuits (Figure 6-27)

19. A clipper circuit is used to cut off or eliminate an unwanted section of a waveform. It is either used to remove a natural part of the waveform or to prevent a voltage from exceeding a certain value. In the latter, clipper circuits are often called limiter circuits because they will limit a high-amplitude voltage pulse or other signal.
20. There are two types of clipper circuits: the series clipper, which contains a diode that is in series with the load, and the shunt clipper, which has a diode that is in shunt or in parallel with the load.
21. A negative series clipper circuit has a diode connected in *series* with the load or output, and its orientation is such that it will *clip* off the *negative* alternation of the ac input.
22. A positive series clipper circuit has a diode connected in *series* with the load or output, and its orientation is such that it will *clip* off the *positive* alternation of the ac input.
23. With biased series clipper circuits, the clipping level can be changed by applying a bias voltage to the series connected diode.

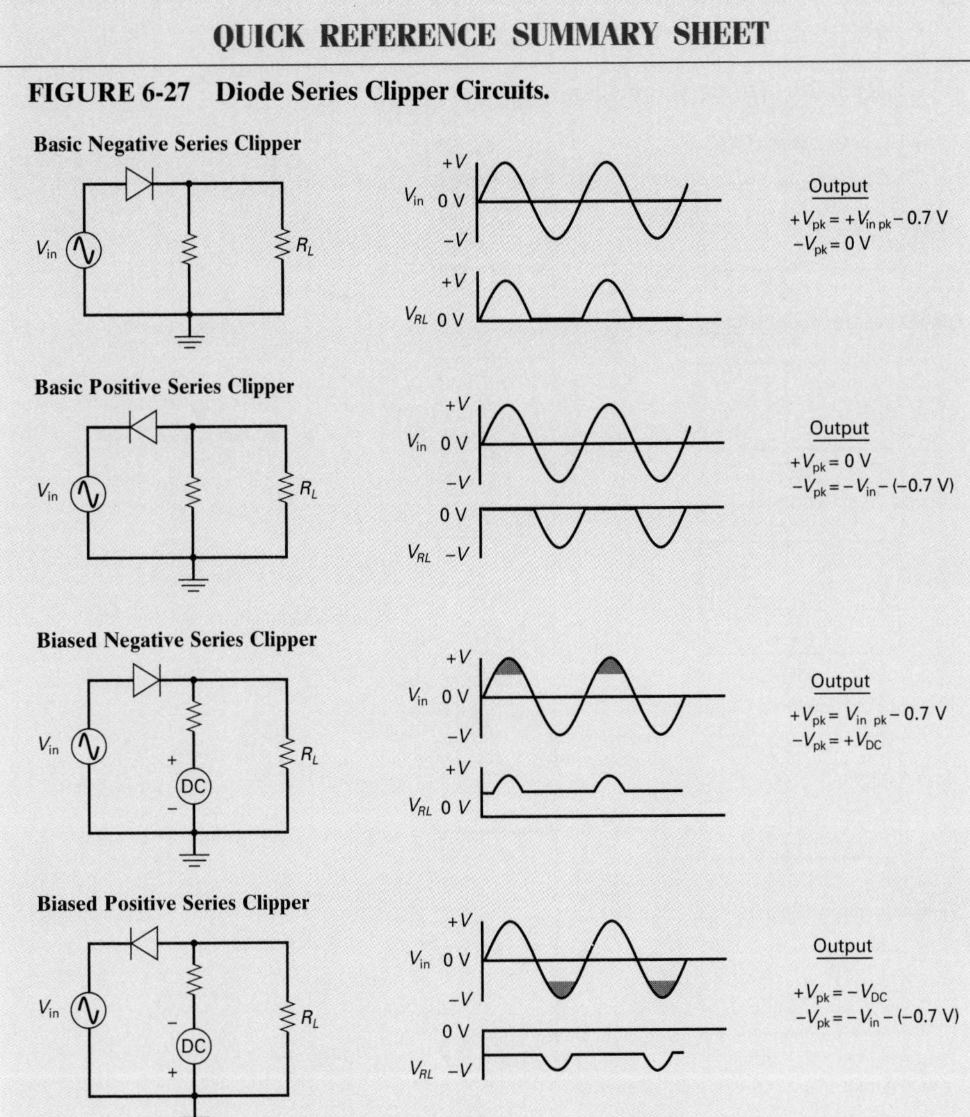

FIGURE 6-27 Diode Series Clipper Circuits.

24. Communication is defined as "a process by which information is exchanged." The two basic methods of transferring information are the spoken word and the written word. Communication of the spoken word began with face-to-face contact and then evolved into telephone and radio communications. Communication of the written word began with hand-carried letters and evolved into newspapers, the mail system, the telegraph, and, now, electronic mail.

Shunt Clipper Circuits (Figure 6-28)

25. Shunt clipper circuits achieve exactly the same results as the series clipper circuits but operate in almost an opposite way. The series clipper produced an output when the diode was forward biased and no output when the

QUICK REFERENCE SUMMARY SHEET

FIGURE 6-28 Diode Shunt Clipper Circuits.

Basic Negative Shunt Clipper

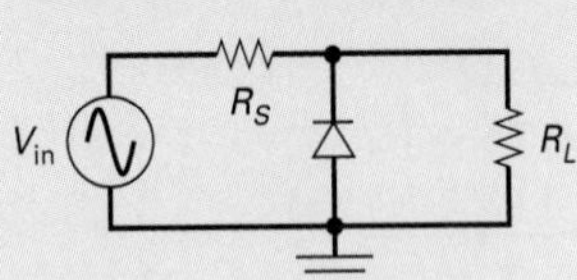

V_{in} 0 V

+V
V_{RL} 0 V

Output

$$+V_{pk} = \frac{R_L}{R_S + R_L} \infty V_{in\,pk}$$

$$-V_{pk} = -0.7\text{ V}$$

Basic Positive Shunt Clipper

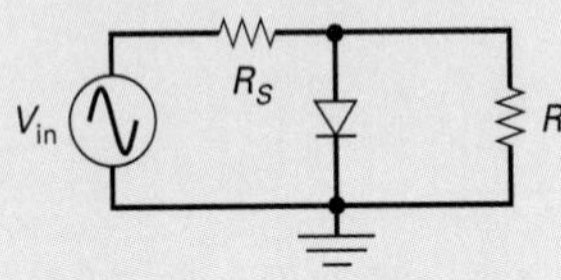

+V
V_{in} 0 V

V_{RL} 0 V
−V

Output

$$+V_{pk} = +0.7\text{ V}$$

$$-V_{pk} = \frac{}{R_S + R_L} \infty (-V_{in\,pk})$$

Biased Negative Shunt Clipper

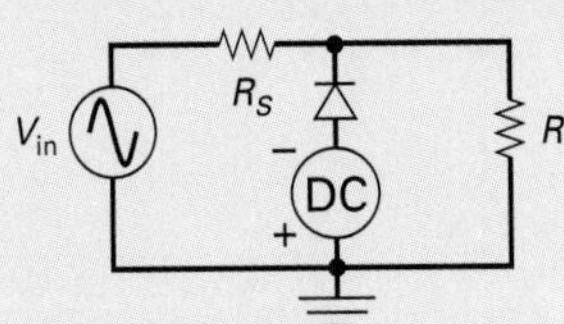

V_{in} 0 V

V_{RL} 0 V

Output

$$+V_{pk} = \frac{R_L}{R_L + R_S} \infty V_{in\,pk}$$

$$-V_{pk} = (-V_{DC}) + (-0.7\text{ V})$$

Biased Positive Shunt Clipper

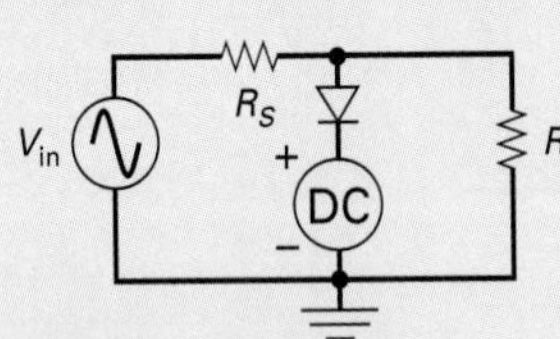

V_{in} 0 V

V_{RL} 0 V

Output

$$+V_{pk} = V_{DC} + 0.7\text{ V}$$

$$-V_{pk} = \frac{R_L}{R_L + R_S} \infty (-V_{in\,pk})$$

Variable Shunt Clipper

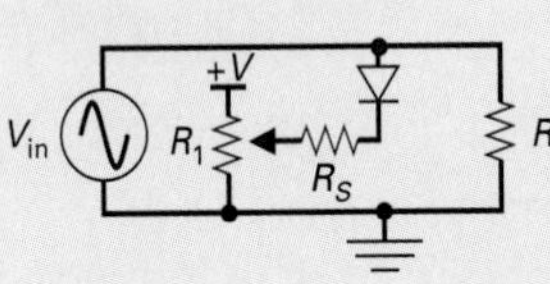

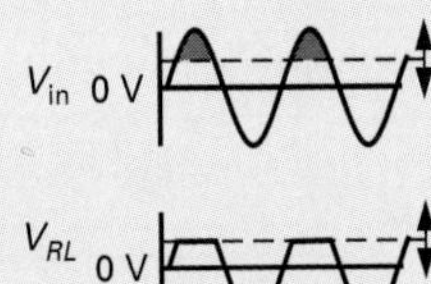

Output

$+V_{pk} = R_1$ adjustable

$$-V_{pk} = -V_{in\,pk}$$

Zener Shunt Clipper (−Clipped, +Zenered)

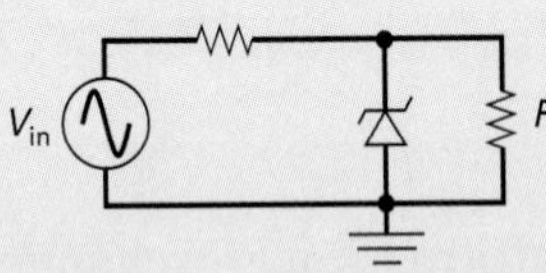

V_{in} 0 V

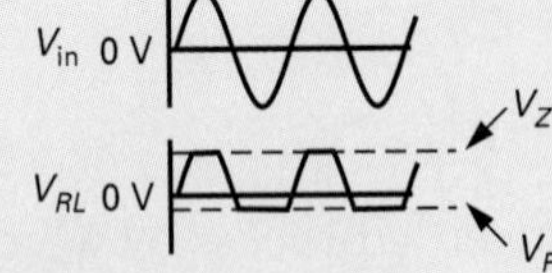

Output

$$+V_{pk} = +V_Z$$

$$-V_{pk} = -0.7\text{ V}$$

Zener Shunt Clipper (+Clipped, −Zenered)

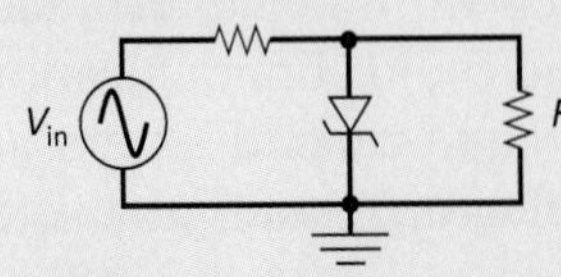

V_{in} 0 V

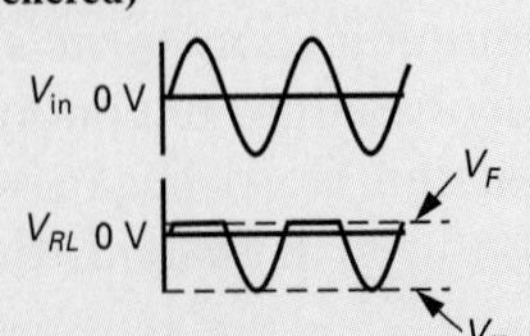

Output

$$+V_{pk} = +0.7\text{ V}$$

$$-V_{pk} = -V_Z$$

Symmetrical Zener Shunt Clipper

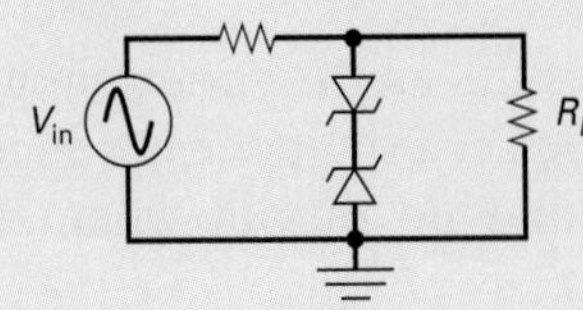

V_{in} 0 V

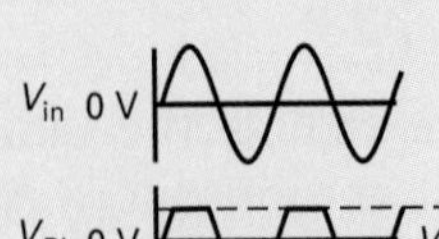

Output

$$+V_{pk} = V_Z + 0.7\text{ V}$$

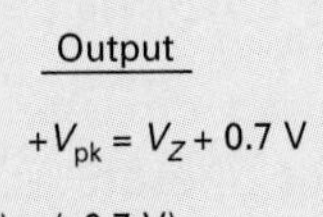

diode was reverse biased, whereas the shunt clipper will short the input signal to ground when it is forward biased and produce an output when the diode is reverse biased.

26. The negative shunt clipper circuit has a diode connected in *shunt* with the load or output, and its orientation is such that it will *clip* off the *negative* alternation of the ac input by shunting it to ground.
27. The positive shunt clipper circuit has a diode connected in *shunt* with the load or output, and its orientation is such that it will *clip* off the *positive* alternation of the ac input by shunting it to ground.
28. Biased shunt clipper circuits have a shunt connected diode and a bias voltage connected to clip off or shunt to ground a section of the input alternation.
29. A variable shunt clipper circuit has a potentiometer connected to adjust the positive shunt clipping level.
30. A zener shunt clipper circuit makes use of the zener diode's forward biased switching action and its reverse bias zener action to clip both the positive and negative alternation of the input signal.
31. A symmetrical zener shunt clipper circuit contains two back-to-back zener diodes that can be used to limit both of the input signal's peaks to produce a clipped symmetrical output.
32. In most applications, shunt clipper circuits are used to prevent a voltage from exceeding a certain value. In these instances, the shunt clipper circuit is often called a limiter because it will limit an unwanted high amplitude voltage or current pulse or spike. These voltage spikes or current surges are often referred to as transient voltages or currents, which by definition are pulses, or sudden momentary cycles that occur in a circuit because of a sudden change in voltage or load.

Clamper Circuits (Figure 6-29)

33. Unlike the clipper circuit, the clamper circuit does not change the shape of the waveform. It simply shifts a waveform's dc reference so that it alternates above and below a dc reference voltage instead of alternating above and below 0 V.
34. A clamper circuit contains a series connected capacitor and a shunt connected diode. The diode's orientation determines whether the circuit is a positive clamper or a negative clamper.
35. A negative clamper will shift the reference or start point of the input waveform to a negative dc voltage, whereas a positive clamper will shift the reference or start point of the input waveform to a positive dc voltage.
36. A television system converts a succession of visual images into corresponding electric signals and then transmits these signals by radio or over wires (cable) to distant receivers (TV sets), which will respond to these signals and reproduce the original images. Every television broadcast consists of two separate signals: an audio (sound) signal generated by a microphone, and a video (picture) signal generated by a camera tube. To explain the video in more detail, one TV picture is made up of about 150,000 small dots

QUICK REFERENCE SUMMARY SHEET

FIGURE 6-29 Diode Clamper Circuits.

Negative Clamper

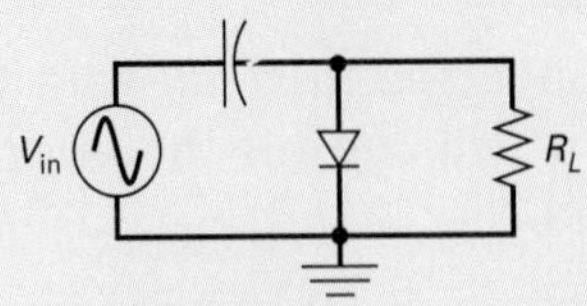

+V
V_{in} 0 V
−V
0 V
V_{RL}
−V

Output

$+V_{pk} = +0.7$ V

$-V_{pk} = -V_{in\ pk\text{-}pk}$

Positive Clamper

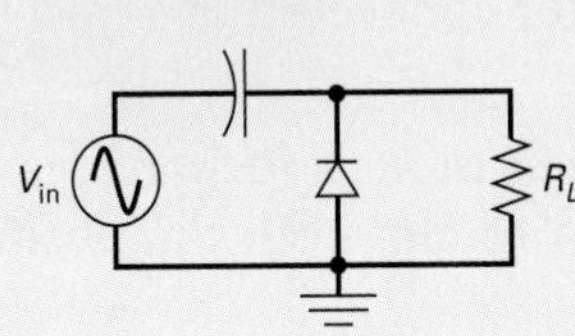

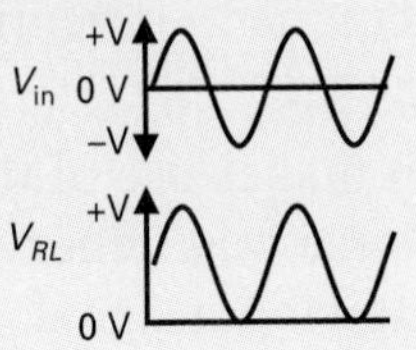

Output

$+V_{pk} = +V_{in\ pk\text{-}pk}$

$-V_{pk} = -0.7$ V

or picture elements called "pixels." The video camera contains a special camera tube that generates a video signal by scanning the screen horizontally line by line, covering 525 horizontal lines in succession from the top to the bottom of the picture in 1/30th of a second. These 525 lines make up one "frame" of a TV picture.

37. A positive dc clamper circuit is included in the final stage of a television receiver to restore the dc component of the video signal which is generally lost due to several capacitor coupled amplifier stages. The resulting dc voltage serves as the bias voltage for the grid of the television set's cathode ray tube.

NEW TERMS

Amplitude Modulation (AM)
Audio Signal
Biased Negative Series Clipper
Biased Negative Shunt Clipper Circuit
Biased Positive Series Clipper
Biased Positive Shunt Clipper Circuit
Biased Series Clipper Circuit
Biased Shunt Clipper Circuit
Carrier Frequency
Carrier Wave
Clamper Circuit
Clipper Circuit
DC Restorer Circuit
Demodulator Circuit
Detector Circuit
Diode AM Demodulator Circuit
Diode AM Detector Circuit
Diode AM Modulator Circuit
Electronic Communication Circuits
Full-Wave Voltage-Doubler Circuit
Half-Wave Voltage-Doubler Circuit
Limiter Circuit
Modulation
Modulator Circuit
Negative Clamper Circuit
Negative Edge Triggering
Negative Half-Wave Voltage Doubler
Negative Series Clipper

Negative Shunt Clipper
Pixels
Positive Clamper Circuit
Positive Edge Triggering
Positive Half-Wave Voltage Doubler
Positive Series Clipper
Positive Shunt Clipper
Series-Aiding Voltage Sources
Series Clipper Circuit
Series-Opposing Voltage Sources
Shunt Clipper Circuit
Surge Protection Power Strip
Symmetrical Zener Shunt Clipper Circuit
Transient Suppressor Diode
Transient Voltages or Currents
Variable Shunt Clipper Circuit
Video Signal
Voltage-Doubler Circuit
Voltage-Multiplier Circuit
Voltage-Quadrupler Circuit
Voltage-Tripler Circuit
Zener Shunt Clipper Circuit

REVIEW QUESTIONS

Multiple-Choice Questions

1. Which of the following diode circuits is used to change or restore the dc reference of an alternative waveform?
   a. The voltage-multiplier circuit
   b. The diode clipper circuit
   c. The diode clamper circuit
   d. The rectifier circuit
2. Which of the following diode circuits is used to generate a large dc supply voltage that is some multiple of the ac peak input voltage?
   a. The voltage-multiplier circuit
   b. The diode clipper circuit
   c. The diode clamper circuit
   d. The rectifier circuit
3. Which of the following diode circuits is used to eliminate an unwanted portion of the ac input signal?
   a. The voltage multiplier circuit
   b. The diode clipper circuit
   c. The diode clamper circuit
   d. The rectifier circuit
4. Without a step-up transformer, rectifier circuits can only produce an output voltage that is equal to or less than the peak value of the ac input. Which circuits can be used in place of rectifiers to generate higher dc supply voltages?
   a. Voltage multiplier circuits
   b. Clipper circuits
   c. Clamper circuits
   d. Encoder circuits
5. What key advantage does the full-wave voltage doubler have over the half-wave voltage doubler?
   a. Its output ripple frequency is lower
   b. Its output ripple frequency is higher
   c. It uses less components
   d. Both (a) and (b) are true
6. If a 115 V rms, 60 Hz input voltage was connected to a half-wave doubler circuit, what would be the peak output voltage and ripple frequency?
   a. 230 V, 60 Hz
   b. 230 V, 120 Hz
   c. 325.2 V, 60 Hz
   d. 325.2 V, 120 Hz

7. If a 115 V rms, 60 Hz input voltage was connected to a full-wave doubler circuit, what would be the peak output voltage and ripple frequency?

   a. 230 V, 60 Hz **c.** 325.2 V, 60 Hz
   b. 230 V, 120 Hz **d.** 325.2 V, 120 Hz

8. If a 115 V rms, 60 Hz input voltage was connected to a voltage tripler circuit, what would be the peak output voltage?

   a. 487.8 V **c.** 406 V
   b. 345 V **d.** 325.2 V

9. The ____________ clipper contains a diode that is connected end to end with the load, whereas the ____________ clipper has a diode that is connected in parallel with the load.

   a. Series, biased **c.** Series, shunt
   b. Shunt, series **d.** Shunt, biased

10. A positive series clipper circuit will pass the ____________ alternation of the ac input.

    a. Positive **b.** Negative

11. A negative series clipper circuit will eliminate the ____________ alternation of the ac input.

    a. Positive **b.** Negative

12. Which of the following circuits would be used in modulator communication circuits?

    a. Shunt clipper **c.** Positive clamper
    b. Series clipper **d.** Symmetrical shunt clipper

13. Which of the following shunt clipper circuits is generally used at the front end of a dc power supply circuit to protect against transients from the 115 V ac input?

    a. Positive shunt clipper **c.** Variable shunt clipper
    b. Negative shunt clipper **d.** Symmetrical zener shunt clipper

14. A shunt clipper diode is generally always connected across a coil to protect the drive circuit from ____________ .

    a. Electromagnetic interference **c.** The energizing voltage
    b. Modulation **d.** The coil's counter emf

15. Unlike the clipper circuit, the ____________ circuit does not change the shape of the waveform.

    a. Clamper **c.** Rectifier
    b. Voltage multiplier **d.** Both (b) and (c) are true

Essay Questions

16. Define the function of the following diode circuits: (Intro.)

    a. A multiplier circuit **b.** A clipper circuit **c.** A clamper circuit

17. What is the difference between the peak and the rms of an ac input voltage? (basic ac theory)

18. Sketch a half-wave voltage doubler circuit, and show its input/output waveforms. (6-1-1)

19. What advantage(s) does the full-wave voltage-doubler circuit have over the half-wave voltage doubler? (6-1-1)
20. Why are voltage triplers and quadruplers seldom used? (6-1-3)
21. What is a limiter circuit? (6-2)
22. Describe and sketch the difference(s) between a series and shunt clipper circuit. (6-2-1, 6-2-2)
23. What is the difference between a basic series clipper circuit and a half-wave rectifier circuit? (6-2-1)
24. What is a biased series clipper circuit? (6-2-1)
25. Briefly describe how series clipper circuits are used to demodulate a carrier frequency. (6-2-1)
26. What is amplitude modulation? (6-2-1)
27. Would the output waveshape from a basic positive series clipper be almost identical to that of a basic negative shunt clipper? (6-2-1 and 6-2-2)
28. In what applications are shunt clippers most often used? (6-2-2)
29. What is a clamper circuit? (6-3)
30. In what applications are clamper circuits used? (6-3-2)

Practice Problems

31. Identify the circuits shown in Figure 6-30, and then calculate the output voltage from each circuit.
32. Identify the circuit shown in Figure 6-31, and then calculate the voltage measured by voltmeters *A*, *B*, and *C*.
33. Why would the circuit in Figure 6-31 be called a dual output power supply circuit?
34. Identify the circuit shown in Figure 6-32, and indicate which of the output waveforms shown in Figure 6-32(c), (d), or (e) will be present on the oscilloscope for the function generator input signal shown in Figure 6-32(b).
35. Calculate the positive and negative peak of the output from the circuit in Figure 6-32.

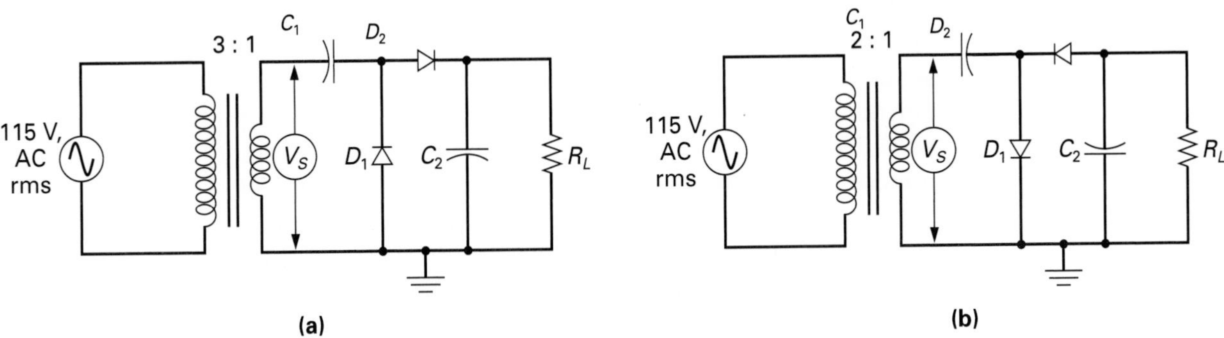

FIGURE 6-30 Multiplier Circuits.

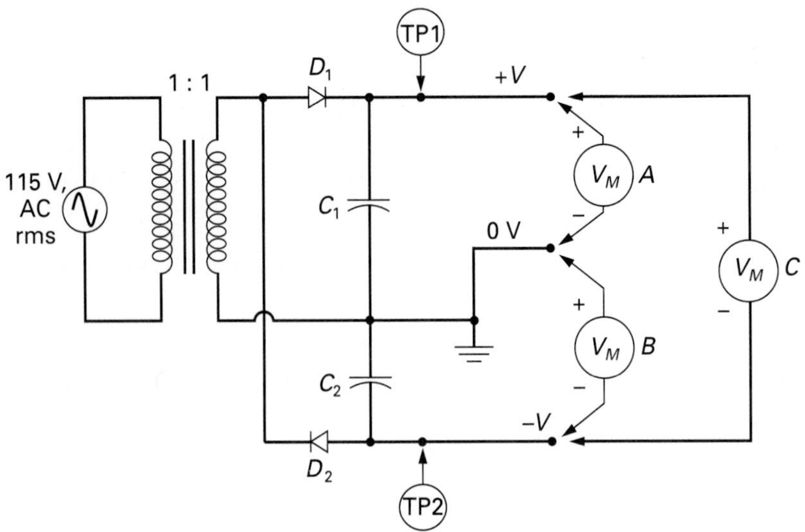

FIGURE 6-31 A Dual-Polarity Power Supply Circuit.

36. What circuits would generate the outputs shown in Figure 6-32(c), (d), and (e), for the input shown in Figure 6-32(b)?
37. Identify the circuit shown in Figure 6-33, and calculate the positive and negative peak of the output signal.
38. Figure 6-34 shows the basic blocks for an ultrasonic motion detector circuit. The circuit operates by flooding an area with ultrasonic (high frequency sound waves) signals and then sensing the changes in amplitude of the reflected signal. Looking at the schematic diagram, you can see that the circuit is powered by a +8 V dc supply. This 8 V supply voltage is connected to two amplifiers, a comparator, and a 35 kHz oscillator or sine wave generator. The 35 kHz sine wave signal from the oscillator causes a transmit crystal to expand and contract, and this movement will disrupt the surrounding air sending out 35 kHz sound waves. With no motion, the signal sent back to the receive crystal will be of a constant amplitude and the "motion indicator LED" (D_2) will be OFF. When there is motion in the transmit area, the signal reflected back to the receive crystal will be amplitude modulated (AM), as shown in the waveforms in Figure 6-34. The detector circuit (D_1 and R_1) will demodulate this signal. It will then be amplified and filtered and amplified by an active band-pass filter (which is included to only pass signals centered at 35 kHz so that only our motion detector signals will trigger the circuit). Finally, a comparator circuit compares a reference voltage provided by R_4 and R_5 with the motion signal voltage (which can be adjusted in sensitivity by R_2) to produce a MOTION/NO MOTION signal that either turns ON or OFF the LED D_2. Calculate the positive and negative voltage peak ($+V_{pk}$ and $-V_{pk}$) and the NO MOTION peak ($+V_{NM}$) of the signal at the output of the diode detector (D_1, R_1).
39. Identify the D_1, R_1 clipper circuit in Figure 6-34.
40. Identify the circuit shown in Figure 6-35 (p. 283), and indicate which of the

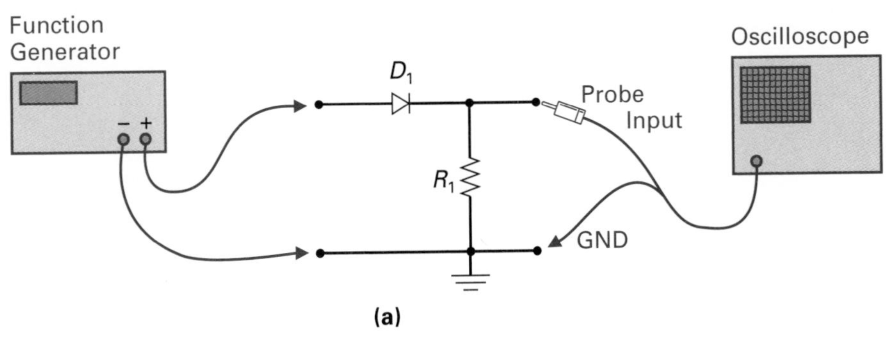

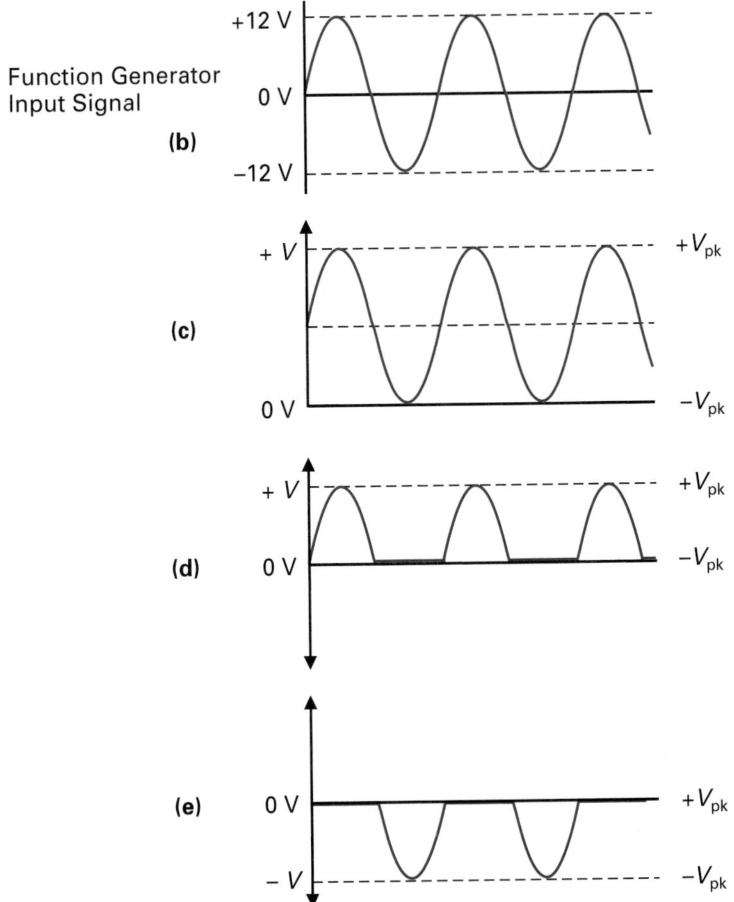

FIGURE 6-32 A Circuit's Input/Output Signals.

output waveforms shown in Figure 6-35(c), (d), or (e) will be present on the oscilloscope for the function generator input signal shown in Figure 6-35(b).

41. Calculate the positive and negative peak of the output from the circuit in Figure 6-35.

42. What circuits would generate the outputs shown in Figure 6-35(c), (d), and (e), for the input shown in Figure 6-35(b)?

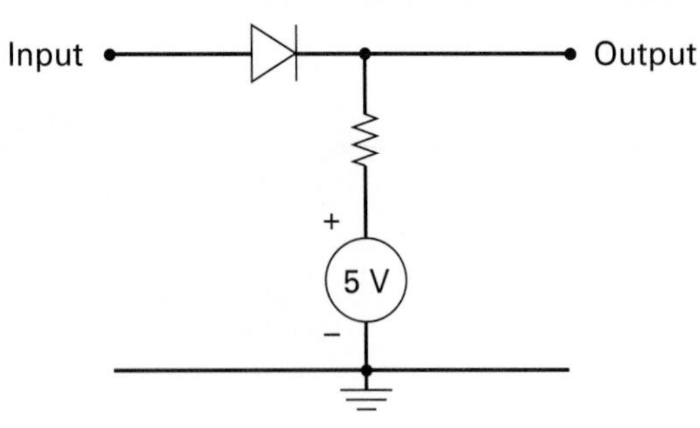

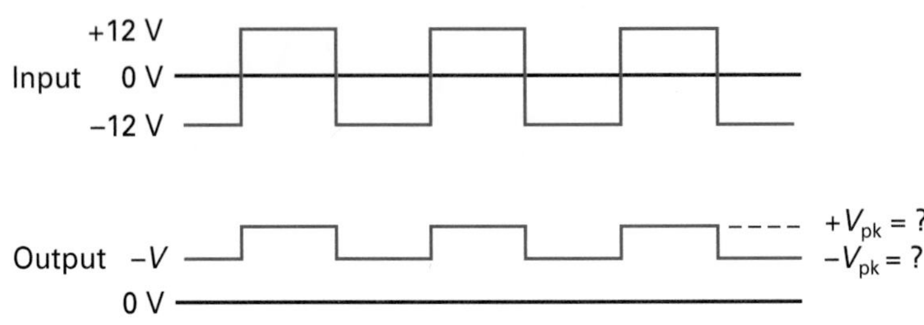

FIGURE 6-33 A Biased Negative Series Clipper Circuit.

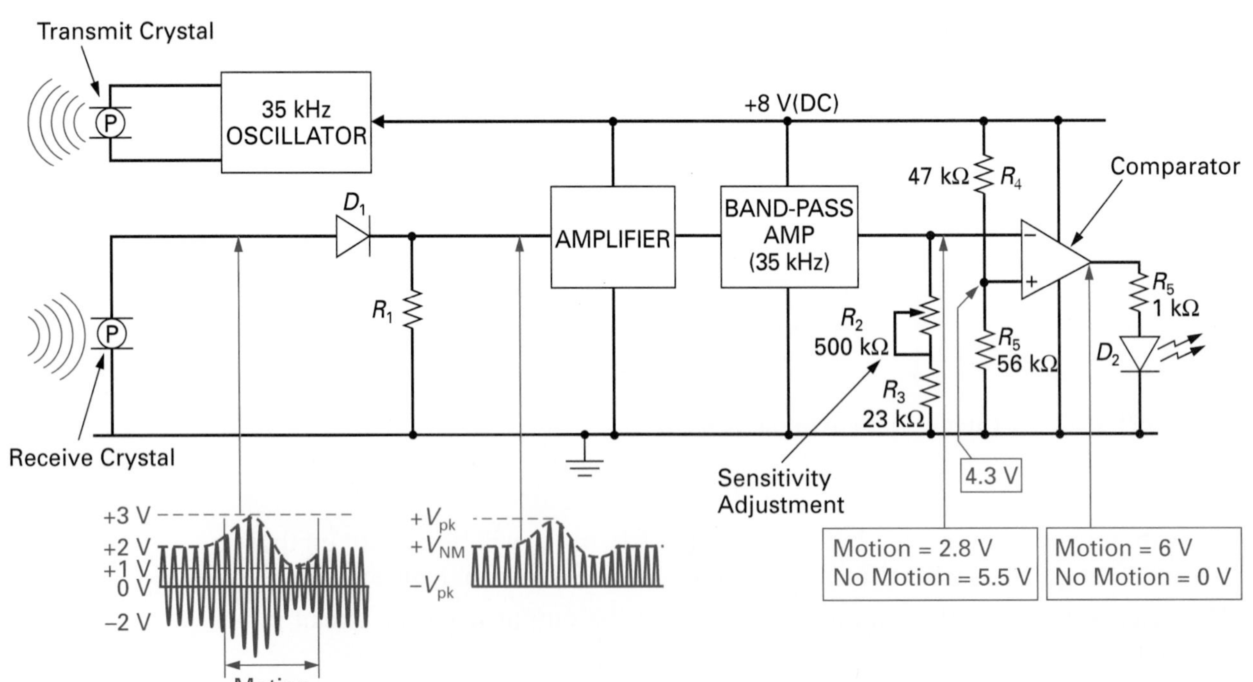

FIGURE 6-34 Motion Detector Circuit.

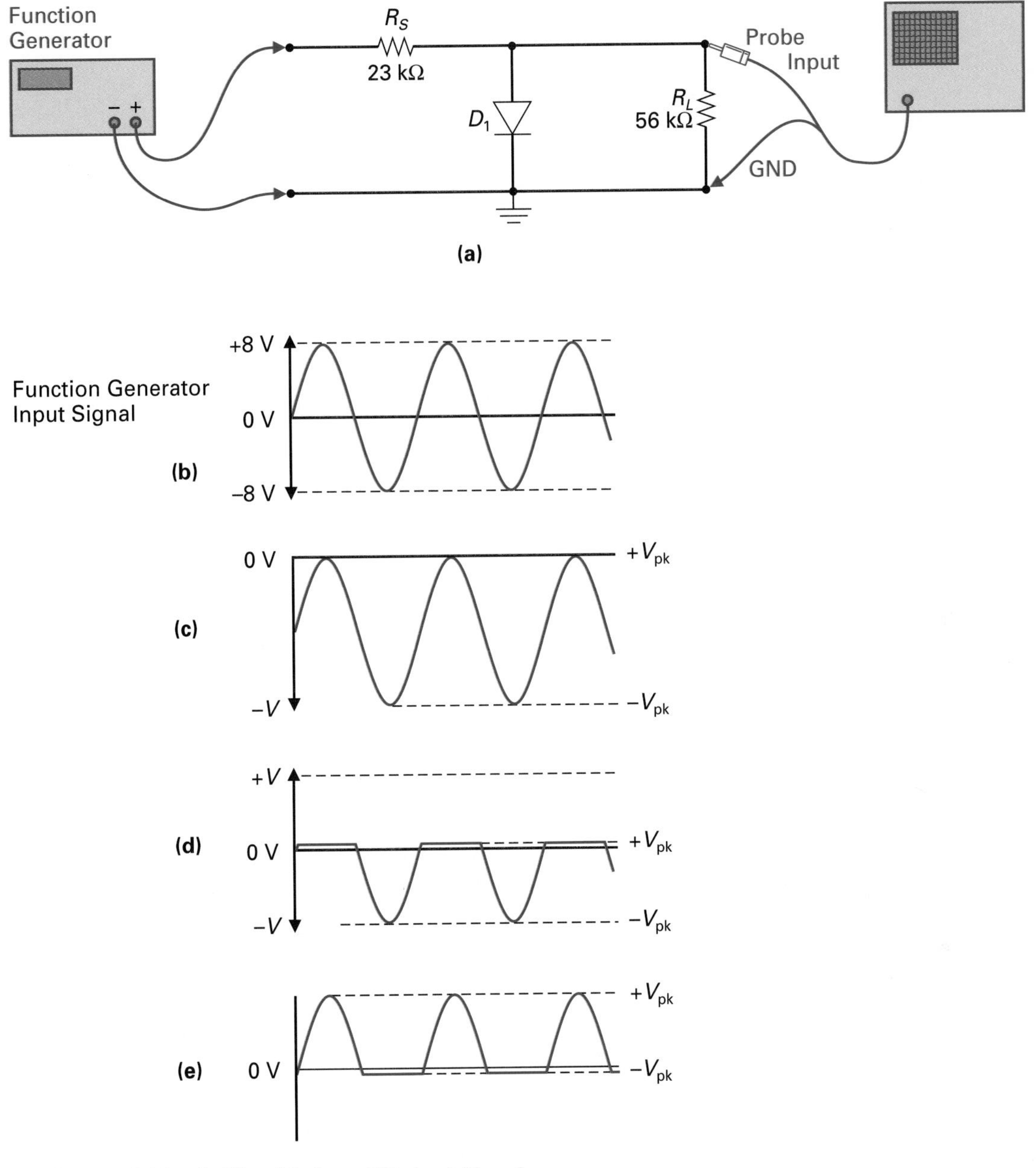

FIGURE 6-35 A Circuit's Input/Output Signals.

43. Some computers save programs (which are instructions and data) that do not presently need to be used on the tape of an audio cassette. These programs are saved in exactly the same way that you save or record a piece of music on an audio cassette. When a computer wants to use this program, the cassette will have to be played and the program loaded back into the computer's memory. Figure 6-36 shows how a retrieved computer program that has been saved on an audio cassette is limited and amplified before being put back into the computer's memory. Identify the circuit formed by D_1 and D_2, and describe why you think this circuit is included.

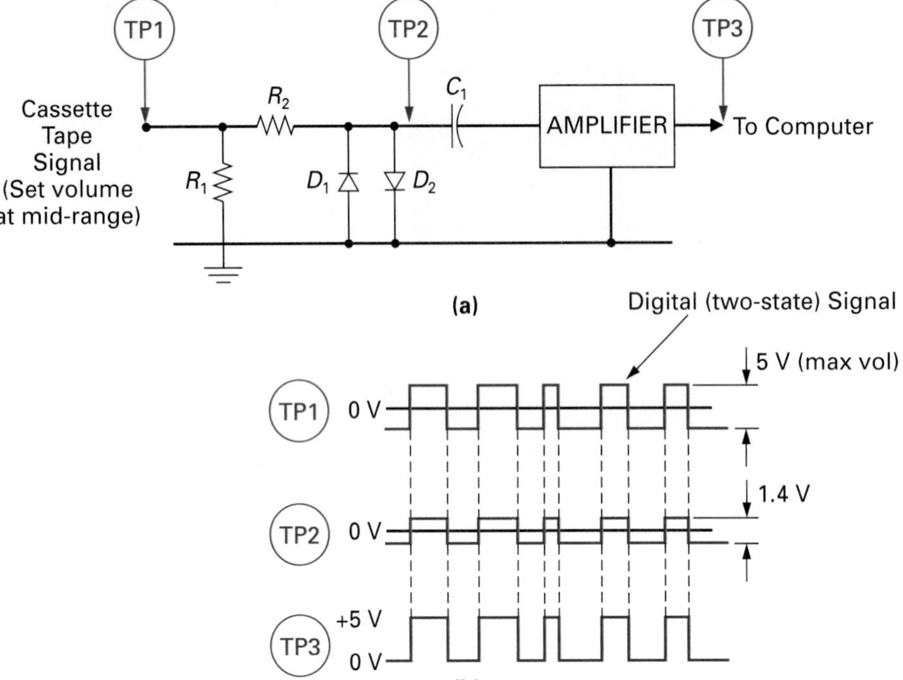

FIGURE 6-36 Digital Cassette Tape Signal Processing.

44. The four drive circuits in Figure 6-37 are used to produce a sequence of 0 V outputs to the four windings of a stepper motor. For example, first drive circuit *A* will switch 0 *V* to its output and cause winding *A* to energize, then drive circuit *B* will produce a 0 V output and energize winding *B*, and then *C*, and then *D*, and then the cycle will repeat. The result is that the rotor of the motor will step round in a clockwise direction. Why are diodes D_1 through D_4 included in this circuit, and what forward current rating should these diodes have?

45. Identify the circuit shown in Figure 6-38, and calculate the positive and negative peak of the output signal.

Troubleshooting Questions

46. Figure 6-39(a) shows a voltage-quadrupler circuit. Determine whether or not there is a problem with the circuit, based on the capacitor voltage readings given in the three examples in Figure 6-39(b), (c), and (d). Indicate which components you would suspect, and therefore check, in each example.

47. Determine whether or not there is a problem with the series clipper circuits shown in Figure 6-40.

48. Determine whether or not there is a problem with the shunt clipper circuits shown in Figure 6-41 (p. 287).

49. Determine whether or not there is a problem with the shunt clamper circuits shown in Figure 6-42 (p. 287).

50. What would happen to the output waveform in Figure 6-42(b) if the diode D_1 were to open?

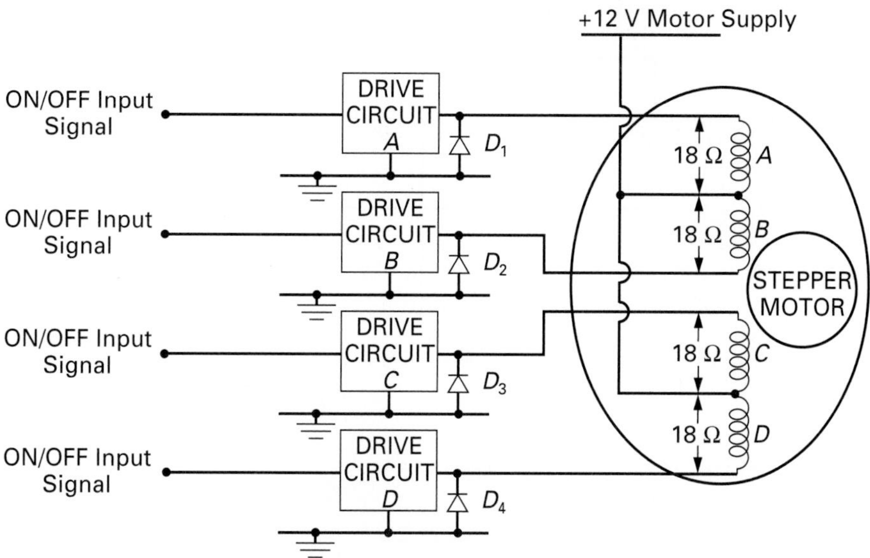

FIGURE 6-37 A Stepper Motor Drive Circuit.

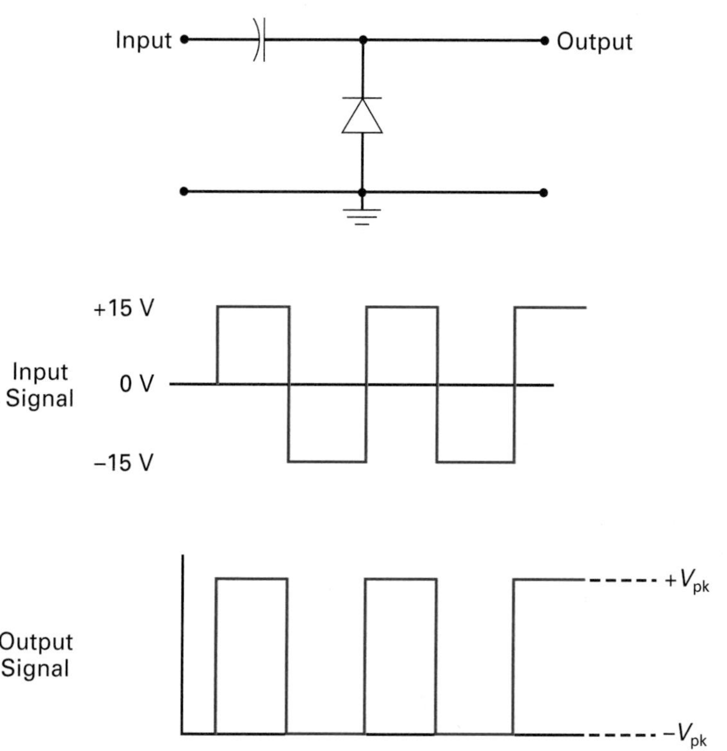

FIGURE 6-38 A DC Restorer Circuit.

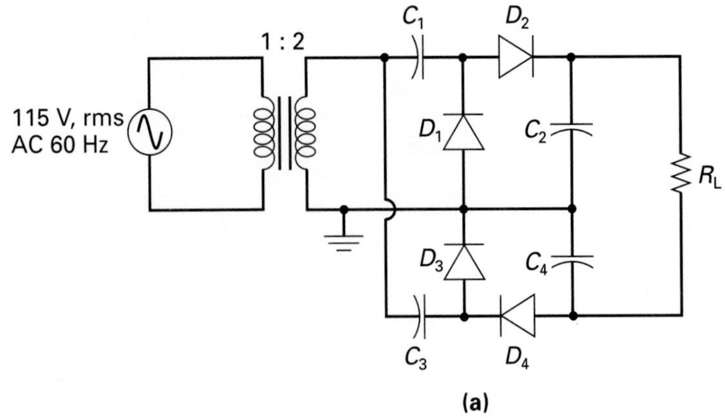

(a)

(b)	(c)	(d)
$V_p = 162.6\ V_{pk}$	$V_p = 162.6\ V_{pk}$	$V_p = 162.6\ V_{pk}$
$V_S = 325.2\ V_{pk}$	$V_S = 325.2\ V_{pk}$	$V_S = 325.2\ V_{pk}$
$V_{C1} = 325.2$ V	$V_{C1} = 325.2$ V	$V_{C1} = 325.2$ V
$V_{C2} = 325.2$ V	$V_{C2} = 650.4$ V	$V_{C2} = 650.4$ V
$V_{C3} = 325.2$ V	$V_{C3} = 325.2$ V	$V_{C3} = 50.0$ V
$V_{C4} = 650.4$ V	$V_{C4} = 650.4$ V	$V_{C4} = 50.0$ V
$V_{RL} = 975.6$ V	$V_{RL} = 1300.8$ V	$V_{RL} = 200.4$ V

FIGURE 6-39 Troubleshooting a Voltage Multiplier Circuit.

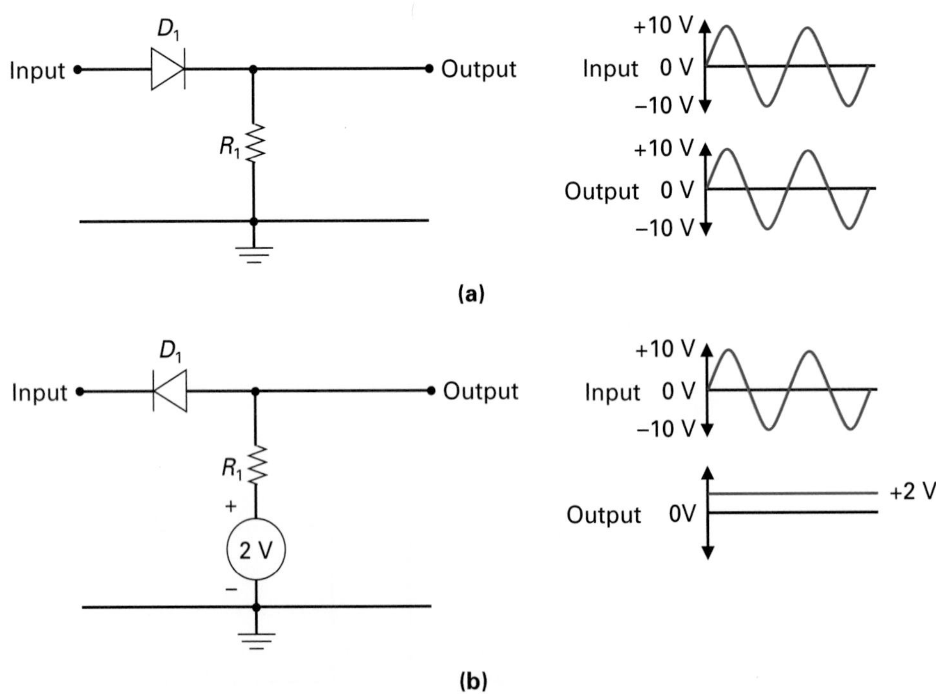

FIGURE 6-40 Troubleshooting Series Clipper Circuits.

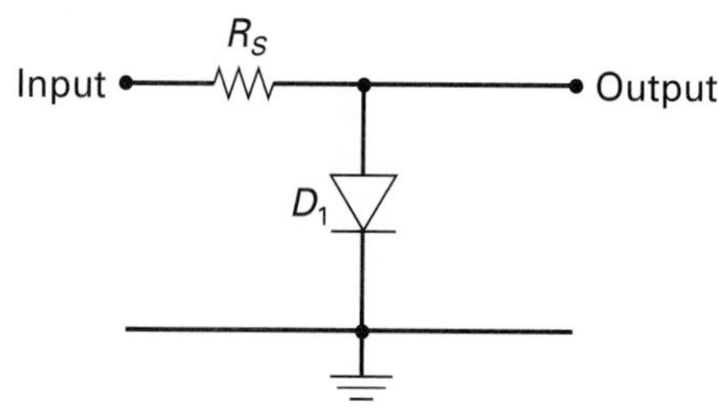

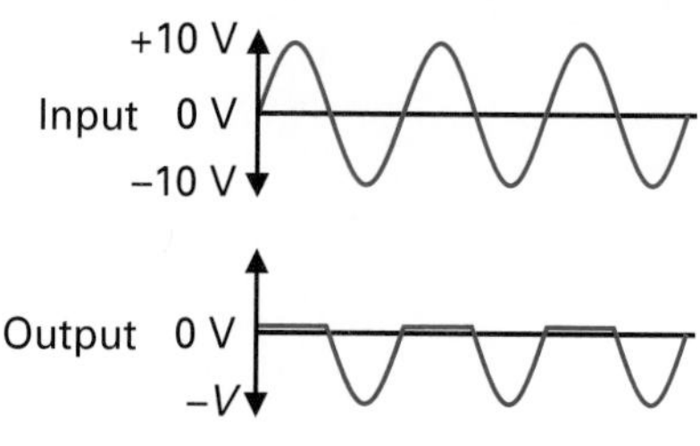

(a)

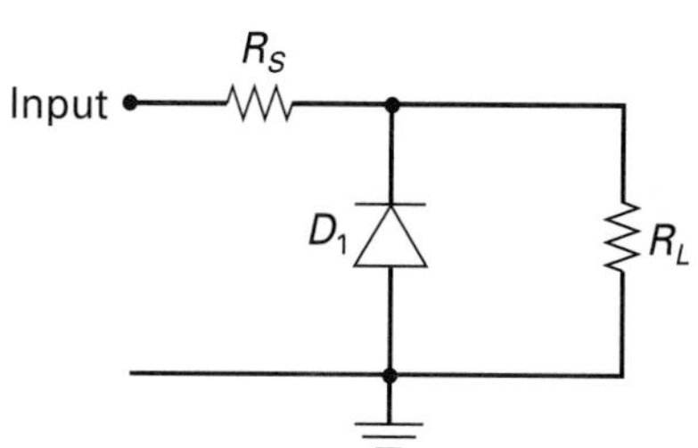

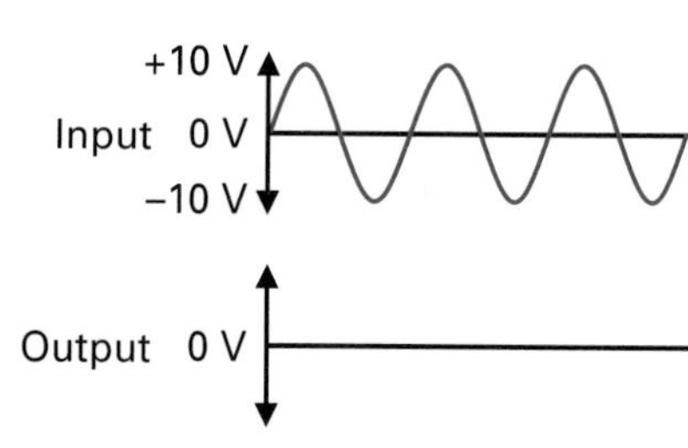

(b)

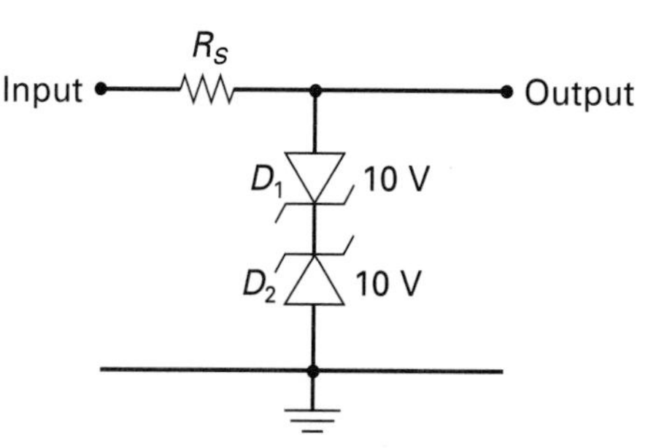

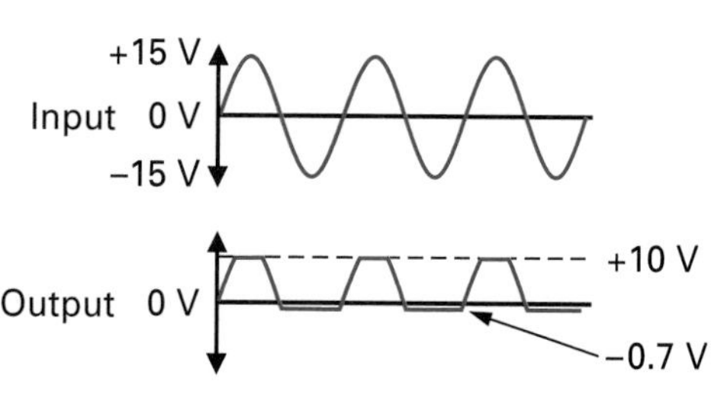

(c)

FIGURE 6-41 Troubleshooting Shunt Clipper Circuits.

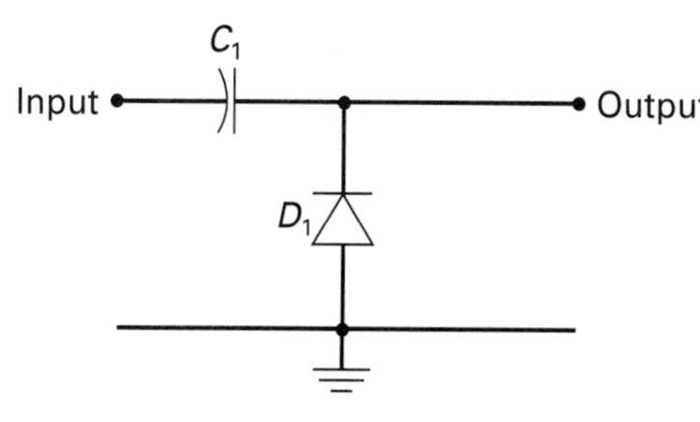

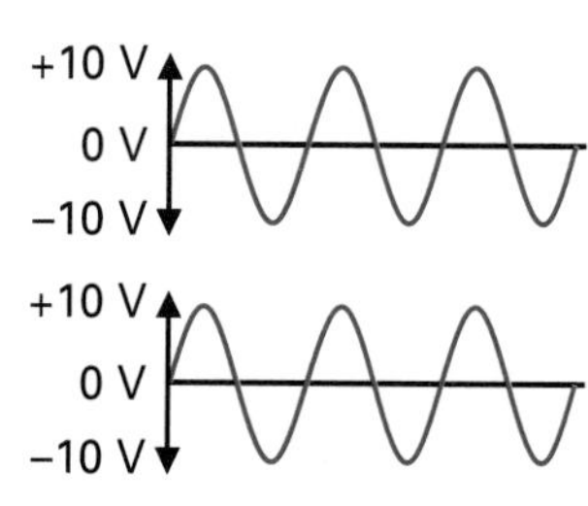

(a)

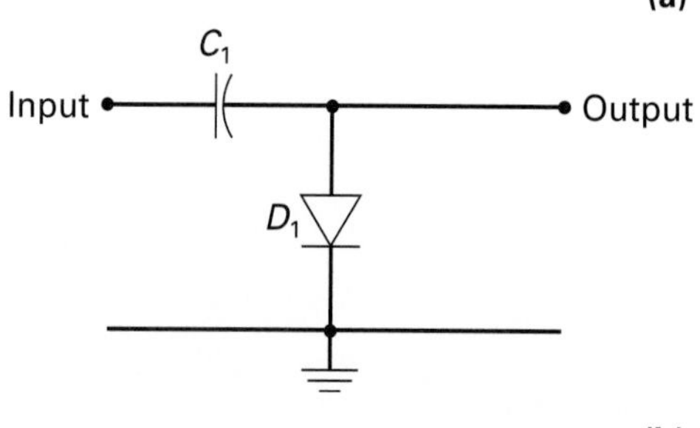

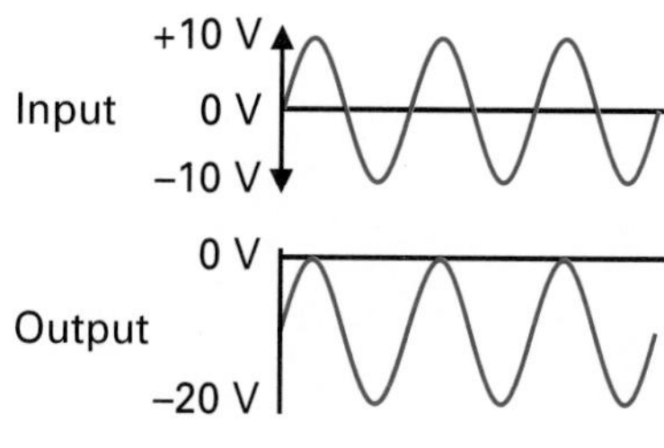

(b)

FIGURE 6-42 Troubleshooting Clamper Circuits.

CHAPTER

7

OBJECTIVES

After completing this chapter, you will be able to:

1. List the three bands within the optical frequency spectrum.
2. Identify the schematic symbol of the photodiode, and describe its characteristics and operation.
3. Determine the key ratings of a photodiode from a manufacturer's data sheet.
4. Explain how a photodiode is used in fiber optic communications.
5. List the advantages that a fiber optic communication link has over a copper conductor communication link.
6. Identify the schematic symbol of the varactor diode, and describe its characteristics and operation.
7. Determine the key ratings of a varactor diode from a manufacturer's data sheet.
8. Explain how a varactor diode can be used in a receiver's tuned circuit.
9. Identify the schematic symbol of the transient suppressor diode, and describe its characteristics and operation.
10. Determine the key ratings of a transient suppressor diode from a manufacturer's data sheet.
11. Identify the schematic symbol of the constant-current diode, and describe its characteristics and operation.
12. Determine the key ratings of a constant-current diode from a manufacturer's data sheet.
13. Identify the schematic symbol of the schottky and tunnel diode, and describe their characteristics and operation.
14. List and basically describe the typical applications of the schottky and tunnel diode.

Special Application Diodes

OUTLINE

V I G N E T T E

Magnetic Attraction

Andre Ampere was born near Lyon, France, in 1775. His father was a rich silk merchant who tutored him privately; however, by the time the boy had reached his teens he had read the works of many of the great mathematicians. Ampere possessed a photographic memory which, in conjunction with a precocious ability in mathematics, made him a challenge for any professor of science.

In 1796 he began giving private lessons in mathematics, chemistry, and languages. It was in this capacity that he met his wife and was married in 1799. In 1801 he was offered a position as the professor of physics at Bourg, and because his wife was sick at the time, he traveled on ahead. His wife died a few days later and he never recovered from the blow. In fact, in later life he confided to a friend that he realized at the time of his wife's death that he could love nothing else except his work. In 1809 he became professor of mathematics at École Polytechnique in Paris, a position he held for the remainder of his life.

In 1820 Ampere witnessed Hans Christian Oersted's discovery that a compass needle could be deflected by a current-carrying wire. Inspired by this first basic step in electromagnetism, Ampere began to experiment with characteristic industry and care. In only a few weeks he advanced Oersted's discovery by leaps and bounds, developing several mathematical laws of electromagnetism. He also discovered that a coil of wire carrying a current would act like a magnet, and if an iron bar were placed in its center, it would become magnetized. He called this device a solenoid, a name that is still given to electromagnets containing a movable element. In honor of his achievements in electricity, the basic unit of electric current, the ampere, is named after him. On June 10, 1836, Ampere died in Marseille of what he told a friend just moments before his death was "a broken heart."

I N T R O D U C T I O N

Up to this point we have seen how the semiconductor P-N junction can be used to construct a junction diode, zener diode, and light-emitting diode, and how these devices are used in a variety of circuit applications. In this chapter, we will study some other types of diodes that have unique characteristics that make them well suited for special circuit application. These diodes include: the photodiode, varactor diode, transient suppressor diode, constant-current diode, and two high-frequency diodes called the schottky diode and tunnel diode.

7-1 THE PHOTODIODE

In Chapter 4 we discussed in detail the operation and characteristics of the light-emitting diode or LED. The LED is a photo-emitting or light-transmitting device. In this section we will be discussing in detail the operation and characteristics of its counterpart, the photodiode, which is a photo-detecting or light-receiving device. The LED and photodiode are referred to as "optoelectronic devices" because their operation combines both optics and electronics. Like most areas of science, optoelectronics has its own set of terms and units. Before discussing the photodiode, we will briefly examine the basic principles of light.

7-1-1 The Optical Electromagnetic Spectrum

Like radio waves, light is electromagnetic radiation—but at a different frequency. The light, or **optical electromagnetic spectrum,** is midway between the microwave frequency band and the X-ray frequency band, as shown in Figure 7-1(a). This optical light spectrum extends from 300 GHz to about 9,300,000 GHz and contains three bands called the **infrared band, visible light band,** and **ultraviolet band.** The human eye can only detect the narrow band of frequencies in the visible light band between about 428,000 GHz and 750,000 GHz. Light waves cannot be seen below this visible region in the infrared band and above this visible region in the ultraviolet band. The frequencies in the visible light band can be divided into the colors of the rainbow: red, orange, yellow, green, blue, and violet, with each color corresponding to a narrow range of frequencies.

As you can see in Figure 7-1(a), the frequency values in the optical spectrum are large and difficult to work with. To make these values easier to handle, most optical frequencies are converted to their wavelength equivalent. To review, wavelength is symbolized by the Greek letter lambda (λ), and is measured in meters. As can be seen in Figure 7-1(b), wavelength is as its name states, the physical length of one cycle of a transmitted electromagnetic wave. Like radio waves, the wavelength of a light wave is calculated by dividing the light wave's velocity by its frequency, as shown in the following formula:

Optical Electromagnetic Spectrum
That part of the electromagnetic spectrum that encompasses infrared, visible, and ultraviolet light.

Infrared Band
The range of frequencies below that which the human eye can detect.

Visible Light Band
The range of frequencies the human eye can detect.

Ultraviolet Band
The range of frequencies above that which the human eye can detect.

$$\lambda = \frac{c}{f}$$

where λ = the length of one cycle of the electromagnetic wave in meters;
c = the speed of the electromagnetic wave, which is the speed of light or 3×10^8 meters/second; and
f = the frequency of the electromagnetic wave in hertz.

■ **EXAMPLE:**

Calculate the wavelength of the two frequencies at either end of the microwave frequency band, which extends from 300 MHz to 300 GHz.

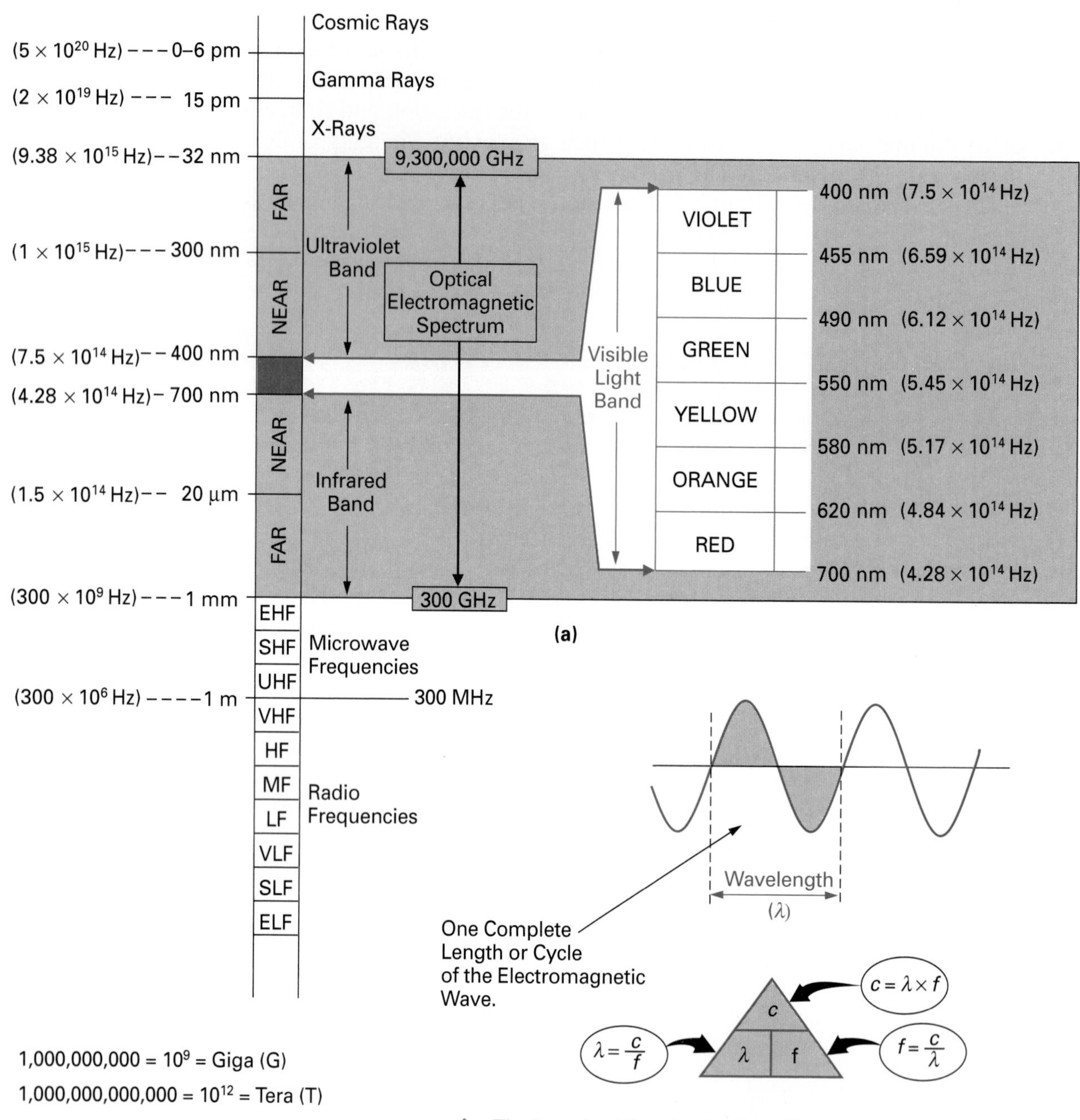

λ = The Length of One Cycle of the Electromagnetic Wave in meters.

c = The Speed of the Electromagnetic Wave, which is the Speed of Light, or 3×10^8 meters/second.

f = The Frequency of the Electromagnetic Wave in Hertz.

(b)

FIGURE 7-1 The Optical Electromagnetic Spectrum.

■ ***Solution:***

For 300 MHz:

$$\lambda = \frac{c}{f} = \frac{3 \times 10^8 \text{ m/s}}{300 \times 10^6 \text{ Hz}} = 1 \text{ meter}$$

For 300 GHz:

$$\lambda = \frac{c}{f} = \frac{3 \times 10^8 \text{ m/s}}{300 \times 10^9 \text{ Hz}} = 0.001 \text{ meter}$$

These wavelengths are shown alongside their frequency equivalents in Figure 7-1(a).

As you can see from the previous example, because wavelength is inversely proportional to frequency, the higher the frequency the smaller the wavelength ($f\uparrow, \lambda\downarrow$). In fact, at light frequencies the calculated wavelengths are so small, we have to continually convert the answer in meters to a smaller unit such as micrometers or nanometers. For example, the visible light band begins at 4.28×10^{14} Hz. This frequency is equivalent to the following wavelength:

$$\lambda = \frac{c}{f} = \frac{3 \times 10^8 \text{ m/s}}{4.28 \times 10^{14} \text{ Hz}} = 0.0000007 \text{ meter}$$

To add a prefix to meter:

$$0.0000007 \text{ m} = 0.7 \times 10^{-6} \text{ m} = 0.7 \text{ micrometers or } 0.7 \text{ microns}$$

$$\text{or } 0.0000007 \text{ m} = 700 \times 10^{-9} \text{ m} = 700 \text{ nanometers}$$

To save a step, we could use a smaller light-speed unit and convert frequency directly to micrometers (more commonly called microns) or nanometers. For example:

Speed of Light (↑)	*Wavelength* (↓)
3×10^8 meters/second	= meters
3×10^{10} centimeters/second	= centimeters
3×10^{11} millimeters/second	= millimeters
3×10^{14} micrometers/second	= micrometers or microns
3×10^{17} nanometers/second	= nanometers

Do not think that these speed of light values are all different. Light will simply travel one centimeter 100 times faster than it will travel one meter.

To apply these to the visible light band starting frequency of 4.28×10^{14} Hz, the wavelength in microns and nanometers would be:

$$\lambda = \frac{c}{f} = \frac{3 \times 10^{14} \ \mu\text{m/s}}{4.28 \times 10^{14} \text{ Hz}} = 0.7 \text{ micrometers or } 0.7 \text{ microns}$$

$$\lambda = \frac{c}{f} = \frac{3 \times 10^{17} \text{ nm/s}}{4.28 \times 10^{14} \text{ Hz}} = 700.9 \text{ nanometers}$$

Another unit that is commonly used is the **angstrom,** which is symbolized by Å, and is equal to 1×10^{-10} meters. Therefore, our example of 0.0000007009

Angstrom (Å) 1×10^{-10} meters

meters would be equal to 7009×10^{-10} meters, or 7009 angstroms (7009 Å). To convert to angstroms, simply use 3×10^{18} for the speed of light:

$$\lambda = \frac{c}{f} = \frac{3 \times 10^{18}}{4.28 \times 10^{14}} = 7009 \text{ Angstroms}$$

It is important to understand the relationship between frequency and wavelength, because these two terms are used interchangeably when we discuss electromagnetic radio or light waves. In fact, most manufacturer's data sheets will only list an LED's output frequency or a photodiode's best input frequency by wavelength.

■ **EXAMPLE:**

A manufacturer's data sheet lists an LED as producing the highest output power at a wavelength of 800 nm or 8000 Å. In what light wave band is this wavelength?

■ ***Solution:***

Referring to Figure 7-1, you can see that a wavelength of 600 nanometers is within the near infrared light wave band. This band, which is close to the visible band (hence the name "near infrared"), extends in wavelength from 20 microns to 700 nanometers. In frequency, these wavelengths are equivalent to 150 terahertz (150×10^{12} Hz or 1.5×10^{14} Hz) to 428 terahertz (4.28×10^{14} Hz).

Wave Theory of Light
Assumes that light propagates or travels as electromagnetic waves.

Quantum Theory or Particle Theory
States that light consists of tiny particles.

Photon
A discrete particle or quantum of light.

Before we begin our discussion on the photodiode, there is one other aspect of light that we should address. The **wave theory** of light assumes that light propagates or travels as electromagnetic waves. This theory is ideal for explaining why light travels at the speed of light through a vacuum (186,282.4 miles/s or 3×10^8 meters/s), at a slightly slower velocity in air, at even slower speeds in other materials such as glass and water, and the light bending action that occurs as light passes through these materials. The wave theory, however, cannot explain the interaction that occurs between light and semiconductor materials. To explain this action, we must use the **quantum theory** or **particle theory** of light. The quantum theory states that light consists of tiny particles, and each of these discrete quanta or individual packets of energy is called a **photon.** These photons are uncharged particles that have wave-like characteristics. The photon's energy is determined by its frequency, with a higher frequency photon having more energy than a lower frequency photon. Therefore, ultraviolet photons will have a higher energy content than a visible light-wave photon.

To understand and explain the behavior of light waves, we will have to think of them as electromagnetic waves that contain many tiny particles.

7-1-2 Photodiode Construction and Symbol

Photodiode
A photo-detecting or light-receiving device that contains a semiconductor P-N junction.

A **photodiode** is a photo-detecting or light-receiving device that contains a semiconductor P-N junction.

Figure 7-2(a) shows a typical photodiode package. A glass window or convex lens allows light to enter the case and strike the semiconductor photodiode that is mounted within the metal case.

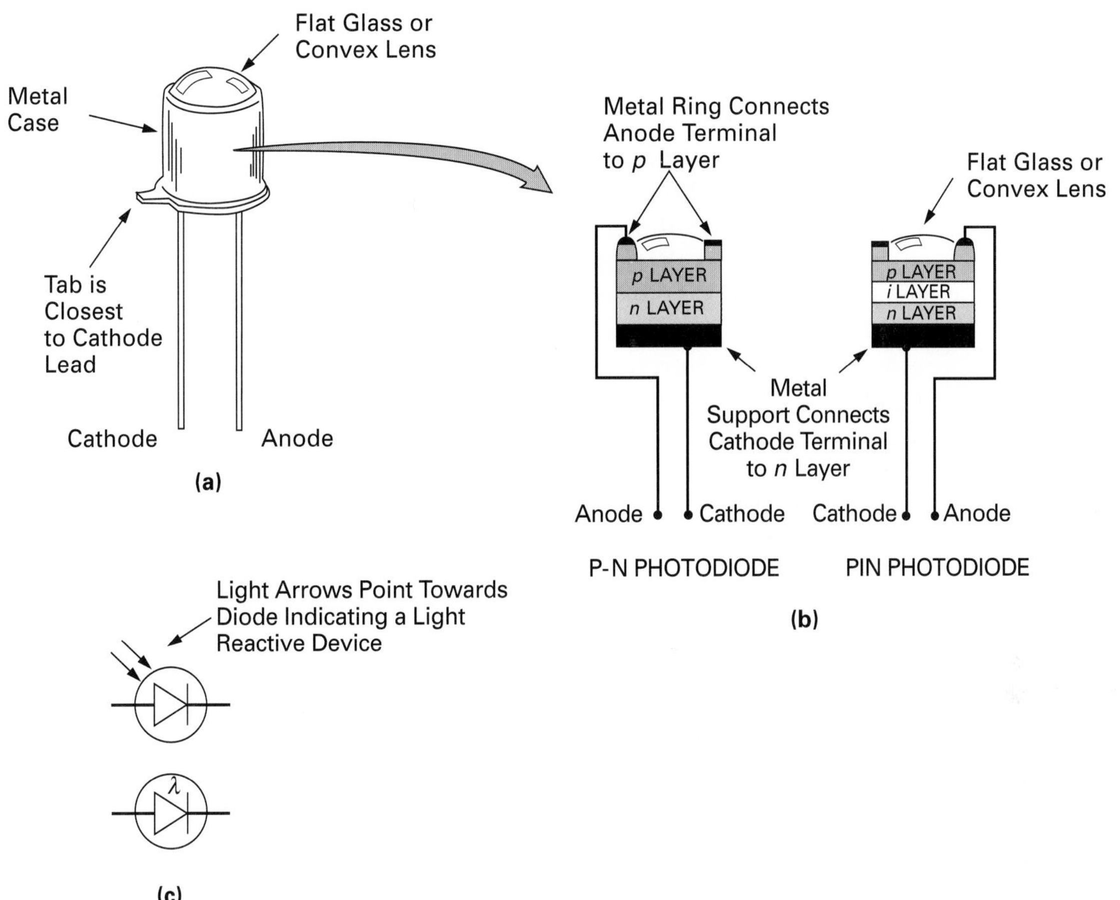

FIGURE 7-2 Photodiodes. (a) Typical Package. (b) P-N and PIN Photodiode Construction. (c) Schematic Symbols. (d) Physical Appearance.

Figure 7-2(b) shows how the photodiode is constructed in basically one of two ways. The **P-N photodiode** contains a *p*-type region that is diffused into an *n*-type substrate. A metal base makes the connection between the cathode terminal and the *n*-type region, while a metal ring makes contact between the anode terminal and the *p*-type region. Light enters the photodiode through the hole in the metal ring. The **PIN photodiode** is constructed in almost exactly the same way, except that the device has an intrinsic layer between the *p* and *n* regions, hence the name PIN (*p*-type layer, intrinsic layer, *n*-type layer). The addition of the intrinsic layer, which is a pure semiconductor having no impurities, makes the photodiode respond better to low-frequency (infrared) photons which tend to penetrate deeper into the diode's regions. The intrinsic layer also creates a larger depletion region, which causes the photodiode to produce a more linear change in current in response to light intensity changes.

P-N Photodiode
Contains a *p*-type region that is diffused into an *n*-type substrate.

PIN Photodiode
Has an intrinsic layer between the *p* and *n* regions.

Figure 7-2(c) shows the two commonly used photodiode schematic symbols. With the LED, the two arrows pointed away from the diode to indicate that it generated a light output. With the photodiode, the two arrows point towards the diode to indicate that it responds to a light input.

7-1-3 Photodiode Operation

Photodiodes can be operated in one of two modes, as shown in Figure 7-3.

Photovoltaic Mode When the photodiode generates an output voltage in response to a light input.

When used in the **photovoltaic mode,** the photodiode will generate an output voltage (voltaic) in response to a light (photo) input. Figure 7-3(a) shows how the photovoltaic photodiode cell, or solar cell, will operate in this mode. When light passes through the photodiode's window, the light's photons are absorbed at different depths in the semiconductor material, depending on their energy content (wavelength). High-energy photons will collide with semiconductor atoms and transfer energy to the atoms. If enough energy is transferred from photon to atom, a valence electron will be released from the valence band to the conduction band. This will result in a free electron (negative charge) and a positively charged atom (electron-hole pair). Negative free electrons in the depletion region will be attracted to the positive ions in the *n*-type region, while positive holes in the depletion region will be attracted to the negative ions in the *p*-type region. This separation of charges will generate a potential difference or small voltage across the P-N junction that is typically about 0.45 V. If a load were connected across the photodiode, a small electron current would flow from the *n*-type region (cathode) to the *p*-type region (anode). Photovoltaic photodiodes can be used as a light meter in a camera. They can also be arranged into banks or arrays, where they charge batteries in remote locations such as communications satellites, freeway emergency telephones, and portable equipment such as calculators.

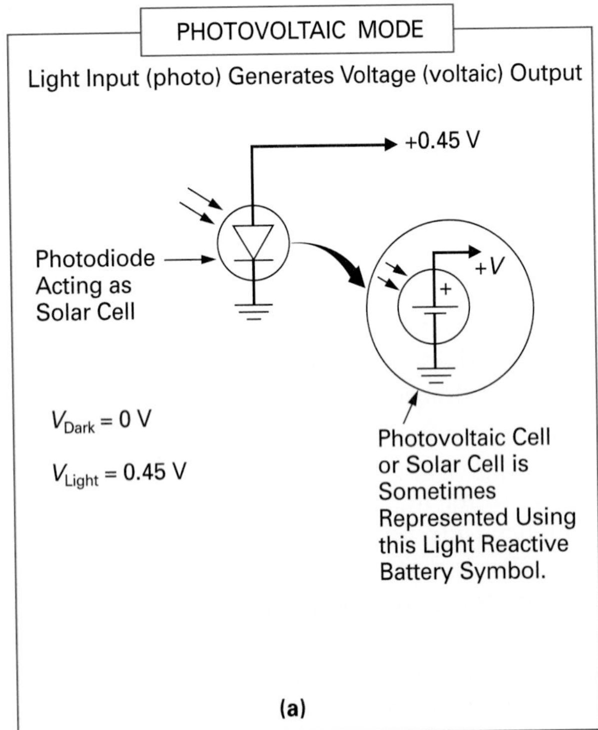

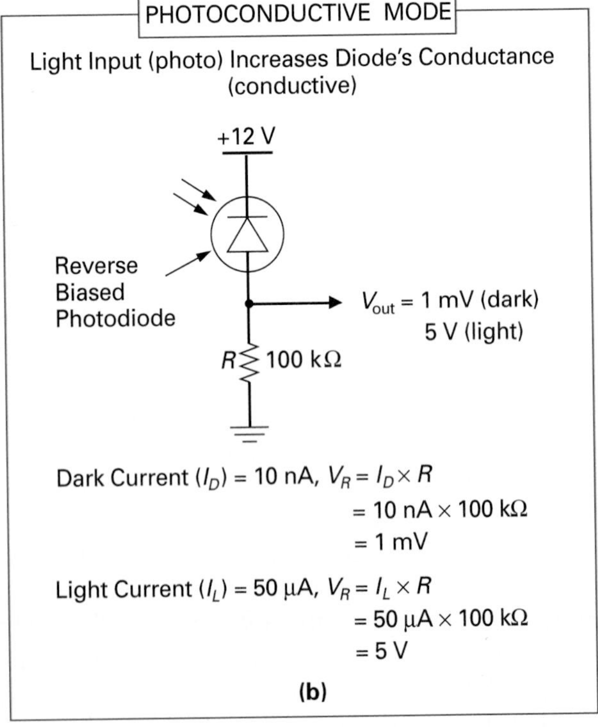

FIGURE 7-3 Photodiode Operation.

Photodiodes are most widely used in the **photoconductive mode,** in which they will change their conductance (conductive) when light (photo) is applied. In this mode, the photodiode is reverse biased (n-type region is made positive, p-type region is made negative), as shown in the example circuit in Figure 7-3(b). The reverse biased photodiode will have a wide depletion region, and only a small reverse current will pass through the diode. The reverse current that passes through the photodiode when no light is being applied is called the **dark current (I_D),** and in the example in Figure 7-3(b) this is equal to 10 nA. When light is applied, photons enter the depletion region and create electron-hole pairs. The electrons are attracted to the positive bias voltage (+12 V), and the holes are attracted to the negative bias voltage (ground). This movement of separated electrons and holes makes up a reverse current through the photodiode. An increase in the light intensity will result in an increase in the reverse current, and in the photodiode's conductivity. The reverse current that passes through the photodiode when light is being applied is called the **light current (I_L),** and in the example in Figure 7-3(b) this is equal to 50 μA.

Photoconductive Mode
When the photodiode changes conductance in response to a light input.

Dark Current (I_D)
The reverse current that passes through the photodiode when no light is being applied.

Light Current (I_L)
The reverse current that passes through the photodiode when light is being applied.

■ **EXAMPLE:**

Referring to Figure 7-3(b), calculate:

a. The ratio of light current to dark current.
b. The output voltage when no light is applied.
c. The output voltage when light is applied.

■ ***Solution:***

a. $$\text{Ratio} = \frac{I_L}{I_D} = \frac{50\ \mu\text{A}}{10\ \mu\text{A}} = 5000$$

Therefore, the photodiode's light current is 5000 times larger than the photodiode's dark current.

b. $$V_R = I_D \times R = 10\ \text{nA} \times 100\ \text{k}\Omega = 1\ \text{mV}$$

c. $$V_R = I_L \times R = 50\ \mu\text{A} \times 100\ \text{k}\Omega = 5\ \text{V}$$

The PIN photodiode is more widely used than the P-N photodiode due to its greater sensitivity. This is because the intrinsic layer adds to the photodiode's depletion region and makes a much wider depletion region for a given reverse bias voltage. This wider depletion region increases the chance that electron-hole pairs will be generated by photons and increases the conductance of the photodiode.

7-1-4 Photodiode Data Sheet

Figure 7-4 shows a typical manufacturer's data sheet for a PIN silicon photodiode. Using this data sheet, you can find the photodiode's key characteristics. The most important ones have been explained in the notes in this figure. As you can

DEVICE: Photo Detector Diode — MRD500 and MRD510

Features:

- Ultra Fast Response — (<1 ns Typ)
- High Sensitivity — MRD500 (1.2 μA/(mW/cm^2) Min)
  MRD510 (0.3 μA/(mW/cm^2) Min)
- Available with Convex Lens (MRD500) or Flat Glass (MRD510) for Design Flexibility
- Popular TO-18 Type Package for Easy Handling and Mounting
- Sensitive Throughout Visible and Near Infrared Spectral Range for Wide Application
- Annular Passivated Structure for Stability and Reliability

Applications:

- Industrial Processing and Control
- Shaft or Position Readers
- Optical Switching
- Remote Control
- Laser Detection
- Light Modulators
- Logic Circuits
- Light Demodulation/Detection
- Counters
- Sorters

PHOTO DETECTORS
DIODE OUTPUT
PIN SILICON
250 MILLIWATTS
100 VOLTS

Convex lens is used when we need to capture surrounding light.

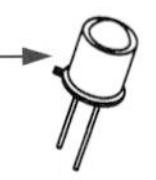

CASE 209-01, Style 1
MRD500
(CONVEX LENS)

Flat glass can be used when light is applied directly into photodiode window.

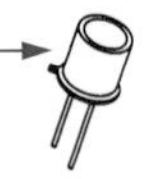

CASE 210-01, Style 1
MRD510
(FLAT GLASS)

MAXIMUM RATINGS (T_A = 25°C unless otherwise noted)

Rating	Symbol	Value	Unit
Reverse Voltage	V_R	100	Volts
Total Power Dissipation @ T_A = 25°C Derate above 25°C	P_D	250 2.27	mW mW/°C
Operating Temperature Range	T_{op}	−55 to +125	°C
Storage Temperature Range	T_{stg}	−65 to +200	°C

STATIC ELECTRICAL CHARACTERISTICS (T_A = 25°C unless otherwise noted)

Characteristic	Fig. No.	Symbol	Min	Typ	Max	Unit
Dark Current (V_R = 20 V, R_L = 1 megohm)(2) T_A = 25°C T_A = 100°C	2 and 3	I_D	 — —	 — 14	 2 —	nA
Reverse Breakdown Voltage (I_R = 10 μA)	—	$V_{(BR)R}$	100	200	—	Volts
Forward Voltage (I_F = 50 mA)	—	V_F	—	—	1.1	Volts
Series Resistance (I_F = 50 mA)	—	R_s	—	—	10	Ohms
Total Capacitance (V_R = 20 V, f = 1 MHz)	5	C_T	—	—	4	pF

Typical Dark Current (I_D) = 14 nA.

OPTICAL CHARACTERISTICS (T_A = 25°C unless otherwise noted)

Characteristic	Fig. No.	Symbol	Min	Typ	Max	Unit
Light Current (V_R = 20 V)(1) MRD500 MRD510	1	I_L	6 1.5	9 2.1	— —	μA
Sensitivity at 0.8 μm (V_R = 20 V)(3) MRD500 MRD510	—	$S(\lambda = 0.8\ \mu m)$	— —	6.6 1.5	— —	μA/(mW/cm^2)
Response Time (V_R = 20 V, R_L = 50 Ohms)	—	$t_{(resp)}$	—	1	—	ns
Wavelength of Peak Spectral Response	5	λ_s	—	0.8	—	μm

Typical Light Current (I_L) = 9 μA.

If 1 mW per cm^2 of light is applied to photodiode at a wavelength of 0.8 μm (800 nm), light current will be 6.6 μA. For 2 mW/cm^2, I_L will be 2 × 6.6 μA = 13.2 μA. This rating indicates photodiode's sensitivity.

Relative Spectral Response

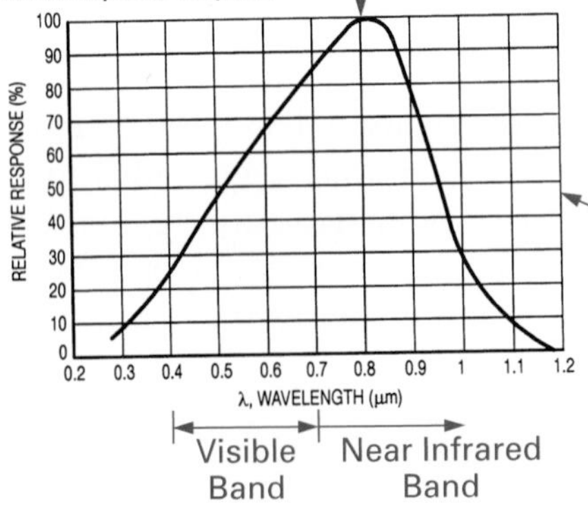

Photodiode's response is best when LED or light source is emitting 0.8 μm or 800 nm. Referring to Figure 7-1, you will find this wavelength in the near infrared band.

Relative Spectral Response Curve shows photodiode's response to different input wavelengths.

FIGURE 7-4 Device Data Sheet for a Photodiode. (Copyright of Motorola. Used by permission.)

see, this device has a typical dark current of 14 nA, a typical light current of 9 μA, and a typical sensitivity of 6.6 $\mu A/mW/cm^2$. In a perfect photodiode, each photon will generate an electron-hole pair; however, in most cases only 60% of the photons reaching the photodiode will cause an electron-hole pair or reverse current. The "relative spectral response curve," and the "wavelength of peak spectral response" rating indicate that this photodiode is most sensitive to a light input wavelength of 800 nm.

■ **EXAMPLE:**

Referring to the data sheet in Figure 7-4, calculate:

a. The light input frequency that this photodiode is most sensitive to.
b. The optical frequency band that a 800 nm wavelength is in.

■ ***Solution:***

a. By transposing the wavelength formula, we can calculate frequency from wavelength.

$$\lambda = \frac{c}{f} \text{ therefore,}$$

$$f = \frac{c}{\lambda} = \frac{3 \times 10^{17} \text{ nanometers/s}}{800 \text{ nanometers}} = \frac{3 \times 10^{17}}{800} = 3.75 \times 10^{14} \text{ or } 375 \text{ THz}$$

b. 3.75×10^{14} Hz is in the near infrared band.

7-1-5 ***Photodiode Application: Fiber Optic Communications***

The PIN photodiode is one of the most frequently used light detectors in fiber optic communication links. Fiber optics is a technology in which light is transmitted along the inside of a thin flexible glass or plastic fiber. The light signal transmitted down an optical fiber is equivalent to an electrical signal passing down a copper wire.

To examine the advantages of fiber optic communications, let us compare the copper wire communication link in Figure 7-5(a) with the fiber optic communication link in Figure 7-5(b). In both of these example circuits, we will be transmitting digital data from a computer to a printer. At first glance, the copper wire link seems to be a much better system to use because of its simplicity since the fiber optic link has to convert the electrical signal to a light signal, transmit the light through a fiber, and then convert the light back into an electrical signal. However, the fiber optic link has several key advantages over the copper wire link. Let us examine these advantages in more detail.

Large Bandwidth

With communications, the goal is to send more information, more efficiently, over a medium requiring less space. The amount of information that can be sent over a communication channel increases with frequency. It is no sur-

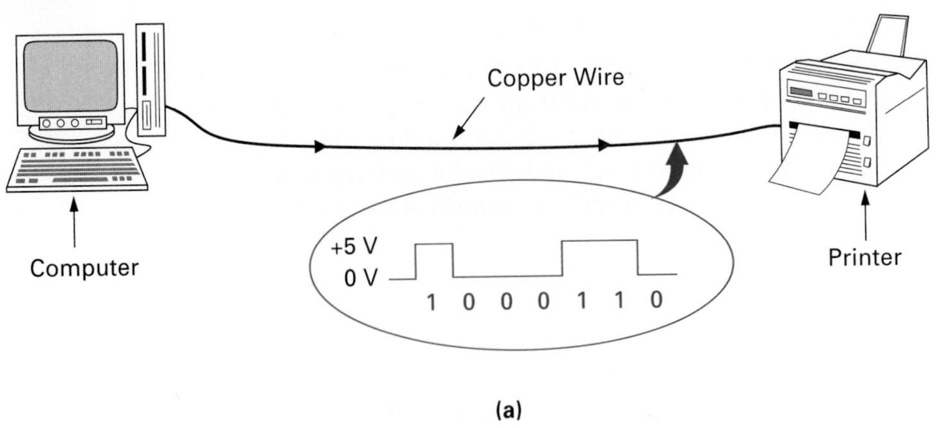

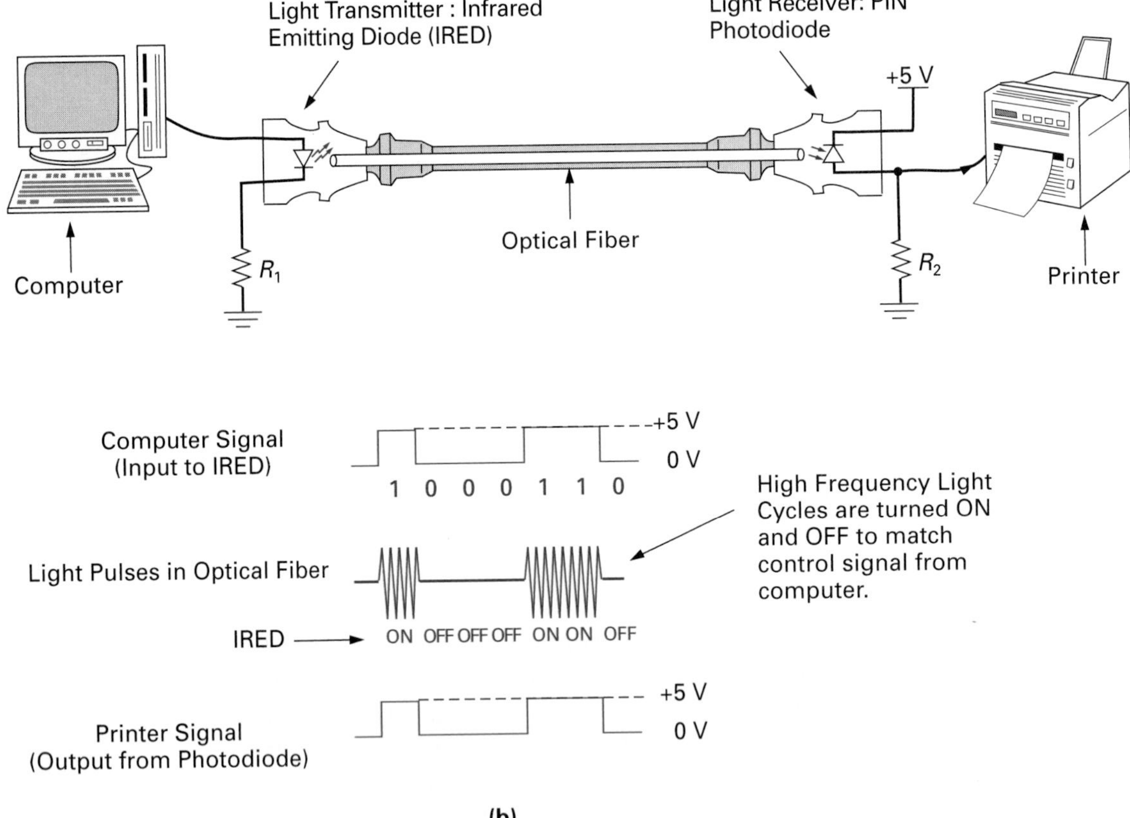

FIGURE 7-5 Fiber Optic Communications. (a) Basic Cable Communication Link. (b) Basic Fiber Optic Communication Link.

prise that an optical fiber has the potential to carry a great deal more information—light frequencies are several thousand times higher than radio frequencies. For example, present-day single-fiber telephone communication links can carry up to 10,000 voices, compared to a single copper wire that can only carry a few hundred voices.

Low Attenuation

Whether the transmission medium is a wire, space, or an optical fiber, a signal loses power (is attenuated) as it travels from one point to another. Optical fibers, however, offer very low losses compared to their copper wire counterparts. This is because, as frequency increases, the changes in current in a wire increase, causing the wire's inductance or current opposition to increase. This self-inductance (which is also called skin effect) is greatest at the center of the wire; therefore, current is forced to travel on the outside of the conductor. The result is that copper conductors carrying high frequency signals have less cross-sectional area for the signal current, which means they have a greater resistance and will heavily attenuate the signal. Also, because the signal energy is at the outside of the conductor, it will radiate more easily causing a further loss in signal power.

Optical fibers are insulators and so they do not experience inductance. This means that their signal attenuation is not frequency dependent. As a result, fewer repeaters are required for long distance communication. A **repeater** is a system that is included at certain intervals in a long distance communication link to rebuild a distorted and attenuated signal. For example, a coaxial cable telephone system has a standard repeater spacing of 1 km, whereas its optical counterpart uses spaces of 11 km.

Repeater
A system that is included at certain intervals in a long distance communication link to rebuild a distorted and attenuated signal.

Electromagnetic Immunity

Electromagnetic interference (EMI) is unwanted energy that is given off by electronic devices and circuits that contain rapidly alternating currents, current-carrying conductors, transmitting antennas, and almost every electrical and electronic system in the home, office, and factory. Any metal conductor will radiate and receive energy. To protect a signal from getting out, and outside interference from getting in, a copper cable is often shielded. At low frequencies, twisting wires together can provide adequate protection. At medium frequencies, shielded or coaxial cables have to be used, which raises the cable costs. At higher frequencies, the attenuation in coaxial cable becomes so severe that only short runs of cable are practical. At even higher microwave frequencies, wire can no longer be used. Signals have to be guided through hollow metal tubes called waveguides.

Fiber optic communication links are immune to the electromagnetic interference problems associated with wires because the optical fiber is made of an insulator material. The possibility of EMI causing cross-talk, or injecting an error signal in a wire, is eliminated.

Small Size and Light Weight

Optical fibers are considerably smaller than copper cables. This means that in areas where space is a problem, such as overcrowded telephone conduits and ever-expanding office and local area network conduits, the fiber optic cable is an obvious advantage. For example, a single small fiber will have the same capacity as a 900 pair copper cable that is 27 times as thick.

Also, because glass weighs less than copper, and less fiber optic cable is needed for the same information capacity, a weight advantage makes fiber optics ideal in applications such as aircraft and cars.

DEVICE: Snap-In Fiber Optic Link—HFBR-0500 Series

Features

- GUARANTEED LINK PERFORMANCE OVER TEMPERATURE
  High Speed Links: dc to 5 MBd
  Extended Distance Links up to 111 m
  Low Current Links: 6 mA Peak Supply Current for an 10 m Link
  Photo Interrupters
- LOW COST PLASTIC DUAL-IN-LINE PACKAGE
- EASY FIELD CONNECTORING
- EASY TO USE RECEIVERS:
  Logic Compatible Output Level
  Single +5 V Receiver Power Supply
  High Noise Immunity
- LOW LOSS PLASTIC CABLE:
  Simplex and Zip Cord Style Duplex Cable
  Extra Low Loss Simplex and Duplex

Applications

- HIGH VOLTAGE ISOLATION
- SECURE DATA COMMUNICATIONS
- REMOTE PHOTO INTERRUPTER
- LOW CURRENT LINKS
- INTER/INTRA-SYSTEM LINKS
- STATIC PROTECTION
- EMC REGULATED SYSTEMS (FCC, VDE)

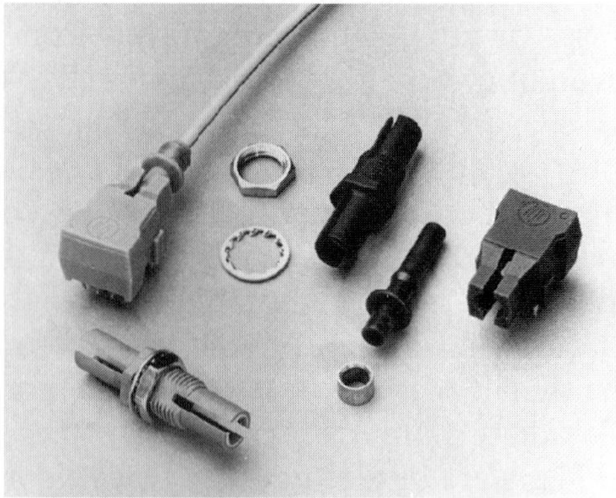

Description

The HFBR-0500 series is a complete family of fiber optic link components for configuring low-cost control, data transmission, and photo interrupter links. These components are designed to mate with plastic snap-in connectors and low-cost plastic cable.* Link design is simplified by the logic compatible receivers and the ease of connectoring the plastic fiber cable. The key parameters of links configured with the HFBR-0500 family are fully guaranteed.

*Cable is available in standard low loss and extra low loss varieties.

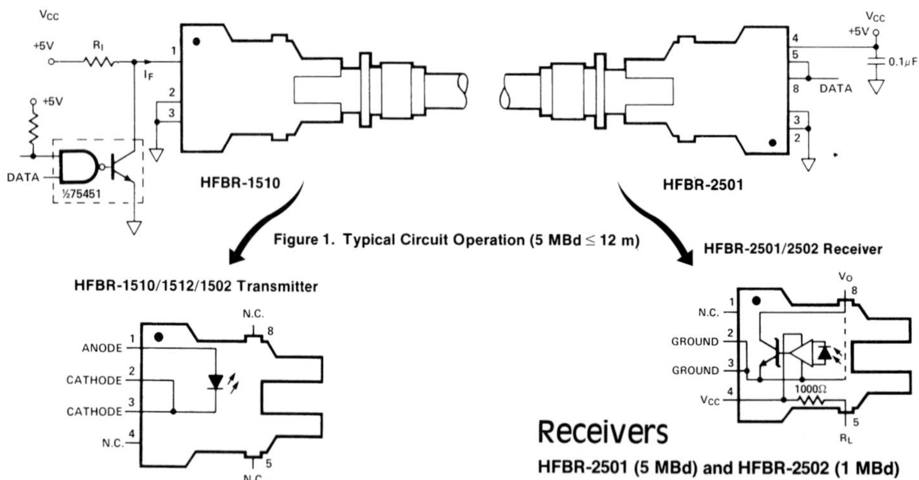

Receivers

HFBR-2501 (5 MBd) and HFBR-2502 (1 MBd)

The HFBR-2501/2502 Receiver modules feature a shielded integrated photodetector and wide bandwidth DC amplifier for high EMI immunity. A Schottky clamped open-collector output transistor allows interfacing to common logic families and enables "wired-OR" circuit designs. The open collector output is specified up to 18V. An integrated 1000 ohm resistor internally connected to Vcc may be externally jumpered to provide a pull-up for ease-of-use with +5V logic. The combination of high optical power levels and fast transitions falling edge could result in distortion of the output signal (HFBR-2502 only), that could lead to multiple triggering of following circuitry.

665 nm Transmitters

HFBR-1502/HFBR-1510 and HFBR-1512

The HFBR-1510/1502/1512 Transmitter modules incorporate a 665 nm LED emitting at a low attenuation wavelength for the HFBR-R/E plastic fiber optic cable. The transmitters can be easily interfaced to standard TTL logic. The optical power output of the HFBR-1510/1512/1502 is specified at the end of 0.5 m of cable. The HFBR-1512 output optical power is tested and guaranteed at low drive currents.

FIGURE 7-6 Device Data Sheet for a Fiber Optic Link. (Courtesy of Hewlett-Packard Company.)

Security and Safety

It is difficult to tap into fiber optic communication links because the fiber does not radiate energy. This makes it a very private system.

Also, the fiber will not generate any sparks if its line is damaged, making it ideal in environments containing flammable materials.

Applications

The many advantages of fiber optics makes it ideal for many applications in telecommunications, computer communications, test equipment, medical, automotive, military, and industrial control. The telephone industry, at present the largest user of fiber optics, makes use of its wide bandwidth and low losses, which allow for high speed, long distance links. The computer industry uses fiber optics because of its electromagnetic noise immunity and bandwidth, which allows for error free high-speed data transfer. As factories become more automated, industrial systems are beginning to switch to fiber optics because of its electromagnetic immunity and the safety offered by electrical isolation and the absence of a spark. The medical field needs its electromagnetic immunity to isolate data lines from interference generated by an ever increasing number of systems, such as Magnetic Resonance Imaging (MRI), and so on. The security of a transmission medium that does not radiate appeals to both the military and consumers.

As an example, Figure 7-6 shows the data sheet for a short distance fiber optic link. The transmitter module contains a 665 nm (red) LED, and the receiver module contains a PIN photodiode. A short link system such as this would typically be less than 1 kilometer. It would be used in low-speed data communication links between a computer and printer or between computer and computer (a network).

The receiver module is called an **integrated detector and preamplifier (IDP)** because it contains an amplifier along with the PIN photodiode detector. The advantage of this system is that the received signal is immediately amplified or strengthened to a good usable output voltage before it meets the noise associated with the following circuits.

Integrated Detector and Preamplifier (IDP)
A receiver module that contains an amplifier along with the PIN photodiode detector.

SELF-TEST REVIEW QUESTIONS FOR SECTION 7-1

1. What is the difference between an LED and a photodiode?
2. Define the difference between photoconduction and photovoltaic action.
3. A photodiode has a peak spectral response at 615 nm. This wavelength is in what optical electromagnetic band?
4. Give the full names of the following abbreviations:
   - a. PIN
   - b. IDP
   - c. IRED
5. An increase in light will cause a/an ____________ in the conduction of a photodiode.

7-2 THE VARACTOR DIODE

When any diode is reverse biased, a depletion region is formed, as shown in Figure 7-7(a). Increasing the reverse bias voltage applied across the diode increases the width of the depletion layer. Conversely, when the reverse bias voltage is decreased, the depletion region width becomes narrower. This depletion region has an absence of majority carriers and acts like an insulator, preventing conduction between the *n* and *p* regions of the diode—just as a dielectric separates the two plates of a capacitor. In fact, the similarities between a reverse biased diode and a capacitor are many. When reverse biased, a diode exhibits a small value of capacitance that can be varied by varying the reverse bias voltage, and therefore the width of the depletion region. The basic P-N junction diode has only a very small amount of internal junction capacitance; however, special diodes can be constructed to have a larger value of internal capacitance. These special diodes are called **varactor diodes** because, by varying the reverse bias voltage, they can be made to operate as *voltage-controlled variable capacitors.*

Varactor Diodes
Diodes that operate as voltage-controlled variable capacitors.

7-2-1 Varactor Diode Operation and Characteristics

Figure 7-7(b) shows the operation of the varactor diode, which is operated in its reverse region. The varactor diode is a specially constructed diode with a small impurity dose at its junction. The impurity level increases as you travel away from the junction, resulting in a greater "capacitance to reverse bias voltage" change than that of a conventional junction diode. The reverse bias voltage (V_R) is varied to control the width of the depletion region and is always less than the reverse breakdown voltage rating of the diode. As an example, Figure 7-7(b) shows how a small reverse bias voltage will produce a small depletion region, while a large reverse bias voltage will produce a large depletion region. Remembering the capacitance formula, we know that capacitance is inversely proportional to the distance between the plates ($C \propto 1/d$). Therefore, a small reverse voltage will produce a small depletion region, or dielectric, and a large capacitance ($d\downarrow$, $C\uparrow$). On the other hand, an increase in reverse bias voltage will result in a large depletion region or dielectric and a small capacitance ($d\uparrow$, $C\downarrow$).

Figure 7-7(c) illustrates the typical "capacitance versus reverse bias voltage" curves produced by a conventional junction and varactor diode. As you can see in the curve in Figure 7-7(c), the varactor diode's capacitance varies inversely with the applied reverse bias voltage. For example, at –2 V, the varactor's depletion region will be small and the capacitance large at approximately 80 pF. On the other hand, as the reverse bias voltage across the varactor is increased toward –20 V, the varactor's capacitance decreases rapidly to approximately 25 pF. Figure 7-7(c) also shows the very small change in a basic P-N junction diode's internal capacitance over the same reverse bias voltage range.

The schematic symbols used to represent a varactor (variable-capacitor) diode are shown in Figure 7-7(d). Figure 7-7(e) shows the typical low-frequency (below 500 mHz) and high-frequency (above 500 mHz) packages. A wide assortment of varactor diodes are available with capacitance values that range from 1 pF to 2000 pF, and with power ratings that range from 500 mW to 35 W.

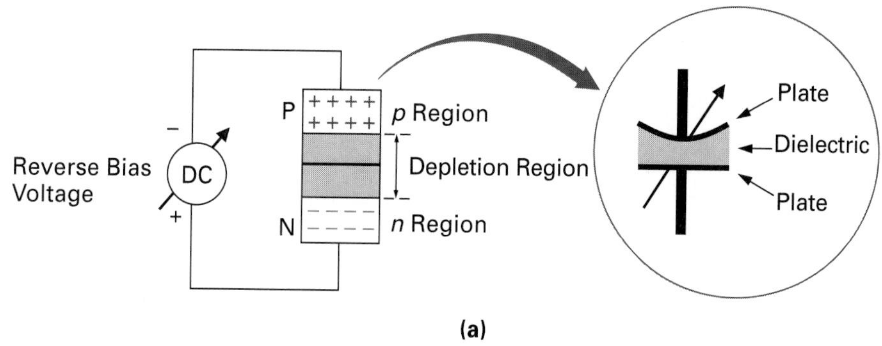

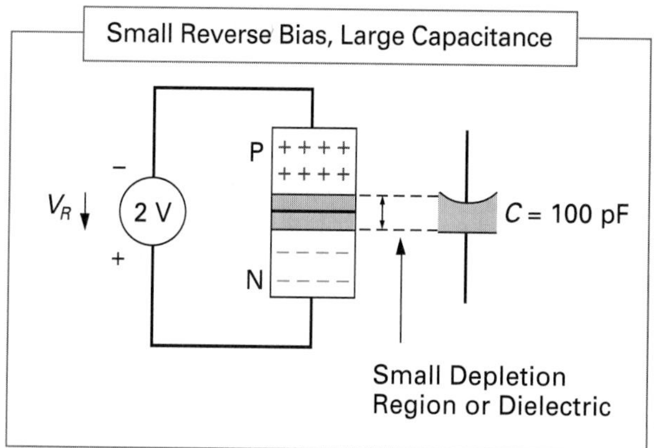

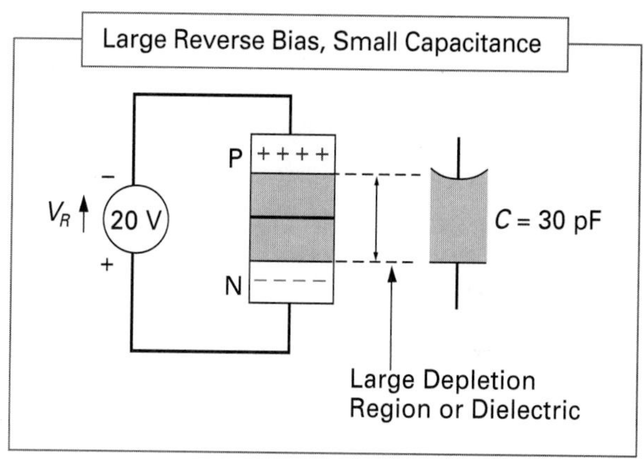

(b)

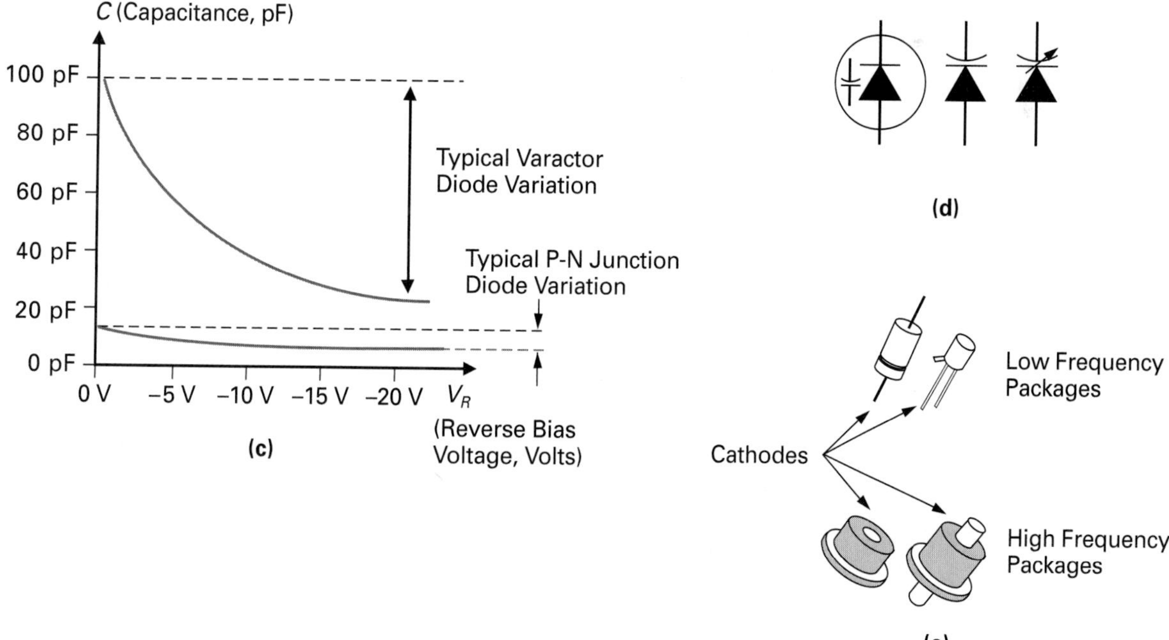

FIGURE 7-7 **The Varactor Diode. (a) Reverse Bias Capacitance Effect. (b) Operation. (c) Typical Capacitance versus Reverse Bias Voltage Characteristics. (d) Schematic Symbols. (e) Low Frequency and High Frequency Packages.**

7-2-2 *Varactor Diode Data Sheet*

Figure 7-8 is an example of a typical manufacturer's data sheet for two varactor diodes that, due to their different package styles, have different forward power dissipation ratings. All other diode ratings apply to both types of varactor diodes, and any new terms are explained in Figure 7-8.

7-2-3 *Varactor Diode Application: Tuned Circuit*

Varactor diodes are used in place of variable capacitors in many applications. One such application is shown in Figure 7-9(a), in which a varactor diode will be used in a tuned circuit in a radio receiver. The tuned circuit in a radio receiver

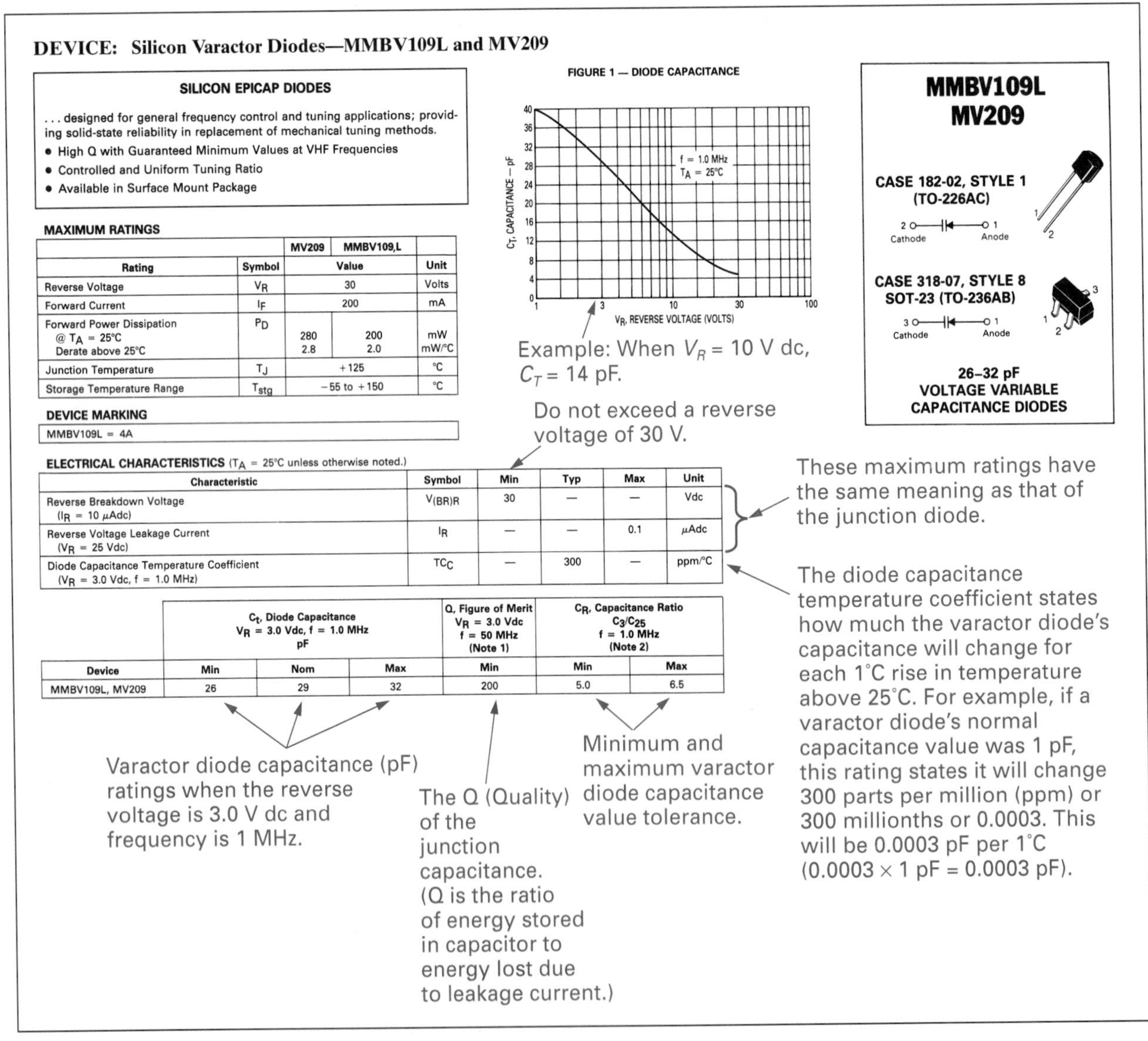

DEVICE: Silicon Varactor Diodes—MMBV109L and MV209

SILICON EPICAP DIODES

. . . designed for general frequency control and tuning applications; providing solid-state reliability in replacement of mechanical tuning methods.

- High Q with Guaranteed Minimum Values at VHF Frequencies
- Controlled and Uniform Tuning Ratio
- Available in Surface Mount Package

MAXIMUM RATINGS

Rating	Symbol	MV209 Value	MMBV109,L Value	Unit
Reverse Voltage	V_R	30		Volts
Forward Current	I_F	200		mA
Forward Power Dissipation @ T_A = 25°C Derate above 25°C	P_D	280 2.8	200 2.0	mW mW/°C
Junction Temperature	T_J	+125		°C
Storage Temperature Range	T_{stg}	−55 to +150		°C

DEVICE MARKING

MMBV109L = 4A

ELECTRICAL CHARACTERISTICS (T_A = 25°C unless otherwise noted.)

Characteristic	Symbol	Min	Typ	Max	Unit
Reverse Breakdown Voltage (I_R = 10 μAdc)	$V_{(BR)R}$	30	—	—	Vdc
Reverse Voltage Leakage Current (V_R = 25 Vdc)	I_R	—	—	0.1	μAdc
Diode Capacitance Temperature Coefficient (V_R = 3.0 Vdc, f = 1.0 MHz)	TC_C	—	300	—	ppm/°C

Device	C_t, Diode Capacitance V_R = 3.0 Vdc, f = 1.0 MHz pF Min	Nom	Max	Q, Figure of Merit V_R = 3.0 Vdc f = 50 MHz (Note 1) Min	C_R, Capacitance Ratio C_3/C_{25} f = 1.0 MHz (Note 2) Min	Max
MMBV109L, MV209	26	29	32	200	5.0	6.5

FIGURE 7-8 Device Data Sheet For a Varactor Diode. (Copyright of Motorola. Used by permission.)

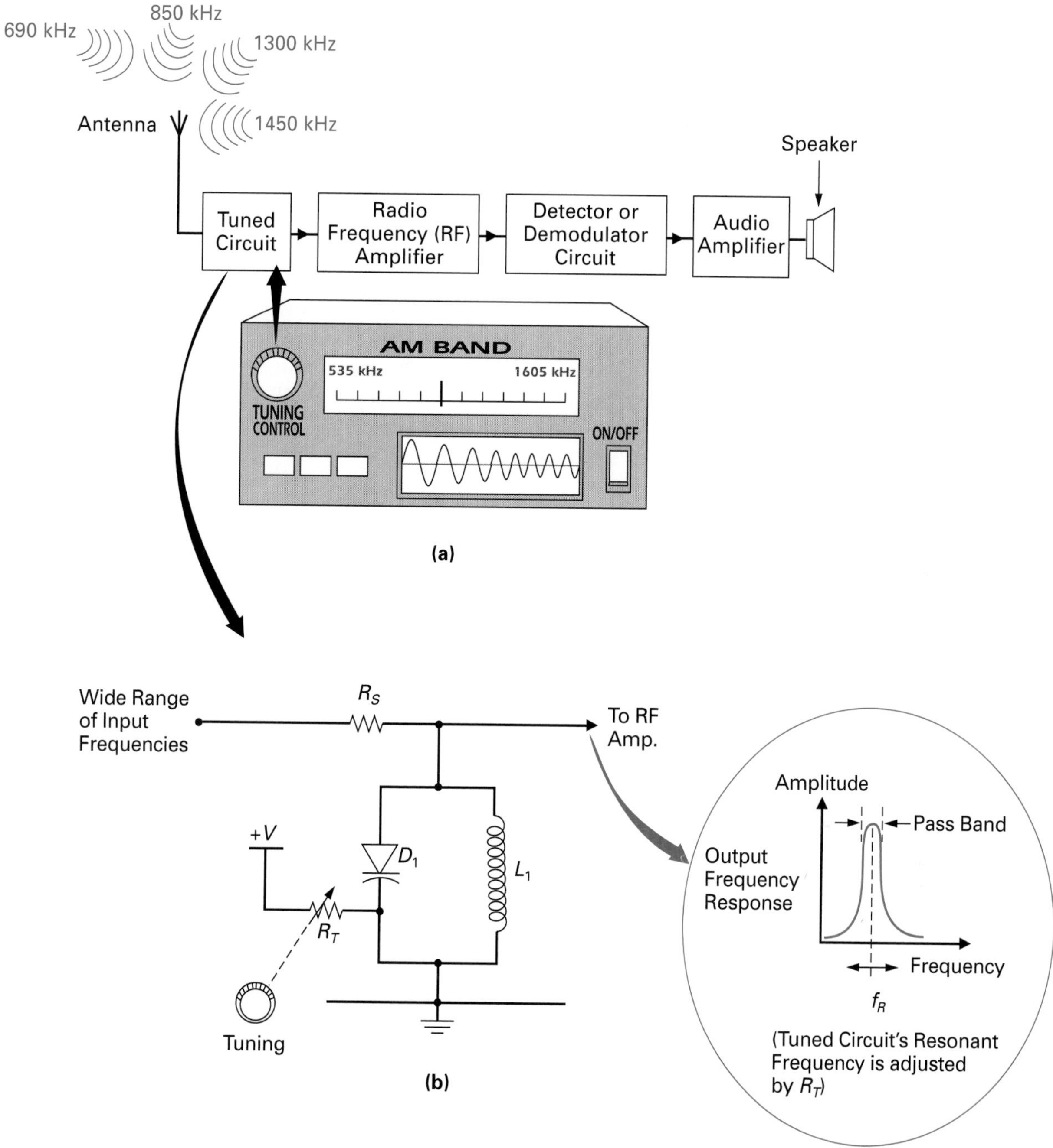

FIGURE 7-9 The Varactor Diode In a Tuned Circuit.

acts as a filter to pass a selected station, or frequency, and block all other radio stations, or frequencies. To explain this further, Figure 7-9(a) shows the basic block diagram of an AM (amplitude modulated) radio receiver. As you can see, the tuned circuit is at the front end of a radio receiver, and its operation will be controlled by the radio's tuning control to select one of the AM radio stations present in the AM radio band between 535 kHz and 1605 kHz.

Figure 7-9(b) shows the schematic diagram of a basic tuned circuit, which contains a varactor diode. Notice that the varactor diode is reverse biased because its cathode is connected to the positive source voltage ($+V$) via the tuning resistor (R_T). As R_T is adjusted, the capacitance of the varactor D_1 will be changed, changing the resonant frequency of the tank circuit made up of D_1's capacitance and L_1's inductance. Because the varactor diode acts as a variable capacitor, it can be used in place of the more expensive mechanically-variable air capacitors. Let us review how this parallel resonant band-pass tuning circuit will operate. At resonance (which is determined by the capacitance of D_1 and the inductance of L_1), the tank has a high impedance, so very little of the input current at this resonant frequency will be shunted away from the output. At frequencies above resonance, X_C will be low, and most of the input signal at frequencies above resonance will be shunted away from the output by the capacitance of D_1. At frequencies below resonance, X_L will be low, and the shunting action of L_1 will again prevent any frequencies below resonance from appearing at the output.

This band-pass filter circuit will therefore tune-in (select or pass) one frequency that contains the information we desire and allow this signal to proceed to the other circuits in the radio or television receiver. All of the other millions of information carrying frequencies, however, will be blocked by the resonant action of the filter. Varactor diodes that are used in these circuits are often called *varicap diodes* or *tuning diodes.*

■ **EXAMPLE:**

Calculate the resonant frequency of the tuned circuit in Figure 7-9(b), if:

D_1 Capacitance = 14 pF

L_1 Inductance = 5 mH

■ ***Solution:***

$$f_R = \frac{1}{2\pi\sqrt{L \times C_{VD1}}} = \frac{1}{2\pi\sqrt{(5 \times 10^{-3}) \times (14 \times 10^{-12})}}$$

$$= 601.5 \text{ kHz}$$

SELF-TEST REVIEW QUESTIONS FOR SECTION 7-2

1. True or False: The varactor diode is normally always operated in its forward region.
2. In what application can varactors normally be found?
3. True or False: As the reverse bias is increased, the capacitance of the varactor will decrease.
4. If a silicon varactor diode was forward biased by an applied voltage of +0.7 V, what would be its value of capacitance?

7-3 TRANSIENT SUPPRESSOR DIODE

Lightning, power line faults, and the switching on and off of motors, air-conditioners, and heaters can cause the normal 115 V rms ac line voltage at the wall outlet to contain under-voltage dips and over-voltage spikes. Although these *transients* only last for a few microseconds, the over-voltage spikes can cause the input line voltage to momentarily increase by 1000 V or more. In sensitive equipment, such as televisions and computers, shunt filtering devices are connected between the ac line input and the primary of the dc power supply's transformer to eliminate these transients before they get into, and possibly damage, the system.

One such device that can be used to filter the ac line voltage is the **transient suppressor diode.** Referring to Figure 7-10(a), you can see that this diode contains two zener diodes that are connected back-to-back. The schematic symbol for this diode is shown in Figure 7-10(b).

Transient Suppressor Diode
A device used to protect voltage sensitive electronic devices in danger of destruction by high energy voltage transients.

Figure 7-10(c) shows how a transorb would be connected across the ac power line input to a dc power supply. Because the zeners within the transient suppressor diode are connected back-to-back, they will operate in either direction (the device is "bi-directional") and monitor both alternations of the ac input. If a voltage surge occurs that exceeds the V_Z (zener voltage) of the diodes, they will break down and shunt the surge away from the power supply.

Transient suppressor diodes are also called **transorbs** because they "absorb transients."

Transorb
Absorb transients. Another name for transient suppressor diode.

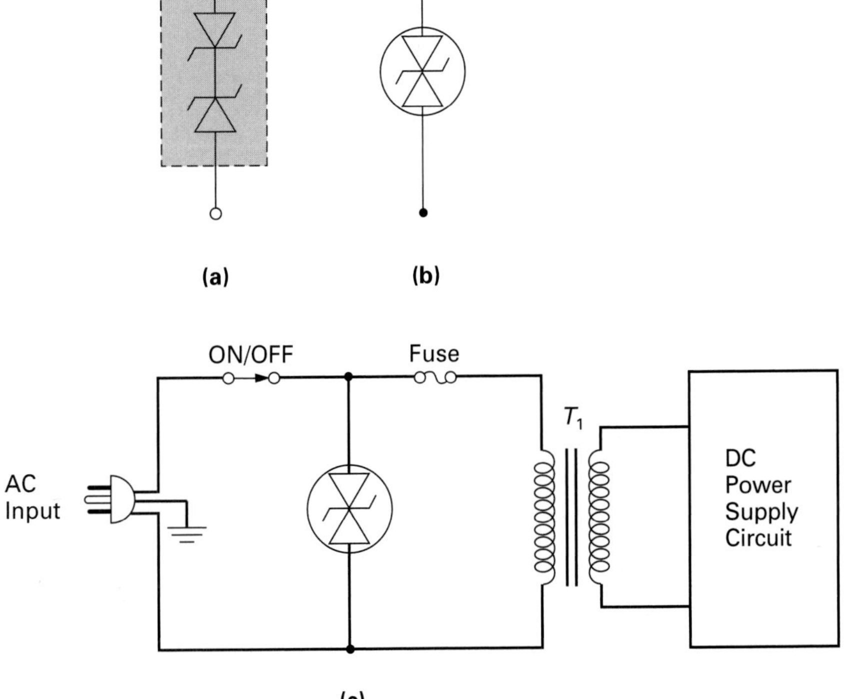

FIGURE 7-10 Transient Suppressor Diodes. (a) Construction. (b) Schematic Symbol. (c) Bidirectional Circuit Application (AC Line Voltage).

DEVICE: Undirectional Transcient Suppressor Diode

Zener Transient Voltage Suppressors Undirectional and Bidirectional

The P6KE6.8A series is designed to protect voltage sensitive components from high voltage, high energy transients. They have excellent clamping capability, high surge capability, low zener impedance and fast response time. The P6KE6.8A series is supplied in Motorola's exclusive, cost-effective, highly reliable Surmetic axial leaded package and is ideally-suited for use in communication systems, numerical controls, process controls, medical equipment, business machines, power supplies and many other industrial/ consumer applications.

Specification Features:

- Standard Zener Voltage Range — 6.8 to 200 Volts
- Peak Power — 600 Watts @ 1 ms
- Maximum Clamp Voltage @ Peak Pulse Current
- Low Leakage < 5 μA Above 10 Volts
- Maximum Temperature Coefficient Specified

MAXIMUM RATINGS

Rating	Symbol	Value	Unit
Peak Power Dissipation (1) @ $T_L \leq 25°C$	P_{PK}	600	Watts
Steady State Power Dissipation @ $T_L \leq 75°C$, Lead Length = 3/8″ Derated above $T_L = 75°C$	P_D	5 50	Watts mW/°C
Forward Surge Current (2) @ $T_A = 25°C$	I_{FSM}	100	Amps
Operating and Storage Temperature Range	T_J, T_{stg}	– 65 to +175	°C

Lead Temperature not less than 1/16″ from the case for 10 seconds: 230°C

Mechanical Characteristics:

CASE: Void-free, transfer-molded, thermosetting plastic
FINISH: All external surfaces are corrosion resistant and leads are readily solderable
POLARITY: Cathode indicated by polarity band. When operated in zener mode, will be positive with respect to anode
MOUNTING POSITION: Any

NOTES: 1. Nonrepetitive current pulse per Figure 4 and derated above $T_A = 25°C$ per Figure 2.
2. 1/2 sine wave (or equivalent square wave), PW = 8.3 ms, duty cycle = 4 pulses per minute maximum.

P6KE6.8A through P6KE200A

ZENER OVERVOLTAGE TRANSIENT SUPPRESSORS 6.8–200 VOLT 600 WATT PEAK POWER 5 WATTS STEADY STATE

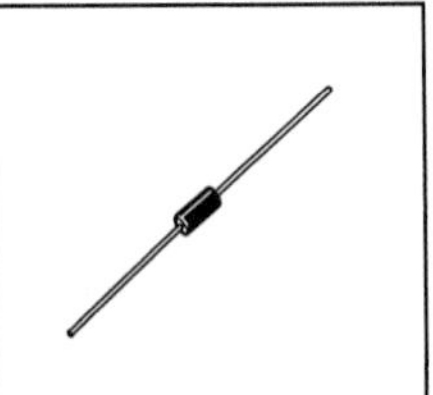

APPLICATION NOTES

RESPONSE TIME

In most applications, the transient suppressor device is placed in parallel with the equipment or component to be protected. In this situation, there is a time delay associated with the capacitance of the device and an overshoot condition associated with the inductance of the device and the inductance of the connection method. The capacitance effect is of minor importance in the parallel protection scheme because it only produces a time delay in the transition from the operating voltage to the clamp voltage as shown in Figure A.

TYPICAL PROTECTION CIRCUIT

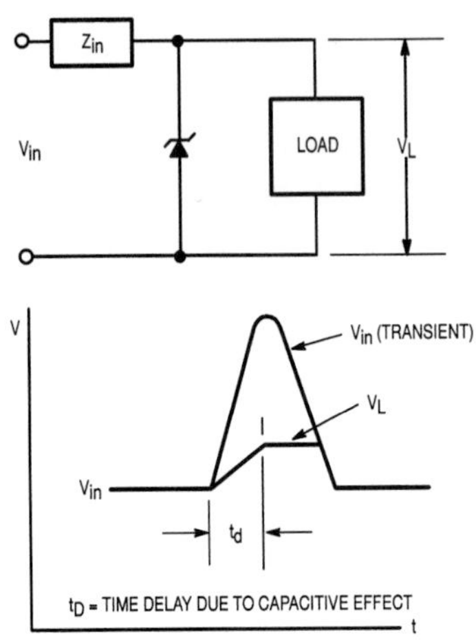

Figure A.

The inductive effects in the device are due to actual turn-on time (time required for the device to go from zero current to full current) and lead inductance. This inductive effect produces an overshoot in the voltage across the equipment or component being protected as shown in Figure B. Minimizing this overshoot is very important in the application, since the main purpose for adding a transient suppressor is to clamp voltage spikes. The P6KE6.8A series has very good response time, typically < 1 ns and negligible inductance. However, external inductive effects could produce unacceptable overshoot. Proper circuit layout, minimum lead lengths and placing the suppressor device as close as possible to the equipment or components to be protected will minimize this overshoot.

Some input impedance represented by Z_{in} is essential to prevent overstress of the protection device. This impedance should be as high as possible, without restricting the circuit operation.

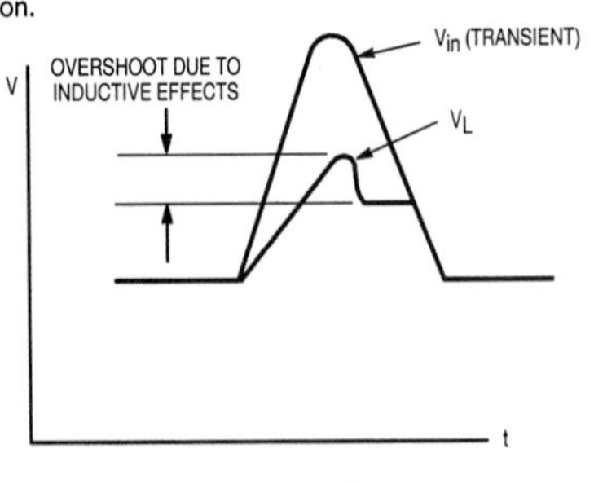

Figure B.

FIGURE 7-11 Data Sheet for a Unidirectional Transient Suppressor Diode. (Copyright of Motorola. Used by permission.)

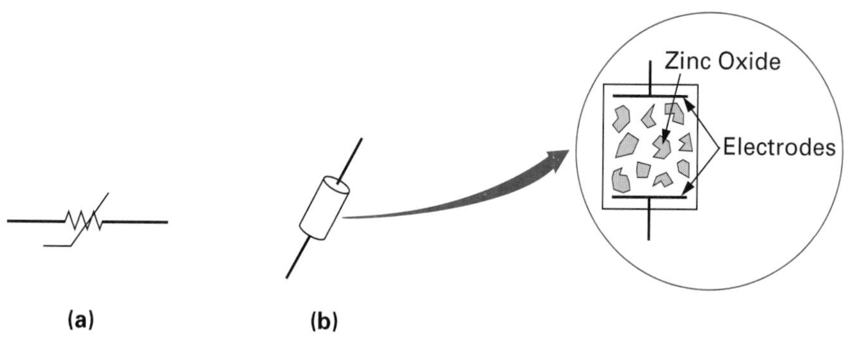

FIGURE 7-12 **Metal Oxide Varistors (MOVs). (a) Schematic Symbol. (b) Physical Appearance and Construction.**

Most manufacturers' transorbs have a high power dissipation rating because they may have to handle momentary power line surges in the hundreds of watts. For example, the Motorola 1N5908-1N6389 series of transorbs can dissipate 1.5 kW for a period of approximately 10 ms (most surges last for a few milliseconds). The devices must also have a fast turn-on time so that they can limit or clamp any voltage spikes. For example, the Motorola P6KE6.8 series has a response time of less than 1 ns.

In dc applications, a single unidirectional (one-direction) transient suppressor can be used instead of a bidirectional (two-direction) transient suppressor. These single transorbs have the same schematic symbol as a zener. The data sheet and application notes for a typical single transorb is shown in Figure 7-11. As you can see in the typical protection circuit diagram in Figure 7-11, the transorb would be connected in shunt with the dc input and reverse biased (cathode to positive dc).

Metal oxide varistors (MOVs) are currently replacing zener-diode and transient-diode suppressors because they are able to shunt a much higher current surge and are cheaper. These are not semiconductor devices; in fact, they contain a zinc-oxide and bismuth-oxide compound in a ceramic body but are connected in the same way as a transient suppressor diode. They are called **varistors** because they operate as a "voltage dependent resistor" that will have a very low resistance at a certain breakdown voltage. The MOV's schematic symbol, typical appearance, and construction is shown in Figure 7-12.

Metal Oxide Varistors (MOVs)
Devices that are replacing zener-diode and transient-diode suppressors because they are able to shunt a much higher current surge and are cheaper.

Varistor
Voltage dependent resistor.

SELF-TEST REVIEW QUESTIONS FOR SECTION 7-3

1. True or False: A bidirectional transient suppressor diode would be used to suppress ac power surges.
2. True or False: A unidirectional transient suppressor diode would be used to suppress dc power surges.
3. Besides the ability to dissipate the large burst of power in a surge, what other important feature should a transorb have?
4. What device is largely replacing the transorb in transient protection applications?

7-4 CONSTANT-CURRENT DIODE

The constant-current diode does what its name states: *maintains a constant output current despite variations in the input voltage.* The schematic symbols used for this current regulating diode are shown in Figure 7-13(a).

To understand how this diode operates, refer to the forward voltage-current characteristics shown in Figure 7-13(b), which shows the *V-I* characteristics for a few example constant-current diodes. As we know, when a basic P-N junction diode is forward biased, the anode-to-cathode forward voltage drop will be almost constant at 0.7 V, as shown by the dashed curve in Figure 7-13(b). When a constant-current diode is forward biased, however, the anode-to-cathode forward voltage drop will not be constant. In fact, it can be anywhere between 0.1 V and 100 V. The constant-current diode's forward current (I_F), on the other hand, will remain constant. For example, the 1N5298 constant current diode curve shows that forward current (I_F) will increase as forward voltage (V_F) is increased from 0 V to 1 V. Any increase in V_F beyond 1 V, however, will not cause any further increase in forward current. At this point, the current regulator diode is regulating the forward current through the diode to 1 mA, even if V_F were increased to 100 V. As another example, from about 2 V to 100 V, the 1N5309 will maintain the forward current through the diode to 3 mA.

Series Current Regulator Circuit A circuit that utilizes a forward biased constant-current diode connected in series with the load.

Regulator Current (I_P) The constant current maintained by the regulator circuit.

Shunt Current Regulator Circuit A circuit that utilizes a forward biased constant-current diode connected in shunt, or parallel, with the load.

Figure 7-13(c) shows how the constant-current diode would be connected to regulate, or maintain constant, the current delivered to a circuit (which is represented as R_L). With the **series current regulator** circuit, the forward biased D_1 is connected in *series* with the load. Even though the input voltage has a slight fluctuation above and below 3 V, D_1 will maintain the **regulator current (I_P),** and therefore the load current, constant at 3 mA. This constant current will develop a constant voltage across the load (assuming R_L does not change) of 3 V.

$$V_{R_L} = I_P \times R_L = 3\text{ mA} \times 1\text{ k}\Omega = 3\text{ V}$$

With the **shunt current regulator** circuit, the forward biased D_2 is connected in *shunt,* or parallel, with the load. In this arrangement, a constant 1 mA of current will be shunted away from the load by D_2. Therefore, the load current (I_L) will be equal to the input current (I_{in}) minus the regulator shunt current (I_P).

$$I_L = I_{in} - I_P = 5\text{ mA} - 1\text{ mA} = 4\text{ mA}$$

The output voltage (V_{R_L}) in this example will therefore equal:

$$V_{R_L} = I_L \times R_L = 4\text{ mA} \times 1\text{ k}\Omega = 4\text{ V}$$

Figure 7-14 shows the data sheet for a series of current regulator diodes. The constant-current diode should always be forward biased, and once the maximum limiting voltage (V_L) is exceeded, the diode will produce a constant regulator current (I_P). The electrical characteristics table in Figure 7-14 lists the voltage at which current regulation begins and the regulated forward current for all the constant current diodes in this series.

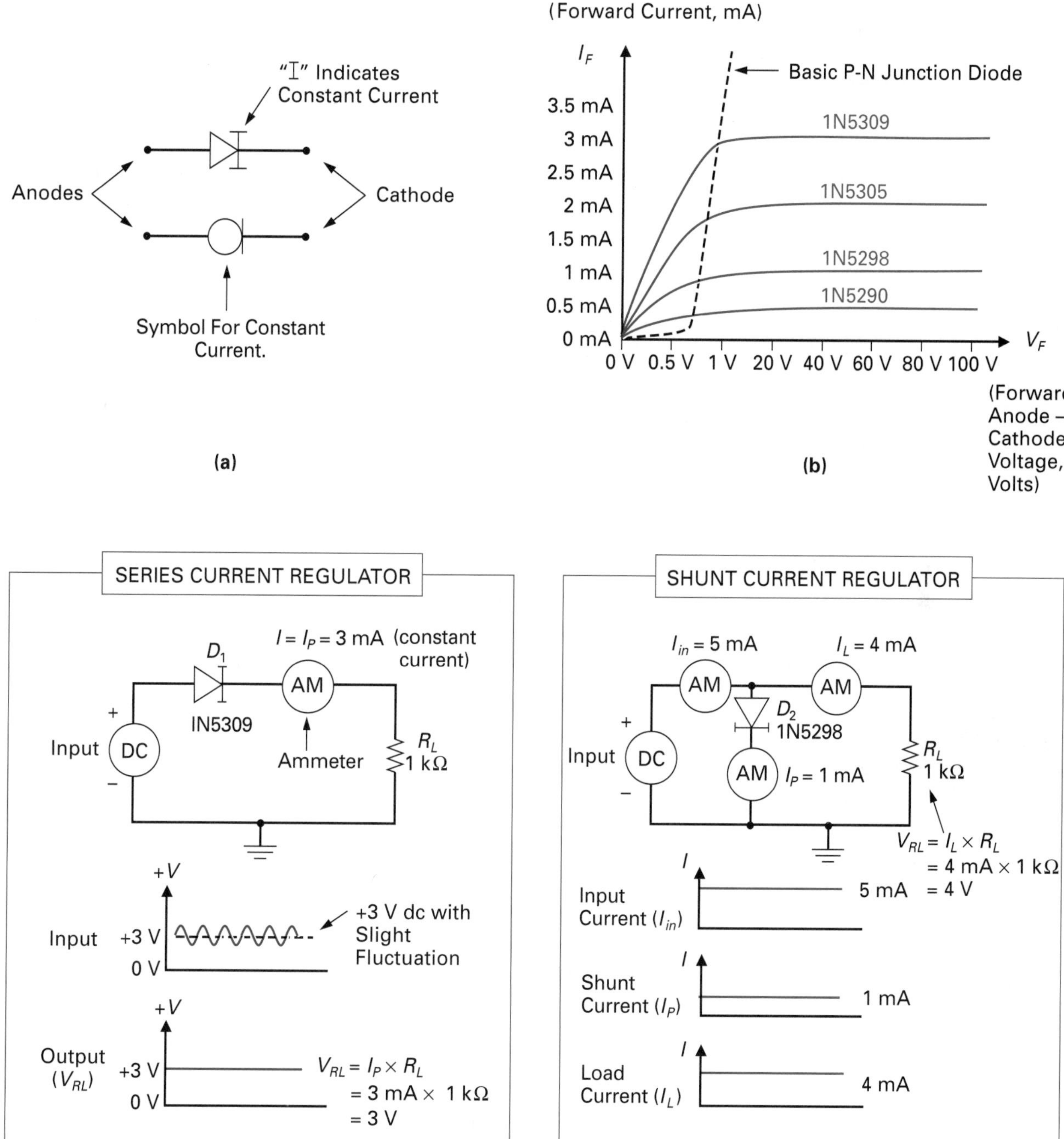

FIGURE 7-13 **Constant Current Diodes. (a) Schematic Symbols. (b) Forward Voltage—Current Characteristics. (c) Application—Current Regulator Circuits.**

DEVICE: Current Regulator Diodes—1N5283 Through 1N5314

Current Regulator Diodes

1N5283 through 1N5314

CURRENT REGULATOR DIODES

Field-effect current regulator diodes are circuit elements that provide a current essentially independent of voltage. These diodes are especially designed for maximum impedance over the operating range. These devices may be used in parallel to obtain higher currents.

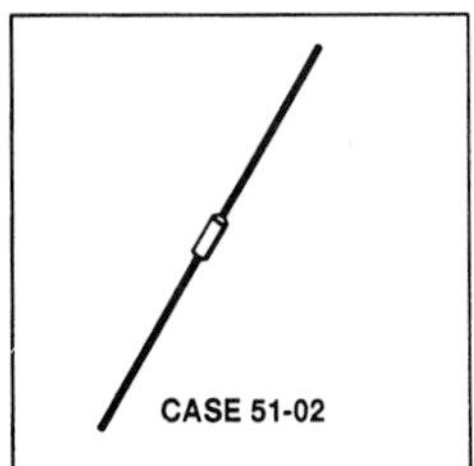

MAXIMUM RATINGS

Rating	Symbol	Value	Unit
Peak Operating Voltage (T_J = −55°C to +200°C)	POV	100	Volts
Steady State Power Dissipation @ T_L = 75°C Derate above T_L = 75°C Lead Length = 3/8″ (Forward or Reverse Bias)	P_D	600 4.8	mW mW/°C
Operating and Storage Junction Temperature Range	T_J, T_{stg}	−55 to +200	°C

Forward voltage (V_F) must not exceed this maximum rating.

ELECTRICAL CHARACTERISTICS (T_A = 25°C unless otherwise noted)

Type No.	Regulator Current Ip (mA) @ V_T = 25 V Nom	Min	Max	Maximum Limiting Voltage @ I_L = 0.8 Ip (min) V_L (Volts)
1N5283	0.22	0.198	0.242	1.00
1N5284	0.24	0.216	0.264	1.00
1N5285	0.27	0.243	0.297	1.00
1N5286	0.30	0.270	0.330	1.00
1N5287	0.33	0.297	0.363	1.00
1N5288	0.39	0.351	0.429	1.05
1N5289	0.43	0.387	0.473	1.05
1N5290	0.47	0.423	0.517	1.05
1N5291	0.56	0.504	0.616	1.10
1N5292	0.62	0.558	0.682	1.13
1N5293	0.68	0.612	0.748	1.15
1N5294	0.75	0.675	0.825	1.20
1N5295	0.82	0.738	0.902	1.25
1N5296	0.91	0.819	1.001	1.29
1N5297	1.00	0.900	1.100	1.35
1N5298	1.10	0.990	1.21	1.40
1N5299	1.20	1.08	1.32	1.45
1N5300	1.30	1.17	1.43	1.50
1N5301	1.40	1.26	1.54	1.55
1N5302	1.50	1.35	1.65	1.60
1N5303	1.60	1.44	1.76	1.65
1N5304	1.80	1.62	1.98	1.75
1N5305	2.00	1.80	2.20	1.85
1N5306	2.20	1.98	2.42	1.95
1N5307	2.40	2.16	2.64	2.00
1N5308	2.70	2.43	2.97	2.15
1N5309	3.00	2.70	3.30	2.25
1N5310	3.30	2.97	3.63	2.35
1N5311	3.60	3.24	3.96	2.50
1N5312	3.90	3.51	4.29	2.60
1N5313	4.30	3.87	4.73	2.75
1N5314	4.70	4.23	5.17	2.90

Voltage at which diode begins to regulate current.

Regulated Forward Current (IP)

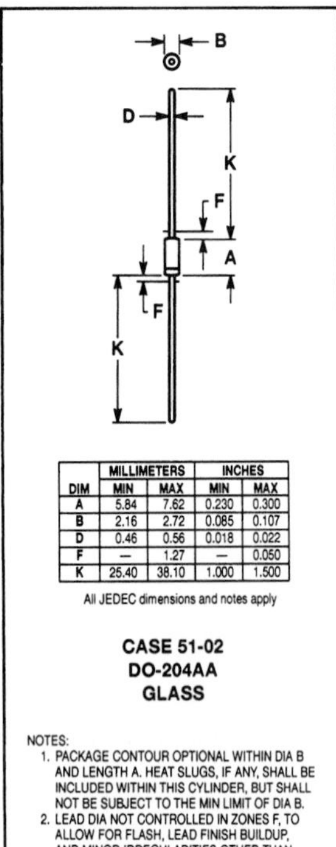

DIM	MILLIMETERS MIN	MAX	INCHES MIN	MAX
A	5.84	7.62	0.230	0.300
B	2.16	2.72	0.085	0.107
D	0.46	0.56	0.018	0.022
F	—	1.27	—	0.050
K	25.40	38.10	1.000	1.500

All JEDEC dimensions and notes apply

CASE 51-02
DO-204AA
GLASS

NOTES:
1. PACKAGE CONTOUR OPTIONAL WITHIN DIA B AND LENGTH A. HEAT SLUGS, IF ANY, SHALL BE INCLUDED WITHIN THIS CYLINDER, BUT SHALL NOT BE SUBJECT TO THE MIN LIMIT OF DIA B.
2. LEAD DIA NOT CONTROLLED IN ZONES F, TO ALLOW FOR FLASH, LEAD FINISH BUILDUP, AND MINOR IRREGULARITIES OTHER THAN HEAT SLUGS.

FIGURE 7-14 **Data Sheet For a Constant Current Diode. (Copyright of Motorola. Used by permission.)**

SELF-TEST REVIEW QUESTIONS FOR SECTION 7-4

1. Just as the zener diode regulates ____________, the constant-current diode regulates ____________.
2. The constant-current diode is always connected in circuit so that it is ____________ biased.

7-5 HIGH-FREQUENCY DIODES

In the following section, you will be introduced to two diodes that are used in high frequency communication and digital computer circuit applications. The characteristics of these two diodes in this section were chosen to give you a good understanding of special diode characteristics. Although there are many other types of special application diodes, these are best covered in conjunction with their circuit application at a later time.

7-5-1 *The Schottky Diode*

The schottky diode was named after its German inventor who discovered the operating principle in 1938. Figure 7-15 shows the schottky diode's construction, schematic symbol, voltage-current characteristics, and packaging. Referring to the construction of this diode in Figure 7-15(a), you can see that the diode is formed by joining an *n-type semiconductor region* with a *metal region* such as gold or silver. This unique construction gives the schottky diode a few distinct characteristics.

When unbiased, the metal region (like all conductors) will have more electrons traveling in larger orbits than the *n*-type semiconductor region. This difference in energy levels between the two regions will develop a *schottky barrier voltage* within the device that is approximately equal to 0.4 V.

When the schottky diode is forward biased ($-V \rightarrow n$ region, $+V \rightarrow$ metal region), the free electrons in the *n* region are injected into the metal region. Due to the very fast response of the metal conductor region, the schottky diode will turn ON at about +0.4 V, which is almost one-half that of an ordinary P-N silicon diode, as shown in Figure 7-15(c). This quick turn ON, and therefore OFF time, makes the schottky diode ideal for fast-switching or high-frequency circuits operating at 20 GHz or more.

Most of the other special diodes that are capable of switching at this fast rate or frequency cannot handle high values of current. The schottky diode is a relatively high current device that can provide forward currents of typically 50 A, making it ideal as a fast-switching rectifier diode. Schottky diodes are often used in low-voltage rectifier circuits because of their low voltage drop per diode (0.4 V) compared to a P-N junction diode (0.7 V).

Figure 7-15(c) shows that the basic P-N junction diode will not be so easily damaged by a reverse voltage as the schottky diode. For example, a schottky

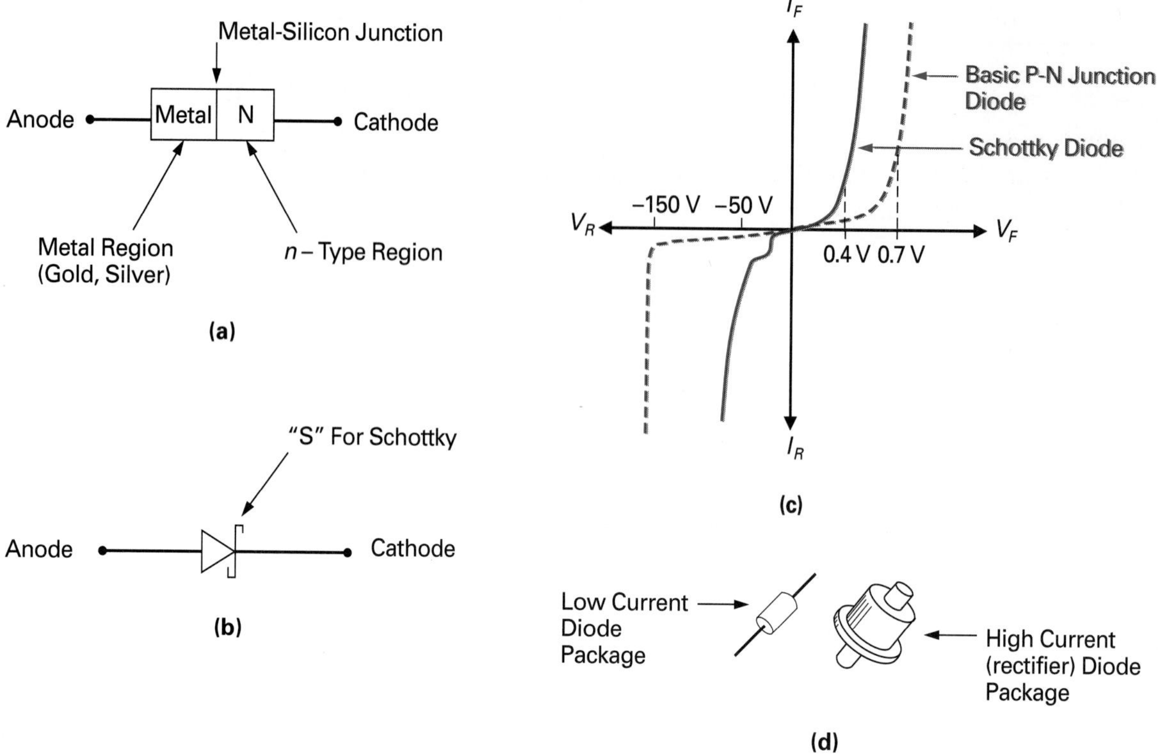

FIGURE 7-15 Schottky Diodes. (a) Construction. (b) Schematic Symbol. (c) Voltage-Current Characteristics. (d) Package Types.

diode will typically have a reverse breakdown voltage rating of about –50 V compared to a typical P-N junction diode's reverse breakdown rating of –150 V.

Hot Carrier Diodes (HCD)
Another name for schottky diodes, so called because the speed with which electrons move from the *n* region to the anode is similar to the speed with which electrons move from the heated (hot) cathode of a vacuum tube.

Schottky diodes are sometimes also called **hot carrier diodes (HCD)** because of the speed at which electrons move from the *n* region across the metal region to the anode. This speed is similar to the speed at which electrons move from the heated (hot) cathode of a vacuum tube.

The ability of the schottky to change operating states (between ON and OFF) at a faster rate than that of a basic junction diode accounts for its use in microwave mixers (which are circuits that combine two microwave frequency signals), detectors or demodulators (which are circuits used to extract information from a carrier), and high-speed digital computer circuits (because a faster circuit switching speed leads to a faster computer operating speed).

Figure 7-15(d) illustrates the typical low-current and high-current schottky diode package.

7-5-2 *The Tunnel Diode*

The schematic symbols used to represent the tunnel diode are shown in Figure 7-16(a). This small two-lead germanium or gallium-arsenide device is formed by doping the *p* and *n* regions one hundred to several thousand times heavier than

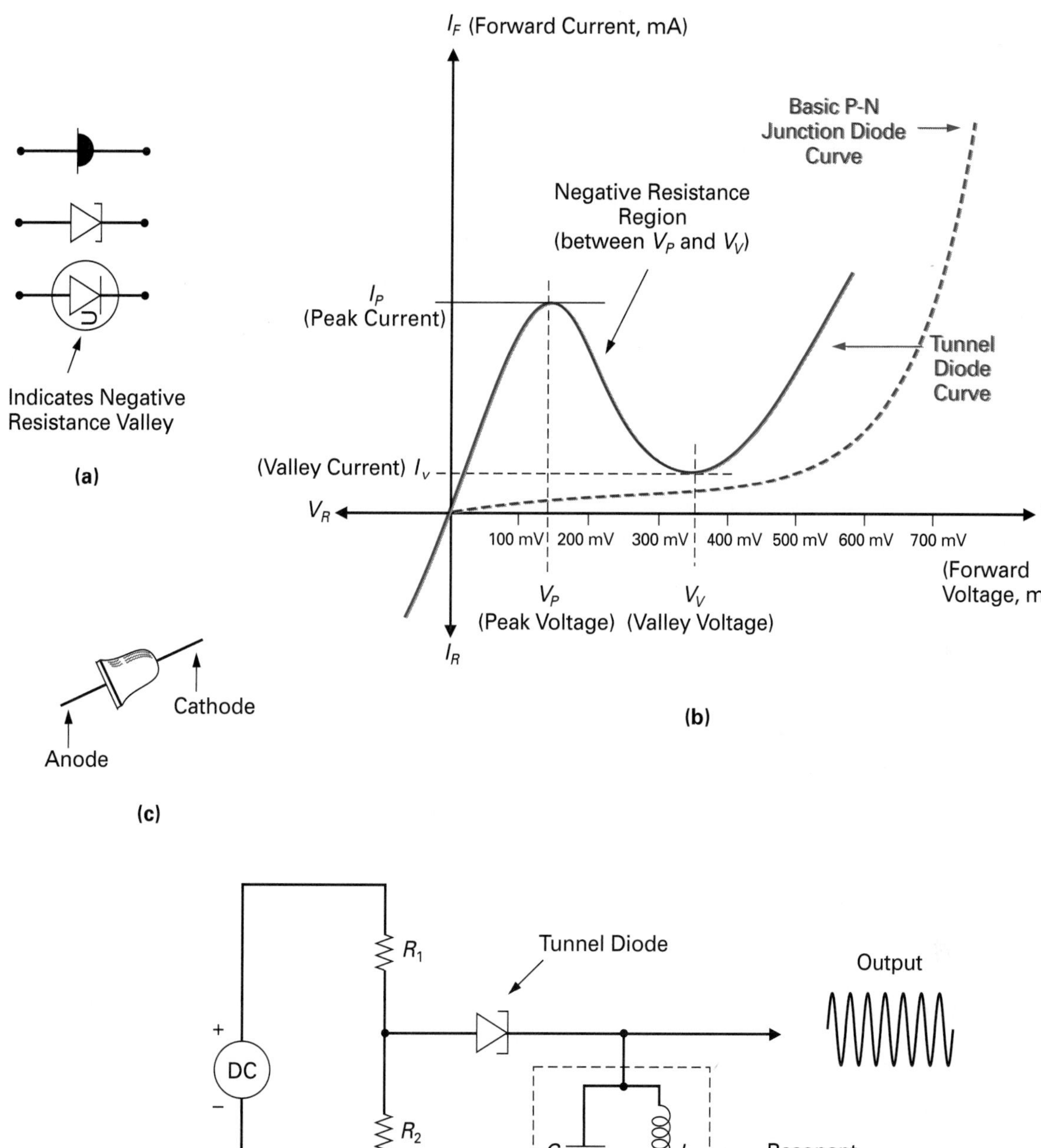

FIGURE 7-16 The Tunnel Diode. (a) Schematic Symbols. (b) Forward Voltage-Current Characteristics. (c) Physical Appearance. (d) Application—Tunnel Diode Oscillator.

that of an ordinary P-N junction diode. This heavy doping is used to create a high internal barrier voltage and a very narrow depletion region.

Figure 7-16(b) shows the *V-I* characteristics for a typical tunnel diode. When reverse biased, the tunnel diode has a low reverse breakdown voltage and will therefore conduct a high reverse current when it is connected to even a small reverse voltage. The tunnel diode is not normally operated in the reverse region because its key characteristic occurs when it is forward biased. Referring to the forward region in Figure 7-16(b), you can see that a forward voltage increase will cause a corresponding increase in forward current. It appears, therefore, that at low forward bias voltages, electrons manage to travel or "tunnel" through the narrow depletion region despite the tunnel diode's high internal barrier voltage. As the applied forward bias voltage is increased, a point is reached where the forward current through the tunnel diode reaches a peak value. This peak value of current is called the tunnel diode's **peak current (I_P),** and the applied forward voltage that causes this peak current value is called the **peak voltage (V_P).** At this point in the curve, a strange tunnel diode characteristic called **negative resistance** occurs. This portion of the curve is called the negative resistance region because the current through the diode is decreasing ($I\downarrow$) as the applied voltage is increasing ($V\uparrow$). This region of the curve seems to go against Ohm's law which states that a voltage increase ($V\uparrow$) usually results in a current increase ($I\uparrow$). The tunnel diode's current continues to decrease until a minimum value of forward diode current is reached. This minimum value of current is called the tunnel diode's **valley current (I_V),** and the applied forward voltage that causes this minimum current value is called the **valley voltage (V_V).** If the forward bias voltage across the tunnel diode is further increased, the forward current will again begin to increase and the remainder of the curve will resemble that of a normal P-N junction diode.

Peak Current (I_P)
The peak value of the current for tunnel diodes.

Peak Voltage (V_P)
The applied forward voltage that causes the peak current for a tunnel diode.

Negative Resistance
The point, for a diode, at which the current decreases as the applied voltage is increasing.

Valley Current (I_V)
The minimum value of current for a tunnel diode.

Valley Voltage (V_V)
The applied forward voltage that causes the valley current for a tunnel diode.

Figure 7-16(c) shows the typical physical appearance of a tunnel diode, while Figure 7-16(d) shows how a tunnel diode can be used in an "oscillator" or sine wave generator circuit. Oscillator circuits will be covered in more detail in a later chapter. For now, let us just basically understand how this circuit operates using a tunnel diode. Manufacturers of tunnel diodes usually specify the values of I_P, V_P, I_V, and V_V on device data sheets since most tunnel diodes are biased to operate in their forward negative resistance region. With the circuit shown in Figure 7-16(d), the voltage divider circuit formed by R_1 and R_2 is used to bias the tunnel diode in the negative resistance region. When power is first applied to the circuit, a surge of dc power from the dc source causes a ringing action in the *LC* tank at its resonant frequency. These sine wave oscillations are applied to the output and also to the cathode of the tunnel diode, causing the diode's applied bias voltage to change in sync with the *LC* tank's oscillations. Power losses in the *LC* tank cause a decrease in the sine wave's amplitude, which controls the tunnel diode's negative resistance to allow more power through to compensate for the original loss. The tunnel diode therefore sustains oscillations.

Tunnel diodes are also used as high-speed switches in which they are switched between the two states of high forward current (I_P) and low forward current (I_V).

SELF-TEST REVIEW QUESTIONS FOR SECTION 7-5

1. A ____________ has a metal-to-semiconductor junction.
   a. Tunnel diode
   b. Hot carrier diode
   c. PIN photodiode
   d. Varactor diode
2. The barrier voltage of a schottky diode is approximately equal to ____________ volts.
3. Tunnel diodes are normally always ____________ biased, and operated in the ____________ region.
4. True or False: A tunnel diode is normally biased between V_P and V_V.

SUMMARY

1. The LED is a photo-emitting or light-transmitting device, while the photodiode is a photo-detecting or light-receiving device.
2. The LED and photodiode are referred to as "optoelectronic devices" because their operation combines both optics and electronics.

The Optical Electromagnetic Spectrum (Figure 7-17)

3. Like radio waves, light is electromagnetic radiation. Its only difference from radio waves is its frequency.
4. The light or optical electromagnetic spectrum is midway between the microwave frequency band and the X-ray frequency band.
5. The optical light spectrum extends from 300 GHz to about 9,300,000 GHz and contains three bands called the infrared band, visible light band, and ultraviolet band.
6. The human eye can only detect the narrow band of frequencies in the visible light band between about 428,000 GHz and 750,000 GHz.
7. Light waves cannot be seen below the visible region in the infrared band and above the visible region in the ultraviolet band.
8. The frequencies in the visible light band can be divided into the colors of the rainbow: red, orange, yellow, green, blue, and violet, with each color corresponding to a narrow range of frequencies.
9. The frequency values in the optical spectrum are large and difficult to work with. To make these values easier to handle, most optical frequencies are converted to their wavelength equivalent.
10. Wavelength, the physical length of one cycle of a transmitted electromagnetic wave, is symbolized by the Greek letter lambda (λ) and is measured in meters.
11. Like radio waves, the wavelength of a light wave is calculated by dividing the light wave's velocity by its frequency.

QUICK REFERENCE SUMMARY SHEET

FIGURE 7-17 The Optical Electromagnetic Spectrum.

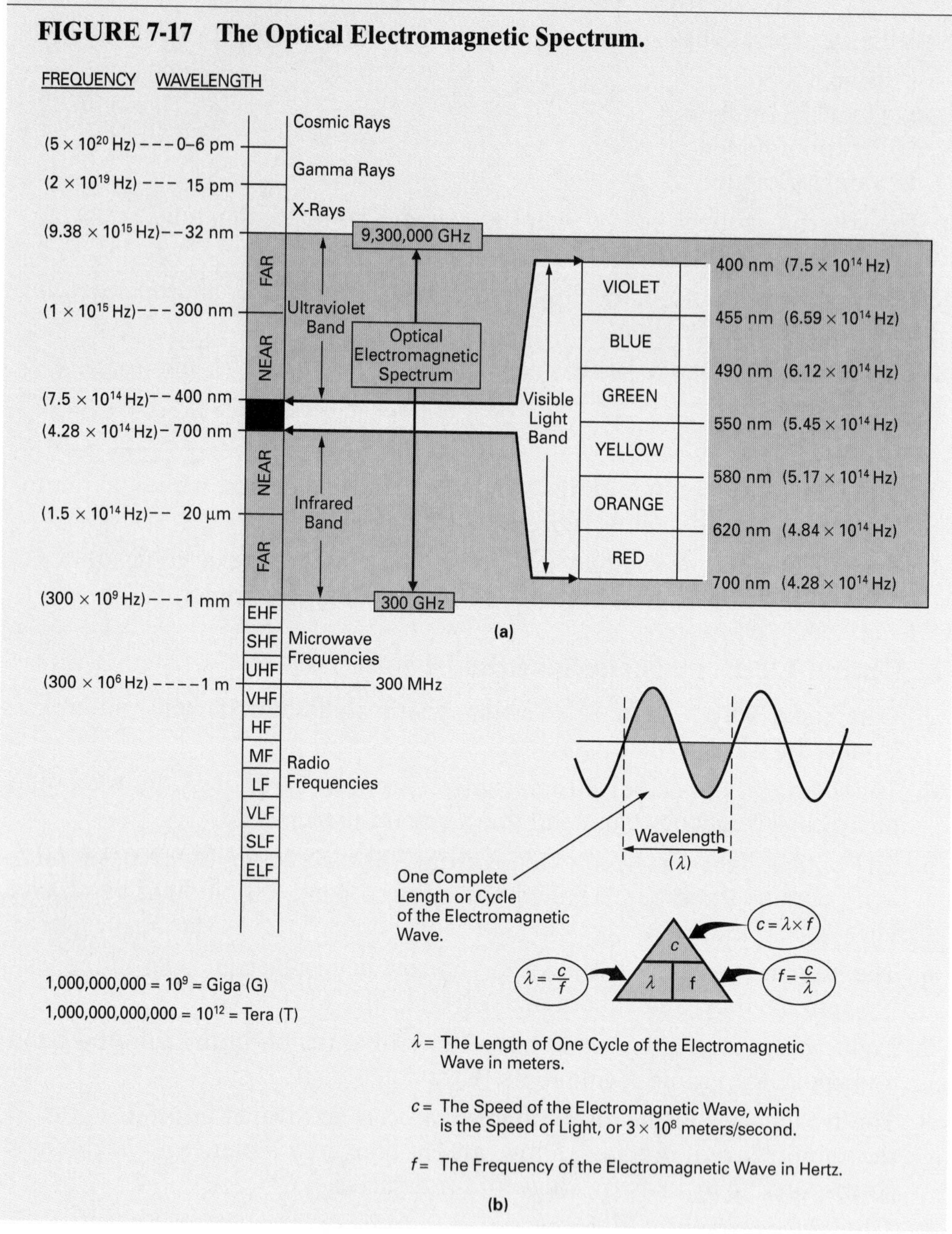

12. Use a smaller light speed unit to convert frequency directly to micrometers (more commonly called microns) or nanometers.

Speed of Light (↑)	*Wavelength* (↓)
3×10^8 meters/second	= meters
3×10^{10} centimeters/second	= centimeters
3×10^{11} millimeters/second	= millimeters
3×10^{14} micrometers/second	= micrometers or microns
3×10^{17} nanometers/second	= nanometers

13. The angstrom, symbolized by Å and equal to 1×10^{-10} meters, is also used as a wavelength unit.
14. The wave theory of light assumes that light propagates or travels as electromagnetic waves. This theory is ideal for explaining why light travels at the speed of light through a vacuum (186,282.4 miles/s or 3×10^8 meters/s), at a slightly slower velocity in air, at even slower speeds in other materials such as glass and water, and the light bending action that occurs as light passes through these materials.
15. The wave theory, however, cannot explain the interaction that occurs between light and semiconductor materials. To explain this action, we must use the quantum theory of light. The quantum theory states that light consists of tiny particles, and each of these discrete quanta or individual packets of energy is called a photon. These photons are uncharged particles that have wave-like characteristics. The photon's energy is determined by its frequency, with a higher frequency photon having more energy than a lower frequency photon.
16. To understand and explain the behavior of light waves, we will have to think of them as electromagnetic waves that contain many tiny particles.

Special Application Diodes (Figure 7-18)

17. The photodiode contains a glass window or convex lens that allows light to enter the case and strike the semiconductor photodiode that is mounted within the metal case.
18. There are two basic types of photodiodes: the basic P-N junction photodiode and the PIN photodiode.
19. The PIN photodiode is constructed in almost exactly the same way as the photodiode, except that the device has an intrinsic layer between the *p* and *n* regions, hence the name PIN (*p*-type layer, intrinsic layer, *n*-type layer). The PIN photodiode is more widely used than the P-N photodiode due to its greater sensitivity. This is because the intrinsic layer adds to the photodiode's depletion region and makes a much wider depletion region for a given reverse bias voltage. This wider depletion region increases the chance that electron-hole pairs will be generated by photons and increases the conductance of the photodiode.
20. The PIN photodiode is one of the most frequently used light detectors in fiber optic communication links. Fiber optics is a technology in which light is transmitted along the inside of a thin flexible glass or plastic fiber.
21. Fiber optic communication links have several key advantages over the copper wire communication links. These advantages are:

    a. Large bandwidth
    b. Low attenuation
    c. Electromagnetic immunity
    d. Small size and light weight
    e. Security and safety

22. Fiber optic communications is used in telecommunications, computer communications, test equipment, medical, automotive, military, and industrial control.

QUICK REFERENCE SUMMARY SHEET

FIGURE 7-18 Special Application Diodes.

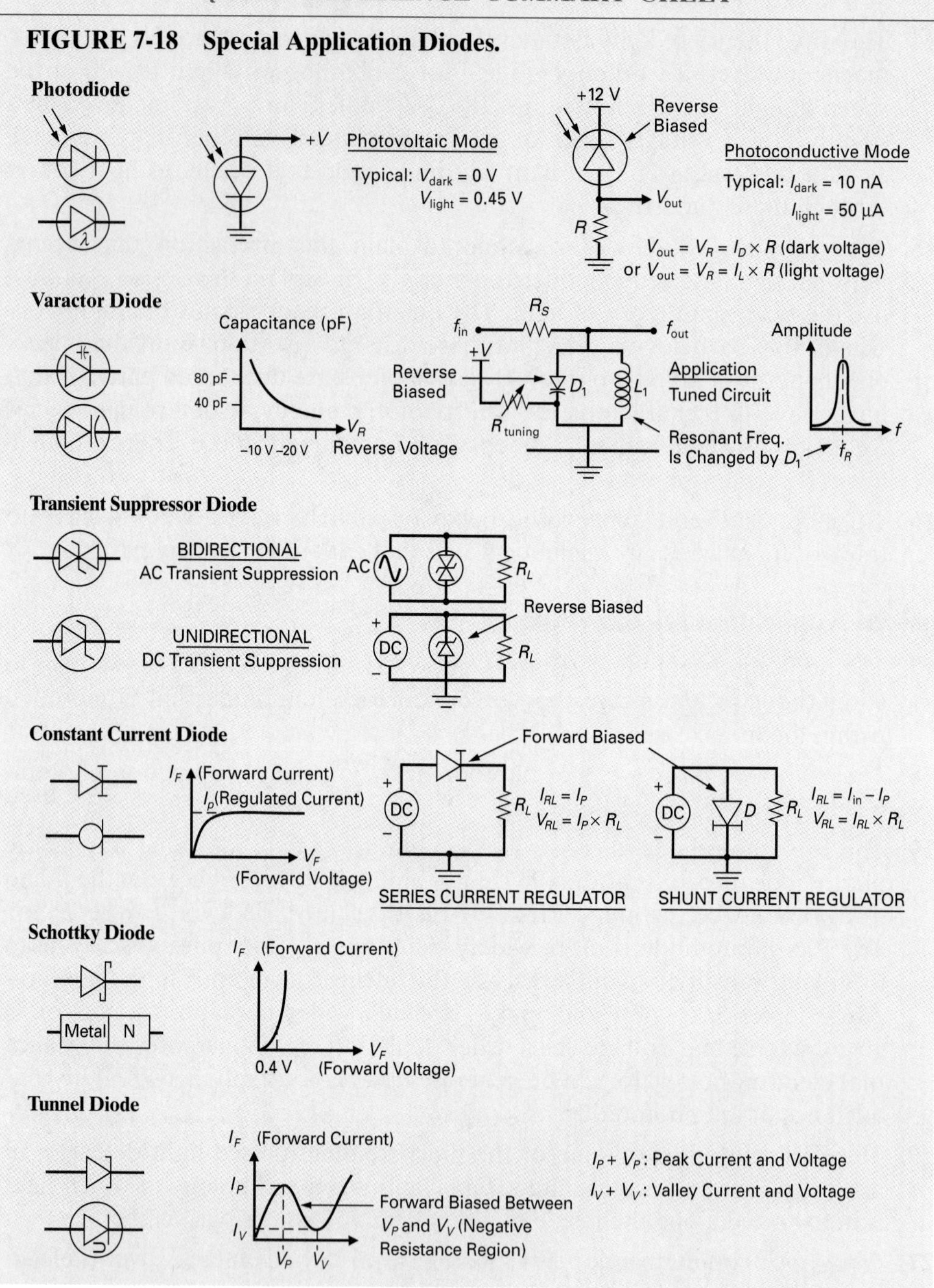

23. The basic P-N junction diode has only a very small amount of internal junction capacitance; however, special diodes can be constructed to have a larger value of internal capacitance. These special diodes are called varactor diodes because, by varying the reverse bias voltage, they can be made to operate as voltage-controlled variable capacitors.

24. With a varactor diode, the reverse bias voltage (V_R) is varied to control the width of the depletion region and the diode's junction capacitance.
25. The varactor diode's capacitance varies inversely with the applied reverse bias voltage.
26. A wide assortment of varactor diodes are available, with capacitance values that range from 1 pF to 2000 pF and with power ratings that range from 500 mW to 35 W.
27. Varactor diodes are used in place of variable capacitors in many applications such as a radio or television receiver. Varactor diodes that are used in these circuits are often called varicap diodes or tuning diodes.
28. Lightning, power-line faults, and the switching on and off of motors, air-conditioners, and heaters can cause the normal 115 V rms ac line voltage at the wall outlet to contain under-voltage dips and over-voltage spikes. Although these transients only last for a few microseconds, the over-voltage spikes can cause the input line voltage to momentarily increase by 1000 V or more. In sensitive equipment, such as televisions and computers, shunt filtering devices are connected between the ac line input and the primary of the dc power supply's transformer to eliminate these transients before they get into, and possibly damage, the system.
29. The transient suppressor diode is used to remove or absorb transients, which accounts for why it is also called a transorb.
30. The bidirectional transorb contains two zener diodes that are connected back-to-back and will therefore monitor both alternations of the ac input in exactly the same way as a symmetrical zener shunt clipper circuit.
31. In dc applications, a single unidirectional transient suppressor can be used. These single transorbs have the same schematic symbol as a zener diode.
32. Metal oxide varistors (MOVs) are currently replacing zener-diode and transient-diode suppressors because they are able to shunt a much higher current surge and are cheaper. They are called varistors because they operate as a voltage dependent resistor that will have a very low resistance at a certain breakdown voltage.
33. The constant current diode does what its name states—maintains a constant output current despite variations in the input voltage.
34. When a constant current diode is forward biased, the anode-to-cathode forward voltage drop will be anywhere between 0.1 V and 100 V. The constant current diode's forward current (I_F), on the other hand, will remain constant.
35. With the series current regulator circuit, the forward biased constant current diode is connected in series with the load, whereas with the shunt current regulator circuit, the forward biased constant current diode is connected in shunt, or parallel, with the load.
36. The schottky diode was named after its German inventor, who discovered the operating principle in 1938. The diode is formed by joining an *n*-type semiconductor region with a metal region such as gold or silver.

37. Due to the very fast response of the metal conductor region, the schottky diode will turn ON at about +0.4 V, which is almost one-half that of an ordinary P-N silicon diode.
38. The schottky diode is a relatively high current device that can provide forward currents of typically 50 A, making it ideal as a fast-switching rectifier diode.
39. Schottky diodes are sometimes also called hot carrier diodes (HCD) because of the speed at which electrons move from the *n* region across the metal region to the anode.
40. The ability of the schottky to change operating states (between ON and OFF) at a faster rate than that of a basic junction diode accounts for its use in microwave mixers, detectors or demodulators, and high-speed digital computer circuits.
41. The P-N junction of a tunnel diode is heavily doped to produce a high internal barrier voltage and very narrow depletion region.
42. The key characteristic of a tunnel diode is its negative resistance region, in which an applied forward voltage increase causes a forward current decrease. Most tunnel diodes are biased to operate in their forward negative resistance region.

NEW TERMS

Angstrom
Constant-Current Diode
Dark Current
Fiber Optic Communications
Hot Carrier Diode
Infrared Band
Integrated Detector and Preamplifier
Light Current
Metal Oxide Varistor
Negative Resistance
Optical Electromagnetic Spectrum
Optical Frequency Spectrum
Particle Theory
Peak Current
Peak Voltage
Photoconductive Mode
Photodiode
Photon
Photovoltaic Mode
PIN Photodiode
P-N Photodiode
Quantum Theory
Regulator Current
Repeater
Schottky Barrier Voltage
Schottky Diode
Series Current Regulator
Shunt Current Regulator
Transient Suppressor Diode
Transorb
Tuned Circuit
Tuning Diode
Tunnel Diode
Ultraviolet Band
Valley Current
Valley Voltage
Varactor Diode
Varicap Diode
Varistor
Visible Light Band
Voltage Regulator
Wave Theory
Wavelength

REVIEW QUESTIONS

Multiple-Choice Questions

1. Which of the following wavelengths is equivalent to the frequency 545 THz?

   a. 580 nm **b.** 550 nm **c.** 400 nm **d.** 700 nm

2. When operated in the photoconductive mode, the photodiode is normally:

   a. Forward biased **c.** Acting as a voltage source
   b. Reverse biased **d.** None of the above

3. When light increases, the conductance of a photodiode will ________.

   a. Increase **b.** Decrease **c.** Remain the same **d.** Be unaffected

4. The LED is classified as a photo- ________ device, while the photodiode is classified as a photo- ________ device.

   a. Transmitting, emitting **c.** Receiving, transmitting
   b. Detecting, emitting **d.** Emitting, detecting

5. A varactor diode is generally operated in its ________ biased mode.

   a. Forward **b.** Reverse **c.** Photo **d.** Un-

6. As the varactor diode's reverse bias voltage increases, the width of a varactor's depletion region ________, and therefore the varactor's capacitance ________.

   a. Increases, decreases **c.** Decreases, increases
   b. Increases, increases **d.** Both (a) and (c) are true

7. In what application are varactor diodes typically used?

   a. Voltage Regulator Circuits **c.** Transient Suppressor Applications
   b. Constant-Current Circuits **d.** Tuned Circuits

8. Which diode is used to shunt high voltage surges?

   a. P-N junction Diode **c.** Transorb Diode
   b. Constant Current Diode **d.** Varactor Diode

9. Which of the following diodes are normally always reverse biased?

   a. Varicap **c.** PIN Photodiode
   b. Transorb **d.** All of the above

10. Which of the following devices could be used in place of a unidirectional transorb?

    a. Tuning Diode **c.** Metal Oxide Varistor
    b. Zener Diode **d.** Both (b) and (c) are true

11. With the constant current diode, a forward voltage increase will cause the voltage drop across the diode to ________ and the forward current to ________.

    a. Increase, remain constant **c.** Remain constant, increase
    b. Decrease, increase **d.** Decrease, remain constant

12. A ________ regulates voltage while a ________ regulates current.

    a. Zener, varactor **c.** Varactor, constant-current diode
    b. Transorb, zener **d.** Zener, constant-current diode

13. Negative resistance, by definition, occurs when a voltage ____________ causes a current ____________.

a. Increase, increase c. Increase, decrease
b. Decrease, decrease d. Both (a) and (b) are true

14. Which of the following diodes exhibit negative resistance?

a. Varactor b. Tunnel c. Schottky d. All of the above

15. Which of the following diodes has a metal-semiconductor junction?

a. Varactor b. Tunnel c. Schottky d. Zener

Essay Questions

16. Sketch the schematic symbols for the following diodes:

a. Junction diode (Chapter 3)
b. Zener diode (Chapter 4)
c. Photodiode (7-1)
d. Varactor diode (7-2)
e. Transient suppresser diode (bidirectional & unidirectional) (7-3)
f. Constant-Current diode (7-4)
g. Schottky diode (7-5-1)
h. Tunnel diode (7-5-2)

17. Describe the relationship between the frequency and wavelength of an electromagnetic wave. (7-1-1)

18. Give the full names of the following abbreviations:

a. PIN (7-1-2)
b. EMI (7-1-5)
c. Varicap (7-2-3)
d. Transorb (7-3)
e. MOV (7-3)

19. Why are PIN photodiodes more frequently used than the P-N photodiode? (7-1-3)

20. Describe the difference between the photodiode's photovoltaic mode and photoconductive mode of operation. (7-1-3)

21. Briefly describe how photodiodes are used in fiber optic communications. (7-1-5)

22. What advantages does a fiber optic communication link have over a copper conductor communication link? (7-1-5)

23. Describe the basic operation of a varactor diode. (7-2)

24. Sketch a circuit, and describe how the varactor can be used in a tuned circuit. (7-2-3)

25. What advantage does a varactor diode have over a variable capacitor? (7-2-3)

26. Briefly describe how the transorb diode can be used to suppress a voltage surge. (7-3)

27. What is the difference between a bidirectional and unidirectional transient suppressor diode? (7-3)

28. Describe the basic operation of the constant current diode. (7-4)
29. In what application would the current-regulator diode be used? (7-4)
30. What is the typical forward voltage drop across a schottky diode, and how does its construction differ from other diodes? (7-5-1)
31. What are the two key advantages of the schottky diode? (7-5-1)
32. What is negative resistance? (7-5-2)
33. In what region will the tunnel diode generally be biased? (7-5-2)
34. In what applications will the schottky and tunnel diode be used? (7-5)
35. Indicate which of the following diodes are generally operated in the forward bias region and which are generally operated in the reverse bias region.
    - **a.** Zener diode (Chapter 4)
    - **b.** Photodiode (7-1)
    - **c.** Varactor diode (7-2)
    - **d.** Transorb diode (7-3)
    - **e.** Constant-Current diode (7-4)
    - **f.** Schottky diode (7-5-1)
    - **g.** Tunnel diode (7-5-2)

Practice Problems

36. Convert the following optical frequencies to their wavelength equivalents:
    - **a.** 6×10^{14} Hz to a wavelength in nanometers.
    - **b.** 1×10^{15} Hz to a wavelength in angstroms.
    - **c.** 1×10^{14} Hz to a wavelength in microns.
37. Convert the following wavelengths to their frequency equivalents:
    a. 500 nm **b.** 4550 Å **c.** 850 μm
38. Calculate the ratio of light current to dark current for the photodiodes shown in Figure 7-19.
39. Which of the photoconductive photodiodes in Figure 7-19 are correctly biased?

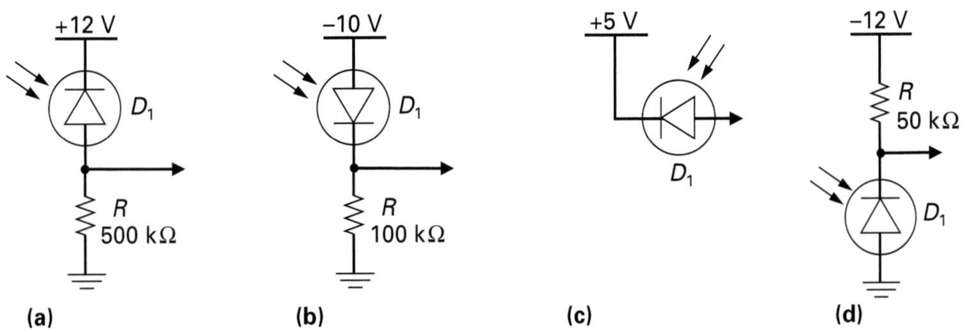

FIGURE 7-19 Photodiode Biasing.

40. Calculate the dark and light voltage developed across the resistor in Figure 7-19(a).
41. What resistance does the photodiode have in Figure 7-19(a) when dark and light?
42. Calculate the dark and light voltage developed across the resistor in Figure 7-19(b).
43. Calculate the dark and light output voltage developed across the photodiode in Figure 7-19(d).
44. In Figure 7-20, a small circular disk containing 12 holes is attached to a dc (toy) motor. The electronic circuit surrounding this motor is designed to function as a "tachometer," which is an instrument that measures angular speed in revolutions per minute (rpm). The input control resistor R_1 is used to adjust the speed of the motor, while the output LED D_2 and the oscilloscope are used to calculate the motor's angular speed in rpm. This circuit operates in the following way. LED D_1 is permanently ON, and its light is applied to the 12 hole rotating disc that is attached to the motor. As the motor rotates, light passes through the holes and is applied to a PIN photo-

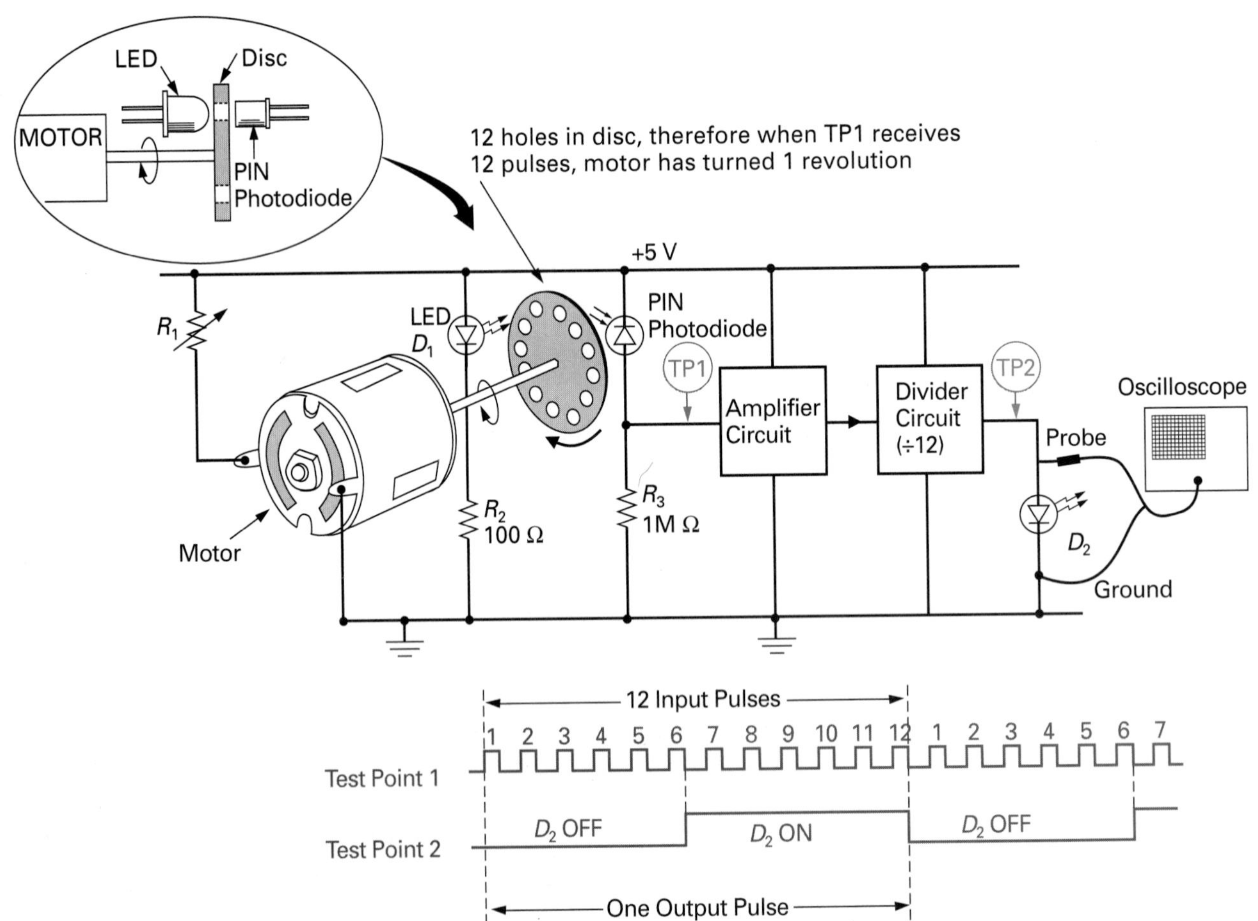

FIGURE 7-20 Tachometer Circuit.

diode. The photodiode will conduct more and less to develop a series of pulses across the resistor R_3 at test point 1, as seen in the TP1 waveform. Because there are 12 holes in the disc, 12 pulses will be generated for each revolution of the disc, and therefore each revolution of the motor. These pulses are amplified. Then a divide-by-12 circuit will generate 1 pulse for every 12 input pulses at TP2, as shown in the TP2 waveform. Because the disc generates 12 pulses/revolution of the motor (×12), and the divider circuit generates 1 pulse for every 12 input pulses (÷12), the LED (D_2) and the oscilloscope will receive 1 pulse/revolution of the motor. Measuring the pulses/second received at D_2 and the oscilloscope will enable us to calculate the motor's angular speed in revolutions/minute.

a. If the motor is turning at a speed of 3300 rpm, what will be the frequency of the pulses at TP1 and TP2?

b. If the oscilloscope is measuring 24 pps at TP2, what is the motor's rpm and the pps rate at TP1?

45. Are the varactor diodes in Figure 7-21 correctly or incorrectly biased?

46. Calculate the frequency range of the tuned circuit shown in Figure 7-21.

47. Which of the transient suppressor diodes in Figure 7-22(c), (d), (e), and (f) should be used to protect the circuits in Figure 7-22(a) and (b)?

48. Referring to the circuit in Figure 7-23, calculate:

a. The current through R_L

b. The voltage developed across R_L

49. In Figure 7-24, the schottky diodes are being used in this low-voltage bridge rectifier circuit due to their smaller voltage drop. If the input ac is 6 V rms, what will be the peak of the pulsating dc from the rectifier (taking into account an individual schottky diode drop of 0.4 V)?

50. Is the tunnel diode in Figure 7-25 biased approximately midway in its negative resistance region by R_1 and R_2?

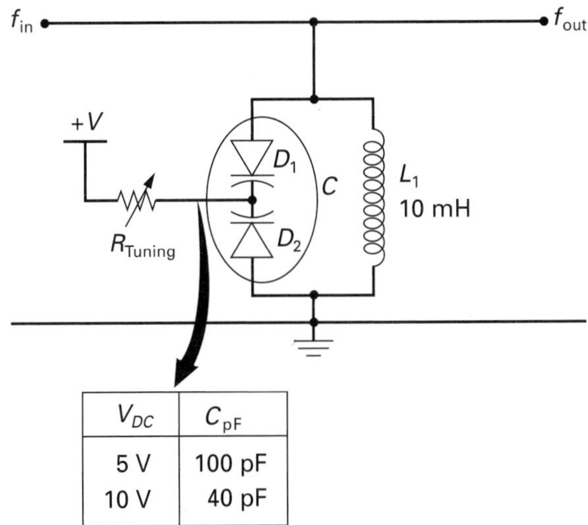

V_{DC}	C_{pF}
5 V	100 pF
10 V	40 pF

FIGURE 7-21 Tuned Varactor Diode Circuit.

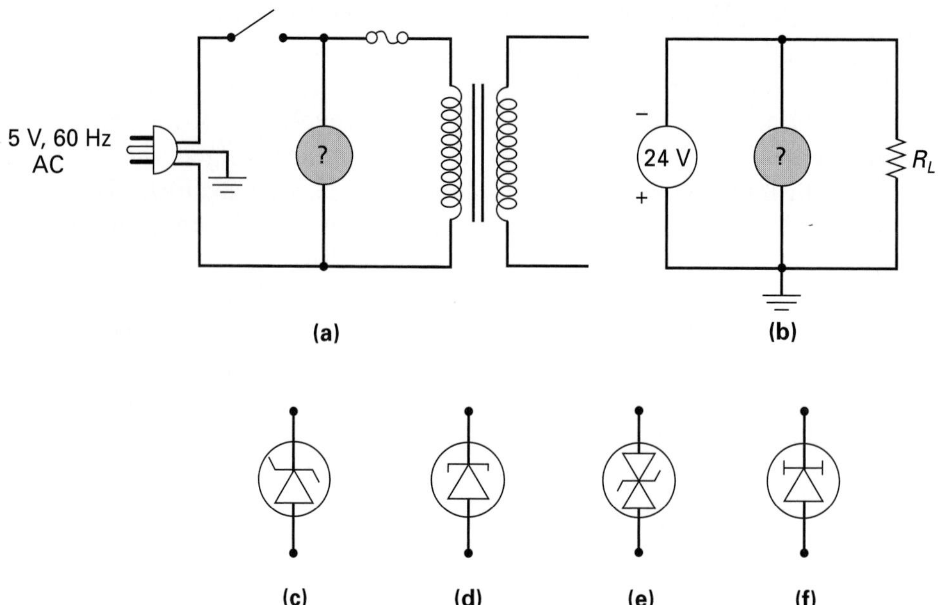

FIGURE 7-22 Transient Voltage Protection.

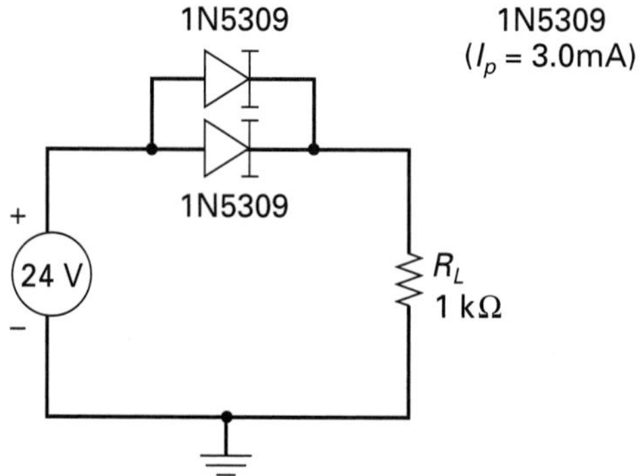

FIGURE 7-23 Series Current Regulator Circuit.

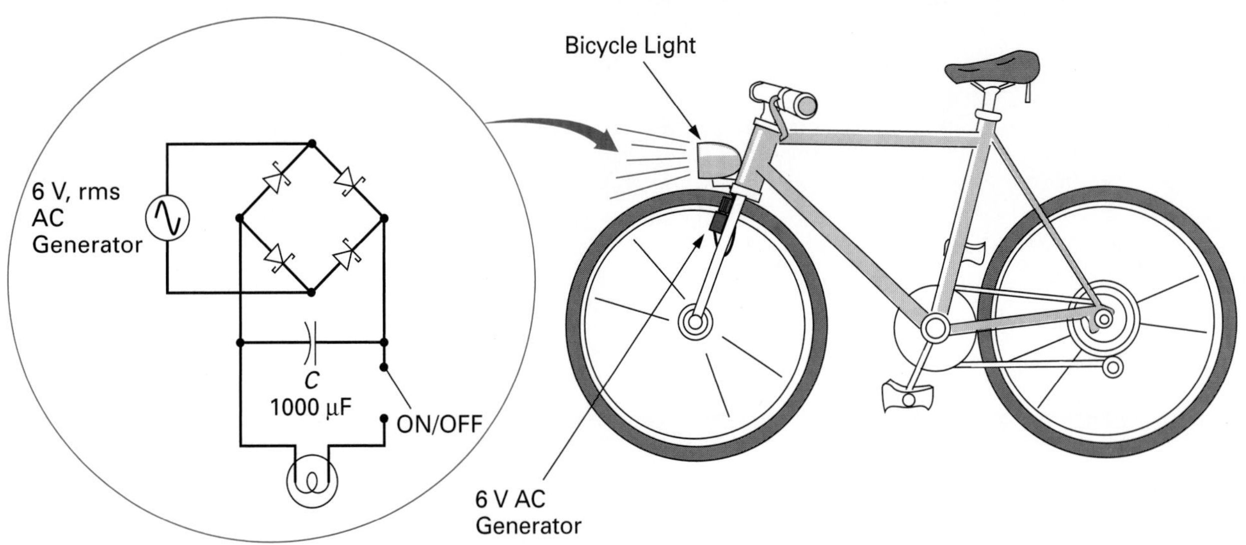

FIGURE 7-24 Schottky Diode Application.

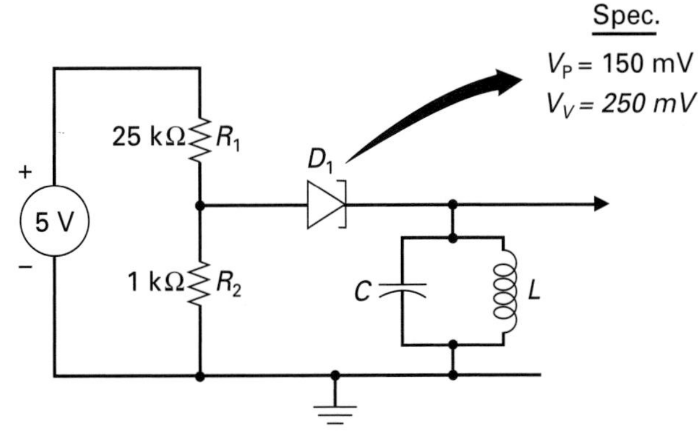

FIGURE 7-25 A Tunnel Diode Oscillator Circuit.

CHAPTER 8

OBJECTIVES

After completing this chapter, you will be able to:

1. Name the three terminals of the bipolar junction transistor.
2. Describe the difference between the construction and schematic symbol of an NPN and PNP bipolar transistor.
3. Identify the base, collector, and emitter terminals of typical low-power and high-power transistor packages.
4. Describe the two basic actions of a bipolar transistor:
   - **a.** ON/OFF switching action
   - **b.** Variable-Resistor action
5. Define the terms transistor and transistance.
6. Explain how the transistor can be used in the following basic applications:
   - **a.** A digital (two-state) circuit, such as a logic gate
   - **b.** An analog (linear) circuit, such as an amplifier
7. Describe in more detail the bipolar junction transistor action.
8. Explain in more detail how the transistor can be used as an amplifier.
9. Identify and list the characteristics of the three transistor circuit configurations, namely:
   - **a.** Common base
   - **b.** Common emitter
   - **c.** Common collector
10. Explain the meaning of the following:
    - **a.** Transistor voltage and current abbreviations
    - **b.** DC alpha
    - **c.** DC beta
    - **d.** Collector characteristic curve
    - **e.** AC beta
    - **f.** Input resistance or impedance
    - **g.** Output resistance or impedance
11. Calculate a transistor circuit's
    - **a.** DC current gain **c.** Voltage gain
    - **b.** AC current gain **d.** Power gain
12. Interpret the specifications given in a manufacturer's data sheet for the following:
    - **a.** A general purpose transistor
    - **b.** A high-voltage transistor
    - **c.** A high-current transistor
    - **d.** A high-power transistor
13. Describe the function of a transistor tester.
14. Explain how an ohmmeter can be used to check the P-N junctions of a transistor for opens and shorts.
15. Calculate the different values of circuit voltage and current for the following transistor biasing methods:
    - **a.** Base biasing **b.** Voltage-Divider biasing
16. Define the following terms:
    - **a.** DC load line **c.** Saturation point
    - **b.** Cutoff point **d.** Quiescent point

Bipolar Junction Transistors (BJTs)

OUTLINE

VIGNETTE

The Fairchildren

On December 23, 1947, John Bardeen, Walter Brattain, and William Shockley first demonstrated how a semiconductor device, named the transistor, could be made to amplify. However, the device had mysterious problems and was very unpredictable. Shockley continued his investigations, and in 1951 he presented the world with the first reliable junction transistor. In 1956, the three shared the Nobel Prize in physics for their discovery. In 1972, Bardeen would win a rare second Nobel Prize for his research at the University of Illinois in the field of superconductivity.

Shockley left Bell Labs in 1955 to start his own semiconductor company near his home in Palo Alto and began recruiting personnel. He was, however, very selective and only hired those who were bright, young, and talented. The company was a success, although many of the employees could not tolerate Shockley's eccentricities, such as posting everyone's salary and requiring that the employees rate one another. Two years later, eight of Shockley's most talented employees defected. The "traitorous eight," as Shockley called them, started their own company named Fairchild Semiconductor only a dozen blocks away.

More than fifty companies would be founded by former Fairchild employees. One of the largest was started by Robert Noyce and two other colleagues from the group of eight Shockley defectors; they named their company Intel, which was short for "intelligence."

INTRODUCTION

In 1948, a component known as a transistor sparked a whole new era in electronics, the effects of which have not been fully realized even to this day. A transistor is a three-element device made of semiconductor materials used to control electron flow, the amount of which can be controlled by varying the voltages applied to its three elements. Having the ability to control the amount of current through the transistor allows us to achieve two very important applications: switching and amplification.

Like the diode, transistors are formed by p and n regions and, as we are already aware, the point at which a p and an n-region join is known as a junction. Transistors in general are classified as being either the *bipolar* or *unipolar* type. The bipolar type has two P-N junctions, while unipolar transistors have only one P-N junction. In this chapter we will study all of the details relating to the *bipolar* transistor, or as it is also known, the *bipolar junction transistor* or *BJT*.

8-1 FIRST APPROXIMATION DESCRIPTION OF A BIPOLAR TRANSISTOR

In most cases it is easier to build a jigsaw puzzle when you can refer to the completed picture on the box. The same is true whenever anyone is trying to learn anything new, especially a science that contains many small pieces. These first approximation descriptions are a means for you to quickly see the complete picture without having to wait until you connect all of the pieces. Like the diode's first approximation description, this general overview will cover the transistor's basic construction, schematic symbol, physical appearance, basic operation, and main applications.

8-1-1 *Transistor Types (NPN and PNP)*

Like the diode, a bipolar transistor is constructed from a semiconductor material. However, unlike the diode, which has two oppositely doped regions and one P-N junction, the transistor has three alternately doped semiconductor regions and two P-N junctions. These three alternately doped regions are arranged in one of two different ways, as shown in Figure 8-1.

With the **NPN transistor** shown in Figure 8-1(a), a thin lightly doped *p*-type region known as the **base** (symbolized *B*) is sandwiched between two *n*-type regions called the **emitter** (symbolized *E*) and the **collector** (symbolized *C*). Looking at the NPN transistor's schematic symbol in Figure 8-1(b), you can see that an arrow is used to indicate the emitter lead. As a memory aid for the NPN transistor's schematic symbol, you may want to remember that when the emitter arrow is "**N**ot **P**ointing i**N**" to the base, the transistor is an **"NPN."** An easier method is to think of the arrow as a diode, with the tip of the arrow or cathode pointing to an *n* terminal and the back of the arrow or anode pointing to a *p* terminal, as seen in the inset in Figure 8-1(b).

The **PNP transistor** can be seen in Figure 8-1(c). With this transistor type, a thin, lightly doped *n*-type region (base) is placed between two *p*-type regions (emitter and collector). Figure 8-1(d) illustrates the PNP transistor's schematic symbol. Once again, if you think of the emitter arrow as a diode, as shown in the inset in Figure 8-1(d), the tip of the arrow or cathode is pointing to an *n* terminal and the back of the arrow or anode is pointing to a *p* terminal.

NPN Transistor
A thin, lightly doped *p*-type region (base) is sandwiched between two *n*-type regions (emitter and collector).

Base
The region that lies between an emitter and a collector of a transistor and into which minority carriers are injected.

Emitter
A transistor region from which charge carriers are injected into the base.

Collector
A semiconductor region through which a flow of charge carriers leaves the base of the transistor.

PNP Transistor
A thin, lightly doped *n*-type region (base) is placed between two *p*-type regions (emitter and collector).

8-1-2 *Transistor Construction and Packaging*

Like the diode, the three layers of an NPN or PNP transistor are not formed by joining three alternately doped regions. These three layers are formed by a "diffusion process," which first melts the base region into the collector region, and then melts the emitter region into the base region. For example, with the NPN transistor shown in Figure 8-2(a), the construction process would begin by diffusing or melting a *p*-type base region into the *n*-type collector region. Once this *p*-type base region is formed, an *n*-type emitter region is diffused or melted into the newly diffused *p*-type base region to form an NPN transistor. Keep in mind

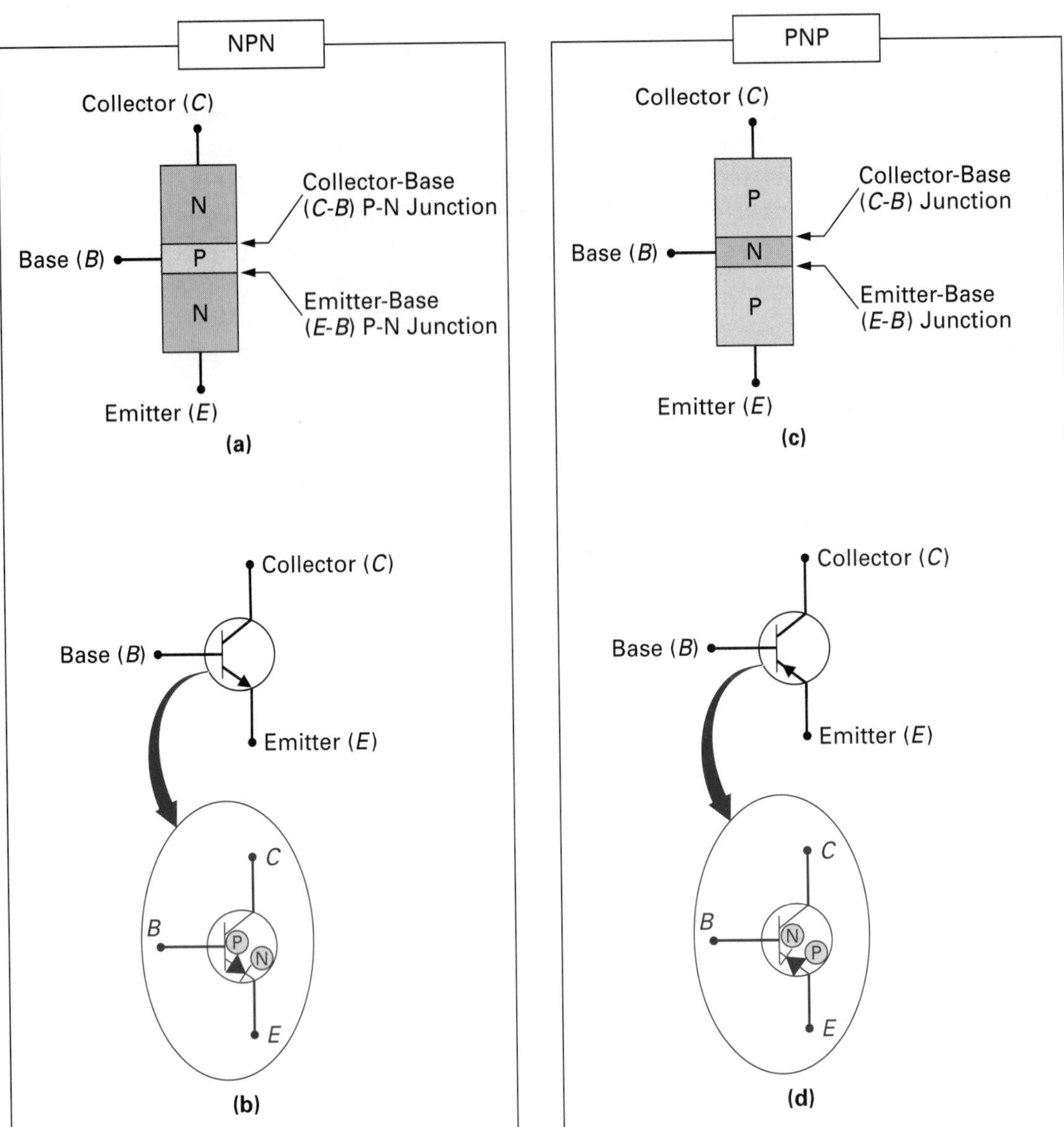

FIGURE 8-1 Bipolar Junction Transistor (BJT) Types.

that manufacturers will generally construct thousands of these transistors simultaneously on a thin semiconductor wafer or disc, as shown in Figure 8-2(a). Once tested, these discs, which are about 3 inches in diameter, are cut to separate the individual transistors. Each transistor is placed in a package, as shown in Figure 8-2(b). The package will protect the transistor from humidity and dust, provide a means for electrical connection between the three semiconductor regions and the three transistor terminals, and serve as a heat sink to conduct away any heat generated by the transistor.

Figure 8-3 illustrates some of the typical low-power and high-power transistor packages. Most low-power, small-signal transistors are hermetically sealed in a metal, plastic, or epoxy package. Four of the low-power packages shown in Fig-

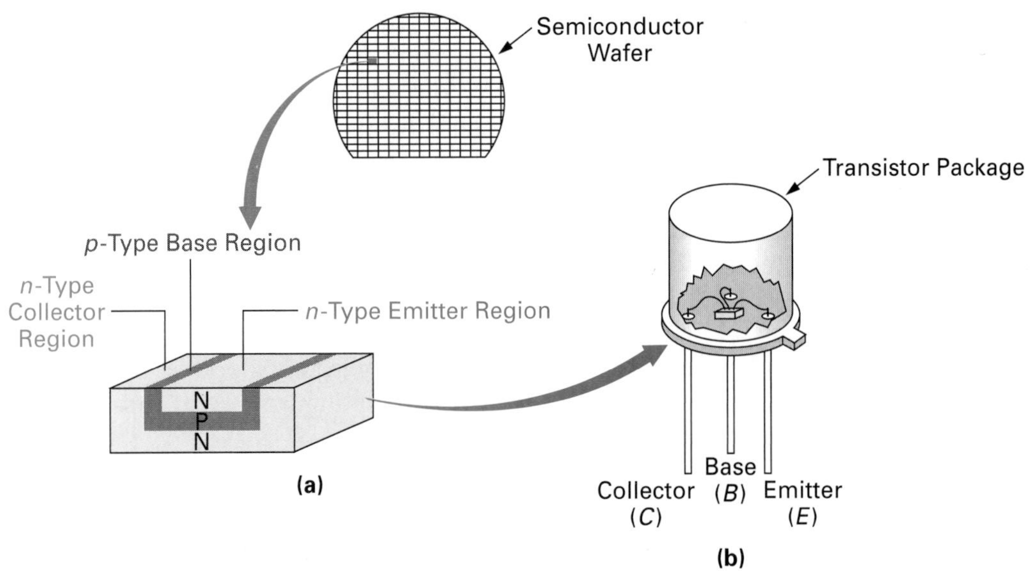

FIGURE 8-2 **Bipolar Junction Transistor Construction and Packaging.**

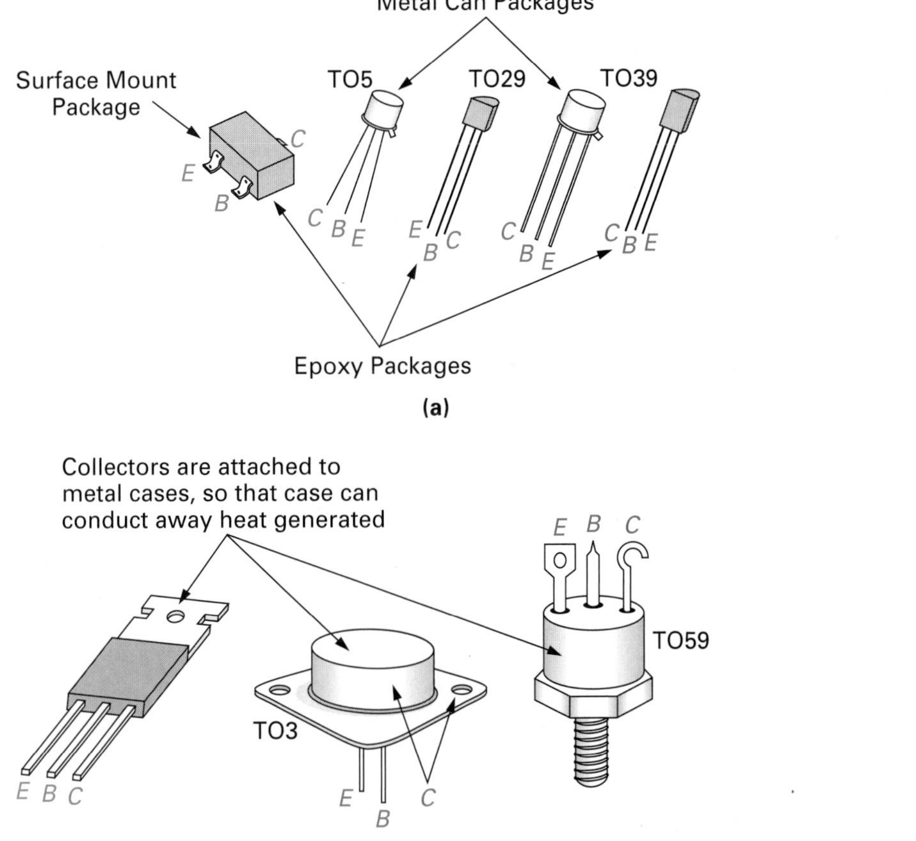

FIGURE 8-3 **Bipolar Junction Transistor Package Types. (a) Low-Power. (b) High-Power.**

ure 8-3(a) have their three leads protruding from the bottom of the package because these package types are usually inserted and soldered into holes in printed circuit boards (PCBs). The surface mount technology (SMT) low-power transistor package, on the other hand, has flat metal legs that mount directly onto the surface of the PCB. These transistor packages are generally used in high component density PCBs because they use less space than a "through-hole" package. To explain this in more detail, a through-hole transistor package needs a hole through the PCB and a connecting pad around the hole to make a connection to the circuit. With an SMT package, however, no holes are needed, only a small connecting pad. Without the need for holes, pads on printed circuit boards can be smaller and placed closer together, resulting in considerable space saving.

The high-power packages, shown in Figure 8-3(b), are designed to be mounted onto the equipment's metal frame or chassis so that the additional metal will act as a heat sink and conduct the heat away from the transistor. With these high-power transistor packages, two or three leads may protrude from the package. If only two leads are present, the metal case will serve as a collector connection, and the two pins will be the base and emitter.

Transistor package types are normally given a reference number. These designations begin with the letters "TO," which stands for transistor outline and are followed by a number. Figure 8-3 includes some examples of TO reference designators.

8-1-3 *Transistor Operation*

Figure 8-4 shows an NPN bipolar transistor, and the inset shows how a transistor can be thought of as containing two diodes: a *base-to-collector diode* and a *base-to-emitter diode.* With an NPN transistor, both diodes will be back-to-back and

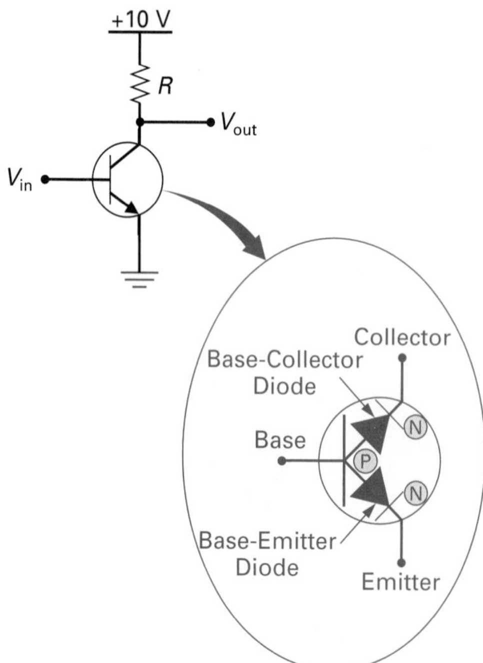

FIGURE 8-4 The Base-Collector and Base-Emitter Diodes Within a Bipolar Transistor.

"**N**ot be **P**ointing i**N**" (NPN) to the base, as shown in the inset in Figure 8-4. For a PNP transistor, the base-collector and base-emitter diode will both be pointing into the base.

Transistors are basically controlled to operate as a switch, or they are controlled to operate as a variable resistor. Let us now examine each of these operating modes.

The Transistor's ON/OFF Switching Action

Figure 8-5 illustrates how the transistor can be made to operate as a switch. This ON/OFF switching action of the transistor is controlled by the transistor's base-to-emitter (*B-E*) diode. If the *B-E* diode of the transistor is forward biased, the transistor will turn ON; if the *B-E* diode of the transistor is reverse biased, the transistor will turn OFF.

To begin with, let us see how the transistor can be switched ON. In Figure 8-5(a), the *B-E* diode of the transistor is forward biased (anode at base is +5 V, cathode at emitter is 0 V), and the transistor will turn ON. Its collector and emitter output terminals will be equivalent to a closed switch, as shown in Figure 8-5(b). This low resistance between the transistor's collector and emitter will cause a current (*I*), as shown in Figure 8-5(b). The output voltage in this condition will be zero volts because all of the +10 V supply voltage will be dropped across *R*. Another way to describe this would be to say that the low resistance path between the transistor's emitter and collector connects the zero volt emitter potential through to the output.

Now let us see how the transistor can be switched OFF. In Figure 8-5(c), the transistor has 0 V being applied to its base input. In this condition, the *B-E* diode of the transistor is reverse biased (anode at base is 0 V, cathode at emitter is 0 V), and so the transistor will turn OFF and its collector and emitter output

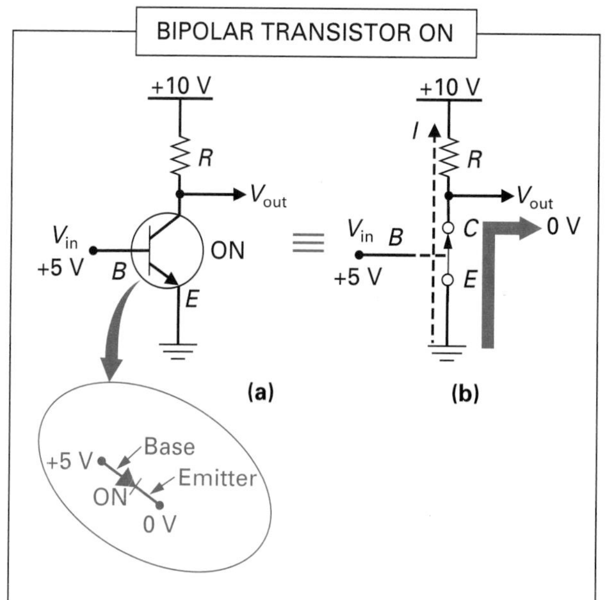

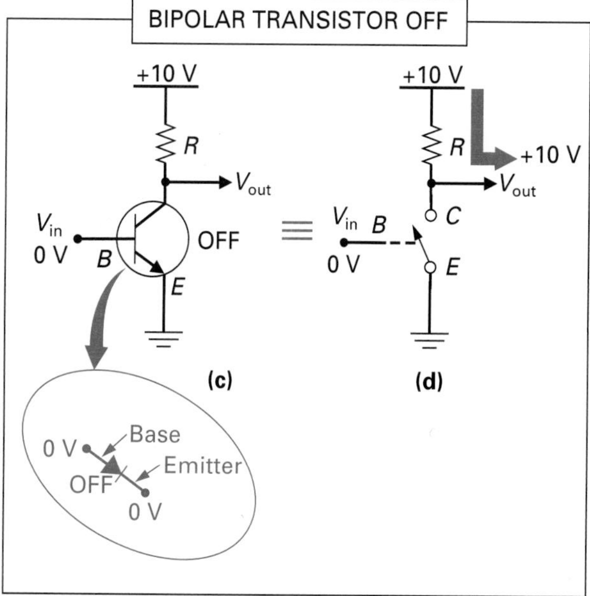

FIGURE 8-5 The Bipolar Transistor's ON/OFF Switching Action.

terminals will be equivalent to an open switch, as shown in Figure 8-5(d). This high resistance between the transistor's collector and emitter will prevent any current and any voltage drop, resulting in the full +10 V supply voltage being applied to the output, as shown in Figure 8-5(d).

The Transistor's Variable-Resistor Action

In the previous section we saw how the transistor can be biased to operate in one of two states: ON or OFF. When operated in this two-state way, the transistor is being switched ON and OFF in almost the same way as a junction diode. The transistor, however, has another ability that the diode does not have—it can also function as a variable resistor, as shown in the equivalent circuit in Figure 8-6(a). In Figure 8-5 we saw how +5 V base input bias voltage would result in a low resistance between emitter and collector (closed switch) and how a 0 V base input bias would result in a high resistance between emitter and collector (open switch). The table in Figure 8-6(b) shows an example of the relationship between base input bias voltage (V_B) and emitter-to-collector resistance ($R_{C\text{-}E}$). In this table, you can see that the transistor is not only going to be driven between the two extremes of fully ON and fully OFF. When the base input voltage is at some voltage level between +5 V and 0 V, the transistor is partially ON; therefore, the transistor's emitter-to-collector resistance is somewhere between 0 Ω and maximum Ω. For example, when V_B = +4 V, the transistor is not fully ON, and its emitter-to-collector resistance will be slightly higher, at 100 Ω. If the base input bias voltage is further reduced to +3 V, for example, you can see in the table that the emitter-to-collector resistance will further increase to 10 kΩ.

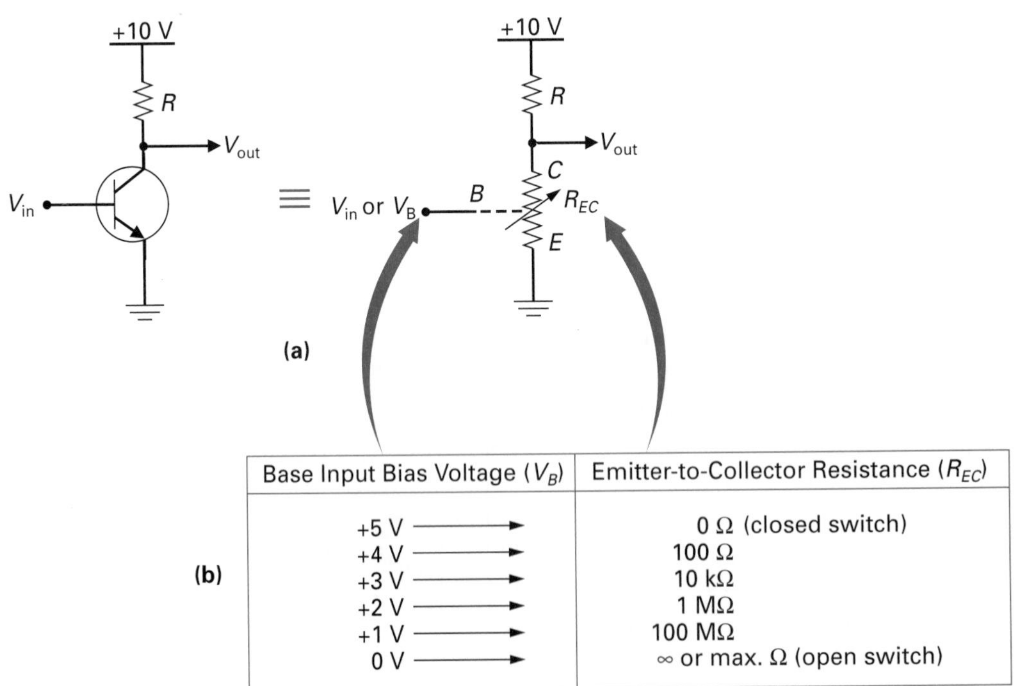

Base Input Bias Voltage (V_B)	Emitter-to-Collector Resistance (R_{EC})
+5 V →	0 Ω (closed switch)
+4 V →	100 Ω
+3 V →	10 kΩ
+2 V →	1 MΩ
+1 V →	100 MΩ
0 V →	∞ or max. Ω (open switch)

FIGURE 8-6 The Bipolar Transistor's Variable-Resistor Action.

Further decreases in base input voltage ($V_B\downarrow$) will cause further increases in emitter-to-collector resistance ($R_{C\text{-}E}\uparrow$) until $V_B = 0$ V and $R_{C\text{-}E}$ = maximum Ω.

As a matter of interest, the name transistor was derived from the fact that through base control we can "transfer" different values of "resistance" between the emitter and collector. This effect of "transferring resistance" is known as **transistance** and the component that functions in this manner is called the transistor.

Transistance The effect of transferring resistance.

Now that we have seen how the transistor can be made to operate as either a switch or as a variable resistor, let us see how these characteristics can be made use of in circuit applications.

8-1-4 Transistor Applications

The transistor's impact on electronics has been phenomenal. It initiated the multi-billion dollar semiconductor industry and was the key element behind many other inventions, such as integrated circuits (*IC*s), optoelectronic devices, and digital computer electronics. In all of these applications, however, the transistor is basically made to operate in one of two ways: as a switch or as a variable resistor. Let us now briefly examine an example of each.

Digital Logic Gate Circuit

A digital logic gate circuit makes use of the transistor's ON/OFF switching action. Digital circuits are often referred to as "switching" or "two-state" circuits because their main control device (the transistor) is switched between the two states of ON and OFF. The transistor is at the very heart of all digital electronic circuits. For example, transistors are used to construct logic gate circuits, gates are used to construct flip-flop circuits, flip-flops are used to construct register and counter circuits, and these circuits are used to construct microprocessor, memory, and input/output circuits—the three basic blocks of a digital computer.

In Chapter three you were introduced to the OR gate, which would produce a HIGH (logic 1) output when either input *A* OR *B* were HIGH. Figure 8-7(a) shows how the transistor can be used to construct another type of digital logic gate. This gate is called the NOT gate, or more commonly, the INVERTER gate. The basic NOT gate circuit is constructed using one NPN transistor and two resistors. This logic gate has only one input (*A*) and one output (*Y*), and its schematic symbol is shown in Figure 8-7(b). Figure 8-7(c) shows how this logic gate will react to the two different input possibilities. When the input is 0 V (logic 0), the transistor's base-emitter P-N diode will be reverse biased and so the transistor will turn OFF. Referring to the inset for this circuit condition in Figure 8-7(c), you can see that the OFF transistor is equivalent to an open switch between emitter and collector, and therefore the +5 V supply voltage will be connected to the output. In summary, a logic 0 input (0 V) will be converted to a logic 1 output (+5 V). On the other hand, when the input is +5 V (logic 1), the transistor's base-emitter P-N diode will be forward biased and so the transistor will turn ON. Referring to the inset for this circuit condition in Figure 8-7(c), you can see that the ON transistor is equivalent to a closed switch between emitter and collector, and therefore 0 V will be connected to the output. In summary, a logic 1 input (+5 V) will be converted to a logic 0 output (0 V).

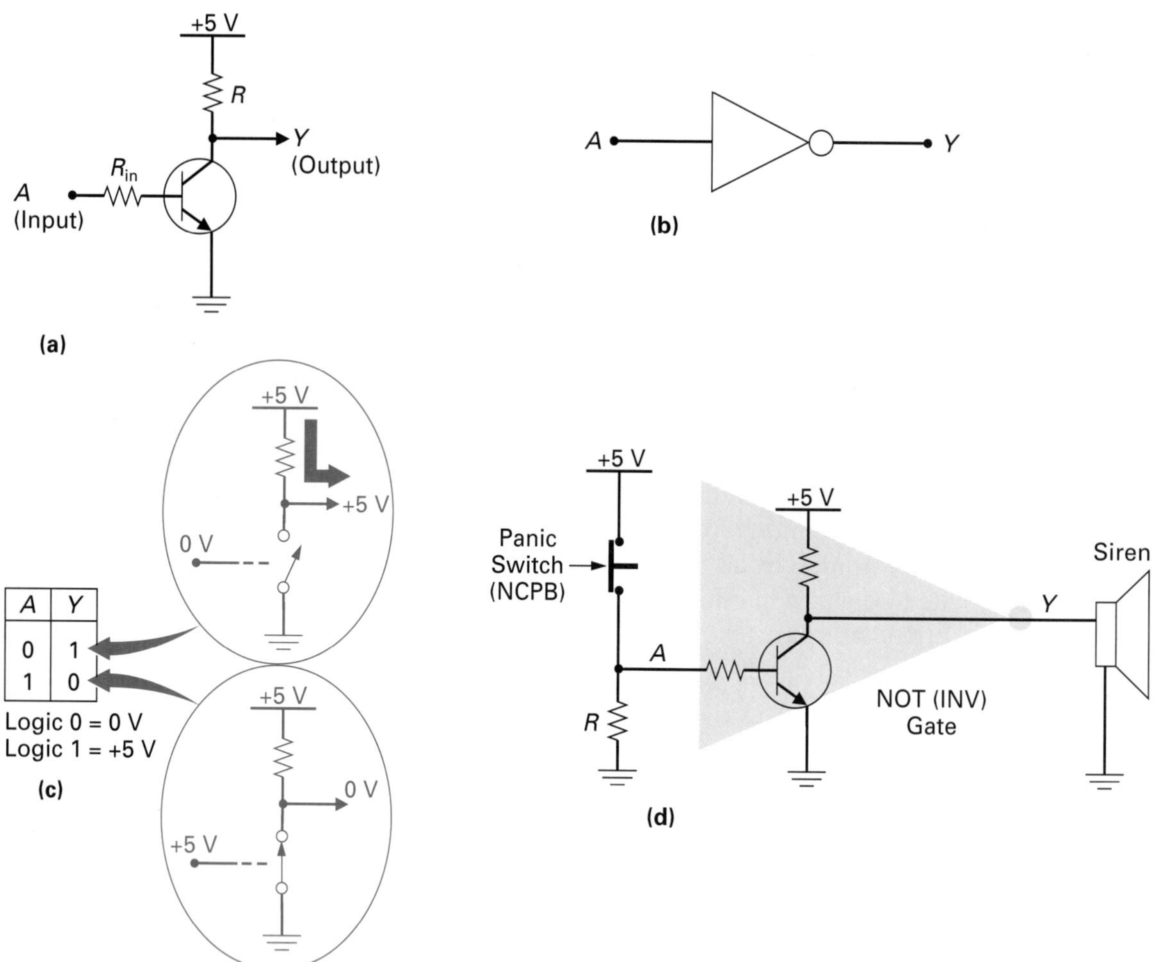

FIGURE 8-7 The Transistor Being Used in a Digital Electronic Circuit. (a) Basic NOT or INVERTER Gate Circuit. (b) NOT Gate Schematic Symbol. (c) NOT Gate Function Table. (d) NOT Gate Security System Application.

Referring to the function table in Figure 8-7(c), you can see that the output logic level is "NOT" the same as the input logic level—hence the name NOT gate.

As an application, Figure 8-7(d) shows how a NOT or INV gate can be used to invert an input control signal. In this circuit, you can see that a normally closed push button (NCPB) switch is used as a panic switch to activate a siren in a security system. Because the push button is normally closed, it will produce +5 V at *A* when it is not in alarm. If this voltage were connected directly to the siren, the siren would be activated incorrectly. By including the NOT gate between the switch circuit and the siren, the normally HIGH output of the NCPB will be inverted to a LOW, and not activate the siren when we are not in alarm. When the panic switch is pressed however, the NCPB contacts will open producing a LOW input voltage to the NOT gate. This LOW input will be inverted to a HIGH output and activate the siren.

The NOT gate, and other digital logic gate circuits will be discussed in more detail in Chapter 9.

Analog Amplifier Circuit

When used as a variable resistor, the transistor is the controlling element in many analog or linear circuits applications such as amplifiers, oscillators, modulators, detectors, regulators, and so on. The most important of these applications is **amplification,** which is the boosting in strength or increasing in amplitude of electronic signals.

Amplification Boosting in strength, or increasing amplitude, of electronic signals.

Figure 8-8(a) shows a simplified transistor amplifier circuit, while Figure 8-8(b) shows the voltage waveforms present at different points in the circuit. As you can see, the transistor is labeled Q_1 because the letter "Q" is the standard letter designation used for transistors.

Before applying an ac sine wave input signal, let us determine the dc voltage levels at the transistor's base and collector. The 6.3 kΩ/3.7 kΩ resistance ratio of the voltage divider R_1 and R_2 causes the +10 V supply voltage to be proportionally divided, producing +3.7 V dc across R_2. This +3.7 V dc will be applied to the base of the transistor, causing the base-emitter junction of Q_1 to be forward biased and Q_1 to turn ON. With transistor Q_1 ON, a certain value of resistance will exist between the transistor's collector and emitter (R_{CE}), and this resistance will form a voltage divider with R_E and R_C, as seen in the inset in Figure 8-8(a). The dc voltage at the collector of Q_1 relative to ground (V_C) will be equal to the voltage developed across Q_1's collector-emitter resistance (R_{CE}) and R_E. In this example circuit, with no ac signal applied, a V_b of +3.7 V dc will cause R_{CE} and R_E to cumulatively develop +8 V at the collector of Q_1. The transistor has a base bias voltage (V_b) that is +3.7 V dc relative to ground and a collector voltage (V_C) that is +8 V dc relative to ground. Capacitors C_1 and C_2 are included to act as dc blocks, with C_1 preventing the +3.7 V dc base bias voltage (V_b) from being applied back to the input (V_{in}) and C_2 preventing the +8 V dc collector reference voltage (V_C) from being applied across the output (V_{R_L} or V_{out}).

Let us now apply an input signal and see how it is amplified by the amplifier circuit in Figure 8-8(a). The alternating input sine wave signal (V_{in}) is applied to the base of Q_1 via C_1, which, like most capacitors, offers no opposition to this ac signal. This input signal, which has a peak-to-peak voltage change of 200 mV, is shown in the first waveform in Figure 8-8(b). The alternating input signal will be superimposed on the +3.7 V dc base bias voltage and cause the +3.7 V dc at the base of Q_1 to increase by 100 mV (3.7 V +100 mV = 3.8 V), and decrease by 100 mV (3.7 V – 100 mV = 3.6 V), as seen in the second waveform in Figure 8-8(b). An increase in the input signal ($V_{in}\uparrow$), and therefore the base voltage ($V_B\uparrow$), will cause an increase in the emitter diode's forward bias, causing Q_1 to turn more ON and the emitter-to-collector resistance of Q_1 to decrease ($R_{EC}\downarrow$). Because voltage drop is always proportional to resistance, a decrease in $R_{EC}\downarrow$ will cause a decrease in the voltage drop across R_{CE} and R_E ($V_C\downarrow$), and this decrease in V_C will be coupled to the output via C_2, causing a decrease in the output voltage developed across the load ($V_{out}\downarrow$).

Now let us examine what will happen when the sine wave input signal decreases. A decrease in the input signal ($V_{in}\downarrow$), and therefore the base voltage ($V_B\downarrow$), will cause a decrease in the emitter diode's forward bias, causing Q_1 to turn less ON and the emitter-to-collector resistance of Q_1 to increase ($R_{EC}\uparrow$).

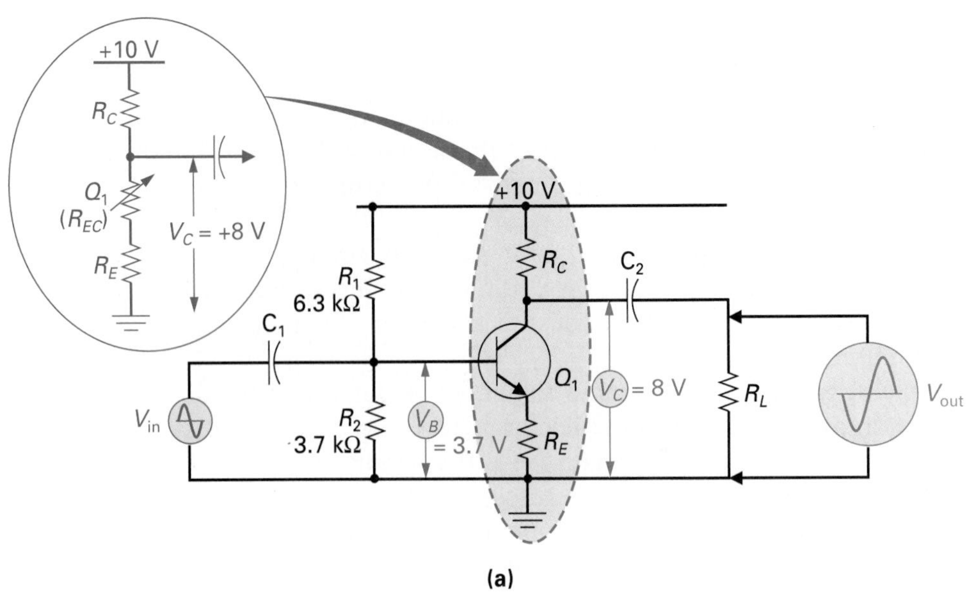

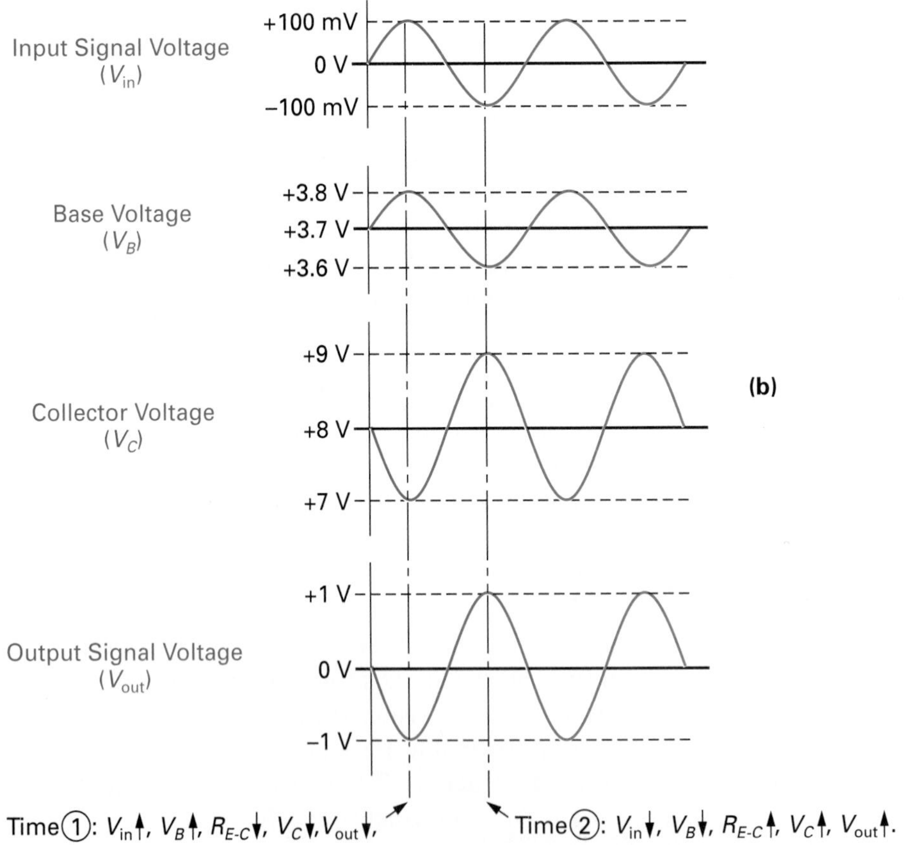

FIGURE 8-8 The Transistor Being Used in an Analog Electronic Circuit. (a) Basic Amplifier Circuit. (b) Input/Output Voltage Waveforms.

Because voltage drop is always proportional to resistance, an increase in $R_{EC}\uparrow$ will cause an increase in the voltage drop across R_{EC} and R_E ($V_C\uparrow$). This increase in V_C will be coupled to the output via C_2, causing an increase in the output voltage developed across the load ($V_{out}\uparrow$).

Comparing the input signal voltage (V_{in}) to the output signal voltage (V_{out}) in Figure 8-8(b), you can see that a change in the input signal voltage produces a corresponding greater change in the output signal voltage. The ratio (comparison) of output signal voltage change to input signal voltage change is a measure of this circuit's **voltage gain (A_v).** In this example, the **output signal voltage change (ΔV_{out})** is between +1 V and –1 V, and the **input signal voltage change (ΔV_{in})** is between +100 mV and –100 mV. The circuit's voltage gain between input and output will therefore be:

Voltage Gain (A_V)
The ratio of the output signal voltage change to input signal voltage change.

Output Signal Voltage Change
Change in output signal voltage in response to a change in the input signal voltage.

Input Signal Voltage Change
The input voltage change that causes a corresponding change in the output voltage.

$$\text{Voltage Gain } (A_v) = \frac{\text{Output Voltage Change } (\Delta V_{out})}{\text{Input Voltage Change } (\Delta V_{in})}$$

$$A_v = \frac{+1 \text{ to } -1 \text{ V}}{+100 \text{ mV to } -100 \text{ mV}} = \frac{2 \text{ V}}{200 \text{ mV}} = 10$$

A voltage gain of 10 means that the output voltage is ten times larger than the input voltage. The transistor does not produce this gain magically within its NPN semiconductor structure. The gain or amplification is achieved by the input signal controlling the conduction of the transistor, which takes energy from the collector supply voltage and develops this energy across the load resistor. Amplification is achieved by having a small input voltage control a transistor and its large collector supply voltage, so that a small input voltage change results in a similar but larger output voltage change.

Comparing the input signal to the output signal at time 1 and time 2 in Figure 8-8(b), you can see that this circuit will invert the input signal voltage in the same way that the NOT gate inverts its input voltage (positive input voltage swing produces a negative output voltage swing, and vice versa). This inversion always occurs with this particular transistor circuit arrangement; however, it is not a problem since the shape of the input signal is still preserved at the output (both input and output signals are sinusoidal).

This and other bipolar junction transistor amplifiers will be covered in greater detail in Chapter 10.

SELF-TEST REVIEW QUESTIONS FOR SECTION 8-1

1. What are the two basic types of bipolar transistor?
2. Name the three terminals of a bipolar transistor.
3. What are the two basic ways in which a transistor is made to operate?
4. Which of the modes of operation mentioned in question 3 is made use of in digital circuits and which is made use of in analog circuits?

8-2 SECOND APPROXIMATION DESCRIPTION OF A BIPOLAR TRANSISTOR

Now that we have a good understanding of the bipolar junction transistor's (BJTs) general characteristics, operation and applications, let us examine all of these aspects in a little more detail.

8-2-1 *Basic Bipolar Transistor Action*

When describing diodes previously, we saw how the P-N junction of a diode could be either forward or reverse biased to either permit or block the flow of current through the device. The transistor must also be biased correctly; however, in this case, two P-N junctions rather than one must have the correct external supply voltages applied.

A Correctly Biased NPN Transistor Circuit

Figure 8-9(a) shows how an NPN transistor should be biased for normal operation. In this circuit, a +10 V supply voltage is connected to the transistor's collector (C) via a 1 kΩ collector resistor (R_C). The emitter (E) of the transistor is connected to ground via a 1.5 kΩ emitter resistor (R_E), and, as an example, an input voltage of +3.7 V is being applied to the base (B). The output voltage (V_{out}) is taken from the collector, and this collector voltage (V_C) will be equal to the voltage developed across the transistor's collector-to-emitter and the emitter resistor R_E.

As previously mentioned in the first approximation description of the transistor, the transistor can be thought of as containing two diodes, as shown in Figure 8-9(b). In normal operation, *the transistor's emitter diode or junction is forward biased, while the transistor's collector diode or junction is reverse biased.* To explain how these junctions are biased ON and OFF simultaneously, let us see how the input voltage of +3.7 V will affect this transistor circuit. An input voltage of +3.7 V is large enough to overcome the barrier voltage of the emitter diode (base-emitter junction), and so it will turn ON (base or anode is +, emitter or cathode is connected to ground or 0 V). Like any forward biased silicon diode, the emitter diode will drop 0.7 V between base and emitter, and so the +3.7 V at the base will produce +3.0 V at the emitter. Knowing the voltage drop across the emitter resistor (V_{R_E} = 3 V) and the resistance of the emitter resistor (R_E = 1.5 kΩ), we can calculate the value of current through the emitter resistor.

$$I_{R_E} = \frac{V_{R_E}}{R_E} = \frac{3\text{ V}}{1.5\text{ k}\Omega} = 2\text{ mA}$$

This emitter resistor current of 2 mA will leave ground, travel through R_E, and then enter the transistor's *n*-type emitter region. This current at the transistor's emitter terminal is called the **emitter current (I_E).** The forward biased emitter diode will cause the steady stream of electrons entering the emitter to head toward the base region, as shown in the inset in Figure 8-9(b). The base is a very thin, lightly doped region with very few holes in relation to the number of elec-

Emitter Current (I_E) The current at the transistor's emitter terminal.

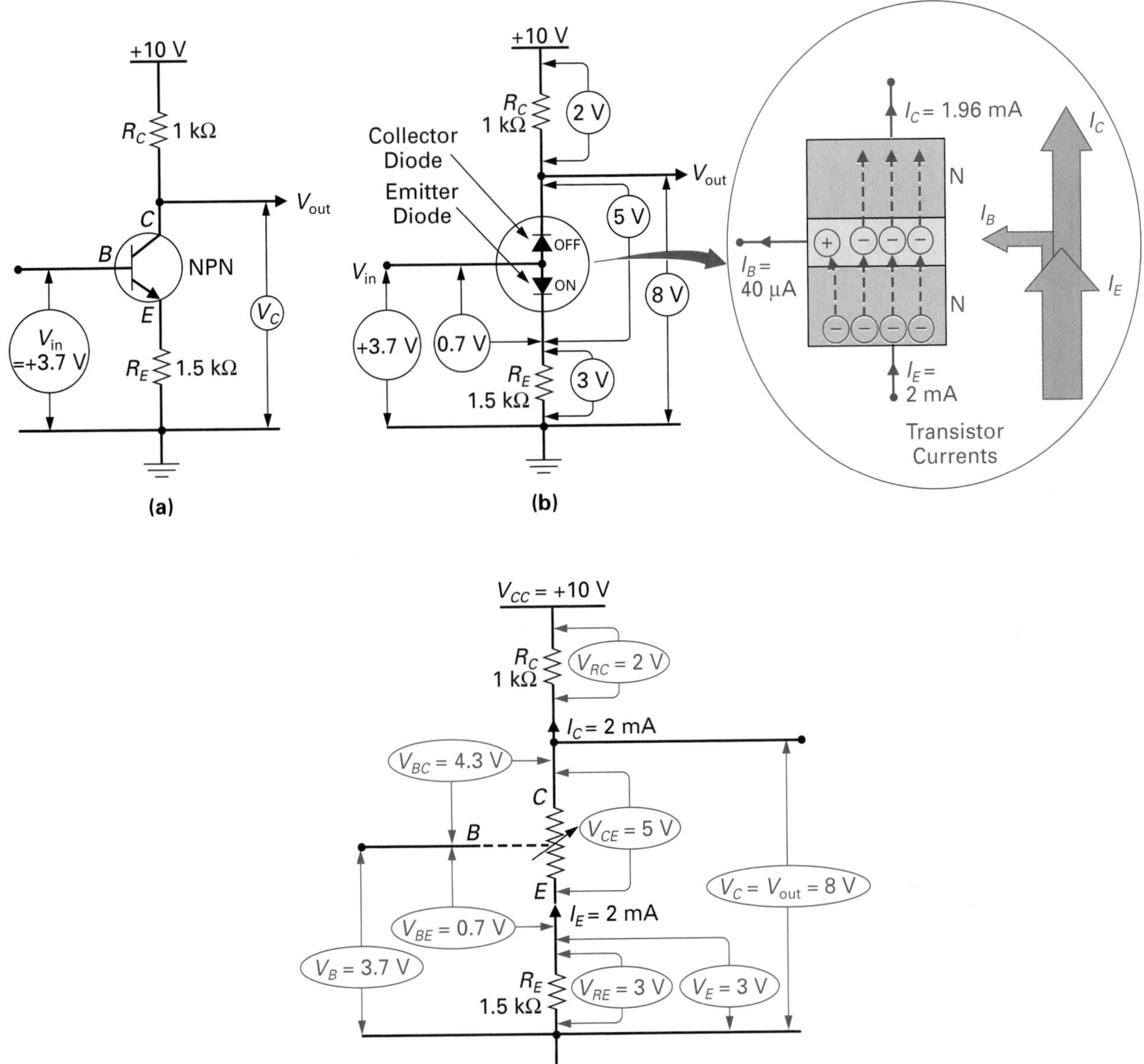

FIGURE 8-9 **A Correctly Biased NPN Transistor Circuit.**

trons entering the transistor from the emitter. Consequently, only a few electrons combine with the holes in the base region and flow out of the base region. This relatively small current at the transistor's base terminal is called the **base current (I_B).** Because only a few electrons combine with holes in the base region, there is an accumulation of electrons in the base's *p* layer. These free electrons, feeling the attraction of the large positive collector supply voltage (+10 V), will travel through the *n*-type collector junction and out of the transistor to the positive external collector supply voltage. The current emerging out of

Base Current (I_B) The relatively small current at the transistor's base terminal.

Collector Current (I_C)
The current emerging out of the transistor's collector.

the transistor's collector is called the **collector current (I_C).** Because both the collector current and base current are derived from the emitter current, we can state that:

$$I_E = I_B + I_C$$

In the example in the inset in Figure 8-9(b), you can see that this is true because

$$I_E = I_B + I_C$$
$$I_E = 40\ \mu\text{A} + 1.96\text{ mA} = 2\text{ mA}\quad (40\ \mu\text{A} = 0.04\text{ mA})$$

Stated another way, we can say that the collector current is equal to the emitter current minus the current that is lost out of the base.

$$I_C = I_E - I_B$$
$$I_C = 2\text{ mA} - 40\ \mu\text{A} = 1.96\text{ mA}$$

Approximately 98% of the electrons entering the emitter of a transistor will arrive at the collector. Because of the very small percentage of current flowing out of the base (I_B equals about 2% of I_E), we can approximate and assume that I_C is equal to I_E.

$$I_C \cong I_E$$

(I_C approximately equals I_E)

The Current-Controlled Transistor

In the previous section, we discovered that because the collector and base currents (I_C and I_B) are derived from the emitter current (I_E), an increase in the emitter current ($I_E\uparrow$), for example, will cause a corresponding increase in collector and base current ($I_C\uparrow$, $I_B\uparrow$). Looking at this from a different angle, an increase in the applied base voltage (base input increases to +3.8 V) will increase the forward bias applied to the emitter diode of the transistor, which will draw more electrons up from the emitter and cause an increase in I_E, I_B, and I_C. Similarly, a decrease in the applied base voltage (base input decreases to +3.6 V) will decrease the forward bias applied to the emitter diode of the transistor, which will decrease the number of electrons being drawn up from the emitter and cause a decrease in I_E, I_B, and I_C. The applied input base voltage will control the amount of base current, which will in turn control the amount of emitter and collector current, and therefore the conduction of the transistor. This is why *the bipolar transistor is known as a current-controlled device.*

Continuing our calculations for the example circuit in Figure 8-9(b), let us apply this current relationship and assume that I_C is equal to I_E, which, as we previously calculated, is equal to 2 mA. Knowing the value of current for the collector resistor (I_{R_C} = 2 mA) and the resistance of the collector resistor (R_C = 1 kΩ), we can calculate the voltage drop across the collector resistor.

$$V_{R_C} = I_{R_C} \times R_C = 2\text{ mA} \times 1\text{ k}\Omega = 2\text{ V}$$

With 2 V being dropped across R_C, the voltage at the transistor's collector (V_C) will be:

$$V_C = +10\text{ V} - V_{R_C} = 10\text{ V} - 2\text{ V} = 8\text{ V}$$

Because the voltage at the transistor's collector relative to ground is applied to the output, the output voltage will also be equal to 8 V.

$$V_C = V_{out} = 8\text{ V}$$

At this stage, we can determine a very important point about any correctly biased NPN transistor circuit. *A properly biased transistor will have a forward biased base-emitter junction (emitter diode is ON), and a reverse biased base-collector junction (collector diode is OFF).* We can confirm this with our example circuit in Figure 8-9(b), because we now know the voltages at each of the transistor's terminals.

Emitter diode (base-emitter junction) is forward biased (ON) because
Anode (base) is connected to +3.7 V (V_{in})
Cathode (emitter) is connected to 0 V via R_E.

Cathode diode (base-collector junction) is reverse biased (OFF) because
Anode (base) is connected to +3.7 V (V_{in})
Cathode (collector) is at +8 V (due to 2 V drop across R_C)

Keep in mind that even though the collector diode (base-collector junction) is reverse biased, current will still flow through the collector region. This is because most of the electrons traveling from emitter-to-base (through the forward biased emitter diode) do not find many holes in the thin, lightly doped base region, and therefore the base current is always very small. Almost 98% of the electrons accumulating in the base region feel the strong attraction of the positive collector supply voltage and flow up into the collector region and then out of the collector as collector current.

With the example circuit in Figure 8-9(a) and (b), the emitter diode is ON and the collector diode is OFF, and the transistor is said to be operating in its normal, or *active region.*

Operating a Transistor in the Active Region

A transistor is said to be in **active operation,** or in the **active region,** when its base-emitter junction is forward biased (emitter diode is ON), and the base-collector junction is reverse biased (collector diode is OFF). In this mode, the transistor is equivalent to a variable-resistor between collector and emitter.

Active Operation or in the Active Region
When the base-emitter junction is forward biased and the base-collector junction is reverse biased. In this mode, the transistor is equivalent to a variable-resistor between collector and emitter.

In Figure 8-9(c), our transistor circuit example has been redrawn with the transistor this time being shown as a variable-resistor between collector and emitter and with all of our calculated voltage and current values inserted. Before we go any further with this circuit, let us discuss some of the letter abbreviations used in transistor circuits. To begin with, the term V_{CC} is used to denote the "stable collector voltage" and this dc supply voltage will typically be positive for an NPN transistor. Two *C*s are used in this abbreviation ($+V_{CC}$) because V_C (V sub single C) is used to describe the voltage at the transistor's collector relative to ground. The doubling up of letters such as V_{CC}, V_{EE}, or V_{BB} is used to denote a constant dc bias voltage for the collector (V_{CC}), emitter (V_{EE}), and base (V_{BB}). A single sub letter abbreviation such as V_C, V_E, or V_B is used to denote a transistor terminal voltage relative to ground. The other voltage abbreviations, V_{CE}, V_{BE}, and V_{CB}, are used for the voltage difference between two terminals of the transistor. For example, V_{CE} is used to denote the potential difference between the transistor's collector and emitter terminals. Finally, I_E, I_B, and I_C are, as previously stated, used to denote the transistor's emitter current (I_E), base current (I_B), and collector current (I_C).

Because the transistor's resistance between emitter and collector in Figure 8-9(c) is in series with R_C and R_E, we can calculate the voltage drop between collector and emitter (V_{CE}) because V_{RC} and V_{RE} are known.

$$V_{CE} = V_{CC} - (V_{R_E} + V_{R_C})$$

$$V_{CE} = 10\text{ V} - (3\text{ V} + 2\text{ V}) = 10\text{ V} - 5\text{ V} = 5\text{ V}$$

Now that we know the voltage drop between the transistor's collector and emitter (V_{CE}), we can calculate the transistor's equivalent resistance between collector and emitter (R_{C_E}) because we know that the current through the transistor is 2 mA.

$$R_{CE} = \frac{V_{CE}}{I_C} = \frac{5\text{ V}}{2\text{ mA}} = 2.5\text{ k}\Omega$$

Operating the Transistor in Cutoff and Saturation

Figure 8-10 shows the three basic ways in which a transistor can be operated. As we have already discovered, the bias voltages applied to a transistor control the transistor's operation by controlling the two P-N junctions (or diodes) in a bipolar transistor. For example, the center column reviews how a transistor will operate in the active region. As you can see, our previous circuit example with all of its values has been used. To summarize: *when a transistor is operated in the active region, its emitter diode is biased ON, its collector diode is biased OFF, and the transistor is equivalent to a variable-resistor between the collector and the emitter.*

Cutoff
A transistor is in cutoff when the bias voltage is reduced to a point that it stops current in the transistor.

The left column in Figure 8-10 shows how the same transistor circuit can be driven into **cutoff.** A transistor is in cutoff when the bias voltage is reduced to a point that it stops current in the transistor. In this example circuit, you can see that when the base input bias voltage (V_B) is reduced to 0 V, the transistor is cut off. *In cutoff, both the emitter and the collector diode of the transistor will be biased OFF, the transistor is equivalent to an open switch between the collector and the emitter, and the transistor current is zero.*

Saturation
A transistor is in saturation when the bias voltage is increased to such a point that further increase will not cause any increase in current through the transistor.

The right column in Figure 8-10(c) shows how the same transistor circuit can be driven into **saturation.** A transistor is in saturation when the bias voltage is increased to such a point that any further increase in bias voltage will not cause any further increase in current through the transistor. In the equivalent circuit in Figure 8-10(c), you can see that when the base input bias voltage (V_B) is increased to +6.7 V, the emitter diode of the transistor will be heavily forward biased and the emitter current will be large.

$$I_E = \frac{V_B - V_{BE}}{R_E} = \frac{6.7\text{ V} - 0.7\text{ V}}{1.5\text{ k}\Omega} = 4\text{ mA}$$

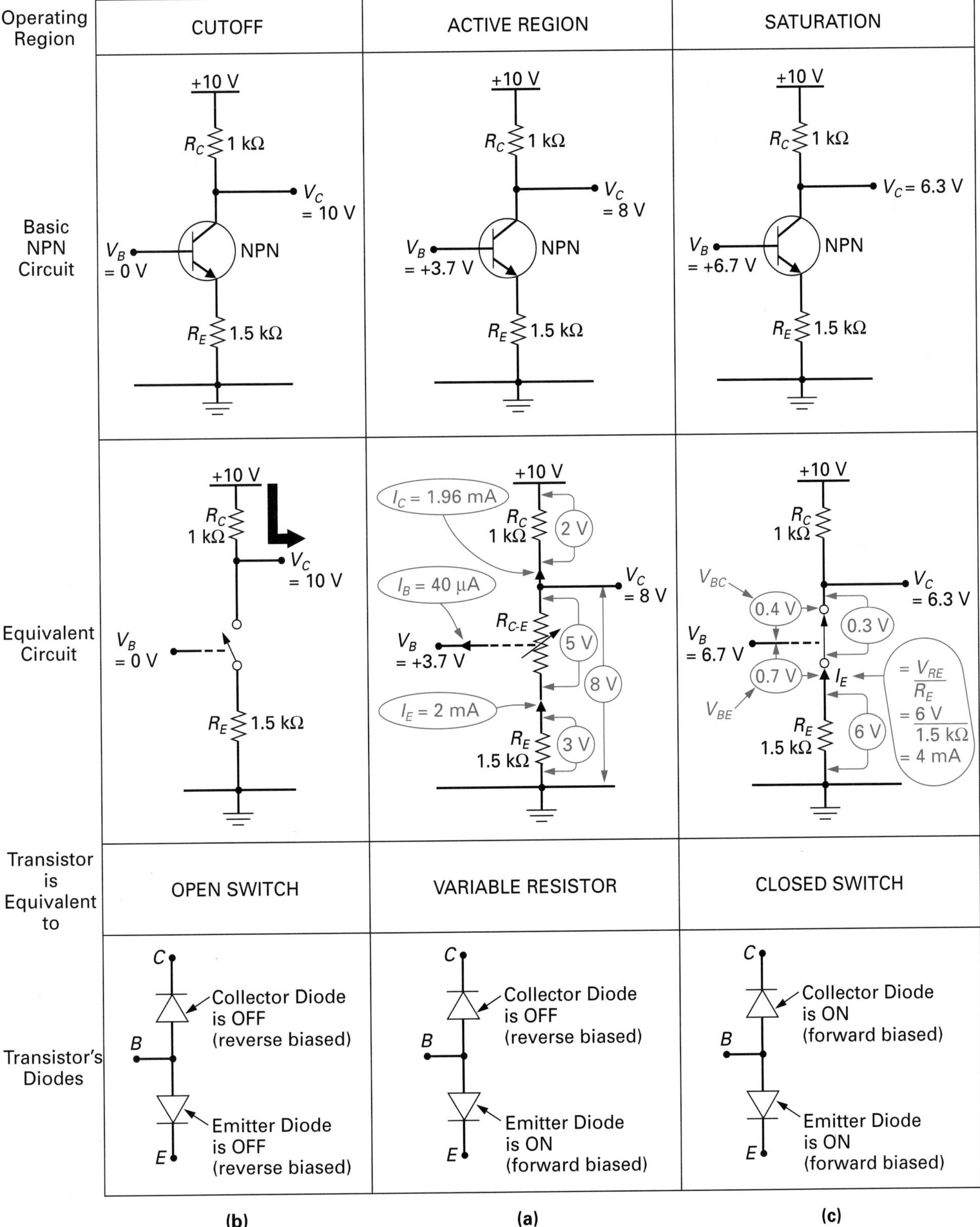

FIGURE 8-10 The Three Bipolar Transistor Operating Regions.

Because I_B and I_C are both derived from I_E, an increase in I_E will cause a corresponding increase in both I_B and I_C. These high values of current through the transistor account for why a transistor operating in saturation is said to be equivalent to a closed switch (high conductance, low resistance). Although the transistor's resistance between the collector and emitter (R_{CE}) is assumed to be 0 Ω, there is still some small value of R_{CE}. Typically, a saturated transistor will have a 0.3 V drop between the collector and emitter (V_{CE} = 0.3 V), as shown in the equivalent circuit in Figure 8-10(c). If V_{BE} = 0.7 V and V_{CE} = 0.3 V, then the voltage drop across the collector diode of a saturated transistor (V_{BC}) will be 0.4 V. This means that the base of the transistor (anode) is now +0.4 V relative to the collector (cathode), and there is not enough reverse bias voltage to turn OFF the collector diode. *In saturation both the emitter and collector diodes are said to be forward biased, the transistor is equivalent to a closed switch, and any further increase in bias voltage will not cause any further increase in current through the transistor.*

Biasing PNP Bipolar Transistors

Generally the PNP transistor is not employed as much as the NPN transistor in most circuit applications. The only difference that occurs with PNP transistor circuits is that the polarity of V_{CC} and the base bias voltage (V_B) need to be reversed to a negative voltage, as shown in Figure 8-11. The PNP transistor has the same basic operating characteristics as the NPN transistor, and all of the previously discussed equations still apply. Referring to the inset in Figure 8-11, you will see that the –3.7 V base bias voltage will forward bias the emitter diode, and the –10 V V_{CC} will reverse bias the collector diode, so that the transistor is operating in the active region. Also, the electron transistor currents are in the opposite direction. This, however, makes no difference because the fact that the sum of the collector current entering the collector and base current entering the base is equal to the value of emitter current leaving the emitter, so $I_E = I_B + I_C$ still applies.

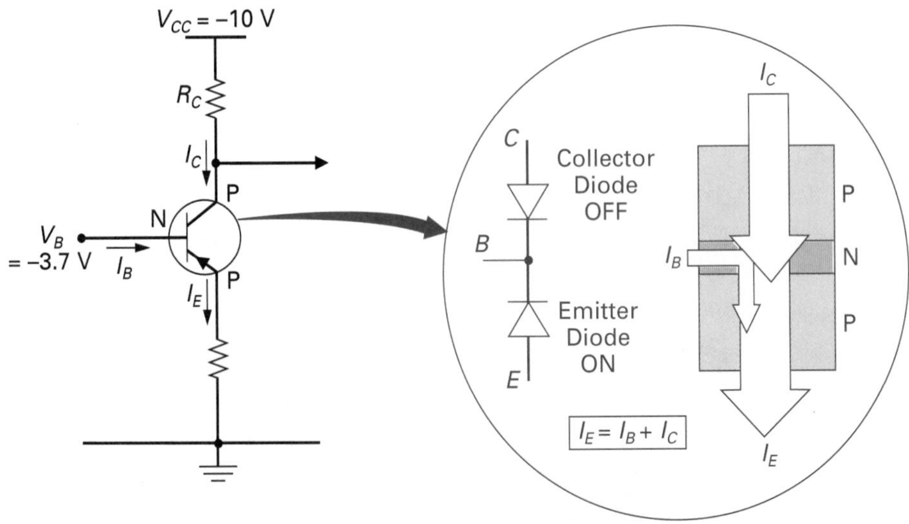

FIGURE 8-11 A Correctly Biased PNP Transistor Circuit.

8-2-2 *Bipolar Transistor Circuit Configurations and Characteristics*

In the previous sections, we have seen how the bipolar junction transistor can be used in digital two-state switching circuits and analog or linear circuits such as the amplifier. In all of these different circuit interconnections or **configurations,** the bipolar transistor was used as the main controlling element, with one of its three leads being used as a common reference and the other two leads being used as an input and an output. Although there are many thousands of different bipolar transistor circuit applications, all of these circuits can be classified in one of three groups based on which of the transistor's leads is used as the **common** reference. These three different circuit configurations are shown in Figure 8-12. With the *common-emitter (C-E)* bipolar transistor circuit configuration, shown in Figure 8-12(a), the input signal is applied between the base and emitter, while the output signal appears between the transistor's collector and emitter. With this circuit arrangement, the input signal controls the transistor's base current, which in turn controls the transistor's output collector current, and the emitter lead is common to both the input and output. Similarly, with the *common-base (C-B)* circuit configuration shown in Figure 8-12(b), the input signal is applied between the transistor's emitter and base, the output signal is developed across the transistor's collector and base, and the base is common to both input and output. Finally, with the *common-collector (C-C)* circuit configuration shown in Figure 8-12(c), the input is applied between the base and collector, the output is developed across the emitter and collector, and the collector is common to both the input and output.

Configurations
Different circuit interconnections.

Common
Shared by two or more services, circuits, or devices. Although the term "common ground" is frequently used to describe two or more connections sharing a common ground, the term "common" alone does not indicate a ground connection, only a shared connection.

To begin with, we will discuss the common-emitter circuit configuration characteristics because it has been this circuit arrangement that we have been using in all of the circuit examples in this chapter.

Common-Emitter Circuits

With the **common-emitter circuit,** the transistor's emitter lead is common to both the input and output signal. In this circuit configuration, the base serves as the input lead, and the collector serves as the output lead. Figure 8-13 con-

Common-Emitter (*C-E*) Circuit
Configuration in which the input signal is applied between the base and the emitter, while the output signal appears between the transistor's collector and emitter.

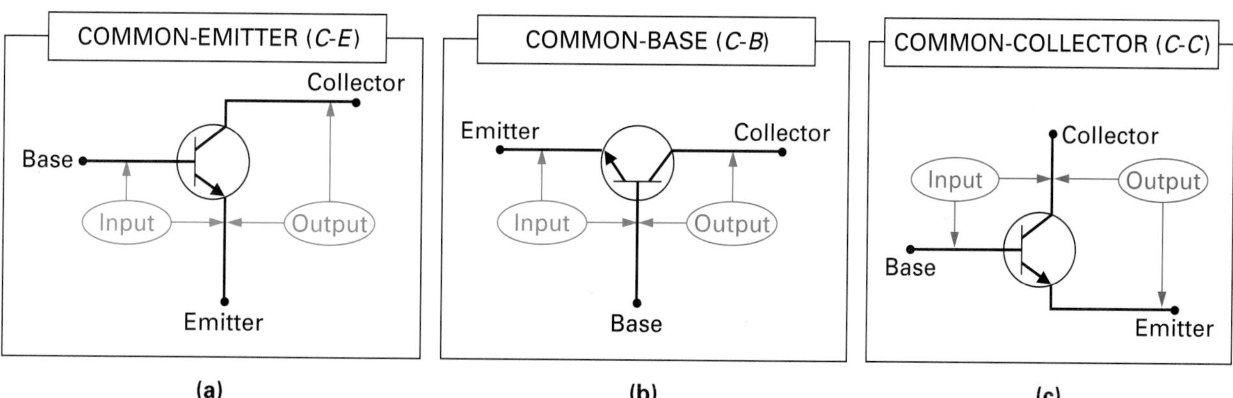

FIGURE 8-12 **Bipolar Transistor Circuit Configurations.**

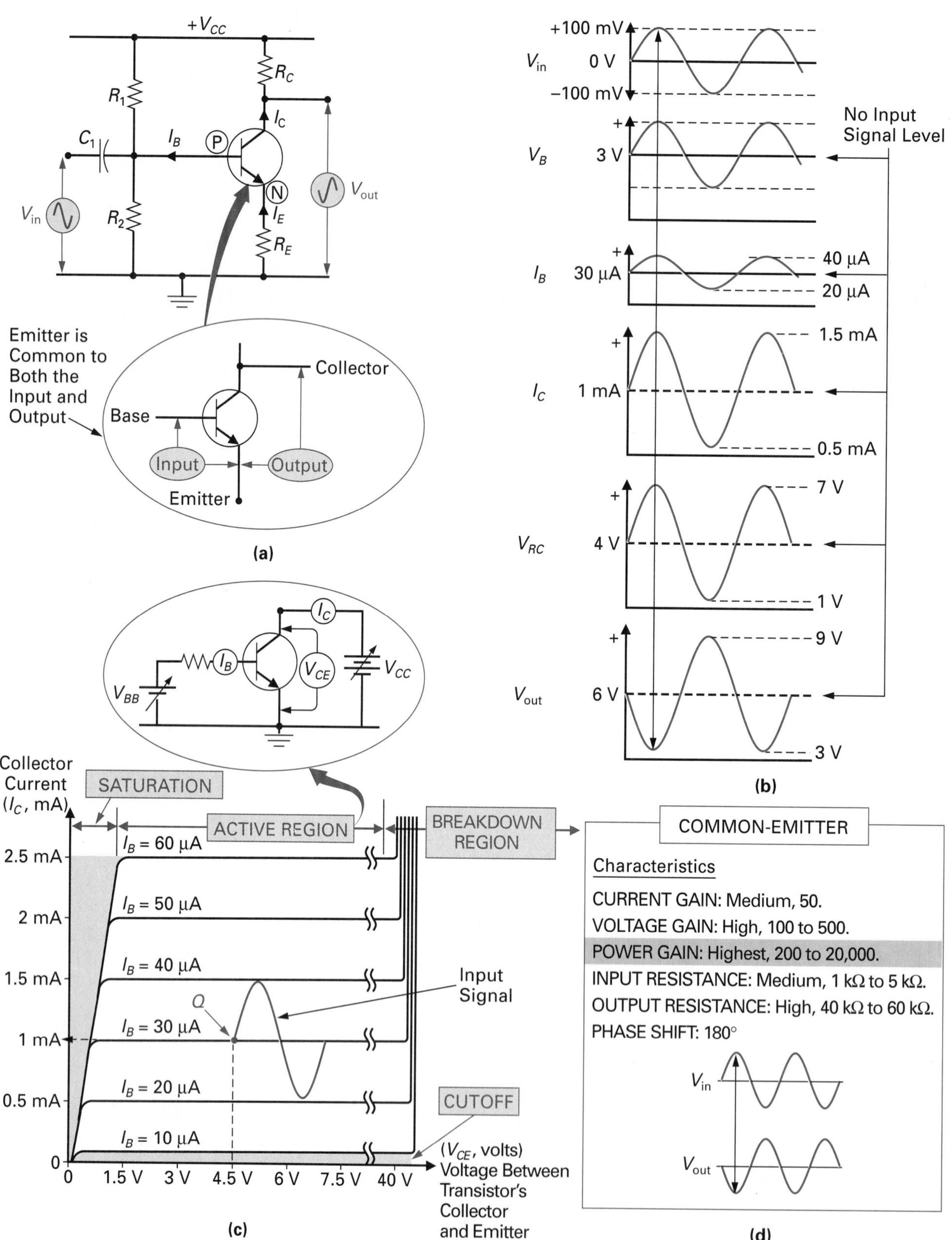

FIGURE 8-13 Common-Emitter (C-E) Circuit Configuration Characteristics.

tains a basic common-emitter (*C-E*) circuit, its associated input/output voltage and current waveforms, characteristic curves, and table of typical characteristics. Using this illustration, we will examine the operation and characteristics of the *C-E* circuit configuration.

DC current gain Referring to the C-E circuit in Figure 8-13(a), and its associated waveforms in Figure 8-13(b), let us now examine this circuit's basic operation.

Before applying the ac sine wave input signal (V_{in}), let us assume that V_{in} = 0 V and examine the transistor's dc operating characteristics, or the "no input signal" condition. The voltage divider R_1 and R_2 will divide the V_{CC} supply voltage, producing a positive dc base bias voltage across R_2. This base bias voltage will be applied to the base of Q_1, and because it is generally greater than 0.7 V, it will forward bias the transistor's base-emitter junction, turning Q_1 ON (in the example in Figure 8-13, V_B = 3 V dc). Capacitor C_1 is included to act as a dc block, preventing the base bias voltage (V_B) from being applied back to the input (V_{in}). The value of dc base bias voltage will determine the value of base current (I_B) flowing out of the transistor's base, and this value of I_B will in turn determine the value of collector current (I_C) flowing out of the transistor's collector and through R_C. Because the transistor's output current (I_C) is so much larger than the transistor's very small input current (I_B), the circuit produces an increase in current, or a **current gain.** The current gain in a common-emitter configuration is called the transistor's **beta** (symbolized β). A transistor's **dc beta** (β_{DC}) indicates a common-emitter transistor's "dc current gain," and it is the ratio of its output current (I_C) to its input current (I_B). This ratio can be expressed mathematically as:

Current Gain The increase in current produced by the transistor circuit.

Beta (β) The transistor's current gain in a common-emitter configuration.

DC Beta (β_{DC}) The ratio of a transistor's dc output current to its input current.

$$\beta_{DC} = \frac{I_C}{I_B}$$

■ EXAMPLE:

As can be seen in Figure 8-13(b), the "no input signal" level of I_C = 1 mA, and I_B = 30 μA. What is the transistor's dc beta?

■ *Solution:*

$$\beta_{DC} = \frac{I_C}{I_B}$$

$$= \frac{1 \text{ mA}}{30 \text{ μA}}$$

$$= 33.3$$

This value indicates that I_C is 33.3 times greater than I_B, therefore the dc current gain between input and output is 33.3.

Another form of system analysis, known as "hybrid parameters," uses the term *hfe* instead of β_{DC} to indicate a transistor's dc current gain.

AC current gain When amplifying an ac waveform, a transistor has applied to it both dc voltages to make it operational and the ac signal voltage to be amplified that varies the base bias voltage and the base current. Referring to the first four waveforms in Figure 8-13(b), let us now examine what will happen when we apply a sine wave input signal. As V_{in} increases above 0 V to a peak positive voltage ($V_{in}\uparrow$), it will cause the forward base-bias voltage applied to the transistor to increase ($V_B\uparrow$) above its dc reference or "no input signal level." This increase in V_B will increase the forward conduction of the transistor's emitter diode, resulting in an increase in the input base current ($I_B\uparrow$) and a corresponding larger increase in the output collector current ($I_C\uparrow$). The ratio of output collector current change (ΔI_C) to input base current change (ΔI_B) is the transistor's ac current gain or **AC beta (β_{AC})**. The formula for calculating a transistor's ac current gain is:

AC Beta (β_{AC}) The ratio of a transistor's ac output current to input current.

$$\beta_{AC} = \frac{\Delta I_C}{\Delta I_B}$$

■ EXAMPLE:

Calculate the ac current gain of the example in Figure 8-13(b).

■ *Solution:*

$$\beta_{AC} = \frac{\Delta I_C}{\Delta I_B}$$

$$= \frac{1.5 \text{ to } 0.5 \text{ mA}}{40 \text{ to } 20\ \mu\text{A}}$$

$$= \frac{1 \text{ mA}}{20\ \mu\text{A}}$$

$$= 50$$

This means that the alternating collector current at the output is 50 times greater than the alternating base current at the input.

Common-emitter transistors will typically have beta, or current gain values, of 50.

Voltage gain The common-emitter circuit is not only used to increase the level of current between input and output. It can also be used to increase the amplitude of the input signal voltage, or produce a **voltage gain.** This action can be seen by examining the C-E circuit in Figure 8-13(a) and by following the changes in the associated waveforms in Figure 8-13(b). An input signal voltage increase from 0 V to a positive peak ($V_{in}\uparrow$) causes an increase in the dc base bias voltage ($V_B\uparrow$), causing the emitter diode of Q_1 to turn more ON and result in an increase in both I_B and I_C. Because the I_C flows through R_C, and because voltage drop is proportional to current, an increase in I_C will cause an increase in the voltage drop across R_C ($V_{R_C}\uparrow$). The output voltage is equal to the voltage developed across Q_1's collector-to-emitter resistance (R_{CE}) and R_E. Because R_C is in

Voltage Gain (A_V) The ratio of the output signal voltage change to input signal voltage change.

series with Q_1's collector-to-emitter resistance (R_{CE}) and R_E, an increase in $V_{R_C}\uparrow$ will cause a decrease in the voltage developed across R_{CE} and R_E, which is $V_{out}\downarrow$. This action is summarized with Kirchhoff's voltage law which states that the sum of the voltages in a series circuit is equal to the voltage applied ($V_{R_C} + V_{out} = V_{CC}$). Using the example in Figure 8-13(b), you can see that when:

$V_{R_C}\uparrow$ to 7 V, $V_{out}\downarrow$ to 3 V. $\quad (V_{R_C} + V_{out} = V_{CC}, 7\text{ V} + 3\text{ V} = 10\text{ V})$

$V_{R_C}\downarrow$ to 1 V, $V_{out}\uparrow$ to 9 V. $\quad (V_{R_C} + V_{out} = V_{CC}, 1\text{ V} + 9\text{ V} = 10\text{ V})$

Although the input voltage (V_{in}) and output voltage (V_{out}) are out of phase with one another, you can see from the example values in Figure 8-13(b) that there is an increase in the signal voltage between input and output. This voltage gain between input and output is possible because the output current (I_C) is so much larger than the input current (I_B). The amount of **voltage gain** (which is symbolized $\boldsymbol{A_V}$) can be calculated by comparing the output voltage change (ΔV_{out}) to the input voltage change (ΔV_{in}).

$$A_V = \frac{\Delta V_{out}}{\Delta V_{in}}$$

■ **EXAMPLE:**

Calculate the voltage gain of the circuit and its associated waveforms in Figure 8-13(a) and (b).

■ ***Solution:***

$$A_V = \frac{\Delta V_{out}}{\Delta V_{in}} = \frac{+9\text{ V } to +3\text{ V}}{+100\text{ mV } to - 100\text{ mV}} = \frac{6\text{ V}}{200\text{ mV}} = 30$$

This value indicates that the output ac signal voltage is 30 times larger than the ac input signal voltage.

Most common-emitter transistor circuits have high voltage gains between 100 to 500.

Power gain As we have seen so far, the common-emitter circuit provides both current gain and voltage gain. Because power is equal to the product of current and voltage ($P = V \times I$), it is not surprising that the C-E circuit configuration also provides **power gain ($\boldsymbol{A_P}$)**. The power gain of a circuit can be calculated by dividing the output signal power (P_{out}) by the input signal power (P_{in}).

Power Gain ($\boldsymbol{A_P}$) The ratio of the output signal power to the input signal power.

$$A_P = \frac{P_{out}}{P_{in}}$$

To calculate the amount of input power (P_{in}) applied to the C-E circuit, we will have to multiply the change in input signal voltage (ΔV_{in}) by the accompanying change in input signal current (ΔI_{in} or ΔI_B).

$$P_{in} = \Delta V_{in} \times \Delta I_{in}$$

To calculate the amount of output power (P_{out}) delivered by the C-E circuit, we will have to multiply the change in output signal voltage (ΔV_{out}) produced by the change in output signal current (ΔI_{out} or ΔI_C).

$$P_{out} = \Delta V_{out} \times \Delta I_{out}$$

The power gain of a common-emitter circuit is therefore calculated with the formula:

$$A_P = \frac{P_{out}}{P_{in}} = \frac{\Delta V_{out} \times \Delta I_{out}}{\Delta V_{in} \times \Delta I_{in}}$$

■ **EXAMPLE:**

Calculate the power gain of the example circuit in Figure 8-13.

■ ***Solution:***

$$A_P = \frac{P_{out}}{P_{in}} = \frac{\Delta V_{out} \times \Delta I_{out}}{\Delta V_{in} \times \Delta I_{in}}$$

$$= \frac{(9\text{ V to }3\text{ V}) \times (1.5\text{ mA to }0.5\text{ mA})}{(+100\text{ mV to }-100\text{ mV}) \times (40\ \mu\text{A to }20\ \mu\text{A})}$$

$$= \frac{6\text{ V} \times 1\text{ mA}}{200\text{ mV} \times 20\ \mu\text{A}} = \frac{6\text{ mW}}{4\ \mu\text{W}} = 1500$$

In this example, the common-emitter circuit has increased the input signal power from 4 μW to 6 mW—a power gain of 1500.

The power gain of the circuit in Figure 8-13 can also be calculated by multiplying the previously calculated C-E circuit voltage gain (A_V) by the previously calculated C-E circuit current gain (β_{AC}).

$$\text{Since } A_V = \frac{\Delta V_{out}}{\Delta V_{in}},\ and\ \beta_{AC} = \frac{\Delta I_{out}\ (or\ \Delta I_C)}{\Delta I_{in}\ (or\ \Delta I_B)}$$

$$A_P = V \times I = \frac{\Delta V_{out}}{\Delta V_{in}} \times \frac{\Delta I_{out}\ (or\ \Delta I_C)}{\Delta I_{in}\ (or\ \Delta I_B)} \text{ or } A_P = A_V \times \beta_{AC}$$

$$A_P = A_V \times \beta_{AC}$$

For the example circuit in Figure 8-13, this will be:

$$A_P = A_V \times \beta_{AC} = 30 \times 50 = 1500$$

indicating that the common-emitter circuit's output power in Figure 8-13 is 1500 times larger than the input power.

The power gain of common-emitter transistor circuits can be as high as 20,000, making this characteristic the circuit's key advantage.

Collector characteristic curves One of the easiest ways to compare several variables is to combine all of the values in a graph. Figure 8-13(c) shows a special graph for a typical common-emitter transistor circuit called the **collector characteristic curves.** The data for this graph is obtained by using the transistor test circuit, shown in the inset in Figure 8-13(c), which will apply different values of base bias voltage (V_{BB}) and collector bias voltage (V_{CC}) to an NPN transistor. The two ammeters and one voltmeter in this test circuit are used to measure the circuit's I_B, I_C, and V_{CE} response to each different circuit condition. The values obtained from this test circuit are then used to plot the transistor's collector current (I_C in mA) in the vertical axis against the transistor's collector-emitter voltage (V_{CE}) drop in the horizontal axis for various values of base current (I_B in μA). This graph shows the relationship between a transistor's input base current, output collector current, and collector-to-emitter voltage drop.

Collector Characteristic Curves
Graph for a typical common-emitter transistor circuit.

Let us now examine the typical set of collector characteristic curves shown in Figure 8-13(c). When V_{CE} is increased from zero, by increasing V_{CC}, the collector current rises very rapidly, as indicated by the rapid vertical rise in any of the curves. When the collector diode of the transistor is reverse biased by the voltage V_{CE}, the collector current levels off. At this point, any one of the curves can be followed based on the amount of base current, which is determined by the value of base bias voltage (V_{BB}) applied. This flat part of the curve is known as the transistor's **active region.** The transistor is normally operated in this region, where it is equivalent to a variable-resistor between the collector and emitter.

Active Region
Flat part of the collector characteristic curve. A transistor is normally operated in this region, where it is equivalent to a variable-resistor between the collector and emitter.

As an example, let us use these curves in Figure 8-13(c) to calculate the value of output current (I_C) for a given value of input current (I_B) and collector-to-emitter voltage (V_{CE}). If V_{BB} is adjusted to produce a base current of 30 μA, and V_{CC} is adjusted until the voltage between the transistor's collector and emitter (V_{CE}) is 4.5 V, the output collector current (I_C) will be equal to approximately 1 mA. This is determined by first locating 4.5 V on the horizontal axis (V_{CE} = 4.5 V), following this point directly up to the I_B = 30 μA curve, and then moving directly to the left to determine the value of output current on the vertical axis (I_C = 1 mA). When these values of V_{CE} and I_B are present, the transistor is said to be operating at point "*Q*." This dc operating point is often referred to as a **quiescent operating point (*Q* point),** which means a dc steady-state or no-input signal operating point. The *Q* point of a transistor is set by the circuit's dc bias components and supply voltages. For instance, in the circuit in Figure 8-13(a), R_1 and R_2 were used to set the dc base bias voltage (V_B, and therefore I_B), and the value of V_{CC}, R_C, and R_E were chosen to set the transistor's dc collector-emitter voltage (V_{CE}). At this dc operating point, we can calculate the transistor's dc current gain *(β_{DC}),* since both I_B and I_C are known:

Quiescent Operating Point (*Q* point)
The voltage or current value that sets up the no-signal input or operating point bias voltage.

$$\beta_{DC} = \frac{I_C}{I_B} = \frac{1\text{ mA}}{30\text{ μA}} = 33.3$$

If a sine wave signal was applied to the circuit, as shown in the waveforms in Figure 8-13(b) and the characteristic curves in Figure 8-13(c), it would cause the transistor's input base current, and therefore output collector current, to alternate above and below the transistor's *Q* point (dc operating point). This ac input signal voltage will cause the input base current (I_B) to increase between 20 μA and 40 μA, and this input base current change will generate an output collector current change of 0.5 mA to 1.5 mA. The transistor's ac current gain (β_{AC}) will be:

$$\beta_{AC} = \frac{\Delta I_C}{\Delta I_B}$$

$$= \frac{1.5 \text{ to } 0.5 \text{ mA}}{40 \text{ to } 20 \text{ μA}}$$

$$= \frac{1 \text{ mA}}{20 \text{ μA}}$$

$$= 50$$

Breakdown Region
The point at which the collector supply voltage will cause a damaging value of current through the transistor.

Saturation Region
The point at which the collector supply voltage has the transistor operating at saturation.

Cutoff Region
The point at which the collector supply voltage has the transistor operating in cutoff.

Returning to the collector characteristic curve in Figure 8-13(c), you can see that if the collector supply voltage (V_{CC}) is increased to an extreme, a point will be reached where the V_{CE} voltage across the transistor will cause the transistor to break down, as indicated by the rapid rise in I_C. This section of the curve is called the **breakdown region** of the graph, and the damaging value of current through the transistor will generally burn out and destroy the device. As an example, for the 2N3904 bipolar transistor, breakdown will occur at a V_{CE} voltage of 40 V.

There are two shaded sections shown in the set of collector characteristic curves in Figure 8-13(c). These two shaded sections represent the other two operating regions of the transistor. To begin with, let us examine the vertically shaded **saturation region.** If the base bias voltage (V_{BB}) is increased to a large positive value, the emitter diode of the transistor will turn ON heavily, I_B will be a large value, and the transistor will be operating in saturation. In this operating region, the transistor is equivalent to a closed switch between its collector and emitter (both the emitter and collector diode are forward biased), and therefore the voltage drop between collector and emitter will be almost zero (V_{CE} = typically 0.3 V, when transistor is saturated), and I_C will be a large value that is limited only by the externally connected components. The horizontally shaded section represents the **cutoff region** of the transistor. If the base bias voltage (V_{BB}) is decreased to zero, the emitter diode of the transistor will turn OFF, I_B will be zero, and the transistor will be operating in cutoff. In this operating region, the transistor is equivalent to an open switch between its collector and emitter (both the emitter and collector diode are reverse biased), and the voltage drop between collector and emitter will be equal to V_{CC} and I_C will be zero.

A set of collector characteristic curves are therefore generally included in a manufacturer's device data sheet, and can be used to determine the values of I_B, I_C and V_{CE} at any operating point.

Input Resistance (R_{in})
The amount of opposition offered to an input signal by the input base-emitter junction (emitter diode).

Input resistance The **input resistance (R_{in})** of a common-emitter transistor is the amount of opposition offered to an input signal by the input base-emitter junction (emitter diode). Because the base-emitter junction is normally forward biased when the transistor is operating in the active region, the opposition to input current is relatively small. However, the extremely small base region will only support a very small input base current. On average, if no additional components are connected in series with the transistor's base-emitter junction, the input resistance of a C-E transistor circuit is typically a medium value between 1 kΩ to 5 kΩ. This typical value is an average because the transistor's input resistance is a "dynamic or changing quantity" that will vary slightly as the input signal changes the conduction of the C-E transistor's emitter diode, and this changes I_B ($R\updownarrow = V/I\updownarrow$).

Because the transistor has a small value of input P-N junction capacitance and input terminal inductance, the opposition to the input signal is not only resistive but, to a small extent, reactive. For this reason, the total opposition offered by the transistor to an input signal is often referred to as the **input impedance (Z_{in})** because impedance is the total combined resistive and reactive input opposition.

Input Impedance (Z_{in})
The total opposition offered by the transistor to an input signal.

Output resistance The **output resistance (R_{out})** of a common-emitter transistor is the amount of opposition offered to an output signal by the output base-collector junction (collector diode). This junction is normally reverse biased when the transistor is operating in the active region, and therefore the C-E transistor's output resistance is relatively high. However, because a unique action occurs within the transistor and allows current to flow through this reverse biased junction (electron accumulation at the base and then conduction through collector diode due to attraction of $+V_{CC}$), the output current (I_C) is normally large, and so the output resistance is not an extremely large value. On average, if no load resistor is connected in series with the transistor's collector diode, the output resistance of a C-E transistor circuit is typically a high value between 40 kΩ to 60 kΩ.

Output Resistance (R_{out})
The amount of opposition offered to an output signal by the output base-collector junction (collector diode).

Because the transistor has a small value of output P-N junction capacitance and output terminal inductance, the opposition to the output signal is not only resistive but also reactive. For this reason, the total opposition offered by the transistor to the output signal is often referred to as the **output impedance (Z_{out}).**

Output Impedance (Z_{out})
The total opposition offered by the transistor to the output signal.

Common-Base Circuits

With the **common-base circuit,** the transistor's base lead is common to both the input and output signal. In this circuit configuration, the emitter serves as the input lead, and the collector serves as the output lead. Figure 8-14 contains a basic common-base (*C-B*) circuit, its associated input/output voltage and current waveforms, and table of typical characteristics. Using this illustration, we will examine the operation and characteristics of the *C-B* circuit configuration.

Common-Base (*C-B*) Circuit
Configuration in which the input signal is applied between the transistor's emitter and base, while the output is developed across the transistor's collector and base.

DC current gain Referring to the *C-B* circuit in Figure 8-14(a), and its associated waveforms in Figure 8-14(b), let us now examine this circuit's basic operation.

Before applying the ac sine wave input signal (V_{in}), let us assume that V_{in} = 0 V and examine the transistor's dc operating characteristics, or the "no input signal" condition. The voltage divider R_1 and R_2 will divide the V_{CC} supply voltage, producing a positive dc base bias voltage across R_2. This base bias voltage will be applied to the base of Q_1. Because it is generally greater than 0.7 V, it will forward bias the transistor's base-emitter junction, turning Q_1 ON. Because the common-base circuit's input current is I_E and its output current is I_C, the current gain between input and output will be determined by the ratio of I_C to I_E. This

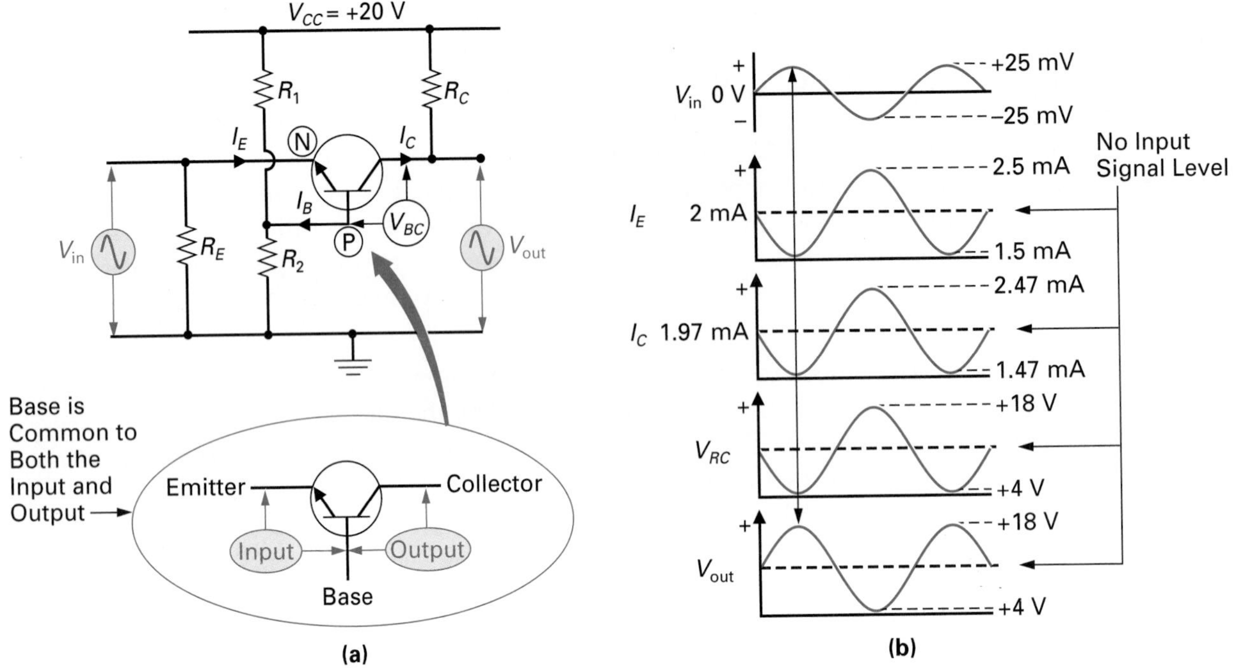

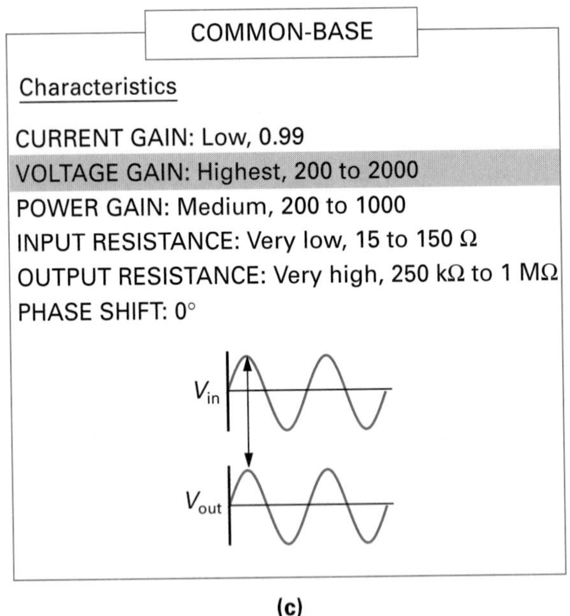

(c)

FIGURE 8-14 Common-Base (*C-B*) Circuit Configuration Characteristics.

ratio for calculating a *C-B* transistor's dc current gain is called the transistor's **DC alpha (α_{DC})** and is equal to

DC Alpha (α_{DC})
The ratio for calculating a *C-B* transistor's dc current gain.

$$\alpha_{DC} = \frac{I_C}{I_E}$$

The "no input signal," or "steady state" dc levels of I_E and I_C are determined by the value of voltage developed across R_2, which is controlling the conduction of the transistor's forward biased base-emitter junction (emitter diode). Because the output current I_C is always slightly lower than the input current I_E (due to the small I_B current flow out of the base), the C-B transistor circuit does not increase current between input and output. In fact, there is a slight loss in current between input and output, which is why the C-B circuit configuration is said to have a current gain that is less than 1.

■ **EXAMPLE:**

Calculate the dc alpha of the circuit in Figure 8-14(a), if I_C = 1.97mA and I_E = 2 mA.

■ ***Solution:***

$$\alpha_{DC} = \frac{I_C}{I_E} = \frac{1.97 \text{ mA}}{2 \text{ mA}} = 0.985$$

This value of 0.985 indicates that I_C is 98.5% of I_E ($0.985 \times 100 = 98.5$).

As you can see from this example, the difference between I_C and I_E is generally so small that we always assume that the dc alpha is 1, which means that $I_C = I_E$.

AC current gain When amplifying an ac waveform, a transistor has applied to it both dc voltages to make it operational and the ac signal voltage to be amplified that varies the base-emitter bias and the input emitter current. Referring to the waveforms in Figure 8-14(b), let us now examine what will happen when we apply a sine wave input signal. As mentioned previously, the positive voltage developed across R_2 will make the NPN transistor's base positive with respect to the emitter and forward bias the P-N base-emitter junction.

As the input voltage swings positive ($V_{in}\uparrow$), it will reduce the forward bias across the transistor's P-N base-emitter junction. For example, if V_B = +5 V and the transistor's *n*-type emitter is made positive, the P-N base-emitter diode will be turned more OFF. Turning the transistor's emitter diode less ON will cause a decrease in emitter current ($I_E\downarrow$), a decrease in collector current ($I_C\downarrow$), and a decrease in the voltage developed across R_C ($V_{R_C}\downarrow$). Because V_{R_C} and V_{out} are connected in series across V_{CC}, a decrease in $V_{R_C}\downarrow$ must be accompanied by an increase in $V_{out}\uparrow$. To explain this another way, the decrease in I_E and the subsequent decrease in both I_C and I_B means that the conduction of the transistor has decreased. This decrease in conduction means that the normally forward biased base-emitter junction has turned less ON, and the normally reverse biased base-collector junction has turned more OFF. Because the transistor's base-collector junction is in series with R_C and R_2 across V_{CC}, an increase in the transistor's base-collector resistance ($R_{BC}\uparrow$) will cause an increase in the voltage developed across the transistor's base-collector junction ($V_{BC}\uparrow$), which will cause an increase in $V_{out}\uparrow$.

Similarly, as the input voltage swings negative ($V_{in}\downarrow$), it will increase the forward bias across the transistor's P-N base-emitter junction. For example, if $V_B = -5$ V and the transistor's n-type emitter is made negative, the P-N base-emitter diode will be turned more ON. Turning the transistor's emitter diode more ON will cause an increase in emitter current ($I_E\uparrow$), an increase in collector current ($I_C\uparrow$), and an increase in the voltage developed across R_C ($V_{R_C}\uparrow$). Because V_{R_C} and V_{out} are connected in series across V_{CC}, an increase in $V_{R_C}\uparrow$ must be accompanied by a decrease in $V_{out}\downarrow$.

Now that we have seen how the input voltage causes a change in input current (I_E) and output current (I_C), let us examine the *C-B* circuit's ac current gain. The ratio of input emitter current change (ΔI_E) to output collector current change (ΔI_C) is the *C-B* transistor's ac current gain or **AC alpha (α_{AC})**. The formula for calculating a *C-B* transistor's ac current gain is:

AC Alpha (α_{AC}) The ratio of input emitter current change to output collector current change.

$$\alpha_{AC} = \frac{\Delta I_C}{\Delta I_E}$$

■ **EXAMPLE:**

Calculate the ac current gain of the example in Figure 8-14(b).

■ ***Solution:***

$$\begin{aligned}\alpha_{AC} &= \frac{\Delta I_C}{\Delta I_E} \\ &= \frac{2.47 \text{ to } 1.47 \text{ mA}}{2.5 \text{ to } 1.5 \text{ mA}} \\ &= \frac{1 \text{ mA}}{1 \text{ mA}} \\ &= 1\end{aligned}$$

This means that the change in output collector current is equal to the change in input emitter current, and therefore the ac current gain is 1. (The output is 1 times larger than the input, 1 mA × 1 = 1 mA.)

Common-base transistors will typically have an ac alpha, or ac current gain, of 0.99.

Voltage gain Although the common-base circuit does not achieve any current gain, it does make up for this disadvantage by achieving a very large voltage gain between input and output. Returning to the *C-B* circuit in Figure 8-14(a) and its waveforms in Figure 8-14(b), let us see how this very high voltage gain is obtained. Only a small input voltage (V_{in}) is needed to control the conduction of the transistor's emitter diode, and therefore the input emitter current (I_E) and output collector current (I_C). Even though I_C is slightly lower than I_E, it is still a relatively large value of current and will develop a large voltage change across R_C for a very small change in V_{in}. Because V_{R_C} and V_{out} are in series and connected across V_{CC}, a large change in voltage across R_C will cause a large change in the voltage developed across the transistor output (V_{BC}) and V_{out}. As before, the amount of **voltage gain** (which is symbolized $\boldsymbol{A_V}$) can be calculated by comparing the output voltage change (ΔV_{out}) to the input voltage change (ΔV_{in}).

$$A_V = \frac{\Delta V_{out}}{\Delta V_{in}}$$

■ **EXAMPLE:**

Calculate the voltage gain of the circuit and its associated waveforms in Figure 8-14(a) and (b).

■ ***Solution:***

$$A_V = \frac{\Delta V_{out}}{\Delta V_{in}} = \frac{+18\text{ V } \textit{to} +4\text{ V}}{+25\text{ mV } \textit{to} -25\text{ mV}} = \frac{14\text{ V}}{50\text{ mV}} = 280$$

This value indicates that the output ac signal voltage is 280 times larger than the ac input signal voltage.

Most common-base transistor circuits have very high voltage gains between 200 to 2000. While on the topic of comparing the input voltage to output voltage, you can see by looking at Figure 8-14(b) that, unlike the *C-E* circuit, the common-base circuit has no phase shift between input and output (V_{out} is in phase with V_{in}).

Power gain Although the common-base circuit achieves no current gain, it does have a very high voltage gain and therefore can provide a medium amount of power gain ($P\uparrow = V\uparrow \times I$). The power gain of a circuit can be calculated by dividing the output signal power (P_{out}) by the input signal power (P_{in}).

$$A_P = \frac{P_{out}}{P_{in}}$$

To calculate the amount of input power (P_{in}) applied to the *C-B* circuit, we will have to multiply the change in input signal voltage (ΔV_{in}) by the accompanying change in input signal current (ΔI_{in} or ΔI_E).

$$P_{in} = \Delta V_{in} \times \Delta I_{in}$$

$$P_{in} = 50\text{ mV} \times 1\text{ mA} = 50\ \mu\text{W}$$

To calculate the amount of output power (P_{out}) delivered by the C-B circuit, we will have to multiply the change in output signal voltage (ΔV_{out}) produced by the change in output signal current (ΔI_{out} or ΔI_C).

$$P_{out} = \Delta V_{out} \times \Delta I_{out}$$

$$P_{out} = 14 \text{ V} \times 1 \text{ mA} = 14 \text{ mW}$$

The power gain of a common-base circuit is calculated with the formula:

$$A_P = \frac{P_{out}}{P_{in}} = \frac{\Delta V_{out} \times \Delta I_{out}}{\Delta V_{in} \times \Delta I_{in}}$$

■ **EXAMPLE:**

Calculate the power gain of the example circuit in Figure 8-14.

■ ***Solution:***

$$A_P = \frac{P_{out}}{P_{in}} = \frac{\Delta V_{out} \times \Delta I_{out}}{\Delta V_{in} \times \Delta I_{in}}$$

$$= \frac{14 \text{ V} \times 1 \text{ mA}}{50 \text{ mV} \times 1 \text{ mA}} = \frac{14 \text{ mW}}{50 \text{ μW}} = 280$$

In this example, the common-base circuit has increased the input signal power from 50 μW to 14 mW—a power gain of 280.

The power gain of the circuit in Figure 8-14 can also be calculated by multiplying the previously calculated *C-B* circuit voltage gain (A_V) by the previously calculated *C-B* circuit current gain (α_{AC}).

$$A_P = A_V \times \alpha_{AC}$$

For the example circuit in Figure 8-14, this will be

$$A_P = A_V \times \alpha_{AC} = 280 \times 1 = 280$$

indicating that the common-base circuit's output power in Figure 8-14 is 280 times larger than the input power.

Typical common-base circuits will have power gains from 200 to 1000.

Input resistance The input resistance (R_{in}) of a common-base transistor is the amount of opposition offered to an input signal by the input base-emitter junction (emitter diode). Because the base-emitter junction is normally forward biased and the input emitter current (I_E) is relatively large, the input signal sees a very low input resistance. On average, if no additional components are connected in series with the transistor's base-emitter junction, the input resistance of a *C-B* transistor circuit is typically a low value between 15Ω to 150Ω. This typical value is an average because the transistor's input resistance is a dynamic or changing quantity that will vary slightly as the input signal changes the conduction of the *C-B* transistor's emitter diode, and this changes I_E ($R\updownarrow = V/I\updownarrow$).

Output resistance The output resistance (R_{out}) of a common-base transistor is the amount of opposition offered to an output signal by the output base-collector junction (collector diode). This junction is normally reverse biased when the transistor is operating in the active region, and therefore the *C-B* transistor's output resistance is relatively high. On average, if no load resistor is connected in series with the transistor's collector diode, the output resistance of a *C-B* transistor circuit is typically a very high value between 250 kΩ to 1 MΩ.

Common-Collector Circuits

Common-Collector (*C-C*) Circuit Configuration in which the input signal is applied between the transistor's base and collector, while the output is developed across the transistor's collector and emitter.

With the **common-collector circuit,** the transistor's collector lead is common to both the input and output signal. In this circuit configuration therefore, the base serves as the input lead, and the emitter serves as the output lead. Figure 8-15 contains a basic common-collector (*C-C*) circuit, its associated input/output voltage and current waveforms, and table of typical characteristics. Using this illustration, we will examine the operation and characteristics of the *C-C* circuit configuration.

DC current gain Referring to the *C-C* circuit in Figure 8-15(a) and its associated waveforms in Figure 8-15(b), let us now examine this circuit's basic operation.

Before applying the ac sine wave input signal (V_{in}), let us assume that V_{in} = 0 V and examine the transistor's dc operating characteristics, or the "no input signal" condition. The voltage divider R_1 and R_2 will divide the V_{CC} supply voltage, producing a positive dc base bias voltage across R_2. This base bias voltage will be applied to the base of Q_1 and, because it is generally greater than 0.7 V, it will forward bias the transistor's base-emitter junction, turning Q_1 ON. Because the common-collector (*C-C*) circuit's input current I_B is much smaller than the output current I_E, the circuit provides a high current gain. In fact, the common-collector circuit provides a slightly higher gain than the *C-E* circuit because the common-collector's output current (I_E) is slightly higher than the *C-E*'s output current (I_C).

Like any circuit configuration, the dc current gain is equal to the ratio of output current to input current. For the *C-C* circuit, this is equal to the ratio of I_E to I_B:

$$\text{DC Current Gain} = \frac{I_E}{I_B}$$

Transistor manufacturers will generally not provide specifications for all three circuit configurations. In most cases, because the common-emitter (*C-E*) circuit configuration is most frequently used, manufacturers will give the transistor's characteristics for only the *C-E* circuit configuration. In these instances, we will have to convert this *C-E* circuit data to equivalent specifications for other

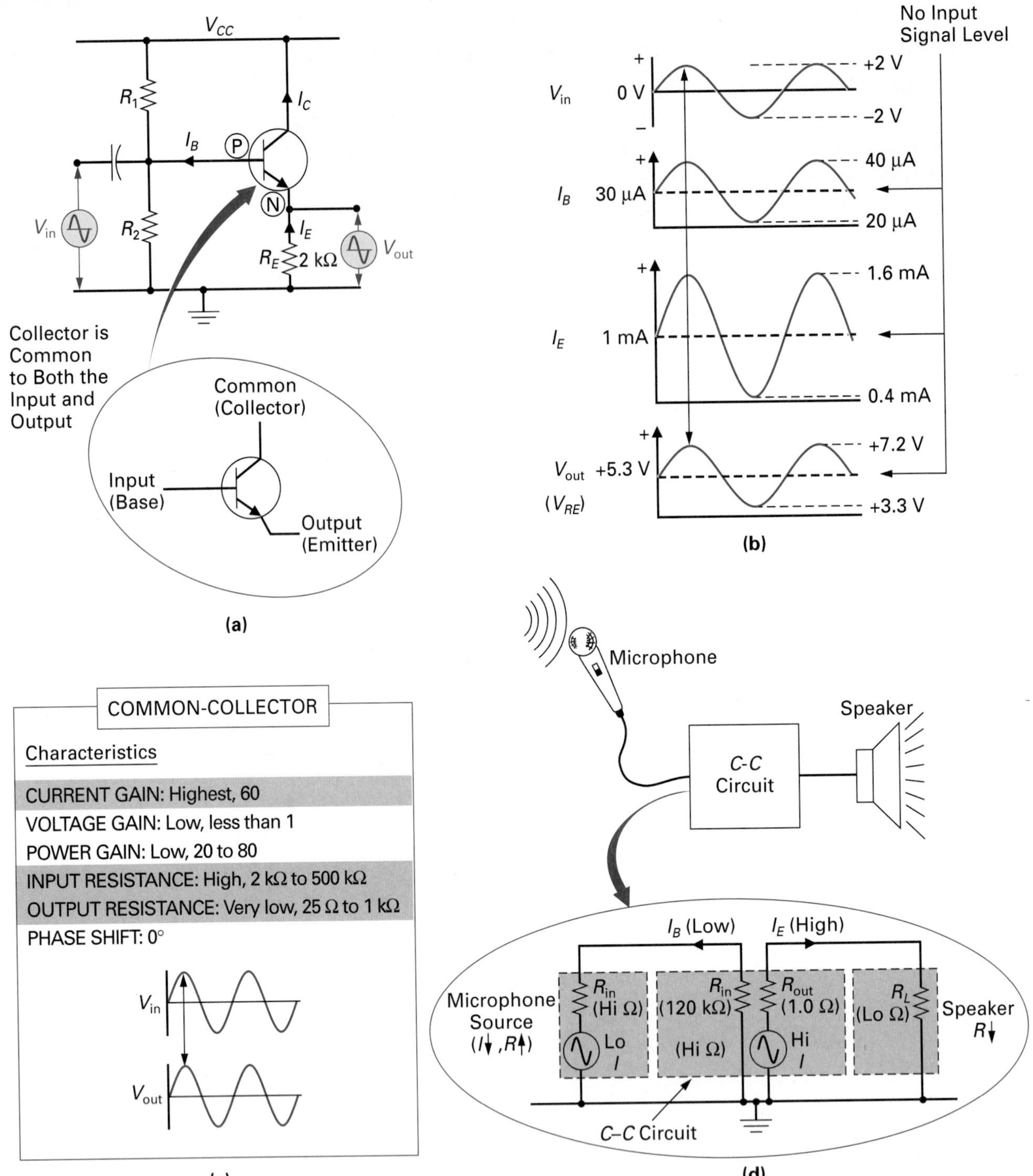

FIGURE 8-15 Common-Collector (C-C) Circuit Configuration Characteristics.

configurations. For example, in most data sheets the transistor's dc current gain will be listed as β_{DC}. As we know, dc beta is the measure of a *C-E* circuit's current gain because it compares input current I_B to output current I_C. How then, can we convert this value so that it indicates the dc current gain of a common-collector circuit? The answer is as follows:

$$\text{Common} = \text{Collector DC Current Gain} = \frac{\text{Output Current}}{\text{Input Current}} = \frac{I_E}{I_B}$$

Since $I_E = I_B + I_C$,

$$\text{DC Current Gain} = \frac{I_E}{I_B} = \frac{(I_B + I_C)}{I_B}$$

Since $I_B \div I_B = 1$,

$$\text{DC Current Gain} = 1 + \frac{I_C}{I_B}$$

Since $\frac{I_C}{I_B} = \beta_{DC}$,

$$\text{DC Current Gain} = 1 + \frac{I_C}{I_B} = 1 + \beta_{DC}$$

$$\text{DC Current Gain} = 1 + \beta_{DC}$$

■ **EXAMPLE:**

Calculate the dc current gain of the *C-C* circuit in Figure 8-15, if the transistor's $\beta_{DC} = 32.33$.

■ ***Solution:***

$$\text{DC Current Gain} = 1 + \frac{I_C}{I_B} = 1 + \frac{(I_E - I_B)}{I_B} = 1 + \frac{1\text{ mA} - 30\ \mu\text{A}}{30\ \mu\text{A}}$$

$$= 1 + \frac{970\ \mu\text{A}}{30\ \mu\text{A}} = 1 + 32.33 = 33.33$$

or,

$$\text{DC Current Gain} = 1 + \beta_{DC} = 1 + 32.33 = 33.33$$

As you can see in this example, the dc current gain of a common-collector circuit ($\beta_{DC} + 1$) is slightly higher than the dc current gain of a *C-E* circuit (β_{DC}). In most instances, the extra 1 makes so little difference when the transistor's dc current gain is a large value of about 30, as in this example, that we assume that the current gain of a common-collector circuit is equal to the current gain of a *C-E* circuit.

$$\text{C-C DC Current Gain} \cong \textit{C-E}\text{ DC Current Gain } (\beta_{DC})$$

AC current gain When amplifying an ac waveform, a transistor has applied to it both dc voltages to make it operational and the ac signal voltage that varies the base-emitter bias and the input base current. Referring to the waveforms in Figure 8-15(b), let us now examine what will happen when we apply a sine wave input signal. As mentioned previously, the positive voltage developed across R_2 will make the NPN transistor's base positive with respect to the emitter, and therefore forward bias the P-N base-emitter junction.

As the input voltage swings positive ($V_{in}\uparrow$), it will add to the forward bias applied across the transistor's P-N base-emitter junction. This means that the transistor's emitter diode will turn more ON, cause an increase in the $I_B\uparrow$, and therefore a proportional but much larger increase in the output current $I_E\uparrow$.

Similarly, as the input voltage swings negative ($V_{in}\downarrow$), it will subtract from the forward bias applied across the transistor's P-N base-emitter junction. This means that the transistor's emitter diode will turn less ON, cause a decrease in the $I_B\downarrow$, and therefore a proportional but larger decrease in the output current $I_E\downarrow$.

The ac current gain of a common-collector transistor is calculated using the same formula as dc current gain. However, with an ac current, we will compare the output current change (ΔI_E) to the input current change (ΔI_B).

$$\text{AC Current Gain} = \frac{\Delta I_E}{\Delta I_B}$$

■ **EXAMPLE:**

Calculate the ac current gain of the circuit in Figure 8-15(a), using the values in Figure 8-15(b).

■ ***Solution:***

$$\text{AC Current Gain} = \frac{\Delta I_E}{\Delta I_B} = \frac{1.6\text{ mA} - 0.4\text{ mA}}{40\ \mu\text{A} - 20\ \mu\text{A}} = 60$$

Like the common-collector's dc current gain, because there is so little difference between I_E and I_C, we can assume that the ac current gain is equivalent to β_{AC}.

$$C\text{-}C\text{ AC Current Gain} \cong C\text{-}E\text{ AC Current Gain } (\beta_{AC})$$

Common-collector transistor circuit configurations can have current gains as high as 60, indicating that I_E is sixty times larger than I_B.

Voltage gain Although the common-collector circuit has a very high current gain rating, it cannot increase voltage between input and output. Returning to the *C-C* circuit in Figure 8-15(a) and its waveforms in Figure 8-15(b), let us see why this circuit has a very low voltage gain.

As the input voltage swings positive ($V_{in}\uparrow$), it will add to the forward bias applied across the transistor's P-N base-emitter junction ($V_{BE}\uparrow$). As the transistor's emitter diode turns more ON, it will cause an increase in $I_B\uparrow$, a proportional but larger increase in $I_E\uparrow$, and therefore an increase in the voltage developed across R_E (V_{R_E}, V_{out}, or $V_E\uparrow$). This increase in the voltage developed across R_E has a **degenerative effect** because an increase in the emitter voltage ($V_E\uparrow$) will counter the initial increase in base voltage ($V_B\uparrow$), and therefore the voltage difference between the transistor's base and emitter will remain almost constant (V_{BE} is almost constant). In other words, if the base goes positive and then the emitter goes positive, there is almost no increase in the potential difference between the base and the emitter and so the change in forward bias is almost zero. There is, in fact, a very small change in forward bias between base and emitter, and this will cause a small change in I_B and I_E, and therefore a small output voltage will be developed across R_E. Comparing the input and output voltage signals in Figure 8-15(b), you can see that both are about 4 V pk-pk, and both are in phase with one another. The common-collector circuit is often referred to as an **emitter-follower** or **voltage-follower** because the emitter output voltage seems to track or follow the phase and amplitude of the input voltage.

Degenerative Effect
An effect that causes a reduction in amplification due to negative feedback.

Emitter-Follower or Voltage-Follower
The common-collector circuit in which the emitter output voltage seems to track or follow the phase and amplitude of the input voltage.

As with all circuit configurations, the amount of **voltage gain (A_V)** can be calculated by comparing the output voltage change (ΔV_{out}) to the input voltage change (ΔV_{in}).

$$A_V = \frac{\Delta V_{out}}{\Delta V_{in}}$$

■ **EXAMPLE:**

Calculate the voltage gain of the circuit and its associated waveforms in Figure 8-15(a) and (b).

■ ***Solution:***

$$A_V = \frac{\Delta V_{out}}{\Delta V_{in}} = \frac{+7.2\text{ V to }+3.3\text{ V}}{+2\text{ V to }-2\text{ V}} = \frac{3.9\text{ V}}{4\text{ V}} = 0.975$$

This value indicates that the output ac signal voltage is 0.975 or 97.5% of the ac input signal voltage ($0.975 \times 4\text{ V} = 3.9\text{ V}$).

Most common-collector transistor circuits have a voltage gain that is less than 1. However, in most circuit examples it is assumed that output voltage change equals input voltage change.

Power gain Although the common-collector circuit achieves no voltage gain, it does have a very high current gain, and therefore can provide a medium amount of power gain ($P\uparrow = V \times I\uparrow$). As before, the power gain of a circuit can be calculated by dividing the output signal power (P_{out}) by the input signal power (P_{in}).

$$A_P = \frac{P_{out}}{P_{in}}$$

To calculate the amount of input power (P_{in}) applied to the *C-C* circuit, we will have to multiply the change in input signal voltage (ΔV_{in}) by the accompanying change in input signal current (ΔI_{in} or ΔI_B).

$$P_{in} = \Delta V_{in} \times \Delta I_{in}$$

$$P_{in} = 4 \text{ V} \times 20 \text{ µA} = 80 \text{ µW}$$

To calculate the amount of output power (P_{out}) delivered by the *C-C* circuit, we will have to multiply the change in output signal voltage (ΔV_{out}) produced by the change in output signal current (ΔI_{out} or ΔI_E).

$$P_{out} = \Delta V_{out} \times \Delta I_{out}$$

$$P_{out} = 3.9 \text{ V} \times 1.2 \text{ mA} = 4.68 \text{ mW}$$

The power gain of a common-collector circuit is therefore calculated with the formula:

$$A_P = \frac{P_{out}}{P_{in}} = \frac{\Delta V_{out} \times \Delta I_{out}}{\Delta V_{in} \times \Delta I_{in}}$$

■ **EXAMPLE:**

Calculate the power gain of the example circuit in Figure 8-15.

■ ***Solution:***

$$A_P = \frac{P_{out}}{P_{in}} = \frac{\Delta V_{out} \times \Delta I_{out}}{\Delta V_{in} \times \Delta I_{in}}$$

$$= \frac{3.9 \text{ V} \times 1.2 \text{ mA}}{4 \text{ V} \times 20 \text{ µA}} = \frac{4.68 \text{ mW}}{80 \text{ µW}} = 58.5$$

In this example, the common-collector circuit has increased the input signal power from 80 µW to 4.68 mW—a power gain of 58.5.

The power gain of the circuit in Figure 8-15 can also be calculated by multiplying the previously calculated C-C circuit voltage gain (A_V) by the previously calculated C-C circuit current gain.

$$A_P = A_V \times \text{AC Current Gain}$$

For the example circuit in Figure 8-15, this will be

$$A_P = A_V \times \text{AC Current Gain} = 0.975 \times 60 = 58.5$$

indicating that the common-collector circuit's output power in Figure 8-15 is 58.5 times larger than the input power.

Typical common-collector circuits will have power gains from 20 to 80.

Input resistance An input signal voltage will see a very large input resistance when it is applied to a common-collector circuit. This is because the input signal sees the very large emitter connected resistor ($R_E\uparrow\uparrow$) and, to a smaller extent, the resistance of the forward biased base-emitter junction ($R_{in}\uparrow = V_{in}/I_B\downarrow$: R_{in} is large because I_{in} or I_B is small). Using these two elements, we can derive a formula for calculating the input resistance of a C-C transistor circuit.

$$R_{in} = R_E \times \text{AC Current Gain}$$

Since

$$C\text{-}C \text{ AC Current Gain} \cong C\text{-}E \text{ AC Current Gain } (\beta_{AC})$$

the input resistance can also be calculated with the formula:

$$R_{in} = R_E \times \beta_{AC}$$

■ **EXAMPLE:**

Calculate the input resistance of the circuit in Figure 8-15, assuming $\beta_{AC} = 60$.

■ ***Solution:***

$$R_{in} = R_E \times \beta_{AC} = 2\text{ k}\Omega \times 60 = 120\text{ k}\Omega$$

This means that an input voltage signal will see this *C-C* circuit as a resistance of 120 kΩ.

The input resistance of a C-C transistor circuit is typically a very large value between 2 kΩ to 500 kΩ.

Output resistance The output signal from a common-collector circuit sees a very low output resistance, as proved by this circuit's very high output current gain. Like this circuit's input resistance, the output resistance is largely dependent on the value of the emitter resistor R_E.

The output resistance of a typical C-C transistor circuit is a very low value between 25 Ω to 1 kΩ.

Impedance or resistance matching Do not be misled into thinking that the very high input resistance and low output resistance of the common-collector transistor circuit are disadvantages. On the contrary, the very high input resistance and low output resistance of this configuration are made use of in many circuit applications, along with the *C-C* circuit's other advantage of high current gain.

To explain why a high input resistance and a low output resistance are good circuit characteristics, refer to the application circuit in Figure 8-15(d). In this example, a microphone is connected to the input of a *C-C* amplifier, and the output of this circuit is applied to a speaker. As we know, the sound wave input to the microphone will physically move a magnet within the microphone, which will in turn interact to induce a signal voltage into a stationary coil. This voice signal voltage from the microphone, which is our source, is then applied across the input resistance of our example *C-C* circuit, which is our load.

In the inset in Figure 8-15(d), you can see that the microphone has been represented as a low-current ac source with a high internal resistance, and the input resistance of the *C-C* circuit is shown as a high value (in the previous example, 120 kΩ) resistor. Remembering our previous discussion on sources and loads from Chapter 5, we know that a small load resistance will cause a large current to be drawn from the source, and this large current will drain or pull down the source voltage. Many signal sources, such as microphones, can only generate a small signal source voltage because they have a high internal resistance. If this small signal source voltage is applied across an amplifier with a small input resistance, a large current will be drawn from the source. This heavy load will pull the signal voltage down to such a small value that it will not be large enough to control the amplifier. A large amplifier input resistance ($R_{in}\uparrow$), on the other hand, will not load the source. Therefore the input voltage applied to the amplifier will be large enough to control the amplifier circuit, to vary its transistor currents, and achieve the gain between amplifier input and output. In summary, the high input resistance of the *C-C* circuit can be connected to a high resistance source because it will not draw an excessive current and pull down the source voltage.

Referring again to the inset in Figure 8-15(d), you can see that at the output end, the *C-C*'s output circuit has been represented as a high current source with a low value internal output resistor, and the speaker has been represented as a low resistance load. The low output resistance of the *C-C* circuit means that this circuit can deliver the high current output that is needed to drive the low resistance load.

Impedance Matching Circuit
A circuit that can match, or isolate, a high resistance (low current) source.

Buffer Current Amplifier
The *C-C* circuit that can ensure that power is efficiently transferred from source to load.

As you will see later in application circuits, most *C-C* circuits are used as a resistance or **impedance matching circuit** that can match, or isolate, a high resistance (low current) source, such as a microphone, to a low resistance (high current) load, such as the speaker. By acting as a **buffer current amplifier,** the *C-C* circuit can ensure that power is efficiently transferred from source to load.

8-2-3 Bipolar Junction Transistor Data Sheets

Like the diode's data sheets, manufacturer's bipolar transistor data sheets list the typical dc and ac operating characteristics of the device. To illustrate some of the many different types of transistors available,

a. Figure 8-16 shows the data sheet of a typical *general purpose switching or amplifying bipolar transistor.*
b. Figure 8-17 shows the data sheet of a typical *high-voltage bipolar transistor.*
c. Figure 8-18 shows the data sheet of a typical *high-current bipolar transistor.*
d. Figure 8-19 (p. 377) shows the data sheet of a typical *high-power bipolar transistor.*

As before, notes are included in these data sheets to call out important characteristics and to explain some of the terms that have not been previously used.

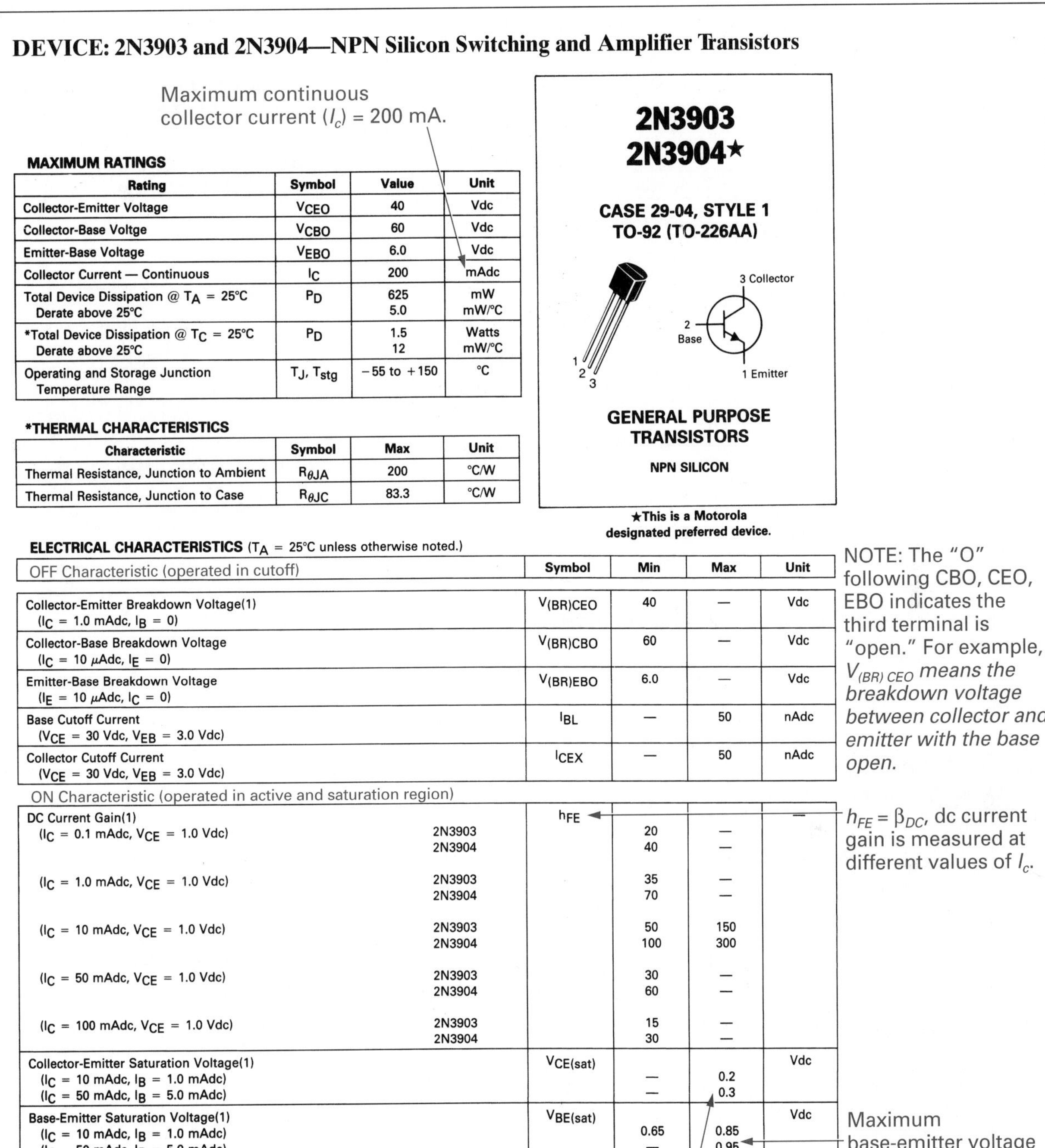

DEVICE: 2N3903 and 2N3904—NPN Silicon Switching and Amplifier Transistors

Maximum continuous collector current (I_c) = 200 mA.

MAXIMUM RATINGS

Rating	Symbol	Value	Unit
Collector-Emitter Voltage	V_{CEO}	40	Vdc
Collector-Base Voltge	V_{CBO}	60	Vdc
Emitter-Base Voltage	V_{EBO}	6.0	Vdc
Collector Current — Continuous	I_C	200	mAdc
Total Device Dissipation @ T_A = 25°C Derate above 25°C	P_D	625 5.0	mW mW/°C
*Total Device Dissipation @ T_C = 25°C Derate above 25°C	P_D	1.5 12	Watts mW/°C
Operating and Storage Junction Temperature Range	T_J, T_{stg}	−55 to +150	°C

2N3903
2N3904★

CASE 29-04, STYLE 1
TO-92 (TO-226AA)

GENERAL PURPOSE TRANSISTORS

NPN SILICON

★This is a Motorola designated preferred device.

***THERMAL CHARACTERISTICS**

Characteristic	Symbol	Max	Unit
Thermal Resistance, Junction to Ambient	$R_{\theta JA}$	200	°C/W
Thermal Resistance, Junction to Case	$R_{\theta JC}$	83.3	°C/W

ELECTRICAL CHARACTERISTICS (T_A = 25°C unless otherwise noted.)

OFF Characteristic (operated in cutoff)		Symbol	Min	Max	Unit
Collector-Emitter Breakdown Voltage(1) (I_C = 1.0 mAdc, I_B = 0)		$V_{(BR)CEO}$	40	—	Vdc
Collector-Base Breakdown Voltage (I_C = 10 μAdc, I_E = 0)		$V_{(BR)CBO}$	60	—	Vdc
Emitter-Base Breakdown Voltage (I_E = 10 μAdc, I_C = 0)		$V_{(BR)EBO}$	6.0	—	Vdc
Base Cutoff Current (V_{CE} = 30 Vdc, V_{EB} = 3.0 Vdc)		I_{BL}	—	50	nAdc
Collector Cutoff Current (V_{CE} = 30 Vdc, V_{EB} = 3.0 Vdc)		I_{CEX}	—	50	nAdc
ON Characteristic (operated in active and saturation region)					
DC Current Gain(1) (I_C = 0.1 mAdc, V_{CE} = 1.0 Vdc)	2N3903 2N3904	h_{FE}	20 40	— —	—
(I_C = 1.0 mAdc, V_{CE} = 1.0 Vdc)	2N3903 2N3904		35 70	— —	
(I_C = 10 mAdc, V_{CE} = 1.0 Vdc)	2N3903 2N3904		50 100	150 300	
(I_C = 50 mAdc, V_{CE} = 1.0 Vdc)	2N3903 2N3904		30 60	— —	
(I_C = 100 mAdc, V_{CE} = 1.0 Vdc)	2N3903 2N3904		15 30	— —	
Collector-Emitter Saturation Voltage(1) (I_C = 10 mAdc, I_B = 1.0 mAdc) (I_C = 50 mAdc, I_B = 5.0 mAdc)		$V_{CE(sat)}$	— —	0.2 0.3	Vdc
Base-Emitter Saturation Voltage(1) (I_C = 10 mAdc, I_B = 1.0 mAdc) (I_C = 50 mAdc, I_B = 5.0 mAdc)		$V_{BE(sat)}$	0.65 —	0.85 0.95	Vdc

NOTE: The "O" following CBO, CEO, EBO indicates the third terminal is "open." For example, $V_{(BR)\,CEO}$ *means the breakdown voltage between collector and emitter with the base open.*

$h_{FE} = \beta_{DC}$, dc current gain is measured at different values of I_c.

Maximum base-emitter voltage (V_{BE}) when transistor is saturated

Maximum value of voltage between collector and emitter (V_{CE}) when transistor is in saturation.

FIGURE 8-16 A General Purpose NPN Silicon Transistor. (Copyright of Motorola. Used by permission.)

DEVICE: BFW43—PNP Silicon High Reverse Voltage Transistor

MAXIMUM RATINGS

Rating	Symbol	Value	Unit
Collector-Emitter Voltage	V_{CEO}	150	Vdc
Collector-Base Voltage	V_{CBO}	150	Vdc
Emitter-Base Voltage	V_{EBO}	6.0	Vdc
Collector Current — Continuous	I_C	0.1	Adc
Total Device Dissipation @ T_A = 25°C Derate above 25°C	P_D	0.4 2.28	Watt mW/°C
Total Device Dissipation @ T_C = 25°C Derate above 25°C	P_D	1.4 8.0	Watt mW/°C
Operating and Storage Junction Temperature Range	T_J, T_{stg}	−65 to +200	°C

High Reverse Voltage Ratings

APPLICATIONS: High voltage circuits found in televisions and computer monitors.

BFW43

CASE 22-03, STYLE 1
TO-18 (TO-206AA)

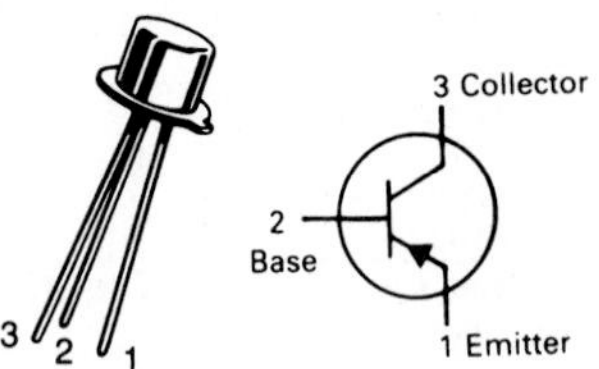

HIGH VOLTAGE TRANSISTOR

PNP SILICON

FIGURE 8-17 A High-Voltage Transistor. (Copyright of Motorola. Used by permission.)

DEVICE: MPS6714—NPN Silicon High Current (I_C) Transistor

MAXIMUM RATINGS

Rating	Symbol	Value	Unit
Collector-Emitter Voltage MPS6714 MPS6715	V_{CEO}	 30 40	Vdc
Collector-Base Voltage MPS6714 MPS6715	V_{CBO}	 40 50	Vdc
Emitter-Base Voltage	V_{EBO}	5.0	Vdc
Collector Current — Continuous	I_C	1.0	Adc
Total Device Dissipation @ T_A = 25°C Derate above 25°C	P_D	1.0 8.0	Watt mW/°C
Total Device Dissipation @ T_C = 25°C Derate above 25°C	P_D	2.5 20	Watts mW/°C
Operating and Storage Junction Temperature Range	T_J, T_{stg}	−55 to +150	°C

High I_C Rating

Applications: Current Regulator Circuits

MPS6714
MPS6715

CASE 29-05, STYLE 1
TO-92 (TO-226AE)

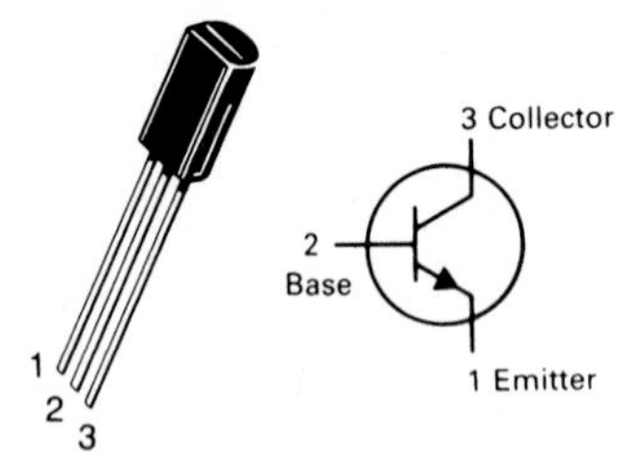

ONE WATT AMPLIFIER TRANSISTORS

NPN SILICON

FIGURE 8-18 A High-Current Transistor. (Copyright of Motorola. Used by permission.)

DEVICE: BUX48—NPN Silicon High Power Dissipation Transistor

SWITCHMODE II SERIES
NPN SILICON POWER TRANSISTORS

The BUX 48/BUX 48A transistors are designed for high-voltage, high-speed, power switching in inductive circuits where fall time is critical. They are particularly suited for line-operated switchmode applications such as:

- Switching Regulators
- Inverters
- Solenoid and Relay Drivers ◄— Applications
- Motor Controls
- Deflection Circuits

Fast Turn-Off Times
 60 ns Inductive Fall Time — 25°C (Typ)
 120 ns Inductive Crossover Time — 25°C (Typ)

Operating Temperature Range -65 to +200°C

100°C Performance Specified for:
 Reverse-Biased SOA with Inductive Loads
 Switching Times with Inductive Loads
 Saturation Voltage
 Leakage Currents (125°C)

15 AMPERES
NPN SILICON
POWER TRANSISTORS
400 AND 450 VOLTS
V(BR)CEO
850 – 1000 VOLTS
V(BR)CEX
175 WATTS

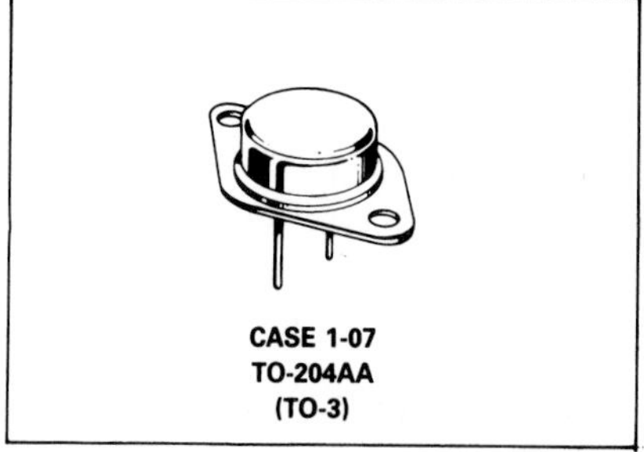

CASE 1-07
TO-204AA
(TO-3)

MAXIMUM RATINGS

Rating	Symbol	BUX48	BUX48A	Unit
Collector-Emitter Voltage	$V_{CEO(sus)}$	400	450	Vdc
Collector-Emitter Voltage (V_{BE} = -1.5V)	V_{CEX}	850	1000	Vdc
Emitter Base Voltage	V_{EB}	7		Vdc
Collector Current — Continuous — Peak (1) — Overload	I_C I_{CM} I_{OI}	15 30 60		Adc
Base Current — Continuous — Peak (1)	I_B I_{BM}	5 20		Adc
Total Power Dissipation — T_C = 25°C — T_C = 100°C Derate above 25°C (High power dissipation rating.)	P_D	175 100 1		Watts W/°C
Operating and Storage Junction Temperature Range	T_J, T_{stg}	-65 to +200		°C

THERMAL CHARACTERISTICS

Characteristic	Symbol	Max	Unit
Thermal Resistance, Junction to Case	$R_{\theta JC}$	1	°C/W
Maximum Lead Temperature for Soldering Purposes: 1/8″ from Case for 5 Seconds	T_L	275	°C

FIGURE 8-19 A High-Power Transistor. (Copyright of Motorola. Used by permission.)

8-2-4 *Testing Bipolar Junction Transistors*

Although transistors are exceptionally more reliable than their counterpart, the vacuum tube, they still will malfunction. These failures are normally the result of excessive temperature, current, or mechanical abuse and generally result in one of three problems:

1. An open between two or three of the transistor's leads
2. A short between two or three of the transistor's leads
3. A change in the transistor's characteristics

Transistor Tester

Transistor Tester
Special test instrument that can be used to test both NPN and PNP bipolar transistors.

The **transistor tester** shown in Figure 8-20 is a special test instrument that can be used to test both NPN and PNP bipolar transistors. This special meter can be used to determine whether an open or short exists between any of the transistor's three terminals, the transistor's dc current gain (β_{DC}), and whether an undesirable value of leakage current is present through one of the transistor's junctions.

Ohmmeter Transistor Test

If the transistor tester is not available, the ohmmeter can be used to detect open and shorted junctions, which are the most common transistor failure. Figure 8-21 illustrates how the ohmmeter can be used to check the emitter and collector diode of an NPN or PNP transistor. Referring to this diagram, notice that reverse biasing either the collector or emitter diode of any good transistor

FIGURE 8-20 Transistor Tester. (Courtesy of Sencore, Inc.—Test Equipment for the Professional Servicer. 1-800-SENCORE.)

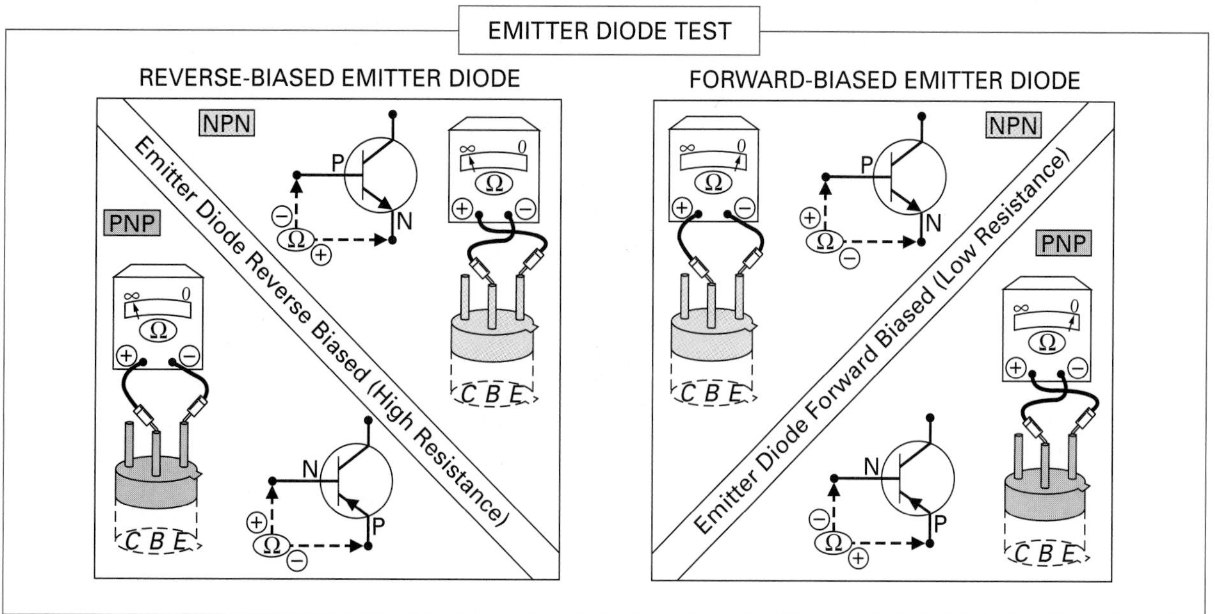

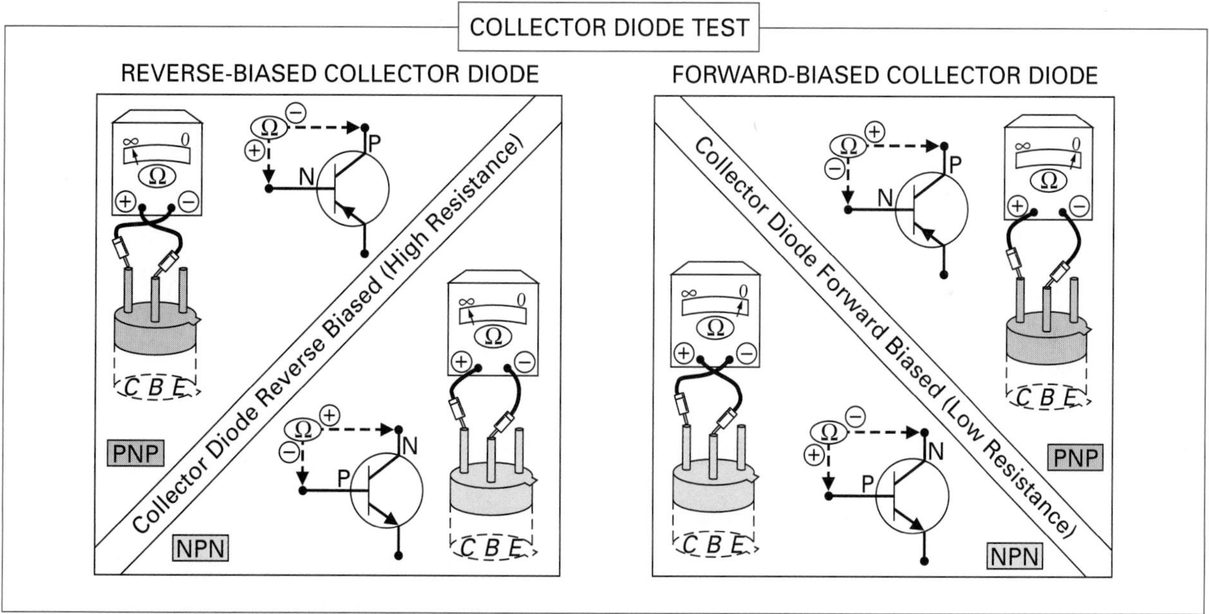

FIGURE 8-21 Using the Ohmmeter to Test a Bipolar Junction Transistor.

should cause the ohmmeter to display a relatively high resistance (several hundred thousand ohms or more). If a reverse biased emitter or collector diode has a low ohms reading, the respective transistor junction can be assumed to be shorted. Conversely, forward biasing either the collector or emitter diode of any good transistor should cause the ohmmeter to exhibit a low resistance (several hundred ohms or less). If a forward biased emitter or collector diode has a high ohms reading, the respective transistor junction can be assumed to be an open.

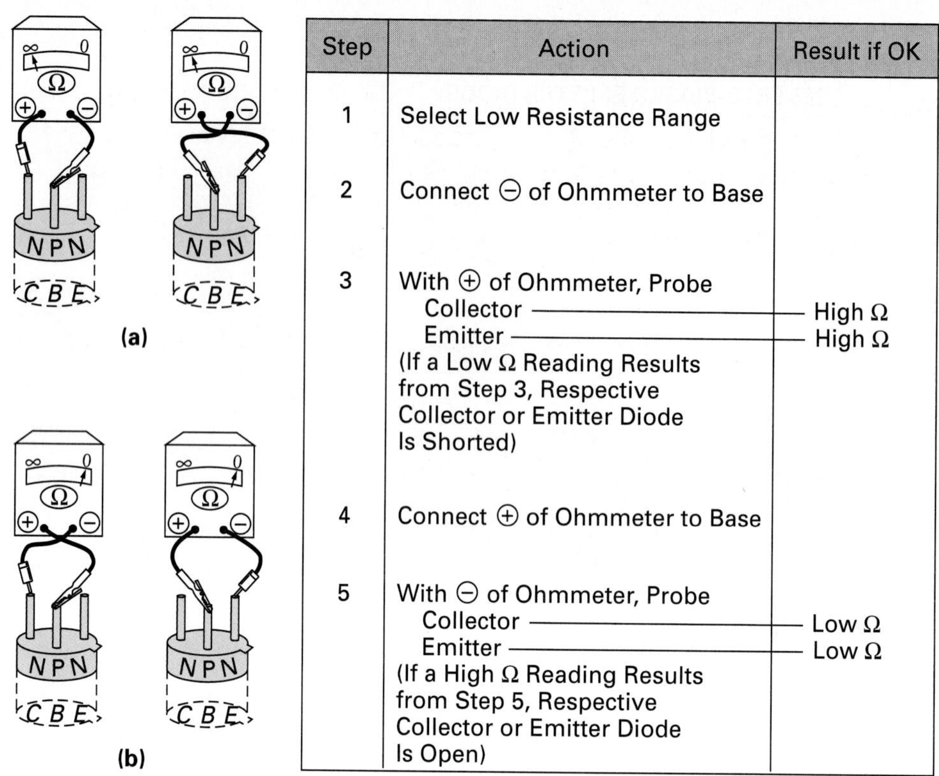

Step	Action	Result if OK
1	Select Low Resistance Range	
2	Connect ⊖ of Ohmmeter to Base	
3	With ⊕ of Ohmmeter, Probe Collector Emitter (If a Low Ω Reading Results from Step 3, Respective Collector or Emitter Diode Is Shorted)	 High Ω High Ω
4	Connect ⊕ of Ohmmeter to Base	
5	With ⊖ of Ohmmeter, Probe Collector Emitter (If a High Ω Reading Results from Step 5, Respective Collector or Emitter Diode Is Open)	 Low Ω Low Ω

FIGURE 8-22 NPN Ohmmeter Test Procedure. (a) Reverse-Biasing Collector, Then Emitter Diode. (b) Forward-Biasing Collector, Then Emitter Diode.

Transistor Testing Procedure

Figure 8-22 shows the step-by-step procedure for testing an NPN transistor. Following through this test procedure, we begin by reverse biasing the collector diode and then the emitter diode, and then forward biasing the collector diode and then the emitter diode. The table in Figure 8-22 shows the order and action to be performed for each step and the reading that should result if the NPN transistor junction is operating correctly.

Transistor Lead Identification

In most instances, to identify the bipolar transistor's leads, we would turn the transistor upside down, and then locate the emitter lead by determining which lead is closest to the side tag, as seen in Figure 8-22. Once the emitter is located, we would then assume that the lead at the other end of the case is the collector and that the center lead is the base. Although this procedure will most often be true, in some instances the base lead is not always between the emitter and the collector. For example, the transistor leads may be organized as *ECB* (emitter-collector-base) instead of *EBC* (emitter-base-collector). To be sure which lead is which, you should look up the device in the manufacturer's data book. If this is not available, the ohmmeter can be used to determine the transistor type (NPN or PNP) and the transistor's lead orientation *(EBC, ECB)*. The procedure, which will only work if the transistor is fully operational, is as follows.

Choosing the center lead, and assuming that it is the base, connect the negative lead of the ohmmeter and touch the two other transistor leads with the positive meter lead. If a low-resistance reading is observed in both cases, the center lead is definitely the base and the transistor is of the PNP type. The emitter lead is nearest to the tag and the collector is the remaining lead. If a high-resistance reading is observed in both cases, place the positive lead of the meter on the center lead of the transistor and probe the other two. If a low reading results in both instances, the center lead is the base and the transistor is an NPN type.

If a low resistance does not occur in both cases, the center lead is not the base and another lead should be selected to act as the assumed base. Repeating this procedure and observing the results will eventually reveal the lead configuration and transistor type.

8-2-5 Bipolar Transistor Biasing Circuits

As we discovered in the previous discussion on transistor circuit configurations, the ac operation of a transistor is determined by the "DC bias level," or "no-input signal level." This steady-state or dc operating level is set by the value of the circuit's dc supply voltage (V_{CC}) and the value of the circuit's biasing resistors. This single supply voltage and the one or more biasing resistors set up the initial dc values of transistor current (I_B, I_E and I_C) and transistor voltage (V_{BE}, V_{CE} and V_{BC}).

In this section, we will examine some of the more commonly used methods for setting the "initial dc operating point" of a bipolar transistor circuit. As you encounter different circuit applications in later chapters, you will see that many of these circuits include combinations of these basic biasing techniques and additional special purpose components for specific functions. Because the common-emitter (*C-E*) circuit configuration is used more extensively than the *C-B* and the *C-C*, we will use this configuration in all of the following basic biasing circuit examples.

Base Biasing

Figure 8-23(a) shows how a common-emitter transistor circuit could be base biased. With **base biasing,** the emitter diode of the transistor is forward biased by applying a positive base-bias voltage ($+V_{BB}$) via a current-limiting resistor (R_B) to the base of Q_1. In Figure 8-23(b), the transistor circuit from Figure 8-23(a) has been redrawn so as to simplify the analysis of the circuit. The transistor is now represented as a diode between base and emitter (emitter diode), and the transistor's emitter to collector has been represented as a variable resistor. Assuming Q_1 is a silicon bipolar transistor, the forward biased emitter diode will have a standard base-emitter voltage drop of 0.7 V (emitter diode drop = 0.7 V).

Base Biasing
A transistor biasing method in which the dc supply voltage is applied to the base of the transistor via a base bias resistor.

$$V_{BE} = 0.7 \text{ V}$$

The base bias resistor (R_B) and the transistor's emitter diode form a series circuit across V_{BB}, as seen in Figure 8-23(b). Therefore, the voltage drop across R_B (V_{R_B}) will be equal to the difference between V_{BB} and V_{BE}.

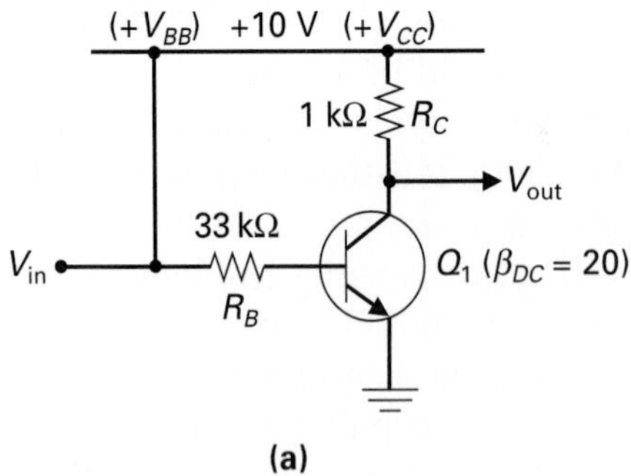

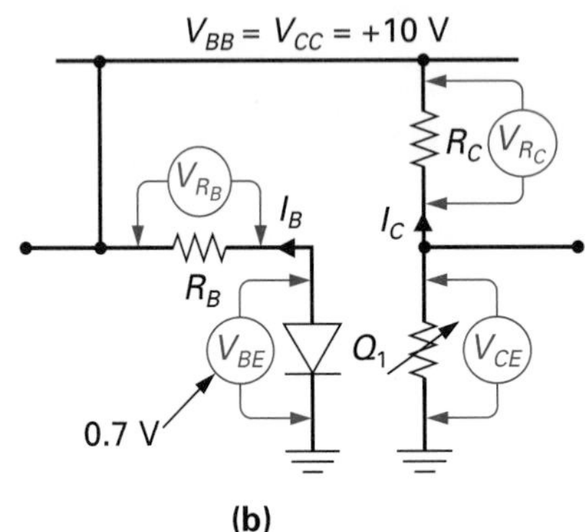

FIGURE 8-23 A Base Biased Common Emitter Circuit. (a) Basic Circuit. (b) Simplified Equivalent Circuit.

$$
\begin{aligned}
V_{R_B} &= V_{BB} - V_{BE} \\
&= V_{BB} - 0.7\text{ V}
\end{aligned}
$$

$$V_{R_B} = V_{BB} - V_{BE} = 10\text{ V} - 0.7\text{ V} = 9.3\text{ V}$$

Now that the resistance and voltage drop across R_B are known, we can calculate the current through R_B (I_{R_B}). Because a series circuit is involved, the current through R_B (I_{R_B}) will also be equal to the transistor base current I_B.

$$I_B = \frac{V_{R_B}}{R_B}$$

$$I_B = \frac{V_{R_B}}{R_B} = \frac{9.3\text{ V}}{33\text{ k}\Omega} = 282\ \mu\text{A}$$

Because the transistor's dc current gain (β_{DC}) is given in Figure 8-23(a), we can calculate I_C because β_{DC} tells us how much greater the output current I_C is compared to the input current I_B.

$$I_C = I_B \times \beta_{DC}$$

$$I_C = I_B \times \beta_{DC} = 282\ \mu\text{A} \times 20 = 5.6\text{ mA}$$

Because the current through R_C is I_C, we can now calculate the voltage drop across R_C (V_{R_C}).

$$V_{R_C} = I_C \times R_C$$

$$V_{R_C} = I_C \times R_C = 5.6 \text{ mA} \times 1 \text{ k}\Omega = 5.6 \text{ V}$$

Now that V_{R_C} is known, we can calculate the voltage drop across the transistor's collector-to-emitter because V_{CE} and V_{R_C} are in series and will be equal to the applied voltage V_{CC}.

$$V_{CE} = V_{CC} - V_{R_C}$$

$$V_{CE} = V_{CC} - V_{R_C} = 10 \text{ V} - 5.6 \text{ V} = 4.4 \text{ V}$$

Combining the previous two equations, we can obtain the following V_{CE} formula:

$$V_{CE} = V_{CC} - V_{R_C}$$

Since

$$V_{R_C} = I_C \times R_C$$

$$V_{CE} = V_{CC} - (I_C \times R_C)$$

$$V_{CE} = V_{CC} - (I_C \times R_C) = 10 \text{ V} - (5.6 \text{ mA} \times 1 \text{ k}\Omega) = 4.4 \text{ V}$$

Using the above formulas, which are all basically Ohms' law, you can calculate the current and voltage values in a base biased circuit.

DC load line In a transistor circuit, such as the example in Figure 8-23, V_{CC} and V_{R_C} are constants. On the other hand, the input current I_B and the output current I_C are variables. Using the example circuit in Figure 8-23, let us calculate what collector-to-emitter voltage drops (V_{CE}) will result for different values of I_C.

a. When Q_1 is OFF, $I_C = 0$ mA, and therefore V_{CE} equals:

$$V_{CE} = V_{CC} - (I_C \times R_C) = 10 \text{ V} - (0 \text{ mA} \times 1 \text{ k}\Omega) = 10 \text{ V} - 0 \text{ V} = 10 \text{ V}$$

This would make sense because Q_1 would be equivalent to an open switch between collector and emitter when it is OFF, and therefore all of the 10 V V_{CC} supply voltage would appear across the open. Figure 8-24 shows how this point would be plotted on a graph (point A).

b. When $I_C = 1$ mA,

$$V_{CE} = 10 \text{ V} - (1 \text{ mA} \times 1 \text{ k}\Omega) = 10 \text{ V} - 1 \text{ V} = 9 \text{ V (point } B)$$

c. When $I_C = 2$ mA,

$$V_{CE} = 10 \text{ V} - (2 \text{ mA} \times 1 \text{ k}\Omega) = 10 \text{ V} - 2 \text{ V} = 8 \text{ V (point } C)$$

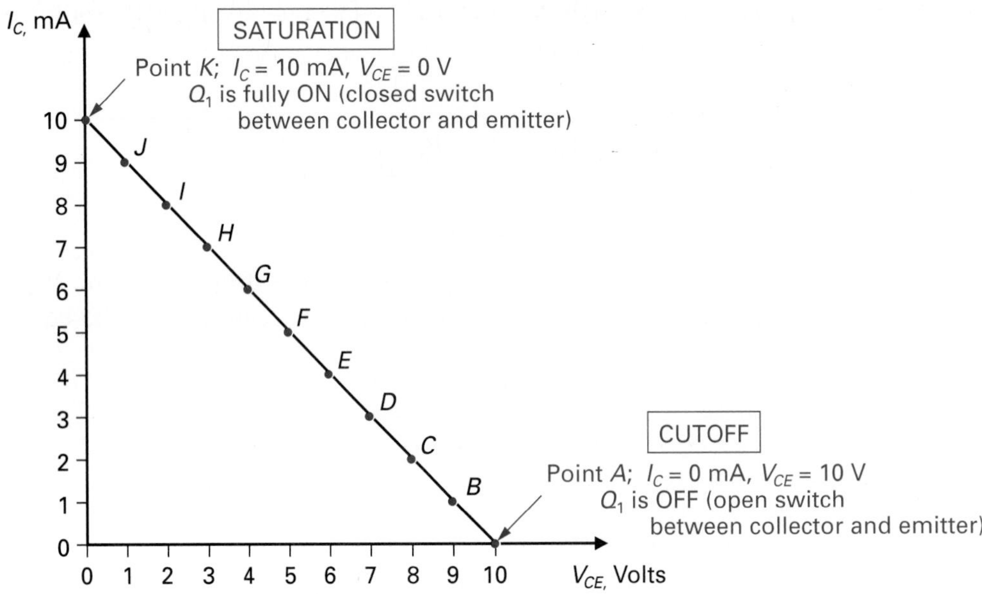

FIGURE 8-24 A Transistor DC Load Line with Cutoff and Saturation Points.

d. When $I_C = 3$ mA,

$$V_{CE} = 10\text{ V} - (3\text{ mA} \times 1\text{ k}\Omega) = 10\text{ V} - 3\text{ V} = 7\text{ V (point } D)$$

e. When $I_C = 4$ mA, $V_{CE} = 6$ V (point *E*)

f. When $I_C = 5$ mA, $V_{CE} = 5$ V (point *F*)

g. When $I_C = 6$ mA, $V_{CE} = 4$ V (point *G*)

h. When $I_C = 7$ mA, $V_{CE} = 3$ V (point *H*)

i. When $I_C = 8$ mA, $V_{CE} = 2$ V (point *I*)

j. When $I_C = 9$ mA, $V_{CE} = 1$ V (point *J*)

k. When $I_C = 10$ mA, the only resistance is that of R_C because Q_1 is fully ON and is equivalent to a closed switch between collector and emitter. It is not a surprise that the voltage drop across Q_1's collector-to-emitter is almost 0 V.

$$V_{CE} = 10\text{ V} - (10\text{ mA} \times 1\text{ k}\Omega) = 10\text{ V} - 10\text{ V} = 0\text{ V (point } K)$$

DC Load Line
A line representing all the dc operating points of the transistor for a given load resistance.

The line drawn in the graph in Figure 8-24 is called the **DC load line** because it is a line representing all the dc operating points of the transistor for a given load resistance. In this example, the transistor's load was the 1 kΩ collector connected resistor R_C in Figure 8-23.

Cutoff and saturation points Let us now examine the two extreme points in a transistor's dc load line, which in the example in Figure 8-24 were points *A* and *K*. If a transistor's base input bias voltage is reduced to zero, its input current I_B will be zero, Q_1 will turn OFF and be equivalent to an open switch between the collector and emitter, the output current I_C will be 0 mA, and a V_{CE}

will be 10 V. This point in the transistor dc load line is called **cutoff** (point A in Figure 8-24) because the output collector current is reduced to zero, or cut off. In summary, at cutoff:

$$I_{C(Cutoff)} = 0 \text{ mA}$$
$$V_{CE(Cutoff)} = V_{CC}$$

In the example circuit in Figure 8-23 and its dc load line in Figure 8-24, with Q_1 cut OFF:

$$I_{C(Cutoff)} = 0 \text{ mA}, V_{CE(Cutoff)} = V_{CC} = 10 \text{ V}$$

If the base input bias voltage is increased to a large positive value, the transistor's collector diode (which is normally reverse biased) will be forward biased. In this condition, I_B will be at its maximum, Q_1 will be fully ON and equivalent to a closed switch between the collector and emitter, I_C will be at its maximum of 10 mA, and V_{CE} will be 0 V. This point in the transistor's dc load line is called **saturation** (point K in Figure 8-24) because, just as a point is reached where a wet sponge is saturated and cannot hold any more water, the transistor at saturation cannot increase I_C beyond this point. In summary, at saturation:

$$I_{C(Sat.)} = \frac{V_{CC}}{R_C}$$
$$V_{CE(Sat.)} = 0 \text{ V}$$

In the example circuit in Figure 8-23 and its dc load line in Figure 8-24, with Q_1 saturated:

$$I_{C(Sat.)} = \frac{V_{CC}}{R_C} = \frac{10 \text{ V}}{1 \text{ k}\Omega} = 10 \text{ mA}$$
$$V_{CE(Sat.)} = 0 \text{ V}$$

Rearranging the formula $\beta_{DC} = I_C/I_B$, we can calculate the value of input base current that causes the output saturation current:

$$\beta_{DC} = \frac{I_C}{I_B}, \text{ therefore,}$$

$$I_{B(Sat.)} = \frac{I_{C(Sat.)}}{\beta_{DC}}$$

In the example circuit in Figure 8-23 and its dc load line in Figure 8-24, the input current that will cause saturation will be

$$I_{B(Sat.)} = \frac{I_{C(Sat.)}}{\beta_{DC}} = \frac{10 \text{ mA}}{20} = 500 \text{ µA}$$

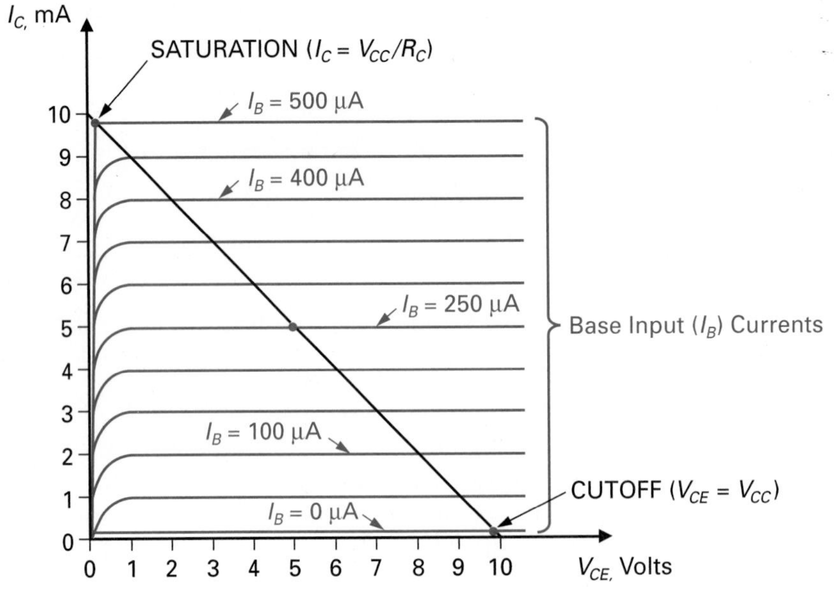

FIGURE 8-25 Transistor Input/Output Characteristic Graph.

Figure 8-25 summarizes all of our base bias circuit calculations so far by including the dc load line from Figure 8-24 in a set of collector characteristic curves for the transistor circuit example in Figure 8-23. As you can see in the graph in Figure 8-24, at cutoff: I_B = 0 μA, I_C = 0 mA, and $V_{CE} = V_{CC}$ which is 10 V. On the other hand, at saturation: I_B = 500 μA, I_C = 10 mA, and V_{CE} = 0 V.

Quiescent point Generally, the value of the base biased resistor (R_B) is chosen so that the value of base current (I_B) is near the middle of the dc load line. For example, if a base bias resistance of 37.2 kΩ was used in the example circuit in Figure 8-23 (R_B = 37.2 kΩ), it would produce a base current of 250 μA (I_B = 9.3 V/37.2 kΩ = 250 μA). Referring to the dc load line in Figure 8-25, you can see that this value of base current is half-way between cutoff at 0 μA, and saturation at 500 μA. This point is called the **quiescent** (at rest) or ***Q* point** and is defined as *the dc bias point at which the circuit rests when no ac input signal is applied.* An ac input signal voltage will vary I_B above and below this Q point, resulting in a corresponding but larger change in I_C.

■ **EXAMPLE:**

Complete the following for the circuit shown in Figure 8-26.

a. Calculate I_B
b. Calculate I_C
c. Calculate V_{CE}
d. Sketch the circuit's dc load line with saturation and cutoff points
e. Indicate where the Q point is on the circuit's dc load line

■ ***Solution:***

a. Since V_{BE} = 0.7 V and V_{BB} = 12 V,

$$V_{R_B} = 12\text{ V} - 0.7\text{ V} = 11.3\text{ V}$$

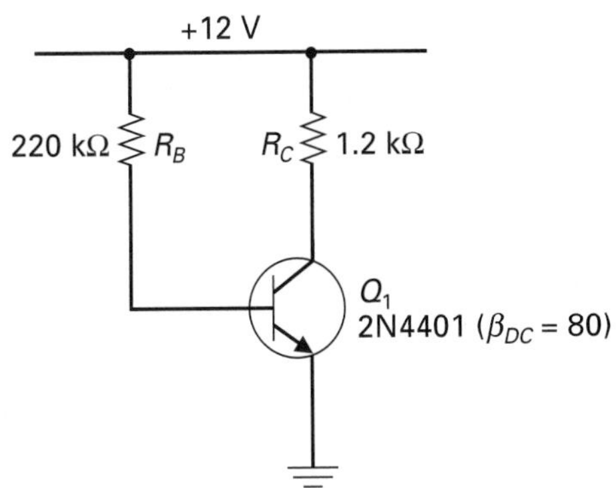

FIGURE 8-26 Bipolar Transistor Example.

$$I_B = \frac{V_{R_B}}{R_B} = \frac{11.3\text{ V}}{220\text{ k}\Omega} = 51.4\ \mu\text{A}$$

b. $I_C = I_B \times \beta_{DC} = 51.4\ \mu\text{A} \times 80 = 4.1\text{ mA}$

c. $V_{R_C} = I_C \times R_C = 4.1\text{ mA} \times 1.2\text{ k}\Omega = 4.92\text{ V}$

$$V_{CE} = V_{CC} - V_{R_C} = 12\text{ V} - 4.92\text{ V} = 7.08\text{ V}$$

d. At cutoff, the transistor is OFF and therefore equivalent to an open switch between collector and emitter. All of the V_{CC} supply voltage will therefore be across Q_1.

At cutoff, $V_{CE} = V_{CC} = 12$ V (see cutoff in the dc load line in Figure 8-27).

At saturation, the transistor is fully ON and therefore equivalent to a closed switch between the collector and emitter. The only resistance is that of R_C, and so:

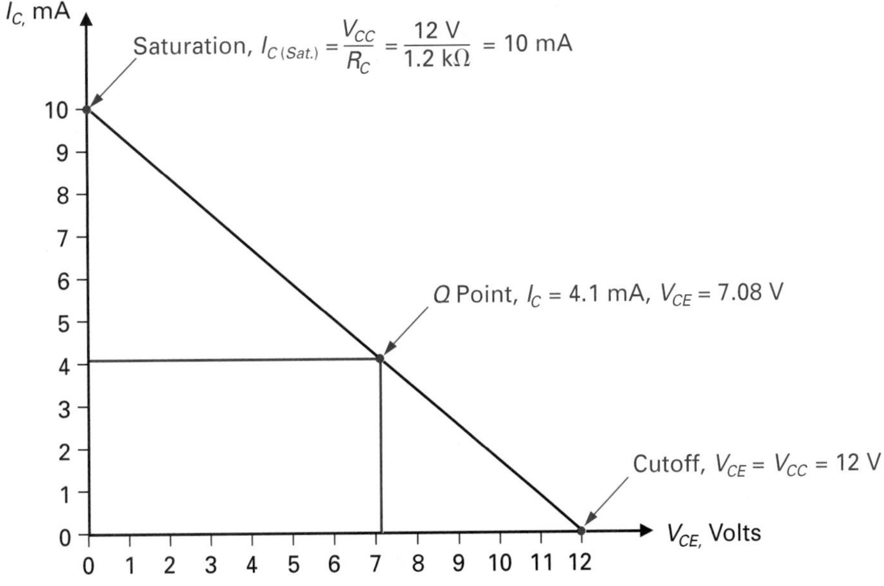

FIGURE 8-27 The DC Load Line for the Circuit in Figure 8-26.

At saturation,

$$I_{C(Sat.)} = \frac{V_{CC}}{R_C} = \frac{12 \text{ V}}{1.2 \text{ k}\Omega} = 10 \text{ mA}$$

(see saturation in the dc load line in Figure 8-27).

e. The operating point or Q point of this circuit is set by the base bias resistor R_B. This Q point will be at

$$I_C = 4.1 \text{ mA}$$

which produces a

$$V_{CE} = 7.08 \text{ V}$$

This quiescent (Q) point is also shown on Figure 8-27.

■ **EXAMPLE:**

Calculate the current through the lamp in Figure 8-28.

■ ***Solution:***

$$V_{BE} = 0.7 \text{ V}, V_{in} = +5 \text{ V, therefore,}$$

$$V_{R_B} = V_{in} - 0.7 \text{ V}$$

$$= 5 \text{ V} - 0.7 \text{ V} = 4.3 \text{ V}$$

$$I_B = \frac{V_{R_B}}{R_B} = \frac{4.3 \text{ V}}{10 \text{ k}\Omega} = 430 \text{ }\mu\text{A}$$

$$I_C = I_B \times \beta_{DC}$$

$$= 433 \text{ }\mu\text{A} \times 125 = 53.75 \text{ mA}$$

An input of zero volts ($V_{in} = 0$ V) will turn OFF Q_1, and therefore lamp L_1. On the other hand, an input of +5 V will turn ON Q_1 and permit a collector current, and therefore lamp current, of 53.75 mA.

Base-biasing applications Base-bias circuits are used in switching circuit applications like the two-state ON/OFF lamp circuit discussed in the previous example. In these circuits, the bipolar transistor is equivalent to a switch and is

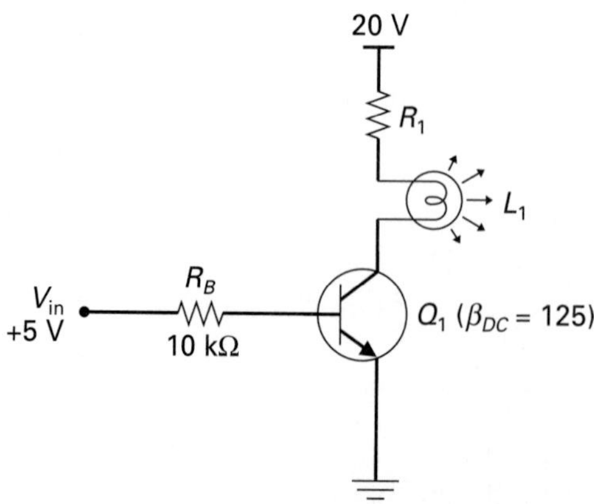

FIGURE 8-28 Two-State Lamp Circuit.

controlled by a HIGH/LOW input voltage that drives the transistor between the two extremes of cutoff and saturation.

The advantage of this biasing technique is circuit simplicity because only one resistor is needed to set the base bias voltage. The disadvantage of the base-biased circuit is that it cannot compensate for changes in its dc bias current due to changes in temperature. To explain this in more detail, a change in temperature will result in a change in the internal resistance of the transistor (all semiconductor devices have a negative temperature coefficient of resistance—temperature ↑ causes internal resistance ↓). This change in the transistor's internal resistance will change the transistor's dc bias currents (I_B and I_C), which will change or shift the transistor's dc operating point or Q point away from the desired midpoint.

Base-bias troubleshooting Figure 8-29 repeats our original base-bias circuit example, and the previously calculated values of circuit voltage and current under normal operating conditions. As in any troubleshooting exercise, remember to apply the three step troubleshooting procedure of "diagnose, isolate, and repair."

Before beginning with any circuit troubleshooting, let us first list some of the obvious errors that should not be overlooked:

a. Is the power supplied to the circuit (+10 V) present and of the correct value?

b. Is ground connected to the circuit?

c. If this is a newly constructed circuit, are all of the components connected correctly, especially the three terminals of the transistor?

Once you have determined that the problem is, in fact, within the transistor circuit, perform a visual check of the circuit to look for

a. Shorts caused by badly connected components, loose clippings of wire, solder bridges, and so on.

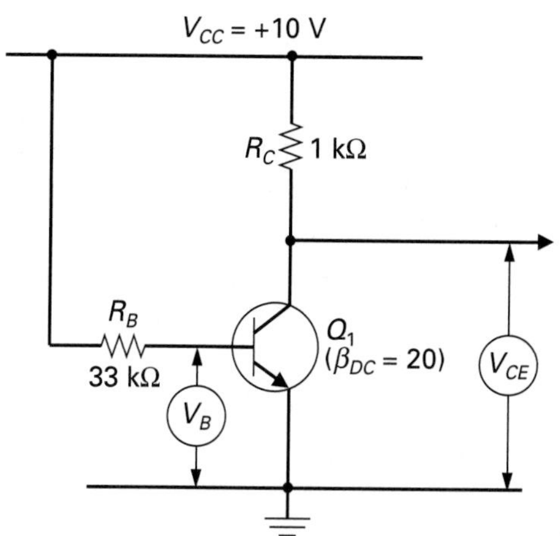

$V_{BE} = 0.7$ V	$I_B = 282$ μA
$V_{R_B} = 9.3$ V	$I_C = 5.6$ mA
$V_{R_C} = 5.6$ V	$I_{C(Sat.)} = 10$ mA
$V_{CE} = 4.4$ V	$I_{B(Sat.)} = 500$ μA

FIGURE 8-29 Troubleshooting a Base-Biased Circuit.

b. Incorrect component values, such as resistors.

c. Sign of excessive heat on wire insulation, or circuit components.

The next step would be to use test equipment to determine what "effect" the problem is causing, and what could be the "cause." To give you some examples, some of the more common problems that can occur within a base-biased common-emitter bipolar transistor circuit are listed below.

CAUSE	EFFECT
R_C Open	Collector resistor opens—there is no path for current in collector circuit. Transistor is cutoff. V_{CC} will appear across open. $V_{R_C} = 10$ V, $V_{CE} = 0$ V.
R_C Short	Collector resistor short—value of R_C is chosen so that I_C is kept within safe limits: A R_C short will cause excessive collector current and burn out Q_1.
R_B Open	When base resistor opens, no biasing current can flow in base. Transistor will cut off. $V_B = 0$ V, $V_{CE} = 10$ V.
R_B Short	Base resistor short—base current increases to a maximum, transistor goes into heavy saturation (fully ON). Base will be 0.7 V above emitter voltage ($V_B = +0.7$ V). $V_{CE} = 0$ V. Transistor could burn out due to heavy internal transistor currents.
$Q_{1\,(B\text{-}E)}$ Open	Transistor cuts OFF—no circuit current. $V_B = V_{CC}$, $V_C = V_{CC}$. To verify, disconnect power, perform ohmmeter check on transistor and R_B to isolate problem.
Q_1 Leaky	Transistor is leaky, which means partially shorted. Transistor appears to be in saturation, with symptoms the same as R_B short. Disconnect power, perform ohmmeter check on transistor and R_B to isolate problem.

Voltage-Divider Biasing

Voltage-Divider Biasing
A biasing method used with amplifiers in which a series arrangement of two fixed-value resistors are connected across the voltage source. The result is that a desired fraction of the total voltage is obtained at the center of the two resistors and is used to bias the amplifier.

Figure 8-30(a) shows how a common-emitter transistor circuit could be **voltage-divider biased.** The name of this biasing method comes from the two resistor series-voltage divider (R_1 and R_2) connected to the transistor's base. In this most widely used biasing method, the emitter diode of Q_1 is forward biased by the voltage developed across R_2 (V_{R_2}), as seen in the simplified equivalent circuit in Figure 8-30(b). To calculate the voltage developed across R_2, and therefore the voltage applied to Q_1's base, we can use the voltage divider formula.

$$V_{R_2} \text{ or } V_B = \frac{R_2}{R_1 + R_2} \times V_{CC}$$

$$V_{R_2} \text{ or } V_B = \frac{R_2}{R_1 + R_2} \times V_{CC} = \frac{10\text{ k}\Omega}{20\text{ k}\Omega + 10\text{ k}\Omega} \times 20\text{ V} = 0.333 \times 20\text{ V} = 6.7\text{ V}$$

Because the current through R_1 and R_2 (from ground to $+V_{CC}$) is generally more than 10 times greater than the base current of Q_1 (I_B), it is nor-

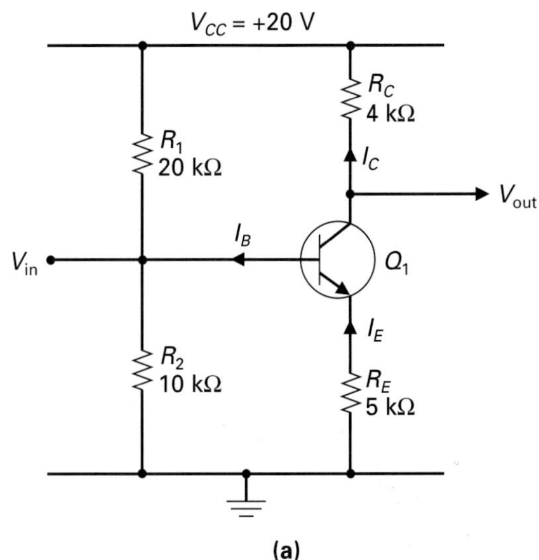

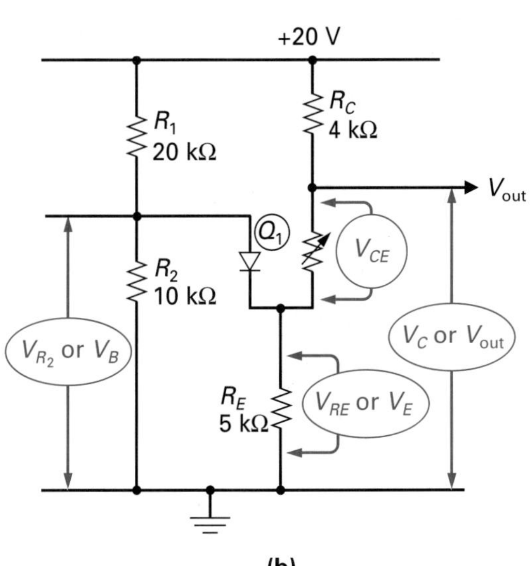

FIGURE 8-30 A Voltage Divider Biased Common Emitter Circuit. (a) Basic Circuit. (b) Simplified Equivalent Circuit.

mally assumed that I_B will have no effect on the voltage divider current through R_1 and R_2. The R_1 and R_2 voltage divider can be assumed to be independent of Q_1, and the previous voltage divider formula can be used to calculate V_{R_2} or V_B.

Because $V_B = 6.7$ V, the emitter diode of Q_1 will be forward biased. Assuming a 0.7 V drop across the transistor's base-emitter junction ($V_{BE} = 0.7$ V), the voltage at the emitter terminal of Q_1 (V_E) will be:

$$V_{R_E} \text{ or } V_E = V_B - 0.7 \text{ V}$$

$$V_{R_E} \text{ or } V_E = V_B - 0.7 \text{ V} = 6.7 \text{ V} - 0.7 \text{ V} = 6 \text{ V}$$

Now that the voltage drop across R_E (V_{R_E}) is known, along with its resistance, we can calculate the current through R_E and the value of current being injected into the transistor's emitter.

$$I_{R_E} = I_E = \frac{V_{R_E}}{R_E}$$

$$I_{R_E} = I_E = \frac{V_{R_E}}{R_E} = \frac{6\text{ V}}{5\text{ k}\Omega} = 1.2\text{ mA}$$

Because we know that a transistor collector current (I_C) is approximately equal to the emitter current (I_E), we can state that

$$I_E \cong I_C$$

$$I_E \cong I_C = 1.2\text{ mA}$$

Now that I_C is known, we can calculate the voltage drop across R_C (V_{R_C}) because both its resistance and current are known.

$$V_{R_C} = I_C \times R_C$$

$$V_{R_C} = I_C \times R_C = 1.2\text{ mA} \times 4\text{ k}\Omega = 4.8\text{ V}$$

The dc quiescent voltage at the collector of Q_1 with respect to ground (V_C), which is also V_{out}, will be equal to the dc supply voltage (V_{CC}) minus the voltage drop across R_C.

$$V_C \text{ or } V_{out} = V_{CC} - V_{R_C}$$

$$V_C \text{ or } V_{out} = V_{CC} - V_{R_C} = 20\text{ V} - 4.8\text{ V} = 15.2\text{ V}$$

Because V_{CC} is connected across the series voltage divider formed by R_C, Q_1's collector-to-emitter resistance (R_{CE}), and R_E, we can calculate V_{CE} if both V_{R_C} and V_E are known:

$$V_{CE} = V_{CC} - (V_{R_C} + V_E)$$

$$V_{CE} = V_{CC} - (V_{R_C} + V_E) = 20\text{ V} - (4.8\text{ V} + 6\text{ V}) = 20\text{ V} - 10.8\text{ V} = 9.2\text{ V}$$

DC load line Figure 8-31 shows the dc load line for the example circuit in Figure 8-30. Referring to the dc load line's two end points, let us examine this circuit's saturation and cutoff points.

When transistor Q_1 is fully ON or saturated, it will have approximately 0 Ω of resistance between its collector and emitter. As a result, R_C and R_E determine the value of I_C when Q_1 is saturated.

$$I_{C(Sat.)} = \frac{V_{CC}}{R_C + R_E}$$

$$I_{C(Sat.)} = \frac{V_{CC}}{R_C + R_E} = \frac{20 \text{ V}}{4 \text{ k}\Omega + 5 \text{ k}\Omega} = \frac{20 \text{ V}}{9 \text{ k}\Omega} = 2.2 \text{ mA}$$

As you can see in Figure 8-31, at saturation, I_C is maximum at 2.2 mA, and V_{CE} is 0 V because Q_1 is equivalent to a closed switch (0 Ω) between Q_1's collector and emitter.

$$V_{CE(Sat.)} = 0 \text{ V}$$

At the other end of the dc load line in Figure 8-31, we can see how the transistor's characteristics are plotted when it is cut off. When Q_1 is cut OFF, it is equivalent to an open switch between collector-to-emitter. Therefore all of the V_{CC} supply voltage will appear across the series circuit open.

$$V_{CE(Cutoff)} = V_{CC}$$

$$V_{CE(Cutoff)} = V_{CC} = 20 \text{ V}$$

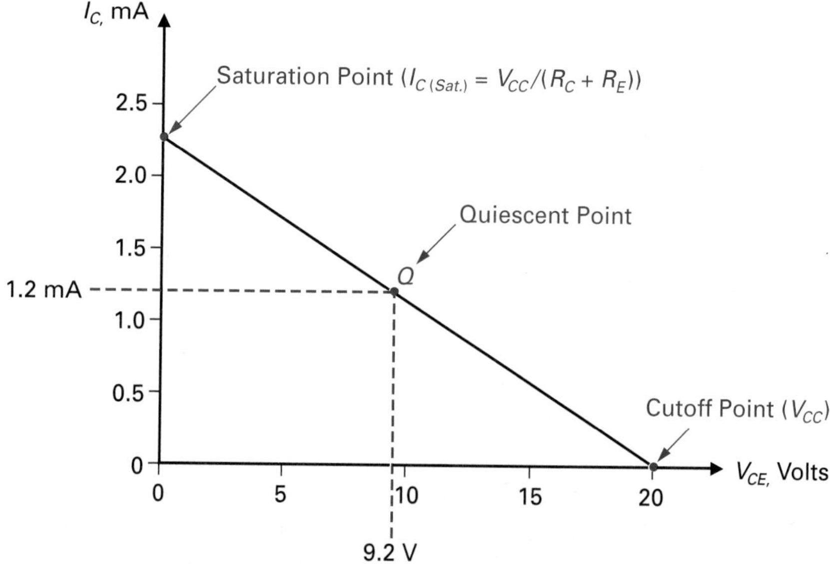

FIGURE 8-31 The DC Load Line for the Circuit in Figure 8-30.

As you can see in Figure 8-31, when Q_1 is cut OFF, all of the V_{CC} supply voltage will appear across Q_1's collector-to-emitter terminals, and I_C will be blocked and equal to zero.

$$I_{C(Cutoff)} = 0 \text{ mA}$$

Generally, the value of the voltage divider resistors R_1 and R_2 are chosen so that the value of base current (I_B) is near the middle of the dc load line. Referring to Figure 8-31, you can see that by plotting our previously calculated values of I_C (which at rest was 1.2 mA) and V_{CE} ($V_{CE} = 9.2$ V), we obtain a Q point that is near the middle of the dc load line.

■ **EXAMPLE:**

Calculate the following for the circuit shown in Figure 8-32.

a. V_B and V_E
b. Determine whether C_E will have any effect on the dc operating voltages.
c. I_C
d. V_C and V_{CE}
e. Sketch the circuit's dc load line and include the saturation, cutoff, and Q points.

■ ***Solution:***

a. $V_B = \dfrac{R_2}{R_1 + R_2} \times V_{CC} = \dfrac{2.2 \text{ k}\Omega}{10 \text{ k}\Omega + 2.2 \text{ k}\Omega} \times 12 \text{ V} = 2.16 \text{ V}$

$V_E = V_B - 0.7 \text{ V} = 2.16 \text{ V} - 0.7 \text{ V} = 1.46 \text{ V}$

b. Since all capacitors can be thought of as a dc block, C_E will have no effect on the circuit's dc operating voltages.

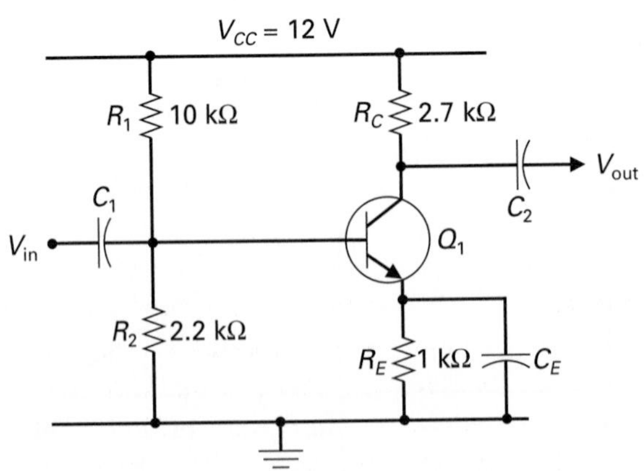

FIGURE 8-32 A Common-Emitter Amplifier Circuit Example.

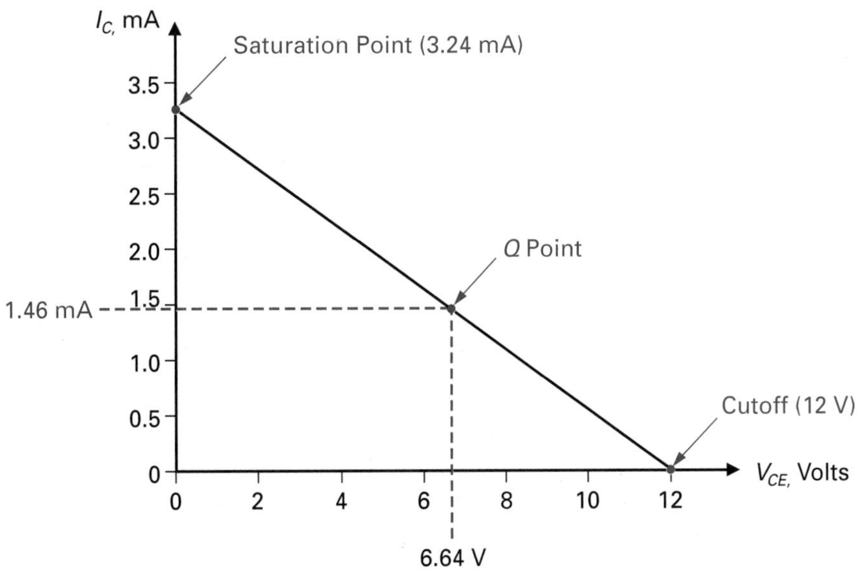

FIGURE 8-33 The DC Load Line for the Circuit in Figure 8-32.

c. $$I_E = \frac{V_E}{R_E} = \frac{1.46\text{ V}}{1\text{ k}\Omega} = 1.46\text{ mA}$$

$$I_C \cong I_E = 1.46\text{ mA}$$

d. $V_{R_C} = I_C \times R_C = 1.46\text{ mA} \times 2.7\text{ k}\Omega = 3.9\text{ V}$

V_{out} or $V_C = V_{CC} - V_{R_C} = 12\text{ V} - 3.9\text{ V} = 8.1\text{ V}$

$V_{CE} = V_{CC} - (V_{R_C} + V_E) = 12\text{ V} - (3.9\text{ V} + 1.46\text{ V}) = 12\text{ V} - 5.36\text{ V} = 6.64\text{ V}$

e. $$I_{C(Sat.)} = \frac{V_{CC}}{R_C + R_E} = \frac{12\text{ V}}{2.7\text{ k}\Omega + 1\text{ k}\Omega} = \frac{12\text{ V}}{3.7\text{ k}\Omega} = 3.24\text{ mA}$$

$V_{CE(Cutoff)} = V_{CC} = 12\text{ V}$

Q Point, $I_C = 1.46$ mA *and* $V_{CE} = 6.64$ V

(This information is plotted on the graph in Figure 8-33)

Voltage-divider bias applications Voltage-divider bias circuits are used in analog or linear circuit applications such as the amplifier circuit discussed in the previous example. In these circuits, the bipolar transistor is equivalent to a variable-resistor and is controlled by an alternating input signal voltage.

Unlike the base-biased circuit, the voltage-divider biased circuit has very good temperature stability due to the emitter resistor R_E. To explain this in more detail, let us assume that there is an increase in the temperature surrounding a voltage-divider circuit, such as the example circuit in Figure 8-32. As temperature increases, it causes an increase in the transistor's internal currents ($I_B\uparrow$, $I_E\uparrow$, $I_C\uparrow$) because all semiconductor devices have a negative temperature coefficient of resistance (temperature $\uparrow$, $R\downarrow$, $I\uparrow$). An increase in $I_E\uparrow$ will cause an increase in the voltage drop across $R_E\uparrow$, which will decrease the voltage difference between the transistor's base and emitter ($V_{BE}\downarrow$). Decreasing the forward bias applied to the transistor's emitter diode will decrease all of the transistor's internal currents ($I_B\downarrow$, $I_E\downarrow$, $I_C\downarrow$) and return them to their original values. Therefore, a change in output current (I_C) due to temperature will effectively be fed back to the input and change the input current (I_B), which is why a circuit containing an emitter resistor is said to have **emitter feedback** for temperature stability.

Emitter Feedback
The coupling from the emitter output to the base input in a transistor amplifier.

Voltage-divider bias troubleshooting Figure 8-34 repeats our original voltage-divider biased circuit example and the previously calculated values of circuit voltage and current under normal operating conditions. As in any troubleshooting exercise, remember to apply the three step troubleshooting procedure of "diagnose, isolate, and repair."

Once again, begin by looking for the obvious errors:

a. Is the power supplied to the circuit (+20 V) present and of the correct value?

b. Is ground connected to the circuit?

c. If this is a newly constructed circuit, are all of the components connected correctly, especially the three terminals of the transistor?

Once you have determined that the problem is, in fact, within the transistor circuit, perform a visual check of the circuit to look for anything out of the normal.

The next step would be to use test equipment to determine what "effect" the problem is causing and what could be the "cause." To give you some examples, some of the more common problems that can occur within a voltage-divider biased common-emitter bipolar transistor circuit are listed below. Keep in mind that if a component problem turns the transistor fully OFF or ON, you

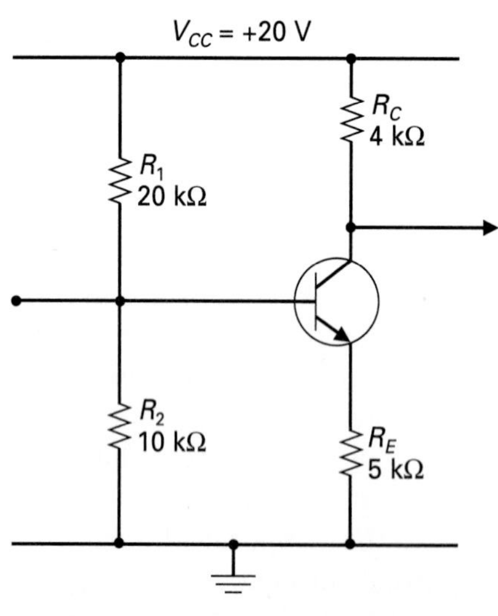

FIGURE 8-34 Troubleshooting a Voltage-Divider Biased Circuit.

may have to disconnect power and perform the ohmmeter test on the transistor to determine whether the problem is the transistor or another component in the circuit affecting the transistor.

CAUSE	EFFECT
R_1 Open	With R_1 open, there is no path for base current to $+V_{CC}$, no voltage developed across R_2, and therefore $V_B = 0$ V. Transistor is OFF, and equivalent to open switch between collector and emitter, therefore; $V_E = 0$ V, $V_C = +V_{CC}$.
R_2 Open	If R_2 opens, base of transistor will be a large positive voltage and so it will turn heavily ON (saturate). Transistor is equivalent to a closed switch between collector and emitter, and therefore; $V_{CE} = 0$ V, $V_E = R_E \times I_{C(Sat)}$, $V_C = V_E$, $V_B = V_E + 0.7$ V.
R_E Open	With R_E open, there can be no emitter current, and therefore there can be no input base current or output collector current. Transistor will be OFF, therefore V_C will equal $+V_{CC}$.
R_C Open	With R_C open, there can be no path for the output collector current, $I_C = 0$. Since B-E junction is forward biased I_E will equal I_B, and so; V_E will be a small voltage, $V_B = V_E + 0.7$ V, $V_C = V_E$ (all three transistor terminal voltages will be low).

SELF-TEST REVIEW QUESTIONS FOR SECTION 8-2

1. The bipolar transistor is a ____________ (voltage/current) controlled device.

2. When a bipolar transistor is being operated in the active region, its emitter diode is ____________ biased and its collector diode is ____________ biased.

3. Which of the following is correct:

a. $I_E = I_C - I_B$

b. $I_C = I_E - I_B$

c. $I_B = I_C - I_E$

4. When a transistor is in cutoff, it is equivalent to a/an ____________ between its collector and emitter.

5. When a transistor is in saturation, it is equivalent to a/an ____________ between its collector and emitter.

6. Which of the bipolar transistor circuit configurations has the best

a. Voltage gain

b. Current gain

c. Power gain

7. Which biasing method makes use of two series connected resistors across the V_{CC} supply voltage?

8. Which biasing technique has a single resistor connected in series with the base of the transistor?

SUMMARY

Bipolar Junction Transistors (Figure 8-35)

1. A transistor is a three-element device made of semiconductor materials used to control electron flow, the amount of which can be controlled by varying the voltages applied to its three elements. Having the ability to control the amount of current through the transistor allows us to achieve two very important applications: switching and amplification.
2. Transistors are generally classified as being either bipolar or unipolar. The bipolar junction transistor (BJT) has two P-N junctions whereas the unipolar transistor has only one P-N junction.
3. The bipolar transistor has three alternately doped semiconductor regions called the emitter (E), base (B), and collector (C).
4. The NPN transistor type has a thin, lightly doped *p*-type semiconductor layer sandwiched between two *n*-type semiconductor regions.
5. The PNP transistor type has a thin, lightly doped *n*-type region sandwiched between two *p*-type regions.
6. As a memory aid for the NPN transistor's schematic symbol, you may want to remember that when the emitter arrow is "**N**ot **P**ointing i**N**" to the base, the transistor is an **"NPN."**
7. The package will protect the transistor from humidity and dust, provide a means for electrical connection between the three semiconductor regions and the three transistor terminals, and serve as a heat sink to conduct away any heat generated by the transistor.
8. Transistor package types are normally given a reference number. These designations begin with the letters "TO," which stands for transistor outline and are followed by a number.
9. A transistor can be thought of as containing two diodes: a base-to-collector diode and a base-to-emitter diode. With an NPN transistor, both diodes will be back-to-back and pointing away from the base. For a PNP transistor, the base-collector and base-emitter diode will both be pointing into the base.
10. Transistors are basically controlled to operate as a switch, or as a variable resistor:
    a. The ON/OFF switching action of the transistor is controlled by the transistor's base-to-emitter (*B-E*) diode. If the *B-E* diode of the transistor is forward biased, the transistor will turn ON and be equivalent to a closed switch between its collector and emitter. If the *B-E* diode of the transistor is reverse biased, the transistor will turn OFF and be equivalent to an open switch between its collector and emitter.
    b. The variable resistor action of the transistor is also controlled by the transistor's base-to-emitter (*B-E*) diode. When the base input control voltage is between the OFF input voltage and the fully ON input voltage, the transistor is partially ON, and therefore the transistor's emitter-to-collector resistance is somewhere between maximum ohms and zero ohms.
11. The name transistor was derived from the fact that through base control we can "transfer" different values of "resistance" between the emitter and col-

QUICK REFERENCE SUMMARY SHEET

FIGURE 8-35 Bipolar Junction Transistors (BJTs).

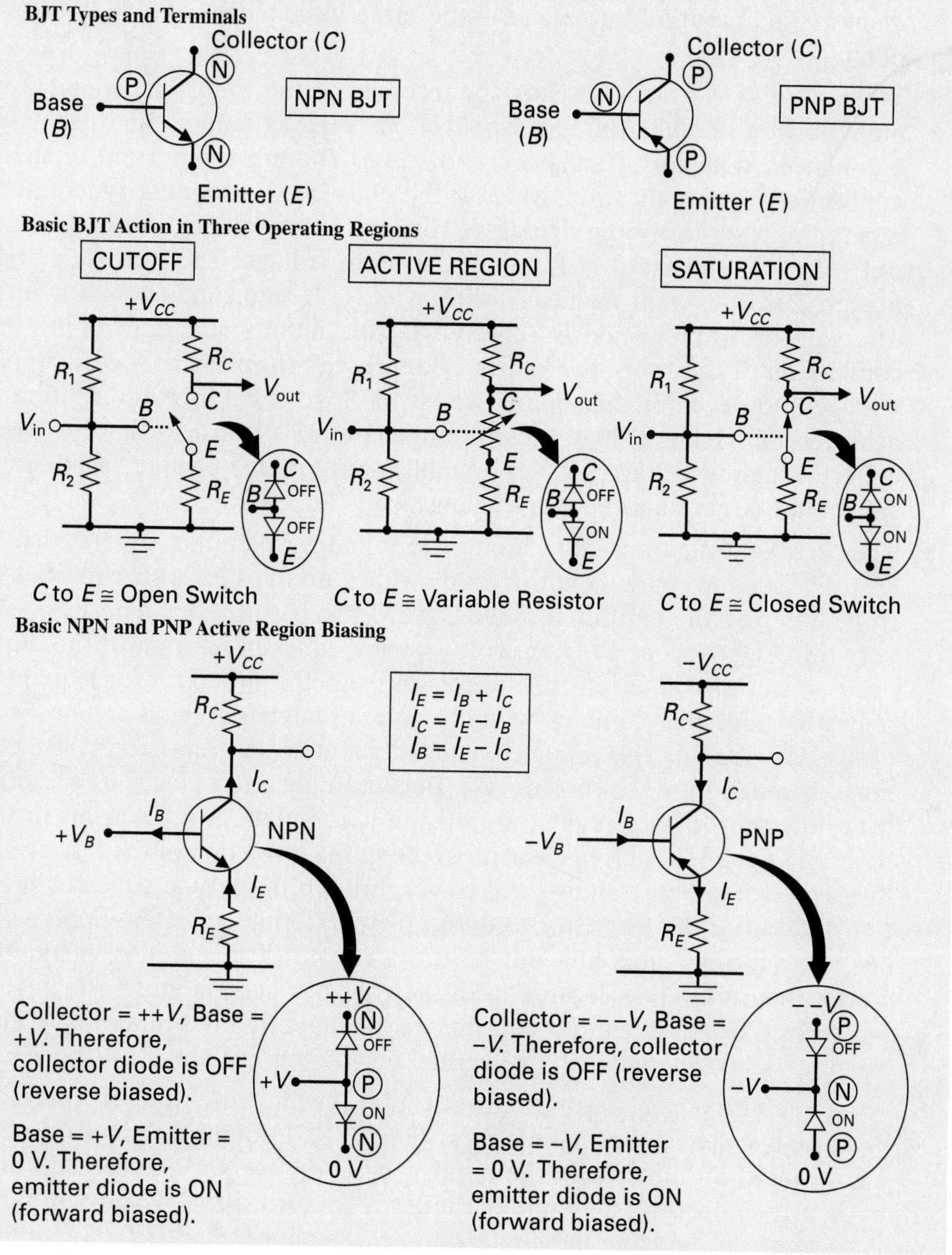

BJT Types and Terminals

Collector (C)
Base (B)
Emitter (E)
NPN BJT

Collector (C)
Base (B)
Emitter (E)
PNP BJT

Basic BJT Action in Three Operating Regions

CUTOFF
C to $E \cong$ Open Switch

ACTIVE REGION
C to $E \cong$ Variable Resistor

SATURATION
C to $E \cong$ Closed Switch

Basic NPN and PNP Active Region Biasing

$$I_E = I_B + I_C$$
$$I_C = I_E - I_B$$
$$I_B = I_E - I_C$$

NPN

Collector = $++V$, Base = $+V$. Therefore, collector diode is OFF (reverse biased).

Base = $+V$, Emitter = 0 V. Therefore, emitter diode is ON (forward biased).

PNP

Collector = $--V$, Base = $-V$. Therefore, collector diode is OFF (reverse biased).

Base = $-V$, Emitter = 0 V. Therefore, emitter diode is ON (forward biased).

lector. This effect of "transferring resistance" is known as transistance, and the component that functions in this manner is called the transistor.

12. The transistor initiated the multi-billion dollar semiconductor industry and was the key element behind many other inventions, such as integrated circuits (ICs), optoelectronic devices, and digital computer electronics.

13. A digital logic gate circuit makes use of the transistor's ON/OFF switching action. Transistors are used to construct logic gate circuits; gates are used to construct flip-flop circuits; flip-flops are used to construct register and counter circuits, and these circuits are used to construct microprocessor, memory and input/output circuits—the three basic blocks of a digital computer.
14. When used as a variable resistor, the transistor is the controlling element in many analog or linear circuits applications such as amplifiers, oscillators, modulators, detectors, regulators, and so on. The most important of these applications is amplification, which is the boosting in strength or increasing in amplitude of electronic signals. A voltage gain of 10 means that the output voltage is ten times larger than the input voltage. The transistor does not produce this gain magically within its NPN semiconductor structure. The gain or amplification is achieved by the input signal controlling the conduction of the transistor, which takes energy from the collector supply voltage and develops this energy across the load resistor. Amplification is achieved by having a small input voltage control a transistor and its large collector supply voltage so that a small input voltage change results in a similar but larger output voltage change.
15. The current at the transistor's emitter terminal is called the emitter current (I_E). The forward biased emitter diode will cause the steady stream of electrons entering the emitter to head toward the base region. The base is a very thin, lightly doped region with very few holes in relation to the number of electrons entering the transistor from the emitter. Consequently, only a few electrons combine with the holes in the base region and flow out of the base region. This relatively small current at the transistor's base terminal is called the base current (I_B). Because only a few electrons combine with holes in the base region, there is an accumulation of electrons in the base's *p* region. These free electrons, feeling the attraction of the large positive collector supply voltage, will travel through the *n*-type collector junction and out of the transistor to the positive external collector supply voltage. The current emerging out of the transistor's collector is called the collector current (I_C). Because both the collector current and base current are derived from the emitter current, we can state that the sum of base current and the collector current will equal the emitter current.
16. An increase in the emitter current ($I_E\uparrow$) will cause a corresponding increase in collector and base current ($I_C\uparrow$, $I_B\uparrow$). Looking at this from a different angle, an increase in the applied base voltage will increase the forward bias applied to the emitter diode of the transistor, which will draw more electrons up from the emitter and cause an increase in I_E, I_B, and I_C. Similarly, a decrease in the applied base voltage will decrease the forward bias applied to the emitter diode of the transistor, which will decrease the number of electrons being drawn up from the emitter and cause a decrease in I_E, I_B, and I_C. The applied input base voltage will control the amount of base current, which will in turn control the amount of emitter and collector current and the conduction of the transistor. This is why the bipolar transistor is known as a current-controlled device.

17. A transistor is said to be in the active region when its base-emitter junction is forward biased (emitter diode is ON), and the base-collector junction is reverse biased (collector diode is OFF). In this mode, the transistor is equivalent to a variable-resistor between the collector and emitter.
18. A transistor is in cutoff when the bias voltage is reduced to a point that it stops current in the transistor. In cutoff, both the emitter and collector diode of the transistor will be biased OFF, the transistor is equivalent to an open switch between the collector and emitter, and transistor current is zero.
19. A transistor is in saturation when the bias voltage is increased to such a point that any further increase in bias voltage will not cause any further increase in current through the transistor. In saturation, both the emitter and collector diodes are said to be forward biased, and the transistor is equivalent to a closed switch.
20. Generally the PNP transistor is not employed as much as the NPN transistor in most circuit applications. The only difference that occurs with PNP transistor circuits is that the polarity of V_{CC} and the base bias voltage (V_B) need to be reversed to a negative voltage.

BJT Circuit Configuration Characteristics (Figure 8-36)

21. Although there are many thousands of different bipolar transistor circuits, all of these circuits can be classified into one of three groups based on which of the transistor's leads is used as the "common reference." These three different circuit configurations are the:
    - **a.** Common-emitter (*C-E*), in which the input signal is applied between the base and emitter, while the output signal appears between the transistor's collector and emitter. With this circuit arrangement, the input signal controls the transistor's base current, which in turn controls the transistor's output collector current, and the emitter lead is common to both the input and output.
    - **b.** Common-base (*C-B*), in which the input signal is applied between the transistor's emitter and base, the output signal is developed across the transistor's collector and base, and the base is common to both input and output.
    - **c.** Common-collector (*C-C*), in which the input is applied between the base and the collector, the output is developed across the emitter and the collector, and the collector is common to both the input and output.
22. The current gain in a common-emitter configuration is called the transistor's beta (symbolized β) and is the ratio of output collector current to input base current. The term *hfe* is often used instead of dc current gain (β_{DC}) in manufacturer data sheets.
23. The common-emitter circuit is not only used to increase the level of current between input and output. It can also be used to provide an increase in the amplitude of the input signal voltage or produce a voltage gain (A_V).
24. The common-emitter circuit provides both current gain and voltage gain. Because power is equal to the product of current and voltage ($P = V \times I$), it

QUICK REFERENCE SUMMARY SHEET

FIGURE 8-36 BJT Circuit Configuration Characteristics.

Common-Emitter (C-E) Circuit

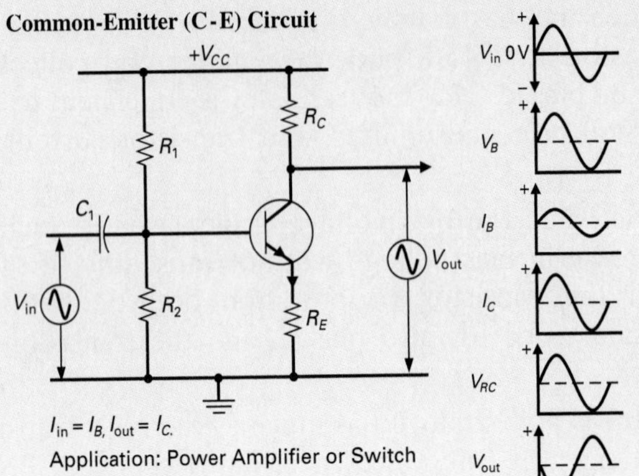

CHARACTERISTICS

- Current Gain: Medium, 50
- Voltage Gain: High, 100 to 500
- **Power Gain: Highest, 200 to 20,000**
- Input Resistance: Medium, 1 kΩ to 5 kΩ
- Output Resistance: High, 40 kΩ to 60 kΩ
- Phase Shift: 180°

Current Gain: (DC)$\beta_{DC} = \frac{I_C}{I_B}$, (AC)$\beta_{AC} = \frac{\Delta I_C}{\Delta I_B}$

Voltage Gain: $A_V = \frac{\Delta V_{out}}{\Delta V_{in}}$

Power Gain:

$$A_P = \frac{P_{out}}{P_{in}} = \frac{\Delta V_{out} \times \Delta I_{out}}{\Delta V_{in} \times \Delta I_{in}} = A_V \times \beta_{AC}$$

Common-Base (C-B) Circuit

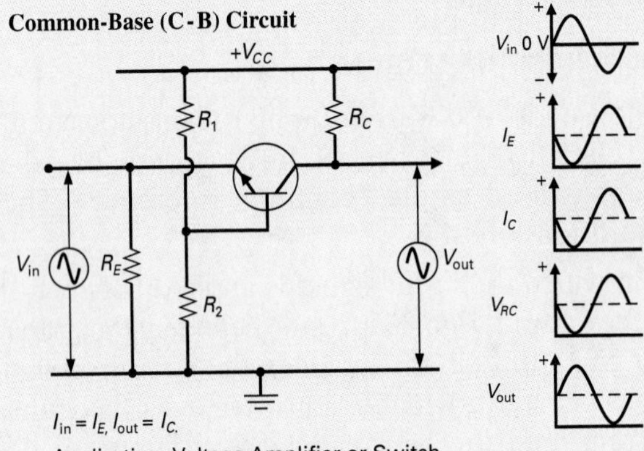

CHARACTERISTICS

- Current Gain: Low, 0.99
- **Voltage Gain: Highest, 200 to 2000**
- Power Gain: Medium, 200 to 1000
- Input Resistance: Very low, 15 to 150 Ω
- Output Resistance: Very high, 250 kΩ to 1 MΩ
- Phase Shift: 0°

Current Gain: (DC)$\alpha_{DC} = \frac{I_C}{I_E}$, (AC)$\alpha_{AC} = \frac{\Delta I_C}{\Delta I_E}$

Voltage Gain: $A_V = \frac{\Delta V_{out}}{\Delta V_{in}}$

Power Gain:

$$A_P = \frac{P_{out}}{P_{in}} = \frac{\Delta V_{out} \times \Delta I_{out}}{\Delta V_{in} \times \Delta I_{in}} = A_V \times \alpha_{AC}$$

Common-Collector (C-C) Circuit

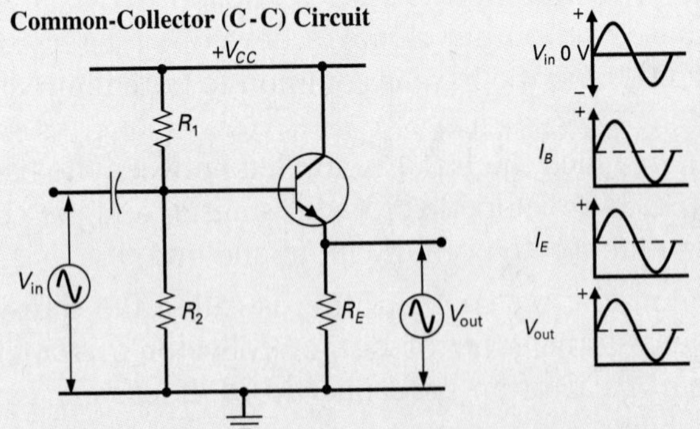

$I_{in} = I_B$, $I_{out} = I_E$.

Application: Current Amplifier or Switch, and Impedance or Resistance Matching Device

CHARACTERISTICS

- **Current Gain: Highest, 60**
- Voltage Gain: Low, less than 1
- Power Gain: Low, 20 to 80
- **Input Resistance: High, 2 kΩ to 500 kΩ**
- **Output Resistance: Very low, 25 Ω to 1 kΩ**
- Phase Shift: 0°

Current Gain: (DC) $= \frac{I_E}{I_B} = 1 + \beta$, AC $= \frac{\Delta I_E}{\Delta I_B}$

Voltage Gain: $A_V = \frac{\Delta V_{out}}{\Delta V_{in}}$

Power Gain:

$$A_P = \frac{P_{out}}{P_{in}} = \frac{\Delta V_{out} \times \Delta I_{out}}{\Delta V_{in} \times \Delta I_{in}}$$

$$= A_V \times \text{AC Current Gain}$$

Input Resistance: $R_E \times \beta_{AC}$

is not surprising that the *C-E* circuit configuration also provides a high value of power gain (A_P).

25. The set of collector characteristic curves graph shows the relationship between a transistor's input base current, output collector current, and collector-to-emitter voltage drop.
26. The input resistance (R_{in}), or input impedance (Z_{in}), of a transistor is the amount of opposition offered to an input signal by the transistor's input junction.
27. The output resistance (R_{out}), or output impedance (Z_{out}), of a transistor is the amount of opposition offered to an output signal by the transistor's output junction.
28. The current gain in a common-base configuration is called the transistor's alpha (symbolized α) and is the ratio of output collector current to input emitter current.
29. Although the common-base circuit configuration does not achieve any current gain, it does have a very high voltage gain.
30. The common-collector circuit provides a slightly higher gain than the *C-E* circuit because the common-collector's output current (I_E) is slightly higher than the *C-E*'s output current (I_C).
31. The common-collector circuit is often referred to as an emitter-follower or voltage-follower because the emitter output voltage seems to track or follow the phase and amplitude of the input voltage.
32. The very high input resistance and low output resistance of the common-collector configuration can be used as a resistance or impedance matching device that can match, or isolate, a high resistance (low current) source to a low resistance (high current) load. By acting as a buffer current amplifier, the *C-C* circuit can ensure that power is efficiently transferred from source to load.

BJT Testing (Figure 8-37)

33. The transistor tester is a special test instrument that can be used to test both NPN and PNP bipolar transistors. This special meter can be used to determine whether an open or short exists between any of the transistor's three terminals, the transistor's dc current gain (β_{DC}), and whether an undesirable value of leakage current is present through one of the transistor's junctions.
34. If the transistor tester is not available, the ohmmeter can be used to detect open or shorted junctions, which are the most common transistor failures.
35. Reverse biasing either the collector or emitter diode of any good transistor should cause the ohmmeter to display a relatively high resistance (several hundred thousand ohms or more). If a reverse biased emitter or collector diode has a low ohms reading, the respective transistor junction can be assumed to be shorted.
36. Forward biasing either the collector or emitter diode of any good transistor should cause the ohmmeter to exhibit a low resistance (several hundred

QUICK REFERENCE SUMMARY SHEET

FIGURE 8-37 BJT Ohmmeter Testing.

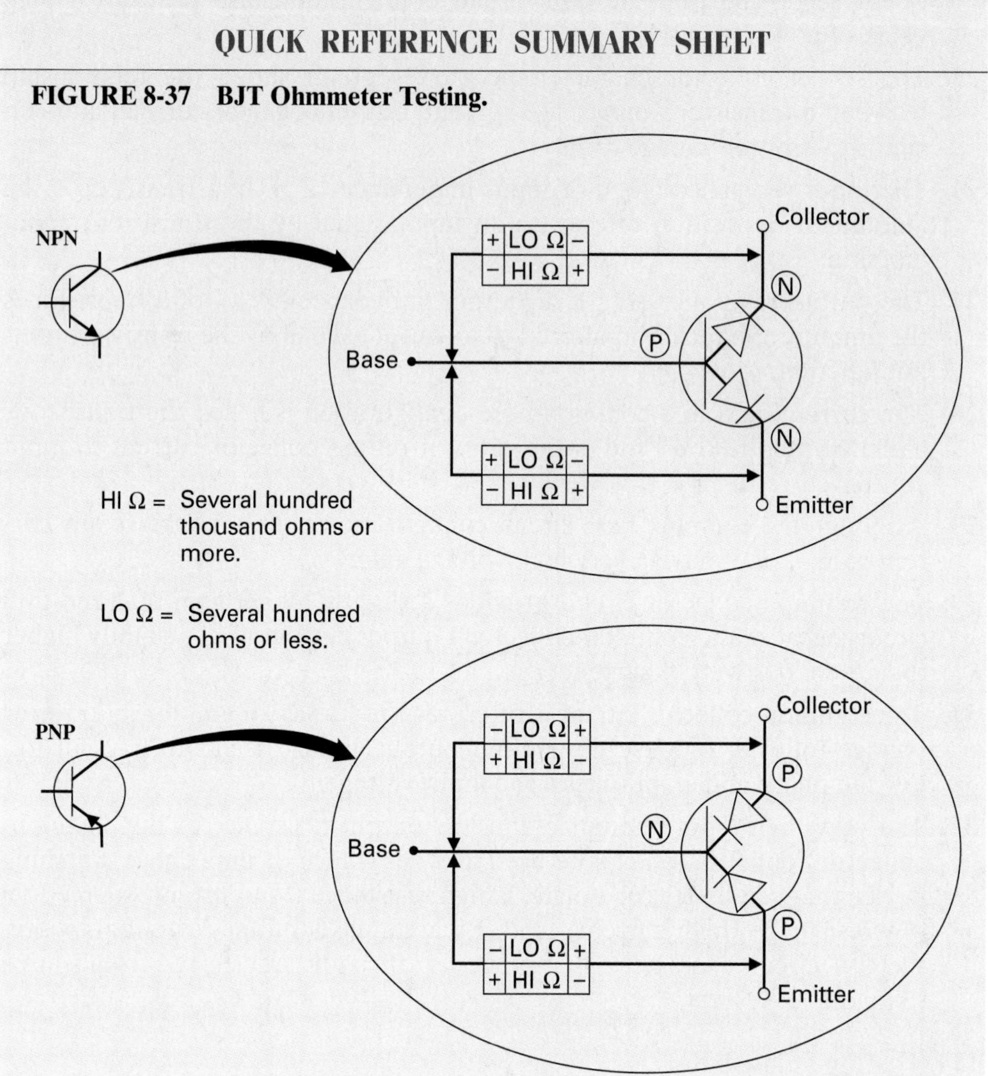

ohms or less). If a forward biased emitter or collector diode has a high ohms reading, the respective transistor junction can be assumed to be open.

BJT Biasing (Figure 8-38)

37. The ac operation of a transistor is determined by the "dc bias level," or "no-input signal level." This steady-state or dc operating level is set by the value of the circuit's dc supply voltage (V_{CC}) and the value of the circuit's biasing resistors. This single supply voltage, and the one or more biasing resistors, set up the initial dc values of transistor current (I_B, I_E and I_C) and transistor voltage (V_{BE}, V_{CE} and V_{BC}).

38. With base biasing, the emitter diode of the transistor is forward biased by applying a positive base-bias voltage ($+V_{BB}$) via a current-limiting resistor (R_B) to the base of Q_1.

QUICK REFERENCE SUMMARY SHEET

FIGURE 8-38 BJT Biasing.

Base-Biasing

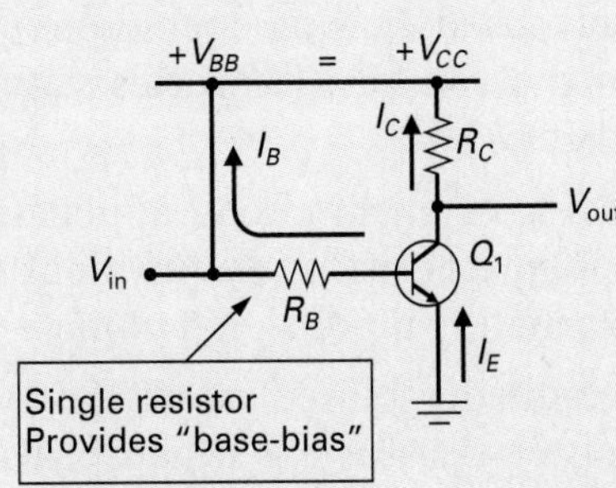

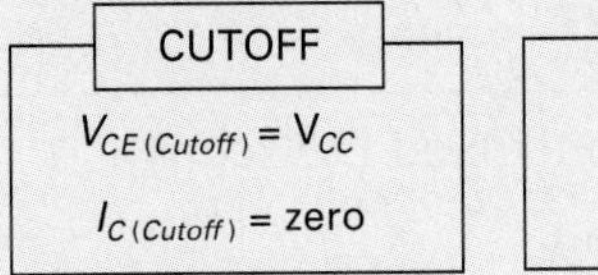

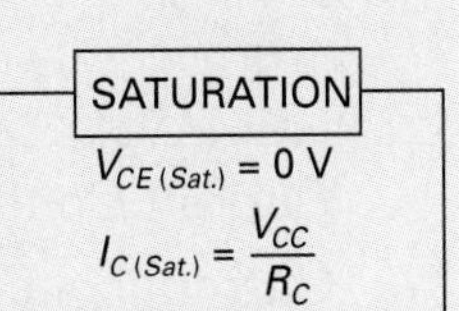

Q-POINT FORMULAS

$V_{BE} = 0.7$ V

$V_{R_B} = V_{BB} - V_{BE} = V_{BB} - 0.7$ V

$I_B = \frac{V_{R_B}}{R_B}$

$I_C = I_B \times \beta_{DC}$

$V_{R_C} = I_C \times R_C$

$V_{CE} = V_{CC} - V_{R_C} = V_{CC} - (I_C \times R_C)$

Advantage: Circuit Simplicity

Disadvantage: Temperature instability causes Q-point shift

Application: Switching (ON/OFF) Circuits.

Voltage-Divider Biasing

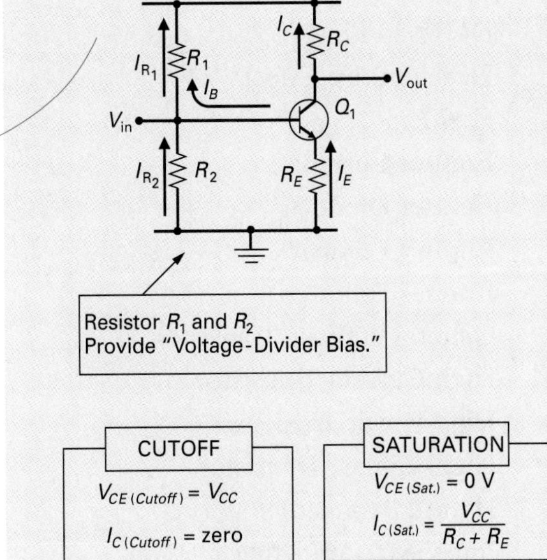

Q-POINT FORMULAS

V_{R_2} or $V_B = \frac{R_2}{R_1 + R_2} \times V_{CC}$

V_{RE} or $V_E = V_B - 0.7$ V

$I_E = \frac{V_E}{R_E}$

$I_C \cong I_E$

$V_{R_C} = I_C \times R_C$

V_{out} or $V_C = V_{CC} - V_{R_C}$

$V_{CE} = V_{CC} - (V_{R_C} + V_{CE})$

Advantage: Good temperature stability

Disadvantage: Needs additional bias resistor

Application: Analog or linear circuits, such as the amplifier

39. The dc load line is a line plotted on a graph representing all the dc operating points of the transistor for a given load resistance.

40. Base-bias circuits are used in switching circuit applications like the two-state ON/OFF lamp circuit discussed in the previous example. In these circuits, the bipolar transistor is equivalent to a switch and controlled by a

HIGH/LOW input voltage that drives the transistor between the two extremes of cutoff and saturation.

41. The advantage of this biasing technique is circuit simplicity because only one resistor is needed to set the base bias voltage. The disadvantage of the base-biased circuit is that it cannot compensate for changes in its dc bias current due to changes in temperature.

42. Voltage-divider biasing is the most widely used biasing technique. The name of this biasing method comes from the two resistor series voltage-divider (R_1 and R_2) connected to the transistor's base.

43. Voltage-divider bias circuits are used in analog or linear circuit applications like the amplifier. In these circuits, the bipolar transistor is equivalent to a variable-resistor and is controlled by an alternating input signal voltage.

44. Unlike the base-biased circuit, the voltage-divider biased circuit has very good temperature stability due to the emitter resistor R_E which provides emitter feedback.

NEW TERMS

AC Alpha
AC Beta
Active Region
Amplification
Amplifier Circuit
Base
Base Biasing
Base-Collector Diode
Base Current
Base-Emitter Diode
Beta
Bias Voltages
Bipolar Junction Transistor
Bipolar Transistor
BJT
Breakdown Region
Buffer Current Amplifier
Collector
Collector Characteristic Curves
Collector Current
Collector Diode
Common-Base Circuit
Common-Collector Circuit
Common-Emitter Circuit
Configuration
Current Controlled Device
Current Gain
Cutoff Point
Cutoff Region
DC Alpha
DC Beta
DC Load Line
Degenerative Effect
Emitter
Emitter Current
Emitter Diode
Emitter Feedback
Emitter Follower
General Purpose Transistor
High-Current Transistor
High-Power Transistor
High-Voltage Transistor
Hybrid Parameters
Impedance Matching
Input Impedance
Input Resistance
Input Signal Voltage Change
Inverter Gate
NOT Gate
NPN Transistor

Output Impedance
Output Resistance
Output Signal Voltage Change
PNP Transistor
Power Gain
Quiescent Point
Saturation Point
Saturation Region
Transistance
Transistor
Transistor Outline Package
Transistor Tester
Voltage Controlled Switch
Voltage Controlled Variable Resistor
Voltage-Divider Biasing
Voltage Follower
Voltage Gain

REVIEW QUESTIONS

Multiple-Choice Questions

1. The bipolar junction transistor has three terminals called the:
   a. Drain, source, gate **c.** Main terminal 1, main terminal 2, gate
   b. Anode, cathode, gate **d.** Emitter, base, collector
2. The term bipolar junction transistor was given to the device because it has:
   a. Two P-N junctions **c.** One *p* region and one *n* region
   b. Two magnetic poles **d.** Two magnetic junctions
3. An NPN transistor is normally biased so that its base is ____________.
   a. Positive **b.** Negative
4. Which is considered the most common bipolar junction transistor configuration?
   a. Common base **c.** Common emitter
   b. Common collector **d.** None of the above
5. A common-collector circuit is often called a/an ____________.
   a. Base follower **c.** Collector follower
   b. Emitter follower **d.** None of the above
6. With the NPN transistor schematic symbol, the emitter arrow will point ____________ the base, whereas with the PNP transistor schematic symbol, the emitter arrow will point ____________ the base.
   a. Towards, away from **b.** Away from, towards
7. The transistor's ON/OFF switching action is made use of in ____________ circuits.
   a. Analog **b.** Digital **c.** Linear **d.** Both (a) and (c) are true
8. The transistor's variable resistor action is made use of in ____________ circuits.
   a. Analog **b.** Digital **c.** Linear **d.** Both (a) and (c) are true
9. Approximately 98 percent of the electrons entering the ____________ of a bipolar transistor will arrive at the ____________, and the remainder will flow out of the ____________.
   a. Emitter, collector, base **c.** Collector, emitter, base
   b. Base, collector, emitter **d.** Emitter, base, collector

10. The common-base circuit configuration achieves the highest _______________ gain, the common-emitter achieves the highest _______________ gain, and the common-collector achieves the highest _______________ gain.

a. Voltage, current, power c. Voltage, power, current
b. Current, power, voltage d. Power, voltage, current

11. Which of the following abbreviations is used to denote the voltage drop between a transistor's base and emitter?

a. I_{BE} b. V_{CC} c. V_{CE} d. V_{BE}

12. Which of the following abbreviations is used to denote the voltage drop between a transistor's collector and emitter?

a. V_C b. V_{CE} c. V_E d. V_{CC}

13. A transistor's _______________ specification indicates the gain in dc current between the input and output of a common emitter circuit.

a. α_{AC} b. α_{DC} c. β_{AC} d. β_{DC}

14. Consider the following for a base-biased bipolar transistor circuit: R_B = 33 kΩ, R_C = 560 Ω, Q_1 (β_{DC}) = 25, V_{CC} = +10 V. What is V_{BE}?

a. 1.43 mV c. 0.7 V
b. 25 × 33 kΩ d. Not enough information given to calculate.

15. Which point on the dc load line results in an $I_C = V_{CC}/R_C$ and a V_{CE} = 0 V?

a. Saturation point b. Cutoff point c. Q point d. None of the above

16. Which point on the dc load line results in a $V_{CE} = V_{CC}$, and an I_C = 0?

a. Saturation point b. Cutoff point c. Q point d. None of the above

17. The midway point on the dc load line at which a transistor is biased with dc voltages when no signal input is applied is called the:

a. Saturation point b. Cutoff point c. Q point d. None of the above

18. Which transistor biasing method makes use of one current limiting resistor in the base circuit?

a. Base-Bias c. Emitter follower bias
b. Voltage-Divider bias d. Current divider bias

19. A forward biased transistor emitter or collector diode should have a _______________ resistance, while a reverse biased emitter and collector diode should have a _______________ resistance.

a. Low, low b. High, low c. High, high d. Low, high

20. A transistor tester will check a transistor's:

a. Opens or shorts between any of the terminals
b. Gain
c. Reverse leakage current value
d. All of the above

Essay Questions

21. What are the two basic transistor types? (8-1-1)

22. Give the full names of the following abbreviations. (Chapter 8)

a. BJT	**e.** V_{CE}	**i.** C-E	**m.** R_{in}
b. TO Package	**f.** I_C	**j.** β_{DC}	**n.** α_{AC}
c. SMT	**g.** I_B	**k.** A_P	**o.** ΔV_{out}
d. A_V	**h.** V_C	**l.** Q point	**p.** $I_{C(Sat.)}$

23. Briefly describe how transistors are constructed and packaged. (8-1-2)

24. Define transistor and transistance. (8-1-3)

25. Briefly describe how the bipolar transistor can be made to operate as a: (8-1-3)

a. Switch **b.** Variable resistor

26. Briefly describe how the BJT is used in: (8-1-4)

a. Digital circuit applications **b.** Analog circuit applications

27. Sketch and briefly describe the operation of (8-1-4)
a. A transistor logic gate **b.** A transistor amplifier

28. Briefly describe how a transistor achieves a gain in voltage between input and output. (8-1-4)

29. What is the relationship between a bipolar transistor's emitter current, base current, and collector current? (8-2-1)

30. Why is the bipolar transistor known as a current controlled device? (8-2-1)

31. In normal operation, which of the bipolar transistor's P-N junctions or diodes is forward biased, and which is reverse biased? (8-2-1)

32. What is the bipolar transistor equivalent to when it is operated in: (8-2-1)

a. Saturation **b.** The active region **c.** Cutoff

33. What are the three bipolar transistor circuit configurations? (8-2-2)

34. Briefly describe the following terms: (8-2-2)

a. DC beta **b.** AC beta **c.** DC alpha **d.** AC alpha

35. What is a collector characteristic curve? (8-2-2)

36. List the key characteristics of each of the three transistor configurations. (8-2-2)

37. Why is the common-emitter circuit configuration the most widely used? (8-2-2)

38. Why is the common-collector circuit configuration well suited as an impedance matching device? (8-2-2)

39. Why is impedance matching a desirable condition? (8-2-2)

40. What characteristics of a transistor will a transistor tester check? (8-2-4)

41. Use a sketch to show how an ohmmeter can be used to check a transistor's P-N junctions for opens and shorts. (8-2-4)

42. Sketch a base-biased common-emitter circuit. (8-2-5)

43. What is the disadvantage of a base-biased circuit? (8-2-5)

44. Sketch a voltage-divider biased: (8-2-5)

a. Common-emitter circuit
b. Common-base circuit
c. Common-collector circuit

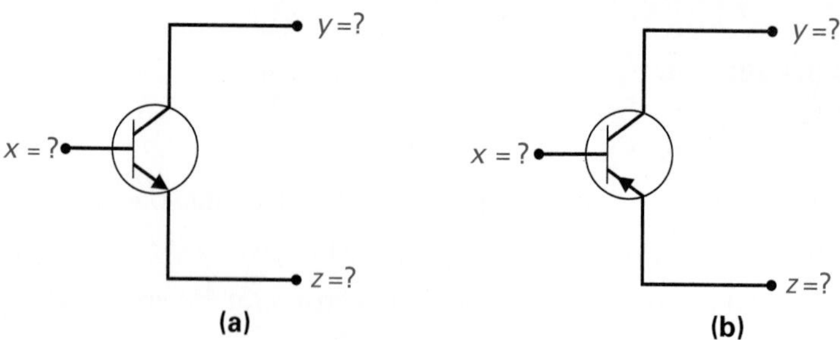

FIGURE 8-39 **Identify the Transistor Type and Terminals.**

45. How does an emitter resistor prevent Q point shift due to temperature change? (8-2-5)

Practice Problems

46. Identify the type and terminals of the transistors shown in Figure 8-39.

47. A bipolar transistor is correctly biased for operation in the active region when its emitter diode is forward biased and its collector diode is reverse biased. Referring to Figure 8-40, which of the bipolar transistor circuits is correctly biased?

48. Identify which of the leads of the bipolar transistor packages in Figure 8-41 is the emitter.

49. Digital circuits are often referred to as switching circuits because their control devices (diodes or transistors) are switched between the two extremes of ON and OFF. These digital circuits are also called two-state circuits because their control devices are driven either into the saturation state (ON) or cutoff state (OFF). Figure 8-42 illustrates three digital logic gates.

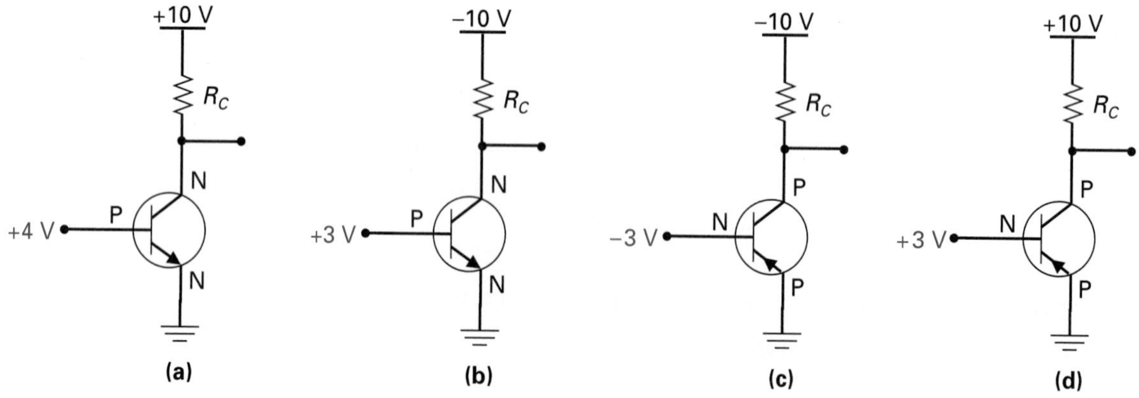

FIGURE 8-40 **Identifying the Correctly Biased (Active Region) Bipolar Transistors.**

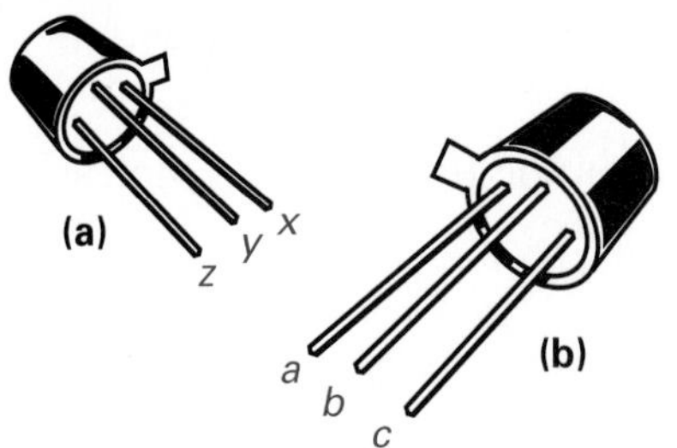

FIGURE 8-41 **Identifying the Bipolar Transistor's Emitter Lead.**

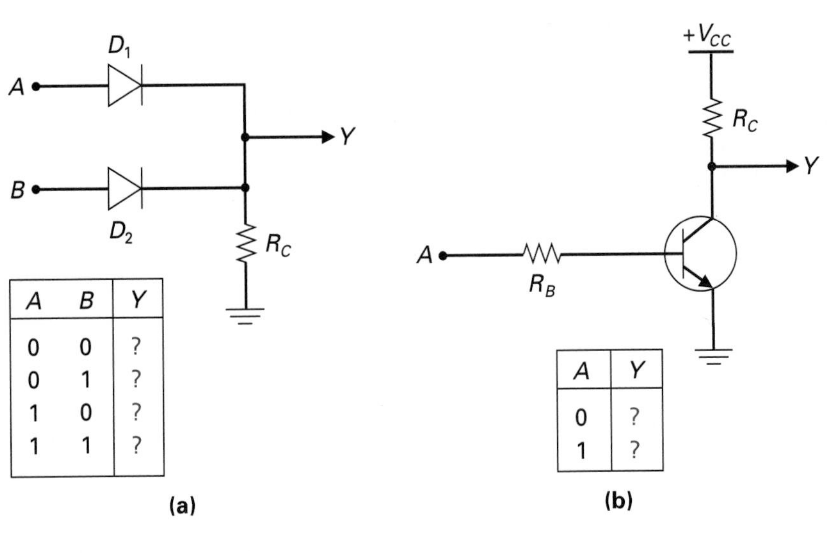

A	B	Y
0	0	?
0	1	?
1	0	?
1	1	?

(a)

A	Y
0	?
1	?

(b)

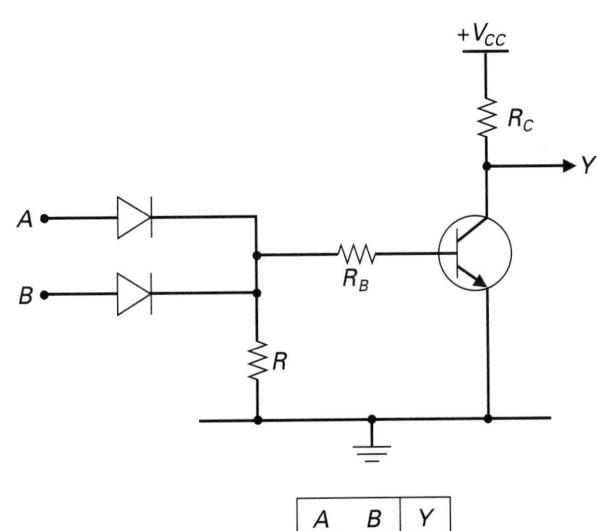

A	B	Y
0	0	?
0	1	?
1	0	?
1	1	?

(c)

FIGURE 8-42 **Digital Logic Gate Circuits (Logic 0 = 0 V, Logic 1 = +5 V).**

Complete their function tables by indicating what logic level will be at the output Y for each of the input combinations.

50. Calculate the value of the missing current in the following examples:

a. $I_E = 25$ mA, $I_C = 24.6$ mA, $I_B = ?$
b. $I_B = 600$ μA, $I_C = 14$ mA, $I_E = ?$
c. $I_E = 4.1$ mA, $I_B = 56.7$ μA, $I_C = ?$

51. Calculate the value of the missing current in the examples in Figure 8-43.

52. Calculate the dc beta for the transistor circuits in Figure 8-43.

53. Calculate the voltage gain (A_V) of the transistor amplifier whose input/output waveforms are shown in Figure 8-44.

54. Identify the configuration of the actual bipolar transistor electronic system circuits shown in Figure 8-45.

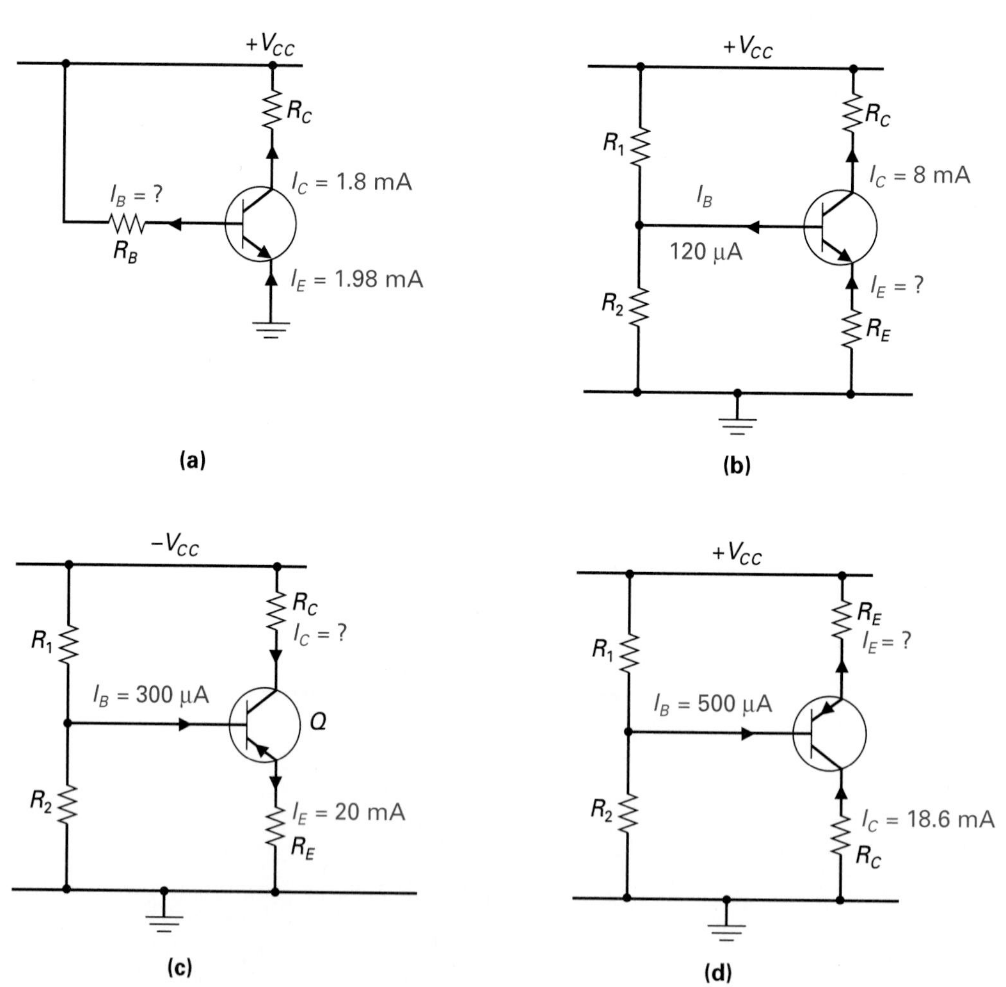

FIGURE 8-43 Calculating Transistor Current Values.

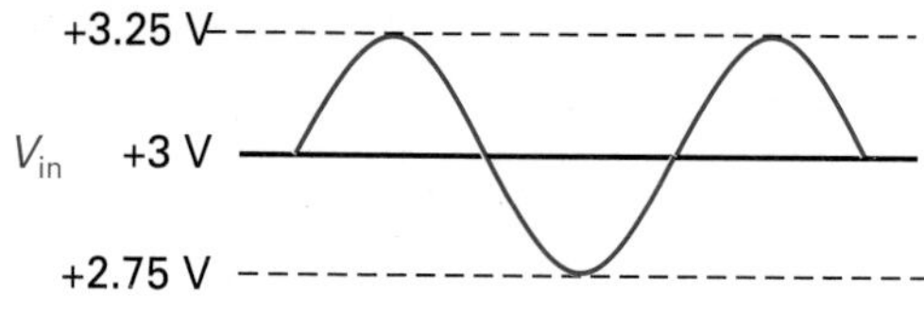

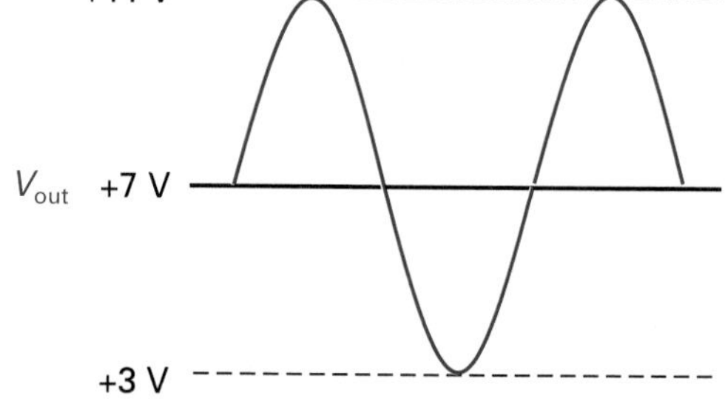

FIGURE 8-44 Transistor Amplifier Input/Output Waveforms.

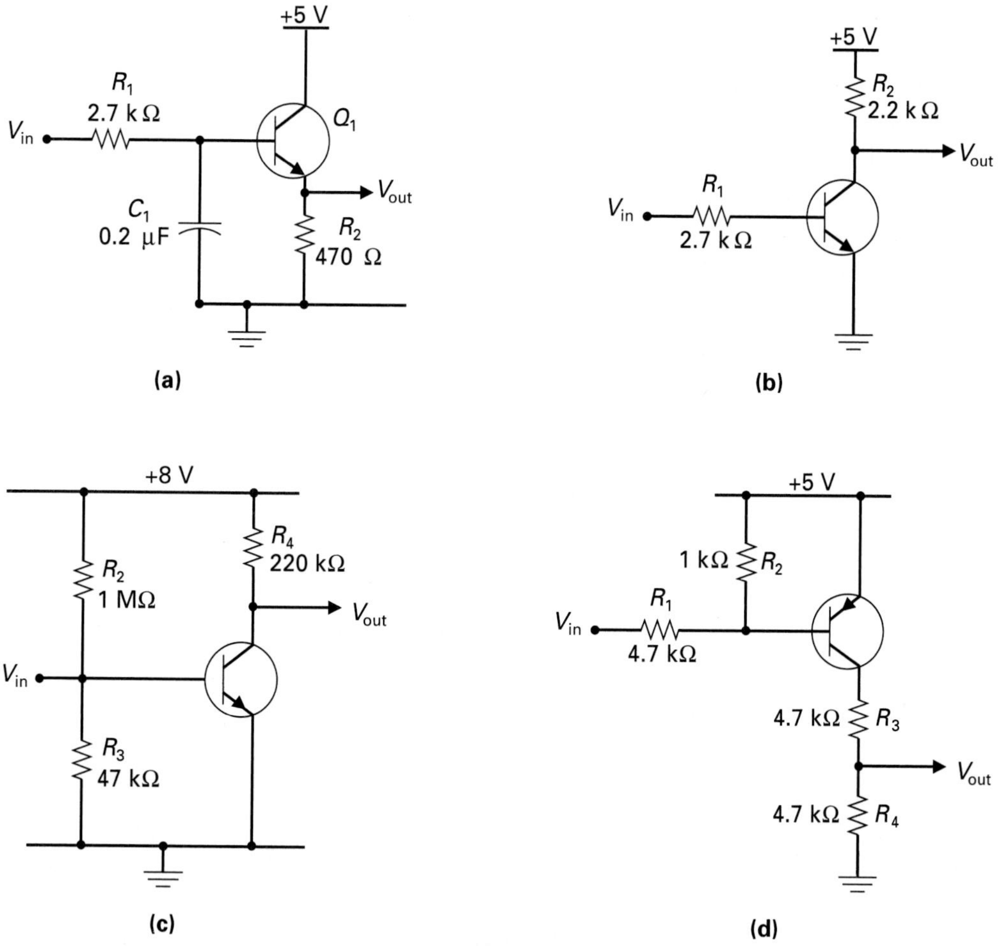

(a)

(b)

(c)

(d)

FIGURE 8-45 Identifying the Configuration of Actual BJT Circuits.

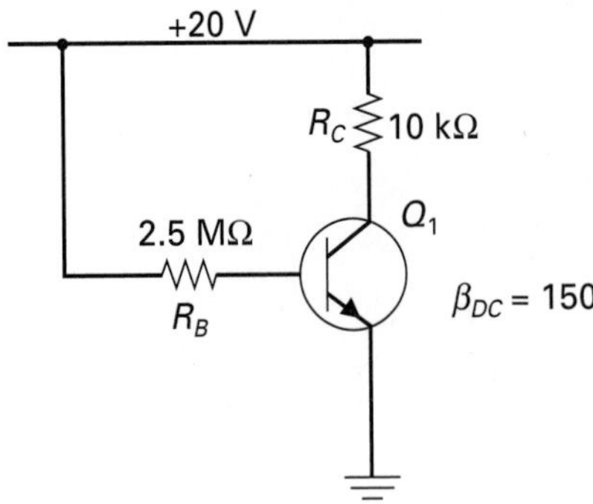

FIGURE 8-46 Base Biased Transistor Circuit.

55. Identify the bipolar transistor type, and the biasing technique used in Figure 8-45.

56. Calculate the following for the base biased transistor circuit shown in Figure 8-46:

a. I_B **b.** I_C **c.** V_{CE}

57. Sketch the dc load line for the circuit in Figure 8-46, showing the saturation, cutoff, and *Q* points.

58. Calculate the following for the voltage-divider biased transistor circuit shown in Figure 8-47:

a. V_B and V_E **b.** I_C **c.** V_C **d.** V_{CE}

59. Sketch the dc load line for the circuit in Figure 8-47, showing the saturation, cutoff, and *Q* points.

60. In Figure 8-47, does Kirchhoff's voltage law apply to the voltage-divider made up of R_C, R_{CE}, and R_E?

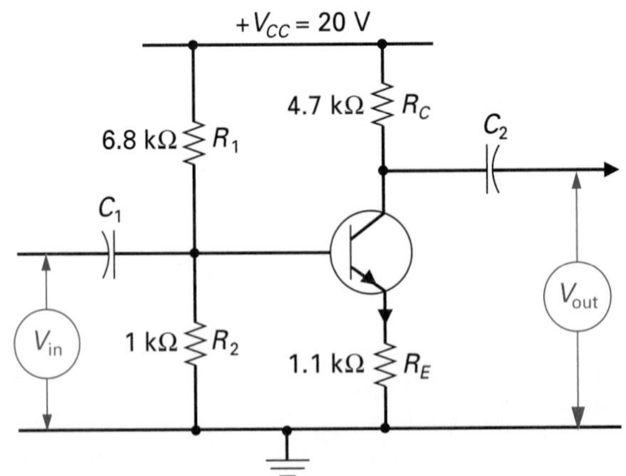

FIGURE 8-47 Voltage-Divider Biased Transistor Circuit.

Troubleshooting Questions

61. Briefly describe what characteristics a transistor tester can check.
62. How can an ohmmeter be used to test transistors?
63. Which of the transistors being tested in Figure 8-48 are good or bad? If bad, state the suspected problem.
64. Figure 8-49 shows a base biased circuit with its normal circuit values. Determine what "effect" the two faults in Figure 8-49(a) and (b) will have on this circuit. Describe the logic behind the cause and effect, and list other circuit faults that could possibly have the same effect.
65. Figure 8-50 shows a voltage-divider biased circuit with its normal circuit values. Determine what you think is the "root cause" of the three faults in Figure 8-50 (a), (b), and (c), based on the voltage readings given. Describe the logic behind the cause and effect, and list other circuit faults that could possibly have the same effect.

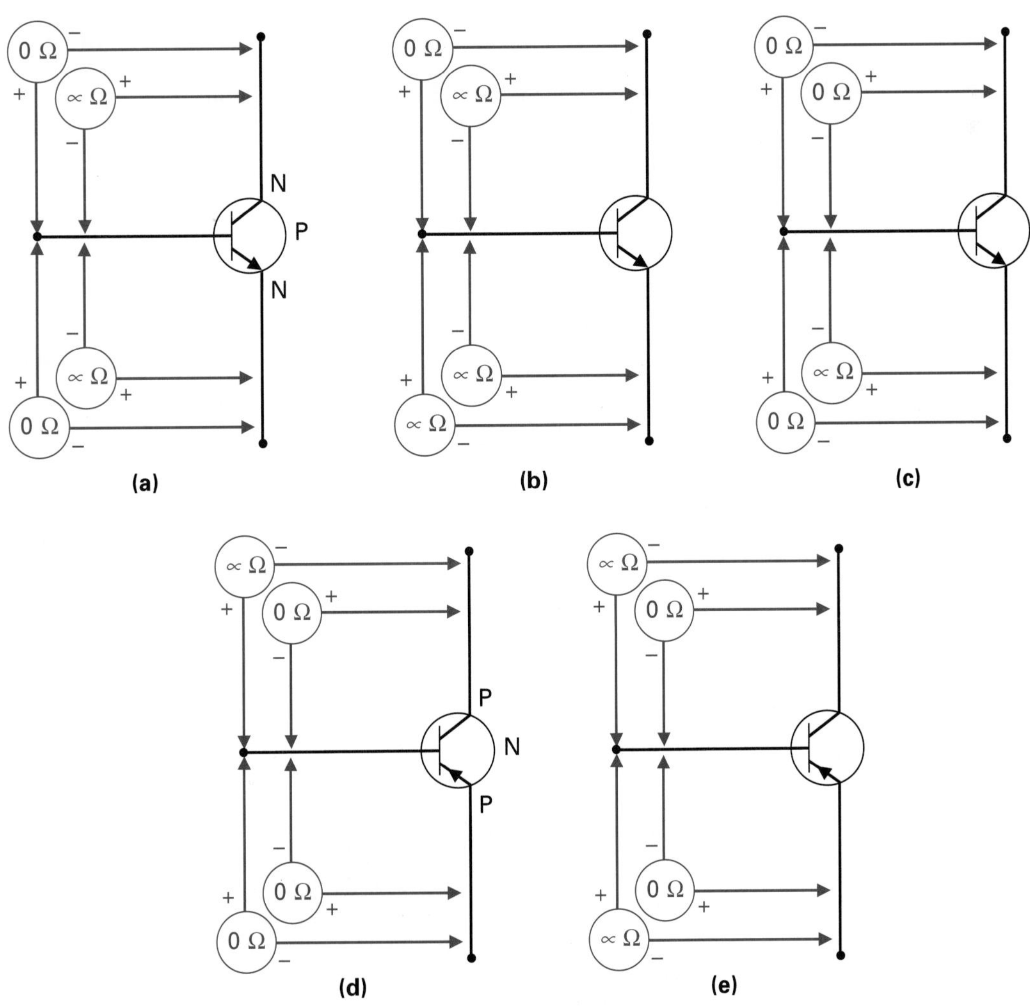

FIGURE 8-48 **Testing Transistors with the Ohmmeter.**

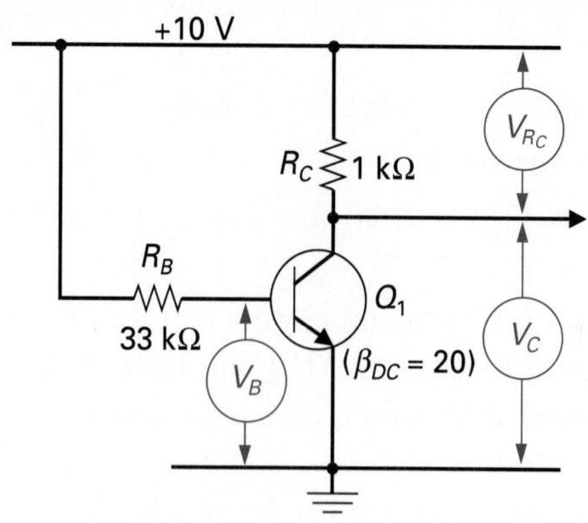

Under Normal Operating Conditions:

$V_{BE} = 0.7$ V	$I_B = 282$ μA
$V_{R_B} = 9.3$ V	$I_C = 5.6$ mA
$V_{R_C} = 5.6$ V	$I_{C(Sat.)} = 10$ mA
$V_{CE} = 4.4$ V	$I_{B(Sat.)} = 500$ μA

Cause—Q_1 Base resistor is open

Effect— Q_1 =
V_B =
V_C =
V_{R_C} =

(a)

Cause—Emitter diode is open

Effect— Q_1 =
V_B =
V_C =
V_{R_C} =

(b)

FIGURE 8-49 Troubleshooting Base Bias Circuit Problems.

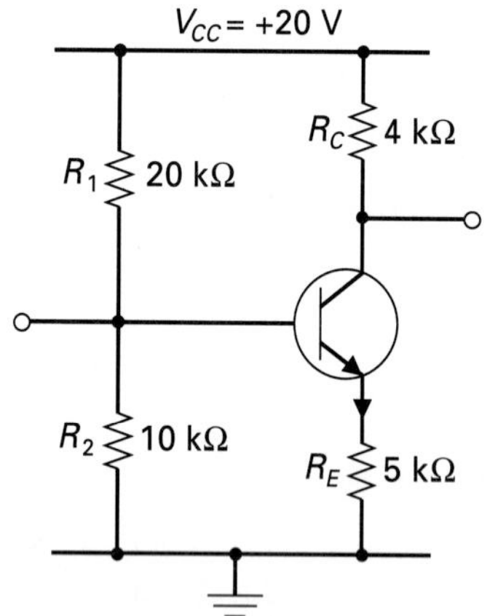

Under Normal Operating Conditions:

$V_B = 6.7$ V	$I_E = 1.2$ mA
$V_E = 6.0$ V	$I_C = 1.2$ mA
$V_{R_C} = 4.8$ V	$I_{C(Sat.)} = 2.2$ mA
$V_C = V_{out} = 15.2$ V	
$V_{CE} = 9.2$ V	

Effect— $V_B = 0$ V
$V_E = 0$ V
$V_C = +20$ V

Root Cause—?

(a)

Effect— $V_B = 11.7$ V
$V_E = 11.0$ V
$V_C = 11.0$ V

Root Cause—?

(b)

Effect— $V_B = 6.7$ V
$V_E = 0$ V
$V_C = +20$ V

Root Cause—?

(c)

FIGURE 8-50 Troubleshooting Voltage-Divider Bias Circuits.

CHAPTER 9

OBJECTIVES

After completing this chapter, you will be able to:

1. Describe how the bipolar transistor is used in digital or two-state switching circuit applications.
2. Explain how binary data is represented with digital hardware.
3. Describe the evolution of the two-state or digital switch.
4. List and describe the action of the five basic logic gates.
5. Describe the operation of a bipolar transistor standard TTL (transistor-transistor logic) gate circuit.
6. List and describe the standard bipolar transistor TTL logic gate characteristics.
7. Describe the circuit operation, characteristics, and applications of other bipolar transistor TTL logic gate circuits, such as:
   a. Low-power TTL logic gates
   b. Schottky TTL logic gates
   c. Open collector TTL logic gates
   d. Three-state output TTL logic gates
   e. Buffer/driver TTL logic gates
   f. Schmitt trigger TTL logic gates
8. Describe the circuit operation, characteristics, and applications of other non-standard bipolar transistor logic gate circuits, such as:
   a. Emitter-coupled logic (ECL) gates
   b. Integrated injection (I^2L) logic gates
9. Explain the operation, characteristics, and application of the following bipolar transistor digital timer and control circuits:
   a. The astable multivibrator circuits
   b. The monostable multivibrator circuit
   c. The bistable multivibrator circuit
   d. The 555 8-pin IC timer circuit
10. Describe how the 555 IC timer can be used in various applications, such as:
    a. A square wave or pulse wave generator
    b. A one-shot timer
11. Describe how traditional analog test equipment, such as the multimeter and oscilloscope, can be used to troubleshoot digital circuits.
12. Describe the operation and application of digital test equipment, such as the:
    a. Logic clip **c.** Logic pulser
    b. Logic probe **d.** Current tracer
13. Describe how test equipment and trouble-shooting techniques can be used to isolate digital circuit problems, such as:
    a. Internal I_C opens
    b. Internal I_C shorts
    c. External I_C opens and shorts
    d. Power supply problems
    e. Clock signal problems

Bipolar Transistor Digital Circuits

OUTLINE

V I G N E T T E

Moon Walk

In July of 1969, almost everyone throughout the world was caught at one stage or another gazing up to the stars and wondering. For American astronaut Neil Armstrong, the age-old human dream of walking on the moon was close to becoming a reality. On earth, millions of people waited anxiously to celebrate a successful moon landing. Then, the rather brief but eloquent message from Armstrong was transmitted 240,000 miles to Houston, Texas, where it was immediately retransmitted to a waiting world. This message was, "That's one small step for a man, one giant leap for mankind." These words were received by many via television, but for magazines and newspapers this entire mission, including the speech from Armstrong, was converted into a special code made up of ON-OFF pulses that traveled from computer to computer. Every letter of every word was converted into a code which used the two symbols of the binary number system, zero and one. This code is still used extensively by all modern computers and is called American Standard Code for Information Interchange, or ASCII (pronounced "askey").

It is only fitting that these codes made up of zeros and ones conveyed the finale to this historic mission because they played such an important role throughout. Commands encoded into zeros and ones were used to control almost everything, from triggering the take-off to keeping the spacecraft at the proper angle for re-entry into the earth's atmosphere.

No matter what its size or application, a digital electronic computer is, quite simply, a system that manages the flow of information in the form of zeros (0) and ones (1). Referring to the ASCII code table in Figure 9-1 (p. 423), see if you can decode the following famous Armstrong message, running from left to right.

0100010	1010100	1101000	1100001	1110100	0100111	1110011	0100000	1101111
1101110	1100101	0100000	1110011	1101101	1100001	1101100	1101100	0010000
1110011	1110100	1100101	1110000	0100000	1100110	1101111	1110010	0100000
1100001	0100000	1101101	1100001	0110110	0101100	0100000	1101111	1101110
1100101	0100000	1100111	1101001	1100001	1101110	0111010	0100000	1101100
1100101	1100001	1110000	0100000	1100110	1101111	1110010	0100000	1101101
1100001	1101110	1100100	0101110	0100010	0100000	0100000	0101101	1001110
1100101	1101001	1101100	0100000	1000001	1110010	1101101	1110011	1110100
1110010	1101111	1101110	1100111	0101100	0100000	1000001	1110000	1101111
1101100	1101100	1101111	0100000	0110001	0110001			

INTRODUCTION

In the previous chapter, we discovered how the bipolar junction transistor could be made to function as:

a. A two-state switch, or as

b. A variable resistor.

It is the first of these, the bipolar transistor's switching action, that is made use of in **digital electronic circuits.** These digital electronic circuits dominate almost every electronic system, which is why you need to be introduced to them at this point in your studies. Their influence has greatly improved the ability of electronic equipment to perform its primary function, which is to manage the flow of information. In fact, there are probably very few people today who have not been affected in some way by the biggest application of digital electronics: the **computer** or **data processor.**

In this chapter, you will be introduced to digital electronic **hardware,** which is used to represent and manipulate binary information. The hardware of a digital computer is all of the electronic, magnetic, and mechanical devices of a digital system. The **software** of a computer is the binary data (like ASCII codes) and binary instructions that are processed by the digital system hardware. To use an analogy, we could say that a compact disc player is hardware, and the music information stored on a compact disc is software. Just as a CD player is useless without CDs, a computer's hardware is useless without software. In other words, it is the information on the CD that determines what music is played, and, similarly, it is the computer's software that determines the actions of the computer hardware.

The first section in this chapter will discuss the evolution of the electronic switch, and then show how the bipolar transistor switch is used to construct basic **digital logic gate circuits.** These logic gates are used as the basic building blocks for digital circuits, and their function is to control the movement of binary data and binary instructions.

In the second section of this chapter, we will see how the bipolar transistor is used to construct basic **digital timer and control circuits.** These circuits generate, manage, and modify digital signals.

In the final section of this chapter, we will discuss **digital circuit troubleshooting,** including the use of digital test equipment, typical digital circuit failures, and digital troubleshooting techniques.

Digital Electronic Circuits
Circuits that have outputs of only two distinct levels or steps, for example on-off, open-closed, and so on.

Computer or Data Processor
Piece of equipment used to process information or data.

Hardware
Electrical, electronic, mechanical, or magnetic devices or components. The physical equipment.

Software
Program of instructions that directs the operation of a computer.

Digital Logic Gate Circuit
Circuits that employ digital logic gates to control the movement of binary data and binary instructions.

Digital Timer and Control Circuits
Circuits that generate, manage, and modify digital signals.

Digital Circuit Troubleshooting
The use of digital test equipment, typical digital circuit failures, and digital troubleshooting techniques.

9-1 DIGITAL LOGIC GATE CIRCUITS

Digital circuits are often referred to as "switching circuits," because their control devices (diodes or transistors) are switched between the two extremes of ON and OFF. These digital circuits are also called "two-state circuits" because their control devices are driven either into the saturation state (fully ON) or cutoff state (fully OFF). With digital electronic circuits, information is first encoded into a series of "1s" (ones) and "0s" (zeros), such as the Armstrong message discussed in this chapter's vignette. Figure 9-1(a) lists the "American Standard Code for Information Interchange," which is abbreviated ASCII (pronounced "askey"). For example, if you were to look up the code shown in Figure 9-1(b), you will find that the "1101001" information or data stream code corresponds to the lower case "i" in the ASCII table in Figure 9-1(a). An example of a circuit that makes use of this two-state ASCII code is the digital computer keyboard, shown in Figure 9-1(b). Whenever the lower case "i" key on a computer keyboard is pressed, a "1101001" code is generated, thereby encoding the information "i" into a group of pulses.

The next question you may have is: why do we go to all of this trouble to encode all of our data or information into these two-state codes? The answer can best be explained by examining history. The early digital systems constructed in the 1950s made use of ten levels or voltages. Each of these voltages corresponded to one of the ten digits in the decimal number system (0 = 0 V, 1 = 1 V, 2 = 2 V, 3 = 3 V, up to 9 = 9 V). These circuits, however, were very complex because they had to generate one of ten voltage levels and sense the difference between all ten voltage levels. This complexity led to inaccuracy because some circuits would periodically confuse one voltage level for a different voltage level. *The solution to the problem of circuit complexity and inaccuracy was solved by adopting a two-state system instead of a ten-state system.* Using a two-state or two-digit system, you can generate codes for any number, letter, or symbol, as we have seen in the ASCII table in Figure 9-1(a). However, the electronic circuits that manage the two-state codes are less complex because they only have to generate and sense a HIGH and LOW voltage, and they are much more accurate because there is little room for error between the two extremes of ON and OFF, or HIGH voltage and LOW voltage.

Abandoning the ten-state system and adopting the two-state system for the advantages of circuit simplicity and accuracy means that we are no longer dealing with the decimal number system. Having only two digits (0 and 1) means that we are now operating in the two-state number system, which is called **binary.** Figure 9-2(a) shows how the familiar decimal system has ten digits or levels labeled 0 through 9, and Figure 9-2(b) shows how we could electronically represent each decimal digit. With the base 2, or binary number system, we only have two digits in the number scale. Therefore, only the digits "0" and "1" exist in binary, as shown in Figure 9-2(c). These two states are typically represented in an electronic circuit as two different values of voltage (0 = LOW voltage, 1 = HIGH voltage), as shown in Figure 9-2(d). Using combinations of **binary digits** (abbreviated **"bits"**), we can represent information as a binary code. This code is called a **digital signal** because it is an *information signal* that makes use of *binary digits.* Today, almost all information, from your voice telephone conversations to the music on your compact discs, is **digitized** or converted to a binary data form.

Binary
The two-state number system.

Binary Digits or Bits
Used to represent information in binary code.

Digital Signal
An information signal that makes use of binary digits.

Digitized
Information that is converted to a binary data form.

To develop a digital electronic system that could manipulate binary information, inventors needed a two-state electronic switch. Early machines used mechanical switches and electromechanical relays, such as the examples seen in Figure 9-3(a) and (b) (p. 425), to represent binary data by switching current ON and OFF. These mechanical devices were eventually replaced by the vacuum

The American Standard Code for Information Interchange (ASCII)

Character								Character							
•	0	1	0	0	0	0	0	P	1	0	1	0	0	0	0
!	0	1	0	0	0	0	1	Q	1	0	1	0	0	0	1
"	0	1	0	0	0	1	0	R	1	0	1	0	0	1	0
#	0	1	0	0	0	1	1	S	1	0	1	0	0	1	1
$	0	1	0	0	1	0	0	T	1	0	1	0	1	0	0
%	0	1	0	0	1	0	1	U	1	0	1	0	1	0	1
&	0	1	0	0	1	1	0	V	1	0	1	0	1	1	0
'	0	1	0	0	1	1	1	W	1	0	1	0	1	1	1
(	0	1	0	1	0	0	0	X	1	0	1	1	0	0	0
)	0	1	0	1	0	0	1	Y	1	0	1	1	0	0	1
*	0	1	0	1	0	1	0	Z	1	0	1	1	0	1	0
+	0	1	0	1	0	1	1	[	1	0	1	1	0	1	1
,	0	1	0	1	1	0	0	/	1	0	1	1	1	0	0
–	0	1	0	1	1	0	1	]	1	0	1	1	1	0	1
.	0	1	0	1	1	1	0	^	1	0	1	1	1	1	0
/	0	1	0	1	1	1	1	_	1	0	1	1	1	1	1
0	0	1	1	0	0	0	0	'	1	1	0	0	0	0	0
1	0	1	1	0	0	0	1	a	1	1	0	0	0	0	1
2	0	1	1	0	0	1	0	b	1	1	0	0	0	1	0
3	0	1	1	0	0	1	1	c	1	1	0	0	0	1	1
4	0	1	1	0	1	0	0	d	1	1	0	0	1	0	0
5	0	1	1	0	1	0	1	e	1	1	0	0	1	0	1
6	0	1	1	0	1	1	0	f	1	1	0	0	1	1	0
7	0	1	1	0	1	1	1	g	1	1	0	0	1	1	1
8	0	1	1	1	0	0	0	h	1	1	0	1	0	0	0
9	0	1	1	1	0	0	1	i	1	1	0	1	0	0	1
:	0	1	1	1	0	1	0	j	1	1	0	1	0	1	0
;	0	1	1	1	0	1	1	k	1	1	0	1	0	1	1
<	0	1	1	1	1	0	0	l	1	1	0	1	1	0	0
=	0	1	1	1	1	0	1	m	1	1	0	1	1	0	1
>	0	1	1	1	1	1	0	n	1	1	0	1	1	1	0
?	0	1	1	1	1	1	1	o	1	1	0	1	1	1	1
@	1	0	0	0	0	0	0	p	1	1	1	0	0	0	0
A	1	0	0	0	0	0	1	q	1	1	1	0	0	0	1
B	1	0	0	0	0	1	0	r	1	1	1	0	0	1	0
C	1	0	0	0	0	1	1	s	1	1	1	0	0	1	1
D	1	0	0	0	1	0	0	t	1	1	1	0	1	0	0
E	1	0	0	0	1	0	1	u	1	1	1	0	1	0	1
F	1	0	0	0	1	1	0	v	1	1	1	0	1	1	0
G	1	0	0	0	1	1	1	w	1	1	1	0	1	1	1
H	1	0	0	1	0	0	0	x	1	1	1	1	0	0	0
I	1	0	0	1	0	0	1	y	1	1	1	1	0	0	1
J	1	0	0	1	0	1	0	z	1	1	1	1	0	1	0
K	1	0	0	1	0	1	1	(	1	1	1	1	0	1	1
L	1	0	0	1	1	0	0	¦	1	1	1	1	1	0	0
M	1	0	0	1	1	0	1	)	1	1	1	1	1	0	1
N	1	0	0	1	1	1	0	~	1	1	1	1	1	1	0
O	1	0	0	1	1	1	1	DEL	1	1	1	1	1	1	1

(a)

(b)

FIGURE 9-1 The Two-State (Binary) ASCII Code.

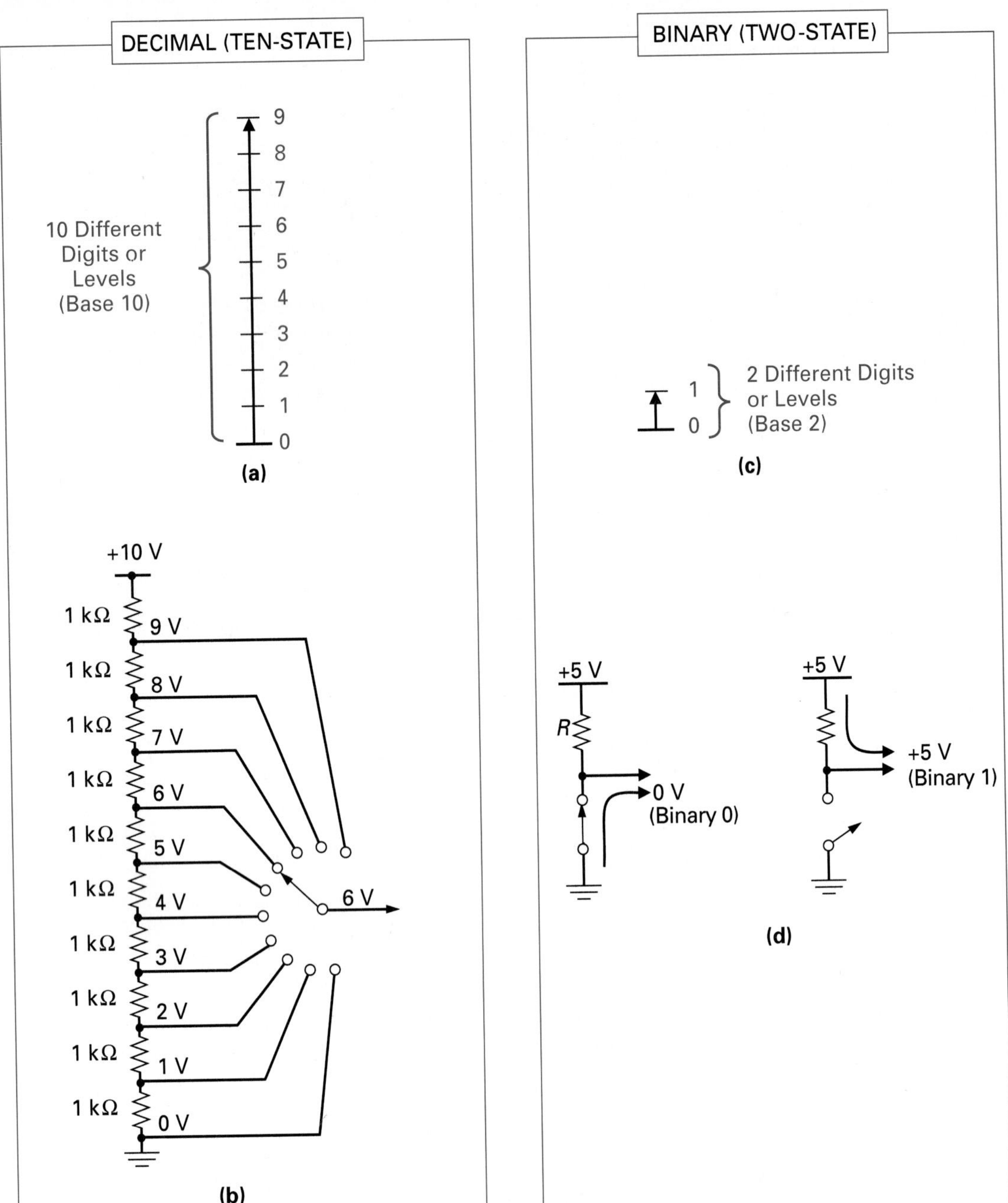

FIGURE 9-2 Electronically Representing the Digits of a Number System.

tube, shown in Figure 9-3(c), which, unlike switches and relays, had the advantage of no moving parts. The vacuum tube was bulky, fragile, had to warm up, and consumed an enormous amount of power. Finally, compact and low-power digital electronic circuits and systems became a reality with the development of a new kind of switch called the bipolar junction transistor, which is shown in Figure 9-3(d).

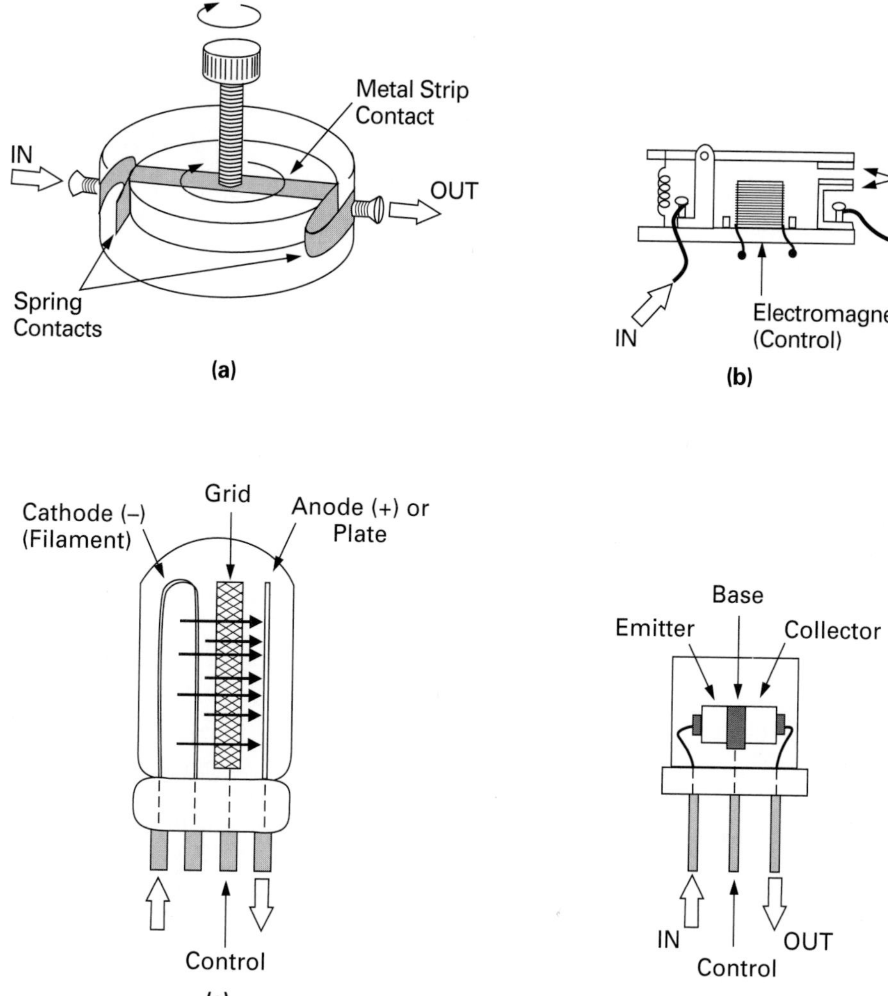

FIGURE 9-3 The Evolution of the Switch for Digital Electronic Circuits. (a) A 19th Century Mechanical Turn Switch. (b) An Electromechanical Relay. (c) A 1906 Vacuum Tube. (d) A 1948 Bipolar Junction Transistor.

9-1-1 *Basic Logic Gates*

Every digital electronic circuit uses **logic gate circuits** to manipulate the coded pulses of binary language. These logic gate circuits are constructed using transistors and are the basic "decision-making elements" in all digital circuits.

Logic Gate Circuits
Circuits constructed using transistors. They are the basic "decision-making elements" in all digital circuits.

Figure 9-4 shows the five basic logic gates. These logic gates accept HIGH or LOW voltage signals at their inputs, judge these input levels based on a predetermined set of rules, and then produce a single HIGH or LOW voltage signal at their output. You have already been introduced to most of these logic gates in previous chapters where they have been used as examples of digital circuit applications for the diode and transistor. A brief review of logic gates follows.

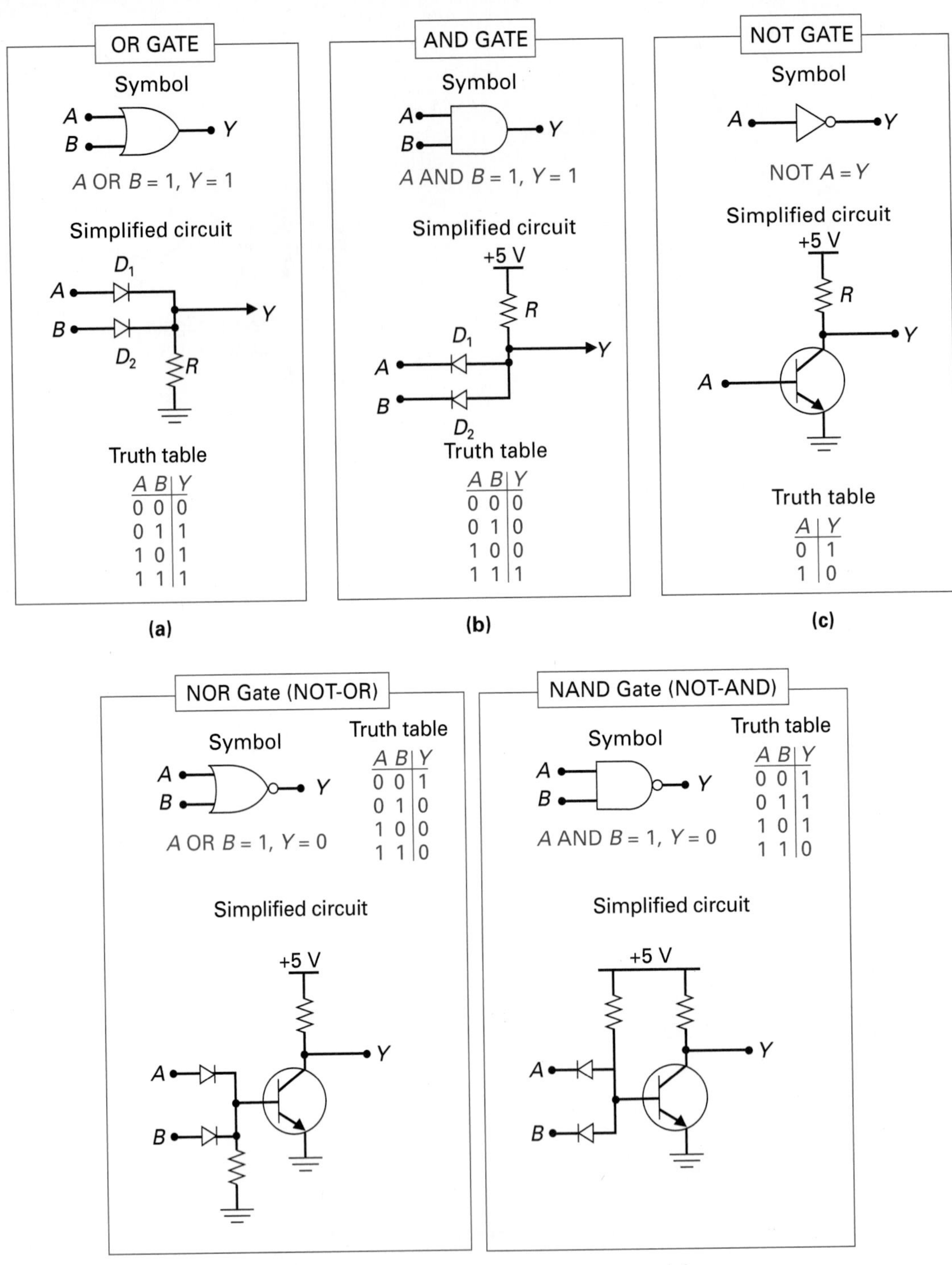

FIGURE 9-4 The Five Basic Logic Gates.

OR gate Figure 9-4(a) shows the symbol, simplified circuit, and truth table or function table for the OR gate. As illustrated, an OR gate will deliver a binary 1 or logical TRUE (+5 V) output, if any one of its inputs is TRUE. An OR gate will only generate a binary 0 or logical FALSE (0 V) output when all of its inputs are FALSE.

The OR gate could have more than two inputs; however, it will always only have one output. More inputs do not change the basic action of the OR gate because it will still generate a HIGH output if any of its inputs are HIGH and only generate a LOW output when all of its inputs are LOW.

AND gate Figure 9-4(b) shows the symbol, simplified circuit, and truth table or function table for the AND gate. As illustrated, an AND gate will deliver a binary 1 or logical TRUE output only when all of its inputs are logically TRUE. Explained another way, we can say that the output of an AND gate will be LOW if any of its inputs are LOW.

NOT gate Figure 9-4(c) shows the symbol, simplified circuit, and truth table or function table for the NOT or INVERTER (INV.) gate. As illustrated, a NOT gate accepts just one input which it then reverses. For example, a NOT gate will turn a binary 0 input into binary 1 output and binary 1 input into binary 0 output.

NOR and NAND gates NOT gates are often combined with OR gates and AND gates to form NOR (NOT-OR) and NAND (NOT-AND) gates as shown in Figure 9-4(d) and (e). These combined devices process the inputs using the usual AND/OR rules and then automatically invert the output. For example, comparing the truth table of the NOR gate and the OR gate, you can see that the *Y* output is simply opposite. The same is true if you compare the NAND gate *Y* output with the AND gate *Y* output. The circle or "bubble" at the output of the NOR and NAND schematic symbol is used to indicate the NOT or inverting action that occurs with these gates. Studying the simplified circuit of the NOR and NAND gate, you can see that they are combinations of an OR followed by a NOT and an AND followed by a NOT.

These logic gates are the basic elements in all digital electronic circuits. The rules that control the operation of each logic gate are used to control the movement of data in a digital circuit. For example, a certain input signal would only be allowed to pass to another circuit if both the control inputs to an AND gate are true.

Let us now examine the basic construction of these logic gate devices.

9-1-2 Basic Logic Gate Construction

The simplified logic gate circuits shown in Figure 9-4 only serve to show how each logic gate will basically operate. In reality, many more components are included in a typical logic gate circuit to obtain better input/output characteristics. For example, Figure 9-5 shows a typical logic gate integrated circuit (IC). Referring to the inset in this figure, you can see that a logic gate is actually constructed using a number of bipolar transistors, diodes, and resistors. All of these components are formed and interconnected on one side of a silicon chip. This

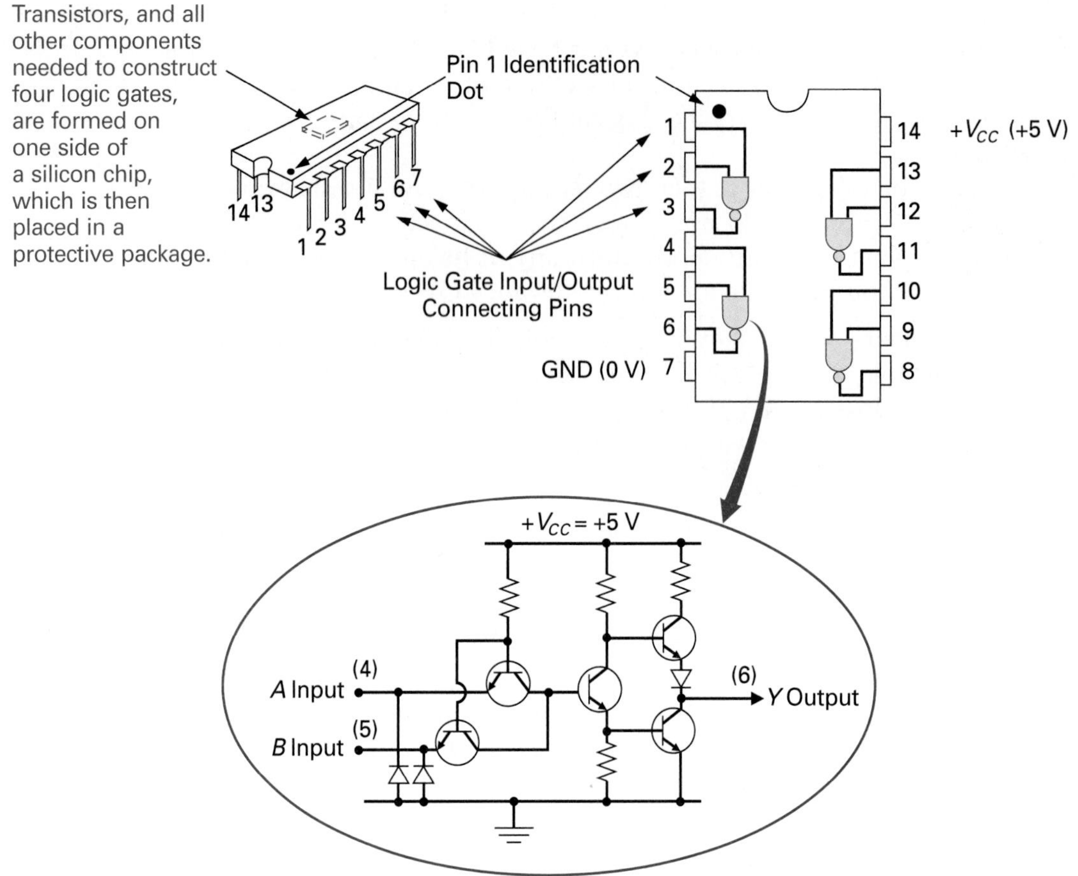

FIGURE 9-5 **Basic Logic Gate IC.**

single piece of silicon actually contains four logic gate transistor circuits, with the inputs and output of each logic gate connected to the external pins of the IC package. For example, looking at the top view of the IC in Figure 9-5, you can see that one logic gate has its input on pin 1 and 2 and its output on pin 3. Because all four logic gate circuits will need a $+V_{CC}$ supply voltage (typically +5 V) and a ground (0 V), two pins are assigned for this purpose (pin 14 = $+V_{CC}$, pin 7 = ground). In the example in Figure 9-5, the IC contains four two-input NAND gates. This IC package is called a dual-in-line package (DIP) because it contains two rows of connecting pins.

9-1-3 TTL Series of Logic Gates

Transistor-Transistor Logic (TTL) ICs Circuit used to construct the logic gates that contain several interconnected transistors.

In 1964, Texas Instruments introduced a complete range of logic gate integrated circuits. They called this line of products **transistor-transistor logic (TTL) ICs** because the circuit used to construct the logic gates contained several interconnected transistors (**T**ransistor-connected-to-**T**ransistor **L**ogic circuit), as shown in the example in the inset in Figure 9-5. These ICs became the building blocks for all digital circuits. They were building blocks in the true sense of the word.

TABLE 9-1 7400 Standard TTL Series

DEVICE NUMBER	LOGIC GATE CONFIGURATION
7400	Quad (four) 2-input NAND gates
7402	Quad 2-input NOR gates
7404	Hex (six) NOT or Inverter gates
7408	Quad 2-input AND gates
7410	Triple (three) 3-input NAND gates
7411	Triple 3-input AND gates
7420	Dual (two) 4-input NAND gates
7421	Dual 4-input AND gates
7427	Triple 3-input NOR gates
7430	Single (one) 8-input NAND gate
7432	Quad 2-input OR gates

No knowledge of electronic circuit design was necessary because the electronic circuits had already been designed and fabricated on a chip. All that had to be done was to connect all of the individual ICs in a combination that achieved the specific or the desired circuit operation. All of these TTL ICs were compatible, which meant they all responded and generated the same logic 1 and 0 voltage levels, they all used the same value of supply voltage, and therefore the output of one TTL IC could be directly connected to the input of one or more other TTL ICs in any combination.

By modifying the design of the bipolar transistor circuit within the logic gate, TTL IC manufacturers can change the logic function performed. The circuit can also be modified to increase the number of inputs to the gate. The most common type of TTL circuits are the 7400 (seventy-four hundred series originally developed by Texas Instruments and now available from almost every digital IC manufacturer). As an example, Table 9-1 lists a few of the frequently used 7400 series TTL ICs.

The TTL IC example shown in Figure 9-5 would have a device part number of "7400" since it contains four (quad) 2-input NAND gates.

■ **EXAMPLE:**

Figure 9-6(a) shows how a 7432 (quad 2-input OR gate TTL IC) could be connected to test the operation of an OR gate. Sketch a timing diagram showing a 0 V to 5 V square wave input to the OR gate from the function generator, a randomly selected HIGH and LOW control voltage input from the ENABLE/DISABLE switch, and the resulting OR gate output displayed on the oscilloscope.

■ ***Solution:***

Referring to the waveforms in Figure 9-6(b), you can see that any HIGH input will produce a HIGH output. Therefore, the alternating HIGH and LOW square wave signal will only be present at the output if the ENABLE/DISABLE switch is LOW (enabling the square wave to reach the output).

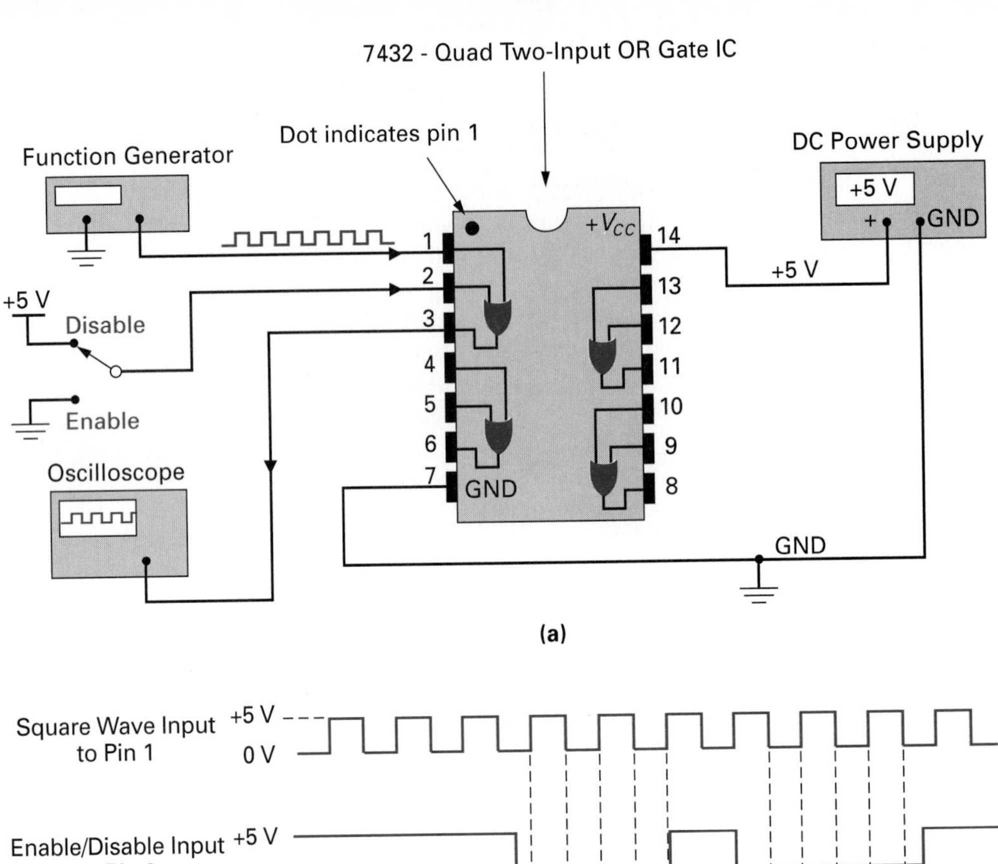

FIGURE 9-6 OR Gate Test Circuit.

■ **EXAMPLE:**

Figure 9-7(a) shows how a 7408 (quad 2-input AND gate TTL IC) could be connected to test the operation of an AND gate. Sketch a timing diagram showing a 0 V to 5 V square-wave input to the AND gate from the function generator, a randomly selected HIGH and LOW control voltage input from the ENABLE/DISABLE switch, and the resulting AND gate output displayed on the oscilloscope.

■ ***Solution:***

Referring to the waveforms in Figure 9-7(b), you can see that only when both inputs are HIGH will the output be HIGH. Therefore, the alternating HIGH and LOW square wave signal will only be present at the output if the ENABLE/DISABLE switch is HIGH (enabling the square wave so that it will reach the output).

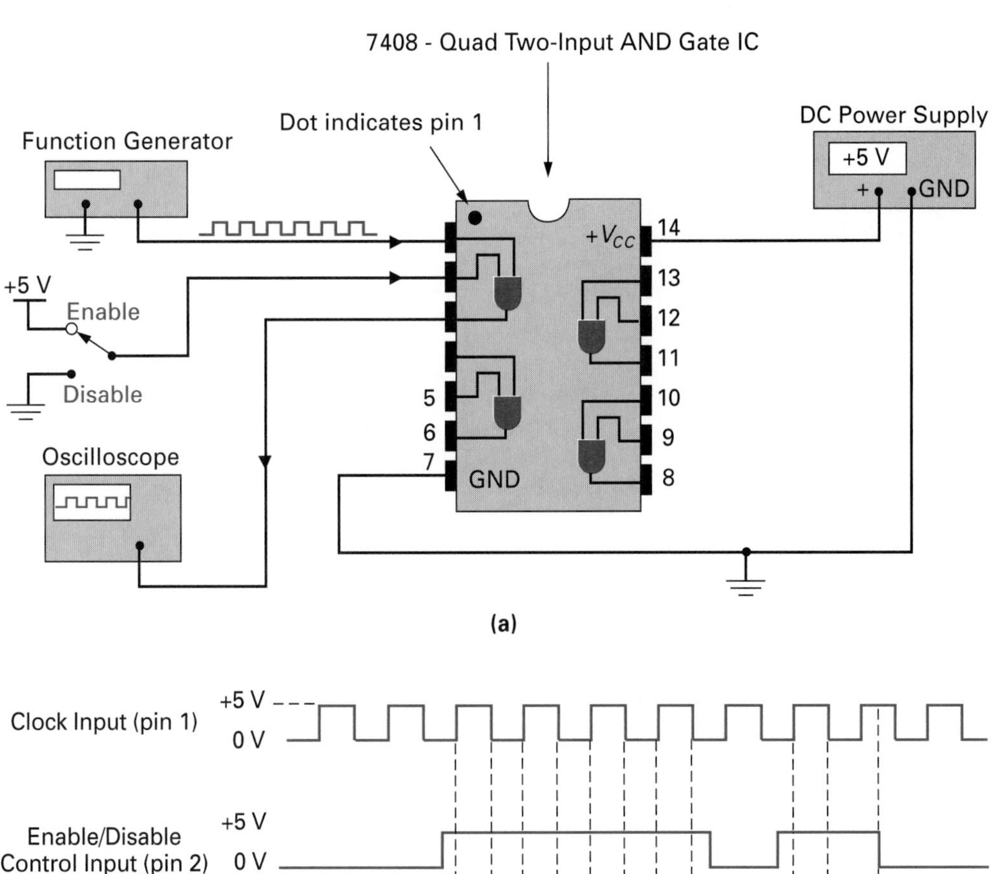

FIGURE 9-7 AND Gate Test Circuit.

9-1-4 Standard TTL Logic Gate Circuits

It is important that you understand the inner workings of a logic gate circuit. You will only be able to diagnose a malfunctioning logic gate if you know how a logic gate is supposed to operate. In this section, we will examine the bipolar transistor circuits within two typical standard TTL ICs.

Bipolar Transistor NOT Gate Circuit

Figure 9-8(a) shows the pin assignment for the six NOT gates within a 7404 TTL IC. All six NOT gates are supplied power by connecting the $+V_{CC}$ terminal (pin 14) to +5 V and the ground terminal (pin 7) to 0 V. Connecting power and ground to the 7404 will activate all six of the NOT gate circuits so that they are ready to function as inverters.

The inset in Figure 9-8(a) shows the circuit for a bipolar TTL NOT or INVERTER gate. As expected, the INVERTER has only one input, which is

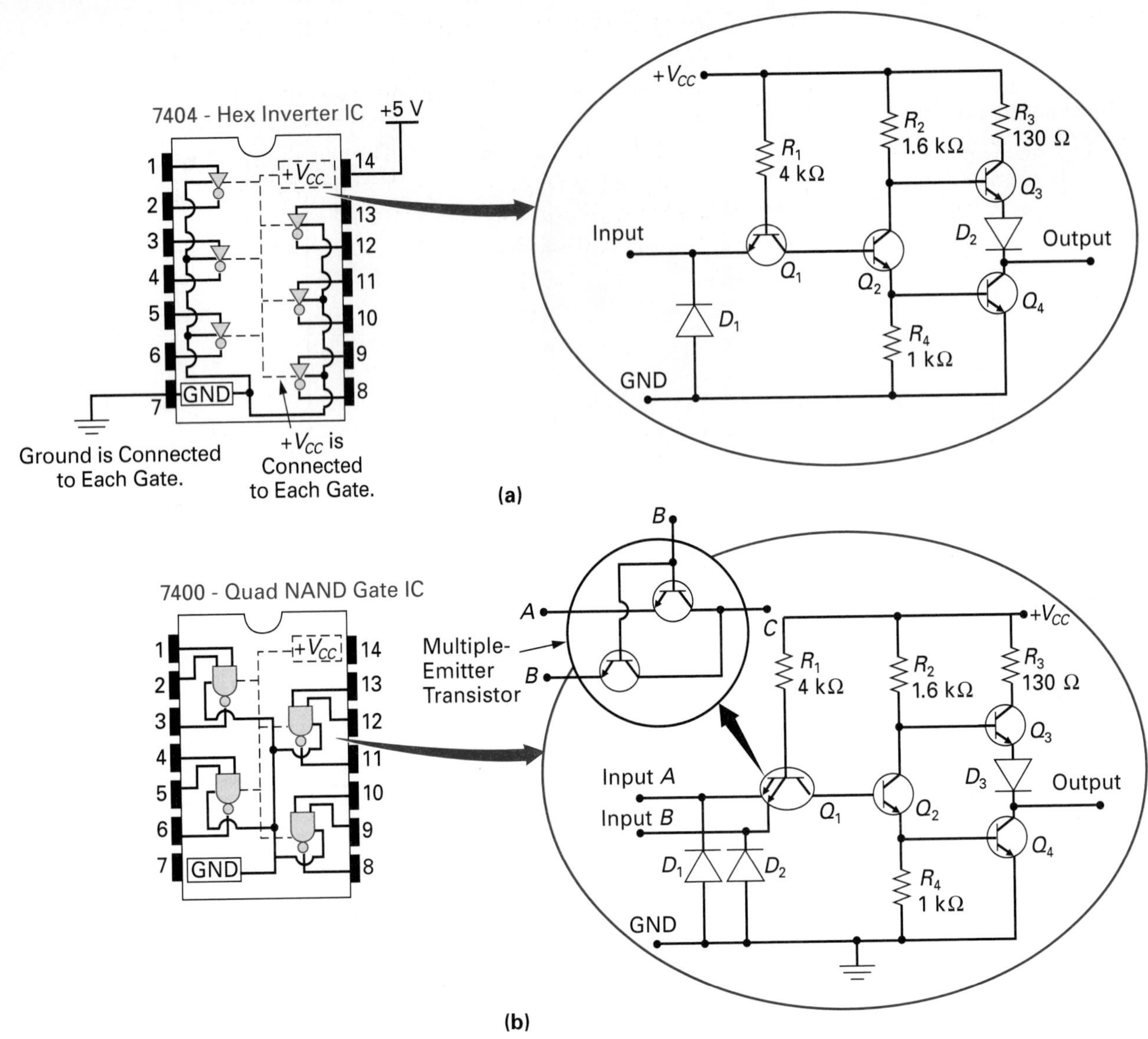

FIGURE 9-8 Standard TTL Logic Gate Circuits. (a) NOT (Inverter) Gate. (b) NAND Gate.

Totem Pole Circuit A transistor circuit containing two transistors connected one on top of the other with two inputs and one output.

Phase Splitter Circuit elements whose collector and emitter outputs are out of phase with one another.

applied to the *coupling transistor*, Q_1, and one output, which is developed by the **totem pole circuit** made up of Q_3 and Q_4. This totem pole circuit derives its name from the fact that Q_3 sits on top of Q_4 like the elements of a Native American totem pole. Transistor Q_2 acts as a **phase splitter,** because its collector and emitter outputs will be out of phase with one another (base-to-emitter = 0° phase shift, base-to-collector = 180° phase shift), and it is the two opposite outputs that will drive the bases of Q_3 and Q_4.

Let us now see how this circuit will respond to a LOW input. A 0 volt (binary 0) input to the NOT gate will apply 0 V to the emitter of Q_1. This will forward bias the P-N base-emitter junction of Q_1 (base is connected to +5 V, emitter is at 0 V), causing electron flow from the 0 V input through Q_1's base emitter through R_1 to $+V_{CC}$. Because the base of Q_1 is only 0.7 V above the emitter voltage, the collector diode of Q_1 will be reversed biased; the base of Q_2 will receive no input voltage, and therefore turn OFF. With Q_2 cut off, all of the +5 V supply voltage will be applied via R_2 to the base of Q_3, sending Q_3 into saturation. With Q_3 saturated, its collector to emitter junction is equivalent to a closed switch between collector and emitter. The +5 V supply voltage ($+V_{CC}$) will be connected to the output via the low output resistance path of R_3, Q_3's collector-to-emitter, and D_2. A LOW input, therefore, will result in a HIGH output. In this condition, transistor Q_4 is cut off and equivalent to an open switch between collector and emitter. Q_2 is OFF, and therefore the emitter of Q_2 and the base of Q_4 are at 0 V.

A HIGH input (+5 V) to the emitter of Q_1 will reverse bias the P-N base-emitter junction of Q_1 (base is connected to +5 V, emitter = +5 V). The +5 V at the base of Q_1 will forward bias Q_1's base collector junction, applying +5 V to the base of Q_2 sending it into saturation. When Q_2 was cut off, all of the $+V_{CC}$ supply voltage was applied to the base of Q_3. In this condition, Q_2 is heavily ON and equivalent to a closed switch between the collector and emitter. A proportion of the $+V_{CC}$ supply voltage will be developed across R_4 and applied to the base of Q_4. Q_4 will turn ON and switch its emitter voltage of 0 V to the collector, and therefore the output. A HIGH input will result in a LOW output. In this condition, Q_3 is kept OFF by a sufficiently low voltage level on Q_2's collector, and by diode D_2, which adds an additional diode to the emitter diode of Q_3 ensuring Q_3 stays OFF when Q_4 is ON.

The only other component that has not been discussed is the input diode D_1, which is included to prevent any negative input voltage spikes (negative transient) from damaging Q_1. D_1 will conduct these negative input voltages to ground.

Bipolar Transistor NAND Gate Circuit

Figure 9-8(b) shows the pinout for the standard TTL 7400 integrated circuit and the bipolar transistor circuitry needed for each of the four NAND gates within this chip. The circuit is basically the same as the INVERTER or NOT gate circuit previously discussed, except for the **multiple-emitter input transistor,** Q_1. This transistor is used in all two-input logic gates, and if three inputs were needed, you would see a three-emitter input transistor. Q_1 can be thought of as having two transistors with separate emitters, with the bases and emitters connected, as shown adjacent to the NAND circuit in Figure 9-8(b). Any LOW input to either A or B will cause the respective emitter diode to conduct current from the LOW input through Q_1's base emitter and R_1 to $+ V_{CC}$. The LOW voltage at the base of Q_1 will reverse bias Q_1's collector diode, preventing any voltage from reaching Q_2's base, and so Q_2 will cut OFF. The remainder of the circuit will operate in exactly the same way as the NOT gate circuit. With Q_2 OFF, Q_3 will turn ON and connect $+V_{CC}$ through to the output. A LOW input will generate a HIGH output, which is in keeping with the NAND gate's truth table given in Figure 9-4(e). Only when both inputs are HIGH will both of Q_1's emitter diodes be OFF, and the HIGH at the base of Q_1 will be applied via the forward bias collector diode of Q_1 to the base of Q_2. With Q_2 ON, Q_4 will receive a positive base voltage and switch ground, or a LOW voltage, through to the output.

Multiple-Emitter Input Transistor
A transistor, specially constructed to have more than one emitter.

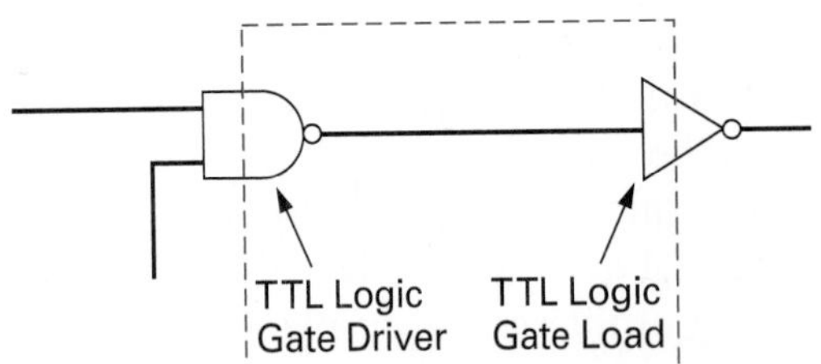

FIGURE 9-9 Connecting Logic Gates.

TTL Logic Gate Characteristics

Almost every digital IC semiconductor manufacturer has their own version of the standard TTL logic gate family or series of ICs. For instance, Fairchild has its 9300 series of TTL ICs, Signetics has the 8000 series, and so on. No matter what manufacturer is used, all of the TTL circuits are compatible in that they all have the same characteristics. To examine the compatibility of these bipolar transistor logic gate circuits, we will have to connect a *TTL driver*, or logic source, to a *TTL load* and see how these two operate together, as shown in Figure 9-9.

Figure 9-10 lists all of the standard TTL logic gate characteristics. Let us now step through the items mentioned in this list of characteristics and understand their meaning.

For TTL circuits, the *valid input logic levels* to a gate and *valid output logic levels* from a gate are shown in the blocks in Figure 9-10. Looking at the block on the left, you can see that a gate must generate an output voltage that is between the minimum and maximum points in order for that output to be recognized as a valid logical 1 or 0. Similarly, as long as the input voltages to a gate are between the minimum and maximum points, the gate will recognize the input as a valid logical 1 or 0.

Power Dissipation Amount of heat energy generated by a device in one second when current flows through it.

The second item on our list of TTL characteristics in Figure 9-10 is **power dissipation.** A standard TTL circuit will typically dissipate 10 mW per gate.

Propagation Delay Time The time it takes for the output of a gate to change after the inputs have changed.

The third item on our list of TTL characteristics in Figure 9-10 is the **propagation delay time** of a gate, which is the time it takes for the output of a gate to change after the inputs have changed. This delay time is typically about 10 nanoseconds for a standard TTL gate. For example, it would normally take 10 ns for the output of an AND gate to go HIGH after its two inputs go HIGH.

As you can see from the next two items mentioned in Figure 9-10, there are two TTL series available. The 7400 TTL series is used for all commercial applications, and it will operate reliably when the supply voltage is between 4.75 V to 5.25 V and the ambient temperature is between 0°C and 70°C. The *5400 TTL series* is a more expensive line used for military applications because of its increased supply voltage tolerance (4.5 V to 5.5 V) and temperature tolerance (–55°C to 125°C). A 5400 series TTL IC will have the same pin configuration and perform exactly the same logic function as a 7400 series TTL device. For example, a 5404 (hex INVERTER) has the same gate input, gate output, $+V_{CC}$, and ground pin numbers as a 7404.

Noise Immunity Unaffected by noise fluctuations at the gate's input.

Noise is an unwanted voltage signal that is induced into conductors due to electromagnetic radiation from adjacent current carrying conductors. All TTL logic circuits have a high **noise immunity,** which means they will be unaffected

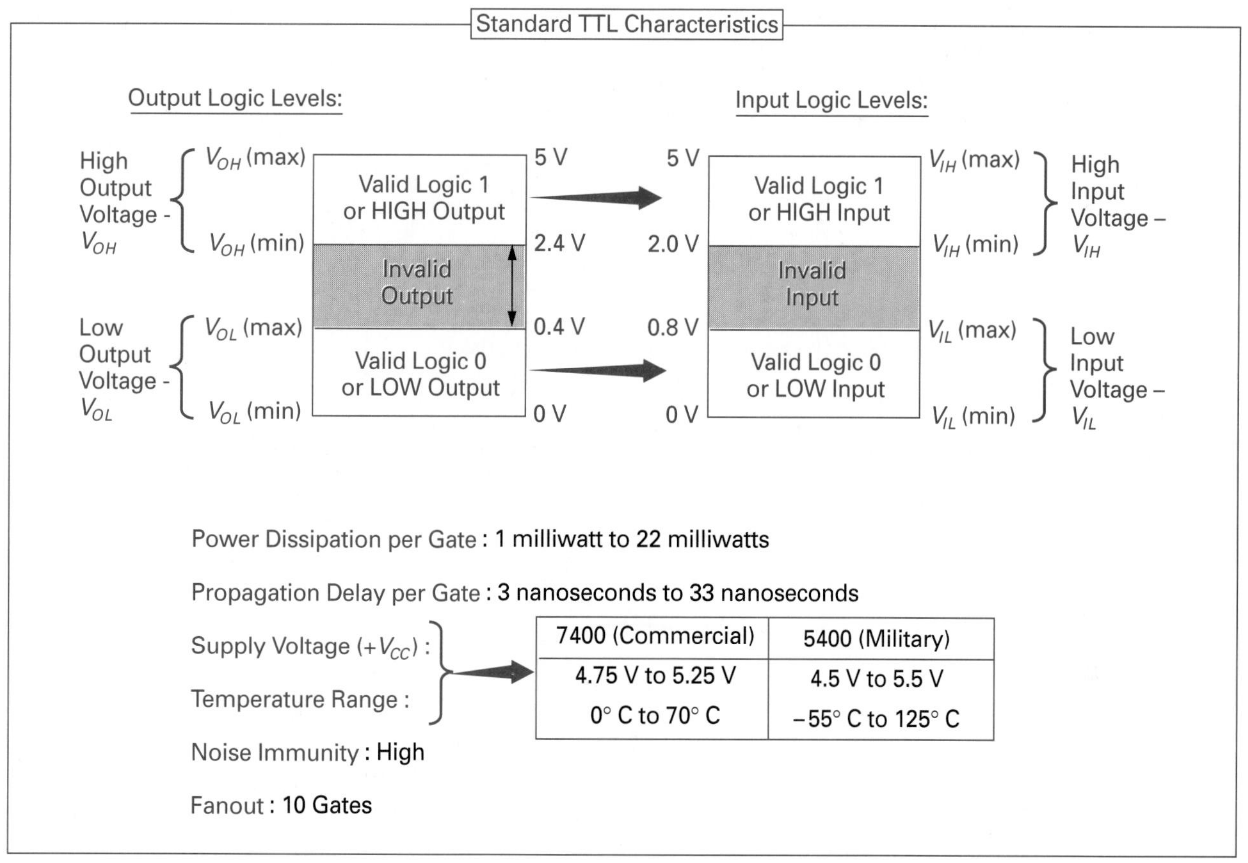

FIGURE 9-10 **Standard TTL Logic Gate Characteristics.**

by noise fluctuations at the gate's inputs. Referring back to the TTL specifications in Figure 9-10, let us see how a TTL logic gate has a *noise margin* of 0.4 V between a TTL driver and TTL load. If a TTL driver were to output its worse case logic 0 V (0.4 V), and induced noise increased this by 0.4 V to 0.8 V, the TTL load would still accept this input as a valid logic 0 because it is between 0 V and 0.8 V. This is shown in the simplified diagram in Figure 9-11(a). On the other hand, if a TTL driver were to output its worse case logic 1 voltage (2.4 V), and induced noise decreased this by 0.4 V to 2.0 V, the TTL load would still accept this input as a valid logic 1 because it is between 2.0 V and 5 V. This is shown in the simplified diagram in Figure 9-11(b).

Figure 9-12 illustrates how a TTL driver has to act as both a **current sink** and a **current source.** Looking at the direction of current in this illustration, you can see that this characteristic is described using *conventional current flow* (+ → –) instead of *electron current flow* (– → +). Whether the current flow is from ground to $+V_{CC}$ (electron flow) or from $+V_{CC}$ to ground (conventional flow), the value of current is always the same if the same circuit exists between these two points. We have changed to conventional at this point only because the manufacturer's specifications for the terms "sinking" and "sourcing" are for conventional current flow.

Current Sink A circuit or device that absorbs current from a load circuit.

Current Source A circuit or device that supplies current to a load circuit.

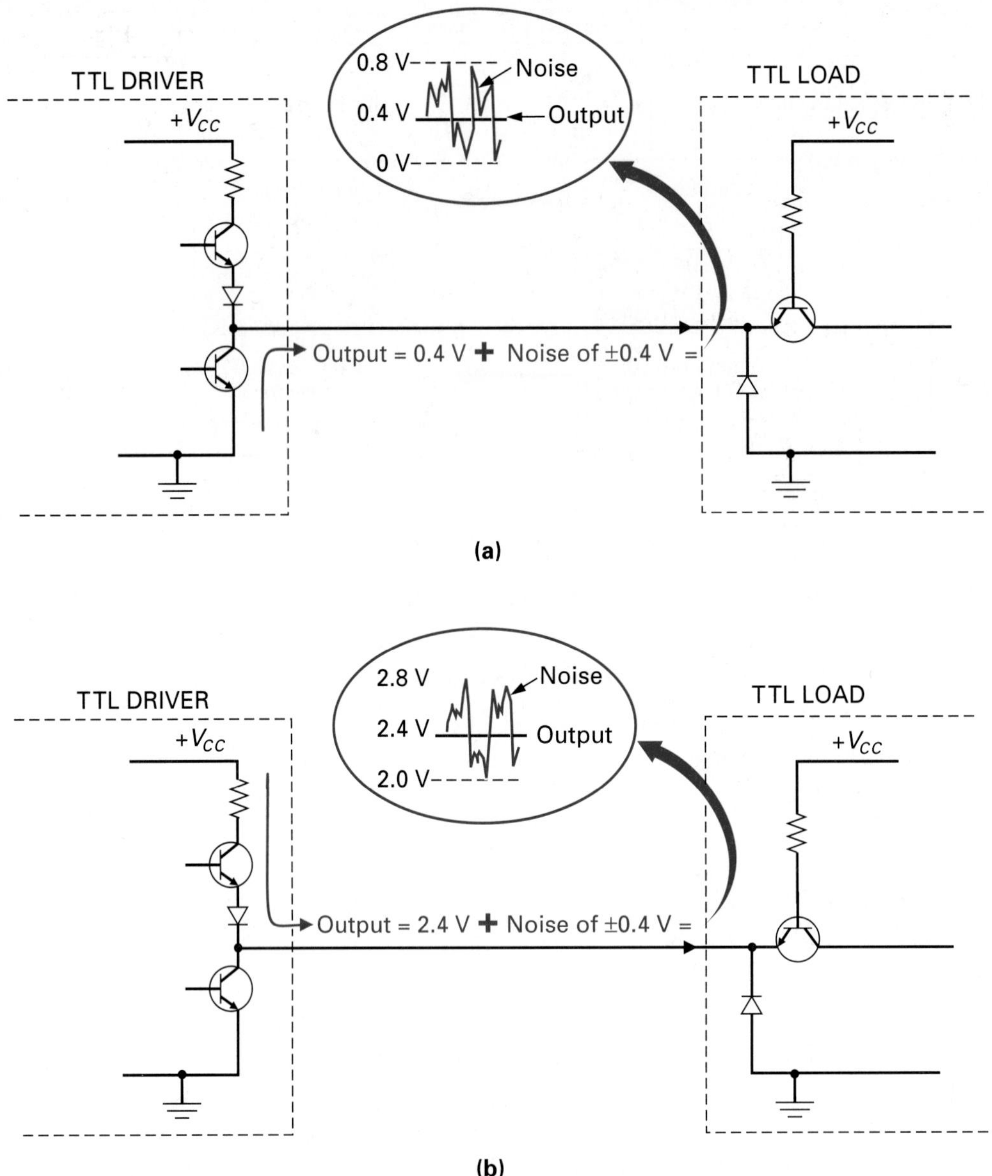

FIGURE 9-11 TTL Noise Margin.

In Figure 9-12(a), a single logic gate load has been connected to a TTL logic gate driver which is producing a LOW output (Q_4 in driver's totem pole is ON). In this instance, Q_4 will "sink" the single gate load current of 1.6 mA to ground. Manufacturer's data sheets state that Q_4 can sink a maximum of 16 mA and still produce a LOW output voltage of 0.4 V or less. Because the maximum sinking current of 16 mA is ten times larger than the single gate load current of 1.6 mA, we can connect ten TTL loads to each TTL driver, as seen in Figure 9-12(b), and still stay within spec.

In Figure 9-12(c), a single TTL load has been connected to a TTL driver which is producing a HIGH output (Q_3 in the driver's totem pole is ON). In this instance, Q_3 will "source" the single gate load current of 40 μA. Manufacturer's

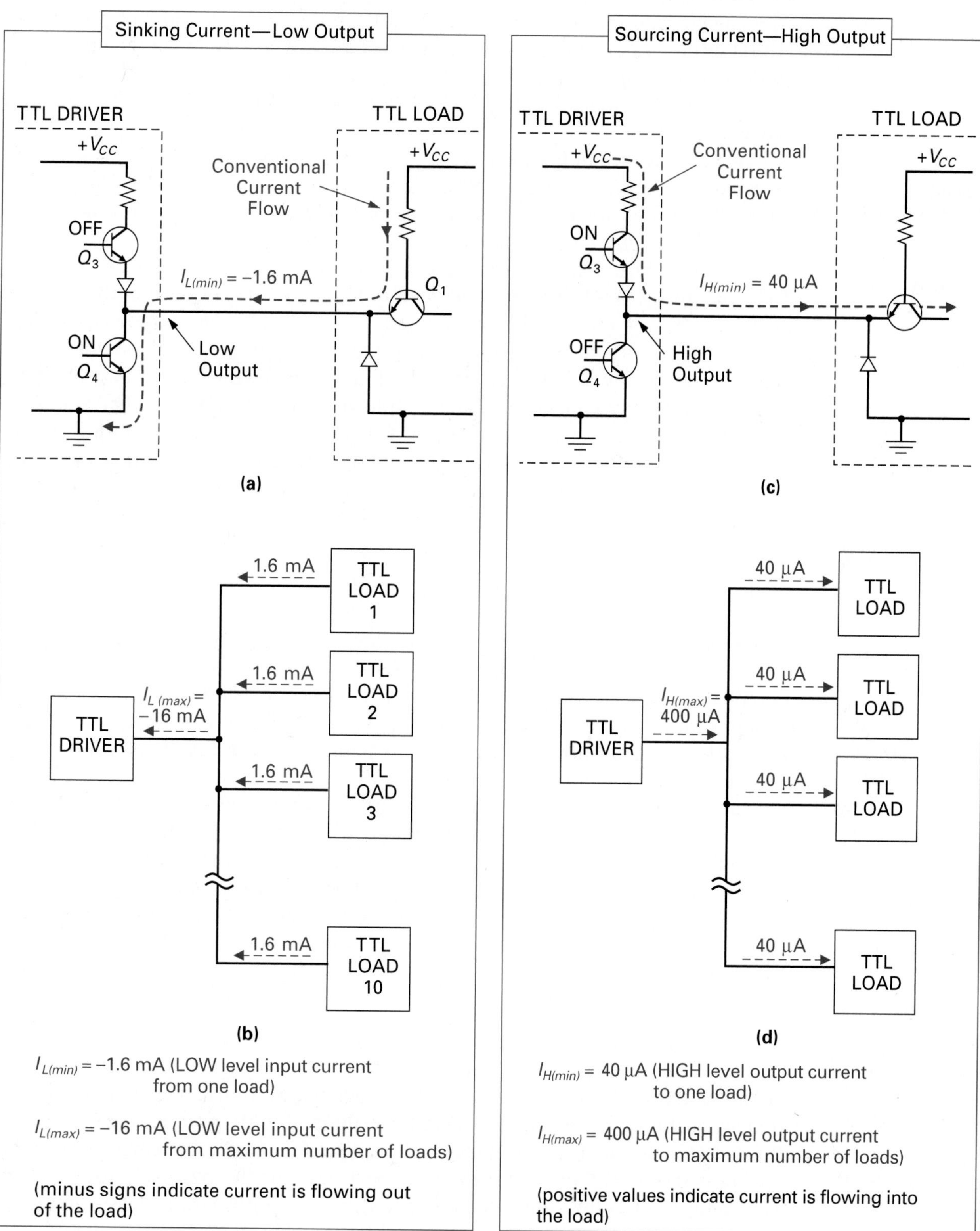

FIGURE 9-12 Sinking and Sourcing Current.

data sheets state that Q_3 can source a maximum of 400 μA and still produce a HIGH output voltage of 2.4 V or more. Because the maximum source current of 400 μA is ten times larger than the single gate load current of 40 μA, we can connect ten TTL loads to each TTL driver, as seen in Figure 9-12(d), and still stay within spec.

Fanout
The maximum number of TTL loads that can be reliably driven by a single TTL driver.

As you can see in Figure 9-12, the maximum number of TTL loads that can be reliably driven by a single TTL driver is called the **fanout.** Referring back to the standard TTL logic gate characteristics in Figure 9-10, you will now understand why a standard TTL gate has a fanout of 10.

Floating Input
An unconnected input.

Using Figure 9-12, we can describe another TTL characteristic. When a TTL load receives a LOW input, as seen in Figure 9-12(a), a large current exists in the emitter of Q_1. On the other hand, when a TTL load receives a HIGH input, as seen in Figure 9-12(c), the emitter current in Q_1 is almost zero. This action explains why a TTL logic gate acts like it is receiving a HIGH input when its input lead is not connected. For example, if the input lead to a NOT gate was not connected to a circuit, and power was applied to the gate, the output of the NOT gate would be LOW. An unconnected input is called a **floating input.** In this condition, no emitter current will exist in Q_1 because the input is disconnected, and so *a floating TTL input is equivalent to a HIGH input.*

9-1-5 *Low-Power TTL Logic Gates*

A low-power TTL circuit is basically the same as the standard TTL circuit previously discussed except that all of the internal resistance values in the logic gate circuit are ten times larger. This means that the power dissipation of a low-power TTL circuit is one-tenth that of a standard TTL circuit (low-power TTL logic gates will typically have a power dissipation of about 1 mW per gate).

Increasing the internal resistance, however, will decrease the circuit's internal currents and, because the bipolar transistor is a current-operated device, the response or switching time of a logic gate will decrease (lowpower TTL logic gate will typically have a propagation delay time of 33 ns per gate).

With low-power TTL circuits, high speed is sacrificed for low power consumption. Low power logic gate ICs can be identified because the device number will have the letter "L" following the series number 74 or 54. For example, 74L00, 74L01, 74L02, and so on.

Saturation Delay Time
The short time it takes for extra carriers to leave the transistor once it has been switched off.

Schottky TTL Logic Gate
A device that employs a schottky transistor.

Schottky Transistor
A device that incorporates a schottky diode between the collector and base of each bipolar transistor.

9-1-6 *Schottky TTL Logic Gates*

With standard TTL and low-power TTL logic gates, transistors go into saturation when they are switched ON and are flooded with extra carriers. When a transistor is switched OFF, it takes a short time for these extra carriers to leave the transistor, and therefore it takes a short time for the transistor to cut OFF. This **saturation delay time** accounts for the slow propagation delay times, or switching times, of standard and low power TTL logic gates.

Schottky TTL logic gates overcome this problem by fabricating a schottky diode between the collector and base of each bipolar transistor, as seen in Figure 9-13(a). This transistor is called a **schottky transistor** and uses the symbol shown

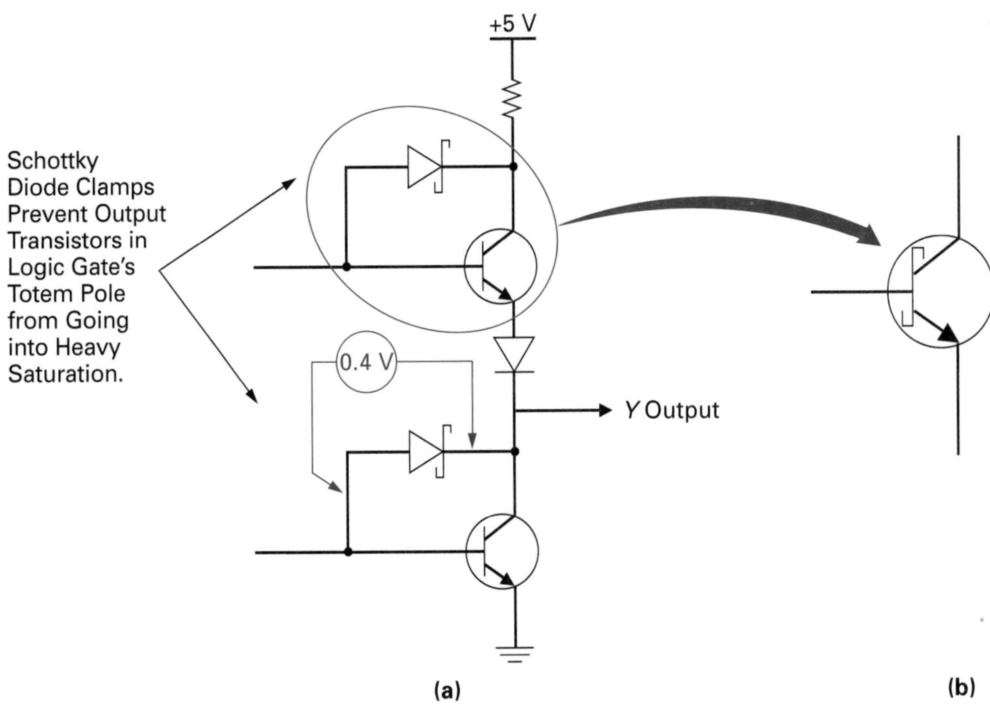
+5 V
Schottky Diode Clamps Prevent Output Transistors in Logic Gate's Totem Pole from Going into Heavy Saturation.
0.4 V
Y Output
(a)
(b)

FIGURE 9-13 Schottky Clamped Bipolar Transistors. (a) Connecting a Schottky Diode Across a Transistor Collector-Base Junction. (b) Schottky Transistor.

in Figure 9-13(b). Let us now see how this schottky diode will affect the operation of the transistor and decrease saturation delay time. When the transistor begins to turn ON, its collector voltage will fall. When the collector drops below 0.4 V of the base voltage, the schottky diode will conduct and shunt current away from the collector junction. This action effectively clamps the collector to 0.4 V of the base voltage, keeping the collector diode of the transistor reversed biased. Keeping the collector diode reverse biased prevents the transistor from slipping into heavy saturation, and therefore decreasing saturation delay time.

Schottky TTL devices are fast, with propagation delay times of typically 3 ns. This fast switching time means that the input signals can operate at extremely high frequencies of typically 100 mHz. Schottky TTL ICs are labeled with an "S" after the first two digits of the series number, for example, 74S04, 74S08, and so on.

By using schottky transistors and increasing the internal resistance in a logic gate circuit, manufacturers compromised the speed of the schottky to obtain a lower dissipation rating. Schottky TTL devices will typically have a power dissipation of 19 mW per gate, whereas **low-power schottky TTL devices** typically have a power dissipation of 2 mW per gate and propagation delay times of 9.5 ns. These low power schottky TTL devices include the letters "LS" in the device number, for example, 74LS32, 74LS123, and so on.

Special schottky TTL devices are available that use a new integration process called **oxide isolation,** in which the transistor's collector diode is isolated by a thin oxide layer instead of a reverse bias junction. These **advanced schottky**

Low-Power Schottky TTL Devices
Schottky devices that typically have a power dissipation of 2 mW per gate and propagation delay times of 9.5 ns.

Oxide Isolation
New integration process in which the transistor's collector diode is isolated by a thin oxide layer instead of a reverse bias junction.

Advanced Schottky TTL Devices
Devices that employ oxide isolation.

TTL devices (labeled 74AS00) have a typical power dissipation of 8.5 mW per gate and a propagation delay time of 1.5 ns. By increasing internal resistance once again manufacturers came up with an **advanced low-power schottky TTL** line of products (labeled 74ALS00), which have a power dissipation of 1 mW and propagation time of 4 ns. Some manufacturers, such as Fairchild, use the letter “F” for fast in their AS series of ICs, for example, 74F10, 74F14, and so on.

Advanced Low-Power Schottky TTL Devices
Schottky devices that typically have a power dissipation of 1 mW per gate and propagation delay times of 4 ns.

9-1-7 Open Collector TTL Gates

Some TTL devices have an **open collector output** instead of the totem pole output. Figure 9-14(a) shows a standard TTL INVERTER circuit with an open collector output, and Figure 9-14(b) shows the symbol used to represent an open collector output gate.

Open Collector Output
A type of output structure found in certain bipolar logic families. Resistive pull-ups are generally added to provide the high level output voltage.

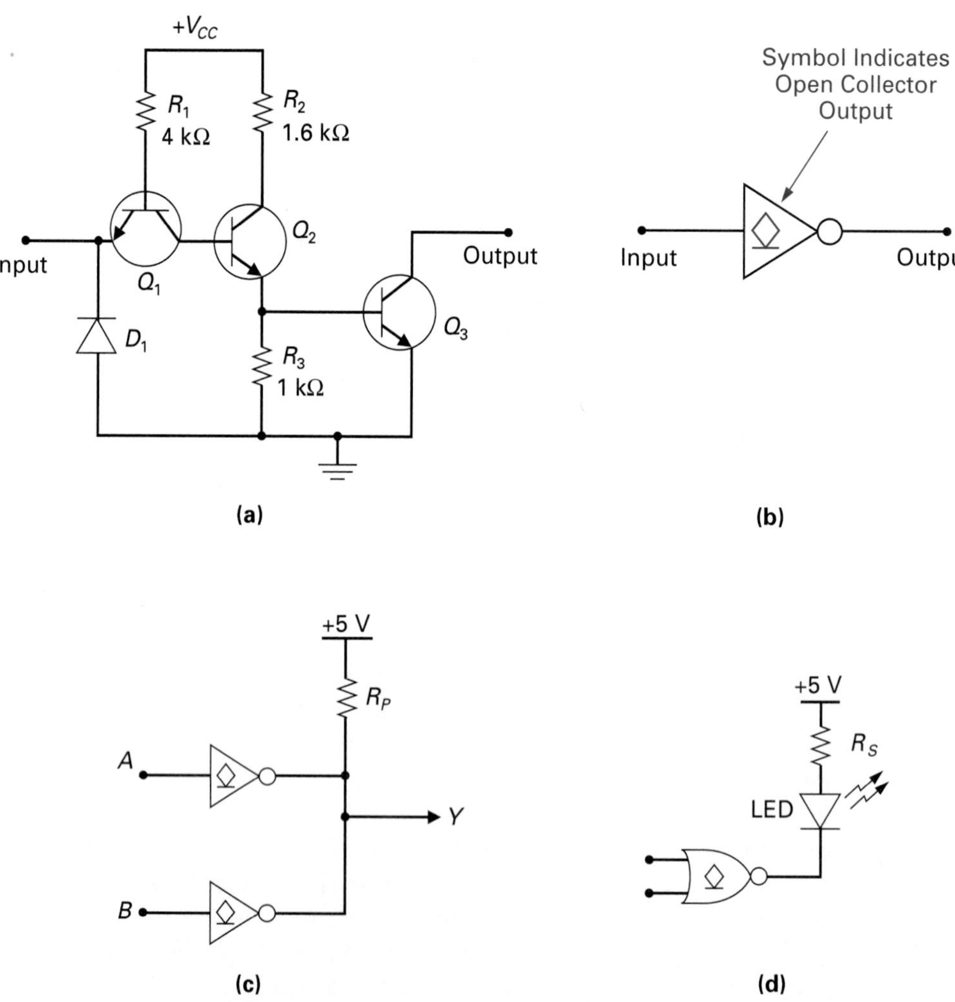

FIGURE 9-14 Open Collector Gates. (a) Bipolar Transistor Open Collector INVERTER Circuit. (b) Open Collector Symbol. (c) Wired INVERTER Open Collector Gates. (d) Open Collector NOR Driver Gate.

In order to get a HIGH or LOW output from an open collector gate, an external **pull-up resistor** (R_P) must be connected between $+V_{CC}$ and the collector of Q_3 (which is the output of the logic gate) as seen in Figure 9-14(c). In this circuit, two open collector inverters are driving the same output Y. If two standard TTL logic gates with totem pole outputs were wired together in this way and the A INVERTER produced a HIGH output and the B INVERTER a LOW output, there would be a direct short between the two causing both gates to burn out. With open collector gates, the outputs can be wired together without causing any HIGH-LOW conflicts between gates. With the circuit in Figure 9-14(c), only when both inputs are LOW will the output transistors of both inverters be OFF and the output be pulled up by R_P to +5 V ($A = 0$, $B = 0$, $Y = 1$). If either A input or B input is HIGH, one of the output transistors will turn ON and pull down or ground the output Y ($A = 1$ or $B = 1$, $Y = 0$). You may recognize this logic operation as that of a NOR gate.

Pull-Up Resistor
A resistor connected between a signal output and positive V_{CC} to pull the signal line HIGH when it is not being driven LOW.

Open collector gates can sink up to 40 mA, which is a vast improvement over standard TTL gates that only sink 16 mA (74ASXX gates can sink 20 mA). In Figure 9-14(d), you can see how the high sinking current ability of an open collector NOR gate can be used to drive the high current demand of a light emitting diode (LED). The current limiting resistor (R_S) is being used to limit the current through the LED to about 25 mA, which produces a good level of brightness from an LED. In many applications, open collector gates are used to drive external loads that require increased current such as LEDs, lamps, relays, and so on.

9-1-8 *Three-State (Tri-State) Output TTL Gates*

The **three-state output gate** has, as its name implies, three output states or conditions. These three output states are LOW, HIGH, and FLOATING, or high-impedance output state. Before we discuss why we need this third output state, let us see how this circuit operates.

Three-State Output Gate
A gate that had three output states or conditions: LOW, HIGH, and FLOATING.

Figure 9-15(a) illustrates the internal bipolar transistor circuit for a three-state NOT gate. The circuit is basically the same as the previously discussed INVERTER gate, except for the additional control input applied to transistor Q_4. When the enable input is LOW, Q_4 is OFF and has no effect on the normal inverting operation of the circuit. However, when the enable input is HIGH, Q_4 is turned ON, and the ground on its emitter will be switched through to the second emitter of Q_1 and the base of Q_3 via D_1. The ON Q_1 will switch ground from its emitter through to the base of Q_2, turning Q_2 OFF and therefore Q_5 OFF. Transistor Q_4 will also ground the base of Q_3, turning it OFF. The result will be that both totem pole transistors are cut OFF, as seen in the equivalent circuit shown in the inset in Figure 9-15(a). With the output of the gate completely disconnected from the circuit, this lead is said to be FLOATING and can either be pulled up to $+V_{CC}$ or pulled down to ground depending on what is connected externally.

Figure 9-15(b) shows the schematic symbol for this three-state output NOT gate. This circle or bubble attached to the enable input is used to indicate an **active-LOW control line,** which means that if you want to activate the enable

Active-LOW Control Line
The control line is taken LOW to activate the line.

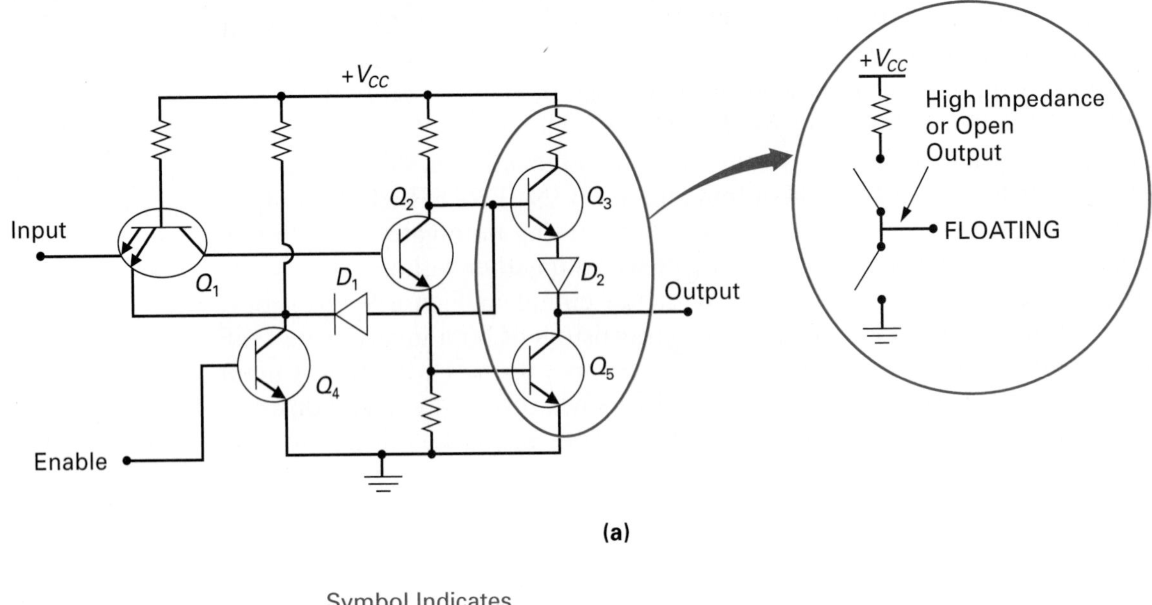

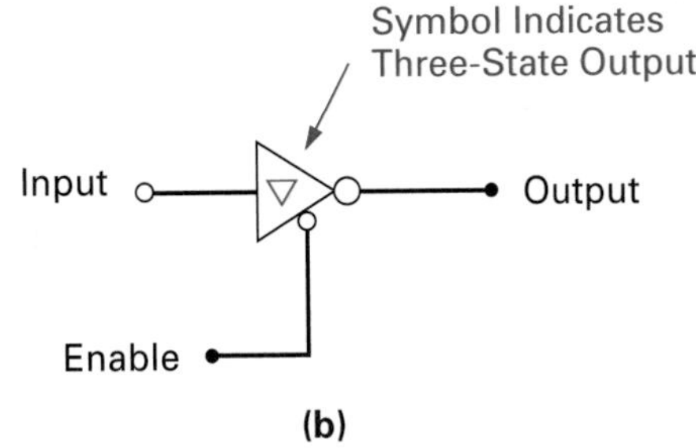

(b)

Enable	Input	Output
LOW (enabled)	LOW	HIGH
LOW (enabled)	HIGH	LOW
HIGH (disabled)	LOW	FLOATING
HIGH (disabled)	HIGH	FLOATING

(c)

FIGURE 9-15 Three-State Output Gates. (a) Bipolar Transistor Circuits. (b) Symbol. (c) Truth Table.

line you must make this control line LOW. In other words, to disable the gate we would not make this line active, and so the enable line would be HIGH. Studying the truth table or function table in Figure 9-15(c), you can see this gate's operation summarized. When the enable control line is LOW, the circuit is enabled for normal INVERTER operation. When the enable control line is HIGH, however, the output is always FLOATING, no matter what input logic level is present.

Like the previously discussed open collector gates, these devices are used in applications where two or more gate outputs are connected to a single output line. To avoid HIGH/LOW conflicts between gates, only one gate is enabled at one time to drive the output line. All other gates being disabled and their outputs are FLOATING.

9-1-9 Buffer/Driver TTL Gates

Buffer
A device that isolates one circuit from another.

A **buffer** is a device that isolates one circuit from another. In order to achieve this isolation, a buffer should have a high input impedance (low input current) and a low output impedance (high output current). A standard TTL gate provides a certain amount of isolating or buffering since the output current is ten times that of the input current.

	Low Output	High Output
Standard TTL	$I_{L(min)} = -1.6$ mA	$I_{H(min)} = 40$ μA
	$I_{L(max)} = -16$ mA	$I_{H(max)} = 400$ μA

A buffer or driver is a basic gate that has been slightly modified to increase the output current. For example, a 7428 is a "quad 2-input NOR gate buffer" whose maximum LOW and HIGH currents are thirty times greater than the minimum LOW and HIGH current, enabling it to drive heavier loads.

	Low Output	High Output
Buffer/Driver TTL	$I_{L(min)} = -1.6$ mA	$I_{H(min)} = 40$ μA
	$I_{L(max)} = -48$ mA	$I_{H(max)} = 1.2$ mA

9-1-10 Schmitt Trigger TTL Gates

Schmitt Trigger Circuit
A two-state device used for pulse shaping.

The **schmitt trigger circuit,** named after its inventor, is a two-state device that is used for pulse shaping. The typical schmitt trigger circuit can be seen in Figure 9-16(a), while the schematic symbol for the schmitt trigger is shown in Figure 9-16(b). The waveforms shown in Figure 9-16(c) and (d) illustrate the two basic applications of the schmitt trigger which are to convert a sine wave into a rectangular wave and to sharpen the rise and fall times of a rectangular wave.

Like all of the previously discussed digital switching circuits, the schmitt trigger circuit may be constructed with discrete or individual components, or purchased as an integrated circuit (IC) containing several complete schmitt trigger circuits. Using the circuit shown in Figure 9-16(a) and the example input/output waveforms shown in Figure 9-16(c) and (d), let us see how this bipolar transistor circuit will operate. The input signal is applied to the base of Q_1 and, if this signal is below the ON voltage (V_{ON}), Q_1 will be cut off and its collector voltage will be HIGH. The HIGH Q_1 collector voltage is coupled to the base of Q_2, causing it to saturate (to turn heavily ON). As a result, Q_2 conducts a large current (shown as electron flow) through R_6, Q_2 emitter-to-collector, and R_2 to $+V_{CC}$. The voltage developed across R_6 (V_{R6}) establishes the V_{ON} voltage of the schmitt trigger because the base of Q_1, and therefore the input, will have to be 0.7 V greater than the emitter of Q_1, or V_{R6}, if Q_1 is to turn ON.

When the input does reach a value that is 0.7 V above V_{R6}, Q_1 will turn ON and its collector voltage will fall. This decrease is coupled to the base of Q_2, turning it OFF and causing its collector voltage, and therefore the output, to rise. With Q_1 now ON, and Q_2 OFF, the voltage across R_6 is now being established by the current path through R_6, Q_1 emitter-to-collector, and R_1 to $+V_{CC}$. This will set up a different V_{R6} voltage, and therefore a different V_{OFF} voltage since the input must fall 0.7 V below V_{R6} for Q_1 to turn OFF. When the input

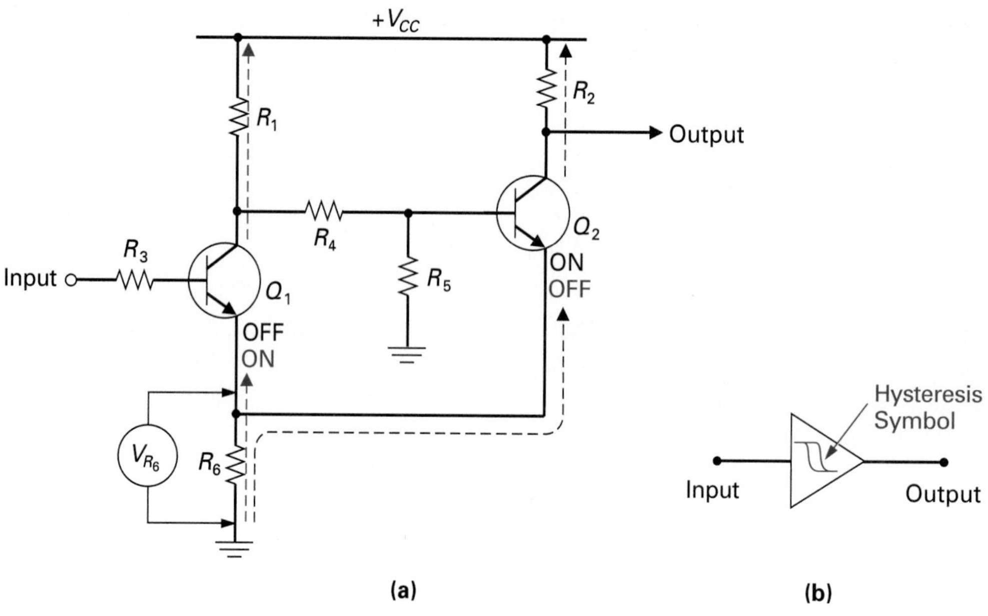

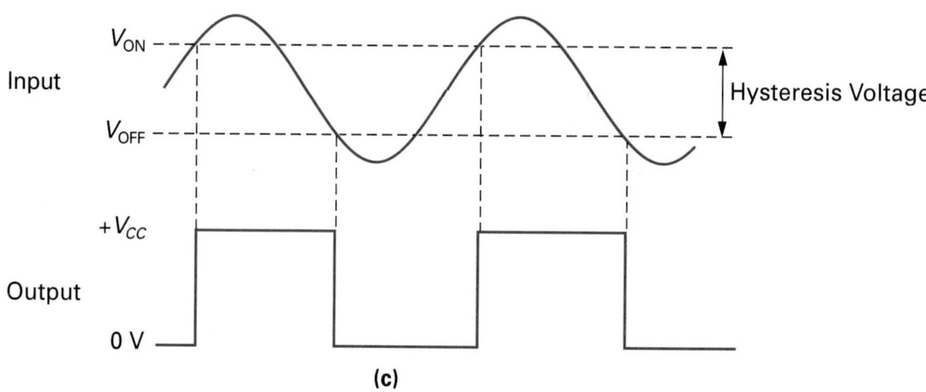

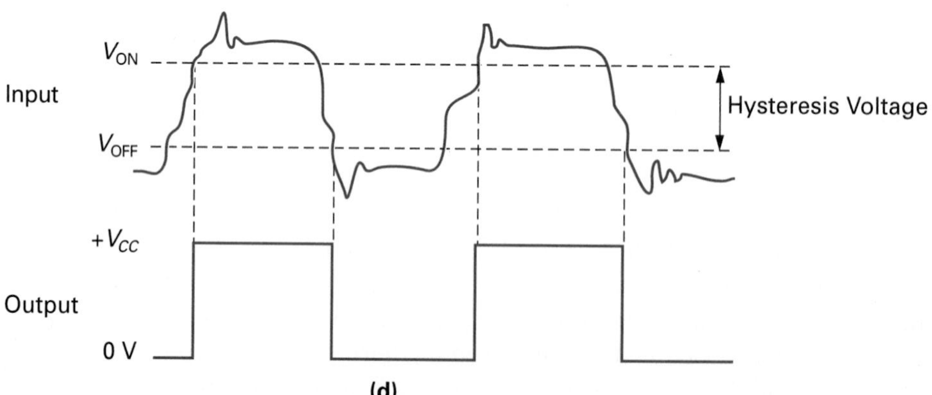

FIGURE 9-16 The Schmitt Trigger (a) Bipolar Transistor Circuit. (b) Logic Gate Symbol. (c) Converting a Sine Wave into a Rectangular Wave. (d) Sharpening the Rise and Fall Times of a Rectangular Wave.

falls below this voltage, Q_1 will turn OFF. Its collector voltage will rise and turn ON Q_2, causing its collector voltage and the output voltage to fall. The circuit is now returned to its original condition, awaiting the input to once again exceed the V_{ON} threshold voltage.

The symbol for any schmitt trigger gate will contain the distinctive hysteresis symbol, as shown in Figure 9-16(b). Hysteresis can be loosely defined as the lag between cause and effect. Referring to the waveforms in Figure 9-16(c) and (d), you can see that there is a lag between the input (cause) and output (effect). The input has to rise to an "ON voltage" (V_{ON}) before the output will rise to its positive peak, and drop below an "OFF voltage" (V_{OFF}) before the output will drop to its negative peak. This lag is desirable because it produces a sharp, well defined output signal. The difference between the V_{ON} and V_{OFF} voltage is called the **hysteresis voltage.**

Hysteresis Voltage
The difference between the V_{ON} and V_{OFF} voltage.

Several schmitt trigger TTL ICs are available, including 7413 (dual 4-input NAND schmitt trigger), 7414 (hex schmitt trigger inverters), 74132 (quad 2-input NAND schmitt trigger) and so on. These logic circuits will have a schmitt trigger circuit connected to the output of each logic gate circuit so that the gate will perform its normal function, and then sharpen up the rise and fall times of pulses that have been corrupted by noise and attenuation.

9-1-11 *Emitter-Coupled Logic (ECL) Gate Circuits*

Schottky TTL circuits improved the switching speed of digital logic circuits by preventing the bipolar transistor from saturating. **Emitter-coupled logic,** or **ECL,** is another nonsaturating bipolar transistor logic circuit arrangement. However, ECL is the fastest of all logic circuits. It is used in large digital computer systems that require fast propagation delay times (typically 0.8 ns per gate) and are willing to sacrifice power dissipation (typically 40 mW per gate).

Emitter Coupled Logic (ECL)
A nonsaturating bipolar transistor logic circuit arrangement that is the fastest of all logic circuits.

Figure 9-17(a) shows a typical ECL gate circuit. The two inputs to this circuit, *A* and *B,* are applied to the transistors Q_1 and Q_2, which are connected in parallel. If additional inputs are needed, more input transistors can be connected in parallel across Q_1 and Q_2, as shown in the dashed lines in Figure 9-17(a). The two collector outputs of Q_2 and Q_3 drive the two low output impedance emitter followers Q_5 and Q_6. These emitter follower outputs provide an OR output from Q_6 and its complement or opposite, which is a NOR output from Q_5. This dual output is one of the advantages of ECL gates. External *pull-down resistors* are required, as seen in Figure 9-17(b) (which also shows the logic symbol for this OR/NOR ECL gate). The open outputs of ECL gates means that two or more gate outputs can drive the same line without causing any HIGH/LOW voltage conflicts. The V_{EE} pin of an ECL IC is typically connected to –5.2 V, and the V_{CC} pin is connected to ground. Although these dc supply values do not seem to properly bias the circuit, remember that the V_{CC} power line is still positive (since it is at ground) relative to the V_{EE} power line (which is at –5.2 V). The transistor Q_4 has a constant base bias voltage and an emitter resistor and will produce a temperature stabilized constant voltage, or reference voltage, of approximately –1.3 V to the base of Q_3. The high speed of any ECL logic gate is achieved by ensuring that none of the transistors saturate. The bipolar

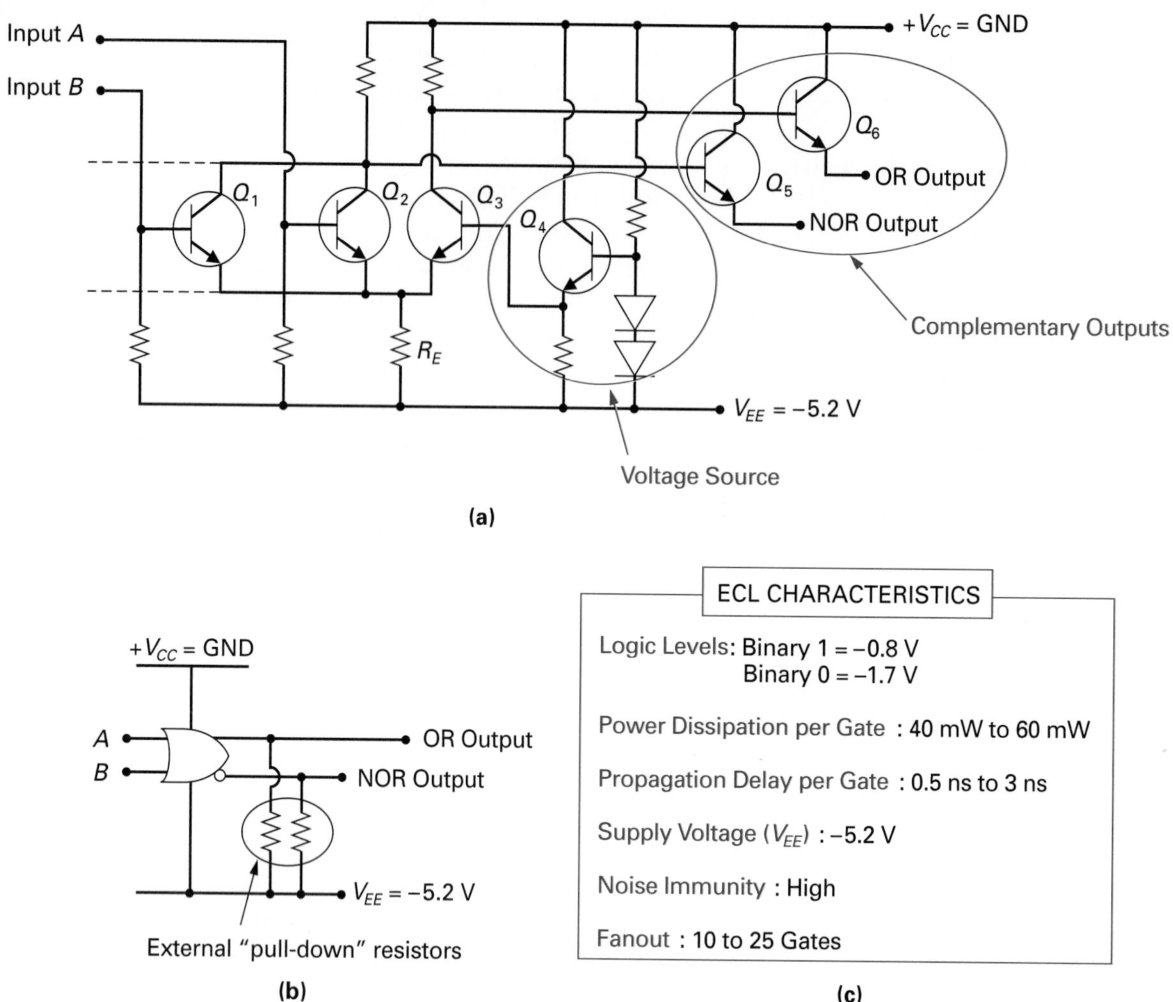

FIGURE 9-17 Emitter-Coupled Logic (ECL) Gates. (a) Bipolar Transistor OR/NOR Circuit. (b) Logic Symbol. (c) Characteristics.

transistors in the circuit simply conduct more or conduct less depending on whether they represent a LOW or HIGH. Preventing transistors from going into saturation accounts for the fast speed of ECL logic gate circuits. The fact that transistors, for the most part, are conducting and drawing a current accounts for the ECL's increase in power dissipation. The typical characteristics of ECL gates are listed in Figure 9-17(c). Notice that the binary 0 and binary 1 voltage levels are slightly different from TTL circuits. However, the binary 1 voltage level (–0.8 V) is still positive relative to binary 0 (–1.7V).

Let us now examine this ECL circuit's operation. When either one or both of the inputs go to a HIGH (–0.8 V), that transistor will conduct more (since its base is now positive with respect to the emitter). In so doing, the input transistor will make transistor Q_3 turn more OFF (because the base-emitter junction of the input transmitter will make Q_3's emitter positive with respect to its large negative base voltage). The conducting input transistor will switch its negative emitter voltage through to the collector and therefore to the base of Q_5. Q_5 will turn more OFF, and its output will be pulled LOW. This is what should happen to a NOR gate output; any binary 1 input will produce a binary 0 output. On the other hand, with Q_3 more OFF, its collector will be more positive (closer to 0 V) and so Q_6 will turn more ON and connect the more positive V_{CC} to the OR output. This is what should happen to an OR gate output; any binary 1 input will produce a binary 1 output. When all inputs are LOW (–1.7 V), all of the input transistors will be more OFF, producing a HIGH output to Q_5 (all inputs are 0, output from NOR is 1). With Q_1 and Q_2 more OFF, Q_3 will conduct more, producing a LOW out to Q_6 (all inputs are 0, output of OR is 0).

The high speed, high power consumption, and high cost of these logic circuits are the reasons why they are only used when absolutely necessary. The most widely used ECL family of ICs is Motorola's ECL (MECL) 10K and 100K series, in which devices are numbered MC10XXX or MC100XXX.

9-1-12 *Integrated-Injection Logic (I^2L) Gate Circuits*

Integrated-injection logic (IIL), or as they are sometimes abbreviated I^2L (pronounced I squared, L), is a series of digital bipolar transistor circuits that have good speed, low power consumption, and very good *circuit packing density.* In fact, one I^2L gate is one tenth the size of a standard TTL gate, making it ideal for compact or high density circuit applications.

Integrated-Injection Logic (IIL)
A series of digital bipolar transistor circuits that have good speed, low power consumption, and very good circuit packing density.

A typical I^2L logic NOT gate is shown in Figure 9-18(a). Transistor Q_1 acts as a current source, and the multiple collector transistor Q_2 acts as the INVERTER with two outputs. Studying the circuit, you can see that the base of Q_1 is common to the emitter of Q_2, and the collector of Q_1 is common to the base of Q_2. Because of these common connections, the entire I^2L gate can be constructed as one transistor with two emitters and two collectors, and it uses the same space on a silicon chip as one standard TTL multiple-emitter transistor.

The circuit in Figure 9-18(a) operates in the following way. The emitter of Q_1 is connected to an external supply voltage ($+V_S$) which can be anywhere from 1 V to 15 V depending on the *injector current* required. The constant base emitter bias of Q_1 means that a constant current is available at the collector of Q_1 (called the *current injector transistor*) and this current will either be applied to the base of Q_2 or out of the input, depending on the input voltage logic level. To be specific, a LOW input will pull the injector current out of the input and away from the base of Q_2. This will turn Q_2 OFF, and its outputs will be open, which is equivalent to a HIGH output. On the other hand, a HIGH input will cause the injector current of Q_1 to drive the base of Q_2. This will turn Q_2 ON and switch the LOW on its emitter through to the collector outputs.

Figure 9-18(b) illustrates an I^2L NAND gate. As you can see, the only difference between this circuit and the INVERTER is the two inputs *A* and *B*. Transistor Q_3 and Q_4 are shown in this circuit; however, they are the multiple-collector transistors of the gates driving input *A* and *B* of this NAND gate circuit. If either one of these transistors is ON (a low input to the NAND), the injector current of Q_1 will be pulled away from Q_2's base, turning OFF Q_2 and causing a HIGH output. Only when both inputs are HIGH (Q_3 and Q_4 are OFF), will the injector current from Q_1 turn ON Q_2 and produce a LOW output.

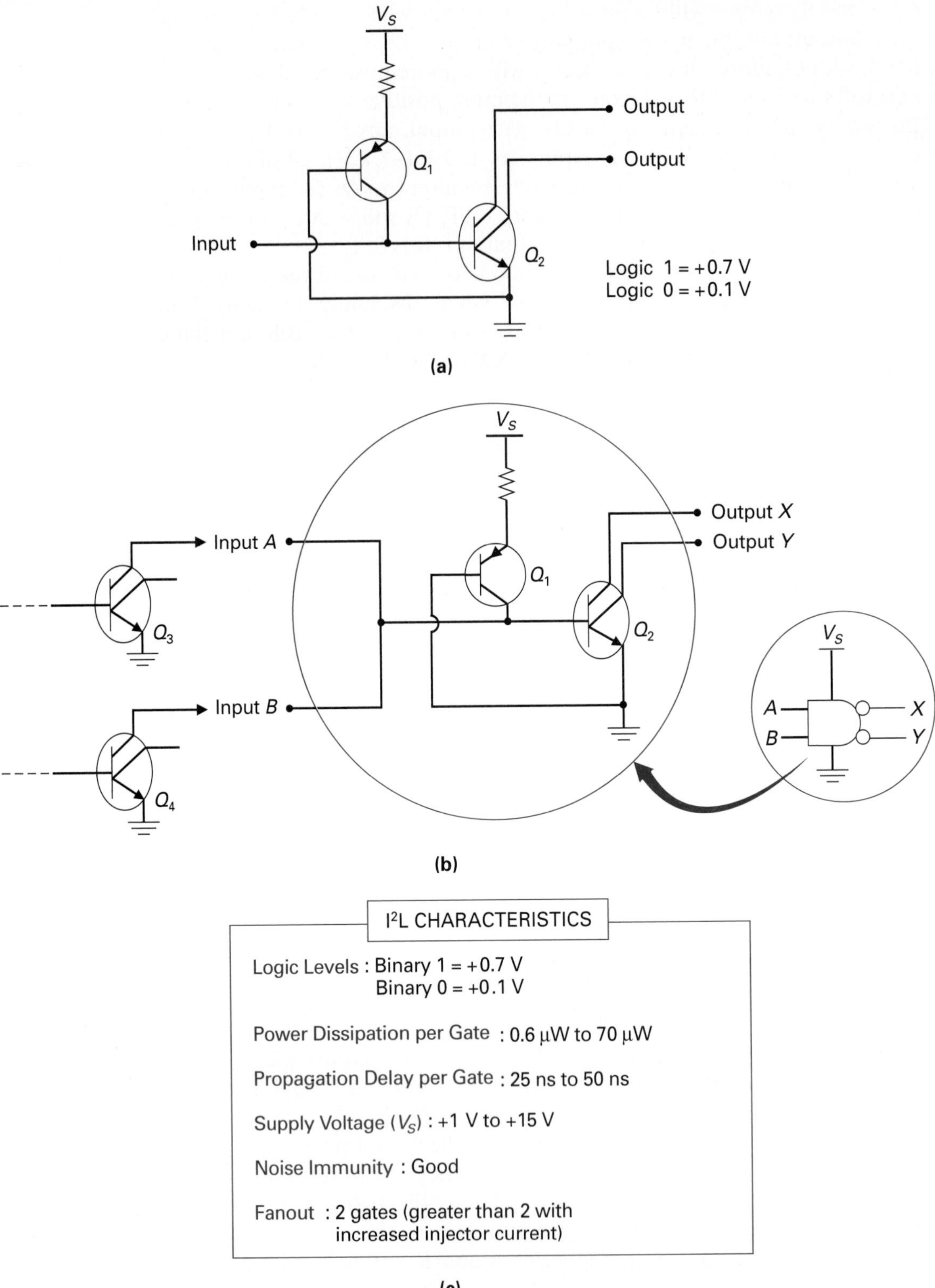

FIGURE 9-18 Integrated-Injection Logic (IIL or I²L) Gates. (a) Bipolar Transistor INVERTER (NOT) Gate Circuit. (b) Bipolar Transistor NAND Gate Circuit. (c) Characteristics.

Figure 9-18(c) lists the characteristics of I^2L logic-gate circuits. These circuits are used almost extensively in digital watches, cameras, and all compact, battery-powered circuit applications. As mentioned earlier, its good speed, low power consumption, and small size make it ideal for any portable system. Also, the simplicity of the I^2L circuits make it is easy to combine analog or linear circuits (such as amplifiers and oscillators) with digital circuits to create complete electronic systems on one chip. The circuit complexity of standard TTL digital circuits means that it is normally too difficult to place digital and linear circuits on the same IC chip. It is also not cost effective due to fabrication difficulties and therefore high cost.

SELF-TEST REVIEW QUESTIONS FOR SECTION 9-1

1. What are the two basic actions of the bipolar junction transistor, and which action is made use of in digital electronic circuits?
2. List the five basic logic gates.
3. TTL is an abbreviation for ____________?
4. What is the fanout of a standard TTL gate?
5. A floating TTL input is equivalent to a ____________ (HIGH/LOW) input.
6. The three output states of a tri-state logic gate are ____________, ____________, and ____________.
7. In what application would ECL circuits be used?
8. What is the advantage of I^2L logic circuits?

Clock Signal
Generally a square wave used for the synchronization and timing of several circuits.

Clock Oscillator
A device for generating a clock signal.

Astable Multivibrator
A device commonly used as a clock oscillator.

Monostable Multivibrator
A device that when triggered will generate a rectangular pulse of fixed duration.

Bistable Multivibrator
A digital control device that can be either set or reset.

Free-Running Multivibrator
A circuit that requires no input signal to start its operation, but simply begins to oscillate the moment the dc supply voltage is applied.

9-2 DIGITAL TIMER AND CONTROL CIRCUITS

Timing is everything in digital logic circuits. To control the timing of digital circuits, a **clock signal** is distributed throughout the digital system. This square wave clock signal is generated by a **clock oscillator,** and its sharp positive (leading) and negative (trailing) edges are used to control the sequence of operations in a digital circuit. In this section we will discuss the **astable multivibrator,** which is commonly used as a clock oscillator, and the **monostable multivibrator,** which when triggered will generate a rectangular pulse of a fixed duration. To complete our discussion on multivibrator circuits, we will also discuss the **bistable multivibrator,** which is a digital control device that can be either set or reset.

9-2-1 *The Astable Multivibrator Circuit*

The astable multivibrator circuit, seen in Figure 9-19(a), is used to produce an alternating two-state square or rectangular output waveform. This circuit is often called a **free-running multivibrator** because the circuit requires no input

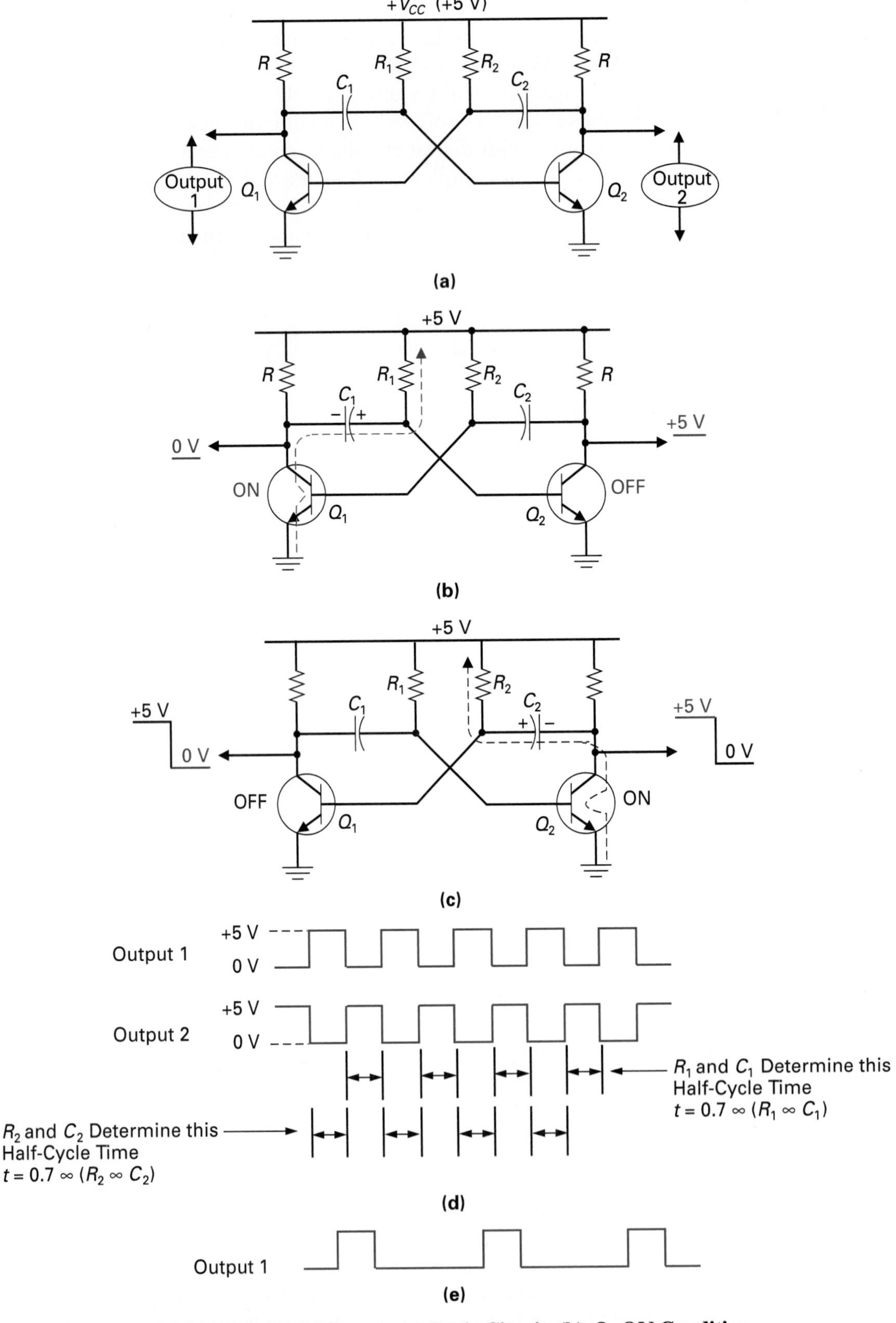

FIGURE 9-19 The Astable Multivibrator. (a) Basic Circuit. (b) Q_1 ON Condition. (c) Q_2 ON Condition. (d) Square Wave Mode. (e) Rectangular Wave Mode.

signal to start its operation. It will simply begin oscillating the moment the dc supply voltage is applied.

The circuit consists of two *cross-coupled bipolar transistors,* which means that there is a cross connection between the base and the collector of the two transistors Q_1 and Q_2. This circuit also contains two RC timing networks: R_1/C_1 and R_2/C_2.

Let us now examine the operation of this astable multivibrator circuit. When no dc supply voltage is present ($V_{CC} = 0$ V), both transistors are OFF, and therefore there is no output. When a V_{CC} supply voltage is applied to the circuit (for example, +5 V), both transistors will receive a positive bias base voltage via R_1 and R_2. Although both Q_1 and Q_2 are matched bipolar transistors, which means that their manufacturer ratings are identical, no two transistors are ever the same. This difference, and the differences in R_1 and R_2 due to resistor tolerances, means that one transistor will turn ON faster than the other. Let us assume that Q_1 turns on first, as seen in Figure 9-19(b). As Q_1 conducts, its collector voltage decreases because it is like a closed switch between the collector and emitter. This decrease in collector voltage is coupled through C_1 to the base of Q_2, causing it to conduct less and eventually turn OFF. With Q_2 OFF, its collector voltage will be high (+5 V) because Q_2 is equivalent to an open switch between the collector and emitter. This increase in collector voltage is coupled through C_2 to the base of Q_1 causing it to conduct more and eventually turn fully ON. The cross coupling between these two bipolar transistors will reinforce this condition with the LOW Q_1 collector voltage keeping Q_2 OFF and the HIGH Q_2 collector voltage keeping Q_1 ON. With Q_1 equivalent to a closed switch, a current path now exists for C_1 to charge as seen in Figure 9-19(b). As soon as the charge on C_1 reaches about 0.7 V, Q_2 will conduct because its base-emitter junction will be forward biased. This condition is shown in Figure 9-19(c).When Q_2 conducts, its collector voltage will drop, cutting OFF Q_1 and creating a charge path for C_2. As soon as the charge on C_2 reaches 0.7 V, Q_1 will conduct again and the cycle will repeat.

The output waveforms switch between the supply voltage ($+V_{CC}$) when a transistor is cut off, and zero volts when a transistor is saturated (ON). The result is two square-wave outputs that are out of phase with one another, as seen in the waveforms in Figure 9-19(d). Referring to the output waveforms in Figure 9-19(d), you can see that the time constant of R_1 and C_1, and R_2 and C_2 determine the complete cycle time. If the R_1/C_1 time constant is equal to the R_2/C_2 time constant, both halves of the cycle will be equal (50% duty cycle) and the result will be a square wave. Referring to Figure 9-19(d), you can see that the R_1/C_1 time constant will determine the time of one half-cycle, while the R_2/C_2 time constant will determine the time of the other half-cycle. The formula for calculating the time of one half-cycle is equal to:

$$t = 0.7 \times (R_1 \times C_1) \quad or \quad T = 0.7 \times (R_2 \times C_2)$$

The frequency of this square wave can be calculated by taking the reciprocal of both half-cycles, which will be

$$f = \frac{1}{1.4 \times RC}$$

■ **EXAMPLE:**

Calculate the positive and negative cycle time and circuit frequency of the astable multivibrator circuit in Figure 9-19, if

$$R_1 \text{ and } R_2 = 100\ \text{k}\Omega \qquad C_1 \text{ and } C_2 = 1\ \mu\text{F}$$

■ ***Solution:***

Because the time constant of R_1/C_1 and R_2/C_2 are the same, each half-cycle time will be the same and equal to

$$t = 0.7 \times (R \times C)$$
$$= 0.7 \times (100\ \text{k}\Omega \times 1\ \mu\text{F}) = 0.07\ \text{s or } 70\ \text{ms}$$

The frequency of the astable circuit will be equal to the reciprocal of the complete cycle, or the reciprocal of twice the half-alternation time.

$$f = \frac{1}{1.4 \times RC} = \frac{1}{1.4 \times (100\ \text{k}\Omega \times 1\ \mu\text{F})} = 7.14\ \text{Hz}$$

$$\text{or } f = \frac{1}{2 \times t} = \frac{1}{2 \times 70\ \text{ms}} = \frac{1}{0.14} = 7.14\ \text{Hz}$$

If the time constants of the two *RC* timing networks in the astable circuit are different, however, the result will be a rectangular or pulse waveform, as seen in Figure 9-19(e). In this instance, the same formula can be used to calculate the time for each alternation, and the frequency will be equal to the reciprocal of the time for both alternations.

9-2-2 *The Monostable Multivibrator Circuit*

The astable multivibrator is often referred to as an "unstable multivibrator" because it is continually alternating or switching back and forth, and therefore it has no stable condition or state. The monostable multivibrator has, as its name implies, one (mono) stable state. The circuit will remain in this stable state indefinitely until a trigger is applied and forces the monostable multivibrator into its unstable state. It will remain in its unstable state for a small period of time and then switch back to its stable state and await another trigger. The monostable multivibrator is often compared to a gun and is called a **one-shot multivibrator** because it will produce one output pulse or shot for each input trigger.

One-Shot Multivibrator
Produces one output pulse or shot for each input trigger.

Referring to the monostable multivibrator circuit in Figure 9-20(a), you can see that the monostable is similar to the astable except for the trigger input circuit and for the fact that it has only one *RC* timing network. To begin with, let us consider the stable state of the monostable. Components R_2, D_1, and R_5 form a voltage divider, the values of which are chosen to produce a large positive Q_2 base voltage. This large positive base bias voltage will cause Q_2 to saturate (turn heavily ON), which in turn will produce a LOW Q_2 collector voltage, which will be coupled via R_4 to the base of Q_1, cutting it OFF. The circuit remains in this stable state (Q_2 ON, Q_1 OFF) until a **trigger input** is received.

Trigger Input
Pulse used to initiate a circuit action.

Referring to the timing waveforms in Figure 9-20(b), you can see how the circuit reacts when a positive input trigger is applied. The pulse is first applied to

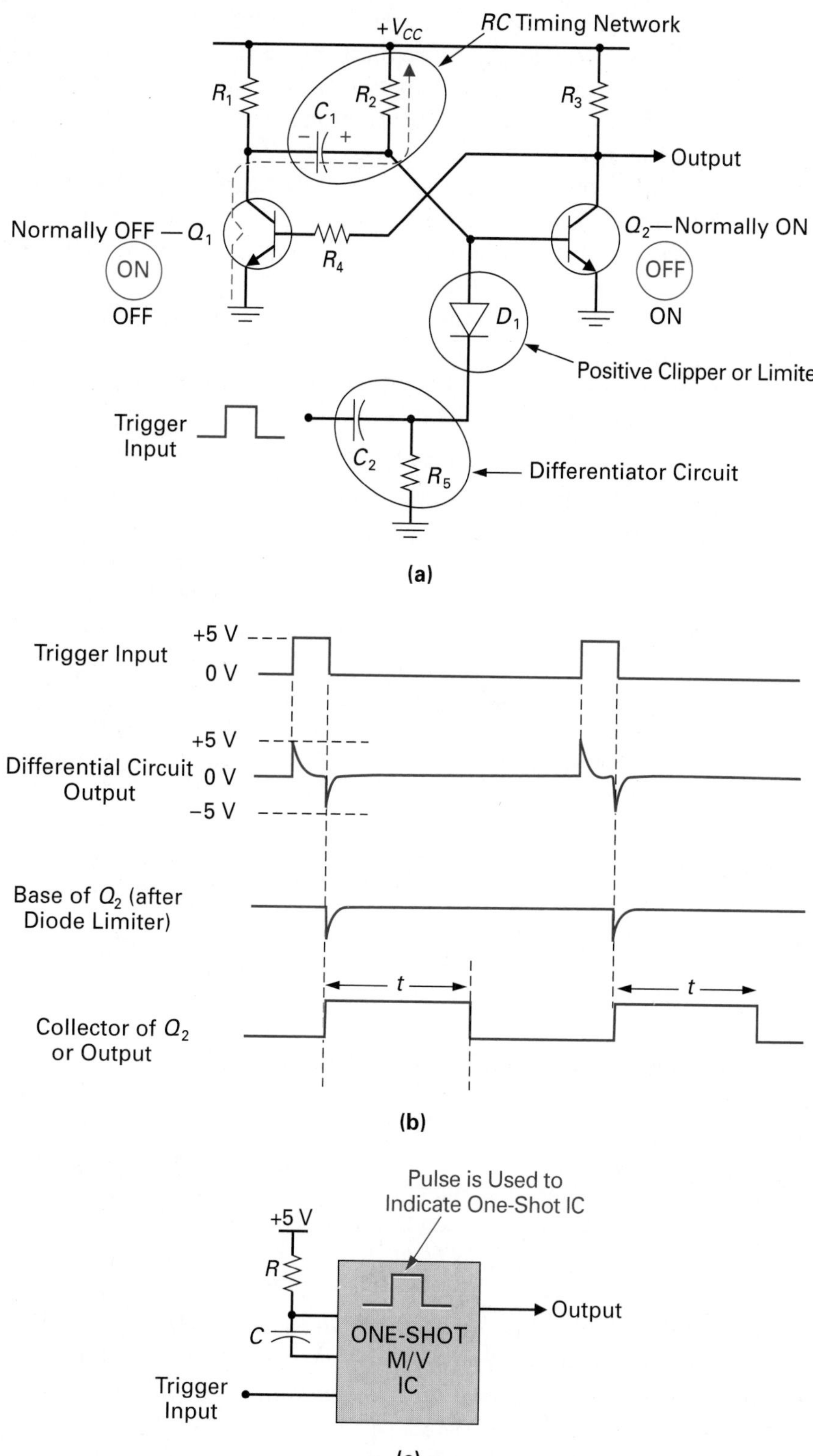

FIGURE 9-20 The Monostable Multivibrator Circuit. (a) Bipolar Transistor Circuit. (b) Input/Output Timing Waveforms. (c) One-Shot IC

a differentiator circuit (C_2 and R_5) that converts the pulse into a positive and a negative spike. These spikes are then applied to the positive clipper diode D_1, which only allows the negative spike to pass to the base of Q_2. This negative spike will reverse bias Q_2's base-emitter junction, turning Q_2 OFF and causing its collector voltage to rise to $+V_{CC}$, as seen in the waveforms. This increased Q_2 collector voltage will be coupled to the base of Q_1, turning it ON. The monostable multivibrator is now in its unstable state, which is indicated in the second color in Figure 9-20(a). In this condition, C_1 will charge as shown by the dashed current line. However, as soon as the voltage across C_1 reaches 0.7 V (which is dependent on the R_2/C_2 time constant), it will force Q_2 to conduct, which in turn will cause Q_1 to cut OFF and the monostable to return to its stable state. The output pulse width or pulse time (t) seen in Figure 9-20(b), can be calculated with the same formula used for the astable multivibrator:

$$t = 0.7 \times (R_2 \times C_1)$$

The one-shot multivibrator is sometimes used in *pulse-stretching* applications. For example, referring to the waveforms in Figure 9-20(b), imagine the input positive trigger pulse is 1µs in width, and the RC time constant of R_2 and C_1 is such that the output pulse width (t) is 500 µs. In this example, the input pulse would be effectively stretched from 1 µs to 500 µs. The monostable multivibrator, or one-shot timer circuit, is also used to introduce a *time delay*. Referring again to the waveforms in Figure 9-20(b), imagine a differentiator circuit connected to the output of the monostable circuit. If the output pulse width was again set to 500 µs, there would be a 500 µs delay between the differentiated negative edge of the input pulse and the differentiated negative edge of the output pulse.

Figure 9-20(c) shows the logic symbol for a monostable (one-shot) multivibrator. Nearly all one-shot circuits in use today are in integrated form. These IC one-shots operate in exactly the same way as their discrete component counterparts. For example, the 74LS123 IC contains two fully independent monostable multivibrators. Like the symbol in Figure 9-20(c), the 74LS123 has pins for connecting external timing resistors and capacitors.

9-2-3 The Bistable Multivibrator Circuit

S-R (Set-Reset) Flip-Flop
A multivibrator circuit in which a pulse on the SET input will "flip" the circuit into the set state while a pulse on the RESET input will "flop" the circuit into its reset state.

The bistable multivibrator has two (bi) stable states, and its bipolar transistor circuit is illustrated in Figure 9-21(a). The circuit has two inputs called the "SET" and "RESET" inputs, and these inputs drive the base of Q_1 and Q_2. The two outputs from this circuit are taken from the collectors of Q_1 and Q_2 and are called "Q" and "$\overline{Q}$" (pronounced "Q not"). The $\overline{Q}$ output derives its name from the fact that its voltage level is always the opposite of the Q output. For example, if Q is HIGH, $\overline{Q}$ will be LOW, and if Q is LOW, $\overline{Q}$ will be HIGH. The bistable multivibrator circuit is often called an ***S-R* (set-rest) flip-flop** because

*A pulse on the SET input will "flip" the circuit into the set state (*Q *output is set HIGH), while*

*A pulse on the RESET input will "flop" the circuit into its reset state (*Q *output is reset LOW).*

Figure 9-21(b) shows the logic symbol for a *S-R* or *R-S* flip-flop, or bistable multivibrator circuit.

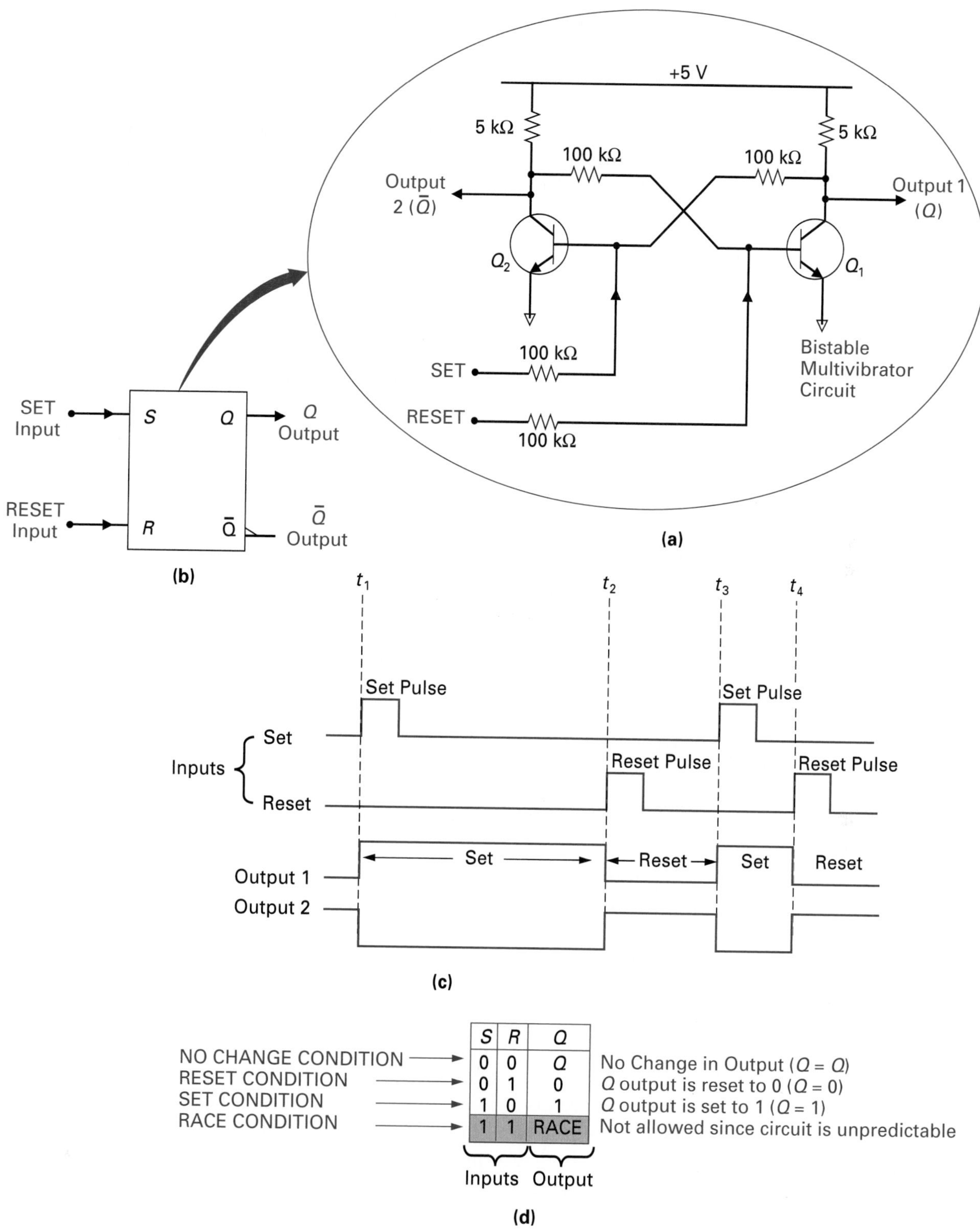

S	R	Q
0	0	Q
0	1	0
1	0	1
1	1	RACE

FIGURE 9-21 The Bistable Multivibrator or Set-Rest ($S = R$ or $R = S$) Flip-Flop. (a) Bipolar Transistor Circuit. (b) SR Symbol. (c) Timing Waveforms. (d) Truth Table.

To fully understand the operation of the circuit, refer to the waveforms in Figure 9-21(c). When power is first applied, one of the transistors will turn "ON" first and, because of the cross coupling, turn the other transistor "OFF." Let us assume that the circuit in Figure 9-21(a) starts with Q_1 ON and Q_2 OFF. The low voltage (approximately 0.3 V) on Q_1's collector will be coupled to Q_2's base, thus keeping it OFF. The high voltage on Q_2's collector (approximately +5 V) will be coupled to the base of Q_1, keeping it ON. This condition is called the **reset state** because the primary output (output 1, or Q) has been reset to binary 0, or 0 V. The cross-coupling action between the transistors will keep the transistors in the reset state (output 1 or Q = 0 V, and output 2 or $\overline{Q}$ = 5 V) until an input appears.

Reset State
Primary output set LOW.

Following the waveforms in Figure 9-21(c), you can see that the first input to go active is the SET input at time "t_1." This positive pulse will be applied to the base of Q_2 and forward bias its base-emitter junction. As a result, Q_2 will go ON and its LOW collector voltage will be applied to output 2 ($\overline{Q}$). This LOW on Q_2's collector will also be cross-coupled to the base of Q_1, turning it OFF and therefore making output 1 (Q) go HIGH. When a pulse appears on the SET input, the circuit will be put in its **set state** which means that the primary output (output 1 or Q) will be set HIGH. Studying the waveforms in Figure 9-21(c) once again, you will notice that after the positive SET input pulse has ended, the bistable will still remain **latched** or held in its last state, due to the cross-coupling between the transistors. This ability of the bistable multivibrator to remain in its last condition or state explains why the *S-R* flip-flop is also called an ***S-R* Latch.**

Set State
Primary output set HIGH.

Latched
Held in the last state.

***S-R* Latch**
Another name for *S-R* flip-flop, so called because the output remains latched in the set or reset state even though the input is removed.

Following the waveforms you can see that a RESET pulse occurs at time t_2, and resets the primary output (output 1 or Q) LOW. The flip-flop then remains latched in its reset state, until a SET pulse is applied to the set input at time t_3, setting Q HIGH. Finally, a positive RESET pulse is applied to the reset input at time t_4, and Q is reset LOW.

The operation of the *S-R* flip-flop is summarized in the truth table or function table shown in Figure 9-21(d). When only the R input is pulsed HIGH (reset condition), the Q output is reset to a binary 0 or reset LOW ($\overline{Q}$ will be the opposite, or HIGH). On the other hand, when only the S input is pulsed HIGH (set condition), the Q output is set to a binary 1 or set HIGH ($\overline{Q}$ will be the opposite, or LOW). When both the S and R inputs are LOW, the *S-R* flip-flop is said to be in the **no-change** or latch condition because there will be no change in the output Q. For example, if the output Q is SET, and then the S and R inputs are made LOW, the Q output will remain SET, or HIGH. On the other hand, if the output Q is RESET, and then the S and R inputs are made LOW, the Q output will remain RESET, or LOW.

No-Change or Latch Condition
When both inputs are LOW, the *S-R* flip-flop is said to be in the no-change or latch condition because there will be no change in the output.

The external circuits driving the S and R inputs will be designed so that these inputs are never both HIGH, as shown in the last condition in the table in Figure 9-21(d). This is called the *race condition* because both bipolar transistors will have their bases made positive, and therefore they will race to turn ON, and then shut the other transistor OFF via the cross-coupling. This input condition is not normally applied because the output condition is unpredictable.

The schmitt trigger circuit discussed previously is a bistable multivibrator circuit because its output voltage can be either one of two states. That is, its output can be either SET HIGH or RESET LOW, based on the voltage level of the input control voltage.

The bistable multivibrator *S-R* flip-flop or *S-R* latch has become one of the most important circuits in digital electronics. It is used in a variety of applications ranging from data storage to counting and frequency division. These applications will be covered later in the digital circuits chapter of this textbook. You will, however, see how this *S-R* flip-flop circuit is made use of in the following 555 timer circuit.

9-2-4 *The 555 Timer Circuit*

One of the most frequently used low-cost integrated circuit timers is the *555 Timer*. Its IC package consists of 8 pins, as seen in Figure 9-22(a), and derives its number identification from the distinctive voltage divider circuit, seen in Figure 9-22(b), consisting of three 5 kΩ resistors. It is a highly versatile timer that can be made to function as an astable multivibrator, monostable multivibrator, frequency divider, or modulator depending on the connection of external components.

Nearly all the IC manufacturers produce a version of the 555 timer, which can be labeled in different ways, for example: SN72 555, MC14 555, SE 555, and so on. Two 555 timers are also available in a 16 pin dual IC package that is labeled with the numbers 556.

Basic 555 Timer Circuit Action

Referring to the basic block diagram in Figure 9-22(b), let us examine the basic action of all the devices in a 555 timer circuit. The three 5 kΩ resistors develop reference voltages at the inputs of the two comparators *A* and *B*. As previously mentioned, a comparator is a circuit that compares an input signal voltage to an input reference voltage and then produces a YES/NO or HIGH/LOW decision output. The negative input of comparator *A* will have a reference that is 2/3 of V_{CC}, and therefore the positive input (pin 6, threshold) will have to be more positive than 2/3 of V_{CC} for the output of comparator *A* to go HIGH. With comparator *B*, the positive input has a reference that is 1/3 of V_{CC}, and therefore the negative input (pin 2, trigger) will have to be more negative, or fall below, 1/3 of V_{CC} for the output of comparator *B* to go HIGH. If the output of comparator *A* were to go HIGH, the set/reset flip-flop output would be reset LOW to 0 V. This LOW output would be inverted by the INVERTER to a HIGH, and then inverted and buffered (boosted in current) by the final INVERTER to appear as a LOW at the output pin 3. If the output of comparator *B* were to go high, the set/reset flip-flop would be set HIGH to +5 V. This HIGH output would be inverted by the INVERTER to a LOW, and then inverted and buffered by the final INVERTER to appear as a HIGH at the output pin 3. When the output of the set/reset flip-flop is LOW (reset), the input at the base of the discharge transistor will be HIGH. The transistor will therefore turn ON, and its low emitter-to-collector resistance will ground pin 7. When the output of the set/reset flip-flop is HIGH (set), the input at the base of the dis-

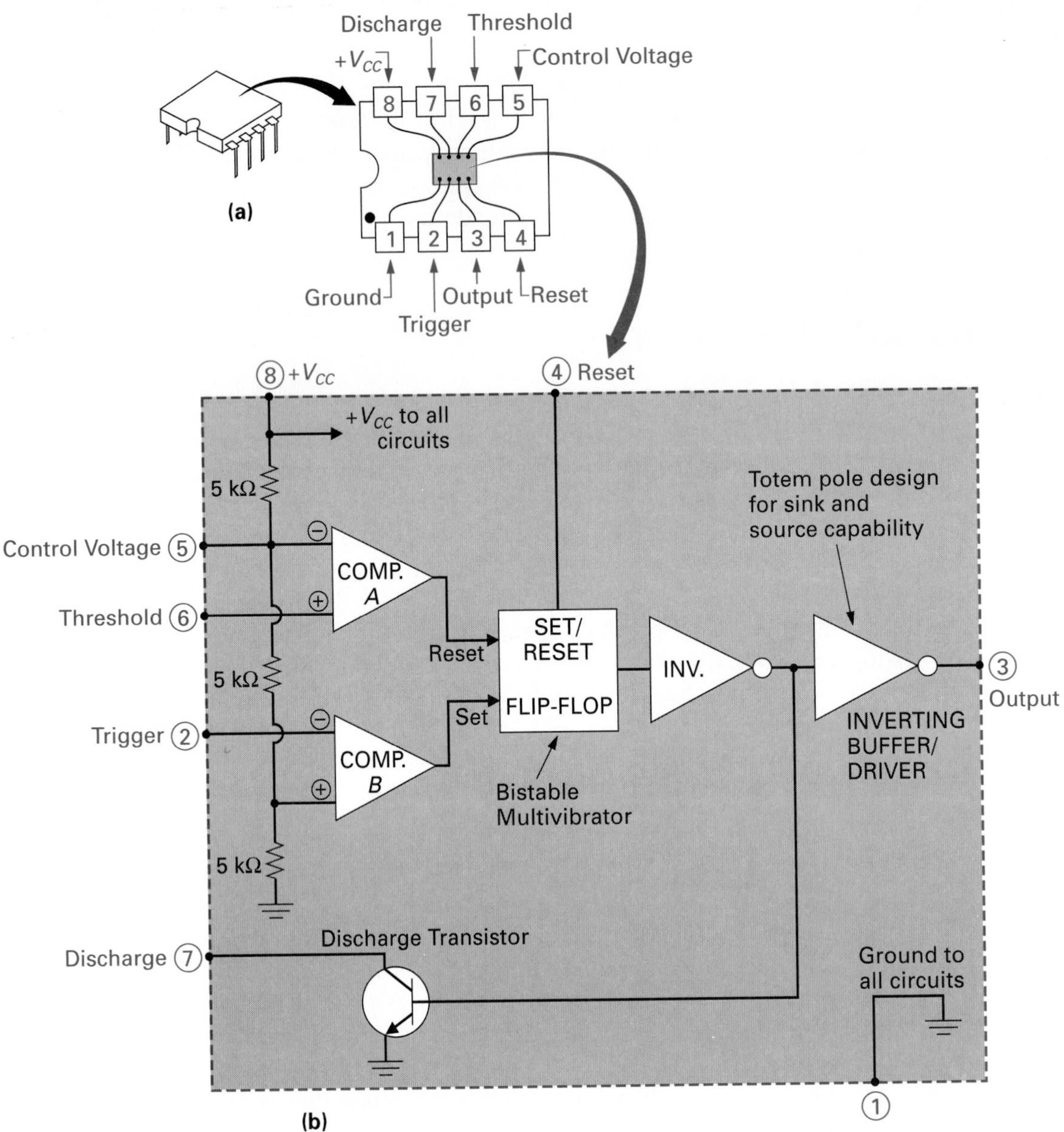

FIGURE 9-22 **The 555 Timer. (a) IC Pin Layout. (b) Basic Block Diagram.**

charge transistor will be LOW. The transistor will therefore turn OFF and its high emitter-to-collector resistance will cause pin 7 to ground.

The 555 Timer as an Astable Multivibrator

Figure 9-23(a) shows how the 555 timer can be connected to operate as an astable or free-running multivibrator. The waveforms in Figure 9-23(b) show how the externally connected capacitor C will charge and discharge and how the output will continually switch between its positive ($+V_{CC}$) and negative (0 V) peaks.

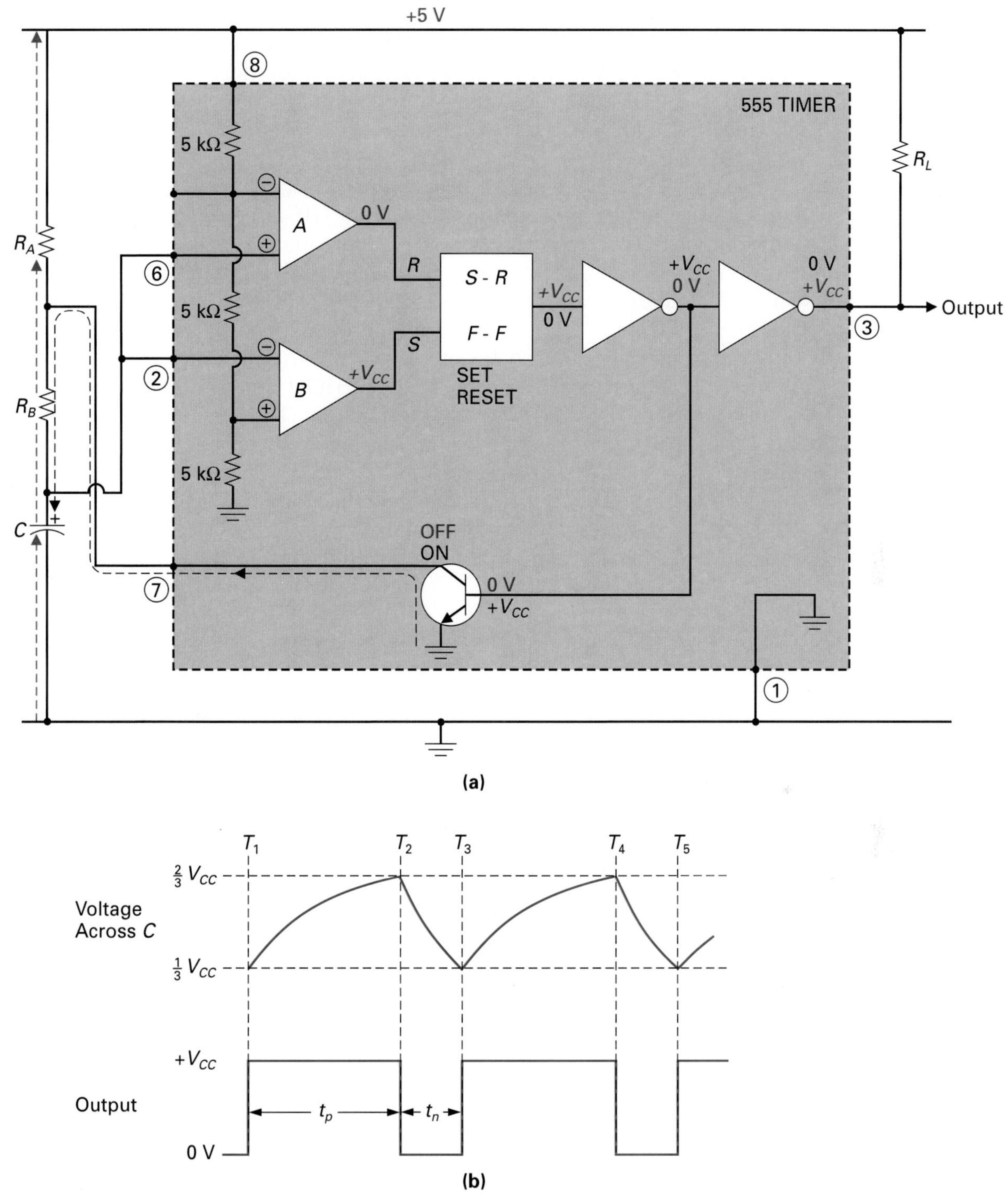

(a)

(b)

FIGURE 9-23 The 555 Timer as an Astable Multivibrator. (a) Circuit. (b) Waveforms.

To begin with, let us assume that the output of the S-R flip-flop is set HIGH, and therefore the output will be HIGH (blue condition in Figure 9-23(a), time T_1 in Figure 9-23(b)). The HIGH output of the S-R flip-flop will be inverted to a LOW and turn OFF the 555's internal discharge transistor. With this transistor OFF, the external capacitor C can begin to charge towards $+V_{CC}$ via R_A and R_B.

At time T_2, the capacitor's charge has increased beyond 2/3 of V_{CC}, and therefore the output of comparator A will go HIGH and RESET the S-R flip-flop's output LOW. This will cause the output (pin 3) of the 555 to go LOW. However, the discharge transistor's base will be HIGH, and so it will turn ON. With the discharge transistor ON, the capacitor can begin to discharge (black condition in Figure 9-23(a), time T_2 in Figure 9-23(b)). At time T_3, the capacitor's charge has fallen below 1/3 of V_{CC}, or the trigger level of comparator B. As a result, comparator B's output will go HIGH and SET the output of the S-R flip-flop HIGH, or back to its original state. The discharge transistor will once again be cut OFF, allowing the capacitor to charge and the cycle to repeat.

As you can see in Figure 9-23(a), the capacitor charges through R_A and R_B to 2/3 of V_{CC}, and then discharges through R_B to 1/3 of V_{CC}. As a result, the positive half-cycle time (t_p) can be calculated with the formula:

$$t_p = 0.7 \times C \times (R_A + R_B)$$

The negative half-cycle time (t_n) can be calculated with the formula:

$$t_n = 0.7 \times C \times R_B$$

The total cycle time will equal the sum of both half-cycles ($t = t_p + t_n$), and the frequency will equal the reciprocal of time ($f = 1/t$).

■ **EXAMPLE:**

Calculate the positive half-cycle time, negative half-cycle time, complete cycle time, and frequency of the 555 astable multivibrator circuit in Figure 9-23, if

$$R_A = 1\ \text{k}\Omega \quad R_B = 2\ \text{k}\Omega \quad C = 1\ \mu\text{F}$$

■ ***Solution:***

The positive half-cycle will last for

$$\begin{aligned} t_p &= 0.7 \times C \times (R_A + R_B) \\ &= 0.7 \times 1\ \mu\text{F} \times (1\ \text{k}\Omega + 2\ \text{k}\Omega) = 2.1\ \text{ms} \end{aligned}$$

The negative half-cycle will last for

$$\begin{aligned} t_n &= 0.7 \times C \times R_B \\ &= 0.7 \times 1\ \mu\text{F} \times 2\ \text{k}\Omega = 1.4\ \text{ms} \end{aligned}$$

The complete cycle time will be

$$t = t_p + t_n$$
$$= 2.1 \text{ ms} + 1.4 \text{ ms} = 3.5 \text{ ms}$$

The frequency of this 555 astable multivibrator will be

$$f = \frac{1}{t}$$
$$= \frac{1}{3.5 \text{ ms}} = 285.7 \text{ Hz}$$

The 555 Timer as a Monostable Multivibrator

Figure 9-24(a) shows how the 555 timer can be connected to operate as a monostable or one-shot multivibrator. The waveforms in Figure 9-24(b) show the time relationships between the input trigger, the charge and discharge of the capacitor, and the output pulse. The width of the output pulse (P_W) is dependent on the values of the external timing components R_A and C.

At time T_1 in Figure 9-24(b), the set-reset flip-flop (*S-R F-F*) is in the reset condition and is therefore producing a LOW output. This LOW from the *S-R F-F* is inverted by the INVERTER, and then inverted and buffered by the final stage to produce a LOW (0 V) output from the 555 timer at pin 3. The LOW output from the *S-R F-F* will be inverted and appear as a HIGH at the base of the discharge transistor, turning it ON and providing a discharge path for the capacitor to ground.

At time T_2, a trigger is applied to pin 2 of the 555 monostable multivibrator. This negative trigger will cause negative input of comparator *B* to fall below 1/3 of V_{CC}, and so the output of comparator *B* will go HIGH and SET the output of the *S-R F-F* HIGH. This HIGH from the *S-R F-F* will send the output of the 555 timer (pin 3) HIGH, and turn OFF the discharge transistor. Once the path to ground through the discharge transistor has been removed from across the capacitor, the capacitor can begin to charge via R_A to $+ V_{CC}$, as seen in the waveforms in Figure 9-24(b). The output of the 555 timer remains HIGH until the charge on the capacitor exceeds 2/3 of V_{CC}. At this time (T_3), the output of comparator *A* will go HIGH, resetting the *S-R F-F* and causing the output of the 555 timer to go LOW and also turning ON the discharge transistor to discharge the capacitor. The circuit will then remain in this stable condition until a new trigger arrives to initiate the cycle once again.

The leading edge of the positive output pulse is initiated by the input trigger, while the trailing edge of the output pulse is determined by the R_A and C charge time, which is dependent on their values. Because the capacitor can charge to 2/3 of V_{CC} in a little more than one time constant (1 time constant = 0.632, 2/3 = 0.633), the following formula can be used to calculate the pulse width (P_W):

$$P_W = 1.1 \times (R_A \times C)$$

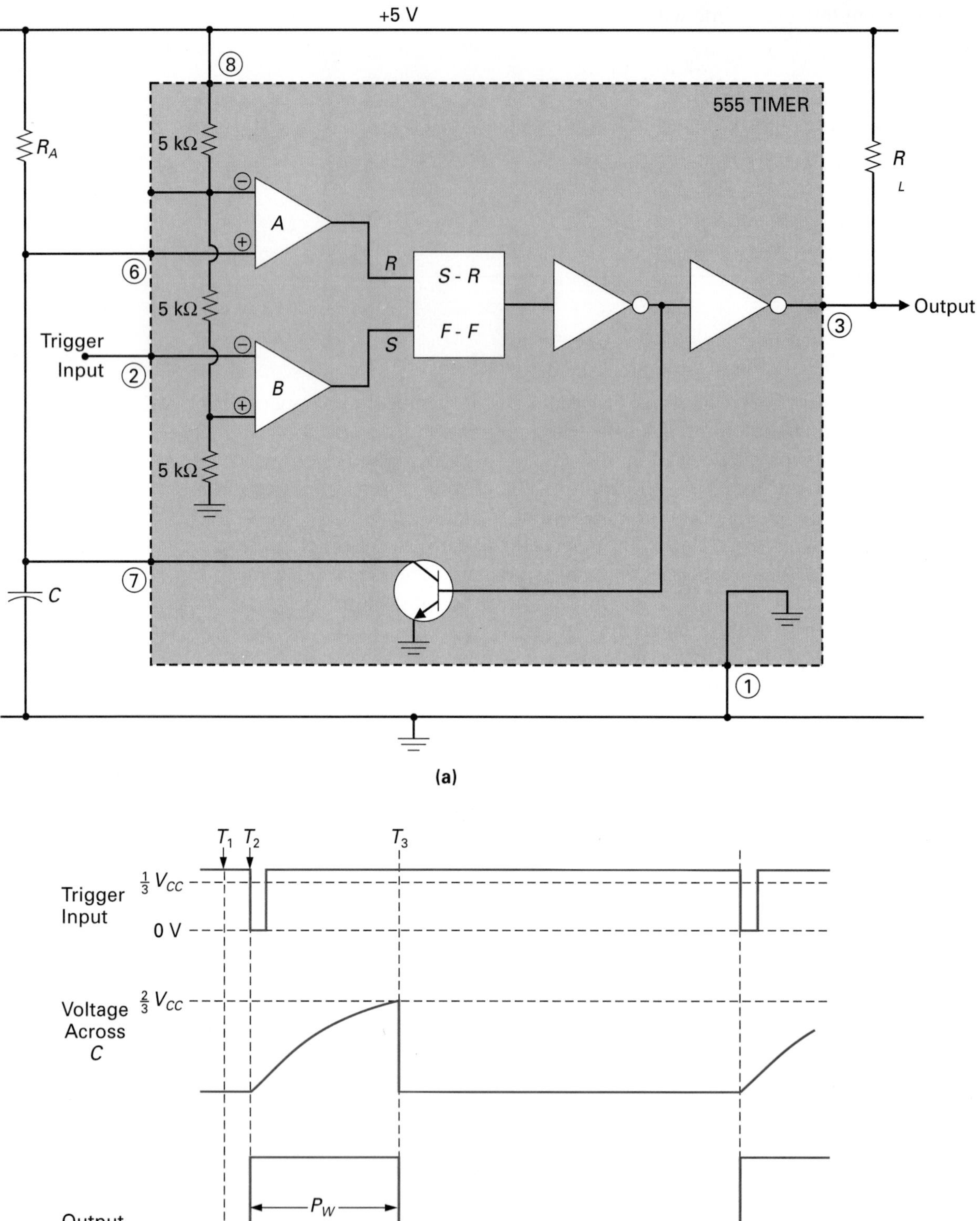

FIGURE 9-24 The 555 Timer as a Monostable Multivibrator. (a) Circuit. (b) Waveforms.

■ **EXAMPLE:**

Calculate the pulse width of a 555 monostable multivibrator, if

$$R_A = 2\ \text{M}\Omega, \text{ and } C = 1\ \mu\text{F}.$$

■ ***Solution:***

The width of the output pulse will be

$$P_W = 1.1 \times (R_A \times C)$$
$$= 1.1 \times (2\ \text{M}\Omega \times 1\ \mu\text{F}) = 2.2$$

SELF-TEST REVIEW QUESTIONS FOR SECTION 9-2

1. The astable multivibrator is also called the ____________ multivibrator.
2. Which of the multivibrator circuits has only one stable state?
3. Which multivibrator is also called a set-reset flip-flop?
4. Why is the set-reset flip-flop also called a latch?
5. The 555 timer derived its number identification from ____________.
6. List two applications of the 555 timer.

9-3 TROUBLESHOOTING DIGITAL CIRCUITS

To be an effective electronic technician or troubleshooter, you must have a thorough knowledge of electronics, test equipment, troubleshooting techniques, and equipment repair. Like analog circuits, digital circuits do occasionally fail. In most cases, a technician is required to quickly locate the problem within the system and then make the repair. The procedure for fixing a failure can be broken down into three basic steps.

Step 1: DIAGNOSE

The first step is to determine whether a problem really exists. To carry out this step, a technician must collect as much information as possible about the system, circuit, and components used, and then diagnose the problem.

Step 2: ISOLATE

The second step is to apply a logical and sequential reasoning process to isolate the problem. In this step, a technician will operate, observe, test, and apply troubleshooting techniques in order to isolate the malfunction.

Step 3: REPAIR

The third and final step is to make the actual repair, and then final test the circuit.

In this section, you will first be introduced to the tools of the trade: digital test equipment. Following this section, we will examine some of the typical internal and external digital IC failures and how these failures can be recognized.

9-3-1 *Digital Test Equipment*

There is a variety of test equipment to help you troubleshoot digital circuits and systems. Some are standard, such as the *multimeter* and *oscilloscope,* which can be used for either analog or digital circuits. Other test instruments such as the *logic clip, logic probe, logic pulser,* and *current tracer* have been designed specifically to test digital logic circuits. To begin with, let us see how the multimeter can be used to test digital circuits.

Testing with the Multimeter

Both analog and digital type multimeters can be used to troubleshoot digital circuits. In most cases, you will only use the multimeter to measure voltage and resistance. The current settings are rarely used because a path needs to be opened to measure current. Most circuits are soldered on to a printed circuit board, making current measurement impractical.

Some of the more common tests include using the multimeter on the **AC VOLTS** setting to check the 120 V ac input into the equipment's dc power supply and using the **DC VOLTS** setting to check all of the dc power supply voltages out of the power supply circuit, as seen in Figure 9-25(a). The multimeter can also be used to check that these dc supply voltages are present at each printed circuit board within the system, as seen in Figure 9-25(b). You can also use the multimeter on the DC VOLTS setting to test the binary 0 and binary 1 voltages throughout the digital circuit, as seen in Figure 9-25(b). The multimeter is also frequently used on the **OHMS** setting to test components such as resistors, fuses, diodes, transistors, and so on. It can also be used on the OHMS setting to test wires and cables for opens and shorts as shown in Figure 9-25(c).

Testing with the Oscilloscope

To review, the **pulse or rectangular wave** alternates between two peak values. However, unlike the square wave, the pulse wave does not remain at the two peak values for equal lengths of time. The **positive pulse waveform** seen in Figure 9-26(a), for instance, remains at its negative value for long periods of time and only momentarily pulses positive. On the other hand, the **negative pulse waveform** seen in Figure 9-26(b) remains at its positive value for long periods of time and momentarily pulses negative.

When referring to pulse waveforms, the term **pulse repetition frequency (PRF)** is used to describe the frequency or rate of the pulses, and the term **pulse repetition time (PRT)** is used to describe the period or time of one complete cycle, as seen in Figure 9-26(c). For example, if 1000 pulses are generated every second (PRF = 1000 pulses per second, pps or 1 kHz) each cycle will last 1/1000th of a second or 1 ms (PRT = 1/PRF = 1/kHz = 1 ms or 1000 μs).

The **duty cycle** of a pulse waveform indicates the ratio of pulse width time to the complete cycle time and is calculated with the formula

$$\text{Duty Cycle} = \frac{\text{Pulse Width}(P_W)}{\text{Period}(t)} \times 100\%$$

Pulse or Rectangular Wave
A wave that alternates between two peak values; however, unlike the square wave the pulse wave does not remain at the two peak values for an equal length of time.

Positive Pulse Waveform
A wave that remains at its negative value for long periods of time and only momentarily pulses positive.

Negative Pulse Waveform
A wave that remains at its positive value for long periods of time and only momentarily pulses negative.

Pulse Repetition Frequency (PRF)
The frequency or rate of the pulses.

Pulse Repetition Time (PRT)
The period or time of one complete cycle.

Duty Cycle
The ratio of the pulse width time to the complete cycle time.

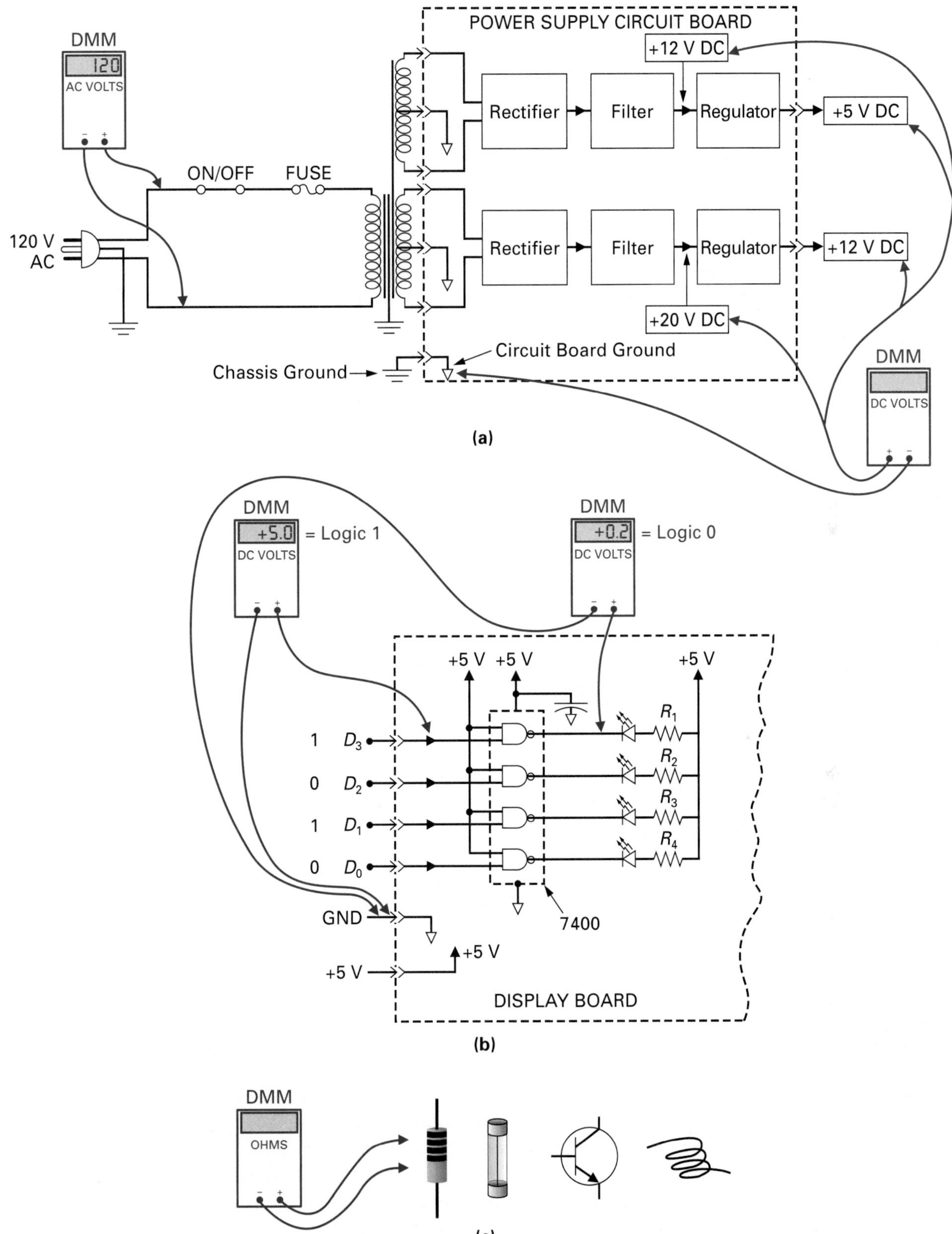

FIGURE 9-25 Testing With the Multimeter.

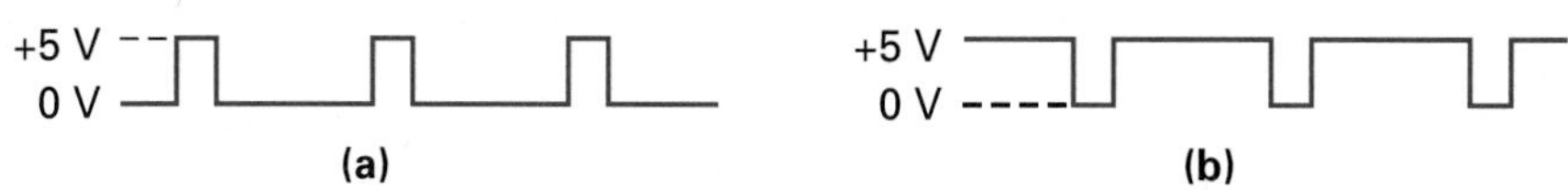

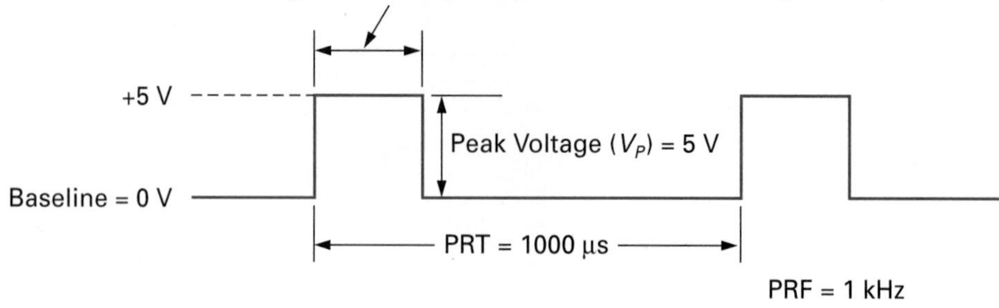

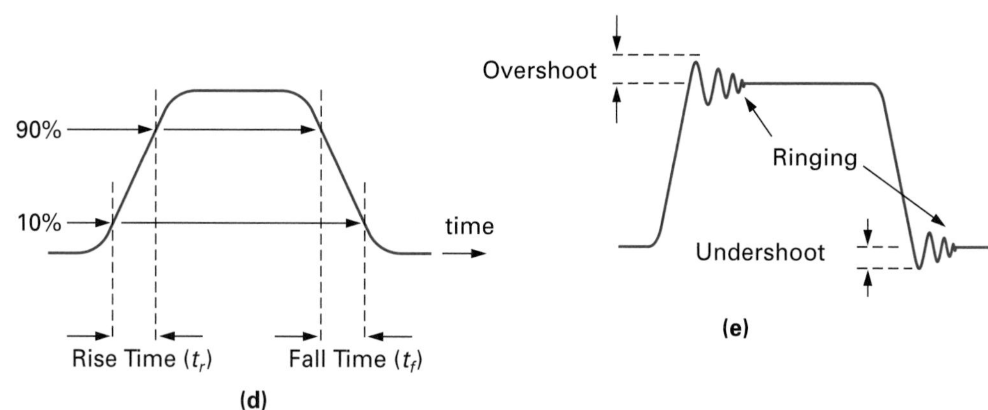

FIGURE 9-26 The Pulse Waveform. (a) Positive Pulse Waveform. (b) Negative Pulse Waveform. (c) Peak Voltage, Pulse Width, PRF and PRT. (d) Rise Time and Fall Time. (e) Overshoot, Undershoot and Ringing.

In the example in Figure 9-26(c), a pulse width of 100 μs and a period of 1000 μs will produce a duty cycle of

$$\text{Duty Cycle} = \frac{\text{Pulse Width}(P_W)}{\text{Period}(t)} \times 100\%$$

$$= \frac{100\ \mu\text{s}}{1000\ \mu\text{s}} \times 100\% = 10\%.$$

A duty cycle of 10 percent means that the positive pulse lasts for 10 percent of the complete cycle time, or PRT.

Rise Time
The time needed for the pulse to rise from 10% to 90% of the peak amplitude.

Fall Time
The time needed for the pulse to fall from 90% to 10% of the peak amplitude.

The pulse waveforms seen in Figure 9-26(a, b, and c) all rise and fall instantly between their two peak values. In reality, there is a time lapse referred to as the **rise time** and **fall time** as seen in Figure 9-26(d). The rise time is defined as the time needed for the pulse to rise from 10 percent to 90 percent of the peak amplitude, while the fall time is the time needed for the pulse to fall from 90 percent to 10 percent of the peak amplitude.

When carefully studying pulse waveforms on the oscilloscope, you will be able to measure pulse width, PRT, rise time, and fall time. You will also see a few other conditions known as *overshoot, undershoot,* and *ringing*. These unwanted conditions tend to always accompany high frequency pulse waveforms due to imperfections in the circuit, as seen in Figure 9-26(e). As you can see, the positive rising edge of the waveform "overshoots" the peak, and then a series of sine wave oscillations or "ringing" occur that slowly decrease in size. At the end of the pulse, you can see that the negative falling edge "undershoots" the negative peak value, and then this is also followed by some ringing.

The vertical graticules on an oscilloscope's display can be used as a scale to measure amplitude, while the horizontal graticules can be used as a scale to measure time, as seen in Figure 9-27(a). This ability of the oscilloscope enables us to measure almost every characteristic of a digital signal. To begin with, the vertical scale can be used to measure HIGH/LOW logic levels, as shown in Figure 9-27(b). On the other hand, the horizontal scale can be used to measure the period, rise time, fall time, and pulse width of a digital signal as seen in Figure 9-27(c). Once the period of a cycle has been calculated, the frequency can be checked because $f = 1/t$.

In nearly all digital systems, you will use the oscilloscope to check the clock or master timing signal and to see that it is not only present at every point in the circuit but also that its frequency is correct. You will also use the oscilloscope frequently to monitor two or more signals simultaneously, as shown in Figure 9-27(d). In this instance, an input and output waveform are being compared to determine the NAND gate's propagation delay time. In the oscilloscope display in Figure 9-27(d), you can also see that the oscilloscope displays noise pulses within a signal, such as overshoot, undershoot, and ringing. The oscilloscope can be used to measure these waveform distortions to see if they are causing any false operations, such as incorrectly triggering a gate. A noise problem in a digital circuit is often called a **glitch,** which stands for "gremlins loose in the computer housing."

Glitch
Noise problem in a digital circuit.

From the previous discussion, you can see that the oscilloscope is a very versatile test instrument, able to measure almost every characteristic of a digital signal. Its accuracy, however, does not match either the multimeter for measuring logic level voltages or the frequency counter for measuring frequency. However, the scope is a more visual instrument, able to display a picture image of signals at each point in a circuit and enabling the operator to see phase relationships, wave shape, pulse widths, rise and fall times, distortion, and so on.

When using an oscilloscope to troubleshoot a digital circuit, try to

a. Always use a *multi-trace oscilloscope* when possible, so that you can compare input and output waveforms. Remember that a chopped or multiplexed display will not be accurate at high frequencies because one trace is trying to display two waveforms. A dual beam oscilloscope is best for high frequency digital circuit troubleshooting.

b. Trigger the horizontal sweep of the oscilloscope with the input signal, or a frequency related signal. Having a *triggered sweep* will ensure more accurate timing measurements and enable the oscilloscope to more easily lock on to your input.

c. Use an oscilloscope with a large *bandwidth.* The input signals are fed to vertical amplifiers within the oscilloscope, and these circuits need to have a

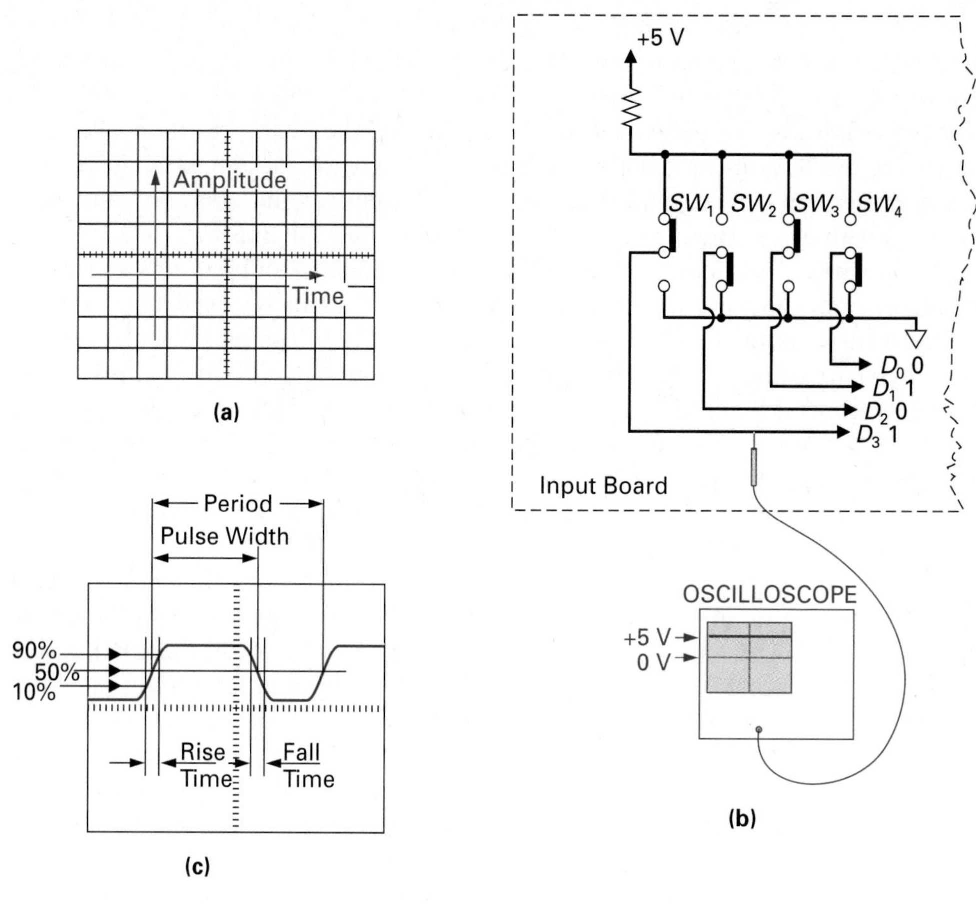

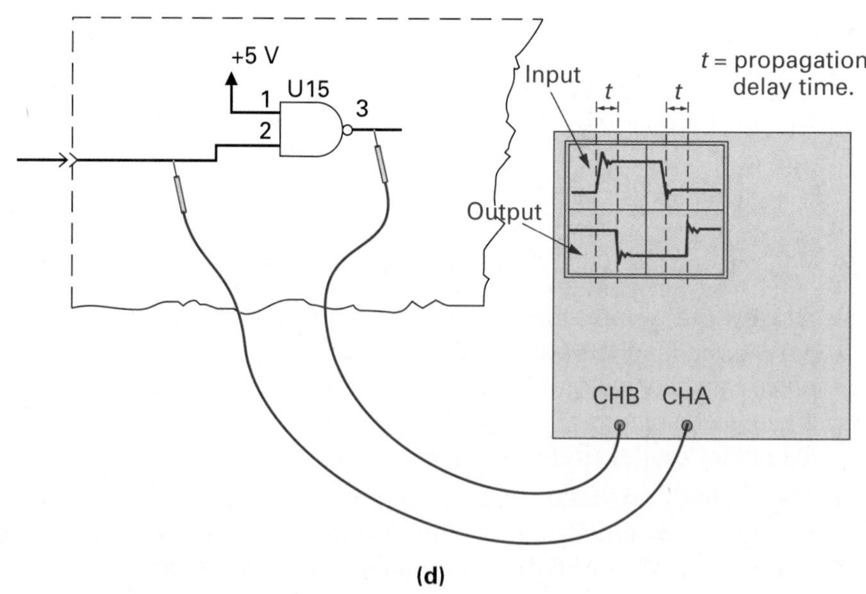

FIGURE 9-27 Testing With the Oscilloscope.

high upper frequency limit. Since all scopes can measure dc signals (0 Hz), this lower frequency is generally not listed. The upper frequency limit, however, is normally printed somewhere on the front panel of the scope. It should be 20 MHz for most digital measurements, 40 to 60 MHz for high speed TTL, and between 100 to 1000 MHz for high-speed ECL circuits.

Testing with the Logic Clip

The **logic clip,** which is shown in Figure 9-28(a), was specially designed for troubleshooting digital circuits. It consists of a spring loaded clip that clamps onto a standard IC package, where it makes contact with all the pins of the IC, as shown in Figure 9-28(b). These contacts are then available at the end of the clip, so that a multimeter or oscilloscope probe can easily be connected.

Logic Clip
A spring loaded clip that clamps onto a standard IC package, where it makes contact with all the pins. It was designed for troubleshooting digital circuits.

Some logic clips simply have connections on the top of the clip, while others have light emitting diodes (LEDs) and test points on the top of the clip, as seen in Figure 9-28(a). The LEDs give a quick indication of the binary level on the pins of the IC, with a logic 0 turning an LED OFF and a binary 1 turning an LED ON.

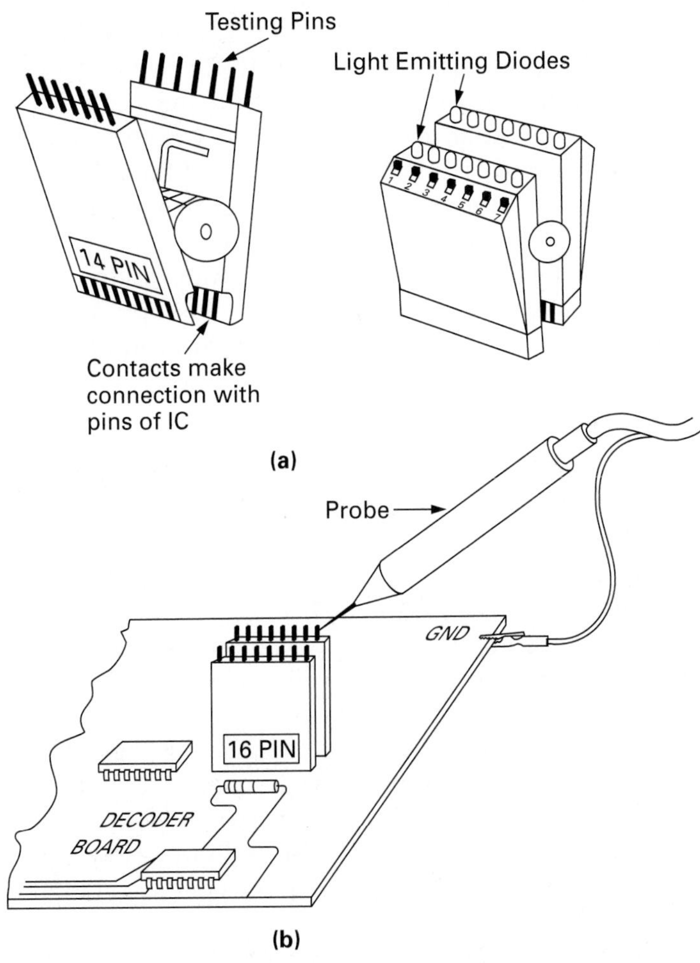

FIGURE 9-28 Testing With the Logic Clip.

Testing with the Logic Probe

Logic Probe An instrument designed specifically for digital circuit troubleshooting.

Figure 9-29(a) shows a typical **logic probe,** which is an instrument designed specifically for digital circuit troubleshooting. In order for the probe to operate, its red and black power leads must be connected to a power source. This power source can normally be obtained from the circuit being tested, as shown in Figure 9-29(b). Once the power leads have been connected, the logic probe is ready to sense the voltage at any point in a digital circuit and give an indication as to whether that voltage is a valid binary 1 or valid binary 0. The "HI" and "LO" LEDs on the logic probe are used to indicate the logic level as seen in Figure 9-29(b). Because the voltage thresholds for a binary 0 and binary 1 can be different, most logic probes have a selector switch to select different internal threshold detection circuits for TTL, ECL, or another digital family of ICs that will be discussed later called CMOS. In addition to being able to perform these **static tests** (constant or nonchanging) like the multimeter, the logic probe, like the oscilloscope, can also perform **dynamic tests,** which detect a changing signal such as a single momentary pulse or a clock-signal pulse train (sequence of pulses). Nearly all logic probes have an internal low-frequency oscillator that will flash the "PULSE" LED ON and OFF if a pulse train is detected at a frequency of up to 100 mHz. If a single pulse occurs, the width of the pulse may be so small that the operator may not be able to see the PULSE LED blink. In this instance, a memory circuit can be enabled using the MEMORY switch so that any pulse (even as narrow as 10 nanoseconds) will turn ON, and keep ON, the PULSE LED. This memory circuit within the logic probe senses the initial voltage level present at the test point (either HIGH or LOW) and then turns ON the pulse light if there is a logic transition or change. To reset this memory circuit, simply toggle (switch OFF and then ON) the memory switch.

Static Tests Tests on nonchanging signals.

Dynamic Tests Tests on changing signals such as a single momentary pulse or a clock-signal pulse train.

Even though the logic probe is not as accurate as the multimeter and cannot determine as much about a signal as the oscilloscope, it is ideal for digital circuit troubleshooting because of its low cost, small size, and versatility.

Testing with the Logic Pulser

Using the logic probe to test the static HIGH and LOW inputs and output of a gate does not fully test the operation of that logic gate. In most instances, it is necessary to trigger the input of a circuit and then monitor its response to this dynamic test. A **logic pulser** is a signal generator designed to produce either a single pulse or a pulse train. Its appearance is similar to the logic probe shown in Figure 9-30(a) in that it requires a supply voltage to operate and has a probe which is used to apply pulses to the circuit under test, as shown in Figure 9-30(b).

Logic Pulser A signal generator designed to produce either a single pulse or a pulse train.

The logic pulser is operated by simply connecting the probe to any conducting point on a circuit, and then pressing the SINGLE PULSE button for a single pulse or selecting the PULSE TRAIN switch position for a constant sequence of pulses. The frequency of the pulse train in this example model can be either one pulse per second (1 pps) or 500 pulses per second (500 pps), based on the pps selector switch.

The logic probe is generally used in conjunction with the logic pulser to sense the pulse or pulses generated by the logic pulser, as shown in Figure

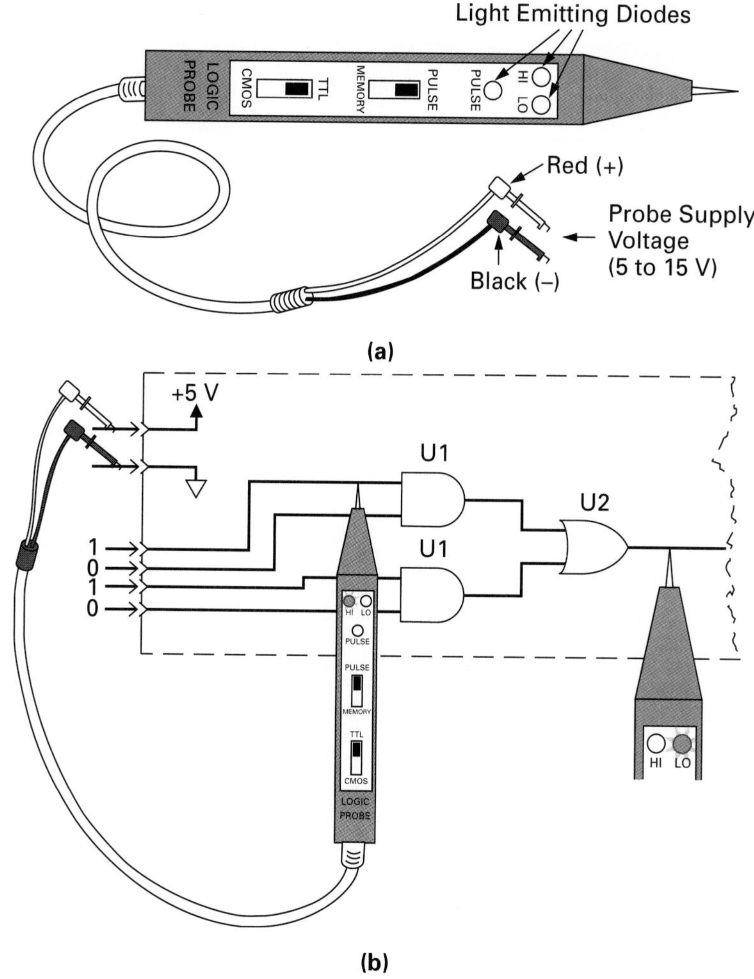

FIGURE 9-29 Testing With the Logic Probe.

9-30(b). In this example, the logic pulser is being used to inject a single pulse into the input of an OR gate, and a logic probe is being used to detect or sense this pulse at the output of the OR gate. If the input to the OR gate was HIGH, the logic pulser would have automatically sensed the binary 1 voltage level and then driven the input to the OR gate LOW when the pulse button was pressed. In this example, however, the input logic level to the OR gate was LOW, and so the logic pulser sensed the binary 0 voltage level and then pulsed the OR gate input HIGH. This ability of the logic pulser to sense the logic level present at any point and then pulse the line to its opposite state means that the operator does not have to determine these logic levels before a point is tested, making the logic pulser a fast and easy instrument to use. The logic pulser achieves this feature with an internal circuit that will override the logic level in the circuit under test, and either source current or sink current when a pulse needs to be generated. This complete *in-circuit testing* ability means that components do not need to be removed, and input and output paths opened in order to carry out a test.

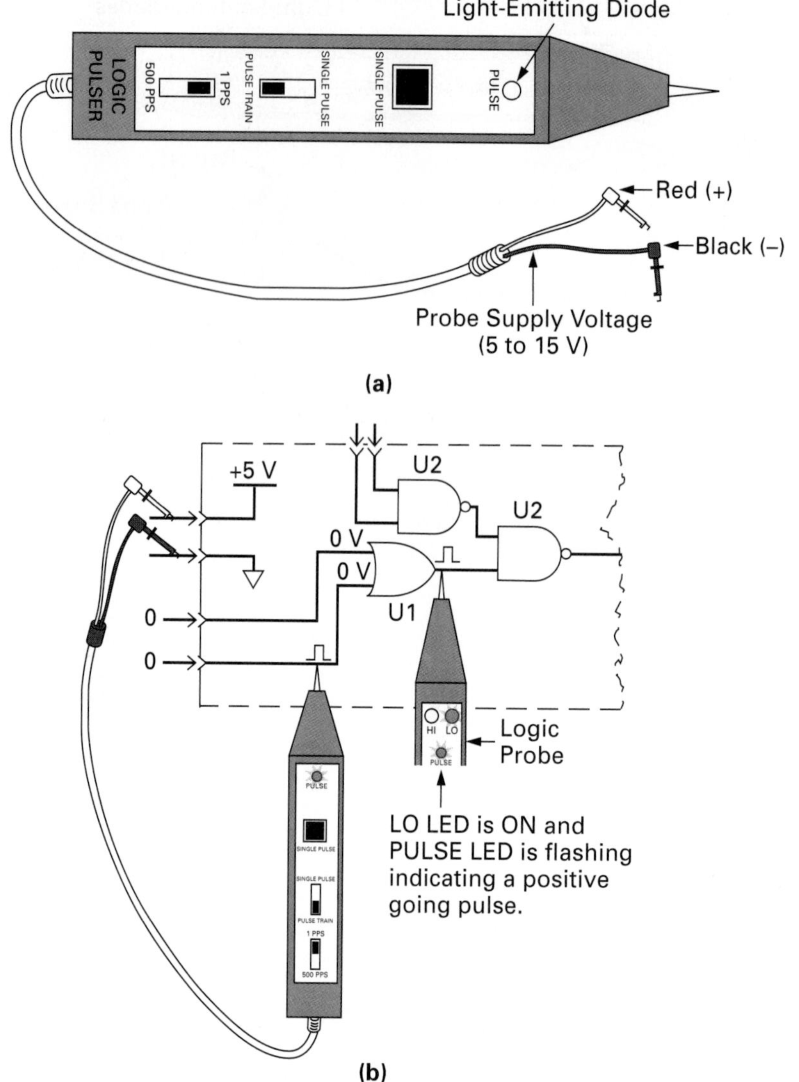

FIGURE 9-30 Testing With the Logic Pulser.

Since TTL, ECL, and CMOS (which will be discussed later) all have different input and output voltages and currents, logic pulsers will need different internal circuits for each family of logic ICs. Typical pulse widths for logic pulsers are between 500 nanoseconds to 1 microsecond for TTL logic pulsers and 10 microseconds for CMOS logic pulsers.

Current Tracer
A sensing test instrument that senses the relative values of current in a conductor.

Testing with the Current Tracer

The logic pulser can also be used in conjunction with a **current tracer,** which is shown in Figure 9-31(a). Like the logic probe, the current tracer is also a sensing test instrument. Unlike the logic probe, the current tracer senses the relative values of current in a conductor. It achieves this by using an insulated inductive

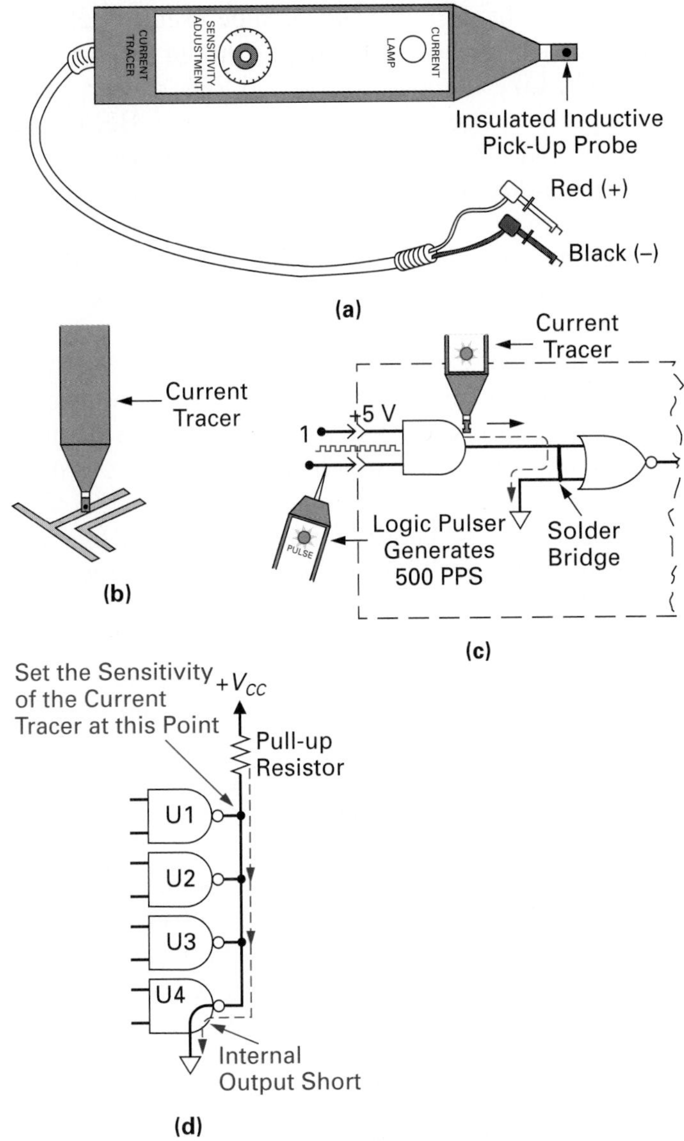

FIGURE 9-31 Testing With the Current Tracer.

pick-up tip, which senses the magnetic field generated by the current in a conductor. By adjusting the sensitivity control on the current tracer and observing the lamp's intensity when the probe is placed on a pulsating logic signal line, a shorted path can be found by simply tracing the path of high current.

To troubleshoot a short in an IC's input or output circuit, a power supply, or a printed circuit board (PCB) track or cable, we would traditionally have to cut PCB tracks, snip pins, or open component leads. This would isolate the short because the excessive current path or short would be broken when the path was opened. The current tracer allows the troubleshooter to isolate the shorted path without tampering with the circuit's construction.

Let us now see how the current tracer can be used to detect shorted paths. Figure 9-31(b) illustrates how the current tracer should be correctly aligned on a conducting PCB track. Like the logic probe and logic pulser, the current tracer can use the same supply voltage as the circuit under test. Because the probe's tip is insulated, it can be placed directly on the track. However, be sure that the probe is always perpendicular to the board. The other important point to remember is to ensure that the small holes on either side of the tip are aligned so that they face up-and-down the track being traced at all times. This is so that the inductive coil in the insulated tip is oriented to pick up the maximum amount of magnetic flux.

Figure 9-31(c) shows an example of how a current tracer and logic pulser can be used to locate a short caused by a solder bridge. In this example, we will first connect the logic pulser so that it will generate a pulsating current, and therefore a changing magnetic field, which can be detected by the inductive tip of the current tracer. The next step is to touch the tip of the current tracer on the output of the AND gate and adjust the sensitivity level of the current tracer to light the current lamp (the sensitivity adjustment allows the current tracer to sense a wide range of current levels, from 1 mA to 1 A typically). Once this reference level of brightness has been set, trace along the path between the output of the AND gate and the input of the NOR gate. Because the current value should be the same at all points along this track, the lamp should glow at the same level of brightness. Once the current tracer moves past the solder bridge, the lamp will go out because no current is present beyond this point. At this point you should, after a visual inspection, be able to locate the short.

The current tracer is also an ideal tool for troubleshooting a circuit in which many outputs are tied to a common point (such as open collector gates), as seen in the example in Figure 9-31(d). If any of these gates were to develop an internal short, it would be difficult to isolate which of the parallel paths is causing the problem. In the past, you would have had to remove each gate from the circuit (which is difficult with a soldered PCB), or clip the output IC pin of each gate until the short was removed and the faulty gate isolated. The current tracer can quickly isolate faults of this nature because it will highlight the current path provided by the short and lead you directly to the faulty gate.

9-3-2 *Digital Circuit Problems*

Because ICs account for almost 85 percent of the components within digital systems, it is highly likely that most digital circuit problems will be caused by a faulty IC. Digital circuit failures can be basically divided into two categories: *digital IC problems* which are failures within the IC or *other digital circuit device problems.* When troubleshooting digital circuits, you will have to first isolate the faulty IC, and then determine whether the problem is internal or external to the IC. An internal IC problem cannot be fixed and the IC will have to be replaced. An external IC problem can be caused by electronic devices, connecting devices, mechanical or magnetic devices, power supply voltages, clock signals, and so on. To begin with, let us discuss some of the typical internal IC failures.

Digital IC Problems

Digital IC logic gate failures can basically be classified as either opens or shorts in either the inputs or output, as summarized in Figure 9-32. In all cases, these failures will result in the IC having to be discarded and replaced. The following examples will help you to isolate an internal IC failure.

Internal IC opens An internal open gate input or open gate output are very common internal IC failures. These failures are generally caused by the very thin wires connecting the IC to the pins of the package coming loose or burning out due to excessive values of current or voltage.

Figure 9-33 illustrates how a logic probe and logic pulser can be used to isolate an internal open NAND gate input. In Figure 9-33(a), the open input on pin 10 has been connected to $+V_{CC}$, and the other gate input (pin 9) is being driven by the logic pulser. As you know, any LOW input to a NAND gate will produce a HIGH output, and only when both inputs are HIGH will the output be LOW, as shown in the inset in Figure 9-33(a). The logic probe, which is monitoring the gate on pin 8, indicates that a negative going pulse waveform is present at the output and the gate seems to be functioning normally. When the inputs are reversed, however, as seen in Figure 9-33(b), the output remains permanently LOW. These conditions could only occur if the pin 10 input was permanently

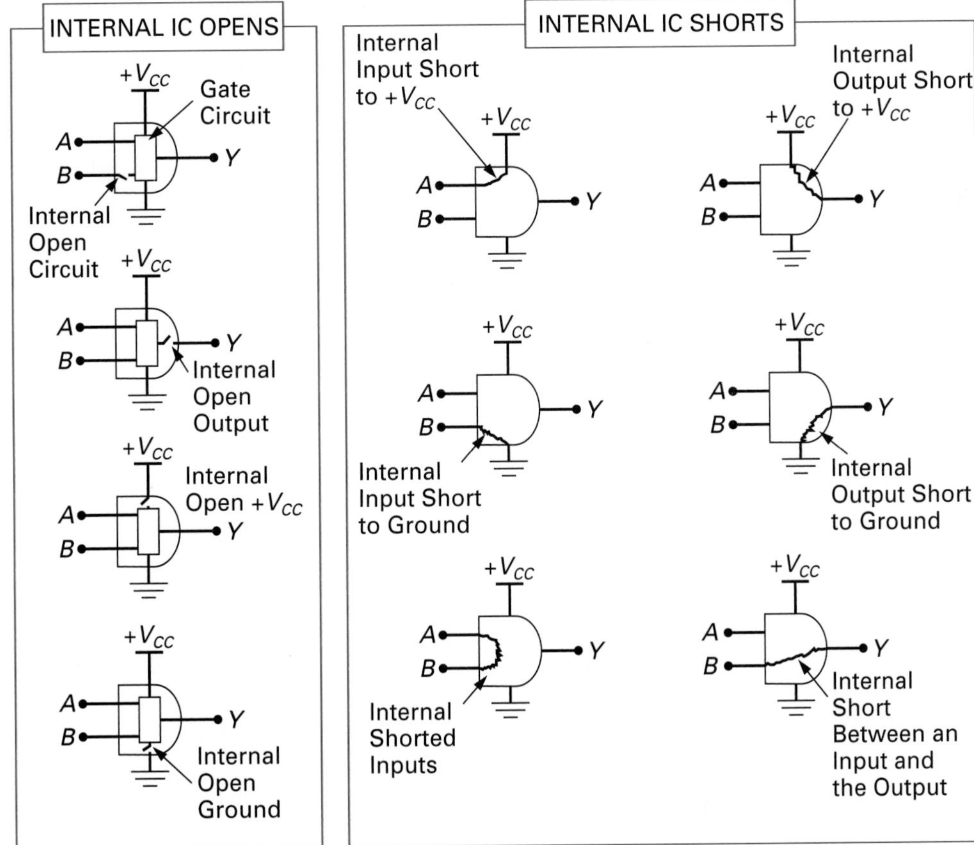

FIGURE 9-32 Internal IC Failures.

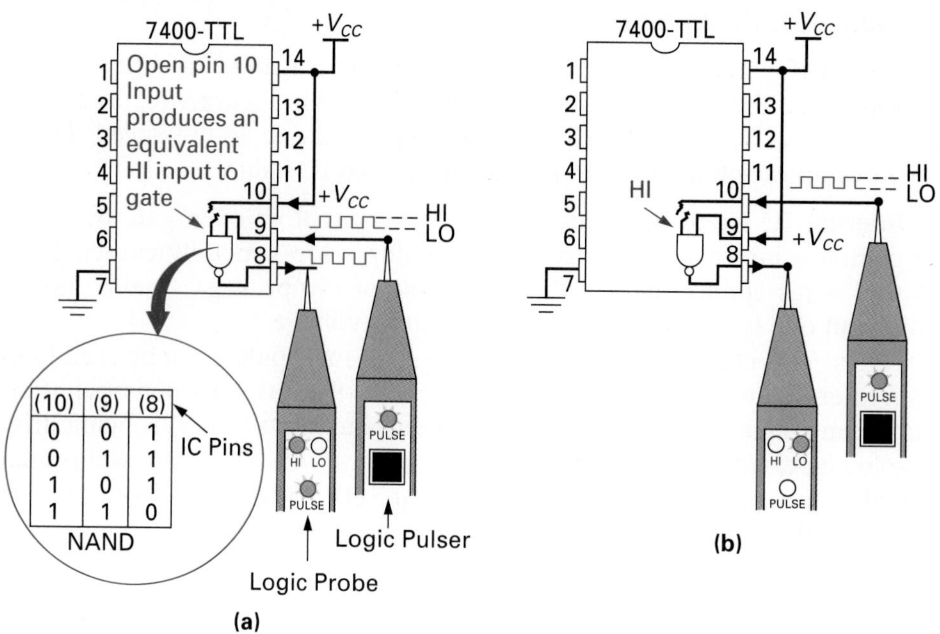

FIGURE 9-33 An Internal Open Gate Input.

HIGH. As discussed previously, an open or floating input is equivalent to HIGH input because both have next to no input current. The cause must be either an internal open at pin 10 (because this is equivalent to a HIGH input) or a $+V_{CC}$ short to the pin 10 input of this gate (which is less common than an open). In either case, the logic gate IC has an internal failure and will have to be replaced.

Figure 9-34 illustrates how a logic pulser and logic probe can be used to isolate an internal open gate output. In Figure 9-34(a), the logic gate's input at pin

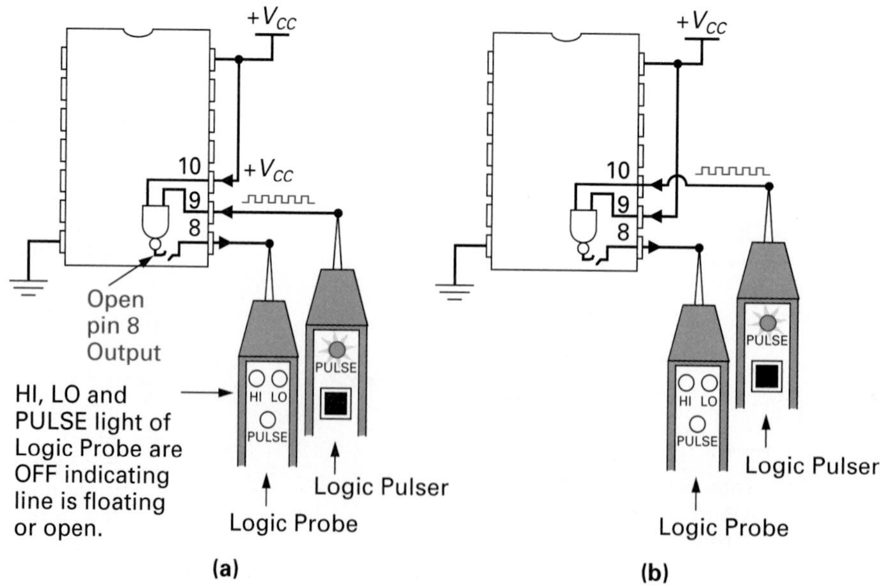

FIGURE 9-34 An Internal Open Gate Output.

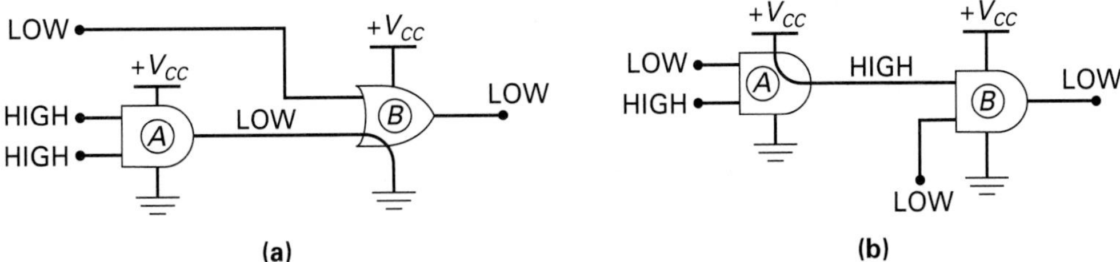

FIGURE 9-35 Internal IC Shorts.

10 has been connected to $+V_{CC}$, and the other input at pin 9 is being driven by the logic pulser. The logic probe, which is monitoring the output on pin 8, has none of its indicator LED's ON, which indicates that the logic level is neither a valid logic 1 or a valid logic 0 (line is probably floating). In Figure 9-34(b), the inputs have been reversed; however, the logic probe is still indicating the same effect. This test highlights the failure which is an open gate output.

Internal IC shorts Internal logic gate shorts to either $+V_{CC}$ or ground will cause the inputs or output to be either stuck HIGH or LOW. These internal shorts are sometimes difficult to isolate, and so let us examine a couple of typical examples.

Figure 9-35(a) shows the effect an internal input lead short to ground will have on a digital logic gate circuit. The input to gate *B* is shorted to ground. This LOW between the *A* output and *B* input makes it appear as though gate *A* is malfunctioning because it has two HIGH inputs and therefore should be giving a LOW output. The problem is to determine whether the fault is in gate *A* or gate *B*. This can best be achieved by disconnecting gate *A* from gate *B* because gate *A*'s output will go HIGH and indicate it is functioning normally when the short in gate *B* is removed.

In Figure 9-35(b), the output of gate *A* has shorted to $+V_{CC}$. This circuit condition gives the impression that gate *A* is malfunctioning. However, if the *B* gate input had shorted to $+V_{CC}$, we would get the same symptoms. Once again, we will have to isolate the *A* output from the *B* input to determine which gate is faulty.

Troubleshooting internal IC failures Let us now try a few examples to test our understanding of digital test equipment and troubleshooting techniques.

■ **EXAMPLE:**

Determine whether a problem exists in the circuit in Figure 9-36(a), based on the logic probe readings indicated. If a problem does exist, indicate what you think could be the cause.

■ ***Solution:***

Gates *B* and *C* seem to be producing the correct outputs based on their input logic levels. Gate *A*'s output, on the other hand, should be LOW, and it is reading a HIGH. To determine whether the problem is the source gate *A* or the load gate *B*, the two gates have been isolated as shown in Figure 9-36(b). After isolating gate *A* from gate *B*, you can see that the pin 10 input to gate *C* seems to be stuck HIGH probably due to an internal short to $+V_{CC}$.

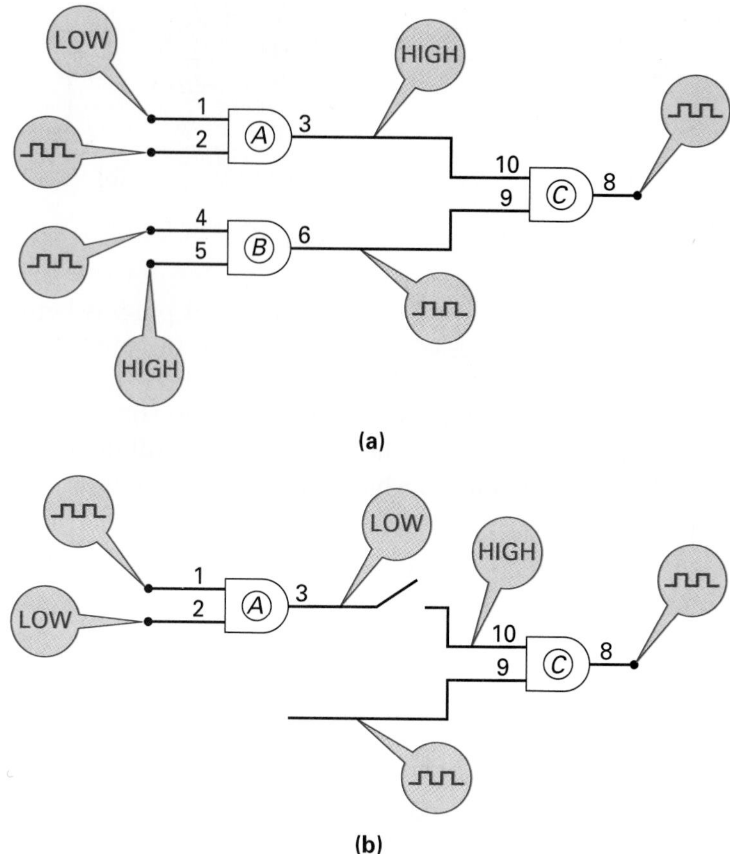

FIGURE 9-36 Troubleshooting Gate Circuits.

Even though gate *C* seems to be the only gate malfunctioning, the whole 7408 IC will have to be replaced. In situations like this, it is a waste of time isolating the problem to a specific gate when the problem has already been isolated down to a single component. Use your understanding of internal logic gate circuits to isolate the problem to a specific component and not to a device within that component.

■ **EXAMPLE:**

What could be the possible fault in the circuit shown in Figure 9-37?

■ ***Solution:***

The output of NOT gate *A* should be HIGH; however, it is pulsating. This pulsating signal can only have come from the pin 2 of the NAND gate. By disconnecting the pulsating input at pin 2 of the NAND gate, and then single pulsing this input using the logic pulser, you can use the logic probe to see if the pin 1 input of the NAND gate always follows what is on pin 2. Once an input short has been determined, visually inspect the NAND gate IC to be sure that there is no external short between the pins 1 and 2.

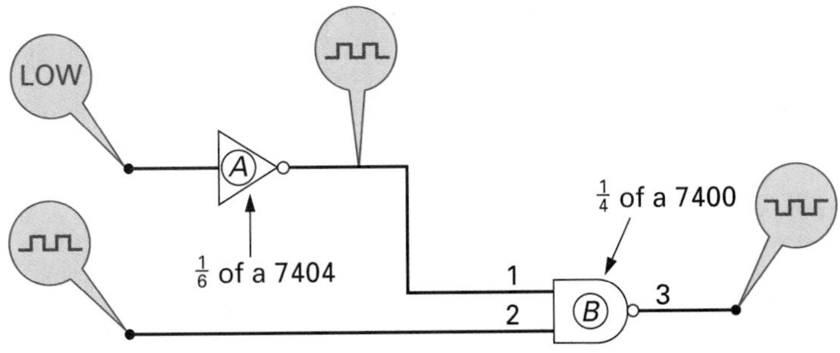

FIGURE 9-37 Troubleshooting Gate Circuits.

Other Digital Circuit Device Problems

Failures external to ICs can produce almost exactly the same symptoms as an internal IC failure, and therefore your task is to determine whether the problem is internal or external to the IC. The external IC circuit problems can be wide and varied; however, let us discuss some of the more typical failures.

External IC shorts and opens Most digital signal line shorts can be isolated by injecting a signal on to the line using the logic pulser, and then using the current tracer to find the shorted path. Here are some examples of typical external IC short circuit problems.

a. A short between the pins of the IC package due to a solder bridge, sloppy wiring, wire clippings, or an improperly etched printed circuit board.

b. A bending in or out of the IC pins as they are inserted into a printed circuit board.

c. The shorting of an externally connected component, such as a shorted capacitor, resistor, diode, transistor, LED, switch, and so on.

d. A shorted connector such as an IC socket, circuit board connector, or cable connector.

Most digital signal line opens are easily located because the digital signal does not arrive at its destination. The logic probe is ideal for tracing these open paths. Here are some examples of typical external IC open circuit problems.

a. An open in a signal line due to an improperly soldered pin, a deep scratch in the etched printed circuit board, a bent or broken IC pin, a broken wire, and so on.

b. An IC that is placed into the printed circuit board backwards.

c. The failure of externally connected components such as resistors, diodes, transistors, and so on.

d. An open connector such as an IC socket, circuit board connector, or cable connector.

Power supply problems Like analog circuits, digital circuits suffer from more than their fair share of power supply problems. When a digital circuit seems to have a power supply problem, remember to use the half-split method of troubleshooting first, as seen in Figure 9-38(a). First, check the dc supply volt-

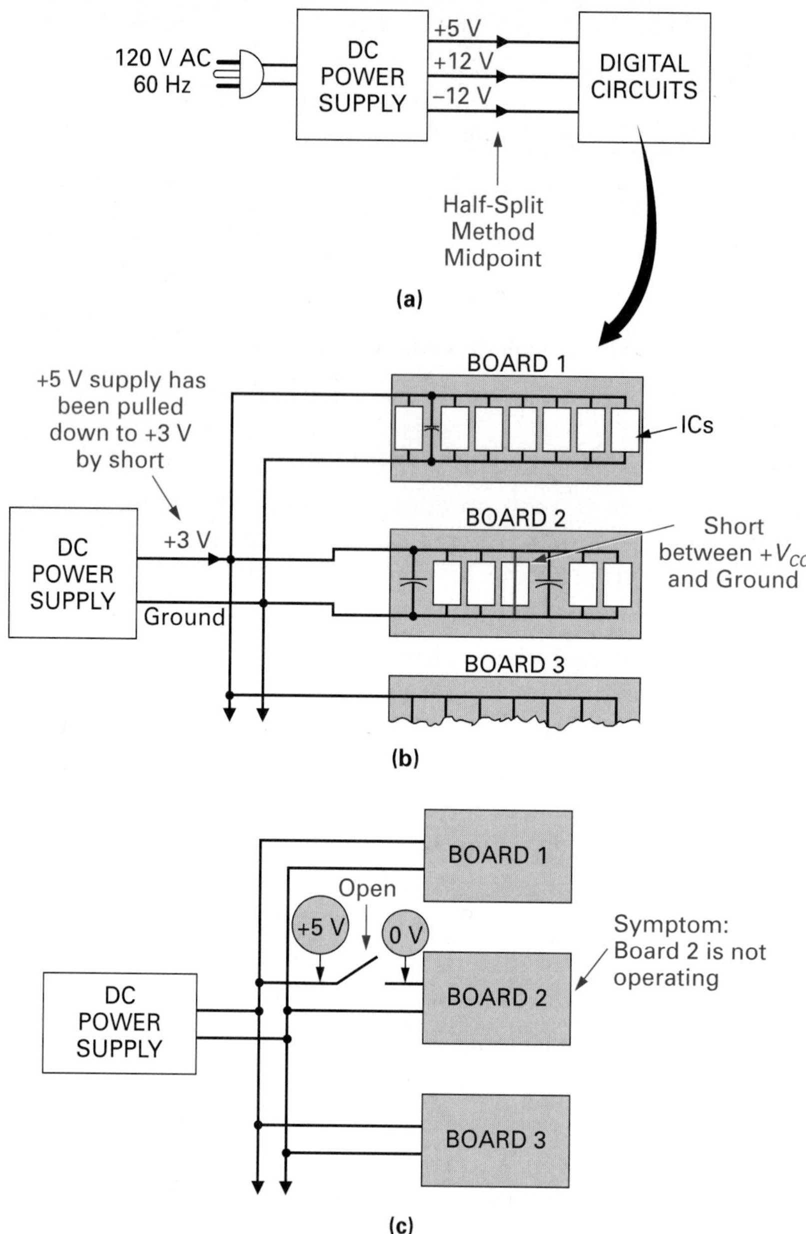

FIGURE 9-38 Power Supply Problems in Digital Circuits. (a) Using the Half-Split Method. (b) Tracing Shorts. (c) Tracing Opens.

ages from the dc power supply unit to determine whether the power problem is in the power supply (source) or in the digital circuits (load).

If the dc output voltages are not present at the output of the dc power supply, then the fault probably exists before this test point. In this instance, you should test the dc power supply circuit, applying the troubleshooting techniques outlined in Chapter 5. However, a short in the digital circuits can pull down the dc supply voltage and make it appear as though the problem is in the dc power supply, as shown in Figure 9-38(b). Once again, the current tracer is ideal for tracing the faulty board and then the faulty device in that printed circuit board. If you cannot get power to the digital circuits because the short causes the dc power supply fuse to continually blow, use the logic pulser to inject a current into the supply line and then use the current tracer to follow the current path to the short.

Open power supply lines are easy to locate because power is prevented from reaching its destination. The logic probe is ideal for tracking down power line opens, as shown in Figure 9-38(c).

Clock signal problems All digital systems use a master timing signal called a clock. This two-state timing signal is generated by a digital oscillator circuit, like the previously discussed astable multivibrator circuit. This signal is distributed throughout the digital electronic system to ensure that all circuits operate in sync, or at their correct time. The presence or absence of this clock signal can be checked with a logic probe. However, in most cases it is best to check the quality of the signal with an oscilloscope. Even small variations in frequency, pulse width, amplitude, and other characteristics can cause problems throughout the digital system.

To isolate a clock signal problem, first apply the half-split method to determine whether the fault is in the multivibrator circuit (source) or somewhere in the digital circuits (load), as shown in Figure 9-39(a). If the clock signal from the

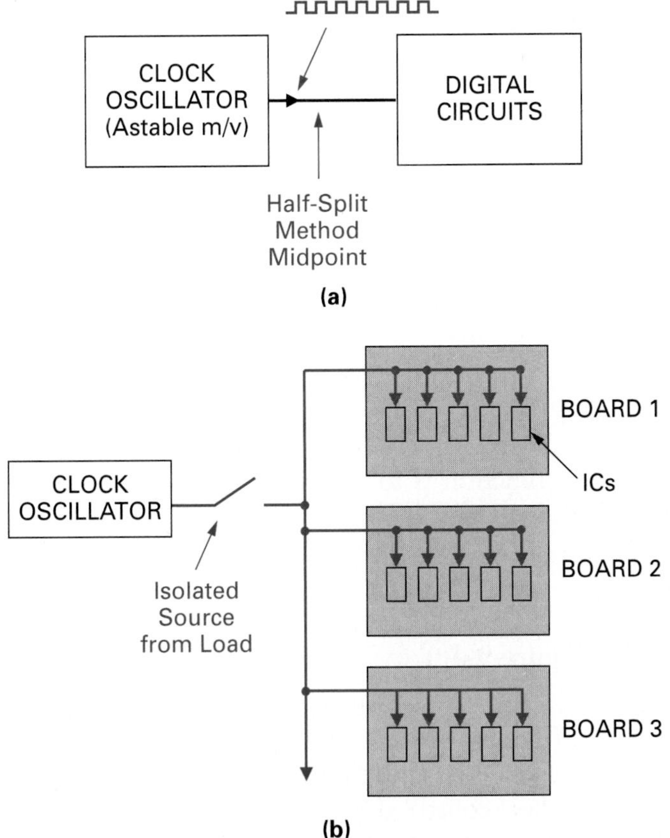

FIGURE 9-39 Clock Signal Problems.

multivibrator is incorrect, or if no signal exists, the problem will generally be in the multivibrator clock circuit. However, like the dc supply voltage, the clock signal is distributed throughout the digital system. You may have to disconnect the clock signal output from the multivibrator to determine whether a short is pulling down the signal or whether the multivibrator clock circuit is not generating the proper signal, as shown in Figure 9-39(b). If there is a short somewhere in one of the digital circuits, the current tracer can be used to find the shorted path. An open in a clock signal line can easily be traced with a logic probe.

SELF-TEST REVIEW QUESTIONS FOR SECTION 9-3

1. What would be the most accurate instrument for measuring dc voltages?
2. Which test instrument would be best for measuring gate propagation delay time?
3. A logic ____________ is used to generate a single or train of test pulses.
4. A ____________ detects current pulses, while a ____________ detects voltage levels and pulses.
5. How are most internal IC failures repaired?
6. An ____________ is easier to isolate and will generally not affect other components while a ____________ is harder to isolate and can often damage other components (short or open).

SUMMARY

1. It is the bipolar transistor's switching action that is made use of in digital electronic circuits.
2. The hardware of a digital computer is all of the electronic, magnetic, and mechanical devices of a digital system. The software of a computer is the binary data and binary instructions that are processed by the digital system hardware.
3. Digital logic gate circuits are constructed using switching bipolar junction transistors. These logic gates are used as the basic building blocks for digital circuits, and their function is to control the movement of binary data and binary instructions.
4. Digital timer and control circuits are used to generate, manage, and modify digital signals.

The ASCII Code (Figure 9-40)

5. Digital circuits are often referred to as "switching circuits" because their control devices (diodes or transistors) are switched between the two extremes of ON and OFF. These digital circuits are also called "two-state circuits" because their control devices are driven either into the saturation state (fully ON) or cutoff state (fully OFF). With digital electronic circuits, information is first encoded into a series of "1s" (ones) and "0s" (zeros).

QUICK REFERENCE SUMMARY SHEET

FIGURE 9-40 The ASCII Code.

	Column	0	1	2	3	4	5	6	7
Row	Bits 4321 765 →	000	001	010	011	**100**	101	110	111
0	0000	NUL	DLE	SP	0	@	P	\	p
1	0001	SOH	DC1	!	1	A	Q	a	q
2	0010	STX	DC2	"	2	B	R	b	r
3	0011	ETX	DC3	#	3	C	S	c	s
4	0100	EOT	DC4	$	4	D	T	d	t
5	0101	ENQ	NAK	%	5	E	U	e	u
6	0110	ACK	SYN	&	6	F	V	f	v
7	0111	BEL	ETB	'	7	G	W	g	w
8	1000	BS	CAN	(	8	H	X	h	x
9	1001	HT	EM	)	9	I	Y	i	y
10	1010	LF	SUB	*	:	J	Z	j	z
11	**1011**	VT	ESC	+	;	**K**	[	k	{
12	1100	FF	FS	,	<	L	\	l	¦
13	1101	CR	GS	-	=	M	]	m	}
14	1110	SO	RS	.	>	N	⌒	n	~
15	1111	SI	US	/	?	O	—	o	DEL

Example: Upper Case "K" = Column 4 (100), Row 11 (1011)

Bits	7	6	5	4	3	2	1
ASCII Code	1	0	0	1	0	1	1

NUL Null
SOH Start of Heading
STX Start of Text
ETX End of Text
EOT End of Transmission
ENQ Enquiry
ACK Acknowledge
BEL Bell (audible signal)
BS Backspace
HT Horizontal Tabulation (punched card skip)
LF Line Feed
VT Vertical Tabulation

FF Form Feed
CR Carriage Return
SO Shift Out
SI Shift In
SP Space (blank)
DLE Data Link Escape
DC1 Device Control 1
DC2 Device Control 2
DC3 Device Control 3
DC4 Device Control 4
NAK Negative Acknowledge
SYN Synchronous Idle
ETB End of Transmission Block

CAN Cancel
EM End of Medium
SUB Substitute
ESC Escape
FS File Separator
GS Group Separator
RS Record Separator
US Unit Separator
DEL Delete

ASCII Abbreviations

The "American Standard Code for Information Interchange" is abbreviated as ASCII (pronounced "askey").

6. The early digital systems constructed in the 1950s made use of ten levels or voltages. Each of these voltages corresponded to one of the ten digits in the decimal number system (0 = 0 V, 1 = 1 V, 2 = 2 V, 3 = 3 V, up to 9 = 9 V).

These circuits, however, were very complex because they had to generate one of ten voltage levels and sense the difference between all ten voltage levels. This complexity led to inaccuracy because some circuits would periodically confuse one voltage level for a different voltage level.

7. Using a two-state or two-digit system, you can generate codes for any number, letter, or symbol, as we have seen in the ASCII table. The electronic circuits that manage these two-state codes are less complex because they only have to generate and sense a HIGH and LOW voltage and much more accurate because there is little room for error between the two extremes of ON and OFF, or HIGH voltage and LOW voltage.

8. Abandoning the ten-state system and adopting the two-state system for the advantages of circuit simplicity and accuracy means that we are no longer dealing with the decimal number system. Having only two digits (0 and 1) means that we are now operating in the two-state number system which is called binary. With the base 2, or binary number system, we only have two digits in the number scale, and therefore only the digits "0" and "1" exist in binary. These two states are typically represented in an electronic circuit as two different values of voltage (0 = LOW voltage, 1 = HIGH voltage). Using combinations of binary digits (abbreviated "bits"), we can represent information as a binary code. This code is called a digital signal because it is an information signal that makes use of binary digits. Today, almost all information, from your voice telephone conversations to the music on your compact discs, is digitized or converted to binary data form.

9. To develop a digital electronic system that could manipulate binary information, inventors needed a two-state electronic switch. Early machines used mechanical switches, electromechanical relays, and vacuum tubes. Finally, compact and low-power digital electronic circuits and systems became a reality with the development of a new kind of switch called the bipolar junction transistor.

The Five Basic Logic Gates (Figure 9-41)

10. Every digital electronic circuit uses logic-gate circuits to manipulate the coded pulses of binary language. These logic-gate circuits are constructed using transistors and are the basic "decision-making elements" in all digital circuits.

11. Logic gates accept HIGH or LOW voltage signals at their inputs, judge these input levels based on a predetermined set of rules, and then produce a single HIGH or LOW voltage signal at their output.

12. An OR gate will deliver a binary 1 or logical TRUE (+5 V) output, if any one of its inputs is TRUE. An OR gate will generate a binary 0 or logical FALSE (0 V) output only when all of its inputs are FALSE.

13. An AND gate will deliver a binary 1 or logical TRUE output only when all of its inputs are logically TRUE. Explained another way, we can say that the output of an AND gate will be LOW if any of its inputs are LOW.

QUICK REFERENCE SUMMARY SHEET

FIGURE 9-41 The Five Basic Logic Gates.

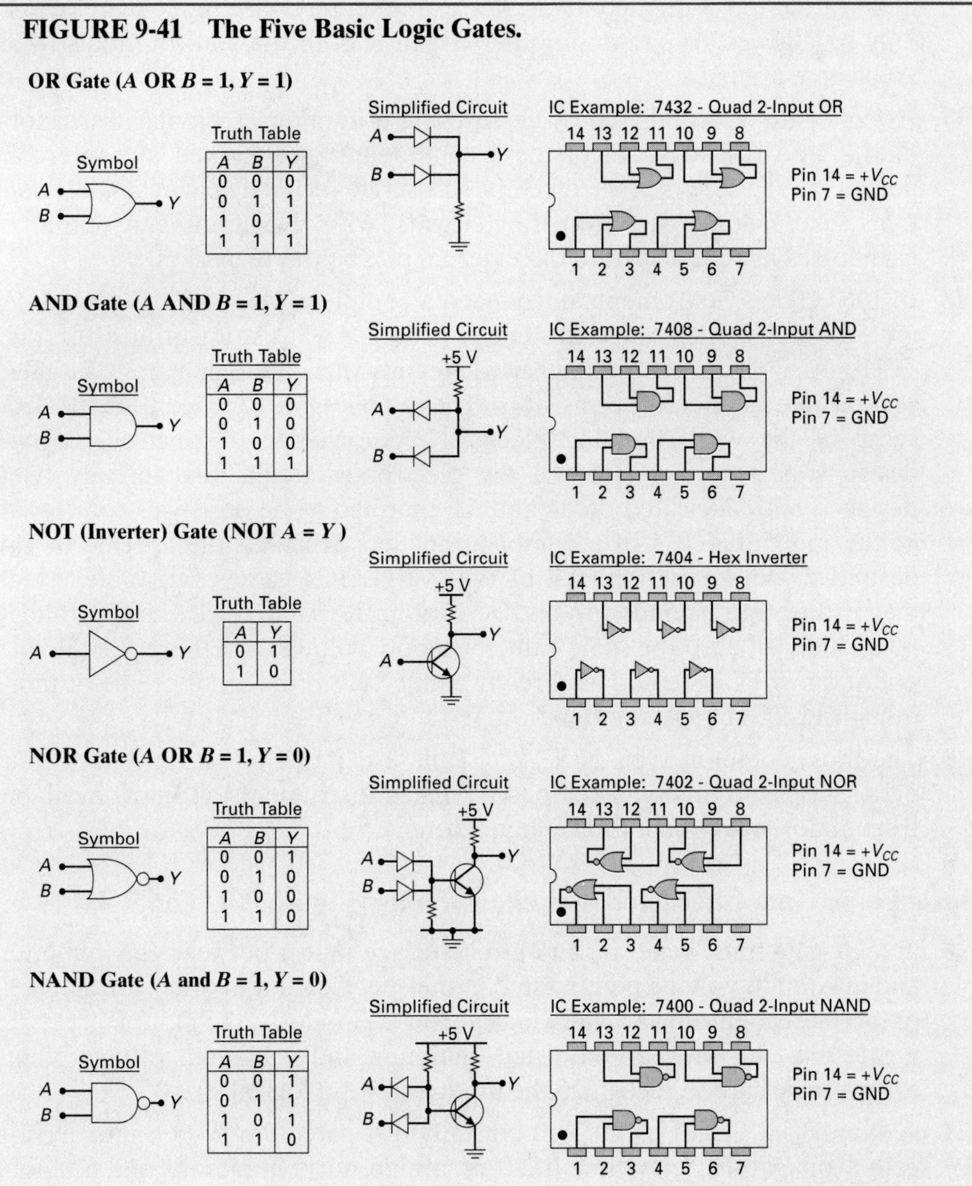

OR Gate truth table:

A	B	Y
0	0	0
0	1	1
1	0	1
1	1	1

AND Gate truth table:

A	B	Y
0	0	0
0	1	0
1	0	0
1	1	1

NOT Gate truth table:

A	Y
0	1
1	0

NOR Gate truth table:

A	B	Y
0	0	1
0	1	0
1	0	0
1	1	0

NAND Gate truth table:

A	B	Y
0	0	1
0	1	1
1	0	1
1	1	0

14. A NOT or INVERTER gate accepts just one input which it then reverses, turning binary 0 input into binary 1 output, or a binary 1 input into binary 0 output.
15. NOT gates are often combined with OR gates and AND gates to form NOR (NOT-OR) and NAND (NOT-AND) gates. These combined devices process the inputs using the usual AND/OR rules, and then automatically invert the output.

16. Logic gates are the basic elements in all digital electronic circuits. The rules that control the operation of each logic gate are used to control the movement of data in a digital circuit. For example, a certain input signal would only be allowed to pass to another circuit if both the control inputs to an AND gate are true.
17. A logic gate is actually constructed using a number of bipolar transistors, diodes, and resistors. All of these components are formed and interconnected on one side of a silicon chip. This single piece of silicon actually contains several logic gate transistor circuits, with the inputs and output of each logic gate connected to the external pins of the IC package.
18. In 1964, Texas Instruments introduced a complete range of logic gate integrated circuits. They called this line of products transistor-to-transistor logic (TTL) ICs, because the circuit used to construct the logic gates contained several interconnected transistors. These logic gate ICs were in the true sense of the word building blocks. No knowledge of electronic circuit design was necessary because the electronic circuits had already been designed and fabricated on a chip. All that had to be done was connect all of the individual ICs in a combination that achieved the specific or the desired circuit operation. All of these TTL ICs were compatible, which meant they all responded to and generated the same logic 1 and 0 voltage levels, they all used the same value of supply voltage, and therefore the output of one TTL IC could be directly connected to the input of one or more other TTL ICs in any combination.
19. The most common type of TTL circuits are the 7400 (seventy-four hundred) series, originally developed by Texas Instruments and now available from almost every digital IC manufacturer.

Basic Logic Gate Characteristics and Circuits (Figure 9-42 and 9-43)

20. A logic gate must generate an output voltage that is between the minimum and maximum voltage points listed in the specification in order for that output to be recognized as a valid logical 1 or 0. Similarly, the input voltages to a gate must be between the listed minimum and maximum voltage points for the logic gate to recognize the input as a valid logical 1 or 0.
21. A standard TTL logic gate will typically dissipate 10 mW per gate, have a gate propagation delay time of 10 ns, a high noise immunity, and a fanout of 10.
22. The 7400 TTL series is used for all commercial applications and it will operate reliably when the supply voltage is between 4.75 V to 5.25 V, and when the ambient temperature is between 0°C and 70°C. The 5400 TTL series is a more expensive line used for military applications because of its increased supply voltage tolerance (4.5 V to 5.5 V) and temperature tolerance (–55°C to 125°C).
23. Noise is an unwanted voltage signal that is induced into conductors due to electromagnetic radiation from adjacent current carrying conductors. All TTL logic circuits have a high noise immunity, which means they will be unaffected by noise fluctuations at the gate's inputs.

QUICK REFERENCE SUMMARY SHEET

FIGURE 9-42 Logic Gate Characteristics (Bipolar Transistor).

Transistor-Transistor Logic (TTL or T²L) Advantage: Low Cost

Standard TTL: (74XX)
Output HIGH = 2.4 V to 5.0 V
Output LOW = 0 V to 0.4 V
Input HIGH = 2.0 V to 5.0 V
Input LOW = 0 V to 0.8 V
Power Dissipation/gate = 1 mW to 22 mW (10 mW typ.)
Propagation Delay/gate = 3 ns to 33 ns (10 ns typ.)

	7400 (Commercial)	5400 (Military)
Supply Voltage ($+V_{CC}$):	4.75 V to 5.25 V	4.5 V to 5.5 V
Temperature Range:	0°C to 70°C	−55°C to 125°C

Noise Immunity: High
Fanout: 10 Gates
LOW level Input Current (1 load) = −1.6 mA
LOW level Input Current (10 loads) = −16 mA } Sinking
HIGH level Output Current (1 load) = 40 μA
HIGH level Output Current (10 loads) = 400 μA } Sourcing

Low-Power TTL: (74LXX) Power: 1 mW/gate typ., Speed: 33 ns/gate typ.
Schottky TTL: (74SXX) Power: 19 mW/gate typ., Speed: 3 ns/gate typ.
Low-Power Schottky TTL: (74LSXX) Power: 2 mW/gate typ., Speed: 9.5 ns/gate typ.
Advanced Schottky TTL: (74ASXX) Power: 8.5 mW/gate typ., Speed: 1.5 ns/gate typ.
Advanced Low-Power Schottky TTL: (74ALSXX or 74FXX) Power: 1 mW/gate typ., Speed: 4 ns/gate typ.

Open Collector Gates: Can sink up to 40 mA/gate.

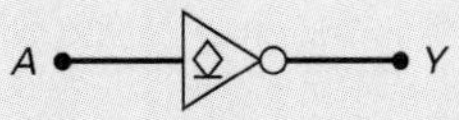

Three-State Output Gates:

Enable	Input	Output
LOW	0	0
LOW	1	1
HIGH	0	Floating
HIGH	1	Floating

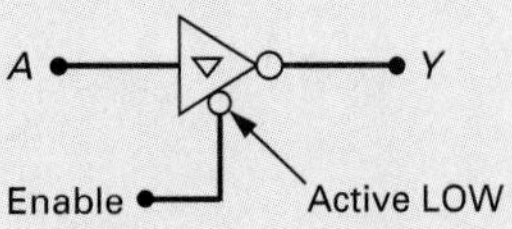

Buffer/Driver Gates: Output current is 30 times greater than standard TTL.

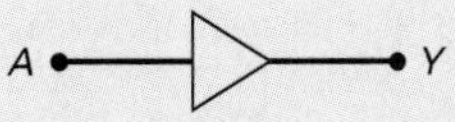

Schmitt Trigger Gates: Used to sharpen rise and fall times of pulses.

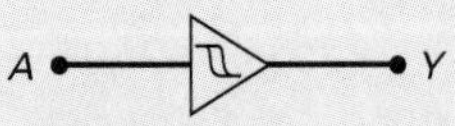

Emitter-Coupled Logic (ECL) Advantage: High Speed

Logic Levels: Binary 1 = −0.8 V
Binary 0 = −1.7 V
Power Dissipation/gate: 40 mW to 60 mW
Propagation Delay/gate: 0.5 ns to 3 ns
Supply Voltage (V_{EE}): −5.2 V
Noise Immunity : HIGH
Fanout : 10 to 25 gates

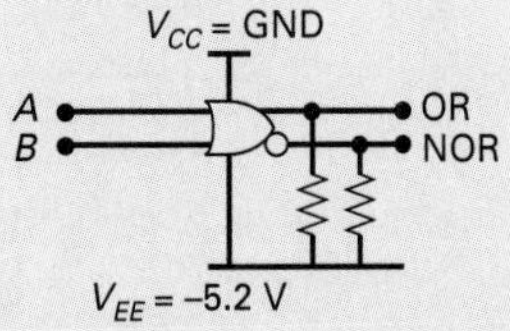

Integrated-Injection Logic (IIL or I²L) Advantage: Small Size

Logic Levels: Binary 1 = +0.7 V
Binary 0 = +0.1 V
Power Dissipation/gate: 0.6 μW to 70 μW
Propagation Delay/gate: 25 ns to 50 ns
Supply Voltage (V_S): +1 V to +15 V
Noise Immunity: Good
Fanout : 2 gates (greater with increased injector current)

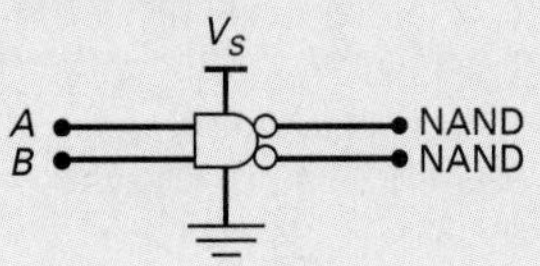

QUICK REFERENCE SUMMARY SHEET

FIGURE 9-43 Basic Digital Logic Gate Circuits (Bipolar Transistor).

Standard TTL Logic Gate Circuit

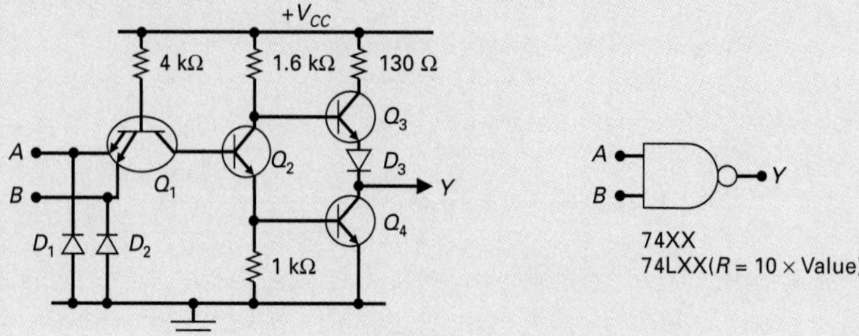

Schottky TTL Logic Gate Circuit

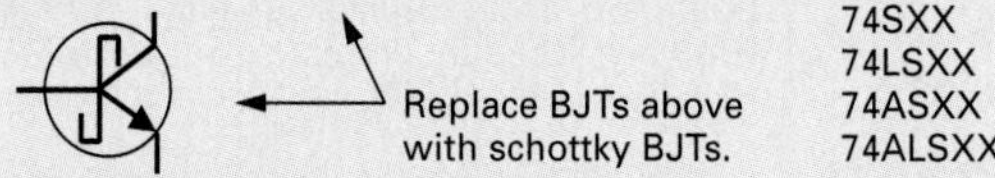

Open Collector TTL Logic Gate Circuits

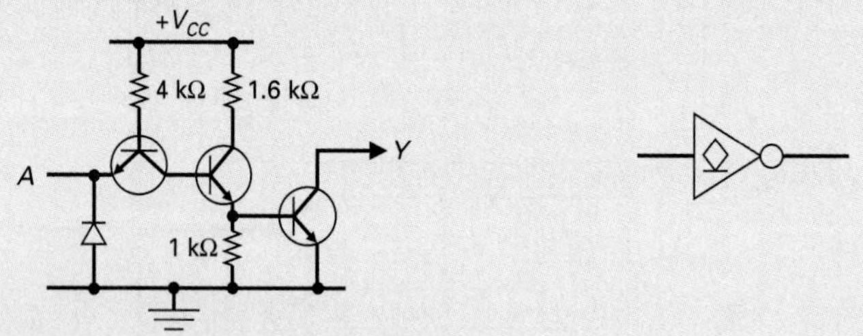

Three-State Output TTL Logic Gate Circuits

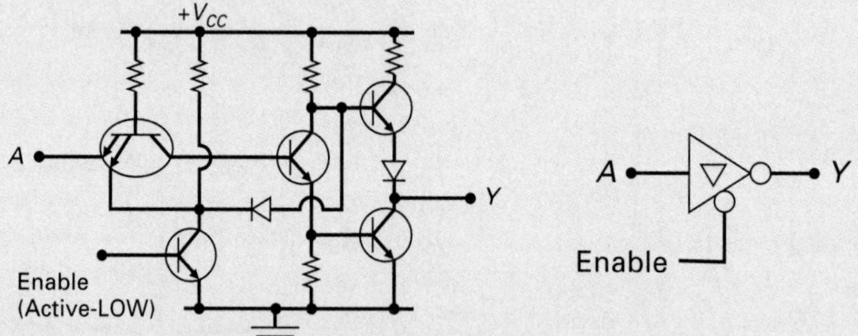

Schmitt Trigger TTL Logic Gate Circuits

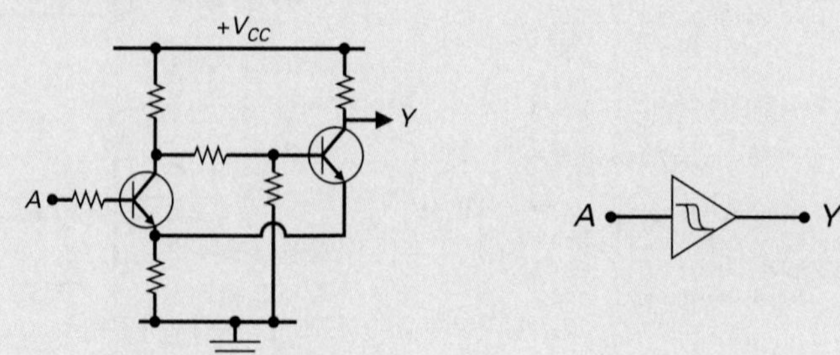

QUICK REFERENCE SUMMARY SHEET

FIGURE 9-43 (continued).

Emitter-Coupled Logic Gate Circuits (ECL)

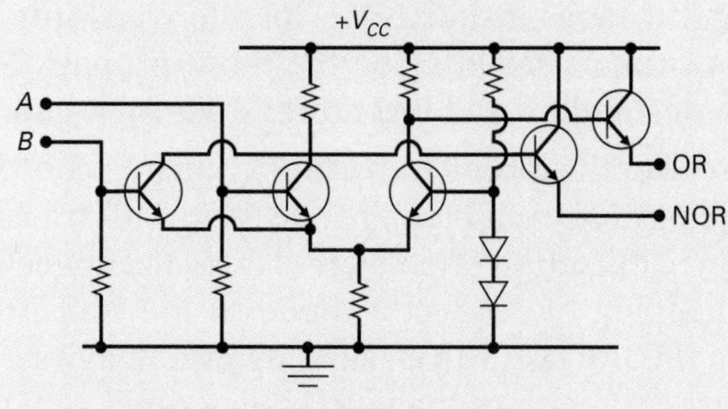

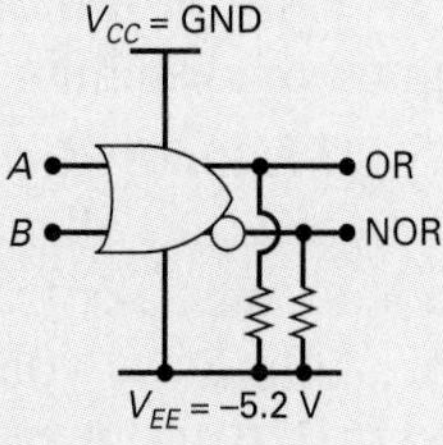

Integrated-Injection Logic Gate Circuits (I²L)

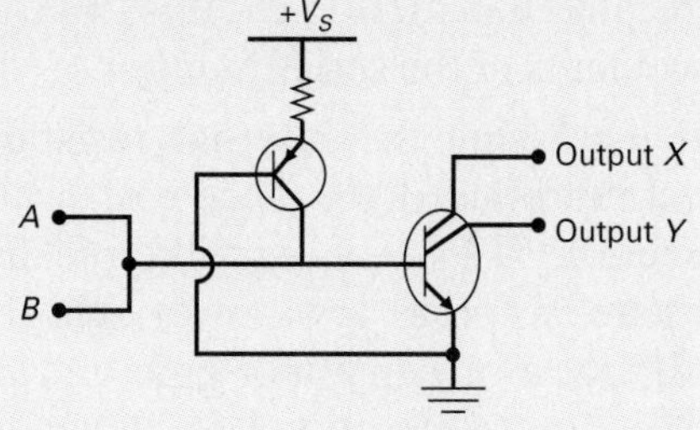

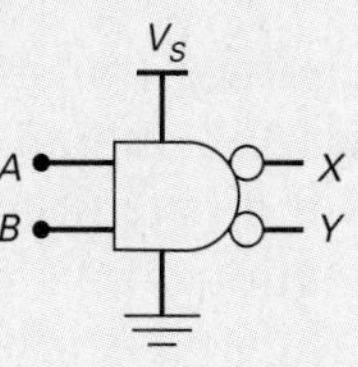

24. A standard TTL gate can sink a maximum of 16 mA and still produce a LOW output voltage of 0.4 V or less. Because the maximum sinking current of 16 mA is ten times larger than the single gate load current of 1.6 mA, we can connect ten TTL loads to each TTL driver. Similarly, a standard TTL gate can source a maximum of 400 μA and still produce a HIGH output voltage of 2.4 V or more. Because the maximum source current of 400 μA is ten times larger than the single gate load current of 40 μA, we can again connect ten TTL loads to each TTL driver. Therefore, the maximum number of TTL loads that can be reliably driven by a single TTL driver is called the fanout.

25. An unconnected logic gate input is called a floating input. In this condition, no emitter current will exist in Q_1 because the input is disconnected, and so a floating TTL input is equivalent to a HIGH input.

26. A low-power TTL circuit is basically the same as the standard TTL circuit previously discussed except that all of the internal resistance values in the logic gate circuit are ten times larger. This means that the power dissipation of a low-power TTL circuit is one-tenth that of a standard TTL circuit (typically 1 mW/gate). Increasing the internal resistance, however, will decrease the circuit's internal currents. Because the bipolar transistor is a current operated device, the response or switching time of a logic gate will decrease (typical propagation delay time is 33 ns per gate). With low-power TTL circuits, high speed is sacrificed for low power consumption. Low power logic

gate ICs can be identified because the device number will have the letter "L" following the series number 74 or 54.

27. With standard TTL and low-power TTL logic gates, transistors go into saturation when they are switched ON and are flooded with extra carriers. When a transistor is switched OFF, it takes a short time for these extra carriers to leave the transistor, and it takes a short time for the transistor to cut OFF. This saturation delay time accounts for the slow propagation delay times, or switching times, of standard and low power TTL logic gates.

28. Schottky TTL logic gates overcome this problem by fabricating a schottky diode between the collector and base of each bipolar transistor. This transistor is called a schottky transistor. This action effectively clamps the collector to 0.4 V of the base voltage, keeping the collector diode of the transistor reversed biased and preventing the transistor from slipping into heavy saturation. Schottky TTL devices are fast, with propagation delay times of typically 3 ns. This fast switching time means that the input signals can operate at extremely high frequencies of typically 100 mHz. Schottky TTL ICs are labeled with an "S" after the first two digits of the series number.

29. By using schottky transistors and increasing the internal resistance in a logic gate circuit, manufacturers compromised the speed of schottky to obtain a lower dissipation rating. Schottky TTL devices will typically have a power dissipation of 19 mW per gate, whereas low-power schottky TTL devices typically have a power dissipation of 2 mW per gate and propagation delay times of 9.5 ns. These low-power schottky TTL devices include the letters "LS" in the device number, for example 74LS32, 74LS123, and so on.

30. Special schottky TTL devices are available that use a new integration process called oxide isolation, in which the transistor's collector diode is isolated by a thin oxide layer instead of a reverse bias junction. These advanced schottky TTL devices (labeled 74AS00) have a typical power dissipation of 8.5 mW per gate and a propagation delay time of 1.5 ns. By increasing internal resistance, once again manufacturers came up with an advanced low-power schottky TTL line of products (labeled 74ALS00), which have a power dissipation of 1 mW and propagation time of 4 ns.

31. In order to get a HIGH or LOW output from an open collector gate, an external pull-up resistor must be connected between $+V_{CC}$ and the gate output. With open collector gates, the outputs can be wired together without causing any HIGH-LOW conflicts between gates. Open collector gates can sink up to 40 mA. This high sinking current ability can be used to drive high current demand loads.

32. The three-state output gate has three output states: LOW, HIGH, and FLOATING, or high-impedance output state. Like the previously discussed open collector gates, these devices are used in applications where two or more gate outputs are connected to a single output line. To avoid HIGH/LOW conflicts between gates, only one gate is enabled at one time to drive the output line, with all other gates being disabled and their outputs FLOATING.

33. A buffer is a device that isolates one circuit from another. In order to achieve this isolation, a buffer should have a high input impedance (low input current) and a low output impedance (high output current). A standard TTL gate provides a certain amount of isolating or buffering since the output current is ten times that of the input current. A buffer/driver is a basic gate that has been slightly modified to increase the output current so that it is thirty times greater than the input currents, enabling it to drive heavier loads.

34. The schmitt trigger circuit, named after its inventor, is a two-state device that is used to convert a sine wave into a rectangular wave, and to sharpen the rise and fall times of a square or rectangular wave. The symbol for any schmitt trigger gate contains the hysteresis symbol. Hysteresis can be loosely defined as the lag between cause and effect. This lag is desirable since it produces a sharp, well-defined, output signal.

35. Emitter-coupled logic, or ECL, is another nonsaturating bipolar transistor logic circuit arrangement; however, ECL is the fastest of all logic circuits. It is used in large digital computer systems that require fast propagation delay times (typically 0.8 ns per gate) and are willing to sacrifice power dissipation (typically 40 mW per gate).

36. Integrated injection logic (IIL), or as they are sometimes abbreviated I^2L (pronounced I squared L), is a series of digital bipolar transistor circuits that have good speed, low power consumption, and very good circuit packing density. These circuits are used almost extensively in digital watches, cameras, and all compact battery-powered circuit applications.

Basic Digital Timer and Control Circuits (Figure 9-44)

37. To control the timing of digital circuits, a clock signal is distributed throughout the digital system. This square wave clock signal is generated by a clock oscillator, and its sharp positive (leading) and negative (trailing) edges are used to control the sequence of operations in a digital circuit.

38. The astable multivibrator circuit is used to produce an alternating two-state square or rectangular output waveform. This circuit is often called a free-running multivibrator because the circuit requires no input signal to start its operation. It will simply begin oscillating the moment the dc supply voltage is applied. If the time constants of the two *RC* timing networks in the astable circuit are different, the result will be a rectangular or pulse waveform.

39. The astable multivibrator is often referred to as an "unstable multivibrator" because it is continually alternating or switching back and forth, and therefore it has no stable condition or state. The monostable multivibrator has, as its name implies, one (mono) stable state. The circuit will remain in this stable state indefinitely until a trigger is applied and forces the monostable multivibrator into its unstable state. It will then remain in its unstable state for a small period of time, and then switch back to its stable state and await another trigger. The monostable multivibrator is often compared

QUICK REFERENCE SUMMARY SHEET

FIGURE 9-44 Basic Digital Timer and Control Circuits (Bipolar Transistor).

Astable Multivibrator Circuit (Square-Wave Oscillator)

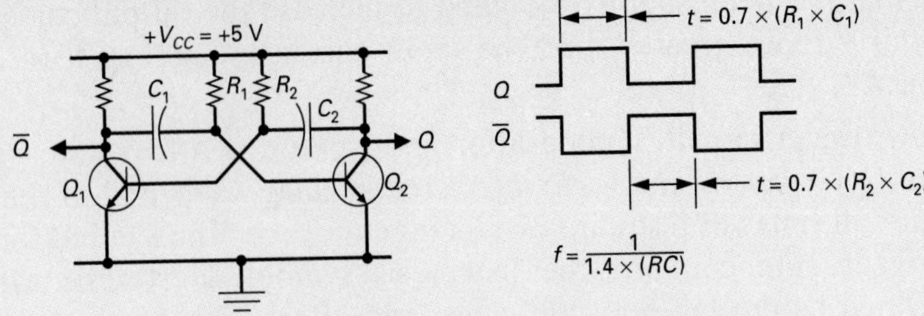

Monostable Multivibrator Circuit (One-Shot Pulse Generator)

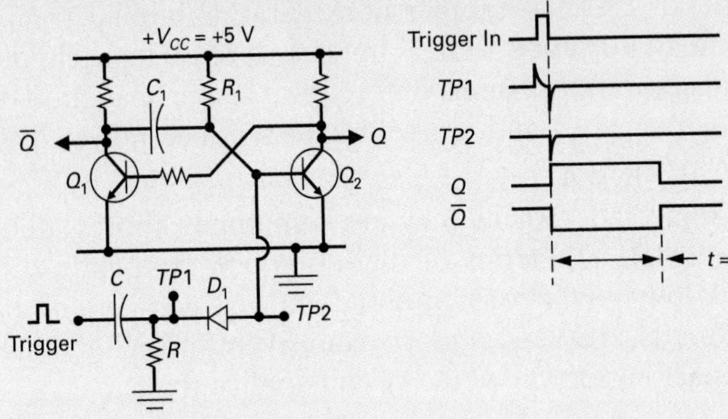

Bistable Multivibrator Circuit (Set-Reset Flip-Flop or Latch)

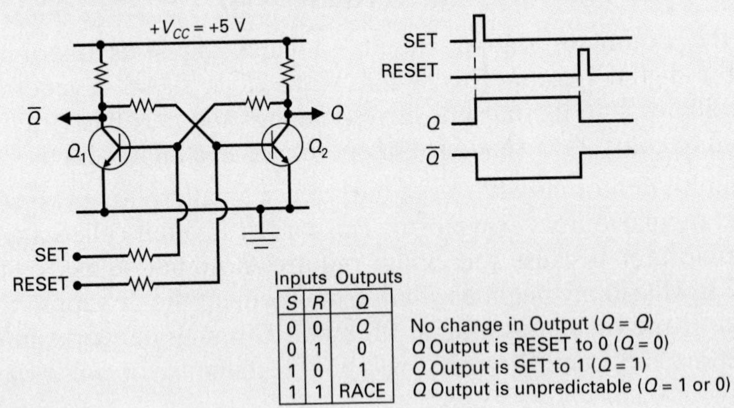

Inputs		Outputs	
S	R	Q	
0	0	Q	No change in Output (Q = Q)
0	1	0	Q Output is RESET to 0 (Q = 0)
1	0	1	Q Output is SET to 1 (Q = 1)
1	1	RACE	Q Output is unpredictable (Q = 1 or 0)

to a gun and called a one-shot multivibrator because it will produce one output pulse or shot for each input trigger. This one-shot timer circuit is often used in pulse-stretching and time delay applications.

40. The bistable multivibrator has two (bi) stable states, two inputs called the "SET" and "RESET," and two outputs called "Q" and "$\overline{Q}$." The bistable multivibrator circuit is often called an *S-R* (set-reset) flip-flop because a pulse on the SET input will "flip" the circuit into the set state (Q output is set HIGH), while a pulse on the RESET input will "flop" the circuit into its reset state (*Q* output is reset LOW). The ability of the bistable multivibrator to remain in its last condition or state accounts for why the *S-R* flip-flop is also called an *S-R* latch. The bistable multivibrator *S-R* flip-flop or *S-R* latch has become one of the most important circuits in digital electronics. It is used in a variety of applications ranging from data storage to counting and frequency division.

The 555 Timer Circuit (Figure 9-45)

41. One of the most frequently used low-cost integrated circuit timers is the 555 timer. Its IC package consists of 8 pins, and it derives its number identification from the distinctive input voltage divider circuit consisting of three 5 kΩ resistors. It is a highly versatile timer that can be made to function as an astable multivibrator, monostable multivibrator, frequency divider, or modulator depending on the connection of external components.

Basic Digital Test Equipment (Figure 9-46)

42. To be an effective electronic technician or troubleshooter, you must have a thorough knowledge of electronics, test equipment, troubleshooting techniques, and equipment repair.
43. There is a variety of test equipment to help you troubleshoot digital circuits and systems. Some are standard, such as the multimeter and oscilloscope, which can be used for either analog or digital circuits. Other test instruments such as the logic clip, logic probe, logic pulser, and current tracer have been designed specifically to test digital logic circuits.
44. The pulse or rectangular wave alternates between two peak values. However, unlike the square wave, the pulse wave does not remain at the two peak values for equal lengths of time. The positive pulse waveform remains at its negative value for long periods of time and only momentarily pulses positive. On the other hand, the negative pulse waveform remains at its positive value for long periods of time and momentarily pulses negative.
45. When referring to pulse waveforms, the term pulse repetition frequency (PRF) is used to describe the frequency or rate of the pulses, and the term pulse repetition time (PRT) is used to describe the period or time of one complete cycle. The duty cycle of a pulse waveform indicates the ratio of pulse width time to the complete cycle time.
46. The rise time of a pulse waveform is defined as the time needed for the pulse to rise from 10 percent to 90 percent of the peak amplitude, while the fall time of a pulse waveform is the time needed for the pulse to fall from 90 percent to 10 percent of the peak amplitude.
47. When carefully studying pulse waveforms on the oscilloscope, you will be able to measure pulse width, PRT, rise time, and fall time. You will also see a few other conditions known as overshoot, undershoot, and ringing. These unwanted conditions tend to always accompany high frequency pulse waveforms due to imperfections in the circuit.

QUICK REFERENCE SUMMARY SHEET

FIGURE 9-45 The 555 Timer Circuit.

The 555 Timer IC Pins and Basic Circuit

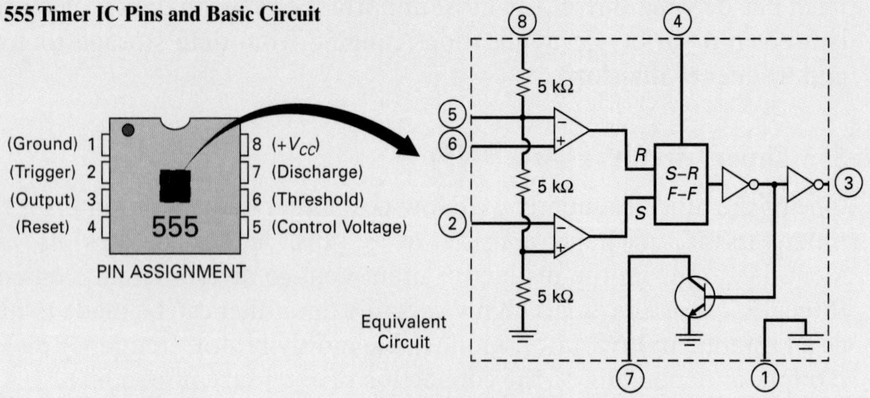

PIN ASSIGNMENT

Equivalent Circuit

555 Timer Example Circuits

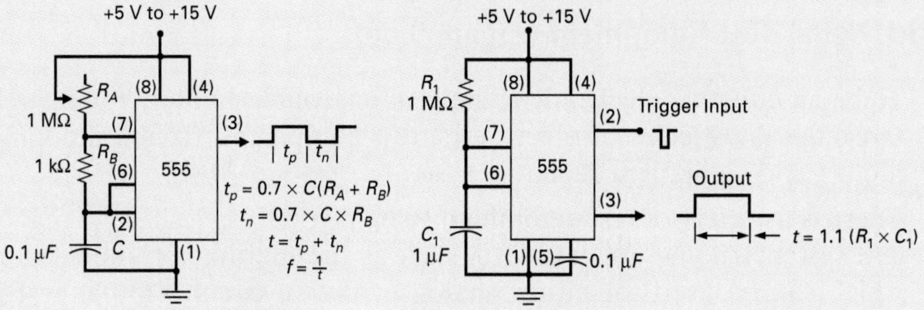

$t_p = 0.7 \times C(R_A + R_B)$

$t_n = 0.7 \times C \times R_B$

$t = t_p + t_n$

$f = \frac{1}{t}$

Astable Multivibrator (Pulse Generator)

$t = 1.1\,(R_1 \times C_1)$

Monostable Multivibrator (One-Shot Timer)

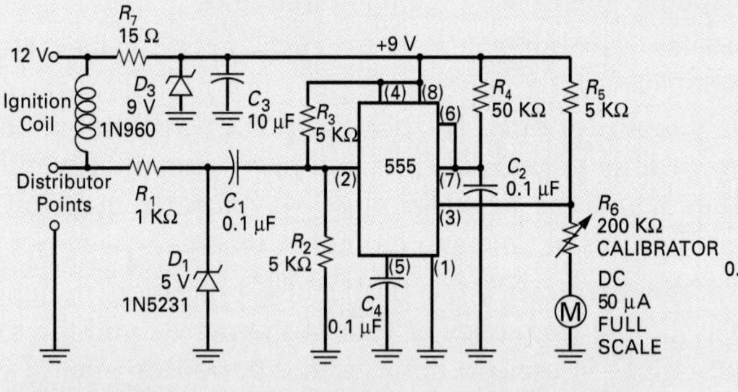

Car Tachometer: 555 Timer receives pulses from distributor points. Meter receives a current through R_6 when output is HIGH, no current when output is LOW. Ratio of ON/OFF current causes meter to provide indication of engine speed.

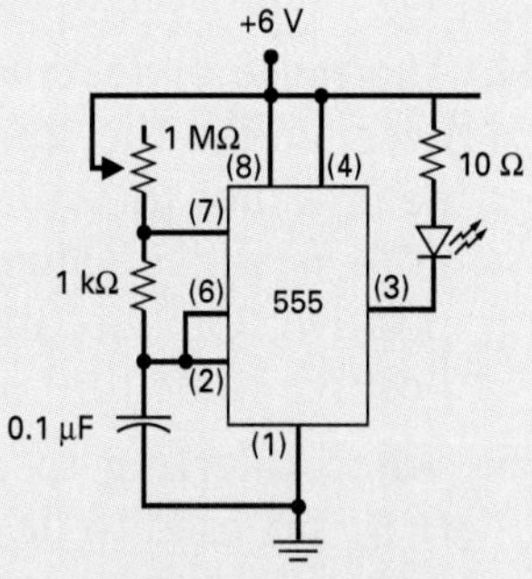

LED Transmitter Circuit

48. The logic clip was specially designed for troubleshooting digital circuits. It consists of a spring loaded clip that clamps on to a standard IC package where it makes contact with all the pins of the IC. These contacts are then available at the end of the clip so that a test instrument probe can easily be connected.

49. The logic probe is an instrument designed specifically for digital circuit troubleshooting. In order for the probe to operate, its red and black power

QUICK REFERENCE SUMMARY SHEET

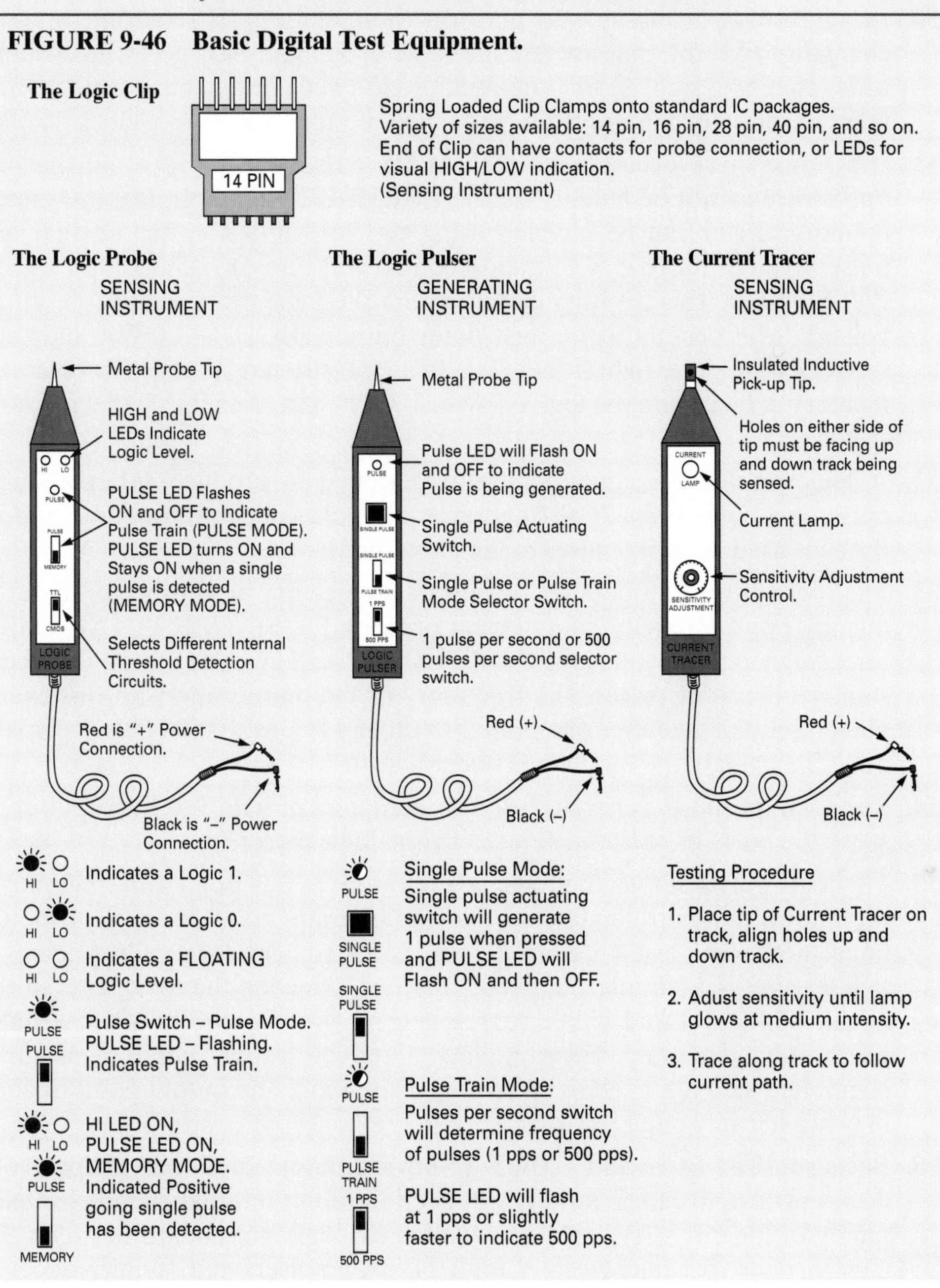
The Logic Clip
14 PIN
Spring Loaded Clip Clamps onto standard IC packages.
Variety of sizes available: 14 pin, 16 pin, 28 pin, 40 pin, and so on.
End of Clip can have contacts for probe connection, or LEDs for visual HIGH/LOW indication.
(Sensing Instrument)
The Logic Probe
SENSING INSTRUMENT
Metal Probe Tip
HIGH and LOW LEDs Indicate Logic Level.
PULSE LED Flashes ON and OFF to Indicate Pulse Train (PULSE MODE). PULSE LED turns ON and Stays ON when a single pulse is detected (MEMORY MODE).
Selects Different Internal Threshold Detection Circuits.
HI LO
PULSE
MEMORY
TTL
CMOS
LOGIC PROBE
Red is "+" Power Connection.
Black is "–" Power Connection.
HI LO Indicates a Logic 1.
HI LO Indicates a Logic 0.
HI LO Indicates a FLOATING Logic Level.
PULSE
PULSE
Pulse Switch – Pulse Mode.
PULSE LED – Flashing.
Indicates Pulse Train.
HI LO
PULSE
MEMORY
HI LED ON,
PULSE LED ON,
MEMORY MODE.
Indicated Positive going single pulse has been detected.
The Logic Pulser
GENERATING INSTRUMENT
Metal Probe Tip
Pulse LED will Flash ON and OFF to indicate Pulse is being generated.
Single Pulse Actuating Switch.
Single Pulse or Pulse Train Mode Selector Switch.
1 pulse per second or 500 pulses per second selector switch.
LOGIC PULSER
Red (+)
Black (–)
PULSE
SINGLE PULSE
SINGLE PULSE
Single Pulse Mode:
Single pulse actuating switch will generate 1 pulse when pressed and PULSE LED will Flash ON and then OFF.
PULSE
PULSE TRAIN
1 PPS
500 PPS
Pulse Train Mode:
Pulses per second switch will determine frequency of pulses (1 pps or 500 pps).
PULSE LED will flash at 1 pps or slightly faster to indicate 500 pps.
The Current Tracer
SENSING INSTRUMENT
Insulated Inductive Pick-up Tip.
Holes on either side of tip must be facing up and down track being sensed.
Current Lamp.
Sensitivity Adjustment Control.
CURRENT
LAMP
SENSITIVITY ADJUSTMENT
CURRENT TRACER
Red (+)
Black (–)
Testing Procedure
1. Place tip of Current Tracer on track, align holes up and down track.
2. Adust sensitivity until lamp glows at medium intensity.
3. Trace along track to follow current path.

FIGURE 9-46 Basic Digital Test Equipment.

leads must be connected to a power source. Once the power leads have been connected, the logic probe is ready to sense the voltage at any point in a digital circuit and give an indication as to whether that voltage is a valid binary 1 or valid binary 0. The "HI" and "LO" LEDs on the logic probe are used to indicate the logic level. Because the voltage thresholds for a binary 0 and binary 1 can be different, most logic probes have a selector switch to select different internal threshold detection circuits for TTL,

ECL, or another digital family of ICs that will be discussed later called CMOS.

50. In addition to being able to perform these static tests (constant or non-changing) like the multimeter, the logic probe can also perform dynamic tests (changing signals) like the oscilloscope, such as detecting a single momentary pulse or a pulse train (sequence of pulses) such as a clock signal.

51. If a single pulse occurs, the width of the pulse may be so small that the operator may not be able to see the PULSE LED blink. In this instance, a memory circuit can be enabled using the MEMORY switch so that any pulse (even as narrow as 10 nanoseconds) will turn ON, and keep ON, the PULSE LED. This memory circuit within the logic probe senses the initial voltage level present at the test point (either HIGH or LOW), and then turns ON the pulse light if there is a logic transition or change. To reset this memory circuit, simply toggle (switch OFF and then ON) the memory switch.

52. A logic pulser is a signal generator designed to produce either a single pulse or a pulse train. It also requires a supply voltage to operate, and its probe is used to apply pulses to the circuit under test.

53. The logic pulser is operated by simply connecting the probe to any conducting point on a circuit, and then pressing the SINGLE PULSE button for a single pulse or selecting the PULSE TRAIN switch position for a constant sequence of pulses. The frequency of the pulse train, with this example model, can be either one pulse per second (1 pps) or 500 pulses per second (500 pps), based on the pps selector switch.

54. The logic probe is generally used in conjunction with the logic pulser to sense the pulse or pulses generated by the logic pulser.

55. The logic pulser senses the logic level present at any point and then pulses the line to its opposite state so that the operator does not have to determine these logic levels before a point is tested. The logic pulser achieves this feature with an internal circuit that will override the logic level in the circuit under test and either source current or sink current when a pulse needs to be generated. This complete "in-circuit testing" ability means that components do not need to be removed and input and output paths opened in order to carry out a test.

56. Because TTL, ECL, and CMOS all have different input and output voltages and currents, logic pulsers will need different internal circuits for each family of logic ICs. Typical pulse widths for logic pulsers are between 500 nanoseconds to 1 microsecond for TTL logic pulsers and 10 microseconds for CMOS logic pulsers.

57. The logic pulser can also be used in conjunction with a current tracer, which is used to sense the relative values of current in a conductor. It achieves this by using an insulated inductive pick-up tip, which senses the magnetic field generated by the current in a conductor. By adjusting the sensitivity control on the current tracer and observing the lamp's intensity when the probe is placed on a pulsating logic signal line, a shorted path can be found by simply tracing the path of high current.

58. Because ICs account for almost 85 percent of the components within digital systems, it is highly likely that most digital circuit problems will be caused by a faulty IC. Digital circuit failures can be basically divided into two categories: "digital IC problems," which are failures within the IC, or "other digital circuit device problems." When troubleshooting digital circuits, you will have to first isolate the faulty IC, and then determine whether the problem is internal or external to the IC. An internal IC problem cannot be fixed and the IC will have to be replaced. An external IC problem can be caused by electronic devices, connecting devices, mechanical or magnetic devices, power supply voltages, clock signals, and so on.

NEW TERMS

555 Timer
Active LOW Control Line
Advanced Low-Power Schottky
Advanced Schottky
American Standard Code for Information Interchange (ASCII)
AND Gate
Astable Multivibrator
Bistable Multivibrator
Buffer/Driver TTL Gate
Clipper Circuit
Coupling Transistor
Cross Coupled Transistors
Current Source
Current Sink
Current Tracer
Data Processor
Digital Circuit Troubleshooting
Digital Control Circuits
Digital Logic Gate Circuits
Digital Timer Circuits
Digitized
Emitter-Coupled Logic (ECL)
Fall Time
Floating Input
Free-running Multivibrator
Glitch
Hardware
Hysteresis
Hysteresis Voltage
Injector Current
Integrated-Injection Logic (I^2L)
Latch
Logic Clip
Logic Gate Circuits
Logic Probe
Logic Pulser
Low-Power Schottky
Monostable Multivibrator
Multiple-Emitter Input Transistor
NAND Gate
New Dynamic Tests
New In-Circuit Testing
Noise Immunity
Noise Margin
NOR Gate
NOT Gate
One-Shot Multivibrator
Open Collector TTL Gates
OR Gate
Overshoot
Oxide Isolation
Phase Splitter
Pull-down Resistor
Pull-up Resistor
Pulse Stretching
Pulse Waveforms
Ringing
Rise Time
R-S Flip-Flop
R-S Latch
Schmitt Trigger Circuit
Schmitt Trigger TTL Gate
Schottky TTL
Software
S-R Flip-Flop
S-R Latch
Standard TTL Logic Gate
Static Tests
Three-State Output TTL Gates
Timer
Totem Pole Circuit
Transistor-Transistor Logic (TTL)
Tri-State Logic Gates
TTL Driver
TTL Load
Undershoot

REVIEW QUESTIONS

Multiple-Choice Questions

1. Which of the following ASCII codes is used to represent the number 6?

   a. 0101011 **b.** 1100010 **c.** 0110110 **d.** 1110001

2. Which device is currently being used extensively as a two-state switch in digital circuits?

   a. The BJT **b.** The vacuum tube **c.** The relay **d.** The toggle switch

3. If the inputs to a 3-input OR gate were 010, the output would be:

   a. 0 **b.** 1 **c.** Floating **d.** None of the above

4. Which logic gate does the following expression describe? Only when both inputs are 1 will the output be 1.

   a. OR **b.** NOT **c.** NOR **d.** NAND **e.** AND

5. Which logic gate does the following expression describe? Any HIGH input will produce a HIGH output.

   a. OR **b.** NOT **c.** NOR **d.** NAND **e.** AND

6. Which of the following produces a LOW output impedance for both a HIGH and LOW output logic level?

   a. Phase splitter transistor **c.** Multiple-emitter transistor
   b. Totem-pole transistor circuit **d.** Cross coupled transistor circuit

7. A floating TTL input has exactly the same effect as a/an:

   a. LOW input **c.** Invalid input
   b. HIGH input **d.** Both (a) and (c)

8. A standard TTL logic gate can sink a maximum of ____________ and source a maximum of ____________.

   a. 400 μA, 1.6 mA **c.** 16 mA, 400 μA
   b. 40 μA, 16 mA **d.** 1.6 mA, 40 μA

9. Which logic gates require an external pull-up resistor?

   a. ECL **c.** Schottky TTL
   b. Open collector **d.** Both (a) and (b)

10. The output current of a standard TTL gate is ____________ times greater than the input current, while the output current of a buffer/driver TTL gate is ____________ times greater than the input.

    a. 10, 30 **b.** 10, 15 **c.** 30, 10 **d.** 15, 30

11. ____________ logic gates are generally incorporated in digital circuits to sharpen the rise and fall times of input pulses.

    a. Buffer/driver **c.** Open collector
    b. Schmitt trigger **d.** Tri-state

12. The ____________ multivibrator will produce a continuously alternating square wave or pulse wave output.

    a. Astable **b.** Monostable **c.** Bistable **d.** Schmitt

13. The ____________ multivibrator is also called a one-shot.

    a. Astable **b.** Monostable **c.** Bistable **d.** Schmitt

14. Which multivibrator is also called a set-reset flip-flop?

a. Astable **b.** Monostable **c.** Bistable **d.** Schmitt

15. The output pulse width of a 555 timer is determined by the externally connected ____________ and ____________.

a. Power supply, resistor **c.** Capacitor, load resistor
b. Load resistance, capacitor **d.** Input Resistor, capacitor

16. The 555 timer consists of a ____________ resistor voltage divider, ____________ comparator(s), ____________ *R-S* flip-flop, an INVERTER, output stage and discharge transistor on a single IC.

a. 2, 2, 2 **b.** 3, 2, 1 **c.** 1, 2, 3 **d.** 3, 1, 2

17. A monostable multivibrator will generally make use of a ____________ and ____________ circuit on the trigger input.

a. Integrator, Clipper **c.** Schmitt, Clipper
b. Differentiator, Schmitt **d.** Differentiator, Clipper

18. Which of the two-state circuits could be used to convert a sine wave into a square wave?

a. Schmitt **b.** Bistable **c.** Astable **d.** Monostable

19. Which of the multivibrators could be used to stretch the width of a pulse?

a. Schmitt **b.** Bistable **c.** Astable **d.** Monostable

20. A ____________ wave has a 50 percent duty cycle.

a. Triangular **b.** Sawtooth **c.** Rectangular **d.** Square

21. If a rectangular wave had a PRF of 1000 pps, what would be its PRT?

a. 1/1000 s **b.** 1 ms **c.** 1000 μs **d.** All of the above

22. The voltmeter can be used to make ____________ tests, while the oscilloscope can be used to make ____________ tests.

a. Dynamic, static **b.** Static, dynamic

23. The logic pulser

a. Has an insulated tip.
b. Senses current pulses at different points in a circuit.
c. Is a signal generator.
d. Is a signal sensing instrument.

24. Which digital test instrument can be used to sense the logic level at different points in a circuit?

a. Multimeter **b.** Logic probe **c.** Oscilloscope **d.** All of the above

25. The current tracer is ideal for locating what type of circuit faults?

a. Shorts **b.** Opens **c.** Invalid logic levels **d.** Rise/fall signals problems

Essay Questions

26. What is the ASCII code? (9-1)

27. Which basic bipolar transistor action is made use of in digital electronic circuits? (Intro.)

28. Define the following terms: (Intro.)

    a. Data processor b. Hardware c. Software

29. Why was the two-state system of logic used instead of the ten-state system of logic? (9-1)
30. Sketch the schematic symbol of the five basic logic gates, and show their 2-input truth tables. (9-1-1)
31. What is transistor-transistor logic? (9-1-3)
32. Sketch a bipolar transistor NAND gate standard TTL circuit, and describe the operation of the circuit for all input combinations. (9-1-4)
33. Define the purpose of the following; (9-1-4)
    a. A bipolar multiple-emitter input transistor
    b. A totem-pole circuit
    c. A bipolar phase splitter transistor
34. List and describe the following standard TTL logic gate characteristics: (9-1-4)
    a. The valid logic 1 output voltage range
    b. The valid logic 1 input voltage range
    c. The valid logic 0 output voltage range
    d. The valid logic 0 input range
    e. The typical power dissipation per gate
    f. The typical propagation delay per gate
    g. The logic gate fanout
    h. The $+V_{CC}$ commercial supply voltage range
35. What is the maximum and minimum current a standard TTL logic gate can source and sink? (9-1-4)
36. Why is a floating input equivalent to a HIGH input? (9-1-4)
37. Describe the basic operation, characteristics and applications of the following bipolar transistor logic gate circuits:
    a. Low-power TTL (9-1-5)
    b. Schottky TTL (9-1-6)
    c. Open collector TTL (9-1-7)
    d. Three-state output TTL (9-1-8)
    e. Buffer/driver TTL (9-1-9)
    f. Schmitt trigger (9-1-10)
    g. Emitter-coupled logic (9-1-11)
    h. Integrated injection logic (9-1-12)
38. What is a schottky transistor? (9-1-6)
39. Why is the astable multivibrator also known as the free-running multivibrator? (9-2-1)
40. Why is the sine wave considered an analog signal and the square or rectangular wave considered a digital signal? (Chapter 1)
41. What is the other name given to the monostable multivibrator? (9-2-2)
42. Sketch an astable multivibrator circuit, and briefly describe its operation. (9-2-1)

43. What is the purpose of the differentiator and clipper on the trigger input of a monostable multivibrator? (9-2-2)

44. How can the monostable multivibrator be used to stretch a pulse? (9-2-2)

45. Sketch a monostable multivibrator circuit, and briefly describe its operation. (9-2-2)

46. What is the similarity between the bistable multivibrator and the schmitt trigger? (9-2-3)

47. Sketch the bistable multivibrator circuit, and briefly describe its operation. (9-2-3)

48. Define hysteresis. (9-1-10)

49. Why is the schmitt trigger logic gate symbolized with the hysteresis curve? (9-1-10)

50. Why is the 555 timer so popular, and where were the numbers 555 derived from? (9-2-4)

51. Briefly describe how the 555 timer can be made to function as (9-2-4)

    a. A free-running multivibrator
    b. A one-shot multivibrator

52. Define the following, as they relate to two-state waveforms: (9-3-1)

    a. Rise time **d.** Duty cycle
    b. Fall time **e.** PRT
    c. Pulse width **f.** PRF

53. What is the difference between a pulse waveform and a square waveform? (9-3)

54. Show the schematic symbol and describe the operation of the set-reset flip-flop. (9-2-3).

55. Using the block diagram of the 555 timer, describe the function of each block. (9-2-4)

Practice Problems

56. Convert the following letter, number or command to its ASCII code equivalent (encode):

    a. A **b.** 4 **c.** % **d.** r **e.** +

57. Convert the following ASCII codes to their letter, number or command equivalent (decode);

    a. 1011010 **b.** 0111111 **c.** 0111101 **d.** 0110000 **e.** 0101110

58. Figure 9-47 shows an ASCII encoder, or code generator circuit. Which three ASCII codes will be generated for the three positions of the rotary switch, and what letter, number, or command do these ASCII codes represent?

59. In the home security system shown in Figure 9-48(a), a two-input logic gate is needed to actuate an alarm (output *Y*) if either the window is opened (input *A*) or the door is opened (input *B*). Which decision-making logic gate should be used in this application?

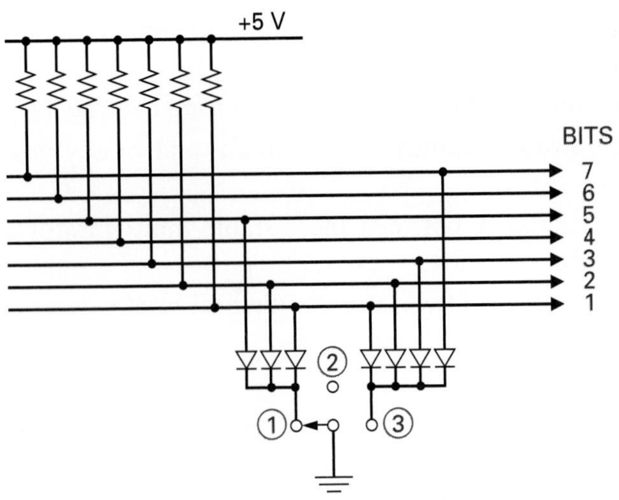

FIGURE 9-47 An ASCII Encoder (Code Generator) Circuit.

60. In the office building cooling system shown in Figure 9-48(b), a thermostat (input A) is used to turn ON and OFF a fan (output Y). A thermostat enable switch is connected to the other input of the logic gate (input B), and it will either enable thermostat control of the fan, or disable thermostat control of the fan. Which decision making logic gate should be used in this application?

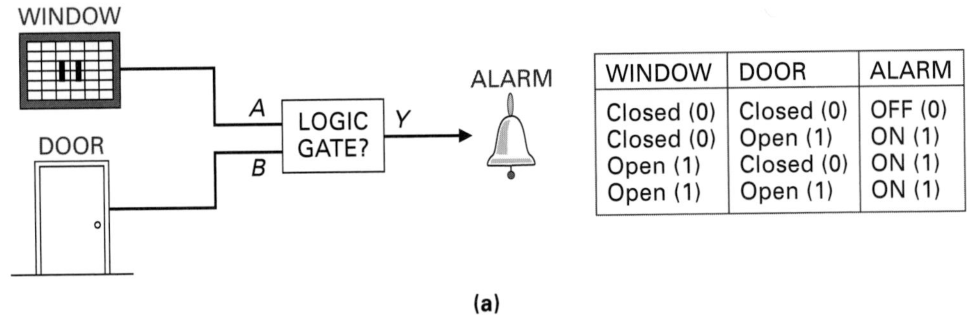

WINDOW	DOOR	ALARM
Closed (0)	Closed (0)	OFF (0)
Closed (0)	Open (1)	ON (1)
Open (1)	Closed (0)	ON (1)
Open (1)	Open (1)	ON (1)

(a)

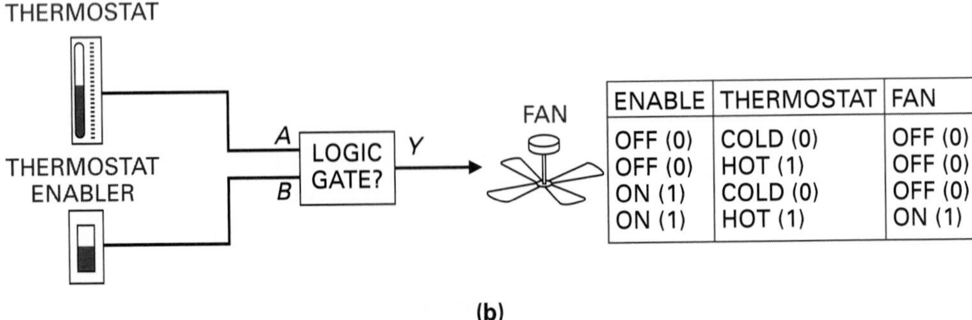

ENABLE	THERMOSTAT	FAN
OFF (0)	COLD (0)	OFF (0)
OFF (0)	HOT (1)	OFF (0)
ON (1)	COLD (0)	OFF (0)
ON (1)	HOT (1)	ON (1)

(b)

FIGURE 9-48 Decision Making Logic Gates.

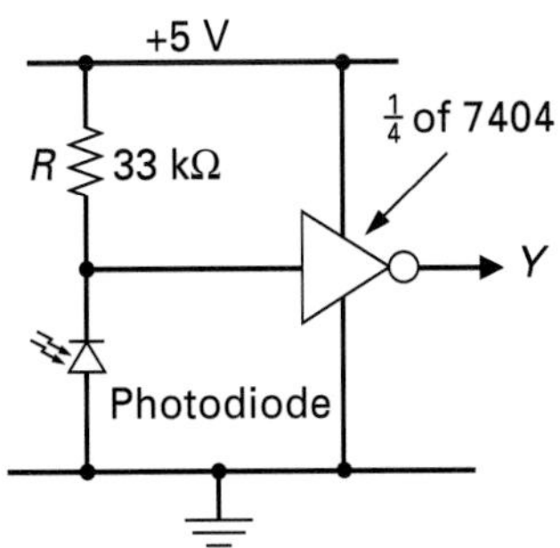

FIGURE 9-49 **Photodiode Receiver.**

61. Briefly describe the operation of the photodiode receiver circuit in Figure 9-49. What will be the output at *Y* when light is being applied to the photodiode and when light is not being applied?

62. Briefly describe the operation of the gated LED flasher circuit shown in Figure 9-50.

63. Which logic gate would produce the output waveform shown in Figure 9-51?

64. Indicate which codes would be generated for the OR gate encoder circuit shown in Figure 9-52.

65. Do the codes generated by the OR gate encoder circuit in Figure 9-52 have any similarity to the ASCII codes for the same decimal digits?

66. Identify the circuit shown in Figure 9-53.

67. Calculate the frequency of the output for the circuit shown in Figure 9-53.

68. Identify the circuit shown in Figure 9-54.

69. Calculate the width of the output pulse (t_2) from the circuit shown in Figure 9-54.

70. Comparing the input pulse to the output pulse in Figure 9-54, how much pulse stretching has actually occured?

71. What would be the approximate width of the trigger pulse at *TP*2 in Figure 9-54?

72. Identify the circuit shown in Figure 9-55 (p. 506).

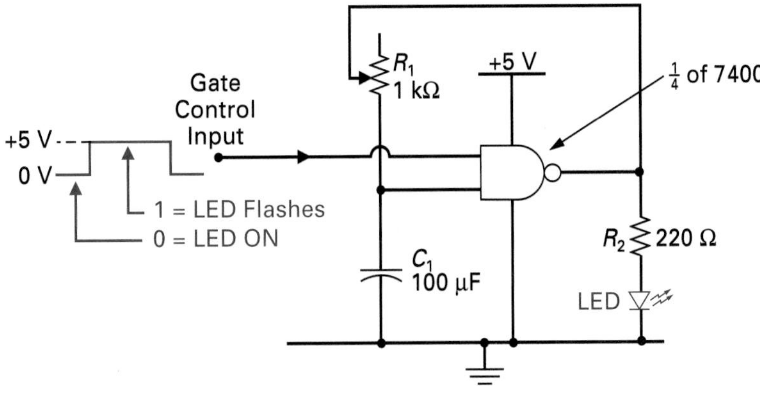

FIGURE 9-50 **Gated LED Flasher Circuit.**

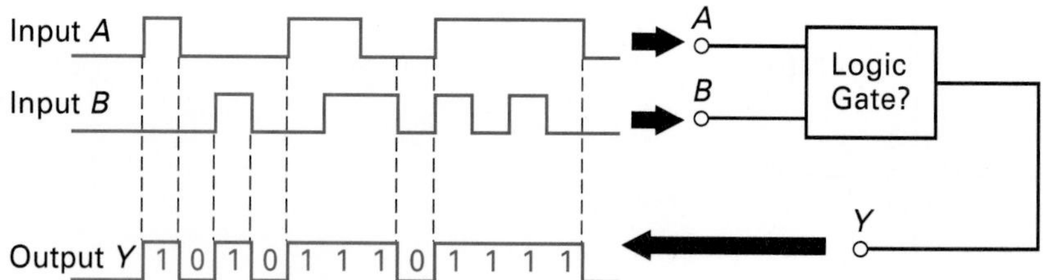

FIGURE 9-51 **Timing Analysis of a Logic Gate.**

73. Calculate the following in relation to the circuit shown in Figure 9-55:

a. The positive half-cycle time **c.** The cycle's period
b. The negative half-cycle time **d.** The cycle's frequency

74. What would be the duty cycle of the output waveform generated by the 555 timer circuit in Figure 9-55?

75. Referring to Figure 9-56, what is the waveform's:

a. Pulse width **c.** Frequency
b. Period **d.** Duty cycle

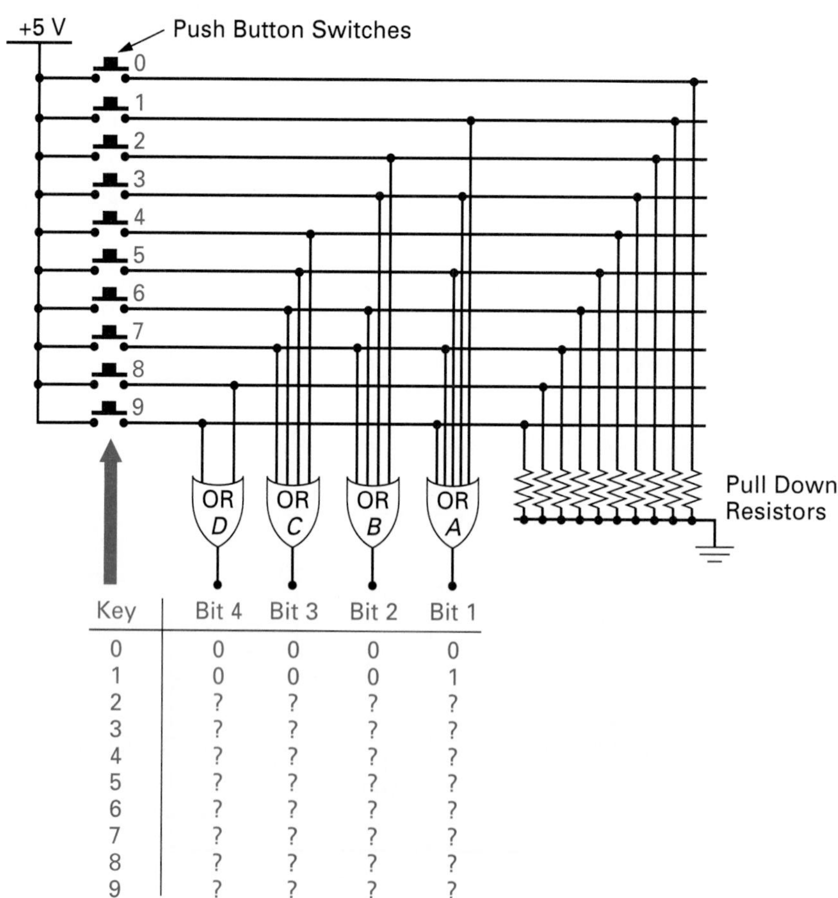

Key	Bit 4	Bit 3	Bit 2	Bit 1
0	0	0	0	0
1	0	0	0	1
2	?	?	?	?
3	?	?	?	?
4	?	?	?	?
5	?	?	?	?
6	?	?	?	?
7	?	?	?	?
8	?	?	?	?
9	?	?	?	?

FIGURE 9-52 **An OR Gate Encoder Circuit.**

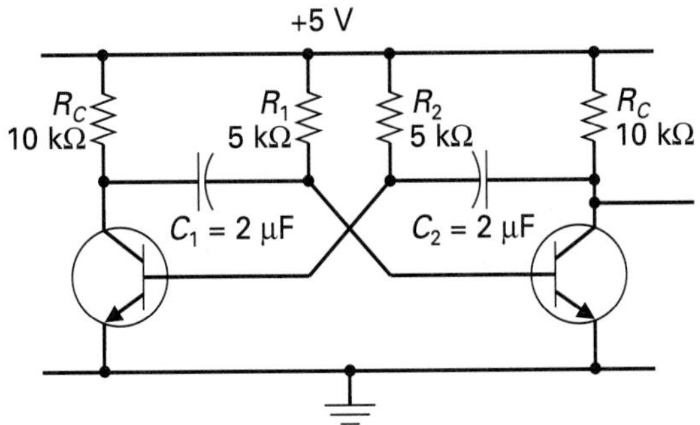

FIGURE 9-53 A Two-State Switching Circuit.

Troubleshooting Questions

76. Briefly describe how the multimeter can be used to test digital circuits. (9-3-1)

77. If a logic probe can be used for both static and dynamic testing, why would you still use a multimeter and oscilloscope? (9-3-1)

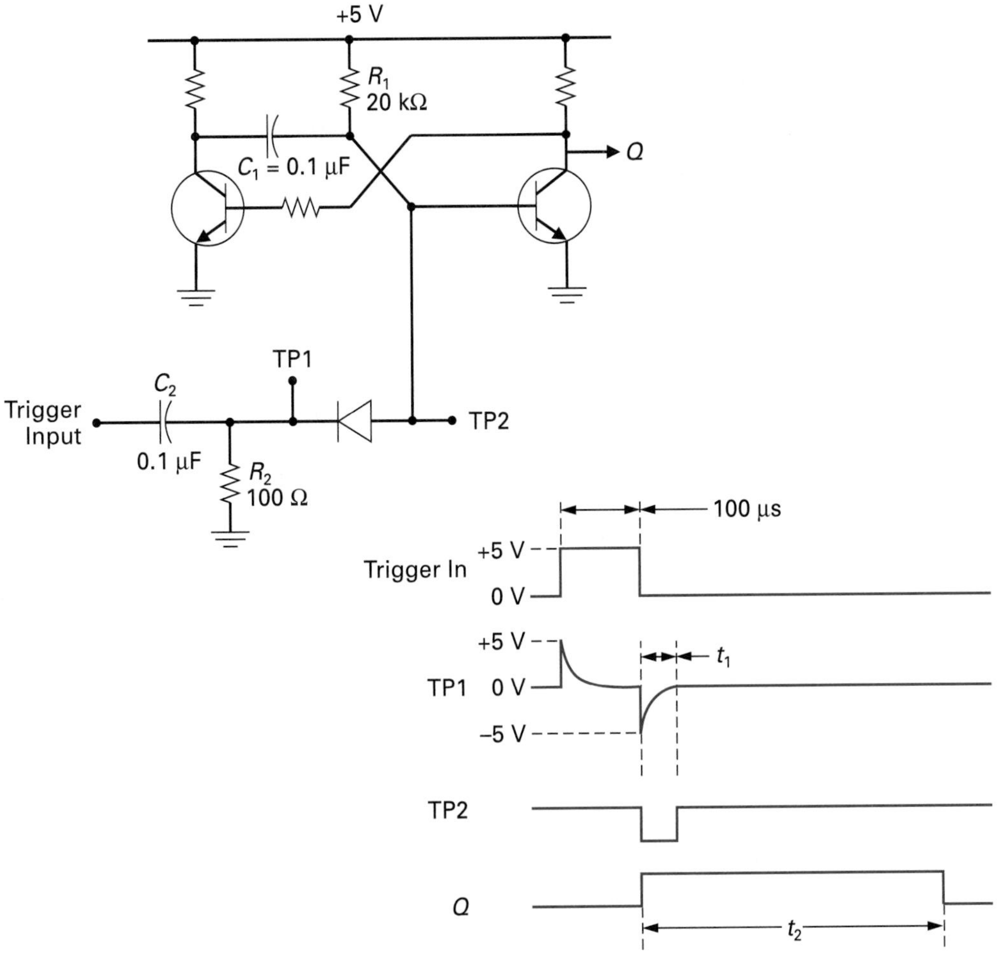

FIGURE 9-54 A Pulse Stretching Circuit.

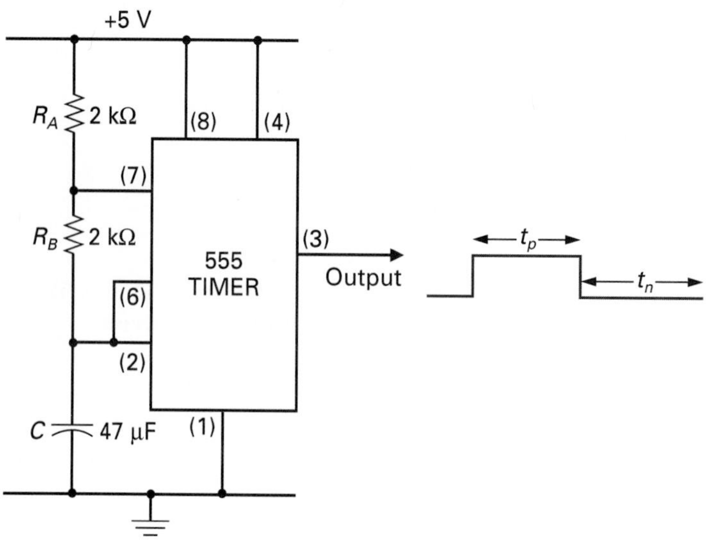

FIGURE 9-55 **A 555 Timer Circuit.**

78. What is a logic clip? (9-3-1)

79. Briefly describe how a logic probe and logic pulser can be used in conjunction with one another. (9-3-1)

80. Briefly describe how a logic pulser and current tracer can be used in conjunction with one another. (9-3-1)

81. List some of the typical internal IC problems. (9-3-2)

82. How are opens and shorts isolated in power supply lines? (9-3-2)

83. Why is it important to be able to recognize different types of internal IC failures if the ICs themselves cannot be repaired? (9-3-2)

84. In Figure 9-57(a), a logic probe has been used to check the circuit since it is not functioning as it should. In Figure 9-57(b), you can see what logic levels were obtained when the output of the NAND gate was isolated. Which IC should be replaced, and what do you suspect is wrong?

85. In Figure 9-58, an oscilloscope has been used to test several points on the digital circuit. Which IC should be replaced, and what do you think is the problem?

86. Are the logic gates in Figure 9-59 operating as they should?

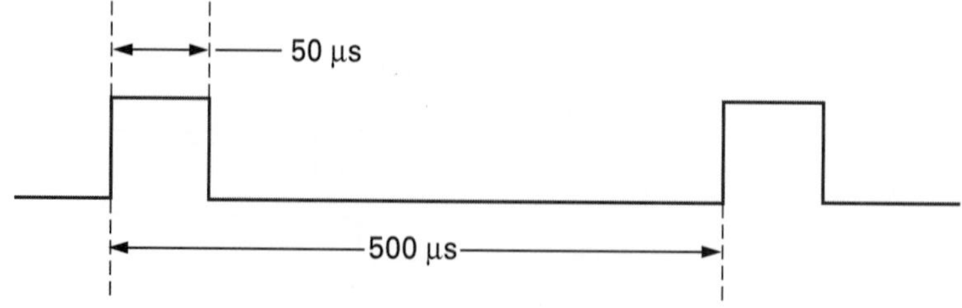

FIGURE 9-56 **A Rectangular or Pulse Waveform.**

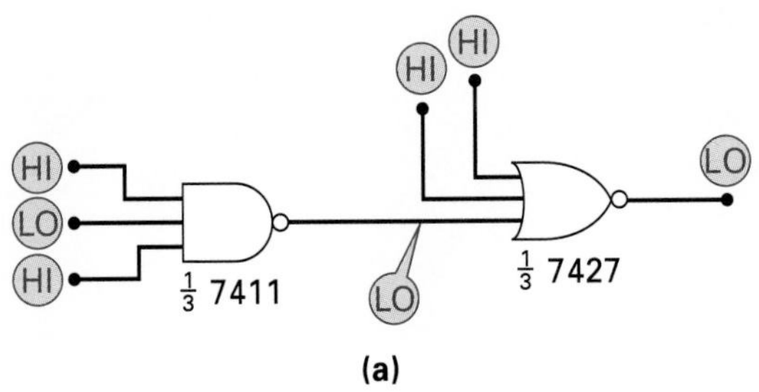

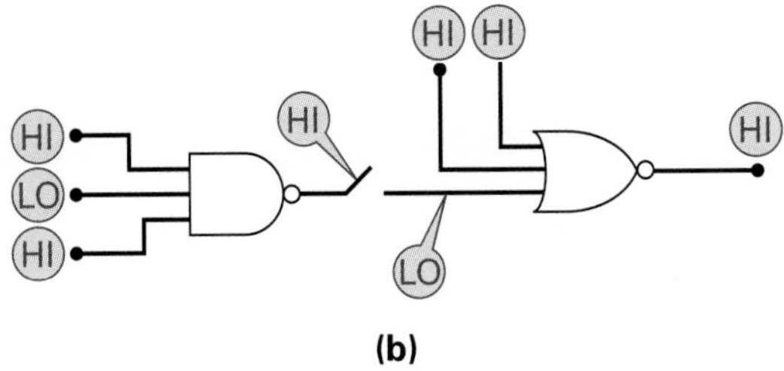

FIGURE 9-57 **Troubleshooting Exercise 1.**

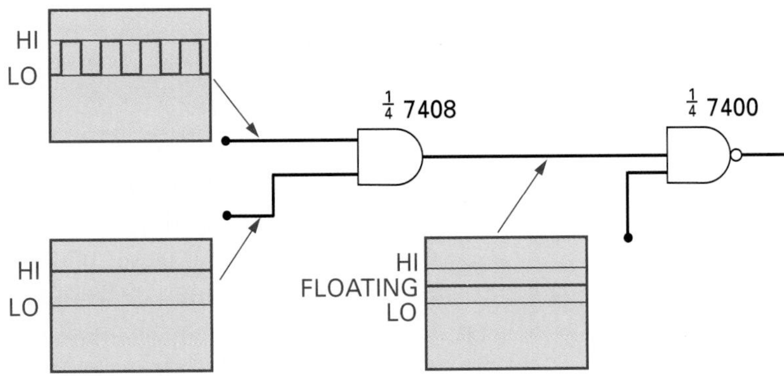

FIGURE 9-58 **Troubleshooting Exercise 2.**

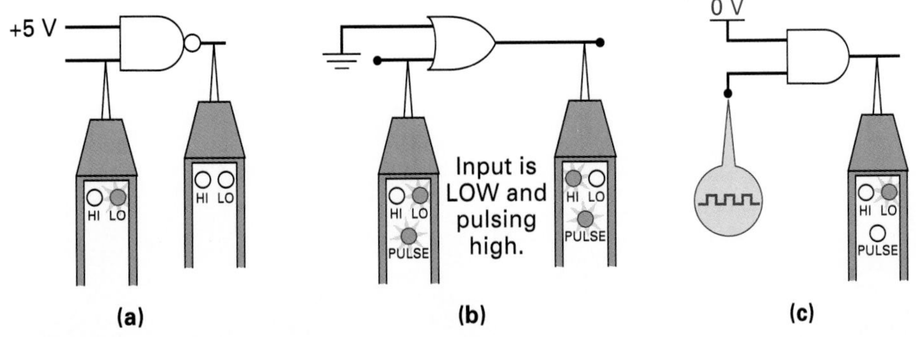

FIGURE 9-59 **Troubleshooting Exercise 3.**

CHAPTER

10

OBJECTIVES

After completing this chapter, you will be able to:

1. List the three basic amplifier types.
2. Explain how the gain of a voltage, current, and power amplification can be calculated.
3. Define the term *decibels.*
4. Describe all of the mathematical details relating to logarithms.
5. Explain how gain and loss can be expressed in decibels and how to convert between arithmetic gain and decibel gain and vice versa.
6. Describe the different amplifier classes of operation.
7. Explain how frequency response, bandwidth, cutoff frequency, and bode diagram relate to amplifiers.
8. Describe the operation, characteristics, and application of the following direct current (dc) amplifier circuits:
   - **a.** Single-stage dc amplifier
   - **b.** Direct-coupled multiple-stage dc amplifier
   - **c.** Resistive-coupled multiple-stage dc amplifier
   - **d.** Zener-coupled multiple-stage amplifier
   - **e.** Complimentary dc amplifier
   - **f.** Darlington pair dc amplifier
   - **g.** Differential dc amplifier
9. Describe the operation characteristics and application of the following audio frequency (AF) amplifier circuits:
   - **a.** Single-stage AF voltage amplifier
   - **b.** Multiple-stage AF voltage amplifier
   - **c.** Variable-gain and variable-frequency response AF voltage amplifier
   - **d.** Push-pull class-B AF power amplifier
   - **e.** Complimentary AF power amplifier
   - **f.** Quasi-complimentary power AF amplifier
10. Describe how to isolate whether a circuit malfunction is within an amplifier circuit or within one of the circuits external to the amplifier.
11. Explain how to troubleshoot a voltage and power amplifier circuit.

Bipolar Transistor Low-Frequency Amplifier Circuits

OUTLINE

VIGNETTE

The First Pocket Calculator

During the seventeenth century, European thinkers were obsessed with any device that could help them with mathematical calculation. Scottish mathematician John Napier decided to meet this need, and in 1614 he published his new discovery of logarithms. In this book, consisting mostly of tediously computed tables, Napier stated that a logarithm is the exponent of a base number. For example, the common logarithm (base 10) of 100 is 2 ($100 = 10^2$), the common logarithm of 10 is 1 ($10 = 10^1$), the common logarithm of 27 is 1.43136 ($27 = 10^{1.43136}$), the common logarithm of 6 is 0.77815 ($6 = 10^{0.77815}$). Any number, no matter how large or small, can be represented by or converted to a logarithm. He also outlined how the multiplication of two numbers could be achieved by simply adding the numbers' logarithms. For example, if the logarithm of 2 (which is 0.30103) is added to the logarithm of 4 (which is 0.60206), the result will be 0.90309, which is the logarithm of the number 8 ($0.30103 + 0.60206 = 0.90309$, $2 \times 4 = 8$). Therefore, the multiplication of two large numbers can be achieved by looking up the logarithm of the two numbers in a log table, adding them together, and then finding the number that corresponds to the sum in an antilog (reverse log) table. In this example, the antilog of 0.90309 is 8.

Napier's table of logarithms was used by William Oughtred, who, just ten years after Napier's death in 1617, developed a handy mechanical device that could be used for rapid calculation. This device, considered the first pocket calculator, was the slide rule.

As well as being a brilliant mathematician, Napier was also interested in designing military weapons. One such unfinished project was a death ray system consisting of an arrangement of mirrors and lenses that would produce a concentrated lethal beam of sunlight.

INTRODUCTION

The amplifier is probably one of the most widely used electronic circuits. Its function is to increase the amplitude of either an ac or dc input signal. Although this function is quite basic, it is vital and needed in almost every electronic system. Amplification is achieved by having a small signal at the input control a large amount of power at the output. The output signal should be a direct copy of the input signal, except larger in amplitude.

In Chapter 8, we discussed the two basic applications of the bipolar junction transistor (BJT). In Chapter 9 we examined the first of these two applications: *the bipolar transistor's switching action.* In this chapter, we will see how the second of these two applications, *the bipolar transistor's variable resistor action,* can be put to good use.

As an electronics technician, it is important for you to be able to recognize the appearance, understand the operation, and know the characteristics of various bipolar transistor amplifier circuits. The amplifier circuits in this chapter have been organized by the frequency of the signal they will amplify. In this chapter we will examine *direct current (dc) amplifiers,* which amplify dc and low frequency ac signals, and *audio frequency (AF) amplifiers,* which amplify ac signals between 20 Hz and 20,000 Hz.

In the following chapter, we will continue the coverage started in this chapter, examining bipolar transistor amplifier circuits that operate at progressively higher frequencies, such as *radio frequency (RF) amplifiers, intermediate frequency (IF) amplifiers,* and *video frequency (VF) amplifiers.*

10-1 AMPLIFIER PRINCIPLES

Before we examine how the bipolar transistor can be used as the controlling element in a direct-current (dc) or audio-frequency (AF) amplifier, let us begin by discussing a few basic amplifier principles.

10-1-1 *Basic Amplifier Types and Gain*

The amplifier's main objective is to produce a **gain,** which is symbolized **A.** Gain is the ratio of the amplitude of the output signal to the amplitude of the input signal. This can be stated mathematically as:

Gain, *A*
The ratio of the amplitude of the output signal to the amplitude of the input signal.

$$\text{Gain}(A) = \frac{\textit{Amplitude of Output Signal}}{\textit{Amplitude of Input Signal}}$$

There are three basic amplifier types, and these are shown in Figure 10-1. As you can see from this illustration, and as shown with the buffer amplifier in the previous chapter, an amplifier is often symbolized as a triangle with an input terminal on the left and an output terminal on the right.

Voltage Amplifier
An amplifier designed to produce an output signal voltage that is greater than the input signal voltage.

Voltage Gain

Figure 10-1(a) shows the **voltage amplifier,** which is designed to produce an output signal voltage that is greater than the input signal voltage. This type of amplifier will typically receive an input signal (V_{IN}) measured in millivolts and produce an output signal (V_{OUT}) normally measured in volts. The amount of voltage amplification, or **voltage gain (A_V),** is a ratio of the output signal volt-

Voltage Gain (A_V)
The ratio of the output signal voltage to the input signal voltage.

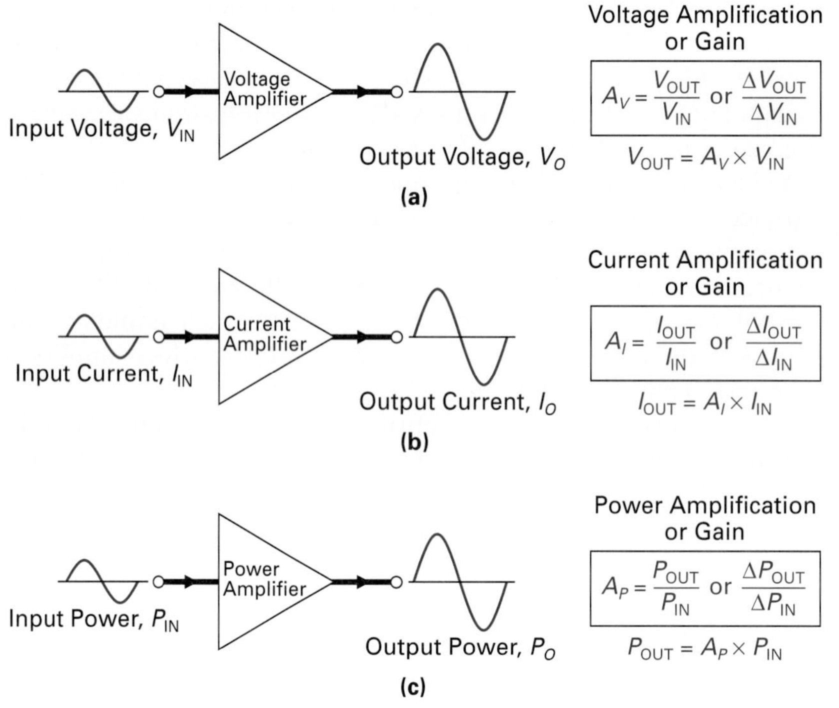

FIGURE 10-1 Basic Amplifier Types and Gain. (a) Voltage Amplifier. (b) Current Amplifier. (c) Power Amplifier.

age to the input signal voltage. If a dc voltage signal is applied, the following formula can be used:

$$A_V = \frac{V_{OUT}}{V_{IN}}$$

DC Voltage Amplifier Gain

■ **EXAMPLE:**

A photodiode is being used to sense the light level in a room. Based on the amount of light present, the photodiode will produce a dc signal voltage of between 0.2 V and 5 V. This dc voltage signal is applied to a direct current (dc) voltage amplifier, which is included to boost the dc voltage signal to a more usable voltage level. If an input voltage signal of 200 mV is applied to the voltage amplifier, and the amplifier produces an output voltage of 2 V, what is the amplifier's voltage gain?

■ ***Solution:***

$$A_V = \frac{V_{OUT}}{V_{IN}} = \frac{2\text{V}}{200\text{mV}} = 10$$

If an ac signal is applied to a voltage amplifier, the Greek capital letter delta (Δ) precedes the V_{OUT} and V_{IN} in the gain formula to indicate the ac signal voltage change.

$$A_V = \frac{\Delta V_{\text{OUT}}}{\Delta V_{\text{IN}}}$$

AC Voltage Amplifier Gain

■ **EXAMPLE:**

A microphone is being used to sense the loudness level in a room. The ac signal voltage produced by the microphone is applied to an audio frequency (AF) voltage amplifier. If an input voltage signal of 400 mV peak-peak is applied to the voltage amplifier, and the amplifier produces an output voltage of 8 V peak-peak, what is the amplifier's voltage gain?

■ ***Solution:***

$$A_V = \frac{\Delta V_{\text{OUT}}}{\Delta V_{\text{IN}}} = \frac{8\text{V}}{400\text{mV}} = 20$$

By transposing the gain formula, we can obtain a formula for calculating the output of a voltage amplifier if the voltage gain and input voltage are known.

$$V_{\text{OUT}} = A_V \times V_{\text{IN}}$$

■ **EXAMPLE:**

What would be the output from a common-base voltage amplifier with a gain of 50 if a multimeter on the ac-volt's setting measures an input signal of 18 mV?

■ ***Solution:***

A multimeter on the ac-volt's setting is calibrated to display rms voltage. If the input voltage is an rms value, the output voltage value will also be an rms value. Therefore:

$$V_{\text{OUT}(rms)} = A_V \times V_{\text{IN}(rms)} = 50 \times 18\text{ mV} = 900\text{ mV}_{rms}$$

Current Gain

Figure 10-1(b) shows the **current amplifier,** which is designed to produce an output signal current (I_{OUT}) that is greater than the input signal current (I_{IN}). The amount of current amplification, or **current gain (A_I),** can be calculated by using the same gain ratio formula of output over input.

Current Amplifier An amplifier designed to produce an output signal current (I_{OUT}) that is greater than the input signal current (I_{IN}).

Current Gain (A_I) The ratio of the output signal current to the input signal current.

$$A_I = \frac{I_{\text{OUT}}}{I_{\text{IN}}} \qquad A_I = \frac{\Delta I_{\text{OUT}}}{\Delta I_{\text{IN}}}$$

DC Current Amplifier Gain   AC Current Amplifier Gain

■ EXAMPLE:

A common-collector amplifier has an input of 0.25 mA and an output of 0.37 mA. What is its current gain?

■ *Solution:*

$$A_I = \frac{I_{\text{OUT}}}{I_{\text{IN}}} = \frac{0.37\,\text{mA}}{0.25\,\text{mA}} = 1.5$$

By transposing these gain formulas, we can calculate the output of a current amplifier by multiplying the amplifier's current gain by the input current.

$$I_{\text{OUT}} = A_I \times I_{\text{IN}}$$

■ EXAMPLE:

If a 150 μA input signal is applied to a current amplifier with a gain of 35, what will the output signal current be?

■ *Solution:*

$$I_{\text{OUT}} = A_I \times I_{\text{IN}} = 35 \times 150\ \mu\text{A} = 5.25\ \text{mA}$$

Power Gain

Power Amplifier An amplifier designed to produce an output signal power that is greater than the input signal power.

Power Gain (A_P) The ratio of the output signal power to the input signal power.

Figure 10-1(c) shows the **power amplifier,** which is designed to produce an output signal power that is greater than the input signal power. This type of amplifier will typically receive an input signal (P_{IN}) measured in milliwatts and produce an output signal (P_{OUT}) normally measured in watts. The amount of power amplification, or **power gain (A_P),** can be calculated by using the same gain-ratio formula of output over input.

$$A_P = \frac{P_{\text{OUT}}}{P_{\text{IN}}} \qquad\qquad A_P = \frac{\Delta P_{\text{OUT}}}{\Delta P_{\text{IN}}}$$

DC Power Amplifier Gain AC Power Amplifier Gain

As before, we can transpose the power gain formula to calculate the output from a power amplifier when the input power is known.

$$P_{\text{OUT}} = A_P \times P_{\text{IN}}$$

Because power is equal to the product of voltage and current ($P = V \times I$), the power gain of an amplifier can also be calculated by multiplying the amount of voltage gain by the amount of current gain.

$$A_P = A_V \times A_I$$

■ **EXAMPLE:**

Calculate the gain of a common-emitter power amplifier if it has a voltage gain of 43 and a current gain of 10. Also, determine the output power from the amplifier when an input of 53 μW is applied.

■ ***Solution:***

$$A_P = A_V \times A_I = 43 \times 10 = 430$$

$$P_{\text{OUT}} = A_P \times P_{\text{IN}} = 430 \times 53\ \mu\text{W} = 22.8\ \text{mW}$$

10-1-2 *Bipolar Transistor Circuit Configurations*

The voltage, current, and power amplifier block symbols shown in Figure 10-1 are used to represent a complete electronic amplifier circuit, as shown in Figure 10-2. This amplifier circuit generally contains one or more transistors, which act as the circuit's controlling element, and several associated components, such as resistors and capacitors, which control the characteristics and stability of the transistor. As we discussed in Chapter 8, by connecting the bipolar junction transistor in one of three different configurations, we could have it function as a power amplifier, voltage amplifier, or current amplifier. Let us briefly review these three circuit configurations and their characteristics, which are shown in Figure 10-3.

1. The common-emitter circuit configuration, shown in Figure 10-3(a), provides a high current gain, high voltage gain, and therefore the combined

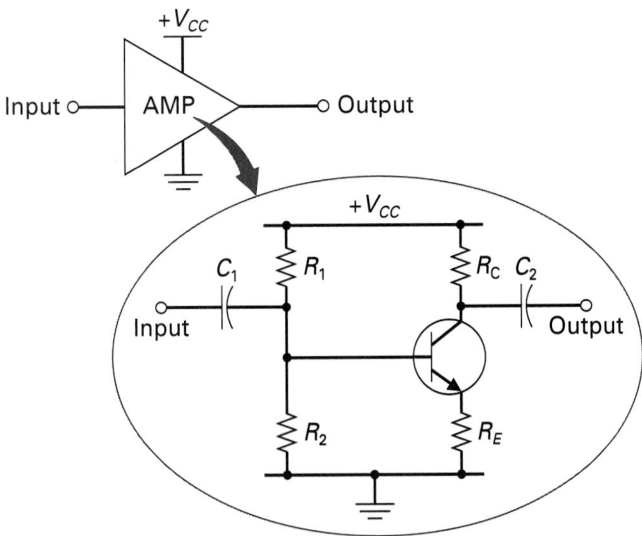

FIGURE 10-2 The Amplifier Circuit's Components.

(a) Common-Emitter (C-E) Circuit

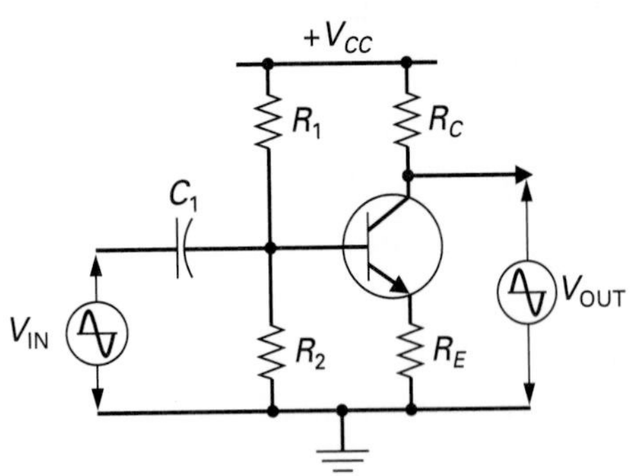

$I_{IN} = I_B$, $I_{OUT} = I_C$
Application: Power Amplifier or Switch

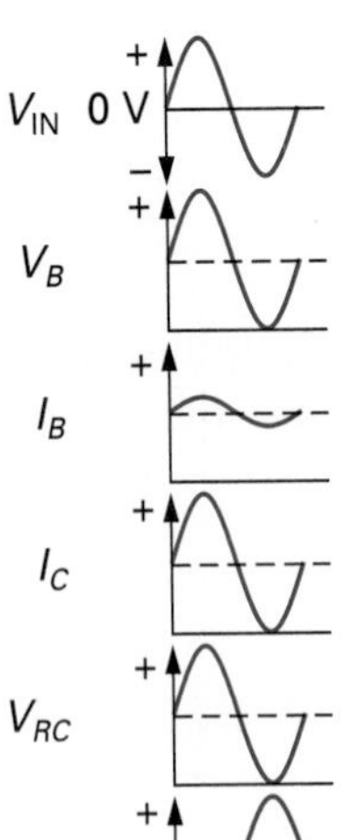

CHARACTERISTICS
Current Gain: Medium, 50
Voltage Gain: High, 100 to 500
Power Gain: Highest, 200 to 20,000
Input Resistance: Medium, 1 kΩ to 5 kΩ
Output Resistance: High, 40 kΩ to 60 kΩ
Phase Shift: 180°

Current Gain: (DC)$\beta_{DC} = \dfrac{I_C}{I_B}$, (AC)$\beta_{AC} = \dfrac{\Delta I_C}{\Delta I_B}$

Voltage Gain: $A_V = \dfrac{\Delta V_{OUT}}{\Delta V_{IN}}$

Power Gain: $A_P = \dfrac{P_{OUT}}{P_{IN}} = \dfrac{\Delta V_{OUT} \times \Delta I_{OUT}}{\Delta V_{IN} \times \Delta I_{IN}} = A_V \times \beta_{AC}$

(b) Common-Base (C-B) Circuit

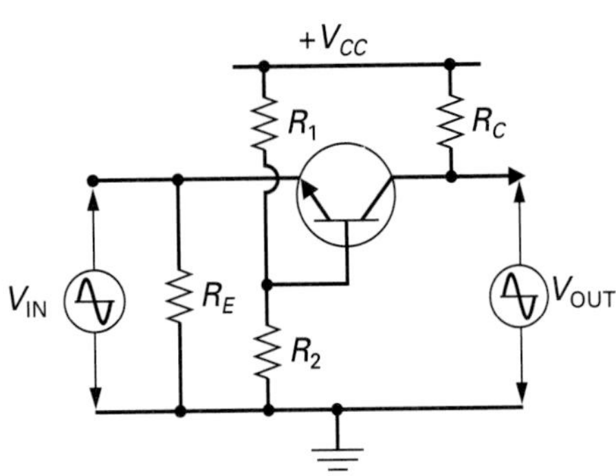

$I_{IN} = I_E$, $I_{OUT} = I_C$
Application: Voltage Amplifier or Switch

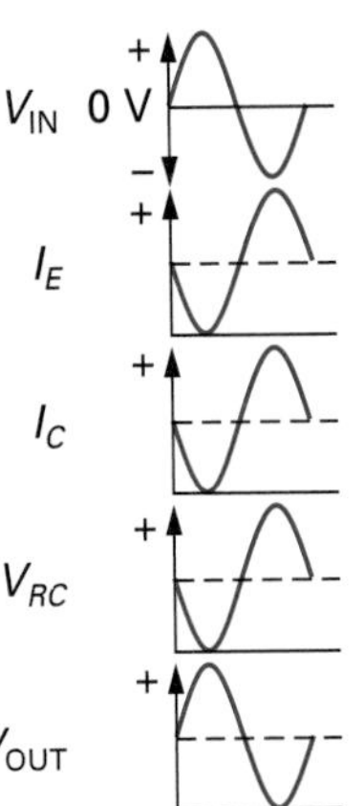

CHARACTERISTICS
Current Gain: Low, 0.99
Voltage Gain: Highest, 200 to 2000
Power Gain: Medium, 200 to 1000
Input Resistance: Very low, 15 Ω to 150 Ω
Output Resistance: Very high, 250 Ω k to 1 MΩ
Phase Shift: 0°

Current Gain: (DC)$\alpha_{DC} = \dfrac{I_C}{I_E}$, (AC)$\alpha_{AC} = \dfrac{\Delta I_C}{\Delta I_E}$

Voltage Gain: $A_V = \dfrac{\Delta V_{OUT}}{\Delta V_{IN}}$

Power Gain: $A_P = \dfrac{P_{OUT}}{P_{IN}} = \dfrac{\Delta V_{OUT} \times \Delta I_{OUT}}{\Delta V_{IN} \times \Delta I_{IN}} = A_V \times \alpha_{AC}$

(c) Common-Collector (C-C) Circuit

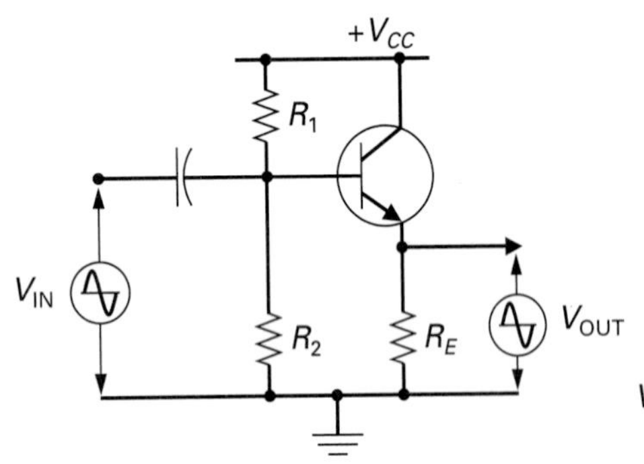

$I_{IN} = I_B$, $I_{OUT} = I_E$
Application: Current Amplifier or Switch, and Impedance or Resistance Matching Device.

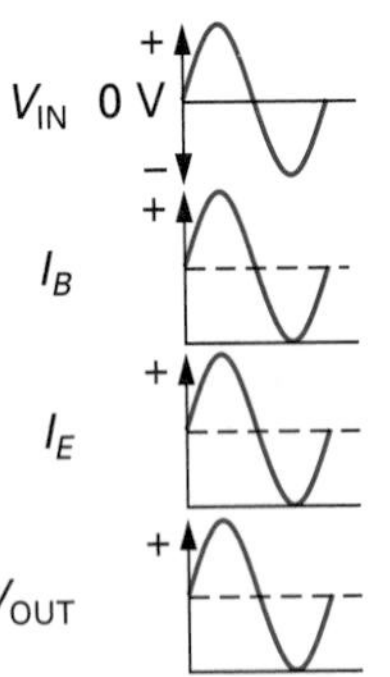

CHARACTERISTICS
Current Gain: Highest, 60
Voltage Gain: Low, less than 1
Power Gain: Low, 20 to 80
Input Resistance: High, 2 kΩ to 500 kΩ
Output Resistance: Very Low, 25 Ω to 1 kΩ
Phase Shift: 0°

Current Gain: DC $= \dfrac{I_E}{I_B} = 1 + \beta$, AC $= \dfrac{\Delta I_E}{\Delta I_B}$

Voltage Gain: $A_V = \dfrac{\Delta V_{OUT}}{\Delta V_{IN}}$

Power Gain: $A_P = \dfrac{P_{OUT}}{P_{IN}} = \dfrac{\Delta V_{OUT} \times \Delta I_{OUT}}{\Delta V_{IN} \times \Delta I_{IN}} = A_V \times$ AC Current Gain

Input Resistance: $R_E \times \beta_{AC}$

FIGURE 10-3 Bipolar Transistor Circuit Configuration Characteristics.

power gain is high. It also has a medium input impedance, a high output impedance, and a 180° phase shift between input and output. All of these generally good characteristics make the common-emitter transistor circuit configuration the most widely used.

2. The common-base circuit configuration, shown in Figure 10-3(b), provides a high voltage gain. However, its current output is slightly less than the current input, so the combined power gain is medium. It also has a very low input impedance, which will load a source, and a high output impedance, which accounts for the low output current. This configuration does not invert the signal between input and output. All of these generally poor characteristics make the common-base transistor circuit configuration the most infrequently used.
3. The common-collector circuit configuration, shown in Figure 10-3(c), provides a high current gain, however its voltage output is slightly less than the voltage input, so the combined power gain is low. The circuit's key advantage is its high input impedance and very low output impedance, making it ideal as an impedance matching device between a low-current (high-impedance) source and a high-current (low-impedance) load. This circuit is also referred to as an emitter follower.

All of the direct-current, audio-frequency amplifiers discussed in this chapter, and all of the radio-frequency, intermediate-frequency, and video-frequency amplifiers discussed in the next chapter, will use one of these three circuit configurations in order to take advantage of that configuration's characteristics.

10-1-3 **Decibels**

So far we have discussed the three basic amplifier types, how the bipolar transistor can be used in different configurations to give voltage, current, or power gain, and the basic amplifier gain formulas. While on the topic of gain, we next need to examine the decibel, a term you will encounter frequently in electronics. For example, the gain of an amplifier is normally given in decibels, the noise level in an amplifier circuit is generally described as the number of decibels it is below the level of the desired signal, an amplifier's response to frequency is generally graphed in decibels, and many other devices, such as antennas and microphones, are also rated in decibels. Because decibels are logarithmic, you should first read the following mini-math review on logarithms.

MINI-MATH REVIEW–LOGARITHMS

This mini-math review is designed to remind you of the mathematical details relating to **logarithms** or **logs,** which will be used in following sections of this text.

Logarithms
The exponent that indicates the power to which a number is raised to produce a given number.

As discussed in this chapter's vignette, logarithms can be used to simplify problem solving by

1. simplifying multiplication to addition
2. simplifying division to subtraction
3. simplifying raising to a power to simple multiplication
4. simplifying extracting square roots to simple division.

Today, multiplication, division, raising a number to some power, and extracting square roots can be accomplished more easily with a calculator. The use of logarithms to simplify problem solving has dramatically decreased. In what application are logarithms used in electronics? We need an understanding of logarithms primarily because the difference in power at the output of an amplifier compared to the input is given in the logarithmic unit **decibels** or **dBs.** To properly understand decibels, we will need to first have a clear understanding of logarithms.

Decibels (dBs)
One tenth of a bel (1×10^{-1} bel). The logarithmic unit for the difference in power at the output of an amplifier compared to the input.

There are two basic types of logarithms: **natural logarithms,** which use a base of 2.71828, and **common logarithms,** which use a base of 10. We will be concentrating on common logarithms because they are used in association with amplifiers.

Natural Logarithms
A base 2.71828 exponent.

Common Logarithms
A base 10 exponent.

Powers of 10
Base 10 exponents.

Common logarithms. Stated simply, a common logarithm is a base 10 exponent. So far, you have already had considerable experience with base 10 exponents or, as they are officially called, **powers of 10.** For example, you know that 1 megavolt is equal to 1×10^6 V (or 1,000,000 V) and that 1 kilohm is equal to $1 \times 10^3\ \Omega$ (or 1000Ω). The power of 10 exponent used in these examples is the logarithm. For the first example, "*the common logarithm (base 10) of 1,000,000 is 6,*" and for the other example "*the common logarithm (base 10) of 1,000 is 3.*" If we were to state the first of these examples in mathematical form, it would appear as follows:

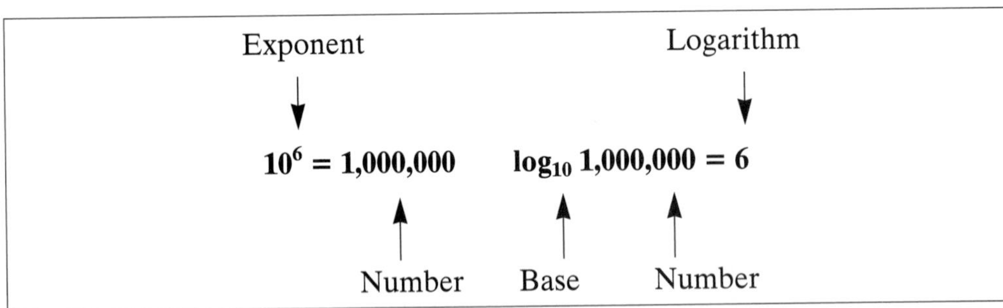

Exponential Expression
An expression in which the quantity is expressed in exponential form.

Logarithmic Expression
An expression in which the quantity is expressed in logarithmic form.

The first expression ($10^6 = 1{,}000{,}000$) is an **exponential expression** because the quantity is expressed in **exponential form.** The second expression ($\log_{10} 1{,}000{,}000 = 6$) is a **logarithmic expression** because the quantity is expressed in **logarithmic form.**

The base of a logarithm should be indicated when there is any chance of confusion; however, when using common logs it is not necessary to indicate or say the base because it is assumed to be 10. Therefore, "$\log_{10} 1{,}000{,}000 = 6$," can be simplified to "$\log 1{,}000{,}000 = 6$."

As another example, what is the log of 1,000? To find this, we simply convert 1,000 to its power of ten equivalent, which in exponential form will be 10^3.

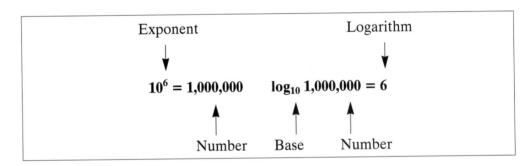

Because the common logarithm of any number is its base 10 exponent, in logarithmic form, log 1000 = 3.

Here is a list of logs for numbers that are greater than 1.

EXPONENTIAL FORM	LOGARITHMIC FORM
$10^0 = 1$	log 1 = 0
$10^1 = 10$	log 10 = 1
$10^2 = 100$	log 100 = 2
$10^3 = 1{,}000$	log 1,000 = 3
$10^4 = 10{,}000$	log 10,000 = 4
$10^5 = 100{,}000$	log 100,000 = 5
$10^6 = 1{,}000{,}000$	log 1,000,000 = 6

What do we do when we are dealing with numbers that are a fraction of 1? For example, 0.001 mV, 0.000001 μA, and so on. The answer is to follow exactly the same procedure. First, express the number or value in exponential form ($10^{-3} = 0.001$). Then, because the logarithm is the base 10 exponent, express the number in logarithmic form (log 0.001 = –3).

Here is a list of logs for numbers that are a fraction of 1.

EXPONENTIAL FORM	LOGARITHMIC FORM
$10^0 = 1$	log 1 = 0
$10^{-1} = 0.1$	log 0.1 = –1
$10^{-2} = 0.01$	log 0.01 = –2
$10^{-3} = 0.001$	log 0.001 = –3
$10^{-4} = 0.0001$	log 0.0001 = –4
$10^{-5} = 0.00001$	log 0.00001 = –5
$10^{-6} = 0.000001$	log 0.000001 = –6

The calculator's table of common logarithms. Your calculator contains a table of common logarithms. For example, to find the logarithms of the previous numbers, we would press the following keys.

[log] Common Logarithm Key—calculates the common logarithm (base 10) of the number in the display.

To calculate the log of 1,000, you simply input the value 1,000, and then press the log calculator key, as follows:

Log 1,000 = ?

Press Keys: [1] [0] [0] [0] [log]

Display Shows: 3

As another example, to calculate the log of 0.00001, press the following calculator keys:

Log 0.00001 = ?

Press Keys: [0] [.] [0] [0] [0] [0] [1] [log]

Display Shows: –5

Numbers that are not multiples of ten. Only numbers that are multiples of 10 will have an exponent or logarithm that is a whole number. For example, 0.001, 0.1, 100, 1,000,000,000, and so on. In the following examples, you will see that any number between two multiples of 10 will have a decimal fraction log. Using your calculator, check the logs (which have been rounded off to 4 decimal places) for the following numbers.

LOGARITHMIC FORM	EXPONENTIAL FORM
$\log 0.000089 = -4.0506$	$10^{-4.0506} = 0.000089$
$\log 0.28 = -0.5528$	$10^{-0.5528} = 0.28$

Because the example numbers are a fraction of 1, the log or exponent will be negative.

$\log 2.5 = 0.3979$	$10^{0.3979} = 2.5$
$\log 6 = 0.7782$	$10^{0.7782} = 6$
$\log 9.75 = 0.9890$	$10^{0.9890} = 9.75$

Because the log of 1 is 0 and the log of 10 is 1, all of the example numbers between 1 and 10 will have logs or exponents that are between 0 and 1.

$\log 12 = 1.0792$	$10^{1.0792} = 12$
$\log 76.4 = 1.8831$	$10^{1.8831} = 76.4$

Because the log of 10 is 1 and the log of 100 is 2, all of the example numbers between 10 and 100 will have logs or exponents that are between 1 and 2.

$\log 135.54 = 2.132$	$10^{2.132} = 135.54$
$\log 873 = 2.941$	$10^{2.941} = 873$

Because the log of 100 is 2 and the log of 1000 is 3, all of the example numbers between 100 and 1,000 will have logs or exponents that are between 2 and 3.

Using logarithms to simplify multiplication. Now that we know what a logarithm is and how to find them, let us see what we can do with them. One reason for converting numbers to logarithms is to simplify calculations. For example, using logarithms, *the multiplication of two numbers can be reduced to simple addition.* This advantage was first shown in this chapter's opening vignette; however, let us study the steps involved in a little more detail. To find the product of two numbers using logs, you have to follow these four steps:

1. Find the log of each number,
2. Change each number to base 10,
3. Add the base 10 exponents,
4. Convert the log answer back to a number using antilog.

As an example, let us use logarithms to calculate the product of 26×789. The first step calls for us to find the log of each number:

$$\log 26 = 1.41497$$
$$\log 789 = 2.89708$$

The second step states that we should change each number to base 10. Because the log of a number is the base 10 exponent, this step will result in

$$26 = 10^{1.41497}$$

$$789 = 10^{2.89708}$$

The third step says that the answer to 26 × 789 can be obtained by simply adding the exponents from the second step.

$$10^{1.41497} \times 10^{2.89708} = 10^{1.41497+2.89708} = 10^{4.31205}$$

(Note: This law has always applied to exponents. For example, $100 \times 1000 = 10^2 \times 10^3 = 100{,}000 = 10^5$. When multiplying like bases, the exponents or number of zeroes are added.)

Once you have converted all your numbers to logarithms and done all your calculations, you then have to convert the log answer back to a number using a **common antilogarithm** table. This reverse log table is also stored in your calculator and operates as follows,

Common Antilogarithm Reverse common logarithm.

[INV][log]	Common Antilogarithm key—calculates the common antilogarithm of the displayed value (raises 10 to the
or [10^x]	displayed power).

Completing the fourth step in this multiplication example, let us now find the antilog of our answer $10^{4.31205}$. To reverse the log 4.31205, we would press the following calculator keys:

Press the Keys: [4][.][3][1][2][0][5][INV][log]

Display Shows: 20513.98

Using the calculator to double check this answer, we find that 26 × 789 = 20,514.

Using logarithms to simplify division. Using logarithms, *the division of one number by another number can be reduced to simple subtraction.* The steps to follow to calculate the result of a division using logs are

1. Find the log of the dividend and divisor,
2. Change each number to base 10,
3. Subtract the log of the divisor from the log of the dividend, and
4. Convert the log answer back to a number using antilog.

As an example, let us use logarithms to calculate the quotient of 8992.5 ÷ 165. The first step calls for us to find the log of both the dividend (8992.5) and the divisor (165).

$$\log 8992.5 = 3.95388$$

$$\log 165 = 2.21748$$

The second step states that we should change each number to base 10. Because the log of a number is the base 10 exponent, this step will result in

$$8992.5 = 10^{3.95388}$$

$$165 = 10^{2.21748}$$

The third step says that the answer to 8992.5 ÷ 165 can be obtained by simply subtracting the log or exponent of the divisor from the log of the dividend.

$$10^{3.95388} \div 10^{2.21748} = 10^{3.95388 - 2.21748} = 10^{1.7364}$$

(Note: This law has always applied to exponents. For example, $10{,}000 \div 10 = 10^4 \div 10^1 = 10^{4-1} = 10^3 = 1{,}000$. When like bases are divided, the exponents are subtracted)

In the fourth step, we convert the log answer back to a number using the antilog table in your calculator.

$$\text{antilog } 1.7364 = 54.5$$

Using the calculator to double check this answer, we find that 8992.5 ÷ 165 = 54.5.

Summary. In this mini-math review, we discussed how any number can be expressed as a power of 10 and that the exponent of this power of 10 is called the common logarithm. Once a number has been converted to its power of 10 equivalent, certain mathematical operations, such as multiplication and division, are simplified. Today, multiplication, division, and other mathematical operations such as raising a number to some power or extracting square roots, can be accomplished more easily with a calculator. The use of logarithms to simplify problem solving has dramatically decreased. Logarithms, however, are still used in a variety of applications in most of the sciences. In electronics, one of the primary uses of logarithms is in power-to-decibel conversions, a topic that will be discussed in the following section.

The Human Ear's Response to Sound

It has been proven through experimentation that the human ear responds logarithmically to changes in sound level. To explain this in more detail, refer to the illustration in Figure 10-4(a), which shows a sine-wave generator connected to a speaker. As the volume control of the generator is increased, the power output of the speaker is gradually increased from 1 watt to 100 watts, and the listener is asked to indicate the points at which a change in volume is detected. These points are listed in the table in Figure 10-4(b). Studying this table, you can see that the power has to be increased from its start point of 1 W to approximately 1.26 W before the ear can detect the increase in volume. As the power is further increased, a change in volume is detected at 1.58 W, 2 W, 2.5 W, 3.17 W, and so on, as shown in the table. This table demonstrates an important point: *The human ear is not sensitive to the amount of change that occurs, but rather to the ratio of change that occurs.*

In other words, at a low-power level the amount of power needed to sense a change in sound is small (between 1 W and 1.26 W there is an increase of 0.26 W), whereas at a high-power level, the amount of power needed to sense a change in sound is high (between 79.4 W and 100 W there is an increase of 20.6 W). Although the amount of change between any two points is not the same, the ratio of change between any two powers is the same:

$$\frac{1.26}{1} = 1.26 \qquad \frac{100}{79.4} = 1.26$$

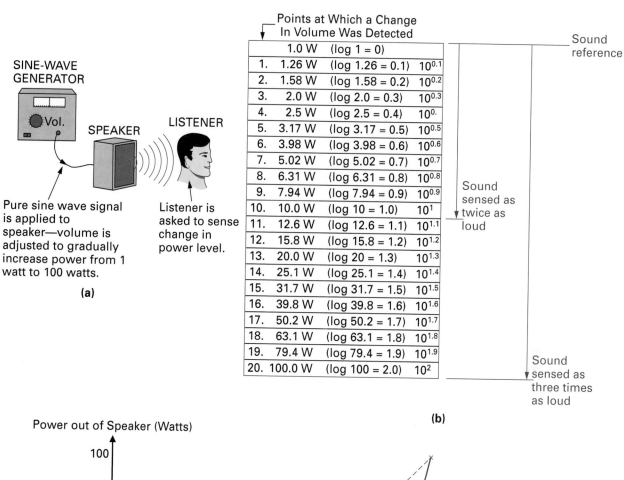

SINE-WAVE GENERATOR

Vol.

SPEAKER

LISTENER

Pure sine wave signal is applied to speaker—volume is adjusted to gradually increase power from 1 watt to 100 watts.

Listener is asked to sense change in power level.

(a)

Points at Which a Change In Volume Was Detected

	1.0 W	(log 1 = 0)	
1.	1.26 W	(log 1.26 = 0.1)	$10^{0.1}$
2.	1.58 W	(log 1.58 = 0.2)	$10^{0.2}$
3.	2.0 W	(log 2.0 = 0.3)	$10^{0.3}$
4.	2.5 W	(log 2.5 = 0.4)	$10^{0.}$
5.	3.17 W	(log 3.17 = 0.5)	$10^{0.5}$
6.	3.98 W	(log 3.98 = 0.6)	$10^{0.6}$
7.	5.02 W	(log 5.02 = 0.7)	$10^{0.7}$
8.	6.31 W	(log 6.31 = 0.8)	$10^{0.8}$
9.	7.94 W	(log 7.94 = 0.9)	$10^{0.9}$
10.	10.0 W	(log 10 = 1.0)	10^{1}
11.	12.6 W	(log 12.6 = 1.1)	$10^{1.1}$
12.	15.8 W	(log 15.8 = 1.2)	$10^{1.2}$
13.	20.0 W	(log 20 = 1.3)	$10^{1.3}$
14.	25.1 W	(log 25.1 = 1.4)	$10^{1.4}$
15.	31.7 W	(log 31.7 = 1.5)	$10^{1.5}$
16.	39.8 W	(log 39.8 = 1.6)	$10^{1.6}$
17.	50.2 W	(log 50.2 = 1.7)	$10^{1.7}$
18.	63.1 W	(log 63.1 = 1.8)	$10^{1.8}$
19.	79.4 W	(log 79.4 = 1.9)	$10^{1.9}$
20.	100.0 W	(log 100 = 2.0)	10^{2}

Sound reference

Sound sensed as twice as loud

Sound sensed as three times as loud

(b)

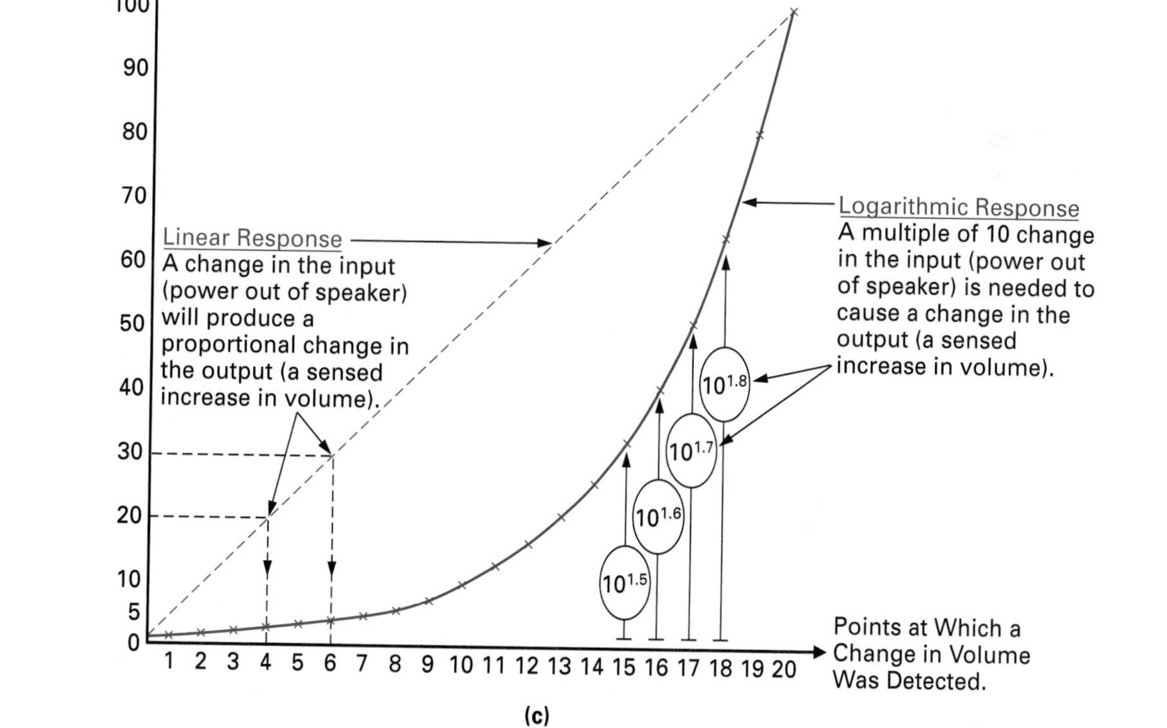

(c)

FIGURE 10-4 The Human Ear's Logarithmic Response to Changes in Sound Level.

The Bel and Decibel

Studying the table in Figure 10-4(b), you can see that the human ear is rather slow to respond to increased stimulation. For example, referring to Figure 10-4(b), you can see that as the sound is increased in intensity to twice its original value (1 W to 2 W), it is not heard by the listener as twice as loud. In fact, the listener seems to only hear a few small changes in loudness. To make the sound twice as loud, the power must be increased by ten times (1 W to 10 W), and to make the sound three times as loud, the power must be increased by one hundred times (1 W to 100 W). To connect this point to something we all understand, let us imagine that you are about to purchase an audio amplifier for your home music system, and you cannot decide whether to buy a 50-watt amplifier or spend twice as much and buy a 100-watt amplifier. Although the 100 W amplifier delivers twice the output power, our ears will only perceive this as a small difference in volume. For your ears to detect a significant difference in volume, you would need an amplifier that could deliver ten times the power. Only a 500-W amplifier will sound as though you are getting twice your money's worth, compared to a 50-W amplifier. This relationship between detected sound and power output is logarithmic because the sound detected goes up in multiples of 10. Because the human ear responds to sound intensity in this way, it is convenient for us to measure sound intensity in a **logarithmic unit,** which is a unit that goes up in equal steps for every tenfold increase in intensity. This unit is called the **bel,** in honor of the great American scientist Alexander Graham Bell. The bel was originally used to indicate the amount of signal loss in two miles of telephone wire. When dealing with small electronic circuits like amplifiers, we need to use a smaller unit of measure. A **decibel (dB)** is equal to 1/10th of a bel (1×10^{-1} bel). To understand the relationship between bel and decibel, consider

Logarithmic Unit
A unit that goes up in equal steps for every tenfold increase in intensity.

Bel
The logarithmic unit of sound.

Decibel (dB)
One tenth of a bel (1×10^{-1} bel). The logarithmic unit for the difference in power at the output of an amplifier compared to the input.

$$0.1 \text{ amps} = 1 \text{ deciamp } (1 \times 10^{-1} \text{ A, or } 1 \text{ dA}) = 10 \text{ centiamps}$$
$$(10 \times 10^{-2} \text{ A}) = 100 \text{ milliamps } (100 \times 10^{-3} \text{ A}).$$

The same is true for bels.

$$0.1 \text{ bels} = 1 \text{ decibel } (1 \times 10^{-1} \text{ B, or } 1 \text{ dB}).$$

Referring again to the table in Figure 10-4(b), you can see that it requires a 1 decibel (0.1 bel) change in sound level before we are aware that a change in sound intensity has taken place. Figure 10-4(c) shows the logarithmic response curve of the human ear by plotting the information in the table in Figure 10-4(b). Comparing this logarithmic response to the linear response, which is also shown in Figure 10-4(c), you can see that a linear response increases by 1, 2, 3, 4, and so on, whereas a logarithmic response increases by 0.1, 1, 10, 100, and so on.

To show how the dB is used to indicate the level of sound, here are some typical dB ratings.

Whisper	0 dB
Normal Conversation	45 dB
Industrial Equipment	90 dB
Very Loud Music	110 dB
Threshold of Pain	130 dB

Power Gain or Loss in Decibels

By definition the bel is the unit of sound level and is equal to the logarithm (base 10) of the ratio of output power to input power. Stated mathematically:

$$10^{\text{bel}} = \frac{P_{\text{OUT}}}{P_{\text{IN}}} \qquad \text{bel} = \log_{10} \frac{P_{\text{OUT}}}{P_{\text{IN}}}$$

In Exponential Form   In Logarithmic Form

Power gains in bels are expressed as the logarithm of the ratio of power levels and are equal to

$$A_P(\text{power gain in bels}) = \log_{10} \frac{P_{\text{OUT}}}{P_{\text{IN}}}$$

However, as stated earlier, the bel is too large a unit, and so we will generally use the decibel, which is 1/10th of a bel. Therefore, power gain in decibels will be equal to

$$A_{P(db)}(\text{power gain in decibels}) = 10 \times \log_{10} \frac{P_{\text{OUT}}}{P_{\text{IN}}}$$

This formula can be used to convert arithmetic power ratios to logarithmic decibel units.

Watts to decibels. Let us now apply this formula to a few examples to see how the decibel can give us an indication of power gain.

■ **EXAMPLE:**

Calculate the power gain in decibels of an amplifier that has an input of 10 mW and an output of 1 W.

■ ***Solution:***

$$A_{P(dB)} = 10 \times \log \frac{P_{\text{OUT}}}{P_{\text{IN}}} = 10 \times \log \frac{1000 \text{ mW}}{10 \text{ mW}} = 10 \times \log 100 = 10 \times 2 = +20 \text{ dB}$$

Remember that a power gain of 20dBs does not mean that the output of the amplifier is 20 times greater than the input. From the values given, we can see that 1 W (or 1000 mW) is 100 times greater than 10 mW. To convert the dB power gain value ($A_{P(dB)}$) to a normal power gain value (A_P), we simply reorganize the original power formula as follows.

$$A_{P(dB)} = 10 \times \log \frac{P_{\text{OUT}}}{P_{\text{IN}}} = 10 \times \log A_P$$

To solve for A_P, divide both sides by 10

$$\frac{A_{P(dB)}}{10} = \frac{10 \times \log A_P}{10}$$

and take the antilog of both sides.

$$\text{antilog}\,\frac{A_{P(dB)}}{10} = A_P$$

$$A_P = \text{antilog}\,\frac{A_{P(dB)}}{10}$$

For the previous example,

$$A_P = \text{antilog}\,\frac{A_{P(dB)}}{10} = \text{antilog}\,\frac{+20\text{ dB}}{10} = \text{antilog } 2 = 100$$

■ **EXAMPLE:**

A cable has an input of 1.5 kW, and an output of 1.2 kW. Calculate the power loss in decibels.

■ ***Solution:***

$$A_{P(dB)} = 10 \times \log \frac{P_{\text{OUT}}}{P_{\text{IN}}} = 10 \times \log \frac{1.2\text{ kW}}{1.5\text{ kW}} = 10 \times \log 0.8 = 10 \times 0.0969 = -1\text{ dB}$$

Once again, a –1 dB loss does not mean that the output from the cable is equal to the input minus 1. To convert this loss in dBs to a normal gain figure, simply reverse the dB formula as before.

$$A_P = \text{antilog}\,\frac{A_{P(dB)}}{10} = \text{antilog}\,\frac{-1\text{ dB}}{10} = \text{antilog } -0.1 = 0.79$$

The output from the cable is equal to 0.79, or 79%, of the input.

■ **EXAMPLE:**

Using the following table, calculate the total dB loss if 38 feet of RG 58 coaxial cable is used to connect a factory's computer controller to a manufacturing robot.

RG/U TYPE COAXIAL CABLE	IMPEDANCE (Ω)	DIAMETER (IN.)	dB/100 FT
RG8A/U	52	0.405	9.0
RG58C/U	50	0.195	17.5
RG108A/U	78	0.235	26.2
RG179B/U	75	0.100	25.0
RG188A/U	50	0.110	30.0
RG196A/U	50	0.08	45.0
RG213/U	50	0.405	9.0

■ ***Solution:***

The RG 58 coaxial cable has a loss of 17.5 dBs per 100 ft of cable. For 38 feet, the loss will be

$$\text{Cable loss} = \frac{38}{100} \times 17.5 = 38\% \text{ of } 17.5 \text{ dB} = 6.65 \text{ dB}$$

The previous examples illustrate an important point regarding decibels. If there is a power gain (the power out is greater than the power in), the dB value is positive. On the other hand, if there is a power loss (the power out is less than the power in), the dB value is negative. Table 10-1 lists some of the more common dB power gain and loss ratios.

Quickly converting between power and decibels. By applying the following two relationships, you can quickly convert power ratios to dBs and dB values to power ratios.

1. An arithmetic power ratio of 2:1 is equivalent to +3 dB, which means if power is doubled the ratio increases by +3 dB. Using a 2:1 power ratio and +3 dB as a base, we can say that a 4:1 power ratio (2 × 2) will be equal to +6 dB (3 dB + 3 dB), an 8:1 ratio (2 × 2 × 2) will be equal to +9 dB (3 dB + 3 dB + 3 dB), and so on.

   Similarly, a power ratio of 1:2 is equivalent to –3 dB, which means if power is halved, the ratio is decreased by –3 dB. Using a 1:2 power ratio and –3 dB as a base, we can say that a 6 dB loss (–6 dB) is equivalent to one-fourth (1/4) the power, a 9 dB loss (–9 dB) is equal to 1/8th the power, and so on.
2. A power ratio of 10:1 is equivalent to +10 dB, which means if power is increased by 10, the ratio increases by 10 dB. Using a 10:1 ratio and 10 dB as a base, we can say that a 100:1 power ratio (10 × 10) will be equal to +20 dB (10 dB + 10 dB), a 1000:1 power ratio (10 × 10 × 10) is equal to +30 dB (10dB + 10dB + 10dB), and so on.

TABLE 10-1 Common Decibel Power Gain and Loss Ratios

GAIN (+dB)	LOSS (–dB)
+3dB = × 2	–3dB = ÷ 2 or 1/2
+6dB = × 4	–6dB = ÷ 4 or 1/4
+9dB = × 8	–9dB = ÷ 8
+10dB = × 10	–10dB = ÷ 10
+12dB = × 16	–12dB = ÷ 16
+15dB = × 32	–15dB = ÷ 32
+20dB = × 100	–20dB = ÷ 100
+30dB = × 1,000	–30dB = ÷ 1,000
+40dB = × 10,000	–40dB = ÷ 10,000
+50dB = × 100,000	–50dB = ÷ 100,000

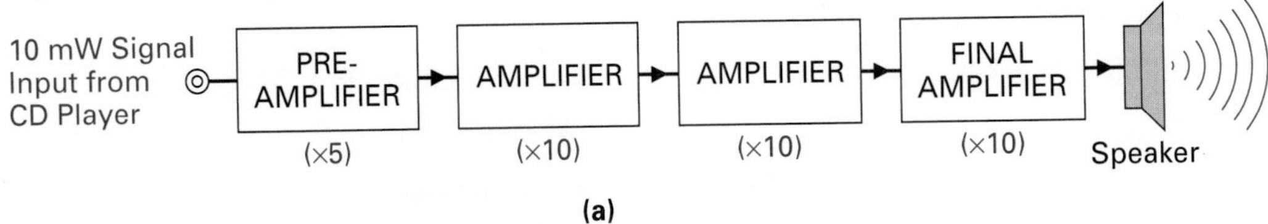

(a)

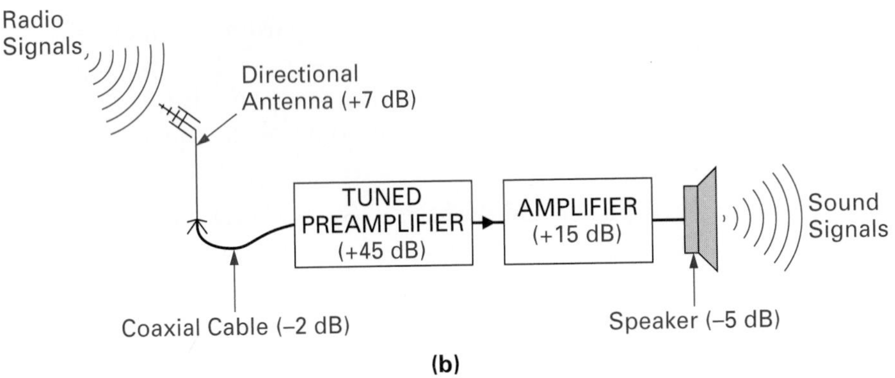

(b)

FIGURE 10-5 Calculating Multiple Stage Gain.

Similarly, a power ratio of 1:10 is equivalent to –10 dB, which means if power is reduced to one-tenth, the ratio is decreased by –10 dB. Using a 1:10 power ratio and –10 dB as a base, we can say that a 20 dB loss (–20 dB) is equivalent to one-hundredth (1/100) the power, a 30 dB loss (–30 dB) is equal to 1/1000th the power, and so on.

Multiple-stage gain. Up until now we have been dealing with only a single device. How would we calculate the total gain or loss of a system that contains several stages? If the gain of several stages was given as an arithmetic power ratio, as seen in Figure 10-5(a), the total gain is equal to the product of all of the stages. Therefore, the total gain in this system will be

$$A_{\text{Total}} = A_1 \times A_2 \times A_3 \times \cdots$$

■ **EXAMPLE:**

Calculate the total gain of the amplifier in Figure 10-5(a) and its final output power.

■ ***Solution:***

$$A_{\text{Total}} = A_1 \times A_2 \times A_3 \times A_4 = 5 \times 10 \times 10 \times 10 = 5000$$

Once the multi-stage gain of the amplifier is known, the output from this four-stage amplifier can be calculated because we know that the output power is 5,000 times larger than the input power. An input of 10 mW, therefore, would produce an output of

$$P_{\text{OUT}} = A_P \times P_{\text{IN}} = 5000 \times 10 \text{ mW} = 50 \text{ W}$$

The next question is: How do we deal with a multiple-stage gain or loss when the ratio of change for each stage is given in decibels? To answer this question, let us use an example.

■ **EXAMPLE:**

Calculate the gain of the radio receiver shown in Figure 10-5(b).

■ ***Solution:***

Because decibels are logarithmic, the total gain or loss of the circuit shown in Figure 10-5(b) can be calculated by simply adding all of the individual dB values. The directional antenna captures a signal of +7 dB, there is a –2 dB loss in the coaxial cable between the antenna and the tuned preamplifier, the preamplifier has a gain of +45 dB, the amplifier a gain of 15 dB, and the speaker has a 5 dB loss. The radio receiver gain will be;

$$A_{\text{Total}} = (+7\text{ dB}) + (-2\text{ dB}) + (+45\text{ dB}) + (+15\text{ dB}) + (-5\text{ dB}) = +60\text{ dB}$$

When the gain of a multi-stage circuit or system is measured in dBs the total gain or loss of the amplifier can be calculated by simply adding all of the individual dB values.

$$A_{\text{Total}} = A_{1(dB)} + A_{2(dB)} + A_{3(dB)} + \cdots$$

Scaling power in dBm. In the preceding examples, both the input power and the output power were given. In some applications, the power at a circuit test point is given with respect to a reference level of power. For example, cable and telecommunication companies have established a standard of 1 mW as a reference power. With this **decibels relative to 1 milliwatt** or **dBm** system, power levels below 1 mW are indicated as –dBm values, and power levels above 1 mW are shown as +dBm values. To calculate dBm, the formula will be

dBm
Decibels relative to 1 milliwatt.

$$A_{P(dBm)} = \log \frac{P_{\text{OUT}}}{1\text{ mW}}$$

■ **EXAMPLE:**

An amplifier has an output of 1.5 watts; what is this value in dBm?

■ ***Solution:***

$$A_{P(dBm)} = \log \frac{P_{\text{OUT}}}{1\text{ mW}} = \log \frac{1.5\text{ W}}{1\text{ mW}} = \log 1500 = 3.18\text{ dBm}$$

Decibels to watts. Transposing the original dB power formula, we can derive formulas for calculating power out and power in.

$$A_{P(dB)} = 10 \times \log \frac{P_{\text{OUT}}}{P_{\text{IN}}} \text{ (original dB formula)}$$

$$P_{OUT} = P_{IN} \times \text{antilog}\,\frac{dB}{10}$$

$$P_{IN} = \frac{P_{OUT}}{\text{antilog}\,\frac{dB}{10}}$$

■ **EXAMPLE:**

Calculate the power out of a 35 dB amplifier for an input of 4 mW.

■ ***Solution:***

$$P_{OUT} = P_{IN} \times \text{antilog}\,\frac{dB}{10} = 4\text{ mW} \times \text{antilog}\,\frac{35}{10} = 12.65\text{ W}$$

■ **EXAMPLE:**

Calculate the input power needed for a 30 dB amplifier to produce a 2 watt output.

■ ***Solution:***

$$P_{IN} = \frac{P_{OUT}}{\text{antilog}\,\frac{dB}{10}} = \frac{2\text{ W}}{\text{antilog}\,\frac{30}{10}} = \frac{2\text{ W}}{\text{antilog } 3} = 2\text{ mW}$$

Voltage Gain or Loss in Decibels

A voltage gain or loss can also be expressed in decibels by substituting V^2/R for power in the decibel power formula.

$$A_{P(dB)} = 10 \times \log\,\frac{P_{OUT}}{P_{IN}}$$

Since $$P_{OUT} = \frac{V_{OUT}^2}{R_{OUT}} \quad \text{and} \quad P_{IN} = \frac{V_{IN}^2}{R_{IN}}$$

We can write the formula for voltage gain in decibels.

$$A_{V(dB)} = 10 \times \log\left(\frac{\frac{V_{OUT}^2}{R_{OUT}}}{\frac{V_{IN}^2}{R_{IN}}}\right)$$

$$= 10 \times \log\left(\frac{V^2_{OUT}}{V^2_{IN}} \times \frac{R_{IN}}{R_{OUT}}\right)$$

$$= 10 \times \log \frac{V^2_{\text{OUT}}}{V^2_{\text{IN}}} + 10 \times \log \frac{R_{\text{IN}}}{R_{\text{OUT}}}$$

If $R_{\text{OUT}} = R_{\text{IN}}$,

$$= 10 \times \log \left(\frac{V_{\text{OUT}}}{V_{\text{IN}}}\right)^2 + (10 \times \log 1)$$

$$= 20 \times \log \frac{V_{\text{OUT}}}{V_{\text{IN}}} + 0 \qquad \text{therefore} \qquad A_{V(dB)} = 20 \times \log \frac{V_{\text{OUT}}}{V_{\text{IN}}}$$

$$A_{V(dB)} = 20 \times \log \frac{V_{\text{OUT}}}{V_{\text{IN}}}$$

■ **EXAMPLE:**

When a dc voltage signal of 60 mV is applied to an amplifier, the output is 1.45V. If $R_{\text{OUT}} = R_{\text{IN}}$, what is the voltage gain in decibels?

■ ***Solution:***

$$A_{V(dB)} = 20 \times \log \frac{V_{\text{OUT}}}{V_{\text{IN}}} = 20 \times \log \frac{1.45\text{ V}}{60\text{ mV}} = 20 \times \log 24.167 = 20 \times 1.383 = 27.66\text{ dB}$$

Remember that a voltage gain of 27.66 dBs does not mean that the output voltage of the amplifier is 27.66 times greater than the input. From the values given, we can see that 1.45 V is 24.167 times greater than 60 mV. To convert the dB voltage gain value ($A_{V(dB)}$) to a normal voltage gain value (A_V), we simply reorganize the original voltage formula as follows.

$$A_{V(dB)} = 20 \times \log \frac{V_{\text{OUT}}}{V_{\text{IN}}} = 20 \times \log A_V$$

To solve for A_V, divide both sides by 20

$$\frac{A_{V(dB)}}{20} = \frac{20 \times \log A_V}{20}$$

and take the antilog of both sides.

$$\text{antilog} \frac{A_{V(dB)}}{20} = A_V$$

$$A_V = \text{antilog} \frac{A_{V(dB)}}{20}$$

For the previous example,

$$A_V = \text{antilog} \frac{A_{V(dB)}}{20} = \text{antilog} \frac{27.66\text{ dB}}{20} = \text{antilog } 1.383 = 24.15$$

Current Gain or Loss in Decibels

A current gain or loss can also be expressed in decibels by substituting $I^2 \times R$ for power in the decibel power formula to obtain

$$A_{I(dB)} = 20 \times \log \frac{I_{OUT}}{I_{IN}}$$

■ **EXAMPLE:**

A current amplifier produces a 400 mA signal output when an input signal current of 25 mA is applied. If $R_{OUT} = R_{IN}$, what is the current gain in decibels?

■ ***Solution:***

$$A_{I(dB)} = 20 \times \log \frac{I_{OUT}}{I_{IN}} = 20 \times \log \frac{400 \text{ mA}}{25 \text{ mA}} = 20 \times \log 16 = 20 \times 1.204 = 24.08 \text{ dB}$$

Once again, remember that a current gain of 24 dBs does not mean that the output current of the amplifier is 24 times greater than the input. From the values given, we can see that 400 mA is 16 times greater than 25 mA. To convert the dB current gain value ($A_{I(dB)}$) to a normal current gain value (A_I), we simply reorganize the original current formula to obtain

$$A_I = \text{antilog} \frac{A_{I(dB)}}{20}$$

For the previous example,

$$A_I = \text{antilog} \frac{A_{I(dB)}}{20} = \text{antilog} \frac{24.08 \text{ dB}}{20} = \text{antilog } 1.204 = 16$$

Common Power and Voltage/Current Ratios

In summary, the decibel is a measure of change, and its value gives us an indication of the performance of a device, circuit, or system. Table 10-2 makes a comparison between the common decibel values for power and voltage/current ratios. Looking at this table, you can see that the corresponding voltage and current decibel values are twice that of the same power ratio. For example, if power is doubled (× 2), the decibel rating is 3 dB. If voltage or current is doubled, the decibel rating is 6 dB. Similarly, 100 times the power is 20 dB, whereas 100 times the voltage or current is 40 dB. This difference occurs because power is proportional to voltage squared ($P = V^2/R$) and current squared ($P = I^2R$), and so decibel formulas for voltage and current have a multiplying factor of 20 rather than 10.

TABLE 10-2 Common Power and Voltage/Current Ratios

	DECIBELS	POWER RATIO	VOLTAGE/CURRENT RATIO
Gain:	40	× 10,000	× 100
	20	× 100	× 10
	6	× 4	× 2
	3	× 2	× 1.4
	1	× 1.2	× 1.1
Loss:	–1	÷ 1.2	÷ 1.1
	–3	÷ 2	÷ 1.4
	–6	÷ 4	÷ 2
	–20	÷ 100	÷ 10
	–40	÷ 10,000	÷ 100

10-1-4 Amplifier Class of Operation

Before we discuss the four basic amplifier classes of operation, or modes of operation, let us briefly review the basic dc biasing of a transistor amplifier by referring to Figure 10-6. This group of characteristic curves was discussed previously in Chapter 8. The key point to remember with regard to this section is that the transistor amplifier is generally biased at a dc operating point, which is called

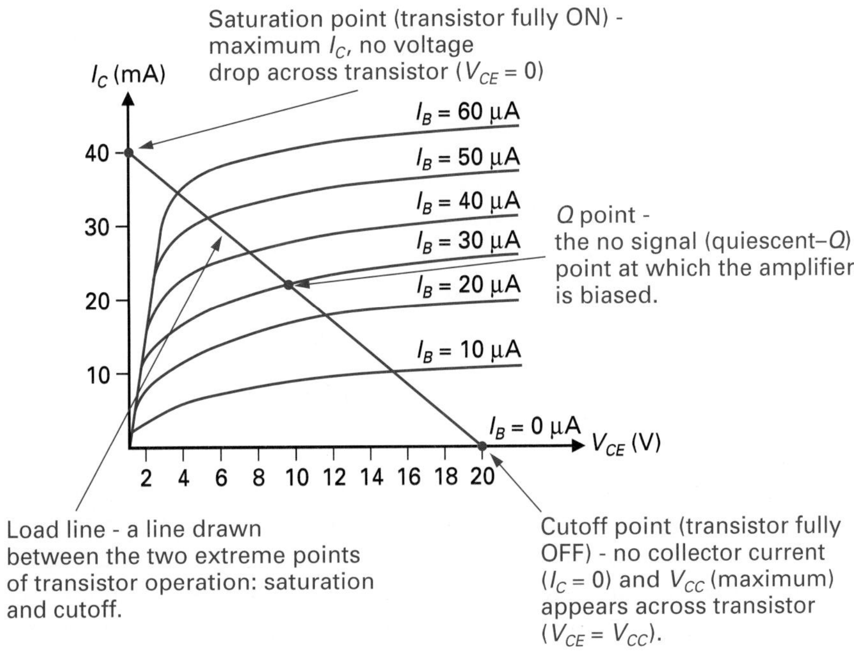

FIGURE 10-6 The Bipolar Transistor's Load Line and *Q* Point in Its Family of Characteristic Curves.

the quiescent or Q point. The input signal will vary the transistor's dc bias voltage above and below its Q point, causing a change in base current (I_B), collector current (I_C), and output voltage (V_{CE}), as shown in Figure 10-7(a). All of the bipolar transistor amplifier circuits discussed in Chapter 8 were operated in this way. However, as you will see in the different application amplifier circuits discussed in this chapter, by biasing an amplifier at different points on the dc load line, we can obtain some circuit characteristics that are ideal for certain applications.

Class-A Amplifier Operation

Class-A Amplifier A transistor amplifier that has its dc operating point set near the center of the load line, so that the output current (I_C) flows during the entire cycle of the ac input signal.

Any transistor amplifier that has its dc operating point set near the center of the load line, so that the output current (I_C) flows during the entire cycle of the ac input signal, is said to be operating as a **class-A amplifier.** All of the amplifier circuits discussed in Chapter 8 were of the class A type because they were biased at a mid-point on the dc load line so that any change in the input current would produce a proportional but amplified change in the output current, as shown in Figure 10-7(a). Class A amplifiers are said to operate in a linear manner because a change in the input signal produces a proportional change in the output signal.

Class-B Amplifier Operation

Class-B Amplifier An amplifier biased so that its output current will flow for only half of the complete ac input signal cycle.

When an amplifier is biased so that its output current will flow for only half of the complete ac input signal cycle, the circuit is said to be operating as a **class-B amplifier,** as can be seen in Figure 10-7(b). To achieve this mode of operation, the amplifier is biased so that its Q point is at cutoff; therefore, the output current will only flow during the time that the input alternation forward biases the transistor. The advantage of this amplifier class of operation is that the circuit will only consume half the power of an equivalent class A amplifier because the amplifier is only turned ON for 50% of the time. Stated another way, a class B amplifier works for half of the input cycle and then rests for the other half of the input cycle, as opposed to a class A amplifier, which is working all of the time. Class B amplifiers can be pushed to deliver a lot more power than a class A amplifier because of their 50% work ratio. A simple analogy is that if a person who normally works a full eight-hour day was instructed to work for only half that time, he or she could be pushed to deliver a lot more in four hours than in the usual eight-hour day.

Class-C Amplifier Operation

Class-C Amplifier An amplifier biased so that its output current will flow for less than one half of the ac input cycle.

When an amplifier is biased so that its output current will flow for less than one half of the ac input cycle, it is said to be operating as a **class-C amplifier,** as seen in Figure 10-7(c). This is achieved by biasing the transistor amplifier operating point below the cutoff point so that for most of the time the amplifier is reverse biased or cutoff. For only a small period of time will it produce an output pulse of current. This type of amplifier is therefore very efficient because it only consumes power for a very short time, compared to the complete cycle of the ac input signal.

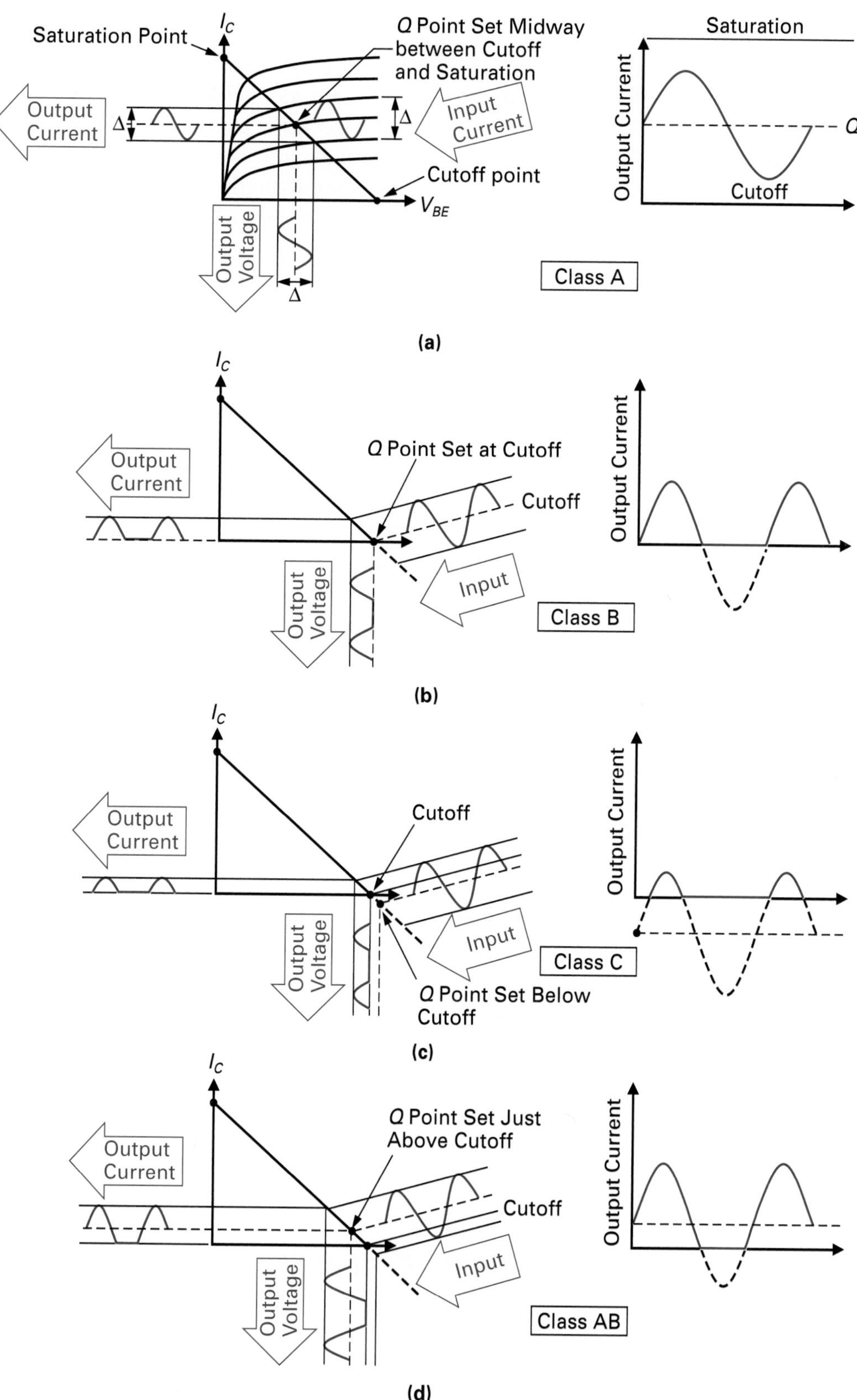

FIGURE 10-7 Transistor Classes of Amplification. (a) Class A Amplifier. (b) Class B Amplifier. (c) Class C Amplifier. (d) Class AB Amplifier.

Class-AB Amplifier Operation

Class-AB Amplifier An amplifier biased so that its output current will flow for slightly less than one full cycle of the ac input (class A), but more than one half of the ac input cycle (class B).

When an amplifier is biased so that its output current will flow for slightly less than one full cycle of the ac input (class A), but more than one half of the ac input cycle (class B), the amplifier is said to be operating as a **class-AB amplifier,** as shown in Figure 10-7(d). This type of amplifier is a compromise between a class A and a class B, and its operating point is set at slightly above cutoff causing an output current to be produced for slightly more than one half of the input cycle.

Amplifier Efficiency

The Greek letter *eta,* η, (pronounced "eat-eh"), is used to designate an amplifier's efficiency, which is the ratio of the power delivered to the load by the amplifier to the power consumed by the amplifier. As a formula, it would appear as follows:

$$\eta = \frac{\textit{Power Delivered to Load by Amplifier } (P_L)}{\textit{DC Power Consumed by Amplifier } (P_S)} \times 100\%$$

Class-A amplifiers are constantly drawing power from the $+V_{CC}$ dc power supply the whole time that they are supplying power to the load. As a result, class A amplifiers tend to be fairly inefficient. On the other hand, class-B amplifiers are OFF for 50% of the input cycle, and, because they consume no power during this time, they tend to be more efficient than their class-A counterparts. Class-C amplifiers are OFF, and therefore not consuming power, for more than 50% of the input cycle. A class-C amplifier is therefore more efficient than a class-B amplifier. The class-AB amplifier's power consumption is midway between the class A and the class B and so is its efficiency.

As we proceed through this chapter, you will see how the efficiency of these different classes of amplifiers can be calculated and how their efficiency is generally at best as follows:

AMPLIFIER CLASS	EFFICIENCY
A	35%
AB	50%
B	75%
C	99%

10-1-5 Amplifier Frequency Response

By changing the design of an amplifier circuit, you can change its characteristics. One of the key characteristics of an amplifier circuit is its response to different input signal frequencies. Some amplifier circuits are best suited to amplify high frequency signals, while other amplifier circuits respond better to low frequency signals. For example, audio frequency amplifiers provide a high gain to any audio frequency signals between 20 Hz and 20 kHz, whereas a radio frequency amplifier would provide almost no gain to any input signal in this range. A **frequency response curve** is used to show the response of an amplifier circuit to different signal input frequencies by plotting gain against frequency. As an example, Figure 10-8 shows the frequency response curve for an audio frequency power amplifier. The amplifier's **bandwidth** is the group or band of frequencies between the half-power points. These half-power points are the points at which the gain has fallen below 50% of the maximum power gain (–3dB). Referring to the example in Figure 10-8, you can see that the audio frequency power amplifier's gain increases above half-power at 20 Hz and falls below half-power at 20,000 Hz. The bandwidth for this amplifier will be 19,980 Hz (19.98 kHz), which is the difference between 20,000 Hz and 20 Hz.

Frequency Response Curve
A plot of gain versus frequency.

Bandwidth
The group or band of frequencies between the half-power points.

$$BW = f_{HI} - f_{LO}$$

BW = Bandwidth, f_{HI} = Upper frequency limit, f_{LO} = Lower frequency limit

$$BW = f_{HI} - f_{LO} = 20{,}000 \text{ Hz} - 20 \text{ Hz} = 19{,}980 \text{ Hz}$$

The upper frequency limit at which the gain of the amplifier falls below half power is also called the **cutoff frequency.**

Cutoff Frequency
The upper frequency limit at which the gain of the amplifier falls below half power.

Figure 10-8(b) shows how the bandwidth of a current amplifier is determined. In this diagram, you may have noticed that 70.7%, or 0.707 of the maximum gain, is used to determine the bandwidth. This is because 70.7% of the maximum current gain is equivalent to the half-power points. To prove this, let us use an example.

■ **EXAMPLE:**

An amplifier is delivering 100 mA to a 2 kΩ load. Therefore, the power delivered is

$$P = I^2 \times R = 100 \text{ mA} \times 2 \text{ k}\Omega = 20 \text{ W}$$

If the input signal frequency were to increase and cause the amplifier's current output to drop to 70.7 mA, what would be the power delivered to the same load?

■ ***Solution:***

$$P = I^2 \times R = 70.7 \text{ mA} \times 2 \text{ k}\Omega = 10 \text{ W}$$

As you can see from this example, the 70.7% current points are equal to the half-power points; an amplifier's bandwidth exists between the 70.7% current points, or half-power points.

The same is true for a voltage amplifier, as shown in Figure 10-8(c), because 70.7% of the maximum voltage gain is equivalent to the half-power points. Let us again prove this with an example.

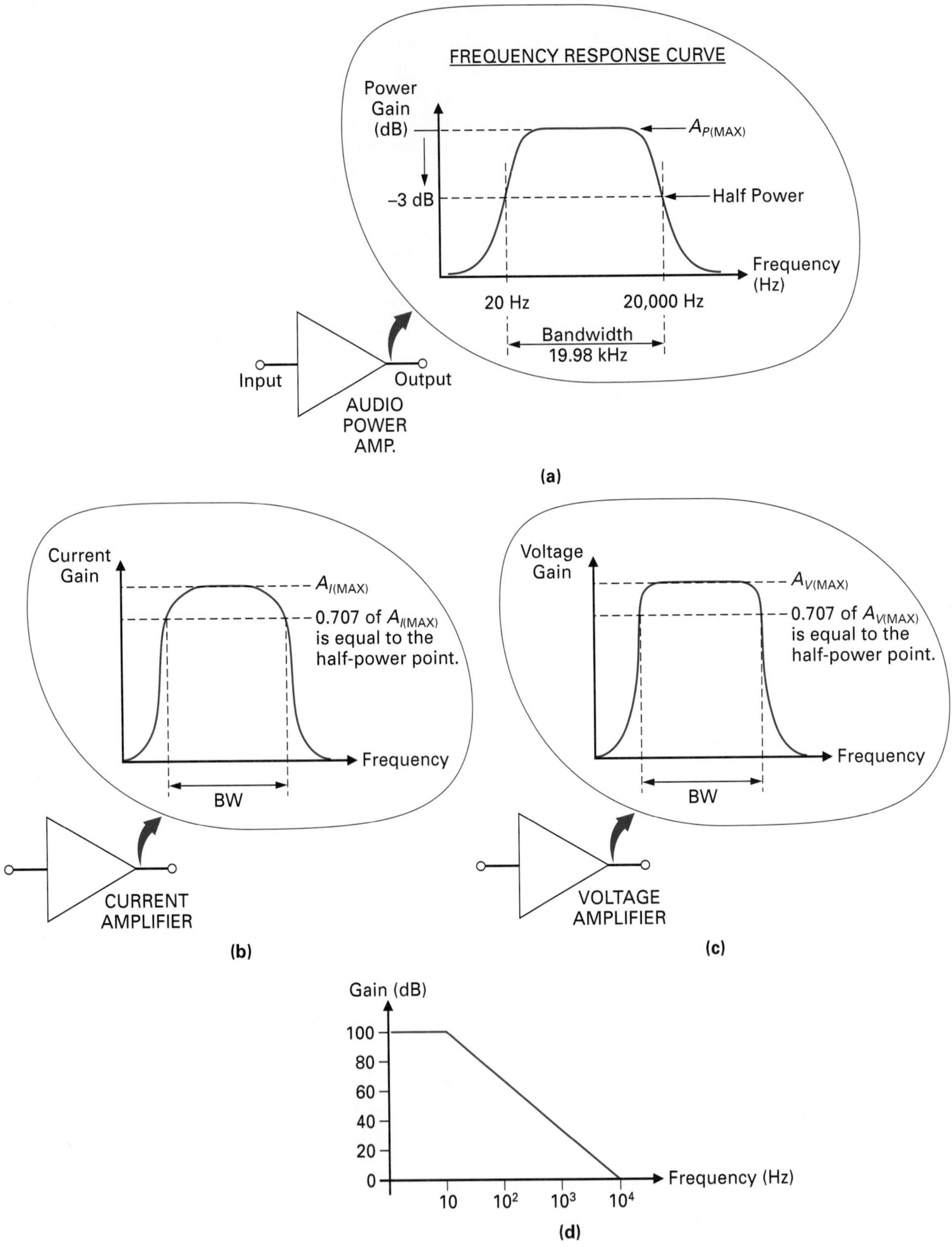

FIGURE 10-8 An Amplifier's Frequency Response and Bandwidth.

■ **EXAMPLE:**

An amplifier is delivering 10 V to a 1 kΩ load. Therefore, the power delivered is

$$P = \frac{V^2}{R} = \frac{10\ V^2}{1\ \text{k}\Omega} = 100\ \text{mW}$$

If the input signal frequency were to increase and cause the amplifier's voltage output to drop to 7.07V, what would be the power delivered to the same load?

■ ***Solution:***

$$P = \frac{V^2}{R} = \frac{7.07\ V^2}{1\ \text{k}\Omega} = 50\ \text{mW}$$

Once again, as you can see from this example, the 70.7% voltage points are equal to the half-power points. An amplifier's bandwidth therefore exists between the 70.7% voltage or current points, or half-power points.

A **bode diagram** is another form of frequency response curve. As you can see from the example in Figure 10-8(d), this diagram is an approximation using straight lines and plots a circuit's gain in decibels against frequency.

Bode Diagram
A frequency response curve that plots gain in decibels versus frequency.

SELF-TEST REVIEW QUESTIONS FOR SECTION 10-1

1. How is the gain of an amplifier calculated?
2. If a voltage amplifier has a gain of 75, and an input of 0.5 mV, what will be the output voltage?
3. The common-base configuration provides a good ____________ gain, the common-emitter provides a good ____________ gain, and the common-collector provides a good ____________ gain.
4. A common logarithm is a base 10 ____________ .
5. A power amplifier has an input of 10 mW and an output of 15W. What is the amplifier's gain in decibels?
6. A power gain of 40 dB means that the output of the amplifier is ____________ times greater than the input.
7. Which amplifier class of operation has its Q point set near the center of the load line so that output current is present for the entire cycle of the ac input signal?
8. A ____________ is used to show the response of an amplifier circuit to different input signal frequencies.

10-2 DIRECT CURRENT (DC) AMPLIFIERS

Now that the principles of amplifier circuits have been discussed, let us discuss some specific amplifier types and their applications. In this section we will examine the operation and characteristics of **direct current or dc amplifiers,**

Direct Current (DC) Amplifiers
Amplifiers for dc and low frequency ac signals.

which are normally always class A biased (Q point is midway between saturation and cutoff); therefore, the transistor is always conducting. Direct current amplifiers are used to amplify dc voltages or currents, or low-frequency ac voltages or currents, produced by transducers or sensors. These transducers will sense heat, light, pressure, or vibration, and produce a corresponding electrical signal that is usually extremely weak in amplitude and must be amplified up to a more usable level. To begin with, we will study a basic single-stage dc amplifier.

10-2-1 Single-Stage DC Amplifier Circuits

Figure 10-9(a) shows how a direct current (dc) amplifier can be used to amplify a +0.2 V to +5 V dc signal from a light dependent resistor (LDR), which is being used to sense the ambient light level. Resistors R_1 and R_2 provide voltage-divider bias for the common-emitter transistor amplifier. Collector resistor R_3 operates in conjunction with the transistor to develop an output signal voltage. Resistor R_E provides "emitter feedback" for temperature stability. To review, resistor R_E achieves circuit thermal stability through emitter feedback. With R_E included in circuit, V_{CC} is developed across R_C, Q_1's collector-emitter, and R_E. An increase in temperature will cause an increase in I_C, I_E, and therefore the voltage developed across R_E. An increase in V_{RE} will decrease the base-emitter bias voltage applied, causing Q_1's internal currents to decrease back to their original values. A change in the output current (I_C) due to temperature will cause the emitter resistor to feedback a control voltage to the input. Because the input signal is a dc voltage, capacitors cannot be connected in the signal path (a capacitor will act as a block to any dc signals). The lack of an input coupling capacitor means that the base bias of Q_1 is not only set by the resistance values of R_1 and R_2 but also by the internal resistance of the LDR source in parallel with R_1. Similarly, the lack of an output coupling capacitor means that the resistance of the load is connected in parallel with the resistance of transistor Q_1 and R_4. When designing a dc amplifier, the resistance of the input source and resistance of the load will have to be taken into account because these values will affect the bias network. The bias network will govern base voltage, base current, and, in turn, control collector current and output voltage. As you know, the load resistance shown in Figure 10-9(a) merely represents the resistance of the device connected to the output of this amplifier. In most instances, this load resistance will represent the input resistance of the next amplifier stage.

The common-emitter dc amplifier shown in Figure 10-9(a) will provide both voltage and current amplification; however, its main purpose is as a voltage amplifier. As you can see in Figure 10-9(b), which shows the typical frequency response of a dc amplifier, the voltage gain of a dc amplifier remains almost constant from 0 Hz (dc) to about 2000 Hz (ac). Beyond 2000 Hz or 2 kHz, the voltage gain of a dc amplifier drops off rapidly. The frequency at which a transistor's gain falls below 70.7% is called the transistor's "cutoff frequency." Consider this frequency cutoff action in more detail: as the ac input signal frequency increases, its cycle time or period decreases, and eventually

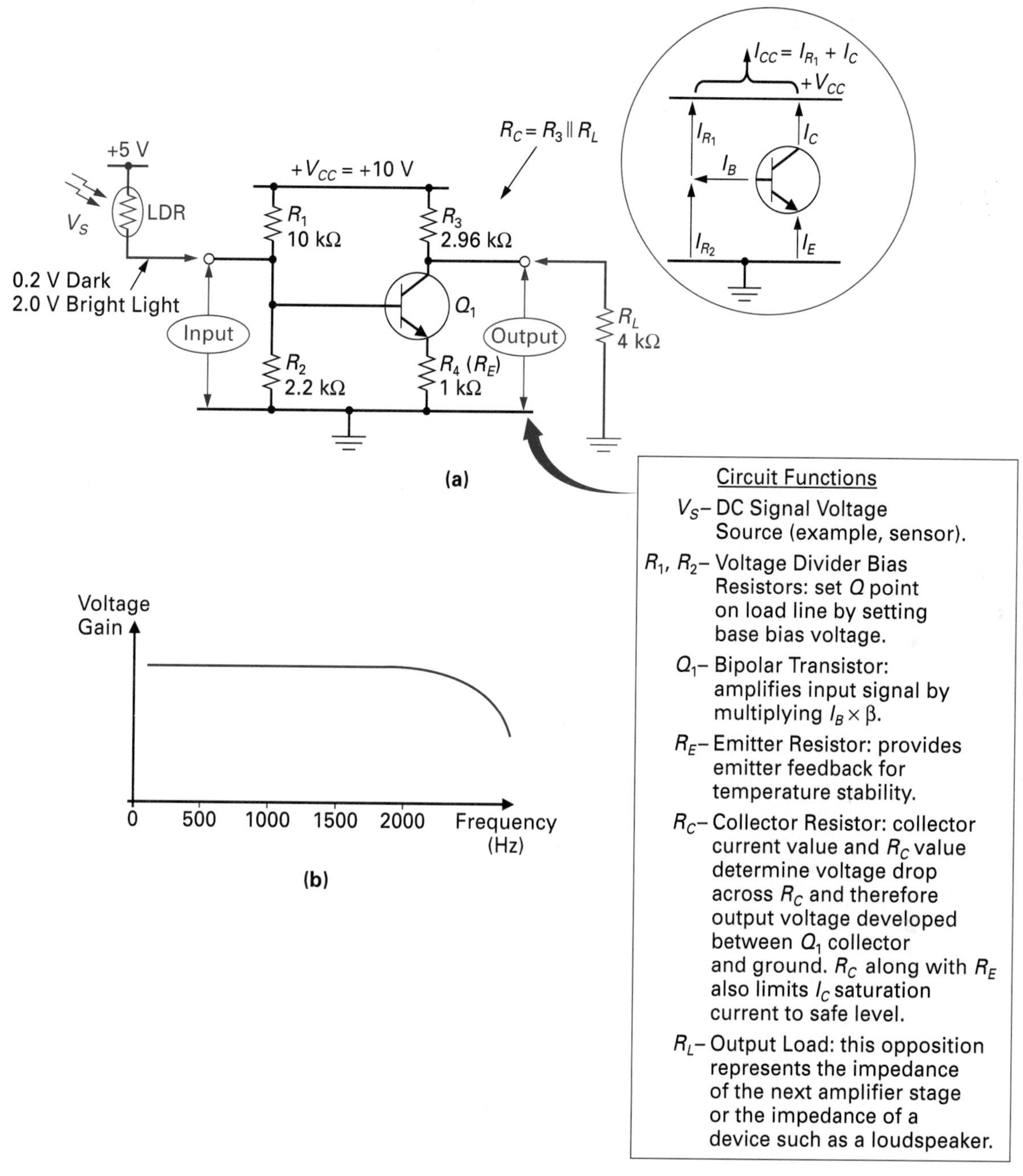

FIGURE 10-9 Single-Stage Direct Current (DC) Amplifier. (a) Basic Circuit. (b) Frequency Response.

this cycle time is faster than the transit time needed for a charge carrier to pass through the transistor. At this time, the gain of the transistor will fall to almost zero because the transistor's internal currents do not have enough time to respond to the high-frequency input signal changes. Different types of bipolar transistors will have different cutoff frequency ratings, which will be specified in the manufacturer's data sheet.

From a dc input signal of 0 Hz up to an ac input signal frequency of about 2 kHz, the voltage gain of a dc amplifier is approximately equal to the ratio of the transistor's collector resistance R_C to the emitter resistor R_E.

$$A_V = \frac{R_C}{R_E}$$

No Load Connected

With no load connected, the voltage gain of the dc amplifier in Figure 10-9(a) will be equal to R_3 divided by R_4.

However, if a load is connected to the output, the collector current out of the transistor will split and pass through R_3 and R_L; therefore, these two resistors are effectively in parallel. The voltage gain of an amplifier in this instance is equal to the equivalent parallel resistance of R_3 and R_L divided by the emitter resistance (R_4 or R_E).

$$A_V = \frac{R_3 \| R_L}{R_E}$$

Load Connected

To calculate the transistor's equivalent parallel collector resistance, we will use the standard parallel resistance formula: product over sum.

$$R_C = R_3 \| R_L = \frac{R_3 \times R_L}{R_3 + R_L}$$

Let us now practice using these formulas with an example.

■ **EXAMPLE:**

Calculate the voltage gain of the circuit in Figure 10-9(a). For the dc input signal voltage range given, what would be the output dc signal voltage range?

■ ***Solution:***

$$R_C = \frac{R_3 \times R_L}{R_3 + R_L} = \frac{2.96\ \text{k}\Omega \times 4\ \text{k}\Omega}{2.96\ \text{k}\Omega + 4\ \text{k}\Omega} = \frac{11.84\ \text{k}\Omega}{6.96\ \text{k}\Omega} = 1.7\ \text{k}\Omega$$

$$A_V = \frac{R_C}{R_E} = \frac{1.7\ \text{k}\Omega}{1\ \text{k}\Omega} = 1.7$$

The output signal voltage will be 1.7 times greater than the input signal voltage. Therefore, because the dc input signal voltage can be anywhere from 0.2 V to 2 V, the output signal voltage will be anywhere from

$$V_{OUT} = V_{IN} \times A_V = 0.2\text{ V} \times 1.7\text{ V} = 0.34\text{ V}$$

to

$$V_{OUT} = V_{IN} \times A_V = 2\text{ V} \times 1.7\text{ V} = 3.4\text{ V}$$

As mentioned previously, an amplifier's efficiency is equal to the ratio of the power delivered to the load by the amplifier to the power consumed by the amplifier.

$$\eta = \frac{\textit{Power Delivered to Load by Amplifier } (P_L)}{\textit{DC Power Consumed by Amplifier } (P_S)} \times 100\%$$

To calculate the efficiency of the dc amplifier in Figure 10-9, we will have to know how much power is being delivered to the load and how much dc power is being consumed by the amplifier. To calculate the amount of power delivered to the load (P_L), we can use the standard V^2/R power formula.

$$P_L = \frac{V_L^2}{R_L}$$

P_L = Power delivered to the load
V_L = Voltage developed across load
R_L = Load resistance

To calculate the amount of power consumed by the amplifier circuit, or the power supplied to the amplifier by the dc power supply (P_S), we will use the standard *VI* power formula.

$$P_S = V_{CC} \times I_{CC}$$

P_S = power supplied to the amplifier by the dc power supply
V_{CC} = Circuit's dc supply voltage
I_{CC} = Total current drawn by the amplifier circuit

To see how the efficiency of an amplifier is calculated, let us apply these formulas to the example circuit in Figure 10-9(a).

■ **EXAMPLE:**

Calculate the efficiency of the class A amplifier in Figure 10-9(a).

■ ***Solution:***

Because we know from the previous example that the output voltage is between 0.34 V to 3.4 V, the average output voltage will be

$$V_{AVG} = \frac{V_1 + V_2}{2} = \frac{0.34\text{ V} + 3.4\text{ V}}{2} = 1.87\text{ V}$$

This voltage will be applied across a 4 kΩ load, so the power delivered to the load will be

$$P_L = \frac{V^2{}_L}{R_L} = \frac{1.87\ \text{V}^2}{4\ \text{k}\Omega} = 874.2\ \mu\text{W}$$

To calculate the amount of power consumed by the amplifier, we will need to know the total amount of current being drawn by the amplifier circuit (I_{CC}). This total value of current is equal to the value of current flowing through the voltage divider resistor R_1 (I_{R_1}) and the value of collector current, (I_C), as shown in the inset in Figure 10-9(a). To calculate these values, we will have to do the following calculations.

$$V_{R_2} = \frac{R_2}{R_1 + R_2} \times V_{CC} = 1.8\ \text{V}$$

$$I_{R_2} = \frac{V_{R_2}}{R_2} = \frac{1.8\ \text{V}}{2.2\ \text{k}\Omega} = 818\ \mu\text{A}$$

$I_{R_2} \cong I_{R_1}$ (I_B can be ignored), therefore $I_{R_1} = 818\ \mu\text{A}$

$$V_E = V_B - 0.7\ \text{V} = 1.8\ \text{V} - 0.7\ \text{V} = 1.1\ \text{V}$$

$$I_E = \frac{V_E}{R_E} = \frac{1.1\ \text{V}}{1\ \text{k}\Omega} = 1.1\ \text{mA}$$

$\text{I}_\text{E} \cong I_C$, therefore $\text{I}_\text{C} = 1.1\text{mA}$

Now that I_{R_1} and I_C are known, we can calculate I_{CC}.

$$I_{CC} = I_{R_1} + I_C = 818\ \mu\text{A} + 1.1\ \text{mA} = 1.92\ \text{mA}$$

Now that I_{CC} is known, we can calculate the power supplied to the amplifier by the dc power supply.

$$P_S = V_{CC} \times I_{CC} = 10\ \text{V} \times 1.92\ \text{mA} = 19.2\ \text{mW}$$

If the power delivered to the load is equal to 874.2 μW, and the power supplied to the amplifier by the dc power supply is equal to 19.2 mW, the class A amplifier's efficiency in Figure 10-9(a) will be

$$\eta = \frac{P_L}{P_S} \times 100\% = \frac{874.2\ \mu\text{W}}{19.2\ \text{mW}} \times 100\% = 4.5\%$$

With the common-emitter circuit configuration used in Figure 10-9(a), there will be a 180° phase difference between the input signal voltage and the output signal voltage. For example, when the input signal voltage goes more positive, the output voltage will go less positive, and when the input signal voltage goes less positive, the output voltage will go more positive. This inverting action of the amplifier is not a problem because the output voltage signal still changes in accordance with the input signal change. For example, if the light signal increased in Figure 10-9(a), the positive voltage change at the input of Q_1 would also increase, causing the voltage signal at the collector of Q_1 to decrease proportionally, so that the drop in voltage accurately represents the signal increase.

10-2-2 *Multiple-Stage DC Amplifier Circuits*

In some applications, a single dc amplifier stage may not provide enough gain. In these instances, several dc amplifier stages may be needed to increase the input signal up to a desired amplitude.

Direct-Coupled DC Amplifier Circuits

Figure 10-10(a) shows how we can couple the output of one dc amplifier stage into the input of another dc amplifier stage, so that we can achieve a higher overall gain. This two-stage dc amplifier circuit uses **direct coupling,** which means that the output of the first stage is coupled directly into the input of the second stage. The first stage of this multiple-stage dc amplifier is identical to the single-stage dc amplifier discussed previously in Figure 10-9(a) in that it contains four resistors (R_1 to R_4) and a transistor (Q_1). The second-stage dc amplifier circuit is made up of transistor Q_2 and resistors R_5 and R_6. The base bias for Q_2 is provided by the dc collector voltage of Q_1. The input signal is applied to the base of Q_1 and, in controlling Q_1, this signal will also control Q_2. Because the input signal may be a dc voltage or current, dc blocking components such as capacitors and transformers cannot be connected in the signal path.

Direct Coupling
The output of the first stage is coupled directly into the input of the second stage.

The dc input signal, or ac input signal, is amplified by the first stage and then further amplified by the second stage. When two or more dc amplifiers are connected end-to-end, or **cascaded,** in this way the overall dc amplifier voltage gain is equal to the product of all the individual stage gains.

Cascaded
Two or more dc amplifiers connected end-to-end.

$$A_{\text{Total}} = A_1 \times A_2 \times A_3 \times \cdots$$

The disadvantage of this direct-coupled dc amplifier is its lack of isolation between stages. As the number of direct-coupled stages increases, the collector bias voltage applied to the next stage become progressively larger. Because it is this collector voltage that controls the dc bias of the next stage, only a few direct-coupled stages can be cascaded before the collector voltage becomes so large that it will drive the next amplifier stage's output beyond the $+V_{CC}$ supply voltage. Referring to the inset in Figure 10-10(a), you can see that the Q point (dc base bias point) on the load line is continually increased by each stage until the signal input drives the transistor into saturation, and the upper part of the input signal is clipped. The output signal of an amplifier should in most cases be a direct copy of the input signal, only larger in amplitude. Any differences or irregularities introduced unintentionally into the signal are unwanted and called **signal distortion.** To prevent this distortion from occuring with direct coupled dc amplifiers, it is generally necessary to decrease the gain of the first stage, so that we can control the gain of the second stage. This compromise of sacrificing gain for circuit control is typical with most amplifier circuits.

Signal Distortion
Any unwanted differences or irregularities introduced unintentionally into the signal.

■ **EXAMPLE:**

Both of the stages of the dc amplifier in Figure 10-10(a) have a voltage gain of 10. If the gain of the first stage is decreased to 80%, what is the overall gain of this two-stage amplifier?

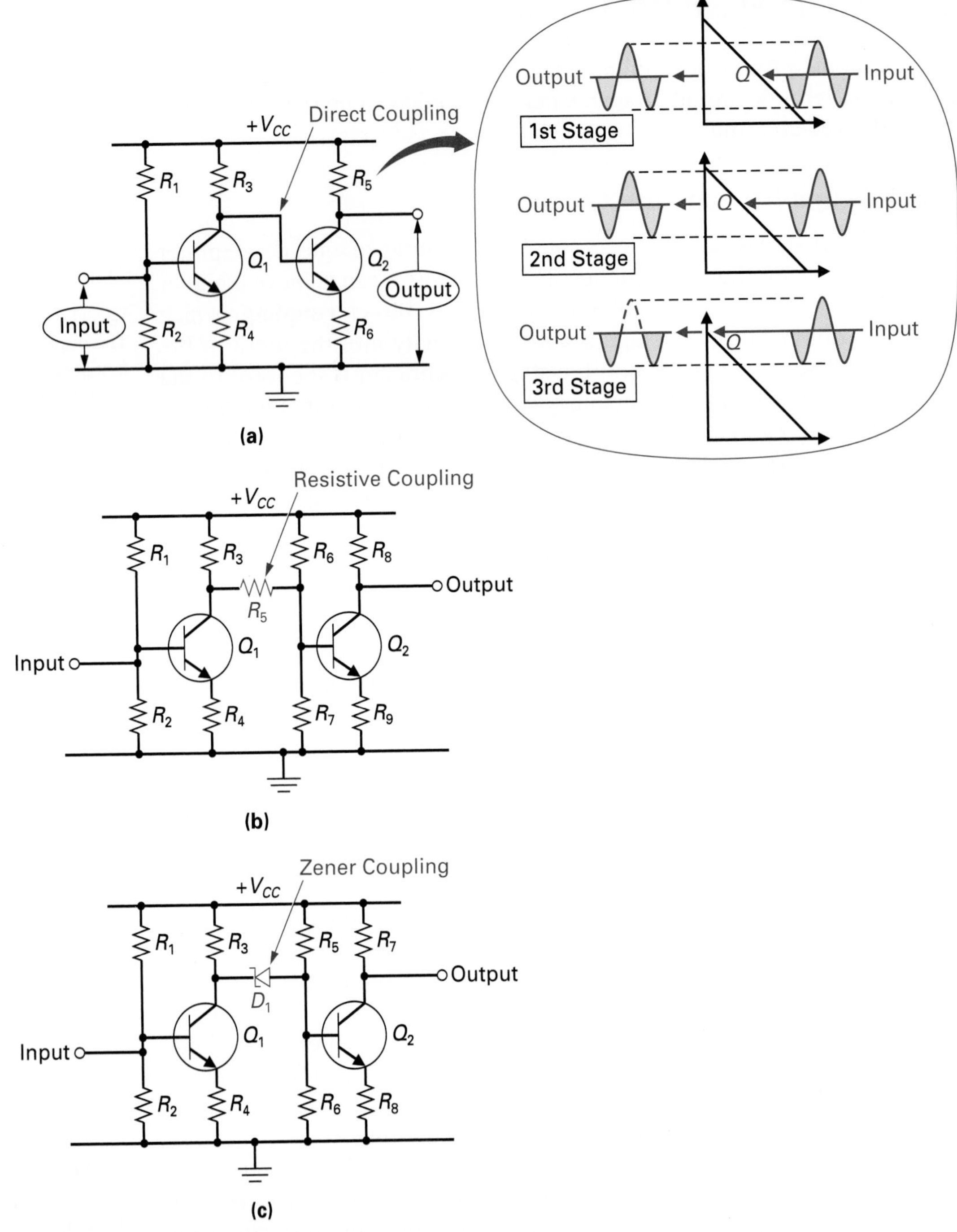

FIGURE 10-10 Multiple Stage DC Amplifiers. (a) Direct Coupling. (b) Resistive Coupling. (c) Zener Coupling.

■ ***Solution:***

$$80\% \text{ of } 10 = 8$$

$$A_{V(\text{Total})} = A_{V_1} \times A_{V_2} = 8 \times 10 = 80$$

The output signal voltage of the amplifier in Figure 10-10(a) will be 80 times greater than the input signal voltage.

Resistive-Coupled DC Amplifier Circuits

Figure 10-10(b) shows how we can overcome the lack of isolation (and therefore lack of control) of a direct-coupled dc amplifier by connecting a large-value resistor (R_5) between two complete dc amplifier stages. This **resistive coupling** technique means that the bias voltages and currents of each stage are somewhat independent of one another, so we can control the gain of each stage to produce a higher gain output than a direct-coupled amplifier. The disadvantage of this circuit is that, because the resistor is connected in the signal path, there is a voltage drop and therefore signal voltage loss across the coupling resistor.

Resistive Coupling
A connection made by a large-value resistor.

Zener-Coupled DC Amplifier Circuits

Figure 10-10(c) shows how we can achieve circuit control (through stage isolation) and reduce coupling loss by **zener coupling** two dc amplifier stages. The zener diode (D_1) is reverse biased and operating in its zener region because the voltage at the collector of Q_1 is generally several volts more positive than the voltage at the base of Q_2.

Zener Coupling
A connection made by a zener diode.

An applied input signal will be amplified by the first stage as before, and then be applied to the zener. This changing signal voltage at the collector of Q_1 will cause the zener diode's internal resistance to continually change, resulting in a change in reverse current through the zener but an almost constant voltage drop across the zener. Because the signal voltage at the collector of Q_1 will be developed across the zener diode and R_6, the voltage across the zener diode will remain almost constant, and the changing signal voltage will be developed across R_6. The changing signal voltage developed across R_6 will be applied to the base of Q_2, controlling the second stage dc amplifier, and therefore causing additional gain. Zener coupling two dc amplifier stages in this way causes the overall voltage gain to be almost equal to the calculated voltage gain of stage-1 gain times stage-2 gain.

Complementary DC Amplifier Circuits

Figure 10-11(a) shows the basic **complementary DC amplifier,** which makes use of an NPN transistor and a PNP transistor. This circuit operates in almost the same way as a direct coupled DC amplifier, except for the use of complementary or opposite bipolar transistor types. An applied input signal will be amplified by the first stage as before. To properly bias the second PNP stage, Q_2's emitter and collector leads are connected in reverse so that its emitter is connected to $+V_{CC}$ via R_5, and its collector lead is connected to ground via R_6. As we know, the base of a PNP transistor must be more positive than its collector, while the opposite is true for the NPN transistor, whose base must be more negative than its collector. Because the input signal is applied to the base of Q_2

Complementary DC Amplifier
An amplifier in which NPN and PNP transistors are used in an alternating sequence.

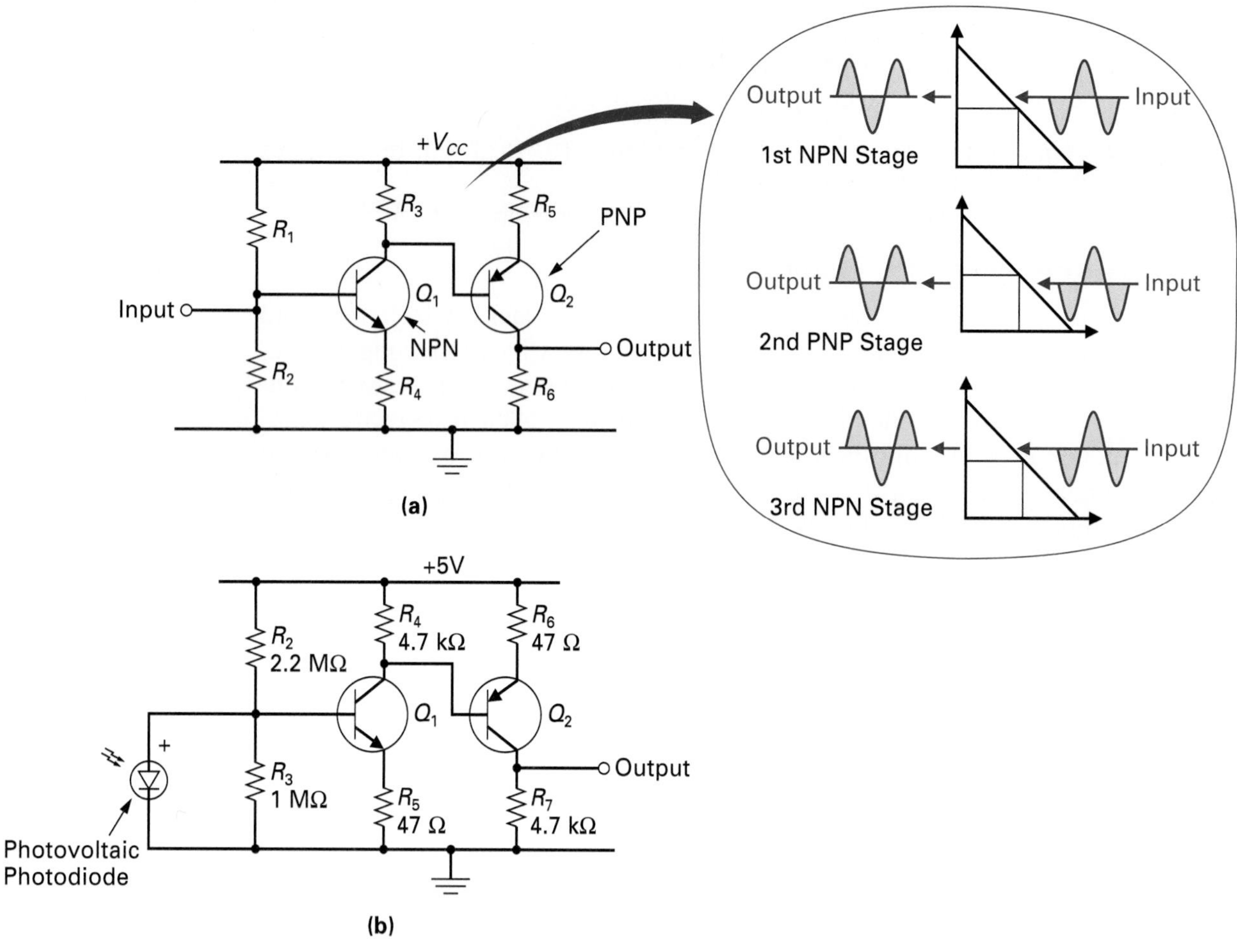

FIGURE 10-11 Complementary DC Amplifier Circuits. (a) Basic Circuit. (b) Application: Light-Sensing Amplifier.

and the output signal is taken from the collector, this second stage dc amplifier is also a common-emitter amplifier. It is the voltage developed across R_3 (between Q_1's collector and + V_{CC}) that is applied to Q_2's base-emitter and controls Q_2. It is the voltage developed across R_6 (between Q_2's collector and ground) that is applied to the output and is used to control the following stage.

As mentioned previously, with a standard direct-coupled dc amplifier, as the number of direct coupled stages increases, the collector voltage of each stage becomes progressively larger. As a result, only a few stages can be cascaded before the collector bias voltage for the next stage becomes so large that the input signal drives the transistor into saturation, and signal distortion occurs. A solution to this problem is to cascade complementary transistors (have an NPN followed by a PNP, followed by an NPN, followed by a PNP, and so on). Because the collector of an NPN stage is connected to $+V_{CC}$ via a resistor, its output collector bias voltage will be more positive. On the other hand, since the collector of a PNP transistor is connected to ground via a resistor, its output collector bias voltage will be less positive. As can be seen in the inset in Figure 10-11(a), this complementary dc amplifier arrangement enables us to cascade several stages to achieve high signal gain without distortion because the Q point, which is set by the previous stage, continually alternates between being either more positive (slightly above the mid-point) or less positive (slightly below the mid-point).

Figure 10-11(b) shows a typical two-stage complementary dc amplifier circuit. This circuit is being used to amplify a small voltage input signal being generated by a photovoltaic cell. The voltage gain of each stage can be calculated with the previously discussed formula and is approximately equal to the ratio of the collector resistance to the emitter resistance.

$$A_V = \frac{R_C}{R_E}$$

■ **EXAMPLE:**

Calculate the approximate individual-stage gain and overall dc amplifier gain of the circuit in Figure 10-11(b).

■ ***Solution:***

The gain of the first NPN stage will be approximately equal to the ratio of R_4 to R_5 (ignoring the input resistance of Q_2).

$$A_V = \frac{R_C}{R_E} = \frac{R_4}{R_5} = \frac{4.7\ \text{k}\Omega}{47\ \Omega} = 100$$

The gain of the second PNP stage will be equal to the ratio of R_7 to R_6.

$$A_V = \frac{R_C}{R_E} = \frac{R_7}{R_6} = \frac{4.7\ \text{k}\Omega}{47\ \Omega} = 100$$

The overall gain of the circuit is equal to the product of the first-stage gain and the second-stage gain.

$$A_{\text{Total}} = A_1 \times A_2 = 100 \times 100 = 10{,}000$$

Darlington Pair DC Amplifier Circuits

Figure 10-12(a) shows how the collectors of two transistors can be tied together, and the emitter of one transistor can be direct coupled to the base of the other to form a very useful dc amplifier arrangement called a **darlington pair.** Named after its creator, this pair of bipolar transistors function as one and are packaged no differently than an ordinary transistor, as shown in the inset in Figure 10-12(a). Referring to the circuit in Figure 10-12(a), you can see that resistors R_1 and R_2 provide voltage divider bias for the first transistor Q_1, while resistor R_3 serves as a collector resistor for both transistors. An emitter resistor is generally not included in the circuit because it would reduce the amount of current gain (I_E and therefore I_C), which in turn would reduce voltage gain (V_{OUT} or V_{CE}). An applied input signal voltage will cause a variation in the base current of Q_1, a corresponding change in Q_1's emitter current, and therefore a variation in Q_2's base current and collector current. By controlling Q_1, you can control the conduction of Q_2. In other words, turning Q_1 more ON will turn Q_2 more ON, and turning Q_1 more OFF will turn Q_2 more OFF. The overall common-emitter current gain of this circuit arrangement is effectively equal to the

Darlington Pair
A dc amplifier arrangement wherein the collectors of two transistors are tied together with the emitter of one transistor direct coupled to the base of the other.

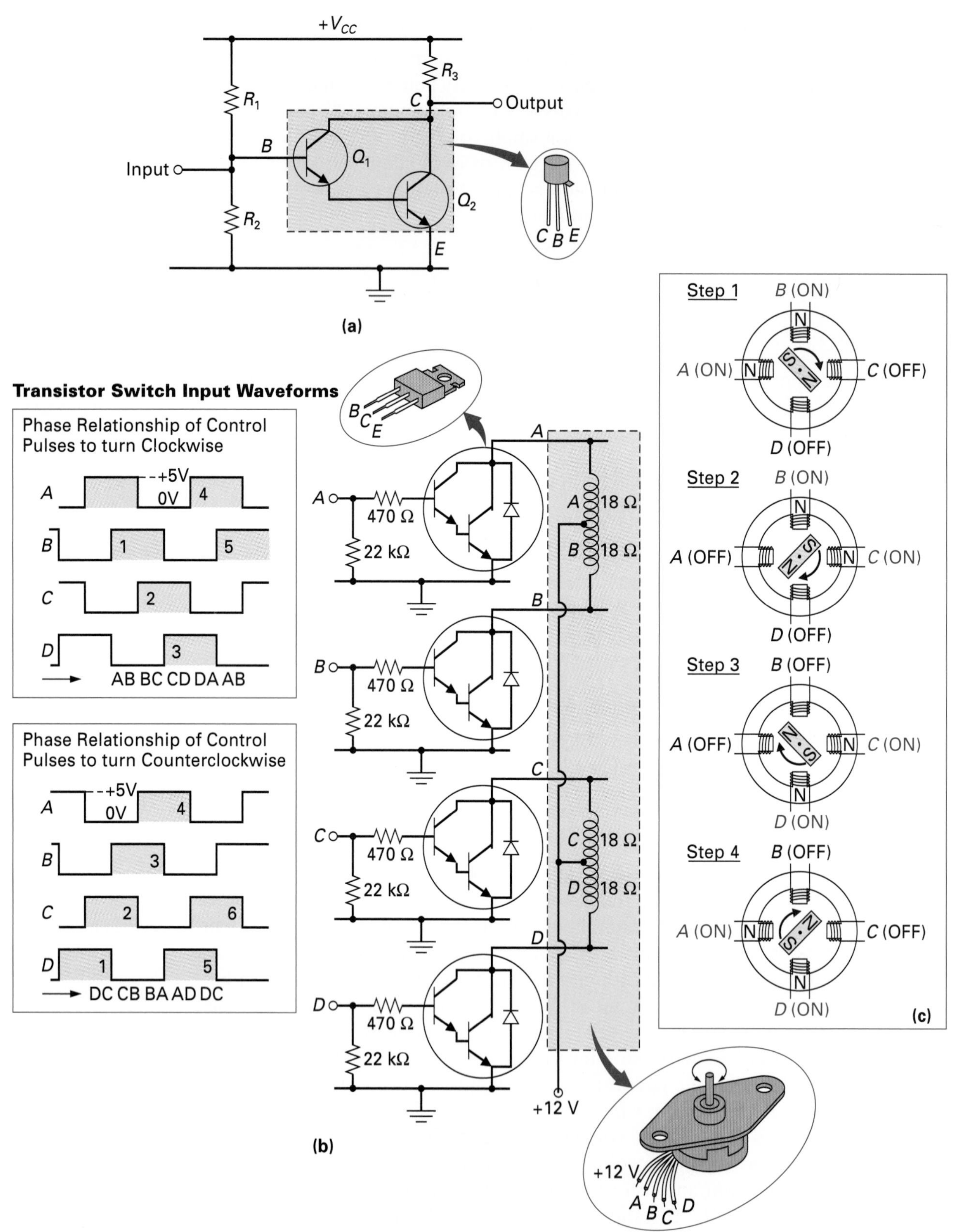

FIGURE 10-12 Darlington Pair DC Amplifier Circuits. (a) Basic Circuit. (b) Application: DC Stepper Motor Control. (c) Stepper Motor Clockwise Operation.

current gain of the first transistor (beta Q_1) multiplied by the current gain of the second transistor (beta Q_2). In practice, however, because the emitter current of Q_1 drives the base of Q_2, the base bias of Q_1 is generally made lower than normal to keep Q_1's emitter current from driving Q_2 into saturation.

Figure 10-12(b) shows how the high-current switching ability of a darlington pair can be made use of to control the operation of a **stepper motor.** The dc stepper motor is ideally suited for computer controlled industrial applications because its rotor's position, speed, and direction can be controlled by a computer's two-state output control signals. The stepper motor derives its name from the fact that its rotor rotates in uniform angular steps. A variety of different types of stepper motors are available with a different number of steps per revolution. For example, if a stepper motor has 48 steps per revolution, each step will cause the rotor to turn 7.5°.

Stepper Motor
A motor that rotates in small angular steps.

$$\text{Degrees per step} = \frac{360°}{\text{Number of steps}}$$

$$= \frac{360°}{48} = 7.5° \text{ per step}$$

Table 10-3 lists some of the more frequently used steps per revolution sizes of stepper motors.

Let us now examine how the darlington pair circuit in Figure 10-12(b) operates. A pulse pattern is applied to the inputs of the four darlington pair transistors when the computer controlling this circuit wants to set the motor in motion. A HIGH to any of the inputs will turn ON both of the transistors in the darlington pair and switch ground from the emitter of the transistor through to the end of its connected winding. Explained another way, when a darlington is turned ON, its respective coil will be energized because the coil's center tap is connected to +12V, and its end is connected to ground via the darlington. The internal shunt connected clipper diodes within the darlington pair transistors are used to protect the driver's circuit from the counter emf transient that is generated by the stator coils, as discussed in Section 6-2-2. The basic operation of the stepper motor is shown in Figure 10-12(c). For ease of explanation, only four stator poles and one rotor are shown, providing 4 steps. In reality, there would be many more stator and rotor poles providing the number of steps required. The 4-step clockwise excitation sequence is shown in the simplified diagram in Figure 10-12(c). In step 1, stator poles A and B are energized to produce two

TABLE 10-3 Typical Stepper Motor Sizes

STEPS PER REVOLUTION	ROTOR ANGLE MOVEMENT PER STEP
240	1.5°
180	2.0°
144	2.5°
72	5.0°
48	7.5°
24	15°
12	30°

north poles so the rotor's south pole will align itself between stator poles A and B. In the second step, stator poles B and C are energized so the rotor's south pole will move clockwise and align itself between B and C. In step three, stator poles C and D are energized so the rotor's south pole will align itself between these two stator poles. In step four, stator poles D and A are energized so the rotor aligns itself between these two stators. The next stepping action would energize stator poles A and B, which puts us back to our original position. The signal sequence needed to control this stepper motor is shown in the timing diagrams in Figure 10-12(b). To drive the stepper motor clockwise, the stator coils are energized in the forward sequence AB, BC, CD, DA, AB, and so on, which was the sequence shown in Figure 10-12(c). To drive the stepper motor counterclockwise, the stator coils are energized in the reverse sequence DC, CB, BA, AD, DC, and so on.

Differential DC Amplifier Circuits

Figure 10-13(a) illustrates a basic differential dc amplifier circuit. This circuit contains two transistors (Q_1 and Q_2) which are matched to have identical characteristics, each of which has its own collector resistor and voltage divider bias. The distinguishing feature of this circuit is the sharing of an emitter resistor by both transistors. Most differential amplifiers are not constructed using the discrete or individual components used in Figure 10-13(a). Differential amplifiers are available in integrated circuit (IC) form as shown in Figure 10-13(b). In this circuit, the voltage divider bias resistors are not needed since the base-emitter junction of the transistor is forward biased by applying a negative emitter supply voltage ($-V_{\mathrm{EE}}$). The symbol for this circuit is shown in Figure 10-13(c).

With two inputs and two outputs, the differential amplifier can have many different input/output combinations. Figure 10-13(d), (e), and (f) illustrate the more common differential amplifier circuit arrangements. Let us discuss each of these circuit arrangements in detail.

Single-input single-output connection. In this arrangement the input signal is applied to the base of Q_1 (V_{B_1}), and the output is taken from the collector of Q_2(V_{C_2}), as shown in Figure 10-13(d). Transistor Q_1 acts as an emitter follower (common collector), and therefore amplifies the signal input producing an output emitter voltage (V_E) that will follow the input signal applied to the base. This signal voltage developed across the emitter resistor will be applied to the emitter of Q_2, which will function as a common-base amplifier stage, developing an output voltage at its collector (V_{C_2}). The amount of voltage gain achieved by this circuit is equal to the voltage gain of the common base amplifier stage only (Q_2) because Q_1 will have no voltage gain, only current gain and impedance isolation.

Single-input, differential-output connection. In this arrangement, the input signal is applied to the base of Q_1 (V_{B_1}), and two outputs are available at each of the transistor's collectors (V_{C_1} and V_{C_2}), as shown in Figure 10-13(e). Transistor Q_1 amplifies the input signal and produces an output at its collector (V_{C_1}) that is out of phase with the input signal because Q_1 is operating as a common-emitter amplifier stage. The changes in Q_1's emitter current will be developed across the emitter resistor R_2 (V_E), and this voltage variation will be in phase with the input signal at the base of Q_1 (V_{B_1}). Transistor Q_2 amplifies the emitter

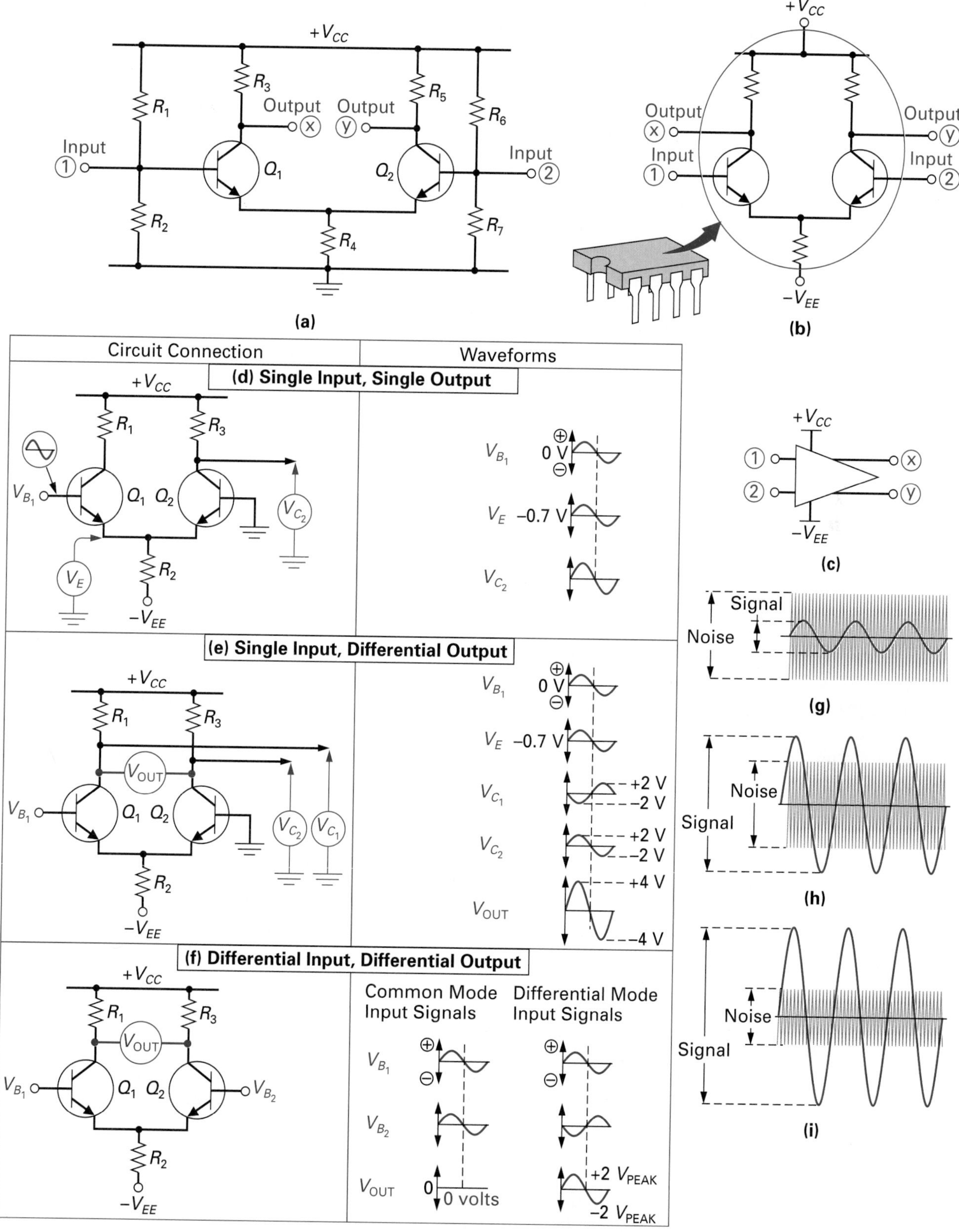

FIGURE 10-13 Differential Amplifier Circuits. (a) Basic Discrete Circuit. (b) Basic IC Circuit. (c) Symbol. (d), (e), (f) Modes of Operation. (g), (h), (i) Noise.

voltage signal from Q_1, acting as a common-base amplifier stage, so there is no inversion between the emitter input of Q_2 and the collector output of Q_2 (V_{C_2}). The two separate outputs at the collector of Q_1 relative to ground, and at the collector of Q_2 relative to ground, are 180° out of phase. In most applications, the differential amplifier would not be used to provide two outputs with respect to ground in this way. Instead, the output signal voltage would be equal to the voltage difference between the collector of Q_1 and the collector of Q_2 (V_{OUT}), as shown in the single-input differential-output circuit in Figure 10-13(e). Because the voltages at the collectors of Q_1 and Q_2 are 180° out of phase, a very large voltage difference—and therefore voltage gain—will exist between these two points.

Differential-input, differential-output connection. In this arrangement, the differential amplifier is used to its full advantage, and its action demonstrates how the differential, or difference, amplifier derived its name. Referring to the differential-input, differential-output circuit in Figure 10-13(f), you can see that two input signals are applied to the base of Q_1 and Q_2. The output is equal to the voltage difference between the collector of Q_1 and the collector of Q_2.

Common Mode Input Signals
Two input signals in phase with one another.

When both input signals are in phase with one another, they are referred to as **common mode input signals,** as seen in the waveforms in Figure 10-13(f). If both input signal voltages are of the same amplitude and phase ($V_{B_1} = V_{B_2}$), the difference between these two inputs is zero, and this is the voltage that appears at the output ($V_{OUT} = 0$ V). This occurs because the two identical input signal voltages turn both Q_1 and Q_2 ON by the same amount. Therefore, their emitter currents, collector currents, and collector output voltages are the same ($V_{C_1} = V_{C_2}$). If the collector voltage at Q_1 is equal to the collector voltage of Q_2, then the output voltage, which is equal to the potential difference between V_{C_1} and V_{C_2}, will be 0 V.

Differential Mode Input Signals
Two input signals out of phase with one another.

If both of the input signals are out of phase with one another, they are referred to as **differential mode input signals,** as seen in the waveforms in Figure 10-13(f). If both input signal voltages are out of phase with one another, the output voltage will be equal to the difference between the two input signal voltages. Let us see why this happens by examining each transistor's operation separately. When V_{B_1} swings positive, Q_1 will turn more ON. Due to Q_1's common-emitter action, its collector voltage will swing negative. The emitter of Q_1, on the other hand, will swing positive, and this will be applied to Q_2's emitter, which will turn Q_2 more OFF. At the same time, Q_2's base is being driven negative by its input signal, and so its collector voltage will swing positive, and its emitter voltage will swing negative. This negative voltage at the emitter of Q_2 will be applied to the emitter of Q_1, and turn Q_1 more ON. In summary, the positive input voltage at the base of Q_1 and the negative voltage at the emitter of Q_1 from Q_2 will work together to turn Q_1 more ON. On the other hand, the negative input voltage at the base of Q_2 and the positive voltage at the emitter of Q_2 from Q_1 will work together to turn Q_2 more OFF. The result is that V_{B_1} and V_{B_2} work together to produce two amplified output signals that are 180° out of phase. This difference between the two collector signal voltages means that there will be a large difference in potential across the output voltage terminals.

Common-mode rejection. This ability of the differential amplifier to reject common mode input signals is its key advantage and accounts for why it is perhaps the most important of all the dc amplifiers. The question you may be asking at this stage is: What are common mode input signals, and why should we want to reject them? The answer is that temperature changes and noise are common-mode input signals, and they are unwanted signals.

Removing Amplifier Temperature Instability. As you know, temperature variations within electronic equipment affect the operation of semiconductor materials and therefore semiconductor devices. These temperature variations can cause the dc output voltage of the first stage to drift away from its normal Q point, or no-input-signal bias level. The second stage amplifier will amplify this voltage change in the same way as it would amplify any dc input signal and so would all of the following amplifier stages. The increase or decrease in the normal Q point bias for all of the amplifier stages will get progressively worse due to this thermal instability. The final stage may have a Q point that is so far off of its mid position that an input signal may drive it into saturation or cutoff, causing signal distortion. With the difference amplifier, the voltages between the collector of Q_1 and ground, and the collector of Q_2 and ground will tend to increase and decrease as temperature changes by exactly the same amount. This common mode change that occurs within every amplifier due to temperature changes will not appear at the output of the differential amplifier due to its **common mode rejection.**

Removing Amplifier Noise. The second common-mode input signal that the differential amplifier removes is noise. It is often necessary to amplify low-level signals from low-sensitivity sources such as microphones, light detectors, and other transducers. High-gain amplifiers are used to increase the amplitude of the input signal until it is large enough to drive or control a load such as a loudspeaker. The 60 Hz ac power line, or any other electrical variation, can induce a noise signal along with the input signal at the input of this high-gain multiple-stage amplifier. If this noise signal is larger than the desired signal, as shown in Figure 10-13(g), the signal will never be able to be extracted from the noise. Under normal conditions, the noise from a microphone should be at least 30 dB below the signal level for voice communication, as shown in Figure 10-13(h), and 60 dB below the signal level for quality music recording and broadcasting, as shown in Figure 10-13(i). These noise signals will be induced at all points in the circuit and be identical in amplitude and phase. The differential amplifier will block these unwanted signals because they will be present at both inputs of the differential amplifier (noise will be a common-mode input). A true input signal, on the other hand, will appear at the two inputs of the differential amplifier as a differential input signal and be amplified.

To summarize, the desired signal that appears on only one input (single-input mode) or both inputs (differential-input mode) of a differential amplifier will be amplified and applied to the output. On the other hand, unwanted signals caused by temperature variations or noise will appear as common-mode input signals and be rejected. A differential amplifier's ability to provide a high **differential gain (A_{VD})** and a low **common-mode gain (A_{CM})** is a measure of a differential amplifier's performance. This ratio is called the **common-mode rejection ratio (CMRR)** and is calculated with the following formula:

Common Mode Rejection
Rejection, by a differential amplifier, of the common mode change that occurs within every amplifier due to temperature changes.

Differential Gain (A_{VD})
The amplification of differential-mode input.

Common-Mode Gain (A_{CM})
The amplification of common-mode input. This value is typically less than 1.

Common-Mode Rejection Ratio (CMRR)
The ratio of an operational amplifier's differential gain to the common-mode gain.

$$\text{CMRR} = \frac{A_{VD}}{A_{CM}}$$

Looking at this formula, you can see that the higher the A_{VD} (differential gain), or the smaller the A_{CM} (common-mode gain), the higher the CMRR value, and therefore the better the differential amplifier. This ratio can also be expressed in dBs by using the following formula:

$$\text{CMRR} = 20 \times \log \frac{A_{VD}}{A_{CM}}$$

■ **EXAMPLE:**

If a differential amplifier has a differential gain of 5,000 and a common-mode gain of 0.5, what is the amplifier's CMRR? Express the answer in standard gain and dBs.

■ ***Solution:***

$$\text{CMRR} = \frac{A_{VD}}{A_{CM}} = \frac{5000}{0.5} = 10{,}000$$

$$\text{CMRR} = 20 \times \log \frac{A_{VD}}{A_{CM}} = 20 \times \log 10{,}000 = 20 \times 4 = 80 \text{ dB}$$

A CMRR of 10,000 or 80 dB means that the differential desired input signals will be amplified 10,000 times more than the unwanted common-mode input signals.

The versatility, temperature stability, and noise rejection of the differential amplifier make it an ideal candidate for a variety of applications. One such application is the "operational amplifier," an IC amplifier that has almost completely replaced the discrete, or individual-component, amplifier circuits. The operational amplifier, which is covered in Chapter 14, contains several differential-amplifier stages, and so we will be discussing the differential amplifier's characteristics in more detail along with this important application.

SELF-TEST REVIEW QUESTIONS FOR SECTION 10-2

1. Direct current amplifiers are generally used to boost a weak signal from a transducer. (true/false)
2. How does the inclusion of an emitter resistor improve the amplifier's temperature stability?
3. What is the advantage of the complementary dc amplifier?
4. The gain of a multi-stage amplifier is always slightly less than the ____________ of all the stage gains.
5. The darlington pair provides a very high ____________ gain.
6. The advantage of the differential dc amplifier is that it provides a very ____________ differential gain and a very ____________ common-mode gain.

10-3 AUDIO FREQUENCY (AF) AMPLIFIERS

Audio is a Greek word meaning "I hear," which is why the term audio frequency range includes all of the frequencies that the human voice can produce and the human ear can respond to. This audio frequency range is from about 20 Hz to 20,000 Hz. An **audio frequency (AF) amplifier** is a circuit containing one or more amplifier stages designed to amplify an audio-frequency signal. These audio amplifiers are used within any electronic system that processes, transmits, or receives audio signals, such as radio transmitters and receivers, television transmitters and receivers, telephone communication systems, music systems, multimedia computer systems, and so on. The two basic types of audio amplifiers are the AF voltage amplifier, which is generally class A biased, and AF power amplifier, which is generally class B or class AB biased. Let us begin by discussing the appearance, operation, and characteristics of the AF voltage amplifier.

Audio Frequency (AF) Amplifier
Amplifiers for ac signals between 20 Hz and 20,000 Hz.

10-3-1 *Audio Frequency (AF) Voltage Amplifiers*

A voltage amplifier is used to boost the voltage level of a weak input signal, so that it is large enough in amplitude to control a power amplifier. When used in this manner, voltage amplifiers are often referred to as **preamplifiers** or preamps because they precede the main power amplifier, which is used to boost signal current so that the final output can control a low-impedance load, such as a loudspeaker.

Preamplifiers
Amplifiers that precede the main power amplifier and boost signal current so that the final output can control a low impedance load.

Single-Stage AF Voltage Amplifier Circuits

Figure 10-14(a) shows a basic audio-frequency common-emitter voltage amplifier circuit. The transistor Q_1 is class A biased by the voltage divider network R_1 and R_2. The input coupling capacitor (C_1) prevents Q_1's dc base bias voltage from being applied back to the source, while the output coupling capacitor (C_2) prevents Q_1's dc collector bias voltage from being applied to the load. Collector resistor R_3 operates in conjunction with the transistor to develop an output signal voltage, and resistor R_4 provides emitter feedback for temperature stability. Capacitor C_E has been added to remedy a problem that occurs when an amplifier is amplifying an ac signal. To explain this in detail, a signal variation at the base causes a variation in base current (I_B) and a corresponding change in output current (I_C). These variations in transistor current produce a varying voltage drop across R_3 but because R_4 is now included in the circuit for temperature stability, a similar voltage swing is also developed across R_4. Therefore, as the input signal swings positive, the transistor's V_{BE} bias voltage is increased, causing an increase in I_B and I_E and therefore an increase in the voltage drop across R_4. This increase in the voltage developed across R_4 will increase the positive potential at the transistor's emitter and therefore decrease or degenerate the original forward bias of the transistor's base-emitter (PN) junction. This **degenerative feedback** dramatically reduces the gain of any ac signal amplifier but can be remedied by including a capacitor in parallel with R_4, as shown in Figure 10-14(a). Because a capacitor basically operates as a closed switch to ac and an

Degenerative Feedback
Also called negative feedback, degenerative feedback is a method for coupling a portion of the output of a circuit back so that it is 180° out of phase with the input signal. This reduces amplification and stabilizes the circuit to reduce distortion and noise.

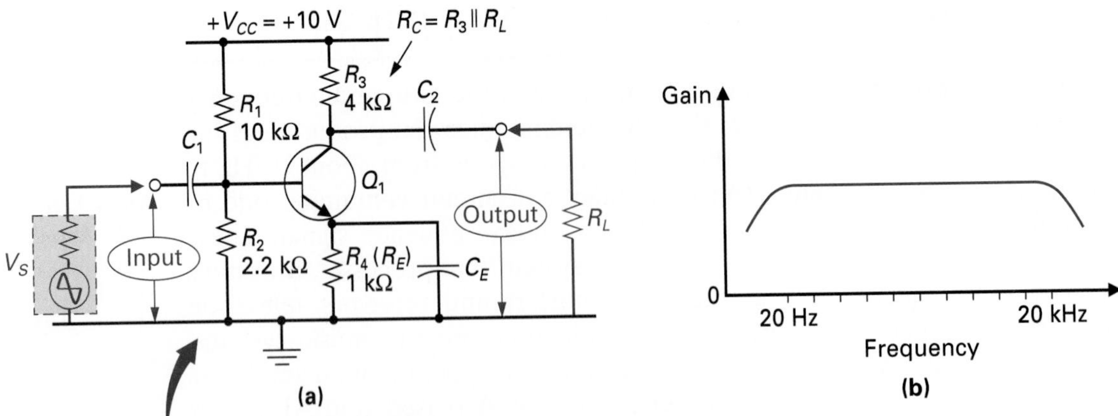

Circuit Functions

V_S – Audio frequency voltage source (example, microphone).

C_1 – Input Coupling capacitor: blocks DC from passing back to source, passes AC audio signal.

R_1, R_2 – Voltage divider bias resistors: sets Q point on load line by setting base voltage.

Q_1 – Bipolar Transistor: amplifies the input signal by multiplying I_B by β.

R_C – Collector Resistor: collector current value and R_C value determine voltage drop across R_C and therefore output voltage developed between Q_1 collector and ground. R_C along with R_E also limits I_C saturation current to safe level.

R_E – Emitter Resistor: provides emitter feedback for temperature stability.

C_E – Emitter Bypass Capacitor: prevents degenerative feedback.

R_L – Output Load: this opposition represents the impedance of the next amplifier stage or the impedance of a device such as a speaker.

FIGURE 10-14 Single Stage Audio Frequency (AF) Voltage Amplifier. (a) Basic Circuit. (b) Frequency Response.

open switch to dc, the ac signal appearing at the emitter will now be shunted around R_4, preventing the degenerative feedback. The capacitor will not interfere with the dc feedback signal developed across R_4 to compensate for changes in temperature. Although the temperature stability feedback provided by R_4 varies up and down and is also degenerative, its changes occur at a much slower rate and can be considered dc and will suffer no interference from the capacitor.

Like the dc amplifier, the voltage gain is equal to the ratio of the transistor's total collector resistance to emitter resistance. Because the audio amplifier is dealing with ac signals, the voltage gain of an audio frequency amplifier will be equal to the ratio of total ac collector resistance to total ac emitter resistance.

$$A_{V(ac)} = \frac{R_{C(ac)}}{R_{E(ac)}}$$

Because the collector current from the transistor has two paths—one through the collector resistor R_3 and the second through the load resistance R_L—R_3 and R_L are effectively in parallel with one another. Therefore, the total ac collector resistance is equal to

$$R_{C(ac)} = \frac{R_3 \times R_L}{R_3 + R_L}$$

The ac emitter resistance of a small-signal voltage amplifier will be a lot lower than the value of the emitter resistor due to the bypass capacitor C_E and can be calculated with the formula

$$R_{E(ac)} = \frac{25 \text{ mV}}{I_E}$$

Let us now see how we can use these formulas to calculate the voltage gain of a typical AF voltage amplifier.

■ **EXAMPLE:**

Referring to the class-A biased AF voltage amplifier in Figure 10-14(a), calculate

1. Its voltage gain
2. The signal voltage output if the input was 125 μV

■ ***Solution:***

1. To begin with we will have to use our voltage-divider bias formulas to determine V_B, V_E, and I_E.

$$V_B = \frac{R_2}{R_1 + R_2} \times V_{CC} = \frac{2.2 \text{ k}\Omega}{10 \text{ k}\Omega + 2.2 \text{ k}\Omega} \times 10 \text{ V} = 1.8 \text{ V}$$

$$V_E = V_B - 0.7 \text{ V} = 1.8 \text{ V} - 0.7 \text{ V} = 1.1 \text{ V}$$

$$I_E = \frac{V_E}{R_E} = \frac{1.1 \text{ V}}{1 \text{ k}\Omega} = 1.1 \text{ mA}$$

Now that V_B, V_E, and I_E are known, we can calculate ac collector resistance, ac emitter resistance, and, finally, ac voltage gain.

$$R_{C(ac)} = \frac{R_3 \times R_L}{R_3 + R_L} = \frac{4 \text{ k}\Omega \times 15 \text{ k}\Omega}{4 \text{ k}\Omega + 15 \text{ k}\Omega} = \frac{60 \text{ M}\Omega}{19 \text{ k}\Omega} = 3.16 \text{ k}\Omega$$

$$R_{E(ac)} = \frac{25 \text{ mV}}{I_E} = \frac{25 \text{ mV}}{1.1 \text{ mA}} = 22.73 \ \Omega$$

$$A_{V(ac)} = \frac{R_{C(ac)}}{R_{E(ac)}} = \frac{3.16 \text{ k}\Omega}{22.73 \ \Omega} = 139$$

2. The ac signal voltage at the output is 139 times greater than the input voltage. For an input signal voltage that is 125 μV, the output voltage will be

$$V_{OUT(ac)} = V_{IN(ac)} \times A_{V(ac)} = 125\ \mu V \times 139 = 17.4\ mV$$

■ **EXAMPLE:**

Calculate the efficiency of an AF voltage amplifier if it has the following circuit conditions:

$$V_{OUT} = 2\ V$$
$$R_L = 1\ k\Omega$$
$$V_{CC} = 12\ V$$
$$I_{CC} = 1\ mA$$

■ ***Solution:***

As discussed previously, the efficiency of an amplifier is equal to

$$\eta = \frac{\textit{Power Delivered to Load by Amplifier } (P_L)}{\textit{DC Power Consumed by Amplifier } (P_S)} \times 100\%$$

Because the voltage supplied to the 1 kΩ load is 2 V, the power delivered to the load will be

$$P_L = \frac{V_L^2}{R_L} = \frac{2V^2}{1\ k\Omega} = 4\ mW$$

The dc power supplied to the amplifier is equal to

$$P_S = V_{CC} \times I_{CC} = 12\ V \times 1\ mA = 12\ mW$$

If the power delivered to the load is equal to 4 mW, and the power supplied to the amplifier by the dc power supply is equal to 12 mW, the class A amplifier's efficiency in Figure 10-14(a) will be

$$\eta = \frac{P_L}{P_S} \times 100\% = \frac{4\ mW}{12\ mW} \times 100\% = 33.3\%$$

The single-stage class-A AF voltage amplifier shown in Figure 10-14(a) will provide a large voltage gain for a wide range of input signal frequencies. The frequency response of this audio frequency amplifier is shown in Figure 10-14(b). Looking at this curve, you can see that the amplifier cannot be used to amplify dc, or very low frequency ac, signals because of the opposition or reactance of the input and output coupling capacitors. To amplify dc, or very low frequency ac, signals a dc amplifier would have to be used, which, as you know, does not include any dc blocking components in the signal path, such as capacitors and transformers. As the ac input signal frequency increases, its cycle time or period decreases, and eventually this cycle time is faster than the transit time needed for a charge carrier to pass through the transistor. At this time, the gain of the transistor will fall to almost zero, as shown at the upper end of the frequency response curve in Figure 10-14(b), because the transistor's internal currents do not have enough time to respond to the high-frequency input signal changes.

Multiple-Stage AF Voltage Amplifier Circuits

If a single-stage AF amplifier cannot provide the amount of gain needed, a multiple stage AF circuit can be used that includes two or more cascaded amplifier stages. Figure 10-15(a) shows a two-stage ***RC* coupled AF voltage amplifier.** A capacitor is commonly used to connect one amplifier stage to another. When used in this way, the capacitor is called a **coupling capacitor,** which means that it connects two circuits together electronically so that the signal from one circuit can pass to the next circuit with no distortion. You may remember from your introductory dc/ac electronics that the reactance or opposition of a capacitor is inversely proportional to frequency.

***RC* Coupled AF Voltage Amplifier** A multiple-stage AF voltage amplifier coupled by a capacitor.

Coupling Capacitor A capacitor used to connect one amplifier stage to another.

$$X_C = \frac{1}{2\pi fC}$$

The coupling capacitor actually performs two functions:

1. Due to its low reactance to ac signals, it will pass the audio-frequency information signal from the first stage to the second stage with no distortion, and
2. Due to its high reactance to dc, the coupling capacitor will isolate the dc bias voltages of one stage from the dc bias voltages of another.

Referring to Figure 10-15(a), you can see that coupling capacitor C_2 performs these two functions of coupling the ac signal from the collector of Q_1 to the base of Q_2, but isolating the dc collector bias voltage of Q_1 (+5.6V) from the dc base bias of Q_2 (+1.8V).

The disadvantage of the *RC* coupled amplifier is that maximum power is not transferred between one stage and the next. In the *RC*-coupled amplifier's first stage, the collector current has two paths: one through the collector resistor R_3 and the second through the low-input impedance of the second amplifier stage. The low-input impedance of the second stage and the first stage's collector resistor R_3 are therefore effectively in parallel with one another. This means that the overall load resistance of the first stage is reduced, and the voltage developed at the output and applied to the next stage is also reduced. Explained another way, the low-input impedance of the second stage (caused by the parallel connection of R_3 and the input impedance of Q_2) will "load" the first stage, causing a high-output current ($I\uparrow$) due to the low-equivalent resistance ($R\downarrow$). A small voltage will be developed across the output of the first stage ($V_{OUT}\downarrow$). With the output voltage of the first stage low, maximum power is not being transferred from one stage to the next. To achieve an overall high voltage gain, several *RC*-coupled AF voltage amplifier stages may be needed to compensate for the individual-stage gains, which will be slightly less than expected.

The **transformer coupled AF voltage amplifier** is shown in Figure 10-15(b). As you can see, the transformer T_1 is used to couple the output of the first stage to the input of the second stage. The ac audio signal is amplified by Q_1 and then developed across the primary of T_1, which acts as Q_1's collector load. Through electromagnetic induction, the ac audio signal is coupled to the secondary of T_1, where it is applied to the base of Q_2, causing a variation in the base bias and therefore base current. Like the first stage, the primary of T_2 serves as a collec-

Transformer Coupled AF Voltage Amplifier A multiple-stage AF voltage amplifier coupled by a transformer.

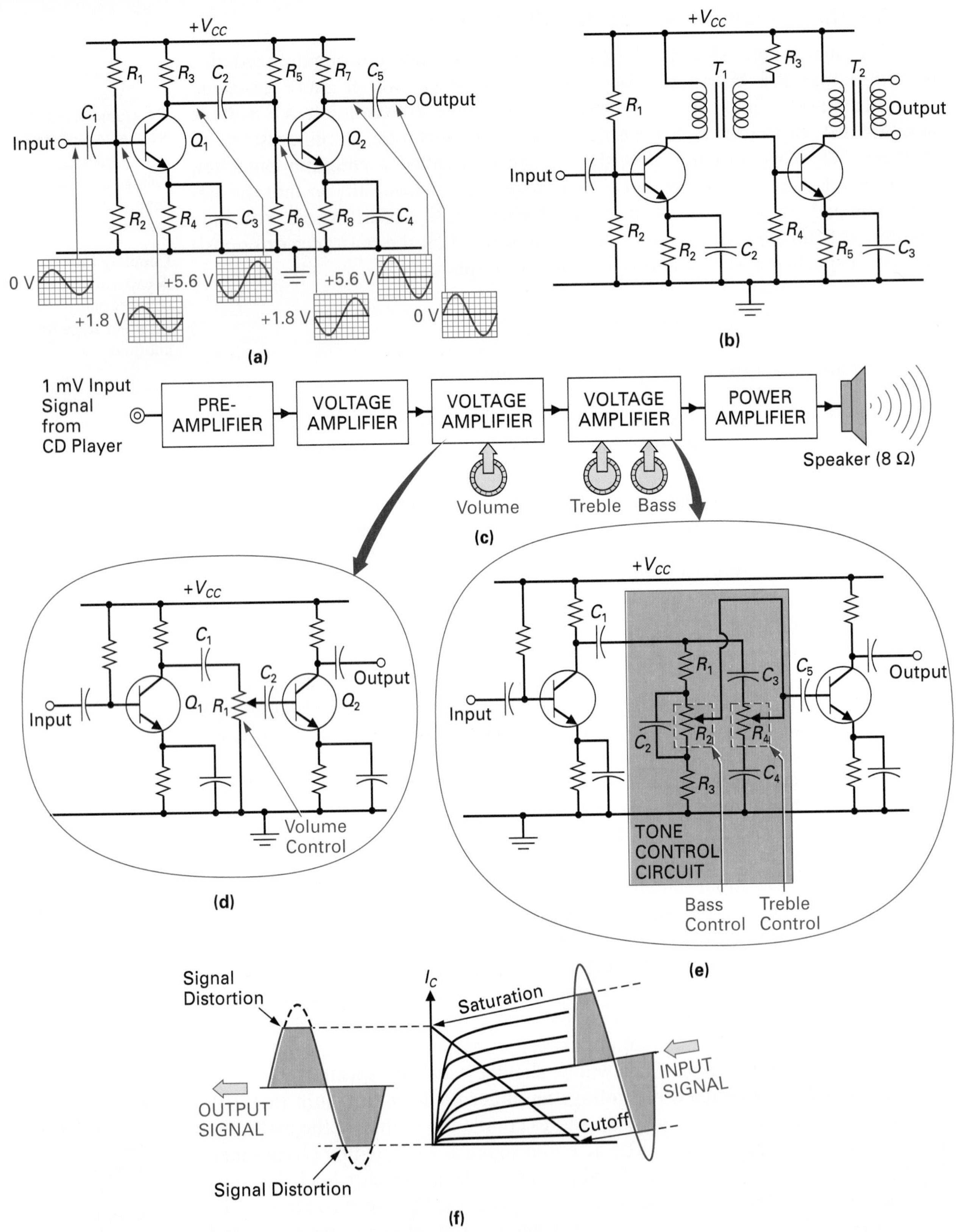

FIGURE 10-15 Multiple Stage AF Voltage Amplifiers. (a) *RC* Coupled. (b) Transformer Coupled. (c) Application: CD Music Amplifier System. (d) Volume Control. (e) Tone Control. (f) Class A Distortion.

tor load, and the output of this amplifier stage is developed across the secondary of T_2. As well as providing dc isolation between amplifier stages, the transformers can also match the output impedance of one amplifier stage to the input impedance of the following stage. Using the different inductive reactance values of the transformer's primary and secondary, the low-input impedance of the second amplifier stage (typically 4 kΩ to 6 kΩ) can appear to equal the higher output impedance of the first amplifier stage (typically 40 kΩ to 60 kΩ). By matching input and output impedances in this way, the low-input impedance of the second amplifier stage will not load down the first amplifier stage, and so a high value of signal current and signal voltage (and therefore signal power) will be applied to the second stage. Transformer coupling is often used in audio frequency amplifiers because of its efficient transfer of power between stages. However, the transformer is a more expensive coupling method than the less efficient *RC* coupling method.

Volume and Tone Control of AF Voltage Amplifiers

Figure 10-15(c) shows the basic block diagram of a music system's amplifier. A 1 mV music signal from a compact disc player is applied first to a preamplifier, which would be similar to the single-stage voltage amplifier circuit discussed previously in Figure 10-14. The output signal of this preamplifier is then applied to a two-stage voltage amplifier, which is shown in Figure 10-15(d). Unlike the previously discussed multiple-stage amplifiers, the overall gain of this amplifier can be controlled by the variable resistor R_1. This variable resistor will control the volume level or loudness of the music system by controlling the voltage developed across R_{C_1}, and therefore the voltage applied to the second stage. For example, because C_1 has a very low capacitive reactance, nearly all of the signal voltage from Q_1 will be developed across the variable resistor R_1. If the volume control is turned "down" (wiper is moved down), a smaller resistance exists between the base and emitter of the following stage (Q_2). A smaller signal voltage will be developed across this resistance and applied to the base of Q_2 as an input. On the other hand, as the volume control is turned "up" (wiper is moved up), a greater resistance exists between the base and emitter of the following stage "Q_2." A greater signal voltage will be developed across this resistance and applied to the base of Q_2 as an input.

Figure 10-15(e) shows how a circuit can be inserted between a two-stage voltage amplifier to control the **treble** and **bass** of the music, which combined is called the **tone** of the music. Because people hear the same music in different ways, music systems will generally provide these controls so that the amplifier's sensitivity to certain frequencies is either emphasized or de-emphasized. If the bass and treble controls are placed in their mid position, the frequency response of an audio frequency amplifier is generally flat, as was shown previously in Figure 10-14(b). If these controls are adjusted, however, the frequency response, or audio frequency amplifier gain at certain frequencies, will be modified. For example, the bass control will either increase or decrease the gain (or response) of the audio frequency amplifier to low-end bass frequencies, while the treble control will either increase or decrease the response of the audio frequency amplifier to high-end treble frequencies. Referring to the circuit in Figure 10-15(e), let us now examine how this overall music signal tone con-

Treble
High audio frequencies normally handled by a tweeter in a sound system.

Bass
Low audio frequencies normally handled by a woofer in a sound system.

Tone
A term describing both the bass and treble of a sound signal.

trol is achieved. The large value coupling capacitors C_1 and C_5 will have a very low reactance, and therefore the audio frequency signal will pass through these capacitors with very little opposition. Capacitor C_2 will shunt the high audio frequencies around the bass control resistor R_2. Low audio frequencies (bass frequencies), however, will be developed across the voltage divider made up of R_1, R_2, and R_3. Moving the wiper of R_2 up will allow all of the bass frequencies to pass to the second stage with only little opposition, whereas moving the wiper of R_2 down will force the bass frequencies to pass through a greater resistance and therefore decrease their amplitude before they are passed to the following stage. With the treble control, capacitors C_3 and C_4 will have a very low reactance to high (treble) audio frequencies and a high reactance to the low (bass) audio frequencies. As a result, the high treble frequencies will be developed across the variable resistor R_4. Moving the wiper of R_4 up will allow all of the treble frequencies to pass to the second stage with only little opposition, whereas moving the wiper of R_4 down will force the treble frequencies to pass through a greater resistance and therefore decrease their amplitude before they are passed to the following stage.

A typical large amplifier system, like the one seen in Figure 10-15(c), may have several amplifier stages. The volume control and input signal level should be carefully adjusted so as not to heavily drive the amplifiers. For example, if the input signal level or volume is set too high, the signal will try to drive an amplifier beyond the limits of the power supply, as shown in Figure 10-15(f). This **class-A signal distortion** causes the top and bottom of the audio signal to be clipped when the signal amplitude reaches $+V_{CC}$ (transistor saturates) and 0 V (transistor is cut off). In practice, signal swings must be limited to about 80% of the power supply voltage for minimal distortion.

Class-A Signal Distortion
The condition that occurs when the input signal level or volume is set too high and the signal tries to drive an amplifier beyond the limits of the power supply.

10-3-2 *Audio Frequency (AF) Power Amplifiers*

Figure 10-16(a) repeats the simple block diagram of an audio frequency music amplifier. Up until this point, we have only discussed the voltage amplifier stages, which are used to increase the ac audio signal input of 1 mV to a large enough voltage to control the final stage power amplifier. Although the **AF power amplifier** does not increase the signal voltage, it does supply the high-output signal current (and therefore power) that is needed to operate a low-impedance load, such as a loudspeaker.

AF Power Amplifier
An amplifier that supplies the high output signal current (and therefore power) that is needed to operate a low-impedance load such as a loudspeaker.

The power delivered to a load can be calculated using the standard V^2/R formula, as follows.

$$P_{RL} = \frac{V_{RL(rms)}^2}{R_L}$$

P_{RL} = The ac power supplied to the load
V_{RL} = The rms voltage developed across the load
R_L = The load resistance

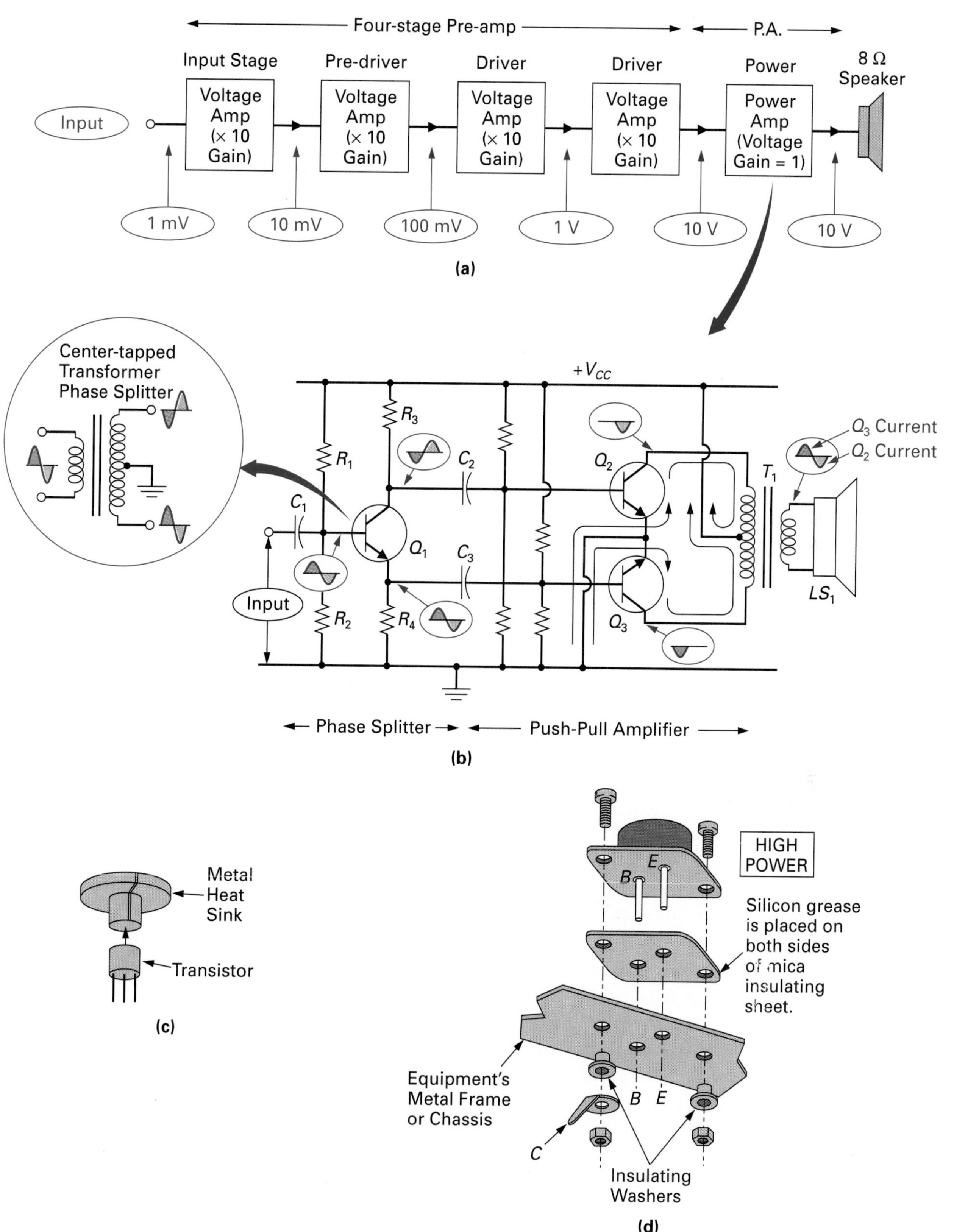

FIGURE 10-16 **Basic Audio Frequency (AF) Power Amplifiers. (a) Music Amplifier Block Diagram. (b) A Push-Pull Power Amplifier with a Transistor Phase Splitter Circuit. (c) Low-Power Heat Sink. (d) High-Power Heat Sink.**

■ **EXAMPLE:**

Using the ac voltmeter readings given in Figure 10-16(a), calculate the power delivered by the audio frequency amplifier to the 8 Ω loudspeaker load.

■ ***Solution:***

Because a voltmeter is calibrated to display ac volts in rms, the power amplifier output voltage of 10 V can be entered directly into the formula.

$$P_{RL} = \frac{V^2_{RL(rms)}}{R_L} = \frac{(10\text{ V})^2}{8\ \Omega} = 12.5\text{ watts}$$

If a 4 Ω loudspeaker were connected to the output of the amplifier in Figure 10-16(a), instead of the 8 Ω loudspeaker, the output power delivered should be doubled.

$$P_{RL} = \frac{V^2_{RL(rms)}}{R_L} = \frac{(10\text{ V})^2}{4\ \Omega} = 25\text{ W}$$

However, the output power delivered could only be doubled if the amplifier circuit could supply the extra current without the lower resistance load pulling down (or loading) the signal voltage. In reality, the power delivered would be less than double since the increase in circuit current will cause an increase in the heat power generated by the circuit, and therefore a loss in the signal power delivered to the load. Most amplifiers are rated for a specific load resistance, and you should always check to see that the load is within specification. For example, most audio frequency music system amplifiers are designed to drive loudspeaker loads of between 8 Ω to 15 Ω. Using a load resistance outside of the rated range can cause the amplifier to overdrive the speaker or the speaker to overload the amplifier.

Push-Pull AF Power Amplifier Circuits

Figure 10-16(b) shows how two transistors (Q_2 and Q_3) can be connected to form a special type of power amplifier circuit called a **push-pull power amplifier.** Before we discuss these two transistors, we will first need to examine the transistor that is driving the push-pull power amplifier, Q_1. The class A biased transistor Q_1 operates as a **phase splitter,** providing two signal outputs: an inverted output from the collector that is 180° out of phase with the base input signal and an emitter signal that is in phase with the base input signal. The resistance values of R_3 and R_4 are chosen to compensate for the transistor's different emitter and collector characteristics to produce two **balanced output signal voltages** that are equal in amplitude but opposite in phase. As you can see in the inset in Figure 10-16(b), a center-tapped transformer could also be used to supply this two-phase output signal. However, a transistor phase splitter has a wider frequency response than a transformer. The complementary output signals from the phase splitter drive the inputs of the two class-B biased push-pull power transistors Q_2 and Q_3. Biased in class B means that each transistor will conduct for only half of the input signal cycle, as shown in the waveforms in Figure 10-16(b). To explain how and when each of the power transistors will conduct, let us examine each half cycle of the input signal in more detail.

Push-Pull Power Amplifier
A balanced amplifier that uses two similar equivalent amplifying transistors working in phase opposition.

Phase Splitter
A circuit that takes a single input signal and produces two output signals that are 180° apart in time.

Balanced Output Signal Voltages
Signal voltages that are equal in amplitude but opposite in phase.

1. When the input signal swings positive, the emitter of Q_1 also swings positive, while its collector swings negative. Because both Q_2 and Q_3 are NPN transistors with their emitters at 0 V, a positive input voltage is needed to turn ON the transistor. Transistor Q_3 will conduct during this alternation of the input signal. Current will pass through Q_3's emitter-to-collector, through the lower half of the output transformer, and then to $+V_{CC}$, which is at the transformer's center tap. Transistor Q_2 remains cut OFF during this half of the input cycle.
2. When the input signal swings negative, the emitter of Q_1 also swings negative, while its collector swings positive. The positive signal voltage at the base of Q_2 will turn Q_2 ON, and current will pass through Q_2's emitter-to-collector, through the upper half of the output transformer, and then to $+V_{CC}$, which is at the transformer's center tap. Transistor Q_3 remains cut OFF during this half of the input cycle.

When Q_2 conducts, current passes through the upper half of the output transformer (which acts as Q_2's collector load), and a magnetic field is generated and coupled to the secondary. When Q_3 conducts, current passes through the lower half of the output transformer (which acts as Q_3's collector load), and a magnetic field of the opposite polarity will build up and be coupled to the secondary. Therefore, although only half of the cycle appears at the output of each transistor, the complete cycle is reconstructed at the secondary of the output transformer T_1, so a complete high-power ac audio frequency signal drives the loudspeaker. The name "push-pull" is derived from this action because one transistor is responsible for driving current through the load in one direction (pushing), while the other transistor is responsible for driving current through the load in the opposite direction (pulling).

■ EXAMPLE:

If an oscilloscope were connected across the secondary of the output transformer T_1 and the ac audio frequency signal applied to the 8 Ω speaker had a peak voltage of 25 V, what would be the power delivered to the speaker?

■ ***Solution:***

To calculate the power delivered to the speaker, we will have to convert the peak voltage reading taken from the oscilloscope to an rms value.

$$\text{rms} = V_{PK} \times 0.707 = 25\text{ V} \times 0.707 = 17.7\text{ V}$$

$$P_{RL} = \frac{V_{RL(rms)}^2}{R_L} = \frac{(17.7\text{ V})^2}{8\ \Omega} = 39.2\text{ W}$$

Matched Transistors
Two output transistors that have the same part number and therefore the same operating characteristics.

To ensure that the two output half-cycles from the push-pull amplifier are identical, the two power transistors should be **matched transistors.** This means that the two output transistors must have the same part number and the same operating characteristics. For example, both Q_2 and Q_3 in Figure 10-16(b) should be 2N3902s, which are typical NPN silicon power transistors (discussed previously in Figure 8-19). If one of the transistors in a class-B amplifier were to fail, both transistors are normally replaced because a failure of one transistor generally damages or changes the characteristics of the other transistor.

Class-B Power Amplifier Efficiency

A transistor's performance, or operating characteristics, is closely related to the temperature conditions. With power transistors, you have seen that the object is not to increase the voltage output because most power amplifiers have a voltage gain of 1, which means that the output voltage equals input voltage. The primary function of a power amplifier is to boost the output signal current so that the signal can operate a low-impedance (high-power) load. However, high current always goes hand-in-hand with high heat ($P = I^2R$). A class-B operated transistor is much more efficient than a class-A power amplifier because it is not generating heat during the time intervals when it is cut off. This rest period between heavy work loads means that the class-B amplifier will work a lot more efficiently than a class-A power amplifier that is ON, or working, all of the time.

■ **EXAMPLE:**

Calculate the efficiency of an AF push-pull power amplifier under the following circuit conditions:

$$V_{OUT} = 2\text{ V}$$
$$R_L = 16\ \Omega$$
$$V_{CC} = 20\text{ V}$$
$$I_{CC} = 16\text{ mA}$$

■ ***Solution:***

As discussed previously, the efficiency of an amplifier is equal to

$$\eta = \frac{\textit{Power Delivered to Load by Amplifier } (P_L)}{\textit{DC Power Consumed by Amplifier } (P_S)} \times 100\%$$

Because the voltage supplied to the 16 Ω load is 2 V, the power delivered to the load will be

$$P_L = \frac{V_L^2}{R_L} = \frac{(2\text{ V})^2}{16\ \Omega} = 250\text{ mW}$$

The dc power supplied to the amplifier is equal to

$$P_S = V_{CC} \times I_{CC} = 20\text{ V} \times 16\text{ mA} = 320\text{ mW}$$

If the power delivered to the load is equal to 250 mW, and the power supplied to the amplifier by the dc power supply is equal to 320 mW, the class-B push-pull amplifier's efficiency will be

$$\eta = \frac{P_L}{P_S} \times 100\% = \frac{250\text{ mW}}{320\text{ mW}} \times 100\% = 78.125\%$$

Power Amplifier Heat Sinks

The high levels of current handled by power transistors are the reason they are usually connected to **heat sinks,** which are metal extensions of the transistor's heat radiating package. Figure 10-16(c) shows how a metal cap can be attached to the low-power transistor to increase the metal heat-radiating surface of the transistor. Figure 10-16(d) shows how high-power transistors can be mounted to a system's metal rear panel or chassis in order to increase the transistor's heat radiating metal surface area. Unless the transistor's collector is designed to be at ground potential, a mica insulating sheet is needed to isolate the transistor's collector (the case) from the grounded chassis. When an insulating sheet is used to isolate the collector case from the chassis in this way, both sides of the mica sheet are coated with silicon grease (a bad electrical conductor but good heat conductor), which will thermally connect the transistor's case to the chassis.

Heat Sinks
Metal extensions of the transistor's heat radiating package.

Complementary AF Power Amplifier Circuits

Figure 10-17(a) shows another type of push-pull amplifier that uses the complementary nature of an NPN and a PNP transistor to perform the push-pull action. Because the NPN transistor (Q_1) will conduct when the input swings positive and the PNP transistor (Q_2) will conduct when the input swings negative, a phase splitter is not required, and a power amplifier circuit can be constructed with fewer components. Transistors Q_1 and Q_2 are connected in series between $+V_{CC}$ and ground. Resistors R_1 and R_2, and diodes D_1 and D_2 form a voltage divider circuit that biases Q_1 and Q_2 so that the $+V_{CC}$ supply voltage is divided equally across Q_1's collector-emitter and R_3 and Q_2's collector-emitter and R_4. Capacitor C_2 is included to isolate this dc bias voltage (appearing at the junction of R_3 and R_4) from the speaker. To be properly biased for class-B operation, each transistor's base should be 0.7 V with respect to its emitter, and a total of 1.4 V (2 × 0.7 V) should exist between the base of Q_1 and the base of Q_2. This voltage difference is achieved by the diodes D_1 and D_2. Diodes D_1 and D_2 are also included to prevent thermal runaway, which occurs when a temperature increase in the transistor causes a corresponding current increase in the transistor, which causes a corresponding temperature increase, and so on. With D_1 and D_2 included in the circuit, a temperature increase causes an increasing current through Q_1 and Q_2 and D_1 and D_2 because all these devices are semiconductor components. The increase in current through D_1 and D_2 will decrease the voltage drop developed across the diodes, causing a decrease in the voltage difference between Q_1 and Q_2's base and therefore a decrease in the forward bias being applied to the transistors. As a result, an increase in current due to temperature will be counteracted by a decrease in forward bias and therefore a decrease in current. Emitter feedback resistors R_3 and R_4 also assist in this temperature stabilization by providing degenerative feedback.

Figure 10-17(b) shows how **crossover distortion** can occur if diodes D_1 and D_2 are not included to keep Q_1 and Q_2 biased correctly despite variations in temperature. The addition of these two diodes in the input section ensures that each transistor is biased at exactly 0.7 V relative to its emitter (class B). Therefore, Q_1 is biased just OFF and ready to conduct the moment the input signal

Crossover Distortion
Distortion of a signal at the zero crossover point.

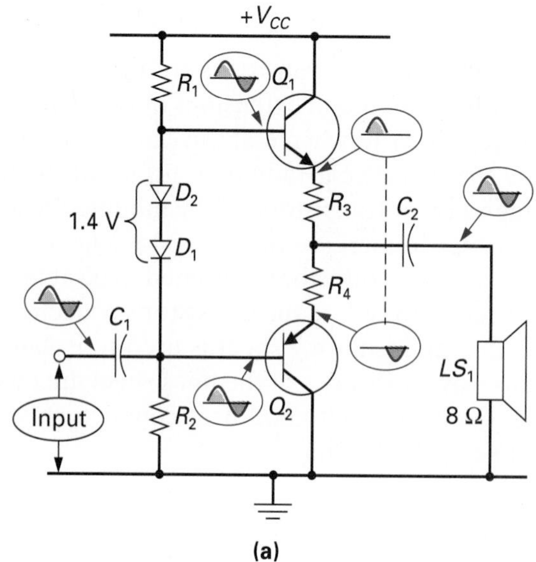

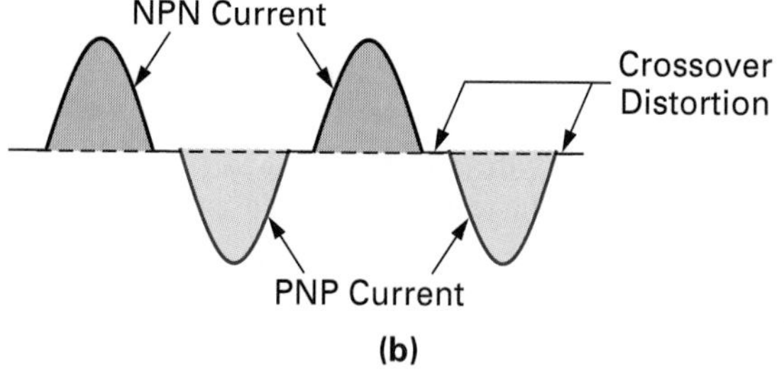

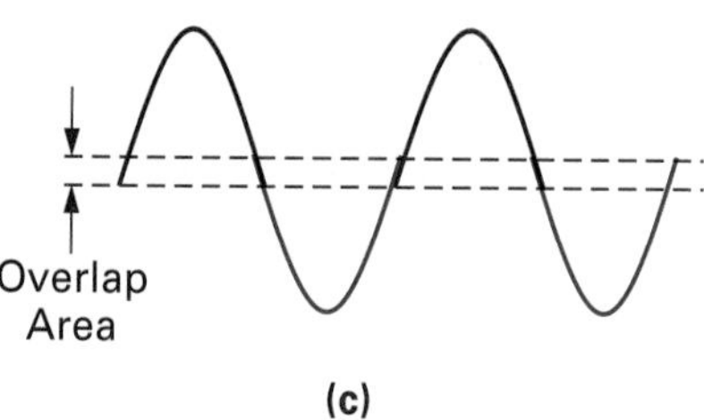

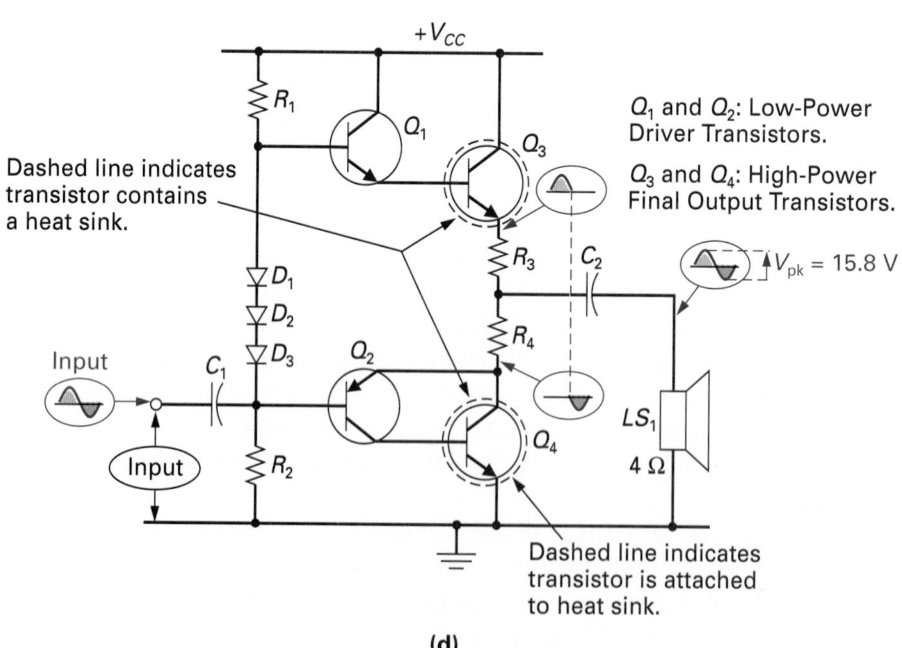

FIGURE 10-17 Complementary Audio Frequency (AF) Power Amplifiers. (a) Complementary Power Amplifier Circuit. (b) Crossover Distortion. (c) Class AB Biased. (d) Quasi-Complementary Power Amplifier Circuit.

crosses zero and swings positive, and Q_2 is also biased just OFF and ready to conduct the moment the input signal swings negative.

In most practical push-pull power amplifiers, the two output transistors are class AB biased to ensure that the output signal is not affected by crossover distortion. Unlike the class-B transistors that are biased OFF when no input signal is present, the class-AB amplifiers will have a small value of current flowing through both transistors when no input signal is present, as shown in Figure 10-17(c).

Once again, the two power transistors should be matched. In this case, however, one is an NPN and the other is a PNP, and you will have to check the device data sheet to find an NPN transistor-matched PNP counterpart. For example, the NPN 2N3904 transistor has the same operating characteristics and specifications as the PNP 2N3906. As before, if one transistor fails, you should replace both transistors just in case the other transistor has been damaged by the failure or its characteristics have been altered by the fault, and will therefore damage the replacement, .

Quasi-Complementary AF Power Amplifier Circuits

Figure 10-17(d) shows how a higher power output (10 watts or more) can be obtained by having a low-power push-pull transistor pair (Q_1 and Q_2) drive a high-power push-pull transistor pair (Q_3 and Q_4). This circuit is called **quasi-complementary** because the circuit action of two final output NPN transistors (Q_3 and Q_4) operate like (quasi) the NPN and PNP output transistors in a complementary amplifier circuit. The advantage of this circuit is not only its higher output power due to the two stages of power amplification, but also the fact that it uses two high-power final stage NPN transistors, which are much cheaper than high-power PNP transistors.

Quasi-complementary A circuit of two final output NPN transistors that operate like the NPN and PNP output transistors in a complementary amplifier circuit.

When the input signal swings positive, NPN transistor Q_1 turns ON, while PNP transistor Q_2 turns OFF. With Q_2 OFF, Q_4 will receive no input base current, and so it will also cut OFF. Transistors Q_1 and Q_3 form an NPN darlington pair, and so this high current arrangement provides an AF emitter-follower output signal to the load when the input signal is positive.

When the input signal swings negative, NPN transistor Q_1 turns OFF, while PNP transistor Q_2 turns ON. With Q_1 OFF, Q_3 will receive no input base current, and so it will also cut OFF. The negative signal alternation at the base of Q_2 will be inverted to a positive signal alternation, due to Q_2's base-collector inversion. This positive signal alternation at the collector of Q_2 will turn ON Q_4, which will again invert its positive input signal alternation to a negative signal alternation, which will then drive the loudspeaker. Like Q_1 and Q_3, Q_2 and Q_4 will operate together as a two-stage power amplifier.

The other difference with this circuit is that it has three diodes between the base of Q_1 and Q_2. This extra diode is needed because there are now three base-emitter junctions between the Q_1 and Q_2's base (B-E of Q_1, B-E of Q_3, and B-E of Q_2).

■ **EXAMPLE:**

Calculate the power delivered to the load by the power amplifier circuit shown in Figure 10-17(d).

■ ***Solution:***

$$V_{rms} = V_{PK} \times 0.707 = 15.8\ \text{V} \times 0.707 = 11.17\ \text{V}$$

$$P_{RL} = \frac{V_{RL(rms)}^{\ 2}}{R_L} = \frac{(11.17\ \text{V})^2}{4\ \Omega} = 31.2\ \text{W}$$

SELF-TEST REVIEW QUESTIONS FOR SECTION 10-3

1. Audio frequency amplifiers are designed to amplify all of the frequency that the human voice can produce and the human ear can respond to. This audio frequency range is from about __________ Hz to __________ Hz.
2. What are the two basic types of audio frequency amplifiers?
3. A __________ amplifier is used to boost the voltage level of a weak input signal so that it is large enough to control a __________ amplifier.
4. Why is an emitter capacitor generally included in parallel with an ac amplifier's emitter resistor?
5. A coupling capacitor is inserted between amplifier stages to provide __________ and __________ between amplifier stages.
6. A push-pull amplifier contains two transistors that are both class __________ bias.
7. What advantage(s) does the complementary power amplifier have over the basic push-pull power amplifier?

10-4 TROUBLESHOOTING AMPLIFIER CIRCUITS

Now that we know how these amplifier circuits are supposed to work, let us see how we can troubleshoot a problem when they do not work. First, let us review our basic troubleshooting procedure.

Step 1: DIAGNOSE

The first step is to determine whether a problem really exists. To carry out this step, a technician must collect as much information as possible about the system, circuit, and components used, and then diagnose the problem.

Step 2: ISOLATE

The second step is to apply a logical and sequential reasoning process to isolate the problem. In this step, a technician will operate, observe, test, and apply troubleshooting techniques in order to isolate the malfunction.

Step 3: REPAIR

The third and final step is to make the actual repair, then final test the circuit.

10-4-1 *Isolating the Problem to the Amplifier*

Once you have diagnosed that an amplifier circuit is malfunctioning, the next step is to isolate whether the problem is within the amplifier circuit or external to the amplifier circuit. For example, Figure 10-18(a) shows a typical two-stage audio-frequency voltage amplifier. Referring to the inset in this figure, you can see that the amplifier receives power from the dc power supply, an input signal from the input signal source, and supplies an output signal to a load. Our first step is to isolate whether the problem is within the amplifier circuit or within one of the external circuits. To achieve this, the following checks should be made.

1. Power Supply or Amplifier? If the amplifier is not functioning as it should, first check the power ($+V_{CC}$) and ground connection to the amplifier circuit. If the dc power supply voltage to the circuit is not correct, you will next need to determine whether the problem is in the source (the dc power supply circuit) or in the load (the amplifier circuit). If the amplifier is disconnected from the dc power supply and the voltage returns to its normal potential, the amplifier has developed some short within its circuit and this is pulling down the $+V_{CC}$ supply voltage. If, on the other hand, the amplifier is disconnected from the dc power supply and the supply voltage still remains at its incorrect value, the problem is within the dc power supply circuit.
2. Amplifier Source or Output Load? If the output signal from the multiple-stage amplifier is incorrect, remember that a bad load can pull down a voltage signal and make it appear as though the amplifier is faulty. When you have no output, or a bad output, you will first need to disconnect the output of the amplifier from the load to determine whether the problem is in the source (amplifier) or the load. If the load is disconnected and the signal is good, the impedance of the load is pulling down the output of the amplifier. On the other hand, if the load is disconnected and the signal output of the amplifier is still bad, the problem is in the amplifier.
3. Signal Source or Amplifier Load? If the output signal from a multiple-stage amplifier is incorrect, do not backtrack through the amplifier. Go immediately to the multiple-stage amplifier's input, and check the input signal to the first stage. Once again, if the input signal to the amplifier is incorrect, you will first have to isolate whether the problem exists in the signal source or in the amplifier. If the input signal source is disconnected from the amplifier and the signal generated by V_S returns to its normal level, a fault in the amplifier input is pulling down the input signal. On the other hand, if the input signal source is disconnected from the amplifier, and the signal does not return to its normal level, a fault exists in the input signal source circuit.

10-4-2 *Troubleshooting a Voltage Amplifier Circuit*

Once you have isolated the problem to the amplifier circuit, you will next need to test the amplifier's dc bias and ac signal voltages at different points in the circuit to isolate the faulty component. Figure 10-18(a) shows how a fully operational two-stage AF voltage amplifier should test at different points in the circuit

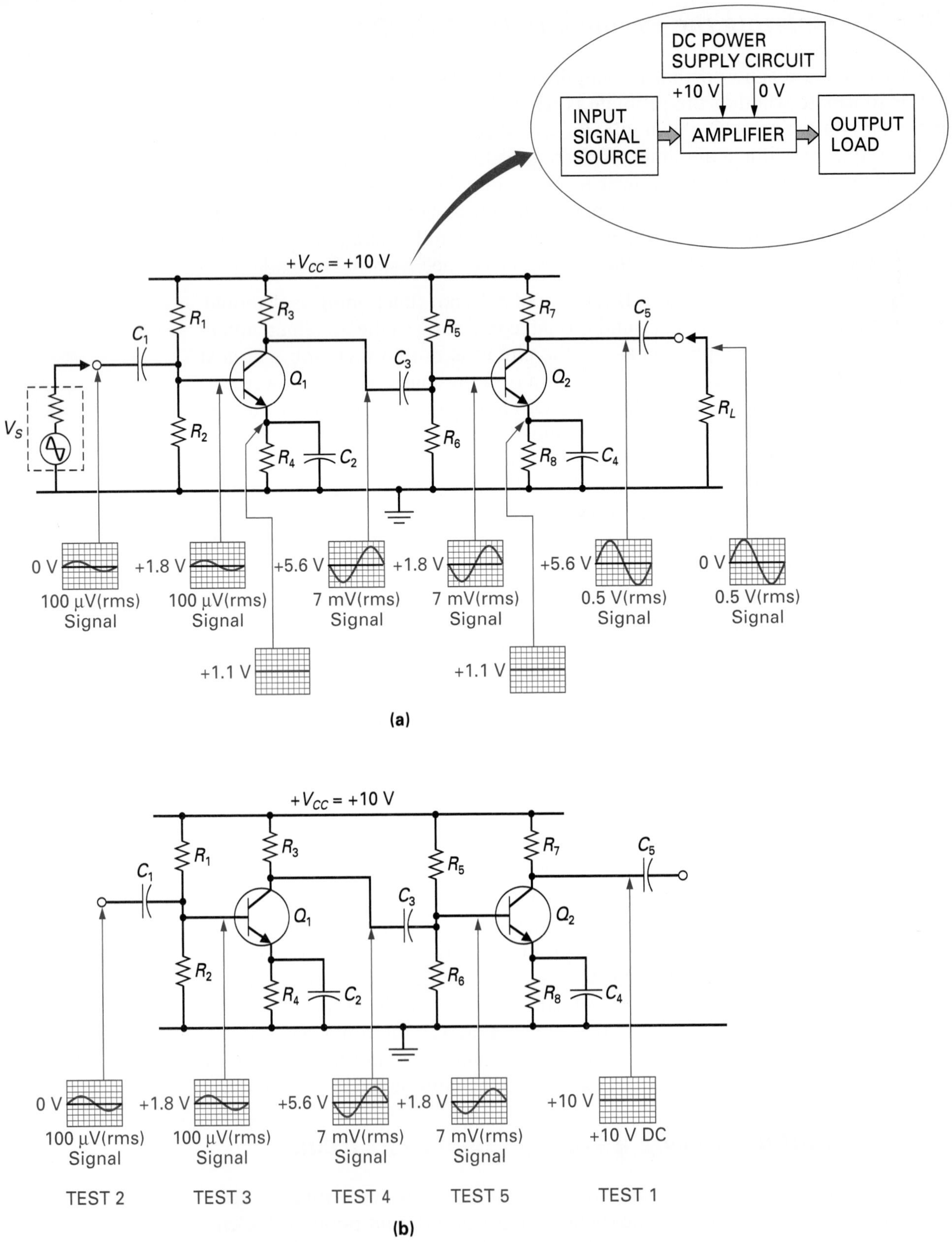

DC POWER SUPPLY CIRCUIT
+10 V
0 V
INPUT SIGNAL SOURCE
AMPLIFIER
OUTPUT LOAD
+V_{CC} = +10 V
R_1
R_2
R_3
R_4
R_5
R_6
R_7
R_8
R_L
C_1
C_2
C_3
C_4
C_5
Q_1
Q_2
V_S
0 V
100 μV(rms) Signal
+1.8 V
100 μV(rms) Signal
+5.6 V
7 mV(rms) Signal
+1.8 V
7 mV(rms) Signal
+5.6 V
0.5 V(rms) Signal
0 V
0.5 V(rms) Signal
+1.1 V
+1.1 V
(a)
+V_{CC} = +10 V
0 V
100 μV(rms) Signal
+1.8 V
100 μV(rms) Signal
+5.6 V
7 mV(rms) Signal
+1.8 V
7 mV(rms) Signal
+10 V
+10 V DC
TEST 2
TEST 3
TEST 4
TEST 5
TEST 1
(b)

FIGURE 10-18 Troubleshooting Voltage Amplifier Circuits.

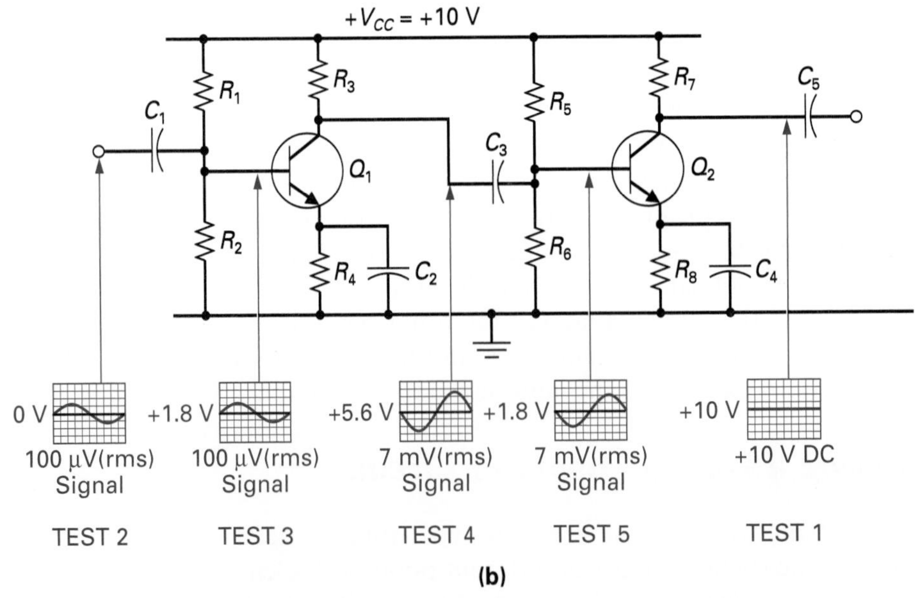

using the oscilloscope. The oscilloscope has been used on its dc setting to make these tests because it will allow you to check both the dc bias and ac signal voltages at various test points throughout the circuit. When using the oscilloscope to test several amplifier stages, remember that the input signal will be much smaller in amplitude than the final output signal. Therefore, you may have to reduce the oscilloscope's volts/division setting as you go back to test the input signal. For example, if the amplifier's final output signal is 1000 times greater than the amplifier's input, you may find that the best volts/division oscilloscope setting for viewing the output waveform will be so large that the input signal will not even be visible. This could mislead you into believing that the amplifier does not have an input signal.

As an example, the amplifier circuit in Figure 10-18(b) shows the procedure you should follow to find the problem in this circuit. Testing the output of the amplifier, you can see that we are not getting an ac output signal. The output at the collector of Q_2 is only a steady +10 V dc voltage (TEST 1). Knowing that there is a problem at some preceding point in the circuit, the next test should be to check the input signal because an amplifier cannot give a good output unless it has a good input (TEST 2). Test 2 shows that the input signal is present, and test 3 indicates that the dc base bias and ac input signal is present at the base of Q_1. The signal is then traced through the amplifier until you find the stage that has a good input signal but a bad output signal. The malfunctioning stage in Figure 10-18(b) is Q_2. Test 5 shows that the dc base bias voltage and ac signal voltage is present at the base of Q_2, and therefore Q_2 should be ON. The +10 V on the collector of Q_2, however, indicates that the transistor is OFF. Because $V_{C(Q2)} = V_{CC}$, a collector-emitter open should be suspected, and you should replace Q_2. Although Q_2 is the most likely cause of the problem in this circuit, remember that a transistor will not operate correctly if one of its associated components is faulty. For example, Q_2 will be cut OFF if either R_7 or R_8 are open. To be sure that Q_2 is the fault, remove the transistor from the circuit and test its P-N junctions with an ohmmeter.

10-4-3 Troubleshooting a Power Amplifier Circuit

In Figure 10-19(a), a class-A voltage amplifier is driving the input of a diode-biased complementary push-pull power amplifier. The oscilloscope displays show the dc bias voltages and ac signal voltages obtained at different points in this operational circuit. Like the previous voltage amplifier troubleshooting procedure, you must first determine that the problem is within the amplifier circuit by checking

1. the $+V_{CC}$ voltage applied to the amplifier and the ground connection to the amplifier
2. that the load is not loading the output amplifier signal
3. that the amplifier is receiving a good input signal

Once the problem is isolated to the two-stage amplifier in Figure 10-19(a), the next step would be to use the oscilloscope to trace the signal through the amplifier circuit. If an incorrect signal is obtained at the output of Q_1, do not

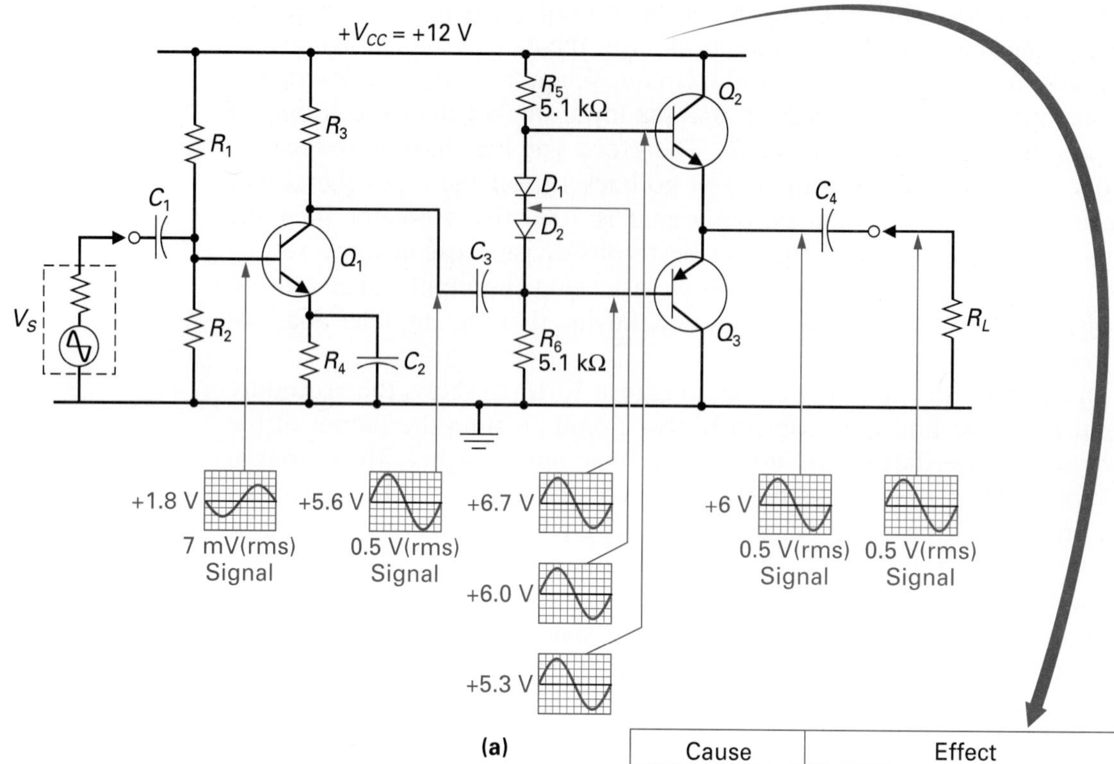

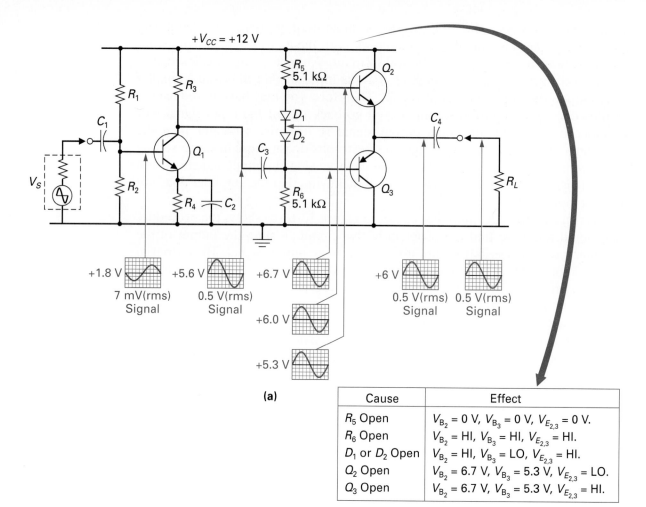

(a)

Cause	Effect
R_5 Open	V_{B_2} = 0 V, V_{B_3} = 0 V, $V_{E_{2,3}}$ = 0 V.
R_6 Open	V_{B_2} = HI, V_{B_3} = HI, $V_{E_{2,3}}$ = HI.
D_1 or D_2 Open	V_{B_2} = HI, V_{B_3} = LO, $V_{E_{2,3}}$ = HI.
Q_2 Open	V_{B_2} = 6.7 V, V_{B_3} = 5.3 V, $V_{E_{2,3}}$ = LO.
Q_3 Open	V_{B_2} = 6.7 V, V_{B_3} = 5.3 V, $V_{E_{2,3}}$ = HI.

Normal Voltages: V_{B_2} = 6.7 V, V_{B_3} = 5.3 V, $V_{E_{2,3}}$ = 6.0 V.

LO = Voltages lower than normal.
HI = Voltages higher than normal.

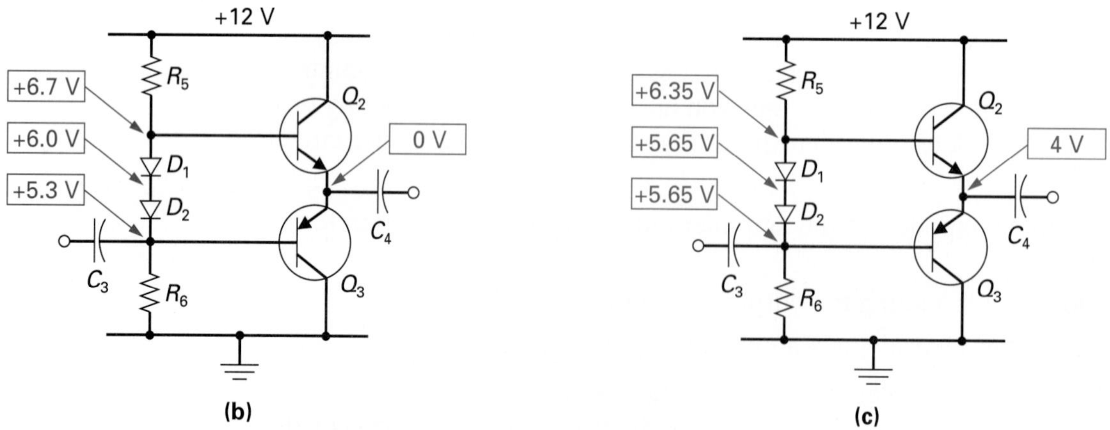

FIGURE 10-19 Troubleshooting Power Amplifier Circuits.

assume that Q_1 is the problem because a short or low-resistance path in the input of the power amplifier stage can pull down the output signal from Q_1. In this case, you will have to isolate the voltage amplifier stage from the power amplifier stage. Because these two amplifiers are probably mounted on a printed circuit board, the best way to isolate Q_1 from Q_2 and Q_3 is to unsolder one lead of the coupling capacitor C_3. If the signal at the output of the voltage amplifier returns to its normal level, the problem is in the power amplifier. On the other hand, if after Q_1 is isolated from the power amplifier, the signal voltage from the Q_1 is still not accurate, the problem is within the voltage amplifier.

Because voltage amplifier troubleshooting was discussed in the previous section, we will now concentrate on troubleshooting the power amplifier circuit. The oscilloscope tests in Figure 10-19(a) show how the power amplifier should operate under normal conditions, while the table in Figure 10-19(a) lists the effect you should expect for a given cause. Let us apply our troubleshooting understanding to two examples. After isolating the fault to the power amplifier and disconnecting the input signal, the dc bias voltage readings shown in Figure 10-19(b) and (c) were obtained. In Figure 10-19(b), the base bias voltages are all normal, and therefore 6 V should be at the emitter junction of Q_2 and Q_3. The O_V reading appears to indicate that Q_3 has developed either an internal short or an internal open. In either case, Q_2 may have damaged Q_3, and so, you should replace both Q_2 and Q_3 with matched transistors. In Figure 10-19(c), the readings indicate that D_2 has shorted. Because D_1 and D_2 are also normally matched to have identical characteristics, both diodes should be replaced. D_1 may have been damaged or altered by the D_2 short.

SELF-TEST REVIEW QUESTIONS FOR SECTION 10-4

1. When troubleshooting an amplifier circuit, the first step is to isolate whether the problem is within the amplifier circuit or within the ____________, ____________, or ____________.
2. How can we best isolate whether a problem is in a source or in a load?
3. Why is the oscilloscope best suited for testing an amplifier circuit?
4. Why should both transistors in a push-pull amplifier be replaced if only one transistor has an internal fault?

SUMMARY

Basic Amplifier Types (Figure 10-20)

1. The amplifier's main objective is to produce a gain (symbolized A) which is the ratio of the amplitude of the output signal to the amplitude of the input signal.
2. An amplifier is often symbolized as a triangle with an input terminal on the left and an output terminal on the right.
3. The voltage amplifier, which is designed to produce an output signal voltage that is greater than the input signal voltage, will typically receive an

QUICK REFERENCE SUMMARY SHEET

FIGURE 10-20 Basic Amplifier Types.

Voltage Amplifier

V_{IN} → V → V_{OUT}

$$A_V = \frac{V_{OUT}}{V_{IN}} \text{ or } \frac{\Delta V_{OUT}}{\Delta V_{IN}}$$

$$V_{OUT} = A_V \times V_{IN}$$

Current Amplifier

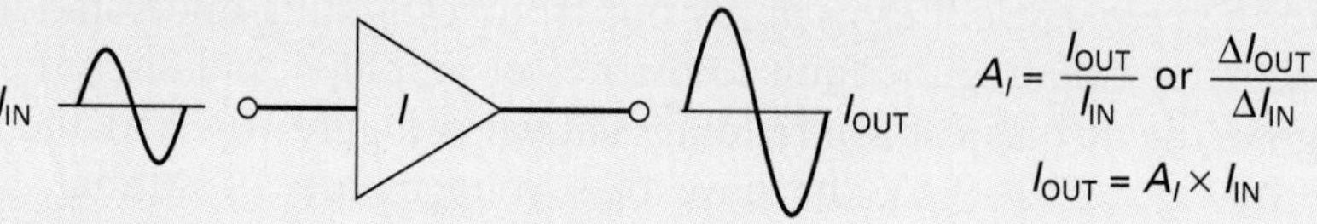

I
I_{IN}
I_{OUT}
$A_I = \frac{I_{OUT}}{I_{IN}}$ or $\frac{\Delta I_{OUT}}{\Delta I_{IN}}$
$I_{OUT} = A_I \times I_{IN}$

Power Amplifier

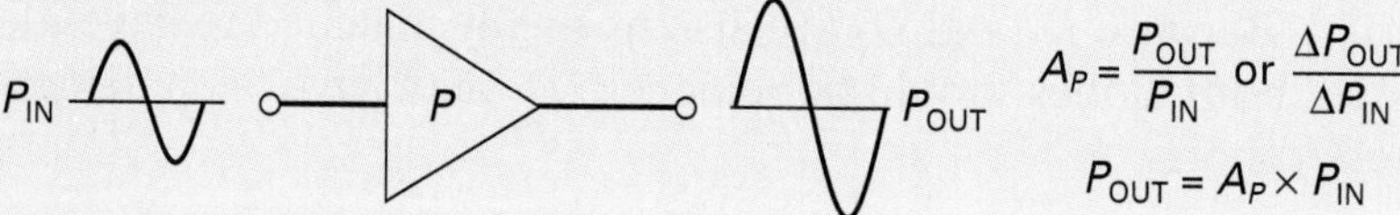

input signal (V_{IN}) measured in millivolts and produce an output signal (V_{OUT}) that is normally measured in volts. The amount of voltage amplification, or voltage gain (A_V), is a ratio of the output signal voltage to the input signal voltage.

4. The current amplifier is designed to produce an output signal current (I_{OUT}) that is greater than the input signal current (I_{IN}). The amount of current amplification, or current gain (A_I), can be calculated by using the same gain ratio formula of output over input.
5. The power amplifier is designed to produce an output signal power that is greater than the input signal power. This type of amplifier will typically receive an input signal (P_{IN}) measured in milliwatts and produce an output signal (P_{OUT}) normally measured in watts. The amount of power amplification, or power gain (A_P), can be calculated by using the same gain ratio formula of output over input. Because power is equal to the product of voltage and current ($P = V \times I$), the power gain of an amplifier can also be calculated by multiplying the amount of voltage gain by the amount of current gain.
6. The voltage, current, and power amplifier block symbols are used to represent a complete electronic amplifier circuit. This amplifier circuit generally contains one or more transistors that act as the circuit's controlling element

and several associated components, such as resistors and capacitors, that control the characteristics and stability of the transistor.

7. The bipolar junction transistor can be connected in one of three different configurations to have it function as a power amplifier, voltage amplifier, or current amplifier.
8. The common-emitter circuit configuration provides a high current gain and a high voltage gain. Therefore the combined power gain is high. It also has a medium input impedance, a high output impedance, and a 180° phase shift between input and output. All of these generally good characteristics make the common-emitter transistor circuit configuration the most widely used.
9. The common-base circuit configuration provides a high voltage gain. However, its current output is slightly less than the current input, so the combined power gain is medium. It also has a very low input impedance, which will load a source, and a high output impedance, which accounts for the low output current. This configuration does not invert the signal between input and output. All of these generally poor characteristics make the common-base transistor circuit configuration the most infrequently used.
10. The common-collector circuit configuration provides a high current gain. However, its voltage output is slightly less than the voltage input, so the combined power gain is low. The circuit's key advantage is its high input impedance and very low output impedance, making it ideal as an impedance matching device between a low-current (high-impedance) source and a high-current (low-impedance) load. This circuit is also referred to as an emitter follower.

Decibels (dBs) and Common Power and Voltage/Current Ratios (Figure 10-21)

11. The gain of an amplifier is normally given in decibels, the noise level in an amplifier circuit is generally described as the number of decibels it is below the level of the desired signal, an amplifier's response to frequency is generally graphed in decibels, and many other devices, such as antennas and microphones, are also rated in decibels.
12. Logarithms can be used to simplify problem solving, by
    - **a.** simplifying multiplication to addition,
    - **b.** simplifying division to subtraction,
    - **c.** simplifying raising to a power to simple multiplication, and
    - **d.** simplifying extracting square roots to simple division.
13. One of the primary reasons we need an understanding of logarithms is because the difference in power at the output of an amplifier compared to the input is given in the logarithmic unit decibels or dBs. In order to properly understand decibels, we will need to first have a clear understanding of logarithms.
14. There are two basic types of logarithms: natural logarithms, which use a base of 2.71828, and common logarithms, which use a base of 10.

QUICK REFERENCE SUMMARY SHEET

FIGURE 10-21 Decibels (dBs), Common Power, and Voltage/Current Ratios.

	Decibels	Power Ratio	Voltage/Current Ratio
Gain:	40	× 10,000	× 100
	20	× 100	× 10
	6	× 4	× 2
	3	× 2	× 1.4
	1	× 1.2	× 1.1
Loss:	−1	÷ 1.2	÷ 1.1
	−3	÷ 2	÷ 1.4
	−6	÷ 4	÷ 2
	−20	÷ 100	÷ 10
	−40	÷ 10,000	÷ 100

$$A_{P(dB)} = 10 \times \log_{10} \frac{P_{OUT}}{P_{IN}}$$

$$A_P = \text{antilog} \frac{A_{P(dB)}}{10}$$

$$A_{TOTAL} = A_1 \times A_2 \times A_3 \times \cdots$$

$$A_{TOTAL} = A_{1(dB)} + A_{2(dB)} + A_{3(dB)} + \cdots$$

$$A_{P(dBm)} = \log \frac{P_{OUT}}{1 \text{ mW}}$$

$$P_{OUT} = P_{IN} \times \text{antilog} \frac{A_{P(dB)}}{10}$$

$$P_{IN} = \frac{P_{OUT}}{\text{antilog} \frac{A_{P(dB)}}{10}}$$

$$A_{V(dB)} = 20 \times \log \frac{V_{OUT}}{V_{IN}} \qquad A_V = \text{antilog} \frac{A_{V(dB)}}{20}$$

$$A_{I(dB)} = 20 \times \log \frac{I_{OUT}}{I_{IN}} \qquad A_I = \text{antilog} \frac{A_{I(dB)}}{10}$$

15. A common logarithm is a base 10 exponent. For example, you know that 1 megavolt is equal to 1×10^6 V (or 1,000,000 V) and that 1 kilohm is equal to $1 \times 10^3\ \Omega$ (or 1000 Ω). The power of 10 exponent used in these examples is the logarithm. For example, the common logarithm (base 10) of 1,000,000 is 6, and the common logarithm (base 10) of 1,000 is 3.

16. The expression $10^6 = 1{,}000{,}000$ is an exponential expression because the quantity is expressed in exponential form. The expression $\log_{10} 1{,}000{,}000 = 6$ is a logarithmic expression because the quantity is expressed in logarithmic form.

17. The human ear is not sensitive to the amount of change that occurs but rather to the ratio of change that occurs.

18. To make the sound twice as loud, the power must be increased by ten times (1 W to 10 W); to make the sound three times as loud, the power must be increased by one hundred times (1 W to 100 W).

19. This relationship between detected sound and power output is logarithmic because the sound detected goes up in multiples of 10. Because the human ear responds to sound intensity in this way, it is convenient for us to measure sound intensity in a logarithmic unit, which is a unit that goes up in equal steps for every tenfold increase in intensity. This unit is called the bel, in honor of the great American scientist Alexander Graham Bell. The bel was originally used to indicate the amount of signal loss in two miles of telephone wire. When dealing with small electronic circuits, such as amplifiers, we will use a smaller unit of measure, the decibel (dB), which is equal to 1/10th of a bel (1×10^{-1} bel).

20. If the gain of several stages is given as an arithmetic power ratio, the total gain is equal to the product of all of the stages.
21. When the gain of a multi-stage circuit or system is measured in dBs, the total gain or loss of the amplifier can be calculated by simply adding all of the individual dB values.
22. Cable and telecommunication companies have established a standard of 1 mW as a reference power. In this decibels-relative-to 1-milliwatt, or dBm, system, power levels below 1 mW are indicated as –dBm values, and power levels above 1 mW are shown as +dBm values.
23. The decibel is a measure of change, and its value gives us an indication of device, circuit, or system performance.

Amplifier Classes of Operation (Figure 10-22)

24. The transistor amplifier is generally biased at a dc operating point that is called the quiescent or *Q* point. The input signal will vary the transistor's dc bias voltage above and below its *Q* point, causing a change in base current (I_B), collector current (I_C), and therefore output voltage (V_{CE}). By biasing an amplifier at different points on the dc load line, we can obtain some circuit characteristics that are ideal for certain applications.
25. Any transistor amplifier that has its dc operating point set near the center of the load line so that the output current (I_C) flows during the entire cycle of the ac input signal is said to be operating as a class-A amplifier. Class-A amplifiers are said to operate in a linear manner because a change in the input signal produces a proportional change in the output signal.
26. When an amplifier is biased so that its output current will flow for only half of the complete ac input signal cycle, the circuit is said to be operating as a class-B amplifier. To achieve this mode of operation, the amplifier is biased

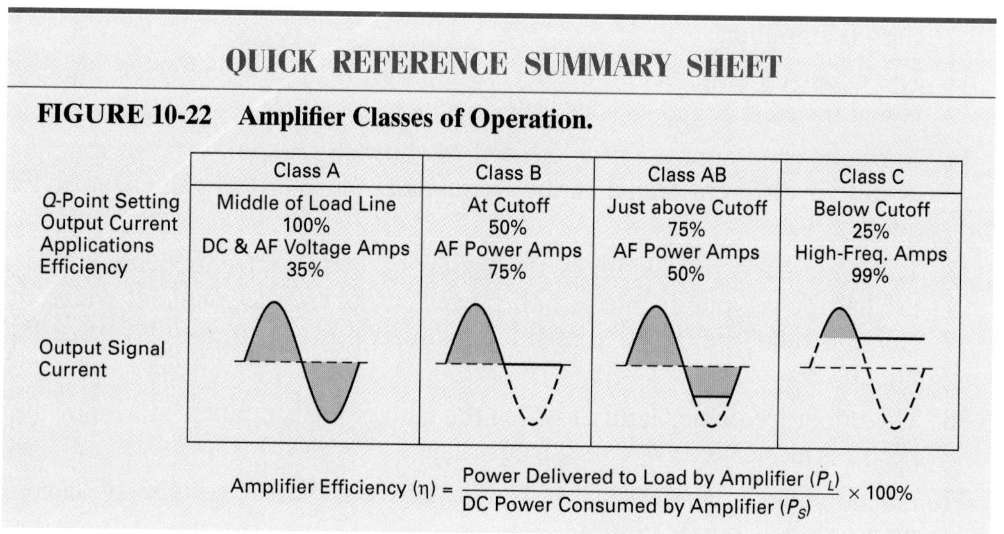

QUICK REFERENCE SUMMARY SHEET

FIGURE 10-22 Amplifier Classes of Operation.

	Class A	Class B	Class AB	Class C
Q-Point Setting	Middle of Load Line	At Cutoff	Just above Cutoff	Below Cutoff
Output Current	100%	50%	75%	25%
Applications	DC & AF Voltage Amps	AF Power Amps	AF Power Amps	High-Freq. Amps
Efficiency	35%	75%	50%	99%
Output Signal Current				

$$\text{Amplifier Efficiency } (\eta) = \frac{\text{Power Delivered to Load by Amplifier } (P_L)}{\text{DC Power Consumed by Amplifier } (P_S)} \times 100\%$$

so that its Q point is at cutoff, and therefore the output current will only flow during the time that the input alternation forward biases the transistor. The advantage of this amplifier class of operation is that, because the amplifier is only turned ON for 50% of the time, the circuit will only consume half the power of an equivalent class-A amplifier.

27. When an amplifier is biased so that its output current will flow for less than one-half of the ac input cycle, it is said to be operating as a class-C amplifier. This is achieved by biasing the transistor amplifier's dc operating point below cutoff so that for most of the time the amplifier is reverse biased or cutoff, and for only a small period of time will it produce an output pulse of current. This type of amplifier is very efficient because it only consumes power for a very short time compared to the complete cycle of the ac input signal.

28. When an amplifier is biased so that its output current will flow for slightly less than one full cycle of the ac input (class A), but more than one-half of the ac input (class-B), the amplifier is said to be operating as a class-AB amplifier. This type of amplifier is a compromise between a class A and a class B. Its operating point is set at slightly above cutoff, causing an output current to be produced for slightly more than one-half of the input cycle.

29. The Greek letter "eta" (η, pronounced "eat-eh"), is used to designate an amplifier's efficiency, which is the ratio of the power delivered to the load by the amplifier to the power consumed by the amplifier.

30. Class-A amplifiers are constantly drawing power from the $+V_{CC}$ dc power supply the whole time that they are supplying power to the load. As a result, class-A amplifiers tend to be fairly inefficient.

31. Class-B amplifiers are OFF for 50% of the input cycle and, because they consume no power during this time, they tend to be more efficient than their class-A counterparts.

32. Class-C amplifiers are OFF, and therefore not consuming power, for more than 50% of the input cycle. A class-C amplifier is more efficient than a class-B amplifier.

33. The class-AB amplifier's power consumption is midway between the class A and the class B and so is its efficiency.

34. A frequency response curve is used to show the response of an amplifier circuit to different signal input frequencies by plotting gain against frequency.

35. The amplifier's bandwidth is the group or band of frequencies between the half-power points. These half-power points are the points at which the gain has fallen below 50% of the maximum power gain (–3dB). (See Figure 10-23.)

36. The upper frequency limit at which the gain of the amplifier falls below half power is also called the cutoff frequency.

37. An amplifier's bandwidth exists between the 70.7% voltage or current points, or half-power points.

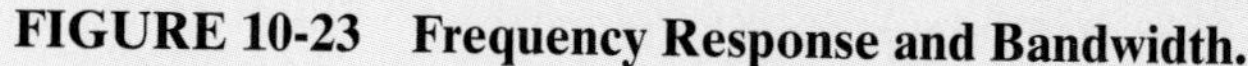

QUICK REFERENCE SUMMARY SHEET

FIGURE 10-23 Frequency Response and Bandwidth.

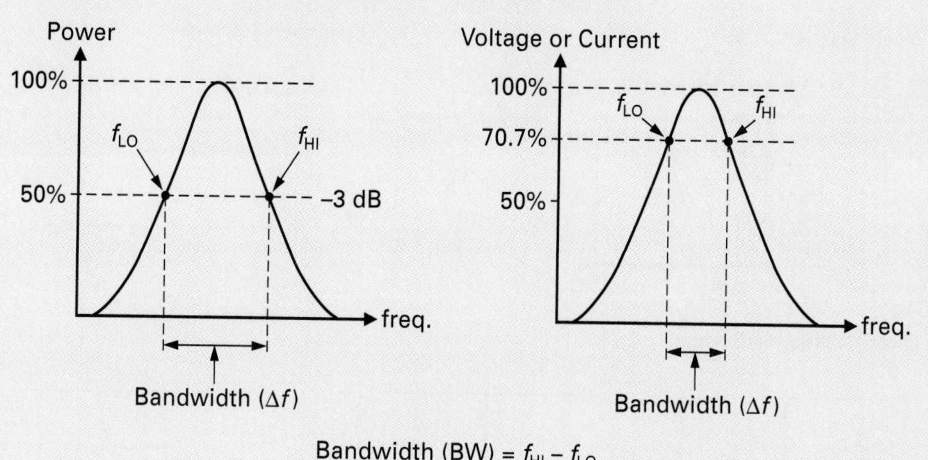

38. A bode diagram is another form of frequency response curve that plots a circuit's gain in decibels against frequency.

Direct Current (DC) Amplifiers (Figure 10-24)

39. Direct current amplifiers, which are normally always class A biased, are used to amplify dc or low frequency ac voltages or currents produced by transducers or sensors. These transducers will sense heat, light, pressure, or vibration and produce a corresponding electrical signal that is usually extremely weak in amplitude and must be amplified up to a more usable level.
40. Emitter resistor R_E provides "emitter feedback" for temperature stability.
41. Because the input signal to a dc amplifier is a dc voltage, capacitors cannot be connected in the signal path. A capacitor will act as a block to any dc signals.
42. The voltage gain of a dc amplifier remains almost constant from 0 Hz (dc) to about 2000 Hz (ac). Beyond 2000 Hz or 2 kHz, the voltage gain of a dc amplifier drops off rapidly.
43. From a dc input signal of 0 Hz up to an ac input signal frequency of about 2 kHz, the voltage gain of a dc amplifier is approximately equal to the ratio of the transistor's collector resistance R_C to the emitter resistor R_E.
44. From a dc input signal of 0 Hz up to an ac input signal frequency of about 2 kHz, the voltage gain of a dc amplifier is approximately equal to the ratio of the transistor's collector resistance R_C to the emitter resistor R_E. However, if a load is connected to the output, the voltage gain of an amplifier is equal to the equivalent parallel resistance of R_C and R_L divided by R_E.
45. To calculate the efficiency of a dc amplifier, we will have to know how much power is being delivered to the load and how much dc power is being consumed by the amplifier.

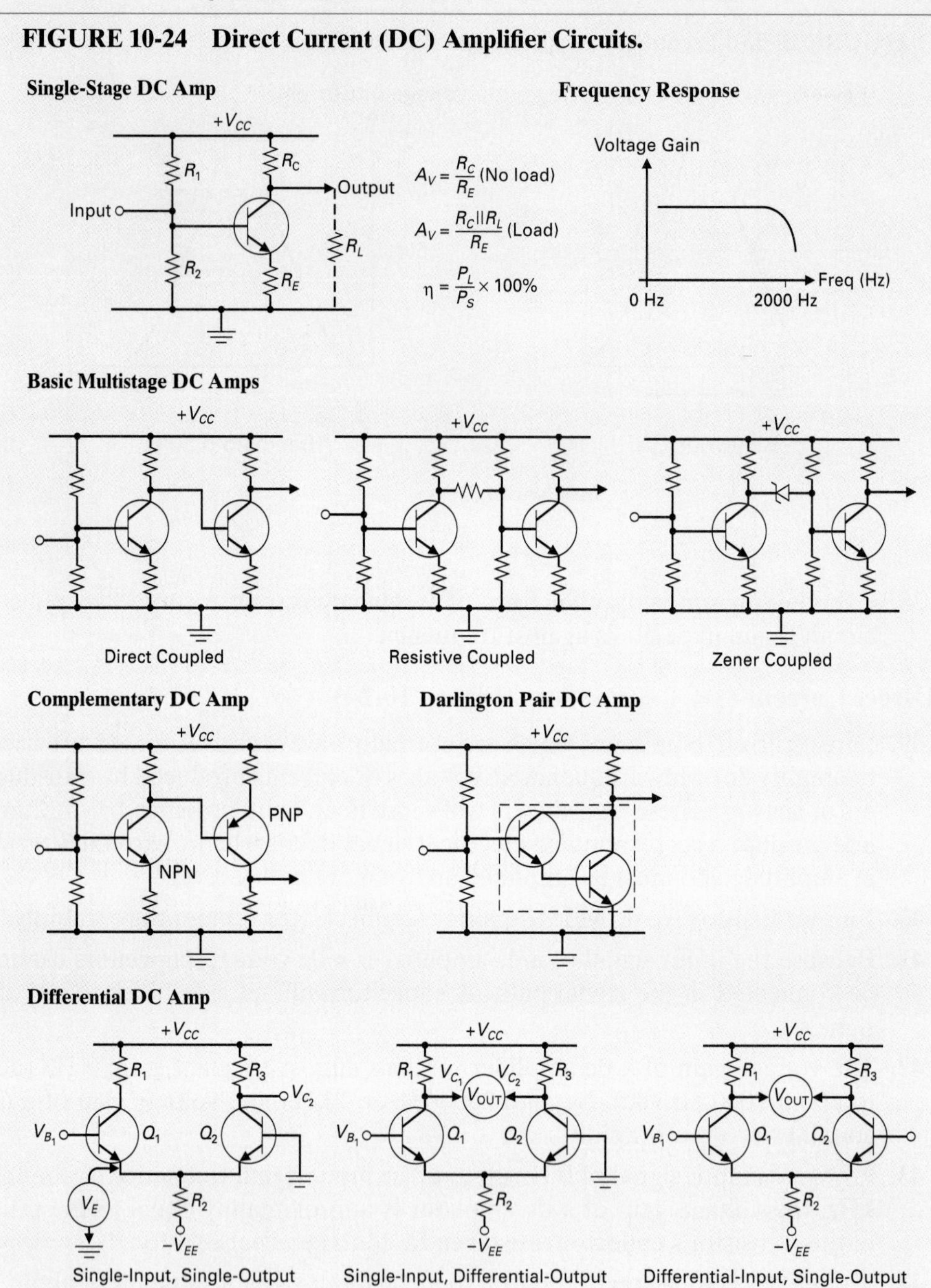

FIGURE 10-24 Direct Current (DC) Amplifier Circuits.

46. With the common-emitter dc amplifier circuit, there will be a 180° phase difference between the input signal voltage and the output signal voltage. This inverting action of the amplifier is not a problem because the output voltage signal still changes in accordance with the input signal change.

47. In some applications, a single dc amplifier stage may not provide enough gain. In these instances, several dc amplifier stages may be needed to increase the input signal up to a desired amplitude.

48. A direct coupled dc amplifier has the output of the first stage coupled directly into the input of the second stage. The disadvantage of this direct coupled dc amplifier is its lack of isolation between stages.

49. When two or more dc amplifiers are connected end-to-end, or cascaded, the overall dc amplifier voltage gain is equal to the product of all the individual stage gains.

50. To prevent distortion from occurring in a direct-coupled dc amplifier, it is generally necessary to decrease the gain of the first stage, so that we can control the gain of the second stage. This sacrifice of gain for circuit control is typical with most amplifier circuits.

51. A resistive-coupled dc amplifier has a large value resistor connected between stages so that the bias voltages and currents of each stage are somewhat independent of one another, and we can control the gain of each stage to produce a higher gain output than a direct-coupled amplifier. The disadvantage of this circuit is that, because the resistor is connected in the signal path, there is a voltage drop and signal voltage loss across the coupling resistor.

52. A zener coupled dc amplifier has a reverse-biased zener diode connected between stages. Zener coupling two dc amplifier stages causes the overall voltage gain to be almost equal to the calculated voltage gain of stage 1 times stage-2 gain.

53. The basic complementary dc amplifier makes use of an NPN transistor and a PNP transistor. The complementary dc amplifier arrangement enables us to cascade several stages to achieve high signal gain without distortion because the Q point, which is set by the previous stage, continually alternates between being either more positive (slightly above the mid-point) or less positive (slightly below the mid-point).

54. With the darlington pair dc amplifier, the collectors of two transistors can be tied together, and the emitter of one transistor is direct-coupled to the base of the other. The overall common-emitter current gain of this circuit arrangement is effectively equal to the current gain of the first transistor (beta Q_1) multiplied by the current gain of the second transistor (beta Q_2).

55. The differential dc amplifier circuit contains two transistors (Q_1 and Q_2) which are matched to have identical characteristics, each of which has its own collector resistor and voltage divider bias. The distinguishing feature of this circuit is the sharing of an emitter resistor by both transistors.

56. The ability of the differential amplifier to reject common mode input signals caused by noise or temperature instability is its key advantage and accounts for why it is perhaps the most important of all the dc amplifiers. A differential amplifier's ability to provide a high differential gain (A_{VD}) and a low common-mode gain (A_{CM}), is a measure of a differential amplifier's performance, and this ratio is called the common-mode rejection ratio (CMRR).

57. The key application of the differential amplifier is in the "operational amplifier," which contains several differential amplifier stages. The operational amplifier is an IC amplifier that has almost completely replaced the discrete, or individual-component, amplifier circuits.

Audio Frequency (AF) Amplifiers (Figure 10-25)

58. *Audio* is a Greek word meaning "I hear," which is why the term *audio frequency range* includes all of the frequencies that the human voice can produce and the human ear can respond to. This audio frequency range is from about 20 Hz to 20,000 Hz. An audio frequency (AF) amplifier is a circuit containing one or more amplifier stages designed to amplify an audio-frequency signal. These audio amplifiers are used within any electronic system that processes, transmits, or receives audio signals such as radio transmitters and receivers, television transmitters and receivers, telephone communication systems, music systems, multimedia computer systems, and so on.
59. The two basic types of audio amplifiers are the AF voltage amplifier, which is generally class A biased, and AF power amplifier, which is generally class B or class AB biased.
60. A voltage amplifier is used to boost the voltage level of a weak input signal so that it is large enough in amplitude to control a power amplifier. When used in this manner, voltage amplifiers are often referred to as preamplifiers or preamps because they precede the main power amplifier, which is used to boost signal current so that the final output can control a low-impedance load, such as a loudspeaker.
61. Degenerative feedback dramatically reduces the gain of any ac signal amplifier, but can be remedied by including a capacitor in parallel with the emitter resistor.
62. With an *RC*-coupled AF voltage amplifier, a capacitor is used to connect one amplifier stage to another. This coupling capacitor performs two functions:
    1. Due to its low reactance to ac signals, it will pass the audio frequency information signal from the first stage to the second stage with no distortion, and
    2. Due to its high reactance to dc, the coupling capacitor will isolate the dc bias voltages of one stage from the DC bias voltages of another.
63. The disadvantage of the *RC* coupled amplifier is that maximum power is not transferred between one stage and the next due to the lack of impedance matching between stages.
64. With a transformer-coupled dc amplifier, the transformer is used to match the output impedance of the first amplifier stage to the low input impedance of the second amplifier stage. Impedance matching the amplifier stages will not load down the first amplifier stage, so a high value of signal current and signal voltage (and therefore signal power) will be applied to the second stage. Transformer coupling is often used in audio frequency amplifiers because of its efficient transfer of power between stages. How-

QUICK REFERENCE SUMMARY SHEET

FIGURE 10-25 Audio Frequency (AF) Amplifier Circuits.

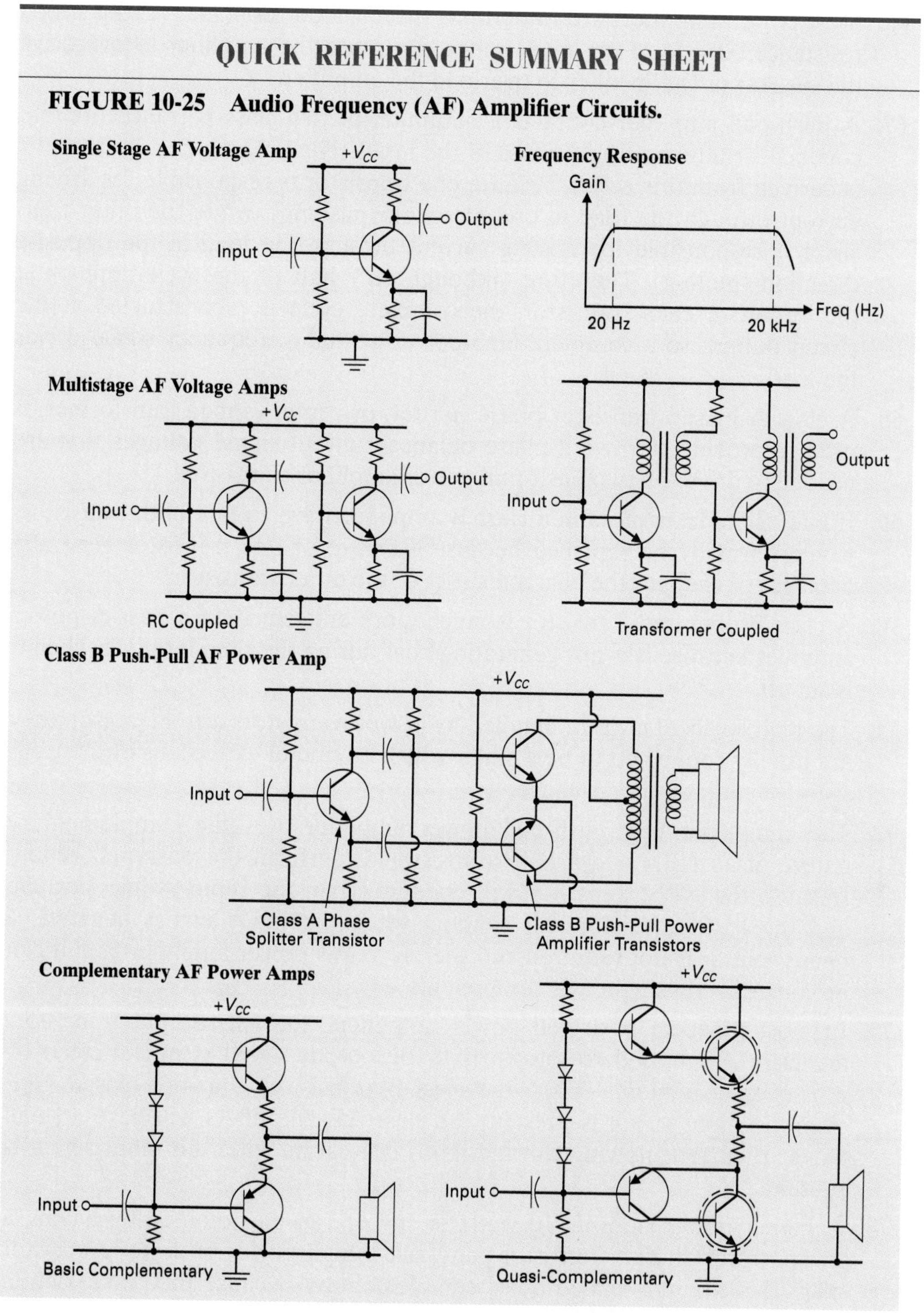

Single Stage AF Voltage Amp

$+V_{CC}$ Input Output

Frequency Response

Gain Freq (Hz) 20 Hz 20 kHz

Multistage AF Voltage Amps

$+V_{CC}$ Input Output

RC Coupled

Input Output

Transformer Coupled

Class B Push-Pull AF Power Amp

$+V_{CC}$ Input

Class A Phase Splitter Transistor

Class B Push-Pull Power Amplifier Transistors

Complementary AF Power Amps

$+V_{CC}$ Input

Basic Complementary

$+V_{CC}$ Input

Quasi-Complementary

ever, the transformer is a more expensive coupling method than the less efficient *RC* coupling method.

65. Although the AF power amplifier does not increase the signal voltage, it does supply the high output signal current (and therefore power) that is needed to operate a low-impedance load such as a loudspeaker.

66. Most power amplifiers are rated for a specific load resistance. Using a load resistance outside of the rated range can cause the amplifier to overdrive the speaker or the speaker to overload the amplifier.

67. A push-pull amplifier contains a minimum of two class-B transistors that conduct for alternate half cycles of the input signal. The name "push-pull" is derived from this action because one transistor is responsible for driving current through the load in one direction (pushing), while the other transistor is responsible for driving current through the load in the opposite direction (pulling). Therefore, although only half of the cycle appears at the output of each transistor, the complete cycle is reconstructed at the circuit output, so a complete high-power ac audio frequency signal drives the load.

68. A class A biased transistor phase splitter, or center-tapped transformer, is used to provide the two 2-phase balanced output signal voltages that are needed to drive the two push-pull power amplifier transistors.

69. If one of the transistors in a class-B amplifier were to fail, both transistors would normally be replaced because a failure of one transistor generally damages or changes the characteristics of the other transistor.

70. A class-B operated transistor is much more efficient than a class-A power amplifier because it is not generating heat during the time intervals when it is cut off.

71. The high levels of current handled by power transistors are the reason they are usually connected to heat sinks, which are metal extensions of the transistor's heat radiating package.

72. The complementary push-pull power amplifier uses the complementary nature of an NPN and a PNP transistor to perform the push-pull action. Because the NPN transistor will conduct when the input swings positive and the PNP transistor will conduct when the input swings negative, a phase splitter is not required and therefore this power amplifier circuit can be constructed using fewer components.

73. In most practical push-pull power amplifiers, the two output transistors are class AB biased to ensure that the output signal is not affected by crossover distortion. Unlike the class-B transistors that are biased OFF when no input signal is present, the class-AB amplifiers will have a small value of current flowing through both transistors when no input signal is present.

74. A higher power output (10 watts or more) can be obtained by having a low-power push-pull transistor pair drive a high-power push-pull transistor pair. This circuit is called quasi-complementary because the circuit action of two final output NPN transistors (Q_3 and Q_4) operate like (quasi) the NPN and PNP output transistors in a complementary amplifier circuit. The advantage of this circuit is not only its higher output power due to the two stages of power amplification, but also the fact that it uses two high-power final stage NPN transistors, which are much cheaper than high-power PNP transistors.

Troubleshooting Amplifier Circuits

75. Once you have diagnosed an amplifier circuit as malfunctioning, the next step is to isolate the problem to the amplifier circuit or one of the external circuits. To achieve this, the following checks should be made.

1. Power Supply or Amplifier? If the amplifier is not functioning as it should, first check the power ($+V_{CC}$) and ground connection to the amplifier circuit. If the dc power supply voltage to the circuit is not correct, you will need to determine whether the problem is in the source (the dc power supply circuit) or in the load (the amplifier circuit). Disconnect to isolate.
2. Amplifier Source or Output Load? If the output signal from the multiple stage amplifier is incorrect, remember that a bad load can pull down a voltage signal and make it appear as though the amplifier is faulty. Disconnect to isolate.
3. Signal Source or Amplifier Load? If the output signal from a multiple stage amplifier is incorrect, do not backtrack through the amplifier. Go immediately to the multiple stage amplifier's input and check the input signal to the first stage. Once again, if the input signal to the amplifier is incorrect, you will first have to isolate whether the problem exists in the signal source or in the amplifier. Disconnect to isolate.

76. Once you have isolated the problem to the amplifier circuit, you will next need to test the amplifier's dc bias and ac signal voltages at different points in the circuit to isolate the faulty component.

77. The oscilloscope on its dc setting is ideal for testing an amplifier circuit because it will allow you to check both the dc bias and ac signal voltages at various test points throughout the circuit. When using the oscilloscope to test several amplifier stages, remember that the input signal will be much smaller in amplitude than the final output signal, and therefore you may have to reduce the oscilloscope's volts/division setting as you go back to test the input signal.

NEW TERMS

Audio Frequency (AF) Amplifier
Audio Frequency (AF) Power Amplifier
Audio Frequency (AF) Voltage Amplifier
Balanced Power Signal Voltages
Bandwidth
Bass
Bel
Bode Diagram
Cascaded Amplifier
Class-A Amplifier
Class-A Signal Distortion
Class-AB Amplifier
Class-B Amplifier
Class-C Amplifier
Class of Operation
Common Antilogarithm
Common Logarithm
Common Mode Gain
Common Mode Input Signals
Common Mode Rejection
Common Mode Rejection Ratio
Complementary AF Power Amplifier

Complementary DC Amplifier
Coupling Capacitor
Crossover Distortion
Current Amplifier
Current Gain
Cutoff Frequency
Darlington Pair
Darlington Pair DC Amplifier
dB
dBm
DC Amplifier
Decibels
Degenerative Feedback
Differential DC Amplifier
Differential Gain
Differential Input Single-Output Connection
Differential Mode Input Signals
Direct Coupled
Direct Coupling
Direct Current (DC) Amplifier
Exponential Expression
Exponential Form
Frequency Response Curve
Gain
Heat Sinks
Log
Logarithm
Logarithmic Expression
Logarithmic Form
Logarithmic Unit
Matched Transistors
Multiple Stage Amplifier
Multistage AF Voltage Amplifier
Natural Logarithm
Phase Splitter
Power Amplifier
Power Gain
Powers of 10
Preamplifier
Push-Pull Power Amplifier
Quasi-Complementary AF Power Amplifier
RC Coupled AF Voltage Amplifier
Resistive Coupled DC Amplifier
Resistive Coupling
Signal Distortion
Single-Input Differential-Output Connection
Single-Input Single-Output Connection
Single-Stage AF Voltage Amplifier
Stepper Motor
Tone
Tone Control
Transformer Coupled AF Voltage Amplifier
Treble
Voltage Amplifier
Voltage Gain (A_V)
Volume Control
Zener Coupled DC Amplifier
Zener Coupling

REVIEW QUESTIONS

Multiple-Choice Questions

1. A class ____________ amplifier has its dc operating point set at cutoff.

a. A **b.** B **c.** C **d.** AB

2. A class ____________ amplifier has its dc operating point set below cutoff.

a. A **b.** B **c.** C **d.** AB

3. A class ____________ amplifier has its DC operating point set slightly above cutoff.

   a. A **b.** B **c.** C **d.** AB

4. A class ____________ amplifier has its operating point set near the center of the load line.

   a. A **b.** B **c.** C **d.** AB

5. The logarithm of 10,000 is

   a. 2 **b.** 40 **c.** 4 **d.** 100

6. If an amplifier has an input of 25 mW and an output of 25 W, its gain in dBs will be

   a. +30 dB **b.** 1000 dB **c.** −20 dB **d.** +10 dB

7. If a cable has a loss of 4 dBs, the output from the cable will equal ____________% of the input.

   a. 60.2 **b.** 39.8 **c.** 70.7 **d.** 63.2

8. If the first stage of a multi-stage amplifier has a voltage gain of 10 and the second stage has a voltage gain of 2, what will be the overall voltage gain?

   a. 12 **b.** 6 **c.** 20 **d.** 8

9. If an amplifier has a gain of 35 dB and the speaker it is driving has an 8 dB loss, what will be the overall gain?

   a. +43 dB **b.** 4.38 dB **c.** 35 dB **d.** +27 dB

10. A common-emitter power amplifier has a gain of 30 dB and an input of 20 mW. What will be the power out of this amplifier?

    a. 20 W **b.** 10 W **c.** 75 W **d.** 100 W

11. A common collector current amplifier produces a 200 mA output signal when a 30 mA input signal is applied. What is the current gain of this amplifier in decibels?

    a. 16.5 dB **b.** 66.7 dB **c.** 24.3 dB **d.** 12.6 dB

12. Any differences or irregularities introduced unintentionally into the signal are called

    a. cascading **b.** complementary **c.** distortion **d.** darlington effect

13. A darlington pair has

    a. three transistors
    b. a single base-emitter voltage drop
    c. a very low input impedance
    d. a very high current gain

14. The versatility, temperature stability, and noise rejection of the ____________ dc amplifier make it an ideal candidate for many applications.

    a. darlington pair
    b. zener coupled
    c. complementary
    d. differential

15. A capacitor acts as a/an

    a. open switch to dc
    b. block to ac
    c. closed switch to dc
    d. open switch to ac

16. The efficiency of an amplifier is equal to the ratio of

a. R_C to R_E **c.** P_L to P_S
b. V_{OUT} to V_{IN} **d.** All of the above

17. Which dc amplifier would be best suited for switching ON and OFF a high-current relay coil?

a. *C-B* **c.** Zener coupled
b. Complementary **d.** Darlington pair

18. A stepper motor with 72 steps/revolution will cause the rotor to turn ____________ for each step.

a. 5° **c.** 2°
b. 7.5° **d.** 15.5°

19. A differential amplifier should have a ____________ differential gain and a ____________ common-mode gain.

a. high, high **c.** low, low
b. low, high **d.** high, low

20. What are common-mode input signals?

a. voice signal **c.** transducer input signal
b. noise **d.** none of the above

21. Audio frequency voltage amplifiers are often referred to as ____________ amplifiers.

a. dc **b.** pre- **c.** power **d.** IF

22. Audio frequency power amplifiers are usually class ____________ biased.

a. A **b.** B or AB **c.** C **d.** D

23. Audio frequency voltage amplifiers are usually class ____________ biased.

a. A **b.** B or AB **c.** C **d.** D

24. The capacitor that connects an ac ground to the emitter of an ac signal amplifier

a. prevents degenerative feedback
b. acts as an ac bypass for R_E
c. performs the opposite of a coupling capacitor
d. all of the above

25. Because power transistors are designed to boost ____________, a large amount of ____________ is generated. To ensure their operating characteristics are stable and their efficiency good, they are operated with class ____________ bias.

a. voltage, heat, A **c.** impedance, power, A
b. voltage, current, C **d.** current, heat, B

26. Audio frequency power amplifiers are normally operated in class AB to prevent

a. temperature stability **c.** crossover distortion
b. degenerative feedback **d.** all of the above

27. If $R_C = 2.7\ k\Omega$ and $R_L = 10\ k\Omega$, the equivalent collector load resistance will be

a. 12.7 kΩ **b.** 2.1 kΩ **c.** 2.7 kΩ **d.** 10 kΩ

28. The power gain of a class-B push-pull power amplifier is

a. equal to the voltage gain
b. equal to P_{OUT} divided by P_{IN}
c. less than the voltage gain
d. equal to twice the amplifier's current gain

29. If the coupling capacitor were to open, the ac input signal voltage at the transistor's base would

a. be zero **c.** be unchanged
b. decrease **d.** increase

30. The advantage of a complementary push-pull amplifier is that it

a. is class-A biased **c.** contains a phase splitter
b. uses fewer components **d.** uses diode biasing

Essay Questions

31. What is an amplifier, and why is it such an important electronic circuit? (Introduction and 10-1)
32. What are the three basic amplifier types? (10-1-1)
33. List the three bipolar transistor circuit configurations and their key advantages. (10-1-2)
34. What is amplifier gain, and how is it calculated? (10-1-1)
35. What is the decibel, and how is it related to gain? (10-1-3)
36. What is a common logarithm? (10-1-3)
37. What is the difference between an exponential expression and a logarithmic expression? (10-1-3)
38. What function is performed by the log and antilog keys on a standard scientific calculator? (10-1-3)
39. What is the relationship between logarithms and the human ears' response to increases in sound? (10-1-3)
40. Why would gain or loss be better expressed in decibels rather than an arithmetic ratio? (10-1-3)
41. State how power, voltage, and current gain or loss can be calculated in decibels. (10-1-3)
42. How do we convert a decibel gain or loss figure back to an arithmetic ratio figure? (10-1-3)
43. List and briefly describe the four basic amplifier classes of operation. (10-1-4)
44. How is the efficiency of an amplifier calculated? (10-1-4)
45. What is a frequency response curve? (10-1-5)
46. Define the term "amplifier bandwidth." (10-1-5)

47. What is a bode diagram? (10-1-5)
48. In what application would a dc amplifier be used? (10-2)
49. Why is an emitter resistor included in most amplifier circuits? (10-2-1)
50. Describe how the gain of a dc amplifier is calculated. (10-2-1)
51. Most dc amplifiers are biased to operate in what class of operation? (10-2)
52. How is the gain of a multiple stage amplifier calculated? (10-2-2)
53. Define the following terms (10-2-2)
    a. Cascaded amplifier
    b. Signal distortion
    c. Direct coupling
    d. Complementary Amplifier
54. What advantage does the multi-stage complementary dc amplifier have over a direct coupled multi-stage amplifier? (10-2-2)
55. Briefly describe the operation and advantages of the darlington pair dc amplifier. (10-2-2)
56. Sketch the following dc amplifier circuits (10-2-2)
    a. Single-stage voltage divider biased direct coupled
    b. Complementary dc amplifier
    c. Darlington pair dc amplifier
    d. Differential dc amplifier
57. Describe the basic operation of the dc stepper motor and how a darlington pair can be used as a stepping control device. (10-2-2)
58. Briefly describe the operation and versatility of the differential dc amplifier. (10-2-2)
59. What are common-mode input signals? (10-2-2)
60. Define the term *common-mode rejection ratio.* (10-2-2)
61. What is an audio-frequency amplifier, and what are the two basic types? (10-3)
62. Why are AF voltage amplifiers often called preamplifiers? (10-3-1)
63. What is degenerative feedback, and how can an emitter-bypass capacitor prevent it? (10-3-1)
64. What advantage does a transformer coupled AF amplifier have over a capacitor coupled AF amplifier? (10-3-1)
65. Briefly describe and sketch how an AF voltage amplifier can have its gain adjusted in order to control audio signal volume. (10-3-1)
66. Briefly describe and sketch how an AF voltage amplifier can have its frequency response adjusted in order to control audio signal treble and bass. (10-3-1)
67. What is class-A signal distortion? (10-3-1)
68. What is a power amplifier? (10-3-2)
69. Why are power amplifiers generally class-B biased? (10-3-2)
70. Why is a phase splitter needed with some push-pull power amplifiers? (10-3-2)

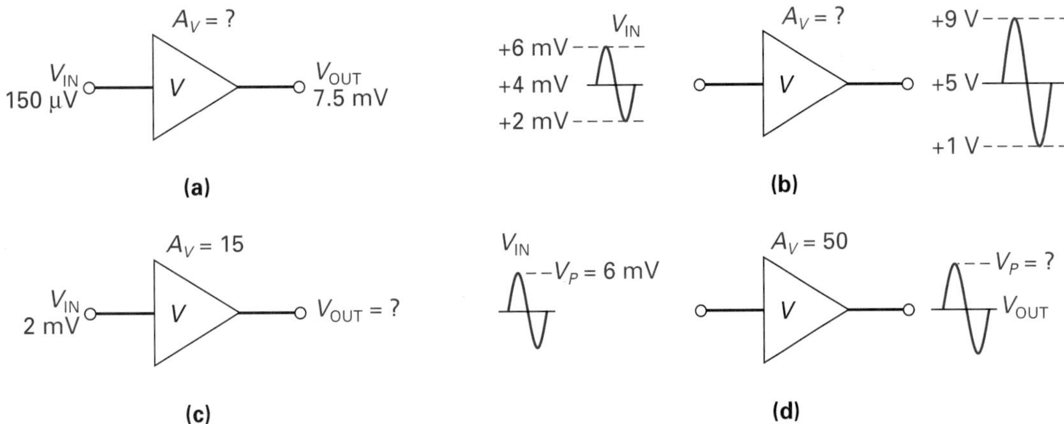

FIGURE 10-26 Voltage Amplifiers.

71. Sketch a typical complementary push-pull amplifier and briefly describe the operation of the circuit. (10-3-2)

72. What advantage(s) does the complementary power amplifier have over the standard push-pull power amplifier? (10-3-2)

73. Define the term "crossover distortion." (10-3-2)

74. To be more specific, why are most power amplifiers class AB biased instead of class B biased? (10-3-2)

75. Sketch and describe the operation of the quasi-complementary power amplifier circuit. (10-3-2)

76. Why do most power transistors have heat sinks? (10-3-2)

Practice Problems

77. Referring to the voltage amplifiers in Figure 10-26, calculate the unknown.

78. Referring to the current amplifiers in Figure 10-27, calculate the unknown.

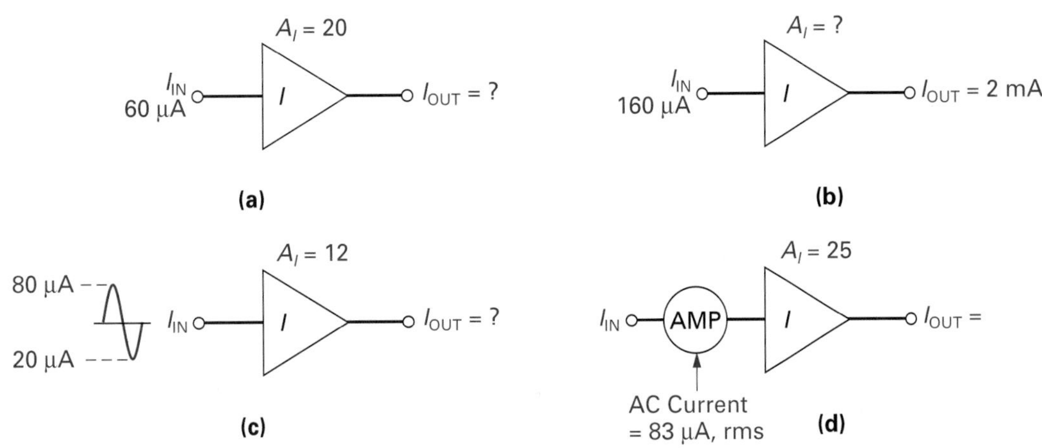

FIGURE 10-27 Current Amplifiers.

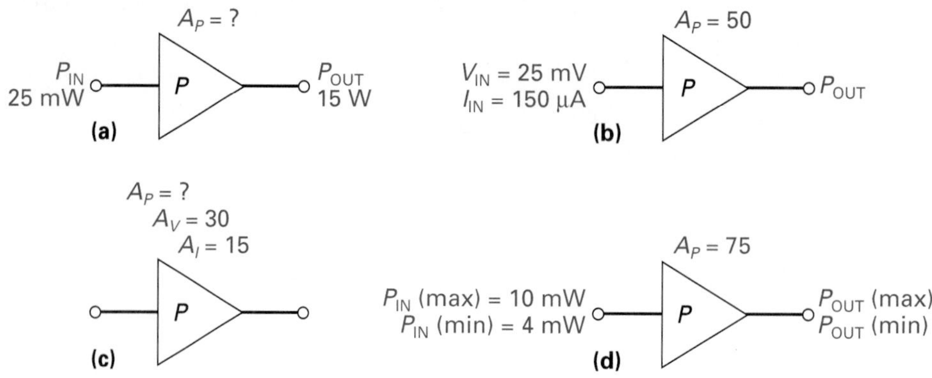

FIGURE 10-28 Power Amplifiers.

79. Referring to the power amplifiers in Figure 10-28, calculate the unknown.

80. Referring to the actual system-amplifier circuits shown in Figure 10-29

a. Identify the bipolar transistor circuit configurations.
b. Which of the circuits would provide a high voltage gain?
c. Which of the circuits would provide a high current gain?
d. Which of the circuits would provide a high power gain?

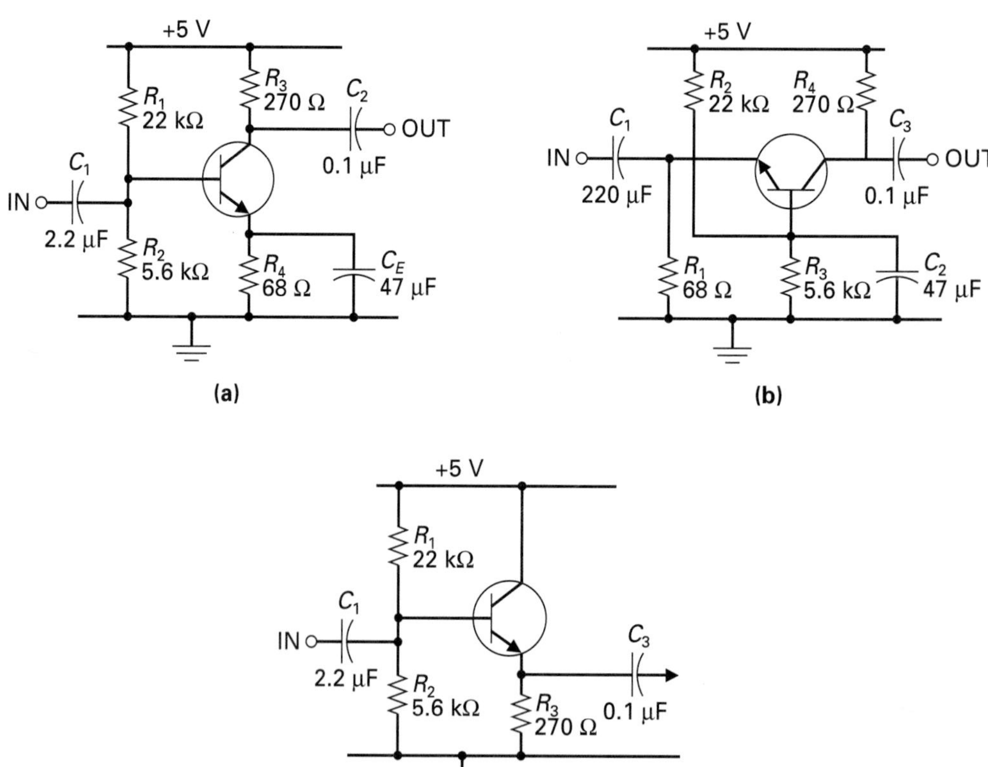

FIGURE 10-29 Amplifier Circuits.

e. Which of the circuits would be best suited as an impedance-matching circuit?
f. Are these circuits designed for AF or dc signals?
g. What type of biasing method is used for all three circuits?

81. Determine the log of the following:

a. 10	e. 10,500
b. 100	f. 3746
c. 25	g. 0.35
d. 150	h. 1.75

82. Determine the antilog of the following:

a. 1.75	e. 5.5
b. 2.35	f. 0.25
c. 5	g. 0.1
d. 1	h. 10

83. Calculate the voltage gain in decibels of the amplifiers shown in Figure 10-26(a) and (b).

84. Calculate the current gain in decibels of the amplifiers shown in Figure 10-27(a) and (b).

85. Calculate the power gain in decibels of the amplifiers shown in Figure 10-28(a) and (b).

86. A 12-foot coaxial cable has an input of 25 W and an output of 23.7 W. What is the cable loss in decibels?

87. Calculate the arithmetic gain or loss of the following:

a. Power gain = 20 dB	c. Current gain = 35.8 dB
b. Voltage loss = −1.5 dB	d. Power gain = 1.75 dB

88. The power output from an audio frequency amplifier is greater than 50% when the input frequencies are between 30 Hz to 3000 Hz. What is the amplifier's bandwidth?

89. Referring to Figure 10-30, calculate the
   a. Amplifier's voltage gain
   b. Amplifier's output voltage
   c. Amplifier's efficiency

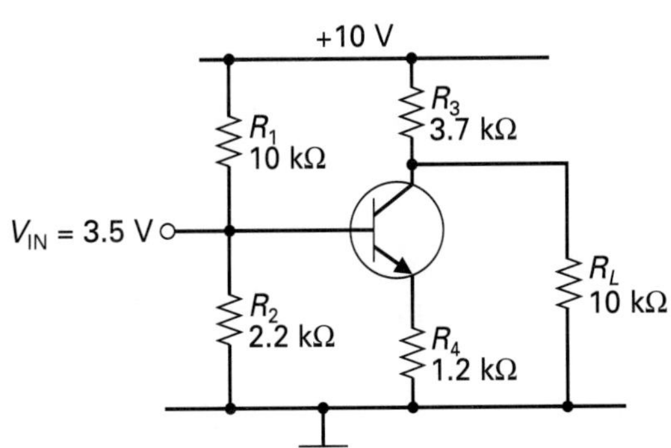

FIGURE 10-30 A Direct Current Amplifier.

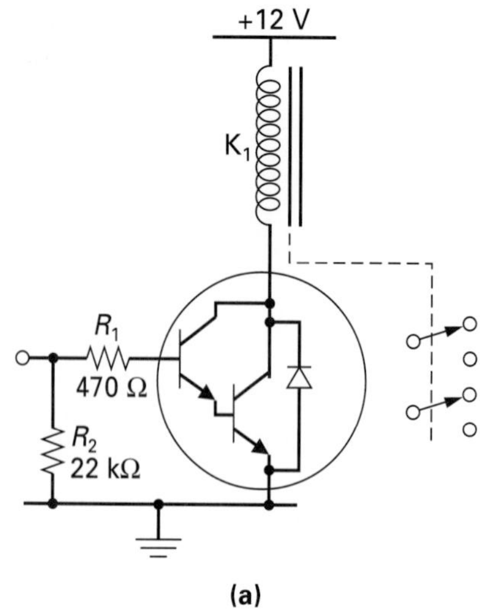

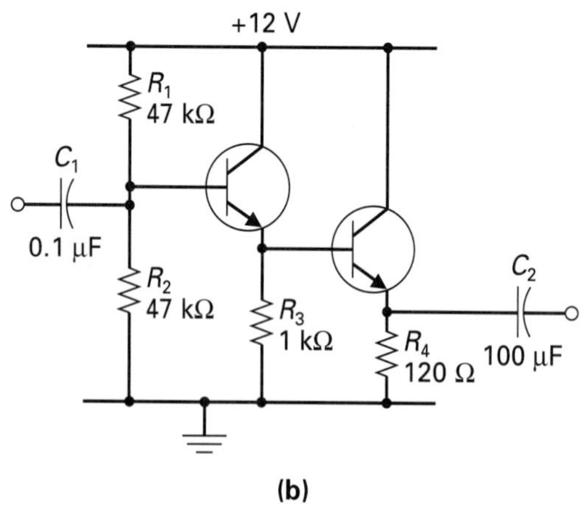

FIGURE 10-31 **Transistor Circuits.**

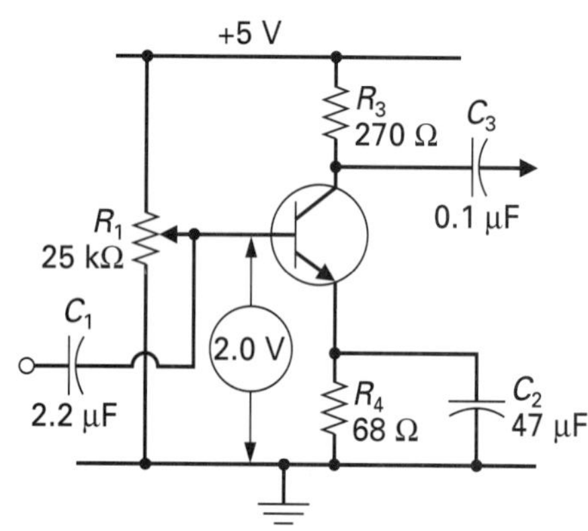

FIGURE 10-32 **An AF Voltage Amplifier.**

90. Identify the circuits shown in Figure 10-31, and briefly describe the circuit's function.

91. If a differential amplifier has a differential gain of 6000 and a common-mode gain of 0.33, what would be the amplifier's common-mode rejection ratio?

92. Referring to the audio frequency amplifier circuit in Figure 10-32, calculate the

a. voltage gain

b. output signal voltage for an input signal of 2 mV

93. Calculate the efficiency of the amplifier shown in Figure 10-32 if a 4.2 V signal is applied to a 1 kΩ load, and $I_{CC} = 20$ mA.

94. What would be the power delivered to an 8 Ω speaker if a power amplifier outputs a 23 V rms signal?

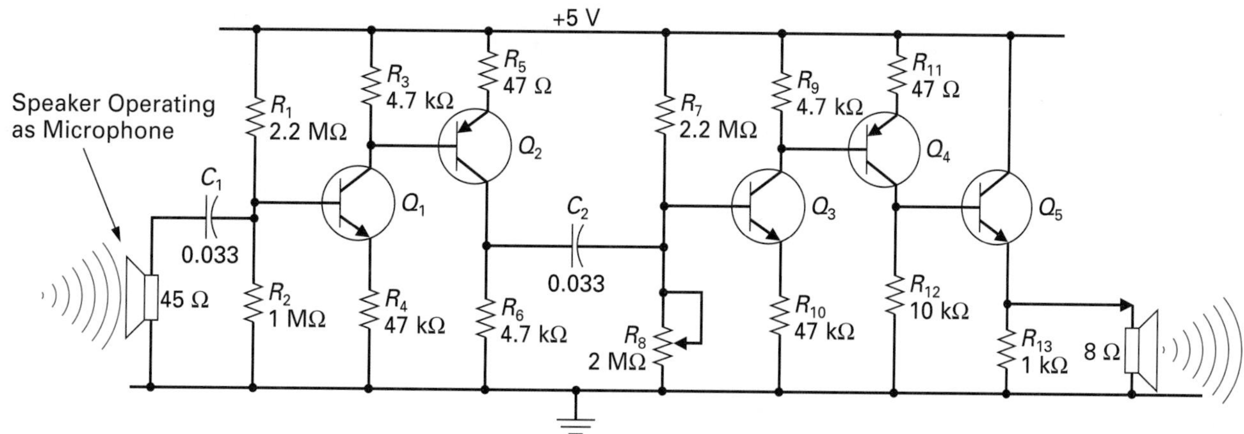

FIGURE 10-33 Troubleshooting an Amplifier Circuit.

Troubleshooting Questions

95. Why is the oscilloscope ideal for testing amplifier circuits? (10-4-2)

96. If a problem existed in the amplifier circuit in Figure 10-33, briefly describe how you could isolate whether the problem is within the amplifier circuit or within the signal source, load, or power supply. (10-4-1)

97. Briefly describe the test points you would check using the oscilloscope if a problem existed in the amplifier circuit in Figure 10-33. (10-4-2)

98. How would you isolate the different stages of the amplifier shown in Figure 10-33? (10-4-2, 10-4-3)

99. Referring to Figure 10-33, describe the effect you think you would obtain from the following circuit malfunctions:

a. C_1 open **b.** R_8 short **c.** Q_1 C-E open **d.** R_{12} short

100. Referring to Figure 10-34, describe the effect you think you would obtain from the following circuit malfunctions:

a. C_1 open **b.** R_3 short **c.** Q_2 C-E open **d.** C_2 short

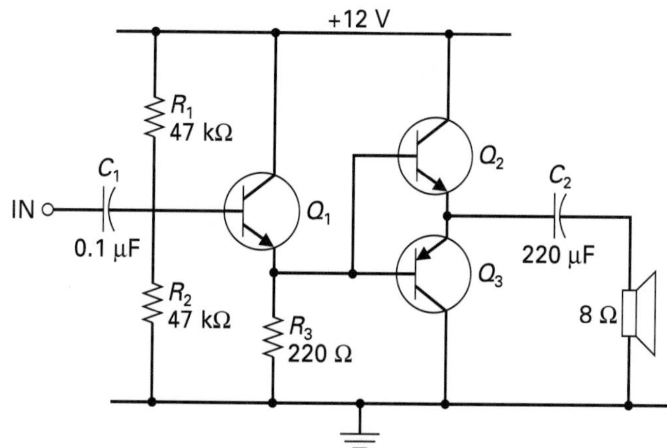

FIGURE 10-34 A Power Amplifier Circuit.

CHAPTER

11

OBJECTIVES

After completing this chapter, you will be able to:

1. Describe the need for high-frequency amplifier circuits.
2. Explain the operation of a basic amplitude modulated radio communication transmitter.
3. Describe the function and operation of an RF amplifier/multiplier circuit.
4. Explain why the resonant action of an LC tank circuit is so useful in high frequency amplifier circuits.
5. Describe the function and operation of an RF amplifier/modulator circuit.
6. Define these terms as they relate to an amplitude modulated waveform: carrier, sidebands, and bandwidth.
7. Describe the function and operation of an RF fixed-tuned amplifier circuit.
8. Describe how the characteristics of a tuned circuit will determine a high-frequency amplifier's bandwidth and selectivity.
9. Explain how a transformer's coefficient of coupling can determine a tuned circuit's frequency response.
10. Define the term "Miller effect capacitance," and describe how it can be neutralized.
11. Describe the function and operation of an RF variable-tuned amplifier circuit.
12. Explain the operation of a basic amplitude modulated radio communication tuned receiver.
13. Explain the operation of a basic amplitude modulated radio communication super-heterodyne receiver.
14. Describe the function and operation of an RF amplifier/mixer circuit.
15. Describe the function, operation, and purpose of intermediate-frequency (IF) amplifier circuits.
16. Describe the purpose of video frequency (VF) amplifiers, and describe how the high cutoff frequency of these amplifiers can be increased.
17. Describe the function and operation of a basic video amplifier circuit and typical television receiver circuits.
18. List the procedure and describe the method for tuning a high-frequency amplifier's tuned circuit.
19. Describe how to troubleshoot circuit malfunctions in high-frequency amplifier circuits.

Bipolar Transistor High-Frequency Amplifier Circuits

OUTLINE

VIGNETTE

Boole-Headed

George Boole was born in the industrial town of Lincoln in eastern England in 1815. His parents were poor tradespeople and even though there was a school for boys in Lincoln, there is no record of his ever attending. In those hard times, children of the working class had no hope of receiving any form of education, and their lives generally followed the pattern of their parents. George Boole, however, broke the mold. He rose from these humble beginnings to become one of the most respected mathematicians of his day.

Boole's father had taught himself a small amount of mathematics, and because his son of six seemed to have a thirst for learning, he began to pass on his knowledge. At eight years old, Boole had surpassed his father's understanding and craved more. He quickly realized that his advancement was heavily dependent on understanding Latin. Luckily, a family friend who owned a local book shop knew enough about the basics of Latin to get Boole started. Once this friend had taught Boole all he knew, Boole continued with the books at his disposal. By the age of twelve he had mastered Latin and by fourteen he had added Greek, French, German, and Italian to his repertoire.

At the age of sixteen, however, poverty stood in his way. Because his parents could no longer support him, he was forced to take a job as a poorly paid teaching assistant. After studying all the material in the entire school system, he left four years later and opened his own school in which he taught all subjects. He discovered his command of mathematics was weak, so he began studying the mathematical journals at the local library in an attempt to stay ahead of his students. He quickly discovered that he had a talent for mathematics. As well as mastering all the present day ideas, he began to develop some of his own, which were later accepted for publication. After a stream of articles, he became so highly regarded that he was asked to join the mathematics faculty at Queens College in 1849.

After accepting the position, Boole concentrated more on his ideas, one of which was to develop a system of symbolic logic. In this system he created a form of algebra that had its own set of symbols and rules. Using this system, Boole could encode any statement that had to be proved (a proposition) into his symbolic language, then manipulate it to determine whether it was true or false. Boole's algebra has three basic operations that are simulated electronically by the logic gates AND, OR, and NOT. Using these three operations, Boole could add, subtract, multiply, divide, and compare. Boole's theory was that if all logical arguments could be reduced to one of two basic levels, the questionable middle ground would be removed, making it easier to arrive at a valid conclusion. Therefore logic gates are binary in nature, dealing with only two entities: TRUE or FALSE, YES or NO, OPEN or CLOSED, ZERO or ONE, and so on.

At the time, Boole's system, later called *Boolean algebra,* was either ignored or criticized by colleagues who called it a folly with no practical purpose. Almost a century later, however, scientists would combine George Boole's Boolean algebra with binary numbers and make possible the digital electronic computer.

INTRODUCTION

Communication is defined as *a process by which information is exchanged.* The two basic methods of transferring information are the *spoken word,* and the *written word.* Communication of the spoken word began face-to-face, then evolved into radio communications, telephone communications, and now video telecommunications. Communication of the written word began with hand-carried letters, then evolved into newspapers, the mail system, the telegraph, and now electronic mail.

At a very early stage, it was found that the distance we could communicate was important. Systems of long distance communication developed, beginning with smoke signals and beating drums. However, it was not until the late 1800s that scientists discovered that a high-frequency electrical current passing through a conductor would radiate an energy wave that could be modified or altered to carry information. These radiated *electromagnetic waves,* or *radio waves,* could be used to communicate information of different types. For example (to name but a few applications), radio stations transmit audio information such as voice and music, television stations transmit moving-picture information, and cellular telephones transmit voice or facsimile (graphic) data. All of these radio communication transmitters and receivers need *high-frequency amplifiers* to boost the amplitude of the high-frequency information signals in both the transmitting circuits and receiving circuits.

In the previous chapter, we studied basic amplifier principles, and the operation, characteristics, and testing of direct current (dc) amplifier circuits and audio frequency (AF) amplifier circuits. In this chapter, we will continue our coverage of amplifiers, concentrating on bipolar-transistor amplifier circuits that operate at progressively higher frequencies. These circuits are named after the particular frequency that they are designed to amplify: *radio frequency (RF) amplifiers*, *intermediate frequency (IF) amplifiers*, and *video frequency (VF) amplifiers*.

11-1 RADIO FREQUENCY (RF) AMPLIFIER CIRCUITS

Radio frequency (RF) amplifiers are mainly used in communication systems, so we will first need to understand the basics of a simple communication system if we are to see why RF amps are so important. As an example, Figure 11-1 shows the block diagram of a basic AM (amplitude modulated) transmitter, with each block representing a bipolar transistor circuit. The AF voltage amplifier and AF

Radio Frequency (RF) Amplifiers
An amplifier that has one or more transistor stages designed to amplify radio-frequency signals.

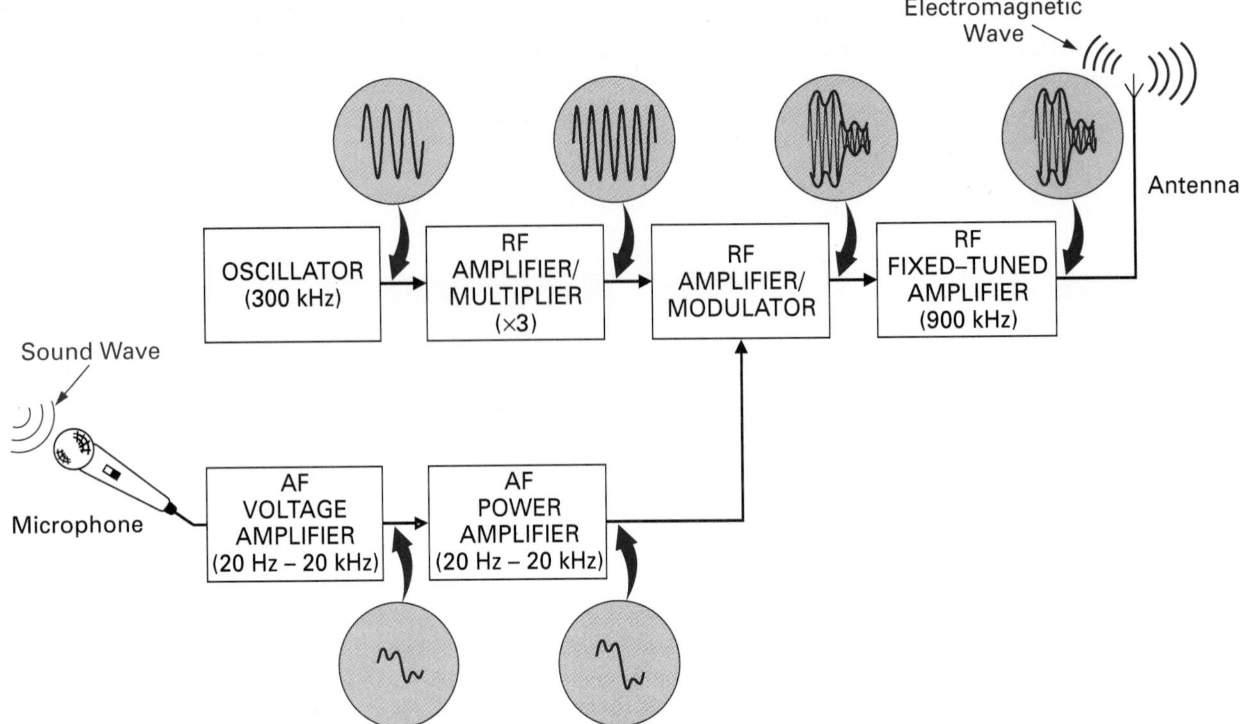

FIGURE 11-1 Basic AM (Amplitude Modulated) Transmitter.

power amplifier circuits were discussed in the previous chapter and, although oscillator circuits are due to be covered in the next chapter, we already know that these circuits are used to generate a continually repeating ac signal. This leaves us with three remaining circuits, all of which contain an RF amplifier.

Before proceeding to the operation of these RF amplifier circuits, let us examine the basic function of each of the blocks or circuits in the AM transmitter shown in Figure 11-1. The high-frequency oscillator supplies a repeating, high-frequency, constant-amplitude, sine wave signal to the RF amplifier/multiplier circuit, that triples the input frequency and applies its output to the RF amplifier/modulator. Using the example values given in Figure 11-1, the oscillator applies a 300 kHz sine wave to the frequency multiplier, which converts this input to a 900 kHz sine wave (300 kHz × 3 = 900 kHz). The radio frequency sine wave from the RF frequency multiplier circuit is applied to the RF amplifier/modulator along with an audio frequency input signal from the microphone, which is amplified by an AF voltage and AF power amplifier. The RF amplifier/modulator combines the low-AF information signal and the high-RF wave, producing an amplitude modulated (AM) output signal. The basics of amplitude modulation were first introduced in Chapter 6. To review, each electromagnetic transmitting station is assigned its own transmit frequency by the FCC, or Federal Communications Commission, so that its radiated signal will not interfere with other stations. The frequency at which a station transmits is called its **carrier frequency.** For example, your local AM radio stations each have their own assigned carrier frequency in the AM radio band, and this high radio

Carrier Frequency
The frequency generated by a radio transmitter, or the average frequency of the emitted wave when modulated by an information signal.

frequency acts as a carrier for the station's audio frequency (AF) information signals (speech and music). Each station will have to convert their lower frequency audio signals to their assigned higher carrier frequency before transmitting. This process of converting a low-frequency information signal to a high-frequency carrier signal is called **modulation.** In Figure 6-14, we saw how the diode could be used in a series clipper circuit to achieve **amplitude modulation (AM),** the process by which the amplitude of the high-frequency carrier is changed or varied to follow the amplitude of the low-frequency information signal. This action is best seen by looking at the change in the waveforms in Figure 11-1. Studying the RF amplifier/modulator's output waveform, you can see that the amplitude of the carrier, which is the high-frequency sine wave from the frequency multiplier, varies at the same rate as the audio signal, which originated from the microphone. In fact, the envelope of the carrier is an exact replica of the audio information signal. The final RF fixed-tuned amplifier, which in our example would be tuned to 900 kHz, is included to boost the amplitude-modulated or amplitude-varied carrier signal before it is applied to the antenna. It is, in fact, the voltage that is applied across the antenna by the final-stage RF amp that generates an **electric (electro) or voltage field.** The current that is generated by the final-stage RF amp and passes through the antenna generates a **magnetic or current field.** These combined fields make up an **electromagnetic field** that radiates away from the antenna in all directions at the speed of light (approximately 186,000 miles per second).

The radio frequency band includes any frequencies from about 10 kHZ to about 30,000 MHz. Let us now examine, in more detail, the three bipolar transistor RF amplifier circuits discussed in Figure 11-1.

Modulation
The process of converting a low-frequency information signal to a high-frequency carrier signal.

Amplitude Modulation
The process by which the amplitude of the high-frequency carrier is changed or varied to follow the amplitude of the low-frequency information signal.

Electric or Voltage Field
A field or force that exists in the space between two different potentials or voltages.

Magnetic or Current Field
Magnetic lines of force traveling from the north to the south pole of a magnet.

Electromagnetic Field
Field having both an electric (voltage) and magnetic (current) field.

11-1-1 An RF Amplifier/Multiplier Circuit

Figure 11-2 repeats the basic AM transmitter block diagram and focuses in on the operation of the **RF amplifier/multiplier circuit.** This block is included in the basic AM transmitter to convert the stable low-frequency oscillation from the 300 kHz oscillator circuit to a stable high-frequency signal that is three times (3×) the input frequency (input = 300 kHz, output = 3 × 300 kHz = 900 kHz).

Figure 11-2(a) shows a basic **RF amplifier/frequency tripler circuit,** and Figure 11-2(b) shows the circuit's associated input/output waveforms. The bipolar transistor Q_1 is class C biased by the negative dc bias voltage ($-V_{BB}$) and resistor R_1. This means that Q_1 will only conduct when the input voltage is large enough to overcome the negative reverse bias voltage at the base of Q_1 . Referring to the waveforms in Figure 11-2(b), you can see that Q_1 conducts only for a short time during the positive peak of the input signal voltage. During these positive peaks, Q_1 will turn ON and send a pulse of current into the LC tank circuit in the collector of Q_1 made up of capacitor C_2 and the inductance of T_1's primary (L_1). If this tank circuit was tuned to the same frequency as the input signal, the natural flywheel action of the LC tank would reproduce a copy of the input sine wave at the output that is of the same frequency but slightly larger in amplitude due to the amplifying action of Q_1. However, with the frequency tripler circuit in Figure 11-2(a), the LC tank is tuned to three times the input signal frequency,

RF Amplifier/ Multiplier Circuit
A radio frequency amplifier circuit in which the frequency of the output is an exact multiple of the input frequency.

RF Amplifier/ Frequency Tripler Circuit
A radio frequency amplifier circuit in which the frequency of the output is three times that of the input frequency.

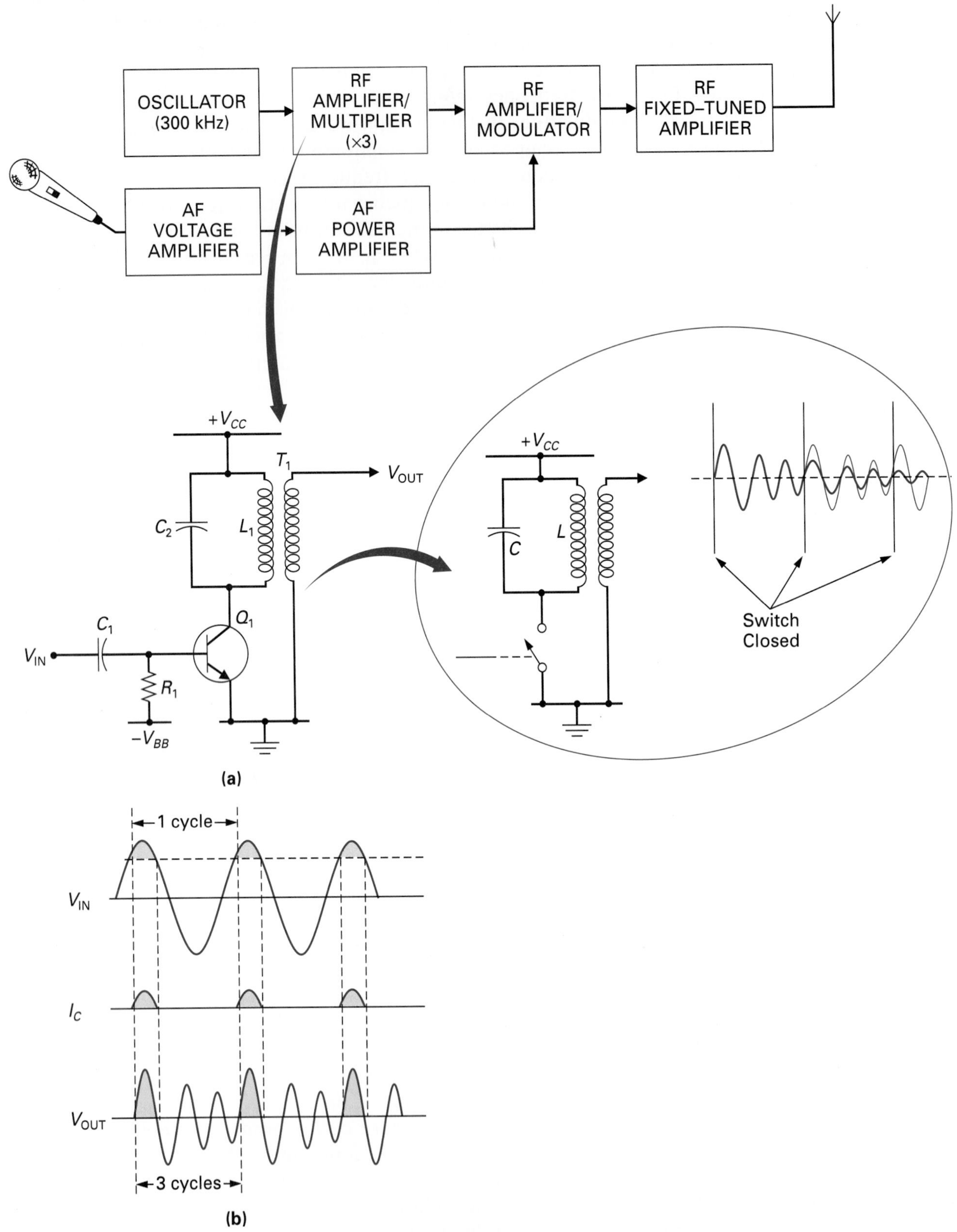

FIGURE 11-2 **Radio Frequency (RF) Amplifier/Multiplier Circuits. (a) Frequency Tripler Circuit. (b) Frequency Tripler Input/Output Waveforms.**

and therefore the tank will produce three sine wave output cycles for every one cycle of the input. The pulse of collector current produced by Q_1 will *shock excite* the tank into oscillation and, due to the tank's natural flywheel action, the parallel LC circuit will oscillate at its natural resonant frequency. When operated with class C bias, it would seem that a transistor would introduce signal distortion because the full sine wave input signal is distorted to a small pulse of current. Signal distortion, however, does not occur because the rest of the positive half-cycle, and the full negative half-cycle, is recreated by the LC tank's flywheel action. The inset in Figure 11-2(a) shows transistor Q_1 simplified to a switch that, when closed, will cause the LC tank to oscillate at its resonant frequency. If the transistor switch did not close again and produce additional current pulses, the oscillations would slowly decrease in size and eventually fall to zero. However, if additional current pulses are produced at exactly three times the input frequency, these current pulses will continually re-energize the tank circuit and maintain the oscillations.

To review, the name "flywheel action" is derived from the fact that this circuit's action resembles a mechanical flywheel, which, once started, will keep spinning back and forward until friction reduces the magnitude of the rotations to zero. The electronic circuit equivalent of a mechanical flywheel is a parallel resonant LC circuit. Supplying this circuit with a pulse of current will cause a circulating current within the LC tank that alternates at the circuit's frequency of resonance, calculated by the formula

$$f_R = \frac{1}{2\pi\sqrt{LC}}$$

■ **EXAMPLE:**

Determine the center resonant frequency of a tuned amplifier circuit if the parallel tank has a capacitor of 80 pF and an inductor of 1 mH.

■ ***Solution:***

$$f_R = \frac{1}{2\pi\sqrt{LC}} = \frac{1}{2\pi\sqrt{(1\ \text{mH}) \times (80\ \text{pF})}} = 562.7\ \text{kHz}$$

Energy is stored in the capacitor in an electric field between its plates on one half-cycle. Then the capacitor discharges, supplying current to the inductor, causing it to store energy in a magnetic field during the other half-cycle. Once the inductor's magnetic field has built up to a maximum, the magnetic field will begin to collapse and supply a charge current to the capacitor. After the capacitor has charged to a maximum, it will discharge, supplying a current back to the inductor, and so on. The oscillations within a tank circuit will eventually fall to zero, due mainly to the coil resistance of the inductor. The energy "storing action" accounts for why this circuit is also called a "tank."

The circuit in Figure 11-2(a) is called a frequency tripler because the output frequency is three times the input frequency as seen in Figure 11-2(b). To ensure that the pulse of current occurs at the right point to sustain oscillations, the tank must be "tuned" to an exact multiple of the input frequency, so that the current pulse from Q_1 will reinforce the oscillations. Frequency doubler and frequency quadrupler circuits are often used in RF circuits to convert a stable low-frequency oscillation to a stable high-frequency oscillation. Operating with class C bias means that Q_1 is conducting (and consuming power) for only a small portion of time; therefore, the circuit is more efficient than a class-A circuit. Signal distortion does not occur with a class-C biased transistor if an LC tank is included in the circuit because the natural flywheel action of a parallel resonant circuit will recreate the entire sine wave cycle at the output.

11-1-2 An RF Amplifier/Modulator Circuit

Figure 11-3 again repeats the basic AM transmitter block diagram and, in this case, focuses on the operation of a bipolar transistor RF amplifier/modulator circuit. Most practical amplitude modulators include an amplifier stage—an advantage that the diode modulator discussed previously in Chapter 6 did not have.

Referring to the circuit in Figure 11-3, you can see that the RF carrier from the frequency multiplier circuit is applied through coupling capacitor C_1 to the base of the class-A voltage-divider biased transistor Q_1. The AF modulating signal from the AF amplifiers is applied to the emitter circuit of Q_1 via the transformer T_1. Remember that there is a very large frequency difference between the RF carrier input and the AF modulating signal input. For example, using the values chosen in our basic AM-transmitter block diagram, the RF carrier input frequency is 900 kHz, while the AF modulating signal frequency could typically be about 900 Hz, making the RF carrier frequency 1,000 times higher than the AF modulating signal frequency.

$$\frac{900\text{ kHz}}{900\text{ Hz}} = 1000$$

Let's see how this circuit combines these two input signals to produce a 900 kHz RF sine wave output that varies in amplitude in accordance with the AF modulating signal input. Coupling capacitors C_1 and C_3 and emitter-bypass capacitor C_2 have values that will offer no opposition to the 900 kHz RF carrier signal. So as far as the RF carrier input signal is concerned, this circuit is simply a common-emitter amplifier stage. With regard to the AF modulating signal, the emitter current of Q_1 is determined by the resistance of R_4 and the voltage developed across R_4. The voltage on the top of R_4 is determined by the voltage divider bias resistors R_1 and R_2 and the 0.7 V drop between Q_1's base and emitter. When the AF modulating signal input swings positive, the voltage at the bottom of R_4 increases, and the voltage difference (or drop) across R_4 decreases, causing a decrease in I_E. On the other hand, when the AF modulating signal input swings negative, the voltage at the bottom of R_4 decreases, and the voltage difference or drop across R_4 increases, causing an increase in I_E. In the previous chapter, it was stated that the voltage gain of an amplifier stage is equal to the ratio of the transistor's ac collector resistance to ac emitter resistance.

$$A_V = \frac{R_{C(ac)}}{R_{E(ac)}}$$

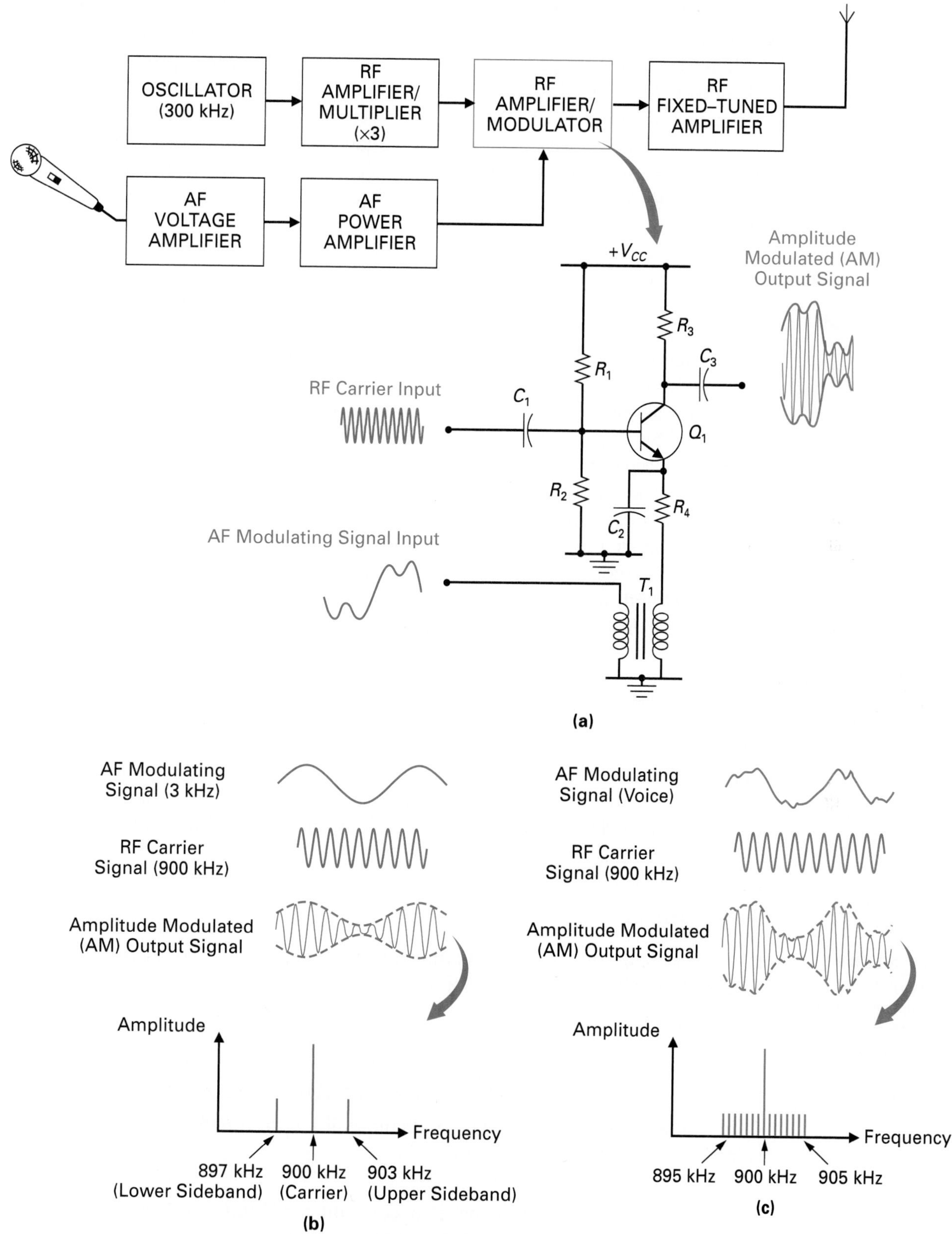

FIGURE 11-3 Radio Frequency (RF) Amplifier/Modulator Circuit.

As the AF modulating input signal increases and decreases, the ac emitter resistance of Q_1 is increased and decreased, and the gain of Q_1 is increased and decreased. To be specific, as the AF modulating signal swings positive, the voltage across R_4 decreases, causing I_E to decrease, the ac emitter resistance to increase, and the transistor's gain to decrease. On the other hand, as the AF modulating signal swings negative, the voltage across R_4 increases, causing I_E to increase, the ac emitter resistance to decrease, and the transistor's gain to increase. The AF modulating signal increases and decreases the gain of Q_1, and therefore the amplitude of the signal being amplified, which is the RF carrier signal. This **emitter modulator** circuit shows how an RF carrier signal can be amplitude modulated so that its envelope will follow the changes in the AF information signal.

Emitter Modulator
A modulating circuit in which the modulating signal is applied to the emitter of a bipolar transistor.

Although communication techniques will be discussed in a later chapter, we will need to examine our amplitude modulated output wave in a little more detail so that we can better understand RF amplifier circuits. Referring to the waveforms shown in Figure 11-3(b), you can see that if a 900 kHz RF carrier is modulated by a constant 3 kHz AF tone, a 900 kHz signal is produced that has a 3 kHz envelope. If we were to study the frequency components within this amplitude modulated wave, we would find that it is made up of a 900 kHz center carrier frequency with two other frequency components at 897 kHz and 903 kHz, as seen in the frequency spectrum shown in Figure 11-3(b). These two additional frequencies are called **sidebands:** the 897 kHz signal is called the **lower sideband,** and the 903 kHz signal is called the **upper sideband.** In this example, the AM radio frequency signal would have a bandwidth of 6 kHz.

Sidebands
A band of frequencies on both sides of the carrier frequency of a modulated signal produced by modulation.

$$BW = f_{HI} - f_{LO} = 903\text{ kHz} - 897\text{ kHz} = 6\text{ kHz}$$

In most applications, the RF carrier is modulated by a voice or music signal instead of a constant tone, as shown in Figure 11-3(c). In this case, the modulating signal contains many different frequencies which are constantly changing. Because one set of sidebands (upper and lower) is produced for each modulating frequency, the resulting AM waveform will have the frequency spectrum shown in Figure 11-3(c). In most applications, the bandwidth is always twice the highest modulating frequency.

Lower Sideband
A group of frequencies that are equal to the differences between the carrier and modulation frequencies.

Upper Sideband
A group of frequencies that are equal to the sums of the center and modulation frequencies.

■ **EXAMPLE:**

If a 690 kHz carrier is voice modulated by a signal that is between 500 Hz to 5 kHz, what will be the transmission's bandwidth?

■ ***Solution:***

Because the bandwidth is always twice the highest modulating frequency

$$BW = 2 \times f_{AF(HI)} = 2 \times 5\text{ kHz} = 10\text{ kHz}$$

Centered at 690 kHz, the lower sideband will begin at 685 kHz, and the upper sideband will extend up to 695 kHz.

The carrier, therefore, remains at a constant amplitude and frequency at all times; however, the sidebands are constantly changing amplitude and frequency in accordance with the modulating signal.

11-1-3 *An RF Fixed-Tuned Amplifier Circuit*

Figure 11-4(a) continues our coverage of the basic AM transmitter by showing a final stage **RF fixed-tuned amplifier circuit.** This circuit contains an *LC* tuned circuit in both the input and the output that will pass a select band of frequencies and reject all other frequencies outside the band. In this basic AM transmitter example, the tuned circuits will both be tuned to our transmit frequency of 900 kHz, and therefore our amplitude modulated RF carrier will be amplified and passed onto the transmitting antenna.

RF Fixed-Tuned Amplifier Circuit A circuit that contains an *LC* tuned circuit in both the input and the output and will pass a select band of frequencies and reject all other frequencies outside the band.

A Tuned Circuit's Bandwidth and Selectivity

At resonance, a series *LC* tank has a minimum impedance and therefore a maximum current, whereas a parallel *LC* tank circuit has a maximum impedance and minimum current at resonance. Since the characteristics of the *LC* tuned circuit are so important to the operation of the RF amplifier, let us briefly review the tuned circuit material discussed previously in dc/ac electronics. Referring to Figure 11-4(b), you can see that a parallel resonant *LC* circuit has a maximum impedance at the resonant frequency. At frequencies above and below resonance, the impedance falls off to a low value. This high impedance at resonance is important in a tuned RF amplifier circuit because the voltage developed across a parallel resonant circuit is determined by the impedance of the resonant circuit. This impedance determines the amplitude of the signal voltage delivered to the output.

$$V_{OUT} = I \times Z$$

Figure 11-4(b) shows a parallel resonant *LC* circuit and its frequency response curve in the inset. At resonance, the parallel circuit has a maximum impedance ($Z\uparrow$), and therefore the voltage applied to the output will also be maximum at this frequency ($V_{OUT}\uparrow$). At frequencies above and below resonance, the impedance falls to a lower value ($Z\downarrow$), and therefore the voltage applied to the output also falls to a lower value ($V_{OUT}\downarrow$).

As mentioned in the previous chapter, the bandwidth of a tuned circuit is the band of frequencies that will be passed through to the output and that have a usable power level of 50% or greater. These frequencies that are above the half-power points correspond to the 70.7% points on the impedance axis of the frequency response curve shown in Figure 11-4(b). Between frequencies f_1 and f_2, the amplitude of the output voltage signal will be greater than 70.7% of the maximum voltage. The bandwidth of a standard AM radio station is generally always twice the highest modulating frequency. Therefore, if the highest AF modulating frequency is 3 kHz, the bandwidth will be 6 kHz. To properly pass an AM carrier of 900 kHz and its sidebands within a 6 kHz bandwidth, we would have to make sure that frequencies between 897 kHz and 903 kHz are coupled to the output with a usable power level of 50% or greater.

The bandwidth of the parallel *LC* circuit is a very important characteristic because it determines the **selectivity** of the tuned circuit and the selectivity of the RF amplifier. By definition, selectivity is the ability of a circuit to select the wanted signal(s) and reject the unwanted signal(s). For example, a circuit with a

Selectivity The ability of a circuit to select the wanted signal(s) and reject the unwanted signal(s).

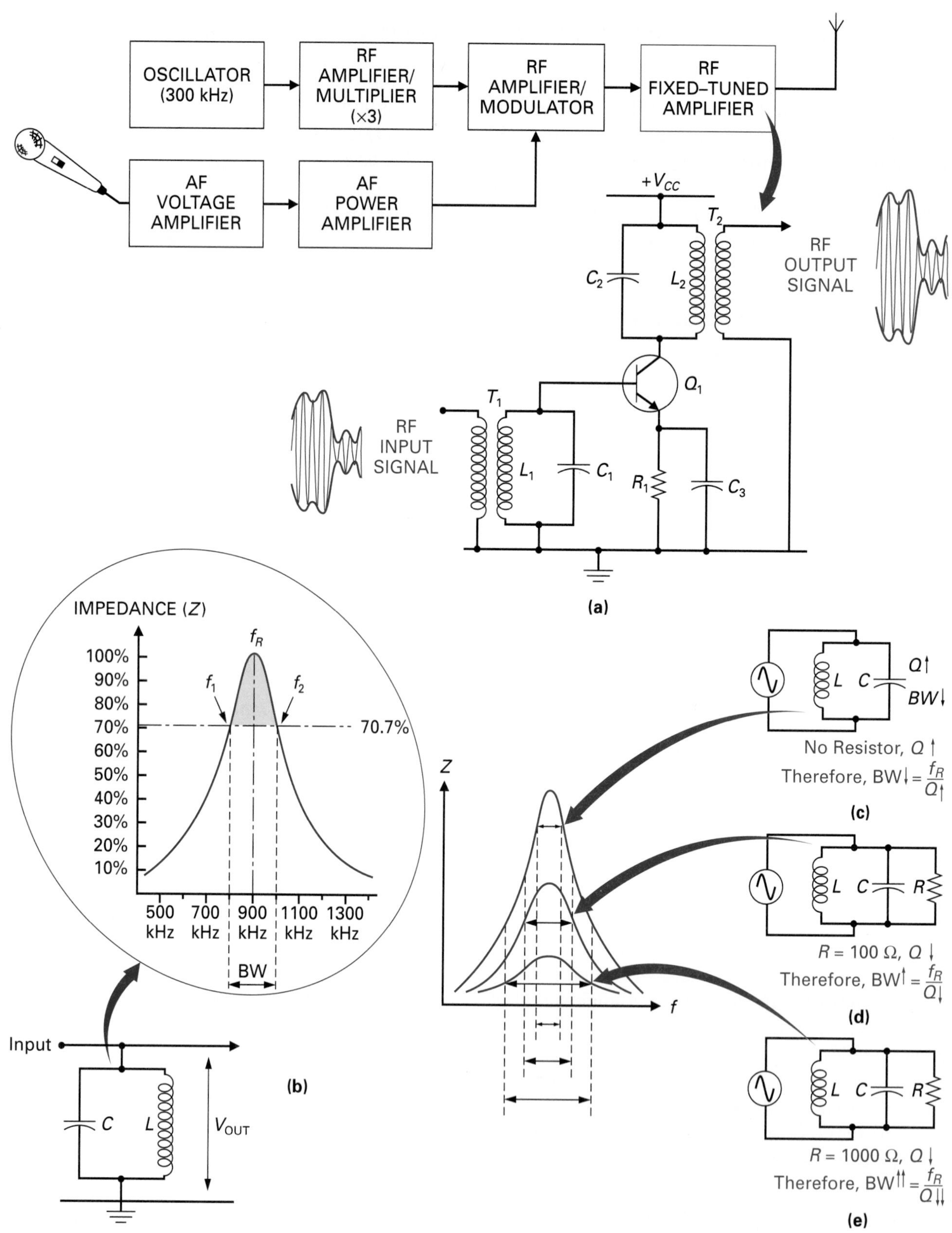

FIGURE 11-4 **Fixed-Tuned Radio Frequency (RF) Amplifier Circuit. (a) Basic Circuit. (b) Parallel Resonant Circuit's Response Curve. (c), (d), (e) Bandwidth Increase as *Q* Decreases.**

wide bandwidth will pass a wide band of frequencies through to the output with a usable output voltage, and this circuit would be said to have a "poor selectivity." On the other hand, a circuit with a very narrow bandwidth will pass very few frequencies through to the output at a usable voltage level, and therefore this circuit is said to be "highly selective." Bandwidth and selectivity are inversely proportional: a circuit with a large bandwidth is not very selective, while a circuit with a small bandwidth is very selective.

$$\text{Selectivity} \propto \frac{1}{\text{Bandwidth}}$$

The bandwidth of a tuned circuit, and therefore the tuned circuit's selectivity, is largely determined by the Q of the circuit.

$$\text{BW} = \frac{f_R}{Q}$$

This Q factor indicates the quality of a parallel resonance circuit and is a ratio of the tuned circuit's reactance to resistance.

$$Q = \frac{X}{R}$$

Stated another way, the Q of a tank circuit is equal to the ratio of energy stored in the tank to energy lost in the tank. Because the only resistance in the tank is that of the inductor's winding resistance, the Q of a tuned circuit is more specifically equal to the inductive reactance of the coil divided by the coil's winding resistance.

$$Q = \frac{X_L}{R_W}$$

X_L = Reactance of the Coil at Resonance

R_W = Resistance of the Winding

■ **EXAMPLE:**

An amplifier's tuned circuit is tuned to 800 kHz and has a reactance-to-resistance ratio 6.2 to 1. Calculate

1. The tuned circuit's Q,
2. The bandwidth of the tuned circuit at the ratio of 6.2 to 1, and
3. The bandwidth of the tuned circuit if a resistor is added to the tuned circuit, doubling the tuned circuit's resistance.

■ ***Solution:***

1. As stated in the question, the reactance is 6.2 times greater than the resistance, so the Q of the tuned circuit will be

$$Q = \frac{X}{R} = \frac{6.2}{1} = 6.2$$

2.

$$\text{BW} = \frac{f_R}{Q} = \frac{800\text{ kHz}}{6.2} = 129\text{ kHz}$$

3. If the resistance of the tuned circuit is doubled, the Q will be halved to 3.1. Therefore, the bandwidth will be

$$\text{BW} = \frac{f_R}{Q} = \frac{800\text{ kHz}}{3.1} = 258\text{ kHz}$$

The bandwidth, and therefore selectivity, of a tuned circuit can be controlled by simply increasing or decreasing the tuned circuit's resistance, as shown in Figure 11-4(c), (d), and (e). For example, no additional resistance other than that of the inductor's coil resistance (capacitors have almost no resistance) will result in a low value of tuned circuit resistance ($R\downarrow$), a high circuit Q ($Q\uparrow$), small bandwidth (BW$\downarrow$), and the circuit will have a very high circuit selectivity. On the other hand, by increasing the tuned circuit's resistance ($R\uparrow$), we will decrease the circuit's Q ($Q\downarrow$), increasing the circuit's bandwidth (BW$\uparrow$), and the tuned circuit will have a very broad selectivity.

Comparing the three frequency response curves in Figure 11-4(c), (d), and (e), it would seem that a high Q is always the best situation because it yields a higher impedance at resonance, and therefore a higher gain and selectivity. However, this is not always the case. For example, radio frequency amplifiers in television receivers must have a broad bandwidth of 6 MHz, whereas radio frequency amplifiers in a transmitter will have a narrow bandwidth to prevent unwanted signals from being transmitted.

Typically, the Q of an RF amplifier's tuned circuit is adjusted by adding in a parallel resistance, called a **swamping resistor,** to increase the bandwidth.

Swamping Resistor
A parallel resistance added to the RF amplifier tuned circuit to increase the bandwidth.

Transformer Coupling and Frequency Response

Figure 11-5 concentrates on the characteristics of the RF amplifier's tuned-input and tuned-output transformers. By including step-up voltage transformers in an RF amplifier, additional voltage gain can be achieved. These transformers can also determine the tuned circuit's frequency response by controlling the transformer's **coefficient of coupling.** To review, the coefficient of coupling for a transformer is the ratio of the number of magnetic lines of force that cut the secondary compared to the number of magnetic flux lines being produced by the primary. For example, Figure 11-5(a), (b), and (c) show how the coefficient of coupling between a transformer's primary and secondary can be changed to alter a tuned circuit's bandwidth. As an example, let us imagine that we wish to pass all of the upper and lower frequency components of a station between f_1 and f_2 and block all other frequencies.

Coefficient of Coupling
The degree of coupling that exists between two circuits.

In Figure 11-5(a), the coefficient is low because the transformer coils are far apart and less signal energy is transferred between primary to secondary. Looking at the frequency response for this **loose-coupled transformer,** you can see that the transformer passes the higher amplitude center frequency (or resonant frequency). However, frequencies either side of resonance (f_1 and f_2) are not coupled to the secondary with sufficient amplitude. This means that the lower frequency (f_1) elements of our modulating signal and the higher frequency (f_2) elements of our modulating signal will not be passed, causing signal frequency distortion.

Loose-Coupled Transformer
A transformer in which the coils are far apart and less signal energy is transferred between primary and secondary.

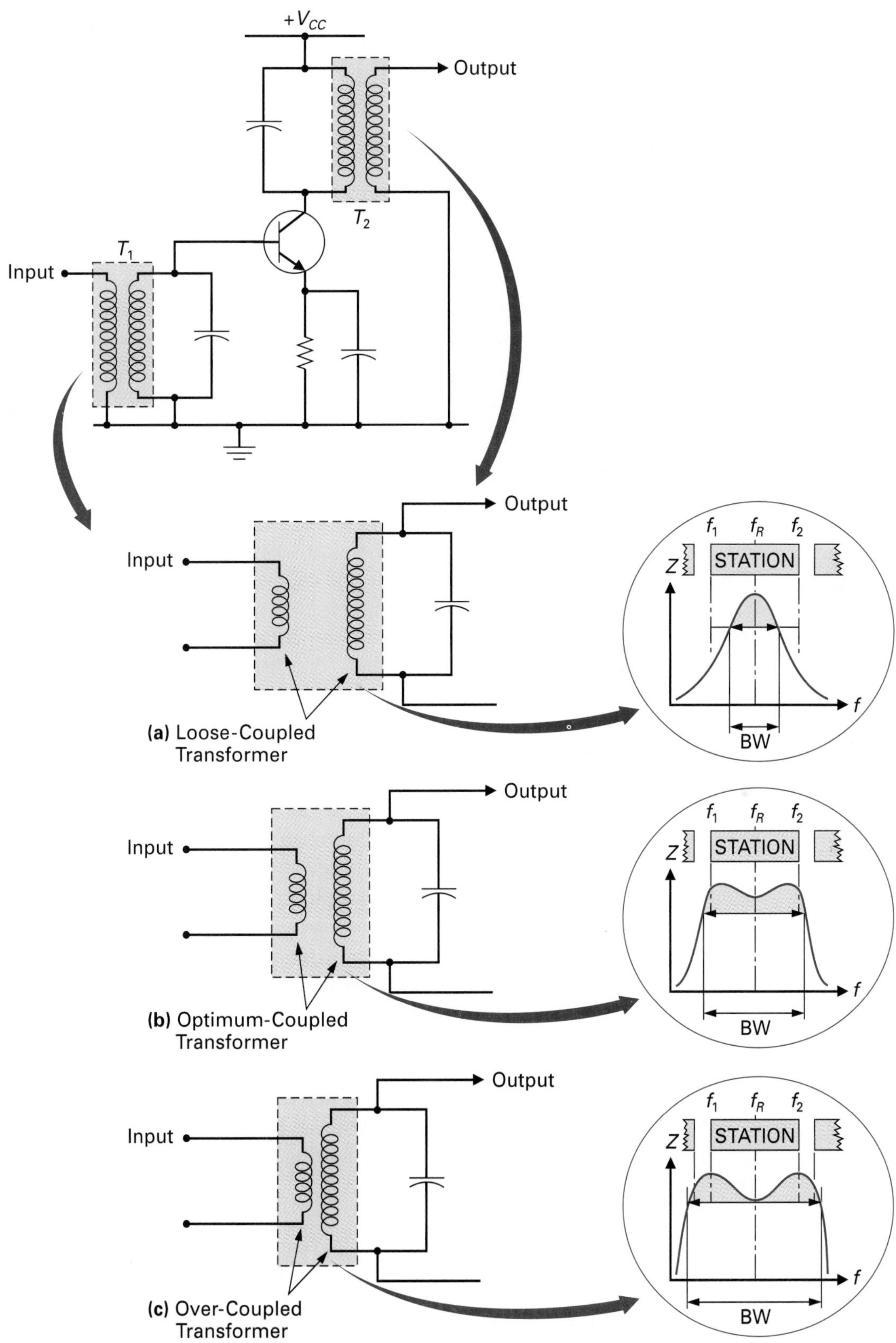

FIGURE 11-5 Transformer Coupling and Frequency Response.

In Figure 11-5(b), the coefficient of coupling has been increased so that more signal energy is transferred, causing an increase in the bandwidth and therefore a better response to a signal's resonant frequency and associated higher and lower frequency components.

Overcoupled
A transformer in which the coils are so close together that a dip appears in the center of the response curve causing the resonant frequency component to be of a lower amplitude than the upper and lower frequency components of the signal.

In Figure 11-5(c), the transformer's primary and secondary have been **overcoupled** and, as a result, a dip appears in the center of the response curve causing the resonant frequency component to be of a lower amplitude than the upper and lower frequency components of the signal. The other disadvantage with an overcoupled transformer is that its extremely wide bandwidth will permit unwanted outside signals to pass through to the next stage.

Ideally, transformers have optimum coupling, as shown in Figure 11-5(b), so that the circuit provides an equal response to all the desired frequencies between f_1 and f_2. Frequencies outside the f_1 to f_2 band will meet with a reduced impedance, and the voltage gain for these circuits will be low and outside the passed band of the amplifier.

Neutralizing Miller-Effect Capacitance in Transistors

Miller-Effect Capacitance (C_M)
An undesirable inherent capacitance that exists between the junctions of transistors.

Junction Capacitance
The small value of capacitance that exists between the junctions of all transistors.

Negative Feedback
An out-of-phase feedback in which the feedback voltage is opposite in polarity to the input signal voltage, causing the input signal to be decreased.

Neutralizing Capacitor (C_N)
A capacitor that couples an in-phase signal back to the base of the RF transistor to compensate for the out-of-phase signal coupled back to the base of the transistor by the Miller capacitance.

Figure 11-6 shows our fixed-tuned RF amplifier circuit and details the effects of an undesirable inherent capacitance within transistors called **Miller-effect capacitance (C_M).** A small value of capacitance exists between the junctions of all transistors as shown in Figure 11-6(a). These **junction capacitances** are determined by the physical size of the junction, the spacing between the transistors' leads, and the junction bias condition. At low frequencies these capacitances have very little effect because X_C is large. However, when frequency increases, X_C decreases, and the capacitance between a transistor's collector and base, which is known as the transistor's Miller capacitance [as shown in Figure 11-6(b)], becomes a problem: Its low-impedance path will couple a portion of the output signal back to the input. Because a phase inversion exists between the base and collector of a common emitter amplifier, this feedback voltage will be opposite in polarity to the input signal voltage, causing the input signal voltage to be decreased and the amplifier's gain to also be decreased. This type of out-of-phase feedback is called **negative feedback** because it has a degenerative effect.

Figure 11-6(c) shows how the negative feedback caused by Miller capacitance can be neutralized. To counter the effect of Miller capacitance in high-frequency amplifiers, a **neutralizing capacitor (C_N)** is included to couple an in-phase signal back to the base of the RF transistor to compensate for the out-of-phase signal coupled back to the base of the transistor by the Miller capacitance C_M. Coupling an in-phase voltage back to the base of Q_1 in this way effectively neutralizes the out-of-phase feedback voltage also being fed back to the base of Q_1.

11-1-4 An RF Variable-Tuned Amplifier Circuit

Up until this point, we have seen how the RF amplifier can be used in a basic AM transmitter. In this section, we will see how the RF amplifier finds application in a basic AM receiver.

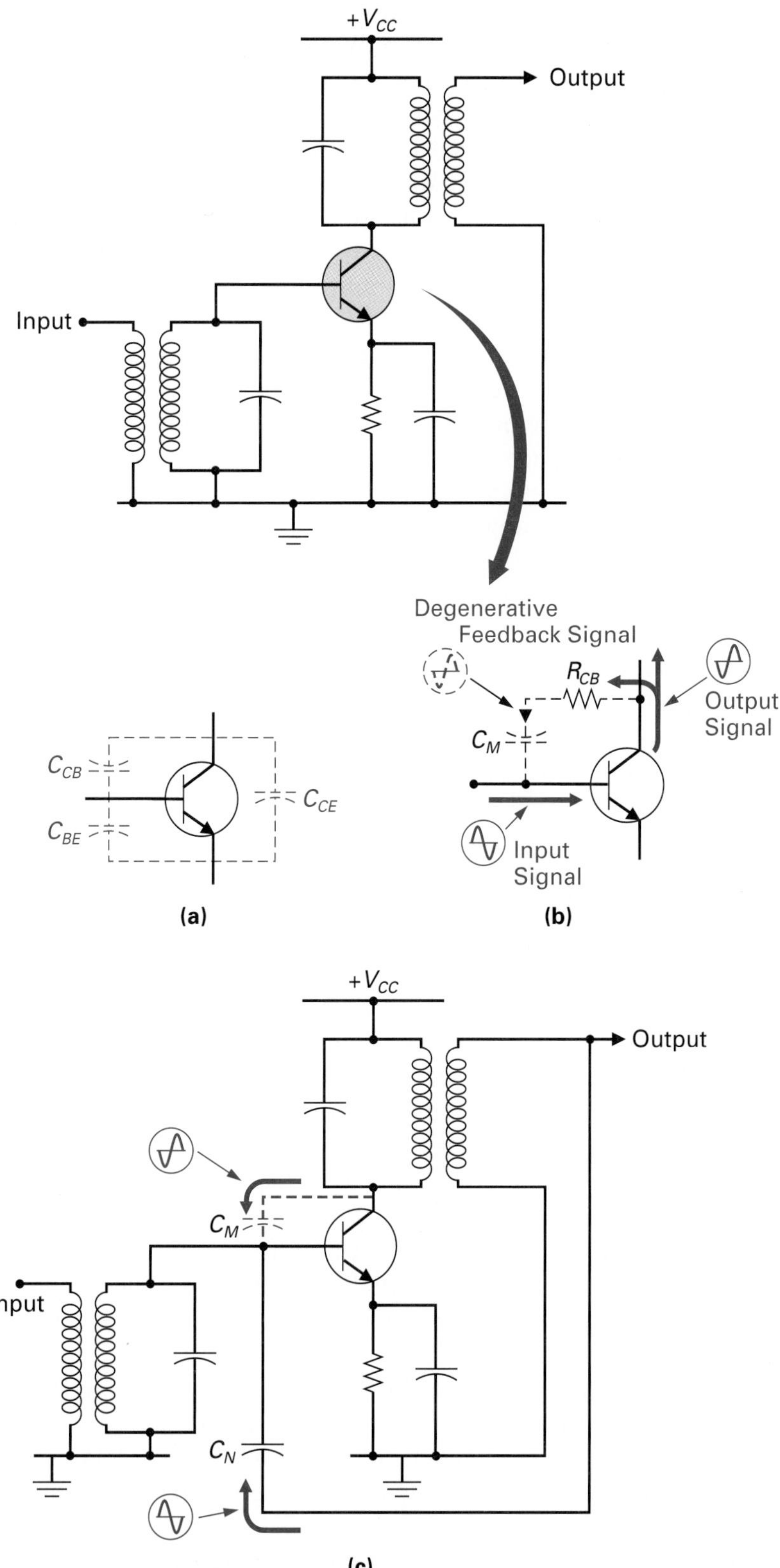

FIGURE 11-6 Neutralizing Miller-Effect Capacitance in Transistors.

Tuned AM Receiver An AM receiver circuit that can have its components varied so that the circuit responds to one selected frequency yet heavily attenuates all other frequencies.

Figure 11-7(a) shows the simplified block diagram of a **tuned AM receiver.** The electromagnetic waves transmitted by all of the different local AM radio stations will be present at the receiving antenna, which is simply a piece of wire. These electromagnetic signals will cut through the receiving antenna and induce an amplitude modulated signal that corresponds in wave shape to the original AM transmitted signal; however, its amplitude will now be considerably lower, typically only a few hundred microvolts. To boost this very weak signal, receiver circuits will generally contain several front-end RF amplifier stages to boost the received signal to a more usable amplitude before it is applied to the detector.

Variable-Tuned RF Amplifier A tuned radio frequency amplifier in which the tuned circuit(s) can be adjusted to select the desired station carrier frequency.

Ganged Mechanically linked by a common shaft so that adjustments can be made simultaneously.

Before discussing the detector circuit, let us examine these **variable-tuned RF amplifier** stages in more detail. With every local AM radio station signal at the input of the receiver, the question is: how do we tune-in or select the desired station? The answer is to once again include *LC* tank circuits as band-pass filters in the RF amplifier circuit, as shown in Figure 11-7(b). Unlike the previously discussed fixed-tuned RF amplifier circuit, however, these tank circuits contain variable capacitors that can be adjusted to control the resonant frequency of the *LC* tuned circuit and therefore the band of frequencies or station that will be passed onto the next stage. The dashed line between the two capacitors C_{1A} and C_{1B} is used to indicate that these variable capacitors are **ganged,** or mechanically linked by a common shaft, so that the operator can adjust the "tuning control" to simultaneously set the band-pass frequency for both tuned circuits, and therefore select which station is passed or filtered-out, or blocked. These tuned circuits are tunable over a wide range of frequencies, which for the AM band would be between 535 kHz to 1605 kHz.

Let us now examine the operation of the variable-tuned RF amplifier circuit shown in Figure 11-7(b). The amplitude modulated induced signals from the antenna are transformer coupled to the first parallel *LC* tank circuit made up of capacitor C_{1A} and the secondary of T_1. In the example in Figure 11-7(b), the tuning control is set to 900 kHz, and so the input tuned circuit will pass this frequency to the base of Q_1 and reject all others. To be more specific, the 900 kHz signal will see the parallel tuned circuit as a very high impedance, and therefore nearly all of the signal voltage will be developed across the tank and applied to the base of Q_1. All other frequencies will see a very low tank impedance and be shunted to ground. The selected RF signal is then amplified by Q_1 and transformer-coupled to the output tuned circuit made up of T_2's secondary and capacitor C_{1B}. The output tank, which is tuned to the exact same frequency as the input tank circuit, is included to make the RF amplifier highly selective. Transformer input and output coupling is used to both isolate amplifier stages and match impedances for a high gain.

Detector Circuit The circuit that recovers audio frequency information signal from the radio frequency carrier.

Let us now return to the basic block diagram of the tuned AM receiver block diagram shown in Figure 11-7(a) to discuss the function of the remaining blocks. As discussed previously, each information signal is placed on its own unique carrier frequency to prevent its transmission from interfering with other station transmissions. However, the human ear cannot hear high radio frequency signals, and so a demodulator, or **detector circuit,** is included in every receiver to recover the AF information signal from the radio frequency (RF) carrier. The

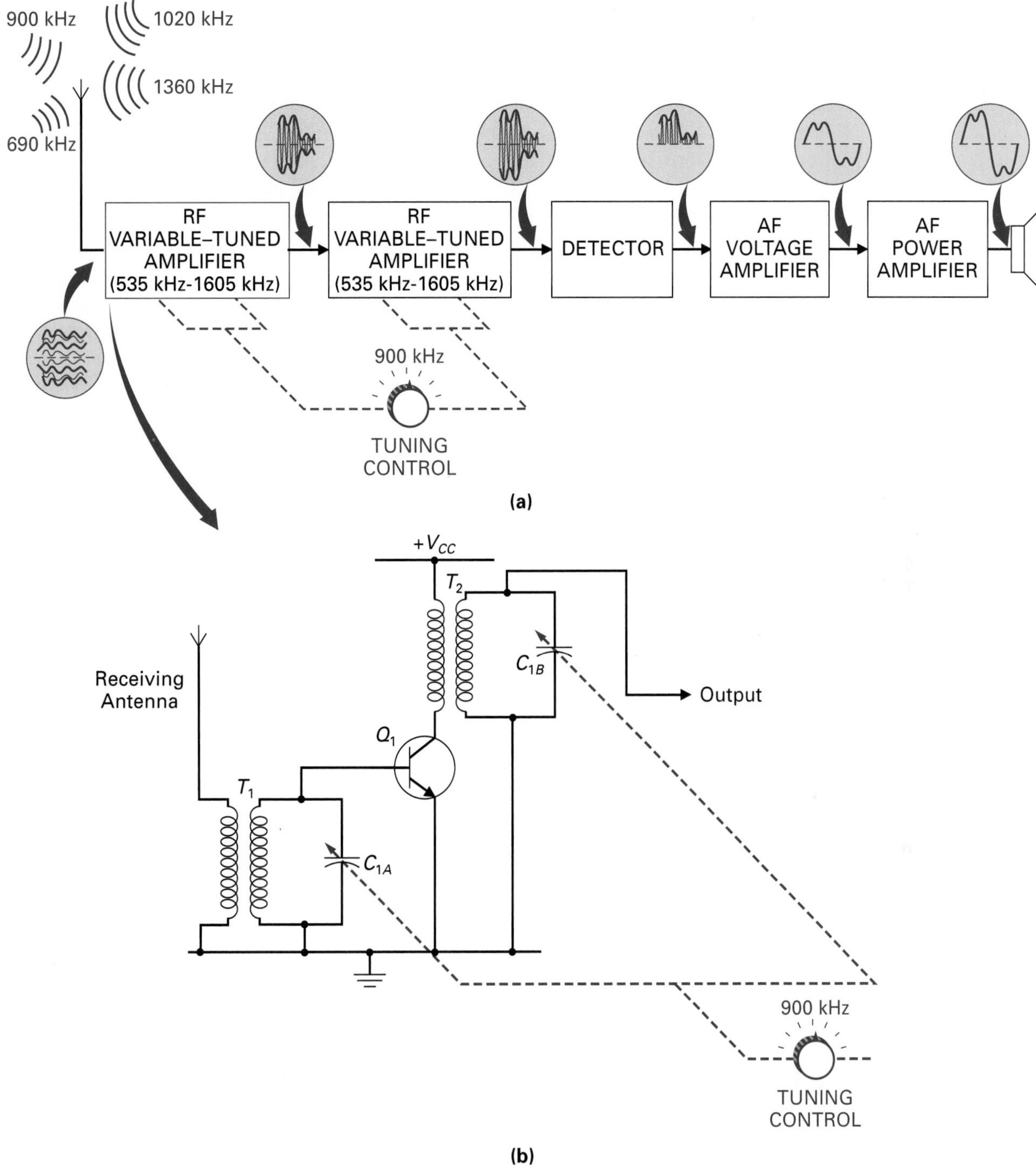

FIGURE 11-7 Tuned AM Radio Frequency (RF) Receiver. (a) Basic Block Diagram of RF Receiver. (b) Variable-Tuned RF Amplifier Circuit.

recovered audio frequency information signal is then amplified by an AF voltage and power amplifier so that the signal strength is large enough to drive the speaker.

The tuned AM radio frequency receiver, shown in Figure 11-7(a), had two key disadvantages. The first was that several ganged RF variable-tuned amplifier stages were needed to increase the signal level to a usable voltage level before it could be applied to the demodulator. This large number of RF amplifiers was necessary because negative feedback is always present when amplifying high-frequency signals. The second disadvantage of this receiver was that, because all of the RF amplifier stages had to be ganged, the capacitor was both bulky and expensive. As a result, the tuned AM receiver circuit was replaced by the "superheterodyne receiver," which will be discussed in the following section.

SELF-TEST REVIEW QUESTIONS FOR SECTION 11-1

1. Radio frequency amplifiers are used to amplify radio frequency signals from about
   a. 20 Hz to 20 kHz
   b. 10 Hz to 35 MHz
   c. 10 kHz to 30 GHz
   d. 20 kHz to 20 MHz
2. Give the full names for the following abbreviations:
   a. RF **b.** AM **c.** FCC
3. Which RF amplifier circuit has a collector-tuned circuit that is tuned to an exact multiple of the input frequency?
4. What would be the bandwidth of an AM 1500 kHz carrier if it were modulated by a 3.5 kHz AF tone?
5. An RF amplifier with a narrow bandwidth has a ____________ selectivity.
6. When a swamping resistor is added to an amplifier's tuned circuit, its bandwidth is ____________, and its gain is ____________.

11-2 INTERMEDIATE-FREQUENCY (IF) AMPLIFIER CIRCUITS

Superheterodyne Receiver
RF receiver that converts all RF inputs to a common intermediate frequency (IF) before demodulation.

In the previous section, we discovered that the "tuned AM receiver" needed several high-frequency variable-tuned RF amplifier stages to boost the weak input signal to a usable voltage level and an expensive ganged capacitor for each tuned circuit in each RF amplifier stage. These two disadvantages were overcome by using a radio receiving technique that has been so successful it is used in nearly every AM radio, FM radio, and television receiver. This special type of receiver is called a **superheterodyne receiver,** with the word "heterodyning" meaning "frequency converting."

11-2-1 *The Superheterodyne Receiver*

To understand the operation of a superheterodyne receiver, Figure 11-8(a) shows the block diagram of a basic superhet AM receiver. As before, the received AM signal is first amplified by a variable-tuned RF amplifier stage. The output of this stage is then applied to the distinct feature of this type of receiver, its **RF amplifier/mixer circuit.** This mixer circuit is a special circuit designed to combine, or mix, two input frequencies. The two input frequencies that are combined in the mixer circuit are the received AM signal from the RF variable-tuned amplifier and the RF sinewave signal from the **RF local oscillator circuit.** Referring to the tuning control in Figure 11-8(a), you can see that it is ganged to control both the tuning of the front-end RF amplifier and the frequency of the RF local oscillator. The frequency of the RF local oscillator is controlled so that its frequency is 455 kHz greater than the received RF input frequency. For example, let us assume that the tuning control is set to 900 kHz. The mechanical linkage from the tuning control to the RF amplifier will vary its internal tuned circuit capacitors so that they will pass the band of frequencies centered at 900 kHz. At the same time, the mechanical linkage from the tuning control to the RF local oscillator will vary its internal tuned circuits so that it will produce an RF sine-wave frequency that is 455 kHz greater than the received RF signal.

RF Amplifier/Mixer Circuit
A circuit designed to combine or mix two input frequencies.

RF Local Oscillator Circuit
The radio frequency oscillator in a superhetrodyne receiver.

$$f_{RF\,\text{Amp}} = 900\text{ kHz}$$

$$f_{\text{LO}} = \text{RF} + 455\text{ kHz} = 1355\text{ kHz}$$

These two input signals are combined in the RF amplifier/mixer circuit, producing four signals at the output. These output signals are

1. The original RF input signal from the RF amplifier (900 kHz),
2. The original RF signal from the local oscillator (1355 kHz),
3. The sum of the received RF signal and the local oscillator signal (900 kHz + 1355 kHz = 2255 kHz), and
4. The difference between the received RF signal and the local oscillator signal (1355 kHz − 900 kHz = 455 kHz).

11-2-2 *An RF Amplifier/Mixer Circuit*

Figure 11-8(b) shows an example of a typical RF amplifier/mixer circuit. The received AM signal from the RF amplifier and the RF sine-wave signal from the RF local oscillator are both coupled to the base of Q_1, where they are amplified and combined to produce four signal combinations at the transistor's collector. The **double-tuned transformer** in the collector of Q_1 is fixed-tuned to the difference frequency of 455 kHz; therefore, only this signal is passed to the output.

Double-Tuned Transformer
A transformer having both a tuned primary and tuned secondary.

Because the frequency of the RF local oscillator is controlled to always be 455 kHz above the tuned frequency of the RF amplifier, the difference output signal from the mixer circuit will always be fixed at 455 kHz. For example, if the tuning control were set to 690 kHz, the RF amplifier's internal tuned circuits will pass the band of frequencies centered at 690 kHz, and the RF local oscillator

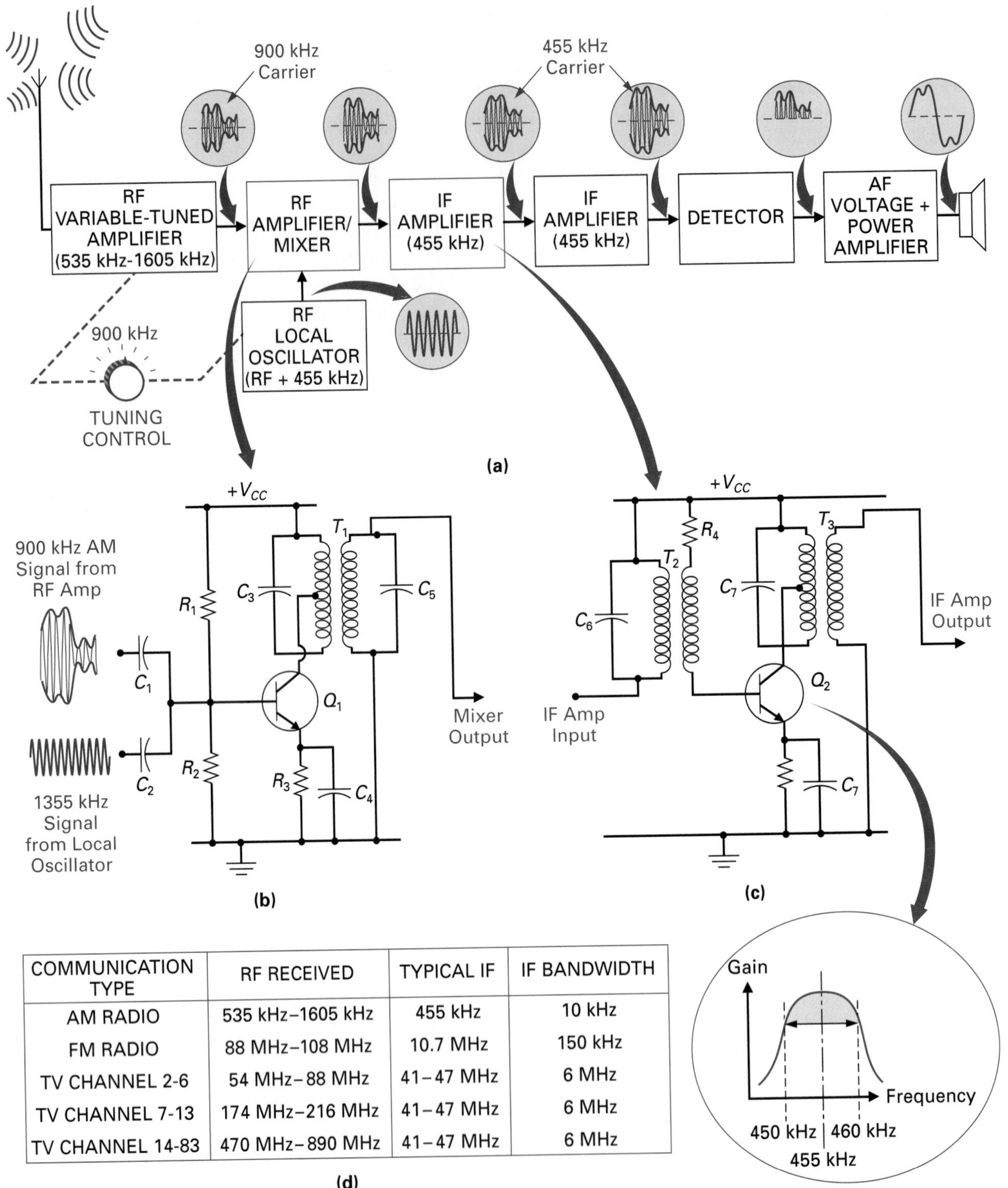

COMMUNICATION TYPE	RF RECEIVED	TYPICAL IF	IF BANDWIDTH
AM RADIO	535 kHz–1605 kHz	455 kHz	10 kHz
FM RADIO	88 MHz–108 MHz	10.7 MHz	150 kHz
TV CHANNEL 2-6	54 MHz–88 MHz	41–47 MHz	6 MHz
TV CHANNEL 7-13	174 MHz–216 MHz	41–47 MHz	6 MHz
TV CHANNEL 14-83	470 MHz–890 MHz	41–47 MHz	6 MHz

FIGURE 11-8 Superheterodyne AM Radio Frequency Receivers. (a) Basic Block Diagram of RF Receiver. (b) RF Amplifier/Mixer Circuit. (c) Intermediate Frequency (IF) Amplifier Circuit. (d) Typical IF Frequencies.

will produce an RF sine-wave frequency that is 455 kHz greater than the received RF signal.

$$f_{\text{RF Amp}} = 690 \text{ kHz}$$

$$f_{\text{LO}} = \text{RF} + 455 \text{ kHz} = 690 \text{ kHz} + 455 \text{ kHz} = 1145 \text{ kHz}$$

These two input signals of 690 kHz and 1145 kHz are then combined in the RF amplifier/mixer circuit, and the difference of 455 kHz is extracted.

$$f_{\text{Mixer}} = f_{\text{LO}} - f_{\text{RF Amp}} = 1145 \text{ kHz} - 690 \text{ kHz} = 455 \text{ kHz}$$

Returning to the block diagram's waveforms in Figure 11-8(a), you can see that the 455 kHz signal from the RF amplifier mixer is amplitude modulated in exactly the same way as the 900 kHz RF input signal from the antenna. Our AF modulating information signal is now simply being carried on the lower 455 kHz carrier instead of the higher 690 kHz radio frequency carrier. This system of transferring the information from a high radio frequency to a lower intermediate frequency is the key advantage of this superheterodyne receiver. Now that we have the information signal on a lower fixed frequency, it will be easier to amplify it up to a more usable voltage level.

11-2-3 *An IF Amplifier Circuit*

Returning to the block diagram in Figure 11-8(a), you can see that the 455 kHz difference signal output from the RF amplifier/mixer is next applied to a two-stage amplifier. The difference frequency from the RF amplifier/mixer is about midway between the RF signal received at the antenna and the AF final output signal applied to the speaker. It is called an **intermediate frequency** or IF (pronounced "eye-eff"). The tuned circuits in these intermediate frequency (IF) amplifier circuits are also fixed-tuned to 455 kHz. Figure 11-8(c) shows an example of a typical IF amplifier circuit and its frequency response curve. Capacitor C_6 and the primary of T_2 are fixed-tuned to the IF frequency of 455 kHz, along with the output tank circuit made up of C_7 and the primary of T_3.

Intermediate Frequency
The difference frequency from the RF amplifier/mixer that is about midway between the RF signal received at the antenna and the AF final output signal applied to the speaker.

As mentioned previously, looking at the waveforms in Figure 11-8(a), you can see that the 455 kHz signal being amplified by the IF amplifiers is amplitude modulated in exactly the same way as the 900 kHz RF input signal from the antenna. Our AF modulating information signal is now simply being carried on a lower intermediate frequency instead of the higher radio frequency. This system of transferring the information from a high radio frequency to a lower intermediate frequency is the key advantage of this superheterodyne receiver. Fixed-tuned intermediate frequency amplifiers are less expensive, and have less high-frequency losses, than variable-tuned radio frequency amplifiers. In addition, as we discovered earlier, narrow band high-Q amplifier circuits, such as the IF amp, have higher gains.

As seen in Figure 11-8(a), the remainder of the superhet receiver following the IF amplifiers is much the same as the tuned receiver. The 455 kHz signal from the final IF amplifier stage is applied to a detector, which extracts the AF information signal from the 455 kHz carrier. Then the AF signal is applied to an amplifier and boosted in amplitude before being applied to the speaker.

The table in Figure 11-8(d) lists some of the typical IF frequencies and bandwidths used in several different types of communication circuits.

SELF-TEST REVIEW QUESTIONS FOR SECTION 11-2

1. In a superheterodyne receiver, the RF input signal is converted to a lower ___________.
   a. RF
   b. VF
   c. AF
   d. IF
2. The ___________ circuit in a superhet receiver combines the RF input frequency and the RF signal from the local oscillator.
3. What are the two advantages for transferring an input signal onto an IF carrier instead of an RF carrier?
4. Are IF amplifiers fixed-tuned or variable-tuned?

11-3 VIDEO FREQUENCY (VF) AMPLIFIER CIRCUITS

Video Frequency (VF) Amplifiers Amplifiers used in televisions, computer monitors, video games, virtual reality systems, and radar systems to amplify picture information signals (video signals).

Video Signals A signal that contains visual information for television or radar systems.

Video frequency (VF) amplifiers are used in televisions, computer monitors, video games, virtual reality systems, and radar systems to amplify picture-information signals, which are also called **video signals.** These video signals include a very wide range of frequencies, spanning from about 10 Hz to 5 MHz; therefore, the video amplifier must provide a high gain over a very wide bandwidth. To begin with, let us discuss how we can modify a basic amplifier circuit in order to get it to respond to this wide band of frequencies.

11-3-1 Video Amplifier Shunt Capacitance

As discussed previously in the RF amplifier section, a small value of capacitance exists between all the junctions of a transistor. These junction capacitances are determined by the physical size of the junction, the spacing between the transistor's leads, and the junction bias condition. In addition, because many of the tracks on a printed circuit board are in close proximity to one another, they possess a certain value of capacitance that is inherent in every circuit, called stray wire capacitance.

Output Stray-Wire Capacitance (C_{SO}) The capacitance that is formed between any conductor and an adjacent conductor.

Referring to Figure 11-9(a), you can see a basic *RC* coupled amplifier with all of the circuit's inherent shunt capacitances drawn in using dashed lines. The shunt output capacitance of transistor Q_1 ($C_{\text{OUT}(Q_1)}$) is made up of the transistor's collector-to-emitter capacitance (C_{CE}) and the **output stray-wire capacitance (C_{SO}).** At the other end, we have the input capacitance of Q_2, which is made up of Q_2's collector-to-base Miller capacitance (C_M), base-to-emitter capacitance (C_{BE}), and the input stray-wire capacitance (C_{SI}). Because all of these capacitance values are in parallel, we can add them together to obtain a total shunt capacitance value (C_T). This total shunt capacitance value is typically

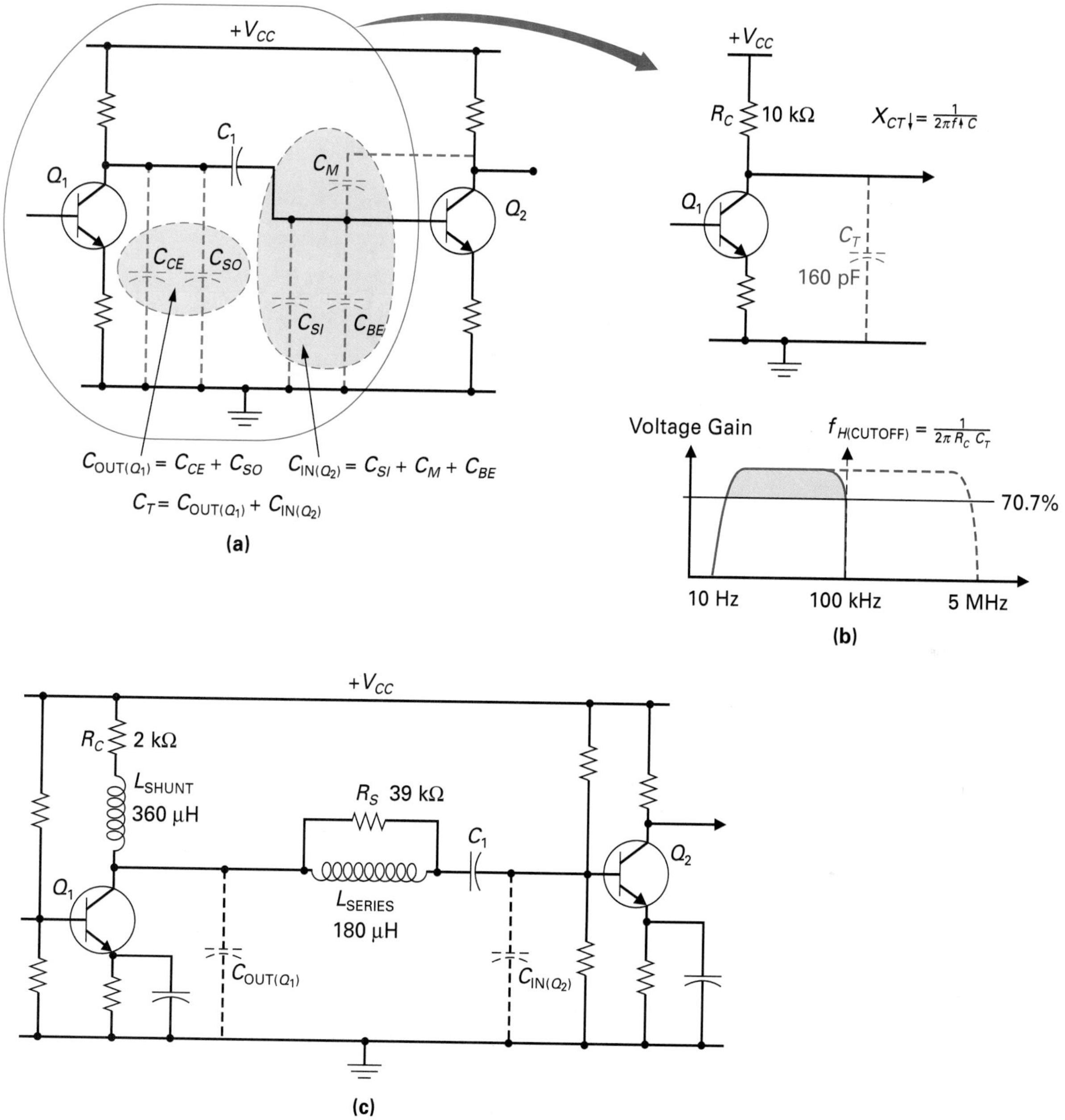

+V_{CC}
C_1
C_M
Q_1
Q_2
C_{CE}
C_{SO}
C_{SI}
C_{BE}
$C_{OUT(Q_1)} = C_{CE} + C_{SO}$
$C_{IN(Q_2)} = C_{SI} + C_M + C_{BE}$
$C_T = C_{OUT(Q_1)} + C_{IN(Q_2)}$
(a)
+V_{CC}
R_C 10 kΩ
$X_{CT}\downarrow = \frac{1}{2\pi f\uparrow C}$
Q_1
C_T
160 pF
Voltage Gain
$f_{H(CUTOFF)} = \frac{1}{2\pi R_C C_T}$
70.7%
10 Hz
100 kHz
5 MHz
(b)
+V_{CC}
R_C 2 kΩ
L_{SHUNT}
360 μH
R_S 39 kΩ
C_1
Q_1
Q_2
L_{SERIES}
180 μH
$C_{OUT(Q_1)}$
$C_{IN(Q_2)}$
(c)

FIGURE 11-9 Video Amplifier Circuits. (a) Shunt Capacitances. (b) High Frequency Cutoff. (c) Frequency Compensation.

160 pF and is shown in Figure 11-9(b). When the amplifier is amplifying low-frequency signals, this capacitance has very little effect because X_C is large. However, as the frequency of the signal being amplified increases, X_C decreases, and the total shunt capacitance becomes a problem because it shunts some of the signal voltage to ground. The point at which the voltage of the signal frequency drops below 70.7% is called the **high-frequency cutoff** ($f_{H(\text{CUTOFF})}$) and can be calculated with the formula

High-Frequency Cutoff ($F_{H(\text{CUTOFF})}$)
The point at which the voltage of the signal frequency drops below 70.7%.

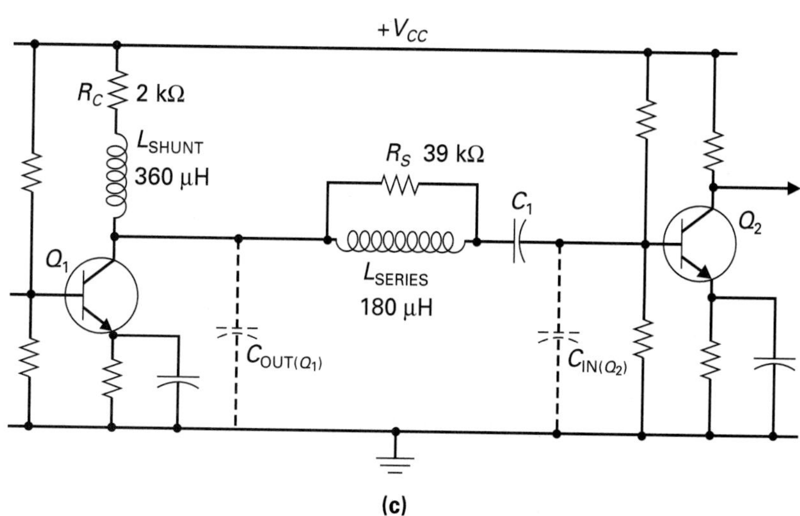

$$f_{H(\text{CUTOFF})} = \frac{1}{2\pi R_C C_T}$$

■ **EXAMPLE:**

Calculate the high-frequency cutoff for the values given in Figure 11-9(b).

■ ***Solution:***

$$f_{H(\text{CUTOFF})} = \frac{1}{2\pi R_C C_T} = \frac{1}{2 \times \pi \times 10\text{ k}\Omega \times 160\text{ pF}}$$

$$= \frac{0.159}{(10 \times 10^3) \times (160 \times 10^{-12})} = 99.47\text{ kHz}$$

This high-frequency cutoff at 100 kHz is shown in the frequency response curve in Figure 11-9(b). Unfortunately, the needed frequency response for video signals is up to 5 MHz, as shown by the dashed line. To increase the frequency response of a video amplifier we could decrease the value of R_C.

■ **EXAMPLE:**

What would be the high-frequency cutoff if R_C were reduced to 2 kΩ, as shown in Figure 11-9(c)?

■ ***Solution:***

This would extend our frequency range to

$$f_{H(\text{CUTOFF})} = \frac{1}{2\pi R_C C_T} = \frac{1}{2 \times \pi \times 2\text{ k}\Omega \times 160\text{ pF}} = 497.4\text{ kHz}$$

However, decreasing R_C will also reduce the voltage gain of the amplifier ($A_V = R_C/R_E$). Therefore, more video amplifier stages will be needed to achieve the same overall gain.

11-3-2 *Video Amplifier Peaking Coils*

Peaking Coils
Coils included in a circuit to compensate for signal loss at high frequencies.

Another way to compensate for this signal loss at high frequencies is to somehow cancel the effect of the circuit's inherent shunt capacitances. This is achieved by using the **peaking coils** shown in the video amplifier circuit in Figure 11-9(c). The *shunt peaking coil* (L_{SHUNT}) included in the collector of Q_1 will compensate for the output capacitance of Q_1 ($C_{\text{OUT}(Q_1)}$). By carefully choosing the value of the shunt peaking coil, we can cause L_{SHUNT} and $C_{\text{OUT}(Q_1)}$ to resonate at the higher frequencies. Because L_{SHUNT} and $C_{\text{OUT}(Q_1)}$ are in parallel with one another (collector current will split), this parallel resonance circuit will have a high impedance at resonance and therefore develop a larger output voltage. A *series peaking coil* (L_{SERIES}) can also be used to increase the high-frequency response of a video amplifier by compensating for the input capacitance of Q_2 ($C_{\text{IN}(Q_2)}$), as seen in Figure 11-9(c). Because the inductive reactance of this series peaking coil will increase with frequency ($X_L \propto f$), this coil will isolate the output capacitance of

Q_1 from the input capacitance of Q_2, and therefore the large total shunt capacitance is now reduced to two smaller value capacitances. Because $C_{OUT(Q_1)}$ is smaller than C_T, a larger value of R_C can be used, and therefore the amplifier will have a larger voltage gain. The swamping resistor R_S in parallel with the series peaking coil reduces the Q of the inductor to widen the response, or bandwidth, of the coil so that it will respond well to all video frequencies.

11-3-3 Basic Television Receiver Circuits

Video amplifiers find their largest application in television receivers. Figure 11-10 shows the basic block diagram of a television receiver and examples of some typical television receiver circuits.

Figure 11-10(a) shows a typical VHF fixed-tuned RF amplifier circuit. The tuned circuits, made up of L_1 with C_1 and C_3 with the primary of T_1, reject frequencies that are outside of the selected band. Transistor Q_1 operates as a class-A amplifier, with its gain controlled by an **automatic gain control (AGC)** input. This control voltage originates from a circuit that automatically senses the strength of the input signal, then increases or decreases the gain of the amplifier to ensure that the gain is almost constant over a wide range of input signal voltages. For example, when the input signal is too strong, the AGC circuit uses degenerative feedback to reduce the input signal strength, whereas if the input signal is weak, the degenerative feedback is reduced to increase the amplifier gain. By ground-shielding the transistor Q_1, any stray RF signals will be shunted to ground instead of being introduced into the signal as noise. The noise level at this early amplifier stage should be kept to a minimum because any noise introduced will be amplified along with the desired signal by all of the following amplifier stages. Resistor R_2 and capacitor C_4 form a decoupling filter circuit that will prevent any of the RF signal from reaching the +10 V power supply line and then interfering with other circuits, while the neutralizing capacitor (C_N) compensates for Miller capacitance.

Automatic Gain Control (AGC)
The control voltage from a circuit that automatically senses the strength of the input signal then increases or decreases the gain of the amplifier to ensure that the gain is almost constant over a wide range of input signal voltages.

Figure 11-10(b) shows a typical IF amplifier for a TV receiver, which is double transformer tuned to the intermediate frequency of 45 MHz. Once again, the gain of Q_1 is controlled by the AGC control input, and capacitor C_N is included to neutralize the Miller effect.

Figure 11-10(c) shows a typical video frequency amplifier for a television receiver. This black and white television amplifier will provide a voltage gain of about 30 for video signals from about 10 Hz to 3.5 MHz (for color, the bandwidth will have to be extended to 5 MHz). Transistor Q_1 is an emitter follower, included to act as a high-impedance load for the previous detector stage and to act as a low-impedance source for the following stage. The video output signal from Q_1 is developed across R_1 and then coupled to the base of Q_2 via the large value capacitor C_1, which will have a low reactance to low-frequency signals. The common-emitter amplifier stage (Q_2) makes use of a very high V_{CC} supply voltage, which is needed to drive the television's cathode ray tube (CRT). Resistor R_5 is not bypassed and will be degenerative, controlling the gain of this video amplifier, and as a result, controlling the **contrast** of the picture. For example, by adjusting the TV's contrast control to decrease R_5's resistance (moving R_5's wiper up), the gain of Q_2 will be increased, and the difference between the black and white dots on the TV screen will increase, causing an increase in contrast.

Contrast
The difference between the light and dark areas in a video picture. High contrast pictures have dark blacks and brilliant whites. Low contrast pictures have an overall gray appearance.

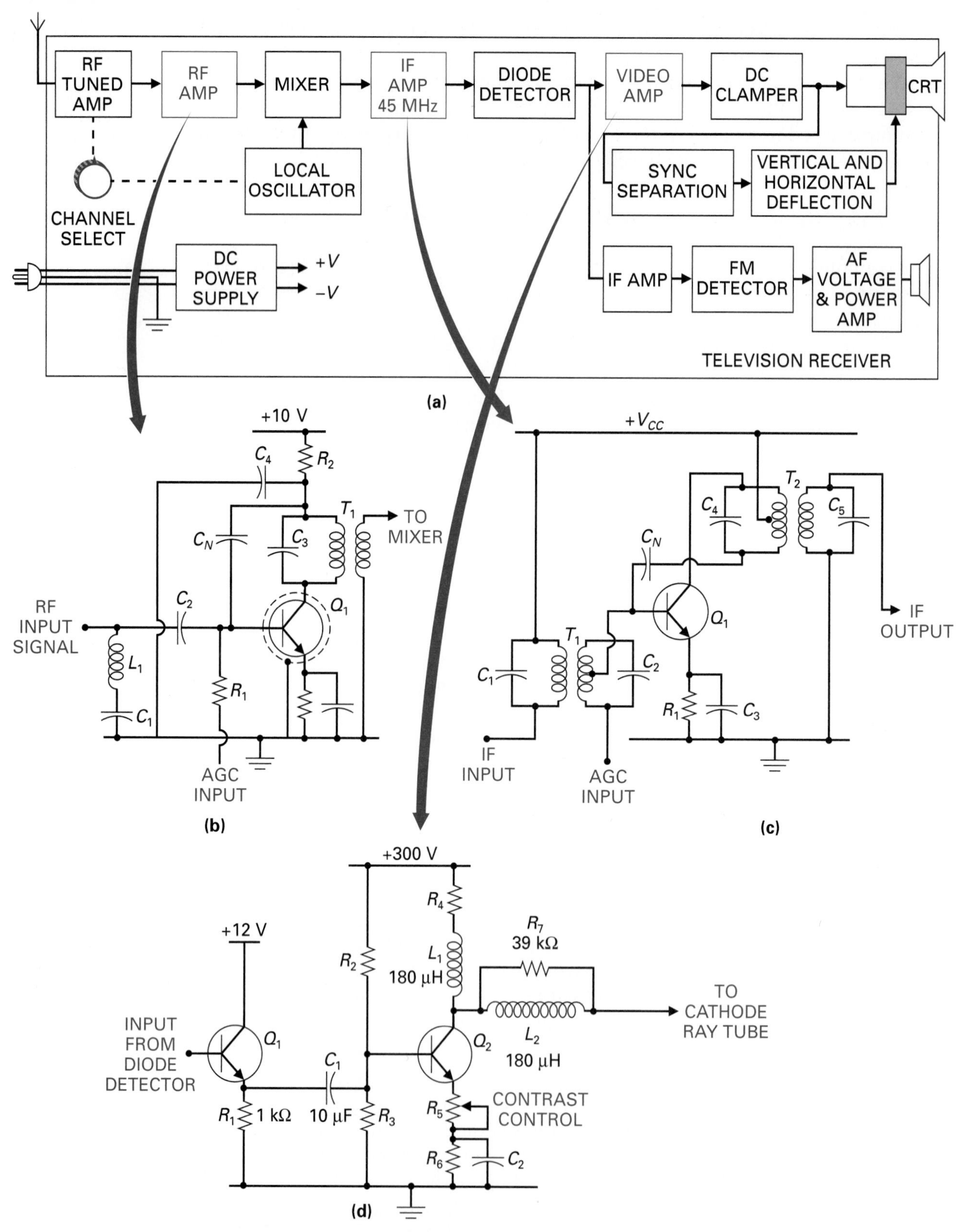

FIGURE 11-10 Television Receiver Circuits. (a) Basic Television Receiver Block Diagram. (b) RF Amplifier. (c) IF Amplifier. (d) VF Amplifier.

Looking at the collector of the video amplifier Q_2, you can see that a shunt peaking coil (L_1), a series peaking coil (L_2), and a swamping resistor (R_7) have been added to extend the amplifier's high-frequency cutoff point.

SELF-TEST REVIEW QUESTIONS FOR SECTION 11-3

1. Where are video frequency amplifiers used?
2. Video frequency amplifiers are used to amplify video frequency signals from about
   a. 20 Hz to 20 kHz
   b. 10 Hz to 5 MHz
   c. 10 kHz to 30 GHz
   d. 20 kHz to 20 MHz
3. Why are peaking coils included in most VF amplifier circuits?
4. Decreasing the value of R_C in a video amplifier will increase the amplifier's high-frequency cutoff and reduce the amplifier's gain. (True/False)

11-4 TUNING AND TROUBLESHOOTING HIGH-FREQUENCY AMPLIFIERS

In this section we will examine the details of tuning and troubleshooting of high-frequency amplifier circuits.

11-4-1 *Tuning a Tuned Amplifier Circuit*

One of the biggest problems with tuned circuit amplifiers is that their tuned circuits are not tuned to the desired center frequency. The reasons for this problem can range from capacitor and inductor tolerances to the inherent stray and junction circuit capacitance values that are always present. To compensate for these problems, most tuned amplifier circuits either have a built-in variable inductor as seen in Figure 11-11(a) or a built-in variable capacitor as seen in Figure 11-11(b). A variable inductor is more frequently used because it is less expensive than a variable capacitor.

The next question is: How do we **tune** the amplifier? A simple and accurate method is to

Tune
To adjust the resonance of a circuit so that it will select the desired frequency.

1. First connect an ac current meter between V_{CC} and the tuned circuit, as shown in Figure 11-11(c),
2. Next, connect a frequency generator to the input of the amplifier, and set it to the desired frequency, and
3. Finally, adjust the variable inductor (L_1) until the meter shows a large dip or null in current.

Remember that because a parallel tuned circuit will have a maximum impedance at resonance, all we are doing is adjusting the value of inductance (and

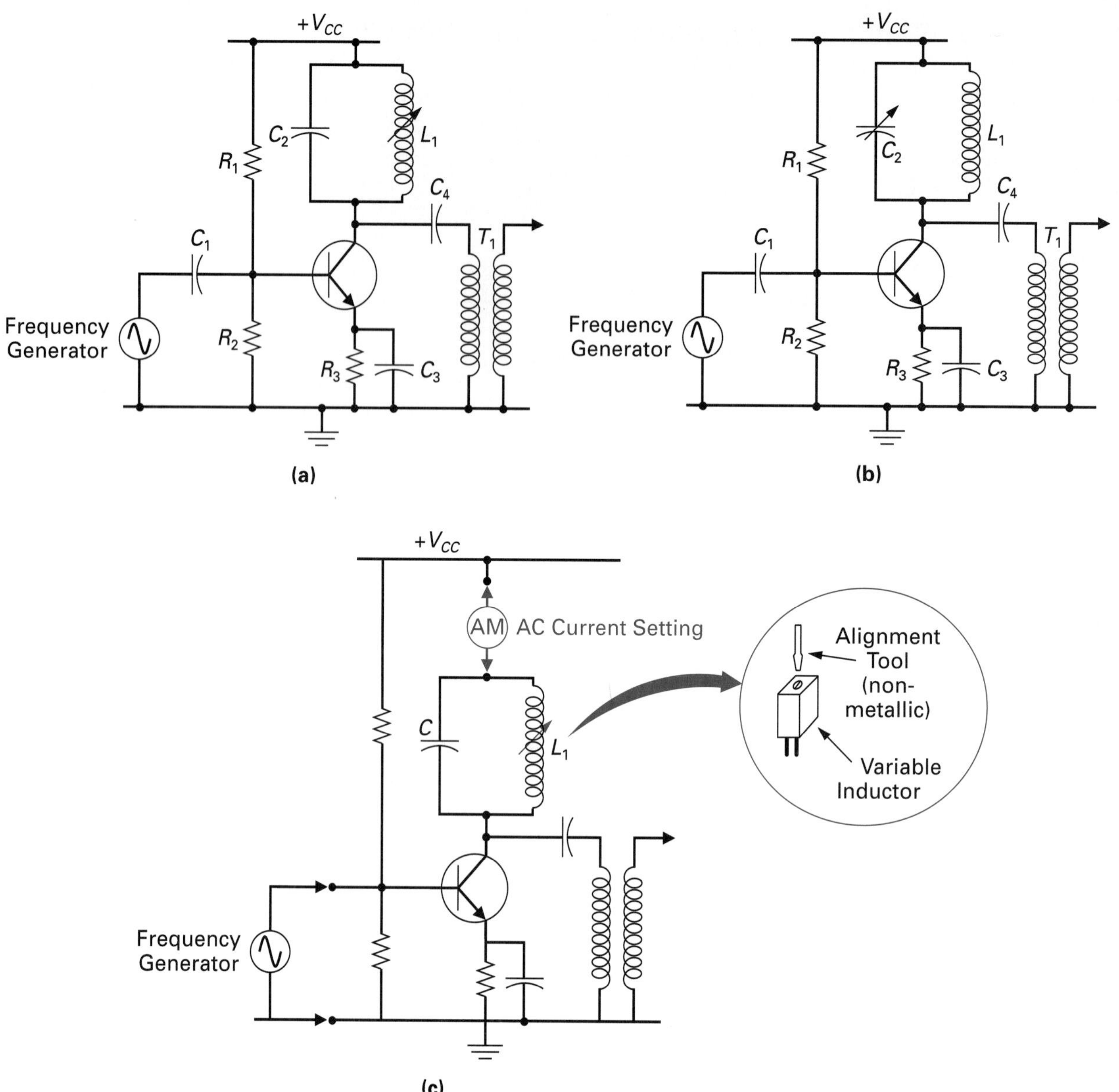

FIGURE 11-11 Tuning a Tuned Amplifier Circuit.

therefore the tuned circuit resonant frequency) until the current drops to a minimum. At frequencies above and below resonance, a small value of current will pass through the tank; however, when current drops to a minimum, we know that the tuned circuit's impedance is at a maximum, and therefore the tank is tuned to the same frequency as the frequency generator's input frequency.

Another important point to remember with this procedure is to always use a nonmetallic screwdriver because the proximity of any metal to the electric and magnetic fields of an adjustable capacitor or inductor will change its operation, and therefore change the frequency to which the tank is tuned.

11-4-2 *Troubleshooting a Tuned Amplifier Circuit*

As discussed in detail in the previous chapter, to troubleshoot an amplifier circuit you should start by making sure that the tuned high-frequency amplifier is, in fact, the cause of the problem. Follow the troubleshooting procedures outlined in the low-frequency amplifier chapter in exactly the same way to determine whether the problem is within the dc power supply, the amplifier's load, the amplifier's signal source, or the amplifier circuit itself.

The typical problem within tuned amplifier circuits is **frequency drift.** This is caused by component aging, changing temperature conditions, and a variety of other elements. This problem can usually be easily remedied by simply retuning the amplifier as discussed in the previous section. If tuning the amplifier does not cure the selectivity problem with the amplifier, you will have to test the amplifier's tuned circuit components. Figure 11-12 describes the effects of a tuned circuit's inductor and capacitor shorts and opens. If either the inductor or

Frequency Drift
A slow change in the frequency of a circuit due to temperature or frequency determining component value changes.

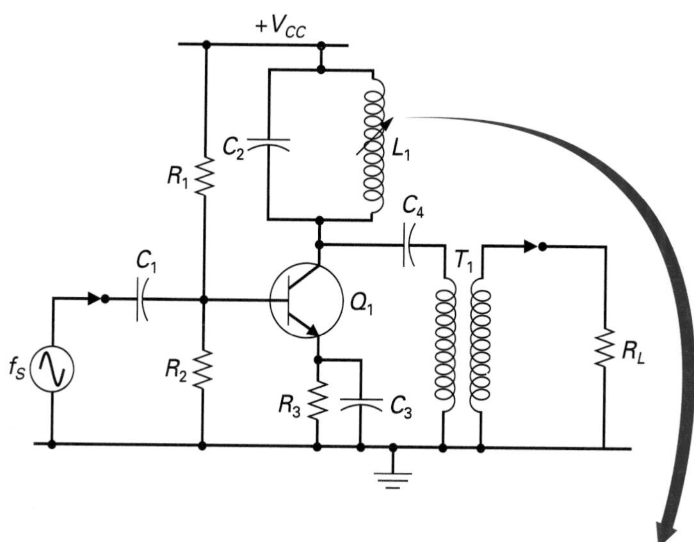

Cause	Effect
L_1 Short	With L_1 shorted, the tank circuit is bypassed and therefore the circuit will function as an emitter follower, however V_C will remain at V_{CC}, and therefore there will be no output signal.
C_2 Short	Same as above.
L_1 Open	The V_{CC} dc supply is disconnected from Q_1 collector since capacitor C_2 acts as a dc block. Collector current and therefore V_C drops to zero.
C_2 Open	A capacitor is usually an open to the dc supply voltage, however with it open the tuned circuit becomes inductive and so tuning is lost, and circuit will simply amplify and pass all frequencies.

FIGURE 11-12 Troubleshooting a Tuned Amplifier Circuit.

capacitor in the tuned circuit were to short, the circuit effects will be almost identical, due to the parallel connection. Therefore, the best course of action would be to disconnect power from the circuit and test the resistance of the tuned circuit. If the capacitor begins to charge (resistance rises), the inductor is probably at fault; however, if the resistance remains at zero, the capacitor is probably shorted.

SELF-TEST REVIEW QUESTIONS FOR SECTION 11-4

1. How is a high-frequency amplifier's tank circuit tuned?
2. What is frequency drift?

SUMMARY

1. Communication is a process by which information is exchanged.
2. The two basic methods of transferring information are the spoken word and the written word. Communication of the spoken word began face-to-face, then evolved into radio communications, telephone communications, and now video telecommunications. Communication of the written word began with hand-carried letters, and evolved into newspapers, the mail system, the telegraph, and now electronic mail.
3. At a very early stage, it was found that the distance we could communicate was important; therefore, systems of long distance communication were developed, beginning with smoke signals and beating drums. It was not until the late 1800s, that scientists discovered that a high-frequency electrical current passing through a conductor would radiate an energy wave that could be modified or altered to carry information. These radiated electromagnetic waves, or radio waves, could be used to communicate information of different types.
4. All radio communication transmitters and receivers need high-frequency amplifiers to boost the amplitude of the high-frequency information signals in both the transmitting circuits and receiving circuits.

Radio Frequency (RF) Amplifiers (Figure 11–13)

5. The radio frequency band includes any frequencies from about 10 kHZ to about 30,000 MHz.
6. An RF amplifier/multiplier circuit converts a stable low-frequency oscillation input into a stable high-frequency output signal that is some multiple of the input frequency.
7. When operated with class-C bias, it would seem that an RF amplifier transistor would introduce signal distortion because the full sine wave input signal is distorted to a small pulse of current. However, signal distortion does not occur because the rest of the positive half-cycle, and the full negative half-cycle, is recreated by the *LC* tank's flywheel action.

QUICK REFERENCE SUMMARY SHEET

FIGURE 11-13 Radio Frequency (RF) Amplifiers (Bipolar Transistor).

RF Amplifier/Multiplier Circuit

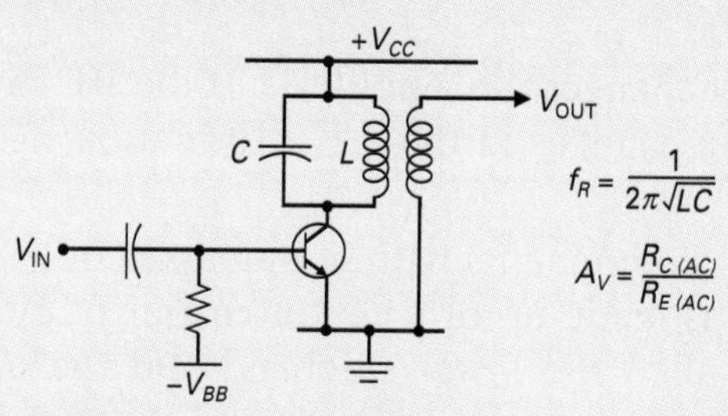

$f_R = \frac{1}{2\pi\sqrt{LC}}$

$A_V = \frac{R_{C(AC)}}{R_{E(AC)}}$

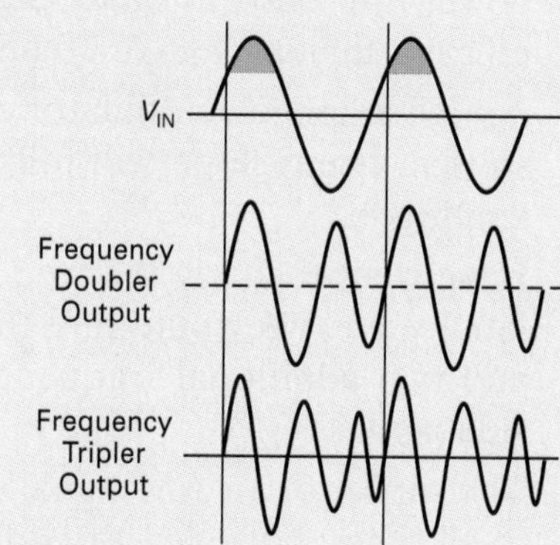

RF Amplifier/Modulator Circuit

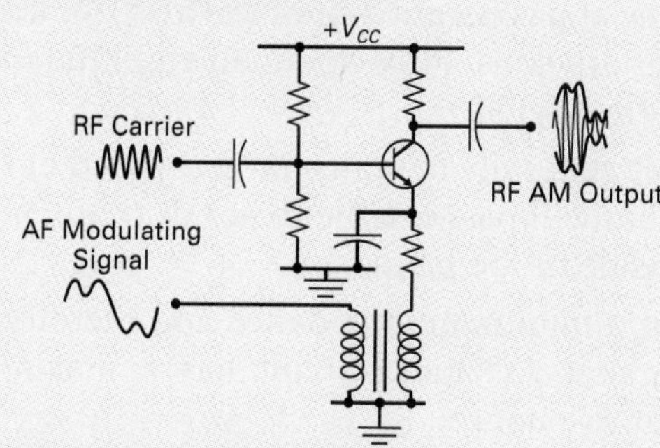

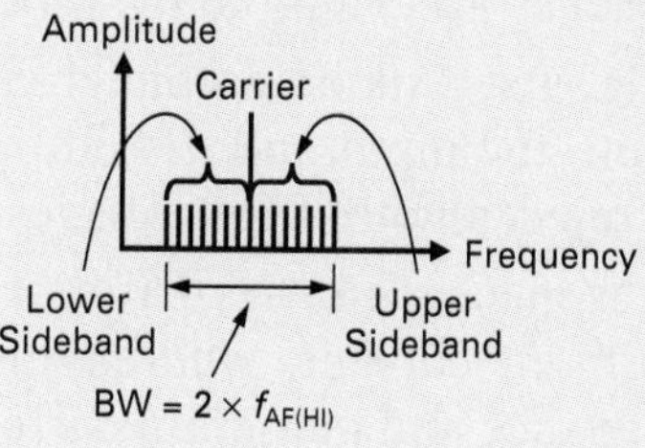

RF Fixed-Tuned Amplifier Circuit

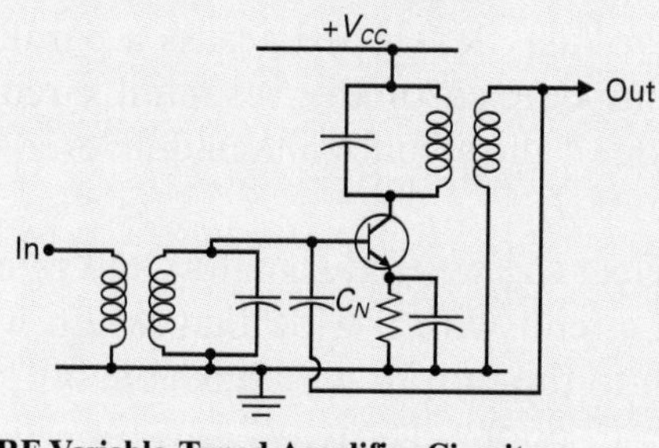

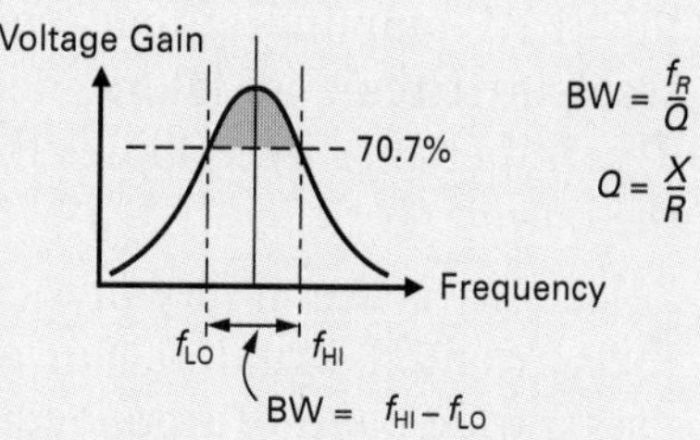

$BW = \frac{f_R}{Q}$

$Q = \frac{X}{R}$

RF Variable-Tuned Amplifier Circuit

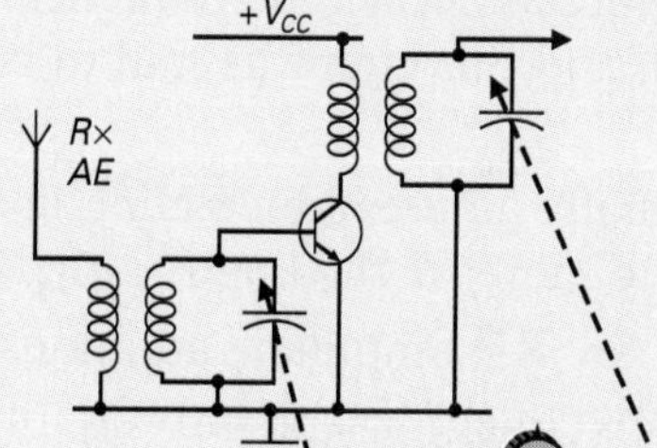

RF Amplifier/Mixer Circuit (for Superhet Receiver)

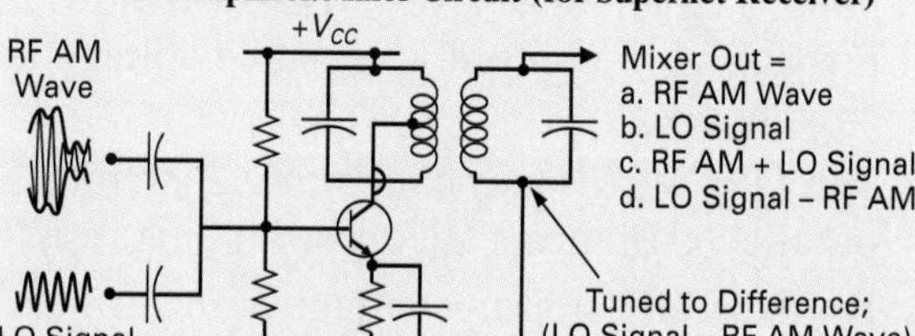

8. Frequency doubler, tripler, and quadrupler circuits are often used in RF circuits to convert a stable low-frequency oscillation to a stable high-frequency oscillation.
9. Operating with class-C bias means that Q_1 is conducting and consuming power for only a small portion of time, therefore the circuit is more efficient than a class-A circuit.
10. An RF amplifier/modulator circuit changes the amplitude of the RF carrier so that its envelope will follow the changes in the AF information signal's amplitude.
11. If we were to study the frequency components within an amplitude modulated wave, we would find that it is made up of a center carrier frequency and two additional frequencies called the lower sideband and the upper sideband.
12. The carrier of an AM signal remains at a constant amplitude and frequency at all times; however, the sidebands are constantly changing amplitude and frequency according to the modulating signal.
13. An RF fixed-tuned amplifier circuit generally contains an LC tuned circuit in both the input and the output that will pass a select band of frequencies and reject all other frequencies outside of the band.
14. At resonance, a series LC tank has a minimum impedance and therefore a maximum current, whereas a parallel LC tank circuit has a maximum impedance and minimum current at resonance.
15. The high impedance of a parallel LC tank at resonance is important in a tuned RF amplifier circuit because the voltage developed across a parallel resonant circuit is determined by the impedance of the resonant circuit. This impedance determines the amplitude of the signal voltage delivered to the output.
16. Selectivity is the ability of a circuit to select the wanted signal(s) and reject the unwanted signal(s). For example, a circuit with a wide bandwidth will pass a wide band of frequencies through to the output with a usable output voltage, and this circuit is said to have a "poor selectivity." On the other hand, a circuit with a very narrow bandwidth will pass very few frequencies through to the output at a usable voltage level, and this circuit is said to be "highly selective."
17. The Q of an RF amplifier's tuned circuits is usually adjusted by adding in a parallel resistance, called a swamping resistor, to increase the bandwidth.
18. By including step-up voltage transformers in an RF amplifier, additional voltage gain can be achieved. These transformers can also determine the tuned circuit's frequency response by controlling the transformer's coefficient of coupling.
19. A small value of capacitance exists between the junctions of all transistors. These junction capacitances are determined by the physical size of the junction, the spacing between the transistor's leads, and the junction bias condition. At low frequencies, these capacitances have very little effect since X_C is large. However, when frequency increases, X_C decreases, and the capacitance between a transistor's collector and base, which is known as the tran-

sistor's Miller capacitance, becomes a problem because its low-impedance path will couple a portion of the output signal back to the input. Because a phase inversion exists between the base and collector of a common emitter amplifier, this feedback voltage will be opposite in polarity to the input signal voltage, causing the input signal voltage to be decreased and the amplifier's gain to also be decreased. This type of out-of-phase feedback is called negative feedback because it has a degenerative effect.

20. Negative feedback caused by Miller capacitance can be neutralized by including a neutralizing capacitor (C_N) to couple an in-phase signal back to the base of the RF transistor to compensate for the out-of-phase signal coupled back to the base of the transistor by the Miller capacitance C_M. Coupling an in-phase voltage back to the base of Q_1 in this way effectively neutralizes the out-of-phase feedback voltage also being fed back to the base of Q_1.

21. Unlike the fixed-tuned RF amplifier circuit, the variable-tuned RF amplifier has tank circuits that contain variable capacitors that can be adjusted to control the resonant frequency of the *LC* tuned circuit and the band of frequencies or station that will be passed onto the next stage. The dashed line between the two capacitors is used to indicate that these variable capacitors are ganged, or mechanically linked by a common shaft, so that the operator can adjust the "tuning control" to simultaneously set the band-pass frequency for both tuned circuits, and select which station is passed and which are filtered-out or blocked.

22. The RF amplifier/mixer circuit is a special circuit designed to combine, or mix, two input frequencies. The two input frequencies that are combined in the mixer circuit are the received AM signal from the RF variable-tuned amplifier and the RF sine-wave signal from an RF local oscillator circuit. These two input signals are combined in the RF amplifier/mixer circuit, producing four signals at the output. These output signals are

a. The original RF input signal from the RF amplifier,
b. The original RF signal from the local oscillator,
c. The sum of the received RF signal and the local oscillator signal, and
d. The difference between the received RF signal and the local oscillator signal.

Because the output of the RF amplifier/mixer is fixed-tuned to the difference frequency, only this signal will be passed through to the output.

Intermediate Frequency (IF) Amplifiers (Figure 11-14)

23. The difference frequency from the RF amplifier/mixer is about midway between the RF signal received at the antenna and the AF final output signal applied to the speaker; it is called an intermediate frequency or IF.

24. This system of transferring the information signal from a high radio frequency to a lower intermediate frequency is the key advantage of the superheterodyne receiver. Fixed-tuned intermediate frequency amplifiers are less expensive, and have less high frequency losses, than variable-tuned radio frequency amplifiers. In addition, as we discovered earlier, narrow-band high-*Q* amplifier circuits, like the IF amp, have higher gains.

QUICK REFERENCE SUMMARY SHEET

FIGURE 11-14 Intermediate Frequency (IF) Amplifiers (Bipolar Transistor).

Radio IF Amplifier Circuit

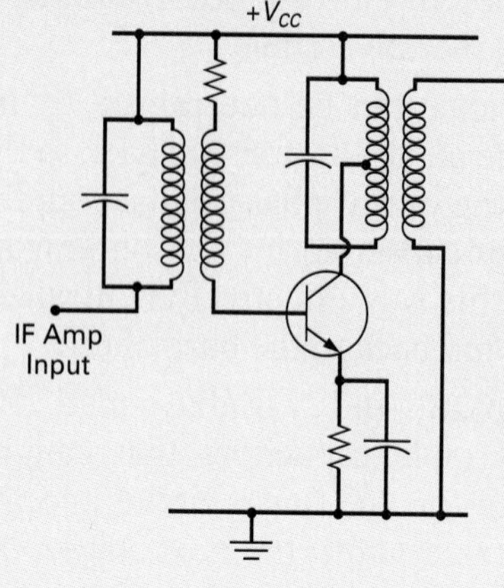

COMMUNICATION TYPE	RF RECEIVED	TYPICAL IF	IF BW
AM Radio	535 kHz-1605 kHz	455 kHz	10 kHz
FM Radio	88 MHz-108 MHz	10.7 MHz	150 kHz
TV Channel 2-6	54 MHz-88 MHz	41-47 MHz	6 MHz
TV Channel 7-13	174 MHz-216 MHz	41-47 MHz	6 MHz
TV Channel 14-83	470 MHz-890 MHz	41-47 MHz	6 MHz

Television IF Amplifier Circuit

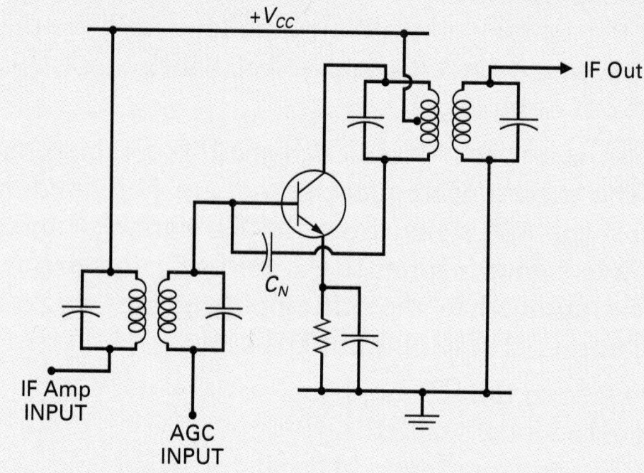

Video Frequency (VF) Amplifiers (Figure 11-15)

25. Video frequency (VF) amplifiers are used in televisions, computer monitors, video games, virtual reality systems, and radar systems to amplify picture-information signals, which are also called video signals. These video signals include a very wide range of frequencies, spanning from about 10 Hz to 5 MHz, and therefore the video amplifier must provide a high gain over a very wide bandwidth.

26. The shunt output capacitance of a transistor is made up of the transistor's collector-to-emitter capacitance and the output stray-wire capacitance. At the other end, we have the input capacitance of the second transistor stage, which is made up of Q_2's collector-to-base Miller capacitance, base-to-emitter capacitance, and the input stray-wire capacitance. Because all of these capacitance values are in parallel, we can add them together to obtain a total shunt capacitance value (C_T). This total shunt capacitance value is typically about 160 pF.

QUICK REFERENCE SUMMARY SHEET

FIGURE 11-15 Video Frequency (VF) Amplifiers (Bipolar Transistor).

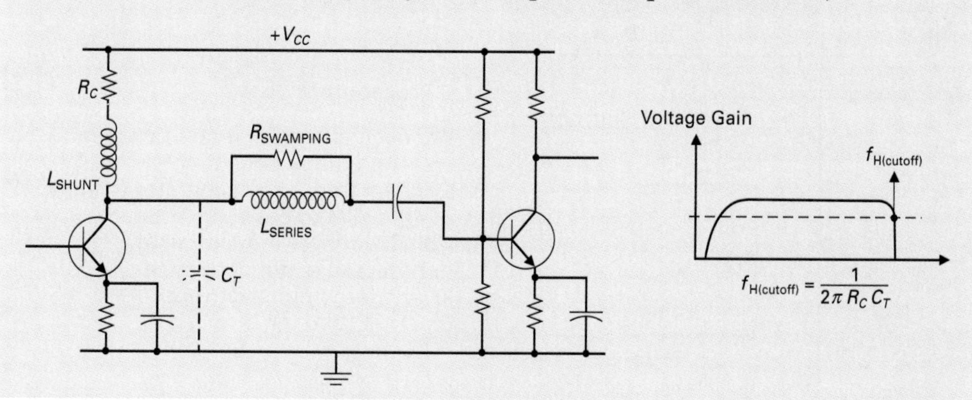

27. When the amplifier is amplifying low-frequency signals, the total shunt capacitance has very little effect because X_C is large. However, as the frequency of the signal being amplified increases, X_C decreases. The total shunt capacitance becomes a problem because it shunts some of the signal voltage to ground. The point at which the voltage of the signal frequency drops below 70.7% is called the high frequency cutoff.

28. To compensate for signal loss at high frequencies, peaking coils are often included in video amplifier circuits.

29. A shunt peaking coil is included in the collector of a transistor to compensate for output capacitance. By carefully choosing the value of the shunt peaking coil, we can cause L_{SHUNT} and C_{OUT} to resonate at the higher frequencies and develop a larger output voltage.

30. A series peaking coil can also be used to increase the high-frequency response of a video amplifier by compensating for input capacitance. Because the inductive reactance of this series peaking coil will increase with frequency ($X_L \propto f$), this coil will isolate the output capacitance of Q_1 from the input capacitance of Q_2, and therefore the large total shunt capacitance is now reduced to two smaller value capacitances. A swamping resistor is sometimes connected in parallel with the series peaking coil to reduce the Q of the inductor, which widens the response or bandwidth of the coil so that it will respond well to all video frequencies.

31. The automatic gain control (AGC) signal is a control voltage that originates from a circuit that automatically senses the strength of the input signal, then increases or decreases the gain of the amplifier to ensure that the gain is almost constant over a wide range of input signal voltages.

Tuning a Tuned Circuit (Figure 11-16)

32. One of the biggest problems with tuned circuit amplifiers is that their tuned circuits are not tuned to the desired center frequency. The reasons for this problem range from capacitor and inductor tolerances to the inherent stray

QUICK REFERENCE SUMMARY SHEET

FIGURE 11-16 Tuning a Tuned Circuit.

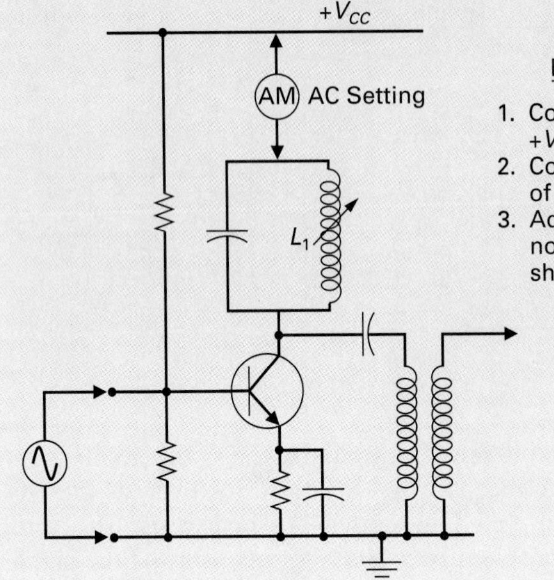

Procedure:

1. Connect ac current meter between +V_{CC} and tuned circuit.
2. Connect frequency generator to input of amp. Set to desired frequency.
3. Adjust variable inductor with nonmagnetic tool until meter shows large dip in current.

and junction circuit capacitance values that are always present. To compensate for these problems, most tuned amplifier circuits either have a built-in variable inductor or variable capacitor. In most tuned circuits, a variable inductor is used because it is less expensive than a variable capacitor.

33. To tune an amplifier

   1. First connect an ac current meter between V_{CC} and the tuned circuit, as shown in Figure 11-16,

   2. Next, connect a frequency generator to the input of the amplifier and set it to the desired frequency, and

   3. Finally, adjust the variable inductor (L_1) until the meter shows a large dip or null in current.

34. Always use a nonmetallic screwdriver to adjust the tuning of a tank because the proximity of any metal to the electric and magnetic fields of an adjustable capacitor or inductor will change its operation and the frequency to which the tank is tuned.

35. The typical problem within tuned amplifier circuits is frequency drift. This is caused by component aging, changing temperature conditions, and a variety of other elements. This problem can usually be easily remedied by simply retuning the amplifier as discussed in the previous section. If tuning the amplifier does not cure the selectivity problem with the amplifier, you will have to test the amplifier's tuned circuit components.

NEW TERMS

Amplitude Modulation
Automatic Gain Control
Bandwidth
Carrier Frequency
Coefficient of Coupling
Contrast Control
Detector Circuit
Double Tuned Transformer
Electric or Voltage Field
Electromagnetic Field
Emitter Modulator Circuit
Frequency Doubler
Frequency Drift
Frequency Quadrupler
Frequency Tripler
Ganged Capacitors
High-Frequency Amplifier
High-Frequency Cutoff
Intermediate Frequency
Intermediate Frequency (IF) Amplifier
Junction Capacitance
Loose-Coupled Transformer
Lower Sideband
Magnetic or Current Field
Miller-Effect Capacitance
Modulation
Negative Feedback
Neutralizing Capacitor
Output Stray-Wire Capacitance
Overcoupled Transformer
Peaking Coil
Radio-Frequency (RF) Amplifier
RF Amplifier/Frequency Tripler Circuit
RF Amplifier/Mixer Circuit
RF Amplifier/Modulator Circuit
RF Amplifier/Multiplier Circuit
RF Fixed-Tuned Amplifier Circuit
RF Local Oscillator Circuit
RF Variable-Tuned Amplifier Circuit
Selectivity
Series Peaking Coil
Shunt Peaking Coil
Sideband
Stray-Wire Capacitance
Superheterodyne Receiver
Swamping Resistor
Transistor Junction Capacitance
Tune
Tuned AM Receiver
Tuned Amplifier
Tuning
Upper Sideband
Video-Frequency (VF) Amplifier
Video Signals

REVIEW QUESTIONS

Multiple-Choice Questions

1. The radio frequency at which a station transmits is called its ______________ frequency.
   a. Intermediate
   b. Superheterodyne
   c. Carrier
   d. Neutralizing
2. The voltage applied across a transmitting antenna generates a/an ______________ field, while the current passing through the antenna generates a/an ______________ field.
   a. electric, voltage
   b. electric or voltage, current
   c. current, electric
   d. electron, electric

3. Which of the following circuits could be used to vary the amplitude of an RF carrier in accordance with the AF signal?
   a. RF Amplifier/Multiplier Circuit
   b. RF Fixed-Tuned Amplifier Circuit
   c. RF Variable-Tuned Amplifier Circuit
   d. RF Amplifier/Modulator Circuit
4. Which of the following circuits could be used to double the frequency of an RF input sine wave signal?
   a. RF Amplifier/Multiplier Circuit
   b. RF Fixed-Tuned Amplifier Circuit
   c. RF Variable-Tuned Amplifier Circuit
   d. RF Amplifier/Modulator Circuit
5. Which of the following circuits is used in a receiver to tune in or select a desired station?
   a. RF Amplifier/Multiplier Circuit
   b. RF Fixed-Tuned Amplifier Circuit
   c. RF Variable-Tuned Amplifier Circuit
   d. RF Amplifier/Modulator Circuit
6. Which of the following circuits is used to boost the amplitude of an RF signal of a specific frequency?
   a. RF Amplifier/Multiplier Circuit
   b. RF Fixed-Tuned Amplifier Circuit
   c. RF Variable-Tuned Amplifier Circuit
   d. RF Amplifier/Modulator Circuit
7. An amplifier with a high Q will have a _____________ bandwidth, and therefore a _____________ selectivity.
   a. small, good **c.** small, poor
   b. large, good **d.** large, poor
8. To neutralize the _____________ feedback caused by Miller effect in high-frequency amplifiers, a _____________ is usually included to couple an _____________ signal back to the base.
   a. negative, resistor, out-of-phase
   b. positive, capacitor, in-phase
   c. negative, capacitor, in-phase
   d. positive, resistor, out-of-phase
9. With a superheterodyne receiver, the received RF signal is converted to a lower frequency called a/an _____________ before the information signal is extracted.
   a. VF **c.** RF
   b. IF **d.** MF
10. Intermediate frequency amplifiers are always
   a. Fixed-Tuned **b.** Variable Tuned
11. The standard IF frequency for an AM radio receiver is
   a. 41 MHz **c.** 10 kHz
   b. 10.7 MHz **d.** 455 kHz

12. Video amplifiers should ideally have a frequency response of ____________ to ____________.

   a. 10 kHz to 30,000 MHz c. 10 Hz to 5 MHz
   b. 41 MHz to 47 MHz d. 20 Hz to 20 kHz

13. To increase the high-frequency gain of a video amplifier, we could ____________ the value of R_C, and/or use ____________.

   a. decrease, peaking coils
   b. decrease, swamping resistors
   c. increase, shunt capacitances
   d. increase, peaking capacitors

14. An amplifier's low-frequency response will not be affected by the circuit's inherent shunt capacitance.

   a. True b. False

15. A tuned amplifier has a tuned circuit load, therefore load ____________ and amplifier ____________ will vary with frequency.

   a. center frequency, power rating
   b. capacitance, frequency
   c. impedance, gain
   d. both (a) and (b) are true

Essay Questions

16. Define the following terms (Chapter 11)

   a. Communication b. Carrier Frequency c. Amplitude Modulation

17. What is a high-frequency amplifier circuit? (Intro.)
18. Briefly describe how the electrical signal applied to an antenna is converted to an electromagnetic wave. (11-1)
19. Describe the function of an RF amplifier/multiplier circuit. (11-1-1)
20. Why are RF amplifiers class-C bias? (11-1-1)
21. Why are *LC* tank circuits included in high-frequency amplifiers? (11-1-1)
22. Briefly describe the function of an RF amplifier/modulator circuit. (11-1-2)
23. Define the term modulation. (11-1)
24. What are sidebands? (11-1-2)
25. Briefly describe the function of an RF fixed-tuned amplifier circuit. (11-1-3)
26. Explain the difference between a series resonance circuit and a parallel resonance circuit. (11-1-3)
27. Why are parallel tuned circuits ideally suited for high frequency amplifiers? (11-1-3)
28. Define the terms bandwidth and selectivity, and describe their relationship. (11-1-3)
29. What is a swamping resistor and how is it used to control a tuned circuit's bandwidth? (11-1-3)

30. Why are transformers often used for coupling in high-frequency amplifiers, and how does the coefficient of coupling alter an amplifier's characteristics? (11-1-3)
31. What is Miller-effect capacitance? (11-1-3)
32. What is the purpose of a neutralizing capacitor? (11-1-3)
33. Describe the function of an RF variable-tuned amplifier circuit. (11-1-4)
34. How are ganged capacitors used in a tuned radio receiver circuit? (11-1-4)
35. Define the term heterodyning, and describe the operation of a superheterodyne receiver. (11-2)
36. What advantages does the superheterodyne receiver have over the basic tuned receiver? (11-2)
37. What is an intermediate frequency? (11-2-3)
38. Describe the function of an RF amplifier/mixer circuit. (11-2-2)
39. Are RF amplifier circuits fixed-tuned or variable-tuned? (11-2-3)
40. Where are video frequency amplifiers used? (11-3)
41. What is the frequency range of video signals? (11-3)
42. Why is shunt capacitance a problem in high-frequency amplifiers? (11-3-1)
43. How can we extend the high-frequency cut-off of amplifiers? (11-3-1)
44. What is a peaking coil? (11-3-2)
45. Describe how shunt and series peaking coils can be used in video amplifier circuits. (11-3-2)
46. What is automatic gain control? (11-3-3)

Practice Problems

47. Referring to Figure 11-17, calculate the resonant frequency of the amplifier's tank circuits.
48. Calculate the Q of the tuned circuits shown in figure 11-17.
49. Calculate the bandwidth of the tuned circuits shown in figure 11-17.
50. Calculate the high-frequency cutoff of the video amplifier circuit shown in Figure 11-18.
51. If a 1360 kHz carrier is voice modulated by a signal that is between 200 Hz to 20 kHz, what will be the transmission's upper and lower sidebands and bandwidth?

Troubleshooting Questions

52. How do we tune an amplifier? (11-4-1)
53. Why is it so important that an amplifier is tuned to the desired frequency? (11-4-1)
54. What is frequency drift? (11-4-2)

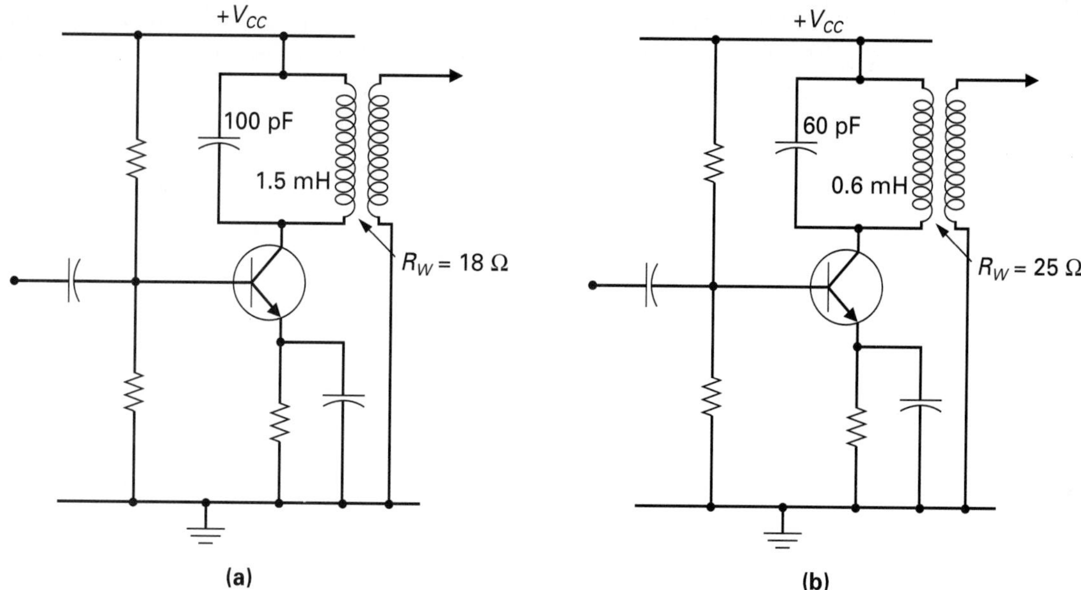

FIGURE 11-17 Tuned Amplifier Circuits.

55. Describe some of the typical amplifier tuned circuit problems, and how these failures can be recognized and isolated. (11-4-2)
56. The tuned amplifier circuit shown in Figure 11-19 has a collector voltage that is stuck at $+V_{CC}$. What do you think the possible circuit problem could be? (11-4-2)

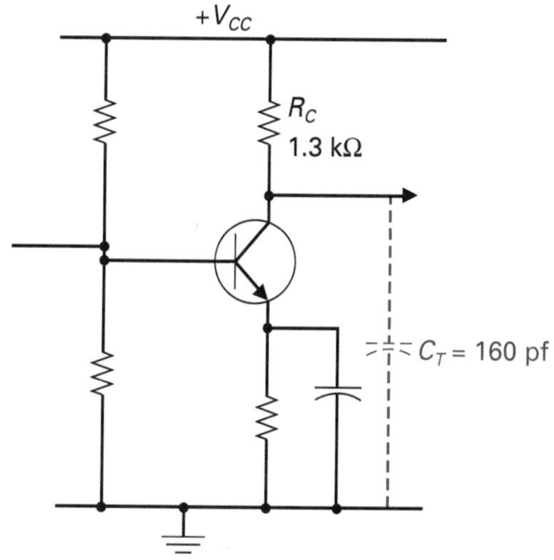

FIGURE 11-18 Video Amplifier Circuit.

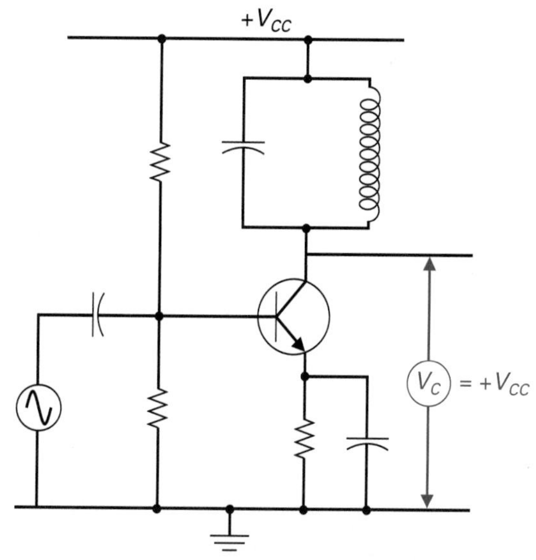

FIGURE 11-19 Troubleshooting Tuned Amplifier Circuits.

CHAPTER

12

OBJECTIVES

After completing this chapter, you will be able to:

1. State the purpose of the oscillator.
2. Describe how oscillator circuits are generally classified into three basic groups.
3. Describe the basic *LC* oscillator or flywheel action.
4. Identify, explain the operation of, and calculate the frequency of the following oscillator circuits:
   - **a.** The Armstrong Oscillator
   - **b.** The Hartley Oscillator
   - **c.** The Colpitts Oscillator
   - **d.** The Clapp Oscillator
5. Explain why the Barkhausen criterion is important to an oscillator's operation.
6. Describe how transistors can generate parasitic oscillations due to a circuit's inherent *LC* feedback paths.
7. Describe the basic characteristics of crystals and how these characteristics can be made use of in oscillator circuits.
8. Identify, explain the operation of, and calculate the frequency of the following crystal oscillator circuits:
   - **a.** The Hartley Crystal Oscillator
   - **b.** The Colpitts Crystal Oscillator
   - **c.** The Pierce Crystal Oscillator
9. Explain the characteristics of an *RC* phase-shift network.
10. Identify, explain the operation of, and calculate the frequency of the following crystal oscillator circuits:
    - **a.** The Phase-Shift Oscillator
    - **b.** The Wien-Bridge Oscillator
11. Describe the procedure to be used when troubleshooting an oscillator circuit.

Bipolar Transistor Oscillator Circuits

OUTLINE

VIGNETTE

Making an Impact

John Von Neuman, a mathematics professor at the Institute of Advanced Studies, delighted in amazing his students by performing complex computations in his head faster than they could with pencil, paper, and reference books. He possessed a photographic memory and at parties in his home in Princeton, New Jersey, he gladly occupied center stage to recall from memory entire pages of books read years previously, the lineage of European royal families, and a store of controversial limericks. His memory, however, failed him in his search for basic items in a house he had lived in for seventeen years. On many occasions when traveling, he would become so completely absorbed in mathematics that he would have to call his office to find out where he was going and why.

Born in Hungary, he was quick to demonstrate his genius. At the age of six, he would joke with his father in classical Greek. At the age of eight, he had mastered calculus, and in his midtwenties, he was teaching and making distinct contributions to the science of quantum mechanics, which is the cornerstone of nuclear physics.

Next to fine clothes, expensive restaurants, and automobiles which he had to replace annually due to smashups, he loved his work. His interest in computers began when he worked on the top secret Manhattan Project at Los Alamos, New Mexico, where he proved mathematically the implosive method of detonating an atom bomb. Working with the then available computers he became aware that they could become much more than a high speed calculator. He believed that they could be an all-purpose scientific research tool, and he published these ideas in a paper. This was the first document to outline the logical organization of the electronic digital computer, and it was widely circulated to all scientists throughout the world. In fact, even to this day, scientists still refer to computers as "Von Neuman machines."

Von Neuman collaborated on a number of computers of advanced design for military applications such as the development of the hydrogen bomb and ballistic missiles.

In 1957, at the age of 54, he lay in the hospital dying of bone cancer. Under the stress of excruciating pain, his brilliant mind began to break down. Because he had been privy to so much highly classified information, the Pentagon had him surrounded with only medical orderlies specially cleared for security for fear he might, in pain or sleep, give out military secrets.

INTRODUCTION

An oscillator is an electronic circuit that draws power from its dc supply to generate a continuously repeating ac output signal. Stated another way, the oscillator circuit is simply a signal generator converting its V_{CC} supply voltage input into a continuously repeating ac signal output.

Oscillators are used in a variety of applications. For example, in the previous chapter an oscillator was used in a transmitter to generate the carrier frequency of the station. In a superheterodyne receiver, an oscillator was used to generate an RF signal of the correct frequency to mix with the incoming signal to produce the correct IF signal. In other applications, the oscillator's ac signal is distributed throughout an electronic system to time or synchronize operations. When used in this way, the electronic oscillator can be compared to the pendulum of a clock. Just as the pendulum oscillates back and forth timing the minutes, the continuously repeating electronic oscillator's output signal is used to control the timing of circuit operations.

Like the amplifier circuit, some oscillator circuits are designed to generate low-frequency oscillations of only a few hertz, while other high-frequency oscillators generate oscillations at several hundred gigahertz. Oscillators are generally classified into one of three basic groups based on the frequency determining components used within the circuit. These three groups are

LC Oscillators
Crystal Oscillators
RC Oscillators

In this chapter, we will study the operation and characteristics of typical bipolar transistor *LC*, *RC*, and crystal oscillator circuits.

12-1 *LC* OSCILLATOR CIRCUITS

As stated in the introduction, the oscillator circuit is simply a signal generator, converting its V_{CC} or dc supply voltage input into a continuously repeating ac signal output. In this section, we will discuss several different bipolar transistor *LC* oscillator circuits, which use an *LC* tuned circuit to set the frequency of oscillation and are all named after their inventors.

12-1-1 Basic LC Oscillator Action

To explain the basic action of an *LC* oscillator circuit, Figure 12-1(a) shows how an *LC* tank circuit can be "shock excited" into oscillation by a momentary surge of current from a dc power switch. Looking at the output of this circuit, you can see that the back-and-forth action of the circulating current within the tank generates a sine wave at the output. This sine-wave output, however, will slowly decrease in amplitude due to the energy lost in the coil's resistance. To prevent the oscillations from the tank from being "damped" in this way, we need to

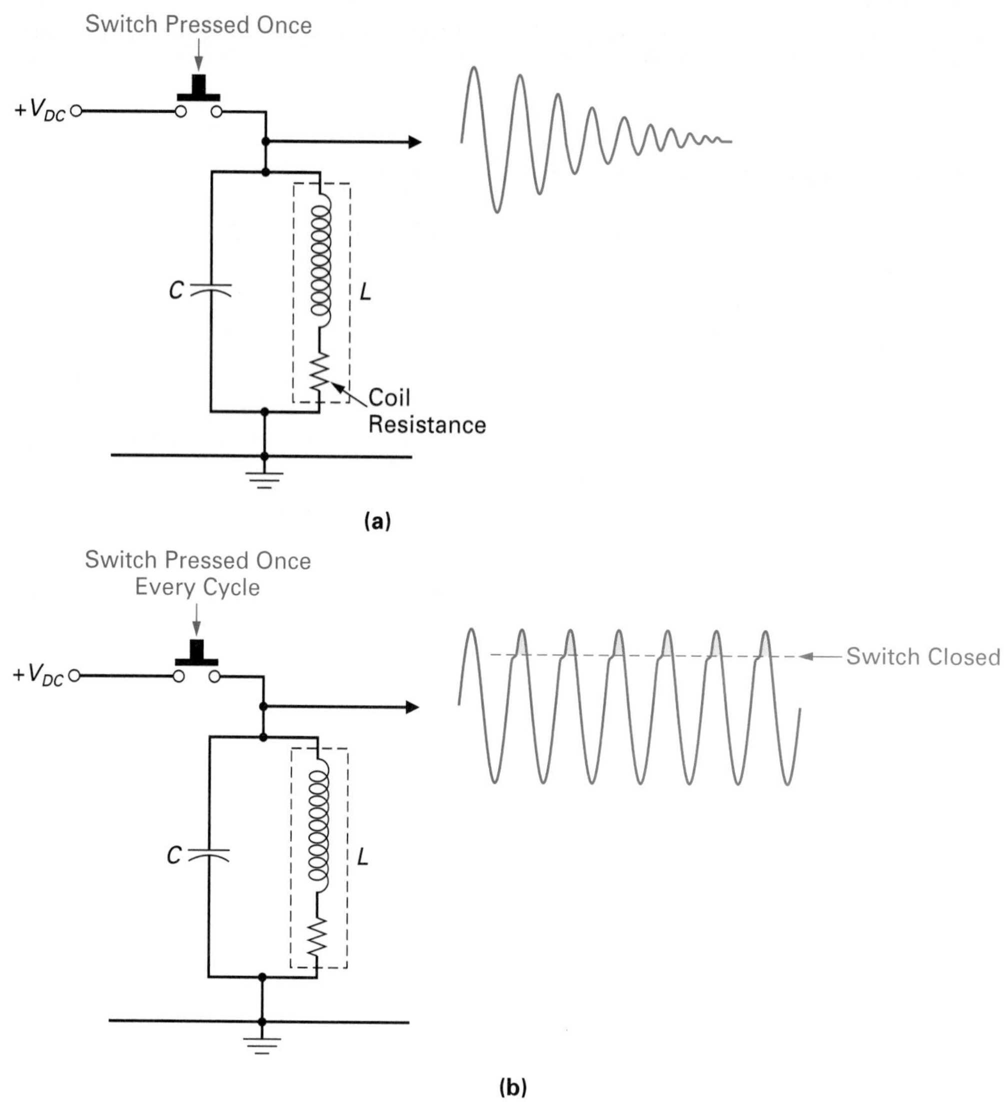

FIGURE 12-1 **Basic *LC* Oscillator Action.**

Regenerative (Positive) Feedback Reinforcing energy that regenerates or adds to the sine-wave oscillations of an oscillator circuit.

somehow replace the energy lost. Figure 12-1(b) shows how this can be achieved by closing the dc power switch once every cycle. The timing of this switch closure is crucial because the pulse of dc current must reinforce the tank's circulating current, and therefore be in phase with the tank's output sine wave, as seen in the waveform in Figure 12-1(b). Reinforcing energy that "regenerates" or "adds to" the sine wave oscillations in this way is called **regenerative feedback** or **positive feedback.** Most oscillator circuits operate on this positive feedback principle, which means that an in-phase signal must be constantly injected into the oscillator circuit in order to replace the energy lost and maintain oscillations.

Although the oscillator circuit shown in Figure 12-1(b) would work, it is obvious that some sort of automatic electronic positive feedback system is needed instead of the very impractical manual mechanical positive feedback system. One of the simplest oscillator circuits that makes use of a tuned *LC* tank and has an automatic electronic positive feedback loop is the Armstrong oscillator, which we will now examine.

12-1-2 *The Armstrong Oscillator*

Armstrong Oscillator A tuned transformer oscillator developed by E. H. Armstrong.

Figure 12-2(a) shows the **Armstrong oscillator** circuit. Transistor Q_1 has the usual voltage divider base bias (R_1 and R_2), emitter resistor and capacitor (R_3 and C_3), and input coupling capacitor (C_2). The frequency at which this circuit will oscillate is determined by the parallel resonant circuit C_1 and L_1, which is connected to the collector of transistor Q_1. Transformer T_1's primary coil (L_1) and secondary coil (L_2) are wound so that the voltage induced at the top of L_2 will be 180° out of phase with the voltage present at the bottom of L_1 (or collector of Q_1). In addition, this circuit includes a positive feedback loop to sustain oscillations between the circuit's output and the transistor's input.

In Figures 12-2(b), (c), (d), and (e), the bias components and coupling capacitors have been removed so that we can see more clearly the steps involved in this circuit's oscillating cycle.

Step 1: FIGURE 12-2(b)

When circuit power ($+V_{CC}$) is first applied, the voltage divider bias resistors R_1 and R_2 cause Q_1 to turn ON and collector current to flow through Q_1 and the tank circuit to $+V_{CC}$. With Q_1 ON, its collector voltage will be LOW, applying a LOW to the bottom of L_1. Due to the phase inversion of T_1, the top of L_2 will go HIGH. This HIGH positive voltage at the top of L_2 will be fed back to the base of Q_1 via the feedback loop. The conduction of Q_1 into saturation is increased, resulting in an increase in collector current and the charging of C_1 to the polarity shown.

Step 2: FIGURE 12-2(c)

When capacitor C_1 is fully charged, current through the tank circuit will drop to zero, and this lack of changing current will cause the transformer's induced positive voltage at the top of L_2 to drop to zero. Without this additional base-emitter forward bias from the feedback loop, Q_1 will turn OFF. With Q_1 OFF, capacitor C_1 will begin to discharge, transferring all of its energy to L_1 and initiating the tank's flywheel action. When C_1 is fully discharged, current through L_1 will cease and inductor L_1's magnetic field will begin to collapse.

Step 3: FIGURE 12-2(d)

As inductor L_1's magnetic field collapses, an opposite polarity is induced into the coil. Capacitor C_1 begins to charge as energy is transferred from the tank's coil to the capacitor. The positive polarity at the bottom of C_1 is applied to L_1, inducing a negative polarity at the top of L_2, driving Q_1 further into cutoff.

Step 4: FIGURE 12-2(e)

Continuing the flywheel action, capacitor C_1, which is now fully charged, begins to discharge through L_1 in the opposite direction. The coil's magnetic field once again begins to build. When capacitor C_1 is fully discharged,

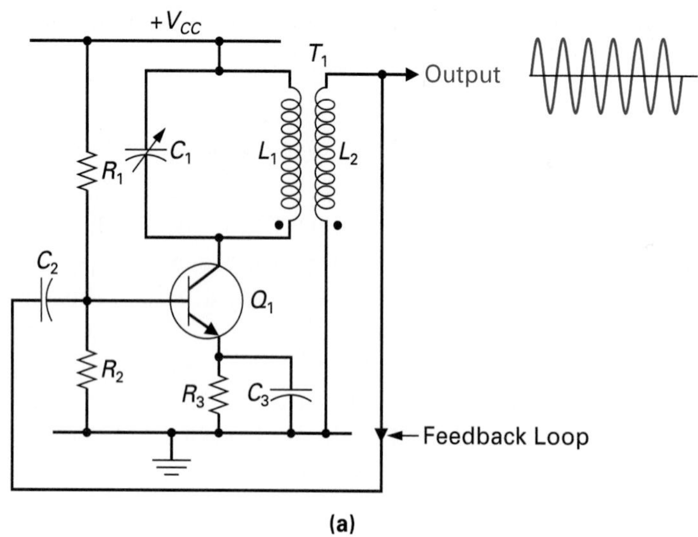

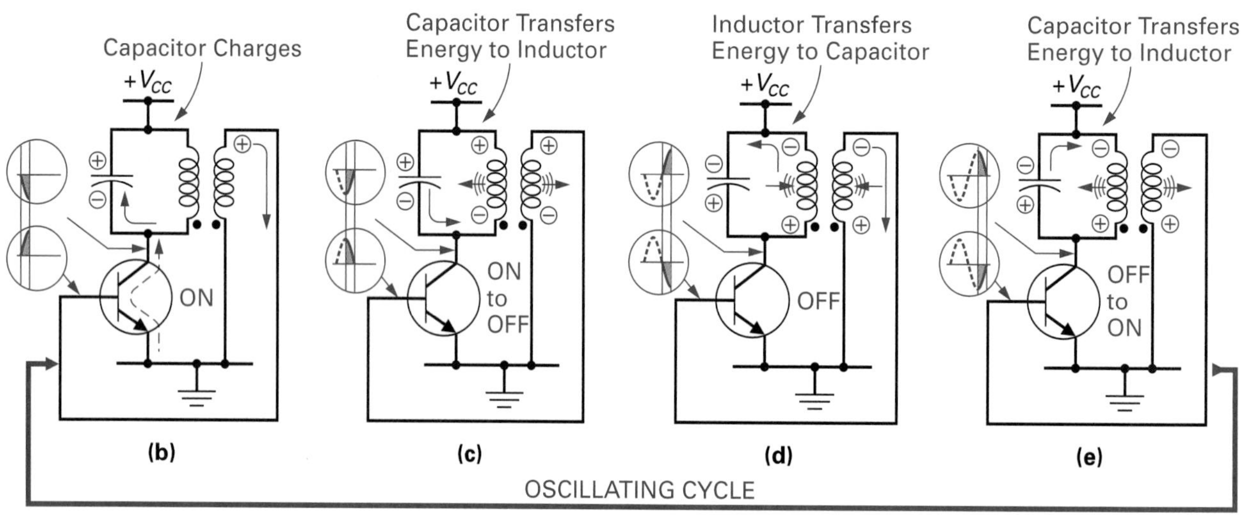

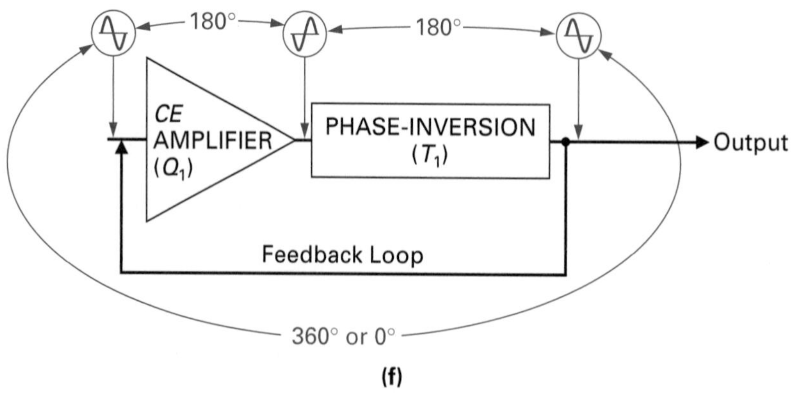

FIGURE 12-2 The Armstrong Oscillator.

circuit current once again drops to zero and therefore the magnetic field around L_1 starts to collapse and C_1 begins to recharge.

Step 5: FIGURE 12-2(b)

The collapsing field of L_1 charges capacitor C_1 to the polarity shown. At the same time, the negative potential at the bottom of L_1 induces a positive potential at the top of L_2, and this potential forward biases Q_1, which turns ON. The transistor current assists the charging of capacitor C_1 and replaces the energy lost in the tank. Transistor Q_1 is therefore acting like an automatic switch, similar to the manual switch in Figure 12-1(b), replacing the energy lost in the tank circuit with synchronized positive feedback.

As you have seen in stepping through the oscillating cycle, capacitor C_1 and inductor L_1 set the frequency at which this circuit oscillates because their values control the rate at which energy is transferred back and forth within the tank. In many applications, the capacitor in the tuned circuit will be made variable so that the frequency of oscillation can be changed. The frequency at which this circuit will oscillate, and therefore the frequency of the sine-wave output, can be calculated with the formula

$$f_R = \frac{1}{2\pi\sqrt{L_1 \times C_1}}$$

■ **EXAMPLE:**

Calculate the frequency at which the circuit shown in Figure 12-2(a) will oscillate if $L_1 = 47$ mH and $C_1 = 0.033$ μF.

■ ***Solution:***

$$f_R = \frac{1}{2\pi\sqrt{L_1 \times C_1}} = \frac{1}{2\pi\sqrt{47\text{ mH} \times 0.033\ \mu\text{F}}} = 4\text{ kHz}$$

Another important point to stress is that the transistor Q_1 acts as a switch and is only turned ON when the output signal is positive and only when the transistor's collector current will assist the tank's circulating current in the charging of the capacitor. This "topping off" action is similar to someone giving a little push at the right time to a person on a playground swing.

To stress the idea of positive feedback in more detail, Figure 12-2(f) shows a block diagram of the Armstrong oscillator. Because a common-emitter amplifier has a 180° phase inversion between input and collector output, an additional 180° phase shift device is needed between the output and the input so that the feedback signal will be positive or reinforcing, and the circuit will oscillate. The additional 180° phase shift is provided by the transformer T_1. When the base of Q_1 swings positive, the collector of Q_1 swings negative, the bottom of L_1 swings negative, the top of L_2 swings positive. This positive swing at the output is fed back to the input to reinforce the original positive input swing and sustain oscillations. As a result, the signal is shifted 360° as it is passed from input to output and then back to input. Because a phase shift of 360° is the same as not shifting the signal at all, the feedback signal is in phase with the input. If the phase inverting transformer T_1 was not included and the collector output was fed directly back to the input, the feedback signal would oppose the input signal. This type of out-of-phase feedback is known as **degenerative feedback** or **negative feedback** and would result in a canceling of the input and therefore no output oscillations.

Degenerative (Negative) Feedback Feedback in which a portion of the output signal is fed back 180° out of phase with the input signal. Also called degenerative feedback, inverse feedback, or stabilized feedback.

12-1-3 The Hartley Oscillator

Hartley Oscillator An *LC* tuned oscillator circuit in which the tank coil has an intermediate tap.

Figure 12-3(a) shows the schematic of a **Hartley oscillator,** which, unlike the Armstrong oscillator, does not require a transformer. The 180° phase inversion needed for positive feedback is achieved by this circuit's identifying feature: the tapped inductor L_1. Transistor Q_1 has the usual bias and stability components; however, a radio frequency choke (RFC) is connected in the collector to connect the dc power to the transistor while preventing the high frequency oscillations at the output from passing to the $+V_{CC}$ line and back to the power supply. Coupling capacitor C_3 is included to block the dc collector bias from the output but couple ac collector variations to the tank. Coupling capacitor C_2 will block the dc base bias from the tank circuit, while coupling the resonant tank variations to the base of Q_1.

To describe this circuit's operation, we will once again break down the oscillating cycle to four steps, which are shown in Figures 12-3(b), (c), (d), and (e).

Step 1: FIGURE 12-3(b)

When circuit power ($+V_{CC}$) is first applied, the voltage divider bias resistors R_1 and R_2 cause Q_1 to turn ON, and collector current will flow through Q_1 to the tank circuit. This current causes the magnetic field around L_1 to expand, and because the center tap of L_1 is grounded, the opposite ends of L_1 will be at different potentials. Making the top of L_1 LOW will make the bottom of L_1 HIGH, and this HIGH will be applied to the base of Q_1 causing it to quickly saturate.

Step 2: FIGURE 12-3(c)

When Q_1 is saturated, there will be no further increase in current, and the magnetic field produced by L_1 will collapse. The energy in the inductor will be transferred to the capacitor. As the bottom of C_1 is now charging to a negative potential—and this point is connected to the base of Q_1—the transistor will begin to feel a reverse bias voltage and start to cut OFF.

Step 3: FIGURE 12-3(d)

With Q_1 OFF, the tank's flywheel action takes over, and C_1 begins to discharge and transfer its energy from the capacitor to the inductor.

Step 4: FIGURE 12-3(e)

When capacitor C_1 is fully discharged, the magnetic field around L_1 starts to collapse and C_1 begins to recharge. Because the bottom of C_1 is charging

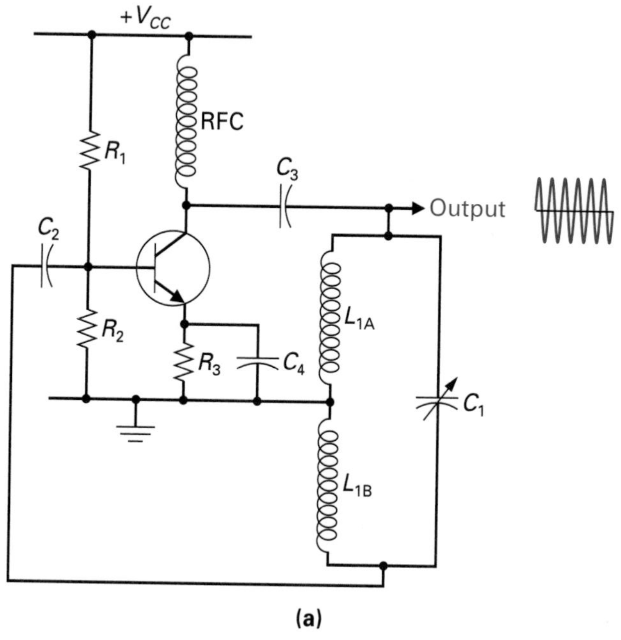

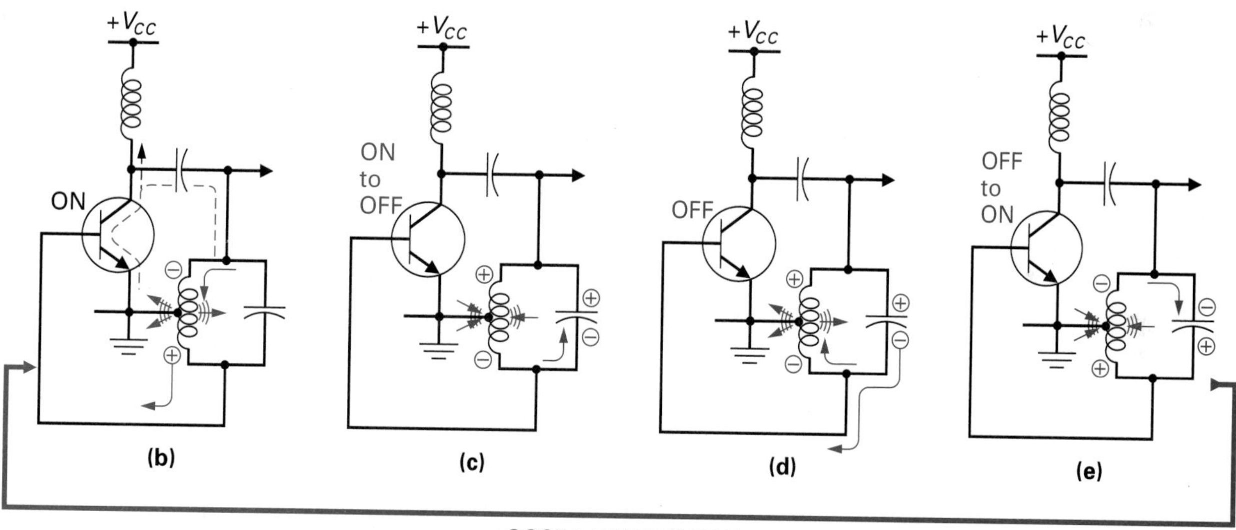

FIGURE 12-3 The Hartley Oscillator.

to a positive potential—and this is connected to the base of Q_1—the transistor will begin to receive a forward bias and therefore start to turn ON. The transistor current assists the charging of capacitor C_1 and replaces the energy lost in the tank. Once C_1 is fully charged, flywheel action will take over, the capacitor's energy will be transferred to L_1 as shown in Figure 12-3(b), and the oscillating cycle will repeat.

The frequency at which this circuit oscillates is set by the tank components C_1 and L_T, and the frequency of the sine wave output can be calculated with the formula

$$f_R = \frac{1}{2\pi\sqrt{L_T \times C_1}}$$
$$L_T = L_{1A} + L_{1B}$$

■ **EXAMPLE:**

Calculate the frequency at which the circuit shown in Figure 12-3(a) will oscillate if $L_{1A} = 47\ \mu H$, $L_{1B} = 22\ \mu H$ and $C_1 = 22$ nF.

■ ***Solution:***

$$L_T = L_{1A} + L_{1B} = 47\ \mu H + 22\ \mu H = 69\ \mu H$$

$$f_R = \frac{1}{2\pi\sqrt{L_T \times C_1}} = \frac{1}{2\pi\sqrt{69\ \mu H \times 22\ nF}} = 129.2\text{ kHz}$$

12-1-4 *The Colpitts Oscillator*

Colpitts Oscillator An *LC* tuned oscillator circuit in which two tank capacitors are used instead of a tapped coil.

The **Colpitts oscillator,** shown in Figure 12-4(a), is similar to the Hartley; however, in this circuit two capacitors are used instead of a center tapped coil. These two capacitors, which are grounded at the center, are the identifying feature of this oscillator and will provide the needed 180° of phase shift for positive feedback.

To describe this circuit's operation, we will once again break down the oscillating cycle to four steps, which are shown in Figures 12-4(b), (c), (d), and (e).

Step 1: FIGURE 12-4(b)

When dc power is first applied to this circuit, the voltage divider bias provided by R_1 and R_2 will cause Q_1 to conduct. With Q_1 ON, dc collector current will pass from Q_1's emitter to collector, through the radio frequency choke and to $+V_{CC}$. The drop in voltage at the collector of Q_1 will be applied to the top plate of C_1, causing C_1 to charge. Due to the ground at the center of C_1 and C_2, the bottom plate of C_2 will charge to a positive potential through L_1. This positive potential will be applied via the feedback loop to Q_1, sending it into saturation.

Step 2: FIGURE 12-4(c)

When Q_1 saturates, there is no further voltage change at the collector of Q_1, so C_1 and C_2 receive no further charge, and the flywheel action of the tank circuit takes over. Capacitors C_1 and C_2 act together as they discharge through L_1, causing L_1's magnetic field to build up.

Step 3: FIGURE 12-4(d)

When capacitors C_1 and C_2 have fully discharged, current decreases to zero and the magnetic field around L_1 begins to collapse, inducing an opposite

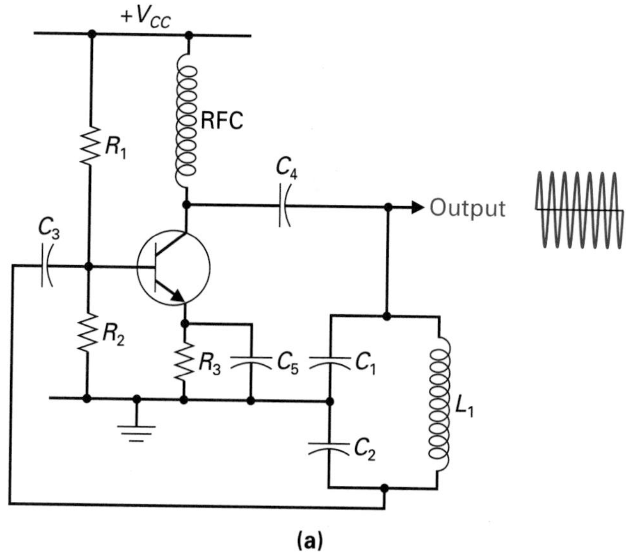

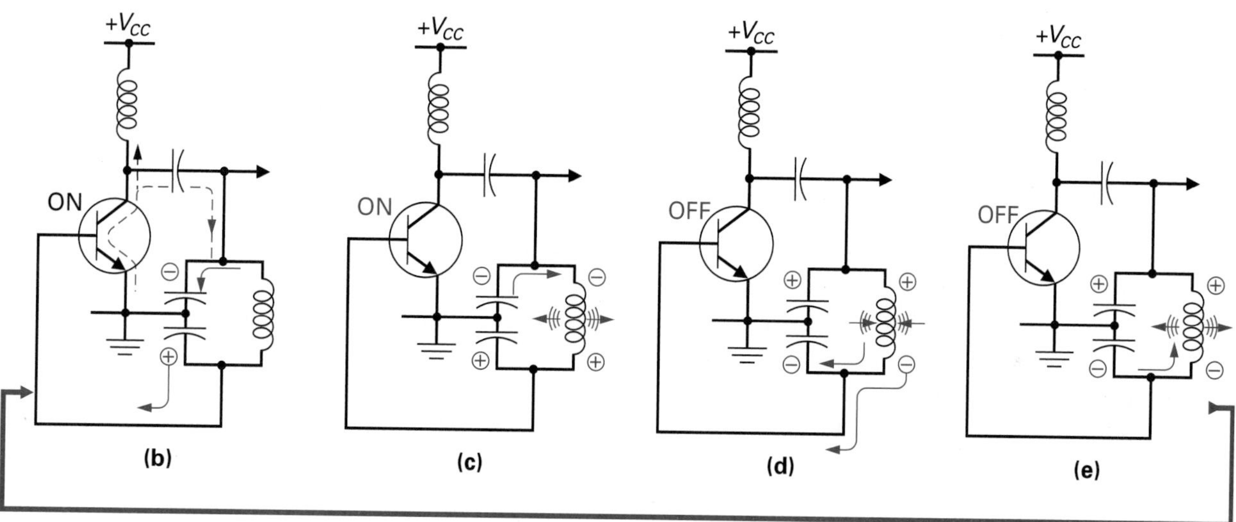

FIGURE 12-4 The Colpitts Oscillator.

voltage into L_1, which then charges C_1 and C_2. Because the bottom plate of C_2 charges to a negative potential, and this is applied to the base of Q_1, the transistor will be driven into cutoff. The capacitors will then charge as L_1 transfers all of its energy into C_1 and C_2.

Step 4: FIGURE 12-4(e)

When capacitors C_1 and C_2 have received all of the energy from L_1, the flywheel action will again reverse causing C_1 and C_2 to discharge through L_1. When all energy has been transferred to L_1, current will again drop to zero and the inductor's magnetic field will collapse. This induces a voltage into L_1 that will charge capacitors C_1 and C_2 to the polarity shown in Figure 12-4(b), and the cycle will repeat.

As with all of the previous *LC* oscillators, the frequency of oscillation is determined by the tank's inductive and capacitive values, which in this case are equal to L_1 and the series combination of C_1 and C_2.

$$f_R = \frac{1}{2\pi\sqrt{L_1 \times C_T}}$$

$$C_T = \frac{C_1 \times C_2}{C_1 + C_2}$$

■ **EXAMPLE:**

Calculate the frequency at which the circuit shown in Figure 12-4(a) will oscillate if $C_1 = 0.27\ \mu H$, $C_2 = 0.47\ \mu H$, and $L_1 = 0.6$ mH.

■ ***Solution:***

$$C_T = \frac{C_1 \times C_2}{C_1 + C_2} = \frac{0.27\ \mu F \times 0.47\ \mu F}{0.27\ \mu F + 0.47\ \mu F} = 0.17\ \mu F$$

$$f_R = \frac{1}{2\pi\sqrt{L_1 \times C_T}} = \frac{1}{2\pi\sqrt{0.6\ mH \times 0.17\ \mu F}} = 15.76\ kHz$$

12-1-5 The Feedback Factor

Feedback Factor (B_F)
The percentage of output returned to the input.

Feedback Voltage (V_F)
The output feedback voltage that is applied to the input.

Now that you have a good understanding of oscillator operation, let's examine how the amount of feedback can affect the output of an oscillator. To illustrate this point, Figure 12-5 includes three examples in which the **feedback factor (B_F)** has been changed. The feedback factor describes the percentage of output returned to the input. For example, in Figure 12-5(a), the feedback factor is 0.001 or 0.1%, which means that 0.1% of the output voltage is returned as positive feedback to the input. This **feedback voltage (V_F)** is applied to the input of the amplifier (V_{IN}) and can be calculated using the formula

$$V_{IN} \text{ or } V_F = V_{OUT} \times B_F$$

Once we know the value of the input voltage or feedback voltage, we can then calculate the output voltage if the amplifier's voltage gain is known

$$V_{OUT} = A_V \times V_F$$

Using the example values given in Figure 12-5(a), you can see that the amplifier has a voltage gain of 100, a feedback factor of 0.001, and, when the oscillator is first turned ON, an input of 0.1 V_{pk}.

Looking at the table in Figure 12-5(a), you can see that

1. For the first cycle of the output, an input of 0.1 V_{pk} will produce an output voltage of 10 V_{pk} ($V_{OUT} = A_V \times V_F = 100 \times 0.1 = 10\ V_{pk}$). Because 0.1% of this output will be fed back to the input, the feedback voltage for the second cycle will be 0.01 V_{pk} ($V_F = V_{OUT} \times B_F = 10 \times 0.001 = 0.01\ V_{pk}$).

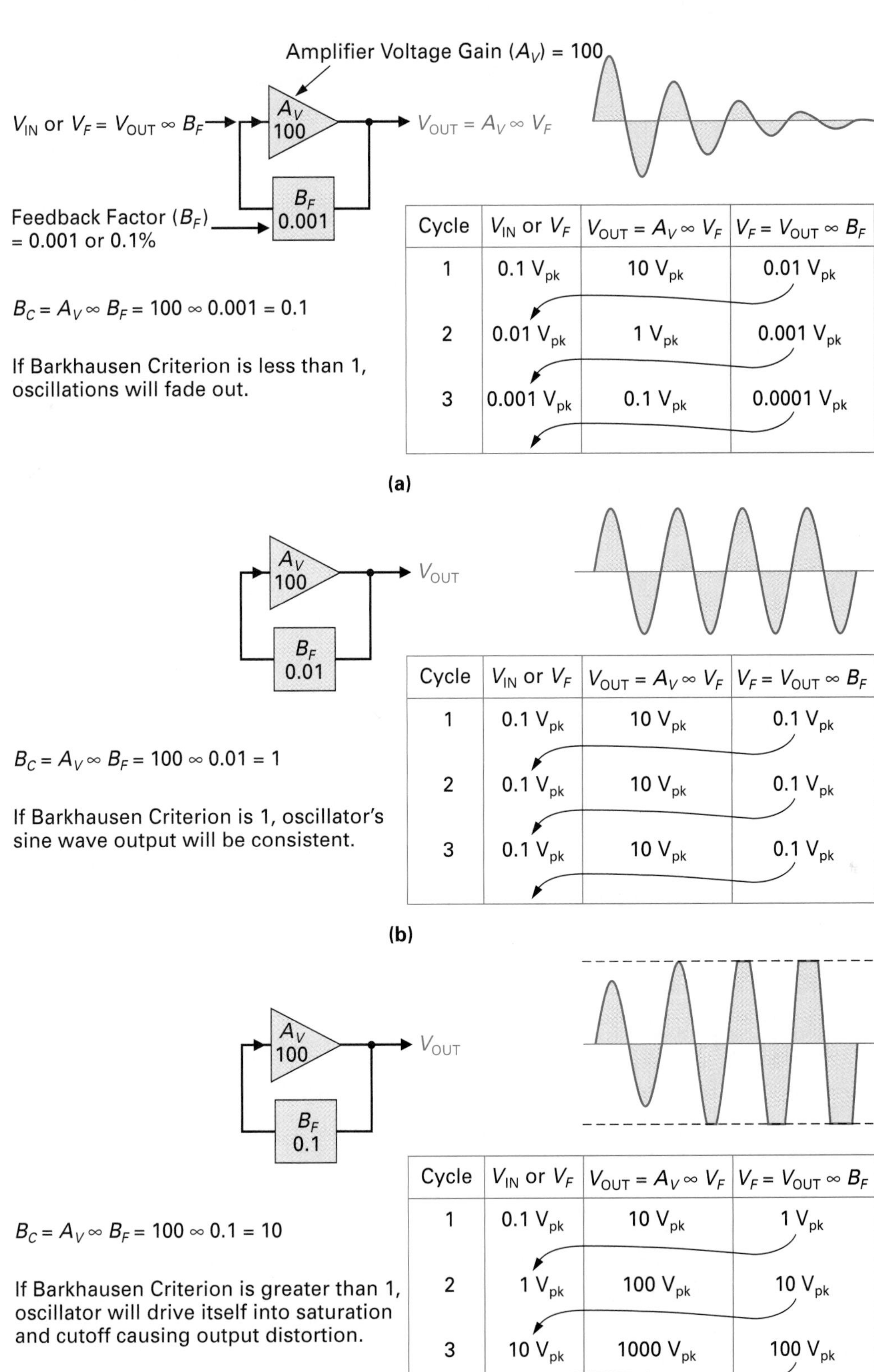

Amplifier Voltage Gain (A_V) = 100
V_{IN} or $V_F = V_{OUT} \infty B_F$
A_V 100
$V_{OUT} = A_V \infty V_F$
B_F 0.001
Feedback Factor (B_F) = 0.001 or 0.1%
$B_C = A_V \infty B_F = 100 \infty 0.001 = 0.1$
If Barkhausen Criterion is less than 1, oscillations will fade out.
(a)
A_V 100
V_{OUT}
B_F 0.01
$B_C = A_V \infty B_F = 100 \infty 0.01 = 1$
If Barkhausen Criterion is 1, oscillator's sine wave output will be consistent.
(b)
A_V 100
V_{OUT}
B_F 0.1
$B_C = A_V \infty B_F = 100 \infty 0.1 = 10$
If Barkhausen Criterion is greater than 1, oscillator will drive itself into saturation and cutoff causing output distortion.
(c)

(a)

Cycle	V_{IN} or V_F	$V_{OUT} = A_V \infty V_F$	$V_F = V_{OUT} \infty B_F$
1	0.1 V_{pk}	10 V_{pk}	0.01 V_{pk}
2	0.01 V_{pk}	1 V_{pk}	0.001 V_{pk}
3	0.001 V_{pk}	0.1 V_{pk}	0.0001 V_{pk}

(b)

Cycle	V_{IN} or V_F	$V_{OUT} = A_V \infty V_F$	$V_F = V_{OUT} \infty B_F$
1	0.1 V_{pk}	10 V_{pk}	0.1 V_{pk}
2	0.1 V_{pk}	10 V_{pk}	0.1 V_{pk}
3	0.1 V_{pk}	10 V_{pk}	0.1 V_{pk}

(c)

Cycle	V_{IN} or V_F	$V_{OUT} = A_V \infty V_F$	$V_F = V_{OUT} \infty B_F$
1	0.1 V_{pk}	10 V_{pk}	1 V_{pk}
2	1 V_{pk}	100 V_{pk}	10 V_{pk}
3	10 V_{pk}	1000 V_{pk}	100 V_{pk}

FIGURE 12-5 **The Feedback Factor.**

2. For the second cycle of the output, an input of 0.01 V_{pk} will produce an output voltage of 1 V_{pk} ($V_{OUT} = A_V \times V_F = 100 \times 0.01 = 1\ V_{pk}$). Because 0.1% of this output will be fed back to the input, the feedback voltage for the third cycle will be 0.001 V_{pk} ($V_F = V_{OUT} \times B_F = 1 \times 0.001 = 0.001\ V_{pk}$).
3. For the third cycle of the output, an input of 0.001 V_{pk} will produce an output voltage of 0.1 V_{pk}. Because 0.1% of this output will be fed back to the input, the feedback voltage for the fourth cycle will be 0.0001 V_{pk}.

As you can see from this example, if the feedback factor is too small, the feedback voltage will be small, and the amplifier's output voltage will decrease with each cycle until the oscillations fade to zero.

In Figure 12-5(b), the feedback factor has been increased to 0.01 or 1%. Using the initial circuit turn ON voltage input of 0.1 V_{pk}, you can see by looking at the table in Figure 12-5(b) that

1. For the first cycle of the output, an input of 0.1 V_{pk} will produce an output voltage of 10 V_{pk}. Because 1% of this output will be fed back to the input, the feedback voltage for the second cycle will be 0.1 V_{pk}.
2. For the second cycle of the output, an input of 0.1 V_{pk} will produce an output voltage of 10 V_{pk}. Because 1% of this output will be fed back to the input, the feedback voltage for the third cycle will be 0.1 V_{pk}.
3. For the third cycle of the output, an input of 0.1 V_{pk} will produce an output voltage of 10 V_{pk}. Because 1% of this output will be fed back to the input, the feedback voltage for the fourth cycle will be 0.1 V_{pk}.

As you can see from this example, if the feedback factor is of the right value, the oscillator will continuously produce a consistent sine-wave output.

In Figure 12-5(c), the feedback factor has been increased to 0.1 or 10%. Using the initial circuit turn ON voltage input of 0.1 V_{pk}, you can see by looking at the table in Figure 12-5(c) that

1. For the first cycle of the output, an input of 0.1 V_{pk} will produce an output voltage of 10 V_{pk}. Because 10% of this output will be fed back to the input, the feedback voltage for the second cycle will be 1 V_{pk}.
2. For the second cycle of the output, an input of 1 V_{pk} will produce an output voltage of 100 V_{pk}. Because 10% of this output will be fed back to the input, the feedback voltage for the third cycle will be 10 V_{pk}.
3. For the third cycle of the output, an input of 10 V_{pk} will produce an output voltage of 100 V_{pk}. Because 10% of this output will be fed back to the input, the feedback voltage for the fourth cycle will be 100 V_{pk}.

The transistor's output will, of course, never reach 100 V_{pk} because the output cannot exceed the $+V_{CC}$ supply voltage. As shown in this example, if the feedback factor is too large, the transistor will soon be driven into saturation and cutoff for long periods of time, and therefore the output sine wave will become distorted due to clipping.

From the examples shown in Figure 12-5, you can see that the feedback factor is vital if an oscillator is going to oscillate. This point was initially discovered in 1919 by Barkhausen, who studied in detail the characteristics of oscillator feedback. The **Barkhausen criterion (B_C)** is the relationship between the oscillator's feedback factor (B_F) and the amplifier's voltage gain (A_V) and should be equal to 1 for proper oscillator operation.

Barkhause Criterion (B_C)
The relationship between the oscillator's feedback factor (B_F) and the amplifier's voltage gain (A_V).

$$B_C = A_V \times B_F$$

In the example in Figure 12-5(a), the Barkhausen criterion was less than 1, and, as we discovered, there was not enough feedback to sustain oscillations.

$$B_C = A_V \times B_F = 100 \times 0.001 = 0.1$$

In the example in Figure 12-5(b), the Barkhausen criterion was a perfect 1, and, as we discovered, the oscillator's sine wave output was continually repeating and consistent in amplitude.

$$B_C = A_V \times B_F = 100 \times 0.01 = 1$$

In the example in Figure 12-5(c), the Barkhausen criterion was greater than 1, and, as we discovered, the oscillator was driven into saturation and cutoff, causing excessive sine-wave output distortion.

Let us now examine in more detail the feedback factor as it applies to the previously discussed Colpitts oscillator. By adjusting the value of components in an oscillator's feedback tank circuit, we can control the amount of feedback from an oscillator's output to input, and therefore maintain the Barkhausen criterion at 1. Referring again to the circuit in Figure 12-4, you will remember that it is the series combination of capacitors C_1 and C_2 and the inductor L_1 that determine the Colpitts frequency of oscillation. The amount of feedback, however, is controlled by adjusting the value of C_2. The voltage developed between this capacitor's bottom plate and its other plate, which is connected to ground, is the feedback voltage applied to the base of Q_1. For example, if we were to increase the value of C_2 ($C\uparrow$), its capacitive reactance would decrease ($X_C\downarrow$), and therefore a smaller feedback voltage would be developed across C_2 and applied to the base of Q_1. The feedback factor of a Colpitts oscillator will typically be between 0.1 (10%) to 0.5 (50%) and will be controlled by adjusting the ratio of C_1 to C_2.

$$\text{Feedback factor } (B_F) = \frac{C_2}{C_1 + C_2}$$

This feedback factor is adjusted so that it is large enough to sustain oscillations but not too large to distort the output signal.

■ **EXAMPLE:**

Calculate the feedback factor of the Colpitts oscillator shown in Figure 12-4(a), if $C_1 = 0.47\ \mu F$ and $C_2 = 0.33\ \mu F$.

■ ***Solution:***

$$\text{Feedback factor } (B_F) = \frac{C_2}{C_1 + C_2} = \frac{0.33\ \mu\text{F}}{0.47\ \mu\text{F} + 0.33\ \mu\text{F}} = \frac{0.33\ \mu\text{F}}{0.8\ \mu\text{F}}$$

$$= 0.4125 \text{ or } 41.25\%$$

12-1-6 *The Clapp Oscillator*

Clapp Oscillator
A series tuned colpitts oscillator circuit.

The **Clapp oscillator** shown in Figure 12-6 is a variation on the Colpitts oscillator. Studying the schematic you can see that the only circuit difference is the additional capacitor C_3. This capacitor forms a series resonance circuit with inductor L_1, and if it is made variable, it can adjust the oscillator's frequency without interfering with the circuit's feedback factor, which is controlled by the ratio of C_1 to C_2. Other than this small modification, the operation of this circuit is the same as the previous Colpitts oscillator.

Since the value of capacitor C_3 is generally much lower than C_1 and C_2, it becomes the dominant capacitance and is used in oscillator frequency calculations.

$$f_R = \frac{1}{2\pi\sqrt{L_1 \times C_3}}$$

■ **EXAMPLE:**

Calculate the frequency range at which the circuit shown in Figure 12-6 will oscillate if $L_1 = 0.47$ mH and $C_3 = 0.33$ μF to 0.47 μF.

■ ***Solution:***

$$f_{R(\text{Low})} = \frac{1}{2\pi\sqrt{L_1 \times C_3}} = \frac{1}{2\pi\sqrt{0.47\ \text{mH} \times 0.47\ \mu\text{H}}} = 10.7\ \text{kHz}$$

$$f_{R(\text{High})} = \frac{1}{2\pi\sqrt{L_1 \times C_3}} = \frac{1}{2\pi\sqrt{0.47\ \text{mH} \times 0.33\ \mu\text{H}}} = 12.8\ \text{kHz}$$

12-1-7 *High-Frequency Oscillations*

Parasitic Oscillations
An unwanted self-sustaining oscillation or self-generated random pulse.

As discussed in the previous chapter, transistors have an inherent collector to base capacitance and transistor lead inductance, as shown in Figure 12-7. At low frequencies, this small value of capacitance and inductance does not affect the circuit. At high frequencies, however, these elements combine to produce series resonant and parallel resonant circuits that will cause unwanted **parasitic oscillations.** In amplifier circuits, these high-frequency oscillations are undesirable and circuit modifications are made to counter the effects. High frequency oscillators, on the other hand, make use of the transistor circuit's inherent resonant tanks, which function as feedback networks. For example, the circuit shown in Figure 12-7 would not include the transistor lead inductance and junction capacitance devices on a schematic diagram because they are not physically on the printed circuit board. This circuit, however, could be listed as a high-frequency oscillator

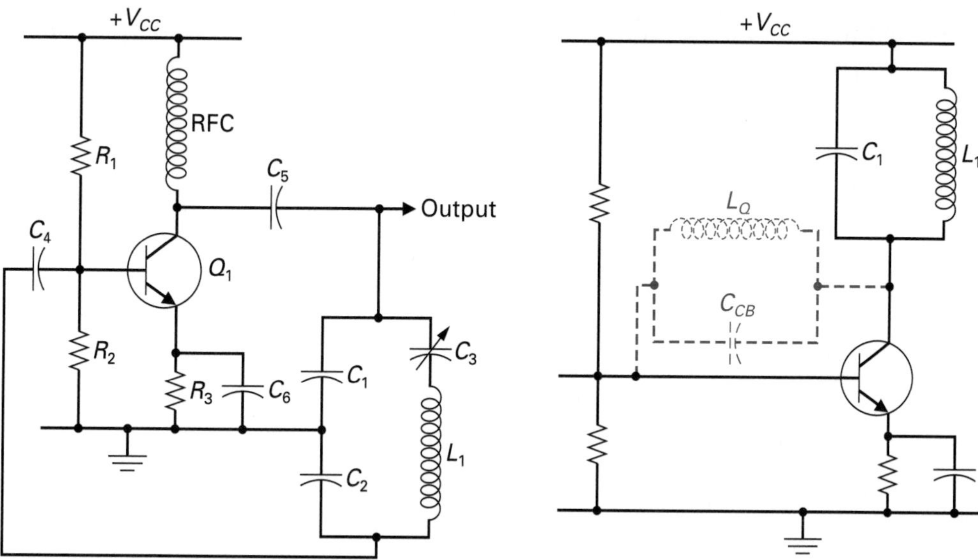

FIGURE 12-6 The Clapp Oscillator.

FIGURE 12-7 High-Frequency Transistor Oscillations.

even though no feedback path is evident. The frequency of this oscillator would be controlled by the collector's tank circuit made up of L_1 and C_1.

SELF-TEST REVIEW QUESTIONS FOR SECTION 12-1

1. What are the three basic oscillator classifications?
2. What type of feedback is needed for an oscillator to continually oscillate?
3. Which component(s) provide the 180° phase shift in the following oscillator circuits?
   a. Armstrong
   b. Hartley
   c. Colpitts
4. What value should the Barkhausen criterion be if an oscillator's sine-wave output is to be consistent in amplitude and free of distortion?
5. What are parasitic oscillations?
6. An oscillator replaces lost tank energy when the transistor is:
   a. ON
   b. OFF

12-2 CRYSTAL OSCILLATOR CIRCUITS

The *LC* oscillator circuits discussed in the previous section would typically have a 0.8% frequency drift. This frequency drift is due to circuit temperature changes, the aging of the components, and changes in the resistance of the load connected to the oscillator. If an *LC* oscillator were used to generate the timing signal for a wristwatch, a 0.8% frequency drift would mean that the watch may gain or lose approximately eleven and a half minutes in a day.

$$60 \text{ minutes} \times 24 \text{ hours} = 1440 \text{ minutes per day}$$

$$0.8\% \text{ of } 1440 = 11.5 \text{ minutes per day}$$

If a high degree of oscillator stability is needed, crystal oscillators are used to generate the timing signal. Crystal controlled oscillators will typically have a frequency drift of 0.0001%, and, if used in a wristwatch, may gain or lose a maximum of half a minute a year.

$$60 \text{ minutes} \times 24 \text{ hours} \times 365 \text{ days} = 525{,}600 \text{ minutes per year}$$

$$0.0001\% \text{ of } 525{,}600 = 0.5 \text{ minutes per year}$$

Therefore, in applications where a high degree of frequency stability is needed, such as communication and computer systems, crystal-controlled oscillators are used. Crystal-controlled oscillators make use of a quartz crystal to control the circuit's frequency of operation. Before we discuss the operation of various crystal oscillator circuits, let us review the basic characteristics of a crystal.

12-2-1 The Characteristics of Crystals

The quartz crystal is made of silicon dioxide and is naturally a six-sided (hexagonal) compound with pyramids at either end, as seen in Figure 12-8(a). To construct an electronic component, a thin slice or slab of crystal is cut from the mother stone, mounted between two metal plates that make electrical contact, then placed in a protective holder as seen in Figure 12-8(b). On a schematic diagram, a crystal is generally labeled either "XTAL" or "Y," and has the symbol shown in Figure 12-8(c).

The crystal is basically operated as a transducer, or energy converter, transforming mechanical energy to electrical energy as shown in Figure 12-8(d), (e), and (f), or electrical energy to mechanical energy as shown in Figure 12-8(g).

Mechanical-to-Electrical Conversion

In Figure 12-8(d), you can see that a crystal normally has its internal charges evenly distributed throughout, and the potential difference between its two plates is zero. If the crystal is compressed by applying pressure to either side, as shown in Figure 12-8(e), opposite charges accumulate on either side of the crystal and a potential difference is generated. Similarly, if the crystal is expanded by applying pressure to the top and bottom, as shown in Figure 12-8(f), opposite charges accumulate on either side of the crystal, and a potential difference of the opposite polarity is generated. If a crystal were subjected to an alternating pressure that caused it to continually expand and compress, an alternating, or ac, voltage would be generated. Crystal microphones make use of this principle. Sound waves are applied to the microphone, and these mechanical waves continually compress and expand the crystal, generating an electrical wave that is equivalent to the original sound wave.

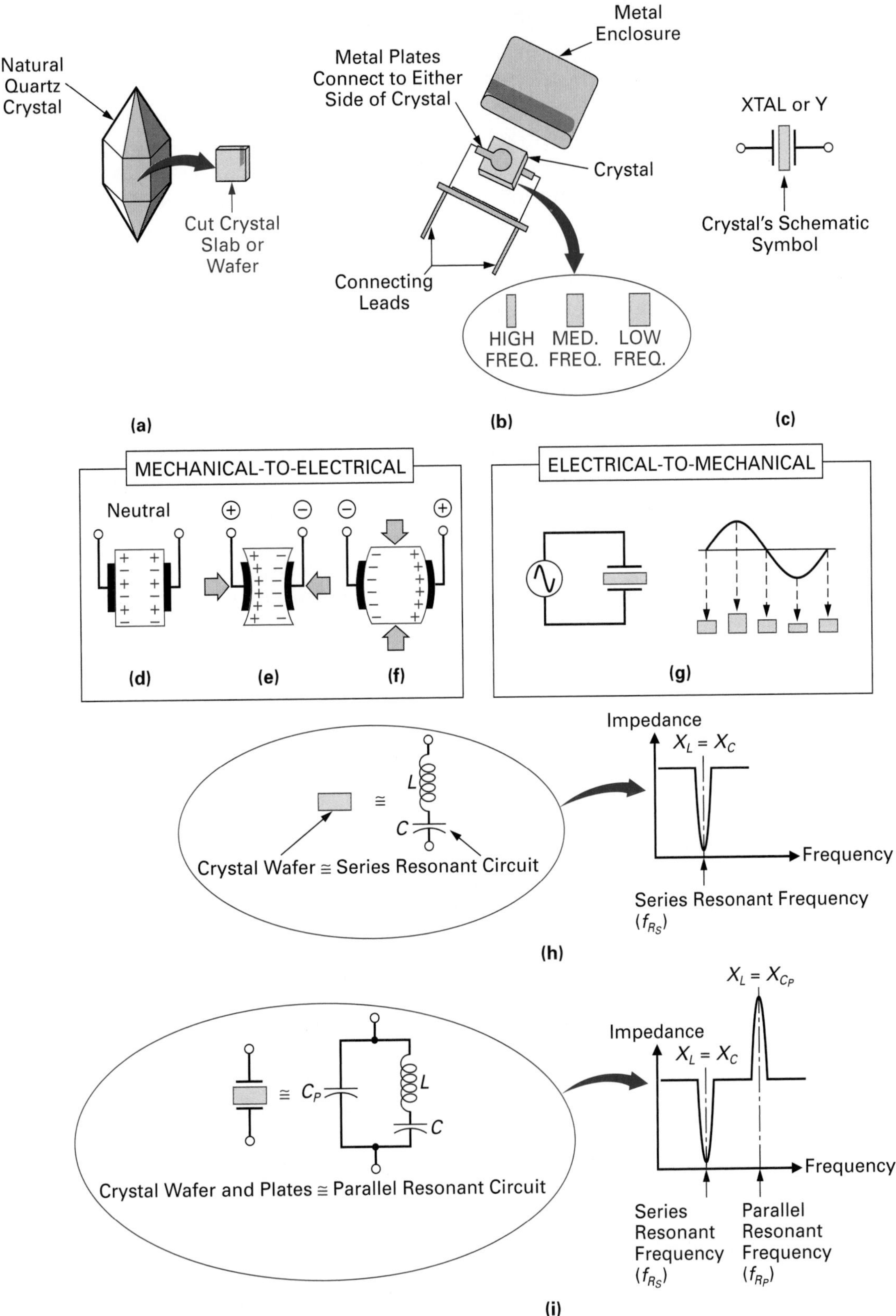

FIGURE 12-8 The Characteristics of Crystals.

Electrical-to-Mechanical Conversion

Due to their composition, crystals have a "natural frequency of vibration." This means that if the ac voltage applied across a crystal matches the crystal's natural frequency of vibration as shown in Figure 12-8(g), the crystal will physically expand and contract by a relatively large amount. If, on the other hand, the frequency of the applied ac voltage is either above or below the crystal's natural frequency, the vibration is only slight. Crystals are, therefore, very frequency selective—a characteristic we can make use of in filter circuits and oscillator circuits. This action is called **piezoelectric effect,** which by definition is the tendency of a crystal to vibrate at a constant rate when it is subjected to a changing electric field produced by an applied ac voltage. The crystal's natural frequency of vibration is dependent on the thickness of the crystal between the two plates, as seen in the inset in Figure 12-8(b). By cutting a crystal to the right size, we can obtain a crystal that will naturally vibrate at an exact frequency. This rating is usually printed on the crystal's package.

Piezoelectric Effect
The tendency of a crystal to vibrate at a constant rate when subjected to a changing electric field.

When the frequency of the ac voltage applied to a crystal matches the crystal's natural frequency of vibration, the impedance or current opposition of the crystal drops to a minimum, as seen in Figure 12-8(h). This makes the crystal wafer or slab equivalent to a series resonant circuit, and, as you can see in the frequency response curve in Figure 12-8(h), the very sharp response at the crystal's series resonant frequency (f_{R_S}) means that the crystal is extremely frequency selective.

When a crystal is mounted between two connecting plates, the device is equivalent to the crystal's original series resonant circuit in parallel with a small value of capacitance due to the connecting plates (C_P), as seen in Figure 12-8(i). Referring to the frequency response curve in Figure 12-8(i), you can see that the crystal component will still respond in exactly the same way when the applied ac frequency is equal to the crystal's natural series resonant frequency. As frequency is increased beyond the series resonant frequency of the crystal, however, the reactance of C will decrease, and the crystal will become inductive. At a higher frequency, there will be a point at which the inductive reactance of the crystal is equal to the capacitive reactance of the connecting plate's ($X_L = X_{C_P}$). Because the crystal's inductance is in parallel with the mounting plate's capacitance, the two form a parallel resonant circuit. Therefore, the device's impedance will be maximum at this parallel resonant frequency (f_{R_P}), as shown in Figure 12-8(i). Circuit applications can make use of either the crystal's series resonant selectivity or parallel resonant selectivity.

Most crystal-oscillator circuits will make use of the crystal's series resonant frequency response to feed back only the desired frequency to the input and tightly maintain the frequency stability of the oscillator. A crystal's series resonant tuned circuit will typically have a very high Q of 40,000. If you compare this to an LC tuned circuit that would typically have a Q of 200, it is easy to see why crystal controlled oscillators are much more stable than LC controlled oscillators.

12-2-2 The Hartley Crystal Oscillator

As you can see in Figure 12-9, the Hartley crystal oscillator is identical to the previously discussed LC Hartley oscillator, except for the crystal inserted in series with the feedback path. When the oscillator generates a frequency that is

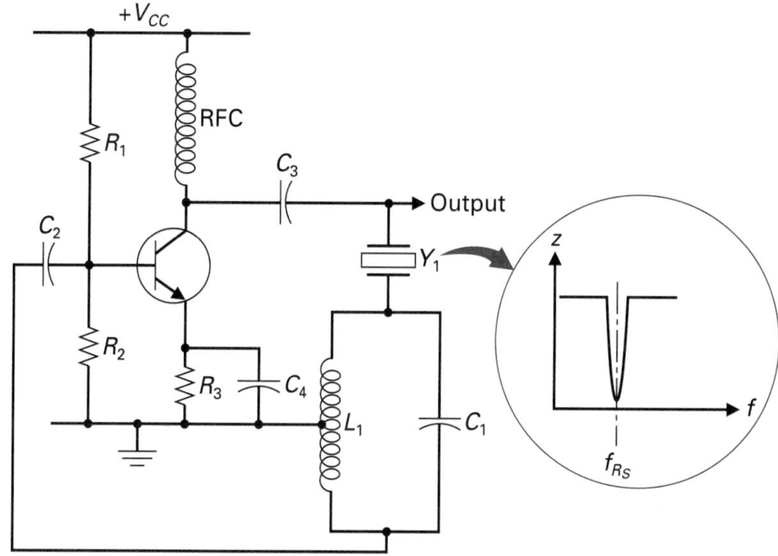

FIGURE 12-9 The Hartley Crystal Oscillator.

identical to the series resonant frequency of the crystal, the crystal will offer almost no opposition and the feedback signal will be maximum. If the oscillator's frequency begins to drift above or below the crystal's natural series resonant frequency, the impedance of the crystal will increase, causing the feedback to decrease. Because the frequency of the feedback signal controls the circuit's frequency of oscillation, the circuit is forced to return to the natural frequency of the crystal. Including a highly selective crystal in the feedback path means that this crystal-controlled Hartley oscillator will have a frequency drift rating that is almost 1,000 times better than an *LC* controlled Hartley oscillator.

12-2-3 *The Colpitts Crystal Oscillator*

Referring to Figure 12-10, you can see that the Colpitts crystal oscillator is identical to the previously discussed *LC* Colpitts oscillator, except for the crystal inserted in series with the feedback path. This circuit will operate in exactly the same way as the previously discussed *LC* Colpitts oscillator; however, the circuit will have less frequency drift due to the inclusion of a crystal.

12-2-4 *The Pierce Crystal Oscillator*

The **Pierce crystal oscillator** is shown in Figure 12-11, and, as you can see, it is almost identical to the Colpitts oscillator except for crystal Y_1, which replaces the tank's coil. Crystal Y_1 will be operating at its parallel resonant frequency, which means that it will offer a very high impedance to the oscillator's generated frequency. This action is not any different from the previous *LC* parallel resonant tank, which would also have a high impedance at resonance. However, now that a crystal controls the tank's impedance, the tuned circuit will be much more

Pierce Crystal Oscillator
An oscillator circuit in which a piezoelectric crystal is connected in a tank between output and input.

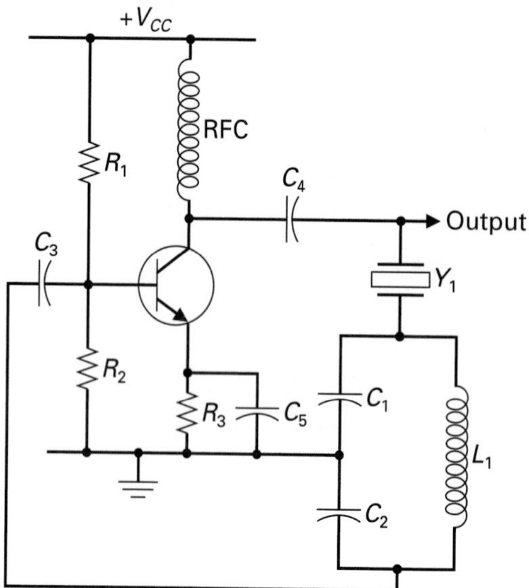

FIGURE 12-10 The Colpitts Crystal Oscillator.

selective, and the oscillator's output will be more stable. At resonance, the tank circuit will have a very large impedance, and therefore a large feedback voltage will be developed across C_2 and applied to the base of Q_1. If the oscillator's frequency drifts above or below resonance, the crystal's impedance will quickly decrease, decreasing the voltage developed across the tank and the amplitude of the feedback signal.

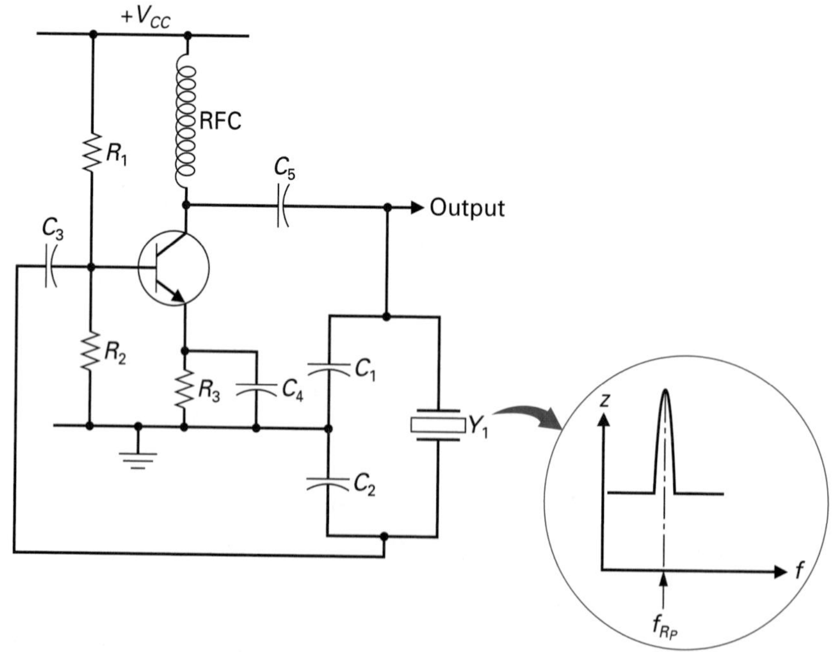

FIGURE 12-11 The Pierce Crystal Oscillator.

SELF-TEST REVIEW QUESTIONS FOR SECTION 12-2

1. What key advantage does a crystal oscillator circuit have over an *LC* oscillator circuit?
2. The physical size of the crystal determines its natural frequency of vibration, or resonant, frequency. A thinly cut crystal will resonate at a ____________ frequency, whereas a thickly cut crystal will resonate at a ____________ frequency.
3. At a crystal's series resonant frequency, impedance will be ____________, whereas at the parallel resonant frequency impedance will be ____________.
4. Both the Hartley and Colpitts crystal oscillator circuits make use of the crystal's ____________ resonant frequency, whereas the Pierce crystal oscillator makes use of the crystal's ____________ resonant frequency.

12-3 *RC* OSCILLATOR CIRCUITS

In Chapter 9, we saw how *RC* networks can be used within multivibrator circuits to generate a single or repeating square wave or rectangular wave output. In this section we will see how *RC* networks can be used within oscillator circuits to generate a sine-wave output. Before we begin, however, let us examine why we need *RC* oscillators. Within electronic systems, you will see *LC* and crystal oscillators used to generate medium-frequency and high-frequency sine-wave oscillations. As previously stated, the choice between these two oscillator types is basically governed by two factors: frequency stability and price. If frequency stability is of prime importance, a crystal oscillator is used; if frequency stability is not crucial, the lower priced *LC* oscillator can be used. The next question to answer is: Is it practical to use these oscillator circuits to generate low audio-frequency oscillations? The answer is no. At low frequencies, the size of the inductor needed for an AF tuned tank is very large, and therefore a low-frequency *LC* oscillator would be large in size and expensive. As far as the crystal oscillator is concerned, the physical size of the crystal makes it only practical for crystals to be used in applications requiring an output frequency of 50 kHz or greater. Although *RC* oscillators are not as stable as *LC* or crystal oscillators, they are inexpensive and are ideal in applications where we need to generate low-frequency oscillations of less than 10 kHz. In this section we will discuss the operation and characteristics of two basic *RC* oscillator circuits: the phase-shift oscillator and the Wien-Bridge oscillator.

12-3-1 ***The Phase-Shift Oscillator***

Before we examine the phase-shift oscillator circuit, let's review the basic theory of an *RC* phase-shift network. In your previous introductory dc/ac electronics course, you found that *when a voltage was applied to a purely capacitive circuit, the current led the voltage by 90°.* When a voltage source is applied across a capacitor with no charge, the circuit current is initially maximum as the capacitor charges, while the voltage across the capacitor is initially zero. As the voltage across the capacitor slowly increases, the circuit current slowly decreases until the voltage across the capacitor is equal but opposite to the source voltage, and therefore the circuit current is zero. In a purely capacitive circuit, circuit current leads the capacitor voltage by 90°. On the other hand, *in a purely resistive circuit the circuit current and voltage developed across a resistor are always in phase.* This means that when a source voltage is applied across a resistor, the voltage developed across the resistor and current passing through the resistor will immediately rise to their Ohm's law values. The next question is: What happens to voltage and current in a circuit or network that contains both capacitance and resistance? The answer is shown in Figure 12-12(a). *In a resistive-capacitive (RC) network, the circuit current leads the voltage by some value between 0° and 90° depending on the values of resistance and capacitance.* For the circuit shown in Figure 12-12(a), the input voltage is applied across the series connected capacitor and resistor, and the output voltage is taken across the resistor. To review the formulas involved, the phase difference between the output voltage (which is the voltage developed across the resistor, V_R or V_{OUT}) and the input voltage (V_{IN}) is dependent on the capacitive reactance (X_C) and resistance (R) of the RC network. For the values given in Figure 12-12(a), the capacitor's capacitive reactance will be

$$X_C = \frac{1}{2\pi fC} = \frac{1}{2 \times \pi \times 100\ \text{Hz} \times 0.6\ \mu\text{F}} = 2.65\ \text{k}\Omega$$

and therefore the degree of phase shift will be

$$\theta = \text{inv tan}\ \frac{X_C}{R} = \text{inv tan}\ \frac{2.65\ \text{k}\Omega}{1.53\ \text{k}\Omega} = 60°$$

Comparing the V_{IN} waveform to the V_{OUT} waveform in Figure 12-12(a) or looking at the vector diagram, you can see that the output voltage leads the input voltage by 60°. As you have seen from the previous formulas, this phase shift is dependent on the values of R and C and the frequency of the applied ac input voltage. If the frequency of the input voltage were to change, the phase shift would also change.

Looking at Figure 12-12(b), you can see that if three 60° phase-shift networks were connected in series, the combined phase shift would be 180°. This 180° phase shift is exactly what is needed for positive feedback in an oscillator circuit. Therefore, if this network were placed between the collector output and base input of a common-emitter amplifier, the result would be a **phase-shift oscillator** circuit, as shown in Figure 12-12(c). Transistor Q_1 will typically have a voltage gain of between 40 and 50 so that it can compensate for the power loss in the RC phase-shift network. The circuit operates in much the same way as all of the previously discussed oscillator circuits. For instance, when power is first applied to the circuit, the bias resistors R_4 and R_5 ensure that Q_1 turns ON, resulting in a drop in voltage at the collector of Q_1. This decrease in voltage at the collector of Q_1 is inverted to an increase in voltage at the base of Q_1 by the 180° RC phase-shift network, which increases the forward bias applied to Q_1, sending it into saturation. When Q_1 saturates, there is no further change in the

Phase-Shift Oscillator
An RC oscillator circuit in which the 180° phase shift is achieved with several RC networks.

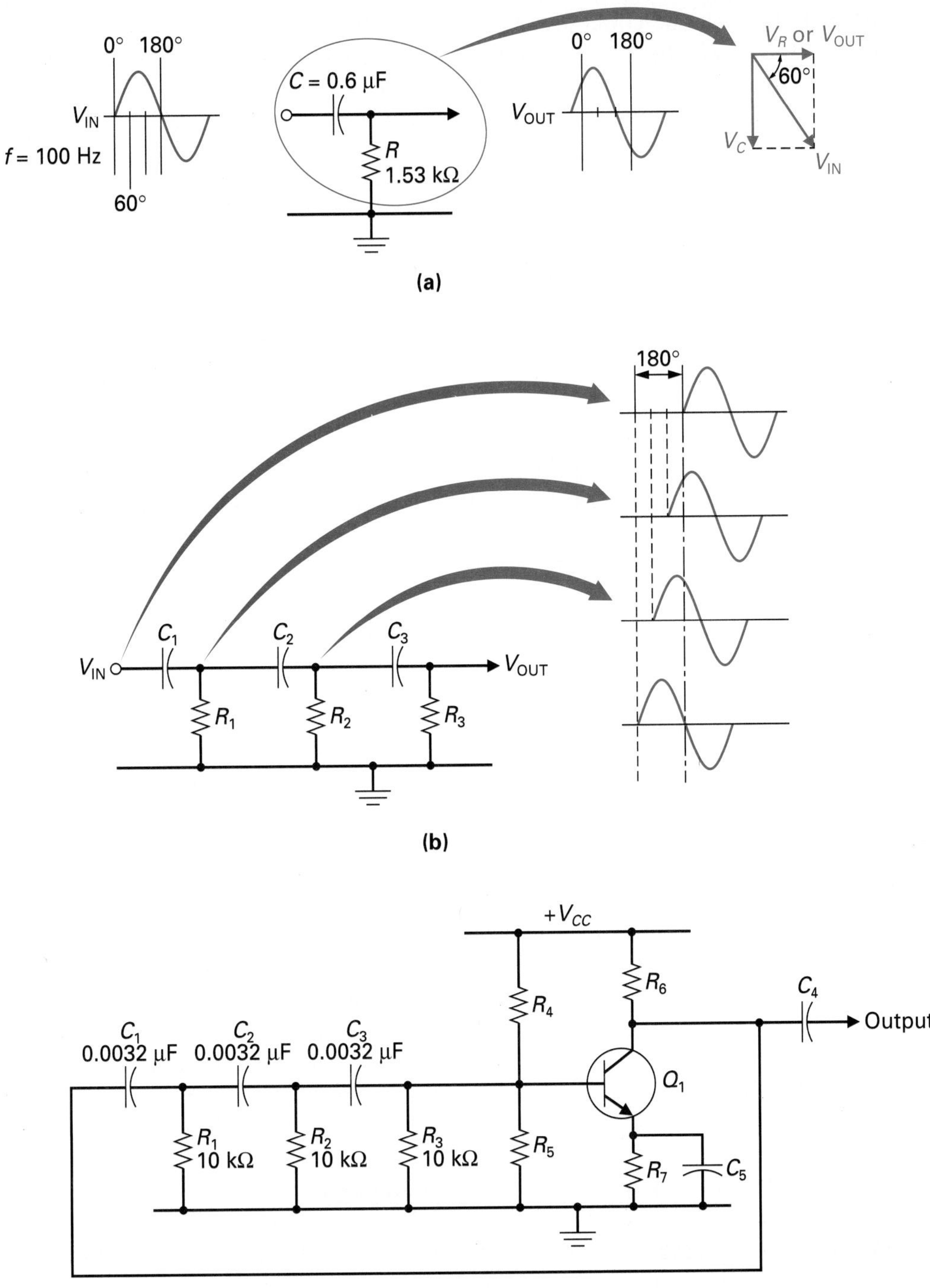

FIGURE 12-12 **The *RC* Phase Shift Oscillator.**

voltage at the collector of Q_1, and therefore the base of Q_1. Q_1 begins to cut off, causing its collector voltage to rise. This increase in voltage at the collector of Q_1 is inverted to a decrease in voltage at the base of Q_1 by the 180° *RC* phase-shift network, which decreases the forward bias applied to Q_1, sending it into cutoff.

When the voltage at the collector of Q_1 reaches $+V_{CC}$, the voltage at the base of Q_1 will be the opposite, or 0 V, and the bias resistors R_4 and R_5 will provide sufficient base voltage to turn Q_1 ON. As Q_1 turns ON, there will be a drop in voltage at the collector of Q_1, which will be inverted to an increase in voltage at the base of Q_1 by the 180° *RC* phase-shift network. This increases the forward bias applied to Q_1, sending it into saturation. The oscillator action will then repeat with the continual rise and fall in voltage at the collector of Q_1, producing a sine wave at the output of the oscillator. The frequency of oscillation can be approximately calculated using the following formula:

$$f_R = \frac{1}{2 \times \pi \times \sqrt{6} \times R \times C}$$

R is the value of one of the resistors in the phase-shift network
C is the value of one of the capacitors in the phase-shift network

■ **EXAMPLE:**

Calculate the frequency at which the phase-shift oscillator in Figure 12-12(c) will oscillate.

■ ***Solution:***

$$f_R = \frac{1}{2 \times \pi \times \sqrt{6} \times R \times C} = \frac{1}{2 \times \pi \times 2.45 \times 10\ \text{k}\Omega \times 0.0032\ \mu\text{F}} = 2\ \text{kHz}$$

The *RC* phase-shift oscillator is typically only used as a fixed-frequency oscillator because any changes in the resistance or capacitance in the feedback network will interfere with the regenerative feedback. To improve oscillator frequency stability, many *RC* phase-shift oscillator circuits will have four 45°-*RC* networks, or six 30°-*RC* networks, which combined will still provide the 180° phase shift that is needed. Increasing the number of *RC* phase-shift networks will mean that each network will be responsible for less phase shift; therefore, frequency drift due to component aging and temperature will be reduced.

12-3-2 The RC Wien-Bridge Oscillator

Wien-Bridge Oscillator
An *RC* oscillator circuit in which a Wien Bridge determines frequency.

The **Wien-Bridge oscillator,** which also makes use of *RC* networks in the feedback path, can be used to generate sine-wave signal outputs up to 1 MHz. Before we begin with the oscillator circuit, let us first examine the two *RC* networks used in the Wien-Bridge oscillator, which are shown in Figure 12-13(a). Series connected C_1 and R_1 form a high-pass filter because the reactance of C_1 will be HIGH at low frequencies and LOW at high frequencies. Low frequencies will be blocked and high frequencies will be passed. Parallel connected C_2 and R_2 form a low-pass filter because the reactance of C_2 will be HIGH at low frequencies and LOW at high frequencies. High frequencies will be shunted to

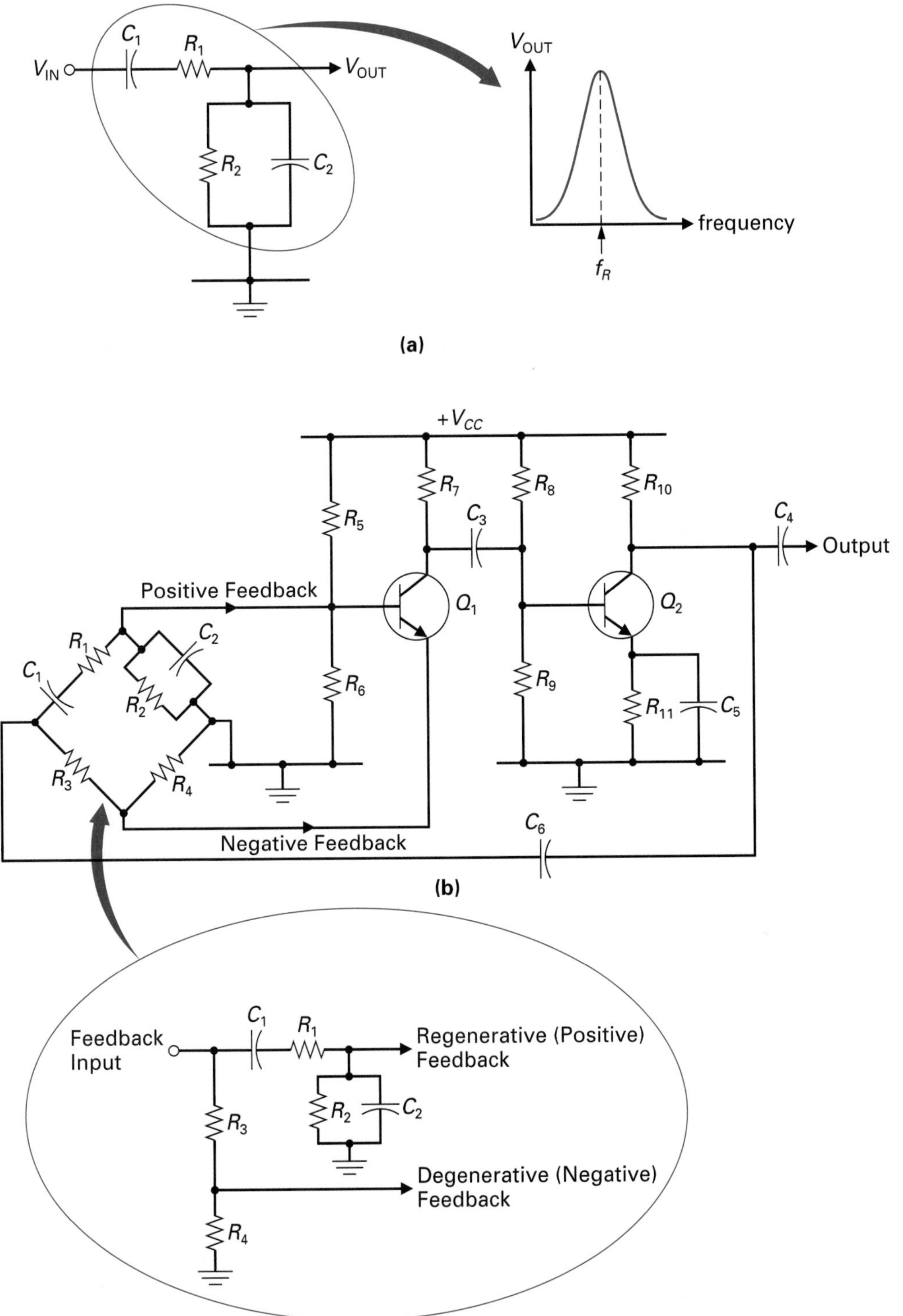

FIGURE 12-13 **The Wien-Bridge Oscillator.**

ground and low frequencies passed. The combined effect of this series high-pass filter and parallel low-pass filter is a band-pass filter that will produce a maximum output voltage when the input voltage is at the circuit's resonant frequency. The resonant frequency of these combined RC networks can be calculated using the following formula:

$$f_R = \frac{1}{2\pi\sqrt{C_1 \times R_1 \times C_2 \times R_2}}$$

To review, a band-pass filter does not introduce a phase shift when it is operated at its resonant frequency. This means that when a low-input frequency is applied to a band-pass filter, the output voltage will lead the input voltage (the phase angle is positive). When a high-input frequency is applied, the output voltage will lag the input voltage (the phase angle is negative). Only when the frequency of the input signal matches the band-pass filter's resonant frequency will the series and parallel circuit phase shifts be equal but opposite—and therefore cancel—causing the output signal voltage to be in phase with the input signal voltage. In summary, *there is no phase shift between the input and output of a band-pass filter when the input signal frequency is equal to the band-pass filter's resonant frequency.*

Figure 12-13(b) shows how this RC band-pass filter network can be connected to a Wien-Bridge oscillator circuit. In this circuit, transistor Q_1 operates as the oscillating transistor. Studying the inset in Figure 12-13(b), you can see that the band-pass filter network (C_1, R_1, C_2, R_2) is used to provide "positive feedback signal," which produces oscillations. A voltage-divider network (R_3, R_4) is used to provide "negative feedback signal," which controls the voltage gain of transistor Q_1. The positive feedback signal is applied to the base of Q_1, while the negative feedback signal is applied to the emitter of Q_1. Because the band-pass filter in the bridge network will not provide any phase shift, the common-emitter signal inverting transistor Q_2 is included. It will provide the 180° of phase shift that is needed for positive feedback. Capacitor C_4 couples the circuit's oscillations to the output, while the feedback signal is coupled to the input of the bridge network by C_6. Let us now examine the action of the band-pass filter circuit and voltage-divider circuit within the bridge network in more detail.

Positive Feedback Path

As stated earlier, the band-pass filter will not introduce a phase shift when it is operated at its resonant frequency. If the oscillator's frequency begins to increase, the reactance of C_2 will begin to decrease and less positive feedback will be applied to Q_1. On the other hand, if the oscillator's frequency begins to increase, the reactance of C_1 will begin to increase, blocking the positive feedback from being applied to Q_1. Only when the oscillator's output frequency is within the narrow band of the band-pass filter will the full positive feedback signal be applied to Q_1 to sustain oscillations.

Negative Feedback Path

The voltage divider, R_3 and R_4, is included to control the voltage gain of Q_1 so that the output signal is not clipped due to the transistor's being driven too heavily into cutoff and saturation. If the amplitude of the output signal begins to increase, there will be an increase in the feedback signal, which will cause an increase in the voltage developed across R_4 and therefore applied to the emitter of Q_1. Increasing the voltage at the emitter of Q_1 will decrease its base-emitter forward bias and therefore decrease the gain of Q_1, counteracting the original increase. Similarly, if the amplitude of the output signal begins to decrease, there will be a decrease in the feedback signal, which will cause a decrease in the voltage developed across R_4, and therefore applied to the emitter of Q_1. Decreasing the voltage at the emitter of Q_1 will increase its base-emitter forward bias and increase the gain of Q_1, counteracting the original decrease. It is this gain-control feedback signal that makes the Wien-Bridge oscillator circuit very popular because it is able to generate stable sine-wave oscillations up to 10 MHz.

The frequency of the Wien-Bridge oscillator can be adjusted by varying resistors R_1 and R_2, which are typically ganged. Like all *RC* oscillators, the circuit's gain must be quite high in order to compensate for the power losses in the bridge network's resistors.

SELF-TEST REVIEW QUESTIONS FOR SECTION 12-3

1. *RC* oscillators are generally used in applications requiring a sine wave of less than __________.
2. For positive feedback, several *RC* phase-shift networks are needed to shift the feedback signal by __________ degrees.
3. Which *RC* oscillator circuit makes use of both positive and negative feedback?
4. At resonance, the Wien-Bridge oscillator's band-pass filter will shift the feedback signal by __________ degrees.

12-4 TROUBLESHOOTING OSCILLATOR CIRCUITS

Now that we know how these oscillator circuits are supposed to work, let us see how we can troubleshoot a problem when they do not work. First, let us review our basic troubleshooting procedure.

Step 1: DIAGNOSE

The first step is to determine whether a problem really exists. To carry out this step, a technician must collect as much information as possible about the system, circuit, and components used, and then diagnose the problem.

Step 2: ISOLATE

The second step is to apply a logical and sequential reasoning process to isolate the problem. In this step, a technician will operate, observe, test, and apply troubleshooting techniques in order to isolate the malfunction.

Step 3: REPAIR

The third and final step is to make the actual repair, then final test the circuit.

12-4-1 *Isolating the Problem to the Oscillator*

Once you have diagnosed that an oscillator circuit is malfunctioning, the next step is to isolate whether the problem is within the oscillator circuit, or external to the oscillator circuit. For example, Figure 12-14 shows the previously discussed Armstrong oscillator circuit, which receives power from the dc power supply and supplies an output signal to a load. Our first step is to isolate whether the problem is within the amplifier circuit or within one of these external circuits. Unlike the previous amplifier circuits, we will not need to isolate whether or not there is a problem in the signal source because the only input to an oscillator is dc power.

1. Power Supply or Oscillator? If the oscillator is not functioning as it should, first check the power ($+V_{CC}$) and ground connection to the circuit. If the dc power supply voltage to the circuit is not correct, you will next need to determine whether the problem is in the source (the dc power supply cir-

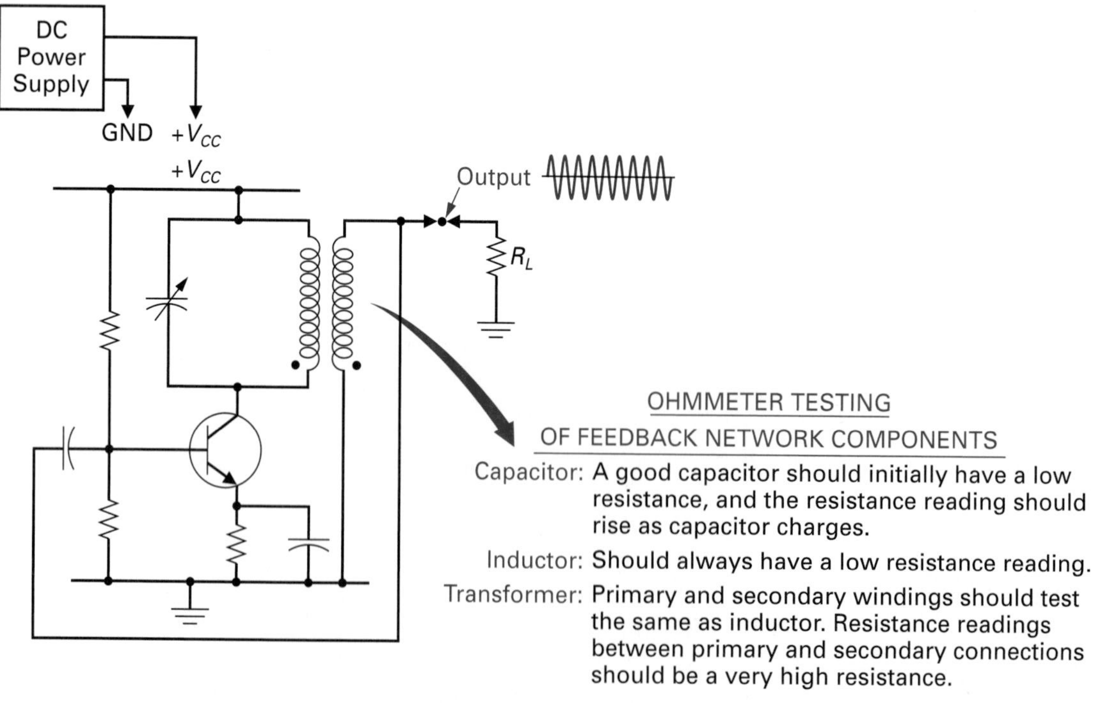
DC Power Supply
GND
$+V_{CC}$
$+V_{CC}$
Output
R_L
OHMMETER TESTING
OF FEEDBACK NETWORK COMPONENTS
Capacitor: A good capacitor should initially have a low resistance, and the resistance reading should rise as capacitor charges.
Inductor: Should always have a low resistance reading.
Transformer: Primary and secondary windings should test the same as inductor. Resistance readings between primary and secondary connections should be a very high resistance.

FIGURE 12-14 Troubleshooting Oscillator Circuits.

cuit) or in the load (the oscillator circuit). If the oscillator is disconnected from the dc power supply, and the voltage returns to its normal potential, the oscillator has developed some short within its circuit and this is pulling down the $+V_{CC}$ supply voltage. If on the other hand, the oscillator is disconnected from the dc power supply and the voltage supply voltage still remains at its incorrect value, the problem is within the dc power supply circuit.

2. Oscillator Source or Output Load? If the output signal from the oscillator is incorrect, remember that a bad load can pull down a voltage signal and make it appear as though the oscillator is faulty. When you have no output, or a bad output, you will first need to disconnect the output of the oscillator from the load to determine whether the problem is in the source (oscillator) or the load. If the load is disconnected and the signal is good, the impedance of the load is pulling down the output of the oscillator. On the other hand, if the load is disconnected and the signal output of the oscillator is still bad, the problem is in the oscillator.

12-4-2 *Troubleshooting the Oscillator Circuit*

Once you have isolated the problem to the oscillator circuit, you will next need to test the oscillator's dc bias and ac voltage signals at different points in the circuit to isolate the faulty component. Oscillator circuits are notorious as being difficult circuits to troubleshoot due to their feedback system of generating an output and then feeding back part of that output to generate an output, and so on. Because every component plays such an important role, any break due to a component failure in the loop will shut down the whole circuit. In most cases, a lack of output is generally caused by a failure of the transistor or feedback network components, which can be tested with the ohmmeter in the usual way detailed in Figure 12-14. If you do have to replace one of the feedback network's reactive components, remember that you will probably need to test the output frequency and retune if necessary.

SELF-TEST REVIEW QUESTIONS FOR SECTION 12-4

1. What test(s) can be performed on an oscillator circuit to isolate whether the fault is within the oscillator circuit or in the signal input?
2. Why are oscillator circuits difficult to troubleshoot?

SUMMARY

1. An oscillator is an electronic circuit that draws power from its dc supply to generate a continuously repeating ac output signal.
2. Oscillators are used to generate a transmitter's carrier frequency, to generate an RF signal within a receiver, or to generate an ac signal that can be distributed throughout an electronic system to time or synchronize operations.

3. Some oscillator circuits are designed to generate low-frequency oscillations of only a few hertz, while other high-frequency oscillators generate oscillations at several hundred gigahertz.
4. Oscillators are generally classified in one of three basic groups based on the frequency determining components used within the circuit. These three groups are

   LC Oscillators
   Crystal Oscillators
   RC Oscillators

LC Oscillators (Figure 12-15)

5. An *LC* tank circuit can be "shock excited" into oscillation by a momentary surge of current from a dc power switch. However, the oscillations will slowly decrease in amplitude due to the energy lost in the coil's resistance. To prevent the oscillations from the tank from being "damped" in this way, we need to somehow replace the energy lost.
6. Reinforcing energy that "regenerates" or "adds to" the sine-wave oscillations is called regenerative feedback or positive feedback. Most oscillator circuits operate on this positive feedback principle, which means that an in-phase signal must be constantly injected into the oscillator circuit in order to replace the energy lost and maintain oscillations.
7. With the Armstrong oscillator circuit, the 180° phase shift is provided by a transformer. If the phase-inverting transformer were not included and the collector output were fed directly back to the input, the feedback signal would oppose the input signal. This type of out-of-phase feedback is known as degenerative feedback or negative feedback and would result in a canceling of the input and therefore no output oscillations.
8. An oscillator's tuned tank, formed from a capacitor and inductor, sets the frequency at which the circuit oscillates. The capacitor's and inductor's values control the rate at which energy is transferred back and forth within the tank.
9. The transistor within an oscillator acts as a switch, which is only turned ON when the output signal is positive and when the transistor's collector current will assist the tank's circulating current in the charging of the capacitor.
10. Unlike the Armstrong oscillator, the Hartley oscillator does not require a transformer. The 180° phase inversion needed for positive feedback is achieved by this circuit's identifying feature, the tapped inductor L_1.
11. The Colpitts oscillator is similar to the Hartley; however, in this circuit two capacitors, which are grounded at the center, are the identifying feature of the oscillator. They provide the needed 180° phase shift for positive feedback.
12. The feedback factor describes the percentage of output returned to the input.

QUICK REFERENCE SUMMARY SHEET

FIGURE 12-15 ***LC* Oscillators.**

Armstrong Oscillator Circuit

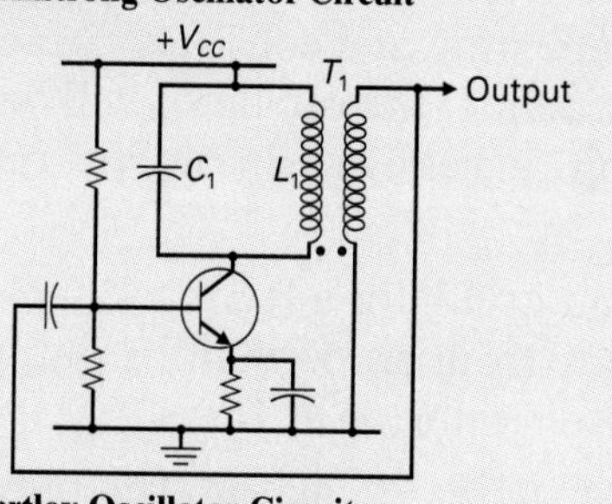

$$f_R = \frac{1}{2\pi\sqrt{L_1 \times C_1}}$$

Identifying Feature

Single transformer with parallel capacitor.

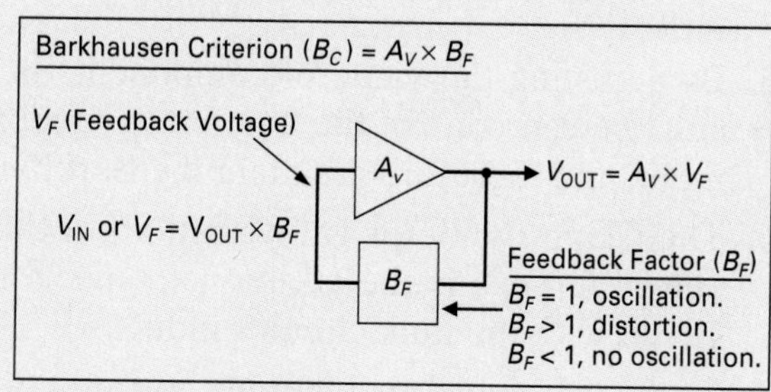

Hartley Oscillator Circuit

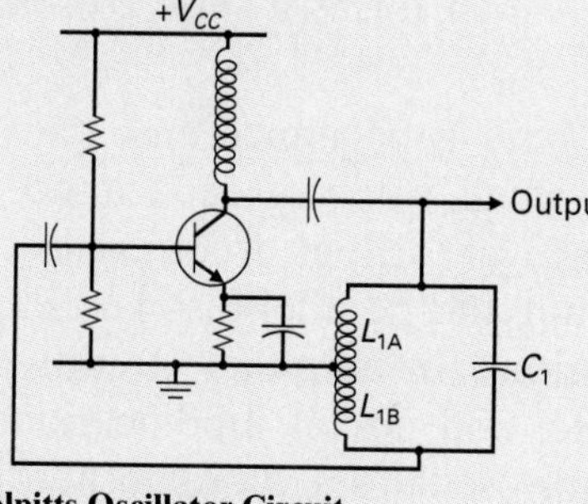

$$f_R = \frac{1}{2\pi\sqrt{L_T \times C_1}}$$

$$L_T = L_{1A} + L_{1B}$$

Identifying Feature

Tapped inductor with parallel capacitor.

Colpitts Oscillator Circuit

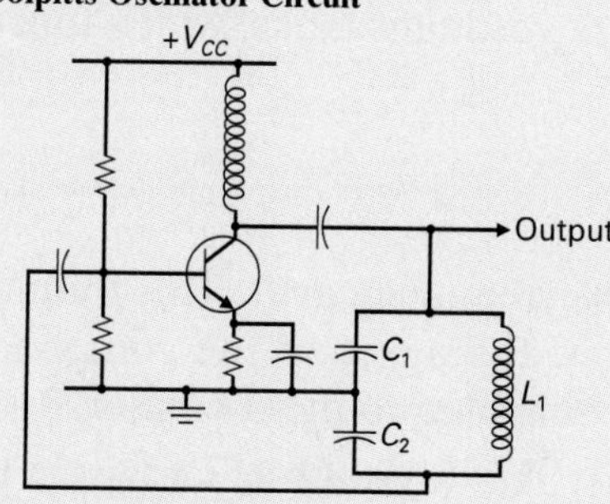

$$f_R = \frac{1}{2\pi\sqrt{L_1 \times C_T}} \qquad B_F = \frac{C_2}{C_1 + C_2}$$

$$C_T = \frac{C_1 \times C_2}{C_1 + C_2}$$

Identifying Feature

Two series connected capacitors, grounded at center, and a single inductor.

Clapp Oscillator Circuit

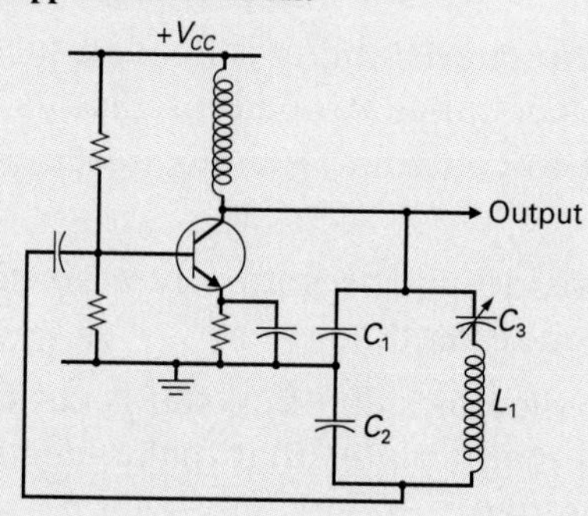

$$f_R = \frac{1}{2\pi\sqrt{L_1 \times C_3}}$$

Identifying Feature

Same circuit as Colpitts, except for capacitor in series with inductor.

13. The Barkhausen criterion (B_C) is the relationship between the oscillator's feedback factor (B_F) and the amplifier's voltage gain (A_V). This should be equal to 1 for proper oscillator operation.
14. If the Barkhausen criterion is less than 1, there will not be enough feedback to sustain oscillations. If the Barkhausen criterion is a perfect 1, the oscillator's sine-wave output will be continually repeating and consistent in amplitude. If the Barkhausen criterion is greater than 1, the oscillator will be driven into saturation and cutoff, causing excessive sine-wave output distortion.
15. By adjusting the value of components in an oscillator's feedback tank circuit, we can control the amount of feedback from an oscillator's output to input, and therefore maintain the Barkhausen criterion at 1.
16. The Clapp oscillator is a variation on the Colpitts oscillator. The only circuit difference is the additional capacitor, which forms a series resonance circuit with the tank circuit's inductor.
17. Transistors have an inherent collector to base capacitance and transistor lead inductance. At low frequencies, this small value of capacitance and inductance does not affect the circuit. However, at high frequencies these elements combine to produce series resonant and parallel resonant circuits that will cause unwanted parasitic oscillations. In amplifier circuits these high-frequency oscillations are undesirable, and circuit modifications are made to counter the effects. High-frequency oscillators, on the other hand, make use of the transistor circuit's inherent resonant tanks, which function as feedback networks.

Crystal Oscillators (Figure 12-16)

18. *LC* oscillator circuits typically have a 0.8% frequency drift. This frequency drift is due to circuit temperature changes, the aging of the components, and changes in the resistance of the load connected to the oscillator.
19. If a high degree of oscillator stability is needed, crystal oscillators are used because they will typically have a frequency drift of 0.0001%.
20. In applications where a high degree of frequency stability is needed such as communication and computer systems, crystal-controlled oscillators are used, which make use of a quartz crystal to control the circuit's frequency of operation.
21. The quartz crystal is made of silicon dioxide and is naturally a six-sided (hexagonal) compound with pyramids at either end.
22. To construct a crystal component, a thin slice or slab of crystal is cut from the mother stone, mounted between two metal plates that make electrical contact, and then placed in a protective holder.
23. On a schematic diagram, a crystal is generally labeled either "XTAL" or "*Y*."
24. The crystal is basically operated as a transducer, or energy converter, transforming mechanical energy to electrical energy, or electrical energy to mechanical energy.

QUICK REFERENCE SUMMARY SHEET

FIGURE 12-16 Crystal Oscillators.

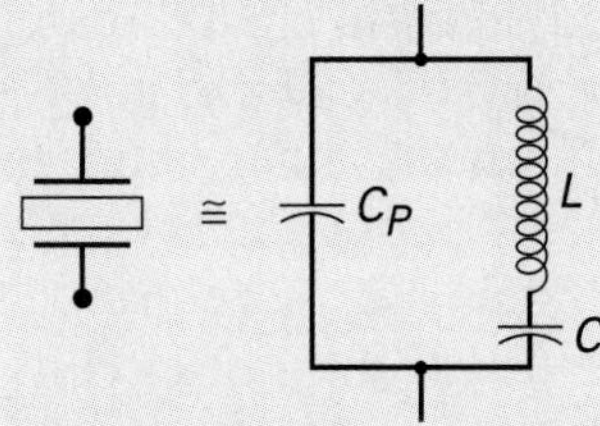

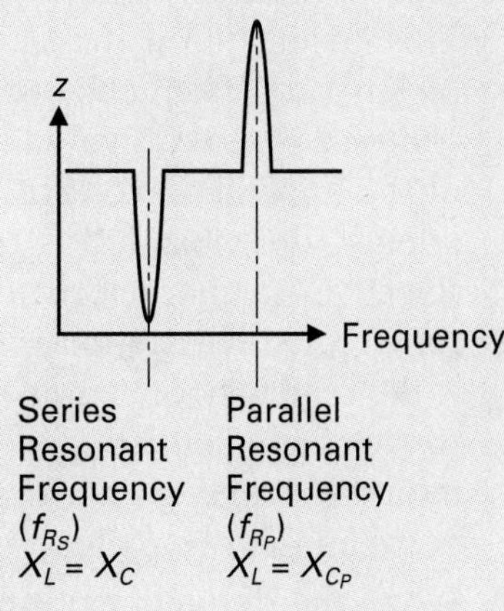

Hartley Crystal Oscillator Circuit

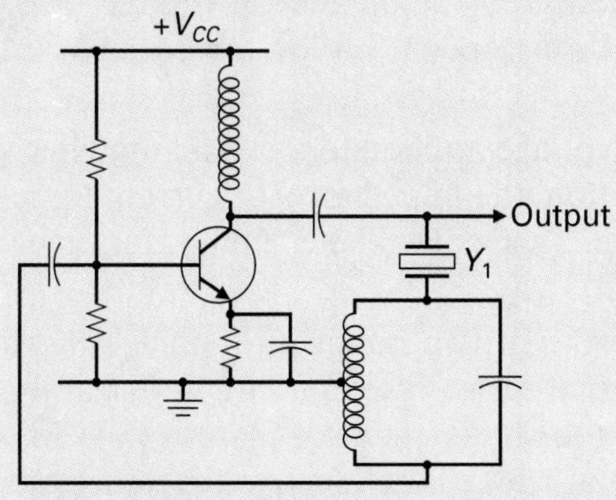

Colpitts Crystal Oscillator Circuit

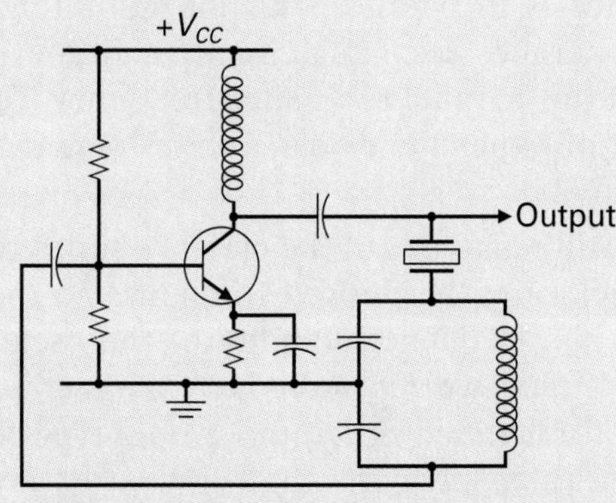

Pierce Crystal Oscillator Circuit

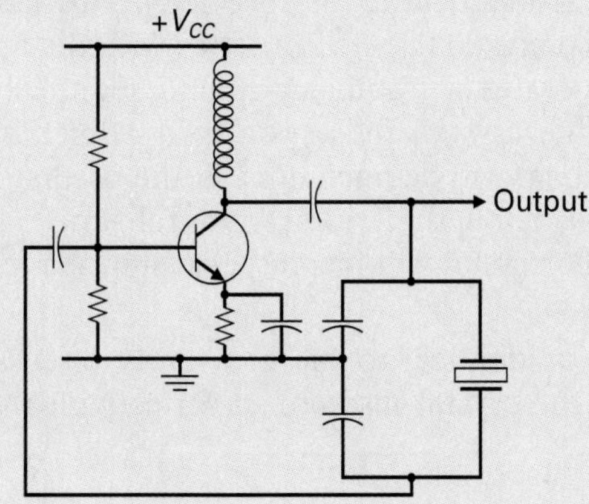

25. If a crystal were subjected to an alternating pressure that caused it to continually expand and compress, an alternating or ac voltage would be generated.

26. Due to their composition, crystals have a "natural frequency of vibration." This means that if the ac voltage applied across a crystal matches the crystal's natural frequency of vibration, the crystal will physically expand and contract by a relatively large amount. If, on the other hand, the frequency of the applied ac voltage is either above or below the crystal's natural frequency, the vibration is only slight. Crystals are therefore very frequency selective—a characteristic we can make use of in filter circuits and oscillator circuits. This action is called the piezoelectric effect, which by definition is the tendency of a crystal to vibrate at a constant rate when it is subjected to a changing electric field produced by an applied ac voltage. The crystal's natural frequency of vibration is dependent on the thickness of the crystal between the two plates.

27. When the frequency of the ac voltage applied to a crystal matches the crystal's natural frequency of vibration, the impedance, or current opposition of the crystal, drops to a minimum. This makes the crystal wafer or slab equivalent to a series resonant circuit.

28. When a crystal is mounted between two connecting plates, the device is equivalent to the crystal's original series resonant circuit in parallel with a small value of capacitance due to the connecting plates (C_P). At a higher frequency, there will be a point at which the inductive reactance of the crystal is equal to the capacitive reactance of the connecting plates ($X_L = X_{C_P}$). Because the crystal's inductance is in parallel with the mounting plate's capacitance, the two form a parallel resonant circuit, and the device's impedance will be maximum at this parallel resonance frequency (f_{R_P}). Circuit applications can make use of either the crystal's series resonant selectivity or parallel resonant selectivity.

29. Most crystal-oscillator circuits will make use of the crystal's series resonant frequency response to feed back only the desired frequency to the input, and therefore tightly maintain the frequency stability of the oscillator. A crystal's series resonant tuned circuit will typically have a very high Q of 40,000. If you compare this to an LC tuned circuit that would typically have a Q of 200, it is easy to see why crystal-controlled oscillators are much more stable than LC-controlled oscillators.

30. The Hartley crystal oscillator is identical to the previously discussed LC Hartley oscillator, except for the crystal inserted in series with the feedback path. When the oscillator generates a frequency that is identical to the series resonant frequency of the crystal, the crystal will offer almost no opposition and the feedback signal will be maximum. If the oscillator's frequency begins to drift above or below the crystal's natural series resonant frequency, the impedance of the crystal will increase, causing the feedback to decrease.

31. The Colpitts crystal oscillator is identical to the previously discussed LC Colpitts oscillator, except for the crystal inserted in series with the feedback path.

32. The Pierce crystal oscillator is almost identical to the Colpitts oscillator except for crystal Y_1 which replaces the tank's coil. Crystal Y_1 will be operating at its parallel resonant frequency, which means that it will offer a very high impedance to the oscillator's generated frequency. This action is not any different from an *LC* parallel resonant tank, which would also have a high impedance at resonance. However, now that a crystal controls the tank's impedance, the tuned circuit will be much more selective, and the oscillator's output will be more stable.

RC Oscillators (Figure 12-17)

33. Although *RC* oscillators are not as stable as *LC* or crystal oscillators, they are inexpensive and are ideal in applications where we need to generate low-frequency oscillations of less than 10 kHz.
34. When a voltage is applied to a purely capacitive circuit, the current leads the voltage by 90°. In a purely resistive circuit, the circuit current and voltage developed across a resistor are always in phase. In a resistive-capacitive (*RC*) network, the circuit current leads the voltage by some value between 0° and 90°, depending on the values of resistance and capacitance.
35. If three 60° phase-shift networks were connected in series, the combined phase-shift would be 180°. This 180° phase-shift is exactly what is needed for positive feedback in an oscillator circuit. Therefore, if this network were placed between the collector output and base input of a common-emitter amplifier, the result would be a phase-shift oscillator circuit.
36. The *RC* phase-shift oscillator is typically only used as a fixed-frequency oscillator because any changes in the resistance or capacitance in the feedback network will interfere with the regenerative feedback.
37. The Wien-Bridge oscillator, which also makes use of *RC* networks in the feedback path, can be used to generate sine-wave signal outputs of up to 1 MHz.
38. The band-pass filter network in a Wien-Bridge oscillator is used to provide a "positive feedback signal," which produces oscillations, while a voltage-divider network is used to provide a "negative feedback signal," which controls the voltage gain of transistor Q_1. The positive feedback signal is applied to the base of the oscillating transistor, while the negative feedback signal is applied to the emitter of the oscillating transistor. Because the band-pass filter in the bridge network will not provide any phase shift, a common-emitter signal inverting transistor is included to provide the 180° phase-shift that is needed for positive feedback.

Troubleshooting Oscillator Circuits

39. An oscillator receives power from the dc power supply and supplies an output sine-wave signal to a load. Our first step is to isolate whether the problem is within the amplifier circuit or within one of these external circuits. Unlike the previous amplifier circuits, we will not need to isolate whether or not there is a problem in the signal source because the only input to an oscillator is dc power.

QUICK REFERENCE SUMMARY SHEET

FIGURE 12-17 ***RC* Oscillators.**

The Phase-Shift Oscillator Circuit

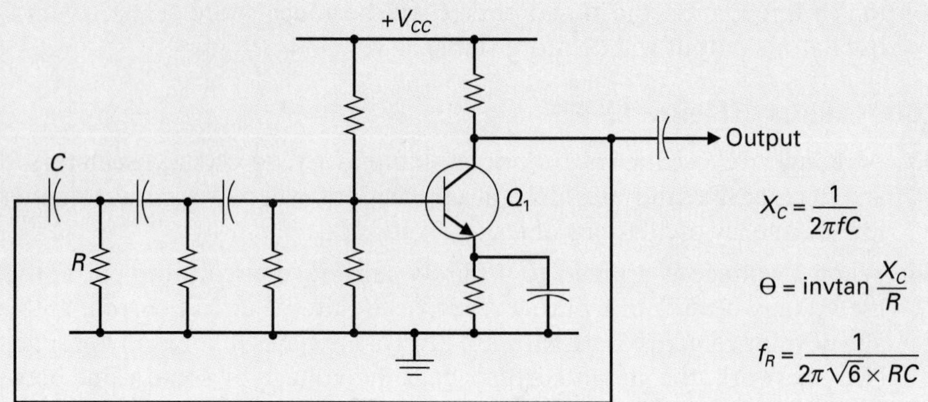

The Wein-Bridge Oscillator Circuit

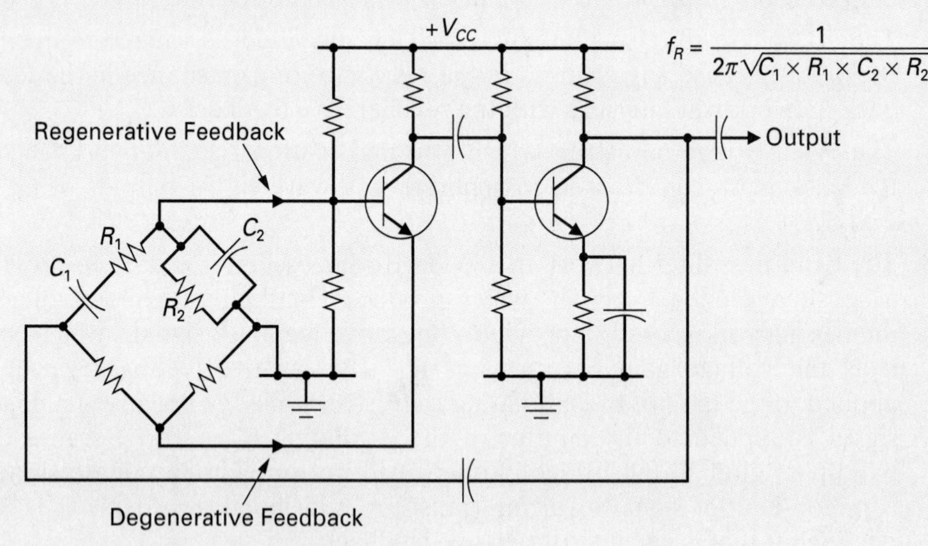

40. Oscillator circuits are notorious as being difficult circuits to troubleshoot, due to their feedback system. Because every component plays such an important role, any break due to a component failure in the loop will shut down the whole circuit.

NEW TERMS

Armstrong Oscillator
Barkhausen Criterion
Clapp Oscillator
Colpitts Crystal Oscillator
Colpitts Oscillator
Crystal Oscillator
Degenerative Feedback
Feedback Factor
Feedback Voltage
Frequency Drift
Frequency of Oscillation
Frequency Stability
Hartley Crystal Oscillator
Hartley Oscillator
LC Oscillator
Negative Feedback
Oscillator Circuit
Parasitic Oscillations
Phase-Shift Network
Phase-Shift Oscillator
Pierce Crystal Oscillator
Piezoelectric Effect
Positive Feedback
Positive Feedback Signal
Quartz Crystal
RC Oscillator
Regenerative Feedback
Wien-Bridge Oscillator

REVIEW QUESTIONS

Multiple-Choice Questions

1. An oscillator's circuit basically converts

a. ac to dc **c.** ac to ac
b. dc to ac **d.** dc to dc

2. To oscillate, a tuned circuit will need an amplifier with

a. Regenerative feedback **c.** Negative feedback
b. Degenerative feedback **d.** Both (a) and (c) are true

3. Which of the *LC* oscillators makes use of a tuned transformer?

a. Hartley oscillator **c.** Armstrong oscillator
b. Colpitts oscillator **d.** Clapp oscillator

4. Which *LC* oscillator type makes use of a tapped inductor in the tuned circuit?

a. Hartley oscillator **c.** Armstrong oscillator
b. Colpitts oscillator **d.** Clapp oscillator

5. The ____________ has an additional series resonant circuit in parallel with two series connected capacitors.

a. Hartley oscillator **c.** Armstrong oscillator
b. Colpitts oscillator **d.** Clapp oscillator

6. The ____________ will produce an output sine-wave frequency that is determined by the values of an inductor in parallel with two series connected capacitors.

a. Hartley oscillator **c.** Armstrong oscillator
b. Colpitts oscillator **d.** Clapp oscillator

7. An amplifier is necessary in an oscillator circuit to replace circuit losses.

a. True **b.** False

8. In order to produce a sine wave output that is consistent in amplitude, the oscillator's Barkhausen criterion must be

a. Less than 1 **c.** Greater than 1
b. 23 **d.** 1

9. If the feedback in an oscillator circuit is too little, the output waveform will be

a. Constant **c.** Damped
b. Distorted **d.** All of the above

10. If the feedback in an oscillator circuit is too much, the output waveform will be

a. Constant **c.** Damped
b. Distorted **d.** All of the above

11. When an oscillator circuit is first turned ON, the transistor will

a. Conduct **c.** Block current to the tank
b. Turn OFF **d.** All of the above

12. An oscillator circuit replaces the energy lost in the tank (due to its resistance) during the time the transistor is

a. OFF **b.** ON

13. If a high-frequency stable output is needed in a circuit application, the ideal choice is a(an) ____________ oscillator.

a. *LC* **b.** Crystal **c.** *RC*

14. If a low frequency oscillation of approximately 120 Hz is needed, the ideal choice would be a(an) ____________ oscillator.

a. *LC* **b.** Crystal **c.** *RC*

15. In applications where a high-frequency sine wave oscillation is needed, and the frequency stability is not important, a(an) ____________ oscillator is used.

a. *LC* **b.** Crystal **c.** *RC*

16. If the ac signal frequency applied to a crystal matches the crystal's natural series resonant frequency, the crystal's impedance will be

a. maximum **b.** minimum **c.** unaltered **d.** zero

17. A crystal oscillator's high-frequency stability is due to the

a. crystal's high *Q* **c.** crystal's selectivity
b. transistor's gain **d.** Both (a) and (c) are true

18. Which of the following is an *RC* oscillator?

a. Hartley **c.** Phase-shift
b. Pierce **d.** Colpitts

19. The Wien-Bridge oscillator uses a ____________ to provide the 180° phase shift needed for positive feedback.

a. *RC* network **c.** tapped inductor
b. tuned tank **d.** common-emitter transistor

20. When a phase-shift oscillator is producing an output that matches the circuit's resonant frequency, the feedback voltage is

a. maximum **c.** zero
b. minimum **d.** both (b) and (c) are true

Essay Questions

21. What is an oscillator? (Intro.)

22. List the three basic oscillator groups. (Intro.)

23. Using a sketch, describe how an *LC* tank circuit can be shock excited into oscillation. (12-1-1)

24. What is positive feedback? (12-1-1)

25. Sketch the Armstrong *LC* oscillator's circuit, and then briefly describe how the circuit operates. (12-1-2)

26. How does an oscillator circuit synchronize the turning ON and OFF of its transistor? (12-1-2)

27. Why does the tank circuit in an *LC* oscillator set the operating frequency? (12-1-2)

28. Describe how the transistor in an oscillator circuit operates. (12-1-2)

29. What is negative feedback? (12-1-2)

30. Sketch the Hartley *LC* oscillator circuit, and describe its operation. (12-1-3)

31. Sketch the Colpitts *LC* oscillator circuit, and describe its operation. (12-1-4)

32. Why is the feedback factor crucial to the correct operation of an oscillator circuit? (12-1-5)

33. Describe the Barkhausen criterion. (12-1-5)

34. Sketch the Clapp oscillator circuit, and describe its operation. (12-1-6)

35. What are parasitic oscillations? (12-1-7)

36. What advantage does a crystal oscillator have over an *LC* oscillator? (12-2)

37. Define the term "frequency stability." (12-2)

38. What is piezoelectric effect? (12-2-1)

39. What type of *LC* tank circuit is a crystal component equivalent to? (12-2-1)

40. Sketch and describe the operation of the following crystal oscillator circuits:

- **a.** The Hartley crystal oscillator (12-2-2)
- **b.** The Colpitts crystal oscillator (12-2-3)
- **c.** The Pierce crystal oscillator (12-2-4)

41. What is the key advantage of *RC* oscillators? (12-3)

42. Sketch the phase-shift *RC* oscillator circuit, and describe its operation. (12-3-1)

43. What is the phase shift between input and output of a band-pass filter when the input signal frequency is equal to the band-pass filter's resonant frequency? (12-3-2)

44. Sketch the *RC* Wien-Bridge oscillator circuit, and describe its operation. (12-3-2)

45. Why does the Wien-Bridge *RC* oscillator have both a positive and negative feedback path? (12-3-2)

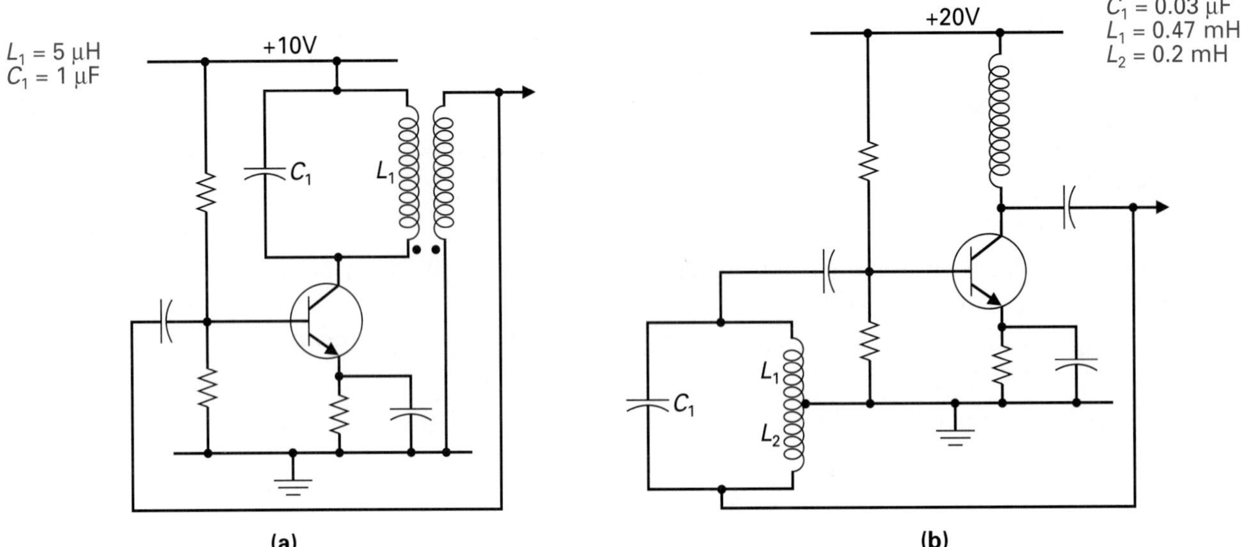

FIGURE 12-18 ***LC* Oscillator Circuits.**

Practice Problems

46. Identify the circuits shown in Figure 12-18, and then calculate their operating frequency.

47. Identify the circuit shown in Figure 12-19, and then calculate the circuit's

  a. Operating frequency

  b. Feedback factor

  c. Barkhausen criterion

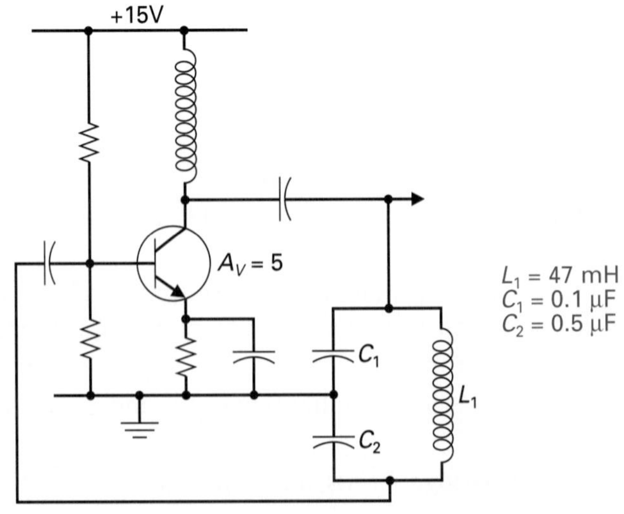

FIGURE 12-19 **An *LC* Oscillator.**

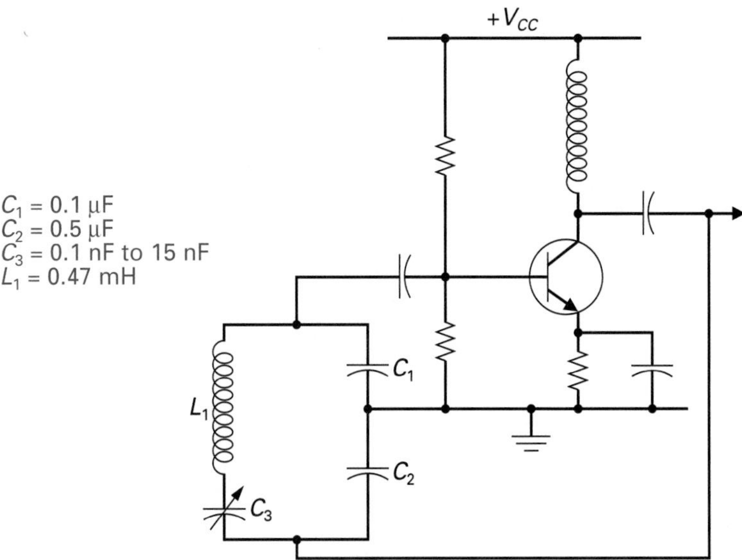

FIGURE 12-20 An *LC* Oscillator.

48. Identify the circuit shown in Figure 12-20, and then calculate its operating frequency range.

49. Identify the circuits shown in Figure 12-21, and then determine the operating frequency of these oscillator circuits.

50. Identify the circuits shown in Figure 12-22, and then calculate their operating frequency.

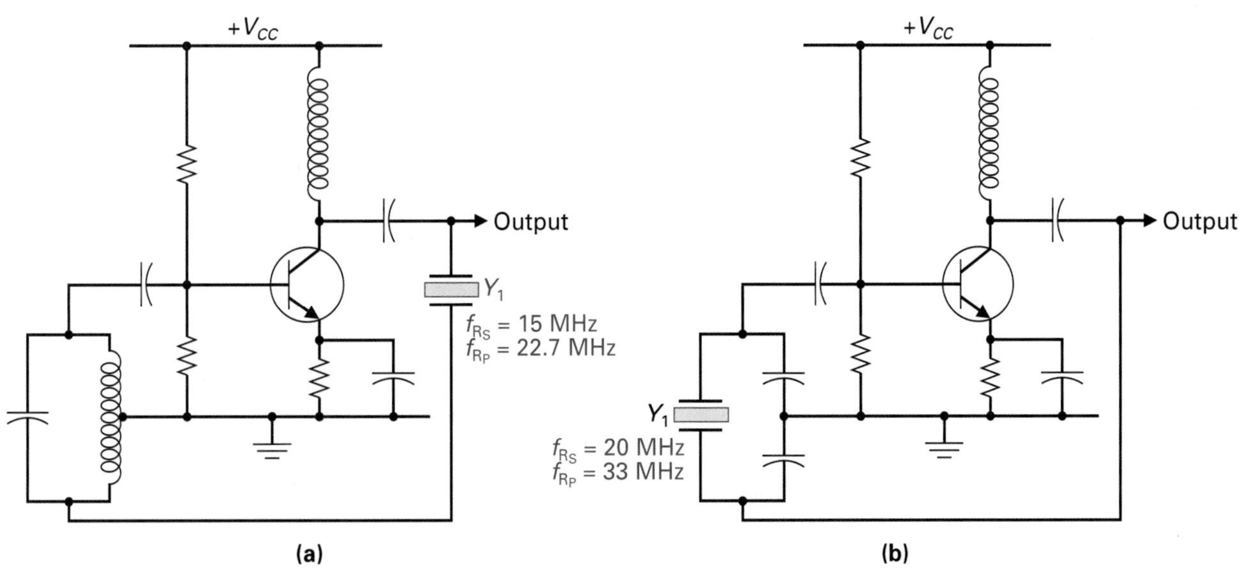

FIGURE 12-21 Crystal Oscillator Circuits.

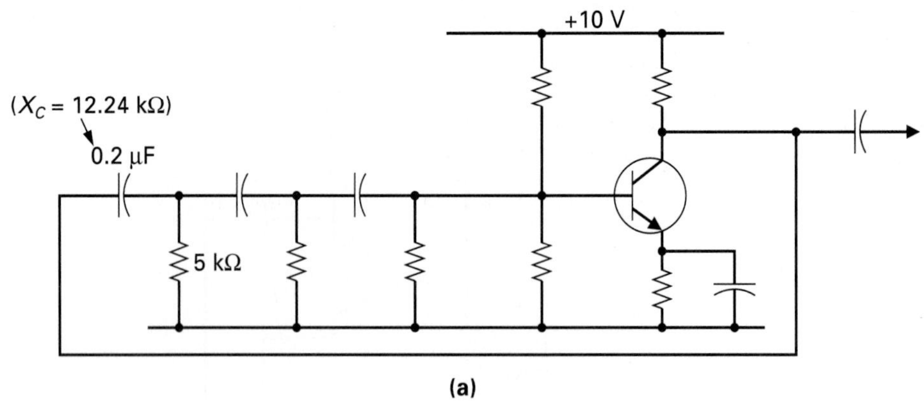

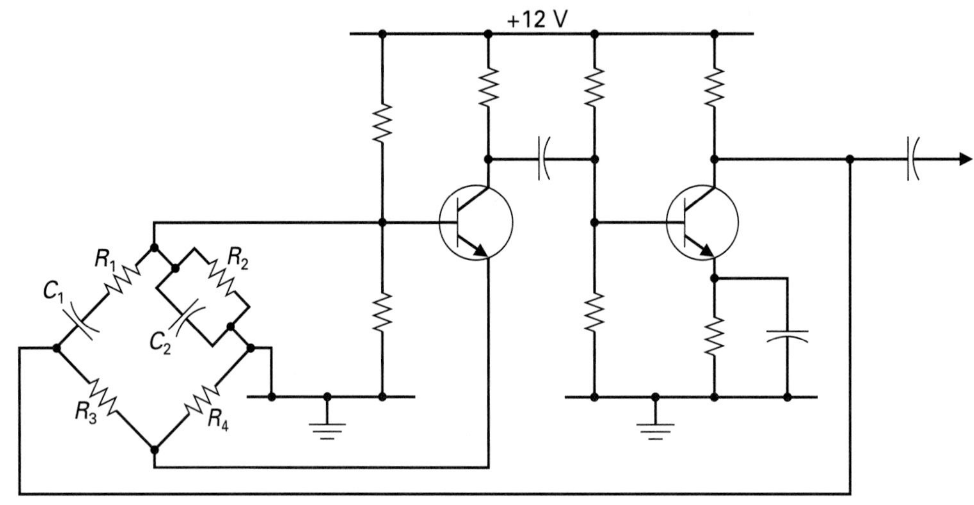

FIGURE 12-22 ***RC* Oscillator Circuits.**

Troubleshooting Questions

51. What is the three step troubleshooting procedure? (12-4)
52. What test should be first performed to determine whether the fault is within the oscillator circuit or external to the oscillator circuit? (12-4-1)
53. If a short existed in the load of an oscillator, what would be the output of the oscillator circuit, and how could this problem be isolated? (12-4-1)
54. Why are oscillator circuits normally difficult to troubleshoot? (12-4-2)
55. Identify the oscillator circuit in Figure 12-23, and then explain what circuit malfunction could be causing no sine-wave output, based on the voltmeter reading given relative to ground.

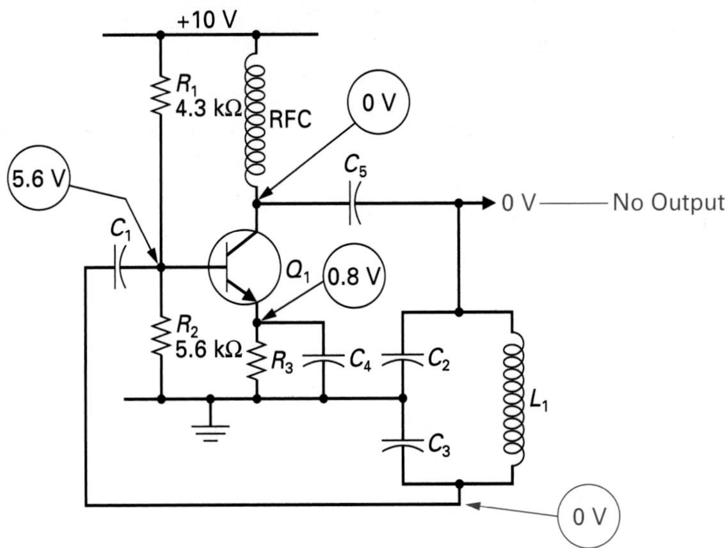

FIGURE 12-23 Troubleshooting Oscillator Circuits.

CHAPTER

13

OBJECTIVES

After completing this chapter, you will be able to:

1. Name the two different types of field effect transistors.
2. Describe the physical construction and operation of the junction FET (JFET), and name and identify its three terminals.
3. Describe why the term "field effect" is used in the name of an FET and why the term "junction" is used in JFET.
4. Explain how the JFET operates.
5. Explain the JFET operation, and the following characteristics:
   a. V_P, V_{BR}, I_{DSS}, $V_{GS(\mathrm{OFF})}$
   b. Transconductance
   c. High-input impedance
6. Interpret the JFET specifications given in a typical manufacturer's data sheet.
7. Calculate the different values of circuit voltage and current for the following transistor biasing methods:
   a. Gate Biasing
   b. Self Biasing
   c. Voltage-Divider Biasing
8. Identify and list the characteristics of the following three FET circuit configurations:
   a. Common Source
   b. Common Gate
   c. Common Drain
9. Explain how the JFET's characteristics can be used in digital and analog circuit applications.
10. Describe how to test a JFET and how to troubleshoot JFET circuits.

Junction Field Effect Transistors (JFETs)

OUTLINE

VIGNETTE

Spitting Lighting Bolts

Nikola Tesla was born in Yugoslavia in 1856. He studied mathematics and physics in Prague, and in 1884 he emigrated to the United States.

In New York he met Thomas Edison, the self-educated inventor who is best known for his development of the phonograph and the incandescent light bulb. Both men were gifted and eccentric and, due to their common interest in "invention," they got along famously. Because Tesla was unemployed, Edison offered him a job. In his lifetime, Edison would go on to take out 1,033 patents and become one of the most prolific inventors of all time. Tesla would go on to invent many different types of motors, generators, and transformers, one of which is named the "Tesla coil" and produces five-foot lighting bolts. With this coil, Tesla investigated "wireless power transmission," the only one of his theories that has not come into being.

Both Tesla and Edison had very strong views on different aspects of electricity and, as time passed, the two men began to engage in very long, loud, and angry arguments. One such discussion concerned whether power should be distributed as alternating current or direct current. Eventually, the world would side with Tesla and choose ac. At the time, this topic, like many others, would cause a hatred to develop between the two men. Eventually Tesla left Edison and started his own company. However, the anger remained, and on one occasion when they were both asked to attend a party for their friend Mark Twain, both refused to come because the other had been invited.

In 1912, Tesla and Edison were both nominated for the Nobel prize in physics, but because neither one would have anything to do with the other, the prize went to a third party—proving that bitterness really will cause a person to cut off his nose to spite his face.

When angry count up to four; when very angry, swear.
Mark Twain

INTRODUCTION

In the five previous chapters, we have concentrated on all aspects of the bipolar junction transistor or BJT. We have examined its construction, operation, characteristics, testing, analog and digital circuit applications such as logic gates, amplifiers, and oscillators, and the troubleshooting of these circuits. In the following two chapters we will examine another type of transistor called the *field effect transistor,* which is more commonly called an FET (pronounced "eff-ee-tee"). Like the BJT, the FET has three terminals and can operate as a switch and can be used in digital circuit applications. It can also operate as a variable resistor and be used in analog or linear circuit applications. In fact, as we step through these next two chapters, you will see many similarities between the BJT and FET. You will also notice a few distinct differences between these two transistor types, and these differences are what make the BJT ideal in some applications and the FET ideal in other applications.

There are two types of field effect transistors or FETs. One type is the *junction field effect transistor,* which is more typically called a JFET (pronounced "jay-fet"). The other type is the *metal oxide semiconductor field effect transistor,* which is more commonly called a MOSFET (pronounced "moss-fet"). In this chapter we will concentrate on the JFET, and in the following chapter we will examine the MOSFET.

Junction Field Effect Transistor (JFET)
A field-effect transistor made up of a gate region diffused into a channel region. When a control voltage is applied to the gate, the channel is depleted or enhanced, and the current between source and drain is thereby controlled.

13-1 JUNCTION FIELD EFFECT TRANSISTOR (JFET) BASICS

Like the bipolar junction transistor, the **junction field effect transistor** or **JFET** is constructed from *n*-type and *p*-type semiconductor materials. However, the JFET's construction is very different from the BJT's construction, and therefore we will need to first see how the JFET device is built before we can understand how it operates.

13-1-1 *JFET Construction*

n-Channel JFET
A junction field effect transistor having an *n*-type channel between source and drain.

p-Channel JFET
A junction field effect transistor having a *p*-type channel between source and drain.

Gate
One of the field effect transistor's electrodes (also used for thyristor devices).

Source
One of the field effect transistor's electrodes.

Drain
One of the field effect transistor's electrodes.

Channel
A path for a signal.

Just as the bipolar junction transistor has two basic types (NPN BJT or PNP BJT), there are two types of junction field effect transistor called the ***n*-channel JFET** and ***p*-channel JFET.** The construction and schematic symbol for these two JFET types are shown in Figure 13-1.

To begin with, let us examine the construction of the more frequently used *n*-channel JFET, shown in Figure 13-1(a). This type of JFET basically consists of an *n*-type block of semiconductor material on top of a *p*-type substrate, with a "U" shaped *p*-type section attached to the surface of a *p*-type substrate. Like the BJT, the JFET has three terminals called the **gate, source,** and **drain.** The gate lead is attached to the *p*-type substrate, and the source and drain leads are attached to either end of an *n*-type **channel** that runs through the middle of the "U" shaped *p*-type section. In the simplified two-dimensional view in the inset in Figure 13-1(a), you can see the *n*-type channel that exists between the *n*-channel JFET's source and drain. The schematic symbol for the *n*-channel JFET is shown in Figure 13-1(b), and, as you can see, the gate lead's arrowhead points into the device. To aid your memory, you can imagine this arrowhead as a P-N junction diode as shown in the inset in Figure 13-1(b). The gate lead is connected to the diode's anode, which is a *p*-type material, and the source and drain leads are connected to either end of the diode's cathode, which is an *n*-type material. Because the source-to-drain channel is made from an *n*-type material, this is an *n*-channel JFET. The *p*-type gate and *n*-type source and drain makes this *n*-channel JFET equivalent to an NPN BJT, which has a *p*-type base and *n*-type emitter and collector, as shown in the inset in Figure 13-1(b).

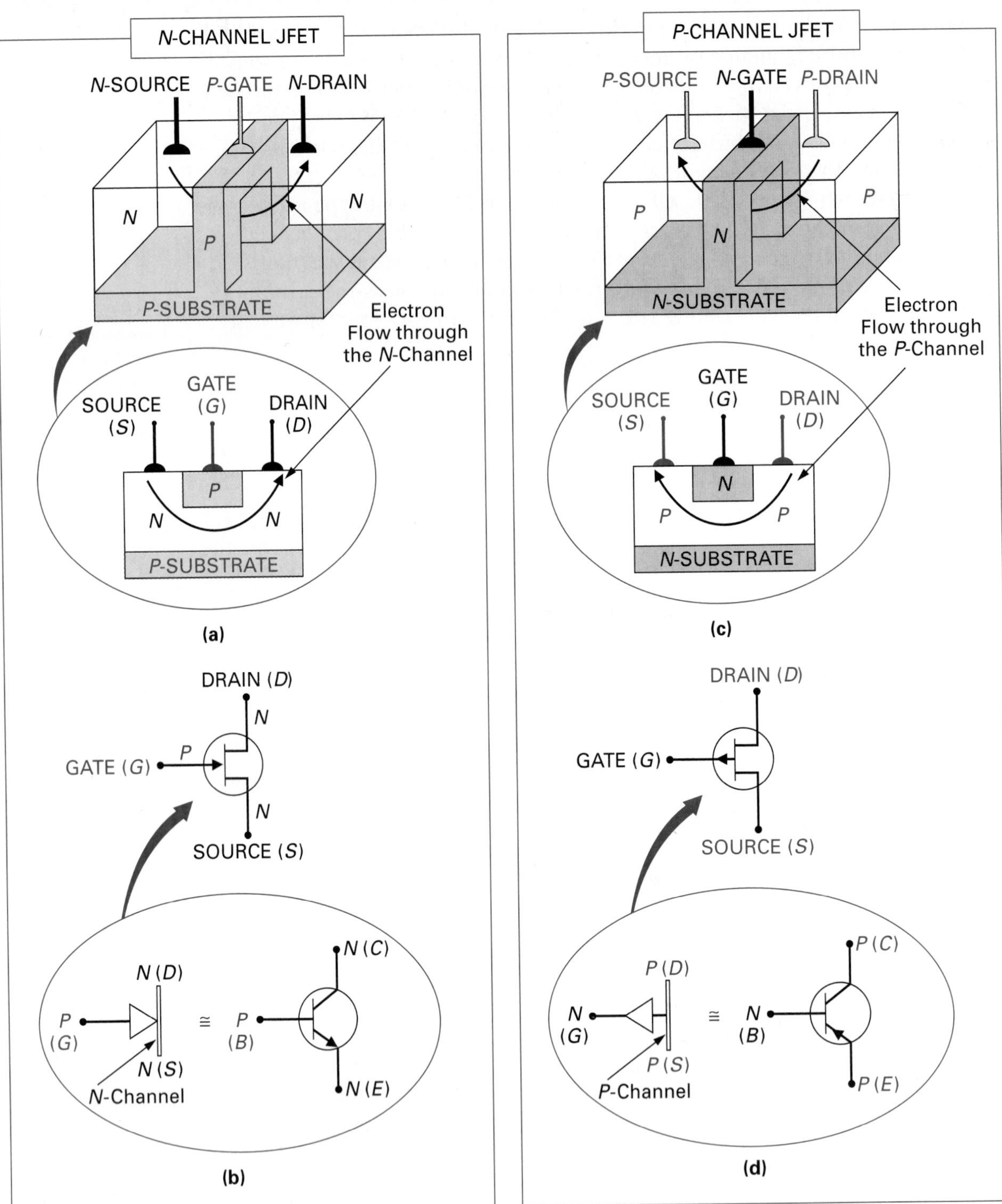

FIGURE 13-1 The Junction Field Effect Transistor (JFET) Types.

Figure 13-1(c) shows the construction of the *p*-channel JFET. This JFET type is constructed in exactly the same way as the *n*-channel JFET except that the gate lead is attached to an *n*-type substrate, and the source and drain leads are attached to either end of a *p*-type channel. Looking at the schematic symbol for the *p*-channel JFET in Figure 13-1(d), you can see that the gate's arrowhead points out of the device. To help you distinguish the symbols used for the *n*-channel JFET from the *p*-channel JFET, once again imagine the arrowhead as a P-N junction diode as shown in the inset in Figure 13-1(d). Because the gate lead is connected to the diode's cathode, the gate must be an *n*-type material. Because the source and drain leads are connected to either end of the diode's cathode, the source-to-drain channel is therefore made from a *p*-type material, and this is a *p*-channel JFET. The *n*-type gate and *p*-type source and drain makes this *p*-channel JFET equivalent to a PNP BJT, which has an *n*-type base and *p*-type emitter and collector, as shown in the inset in Figure 13-1(d).

13-1-2 *JFET Operation*

As you know, an NPN bipolar transistor needs both a collector supply voltage ($+V_{CC}$) and a base-emitter bias voltage (V_{BE}) in order to operate correctly. The same is true for the JFET, which requires both a **drain supply voltage ($+V_{DD}$)** and a **gate-source bias voltage (V_{GS})**, as shown in Figure 13-2(a). The $+V_{DD}$ bias voltage is connected between the drain and source of the *n*-channel JFET and will cause a current to flow through the *n*-channel. This source-to-drain current—which is made up of electrons because they are the majority carriers within an n-type material—is called the JFET's **drain current (I_D)**. The value of drain current passing through a JFET's channel is dependent on two elements: the value of $+V_{DD}$ applied between the drain and source, and the value of V_{GS} applied between gate and source. Let us examine in more detail why these applied voltages control the value of drain current passing through the JFET's channel.

Drain-Supply Voltage ($+V_{DD}$)
The bias voltage connected between the drain and source of the JFET, which causes current to flow.

Gate-Source Bias Voltage (V_{GS})
The bias voltage applied between the gate and source of a field effect transistor.

Drain Current (I_D)
A JFET's source-to-drain current.

The Relationship between + V_{DD} and I_D (Figure 13-2)

The value of $+V_{DD}$ controls the amount of drain current between source and drain because it is this supply voltage that controls the potential difference applied across channel, as seen in Figure 13-2(a). Therefore, an increase in the voltage applied across the JFET's drain and source ($+V_{DD}\uparrow$) will increase the amount of drain current ($I_D\uparrow$) passing through the channel. Similarly, a decrease in the voltage applied across the JFET's drain and source ($+V_{DD}\downarrow$) will decrease the amount of drain current ($I_D\downarrow$) passing through the channel. In most cases, a schematic diagram will show the V_{DD} supply voltage connection to the JFET as a source connection to ground and a drain connection up to $+V_{DD}$, as shown in Figure 13-2(b).

The Relationship between V_{GS} and I_D (Figure 13-3)

The value of V_{GS} controls the amount of drain current between source and drain because it is this voltage that controls the resistance of the channel. Figure 13-3(a) shows that when the V_{GS} bias voltage is 0 volts (which is the same as

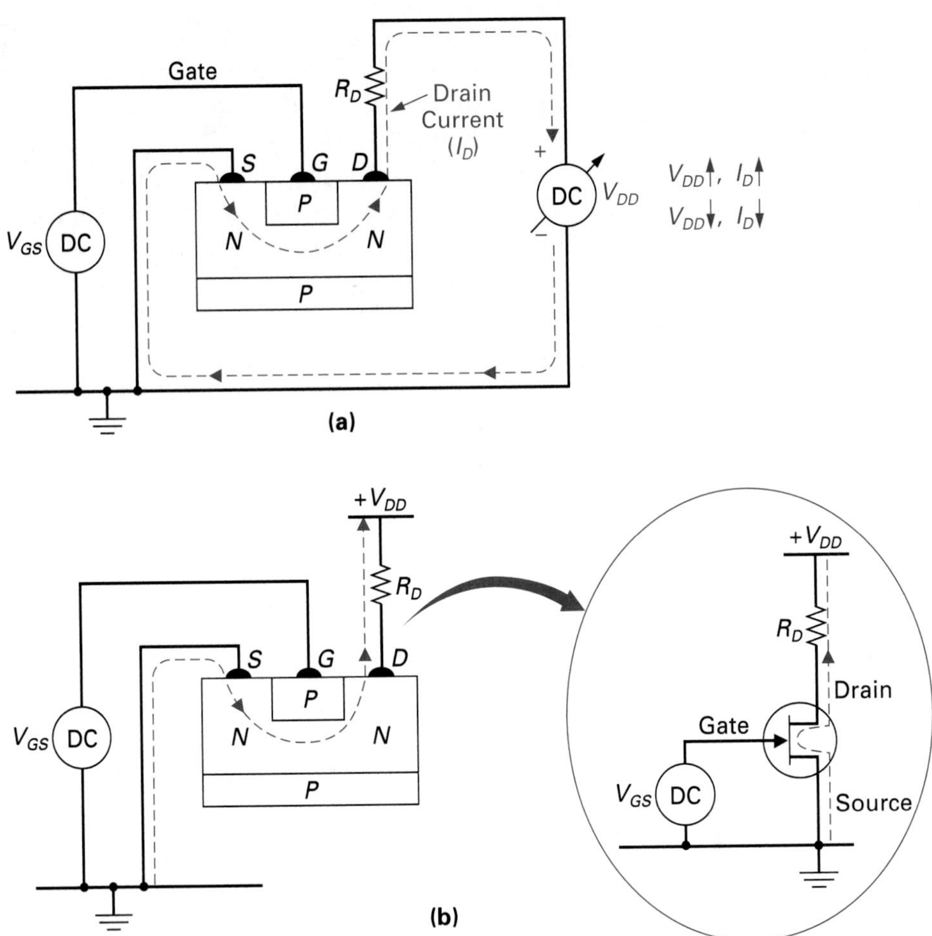

FIGURE 13-2 The Relationship Between a JFET's DC Supply Voltage ($+V_{DD}$) and Output Current (I_D).

connecting the gate to ground), there is no potential difference between the gate and source, and so the gate-to-source P-N junction will be reverse biased. As with all reverse-biased junctions, a small depletion layer will form and spread into the channel. Although it appears as though two depletion regions exist, in fact they are both part of the same depletion region that extends around the wall of the *n*-channel. This extremely small depletion region will offer very little opposition to I_D, and so drain current will be large. In Figure 13-3(b), V_{GS} is increased to –2 V (made more negative), and therefore the gate-to-source P-N junction will be further reverse biased. This causes an increase in the depletion region, decrease in the channel's width, and a decrease in drain current. In Figure 13-3(c), V_{GS} is increased to –4 V, and therefore the gate-to-source P-N junction will be further reverse biased, causing an increase in the depletion region, decrease in the channel's width, and a further decrease in drain current.

In most circuit applications, the $+V_{DD}$ supply voltage is maintained constant and the V_{GS} input voltage is used to control the resistance of the channel and the value of the output current, I_D. This can be seen more clearly in the insets in Figure 13-3. Because the output voltage (V_D or V_{OUT}) is dependent on

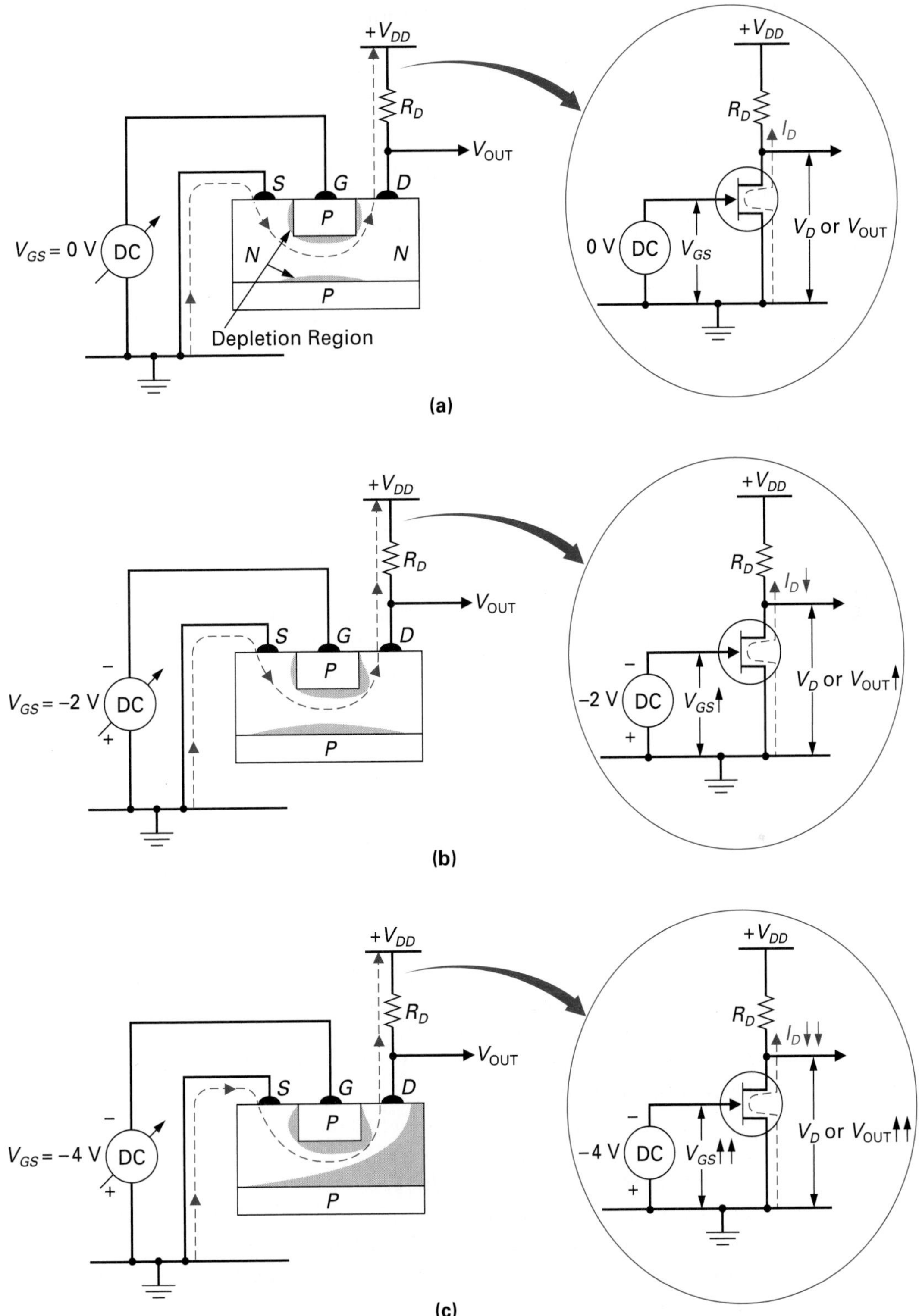

FIGURE 13-3 The Relationship Between a JFET's Input Voltage (V_{GS}) and Output Current (I_D).

the resistance between the JFET's source and drain, by controlling the value of I_D, we can control the output voltage. For instance, if the input voltage V_{GS} is made more negative, the resistance of the channel will be increased, causing the output current I_D to decrease, and therefore the voltage developed between drain and ground (V_D or V_{OUT}) to increase. The gate-to-source junction of an FET is normally always reverse biased by the input voltage (V_{GS}), and it is this input voltage that controls the output current (I_D) and output voltage (V_{OUT}). Because the gate-to-source junction of an FET is normally always reverse biased by the input voltage (V_{GS}), there will be no input current. This characteristic accounts for the FET's naturally high input impedance. The operation of an FET is very different from the BJT, which normally uses an input voltage to forward bias the base-to-emitter junction and vary the input current (which varies the output current), and therefore the output voltage. This is the distinct difference between an FET and a BJT. An FET's input junction is normally reverse biased, and therefore the input voltage controls the output current. A BJT's input junction is normally forward biased and therefore the input current controls the output current. This difference is why BJTs are known as **current-controlled devices** and FETs are known as **voltage-controlled devices.** In fact, the name "field-effect transistor" is derived from this voltage control action because the applied input voltage will generate an electric field. It is this electric field that varies the size of the depletion region, and therefore the resistance of the channel between the FET's drain and source output terminals. In other words, the "effect" of the electric "field" causes "transistance," which is the transferring of different values of resistance between the output terminals.

Current-Controlled Devices
A device in which the input junction is normally forward biased and the input current controls the output current.

Voltage-Controlled Devices
A device in which the input junction is normally reverse biased and the input voltage controls the output current.

The term "junction" is attached to this type of FET because of the single P-N junction formed between the gate and the source-to-drain channel. Therefore, an *n*-channel JFET has a single P-N junction between gate to channel, and a *p*-channel JFET has a single N-P junction between gate to channel.

The field effect transistor is also often referred to as a **unipolar device** because only one type of semiconductor material exists between the output terminals (*n*-type or *p*-type channel between source and drain), and therefore the charge carriers have only one polarity (unipolar). Compare this to a BJT, which is a **bipolar device** because there is a change in semiconductor material between the output terminals (NPN or PNP between emitter and collector), and the charge carriers can be one of two polarities (bipolar, because both majority and minority carriers are used).

Unipolar Device
A device in which only one type of semiconductor material exists between the output terminals and therefore the charge carriers have only one polarity (unipolar).

Bipolar Device
A device in which there is a change in semiconductor material between the output terminals (NPN or PNP between emitter and collector), so the charge carriers can be one of two polarities (bipolar).

13-1-3 JFET Characteristics

Like the BJT, the JFET's response to certain variables is best described by using a graph. Figure 13-4(a) shows a graph plotting drain current (I_D) against drain-to-source voltage (V_{DS}). As you have probably already observed, this **drain characteristic curve** is very similar to a bipolar transistor's collector characteristic curve. Starting at 0 V and moving right along the horizontal axis, you can see that an increase in the drain supply voltage ($+V_{DD}$), and therefore an increase in V_{DS}, will result in a continual increase in I_D. At a certain V_{DS} voltage (in this

Drain Characteristic Curve
A plot of the drain current (I_D) versus the drain-to-source voltage (V_{DS}).

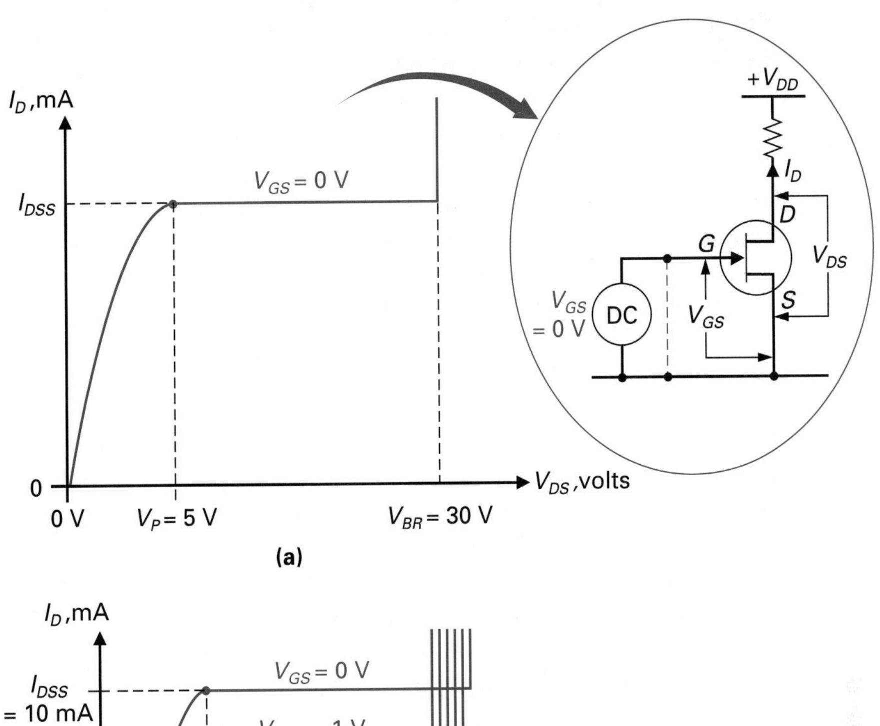

(a)

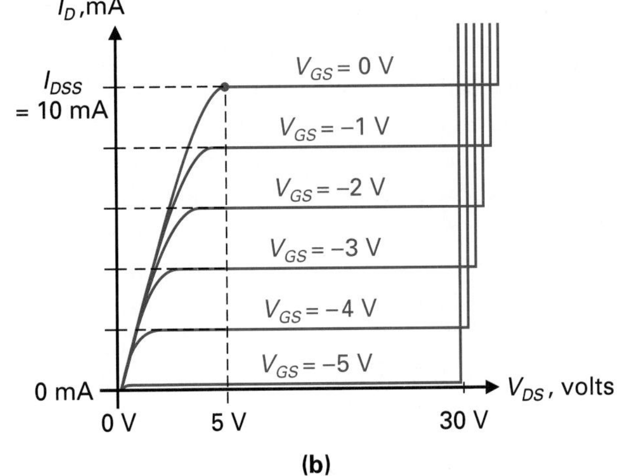

(b)

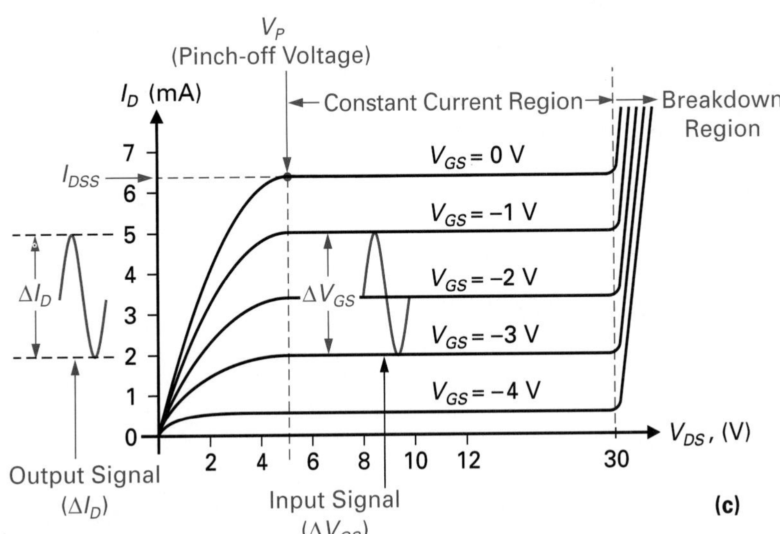

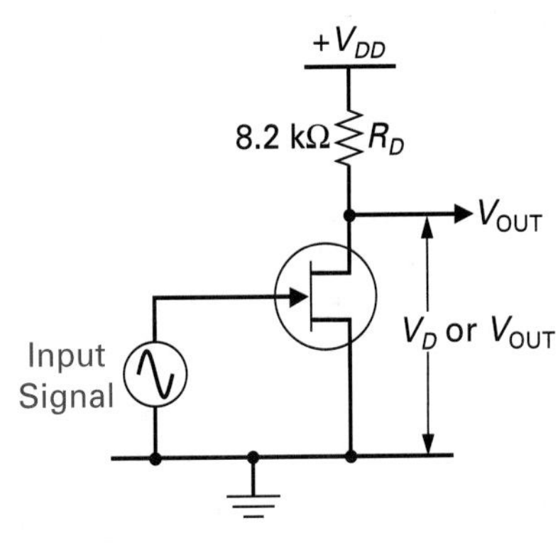

(c)

FIGURE 13-4 JFET Characteristics.

Pinch-Off Voltage (V_P)
The value of V_{DS} at which further increases in V_{DS} will cause no further increase in I_D.

Constant-Current Region
The flat portion of the drain characteristic curve. In this region I_D remains constant despite changes in V_{DS}.

Breakdown Voltage (V_{BR})
The voltage at which a damaging value of I_D will pass through the JFET.

Drain-to-Source Current with Shorted Gate (I_{DSS})
The maximum value of drain current, achieved by holding V_{GS} at 0 V.

Gate-to-Source Cutoff Voltage or $V_{GS(OFF)}$.
The negative V_{GS} bias voltage that causes I_D to drop to approximately zero.

example 5 V), further increases in V_{DS} will cause no further increase in I_D. This value of V_{DS} is called the **pinch-off voltage (V_P)** because it is the point at which the bias voltage has caused the depletion region to pinch-off or restrict drain current. From this point on, further increases in V_{DS} are counteracted by increases in the resistance of the channel, and therefore I_D remains constant. This is shown by the flat portion of the graph in Figure 13-4(a) and is called the **constant-current region** because I_D remains constant despite changes in V_{DS}. If V_{DS} is further increased (by increasing $+V_{DD}$), the JFET will eventually reach its **breakdown voltage (V_{BR})**, at which time a damaging value of I_D will pass through the JFET.

In the example graph in Figure 13-4(a), we plotted what would happen to I_D as V_{DS} increased with V_{GS} at 0 V. In Figure 13-4(b) we will examine what will happen to an *n*-channel JFET when the gate-source junction is reverse biased by several negative voltages. As previously described in the JFET operation section, a negative voltage is normally applied to reverse-bias the gate and set up a depletion region. As V_{GS} is made more negative, the gate will be further reverse biased, and the corresponding I_D value will be smaller. Therefore, when V_{GS} is at 0 V, a maximum value of drain current is passing through the JFET's channel. This maximum value of drain current is called the **drain-to-source current with shorted gate (I_{DSS}).** This name is derived from the fact that when $V_{GS} = 0$ V, as shown in the inset in Figure 13-4(a), the gate and source terminals of the JFET are at the same potential of zero volts, and therefore the gate is effectively shorted to the source as shown by the dashed line. The drain-to-source current with shorted gate (I_{DSS}) rating is therefore the maximum current that can pass through the channel of a given JFET. When given on a specification sheet, this rating is equivalent to a bipolar transistor's $I_{C(SAT)}$ rating.

Returning to Figure 13-4(b), you can see that if V_{GS} is made more negative, the depletion regions within the JFET will get closer and closer and eventually touch, cutting off drain current. This negative V_{GS} bias voltage that causes I_D to drop to approximately zero is called the **gate-to-source cutoff voltage or $V_{GS(OFF)}$.** In the example in Figure 13-4(b), when $V_{GS} = -5$ V, I_D is almost zero and therefore $V_{GS(OFF)} = -5$ V. When cut OFF, the JFET will be equivalent to an open circuit between drain and source, and subsequently all of the drain supply voltage (V_{DD}) will appear across the open JFET ($V_{DS} = V_{DD}$).

To summarize the specifications in Figure 13-4(b),

When $V_{GS} = 0$ V, $I_D = I_{DSS} = 10$ mA
$V_P = 5$ V
$V_{BR} = 30$ V
$V_{GS(OFF)} = -5$ V
Constant Current Region = V_P to V_{BR} = 5 V to 30 V

13-1-4 Transconductance

Figure 13-4(c) illustrates a JFET test circuit and its associated characteristic graph. Before we see how a JFET can be made to amplify, let's first summarize the details given in this graph. If we first consider the curve when $V_{GS} = 0$ V, you can see that up to V_P, I_D increases in almost direct proportion to V_{DS}. This is because the depletion region is not sufficiently large enough to affect I_D, so the channel is simply behaving as a semiconductor with a fixed resistance value between source and drain.

When V_{DS} is equal to V_P, the drain current (I_D) will be pinched into an extremely narrow channel between the wedge-shaped depletion region. Any further increase in V_{DS} will have two effects:

1. Increase the pinching effect on the channel, which will resist current flow, and
2. Increase the potential between the drain and source, which will encourage current flow.

The net result is that channel resistance increases in direct proportion with V_{DS} and consequently I_D remains constant, as shown by the flat portion of the characteristic curve.

Assuming a fixed value of V_{DD}, any increase in the negative voltage of V_{GS} will cause a corresponding decrease in I_D. Therefore, beyond V_P, I_D is controlled by small-signal changes (such as the input signal) in V_{GS} and is independent of changes in V_{DS}. This section of the curve between V_P and V_{BR} is called the constant-current region.

Like the bipolar transistor, an FET can be used to amplify a signal, as shown in Figure 13-4(c). As before, the amount of amplification achieved is a ratio between output and input. For a bipolar transistor, the amount of gain is equal to the ratio of input current to output current (beta). For an FET, there is no input current, and therefore an FET's gain is equal to the ratio of output current change (ΔI_D) to input voltage change (ΔV_{GS}). This ratio is called the FET's **transconductance** (symbolized δ_m).

Transconductance Also called mutual conductance, it is the ratio of a change in output current to the initiating change in input voltage.

$$\delta_m = \frac{\Delta I_D}{\Delta V_{GS}}$$

δ_m = transconductance in siemens (S)
ΔI_D = change in drain current
ΔV_{GS} = change in gate-source voltage

■ **EXAMPLE:**

Calculate the transconductance of the FET for the example shown in Figure 13-4(c).

■ ***Solution:***

$$\delta_m = \frac{\Delta I_D}{\Delta V_{GS}} = \frac{5\text{ mA} - 2\text{ mA}}{-1\text{ V} - (-3\text{ V})} = \frac{3\text{ mA}}{2\text{ V}} = 1.5\text{ millisiemens}$$

A high-gain FET will produce a large change in I_D for a small change in V_{GS}, resulting in a high-transconductance figure ($\delta_m\uparrow$).

13-1-5 *Voltage Gain*

Because transconductance is the ratio of output current change (ΔI_D) to input voltage change (ΔV_{GS}), it is no surprise that this ratio is used to determine a JFET's voltage gain. The voltage gain formula is as follows

$$A_V = \delta_m \times R_D$$

■ **EXAMPLE:**

Calculate the voltage gain for the circuit example shown in Figure 13-4(c).

■ ***Solution:***

$$A_V = \delta_m \times R_D = 1.5\ \text{mS} \times 8.2\ \text{k}\Omega = 12.3$$

This means that the output voltage will be 12.3 times greater than the input voltage.

13-1-6 *JFET Data Sheets*

Throughout this section, we have used certain JFET specifications in our calculations, such as $V_{GS(\text{OFF})}$ and I_{DSS}. As an example, Figure 13-5 shows the data sheet for a typical *n*-channel JFET. As before, notes have been inserted within the data sheet to describe any confusing ratings; however, most of these ratings are self-explanatory.

SELF-TEST REVIEW QUESTIONS FOR SECTION 13-1

---

1. Give the full names of the following abbreviations:
   - **a.** BJT
   - **b.** JFET
   - **c.** MOSFET
2. Name the two different types of JFETs.
3. The BJT is a ____________ operated device while the FET is a ____________ operated device.
4. The gate-source junction of a JFET is always ____________ biased.
5. When $V_{GS} = 0$ V, $I_D = I_{DSS}$. (True/False)
6. When $V_{GS} = V_{GS(\text{OFF})}$, $I_D = \text{max}$. (True/False)
7. ____________ is a ratio of an FET's output current change to input voltage change.
8. A JFET has a ____________ input impedance due to its ____________ biased gate-source junction.

DEVICE: 2N5484 Through 2N5486—N-Channel JFET

Most of these specifications are self-explanatory. $V_{GS\,(off)}$ which has been given throughout is listed in the OFF characteristics, while I_{DSS} is listed in the ON characteristics. The only confusing maximum rating is "Forward Gate Current" because the gate is never normally forward biased (V_{GS} = a negative voltage or zero volts). This rating indicates that if the gate accidentally becomes forward biased, gate current must not exceed 10 mA dc or the JFET will be destroyed.

2N5484
2N5486

CASE 29-04, STYLE 5
TO-92 (TO-226AA)

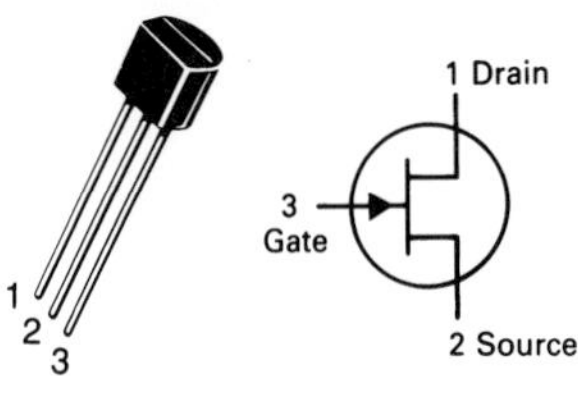

JFET
VHF/UHF AMPLIFIERS
N-CHANNEL — DEPLETION

MAXIMUM RATINGS

Rating	Symbol	Value	Unit
Drain-Gate Voltage	V_{DG}	25	Vdc
Reverse Gate-Source Voltage	V_{GSR}	25	Vdc
Drain Current	I_D	30	mAdc
Forward Gate Current	$I_{G(f)}$		mAdc
Total Device Dissipation @ T_C = 25°C Derate above 25°C	P_D	310 2.82	mW mW/°C
Operating and Storage Junction Temperature Range	T_J, T_{stg}	−65 to +150	°C

ELECTRICAL CHARACTERISTICS (T_A = 25°C unless otherwise noted.)

Characteristic		Symbol	Min	Typ	Max	Unit
OFF CHARACTERISTICS						
Gate-Source Breakdown Voltage (I_G = −1.0 μAdc, V_{DS} = 0)		$V_{(BR)GSS}$	−25	—	—	Vdc
Gate Reverse Current (V_{GS} = −20 Vdc, V_{DS} = 0) (V_{GS} = −20 Vdc, V_{DS} = 0, T_A = 100°C)		I_{GSS}	— —	— —	−1.0 −0.2	nAdc μAdc
Gate Source Cutoff Voltage (V_{DS} = 15 Vdc, I_D = 10 nAdc)	2N5484 2N5485 2N5486	$V_{GS(off)}$	−0.3 −0.5 −2.0	— — —	−3.0 −4.0 −6.0	Vdc
ON CHARACTERISTICS						
Zero-Gate-Voltage Drain Current (V_{DS} = 15 Vdc, V_{GS} = 0)	2N5484 2N5485 2N5486	I_{DSS}	1.0 4.0 8.0	— — —	5.0 10 20	mAdc

$V_{GS\,(off)}$

I_{DSS}

FIGURE 13-5 Data Sheet for an *n*-Channel JFET. (Copyright of Motorola. Used by permission.)

13-2 JFET BIASING

The biasing methods used in FET circuits are very similar to those employed in BJT circuits. In this section we will examine the circuit calculations for the three most frequently used JFET biasing methods: gate-biasing, self-biasing and voltage-divider biasing.

13-2-1 *Gate Biasing*

Figure 13-6(a) shows a gate-biased JFET circuit. The gate supply voltage ($-V_{GG}$) is used to reverse bias the gate-source junction of the JFET. With no gate current, there can be no voltage drop across R_G and therefore the voltage at the gate of the JFET will equal the dc gate-supply voltage.

$$V_{GS} = V_{GG}$$

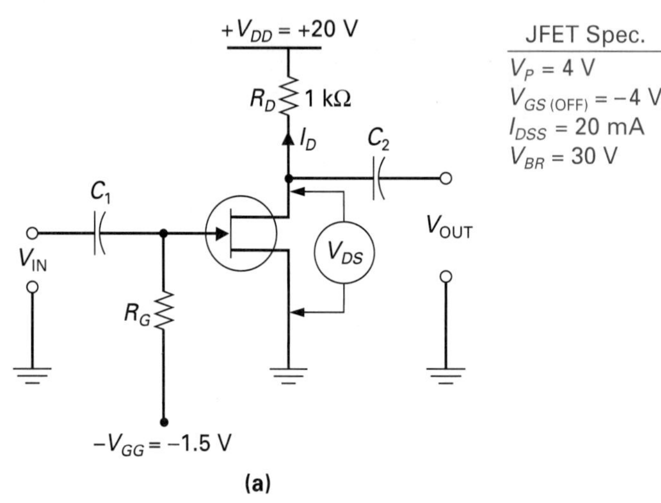

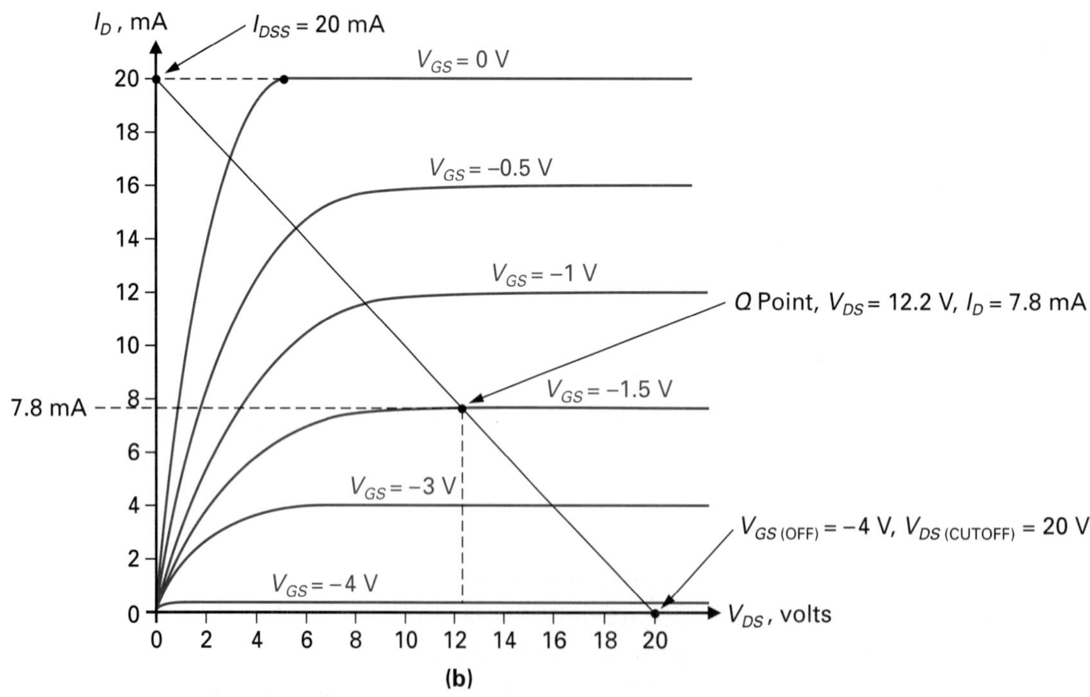

FIGURE 13-6 A Gate-Biased JFET Circuit. (a) Basic Circuit. (b) Drain Characteristic Curves and DC Load Line.

In the example in Figure 13-6,

$$V_{GS} = V_{GG} = -1.5 \text{ V}$$

Knowing V_{GS}, we can calculate I_D if the JFET's current (I_{DSS}) and voltage ($V_{GS(\text{OFF})}$) specification limits are known by using the following formula.

$$I_D = I_{DSS}\left(1 - \frac{V_{GS}}{V_{GS(\text{OFF})}}\right)^2$$

In the example in Figure 13-6,

$$\begin{aligned} I_D &= 20 \text{ mA}\left(1 - \frac{-1.5 \text{ V}}{-4 \text{ V}}\right)^2 \\ &= 20 \text{ mA } (1 - 0.375)^2 \\ &= 20 \text{ mA} \times 0.625^2 \\ &= 20 \text{ mA} \times 0.39 = 7.8 \text{ mA} \end{aligned}$$

Now that I_D is known, we can calculate the voltage drop across R_D using Ohm's law.

$$V_{RD} = I_D \times R_D$$

In the example in Figure 13-6,

$$V_{RD} = I_D \times R_D = 7.8 \text{ mA} \times 1 \text{ k}\Omega = 7.8 \text{ V}$$

Because V_{RD} plus V_{DS} will equal V_{DD}, we calculate V_{DS} once V_{RD} is known with the following formula.

$$V_{DS} = V_{DD} - V_{RD}$$

In the example in Figure 13-6,

$$V_{DS} = V_{DD} - V_{RD} = 20 \text{ V} - 7.8 \text{ V} = 12.2 \text{ V}$$

Figure 13-6(b) shows the drain characteristic curve and dc load line for the example JFET circuit in Figure 13-6(a). Like the bipolar transistor, the JFET's dc load line extends between the maximum output current point, or saturation point, (when the JFET is fully ON, $I_{DSS} = 20$ mA) to the maximum output voltage point (when the JFET is cut OFF, $V_{DS} = 20$ V). The dc operating point, or *Q* point, which was determined with the previous calculations, is also plotted on the dc load line in Figure 13-6(b).

■ **EXAMPLE:**

Calculate the following for the circuit shown in Figure 13-7.

1. V_{GS}
2. I_D

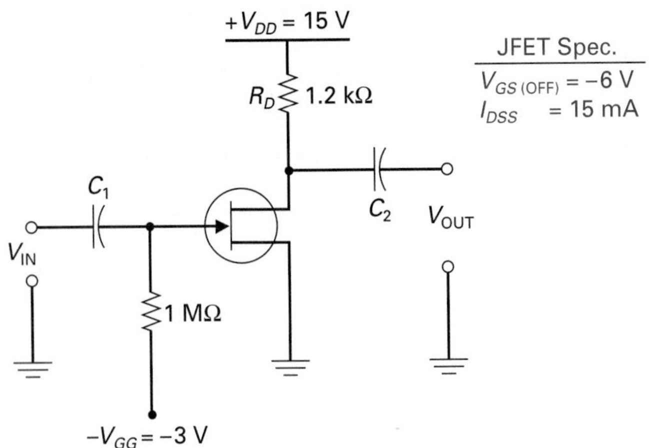

FIGURE 13-7 A Gate-Biased JFET Circuit Example.

3. V_{DS}
4. Maximum value of I_D
5. V_{DS} when $V_{GS} = V_{GS(\text{OFF})}$
6. Q point

■ ***Solution:***

1. $V_{GS} = V_{GG} = -3$ V
2. $I_D = I_{DSS}\left(1 - \frac{V_{GS}}{V_{GS(\text{OFF})}}\right)^2 = 15\text{ mA}\left(1 - \frac{-3\text{ V}}{-6\text{ V}}\right)^2 = 15\text{ mA}\,(1 - 0.5)^2 = 3.75\text{ mA}$
3. $V_{DS} = V_{DD} - V_{RD}$ (since $V_{RD} = I_D \times R_D$, we can substitute)

   $V_{DS} = V_{DD} - (I_D \times R_D) = 15\text{ V} - (3.75\text{ mA} \times 1.2\text{ k}\Omega) = 15\text{ V} - 4.5\text{ V} = 10.5\text{ V}$
4. Maximum value of $I_D = I_{DSS} = 15$ mA
5. When $V_{GS} = V_{GS(\text{OFF})}$, the JFET is cut off and equivalent to an open switch between drain and source. In this condition all of the drain supply voltage will appear across the open JFET.

$$V_{DS(\text{CUTOFF})} = V_{DD} = 15\text{ V}$$

6. The dc operating or Q point is

$$V_{GS} = -3\text{ V}$$
$$I_D = 3.75\text{ mA}$$
$$V_{DS} = 10.5\text{ V}$$

13-2-2 Self Biasing

Figure 13-8(a) shows how to self-bias a JFET circuit. One advantage of this biasing method over gate biasing is that only a single drain supply voltage is needed (V_{DD}) instead of both V_{DD} and a negative gate supply voltage ($-V_{GG}$). The other difference you may have noticed is that a source resistor (R_S) has been included,

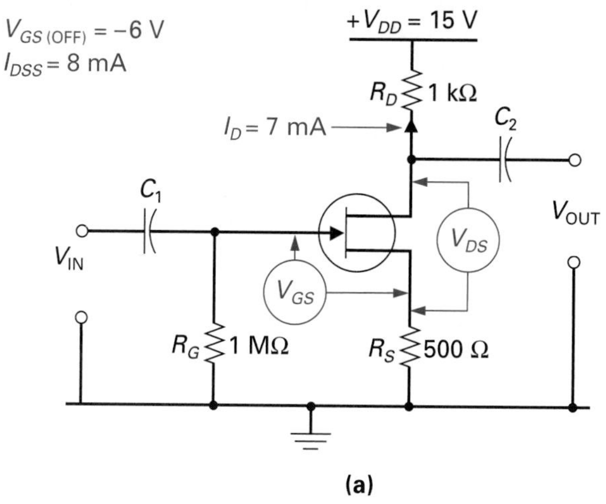

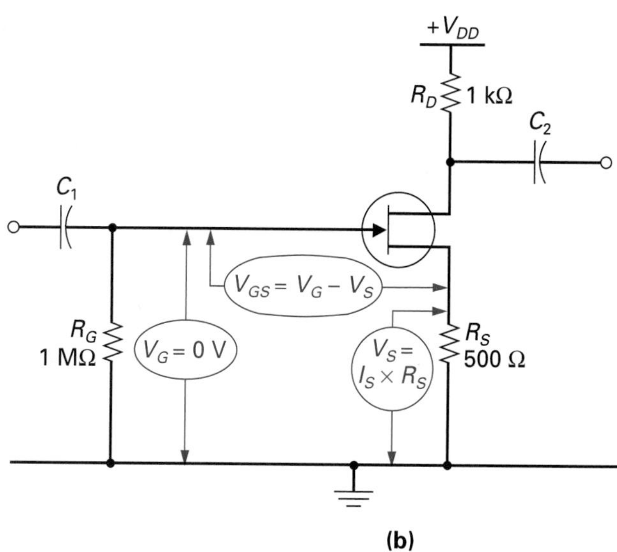

FIGURE 13-8 A Self-Biased JFET Circuit. (a) Basic Circuit. (b) How R_S Develops a $-V_{GS}$.

and R_G has been connected to ground. Although this arrangement seems completely different to the gate bias circuit, the inclusion of R_S and the grounding of R_G will achieve the same result, which is to reverse bias the JFET's gate-source junction. Figure 13-8(b) illustrates how this is achieved. Since there is no gate current in a JFET circuit ($I_G = 0$), all of the current flowing into the source will travel through the channel and flow out of the drain. Therefore

$$I_S = I_D$$

For the example in Figure 13-8,

$$I_S = I_D = 7 \text{ mA}$$

Now that I_S is known, we can calculate the voltage drop across the source resistor (V_{RS}), and therefore the voltage at the JFET's source (V_S).

$$V_{RS} = V_S = I_S \times R_S$$

For the example in Figure 13-8,

$$V_{RS} = V_S = I_S \times R_S = 7 \text{ mA} \times 500\ \Omega = +3.5 \text{ V}$$

Because $I_G = 0$ A, there will be no voltage drop across R_G, and so the voltage at the gate of the JFET will be 0 V.

$$V_G = 0 \text{ V}$$

Now that we know that $V_S = +3.5$ V and $V_G = 0$ V, we can see how the JFET's gate-source junction is reverse biased. To reverse bias a gate-biased JFET, we simply made the gate voltage negative with respect to the source that is at 0 V. With a self-biased JFET, we achieve the same result by making the source voltage positive with respect to the gate that is at 0 V. This makes the gate of the JFET negative with respect to the source. This potential difference from gate-to-source (V_{GS}) is therefore equal to

$$V_{GS} = V_G - V_S$$

Because $V_S = I_S \times R_S$ and $V_G = 0$ V, we can substitute the previous formula to obtain

$$V_{GS} = 0 \text{ V} - (I_S \times R_S)$$

or

$$V_{GS} = -(I_S \times R_S)$$

or because $I_S = I_D$

$$V_{GS} = -(I_D \times R_S)$$

In the example in Figure 13-8,

$$\begin{aligned} V_{GS} &= -(I_S \text{ or } I_D \times R_S) \\ &= -(7 \text{ mA} \times 500\ \Omega) \\ &= -3.5 \text{ V} \end{aligned}$$

If $-V_{GS}$ and R_S are known, we could transpose the above equation to calculate I_D.

$$I_D = \frac{V_{GS}}{R_S}$$

In the example in Figure 13-8,

$$I_D = \frac{V_{GS}}{R_S} = \frac{3.5\text{ V}}{500\ \Omega} = 7\text{ mA}$$

The final calculation is to determine the voltage at the JFET's drain with respect to ground (V_D) and the drain-to-source voltage drop across the JFET.

$$V_D = V_{DD} - V_{RD}$$

Because $V_{RD} = I_D \times R_D$,

$$V_D = V_{DD} - (I_D \times R_D)$$

In the example in Figure 13-8,

$$V_D = V_{DD} - (I_D \times R_D) = 15\text{ V} - (7\text{ mA} \times 1\text{ k}\Omega) = 15\text{ V} - 7\text{ V} = 8\text{ V}$$

Now that the voltage drops across R_D (V_{RD}) and R_S (V_{RS}) are known, we can calculate the voltage drop across the JFET's drain to source (V_{DS}).

$$V_{DS} = V_{DD} - (V_{RD} + V_{RS})$$

In the example in Figure 13-8,

$$V_{DS} = V_{DD} - (V_{RD} + V_{RS}) = 15\text{ V} - (7\text{ V} + 3.5\text{ V}) = 15\text{ V} - 10.5\text{ V} = 4.5\text{ V}$$

■ **EXAMPLE:**

Calculate the following for the circuit shown in Figure 13-9.

1. V_S
2. V_{GS}
3. V_{DS}
4. I_D maximum
5. V_{DS} when the JFET is OFF
6. V_D

■ ***Solution:***

1. Because $I_S = I_D$, $V_S = I_S \times R_S = 4\text{ mA} \times 500\ \Omega = 2\text{ V}$
2. $V_{GS} = V_G - V_S = 0\text{ V} - 2\text{ V} = -2\text{ V}$

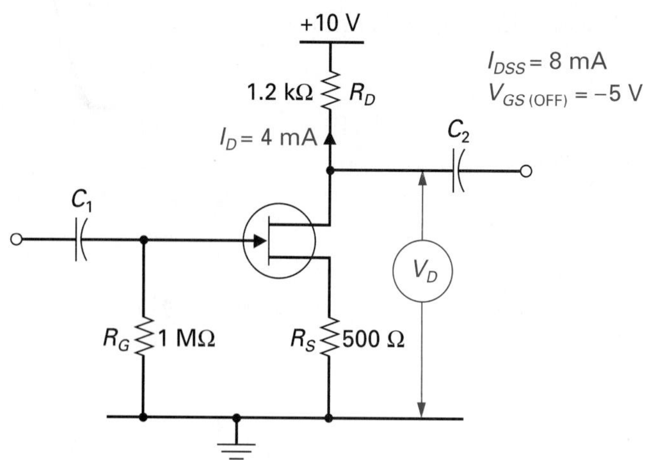

FIGURE 13-9 A Self-Biased JFET Circuit Example.

3. $V_{DS} = V_{DD} - (V_{RD} + V_{RS}) = 10\text{ V} - [(I_D R_D) + 2\text{ V}]$
$= 10\text{ V} - [(4\text{ mA} \times 1.2\text{ k}\Omega) + 2\text{ V}] = 10\text{ V} - (4.8 + 2\text{ V})$
$= 10\text{ V} - 6.8\text{ V} = 3.2\text{ V}$
4. I_D maximum $= I_{DSS} = 8\text{ mA}$
5. $V_{DS(\text{CUTOFF})} = V_{DD} = 10\text{ V}$
6. $V_D = V_{DS} + V_{RS} = 3.2\text{ V} + 2\text{ V} = 5.2\text{ V}$

As previously mentioned, one advantage of this self-biased JFET method is that only a drain supply voltage is needed (V_{DD}). The gate supply voltage (V_{GG}) is not needed due to the inclusion of a source resistor that reverse biases the JFET's gate-source junction by applying a positive voltage to the source with respect to the 0 V on the gate. This method of effectively sending back a negative voltage from the source to the gate is known as "negative feedback." It not only enables us to bias a JFET with one supply voltage, it also provides temperature stability. Any change in the ambient temperature will cause a change in the semiconductor JFET's conduction, which would move the JFET's Q point away from its desired setting. The inclusion of R_S will prevent the Q point from shifting due to temperature in the same way as a BJT's emitter resistor. If temperature were to increase, for instance ($T\uparrow$), the resistance of the semiconductor would decrease ($R\downarrow$) because all semiconductor materials have a negative temperature coefficient of resistance ($T\uparrow$, $R\downarrow$), and this will cause the channel current to increase. If the drain current increases ($I_D\uparrow$), the voltage drop across R_S will increase (V_{RS} or $V_S\uparrow = I_D\uparrow \times R_S$). This increase in V_S will increase the gate-source reverse voltage ($-V_{GS}\uparrow$), causing the JFET's channel to get narrower and the drain current to decrease ($I_D\downarrow$) and counteract the original increase. Similarly, a decrease in temperature will cause a decrease in I_D which will decrease the gate-source reverse bias, resulting in an increase in I_D. The Q point will remain relatively stable despite changes in temperature when a JFET circuit has a source resistor included.

13-2-3 *Voltage-Divider Biasing*

Referring to the voltage-divider biased JFET circuit shown in Figure 13-10, you will probably notice that it is very similar to the voltage-divider biased BJT circuit discussed previously. Like the self-biased circuit, the inclusion of a source resistor stabilizes the Q point despite ambient temperature changes. In addition, using a voltage divider to determine the gate-bias voltage ensures that V_{GS}, and therefore the circuit, has increased stability.

The gate voltage (V_G) is calculated using the following voltage divider formula

$$V_{R_2} \text{ or } V_G = \frac{R_2}{R_1 + R_2} \times V_{DD}$$

For the example in Figure 13-10,

$$V_{R_2} \text{ or } V_G = \frac{R_2}{R_1 + R_2} \times V_{DD} = \frac{5\ \text{M}\Omega}{10\ \text{M}\Omega + 5\ \text{M}\Omega} \times 15\ \text{V} = 5\ \text{V}$$

Because $I_D = I_S$ ($I_G = 0$) and the drain resistance and current are known, we can next calculate the voltage drop across the source resistor (V_{RS}), drain resistor (V_{RD}), and JFET's source-drain junction (V_{DS}).

$$I_S = I_D$$

$$I_S = I_D = 2\ \text{mA}$$

$$V_{RS} = I_S \times R_S$$

$$V_{RS} = 2\ \text{mA} \times 4.3\ \text{k}\Omega = 8.6\ \text{V}$$

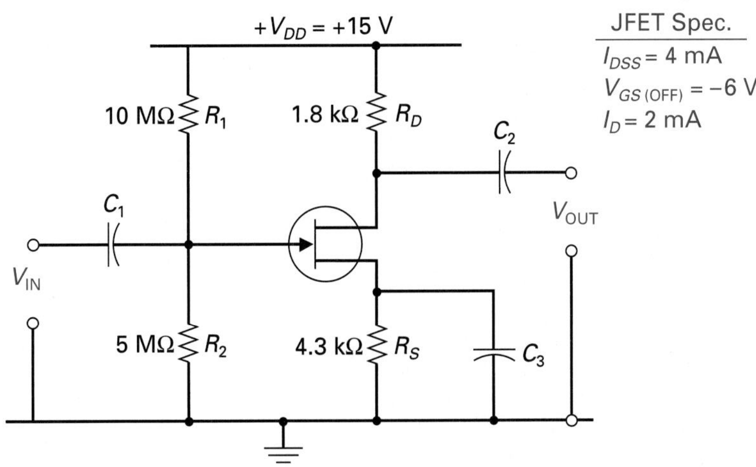

FIGURE 13-10 A Voltage-Divider Biased JFET Circuit.

$$V_{RD} = I_D \times R_D$$

$$V_{RD} = 2 \text{ mA} \times 1.8 \text{ k}\Omega = 3.6 \text{ V}$$

$$V_{DS} = V_{DD} - (V_{RS} + V_{RD})$$

$$V_{DS} = 15 \text{ V} - (8.6 \text{ V} + 3.6 \text{ V}) = 2.8 \text{ V}$$

Now that the JFET's gate and source voltages are known (V_G and V_S), we can calculate the value of gate-source reverse bias ($-V_{GS}$).

$$V_{GS} = V_G - V_S$$

$$(V_G = V_{R_2}, V_S = V_{RS})$$

For the example in Figure 13-10,

$$V_{GS} = 5 \text{ V} - 8.6 \text{ V} = -3.6 \text{ V}$$

■ **EXAMPLE:**

Calculate the following for the voltage-divider biased JFET circuit shown in Figure 13-11:

1. V_G
2. I_S
3. V_S
4. V_{DS}
5. V_{GS}
6. V_D, when $V_{GS} = V_{GS(OFF)}$
7. I_D, when $V_{GS} = 0$ V

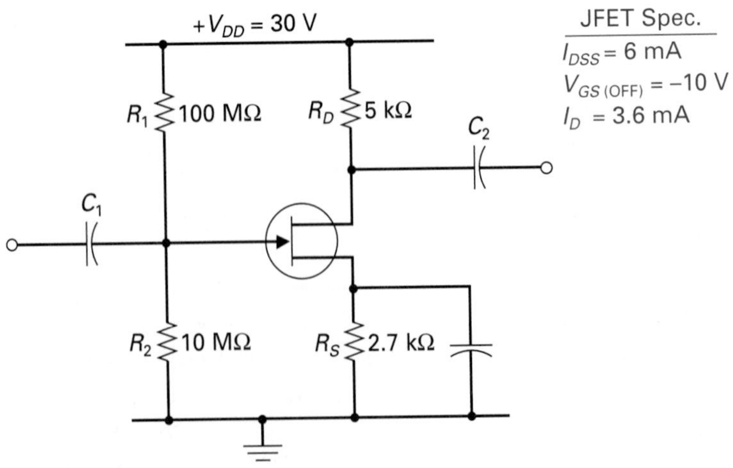

FIGURE 13-11 A Voltage-Divider Biased Circuit Example.

■ ***Solution:***

1. $V_G = \dfrac{R_2}{R_1 + R_2} \times V_{DD} = \dfrac{10\text{ M}\Omega}{100\text{ M}\Omega + 10\text{ M}\Omega} \times 30\text{ V} = 2.7\text{ V}$
2. $I_S = I_D = 3.6\text{ mA}$
3. $V_S = V_{RS} = I_S \times R_S = 3.6\text{ mA} \times 2.7\text{ k}\Omega = 9.7\text{ V}$
4. $V_{DS} = V_{DD} - (V_{RS} + V_{RD}) = 30\text{ V} - [9.7\text{ V} + (I_S \times R_D)]$
   $= 30\text{ V} - [9.7\text{ V} + (3.6\text{ mA} \times 5\text{ k}\Omega)] = 30\text{ V} - (9.7\text{ V} + 18\text{ V}) = 2.3\text{ V}$
5. $V_{GS} = V_G - V_S = 2.7\text{ V} - 9.7\text{ V} = -7\text{ V}$
6. When $V_{GS} = V_{GS(\text{OFF})}$, JFET is OFF and $V_D = V_{DD} = 30\text{ V}$
7. When $V_{GS} = 0\text{ V}$, $I_D = \text{maximum} = I_{DSS} = 6\text{ mA}$

SELF-TEST REVIEW QUESTIONS FOR SECTION 13-2

1. What two advantages does self bias have over gate bias?
2. What component in a JFET circuit provides temperature stability?
3. With a self-biased JFET circuit the source voltage is ____________ with respect to the gate voltage which is ____________. This makes the gate of the JFET ____________ with respect to the source.
4. Like self bias, ____________ bias has negative feedback and therefore maintains the Q point stable.

13-3 JFET CIRCUIT CONFIGURATIONS

The three JFET circuit configurations are illustrated in Figure 13-12 along with their typical circuit characteristics. Like the bipolar transistor configurations, the term "common" is used to indicate which of the JFET's leads is common to both the input and output. In this section we will examine the characteristics of these three configurations: common-source, common-gate, and common-drain.

13-3-1 Common-Source (C-S) *Circuits*

Similar to its bipolar counterpart, the common-emitter configuration, the **common-source configuration** is the most widely used JFET circuit and is detailed in Figure 13-12(a). The input is applied between the gate and source and the output is taken between the drain and source, with the source being common to both input and output. The ac input will pass through the coupling capacitor C_1 and be superimposed on the dc gate-bias voltage provided by resistor R_1, which sets up the dc operating or Q point. As the signal input changes, it will cause a change in gate voltage, which will cause a corresponding change in the output drain current. The output voltage developed between the FET's drain and ground is 180° out of phase with the input because an increase in $V_{IN}\uparrow$ and,

Common-Source Configuration
An FET configuration in which the source is grounded and common to the input and output signal.

	Voltage Gain	Input Impedance	Output Impedance	Circuit Appearance and Application	Waveforms
Common Source (a)	5–10 (Voltage Amp)	Very High 1–15 MΩ	Low 2–10 kΩ	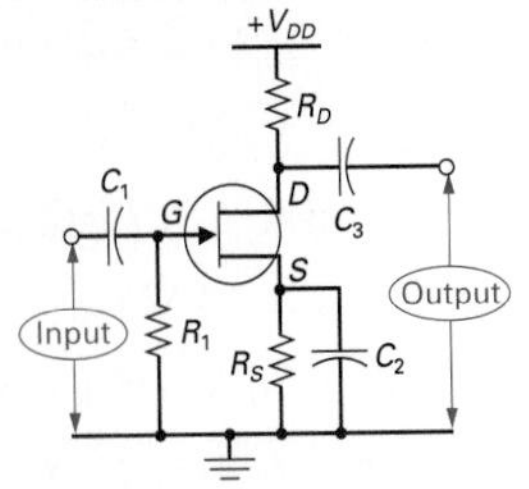 Most widely used FET configuration. It is mainly used as a voltage amplifier, however it is also used as an impedance matching device and can handle the high radio frequency signals.	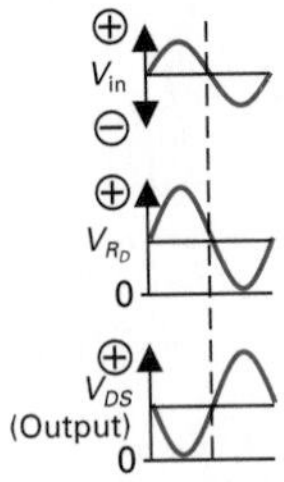 V_{in} and V_{out} are out of phase (180° phase shift)
Common Gate (b)	2–5	Very Low 200–1500 Ω	Medium 5–15 kΩ	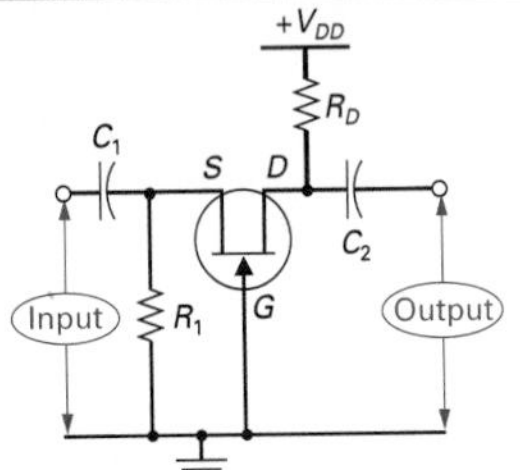 This configuration is used to amplify radio frequency signals due to its very stable nature at high frequencies. It is also used as a buffer to match a low impedance source to a high impedance load.	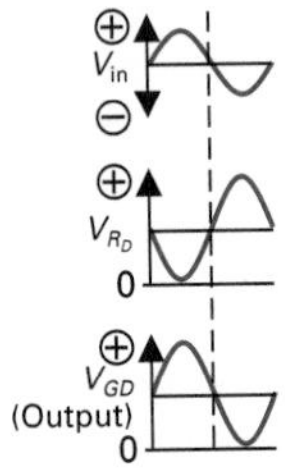 V_{in} and V_{out} are in phase (0° phase shift)
Common Drain (c)	0.98	Very High 1500 MΩ	Low 10 kΩ	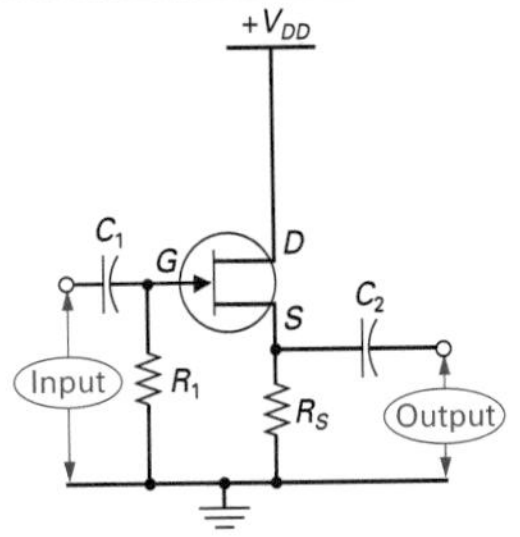 This amplifier is commonly called a source follower as the source follows whatever is applied to the gate. Its very high input impedance will not load down (and therefore not distort) signals from high impedance signal sources, such as a microphone, and its low output impedance is ideal to drive a low impedance load such as an audio amplifier.	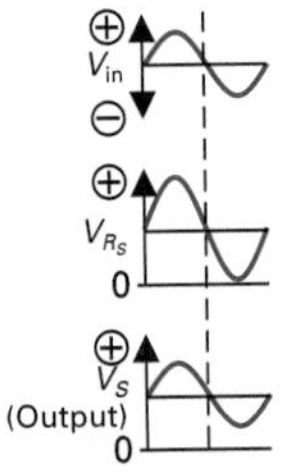 V_{in} and V_{out} are in phase (0° phase shift)

FIGURE 13-12 JFET Circuit Configurations.

therefore $V_{GS}\uparrow$, will cause an increase in $I_D\uparrow$, a decrease in the voltage drop across the FET ($V_{DS}\downarrow$), and a decrease in the output voltage $V_{OUT}\downarrow$. Resistor R_S is included to provide temperature stability and, as with the bipolar transistor, the source decoupling capacitor C_2 is included to prevent degenerative feedback.

When a small ac input signal is applied to the gate of a common-source amplifier, the variations in voltage at the gate control the JFET, which effectively acts as a variable resistor, varying the output drain current. These changes in drain current will vary the voltage drop across R_D and the drain-to-source voltage drop, which, with R_S, determines the output voltage. Referring to the characteristics listed in Figure 13-12(a), you can see that the output voltage (V_{OUT}) of the common-source JFET configuration can be five to ten times larger than the gate control input voltage (V_{IN}). If a high amount of voltage gain is desired, R_D is made relatively large (typically greater than 20 kΩ) and the JFET is biased so that its drain-to-source resistance is also high. A larger resistance will develop a larger voltage.

Also listed in the common-source characteristics in Figure 13-12(a) is the very high input resistance and the relatively low output resistance of this circuit. The high input resistance is due to the JFET's reverse biased gate-source junction, which permits no gate input current, and therefore has a very large resistance. This key characteristic means that the common-source JFET circuit is ideal in applications where we need to provide voltage amplification but do not want to load down a source that can only generate a small input signal. Such applications include

1. Digital circuits in which the outputs of many circuits are connected to one another, and therefore the output resistances of all the circuits load one another. As a result, the signals generated by these circuits are small, and a circuit is needed that will not load the signal source but will still provide voltage gain.
2. Analog circuits in which it can amplify both dc and low- and high-frequency ac input signal voltages. The *C-S* circuit's high input impedance makes it ideal at the front end of systems such as the first RF amplifier stage following the antenna and the first stage in a voltmeter, in which it will not load the source yet will amplify a wide range of input signal voltages.

13-3-2 *Common-Gate* (C-G) *Circuits*

The **common-gate circuit configuration** shown in Figure 13-12(b) is very similar to its bipolar counterpart, the common-base circuit. The input is applied between the source and gate, while the output appears across the drain and gate. Self-bias resistor R_1 sets up the static Q point, and the input is applied through the coupling capacitor C_1 and will cause a change in the JFET's source voltage. An increase in source voltage will cause a decrease in the V_{GS} forward bias (n-type source is driven positive), a decrease in I_D, a decrease in the voltage drop across R_D, and therefore an increase in the voltage dropped between the FET's drain and gate. Because the voltage developed across the JFET's drain and gate is applied to the output, an increase in the input produces an increase in the output, and so the input and output voltage are in-phase with one another. Similarly, as the input voltage decreases, the gate-source forward bias will increase. Therefore, I_D will increase, and there will be more voltage developed across R_D and less voltage developed at the output.

Common-Gate Configuration
An FET configuration in which the gate is grounded and common to the input and output signal.

Referring to the common-gate characteristics listed in Figure 13-12(b), you can see that this circuit can be used to provide a small voltage gain. Because the input is applied to the JFET's high-current source terminal, the input resistance is very low. This low input resistance and relatively high output resistance makes the circuit ideal in applications where we need to efficiently transfer power between a low-resistance source and a high-resistance load.

13-3-3 *Common-Drain* (C-D) *Circuits*

Common-Drain Configuration An FET configuration in which the drain is common to the input and output signal.

Source Follower Another name used for a common-drain circuit configuration.

Comparable to the bipolar transistor's common-collector or emitter follower, the **common-drain configuration** shown in Figure 13-12(c) is sometimes called a **source follower** because the source output voltage follows in polarity and amplitude the input voltage at the gate. Once again, self-bias resistor R_1 sets up the quiescent operating point, and an ac gate input voltage will cause a variation in I_D. When the input voltage at the gate swings positive, the FET will conduct more current, less voltage will be developed across the FET drain to source, and therefore more voltage will be developed across R_S and the output. Similarly, a decrease in the input voltage will cause the resistance of the JFET's drain-source junction to increase. Therefore, V_{DS} will increase and V_{RL}, or V_{OUT}, will decrease.

Referring to the common-drain characteristics listed in Figure 13-12(c), you can see that the output voltage is slightly less than the input voltage (circuit does not provide any voltage gain). The input resistance of the common-drain circuit configuration is extremely high due to the JFET's reverse biased gate and R_S connection, and the output resistance is relatively very low. Inserting a common-drain circuit between a high-resistance source and a low-resistance load will ensure that the two opposite resistances are matched and power is efficiently transferred.

SELF-TEST REVIEW QUESTIONS FOR SECTION 13-3

1. Which FET circuit configuration is most widely used like its BJT common-emitter counterpart?
2. Which FET circuit configuration is also known as a source follower?
3. Which JFET circuit configuration could provide a high input impedance and a good value of voltage gain?
4. Which JFET circuit configuration is best suited for providing a very high input impedance and low output impedance?

13-4 JFET APPLICATIONS

It is the high input impedance of the JFET, and therefore its ability not to load a source, and the voltage amplification ability that are mainly made use of in circuit application. Like the bipolar junction transistor, the JFET can be made to function as a switch or as a variable resistor. Let us begin by examining how the JFET's switching ability can be made use of in digital or two-state circuits.

13-4-1 *Digital (Two-State) JFET Circuits*

As a switch, the JFET makes use of only two points on the load line: saturation (in which it is equivalent to a closed switch between source and drain) and cutoff (in which it is equivalent to an open switch between source and drain). Figure 13-13(a) shows an ON/OFF JFET switch circuit and its associated load line in Figure 13-13(b). Figure 13-13(c) shows the input/output voltages for each of the circuit's two operating states. When $V_{GS} = V_{GS(\text{OFF})}$ (–4 V), the JFET is cut OFF (lower end of the load line) and is equivalent to an open switch between source and drain. With the JFET's drain-source open, $I_D = 0$ mA, and the drain supply voltage will be applied to the output ($V_{DS} = V_{\text{OUT}} = +V_{DD}$). On the other hand, when $V_{GS} = 0$ V, the JFET is saturated (upper end of the load line) and is equivalent to a closed switch between source and drain. With the JFET's drain-source closed, I_D = max. = I_{DSS}, and the 0 V at the source will be applied to the output ($V_{DS} = V_{\text{OUT}} = 0$ V).

A typical FET application in digital circuits would be a buffer circuit, which is used to isolate one device from another. The high input impedance of the FET does not load the input circuit or circuits, while the low output impedance of the FET provides a high output current to the output circuit. The high output current and buffering or isolating characteristics of these circuits account for why they are also called buffer-drivers. The schematic symbol of the buffer-driver is shown in Figure 13-13(d).

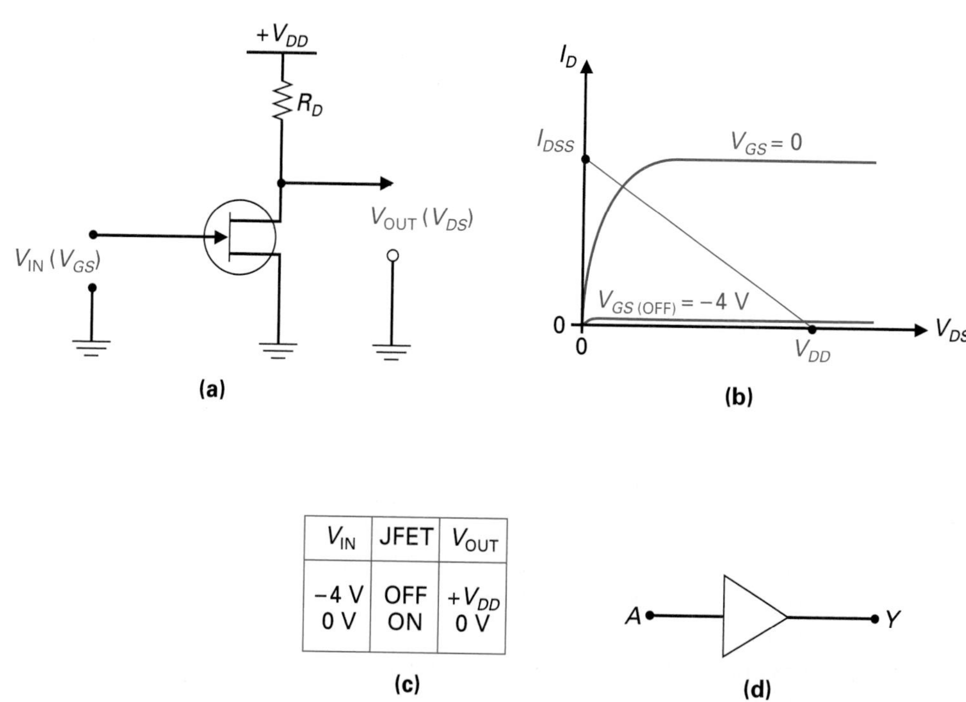

FIGURE 13-13 The JFET's Switching Action—Digital Circuit Applications. (a) Basic Switching Circuit. (b) Load Line. (c) Input/Output Voltages. (d) Digital Buffer/Driver Schematic Symbol.

The high input impedance ($Z_{IN}\uparrow$) of the FET is also made use of in other FET integrated circuits (ICs). When Z_{IN} is high, circuit current is low ($I\downarrow$), and therefore power dissipation is low ($P_D\downarrow$). This condition is ideal for digital integrated circuits (ICs) which contain thousands of transistors all formed onto one small piece of silicon. The low power dissipation of the JFET enables us to densely pack many more components into a very small area.

These and other digital circuits will be discussed in more detail in the digital circuits chapters in this text.

13-4-2 *Analog (Linear) JFET Circuits*

In two-state applications, the JFET is made to operate between the two extreme points of saturation (0 Ω) and cutoff (max. Ω). By controlling the gate-source bias voltage (V_{GS}), the resistance between the JFET's drain and source (R_{DS}) can be changed to be any value between 0 Ω and maximum Ω. The JFET can therefore be made to act as a variable resistor, with an increase in negative V_{GS} causing a larger R_{DS}. In contrast, a decrease in negative V_{GS} causes a smaller R_{DS}. This is illustrated in Figure 13-14(a).

It is the reverse biased gate-source junction of a JFET that gives the JFET its key advantage: *an extremely high input impedance* (typically in the high-MΩ range). In addition, the JFET can provide a small voltage gain and has been found to be a very low noise component. All these characteristics make it an ideal choice as an amplifier.

In previous chapters, we have seen how a light load (large resistance $R_L\uparrow$ and small $I\downarrow$) does not pull down the source voltage by any large amount, whereas a heavy load (small resistance $R_L\downarrow$ and large $I\uparrow$) will pull down the source voltage. A heavy load results in less output voltage ($V_{RL}\downarrow$) and an increased current and heat loss at the source. The overall effect is that a small load resistance or impedance causes less power to be delivered to the load.

The circuit in Figure 13-14(b) shows how a common-source JFET has been connected to function as a preamplifier, which is a circuit that provides gain for a very weak input signal. In this example, the JFET preamplifier matches the high-impedance (small signal) crystal microphone to a low-impedance power amplifier. The reverse biased gate-source junction of a JFET pre-amplifier will offer a large load impedance to the source or microphone. This light load ($R_L\uparrow$) input resistance of the JFET will therefore permit most of the signal voltage being generated by the microphone to be applied to the JFET's gate and then be amplified. In other words, the high input impedance of the JFET amplifier circuit will not pull down the voltage signal being generated by the microphone. Therefore, maximum power will be transferred from source to load.

Figure 13-14(c) shows how a JFET at the front end of a voltmeter or oscilloscope will provide a very high input impedance and therefore not load the circuit under test. In this example, the meter will measure 10 V across R_2, and because the ohms per volt (Ω/V) rating of the voltmeter is 125 kΩ/V, the meter input impedance is

$$Z_{IN} = \Omega/V \times V_{Measured}$$
$$= 125\ k\Omega/V \times 10\ V = 1.25\ M\Omega$$

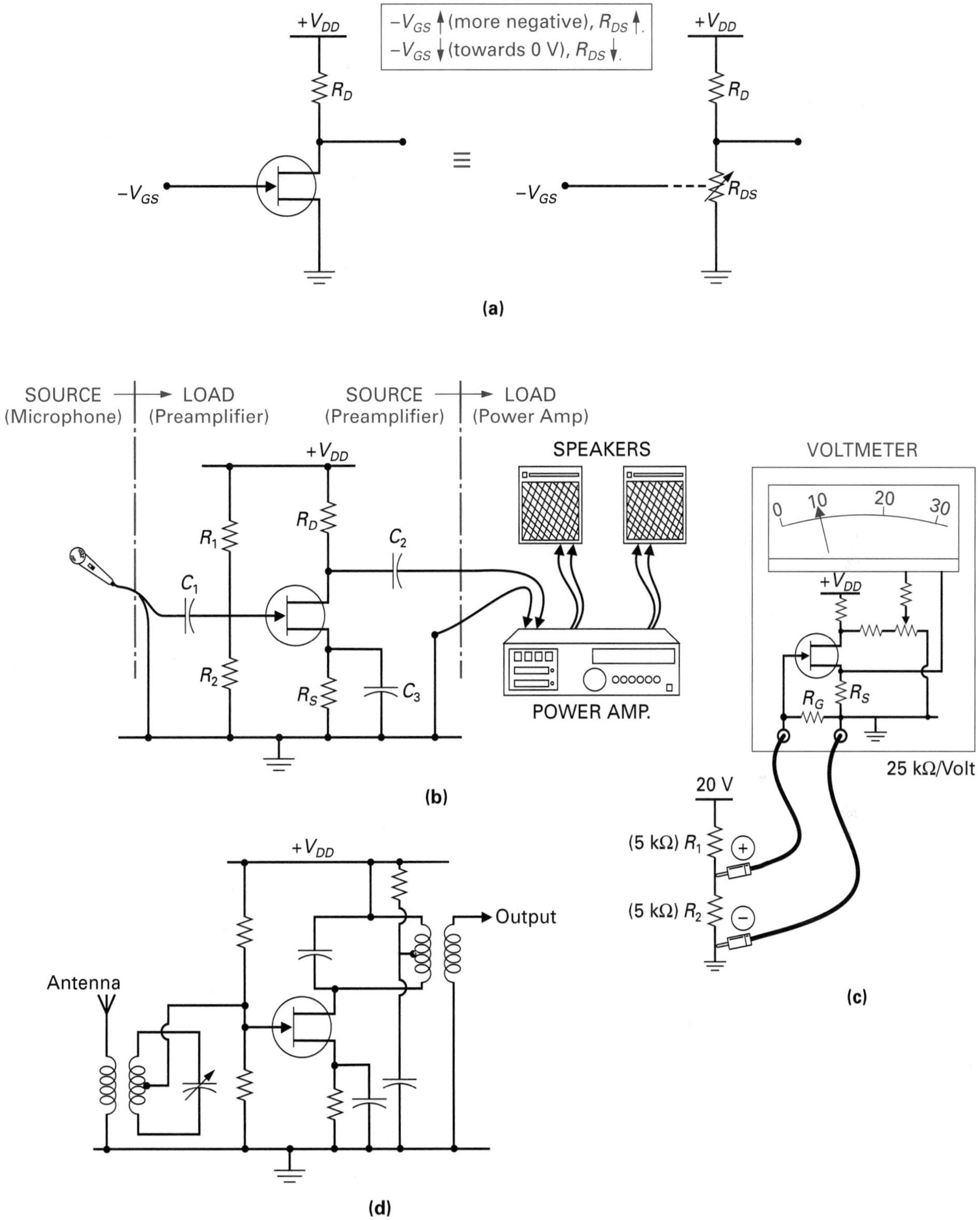

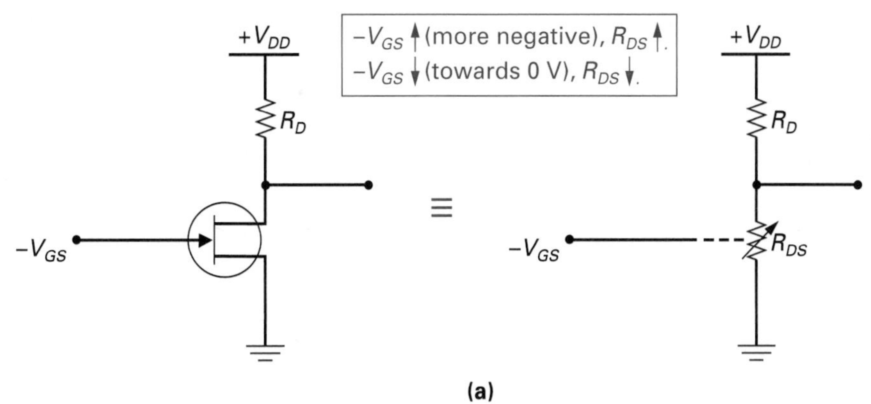

(a)

(b)

(c)

(d)

FIGURE 13-14 The JFET's Variable-Resistor Action—Analog Circuit Applications. (a) Equivalent Circuit. (b) Application 1: An Audio Preamplifier Circuit. (c) Application 2: A Voltmeter High Input Impedance Circuit. (d) Application 3: An RF Amplifier Circuit.

A 1.25 MΩ meter resistance in parallel with the 5 kΩ resistance of R_2 will have very little effect (1.25 MΩ in parallel with 5 kΩ = about 5 kΩ), so an accurate reading will be obtained.

Figure 13-14(d) shows how the JFET can be used as a radio frequency (RF) amplifier. Studying this circuit, you can see that both the gate and drain contain tuned circuits in the same way as the previously discussed BJT RF amplifier circuits. However, there are two advantages that the JFET has over the bipolar transistor as a front end RF amp.

1. The very weak signals injected into the antenna will have a very small value of current. Because the JFET is a voltage operated device, it requires no input current, and it will respond well to the small voltage signal variations picked up by the antenna.
2. The JFET is a very low noise component. Because any noise generated at the front end will be amplified along with the signal at each of the following amplifier stages, this JFET characteristic is ideal in this application.

SELF-TEST REVIEW QUESTIONS FOR SECTION 13-4

1. Which JFET characteristic is made use of in most circuit applications?
2. Why are FET integrated circuits generally four times more densely packed than BJT integrated circuits?
3. Why is the JFET ideal as an RF preamplifier?

13-5 TESTING JFETS AND TROUBLESHOOTING JFET CIRCUITS

One of the nice things about testing JFETs, or troubleshooting JFET circuits, is that they only have one P-N junction. In this section we will examine how to apply our testing and troubleshooting skills to JFET devices and circuits.

13-5-1 *Testing JFETs*

The transistor tester shown in Figure 13-15(a) can be used to test both BJTs and FETs. This tester can be used to determine

1. Whether an open or short exists between any of the terminals,
2. The FET's transconductance/gain, and
3. The FET's value of I_{DSS} and leakage current.

If a transistor tester is not available, the ohmmeter can be used to detect the most common failures: opens and shorts. Figure 13-15(b) indicates what resistance values should be obtained between the terminals of a good *n*-channel and *p*-channel JFET. Looking at these ohmmeter readings, you can see that

(a)

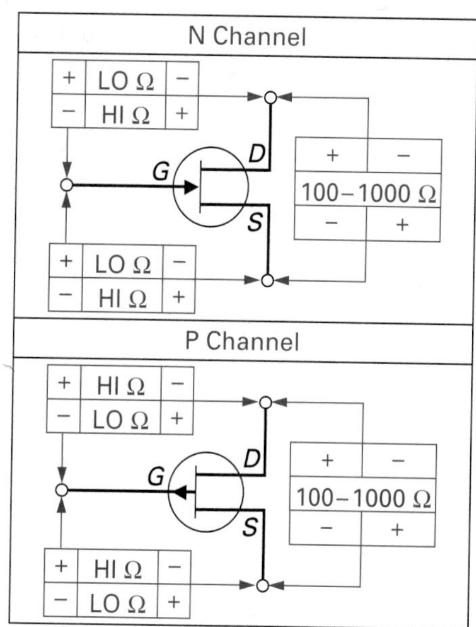

(b)

$+V_{DD}$

I_1 R_1 I_D

I_G

I_2 R_2 R_S I_S

$I_1 = I_2$ $I_S = I_D$

$I_G = 0$

NORMAL CONDITION

(c)

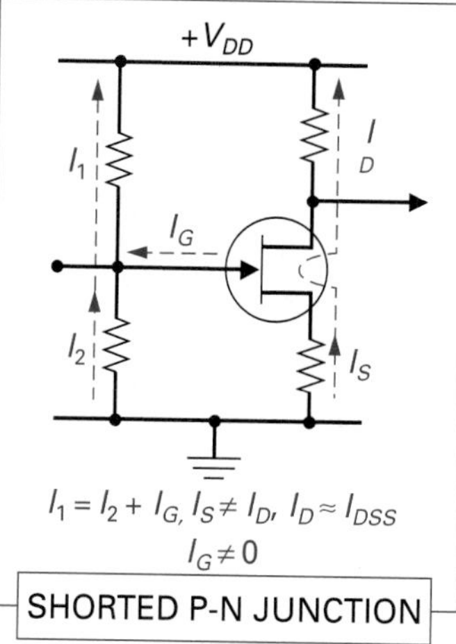

(d)

$+V_{DD}$

I_1 I_D

I_G

I_2 I_S

$I_1 = I_2$, $I_G = 0$, $I_S = I_D \approx I_{DSS}$

I_D is constant despite change in V_{GS}

OPEN P-N JUNCTION

(e)

Cause	Effect
R_2 Open	With R_2 open, $V_2 = +V_{DD}$, and JFET is heavily forward biased and therefore destroyed. Replacement JFET will also be destroyed unless R_2 open is first fixed.
R_1 Open	R_2 provides self-bias. Gate-source bias will increase and output signal may be clipped.
R_D Open	With no drain supply voltage, I_D and I_S will be zero and the JFET will be OFF.
R_S Open	With R_S open, the supply voltage will appear across R_S ($V_S = +V_{DD}$).

(f)

FIGURE 13-15 Testing JFETs and Troubleshooting JFET Circuits. (Photo courtesy of Sencore, Inc.— Test Equipment for the Professional Servicer. 1-800-SENCORE.)

because the JFET has only one P-N junction (gate-to-channel), it is relatively simple to test with an ohmmeter for an open or shorted junction.

13-5-2 *Troubleshooting JFET Circuits*

Let us now examine how to troubleshoot JFET circuits. As with all other circuit troubleshooting, the first step is to isolate whether the problem is within the JFET circuit or external to the JFET circuit. Figure 13-15(c) shows the "normal condition" for a typical voltage divider biased JFET circuit. Because I_G is normally zero, I_1 should equal I_2, and, as we already know, $I_S = I_D$. Because the JFET has only one P-N junction, the two basic circuit problems that can develop are as follows.

1. A shorted gate-source junction will have the effect shown in Figure 13-15(d). With a short between gate and source, there will be no voltage difference between gate and source (V_{GS} will be 0 V), V_G will be a positive voltage, and I_D will be a maximum ($I_D = I_{DSS}$). The gate-source short will allow current to flow out of the gate, and therefore I_S will not equal I_D. Because I_G will combine with I_1, I_1 will not equal I_2. More than likely, the gate current will cause the gate junction to open.
2. An open-gate junction will have the effect shown in Figure 13-15(e). With the gate open, the applied V_{GS} voltage will be present at the gate, and therefore everything will seem normal. However, because V_{GS} is 0 V due to the open gate, I_D will equal maximum ($I_D = I_{DSS}$) and be constant despite changes in the input V_{GS}.

If the JFET circuit component at fault is not the JFET itself, we will need to isolate the effect we are getting, and then try to determine the cause. Figure 13-15(f) lists the symptoms you should get from different JFET circuit component failures.

SELF-TEST REVIEW QUESTIONS FOR SECTION 13-5

1. Using a transistor tester to test a 2N5484 JFET, what typical readings should we obtain for the following:
   a. $V_{GS(OFF)}$
   b. $I_{DSS(MAXIMUM)}$
2. Will the gate-to-source and gate-to-drain of a JFET test with an ohmmeter like any other P-N junction?

SUMMARY

1. The field effect transistor, more commonly called an FET (pronounced "eff-ee-tee"), has three terminals and can operate as a switch—and therefore be used in digital circuit applications—or as a variable resistor—and therefore be used in analog or linear circuit applications.

2. There are two types of field effect transistors or FETs. One type is the junction field effect transistor, which is more typically called a JFET (pronounced "jay-fet"). The other type is the metal oxide semiconductor field effect transistor, which is more commonly called a MOSFET (pronounced "moss-fet").

The Junction Field Effect Transistor—JFET (Figure 13-16)

3. Like the bipolar junction transistor, the junction field effect transistor or JFET is constructed from *n*-type and *p*-type semiconductor materials.
4. Just as the bipolar junction transistor has two basic types (NPN BJT or PNP BJT), there are two types of junction field effect transistor called the *n*-channel JFET and *p*-channel JFET.
5. The more frequently used *n*-channel JFET consists of an *n*-type block of semiconductor material on top of a *p*-type substrate with a "U" shaped *p*-type section attached to the surface of a *p*-type substrate.
6. Like the BJT, the JFET has three terminals called the gate, source, and drain. With the *n*-channel JFET, the gate lead is attached to the *p*-type substrate, and the source and drain leads are attached to either end of an *n*-type channel that runs through the middle of the "U" shaped *p*-type section.
7. With the schematic symbol of the *n*-channel JFET, the gate lead's arrowhead points into the device.
8. The *p*-channel JFET is constructed in exactly the same way as the *n*-channel JFET except that the gate lead is attached to an *n*-type substrate, and the source and drain leads are attached to either end of a *p*-type channel.
9. With the schematic symbol of the *p*-channel JFET, the gate's arrowhead points out of the device.
10. The JFET requires both a drain supply voltage ($+V_{DD}$) and a gate-source bias voltage (V_{GS}) in order to operate. The $+V_{DD}$ bias voltage is connected between the drain and source of the *n*-channel JFET and will cause a current to flow through the *n*-channel. This source-to-drain current, which is made up of electrons because they are the majority carriers within an *n*-type material, is called the JFET's drain current (I_D). The value of drain current passing through a JFET's channel is dependent on two elements: the value of $+V_{DD}$ applied between the drain and source and the value of V_{GS} applied between gate and source.
11. In most circuit applications, the $+V_{DD}$ supply voltage is maintained constant and the V_{GS} input voltage is used to control the resistance of the channel and the value of the output current, I_D.
12. The gate-to-source junction of an FET is normally always reverse biased by the input voltage (V_{GS}), and it is this input voltage that controls the output current (I_D) and therefore output voltage (V_{OUT}). Because the gate-to-source junction of an FET is normally always reverse biased by the input voltage (V_{GS}), there will be no input current. This characteristic accounts for the FET's naturally high input impedance.

QUICK REFERENCE SUMMARY SHEET

FIGURE 13-16 The Junction Field Effect Transistor (JFET).

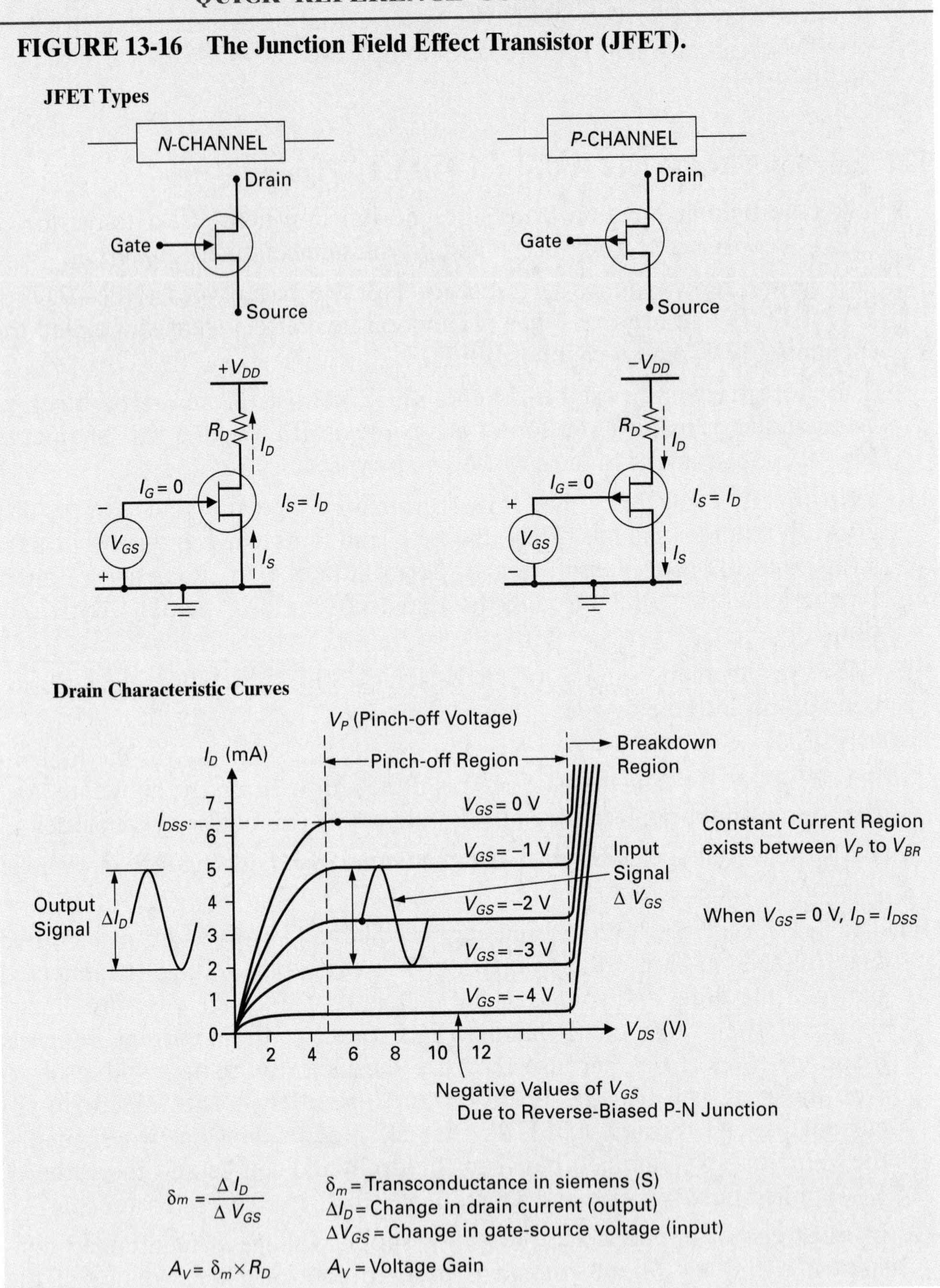

13. The difference between an FET and a BJT is that an FET's input junction is normally reverse biased, and therefore the input voltage controls the output current. A BJT's input junction is normally forward biased, and therefore the input current controls the output current. This is why BJTs are known as current-controlled devices, and FETs are known as voltage-controlled devices.

14. The name "field-effect transistor" is derived from the device's voltage control action because the applied input voltage will generate an electric field. It is this electric field that varies the size of the depletion region, and therefore the resistance of the channel between the FET's drain and source output terminals.
15. The term "junction" is attached to this type of FET because of the single P-N junction formed between the gate and the source-to-drain channel. Therefore, an *n*-channel JFET has a single P-N junction between gate to channel, and a *p*-channel JFET has a single N-P junction between gate to channel.
16. The field effect transistor is also often referred to as a unipolar device since only one type of semiconductor material exists between the output terminals, and therefore the charged carriers have only one polarity (unipolar). Compare this to a BJT, which is a bipolar device because there is a change in semiconductor material between the output terminals, and the charged carriers can be one of two polarities (bipolar, since both majority and minority carriers are used).
17. The drain characteristic curve is a graph plotting drain current (I_D) against drain-to-source voltage (V_{DS}).
18. At a certain V_{DS} voltage, further increases in V_{DS} will cause no further increase in I_D. This value of V_{DS} is called the pinch-off voltage (V_P) because it is the point at which the bias voltage has caused the depletion region to pinch off, or restrict, drain current. From this point on, further increases in V_{DS} are counteracted by increases in the resistance of the channel, and therefore I_D remains constant. This flat portion in the drain characteristic curve is called the constant-current region because I_D remains constant despite changes in V_{DS}. If V_{DS} is further increased (by increasing $+V_{DD}$), the JFET will eventually reach its breakdown voltage (V_{BR}), at which time a damaging value of I_D will pass through the JFET.
19. When V_{GS} is at 0 V, a maximum value of drain current is passing through the JFET's channel. This maximum value of drain current is called the drain-to-source current with shorted gate (I_{DSS}).
20. If V_{GS} is made more and more negative, a point will be reached where the depletion regions within the JFET will get closer and closer—then eventually touch—cutting off drain current. This negative V_{GS} bias voltage that causes I_D to drop to approximately zero is called the gate-to-source cut-off voltage or $V_{GS(OFF)}$.
21. Like the bipolar transistor, an FET can be used to amplify a signal. As before, the amount of amplification achieved is a ratio between output and input. For a bipolar transistor, the amount of gain is equal to the ratio of input current to output current (beta). For an FET, there is no input current, and therefore an FET's gain is equal to the ratio of output current change (ΔI_D) to input voltage change (ΔV_{GS}). This ratio is called the FET's transconductance (symbolized δ_m).
22. Because transconductance is the ratio of output current change (ΔI_D) to input voltage change (ΔV_{GS}), it is no surprise that this ratio is used to determine a JFET's voltage gain.

JFET Biasing Methods (Figure 13-17)

23. The three most frequently used JFET biasing methods are gate biasing, self biasing, and voltage-divider biasing.
24. Gate biasing is the most simple of the three biasing methods.
25. With self biasing, only a single drain supply voltage is needed (V_{DD}), whereas gate-biasing requires both V_{DD} and a negative gate supply voltage ($-V_{GG}$).
26. With a self-biased JFET, we achieve the same result as gate-biasing by making the source voltage positive with respect to the gate, which is at 0 V. This makes the gate of the JFET negative with respect to the source.
27. The Q point will remain relatively stable despite changes in temperature when a JFET circuit has a source resistor included.

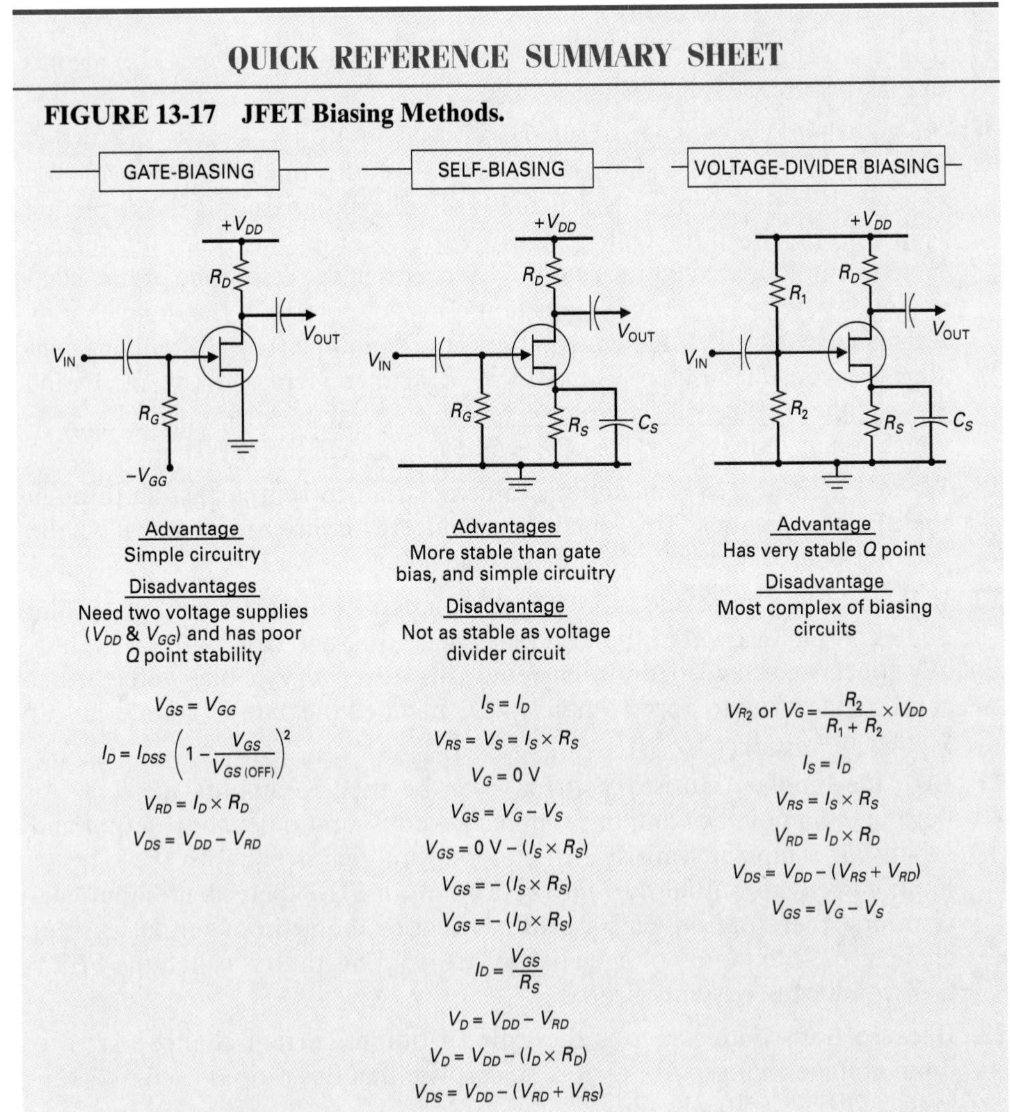

FIGURE 13-17 JFET Biasing Methods.

28. Like the self-biased circuit, the voltage-divider biased circuit includes a source resistor to stabilize the Q point despite ambient temperature changes. In addition, using a voltage divider to determine the gate-bias voltage ensures that V_{GS}, and therefore the circuit, has increased stability.

JFET Circuit Configurations (Figure 13-18)

29. The three JFET circuit configurations are called common-source, common-gate, and common-drain.

30. Similar to its bipolar counterpart, the common-emitter configuration, the common-source configuration is the most widely used JFET circuit. The input is applied between the gate and source and the output is taken between the drain and source, with the source being common to both input and output. Resistor R_S is included to provide temperature stability, and, as

QUICK REFERENCE SUMMARY SHEET

FIGURE 13-18 JFET Circuit Configurations.

	Voltage Gain	Input Impedance	Output Impedance	Circuit Appearance and Application	Waveforms
Common Source (a)	5–10 (Voltage Amp)	Very High 1–15 MΩ	Low 2–10 kΩ	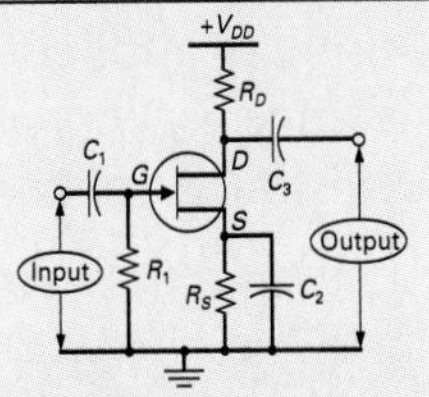 Most widely used FET configuration. It is mainly used as a voltage amplifier, however it is also used as an impedance matching device and can handle the high radio frequency signals.	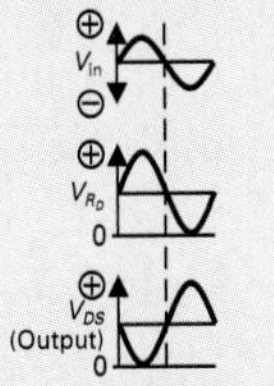 V_{in} and V_{out} are out of phase (180° phase shift)
Common Gate (b)	2-5	Very Low 200–1500 Ω	Medium 5–15 kΩ	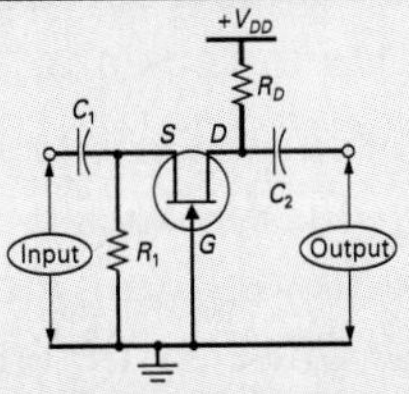 This configuration is used to amplify radio frequency signals due to its very stable nature at high frequencies. It is also used as a buffer to match a low impedance source to a high impedance load.	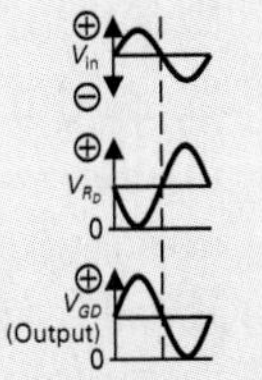 V_{in} and V_{out} are in phase (0° phase shift)
Common Drain (c)	0.98	Very High 1500 MΩ	Low 10 kΩ	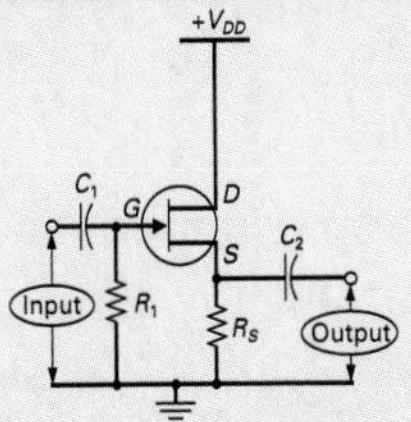 This amplifier is commonly called a source follower as the source follows whatever is applied to the gate. Its very high input impedance will not load down (and therefore not distort) signals from high impedance signal sources, such as a microphone, and its low output impedance is ideal to drive a low impedance load such as an audio amplifier.	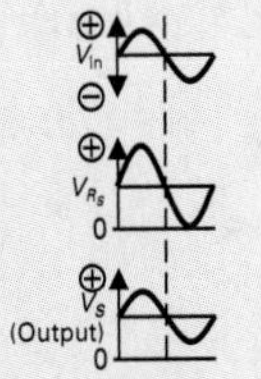 V_{in} and V_{out} are in phase (0° phase shift)

with the bipolar transistor, the source decoupling capacitor C_2 is included to prevent degenerative feedback.

31. When a small ac input signal is applied to the gate of a common-source amplifier, the variations in voltage at the gate control the JFET, which effectively acts as a variable resistor, varying the output drain current. These changes in drain current will vary the voltage drop across R_D and the drain-to-source voltage drop that, with R_S, determines the output voltage.
32. The output voltage (V_{OUT}) of the common-source JFET configuration can be five to ten times larger than the gate-control input voltage (V_{IN}).
33. The common-source circuit has a very high input resistance and relatively low output resistance. The high input resistance is due to the JFET's reverse biased gate-source junction, which permits no gate input current, and therefore has a very large resistance. This key characteristic means that the common-source JFET circuit is ideal in applications where we need to provide voltage amplification but do not want to load down a source that can only generate a small input signal.
34. The common-gate circuit configuration is very similar to its bipolar counterpart, the common-base circuit. The input is applied between the source and gate, while the output appears across the drain and gate.
35. The common-gate circuit can be used to provide a small voltage gain. Because the input is applied to the JFET's high current source terminal, the input resistance is very low. This low input resistance and relatively high output resistance makes the circuit ideal in applications where we need to efficiently transfer power between a low-resistance source and a high-resistance load.
36. Comparable to the bipolar transistor's common-collector, or emitter follower, the common-drain configuration is sometimes called a source-follower because the source output voltage follows in polarity and amplitude the input voltage at the gate.
37. With the common-drain circuit, the output voltage is slightly less than the input voltage (circuit does not provide any voltage gain). The input resistance of the common-drain circuit configuration is extremely high due to the JFET's reverse biased gate and R_L connection, and the output resistance is relatively very low. Inserting a common-drain circuit between a high-resistance source and a low-resistance load will ensure that the two opposite resistances are matched, and therefore power is efficiently transferred.

JFET Applications

38. It is the high input impedance of the JFET, and therefore its ability not to load a source, and the voltage amplification ability that are mainly made use of in circuit application. Like the bipolar junction transistor, the JFET can be made to function as a switch or as a variable resistor.
39. As a switch, the JFET makes use of only two points on the load line: saturation (in which it is equivalent to a closed switch between source and drain) and cutoff (in which it is equivalent to an open switch between source and drain).

40. A typical FET application in digital circuits would be a buffer circuit, which is used to isolate one device from another. The high input impedance of the FET does not load the input circuit or circuits, while the low output impedance of the FET provides a high output current to the output circuit. The high output current and buffering, or isolating, characteristics of these circuits account for why they are also called buffer-drivers.

41. The high input impedance ($Z_{IN}\uparrow$) of the FET is also made use of in other FET integrated circuits (ICs). When Z_{IN} is high, circuit current is low ($I\downarrow$) and therefore power dissipation is low ($P_D\downarrow$). This condition is ideal for digital integrated circuits (ICs) which contain thousands of transistors all formed onto one small piece of silicon. The low power dissipation of the JFET enables us to densely pack many more components into a very small area.

42. In two-state applications, the JFET is made to operate between the two extreme points of saturation (0 Ω) and cutoff (max. Ω). By controlling the gate-source bias voltage (V_{GS}), the resistance between the JFET's drain and source (R_{DS}) can be changed to be any value between 0 Ω and maximum Ω. The JFET can therefore be made to act as a variable resistor with an increase in negative V_{GS} causing a larger R_{DS} and, in contrast, a decrease in negative V_{GS} causing a smaller R_{DS}.

43. It is the reverse-biased gate-source junction of a JFET that gives the JFET its key advantage: an extremely high input impedance (typically in the high MΩ range). In addition, the JFET can provide a small voltage gain and has been found to be a very low noise component. All these characteristics make it an ideal choice as an amplifier.

44. The *C-S* circuit is typically used at the front end of systems such as the first RF amplifier stage following the antenna and the first stage in a voltmeter. It will not load the source, yet it will amplify a wide range of input signal voltages.

Testing JFETs and Troubleshooting JFET Circuits (Figure 13-19)

45. The transistor tester can be used to test both BJTs and FETs. This tester can be used to determine

    a. Whether an open or short exists between any of the terminals,
    b. The FET's transconductance/gain, and
    c. The FET's value of I_{DSS} and leakage current.

46. The ohmmeter can be used to detect the most common JFET failures: opens and shorts.

47. As with all other circuit troubleshooting, the first step is to isolate whether the problem is within the JFET circuit or external to the JFET circuit.

48. Because the JFET has only one P-N junction, the two basic circuit problems that can develop are as follows.

    a. With a shorted gate-source junction there will be no voltage difference between gate and source (V_{GS} will be 0 V), and so V_G will be a positive voltage, and I_D will be a maximum ($I_D = I_{DSS}$). The gate-source short will allow current to flow out of the gate, and therefore I_S will not equal I_D.

QUICK REFERENCE SUMMARY SHEET

FIGURE 13-19 Testing JFETs and Troubleshooting JFET Circuits.

Ohmmeter Test

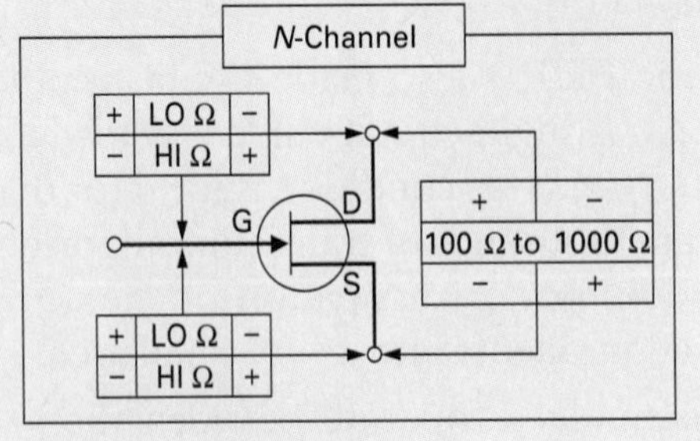

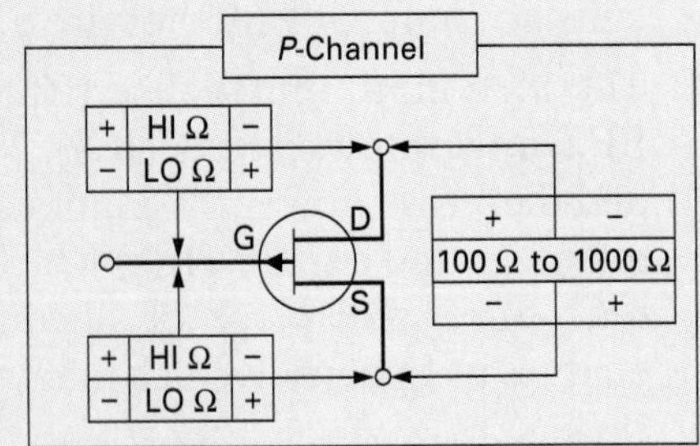

Circuit Troubleshooting

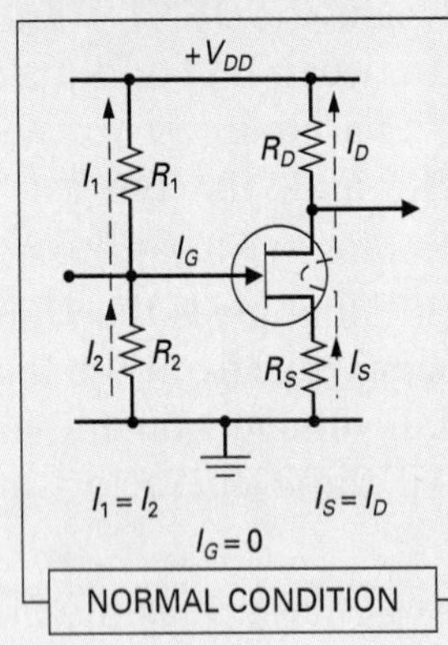

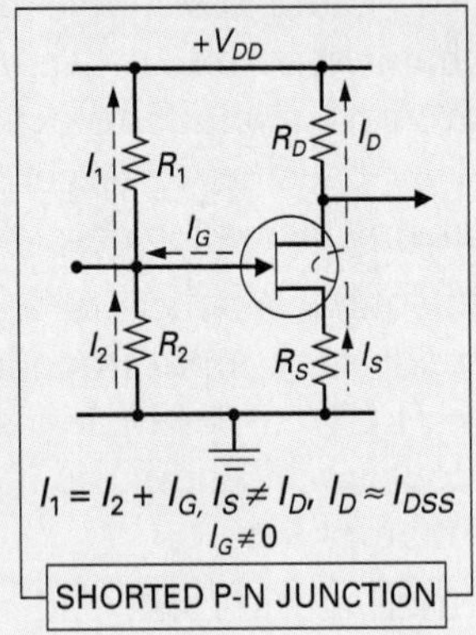

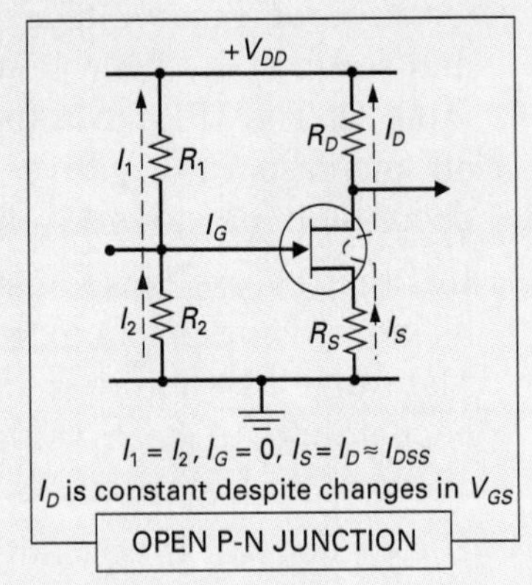

Because I_G will combine with I_1, I_1 will not equal I_2. More than likely, the gate current will cause the gate junction to open.

b. With an open gate junction, the applied V_{GS} voltage will be present at the gate, and therefore everything will seem normal. However, because V_{GS} is 0 V due to the open gate, I_D will equal maximum ($I_D = I_{DSS}$) and be constant despite changes in the input V_{GS}.

NEW TERMS

Bipolar Device
Breakdown Voltage
Buffer-Driver
Channel
Common-Drain Circuit Configuration
Common-Gate Circuit Configuration
Common-Source Circuit Configuration
Constant-Current Region
Current-Controlled Device
Drain
Drain Characteristic Curve
Drain Current
Drain Supply Voltage
Drain-to-Source Current with Shorted Gate (I_{DSS})
Field Effect Transistor (FET)
Gate
Gate Biasing
Gate Source Bias Voltage

Gate-to-Source Cut-Off Voltage ($V_{GS(OFF)}$)
Junction Field Effect Transistor (JFET)
Metal Oxide Semiconductor Field Effect Transistor (MOSFET)
n-Channel JFET
Negative Feedback
p-Channel JFET
Pinch-Off Voltage
Preamplifier
Self Biasing
Source
Source Follower
Temperature Stability
Transconductance
Unipolar Device
Voltage-Controlled Device

REVIEW QUESTIONS

Multiple-Choice Questions

1. A (an) ____________ has three terminals called the gate, source, and drain.

a. BJT **c.** Bipolar Transistor
b. Zener **d.** JFET

2. The ____________ channel JFET schematic symbol has the arrow pointing out while the ____________ channel JFET schematic symbol has the arrow pointing in.

a. *D-S, p* **b.** *p, G-S* **c.** *n, p* **d.** *p, n*

3. The BJT is a ____________ controlled device whereas the FET is a ____________ controlled device.

a. voltage, current **b.** current, voltage

4. The gate-source junction of a JFET is always ____________ biased since the input voltage is normally ____________ or some ____________ voltage.

a. reverse, 0 V, negative **c.** reverse, –4 V, positive
b. forward, 0 V, negative **d.** forward, –4 V, negative

5. When $V_{GS} = V_{GS(OFF)}$, $I_D = ?$

a. I_{DSS} **c.** Maximum
b. Zero **d.** V_P

6. Which JFET circuit configuration is also known as a source follower?

a. Common-Drain **c.** Common-Gate
b. Common-Source **d.** Both (a) and (b) are true

7. Which JFET circuit configuration provides a high input impedance and a good voltage gain?

a. Common-Drain **c.** Common-Gate
b. Common-Source **d.** Both (a) and (c) are true

8. Which biasing method makes use of a $-V_{GG}$ supply voltage?

a. Self Biasing **c.** Base Biasing
b. Gate Biasing **d.** Voltage-Divider Biasing

9. Which biasing method uses a source resistor?

a. Self Biasing **c.** Voltage-Divider Biasing
b. Gate Biasing **d.** Both (a) and (c)

10. Transconductance is a ratio of

a. ΔI_D to ΔV_{DS} c. ΔI_D to ΔV_{GS}
b. ΔV_{GD} to ΔI_D d. ΔV_{GS} to ΔV_{DS}

11. The input resistance of an FET is much higher than the input resistance of a BJT.

a. True b. False

12. With a junction FET, as V_{GS} is made more negative the depletion region will ____________, the channel size will get ____________, and therefore I_D will ____________.

a. decrease, larger, increase c. decrease, smaller, increase
b. increase, smaller, decrease d. increase, larger, decrease

13. Which JFET circuit configuration has phase inversion between input and output?

a. Common-Drain c. Common-Gate
b. Common-Source d. Both (b) and (c)

14. With a *p*-channel JFET, the gate should be ____________ with respect to the source, and the drain should be ____________ with respect to the source.

a. negative, negative c. positive, positive
b. negative, positive d. positive, negative

15. The current between ____________ leads of a JFET is controlled by varying the reverse bias voltage applied to the ____________ leads.

a. gate and source, source and drain c. source and drain, gate and source
b. source and drain, drain and gate d. drain and gate, gate and source

Essay Questions

16. What are the basic differences between a BJT and an FET? (13-1)
17. Identify the schematic symbols shown in Figure 13-20. (13-1-1)
18. Give the full names of the following abbreviations

a. JFET (13-1) d. V_{GS} (13-1-2)
b. MOSFET (Intro.) e. I_{DSS} (13-1-3)
c. BJT (Intro.) f. $V_{GS(OFF)}$ (13-1-3)

19. Name the two different types of JFETs. (13-1-1)
20. Briefly describe the operation of a JFET. (13-1-2)

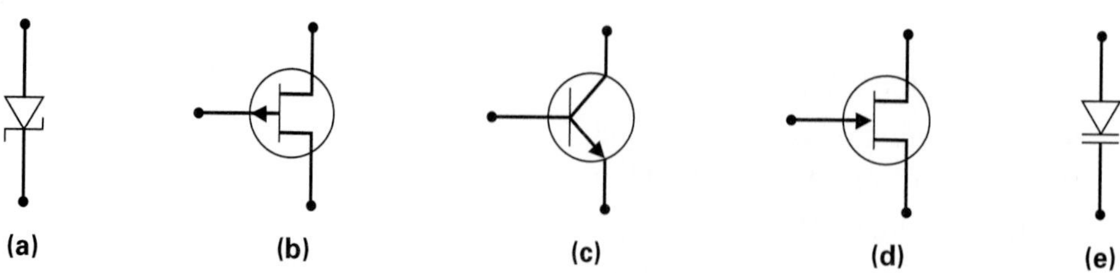

FIGURE 13-20 Schematic Symbols.

21. Why is an FET referred to as a voltage-operated device, and the BJT referred to as a current-operated device? (13-1-2)
22. Name the two different types of FETs. (Intro.)
23. In relation to the drain characteristic curve of a JFET, define the following points: (13-1-3)
    a. Pinch-Off Voltage (V_P)
    b. Constant Current Region
    c. Breakdown Voltage (V_{BR})
    d. Drain-to-Source Current with Shorted Gate (I_{DSS})
    e. Gate-to-Source Cut-Off Voltage ($V_{GS(OFF)}$)
24. What is transconductance? (13-1-4)
25. What key JFET characteristic is made use of in application circuits? (13-4)
26. List the three different JFET circuit configurations, and briefly describe their characteristics. (13-3)
27. Briefly describe the following JFET biasing methods: (13-2)
    a. Gate Biasing **c.** Voltage-Divider Biasing
    b. Self Biasing
28. Why is the FET better suited to miniaturization than the BJT? (13-4-1)
29. Describe some of the typical digital and analog circuit applications of the JFET. (13-4)
30. What two specifications need to be obtained from a manufacturer's data sheet in order to determine JFET circuit current and voltage values? (13-1-6)

Practice Problems

31. Referring to Figure 13-5, you can see that when the gate of a 2N5484 JFET is reverse biased (normal operation), the gate reverse current (I_{GSS}) = 1.0 nA when V_{GS} = –20 V (V_{DS} = 0 V). Calculate the input or gate-source impedance of the JFET.
32. Calculate the following for the amplifier circuit in Figure 13-21:
    a. Transconductance **b.** Voltage Gain
33. Calculate the following for the circuit in Figure 13-22:
    a. V_{GS} **d.** $I_{D(MAXIMUM)}$
    b. I_D **e.** Q point
    c. V_{DS}
34. Calculate the following for the circuit in Figure 13-23:
    a. V_S **b.** V_{GS} **c.** V_{DS}
35. Calculate the following for the circuit in Figure 13-24:
    a. V_G **b.** V_S **c.** V_{GS} **d.** V_{DS}
36. Referring to Figure 13-25,
    a. What is the circuit configuration?
    b. What is the circuit's voltage gain?
    c. What is a circuit like this typically used for?

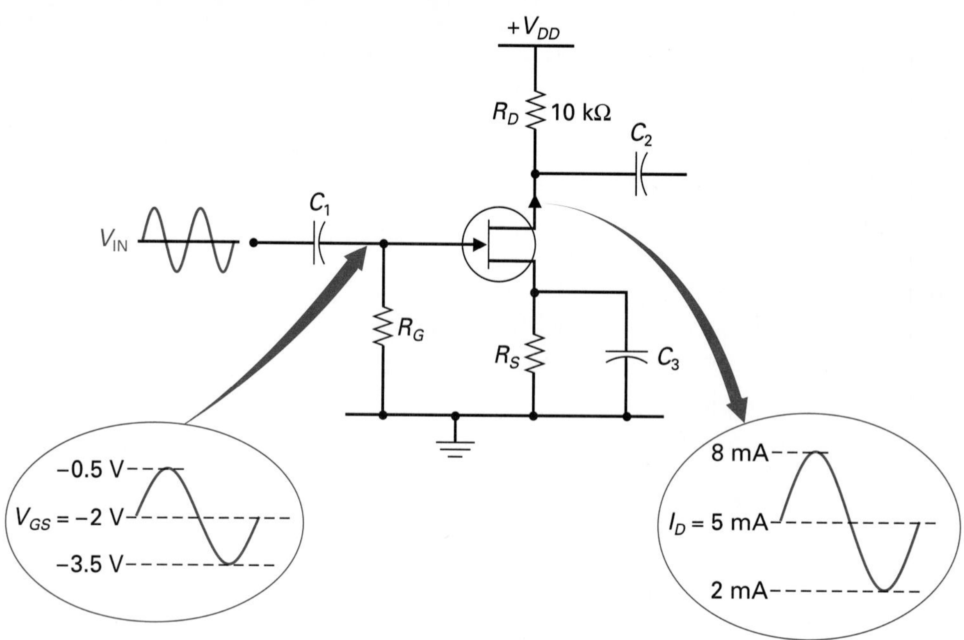

FIGURE 13-21 A Common-Source Amplifier.

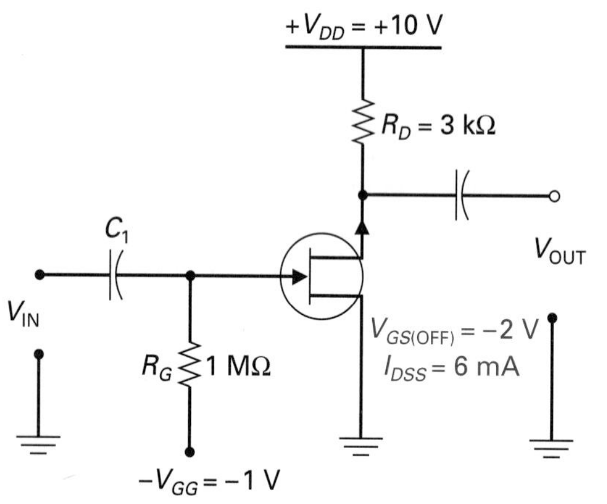

FIGURE 13-22 A Gate-Biased JFET Circuit.

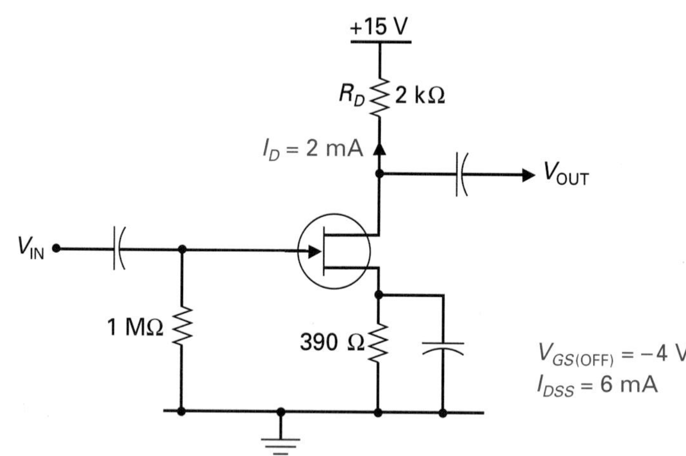

FIGURE 13-23 A Self-biased JFET Circuit.

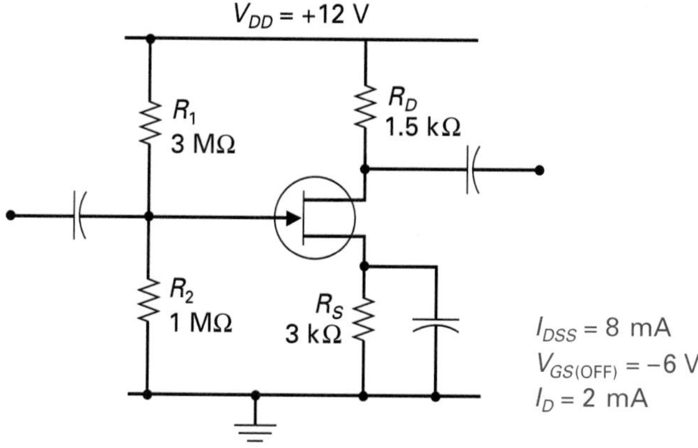

FIGURE 13-24 A Voltage Divider Biased JFET Circuit.

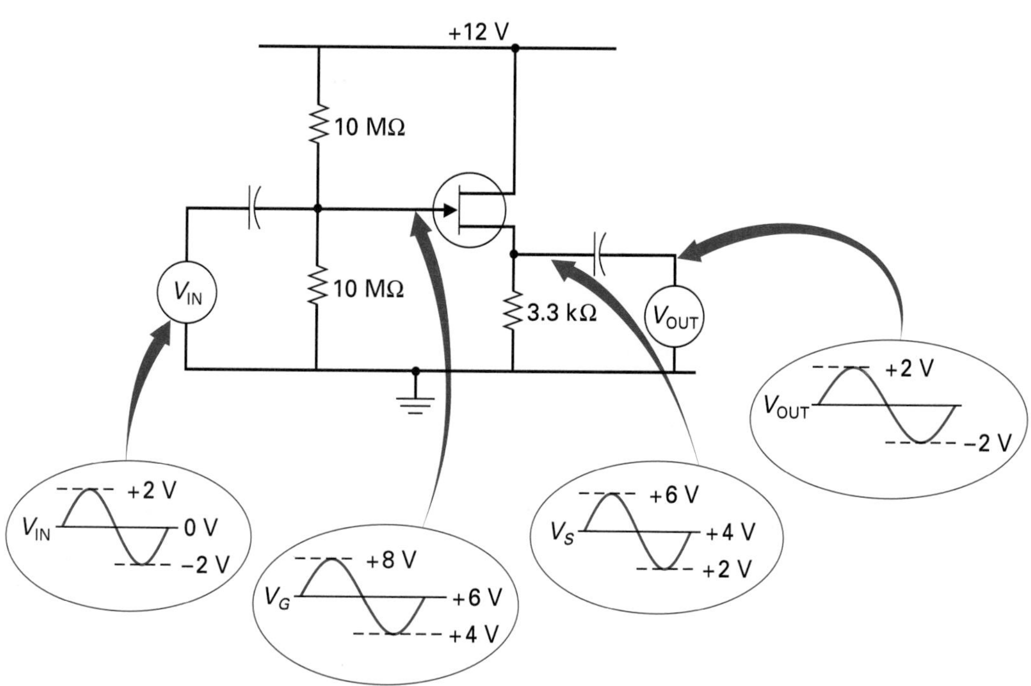

FIGURE 13-25 An Analog JFET Circuit.

37. Referring to the circuit in Figure 13-26,

a. What is the circuit configuration?
b. What will be the output when the input is 0 V?
c. What will be the output when the input is –6 V?
d. Is there phase inversion between input and output?

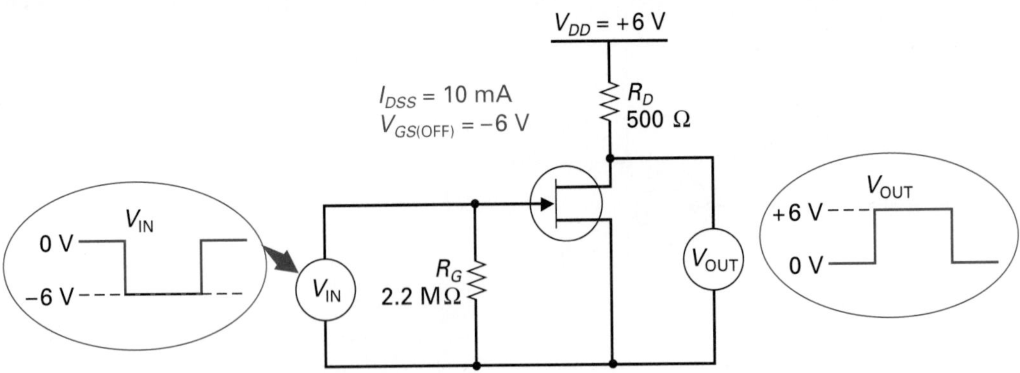

FIGURE 13-26 **A Digital JFET Circuit.**

Troubleshooting Questions

38. Briefly list what characteristics of a JFET can be tested with a transistor tester.

39. How can an ohmmeter test JFETs?

40. Which of the JFETs in Figure 13-27 are good, and which are bad? If bad, state the suspected problem.

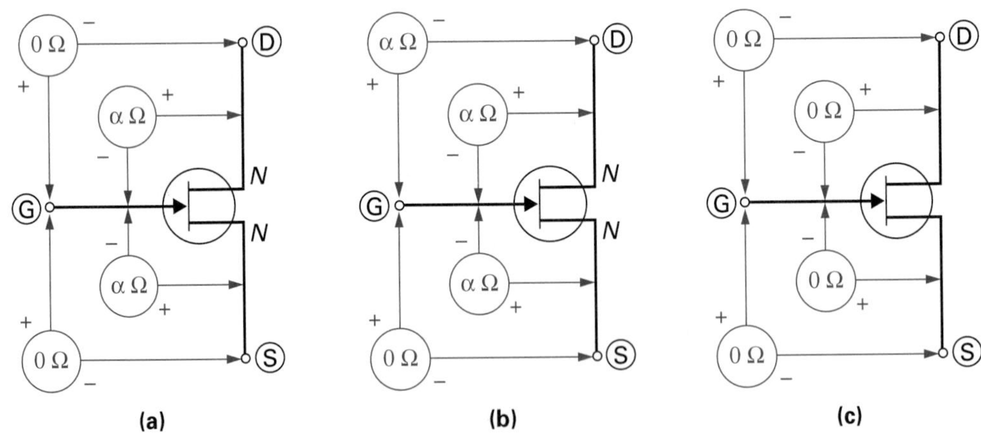

FIGURE 13-27 **Testing JFETs with the Ohmmeter.**

CHAPTER

14

OBJECTIVES

After completing this chapter, you will be able to:

1. Explain the meaning of the term "metal oxide semiconductor field effect transistor."

2. Define the terms "depletion-mode operation" and "enhancement-mode operation."

3. Describe how the D-MOSFET is constructed, how it operates, and its characteristics.

4. Explain how the D-MOSFET is usually biased.

5. Describe the typical applications of the D-MOSFET.

6. Discuss the operation, characteristics, and application of the dual-gate D-MOSFET.

7. Describe how the E-MOSFET is constructed, how it operates, and its characteristics.

8. Explain how the E-MOSFET can be biased.

9. Describe the typical applications of the E-MOSFET.

10. Explain the operation and characteristics of the following digital logic circuits:
   a. PMOS
   b. NMOS
   c. CMOS

11. Discuss the operation and applications of the vertical-channel E-MOSFET.

12. Discuss the specifications listed on a typical MOSFET's data sheet.

13. Describe the precautions that should be observed when handling MOSFETs.

14. Explain how MOSFETs are tested.

Metal Oxide Semiconductor Field Effect Transistors (MOSFETs)

OUTLINE

VIGNETTE

Copy Master

Born in 1906 to two invalid Swedish immigrants, Chester Carlson worked part-time day and night to support his parents and still achieved excellent grades at high school. Sadly, when Carlson was seventeen both of his parents died, so Carlson left New York and went to the California Institute of Technology where he graduated in 1930 with a degree in physics.

Carlson's first job was as a researcher for Bell Telephone Laboratories in New York. It was here that he saw the need for a machine that could copy documents—the method at that time was to have someone in a typing pool retype the original. Carlson realized that available photographic methods were too messy and time consuming. He left Bell Labs in 1935 and began developing a clean and quick copying machine in the small rented bedroom that he had converted into a laboratory.

On the morning of October 23, 1938, Carlson statically charged a metal sulfur-coated plate by rubbing it with his handkerchief. He then exposed the plate to a glass slide that had on it the date and place "10-23-38 ASTORIA." When dry black powder on paper was pressed against the metal plate, the world's first photocopy was created.

Refining the design took Carlson several years because he had to develop more sensitive plates and a powder that would stick to paper. In fact, even after he had a good working prototype, it took him two years to find a company that was interested in manufacturing and selling his machine. However, in January, 1947, Haloid Company of Rochester, New York, which was a small photography firm, signed an agreement with Carlson for what he called his "dry printing machine." They were very dubious about whether it would really catch on and told Carlson that they would not pay him for his invention, but only give him a percentage of the profits. The company called the process "Xerography," which meant dry printing. After an almost overnight success, the company gave up all of their other products and renamed the company after their product, "Xerox Corporation."

The early Xerox machines in 1950 needed an operator to actuate the mechanism. However, by 1960 a fully automatic machine produced perfect copies by pressing one button.

Carlson's invention that nobody wanted transformed his lifestyle from a rented bedroom to royalties that paid him several million dollars a year.

INTRODUCTION

As discussed in the previous chapter, there are two basic types of field effect transistor, or FET. In the previous chapter, we concentrated on all aspects of the junction field effect transistor, or JFET. In this chapter, we will be examining the second type of FET, which is called a "metal oxide semiconductor FET," or MOSFET (pronounced "moss-fet").

With the JFET, an input voltage of zero volts would reverse bias the P-N junction, resulting in a maximum channel size and a maximum value of source to drain current. To decrease the size of the channel, the input voltage was made negative to further reverse bias the gate-source junction. This action would deplete the channel of free carriers, reducing the size of the channel and therefore the source-to-drain current. This type of action is actually called "depletion-mode operation," because an input voltage is used to deplete the channel, and therefore reduce the channel's size and current. The MOSFET does not have a P-N gate-channel junction like the JFET. It has a "metal gate" that is insulated from the "semiconductor channel" by a layer of "silicon dioxide," hence the name "metal oxide semiconductor." Like all "field effect transistors" (FETs), the input voltage will generate an "electric field" that will have the "effect" of changing the channel's size.

The key difference between the JFET and MOSFET is that the JFET's input voltage would always have to be zero or a negative voltage in order to reverse bias the gate-source junction. With the MOSFET, the input voltage can be either a positive or negative voltage since gate current will always be zero because the gate is insulated from the channel. Let us examine each of these input voltage possibilities.

1. If the input voltage is negative, the resulting electric field depletes the channel, reducing its size, and the MOSFET is said to be operating in the depletion mode.
2. If the input voltage is positive, the resulting electric field enhances the channel, increasing its size, and the MOSFET is said to be operating in the enhancement mode.

The MOSFET can therefore be operated in either the depletion or enhancement mode due to its insulated gate. The two different types of MOSFET are given names based on their normal mode of operation. For instance, the *depletion-type MOSFET (D-type MOSFET or D-MOSFET)* should actually be called a DE-MOSFET because it can be operated in both the depletion mode and the enhancement mode, whereas the *enhancement-type MOSFET (E-type MOSFET or E-MOSFET)* is correctly named because it can only be operated in the enhancement mode. In this chapter we will examine the construction, operation, characteristics, circuit biasing, applications, and testing of these two MOSFET types.

14-1 THE DEPLETION-TYPE (D-TYPE) MOSFET

Depletion-Type MOSFET
A field effect transistor with an insulated gate (MOSFET) that can be operated in either the depletion or enhancement mode.

***n*-Channel D-Type MOSFET**
A depletion type MOSFET having an *n*-type channel between its source and drain terminals.

***p*-Channel D-Type MOSFET**
A depletion type MOSFET having a *p*-type channel between its source and drain terminal.

The **depletion-type MOSFET** construction is slightly different from the JFET, and therefore we will need to first see how the D-MOSFET device is built before we can understand how it operates.

14-1-1 D-MOSFET Construction

Like the JFET and BJT, the D-MOSFET has two basic transistor types called the ***n*-channel D-type MOSFET** and ***p*-channel D-type MOSFET.** The construction and schematic symbol for these two D-MOSFET types are shown in Figure 14-1.

To begin with, let us examine the construction of the more frequently used *n*-channel D-type MOSFET, shown in Figure 14-1(a). This type of MOSFET basically consists of an *n*-type channel formed on a *p*-type substrate. A source and drain lead are connected to either end of the *n*-channel, and an additional lead is attached to the substrate. In addition, a thin insulating (silicon dioxide) layer is placed on top of the *n*-channel, and a metal plated area with a gate lead attached is formed on top of this insulating layer. Figure 14-1(b) shows the schematic symbol for an *n*-channel D-MOSFET, and, as you can see, the arrow on the substrate (*SS*) or base (*B*) lead points into the device. As a memory aid, imagine this arrowhead as a P-N junction diode as shown in the inset. The source and drain leads are connected to either end of the diode's *n*-type cathode. Therefore, this device must be an *n*-channel D-MOSFET. The basic difference in the construction and schematic symbol of the *p*-channel D-MOSFET can be seen in Figure 14-1(c) and (d).

Figure 14-1(e) and (f) show how the MOSFET is available as a four-terminal or three-terminal device. In some applications, a separate bias voltage will be applied to the substrate terminal for added control of drain current and the four-terminal device will be used. In most circuit applications, however, the three-terminal device, which has its source and substrate lead internally connected, is all that is needed.

14-1-2 D-MOSFET Operation

Figure 14-2(a) shows the typical drain characteristic curve for an *n*-channel depletion-type MOSFET. As you can see, this set of curves has the same general shape as the JFET's set of drain curves and the BJT's set of collector curves. The key difference is that V_{GS} is plotted for both positive and negative values. This is because the D-MOSFET should actually be called a DE-MOSFET because it can be operated in both the depletion mode (in which V_{GS} is a negative value) and the enhancement mode (in which V_{GS} is a positive value). To best understand the operation of the D-MOSFET, let us examine the three operation diagrams shown in Figure 14-2(b), (c), and (d).

Zero-Volt Operation: The center operation diagram, Figure 14-2(b), shows how the *n*-channel D-MOSFET will respond to a V_{GS} input of zero volts.

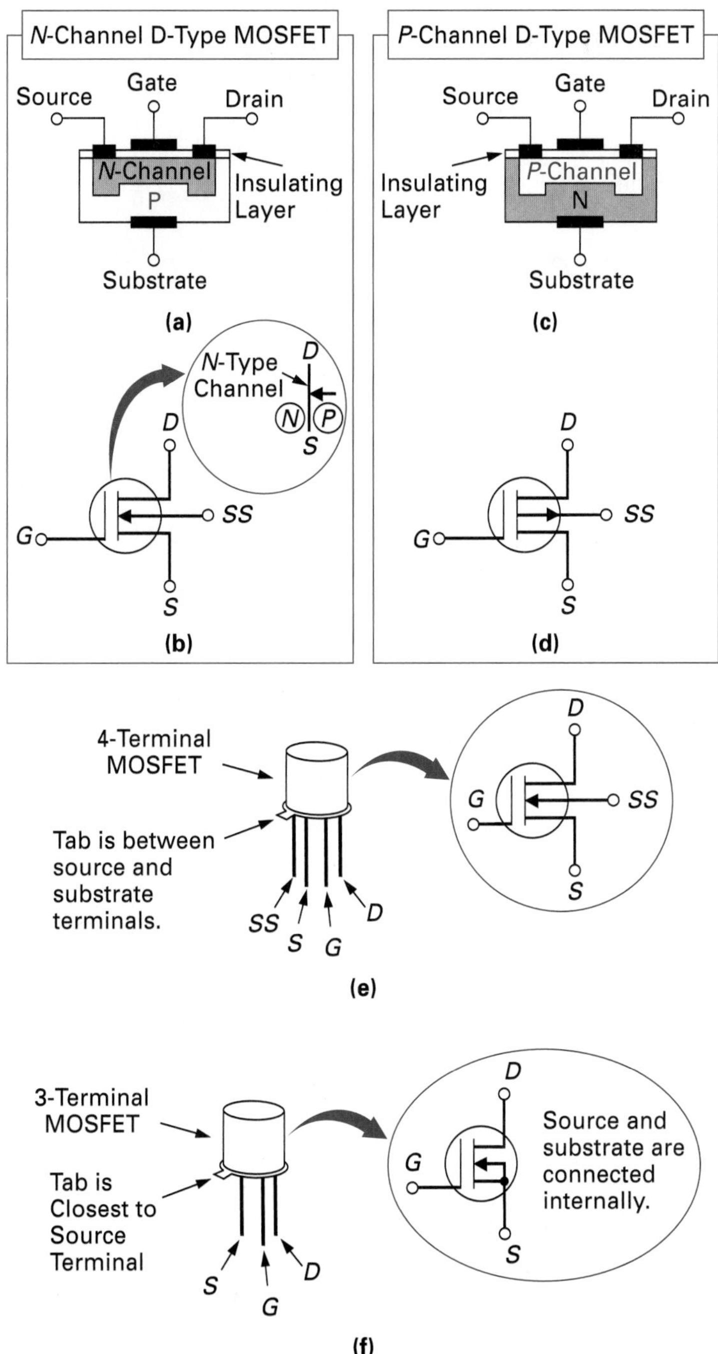

FIGURE 14-1 D-Type MOSFET Construction and Types.

When $V_{GS} = 0$ V, the gate and source terminals are at the same zero volt potential, and therefore the gate is effectively shorted to the source. As in Chapter 13, the value of drain current passing through the channel is called the I_{DSS} value (I_{DSS} is the drain-to-source current passing through the channel when the gate is shorted to the source). Therefore, when

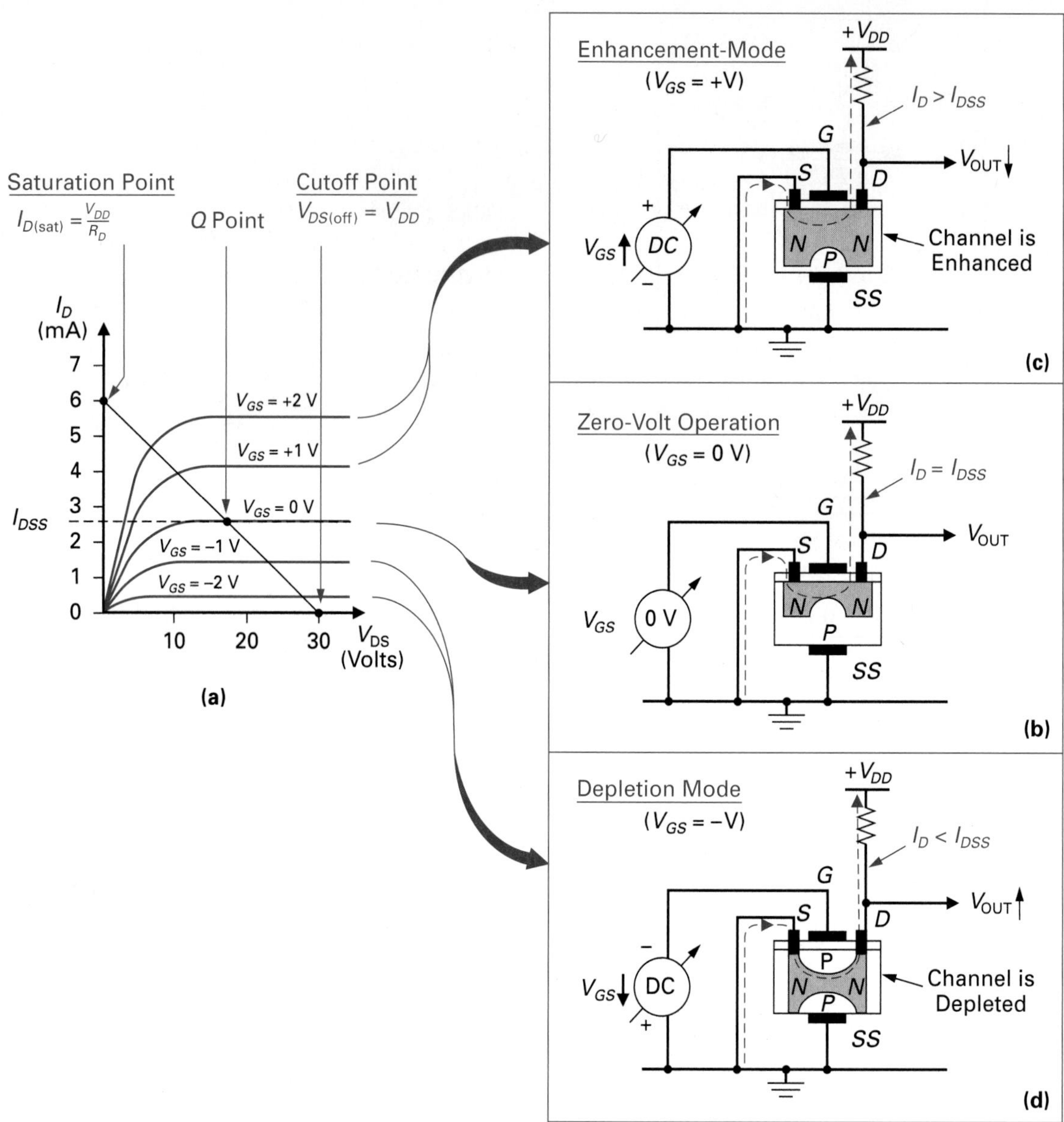

FIGURE 14-2 D-Type MOSFET Operation and Characteristics.

$$V_{GS} = 0 \text{ V}, I_D = I_{DSS}$$

When zero volts is applied to the input of a D-type MOSFET, therefore, it will conduct a value of drain current. With no input, this device is ON, which is why the D-type MOSFET is known as a "normally ON" device.

Enhancement-Mode: The upper operation diagram, Figure 14-2(c), shows how the *n*-channel D-MOSFET will respond when V_{GS} is made positive. In this condition, the channel is enhanced or widened, and the value of I_D is increased above I_{DSS}. Therefore, when

$$V_{GS} = +V, I_D > I_{DSS}$$

Let us examine in more detail why the channel is widened by a positive gate voltage. Because the valence-band holes in the p-type material (majority carriers) will be repelled by a positive gate voltage, and the conduction band electrons in the p-type material (minority carriers) will be attracted to the channel by the positive gate voltage, there will be a build-up of electrons in the p-type material near the channel. This build-up of electrons in the p-type material below the channel will effectively widen the size of the channel, reducing its resistance, and therefore increasing I_D to a value greater than I_{DSS}.

Depletion-Mode: The lower operation diagram, Figure 14-2(d), shows how the n-channel D-MOSFET will respond when V_{GS} is made negative. In this condition, the channel is depleted of free carriers, and therefore the value of I_D is decreased below I_{DSS}. Therefore, when

$$V_{GS} = -V, I_D < I_{DSS}$$

To summarize the n-channel MOSFET's operation, when V_{GS} was either zero volts or a negative voltage, the n-channel D-MOSFET acted in almost exactly the same way as an n-channel JFET. However, unlike the JFET, the D-MOSFET can have a forward-biased gate-to-source P-N junction because the silicon dioxide insulating layer prevents any current from passing through the gate and will still maintain a high input resistance. This dual operating ability is why the depletion-type MOSFET or D-MOSFET should actually be called a depletion-enhancement or DE-MOSFET.

The drain characteristic curves of the D-MOSFET can be used to plot the device's dc load line, as shown in Figure 14-2(a), with

$$I_{D(\text{sat})} = \frac{V_{DD}}{R_D} \text{ at saturation, and}$$

$$V_{DS(\text{OFF})} = V_{DD} \text{ at cutoff.}$$

As with the JFET, the D-MOSFET's transconductance is equal to the ratio of output current change (ΔI_D) to input voltage change (ΔV_{GS}),

$$g_m = \frac{\Delta I_D}{\Delta V_{GS}}$$

and the D-MOSFET's voltage gain is equal to

$$A_V = g_m \times R_D$$

■ **EXAMPLE:**

A D-MOSFET circuit has the following specifications:

$$I_{DSS} = 2 \text{ mA}, V_{GS(\text{OFF})} = -6 \text{ V}, R_D = 3 \text{ k}\Omega, V_{DD} = 12 \text{ V}$$

Calculate the following two extremes on the D-MOSFET's load line:

1. $I_{D(\text{sat})}$
2. $V_{DS(\text{OFF})}$

■ ***Solution:***

1. When the D-MOSFET is saturated, it is equivalent to a closed switch and therefore the only resistance is that of R_D.

$$I_{D(\text{sat})} = \frac{V_{DD}}{R_D} = \frac{12\text{ V}}{3\text{ k}\Omega} = 4\text{ mA}$$

2. When the D-MOSFET is cut off, it is equivalent to an open switch, and therefore the full drain supply voltage will appear across the open between drain and source.

$$V_{DS(\text{OFF})} = V_{DD} = 12\text{ V}$$

14-1-3 D-MOSFET Biasing

Zero Biasing
A configuration in which no bias voltage is applied at all.

Like the JFET, the D-MOSFET can be configured in the same way as a common-drain, common-gate, or common-source circuit, with all of the dc and ac configuration characteristics being the same. As far as biasing, the D-MOSFET is easier to bias than the JFET because of its ability to operate in either the depletion mode ($-V_{GS}$) or the enhancement mode ($+V_{GS}$). In fact, one of the most frequently used D-MOSFET biasing methods is to simply have no biasing at all. This biasing method is called **zero biasing** because the Q point is set at zero volts ($V_{GS} = 0$ V), as seen in Figure 14-3. This makes biasing the D-MOSFET very simple because no gate or source bias voltages are needed. The ac input signal developed across R_G is therefore applied to the extremely high input impedance of the D-MOSFET, causing an increase and decrease in the conduction of the MOSFET above and below the $V_{GS} = 0$ V, Q point.

14-1-4 D-MOSFET Applications

The D-MOSFET is most frequently used in analog or linear circuit applications. This is because the D-MOSFET can be very simply biased at a midpoint in the load line and then have its output current varied above and below this natural Q point in a linear fashion. This, coupled with the D-MOSFET's almost infinite

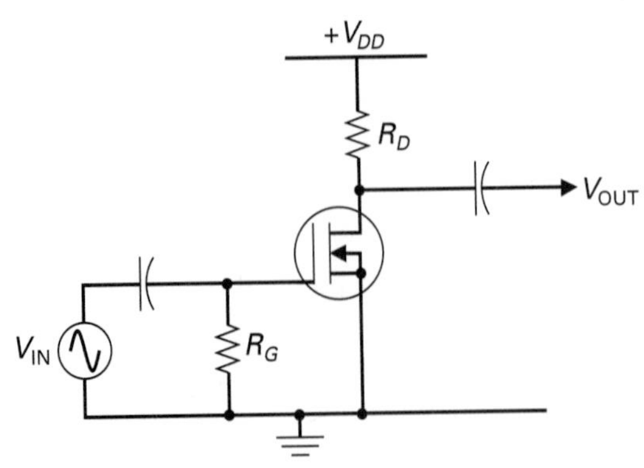

FIGURE 14-3 Zero Biasing a D-MOSFET.

input impedance and low noise properties, makes it ideal as a preamplifier at the front end of a system. Figure 14-4 shows how two D-MOSFETs can be used to construct a typical front-end **cascode amplifier circuit.** This cascode amplifier circuit consists of a self-biased common-source amplifier (Q_1) in series with a voltage-divider biased common-gate amplifier (Q_2). The input signal (V_{IN}) is applied to Q_1's gate, and the amplified output at Q_1's drain is then passed to Q_2's source, where it is further amplified by Q_2 before appearing at Q_2's drain, and therefore at the output (V_{OUT}).

Cascode Amplifier Circuit
An amplifier circuit consisting of a self-biased common-source amplifier in series with a voltage-divider biased common-gate amplifier.

The FET's only limiting factor is that its high input impedance starts to decrease as the input signal's frequency increases. Refer to the inset in Figure 14-4, which shows how the gate, insulator, and channel of a D-MOSFET form a capacitor. This input capacitance of typically 5 pF has very little effect at low input signal frequencies ($X_C\uparrow = 1/2\pi f\downarrow C$) because the input impedance is high ($X_C\uparrow$ therefore $Z_{IN}\uparrow$) and the loading effect is negligible. At higher radio frequency, however, ($X_C\downarrow = 1/2\pi f\uparrow C$) the input impedance is lowered ($X_C\downarrow$ therefore $Z_{IN}\downarrow$) and the D-MOSFET loses its high input impedance advantage. To compensate for this disadvantage, FETs are often connected in series, as in Figure 14-4, so that their input capacitances are also in series. You may recall from your introductory dc/ac electronics theory that series connected capacitors have a lower total capacitance than either of the individual capacitance values. Therefore, the overall input capacitance of two series-connected D-MOSFETs will be less than that of a single D-MOSFET, making this cascode amplifier ideal as a high radio frequency (RF) amplifier: a low input capacitance ($C_{IN}\downarrow$) means a high input reactance ($X_{C(IN)}\uparrow$), and therefore a high input impedance ($Z_{IN}\uparrow$) at high frequencies.

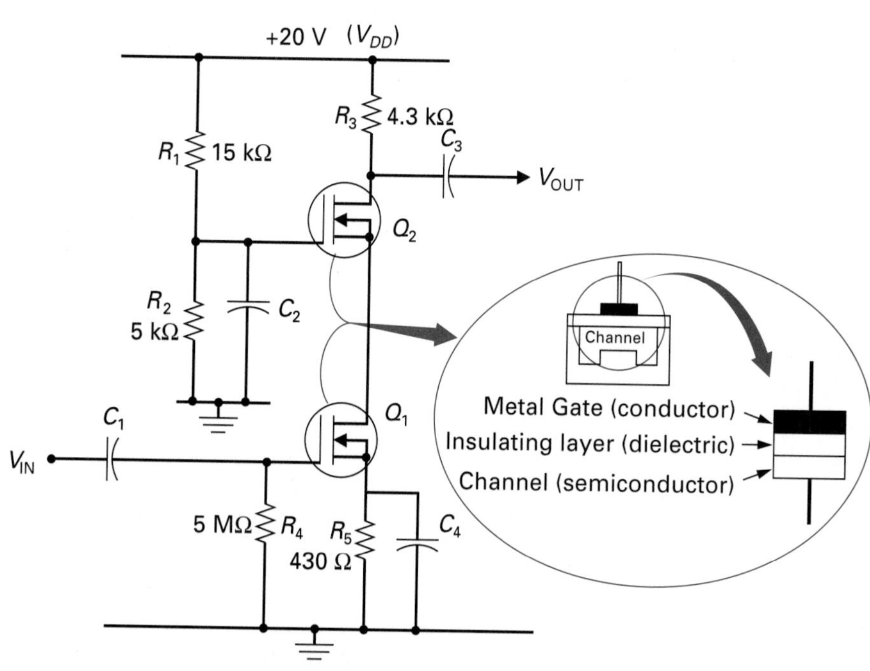

FIGURE 14-4 A D-MOSFET Analog Circuit Application—Cascode Amplifier.

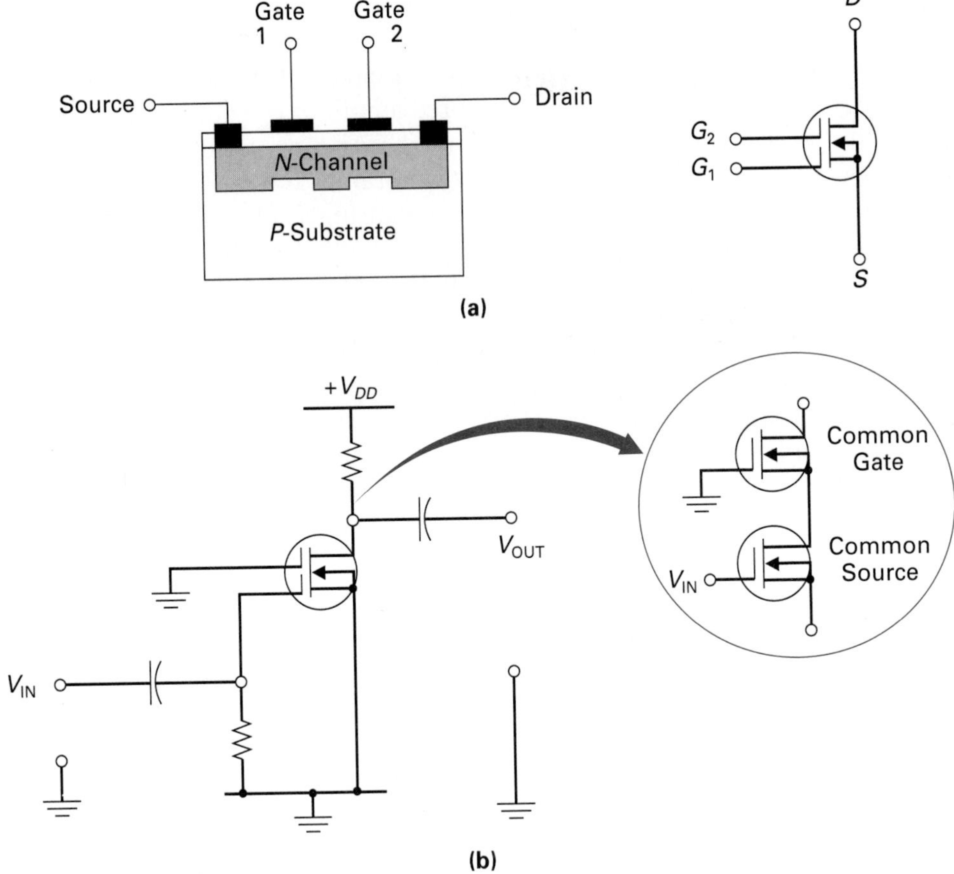

FIGURE 14-5 The Dual-Gate D-MOSFET. (a) Dual-Gate D-MOSFET Construction and Schematic Symbol. (b) Dual-Gate D-MOSFET Application—Cascode Amplifier.

14-1-5 Dual-Gate D-MOSFET

Dual-Gate D-MOSFET
A metal oxide semiconductor FET having two separate gate electrodes.

To compensate for the D-MOSFET's input capacitance problem, the **dual-gate D-MOSFET** was developed. The construction and schematic symbol for the dual-gate D-MOSFET is shown in Figure 14-5(a). In most applications, the dual-gate D-MOSFET is connected so it acts as two series-connected D-MOSFETs, as shown in the cascode amplifier circuit in Figure 14-5(b). With this amplifier, the ac input signal drives the lower gate, which acts like a common-source amplifier. The output of the common-source lower section of the dual-gate D-MOSFET drives the upper half, which acts like a common-gate amplifier. The inset in Figure 14-5(b) shows how the dual-gate D-MOSFET is equivalent to two series-connected D-MOSFETs. As with the previous cascode amplifier, the overall input capacitance of a dual-gate D-MOSFET is less than that of a standard D-MOSFET, and if capacitance is low, X_C, and therefore Z_{IN}, are high.

SELF-TEST REVIEW QUESTIONS FOR SECTION 14-1

1. The two different types of MOSFETs are called the ____________ type MOSFET and ____________ type MOSFET.
2. The D-type MOSFET can be operated in both the depletion and enhancement mode. (True/False)
3. The D-type MOSFET is a normally ____________ (ON/OFF) device.
4. Which FET has a higher input impedance: JFET or MOSFET?
5. Why is the D-MOSFET ideal as a preamplifier?
   a. It can be mid-load-line biased when 0 V is applied.
   b. It has a high input impedance.
   c. It has low noise properties.
   d. All of the above
6. The ____________ MOSFET was developed to lower input capacitance so that it can handle high-frequency signals.

14-2 THE ENHANCEMENT-TYPE (E-TYPE) MOSFET

With an input of zero volts, a D-MOSFET will be ON, and a certain value of current will pass through the channel between source and drain. If the input to the D-MOSFET is made positive, the channel is enhanced, causing the source-to-drain current to increase. If the input is made negative, the channel is depleted, causing the source-to-drain current to decrease. The D-MOSFET can therefore operate in either the enhancement or depletion mode and is called a "normally ON" device because it is ON when nothing (0 V) is applied.

The **enhancement-type MOSFET** or **E-MOSFET** can only operate in the enhancement-mode. In other words, when the input is either zero volts or a negative voltage, the transistor is OFF and there is no source-to-drain current. However, when the input is made positive, the E-MOSFET will turn ON, resulting in a source-to-drain channel current. The E-MOSFET is therefore a "normally OFF" device because it is OFF when nothing (0 V) is applied.

Enhancement-Type MOSFET or E-MOSFET
A field effect transistor with an insulated gate (MOSFET) that can only be turned ON if the channel is enhanced.

_n_-Channel E-Type MOSFET
An enhancement type MOSFET having an _n_-type channel between its source and drain terminals.

_p_-Channel E-Type MOSFET
An enhancement type MOSFET having a _p_-type channel between its source and drain terminals.

14-2-1 *E-MOSFET Construction*

As with all of the other transistor types, it is easier to understand the operation and characteristics of a device once we have seen how the component is constructed. The E-MOSFET has two basic transistor types called the ***n*-channel E-type MOSFET** and ***p*-channel E-type MOSFET.** The construction and schematic symbol for these two E-MOSFET types are shown in Figure 14-6.

To begin with, let us examine the construction of the more frequently used *n*-channel E-type MOSFET, shown in Figure 14-6(a). Studying the construction of this E-MOSFET, notice that no channel exists between the source and drain.

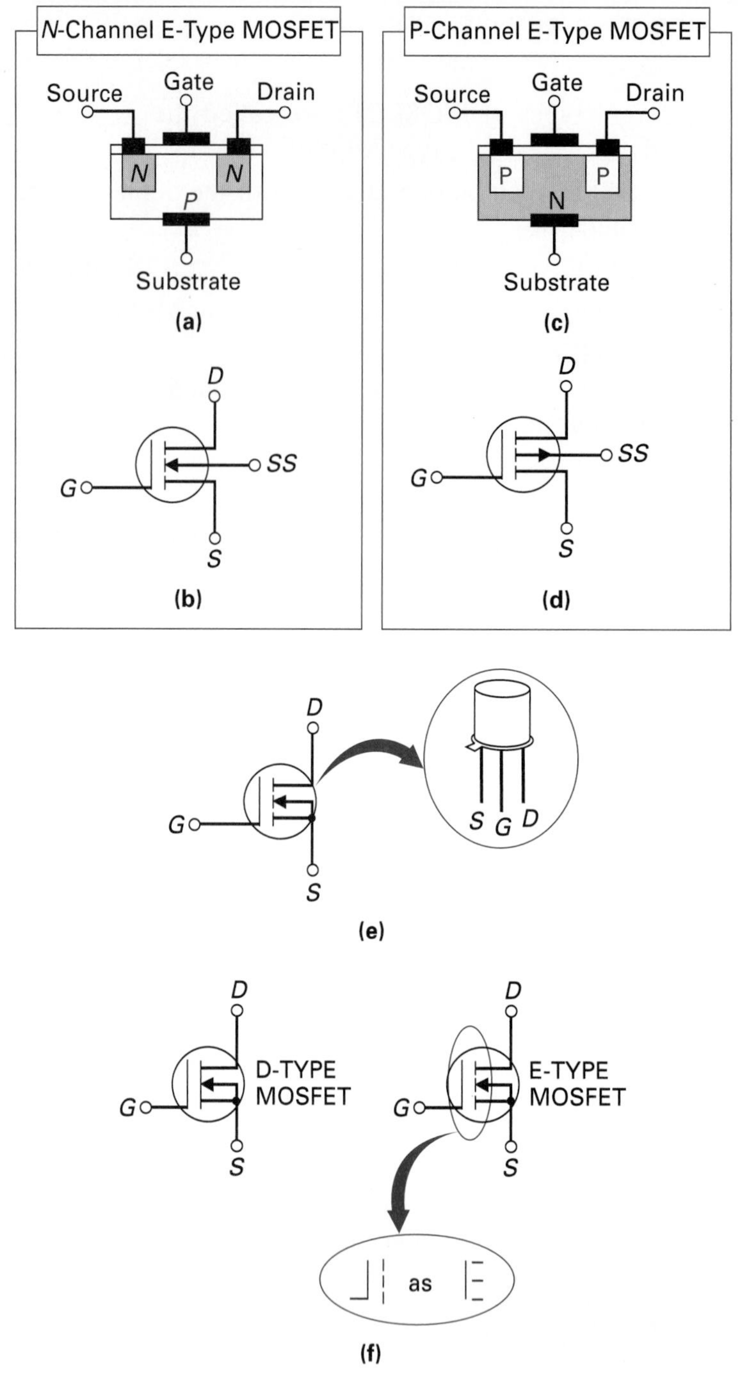

FIGURE 14-6 E-Type MOSFET Construction and Type.

Consequently, with no gate bias voltage, the device will be OFF. When the gate is made positive, however, electrons will be attracted from the substrate, causing a channel to be induced between the source and drain. This enhanced channel will permit drain current to flow, and any further increase in gate voltage will cause a corresponding increase in the size of the channel and therefore the value of I_D.

The schematic symbol for the *n*-channel E-type MOSFET can be seen in Figure 14-6(b). The construction and schematic symbol for the *p*-channel E-MOSFET can be seen in Figure 14-6(c) and (d). As with the D-MOSFET, both 4-terminal and 3-terminal devices are available, with the 3-terminal device having a common connection between source and substrate as shown in Figure 14-6(e).

The only difference between the E-type and D-type MOSFET symbols is the three dashed lines representing the drain, substrate, and source regions. The dashed line is used instead of the solid line to indicate that an E-MOSFET has a normally broken path, or channel, between drain, substrate, and source. To aid memory, the three dashed lines used in the "E"-MOSFET symbol could be thought of as the three horizontal prongs in the capital letter "E," as shown in Figure 14-6(f).

14-2-2 ***E-MOSFET Operation***

Figure 14-7(a) shows the typical drain characteristic curves for an *n*-channel enhancement-type MOSFET. As you can see, this set of drain curves is very similar to the D-MOSFETs set of drain curves; however, in this case only positive values of V_{GS} are plotted. Looking at the relationship between V_{GS} and I_D, you may have noticed that any increase in V_{GS} will cause a corresponding increase in I_D.

To best understand the operation of the E-MOSFET, let us examine the three operation diagrams shown in Figure 14-7(b), (c), and (d).

> **V_{GS} = 0 V Curve:** The lower operation diagram, Figure 14-7(b), shows how the *n*-channel E-MOSFET will respond to a V_{GS} input of zero volts. When $V_{GS} = 0$ V, there is no channel connecting the source and drain, and therefore the drain current will be zero. As a result, $+V_{DD}$ will be present at the output because the E-MOSFET is equivalent to an open switch between drain and source.
>
> **V_{GS} = +5 V Curve:** The center operation diagram, Figure 14-7(c), shows how the *n*-channel E-MOSFET will respond to a V_{GS} input of +5 volts. When $V_{GS} = +5$ V, the E-MOSFET will act in almost exactly the same way as an enhanced D-MOSFET. The positive gate voltage will repel the *p*-type material's majority carriers (holes) away from the gate, while attracting the *p*-type material's minority carriers (electrons) towards the gate. This action will form an *n*-type bridge between the source and drain, and therefore a value of I_D will flow between source and drain, as shown.
>
> **V_{GS} = +10 V Curve:** The upper operation diagram, Figure 14-7(d), shows how the *n*-channel E-MOSFET will respond if the V_{GS} input is further increased to +10 volts. When $V_{GS} = +10$ V, the attraction of electrons and repulsion of holes within the *p*-type material is increased, causing the channel's width to increase and I_D to also increase. As a result, 0 V will be present at the output because the E-MOSFET is equivalent to a closed switch between drain and source.

In summary, when V_{GS} is zero volts, drain current is also zero. As the value of V_{GS} is increased (made more positive), the channel becomes wider, causing I_D

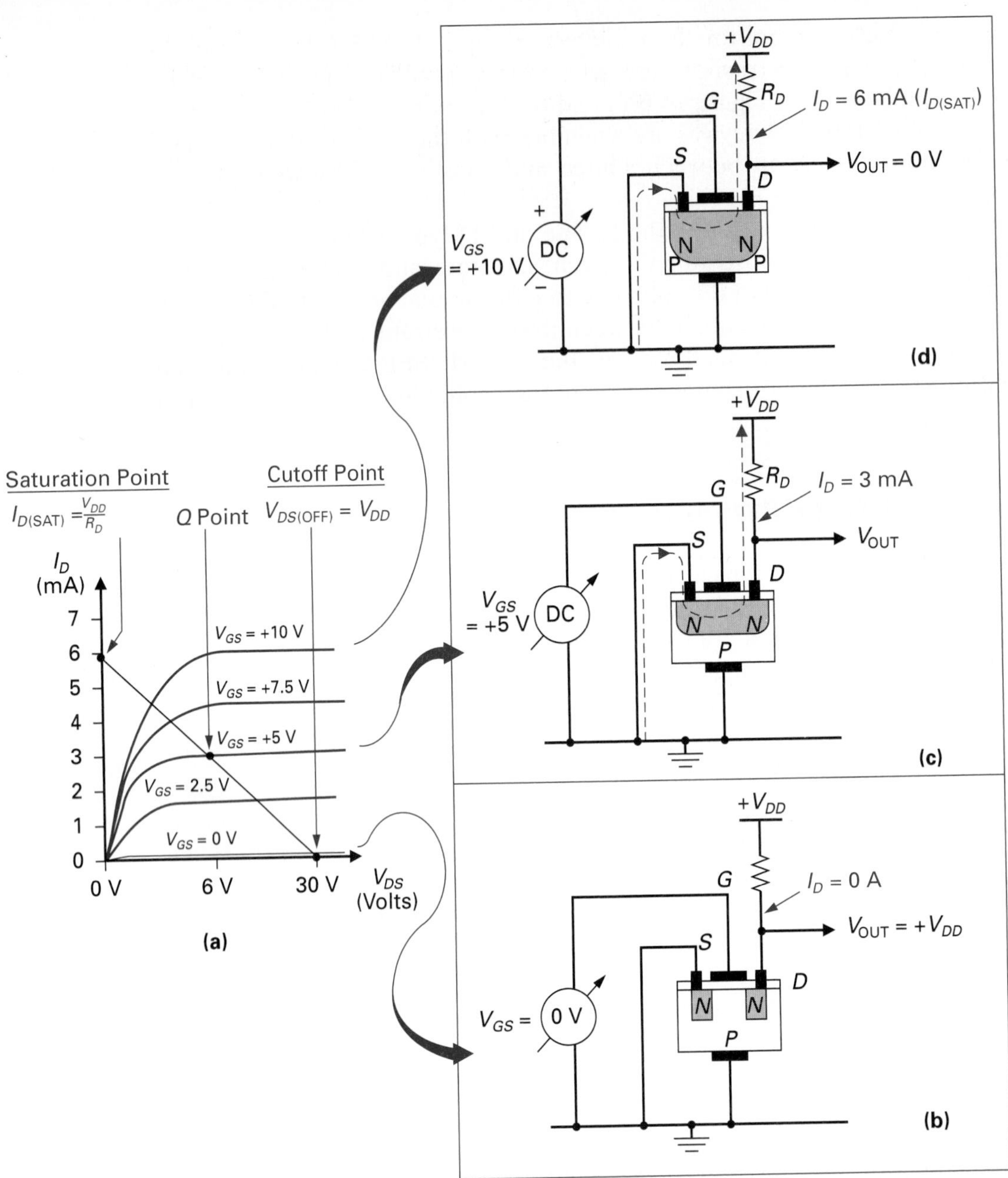

FIGURE 14-7 E-Type MOSFET Operation and Characteristics.

to increase. On the other hand, as the value of V_{GS} is decreased (made less positive), the channel becomes narrower, causing I_D to decrease. In other words, when the input is either zero volts or a negative voltage, the transistor is OFF, and there is no source-to-drain current. However, when the input is made positive, the E-MOSFET will turn ON, resulting in a source-to-drain channel current. The E-MOSFET is therefore a "normally OFF" device because it is OFF when nothing (0 V) is applied.

Although it cannot be seen in the set of drain curves in Figure 14-7(a), V_{GS} will have to increase to a positive threshold voltage of about +1 V before a channel will be induced and a small value of drain current will flow. This "threshold level" is a highly desirable characteristic because it prevents noise or any low-level input signal voltage from turning ON the device. This advantage makes the E-type MOSFET ideally suited as a switch because it can be turned ON by an input voltage and turned OFF once the input voltage falls below the threshold level.

The *p*-channel E-type MOSFET operates in much the same way as the *n*-channel device except that holes are attracted from the substrate to form a *p*-channel and the V_{GS} and V_{DS} bias voltages are reversed.

14-2-3 *E-MOSFET Biasing*

Like the D-MOSFET, the E-MOSFET can be configured as a common-drain, common-gate, or common-source circuit. Unlike the D-MOSFET and JFET, the E-MOSFET cannot be biased using self bias or zero bias because V_{GS} must be a positive voltage. As a result, gate bias and voltage-divider bias can be used; however, more frequently E-MOSFETs are **drain-feedback biased,** as shown in Figure 14-8(a). In this example, R_D equals 8 kΩ, and R_G (which feeds back a positive voltage from the drain, hence the name drain-feedback bias) equals 100 MΩ. Because an E-MOSFET has an extremely high input impedance (due to the insulated gate), no current will flow in the gate circuit. With no gate current, there will be no voltage drop across the gate resistor (V_{RG} = 0 V), and therefore the voltage at the gate will be at the same potential as the voltage at the drain.

Drain-Feedback Biased
A configuration in which the gate receives a bias voltage fed back from the drain.

$$V_{GS} = V_{DS}$$

To help set up the *Q* point, most manufacturer's data sheets specify a load line mid-point drain current $I_{D(\text{ON})}$ and drain voltage $V_{DS(\text{ON})}$. In the example in

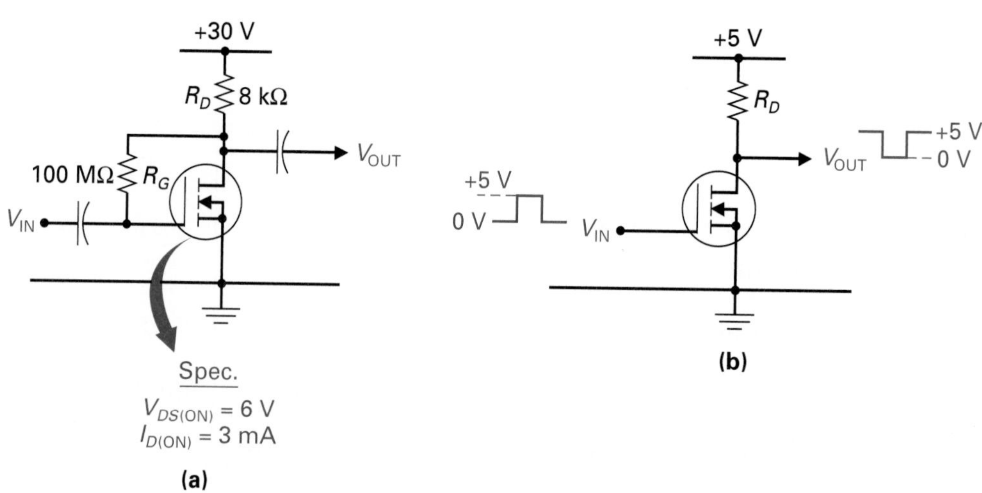

FIGURE 14-8 Biasing E-Type MOSFETs. (a) Drain-Feedback Biasing. (b) Voltage Controlled Switching.

Figure 14-8(a), when the E-MOSFET is ON, or conducting, and $I_D = 3$ mA, the E-MOSFET's drain-to-source voltage drop ($V_{DS(ON)}$) is 6 V. The value of R_D in Figure 14-8(a) has been chosen so that this E-MOSFET circuit will be biased at its specified Q point. To check, we can use the formula

$$V_{DS} = V_{DD} - (I_{D(ON)} \times R_D)$$

In the example in Figure 14-8(a)

$$V_{DS} = V_{DD} - (I_{D(ON)} \times R_D) = 30\text{ V} - (3\text{ mA} \times 8\text{ k}\Omega)$$
$$= 30\text{ V} - 24\text{ V} = 6\text{ V}$$

■ **EXAMPLE:**

A drain-feedback biased E-MOSFET circuit has the following specifications: $R_D = 1\text{ k}\Omega$, $I_{D(ON)} = 10$ mA, $V_{DD} = 20$ V. Calculate V_{DS} and V_{RD}.

■ ***Solution:***

$$V_{DS} = V_{DD} - (I_{D(ON)} \times R_D)$$
$$= 20\text{ V} - (10\text{ mA} \times 1\text{ k}\Omega)$$
$$= 20\text{ V} - 10\text{ V} = 10\text{ V}$$

The constant drain-supply voltage (V_{DD}) will be evenly divided across R_D and the E-MOSFET's drain-to-source ($V_{DS} = 10$ V, $V_{RD} = 10$ V).

In most instances, the E-MOSFET will be used in digital two-state switching circuit applications. In this case, there will be no need for a gate resistor (R_G) because the input voltage (V_{IN}) will either turn ON or OFF the E-MOSFET, as shown in Figure 14-8(b). For example, when $V_{IN} = 0$ V, the E-MOSFET is OFF,

	BIPOLAR FAMILY		
	LS TTL	ECL	IIL
Cost	LOW	HIGH	MEDIUM
Fanout	20	10 to 25	2
Power Dissipation/Gate	2 mW	40 to 60 mW	0.06 to 70 μW
Propagation Delay/Gate	8 ns	0.5 to 3 ns	25 to 50 ns
Typical Clock Frequency	15 to 120 MHz	200 to 1000 MHz	1 to 10 MHz
External Noise Immunity	GOOD	GOOD	FAIR-GOOD
Typical Supply Voltage	+5 V	–5.2 V	1 V to 15 V
Temperature Range	–55 to 125°C 0 to 70°C	–55 to 125°C	0 to 70°C
Internally Generated Noise	MEDIUM–HIGH	LOW–MEDIUM	LOW

(a)

FIGURE 14-9 Comparing Digital IC Logic Gate Families. (a) Bipolar Transistor Family.

therefore $V_{OUT} = +V_{DD} = +5$ V, whereas when $V_{IN} = +5$ V, the E-MOSFET is ON and therefore $V_{OUT} = 0$ V.

14-2-4 E-MOSFET Applications

The E-MOSFET is more frequently used in digital or two-state circuit applications. One reason is that it naturally operates as a "normally OFF-voltage controlled switch" because it can be turned ON when the gate voltage is positive and turned OFF when the gate voltage falls below a threshold level. This threshold level is a highly desirable characteristic because it prevents noise from false triggering, or accidentally turning ON, the device. The other E-MOSFET advantage is its extremely high input impedance, which means that the device's circuit current, and therefore power dissipation, are low. This enables us to densely pack or integrate many thousands of E-MOSFETs onto one small piece of silicon, forming a high component density integrated circuit (IC). These low-power and high-density advantages make the E-MOSFET ideal in battery-powered small-size (portable) applications such as calculators, wristwatches, notebook computers, hand-held video games, digital cellular phones, and so on.

Digital IC Logic Families

Previously in Chapter 9, we saw how the bipolar transistor could be used to construct digital logic gates, which are the basic decision making elements in all digital circuits. A logic family is a group of digital circuits with nearly identical characteristics. The **bipolar family** of digital ICs has three basic types: TTL (transistor-transistor logic), ECL (emitter-coupled logic), and IIL (integrated-injection logic) or I^2L. The characteristics of these three basic bipolar transistor logic gate types are summarized in Figure 14-9(a). For comparison, Figure 14-9(b) shows the three basic E-MOSFET digital IC types, which are more frequently called the **MOS family** of digital ICs. In this section, we will examine the

Bipolar Family
A group of digital logic circuits that make use of the bipolar junction transistor.

MOS Family
A group of digital circuits that make use of metal oxide semiconductor field effect transistors (MOSFETs).

	MOS FAMILY		
	PMOS	NMOS	HCMOS
Cost	HIGH	HIGH	MEDIUM
Fanout	20	20	10
Power Dissipation/Gate	0.2 to 10 mW	0.2 to 10 mW	2.7 nW to 170 μW
Propagation Delay/Gate	300 ns	50 ns	10 ns
Typical Clock Frequency	2 MHz	5 to 10 MHz	5 to 100 MHz
External Noise Immunity	GOOD	GOOD	VERY GOOD
Typical Supply Voltage	−12 V	+5 V	+3 V to +6 V
Temperature Range	−55 to 125°C 0 to 70°C	−55 to 125°C 0 to 70°C	−55 to 125°C −40 to 85°C
Internally Generated Noise	MEDIUM	MEDIUM	LOW–MEDIUM

(b)

FIGURE 14-9 (b) MOS Transistor Family.

three basic types of MOS ICs: *PMOS* (pronounced "pea-moss"), *NMOS* (pronounced "en-moss"), and *CMOS* (pronounced "sea-moss"). To begin with, MOS logic gates use less space due to their simpler construction, have a very high noise immunity, and consume less power than equivalent bipolar logic gates. However, the high input impedance and capacitance due to the E-MOSFET's insulated gate means that time constants are larger and therefore transistor ON/OFF switching speeds are slower than equivalent bipolar gates. This is the trade-off between the two logic families:

1. Bipolar circuits are faster than MOS circuits, but their circuits are larger and consume more power.
2. MOS circuits are smaller and consume less power than bipolar circuits, but they are generally slower.

Previously in Chapter 9, we discussed how a BJT could be used to construct any of the five basic logic gates, reviewed in Figure 14-10(a). In the following sections, we will examine how the E-MOSFET can be used to construct the same gates using MOS logic.

PMOS (*p*-Channel MOS) Logic Circuits

PMOS Logic
Logic circuitry in which logic gates are constructed using *p*-channel E-MOSFETs.

The name **PMOS logic** is used to describe this logic circuitry because the logic gates are constructed using *p*-channel E-MOSFETs. Figure 14-10(b) shows how three *p*-channel E-MOSFETs can be used to construct a NOR gate. Transistor Q_3 can be thought of as a "current limiting resistor" between source and drain because its gate is constantly connected to $-V_{GG}$, and therefore Q_3 is always ON. Any HIGH input will turn OFF its respective E-MOSFET, resulting in a negative or LOW output due to the pull-down action of Q_3. Only when both inputs are LOW will both Q_1 and Q_2 be ON, allowing $+V_{DD}$, which is a HIGH, to be connected through Q_1 and Q_2 to the output.

NMOS (*n*-Channel MOS) Logic Circuits

NMOS Logic
Logic circuitry in which logic gates are constructed using *n*-channel E-MOSFETs.

The name **NMOS logic** is used to describe this logic circuitry because the logic gates are constructed using *n*-channel E-MOSFETs. Figure 14-10(c) shows how three *n*-channel E-MOSFETs can be used to construct a NAND gate. Transistor Q_1 functions as a current-limiting resistor between source and drain because its gate is constantly connected to $+V_{DD}$, and therefore Q_1 is always ON. A LOW on either or both of the inputs will turn OFF one or both of the transistors, Q_2 and Q_3, causing the output to be pulled up to a HIGH by Q_1. Only when both inputs are HIGH will Q_1 and Q_2 turn ON and conduct the ground on Q_3's source through Q_3 and Q_2 to the output.

CMOS Logic
Logic circuitry in which logic gates are constructed using both a *p*-channel E-MOSFET and its "complement" or opposite, an *n*-channel E-MOSFET.

CMOS (Complementary MOS) Logic Circuits

The name **CMOS logic** is used to describe this logic circuitry because the logic gates are constructed using both a *p*-channel E-MOSFET (Q_1) and its "complement" or opposite, an *n*-channel E-MOSFET (Q_2). Figure 14-10(d) shows how a CMOS NOT or inverter gate can be constructed. Notice the *n*-channel E-MOSFET (Q_2) has its source connected to 0 V, and therefore its gate needs to be positive (relative to the 0 V source) for it to turn ON. On the other

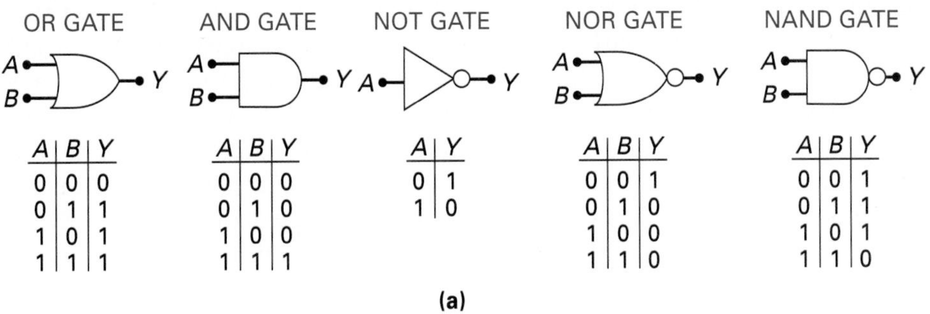

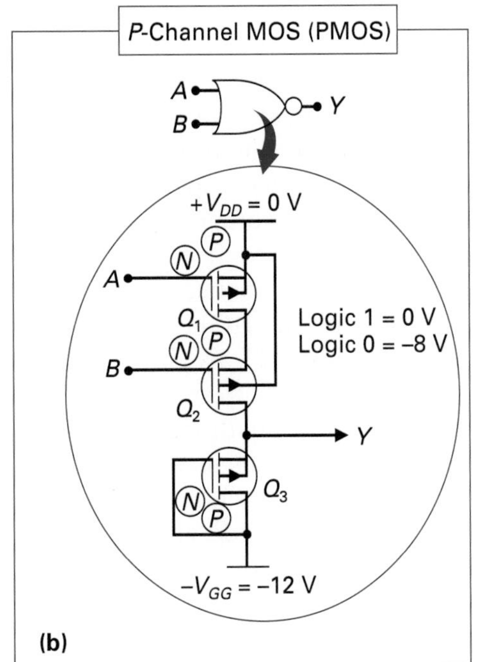

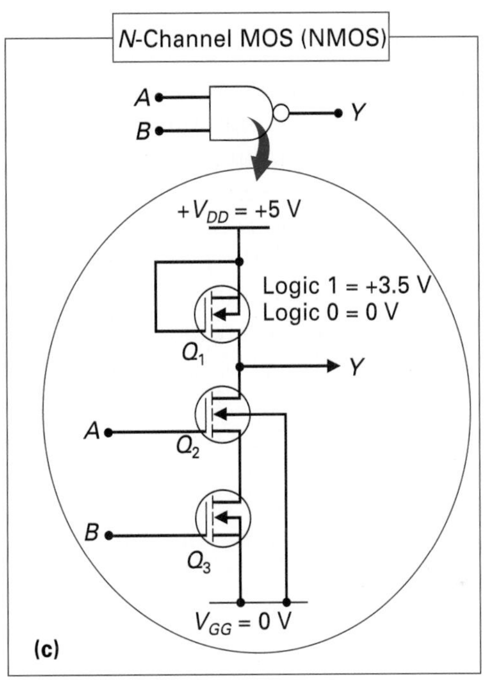

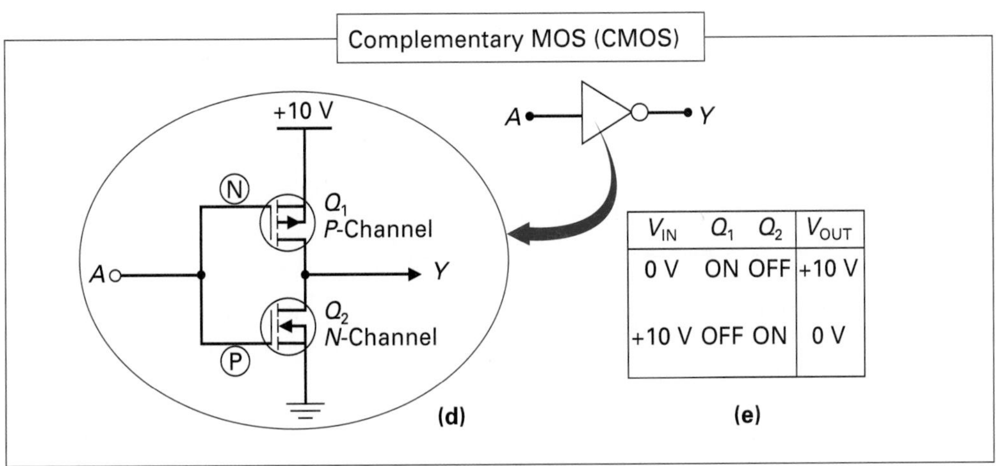

FIGURE 14-10 E-Type MOSFET Digital Application—Logic Gates.

hand, the p-channel E-MOSFET has its source connected to +10 V, and therefore its gate needs to be negative (relative to the +10 V source) for it to turn ON. Figure 14-10(e) gives a table showing how this circuit will function. When V_{IN} is LOW (0 V), Q_1 will turn ON (N-P gate-source is forward biased, $G = 0$ V, $S = +10$ V) and Q_2 will turn OFF (P-N gate-source is reverse biased, $G = 0$ V, $S = 0$ V). A LOW input will therefore turn ON Q_1, creating a channel between source and drain connecting the +10 V supply voltage to the output. On the other hand, when V_{IN} is HIGH (+10 V), Q_1 will turn OFF (N-P gate-source is reverse biased, $G = +10$ V, $S = +10$ V) and Q_2 will turn ON (P-N gate-source is forward biased, $G = +10$ V, $S = 0$ V). A HIGH input will therefore turn ON Q_2, creating a channel between source and drain connecting the 0 V at Q_2's source to the output.

CMOS logic is probably the best all-round logic circuitry because of its low power consumption, very high noise immunity, large power-supply voltage range, and high fanout. These characteristics are given in detail in Figure 14-11. The low power consumption achieved by CMOS circuits is due to the "complementary pairs" present in each CMOS gate circuit. During gate operation there is never a continuous path for current between the supply voltage and ground. For example, when the p-channel device is OFF, $+V_{DD}$ is disconnected from the circuit, and when the n-channel device is OFF, ground is disconnected from the circuit. It is only when the output voltage switches from a HIGH to LOW, or LOW to HIGH, that the p-channel and n-channel E-MOSFET are momentarily ON at the same time. It is during this time of switching over from one state to the next that a small current will flow due to the complete path from $+V_{DD}$ to ground. This is why the power dissipation of CMOS logic gates increases as the operating frequency of the circuit increases.

There is a full line of CMOS ICs available that can provide almost all the same functions as their TTL counterparts. These different types are given differ-

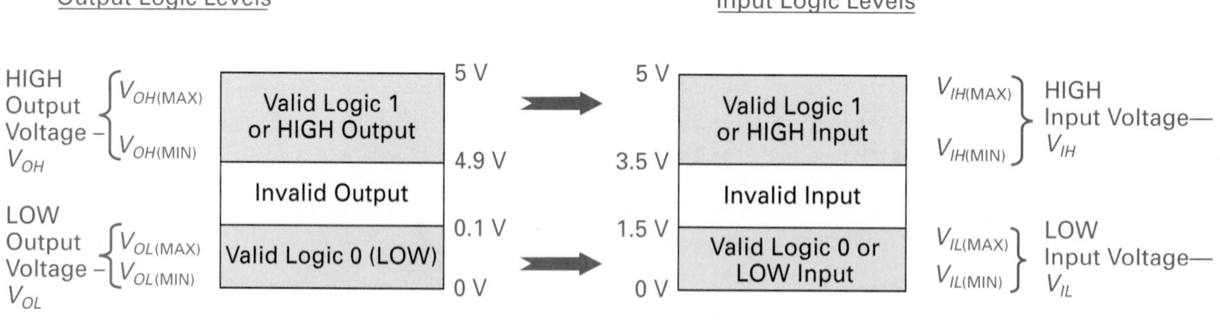

Power Dissipation per Gate: 2.7 nanowatts (static) to 170 μW (at 100 kHz)

Propagation Delay per Gate: 10 ns

Supply Voltage: +3 V to +6 V

Noise Immunity: Very High

Fanout: 10

FIGURE 14-11 CMOS Logic Gate Characteristics (High Speed CMOS—74HCXX).

ent series numbers to distinguish them from others. Some of the more frequently used CMOS ICs are listed here.

1. *4000 Series*—This first line of CMOS digital integrated circuits, originally made by RCA, uses the numbers 40XX, with the number in the "XX" position indicating the circuit type. For example, the 4001 is a "quad 2-input NOR gate," while the 4011 is a "quad 2-input NAND gate." This series was eventually improved and called the 4000B series, and the original was labeled the 4000A series. Both use a V_{DD} supply of between +3 V to +15 V, with a logic 0 (LOW) = 1/3 of V_{DD} and logic 1 (HIGH) = 2/3 of V_{DD}. Propagation delay times for this series are typically 40 ns to 175 ns.
2. *40H00 Series*—This "high-speed" CMOS series improved the propagation delay to approximately 20 ns; however, it was still not able to match the speed of the bipolar TTL 74LS00 series.
3. *74C00 Series*—This series, made by National Semiconductor, was designed to be "pin compatible" with its TTL counterpart. This meant that all of the ICs used the same input and output pin numbers.
4. *74HC00 and 74HCT00 Series*—This high-speed CMOS series of ICs matches the propagation delay times of the 74LS00 bipolar ICs and still maintains the CMOS advantage of low power consumption. The HC and HCT series are both pin compatible with TTL. The HCT (high-speed CMOS TTL compatible) series is a truly compatible CMOS line of ICs because it uses the same input and output voltage levels for a HIGH and LOW. This advantage makes it easier to swap a TTL IC for a CMOS IC and not cause any interfacing problems.
5. *74ACL00 Series*—This "advanced CMOS logic" series has even better characteristics than the 74HC00 and 74HCT00 series.

These and other digital circuits will be discussed in more detail in the digital circuits Chapter 18.

14-2-5 *Vertical-Channel E-MOSFET (VMOS FET)*

As just mentioned, the E-MOSFET is generally used in two-state or digital circuit applications, where it acts as a normally OFF switch. In most digital circuit applications, the channel current is small, and therefore a standard E-MOSFET can be used. However, if a larger current-carrying capability is needed, the **vertical-channel E-MOSFET** or VMOS FET can be used. Figure 14-12(a) shows the construction of a VMOS FET. The gate at the top of the device is insulated from the source (which is also at the top), and as with all E-MOSFETs, no channel exists between the source terminal and the drain terminal with no bias voltage applied. The VMOS FET's semiconductor materials are labeled *P*, *N*+, and *N*– and indicate different levels of doping.

Vertical-Channel E-MOSFET
An enhancement type MOSFET that, when turned ON, forms a vertical channel between source and drain.

When this *n*-channel VMOS FET is biased ON (gate is made positive with respect to source), as seen in Figure 14-12(b), a vertical *n*-type channel is formed between source and drain. This channel is much wider than a standard E-MOSFET's horizontal channel, which is why VMOS FETs can handle a much higher drain current.

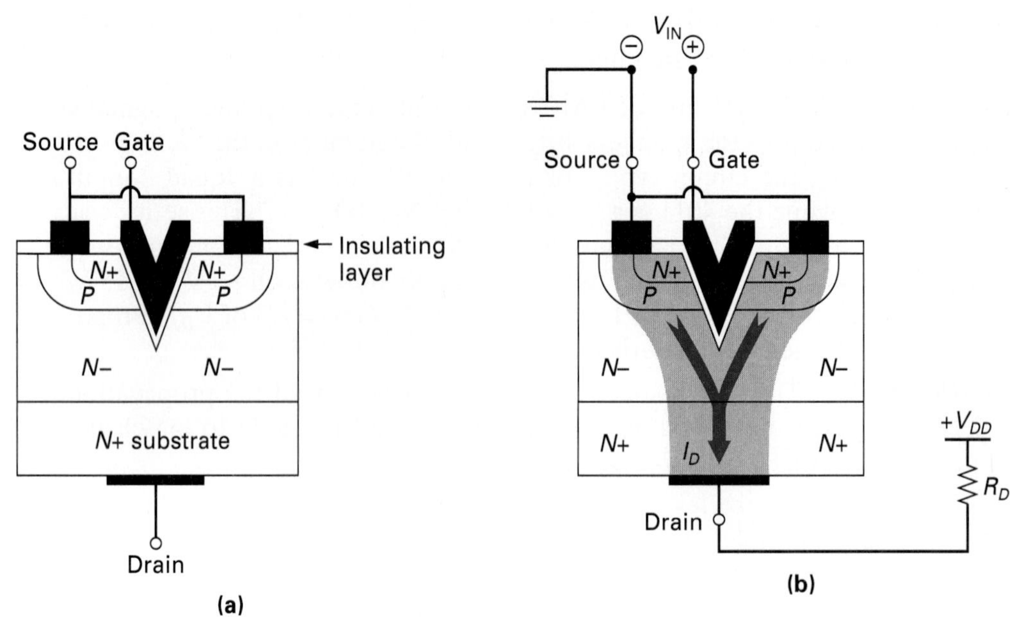

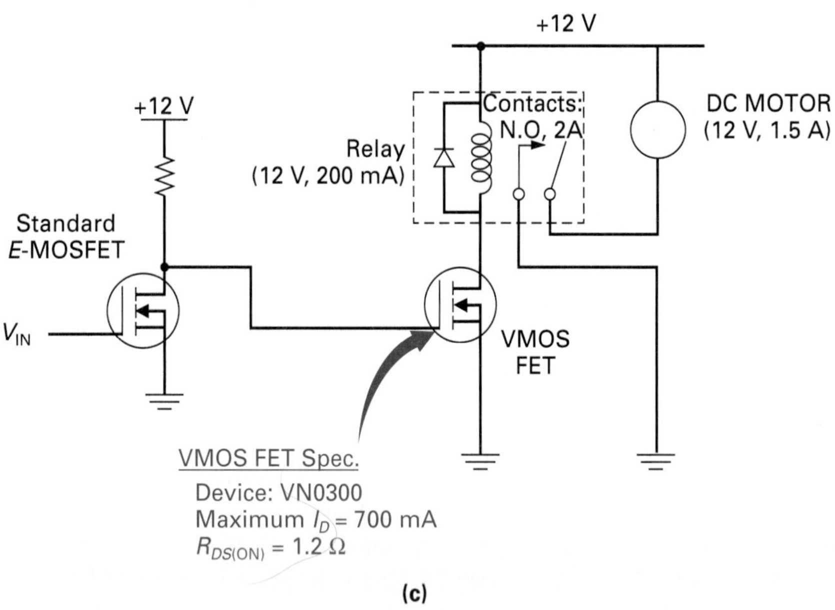

FIGURE 14-12 The Vertical Channel E-MOSFET (VMOS FET). (a) VMOS FET OFF. (b) VMOS FET ON. (c) VMOS FET Application Circuit—Low-Power Signal to High-Power Load Circuit.

Figure 14-12(c) shows how the high-current capability of a VMOS FET can be made use of in an interfacing circuit application. In this circuit, a VMOS FET is used to interface a low-power source input signal to a high-power load. Referring to the current specifications of the VMOS FET, relay, and motor, you can see that each device is used to step up current. The standard E-MOSFET supplies a low-power input signal to the VMOS FET, which can handle enough current to actuate a relay whose contacts can handle enough current to switch power to the dc motor. To be more specific, if V_{IN} is LOW, the standard E-MOSFET will turn OFF, producing a HIGH output to the gate of the VMOS FET, turning it ON. When the VMOS FET turns ON, it effectively switches ground through to the lower end of the relay coil, energizing the relay and closing its normally open contacts. The closed relay contacts switch ground through to the lower end of the motor, which turns ON because it now has the full +12 V supply across its terminals. The motor will stay ON as long as V_{IN} stays LOW. If V_{IN} were to go HIGH, the standard E-MOSFET inverter would produce a LOW output, which would turn OFF the VMOS FET, relay, and motor.

Other than its high-current capability, the VMOS FET also has a positive temperature coefficient of resistance, which means an increase in temperature will cause a decrease in drain current ($T\uparrow$, $R\uparrow$, $I\downarrow$), and this will prevent thermal runaway. This gives the VMOS FET a distinct advantage over the BJT power amplifier, which has a negative temperature coefficient of resistance ($T\uparrow$, $R\downarrow$, $I\uparrow$). This means that if maximum ratings are exceeded, an increase in temperature will cause an increase in current, which will generate a further increase in heat (temperature) and current, and so on.

SELF-TEST REVIEW QUESTIONS FOR SECTION 14-2

1. The E-Type MOSFET can be operated in both the depletion and enhancement mode. (True/False)

2. The E-Type MOSFET is a normally __________ (ON/OFF) device.

3. The E-type MOSFET is generally
a. Zero biased **c.** Drain-Feedback biased
b. Base biased **d.** None of the above

4. The E-MOSFET naturally operates as a voltage controlled switch. (True/False)

5. The E-MOSFET is ideally suited in __________ circuit applications because it is normally __________, whereas the D-MOSFET is better suited for __________ circuit applications because it is normally __________.
a. digital, OFF, analog, ON **c.** analog, OFF, digital, ON
b. digital, ON, analog, OFF **d.** analog, ON, digital, OFF

6. List three advantages that MOS logic gates have over bipolar logic gates.

14-3 MOSFET DATA SHEETS, HANDLING, AND TESTING

In this section we will examine a typical MOSFET manufacturer's data sheet, special MOSFET handling precautions, and MOSFET testing.

14-3-1 *MOSFET Data Sheets*

Throughout this chapter, we have used certain MOSFET specifications in our calculations, such as I_{DSS}, $V_{DS(\text{ON})}$, $I_{D(\text{ON})}$, and so on. As an example, Figure 14-13 shows the data sheet for a typical E-type MOSFET, and, as you can see, these specifications are listed.

14-3-2 *MOSFET Handling Precautions*

Certain precautions must be taken when handling any MOSFET devices. The very thin insulating layer between the gate and the substrate of a MOSFET can easily be punctured if an excessive voltage is applied. Your body can build up extremely large electrostatic charges due to friction. If this charge came in contact with the pins of a MOSFET device, an electrostatic-discharge (ESD) would occur, resulting in a possible arc across the thin insulating layer causing permanent damage. Most MOSFETs presently manufactured have zeners internally connected between gate and source to bypass high voltage static or in-circuit potentials and protect the MOSFET. However, it is important to remember the following.

1. All MOS devices are shipped and stored in a "conductive foam" or "protective foil" so that all of the IC pins are kept at the same potential, and therefore electrostatic voltages cannot build up between terminals.
2. When MOS devices are removed from the conductive foam, be sure not to touch the pins because your body may have built up an electrostatic charge.
3. When MOS devices are removed from the conductive foam, always place them on a grounded surface such as a metal tray.
4. When continually working with MOS devices, use a "wrist grounding strap," which is a length of cable with a 1 MΩ resistor in series. This prevents electrical shock if you come in contact with a voltage source.
5. All test equipment, soldering irons, and work benches should be properly grounded.
6. All power in equipment should be off before MOS devices are removed or inserted into printed circuit boards.
7. Any unused MOSFET terminals must be connected because an unused input left open can build up an electrostatic charge and float to high voltage levels.
8. Any boards containing MOS devices should be shipped or stored with the connection side of the board in conductive foam.

14-3-3 *MOSFET Testing*

Like the BJT and JFET, MOSFETs can be tested with the transistor tester to determine: first, whether an open or short exists between any of the terminals; second, the transistor's transconductance/gain; and third, the value of I_{DSS} and leakage current.

DEVICE: 2N7002LT1—N-Channel Silicon E-MOSFET

2N7002LT1★

CASE 318-07 STYLE 21
SOT-23 (TO-236AB)

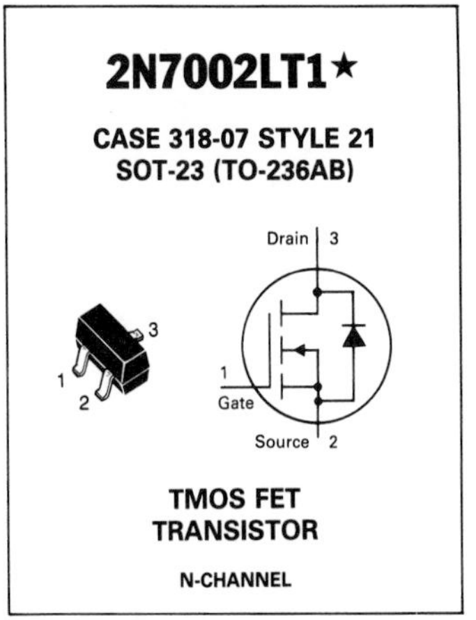

TMOS FET TRANSISTOR

N-CHANNEL

MAXIMUM RATINGS

Rating	Symbol	Value	Unit
Drain-Source Voltage	V_{DSS}	60	Vdc
Drain-Gate Voltage (R_{GS} = 1 MΩ)	V_{DGR}	60	Vdc
Drain Current – Continuous TC = 25°C(1) T_C = 100°C(1) – Pulsed (2)	I_D I_D I_{DM}	±115 ±75 ±800	mA
Gate-Source Voltage – Continuous – Non-repetitive (tp ≤50 μs)	V_{GS} V_{GSM}	±20 ±40	Vdc Vpk
Total Power Dissipation T_C = 25°C T_C = 100°C Derate above 25°C ambient	P_D	200 80 1.6	mW mW/°C

ELECTRICAL CHARACTERISTICS (T_A = 25°C unless otherwise noted.)

Characteristic	Symbol	Min	Typ	Max	Unit
OFF CHARACTERISTICS					
Drain-Source Breakdown Voltage (V_{GS} = 0, I_D = 10 μA)	$V_{(BR)DSS}$	60	—	—	Vdc
Zero Gate Voltage Drain Current (V_{GS} = 0, V_{DS} = 60 V) T_J = 25°C T_J = 125°C	I_{DSS}	— —	— —	1.0 500	μAdc
Gate-Body Leakage Current Forward (V_{GS} = 20 Vdc)	I_{GSSF}	—	—	100	nAdc
Gate-Body Leakage Current Reverse (V_{GS} = −20 Vdc)	I_{GSSR}	—	—	−100	nAdc

(1) The Power Dissipation of the package may result in a lower continuous drain current.
(2) Pulse Width ≤ 300 μs, Duty Cycle ≤ 2.0%.

Characteristic	Symbol	Min	Typ	Max	Unit
ON CHARACTERISTICS*					
Gate Threshold Voltage (V_{DS} = V_{GS}, I_D = 250 μA)	$V_{GS(th)}$	1.0	—	2.5	Vdc
On-State Drain Current ← I_D (on) (V_{DS} ≥ 2.0 $V_{DS(on)}$, V_{GS} = 10 V)	$I_{D(on)}$	500	—	—	mA
Static Drain-Source On-State Voltage (V_{GS} = 10 V, I_D = 500 mA) (V_{GS} = 5.0 V, I_D = 50 mA) ← V_{DS} (on)	$V_{DS(on)}$	— —	— —	3.75 .375	Vdc
Static Drain-Source On-State Resistance (V_{GS} = 10 V, I_D = 500 mA) T_C = 25°C T_C = 125°C (V_{GS} = 5.0 V, I_D = 50 mA) T_C = 25°C T_C = 125°C	$r_{DS(on)}$	— — — —	— — — —	7.5 13.5 7.5 13.5	Ohms
Forward Transconductance (V_{DS} ≥ 2.0 $V_{DS(on)}$, I_D = 200 mA)	g_{FS}	80	—	—	mmhos
DYNAMIC CHARACTERISTICS					
Input Capacitance (V_{DS} = 25 V, V_{GS} = 0, f = 1.0 MHz)	C_{iss}	—	—	50	pF
Output Capacitance (V_{DS} = 25 V, V_{GS} = 0, f = 1.0 MHz)	C_{oss}	—	—	25	pF
Reverse Transfer Capacitance (V_{DS} = 25 V, V_{GS} = 0, f = 1.0 MHz)	C_{rss}	—	—	5.0	pF
SWITCHING CHARACTERISTICS*					
Turn-On Delay Time (V_{DD} = 25 V, I_D ≅ 500 mA,	$t_{d(on)}$	—	—	30	ns
Turn-Off Delay Time R_G = 25 Ω, R_L = 50 Ω)	$t_{d(off)}$	—	—	40	ns
BODY-DRAIN DIODE RATINGS					
Diode Forward On-Voltage (I_S = 11.5 mA, V_{GS} = 0 V)	V_{SD}	—	—	−1.5	V
Source Current Continuous (Body Diode)	I_S	—	—	−115	mA
Source Current Pulsed	I_{SM}	—	—	−800	mA

*Pulse Test: Pulse Width ≤ 300 μs, Duty Cycle ≤ 2.0%.

FIGURE 14-13 An E-Type MOSFET Data Sheet. (Courtesy of Motorola. Used by permission.)

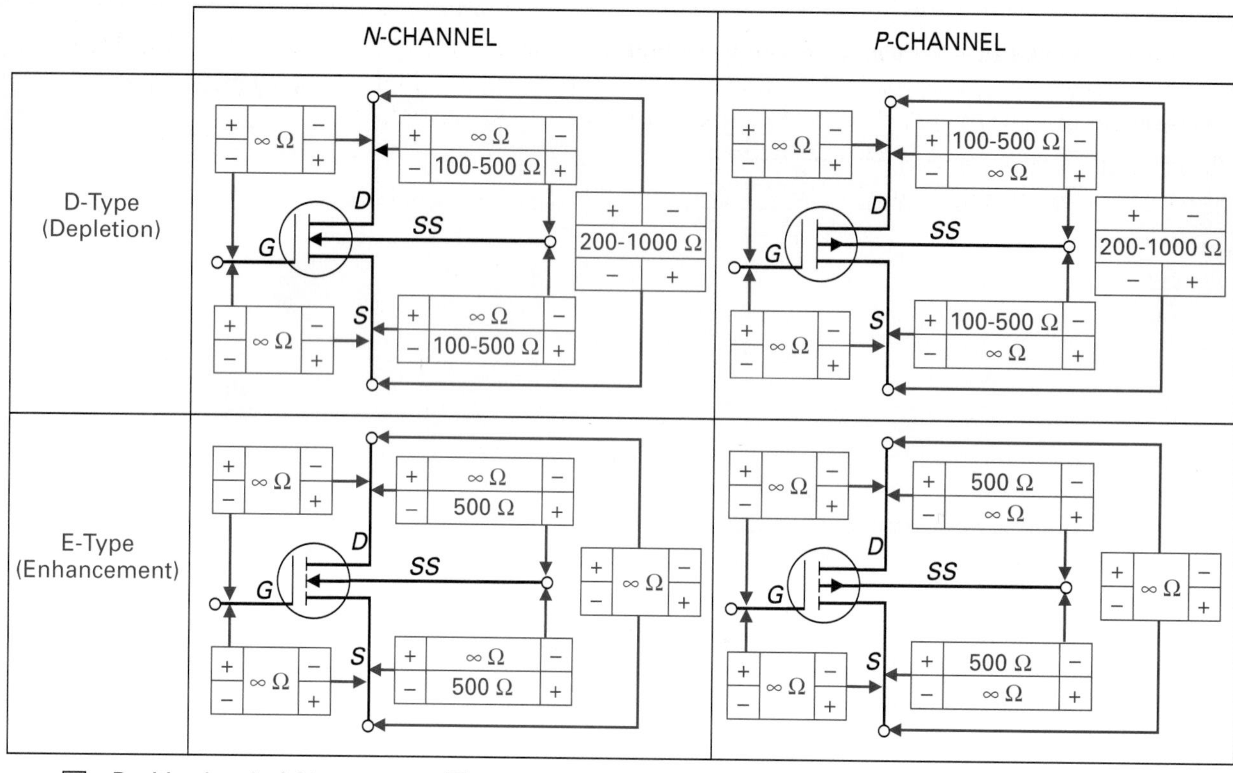

FIGURE 14-14 Testing MOSFETs with the Ohmmeter.

If a transistor tester is not available, the ohmmeter can be used to determine the most common failures: opens and shorts. Figure 14-14 shows what resistance values should be obtained between the terminals of a good *n*-channel or *p*-channel D-MOSFET or E-MOSFET when testing with an ohmmeter. Remember that the gate-to-channel resistance of a MOSFET is always an open due to the insulated gate.

SELF-TEST REVIEW QUESTIONS FOR SECTION 14-3

1. What are the $I_{D(\text{ON})}$ and $V_{DS(\text{ON})}$ values for a 2N4351?
2. If the 2N4351 E-MOSFET was used as a voltage controlled switch, how long would it take
   a. To switch from OFF to ON
   b. To switch from ON to OFF
3. What general precautions should be observed when handling MOSFETs?
4. The gate-to-source or gate-to-drain resistance of a good E-MOSFET should always measure ____________ ohms.

SUMMARY

1. The MOSFET does not have a P-N gate-channel junction like the JFET. It has a "metal gate" that is insulated from the "semiconductor channel" by a layer of "silicon dioxide," hence the name "metal oxide semiconductor."
2. Like all "field effect transistors (FETs)" the MOSFET's input voltage will generate an "electric field" that will have the "effect" of changing the channel's size.
3. The key difference between the JFET and MOSFET is that the JFET's input voltage would always have to be zero or a negative voltage in order to reverse bias the gate-source junction. With the MOSFET, the input voltage can be either a positive or negative voltage because gate current will always be zero (the gate is insulated from the channel).
4. The MOSFET can be operated in either the depletion or enhancement mode due to its insulated gate. The two different types of MOSFET are given names based on their normal mode of operation. The depletion-type MOSFET (D-type MOSFET or D-MOSFET) should actually be called a DE-MOSFET because it can be operated in both the depletion mode and the enhancement mode. The enhancement-type MOSFET (E-type MOSFET or E-MOSFET) is correctly named because it can only be operated in the enhancement mode.

The Depletion-Mode MOSFET (Figure 14-15)

5. The D-MOSFET has two basic transistor types called the *n*-channel D-type MOSFET and *p*-channel D-type MOSFET.
6. This type of MOSFET basically consists of an *n*-type channel formed on a *p*-type substrate. A source and drain lead are connected to either end of the *n*-channel, and an additional lead is attached to the substrate. In addition, a thin insulating (silicon dioxide) layer is placed on top of the *n*-channel, and a metal plated area with a gate lead attached is formed on top of this insulating layer.
7. In some applications, a separate bias voltage will be applied to the substrate terminal for added control of drain current, and the four-terminal device will be used. In most circuit applications, however, the three-terminal device, which has its source and substrate lead internally connected, is all that is needed.
8. The D-MOSFET's set of drain curves have the same general shape as the JFET's set of drain curves and the BJT's set of collector curves. The key difference is that V_{GS} is plotted for both positive and negative values. This is because the D-MOSFET should actually be called a DE-MOSFET because it can be operated in both the depletion mode (in which V_{GS} is a negative value) and the enhancement mode (in which V_{GS} is a positive value).
9. When $V_{GS} = 0$ V, the gate and source terminals are at the same zero-volt potential, and therefore the gate is effectively shorted to the source. The value of drain current passing through the channel in this mode is called the I_{DSS} value (I_{DSS} is the drain-to-source current passing through the channel when the gate is shorted to the source).

QUICK REFERENCE SUMMARY SHEET

FIGURE 14-15 The Depletion-Mode MOSFET (D-MOSFET).

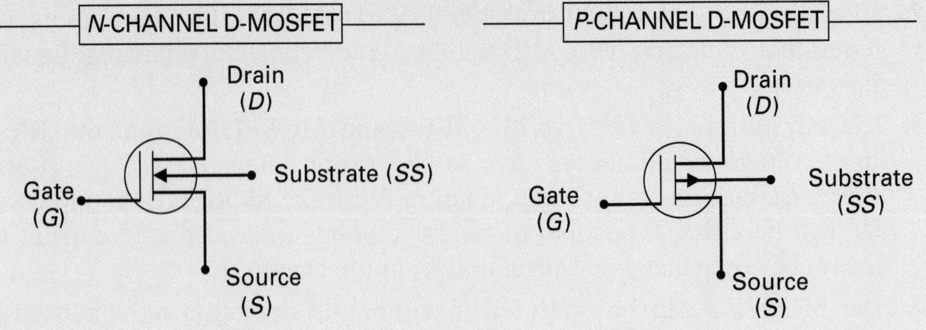

Drain Characteristic Curves

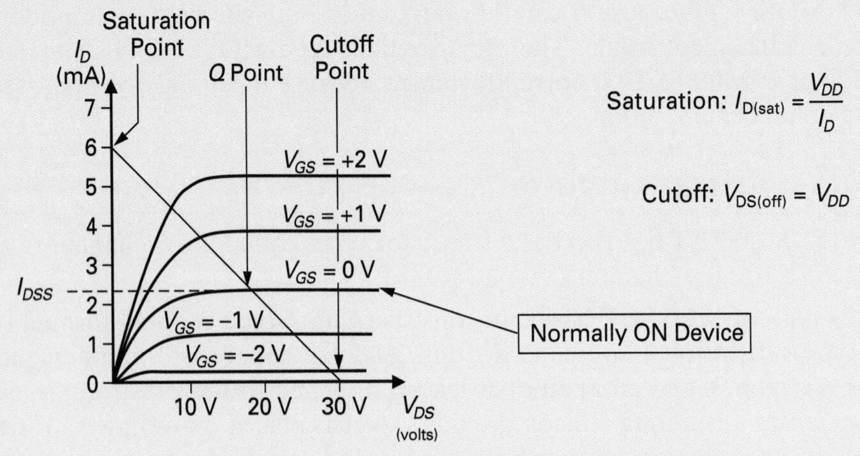

Zero Biasing

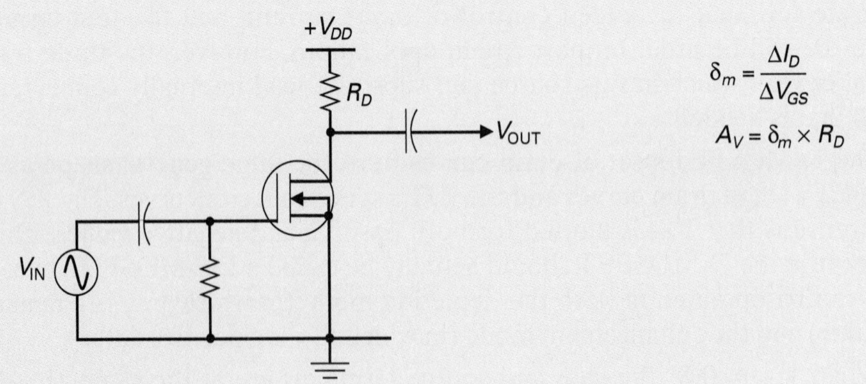

10. When zero volts is applied to the input of a D-type MOSFET, it will conduct a value of drain current. Therefore with no input this device is ON, which is why the D-type MOSFET is known as a "normally ON" device.
11. When V_{GS} is made positive, the channel is enhanced or widened, and therefore the value of I_D is increased above I_{DSS}.
12. When V_{GS} is made negative, the channel is depleted of free carriers, and therefore the value of I_D is decreased below I_{DSS}.
13. As with the JFET, the D-MOSFET's transconductance is equal to the ratio of output current change (ΔI_D) to input voltage change (ΔV_{GS}), and the D-MOSFET's voltage gain is equal to the product of transconductance and the value of drain resistance.
14. Like the JFET, the D-MOSFET can be configured as a common-drain, common-gate or common-source circuit, with all of the dc and ac configuration characteristics being the same.
15. The D-MOSFET is easier to bias than the JFET because of its ability to operate in either the depletion mode ($-V_{GS}$) or the enhancement mode ($+V_{GS}$).
16. A zero-biased D-MOSFET has its Q point set at zero volts ($V_{GS} = 0$ V). Therefore, the ac input signal developed across R_G is applied to the extremely high input impedance of the D-MOSFET, causing an increase and decrease in the conduction of the MOSFET.
17. The D-MOSFET is most frequently used in analog or linear circuit applications. This is because the D-MOSFET can be very simply biased at a mid-point in the load line, and then have its output current varied above and below this natural Q point in a linear fashion. This, coupled with the D-MOSFET's almost infinite input impedance and low-noise properties, makes it ideal as a preamplifier at the front end of a system.
18. The FET's only limiting factor is that its high input impedance starts to decrease as the input signal's frequency increases due to the insulated gate's input capacitance of typically 5 pF. To compensate for this disadvantage, FETs are often connected in series because the overall input capacitance of two series-connected D-MOSFETs will be less than that of a single D-MOSFET.
19. To compensate for the D-MOSFET's input capacitance problem, the dual-gate D-MOSFET was developed.

The Enhancement-Mode MOSFET (Figure 14-16)

20. The enhancement-type MOSFET or E-MOSFET can only operate in the enhancement-mode. In other words, when the input is either zero volts or a negative voltage, the transistor is OFF and there is no source-to-drain current. However, when the input is made positive, the E-MOSFET will turn ON, resulting in a source-to-drain channel current. The E-MOSFET is therefore a "normally OFF" device because it is OFF when nothing (0 V) is applied.
21. The E-MOSFET has two basic transistor types called the *n*-channel E-type MOSFET and *p*-channel E-type MOSFET.

QUICK REFERENCE SUMMARY SHEET

FIGURE 14-16 The Enhancement-Mode MOSFET (E-MOSFET).

N-CHANNEL E-MOSFET

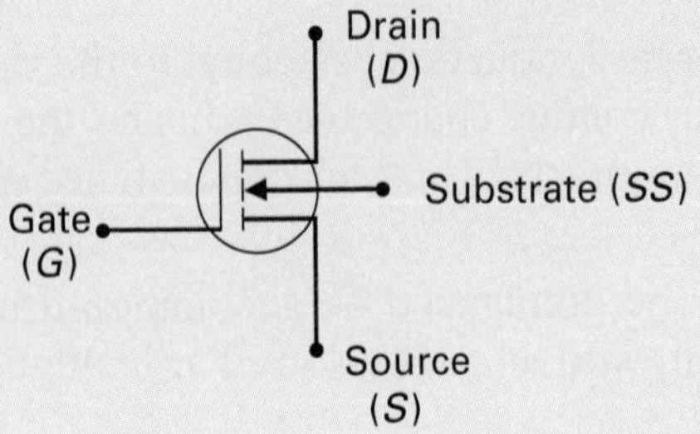

Drain (D)
Gate (G)
Substrate (SS)
Source (S)

P-CHANNEL E-MOSFET

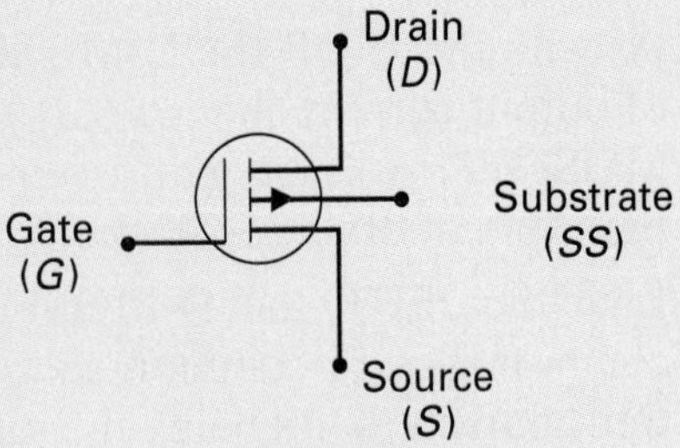

Drain (D)
Gate (G)
Substrate (SS)
Source (S)

Drain Characteristic Curves

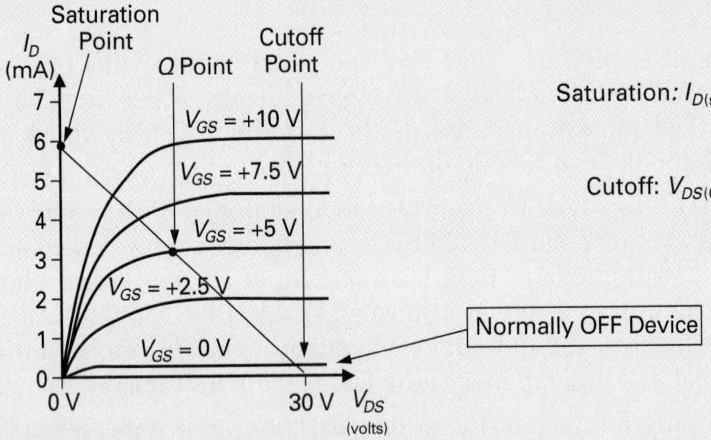

I_D (mA)
Saturation Point
Q Point
Cutoff Point
$V_{GS} = +10$ V
$V_{GS} = +7.5$ V
$V_{GS} = +5$ V
$V_{GS} = +2.5$ V
$V_{GS} = 0$ V
7
6
5
4
3
2
1
0
0 V
30 V V_{DS} (volts)
Normally OFF Device

Saturation: $I_{D(sat)} = \dfrac{V_{DD}}{R_D}$

Cutoff: $V_{DS(OFF)} = V_{DD}$

Drain Feedback Biasing

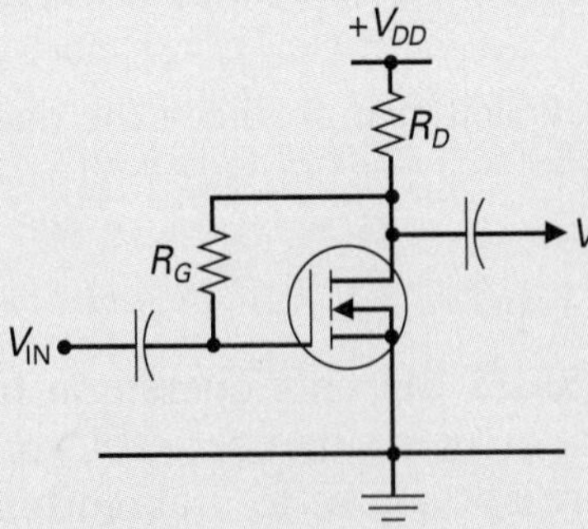

$+V_{DD}$
R_D
R_G
V_{IN}
V_{OUT}

$V_{RG} = 0$ V, $V_{GS} = V_{DS}$

$V_{DS} = V_{DD} - (I_{D(ON)} \times R_D)$

Voltage Controlled Switching

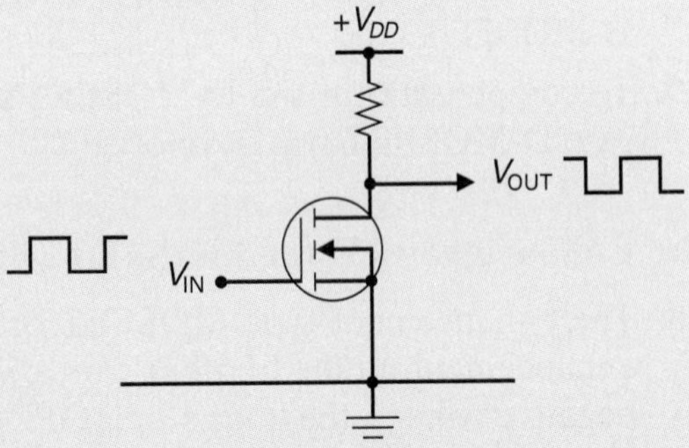

$+V_{DD}$
V_{OUT}
V_{IN}

22. With the E-MOSFET, no channel exists between the source and drain. Consequently, with no gate bias voltage, the device will be OFF. When the gate is made positive, however, electrons will be attracted from the substrate, causing a channel to be induced between the source and drain. This enhanced channel will permit drain current to flow, and any further increase in gate voltage will cause a corresponding increase in the size of the channel and therefore the value of I_D.

23. The only difference between the E-type and D-type MOSFET symbols is the three dashed lines representing the drain, substrate, and source regions. The dashed line is used instead of the solid line to indicate that an E-MOSFET has a normally broken path or channel between drain, substrate, and source.

24. The E-MOSFET's set of drain curves is very similar to the D-MOSFET's set of drain curves; however, in this case only positive values of V_{GS} are plotted, with an increase in V_{GS} causing a corresponding increase in I_D.

25. When $V_{GS} = 0$ V, there is no channel connecting the source and drain, and therefore the drain current will be zero. As a result, $+V_{DD}$ will be present at the output because the E-MOSFET is equivalent to an open switch between drain and source.

26. When $V_{GS} = +5$ V, the E-MOSFET will act in almost exactly the same way as an enhanced D-MOSFET in that an *n*-type bridge will be formed between the source and drain, and a value of I_D will flow between source and drain.

27. When $V_{GS} = +10$ V, the attraction of electrons and repulsion of holes within the *p*-type material is increased, causing the channel's width to increase and I_D to also increase. As a result, 0 V will be present at the output because the E-MOSFET is equivalent to a closed switch between drain and source.

28. Although it cannot be seen in the set of drain curves, V_{GS} will have to increase to a positive threshold voltage of about +1 V before a channel will be induced and a small value of drain current will flow. This "threshold level" is a highly desirable characteristic because it prevents noise or any low-level input signal voltage from turning ON the device. This advantage makes the E-type MOSFET ideally suited as a switch: it can be turned ON by an input voltage and turned OFF once the input voltage falls below the threshold level.

29. Like the D-MOSFET, the E-MOSFET can be configured as a common-drain, common-gate, or common-source circuit. Unlike the D-MOSFET and JFET, the E-MOSFET cannot be biased using self-bias or zero-bias because V_{GS} must be a positive voltage. As a result, gate-bias and voltage-divider bias can be used; however, more frequently E-MOSFETs are drain-feedback biased.

30. In most instances, the E-MOSFET will be used in digital two-state switching circuit applications. In this case, there will be no need for a gate resistor (R_G) because the input voltage (V_{IN}) will either turn ON or OFF the E-MOSFET.

Comparing Digital IC Logic Families (Figure 14-17)

31. The E-MOSFET is more frequently used in digital or two-state circuit applications. One reason is that it naturally operates as a "normally OFF-voltage controlled switch" because it can be turned ON when the gate voltage is positive and turned OFF when the gate voltage falls below a threshold level. This threshold level is a highly desirable characteristic because it prevents noise from false triggering, or accidentally turning ON, the device. The other E-MOSFET advantage is its extremely high input impedance,

QUICK REFERENCE SUMMARY SHEET

FIGURE 14-17 Comparing Digital IC Logic Families.

	BIPOLAR FAMILY		
	LS TTL	ECL	IIL
Cost	LOW	HIGH	MEDIUM
Fanout	20	10 to 25	2
Power Dissipation/Gate	2 mW	40 to 60 mW	0.06 to 70 μW
Propagation Delay/Gate	8 ns	0.5 to 3 ns	25 to 50 ns
Typical Clock Frequency	15 to 120 MHz	200 to 1000 MHz	1 to 10 MHz
External Noise Immunity	GOOD	GOOD	FAIR-GOOD
Typical Supply Voltage	+5 V	–5.2 V	1 V to 15 V
Temperature Range	–55 to 125°C 0 to 70°C	–55 to 125°C	0 to 70°C
Internally Generated Noise	MEDIUM–HIGH	LOW–MEDIUM	LOW

(a)

	MOS FAMILY		
	PMOS	NMOS	HCMOS
Cost	HIGH	HIGH	MEDIUM
Fanout	20	20	10
Power Dissipation/Gate	0.2 to 10 mW	0.2 to 10 mW	2.7 nW to 170 μW
Propagation Delay/Gate	300 ns	50 ns	10 ns
Typical Clock Frequency	2 MHz	5 to 10 MHz	5 to 100 MHz
External Noise Immunity	GOOD	GOOD	VERY GOOD
Typical Supply Voltage	–12 V	+5 V	+3 V to +6 V
Temperature Range	–55 to 125°C 0 to 70°C	–55 to 125°C 0 to 70°C	–55 to 125°C –40 to 85°C
Internally Generated Noise	MEDIUM	MEDIUM	LOW–MEDIUM

(b)

which means that the device's circuit current, and therefore power dissipation, are low. This enables us to densely pack or integrate many thousands of E-MOSFETs onto one small piece of silicon, forming a high component density integrated circuit (IC). These low-power and high-density advantages make the E-MOSFET ideal in battery-powered small-size (portable) applications such as calculators, wristwatches, notebook computers, handheld video games, digital cellular phones, and so on.

32. A logic family is a group of digital circuits with nearly identical characteristics and the bipolar family of digital ICs has three basic types: TTL (transistor-transistor logic), ECL (emitter-coupled logic), and IIL (integrated-injection logic) or I^2L.

33. The three basic E-MOSFET digital IC types (which are more frequently called the MOS family of digital ICs) are: PMOS (pronounced "pea-moss"), NMOS (pronounced "en-moss"), and CMOS (pronounced "sea-moss").

34. MOS logic gates use less space due to their simpler construction, have a very high noise immunity, and consume less power than equivalent bipolar logic gates. However, the high input impedance and capacitance due to the E-MOSFET's insulated gate means that time constants are larger and therefore transistor ON/OFF switching speeds are slower than equivalent bipolar gates. This is the trade-off between the two logic families:

a. Bipolar circuits are faster than MOS circuits, but their circuits are larger and consume more power.

b. MOS circuits are smaller and consume less power than bipolar circuits, but they are generally slower.

MOS Digital Logic Circuits (Figure 14-18)

35. The name PMOS logic is used because the logic gates are constructed using *p*-channel E-MOSFETs, while the name NMOS logic is used because the logic gates are constructed using *n*-channel E-MOSFETs.

36. The name CMOS logic is used because the logic gates are constructed using both a *p*-channel E-MOSFET and its "complement," or opposite, an *n*-channel E-MOSFET.

37. CMOS logic is probably the best all-round logic circuitry because of its low power consumption, very high noise immunity, large power-supply voltage range, and high fanout. The low power consumption achieved by CMOS circuits is due to the "complementary pairs" present in each CMOS gate circuit. During gate operation there is never a continuous path for current between the supply voltage and ground. It is during this time of switching over from one state to the next that a small current will flow due to the complete path from $+V_{DD}$ to ground. This is why the power dissipation of CMOS logic gates increases as the operating frequency of the circuit increases.

38. There is a full line of CMOS ICs available that can provide almost all the same functions as their TTL counterparts. These different types are given different series numbers to distinguish them from others.

QUICK REFERENCE SUMMARY SHEET

FIGURE 14-18 MOS Digital Logic Circuits.

PMOS (*P*-Channel E-MOSFET)

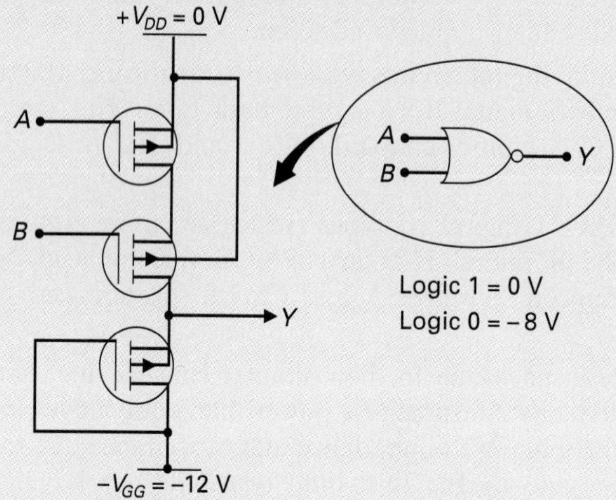

Logic 1 = 0 V
Logic 0 = −8 V

NMOS (*N*-Channel E-MOSFET)

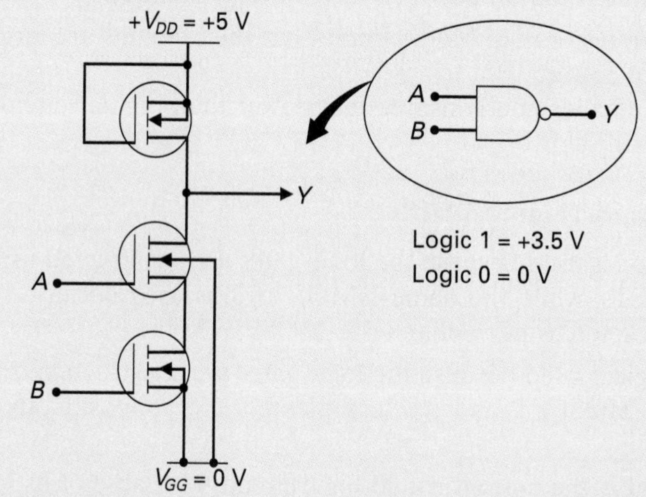

Logic 1 = +3.5 V
Logic 0 = 0 V

CMOS (Complementary MOS — Uses both *N*-Channel and *P*-Channel E-MOSFETs)

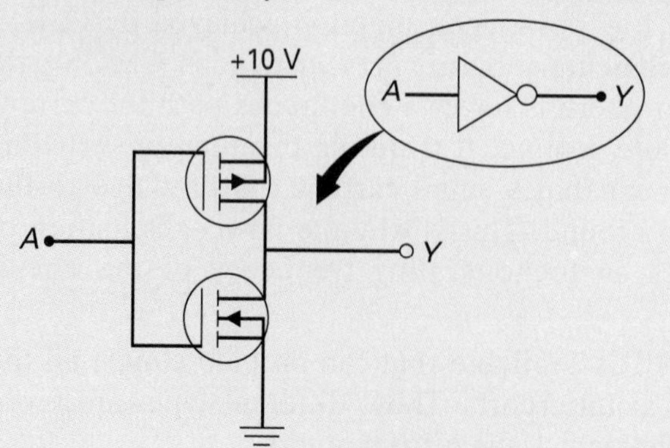

Output HIGH: 4.9 V to 5.0 V
Output LOW: 0 V to 0.1 V
Input HIGH: 3.5 V to 5.0 V
Input LOW: 0 V to 1.5 V
Power Dissipation/gate:
2.7 nW to 170 μW
Propagation Delay/gate: 10 ns
Supply Voltage: +3 V to +6 V
Noise Immunity: Very High
Fanout: 10
(Spec. for 74HC00 Series)

The Vertical Channel MOSFET (VMOS FET)

39. If a larger current-carrying capability than the standard E-MOSFET is needed, the vertical-channel E-MOSFET or VMOS FET can be used.
40. When this *n*-channel VMOS FET is biased ON (gate is made positive with respect to source), a vertical *n*-type channel is formed between source and drain. This channel is much wider than a standard E-MOSFET's horizontal channel, which is why VMOS FETs can handle a much higher drain current.
41. Other than its high current capability, the VMOS FET also has a positive temperature coefficient of resistance, which means an increase in temperature will cause a decrease in drain current ($T\uparrow$, $R\uparrow$, $I\downarrow$), and this will prevent thermal runaway.

MOSFET Handling Precautions

42. The very thin insulating layer between the gate and the substrate of a MOSFET can easily be punctured if an excessive voltage is applied. Your body can build up extremely large electrostatic charges due to friction. If this charge comes in contact with the pins of a MOSFET device, an electrostatic discharge (ESD) would occur, resulting in a possible arc across the thin insulating layer causing permanent damage.
43. Most MOSFETs presently manufactured have zeners internally connected between gate and source to bypass high voltage static or in-circuit potentials and protect the MOSFET.
44. When handling MOSFETs it is important to remember the following.
    - **a.** All MOS devices are shipped and stored in a "conductive foam" so that all of the IC pins are kept at the same potential, and therefore electrostatic voltages cannot build up between terminals.
    - **b.** When MOS devices are removed from the conductive foam, be sure not to touch the pins because your body may have built up an electrostatic charge.
    - **c.** When MOS devices are removed from the conductive foam, always place them on a grounded surface such as a metal tray.
    - **d.** When continually working with MOS devices, use a "wrist grounding strap," which is a length of cable with a 1 MΩ resistor in series. This prevents electrical shock if you come in contact with a voltage source.
    - **e.** All test equipment, soldering irons, and work benches should be properly grounded.
    - **f.** All power in equipment should be off before MOS devices are removed or inserted into printed circuit boards.
    - **g.** Any unused MOSFET terminals must be connected because an unused input left open can build up an electrostatic charge and float to high voltage levels.
    - **h.** Any boards containing MOS devices should be shipped or stored with the connection side of the board in conductive foam.

QUICK REFERENCE SUMMARY SHEET

FIGURE 14-19 Testing MOSFETs with the Ohmmeter.

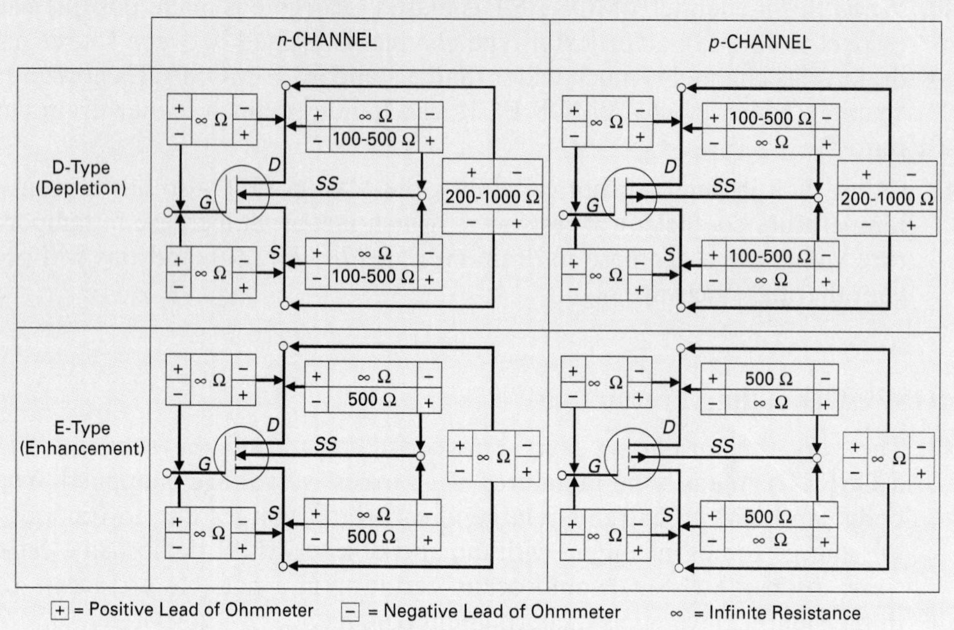

MOSFET Testing (Figure 14-19)

45. Like the BJT and JFET, MOSFETs can be tested with the transistor tester to determine: first, whether an open or short exists between any of the terminals; second, the transistor's transconductance/gain; and third, the value of I_{DSS} and leakage current.

46. If a transistor tester is not available, the ohmmeter can be used to determine the most common *n*-channel and *p*-channel D-MOSFET and E-MOSFET failures: opens and shorts. Remember that the gate-to-channel resistance of a MOSFET is always an open due to the insulated gate.

NEW TERMS

Bipolar Family
Cascade Amplifier
CMOS
CMOS Logic
Complementary MOS (CMOS)
Depletion-Mode Operation
Depletion-Type MOSFET (D-MOSFET)
Drain Feedback Biasing
Dual Gate D-MOSFET
Enhancement-Mode Operation
Enhancement-Type MOSFET (E-MOSFET)
Logic Family
Logic Levels
Metal Oxide Semiconductor FET (MOSFET)
MOS Family

n-Channel MOSFET
NMOS
NMOS Logic
p-Channel MOSFET
PMOS
PMOS Logic
Vertical Channel E-MOSFET
VMOS
Zero Biasing

REVIEW QUESTIONS

Multiple-Choice Questions

1. The _____________ has an insulated gate that allows us to use either a positive or negative gate input voltage.

   a. JFET **c.** BJT
   b. MOSFET **d.** VJT

2. Which type of FET operates in the depletion mode?

   a. JFET **c.** D-MOSFET
   b. E-MOSFET **d.** Both (a) and (c)

3. Which type of FET makes use of an electric field to change the channel's size?

   a. JFET **c.** D-MOSFET
   b. E-MOSFET **d.** All of the above

4. The _____________ has drain current when the gate voltage is zero and is therefore known as a normally _____________ device.

   a. D-MOSFET, OFF **c.** E-MOSFET, OFF
   b. D-MOSFET, ON **d.** E-MOSFET, ON

5. The _____________ has no drain current when gate voltage is zero and is therefore known as a normally _____________ device.

   a. D-MOSFET, OFF **c.** E-MOSFET, OFF
   b. D-MOSFET, ON **d.** E-MOSFET, ON

6. Which type of FET operates in both the depletion and enhancement mode?

   a. JFET **c.** E-MOSFET
   b. D-MOSFET **d.** Both (b) and (c)

7. Which type of MOSFET can use zero biasing to bias it to a mid-load-line point?

   a. D-MOSFET **c.** VMOS FET
   b. E-MOSFET **d.** Both (b) and (c)

8. The dual-gate MOSFET is ideal for interfacing low-voltage devices to high-power loads.

   a. True **b.** False

9. With the _____________, the gate voltage must be greater than the device's threshold voltage to produce drain current.

   a. JFET **c.** E-MOSFET
   b. D-MOSFET **d.** All of the above

10. The vertical channel MOSFET is ideal for interfacing low-voltage devices to high-power loads.

a. True **b.** False

11. Which of the following transistors has the highest input impedance?

a. BJT **c.** MOSFET
b. JFET **d.** Both (a) and (c) are true

12. D-MOSFETs are more frequently used in ____________ circuit applications whereas E-MOSFETs are more often used in ____________ circuit applications.

a. analog, digital **c.** digital, linear
b. linear, analog **d.** two-state, digital

13. A digital code consists of a series of HIGH and LOW voltages called ____________.

a. NOT gates **c.** Logic levels
b. CMOS **d.** Both (a) and (b)

14. The ____________ has made its mark in two-state circuits in which it uses only two points on the load line and acts like a voltage controlled ____________.

a. E-MOSFET, variable resistor **c.** D-MOSFET, variable resistor
b. E-MOSFET, switch **d.** D-MOSFET, switch

15. To avoid possible damage to MOS devices while handling, testing, or in operation, the following precaution should be taken.

a. All leads should be connected except when being tested or in actual operation.
b. Pick up devices by plastic case instead of leads.
c. Do not insert or remove devices when power is applied.
d. All of the above

Essay Questions

16. Define the following terms: (Intro.)

a. Depletion-mode operation **b.** Enhancement-mode operation

17. How does a MOSFET differ from a JFET? (Intro.)
18. Why should a D-MOSFET really be called a DE-MOSFET? (14-1)
19. Identify the device schematic symbols shown in Figure 14-20. (14-1 and 14-2)
20. Briefly describe the operation of the D-MOSFET. (14-1-2)
21. Which MOSFET is called a "normally ON" device and which is called a "normally OFF" device? Describe why. (14-1-2 and 14-2-2)
22. How would a D-MOSFET typically be biased to a mid-point on its load line? (14-1-3)
23. Briefly describe the operation and application of the dual-gate MOSFET. (14-1-5)
24. Briefly describe the operation of the E-MOSFET. (14-2-2)

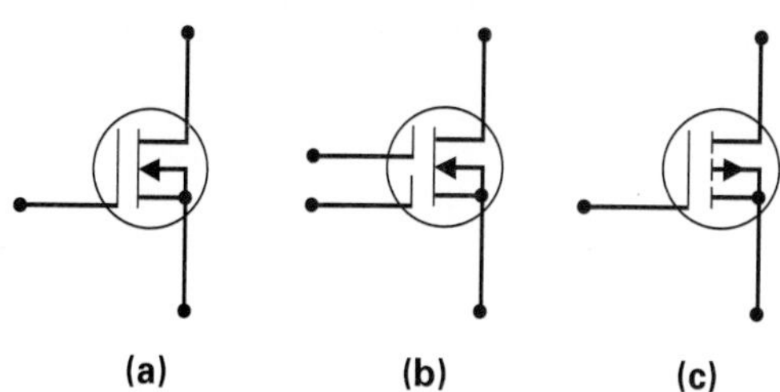

FIGURE 14-20 **MOSFET Schematic Symbols.**

25. How would an E-MOSFET typically be biased to a mid-point on its load line? (14-2-3)
26. List the three basic bipolar digital IC types and the three basic MOS digital IC types. (14-2-4)
27. Briefly describe the operation and application of the vertical channel MOSFET. (14-2-5)
28. Briefly describe the circuit applications for the D-MOSFET and E-MOSFET. (14-1-4 and 14-2-4)
29. Referring to the data sheet in Figure 14-13, determine the following for 2N7002 MOSFET: (14-3-1)
    a. Maximum value of drain current
    b. Drain-to-source reverse breakdown voltage
    c. Input capacitance
30. Why should MOSFETs be handled carefully, and what are some of the procedures to follow? (14-3-2)

Practice Problems

31. Calculate the following for the circuit shown in Figure 14-21:
    a. $I_{D(\text{sat})}$ **b.** $V_{DS(\text{OFF})}$
32. Calculate the following for the circuit shown in Figure 14-22:
    a. I_D **b.** V_{DS}

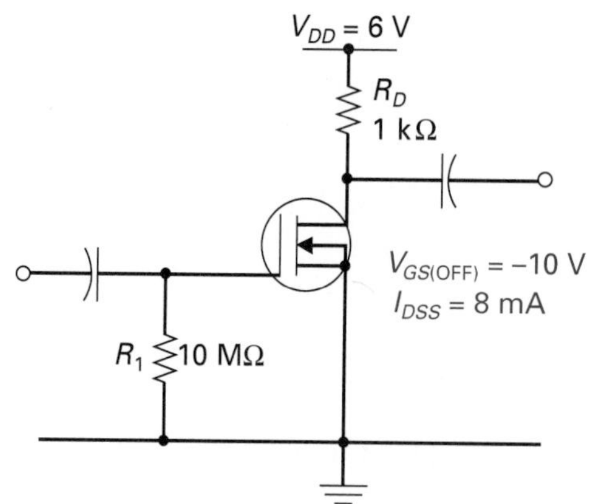

FIGURE 14-21 **A D-MOSFET Amplifier Circuit.**

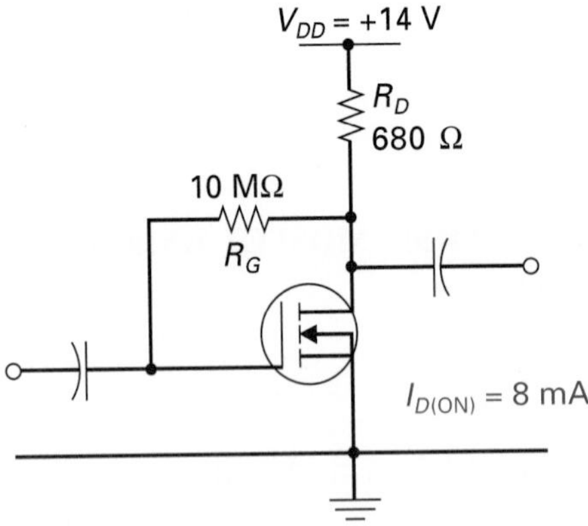

FIGURE 14-22 An E-MOSFET Amplifier Circuit.

33. What biasing method was used in Figure 14-21 and 14-22?

34. Identify the circuit in Figure 14-23. What will the approximate output voltage be if the input equals

a. $V_{IN} = 0$ V **b.** $V_{IN} = +5$ V

35. Referring to the circuit in Figure 14-24, answer the following questions:

a. Is Q_1 ON or OFF, and why?

b. Calculate V_{OUT} when $V_{IN} = +5$ V.

c. Calculate V_{OUT} when $V_{IN} = 0$ V.

Troubleshooting Questions

36. If you had a choice of placing the pins of a good MOSFET transistor on a plastic or metal tray, which would you choose?

37. What resistance would you expect the ohmmeter to show in Figure 14-25?

38. Are the MOSFETs in Figure 14-26 good or bad?

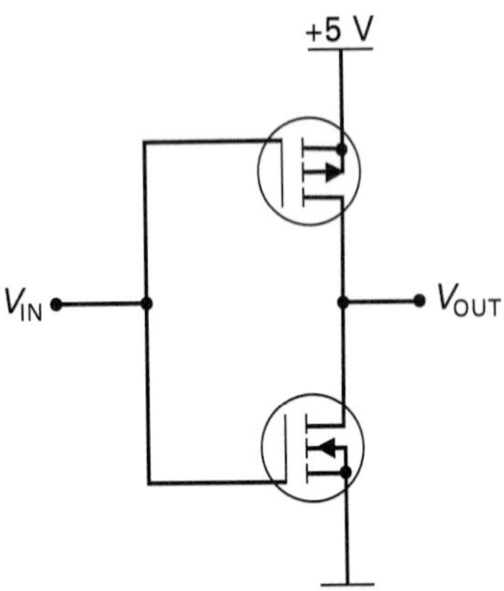

FIGURE 14-23 MOSFET Circuit.

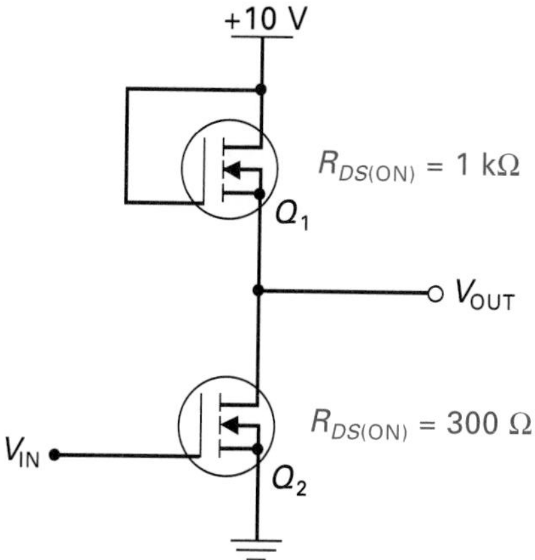

FIGURE 14-24 A MOSFET Acting as a Load Resistor.

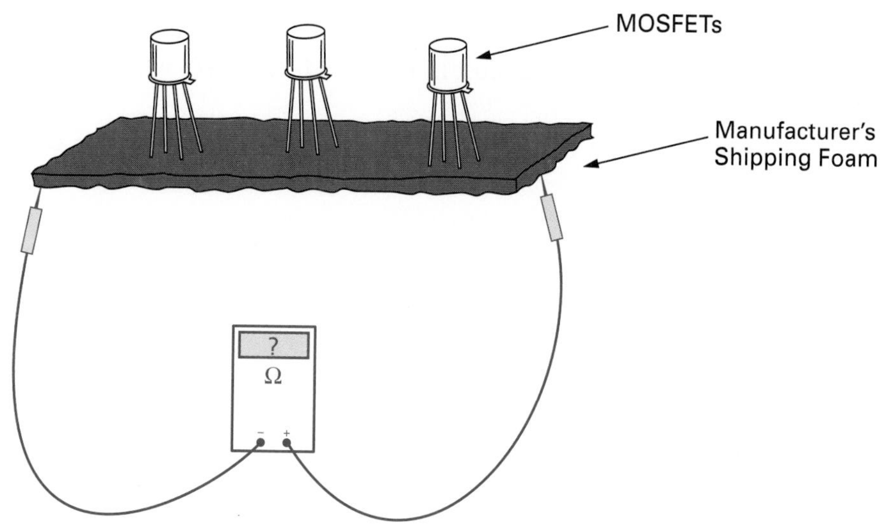

FIGURE 14-25 MOSFETs in Manufacturer's Shipping Foam.

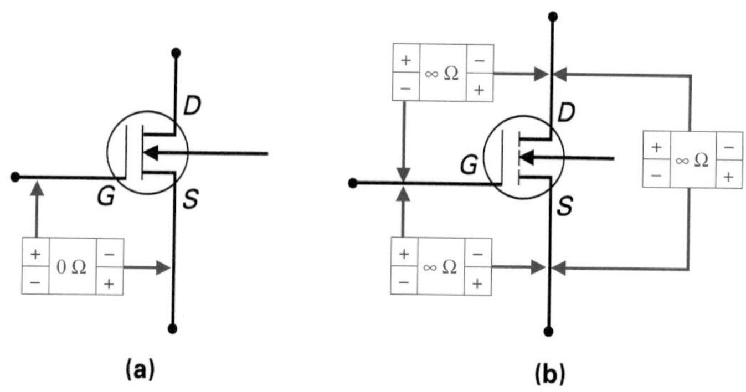

FIGURE 14-26 Testing MOSFETs with the Ohmmeter.

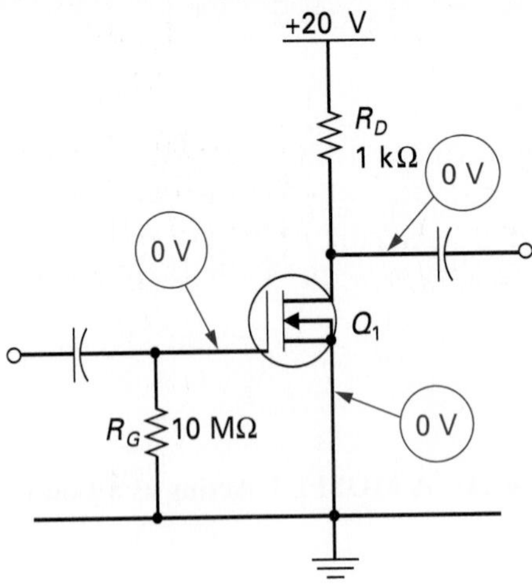

FIGURE 14-27 D-MOSFET Circuit Troubleshooting.

39. Is there a problem with the circuit in Figure 14-27? If so, what do you think it could be?
40. Is there a problem with the circuit in Figure 14-28? If so, what do you think it could be?

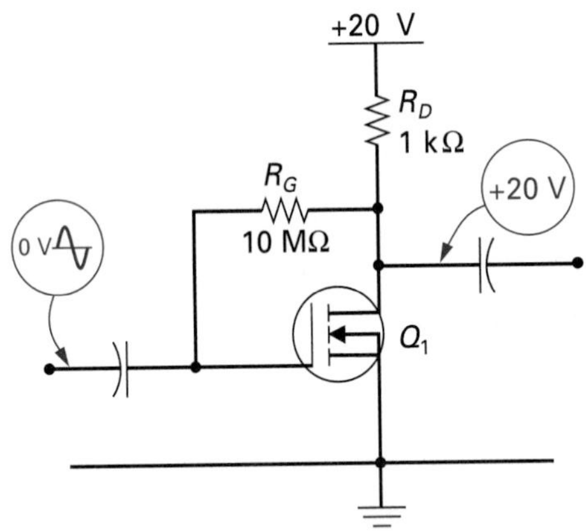

FIGURE 14-28 E-MOSFET Circuit Troubleshooting.

CHAPTER

15

OBJECTIVES

After completing this chapter, you will be able to:

1. Define the following terms:
   - **a.** Discrete components
   - **b.** Integrated circuit
   - **c.** Analog IC
   - **d.** Linear IC
   - **e.** Digital IC
   - **f.** Hybrid IC
   - **g.** Bipolar IC
   - **h.** MOS IC
   - **i.** Small scale integration
   - **j.** Medium scale integration
   - **k.** Large scale integration
   - **l.** Very large scale integration
2. Describe the basic step-by-step construction procedure for an integrated circuit.
3. Describe the different layers of a bipolar and MOS IC.
4. List the advantages of hybrid ICs.
5. Describe some of the IC package types currently available.
6. Explain the classification and some of the applications for SSI, MSI, LSI, and VLSI circuits.

Integrated Circuit Construction

OUTLINE

VIGNETTE

Flying Too High

The term "software"—as opposed to "hardware" or physical components of the computer—was first used in the 1960s to describe the program of instructions that tells a computer what to do. At first, the binary words representing instructions and data were laboriously entered by setting up countless switches and dials. In modern computers, binary coded instructions and data are stored as magnetic signals on disks and then input into the computer's internal memory where the program will tell the computer what to do.

Computer software today is organized into several basic types. There are "translator" programs written to convert high-level computer languages, which are like human speech, into machine language, the ones and zeros understood by the machine or computer. There are also "utility" programs designed to sort and merge data files. Two of the biggest groups of software are "operating systems" and "application programs." An operating system coordinates the operation of the computer system controlling the transfer of data between the computer and its peripheral devices, such as the keyboard, monitor, printer, and disk drive. Application programs are the software that change the personality of a computer from a word processor generating a contract, to an accounting system controlling payroll, to an industrial controller directing a robotics assembly line.

In this vignette, let us examine some of the software-industry folklore relating to operating systems. Most of the 8-bit computers of the late 1970s used an operating system developed by Gary Kildall in 1974 called CP/M, standing for "Control Program for Microcomputers." Just before IBM officially entered into the personal computer market in August of 1981, two IBM executives went to a prearranged meeting with Kildall, who was at the time flying his private plane. After being contacted by radio, Kildall declined to come down and discuss an operating system for their proposed personal computer or "PC." After rescheduling the meeting, the executives were surprised to discover that he had flown off to the Caribbean for a vacation. When he came back, IBM had made a deal with Bill Gates of Microsoft for an operating system. Called MS-DOS, which stood for "Microsoft—Disk Operating System" (DOS means an operating system that is loaded into the computer from a disk), it became an instant success largely due to its blessing from IBM. Hundreds of computer manufacturers started producing computers that were compatible with the IBM PC and therefore used Gates's MS-DOS. Even in an expanding market, Kildall's CP/M could not hold on to any business and is today extinct.

Bill Gate's Microsoft is the largest computer software company in the world, and history is once again repeating itself. Not liking their dependency on Microsoft, IBM has developed their own operating system, which is now in its second version called "OS/2." Another battle begins.

INTRODUCTION

In 1948, scientists discovered that the large vacuum tubes being used in radios, televisions, and other equipment could be replaced by extremely small devices made from semiconductor materials. A few years later, the transistor was born, and the world stepped out of the "vacuum tube era" and into the "solid state era."

At first, a single transistor was built by itself on a single chip or piece of semiconductor material. An electronic circuit was then formed by wiring together all of the individual or "discrete" components, such as the transistors, resistors, capacitors, and so on. In 1959, it was discovered that more than one transistor could be made on a chip of semiconductor material. Soon other components such as resistors, capacitors, and diodes were added with the transistors, and all were interconnected to form a complete circuit. This integrating of various components on a single chip of semiconductor is called an "integrated circuit" or IC.

Today, the IC is extensively used in every branch of electronics with hundreds of thousands of transistors and other components being placed on a chip no bigger than this ∂.

Integrated circuits can be classified as being either "linear" or "digital." A linear IC produces an output signal that varies in proportion to the input signal (linear function), and circuit examples include amplifiers, oscillators, detectors, modulators, regulators, and so on. Because the input and output signals of a linear IC are analog in nature, linear ICs are also called "analog ICs." Digital ICs are the most widely used and are basically two-state switching circuits designed to be used in digital computer and logic circuits.

In previous chapters, you have seen how many semiconductor devices and circuits are within integrated circuit packages. Therefore, it is about time we discussed the integrated circuit in more detail. In this chapter we will discuss the construction of "bipolar ICs," "MOS ICs," and "hybrid ICs," along with the different IC package types available, and the difference between "small scale," "medium scale," "large scale," and "very large scale" integration.

15-1 THE MULTILAYERED INTEGRATED CIRCUIT

When designing a high-rise building, it is important that each floor is designed with connecting horizontal passageways between different points on that floor. It is also important that each floor is connected vertically with stairways and elevators. The procedure for constructing this multi-story building is to first begin with a foundation, then build up floor by floor.

Integrated Circuit (IC)
Two or more components combined into a circuit and then incorporated in one package or housing.

An **integrated circuit** or **IC** is also a multilayered construction that has horizontal connections within each layer and vertical connections between layers. Besides the obvious differences between a high-rise and an integrated circuit, there is a major difference in the construction technique. The fabrication of an integrated circuit begins with a thin slice (4/1000 inch) of silicon called a sub-

Photolithography
A photoengraving process in which each pattern for each layer of an IC is reproduced on a glass plate called a photo mask. This mask then acts as a photographic negative allowing ultraviolet light to imprint the pattern on the silicon substrate.

Photo Mask
The glass plate that acts as a photographic negative allowing ultraviolet light to imprint the pattern on the silicon substrate of an IC.

Ingot
A mass of metal cast into a convenient shape.

Wafers
Very thin (on the order of 4/1000 of an inch thick) slices of ingot.

strate. Then, through a photoengraving process called **photolithography,** each pattern for each layer is reproduced on a glass plate called a **photo mask.** This mask acts as a photographic negative allowing ultraviolet light to imprint the pattern on the silicon substrate. The silicon substrate is then baked, causing the first layer to sink into the silicon substrate. A second pattern or layer is printed onto the substrate, and the baking process is repeated. This causes the second layer to sink into the substrate and the first layer to sink deeper into the substrate. Simple ICs may require a four-layered chip, while complex ICs may require forty layers.

Unlike the high rise building in which we build up from the foundation, with the integrated circuit we build down into the substrate.

15-1-1 Basic Step-by-Step IC Construction Procedure

The basic step-by-step construction of an integrated circuit can be seen in Figure 15-1(a). The process begins with silicon, which is a dull, metallic-looking semiconductor material that is refined from ordinary sand. Through a refining process, the silicon is melted and grown into ingots that are 99.9999999% pure. An **ingot** is shown in step 1 of Figure 15-1(a). It is about six inches in diameter and two feet long.

In step 2, a diamond saw slices the ingot into **wafers** that are 4/1000 of an inch thick. The wafers are then placed in an oven which sterilizes the front and back surface of the wafer and smoothes out any rough edges.

Steps 4, 5, and 6 will be repeated until all of the photo masks have been imprinted, developed, and doped onto the silicon substrate. Figure 15-1(b) shows the photo masks for a four-layered amplifier integrated circuit. These circuit patterns are drawn up on a computer that uses different colors for each layer so one can be superimposed over another to check positioning. Each mask actually contains hundreds of copies of the same pattern on a single glass plate, and these plates are used to imprint hundreds of identical circuits or chips on one wafer.

After hundreds of the same mask pattern have been printed onto the wafer in step 4, the circuit pattern is embedded into the silicon in step 5. Because pure or intrinsic silicon will not support enough current, in step 6 the wafers are doped with hot gaseous ions of phosphorus to create *n*-type sections and boron to create *p*-type sections.

After the final cycle through steps 4, 5, and 6, all of the chips on the wafer are probed by computerized test equipment. Approximately 50% of the ICs fail this initial testing phase due to errors in the previous steps or impurities in the silicon. Due to this low yield, all manufacturing of chips is conducted in clean rooms in which workers are dressed in lint-free caps and smocks and the air is purified. It would only take a speck of dirt, make-up, or graphite from a pencil lead to ruin several ICs.

In step 8, the wafers are sliced into chips, and the defective ICs are discarded. After a final test, the integrated circuit is placed into a protective package, as seen in step 9. Finally, the bonding pads of the circuit are connected to the leads of the package with very thin wires, and the package is sealed and sent for final testing.

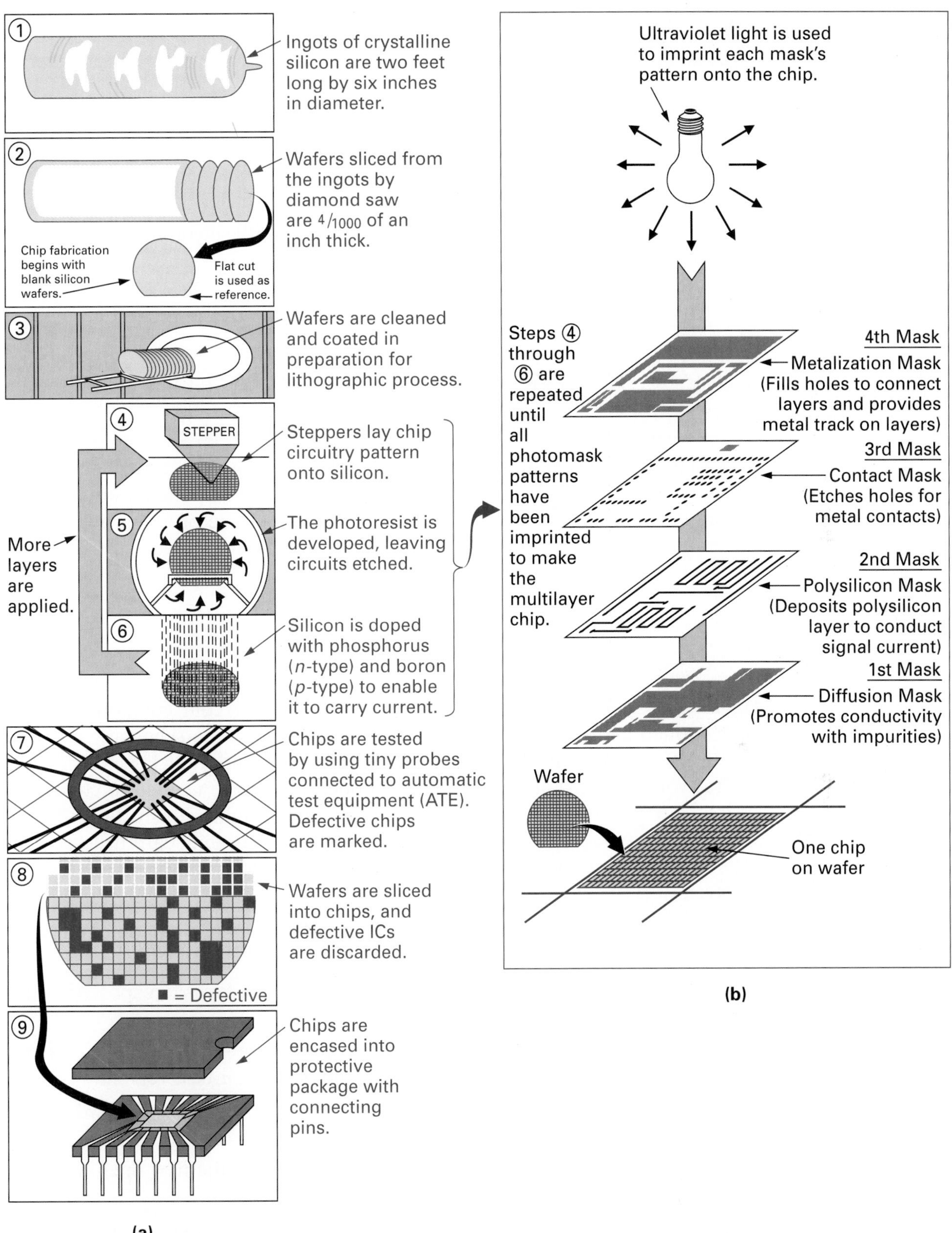

1
Ingots of crystalline silicon are two feet long by six inches in diameter.
2
Wafers sliced from the ingots by diamond saw are 4/1000 of an inch thick.
Chip fabrication begins with blank silicon wafers.
Flat cut is used as reference.
3
Wafers are cleaned and coated in preparation for lithographic process.
4
STEPPER
Steppers lay chip circuitry pattern onto silicon.
5
The photoresist is developed, leaving circuits etched.
More layers are applied.
6
Silicon is doped with phosphorus (n-type) and boron (p-type) to enable it to carry current.
7
Chips are tested by using tiny probes connected to automatic test equipment (ATE). Defective chips are marked.
8
Wafers are sliced into chips, and defective ICs are discarded.
■ = Defective
9
Chips are encased into protective package with connecting pins.
(a)
Ultraviolet light is used to imprint each mask's pattern onto the chip.
Steps 4 through 6 are repeated until all photomask patterns have been imprinted to make the multilayer chip.
4th Mask
Metalization Mask (Fills holes to connect layers and provides metal track on layers)
3rd Mask
Contact Mask (Etches holes for metal contacts)
2nd Mask
Polysilicon Mask (Deposits polysilicon layer to conduct signal current)
1st Mask
Diffusion Mask (Promotes conductivity with impurities)
Wafer
One chip on wafer
(b)

FIGURE 15-1 **The Simplified Step-by-Step Procedure for Constructing an Integrated Circuit.**

SELF-TEST REVIEW QUESTIONS FOR SECTION 15-1

1. IC is an abbreviation for ____________ ____________.
2. ICs are generally classified as being either linear (analog) or ____________.
3. ICs are fabricated using a photoengraving process called ____________.
4. The construction of an integrated circuit begins with silicon which is refined from ____________.

15-2 BIPOLAR AND MOS INTEGRATED CIRCUITS

Now that we have seen how complete integrated circuits are constructed, let us take a closer look at how the individual electronic components are formed. Before we begin, however, we must categorize ICs in one of two groups: those ICs containing bipolar transistors and those ICs containing MOS transistors. Let us start by examining bipolar ICs in more detail.

15-2-1 *Bipolar IC Construction*

Diffusion
A method of producing a junction by diffusing an impurity metal into a semiconductor at a high temperature.

Figure 15-2(a) shows a basic integrated circuit in which four components have been formed and interconnected. Figure 15-2(b) shows the schematic diagram for the IC structure in Figure 15-2(a). Studying the different layers in Figure 15-2(a) you can see that several **diffusion** operations were needed to create the entire chip. All of the components are created simultaneously by diffusing *n*-type and *p*-type impurities into the *p*-type substrate, being careful to control the size, depth, and location of these doped regions. Here is a listing of the steps involved.

1. First a thin insulating layer of silicon oxide is deposited onto the substrate.
2. Then an acid is used to cut windows into the oxide layer to expose the substrate.
3. The next step is to diffuse *n*-type impurities through the open windows and into the substrate. This forms the first, and deepest, *n*-type regions.
4. A layer of oxide is deposited to cover the existing windows, and acid is used once again to open up new windows.
5. A *p*-type impurity is diffused into the substrate via the new open windows to create *p* regions.
6. The process is repeated with another oxide layer, acid to open up windows, and then the final *n* region.

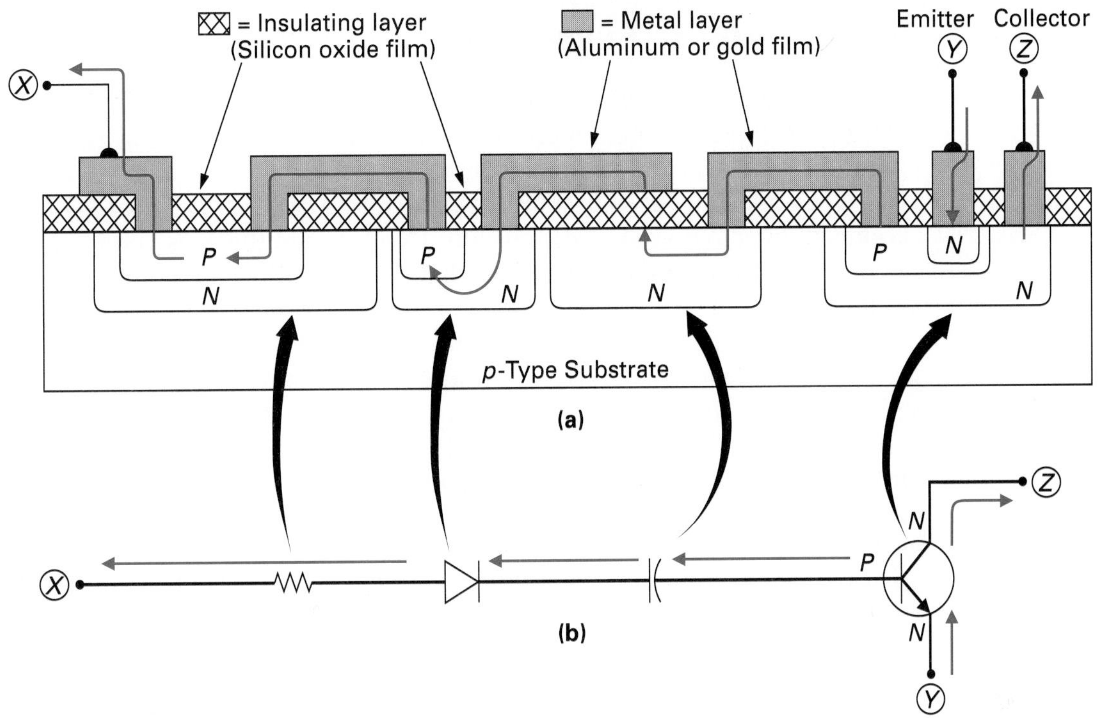

FIGURE 15-2 **Basic Construction of a Bipolar IC. (a) Structure. (b) Schematic.**

7. The last step is to apply a final oxide layer to cover up the windows, and then use acid once again to open up connecting holes. Finally, a coating of aluminum is deposited onto the chip to make the connections to the *n* and *p* regions through the windows and to make the circuit connections between the components.

Studying the components in Figure 15-2(a), you can see that the resistor consists of a *p*-type region on top of an *n*-type region. The resistance can be determined by the length ($L\uparrow = R\uparrow$) and width ($W\uparrow = R\downarrow$) of the *p*-region. The resistance is also determined by the concentration of *p*-type impurities (*p*-type$\uparrow$ = $R\downarrow$, *p*-type $\downarrow = R\uparrow$).

The next component is the P-N junction diode, which is no different from a conventional diode in that it contains a *p*-type anode and *n*-type cathode.

With the capacitor, the layer of metal serves as one plate, the silicon oxide serves as the dielectric and the diffused *n*-type region serves as the other plate. The capacitance is determined by plate area, dielectric thickness, and the dielectric constant.

Finally, the IC NPN transistor is constructed in much the same way as a conventional discreet bipolar NPN transistor. The difference with this transistor is that all three regions have their connections on top of the substrate. Because all of the connections for this transistor lie in a flat plane, the transistor was named a **planar transistor.**

Planar Transistor
A transistor designed so that all the connections lie in a flat plane.

15-2-2 *MOS IC Construction*

Figure 15-3(a) shows the IC structure for an *n*-channel enhancement-type MOSFET (E-MOSFET), and Figure 15-3(b) shows its schematic symbol. As you know, this FET acts as a "voltage-controlled switch" because the device will only turn ON and connect the source to the drain when the input gate voltage exceeds a threshold voltage of typically 1 volt. When this bias voltage is applied, current will flow through an induced channel under the gate between source and drain.

Looking at the construction of the E-MOSFET in Figure 15-3(a), you can see that the gate is isolated from the source and drain regions by an oxide layer, and the source and drain are isolated from one another by P-N junctions. Comparing the construction of the MOSFET in Figure 15-3(a) to the bipolar transistor in Figure 15-2(a), you can see that the construction of the MOS transistor is simpler due to fewer layers. Because of this, MOSFETs can be made smaller than bipolar transistors. MOSFETs can also be used as resistors, as seen in Figure 15-3(c), by simply connecting the gate to the drain and then adjusting the spacing between the source and drain to determine resistance (wide spacing =

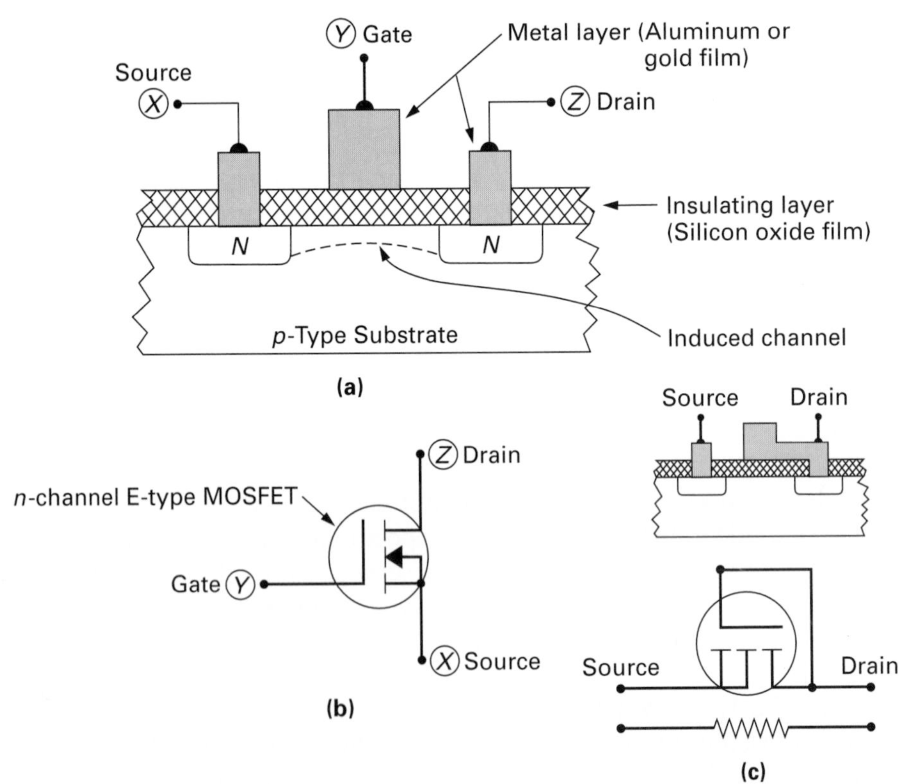

FIGURE 15-3 Basic Construction of a MOS IC. (a) Structure. (b) Schematic. (c) A MOSFET Resistor.

higher resistance). This means that the compact MOSFET can replace the larger planar resistor.

MOS ICs therefore have a higher component density than bipolar circuits, consume less power, have a higher input impedance, noise immunity, and better temperature stability. Bipolar ICs, on the other hand, have one important advantage over MOSFETs: they respond faster and are therefore best suited for high-frequency applications in which circuits need to react to rapidly switching signals.

15-2-3 *Hybrid IC Construction*

Hybrid, by definition, is the blending together of two different elements. In electronics, a **hybrid IC** is a composite or blending together of discrete components and integrated circuits, analog and digital circuits, active and passive components, and standard and custom integrated circuits. All of the components are interconnected with conducting tracks that are printed onto an insulating (generally ceramic) substrate, as seen in Figure 15-4.

Hybrid IC
A composite or blending together of discrete components and integrated circuits, analog and digital circuits, active and passive components, and standard and custom integrated circuits.

The key advantage of a hybrid IC is that if only a few circuits are required, the hybrid circuit does not require the high-cost, extensive circuit layout and elaborate diffusion techniques of an integrated circuit. If a large number of custom circuits are required, the integrated circuit becomes the cheaper solution.

Another advantage to the hybrid circuit is that all of its discrete components (resistors, diodes, transistors, and capacitors) can handle more power than their integrated counterparts due to their larger size. Because of this, hybrid circuits are larger and heavier than integrated circuits.

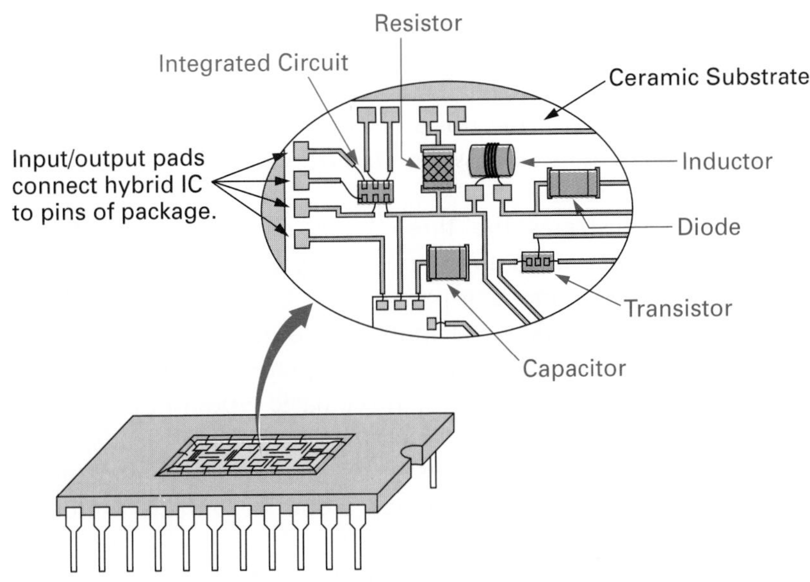

FIGURE 15-4 **Basic Construction of a Hybrid IC.**

SELF-TEST REVIEW QUESTIONS FOR SECTION 15-2

1. Which ICs have a greater component density?
2. Which ICs have a faster response time?
3. A ____________ IC blends together two different devices or technologies.
4. Which of the following would you choose if you only needed a few custom ICs?
   a. MOS ICs
   b. Hybrid ICs
   c. Bipolar ICs

15-3 IC PACKAGE TYPES

Transistor Outline or TO Can Package Cylindrical, metal can type of package for some semiconductor components.

Flat Pack A (generally ceramic) IC package with leads designed to be soldered directly to the tracks of the printed circuit board.

Dual-in-Line Package (DIP) An IC package with two lines of parallel connecting pins. Its leads feed through holes running through the printed circuit board.

Surface-Mount Technology (SMT) Package An IC package that is mounted directly on the printed circuit board and needs no lead feed-through holes.

There are four basic methods of packaging integrated circuits, and these are illustrated in Figure 15-5. Figure 15-5(a) shows the first type package used for ICs called the **transistor outline** or **TO Can package.** This transistor package was modified from the basic 3-lead transistor package to have the extra leads needed and is still used to house linear ICs due to its good heat dissipation ability.

Figure 15-5(b) illustrates another early IC package called the **flat pack.** The leads of the package were designed to be soldered directly to the tracks of the printed circuit board. This package allowed circuits to be placed close together for small-size applications such as avionics and military systems. The package is generally made of ceramic to withstand the high temperatures due to the closely packaged circuits.

Figure 15-5(c) shows the **dual-in-line package** or **DIP,** which has been a standard for many years. Called a DIP because of its two lines of parallel connecting pins, its leads feed through holes running through the printed circuit board. Connection pads circle the holes both on the top and the bottom of the printed circuit board (PCB) so that when the DIP package is soldered in place, it makes a connection.

Figure 15-5(d) illustrates the three types of **surface-mount technology (SMT) packages.** The key advantage of SMT over DIP is that a dip package needs both a hole and a connecting pad around the hole. With SMT, no holes are needed because the package is mounted directly on the printed circuit board. Without the need for holes, pads can be placed closer together resulting in a considerable space saving. There are three basic types of SMT packages: the "small outline IC" (SOIC) package, which uses "L" shaped leads; the "plastic-leaded chip carrier" (PLCC), which uses "J" shaped leads; and the "leadless ceramic chip carrier" (LCCC), which has metallic contacts molded into its ceramic body.

Figure 15-5(e) shows the physical apperance of a wide variety of integrated circuits.

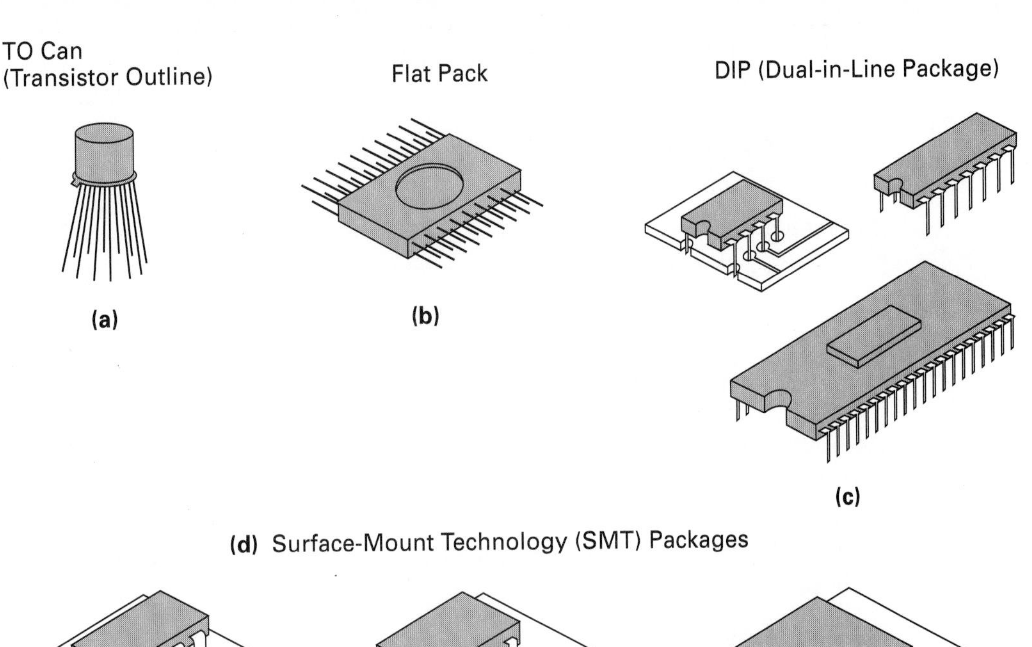

(d) Surface-Mount Technology (SMT) Packages

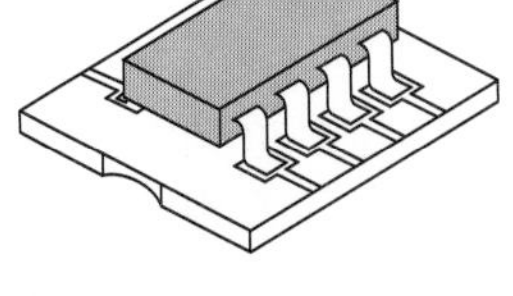

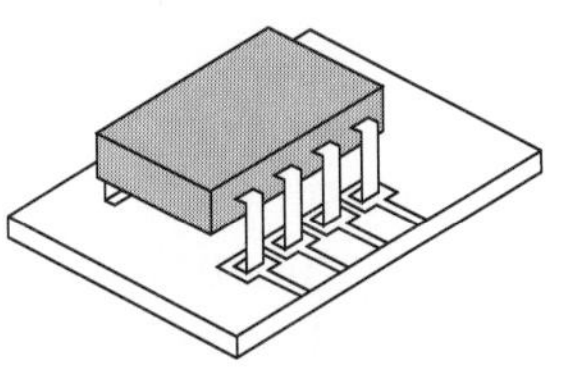

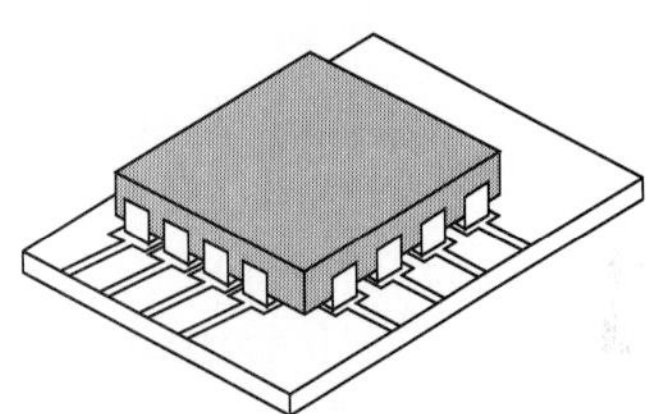

(e)

FIGURE 15-5 Integrated Circuit (IC) Packages. (Photo reprinted with permission of Harris Corporation.)

SELF-TEST REVIEW QUESTIONS FOR SECTION 15-3

1. Give the full names for the following abbreviations:
   a. TO Can **b.** DIP **c.** SMT
2. Which package type saves the most printed circuit board space?

15-4 IC COMPLEXITY CLASSIFICATION

Small Scale Integration (SSI) The simplest form of integrated circuit—such as linear amplifier circuits or single-function digital logic gate circuits—containing less than 180 interconnected components on a single chip.

Medium Scale Integration (MSI) An integrated circuit containing between 180 and 1,500 interconnected components.

Large-Scale Integration (LSI) An integrated circuit containing between 1,500 and 15,000 interconnected components.

Very Large Scale Integration (VLSI) An integrated circuit containing more than 15,000 interconnected components.

Although ICs are generally classified as being either linear (analog) or digital, another method of classification will group ICs depending on their internal circuit complexity. The four categories are called **small scale integration (SSI), medium scale integration (MSI), large scale integration (LSI),** and **very large scale integration (VLSI).**

1. Small Scale Integration (SSI) ICs

   These circuits are the simplest form of integrated circuit, such as linear amplifier circuits or single function digital logic gate circuits. They contain less than 180 interconnected components on a single chip.

2. Medium Scale Integration (MSI) ICs

   These ICs contain between 180 to 1,500 interconnected components, and their advantage over an SSI IC is that the number of ICs needed in a system is less, resulting in a reduced cost and assembly time.

3. Large Scale Integration (LSI) ICs

   These circuits contain several MSI circuits all integrated on a single chip to form larger functional circuits or systems such as memories, microprocessors, calculators, and basic test instruments. These ICs contain between 1,500 and 15,000 interconnected components.

4. Very Large Scale Integration (VLSI) ICs

   VLSI circuits contain extremely complex circuits such as large microprocessors, memories, and single-chip computers. These ICs contain more than 15,000 interconnected components.

SELF-TEST REVIEW QUESTIONS FOR SECTION 15-4

1. An IC containing 4 digital bipolar logic gates with about 44 components would be classified as a/an __________ IC.
   a. SSI **b.** MSI **c.** LSI **d.** VLSI
2. The low circuit current and therefore low heat dissipation of __________ circuits makes it possible to construct small-size circuits, and this is why __________ technology dominates the VLSI circuit market.

SUMMARY

1. In 1948, scientists discovered that the large vacuum tubes being used in radios, televisions, and other equipment could be replaced by extremely small devices made from semiconductor materials. A few years later, the transistor was born, and the world stepped out of the "vacuum tube era" and into the "solid state era."
2. At first, a single transistor was built by itself on a single chip or piece of semiconductor material. An electronic circuit was then formed by wiring together all of the individual or "discrete" components, such as the transistors, resistors, capacitors, and so on.
3. In 1959, it was discovered that more than one transistor could be made on a chip of semiconductor material. Soon other components such as resistors, capacitors, and diodes were added with the transistors, and all were interconnected to form a complete circuit. This integrating of various components on a single chip of semiconductor is called an "integrated circuit" or IC.
4. Integrated circuits can be classified as being either "linear" or "digital."
5. A linear IC produces an output signal that varies in proportion to the input signal (linear function). Circuit examples include amplifiers, oscillators, detectors, modulators, regulators, and so on. Because the input and output signals of a linear IC are analog in nature, linear ICs are also called "analog ICs."
6. Digital ICs are the most widely used and are basically two-state switching circuits designed to be used in digital computer and logic circuits.

Constructing Integrated Circuits

7. The fabrication of an integrated circuit begins with a thin slice (4/1000 inch) of silicon called a substrate. Then through a photoengraving process called photolithography, the pattern for each layer is reproduced on a glass plate called a photo mask. This mask acts as a photographic negative allowing ultraviolet light to imprint the pattern on the silicon substrate. The silicon substrate is then baked, causing the first layer to sink into the silicon substrate. A second pattern or layer is printed onto the substrate, and the baking process is repeated. This causes the second layer to sink into the substrate and the first layer to sink deeper into the substrate.
8. Simple ICs may require a four-layered chip, while complex ICs may require forty layers.

Bipolar, MOS, and Hybrid Integrated Circuits

9. Integrated circuits can basically be categorized as either ICs containing bipolar transistors or ICs containing MOS transistors.
10. A transistor formed in an IC has all connections on top of the substrate. Because all of the connections for this transistor lie in a flat plane, the transistor was named a "planar transistor."
11. A MOS IC transistor has fewer layers than a bipolar IC transistor, and is therefore easier to construct. MOS ICs have a higher component density

than bipolar circuits, consume less power, have a higher input impedance, noise immunity, and better temperature stability.

12. Bipolar ICs, on the other hand, have one important advantage over MOSFETs: they respond faster and are therefore best suited for high-frequency applications in which circuits need to react to rapidly switching signals.
13. A hybrid IC is a composite or blending together of discrete components and integrated circuits, analog and digital circuits, active and passive components, and standard and custom integrated circuits. All of the components are interconnected with conducting tracks that are printed onto an insulating (generally ceramic) substrate.
14. The key advantage of a hybrid IC is that if only a few circuits are required, the hybrid circuit is cheaper to construct. It does not require the extensive circuit layout and elaborate diffusion techniques of an integrated circuit. If a large number of custom circuits are required, however, the integrated circuit becomes the cheaper solution.
15. Another advantage of the hybrid circuit is that all of its discrete components (resistors, diodes, transistors, and capacitors) can handle more power than their integrated counterparts due to their larger size. Because of this, hybrid circuits are larger and heavier than integrated circuits.

Integrated Circuit Packages and Circuit Complexity Classification (Figure 15-6)

16. There are four basic methods of packaging integrated circuits.
17. The transistor outline, or TO Can, was modified from the basic 3-lead transistor package to have the extra leads needed and is still used to house linear ICs due to its good heat dissipation ability.
18. With the flat pack, the leads of the package were designed to be soldered directly to the tracks of the printed circuit board. This package allowed circuits to be placed close together for small size applications, such as in avionics and military systems. The package is generally made of ceramic to withstand the high temperatures due to the closely packaged circuits.
19. The dual-in-line package or DIP has been a standard for many years. Called a DIP because of its two lines of parallel connecting pins, its leads feed through holes running through the printed circuit board. Connection pads circle the holes both on the top and the bottom of the printed circuit board (PCB) so that when the DIP package is soldered in place, it makes a connection.
20. The key advantage SMT (surface mount technology) has over DIP is that a dip package needs both a hole and a connecting pad around the hole. With SMT, no holes are needed because the package is mounted directly on the printed circuit board. Without the need for holes, pads can be placed closer together, resulting in a considerable space saving. There are three basic types of SMT packages:
    a. The "small outline IC" (SOIC) package, which uses "L" shaped leads;

QUICK REFERENCE SUMMARY SHEET

FIGURE 15-6 Integrated Circuit (IC) Packages and Circuit Complexity Classification

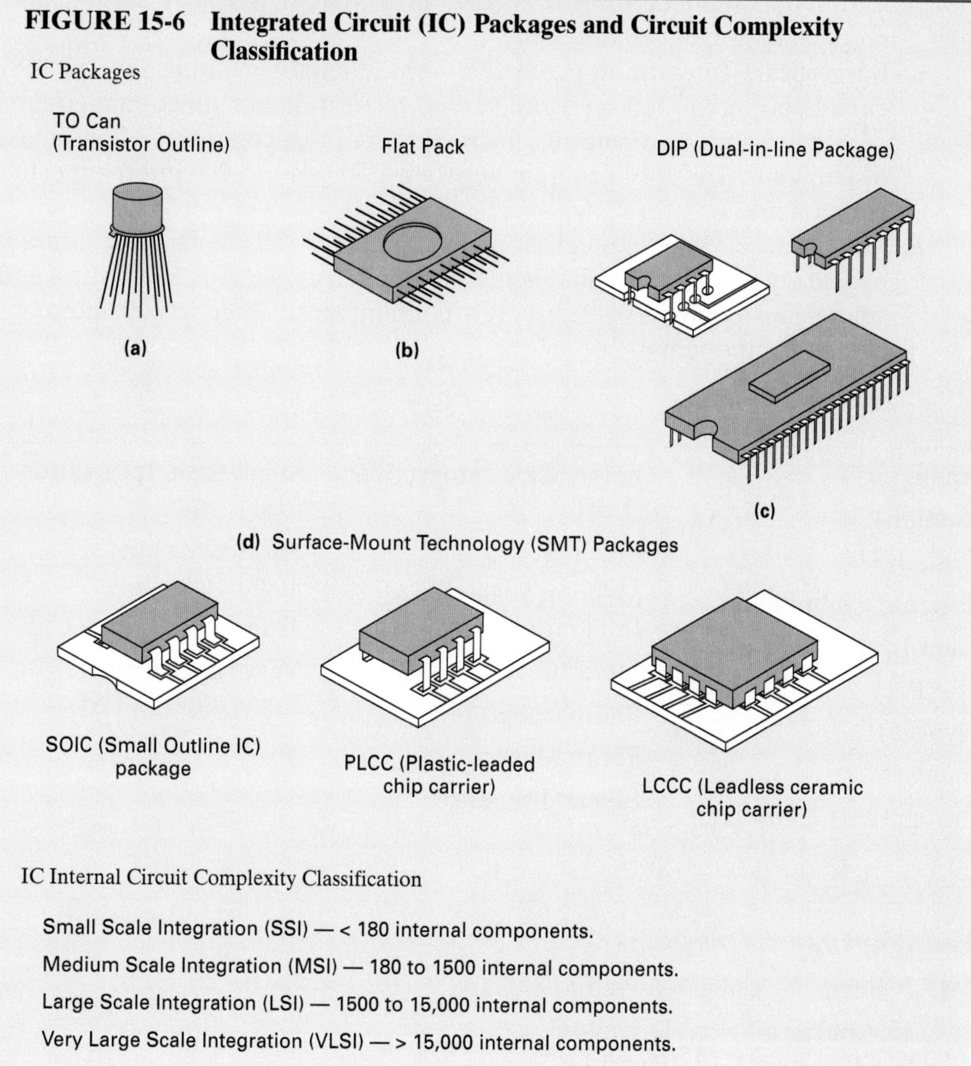

b. The "plastic-leaded chip carrier" (PLCC), which uses "J" shaped leads; and

c. The "leadless ceramic chip carrier" (LCCC), which has metallic contacts molded into its ceramic body.

21. Although ICs are generally classified as being either linear (analog) or digital, another method of classification will group ICs depending on their internal circuit complexity.

a. Small Scale Integration (SSI) ICs—These circuits are the simplest form of integrated circuit, such as linear amplifier circuits or single function digital logic gate circuits. They contain less than 180 interconnected components on a single chip.

b. Medium Scale Integration (MSI) ICs—These ICs contain between 180 to 1500 interconnected components, and their advantage over an SSI IC is that the number of ICs needed in a system is less, resulting in a reduced cost and assembly time.

c. Large Scale Integration (LSI) ICs—These circuits contain several MSI circuits all integrated on a single chip to form larger functional circuits or systems such as memories, microprocessors, calculators, and basic test instruments. These ICs contain between 1,500 and 15,000 interconnected components.

d. Very Large Scale Integration (VLSI) ICs—VLSI circuits contain extremely complex circuits such as large microprocessors, memories, and single-chip computers. These ICs contain more than 15,000 interconnected components.

NEW TERMS

Analog IC
Bipolar IC
Digital IC
Discrete Components
Dual-In-Line Package
Flat Pack
Hybrid IC
Ingots
Intregrated Circuit (IC)
Large Scale Integration (LSI)
Linear IC
Medium Scale Integration (MSI)
MOS IC
Photolithography
Photo Mask
Planar Transistor
Small Scale Integration (SSI)
Surface Mount Technology
Transistor Outline Can
Very Large Scale Integration (VLSI)
Wafers

REVIEW QUESTIONS

Multiple-Choice Questions

1. A linear IC is also referred to as a/an:

   a. Analog IC c. Digital IC
   b. Discrete IC d. Hybrid IC

2. A ____________ IC produces an output signal that varies in proportion to the input signal.

   a. Analog c. Linear
   b. Digital d. Hybrid

3. Digital ICs manage ____________ signals that are basically two-state in nature.

   a. Analog c. Linear
   b. Digital d. Hybrid

4. At first, electronic circuits were formed by wiring together individual or ____________ components.

   a. Chip c. Discrete
   b. Digital d. Passive

5. A ____________ is a glass plate negative that allows ultraviolet light to imprint a circuit pattern onto a silicon substrate.

a. Wafer **c.** Photo mask
b. Ingot **d.** Stepper

6. Through a refining process, silicon is extracted from sand and grown into ____________.

a. Chips **c.** Phosphorus
b. Wafers **d.** Ingots

7. Each circuit pattern, or mask, is actually repeated hundreds of times on a glass plate, and then these plates are used to imprint hundreds of identical chips onto one ____________.

a. Wafer **b.** Ingot **c.** Stepper **d.** ATE

8. A component with all terminals in the same plane is called a ____________ device.

a. Diode **c.** Planar
b. Discrete **d.** MOSFET

9. Which transistor type IC yields the higher component density?

a. Hybrid **c.** Bipolar
b. MOS **d.** Both (a) and (c)

10. Which transistor type IC can be operated at higher frequencies because it has a faster response time?

a. Hybrid **c.** Bipolar
b. MOS **d.** Both (b) and (c)

11. Which IC package achieves the best space saving?

a. DIP **b.** TO **c.** SMT **d.** SIP

12. Which IC classification contains the most components?

a. MSI **b.** VLSI **c.** SSI **d.** LSI

13. Which IC classification contains the least number of components?

a. MSI **b.** VLSI **c.** SSI **d.** LSI

14. ____________ ICs are often used where it is necessary to combine analog (linear) and digital circuits.

a. Bipolar **b.** Hybrid **c.** MOS **d.** SSI

15. A simple operational amplifier would be classified as a/an ____________ circuit.

a. MSI **b.** VLSI **c.** SSI **d.** LSI

Essay Questions

16. Define the following terms: (Intro.)

a. Discrete components **d.** Linear IC
b. Integrated circuit **e.** Digital IC
c. Analog IC

17. Briefly describe the basic IC construction procedure. (15-1)
18. What is photolithography? (15-1)
19. What is the difference between a conventional and planar transistor? (15-2)
20. What is the important advantage(s) of MOS ICs and bipolar ICs? (15-2)
21. What is a hybrid IC, and what are its advantages over an integrated circuit? (15-2-3)
22. What advantages does the SMT package have over the DIP package? (15-3)
23. Give the full names for the following abbreviations: (Chapter 15)

a. SSI	**f.** TO
b. SMT	**g.** PCB
c. ATE	**h.** IC
d. LSI	**i.** MOS
e. DIP	**j.** FET

24. Which FET is commonly called the voltage controlled switch? (15-2-2)
25. How many components can be found in the following IC types? (15-4)

a. SSI	**c.** LSI
b. MSI	**d.** VLSI

CHAPTER

16

OBJECTIVES

After completing this chapter, you will be able to:

1. Describe the symbol, package types, and internal block diagram of the operational amplifier or op-amp.

2. Explain how the op-amp's first differential amplifier stage gives the op-amp its high input impedance, high differential gain, and low common mode gain characteristics.

3. Define common-mode rejection ratio.

4. Describe how the internal second stage voltage amplifier gives the op-amp its very high gain characteristic.

5. Describe how the internal third stage emitter-follower amplifier gives the op-amp its low output impedance and current gain characteristics.

6. Explain how the op-amp operates in the following three basic circuit applications:
  - **a.** Open-loop comparator circuit
  - **b.** Closed-loop inverting amplifier circuit
  - **c.** Closed-loop noninverting amplifier circuit

7. Interpret a typical op-amp data sheet.

8. Compare the characteristics of several different op-amp types.

9. Describe some of the more typical op-amp circuit malfunctions.

10. Identify and describe the operation of the following op-amp circuit applications:
  - **a.** Voltage follower circuit
  - **b.** Summing amplifier circuit
  - **c.** Difference amplifier circuit
  - **d.** Differentiator circuit
  - **e.** Integrator circuit
  - **f.** Signal generator circuits
  - **g.** Active filter circuits

Operational Amplifiers

OUTLINE

VIGNETTE

I See

Standing six feet six inches tall, Jack St. Clair Kilby was a quiet introverted man from Kansas. Excited about the prospect of joining the very well-respected Massachusetts Institute of Technology (MIT) to further his education, he was thoroughly disappointed when he failed the mathematics entrance exam by three points. For the next ten years he worked for a manufacturer of radio and television parts, paying particularly close attention to the new component on the block, the transistor.

In May of 1958, Texas Instruments, a new fast-growing company that developed the first commercially available silicon transistor just four years earlier, offered him a job in their development lab. A project was underway to print electronic components on ceramic wafers and then wire them and stack them together to make a circuit. The more Kilby became involved in the project, the more he realized how complicated and ridiculous the method was. The idea to miniaturize was a good one, but a different solution was needed.

Two months later in July, the company shut down for summer vacation, but Kilby was forced to work because he had not accrued vacation time. This proved to be a blessing in disguise for Kilby who found himself in the lab with a lot of time and resources available to him to develop his idea. His idea was to build resistors and capacitors from the same semiconductor material that was being used to manufacture transistors. This would mean that all of the components that make up a circuit could be manufactured simultaneously on a single slice of semiconductor material.

A few months later Kilby presented his prototype to a very skeptical boss. It contained five components all connected by tiny wires with the complete assembly held together by large blobs of wax. Kilby suddenly found himself the owner of a patent and the richly deserved acclaim that always goes along with being first. The first integrated circuit or IC was born on a thin wafer of germanium just two-fifths of an inch long. Texas Instruments demonstrated its miniaturization advantage by building a computer for the air force using their newly developed technique in a 587 IC. Its rewards were immediately apparent when a 78 cubic foot monster computer was replaced with a more powerful unit measuring only 6.5 cubic inches.

INTRODUCTION

The operational amplifier was initially a vacuum tube circuit used in the early 1940s in analog computers. The name "operational amplifier" or "op-amp" was chosen because the circuit was used as a high-gain DC "amplifier" performing mathematical "operations." These early circuits were expensive and bulky, and they found very little application until the semiconductor integrated circuit was developed in 1958 by Jack Kilby at Texas Instruments. Circuits that once needed hundreds of discrete or individual components can now be integrated into a single IC, making equipment smaller, more energy efficient, cheaper, and easier to design and troubleshoot.

Today's IC op-amp is a very high-gain DC amplifier that can have its operating characteristics changed by connecting different external components. This makes the op-amp very versatile, and it is this versatility that has made the op-amp the most widely used linear IC.

In this chapter, we will be examining the op-amp's operation, characteristics, typical circuit applications, and troubleshooting.

16-1 OPERATIONAL AMPLIFIER BASICS

To begin with, Figure 16-1 introduces the **operational amplifier,** or **op-amp,** by showing its schematic symbol in Figure 16-1(a) and internal circuit in Figure 16-1(b). It would be safe to say that you will not really be learning anything that has not already been covered because the op-amp's internal circuit is simply a combination of three previously covered amplifier circuits. These three circuits are all interconnected and contained within a single IC, and together they function as a "high-gain, high input impedance, low output impedance amplifier."

Operational Amplifier (Op-Amp)
Special type of high-gain amplifier.

16-1-1 Op-Amp Symbol

Referring again to Figure 16-1(a), you can see that the triangle shaped amplifier symbol is used to represent the op-amp in an electronic schematic diagram. Comparing the two symbols, you may have noticed that in some cases the two power supply connections are not shown, even though power is obviously applied.

Let us now examine the op-amp's input and output terminals shown in Figure 16-1. The two op-amp inputs are labeled "−" and "+." The "−" or negative input is called the **inverting input** because any signal applied to this input will be amplified and inverted between input and output (output is 180° out of phase with input). On the other hand, the "+" or positive input is called the **noninverting input** because any signal applied to this input will be amplified but not inverted between input and output (output is in phase with input). An input sig-

Inverting Input
The inverting or negative input of an op-amp.

Noninverting Input
The noninverting or positive input of an op-amp.

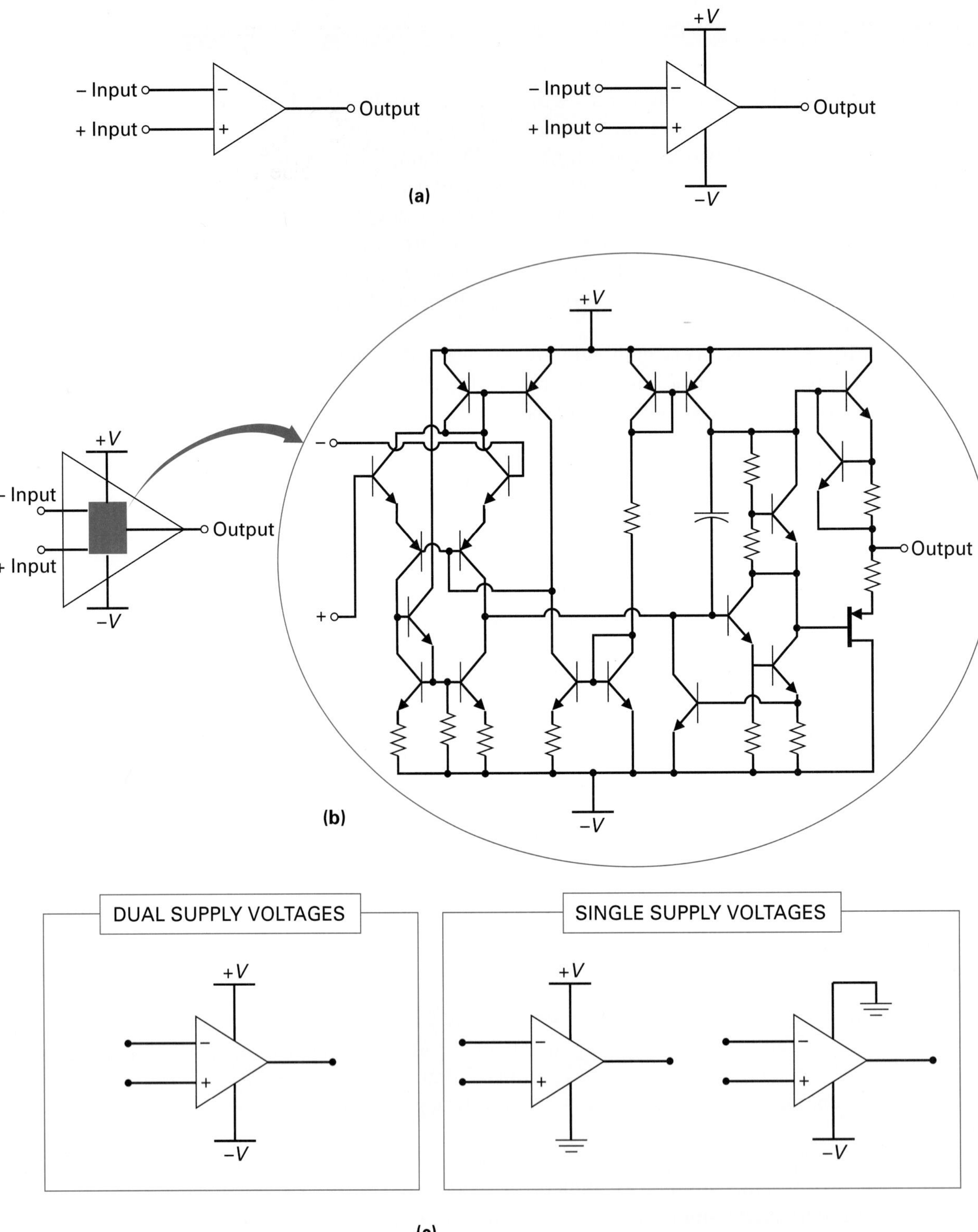

FIGURE 16-1 **The Operational Amplifier. (a) Schematic Symbols. (b) Internal Circuit. (c) Power Supply Connections.**

nal will normally be applied to only one of these inputs, while the other input is used to control the op-amp's operating characteristics.

The two power supply connections to the op-amp are labeled "$+V$" and "$-V$." Figure 16-1(c) shows how power to the op-amp can be supplied by dual supply voltages or by a single supply voltage. When two supply voltages are used (dual supply voltages), the voltage values are of the same value but of opposite polarity (for example, +12 V and −12 V). On the other hand, when only one supply voltage is used (single supply voltage), a positive or negative voltage is applied to its respective terminal while the other terminal is grounded (for example, +5 V and ground or −5 V and ground). Having both a positive and negative power supply voltage will allow the output signal to swing positive and negative, above and below zero. As with all high gain amplifiers, however, the output voltage can never exceed the value of the $+V$ and $-V$ supply voltages.

16-1-2 Op-Amp Packages

The entire op-amp circuit is placed within one of two basic packages, shown in Figure 16-2(a) and (b). The TO-5 metal can package is available with 8, 10, or 12 leads, while the dual in-line through-hole and surface-mount packages typically have 8 or 14 pins.

Like all ICs, an identification code is used to indicate the device manufacturer, device type, and key characteristics. Figure 16-2(c) lists some of the more common manufacturer prefix codes, operating temperature codes, and package codes. In this example, the "MC 741C N" code indicates that the 741 op-amp is made by Motorola, it is designed for commercial application in which the temperature range is between 0 to 70°C, and the package is a through-hole DIP with longer leads.

Referring again to the IC packages in Figure 16-2(a) and (b), you can see that in addition to the two inputs, single output, and two power supply terminals, there are two additional leads labeled offset (also called balance). These two inputs will normally be connected to a potentiometer that can be adjusted to set the output at zero volts when both inverting and noninverting inputs are at zero volts. **Balancing** the op-amp in this way is generally needed due to imbalances within the op-amp's internal circuit.

Balancing
Setting the output of an op-amp to zero volts when both inverting and noninverting inputs are at zero volts.

SELF-TEST REVIEW QUESTIONS FOR SECTION 16-1

1. Is the operational amplifier a discrete component or an integrated circuit?
2. List the names of the op-amp's 5 terminals.
3. Name the two basic op-amp package types.
4. Briefly describe the meaning of each of the three parts in an op-amp's identification code.

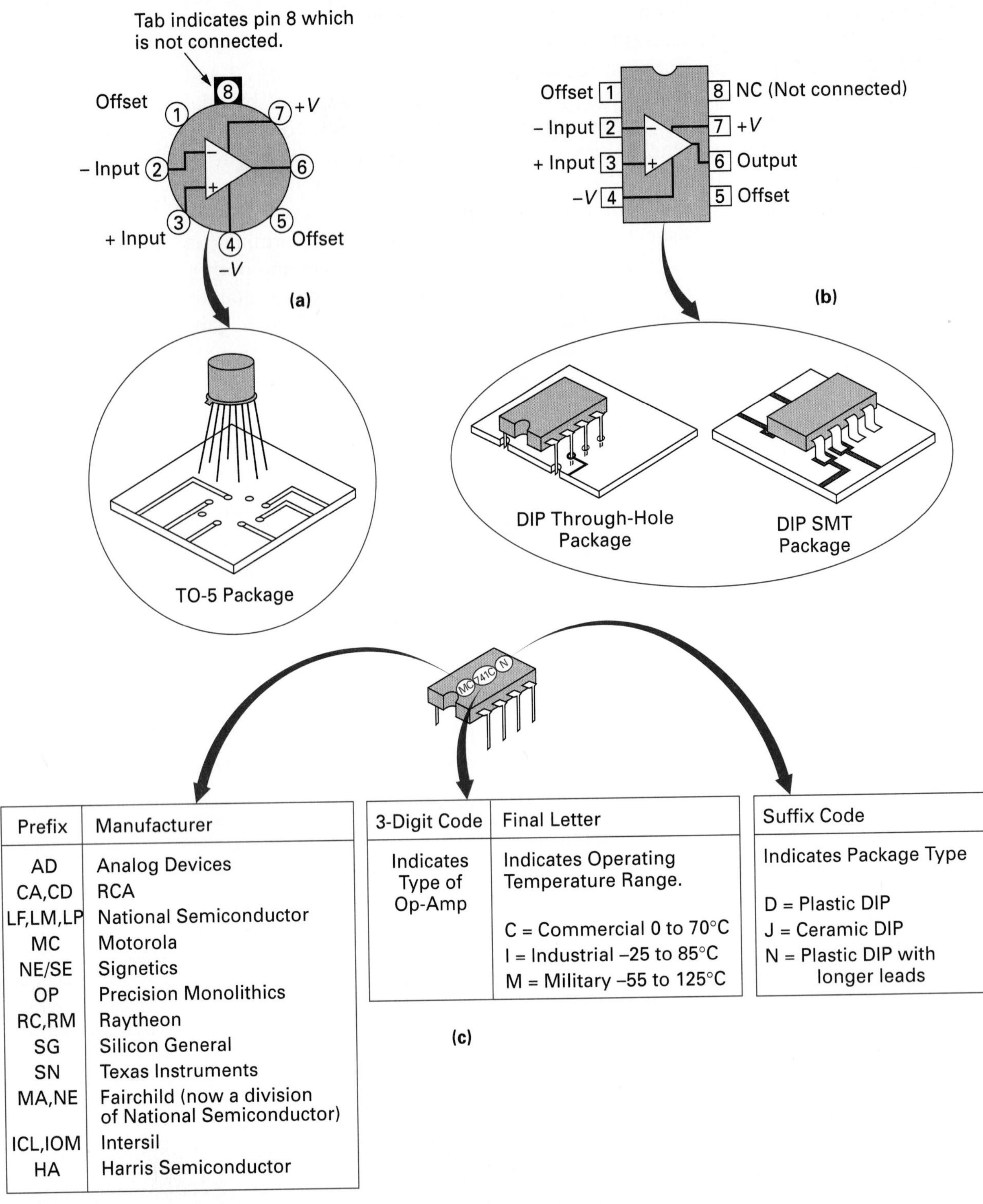

Prefix	Manufacturer
AD	Analog Devices
CA,CD	RCA
LF,LM,LP	National Semiconductor
MC	Motorola
NE/SE	Signetics
OP	Precision Monolithics
RC,RM	Raytheon
SG	Silicon General
SN	Texas Instruments
MA,NE	Fairchild (now a division of National Semiconductor)
ICL,IOM	Intersil
HA	Harris Semiconductor

3-Digit Code	Final Letter
Indicates Type of Op-Amp	Indicates Operating Temperature Range. C = Commercial 0 to 70°C I = Industrial –25 to 85°C M = Military –55 to 125°C

Suffix Code
Indicates Package Type D = Plastic DIP J = Ceramic DIP N = Plastic DIP with longer leads

FIGURE 16-2 Op-Amp Package Types and Identification Codes.

16-2 OP-AMP OPERATION AND CHARACTERISTICS

As mentioned previously, the operational amplifier contains three amplifier circuits, and these three circuits are all interconnected and contained within a single IC. Referring to the block diagram of the op-amp in Figure 16-3(a) you can see that these three circuits are *a differential amplifier, a voltage amplifier,* and *an output amplifier.* Combined, these three circuits give the op-amp its key characteristics, which are *high-gain, high-input impedance,* and *low-output impedance.* All three of these circuits were discussed previously in Chapter 10. However, we will briefly review their characteristics because combined they determine the characteristics of the op-amp.

16-2-1 The Differential Amplifier within the Op-Amp

The differential amplifier within the op-amp is connected to operate in its "differential-input, single-output mode." The operation of the differential amplifier in this mode is reviewed in Figure 16-3(b). When both input signals are equal in amplitude and in phase with one another, they are referred to as **common mode input signals,** as seen in the waveforms in Figure 16-3(b). On the other hand, if the input signals are out of phase with one another, they are referred to as **differential mode input signals,** as seen in the other set of waveforms in Figure 16-3(b). The differential amplifier will amplify differential input signals while rejecting common mode input signals. The questions you may be asking at this stage are what are common mode input signals and why should we want to reject them? The answers are as follows: temperature changes and noise are common-mode input signals, and they are unwanted signals. Let us examine these common-mode signals in more detail.

Common Mode Input Signals
Input signals to an op-amp that are in phase with one another.

Differential Mode Input Signals
Input signals to an op-amp that are out of phase with one another.

1. Temperature variations within electronic equipment affect the operation of semiconductor materials and, therefore, the operation of semiconductor devices. These temperature variations can cause the dc output voltage of the first stage to drift away from its normal Q point. The second-stage amplifier will amplify this voltage change in the same way as it would amplify any dc input signal and so will all of the following amplifier stages. The increase or decrease in the normal Q-point bias for all of the amplifier stages will get progressively worse due to this thermal instability, and the final stage may have a Q point that is so far off of its mid position that an input signal may drive it into saturation or cutoff, causing signal distortion. With the difference amplifier, any change that occurs due to temperature changes will affect both stages and so will not appear at the output of the differential amplifier, due to its **common mode rejection.**

Common Mode Rejection
The ability of a device to reject a voltage signal applied simultaneously to both input terminals.

2. The second common-mode input signal that the differential amplifier removes is noise. It is often necessary to amplify low-level signals from low-sensitivity sources such as microphones, light detectors, and other transducers. High-gain amplifiers are used to increase the amplitude of these small input signals up to a more usable level that is large enough to drive or con-

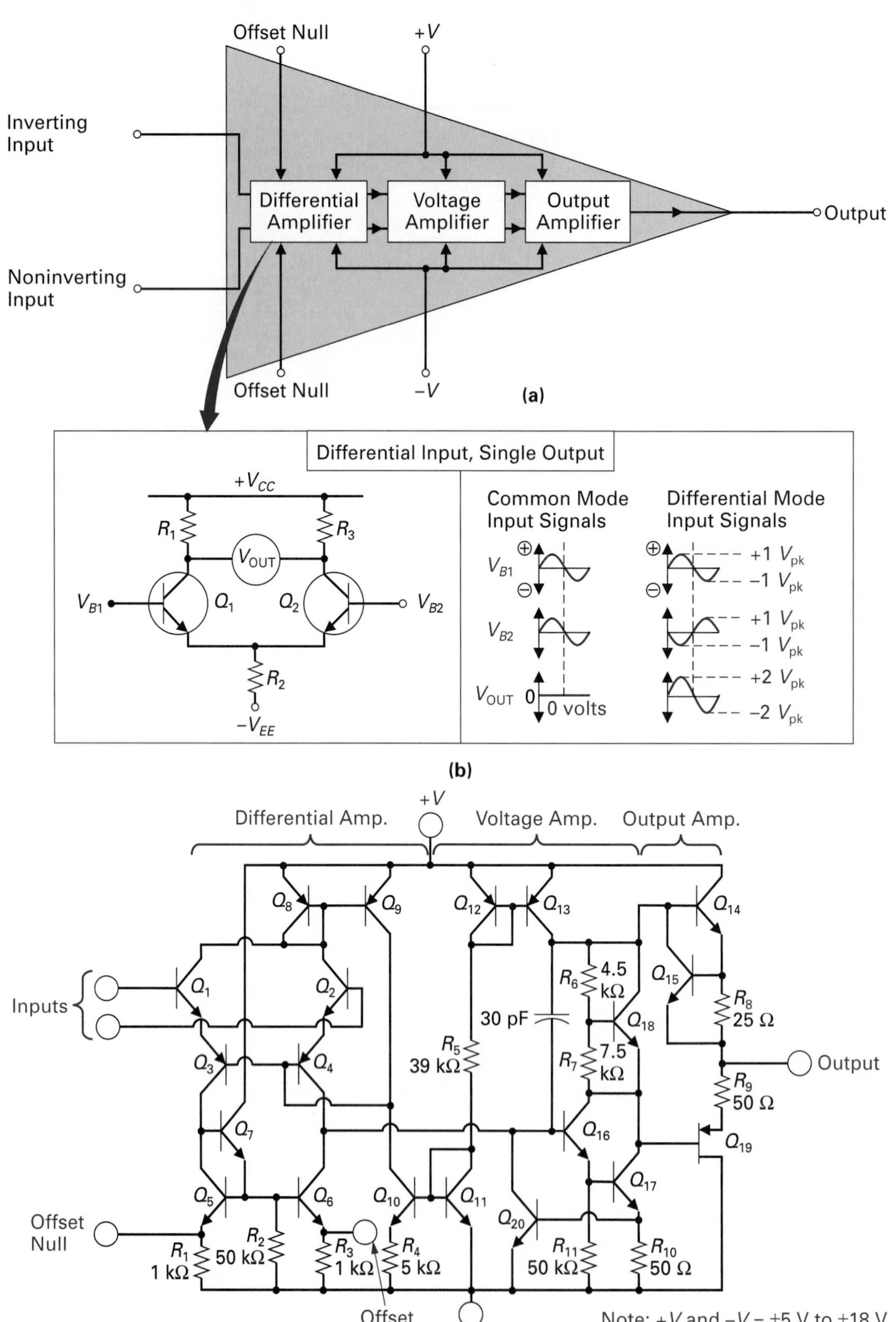

FIGURE 16-3 The Operational Amplifier's Internal Circuit. (a) Block Diagram. (b) The Differential Amplifier's Operation. (c) Circuit Diagram.

trol a load, such as a loudspeaker. The 60 Hz ac power line, or any other electrical variation, can induce a noise signal along with the input signal at the input of this high-gain amplifier. Since these noise signals will be induced at all points in the circuit, and be identical in amplitude and phase, the differential amplifier will block these unwanted signals because they will be present at both inputs of the differential amplifier (noise will be a common-mode input). A true input signal on the other hand, will appear at the two inputs of the differential amplifier as a differential input signal, and therefore be amplified.

16-2-2 *Common-Mode Rejection Ratio*

To summarize, a differential input will be amplified by the op-amp's differential amplifier and passed to the output, while unwanted signals caused by temperature variations or noise will appear as common-mode input signals and therefore be rejected. An op-amp's ability to provide a high **differential gain (A_{VD})** and a low **common-mode gain (A_{CM})** is directly dependent on its internal differential amplifier and is a measure of an op-amp's performance. This ratio is called the **common-mode rejection ratio (CMRR)** and is calculated with the following formula:

Differential Gain (A_{VD})
The amplification of differential-mode input.

Common-Mode Gain (A_{CM})
The amplification of common-mode input.

Common-Mode Rejection Ratio (CMRR)
The ratio of an operational amplifier's differential gain to the common-mode gain.

$$\text{CMRR} = \frac{A_{VD}}{A_{CM}}$$

Looking at this formula, you can see that the higher the A_{VD} (differential gain), or the smaller the A_{CM} (common-mode gain), the higher the CMRR value, and therefore the better the operational amplifier. This ratio can also be expressed in dBs by using the following formula:

$$\text{CMRR} = 20 \times \log \frac{A_{VD}}{A_{CM}}$$

■ **EXAMPLE:**

If a op-amp's differential amplifier has a differential gain of 5,000 and a common-mode gain of 0.5, what is the operational amplifier's CMRR? Express the answer in standard gain and dBs.

■ ***Solution:***

$$\text{CMRR} = \frac{A_{VD}}{A_{CM}} = \frac{5000}{0.5} = 10{,}000$$

$$\text{CMRR} = 20 \times \log \frac{A_{VD}}{A_{CM}} = 20 \times \log 10{,}000 = 20 \times 4 = 80 \text{ dB}$$

A CMRR of 10,000 or 80 dB means that the op-amp's desired input signals will be amplified 10,000 times more than the unwanted common-mode input signals.

16-2-3 *The Op-Amp Block Diagram*

Now that we have reviewed the differential amplifier's circuit characteristics, let us return to the op-amp block diagram in Figure 16-3(a). It is the op-amp's differential-amplifier stage that provides the good common-mode rejection and high differential gain. Because the op-amp's "–" and "+" inputs are applied to either base of the diff-amp, we know that input current will be very small. It is this circuit characteristic that provides the op-amp with another key feature, which is a high input impedance ($I\downarrow$, $Z\uparrow$). The voltage-amplifier stage following the diff-amp usually consists of several darlington-pair stages that provide an overall op-amp voltage gain of typically 50,000 to 200,000. The final output stage consists of a complementary emitter-follower stage to provide a low output impedance and high current gain, so that the op-amp can deliver up to several milliamps, depending on the value of the load.

16-2-4 *The Op-Amp Circuit Diagram*

The complete internal circuit of a typical op-amp can be seen in Figure 16-3(c). With integrated circuits, it is better to have transistors function as resistors wherever possible because they occupy less chip space than actual resistors. This accounts for why the circuit seems to contain many transistors that have their base and collector leads connected. You may also have noticed that no coupling capacitors have been used so that the op-amp can amplify both ac and dc input signals. As with most schematics, the inputs are shown on the left, output on the right, and power is above and below. As discussed previously, the two balancing, or **offset null,** inputs will normally be connected to an external potentiometer that can be adjusted to set the output at zero volts when both the inverting and noninverting inputs are at zero volts. Balancing the op-amp to find the zero-volt output point, or null, in this way is generally needed due to slight imbalances within the op-amp's internal circuit.

Offset Null Inputs
The two balancing inputs used to balance an op-amp.

An important point to realize at this time is that the op-amp is a single component, and up until this time we have concentrated on an understanding of the op-amp's internal circuitry because this helps us to better understand the circuit's normal input/output relationships and characteristics. These operational characteristics are important if we are going to be able to isolate whether a circuit malfunction is internal or external to the op-amp. However, because it is impossible to repair any internal op-amp failures, we will not concentrate on every detail of the op-amp's internal circuit.

16-2-5 *Basic Op-Amp Circuit Applications*

Now that we have an understanding of the op-amp's characteristics, let us now put it to use in some basic circuit applications. To begin with, we will examine the comparator circuit, which was first introduced in the early chapters of this book.

The Open-Loop Comparator Circuit

Figure 16-4 shows how the op-amp can be used to function as a **comparator,** which is a circuit that is used to detect changes in voltage level. In Figure 16-4(a), the inverting input (–) of the op-amp is grounded and the input signal is applied to the op-amp's noninverting input (+). Referring to the associated waveforms, you can see that when the input swings positive relative to the negative input (which is 0), the output of the amplifier goes into immediate saturation due to the very large gain of the op-amp. For example, if the op-amp had a voltage gain of 25,000 ($A_V = 25{,}000$), even a small input of +25 mV would cause the op-amp to try and drive its output to 625 V ($V_{OUT}) = V_{IN} \times A\ V = 25\ mV \times 25{,}000 = 625\ V$). Since the maximum possible positive output voltage cannot exceed the positive supply volt-

Comparator
An op-amp used without feedback to detect changes in voltage level.

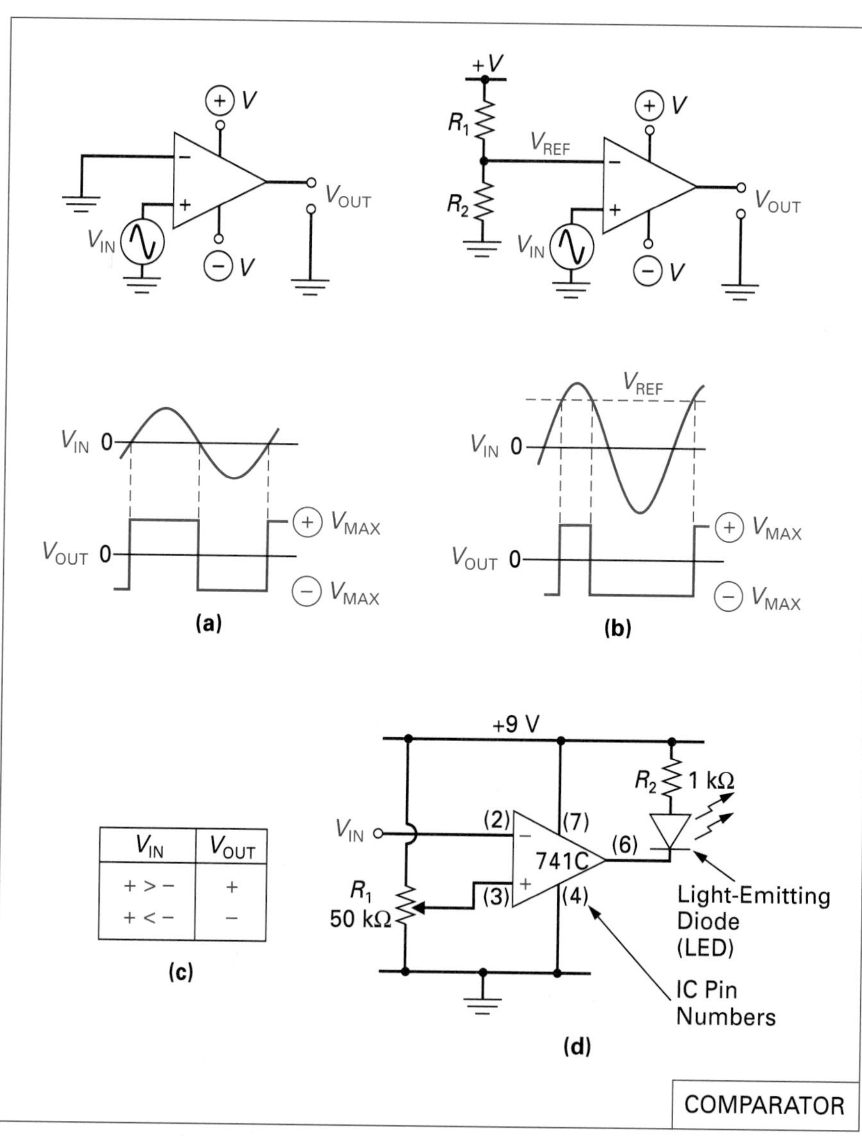

V_{IN}	V_{OUT}
+ > –	+
+ < –	–

FIGURE 16-4 The Open-Loop Comparator Circuit.

age (+V), the output goes to its maximum positive limit, which is equal to the +V supply voltage. When the input swings negative, the amplifier is driven immediately into its opposite state (cut-off), and the output goes to its maximum negative limit, which is equal to the −V supply voltage.

Figure 16-4(b) shows how a voltage divider made up of R_1 and R_2 can be used to supply the inverting input (−) of the op-amp with a reference voltage (V_{REF}) that can be determined by using the voltage divider formula.

$$V_{REF} = \frac{R_2}{R_1 + R_2} \times (+V)$$

Referring to the associated waveforms in Figure 16-4(b), you can see that whenever the ac input signal is more positive than the reference voltage, the output is positive. On the other hand, whenever the ac input signal is less than the reference voltage, the output is negative.

The table in Figure 16-4(c) summarizes the operation of the comparator circuit. In the first line of the table you can see that when the negative input is negative relative to the positive input, the output will go to its positive limit ($V_{OUT} = +V$). In the second line of the table you can see that when the opposite occurs (the negative input is positive, or the positive input is negative), the output will go to its negative limit ($V_{OUT} = -V$).

■ **EXAMPLE:**

Briefly describe the operation of the 741C op-amp comparator circuit shown in Figure 16-4(d).

■ ***Solution:***

Potentiometer R_1 sets up the reference voltage (V_{REF}), which can be anywhere between 0 V and +9 V. When the input voltage (V_{IN}) to the negative input of the op-amp is greater than the positive reference voltage, the op-amp's output will be equal to the −V voltage supply (which is ground), and so the LED will turn ON.

Open-Loop Mode
A control system that has no means of comparing the output with the input for control purposes.

Closed-Loop Mode
A control system containing one or more feedback control loops in which functions of the controlled signals are combined with functions of the commands to tend to maintain prescribed relationships between the commands and the controlled signals.

The Closed-Loop Inverting-Amplifier Circuit

The op-amp is usually operated in either the **open-loop mode** or **closed-loop mode.** With the previously discussed comparator circuit, the op-amp was operating in its open-loop mode because there was no signal feedback from output to input. In most instances, the op-amp is operated in the closed-loop mode, in which there is signal feedback from output back to input. This feedback signal will always be out of phase with the input signal and therefore oppose the original signal, which is why it is called "degenerative or negative feedback." Negative feedback, however, is necessary in nearly all op-amp circuits for the following reasons:

1. Because the op-amp has such an extremely high gain, even a very small input signal will be amplified to a very large signal, which will drive the op-amp out of its linear region and into saturation and cutoff. Negative feedback will lower the op-amp's gain, and therefore control the op-amp to prevent output waveform distortion.

2. Having such a high gain can cause the amplifier to go into oscillation due to positive feedback. Negative feedback prevents an amplifier from going into oscillation by reducing the op-amp's gain.
3. The open-loop gain of an op-amp can have a very large range of value for the same device. For example, the 741's open-loop gain can be anywhere from a minimum of 25,000 to 200,000. Including negative feedback in the op-amp circuit will reduce the gain to a consistent value so that the same part can be relied on to provide the same response.

Figure 16-5(a) shows how an op-amp can be connected as an **inverting amplifier circuit,** which produces an amplified output signal that is 180° out of phase with the input signal. Looking at the output voltage label ($-V_{OUT}$), notice that the negative symbol preceding V_{OUT} is being used to indicate the 180° phase inversion between input and output. In this circuit arrangement, the input signal (V_{IN}) is applied through an input resistor (R_{IN}) to the inverting input (–) of the op-amp, while the non-inverting input (+) is connected to ground. A feedback loop is connected from the output back to the inverting input via the feedback resistor R_F.

Inverting Amplifier Circuit
An op-amp circuit that produces an amplified output signal that is 180° out of phase with the input signal.

Let us now take a closer look at the closed-loop feedback system that occurs within this amplifier circuit. If the applied input voltage was zero volts ($V_{IN} = 0$ V), the differential input signal (which is the difference between the op-amp's "+" and "–" inputs) will be 0 V, because both the inverting and non-inverting inputs will now be at 0 V. A differential input of zero volts therefore will generate an output of zero volts. If the input signal was now to swing positive toward +5 V, the

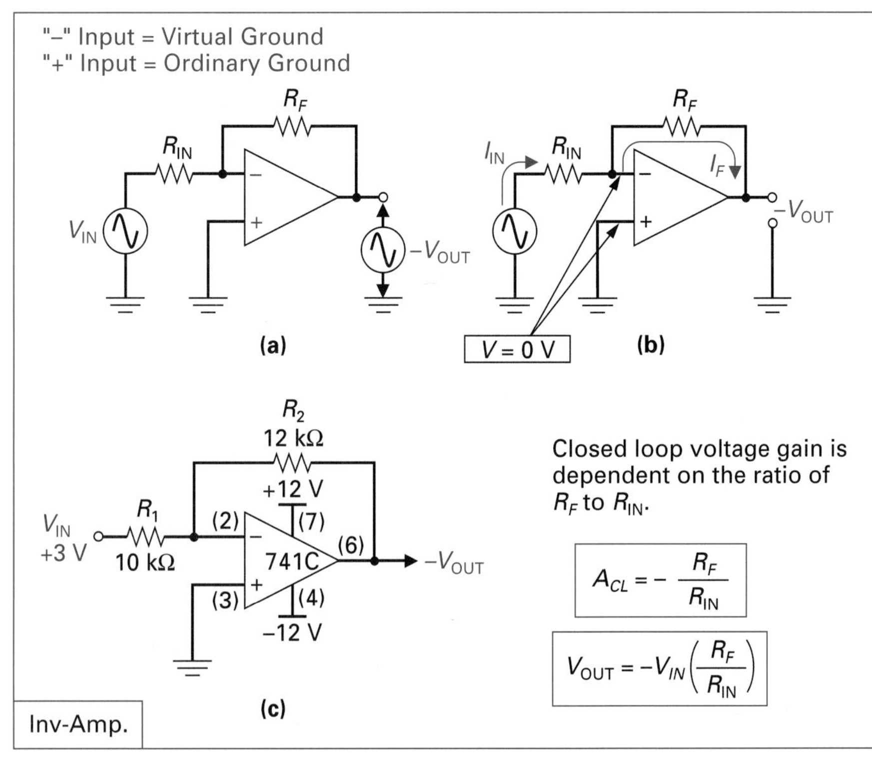

FIGURE 16-5 The Closed-Loop Inverting Amplifier Circuit.

output (V_{OUT}) would swing negative due to the internal op-amp circuit phase inversions. This negative output voltage swing would be applied back to the inverting input via R_F to counteract the original positive input change. The feedback path is designed so that it cannot completely cancel the input signal, for if it did, there would be no input, and therefore no output or feedback. In most instances, the feedback voltage (V_F) will greatly restrain the input voltage change to the point that a +5 V input change at V_{IN} will only be felt as a +5 micro volt change at the op-amp's inverting input. Therefore, even though V_{IN} seems to change in values measured in volts, the inverting input of the op-amp will only change in values measured in micro volts. In fact, if the voltage at the inverting input of the op-amp is measured with a voltmeter as shown in Figure 16-5(b), the "–" input appears to remain at 0 V due to the very minute change at the "–" input. The inverting input of the op-amp in this case would be defined as a **virtual ground,** which is different from an **ordinary ground.** A virtual ground is a voltage ground because this point is at zero volts; however, it is not a current ground because it cannot sink or conduct away any current. An ordinary ground on the other hand, is at zero volts and can sink any amount of current.

Virtual Ground
A ground for voltage but not for current.

Ordinary Ground
A connection in the circuit that is said to be at ground potential or zero volts. Because of its connection to earth it has the ability to conduct electrical current to and from earth.

Returning to the inverting amplifier circuit in Figure 16-5(b), we can now analyze this circuit in a little more detail now that we know its basic operation. To begin with, if the "–" input of the op-amp is at 0 V, then all of the input voltage (V_{IN}) will be dropped across the input resistor (R_{IN}). Therefore, the input current can be calculated if we know the value of V_{IN} and R_{IN}, with the following formula:

$$I_{IN} = \frac{V_{IN}}{R_{IN}}$$

Knowing that the left side of R_F is at 0 V means that all of the output voltage will be developed across R_F, and therefore the value of feedback current (I_F) can be calculated with the formula:

$$I_F = \frac{-V_{OUT}}{R_F}$$

The extremely high input impedance of the op-amp means that only a very small fraction of the input current will enter the inverting input. In fact, nearly all of the current flowing through R_{IN} and reaching the "–" op-amp input will leave this virtual ground point via the easiest path which is through R_F. Therefore, it can be said that the feedback current is equal to the input current, or

$$I_{IN} = I_F$$

If I_{IN} (which equals V_{IN}/R_{IN}) and I_F (which equals $-V_{OUT}/R_F$) are equal, then

$$\frac{V_{IN}}{R_{IN}} = \frac{-V_{OUT}}{R_F}$$

If this is rearranged, we arrive at the following:

$$\frac{V_{\text{OUT}}}{V_{\text{IN}}} = -\frac{R_F}{R_{\text{IN}}}$$

Because the voltage gain of an amplifier is equal to

$$A_V = \frac{V_{\text{OUT}}}{V_{\text{IN}}}$$

and because $V_{\text{OUT}}/V_{\text{IN}} = -R_F/R_{\text{IN}}$, the **closed-loop voltage gain (A_{CL})** of the inverting operational amplifier is equal to the ratio of R_F to R_{IN}.

Closed-Loop Voltage Gain (A_{CL})
The voltage gain of an amplifier when it is operated in the closed loop mode.

$$A_{CL} = -\frac{R_F}{R_{\text{IN}}}$$

(Negative symbol preceding R_F/R_{IN} indicates signal inversion)

By rearranging the previous equation $V_{\text{OUT}}/V_{\text{IN}} = -R_F/R_{\text{IN}}$, we can arrive at the following formula for calculating the output voltage of this circuit.

Since

$$\frac{V_{\text{OUT}}}{V_{\text{IN}}} = -\frac{R_F}{R_{\text{IN}}}$$

$$V_{\text{OUT}} = -V_{\text{IN}}\left(\frac{R_F}{R_{\text{IN}}}\right)$$

The input impedance of this inverting op-amp is equal to the value of R_{IN} because the input voltage (V_{IN}) is developed across the input resistor (R_{IN}).

■ **EXAMPLE:**

Calculate the output voltage of the inverting amplifier shown in Figure 16-5(c).

■ ***Solution:***

$$V_{\text{OUT}} = -V_{\text{IN}}\left(\frac{R_F}{R_{\text{IN}}}\right) = -3\text{ V} \times \frac{12\text{ k}\Omega}{10\text{ k}\Omega} = -3\text{ V} \times 1.2 = -3.6\text{ V}$$

The Closed-Loop Noninverting Amplifier Circuit

Figure 16-6(a) shows how an op-amp can be configured to operate as a **noninverting amplifier circuit.** The input voltage (V_{IN}) is applied to the op-amp's noninverting input (+), and therefore the output voltage (V_{OUT}) will be in phase with the input. To achieve negative feedback, the output is applied back to the inverting input (–) of the op-amp via the feedback network formed by R_F and R_1.

Noninverting Amplifier Circuit
An op-amp circuit that produces an amplified output signal that is in phase with the input signal.

Let us now take a closer look at the closed-loop negative feedback system that occurs within this amplifier circuit. The output voltage (V_{OUT}) is propor-

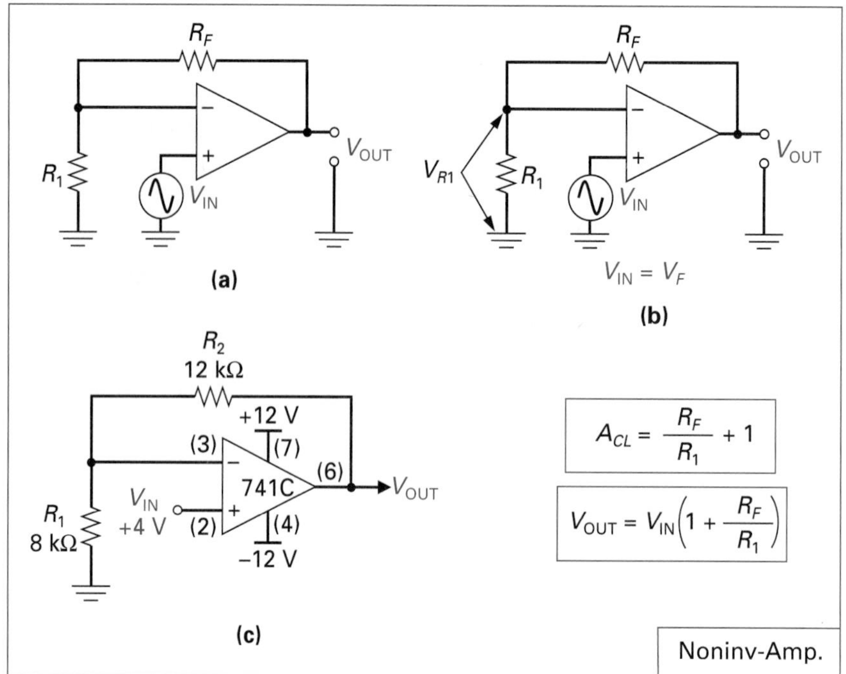

FIGURE 16-6 The Closed-Loop Noninverting Amplifier Circuit.

tionally divided across R_F and R_1, with the feedback voltage (V_F) developed across R_1 being applied to the inverting input (–) of the op-amp, as shown in Figure 16-6(b). Because V_{OUT} is in phase with V_{IN}, the feedback voltage (V_F) will also be in phase with V_{IN}, and therefore these two in-phase inputs to the op-amp will be common-mode input signals. As a result, feedback will be degenerative. However, because V_{OUT} is slightly larger than V_{IN}, there will be a small difference between V_{IN} and V_F, and this differential input will be amplified. To summarize, the noninverting op-amp provides negative feedback by feeding back an in-phase common-mode signal, and this degenerative feedback will lower the op-amp's gain to

1. prevent output waveform distortion,
2. prevent the amplifier from going into oscillation, and
3. reduce the gain of the op-amp to a consistent value.

The very small micro volt difference between V_{IN} and V_F will be amplified; however, since there is such a very small difference between these two input signals, it can be said that

$$V_{IN} = V_F$$

Because the closed-loop voltage gain of any op-amp is equal to

$$A_{CL} = \frac{V_{OUT}}{V_{IN}}$$

the gain of the noninverting amplifier could also be calculated with

$$A_{CL} = \frac{V_{OUT}}{V_F}$$

By using the voltage divider formula, we can develop a formula for calculating V_F.

$$V_F = \frac{R_1}{R_1 + R_F} \times V_{\text{OUT}}$$

By rearranging the above formula as follows

$$\frac{V_{\text{OUT}}}{V_F} = \frac{R_1 + R_F}{R_1} \quad \text{or} \quad \frac{V_{\text{OUT}}}{V_F} = \frac{R_1}{R_1} + \frac{R_F}{R_1} = \frac{R_F}{R_1} + 1$$

and because $V_{\text{OUT}}/V_F = A_{CL}$, the noninverting op-amp's gain can also be calculated with the formula

$$A_{CL} = \frac{R_F}{R_1} + 1$$

Because the output voltage of an amplifier is equal to the product of input voltage and gain ($V_{\text{OUT}} = V_{\text{IN}} \times A_{CL}$), we can add V_{IN} to the previous closed-loop gain formula in order to calculate output voltage.

$$V_{\text{OUT}} = V_{\text{IN}}\left(1 + \frac{R_F}{R_1}\right)$$

With the inverting op-amp, the input impedance is determined by the input resistor (R_{IN}). With the noninverting amplifier, there is no resistor connected because V_{IN} is applied directly into the very high input impedance of the op-amp. As a result, the noninverting amplifier circuit has an extremely high input impedance.

■ **EXAMPLE:**

Calculate the output voltage from the noninverting amplifier shown in Figure 16-6(c).

■ ***Solution:***

$$V_{\text{OUT}} = V_{\text{IN}}\left(1 + \frac{R_F}{R_1}\right) = 4\text{ V}\left(1 + \frac{12\text{ k}\Omega}{8\text{ k}\Omega}\right) = 4\text{ V} \times (1 + 1.5) = 4\text{ V} \times 2.5 = +10\text{ V}$$

The open-loop comparator and closed-loop inverting and noninverting amplifier circuits are just three of many op-amp application circuits. In the final section of this chapter, we will examine many other typical op-amp circuit applications, but for now let us review our understanding of the op-amp by examining a typical manufacturer's data sheet.

DEVICE: µA 747—Dual Linear Op-Amp IC

DESCRIPTION

The 747 is a pair of high-performance monolithic operational amplifiers constructed on a single silicon chip. High common-mode voltage range and absence of "latch-up" make the 747 ideal for use as a voltage-follower. The high gain and wide range of operating voltage provides superior performance in integrator, summing amplifier, and general feedback applications. The 747 is short-circuit protected and requires no external components for frequency compensation. The internal 6dB/octave roll-off insures stability in closed-loop applications. For single amplifier performance, see µA741 data sheet.

PIN CONFIGURATION

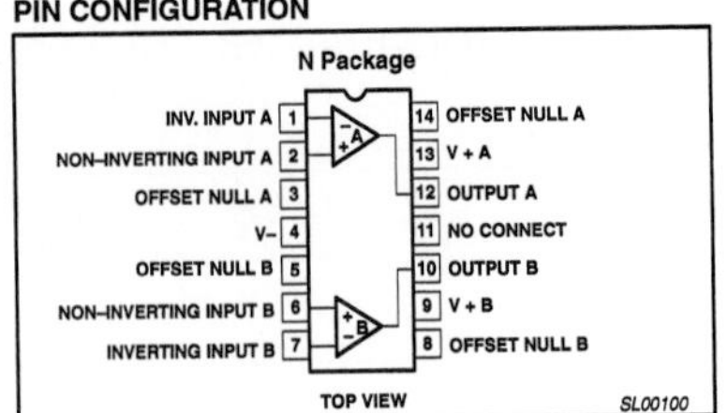

FEATURES

- No frequency compensation required
- Short-circuit protection
- Offset voltage null capability
- Large common-mode and differential voltage ranges
- Low power consumption
- No latch-up

ABSOLUTE MAXIMUM RATINGS

SYMBOL	PARAMETER	RATING	UNIT
V_S	Supply voltage	±18	V
$P_{D\,MAX}$	Maximum power dissipation T_A=25°C (still air)[1]	1500	mW
V_{IN}	Differential input voltage	±30	V
V_{IN}	Input voltage[2]	±15	V
	Voltage between offset null and V-	±0.5	V
T_{STG}	Storage temperature range	-65 to +150	°C
T_A	Operating temperature range	0 to +70	°C
T_{SOLD}	Lead temperature (soldering, 10sec)	300	°C
I_{SC}	Output short-circuit duration	Indefinite	

Explanation of Key Maximum Ratings

Supply Voltage: This is the maximum voltage that can be used to power the op-amp.
Maximum Power Dissipation: This is the maximum power the op-amp can dissipate.
Differential Input Voltage: This is the maximum voltage that can be applied across the + and – inputs.
Input Voltage: This is the maximum voltage that can be applied between an input and ground.
Operating Temperature Range: The temperature range in which the op-amp will operate within the manufacturer's specifications.
Open Short Circuit Duration: This is the amount of time that the op-amp's output can be short circuited to ground or to a supply voltage. This op-amp has an "indefinite" rating since it has an internal circuit that will turn OFF the op-amp's output and protect the internal circuitry if an output short occurs.

FIGURE 16-7 An Op-Amp Data Sheet. (Courtesy of Philips Semiconductors.)

16-2-6 An Op-Amp Data Sheet

To better understand the characteristics of the op-amp, Figure 16-7 shows the data sheet for a "747," which is a 14-pin IC containing two 741 op-amps. Like most data sheets, this one contains a general description of the device, a pin-configuration diagram, a listing of maximum ratings, an internal-circuit diagram, and a listing of input/output characteristics. Notes have been included so that you will be able to understand the meaning of most key terms.

Figure 16-8 compares the characteristics and cost of several different op-amp types to the 741, which is very popular due to its good performance and low price. Referring to the cost-factor column, you can see, for example, that the LF351 is twice the price of the 741. Studying the key characteristics in this figure,

DEVICE: μA 747—Dual Linear Op-Amp IC

EQUIVALENT SCHEMATIC

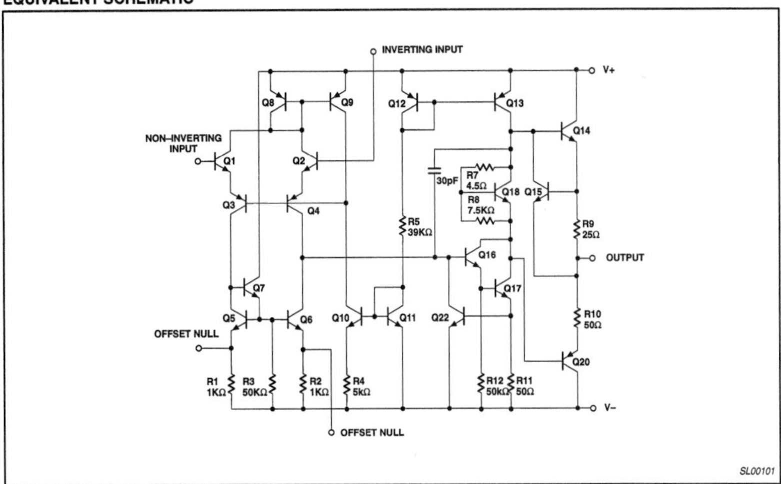

DC ELECTRICAL CHARACTERISTICS

T_A=25°C, V_{CC} = ±15V unless otherwise specified.

SYMBOL	PARAMETER	TEST CONDITIONS	μA747C			UNIT
			Min	Typ	Max	
V_{OS}	Offset voltage	R_S≤10kΩ		2.0	6.0	mV
		R_S≤10kΩ, over temp.		3.0	7.5	mV
$\Delta V_{OS}/\Delta T$				10		μV/°C
I_{OS}	Offset current			20	200	nA
		Over temperature		7.0	300	nA
$\Delta I_{OS}/\Delta T$				200		pA/°C
I_{BIAS}	Input current			80	500	nA
		Over temperature		30	800	nA
$\Delta I_B/\Delta T$				1		nA/°C
V_{OUT}	Output voltage swing	R_L≥2kΩ, over temp.	±10	±13		V
		R_L≥10kΩ, over temp.	±12	±14		V
I_{CC}	Supply current each side			1.7	2.8	mA
		Over temperature		2.0	3.3	mA
P_d	Power consumption			50	85	mW
		Over temperature		60	100	mW
C_{IN}	Input capacitance			1.4		pF
	Offset voltage adjustment range			±15		mV
R_{OUT}	Output resistance			75		Ω
	Channel separation			120		dB
PSRR	Supply voltage rejection ratio	R_S≤10kΩ, over temp.		30	150	μV/V
A_{VOL}	Large-signal voltage gain (DC)	R_L≥2kΩ, V_{OUT}=±10V	25,000			V/V
		Over temperature	15,000			V/V
CMRR	Common-mode rejection ratio	R_S≤10kΩ, V_{CM}=±12V Over temperature	70			dB

Explanation of Key Ratings

Input Offset Voltage: The voltage that must be applied to one input for the output voltage to be zero.
Input Offset Current: The difference of the two input bias currents when the output voltage is zero.
Input Bias Current: The average of the currents flowing into both inputs (ideally input bias currents are equal).
Input Resistance: This is the resistance of either input when the other input is grounded.

Output Resistance: This is the resistance of the op-amp's output.
Output Short-Circuit Current: Maximum output current that the op-amp can deliver to a load.

Supply Current: The current that the op-amp circuit will draw from the power supply.
Slew rate: The maximum rate of change of output voltage under large signal conditions.
Channel Separation: When two op-amps are within one package, there will be a certain amount of interference or "crosstalk" between op-amps.

FIGURE 16-7 (continued)

OP-AMP TYPE	INPUT IMPEDANCE	OUTPUT IMPEDANCE	OPEN-LOOP GAIN (Min.)	CMRR (dB)
741 C	2 MΩ	75Ω	25,000	70
101	800 kΩ	Low	25,000	70
108 A	70 MΩ	Low	80,000	96
351	High	Low	25,000	70
318	High	Low	25,000	70
357	High	Low	50,000	80
363	High	Low	1,000,000	94
356	High	Low	25,000	80

FIGURE 16-8 Comparing the Characteristics of Several Op-Amps.

you can see that the 741, for example, will typically have an input impedance of 2 MΩ, an output impedance of 75Ω, an open-loop gain of 25,000, and a common-mode rejection ratio of 70 dB for any input signal from 0 Hz up to 1 MHz. The gain bandwidth column lists only the upper frequency limit because the op-amp has no internal coupling capacitors, and therefore the lower frequency limit extends down to dc signals, or 0 Hz.

16-2-7 *Troubleshooting Op-Amp Circuits*

Now that we know how operational amplifier circuits are supposed to work, let us see how we can troubleshoot a problem when they do not work. First, let us review our basic troubleshooting procedure.

Step 1: DIAGNOSE

The first step is to determine whether a problem really exists. To carry out this step, a technician must collect as much information as possible about the system, circuit, and components used, and then diagnose the problem.

Step 2: ISOLATE

The second step is to apply a logical and sequential reasoning process to isolate the problem. In this step, a technician will operate, observe, test, and apply troubleshooting techniques in order to isolate the malfunction.

Step 3: REPAIR

The third and final step is to make the actual repair, then test the circuit.

Isolating the Problem to the Op-Amp Circuit

Once you have diagnosed that an op-amp circuit is malfunctioning, the next step is to isolate whether the problem is within the op-amp's internal circuit, or external to the op-amp IC. For example, Figure 16-9(a) shows a typical closed-loop inverting op-amp circuit. As with all amplifiers, the circuit receives power from a dc power supply, an input signal from the input-signal source, and supplies an output signal to a load. Therefore, our first step is to isolate whether the

COST FACTOR	SLEW RATE (V/μs)	GAIN BANDWIDTH	FEATURES
1	0.5	1 MHz	Low cost
1	0.5	1 MHz	Low cost
—	0.3	1 MHz	Precision low drift
2	13	4 MHz	Low bias current
5	70	15 MHz	High slew rate
4	30	20 MHz	High CMRR
45	—	2 MHz	Low noise; high rejection
3	10	5 MHz	Improved 741

FIGURE 16-8 (continued)

problem is within the op-amp circuit or within one of the external circuits. To achieve this, the following checks should be made.

1. Power Supply or Amplifier? If the amplifier is not functioning as it should, first check the power (+12 V and −12 V) and ground connection to the amplifier circuit. If the dc power supply voltage to the circuit is not correct, you will next need to determine whether the problem is in the source (the dc power supply circuit), or in the load (the amplifier circuit). If the amplifier is disconnected from the dc power supply, and the voltage returns to its normal potential, the amplifier has developed some short within its circuit and this is pulling down the $+V_{CC}$ supply voltage. If, on the other hand, the amplifier is disconnected from the dc power supply, and the voltage supply voltage still remains at its incorrect value, the problem is within the dc power supply circuit.
2. Amplifier Source or Output Load? If the output signal from the op-amp is incorrect, remember that a bad load can pull down a voltage signal and make it appear as though the amplifier is faulty. When you have no output, or a bad output, you will first need to disconnect the output of the amplifier from the load to determine whether the problem is in the source (amplifier) or the load. If the load is disconnected and the signal is good, the impedance of the load is pulling down the output of the amplifier. On the other hand, if the load is disconnected and the signal output of the amplifier is still bad, the problem is in the amplifier.
3. Signal Source or Amplifier Load? If the output signal from an op-amp is incorrect, do not backtrack through the amplifier circuit. Go immediately to the amplifier's input and check the input signal to the first stage. Once again, if the input signal to the amplifier is incorrect, we will first have to isolate whether the problem exists in the signal source, or in the amplifier. If the input signal source is disconnected from the amplifier, and the signal generated by V_{IN} returns to its normal level, a fault in the amplifier input is pulling down the input signal. On the other hand, if the input signal source is disconnected from the amplifier, and the signal does

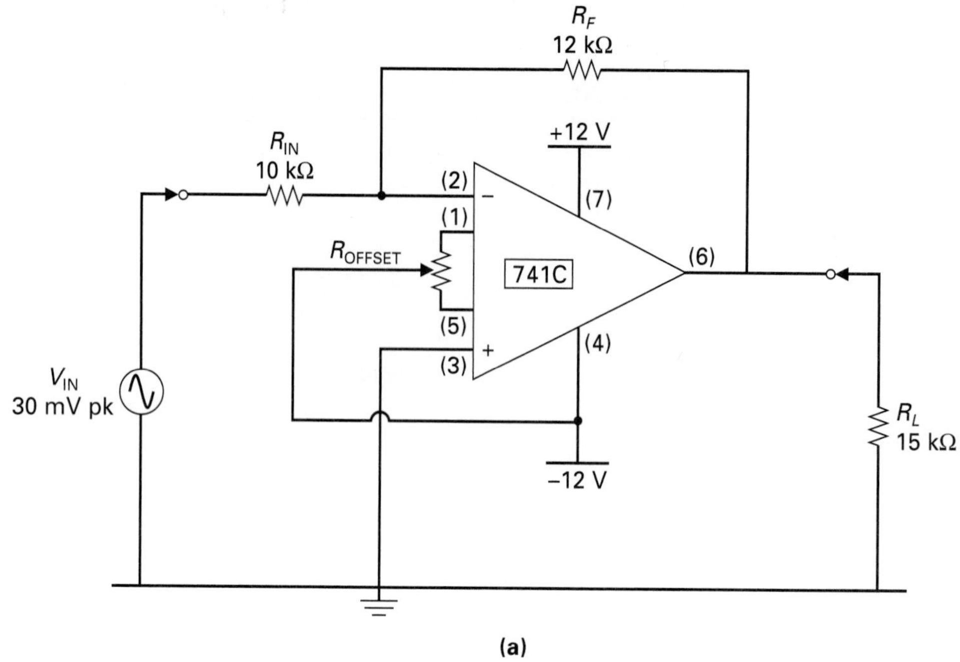

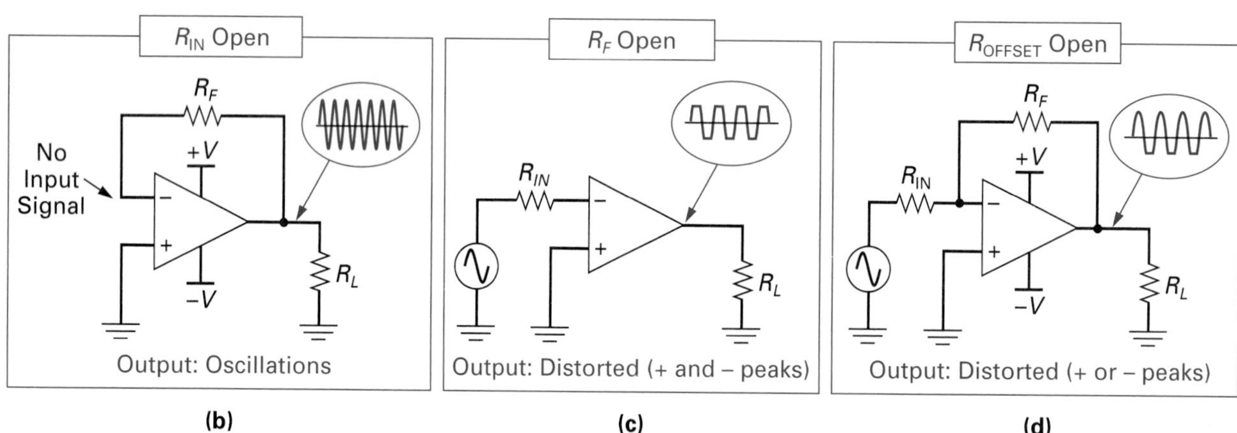

FIGURE 16-9 Troubleshooting Op-Amp Circuits.

not return to its normal level, a fault exists in the input signal source circuit.

Troubleshooting an Op-Amp Circuit

Once you have isolated the problem to the op-amp circuit, you will next need to isolate the faulty component. Because op-amp circuits only contain a handful of components, the isolating process is relatively easy, especially because each component's fault produces its own distinct effect. Figure 16-9(b), (c), and (d) show the effect you should get if each of the circuit resistors were to open.

1. If R_{IN} were to open as shown in Figure 16-9(b), the input signal would be disconnected from the amplifier. Although this condition should result in no output, in most cases it sends the op-amp circuit into oscillation: if the output is positive when R_{IN} opens, a positive voltage will be fed back to the inverting input via R_F. This positive input will be inverted to a negative output, which will be fed back via R_F producing a positive output, which will be fed back via R_F producing a negative output, and so on. The result will be a continuous millivolt oscillation at the output.
2. If R_F were to open as shown in Figure 16-9(c), the feedback path would be disconnected and the gain of the op-amp circuit will increase to its maximum open loop value. For example, if a 30 mV pk input signal were applied to a 741 with a minimum open-loop gain of 25,000, the op-amp would try to produce an output of 750 V pk. Because the output voltage can never exceed the $+V$ and $-V$ supply voltages, the result will be a heavily distorted waveform that has its positive and negative peaks clipped when the op-amp is driven into saturation and cutoff.
3. If the offset resistor were to open, the output of the op-amp will not be zero when the input is zero. In fact, the output will be offset by an amount that is equal to the offset voltage times the op-amp's closed loop gain. As a result, the output waveform will have either its positive or negative peaks clipped, depending on whether the op-amp is unbalanced in the positive or negative direction.
4. As shown previously, the op-amp has an internal circuit that contains a large number of components. If you have isolated the problem to the op-amp's internal circuit, and you have determined that all of the op-amp's externally connected resistors are okay, the next step will be to replace the op-amp IC itself because it is impossible to repair any internal op-amp failures. If only one of the op-amps within a dual op-amp package such as the 747 have gone bad, the whole IC will have to be replaced. When removing an op-amp IC, be sure to note the orientation of the package because a replacement put in backwards will more often than not be destroyed.

SELF-TEST REVIEW QUESTIONS FOR SECTION 16-2

1. Can the op-amp be used to amplify dc as well as ac signal inputs?
2. What is the difference between an open-loop and closed-loop op-amp circuit?
3. List the key characteristics of an op-amp.
4. What are the three basic amplifier types within an op-amp?
5. Which of the following circuits is connected in an open-loop mode?
   a. Comparator
   b. Inverter amplifier
   c. Noninverting amplifier
6. Why is it important for an op-amp circuit to have negative feedback?

16-3 ADDITIONAL OP-AMP CIRCUIT APPLICATIONS

The operational amplifier's flexibility and characteristics make it the ideal choice for a wide variety of circuits applications. In fact, because the op-amp is the most frequently used linear IC, it is safe to say that you will find several op-amps in almost every electronic system. Although it is impossible to cover all of these applications, many circuits predominate, and others are merely variations on the same basic theme. In this section we will concentrate on the operation and characteristics of all of the most frequently used op-amp circuit applications.

16-3-1 *The Voltage Follower Circuit*

Voltage Follower Circuit
An op-amp circuit that has a direct feedback to give unity gain so the output voltage follows the input voltage. Used in applications where a very high input impedance and very low output impedance are desired.

Figure 16-10 shows how an op-amp can be connected to form a noninverting **voltage follower circuit.** Using the noninverting closed-loop gain formula discussed previously, we can calculate the voltage gain of this circuit.

$$A_{CL} = \frac{R_F}{R_1} + 1 = \frac{0\,\Omega}{0\,\Omega} + 1 = 0 + 1 = 1$$

With a gain of 1, the output voltage will be equal to the input voltage—so what is the advantage of this circuit? The answer is the op-amp characteristics of a high-input impedance and a low-output impedance. Similar to the BJT's emitter-follower and the FET's source-follower, the op-amp voltage-follower circuit derives its name from the fact that the output voltage follows the input voltage in both polarity and amplitude. This circuit is therefore ideal as a buffer, interfacing a high-impedance source to a low-impedance load.

16-3-2 *The Summing Amplifier Circuit*

Summing Amplifier Circuit (or Adder Circuit)
An op-amp circuit that will sum or add all of the input voltages.

The **summing amplifier circuit,** or adder amplifier, consists of two or more input resistors connected to the inverting input of an op-amp as shown in Figure 16-11(a). This circuit will sum or add all of the input voltages, and therefore the output voltage will be

$$V_{OUT} = -(V_{IN1} + V_{IN2} + V_{IN3})$$

Using Ohms law ($V = R \times I$), we can also calculate the output voltage with the formula

$$V_{OUT} = -R_4 \times \left(\frac{V_{IN1}}{R_1} + \frac{V_{IN2}}{R_2} + \frac{V_{IN3}}{R_3} \right)$$

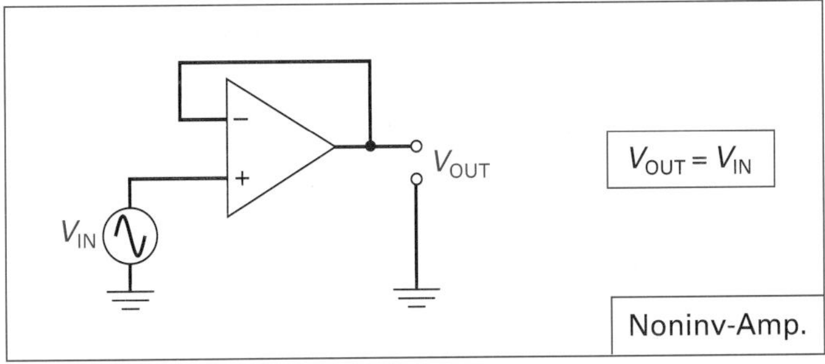

FIGURE 16-10 Voltage Follower Circuit.

Once again, the negative sign preceding the formula indicates that the output signal will be opposite in polarity to the two or three input signals.

■ **EXAMPLE:**

Calculate the output voltage of the summing amplifier circuit shown in Figure 16-11(b).

■ ***Solution:***

$$V_{OUT} = -(V_{IN1} + V_{IN2}) = -(3.7\ \text{V} + 2.6\ \text{V}) = -6.3\ \text{V}$$

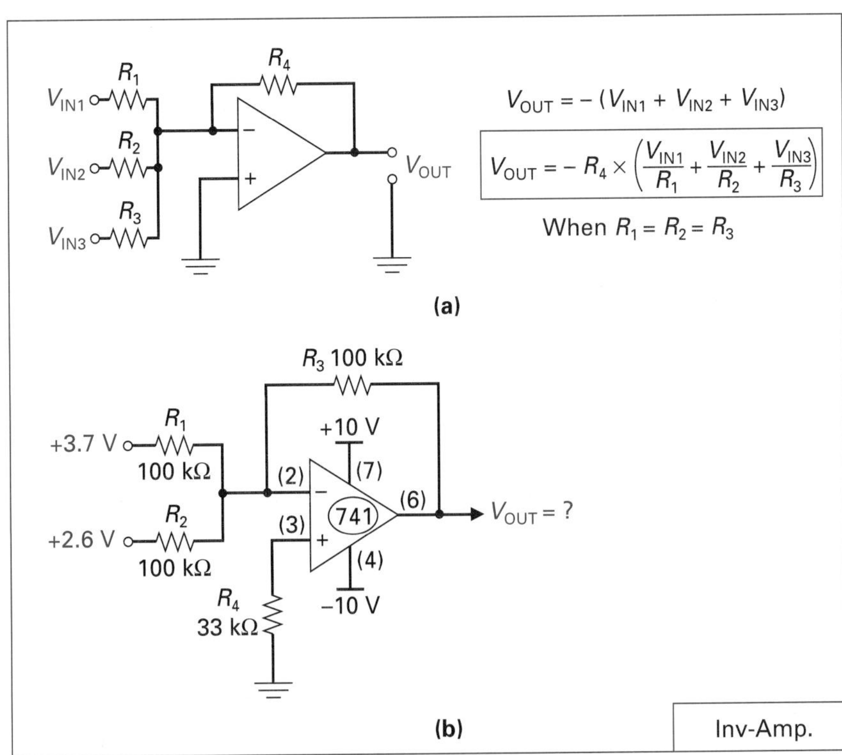

FIGURE 16-11 Summing Amplifier Circuit.

16-3-3 The Difference Amplifier Circuit

Differential Amplifier Circuit An op-amp circuit in which the output voltage is equal to the difference between the two input voltages.

Figure 16-12(a) shows how the op-amp can be connected to operate as a difference or **differential amplifier circuit.** In this circuit application, the op-amp will simply be making use of its first internal amplifier stage, which (as mentioned earlier) is a diff-amp. In this circuit, all four resistors are normally of the same value, and the output voltage is equal to the difference between the two input voltages.

$$V_{OUT} = V_{IN2} - V_{IN1}$$

■ **EXAMPLE:**

Calculate the output voltage of the difference amplifier circuit shown in Figure 16-12(b).

■ ***Solution:***

$$V_{OUT} = V_{IN2} - V_{IN1} = 6\text{ V} - 3.2\text{ V} = 2.8\text{ V}$$

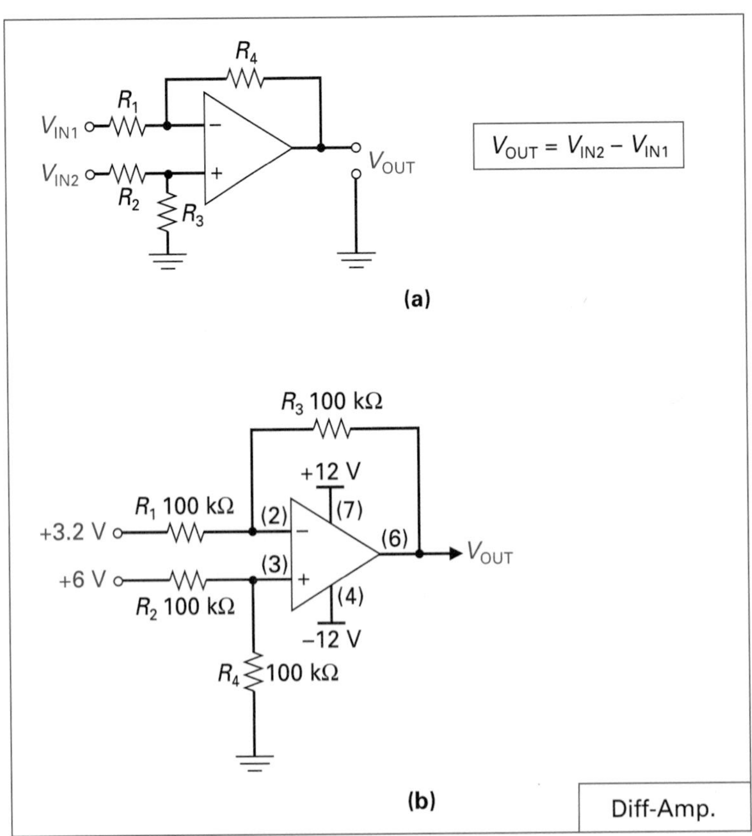

FIGURE 16-12 The Difference Amplifier Circuit.

16-3-4 *The Differentiator Circuit*

The op-amp **differentiator circuit** (not to be confused with the previous differential circuit) is similar to the basic inverting amplifier except that R_{IN} is replaced by a capacitor, as shown in Figure 16-13(a). Including a capacitor in any circuit means that we will develop problems as the frequency of the input signal increases because capacitive reactance is inversely proportional to frequency. This means that the reactance of the input capacitor will decrease for input signals that are higher in frequency, and therefore the input voltage applied to the op-amp and output voltage from the op-amp will increase with frequency. Including an additional resistor (R_S) in series with the input capacitor, as shown in Figure 16-13(b), will decrease the high-frequency gain because gain will now be a ratio of R_F/R_S.

Differentiator Circuit
A circuit whose output voltage is proportional to the rate of change of the input voltage. The output waveform is the time derivative of the input waveform.

Figure 16-13(c) shows the differentiator's input/output waveforms, with the peak of the output square wave being equal to

$$V_{OUT(pk)} = 2\,\pi f \times R_1 \times C_1 \times V_{IN(pk)}$$

■ **EXAMPLE:**

Calculate the peak output voltage of the differentiator circuit shown in Figure 16-13(d).

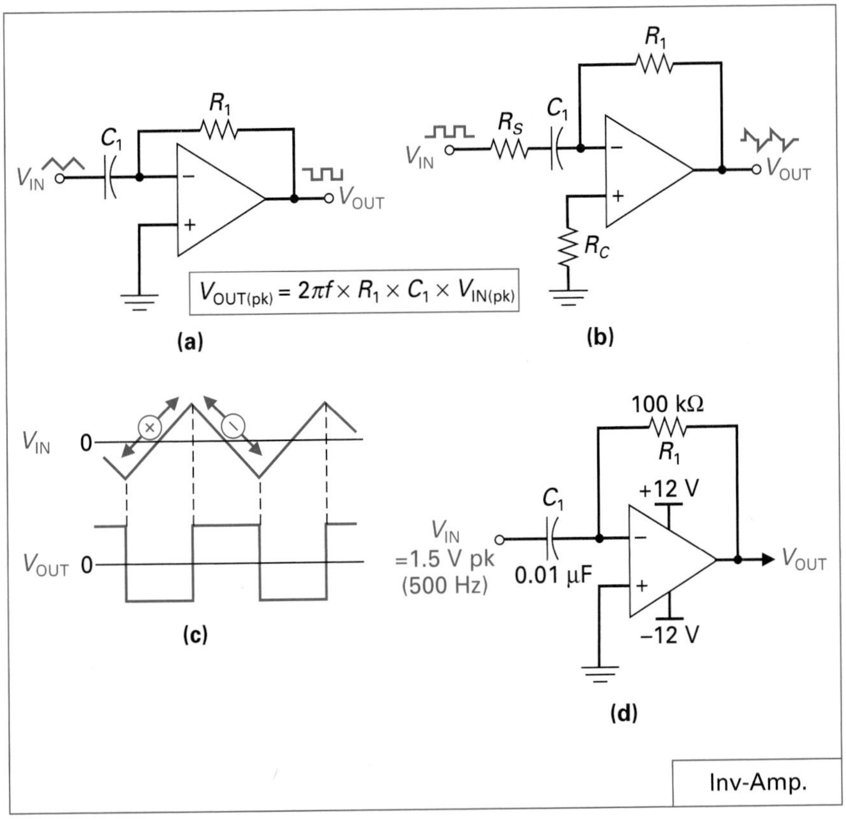

FIGURE 16-13 **Differentiator Circuit.**

■ ***Solution:***

$$V_{OUT(pk)} = 2\,\pi f \times R_1 \times C_1 \times V_{IN(pk)}$$
$$= (2 \times \pi \times 500\text{ Hz}) \times 100\text{ k}\Omega \times 0.01\ \mu\text{F} \times (1.5\text{ V pk}) = 4.7\text{ V}$$

16-3-5 *The Integrator Circuit*

Integrator Circuit
A circuit with an output which is the integral of its input with respect to time.

Figure 16-14(a) shows how the position of the resistor and capacitor in the differentiator circuit can be reversed to construct an op-amp **integrator circuit.** As in the differentiator circuit, the capacitor will alter the gain of the op-amp because its capacitive reactance changes with frequency. To compensate for this effect, a parallel resistor (R_P) is included in shunt with the capacitor, as shown in Figure 16-14(b), to decrease the low-frequency gain: gain will now be a ratio of R_P/R_1.

Figure 16-14(c) shows the integrator's input/output waveforms, with the peak of the output triangular wave being equal to

$$V_{OUT(pk)} = \frac{1}{R_1C_1} \times (\Delta V_{IN(pk)} \times \Delta t)$$

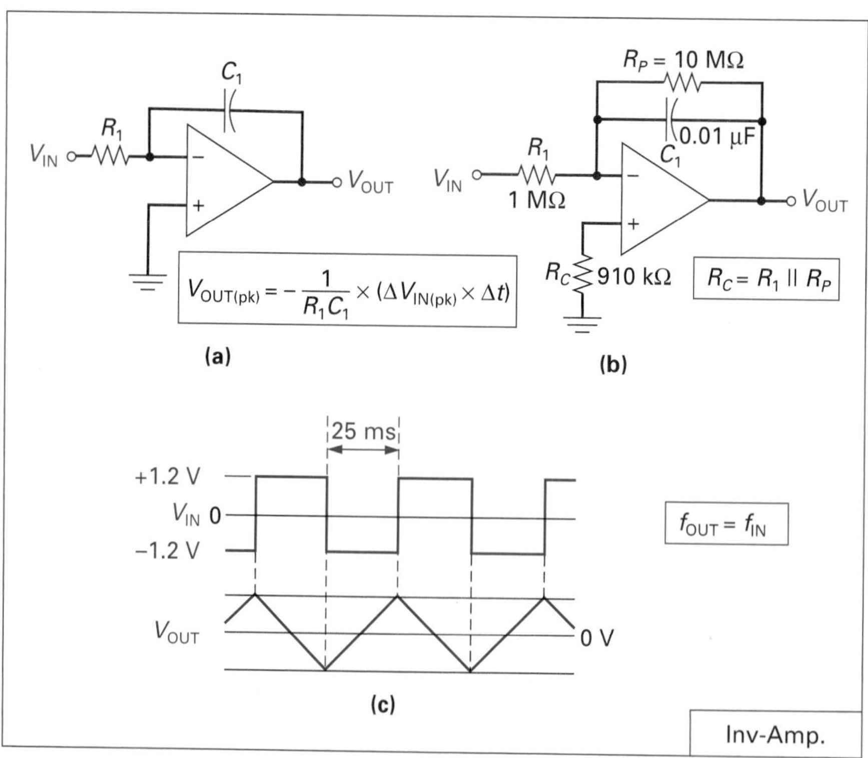

FIGURE 16-14 Integrator Circuit.

■ **EXAMPLE:**

Calculate the peak output voltage of the integrator circuit shown in Figure 16-14(b) considering the input/output waveforms given in Figure 16-14(c).

■ ***Solution:***

$$V_{OUT(pk)} = \frac{1}{R_1C_1} \times (\Delta V_{IN(pk)} \times \Delta t) = \frac{1}{1\ \text{M}\Omega \times 0.01\ \mu\text{F}} \times (0\ \text{V to } 1.2\ \text{V pk} \times 25\ \text{ms})$$

$$= 100 \times (1.2\ \text{V pk} \times 25\ \text{ms}) = 100 \times 0.03 = 3\ \text{V pk}$$

16-3-6 Signal Generator Circuits

Figure 16-15 shows how the op-amp can be connected to act as a signal generator, which is a circuit that will convert a dc supply voltage into a repeating output signal.

In Figure 16-15(a), the **twin-T sine-wave oscillator,** which has two T-shaped feedback networks, will generate a repeating sine-wave output at a frequency equal to

Twin-T Sine-Wave Oscillator
An oscillator circuit that makes use of two T-shaped feedback networks.

$$f_0 = \frac{1}{2\pi RC}$$

Square-Wave Generator
A circuit that generates a continuously repeating square wave.

■ **EXAMPLE:**

Calculate the frequency of the oscillator shown in Figure 16-15(a).

■ ***Solution:***

$$f_0 = \frac{1}{2\pi RC} = \frac{1}{2 \times \pi \times 6.8\ \text{k}\Omega \times 0.033\ \mu\text{F}} = 709\ \text{Hz}$$

Relaxation Oscillator
An oscillator circuit whose frequency is determined by an *RL* or *RC* network, producing a rectangular or sawtooth output waveform.

In Figure 16-15(b), the op-amp has been connected to a **square-wave generator,** or as it is more frequently called a **relaxation oscillator.** The output signal is fed back to both the inverting and noninverting inputs of the op-amp. The capacitor will charge and discharge through *R* controlling the frequency of the output square wave, which is equal to

$$f_0 = \frac{1}{2\ RC \log\left(\frac{2R_1}{R_2} + 1\right)}$$

Triangular-Wave Generator
A signal generator circuit that produces a continuously repeating triangular wave output.

Staircase-Wave Generator
A signal generator circuit that produces an output signal voltage that increases in steps.

Connecting the output of the square-wave generator in Figure 16-15(b) into the inputs of the circuits in Figure 16-15(c) and (d), we can generate a triangular or staircase output waveform. The **triangular-wave generator** in Figure 16-15(c) is simply the integrator circuit discussed previously, with its output frequency equal to the square wave input frequency. When the switch is closed in the **staircase-**

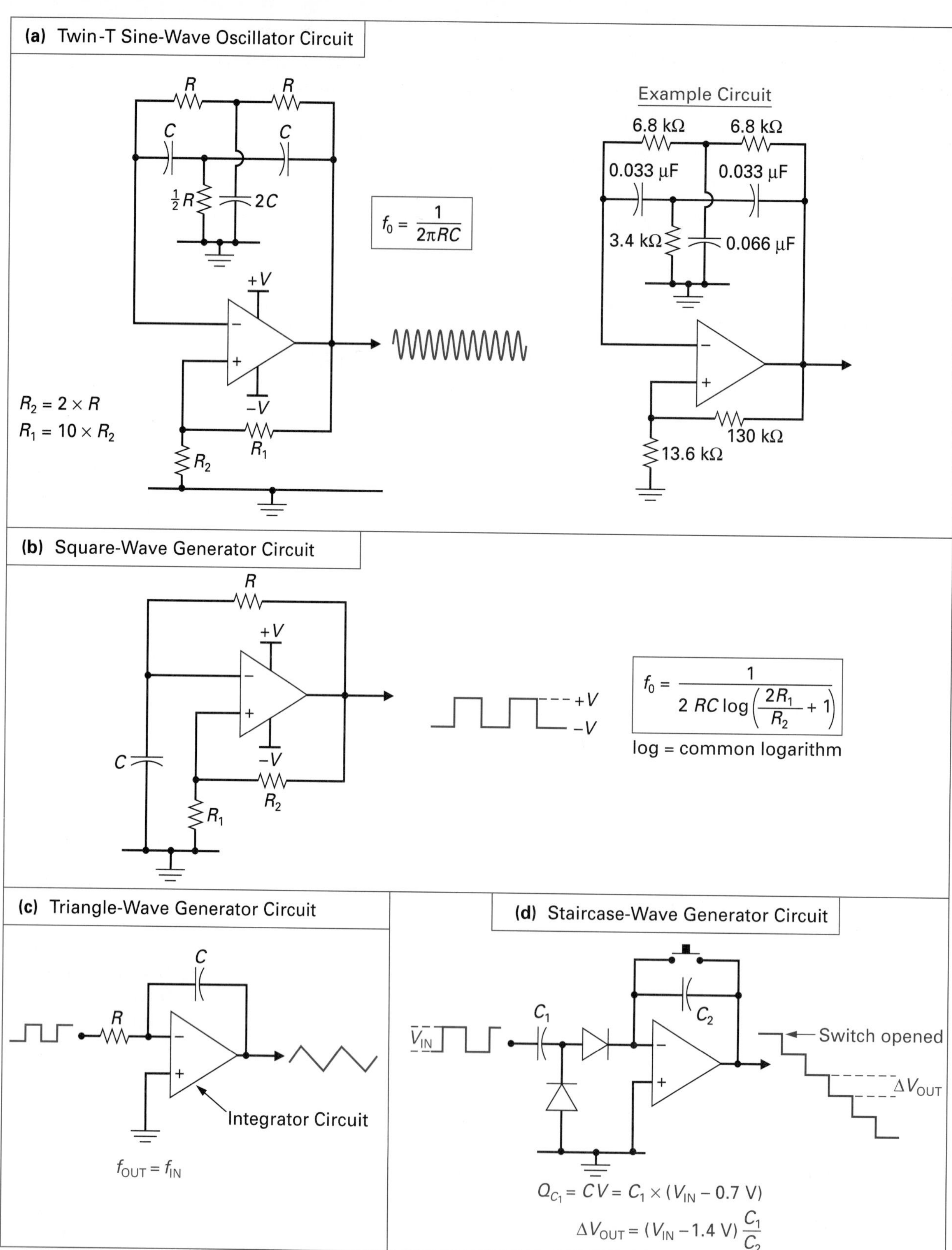

FIGURE 16-15 **Signal Generator Circuits.**

wave generator in Figure 16-15(d) the capacitor is bypassed and will therefore not charge. On the other hand, when the switch is open, C_2 will be charged by each input cycle, producing equal output steps that have the following voltage change

$$\Delta V_{OUT} = (V_{IN} - 1.4\text{ V})\ \frac{C_1}{C_2}$$

16-3-7 *Active Filter Circuits*

Passive filters, which were discussed previously in "Introductory DC/AC Electronics," are circuits that contain passive or nonamplifying components (resistors, capacitors, and inductors) connected in such a way that they will pass certain frequencies while rejecting others. An **active filter** on the other hand, is a circuit that uses an amplifier with passive filter elements to provide frequency paths with rejection characteristics. Active filters, like the op-amp circuits seen in Figure 16-16, have several advantages over passive filters.

Active Filter
A circuit that uses an amplifier with passive filter elements to provide frequency paths with rejection characteristics.

1. Because the op-amp provides gain, the input signal passed to the output will not be attenuated, and therefore better response curves can be obtained.
2. The high input impedance and low output impedance of the op-amp means that the filter circuit does not interfere with the signal source or load.
3. Because active filters provide gain, resistors can be used instead of inductors, and therefore active filters are generally less expensive.

Figure 16-16 illustrates how the op-amp can be connected to form the four basic active-filter types.

Active High-Pass Filter

Figure 16-16(a) illustrates the simple op-amp circuit, frequency response, and relevant formulas for an **active high-pass filter.** As before, the gain of this inverting amplifier is dependent on the ratio of R_F to R_{IN}. When capacitors are included in any circuit, impedance (Z) must be considered instead of simply resistance, and gain is now equal to the ratio of feedback impedance to input impedance.

Active High-Pass Filter
A circuit that uses an amplifier with passive filter elements to pass all frequencies above a cut-off frequency.

$$A_{CL} = -\ \frac{Z_F}{Z_{IN}}$$

The input RC network will offer a high impedance to low frequencies, resulting in a low voltage gain. At high frequencies, the RC network will have a low impedance, causing a high voltage gain. The cutoff frequency for this circuit can be calculated with the following formula when $C_1 = C_2$.

$$f_C = \frac{1}{2\pi RC}$$

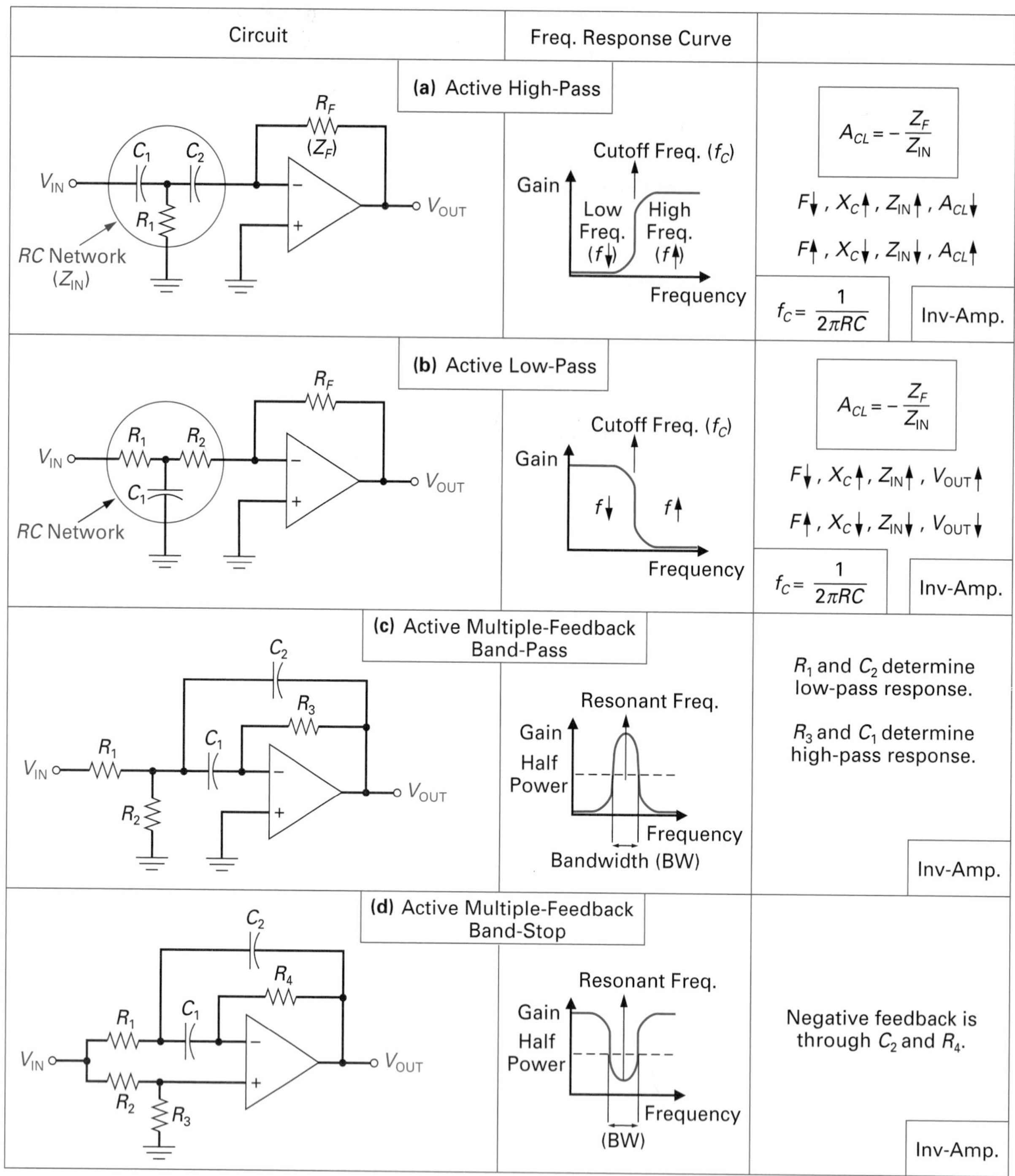

FIGURE 16-16 Active-Filter Circuits.

Active Low-Pass Filter
An amplifier circuit with passive filter elements that pass all frequencies below a cut-off frequency.

Active Low-Pass Filter

Figure 16-16(b) illustrates the op-amp circuit, frequency response curve, and relevant formulas for an **active low-pass filter.** At low frequencies, the capacitor's reactance is high, and low-frequency signals will be passed to the op-amp's input to be amplified and passed to the output. As frequency increases, the capacitive reactance of C_1 will decrease; more of the signal will be shunted away from the op-amp and will not appear at the output. The cutoff frequency for this circuit can be calculated with the following formula when $R_1 = R_2$.

$$f_C = \frac{1}{2\pi RC}$$

Active Band-Pass Filter

Figure 16-16(c) illustrates how the op-amp can be connected to form an **active band-pass filter.** At frequencies outside of the band, V_{OUT} is fed back to the input without being attenuated, and therefore the input signal amplitude is almost equal to the feedback signal amplitude. This results in almost complete cancellation of the signal and therefore a very small output voltage. On the other hand, for the narrow band of frequencies within the band, the feedback network will increase its amount of attenuation. This increase of attenuation means that a very small feedback signal will appear back at the negative input of the op-amp and will have a very small degenerative effect. As a result, the change at the input of the op-amp will be larger when the input signal frequencies are within this band, and the voltage out will also be larger.

Active Band-Pass Filter
A circuit that uses an amplifier with passive filter elements to pass only a band of input frequencies.

Active Band-Stop Filter

Figure 16-16(d) illustrates how the op-amp can be connected to form an **active band-stop filter,** also known as a band-reject or notch filter. The basic operation of this circuit is opposite to that of the previously discussed band-pass filter. At frequencies outside of the band, the feedback signal will be heavily attenuated, and therefore the degenerative effect will be small and the output voltage large. On the other hand, at frequencies within the band, the feedback signal will not be heavily attenuated, and therefore the degenerative effect will be large and the output voltage small.

Active Band-Stop Filter
A circuit that uses an amplifier with passive filter elements to block a band of input frequencies.

SELF-TEST REVIEW QUESTIONS FOR SECTION 16-3

1. Which op-amp circuit provides a voltage gain of 1 and is used as a buffer?
2. Which op-amp circuit will sum all of the input voltages?
3. What is the basic circuit difference and input/output waveform difference between the integrator and differentiator circuit?
4. Which op-amp circuit will generate an output that is equal to the difference between the two inputs?
5. Sketch a circuit showing how the op-amp can be connected to generate a repeating square wave output.
6. What is the difference between an active filter and a passive filter?

SUMMARY

The Operational Amplifier (Figure 16-17)

1. The operational amplifier was initially a vacuum tube circuit used in the early 1940s in analog computers.
2. The name "operational amplifier" or "op-amp" was chosen because the circuit was used as a high-gain dc "amplifier" performing mathematical "operations."
3. The early op-amp circuits were expensive and bulky, and they found very little application until the semiconductor integrated circuit was developed in 1958 by Jack Kilby at Texas Instruments. Circuits that once needed hundreds of discrete or individual components can now be integrated into a single IC, making equipment smaller, more energy efficient, cheaper, and easier to design and troubleshoot.
4. Today's IC op-amp is a very high-gain dc amplifier that can have its operating characteristics changed by connecting different external components. This makes the op-amp very versatile, and it is this versatility that has made the op-amp the most widely used linear IC.
5. The op-amp's internal circuit is simply a combination of three previously covered amplifier circuits. These three circuits are all interconnected and contained within a single IC, and together they function as a "high-gain, high input impedance, low output impedance amplifier."
6. The triangle shaped amplifier symbol is used to represent the op-amp in an electronic schematic diagram.
7. The two op-amp inputs are labeled "–" and "+." The "–" or negative input is called the inverting input because any signal applied to this input will be amplified and inverted between input and output (output is 180° out of phase with input). On the other hand, the "+" or positive input is called the noninverting input because any signal applied to this input will be amplified but not inverted between input and output (output is in phase with input). An input signal will normally be applied to only one of these inputs, while the other input is used to control the op-amp's operating characteristics.
8. The two power supply connections to the op-amp are labeled "$+V$" and "$-V$." When two supply voltages are used (dual supply voltages), the voltage values are of the same value, but of opposite polarity (for example, +12 V and –12 V). On the other hand, when only one supply voltage is used (single supply voltage), a positive or negative voltage is applied to its respective terminal while the other terminal is grounded (for example, +5 V and ground or –5 V and ground). Having both a positive and negative power supply voltage will allow the output signal to swing positive and negative, above and below zero. As with all high-gain amplifiers, however, the output voltage can never exceed the value of the $+V$ and $-V$ supply voltages.
9. The entire op-amp circuit is placed within a TO-5 metal can package or a dual-in-line through-hole or surface-mount package.
10. Like all ICs, an identification code is used to indicate the device manufacturer, device type, and key characteristics.

QUICK REFERENCE SUMMARY SHEET

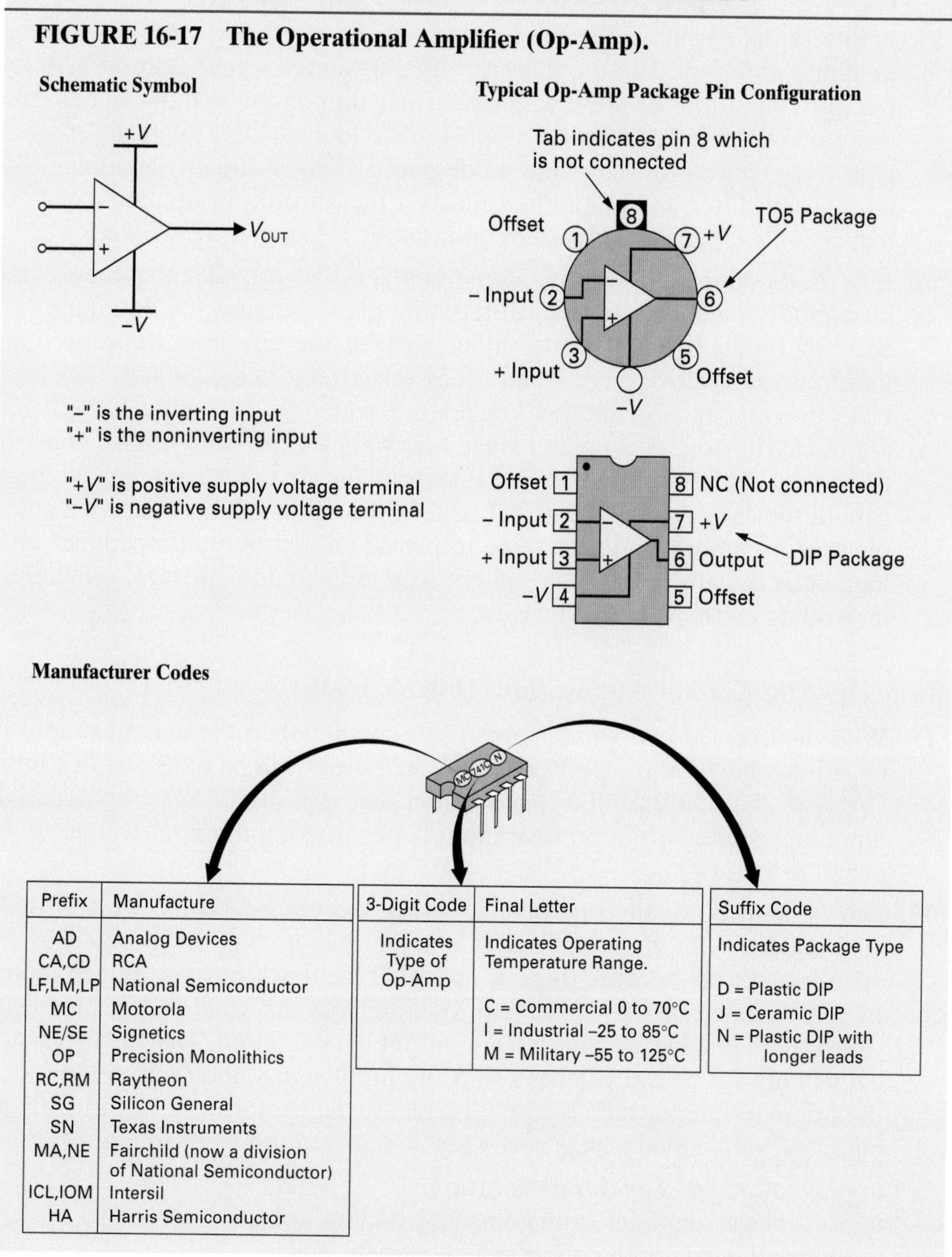

Prefix	Manufacture
AD	Analog Devices
CA,CD	RCA
LF,LM,LP	National Semiconductor
MC	Motorola
NE/SE	Signetics
OP	Precision Monolithics
RC,RM	Raytheon
SG	Silicon General
SN	Texas Instruments
MA,NE	Fairchild (now a division of National Semiconductor)
ICL,IOM	Intersil
HA	Harris Semiconductor

3-Digit Code	Final Letter
Indicates Type of Op-Amp	Indicates Operating Temperature Range. C = Commercial 0 to 70°C I = Industrial −25 to 85°C M = Military −55 to 125°C

Suffix Code
Indicates Package Type D = Plastic DIP J = Ceramic DIP N = Plastic DIP with longer leads

11. The two offset or balancing inputs will normally be connected to a potentiometer that can be adjusted to set the output at zero volts when both the inverting and noninverting inputs are at zero volts. Balancing the op-amp in this way is generally needed due to imbalances within the op-amp's internal circuit.

12. The three circuits all interconnected and contained within the single op-amp IC are a differential amplifier, a voltage amplifier, and an output amplifier.

Combined, these three circuits give the op-amp its key characteristics, which are high-gain, high input impedance, and low output impedance.

13. A differential input will be amplified by the op-amp's first stage differential amplifier and passed to the output, while unwanted signals caused by temperature variations or noise will appear as common-mode input signals and therefore be rejected. An op-amp's ability to provide a high differential gain (A_{VD}) and a low common-mode gain (A_{CM}) is directly dependent on its internal differential amplifier and is a measure of an op-amp's performance. This ratio is called the common-mode rejection ratio (CMRR).

14. It is the op-amp's differential amplifier stage that provides the good common-mode rejection and high differential gain. Because the op-amp's "–" and "+" inputs are applied to either base of the diff-amp, we know that input current will be very small—it is this circuit characteristic that provides the op-amp with another key feature, which is a high input impedance ($I\downarrow$, $Z\uparrow$). The voltage amplifier stage following the diff-amp usually consists of several darlington-pair stages that provide an overall op-amp voltage gain of typically 50,000 to 200,000. The final output stage consists of a complementary emitter-follower stage to provide a low output impedance and high current gain so that the op-amp can deliver up to several milliamps, depending on the value of the load.

Basic Op-Amp Circuit Applications (Figure 16-18)

15. With the open-loop op-amp comparator circuit, when the negative input is negative relative to the positive input, the output will go to its positive limit ($V_{OUT} = +V$). On the other hand, when the opposite occurs (the negative input is positive, or the positive input is negative), the output will go to its negative limit ($V_{OUT} = -V$).

16. The op-amp is usually operated in either the open-loop mode or closed-loop mode. With the comparator circuit, the op-amp is operated in its open-loop mode because there is no signal feedback from output to input. In most instances, the op-amp is operated in the closed-loop mode, in which there is signal feedback from output back to input. This feedback signal will always be out of phase with the input signal and therefore oppose the original signal, which is why it is called "degenerative or negative feedback." Negative feedback is necessary in nearly all op-amp circuits to

a. prevent output waveform distortion,
b. prevent the amplifier from going into oscillation, and
c. reduce the gain of the op-amp to a consistent value.

17. The inverting operational amplifier circuit produces an amplified output signal that is 180° out of phase with the input signal. In this circuit arrangement, the input signal (V_{IN}) is applied through an input resistor to the inverting input (–) of the op-amp, while the noninverting input (+) is connected to ground. A feedback loop is connected from the output back to the inverting input via the feedback resistor.

18. The feedback voltage (V_F) will greatly restrain the input voltage change to the point that a +5 V input change at V_{IN} will only be felt as a +5 micro volt

QUICK REFERENCE SUMMARY SHEET

FIGURE 16-18 Basic Op-Amp Circuit Applications.

Open-Loop Comparator Circuit

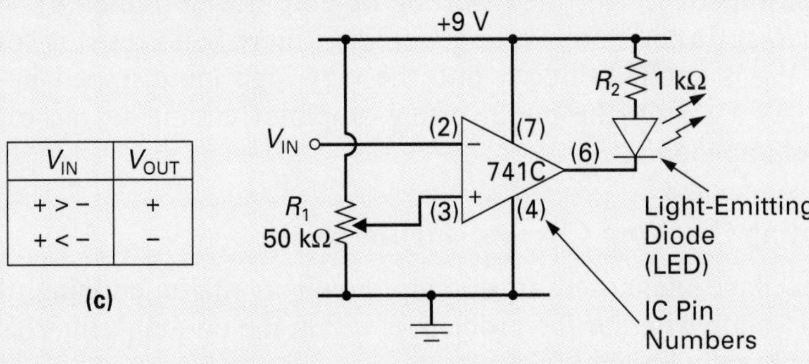

Closed-Loop Inverting Amplifier Circuit

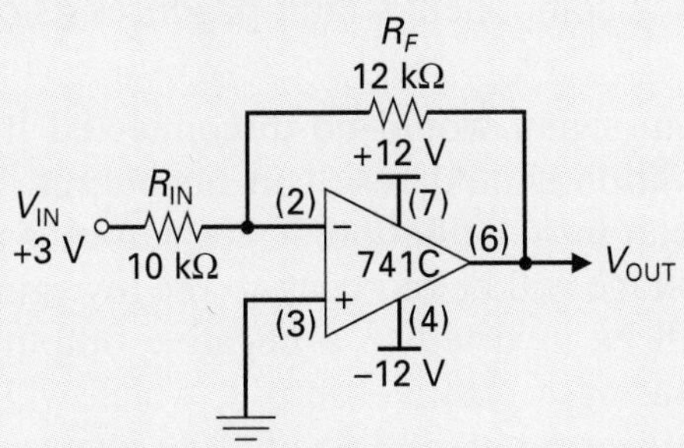

Closed loop voltage gain is dependent on the ratio of R_F to R_{IN}.

$$A_{CL} = -\frac{R_F}{R_{IN}}$$

$$V_{OUT} = -V_{IN}\left(\frac{R_F}{R_{IN}}\right)$$

Closed-Loop Noninverting Amplifier Circuit

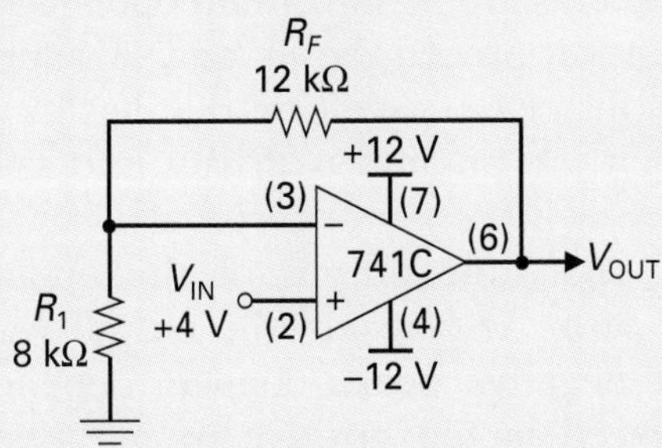

$$A_{CL} = \frac{R_F}{R_1} + 1$$

$$V_{OUT} = V_{IN}\left(1 + \frac{R_F}{R_1}\right)$$

change at the op-amp's inverting input. In fact, if the voltage at the inverting input of the op-amp is measured with a voltmeter, the "−" input appears to remain at 0 V due to the very minute change at the "−" input. The inverting input of the op-amp in this case would be defined as a virtual ground, which is different from an ordinary ground. A virtual ground is a voltage ground because this point is at zero volts; however, it is not a current ground because it cannot sink or conduct away any current. An ordinary ground, on the other hand, is at zero volts and can sink any amount of current.

19. With the noninverting amplifier circuit, the input voltage (V_{IN}) is applied to the op-amp's noninverting input (+), and the output voltage (V_{OUT}) will be in phase with the input. To achieve negative feedback, the output is applied back to the inverting input (–) of the op-amp via the feedback network formed by R_2 and R_1.

20. The input impedance for the inverting op-amp is determined by the input resistor (R_{IN}). In the noninverting amplifier, there is no resistor connected because V_{IN} is applied directly into the very high input impedance of the op-amp. As a result, the noninverting amplifier circuit has an extremely high input impedance.

Troubleshooting Op-Amp Circuits (Figure 16-19)

21. Once you have diagnosed an op-amp circuit as malfunctioning, the next step is to isolate whether the problem is within the op-amp's internal circuit or external to the op-amp IC.

22. The isolating process is relatively easy because op-amp circuits only contain a handful of components and each component's fault produces its own distinct effect.

    a. If R_{IN} were to open, the input signal would be disconnected from the amplifier. Although this condition should result in no output, in most cases it sends the op-amp circuit into oscillation: if the output is positive when R_{IN} opens, a positive voltage will be fed back to the inverting input via R_F, this positive input will be inverted to a negative output, which will be fed back via R_F producing a positive output, which will be fed back via R_F producing a negative output, and so on. The result will be a continuous millivolt oscillation at the output.

    b. If R_F were to open, the feedback path would be disconnected and the gain of the op-amp circuit will increase to its maximum open loop value. Because the output voltage can never exceed the $+V$ and $-V$ supply voltages, the result will be a heavily distorted waveform that has its positive and negative peaks clipped when the op-amp is driven into saturation and cutoff.

    c. If the offset resistor were to open, the output of the op-amp will not be zero when the input is zero. As a result, the output waveform will have either its positive or negative peaks clipped, depending on whether the op-amp is unbalanced in the positive or negative direction.

23. If you have isolated the problem to the op-amp's internal circuit, and you have determined that all of the op-amp's externally connected resistors are okay, the next step will be to replace the op-amp itself because it is impossible to repair any internal op-amp failures.

Additional Op-Amp Circuit Applications (Figure 16-20)

24. The noninverting voltage follower circuit has a voltage gain of 1. Similar to the BJT's emitter-follower and the FET's source-follower, the op-amp voltage-follower circuit derives its name from the fact that the output voltage

QUICK REFERENCE SUMMARY SHEET

FIGURE 16-19 Troubleshooting Op-Amp Circuits.

Basic Inv-Amp Circuit

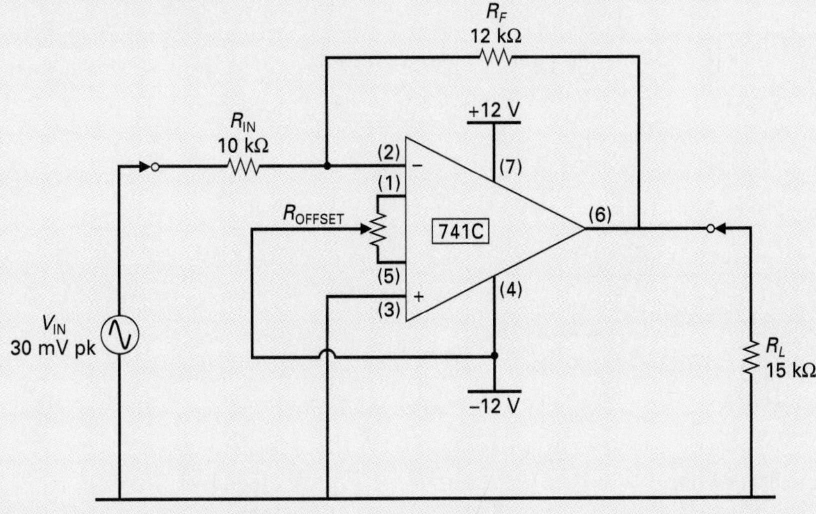

Effect on Output Due to Resistance Opens in Circuit

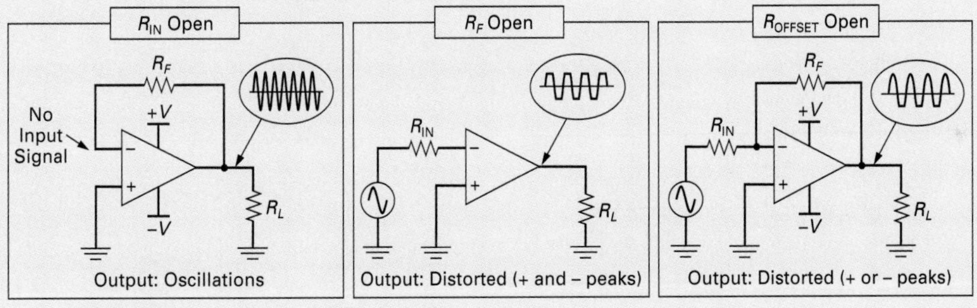

follows the input voltage in both polarity and amplitude. This circuit is ideal as a buffer for interfacing a high-impedance source to a low-impedance load.

25. The inverting summing amplifier circuit, or adder amplifier, will sum or add all of the input voltages.

26. With the difference or differential amplifier circuit, the output voltage is equal to the difference between the two input voltages.

27. The op-amp differentiator circuit (not to be confused with the previous differential circuit) is similar to the basic inverting amplifier except that R_{IN} is replaced by a capacitor. The reactance of the input capacitor will decrease for input signals that are higher in frequency, and therefore the input voltage applied to the op-amp and output voltage from the op-amp will

QUICK REFERENCE SUMMARY SHEET

FIGURE 16-20 Additional Op-Amp Circuit Applications.

Voltage-Follower Circuit

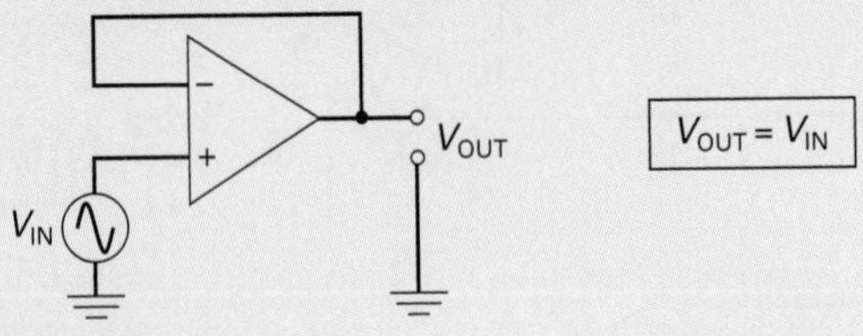

Summing Amplifier Circuit

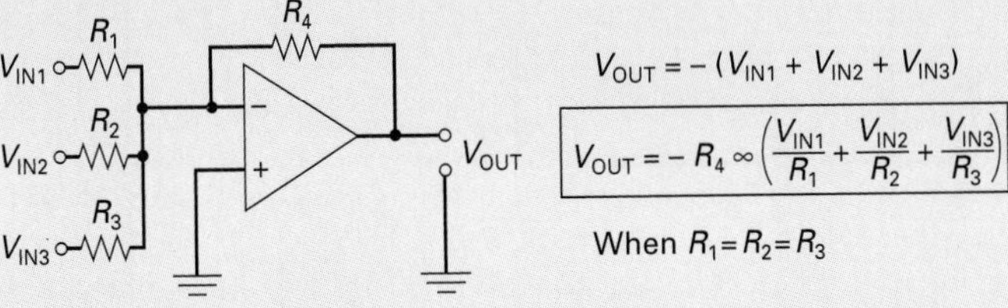

V_{IN1} R_1
V_{IN2} R_2
V_{IN3} R_3
R_4
V_{OUT}
$V_{OUT} = -(V_{IN1} + V_{IN2} + V_{IN3})$
When $R_1 = R_2 = R_3$
$V_{OUT} = -R_4 \infty \left(\frac{V_{IN1}}{R_1} + \frac{V_{IN2}}{R_2} + \frac{V_{IN3}}{R_3}\right)$

Difference Amplifier Circuit

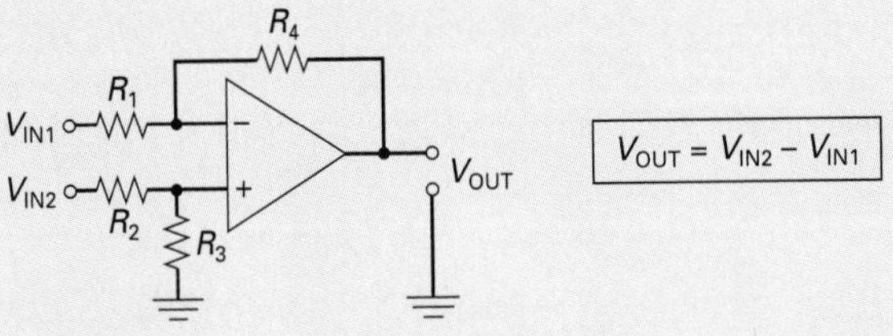

Differentiator Circuit

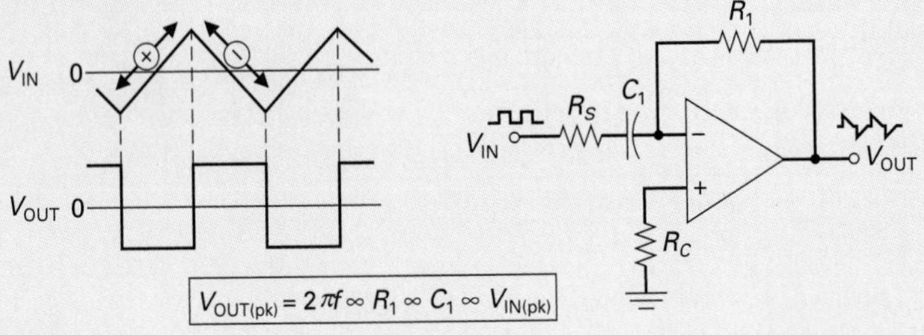

Integrator Circuit

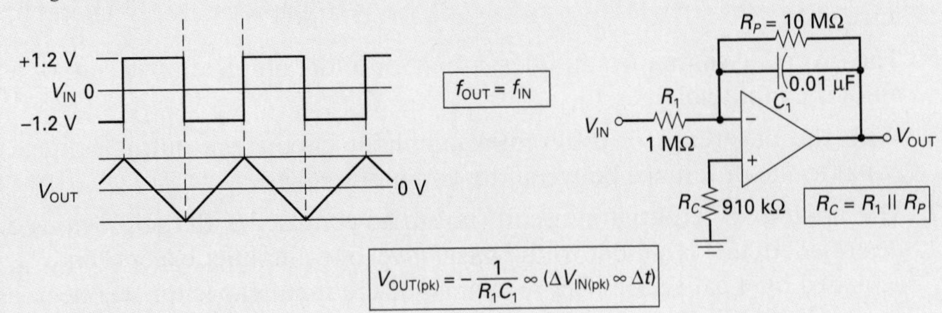

increase with frequency. Including an additional resistor (R_S) in series with the input capacitor will decrease the high-frequency gain, and therefore gain will be a ratio of R_F/R_S.

28. With the op-amp integrator circuit, a resistor is connected in the input and a capacitor is connected in the feedback path. Like the differentiator circuit, the capacitor will alter the gain of the op-amp because its capacitive reactance changes with frequency. To compensate for this effect, a parallel resistor (R_P) is included in shunt with the capacitor to decrease the low-frequency gain because gain will now be a ratio of R_P/R_1.

Op-Amp Signal Generator Circuits (Figure 16-21)

29. A signal generator is a circuit that will convert a dc supply voltage into a repeating output signal.

30. The twin-T sine-wave oscillator, which has two T-shaped feedback networks, will generate a repeating sine wave output.

31. With the square-wave op-amp generator, or relaxation oscillator, the output signal is fed back to both the inverting and noninverting inputs of the op-amp. The capacitor will charge and discharge through R, and this cycle will control the frequency of the output square wave.

32. The triangular-wave generator is simply an integrator circuit with its output frequency equal to the square wave input frequency.

33. When the switch is closed in the staircase-wave generator, the capacitor is bypassed and will not charge. On the other hand, when the switch is open, C_2 will be charged by each input cycle, producing equal output voltage steps.

Op-Amp Active Filter Circuits (Figure 16-22)

34. Passive filters are circuits that contain passive or nonamplifying components (resistors, capacitors, and inductors) connected in such a way that they will pass certain frequencies while rejecting others.

35. An active filter is a circuit that uses an amplifier with passive filter elements to provide frequency paths with rejection characteristics.

36. Active filters have several advantages over passive filters.

- **a.** Because the op-amp provides gain, the input signal passed to the output will not be attenuated, and therefore better response curves can be obtained.
- **b.** The high input impedance and low output impedance of the op-amp means that the filter circuit does not interfere with the signal source or load.
- **c.** Because active filters provide gain, resistors can be used instead of inductors, and therefore active filters are generally less expensive.

37. With the active high-pass filter, the input RC network will offer a high impedance to low frequencies resulting in a low voltage gain, while at high frequencies the RC network will have a low impedance causing a high voltage gain.

QUICK REFERENCE SUMMARY SHEET

FIGURE 16-21 Op-Amp Signal Generator Circuits.

Twin-T Sine-Wave Oscillator Circuit

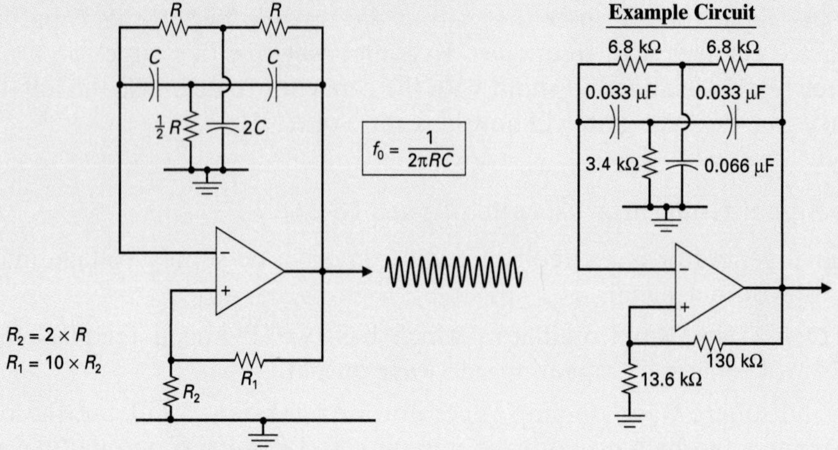

Square-Wave Generator Circuit

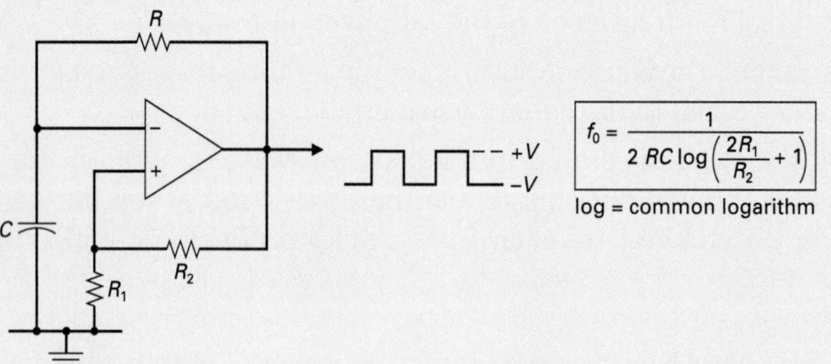

Triangle-Wave Generator Circuit **Staircase-Wave Generator Circuit**

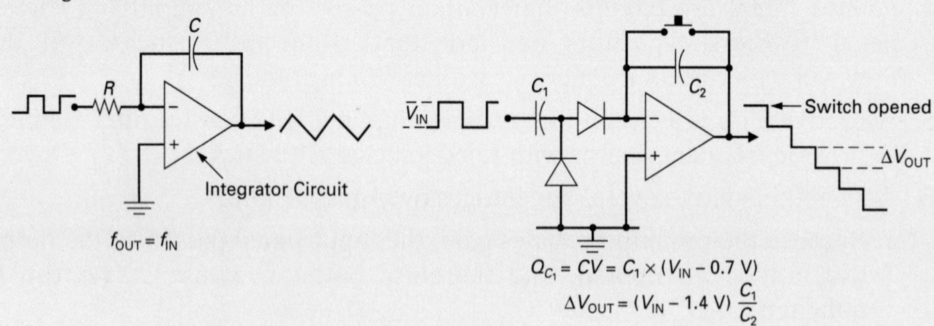

38. With the active low-pass filter, the capacitor's reactance is high at low frequencies, and therefore low-frequency signals will be passed to the op-amp's input to be amplified and passed to the output. As frequency increases, the capacitive reactance of C_1 will decrease, and therefore more of the signal will be shunted away from the op-amp and consequently not appear at the output.

QUICK REFERENCE SUMMARY SHEET

FIGURE 16-22 Op-Amp Active Filter Circuits.

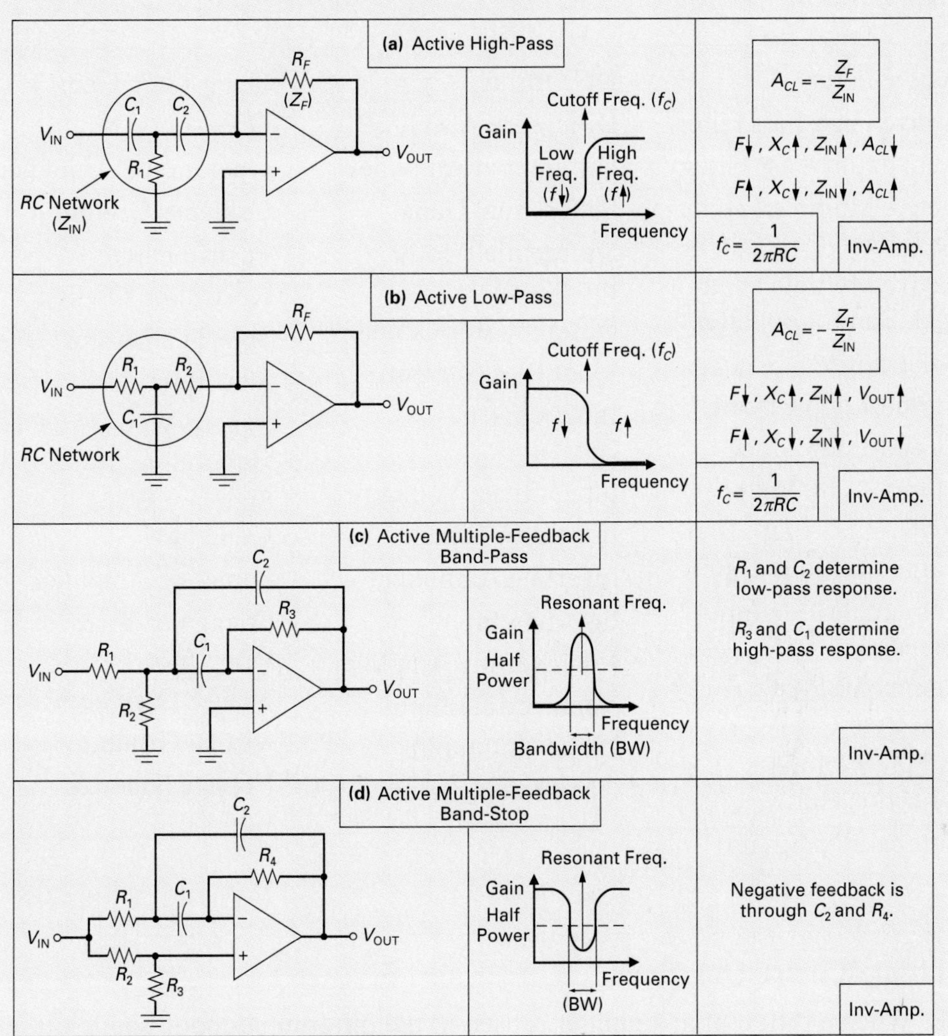

39. At frequencies outside of the band, V_{OUT} is fed back to the input without being attenuated, and therefore the input signal amplitude of the active band-pass filter is almost equal to the feedback signal amplitude. This results in almost complete cancellation of the signal and therefore a very small output voltage. On the other hand, for the narrow band of frequencies within the band, the feedback network will increase its amount of attenuation. This increase of attenuation means that a very small feedback signal will appear back at the negative input of the op-amp and will have a very small degenerative effect.

40. At frequencies outside of the band, the feedback signal will be heavily attenuated, and therefore the degenerative effect of an active band-stop filter will be small and the output voltage large. On the other hand, at frequencies within the band, the feedback signal will not be heavily attenuated, and therefore the degenerative effect will be large and the output voltage small.

NEW TERMS

Active Filter
Active High-Pass Filter
Active Low-Pass Filter
Active Multiple Feedback Band-Pass Filter
Active Multiple Feedback Band-Stop Filter
Balancing
Closed-Loop Circuit
Closed-Loop Mode
Closed-Loop Voltage Gain
Common-Mode Gain
Common-Mode Input Signals
Common Mode Rejection
Common-Mode Rejection Ratio
Comparator Circuit
DC Offsets
Difference Amplifier
Differential Amplifier
Differential Gain
Differential Mode Input Signals
Differentiator Circuit
Input Bias Current
Input Offset Current
Input Offset Voltage
Integrator Circuit
Inverting Amplifier
Inverting Input
Noninverting Amplifier
Noninverting Input
Offset Null Inputs
Open-Loop Circuit
Open-Loop Gain
Open-Loop Mode
Operational Amplifier
Ordinary Ground
Passive Filter
Relaxation Oscillator
Signal Generator
Slew Rate
Square-Wave Generator
Staircase Wave Generator
Summing Amplifier
Triangular Wave Generator
Twin-T Oscillator
Virtual Ground
Voltage Follower

REVIEW QUESTIONS

Multiple-Choice Questions

1. When a differential amplifier is used in the differential-input, single-output mode, it has a ____________ differential gain and a ____________ common mode gain.

a. High, high **c.** Low, low
b. High, low **d.** Low, high

2. The op-amp's internal circuit contains a ____________, ____________, and ____________ amplifier stage.

a. Differentiator, current, power
b. Integrator, voltage, output
c. Darlington-pair, emitter follower, summing
d. A differential, Darlington-pair, emitter follower

3. The op-amp's differential amplifier stage provides the op-amp with a

a. Low common mode gain **c.** High input impedance
b. High differential gain **d.** All of the above

4. Which transistor circuit is used in the op-amp's final output stage to provide a low output impedance and high current gain?

   a. Common emitter **c.** Common collector
   b. Common base **d.** Both (a) and (c)

5. Could an op-amp circuit be constructed using discrete components?

   a. Yes **b.** No

6. The comparator is considered a/an ______________ loop op-amp circuit.

   a. Common **c.** Differential
   b. Open **d.** Closed

7. What is the lower frequency limit of an op-amp?

   a. 20 Hz **b.** 6 Hz **c.** DC **d.** 7.34 Hz

8. A virtual ground is a ground to ______________ but not to ______________.

   a. Current, voltage **b.** Voltage, current

9. The feedback loop in a closed-loop op-amp circuit provides

   a. Positive feedback **d.** Both (a) and (b)
   b. Negative feedback **e.** Both (b) and (c)
   c. Degenerative feedback

10. The ______________ input(s) of an op-amp is used to compensate for slight differences in the transistors in the differential amplifier stage.

    a. Inverting **b.** DC offset **c.** $+V$ **d.** V_{OUT}

Essay Questions

11. Why is the term "operational" used to describe the op-amp? (Intro.)
12. Sketch the op-amp schematic symbol. (16-1-1)
13. Describe some of the different op-amp package types and the meaning of the manufacturer codes. (16-1-2)
14. What are the three basic amplifier blocks within an op-amp? (16-2)
15. In what mode is the differential amplifier used within the op-amp? (16-2-1)
16. What is the difference between common-mode input signals and differential-mode input signals? (16-2-1)
17. Define "common-mode rejection ratio" in relation to the op-amp. (16-2-2)
18. Sketch the basic block diagram of an op-amp's internal circuit, and list the characteristics of each block. (16-2-3)
19. What is the difference between an open-loop and closed-loop op-amp circuit? (16-2-5)
20. Sketch a simple op-amp comparator circuit and briefly describe its operation. (16-2-5)
21. Why is it necessary for an op-amp to have negative feedback? (16-2-5)
22. Sketch a simple inverting operational amplifier circuit, and briefly describe its operation. (16-2-5)
23. How is the gain, and therefore output voltage, of an inverting op-amp circuit determined? (16-2-5)

24. Sketch a simple noninverting operational amplifier circuit, and briefly describe its operation. (16-2-5)
25. How is the gain, and therefore output voltage, of a noninverting op-amp circuit determined? (16-2-5)
26. What is the difference between a virtual ground and an ordinary ground? (16-2-5)
27. What is the difference between a 741 and 747 op-amp IC? (16-2-6)
28. Sketch and describe the operation of the following op-amp application circuits:
    a. Voltage follower circuit (16-3-1)
    b. Summing amplifier circuit (16-3-2)
    c. Difference amplifier circuit (16-3-3)
    d. Differentiator circuit (16-3-4)
    e. Integrator circuit (16-3-5)
29. Sketch a circuit showing how an op-amp can be connected to operate as a sine-wave, square-wave, and triangular-wave generator. (16-3-6)
30. What is a staircase generator circuit? (16-3-6)
31. What is the difference between a passive filter and an active filter, and what are the advantages of active filters? (16-3-7)
32. Show how the op-amp can be used as an active filter by sketching an example circuit and explaining its operation. (16-3-7)

Practice Problems

33. Explain the meaning of the following op-amp manufacturer codes:
    a. LM 318C N **b.** NE 101C D
34. Calculate the common-mode rejection ratio of an op-amp if it has a common-mode gain of 0.8 and a differential gain of 27,000. Also give the answer in dBs.
35. Identify the circuit shown in Figure 16-23. Why must the input to this circuit always be a negative voltage?
36. What would be the output voltage from the circuit in Figure 16-23 if the input voltage were –1.6 V?
37. Identify the circuits shown in Figure 16-24(a) and (b), and then sketch the shape of the output waveform if a square wave were applied to the inputs.

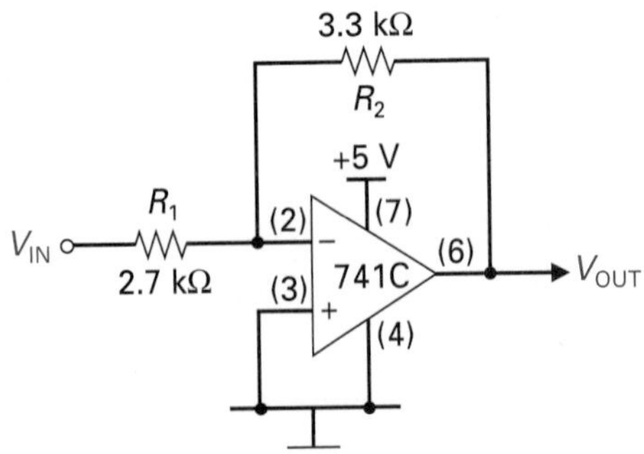

FIGURE 16-23 A 741C Op-Amp Circuit.

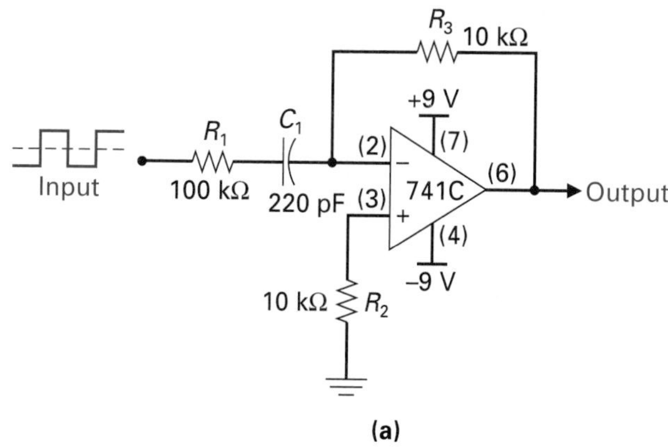

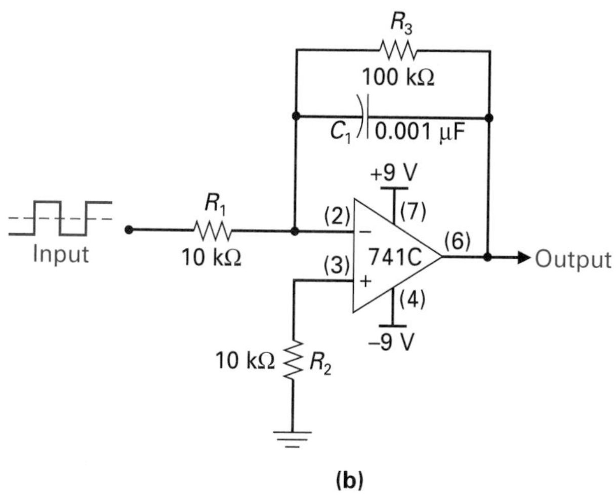

FIGURE 16-24 Two Applications for the 741C Op-Amp.

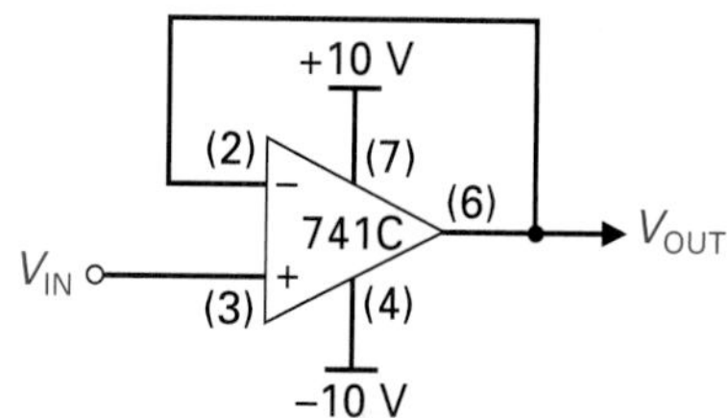

FIGURE 16-25 An Op-Amp Circuit.

38. Referring to the circuit shown in Figure 16-25, what would be the voltage out if a +7.3 V input were applied?
39. Identify the circuits shown in Figure 16-26(a) and (b), and then calculate the output voltages for the given input voltages.
40. Which of the circuits shown in Figure 16-27 is an active high-pass filter and which is an active low-pass filter?
41. Calculate the cutoff frequency of the circuit in Figure 16-27(a) if $C_1 = C_2 = 0.1\ \mu\text{F}$, $R_1 = 10\ \text{k}\Omega$.
42. Calculate the cutoff frequency of the circuit in Figure 16-27(b), if $R_1 = R_2 = 33\ \text{k}\Omega$, $C_1 = 0.33\ \mu\text{F}$.
43. Identify the circuit shown in Figure 16-28 and calculate its closed-loop voltage gain.
44. What would be the output voltage from the circuit in Figure 16-28 for the input voltage given?

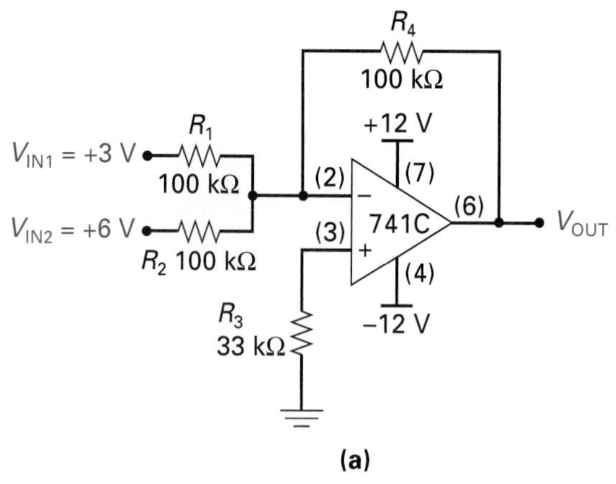

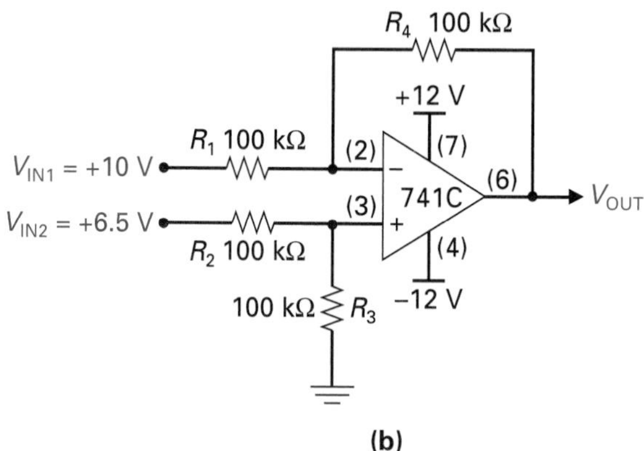

FIGURE 16-26 Op-Amp Circuit Examples.

Troubleshooting Questions

45. If the circuit in Figure 16-28 were not functioning as it should, how would you determine whether the problem is (16-2-7)

a. In the source
b. In the load
c. In the circuit supply voltages

46. If the following faults were introduced to the circuit in Figure 16-28, what would be the circuit symptoms? (16-2-7)

a. R_{IN} open
b. R_F open
c. R_{OFFSET} open

47. As an exercise in troubleshooting, construct the circuit shown in Figure 16-28. Then introduce some of the following circuit problems, note the symptoms, and logically explain why a cause has a certain effect.

a. Disconnect one of the inputs to the op-amp.
b. Disconnect one of the supply voltages.
c. Adjust the offset resistor to one extreme end of its track.
d. Disconnect the load.

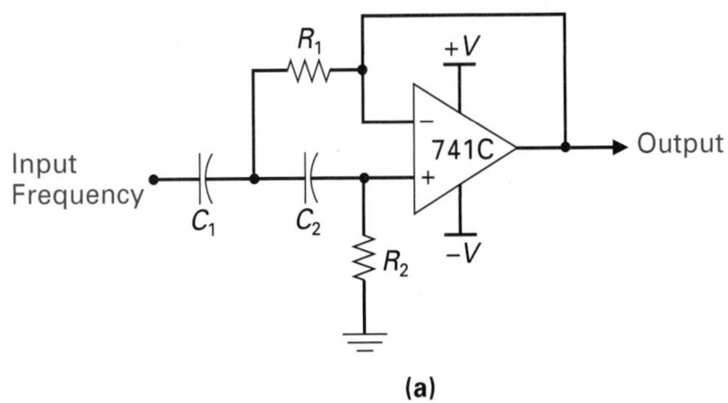

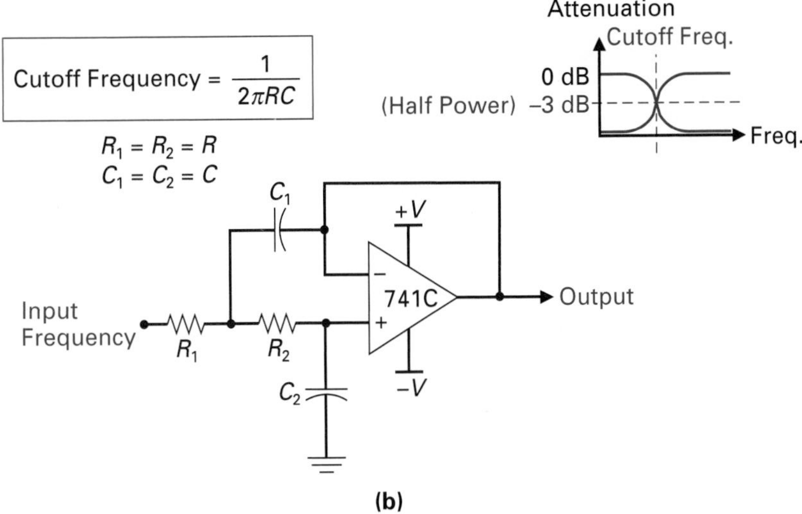

FIGURE 16-27 Op-Amp Active Filter Circuits.

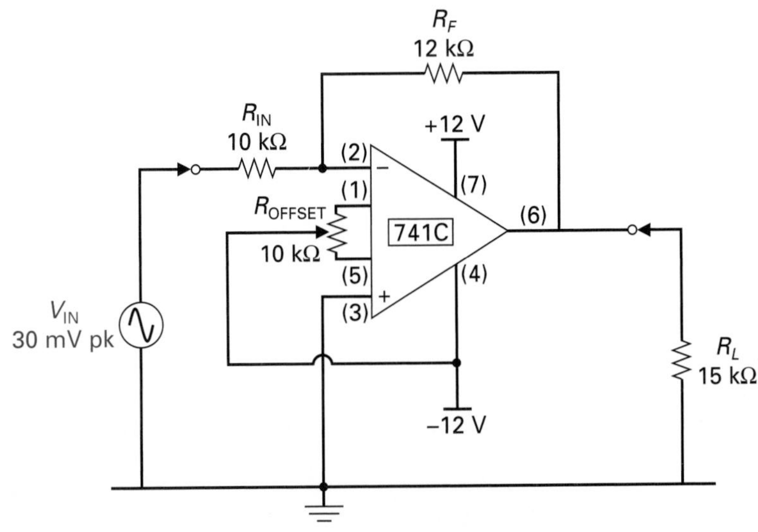

FIGURE 16-28 Op-Amp Amplifier Circuit.

CHAPTER

17

OBJECTIVES

After completing this chapter, you will be able to:

1. Define the difference between a linear IC and a digital IC.
2. List some of the frequently used linear integrated circuits.
3. Define voltage regulation and percent of regulation.
4. Explain the voltage regulator's basic action.
5. List the three types of IC voltage regulators, and describe their operation and application.
6. Describe the difference between series dissipated regulators and series switching regulators.
7. Explain the operation and application of a voltage controlled oscillator and function generator IC.
8. Describe the operation and application of phase-locked loop integrated circuits.

Linear Circuits

O U T L I N E

VIGNETTE

Don't Mention It!

Two years before the Japanese attacked Pearl Harbor, IBM president Thomas J. Watson decided to invest half a million dollars of the company's money in a project conceived by a young Harvard mathematician named Howard Aiken. Aiken's plan was to ignore the architecture of the sorters and calculators that were then available and concentrate on building a machine based on the nineteenth century concepts of Charles Babbage. Using Babbage's original description of his "analytical engine," Aiken set about to build a general-purpose programmable computer. In fact, Aiken stated at the time that "If Babbage had lived seventy-five years later, I would be out of a job."

The Second World War hampered Aiken's development of his machine because he was called to active duty in the navy soon after the attack on Pearl Harbor. In the navy, Aiken distinguished himself by disarming a new type of German torpedo single-handed. Meanwhile, at IBM, Watson did not appreciate the interference of the war and used his influence to get Aiken back on his computer project, arguing that it could greatly help the war effort.

In 1943, the Mark I computer was turned on and underwent its first successful test. It measured 51 feet long and 8 feet high, contained 750,000 devices, and 500 miles of wire. The 3,304 electromechanical relays clicked open and close, and as one observer remarked, "The machine sounded like a room full of old ladies knitting away with steel needles." At this point, personality clashes began to erupt between the Mark I's inventor and the financial backer. Aiken wanted the Mark I's internal circuitry exposed so that scientists could examine it, while Watson wanted the machine encased in glass and stainless steel. Watson won on this point; however, Aiken retaliated by not even mentioning IBM or Watson's involvement in the project when the Mark I was introduced to the press in 1944.

The Mark I was leased to the navy and used to calculate the trajectory of cannon shells. It could add or subtract 23-digit numbers in a few tenths of a second, and multiply them in three seconds. This phenomenal speed—which was only slightly faster than Babbage had imagined 122 years earlier—enabled the navy, in a single day, to speed through calculations that previously took six months.

After the war Aiken stayed at Harvard to develop the second, third, and fourth generations of the Mark I, and had nothing further to do with Watson. Tom Watson was so angered by Aiken's failure to acknowledge him and IBM that he ordered all of his researchers to construct a better and faster system. Revenge launched IBM into the computer business and, when Watson died at the age of 82 in 1956, he had seen his wish fulfilled. IBM had overtaken all opposition, and the name had become so synonymous with computers that most people believed that IBM had invented them.

INTRODUCTION

In 1963, Fairchild Semiconductor introduced the μA702, which was the first commercially available IC op-amp. Realizing the op-amp's ability to adapt to a wide variety of applications, National Semiconductor followed suit and introduced the LM101 op-amp in 1965. In 1967, Fairchild once again made history by unveiling the very popular μA741 op-amp IC, which is still widely used today. From that point on, improvements in technology have refined the op-amp and led to a variety of other spin-off ICs that internally include several op-amps along with their other circuitry. These ICs include "voltage regulators," "timers," "function generators," "phase-locked loops," "analog-to-digital (A/D) converters," "digital-to-analog (D/A) converters," "sample-and-hold amplifiers," and a variety of other specialized ICs. These ICs, along with the op-amp, are collectively called "linear integrated circuits (linear ICs)" because they provide linear signal amplification, in contrast to "digital integrated circuits (digital ICs)" that are primarily used for pulse signal processing.

As you continue your course in electronics, you will see how these ICs can be made use of in a variety of circuit applications. In the previous chapter, we examined the most popular of the linear ICs, the op-amp, and saw how it could be used as a comparator, inverting or noninverting amplifier, signal processor, signal generator, and active filter. In this chapter we will continue our coverage of linear circuits by discussing the operation and application of three frequently used linear IC types: "voltage regulator ICs," "function generator ICs," and "phase-locked loop ICs." As with the op-amp IC, we will be concentrating on the input/output relationships and applications of these linear ICs. These operational characteristics are important if we are going to be able to isolate whether a circuit malfunction is internal or external to the linear IC. However, because it is impossible to repair any internal linear IC failures, we will not need to cover every detail of the device's internal circuit.

17-1 VOLTAGE REGULATOR CIRCUITS

Previously in Chapter 5, we discussed the operation of dc power-supply circuits. Because the linear **voltage regulator IC** is primarily used within dc power-supply circuits, let us begin by first reviewing the operation of the dc power supply.

Voltage Regulator IC
An integrated circuit that maintains the output voltage of a voltage source within required limits despite variations in input voltage and load resistance.

17-1-1 DC Power-Supply Circuit Review

Figure 17-1(a) shows the block diagram of a dc power supply. Because the final output voltage is generally not 120 V, a transformer is usually included to step the ac line voltage up or down to a desired value. The dc power supply's loads, which are electronic circuits, generally require low-voltage dc supply voltages such as 12 V and 5 V, and a step-down transformer is normally used. For example, a step-down transformer with a turns ratio of 10:1 would reduce the 120 V ac input to a 12 V ac output. The output current capability of this transformer will be 1:10, which will be ideal since most electronic circuits require low supply voltages with high current capacity.

As shown by the waveforms in Figure 17-1(a), the rectifier converts the stepped-down ac input from the transformer to a pulsating dc output. This pulsating dc could not be used to power an electronic circuit because of the continuous changes between zero volts and a peak voltage. The filter smoothes out the pulsating dc ripples into a constant dc level, as seen in the waveform following the filter.

The final block is called a voltage regulator, and although there appears to be no difference between the regulator's input and output waveforms, this device performs a very important function. The regulator maintains the dc output voltage from power supply constant, or stable, despite variations in the ac input voltage or variations in the output load resistance.

As an example, Figure 17-1(b) shows the schematic diagram for a dc power supply circuit that can generate a +5 V, +12 V, and –12 V dc output from a 120 V or 240 V ac input. You should recognize most of the pieces of this picture because this circuit contains nearly all of the devices discussed previously in Chapter 5.

The power supply ON/OFF switch (SW_1) switches a 120 V ac input to the 120 V primary winding connection, or a 240 V ac input to the 240 V primary winding connection of the transformer (T_1). By doubling the number of primary turns, we can accept twice the input voltage and deliver the same output voltage. The upper secondary winding of T_1 supplies a bridge rectifier module (BR_1), that generates a +12 V peak full-wave pulsating dc output, which is filtered by C_1 and regulated by the 7805 (U_1) to produce a +5 V dc 750 mA output. The +5 V output will turn on a "power ON" LED (D_3), which is mounted on the power-supply printed circuit board (PCB) along with the filter and regulator. Resistor R_1 is a current-limiting resistor for D_3.

The lower secondary winding of T_1 supplies the two half-wave rectifier diodes D_1 and D_2, which generate –50 V and +50 V peak outputs, respectively. These inputs are filtered by C_2 and C_3, and then regulated by a 7912 (U_2), which generates a –12 V dc 500 mA output, and by a 7812 (U_3), which generates a +12 V dc 500 mA output. The devices C_2, C_3, U_2, and U_3 are also all mounted on the power supply printed circuit board (PCB).

17-1-2 *The Voltage Regulator's Function*

As previously mentioned, the voltage regulator in a dc power supply maintains the dc output voltage constant despite variations in the ac input voltage and the output load resistance. To help explain why these variations occur, let us first examine the relationship between a source and its load.

Ideally, a dc power supply should convert all of its *ac electrical energy input* into a *dc electrical energy output.* However, like all devices, circuits, and systems, a dc power supply is not 100% efficient and along with the electrical energy output, the dc power supply generates wasted heat. This is why a dc power supply can be represented as a voltage source with an internal resistance (R_{INT}), as shown in Figure 17-2(a). The internal resistance, which is normally very small (in this example, 1 Ω), represents the heat energy loss or inefficiency of the dc power supply. The load resistance (R_L) represents the resistance of the elec-

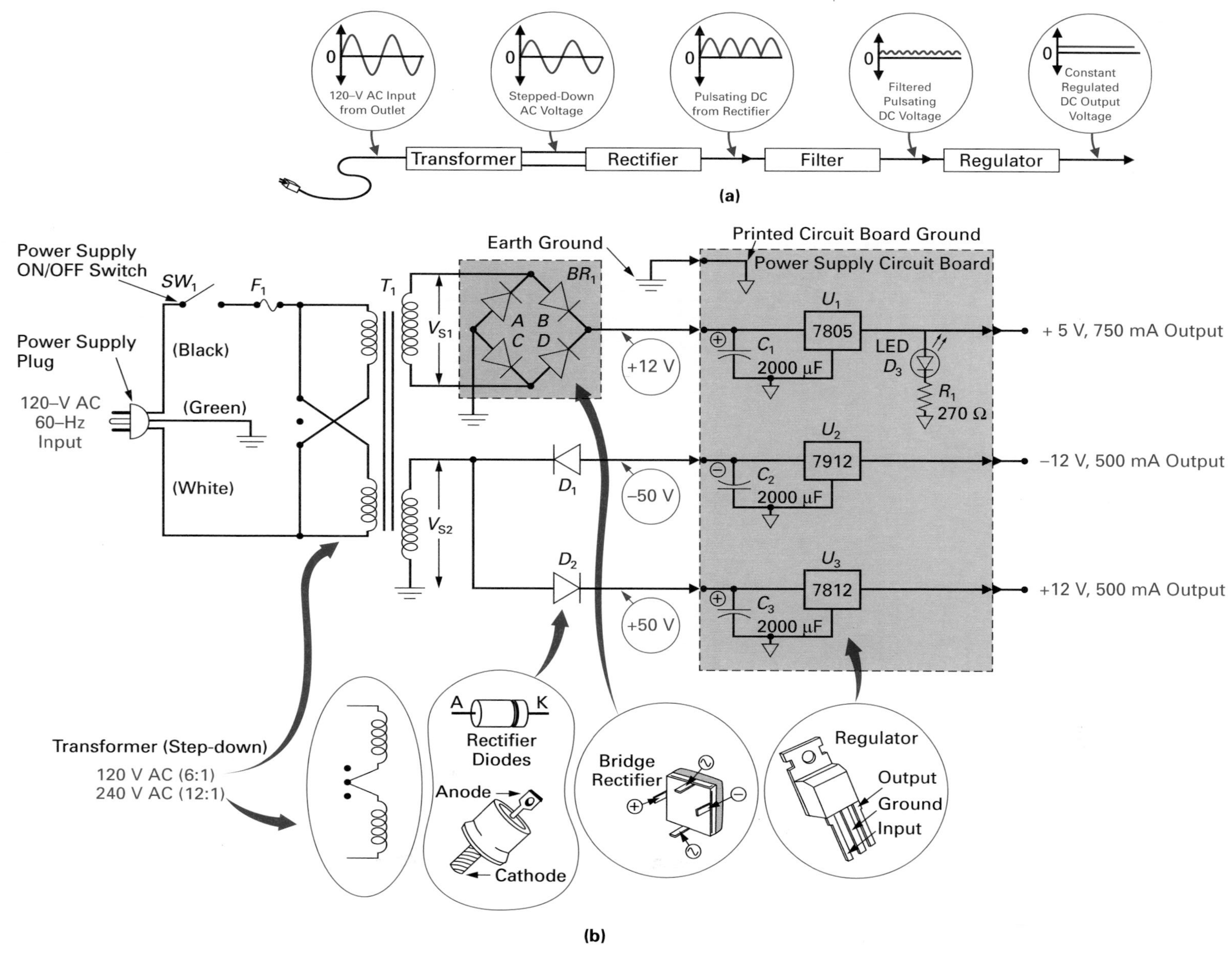

FIGURE 17-1 **DC Power-Supply Circuit. (a) Block Diagram. (b) Circuit Diagram.**

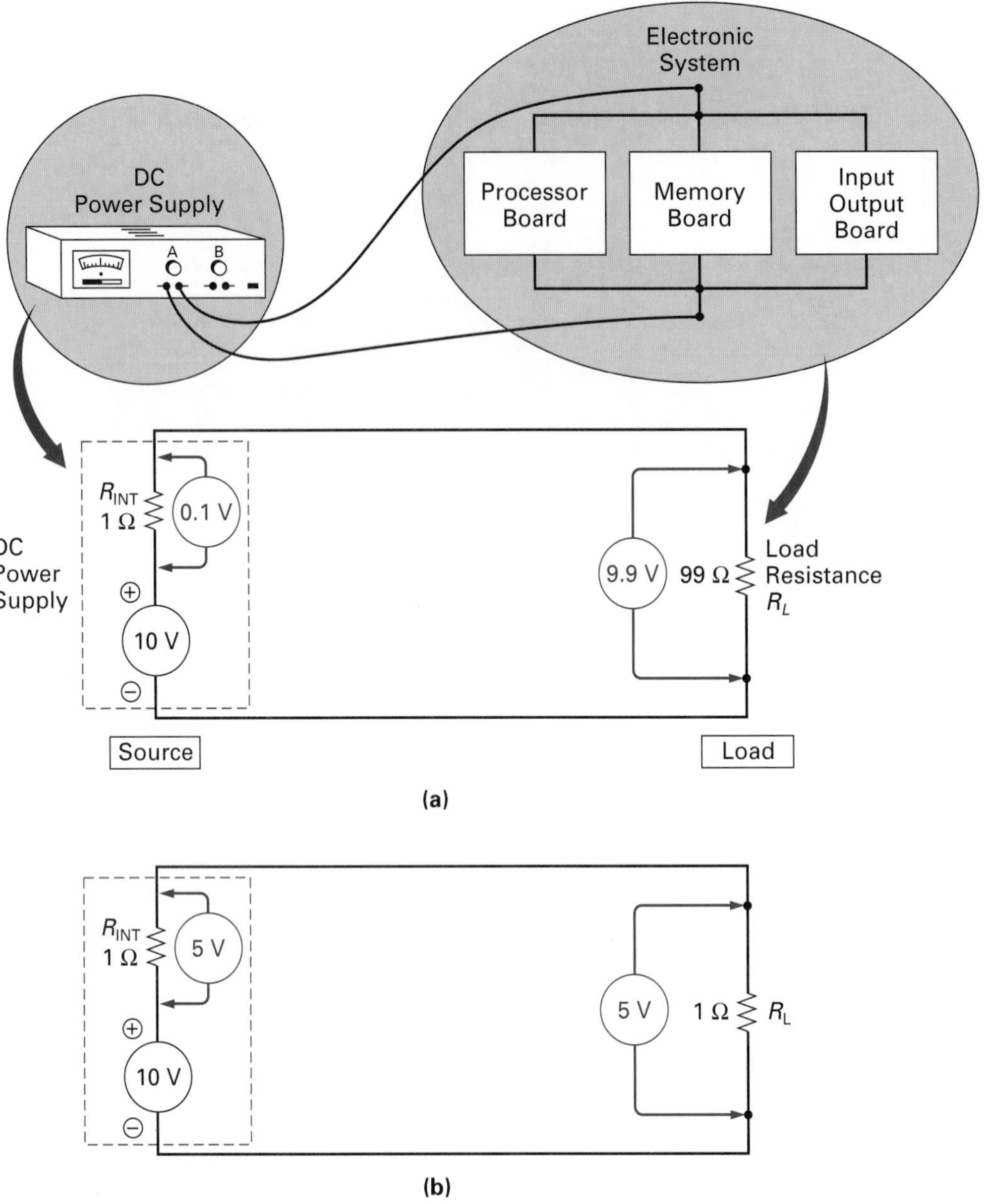

FIGURE 17-2 Source and Load.

tronic circuits in the electronic system. Because R_{INT} and R_L are connected in series with one another, the 10 V supply will be developed proportionally across these resistors, resulting in

$$\begin{aligned} V_{R_{INT}} &= \ \ 0.1\text{ V} \\ V_{R_L} &= \ \underline{\ 9.9\text{ V}\ } \\ &\quad\ 10\text{ V} \quad \text{Total} \end{aligned}$$

If the load resistance (R_L) were to now decrease dramatically to 1 Ω, as shown in Figure 17-2(b), the decrease in load resistance ($R_L\downarrow$) would cause an increase in load current ($I_L\uparrow$). This current increase would cause an increase in the heat generated by the power supply ($P_{R_{INT}}\uparrow = I^2\uparrow \times R$). This loss is seen at the output of the dc power supply, which is now delivering only a 5 V output to the load, because the 10 V source voltage is now being divided equally across R_{INT} and R_L.

$$\begin{aligned} V_{R_{INT}} &= 5\text{ V} \\ V_{R_L} &= \underline{5\text{ V}} \\ & \quad 10\text{ V} \quad \text{Total} \end{aligned}$$

If the load resistance decreases, the load current will increase, and if too much current is drawn from the source, it will pull down the source voltage.

Although a load resistance will generally not change as dramatically as shown in Figure 17-2(b), it will change slightly as different electronic system control settings are selected, and this will *load the source.* For example, the load resistance of the electronic circuits in your music system will change as the volume, bass, treble, and other controls are adjusted. This load resistance change would affect the output of a dc power supply if a regulator were not included to maintain the output voltage constant despite variations in load resistance.

To the electric company, each user appliance is a small part of its load, and it is the source. The ac voltage from the electric company can be anywhere between 105 V and 125 V ac rms (148 to 177 V ac peak), depending on consumer use, which depends on the time of day. When many appliances are in use, the load current will be high and the overall load resistance low, causing the source voltage to be pulled down. If a dc power supply did not have a regulator, these different input ac voltages from the electric company would produce different dc output voltages, when we really want the dc supply voltage to the electronic circuits to always be the same value no matter what ac input is present. This is the other reason that a regulator is included in a dc power supply: it maintains the dc output voltage constant despite variations in the ac input voltage.

In summary, a regulator is included in a dc power supply to maintain the dc output voltage constant despite variations in the output load resistance and the ac input voltage, as shown in Figure 17-3.

17-1-3 Percent of Regulation

The percent of regulation is a measure of the regulator's ability to regulate, or maintain constant, the output dc voltage. It is calculated with the formula

$$\text{percent regulation} = \frac{V_{nl} - V_{fl}}{V_{nl}} \times 100$$

where V_{nl} = no-load voltage, and V_{fl} = full-load voltage.

If we apply this formula to the example in Figure 17-2, the percent of regulation provided by the dc voltage source would be

$$\begin{aligned} \%\text{ regulation} &= \frac{V_{nl} - V_{fl}}{V_{nl}} \times 100 \\ &= \frac{10\text{ V} - 5\text{ V}}{10\text{ V}} \times 100 = 50\% \end{aligned}$$

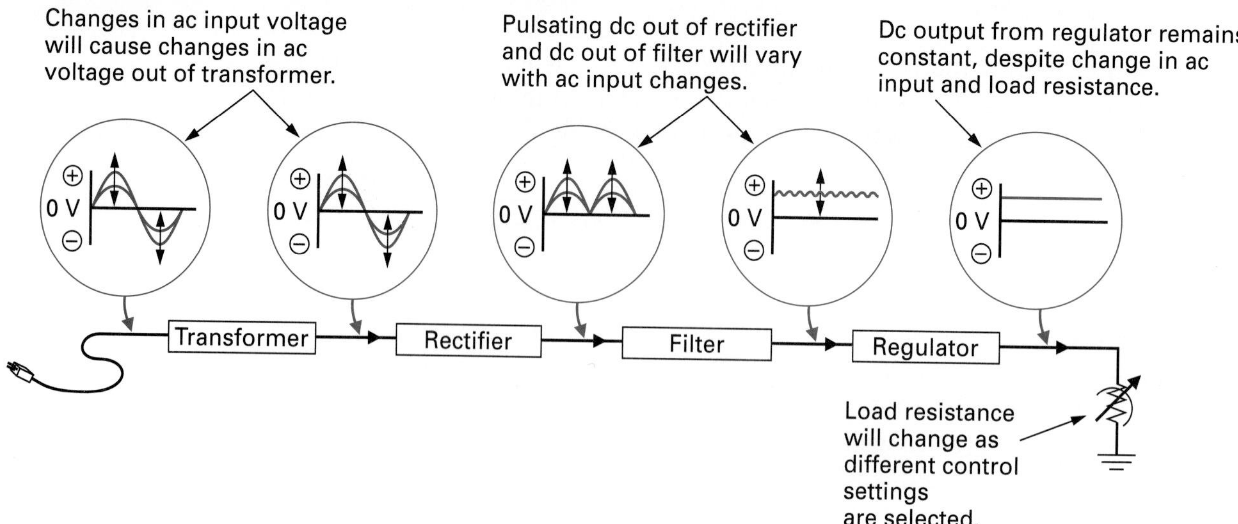

Changes in ac input voltage will cause changes in ac voltage out of transformer.
Pulsating dc out of rectifier and dc out of filter will vary with ac input changes.
Dc output from regulator remains constant, despite change in ac input and load resistance.
0 V
0 V
0 V
0 V
0 V
Transformer
Rectifier
Filter
Regulator
Load resistance will change as different control settings are selected.

FIGURE 17-3 Function of a Regulator.

In the ideal situation, a regulator would maintain the output voltage constant between no-load and full-load, resulting in

$$\% \text{ regulation} = \frac{V_{nl} - V_{fl}}{V_{nl}} \times 100$$

$$= \frac{10\text{ V} - 10\text{ V}}{10\text{ V}} \times 100$$

$$= 0\%$$

Most regulators achieve a percent regulation figure that is not perfect (0%) but is generally in single digits (2% to 8%, typically).

17-1-4 The Zener Regulator

In Chapter 4 it was shown how the zener diode could be used as a voltage regulator. Figure 17-4 shows how a zener diode (D_5) and a series resistor (R_1) would be connected in a dc power supply circuit to provide regulation. An increase in the ac input voltage would cause an increase in the dc output from the filter, which would cause an increase in the current through the series resistor (I_S), the zener diode (I_Z), and the load (I_{R_L}). The reverse-biased zener diode will, however, maintain a constant voltage across its anode and cathode (in this example, 12 V) despite these input voltage and current variations, with the additional voltage being dropped across the series resistor. For example, if the output from the filter were between +15 V and +20 V, the zener diode would always drop 12 V, while the series resistor would drop between 3 V (when input is 15 V) and 8 V (when input is 20 V). Consequently, the zener regulator will maintain a constant output voltage despite variations in the ac input voltage.

From the other standpoint, the zener regulator will also maintain a constant output voltage despite variations in load resistance. The zener diode achieves this by increasing and decreasing its current (I_Z) in response to load resistance changes.

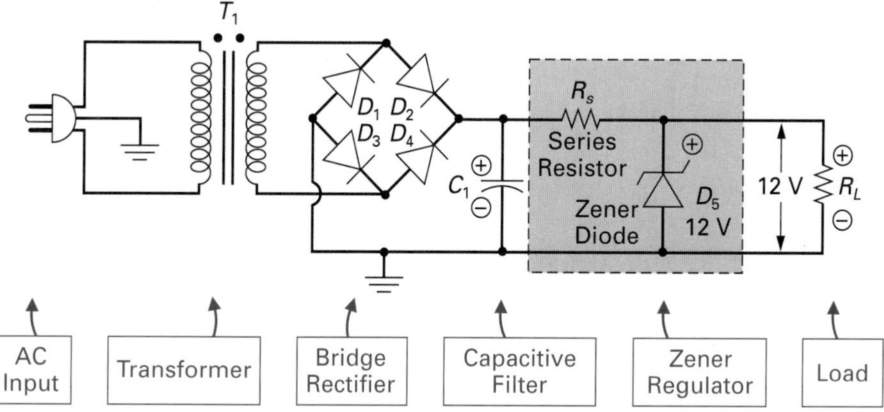

FIGURE 17-4 Zener Diode Regulator.

Despite these changes in zener and load current, however, the zener voltage (V_Z), and therefore the output voltage (V_{OUT} or V_{R_L}), always remains constant. The disadvantage with this regulator is that the series connected resistor will limit load current and, in addition, generate and waste a large amount of power.

17-1-5 IC Voltage Regulators

Most dc power supply circuits today make use of linear integrated circuit (IC) voltage regulators, like the one shown in Figure 17-5. These IC regulators contain about 50 individual or discrete components all integrated together on one silicon semiconductor chip and then encapsulated in a three-pin package. The inset in Figure 17-5 shows the basic block diagram of an IC voltage regulator's internal circuit. The regulator's short-circuit protection and thermal shutdown internal circuits will protect the IC regulator by turning it OFF if the load current drawn exceeds the regulator's rated current—or if the heat sink is too small and the IC regulator is generating heat faster than it can dissipate it.

Although voltage regulators can be constructed using op-amps and other discrete components, complete IC voltage regulators are relatively inexpensive, smaller, and more efficient. Linear voltage regulator ICs can be categorized into one of three basic groups

1. Fixed Output Voltage Regulators
2. Variable Output Voltage Regulators
3. Switching Voltage Regulators

Let us now examine each of these three types in detail, beginning with the fixed output voltage regulator ICs.

Fixed Output Voltage Regulators

As the name implies, a **fixed output voltage regulator** will produce a regulated output voltage that is not variable. These fixed output voltage regulators can be broken down into two basic groups: those producing a positive voltage and those producing a negative output voltage. For example, the 7800 (seventy-

Fixed Output Voltage Regulator
A voltage regulator that produces a regulated voltage that is not variable.

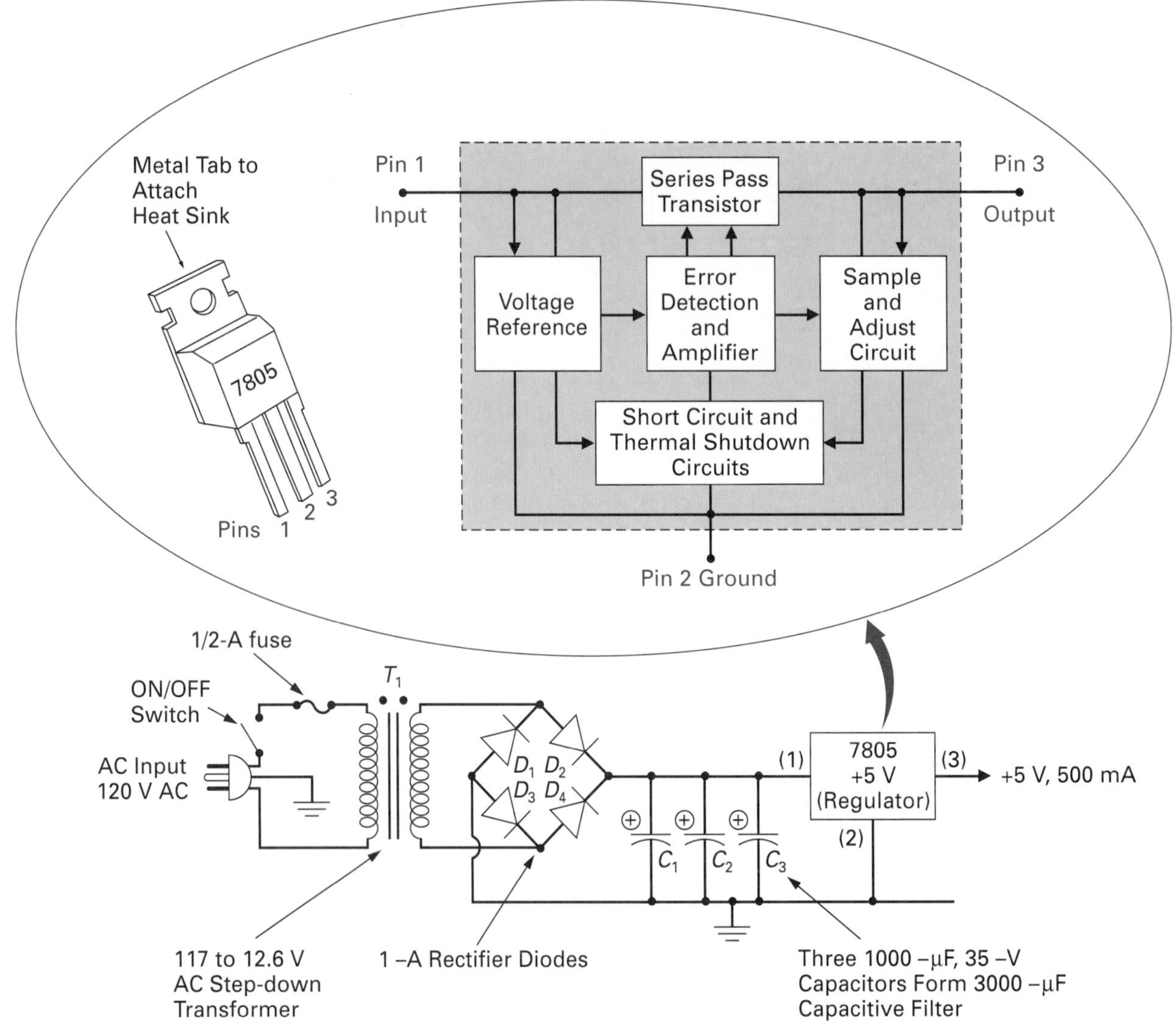

FIGURE 17-5 Integrated Circuit Regulator within a Power Supply Circuit.

eight hundred) series of voltage regulators are detailed in the data sheet in Figure 17-6, and as shown in the "type number/voltage" block, there are seven positive output voltage options. On the other hand, the 7900 series of voltage regulators detailed in the data sheet in Figure 17-7 has nine negative output voltage options. These regulators will deliver a constant regulated output voltage, as long as the input voltage is greater than the regulator's rated output voltage. They can also deliver a maximum output current of up to 1.5 A if properly heat sunk. The three terminals of the IC regulator are labeled input, output, and ground. The package is generally given the identification code of either 78XX or 79XX. The 78XX series of regulators are used to supply a positive output voltage, with the last two digits specifying the output voltage (for example, 7805 = +5 V). On the other hand, the 79XX series of regulators are used to supply a negative output voltage, with the last two digits, once again, specifying the output voltage (for example, 7912 = –12 V).

DEVICE: MC7800 Series—Positive Fixed Voltage Regulators

Three-Terminal Positive Voltage Regulators

These voltage regulators are monolithic integrated circuits designed as fixed-voltage regulators for a wide variety of applications including local, on-card regulation. These regulators employ internal current limiting, thermal shutdown, and safe-area compensation. With adequate heatsinking they can deliver output currents in excess of 1.0 A. Although designed primarily as a fixed voltage regulator, these devices can be used with external components to obtain adjustable voltages and currents.

- Output Current in Excess of 1.0 A
- No External Components Required
- Internal Thermal Overload Protection
- Internal Short Circuit Current Limiting
- Output Transistor Safe-Area Compensation
- Output Voltage Offered in 2% and 4% Tolerance

MC7800 Series

THREE-TERMINAL POSITIVE FIXED VOLTAGE REGULATORS

SEMICONDUCTOR TECHNICAL DATA

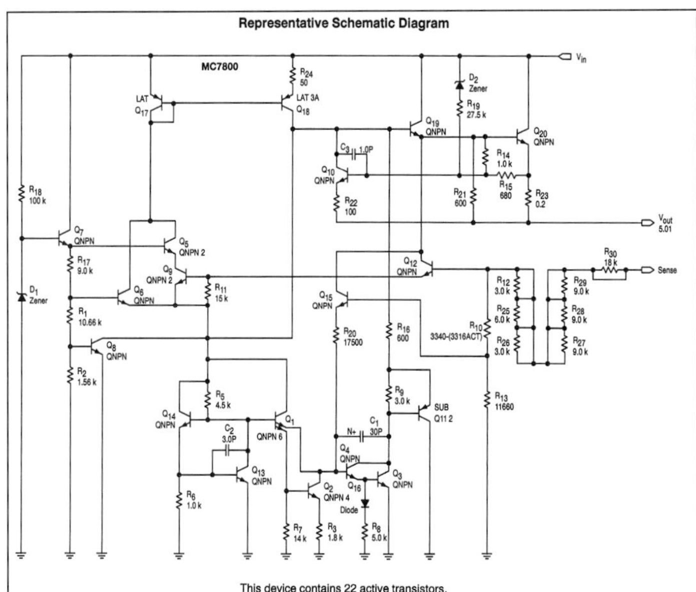

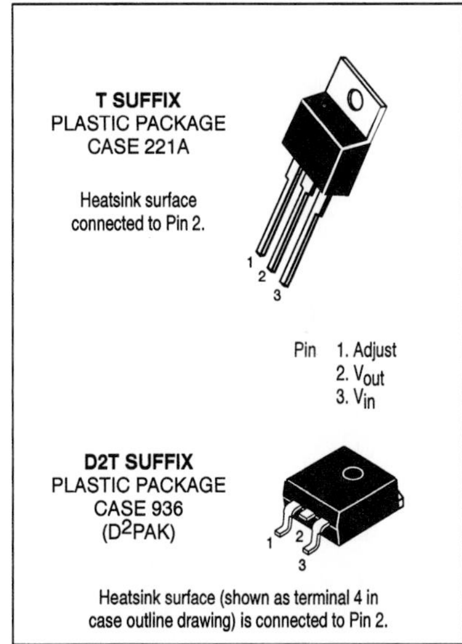

STANDARD APPLICATION

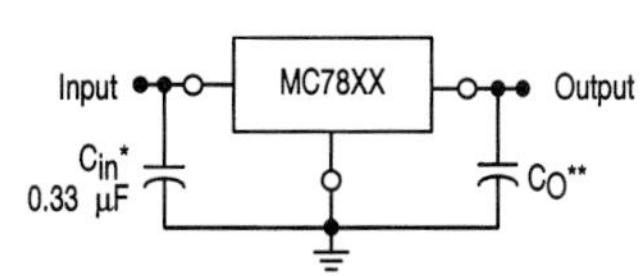

A common ground is required between the input and the output voltages. The input voltage must remain typically 2.0 V above the output voltage even during the low point on the input ripple voltage.

XX, these two digits of the type number indicate nominal voltage.

* C_{in} is required if regulator is located an appreciable distance from power supply filter.

** C_O is not needed for stability; however, it does improve transient response. Values of less than 0.1 µF could cause instability.

ORDERING INFORMATION

Device	Output Voltage Tolerance	Tested Operating Temperature Range	Package
MC78XXACT	2%	T_J = 0° to +125°C	Insertion Mount
MC78XXACD2T			Surface Mount
MC78XXCT	4%		Insertion Mount
MC78XXCD2T			Surface Mount
MC78XXBT		T_J = – 40° to +125°C	Insertion Mount
MC78XXBD2T			Surface Mount

XX indicates nominal voltage.

DEVICE TYPE/NOMINAL OUTPUT VOLTAGE

MC7805	5.0 V	MC7812	12 V
MC7806	6.0 V	MC7815	15 V
MC7808	8.0 V	MC7818	18 V
MC7809	9.0 V	MC7824	24 V

Last two digits indicate output voltage. For example, 7808 will produce a +8 V regulated output.

FIGURE 17-6 Positive Voltage Regulator Data Sheet. (Copyright of Motorola. Used by permission.)

DEVICE: MC7900C Series—Negative Fixed Voltage Regulators

Three-Terminal Negative Voltage Regulators

The MC7900 series of fixed output negative voltage regulators are intended as complements to the popular MC7800 series devices. These negative regulators are available in the same seven-voltage options as the MC7800 devices. In addition, one extra voltage option commonly employed in MECL systems is also available in the negative MC7900 series.

Available in fixed output voltage options from −5.0 V to −24 V, these regulators employ current limiting, thermal shutdown, and safe-area compensation — making them remarkably rugged under most operating conditions. With adequate heatsinking they can deliver output currents in excess of 1.0 A.

- No External Components Required
- Internal Thermal Overload Protection
- Internal Short Circuit Current Limiting
- Output Transistor Safe-Area Compensation
- Available in 2% Voltage Tolerance (See Ordering Information)

MC7900 Series

THREE-TERMINAL NEGATIVE FIXED VOLTAGE REGULATORS

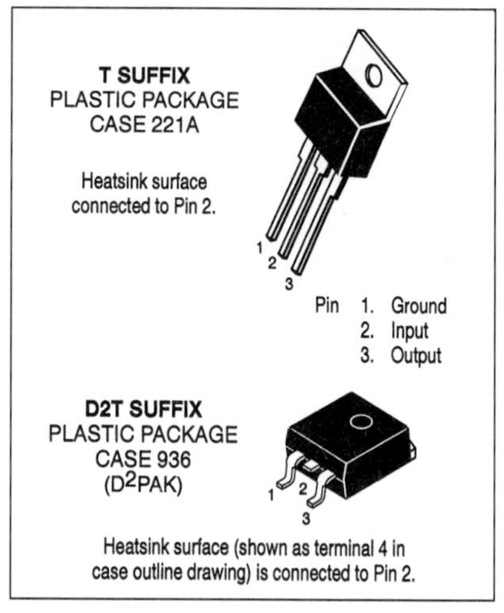

Representative Schematic Diagram

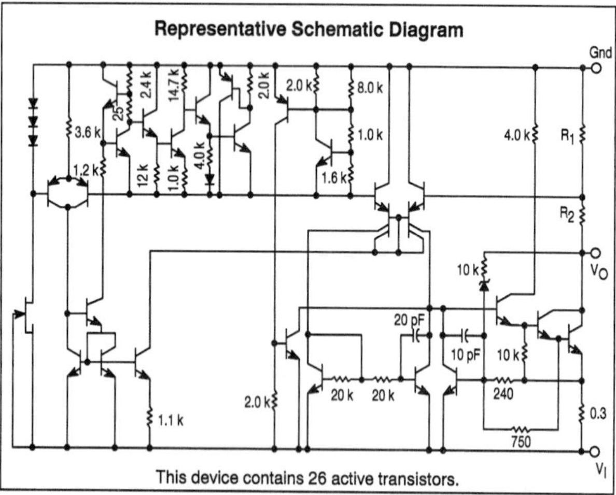

This device contains 26 active transistors.

ORDERING INFORMATION

Device	Output Voltage Tolerance	Tested Operating Temperature Range	Package
MC79XXACD2T	2%	T_J = 0° to +125°C	Surface Mount
MC79XXCD2T	4%		
MC79XXACT	2%		Insertion Mount
MC79XXCT	4%		
MC79XXBD2T	4%	T_J = −40° to +125°C	Surface Mount
MC79XXBT			Insertion Mount

XX indicates nominal voltage.

DEVICE TYPE/NOMINAL OUTPUT VOLTAGE

MC7905	5.0 V	MC7912	12 V
MC7905.2	5.2 V	MC7915	15 V
MC7906	6.0 V	MC7918	28 V
MC7908	8.0 V	MC7924	24 V

STANDARD APPLICATION

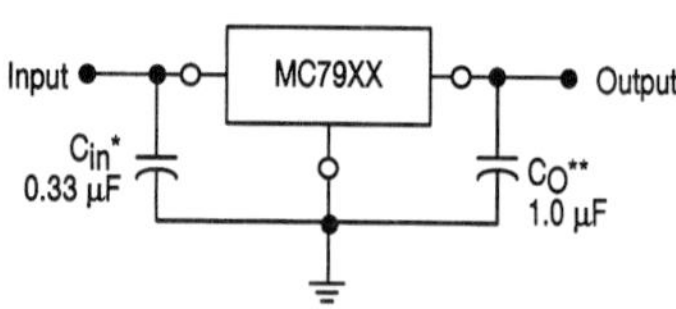

A common ground is required between the input and the output voltages. The input voltage must remain typically 2.0 V above more negative even during the high point of the input ripple voltage.

XX, these two digits of the type number indicate nominal voltage.

* C_{in} is required if regulator is located an appreciable distance from power supply filter.

** C_O improve stability and transient response.

FIGURE 17-7 **Negative Voltage Regulator Data Sheet. (Copyright of Motorola. Used by permission.)**

Variable Output Voltage Regulators

As the name implies, a **variable output voltage regulator** will produce a regulated output voltage that can be adjusted. Like the fixed output voltage regulators, there are positive variable output voltage regulators and negative variable output voltage regulators.

Variable Output Voltage Regulator
A voltage regulator that produces a regulated voltage that can be adjusted.

Figure 17-8 details the LM317 series, which are the most commonly used general-purpose positive variable-output voltage regulators. Figure 17-8(a) shows how the LM317, which has three terminals called V_{IN}, V_{OUT}, and ADJ or adjustment, is normally connected in circuit. Resistors R_1 and R_2 form a voltage divider across the output, with R_2 connected as a rheostat. The output voltage of the LM317 is adjusted by changing the resistance of R_2, and therefore the voltage applied to the ADJ input of the regulator. Figure 17-8(b) shows the typical regulator package types. Figure 17-8(c) lists the different grades of regulators in the series, and, as you can see, the output voltage can be varied from 1.2 V to 37 V for a maximum output current of 0.1 A to 1.5 A.

Figure 17-9 shows the LM317's counterpart, the LM337 series, which are a group of general-purpose negative variable-output voltage regulators. These regulators are available with the same voltage and current options as the LM317 devices.

Switching Voltage Regulators

The fixed and variable output voltage regulators just discussed are all called **series dissipative regulators** because these regulators control the resistance of an internal transistor connected in series between the input voltage and the load. These regulator types maintain a constant output voltage by constantly changing the resistance of the series-connected internal transistor so that variations in input voltage or load current are dissipated as heat. Series dissipative regulators generally have a low "conversion efficiency" of typically 60% to 70% and should only be used in low- to medium-load current applications.

Series Dissipative Regulators
Voltage regulators that maintain a constant output voltage by constantly changing the resistance of the series-connected internal transistor so that variations in input voltage or load current are dissipated as heat.

Series switching regulators, on the other hand, have a conversion efficiency of typically 90%. To explain the operation of these regulator types, refer to the simplified circuit in Figure 17-10(a) (p. 868). To improve efficiency, a series-pass transistor (Q_1) is operated as a switch, rather than as a variable resistor. This means that Q_1 is switched ON and OFF, and therefore either switches the +12 V input at its collector through to its emitter, or blocks the +12 V from passing through to the emitter. These +12 V pulses at the emitter of Q_1 charge capacitor C_1 to an average voltage (which in this example is +5 V) and this voltage is applied to the load (R_L). To explain this in more detail, when Q_1 is turned ON by a HIGH base voltage from the switching regulator IC, the unregulated +12 V at Q_1's collector is switched through to Q_1's emitter, where it reverse biases D_1, and is applied to the series connected inductor L_1 and parallel capacitor C_1. Inductor L_1 and capacitor C_1 act as a low-pass filter because series connected L_1 opposes the ON/OFF changes in current and passes a relatively constant current to the load: shunt connected C_1 opposes the ON/OFF changes in voltage and holds the output voltage relatively constant at +5 V. When Q_1 is turned OFF by a LOW base voltage from the switching regulator IC, the unregulated +12 V input is disconnected from the LC filter, the inductor's magnetic field will col-

Series Switching Regulators
A regulator circuit containing a power transistor in series with the load, that is switched ON and OFF to regulate the dc output voltage delivered to the load.

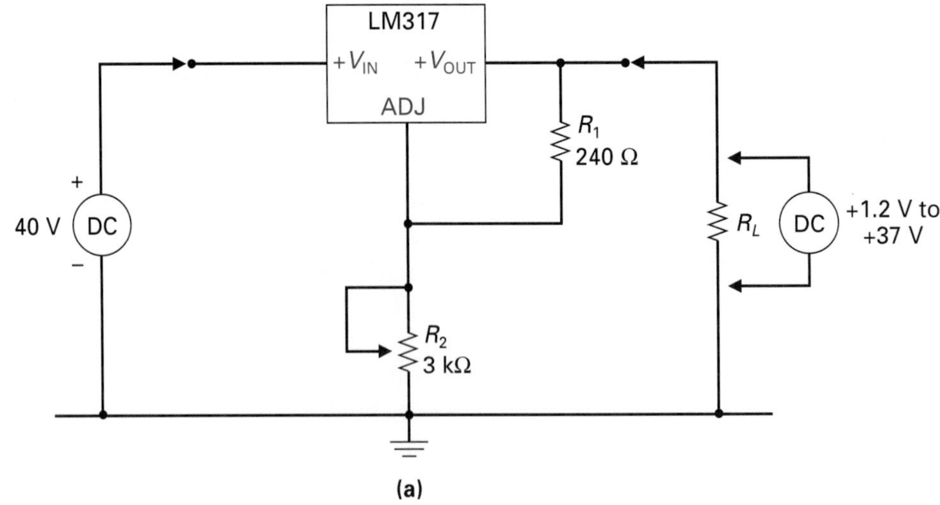

(a)

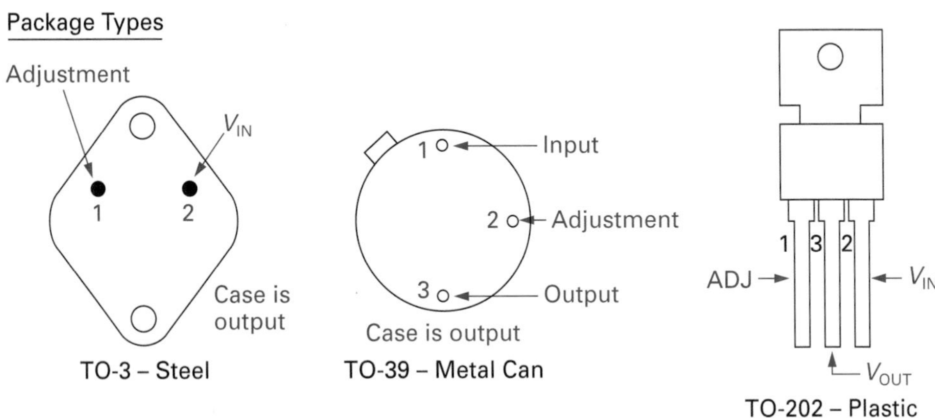

(b)

Regulator Types

Device	Available V_O (V)	Output current (A)	V_{IN} max (V)	Ripple rejection (dB)	Package
LM317	1.2 to 37	1.5	40	80	TO-39
LM317H	1.2 to 37	0.5	40	80	TO-39
LM317HV	1.2 to 57	1.5	60	80	TO-3
LM317HVH	1.2 to 37	0.50	40	80	TO-39
LM317L	1.2 to 37	0.10	40	65	TO-92
LM317M	1.2 to 37	0.50	40	80	TO-202

(c)

FIGURE 17-8 Positive Variable Voltage Regulator (Courtesy of National Semiconductor).

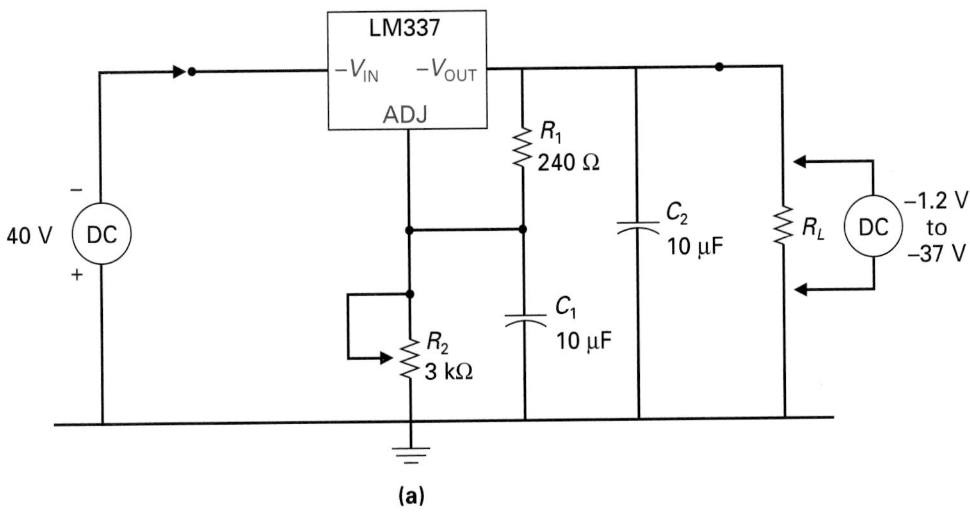

(a)

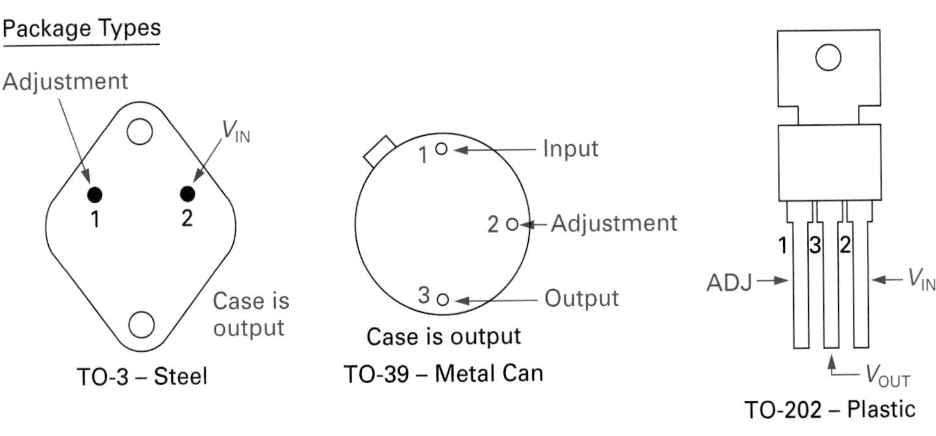

(b)

Regulator Types

Device	Available V_O (V)	Output current (A)	V_{IN} max (V)	Ripple rejection (dB)	Package
LM337	−1.2 to −37	1.5	40	77	TO-39
LM337H	−1.2 to −37	0.5	40	77	TO-39
LM337HV	−1.2 to −47	1.5	50	77	TO-3
LM337HVH	−1.2 to −47	0.5	50	77	TO-39
LM337LZ	−1.2 to −37	0.10	40	65	TO-92
LM337M	−1.2 to −37	0.50	40	77	TO-202

(c)

FIGURE 17-9 Negative Variable Voltage Regulator (Courtesy of National Semiconductor).

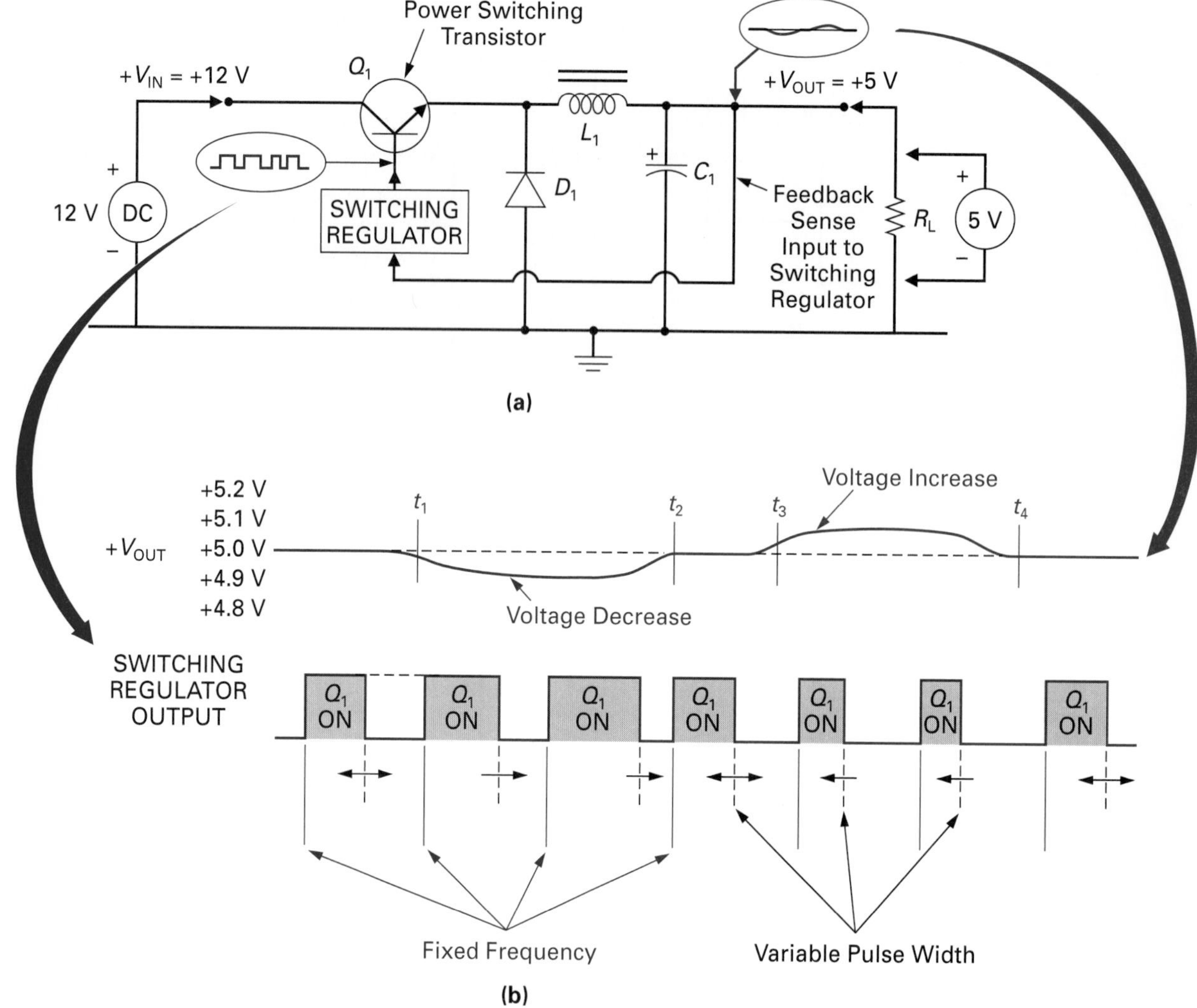

FIGURE 17-10 Basic Switching Regulator Action.

lapse and produce a current through the load, and the +5 V charge held by C_1 will still be applied across the load. Inductor L_1, therefore, smoothes out the current changes, while capacitor C_1 smoothes out the voltage changes caused by the ON/OFF switching of transistor Q_1. The next, and most important, question is: how does this circuit regulate, or maintain constant, the output voltage? The answer is: through a closed-loop "sense and adjust" system controlled by a switching regulator IC. The switching regulator IC operates by comparing an internal fixed reference voltage to a sense input, which is taken from the +5 V output, as is shown in Figure 17-10(a). Referring to the waveforms shown in Figure 17-10(b), you can see that whenever the output voltage falls below +5 V (between time t_1 to t_2), the switching regulator responds by increasing the width of the positive output pulse applied to the base of Q_1. This increases the ON time of Q_1, which raises the average output voltage, bringing the output back up to +5 V. On the other hand, whenever the output voltage rises above +5 V (between time t_3 to t_4), the switching regulator responds by decreasing the width of the positive output pulse applied to the base of Q_1. This decreases the ON time of Q_1, which lowers the average output voltage, bringing the output back down to +5 V. The net result is that the output voltage will remain locked at +5 V despite variations in the input voltage and variations in the load.

A switching regulator, therefore, is a voltage regulator that chops up, or switches ON and OFF (at typically a 20 kHz rate), a dc input voltage to efficiently produce a regulated dc output voltage. A **switching power supply** uses switching regulators and is generally small in size and very efficient. The only disadvantage is that the circuitry is generally a little more complex, and therefore a little more costly.

Switching Power Supply
A dc power supply that makes use of a series switching regulator controlled by a pulse-width-modulator to regulate the output voltage.

Figure 17-11 shows a data sheet for a typical switching regulator IC. This data sheet shows details of the µA78S40 switching regulator IC, which contains an internal voltage reference circuit, oscillator circuit, comparator, set-reset flip-flop or bistable multivibrator, darlington pair, and an uncommitted switching diode and operational amplifier. Figure 17-11 also shows how this IC could be connected to regulate a +5 V output when a +12 V input is applied. The switching frequency is set by the timing capacitor connected between pin 12 and the ground pin 11. The noninverting input (+) of the 78S40's internal comparator (pin 9) is connected to the internal 1.25 V reference voltage at pin 8, while the inverting input (–) of the comparator (pin 10) senses the output voltage that is developed across a voltage divider. The ON/OFF switching of this circuit is opposite to that of the previous example given in Figure 17-10. The 78S40's internal comparator generates an output signal that varies the OFF time of the PNP switching transistor Q_1 instead of varying the ON time of an NPN transistor. To explain in more detail, whenever the output voltage falls below +5 V, the comparator responds by increasing the width of the OFF or LOW output pulse applied to the base of Q_1. Because Q_1 is a PNP transistor, this LOW input increases the ON time of Q_1, which raises the average output voltage, bringing the output back up to +5 V. On the other hand, whenever the output voltage rises above +5 V, the comparator responds by decreasing the width of the LOW output pulse applied to the base of Q_1. This decreases the ON time of Q_1, which lowers the average output voltage, bringing the output back down to +5 V.

SELF-TEST REVIEW QUESTIONS FOR SECTION 17-1

1. List four examples of linear integrated circuits.
2. A voltage regulator will maintain its output voltage constant despite variations in ____________ and ____________.
3. Series dissipative regulators, such as the 7800, 7900, 317, and 377 series, have an internal resistor connected in series between the input voltage and load. This internal resistor is operated as a ____________.
4. Series switching regulators also have an internal transistor connected in series between the input voltage and load; however this transistor is operating as a ____________.

DEVICE: LM78S40 Switching Voltage Regulator

General Description

The LM78S40 is a monolithic regulator subsystem consisting of all the active building blocks necessary for switching regulator systems. The device consists of a temperature compensated voltage reference, a duty-cycle controllable oscillator with an active current limit circuit, an error amplifier, high current, high voltage output switch, a power diode and an uncommitted operational amplifier. The device can drive external NPN or PNP transistors when currents in excess of 1.5A or voltages in excess of 40V are required. The device can be used for step-down, step-up or inverting switching regulators as well as for series pass regulators. It features wide supply voltage range, low standby power dissipation, high efficiency and low drift. It is useful for any stand-alone, low part count switching system and works extremely well in battery operated systems.

Features

- Step-up, step-down or inverting switching regulators
- Output adjustable from 1.25V to 40V
- Peak currents to 1.5A without external transistors
- Operation from 2.5V to 40V input
- Low standby current drain
- 80 dB line and load regulation
- High gain, high current, independent op amp
- Pulse width modulation with no double pulsing

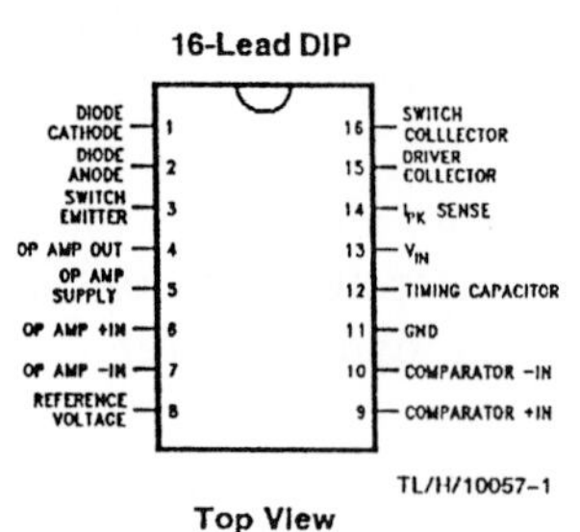

Ordering Information

Part Number	NS Package	Temperature Range
LM78S40J LM78S40J/883	J16A Ceramic DIP J16A Ceramic DIP	−55°C to +125°C
LM78S40N	N16E Molded DIP	−40°C to +125°C
LM78S40CJ LM78S40CN	J16A Ceramic DIP N16E Molded DIP	0°C to +70°C

Block and Connection Diagrams

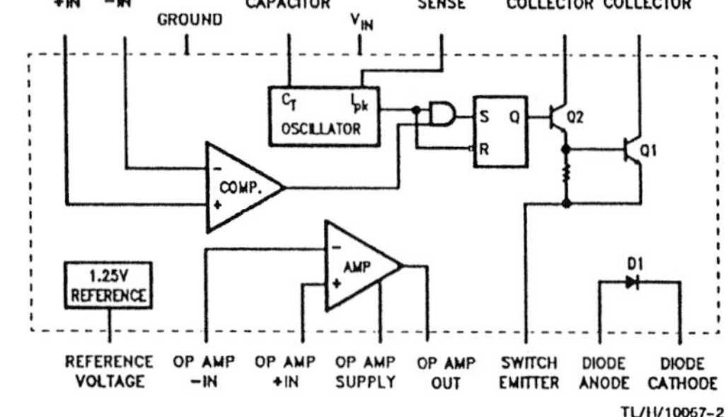

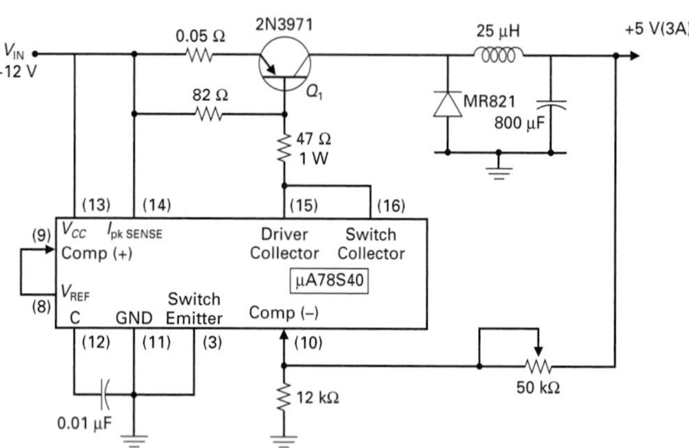

Absolute Maximum Ratings

If Military/Aerospace specified devices are required, please contact the National Semiconductor Sales Office/Distributors for availability and specifications.

Storage Temperature Range	
Ceramic DIP	−65°C to +175°C
Molded DIP	−65°C to +150°C
Operating Temperature Range	
Extended (LM78S40J)	−55°C to +125°C
Industrial (LM78S40N)	−40°C to +125°C
Commercial (LM78S40CN)	0°C to +70°C
Lead Temperature	
Ceramic DIP (Soldering, 60 sec.)	300°C
Molded DIP (Soldering, 10 sec.)	265°C
Internal Power Dissipation (Notes 1, 2)	
16L-Ceramic DIP	1.50W
16L-Molded DIP	1.04W
Input Voltage from V_{IN} to GND	40V
Common Mode Input Range (Comparator and Op Amp)	−0.3 to V+
Differential Input Voltage (Note 3)	±30V
Output Short Circuit Duration (Op Amp)	Continuous
Current from V_{REF}	10 mA
Voltage from Switch Collectors to GND	40V
Voltage from Switch Emitters to GND	40V
Voltage from Switch Collectors to Emitter	40V
Voltage from Power Diode to GND	40V
Reverse Power Diode Voltage	40V
Current through Power Switch	1.5A
Current through Power Diode	1.5A
ESD Susceptibility	(to be determined)

(a)

FIGURE 17-11 A Switching Regulator Data Sheet.

17-2 FUNCTION GENERATOR CIRCUITS

A function generator is by definition a signal generator circuit that delivers a variety of different output waveforms whose frequency can be varied. In this section, we will examine some of the different linear function generator ICs available.

17-2-1 *The Voltage Controlled Oscillator (VCO)*

Previously in the oscillator circuits discussed in Chapter 12, and the multivibrator circuits discussed in Chapter 9, the frequency of the repeating output signal was determined by an *LC, RC,* or crystal network. In many communication circuit applications however, we need an oscillator whose frequency can be controlled by an input "control voltage." The circuit that achieves this function is the **voltage controlled oscillator (VCO)** or, as it is sometimes called, a "voltage-to-frequency converter." The NE/SE 566 shown in Figure 17-12 is a good example of a linear IC VCO. Figure 17-12(a) shows how a 566 can be connected to generate both a square wave and triangular wave output, and Figure 17-12(b) gives the data sheet for this device. The frequency of oscillation is determined by external resistors R_1 and R_2 and capacitor C_1, which determine the **control voltage** applied to pin 5. The triangular wave is generated by linearly charging and discharging the external capacitor C_1, using the 566's internal current source. The charge and discharge levels are determined by the 566's internal Schmitt trigger, which is also used to generate the square wave output. The triangular wave will typically have an amplitude of 2.4 V pk-pk, and the square wave will typically have an amplitude of 5.4 V pk-pk.

Voltage Controlled Oscillator (VCO)
An oscillator whose frequency can be controlled by an input control voltage.

Control Voltage
A voltage signal that starts, stops, or adjusts the operation of a device, circuit, or system.

17-2-2 *Function Generator Circuit*

Function generators are test instruments whose output waveforms are used to test circuits by injecting an input into that circuit and then monitoring its response to the input. For example, sine wave signals are normally used to test

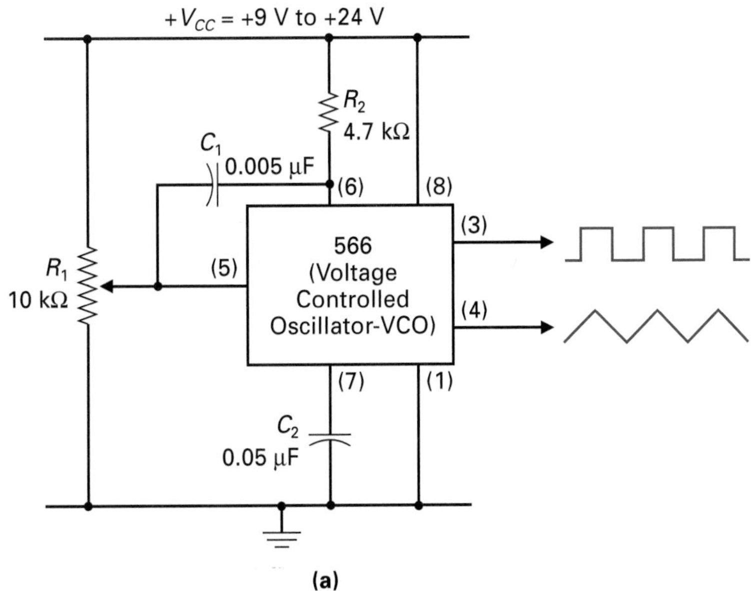

FIGURE 17-12 A Voltage Controlled Oscillator (VCO).

DEVICE: NE/SE 566—Function Generator

DESCRIPTION

The NE/SE566 Function Generator is a voltage-controlled oscillator of exceptional linearity with buffered square wave and triangle wave outputs. The frequency of oscillation is determined by an external resistor and capacitor and the voltage applied to the control terminal. The oscillator can be programmed over a ten-to-one frequency range by proper selection of an external resistance and modulated over a ten-to-one range by the control voltage, with exceptional linearity.

FEATURES

- Wide range of operating voltage (up to 24V; single or dual)
- High linearity of modulation
- Highly stable center frequency (200ppm/°C typical)
- Highly linear triangle wave output
- Frequency programming by means of a resistor or capacitor, voltage or current
- Frequency adjustable over 10-to-1 range with same capacitor

APPLICATIONS

- Tone generators
- Frequency shift keying
- FM modulators
- Clock generators
- Signal generators
- Function generators

PIN CONFIGURATIONS

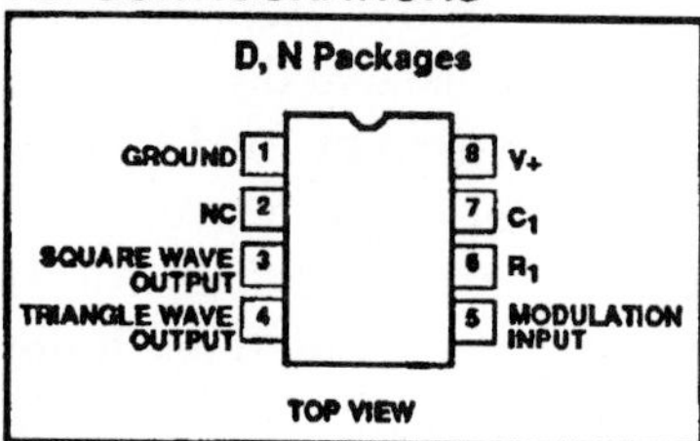

BLOCK DIAGRAM

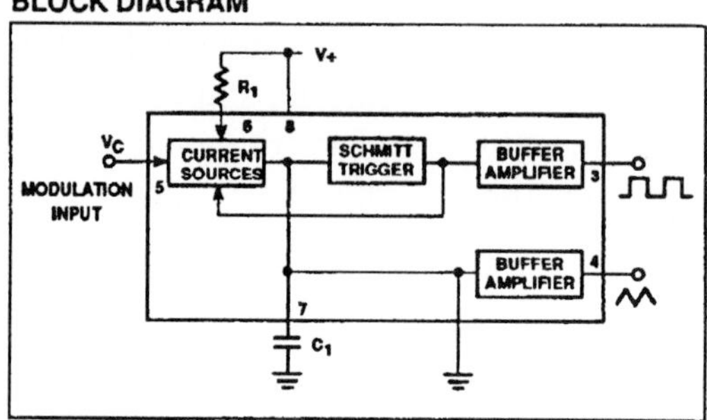

ABSOLUTE MAXIMUM RATINGS

SYMBOL	PARAMETER	RATING	UNIT
V+	Maximum operating voltage	26	V
V_{IN}	Input voltage	3	V_{P-P}
T_{STG}	Storage temperature range	-65 to +150	°C
T_A	Operating ambient temperature range		
	NE566	0 to +70	°C
	SE566	-55 to +125	°C
P_D	Power dissipation	300	mW

(b)

FIGURE 17-12 (continued.)

DEVICE: ICL8038—Waveform Generator/Voltage Controlled Oscillator

Description

The ICL8038 waveform generator is a monolithic integrated circuit capable of producing high accuracy sine, square, triangular, sawtooth and pulse waveforms with a minimum of external components. The frequency (or repetition rate) can be selected externally from 0.001Hz to more than 300kHz using either resistors or capacitors, and frequency modulation and sweeping can be accomplished with an external voltage. The ICL8038 is fabricated with advanced monolithic technology, using Schottky barrier diodes and thin film resistors, and the output is stable over a wide range of temperature and supply variations. These devices may be interfaced with phase locked loop circuitry to reduce temperature drift to less than 250ppm/°C.

Features

- **Low Frequency Drift with Temperature - 250ppm/°C**
- **Simultaneous Sine, Square, and Triangle Wave Outputs**
- **Low Distortion - 1% (Sine Wave Output)**
- **High Linearity - 0.1% (Triangle Wave Output)**
- **Wide Operating Frequency Range - 0.001Hz to 300kHz**
- **Variable Duty Cycle - 2% to 98%**
- **High Level Outputs - TTL to 28V**
- **Easy to Use - Just a Handful of External Components Required**

Absolute Maximum Ratings

Supply Voltage (V- to V+) 36V
Power Dissipation (Note 1) 750mW
Input Voltage (Any Pin) . . . . . V- to V+
Input Current (Pins 4 and 5) 25mA
Output Sink Current (Pins 3 and 9) 25mA
Lead Temperature (Soldering 10 Sec.) . . . . . +300°C

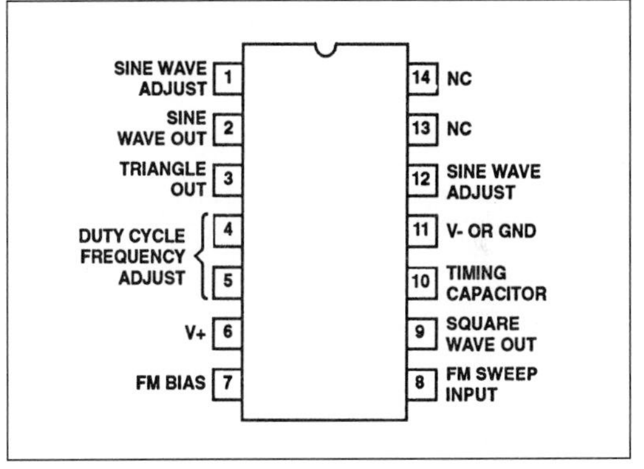

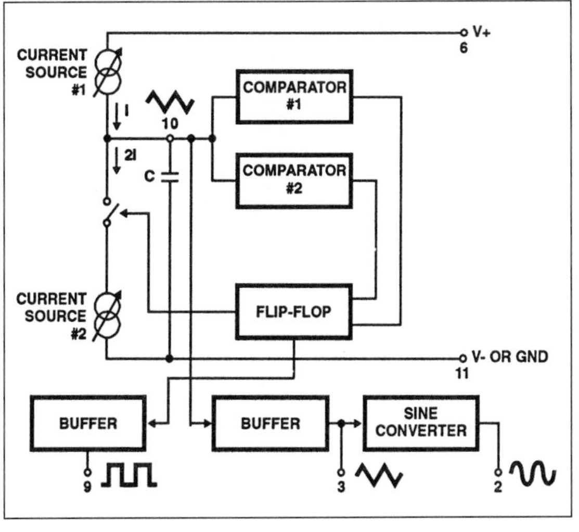

(a)

FIGURE 17-13 Function Generator Circuit.

linear circuits such as amplifiers and filters, whereas square wave signals are often used to test digital circuits such as logic gates. In the past, a function generator circuit was constructed using discrete components; however, today a complete function generator circuit is available in one linear IC. Figure 17-13(a) gives the data sheet for the ICL8038 waveform generator/voltage controlled oscillator linear IC, and Figure 17-13(b) shows how this IC could be connected to form a function generator circuit.

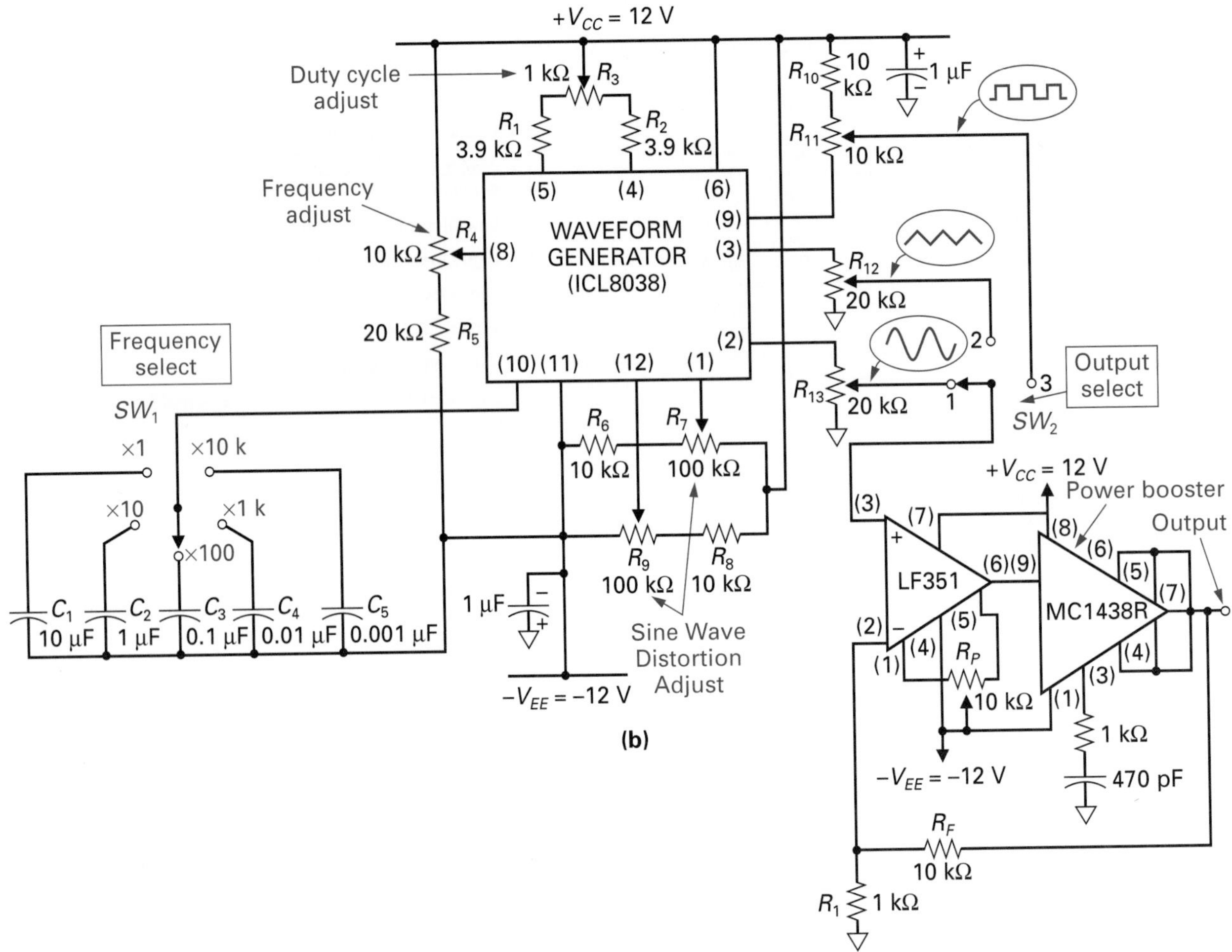

FIGURE 17-13 (continued).

The 8038 function generator circuit in Figure 17-13(b) produces a 20 Hz to 100 kHz sine wave, square wave, or triangular wave output. The frequency select switch (SW_1) is used to select the desired frequency range, and then resistor R_4 is used to adjust the frequency output to the desired frequency within that range. Switch SW_2 is used to select either the square wave, triangular wave, or sine wave output. The position of this output select switch will determine which waveform is applied to the output stage made up of an op amp and power amp. The output amplitude of the square wave can be varied by R_{11}, and the duty cycle of the square wave can be adjusted by R_3. Similarly, the triangular wave's output amplitude can be adjusted by R_{12} and the sine wave's output amplitude can be adjusted by R_{13}. Variable resistors R_7 and R_9 are included to fine tune the sine wave output for minimum distortion.

SELF-TEST REVIEW QUESTIONS FOR SECTION 17-2

1. Give the full name of the abbreviation VCO.
2. A VCO IC generates a repeating output waveform based on a ____________ at its input.
3. Which three waveforms are generated by the 8038 function generator IC?
4. Referring to Figure 17-13(b), which device would be the course frequency adjust, and which device would be the fine frequency adjust?

17-3 PHASE-LOCKED LOOP (PLL) CIRCUITS

Integrated circuit technology has made the **phase-locked loop (PLL) circuit** reasonably priced and therefore widely used in a variety of applications.

Phase-Locked Loop (PLL) Circuit
A circuit consisting of a phase comparator that compares the output frequency of a voltage controlled oscillator with an input frequency. The error voltage out of the phase comparator is then coupled via an amplifier and low-pass filter to the control input of the voltage controlled oscillator to keep it in phase, and therefore at exactly the same frequency as the input.

17-3-1 Phase-Locked Loop Operation

The operation of a phase locked loop is best described by referring to the block diagram of the LM 565 PLL linear IC shown in Figure 17-14(a). As can be seen in this diagram, the PLL contains a phase comparator, amplifier and low-pass filter, and a voltage controlled oscillator. The input frequency (f_{IN}) is applied to one input of the phase comparator and compared with the output frequency (f_{OUT}) from the VCO. If a frequency difference exists between the two comparator inputs, the comparator will detect the difference in phase and generate an **error voltage,** which will be amplified and filtered before it is applied as the control input voltage to the VCO. If the input frequency (f_{IN}) and VCO output frequency (f_{OUT}) differ, the error voltage generated by the phase comparator will shift the VCO frequency until it matches the input frequency, and therefore $f_{IN} = f_{OUT}$. When f_{IN} matches the VCO frequency or f_{OUT}, the PLL is said to be in the **locked** condition. If the input frequency is varied, the VCO output frequency will follow or track f_{IN}, provided that the input frequency is within the capture range of the PLL. Referring to the inset in Figure 17-14(a), you can see how the error voltage from the phase comparator varies while the PLL is trying to **capture** the input frequency. When the input frequency is outside the capture range of the PLL, the VCO will generate a **free-running** or idle frequency.

Error Voltage
A voltage that is proportional to the error that exists between input and output.

Locked
To automatically follow a signal.

Capture
The act of gaining control of a signal.

Free-Running
Operating without any external control.

To summarize: a phase-locked loop circuit consists of a phase comparator that compares the output frequency of a voltage controlled oscillator with an input frequency. The error voltage out of the phase comparator is then coupled via an amplifier and low-pass filter to the control input of the voltage controlled oscillator to keep it in phase—and therefore at exactly the same frequency—as the input frequency. These phase-locked loop circuits will operate in either the free-running state, capture state, or locked state. Figure 17-14(b) shows the data sheet for the 565 PLL linear IC.

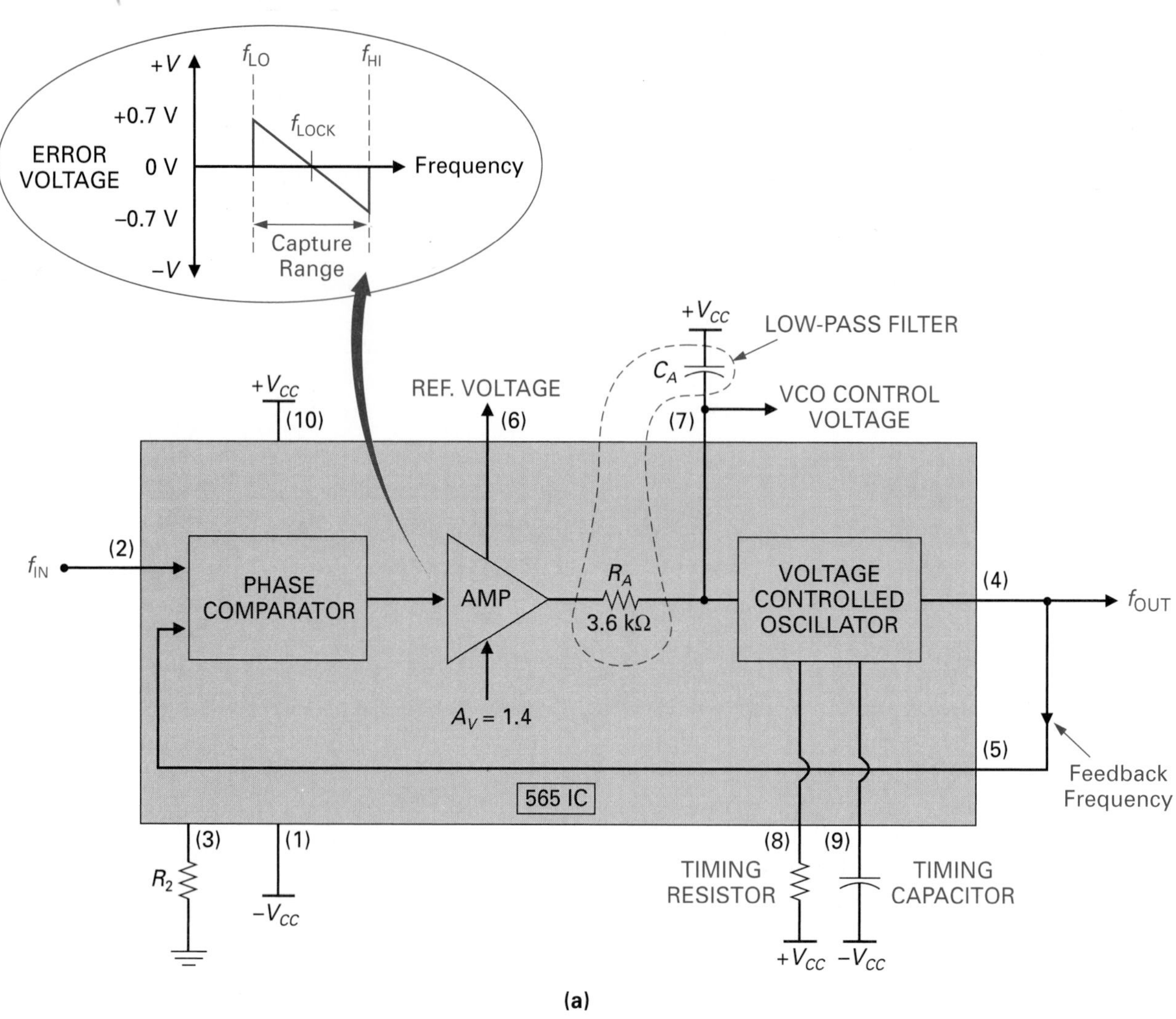

FIGURE 17-14 The Phase Locked Loop (PLL).

17-3-2 *Phase-Locked Loop Applications*

The phase-locked loop circuit can be used in a variety of circuit applications including the demodulation of FM radio signals, frequency multiplication and division, and signal regeneration. Most of these circuits will be covered in your electronic communication course. However, at this time let us examine two of the most important applications: the PLL used with a crystal controlled oscillator as a "frequency multiplier circuit" and the PLL used as a digital demodulator.

PLL Frequency Multiplier Circuit

To generate several stable output frequencies, we would either have to have several crystal controlled oscillators as shown in Figure 17-15(a), or use frequency multiplier circuits (like the one discussed in Chapter 11) to step up

DEVICE: LM565/LM565C Phase Locked Loop

General Description

The LM565 and LM565C are general purpose phase locked loops containing a stable, highly linear voltage controlled oscillator for low distortion FM demodulation, and a double balanced phase detector with good carrier suppression. The VCO frequency is set with an external resistor and capacitor, and a tuning range of 10:1 can be obtained with the same capacitor. The characteristics of the closed loop system—bandwidth, response speed, capture and pull in range—may be adjusted over a wide range with an external resistor and capacitor. The loop may be broken between the VCO and the phase detector for insertion of a digital frequency divider to obtain frequency multiplication.

The LM565H is specified for operation over the −55°C to +125°C military temperature range. The LM565CN is specified for operation over the 0°C to +70°C temperature range.

Features

- 200 ppm/°C frequency stability of the VCO
- Power supply range of ±5 to ±12 volts with 100 ppm/% typical
- 0.2% linearity of demodulated output
- Linear triangle wave with in phase zero crossings available
- TTL and DTL compatible phase detector input and square wave output
- Adjustable hold in range from ±1% to > ±60%

Applications

- Data and tape synchronization
- Modems
- FSK demodulation
- FM demodulation
- Frequency synthesizer
- Tone decoding
- Frequency multiplication and division
- SCA demodulators
- Telemetry receivers
- Signal regeneration
- Coherent demodulators

Connection Diagrams

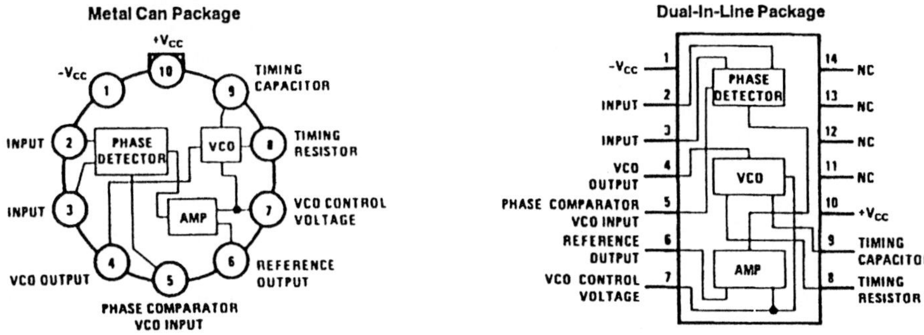

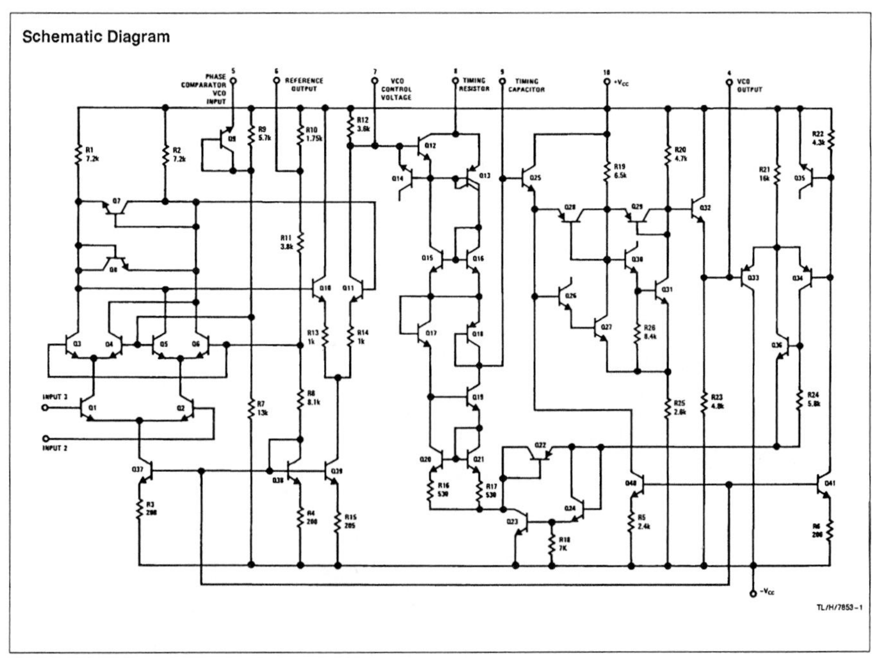

(b)

FIGURE 17-14 (continued).

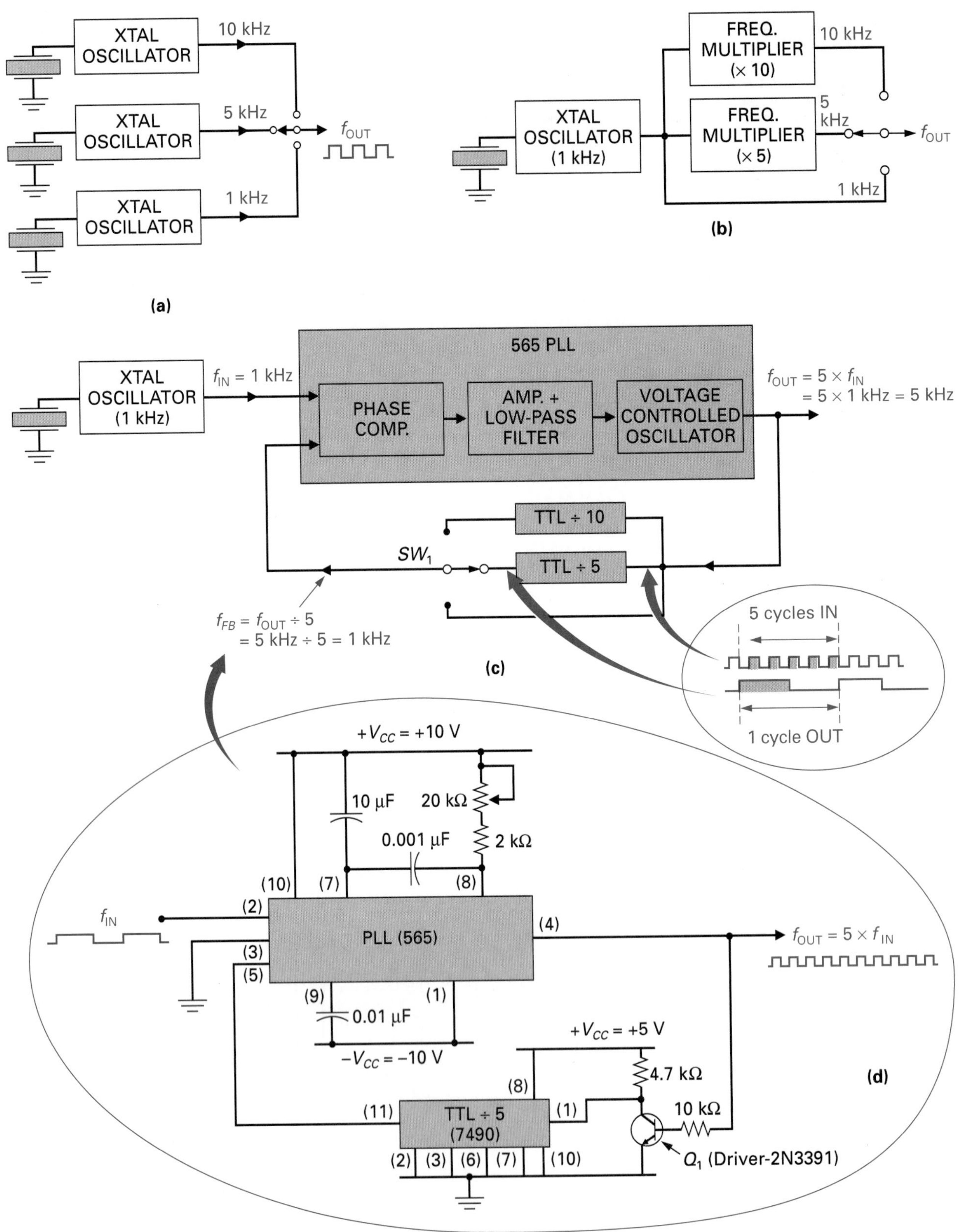

FIGURE 17-15 Frequency Multiplier Using a Phase Locked Loop.

an input frequency as shown in Figure 17-15(b). The disadvantage with the circuit in Figure 17-15(a) is that we will need several complete crystal oscillator circuits, and the disadvantage of the circuit in Figure 17-15(b) is that its frequency characteristics depend on an *LC* tuned circuit that will need frequent alignment. The PLL block diagram shown in Figure 17-15(c) and its associated circuit diagram shown in Figure 17-15(d) will generate several output frequencies that have the same frequency stability as a crystal oscillator's output without the need for several crystal oscillators or frequent *LC* tuning. As can be seen in Figure 17-15(c), a crystal oscillator applies a very stable 1 kHz square wave input to the 565 PLL. When switch 1 is in the lower position, the output of the VCO is fed directly back to the phase comparator, and the circuit will operate in exactly the same way as the circuit in Figure 17-14(a). It will generate an output frequency that tracks the input frequency and is a very stable 1 kHz due to the locking action of the PLL. When switch 1 is in the mid position, a divide-by-five circuit is switched into circuit between the VCO output and the phase comparator input. Referring to the waveforms in Figure 17-15(c), you can see that the divide-by-five circuit will generate one output cycle for every five input cycles. This means that the VCO feedback signal applied to the phase comparator will be one-fifth of the output frequency. Because this signal will not be in phase with the 1 kHz signal from the crystal oscillator, the phase comparator will generate an error voltage instructing the VCO to increase its output frequency by five times. When the VCO generates an output frequency of 5 kHz, the divide-by-five circuit will produce a 1 kHz output (5 kHz ÷ 5 = 1 kHz) to the phase comparator, which will match the phase comparator's other 1 kHz input from the crystal oscillator. As a result, the error voltage from the phase comparator will be zero and the VCO—and therefore the output—will remain locked at 5 kHz. The circuit diagram for this switch position is shown in Figure 17-15(d). If switch SW_1 of the circuit in Figure 17-15(c) is put in the upper position, a divide-by-ten circuit will be included between the VCO output and the phase comparator input. As a result, the PLL will generate an output frequency that is ten times the input frequency ($f_{OUT} = 10 \times f_{IN} = 10 \times 1$ kHz = 10 kHz).

PLL Digital Demodulator Circuit

Telephone lines are often used as a communication link between two computers. Since digital dc HIGH/LOW pulses cannot be sent over telephone lines, a special device called a **modem** is used to connect the computer to the telephone system. The name modem is a contraction of the words "MODulator" and "DEModulator", and this is exactly what a modem contains: a modulator that converts the HIGH and LOW dc digital pulses from the computer to HIGH and LOW frequency pulses that are compatible with the telephone line—and a demodulator that converts the HIGH and LOW frequency pulses from the telephone line to HIGH and LOW dc pulses that are compatible with the computer. One of the modulation techniques used to transfer digital information over telephone lines is called **frequency shift keying (FSK).** A basic block diagram of this communication system is shown in Figure 17-16.

Modem
A special device used to connect the computer to the telephone system.

Frequency Shift Keying (FSK)
A form of frequency modulation in which the modulating wave shifts the output frequency between one of two frequencies.

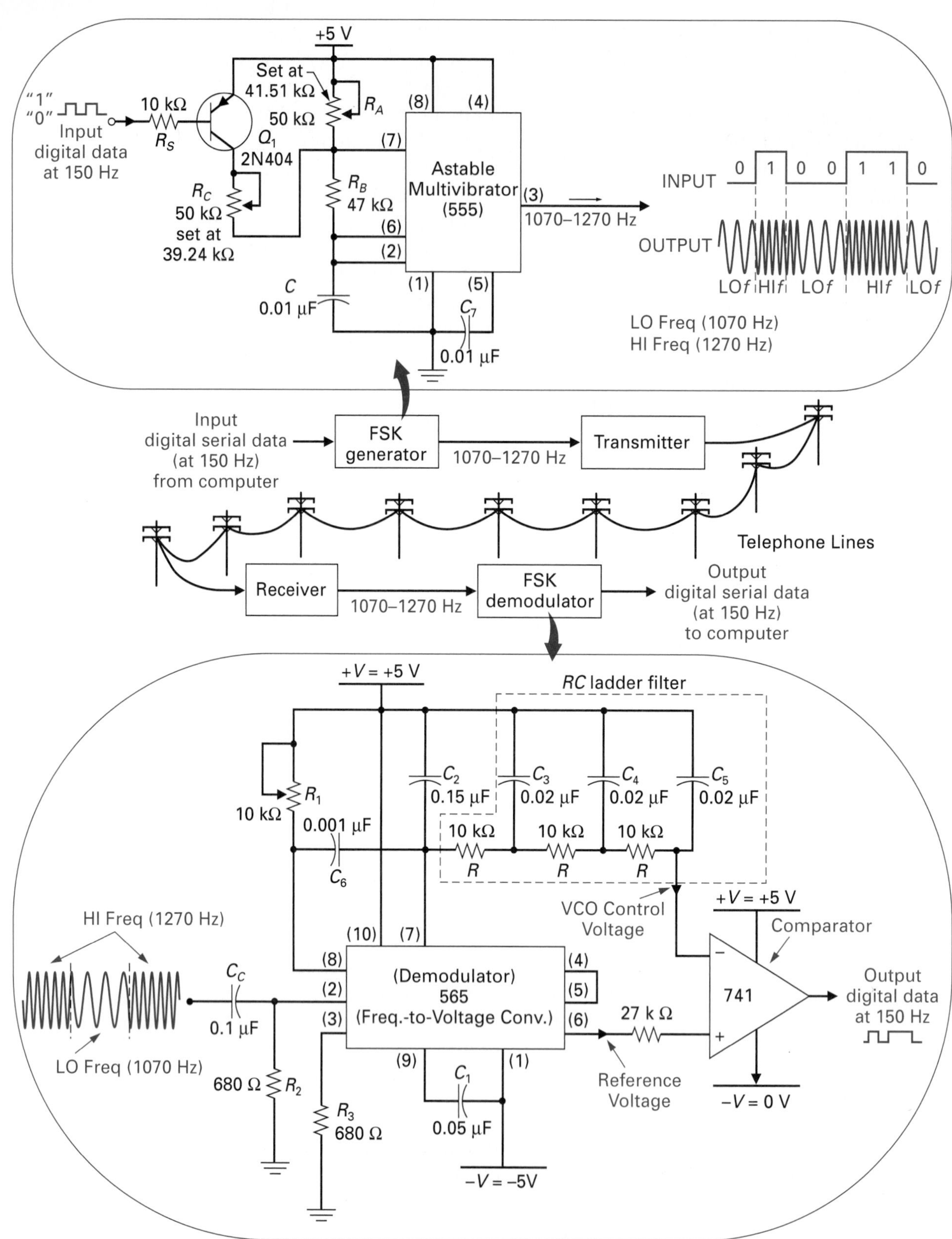

FIGURE 17-16 A Digital Frequency Shift Keying (FSK) Circuit.

Referring to the circuit diagram shown in the inset in the FSK generator block in Figure 17-16, you can see that a 555 astable multivibrator is used to generate one of two output frequencies based on whether a HIGH or LOW input is applied to PNP transistor Q_1. When the digital input from the transmitting computer is HIGH (logical 1), transistor Q_1 will turn OFF, and C will charge via R_A and R_B to 2/3 V_{CC} and discharge via R_B to 1/3 V_{CC}. This RC network will cause the astable multivibrator to generate an output of 1070 Hz for as long as the input to Q_1 is HIGH, or a logical 1. On the other hand, when the input to Q_1 is LOW (logical 0), Q_1 will turn ON and connect R_C in parallel with R_A. This decrease in resistance will decrease the time it takes for C to charge, and therefore cause the output frequency to increase to 1270 Hz. The term "frequency shift keying" is used because the modulating input signal from the computer (or, as it is sometimes called, "keyer") "shifts" the output "frequency" between one of two distinct frequencies.

Referring to the FSK demodulator block in Figure 17-16, you can see in the circuit diagram shown in the inset that a 565 phase-locked loop IC is being used to convert the HIGH and LOW frequency pulses from the telephone line to HIGH and LOW dc pulses that are compatible with the computer. In the previous frequency multiplier application circuit, the output frequency was taken from the PLL's internal VCO at pin 4. In this application, however, we do not need an output frequency; we need an output voltage that follows the input frequency, and this is why the output is taken from pin 7. To explain this in more detail, the PLL will follow the input frequency shift, locking onto each of the two input frequencies and generating a VCO dc control voltage output at pin 7 that follows the input frequency shifts at pin 2. This HIGH/LOW dc digital signal at the PLL output (pin 7) is applied to a comparator via a three-stage RC low-pass ladder filter that is included to filter out any remaining input frequencies. The 741 op-amp comparator compares the reference voltage output from the PLL (pin 6) with the HIGH/LOW dc signal from the PLL (pin 7) and generates an output that shifts between the comparator's two supply voltage extremes of +5 V and 0 V. The communicated digital dc signal information is then passed onto the receiving computer.

SELF-TEST REVIEW QUESTIONS FOR SECTION 17-3

1. What are the three basic blocks in a phase-locked loop circuit?
2. Briefly describe the operation of the phase-locked loop.
3. What is included in the feedback path between the VCO and the phase comparator of a PLL when the PLL is used as a frequency multiplier?
4. In some applications, the PLL will be used to generate an output voltage that follows or tracks changes in frequency at the input. In such an application, which pin of the 565 PLL would be used to obtain this varying voltage?

SUMMARY

1. Improvements in technology have refined the op-amp and led to a variety of other spin-off ICs that internally include several op-amps along with their other circuitry. These ICs include voltage regulators, timers, function generators, phase-locked loops, analog-to-digital (A/D) converters, digital-to-analog (D/A) converters, sample-and-hold amplifiers, and a variety of other specialized ICs. These ICs, along with the op-amp, are collectively called linear integrated circuits (linear ICs) because they provide linear signal amplification, in contrast to digital integrated circuits (digital ICs) that are primarily used for pulse signal processing.

Linear IC Voltage Regulators (Figure 17-17)

2. The regulator maintains the dc output voltage from power supply constant, or stable, despite variations in the ac input voltage or variations in the output load resistance.
3. The percent of regulation is a measure of the regulator's ability to regulate—or maintain constant—the output dc voltage.
4. Most dc power-supply circuits today make use of linear integrated circuit (IC) voltage regulators. These IC regulators contain about 50 individual or discrete components all integrated together on one silicon semiconductor chip and then encapsulated in a three-pin package.
5. These IC regulators include an internal short-circuit protection and thermal shutdown circuit that will protect the IC regulator by turning it OFF if the load current drawn exceeds the regulator's rated current—or if the heat sink is too small and the IC regulator is generating heat faster than it can dissipate it.
6. Although voltage regulators can be constructed using op-amps and other discrete components, complete IC voltage regulators are relatively inexpensive, smaller, and more efficient.
7. Linear voltage regulator ICs can be categorized into one of three basic groups:
   a. Fixed Output Voltage Regulators
   b. Variable Output Voltage Regulators
   c. Switching Voltage Regulators
8. A fixed output voltage regulator will produce a regulated output voltage that is not variable.
9. Fixed output voltage regulators can be broken down into two basic groups: those producing a positive output voltage and those producing a negative output voltage.
10. The 7800 (seventy-eight hundred) series of voltage regulators are used to supply a positive output voltage; the last two digits specifying the output voltage (for example, 7805 = +5 V).
11. The 7900 series of voltage regulators are used to supply a negative output voltage. The last two digits, once again, specifying the output voltage (for example, 7912 = –12 V).

QUICK REFERENCE SUMMARY SHEET

FIGURE 17-17 Linear IC Voltage Regulators.

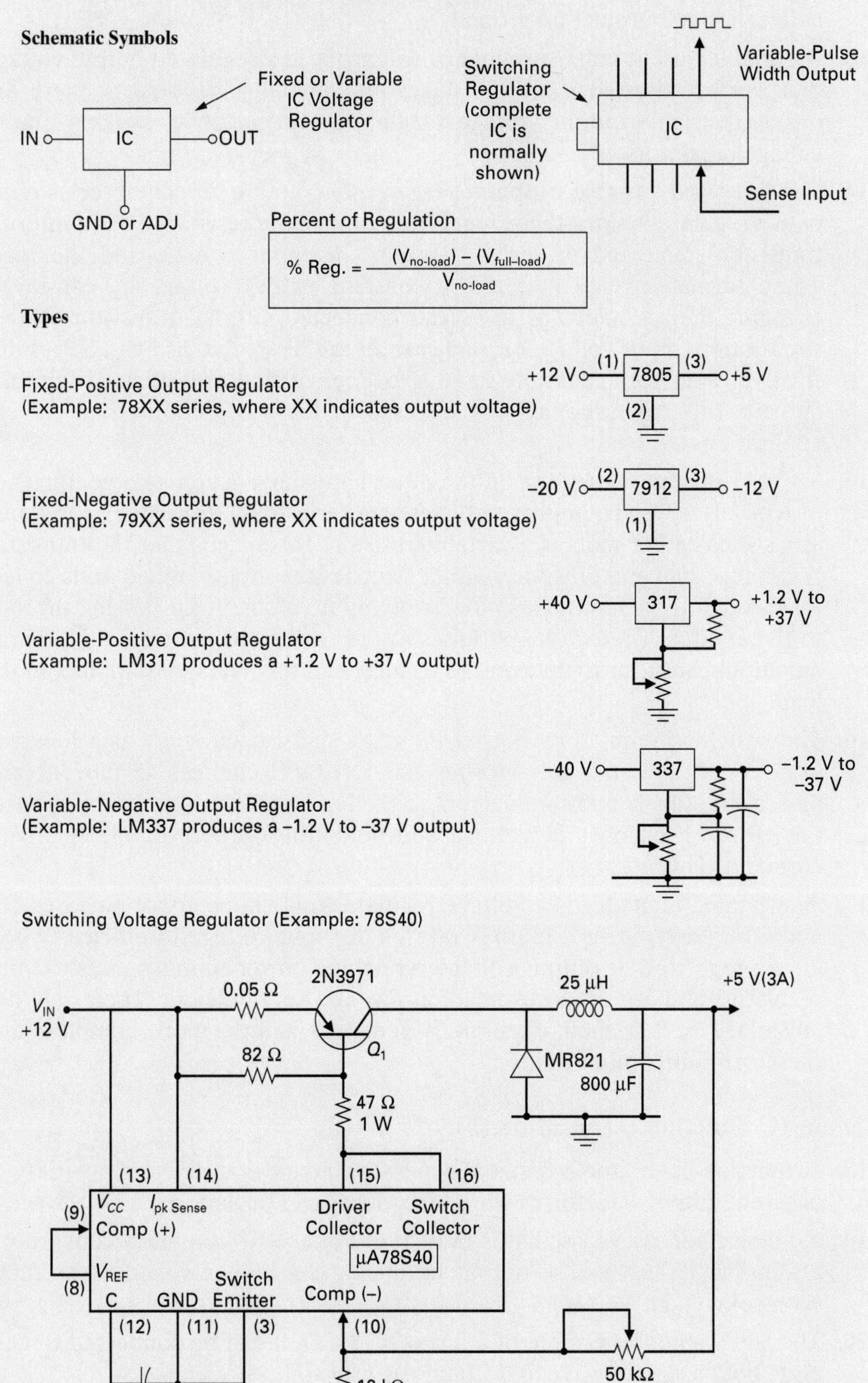

Schematic Symbols
Fixed or Variable IC Voltage Regulator
IN
IC
OUT
GND or ADJ
Switching Regulator (complete IC is normally shown)
IC
Variable-Pulse Width Output
Sense Input
Percent of Regulation
% Reg. = (V no-load) − (V full-load) / V no-load
Types
Fixed-Positive Output Regulator
(Example: 78XX series, where XX indicates output voltage)
+12 V
(1)
7805
(3)
+5 V
(2)
Fixed-Negative Output Regulator
(Example: 79XX series, where XX indicates output voltage)
−20 V
(2)
7912
(3)
−12 V
(1)
Variable-Positive Output Regulator
(Example: LM317 produces a +1.2 V to +37 V output)
+40 V
317
+1.2 V to +37 V
Variable-Negative Output Regulator
(Example: LM337 produces a −1.2 V to −37 V output)
−40 V
337
−1.2 V to −37 V
Switching Voltage Regulator (Example: 78S40)
V IN +12 V
0.05 Ω
2N3971
82 Ω
Q1
47 Ω 1 W
25 μH
+5 V(3A)
MR821
800 μF
(13)
(14)
(15)
(16)
(9)
(8)
V CC
I pk Sense
Comp (+)
Driver Collector
Switch Collector
μA78S40
V REF
C
GND
Switch Emitter
Comp (−)
(12)
(11)
(3)
(10)
0.01 μF
12 kΩ
50 kΩ

12. These fixed output voltage regulators will deliver a constant regulated output voltage as long as the input voltage is greater than the regulator's rated output voltage. They can also deliver a maximum output current of up to 1.5 A if properly heat sunk. The three terminals of the IC regulator are labeled input, output, and ground.
13. A variable output voltage regulator will produce a regulated output voltage that can be adjusted. Like the fixed output voltage regulators, there are positive variable output voltage regulators, and negative variable output voltage regulators.
14. The fixed and variable output voltage regulators are all called series dissipative regulators, since these regulators control the resistance of an internal transistor connected in series between the input voltage and the load. These regulator types maintain a constant output voltage by constantly changing the resistance of the series connected internal transistor so that variations in input voltage or load current are dissipated as heat. Series dissipative regulators generally have a low conversion efficiency of typically 60% to 70%, and should only be used in low- to medium-load current applications.
15. Series switching regulators, on the other hand, have a conversion efficiency of typically 90%. To improve efficiency, a series-pass transistor is operated as a switch rather than as a variable resistor. This means that the transistor is switched ON and OFF, and either switches the input voltage at its collector through to its emitter or blocks the input voltage from passing through to the emitter. These voltage pulses at the emitter of the transistor charge an output capacitor to an average voltage, and this voltage is applied to the load.
16. The switching regulator's output inductor and capacitor act as a low-pass filter, since the inductor opposes the ON/OFF changes in current and passes a relatively constant current to the load, while the capacitor opposes the ON/OFF changes in voltage and holds the output voltage relatively constant.
17. A switching regulator is a voltage regulator that chops up, or switches ON and OFF (at typically a 20 kHz rate), a dc input voltage to efficiently produce a regulated dc output voltage. Switching power supplies use switching regulators and are generally small in size and very efficient. Their only disadvantage is that their circuitry is generally a little more complex, and therefore a little more costly.

Linear IC Function Generators (Figure 17-18)

18. A function generator is a signal generator circuit that delivers a variety of different output waveforms whose frequency can be varied.
19. A voltage controlled oscillator (VCO) or voltage-to-frequency converter is a signal generator whose output frequency can be controlled by an input "control voltage."
20. The 566 is a good example of a linear IC VCO. It can be connected to generate both a square wave and triangular wave output.

QUICK REFERENCE SUMMARY SHEET

FIGURE 17-18 Linear IC Function Generators.

Voltage Controlled Oscillator (VCO) **(Example: 566)**

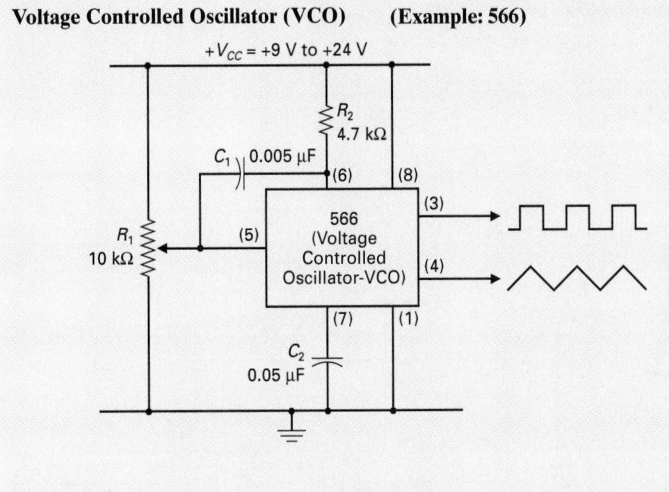

Waveform Generator **(Example: 8038)**

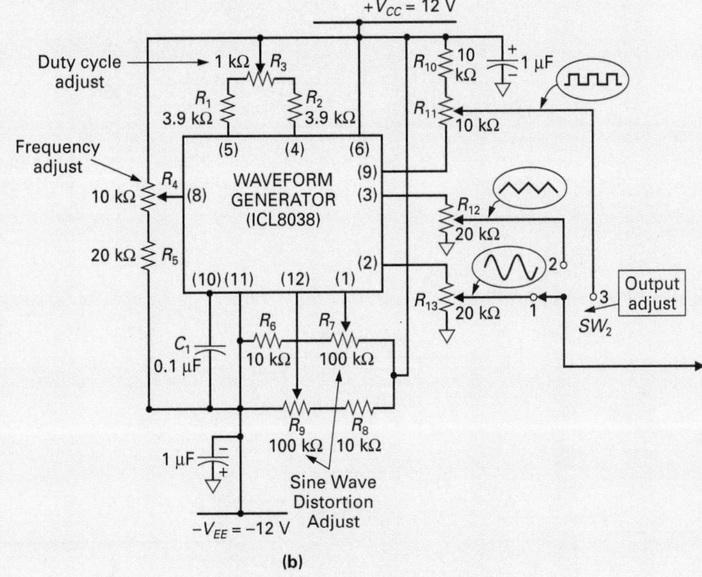

21. Function generators are test instruments whose output waveforms are used to test circuits by injecting an input into that circuit and then monitoring its response to the input. For example, sine wave signals are normally used to test linear circuits such as amplifiers and filters, whereas square wave signals are often used to test digital circuits such as logic gates. In the past, a function generator circuit was constructed using discrete components; however, today a complete function generator circuit is available in one linear IC.

22. The 8038 is an example of a function generator that can generate a 20 Hz to 100 kHz sine wave, square wave, or triangular wave output.

FIGURE 17-19 Linear IC Phase Locked Loops (PLLs).

Basic Block Diagram (Example: 565)

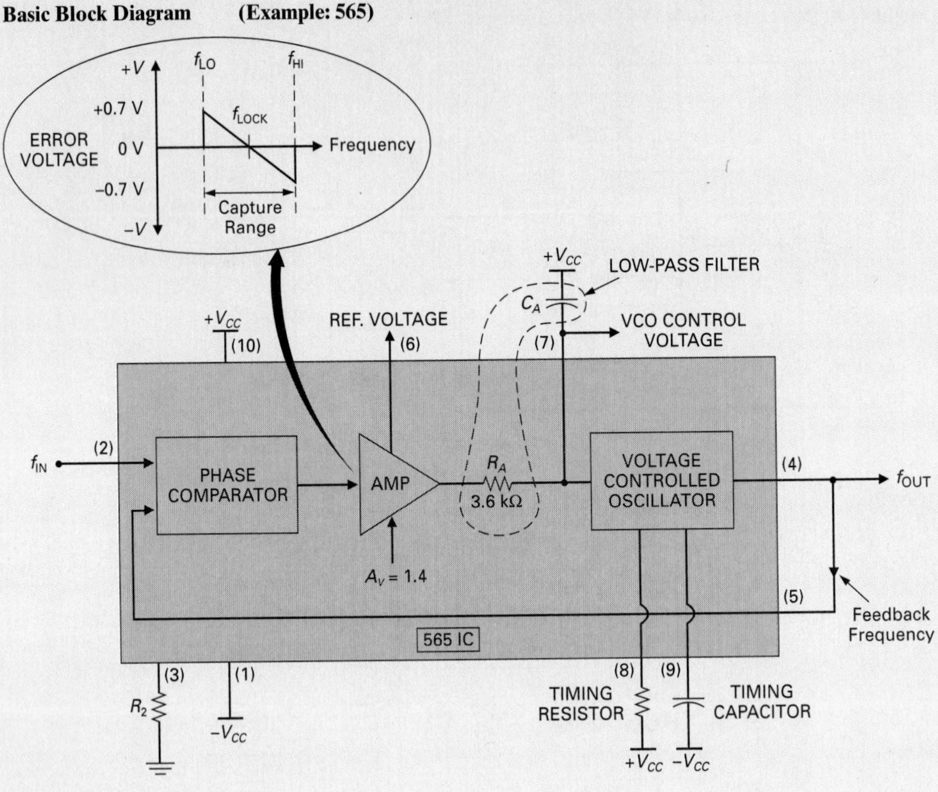

Application: Frequency Multiplier Circuit

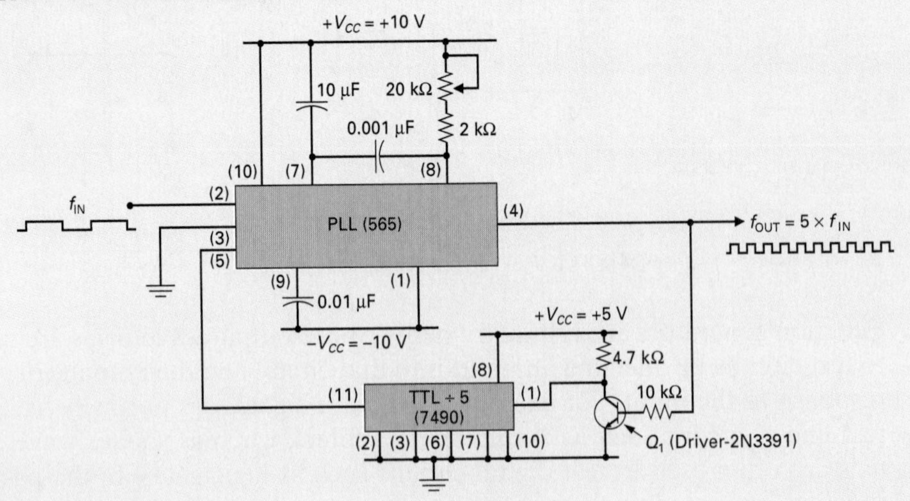

Linear IC Phase-Locked Loops (Figure 17-19)

23. A phase-locked loop circuit consists of a phase comparator that compares the output frequency of a voltage controlled oscillator with an input frequency. The error voltage out of the phase comparator is then coupled via an amplifier and low-pass filter to the control input of the voltage controlled oscillator to keep it in phase, and therefore at exactly the same frequency, as the input frequency. These phase-locked loop circuits will operate in either the free-running state, capture state, or locked state.
24. The 565 is a good example of a linear phase-locked loop IC.
25. The phase-locked loop circuit can be used in a variety of circuit applications, including the demodulation of FM radio signals, frequency multiplication and division, and signal regeneration.
26. When the PLL is used with a crystal-controlled oscillator as a frequency multiplier circuit, divider circuits are included in the feedback path between the VCO and phase comparator to cause the VCO to generate a very stable higher output frequency.

NEW TERMS

Capture State
Control Voltage
Error Voltage
Fixed Output Voltage Regulator
Free-Running State
Frequency Shift Keying
Function Generator IC
Locked State
Phase-Locked Loop IC
Phase-Locked Loop (PLL) Circuit
Modem
Sample-and-Hold Amplifier
Series Dissipative Regulator
Series Switching Regulator
Switching Power Supply
Switching Voltage Regulator
Variable Output Voltage Regulator
Voltage Controlled Oscillator (VCO)
Voltage Regulator IC
Waveform Generator

REVIEW QUESTIONS

Multiple-Choice Questions

1. Which semiconductor manufacturer was the first to develop an op-amp IC?

   a. National Semiconductor **c.** Fairchild
   b. Motorola **d.** Signetics

2. Which of the following is not a linear IC?

   a. Voltage regulator **c.** VCO
   b. Operational amplifier **d.** NAND gate

3. What would be the polarity and voltage output from a 7818 voltage regulator?

   a. –8 V **b.** +18 V **c.** +1.2 V to +37 V **d.** –18 V

4. What is the difference between a 78XX and 79XX voltage regulator?

   a. The 78XX has a variable output **c.** The 78XX has a + output
   b. The 79XX has a + output **d.** The 79XX has a variable output

5. Which of the following is an example of a series dissipative regulator?

   a. Zener regulator **c.** 7912 regulator
   b. 317 regulator **d.** All of the above

6. What advantage does the series dissipative have over the series switching regulator?

   a. Circuit simplicity **b.** Conversion efficiency

7. The frequency output from a VCO is determined by

   a. An *LC* network **c.** A control voltage
   b. An *RC* network **d.** A crystal

8. A PLL circuit contains a

   a. Crystal oscillator **b.** Bypass filter **c.** VCO **d.** VCR

9. What is the similarity between a switching voltage regulator and a phase-locked loop circuit?

   a. Both generate a constant output voltage
   b. Both employ a feedback mechanism to control operation
   c. Both generate a high-frequency output
   d. Both make use of high-pass filters

10. Which is the most frequently used IC?

    a. Voltage regulator **c.** Function generator
    b. Comparator **d.** Op-amp

Essay Questions

11. Define the difference between a linear IC and a digital IC. (Intro.)
12. List five examples of linear ICs. (Intro.)
13. Why are voltage regulators included in dc power-supply circuits? (Section 17-1)
14. Why would the load resistance of a dc power-supply change? (Section 17-1-2)
15. Why would the ac input voltage to a dc power supply vary during the day? (Section 17-1-2)
16. Define the term *percent of regulation.* (Section 17-1-3)
17. What is the disadvantage of the zener regulator? (Section 17-1-4)
18. If a zener regulator and IC voltage regulator are both series dissipative, what advantage does the IC voltage regulator have over the zener regulator? (Section 17-1-5)
19. Using a sketch, show how a 7800 series IC voltage regulator would be connected in circuit. (Section 17-1-5)
20. What is the difference between the three pins of a fixed output voltage regulator and the three pins of a variable voltage regulator? (Section 17-1-5)
21. Using a sketch, show how a 317 variable output voltage regulator would be connected in circuit to generate a variable output voltage. (Section 17-1-5)
22. What is the key advantage of series switching regulators? (Section 17-1-5)

23. Using a block diagram and waveforms, briefly describe the operation of a switching regulator. (Section 17-1-5)

24. What type of filter follows the series switching transistor of a switching regulator? (Section 17-1-5)

25. Give the full names of the following abbreviations:

a. PLL **b.** VCO **c.** IBM **d.** A/D converter

26. What determines the frequency output of a VCO? (Section 17-2-1)

27. What is a function generator? (Section 17-2)

28. For what purpose are function generators used? (Section 17-2-2)

29. Sketch the basic block diagram of a phase-locked loop circuit, and briefly describe its operation and three operating states. (Section 17-3-1)

30. What advantage do PLL circuits have over tuned amplifier circuits in frequency multiplier applications? (Section 17-3-2)

Practice Problems

31. To improve your circuit recognition and operation ability, study the circuit in Figure 17-20 and then answer the following questions:

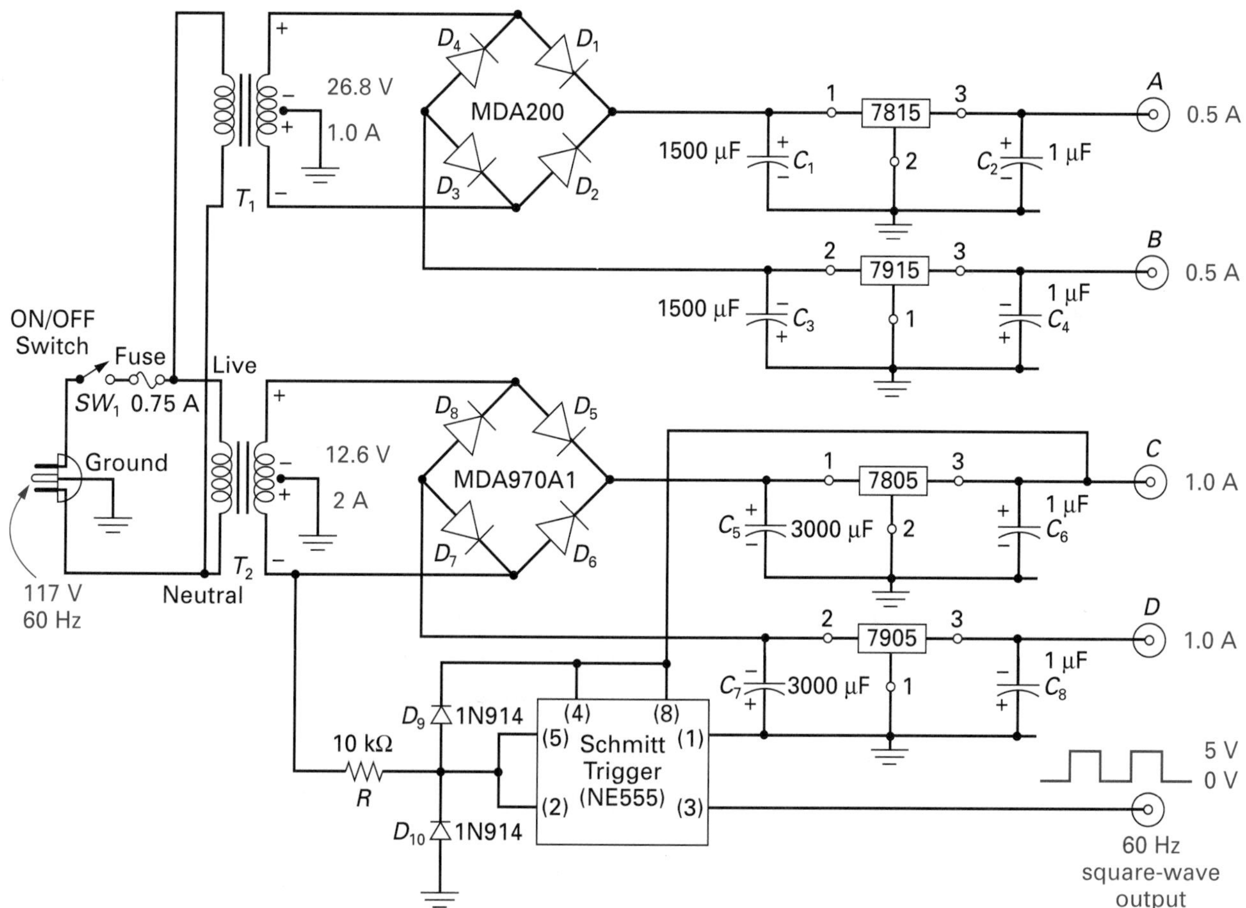

FIGURE 17-20 **Typical DC Power Supply Circuit.**

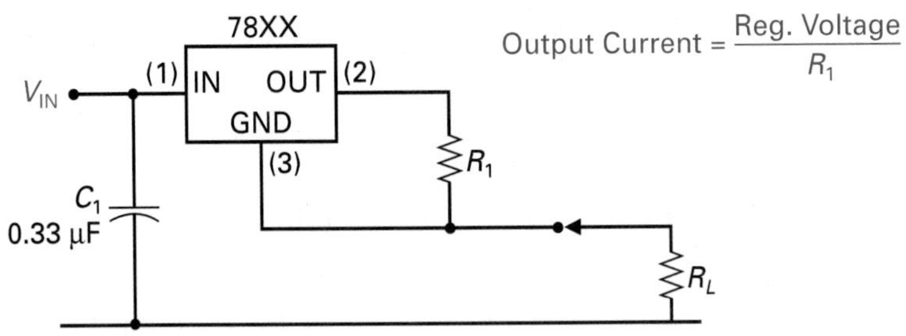

FIGURE 17-21 Constant Current Regulator Circuit.

a. Determine the output voltage and polarity at points *A, B, C* and *D*.
b. Describe whether the regulators used are fixed-positive, fixed-negative, variable-positive, variable-negative, or switching types.
c. What circuit is formed by diodes D_9 and D_{10}?
d. What is the purpose of the 555 timer?

32. Determine the constant current delivered to the load in Figure 17-21 if a 7815 is connected to the load via R_1, which is 1 kΩ.

33. To improve your circuit recognition and operation ability, study the circuit in Figure 17-22 and then answer the following questions:
a. What type of regulator is being used?
b. Why do you think this circuit is called a programmable dc power supply circuit?
c. Briefly describe the operation of this circuit.

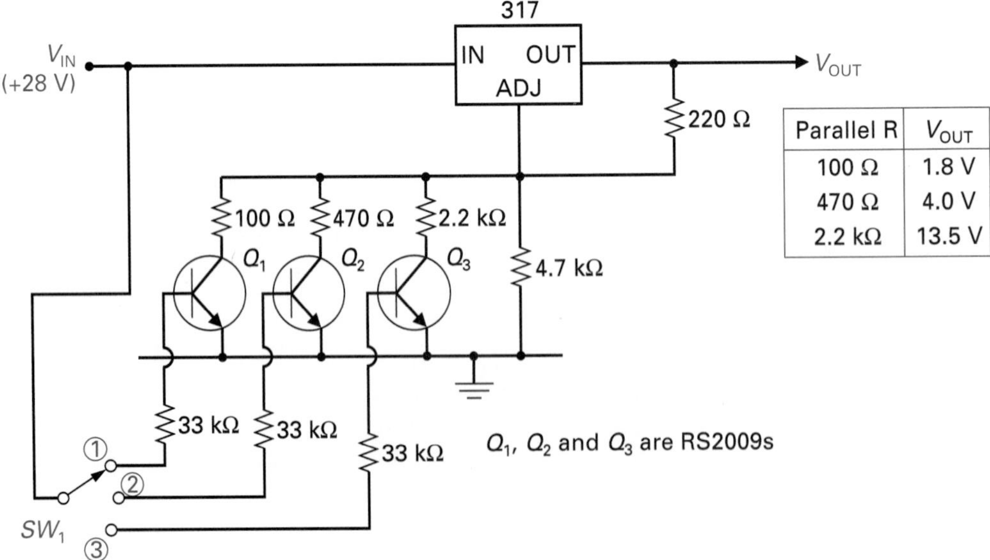

Parallel R	V_{OUT}
100 Ω	1.8 V
470 Ω	4.0 V
2.2 kΩ	13.5 V

FIGURE 17-22 A Programmable DC Power Supply Circuit.

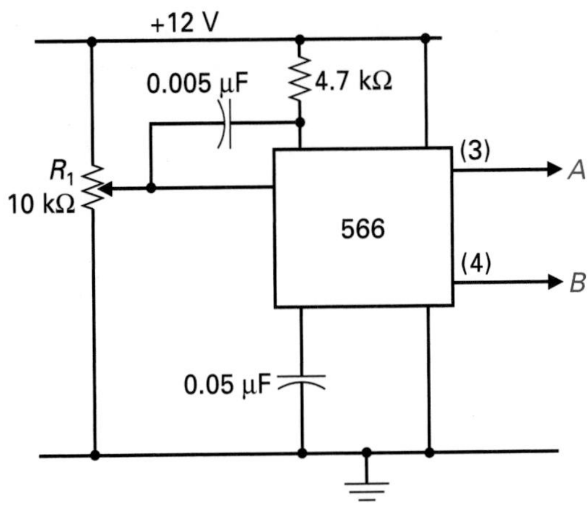

FIGURE 17-23 An Oscillator Circuit.

34. Referring to Figure 17-23,

a. What is the linear IC being used?

b. What would be the waveform output at points *A* and *B*?

c. What is the purpose of R_1?

35. Referring to Figure 17-24,

a. What is the full name of the linear IC being used?

b. What would be the output frequency for each of the switch positions?

c. Why would a PLL be used in this application instead of three crystal oscillator circuits?

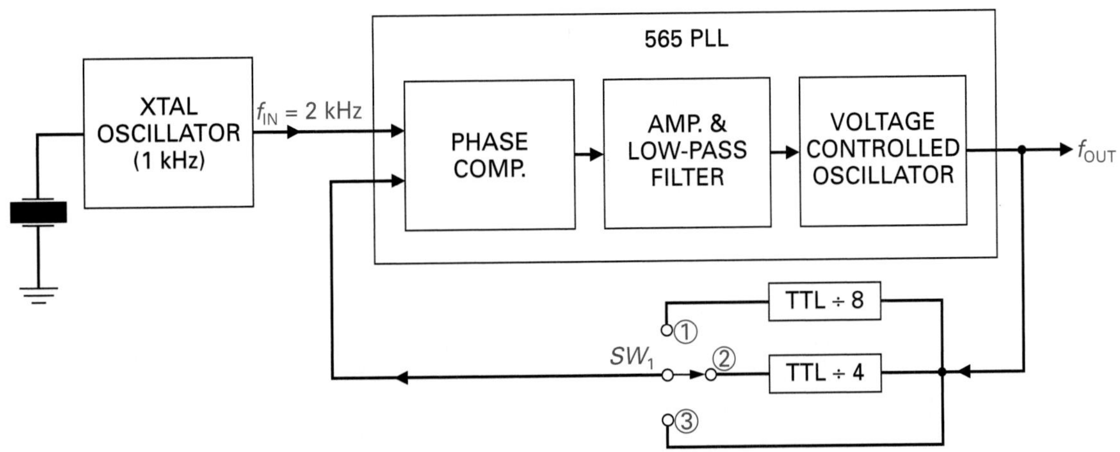

FIGURE 17-24 A Frequency Multiplier Circuit.

CHAPTER

18

OBJECTIVES

After completing this chapter, you will be able to:

1. Explain why digital circuits are two-state instead of ten-state.
2. Describe the differences between the decimal and binary number systems.
3. Explain the positional weight and reset-and-carry action for decimal, binary, and hexadecimal.
4. Explain how to convert numbers between decimal, binary, and hexadecimal.
5. Describe binary coded decimal (BCD) and the American Standard Code for Information Interchange (ASCII) binary codes.
6. Describe the basic operation and typical application for the following digital circuits:
   - **a.** Decoder
   - **b.** Encoder
   - **c.** Comparator
   - **d.** Multiplexer
   - **e.** Demultiplexer
   - **f.** Flip-Flop
7. Explain how a flip-flop can be made to function as a
   - **a.** Frequency divider
   - **b.** Binary counter
   - **c.** Register
   - **d.** One-Shot pulse generator
8. Describe the basic block-diagram operation, circuit-diagram operation, and troubleshooting of a digital frequency counter.

Digital Circuits

OUTLINE

VIGNETTE

Back to the Future

Each and every part of a digital electronic computer is designed to perform a specific task. One would imagine that the function and operation of the computer's basic blocks were first thought of by some pioneer in the twentieth century. In fact, however, two of the key units were first described in 1833 by Charles Babbage.

Born in England in 1791, Charles Babbage became very well known for both his mathematical genius and eccentric personality. For many years Babbage occupied the Cambridge Chair of Mathematics, once held by Isaac Newton. Although he never delivered a single lecture, he wrote a great number of papers on a variety of subjects, ranging from politics to manufacturing techniques. He also helped develop several practical devices including the tachometer and the railroad cowcatcher.

Babbage's ultimate pursuit, however, was that of mathematical accuracy. He delighted in spotting errors in everything from log tables to poetry. In fact, he once wrote to poet Alfred Lord Tennyson pointing out an inaccuracy in his line "every moment dies a man—every moment one is born." Babbage explained to Tennyson that since the world population was actually increasing and not, as he indicated, remaining constant, the line should be rewritten to read "every moment dies a man—every moment one and one-sixteenth is born."

In 1822, Babbage described and built a model of what he called "a difference engine," which could be used to calculate mathematical tables. The Royal Society of Scientists described his machine as "highly deserving of public encouragement," and a year later the government awarded Babbage £1500 for his project. Babbage originally estimated that the project should take three years. But the design had its complications and after ten years of frustrating labor—in which the government grants increased to £17,000—Babbage was still no closer to completion. Finally the money stopped, and Babbage reluctantly decided to let his brainchild go.

In 1833, Babbage developed an idea for a much more practical machine that he named "the analytical engine." It was to be a machine of a more general nature that could be used to solve a variety of problems, depending on instructions supplied by the operator. It would include two units, called the "mill" and the "store," both of which would be made of cogs and wheels. The store, which was equivalent to a modern-day computer memory, could hold up to 100 forty-digit numbers. The mill, which was equivalent to a modern computer's arithmetic and logic unit (ALU), could perform both arithmetic and logic operations on variables or numbers retrieved from the store, and the result could be stored in the store and then acted upon again or printed out. The program of instructions directing these operations would be fed into the analytical engine in the form of punched cards.

Sadly, the analytical engine was never built. All that remains are the volumes of descriptions and drawings and a section of the mill and printer built by Babbage's son, who also had to concede defeat. Unfortunately, for Charles Babbage it was a lifetime of frustration to have conceived the basic building blocks of the modern computer a century before the technology existed to build it.

INTRODUCTION

As discussed previously in Chapter 9, the early digital systems constructed in the 1950s made use of ten levels, or voltages, with each of these voltages corresponding to one of the ten digits in the decimal number system (0 = 0 V, 1 = 1 V, 2 = 2 V, 3 = 3 V, up to 9 = 9 V). However, these circuits were very complex because they had to generate one of ten voltage levels and sense the difference between all ten voltage levels. This complexity led to inaccuracy: some circuits would periodically confuse one voltage level for a different voltage level. *The solution to the problem of circuit complexity and information inaccuracy was solved by adopting a two-state system instead of a ten-state system.* Using a two-state or two-digit system, you can generate codes for any number, letter, or symbol, as we have seen previously in the ASCII table. As a result, the electronic circuits that manage the two-state codes are less complex because they only have to generate and sense a HIGH and LOW voltage. They are more accurate because there is little room for error between the two extremes of ON and OFF or HIGH voltage and LOW voltage.

Abandoning the ten-state system and adopting the two-state system for the advantages of circuit simplicity and accuracy means that we are no longer dealing with the decimal number system. Having only two digits (0 and 1) means that we are now operating in the two-state number system which is called "binary." The binary number system only has two digits in its number scale, and therefore only the digits "0" and "1" exist in binary. These two states are typically represented in an electronic circuit as two different values of voltage (0 = LOW voltage, 1 = HIGH voltage). Using combinations of "binary digits" (abbreviated "bits"), we can represent information as a binary code. This code is called a digital signal because it is an *information signal* that makes use of *binary digits.* Today almost all information—from your voice telephone conversations to the music on your compact disks—is digitized or converted to binary data form.

It would be safe to say that almost every electronic system today employs a combination of both linear and digital circuits. In the past, most electronic systems managed analog information signals, and therefore the majority of their internal circuits were linear. Today, digital electronic circuits tend to dominate most electronic systems because information stored, processed, or communicated in digital form is more accurate because it is less sensitive to noise.

From the beginning of this text, we have concentrated on both linear and digital devices and their applications. In the previous chapter we extended our coverage of linear circuits, and in this chapter we will extend our understanding of digital circuits. In Chapter 9 we examined in detail digital logic gate circuits, digital timer and control circuits, and digital circuit troubleshooting. In this chapter, we will continue our coverage of digital by first covering binary numbers in more detail, and then examining additional digital circuits such as decoders, encoders, comparators, multiplexers, demultiplexers, flip-flops, registers, and counters.

18-1 DIGITAL NUMBER SYSTEMS AND CODES

One of the best ways to understand anything new is to compare it to something that you are already familiar with so that the differences are highlighted. In this section we will be examining the "binary number system," which is the language used by digital electronic circuits. To best understand this new number system, we will compare it to the number system most familiar to us: the "decimal number system."

18-1-1 *The Decimal Number System*

The decimal system of counting and keeping track of items was first created by Hindu mathematicians in India in about 400 AD. Since it involved the use of fingers and thumbs, it was natural that this system would have 10 digits. The system found its way to all of the Arab countries in 800 AD, where it was named the *Arabic number system,* and from there it was eventually adopted by nearly all the European countries in 1200 where it was called the *decimal number system.*

Base or Radix
The base of a number system indicates the number of digits that will be used.

The key feature that distinguishes one number system from another is the number system's **base** or **radix.** This base indicates the number of digits that will be used. The decimal number system, for example, is a base-10 number system, which means that it will use 10 digits (0 through 9) to communicate information about an amount. A subscript is sometimes included after a number when different number systems are being used to indicate the base of the number. For example, $12{,}567_{10}$ would be a base-10 decimal number, whereas 10110_2 would be a base-2 number.

Positional Weight

The position of each digit of a decimal number determines the weight of that digit. A "1" by itself, for instance, is only worth 1, whereas a 1 to the left of three zeros makes the 1 worth 1,000.

Most Significant Digit (MSD)
The left-most digit in a number system.

Least Significant Digit (LSD)
The right-most digit in a number system.

In decimal notation, each position to the left of the decimal point indicates an increased *positive power of ten,* as seen in Figure 18-1(a). The total quantity or amount of the number is therefore determined by the size of each digit and the weighted position each digit is in. For example, the value shown in Figure 18-1(a) has six thousands, zero hundreds, one ten, and nine ones, which combined makes a total of 6019_{10}.

In the decimal number system, the left-most digit is called the **most significant digit (MSD)** while the right-most digit is called the **least significant**

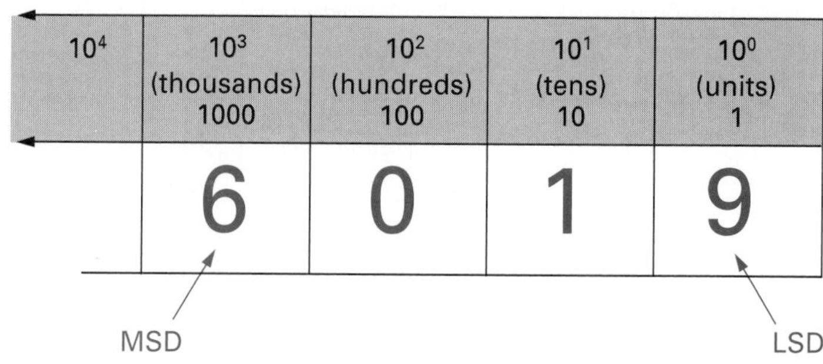

$$
\begin{aligned}
\text{Value of number} &= (6 \times 10^3) + (0 \times 10^2) + (1 \times 10^1) + (9 \times 10^0) \\
&= (6 \times 1000) + (0 \times 100) + (1 \times 10) + (9 \times 1) \\
&= 6000 + 0 + 10 + 9 = 6019_{10}
\end{aligned}
$$

(a)

100	10	1
		0
		1
		2
		3
		4
		5
		6
		7
		8
		9
	1	0
	1	1
	1	2

(b)

100	10	1
	9	4
	9	5
	9	6
	9	7
	9	8
	9	9
1	0	0
1	0	1
1	0	2
1	0	3
1	0	4
1	0	5
1	0	6

(c)

FIGURE 18-1 The Decimal (Base 10) Number System.

digit (LSD). Applying this to the example in Figure 18-1(a), the "6" would be the MSD because its position carries the most weight, while the "9" would be the LSD because its position carries the least weight.

Reset and Carry

Before proceeding to the binary number system, let us just review one other action that occurs when counting in decimal. This action—familiar to us all—is called **reset and carry.** Referring to Figure 18-1(b), you can see that a reset and carry operation occurs after a count of 9. The unit's column, which has reached its maximum count, resets to 0 and carries a 1 to the ten's column, resulting in a final count of "10."

Reset and Carry
An action that occurs in any column that reaches its maximum count.

This reset and carry action will occur in any column that reaches its maximum count. For example, Figure 18-1(c) shows how two reset and carry operations will take place after a count of 99. The unit's column resets and carries a 1 to the ten's column, which in turn resets and carries a 1 to the hundred's column, resulting in a final count of "100."

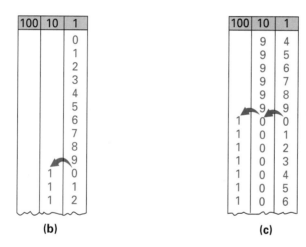

DECIMAL

10^2	10^1	10^0
100	10	1
		0
		1
		2
		3
		4
		5
		6
		7
		8
		9
	1	0
	1	1
	1	2

BINARY

2^3	2^2	2^1	2^0
8	4	2	1
			0
			1
		1	0
		1	1
	1	0	0
	1	0	1
	1	1	0
	1	1	1
1	0	0	0
1	0	0	1
1	0	1	0
1	0	1	1
1	1	0	0

FIGURE 18-2 A Comparison between Decimal and Binary.

18-1-2 The Binary Number System

As in the decimal system, the value of a binary digit is determined by its position relative to the other digits. In the decimal system, each position to the left of the decimal point increases by a power of 10. This case is also true with the binary number system; however, because it is a base-2 (bi) number system, each place to the left of the *binary point* increases by a power of 2. Figure 18-2 shows how columns of the binary and decimal number systems have different weights. For example, with binary, the columns are weighted so that 2^0 is one, 2^1 is two, 2^2 is four, 2^3 is eight ($2 \times 2 \times 2 = 8$), and so on.

As we know, the base or radix of a number system also indicates the maximum number of digits used by the number system. The base-2 binary number system only uses the first two digits on the number scale: 0 (zero) and 1 (one). The 0s and 1s in binary are called "binary digits," or **bits** for short.

Bits
An abbreviation for "binary digits."

Positional Weight

As in the decimal system, each column in binary carries its own weight, as seen in Figure 18-3(a). With the decimal number system, each position to the left increased 10 times. With binary, the weight of each column to the left increases 2 times. The first column therefore has a weight of 1, the second column has a weight of 2, the third column has a weight of 4, the fifth 8, and so on. The value or quantity of a binary number is determined by the digit in each column and the positional weight of the column. For example, in Figure 18-3(a), the binary number 101101_2 is equal in decimal to 45_{10} because we have one $\times$ 32, one $\times$ 8, one $\times$ 4, and one $\times$ 1 ($32 + 8 + 4 + 1 = 45$). The left-most binary digit is called the **most significant bit (MSB)** because it carries the most weight, while the right-most digit is called the **least significant bit (LSB)** because it carries the least weight. Applying this to the example in Figure 18-3(a), the 1 in the thirty-two's column is the MSB, while the 1 in the unit's column is the LSB.

Most Significant Bit (MSB)
The left-most binary digit.

Least Significant Bit (LSB)
The right-most binary digit.

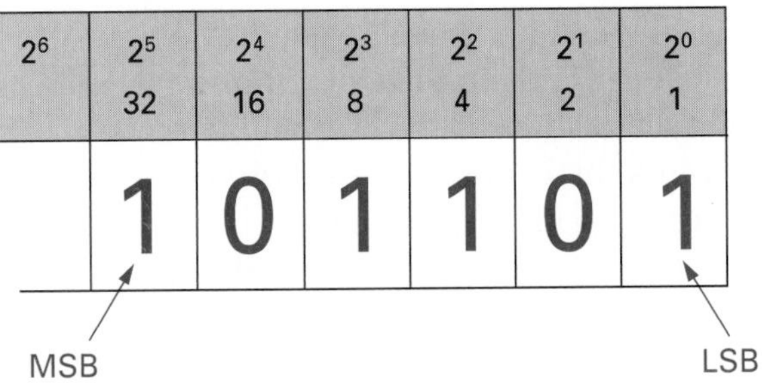

$$
\begin{aligned}
\text{Value of number} &= (1 \times 2^5) + (0 \times 2^4) + (1 \times 2^3) + (1 \times 2^2) + (1 \times 2^1) + (1 \times 2^0) \\
&= (1 \times 32) + (0 \times 16) + (1 \times 8) + (1 \times 4) + (0 \times 2) + (1 \times 1) \\
&= 32 + 0 + 8 + 4 + 0 + 1 \\
&= 45_{10}
\end{aligned}
$$

(a)

DECIMAL

	10^1 10	10^0 1
		0
		1
		2
		3
		4
		5
		6
		7
		8
		9
	1	0
	1	1
	1	2
	1	3
	1	4
	1	5

BINARY

2^3 8	2^2 4	2^1 2	2^0 1
			0
			1
		1	0
		1	1
	1	0	0
	1	0	1
	1	1	0
	1	1	1
1	0	0	0
1	0	0	1
1	0	1	0
1	0	1	1
1	1	0	0
1	1	0	1
1	1	1	0
1	1	1	1

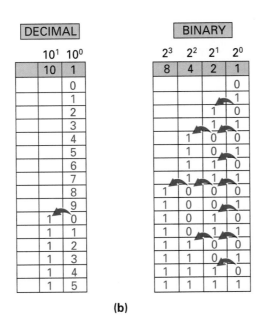

(b)

FIGURE 18-3 The Binary (Base 2) Number System.

Reset and Carry

The reset and carry action will occur in binary in exactly the same way it did in decimal. However, since binary has only two digits, a column will reach its maximum digit a lot sooner, and the reset and carry action in the binary number system will occur much more frequently. Referring to Figure 18-3(b), you can see that the binary counter begins with 0 and then advances to 1. At this stage, the unit's column has reached its maximum, and the next count forces the unit's column to reset and carry into the next column, producing a count of 0010_2 (2_{10}). The unit's column then advances to a count of 0011_2 (3_{10}). At this stage, both the unit's column and two's column have reached their maximum. As the count advances by one, it will cause the unit's column to reset and carry a one into the two's column, which will also have to reset and carry a one into the four's column. This will result in a final count of 0100_2 (4_{10}). The count will then continue to 0101_2 (5_{10}), 0110_2 (6_{10}), 0111_2 (7_{10}), and then 1000_2 (8_{10}), which, as you can see in Figure 18-3(b), is a result of three reset and carries. Comparing the binary reset and carry to the decimal reset and carry in Figure 18-3(b), you can see that because the binary number system uses only two digits, binary numbers quickly turn into multidigit figures. For example, a decimal eight (8) uses only one digit, while a binary eight (1000) uses four digits.

Converting Binary Numbers to Decimal Numbers

Binary numbers can easily be converted to their decimal equivalent by simply adding together all of the column weights that contain a binary 1, as we did previously in Figure 18-3(a).

■ **EXAMPLE:**

Convert the following binary numbers to their decimal equivalents:

a. 1010
b. 101101

■ ***Solution:***

a.

Binary Column Weights	32	16	8	4	2	1
Binary Number			1	0	1	0

Decimal Equivalent $= (1 \times 8) + (0 \times 4) + (1 \times 2) + (0 \times 1)$

$= 8 + 0 + 2 + 0 = 10_{10}$

b.

Binary Column Weights	32	16	8	4	2	1
Binary Number	1	0	1	1	0	1

Decimal Equivalent $=(1\times32)+(0\times16)+(1\times8)+(1\times4)+(0\times2)+(1\times1)$

$= 32 + 0 + 8 + 4 + 0 + 1 = 45_{10}$

■ **EXAMPLE:**

The LEDs in Figure 18-4 are being used as a 4-bit (four binary digit) display. When the LED is OFF it indicates a binary 0, and when the LED is ON it indicates a binary 1. Determine the decimal equivalent of the binary displays shown in Figure 18-4(a), (b), and (c).

■ ***Solution:***

a. $0101_2 = 5_{10}$
b. $1110_2 = 14_{10}$
c. $1001_2 = 9_{10}$

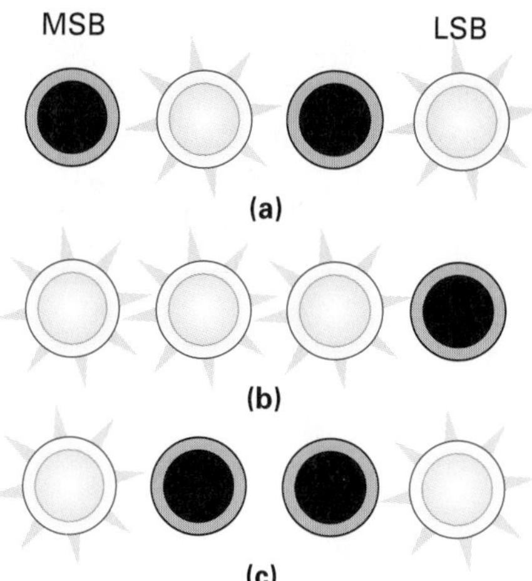

FIGURE 18-4 Using LEDs to Display Binary Numbers.

Converting Decimal Numbers to Binary Numbers

To convert a decimal number to its binary equivalent, continually subtract the largest possible power-of-two until the decimal number is reduced to zero, placing a binary 1 in columns that are used and a binary 0 in the columns that are not used. To explain how simple this process is, refer to Figure 18-5, which shows how decimal 53 can be converted to its binary equivalent. As you can see in this example, the first largest power-of-two that can be subtracted from decimal 53 is 32, therefore a 1 is placed in the 32 column and 21 is remaining. The largest power of 2 that can be subtracted from the remainder 21 is 16. Therefore, a 1 is placed in the 16 column, and 5 is remaining. The next largest power of 2 that can be subtracted from 5 is 4. Therefore, a 1 is placed in the 4 column, and 1 is remaining. The final 1 is placed in the unit's column; therefore, decimal 53 is represented in binary as 11 001, which indicates that the value is 1 × 32, 1 × 16, 1 × 4 and 1 × 1.

■ **EXAMPLE:**

Convert the following decimal numbers to their binary equivalents:

a. 25
b. 55

■ ***Solution:***

a.

	32	16	8	4	2	1
Binary Column Weights	32	16	8	4	2	1
Binary Number		1	1	0	0	1

Decimal Equivalent $25 - 16 = 9, 9 - 8 = 1, 1 - 1 = 0.$

b.

	32	16	8	4	2	1
Binary Column Weights	32	16	8	4	2	1
Binary Number	1	1	0	1	1	1

Decimal Equivalent $55 - 32 = 23, 23 - 16 = 7, 7 - 4 = 3, 3 - 2 = 1, 1 - 1 = 0.$

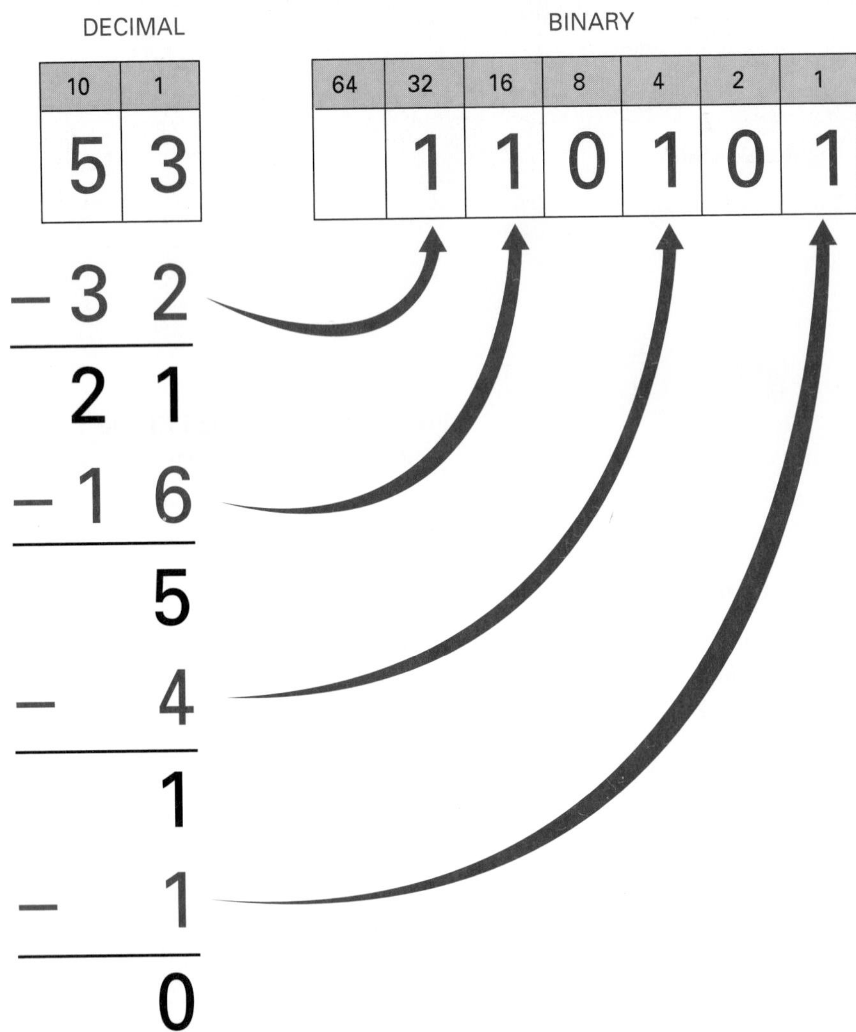

FIGURE 18-5 Decimal to Binary Conversion.

18-1-3 The Hexadecimal Number System

Hexadecimal Number System
A base 16 number system, using the digits 0 through 9 and the six additional digits represented by A, B, C, D, E, and F.

If digital electronic circuits operate using binary numbers and we operate using decimal numbers, why is there any need for us to have the **hexadecimal number system?** Hexadecimal, or hex, is used as a sort of "shorthand" for large strings of binary numbers. To begin with, let us examine the basics of this number system.

Hexadecimal means 16, and this number system has 16 different digits as shown in Figure 18-6(a), which shows a comparison between decimal, hexadecimal, and binary. Looking at the first nine digits in the decimal and hexadecimal columns, you can see that there is no difference between the two. However, beyond 9, hexadecimal makes use of the letters A, B, C, D, E, and F. Digits 0 through 9 and letters A through F make up the total 16 digits of the hexadecimal number system. Comparing hexadecimal to decimal once again, you can see that $A_{16} = 10_{10}$, $B_{16} = 11_{10}$, $C_{16} = 12_{10}$, $D_{16} = 13_{10}$, $E_{16} = 14_{10}$, and $F_{16} = 15_{10}$. Having these extra digits means that a column reset-and-carry will not occur until the

DECIMAL			HEXADECIMAL				BINARY					
10^2	10^1	10^0	16^3	16^2	16^1	16^0	2^5	2^4	2^3	2^2	2^1	2^0
100	10	1	4096	256	16	1	32	16	8	4	2	1
		0				0	0	0	0	0	0	0
		1				1	0	0	0	0	0	1
		2				2	0	0	0	0	1	0
		3				3	0	0	0	0	1	1
		4				4	0	0	0	1	0	0
		5				5	0	0	0	1	0	1
		6				6	0	0	0	1	1	0
		7				7	0	0	0	1	1	1
		8				8	0	0	1	0	0	0
		9				9	0	0	1	0	0	1
	1	0				A	0	0	1	0	1	0
	1	1				B	0	0	1	0	1	1
	1	2				C	0	0	1	1	0	0
	1	3				D	0	0	1	1	0	1
	1	4				E	0	0	1	1	1	0
	1	5				F	0	0	1	1	1	1
	1	6			1	0	0	1	0	0	0	0
	1	7			1	1	0	1	0	0	0	1
	1	8			1	2	0	1	0	0	1	0
	1	9			1	3	0	1	0	0	1	1
	2	0			1	4	0	1	0	1	0	0
	2	1			1	5	0	1	0	1	0	1
	2	2			1	6	0	1	0	1	1	0
	2	3			1	7	0	1	0	1	1	1
	2	4			1	8	0	1	1	0	0	0
	2	5			1	9	0	1	1	0	0	1
	2	6			1	A	0	1	1	0	1	0
	2	7			1	B	0	1	1	0	1	1
	2	8			1	C	0	1	1	1	0	0
	2	9			1	D	0	1	1	1	0	1
	3	0			1	E	0	1	1	1	1	0
	3	1			1	F	0	1	1	1	1	1
	3	2			2	0	1	0	0	0	0	0
	3	3			2	1	1	0	0	0	0	1
	3	4			2	2	1	0	0	0	1	0

(a)

A9	DE7	78	F9	10C
AA	DE8	79	FA	10D
AB	DE9	7A	FB	10E
AC	DEF	7B	FC	10F
AD	DF0	7C	FD	110
AE	DF1	7D	FE	111
AF	DF2	7E	FF	112
B0	DF3	7F	100	
B1		80	101	
		81		

(b)

FIGURE 18-6 Hexadecimal Reset and Carry. (a) Number System Comparison. (b) Hex Counting Examples.

16^4	16^3	16^2	16^1	16^0
65,536	4096	256	16	1
			4	C

Value of Number = (C × 1) + (4 × 16)
= (12 × 1) + (4 × 16)
= 12 + 64 = 76_{10}

FIGURE 18-7 Positional Weight of the Hexadecimal Number System.

count has reached the last and largest digit, F. The hexadecimal column in Figure 18-6(a) shows how reset and carry will occur whenever a column reaches its maximum digit F, and this action is further illustrated in the examples in Figure 18-6(b).

Converting Hexadecimal Numbers to Decimal Numbers

To find the decimal equivalent of a hexadecimal number, simply multiply each hexadecimal digit by its positional weight. Figure 18-7 shows the positional weight of the hexadecimal columns, which are each sixteen times larger. The hexadecimal number 4C therefore indicates that the value has 4 × 16 and C × 1, and because C is equal to 12 in decimal, this is equivalent 12 × 1. The result of 12 + 64 is 76, so hexadecimal 4C is equivalent to decimal 76.

■ **EXAMPLE:**

Convert hexadecimal 8BF to its decimal equivalent.

■ ***Solution:***

Hexadecimal Column Weights	4096	256	16	1
		8	B	F

Decimal Equivalent = (8 × 256) + (B × 16) + (F × 1)
= 2048 + (11 × 16) + (15 × 1)
= 2048 + 176 + 15
= 2239

Converting Decimal Numbers to Hexadecimal Numbers

Decimal to hexadecimal conversion is achieved in the same way as decimal to binary. First subtract the largest possible power of 16, and then keep subtracting the largest possible power of 16 from the remainder. Each time a subtraction takes place, add 1 to the respective column until the decimal value has been reduced to zero. Figure 18-8 illustrates this procedure with an example showing how decimal 425 is converted to its hexadecimal equivalent. To begin with, the largest possible power of 16 (256) is subtracted once from 425 and a 1 is placed in the 256 column, leaving a remainder of 169. The next largest power of 16 is

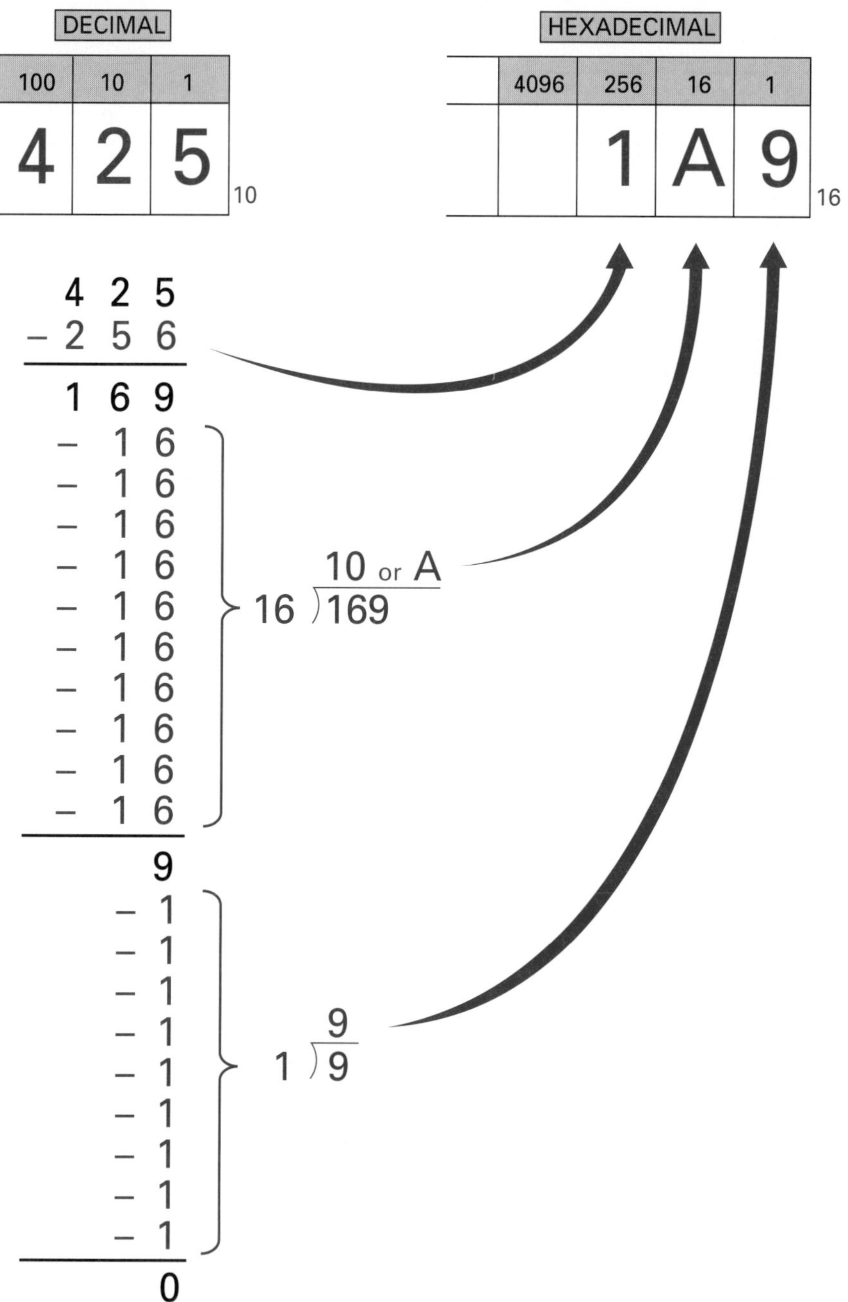

FIGURE 18-8 Decimal to Hexadecimal Conversion.

16, which can be subtracted ten times from 169. Therefore, the hexadecimal equivalent of 10 which is A, is placed in the 16's column, leaving a remainder of 9. Since nine 1s can be subtracted from the remainder of 9, the unit's column is advanced nine times, giving us our final hexadecimal result, 1A9.

■ **EXAMPLE:**

Convert decimal 4525 to its hexadecimal equivalent.

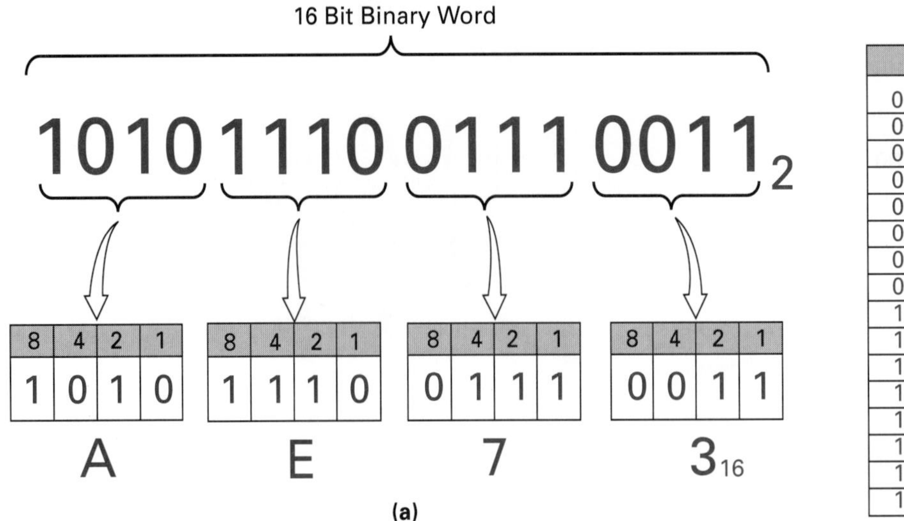

Binary	Hex
0 0 0 0	0
0 0 0 1	1
0 0 1 0	2
0 0 1 1	3
0 1 0 0	4
0 1 0 1	5
0 1 1 0	6
0 1 1 1	7
1 0 0 0	8
1 0 0 1	9
1 0 1 0	A
1 0 1 1	B
1 1 0 0	C
1 1 0 1	D
1 1 1 0	E
1 1 1 1	F

(b)

FIGURE 18-9 Representing Binary Numbers in Hexadecimal.

■ ***Solution:***

Hexadecimal Column Weights	4096	256	16	1
	1	1	A	D

Decimal Equivalent = 4525 – 4096 = 429, 429 – 256 = 173,
173 – 16 – 16 – 16 – 16 – 16 – 16 – 16 – 16 – 16 – 16 = 13,
13 – 1 – 1 – 1 – 1 – 1 – 1 – 1 – 1 – 1 – 1 – 1 – 1 – 1 = 0.

Converting between Binary and Hexadecimal

Binary Word
A numerical value expressed as a group of binary digits.

As mentioned in the beginning of this section, hexadecimal is used as a "shorthand" for representing large groups of binary digits. To illustrate this, Figure 18-9(a) shows how a 16-bit binary number—which is more commonly called a 16-bit **binary word**—can be represented by 4 hexadecimal digits. Figure 18-9(b) shows how a 4-bit binary word can have any value from 0_{10} (0000_2) to 15_{10} (1111_2), and, because hexadecimal has the same number of digits (0 through F), we can use one hexadecimal digit to represent 4 binary bits. As you can see in the example in Figure 18-9(a), it is much easier to work with a number like AE73 than 1010111001110011.

To convert from hexadecimal back to binary, we simply do the opposite, as shown in the example in Figure 18-10. Because each hexadecimal digit represents 4 binary digits, a 4-digit hexadecimal number will convert to a 16-bit binary word.

■ **EXAMPLE:**

Convert hexadecimal 2BF9 to its binary equivalent.

■ ***Solution:***

Hexadecimal Number	2	B	F	9
Binary Equivalent	0010	1011	1111	1001

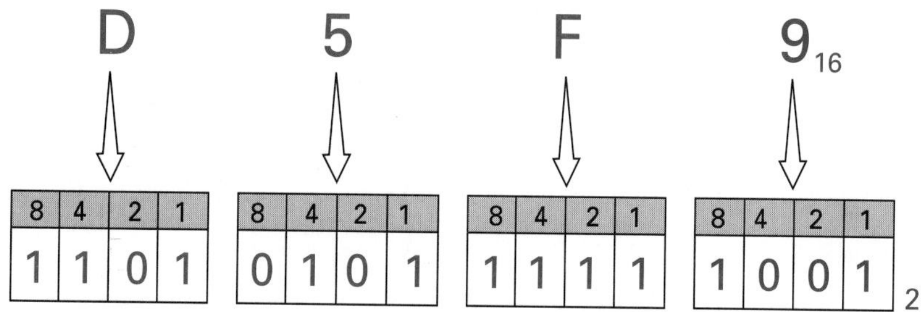

FIGURE 18-10 **Hexadecimal to Binary Conversion.**

■ **EXAMPLE:**

Convert binary 110011100001 to its hexadecimal equivalent.

■ ***Solution:***

Binary Number	1100	1110	0001
Hexadecimal Equivalent	C	E	1

18-1-4 ***Binary Codes***

The process of converting a decimal number to its binary equivalent is called **binary coding.** The result of the conversion is a binary number or code that is called **pure binary.** There are, however, other binary codes used in digital circuits other than pure binary, and we will examine these in this section.

Binary Coding
The process of converting a decimal number to its binary equivalent.

Pure Binary
The binary number that is the result of binary coding.

The Binary Coded Decimal (BCD) Code

No matter how familiar you become with binary, it will always be less convenient to work with than the decimal number system. For example, it will always take us a short time to convert 1111000_2 to 120_{10}. Designers realized this disadvantage early on and developed a binary code that had decimal characteristics, that was appropriately named **binary coded decimal (BCD).** Being a binary code, it had the advantages of a two-state system, but because it has a decimal format it is much easier for an operator to interface (via a decimal keypad or decimal display) to systems such as pocket calculators, wristwatches, and so on.

Binary Coded Decimal (BCD)
A binary code that has decimal characteristics.

The BCD code expresses each decimal digit as a 4-bit word, as shown in the example in Figure 18-11(a). In this example, decimal 1,753 converts to a BCD code of 0001 0111 0101 0011, with the first 4-bit code ($0001_2 = 1_{10}$) representing the 1 in the thousand's column, the second 4-bit code ($0111_2 = 7_{10}$) representing the 7 in the hundred's column, the third 4-bit code ($0101_2 = 5_{10}$) representing the 5 in the ten's column, and the fourth 4-bit code ($0011_2 = 3_{10}$) representing the 3 in the unit's column. As can be seen in Figure 18-11(a), the subscript "BCD" is often used after a BCD code to distinguish it from a pure binary number ($1753_{10} = 0001\ 0111\ 0101\ 0011_{BCD}$).

Figure 18-11(b) compares decimal, binary, and BCD, and, as you can see, the reset-and-carry action occurs in BCD at the same time as it does in decimal. This is because BCD has only ten 4-bit binary codes, 0000, 0001, 0010, 0011,

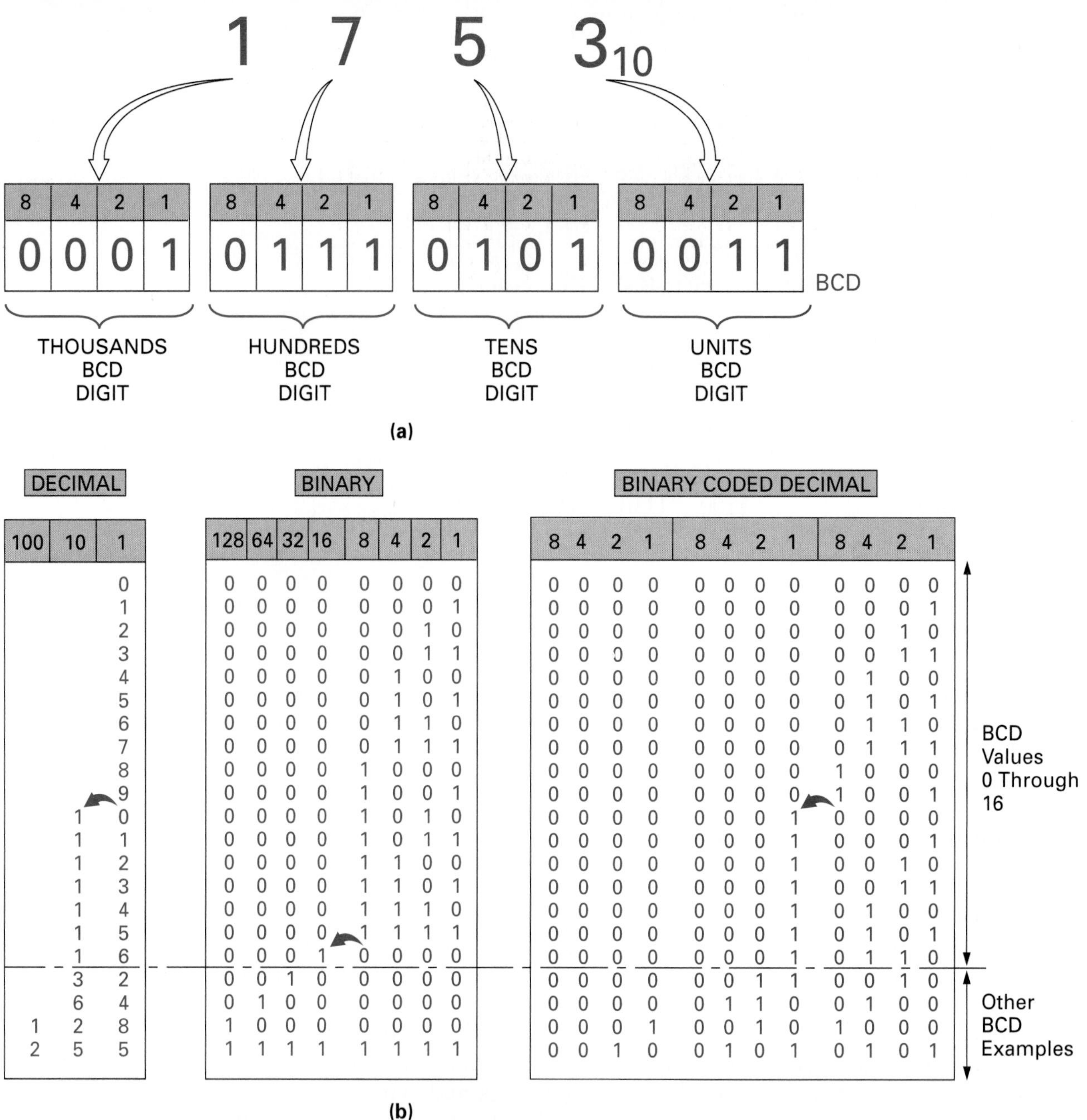

DECIMAL			BINARY								BINARY CODED DECIMAL											
100	10	1	128	64	32	16	8	4	2	1	8	4	2	1	8	4	2	1	8	4	2	1
		0	0	0	0	0	0	0	0	0	0	0	0	0	0	0	0	0	0	0	0	0
		1	0	0	0	0	0	0	0	1	0	0	0	0	0	0	0	0	0	0	0	1
		2	0	0	0	0	0	0	1	0	0	0	0	0	0	0	0	0	0	0	1	0
		3	0	0	0	0	0	0	1	1	0	0	0	0	0	0	0	0	0	0	1	1
		4	0	0	0	0	0	1	0	0	0	0	0	0	0	0	0	0	0	1	0	0
		5	0	0	0	0	0	1	0	1	0	0	0	0	0	0	0	0	0	1	0	1
		6	0	0	0	0	0	1	1	0	0	0	0	0	0	0	0	0	0	1	1	0
		7	0	0	0	0	0	1	1	1	0	0	0	0	0	0	0	0	0	1	1	1
		8	0	0	0	0	1	0	0	0	0	0	0	0	0	0	0	0	1	0	0	0
		9	0	0	0	0	1	0	0	1	0	0	0	0	0	0	0	0	1	0	0	1
	1	0	0	0	0	0	1	0	1	0	0	0	0	0	0	0	0	1	0	0	0	0
	1	1	0	0	0	0	1	0	1	1	0	0	0	0	0	0	0	1	0	0	0	1
	1	2	0	0	0	0	1	1	0	0	0	0	0	0	0	0	0	1	0	0	1	0
	1	3	0	0	0	0	1	1	0	1	0	0	0	0	0	0	0	1	0	0	1	1
	1	4	0	0	0	0	1	1	1	0	0	0	0	0	0	0	0	1	0	1	0	0
	1	5	0	0	0	0	1	1	1	1	0	0	0	0	0	0	0	1	0	1	0	1
	1	6	0	0	0	1	0	0	0	0	0	0	0	0	0	0	0	1	0	1	1	0
	3	2	0	0	1	0	0	0	0	0	0	0	0	0	0	0	1	1	0	0	1	0
	6	4	0	1	0	0	0	0	0	0	0	0	0	0	0	1	1	0	0	1	0	0
1	2	8	1	0	0	0	0	0	0	0	0	0	0	1	0	0	1	0	1	0	0	0
2	5	5	1	1	1	1	1	1	1	1	0	0	1	0	0	1	0	1	0	1	0	1

FIGURE 18-11 Binary Coded Decimal (BCD). (a) Decimal to BCD Conversion Example. (b) A Comparison of Decimal, Binary, and BCD.

0100, 0101, 0110, 0111, 1000, and 1001 (0 through 9), and when the maximum digit is reached, a reset-and-carry will occur.

■ **EXAMPLE:**

Convert the BCD code 0101 1000 0111 0000 to decimal, and the decimal number 369 to BCD.

■ ***Solution:***

BCD Code	0101	1000	0111	0000
Decimal Equivalent	5	8	7	0
Decimal Number	3	6	9	
BCD Code	0011	0110	1001	

The American Standard Code for Information Interchange (ASCII)

Up until this time we have only discussed how we can encode numbers into binary codes. Digital electronic computers must also be able to generate and recognize binary codes that represent letters of the alphabet and symbols. The ASCII code, which was discussed previously in Chapter 9, is the most widely used **alphanumeric code** (alphabet and numeral code). As I write this book, these ASCII codes are being generated by my computer keyboard, and the computer is decoding these ASCII codes and then displaying the alphanumeric equivalent (letter, number, or symbol) on my computer monitor.

Alphanumeric Code
Alphabet and numeral code.

■ **EXAMPLE:**

List the ASCII codes for the message "Digital." Refer to Figure 18-12.

■ ***Solution:***

D = 1000100
i = 1101001
g = 1100111
i = 1101001
t = 1110100
a = 1100001
l = 1101100

SELF-TEST REVIEW QUESTIONS FOR SECTION 18-1

1. What is the base of the
   a. Decimal number system
   b. Binary number system
   c. Hexadecimal number system
2. What is the decimal and hexadecimal equivalent of 10101_2?
3. Convert $1011\ 1111\ 0111\ 1010_2$ to its hexadecimal equivalent.
4. What is the binary and hexadecimal equivalent of 33_{10}?
5. Convert 7629_{10} to BCD.
6. What does the following string of ASCII codes mean?
   1000010 1100001 1100010 1100010 1100001 1100111 1100101

The American Standard Code for Information Interchange (ASCII)

•	0	1	0	0	0	0	0
!	0	1	0	0	0	0	1
"	0	1	0	0	0	1	0
#	0	1	0	0	0	1	1
$	0	1	0	0	1	0	0
%	0	1	0	0	1	0	1
&	0	1	0	0	1	1	0
'	0	1	0	0	1	1	1
(	0	1	0	1	0	0	0
)	0	1	0	1	0	0	1
*	0	1	0	1	0	1	0
+	0	1	0	1	0	1	1
,	0	1	0	1	1	0	0
–	0	1	0	1	1	0	1
.	0	1	0	1	1	1	0
/	0	1	0	1	1	1	1
0	0	1	1	0	0	0	0
1	0	1	1	0	0	0	1
2	0	1	1	0	0	1	0
3	0	1	1	0	0	1	1
4	0	1	1	0	1	0	0
5	0	1	1	0	1	0	1
6	0	1	1	0	1	1	0
7	0	1	1	0	1	1	1
8	0	1	1	1	0	0	0
9	0	1	1	1	0	0	1
:	0	1	1	1	0	1	0
;	0	1	1	1	0	1	1
<	0	1	1	1	1	0	0
=	0	1	1	1	1	0	1
>	0	1	1	1	1	1	0
?	0	1	1	1	1	1	1
@	1	0	0	0	0	0	0
A	1	0	0	0	0	0	1
B	1	0	0	0	0	1	0
C	1	0	0	0	0	1	1
D	1	0	0	0	1	0	0
E	1	0	0	0	1	0	1
F	1	0	0	0	1	1	0
G	1	0	0	0	1	1	1
H	1	0	0	1	0	0	0
I	1	0	0	1	0	0	1
J	1	0	0	1	0	1	0
K	1	0	0	1	0	1	1
L	1	0	0	1	1	0	0
M	1	0	0	1	1	0	1
N	1	0	0	1	1	1	0
O	1	0	0	1	1	1	1

P	1	0	1	0	0	0	0
Q	1	0	1	0	0	0	1
R	1	0	1	0	0	1	0
S	1	0	1	0	0	1	1
T	1	0	1	0	1	0	0
U	1	0	1	0	1	0	1
V	1	0	1	0	1	1	0
W	1	0	1	0	1	1	1
X	1	0	1	1	0	0	0
Y	1	0	1	1	0	0	1
Z	1	0	1	1	0	1	0
[	1	0	1	1	0	1	1
/	1	0	1	1	1	0	0
]	1	0	1	1	1	0	1
^	1	0	1	1	1	1	0
_	1	0	1	1	1	1	1
'	1	1	0	0	0	0	0
a	1	1	0	0	0	0	1
b	1	1	0	0	0	1	0
c	1	1	0	0	0	1	1
d	1	1	0	0	1	0	0
e	1	1	0	0	1	0	1
f	1	1	0	0	1	1	0
g	1	1	0	0	1	1	1
h	1	1	0	1	0	0	0
i	1	1	0	1	0	0	1
j	1	1	0	1	0	1	0
k	1	1	0	1	0	1	1
l	1	1	0	1	1	0	0
m	1	1	0	1	1	0	1
n	1	1	0	1	1	1	0
o	1	1	0	1	1	1	1
p	1	1	1	0	0	0	0
q	1	1	1	0	0	0	1
r	1	1	1	0	0	1	0
s	1	1	1	0	0	1	1
t	1	1	1	0	1	0	0
u	1	1	1	0	1	0	1
v	1	1	1	0	1	1	0
w	1	1	1	0	1	1	1
x	1	1	1	1	0	0	0
y	1	1	1	1	0	0	1
z	1	1	1	1	0	1	0
(	1	1	1	1	0	1	1
¦	1	1	1	1	1	0	0
)	1	1	1	1	1	0	1
~	1	1	1	1	1	1	0
DEL	1	1	1	1	1	1	1

FIGURE 18-12 The American Standard Code for Information Interchange (ASCII).

18-2 DIGITAL CIRCUITS

Previously in Chapter 9, we discussed digital logic gate circuits, digital timer and control circuits, and digital circuit troubleshooting. In this section we extend our understanding of digital circuits, and discuss the operation and application of decoders, encoders, comparators, multiplexers, demultiplexers, flip-flops, registers, and counters.

18-2-1 *Decoder Circuits*

The term decode means to translate coded characters to a more understandable form, or to reverse a previous encoding process. A **digital decoder circuit** is a logic circuit that responds to one specific input word or code while rejecting all others.

Digital Decoder Circuit
A logic circuit that responds to one specific input word or code while rejecting all others.

Basic Decoder Circuits

Figure 18-13(a) shows how an AND gate and two inverters can be connected to act as a basic decoder. As you know, all inputs to an AND gate need to be 1 for the output to be 1. Referring to the function table for this circuit, you can see that only when the input code is 001 will the AND gate have all of its inputs HIGH and therefore produce a HIGH output.

While some decoders are designed to recognize only one input word combination, other decoders are designed to activate one output for every input word

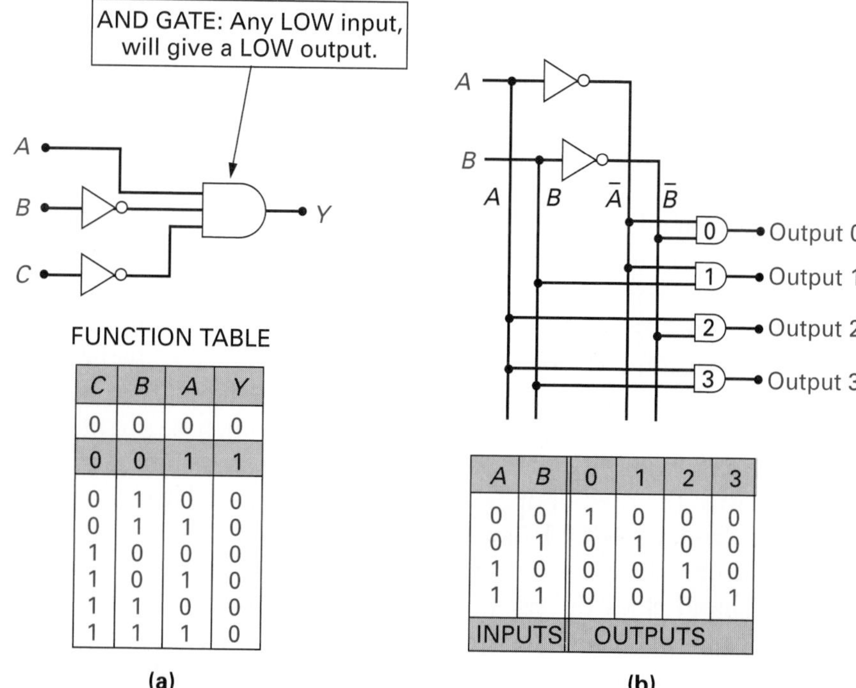

C	B	A	Y
0	0	0	0
0	0	1	1
0	1	0	0
0	1	1	0
1	0	0	0
1	0	1	0
1	1	0	0
1	1	1	0

(a)

A	B	0	1	2	3
0	0	1	0	0	0
0	1	0	1	0	0
1	0	0	0	1	0
1	1	0	0	0	1
INPUTS		OUTPUTS			

(b)

FIGURE 18-13 Basic Decoders.

combination. For example, Figure 18-13(b) shows how four AND gates and two inverters can be connected to activate 1 of 4 outputs based on the binary input word applied to A and B. The two inverters are included to invert the A and B inputs to create two additional input lines called "NOT A" and "NOT B" (symbolized $\overline{A}$ and $\overline{B}$). Looking at the function table for this circuit, you can see that when inputs A and B are 00_2 (zero), only output 0 will be activated or made HIGH. This is because both of the inputs to AND gate 0 are connected to the $\overline{A}$ and $\overline{B}$ input lines, and so when A and B are LOW, $\overline{A}$ and $\overline{B}$ are both HIGH, and therefore the output of AND gate 0 will be HIGH. The inputs of the other AND gates are connected to the input lines so that a 01 (one) input will make output 1 HIGH, a 10 (two) input will make output 2 HIGH, and a 11 (three) input will make output 3 HIGH. This circuit would be described as a **1-of-4 decoder** because an input code is used to select or activate 1 of 4 outputs. If the four outputs were connected to light emitting diodes (LEDs) that were labeled 0 through 3, the decoder would function as a **binary to decimal decoder** because the circuit is translating a binary coded input into a decimal output display.

1-of-4 Decoder
A circuit that uses an input code to select, or activate, 1 of 4 outputs.

Binary-to-Decimal Decoder
A circuit for translating a binary coded input into a decimal output display.

There are a wide variety of digital decoder ICs available that contain a complete internal decoder circuit. Let us now examine the operation and application of a frequently used digital decoder IC, the TTL 7447.

Decoder Application—7-Segment LED Display Decoder Circuit

To review, Figure 18-14(a) shows how the seven segments of an LED display are labeled "a b c d e f g," and Figure 18-14(b) shows how certain ON/OFF segment combinations can be used to display the decimal digits 0–9.

Figure 18-14(c) shows how a 7447 BCD-to-7 segment decoder IC can be connected to drive a common-anode 7-segment LED display. Looking at the 7447 first, you can see that it has four inputs and seven outputs. The input/output combinations for this digital IC are listed in the function table shown in the inset in Figure 18-14(c). Looking at the first line of this table, you can see that if a binary input of 0000 (zero) is applied to the *ABCD* inputs, the 7447 will make LOW outputs *"a b c d e f,"* and make HIGH output *"g."* These outputs are applied to the cathodes of the LEDs within the common-anode display, as seen in the inset. Because all of the anodes are connected to +5 V, a LOW output will turn ON a segment, and a HIGH output will turn OFF a segment. The 7447 therefore generates **active-LOW** outputs, which means that a "LOW" output signal will "activate" its associated output LED. In this example, the active-LOW outputs from the 7447 will turn ON segments *"abcdef,"* while the HIGH output will turn OFF segment *"g,"* causing the decimal digit "0" to be displayed. The remaining lines in the function table indicate which segments are made active for the other binary count inputs 0001 (one) through 1001 (nine).

Active LOW
A signal, input or output, that will activate when LOW.

To summarize the overall operation of the circuit, a 1 Hz clock input signal is applied via a "count ON/OFF" switch to the input of a 7490 digital counter IC. This counter will produce a 4-bit binary word at its outputs *A B C D* based on the number of input clock pulses received at its input (pin 14). This 4-bit binary word at the output of the 7490 is applied to the 7447 7-segment decoder/driver where it will be decoded into a seven-segment code that is applied to a seven-segment common-anode display. When the count ON/OFF switch is closed, the 7490 will count, the 7447 will decode the count, and the display will constantly

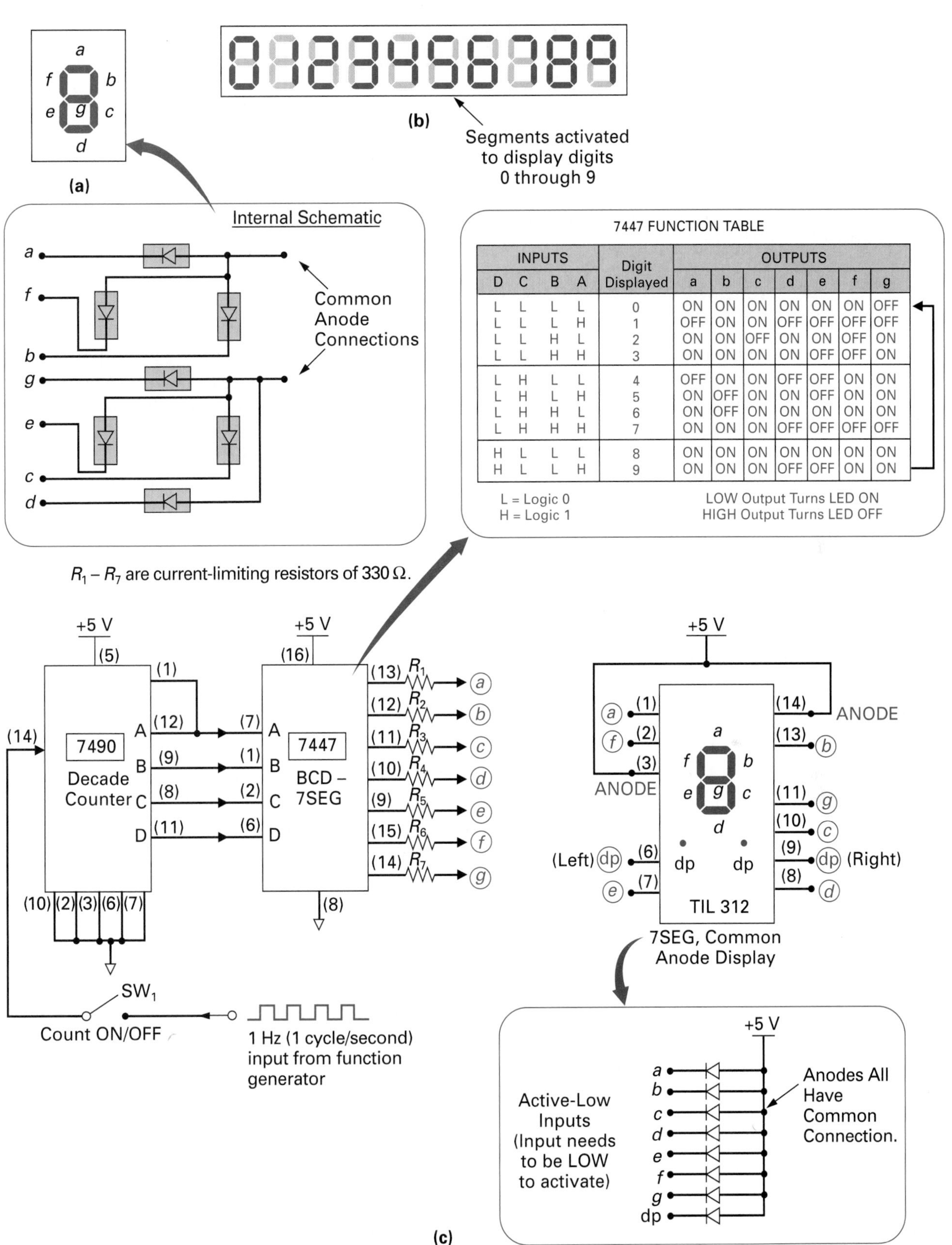

7447 FUNCTION TABLE

INPUTS D	C	B	A	Digit Displayed	OUTPUTS a	b	c	d	e	f	g
L	L	L	L	0	ON	ON	ON	ON	ON	ON	OFF
L	L	L	H	1	OFF	ON	ON	OFF	OFF	OFF	OFF
L	L	H	L	2	ON	ON	OFF	ON	ON	OFF	ON
L	L	H	H	3	ON	ON	ON	ON	OFF	OFF	ON
L	H	L	L	4	OFF	ON	ON	OFF	OFF	ON	ON
L	H	L	H	5	ON	OFF	ON	ON	OFF	ON	ON
L	H	H	L	6	ON	OFF	ON	ON	ON	ON	ON
L	H	H	H	7	ON	ON	ON	OFF	OFF	OFF	OFF
H	L	L	L	8	ON	ON	ON	ON	ON	ON	ON
H	L	L	H	9	ON	ON	ON	OFF	OFF	ON	ON

L = Logic 0
H = Logic 1
LOW Output Turns LED ON
HIGH Output Turns LED OFF

FIGURE 18-14 Decoder Application—7 Segment LED Display/Driver.

cycle through the digits 0 through 9. When the count ON/OFF switch is opened, the counter will freeze at its present count, and so will the number shown on the display.

18-2-2 Encoder Circuits

Encoder Circuit
A circuit designed to generate specific codes.

An **encoder circuit** performs the opposite function of a decoder: a decoder is designed to detect specific codes whereas an encoder is designed to generate specific codes.

Basic Encoder Circuits

Figure 18-15 shows a simple decimal-to-binary encoder circuit using three push buttons, three pull-up resistors, and two NAND gates. The pull-up resistors are included to ensure that the input to the NAND gates are normally HIGH. When button 1 is pressed, the upper input of NAND gate *A* will be pulled LOW, and because any LOW input to a NAND gate will produce a HIGH output, A_0 will be driven HIGH. The two inputs to NAND gate *B* are unaffected by SW_1, and so the two HIGH inputs to this NAND gate will produce a LOW output at A_1. Referring to the table in Figure 18-15, you can see that the two-bit code generated when switch 1 is pressed is 01 (binary one). Studying the circuit and the rest of the function table shown in Figure 18-15, you can see that the code 10 (binary two) will be generated when switch 2 is pressed, and the code 11 (binary three) will be generated when switch 3 is pressed.

There are a wide variety of digital encoder ICs available that contain a complete internal encoder circuit. Let us now examine the operation and application of a frequently used digital encoder IC, the TTL 74147.

Encoder Application—Calculator Keypad Encoder Circuit

Figure 18-16 shows how a 74147 can be used to encode a calculator keypad. The bar over the 74147's $\overline{A}_1$ through $\overline{A}_9$ inputs, and the bubble or small circle at these inputs, are symbols used to indicate that these are active-LOW inputs. This

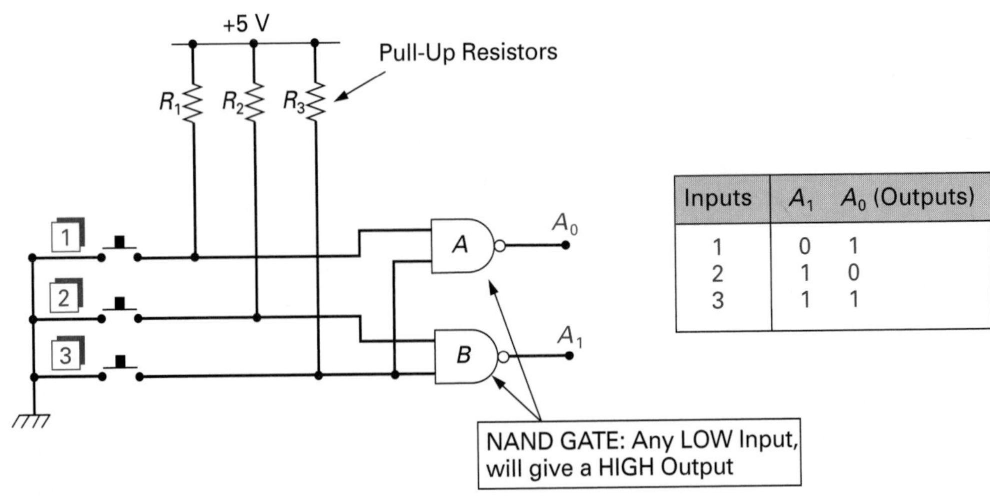

Inputs	A_1	A_0 (Outputs)
1	0	1
2	1	0
3	1	1

FIGURE 18-15 Basic Encoder Circuit.

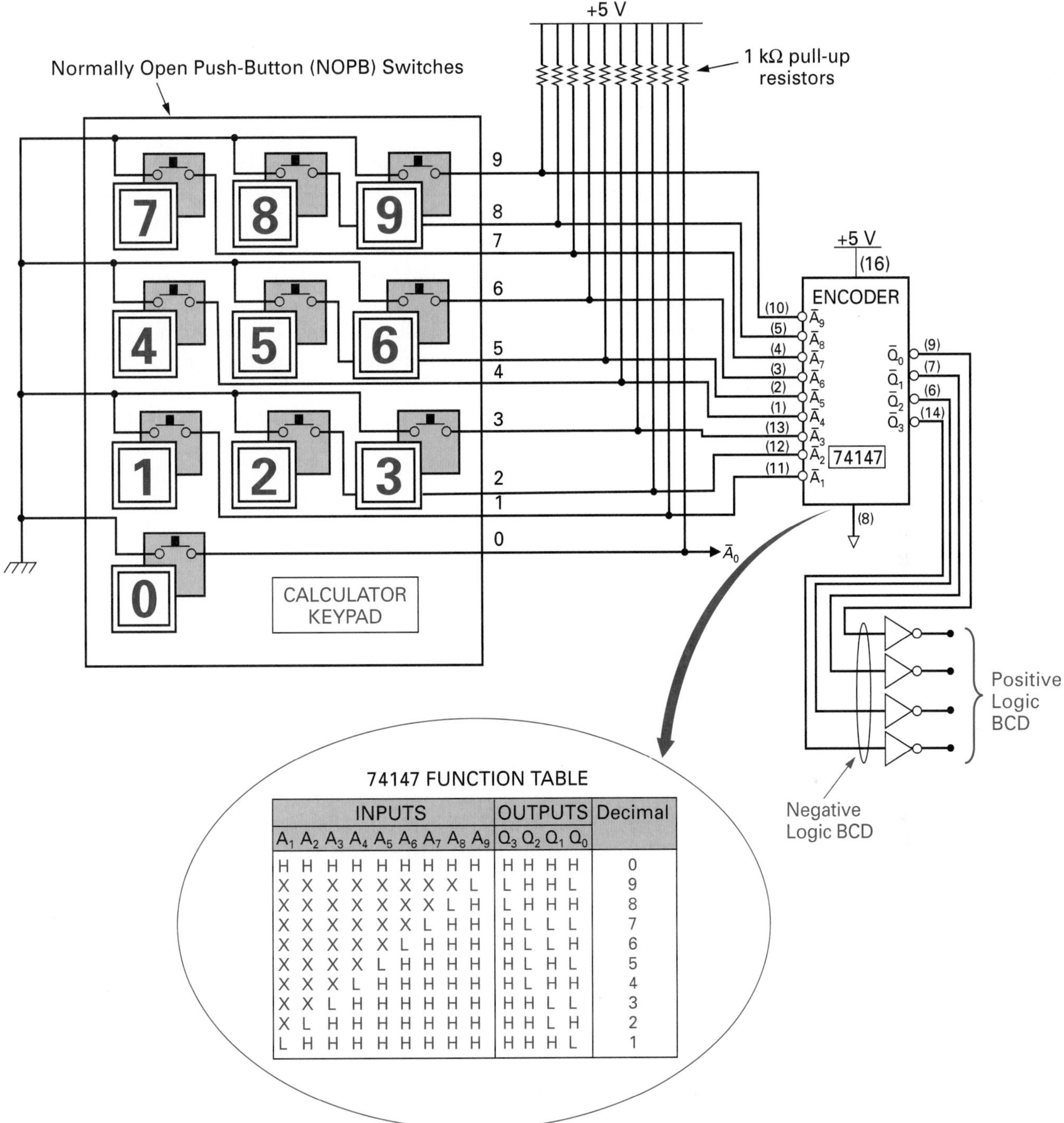

INPUTS									OUTPUTS				Decimal
A_1	A_2	A_3	A_4	A_5	A_6	A_7	A_8	A_9	Q_3	Q_2	Q_1	Q_0	
H	H	H	H	H	H	H	H	H	H	H	H	H	0
X	X	X	X	X	X	X	X	L	L	H	H	L	9
X	X	X	X	X	X	X	L	H	L	H	H	H	8
X	X	X	X	X	X	L	H	H	H	L	L	L	7
X	X	X	X	X	L	H	H	H	H	L	L	H	6
X	X	X	X	L	H	H	H	H	H	L	H	L	5
X	X	X	L	H	H	H	H	H	H	L	H	H	4
X	X	L	H	H	H	H	H	H	H	H	L	L	3
X	L	H	H	H	H	H	H	H	H	H	L	H	2
L	H	H	H	H	H	H	H	H	H	H	H	L	1

FIGURE 18-16 Encoder Application—Keyboard Encoder Circuit.

means that the encoder's inputs will be activated whenever they are made LOW. Referring to the push-button switches controlled by the 1 through 9 keypads, you can see that pull-up resistors are included to ensure that the 74147's inputs are normally HIGH and when a key is pressed, the respective input is taken LOW. The 74147's function table shown in the inset in Figure 18-16 shows how this encoder circuit will respond to different inputs. For example, looking at the first line of the function table you can see that if none of the keys are pressed, all of the 74147 inputs will be HIGH, and it will generate a 1111 output code. This code will then be applied to a set of four inverters that will invert the code to 0000, and so when no key is pressed, the final output code is binary 0000 (zero). You may have noticed that there are also bars over the 74147's $\overline{Q}_0$ through $\overline{Q}_3$ outputs. This is, again, a symbol used to indicate that the output is an active-LOW code (also known as **negative logic**). If you need to convert the output to an active-HIGH code (or **positive logic**), you will need to invert the output, as is being done in Figure 18-16. The rest of the function table shows how a LOW on any of the $\overline{A}_1$ through $\overline{A}_9$ inputs will generate a corresponding negative logic code. For example, a LOW at input $\overline{A}_9$ will cause the 74147 to generate a negative logic code of 0110, which will be inverted to a positive logic code of 1001, which is binary 9.

Negative Logic
Active-LOW code.

Positive Logic
Active-HIGH code.

To explain one other detail, the "Xs" in the function table indicate a "don't care" condition. This is because the 74147 is a "priority encoder," which means that if more than one key is pressed at the same time, the 74147 will only generate the code for the larger digit, or higher priority. For example, if the "9" and "6" keys are pressed at the same time, you can see by the second line of the function table that the 74147 will generate a nine code, responding only to the LOW on the $\overline{A}_9$ input and ignoring the LOW on the $\overline{A}_6$ input.

18-2-3 *Comparator Circuits*

Digital Comparator
A circuit that compares two input words to see if they are equal.

A **digital comparator** compares two binary input words to see if they are equal. There are a wide variety of digital comparator ICs available that contain a complete internal binary comparator circuit. Let us now examine the operation and application of a frequently used digital comparator IC, the TTL 7485.

Comparator Application—A Photocopier Control Circuit

Figure 18-17 shows how a 7485 comparator could be used in a photocopier to control the number of copies made. A storage register *A* holds the 4-bit binary word 0010 (two), indicating the number of copies that have been made. The storage register *B* holds the 4-bit binary word 1000 (eight), indicating the number of copies requested. The 7485 compares these two inputs and generates three output control signals. If the 4-bit input *A* is less than the 4-bit input *B,* the 7485 pin-7 output will be HIGH, signaling that the copier should continue copying. When input *A* is equal to input *B,* the 7485 pin-6 output will be HIGH, signaling that the requested number of copies have been made and the copier should stop copying.

Multiplexer (MUX)
A circuit that is controlled to switch one of several inputs through to one output.

Demultiplexer (DMUX)
A circuit that is controlled to switch a single input through to one of several outputs.

18-2-4 *Multiplexer and Demultiplexer Circuits*

A **digital multiplexer (MUX),** or selector, is a circuit that is controlled to switch one of several inputs through to one output. On the other hand, a **digital demultiplexer (DMUX),** or data distributor, operates in exactly the opposite way: it is a circuit that is controlled to switch a single input through to one of several outputs.

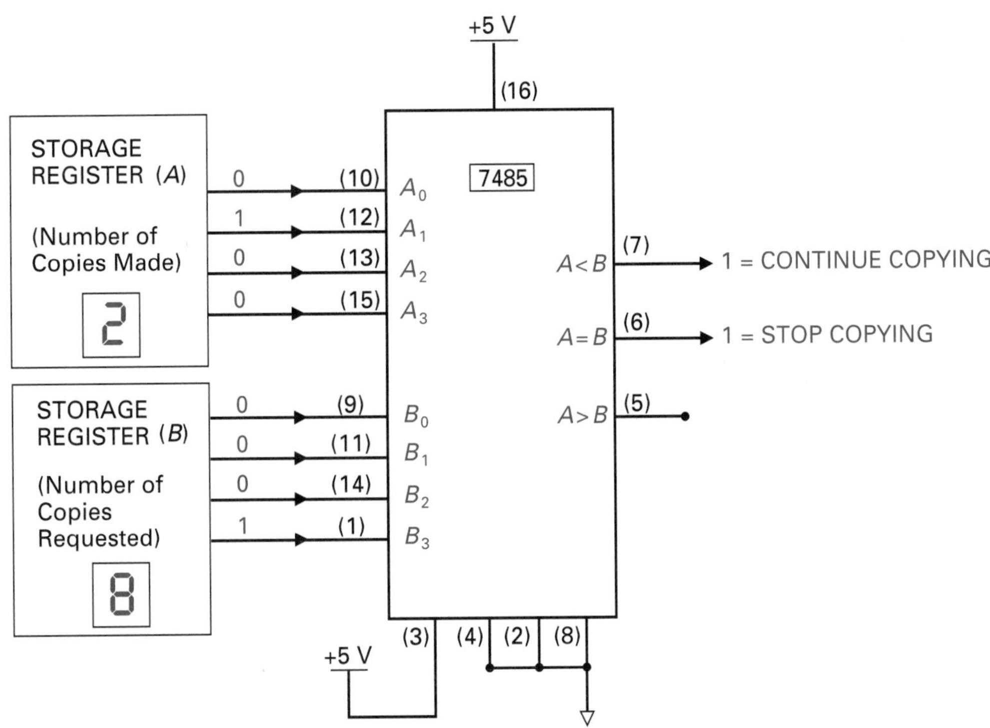

FIGURE 18-17 **Comparator Application—Photocopier Control Circuit.**

Basic Multiplexer and Demultiplexer Circuit

The most basic of all multiplexers is the rotary switch shown in Figure 18-18(a), which will switch any of the six inputs through to the one output based on the switch position control. In most high speed applications, the multiplexer will need to be electronically rather than mechanically controlled, which is why **analog multiplexer** circuits use relays or transistors to electronically switch analog signals. A **digital multiplexer,** on the other hand, uses logic gates to switch one of several digital input signals through to one output.

Analog Multiplexer
A multiplexer that uses relays or transistors to electronically switch analog signals.

Digital Multiplexer
A multiplexer that uses logic gates to switch one of several digital input signals through to one output.

By reversing the direction of the switch, as shown in Figure 18-18(b), you can see how we can construct a basic mechanically controlled demultiplexer circuit. In this instance, the switch is controlled to switch a single input through to one of several outputs.

There are a wide variety of digital multiplexer and demultiplexer ICs available that contain a complete internal circuit. Let us now examine the operation and application of two frequently used digital MUX and DMUX ICs, the TTL 74151 and 74138.

MUX/DMUX Application—A Basic Home Security System

Figure 18-19 shows how a 74151 8-to-1 multiplexer (MUX), and a 74138 1-to-8 demultiplexer (DMUX) could be connected to form a basic home security system. A binary zero (000) to seven (111) counter drives the *A B C* select or control inputs of both the MUX and DMUX. As the counter counts from 0 to 7,

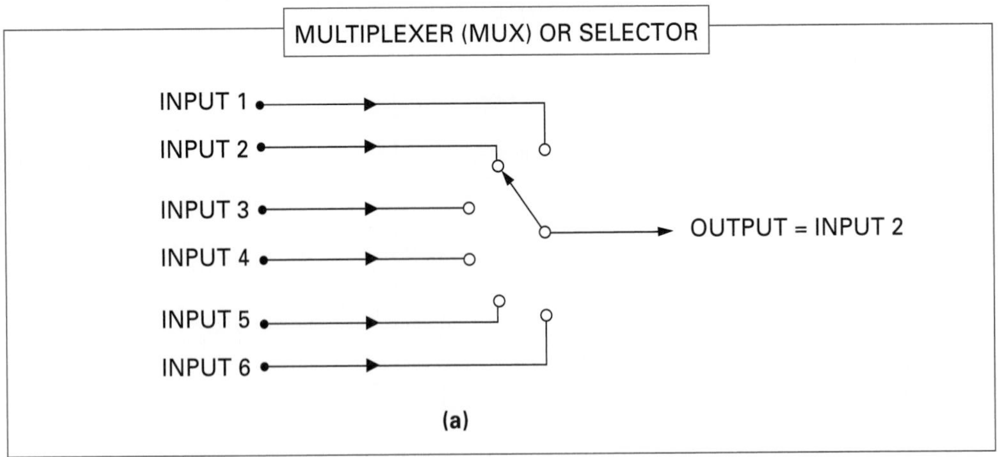

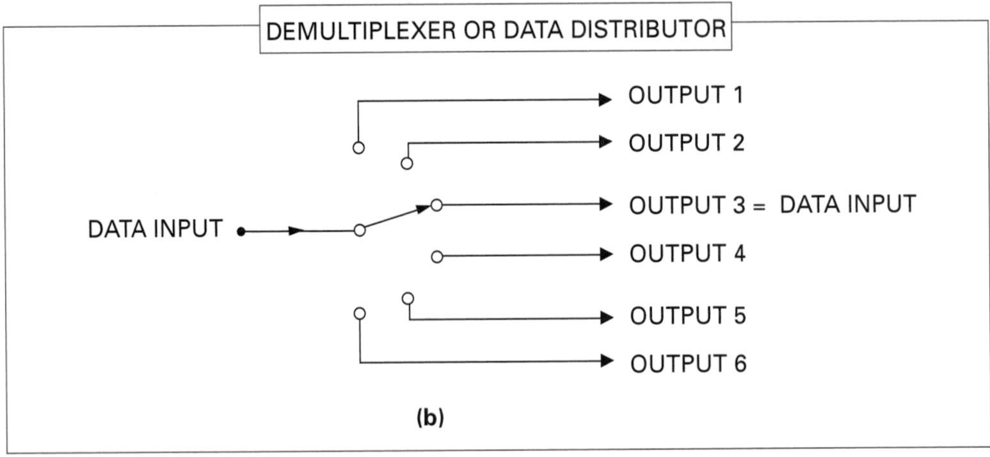

FIGURE 18-18 Basic Multiplexer and Demultiplexer Action.

it "selects" in turn each of the 74151 inputs and switches the HIGH or LOW data through to the *Y* output, as seen in the function table in the inset in Figure 18-19. For example, when the counter is at a count of 011 (three), the HIGH or LOW data present at the 3 input (pin 1) is switched through to the *Y* output. In this example circuit, the 3 input of the 74151 is connected to a window switch. If the window is open, its associated switch will also be open, and the input to the 74151 will be HIGH due to the pull-up resistors. On the other hand, if the window and its associated switch are closed, ground or a LOW logic level will be applied to the input of the 74151.

The 74138 DMUX operates in exactly the opposite way. For example, as we mentioned earlier, when the counter is at a count of 011 (three), the HIGH or LOW data present at the 3 input of the 74151 (pin 1) is switched through to the *Y* output. This HIGH or LOW data is applied to the G_1 input of the 74138, which is also receiving a 011 (three) control-select input from the counter, and so the data at G_1 is switched through to the 3 output (pin 12). Since the 74138 has small bubbles or circles on its outputs, we know that these are active-LOW.

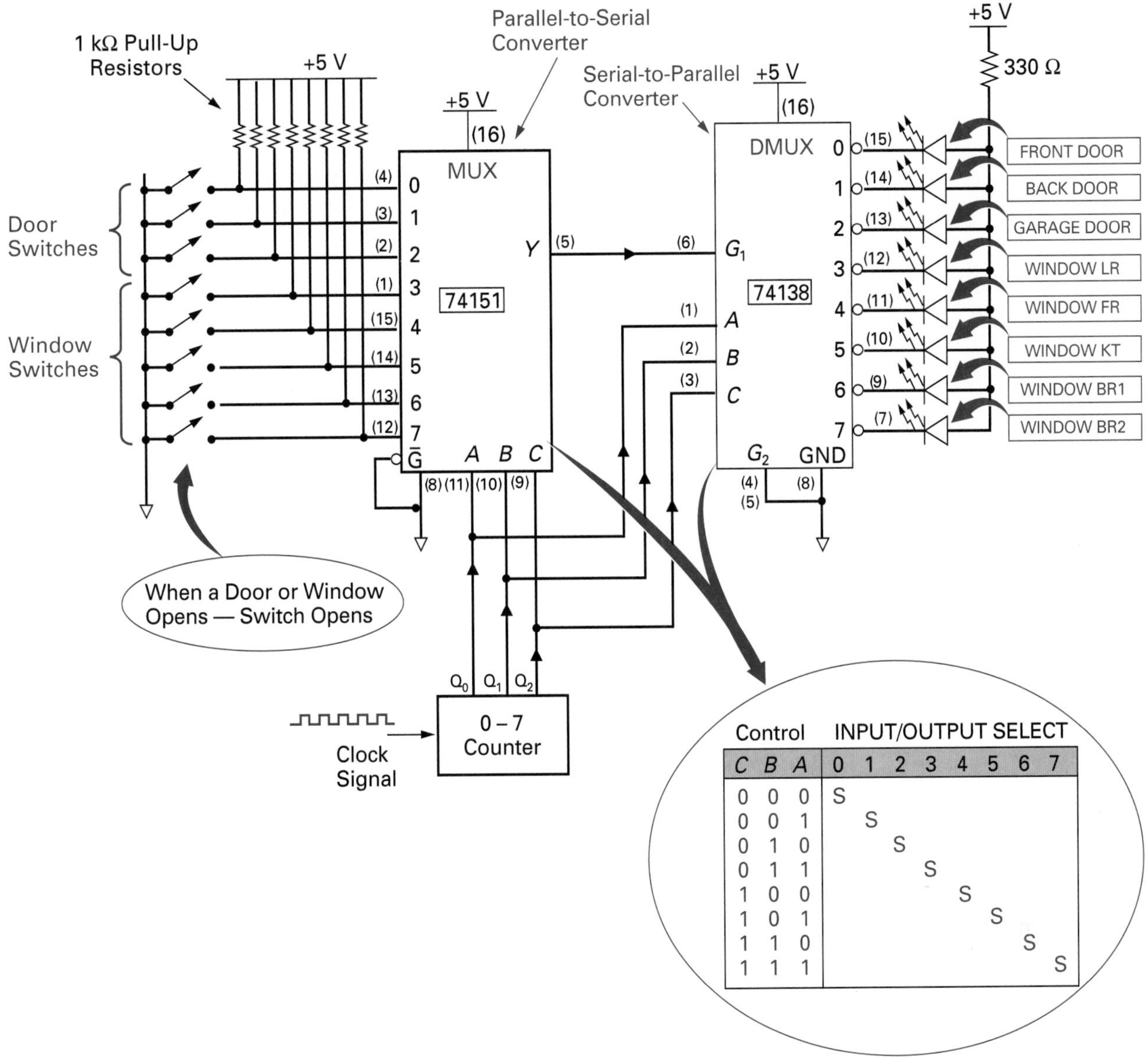

FIGURE 18-19 Multiplexer/Demultiplexer Application—A Basic Home Security System.

This inversion between the 74138's inputs and output is needed since we are driving a common anode LED display, which as we know is also active-LOW (LEDs are activated by a LOW). For example, if the window connected to the 3 input is opened, the switch will open and give a HIGH to the 3 input of the 74151. The 74151 will switch this HIGH data through to the Y output, where it will be applied to the G_1 input of the 74138, which will invert the HIGH to a LOW, then switch it to the 3 output where it will turn ON the living room (LR) window LED. This circuit therefore, senses the condition of switch 0 and then transmits its data to LED 0, senses the condition of switch 1 and then transmits its data to

Parallel-to-Serial Converter
An IC that converts the parallel word applied to its inputs into a serial data stream.

Serial-to-Parallel Converter
An IC that converts a serial data stream applied to its input into a parallel output word.

Serial Data Transmission
The transfer of information sequentially through a single path or channel.

Parallel Data Transmission
The transfer of information simultaneously over a set of parallel paths or channels.

Flip-Flop Circuit
A bistable multivibrator circuit that has two stable states.

Set-Reset Flip-Flop
A bistable multivibrator circuit that has two inputs that are used to either SET or RESET the output.

Reset State
A circuit condition in which the output is reset to binary 0.

Set State
A circuit condition in which the output is set to binary 1.

Latched
A bistable circuit action that causes the circuit to remain held or locked in its last activated state.

LED 1, senses the condition of switch 2 and then transmits its data to LED 2, and so on, until it reaches 7 (111) and then it repeats the process.

The 74151 could be described as a **parallel-to-serial converter** because it converts the parallel 8-bit word applied to its inputs into an 8-bit serial data stream. Conversely, the 74138 could be described as a **serial-to-parallel converter,** since it converts an 8-bit serial data stream applied to its input into a parallel 8-bit output word. One question you may have about this circuit is why don't we simply connect the switches straight to the LEDs and bypass the 74151 and 74138. The answer lies in the basic difference between **serial data transmission** and **parallel data transmission.** With this circuit, the 74151 can be placed close to all the windows and doors, and then only 4 wires (three for the counter and one for data) need to be run to the 74138 and its display. If the switches were connected directly to the display, 8 wires would be needed to connect the switch sensors to the LED display. The advantage of serial data transmission, therefore, is that only one data line is needed. On the other hand, the advantage of parallel data transmission is speed, since the parallel connection will mean that all of the data is transmitted at one time.

18-2-5 Flip-Flop Circuits

A digital **flip-flop circuit** is a "bistable multivibrator" circuit that has two stable states. We discussed the **set-reset flip-flop** circuit previously in Chapter 9, and since the flip-flop is a key device in all digital circuits, let us begin by reviewing the operation of that circuit.

The Set-Reset *(S-R)* Flip-Flop Circuit

The circuit has two inputs called "SET" and "RESET," and two outputs called "Q" and "$\overline{Q}$" (pronounced "queue" and "queue not"). The $\overline{Q}$ output derives its name from the fact that its voltage level is always the opposite of the Q output. For example, if Q is HIGH, $\overline{Q}$ will be LOW, and if Q is LOW, $\overline{Q}$ will be HIGH. The bistable multivibrator circuit is often called an *S-R* (set-reset) flip-flop because *a pulse on the SET input will "flip" the circuit into the set state (Q output is set HIGH), while a pulse on the RESET input will "flop" the circuit into its reset state (Q output is reset LOW).*

Figure 18-20(a) shows the logic symbol for an *S-R* or *R-S* flip-flop circuit. To fully understand the operation of the circuit, refer to the waveforms in Figure 18-20(b). To begin with, let us assume that the Q output is LOW, as shown in the waveforms before time t_1. This condition is called the **reset state** because the primary output *(Q)* has been reset to binary 0, or 0 V. Following the waveforms in Figure 18-20(b), you can see that the first input to go active is the SET input at time "t_1." This positive pulse will put the flip-flop in its **set state,** which means that the primary output *(Q)* will be set HIGH. Studying the waveforms in Figure 18-20(b) once again, you will notice that after the positive SET input pulse has ended, the bistable will still remain **latched** in its last state. This ability of the *S-R* flip-flop to remain in its last condition, or state, accounts for why this circuit is also called an *S-R latch.* Continuing with the waveforms in Figure 18-20(b), you can see that a RESET pulse occurs at time t_2 and resets the primary output *(Q)*

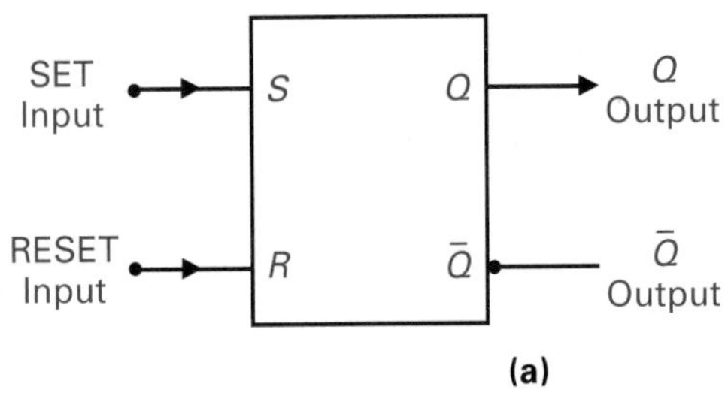

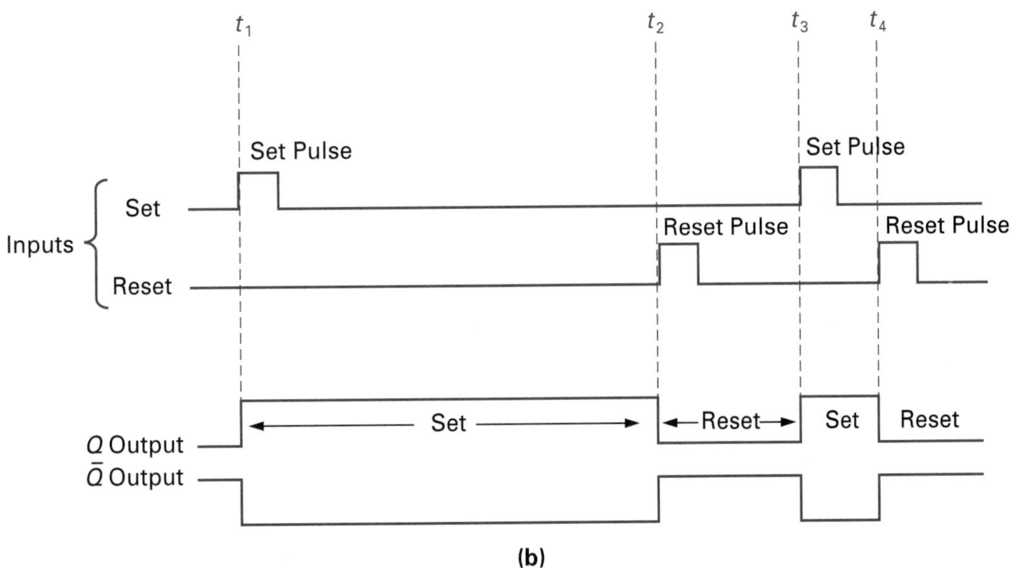

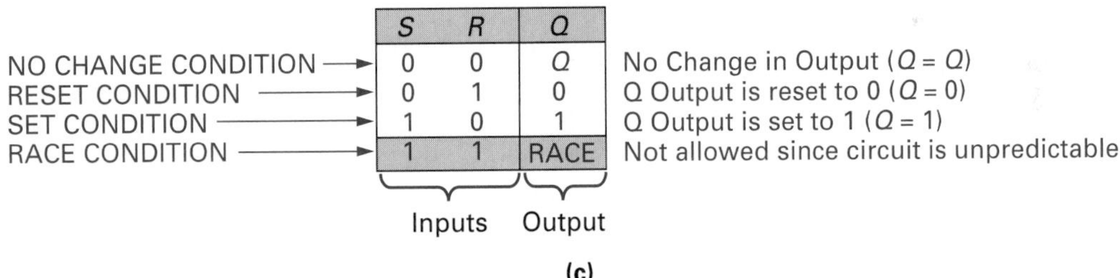

	S	R	Q	
NO CHANGE CONDITION →	0	0	Q	No Change in Output ($Q = Q$)
RESET CONDITION →	0	1	0	Q Output is reset to 0 ($Q = 0$)
SET CONDITION →	1	0	1	Q Output is set to 1 ($Q = 1$)
RACE CONDITION →	1	1	RACE	Not allowed since circuit is unpredictable

FIGURE 18-20 The Set-Reset (*S-R* or *R-S*) Flip-Flop. (a) *S-R* Symbol. (b) Timing Waveforms. (c) Function Table.

LOW. The flip-flop then remains latched in its reset state until a SET pulse is applied to the set input at time t_3, setting Q HIGH. Finally, a positive RESET pulse is applied to the reset input at time t_4, and Q is once again reset LOW.

The operation of the *S-R* flip-flop is summarized in the function or truth table shown in Figure 18-20(c). When only the *R* input is pulsed HIGH (reset condition), the Q output is reset to a binary 0, or reset LOW ($\bar{Q}$ will be the opposite, or HIGH). On the other hand, when only the *S* input is pulsed HIGH (set condition), the Q output is set to a binary 1, or set HIGH ($\bar{Q}$ will be the

No-Change Condition
A latch condition in which there will be no change in the output.

Race Condition
An unpredictable bistable circuit condition.

opposite, or LOW). When both the *S* and *R* inputs are LOW, the *S-R* flip-flop is said to be in the **no-change condition** or latch condition because there will be no change in the output *Q*. For example, if output *Q* is SET, and then the *S* and *R* inputs are made LOW, the *Q* output will remain SET, or HIGH. On the other hand, if output *Q* is RESET, and then the *S* and *R* inputs are made LOW, the *Q* output will remain RESET, or LOW. The external circuits driving the *S* and *R* inputs of this flip-flop type will be designed so that these inputs are never both HIGH, as shown in the last condition in the table in Figure 18-20(c). This is called the **race condition** and should not normally be applied because the output condition is unpredictable.

The *J-K* Flip-Flop Circuit

Clock Signal
A timing signal applied to all of the digital devices to ensure that each device is triggered into operation at the right time.

Triggered
A circuit action which, when initiated, functions for a predetermined time. For example, the duration of one sweep in a cathode-ray tube.

Clock Input (*C*)
An accurate timing signal used for synchronizing the operation of digital circuits.

Negative Edge Triggered Device
A device that will "wake up" and perform when the clock input is making a transition from HIGH to LOW (a negative edge).

Toggle
To switch to the opposite state.

A digital system will typically contain thousands of flip-flop circuits. To coordinate the overall operation of all these digital circuits, a **clock signal** or timing signal is applied to all of the digital devices to ensure that each device is **triggered** into operation at the right time. A clock signal therefore controls when a flip-flop is allowed to operate. Figure 18-21 illustrates the *J-K* flip-flop, whose letters *"J"* and *"K"* were arbitrarily chosen. This flip-flop type is very similar to the *S-R* flip-flop because the *Q* output will be "set HIGH" when the set input *(J)* is active and "reset LOW" when the reset *(K)* input is active. Looking at the *J-K* flip-flop's function table in Figure 18-21(a) and symbol in Figure 18-21(b), you can see that this flip-flop has a **clock input (*C*),** in addition to the standard "set *(J)*" and "reset *(K)*" flip-flop inputs. This clock input will be used to trigger the flip-flop into operation and control the timing of this device in relation to all the other devices in a circuit.

To understand in more detail the operation of this *J-K* flip-flop, refer to the function table shown in Figure 18-21(a). The first three lines of the function table show that if the clock input line is LOW, HIGH, or making a transition from LOW to HIGH (a positive edge), there will be no change at the *Q* output regardless of the *J* and *K* inputs. This is because this *J-K* flip-flop is a **negative edge triggered device,** which means it will only "wake up" and perform the operation applied to the *J* and *K* inputs when the clock input is making a transition from HIGH to LOW (a negative edge). In the last four lines of the function table, you can see that when a negative edge is applied to the clock input, the *J-K* flip-flop will react to its *J* (set) and *K* (reset) inputs. For example, when negative edge triggered, the *J-K* flip-flop will operate in almost the same way as an *S-R* flip-flop: it will reset its *Q* output when the reset input *(K)* is HIGH, set its *Q* output when the set input *(J)* is HIGH, and not change its *Q* output when *J* and *K* are LOW. In the last line of the function table, you can see why the *J-K* flip-flop is an improvement over the *S-R* flip-flop. When both the *S* and *R* inputs of an *S-R* flip-flop are HIGH, it will race and generate an unpredictable *Q* output, which is why this input condition is not used. When both the *J* and *K* inputs of a *J-K* flip-flop are made HIGH, the *Q* output will **toggle** or switch to the opposite state. This means that if *Q* is HIGH it will switch to a LOW, and if *Q* is LOW it will switch to a HIGH. In the function table, it states that if the *J-K* flip-flop is in the toggle condition, and the clock input makes a transition from HIGH to LOW, the *Q* output will switch to the opposite logic level, or the logic level of the $\overline{Q}$ output.

(SET)(RESET)

C	J	K	Q
0	X	X	Q — No Change
1	X	X	Q — No Change
↑	X	X	Q — No Change
↓	0	0	Q — No Change
↓	0	1	0 — Reset
↓	1	0	1 — Set
↓	1	1	$\bar{Q}$ — Toggle

(a)

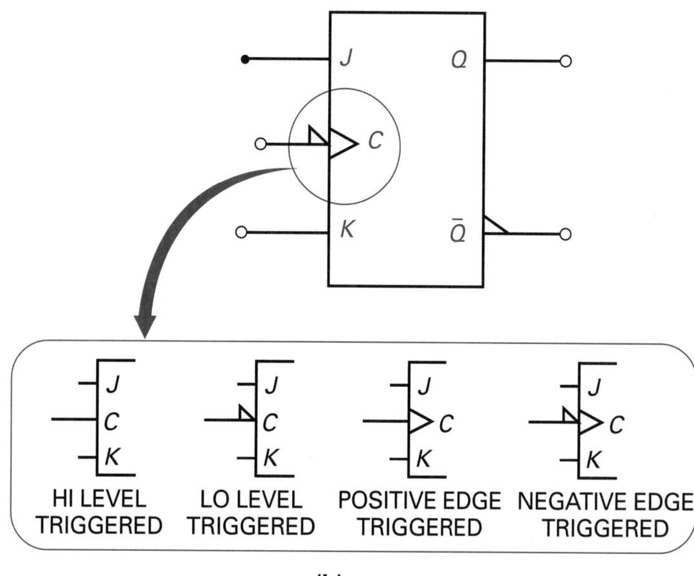

(b)

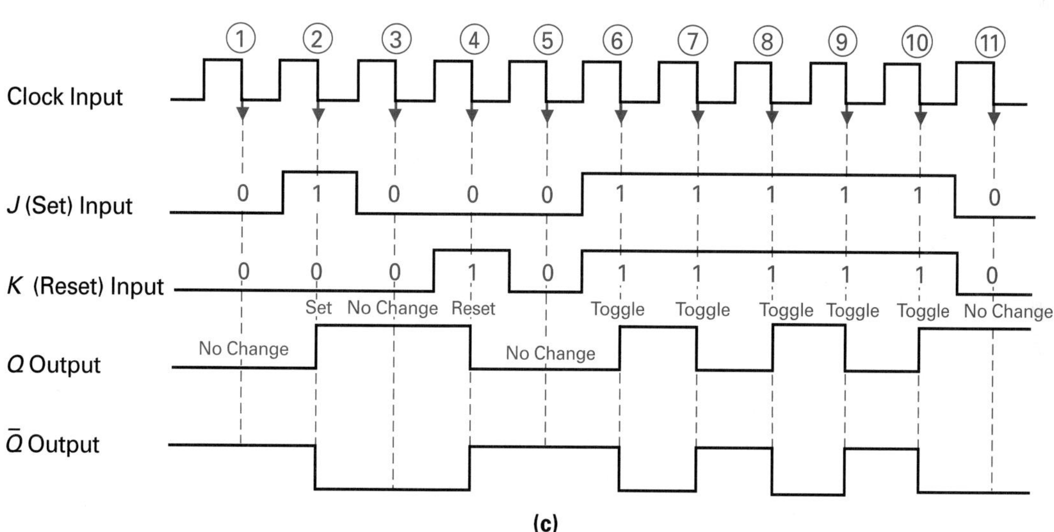

(c)

FIGURE 18-21 Basic *J-K* (Set-Reset) Flip-Flop Operation.

Figure 18-21(b) shows the symbol for a "negative edge triggered *J-K* flip-flop." A full triangle on the clock input inside the rectangular block is used to indicate that this device is **edge triggered,** and the smaller right-angle triangle (or bubble) outside the rectangular block is used to indicate that this input is active-LOW, or negative edge triggered. The other symbols shown in the inset in Figure 18-21(b) show the four possibilities of how *J-K* flip-flops are triggered: by a HIGH-level, LOW-level, positive-edge, or negative-edge.

Edge Triggered
A circuit action which is initiated when the trigger input signal makes a transition from HIGH to LOW or LOW to HIGH.

To reinforce your understanding of the *J-K* flip-flop, Figure 18-21(c) shows how its outputs will respond to a variety of input combinations. Because this *J-K* flip-flop is negative-edge triggered, the device will only respond to the *J* and *K* inputs and change its outputs when the clock signal makes a transition from HIGH to LOW, as indicated by the negative arrows shown on the square-wave clock input waveform. At negative clock edge 1, $J = 0$ and $K = 0$, and so there is no change in the Q output (which stays LOW), and the $\overline{Q}$ output (which stays HIGH). At negative clock edge 2, $J = 1$ and $K = 0$ (set condition), and so the Q output is set HIGH ($\overline{Q}$ is switched to the opposite, LOW). At negative clock edge 3, both *J* and *K* are again logic 0, and so once again the output will not change (Q will remain latched in the set condition). At negative clock edge 4, $J = 0$ and $K = 1$ (reset condition), and the Q output will be reset LOW. At negative edge 5, both *J* and *K* are again LOW, and so there will be no change in the Q output (Q will remain latched in the reset condition). At negative clock edge 6, both *J* and *K* are HIGH (toggle condition), and so the output will toggle or switch to its opposite state (because Q is LOW it will be switched HIGH). At negative clock edges 7, 8, 9, and 10, the *J* and *K* inputs remain HIGH, and therefore the Q output will continually toggle or switch to its opposite logic level. At negative edge 11, both *J* and *K* return to 0, and so the Q output remains in its last state, which in this example is HIGH.

J-K Flip-Flop Application—Frequency Divider/Binary Counter Circuit

Figure 18-22(a) shows how a 555 timer and a 74LS76A *J-K* flip-flop can be connected to form a frequency divider or binary counter circuit. The 555 timer, which was discussed previously in Chapter 9, is connected as an astable multivibrator and will generate the square wave output waveform shown in Figure 18-22(b). The 74LS76A digital IC actually contains two complete *J-K* flip-flop circuits, and because only one is needed for this circuit, the flip-flop is labeled "1/2 of 74LS76A." Because the *J* and *K* inputs of the 74LS76A flip-flop are both connected HIGH, the Q output will continually toggle for each negative edge of the clock-signal input from the 555 timer, as seen in the waveforms in Figure 18-22(b). Because two negative edges are needed at the clock input to produce one cycle at the output, the *J-K* flip-flop is in fact dividing the input frequency by two. An input frequency of 2 kHz from the 555 timer will therefore appear as 1 kHz at the *J-K* flip-flop's Q output because two pulses into the 74LS76A will produce one pulse out.

Now that we have seen how the *J-K* flip-flop can be used as a frequency divider, let us see how it can function as a binary counter. Figure 18-22(c) repeats the output waveforms from the 555 timer and the 74LS76A Q output. If these two outputs were connected to two LEDs, the LEDs would count up in binary as shown in the table in Figure 18-22(c). To explain this in more detail, at time t_0 both the 555 and 74LS76A Q output are LOW, and so our LED binary display shows a count of 00 (binary zero). At time t_1, the binary display will be driven by a HIGH from the 555 timer output, and a LOW from the 74LS76A Q output, and therefore will display 01 (binary one). At time t_2, the display will show 10 (binary two), and at time t_3 the display will show 11 (binary three).

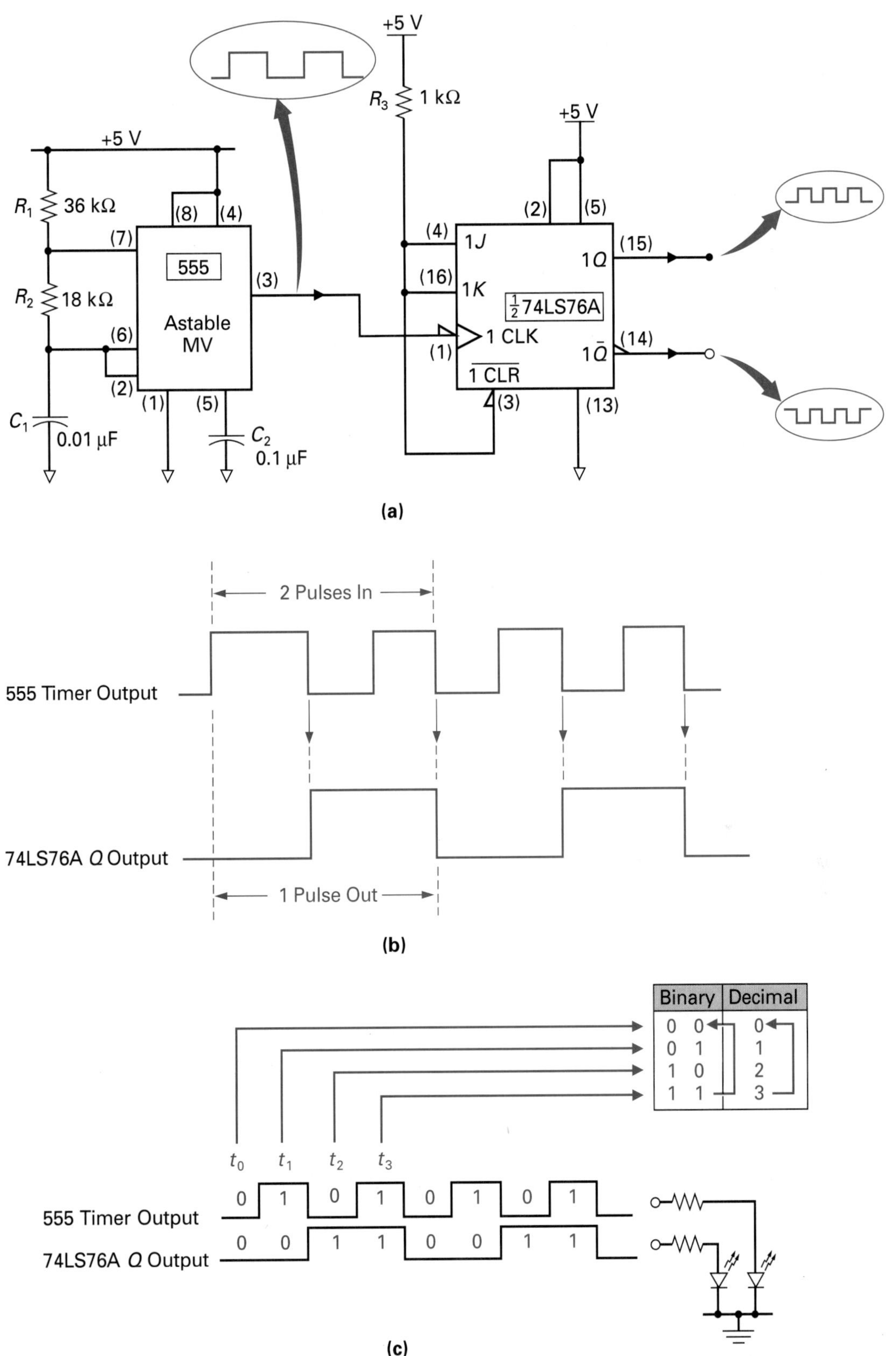

FIGURE 18-22 Flip-Flop Application—Frequency Divider or Binary Counter Circuit.

Combined, therefore, the 555 timer and the divided-by-two output of the 74LS76A can be used to generate a 2-bit word that will count from binary 0 (00) to binary 3 (11), and then continuously repeat the cycle.

Flip-Flop Application—Decade Counter/Divider Circuit

Binary Counter A circuit that contains a group of flip-flops that are all connected to repeatedly divide the input clock frequency by 2.

A **binary counter** circuit contains a group of flip-flops that are all connected to repeatedly divide the input clock frequency by 2 and generate an output code sequence that corresponds to a binary count. Figure 18-23(a) shows how a 74LS160A can be connected as a 0–9 (decade) counter or as a divide-by-10 frequency divider. Referring to the function table in Figure 18-23(a), you can see how the four Q outputs (Q_0, Q_1, Q_2, and Q_3) of the 74LS160A can be triggered

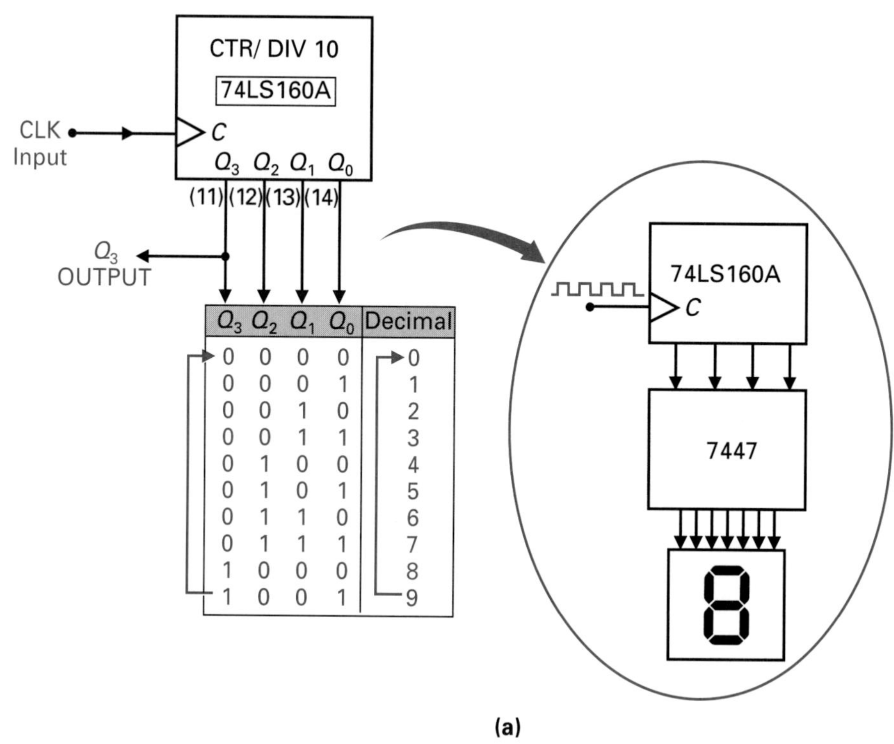

Q_3	Q_2	Q_1	Q_0	Decimal
0	0	0	0	0
0	0	0	1	1
0	0	1	0	2
0	0	1	1	3
0	1	0	0	4
0	1	0	1	5
0	1	1	0	6
0	1	1	1	7
1	0	0	0	8
1	0	0	1	9

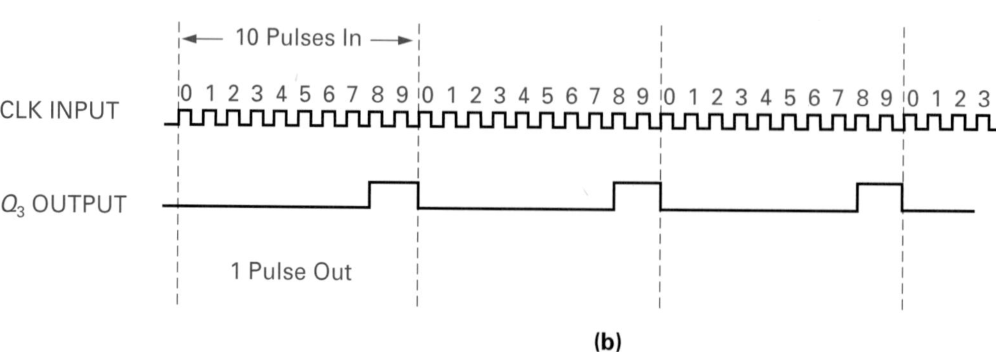

FIGURE 18-23 Flip-Flop Application—0 to 9 Counter (Divide by 10) Circuit.

by a clock input to generate a binary count of 0000 to 1001 (binary zero to nine), and continuously repeat. If the four Q outputs of the 74LS160A were connected to a 7447 BCD-to-7-segment display decoder, as shown in the inset in Figure 18-23(a), the repeating count of 0 to 9 could be seen on the 7-segment LED display.

Figure 18-23(b) shows how the 74LS160A also functions as a divide-by-ten frequency divider. Comparing the clock input to the Q_3 output, you can see that ten input pulses are needed to generate one output pulse. Therefore, if a 10 kHz clock signal were applied to the 76LS160A, a 1 kHz output signal would be present at the Q_3 output.

Flip-Flop Application—Register Circuit

A digital **register** circuit contains a group of flip-flops that act as a memory element to store a group of bits or a binary word. Figure 18-24 shows how a 74LS173 can be connected to form a 4-bit output register. In this application, the "no-change" or latching ability of the flip-flop is made use of to store a 4-bit word generated by a set of four input switches. The 74173 has a HIGH-level clocked input, and an active-LOW LOAD control input ($\overline{\text{LOAD}}$). To explain the operation of this circuit, let us assume that the switches have been open or closed so that a desired 4-bit word is now being applied to the 74173's data inputs (D_1, D_2, D_3, and D_4). To store this 4-bit binary word, the $\overline{\text{LOAD}}$ control line will have to be made LOW, and then we will have to wait until the clock

Register
A digital circuit that contains a group of flip-flops that act as a memory element to store a group of bits or a binary word.

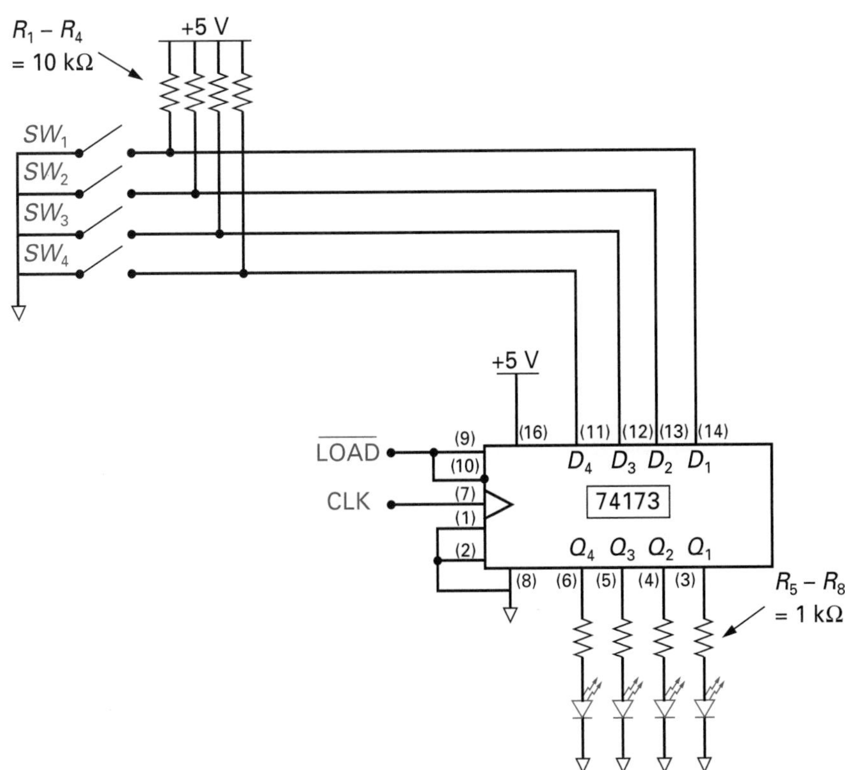

FIGURE 18-24 Flip-Flop Application—Output Storage Register Circuit.

input goes HIGH. When both of these control lines are active, the 4-bit word will be stored in the 74173 and applied to the Q outputs (Q_1, Q_2, Q_3, and Q_4), where it will be displayed on the LEDs. After a 4-bit word has been latched into the 74173 and the $\overline{\text{LOAD}}$ control line is returned HIGH, the contents of the 74173 will not change, even if the ON/OFF condition of the switches are changed. To store a new 4-bit word in the 74173, you would have to once again apply it to the data inputs, take the $\overline{\text{LOAD}}$ control line LOW, and then wait for the clock signal to go HIGH. This new 4-bit data word will then be stored in the 74173 and replace the previously stored 4-bit word.

Flip-Flop Application—One-Shot Pulse Generator Circuit

The astable multivibrator is often referred to as an "unstable multivibrator," because it is continually switching back and forth and has no stable condition or state. The monostable multivibrator on the other hand (discussed previously in Chapter 9), has one (mono) stable state. The circuit will remain in this stable state until a trigger is applied that forces the monostable multivibrator into its unstable state. It will then flip into its unstable state for a small period of time, and then flop back to its stable state and await another trigger. The monostable flip-flop, or multivibrator, is often compared to a gun and called a **one-shot multivibrator** because it will produce one output pulse—or shot—for each input trigger.

One-Shot Multivibrator
A monostable flip-flop, or multivibrator, that will produce one output pulse—or shot—for each input pulse trigger.

Figure 18-25 shows how a 74121 one-shot can be connected as a variable-output pulse generator circuit. The normally open push button switch is used to

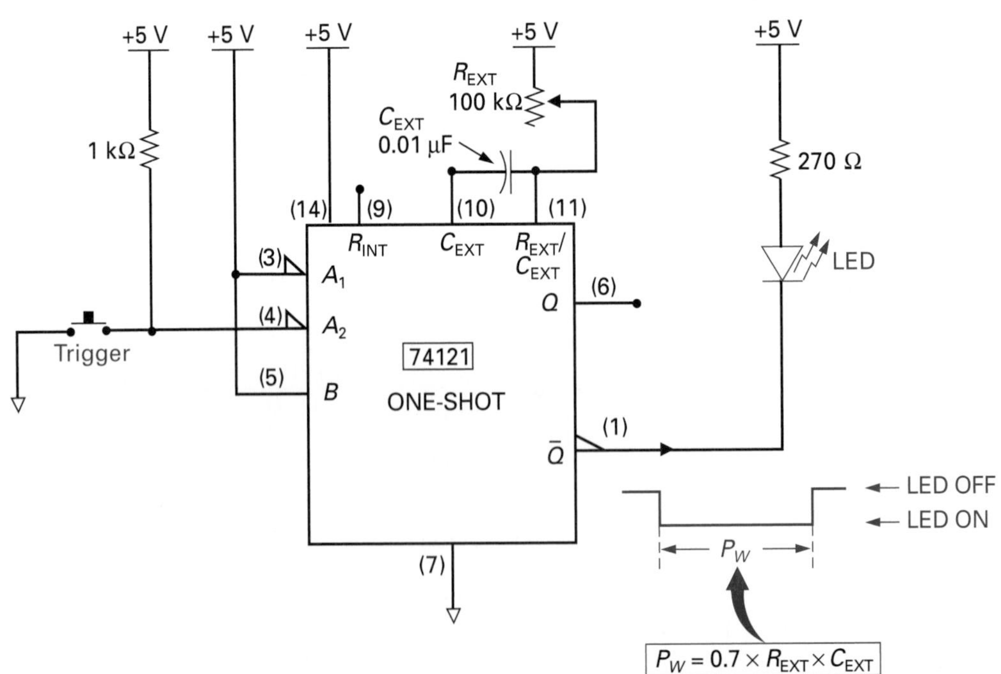

FIGURE 18-25 Flip-Flop Application—One Shot (Monostable MV) Pulse Generator Circuit.

trigger the one-shot, with the output pulse width being determined by R_{EXT} and C_{EXT}. Because the output LED is connected as an active-LOW display (activated or turned ON by a LOW), the $\overline{Q}$ output is used to control the LED.

SELF-TEST REVIEW QUESTIONS FOR SECTION 18-2

1. Why do you think the 7447 decoder/driver is also called a "code converter"?
2. An encoder could also be called a "code generator." (true/false)
3. A multiplexer is a circuit that will switch ____________ input(s) through to ____________ output(s), whereas a demultiplexer is a circuit that will switch ____________ input(s) through to ____________ output(s). ("one of several" or "one")
4. Which of the following input conditions would set HIGH the Q output of a negative edge triggered J-K flip-flop?
   a. $J = 1, K = 0, C = \uparrow$.
   b. $J = 1, K = 0, C = \downarrow$.
   c. $J = 0, K = 1, C = \uparrow$.
   d. $J = 0, K = 1, C = \downarrow$.
5. A digital register circuit makes use of the flip-flop's ____________ ability to store binary data.
6. What are the two main applications for digital counter circuits?

18-3 DIGITAL SYSTEM APPLICATION

As mentioned previously, digital electronic circuits tend to dominate most electronic systems because information stored, processed, or communicated in digital form is more accurate. In this chapter, you have been introduced to some of the many digital electronic devices, and you have seen how these devices can be used in some typical applications. To tie together all of this knowledge, let us see how we can combine many of the previously discussed digital ICs to construct a digital electronic system.

18-3-1 *Basic Frequency Counter Block Diagram*

A frequency counter is a good example of a digital system that makes use of a clock oscillator, frequency dividers, counters, registers, decoders, and a 7-segment display. Before we discuss the frequency counter circuit diagram, let us first examine the basic block diagram of the frequency counter, shown in Figure 18-26. To start at the far left of the timing diagram, you can see that the counter is cleared by a HIGH pulse just before the "count window pulse" begins. The count window pulse and the unknown input frequency are both applied to an

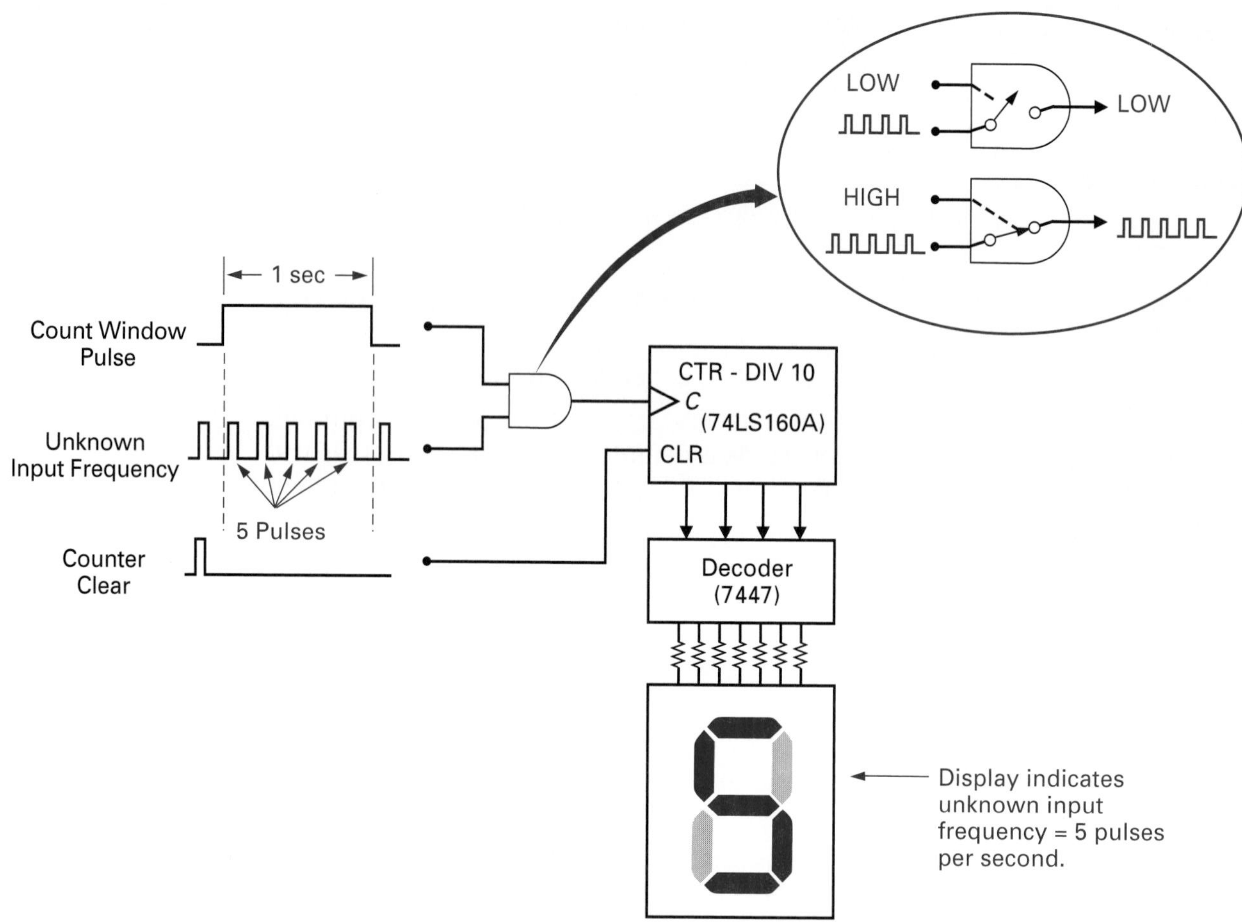

FIGURE 18-26 Simplified Frequency Counter Circuit.

AND gate, which acts as a voltage-controlled switch, as shown in the inset in Figure 18-26. When the count window pulse applies a LOW to the AND gate, it acts as an open switch because any LOW input to an AND gate produces a constant LOW out. On the other hand, when the count window pulse applies a HIGH to the AND gate, it acts as a closed switch connecting the unknown input frequency pulses through to the output. In the example shown in Figure 18-26, five pulses are switched through the AND gate to the clock input of the counter during the 1-second count window pulse. These 5 pulses will trigger the decade counter five times, causing it to count to 5 (0101). This BCD count of 5 from the counter will be decoded by the BCD to 7-segment decoder and displayed on the seven segment LED display. Because 5 pulses occurred in 1 second, the unknown input frequency is 5 pulses per second (5 pps).

The count window pulse is sometimes called the "sample pulse" because the unknown input frequency is sampled during the time that this signal is active. The simple frequency counter circuit shown in Figure 18-26 is extremely limited because it can only determine the unknown input frequency of inputs between 1 to 9 pps. To construct a more versatile frequency counter, we will have to include some additional digital ICs in the circuit.

18-3-2 *A Frequency Counter Circuit Diagram*

Figure 18-27 shows a more complete frequency counter circuit diagram. Although at first glance this appears to be a complex circuit, it can easily be understood by examining one block at a time.

The accuracy of a frequency counter is directly dependent on the accuracy of the count window pulse. To ensure this accuracy, a "100 kHz crystal oscillator" is used to generate a very accurate square wave. The output from the crystal oscillator is used to clock a chain of divide-by-ten ICs (74LS160As) in the "count window generator circuit" block. The 100 kHz input at the right is successively divided by ten to produce a 1 kHz, 100 Hz, 10 Hz, and 1 Hz output. Any one of these count window pulse frequencies can be selected by switch 1 and applied to a divide-by-two stage (1/2 7476) in the "divide-by-two and monostable" block. The waveforms shown in the inset in Figure 18-27 illustrate what occurs when switch SW_1 is placed in the 1 Hz position. The square-wave input to the 7476 completes one cycle in one second; however, the positive half cycle of this waveform only lasts for 0.5 of a second. By including the divide-by-two stage, the 1 Hz input is divided by two to produce an output frequency of 0.5 Hz. This means one complete cycle lasts for two seconds, and therefore the positive alternation or count window pulse lasts for one second. Including a divide-by-two stage in this way means that when SW_1 selects

1. 1 Hz, the count window pulse out of the divide-by-two stage will have a duration of 1 second;
2. 10 Hz, the count window pulse out of the divide-by-two stage will have a duration of 100ms (0.1s);
3. 100 Hz, the count window pulse out of the divide-by-two stage will have a duration of 10ms (0.01s); and
4. 1 kHz, the count window pulse out of the divide-by-two stage will have a duration of 1 ms (0.001 s).

As we will see later, these smaller count window pulses will allow us to measure higher input frequencies.

The count-window pulse out of the divide-by-two stage is applied to an AND gate along with the unknown input frequency. In the example of timing waveforms shown in the inset in Figure 18-27, you can see that ten pulses are switched through the AND gate during the window of one second. These ten pulses are used to clock a four-stage counter in the "counter, register, decoder, and display circuit" block. The "units counter" on the right will count from 0–9, and if the count exceeds 9, it will reset to zero and carry a single tens pulse into the "tens counter." In this example, this is exactly what will happen, and the final display will show decimal 10, meaning that the unknown input frequency is 10 pps. Similarly, the "tens counter" will reset and carry into the "hundreds counter," and the hundreds counter will reset and carry into the "thousands counter," as is expected when counting in decimal. The BCD outputs of the four-stage counter are applied to a set of registers, and the output of these registers drives a four-digit seven-segment display via BCD-to-7-segment display decoders. Referring back to the "divide-by-2 and monostable" block, you can

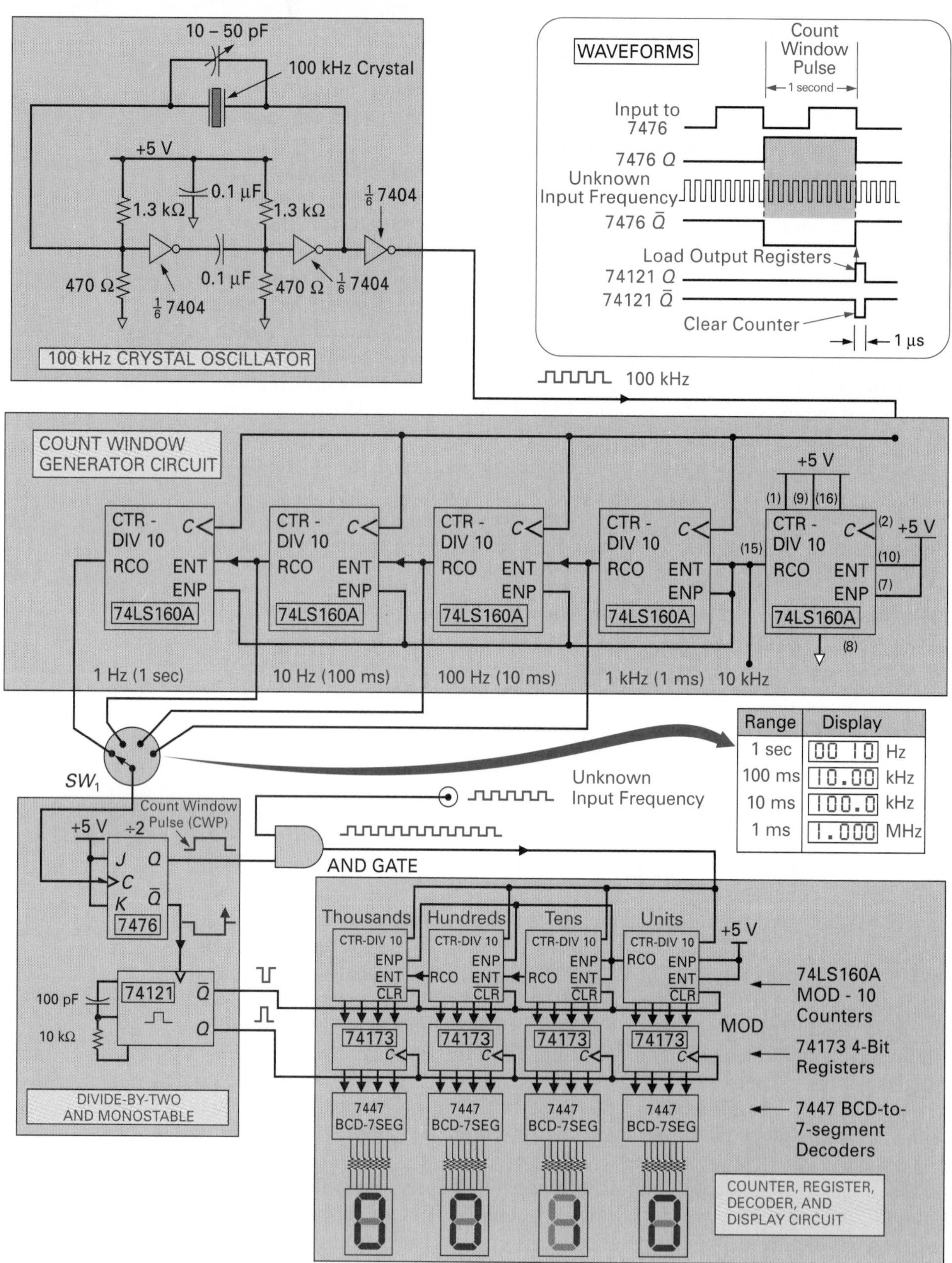

FIGURE 18-27 A Frequency Counter Circuit.

see that the 7476 $\overline{Q}$ output is used to trigger a 74121 one-shot, whose RC network has been chosen to generate a 1 μs pulse. Referring again to the timing waveforms in the inset in Figure 18-27, you can see that the positive edge of the 7476$\overline{Q}$ output triggers the 74121, which generates a positive 1 μs pulse at its Q output and a negative 1 μs pulse at its $\overline{Q}$ output. The positive edge of the 74121 Q output clocks the four 74173 4-bit registers, which store the new frequency count. This stored value appears at the Q outputs of the 74173s where it is decoded by the 7447s and displayed on the seven-segment displays. The 74173s are included to hold the display value constant, except for the few microseconds when the display is being updated to a new value. The negative 1 μs pulse at the 74121 $\overline{Q}$ output is used to clear the 4-stage counter once the new frequency count has been loaded into the 74173 registers.

To explain this circuit in a little more detail, let us examine each of the frequency counter's different range selections (controlled by SW_1) shown in the table in Figure 18-27.

1 Hz or 1 s range. When SW_1 selects this frequency, a count window pulse of 1 second is generated. This window of time and a 4-digit display means that we can measure any input frequency from 0001 Hz to 9999 Hz. If this maximum of 9999 Hz is exceeded, the counter will reset all four digits. For example, an input of 9999 Hz will be displayed as "9999," whereas an input of 10000 Hz will be displayed as "0000." To measure an input frequency in excess of 9999 Hz, we will have to use a smaller window so that fewer pulses are counted.

10 Hz or 100 ms range. When SW_1 selects this frequency, a count window pulse of 0.1 or 1/10th of a second is generated. This means that if the unknown input frequency is 10,000 Hz, only 1/10th of these pulses will be switched through the AND gate to the counter (1/10th of 10,000 = 1000). This value of 1000 is smaller and can be displayed on a 4-digit display; however, a display of 1000 Hz is not accurate because the input frequency is 10,000 Hz. Since 10,000 Hz = 10.00 kHz, on this range we will add a decimal point in the mid position and add a unit label of "kHz" after the value so that the readout is correct, as shown in the table in Figure 18-27. An input of 10,000 Hz therefore will be displayed as "10.00 kHz."

100 Hz or 10 ms range. When SW_1 selects this frequency, a count window pulse of 0.01 or 1/100th of a second is generated. This means that if the unknown input frequency is 100,000 Hz, only 1/100th of these pulses will be switched through the AND gate to the counter (1/100th of 100,000 = 1000). This value of 1000 is smaller and can be displayed on a 4-digit display; however, a display of 1000 Hz is not accurate since the input frequency is 100,000 Hz. Since 100,000 Hz = 100.0 kHz, on this range we will move the position of the decimal point and keep the unit label of "kHz" after the value, so that the readout is correct, as shown in the table in Figure 18-27. An input of 100,000 Hz, therefore, will be displayed as "100.0 kHz."

1 kHz or 1 ms range. When SW_1 selects this frequency, a count window pulse of 0.001 or 1/1000th of a second is generated. This means that if the unknown input frequency is 1,000,000 Hz, only 1/1000th of these pulses will be switched through the AND gate to the counter (1/1000th of 1,000,000 =

1000). This value of 1000 is smaller and can be displayed on a 4-digit display; however, a display of 1000 Hz is not accurate since the input frequency is 1,000,000 Hz. Since 1,000,000 Hz = 1.000 MHz, on this range we will again move the position of the decimal point and add a unit label of "MHz" after the value so that the readout is correct, as shown in the table in Figure 18-27. An input of 1,000,000 Hz therefore, will be displayed as "1.000 MHz."

18-3-3 *Troubleshooting the Frequency Counter Circuit*

To be an effective electronic technician or troubleshooter, you must have a thorough knowledge of electronics, test equipment, troubleshooting techniques, and equipment repair. Like analog circuits, digital circuits do occasionally fail, and in most cases a technician is required to quickly locate the problem and make the repair. The procedure for fixing a failure can be broken down into three basic steps and, to review, these steps are

Step 1: DIAGNOSE

The first step is to determine whether a problem really exists. To carry out this step, a technician must collect information about the system, circuit, and components used, then diagnose the problem.

Step 2: ISOLATE

The second step is to apply a logical and sequential reasoning process to isolate the problem. In this step, a technician will operate, observe, test, and apply troubleshooting techniques in order to isolate the malfunction.

Step 3: REPAIR

The third and final step is to make the actual repair, then final test the circuit.

As an example, we will apply these three steps to the previously discussed frequency counter circuit.

Step 1: Diagnose

It is extremely important that you first understand how a circuit and all of the devices within it are supposed to work so that you can determine whether or not a circuit malfunction exists. If you were preparing to troubleshoot the frequency counter circuit in Figure 18-27, your first step should be to read through the circuit description and review the operation of each integrated circuit until you feel completely confident with the correct operation of the entire circuit. For other digital circuits, refer to service or technical manuals, which generally contain circuit descriptions and troubleshooting guides. As far as the ICs are concerned, refer to the manufacturer's digital data books, which contain a full description of the ICs operation, characteristics, and pin allocations. Referring to all of this documentation before you begin troubleshooting will generally speed up and simplify the isolation step. Once you are fully familiar with the operation of the circuit, you will easily be able to "diagnose the problem" as either

1. operator error, or
2. circuit malfunction.

For example, the following could be interpreted as a circuit malfunction, when in fact it is a simple operator error.

Symptom: SW_1 has selected 1 ms range, input frequency = 100 Hz, display shows 0000.

Diagnosis: Operator error—a 1 millisecond window will not sample any of the input because each cycle of the input lasts for ten milliseconds (1/100 Hz). To determine low input frequencies, operator will need to use a wider window.

Once you have determined that the problem is not an operator error, but is in fact a circuit malfunction, proceed to step two and isolate the circuit failure.

Step 2: Isolate

No matter what circuit or system failure has occurred, you should always follow a logical and sequential troubleshooting procedure. Let us review some of the isolating techniques discussed previously and apply them to our frequency counter circuit.

1. Use a "cause and effect troubleshooting process," which means, study the effects you are getting from the faulty circuit and then logically reason out what could be the cause.
2. Check first for "obvious errors." Is power OFF or not connected to the circuit? Are there wiring errors? Are all of the ICs correctly oriented?
3. Use your "senses" to check for broken wires, loose connectors, overheated or smoking component, pins not making contact, and so on.
4. Using a logic probe or voltmeter test that "power" and "ground" is connected to the circuit and is present at all points requiring power and ground.
5. Test for a "clock signal" using the logic probe or, ideally, the oscilloscope. With the frequency counter circuit, you should first test the 100 kHz output from the crystal oscillator and then test in sequence the 10 kHz, 1 kHz, 100 Hz, 10 Hz, and 1 Hz outputs from the count window generator circuit. Problems in frequency divider chains are generally easily detected because an incorrect frequency output normally indicates that the previous stage has malfunctioned.
6. Perform a "static test" on the circuit. As with our frequency counter circuit example, you should disconnect the inputs to the counter, register, decoder, and display circuit. Then use a logic pulser to clock the counter and a logic probe to test that the valid count is appearing at the outputs of each stage of the counter. For example, after nineteen clock-pulse inputs from the logic pulser, the logic probe should indicate the following logic levels at the counter outputs:

0000	0000	0001	1001_{BCD}
0	0	1	9

   By pulsing the register clock input, this value should appear at the output of the 74173s, be decoded by the 7447s, and displayed. By pulsing the counter clear input, you should be able to clear all of the outputs of the counter. The divide-by-two and monostable circuit can also be tested in this way, by disconnecting the inputs and using the logic pulser as a signal source and the logic probe as a signal detector.

7. With a "dynamic test," the circuit is tested while it is operating at its normal clock frequency. Using an oscilloscope, you should be able to test the frequency of signals at all points throughout the circuit.
8. "Noise" due to electromagnetic interference (EMI) can false-trigger the flip-flops within the counters to a new count, and because of a flip-flop's latching or memory ability, the counter will remain at its new, invalid count. This problem can be overcome by not leaving any of the IC's inputs unconnected and therefore floating. Connect unneeded active-HIGH inputs to ground and active-LOW inputs to $+V_{CC}$.
9. Apply the "half-split method" of troubleshooting first to a system, then to a circuit, then to a section of a circuit to help speed up the isolation process.
10. All digital circuits are subject to the normal malfunctions such as internal and external "shorts" and "opens."
11. "Substitution" can be used to help speed up your troubleshooting process. Once the problem is localized to an area containing only a few ICs, substitute suspect ICs with known working ICs (one at a time) to see if the problem can be quickly remedied.
12. Many electronic system manufacturers provide "troubleshooting trees" in the system technical manual. These charts are a graphical means to show the sequence of tests to be performed on a suspected circuit or system. Figure 18-28 shows a simple troubleshooting tree for our frequency counter circuit.

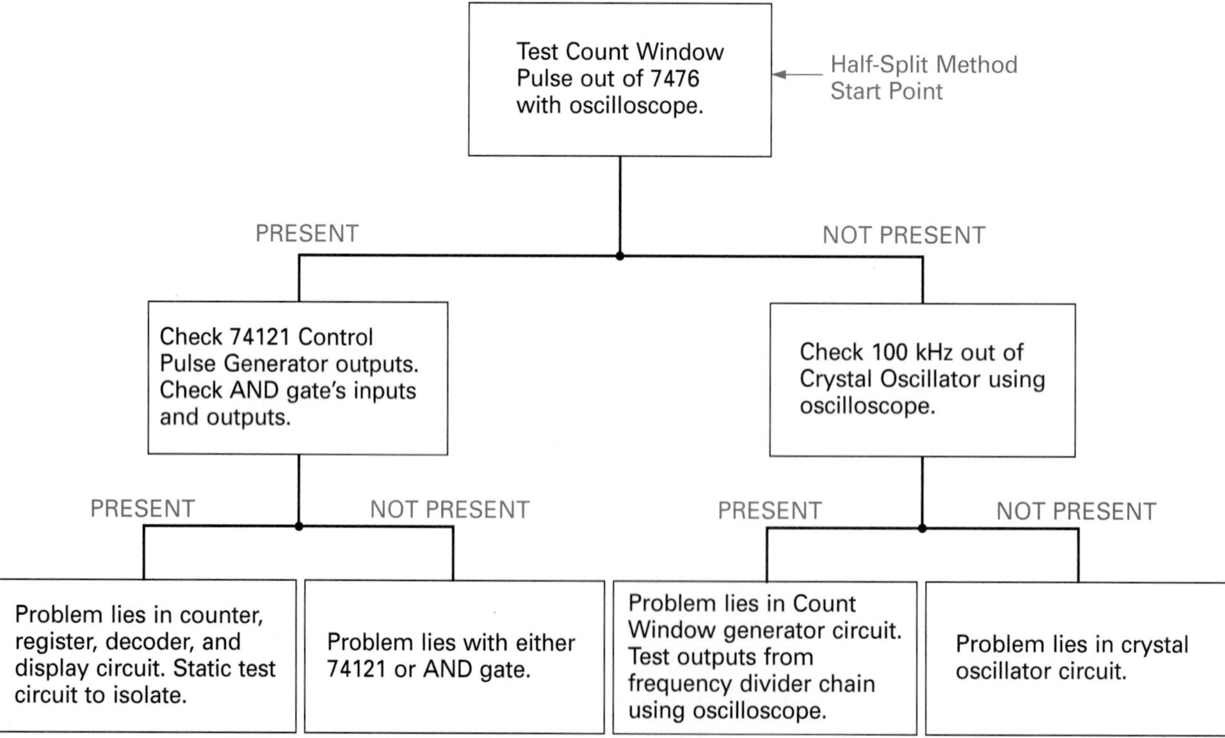

FIGURE 18-28 A Troubleshooting Tree for the Frequency Counter Circuit.

Step 3: Repair

The final step is to repair the circuit, which could involve simply removing an excess piece of wire, resoldering a broken connection, reconnecting a connector, or adjusting the power supply voltage. In most instances, however, the repair will involve the replacement of a faulty component. For a circuit that has been constructed on a prototyping board or bread board, the removal and replacement of a component is simple. However, when a printed circuit board is involved you should make a note of the component's orientation and observe good soldering and desoldering techniques. Also be sure to handle all MOS ICs with care to prevent any damage due to static discharge.

When the circuit has been repaired, always perform a final test to see that the circuit and the system is fully operational.

SELF-TEST REVIEW QUESTIONS FOR SECTION 18-3

1. Which of the 74LS160As in Figure 18-27 function as frequency dividers and which function as counters?
2. To test the "count window generator circuit" in Figure 18-27, the clock oscillator input is disconnected and a logic pulser is used to inject a 500 kHz input. Does a problem exist if the following outputs are measured with the oscilloscope at each of the SW_1 positions: 5 kHz, 500 Hz, 50 Hz, 5 Hz?

SUMMARY

Number Systems (Figure 18-29)

1. The early digital systems constructed in the 1950s made use of ten levels, or voltages, with each of these voltages corresponding to one of the ten digits in the decimal number system. These circuits were very complex since they had to generate one of ten voltage levels and sense the difference between all ten voltage levels. This complexity led to inaccuracy since some circuits would periodically confuse one voltage level for a different voltage level. The solution to the problem of circuit complexity and information inaccuracy was solved by adopting a two-state system instead of a ten-state system.
2. Using a two-state or two-digit system, you can generate codes for any number, letter, or symbol. Electronic circuits that manage two-state codes are less complex because they only have to generate and sense a HIGH and LOW voltage and more accurate because there is little room for error between the two extremes of ON and OFF, or HIGH voltage and LOW voltage.
3. Abandoning the ten-state system and adopting the two-state system meant operating in the two-state number system, which is called "binary."
4. The binary number system only has two digits in its number scale: only the digits "0" and "1" exist in binary.

QUICK REFERENCE SUMMARY SHEET

FIGURE 18-29 Number Systems (Decimal, Binary, and Hexadecimal).

DECIMAL			HEXADECIMAL				BINARY					
10^2	10^1	10^0	16^3	16^2	16^1	16^0	2^5	2^4	2^3	2^2	2^1	2^0
100	10	1	4096	256	16	1	32	16	8	4	2	1
		0				0	0	0	0	0	0	0
		1				1	0	0	0	0	0	1
		2				2	0	0	0	0	1	0
		3				3	0	0	0	0	1	1
		4				4	0	0	0	1	0	0
		5				5	0	0	0	1	0	1
		6				6	0	0	0	1	1	0
		7				7	0	0	0	1	1	1
		8				8	0	0	1	0	0	0
		9				9	0	0	1	0	0	1
	1	0				A	0	0	1	0	1	0
	1	1				B	0	0	1	0	1	1
	1	2				C	0	0	1	1	0	0
	1	3				D	0	0	1	1	0	1
	1	4				E	0	0	1	1	1	0
	1	5				F	0	0	1	1	1	1
	1	6			1	0	0	1	0	0	0	0
	1	7			1	1	0	1	0	0	0	1
	1	8			1	2	0	1	0	0	1	0
	1	9			1	3	0	1	0	0	1	1
	2	0			1	4	0	1	0	1	0	0
	2	1			1	5	0	1	0	1	0	1
	2	2			1	6	0	1	0	1	1	0
	2	3			1	7	0	1	0	1	1	1
	2	4			1	8	0	1	1	0	0	0
	2	5			1	9	0	1	1	0	0	1
	2	6			1	A	0	1	1	0	1	0
	2	7			1	B	0	1	1	0	1	1
	2	8			1	C	0	1	1	1	0	0
	2	9			1	D	0	1	1	1	0	1
	3	0			1	E	0	1	1	1	1	0
	3	1			1	F	0	1	1	1	1	1
	3	2			2	0	1	0	0	0	0	0
	3	3			2	1	1	0	0	0	0	1
	3	4			2	2	1	0	0	0	1	0

5. The decimal system of counting and keeping track of items was first created by Hindu mathematicians in India at about 400 AD. Since it involved the use of fingers and thumbs, it was natural that this system would have 10 digits.
6. The key feature that distinguishes one number system from another is the number system's base or radix. This base indicates the number of digits that will be used. A subscript is sometimes included after a number when different number systems are being used to indicate the base of the number.

7. In decimal notation, each position to the left of the decimal point indicates an increased positive power of ten.
8. In the decimal number system, the left-most digit is called the most significant digit (MSD) while the right-most digit is called the least significant digit (LSD).
9. A reset and carry action will occur in any column that reaches its maximum count.
10. The value of a binary digit is determined by its position relative to the other digits. In the decimal system each position to the left of the decimal point increases by a power of 10. This case is also true with the binary number system; however, since it is a base-2 (bi) number system, each place to the left of the binary point increases by a power of 2.
11. The base-2 binary number system only uses the first two digits on the number scale: 0 (zero) and 1 (one). The 0s and 1s in binary are called "binary digits," or "bits" for short.
12. With binary, the weight of each column to the left increases 2 times. The first column therefore has a weight of 1, the second column has a weight of 2, the third column has a weight of 4, the fourth 8, and so on.
13. The left-most binary digit is called the most significant bit (MSB) since it carries the most weight, while the right-most digit is called the least significant bit (LSB) since it carries the least weight.
14. The reset and carry action will occur in binary in exactly the same way as it did in decimal. However, since binary has only two digits, a column will reach its maximum digit a lot sooner, and the reset and carry action in the binary number system will occur much more frequently.
15. Binary numbers can easily be converted to their decimal equivalent by simply adding together the weights of all the columns that contain a binary 1.
16. To convert a decimal number to its binary equivalent, continually subtract the largest possible power-of-two until the decimal number is reduced to zero, placing a binary 1 in columns that are used and a binary 0 in the columns that are not used.
17. Hexadecimal, or hex, is used as a sort of "shorthand" for large strings of binary numbers.
18. Hexadecimal means 16, and this number system has 16 different digits. Comparing the first nine digits in the decimal and hexadecimal columns, you will find that there is no difference between the two. However, beyond 9 hexadecimal makes use of the letters A, B, C, D, E and F. Digits 0 through 9 and letters A through F make up the total 16 digits of the hexadecimal number system.
19. To find the decimal equivalent of a hexadecimal number, simply multiply each hexadecimal digit by its positional weight.
20. Decimal to hexadecimal conversion is achieved in the same way as decimal to binary. First subtract the largest possible power of 16, and then keep subtracting the largest possible power of 16 from the remainder. Each time a

subtraction takes place add 1 to the respective column until the decimal value has been reduced to zero.

21. A 16-bit binary number, which is more commonly called a 16-bit binary "word," can be represented by 4 hexadecimal digits.

Binary Codes (Figure 18-30)

22. The process of converting a decimal number to its binary equivalent is called binary coding. The result of the conversion is a binary number or code that is pure binary.
23. Using combinations of binary digits, we can represent information as a binary code. There are other binary codes used in digital circuits other than pure binary, such as binary coded decimal (BCD) and the American Standard Code for Information Interchange (ASCII).
24. Binary coded decimal (BCD) is a binary code that has decimal characteristics. Being a binary code it has the advantages of a two-state system, and since it has a decimal format it is also much easier for an operator to interface via a decimal keypad or decimal display to systems such as pocket calculators, wristwatches, and so on.
25. The BCD code expresses each decimal digit as a 4-bit word.
26. BCD has only ten 4-bit binary codes: 0000, 0001, 0010, 0011, 0100, 0101, 0110, 0111, 1000, and 1001 (0 through 9). When the maximum digit (1001) is reached, a reset-and-carry will occur.
27. Digital electronic computers must also be able to generate and recognize binary codes that represent letters of the alphabet and symbols. The ASCII code is the most widely used alphanumeric code (alphabet and numeral code).

Digital Circuits—Decoders and Encoders (Figure 18-31)

28. Almost every electronic system today employs a combination of both linear and digital circuits. In the past, most electronic systems managed analog information signals and therefore the majority of their internal circuits were linear. Today, digital electronic circuits tend to dominate most electronic systems because information stored, processed, or communicated in digital form is more accurate since it is less sensitive to noise.
29. The term decode means to translate coded characters to a more understandable form, or to reverse a previous encoding process. A digital decoder circuit is a logic circuit that responds to one specific input word or code, while rejecting all others.
30. While some decoders are designed to recognize only one input word combination, other decoders are designed to activate one output for every input word combination.
31. A 1 of 4 decoder will select or activate 1 of 4 outputs based on which input code is applied.
32. A binary to decimal decoder translates a binary coded input into a decimal output.

FIGURE 18-30 Binary Codes.

Decimal Number System: 100 Place	10 Place	1 Place	Binary Coded Decimal (BCD) TENS DIGIT: 8	4	2	1	UNITS DIGIT: 8	4	2	1
		0	0	0	0	0	0	0	0	0
		1	0	0	0	0	0	0	0	1
		2	0	0	0	0	0	0	1	0
		3	0	0	0	0	0	0	1	1
		4	0	0	0	0	0	1	0	0
		5	0	0	0	0	0	1	0	1
		6	0	0	0	0	0	1	1	0
		7	0	0	0	0	0	1	1	1
		8	0	0	0	0	1	0	0	0
		9	0	0	0	0	1	0	0	1
	1	0	0	0	0	1	0	0	0	0
	1	1	0	0	0	1	0	0	0	1
	1	2	0	0	0	1	0	0	1	0
	1	3	0	0	0	1	0	0	1	1
	1	4	0	0	0	1	0	1	0	0
	1	5	0	0	0	1	0	1	0	1
	1	6	0	0	0	1	0	1	1	0
	1	7	0	0	0	1	0	1	1	1
	1	8	0	0	0	1	1	0	0	0
	1	9	0	0	0	1	1	0	0	1
	2	0	0	0	1	0	0	0	0	0
	2	1	0	0	1	0	0	0	0	1
	2	2	0	0	1	0	0	0	1	0
	2	3	0	0	1	0	0	0	1	1
	2	4	0	0	1	0	0	1	0	0
	2	5	0	0	1	0	0	1	0	1
	2	6	0	0	1	0	0	1	1	0
	2	7	0	0	1	0	0	1	1	1
	2	8	0	0	1	0	1	0	0	0
	2	9	0	0	1	0	1	0	0	1
	3	0	0	0	1	1	0	0	0	0
	3	1	0	0	1	1	0	0	0	1
	3	2	0	0	1	1	0	0	1	0
	3	3	0	0	1	1	0	0	1	1
	3	4	0	0	1	1	0	1	0	0
	3	5	0	0	1	1	0	1	0	1
	3	6	0	0	1	1	0	1	1	0
	3	7	0	0	1	1	0	1	1	1
	3	8	0	0	1	1	1	0	0	0
	3	9	0	0	1	1	1	0	0	1
	4	0	0	1	0	0	0	0	0	0
	4	1	0	1	0	0	0	0	0	1
	4	2	0	1	0	0	0	0	1	0
	4	3	0	1	0	0	0	0	1	1
	4	4	0	1	0	0	0	1	0	0
	4	5	0	1	0	0	0	1	0	1
	4	6	0	1	0	0	0	1	1	0
	4	7	0	1	0	0	0	1	1	1
	4	8	0	1	0	0	1	0	0	0
	4	9	0	1	0	0	1	0	0	1

The American Standard Code for Information Interchange (ASCII)

Character	Code	Character	Code
•	0100000	P	1010000
!	0100001	Q	1010001
"	0100010	R	1010010
#	0100011	S	1010011
$	0100100	T	1010100
%	0100101	U	1010101
&	0100110	V	1010110
'	0100111	W	1010111
(	0101000	X	1011000
)	0101001	Y	1011001
*	0101010	Z	1011010
+	0101011	[	1011011
,	0101100	/	1011100
–	0101101	]	1011101
.	0101110	^	1011110
/	0101111	_	1011111
0	0110000	'	1100000
1	0110001	a	1100001
2	0110010	b	1100010
3	0110011	c	1100011
4	0110100	d	1100100
5	0110101	e	1100101
6	0110110	f	1100110
7	0110111	g	1100111
8	0111000	h	1101000
9	0111001	i	1101001
:	0111010	j	1101010
;	0111011	k	1101011
<	0111100	l	1101100
=	0111101	m	1101101
>	0111110	n	1101110
?	0111111	o	1101111
@	1000000	p	1110000
A	1000001	q	1110001
B	1000010	r	1110010
C	1000011	s	1110011
D	1000100	t	1110100
E	1000101	u	1110101
F	1000110	v	1110110
G	1000111	w	1110111
H	1001000	x	1111000
I	1001001	y	1111001
J	1001010	z	1111010
K	1001011	(	1111011
L	1001100	¦	1111100
M	1001101	)	1111101
N	1001110	~	1111110
O	1001111	DEL	1111111

QUICK REFERENCE SUMMARY SHEET

FIGURE 18-31 Digital Circuits—Decoders and Encoders.

Decoder Application — BCD-to-7-Segment Decoder (7447)

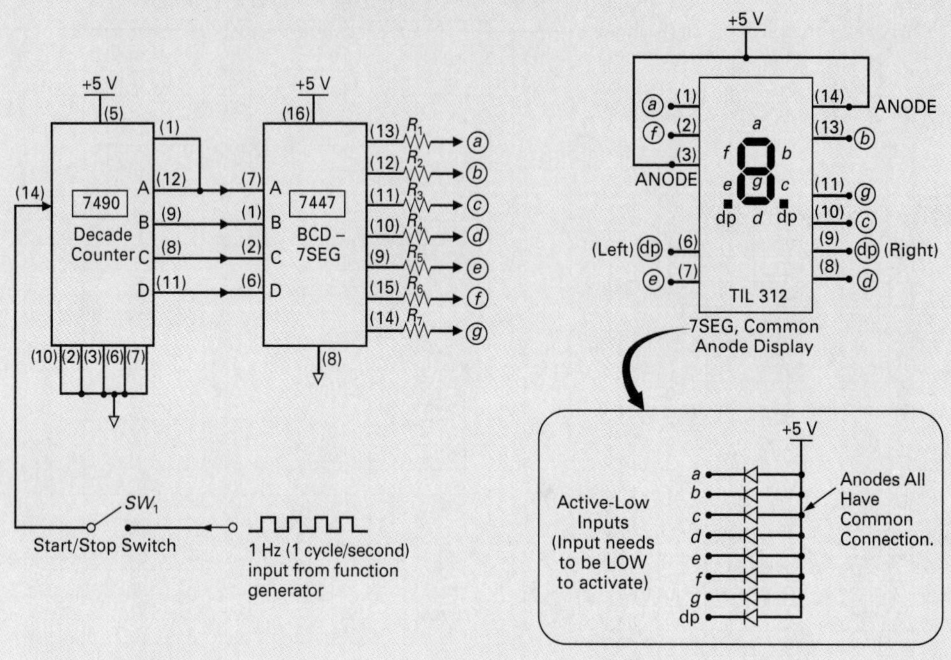

Encoder Application — Keyboard Encoder (74147)

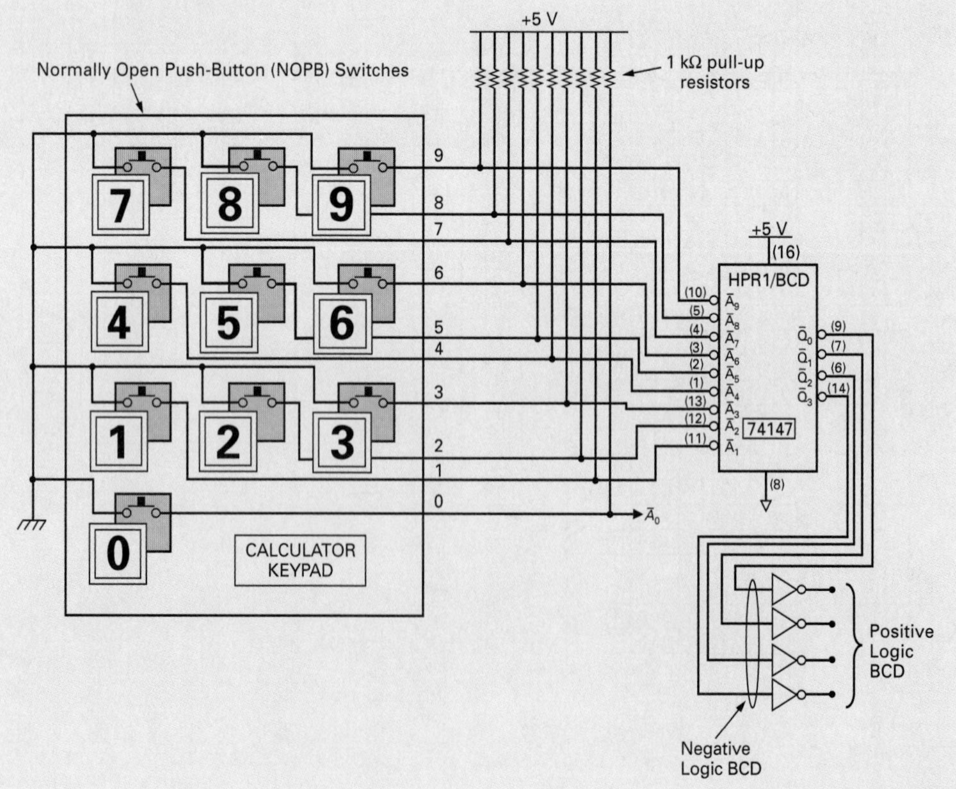

33. A 7447 BCD-to-7-segment decoder IC can be connected to drive a common-anode 7-segment LED display. The 7447 generates active-LOW outputs, which means that a "LOW" output signal will "activate" its associated output LED.
34. An encoder circuit performs the opposite function of a decoder, since a decoder is designed to detect specific codes whereas an encoder is designed to generate specific codes.
35. A 74147 decimal-to-binary encoder can be used to encode a decimal keypad.

Digital Circuits—Comparators, Multiplexers, and Demultiplexers (Figure 18-32)

36. A digital comparator compares two binary input words to see if they are equal.
37. A 7485 compares two 4-bit binary words applied to its inputs, and generates three output control signals.
38. A digital multiplexer (MUX), or selector, is a circuit that is controlled to switch one of several inputs through to one output.
39. A digital demultiplexer (DMUX), or data distributor, is a circuit that is controlled to switch a single input through to one of several outputs.
40. A 74151 is an 8-to-1 multiplexer (MUX), which means that it will switch one of eight inputs through to one output.
41. A 74138 is a 1-to-8 demultiplexer (DMUX), which means that it will switch one input through to one of eight outputs.
42. The 74151 could be described as a parallel-to-serial converter, since it converts the parallel 8-bit word applied to its inputs into an 8-bit serial data stream.
43. The 74138 could be described as a serial-to-parallel converter, since it converts an 8-bit serial data stream applied to its input into a parallel 8-bit output word.
44. The advantage of serial data transmission is that only one data line is needed, whereas the advantage of parallel data transmission is speed because the parallel connection transmits all the data at one time.

Digital Circuits—Set-Reset (*S-R*) Flip-Flop

45. A digital flip-flop circuit is a "bistable multivibrator" circuit that has two stable states.
46. An *S-R* flip-flop has two inputs called "SET" and "RESET" and two outputs called "Q" and "$\overline{Q}$." The $\overline{Q}$ output derives its name from the fact that its voltage level is always the opposite of the Q output.
47. When only the R input is pulsed HIGH (reset condition), the Q output is reset to a binary 0, or reset LOW ($\overline{Q}$ will be the opposite, or HIGH).
48. When only the S input is pulsed HIGH (set condition), the Q output is set to a binary 1, or set HIGH ($\overline{Q}$ will be the opposite, or LOW).

QUICK REFERENCE SUMMARY SHEET

FIGURE 18-32 Digital Circuits—Comparators, Multiplexers, and Demultiplexers.

Comparator Application — Photocopier Control (7485)

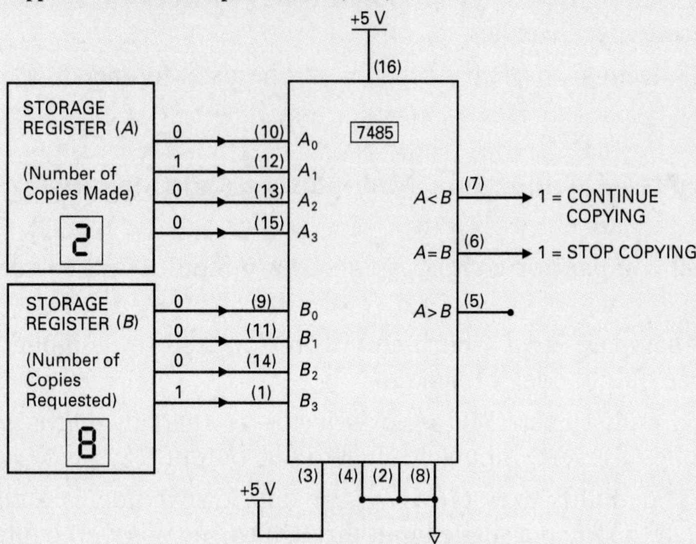

Multiplexer and Demultiplexer Application — Basic Home Security System (74151 and 74138)

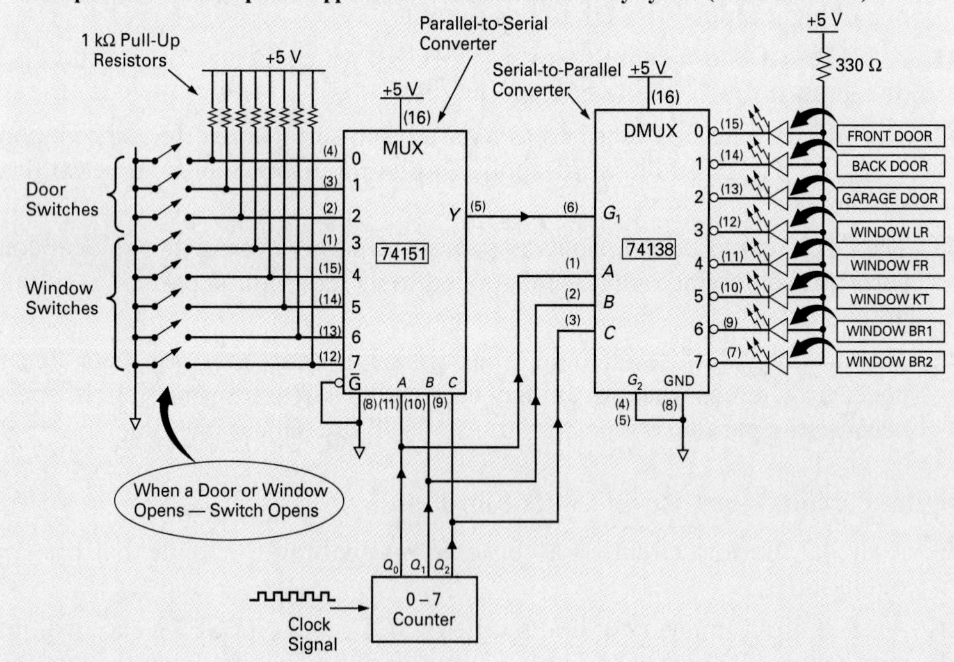

49. When both the S and R inputs are LOW, the S-R flip-flop is said to be in the no-change or latch condition, since there will be no change in the output Q.

50. When both the S and R inputs are HIGH, the S-R flip-flop will race, and the output will be unpredictable.

Digital Circuits—*J-K* Flip-Flop (Figure 18-33)

51. A digital system will typically contain thousands of flip-flop circuits. To coordinate the overall operation of all these digital circuits, a clock or timing signal is applied to all of the digital devices to ensure that each device is triggered into operation at the right time. A clock signal controls when a flip-flop is allowed to operate.

52. The *J-K* flip-flop, (*J* and *K* were arbitrarily chosen) has a clock input *(C),* in addition to the standard "set *(J)*" and "reset *(K)*" flip-flop inputs. This clock input will be used to trigger the flip-flop into operation, and therefore control the timing of this device in relation to all the other devices in a circuit.

53. A negative edge triggered *J-K* flip-flop will "wake up" and perform the operation applied to the *J* and *K* inputs when the clock input is making a transition from HIGH to LOW (a negative edge).

54. When both the *J* and *K* inputs of a *J-K* flip-flop are made HIGH, the *Q* output will toggle or switch to the opposite state. This means that if *Q* is HIGH it will switch to a LOW, and if *Q* is LOW it will switch to a HIGH.

55. A full triangle on the clock input of a flip-flop (inside the rectangular block) is used to indicate that this device is edge triggered, and the smaller right-angle triangle (or bubble) outside the rectangular block is used to indicate that the input is active-LOW, or negative-edge triggered.

56. The 74LS76A digital IC contains two complete negative-edge triggered *J-K* flip-flop circuits.

57. When the *J* and *K* inputs of the 74LS76A flip-flop are both connected HIGH, the *Q* output will continually toggle for each negative edge of the clock-signal input, and therefore divide the input frequency by 2.

58. Combining the input and the divided-by-two output of a 74LS76A set to toggle, we can generate a 2-bit word which will count from binary 0 (00) to binary 3 (11), and then continuously repeat the cycle.

Digital Circuits—Flip-Flop Applications (Figure 18-34)

59. A binary counter circuit contains a group of flip-flops that are all connected to repeatedly divide the input clock frequency by 2 and generate an output code sequence that corresponds to a binary count.

60. A 74LS160A can be connected to function as a 0–9 (decade) counter or as a divide-by-10 frequency divider.

61. A binary register circuit contains a group of flip-flops that act as a memory element to store a group of bits, or a binary word.

62. A 74LS173 can be connected to form a 4-bit register in which the "no-change" or latching ability of the circuit's internal flip-flops are made use of to store a 4-bit word.

63. The monostable multivibrator has (as its name implies) one (mono) stable state. The circuit will remain in this stable state until a trigger is applied and forces the monostable multivibrator into its unstable state. It will then flip into its unstable state for a small period of time, and then flop back to its stable state and await another trigger.

QUICK REFERENCE SUMMARY SHEET

FIGURE 18-33 Digital Circuits—*J-K* Flip-Flop.

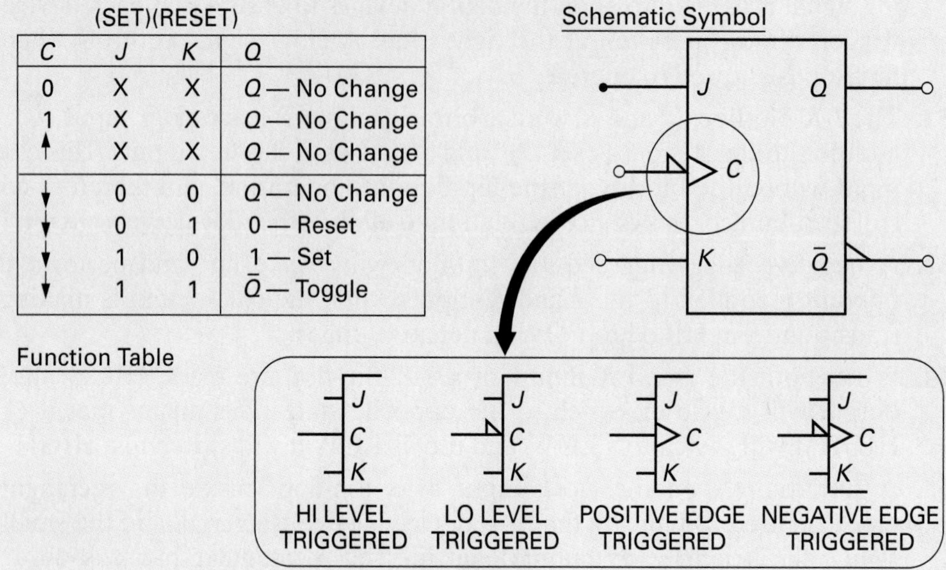

(SET)(RESET)

C	*J*	*K*	*Q*
0	X	X	*Q* — No Change
1	X	X	*Q* — No Change
↑	X	X	*Q* — No Change
↓	0	0	*Q* — No Change
↓	0	1	0 — Reset
↓	1	0	1 — Set
↓	1	1	$\bar{Q}$ — Toggle

Function Table

***J-K* Flip-Flop Application—Frequency Divider or Binary Counter (74LS76A)**

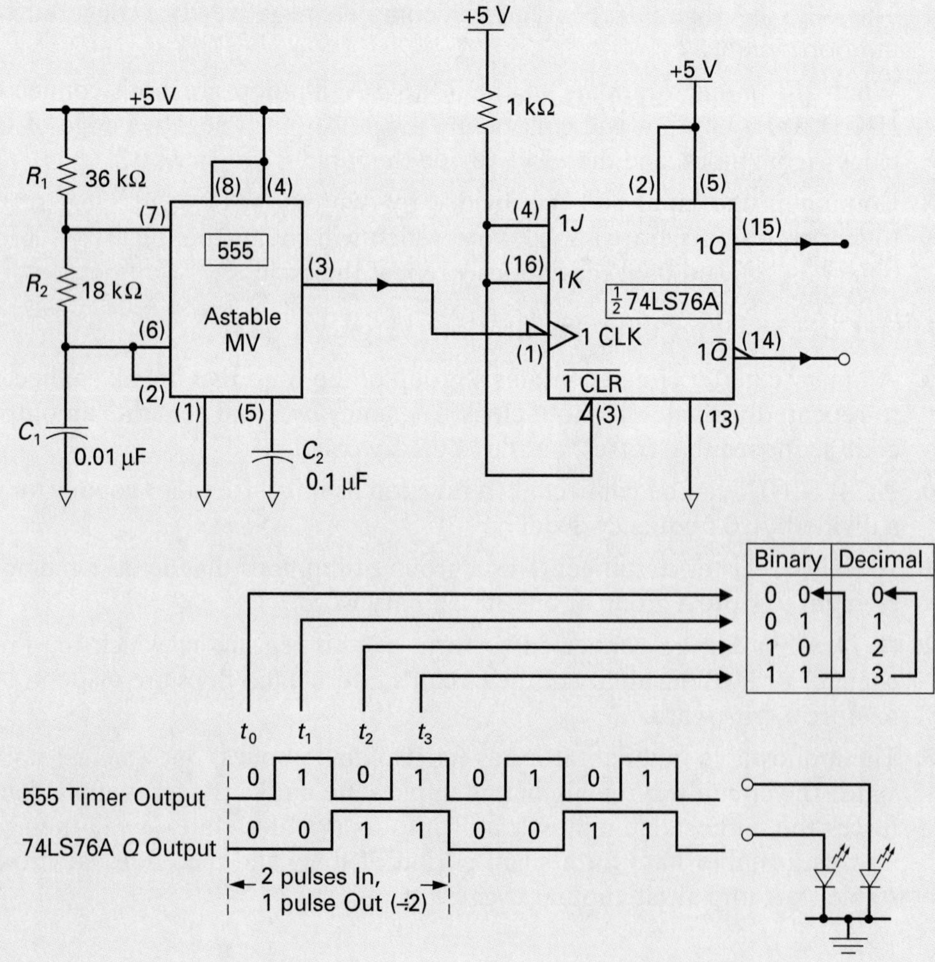

Binary	Decimal
0 0	0
0 1	1
1 0	2
1 1	3

FIGURE 18-34 Digital Circuits—Flip-Flop Applications.

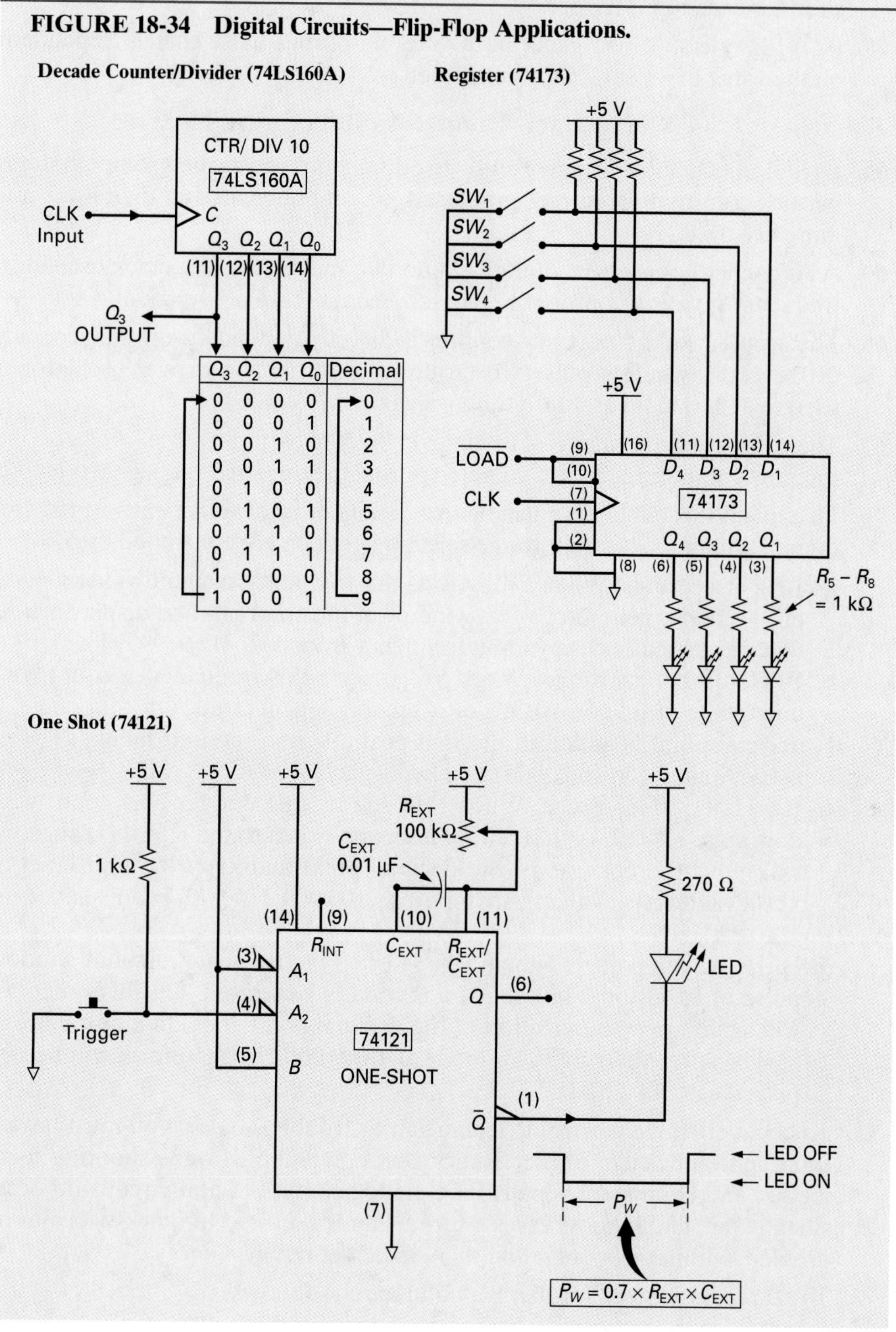

64. The monostable flip-flop, or multivibrator, is often compared to a gun and called a one-shot multivibrator because it will produce one output pulse, or shot, for each input trigger.

65. A 74121 one-shot will generate a variable output pulse that is dependent on the value of an externally connected resistor and capacitor.

Digital Systems—A Frequency Counter Circuit (Figure 18-35)

66. Digital electronic circuits tend to dominate most electronic systems because information stored, processed, or communicated in digital form is more accurate.

67. A frequency counter is a digital system that makes use of a clock oscillator, frequency dividers, counters, registers, decoders, and a 7-segment display.

68. The accuracy of a frequency counter is directly dependent on the accuracy of the count window pulse. To ensure this accuracy a crystal oscillator is used to generate an accurate timing signal.

69. The count window pulse is sometimes called the "sample pulse," because the unknown input frequency is sampled during the time that this signal is active.

70. To explain this circuit in a little more detail, let us examine each of the frequency counter's different range selections, which are controlled by SW_1.
    - **a.** 1 Hz or 1 s range: When SW_1 selects this frequency, a count window pulse of 1 second is generated. This window of time and a 4-digit display means that we can measure any input frequency from 0001 Hz to 9999 Hz.
    - **b.** 10 Hz or 100 ms range: When SW_1 selects this frequency, a count window pulse of 0.1, or 1/10th, of a second is generated. On this range, a decimal point is added in the mid position, and the unit label "kHz" is used. An input of 10,000 Hz will be displayed as "10.00 kHz."
    - **c.** 100 Hz or 10 ms range: When SW_1 selects this frequency, a count window pulse of 0.01, or 1/100th, of a second is generated. On this range we will move the position of the decimal point and keep the unit label of "kHz" after the value. An input of 100,000 Hz will be displayed as "100.0 kHz."
    - **d.** 1 kHz or 1 ms range: When SW_1 selects this frequency, a count window pulse of 0.001, or 1/1000th, of a second is generated. On this range we will again move the position of the decimal point and add a unit label of "MHz" after the value. An input of 1,000,000 Hz, therefore, will be displayed as "1.000 MHz."

71. To be an effective electronic technician or troubleshooter, you must have a thorough knowledge of electronics, test equipment, troubleshooting techniques, and equipment repair. Like analog circuits, digital circuits do occasionally fail, and in most cases a technician is required to quickly locate the problem within the system and then make the repair.

72. The three basic steps for fixing a failure are as follows:
    - **a.** *Step 1:* DIAGNOSE The first step is to determine whether a problem really exists. To carry out this step, a technician must collect information about the system, circuit, and components used, and then diagnose the problem.

QUICK REFERENCE SUMMARY SHEET

FIGURE 18-35 Digital Systems—A Frequency Counter Circuit.

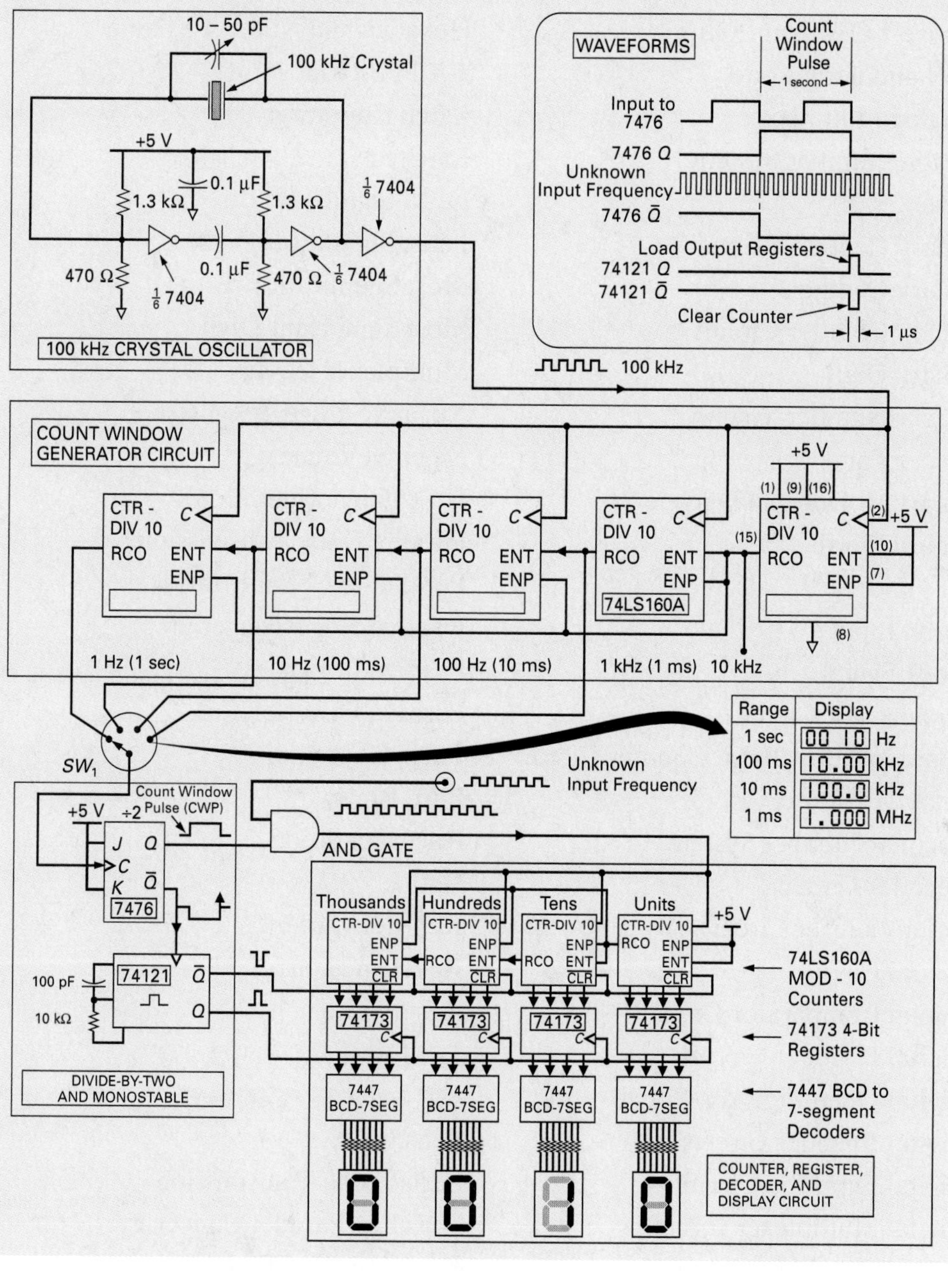

b. *Step 2:* ISOLATE The second step is to apply a logical and sequential reasoning process to isolate the problem. In this step, a technician will operate, observe, test and apply troubleshooting techniques in order to isolate the malfunction.

c. *Step 3:* REPAIR The third and final step is to make the actual repair, then final test the circuit.

NEW TERMS

1-Shot Circuit
1-of-4 Decoder
Active-HIGH Input/Output
Active-LOW Input/Output
Alphanumeric Code
Analog Multiplexer
Arabic Number System
Base
Binary Coded Decimal
Binary Coding
Binary Counter Circuit
Binary Digit
Binary Number System
Binary Point
Binary-to-Decimal Decoder
Binary Word
Bit
Clock Input
Clock Signal
Comparator Circuit
Count Window Pulse Counter Circuit
Data Distributor
Decimal Number System
Decoder Circuit
Demultiplexer Circuit
Digital Circuit
Digital Comparator Circuit
Digital Decoder Circuit
Digital Demultiplexer Circuit
Digital Encoder Circuit
Digital Flip-Flop Circuit
Digital Multiplexer Circuit
Digital One-Shot Circuit
Digital Register Circuit
Edge Triggered
Encoder Circuit
Flip-Flop Circuit
Frequency Counter
Frequency Divider
Hexadecimal Number System
J-K Flip-Flop
Latch Condition
Latched
Least Significant Bit
Least Significant Digit
Most Significant Bit
Most Significant Digit
Multiplexer Circuit
Negative-Edge Triggered
Negative Logic
No-Change Condition
Parallel Data Transmission
Parallel-to-Serial Converter
Positive-Edge Triggered
Positive Logic
Positive Power of Ten
Pulse Generator
Pure Binary
Race Condition
Radix
Register Circuit
Reset and Carry
Reset State
S-R Flip-Flop
S-R Latch
Selector
Serial Data Transmission
Serial-to-Parallel Converter
Set-Reset Flip-Flop
Set State Toggle
Triggered

Multiple-Choice Questions

REVIEW QUESTIONS

1. Binary is a base ____________ number system, decimal a base ____________ number system, and hexadecimal a base ____________ number system.

   a. 1, 2, 3. **c.** 10, 2, 3.
   b. 2, 10, 16. **d.** 8, 4, 2.

2. What is the binary equivalent of decimal 11?

   a. 1010 **c.** 1011
   b. 1100 **d.** 0111

3. What would be displayed on a hexadecimal counter if its count was advanced by one from $39FF_{16}$?

   a. 4000 **c.** 3900
   b. 3A00 **d.** 4A00

4. What is the decimal equivalent of $1001\ 0001\ 1000\ 0111_{BCD}$?

   a. 9187 **b.** 9A56 **c.** 8659 **d.** 254,345

5. An ____________ is a code generator.

   a. Decoder **c.** One-shot
   b. Encoder **d.** MUX

6. A common-cathode display needs to be driven by active ____________ inputs, while a common-anode display needs to be driven by active ____________ inputs.

   a. HIGH, LOW **c.** LOW, HIGH
   b. LOW, LOW **d.** HIGH, HIGH

7. Which segments of a 7-segment display should be active to display the decimal digit nine?

   a. a, b, c, d, f. **c.** a, c, d, f, g.
   b. a, c, d, g, h. **d.** a, b, c, f, g.

8. The negative logic BCD code for 6 is

   a. 1001 **b.** 0110 **c.** 0101 **d.** 1010

9. Multiplexers are often used for ____________ data conversion.

   a. Parallel-to-serial **b.** Serial-to-parallel

10. A digital ____________ is a circuit used to determine whether two parallel input words are equal.

    a. Multiplexer **b.** Decode **c.** Comparator **d.** Counter

11. A ____________ is a digital circuit that can store 1 bit of binary data.

    a. Decoder **c.** Demultiplexer
    b. Flip-flop **d.** Comparator

12. What would be the Q output from a negative-edge triggered J-K flip-flop if $J = 1, K = 0, C = \uparrow, Q = 0$?

    a. No change **c.** LOW
    b. HIGH **d.** Both (a) and (c) are true

13. If a *J-K* flip-flop is connected to toggle, it will
    a. Divide the input frequency by 2.
    b. Function as a 0–3 binary counter.
    c. Produce one output pulse for every two input pulses.
    d. All of the above.
14. Which of the multivibrator types could be used as a clock oscillator?
    a. Astable c. Monostable
    b. Bistable d. Tristable
15. Which of the multivibrator types could be used as a pulse generator?
    a. Astable c. Monostable
    b. Bistable d. Tristable

Essay Questions

16. Why are two-state digital circuits chosen over ten-state digital circuits? (Intro)
17. Give the full names of the following abbreviations (Chapter 18)
    a. Bit c. ASCII e. DMUX
    b. BCD d. MUX f. *S-R F-F*
18. What is the Arabic number system? (18-1-1)
19. Why does reset and carry occur after 9 in decimal and after 1 in binary? (18-1-2)
20. Briefly describe how to (18-1-2)
    a. Convert a binary number to a decimal number
    b. Convert a decimal number to a binary number
21. Why is the hexadecimal number system used in conjunction with digital circuits? (18-1-3)
22. Briefly describe how to (18-1-3)
    a. Convert a hexadecimal number to a decimal number
    b. Convert a decimal number to a hexadecimal number
    c. Convert between binary and hexadecimal
23. What is binary coded decimal, and why is it needed? (18-1-4)
24. Briefly describe how to convert a BCD code to its decimal equivalent. (18-1-4)
25. What is the ASCII code, and in what applications is it used? (18-1-4)
26. What is the difference between a digital decoder circuit and a digital encoder circuit? (18-2-1 and 18-2-2)
27. Using a sketch, describe the operation of a basic decoder and encoder circuit. (18-2-1 and 18-2-2)
28. Describe the difference between an active-LOW output and an active-HIGH output. (18-2-1)
29. What is the difference between positive logic and negative logic? (18-2-2)
30. Define the function of a digital comparator circuit. (18-2-3)

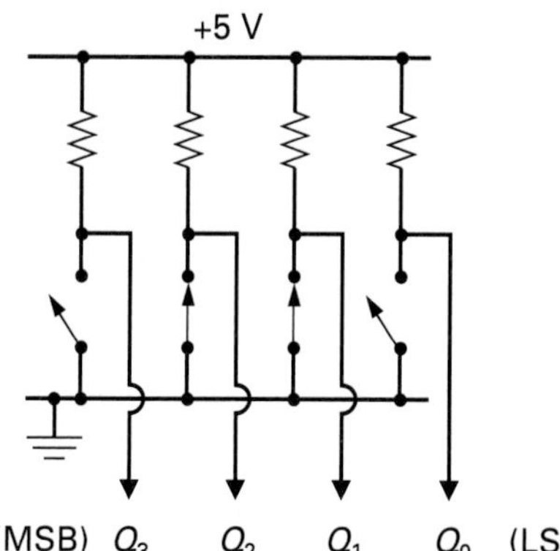

FIGURE 18-36 **A 4-Bit Switch Register.**

31. Using a sketch, describe the operation of a basic multiplexer and demultiplexer circuit. (18-2-4)
32. What are the advantages and disadvantages of serial data transmission and parallel data transmission? (18-2-4)
33. Is the set-reset flip-flop a bistable, astable, or monostable multivibrator? (18-2-5)
34. What advantage does the *J-K* flip-flop have over the *S-R* flip-flop? (18-2-5)
35. Why are flip-flop circuits also called latches? (18-2-5)
36. Why are most flip-flop circuits triggered by a clock signal? (18-2-5)
37. Using a sketch, show how a *J-K* flip-flop can be used as a frequency divider and as a binary counter. (18-2-5)
38. What is a digital register circuit, and in what application would it typically be used? (18-2-5)
39. What is a digital one-shot circuit, and in what application would it typically be used? (18-2-5)
40. Using a sketch, briefly describe the operation of a frequency counter. (18-3)

Practice Problems

41. What is the binary and equivalent decimal output of the switch register shown in Figure 18-36?
42. Convert the following into their decimal equivalent.

    a. 110111_2 **b.** $2F_{16}$ **c.** 10110_{10}
43. What would be the decimal equivalent of the LED display in Figure 18-37 if it were displaying

    a. Pure binary **b.** BCD

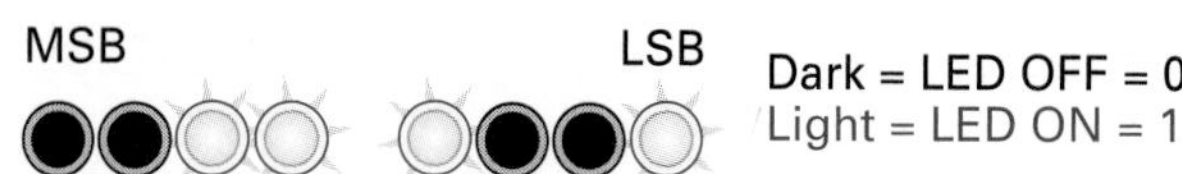

FIGURE 18-37 **An 8-Bit LED Display.**

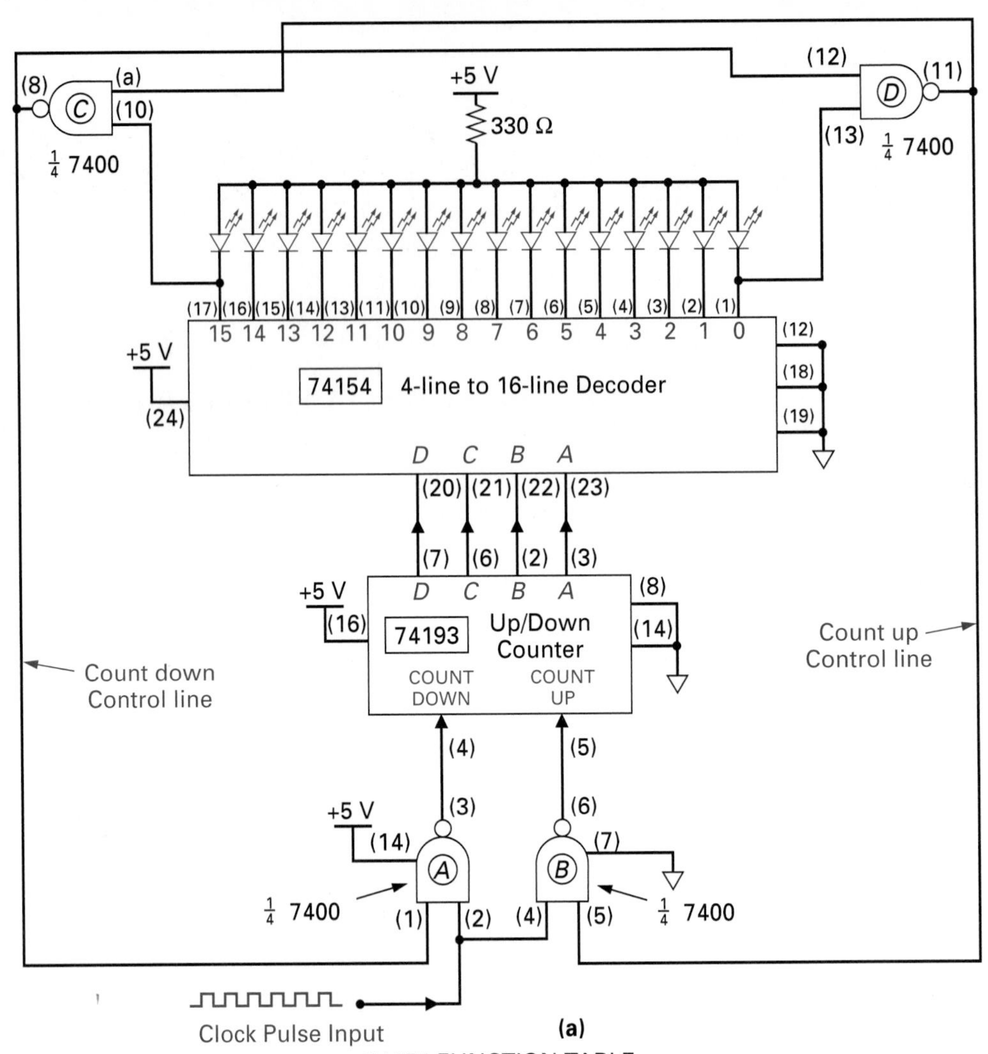

(a)

74154 FUNCTION TABLE

INPUTS				OUTPUTS															
D	*C*	*B*	*A*	0	1	2	3	4	5	6	7	8	9	10	11	12	13	14	15
L	L	L	L	L	H	H	H	H	H	H	H	H	H	H	H	H	H	H	H
L	L	L	H	H	L	H	H	H	H	H	H	H	H	H	H	H	H	H	H
L	L	H	L	H	H	L	H	H	H	H	H	H	H	H	H	H	H	H	H
L	L	H	H	H	H	H	L	H	H	H	H	H	H	H	H	H	H	H	H
L	H	L	L	H	H	H	H	L	H	H	H	H	H	H	H	H	H	H	H
L	H	L	H	H	H	H	H	H	L	H	H	H	H	H	H	H	H	H	H
L	H	H	L	H	H	H	H	H	H	L	H	H	H	H	H	H	H	H	H
L	H	H	H	H	H	H	H	H	H	H	L	H	H	H	H	H	H	H	H
H	L	L	L	H	H	H	H	H	H	H	H	L	H	H	H	H	H	H	H
H	L	L	H	H	H	H	H	H	H	H	H	H	L	H	H	H	H	H	H
H	L	H	L	H	H	H	H	H	H	H	H	H	H	L	H	H	H	H	H
H	L	H	H	H	H	H	H	H	H	H	H	H	H	H	L	H	H	H	H
H	H	L	L	H	H	H	H	H	H	H	H	H	H	H	H	L	H	H	H
H	H	L	H	H	H	H	H	H	H	H	H	H	H	H	H	H	L	H	H
H	H	H	L	H	H	H	H	H	H	H	H	H	H	H	H	H	H	L	H
H	H	H	H	H	H	H	H	H	H	H	H	H	H	H	H	H	H	H	L

(b)

FIGURE 18-38 Decoder Application—Back and Forth Flasher Circuit.

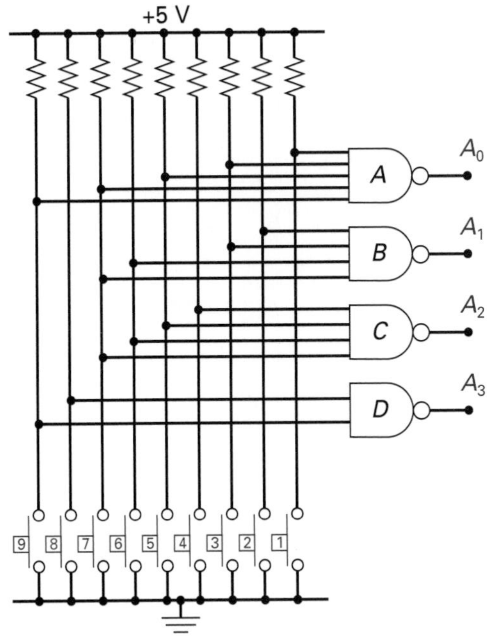

Inputs									Outputs			
1	2	3	4	5	6	7	8	9	A_3	A_2	A_1	A_0
0	1	1	1	1	1	1	1	1	0	0	0	1
1	0	1	1	1	1	1	1	1	0	0	1	0
1	1	0	1	1	1	1	1	1	0	0	1	1
1	1	1	0	1	1	1	1	1	0	1	0	0
1	1	1	1	0	1	1	1	1	0	1	0	1
1	1	1	1	1	0	1	1	1	0	1	1	0
1	1	1	1	1	1	0	1	1	0	1	1	1
1	1	1	1	1	1	1	0	1	1	0	0	0
1	1	1	1	1	1	1	1	0	1	0	0	1
1	1	1	1	1	1	1	1	1	0	0	0	0

FIGURE 18-39 A Basic Encoder Circuit.

44. Convert the following binary numbers to hexadecimal.

a. 111101101001 **c.** 111000
b. 1011 **d.** 1111111

45. Convert the following decimal numbers to BCD.

a. 2,365 **b.** 24

46. Give the ASCII codes for the following:

a. ? **b.** $ **c.** 6

To practice your circuit recognition and understanding of operation, refer to the circuit in Figure 18-38 and answer the following questions.

47. Why would the 74154 also be called a binary-to-hex decoder?

48. Does the 74154 generate active-HIGH or active-LOW outputs?

49. The NAND gates will give a HIGH output for any ____________ input.

50. The rate at which the LEDs will flash backward, then forward, and then repeat, is determined by the

a. 74154 **c.** 7400
b. 74193 **d.** Clock pulse input frequency

To practice your circuit recognition and understanding of operation, refer to the circuit in Figure 18-39 and answer the following questions.

51. What is the function of this circuit?

52. This circuit would best be described as a ____________ to ____________ encoder.

To practice your circuit recognition and understanding of operation, refer to the circuit in Figure 18-40 and answer the following questions.

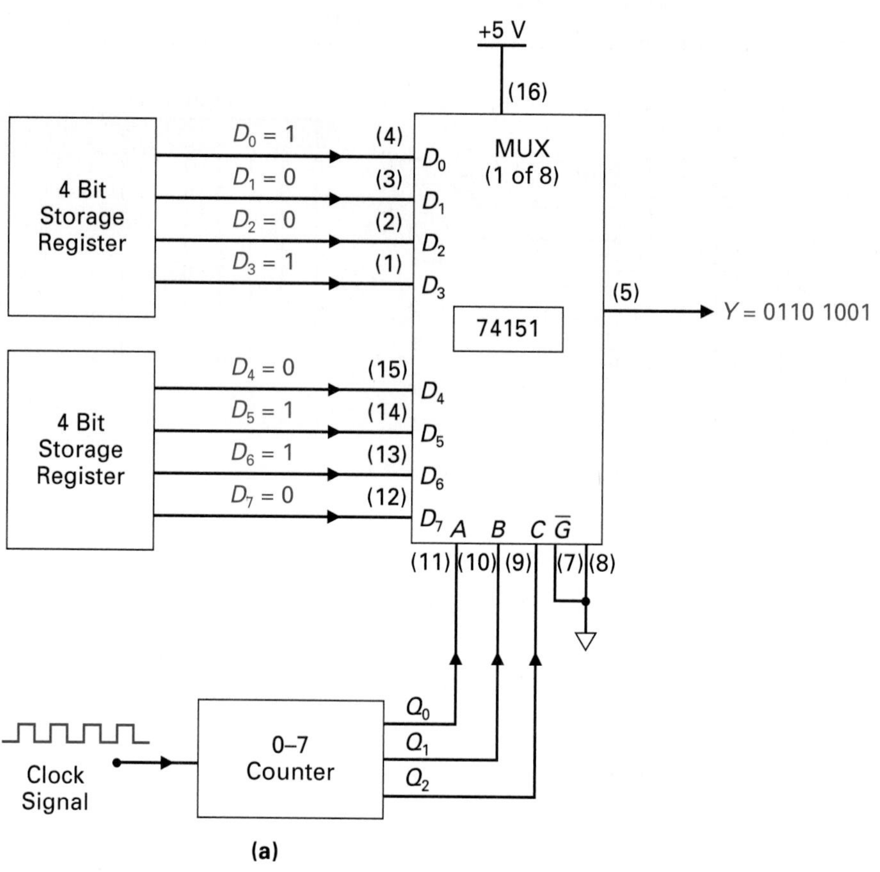

(a)

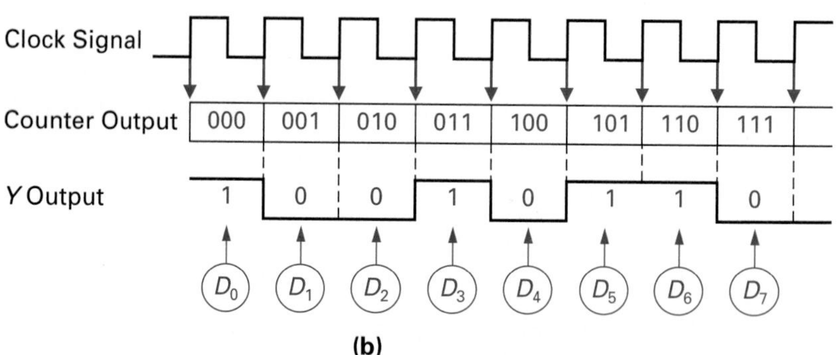

(b)

FIGURE 18-40 Multiplexer Application—Parallel-to-Serial Data Converter.

53. What does "1 of 8 MUX" mean?

54. Briefly describe the operation of the circuit.

55. What is a parallel-to-serial data converter?

To practice your circuit recognition and understanding of operation, refer to the circuit in Figure 18-41 and answer the following questions.

56. The *J-K* flip-flops are____________ edge triggered. (negative/positive)

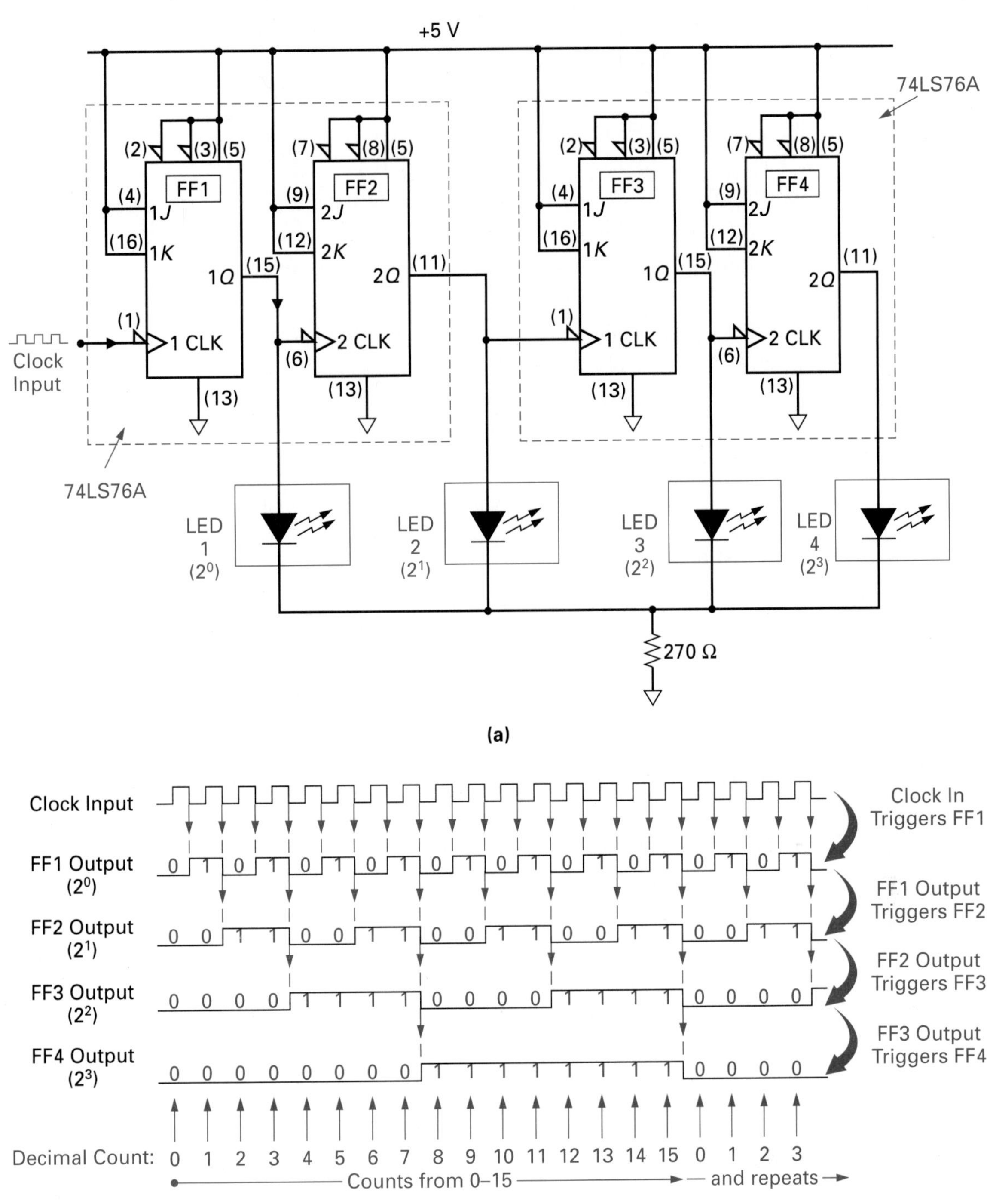

FIGURE 18-41 **Four Bit Binary Up Counter.**

57. The *J-K* flip-flops are all in the ____________ input condition.

   a. Set **c.** No-Change
   b. Reset **d.** Toggle

58. Flip-flop 1 (FF1) will output one cycle for every ____________ input clock cycles and divide the clock input by ____________.

59. Flip-flop 2 (FF2) will output one cycle for every ____________ input clock cycles and divide the clock input by ____________.

60. Flip-flop 3 (FF3) will output one cycle for every ____________ input clock cycles and divide the clock input by ____________.

61. Flip-flop 4 (FF4) will output one cycle for every ____________ input clock cycles and divide the clock input by ____________.

62. As a counter, this circuit will count from binary ____________ to binary ____________, which in decimal is ____________ to ____________.

63. What will be the LED display after 4 clock pulses?

64. How many 74LS76A ICs will be needed to construct this circuit?

Troubleshooting Questions

65. Briefly describe what effect you think the following circuit malfunctions would have on the frequency counter circuit in Figure 18-27.

   a. The crystal oscillator output is permanently HIGH.
   b. The Q output of the 7476 is permanently LOW.
   c. The +5 V supply voltage is not connected to the 74LS160A that generates the 1 Hz output in the "count window generator circuit."
   d. The $\overline{Q}$ output from the 74121 is disconnected from the counter clear inputs.
   e. The Q output from the 74121 is disconnected from the 74173 registers.

CHAPTER

19

OBJECTIVES

After completing this chapter, you will be able to:

1. Define the term thyristor.
2. Describe the operation, characteristics, applications, and testing of the following thyristors:
   - **a.** Silicon controlled rectifier or SCR.
   - **b.** Triode ac semiconductor switch or TRIAC
   - **c.** Diode ac semiconductor switch or DIAC
   - **d.** Unijunction transistor or UJT
   - **e.** Programmable unijunction transistor or PUT
3. Identify the different types of thyristor from their schematic symbols.
4. Define the term transducer.
5. Describe the operation, characteristics, applications, and testing of the following light-sensitive optoelectronic devices:
   - **a.** Photoconductive cell or light-dependent resistor
   - **b.** Photovoltaic cell or solar cell
   - **c.** Photodiode
   - **d.** Phototransistors and photothyristors
6. Describe the operation, characteristics, applications, and testing of the following light-reactive optoelectronic devices:
   - **a.** Light-emitting diode or LED
   - **b.** Injection laser diode or ILD
7. Explain the operation and characteristics of light-emitting diode displays and liquid crystal diode displays.
8. Describe the operation, characteristics, and applications of the following optically coupled isolators:
   - **a.** DIP module
   - **b.** Interrupter module
   - **c.** Reflector module
9. Describe the operation, characteristics, and applications of the following semiconductor sensors:
   - **a.** NTC thermistor
   - **b.** PTC thermistor
   - **c.** Piezoresistive pressure transducer
   - **d.** Hall effect magnetic transducer

Thyristors and Transducers

OUTLINE

VIGNETTE

Leibniz's Language of Logic

Gottfried Wilhelm Leibniz was born in Leipzig, Germany, in 1646. His father was a professor of moral philosophy and spent much of his time discussing thoughts and ideas with his son. Tragically, his father died when he was only six, and from that time on Leibniz spent hour upon hour in his late father's library reading through all of his books.

By the age of twelve he had taught himself history, Latin, and Greek. At fifteen he entered the University of Leipzig, and it was here that he came across the works of scholars such as Johannes Kepler and Galileo. The new frontiers of science fascinated him, so he added mathematics to his curriculum.

In 1666, while finishing his university studies, the twenty-year-old Leibniz wrote what he called modestly a schoolboy essay, "De Arte Combinatoria," which means "On the Art of Combination." In this work, he described how all thinking of any sort on any subject could be reduced to exact mathematical statements. Logic or, as he called it, the laws of thought, could be converted from the verbal realm—which is full of ambiguities—into precise mathematical statements. In order to achieve this, however, Leibniz stated that a "universal language" would be needed. Most of his professors found the paper either baffling or outrageous and this caused Leibniz not to pursue the idea any further.

After graduating from university, Leibniz was offered a professorship at the University of Nuremberg, which he turned down for a position as an international diplomat. This career proved not to be as glamorous as he imagined because most of his time was spent in uncomfortable horse-drawn coaches traveling between the European capitals.

In 1672, his duties took him to Paris where he met Dutch mathematician and astronomer Christian Huygens. After seeing the hours that Huygens spent on endless computations, Leibniz set out to develop a mechanical calculator. A year later Leibniz unveiled the first machine that could add, subtract, multiply, and divide decimal numbers.

In 1676 Leibniz began to concentrate more on mathematics. It was at this time that he invented calculus, which was also independently discovered by Isaac Newton in England. Leibniz's focus, however, was on the binary number system, which occupied him for years. He worked tirelessly to document the long combinations of ones and zeros that make up the modern binary number system and to perfect binary arithmetic. What is ironic is that for all his genius, Leibniz failed to make the connection between his 1666 essay and binary, which was the universal language of logic that he was seeking. It would be a century and a quarter after Leibniz's death (in 1716) when another self-taught mathematician named George Boole would discover it.

INTRODUCTION

In this chapter, we will examine two groups of electronic components called "semiconductor thyristors" and "semiconductor transducers." Thyristors are generally used as dc and ac power control devices, while transducers are generally used as sensing, displaying, and actuating devices.

The semiconductor thyristor acts as an electronically controlled switch, switching power ON and OFF to adjust the average amount of power delivered to a load or connecting or disconnecting power from a load. Some thyristors are "unidirectional," which means that they will only conduct current in one direction (dc), while others are "bi-directional," which means that they can conduct current in either direction (ac). Thyristors are generally used as electronically controlled switches instead of the transistors because they have better power handling capabilities and are more efficient.

A semiconductor transducer is an electronic device that converts one form of energy to another. Electronic transducers can be classified as either input transducers (such as the photodiode) that generate input control signals, or output transducers (such as the LED) that convert output electrical signals to some other energy form. Input transducers are sensors that convert thermal, optical, mechanical, and magnetic energy variations into equivalent voltage and current variations. Output transducers, on the other hand, perform the exact opposite, converting voltage and current variations into optical or mechanical energy variations.

In the first section of this chapter we will examine the different types of semiconductor thyristors, and in the second section we will examine many of the different types of semiconductor transducers.

19-1 THYRISTORS

The most frequently used thyristors are the "silicon controlled rectifier (SCR)," "triode ac semiconductor switch (TRIAC)," "diode ac semiconductor switch (DIAC)," "unijunction transistor (UJT)," and "programmable unijunction transistor (PUT)." In this section, we will examine the operation, characteristics, applications, and testing of all these electronic devices.

Silicon Controlled Rectifier or SCR A three-junction, three-terminal, unidirectional P-N-P-N thyristor that is normally an open circuit. When triggered with the proper gate signal it switches to a conducting state and allows current to flow in one direction.

19-1-1 The Silicon Controlled Rectifier (SCR)

The **silicon controlled rectifier** or **SCR** is the most frequently used of the thyristor family. Figure 19-1(a) shows how this device is a four-layered, alternately-doped component, with three terminals labeled anode (A), cathode (K) and gate (G).

Complementary Latch Circuit A circuit containing an NPN and PNP transistor that once triggered ON will remain latched ON.

SCR Operation

To simplify the operation of the SCR, Figure 19-1(b) shows how the four layers of the SCR can be thought of, when split, as being a PNP and NPN transistor, and Figure 19-1(c) shows how this interconnection forms a **complementary latch**

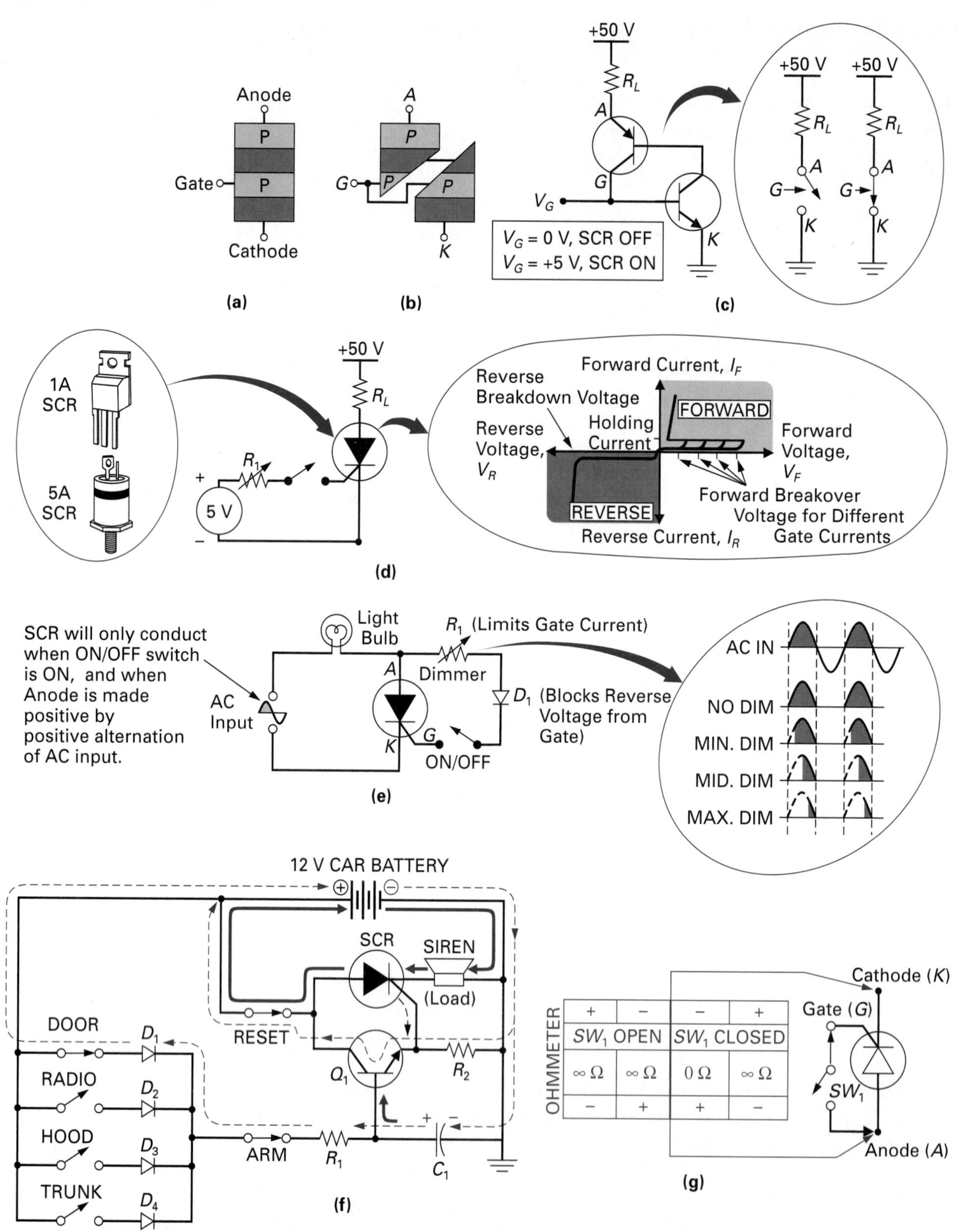

+	−	−	+
SW_1 OPEN		SW_1 CLOSED	
∞ Ω	∞ Ω	0 Ω	∞ Ω
−	+	+	−

FIGURE 19-1 Silicon-Controlled Rectifier. (a) Construction. (b) Complementary Latch. (c) Closed, Open Latch Action. (d) Correctly Biased. (e) SCR Application—Basic Light Dimmer. (f) Application—Alarm System. (g) Testing SCRs.

circuit. To operate as an ON/OFF switch, the SCR must be biased like a diode, with the anode of the SCR made positive relative to the cathode. The gate of the SCR is an active-HIGH input and must be triggered by a voltage that is positive relative to the cathode. In the example in Figure 19-1(c), you can see that the anode of the SCR is connected to +50 V via the load (R_L), the cathode is connected to 0 V, and the gate has a control input that is either 0 V or +5 V. When the gate input is at 0 V, the NPN and PNP transistors are OFF, and the SCR is equivalent to an open switch, as shown in the inset in Figure 19-1(c). On the other hand, when the gate input is +5 V, the NPN transistor's base is made positive with respect to the emitter and so the NPN transistor will turn ON. Turning the NPN transistor ON will connect the 0 V at the emitter of the NPN transistor through to the PNP's base, causing it to also turn ON. Turning the PNP transistor ON will connect the large positive voltage on the PNP's emitter through to the collector and the base of the NPN transistor, keeping it ON even after the +5 V input trigger is removed. As a result, both transistors will be latched ON and held ON by one another, allowing a continuous flow of current from cathode to anode. A momentary positive input trigger, therefore, will cause the SCR to be latched ON and be equivalent to a closed switch, as seen in the inset in Figure 19-1(c).

SCR Characteristics

Figure 19-1(d) shows a correctly biased SCR: the left inset shows a typical low-power and high-power package, and the right inset shows a typical SCR characteristic curve. This graph plots the forward and reverse voltage applied across the SCR's anode-to-cathode against the current through the SCR between cathode and anode. Looking at the forward conduction quadrant, you can see that the voltage needed to turn ON an SCR is called the **forward breakover voltage.** An SCR's forward breakover voltage, or turn ON voltage, is inversely proportional to the value of gate current. For example, a larger gate current will cause the SCR to turn ON when only a small forward voltage is applied between the SCR's anode-to-cathode. If no gate current is applied, the forward voltage applied across the SCR's anode to cathode will have to be very large to make the SCR turn ON and conduct current between cathode and anode. Once the SCR is turned ON, a holding current latches the SCR's complementary latch ON, independent of the gate current. If the forward current between the SCR's cathode and anode falls below this minimum holding current value, the SCR will turn OFF because the NPN/PNP latch will not have enough current to keep each other ON. A gate trigger is therefore used to turn ON the SCR; however, once the SCR is ON, it can only be turned OFF by decreasing the anode-to-cathode voltage so that the cathode-to-anode current passing through the SCR drops below the holding current value.

Forward Breakover Voltage
Voltage needed to turn ON an SCR.

In the reverse direction with the gate switch open, the SCR acts in almost the same way as a diode because a large reverse voltage is needed between anode and cathode to cause the SCR to break down.

SCR Applications

The SCR is used in a variety of electrical power applications, such as light dimmer circuits, motor-speed control circuits, battery charger circuits, temperature control systems, and power regulator circuits. From the characteristic curve

Unidirectional Device A device that will conduct current in only one direction.

in Figure 19-1(d), we discovered that the SCR will only conduct current in the forward direction, which is why it is classed as a **unidirectional device.** This means that if an ac signal is applied across the SCR, it will only respond to a gate trigger during the time that the ac alternation makes the anode positive with respect to the cathode. For example, Figure 19-1(e) shows how the SCR could be connected to form a simple light dimmer circuit. When the ON/OFF switch is open, the SCR will be OFF because its gate current is zero, and the light will be OFF. When the ON/OFF switch is closed, diode D_1 will connect a positive voltage to the gate of the SCR whenever the ac input is positive. The value of gate current applied to the SCR is controlled by the variable dimmer resistor (R_1), and therefore this resistor value will determine the SCR's forward breakover (turn ON) voltage. Referring to the waveforms in the inset in Figure 19-1(e), you can see that when the resistance of R_1 is zero (no dim), gate current will be maximum, the SCR will turn ON for the full positive half cycle of the ac input, and the average power delivered to the light bulb will be HIGH. As the resistance of R_1 is increased, the SCR gate current is decreased, causing the SCR to turn ON for less of the positive alternation, and therefore the average power delivered to the light bulb to decrease. You may ask: why don't we simply connect a variable resistor in series with the light bulb to vary the light bulb's current and therefore brightness? The problem with this arrangement is identical to the disadvantages of the previously discussed "series dissipative regulator." The wattage rating, size, and cost of the variable resistor would be very large, and the circuit would be very inefficient because power is taken away from the light bulb by dissipating it away as heat from the resistor. Varying the SCR's ON/OFF time is a much more efficient system because we will only switch through to the light bulb the power that is desired, and therefore vary the average power applied to the load in almost exactly the same way as the previously discussed "switching regulator circuit."

Now that we have seen the SCR in an ac circuit application, let us now see how it could be used in a dc circuit application. Figure 19-1(f) shows a basic car-alarm system. When both the ARM switch and RESET switch are closed, the alarm system is active, and any of the four sensor switches (door, radio, hood, or trunk) will activate the alarm. For example, Figure 19-1(f) shows what will happen if the car door is opened when the alarm system is armed. Opening the door will cause capacitor C_1 to charge via diode D_1 and resistor R_1. After a short delay, the charge on C_1 is large enough to turn ON Q_1, which will switch the positive potential on its collector from the battery through to its emitter and to the gate of the SCR. This positive gate trigger will turn ON the SCR and activate the siren because the SCR is equivalent to a closed switch when ON and when triggered will connect the full 12-V battery across the siren. Once the SCR is turned ON, it will remain latched ON independent of the ARM switch and the sensor switches. Only by opening the RESET switch, which is hidden within the vehicle, can the siren be shut OFF. The values of R_1 and C_1 are chosen so that a small delay occurs before Q_1 and the SCR are triggered. This delay is included so that the vehicle owner has enough time to enter the car and disarm the alarm system by opening the ARM switch.

SCR Testing

Using the oscilloscope, you can monitor the gate trigger input and ON/OFF switching of an SCR while it is operating in circuit. If you suspect that the SCR is the cause of a circuit malfunction, you should remove the SCR and use the ohmmeter test circuit shown in Figure 19-1(g) to check for terminal-to-terminal opens and shorts. With this test, we will be using the ohmmeter's internal battery to apply different polarities to the different terminals of the SCR and SW_1 to either apply or disconnect gate current. To explain the ohmmeter response table in this illustration, you can see that if SW_1 is open (no gate current), the resistance between anode and cathode should be almost infinite ohms (actually about 250 kΩ), no matter what polarity is applied between anode and cathode. On the other hand, if SW_1 is closed and the anode is made positive with respect to the cathode, the gate will also be made positive due to SW_1, and so the SCR should turn ON and have a very low resistance between anode and cathode. If SW_1 is closed, and a reverse polarity is applied across the SCR (anode is made negative, cathode is made positive), the ohmmeter should once again read infinite ohms.

19-1-2 The Triode AC Semiconductor Switch (TRIAC)

The disadvantage with the SCR is that it is unidirectional, which means that it can only be activated when the applied anode-to-cathode voltage makes the anode positive with respect to the cathode, and it will conduct current in one direction. As a result, the SCR can only control a dc supply voltage, or one-half cycle of the ac supply voltage. To gain control of the complete ac input cycle, we would need to connect two SCRs in parallel, facing in opposite directions, as shown in Figure 19-2(a). This is exactly what was done to construct the **triode ac semiconductor switch** or **TRIAC,** which has three terminals called main terminal 1 (MT_1), main terminal 2 (MT_2), and gate (G). The *P-N* doping for this **bidirectional device** is shown in Figure 19-2(b), and its schematic symbol is shown in Figure 19-2(c).

Triode AC Semiconductor Switch or TRIAC
A bidirectional gate-controlled thyristor that provides full-wave control of ac power.

Bidirectional Device
A device that will conduct current in either direction.

TRIAC Operation and Characteristics

Since the TRIAC is basically two SCRs connected in parallel, back-to-back, it comes as no surprise that its operation and characteristics are very similar to the SCR. The characteristic curve for a typical TRIAC is shown in Figure 19-2(d). Looking at the identical forward and reverse curves, you can see that the key difference with the TRIAC is that it can be triggered or activated by either a positive or negative input gate trigger. This means that the TRIAC can be used to control both the positive and negative alternation of an ac supply voltage. To explain this in more detail, Figure 19-2(e) shows how a TRIAC could be connected across an ac input. When the ON/OFF gate switch is open, the TRIAC will not receive a gate trigger and so it will remain OFF. When the ON/OFF gate switch is closed, the TRIAC will be triggered by the positive and negative cycles of the ac input, via R_1. By adjusting the resistance of R_1, we can control the TRIAC's value of gate current. By controlling gate current, we can

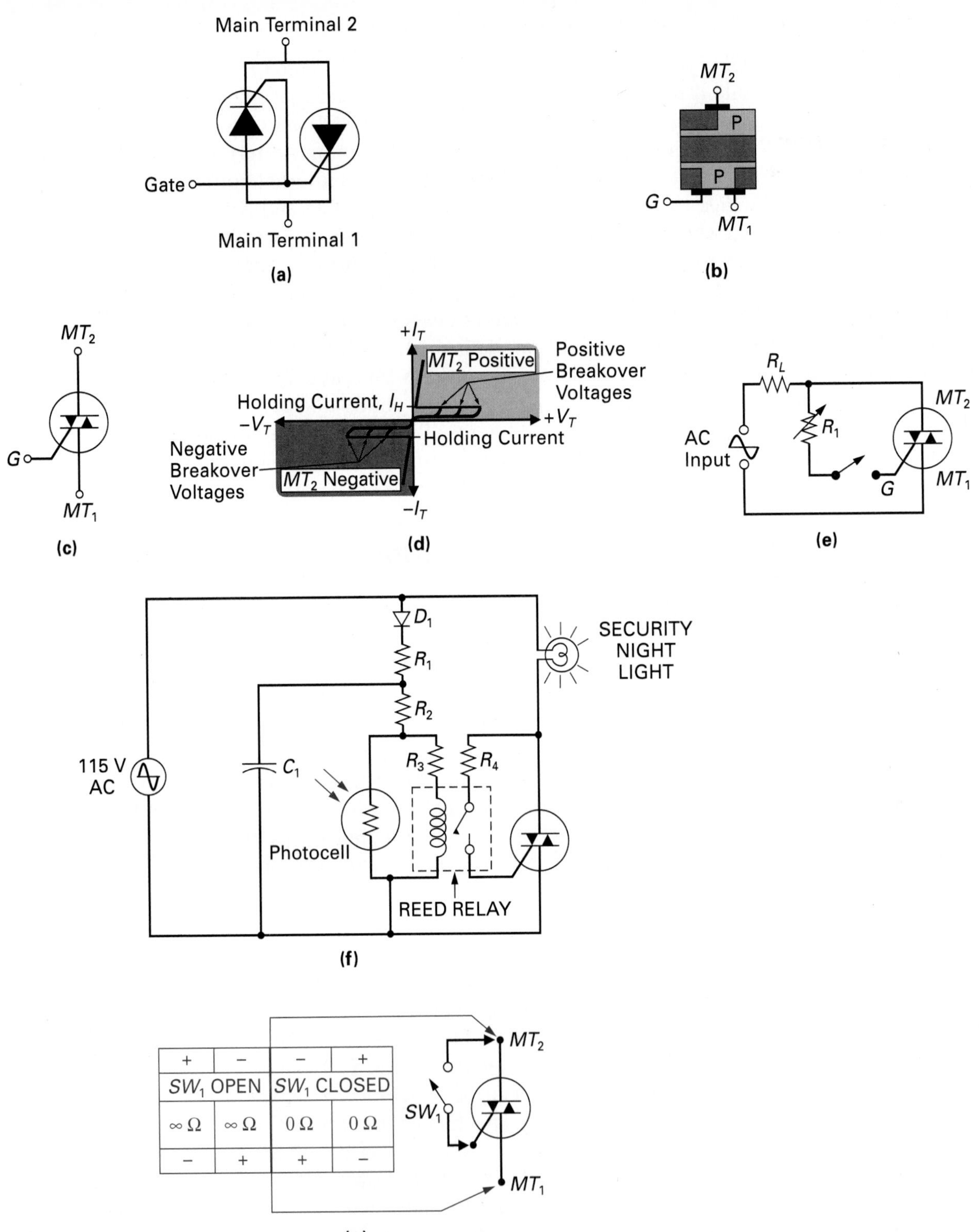

+	−	−	+
SW_1 OPEN		SW_1 CLOSED	
∞ Ω	∞ Ω	0 Ω	0 Ω
−	+	+	−

FIGURE 19-2 TRIAC. (a) TRIAC Equivalent Circuit. (b) Construction. (c) Schematic Symbol. (d) *V-I* Characteristics. (e) Application—AC TRIAC Switch. (f) Application—Automatic Night Light. (g) Testing TRIACs.

control the TRIAC's turn-ON voltage (positive and negative breakover voltage), so that the ac input voltage can be chopped up to adjust the average value of voltage applied to the load.

TRIAC Applications

Figure 19-2(f) shows how a TRIAC could be connected as an automatic night light for home or business security and safety. Later in this chapter, we will discuss the photocell in more detail. However, for now just think of it as a variable resistor that changes its resistance based on the amount of light present. During the day when the photocell is exposed to light, its resistance is less than a few ohms, and since it is in parallel with the energizing coil of a reed relay, most of the current will pass through the photocell keeping the reed relay de-energized. As the sun goes down and the photocell is deprived of light, its resistance increases to a few mega ohms, and this high resistance will cause the current through the reed relay coil to increase, and therefore the reed relay to energize. With the reed-relay switch closed, the TRIAC will be triggered by each half cycle of the ac supply voltage, causing it to turn ON and connect power to the night light.

TRIAC Testing

Using the oscilloscope, you can monitor the gate trigger input, and ON/OFF switching of a TRIAC, while it is operating in-circuit. If you suspect that the TRIAC is the cause of a circuit malfunction, you should remove the TRIAC and use the ohmmeter test circuit shown in Figure 19-2(g), to check for terminal-to-terminal opens and shorts. The ohmmeter response table in this illustration shows that if SW_1 is open (no gate current), the resistance between MT_2 and MT_1 should be almost infinite ohms (actually about 250 kΩ), no matter what polarity is applied. When switch 1 is closed, the gate of the TRIAC will receive a trigger. Since the TRIAC operates on either a positive or negative trigger, it should turn ON no matter what polarity is applied, and therefore have a very low resistance between MT_2 and MT_1.

19-1-3 The Diode AC Semiconductor Switch (DIAC)

One disadvantage with the TRIAC is that its positive breakover voltage is usually slightly different from its negative breakover voltage. This nonsymmetrical trigger characteristic can be compensated for by using a **diode ac semiconductor switch** or **DIAC** to trigger a TRIAC. The DIAC's construction is shown in Figure 19-3(a), its schematic symbol in Figure 19-3(b) and its equivalent circuit in Figure 19-3(c). Equivalent to two back-to-back, series-connected junction diodes, the DIAC has two terminals.

Diode AC Semiconductor Switch or DIAC
A bidirectional diode that has a symmetrical switching mode.

Symmetrical Bidirectional Switch
A device that has the same value of breakover voltage in both the forward and reverse direction.

DIAC Operation and Characteristics

Since the PNP regions of a DIAC are all equally doped, the DIAC will have the same forward and reverse characteristics, as shown in Figure 19-3(d). As a result, the DIAC is classed as a **symmetrical bidirectional switch,** which

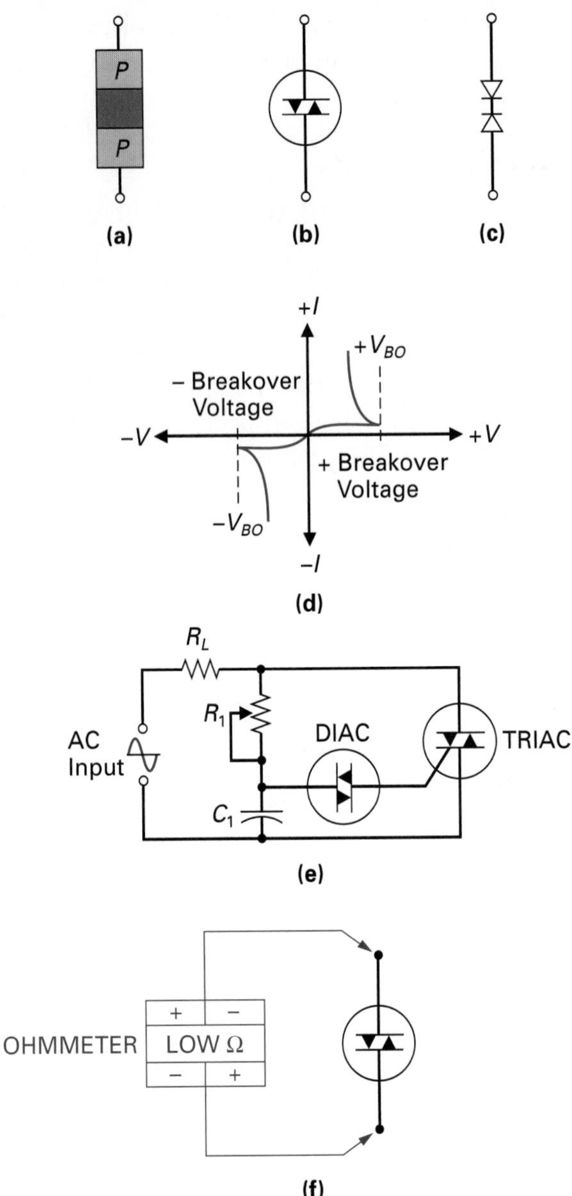

FIGURE 19-3 DIAC. (a) Construction. (b) Schematic Symbol. (c) Equivalent Circuit. (d) *V-I* Characteristics. (e) Application—TRIAC Control. (f) Testing DIACs.

means that it will have the same value of breakover voltage in both the forward and reverse direction.

DIAC Applications

Figure 19-3(e) shows how a DIAC could be connected as a pulse-triggering device in a TRIAC ac power control circuit. The DIAC will turn ON when the capacitor has charged to either the positive or negative breakover voltage ($+V_{BO}$

or $-V_{BO}$). Once this voltage is reached the DIAC turns ON, and the capacitor discharges through the DIAC, triggering the TRIAC into conduction, which then connects the ac supply voltage across the load. The variable resistor R_1 is used to adjust the *RC* charge time constant, so that the DIAC turn-ON time, and therefore TRIAC turn-ON time, can be changed.

DIAC Testing

Using the oscilloscope, you can monitor the ON/OFF switching of a DIAC while it is operating in-circuit. If you suspect that the DIAC is the cause of a circuit malfunction, you should remove the DIAC and check it with the ohmmeter, as seen in Figure 19-3(f). Since the DIAC is basically two diodes connected back-to-back in series, the ohmmeter should show a low resistance reading between its terminals no matter what polarity is applied.

19-1-4 *The Unijunction Transistor (UJT)*

The **unijunction transistor** or **UJT** operates in a very different way to the SCR, TRIAC, and DIAC. Although it is given the name transistor, it is never used as an amplifying device like the BJT and FET: it is only ever used as a voltage-controlled switch. Figure 19-4(a) shows the construction of the UJT and illustrates how the uni (one) junction transistor derives its name from the fact that it has only one P-N junction. Looking at this illustration you can see that the UJT is a three-terminal device with an emitter lead (E) attached to a small p-type pellet, that is fused into a bar of n-type silicon with contacts at either end [labeled base 1 (B_1) and base 2 (B_2)].

Unijunction Transistor or UJT
A P-N device that has an emitter connected to the P-N junction on one side of the bar and two bases at either end of the bar. Used primarily as a switching device.

The schematic symbol for the UJT is shown in Figure 19-4(b). To remember whether the junction is a P-N or N-P type, think of the arrow as a junction diode: the emitter (diode-anode) is positive and the bar (diode-cathode) is negative.

UJT Operation and Characteristics

Figure 19-4(c) illustrates the UJT's equivalent circuit. The emitter-to-bar P-N junction is equivalent to a junction diode, and the bar is equivalent to a two-resistor voltage divider (R_{B1} and R_{B2}). Referring once again to the UJT construction in Figure 19-4(a), you can see that the emitter pellet is closer to terminal B_2 than B_1, and this is why the resistance of R_{B2} is smaller than the resistance of R_{B1}. Figure 19-4(d) shows how a UJT should be correctly biased. The voltage source V_{BB} is connected to make B_2 positive relative to B_1 and the input voltage V_S is connected to make the emitter positive with respect to B_1. The resistor R_E is used to limit emitter current. When V_S is zero, the UJT's emitter diode is OFF, and the resistance between B_2 and B_1 allows only a very small amount of current between ground and $+V_{BB}$. If the emitter supply voltage is increased so that V_E exceeds the voltage at B_1, the emitter diode will turn ON and inject holes into the p region. Flooding the lower half of the UJT with holes increases the amount of current flow through the UJT, dramatically reducing the resistance of R_{B1}. A lower R_{B1} resistance results in a lower V_{B1} voltage drop, and this further increases the E to B_1 P-N forward bias permitting more holes to be injected into the n-type bar between E and B_1. Figure 19-4(e) graphically illustrates this

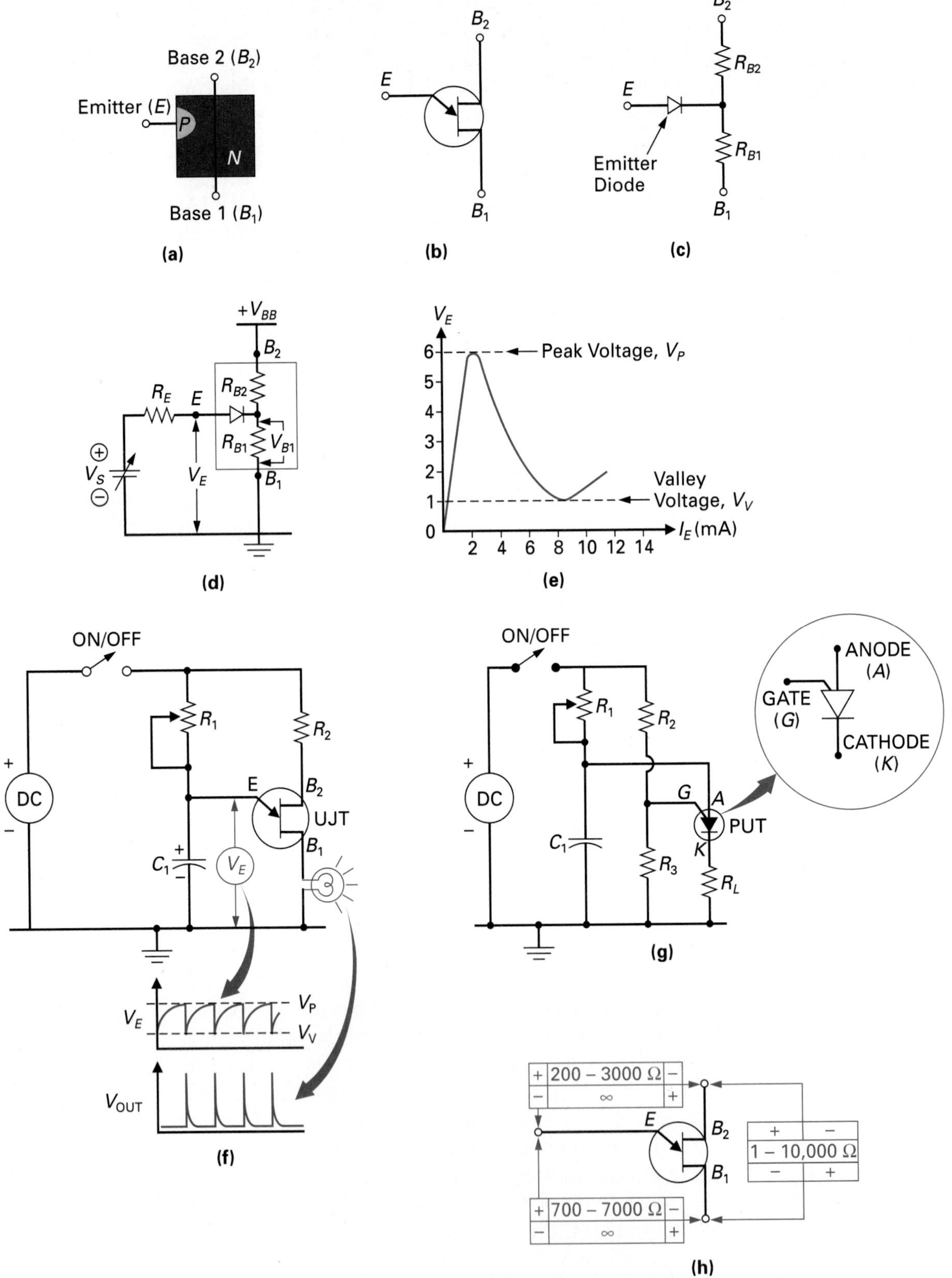

FIGURE 19-4 Unijunction Transistors. (a) Construction. (b) Schematic Symbol. (c) Equivalent Circuit. (d) Correctly Biased. (e) *V-I* Characteristic Curve. (f) Application—Relaxation Oscillator. (g) Programmable Unijunction Transistor (PUT). (h) Ohmmeter Resistances.

action by plotting the UJT's emitter voltage (between E and B_1) against the UJT's emitter current. An increase in V_S, and therefore V_E, produces very little emitter current until the **peak voltage (V_P)** is reached. Beyond V_P, V_E has exceeded V_{B1} and the emitter diode is forward biased, causing I_E to increase and V_E to decrease due to the lower resistance of R_{B1}. This negative resistance region reaches a low point known as the **valley voltage (V_V),** which is a point at which V_E begins to increase and the UJT no longer exhibits a negative resistance. The point at which a UJT turns ON and increases the current between B_1 and B_2 can be controlled, and used in switching applications.

Peak Voltage (V_P)
The maximum value of voltage.

Valley Voltage (V_V)
The voltage at the dip or valley in the characteristic curve.

UJT Applications

The UJT's negative resistance characteristic is useful in switching and timing applications. Figure 19-4(f) shows how a UJT could be connected to form a relaxation oscillator in an emergency flasher circuit. When the ON/OFF switch is closed, capacitor C_1 charges by resistor R_1. When the voltage across C_1 reaches the UJT's V_P value, the UJT will turn ON and its resistance between E and B_1 will drop LOW. This low resistance will allow C_1 to discharge through the UJT's E-to-B_1 junction and into the flasher light bulb, causing it to momentarily flash. As C_1 discharges, its voltage decreases and this causes the UJT to turn OFF. The cycle then repeats since the off UJT will allow capacitor C_1 to begin charging towards V_P, at which time it will trigger the UJT and repeat the process. The circuit's repetition rate, or frequency, is determined by the UJT's V_P rating, the supply voltage, and the RC time constant. To change the flashing rate, the value of R_1 can be changed to vary the rate at which C_1 is charged and therefore how soon the UJT is triggered.

The Programmable UJT (PUT)

The **programmable unijunction transistor** or **PUT** is a variation on the basic UJT thyristor. This four-layer thyristor has three terminals labeled cathode (K), anode (A), and gate (G). The key difference between the basic UJT and the PUT is that the PUT's peak voltage (V_P) can be controlled. Figure 19-4(g) shows how a PUT could also be connected to form a relaxation oscillator circuit. To differentiate the PUT's schematic symbol from the SCR, the gate input is connected into the anode side of the diode symbol instead of the cathode side. This circuit will produce exactly the same output waveform as the circuit shown in Figure 19-4(f). The gate-to-cathode voltage is derived from R_3, which is connected with R_2 to form a voltage divider. This circuit will operate in exactly the same way as the previous relaxation oscillator, in that C_1 will charge via R_1 until the charge across C reaches the V_P value. In this circuit, however, the V_P trigger voltage is set by R_3. When the PUT's anode-to-cathode voltage exceeds the gate voltage by 0.7 V (single diode voltage drop), the PUT will turn ON, and C_1 will discharge through the PUT and develop an output pulse across R_L. To vary the frequency of this circuit, we can change the resistance of R_1 as before, or change the ratio of R_2 to R_3, which controls the V_P value of the PUT. For example, if R_3 is made larger than R_2, the gate voltage and therefore V_P voltage will be larger. A high V_P value will mean that C_1 will have to charge to a larger voltage before the PUT will turn ON. Increasing the time needed for C_1 to charge will decrease the triggering rate of the PUT and therefore decrease the circuit's frequency of operation.

Programmable Unijunction Transistor or PUT
A unijunction transistor that can have its peak voltage controlled.

UJT Testing

Using the oscilloscope, you can monitor the ON/OFF switching of a UJT while it is operating in-circuit. If you suspect that the UJT is the cause of a circuit malfunction, you should remove it and check it with the ohmmeter, as seen in Figure 19-4(h).

SELF-TEST REVIEW QUESTIONS FOR SECTION 19-1

---

1. The SCR's forward breakover voltage will ____________ as gate current is increased.
2. The SCR can only be used to control dc power. (True/False)
3. The TRIAC is a ____________ directional device that can be either positive or negative triggered.
4. Which device can be used to trigger the TRIAC to compensate for the TRIAC's non-symmetrical triggering characteristic?
5. TRIACs should always be used instead of SCRs when we want to control the complete ac power cycle. (True/False)
6. The unijunction transistor has ____________ P-N semiconductor junction(s).
7. The UJT can, like the bipolar and field effect transistors, function as either a switch or an amplifier. (True/False)
8. The PUT's peak voltage can be varied by changing the voltage between ____________ and cathode.

19-2 TRANSDUCERS

Semiconductor Transducer
An electronic device that converts one form of energy to another.

Input Transducers
A transducer that generates input control signals.

Output Transducers
Transducers that convert output electrical signals to some other energy form.

A **semiconductor transducer** is an electronic device that converts one form of energy to another. Electronic transducers can be classified as either **input transducers** (such as the photodiode) that generate input control signals or **output transducers** (such as the LED) that convert output electrical signals to some other energy form. Input transducers are sensors that convert thermal, optical, mechanical, and magnetic energy variations into equivalent voltage and current variations. Output transducers on the other hand, perform the exact opposite, converting voltage and current variations into optical or mechanical energy variations.

19-2-1 *Optoelectronic Transducers*

In Chapter 4 we discussed the light-emitting diode (LED) in detail, and in Chapter 7 we discussed the photodiode in detail. The LED and photodiode are the most frequently used optoelectronic devices; however, they are not the only semiconductor devices in the optoelectronic family. In this section, we will review the operation of the LED and photodiode, along with the operation of other types of semiconductor light transducers.

Light-Sensitive Devices

Figure 19-5 illustrates some of the different types of **light-sensitive semiconductor devices.**

Light-Sensitive Semiconductor Devices
Semiconductor devices that change their characteristics in response to light.

The **photoconductive cell** or **light-dependent resistor (LDR)** shown in Figure 19-5(a) is a two-terminal device that changes its resistance (conductance) when light (photo) is applied. The photoconductive cell is normally mounted in a metal or plastic case with a glass window that allows the sensed light to strike the S-shaped light-sensitive material (typically cadmium sulfide). When light strikes the photoconductive atoms, electrons are released into the conduction band and the resistance between the device's terminals is reduced. When light is not present, the electrons and holes recombine, and the resistance is increased. A photoconductive cell will typically have a "dark resistance" of several hundred mega ohms and a "light resistance" of a few hundred ohms. The photoconductive cell's key advantage is that it can withstand a high operating voltage (typically a few hundred volts). Its disadvantages are that it responds slowly to changes in light level, and that its power rating is generally low (typically a few hundred milliwatts). The schematic symbol for the photoconductive cell is shown in Figure 19-5(b). Figure 19-5(c) shows how the photoconductive cell could be connected to control a street light. During the day, the LDR's resistance is LOW due to the high light levels, and the current through D_1, R_1, R_2, the LDR, and the control relay coil is HIGH. This HIGH value of current will energize the coil and cause the relay's normally closed (*NC*) contacts to open, and therefore the street light to be OFF. When dark, the resistance of the LDR will be HIGH, and the relay will be de-energized, its contacts will return to their normal condition, which is closed, and the street light will be ON.

Photoconductive Cell or Light-Dependent Resistor (LDR)
A two-terminal device that changes its resistance when light is applied.

The **photovoltaic cell** or **solar cell,** shown in Figure 19-5(d), generates a voltage across its terminals that will increase as the light level increases. The solar cell is usually made from silicon, and its schematic symbol is shown in Figure 19-5(e). The solar cell is available as either a discrete device or as a solar panel in which many solar cells are interconnected to form a series-aiding power source, as shown in the calculator application in Figure 19-5(f). The output of a solar cell is normally rated in volts and milliamps. For example, a typical photovoltaic cell could generate 0.5 V and 40 mA. To increase the output voltage simply connect solar cells in series; to increase the output current simply connect solar cells in parallel.

Photovoltaic Cell or Solar Cell
A device that generates a voltage across its terminal that will increase as the light level increases.

A **photodiode** is a photo-detecting or light-receiving device that contains a semiconductor P-N junction. When used in the "photovoltaic mode" as shown in Figure 19-5(g), the photodiode will generate an output voltage (voltaic) in response to a light (photo) input (will operate like a solar cell). Photodiodes are most widely used in the "photoconductive mode," in which they will change their conductance (conductive) when light (photo) is applied (will operate like an LDR). In this mode, the photodiode is reverse biased (*n*-type region is made positive, *p*-type region is made negative).

Photodiode
A photo-detecting or light-receiving device that contains a semiconductor P-N junction.

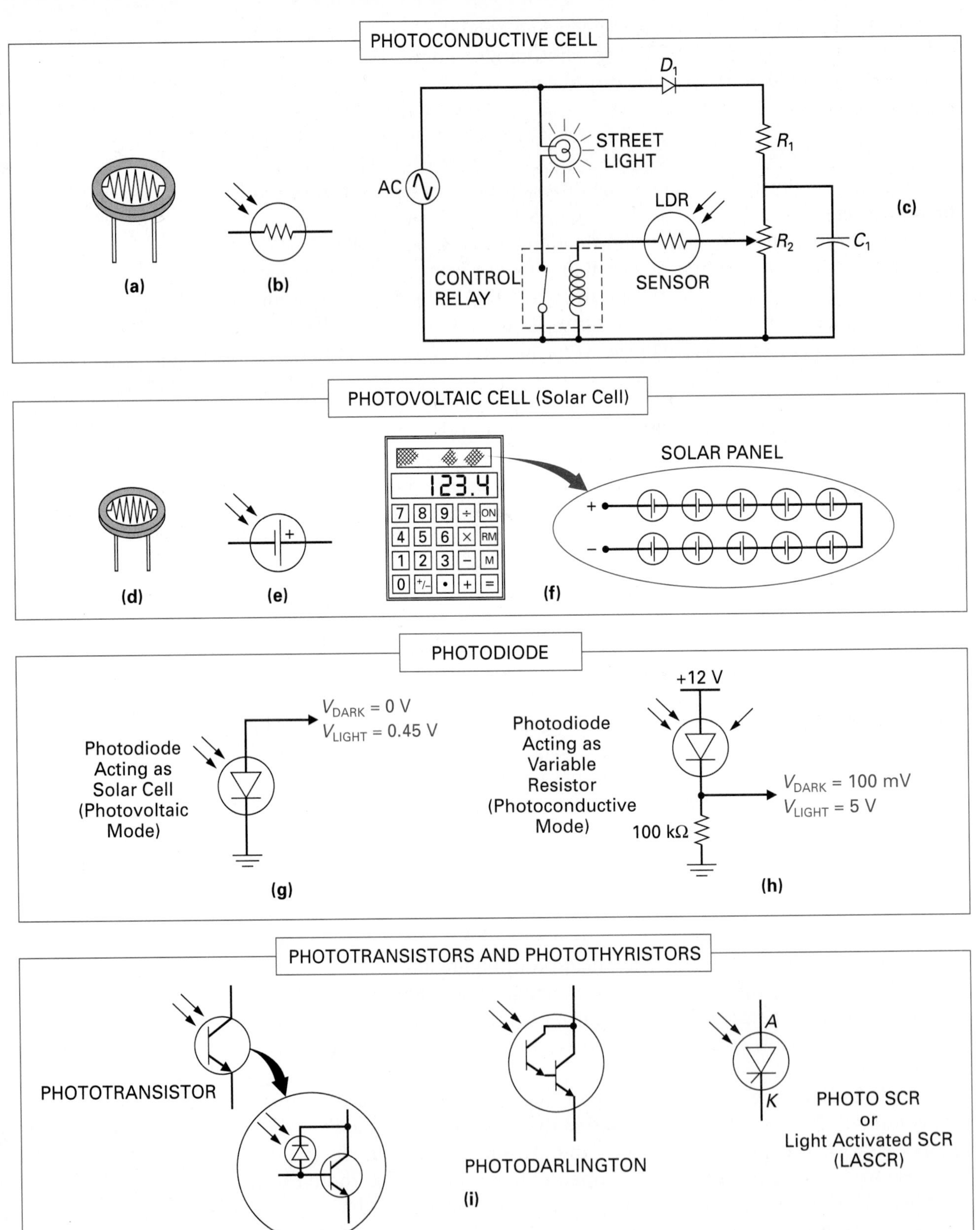

FIGURE 19-5 Light-Sensitive Devices.

The **photo-transistor, photo-darlington,** and **photo-SCR** (or light-activated SCR, LASCR), are all examples of light-reactive devices. As seen in the phototransistor inset, all three devices basically include a photodiode to activate the device whenever light is present. The advantage of the phototransistor is that it can produce a higher output current than the photodiode; however, the photodiode has a faster response time. The advantage of the photodarlington is its high current gain, and therefore it is ideal in low-light applications. When larger current switching is needed (typically a few amps), the LASCR can be used.

Photo-Transistor, Photo-Darlington, and Photo-SCR
Examples of light-reactive devices.

Light-Emitting Devices

Figure 19-6 illustrates some of the different types of **light-emitting semiconductor devices.**

Light-Emitting Semiconductor Devices
Semiconductor devices that will emit light when an electrical signal is applied.

The LED was covered in detail in Chapter 4; however, let us review its basic characteristics. Figure 19-6(a) shows the schematic symbols used to represent a **light-emitting diode** or **LED,** Figure 19-6(b) shows its typical physical appearance, and Figure 19-6(c) shows how the LED can be used as an ON/OFF indicator. The LED is basically a P-N junction diode and like all semiconductor diodes it can be either forward biased or reverse biased. When forward biased it will emit energy in response to a forward current. This emission of energy may be in the form of heat energy, light energy, or both heat and light energy

Light-Emitting Diode or LED
A semiconductor diode that converts electric energy into electromagnetic radiation at visible and near infrared frequencies when its P-N junction is forward biased.

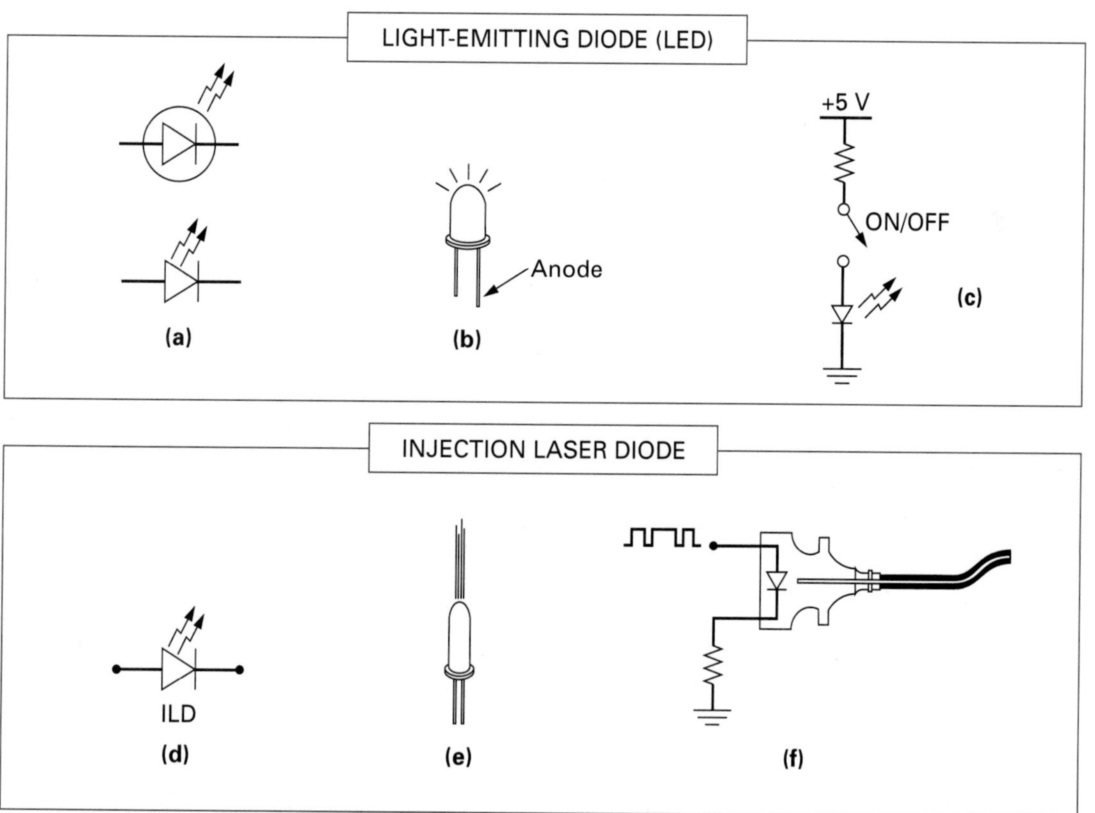

FIGURE 19-6 Light-Emitting Devices.

depending on the type of semiconductor material used. The type of material also determines the color, and therefore frequency, of the light emitted. For example, different compounds are available that will cause the LED to emit red, yellow, green, blue, white, orange, or infrared light when it is forward biased.

Injection Laser Diode or ILD
A semiconductor P-N junction diode that uses a lasing action to increase and concentrate the light output.

Figure 19-6(d) shows the schematic symbol for the **injection laser diode** or **ILD,** Figure 19-6(e) shows its typical physical appearance, and Figure 19-6(f) shows how it is usually used in fiber optic communication. The ILD differs from the LED in that it generates monochromatic (one color, or one frequency) light. Although it appears as though the LED is only generating one color light, it is in fact emitting several wavelengths or different frequencies that combined make up a color. Generating only one frequency is ideal in applications such as fiber optics, where we want to keep the light beam tightly focused so that the beam can travel long distances down a very thin piece of glass or plastic fiber. To explain this point in more detail, let us compare the light from a light bulb to the light produced by an ILD. An ILD generates light of only one frequency (monochromatic), and since all of the small packets of light being generated are of the same frequency, they act like an organized army, marching together in the same direction as a tight concentrated beam. The light bulb, on the other hand, generates white light that is composed of every color or frequency (panchromatic), and since these frequencies are all different, they act completely disorganized, and therefore radiate in all directions.

Optoelectronic Application—Character Displays

One of the biggest applications of LEDs is in the multisegment display. Figure 19-7(a) reviews the variety of LED display types available, which were discussed in detail in Chapter 4. Figure 19-7(b) shows the construction of a common-anode seven segment display. The need to display letters, in addition to numerals, resulted in a display with more segments such as the 14-segment display. For greater character definition, dot matrix displays are used, the most popular of which contains 35 small LEDs arranged in a grid of five vertical columns and seven horizontal rows, forming a 5×7 matrix. The bar display is rapidly replacing analog meter movements for displaying a quantity. For example, a quantity of fuel can be represented by the number of activated segments: no bars lit being zero and all bars lit being maximum.

Liquid Crystal Displays (LCDs)
A digital display having two sheets of glass separated by a sealed quantity of liquid crystal material. When a voltage is applied across the front and back electrodes, the liquid crystal's molecules become disorganized, causing the liquid to darken.

The variety of character displays shown in Figure 19-7(a) are also available as **liquid crystal displays (LCDs),** as shown in Figure 19-7(c). The two key differences between an LED and LCD display are that an LED display generates light while the LCD display controls light, and the LCD display consumes a lot less power than an LED display. The low power consumption feature of the LCD display makes it ideal in portable battery-operated systems such as wristwatches, calculators, video games, portable test equipment, and so on. The LCD's only disadvantage is that the display is hard to see, but this can be compensated for by including back-lighting to highlight the characters. The liquid crystal display contains two pieces of glass that act as a sandwich for a "nematic liquid" or liquid crystal material, as seen in the inset in Figure 19-7(c). The rear piece of glass is completely coated with a very thin layer of transparent metal, while the front piece of glass is coated with the same transparent metal segments in the shape of the desired display. The operation of the liquid crystal display is

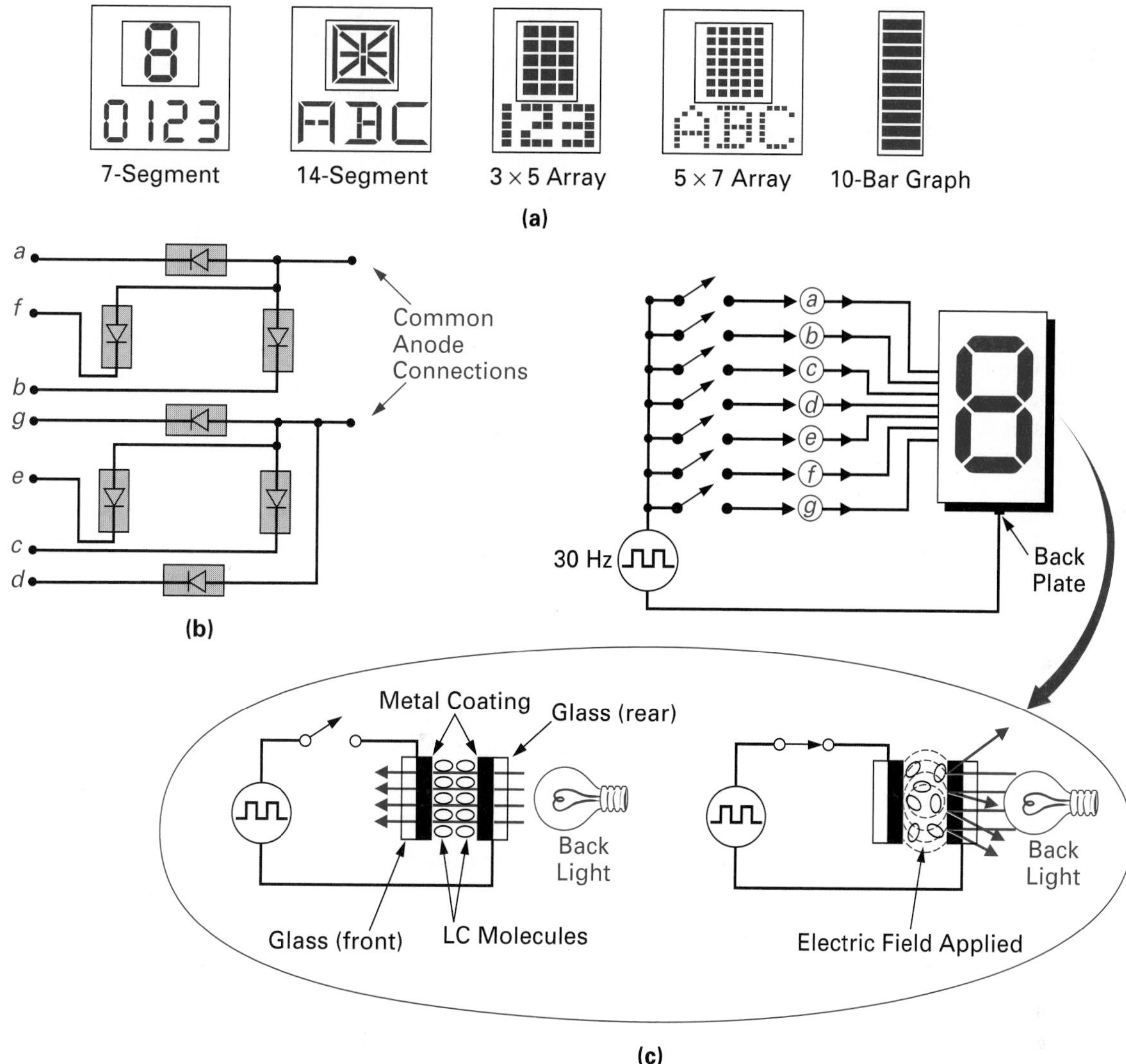

FIGURE 19-7 Optoelectronic Application—Character Displays.

explained in the two illustrations in the inset in Figure 19-7(c). When a segment switch is open, no electric field is generated between the two LCD metal plates, the nematic liquid molecules remain in their normal state which is parallel to the plane of the glass, and so all of the back-lighting passes through to the front display making the segment invisible. When a segment switch is closed, an ac voltage is applied between the two metal plates, generating an electric field between the two LCD metal plates. This electric field will cause the nematic liquid molecules to turn by 90°, and so all of the back-lighting for that segment is blocked making the segment visible.

Optoelectronic Application—Optically Coupled Isolators

Up until this point, we have considered the optoelectronic emitter and detector as discrete or individual devices. There are, however, some devices available that include both a light-emitting and light-sensing device in one pack-

Optically Coupled Isolators
Devices that contain a light-emitting and light-sensing device in one package. They are used to optically couple two electrically isolated points.

Optically Coupled Isolator DIP Module
An optocoupler that contains an infrared-emitting diode and a silicon phototransistor and is generally used to transfer switching information between two electrically isolated points.

Optically Coupled Isolator Interrupter Module
A device that consists of a matched and aligned emitter and detector and that is used to detect opaque or nontransparent targets.

Optically Coupled Isolator Reflector Module
A device that consists of a matched and aligned emitter and detector and that is used to detect targets.

age. These devices are called **optically coupled isolators,** a name that describes their basic function: they are used to *optically couple two electrically isolated points.* To examine these devices in more detail, refer to the three basic types shown in Figure 19-8.

Figure 19-8(a) shows an **optically coupled isolator DIP module.** This optocoupler contains an infrared-emitting diode (IRED) and a silicon phototransistor and is generally used to transfer switching information between two electrically isolated points. Figure 19-8(b) shows how DIP optically coupled isolator ICs can be used in a stepper-motor control circuit (discussed previously in Chapter 10). Since both circuits are identical, let us explain only the upper circuit's operation. Digital logic circuits generate the +5 V peak multiphase ON/OFF switching signal that is applied to IRED in the optocoupler. A HIGH input, for example, will turn ON the IRED, which will emit light to the phototransistor, turning it ON and switching the +12 V on its collector through to the emitter and the base of Q_1. This +12 V input to the base of Q_1 will cause it to turn ON and act as a closed switch, connecting ground to the top of the *A* winding. Since +12 V is connected to the center of the *A/B* winding, the *A* winding will be energized and the motor will move a step. The optocoupler, therefore, couples the ON/OFF switching information from the digital logic circuits and at the same time provides the necessary electrical isolation between the +5 V digital logic circuits and the +12 V motor supply circuits. In the past, isolation was provided by relays or isolation transformers that were larger in size, consumed more power, and were more expensive.

Figure 19-8(c) shows a different type of optocoupler or optoisolator called the **optically coupled isolator interrupter module.** This device consists of a matched and aligned emitter and detector and is used to detect opaque or nontransparent targets. Figure 19-8(d) shows how the interrupter optocoupler module could be used as an optical tachometer. The IRED is permanently ON, as seen in the inset in Figure 19-8(d), and emits a constant light beam towards the phototransistor. A transparent disc mounted to a shaft has opaque targets evenly spaced around the disc. As these opaque targets pass through the infrared beam, they will cause the phototransistor to momentarily turn OFF. On the other hand, a transparent section of the disc will not block the infrared light, and so the phototransistor will remain ON. As a result the phototransistor will turn ON and OFF, generating pulses, the number of which is an indication of the shaft's speed of rotation.

Figure 19-8(e) shows another type of optocoupler called the **optically coupled isolator reflector module.** Like the interrupter module, this device consists of a matched and aligned emitter and detector, and it is also used to detect targets. Figure 19-8(f) shows how the reflector module could also be used as an optical tachometer. The disadvantage with the interrupter module is that the disc has to be exactly aligned between the emitter or detector sections. The reflector module does not have to be positioned so close to the target disc, and therefore the alignment is not so crucial. The target disc used for reflector modules is different in that it is composed of reflective and nonreflective target areas. As the disc rotates, almost no light will be reflected by the dark areas when they are present at the focal point of the IRED and phototransistor. However, when

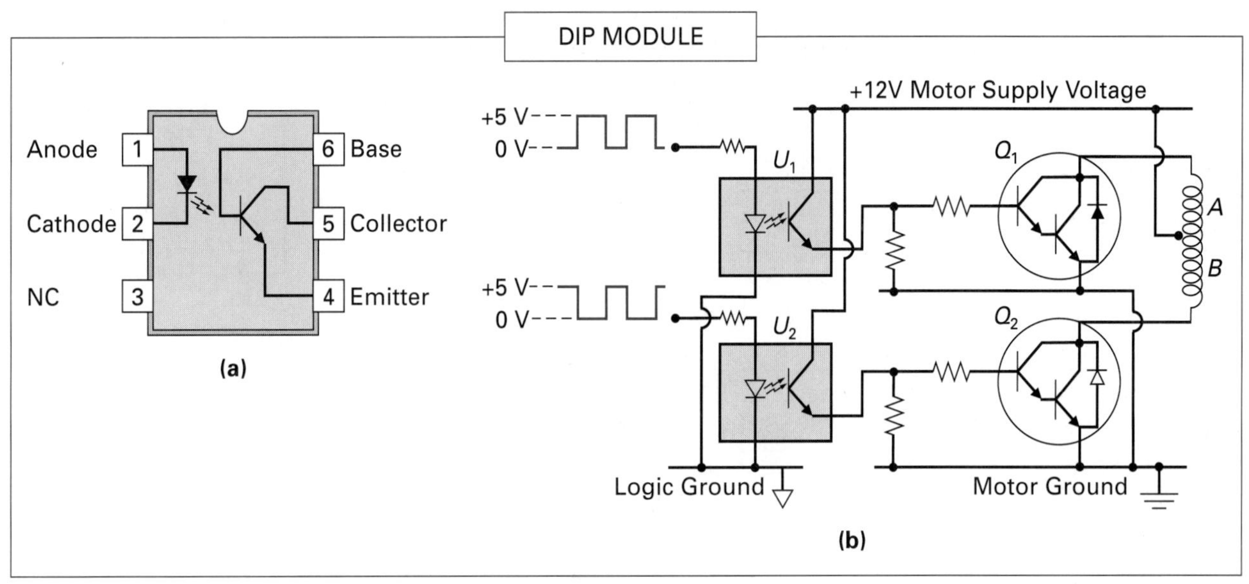

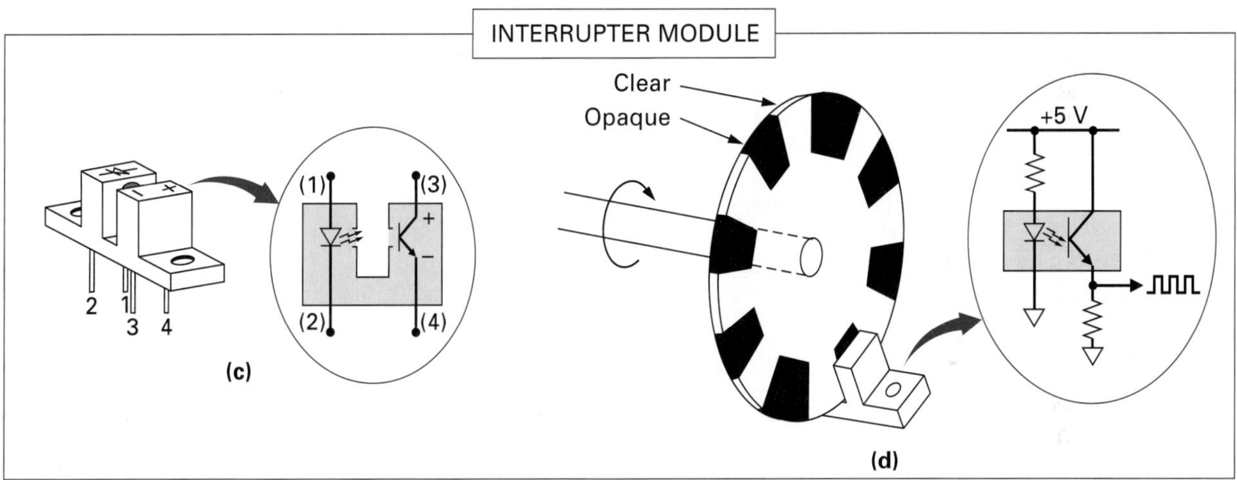

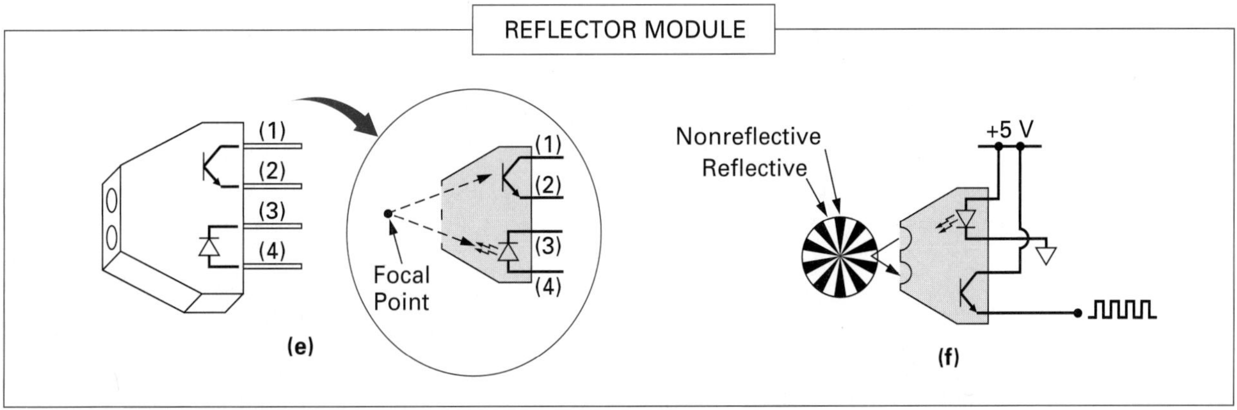

FIGURE 19-8 Optoelectronic Application—Optically Coupled Isolators.

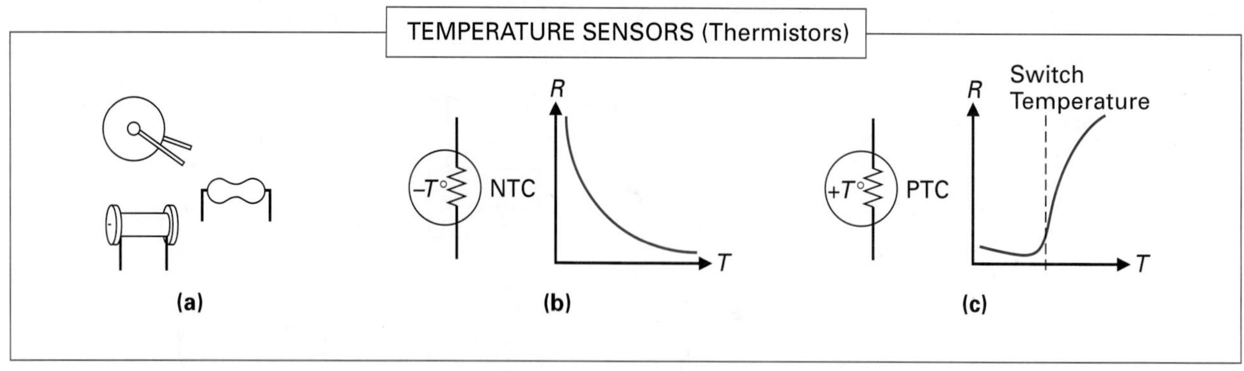
TEMPERATURE SENSORS (Thermistors)
R
−T°
NTC
T
(a)
(b)
R
Switch
Temperature
+T°
PTC
T
(c)

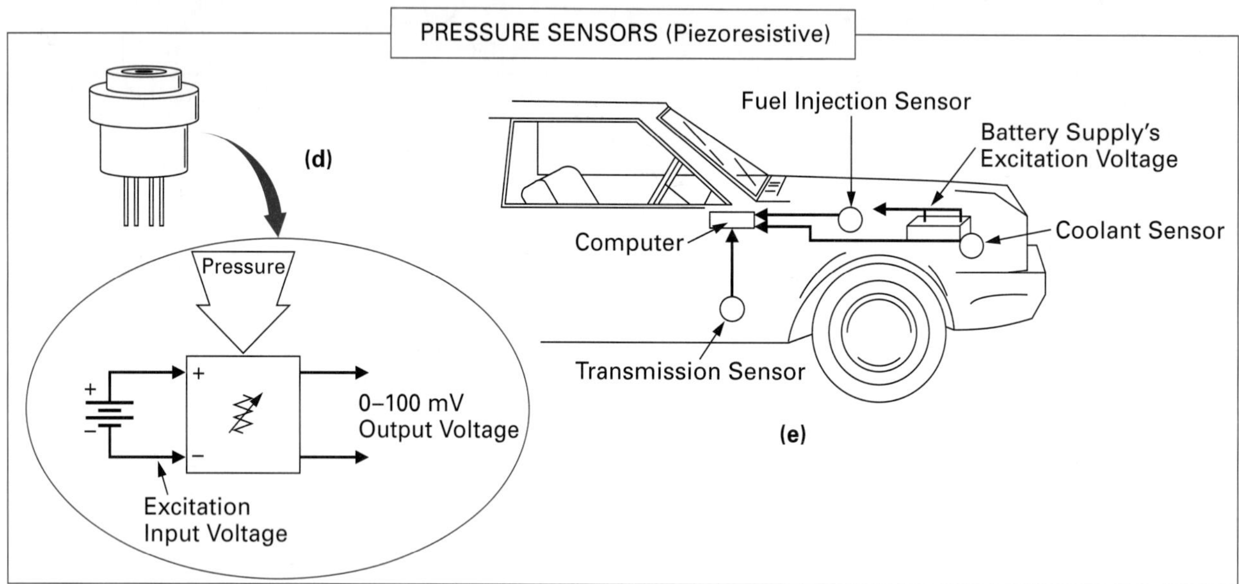
PRESSURE SENSORS (Piezoresistive)
(d)
Pressure
+
−
0–100 mV
Output Voltage
Excitation
Input Voltage
Fuel Injection Sensor
Battery Supply's
Excitation Voltage
Computer
Coolant Sensor
Transmission Sensor
(e)

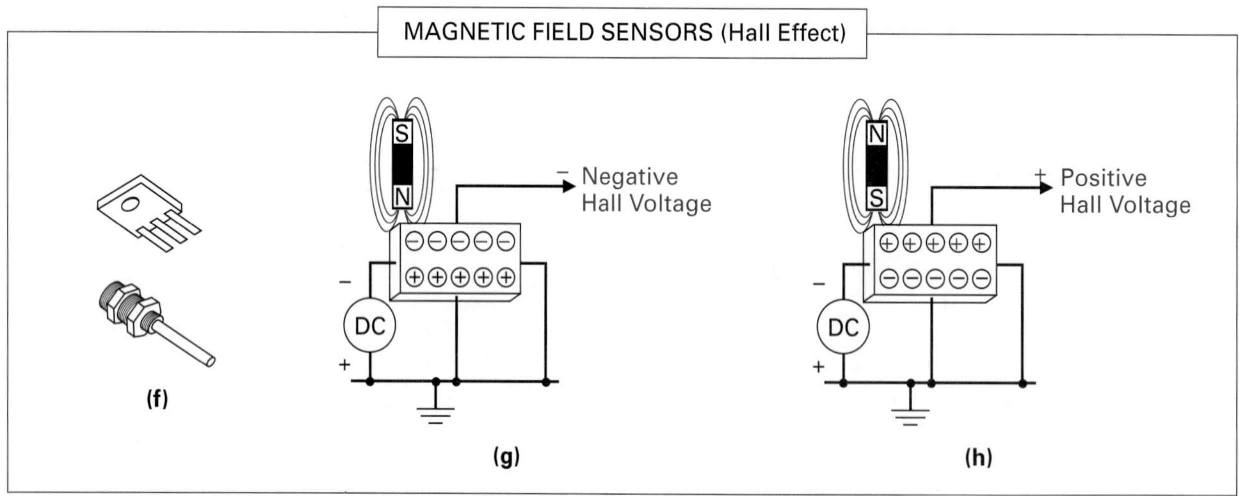
MAGNETIC FIELD SENSORS (Hall Effect)
S
N
− Negative
Hall Voltage
DC
(f)
(g)
N
S
+ Positive
Hall Voltage
DC
(h)

FIGURE 19-9 Semiconductor Transducers.

a reflective target is present at the focal point, light is reflected directly back to the phototransistor, turning it ON and therefore generating an output pulse.

19-2-2 Temperature Transducers

A **thermistor** is a semiconductor device that acts as a temperature sensitive resistor. Figure 19-9(a) shows a few different types of thermistors. There are basically two different types: **positive temperature coefficient (PTC) thermistors** and the more frequently used **negative temperature coefficient (NTC) thermistors.** To explain the difference, Figure 19-9(b) shows how a temperature increase causes the resistance of an NTC thermistor to decrease. On the other hand, Figure 19-9(c) shows how a temperature increase causes the resistance of a PTC thermistor to increase.

Thermistor
A semiconductor device that acts as a temperature sensitive resistor.

Positive Temperature Coefficient (PTC) Thermistors
A thermistor in which a temperature increase causes the resistance to increase.

Negative Temperature Coefficient (NTC) Thermistor
A thermistor in which a temperature increase causes the resistance to decrease.

Thermistors are used in a variety of applications. For example, an NTC could be used in a fire alarm circuit. When the ambient temperatures are LOW, the resistance of the NTC is HIGH, and therefore current cannot energize the alarm. If the ambient temperature increases to a HIGH level, however, the resistance of the thermistor drops LOW, and the alarm circuit is energized.

The PTC, on the other hand, could be used as a sort of circuit protection device (similar to a circuit breaker). With the PTC, the "switch temperature" (which is the temperature at which the resistance rapidly increases) can be varied by different construction techniques from below 0°C to above 160°C. The PTC therefore could be used as a current-limiting circuit protection device because currents lower than a limiting value will not generate enough heat to cause the PTC thermistor to switch to its high resistance state. On the other hand, circuit currents that go above the limiting value will generate enough heat to cause the PTC thermistor to switch to its high-resistance state. Current surges therefore, are limited to a safe value until the surge is over.

19-2-3 Pressure Transducers

A semiconductor pressure transducer will change its resistance in accordance with changes in pressure. Figure 19-9(d) shows a **piezoresistive diaphragm pressure sensor.** Piezoresistance of a semiconductor is described as a change in resistance due to a change in the applied pressure. This device has a dc excitation voltage applied, as seen in the inset in Figure 19-9(d), and will typically generate a 0 to 100 mV output voltage based on the pressure sensed. Figure 19-9(e) shows an application for these devices in which they are used in automobiles to sense the cooling-system pressure, hydraulic-transmission pressure, and fuel-injection pressure.

Piezoresistive Diaphragm Pressure Sensor
A sensor that changes its resistance in response to pressure.

19-2-4 Magnetic Transducers

Hall Effect Sensor
A sensor that generates a voltage in response to a magnetic field.

The **hall effect sensor** was discovered by Edward Hall in 1879 and is used in computers, automobiles, sewing machines, aircraft, machine tools, and medical equipment. Figure 19-9(f) shows a few different types of hall effect sensors. To explain the hall effect principle, Figure 19-9(g) shows that when a magnetic field whose polarity is north is applied to the sensor, it causes a separation of charges, generating in this example a negative hall voltage output. On the other hand, Figure 19-9(h) shows that if the magnetic field polarity is reversed so that a south pole is applied to the sensor, it will cause a polarity separation of charges within the sensor that will result in a positive hall voltage output. The amplitude of the generated positive or negative output voltage from the hall effect sensor is directly dependent on the strength of the magnetic field.

Their small size, light weight, and ruggedness make the hall effect sensors ideal in a variety of commercial and industrial applications. For example, hall effect sensors are embedded in the human heart to serve as timing elements. They are also used to sense shaft rotation, camera shutter positioning, rotary position, flow rate, and so on.

SELF-TEST REVIEW QUESTIONS FOR SECTION 19-2

1. The photoconductive cell basically operates as a light-sensitive ____________.
2. The photovoltaic cell converts light energy into ____________ energy, and is often referred to as a ____________ cell.
3. What are the two modes of operation for the photodiode?
4. With the photoconductive cell, photodiode, phototransistor, photodarlington, photothyristor, and other light-sensitive devices, a light increase always causes a ____________ in conductance.
5. LEDs convert ____________ energy into ____________ energy, whereas an ILD will convert ____________ energy into ____________ energy.
6. LED displays emit light while LCD displays control light. (True/False)
7. An optically coupled isolator DIP module will electrically connect two points, while isolating information from the two points. (True/False)
8. Would a PTC thermistor or NTC thermistor be ideal as a series current limiter?

SUMMARY

1. Thyristors are generally used as dc and ac power control devices, while transducers are generally used as sensing, displaying, and actuating devices.

Thyristors (Figure 19-10)

2. The semiconductor thyristor acts as an electronically controlled switch, connecting or disconnecting power from a load, or switching power ON and OFF to adjust the average amount of power delivered to a load.
3. Some thyristors are "unidirectional," which means that they will only conduct current in only one direction (dc), while others are "bidirectional," which means that they can conduct current in either direction (ac).

QUICK REFERENCE SUMMARY SHEET

FIGURE 19-10 Thyristors.

Silicon Controlled Rectifier (SCR)

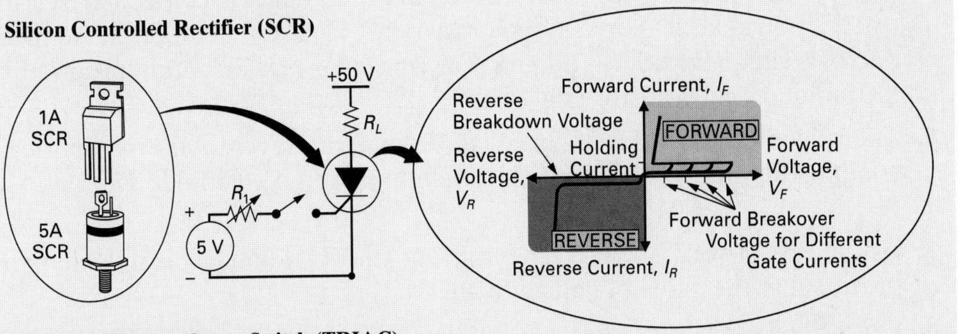

Triode AC Semiconductor Switch (TRIAC)

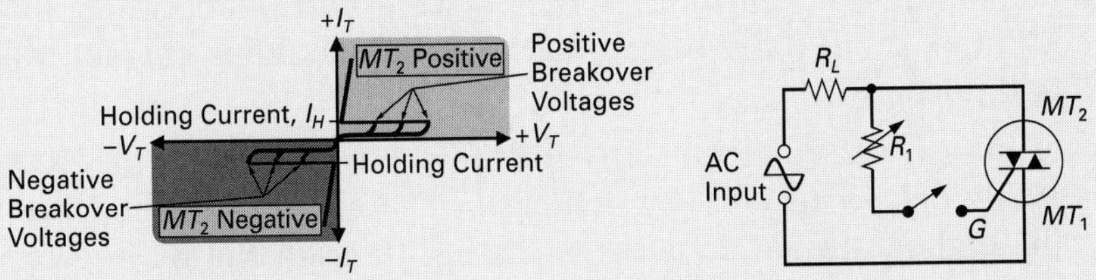

Diode AC Semiconductor Switch (DIAC)

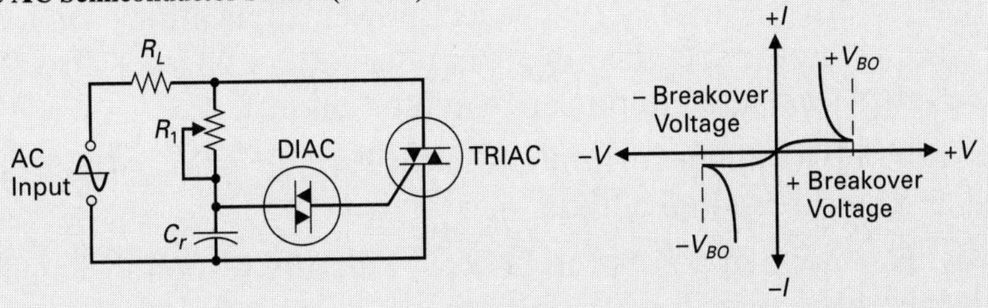

Unijunction Transistors (UJTs)

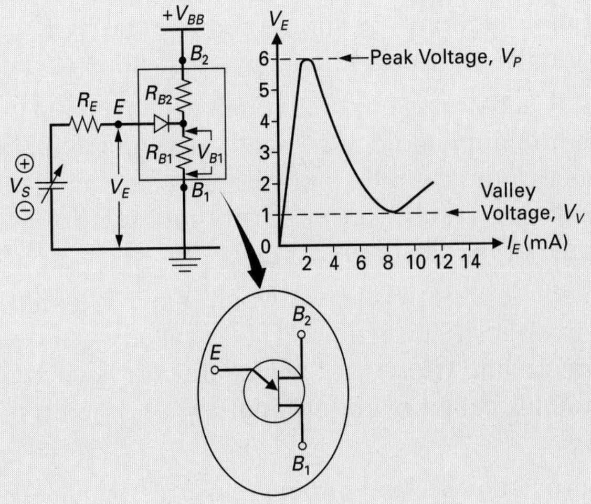

Programmable UJT (PUT)

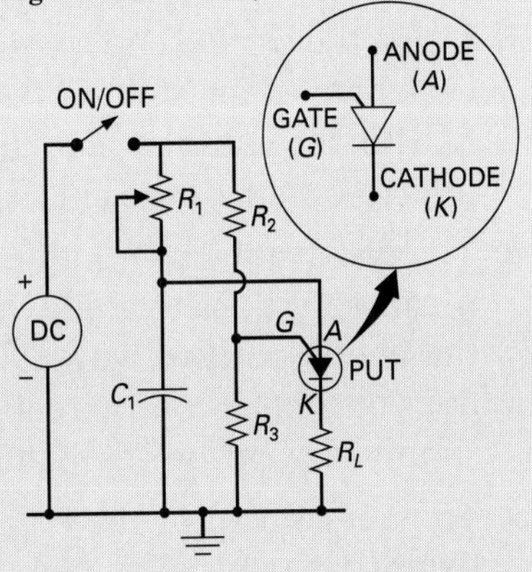

4. Thyristors are generally used as electronically controlled switches instead of the transistors, because they have better power handling capabilities and are more efficient.
5. The most frequently used thyristors are the "silicon controlled rectifier (SCR)," "triode ac semiconductor switch (TRIAC)," "diode ac semiconductor switch (DIAC)", "unijunction transistor (UJT)," and "programmable unijunction transistor (PUT)."
6. The silicon controlled rectifier, or SCR, is a four-layered alternately-doped component with three terminals labeled anode (A), cathode (K) and gate (G).
7. An SCR's forward breakover voltage, or turn-ON voltage, is inversely proportional to the value of gate current.
8. Once the SCR is turned ON, a holding-current latches the SCR ON, independent of the gate current. If the forward current between the SCR's cathode and anode falls below this minimum holding-current value, the SCR will turn OFF.
9. A gate trigger is used to turn ON the SCR; however, once the SCR is ON, it can only be turned OFF by decreasing the anode-to-cathode voltage so that the cathode-to-anode current passing through the SCR drops below the holding-current value.
10. The SCR is used in a variety of electrical power applications, such as light dimmer circuits, motor speed control circuits, battery charger circuits, temperature control systems, and power regulator circuits.
11. The SCR will only conduct current in the forward direction, which is why it is classed as a unidirectional device.
12. If an ac signal is applied across the SCR, it will only respond to a gate trigger during the time that the ac alternation makes the anode positive with respect to the cathode.
13. Using the oscilloscope, you can monitor the gate-trigger input and ON/OFF switching of an SCR while it is operating in circuit. If you suspect that the SCR is the cause of a circuit malfunction, you should remove the SCR and use the ohmmeter to check for terminal-to-terminal opens and shorts.
14. The disadvantage with the SCR is that it is unidirectional, which means that it can only be activated when the applied anode-to-cathode voltage makes the anode positive with respect to the cathode and will therefore only conduct current in one direction. As a result, the SCR can only control a dc supply voltage, or one-half cycle of the ac supply voltage.
15. To gain control of the complete ac input cycle, we would need to connect two SCRs in parallel, facing in opposite directions. This is exactly what was done to construct the bidirectional triode ac semiconductor switch or TRIAC, which has three terminals called main terminal 1 (MT_1), main terminal 2 (MT_2), and gate (G).
16. The TRIAC can be triggered or activated by either a positive or negative input gate trigger. This means that the TRIAC can be used to control both the positive and negative alternation of an ac supply voltage.

17. By controlling gate current, we can control the TRIAC's turn-ON voltage (positive and negative breakover voltage), so that the ac input voltage can be chopped up to adjust the average value of voltage applied to the load.
18. Using the oscilloscope, you can monitor the gate-trigger input and ON/OFF switching of a TRIAC while it is operating in-circuit. If you suspect that the TRIAC is the cause of a circuit malfunction, you should remove the TRIAC and use the ohmmeter to check for terminal-to-terminal opens and shorts.
19. One disadvantage with the TRIAC is that its positive breakover voltage is usually slightly different from its negative breakover voltage. This nonsymmetrical trigger characteristic can be compensated for by using a diode ac semiconductor switch or DIAC to trigger a TRIAC. The DIAC is equivalent to two back-to-back, series connected junction diodes and has two terminals.
20. Since the PNP regions of a DIAC are all equally doped the DIAC will have the same forward and reverse characteristics. As a result, the DIAC is classed as a symmetrical bi-directional switch, which means that it will have the same value of breakover voltage in both the forward and reverse direction.
21. Using the oscilloscope, you can monitor the ON/OFF switching of a DIAC while it is operating in-circuit. If you suspect that the DIAC is the cause of a circuit malfunction, you should remove the DIAC and check it with the ohmmeter.
22. The unijunction transistor or UJT is never used as an amplifying device like the BJT and FET; it is only ever used as a voltage-controlled switch.
23. The UJT has only one P-N junction and is a three-terminal device with an emitter lead (E) attached to a small p-type pellet that is fused into a bar of n-type silicon with contacts at either end labeled base 1 (B_1) and base 2 (B_2).
24. The emitter-to-bar P-N junction is equivalent to a junction diode, and the bar is equivalent to a two-resistor voltage divider (R_{B1} and R_{B1}).
25. An increase in V_S, and therefore V_E, produces very little emitter current until the peak voltage (V_P) is reached. Beyond V_P, V_E has exceeded V_{B1} and the emitter diode is forward biased, causing I_E to increase and V_E to decrease, due to the lower resistance of R_{B1}. This negative resistance region reaches a low point known as the valley voltage (V_V), which is a point at which V_E begins to increase and the UJT no longer exhibits a negative resistance. The point at which a UJT turns ON and increases the current between B_1 and B_2 can be controlled, and therefore used in switching applications.
26. The programmable unijunction transistor or PUT is a variation on the basic UJT thyristor. This four-layer thyristor has three terminals labeled cathode (K), anode (A), and gate (G). The key difference between the basic UJT and the PUT is that the PUT's peak voltage (V_P) can be controlled.
27. Using the oscilloscope, you can monitor the ON/OFF switching of a UJT while it is operating in-circuit. If you suspect that the UJT is the

cause of a circuit malfunction, you should remove it and check it with the ohmmeter.

Transducers

28. A semiconductor transducer is an electronic device that converts one form of energy to another.
29. Electronic transducers can be classified as either input transducers (such as the photodiode) that generate input control signals, or output transducers (such as the LED) that convert output electrical signals to some other energy form.
30. Input transducers are sensors that convert thermal, optical, mechanical and magnetic energy variations into equivalent voltage and current variations.
31. Output transducers on the other hand, perform the exact opposite, converting voltage and current variations into optical or mechanical energy variations.

Optoelectronic Transducers (Figure 19-11)

32. The photoconductive cell, or light-dependent resistor (LDR), is a two-terminal device that changes its resistance (conductance) when light (photo) is applied.
33. When light strikes the photoconductive atoms, electrons are released into the conduction band, and therefore the resistance between the device's terminals is reduced. When light is not present, the electrons and holes recombine, and therefore the resistance is increased.
34. A photoconductive cell will typically have a "dark resistance" of several hundred mega ohms, and a "light resistance" of a few hundred ohms.
35. The photoconductive cell's key advantage is that it can withstand a high operating voltage (typically a few hundred volts). Its disadvantages are that it responds slowly to changes in light level, and that its power rating is generally low (typically a few hundred milliwatts).
36. The photovoltaic cell, or solar cell, generates a voltage across its terminals that will increase as the light level increases.
37. The solar cell is available as either a discrete device, or as a solar panel in which many solar cells are interconnected to form a series-aiding power source.
38. The output of a solar cell is normally rated in volts and milliamps. To increase the output voltage simply connect solar cells in series, and to increase the output current simply connect solar cells in parallel.
39. The phototransistor, photodarlington, and photo SCR (or light-activated SCR, LASCR), are all examples of light-reactive devices. All three devices basically include a photodiode to activate the device whenever light is present.
40. The advantage of the phototransistor is that it can produce a higher output current than the photodiode; however, the photodiode has a faster response time. The advantage of the photodarlington is its high current

QUICK REFERENCE SUMMARY SHEET

FIGURE 19-11 Optoelectronic Transducers.

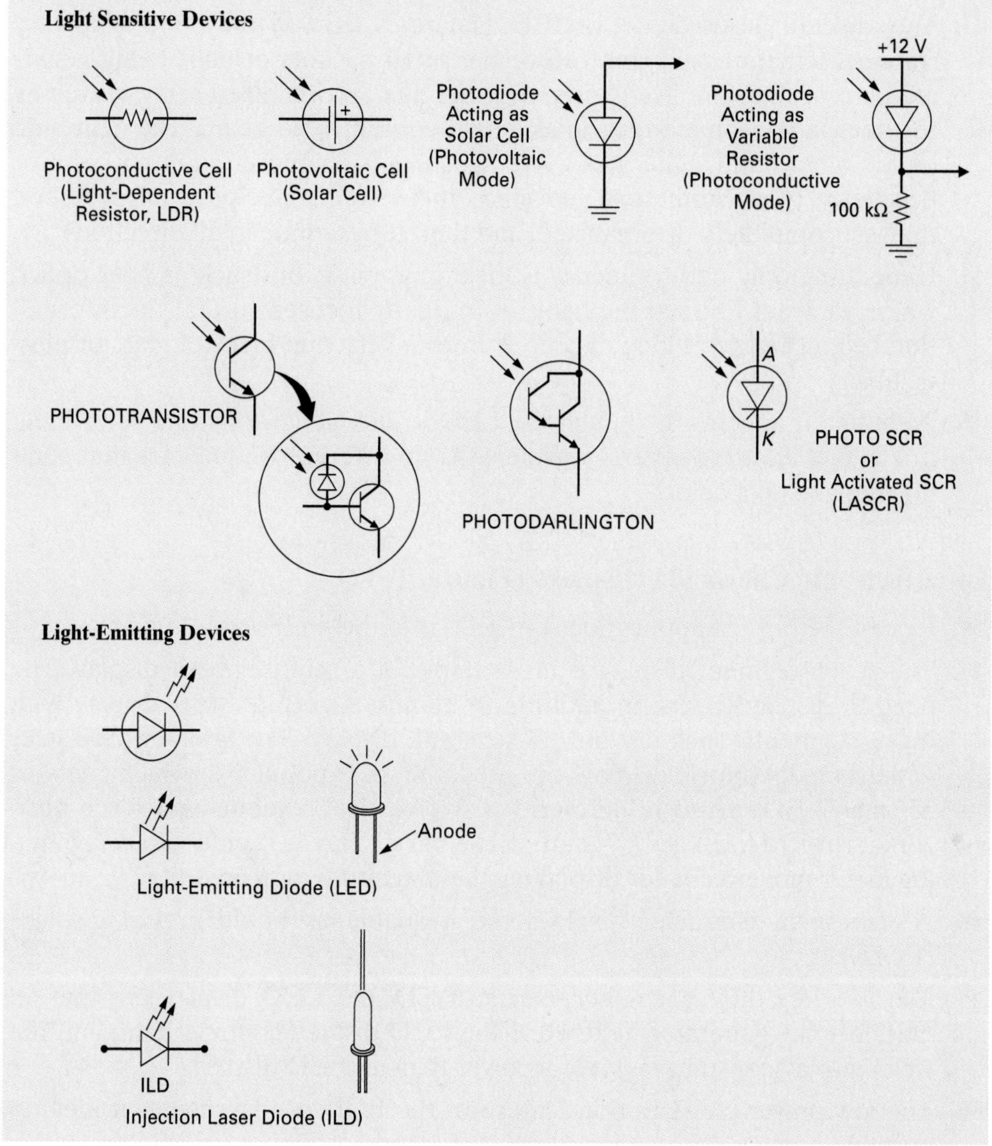

Light Sensitive Devices

Photoconductive Cell (Light-Dependent Resistor, LDR)

Photovoltaic Cell (Solar Cell)

Photodiode Acting as Solar Cell (Photovoltaic Mode)

Photodiode Acting as Variable Resistor (Photoconductive Mode)

+12 V

100 kΩ

PHOTOTRANSISTOR

PHOTODARLINGTON

A

K

PHOTO SCR or Light Activated SCR (LASCR)

Light-Emitting Devices

Anode

Light-Emitting Diode (LED)

ILD

Injection Laser Diode (ILD)

gain, and therefore it is ideal in low-light level applications. When larger current switching is needed (typically a few amps), the LASCR can be used.

41. The LED is basically a P-N junction diode, and like all semiconductor diodes it can be either forward biased or reverse biased. When forward biased it will emit energy in response to a forward current. This emission of energy may be in the form of heat energy, light energy, or both heat and light energy depending on the type of semiconductor material used.

42. The type of material used to construct the LED determines the color, and therefore frequency, of the light emitted. For example, different compounds are available that will cause the LED to emit red, yellow, green, blue, white, orange, or infrared light when it is forward biased.

43. The injection laser diode, or ILD, generates light of only one frequency (monochromatic), and since all of the small packets of light being generated are of the same frequency, they act like an organized army, marching together in the same direction as a tight concentrated beam. The light bulb on the other hand, generates white light that is composed of every color or frequency (panchromatic), and since these frequencies are all different, they act completely disorganized, and therefore radiate in all directions.

44. Generating only one frequency is ideal in applications such as fiber optics, where we want to keep the light beam tightly focused so that the information beam can travel long distances down a very thin piece of glass or plastic fiber.

45. Although it appears as though the LED is only generating one color light, it is in fact emitting several wavelengths or different frequencies that combined make up a color.

Optoelectronic Character Displays (Figure 19-12)

46. One of the biggest applications of LEDs is in the multisegment display.

47. The seven-segment display is an example of a multisegment display. The need to display letters, in addition to numerals, resulted in a display with more segments such as the 14-segment display. For greater character definition, dot matrix displays are used, the most popular of which contains 35 small LEDs arranged in a grid of five vertical columns and seven horizontal rows to form a 5×7 matrix. The bar display is rapidly replacing analog meter movements for displaying the magnitude of a quantity.

48. A variety of character displays are available as liquid crystal displays (LCDs).

49. The two key differences between an LED and LCD display are that an LED display generates light while the LCD display controls light, and the LCD display consumes a lot less power than an LED display.

50. The low power consumption feature of the LCD display makes it ideal in portable battery-operated systems such as wristwatches, calculators, video games, portable test equipment, and so on.

51. The LCD's only disadvantage is that the display is hard to see; however, this can be compensated for by including back-lighting to highlight the characters.

52. When an LCD segment switch is open, no electric field is generated between the two LCD metal plates, the nematic liquid molecules remain in their normal state, which is parallel to the plane of the glass, and all of the back-lighting passes through to the front display making the segment invisible. On the other hand, when a segment switch is closed, an ac voltage is applied between the two metal plates and an electric field is generated

QUICK REFERENCE SUMMARY SHEET

FIGURE 19-12 Optoelectronic Character Displays.

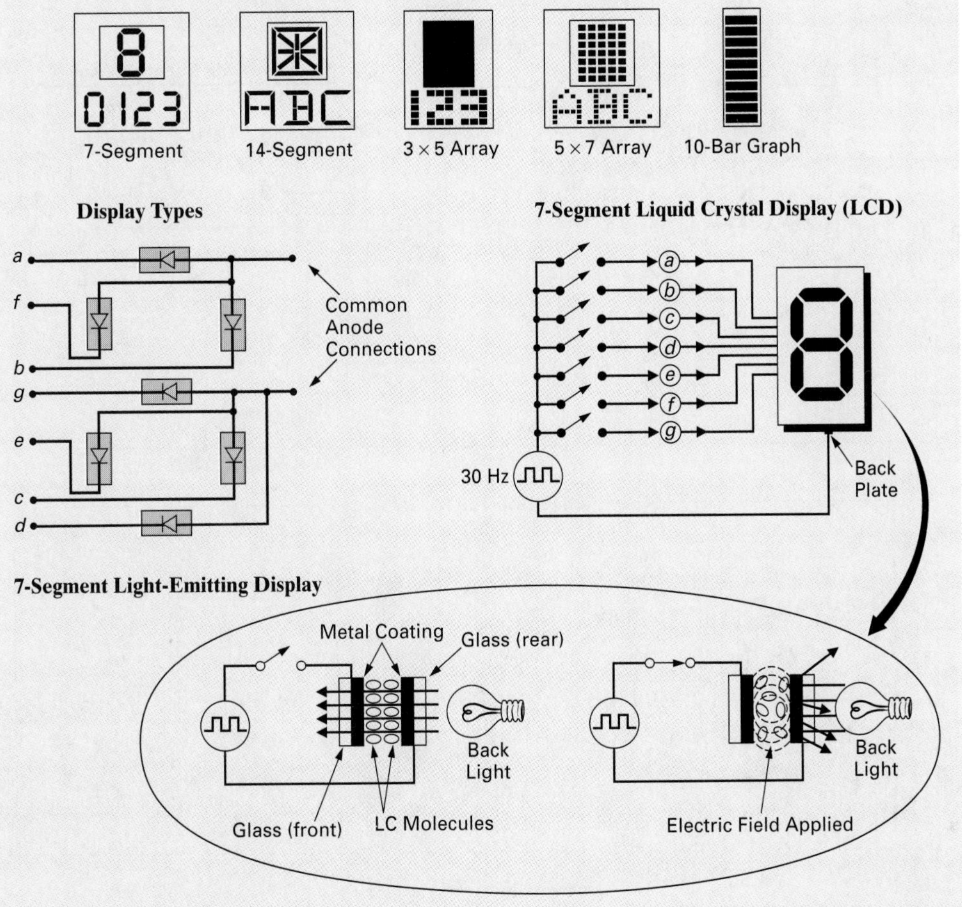

7-Segment 14-Segment 3 × 5 Array 5 × 7 Array 10-Bar Graph

Display Types

7-Segment Liquid Crystal Display (LCD)

7-Segment Light-Emitting Display

between the two LCD metal plates. This electric field will cause the nematic liquid molecules to turn by 90°, so all of the back-lighting for that segment is blocked, making the segment visible.

Optically Coupled Isolators (Figure 19-13)

53. There are devices available that include both a light-emitting and light-sensing device in one package. These devices are called optically coupled isolators, a name that describes their basic function because they are used to optically couple two electrically isolated points.

54. An optically coupled isolator DIP module contains an infrared emitting diode (IRED) and a silicon phototransistor. It is generally used to transfer switching information between two electrically isolated points.

55. The optically coupled isolator interrupter module is a device that contains a matched and aligned emitter and detector and is used to detect opaque or nontransparent targets.

QUICK REFERENCE SUMMARY SHEET

FIGURE 19-13 Optically Coupled Isolators.

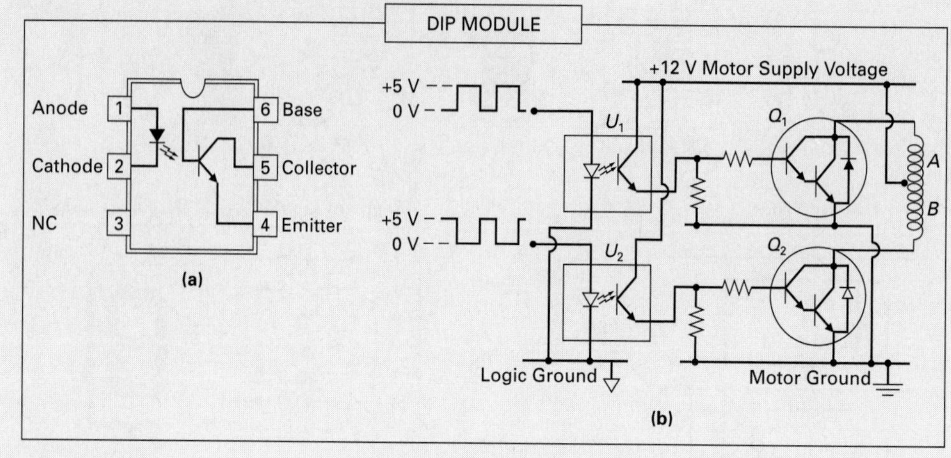

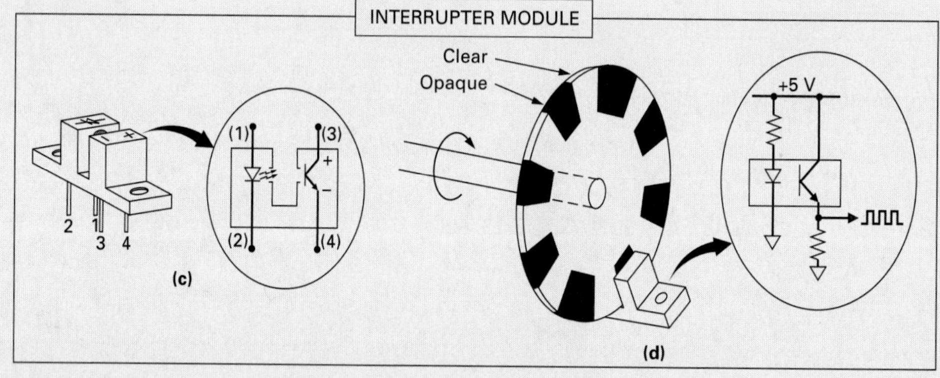

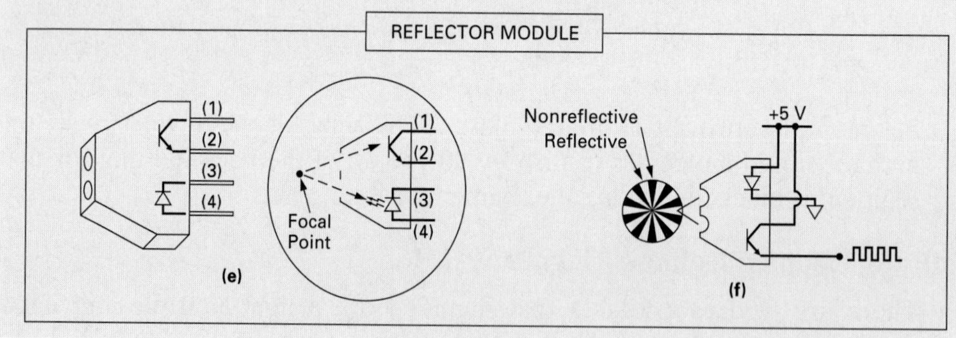

56. The optically coupled isolator reflector module consists of a matched and aligned emitter and detector, and it is also used to detect targets.

57. The disadvantage with the interrupter module is that the disc has to be exactly aligned between the emitter or detector sections. The reflector module does not have to be positioned so close to the target disc, and therefore the alignment is not so crucial.

QUICK REFERENCE SUMMARY SHEET

FIGURE 19-14 Temperature, Pressure, and Magnetic Semiconductor Transducers.

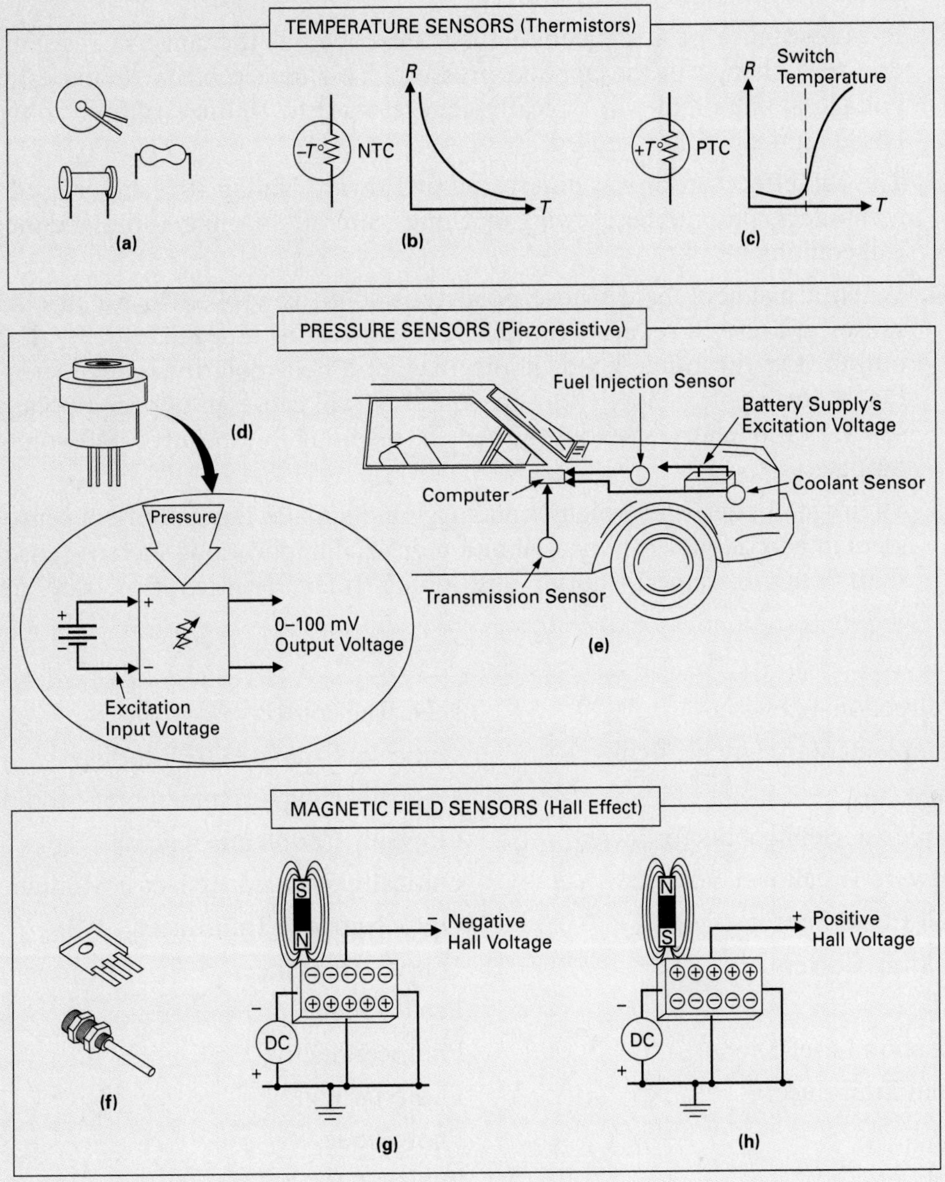

TEMPERATURE SENSORS (Thermistors)

(a) (b) (c)

PRESSURE SENSORS (Piezoresistive)

(d) (e)

MAGNETIC FIELD SENSORS (Hall Effect)

(f) (g) (h)

Temperature, Pressure and Magnetic Semiconductor Transducers (Figure 19-14)

58. A thermistor is a semiconductor device that acts as a temperature sensitive resistor.

59. There are basically two different types of thermistors: positive temperature coefficient (PTC) thermistors and the more frequently used negative temperature coefficient (NTC) thermistors.

60. A temperature increase causes the resistance of an NTC thermistor to decrease. On the other hand, a temperature increase causes the resistance of a PTC thermistor to increase.

61. A semiconductor pressure transducer will change its resistance in accordance with changes in pressure.

62. Piezoresistance of a semiconductor is described as a change in resistance due to a change in the applied pressure. This device has a dc excitation voltage applied and will typically generate a 0 to 100 mV output voltage based on the pressure sensed.

63. The hall effect sensor was discovered by Edward Hall in 1879 and is used in computers, automobiles, sewing machines, aircraft, machine tools, and medical equipment.

64. When a magnetic field whose polarity is south is applied to a hall effect sensor, it causes a separation of charges, generating a negative hall voltage output. On the other hand, if the magnetic field polarity is reversed so that a north pole is applied to the sensor, it will cause an opposite polarity separation of charges within the sensor, resulting in a positive hall voltage output.

65. Their small size, light weight, and ruggedness make the hall effect sensors ideal in a variety of commercial and industrial applications such as sensing shaft rotation, camera shutter positioning, rotary position, flow rate, and so on.

NEW TERMS

Bidirectional Device
Complementary Latch Circuit
DIAC
Diode AC Semiconductor Switch
Forward Breakover Voltage
Hall Effect Sensor
Holding Current
ILD
Injection Laser Diode
Input Transducer
LDR
LED
Light-Dependent Resistor
Light-Emitting Device
Light-Emitting Diode
Light-Emitting Displays
Light-Sensitive Device
Liquid Crystal Displays
Negative Temperature Coefficient
NTC Thermistor
Optically Coupled DIP Module
Optically Coupled Interrupter Module
Optically Coupled Isolators
Optically Coupled Reflector Module
Optoelectronic Transducer
Output Transducer
Peak Voltage
Photoconductive Cell
Photodarlington
Photodiode
Photo-SCR
Phototransistor
Photovoltaic Cell
Piezoresistive Diaphragm Pressure Transducer
Positive Temperature Coefficient
Pressure Transducer
Programmable Unijunction Transistor

PTC Thermistor
PUT
SCR
Semiconductor Transducer
Silicon Control Rectifier
Solar Cell
Symmetrical Bi-Directional Switch
Temperature Transducer
Thermistor
Thyristor
Transducer
TRIAC
Triode AC Semiconductor Switch
UJT
Unidirectional Device
Unijunction Transistor
Valley Voltage

REVIEW QUESTIONS

Multiple-Choice Questions

1. Which of the following thyristors would normally be used to control dc power?
   a. TRIAC **d.** DIAC
   b. SCR **e.** Both (b) and (c)
   c. UJT
2. Which of the following thyristors is a bidirectional device?
   a. TRIAC **c.** UJT
   b. SCR **d.** ENIAC
3. ___________ an SCR's gate current will ___________ its forward breakover voltage.
   a. Increasing, decrease **c.** Decreasing, increase
   b. Increasing, increase **d.** Decreasing, decrease
4. Which of the following thyristors has a symmetrical switching characteristic?
   a. TRIAC **d.** DIAC
   b. SCR **e.** Both (b) and (c)
   c. UJT
5. Which of the following light sensitive devices will generate a voltage when light is applied?
   a. Photoconductive cell **d.** Phototransistor
   b. Photodiode **e.** Both (b) and (c)
   c. Photovoltaic cell
6. Which of the following is a monochromatic light source?
   a. LED **d.** LCD
   b. Light bulb **e.** Both (b) and (c)
   c. ILD
7. Which of the following devices would be best suited to detect whenever a coin has been inserted in a slot?
   a. Piezoresistive sensor **d.** Interrupter optoisolator
   b. Hall Effect sensor **e.** Both (a) and (c)
   c. PTC thermistor

8. What advantage does a liquid crystal display have over an LED display?
   - a. Consumes less power
   - b. Is easier to read
   - c. Is ideal in portable applications
   - d. Both (a) and (c)
   - e. Both (a) and (b)
9. What device could be used as a series connected current limiter?
   - a. Hall effect sensor
   - b. Piezoresistive sensor
   - c. PTC thermistor
   - d. NTC thermistor
10. Which of the following sensors could be used to control the switching of current through a motor's stator windings by sensing the position of the motor's permanent magnet rotor?
    - a. Hall effect sensor
    - b. Piezoresistive sensor
    - c. PTC thermistor
    - d. NTC thermistor

Essay Questions

11. Give the full name for the following abbreviations: (Chapter 19)
    - a. SCR
    - b. TRIAC
    - c. DIAC
    - d. UJT
    - e. PUT
    - f. ILD
    - g. PTC
    - h. NTC
    - i. LCD
12. Explain how the SCR is turned ON, its latching action, and how it is turned OFF. (19-1-1)
13. Why is the SCR classed as a unidirectional device? (19-1-1)
14. Can SCRs be used in ac and dc power control applications? (19-1-1)
15. What is the basic difference between an SCR and a TRIAC? (19-1-2)
16. Why is the TRIAC classed as a bidirectional device? (19-1-2)
17. Which of the thyristors is a symmetrical bidirectional switch? (19-1-3)
18. Briefly describe the operation of the unijunction transistor. (19-1-4)
19. What advantage does a PUT have over the basic UJT? (19-1-4)
20. Define the term "transducer." (19-2)
21. What is the difference between a light-dependent resistor and a solar cell? (19-2-1)
22. Which of the light-sensitive devices would be best suited to detect a very high-speed data transmission from a fiber optic cable? (19-2-1)
23. What advantage does the phototransistor have over the photodiode? (19-2-1)
24. In what application would the LASCR be used? (19-2-1)
25. What is the basic difference between an LED and an ILD? (19-2-1)
26. Briefly describe the operation of a liquid crystal display. (19-2-1)
27. What is an optically coupled isolator? (19-2-1)
28. Describe an application for the following optically coupled isolators: (19-2-1)
    - a. DIP module
    - b. Interrupter module
    - c. Reflector module

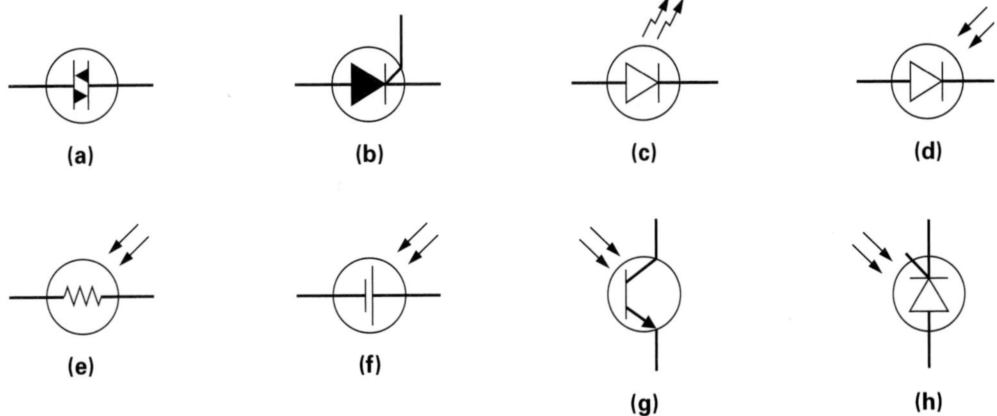

FIGURE 19-15 **Schematic Symbols.**

29. What is the difference between a PTC thermistor and an NTC thermistor? (19-2-2)

30. Briefly describe the operation of the piezoresistive pressure sensor and the hall effect sensor. (19-2-3 and 19-2-4)

Practice Problems

31. Identify the schematic symbols shown in Figure 19-15.

To practice your circuit recognition and operation ability, refer to the circuit in Figure 19-16, and answer the following questions.

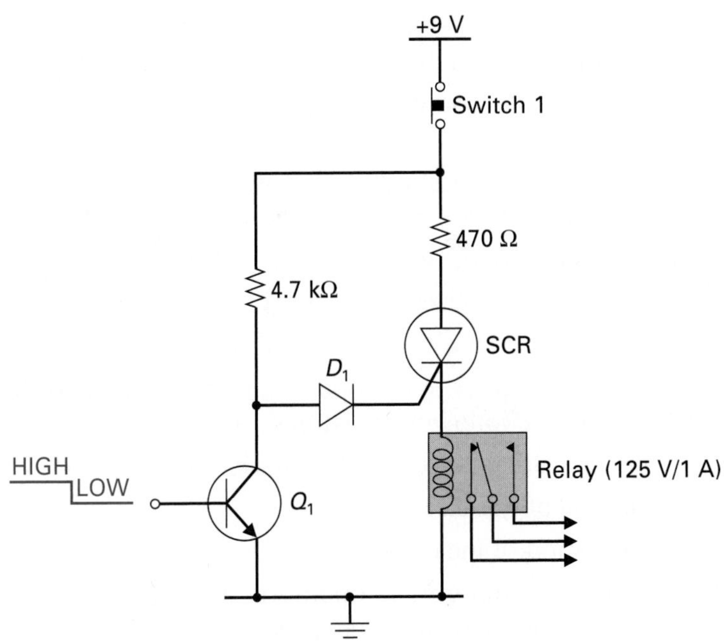

FIGURE 19-16 **Low-Power to High-Power ON/OFF Switching.**

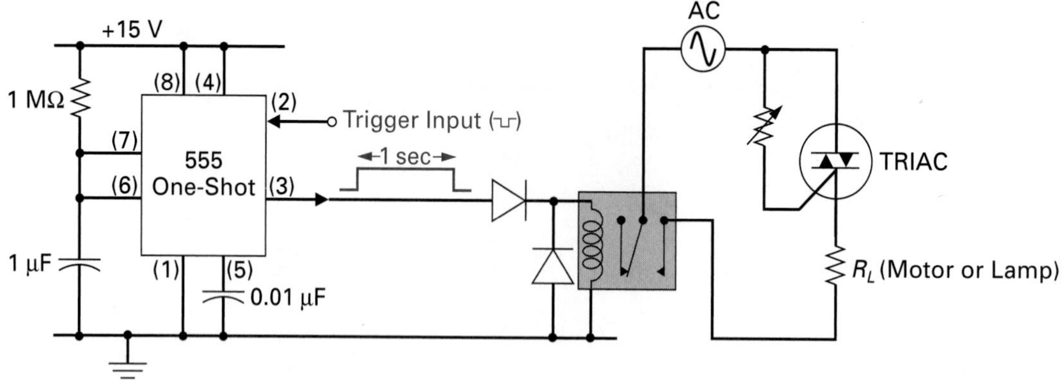

FIGURE 19-17 One-Shot Timer Control of TRIAC.

32. Describe the operation of the circuit when the input is normally HIGH.

33. Describe the operation of the circuit when the input is taken LOW.

34. How is the SCR turned ON, and how is the SCR turned OFF?

To practice your circuit recognition and operation ability, refer to the circuit in Figure 19-17, and answer the following questions.

35. Is the 555 timer connected to function as an astable, monostable, or bistable multivibrator?

36. Describe the basic operation of this circuit.

37. Why do you think a TRIAC is being used in this circuit instead of an SCR?

To practice your circuit recognition and operation ability, refer to the circuit in Figure 19-18(a) and (b), and answer the following questions.

38. Identify the light-emitting and light-sensitive devices used in these circuits.

39. Why was a different light-reactive device used for each of these circuits?

40. Why are these optically coupled isolators needed for computer output device control?

To practice your circuit recognition and operation ability, refer to the circuit in Figure 19-19, and answer the following questions.

41. Identify the light-sensitive device used in this circuit.

42. What is the function of this circuit?

43. Describe the operation of the circuit.

To practice your circuit recognition and operation ability, refer to the circuit in Figure 19-20, and answer the following questions.

44. Identify the light-sensitive device used in this circuit.

45. What is the function of this circuit?

46. Describe the operation of the circuit.

To practice your circuit recognition and operation ability, refer to the circuit in Figure 19-21, and answer the following questions.

47. Identify the light-sensitive device used in this circuit.

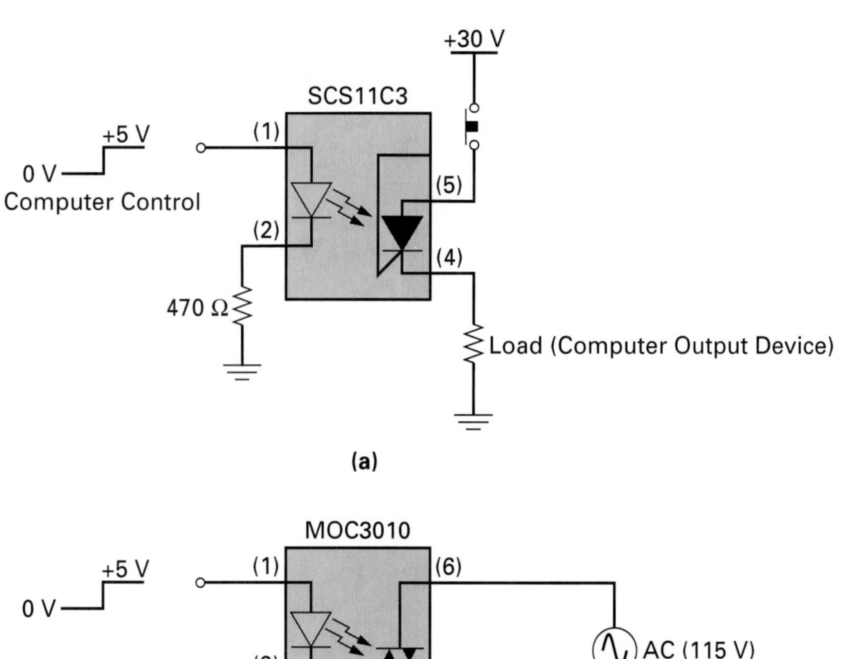

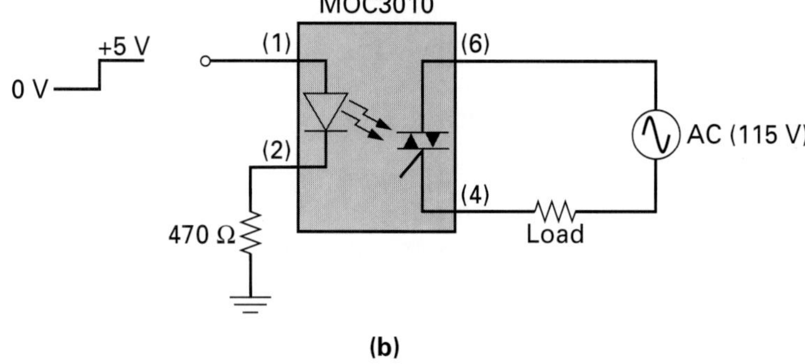

FIGURE 19-18 **Computer Output Device Control.**

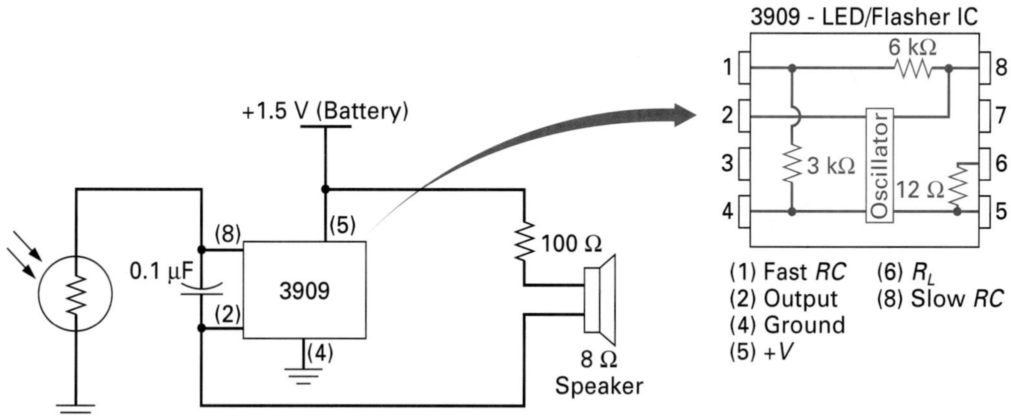

FIGURE 19-19 **Light-Controlled Tone Generator Circuit.**

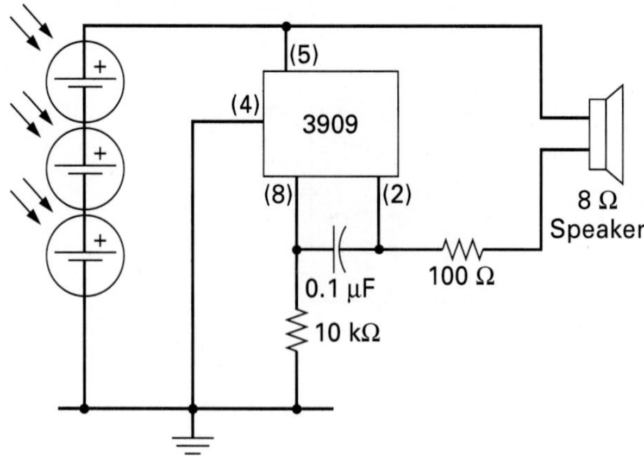

FIGURE 19-20 Solar-Powered Tone Generator.

48. What is the function of this circuit?
49. Describe the operation of the circuit.

To practice your circuit recognition and operation ability, refer to the circuit in Figure 19-22, and answer the following questions.

50. Identify the light-sensitive device used in this circuit.
51. What is the function of this circuit?
52. Briefly describe the function of the 741 op-amp in this circuit.
53. Why is a variable resistor connected to the 741 op-amp?
54. What is the normal voltage drop across an LED?
55. Describe how the bargraph display in this circuit operates.

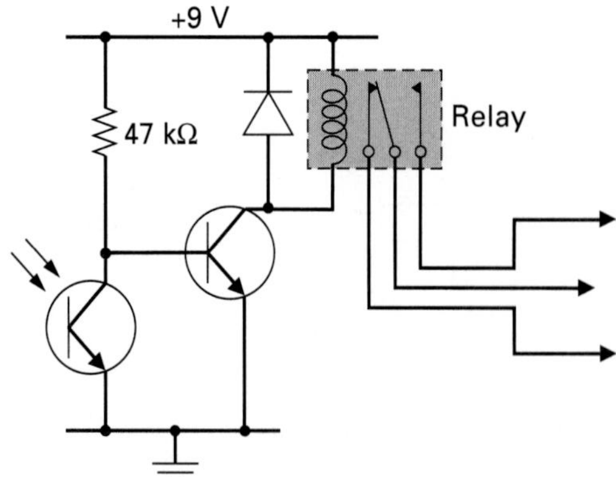

FIGURE 19-21 Optical ON/OFF Relay Control Circuit.

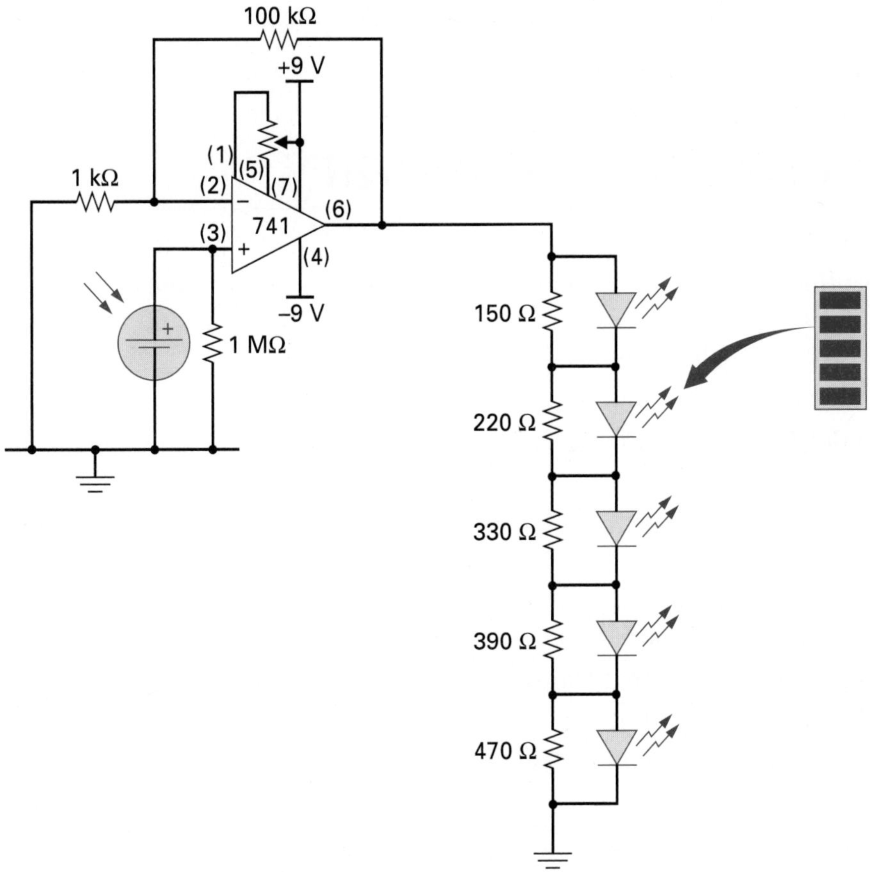

FIGURE 19-22 Lightmeter with Bargraph Display.

APPENDIX

A

Electronics Dictionary

Absorption Loss or dissipation of energy as it travels through a medium. For example, radio waves lose some of their electromagnetic energy as they travel through the atmosphere.

AC Abbreviation for "alternating current."

AC beta (β_{AC}) The ratio of a transistor's ac output current to input current.

Accelerate To go faster.

AC coupling Circuit or component that couples or passes the ac signal yet blocks any dc level.

AC/DC If indicated on a piece of equipment, it means that the equipment will operate from either an ac or dc supply.

AC generator Device that transforms or converts a mechanical input into an ac electrical power output.

Acoustic Relating to sound or the science of sound.

AC power supply Power supply that delivers one or more sources of ac voltage.

Activate To put ac voltage to work or make active. The application of an enabling signal.

Active band-pass filter A circuit that uses an amplifier with passive filter elements to pass only a band of input frequencies.

Active band-stop filter A circuit that uses an amplifier with passive filter elements to block a band of input frequencies.

Active component Component that amplifies a signal or achieves some sort of gain between input and output.

Active equipment Equipment that will transmit and receive, as opposed to passive equipment, which only receives.

Active filter A filter that uses an amplifier with passive filter elements to provide pass or rejection characteristics.

Active high-pass filter A circuit that uses an amplifier with passive filter elements to pass all frequencies above a cut-off frequency.

Active low-pass filter A circuit that uses an amplifier with passive filter elements to pass all frequencies below a cut-off frequency.

AC voltage Alternating voltage.

ADC Abbreviation for "analog-to-digital converter."

Adjustable resistor Resistor whose value can be changed.

Admittance (symbolized *Y*) Measure of how easily ac will flow through a circuit. It is equal to the reciprocal of impedance and is measured in siemens.

Aerial Another term used for antenna.

AF Abbreviation for "audio frequency."

AFC Abbreviation for "automatic frequency control."

AGC Abbreviation for "automatic gain control."

Air-core inductor An inductor that has no metal core.

Aligning tool Small, nonconductive, nonmagnetic screwdriver used to adjust receiver or transmitter tuned circuits.

Alkaline cell Also called an alkaline manganese cell, it is a primary cell that delivers more current than the carbon–zinc cell.

Alligator clip Spring clip normally found on the end of test leads and used to make temporary connections.

Alphanumeric Having numerals, symbols, and letters of the alphabet.

Alpha wave Also called alpha rhythm, it is a brain wave between 9 and 14 Hz.

Alternating current Electric current that rises from zero to a maximum in one direction, falls to zero, and then rises to a maximum in the opposite direction, and then repeats another cycle. The positive and negative alternations being equal.

Alternation One ac cycle consists of a positive and negative alternation.

Alternator Another term used to describe an ac generator.

AM Abbreviation for "amplitude modulation."

Ambient temperature Temperature of the air surrounding the components.

Ammeter Meter placed in the circuit path to measure the amount of current flow.

Ampere (A) Unit of electric current.

Ampere-hour If you multiply the amount of current, in amperes, that can be supplied for a given time frame in hours, a value of ampere-hours is obtained.

Ampere-hour meter Meter that measures the amount of current drawn in a unit of time (hours).

Ampere-turn Unit of magnetomotive force.

Ampere-turn per meter Base unit of magnetic field strength.

Amplification Process of making bigger or increasing the voltage, current, thus increasing the power of a signal.

Amplifier Circuit or device that achieves amplification.

Amplitude Magnitude or size an alternation varies from zero.

Amplitude distortion Changing of a wave shape so that it does not match its original form.

Analog Relating to devices or circuits in which the output varies in direct proportion to the input.

Analog data Information that is represented in a continuously varying form, as opposed to digital data, which have two distinct and discrete values.

Analog display Display in which a moving pointer or some other analog device is used to display the data.

Analog electrical circuits A power circuit designed to control power in a continuously varying form.

Analog electronic circuits Electronic circuits that represent information in a continuously varying form.

Anode Positive electrode or terminal.

Antenna Device that converts an electrical wave into an electromagnetic wave that radiates away from the antenna.

Antenna transmission line System of conductors connecting the transmitter or receiver to the antenna.

Apparent power The power value obtained in an ac circuit by multiplying together effective values of voltage and current, which reach their peaks at different times.

Arc Discharge of electricity through a gas. For example, lightning is a discharge of static electricity buildup through air.

Armature Rotating or moving component of a magnetic circuit.

Armstrong oscillator A tuned transformer oscillator developed by E. H. Armstrong.

Atom Smallest particle of an element.

Atomic number Number of positive charges or protons in the nucleus of an atom.

Atomic weight The relative weight of a neutral atom of an element, based on a neutral oxygen atom having an atomic weight of 16.

Attenuate To reduce in amplitude an action or signal.

Attenuation Loss or decrease in energy of a signal.

Audio Relating to all the frequencies that can be heard by the human ear; from the Latin word meaning "I hear."

Audio frequency Frequency that can be detected by a human ear. From approximately 20 Hz to 20 kHz.

Audio-frequency amplifier An amplifier that has one or more transistor stages designed to amplify audio-frequency signals.

Audio-frequency generator A signal generator that can be set to generate a sinusoidal AF signal voltage at any desired frequency in the audio spectrum.

Autotransformer Single-winding tapped transformer.

Average value Mean value found when the area of a wave above a line is equal to the area of the wave below the line.

Avionics Field of aviation electronics.

AWG Abbreviation for "American wire gauge."

Axial lead Component that has its connecting leads extending from either end of its body.

Balanced bridge Condition that occurs when a bridge is adjusted to produce a zero output.

Ballast resistor A resistor that increases in resistance when current increases. It can therefore maintain a constant current despite variations in line voltage.

Bandpass filter Filter circuit that passes a group or band of frequencies between a lower and an upper cutoff frequency, while heavily attenuating any other frequency outside this band.

Band-stop filter A filter that attenuates alternating currents whose frequencies are between given upper and lower cutoff values while passing frequencies above and below this band.

Bandwidth Width of the group or band of frequencies between the half-power points.

Bar graph display A left-to-right or low-to-high set of bars that are turned on in successive order based on the magnitude they are meant to represent.

Barometer Meter used to measure atmospheric pressure.

Barretter Temperature-sensitive device having a positive temperature coefficient; that is, as temperature increases, resistance increases.

Barrier potential or voltage The potential difference, or voltage, that exists across the junction.

Base The region that lies between an emitter and a collector of a transistor and into which minority carriers are injected.

Base biasing A transistor biasing method in which the dc supply voltage is applied to the base of the transistor via a base bias resistor.

Bass Low audio frequencies normally handled by a woofer in a sound system.

Battery DC voltage source containing two or more cells that convert chemical energy into electrical energy.

Battery charger Piece of electrical equipment that converts ac input power to a pulsating dc output, which is then used to charge a battery.

Baud Unit of signaling speed describing the number of elements per second.

***B-H* curve** Curve plotted on a graph to show successive states during magnetization of a ferromagnetic material.

Bias voltage The dc voltage applied to a semiconductor device to control its operation.

Bidirectional device A device that will conduct current in either direction.

Binary Number system having only two levels and used in digital electronics.

Binary word A numerical value expressed as a group of binary digits.

Bipolar family A group of digital logic circuits that make use of the bipolar junction transistor.

Bipolar transistor A bipolar junction transistor in which excess minority carriers are injected from an emitter region into a base region and from there pass into a collector region.

Bits An abbreviation for "binary digits."

Bleeder current Current drawn continuously from a voltage source. A bleeder resistor is generally added to lessen the effect of load changes or provide a voltage drop across a resistor.

Body resistance The resistance of the human body.

Bolometer Device whose resistance changes when heated.

Branch current A portion of the total current that is present in one path of a parallel circuit.

Breakdown voltage The voltage at which breakdown of a dielectric or insulator occurs.

Bridge rectifier circuit A full-wave rectifier circuit using four diodes that will convert an alternating voltage input into a direct voltage output.

BW Abbreviation for "bandwidth."

Bypass capacitor Capacitor that is connected to provide a low-impedance path for ac.

Byte Group of 8 binary digits or bits.

Cable Group of two or more insulated wires.

CAD Abbreviation for "computer-aided design."

Calculator Device that can come in a pocket (battery-operated) or desktop (ac power) size and is used to achieve arithmetic operations.

Calibration To determine by measurement or comparison with a standard the correct value of each scale reading.

Capacitance (*C*) Measured in farads, it is the ability of a capacitor to store an electrical charge.

Capacitance meter Instrument used to measure the capacitance of a capacitor or circuit.

Capacitive filter A capacitor used in a power supply filter system to supress ripple currents while not affecting direct currents.

Capacitive reactance (X_C) Measured in ohms, it is the ability of a capacitor to oppose current flow without the dissipation of energy.

Capacitor Device that stores electric energy in the form of an electric field that exists within a dielectric (insulator) between two conducting plates, each connected to a connecting lead. This device was originally called a condenser.

Capacitor-input filter Filter in which a capacitor is connected to shunt away low-frequency signals such as ripple from a power supply.

Capacitor microphone Microphone that contains a stationary and movable plate separated by air. The arriving sound waves cause movements in the movable plate, which affects the distance between the plates and therefore the capacitance, generating a varying AF electrical wave. Also called an electrostatic microphone.

Capacitor speaker Also called the electrostatic speaker, its operation depends on mechanical forces generated by an electrostatic field.

Capacity Output current capabilities of a device or circuit.

Capture The act of gaining control of a signal.

Carbon-film resistor Thin carbon film deposited on a ceramic form to create resistance.

Carbon microphone Microphone whose operation depends on the pressure variation in the carbon granules changing their resistance.

Carbon resistor Fixed resistor consisting of carbon particles mixed with a binder, which is molded and then baked. Also called a composition resistor.

Cardiac pacemaker Device used to control the frequency or rhythm of the heart by stimulating it through electrodes.

Carrier Wave that has one of its characteristics varied by an information signal. The carrier will then transport this information between two points.

Cascaded amplifier An amplifier that contains two or more stages arranged in a series manner.

Cassette Thin, flat, rectangular device containing a length of magnetic tape that can be recorded onto or played back.

Cathode Term used to describe a negative electrode or terminal.

Cathode ray tube A vacuum tube in which electrons emitted by a hot cathode are focussed into a narrow beam by an electron gun and then applied to a fluorescent screen. The beam can be varied in intensity and position to produce a pattern or picture on the screen.

Cell Single unit having only two plates that convert chemical energy into a dc electrical voltage.

Celsius-temperature scale Scale that defines the freezing point of water as 0°C and the boiling point as 100°C (named after Anders Celsius).

Center tap Midway connection between the two ends of a winding or resistor.

Center-tapped rectifier Rectifier that makes use of a center-tapped transformer and two diodes to achieve full-wave rectification.

Center-tapped transformer A transformer that has a connection at the electrical center of the secondary winding.

Ceramic capacitor Capacitor in which the dielectric material used between the plates is ceramic.

Cermet Ceramic-metal mixture used in film resistors.

Channel A path for a signal.

Charge Quantity of electrical energy stored in a battery or capacitor.

Charging current Current that flows to charge a capacitor or battery when a voltage is applied.

Chassis Metal box or frame in which components, boards, and units are mounted.

Chassis ground Connection to the metal box or frame that houses the components and associated circuitry.

Choke Inductor used to impede the flow of alternating or pulsating current.

Circuit Interconnection of many components to provide an electrical path between two or more points.

Circuit breaker Reusable fuse. This device will open a current-carrying path without damaging itself once the current value exceeds its maximum current rating.

Circuit failure A malfunction or breakdown of a component in the circuit.

Circular mil (cmil) A unit of area equal to the area of a circle whose diameter is one mil.

Clapp oscillator A series tuned colpitts oscillator circuit.

Class A amplifier An amplifier in which the transistor is in its active region for the entire signal cycle.

Class AB amplifier An amplifier in which the transistor is in its active region for slightly more than half the signal cycle.

Class B amplifier An amplifier in which the transistor is in its active region for approximately half the signal cycle.

Class C amplifier An amplifier in which the transistor is in its active region for less than half the signal cycle.

Clock Generally, a square waveform used for the synchronization and timing of several circuits.

Clock input (C) An accurate timing signal used for synchronizing the operation of digital circuits.

Closed circuit Circuit having a complete path for current to flow.

Closed-loop mode A control system containing one or more feedback control loops, in which functions of the controlled signals are combined with functions of the commands that tend to maintain prescribed relationships between the commands and the controlled signals.

Closed-looped voltage gain (A_{CL}) The voltage gain of an amplifier when it is operated in the closed loop mode.

Coaxial cable Transmission line in which a center signal-carrying conductor is completely covered by a solid dielectric, another conductor, and then an outer insulating sleeve.

Coefficient This is related to the ratio of change under certain conditions.

Coefficient of coupling The degree of coupling that exists between two circuits.

Coercive force (H) Magnetizing force needed to reduce the residual magnetism within a material to zero.

Coil Number of turns of wire wound around a core to produce magnetic flux (an electromagnet) or to react to a changing magnetic flux (an inductor).

Cold resistance The resistance of a device when cold.

Collector A semiconductor region through which a flow of charge carriers leaves the base of the transistor.

Collector characteristic curve A set of characteristic curves of collector voltage versus collector current for a fixed value of transistor base current.

Color code Set of colors used to indicate a component's value, tolerance, and rating.

Colpitts oscillator An *LC* tuned oscillator circuit in which two tank capacitors are used instead of a tapped coil.

Common Shared by two or more services, circuits, or devices. Although the term "common ground" is frequently used to describe two or more connections sharing a common ground, the term "common" alone does not indicate a ground connection, only a shared connection.

Common base A transistor configuration in which the base is common to both the input and output signal.

Common collector A transistor configuration in which the collector is common to both the input and output signal.

Common drain An FET configuration in which the drain is grounded and common to the input and output signal.

Common emitter A transistor configuration in which the emitter is common to both the input and output signal.

Common gate An FET configuration in which the gate is grounded and common to the input and output signal.

Common-mode input An input signal applied equally to both ungrounded inputs of a balanced amplifier. Also called in-phase input.

Common mode rejection The ability of a device to reject a voltage signal applied simultaneously to both input terminals.

Common-mode rejection ratio Abbreviated CMRR, it is the ratio of an operational amplifier's differential gain to common-mode gain.

Common source An FET configuration in which the source is grounded and common to the input and output signal.

Communication Transmission of information between two points.

Comparator An operational amplifier used without feedback to detect changes in voltage level.

Complementary DC amplifier An amplifier in which NPN and PNP transistors are used in an alternating sequence.

Complementary latch circuit A circuit containing a NPN and PNP transistor that once triggered ON will remain latched ON.

Complex numbers Numbers composed of a real number part and an imaginary number part.

Complex plane A plane whose points are identified by means of complex numbers.

Component (1) Device or part in a circuit or piece of equipment. (2) In vector diagrams, it can mean a part of a wave, voltage, or current.

Compound A material composed of united separate elements.

Computer Piece of equipment used to process information or data.

Conductance (*G*) Measure of how well a circuit or path conducts or passes current. It is equal to the reciprocal of resistance.

Conduction Use of a material to pass electricity or heat.

Conductivity Reciprocal of resistivity; it is the ability of a material to conduct current.

Conductors Materials that have a very low resistance and pass current easily.

Connection When two or more devices are attached together so that a path exists between them, there is said to be a connection between the two.

Connector Conductive device that makes a connection between two points.

Constant Fixed value.

Constant current Current that remains at a fixed, unvarying level despite variations in load resistance.

Contact Current-carrying part of a switch, relay, or connector.

Continuity Occurs when a complete path for electric current exists.

Continuity test Resistance test to determine whether a path for electric current exists.

Contrast The difference between the light and dark areas in a video picture. High contrast pictures have dark blacks and brilliant whites. Low contrast pictures have an overall gray appearance.

Control voltage A voltage signal that starts, stops, or adjusts the operation of a device, circuit, or system.

Conventional current flow A current produced by the movement of positive charges toward a negative terminal.

Copper loss Also I^2R loss, it is the power lost in transformers, generators, connecting wires, and other parts of a circuit because of the current flow (I) through the resistance (R) of the conductors.

Core Magnetic material within a coil used to intensify or concentrate the magnetic field.

Coulomb Unit of electric charge. One coulomb equals 6.24×10^{18} electrons.

Counter electromotive force Abbreviated "counter emf," or sometimes called back emf, it is the voltage generated in an inductor due to an alternating or pulsating current and is always of opposite polarity to that of the applied voltage.

Covalent bond A pair of electrons shared by two neighboring atoms.

Crossover distortion Distortion of a signal at the zero crossover point.

CRT Abbreviation for "cathode-ray tube."

Crystal A natural or synthetic piezoelectric or semiconductor material whose atoms are arranged with some degree of geometric regularity.

Current (*I*) Measured in amperes or amps, it is the flow of electrons through a conductor.

Current amplifier An amplifier designed to build up the current of a signal.

Current clamp A device used in conjunction with an ac ammeter, containing a magnetic core in the form of hinged jaws that can be snapped around the current-carrying wire.

Current divider A parallel network designed to proportionally divide the circuit's total current.

Current-limiting resistor Resistor that is inserted into the path of current flow to limit the amount of current to some desired level.

Current sink A circuit or device that absorbs current from a load circuit.

Current source A circuit or device that supplies current to a load circuit.

Current-to-voltage converter An amplifier circuit used for converting a low dc current into a proportional low dc voltage.

Cutoff The minimum value of bias voltage that stops output current in a transistor.

Cutoff frequency Frequency at which the gain of the circuit falls below 0.707 of the maximum current or half-power (–3 dB).

Cycle When a repeating wave rises from zero to some positive value, back to zero, and then to a maximum negative value before returning back to zero, it is said to have completed one cycle. The number of cycles occurring in one second is the frequency measured in hertz (cycles per second).

DAC Abbreviation for "digital-to-analog converter."

Damping Reduction in the magnitude of oscillation due to absorption.

Darlington pair A current amplifier consisting of two separate transistors with their collectors connected together and the emitter of one connected to the base of the other.

D'Arsonval movement When a direct current is passed through a small, lightweight, movable coil, the magnetic field produced interacts with a fixed permanent magnetic field that rotates while moving an attached pointer positioned over a scale.

DC Abbreviation for "direct current."

DC alpha Term for the dc current amplification factor of a transistor.

DC block Component used to prevent the passage of direct current, normally a capacitor.

DC generator Device used to convert a mechanical energy input into a dc electrical output.

DC offsets The change in input voltage that is required to produce an output voltage of zero when no input signal is present.

DC power supply A power line, generator, battery, power pack, or other dc source of power for electrical equipment.

Dead short Short circuit having almost no resistance.

Decibel (dB) One tenth of a bel (1×10^{-1} bel). The logarithmic unit for the difference in power at the output of an amplifier compared to the input.

Degenerative (negative) feedback Feedback in which a portion of the output signal is fed back 180° out of phase with the input signal. Also called degenerative feedback, inverse feedback, or stabilized feedback.

Depletion region A small layer on either side of the junction that becomes empty, or depleted, of free electrons or holes.

Depletion-type MOSFET A field-effect transistor with an insulated gate (MOSFET) that can be operated in either the depletion or enhancement mode.

Design engineer Engineer responsible for the design of a product for a specific application.

Desoldering The process of removing solder from a connection.

Device Component or part.

DIAC A bidirectional diode that has a symmetrical switching mode.

Diagnostic Related to the detection and isolation of a problem or malfunction.

Dielectric Insulating material between two (*di*) plates in which the *electric* field exists.

Dielectric constant (*K*) The property of a material that determines how much electrostatic energy can be stored per unit volume when unit voltage is applied. Also called permittivity.

Dielectric strength The maximum potential a material can withstand without rupture.

Difference amplifier An amplifier whose output is proportional to the difference between the voltages applied to the two inputs.

Differential amplifier An amplifier whose output is proportional to the difference between the voltages applied to its two inputs.

Differential-mode input An input signal applied between the two ungrounded terminals of a balanced three-terminal system.

Differentiator A circuit whose output voltage is proportional to the rate of change of the input voltage. The output waveform is then the time derivative of the input waveform, and the phase of the output waveform leads that of the input by 90°.

Differentiator amplifier A circuit whose output voltage is proportional to the rate of change of the input voltage. The output waveform is the time derivative of the input waveform.

Diffusion A method of producing a junction by diffusing an impurity metal into a semiconductor at a high temperature.

Diffusion current The current that is present when the depletion layer is expanding.

Digital Relating to devices or circuits that have outputs of only two distinct levels or steps, for example, on–off, 0–1, open–closed, and so on.

Digital computer Piece of electronic equipment used to process digital data.

Digital data Data represented in digital form.

Digital display Display in which either light-emitting diodes (LEDs) or liquid crystal diodes (LCDs) are used and each of the segments can be either turned on or off.

Digital electrical (power) circuit An electrical circuit designed to turn power either ON or OFF (two-state).

Digital electronic circuits Electronic circuits that encode information into a group of pulses consisting of HIGH or LOW voltages.

Digital multimeter Multimeter used to measure amperes, volts, and ohms and indicate the results on a digital readout display.

DIP Abbreviation for "dual in-line package."

Direct coupling The coupling of two circuits by means of a wire.

Direct current (dc) Current flow in only one direction.

Direct-current amplifier An amplifier capable of amplifying dc voltages and slowly varying voltages.

Direct voltage (dc voltage) Voltage that causes electrons to flow in only one direction.

Discharge Release of energy from either a battery or capacitor.

Disconnect Breaking or opening of an electric circuit.

Discrete components Separate active and passive devices that were manufactured before being used in a circuit.

Display Visual presentation of information or data.

Display unit Unit designed to display digital data.

Dissipation Release of electrical energy in the form of heat.

Distortion An undesired change in the waveform of a signal.

Distributed capacitance Also known as self-capacitance, it is any capacitance other than that within a capacitor, for example, the capacitance between the coil of an inductor or between two conductors (lead capacitance).

Distributed inductance Any inductance other than that within an inductor, for example, the inductance of a length of wire (line inductance).

Domain Also known as a magnetic domain, it is a moveable magnetized area in a magnetic material.

Doping The process wherein impurities are added to the intrinsic semiconductor material either to increase the number of free electrons or to increase the number of holes.

Dot convention A standard used with transformer symbols to indicate whether the secondary voltage will be in phase or out of phase with the primary voltage.

Double-pole, double-throw switch (DPDT) Switch having two movable contacts (double pole) that can be thrown or positioned in one of two positions (double throw).

Double-pole, single-throw switch (DPST) Switch having two movable contacts (double pole) that can be thrown in only one position (single throw) to either make a dual connection, or opened to disconnect both poles.

Drain One of the field-effect transistor's electrodes.

Drain-feedback biased A configuration in which the gate receives a bias voltage fed back from the drain.

Dropping resistor Resistor whose value has been chosen to drop or develop a given voltage across it.

Dry Cell DC voltage-generating chemical cell using a nonliquid (paste) type of electrolyte.

Dual-gate d-MOSFET A metal oxide semiconductor FET having two separate gate electrodes.

Dual in-line package Package that has two (dual) sets or lines of connecting pins.

Dual-trace oscilloscope Oscilloscope that can simultaneously display two signals.

Dynamic Related to conditions or parameters that change, i.e. motion.

E-core Laminated form, in the shape of the letter E, onto which inductors and transformers are wound.

Eddy currents Small currents induced in a conducting core due to the variations in alternating magnetic flux.

Efficiency Ratio of output power to input power, normally expressed in percent.

Electrical components Components, circuits, and systems that manage the flow of *power*.

Electrical equipment Equipment designed to manage or control the flow of power.

Electrical wave Travelling wave prorogated in a conductive medium that is a variation in voltage or current and travels at slightly less than the speed of light.

Electric charge Electric energy stored on the surface of a material.

Electric current Electron movement or motion.

Electric field Also called a voltage field, it is a field or force that exists in the space between two different potentials or voltages.

Electrician Person involved with the design, repair, and assembly of electrical equipment.

Electricity Science that states that certain particles possess a force field, which with electrons is negative and with protons is positive. Electricity can be divided into two groups: static and dynamic. Static electricity deals with charges at rest, while dynamic electricity deals with charges in motion.

Electric motor Device that converts an electrical energy input into a mechanical energy output.

Electric polarization A displacement of bound charges in a dielectric when placed in an electric field.

Electroacoustic transducer Device that achieves an energy transfer from electric to acoustic (sound), and vice versa. Examples include a microphone and a loudspeaker.

Electroluminescence Transmission or conversion of electrical energy into light energy.

Electrolyte Electrically conducting liquid (wet) or paste (dry).

Electrolytic capacitor Capacitor having an electrolyte between the two plates; due to chemical action, a very thin layer of oxide is deposited on only the positive plate, which accounts for why this type of capacitor is polarized.

Electromagnet A magnet consisting of a coil wound on a soft iron or steel core. When current is passed through the coil a magnetic field is generated and the core is strongly magnetized to concentrate the magnetic field.

Electromagnetic communication Use of an electromagnetic wave to pass information between two points. Also called *wireless* communication.

Electromagnetic energy Radiant electromagnetic energy, such as radio and light waves.

Electromagnetic field Field having both an electric (voltage) and magnetic (current) field.

Electromagnetic induction The voltage produced in a coil due to relative motion between the coil and magnetic lines of force.

Electromagnetic spectrum List or diagram showing the entire range of electromagnetic radiation.

Electromagnetic wave Wave that consists of both an electric and magnetic variation.

Electromagnetism Relates to the magnetic field generated around a conductor when current is passed through it.

Electromechanical transducer Device that transforms electrical energy into mechanical energy (motor), and vice versa (generator).

Electromotive force Force that causes the motion of electrons due to a potential difference between two points.

Electron Smallest subatomic particle of negative charge that orbits the nucleus of the atom.

Electron flow A current produced by the movement of free electrons toward a positive terminal.

Electron-hole pair When an electron jumps from the valence shell or band to the conduction band, it leaves a gap in the covalent bond called a hole. This action creates an electron-hole pair.

Electronic components Components, circuits, and systems that manage the flow of *information*.

Electronic equipment Equipment designed to control the flow of information.

Electronics Science related to the behavior of electrons in devices.

Electron-pair bond or covalent bond A pair of electrons shared by two neighboring atoms.

Electrostatic Related to static electric charge.

Electrostatic field Force field produced by static electrical charges.

Element There are 107 different natural chemical substances, or elements, that exist on earth. These can be categorized as gas, solid, or liquid.

Emitter A transistor region from which charge carriers are injected into the base.

Emitter feedback The coupling from the emitter output to the base input in a transistor amplifier.

Emitter follower A grounded-collector transistor amplifier whose operation provides current gain and impedance matching.

Emitter modulator A modulating circuit in which the modulating signal is applied to the emitter of a bipolar transistor.

Encoder circuit A circuit that produces different output voltage codes, depending on the position of a rotary switch.

Energized Being electrically connected to a voltage source so that the device is activated.

Energy Capacity to do work.

Energy gap The space between two orbital shells.

Engineer Person who designs and develops materials to achieve desired results.

Engineering notation A floating-point system in which numbers are expressed as products consisting of a number that is greater than one multiplied by an appropriate power of ten that is some multiple of three.

Enhancement-mode MOSFET A field-effect transistor in which there are no charge carriers in the channel when the gate-source voltage is zero.

Equipment Term used to describe electrical or electronic units.

Equivalent resistance (R_{eq}) Total resistance of all the individual resistances in a circuit.

Error voltage A voltage that is proportional to the error that exists between input and output.

Excited state An energy level in which a nucleus may exist if given sufficient energy to reach this state from a lower state.

Extrinsic semiconductor A semiconductor whose electrical properties are dependent on impurities added to the semiconductor crystal.

Facsimile Electronic process whereby pictures or images are scanned and the graphical information is converted into electrical signals that can be reproduced locally or transmitted to a remote point, where a likeness or facsimile of the original can be produced.

Fahrenheit temperature scale Temperature scale that indicates the freezing point of water at 32°F and boiling point at 212°F.

Fall time Time it takes a negative edge of a pulse to fall from 90% to 10% of its peak value

Farad (F) Unit of capacitance.

Faraday's law 1. When a magnetic field cuts a conductor, or when a conductor cuts a magnetic field, an electric current will flow in the conductor if a closed path is provided over which the current can circulate. 2. Two other laws related to electrolytic cells.

FCC Abbreviation for "Federal Communications Commission."

Ferrite A powdered, compressed, and sintered magnetic material having high resistivity. The high resistance makes eddy-current losses low at high frequencies.

Ferrite bead Ferrite composition in the form of a bead.

Ferrite core Ferrite core normally shaped like a doughnut.

Ferrite-core inductor An inductor containing a ferrite core.

Ferrites Compound composed of iron oxide, a metallic oxide, and ceramic. The metal oxides include zinc, nickel, cobalt, or iron.

Ferrous Composed of and/or containing iron. A ferrous metal exhibits magnetic characteristics, as opposed to nonferrous metals.

Fiber optics Laser's light output carries information that is conveyed between two points by thin glass optical fibers.

Field-effect transistor Abbreviated FET, it is a transistor in which the resistance of the source to drain current path is changed by applying a transverse electric field to the gate.

Field strength The strength of an electric, magnetic, or electromagnetic field at a given point.

Filament Thin thread of, for example, carbon or tungsten, which when heated by the passage of electric current will emit light.

Filament resistor The resistor in a light bulb or electron tube.

Filter Network composed of resistor, capacitor, and inductors used to pass certain frequencies yet block others through heavy attenuation.

Fission Atomic or nuclear fission is the process of splitting the nucleus of heavy elements such as uranium and plutonium into two parts, which results in large releases of radioactivity and heat (fission or division of a nuclei).

Fixed biasing A constant value of bias for an FET in which the voltage is independent of the input signal strength.

Fixed component Component whose value or characteristics cannot be varied or changed.

Fixed-value capacitor A capacitor whose value is fixed and cannot be varied.

Fixed-value resistor A resistor whose value cannot be changed.

Floating ground Ground potential that is not tied or in reference to earth.

Flow-soldering Flow or wave soldering technique is used in large-scale electronic assembly to solder all the connections on a printed circuit board by moving the board over a wave or flowing bath of molten solder.

Fluorescent lamp Gas-filled glass tube that when bombarded by the flow of electric current causes the gas to ionize and then release light.

Flux A material used to remove oxide films from the surfaces of metals in preparation for soldering.

Flux density A measure of the strength of a wave.

Flywheel effect Sustaining effect of oscillation in an *LC* circuit due to the charging and discharging of the capacitor and the expansion and contraction of the magnetic field around the inductor.

Force Physical action capable of moving a body or modifying its movement.

Forward biased A small depletion region at the junction will offer a small resistance and permit a large current. Such a junction is forward biased.

Forward voltage drop (V_F) The forward voltage drop is equal to the junction's barrier voltage.

Free electrons Electrons that are not in any orbit around a nucleus.

Free-running Operating without any external control.

Frequency Rate or recurrences of a periodic wave normally within a unit of one second, measured in hertz (cycles/second).

Frequency drift A slow change in the frequency of a circuit due to temperature or frequency determining component value changes.

Frequency-division multiplex (FDM) Transmission of two or more signals over a common path by using a different frequency band for each signal.

Frequency-domain analysis A method of representing a waveform by plotting its amplitude versus frequency.

Frequency meter Meter used to measure the frequency or cycles per second of periodic waves.

Frequency multiplier A harmonic conversion circuit in which the frequency of the output signal is an exact multiple of the input signal.

Frequency response Indication of how well a device or circuit responds to the different frequencies applied to it.

Frequency response curve A graph indicating a circuit's response to different frequencies.

Friction The rubbing or resistance to relative motion between two bodies in contact.

Frictional electricity Generation of electric charges by rubbing one material against another.

Full-scale deflection (FSD) Deflection of a meter's pointer to the farthest position on the scale.

Full-wave center-tapped rectifier A rectifier circuit that make use of a center-tapped transformer to cause an output current to flow in the same direction during both half-cycles of the ac input.

Full-wave rectifier Rectifier that makes use of the full ac wave (in both the positive and negative cycle) when converting ac to dc.

Full-wave voltage-doubler circuit A rectifier circuit that doubles the output voltage by charging capacitors during both alternations of the ac input—making use of the full ac input wave.

Function generator Signal generator that can function as a sine, square, rectangular, triangular, or sawtooth waveform generator.

Fundamental frequency This sine wave is always the lowest frequency and largest amplitude component of any waveform shape and is used as a reference.

Fuse This circuit- or equipment-protecting device consists of a short, thin piece of wire that melts and breaks the current path if the current exceeds a rated damaging level.

Fuse holder Housing used to support a fuse with two connections.

Fusion Atomic or nuclear fusion is the process of melting the nuclei of two light atoms together to create a heavier nucleus, which results in a large release of energy (fusion or combining of nuclei).

Gain Increase in power from one point to another. Normally expressed in decibels.

Gamma rays High-frequency electromagnetic radiation from radioactive particles.

Ganged Mechanical coupling of two or more capacitors, switches, potentiometers, or any other components so that the activation of one control will operate all.

Gas Any aeriform or completely elastic fluid, which is not a solid or liquid. All gases are produced by the heating of a liquid beyond its boiling point.

Gate One of the field effect transistor's electrodes (also used for thyristor devices).

Geiger counter Device used to detect nuclear particles.

Generator Device used to convert a mechanical energy input into an electrical energy output.

Giga Prefix for 1 billion (10^9).

Graph origin Center of the graph where the horizontal axis and vertical axis cross.

Greenwich mean time (GMT) Also known as universal time, it is a standard based on the earth's rotation with respect to the sun's position. The solar time at the meridian of Greenwich, England, which is at zero longitude.

Ground An intentional or accidental conducting path between an electrical circuit or system and the earth, or some conducting body acting in the place of the earth.

Gunn diode A semiconductor diode that utilizes the Gunn effect to produce microwave frequency oscillation or to amplify a microwave-frequency signal.

Half-power point A point at which power is 50%. This corresponds to 70.7% of the total current.

Half-split method A troubleshooting technique used to isolate a faulty block in a circuit or system. In this method a mid-point is chosen and tested to determine which half is malfunctioning.

Half-wave rectifier circuit A circuit that converts ac to dc by only allowing current to flow during one-half of the ac input cycle.

Hall effect sensor A sensor that generates a voltage in response to a magnetic field.

Hardware Electrical, electronic, mechanical, or magnetic devices or components. The physical equipment.

Hartley oscillator An *LC* tuned oscillator circuit in which the tank coil has an intermediate tap.

Harmonic Sine wave that is smaller in amplitude and some multiple of the fundamental frequency.

Henry (H) Unit of inductance.

Hertz (Hz) Unit of frequency. One hertz is equal to one cycle per second.

Hexadecimal number system A base 16 number system, using the digits 0 through 9 and the six additional digits represented by A, B, C, D, E, and F.

High fidelity (hi-fi) Sound reproduction equipment that is as near to the original sound as possible.

High-pass filter Network or circuit designed to pass any frequencies above a critical or cutoff frequency and reject or heavily attenuate all frequencies below.

High *Q* Abbreviation for quality and generally related to inductors that have a high value of inductance and very little coil resistance.

High tension Lethal voltage in the kilovolt range and above.

High-voltage probe Accessory to the voltmeter that has added multiplier resistors within the probe to divide up the large potential being measured by the probe.

***H*-lines** Invisible lines of magnetic flux.

Hole A mobile vacancy in the valance structure of a semiconductor. This hole exists when an atom has less than its normal number of electrons, and is equivalent to a positive charge.

Hole flow Conduction in a semiconductor when electrons move into holes when a voltage is applied.

Hologram Three-dimensional picture created with a laser.

Holography Science dealing with three-dimensional optical recording.

Horizontal Parallel to the horizon or perpendicular to the force of gravity.

Horizontally polarized wave Electromagnetic wave that has the electric field lying in the horizontal plane.

Hot resistance The resistance of a device when hot due to the generation of heat by electric current.

Hybrid circuit Circuit that combines two technologies (passive and active or discrete and integrated components) onto one microelectronic circuit. Passive components are generally made by thin-film techniques, while active components are made utilizing semiconductor techniques. Integrated circuits can be mounted on the microelectronic circuit and connected to discrete components also on the small postage-stamp-size boards.

Hydroelectric Generation of electric power by the use of water in motion.

Hysteresis Amount that the magnetization of a material lags the magnetizing force due to molecular friction.

IC Abbreviation for "integrated circuit."

Imaginary number A complex number whose imaginary parts are not zero.

IMPATT diode A semiconductor diode that has a negative resistance characteristic produced by a combination of impact avalanche breakdown and charge carrier transmit time effects in a thin semiconductor chip.

Impedance (*Z*) Measured in ohms, it is the total opposition a circuit offers to current flow (reactive and resistive).

Impedance coupling The coupling of two signal amplifier circuits through the use of an impedance, such as a choke.

Impedance matching Matching of the source impedance to the load impedance causes maximum power to be transferred.

Incandescence State of a material when it is heated to such a high temperature that it emits light.

Incandescent lamp An electric lamp that generates light when an electric current is passed through its filament of resistance, causing it to heat to incandescence.

Induced current Current that flows due to an induced voltage.

Induced voltage Voltage generated in a conductor when it is moved through a magnetic field.

Inductance Property of a circuit or component to oppose any change in current as the magnetic field produced by the change in current causes an induced countercurrent to oppose the original change.

Inductive circuit Circuit that has a greater inductive reactance figure than capacitive reactance figure.

Inductive reactance (X_L) Measured in ohms, it is the opposition to alternating or pulsating current flow without the dissipation of energy.

Inductor Length of conductor used to introduce inductance into a circuit.

Inductor analyzer A test instrument designed to test inductors.

Infinite Having no limits.

Infinity Amount larger than any number can indicate.

Information Data or meaningful signals.

Infrared Electromagnetic heat radiation whose frequencies are above the microwave frequency band and below red in the visible band.

Ingot A mass of metal cast into a convenient shape.

Inhibit To stop an action or block data from passing.

Injection laser diode or ILD A semiconductor P-N junction diode that uses a lasing action to increase and concentrate the light output.

In phase Two or more waves of the same frequency whose maximum positive and negative peaks occur at the same time.

Input impedance Impedance seen when looking into the input terminals of a device.

Insulated When a nonconductive material is used to isolate conducting materials from one another.

Insulating material Material that will in nearly all cases prevent the flow of current due to its chemical composition.

Insulation resistance Resistance of the insulating material. The greater the insulation resistance, the better the insulator.

Insulator Material that has few electrons per atom and those electrons are close to the nucleus and cannot be easily removed.

Integrated When two or more components are combined (into a circuit) and then incorporated in one package or housing.

Integrator amplifier An amplifier whose output is the integral of its input with respect to time.

Integrator circuit A circuit with an output which is the integral of its input with respect to time.

Intermediate-frequency amplifier An amplifier that has one or more transistor stages designed to amplify intermediate frequency signals.

Intermittent Occurring at random intervals of time. An intermittent component, circuit, or equipment problem is undesirable and difficult to troubleshoot, as the problem needs to occur before isolation of the fault can begin.

Internal resistance No voltage source is 100% efficient in that not all the energy in is converted to electrical energy out; some is wasted in the form of heat dissipation. The internal series resistance of a voltage source represents this inefficiency.

Intrinsic semiconductor materials Pure semiconductor materials.

Inverting amplifier An operational amplifier in which the inverting signal or negative input is held virtually at ground by negative feedback and the output is inverted with respect to the input.

Inverting input The inverting or negative input of an operational amplifier.

Ion Atom that has an equal number of protons (+charge) and electrons (−charge) is considered a neutral atom. If more electrons exist in orbit around the atom it is considered a negative ion, whereas if less electrons are in orbit around the atom it is referred to as a positive ion.

Jack Socket or connector into which a plug may be inserted.

JFET A field-effect transistor made up of a gate region diffused into a channel region. When a control voltage is applied to the gate, the channel is depleted or enhanced, and the current between source and drain is thereby controlled.

***j* operator** A prefix used to indicate an imaginary number.

Joule The unit of work and energy.

Junction Contact or connection between two or more wires or cables.

Junction diode A semiconductor diode in which the rectifying characteristics occur at a junction between the *n*-type and *p*-type semiconductor materials.

Kilovolt-ampere 1000 volts at 1 ampere.

Kilowatt-hour 1000 watts for 1 hour.

Kilowatt-hour meter A meter used by electric companies to measure a customer's electric power use in kilowatt-hours.

Kinetic energy Energy associated with motion.

Kirchoff's current law The sum of the currents flowing into a point in a circuit is equal to the sum of the currents flowing out of that same point.

Kirchoff's voltage law The algebraic sum of the voltage drops in a closed-path circuit is equal to the algebraic sum of the source voltage applied.

Knee voltage The voltage at which a curve joins two relatively straight portions of a characteristic curve.

Lag Difference in time between two waves of the same frequency expressed in degrees, i.e. one waveform lags another by a certain number of degrees.

Laminated core Core made up of sheets of magnetic material insulated from one another by an oxide or varnish.

Lamp Device that produces light.

Laser Device that produces a very narrow, intense beam of light. The name is an acronym for "light amplification by stimulated emission of radiation."

Latched A bistable circuit action that causes the circuit to remain held or locked in its last activated state.

***LC* filter** A selective circuit which makes use of an inductance-capacitance network.

Lead The angle by which one alternating signal leads another in time, or a wire that connects two points in a circuit.

Lead-acid cell Cell made up of lead plates immersed in a sulfuric acid electrolyte.

Leakage current or reverse current (I_R) The extremely small current present at the junction.

Least significant bit (LSB) The right-most binary digit.

LED Abbreviation for "light-emitting diode."

Left-hand rule If the fingers of the left hand are placed around the wire so that the thumb points in the direction of the electron flow, the fingers will be pointing in the direction of the magnetic field being produced by the conductor.

Lenz's law The current induced in a circuit due to a change in the magnetic field is so directed as to oppose the change in flux, or to exert a mechanical force opposing the motion.

Lie detector Piece of electronic equipment, also called a polygraph, that determines whether a person is telling the truth by looking for dramatic changes in blood pressure, body temperature, breathing rate, heart rate, and skin moisture in response to certain questions.

Lifetime The time difference between an electron jumping into the conduction band and then falling back into a hole.

Light Electromagnetic radiation in a band of frequencies that can be received by the human eye.

Light-emitting diode A semiconductor diode that converts electric energy into electromagnetic radiation at visible and near infrared frequencies when its P-N junction is forward biased.

Light-emitting semiconductor devices Semiconductor devices that will emit light when an electrical signal is applied.

Limiter Circuit or device that prevents some portion of its input from reaching the output.

Linear Relationship between input and output in which the output varies in direct proportion to the input.

Linear scale A scale whose divisions are uniformly spaced.

Liquid crystal displays (LCDs) A digital display having two sheets of glass separated by a sealed quantity of liquid crystal material. When a voltage is applied across the front and back electrodes, the liquid crystals' molecules become disorganized, causing the liquid to darken.

Live Term used to describe a circuit or piece of equipment that is on and has current flow within it.

Load Source drives a load, and whatever component, circuit, or piece of equipment is connected to the source can be called a load and will have a certain load resistance, which will consequently determine the load current.

Load current The current that is present in the load.

Load impedance Total reactive and resistive opposition of a load.

Loading effect Large load resistance will cause a small load current to flow, and so the loading down of the source or loading effect will be small (light load), whereas a small load resistance will cause a large load current to flow from the source, which will load down the source (heavy load).

Load resistance The resistance of the load.

Locked To automatically follow a signal.

Lodestone A magnetite stone possessing magnetic polarity.

Logarithms The exponent that indicates the power to which a number is raised to produce a given number.

Logic Science dealing with the principles and applications of gates, relays, and switches.

Logic gate circuits Two-state (ON/OFF) circuits used for decision-making functions in digital logic circuits.

Loss Term used to describe a decrease in power.

Lower sideband A group of frequencies that are equal to the differences between the carrier and modulation frequencies.

Low-pass filter Network or circuit designed to pass any frequencies below a critical or cutoff frequency and reject or heavily attenuate all frequencies above.

Magnet Body that can be used to attract or repel magnetic materials.

Magnetic circuit breaker Circuit breaker that is tripped or activated by use of an electromagnet.

Magnetic coil Spiral of a conductor, which is called an electromagnet.

Magnetic core Material that exists in the center of the magnetic coil to either support the windings (nonmagnetic-material) or intensify the magnetic flux (magnetic-material).

Magnetic field Magnetic lines of force traveling from the north to the south pole of a magnet.

Magnetic flux The magnetic lines of force produced by a magnet.

Magnetic leakage The passage of magnetic flux outside the path along which it can do useful work.

Magnetic poles Points of a magnet from which magnetic lines of force leave (north pole) and arrive (south pole).

Magnetism Property of some materials to attract and repel others.

Magnetizing force Also called magnetic field strength, it is the magnetomotive force per unit length at any given point in a magnetic circuit.

Magnetomotive force Force that produces a magnetic field.

Mainframe Large computers that initially were only affordable to medium-sized and large businesses who had the space for them. Mini-computers came after mainframes and were affordable to any business, and now we have microcomputers which are easily affordable to anyone.

Majority carriers The type of carrier that constitutes more than half the total number of carriers in a semiconductor material. In *n*-type materials electrons are the majority carriers, whereas in *p*-type materials holes are the majority carriers.

Matched impedance Condition that occurs when the source impedance is equal to the load impedance, resulting in maximum power being transferred.

Matching Connection of two components or circuits so that maximum energy is transferred or coupled between the two.

Maximum power transfer A theorem that states maximum power will be transferred from source to load when the source resistance is equal to the load resistance.

Maxwell One magnetic line of force or flux is called a maxwell.

Measurement Determining the presence and magnitude of variables.

Medical electronics Branch of electronics involved with therapeutic or diagnostic practices in medicine.

Mercury cell Primary cell that has mercuric oxide cathode, a zinc anode, and a potassium hydroxide electrolyte.

Metal-film resistor A resistor in which a film of metal oxide or alloy is deposited on an insulating substrate.

Metal oxide resistor A metal film resistor in which an oxide of a metal (such as tin) is deposited as a film onto the substrate.

Meter (1) Any electrical or electronic measuring device. (2) In the metric system, it is a unit of length equal to 39.37 inches or 3.28 feet.

Meter FSD current The value of current needed to cause the meter movement to deflect the needle to its full-scale deflection (FSD) position.

Meter resistance The resistance of a meter's armature coil.

Mica capacitor Fixed capacitor that uses mica as the dielectric between its plates.

Microphone Electroacoustic transducer that responds and converts a sound wave input into an equivalent electrical wave out.

Microwave Term used to describe a band of very small wavelength radio waves within the UHF, SHF, and EHF bands.

Mil One thousandth of an inch (0.001 in.).

Miller-effect capacitance (C_M) An undesirable inherent capacitance that exists between the junctions of transistors.

Minority carriers The type of carrier that constitutes less than half the total number of carriers in a semiconductor material. In *n*-type materials holes are the minority carriers, whereas in *p*-type materials electrons are the minority carriers.

Mismatch Term used to describe a difference between the source impedance and load impedance, which will prevent maximum power transfer.

Modulation Process whereby an information signal is used to modify some characteristic of another higher-frequency wave known as a carrier.

Molecule Smallest particle of a compound that still retains its chemical characteristics.

MOS family A group of digital logic circuits that make use of metal oxide semiconductor field effect transistors (MOSFETs).

MOSFET A field-effect transistor in which the insulating layer between the gate electrode and the channel is a metal-oxide layer. Either a *p* or *n* substrate.

Most significant bit (MSB) The left-most binary digit.

Moving-coil microphone Microphone that makes use of a moving coil between a fixed magnetic field. Also called a dynamic microphone.

Moving-coil pickup Dynamic phonograph pickup that uses a coil between a fixed magnetic field, which is moved back and forth by the needle or stylus.

Moving-coil speaker Dynamic speaker that uses a coil placed between a fixed magnetic field and converts the electrical wave input into sound waves.

Multimeter Piece of electronic test equipment that can perform multiple tasks in that it can be used to measure voltage, current, or resistance.

Multiple-emitter input transistor A transistor specially constructed to have more than one emitter.

Multiplier resistor A resistor connected in series with the meter movement of a voltmeter.

Mutual inductance Ability of one inductor's magnetic lines of force to link with another inductor.

Navigation equipment Electronic equipment designed to aid in the direction of aircraft and ships to their destination.

***n*-channel d-type MOSFET** A depletion type MOSFET having an *n*-type channel between its source and drain terminals.

***n*-channel JFET** A junction field effect transistor having an *n*-type channel between source and drain.

Negative (neg.) (1) Some value less than zero. (2) Terminal that has an excess of electrons.

Negative charge An electric charge that has more electrons than protons.

Negative feedback Feedback in which a portion of the output signal is fed back 180° out of phase with the input signal.

Negative ground A system whereby the negative terminal of the voltage source is connected to the system's conducting chassis or body.

Negative ion Atom that has more than the normal neutral amount of electrons.

Negative resistance A resistance such that when the current through it increases, the voltage drop across the resistance decreases.

Negative temperature coefficient Effect that describes that if temperature increases, resistance or capacitance will decrease.

Neon bulb Glass envelope filled with neon gas, which when ionized by an applied voltage will glow red.

Network Combination and interconnection of components, circuits, or systems.

Neutral When an object is neither positive nor negative.

Neutral atom An atom in which the number of positive charges in the nucleus (protons) is equal to the number of negative charges (electrons) that surround the nucleus.

Neutral wire The conductor of a polyphase circuit or a single-phase three-wire circuit that is intended to have a ground potential. The potential differences between the neutral and each of the other conductors are approximately equal in magnitude and are also equally spaced in phase.

Neutron Subatomic particle residing within the nucleus and having no electrical charge.

Nickel-cadmium cell Most popular secondary cell; it uses a nickel oxide positive electrode and cadmium negative electrode.

Node Junction or branch point.

Noise Unwanted electromagnetic radiation within an electrical or mechanical system.

Noninverting amplifier An operational amplifier in which the input signal is applied to the ungrounded positive input terminal to give a gain that is greater than one, and make the output change in phase with the input voltage.

Noninverting input The noninverting or positive input of an operational amplifier.

Nonlinear scale A scale whose divisions are not uniformly spaced.

Normally closed (N.C.) Designation which states that the contacts of a switch or relay are connected normally; however, when activated, these contacts will open.

Normally open (N.O.) Designation which states that the contacts of a switch or relay are normally not connected; however, when activated, these contacts will close.

North pole Pole of a magnet out of which magnetic lines of force are assumed to originate.

Norton's theorem Any network of voltage sources and resistors can be replaced by a single equivalent current source (I_N) in parallel with a single equivalent resistance (R_N).

NPN transistor Negative-positive-negative transistor in which a layer of *p*-type conductive semiconductor is located between two *n*-type regions.

***n*-Type semiconductor** A material that has more valence-band electrons than valence-band holes.

Nuclear energy Atomic energy or power released in a nuclear reaction when either a neutron is used to split an atom into smaller atoms (fission) or when two smaller nuclei are joined together (fusion).

Nuclear reactor Unit that maintains a continuous self-supporting nuclear reaction (fission).

Nucleus Core of an atom; it contains both positive (protons) and neutral (neutrons) subatomic particles.

Octave Interval between two sounds whose fundamental frequencies differ by a ratio of 2 to 1.

Ohm Unit of resistance, symbolized by the Greek capital letter omega (Ω).

Ohmmeter Measurement device used to measure electric resistance.

Ohm's law Relationship between the three electrical phenomena of voltage, current, and resistance, which states that the current flow within a circuit is directly proportional to the voltage applied across the circuit and inversely proportional to its resistance.

Ohms per volt Value that indicates the sensitivity of a voltmeter. The higher the ohms per volt rating, the more sensitive the meter.

Open circuit Break in the path of current flow.

Open collector output A type of output structure found in certain bipolar logic families. Resistive pull-ups are generally added to provide the high level output voltage.

Open-loop mode A control system that has no means of comparing the output with the input for control purposes.

Operational amplifier Special type of high-gain amplifier; also called an op amp.

Operator error An incorrect use of the controls of a circuit or system.

Optimum power transfer Since the ideal maximum power transfer conditions cannot always be achieved, most designers try to achieve maximum power transfer and have the source resistance and load resistance as close in value as possible.

Ordinary ground A connection in the circuit that is said to be at ground potential or zero volts and, because of its connection to earth, has the ability to conduct electrical current to and from the earth.

OR gate When either input *A* OR *B* is HIGH, the output will be HIGH.

Oscillate Continual repetition of or passing through a cycle.

Oscillator Electronic circuit that converts dc to a continuous alternating current out.

Oscilloscope Instrument used to view signal amplitude, period, and shape at different points throughout a circuit.

Out of phase When the maximum and minimum points of two or more waveforms do not occur at the same time.

Output Terminals at which a component, circuit, or piece of equipment delivers current, voltage, or power.

Output impedance Impedance measured across the output terminals of a device without the load connected.

Output power Amount of power a component, circuit, or system can deliver to its load.

Overload Situation that occurs when the load is greater than the component, circuit, or system was designated to handle (load resistance too small, load current too high), resulting in waveform distortion and/or overheating.

Overload protection Protective device such as a fuse or circuit breaker that automatically disconnects or opens a current path when it exceeds an excessive value.

Paper capacitor Fixed capacitor using oiled or waxed paper as a dielectric.

Parallax error The apparent displacement of an object's position caused by a shift in the point of observation of the object.

Parallel Also called shunt; circuit having two or more paths for current flow.

Parallel data transmission The transfer of information simultaneously over a set of parallel paths or channels.

Parallel resonant circuit Circuit having an inductor and capacitor in parallel with one another, offering a high impedance at the frequency of resonance.

Parallel-to-serial converter An IC that converts the parallel word applied to its inputs into a serial data stream.

Parasitic oscillations An unwanted self-sustaining oscillation or self generated random pulse.

Passband Band or range of frequencies that will be passed by a filter circuit.

Passive component Component that does not amplify a signal, such as a resistor or capacitor.

Passive filter A filter that contains only passive (nonamplifying) components and provides pass or rejection characteristics.

Passive system System that emits no energy; in other words it only receives; it does not transmit and consequently reveal its position.

***p*-channel (d-type) MOSFET** A depletion type MOSFET having a *p*-type channel between its source and drain terminal.

***p*-channel JFET** A junction field effect transistor having a *p*-type channel between source and drain.

Peak Maximum or highest-amplitude level.

Peak inverse voltage Abbreviated PIV, it is the maximum rated value of an ac voltage acting in the direction opposite to that in which a device is designed to pass current.

Peak to peak Difference between the maximum positive and maximum negative values.

Pentode A five-electrode electron tube that has an anode, cathode, control grid, and two additional electrodes.

Percent of regulation The change in output voltage that occurs between no-load and full-load in a dc voltage source. Dividing this change by the rated full-load value and multiplying the result by 100 gives percent regulation.

Percent of ripple The ratio of the effective or rms value of ripple voltage to the average value of the total voltage. Expressed as a percentage.

Period Time taken to complete one complete cycle of a periodic or repeating waveform.

Permanence Magnetic equivalent of electrical conductance and consequently equal to the reciprocal of reluctance, just as conductance is equal to the reciprocal of resistance.

Permanent magnet Magnet, normally made of hardened steel, that retains its magnetism indefinitely.

Permeability Measure of how much better a material is as a path for magnetic lines of force with respect to air, which has a permeability of 1 (symbolized by the Greek lowercase letter mu, μ).

Phase Angular relationship between two waves, normally between current and voltage in an ac circuit.

Phase angle Phase difference between two waves, normally expressed in degrees.

Phase-locked loop (PLL) circuit A circuit consisting of a phase comparator that compares the output frequency of a voltage controlled oscillator with an input frequency. The error voltage out of the phase comparator is then coupled via an amplifier and low-pass filter to the control input of the voltage controlled oscillator to keep it in phase, and therefore at exactly the same frequency as the input.

Phase shift Change in phase of a waveform between two points, given in degrees of lead or lag.

Phase-shift oscillator An *RC* oscillator circuit in which the 180° phase shift is achieved with several *RC* networks.

Phase splitter A circuit that takes a single input signal and produces two output signals that are 180° apart in phase.

Phonograph Piece of equipment used to reproduce sound.

Phosphor Luminescent material applied to the inner surface of a cathode ray tube that when bombarded with electrons will emit light.

Photoconduction A process by which the conductance of a material is changed by incident electromagnetic radiation in the light spectrum.

Photoconductive cell Material whose resistance decreases or conductance increases when light strikes it.

Photodetector Component used to detect or sense light.

Photodiode A semiconductor diode that changes its electrical characteristics in response to illumination.

Photometer Meter used to measure light intensity.

Photon Discrete portion of electromagnetic energy. A small packet of light.

Photoresistor Also known as a photoconductive cell or light-dependent resistor, it is a device whose resistance varies with the illumination of the cell.

Photovoltaic action A process by which a device generates a voltage as a result of exposure to radiation.

Photovoltaic cell Component, commonly called a solar cell, used to convert light energy into electric energy (voltage).

Pi Value representing the ratio between the circumference and diameter of a circle and equal to approximately 3.142 (symbolized by the lowercase Greek letter π).

Pierce crystal oscillator An oscillator circuit in which a piezoelectric crystal is connected in a tank between output and input.

Piezoelectric crystal Crystal material that will generate a voltage when mechanical pressure is applied and conversely will undergo mechanical stress when subjected to a voltage.

Piezoelectric effect The operation of a voltage between the opposite sides of a piezoelectric crystal as a result of pressure or twisting. Also, the reverse effect in which the application of a voltage to opposite sides causes deformation to occur at the frequency of the applied voltage.

Piezoresistive diaphragm pressure sensor A sensor that changes its resistance in response to pressure.

Pinch-off region A region on the characteristic curve of an FET in which the gate bias causes the depletion region to extend completely across the channel.

PIN diode A semiconductor diode that has a high-resistance intrinsic region between its low-resistance *p*-type and *n*-type regions.

Pitch Term used to describe the inflection or frequency scale of sounds. When the pitch is increased by one octave, twice the original frequency will be the result.

Plastic film capacitor A capacitor in which alternate layers of metal aluminum foil are separated by thin films of plastic dielectric.

Plate Conductive electrode in either a capacitor or battery.

Plug Movable connector that is normally inserted into a socket.

P-N Junction The point at which two opposite doped materials come in contact with one another.

PNP transistor Positive-negative-positive transistor in which a layer of *n*-type conductive semiconductor is located between two *p*-type regions.

Polar coordinates Either of two numbers that locate a point in a plane by its distance from a fixed point on a line and the angle this line makes with a fixed line.

Polarity Term used to describe positive and negative charges.

Polarized electrolytic capacitor An electrolytic capacitor in which the dielectric is formed adjacent to one of the metal plates, creating a greater opposition to current in one direction only.

Positive Point that attracts electrons, as opposed to negative, which supplies electrons.

Positive charge The charge that exists in a body which has fewer electrons than normal.

Positive ground A system whereby the positive terminal of the voltage source is connected to the system's conducting chassis or body.

Positive ion Atom that has lost one or more of its electrons and therefore has more protons than electrons, resulting in a net positive charge.

Positive shunt clipper A circuit that has a diode connected in shunt with the lead or output, and its orientation is such that it will clip off the positive alternation of the ac input.

Potential difference (pd) Voltage difference between two points, which will cause current to flow in a closed circuit.

Potential energy Energy that has the potential to do work because of its position relative to others.

Potentiometer Three-lead variable resistor that through mechanical turning of a shaft can be used to produce a variable voltage or potential.

Power Amount of energy converted by a component or circuit in a unit of time, normally seconds. It is measured in units of watts (joules/second).

Power amplifier An amplifier designed to deliver maximum output power to a load.

Power dissipation Amount of heat energy generated by a device in one second when current flows through it.

Power factor Ratio of actual power to apparent power. A pure resistor has a power factor of 1 or 100%, while a capacitor has a power factor of 0 or 0%.

Power loss Ratio of power absorbed to power delivered.

Power supply Piece of electrical equipment used to deliver either ac or dc voltage.

Prefix Name used to designate a factor or multiplier.

Pressure The application of force to something by something else in direct contact with it.

Primary First winding of a transformer that is connected to the source, as opposed to the secondary that is connected to the load.

Primary cell Cell that produces electrical energy through an internal electrochemical action; once discharged, it cannot be reused.

Printed circuit board (PCB) Insulating board that has conductive tracks printed onto the board to make the circuit.

Propagation Traveling of electromagnetic, electrical, or sound waves through a medium.

Propagation time Time it takes for a wave to travel between two points.

Proportional Having the same or constant ratio.

Protoboard An experimental arrangement of a circuit on a board. Also called a breadboard.

Proton Subatomic particle within the nucleus that has a positive charge.

***p*-Type Semiconductor** A material that has more valence-band holes than valence-band electrons.

Pull-up resistor A resistor connected between a signal output and positive V_{CC} to pull the signal line HIGH when it is not being driven LOW.

Pulse Rise and fall of some quantity for a period of time.

Pulse fall time Time it takes for a pulse to decrease from 90% to 10% of its maximum value.

Pulse repetition frequency The number of times per second that a pulse is transmitted.

Pulse repetition time The time interval between the start of two consecutive pulses.

Pulse rise time Time it takes for a pulse to increase from 10% to 90% of its maximum value.

Pulse width, pulse length, or pulse duration The time interval between the leading edge and trailing edge of a pulse at which the amplitude reaches 50% of the peak pulse amplitude.

Push-pull amplifier A balanced amplifier that uses two similar equivalent amplifying transistors working in phase opposition.

Pythagorean theorem A theorem in geometry: The square of the length of the hypotenuse of a right triangle equals the sum of the squares of the lengths of the other two sides.

Q Quality factor of an inductor or capacitor; it is the ratio of a component's reactance (energy stored) to its effective series resistance (energy dissipated).

Quiescent point The voltage or current value that sets up the no-signal input or operating point bias voltage.

Race condition An unpredictable bistable circuit condition.

Radar Acronym for "radio detection and ranging"; it is a system that measures the distance and direction of objects.

Radioastronomy Branch of astronomy that studies the radio waves generated by celestial bodies and uses these emissions to obtain more information about them.

Radio broadcast Transmission of music, voice, and other information on radio carrier waves that can be received by the general public.

Radio communication Term used to describe the transfer of information between two or more points by use of radio or electromagnetic waves.

Radio-frequency (RF) amplifier An amplifier that has one or more transistor stages designed to amplify radio-frequency signals.

Radio-frequency generator A generator capable of supplying RF energy at any desired frequency in the radio spectrum.

Radio-frequency probe A probe used in conjunction with an ac meter to measure high-frequency RF signals.

RC Abbreviation for "resistance–capacitance."

***RC* Circuit** Circuit containing both a resistor and capacitor.

***RC* coupling** A coupling method in which resistors are used as the input and output impedances of the two stages. A coupling capacitor is generally used between the stages to couple the ac signal and block the dc supply bias voltages.

***RC* filter** A selective circuit which makes use of a resistance-capacitance network.

***RC* time constant** In one time constant, which is equal to the product of resistance and capacitance in seconds, a capacitor will have charged or discharged 63.2% of the maximum applied voltage.

Reactance (*X*) Opposition to current flow without the dissipation of energy.

Reactive power Also called imaginary power or wattless power, it is the power value obtained by multiplying the effective value of current by the effective value of voltage and the sine of the angular phase difference between current and voltage.

Real number A number that has no imaginary part.

Receiver Unit or piece of equipment used for the reception of information.

Recombination The combination and resultant neutralization of particles or objects having unlike charges. For example, a hole and an electron, or a positive ion and negative ion.

Rectangular coordinates A Cartesian coordinate of a Cartesian coordinate system whose straight-line axes or coordinate planes are perpendicular.

Rectangular wave Also known as a pulse wave; it is a repeating wave that only alternates between two levels or values and remains at one of these values for a small amount of time relative to the other.

Rectification Process that converts alternating current (ac) into direct current (dc).

Rectifier A device that converts alternating current into a unidirectional or dc current.

Rectifier circuit Achieves rectification.

Rectifier diodes or rectifiers Junction diodes that achieve rectification.

Reed relay Relay that consists of two thin magnetic strips within a glass envelope with a coil wrapped around the envelope so that when it is energized the relay's contacts or strips will snap together, making a connection between the leads attached to each of the reed strips.

Regulator A device that maintains a desired quantity at a predetermined voltage.

Relative Not independent; compared with or with respect to some other value of a measured quantity.

Relaxation oscillator An oscillator circuit whose frequency is determined by an *RL* or *RC* network, producing a rectangular or sawtooth output waveform.

Relay Electromechanical device that opens or closes contacts when a current is passed through a coil.

Reluctance Resistance to the flow of magnetic lines of force.

Remanence Amount a material remains magnetized after the magnetizing force has been removed.

Reset and carry An action that occurs in any column that reaches its maximum count.

Reset state A circuit condition in which the output is reset to binary 0.

Residual magnetism Magnetism remaining in the core of an electromagnet after the coil current has been removed.

Resistance Symbolized R and measured in ohms (Ω), it is the opposition to current flow with the dissipation of energy in the form of heat.

Resistive power (True power) The average power consumed by a circuit during one complete cycle of alternating current.

Resistive temperature detector (RTD) A temperature detector consisting of a fine coil of conducting wire (such as platinum) that will produce a relatively linear increase in resistance as temperature increases.

Resistivity Measure of a material's resistance to current flow.

Resistor Component made of a material that opposes the flow of current and therefore has some value of resistance.

Resistor color code Coding system of colored stripes on a resistor that indicates the resistor's value and tolerance.

Resonance Circuit condition that occurs when the inductive reactance (X_L) is equal to the capacitive reactance (X_C).

Resonant circuit Circuit containing an inductor and capacitor tuned to resonate at a certain frequency.

Resonant frequency Frequency at which a circuit or object will produce a maximum amplitude output.

Reverse biased A large depletion region at the junction will offer a large resistance and permit only a small current. Such a junction is reverse biased.

Reverse leakage current (I_R) The undesirable flow of current through a device in the reverse direction.

Reverse voltage drop (V_R) The reverse voltage drop is equal to the source voltage (applied voltage).

RF Abbreviation for "radio frequency."

RF amplifier/frequency tripler circuit A radio frequency amplifier circuit in which the frequency of the output is three times that of the input frequency.

RF amplifier/multiplier circuit A radio frequency amplifier circuit in which the frequency of the output is an exact multiple of the input frequency.

RF local oscillator circuit The radio frequency oscillator in a superheterodyne receiver.

Rheostat Two-terminal variable resistor used to control current.

Right angle triangle A triangle having a 90° or square corner.

Ripple frequency The frequency of the ripple present in the output of a dc source.

Rise time Time it takes a positive edge of a pulse to rise from 10% to 90% of its high value.

***RL* differentiator** An *RL* circuit whose output voltage is proportional to the rate of change of the input voltage.

***RL* filter** A selective circuit of resistors and inductors that offers little or no opposition to certain frequencies while blocking or attenuating other frequencies.

***RL* integrator** An *RL* circuit with an output proportionate to the integral of the input signal.

RMS Abbreviation for "root mean square."

RMS value Rms value of an ac voltage, current, or power waveform is equal to 0.707 times the peak value. The rms value is the effective or dc value equivalent of the ac wave.

Rotary switch Electromechanical device that has a rotating shaft connected to one terminal that is capable of making or breaking a connection.

***R-2R* ladder circuit** A network or circuit composed of a sequence of L networks connected in tandem. This *R*-2*R* circuit is used in digital-to-analog converters.

Saturation The condition in which a further increase in one variable produces no further increase in the resultant effect.

Saturation point The point beyond which an increase in one of two quantities produces no increase in the other.

Sawtooth wave Repeating waveform that rises from zero to a maximum value linearly and then falls to zero and repeats.

Scale Set of markings used for measurement.

Schematic diagram Illustration of the electrical or electronic scheme of a circuit, with all the components represented by their respective symbols.

Schottky diode A semiconductor diode formed by contact between a semiconductor layer and a metal coating. Hot carriers (electrons for *n*-type material and holes for *p*-type material) are emitted from the Schottky barrier of the semiconductor and move to the metal coating. Since majority carriers predominate, there is essentially no injection or storage of minority carriers to limit switching speed.

Scientific notation Numbers are entered and displayed in terms of a power of 10. For example:

Number	Scientific Notation
7642	7.642×10^3
64,000,000	64×10^6
0.0012	1.2×10^{-3}
0.000096	96×10^{-6}

Secondary Output winding of a transformer that is connected across the load.

Secondary cells Electrolytic cells for generating electricity. Once discharged the cell may be restored or recharged by sending an electric current through the cell in the opposite direction to that of the discharge current.

Selectivity Characteristic of a circuit to discriminate between the wanted signal and the unwanted signal.

Self biasing Gate bias for an FET in which a resistor is used to drop the supply voltage and provide gate bias.

Self-inductance The property that causes a counter electromotive force to be produced in a conductor when the magnetic field expands or collapses with a change in current.

Semiconductors Materials that have properties that lie between insulators and conductors.

Serial data transmission The transfer of information sequentially through a single path or channel.

Serial-to-parallel converter An IC that converts a serial data stream applied to its input into a parallel output word.

Series circuit Circuit in which the components are connected end to end so that current has only one path to follow throughout the circuit.

Series clipper A circuit that will clip off part of the input signal. Also known as a limitor since the circuit will limit the ac input. A series clipper circuit has a clipping or limiting device in series with the load.

Series-parallel network Network or circuit that contains components that are connected in both series and parallel.

Series resonance Condition that occurs when the inductive and capacitive reactances are equal and both components are connected in series with one another, and therefore the impedance is minimum.

Series resonant circuit A resonant circuit in which the capacitor and coil are in series with the applied ac voltage.

Series switching regulators A regulator circuit containing a power transistor in series with the load, that is switched ON and OFF to regulate the dc output voltage delivered to the load.

Set-reset flip-flop A bistable multivibrator circuit that has two inputs that are used to either SET or RESET the output.

Set state A circuit condition in which the output is set to binary 1.

Seven-segment display Component that normally has eight LEDs, seven of which are mounted into segments or bars that make up the number 8, and the eighth LED is used as a decimal point.

Shells or bands An orbital path containing a group of electrons that have a common energy level.

Shield Metal grounded cover that is used to protect a wire, component, or piece of equipment from stray magnetic and/or electric fields.

Shock A sudden pain, convulsion, unconsciousness, or death produced by the passage of an electrical current through the body.

Short circuit Also called a short; it is a low-resistance connection between two points in a circuit, typically causing a large amount of current flow.

Shorted out Term used to describe a component that has either internally malfunctioned, resulting in a low-resistance path through the component, or a component that has been bypassed by a low-resistance path.

Shunt clipper A circuit that will clip off part of the input signal. Also known as a limitor since the circuit will limit the ac input. A shunt clipper circuit has a clipping or limiting device in shunt with the load.

Shunt resistor A resistor connected in parallel or shunt with the meter movement of an ammeter.

Side bands A band of frequencies on both sides of the carrier frequency of a modulated signal produced by modulation.

Signal Conveyor of information.

Signal-to-noise ratio Ratio of the magnitude of the signal to the magnitude of the noise, normally expressed in decibels.

Signal voltage Rms or effective voltage value of a signal.

Silicon (Si) Nonmetallic element (atomic number 14) used in pure form as a semiconductor.

Silicon controlled rectifier Abbreviated SCR, it is a three-junction, three-terminal unidirectional P-N-P-N thyristor that is normally an open circuit. When triggered with the proper gate signal it switches to a conducting state and allows current to flow in one direction.

Silicon transistor Transistor using silicon as the semiconducting material.

Silver (Ag) Precious metal that does not easily corrode and is more conductive than copper.

Silvered mica capacitor Mica capacitor with silver deposited directly onto the mica sheets, instead of using conducting metal foil.

Silver solder Solder composed of silver, copper, and zinc with a melting point lower than silver but higher than the standard lead-tin solder.

Simplex Communication in only one direction at a time, for example, facsimile and television.

Simulcast Broadcasting of a program simultaneously in two different forms, for example, a program on both AM and FM.

Sine Sine of an angle of a right-angle triangle is equal to the opposite side divided by the hypotenuse.

Sine wave Wave whose amplitude is the sine of a linear function of time. It is drawn on a graph that plots amplitude against time or radial degrees relative to the angular rotation of an alternator.

Single in-line package (SIP) Package containing several electronic components (generally resistors) with a single row of external connecting pins.

Single-pole, double-throw (SPDT) Three-terminal switch or relay in which one terminal can be thrown in one of two positions.

Single-pole, single-throw (SPST) Two-terminal switch or relay that can either open or close one circuit.

Single sideband (SSB) AM radio communication technique in which the transmitter suppresses one sideband and therefore only transmits a single sideband.

Single-throw switch Switch containing only one set of contacts, which can be either opened or closed.

Sink Device, such as a load, that consumes power, or conducts away heat.

Sintering Process of bonding either a metal or powder by cold-pressing it into a desired shape and then heating to form a strong, cohesive body.

Sinusoidal Varying in proportion to the sine of an angle or time function; for example, alternating current (ac) is sinusoidal.

SIP Abbreviation for "single in-line package."

Skin effect Tendency of high-frequency (rf) currents to flow near the surface layer of a conductor.

Slide switch Switch having a sliding bar, button, or knob.

Slow-acting relay Slow-operating relay that when energized may not pull up the armature for several seconds.

Slow-blow fuse Fuse that can withstand a heavy current (up to ten times its rated value) for a small period of time without opening.

Snap switch Switch containing a spring under tension or compression that causes the contacts to come together suddenly when activated.

SNR Abbreviation for "signal-to-noise ratio."

Soft magnetic material Ferromagnetic material that is easily demagnetized.

Software Program of instructions that directs the operation of a computer.

Solar cell Photovoltaic cell that converts light into electric energy. They are especially useful as a power source for space vehicles.

Solder Metallic alloy that is used to join two metal surfaces.

Soldering Process of joining two metallic surfaces to make an electrical contact by melting solder (usually tin and lead) across them.

Soldering gun Soldering tool having a trigger switch and pistol shape that at its tip has a fast-heating resistive element for soldering.

Soldering iron Soldering tool having an internal heating element that is used for soldering.

Solenoid Coil and movable iron core that when energized by an alternating or direct current will pull the core into a central position.

Solid conductor Conductor having a single solid wire, as opposed to strands.

Solid state Pertaining to circuits and devices that use solid semiconductors such as silicon. Solid-state electronics devices have a solid material between their input and output pins (transistors, diodes), whereas vacuum-tube electronics uses tubes, which have a vacuum between input and output.

Sonar Acronym for "sound navigation and ranging." A system using sound waves to determine a target's direction and distance.

Sonic Pertaining to the speed of sound waves.

Sound wave Traveling wave propagated in an elastic medium that travels at a speed of approximately 1133 ft/s.

Source One of the field effect transistor's electrodes.

Source follower An FET amplifier in which the signal is applied between the gate and drain and the output is taken between the source and drain. Used to handle large-input signals and applications requiring a low-input capacitance.

Source impedance Impedance a source presents to a load.

South pole Pole of a magnet into which magnetic lines of force are assumed to enter.

Spark Momentary discharge of electric energy due to the breakdown of air or some other dielectric material separating two terminals.

SPDT Abbreviation for "single-pole, double-throw."

Speaker Also called a loudspeaker; it is an electroacoustic transducer that converts an electrical wave input into a mechanical sound wave (acoustic) output into the air.

Specification sheet or data sheet Details the characteristics and maximum and minimum values of operation of a device.

Spectrum The frequency spectrum displays all the frequencies and their applications.

Spectrum analyzer Instrument that can display the frequency domain of a waveform, plotting amplitude against frequency of the signals present.

Speed of light Physical constant—the speed at which light travels through a vacuum—equal to 186,282.397 miles/s, 2.997925×10^8 m/s, 161,870 nautical miles/s, or 328 yards/μs.

Speed of sound Speed at which a sound wave travels through a medium. In air it is equal to about 1133 ft/s or 334 m/s, while in water it is equal to approximately 4800 ft/s or 1463 m/s. Also known as sonic speed.

Speedup capacitor Capacitor connected in a circuit to speed up an action due to its inherent behavior.

SPST Abbreviation for "single-pole, single-throw."

Square wave Wave that alternates between two fixed values for an equal amount of time.

Square-wave generator A circuit that generates a continuously repeating square wave.

S-R latch Another name for *S-R* flip-flop, so called because the output remains latched in the set or reset state even though the input is removed.

Staircase-wave generator A signal generator circuit that produces an output signal voltage that increases in steps.

Standard Exact value used as a basis for comparison or calibration.

Static Crackling noise heard on radio receivers caused by electric storms or electric devices in the vicinity.

Static electricity Electricity at rest or stationary.

Stator Stationary part of some rotating device.

Statute mile Distance unit equal to 5280 ft or 1.61 km.

Step-down transformer Transformer in which the ac voltage induced in the secondary is less (due to fewer secondary windings) than the ac voltage applied to the primary.

Stepper motor A motor that rotates in small angular steps.

Step-up transformer Transformer in which the ac voltage induced in the secondary is greater (due to more secondary windings) than the ac voltage applied to the primary.

Stereo sound Sound system in which the sound is delivered through at least two channels and loudspeakers arranged to give the listener a replica of the original performance.

Stranded conductor Conductor composed by a group of twisted wires.

Stray capacitance Undesirable capacitance that exists between two conductors such as two leads or a lead and a metal chassis.

Subassembly Components contained in a unit for convenience in assembling or servicing the equipment.

Subatomic Particles such as electrons, protons, and neutrons that are smaller than atoms.

Substrate The mechanical insulating support on which a device is fabricated.

Superconductor Metal such as lead or niobium that, when cooled to within a few degrees of absolute zero, can conduct current with no electrical resistance.

Superheterodyne receiver Radio-frequency receiver that converts all RF inputs to a common intermediate frequency (IF) before demodulation.

Superhigh frequency (SHF) Frequency band between 3 and 30 GHz, so designated by the Federal Communications Commission (FCC).

Superposition theorem Theorem designed to simplify networks containing two or more sources. It states: In a network containing more than one source, the current at any point is equal to the algebraic sum of the currents produced by each source acting separately.

Supersonic Faster than the speed or velocity of sound (Mach 1).

Supply voltage Voltage produced by a power source or supply.

SW Abbreviation for "shortwave."

Sweep generator Test instrument designed to generate a radio-frequency voltage that continually and automatically varies in frequency within a selected frequency range.

Swing Amount a frequency or amplitude varies.

Switch Manual, mechanical, or electrical device used for making or breaking an electric circuit.

Switching power supply A dc power supply that makes use of a series switching regulator controlled by a pulse-width-modulator to regulate the output voltage.

Switching transistor Transistor designed to switch either on or off.

Synchronization Also called sync; it is the precise matching or keeping in step of two waves or functions.

Synchronous Two or more circuits or devices in step or in phase.

Sync pulse or signal Pulse waveform generated to synchronize two processes.

System Combination or linking of several parts or pieces of equipment to perform a particular function.

Tachometer Instrument that produces an output voltage that indicates the angular speed of the input in revolutions per minute.

Tank circuit Circuit made up of a coil and capacitor that is capable of storing electric energy.

Tantalum capacitor Electrolytic capacitor having a tantalum foil anode.

Tap Electrical connection to some point, other than the ends, on the element of a resistor or coil.

Tapered Nonuniform distribution of resistance per unit length throughout the element.

Technician Expert in troubleshooting circuit and system malfunctions. Along with a thorough knowledge of all test equipment and how to use it to diagnose problems, the technician is also familiar with how to repair or replace faulty components. Technicians basically translate theory into action.

Telegraphy Communication between two points by sending a series of coded current pulses either through wires or by radio.

Telemetry Transmission of instrument reading to a remote location either through wires or by radio waves.

Telephone Apparatus designed to convert sound waves into electrical waves, which are then sent to and reproduced at a distant point.

Telephone line Wires existing between subscribers and central stations.

Telephony Telecommunications system involving the transmission of speech information, therefore allowing two or more persons to converse verbally.

Teletypewriter Electric typewriter that like a teleprinter can produce coded signals corresponding to the keys pressed or print characters corresponding to the coded signals received.

Television (TV) System that converts both audio and visual information into corresponding electric signals that are then transmitted through wires or by radio to a receiver, which reproduces the original information.

Telex Teletypewriter exchange service.

Temperature coefficient of frequency Rate frequency changes with temperature.

Temperature coefficient of resistance Rate resistance changes with temperature.

Tera (T) Prefix that represents 10^{12}.

Terminal Connecting point for making electric connections.

Tesla (T) SI unit of magnetic flux density (1 tesla = 1 Wb/m^2).

Test Sequence of operations designed to verify the correct operation or malfunction of a system.

Tetrode A four-electrode electron tube that has an anode, cathode, control grid, and an additional electrode.

Thermal relay Relay activated by a heating element.

Thermal stability The ability of a circuit to maintain stable characteristics despite changes in the ambient temperature.

Thermistor Temperature-sensitive semiconductor that has a negative temperature coefficient of resistance (as temperature increases, resistance decreases).

Thermocouple Temperature transducer consisting of two dissimilar metals welded together at one end to form a junction that when heated will generate a voltage.

Thermometry Relating to the measurement of temperature.

Thermostat Temperature-sensitive device that opens or closes a circuit.

Thévenin's theorem Theorem that replaces any complex network with a single voltage source in series with a single resistance. It states: Any network of resistors can be replaced with an equivalent voltage source (V_{Th}) and an equivalent series resistance (R_{Th}).

Thick-film capacitor Capacitor consisting of two thick-film layers of conductive film separated by a deposited thick-layer dielectric film.

Thick-film resistor Fixed-value resistor consisting of a thick-film resistive element made from metal particles and glass powder.

Thin-film capacitor Capacitor in which both the electrodes and the dielectric are deposited in layers on a substrate.

Thin-film detector (TFD) A temperature detector containing a thin layer of platinum, and used for very precise temperature readings.

Three-phase supply AC supply that consists of three ac voltages that are 120° out of phase with one another.

Threshold Minimum point at which an effect is produced or indicated.

Thyristor A semiconductor switching device in which bistable action depends on P-N-P-N regenerative feedback. A thyristor can be bidirectional or unidirectional and have from two to four terminals.

Time constant Time needed for either a voltage or current to rise to 63.2% of the maximum or fall to 36.8% of the initial value. The time constant of an *RC* circuit is equal to the product of *R* and *C*, while the time constant of an *RL* circuit is equal to the inductance divided by the resistance.

Time-division multiplex (TDM) Transmission of two or more signals on the same path but at different times.

Time-domain analysis A method of representing a waveform by plotting its amplitude versus time.

Tinned Coated with a layer of tin or solder to prevent corrosion and simplify the soldering of connections.

Toggle switch Spring-loaded switch that is put in one of two positions, either on or off.

Tolerance Permissible deviation from a specified value, normally expressed as a percentage.

Tone A term describing both the bass and treble of a sound signal.

TO package Cylindrical, metal can type of package for some semiconductor components.

Toroidal coil Coil wound on a doughnut-shaped core.

Torque Moving force.

Totem pole circuit A transistor circuit containing two transistors connected one on top of the other with two inputs and one output.

Transconductance Also called mutual conductance, it is the ratio of a change in output current to the initiating change in input voltage.

Transducer Any device that converts energy from one form to another.

Transformer Device consisting of two or more coils that are used to couple electric energy from one circuit to another, yet maintain electrical isolation between the two.

Transformer coupling Also called inductive coupling, it is the coupling of two circuits by means of mutual inductance provided by a transformer.

Transient suppressor diode A device used to protect voltage sensitive electronic devices in danger of destruction by high energy voltage transients.

Transistance The characteristic achieved by a transistor that makes possible the control of voltages and currents so as to achieve gain or switching action.

Transistor (TRANSfer resISTOR) Semiconductor device having three main electrodes called the emitter, base, and collector that can be made to either amplify or rectify.

Transmission Sending of information.

Transmission line Conducting line used to couple signal energy between two points.

Transmitter Equipment used to achieve transmission.

Transorb Absorb transients. Another name for transient suppressor diode.

Treble High audio frequencies normally handled by a tweeter in a sound system.

TRIAC A bidirectional gate-controlled thyristor that provides full-wave control of ac power.

Triangular wave A repeating wave that has equal positive and negative ramps that have linear rates of change with time.

Triangular-wave generator A signal generator circuit that produces a continuously repeating triangular wave output.

Trigger Pulse used to initiate a circuit action.

Triggering Initiation of an action in a circuit which then functions for a predetermined time, for example, the duration of one sweep in a cathode-ray tube.

Trigonometry The study of the properties of triangles and trigonometric functions and of their applications.

Trimmer Small-value variable resistor, capacitor, or inductor.

Triode A three-electrode vacuum tube that has an anode, cathode, and control grid.

Troubleshooting The process of locating and diagnosing malfunctions or breakdowns in equipment by means of systematic checking or analysis.

Tune To adjust the resonance of a circuit so that it will select the desired frequency.

Tuned circuit Circuit that can have its components' values varied so that the circuit responds to one selected frequency yet heavily attenuates all other frequencies.

Tunnel diode A heavily doped junction diode that has negative resistance in the forward direction of its operating range due to quantum mechanical tunnelling.

Turns ratio Ratio of the number of turns in the secondary winding to the number of turns in the primary winding of a transformer.

Twin-T sine-wave oscillator An oscillator circuit that makes use of two T-shaped feedback networks.

Two phase Two repeating waveforms having a phase difference of 90°.

UHF Abbreviation for "ultrahigh frequency."

Ultrasonic Signals that are just above the range of human hearing of approximately 20 kHz.

Uncharged Having a normal number of electrons and therefore no electrical charge.

Unidirectional device A device that will conduct current in only one direction.

Unijunction transistor Abbreviated UJT, it is a P-N device that has an emitter connected to the P-N junction on one side of the bar, and two bases at either end of the bar. Used primarily as a switching device.

Upper sideband A group of frequencies that are equal to the sums of the carrier and modulation frequencies.

VA Abbreviation for "volt-ampere."

Vacuum tube Electron tube evacuated to such a degree that its electrical characteristics are essentially unaffected by the presence of residual gas or vapor. Eventually replaced by the transistor for amplification and rectification.

Valence shell Outermost shell formed by electrons.

Varactor diode A P-N semiconductor diode that is reverse biased to increase or decrease its depletion region width and vary device capacitance.

Variable Quantity that can be altered or controlled to assume a number of distinct values.

Variable capacitor Capacitor whose capacitance can be changed by varying the effective area of the plates or the distance between the plates.

Variable resistor *See* Rheostat *and* Potentiometer.

Variable-tuned RF amplifier A tuned radio frequency amplifier in which the tuned circuit(s) can be adjusted to select the desired station carrier frequency.

Variable-value capacitor A capacitor whose value can be varied.

Variable-value inductor An inductor whose value can be varied.

VCR Abbreviation for "videocassette recorder."

Vector Quantity that has both magnitude and direction. They are normally represented as a line, the length of which indicates magnitude and the orientation of which, due to the arrowhead on one end, indicates direction.

Vector diagram Arrangement of vectors showing the phase relationships between two or more ac quantities of the same frequency.

Vertical-channel E-MOSFET An enhancement type MOSFET that, when turned ON, form a vertical channel between source and drain.

Very high frequency (VHF) Electromagnetic frequency band from 30 to 300 MHz as set by the FCC.

Very low frequency (VLF) Frequency band from 3 to 30 kHz as set by the FCC.

Video Relating to any picture or visual information, from the Latin word meaning "I see."

Video amplifier An amplifier that has one or more transistor stages designed to amplify video signals.

Video signals A signal that contains visual information for television or radar systems.

Virtual ground A ground for voltage but not for current.

Voice coil Coil attached to the diaphragm of a moving coil speaker, which is moved through an air gap between the pole pieces of a permanent magnet.

Voice synthesizer Synthesizer that can simulate speech in any language by stringing together phonemes.

Volt (V) Unit of voltage, potential difference, or electromotive force. One volt is the force needed to produce 1 ampere of current in a circuit containing 1 ohm of resistance.

Voltage (*V* or *E*) Term used to designate electrical pressure or the force that causes current to flow.

Voltage amplifier An amplifier designed to build up the voltage of a signal.

Voltage divider Fixed or variable series resistor network that is connected across a voltage to obtain a desired fraction of the total voltage.

Voltage divider biasing A biasing method used with amplifiers in which a series arrangement of two fixed-value resistors are connected across the voltage source. The result is that a desired fraction of the total voltage is obtained at the center of the two resistors and is used to bias the amplifier.

Voltage drop Voltage or difference in potential developed across a component or conductor due to the loss of electric pressure as a result of current flow.

Voltage follower An operational amplifier that has a direct feedback to give unity gain so that the output voltage follows the input voltage. Used in applications where a very high input impedance and very low output impedance are desired.

Voltage gain Also called voltage amplification, it is equal to the difference between the output voltage level and the input signal voltage level. This value is normally expressed in decibels, which are equal to 20 times the logarithm of the ratio of the output voltage to the input voltage.

Voltage rating Maximum voltage a component can safely withstand without breaking down.

Voltage regulator A device that maintains the output voltage of a voltage source within required limits despite variations in input voltage and load resistance.

Voltage source A circuit or device that supplies voltage to a load circuit.

Voltaic cell Primary cell having two unlike metal electrodes immersed in a solution that chemically interacts with the plates to produce an emf.

Volt-ampere (VA) Unit of apparent power in an ac circuit containing reactance. Apparent power is equal to the product of voltage and current.

Voltmeter Instrument designed to measure the voltage or potential difference. Its scale can be graduated in kilovolts, volts, or millivolts.

Volume Magnitude or power level of a complex audio frequency (AF) wave, expressed in volume units (VU).

VOM meter Abbreviation for volt-ohm-milliamp meter.

VRMS Abbreviation for "volts root-mean-square."

Wall outlet Spring-contact outlet mounted on the wall to which a portable appliance is connected to obtain electric power.

Watt (W) Unit of electric power required to do work at a rate of 1 joule/second. One watt of power is expended when 1 ampere direct current flows through a resistance of 1 ohm. In an ac circuit, the true power is effective volts multiplied by effective amperes, multiplied by the power factor.

Wattage rating Maximum power a device can safely handle continuously.

Watt-hour (Wh) Unit of electrical work, equal to a power of 1 watt being absorbed continuously for 1 hour.

Wattmeter A meter used to measure electric power in watts.

Wave Electric, electromagnetic, acoustic, mechanical, or other form whose physical activity rises and falls or advances and retreats periodically as it travels through some medium.

Waveform Shape of a wave.

Waveguide Rectangular or circular metal pipe used to guide electromagnetic waves at microwave frequencies.

Wavelength (λ) Distance between two points of corresponding phase and is equal to waveform velocity or speed divided by frequency.

Weber (Wb) Unit of magnetic flux. One weber is the amount of flux that when linked with a single turn of wire for an interval of 1 second, will induce an electromotive force of 1 V.

Wet cell Cell using a liquid electrolyte.

Wetting The coating of a contact surface.

Wheatstone bridge A four-arm, generally resistive bridge that is used to measure resistance.

Wideband amplifier Also called broadband amplifier, it is an amplifier that has a flat response over a wide range of frequencies.

Wien-Bridge oscillator An *RC* oscillator circuit in which a Wien Bridge determines frequency.

Winding One or more turns of a conductor wound to form a coil.

Wire Single solid or stranded group of conductors having a low resistance to current flow.

Wire gauge American wire gauge (AWG) is a system of numerical designations of wire sizes, with the first being 0000 (the largest size) and then going to 000, 00, 0, 1, 2, 3, and so on up to the smallest sizes of 40 and above.

Wireless Term describing radio communication that requires no wires between the two communicating points.

Wirewound resistor Resistor in which the resistive element is a length of high-resistance wire or ribbon, usually nichrome, wound onto an insulating form.

Wire-wrapping Method of prototyping in which solderless connections are made by wrapping wire around a rectangular terminal.

Woofer Large loudspeaker designed primarily to reproduce low audio-frequency signals at large power levels.

Work Work is done anytime energy is transformed from one type to another, and the amount of work done is dependent on the amount of energy transformed.

X Symbol for reactance.

***X* axis** Horizontal axis.

Y Symbol for admittance.

***Y* axis** Vertical axis.

***Z* axis** Axis perpendicular to both the *X* and *Y* axes.

Zener diode A semiconductor diode in which a reverse breakdown voltage current causes the diode to develop a constant voltage.

Zeroing Calibrating a meter so that it shows a value of zero when zero is being measured.

APPENDIX

B

Abbreviations

A (amp) Ampere
AC Alternating current
AC/DC Alternating current or direct current
A/D Analog to digital
ADC Analog-to-digital converter
AF Audio frequency
AFC Automatic frequency control
aft Automatic fine tuning
AGC Automatic gain control
Ah Ampere-hour
A_i Current gain
AM Amplitude modulation
AM/FM Amplitude modulation or frequency modulation
AMM Analog multimeter
antilog Antilogarithm
$\mathbf{A}_P$ Power gain
apc Automatic phase control
A_V Voltage gain
AVC Automatic volume control
AWG American wire gauge
B Flux density
BCD Binary coded decimal
bfo Beat frequency oscillator
BJT Bipolar junction transistor
BW Bandwidth
c Centi (10^{-2})
C Capacitance; capacitor
CAD Computer-aided design
CAM Computer-aided manufacturing
CATV Cable TV
CB Common-base configuration
CB Citizen's band
CC Common-collector configuration
CE Common-emitter configuration
cm Centimeter
cmil (cir mil) Circular mil
CPU Central processing unit
C (Q) Coulomb
CR, cr Junction diode
CRO Cathode ray oscilloscope
CRT Cathode ray tube
C_T Total capacitance
cw Continuous wave transmission
d Deci (10^{-1})
D/A or D-A Digital to analog
DC Direct current
d*i* or Δ*i* Change in current
DIP Dual in-line package
DMM Digital multimeter
DPDT Double pole, double throw
DPST Double pole, single throw
d*t* or Δ*t* Change in time
DTL Diode-transistor logic
d*v* or Δ*r* Change in voltage
DVM Digital voltmeter
E DC or rms difference in potential
e Instantaneous difference in potential
ECG Electrocardiogram
ECL Emitter-coupled logic
EHF Extremely high frequency
EHV Extra high voltage
ELF Extremely low frequency
EMF Electromotive force
EMI Electromagnetic interference
EW Electronic warfare
f Frequency
FET Field-effect transistor
FF Flip-flop
fil Filament
FM Frequency modulation
4PDT Four pole, double throw
4PST Four pole, single throw
f_r Frequency at resonance
FSD Full-scale deflection
fsk Frequency-shift keying

G Gravitational force
G Conductance
G Giga (10^9)
H Henry
H Magnetic field intensity
H Magnetizing flux
h Hecto (10^2)
h Hybrid
HF High frequency
hp Horsepower
Hz Hertz
I Current
i Instantaneous current
I_B DC base current
I_C DC collector current
IC Integrated circuit
I_E Total emitter current
I_{eff} Effective current
IF Intermediate frequency
$I_{\max}$ Maximum current
$I_{\min}$ Minimum current
I/O Input/output
IR Infrared
I_R Current through resistance
I_S Secondary current
I_T Total current
JFET Junction FET
K Coefficient of coupling
k Kilo (10^3)
kHz Kilohertz
kV Kilovolt
kVA Kilovoltampere
kW Kilowatt
kWh Kilowatt-hour
L Coil; inductance
LC Inductance–capacitance
LCD Liquid crystal display
L-C-R Inductance–capacitance–resistance
LDR Light-dependent resistor
LED Light-emitting diode
LF Low frequency
L_M Mutual inductance
LNA Low-noise amplifier
LO Local oscillator
LSI Large-scale integration
L_T Total inductance
M Mega (10^6)
M Mutual conductance
M Mutual inductance
m Milli (10^{-3})
mA Milliampere
mag Magnetron
max Maximum
MF Medium frequency
mH Millihenry
MHz Megahertz
min Minimum
mm Millimeter
mmf Magnetomotive force
MOS Metal oxide semiconductor
MOSFET Metal-oxide semiconductor field effect transistor
MPU Microprocessor unit (μP)
MSI Medium scale integrated circuit
mV Millivolt
mW Milliwatt
N Number of turns in an inductor
N Revolutions per minute
n Nano (10^{-9})
N Negative
nA Nanoampere
NC No connection
NC Normally closed
NEG, neg Negative
nF Nanofarad
nH Nanohenry
nm Nanometer
NO Normally open
NPN Negative–positive–negative
ns Nanosecond
nW Nanowatt
OP AMP Operational amplifier
P Pico (1×10^{-12})
P Power
p Instantaneous power
P Positive; peak
PA Public address or power amplifier
pA Picoampere
PAM, pam Pulse-amplitude modulation
P_{ap} Apparent power

P_{av} Average power
PCB Printed circuit board
PCM, pcm Pulse-code modulation
PDM Pulse-duration modulation
pF Picofarad
PLL Phase-locked loop
PM Phase modulation; permanent magnet
PNP Positive–negative–positive
POT, pot Potentiometer
P-P Peak to peak
PPM Pulse-position modulation
PRF Pulse repetition frequency
PRT Pulse repetition time
PW Pulse width
PWM, pwm Pulse-width modulation
Q Charge or quality
q Instantaneous charge
R Resistance
R Reluctance
RAM Random-access memory
RC Resistance-capacitance
rcvr Receiver
rect Rectifier
ref Reference
rf Radio frequencies
RF Radio frequency
RFI Radio frequency interference
R_L Load resistor
RLC Resistance–inductance–capacitance
RMS Root mean square
ROM Read-only memory
rpm Revolutions per minute
SCR Silicon controlled rectifier
SHF Superhigh frequency
SIP Single in-line package
SNR Signal-to-noise ratio
SPDT Single pole, double throw
SPST Single pole, single throw
sq cm Square centimeter
SSB Single sideband
SW Shortwave
SWR Standing-wave ratio
SYNC Synchronous
T Tera (10^{12})
T Torque
T Transformer
t Time (seconds)
TC Time constant; temperature coefficient
TE Transverse electric
temp Temperature
THz Teraherz
TM Transverse magnetic
TR Transmit–receive
TTL Transistor–transistor logic
TV Television
TWT Travelling wave tube
UHF Ultrahigh frequency
UHV Ultrahigh voltage
UJT Unijunction transistor
UV Ultraviolet
V Vacuum tube
V, v Volt
v Instantaneous voltage
VA Voltampere
V_{av} Voltage (average value)
V_{BE} Fixed base-emitter voltage
V_c Capacitive voltage
V_{CE} Fixed collector-emitter voltage
VCO Voltage-controlled oscillator
VHF Very high frequency
V_{in} Input voltage
V_L Inductive voltage
VLF Very low frequency
V_m, V_{max} Maximum voltage
VOM Volt-ohm-milliammeter
V_{out} Output voltage
V_p Primary voltage
V_s Source voltage
***V*SWR** Voltage standing wave ratio
V_T Total voltage
W Watt
X_C Capacitive reactance
X_L Inductive reactance
Y Admittance
Z Impedance
Z_{in} Input impedance
Z_o Output impedance
Z_p Primary impedance
Z_s Secondary impedance
Z_T Total impedance
kΩ Kilohm

MΩ Megohm
μA Microampere
μF (mfd) Microfarad
μH Microhenry
μV Microvolt
μW Microwatt
Ω Ohm
λ Wavelength
° Degrees
°C Degrees Centigrade
°F Degrees Fahrenheit

APPENDIX

Electronic Schematic Symbols

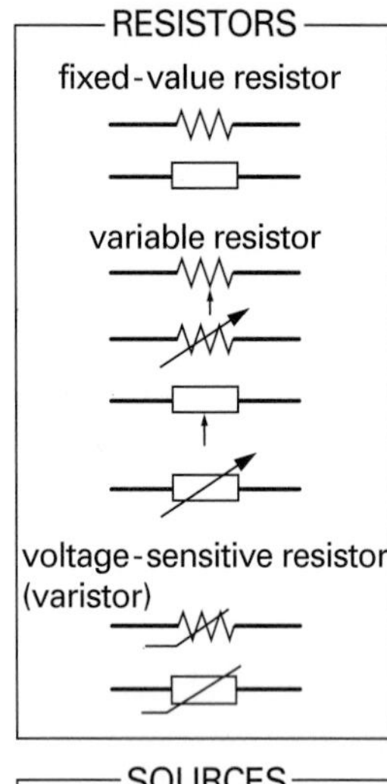
RESISTORS
fixed-value resistor
variable resistor
voltage-sensitive resistor
(varistor)

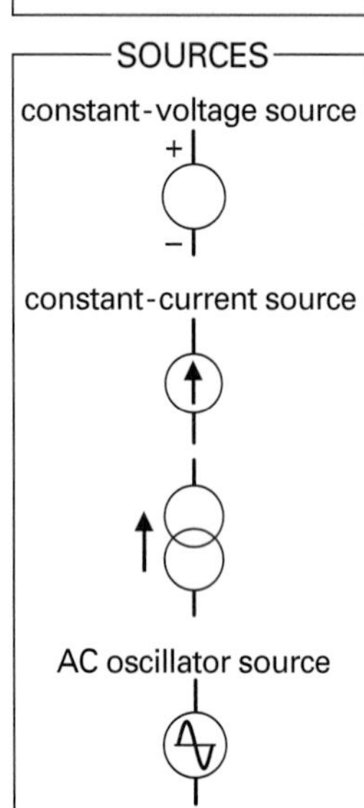
SOURCES
constant-voltage source
+
−
constant-current source
AC oscillator source

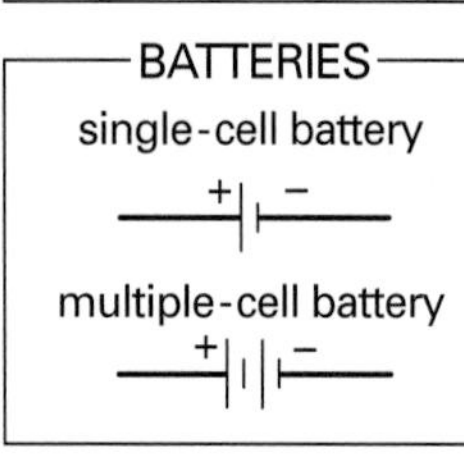
BATTERIES
single-cell battery
+
−
multiple-cell battery
+
−

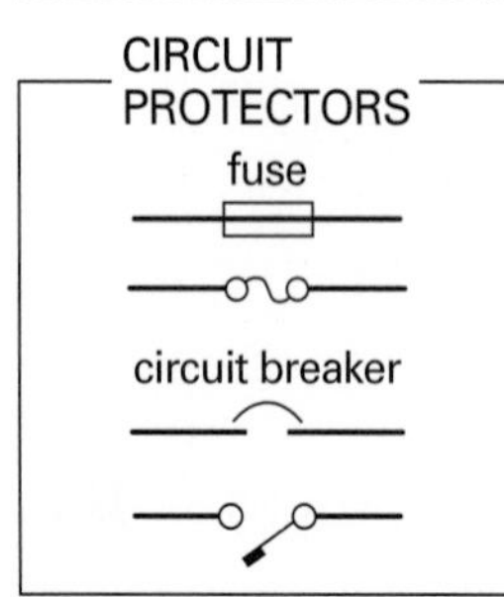
CIRCUIT
PROTECTORS
fuse
circuit breaker

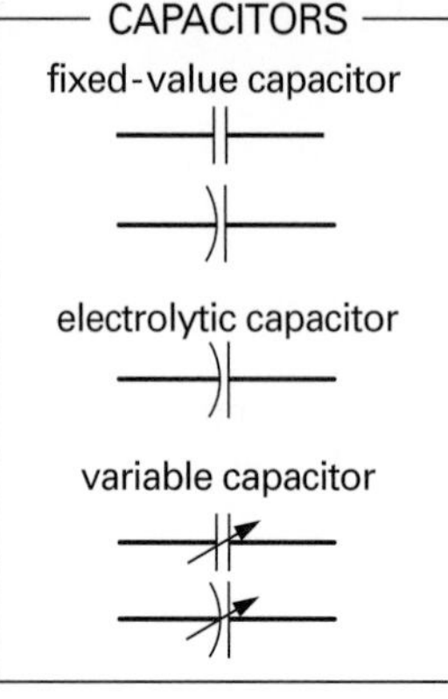
CAPACITORS
fixed-value capacitor
electrolytic capacitor
variable capacitor

CRYSTALS

piezoelectric crystal

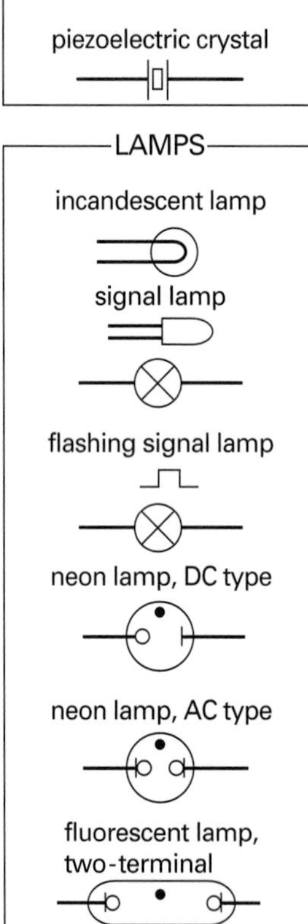
LAMPS
incandescent lamp
signal lamp
flashing signal lamp
neon lamp, DC type
neon lamp, AC type
fluorescent lamp,
two-terminal

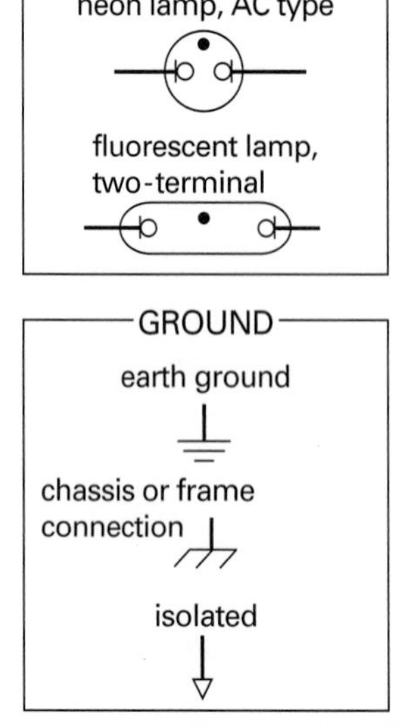
GROUND
earth ground
chassis or frame
connection
isolated

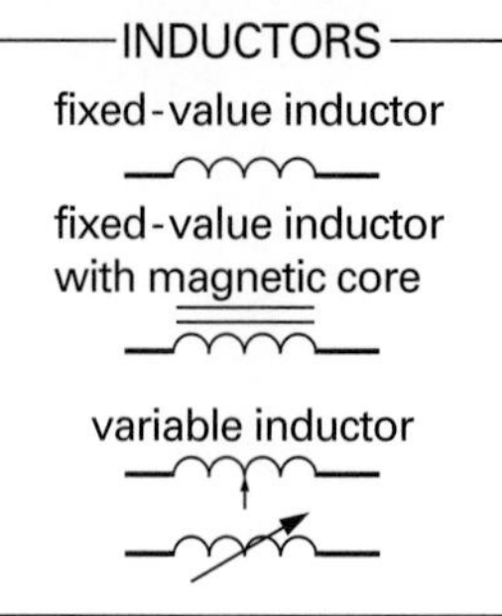
INDUCTORS
fixed-value inductor
fixed-value inductor
with magnetic core
variable inductor

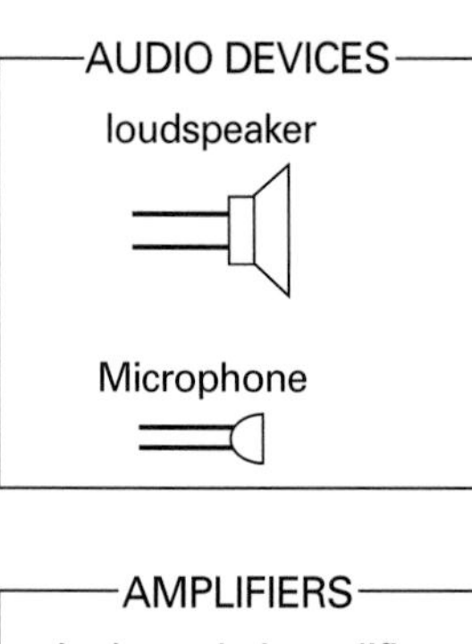
AUDIO DEVICES
loudspeaker
Microphone

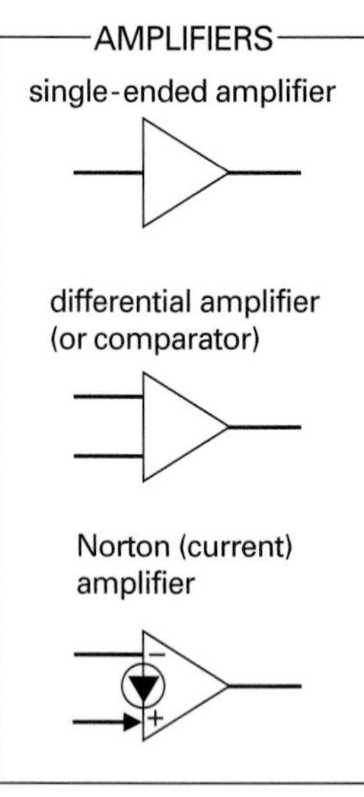
AMPLIFIERS
single-ended amplifier
differential amplifier
(or comparator)
Norton (current)
amplifier
−
+

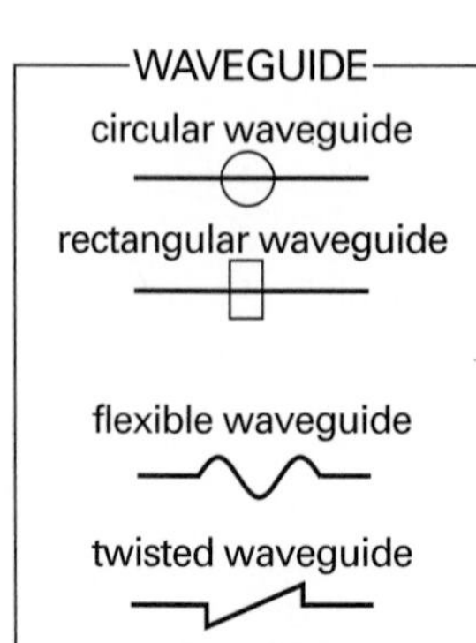
WAVEGUIDE
circular waveguide
rectangular waveguide
flexible waveguide
twisted waveguide

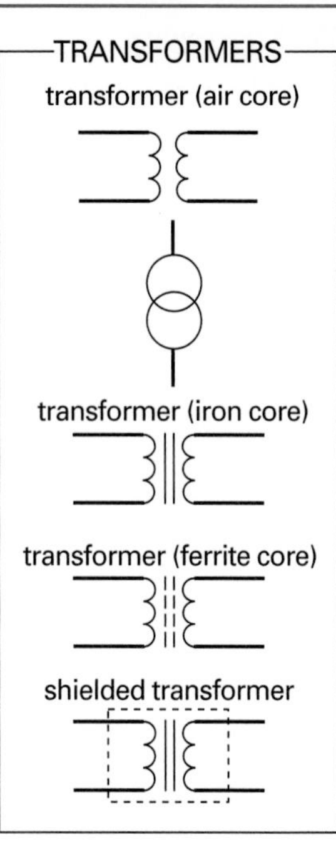
TRANSFORMERS
transformer (air core)
transformer (iron core)
transformer (ferrite core)
shielded transformer

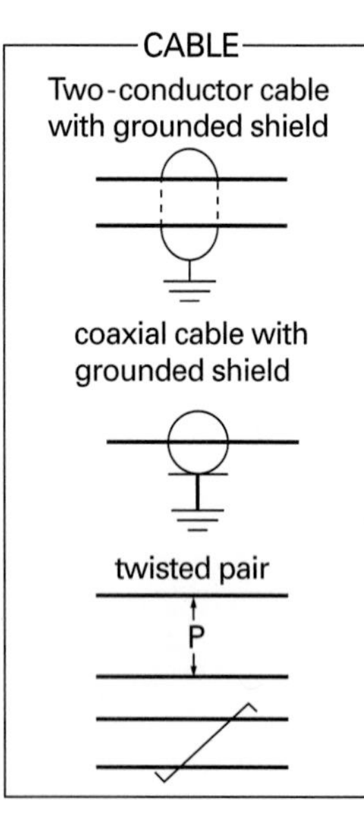
CABLE
Two-conductor cable
with grounded shield
coaxial cable with
grounded shield
twisted pair
P

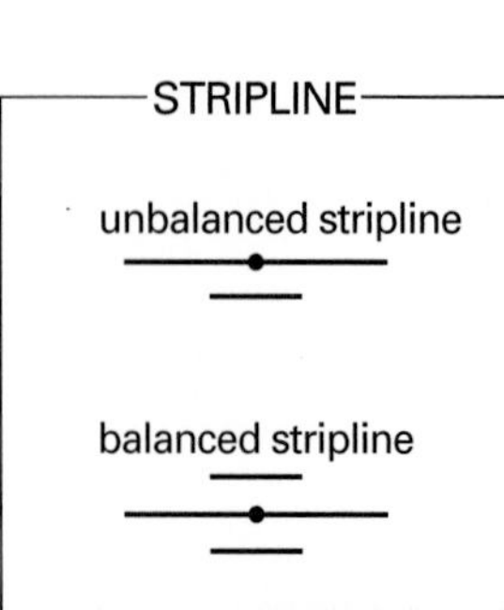
STRIPLINE
unbalanced stripline
balanced stripline

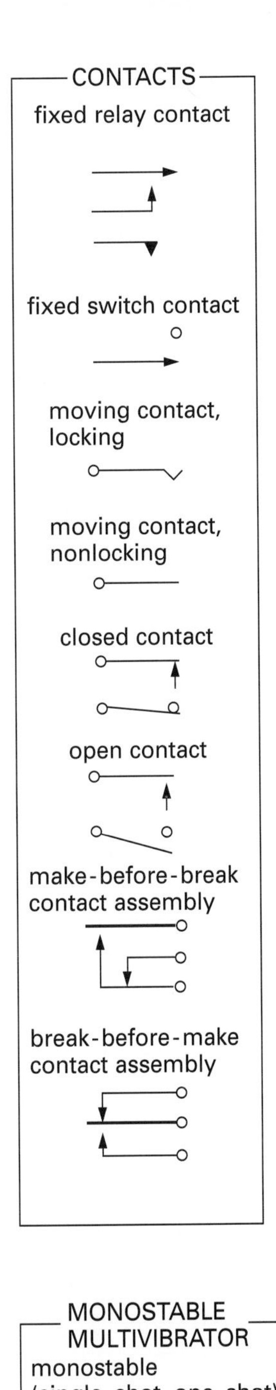
CONTACTS
fixed relay contact
fixed switch contact
moving contact, locking
moving contact, nonlocking
closed contact
open contact
make-before-break contact assembly
break-before-make contact assembly

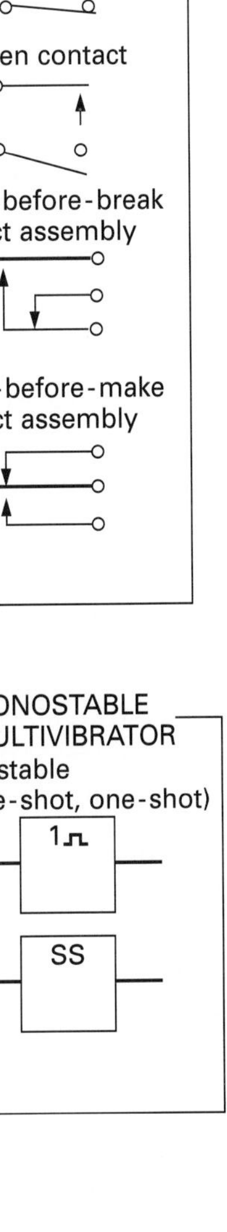
MONOSTABLE MULTIVIBRATOR
monostable (single-shot, one-shot)
1
SS

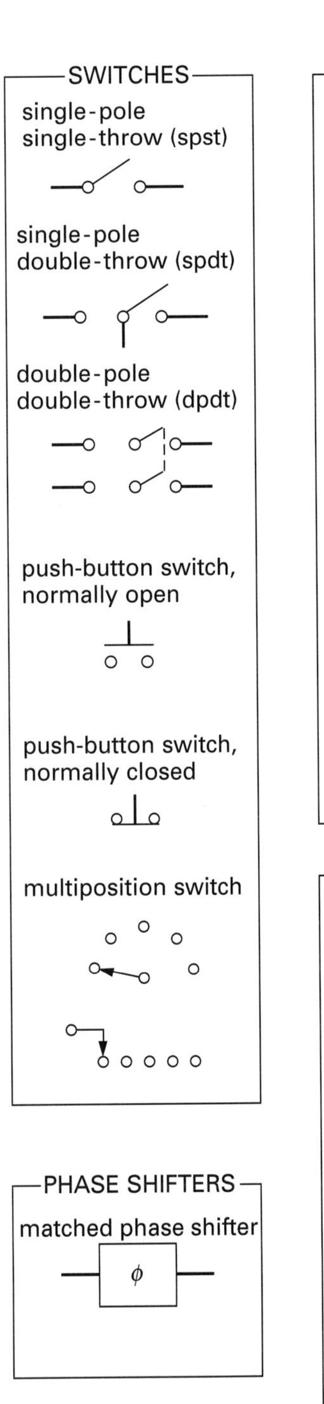
SWITCHES
single-pole single-throw (spst)
single-pole double-throw (spdt)
double-pole double-throw (dpdt)
push-button switch, normally open
push-button switch, normally closed
multiposition switch

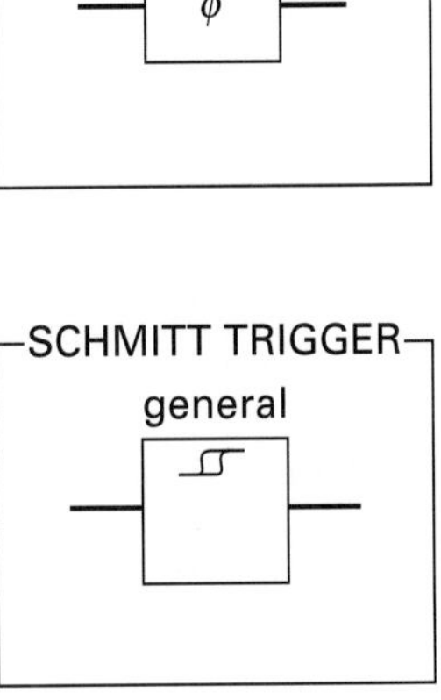
PHASE SHIFTERS
matched phase shifter
ϕ
SCHMITT TRIGGER
general

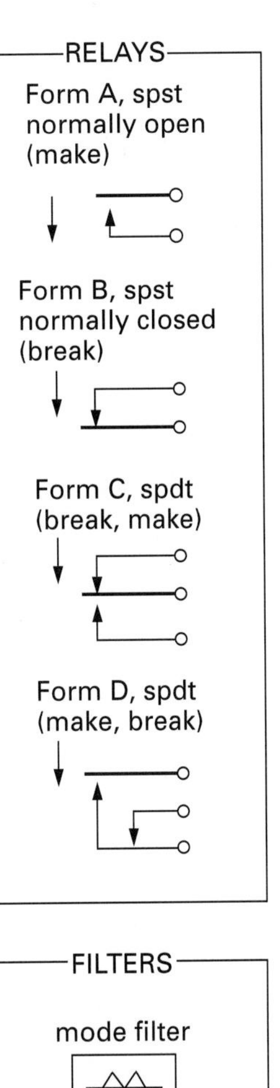
RELAYS
Form A, spst normally open (make)
Form B, spst normally closed (break)
Form C, spdt (break, make)
Form D, spdt (make, break)

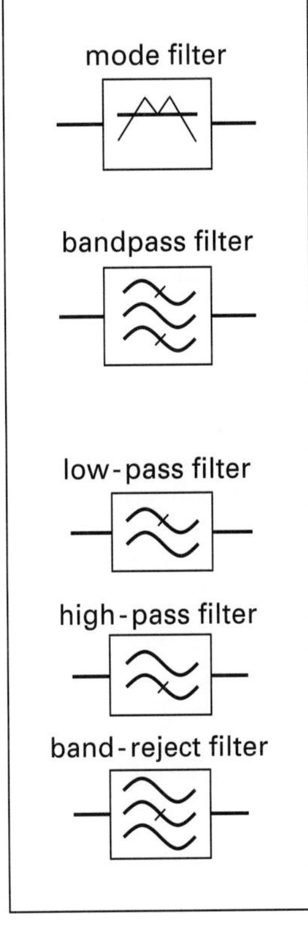
FILTERS
mode filter
bandpass filter
low-pass filter
high-pass filter
band-reject filter

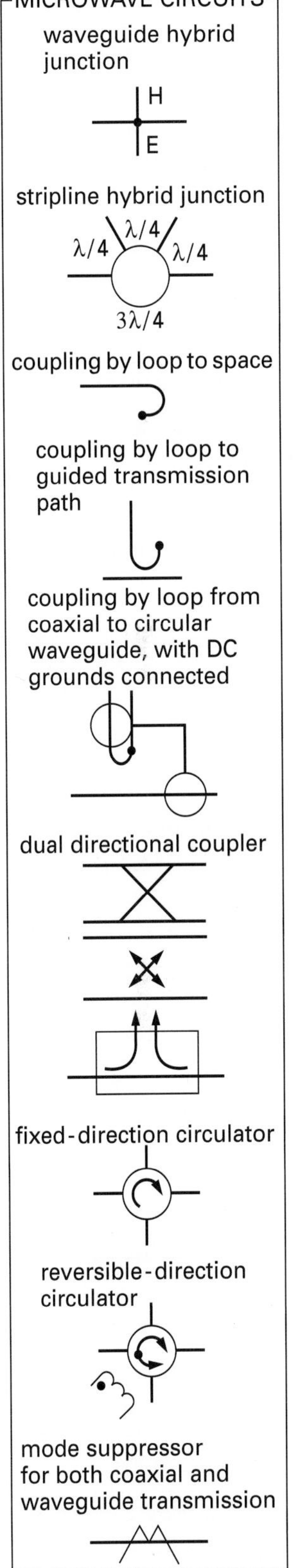
MICROWAVE CIRCUITS
waveguide hybrid junction
H
E
stripline hybrid junction
λ/4
λ/4
λ/4
3λ/4
coupling by loop to space
coupling by loop to guided transmission path
coupling by loop from coaxial to circular waveguide, with DC grounds connected
dual directional coupler
fixed-direction circulator
reversible-direction circulator
mode suppressor for both coaxial and waveguide transmission

DIODES

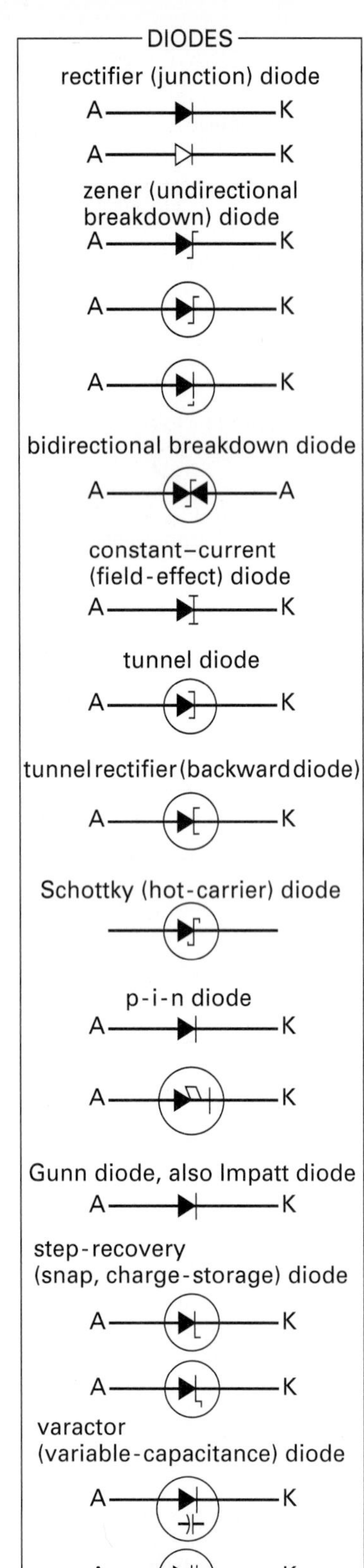
rectifier (junction) diode
A
K
A
K
zener (undirectional breakdown) diode
A
K
A
K
A
K
bidirectional breakdown diode
A
A
constant–current (field-effect) diode
A
K
tunnel diode
A
K
tunnel rectifier (backward diode)
A
K
Schottky (hot-carrier) diode
p-i-n diode
A
K
A
K
Gunn diode, also Impatt diode
A
K
step-recovery (snap, charge-storage) diode
A
K
A
K
varactor (variable-capacitance) diode
A
K
A
K

TRANSISTORS

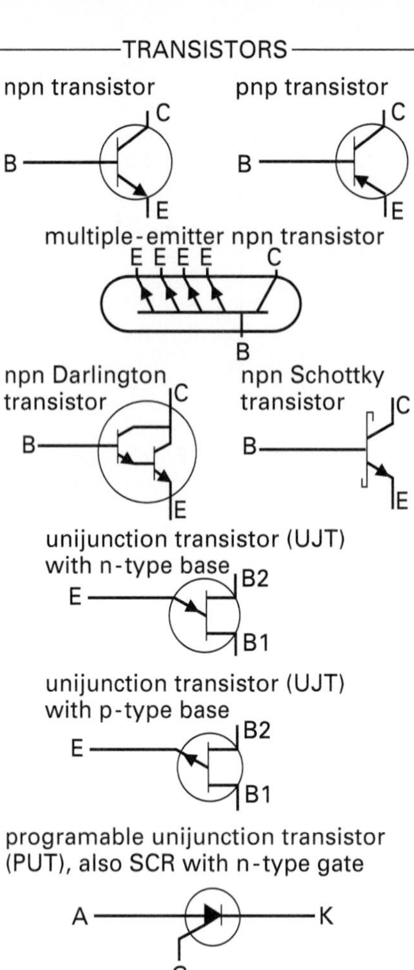
npn transistor
pnp transistor
B
C
E
B
C
E
multiple-emitter npn transistor
E E E E
C
B
npn Darlington transistor
C
B
E
npn Schottky transistor
C
B
E
unijunction transistor (UJT) with n-type base
E
B2
B1
unijunction transistor (UJT) with p-type base
E
B2
B1
programable unijunction transistor (PUT), also SCR with n-type gate
A
K
G

THYRISTORS

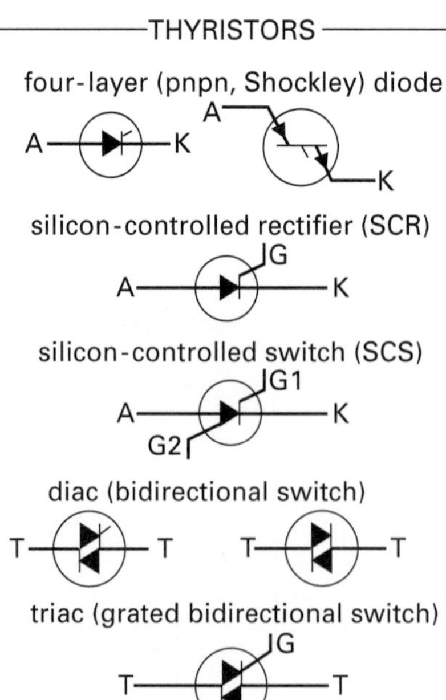
four-layer (pnpn, Shockley) diode
A
K
A
K
silicon-controlled rectifier (SCR)
G
A
K
silicon-controlled switch (SCS)
G1
A
K
G2
diac (bidirectional switch)
T
T
T
T
triac (grated bidirectional switch)
G
T
T

FIELD-EFFECT TRANSISTORS (FETs)

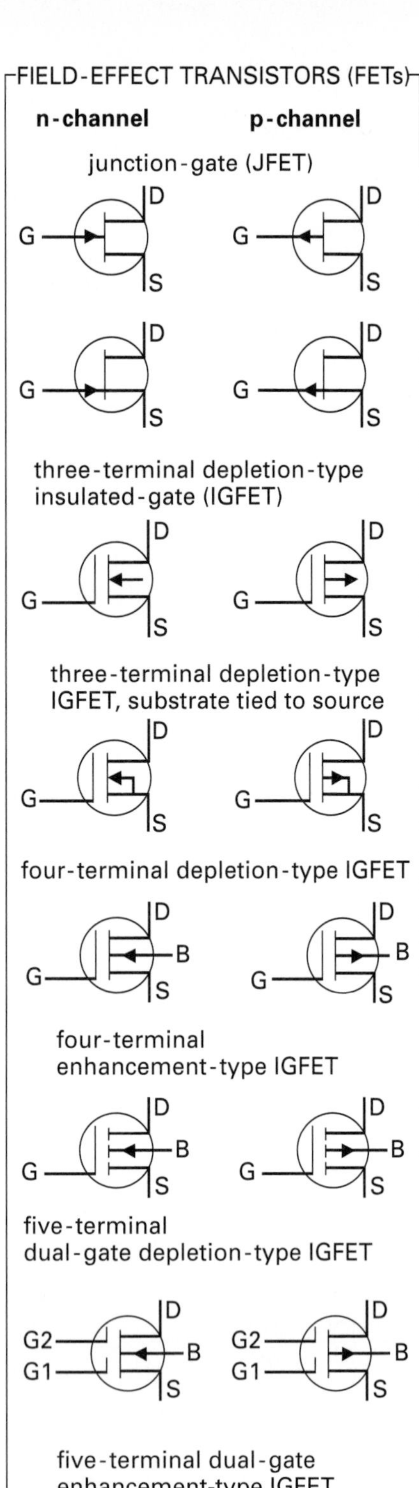
n-channel
p-channel
junction-gate (JFET)
G
D
S
G
D
S
G
D
S
G
D
S
three-terminal depletion-type insulated-gate (IGFET)
G
D
S
G
D
S
three-terminal depletion-type IGFET, substrate tied to source
G
D
S
G
D
S
four-terminal depletion-type IGFET
G
D
B
S
G
D
B
S
four-terminal enhancement-type IGFET
G
D
B
S
G
D
B
S
five-terminal dual-gate depletion-type IGFET
G2
G1
D
B
S
G2
G1
D
B
S
five-terminal dual-gate enhancement-type IGFET

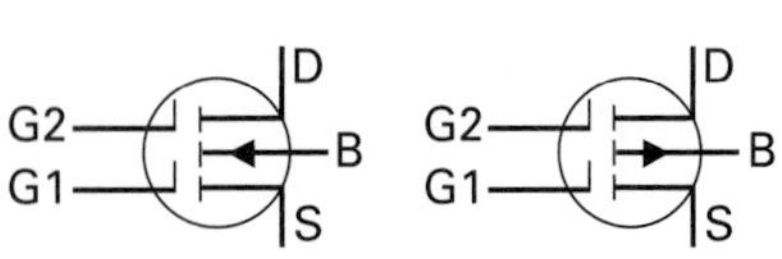
G2
G1
D
B
S
G2
G1
D
B
S

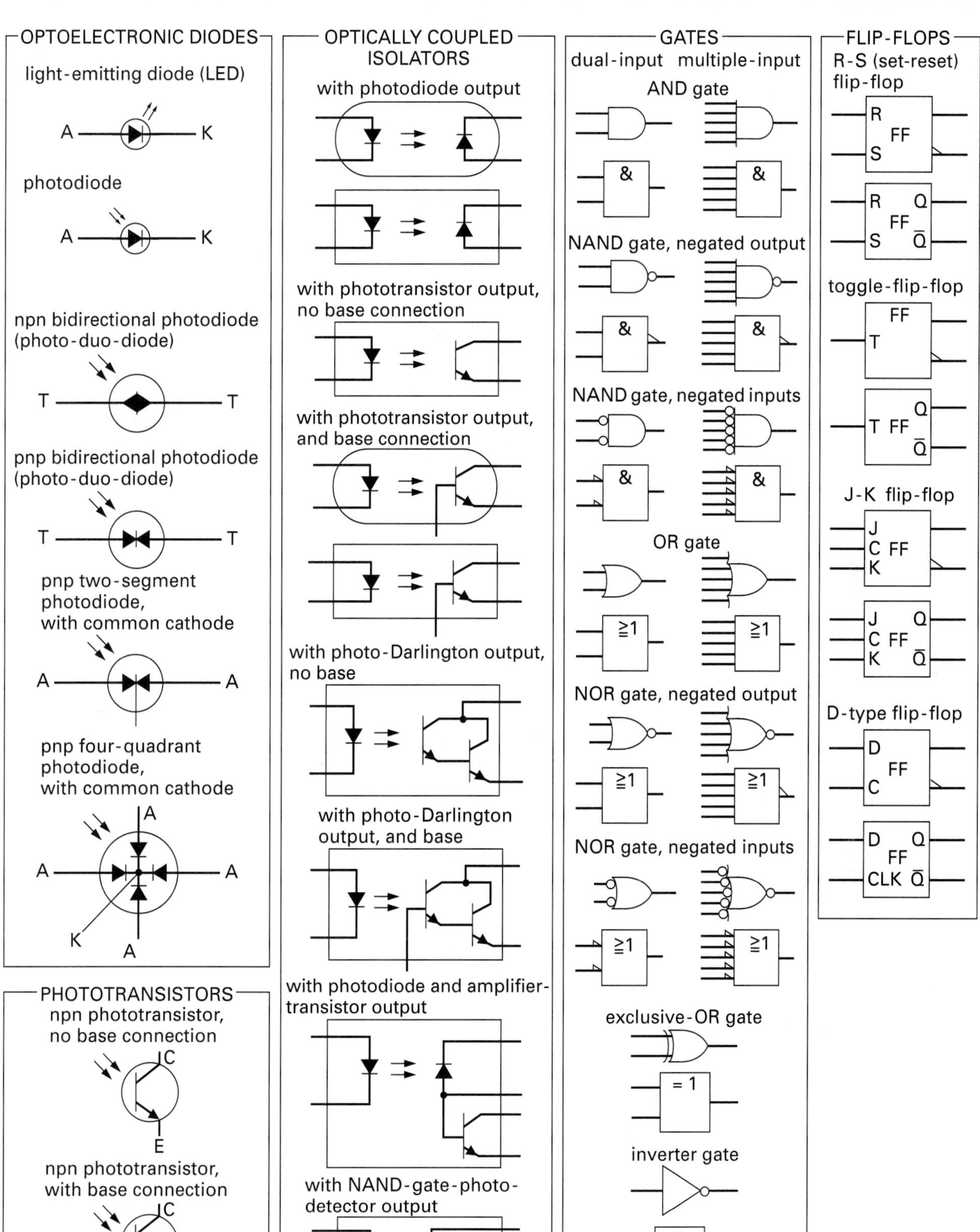
OPTOELECTRONIC DIODES
light-emitting diode (LED)
A
K
photodiode
A
K
npn bidirectional photodiode (photo-duo-diode)
T
T
pnp bidirectional photodiode (photo-duo-diode)
T
T
pnp two-segment photodiode, with common cathode
A
A
pnp four-quadrant photodiode, with common cathode
A
A
A
K
A
PHOTOTRANSISTORS
npn phototransistor, no base connection
C
E
npn phototransistor, with base connection
C
B
E
OPTICALLY COUPLED ISOLATORS
with photodiode output
with phototransistor output, no base connection
with phototransistor output, and base connection
with photo-Darlington output, no base
with photo-Darlington output, and base
with photodiode and amplifier-transistor output
with NAND-gate-photo-detector output
GATES
dual-input
multiple-input
AND gate
&
&
NAND gate, negated output
&
&
NAND gate, negated inputs
&
&
OR gate
≥1
≥1
NOR gate, negated output
≥1
≥1
NOR gate, negated inputs
≥1
≥1
exclusive-OR gate
= 1
inverter gate
1
FLIP-FLOPS
R-S (set-reset) flip-flop
R
FF
S
R
Q
FF
S
Q̄
toggle-flip-flop
FF
T
Q
T FF
Q̄
J-K flip-flop
J
C FF
K
J
Q
C FF
K
Q̄
D-type flip-flop
D
FF
C
D
Q
FF
CLK
Q̄

APPENDIX

D

Device Data Sheets

DEVICE: IN4001 Through IN4007— Silicon Junction Diodes

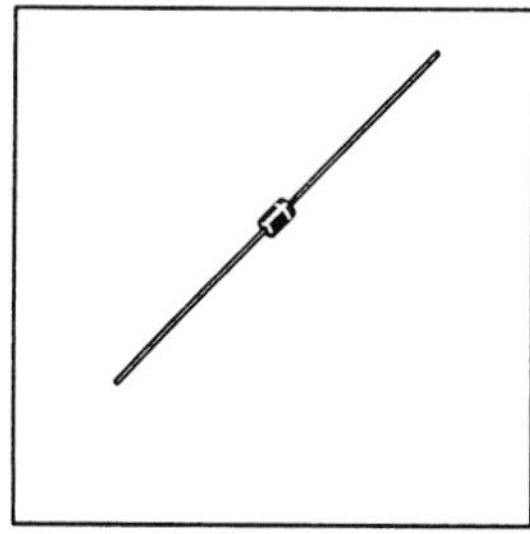

LEAD MOUNTED RECTIFIERS
50–1000 VOLTS
DIFFUSED JUNCTION

Designer's Data Sheet

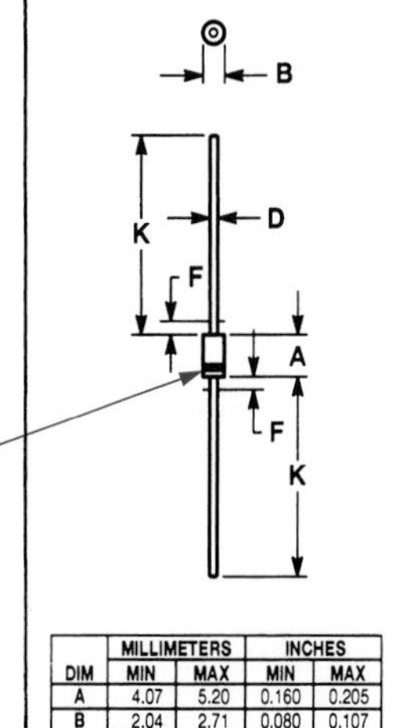

DIM	MILLIMETERS		INCHES	
	MIN	MAX	MIN	MAX
A	4.07	5.20	0.160	0.205
B	2.04	2.71	0.080	0.107
D	0.71	0.86	0.028	0.034
F	—	1.27	—	0.050
K	27.94	—	1.100	—

Mechanical Characteristics

- Case: Epoxy, Molded
- Weight: 0.4 gram (approximately)
- Finish: All External Surfaces Corrosion Resistant and Terminal Leads are Readily Solderable
- Lead and Mounting Surface Temperature for Soldering Purposes: 220°C Max. for 10 Seconds, 1/16″ from case
- Shipped in plastic bags, 1000 per bag.
- Available Tape and Reeled, 5000 per reel, by adding a "RL" suffix to the part number
- Polarity: Cathode Indicated by Polarity Band
- Marking: 1N4001, 1N4002, 1N4003, 1N4004, 1N4005, 1N4006, 1N4007

The **Peak Repetitive Reverse Voltage** is the maximum allowable reverse voltage; i.e., IN4001 will probably break down if the reverse voltage exceeds 50 V. This rating is also called **Working Peak Reverse Voltage** and **DC Blocking Voltage.**

MAXIMUM RATINGS

Rating	Symbol	1N4001	1N4002	1N4003	1N4004	1N4005	1N4006	1N4007	Unit
*Peak Repetitive Reverse Voltage Working Peak Reverse Voltage DC Blocking Voltage	V_{RRM} V_{RWM} V_R	50	100	200	400	600	800	1000	Volts
*Non–Repetitive Peak Reverse Voltage (halfwave, single phase, 60 Hz)	V_{RSM}	60	120	240	480	720	1000	1200	Volts
*RMS Reverse Voltage	$V_{R(RMS)}$	35	70	140	280	420	560	700	Volts
*Average Rectified Forward Current (single phase, resistive load, 60 Hz, see Figure 8, T_A = 75°C)	I_O	1.0							Amp
*Non–Repetitive Peak Surge Current (surge applied at rated load conditions, see Figure 2)	I_{FSM}	30 (for 1 cycle)							Amp
Operating and Storage Junction Temperature Range	T_J T_{stg}	– 65 to +175							°C

Maximum allowable nonrepeating reverse voltage.

The RMS of the Peak Repetitive Reverse Voltage. $V_{rms} = 0.707 \times V_{peak}$.

The maximum average forward current value.

The maximum surge current value.

The diode can be operated and stored at any temperature between –65° to +175° C.

ELECTRICAL CHARACTERISTICS*

Rating	Symbol	Typ	Max	Unit
Maximum Instantaneous Forward Voltage Drop (i_F = 1.0 Amp, T_J = 25°C) Figure 1	v_F	0.93	1.1	Volts
Maximum Full–Cycle Average Forward Voltage Drop (I_O = 1.0 Amp, T_L = 75°C, 1 inch leads)	$V_{F(AV)}$	—	0.8	Volts
Maximum Reverse Current (rated dc voltage) (T_J = 25°C) (T_J = 100°C)	I_R	 0.05 1.0	 10 50	µA
Maximum Full–Cycle Average Reverse Current (I_O = 1.0 Amp, T_L = 75°C, 1 inch leads)	$I_{R(AV)}$	—	30	µA

The maximum voltage drop that will appear across the diode when forward biased.

The maximum average forward voltage drop.

Maximum reverse current at different temps.

This is the maximum average value of reverse current.

DEVICE: IN746 through IN759 and IN957 through IN986—Zener Diodes

500-MILLIWATT HERMETICALLY SEALED GLASS SILICON ZENER DIODES

- Complete Voltage Range — 2.4 to 110 Volts
- DO-35 Package — Smaller than Conventional DO-7 Package
- Double Slug Type Construction
- Metallurgically Bonded Construction
- Oxide Passivated Die

Designer's Data for "Worst Case" Conditions

The Designer's Data sheets permit the design of most circuits entirely from the information presented. Limit curves – representing boundaries on device characteristics – are given to facilitate "worst case" design.

GLASS ZENER DIODES
500 MILLIWATTS
2.4-110 VOLTS

Maximum power dissipation (P_D)

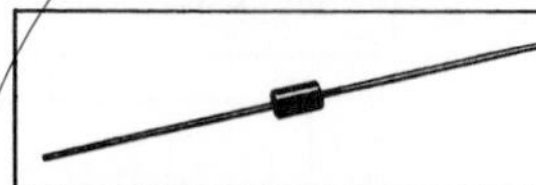

No letter following part number indicates a ± 10% V_z tolerance

1N746 thru 1N759
1N957A thru 1N986A
1N4370 thru 1N4372

Letter "A" following part number indicates a ± 5% V_z tolerance

NOTES:
1. PACKAGE CONTOUR OPTIONAL WITHIN A AND B. HEAT SLUGS, IF ANY, SHALL BE INCLUDED WITHIN THIS CYLINDER, BUT NOT SUBJECT TO THE MINIMUM LIMIT OF B.
2. LEAD DIAMETER NOT CONTROLLED IN ZONE F TO ALLOW FOR FLASH, LEAD FINISH BUILDUP AND MINOR IRREGULARITIES OTHER THAN HEAT SLUGS.
3. POLARITY DENOTED BY CATHODE BAND.
4. DIMENSIONING AND TOLERANCING PER ANSI Y14.5, 1973.

DIM	MILLIMETERS MIN	MILLIMETERS MAX	INCHES MIN	INCHES MAX
A	3.05	5.08	0.120	0.200
B	1.52	2.29	0.060	0.090
D	0.46	0.56	0.018	0.022
F	–	1.27	–	0.050
K	25.40	38.10	1.000	1.500

All JEDEC dimensions and notes apply.

CASE 299-02
DO-204AH
GLASS

MAXIMUM RATINGS

Rating	Symbol	Value	Unit
DC Power Dissipation @ $T_L \le 50^\circ C$, Lead Length = 3/8″	P_D		
*JEDEC Registration		400	mW
*Derate above $T_L = 50^\circ C$		3.2	mW/°C
Motorola Device Ratings		500	mW
Derate above $T_L = 50^\circ C$		3.33	mW/°C
Operating and Storage Junction Temperature Range	T_J, T_{stg}		°C
*JEDEC Registration		-65 to +175	
Motorola Device Ratings		-65 to +200	

*Indicates JEDEC Registered Data.

JEDEC (Joint Electronic Device Engineering Council) values are for military applications. P_D is 100 mW below its actual capability to build in high tolerance for military applications. In non-military applications we use manufacturers specifications:

JEDEC P_D = 400 mW.

Manu. P_D = 500 mW.

ELECTRICAL CHARACTERISTICS ($T_A = 25^\circ C$, $V_F = 1.5$ V max at 200 mA for all types)

Type Number (Note 1)	Nominal Zener Voltage V_Z @ I_{ZT} (Note 2) Volts	Test Current I_{ZT} mA	Maximum Zener Impedance Z_{ZT} @ I_{ZT} (Note 3) Ohms	*Maximum DC Zener Current I_{ZM} (Note 4) mA (JEDEC Spec.)	*Maximum DC Zener Current I_{ZM} (Note 4) mA (Manufacturers Spec.)	Maximum Reverse Leakage Current $T_A = 25^\circ C$ I_R @ $V_R = 1$ V μA	Maximum Reverse Leakage Current $T_A = 150^\circ C$ I_R @ $V_R = 1$ V μA
1N4370	2.4	20	30	150	190	100	200
1N4371	2.7	20	30	135	165	75	150
1N4372	3.0	20	29	120	150	50	100
1N746	3.3	20	28	110	135	10	30
1N747	3.6	20	24	100	125	10	30
1N748	3.9	20	23	95	115	10	30
1N749	4.3	20	22	85	105	2	30
1N750	4.7	20	19	75	95	2	30
1N751	5.1	20	17	70	85	1	20
1N752	5.6	20	11	65	80	1	20
1N753	6.2	20	7	60	70	0.1	20
1N754	6.8	20	5	55	65	0.1	20
1N755	7.5	20	6	50	60	0.1	20
1N756	8.2	20	8	45	55	0.1	20
1N757	9.1	20	10	40	50	0.1	20
1N758	10	20	17	35	45	0.1	20
1N759	12	20	30	30	35	0.1	20

JEDEC Spec. → ← Manufacturers Spec.

Example: IN752

$$I_{ZM} = \frac{P_D(\max)}{V_Z(\max)}$$

$$I_{ZM} = \frac{500 \text{ mW}}{5.6 \text{ V} + 10\%}$$

$$I_{ZM} = \frac{500 \text{ mW}}{6.16 \text{ V}}$$

$$I_{ZM} = 81.2 \text{ mA}$$

Type Number (Note 1)	Nominal Zener Voltage V_Z (Note 2) Volts	Test Current I_{ZT} mA	Maximum Zener Impedance (Note 3) Z_{ZT} @ I_{ZT} Ohms	Z_{ZK} @ I_{ZK} Ohms	I_{ZK} mA	*Maximum DC Zener Current I_{ZM} (Note 4) mA		Maximum Reverse Current I_R Maximum μA	Test Voltage Vdc V_R 5%	Test Voltage Vdc V_R 10%
1N957A	6.8	18.5	4.5	700	1.0	47	61	150	5.2	4.9
1N958A	7.5	16.5	5.5	700	0.5	42	55	75	5.7	5.4
1N959A	8.2	15	6.5	700	0.5	38	50	50	6.2	5.9
1N960A	9.1	14	7.5	700	0.5	35	45	25	6.9	6.6
1N961A	10	12.5	8.5	700	0.25	32	41	10	7.6	7.2
1N962A	11	11.5	9.5	700	0.25	28	37	5	8.4	8.0
1N963A	12	10.5	11.5	700	0.25	26	34	5	9.1	8.6
1N964A	13	9.5	13	700	0.25	24	32	5	9.9	9.4
1N965A	15	8.5	16	700	0.25	21	27	5	11.4	10.8
1N966A	16	7.8	17	700	0.25	19	37	5	12.2	11.5
1N967A	18	7.0	21	750	0.25	17	23	5	13.7	13.0
1N968A	20	6.2	25	750	0.25	15	20	5	15.2	14.4
1N969A	22	5.6	29	750	0.25	14	18	5	16.7	15.8
1N970A	24	5.2	33	750	0.25	13	17	5	18.2	17.3
1N971A	27	4.6	41	750	0.25	11	15	5	20.6	19.4
1N972A	30	4.2	49	1000	0.25	10	13	5	22.8	21.6
1N973A	33	3.8	58	1000	0.25	9.2	12	5	25.1	23.8
1N974A	36	3.4	70	1000	0.25	8.5	11	5	27.4	25.9
1N975A	39	3.2	80	1000	0.25	7.8	10	5	29.7	28.1
1N976A	43	3.0	93	1500	0.25	7.0	9.6	5	32.7	31.0
1N977A	47	2.7	105	1500	0.25	6.4	8.8	5	35.8	33.8
1N978A	51	2.5	125	1500	0.25	5.9	8.1	5	38.8	36.7
1N979A	56	2.2	150	2000	0.25	5.4	7.4	5	42.6	40.3
1N980A	62	2.0	185	2000	0.25	4.9	6.7	5	47.1	44.6
1N981A	68	1.8	230	2000	0.25	4.5	6.1	5	51.7	49.0
1N982A	75	1.7	270	2000	0.25	1.0	5.5	5	56.0	54.0
1N983A	82	1.5	330	3000	0.25	3.7	5.0	5	62.2	59.0
1N984A	91	1.4	400	3000	0.25	3.3	4.5	5	69.2	65.5
1N985A	100	1.3	500	3000	0.25	3.0	4.5	5	76	72
1N986A	110	1.1	750	4000	0.25	2.7	4.1	5	83.6	79.2

DEVICE: Silicon Zener Diodes—1.8 V to 400 V, 225 mW to 5 W

Horizontal listing indicates maximum power dissipation (P_D)

Vertical listing indicates nominal zener voltage (V_Z).

Nominal Zener Breakdown Voltage	225 mW Surface Mount SOT-23		500 mW Surface Mount SOD-123	500 mW Low Level Surface Mount SOD-123	500 mW Surface Mount SOD-123	3 Watt Surface Mount SMB
Volts	Anode, Cathode, No Connection — Plastic Case 318-07 TO-236AB		Plastic Case 425, Style 1			Plastic Case 403A Cathode = Notch
1.8				MMSZ4678T1		
2.0				MMSZ4679T1		
2.2				**MMSZ4680T1**		
2.4	BZX84C2V4LT1	MMBZ5221BLT1	MMSZ2V4T1	MMSZ4681T1	**MMSZ5221BT1**	
2.5		MMBZ5222BLT1			**MMSZ5222BT1**	
2.7	BZX84C2V7LT1	MMBZ5223BLT1	MMSZ2V7T1	MMSZ4682T1	MMSZ5223BT1	
2.8		MMBZ5224BLT1			MMSZ5224BT1	
3.0	BZX84C3V0LT1	MMBZ5225BLT1	MMSZ3V0T1	MMSZ4683T1	**MMSZ5225BT1**	
3.3	BZX84C3V3LT1	MMBZ5226BLT1	MMSZ3V3T1	MMSZ4684T1	MMSZ5226BT1	1SMB5913BT3
3.6	BZX84C3V6LT1	MMBZ5227BLT1	MMSZ3V6T1	**MMSZ4685T1**	MMSZ5227BT1	1SMB5914BT3
3.9	BZX84C3V9LT1	**MMBZ5228BLT1**	MMSZ3V9T1	MMSZ4686T1	MMSZ5228BT1	1SMB5915BT3
4.3	BZX84C4V3LT1	MMBZ5229BLT1	MMSZ4V3T1	MMSZ4687T1	**MMSZ5229BT1**	1SMB5916BT3
4.7	**BZX84C4V7LT1**	MMBZ5230BLT1	MMSZ4V7T1	**MMSZ4688T1**	**MMSZ5230BT1**	1SMB5917BT3
5.1	**BZX84C5V1LT1**	**MMBZ5231BLT1**	**MMSZ5V1T1**	**MMSZ4689T1**	**MMSZ5231BT1**	1SMB5918BT3
5.6	**BZX84C5V6LT1**	**MMBZ5232BLT1**	**MMSZ5V6T1**	**MMSZ4690T1**	**MMSZ5232BT1**	1SMB5919BT3
6.0		MMBZ5233BLT1			MMSZ5233BT1	
6.2	**BZX84C6V2LT1**	**MMBZ5234BLT1**	**MMSZ6V2T1**	MMSZ4691T1	**MMSZ5234BT1**	1SMB5920BT3
6.8	BZX84C6V8LT1	**MMBZ5235BLT1**	MMSZ6V8T1	MMSZ4692T1	**MMSZ5235BT1**	1SMB5921BT3
7.5	BZX84C7V5LT1	MMBZ5236BLT1	MMSZ7V5T1	MMSZ4693T1	MMSZ5236BT1	1SMB5922BT3
8.2	BZX84C8V2LT1	MMBZ5237BLT1	MMSZ8V2T1	MMSZ4694T1	MMSZ5237BT1	1SMB5923BT3
8.7		MMBZ5238BLT1		MMSZ4695T1	MMSZ5238BT1	
9.1	BZX84C9V1LT1	MMBZ5239BLT1	MMSZ9V1T1	MMSZ4696T1	MMSZ5239BT1	1SMB5924BT3
10	BZX84C10LT1	**MMBZ5240BLT1**	MMSZ10T1	MMSZ4697T1	**MMSZ5240BT1**	1SMB5925BT3
11	BZX84C11LT1	MMBZ5241BLT1	MMSZ11T1	MMSZ4698T1	MMSZ5241BT1	1SMB5926BT3
12	**BZX84C12LT1**	**MMBZ5242BL**T1	MMSZ12T1	MMSZ4699T1	**MMSZ5242BT1**	1SMB5927BT3
13	BZX84C13LT1	MMBZ5243BLT1	MMSZ13T1	MMSZ4700T1	MMSZ5243BT1	1SMB5928BT3
14		MMBZ5244BLT1		MMSZ4701T1	MMSZ5244BT1	
15	BZX84C15LT1	**MMBZ5245BLT1**	MMSZ15T1	MMSZ4702T1	**MMSZ5245BT1**	1SMB5929BT3
16	BZX84C16LT1	MMBZ5246BLT1	MMSZ16T1	MMSZ4703T1	**MMSZ5246BT1**	1SMB5930BT3
17		MMBZ5247BLT1		MMSZ4704T1	MMSZ5247BT1	
18	**BZX84C18LT1**	**MMBZ5248BLT1**	**MMSZ18T1**	MMSZ4705T1	**MMSZ5248BT1**	1SMB5931BT3
19		MMBZ5249BLT1		MMSZ4706T1	MMSZ5249BT1	
20	BZX84C20LT1	**MMBZ5250BLT1**	MMSZ20T1	MMSZ4707T1	**MMSZ5250BT1**	1SMB5932BT3
22	BZX84C22LT1	MMBZ5251BLT1	MMSZ22T1	MMSZ4708T1	MMSZ5251BT1	1SMB5933BT3
24	BZX84C24LT1	MMBZ5252BLT1	MMSZ24T1	MMSZ4709T1	**MMSZ5252BT1**	1SMB5934BT3
25		MMBZ5253BLT1		MMSZ4710T1	MMSZ5253BT1	
27	BZX84C27LT1	**MMBZ5254BLT1**	MMSZ27T1	MMSZ4711T1	MMSZ5254BT1	1SMB5935BT3
28		MMBZ5255BLT1		MMSZ4712T1	MMSZ5255BT1	
30	BZX84C30LT1	MMBZ5256BLT1	MMSZ30T1	MMSZ4713T1	**MMSZ5256BT1**	1SMB5936BT3
33	BZX84C33LT1	**MMBZ5257BLT1**	MMSZ33T1	MMSZ4714T1	MMSZ5257BT1	1SMB5937BT3
36	BZX84C36LT1	MMBZ5258BLT1	MMSZ36T1	MMSZ4715T1	MMSZ5258BT1	1SMB5938BT3
39	BZX84C39LT1	MMBZ5259BLT1	**MMSZ39T1**	MMSZ4716T1	MMSZ5259BT1	1SMB5939BT3
43	BZX84C43LT1	MMBZ5260BLT1	MMSZ43T1	MMSZ4717T1	MMSZ5260BT1	1SMB5940BT3
47	BZX84C47LT1	MMBZ5261BLT1	MMSZ47T1		**MMSZ5261BT1**	1SMB5941BT3
51	BZX84C51LT1	MMBZ5262BLT1	MMSZ51T1		MMSZ5262BT1	1SMB5942BT3
56	BZX84C56LT1	MMBZ5263BLT1	MMSZ56T1		MMSZ5263BT1	1SMB5943BT3
60		MMBZ5264BLT1			MMSZ5264BT1	
62	BZX84C62LT1	MMBZ5265BLT1	MMSZ62T1		MMSZ5265BT1	1SMB5944BT3
68	BZX84C68LT1	MMBZ5266BLT1	MMSZ68T1		MMSZ5266BT1	1SMB5945BT3
75	BZX84C75LT1	MMBZ5267BLT1	MMSZ75T1		MMSZ5267BT1	1SMB5946BT3
82		MMBZ5268BLT1			MMSZ5268BT1	1SMB5947BT3
87		MMBZ5269BLT1			MMSZ5269BT1	
91		MMBZ5270BLT1			MMSZ5270BT1	1SMB5948BT3
100					MMSZ5271BT1	1SMB5949BT3
110						1SMB5950BT3
120						1SMB5951BT3
130						1SMB5952BT3
150						1SMB5953BT3
160						1SMB5954BT3
170						
180						1SMB5955BT3
200						1SMB5956BT3

Nominal Zener Breakdown Voltage	1 Watt Cathode = Polarity Band		1.3 Watt Cathode = Polarity Band			1.5 Watt Cathode = Polarity Band	3 Watt Cathode = Polarity Band	5 Watt Cathode = Polarity Band
Volts	Glass Case 59-03 (DO-41)	Plastic Surmetic 30 Case 59-03 (DO-41)	Glass Case 59-03 (DO-41)		Plastic Surmetic 30 Case 59-03 (DO-41)			Plastic Surmetic 40 Case 17
3.3	1N4728A	MZP4728A	BZX85C3V3RL			1N5913B		1N5333B
3.6	1N4729A	MZP4729A	**BZX85C3V6RL**			1N5914B		1N5334B
3.9	1N4730A	MZP4730A	BZX85C3V9RL	MZPY3.9RL	MZD3.9RL	1N5915B	3EZ3.9D5	1N5335B
4.3	1N4731A	MZP4731A	BZX85C4V3RL	MZPY4.3RL	MZD4.3RL	1N5916B	**3EZ4.3D5**	1N5336B
4.7	1N4732A	MZP4732A	BZX85C4V7RL	MZPY4.7RL	MZD4.7RL	1N5917B	3EZ4.7D5	1N5337B
5.1	**1N4733A**	MZP4733A	BZX85C5V1RL	MZPY5.1RL	MZD5.1RL	**1N5918B**	3EZ5.1D5	1N5338B
5.6	**1N4734A**	MZP4734A	BZX85C5V6RL	MZPY5.6RL	MZD5.6RL	1N5919B	3EZ5.6D5	1N5339B
6.0								1N5340B
6.2	**1N4735A**	MZP4735A	BZX85C6V2RL	MZPY6.2RL	MZD6.2RL	**1N5920B**	3EZ6.2D5	1N5341B
6.8	**1N4736A**	MZP4736A	BZX85C6V8RL	MZPY6.8RL	MZD6.8RL	1N5921B	3EZ6.8D5	1N5342B
7.5	1N4737A	MZP4737A	BZX85C7V5RL	MZPY7.5RL	MZD7.5RL	1N5922B	3EZ7.5D5	1N5343B
8.2	**1N4738A**	MZP4738A	BZX85C8V2RL	MZPY8.2RL	MZD8.2RL	1N5923B	3EZ8.2D5	1N5344B
8.7								1N5345B
9.1	1N4739A	MZP4739A	BZX85C9V1RL	MZPY9.1RL	MZD9.1RL	1N5924B	3EZ9.1D5	1N5346B
10	**1N4740A**	MZP4740A	BZX85C10RL	MZPY10RL	MZD10RL	1N5925B	3EZ10D5	1N5347B
11	**1N4741A**	MZP4741A	BZX85C11RL	MZPY11RL	MZD11RL	1N5926B	3EZ11D5	1N5348B
12	**1N4742A**	MZP4742A	BZX85C12RL	MZPY12RL	MZD12RL	1N5927B	3EZ12D5	1N5349B
13	1N4743A	MZP4743A	BZX85C13RL	MZPY13RL	MZD13RL	1N5928B	3EZ13D5	1N5350B
14							3EZ14D5	1N5351B
15	**1N4744A**	MZP4744A	BZX85C15RL	MZPY15RL	MZD15RL	**1N5929B**	3EZ15D5	1N5352B
16	**1N4745A**	MZP4745A	BZX85C16RL	MZPY16RL	MZD16RL	1N5930B	3EZ16D5	1N5353B
17							3EZ17D5	1N5354B
18	**1N4746A**	**MZP4746A**	BZX85C18RL	MZPY18RL	**MZD18RL**	1N5931B	3EZ18D5	1N5355B
19							3EZ19D5	1N5356B
20	**1N4747A**	MZP4747A	BZX85C20RL	MZPY20RL	MZD20RL	1N5932B	3EZ20D5	1N5357B
22	**1N4748A**	MZP4748A	**BZX85C22RL**	MZPY22RL	MZD22RL	1N5933B	3EZ22D5	1N5358B
24	**1N4749A**	**MZP4749A**	**BZX85C24RL**	MZPY24RL	MZD24RL	1N5934B	3EZ24D5	1N5359B
25								1N5360B
27	**1N4750A**	MZP4750A	BZX85C27RL	MZPY27RL	MZD27RL	1N5935B	3EZ27D5	1N5361B
28							3EZ28D5	1N5362B
30	**1N4751A**	**MZP4751A**	BZX85C30RL	MZPY30RL	MZD30RL	**1N5936B**	3EZ30D5	1N5363B
33	**1N4752A**	MZP4752A	BZX85C33RL	MZPY33RL	MZD33RL	1N5937B	3EZ33D5	1N5364B
36	1N4753A	MZP4753A	BZX85C36RL	MZPY36RL	MZD36RL	1N5938B	3EZ36D5	1N5365B
39	1N4754A	MZP4754A	BZX85C39RL	MZPY39RL	MZD39RL	1N5939B	3EZ39D5	1N5366B
43	1N4755A	MZP4755A	BZX85C43RL	MZPY43RL	MZD43RL	1N5940B	3EZ43D5	1N5367B
47	1N4756A	MZP4756A	BZX85C47RL	MZPY47RL	MZD47RL	1N5941B	3EZ47D5	1N5368B
51	1N4757A	MZP4757A	BZX85C51RL	MZPY51RL	MZD51	1N5942B	3EZ51D5	1N5369B
56	1N4758A	MZP4758A	BZX85C56RL	MZPY56RL	MZD56	1N5943B	3EZ56D5	1N5370B
60								1N5371B
62	1N4759A	MZP4759A	BZX85C62RL	MZPY62RL	MZD62	1N5944B	3EZ62D5	1N5372B
68	1N4760A	MZP4760A	BZX85C68RL	MZPY68RL	MZD68	1N5945B	3EZ68D5	1N5373B
75	1N4761A	MZP4761A	BZX85C75RL	MZPY75RL	MZD75	1N5946B	3EZ75D5	1N5374B
82	1N4762A	MZP4762A	BZX85C82RL	MZPY82RL	MZD82	1N5947B	3EZ82D5	1N5375B
87								1N5376B
91	1N4763A	MZP4763A	BZX85C91RL	MZPY91RL	MZD91	1N5948B	3EZ91D5	1N5377B
100	1N4764A	MZP4764A	BZX85C100RL	MZPY100RL	MZD100	1N5949B	3EZ100D5	1N5378B
110		1M110ZS5			MZD110	1N5950B	3EZ110D5	1N5379B
120		1M120ZS5			MZD120	1N5951B	3EZ120D5	1N5380B
130		1M130ZS5			MZD130	1N5952B	3EZ130D5	1N5381B
140							3EZ140D5	1N5382B
150		1M150ZS5			MZD150	1N5953B	3EZ150D5	1N5383B
160		1M160ZS5			MZD160	1N5954B	3EZ160D5	1N5384B
170							3EZ170D5	1N5385B
180		1M180ZS5			MZD180	**1N5955B**	3EZ180D5	1N5386B
190							3EZ190D5	1N5387B
200		1M200ZS5			MZD200	1N5956B	3EZ200D5	1N5388B
220							3EZ220D5	
240							3EZ240D5	
270							3EZ270D5	
300							3EZ300D5	
330							3EZ330D5	
360							3EZ360D5	
400							3EZ400D5	

DEVICE: HLMP- 3000 Series—Red LED Lamps

T-1¾ (5 mm) Red Solid State Lamps

Technical Data

HLMP-3000
HLMP-3001
HLMP-3002
HLMP-3003
HLMP-3050

Features

- **Low Cost, Broad Applications**
- **Long Life, Solid State Reliability**
- **Low Power Requirements: 20 mA @ 1.6 V**
- **High Light Output:** 2.0 mcd Typical for HLMP-3000 4.0 mcd Typical for HLMP-3001
- **Wide and Narrow Viewing Angle Types**
- **Red Diffused and Non-diffused Versions**

Description

The HLMP-3000 series lamps are Gallium Arsenide Phosphide light emitting diodes intended for High Volume/Low Cost applications such as indicators for appliances, smoke detectors, automobile instrument panels, and many other commercial uses.

The HLMP-3000/-3001/-3002/-3003 have red diffused lenses whereas the HLMP-3050 has a red non-diffused lens. These lamps can be panel mounted using mounting clip HLMP-0103. The HLMP-3000/-3001 lamps have 0.025" leads and the HLMP-3002/-3003/-3050 have 0.018" leads.

Package Dimensions

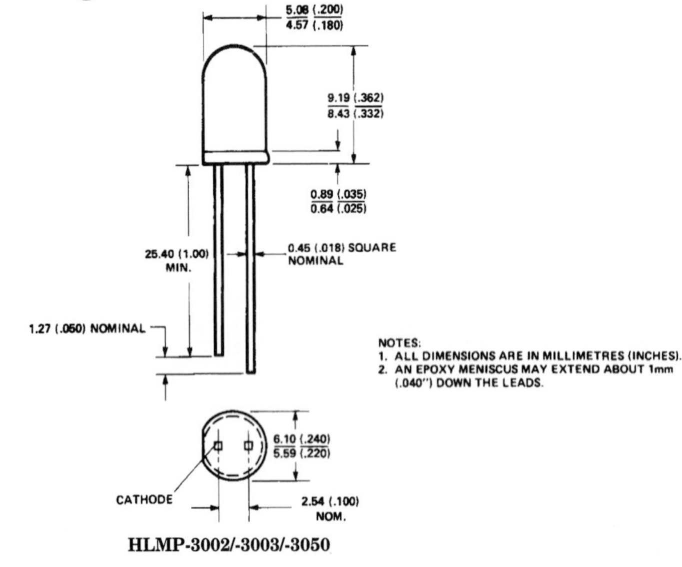

HLMP-3002/-3003/-3050

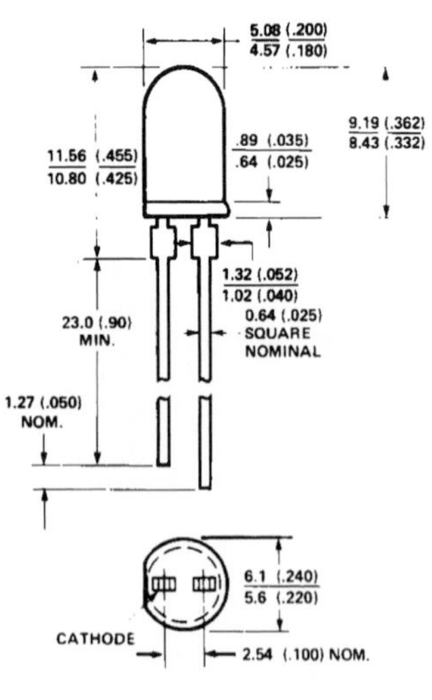

HLMP-3000/-3001

DEVICE: Power Rectifier Diodes

DESCRIPTION

Designed to meet the efficiency demand of switching type power supplies, these devices are useful in many switching applications.
The low thermal resistance and forward voltage drop of this series allows the user to replace DO-5 size devices in many applications.

MECHANICAL SPECIFICATIONS

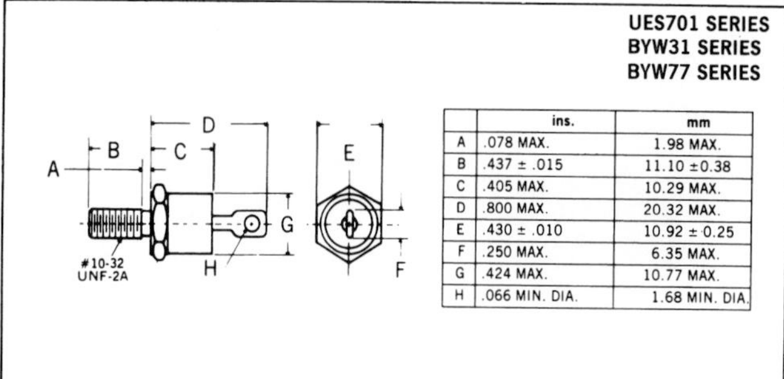

	ins.	mm
A	.078 MAX.	1.98 MAX.
B	.437 ± .015	11.10 ±0.38
C	.405 MAX.	10.29 MAX.
D	.800 MAX.	20.32 MAX.
E	.430 ± .010	10.92 ± 0.25
F	.250 MAX.	6.35 MAX.
G	.424 MAX.	10.77 MAX.
H	.066 MIN. DIA.	1.68 MIN. DIA.

FEATURES

- Low Forward Voltage
- Very Fast Switching
- Low Thermal Resistance
- High Surge Capability
- Mechanically Rugged
- Both Polarities Available

RECTIFIERS
High Efficiency, 25 A

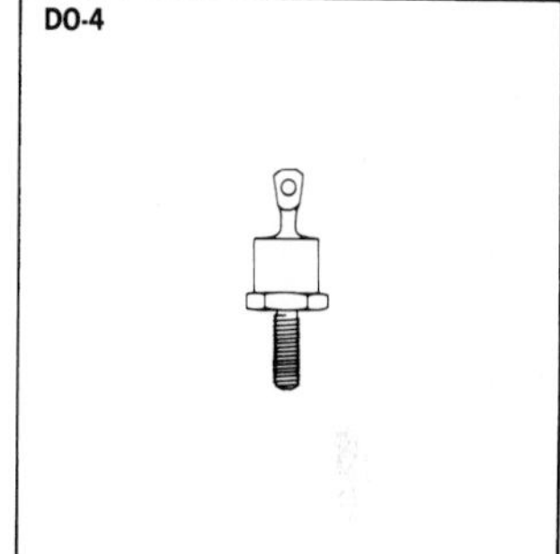

ABSOLUTE MAXIMUM RATINGS

	UES701	UES702	UES703
Peak Inverse Voltage, V_R	50V	100V	150V
Repetitive Peak Inverse Voltag, V_{RRM}	50V	100V	150V
Non-Repetitive Peak Inverse Voltage, V_{RSM}	50V	100V	150V
Maximum Average D.C. Output Current I_O @ T_C	25A @ 100°C		
RMS Forward Current, $I_{F\ (RMS)}$	40A		
Non-Repetitive Sinusoidal Surge Current (8.3mS), I_{FSM}	400A		
Thermal Resistance, Junction to Case, $R_{\theta JC}$	1.5°C/W		
Storage Temperature Range, T_{STG}	−55°C to +175°C		
Maximum Operating Junction Temperature, $T_{J\ MAX}$	+175°C		

ABSOLUTE MAXIMUM RATINGS

	BYW31-50	BYW31-100	BYW31-150	BYW77-50	BYW77-100	BYW77-150
Peak Inverse Voltage, V_R	50V	100V	150V	50V	100V	150V
Repetitive Peak Inverse Voltage, V_{RRM}	50V	100V	150V	50V	100V	150V
Non-Repetitive Peak Inverse Voltage, V_{RSM}	50V	100V	150V	50V	100V	150V
Maximum Average D.C. Output Current, I_O @ T_C = 100°C	25A @ 100°C			30A @ 107°C		
RMS Forward Current, $I_{F\ (RMS)}$	40A			50A		
Non-Repetitive Sinusoidal Surge Current (8.3mS), I_{FSM}	320A			500A		
Thermal Resistance, Junction to Case, $R_{\theta JC}$	1.5°C/W			1.5°C/W		
Storage Temperature Range, T_{STG}	−55°C to +150°C			−55°C to +150°C		
Maximum Operating Junction Temperature, $T_{J\ MAX}$	+150°C			+150°C		

ELECTRICAL SPECIFICATIONS

Type	Maximum Reverse Voltage V_R	Maximum Forward Voltage V_F		Maximum Reverse Current I_R		Maximum Reverse Recovery Time t_{RR}
		T_C = 25°C	T_C = 125°C	T_C = 25°C	T_C = 125°C	
UES701 UES702 UES703	50V 100V 150V	0.95V @ I_F = 25A	0.825V @ I_F = 25A	20μA @ Rated V_R	4mA @ Rated V_R	35ns[1]
		T_C = 25°C	T_C = 100°C	T_C = 25°C	T_C = 100°C	
BYW31-50 BYW31-100 BYW31-150	50V 100V 150V	1.3V @ I_F = 100A	0.85V @ I_F = 20A	20μA @ Rated V_R	2.5mA @ Rated V_R	50ns[2]
		T_C = 25°C	T_C = 100°C	T_C = 25°C	T_C = 100°C	
BYW77-50 BYW77-100 BYW77-150	50V 100V 150V	1.1V @ I_F = 63A	V_F / I_F 0.75V / 10A 0.85V / 20A 1.2V / 100A	25μA @ Rated V_R	2.5mA @ Rated V_R	50ns[2]

DEVICE: GBPC 12, 15, 25 and 35 Series—Full-Wave Bridge Rectifier Modules

GBPC - W Wire leads

GBPC - Standard

FEATURES

- This series is UL recognized under component index, file number E54214
- The plastic package has Underwriters Laboratory flammability recognition 94V-0
- Integrally molded heatsink provide very low thermal resistance for maximum heat dissipation
- Universal 3-way terminals; snap-on, wire wrap-around, or P.C.B. mounting
- High forward surge current capabilities
- Chip junctions are glass passivated
- Typical I_R less than 0.3 μ A
- High temperature soldering guaranteed: 260°C /10 seconds at 5lbs., (2.3 kg) tension

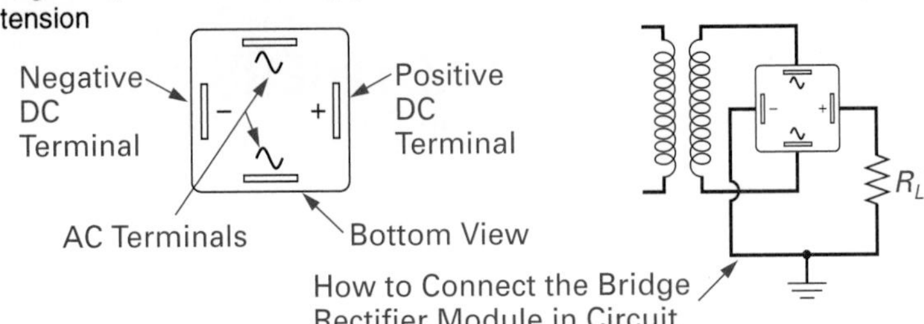

MAXIMUM RATINGS AND ELECTRICAL CHARACTERISTICS

Ratings at 25°C ambient temperature unless otherwise specified.

		SYMBOLS	GBPC12,15,25,35 005	01	02	04	06	08	10	UNITS
Maximum repetitive peak reverse voltage		V_{RRM}	50	100	200	400	600	800	1000	Volts
Maximum RMS voltage		V_{RMS}	35	70	140	280	420	560	700	Volts
Maximum DC blocking voltage		V_{DC}	50	100	200	400	600	800	1000	Volts
Maximum average forward rectified output current (SEE FIG.1)	GBPC12 GBPC15 GBPC25 GBPC35	$I_{(AV)}$	12.0 15.0 25.0 35.0							Amps
Peak forward surge current single sine-wave superimposed on rated load (JEDEC Method)	GBPC12 GBPC15 GBPC25 GBPC35	I_{FSM}	200.0 300.0 300.0 400.0							Amps
Rating (non-repetitive, for t greater than 1 ms and less than 8.3 ms) for fusing	GBPC12 GBPC15 GBPC25 GBPC35	I^2t	160.0 375.0 375.0 660.0							A^2sec
Maximum instantaneous forward voltage drop per leg at	GBPC12 I_F=6.0A GBPC15 I_F=7.5A GBPC25 I_F=12.5A GBPC35 I_F=17.5A	V_F	1.1							Volts
Maximum reverse DC current at rated DC blocking voltage per leg	T_A=25°C T_A=125°C	I_R	5.0 500.0							mA
RMS isolation voltage from case to leads		V_{ISO}	2500.0							Volts
Typical junction capacitance per leg (NOTE 1)		C_J	300.0							pF
Typical thermal resistance per leg (NOTE 2)	GBPC12-25 GBPC35	$R_{\Theta JC}$	1.9 1.4							°C/W
Operating junction storage temperature range		T_J, T_{STG}	-55 to +150							°C

DEVICE: LM78XX Series—IC Voltage Regulators

General Description

The LM78XX series of three terminal regulators is available with several fixed output voltages making them useful in a wide range of applications. One of these is local on card regulation, eliminating the distribution problems associated with single point regulation. The voltages available allow these regulators to be used in logic systems, instrumentation, HiFi, and other solid state electronic equipment. Although designed primarily as fixed voltage regulators these devices can be used with external components to obtain adjustable voltages and currents.

The LM78XX series is available in an aluminum TO-3 package which will allow over 1.0A load current if adequate heat sinking is provided. Current limiting is included to limit the peak output current to a safe value. Safe area protection for the output transistor is provided to limit internal power dissipation. If internal power dissipation becomes too high for the heat sinking provided, the thermal shutdown circuit takes over preventing the IC from overheating.

Considerable effort was expanded to make the LM78XX series of regulators easy to use and mininize the number of external components. It is not necessary to bypass the output, although this does improve transient response. Input bypassing is needed only if the regulator is located far from the filter capacitor of the power supply.

For output voltage other than 5V, 12V and 15V the LM117 series provides an output voltage range from 1.2V to 57V.

Features

- Output current in excess of 1A
- Internal thermal overload protection
- No external components required
- Output transistor safe area protection
- Internal short circuit current limit
- Available in the aluminum TO-3 package

Voltage Range

LM7805C	5V
LM7812C	12V
LM7815C	15V

Absolute Maximum Ratings

Input Voltage (V_O = 5V, 12V and 15V)	35V
Internal Power Dissipation (Note 1)	Internally Limited
Operating Temperature Range (T_A)	0°C to + 70°C
Maximum Junction Temperature	
(K Package)	150°C
(T Package)	150°C
Storage Temperature Range	− 65°C to + 150°C
Lead Temperature (Soldering, 10 sec.)	
TO-3 Package K	300°C
TO-220 Package T	230°C

Metal Can Package
TO-3 (K)
Aluminum

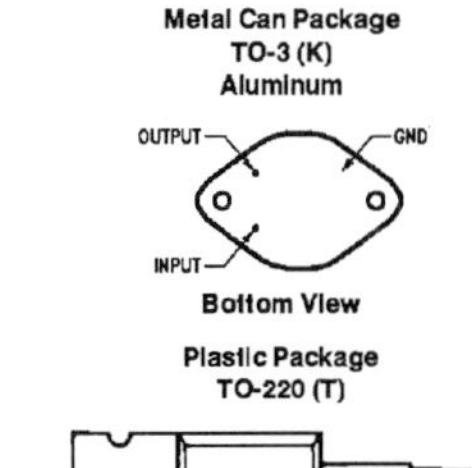

Bottom View

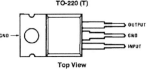

Plastic Package
TO-220 (T)

GND
OUTPUT
GND
INPUT

Top View

Schematic and Connection Diagrams

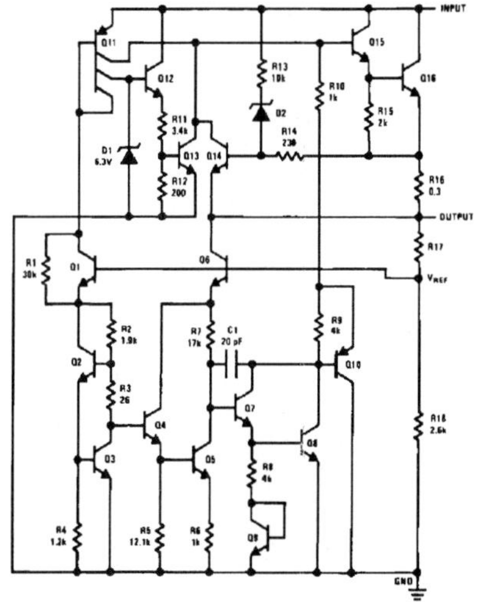

Electrical Characteristics LM78XXC (Note 2) 0°C ≤ Tj ≤ 125°C unless otherwise noted.

Output Voltage				5V			12V			15V			Units
Input Voltage (unless otherwise noted)				10V			19V			23V			
Symbol	**Parameter**	**Conditions**		**Min**	**Typ**	**Max**	**Min**	**Typ**	**Max**	**Min**	**Typ**	**Max**	
V_O	Output Voltage	Tj = 25°C, 5 mA ≤ I_O ≤ 1A		4.8	5	5.2	11.5	12	12.5	14.4	15	15.6	V
		P_D ≤ 15W, 5 mA ≤ I_O ≤ 1A		4.75		5.25	11.4		12.6	14.25		15.75	V
		V_{MIN} ≤ V_{IN} ≤ V_{MAX}		(7.5 ≤ V_{IN} ≤ 20)			(14.5 ≤ V_{IN} ≤ 27)			(17.5 ≤ V_{IN} ≤ 30)			V
ΔV_O	Line Regulation	I_O = 500 mA	Tj = 25°C		3	50		4	120		4	150	mV
			ΔV_{IN}	(7 ≤ V_{IN} ≤ 25)			14.5 ≤ V_{IN} ≤ 30)			(17.5 ≤ V_{IN} ≤ 30)			V
			0°C ≤ Tj ≤ +125°C			50			120			150	mV
			ΔV_{IN}	(8 ≤ V_{IN} ≤ 20)			(15 ≤ V_{IN} ≤ 27)			(16.5 ≤ V_{IN} ≤ 30)			V
		I_O ≤ 1A	Tj = 25°C			50			120			150	mV
			ΔV_{IN}	(7.5 ≤ V_{IN} ≤ 20)			(14.6 ≤ V_{IN} ≤ 27)			(17.7 ≤ V_{IN} ≤ 30)			V
			0°C ≤ Tj ≤ +125°C			25			60			75	mV
			ΔV_{IN}	(8 ≤ V_{IN} ≤ 12)			(16 ≤ V_{IN} ≤ 22)			(20 ≤ V_{IN} ≤ 26)			V
ΔV_O	Load Regulation	Tj = 25°C	5 mA ≤ I_O ≤ 1.5A		10	50		12	120		12	150	mV
			250 mA ≤ I_O ≤ 750 mA			25			60			75	mV
		5 mA ≤ I_O ≤ 1A, 0°C ≤ Tj ≤ +125°C				50			120			150	mV
I_Q	Quiescent Current	I_O ≤ 1A	Tj = 25°C			8			8			8	mA
			0°C ≤ Tj ≤ +125°C			8.5			8.5			8.5	mA
ΔI_Q	Quiescent Current Change	5 mA ≤ I_O ≤ 1A				0.5			0.5			0.5	mA
		Tj = 25°C, I_O ≤ 1A				1.0			1.0			1.0	mA
		V_{MIN} ≤ V_{IN} ≤ V_{MAX}		(7.5 ≤ V_{IN} ≤ 20)			(14.8 ≤ V_{IN} ≤ 27)			(17.9 ≤ V_{IN} ≤ 30)			V
		I_O ≤ 500 mA, 0°C ≤ Tj ≤ +125°C				1.0			1.0			1.0	mA
		V_{MIN} ≤ V_{IN} ≤ V_{MAX}		(7 ≤ V_{IN} ≤ 25)			(14.5 ≤ V_{IN} ≤ 30)			(17.5 ≤ V_{IN} ≤ 30)			V
V_N	Output Noise Voltage	T_A = 25°C, 10 Hz ≤ f ≤ 100 kHz			40			75			90		μV
$\frac{\Delta V_{IN}}{\Delta V_{OUT}}$	Ripple Rejection	f = 120 Hz	I_O ≤ 1A, Tj = 25°C or	62	80		55	72		54	70		dB
			I_O ≤ 500 mA, 0°C ≤ Tj ≤ +125°C	62			55			54			dB
		V_{MIN} ≤ V_{IN} ≤ V_{MAX}		(8 ≤ V_{IN} ≤ 18)			(15 ≤ V_{IN} ≤ 25)			(18.5 ≤ V_{IN} ≤ 28.5)			V
R_O	Dropout Voltage	Tj = 25°C, I_{OUT} = 1A			2.0			2.0			2.0		V
	Output Resistance	f = 1 kHz			8			18			19		mΩ
	Short-Circuit Current	Tj = 25°C			2.1			1.5			1.2		A
	Peak Output Current	Tj = 25°C			2.4			2.4			2.4		A
	Average TC of V_{OUT}	0°C ≤ Tj ≤ +125°C, I_O = 5 mA			0.6			1.5			1.8		mV/°C
V_{IN}	Input Voltage Required to Maintain Line Regulation	Tj = 25°C, I_O ≤ 1A			7.5		14.6			17.7			V

For example, the output voltage of a 7805 could be between 4.8 V to 5.2 V for a 10 V input. The output voltage therefore has a 4% tolerance.

For example, the output of a 7805 will change a maximum of 50 mV as the load current changes from 5 mA to 1.5 A, and will change a maximum of 25 mV as load current changes from 250 mA to 750 mA.

DEVICE: Photo Detector Diode — MRD500 and MRD510

Features:

- Ultra Fast Response — (<1 ns Typ)
- High Sensitivity — MRD500 (1.2 μA/(mW/cm^2) Min)
  MRD510 (0.3 μA/(mW/cm^2) Min)
- Available with Convex Lens (MRD500) or Flat Glass (MRD510) for Design Flexibility
- Popular TO-18 Type Package for Easy Handling and Mounting
- Sensitive Throughout Visible and Near Infrared Spectral Range for Wide Application
- Annular Passivated Structure for Stability and Reliability

Applications:

- Industrial Processing and Control
- Shaft or Position Readers
- Optical Switching
- Remote Control
- Laser Detection
- Light Modulators
- Logic Circuits
- Light Demodulation/Detection
- Counters
- Sorters

**PHOTO DETECTORS
DIODE OUTPUT
PIN SILICON
250 MILLIWATTS
100 VOLTS**

Convex lens is used when we need to capture surrounding light.

**CASE 209-01, Style 1
MRD500
(CONVEX LENS)**

Flat glass can be used when light is applied directly into photodiode window.

**CASE 210-01, Style 1
MRD510
(FLAT GLASS)**

MAXIMUM RATINGS (T_A = 25°C unless otherwise noted)

Rating	Symbol	Value	Unit
Reverse Voltage	V_R	100	Volts
Total Power Dissipation @ T_A = 25°C Derate above 25°C	P_D	250 2.27	mW mW/°C
Operating Temperature Range	T_{op}	−55 to +125	°C
Storage Temperature Range	T_{stg}	−65 to +200	°C

STATIC ELECTRICAL CHARACTERISTICS (T_A = 25°C unless otherwise noted)

Characteristic	Fig. No.	Symbol	Min	Typ	Max	Unit
Dark Current (V_R = 20 V, R_L = 1 megohm)(2) T_A = 25°C T_A = 100°C	2 and 3	I_D	 — —	 — 14	 2 —	nA
Reverse Breakdown Voltage (I_R = 10 μA)	—	$V_{(BR)R}$	100	200	—	Volts
Forward Voltage (I_F = 50 mA)	—	V_F	—	—	1.1	Volts
Series Resistance (I_F = 50 mA)	—	R_s	—	—	10	Ohms
Total Capacitance (V_R = 20 V, f = 1 MHz)	5	C_T	—	—	4	pF

Typical Dark Current (I_D) = 14 nA.

OPTICAL CHARACTERISTICS (T_A = 25°C unless otherwise noted)

Characteristic	Fig. No.	Symbol	Min	Typ	Max	Unit
Light Current (V_R = 20 V)(1) MRD500 MRD510	1	I_L	6 1.5	9 2.1	— —	μA
Sensitivity at 0.8 μm (V_R = 20 V)(3) MRD500 MRD510	—	$S_{(\lambda = 0.8\ \mu m)}$	— —	6.6 1.5	— —	μA/(mW/cm^2)
Response Time (V_R = 20 V, R_L = 50 Ohms)	—	$t_{(resp)}$	—	1	—	ns
Wavelength of Peak Spectral Response	5	λ_s	—	0.8	—	μm

Typical Light Current (I_L) = 9 μA.

If 1 mW per cm^2 of light is applied to photodiode at a wavelength of 0.8 μm (800 nm), light current will be 6.6 μA. For 2 mW/cm^2, I_L will be 2 × 6.6 μA = 13.2 μA. This rating indicates photodiode's sensitivity.

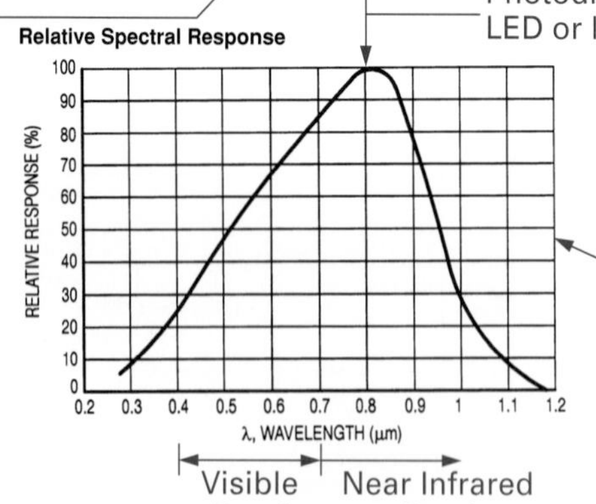

Photodiode's response is best when LED or light source is emitting 0.8 μm or 800 nm. Referring to Figure 7-1, you will find this wavelength in the near infrared band.

Relative Spectral Response Curve shows photodiode's response to different input wavelengths.

DEVICE: Snap-In Fiber Optic Link—HFBR-0500 Series

Features

- GUARANTEED LINK PERFORMANCE OVER TEMPERATURE
  High Speed Links: dc to 5 MBd
  Extended Distance Links up to 111 m
  Low Current Links: 6 mA Peak Supply Current for an 10 m Link
  Photo Interrupters
- LOW COST PLASTIC DUAL-IN-LINE PACKAGE
- EASY FIELD CONNECTORING
- EASY TO USE RECEIVERS:
  Logic Compatible Output Level
  Single +5 V Receiver Power Supply
  High Noise Immunity
- LOW LOSS PLASTIC CABLE:
  Simplex and Zip Cord Style Duplex Cable
  Extra Low Loss Simplex and Duplex

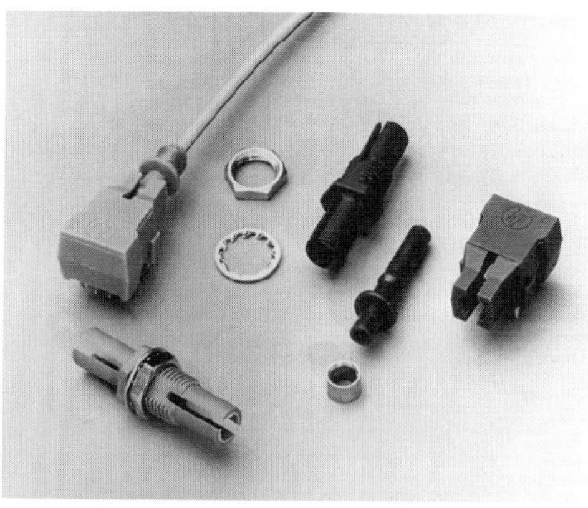

Applications

- HIGH VOLTAGE ISOLATION
- SECURE DATA COMMUNICATIONS
- REMOTE PHOTO INTERRUPTER
- LOW CURRENT LINKS
- INTER/INTRA-SYSTEM LINKS
- STATIC PROTECTION
- EMC REGULATED SYSTEMS (FCC, VDE)

Description

The HFBR-0500 series is a complete family of fiber optic link components for configuring low-cost control, data transmission, and photo interrupter links. These components are designed to mate with plastic snap-in connectors and low-cost plastic cable.* Link design is simplified by the logic compatible receivers and the ease of connectoring the plastic fiber cable. The key parameters of links configured with the HFBR-0500 family are fully guaranteed.

*Cable is available in standard low loss and extra low loss varieties.

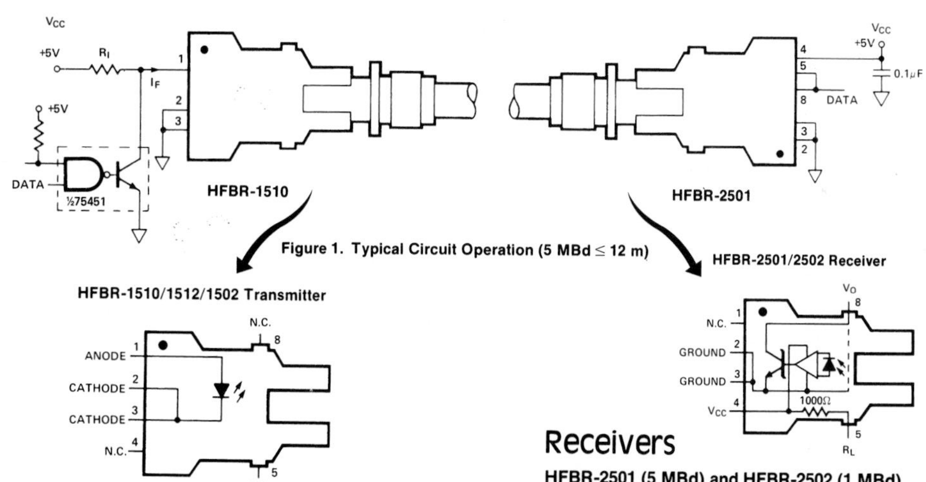

Figure 1. Typical Circuit Operation (5 MBd ≤ 12 m)

665 nm Transmitters

HFBR-1502/HFBR-1510 and HFBR-1512

The HFBR-1510/1502/1512 Transmitter modules incorporate a 665 nm LED emitting at a low attenuation wavelength for the HFBR-R/E plastic fiber optic cable. The transmitters can be easily interfaced to standard TTL logic. The optical power output of the HFBR-1510/1512/1502 is specified at the end of 0.5 m of cable. The HFBR-1512 output optical power is tested and guaranteed at low drive currents.

Receivers

HFBR-2501 (5 MBd) and HFBR-2502 (1 MBd)

The HFBR-2501/2502 Receiver modules feature a shielded integrated photodetector and wide bandwidth DC amplifier for high EMI immunity. A Schottky clamped open-collector output transistor allows interfacing to common logic families and enables "wired-OR" circuit designs. The open collector output is specified up to 18V. An integrated 1000 ohm resistor internally connected to V_{CC} may be externally jumpered to provide a pull-up for ease-of-use with +5V logic. The combination of high optical power levels and fast transitions falling edge could result in distortion of the output signal (HFBR-2502 only), that could lead to multiple triggering of following circuitry.

DEVICE: Silicon Varactor Diodes—MMBV109L and MV209

SILICON EPICAP DIODES

. . . designed for general frequency control and tuning applications; providing solid-state reliability in replacement of mechanical tuning methods.

- High Q with Guaranteed Minimum Values at VHF Frequencies
- Controlled and Uniform Tuning Ratio
- Available in Surface Mount Package

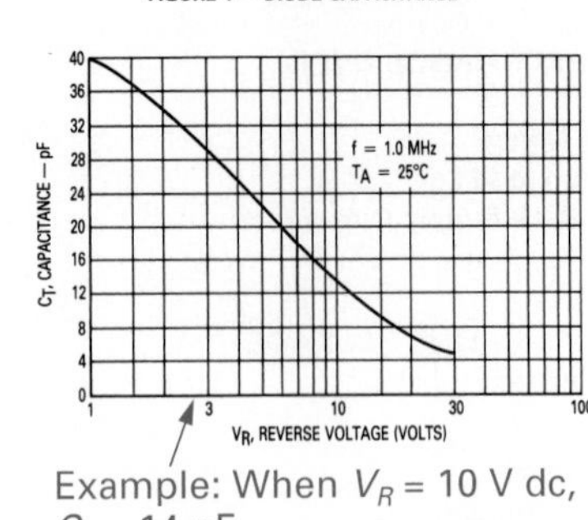

Example: When V_R = 10 V dc, C_T = 14 pF.

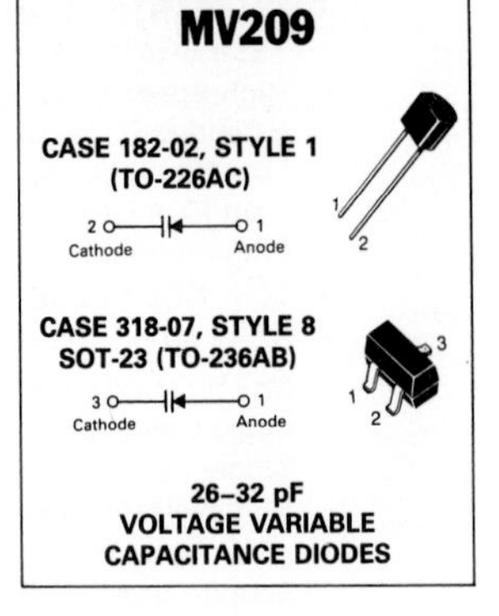

MMBV109L
MV209
CASE 182-02, STYLE 1
(TO-226AC)
Cathode
Anode
CASE 318-07, STYLE 8
SOT-23 (TO-236AB)
Cathode
Anode
26–32 pF
VOLTAGE VARIABLE
CAPACITANCE DIODES

MAXIMUM RATINGS

Rating	Symbol	MV209 Value	MMBV109,L Value	Unit
Reverse Voltage	V_R	30		Volts
Forward Current	I_F	200		mA
Forward Power Dissipation @ T_A = 25°C Derate above 25°C	P_D	280 2.8	200 2.0	mW mW/°C
Junction Temperature	T_J	+125		°C
Storage Temperature Range	T_{stg}	−55 to +150		°C

DEVICE MARKING

MMBV109L = 4A

Do not exceed a reverse voltage of 30 V.

ELECTRICAL CHARACTERISTICS (T_A = 25°C unless otherwise noted.)

Characteristic	Symbol	Min	Typ	Max	Unit
Reverse Breakdown Voltage (I_R = 10 μAdc)	$V_{(BR)R}$	30	—	—	Vdc
Reverse Voltage Leakage Current (V_R = 25 Vdc)	I_R	—	—	0.1	μAdc
Diode Capacitance Temperature Coefficient (V_R = 3.0 Vdc, f = 1.0 MHz)	TC_C	—	300	—	ppm/°C

These maximum ratings have the same meaning as that of the junction diode.

The diode capacitance temperature coefficient states how much the varactor diode's capacitance will change for each 1°C rise in temperature above 25°C. For example, if a varactor diode's normal capacitance value was 1 pF, this rating states it will change 300 parts per million (ppm) or 300 millionths or 0.0003. This will be 0.0003 pF per 1°C (0.0003 × 1 pF = 0.0003 pF).

	C_t, Diode Capacitance V_R = 3.0 Vdc, f = 1.0 MHz pF			Q, Figure of Merit V_R = 3.0 Vdc f = 50 MHz (Note 1)	C_R, Capacitance Ratio C_3/C_{25} f = 1.0 MHz (Note 2)	
Device	**Min**	**Nom**	**Max**	**Min**	**Min**	**Max**
MMBV109L, MV209	26	29	32	200	5.0	6.5

Varactor diode capacitance (pF) ratings when the reverse voltage is 3.0 V dc and frequency is 1 MHz.

The Q (Quality) of the junction capacitance. (Q is the ratio of energy stored in capacitor to energy lost due to leakage current.)

Minimum and maximum varactor diode capacitance value tolerance.

DEVICE: Undirectional Transcient Suppressor Diode

Zener Transient Voltage Suppressors Undirectional and Bidirectional

The P6KE6.8A series is designed to protect voltage sensitive components from high voltage, high energy transients. They have excellent clamping capability, high surge capability, low zener impedance and fast response time. The P6KE6.8A series is supplied in Motorola's exclusive, cost-effective, highly reliable Surmetic axial leaded package and is ideally-suited for use in communication systems, numerical controls, process controls, medical equipment, business machines, power supplies and many other industrial/ consumer applications.

Specification Features:

- Standard Zener Voltage Range — 6.8 to 200 Volts
- Peak Power — 600 Watts @ 1 ms
- Maximum Clamp Voltage @ Peak Pulse Current
- Low Leakage < 5 μA Above 10 Volts
- Maximum Temperature Coefficient Specified

MAXIMUM RATINGS

Rating	Symbol	Value	Unit
Peak Power Dissipation (1) @ $T_L \le 25°C$	P_{PK}	600	Watts
Steady State Power Dissipation @ $T_L \le 75°C$, Lead Length = 3/8″ Derated above $T_L = 75°C$	P_D	5 50	Watts mW/°C
Forward Surge Current (2) @ $T_A = 25°C$	I_{FSM}	100	Amps
Operating and Storage Temperature Range	T_J, T_{stg}	– 65 to +175	°C

Lead Temperature not less than 1/16″ from the case for 10 seconds: 230°C

Mechanical Characteristics:

CASE: Void-free, transfer-molded, thermosetting plastic
FINISH: All external surfaces are corrosion resistant and leads are readily solderable
POLARITY: Cathode indicated by polarity band. When operated in zener mode, will be positive with respect to anode
MOUNTING POSITION: Any

NOTES: 1. Nonrepetitive current pulse per Figure 4 and derated above $T_A = 25°C$ per Figure 2.
2. 1/2 sine wave (or equivalent square wave), PW = 8.3 ms, duty cycle = 4 pulses per minute maximum.

P6KE6.8A through P6KE200A

ZENER OVERVOLTAGE TRANSIENT SUPPRESSORS
6.8–200 VOLT
600 WATT PEAK POWER
5 WATTS STEADY STATE

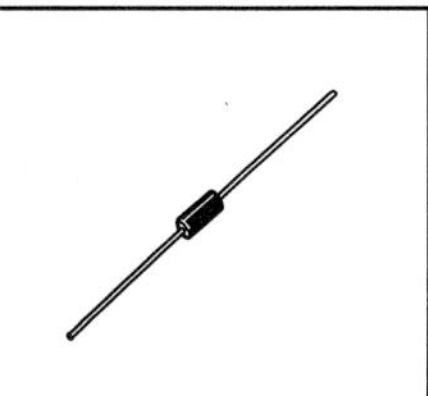

APPLICATION NOTES

RESPONSE TIME

In most applications, the transient suppressor device is placed in parallel with the equipment or component to be protected. In this situation, there is a time delay associated with the capacitance of the device and an overshoot condition associated with the inductance of the device and the inductance of the connection method. The capacitance effect is of minor importance in the parallel protection scheme because it only produces a time delay in the transition from the operating voltage to the clamp voltage as shown in Figure A.

TYPICAL PROTECTION CIRCUIT

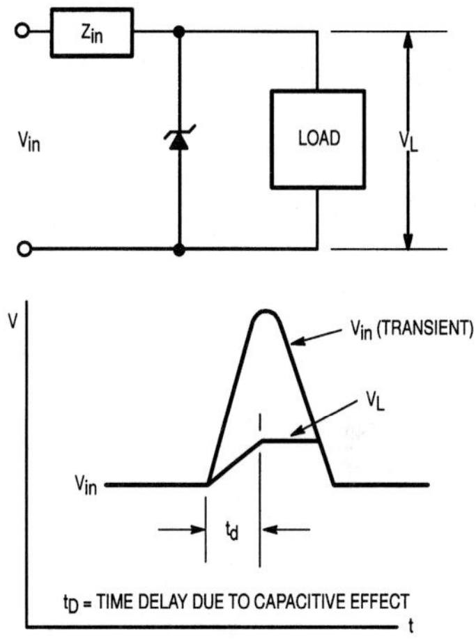

Figure A.

The inductive effects in the device are due to actual turn-on time (time required for the device to go from zero current to full current) and lead inductance. This inductive effect produces an overshoot in the voltage across the equipment or component being protected as shown in Figure B. Minimizing this overshoot is very important in the application, since the main purpose for adding a transient suppressor is to clamp voltage spikes. The P6KE6.8A series has very good response time, typically < 1 ns and negligible inductance. However, external inductive effects could produce unacceptable overshoot. Proper circuit layout, minimum lead lengths and placing the suppressor device as close as possible to the equipment or components to be protected will minimize this overshoot.

Some input impedance represented by Z_{in} is essential to prevent overstress of the protection device. This impedance should be as high as possible, without restricting the circuit operation.

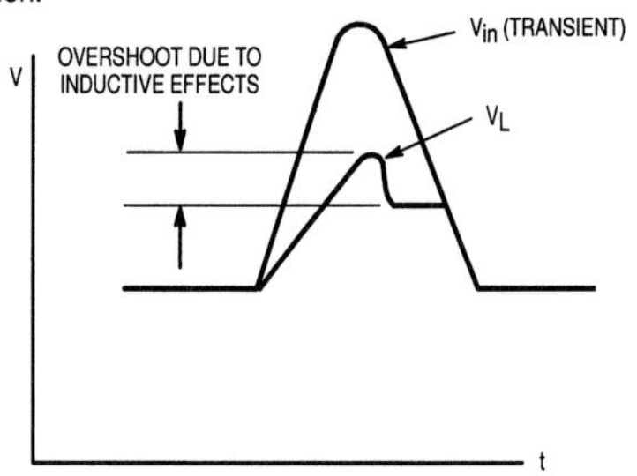

Figure B.

Current Regulator Diodes

1N5283 through 1N5314

CURRENT REGULATOR DIODES

Field-effect current regulator diodes are circuit elements that provide a current essentially independent of voltage. These diodes are especially designed for maximum impedance over the operating range. These devices may be used in parallel to obtain higher currents.

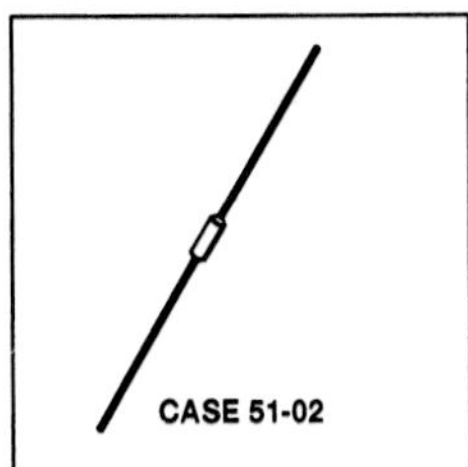

MAXIMUM RATINGS

Rating	Symbol	Value	Unit
Peak Operating Voltage (T_J = –55°C to +200°C)	POV	100	Volts
Steady State Power Dissipation @ T_L = 75°C Derate above T_L = 75°C Lead Length = 3/8″ (Forward or Reverse Bias)	P_D	600 4.8	mW mW/°C
Operating and Storage Junction Temperature Range	T_J, T_{stg}	–55 to +200	°C

Forward voltage (V_F) must not exceed this maximum rating.

ELECTRICAL CHARACTERISTICS (T_A = 25°C unless otherwise noted)

Type No.	Regulator Current Ip (mA) @ V_T = 25 V: Nom	Min	Max	Maximum Limiting Voltage @ I_L = 0.8 Ip (min) V_L (Volts)
1N5283	0.22	0.198	0.242	1.00
1N5284	0.24	0.216	0.264	1.00
1N5285	0.27	0.243	0.297	1.00
1N5286	0.30	0.270	0.330	1.00
1N5287	0.33	0.297	0.363	1.00
1N5288	0.39	0.351	0.429	1.05
1N5289	0.43	0.387	0.473	1.05
1N5290	0.47	0.423	0.517	1.05
1N5291	0.56	0.504	0.616	1.10
1N5292	0.62	0.558	0.682	1.13
1N5293	0.68	0.612	0.748	1.15
1N5294	0.75	0.675	0.825	1.20
1N5295	0.82	0.738	0.902	1.25
1N5296	0.91	0.819	1.001	1.29
1N5297	1.00	0.900	1.100	1.35
1N5298	1.10	0.990	1.21	1.40
1N5299	1.20	1.08	1.32	1.45
1N5300	1.30	1.17	1.43	1.50
1N5301	1.40	1.26	1.54	1.55
1N5302	1.50	1.35	1.65	1.60
1N5303	1.60	1.44	1.76	1.65
1N5304	1.80	1.62	1.98	1.75
1N5305	2.00	1.80	2.20	1.85
1N5306	2.20	1.98	2.42	1.95
1N5307	2.40	2.16	2.64	2.00
1N5308	2.70	2.43	2.97	2.15
1N5309	3.00	2.70	3.30	2.25
1N5310	3.30	2.97	3.63	2.35
1N5311	3.60	3.24	3.96	2.50
1N5312	3.90	3.51	4.29	2.60
1N5313	4.30	3.87	4.73	2.75
1N5314	4.70	4.23	5.17	2.90

Voltage at which diode begins to regulate current.

Regulated Forward Current (IP)

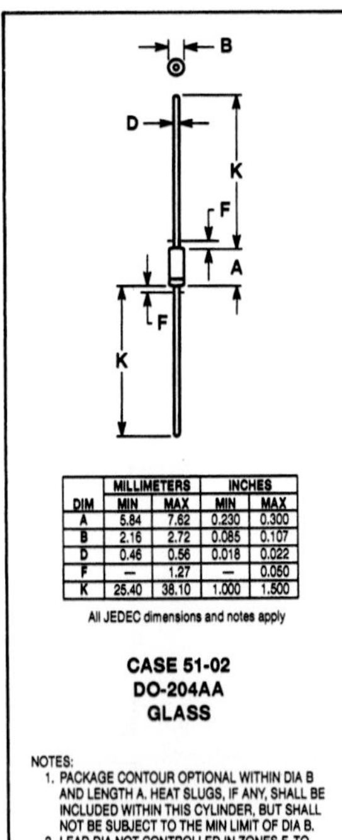

DIM	MILLIMETERS MIN	MILLIMETERS MAX	INCHES MIN	INCHES MAX
A	5.84	7.62	0.230	0.300
B	2.16	2.72	0.085	0.107
D	0.46	0.56	0.018	0.022
F	—	1.27	—	0.050
K	25.40	38.10	1.000	1.500

All JEDEC dimensions and notes apply

CASE 51-02
DO-204AA
GLASS

NOTES:
1. PACKAGE CONTOUR OPTIONAL WITHIN DIA B AND LENGTH A. HEAT SLUGS, IF ANY, SHALL BE INCLUDED WITHIN THIS CYLINDER, BUT SHALL NOT BE SUBJECT TO THE MIN LIMIT OF DIA B.
2. LEAD DIA NOT CONTROLLED IN ZONES F, TO ALLOW FOR FLASH, LEAD FINISH BUILDUP, AND MINOR IRREGULARITIES OTHER THAN HEAT SLUGS.

DEVICE: 2N3903 and 2N3904—NPN Silicon Switching and Amplifier Transistors

Maximum continuous collector current (I_c) = 200 mA.

MAXIMUM RATINGS

Rating	Symbol	Value	Unit
Collector-Emitter Voltage	V_{CEO}	40	Vdc
Collector-Base Voltge	V_{CBO}	60	Vdc
Emitter-Base Voltage	V_{EBO}	6.0	Vdc
Collector Current — Continuous	I_C	200	mAdc
Total Device Dissipation @ T_A = 25°C Derate above 25°C	P_D	625 5.0	mW mW/°C
*Total Device Dissipation @ T_C = 25°C Derate above 25°C	P_D	1.5 12	Watts mW/°C
Operating and Storage Junction Temperature Range	T_J, T_{stg}	−55 to +150	°C

***THERMAL CHARACTERISTICS**

Characteristic	Symbol	Max	Unit
Thermal Resistance, Junction to Ambient	$R_{\theta JA}$	200	°C/W
Thermal Resistance, Junction to Case	$R_{\theta JC}$	83.3	°C/W

2N3903
2N3904★

CASE 29-04, STYLE 1
TO-92 (TO-226AA)

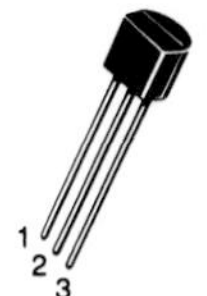

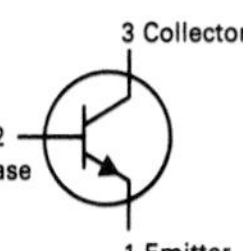

GENERAL PURPOSE TRANSISTORS

NPN SILICON

★This is a Motorola designated preferred device.

ELECTRICAL CHARACTERISTICS (T_A = 25°C unless otherwise noted.)

Characteristic		Symbol	Min	Max	Unit
OFF Characteristic (operated in cutoff)					
Collector-Emitter Breakdown Voltage(1) (I_C = 1.0 mAdc, I_B = 0)		$V_{(BR)CEO}$	40	—	Vdc
Collector-Base Breakdown Voltage (I_C = 10 μAdc, I_E = 0)		$V_{(BR)CBO}$	60	—	Vdc
Emitter-Base Breakdown Voltage (I_E = 10 μAdc, I_C = 0)		$V_{(BR)EBO}$	6.0	—	Vdc
Base Cutoff Current (V_{CE} = 30 Vdc, V_{EB} = 3.0 Vdc)		I_{BL}	—	50	nAdc
Collector Cutoff Current (V_{CE} = 30 Vdc, V_{EB} = 3.0 Vdc)		I_{CEX}	—	50	nAdc
ON Characteristic (operated in active and saturation region)					
DC Current Gain(1) (I_C = 0.1 mAdc, V_{CE} = 1.0 Vdc)	2N3903 2N3904	h_{FE}	20 40	— —	—
(I_C = 1.0 mAdc, V_{CE} = 1.0 Vdc)	2N3903 2N3904		35 70	— —	
(I_C = 10 mAdc, V_{CE} = 1.0 Vdc)	2N3903 2N3904		50 100	150 300	
(I_C = 50 mAdc, V_{CE} = 1.0 Vdc)	2N3903 2N3904		30 60	— —	
(I_C = 100 mAdc, V_{CE} = 1.0 Vdc)	2N3903 2N3904		15 30	— —	
Collector-Emitter Saturation Voltage(1) (I_C = 10 mAdc, I_B = 1.0 mAdc) (I_C = 50 mAdc, I_B = 5.0 mAdc)		$V_{CE(sat)}$	— —	0.2 0.3	Vdc
Base-Emitter Saturation Voltage(1) (I_C = 10 mAdc, I_B = 1.0 mAdc) (I_C = 50 mAdc, I_B = 5.0 mAdc)		$V_{BE(sat)}$	0.65 —	0.85 0.95	Vdc

NOTE: The "O" following CBO, CEO, EBO indicates the third terminal is "open." For example, *$V_{(BR)\ CEO}$ means the breakdown voltage between collector and emitter with the base open.*

$h_{FE} = \beta_{DC}$, dc current gain is measured at different values of I_c.

Maximum base-emitter voltage (V_{BE}) when transistor is saturated

Maximum value of voltage between collector and emitter (V_{CE}) when transistor is in saturation.

DEVICE: BFW43—PNP Silicon High Reverse Voltage Transistor

MAXIMUM RATINGS

Rating	Symbol	Value	Unit
Collector-Emitter Voltage (High Reverse Voltage Ratings)	V_{CEO}	150	Vdc
Collector-Base Voltage (High Reverse Voltage Ratings)	V_{CBO}	150	Vdc
Emitter-Base Voltage (High Reverse Voltage Ratings)	V_{EBO}	6.0	Vdc
Collector Current — Continuous	I_C	0.1	Adc
Total Device Dissipation @ T_A = 25°C Derate above 25°C	P_D	0.4 2.28	Watt mW/°C
Total Device Dissipation @ T_C = 25°C Derate above 25°C	P_D	1.4 8.0	Watt mW/°C
Operating and Storage Junction Temperature Range	T_J, T_{stg}	−65 to +200	°C

APPLICATIONS: High voltage circuits found in televisions and computer monitors.

BFW43

CASE 22-03, STYLE 1
TO-18 (TO-206AA)

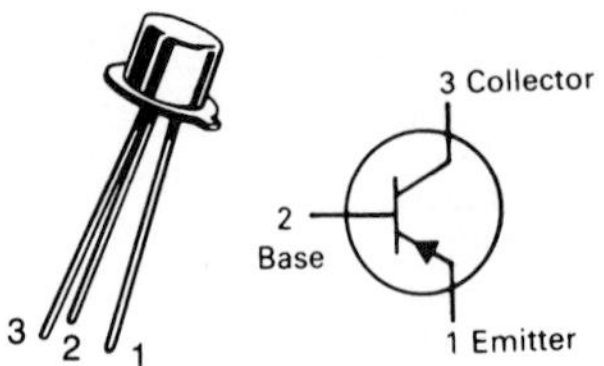

HIGH VOLTAGE TRANSISTOR

PNP SILICON

DEVICE: MPS6714—NPN Silicon High Current (I_C) Transistor

MAXIMUM RATINGS

Rating	Symbol	Value	Unit
Collector-Emitter Voltage MPS6714 MPS6715	V_{CEO}	 30 40	Vdc
Collector-Base Voltage MPS6714 MPS6715	V_{CBO}	 40 50	Vdc
Emitter-Base Voltage	V_{EBO}	5.0	Vdc
Collector Current — Continuous (High I_C Rating)	I_C	1.0	Adc
Total Device Dissipation @ T_A = 25°C Derate above 25°C	P_D	1.0 8.0	Watt mW/°C
Total Device Dissipation @ T_C = 25°C Derate above 25°C	P_D	2.5 20	Watts mW/°C
Operating and Storage Junction Temperature Range	T_J, T_{stg}	−55 to +150	°C

Applications: Current Regulator Circuits

MPS6714
MPS6715

CASE 29-05, STYLE 1
TO-92 (TO-226AE)

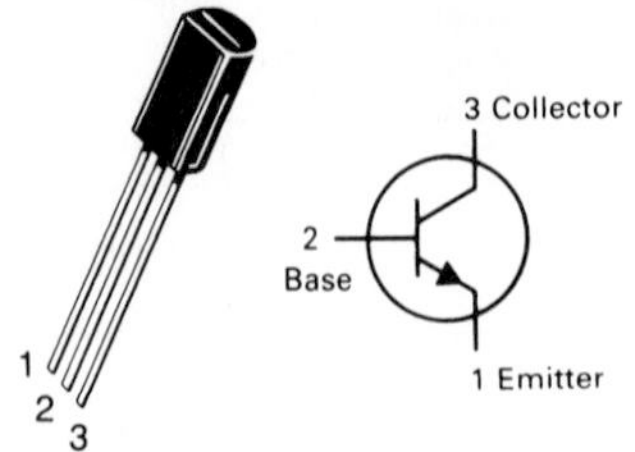

ONE WATT
AMPLIFIER TRANSISTORS

NPN SILICON

DEVICE: BUX48—NPN Silicon High Power Dissipation Transistor

SWITCHMODE II SERIES
NPN SILICON POWER TRANSISTORS

Applications

The BUX 48/BUX 48A transistors are designed for high-voltage, high-speed, power switching in inductive circuits where fall time is critical. They are particularly suited for line-operated switchmode applications such as:

- Switching Regulators
- Inverters
- Solenoid and Relay Drivers
- Motor Controls
- Deflection Circuits

Fast Turn-Off Times
60 ns Inductive Fall Time — 25°C (Typ)
120 ns Inductive Crossover Time — 25°C (Typ)

Operating Temperature Range -65 to +200°C

100°C Performance Specified for:
Reverse-Biased SOA with Inductive Loads
Switching Times with Inductive Loads
Saturation Voltage
Leakage Currents (125°C)

15 AMPERES
NPN SILICON
POWER TRANSISTORS

400 AND 450 VOLTS
V(BR)CEO
850 – 1000 VOLTS
V(BR)CEX
175 WATTS

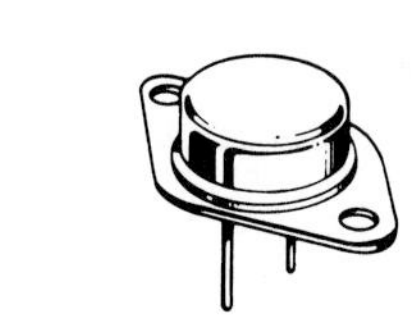

CASE 1-07
TO-204AA
(TO-3)

MAXIMUM RATINGS

Rating	Symbol	BUX48	BUX48A	Unit
Collector-Emitter Voltage	$V_{CEO(sus)}$	400	450	Vdc
Collector-Emitter Voltage (V_{BE} = -1.5V)	V_{CEX}	850	1000	Vdc
Emitter Base Voltage	V_{EB}	7		Vdc
Collector Current — Continuous — Peak (1) — Overload	I_C I_{CM} I_{OI}	15 30 60		Adc
Base Current — Continuous — Peak (1)	I_B I_{BM}	5 20		Adc
Total Power Dissipation — T_C = 25°C — T_C = 100°C Derate above 25°C (High power dissipation rating.)	P_D	175 100 1		Watts W/°C
Operating and Storage Junction Temperature Range	T_J, T_{stg}	-65 to +200		°C

THERMAL CHARACTERISTICS

Characteristic	Symbol	Max	Unit
Thermal Resistance, Junction to Case	$R_{\theta JC}$	1	°C/W
Maximum Lead Temperature for Soldering Purposes: 1/8″ from Case for 5 Seconds	T_L	275	°C

DEVICE: 2N5484 Through 2N5486—N-Channel JFET

Most of these specifications are self-explanatory. $V_{GS\ (\text{off})}$ which has been given throughout is listed in the OFF characteristics, while I_{DSS} is listed in the ON characteristics. The only confusing maximum rating is "Forward Gate Current" because the gate is never normally forward biased (V_{GS} = a negative voltage or zero volts). This rating indicates that if the gate accidentally becomes forward biased, gate current must not exceed 10 mA dc or the JFET will be destroyed.

2N5484
2N5486

CASE 29-04, STYLE 5
TO-92 (TO-226AA)

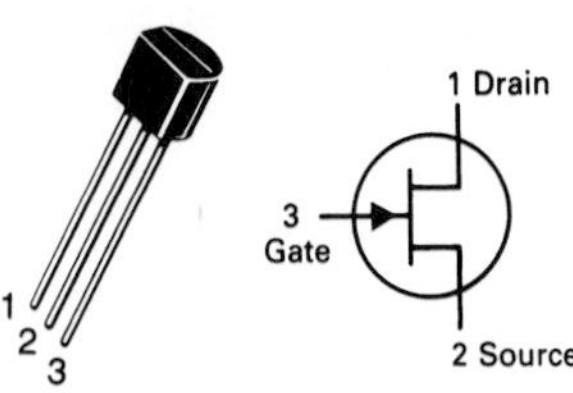

JFET
VHF/UHF AMPLIFIERS
N-CHANNEL — DEPLETION

MAXIMUM RATINGS

Rating	Symbol	Value	Unit
Drain-Gate Voltage	V_{DG}	25	Vdc
Reverse Gate-Source Voltage	V_{GSR}	25	Vdc
Drain Current	I_D	30	mAdc
Forward Gate Current	$I_{G(f)}$		mAdc
Total Device Dissipation @ T_C = 25°C Derate above 25°C	P_D	310 2.82	mW mW/°C
Operating and Storage Junction Temperature Range	T_J, T_{stg}	−65 to +150	°C

ELECTRICAL CHARACTERISTICS (T_A = 25°C unless otherwise noted.)

Characteristic		Symbol	Min	Typ	Max	Unit
OFF CHARACTERISTICS						
Gate-Source Breakdown Voltage (I_G = −1.0 μAdc, V_{DS} = 0)		$V_{(BR)GSS}$	−25	—	—	Vdc
Gate Reverse Current (V_{GS} = −20 Vdc, V_{DS} = 0) (V_{GS} = −20 Vdc, V_{DS} = 0, T_A = 100°C)		I_{GSS}	 — —	 — —	 −1.0 −0.2	 nAdc μAdc
Gate Source Cutoff Voltage (V_{DS} = 15 Vdc, I_D = 10 nAdc)	 2N5484 2N5485 2N5486	$V_{GS(off)}$	 −0.3 −0.5 −2.0	 — — —	 −3.0 −4.0 −6.0	Vdc
ON CHARACTERISTICS						
Zero-Gate-Voltage Drain Current (V_{DS} = 15 Vdc, V_{GS} = 0)	 2N5484 2N5485 2N5486	I_{DSS}	 1.0 4.0 8.0	 — — —	 5.0 10 20	mAdc

$V_{GS\ (\text{off})}$

I_{DSS}

DEVICE: 2N7002LT1—N-Channel Silicon E-MOSFET

2N7002LT1★

CASE 318-07 STYLE 21
SOT-23 (TO-236AB)

Drain 3
3
1
2
1
Gate
Source 2

TMOS FET TRANSISTOR

N-CHANNEL

MAXIMUM RATINGS

Rating	Symbol	Value	Unit
Drain-Source Voltage	V_{DSS}	60	Vdc
Drain-Gate Voltage (R_{GS} = 1 MΩ)	V_{DGR}	60	Vdc
Drain Current – Continuous TC = 25°C(1) T_C = 100°C(1) – Pulsed (2)	 I_D I_D I_{DM}	 ±115 ±75 ±800	mA
Gate-Source Voltage – Continuous – Non-repetitive (tp ≤50 μs)	 V_{GS} V_{GSM}	 ±20 ±40	 Vdc Vpk
Total Power Dissipation T_C = 25°C T_C = 100°C Derate above 25°C ambient	P_D	200 80 1.6	mW mW/°C

ELECTRICAL CHARACTERISTICS (T_A = 25°C unless otherwise noted.)

Characteristic	Symbol	Min	Typ	Max	Unit
OFF CHARACTERISTICS					
Drain-Source Breakdown Voltage (V_{GS} = 0, I_D = 10 μA)	$V_{(BR)DSS}$	60	—	—	Vdc
Zero Gate Voltage Drain Current (V_{GS} = 0, V_{DS} = 60 V) T_J = 25°C T_J = 125°C	I_{DSS}	 — —	 — —	 1.0 500	μAdc
Gate-Body Leakage Current Forward (V_{GS} = 20 Vdc)	I_{GSSF}	—	—	100	nAdc
Gate-Body Leakage Current Reverse (V_{GS} = −20 Vdc)	I_{GSSR}	—	—	−100	nAdc

(1) The Power Dissipation of the package may result in a lower continuous drain current.
(2) Pulse Width ≤ 300 μs, Duty Cycle ≤ 2.0%.

Characteristic	Symbol	Min	Typ	Max	Unit
ON CHARACTERISTICS*					
Gate Threshold Voltage (V_{DS} = V_{GS}, I_D = 250 μA)	$V_{GS(th)}$	1.0	—	2.5	Vdc
On-State Drain Current (V_{DS} ≥ 2.0 $V_{DS(on)}$, V_{GS} = 10 V) [annotation: I_D (on)]	$I_{D(on)}$	500	—	—	mA
Static Drain-Source On-State Voltage (V_{GS} = 10 V, I_D = 500 mA) (V_{GS} = 5.0 V, I_D = 50 mA) [annotation: V_{DS} (on)]	$V_{DS(on)}$	 — —	 — —	 3.75 .375	Vdc
Static Drain-Source On-State Resistance (V_{GS} = 10 V, I_D = 500 mA) T_C = 25°C T_C = 125°C (V_{GS} = 5.0 V, I_D = 50 mA) T_C = 25°C T_C = 125°C	$r_{DS(on)}$	 — — — —	 — — — —	 7.5 13.5 7.5 13.5	Ohms
Forward Transconductance (V_{DS} ≥ 2.0 $V_{DS(on)}$, I_D = 200 mA)	g_{FS}	80	—	—	mmhos
DYNAMIC CHARACTERISTICS					
Input Capacitance (V_{DS} = 25 V, V_{GS} = 0, f = 1.0 MHz)	C_{iss}	—	—	50	pF
Output Capacitance (V_{DS} = 25 V, V_{GS} = 0, f = 1.0 MHz)	C_{oss}	—	—	25	pF
Reverse Transfer Capacitance (V_{DS} = 25 V, V_{GS} = 0, f = 1.0 MHz)	C_{rss}	—	—	5.0	pF
SWITCHING CHARACTERISTICS*					
Turn-On Delay Time (V_{DD} = 25 V, I_D ≅ 500 mA, R_G = 25 Ω, R_L = 50 Ω)	$t_{d(on)}$	—	—	30	ns
Turn-Off Delay Time (V_{DD} = 25 V, I_D ≅ 500 mA, R_G = 25 Ω, R_L = 50 Ω)	$t_{d(off)}$	—	—	40	ns
BODY-DRAIN DIODE RATINGS					
Diode Forward On-Voltage (I_S = 11.5 mA, V_{GS} = 0 V)	V_{SD}	—	—	−1.5	V
Source Current Continuous (Body Diode)	I_S	—	—	−115	mA
Source Current Pulsed	I_{SM}	—	—	−800	mA

*Pulse Test: Pulse Width ≤ 300 μs, Duty Cycle ≤ 2.0%.

DEVICE: µA 747—Dual Linear Op-Amp IC

DESCRIPTION

The 747 is a pair of high-performance monolithic operational amplifiers constructed on a single silicon chip. High common-mode voltage range and absence of "latch-up" make the 747 ideal for use as a voltage-follower. The high gain and wide range of operating voltage provides superior performance in integrator, summing amplifier, and general feedback applications. The 747 is short-circuit protected and requires no external components for frequency compensation. The internal 6dB/octave roll-off insures stability in closed-loop applications. For single amplifier performance, see µA741 data sheet.

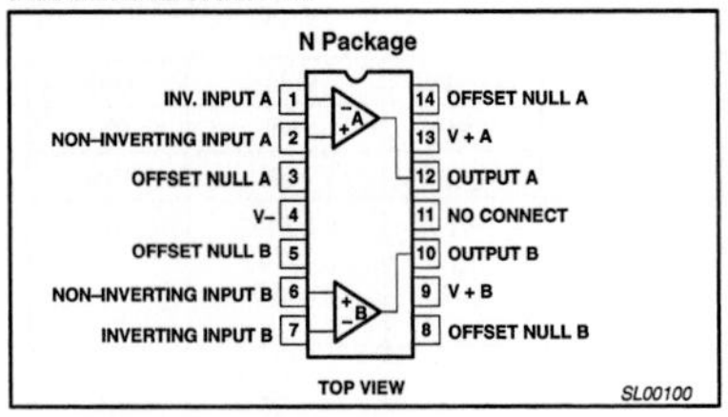

FEATURES

- No frequency compensation required
- Short-circuit protection
- Offset voltage null capability
- Large common-mode and differential voltage ranges
- Low power consumption
- No latch-up

PIN CONFIGURATION

N Package

1	INV. INPUT A	14	OFFSET NULL A
2	NON-INVERTING INPUT A	13	V + A
3	OFFSET NULL A	12	OUTPUT A
4	V–	11	NO CONNECT
5	OFFSET NULL B	10	OUTPUT B
6	NON-INVERTING INPUT B	9	V + B
7	INVERTING INPUT B	8	OFFSET NULL B

TOP VIEW

SL00100

ABSOLUTE MAXIMUM RATINGS

SYMBOL	PARAMETER	RATING	UNIT
V_S	Supply voltage	±18	V
$P_{D\,MAX}$	Maximum power dissipation T_A=25°C (still air)[1]	1500	mW
V_{IN}	Differential input voltage	±30	V
V_{IN}	Input voltage[2]	±15	V
	Voltage between offset null and V-	±0.5	V
T_{STG}	Storage temperature range	-65 to +150	°C
T_A	Operating temperature range	0 to +70	°C
T_{SOLD}	Lead temperature (soldering, 10sec)	300	°C
I_{SC}	Output short-circuit duration	Indefinite	

Explanation of Key Maximum Ratings

Supply Voltage: This is the maximum voltage that can be used to power the op-amp.

Maximum Power Dissipation: This is the maximum power the op-amp can dissipate.

Differential Input Voltage: This is the maximum voltage that can be applied across the + and – inputs.

Input Voltage: This is the maximum voltage that can be applied between an input and ground.

Operating Temperature Range: The temperature range in which the op-amp will operate within the manufacturer's specifications.

Open Short Circuit Duration: This is the amount of time that the op-amp's output can be short circuited to ground or to a supply voltage. This op-amp has an "indefinite" rating since it has an internal circuit that will turn OFF the op-amp's output and protect the internal circuitry if an output short occurs.

DEVICE: µA 747—Dual Linear Op-Amp IC

EQUIVALENT SCHEMATIC

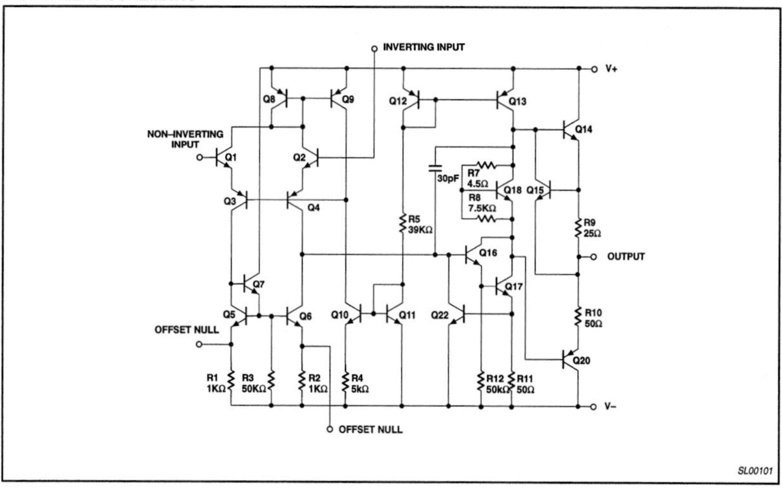

DC ELECTRICAL CHARACTERISTICS

T_A=25°C, V_{CC} = ±15V unless otherwise specified.

SYMBOL	PARAMETER	TEST CONDITIONS	µA747C Min	µA747C Typ	µA747C Max	UNIT
V_{OS}	Offset voltage	R_S≤10kΩ		2.0	6.0	mV
		R_S≤10kΩ, over temp.		3.0	7.5	mV
$\Delta V_{OS}/\Delta T$				10		µV/°C
I_{OS}	Offset current			20	200	nA
		Over temperature		7.0	300	nA
$\Delta I_{OS}/\Delta T$				200		pA/°C
I_{BIAS}	Input current			80	500	nA
		Over temperature		30	800	nA
$\Delta I_B/\Delta T$				1		nA/°C
V_{OUT}	Output voltage swing	R_L≥2kΩ, over temp.	±10	±13		V
		R_L≥10kΩ, over temp.	±12	±14		V
I_{CC}	Supply current each side			1.7	2.8	mA
		Over temperature		2.0	3.3	mA
P_d	Power consumption			50	85	mW
		Over temperature		60	100	mW
C_{IN}	Input capacitance			1.4		pF
	Offset voltage adjustment range			±15		mV
R_{OUT}	Output resistance			75		Ω
	Channel separation			120		dB
PSRR	Supply voltage rejection ratio	R_S≤10kΩ, over temp.		30	150	µV/V
A_{VOL}	Large-signal voltage gain (DC)	R_L≥2kΩ, V_{OUT}=±10V	25,000			V/V
		Over temperature	15,000			V/V
CMRR	Common-mode rejection ratio	R_S≤10kΩ, V_{CM}=±12V Over temperature	70			dB

Explanation of Key Ratings

Input Offset Voltage: The voltage that must be applied to one input for the output voltage to be zero.
Input Offset Current: The difference of the two input bias currents when the output voltage is zero.
Input Bias Current: The average of the currents flowing into both inputs (ideally input bias currents are equal).
Input Resistance: This is the resistance of either input when the other input is grounded.

Output Resistance: This is the resistance of the op-amp's output.
Output Short-Circuit Current: Maximum output current that the op-amp can deliver to a load.

Supply Current: The current that the op-amp circuit will draw from the power supply.
Slew rate: The maximum rate of change of output voltage under large signal conditions.
Channel Separation: When two op-amps are within one package, there will be a certain amount of interference or "crosstalk" between op-amps.

DEVICE: MC7800 Series—Positive Fixed Voltage Regulators

Three-Terminal Positive Voltage Regulators

These voltage regulators are monolithic integrated circuits designed as fixed-voltage regulators for a wide variety of applications including local, on-card regulation. These regulators employ internal current limiting, thermal shutdown, and safe-area compensation. With adequate heatsinking they can deliver output currents in excess of 1.0 A. Although designed primarily as a fixed voltage regulator, these devices can be used with external components to obtain adjustable voltages and currents.

- Output Current in Excess of 1.0 A
- No External Components Required
- Internal Thermal Overload Protection
- Internal Short Circuit Current Limiting
- Output Transistor Safe-Area Compensation
- Output Voltage Offered in 2% and 4% Tolerance

Representative Schematic Diagram

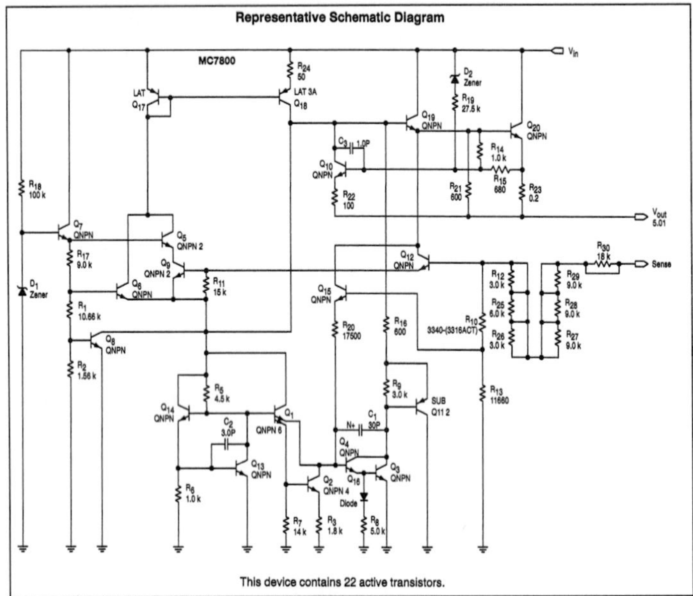

This device contains 22 active transistors.

ORDERING INFORMATION

Device	Output Voltage Tolerance	Tested Operating Temperature Range	Package
MC78XXACT	2%	T_J = 0° to +125°C	Insertion Mount
MC78XXACD2T			Surface Mount
MC78XXCT	4%		Insertion Mount
MC78XXCD2T			Surface Mount
MC78XXBT		T_J = – 40° to +125°C	Insertion Mount
MC78XXBD2T			Surface Mount

XX indicates nominal voltage.

DEVICE TYPE/NOMINAL OUTPUT VOLTAGE

MC7805	5.0 V	MC7812	12 V
MC7806	6.0 V	MC7815	15 V
MC7808	8.0 V	MC7818	18 V
MC7809	9.0 V	MC7824	24 V

Last two digits indicate output voltage. For example, 7808 will produce a +8 V regulated output.

MC7800 Series

THREE-TERMINAL POSITIVE FIXED VOLTAGE REGULATORS

SEMICONDUCTOR TECHNICAL DATA

T SUFFIX
PLASTIC PACKAGE
CASE 221A

Heatsink surface connected to Pin 2.

Pin 1. Adjust
2. V_{out}
3. V_{in}

D2T SUFFIX
PLASTIC PACKAGE
CASE 936
(D^2PAK)

Heatsink surface (shown as terminal 4 in case outline drawing) is connected to Pin 2.

STANDARD APPLICATION

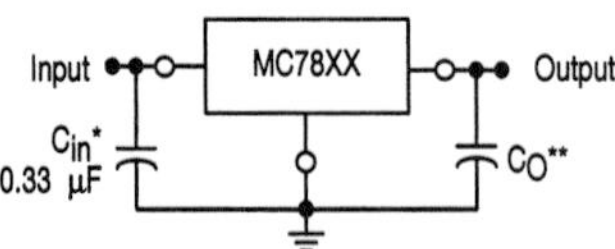

A common ground is required between the input and the output voltages. The input voltage must remain typically 2.0 V above the output voltage even during the low point on the input ripple voltage.

XX, these two digits of the type number indicate nominal voltage.

* C_{in} is required if regulator is located an appreciable distance from power supply filter.

** C_O is not needed for stability; however, it does improve transient response. Values of less than 0.1 µF could cause instability.

DEVICE: MC7900C Series—Negative Fixed Voltage Regulators

Three-Terminal Negative Voltage Regulators

The MC7900 series of fixed output negative voltage regulators are intended as complements to the popular MC7800 series devices. These negative regulators are available in the same seven-voltage options as the MC7800 devices. In addition, one extra voltage option commonly employed in MECL systems is also available in the negative MC7900 series.

Available in fixed output voltage options from −5.0 V to −24 V, these regulators employ current limiting, thermal shutdown, and safe-area compensation — making them remarkably rugged under most operating conditions. With adequate heatsinking they can deliver output currents in excess of 1.0 A.

- No External Components Required
- Internal Thermal Overload Protection
- Internal Short Circuit Current Limiting
- Output Transistor Safe-Area Compensation
- Available in 2% Voltage Tolerance (See Ordering Information)

MC7900 Series

THREE-TERMINAL NEGATIVE FIXED VOLTAGE REGULATORS

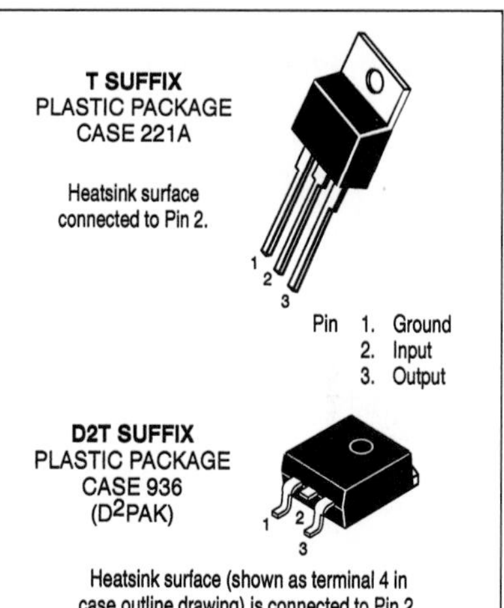

T SUFFIX
PLASTIC PACKAGE
CASE 221A

Heatsink surface connected to Pin 2.

Pin 1. Ground
2. Input
3. Output

D2T SUFFIX
PLASTIC PACKAGE
CASE 936
(D^2PAK)

Heatsink surface (shown as terminal 4 in case outline drawing) is connected to Pin 2.

Representative Schematic Diagram

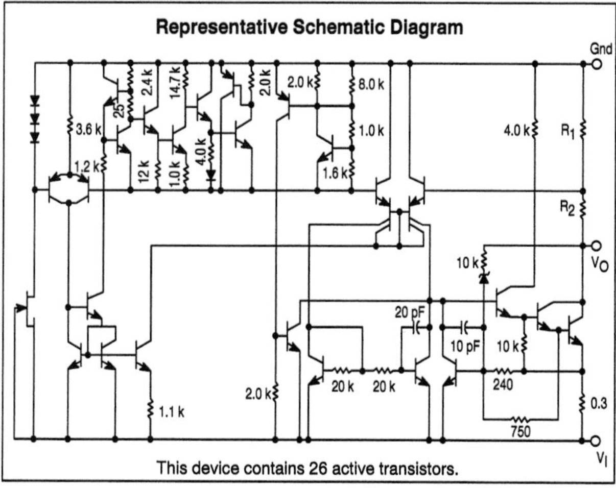

This device contains 26 active transistors.

STANDARD APPLICATION

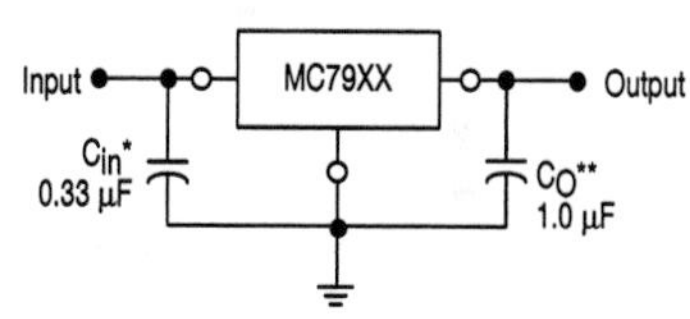

A common ground is required between the input and the output voltages. The input voltage must remain typically 2.0 V above more negative even during the high point of the input ripple voltage.

XX, these two digits of the type number indicate nominal voltage.

* C_{in} is required if regulator is located an appreciable distance from power supply filter.

** C_O improve stability and transient response.

ORDERING INFORMATION

Device	Output Voltage Tolerance	Tested Operating Temperature Range	Package
MC79XXACD2T	2%	T_J = 0° to +125°C	Surface Mount
MC79XXCD2T	4%		
MC79XXACT	2%		Insertion Mount
MC79XXCT	4%		
MC79XXBD2T	4%	T_J = −40° to +125°C	Surface Mount
MC79XXBT			Insertion Mount

XX indicates nominal voltage.

DEVICE TYPE/NOMINAL OUTPUT VOLTAGE

MC7905	5.0 V	MC7912	12 V
MC7905.2	5.2 V	MC7915	15 V
MC7906	6.0 V	MC7918	28 V
MC7908	8.0 V	MC7924	24 V

DEVICE: LM78S40 Switching Voltage Regulator

General Description

The LM78S40 is a monolithic regulator subsystem consisting of all the active building blocks necessary for switching regulator systems. The device consists of a temperature compensated voltage reference, a duty-cycle controllable oscillator with an active current limit circuit, an error amplifier, high current, high voltage output switch, a power diode and an uncommitted operational amplifier. The device can drive external NPN or PNP transistors when currents in excess of 1.5A or voltages in excess of 40V are required. The device can be used for step-down, step-up or inverting switching regulators as well as for series pass regulators. It features wide supply voltage range, low standby power dissipation, high efficiency and low drift. It is useful for any stand-alone, low part count switching system and works extremely well in battery operated systems.

Features

- Step-up, step-down or inverting switching regulators
- Output adjustable from 1.25V to 40V
- Peak currents to 1.5A without external transistors
- Operation from 2.5V to 40V input
- Low standby current drain
- 80 dB line and load regulation
- High gain, high current, independent op amp
- Pulse width modulation with no double pulsing

16-Lead DIP

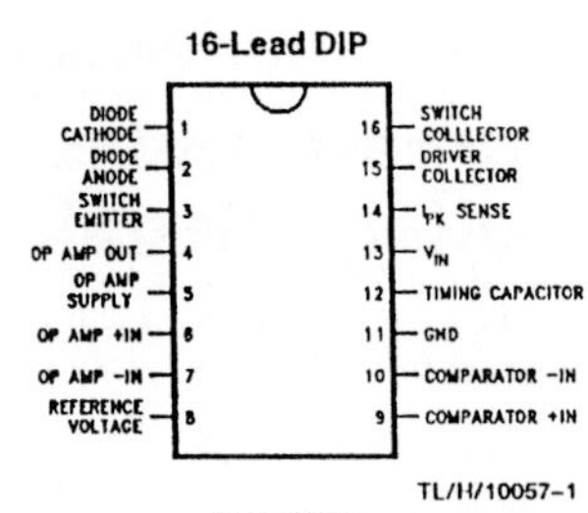

Top View

Ordering Information

Part Number	NS Package	Temperature Range
LM78S40J LM78S40J/883	J16A Ceramic DIP J16A Ceramic DIP	−55°C to +125°C
LM78S40N	N16E Molded DIP	−40°C to +125°C
LM78S40CJ LM78S40CN	J16A Ceramic DIP N16E Molded DIP	0°C to +70°C

Block and Connection Diagrams

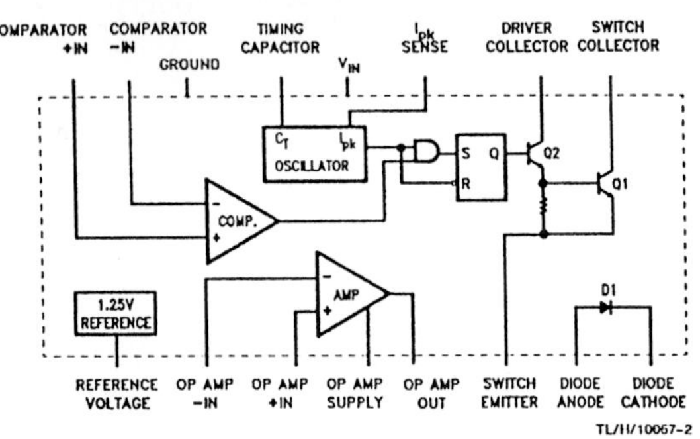

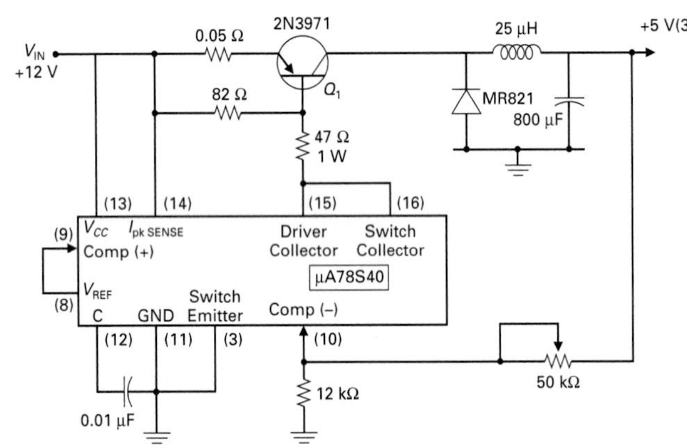

Absolute Maximum Ratings

If Military/Aerospace specified devices are required, please contact the National Semiconductor Sales Office/Distributors for availability and specifications.

Storage Temperature Range	
Ceramic DIP	−65°C to +175°C
Molded DIP	−65°C to +150°C
Operating Temperature Range	
Extended (LM78S40J)	−55°C to +125°C
Industrial (LM78S40N)	−40°C to +125°C
Commercial (LM78S40CN)	0°C to +70°C
Lead Temperature	
Ceramic DIP (Soldering, 60 sec.)	300°C
Molded DIP (Soldering, 10 sec.)	265°C
Internal Power Dissipation (Notes 1, 2)	
16L-Ceramic DIP	1.50W
16L-Molded DIP	1.04W
Input Voltage from V_{IN} to GND	40V
Common Mode Input Range (Comparator and Op Amp)	−0.3 to V+
Differential Input Voltage (Note 3)	±30V
Output Short Circuit Duration (Op Amp)	Continuous
Current from V_{REF}	10 mA
Voltage from Switch Collectors to GND	40V
Voltage from Switch Emitters to GND	40V
Voltage from Switch Collectors to Emitter	40V
Voltage from Power Diode to GND	40V
Reverse Power Diode Voltage	40V
Current through Power Switch	1.5A
Current through Power Diode	1.5A
ESD Susceptibility	(to be determined)

DEVICE: NE/SE 566—Function Generator

DESCRIPTION

The NE/SE566 Function Generator is a voltage-controlled oscillator of exceptional linearity with buffered square wave and triangle wave outputs. The frequency of oscillation is determined by an external resistor and capacitor and the voltage applied to the control terminal. The oscillator can be programmed over a ten-to-one frequency range by proper selection of an external resistance and modulated over a ten-to-one range by the control voltage, with exceptional linearity.

FEATURES

- Wide range of operating voltage (up to 24V; single or dual)
- High linearity of modulation
- Highly stable center frequency (200ppm/°C typical)
- Highly linear triangle wave output
- Frequency programming by means of a resistor or capacitor, voltage or current
- Frequency adjustable over 10-to-1 range with same capacitor

APPLICATIONS

- Tone generators
- Frequency shift keying
- FM modulators
- Clock generators
- Signal generators
- Function generators

PIN CONFIGURATIONS

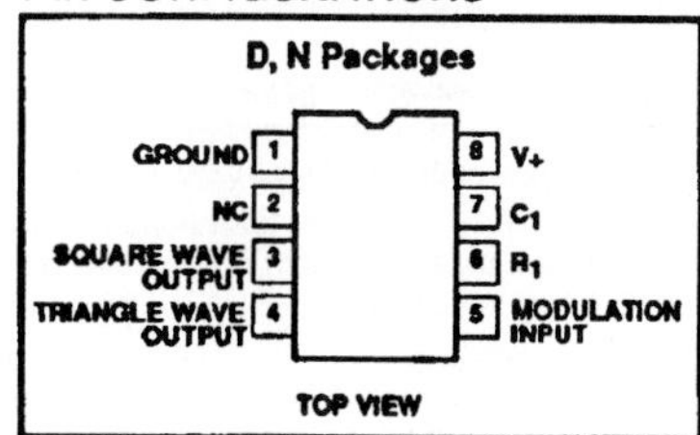

BLOCK DIAGRAM

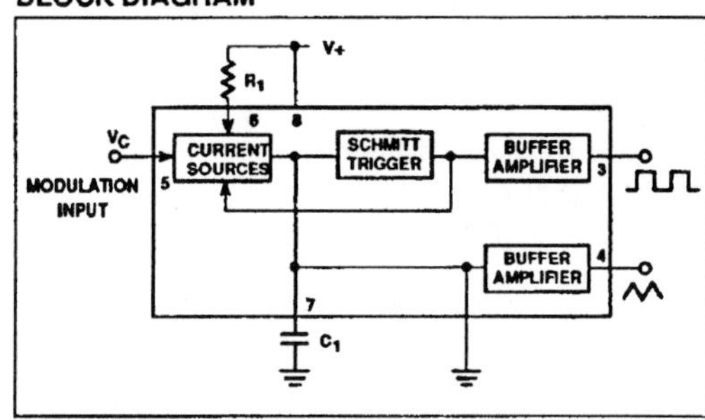

ABSOLUTE MAXIMUM RATINGS

SYMBOL	PARAMETER	RATING	UNIT
V+	Maximum operating voltage	26	V
V_{IN}	Input voltage	3	V_{P-P}
T_{STG}	Storage temperature range	-65 to +150	°C
T_A	Operating ambient temperature range		
	NE566	0 to +70	°C
	SE566	-55 to +125	°C
P_D	Power dissipation	300	mW

DEVICE: ICL8038—Waveform Generator/Voltage Controlled Oscillator

Description

The ICL8038 waveform generator is a monolithic integrated circuit capable of producing high accuracy sine, square, triangular, sawtooth and pulse waveforms with a minimum of external components. The frequency (or repetition rate) can be selected externally from 0.001Hz to more than 300kHz using either resistors or capacitors, and frequency modulation and sweeping can be accomplished with an external voltage. The ICL8038 is fabricated with advanced monolithic technology, using Schottky barrier diodes and thin film resistors, and the output is stable over a wide range of temperature and supply variations. These devices may be interfaced with phase locked loop circuitry to reduce temperature drift to less than 250ppm/°C.

Features

- **Low Frequency Drift with Temperature - 250ppm/°C**
- **Simultaneous Sine, Square, and Triangle Wave Outputs**
- **Low Distortion - 1% (Sine Wave Output)**
- **High Linearity - 0.1% (Triangle Wave Output)**
- **Wide Operating Frequency Range - 0.001Hz to 300kHz**
- **Variable Duty Cycle - 2% to 98%**
- **High Level Outputs - TTL to 28V**
- **Easy to Use - Just a Handful of External Components Required**

Absolute Maximum Ratings

Supply Voltage (V- to V+)	36V
Power Dissipation (Note 1)	750mW
Input Voltage (Any Pin)	V- to V+
Input Current (Pins 4 and 5)	25mA
Output Sink Current (Pins 3 and 9)	25mA
Lead Temperature (Soldering 10 Sec.)	+300°C

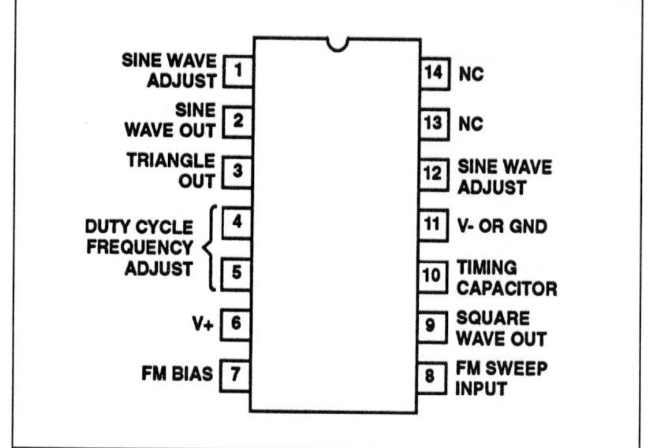

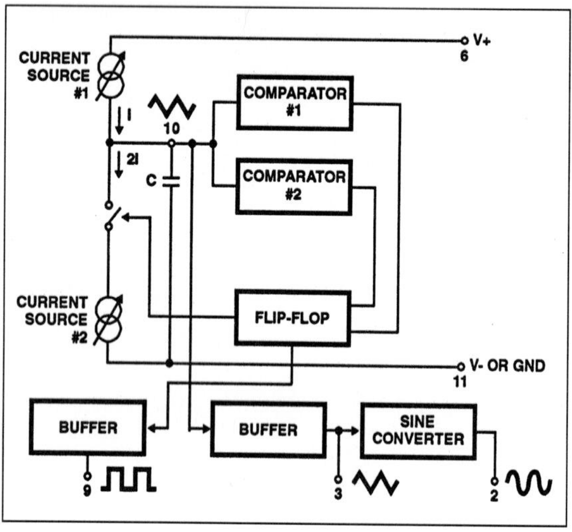

DEVICE: LM565/LM565C Phase Locked Loop

General Description

The LM565 and LM565C are general purpose phase locked loops containing a stable, highly linear voltage controlled oscillator for low distortion FM demodulation, and a double balanced phase detector with good carrier suppression. The VCO frequency is set with an external resistor and capacitor, and a tuning range of 10:1 can be obtained with the same capacitor. The characteristics of the closed loop system—bandwidth, response speed, capture and pull in range—may be adjusted over a wide range with an external resistor and capacitor. The loop may be broken between the VCO and the phase detector for insertion of a digital frequency divider to obtain frequency multiplication.

The LM565H is specified for operation over the −55°C to +125°C military temperature range. The LM565CN is specified for operation over the 0°C to +70°C temperature range.

Features

- 200 ppm/°C frequency stability of the VCO
- Power supply range of ±5 to ±12 volts with 100 ppm/% typical
- 0.2% linearity of demodulated output
- Linear triangle wave with in phase zero crossings available
- TTL and DTL compatible phase detector input and square wave output
- Adjustable hold in range from ±1% to > ±60%

Applications

- Data and tape synchronization
- Modems
- FSK demodulation
- FM demodulation
- Frequency synthesizer
- Tone decoding
- Frequency multiplication and division
- SCA demodulators
- Telemetry receivers
- Signal regeneration
- Coherent demodulators

Connection Diagrams

Metal Can Package

+V_{CC} 10
$-V_{CC}$ 1
TIMING CAPACITOR 9
INPUT 2
PHASE DETECTOR
VCO
TIMING RESISTOR 8
INPUT 3
AMP
VCO CONTROL VOLTAGE 7
VCO OUTPUT 4
5 PHASE COMPARATOR VCO INPUT
REFERENCE OUTPUT 6

Dual-In-Line Package

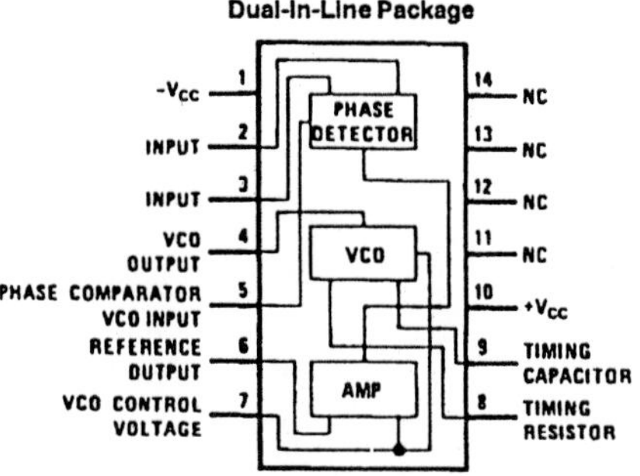

Schematic Diagram

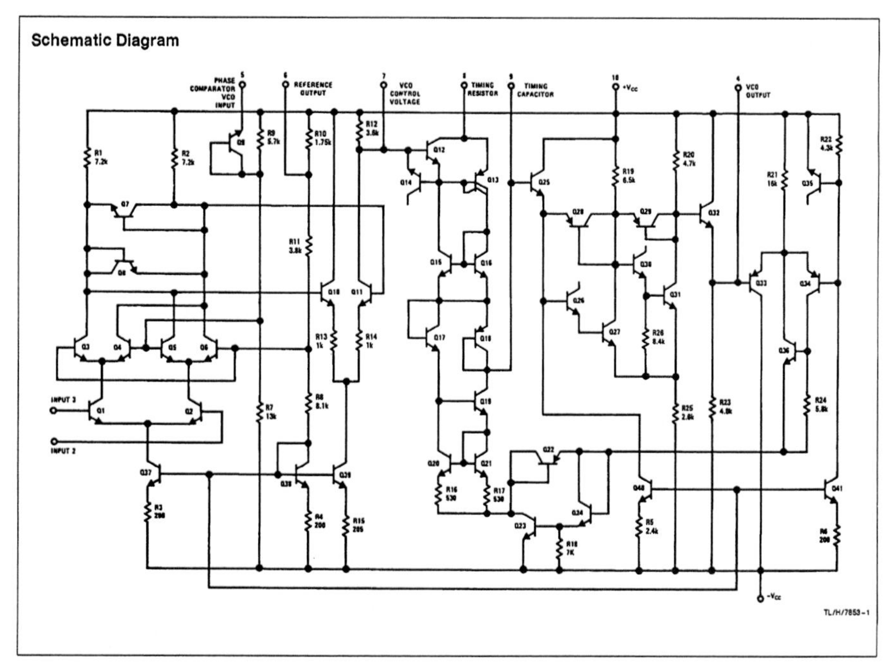

APPENDIX

Semiconductor Devices

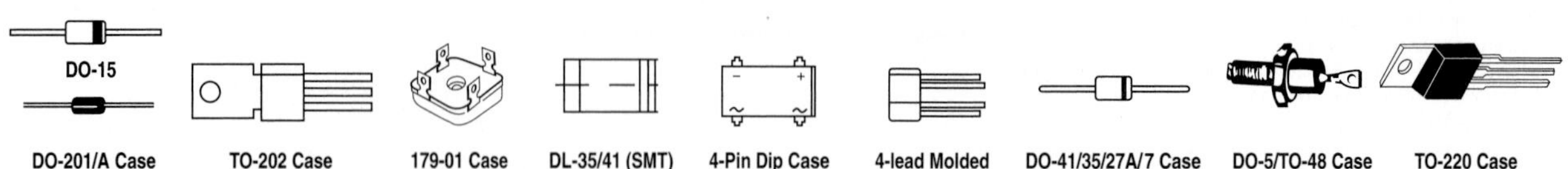

Zener Diodes

PRODUCT NO.	CASE	DESCRIPTION	
1N751	DO-35	5.1 V	500mWatt
1N753	DO-35	6.2 V	500mWatt
1N4732	DO-41	4.7 V	1 Watt
1N4733	DO-41	5.1 V	1 Watt
1N4733SM	DL-41	5.1 V	1 Watt
1N4734	DO-41	5.6 V	1 Watt
1N4735	DO-41	6.2 V	1 Watt
1N4735SM	DL-41	6.2 V	1 Watt
1N4742	DO-41	12.0 V	1 Watt
1N4742SM	DL-41	12.0 V	1 Watt
1N4744	DO-41	15.0 V	1 Watt

Switching Diodes

1N3600	DO-35	50 PRV	200mA
1N3600SM	DL-35	50 PRV	200mA
1N4148	DO-35	75 PRV	10mA (1N 914)
1N4148SM	DL-35	75 PRV	10mA (1N 914)

General Purpose Diode

1N270	DO-7	80 PRV	200mA

Silicon Rectifiers

1N4001	DO-41	50 PRV	1 Amp
1N4001SM	DL-41	50 PRV	1 Amp
1N4002	DO-41	100 PRV	1 Amp
1N4003	DO-41	200 PRV	1 Amp
1N4004	DO-41	400 PRV	1 Amp
1N4004SM	DL-41	400 PRV	1 Amp
1N4005	DO-41	600 PRV	1 Amp
1N4006	DO-41	800 PRV	1 Amp
1N4007	DO-41	1000 PRV	1 Amp
1N5391	DO-15	50 PRV	1.5 Amp
1N5392	DO-15	100 PRV	1.5 Amp
1N5393	DO-15	200 PRV	1.5 Amp
1N5395	DO-15	400 PRV	1.5 Amp
1N5397	DO-15	600 PRV	1.5 Amp
1N5398	DO-15	800 PRV	1.5 Amp
1N5399	DO-15	1000 PRV	1.5 Amp
1N5400	DO-27A	50 PRV	3 Amp
1N5401	DO-27A	100 PRV	3 Amp
1N5402	DO-27A	200 PRV	3 Amp
1N5404	DO-27A	400 PRV	3 Amp

Silicon Rectifiers continued

PRODUCT NO.	CASE	DESCRIPTION	
1N5406	DO-27A	600 PRV	3 Amp
1N5407	DO-27A	800 PRV	3 Amp
1N5408	DO-27A	1000 PRV	3 Amp

Fast Efficient Rectifiers

UF4002	DO41	100 PRV	1 Amp
UF4003	DO41	200 PRV	1 Amp
UF4004	DO41	400 PRV	1 Amp
UF4005	DO41	600 PRV	1 Amp
UF4007	DO41	1000 PRV	1 Amp

Stud Rectifier

1N1188	DO-5	400 PRV	35 Amp

Silicon Controlled Rectifiers (SCR)

C106B1	TO-202	200 PRV 4A (2N6239)*
C38M	TO-48	600 PRV 35A (2N3899)
C122B	TO-220	200 PRV 8A*
C122M	TO-220	600 PRV 8A*

Full Wave Bridge Rectifiers

W02G	4 lead molded	200 PRV	1 Amp
W04G	4 lead molded	400 PRV	1 Amp
W06G	4 lead molded	600 PRV	1 Amp
DF01M	4-Pin DIP case	100 PRV	1 Amp
DF02M	4-Pin DIP case	200 PRV	1 Amp
DF04M	4-Pin DIP case	400 PRV	1 Amp
DF06M	4-Pin DIP case	600 PRV	1 Amp
DF10M	4-Pin DIP case	1000 PRV	1 Amp
MDA990-3	179-01	200 PRV	30 Amp
MDA990-6	179-01	600 PRV	30 Amp

Triacs

SC146B	TO-220	200 PRV10A (2N6342)
SC146M	TO-220	600 PRV10A (2N6344)

Bidirectional Transient Voltage Suppressors

1.5KE6.8CA	DO201	6.8V @10mA	1500W
1.5KE7.5CA	DO201	7.5V @10mA	1500W
1.5KE15CA	DO201	15V @1mA	1500W
1.5KE200CA	DO201	200V @1mA	1500W
1.5KE220CA	DO201	220V @1mA	1500W

Transistors

TO-72 Case TO-3 Case TO-92 Case TO-39 Case SOT-23 (SMT) TO-220 Case 14-pin DIP Case TO-18 Case

PRODUCT NO.	CASE	DESCRIPTION
MPSA06	TO-92	NPN Medium Power
MPSA13	TO-92	NPN Power Darlington
MPSA13SM	SOT-23	NPN Power Darlington
TIP29A	TO-220	NPN Power Transistor
TIP30A	TO-220	PNP Power Transistor
TIP31A	TO-220	NPN Power Transistor
TIP32A	TO-220	PNP Power Transistor
TIP41A	TO-220	NPN Power Transistor
TIP42A	TO-220	PNP Power Transistor
MPSU51	TO-202	PNP Silicon Audio
MPF102	TO-92	JFET (N Channel) RF, VHF, UHF, Amp
TIP102	TO-220	NPN Power Darlington
TIP106	TO-220	PNP Power Darlington
TIP120	TO-220	NPN Power Darlington
TIP125	TO-220	PNP Power Darlington
2N2219A	TO-39	NPN General Purpose
PN2222	TO-92	NPN General Purpose
PN2222SM	SOT-23	NPN General Purpose
2N2222A	TO-18	NPN General Purpose
MPQ2222	14-Pin DIP	NPN Gen. Purp. Small Signal/Quad
PN2907	TO-92	PNP General Purpose

PRODUCT NO.	CASE	DESCRIPTION
PN2907SM	SOT-23	PNP General Purpose
2N2907A	TO-18	PNP General Purpose
MPQ2907	14-Pin DIP	PNP Gen. Purp. Small Signal/Quad
MJE2955T	TO-220	PNP Power Transistor
2N3053	TO-39	NPN General Purpose
MJE3055T	TO-220	NPN Power Transistor
2N3055	TO-3	NPN Power BiPolar, Epitaxial Base
PN3569	TO-92	NPN Medium Power Transistor
2N3772	TO-3	NPN Power BiPolar, Epitaxial Base
2N3823	TO-72	JFET (N Channel) VHF, Amp
2N3904	TO-92	NPN General Purpose
2N3904SM	SOT-23	NPN General Purpose
2N3906	TO-92	PNP General Purpose
2N3906SM	SOT-23	PNP General Purpose
2N4401	TO-92	NPN General Purpose
2N4401SM	SOT-23	NPN General Purpose
2N4403	TO-92	PNP General Purpose
2N4403SM	SOT-23	PNP General Purpose
PN5179	TO-92	NPN High Frequency Transistor
2N5951	TO-92	JFET (N Channel) RF, VHF, UHF, Amp

LEDs and Lamps

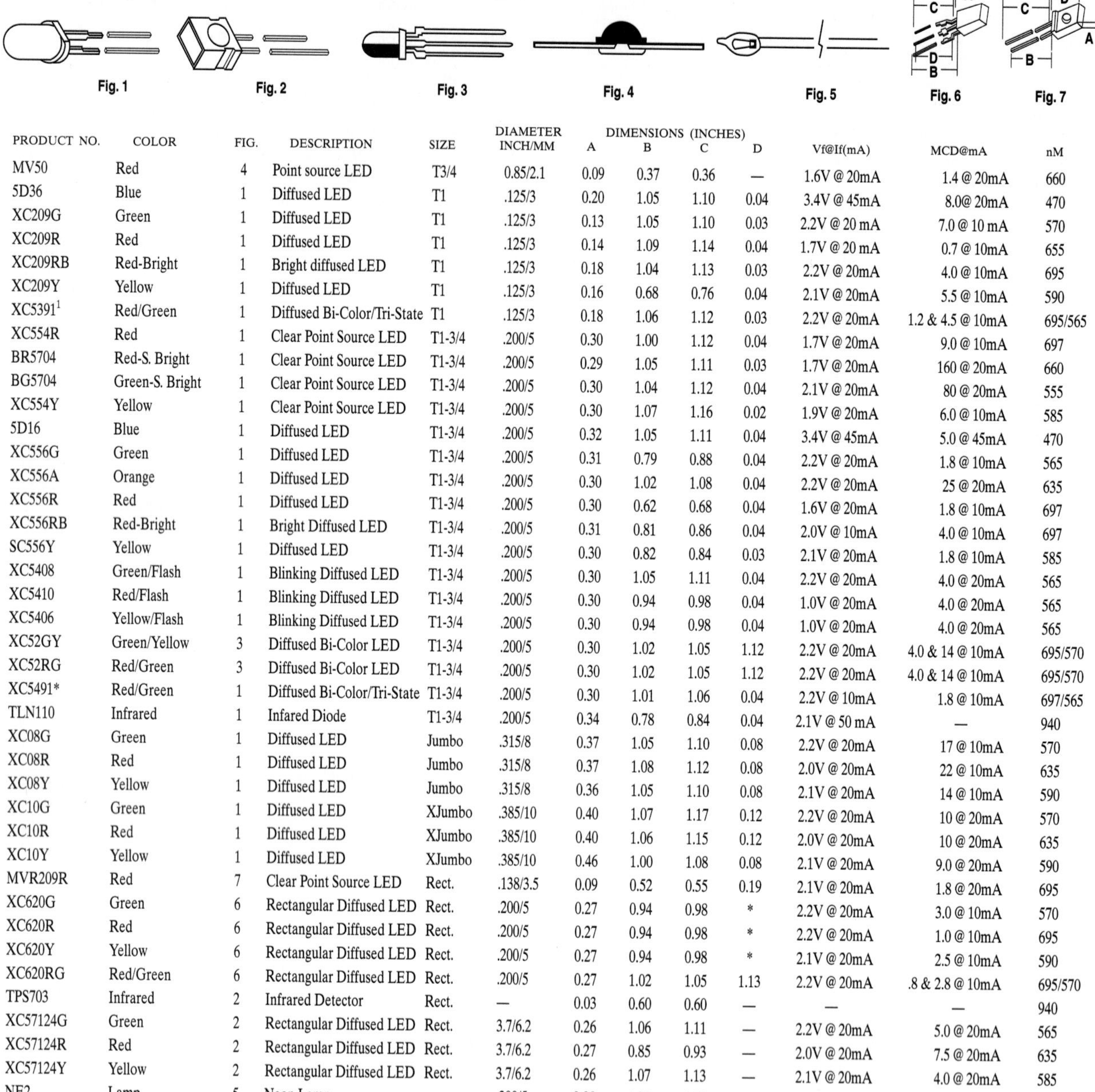

PRODUCT NO.	COLOR	FIG.	DESCRIPTION	SIZE	DIAMETER INCH/MM	DIMENSIONS (INCHES) A	B	C	D	Vf@If(mA)	MCD@mA	nM
MV50	Red	4	Point source LED	T3/4	0.85/2.1	0.09	0.37	0.36	—	1.6V @ 20mA	1.4 @ 20mA	660
5D36	Blue	1	Diffused LED	T1	.125/3	0.20	1.05	1.10	0.04	3.4V @ 45mA	8.0@ 20mA	470
XC209G	Green	1	Diffused LED	T1	.125/3	0.13	1.05	1.10	0.03	2.2V @ 20 mA	7.0 @ 10 mA	570
XC209R	Red	1	Diffused LED	T1	.125/3	0.14	1.09	1.14	0.04	1.7V @ 20 mA	0.7 @ 10mA	655
XC209RB	Red-Bright	1	Bright diffused LED	T1	.125/3	0.18	1.04	1.13	0.03	2.2V @ 20mA	4.0 @ 10mA	695
XC209Y	Yellow	1	Diffused LED	T1	.125/3	0.16	0.68	0.76	0.04	2.1V @ 20mA	5.5 @ 10mA	590
XC5391[1]	Red/Green	1	Diffused Bi-Color/Tri-State	T1	.125/3	0.18	1.06	1.12	0.03	2.2V @ 20mA	1.2 & 4.5 @ 10mA	695/565
XC554R	Red	1	Clear Point Source LED	T1-3/4	.200/5	0.30	1.00	1.12	0.04	1.7V @ 20mA	9.0 @ 10mA	697
BR5704	Red-S. Bright	1	Clear Point Source LED	T1-3/4	.200/5	0.29	1.05	1.11	0.03	1.7V @ 20mA	160 @ 20mA	660
BG5704	Green-S. Bright	1	Clear Point Source LED	T1-3/4	.200/5	0.30	1.04	1.12	0.04	2.1V @ 20mA	80 @ 20mA	555
XC554Y	Yellow	1	Clear Point Source LED	T1-3/4	.200/5	0.30	1.07	1.16	0.02	1.9V @ 20mA	6.0 @ 10mA	585
5D16	Blue	1	Diffused LED	T1-3/4	.200/5	0.32	1.05	1.11	0.04	3.4V @ 45mA	5.0 @ 45mA	470
XC556G	Green	1	Diffused LED	T1-3/4	.200/5	0.31	0.79	0.88	0.04	2.2V @ 20mA	1.8 @ 10mA	565
XC556A	Orange	1	Diffused LED	T1-3/4	.200/5	0.30	1.02	1.08	0.04	2.2V @ 20mA	25 @ 20mA	635
XC556R	Red	1	Diffused LED	T1-3/4	.200/5	0.30	0.62	0.68	0.04	1.6V @ 20mA	1.8 @ 10mA	697
XC556RB	Red-Bright	1	Bright Diffused LED	T1-3/4	.200/5	0.31	0.81	0.86	0.04	2.0V @ 10mA	4.0 @ 10mA	697
SC556Y	Yellow	1	Diffused LED	T1-3/4	.200/5	0.30	0.82	0.84	0.03	2.1V @ 20mA	1.8 @ 10mA	585
XC5408	Green/Flash	1	Blinking Diffused LED	T1-3/4	.200/5	0.30	1.05	1.11	0.04	2.2V @ 20mA	4.0 @ 20mA	565
XC5410	Red/Flash	1	Blinking Diffused LED	T1-3/4	.200/5	0.30	0.94	0.98	0.04	1.0V @ 20mA	4.0 @ 20mA	565
XC5406	Yellow/Flash	1	Blinking Diffused LED	T1-3/4	.200/5	0.30	0.94	0.98	0.04	1.0V @ 20mA	4.0 @ 20mA	565
XC52GY	Green/Yellow	3	Diffused Bi-Color LED	T1-3/4	.200/5	0.30	1.02	1.05	1.12	2.2V @ 20mA	4.0 & 14 @ 10mA	695/570
XC52RG	Red/Green	3	Diffused Bi-Color LED	T1-3/4	.200/5	0.30	1.02	1.05	1.12	2.2V @ 20mA	4.0 & 14 @ 10mA	695/570
XC5491*	Red/Green	1	Diffused Bi-Color/Tri-State	T1-3/4	.200/5	0.30	1.01	1.06	0.04	2.2V @ 10mA	1.8 @ 10mA	697/565
TLN110	Infrared	1	Infared Diode	T1-3/4	.200/5	0.34	0.78	0.84	0.04	2.1V @ 50 mA	—	940
XC08G	Green	1	Diffused LED	Jumbo	.315/8	0.37	1.05	1.10	0.08	2.2V @ 20mA	17 @ 10mA	570
XC08R	Red	1	Diffused LED	Jumbo	.315/8	0.37	1.08	1.12	0.08	2.0V @ 20mA	22 @ 10mA	635
XC08Y	Yellow	1	Diffused LED	Jumbo	.315/8	0.36	1.05	1.10	0.08	2.1V @ 20mA	14 @ 10mA	590
XC10G	Green	1	Diffused LED	XJumbo	.385/10	0.40	1.07	1.17	0.12	2.2V @ 20mA	10 @ 20mA	570
XC10R	Red	1	Diffused LED	XJumbo	.385/10	0.40	1.06	1.15	0.12	2.0V @ 20mA	10 @ 20mA	635
XC10Y	Yellow	1	Diffused LED	XJumbo	.385/10	0.46	1.00	1.08	0.08	2.1V @ 20mA	9.0 @ 20mA	590
MVR209R	Red	7	Clear Point Source LED	Rect.	.138/3.5	0.09	0.52	0.55	0.19	2.1V @ 20mA	1.8 @ 20mA	695
XC620G	Green	6	Rectangular Diffused LED	Rect.	.200/5	0.27	0.94	0.98	*	2.2V @ 20mA	3.0 @ 10mA	570
XC620R	Red	6	Rectangular Diffused LED	Rect.	.200/5	0.27	0.94	0.98	*	2.2V @ 20mA	1.0 @ 10mA	695
XC620Y	Yellow	6	Rectangular Diffused LED	Rect.	.200/5	0.27	0.94	0.98	*	2.1V @ 20mA	2.5 @ 10mA	590
XC620RG	Red/Green	6	Rectangular Diffused LED	Rect.	.200/5	0.27	1.02	1.05	1.13	2.2V @ 20mA	.8 & 2.8 @ 10mA	695/570
TPS703	Infrared	2	Infrared Detector	Rect.	—	0.03	0.60	0.60	—	—	—	940
XC57124G	Green	2	Rectangular Diffused LED	Rect.	3.7/6.2	0.26	1.06	1.11	—	2.2V @ 20mA	5.0 @ 20mA	565
XC57124R	Red	2	Rectangular Diffused LED	Rect.	3.7/6.2	0.27	0.85	0.93	—	2.0V @ 20mA	7.5 @ 20mA	635
XC57124Y	Yellow	2	Rectangular Diffused LED	Rect.	3.7/6.2	0.26	1.07	1.13	—	2.1V @ 20mA	4.0 @ 20mA	585
NE2	Lamp	5	Neon Lamp	—	.200/5	0.38	1.20	1.20	—	65-120VDC @ 130-470mA*/65-120VAC @ 150-500mA[2]		

Optically Coupled Isolators

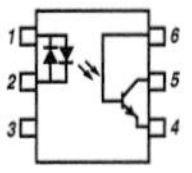

Fig. 1 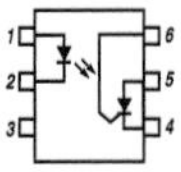 Fig. 2 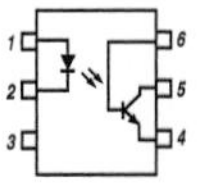 Fig. 3 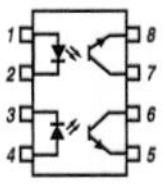 Fig. 4 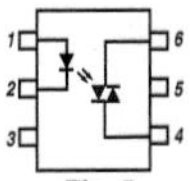 Fig. 5 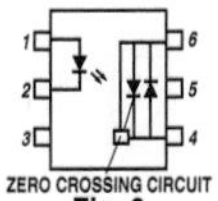 ZERO CROSSING CIRCUIT Fig. 6 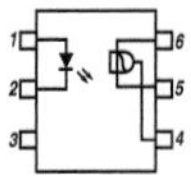 Fig. 7 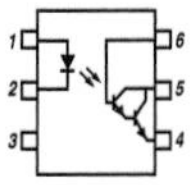 Fig. 8 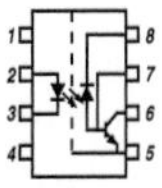 Fig. 9 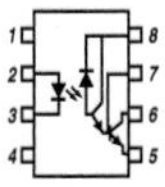 Fig. 10 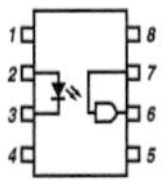 Fig. 11 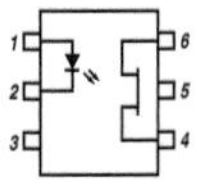 Fig. 12 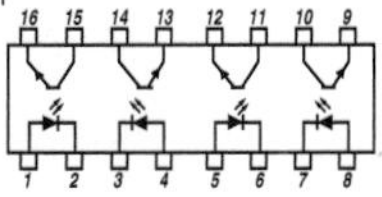 Fig. 13 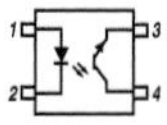 Fig. 14

PRODUCT NO.	FIG.	PINS	DESCRIPTION	INPUT CURRENT (mA)	BLOCKING VOLTAGE (V)	ISOLATION (kV)
H11AA1	1	6	Dual Diode, NPN Phototransistor	20	30	2.50
H11C1	2	6	Photo SCR Output	11	200	5.00
H11C3	2	6	Photo SCR Output	14	200	5.00
H11C4	2	6	Photo SCR Output	11	400	5.00
4N25	3	6	Diode, NPN Phototransistor	10	30	2.50
4N26	3	6	Diode, NPN Phototransistor	10	30	1.50
4N28	3	6	Diode, NPN Phototransistor	10	30	0.50
4N35	3	6	Diode, NPN Phototransistor	10	30	3.50
4N37	3	6	Diode, NPN Phototransistor	10	30	1.50
MCT2	3	6	Diode, NPN Phototransistor	10	30	3.50
IL2	3	6	Diode, NPN Phototransistor	10	70	7.50
IS203	3	6	Diode, NPN Phototransistor	1	30	5.00
IS204	3	6	Diode, NPN Phototransistor	1	70	5.00
SFH600-2	3	6	Diode, NPN Phototransistor	10	70	2.80
ISD203	4	8	Quad Channel NPN Phototransistor(225 CTR)	10	30	5.00
ISD204	4	8	Quad Channel NPN Phototransistor(200 CTR)	10	30	5.00
MCT6	4	8	Dual Channel NPN Phototransistor	10	30	7.50
IS6010	5	6	Bilateral Triac Driver	10	600	7.50
MOC3010	5	6	Bilateral Triac Driver	15	250	7.50
MOC3011	5	6	Bilateral Triac Driver	10	250	7.50
MOC3021	5	6	Bilateral Triac Driver	15	400	7.50
MOC3022	5	6	Bilateral Triac Driver	10	400	7.50
IS622	6	6	Zero Crossing Triac Driver	10	600	7.50
MOC3032	6	6	Zero Crossing Triac Driver	10	250	7.50
MOC3042	6	6	Zero Crossing Triac Driver	10	400	7.50
MOC3031	6	6	Zero Crossing Triac Driver	15	250	7.50
H11L1	7	6	Schmitt Trigger Output Photocoupler	1.6	STTL/TTL	5.00
PC900V	7	6	Schmitt Trigger Output Photocoupler	4	STTL/TTL	5.00
4N32	8	6	Diode, NPN Photo-darlington	10	30	1.50
4N33	8	6	Diode, NPN Photo-darlington	10	30	6.00
6N136	9	8	High Speed NPN Transistor	16	4.5	2.50
6N138	10	8	High Speed, High Gain Darlington	1.6	4.5	2.50
6N139	10	8	High Speed, High Gain Darlington	0.5	4.5	2.50
6N137	11	8	Very High Speed Logic Coupler	5	STTL/TTL	2.50
H11F1	12	6	Isolated FET (200-Ohm ON Resistance)	16	30	2.50
ISQ203	13	16	Quad Channel NPN Phototransistor(225 CTR)	10	30	7.50
ISQ204	13	16	Quad Channel NPN Phototransistor(200 CTR)	10	30	7.50
H21A1	14	4	Single Channel, Slotted Interrupter w/tabs	5	30	—
H22A1	14	4	Single Channel, Slotted Interrupter w/o tabs	5	30	—

Character Displays

24740 17187 24723 104117

PRODUCT NO.	PINS/LEADS	COMPATIBLE WITH JAMECO SOCKET	CHARACTER HEIGHT	COLOR	POLARITY	Vf	If	MCD @ If	WAVELENGTH mM	DEC. PT.
FND10A	9	—	.122"	Red	CC	1.7V	10mA	0.05	665	RHDP
MAN3	9	—	.127"	Red	CC	1.7V	5mA	0.2	665	RHDP
MAN73	8	37161	.294"	Red	CA	1.6V	20mA	0.3	650	±1
MAN71	11	37161	.300"	Red	CA	1.6V	20mA	0.3	650	RHDP
MAN72	11	37161	.300"	Red	CA	1.6V	20mA	0.3	650	LHDP
MAN74	10	37161	.300"	Red	CC	1.6V	20mA	0.3	650	RHDP
NSN373	16	PC board mounted	.300"	Red	2 digit CC	1.7V	10mA	1.6	660	—
TSB3881	16	PC board mounted	.300"	Red	4 digit CC	1.7V	10mA	1.6	660	—
MAN82	11	37161	.300"	Yellow	CA	2.5V	20mA	1.2	590	LHDP
FND357	10	67900	.357"	Red	CC	1.7V	20mA	1.0	655	RHDP
FND350	10	67900	.362"	Red	CA	1.7V	20mA	0.45	665	RHDP
MAN4540	9	37161	.400"	Green	CC	2.5V	20mA	0.51	565	RHDP
MAN4710	11	37161	.400"	Red	CA	1.6V	20mA	0.3	660	RHDP
5082-7653	11	37161	.430"	Orange	CC	2.0V	20mA	1.11	635	RHDP
5082-7750	11	37161	.430"	Red	CA	1.6V	20mA	1.1	655	LHDP
5082-7751	11	37161	.430"	Red	CA	1.6V	20mA	1.1	655	RHDP
5082-7756	12	37161	.430"	Red	Universal ± 1	1.6V	20mA	1.1	655	RHDP
5082-7760	11	37161	.430"	Red	CC	1.6V	20mA	1.1	655	RHDP
FND503	10	102200	.500"	Red	CC	1.7V	20mA	0.6	665	RHDP
FND507	10	102200	.500"	Red	CA	1.7V	20mA	1.2	665	RHDP
MAN6660	10	39335	.560"	Orange	CA	2.5V	20mA	0.5	630	RHDP
LTS5301AP	10	101282	.560"	Red	CA	2.0V	20mA	1.8	660	RHDP
LTD586AG	18	39335	.500"	Green	2 Digit CC	2.0V	20mA	4.0	565	RHDP
MAN6610	18	39335	.560"	Orange	2 Digit CA	2.0V	20mA	4.0	635	RHDP
MAN6640	18	39335	.560"	Orange	2 Digit CC	2.0V	20mA	4.0	635	RHDP
MAN6710	18	39335	.560"	Red	2 Digit CA	2.2V	20mA	1.4	695	RHDP
MAN6740	18	39335	.560"	Red	2 Digit CC	2.2V	20mA	1.4	695	RHDP
FND603	10	102200	.600"	Red	CC	1.7V	20mA	1.0	655	RHDP
HP3401	13	39335	.800"	Red	CA	1.6V	20mA	0.9	655	RHDP
HP3405	13	39335	.800"	Red	CC	1.6V	20mA	0.9	655	LHDP
CSS104H	10	102200	1.0"	Red	CA	2.2V	20mA	3.2	695	RHDP
CSS105H	10	102200	1.0"	Red	CC	2.2V	20mA	3.2	695	RHDP
CSS2314D	10	102200	2.24"	Red	CA	7.0V	20mA	60.0	660	RHDP
CSS2315D	10	102200	2.24"	Red	CC	7.0V	20mA	60.0	660	RHDP
Dot Matrix Displays										
TIL311	11	37161	.270"	Red	4× 7 HEX w/logic	5.0V	5mA	0.1	660	—
LTP1057A	14	101282	1.2" × .85"	Red	CC/AR	1.8V	20mA	3.5	660	—
LTP2158A	14	101282	2.3" × 1.4"	Red	CC/AR	1.8V	20mA	4.0	660	—
Alpha-Numeric Displays										
TSM1416	20	101282	.160"	Red	4 digit 16-segment	5.0V	80mA	0.75	660	—
TSM2416	18	39335/101282	.160"	Red	4 digit 17-segment	5.0V	65mA	1.0	655	—
LT3784E	18	39335	.540"	Orange	2 digit CC	2.1V	20mA	2.6	635	RHDP

Linear Circuits

PRODUCT NO.	PINS	DESCRIPTION
LH0002CN	10	Current Amplifier
OP07	8	Ultra-Low-Offset-Voltage Op Amp
LM10CLN	8	Low Voltage (1.1 volt) Op Amp
MF10CN	20	General Purpose Dual Active Filter
DS0026CN	8	Dual Mos Clock Driver (5MHz) 20V Output
TL061CP	8	Low Power JFET Op Amp
TL062CP	8	Low Power JFET Op Amp
TL064CN	14	Low-Power JFET-Input Op Amp
LH0070-OH	TO-5/3	Precision BCD Buffered Reference
TL071CP	8	Low Noise JFET Op Amp
TL072CP	8	Low Noise JFET Dual Op Amp
TL074CN	14	Low Noise JFET Quad Op Amp
TL081CP	8	JFET-Input Op Amp
SMP81FY	14	Sample and Hold Amplifier
TL082CP	8	JFET-Input Dual Op Amp
TL084CN	14	JFET Input Quad Op Amp
MAX232CPE	16	Dual RS232 Receiver/Transmitter
LM301N	8	Improved Op Amp
LM301H	TO-5/8	Improved Op Amp
LM305H	TO-5/8	Positive Voltage Regulator (4.5V-40V)
LM307N	8	Op Amp (Advanced LM741CN)
LM308N	8	Precision Op Amp
LM308N-14	14	Micro Power Op Amp
LM308H	TO-5/8	Precision Op Amp
LM309K	TO-3	5-Volt Amp Regulator (1 Amp)
LM310N	8	Non-inverting Op Amp/Volt. Follower
LM311N	8	Volt. Comparator
LM317K	TO-3	1.5A Adj. Pos. Regulator (1.2V-37V)
LM317LZ	TO-92/3	100mA Adjustable Pos. Regulator (1.2V-37V)
LM317T	TO-220/3	1.5A Adj. Pos. Reg. (1.2V-37V)
LM318N	8	Precision Hi-Speed Op Amp
LM319N	14	Hi-Speed Dual Comparator
LM323K	TO-3	5-Volt Positive 3-Amp Regulator
LM324N	14	Low Power Quad Op Amp (ULN4336)
LM329DZ	TO-92/3	Precision Zener 100PPM/C°6.9V
LM331N	8	Precision Voltage to Freq. Converter
LM334Z	TO-92/3	Adjustable Current Source
LM335Z	TO-92/3	Precision Temperature Sensor
LM336Z	TO-92/3	Voltage Reference (2.5V)
LM337K	TO-3	3-Terminal Adjustable Negative Regulator
LM337MP	TO-202/3	.5A/3 Term. Adj. Neg. Reg. (1.2V-37V)
LM337T	TO-220/3	1.5A/Adj. Neg. Reg. (-1.2V-37V)
LM338K	TO-3	5A Adjust. Pos. Regulator (1.2V-32V)
LM339N	14	Quad Comparator
LM346N	16	Programmable Quad Op Amp
LF347N	14	Quad JFET Input Op Amp (Wide Band)
LM348N	14	Quad 741 Op Amp
LM350K	TO-3	3A/3 Term. Adj. Pos. Reg. (1.2V-33V)
LM350T	TO-220/3	3 Amp Adjustable Power Regulator
LF351N	8	BIFET Op Amp
LF353N	8	Dual BIFET Op Amp (Compat. w/TL082CP)
LF355N	8	JFET Input Op Amp Low Supply Current
LF356H	TO-5	Mono. JFET Op Amp (Wide Bandwidth)
LF356N	8	JFET Input Op Amp (Wide Band)
LF357N	8	J-FET Input Op Amp (Wide Band Decompens)
LM358N	8	Low Power Dual Op Amp
LM359N	14	High Speed Norton Op Amp
LM360N	8	High Speed Differential Comparator
LM361N	14	Hi Speed Diff. Volt Comp. w/Indep. Strobes (NE529)
LM373N	14	AM/FM/SSB Strip
LM380N	14	2-Watt Audio Power Amp (ULN2280)
LM381N	14	Low Noise Dual Pre-Amp (112dB)
LM383T	TO-220/5	7-Watt Audio High Power Amplifier
LM384N	14	5-Watt Audio Amp (ULN2281)
LM385Z	TO-92/3	Micropower Voltage Reference Diode (2.5V)
LM385Z1.2	TO-92/3	Micropower Voltage Reference Diode (1.2V)
LM386N-1	8	Low Voltage Audio Amp 250Mw/6V
LM386N-3	8	Low Voltage Audio Amp 500Mw/9V
LM387N	8	Low Noise Dual Pre-Amp
LM391N	16	Audio Power Driver (±50V or +100V)
LM393N	8	Dual Comparator
LF398N	8	Sample and Hold Circuit
LM399H	TO-5/4	Temp. Compensated Prec. Ref. (.5 ppm/C°)
LF411CN	8	Low Offset, Low Drift JFET Input Op Amp
LF412CN	8	Low Offset Low Drift JFET Input Dual Op Amp
TL431CLP	8	Programmable Precision Reference
TL494CN	16	Controlled Circuit Pulse Width Modulated
TL497ACN	14	Switching Voltage Regulator
NE555V	8	Timer (MC1455P) (UPC1555C)
XRL555	8	Micro Power 555 Timer
LM556N	14	Dual 555 Timer (MC3456P)
NE558N	16	Quad Timer
NE564N	16	High Frequency Phase Locked Loop
LM565N	14	Phased Locked Loop
LM566CN	8	Function Generator (VCO)
LM567V	8	Tone Decoder
NE570N	16	Compander
NE571N	16	Compander
NE592N	14	Differential Video Amp
TL592P	8	Differential Video Amp
LM710N	14	Voltage Comparator
LM723CH	TO-5/10	Voltage Regulator (2V-37V)
LM723CN	14	Voltage Regulator (2V-37V)
LM725CN	8	Instrumentation Op Amp
LM733CN	14	Differential Amp
LM741CN	8	Op Amp (ULN2151)
LM741CH	TO-5/8	Op Amp
LM741-14N	14	Op Amp
LM747CN	14	Dual 741 Op Amp
LM748CN	8	Frequency Adjusted 741
MC1350P	8	Video IF Amp (12V)
MC1377P	20	Color TV RGB to PAL/NTSC Encoder
LM1456V	8	Op Amp Hi-Performance (NE5556)
LM1458N	8	Dual Op Amp (MC1458P)(RC4558)
LM1488N	14	Quad Line Driver RS232 (75188)
DS14C88N	14	Quad CMOS Line Driver (RS232)
LM1489N	14	Quad Line Receiver RS232 (75189)
DS14C89N	14	Quad CMOS Line Receiver (RS232)
LM1496N	14	Balanced Mod/Demodulator (MC1496P)
MC1648P	14	Voltage Controlled Oscillator
LM1871N	18	Radio Control Encoder/Transmitter
LM1872N	18	Radio Control Receiver/Decoder
LM1875T	TO-220	20-Watt Power Audio Amplifier
LM1877N-9	14	Dual Power Audio Amp
ULN2001	16	Hi-Volt. Hi-Curr. Darlington Transistor Array
ULN2003A	16	Hi-Volt. Hi-Curr. Darlington Transistor Array
ULN2004A	16	High-Volt. Hi-Curr. Darl. Trans. Array (MC1416P)
XR2206	16	Monolithic Function Generator
XR2211	14	FSK Demodulator/Tone Decoder
XR2240	16	Programmable Counter/Timer
26LS29	16	Quad Three-State Single Ended RS-423 Line Drive
26LS30	16	Dual Diff. RS422 Party Line Quad RS423 Line Driver
26LS31	16	Quad Differential Line Driver
26LS32	16	Quad Differential Line Receiver
26LS33	16	Quad Differential Line Receiver
ULN2803A	18	Hi-Volt, Hi-Curr. Darlington Trans. Array
LM2901N	14	Quad Comparator (Low Power) MC3302P
LM2907N	14	Frequency to Voltage Converter
LM2917N-8	8	Frequency to Voltage Converter (CS2917)
LM2917N-14	14	Frequency to Voltage Converter (CS2917)
LM2931CT	TO-220/5	Adjustable Low Dropout Regulator
LM2941T	TO-220/5	1A Low Dropout Adjustable Regulator
MC3470P	18	Floppy Disk Read Amp System
MC3479P	16	Biphase Stepper Motor Driver

PRODUCT NO.	PINS	DESCRIPTION
MC3486P	16	Quad EIA-422/423 Line Rec. w/3-State Outputs
MC3487P	16	Quad EIA-422 Line Driver w/3-State Outputs
SG3524	16	Regulating Pulse Width Modulator
LM3900N	14	Quad Amp
LM3905N	8	Precision Timer
LM3909N	8	LED Flasher/Oscillator
LM3911N	8	Temperature Controller
LM3914N	18	Bar-Graph Display Driver
LM3915N	18	3 DB Steps Bar-Graph Display Driver
LM3916N	18	VU Meter Display Driver
RC4136N	14	Quad Op Amp (XR4136)
RC4151NB	8	Voltage to Freq. Converter (XR4151)
XR4202	14	Quad Programmable Op Amp
NE5532	8	Dual Low-Noise Op Amp
NE5534	8	Low Noise Op Amp
TL7702ACP	8	(-.3 to 6V) Supply Voltage Supervisor
TL7705ACP	8	(-.3 to 10V) Supply Voltage Supervisor
78L05	TO-92	5V Positive Voltage Regulator (100mA)
78L12	TO-92	12V Positive Voltage Regulator (100mA)
78S40	16	Switching Regulator 1.5A (1.25V-40V)
7805K	TO-3	5V Pos. Volt Reg. 1 Amp (LM340K-5)
7805T	TO-220/3	5V Pos. Volt Reg. 1 Amp (LM340T-5)
7806T	TO-220/3	6V Pos. Volt Reg. 1 Amp (LM340T-6)
7808T	TO-220/3	8V Pos. Volt Reg. 1 Amp (LM340T-8)
7812K	TO-3	12V Pos. Volt Reg. 1 Amp (LM340K-12)
7812T	TO-220/3	12V Pos. Volt Reg. 1 Amp (LM340T-12)
7815K	TO-3	15V Pos. Volt Reg. 1 Amp (LM340K-15)
7815T	TO-220/3	15V Pos. Volt Reg. 1 Amp (LM340T-15)
7818T	TO-220/3	18V Pos. Volt Reg. 1 Amp (LM340T-18)
7824T	TO-220/3	24V Pos. Volt Reg. 1 Amp (LM340T-24)
79L05	TO-92	5V Neg. Voltage Reg. (100mA)
79L12	TO-92	12V Negative Voltage Reg. (100mA)
7905T	TO-220/3	5V Neg. Volt Reg. 1 Amp (LM320T-5)
7906K	TO-3	6V Neg. Volt Reg. 1 Amp (LM320K-6)

PRODUCT NO.	PINS	DESCRIPTION
7912T	TO-220/3	12V Neg. Volt Reg. 1 Amp (LM320T-12)
7915T	TO-220/3	15V Neg. Volt Reg. 1 Amp (LM320T-15)
7924K	TO-3	24V Neg. Volt Reg. 1 Amp (LM320K-24)
8038CCPD	14	Precision Wave Form Generator
LF13201N	16	Quad SPST JFET Analog Switch
LM13600N	16	Dual Transconductance Amp
MC14457P	16	CMOS Ultrasonic/Infrared Transmitter
75107	14	Dual Line Receiver (Totem Pole Outputs)
75110	14	Dual Line Driver (12mA Current Switch)
75113	16	Dual Diff. Line Driver w/3-State Outputs
75114	16	Dual Diff. Line Driver (9614)
75150	8	Dual Line Driver (RS232)
75154	16	Quad Line Receiver (9617)
75160	20	Octal Interface Bus Transceiver
75161	20	Octal Interface Bus Transceiver (Single Controller)
75162	22	Octal Interface Bus Transceiver (Multi-Controller)
75172	16	Quadruple Differential Line Driver
75173	16	Quadruple Differential Line Receiver
75174	16	Quad Diff. RS422 Line Driver w/3-State Outputs
75175	16	Quad Diff. EIA 422A/423A Line Rec.w/3-St. Outputs
75176	8	Differential Bus Transceiver
75182	14	Dual Line Receiver (DS8820N)
75183	14	Dual Differential Line Driver (DS8830N)
75188	14	Quad Line Driver RS232 (LM1488N)
75189	14	Quad Line Receiver S232 (LM1489N)
75451	8	Dual Peripheral AND Driver
75452	8	Dual Peripheral NAND Driver
75453	8	Dual Peripheral OR Driver
75477	8	Hi-Speed Switch. Dual Peripheral NAND Driver
75478	8	Hi-Speed Switch. Dual Peripheral OR Drive
75491	14	Quad Segment Driver for LED Readout
75492	14	Hex Digit Driver
MC145106P	18	CMOS Phase Locked Loop Freq. Synthesizer
MC145406P	16	CMOS RS232-C/V.28 Driver/Receiver

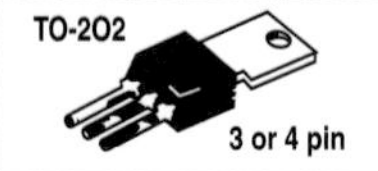

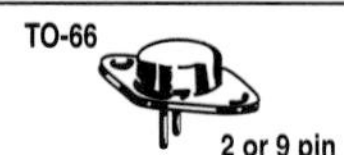

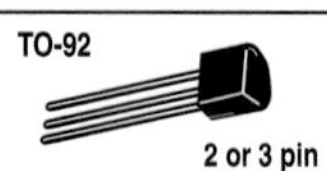

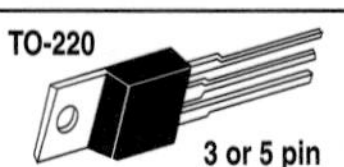

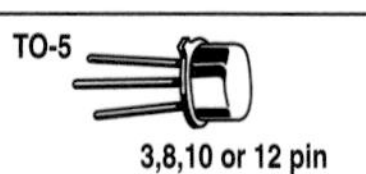

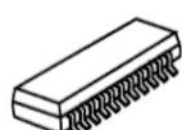

SOIC Package

PRODUCT NO.	PINS	DESCRIPTION
TL071CD	8	Low Noise BIFET Op Amp
TL072CD	8	Low Noise BIFET Dual Op Amp
TL074CD	14	Low Noise BIFET Quad Op Amp
TL082CD	8	Gen. Purpose BIFET Dual. Op Amp
TL084CD	14	Gen. Purpose BIFET Quad Op Amp
LM301D	8	Improved Op Amp
LM311D	8	High Performance Volt Comparator
LM324D	14	Low Power Quad Op Amp
LM336D	8	Voltage Reference (2.5V)
LM339D	14	Quad Comparator
LF351D	8	BIFET Op Amp
LF356D	8	JFET General Purpose Input Op Amp
LM358D	8	Low Power Dual Op Amp
LM385D	8	Micropower Voltage Reference Diode
LM386M-1	8	Low Voltage Audio Amp. 250W/6V
NE555D	8	Timer
LM556D	14	Dual 555 Timer
LM567D	8	Tone Decoder Phased Locked Loop
LM741D	8	Compensated Op Amp
LM747D	14	Dual 741 Op Amp
LM1458D	8	Dual Comp. Op Amp
LM1488D	14	Quad Line Driver RS232
LM1489D	14	Quad Line Receiver RS232
ULN2003AD	16	Hi-Volt. Hi-Curr. Darlington Transistor Array
NE5532D	8	Dual Low-Noise Op Amp
78L05ACM	8	5V Positive Voltage Regulator (100mA)

7400 Series (TTL)

PRODUCT NO.	PINS	DESCRIPTION
7400	14	Quad 2-Input NAND Gate
7401	14	Quad 2-Input NAND Gate (Open Collector)
7402	14	Quad 2-Input NOR Gate
7403	14	Quad 2-Input NAND Gate (Open Collector)
7404	14	Hex Inverter (9016)
7405	14	Hex Inverter (Open Collector)
7406	14	Hex Inverter Buffer/Driver (O.C. Hi-Voltage)
7407	14	Hex Buffer/Driver (O.C. Hi-Voltage)
7408	14	Quad 2-Input AND Gate
7410	14	Triple 3-Input NAND Gate
7411	14	Triple 3-Input AND (O.C.)
7412	14	Triple 3-Input NAND (O.C.)
7413	14	Dual 4-Input Schmitt Trigger
7414	14	Hex Schmitt Trigger Inverter
7416	14	Hex Inverter Buffer/Driver (O.C. Hi-Voltage)
7417	14	Hex Buffer/Driver (O.C. Hi-Voltage)
7420	14	Dual 4-Input NAND Gate
7421	14	Dual 4-Input AND Gate
7422	14	Dual 4-Input NAND Gate (Open Collector)
7425	14	Dual 4-Input NOR Gate
7426	14	Quad 2-Input TTL/MOS Inter. Gate
7427	14	Triple 3-Input NOR Gate
7430	14	8-Input NAND Gate
7432	14	Quad 2-Input OR Gate
7437	14	Quad 2-Input NAND Buffer
7438	14	Quad 2-Input NAND Buffer (Open Collector)
7440	14	Dual 4-Input NAND Buffer
7441	16	BCD-to-Decimal Decoder/Nixie TM Driver
7442	16	BCD-to-Decimal Decoder
7444	16	Excess 3-Gray-to-Decimal
7445	16	BCD-to-Decimal Decoder/Driver
7446	16	BCD-to-7 Segment Decoder/Driver (30V out)
7447	16	BCD-to-7 Segment Decoder/Driver (15V out)
7448	16	BCD-to-7 Segment Decoder/Driver
7451	14	Dual 2-Wide 2-in. AND/OR/Invert Gate
7454	14	4-Wide 2-in. AND/OR/Invert Gate
7470	14	Edge-Triggered JK Flip-Flop
7472	14	JK Master/Slave Flip-Flop
7473	14	Dual JK Master/Slave Flip-Flop (DM8501N)
7474	14	Dual D Flip-Flop
7475	16	4-Bit Bistable Latch
7476	16	Dual JK Master/Slave Flip-Flop
7483	16	4-Bit Binary Full Adder
7485	16	4-Bit Magnitude Comparator
7486	14	Quad 2-Input Exclusive-OR Gate
7489	16	64-Bit RAM (3101)
7490	14	Decade Counter
7492	14	Divide-by-12 Counter
7493	14	4-Bit Binary Counter
7495	14	4-Bit Right Shift/Left Shift Register
7496	16	5-Bit Parallel-IN, Parallel-OUT Shift Register
7497	16	Synchronous 6-Bit Binary Rate Multipliers
74100	24	Dual 4-Bit Bistable Latch
74104	14	Gated J-K Master/Slave Flip-Flop
74105	14	Gated J-K Master/Slave Flip-Flop
74107	14	Dual J-K Master/Slave Flip-Flop
74109	16	Dual J-K Pos-Edge Triggered F-F (9024)
74116	24	Dual 4-Bit Latches w/Clear (9308)
74120	16	Dual Pulse Synchronizer
74121	14	Monostable Multivibrator

PRODUCT NO.	PINS	DESCRIPTION
74122	14	Retriggerable Mono. Multivibrator w/Clear
74123	16	Dual Retriggerable Monostable Multivibrator
74125	14	Tri-State Quad Bus Buffer (DM8093N)
74126	14	Tri-State Quad Bus Buffer (DM8094N)
74128	14	Quad 2-Input NOR Buffer
74132	14	Quad Schmitt Trigger
74136	14	Quad 2-Input Exclusive-OR Gate
74141	16	BCD-to-Decimal Decoder/Driver
74145	16	BCD-to-Decimal Decoder/Driver
74147	16	10 to 4 Line BCD Priority Encoder
74148	16	8 to 3 Line Octal Priority Encoder (9318)
74150	24	16 Line to 1 Line Multiplexer
74151	16	8 Channel Digital Multiplexer
74152	14	8-Channel Data Selector/Multiplexer
74153	16	Dual 4/1 Multiplexer
74154	24	4 to 16 Line Decoder/Demultiplexer (9311)
74155	16	Dual 2/4 Demultiplexer
74156	16	Dual 2/4 Demultiplexer (O.C.)
74157	16	Quad 2/1 Data Selector (9322)
74158	16	Quad 2/1 Multiplexer (Inv. Out)
74159	24	1 of 16 Decoder/Demux O/C Outputs
74160	16	Decade Counter with Asynchronous Clear
74161	16	Synchronous 4-Bit Counter (9316)
74163	16	Synchronous 4-Bit Counter
74164	14	8-Bit Serial Shift Register (DM8570N)
74165	16	Parallel Load 8-bit Serial Shift Reg. (DM8590N)
74166	16	8-Bit Shift Register
74167	16	Rate Multiplier
74170	16	4×4 Register File
74173	16	4-Bit Tri-State Register (DM8551N)
74174	16	Hex D Flip Flop with Clear
74175	16	Quad D Flip Flop with Clear
74177	14	35MHz Presettable Binary Counter (N8281)
74179	16	4-Bit Parallel-Access Shift Register
74180	14	9-Bit Odd/Even Parity Generator/Checker
74181	24	Arithmetic Logic Unit
74182	16	Look-Ahead Carry Generator (9342)
74189	16	Tri-State 64-Bit RAM (DM8599N)
74191	16	Synchronous Binary Up/Down Counter
74192	16	Decade Up/Down Counter (DM8560N)
74193	16	Binary Up/Down Counter (DM8563N)
74194	16	4-Bit Bi-Directional Shift Register
74195	16	4-Bit Parallel-Access Shift Register (DM8300N)
74197	14	Presettable Binary Counter (N8291)
74198	24	8-Bit Shift Register
74199	24	8-Bit Shift Register
74221	16	Dual Monostable Multivibrator
74251	16	Tri-State 8-Channel Multiplexer (DM8121N)
74259	16	8-Bit Addressable Latch (9334)
74265	16	Quad Complementary-Output Elements
74273	20	Octal D-Type Flip Flop with Clear
74279	16	Quad Set-Reset Latch
74283	16	4-Bit Binary Full Address with Fast Carry
74298	16	Quad 2-Input Multiplexer w/Storage
74365	16	Tri-State Hex Buffer (DM8095N)
74366	16	Tri-State Hex Buffer Inverter (DM8096N)
74367	16	Tri-State Hex Buffer (DM8097N)
74368	16	Tri-State Hex Buffer Inverter (DM8098N)
74390	16	Dual 4-Bit Decade Counter
74393	14	Dual 4-Bit Binary Counter w/Individual Clocks

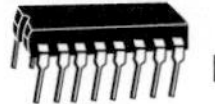 Dual-In-Line Package **74C00 Series (CMOS)**

PRODUCT NO.	PINS	DESCRIPTION
74C00	14	Quad 2-Input NAND Gate
74C02	14	Quad 2-Input NOR Gate
74C04	14	Hex Inverter (CD4069)
74C08	14	Quad 2-Input AND Gate
74C10	14	Triple 3-Input NAND Gate
74C14	14	Hex Inverter Schmitt Trigger (CD40106)
74C32	14	Quad 2-Input OR Gate
74C48	16	BCD to 7-Segment Decoder/Driver
74C73	14	Dual JK Flip-Flop with Clear
74C74	14	Dual D Flip-Flop
74C76	16	Dual J-K Flip-Flop with preset and clear
74C86	14	Quad EXCLUSIVE-OR Gate (4070)
74C90	14	Decade Counter
74C154	24	4 to 16-Line Decoder/Demultiplexer
74C161	16	Synchronous 4-Bit Counter (CD40161)
74C174	16	Hex Flip-Flop (CD40174) (MC14174BPC)
74C192	16	Decade Up/Down Counter (CD40192)
74C193	16	Binary Up/Down Counter (CD40193)
74C221	16	Dual Monostable Multivibrator
74C244	20	Octal Driver Tri-State
74C373	20	Octal D-Type Flip-Flop with Clear Tri-State
74C374	20	Octal D Flip-Flop Tri-State (INS82C06N)
74C911	28	4-Digit Display Controller (INS8247N)
74C912	28	6-Digit BCD Display Controller (INS8248N)
74C917	28	6-Digit Hex Display Controller
74C920	22	256×4 1K Static RAM
74C922	18	16-Key Keyboard Encoder (INS8245N)
74C923	20	20-Key Keyboard Encoder (INS8246N)
74C925	16	4-Digit CTR w/MUX D Segment Driver
74C926	18	4-Digit CTR w/MUX D Segment Driver
80C95	16	Hex Buffer Tri-State (74C365)
80C97	16	Hex Buffer Tri-State (74C367) (CD4503)
80C98	16	Hex Inverter Tri-State (74C368)

Dual-In-Line Package **74F00 Series (TTL Fast)**

PRODUCT NO.	PINS	DESCRIPTION
74F00	14	Quad 2-Input NAND Gate
74F02	14	Quad 2-Input NOR Gate
74F04	14	Hex Inverter
74F08	14	Quad 2-Input AND Gate
74F10	14	Triple 3-Input NAND Gate
74F14	14	Hex Inverter Schmitt Trigger
74F32	14	Quad 2-Input OR Gate
74F74	14	Dual D Flip-Flop
74F86	14	Quad EXCLUSIVE-OR Gate
74F138	16	3 to 8 Decoder/Demultiplexer
74F151	16	8-Input Multiplexer
74F175	16	Quad D Flip-Flop with Clear
74F244	20	Octal Line Driver Tri-State
74F245	20	Octal Bus Transceiver Tri-State
74F251	16	8-Channel Multiplexer Tri-State
74F273	20	Octal D-Type Flip-Flop with Clear
74F373	20	Octal D Flip-Flop Tri-State
74F374	20	Octal D Flip-Flop Edge-Triggered Tri-State
74F379	16	4-Bit Register, Common Enable
74F573	20	Octal D-Type Transparent Latch
74F574	20	Octal D-Type Flip-Flop
74F646	S-24	Octal Bus Trans/Register (Tri-State) .3″ wide

Dual-In-Line Packag **74S00 Series (TTL Schottky)**

PRODUCT NO.	PINS	DESCRIPTION
74S00	14	Quad 2-Input NAND Gate
74S02	14	Quad 2-Input NOR Gate
74S03	14	Quad 2-Input NAND Gate
74S04	14	Hex Inverter
74S05	14	Hex Inverter (O.C.)
74S08	14	Quad 2-Input AND Gate
74S10	14	Triple 3-Input NAND Gate
74S11	14	Triple 3-Input AND Gate
74S20	14	Dual 4-Input NAND Gate
74S32	14	Quad 2-Input OR Gate
74S51	14	Dual 2-Wide 2-Input AND-OR-Invert Gate
74S74	14	Dual D Flip-Flop
74S85	16	4-Bit Magnitude Comparator
74S86	14	Quad EXCLUSIVE-OR Gate
74S124	16	Dual Voltage-Controlled Oscillators
74S138	16	3 to 8 Decoder/Demultiplexer
74S151	16	8-Input Multiplexer
74S174	16	Hex D Flip-Flop with Clear
74S175	16	Quad D Flip-Flop with Clear
74S188	16	256-Bit (32×8) PROM (O.C.) (27S18)
74S189	16	64-Bit (16×4) RAM Tri-State (27S03)
74S240	20	Octal Inverting Line Driver Tri-State
74S241	20	Octal Driver Tri-State
74S244	20	Octal Line Driver Tri-State
74S287	16	1K-Bit (256×4) PROM Tri-State (27S21)
74S288	16	256-Bit (32×8) PROM Tri-State (27S19)
74S373	20	Octal D-Type Flip-Flop Tri-State
74S374	20	Octal D Flip-Flop Edge-Triggered Tri-State
74S387	16	1K-Bit (256×4) PROM (O.C.) (27S20)
74S412	24	8-Bit Latch (8212)
74S472	20	4K-Bit (512×8) PROM Tri-State (27S29)
74S473	20	4K-Bit (512×8) PROM (6348-1J)
74S571	16	2K-Bit (512×4) PROM Tri-State (27S13)
74S573	18	4K-Bit (1024×4) PROM Tri-State (27S33)

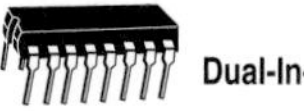

Dual-In-Line Package 74ALS00 Series (TTL Advanced Low-Power Shottky)

PRODUCT NO.	PINS	DESCRIPTION
74ALS00	14	Quad 2-Input NAND Gate
74ALS02	14	Quad 2-Input NOR Gate
74ALS04	14	Hex Inverter
74ALS05	14	Hex Inverter (O.C.)
74ALS08	14	Quad 2-Input AND Gate
74ALS09	14	Quad 2-Input AND Gate (O.C.)
74ALS10	14	Triple 3-Input NAND Gate
74ALS11	14	Triple 3-Input AND Gate
74ALS21	14	Dual 4-Input AND Gate
74ALS30	14	8-Input NAND Gate
74ALS32	14	Quad 2-Input OR Gate
74ALS74	14	Dual D Flip-Flop
74ALS86	14	Quad EXCLUSIVE-OR Gate
74ALS109	16	Dual JK Positive Edge Flip-Flop
74ALS138	16	3 to 8 Decoder/Demultiplexer
74ALS240	20	Octal Inverting Line Driver Tri-State
74ALS244	20	Octal Line Driver Tri-State
74ALS245	20	Octal Bus Transceiver Tri-State
74ALS273	20	Octal D-Type Flip-Flop with Clear
74ALS373	20	Octal D Flip-Flop Tri-State
74ALS374	20	Octal D Flip-Flop Edge-Triggered Tri-State
74ALS573	20	Octal D-Type Transparent Latch
74ALS574	18	Octal D-Type Flip-Flop
74ALS646	S-24	Octal Bus Trans/Register (Tri-State) .3″ wide

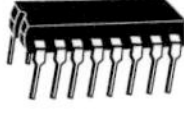

74LS00 Series (Low-Power Schottky)

PRODUCT NO.	PINS	DESCRIPTION
74LS00*	14	Quad 2-Input NAND Gate
74LS02*	14	Quad 2-Input NOR Gate
74LS03	14	Quad 2-Input NAND Gate (O.C.)
74LS04*	14	Hex Inverter
74LS05*	14	Hex Inverter (O.C.)
74LS06*	14	Hex Inverter Buffer/Driver (O.C. Hi-Voltage)
74LS07*	14	Hex Buffer/Driver (O.C. Hi-Voltage)
74LS08*	14	Quad 2-Input AND Gate
74LS09	14	Quad 2-Input AND Gate (O.C.)
74LS10*	14	Triple 3-Input NAND Gate
74LS11*	14	Triple 3-Input AND Gate
74LS13	14	Dual 4-Input NAND Schmitt Trigger
74LS14*	14	Hex Inverter Schmitt Trigger
74LS15	14	Triple 3-Input AND Gate (O.C.)
74LS20*	14	Dual 4-Input NAND Gate
74LS21*	14	Dual 4-Input AND Gate
74LS22	14	Dual 4-Input NAND Gate (O.C.)
74LS26	14	Quad 2-Input NAND Gate (O.C. Hi-Voltage)
74LS27*	14	Triple 3-Input NOR Gate
74LS30*	14	8-Input NAND Gate
74LS32*	14	Quad 2-Input OR Gate
74LS33	14	Quad 2-Input NOR Buffer (O.C.)
74LS37	14	Quad 2-Input NAND Buffer
74LS38	14	Quad 2-Input NAND Buffer (O.C.)
74LS42*	16	BCD-to-Decimal Decoder
74LS47*	16	BCD to 7-Segment Decoder/Driver (O.C. 15V)
74LS48*	16	BCD to 7-Segment Decoder/Driver
74LS51	14	Dual 2-Wide 2-Input AND/OR Invert Gate
74LS54	14	Quad 2-Input AND/OR Invert Gate
74LS55	14	Dual 4-Input AND/OR Invert Gate
74LS73	14	Dual JK Flip-Flop with Clear
74LS74*	14	Dual D Flip-Flop
74LS75*	16	4-Bit Bi-Stable Latch
74LS76*	16	Dual JK Flip-Flop with Preset and Clear
74LS83*	16	4-Bit Binary Full Adder
74LS85*	16	4-Bit Magnitude Comparator
74LS86*	14	Quad EXCLUSIVE-OR Gate
74LS90*	14	Decade Counter
74LS91	14	8-Bit Shift Register
74LS92	14	Divide-by-12 Counter
74LS93*	14	4-Bit Binary Counter
74LS95	14	4-Bit Parallel-Access Shift Register
74LS107	14	Dual JK Flip-Flop with Clear
74LS109	16	Dual JK Positive Edge-Triggered Flip-Flop
74LS112	16	Dual JK Negative Edge-Triggered Flip-Flop
74LS114	14	Dual JK Negative Edge-Triggered Flip-Flop
74LS122	14	Retriggerable Mono Multivibrator Comm CLK
74LS123*	16	Dual Monostable Multivibrator w/Clear
74LS125*	14	Quad Bus Buffer Negative Enable Tri-State
74LS126	14	Quad Bus Buffer Tri-State
74LS132*	14	Quad 2-Input NAND Schmitt Trigger
74LS133	16	13-Input NAND Gate
74LS136	14	Quad EXCLUSIVE-OR Gate (O.C.)
74LS138*	16	3 to 8 Decoder/Demultiplexer
74LS139*	16	2 to 4 Decoder/Demultiplexer
74LS145	16	BCD-to-Decimal Decoder/Driver
74LS147*	16	10 to 4 Priority Encoder
74LS148	16	8 to 3 Priority Encoder
74LS151*	16	8-Input Multiplexer
74LS153*	16	Dual 4 to 1 Selector/Multiplexer
74LS154*	24	4 to 16 Line Decoder/Demultiplexer
74LS155	16	Dual 2/4 Demultiplexer
74LS156	16	Dual 2/4 Demultiplexer (O.C.)
74LS157*	16	Quad 2/1 Data Selector
74LS158	16	Quad 2/1 Multiplexer Inverter
74LS160	16	Decade Counter with Asynchronous Clear
74LS161*	16	Synchronous 4-Bit Counter
74LS163*	16	Fully Synchronous 4-Bit Counter
74LS164*	14	8-Bit Serial Shift Register
74LS165*	16	8-Bit Serial Shift Register, Parallel Load
74LS166	16	8-Bit Shift Register, Parallel Load
74LS169	16	Synch. Binary Up/Down Counter (74LS669)
74LS170	16	4×4 Register File (O.C.)
74LS173	16	4-Bit D-Type Register Tri-State
74LS174*	16	Hex D Flip-Flop with Clear
74LS175*	16	Quad D Flip-Flop with Clear
74LS181	24	Arithmetic Logic Unit/Function Generator
74LS189	16	64-Bit RAM Tri-State
74LS190	16	Up/Down Decade Counter
74LS191*	16	Up/Down Binary Counter
74LS192*	16	Decade Up/Down Counter with Clear
74LS193*	16	Binary Up/Down Counter with Clear
74LS194*	16	4-Bit Bi-Directional Shift Register
74LS195	16	4-Bit Parallel-Access Shift Register
74LS196	14	Presettable Decade Counter
74LS197	14	Presettable Binary Counter
74LS221*	16	Dual Mono Multivibrator Schmitt Trigger
74LS240*	20	Octal Inverting Line Driver Tri-State
74LS241	20	Octal Bus/Line Driver Tri-State
74LS243	14	Quad Transceiver Tri-State
74LS244*	20	Octal Driver Tri-State
74LS245*	20	Octal Bus Transceiver Tri-State
74LS247	16	BCD to 7-Segment Decoder/Driver (O.C.)
74LS251	16	8-Channel Multiplexer Tri-State
74LS253	16	Dual 4 to 1 Multiplexer Tri-State
74LS257	16	Quad 2-Input Multiplexer Tri-State

74LS00 Series (Low-Power Schottky) *Continued*

PRODUCT NO.	PINS	DESCRIPTION
74LS258	16	Quad 2/1 Multiplexer Inverting Tri-State
74LS259	16	8-Bit Addressable Latch
74LS260	14	Dual 5-Input NOR Gate
74LS261	16	2-Bit by 4-Bit Parallel Binary Multipliers
74LS266	14	Quad EXCLUSIVE-NOR Gate (O.C.)
74LS273*	20	Octal D-Type Flip-Flop with Clear
74LS279	16	Quad Set-Reset Latches
74LS280	14	9-Bit Odd/Even Parity Generator/Checker
74LS283	16	4-Bit Binary Full Adder with Fast Carry
74LS293	14	4-Bit Binary Counter
74LS295	14	4-Bit Universal Shift Register
74LS299	20	8-Bit Universal Shift/Storage Register
74LS322	20	8-Bit Serial Reg. w/Sign Extended(25LS22)
74LS323	20	8-Bit Universal Shift/Storage Reg. (25LS23)
74LS365	16	Hex Buffer w/Logical OR Tri-State
74LS366	14	Hex Buffer/Inverter Tri-State
74LS367*	16	Hex Buffer Tri-State
74LS368	16	Hex Inverter Tri-State
74LS373*	20	Octal D Flip-Flop Tri-State
74LS374*	20	Octal Dual Flip-Flop Edge-Triggered Tri-State

PRODUCT NO.	PINS	DESCRIPTION
74LS377	20	Octal D Register Common Enable (25LS07)
74LS390	16	Dual 4-Bit Decade Counter
74LS393*	14	Dual 4-Bit Binary Counter
74LS395	16	4-Bit Cascadable Shift Register Tri-State
74LS399	16	Quad 2-Mux with Storage (25LS09)
74LS534	20	Octal D-Type Flip-Flop Tri-State
74LS540	20	Octal Inverter Buffer/Line Driver Tri-State
74LS541*	20	Octal Buffer/Line Driver Tri-State
74LS573	20	Octal D-Type Flip-Flop Latch Tri-State
74LS574	20	Octal D-Type Flip-Flop Edge-Trig. Tri-State
74LS590	16	8-Bit Binary Counter w/Out. Reg. Tri-State
74LS595	16	8-Bit Serial-to-Parallel Shift Reg. Tri-State
74LS612	40	Memory Mapper Tri-State
74LS629	16	Dual Voltage-Controlled Oscillator
74LS640	20	Octal Inverting Bus Transceiver Tri-State
74LS645	20	Octal Bus Transceiver Tri-State
74LS646	S-24	Octal Bus Trans/Register Tri-State .3″ Wide
74LS670	16	4×4 Register File Tri-State
74LS688*	20	8-Bit Magnitude Comparator (25LS2521)
81LS95	20	Octal Buffer Tri-State (74LS465)

74C00 Series (CMOS Series)

PRODUCT NO.	PINS	DESCRIPTION
74C00	14	Quad 2-Input NAND Gate
74C02	14	Quad 2-Input NOR Gate
74C04	14	Hex Inverter (CD4069)
74C08	14	Quad 2-Input AND Gate
74C10	14	Triple 3-Input NAND Gate
74C14	14	Hex Schmitt Trigger (CD40106)
74C32	14	Quad 2-Input OR Gate
74C48	16	BCD to 7-Segment Decoder/Driver
74C73	14	Dual JK Flip-Flop
74C74	14	Dual D Flip-Flop
74C76	16	Dual J-K Flip-Flop with preset and clear
74C86	14	Quad Exclusive OR Gate (4070)
74C90	14	4-Bit Decade Counter
74C154	24	4-Line to 16-Line Decoder
74C174	16	Hex Flip-Flop (CD40174)(MC14174BPC)
74C192	16	Decade Up/Down Counter (CD40192)
74C193	16	Binary Up/Down Counter (CD40193)
74C221	16	Dual Monostable Multivibrator
74C244	20	Tri-State Octal Buffer (Non-Inverting)
74C373	20	Octal Flow-Thru Latch Tri-State
74C374	20	Octal D Flip-Flop Tri-State (INS82C06N)
74C911	28	4-Digit Display Controller (INS8247N)
74C912	28	6-Digit BCD Display Controller (INS8248N)
74C917	28	6-Digit Hex Display Controller
74C920	22	256×4 1K Static RAM
74C922	18	16-Key Keyboard Encoder (INS8245N)
74C923	20	20-Key Keyboard Encoder (INS8246N)
74C925	16	4-Digit CTR w/MUX D Segment Driver
74C926	18	4-Digit CTR w/MUX D Segment Driver
80C95	16	Tri-State Hex Buffer (74C365)
80C97	16	Tri-State Hex Buffer (74C367) (CD4503)
80C98	16	Tri-State Hex Inverter (74C368)

74HC00 Series (High Speed CMOS)

PRODUCT NO.	PINS	DESCRIPTION
74HC00	14	Quad 2-Input NAND Gate
74HC02	14	Quad 2-Input NOR Gate
74HC04	14	Hex Inverter (Buffered)
74HCU04	14	Hex Inverter (Unbuffered)
74HC05	14	Hex Inverter (Open Collector)
74HC08	14	Quad 2-Input AND Gate
74HC10	14	Triple 3-Input NAND Gate
74HC14	14	Hex Inverter Schmitt Trigger
74HC20	14	Dual 4-Input NAND Gate
74HC30	14	8-Input NAND Gate
74HC32	14	Quad 2-Input OR Gate
74HC42	16	BCD-to-Decimal Decoder
74HC74	14	Dual D Flip Flop with Preset and Clear
74HC75	16	Dual 24-Bit Bistable Latch
74HC76	16	Dual J-K Flip Flop with Preset and Clear
74HC85	16	4-Bit Magnitude Comparator
74HC86	14	Quad 2-Input Exclusive-OR (XOR) Gate
74HC112	16	Dual J-K Flip Flop with Preset and Clear
74HC123	16	Dual Monostable Multivibrator
74HC125	14	Tri-State Quad Buffer
74HC132	14	Quad 2-Input NAND Schmitt Trigger
74HC138	16	3 to 8 Line Decoder
74HC139	16	Expandable Dual 2/4 Line Decoder
74HC153	16	Dual 4/1 Multiplexer
74HC154	S-24	4/16 Line Decoder (use 24SLP Socket)
74HC160	16	Synchronous Decade Counter w/async. clear
74HC161	16	Synchronous Binary Counter w/sync. clear
74HC163	16	Synchronous Binary Counter
74HC164	14	8-Bit Serial-in/Parallel-out Shift Register
74HC165	16	8-Bit Parallel-in/Serial-out Shift Register
74HC174	16	Hex D Flip-Flop with Clear
74HC175	16	Quad D Type Flip-Flop with Clear
74HC191	16	Up/Down Binary Counter
74HC192	16	Synchronous Decade Up/Down Counter
74HC193	16	Synchronous Binary Up/Down Counter
74HC195	16	4-Bit Parallel Access Shift Register
74HC221	16	Dual Monostable Multivibrator
74HC240	20	Inverting Octal Tri-State Buffer
74HC242	14	Inverting Quad Tri-State Transceiver
74HC244	20	Octal Tri-State Buffer

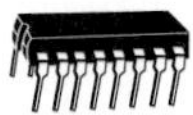

74HC00 Series (High Speed CMOS) *Continued*

PRODUCT NO.	PINS	DESCRIPTION
74HC245	20	Octal Tri-State Transceiver
74HC253	16	Dual 4-Channel Tri-State Multiplexer
74HC257	16	Quad 2-Channel Tri-State Multiplexer
74HC273	20	Octal D Flip-Flop
74HC280	14	9-Bit Odd/Even Parity Generator/Checker
74HC299	20	8-Bit Tri-State Universal Shift Register
74HC365	16	Hex Tri-State Buffer
74HC373	20	Tri-State Octal D Type Latch
74HC374	20	Tri-State Octal D Type Flip-Flop
74HC390	16	Dual 4-Bit Decade Counter
74HC393	14	Dual 4-Bit Binary Counter
74HC541	20	Octal Buffer/Line Driver (Tri-State)
74HC573	20	Octal D-Type Latch Tri-State
74HC574	20	Octal D-Type Flip-Flop Tri-State
74HC590	16	Tri-State 8-Bit Binary Counter with Output Latch Clear
74HC595	16	8-Bit Serial-to-Parallel Shift Register Latch
74HC688	20	8-Bit Mag. Comparator (Equality Detector)

PRODUCT NO.	PINS	DESCRIPTION
74HC942	20	Full Duplex,Low Speed,300Baud Modem Chip
74HC943	20	Full Duplex 300 Baud Modem Chip
74HC4002	14	Dual 4-Input NOR Gate
74HC4017	16	Decade Counter/Div. w/10 Decoded Outputs
74HC4020	16	14-Stage Binary Counter
74HC4024	14	7-Stage Ripple Counter
74HC4040	16	12-Stage Binary Counter
74HC4046	16	CMOS Phase Lock Loop
74HC4049	16	Hex/Buffer/Converter (Inverting)
74HC4050	16	Hex/Buffer/Converter (Non-Inverting)
74HC4051	16	Single 8-Channel Multiplexer/Demultiplexer
74HC4060	16	14-Stage Binary Counter
74HC4066	14	Quad Analog Switch
74HC4075	14	Triple 3-Input OR Gate
74HC4514	S-24	4/16-Line Decoder w/Latch (use 24SLP Socket)
74HC4538	16	Dual Retriggerable Monostable Multivibrator
74HC4543	16	BCD-to-7 Seg. Latch/Decoder for LCDs

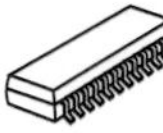

74HC00 (High Speed CMOS-Surface Mount Package)

PRODUCT NO.	PINS	DESCRIPTION
74HC00D	14	Quad 2-Input NAND Gate
74HC02D	14	Quad 2-Input NOR Gate
74HC04D	14	Hex Inverter (Buffered)
74HC05D	14	Hex Inverter (Open Collector)
74HC08D	14	Quad 2-Input NAND Gate
74HC10D	14	Triple 3-Input NAND Gate
74HC14D	14	Hex Inverter Schmitt Trigger
74HC32D	14	Quad 2-Input OR Gate
74HC74D	14	Dual D Flip-Flop with Preset and Clear
74HC76D	16	Dual J-K Flip-Flop with Preset and Clear
74HC86D	14	Quad 2-Input Exclusive-OR (XOR) Gate
74HC123D	16	Dual Monostable Multivibrator
74HC125D	14	Tri-State Quad Buffer
74HC138D	16	3 to 8 Line Decoder
74HC154D	24	4/16 Line Decoder (use 24SLP Socket)
74HC240D	20	Inverting Octal Tri-State Buffer
74HC244D	20	Octal Tri-State Buffer
74HC245D	20	Octal Tri-State Transceiver
74HC273D	20	Octal D Flip-Flop
74HC373D	20	Tri-State Octal D-Type Latch
74HC374D	20	Tri-State Octal D-Type Flip-Flop
74HC573D	20	Tri-State Octal D-Type Latch
74HC574D	20	Octal D-Type Flip-Flop Tri-State
74HC688D	20	8-Bit Mag. Comparator (Equality Detector

APPENDIX

F

Frequency Spectrum

The term "frequency" describes the number of alternations occurring in one second. Direct current (dc) is a steady or constant current that does not alternate and is therefore listed as zero cycles per second or 0 Hz. This appendix illustrates in detail the entire range of frequencies from the lowest subaudible frequency to the highest cosmic rays, along with their applications.

The following page is an overall summary of the complete range of frequencies or spectrum, as it is normally called. The subsequent pages cover each band or section of frequencies in a lot more detail and list the different frequency applications.

This and all of the other appendixes will be useful as references throughout your course of electronic study.

OVERVIEW OF FREQUENCY SPECTRUM

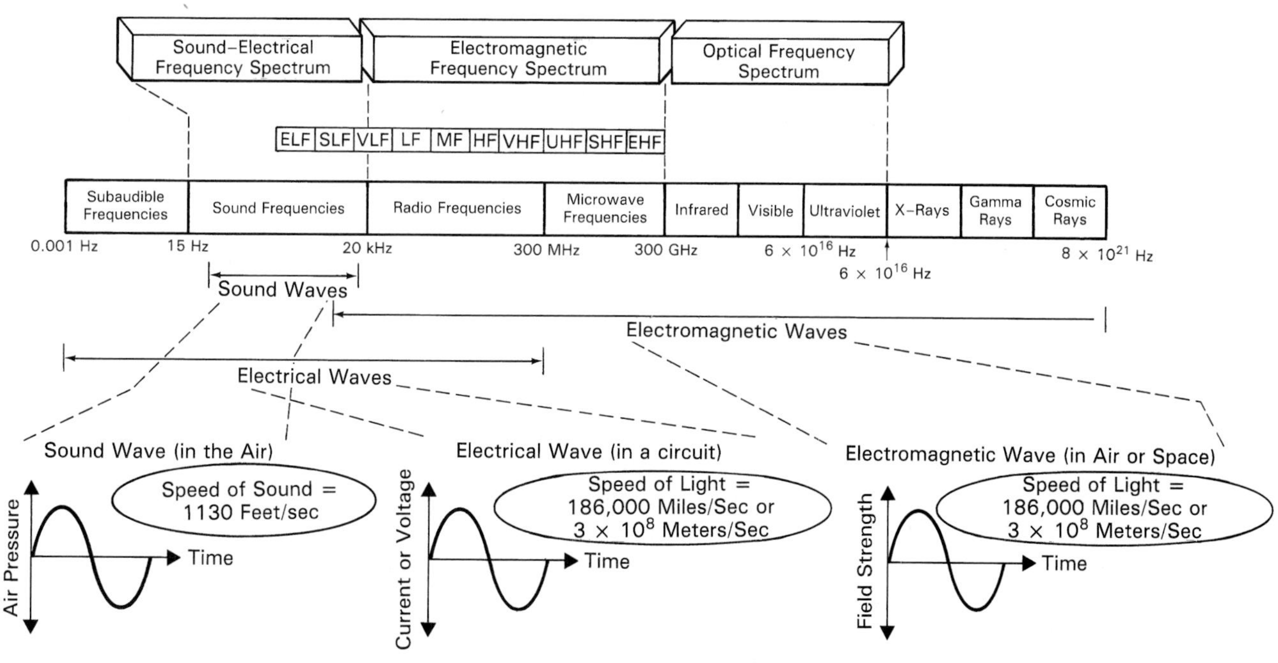

0 Hz	16 → 16 kHz	16 → 30 kHz	10 kHz → 300 GHz
Direct Current (DC) Motors, Relays, Supply Voltages	Audio Frequencies (AC) Motors, Amplifiers, Music Equipment, Speakers, Microphones, Oscillators	Sound (Ultrasonic) Frequencies Sonar, Music, Speech	Electromagnetic (Radio) Frequencies Voice Communications Television Navigation Medical, Scientific, and Military

300 GHz → 4×10^{14}	$4 \times 10^{14} \rightarrow 7.69 \times 10^{14}$	$4 \times 10^{14} \rightarrow 6 \times 10^{16}$	$9.375 \times 10^{15} \rightarrow 3 \times 10^{19}$
Infrared (R) Heating, Photography, Sensing, Military	Visible Color, Photography, Movies, TV	Ultraviolet (UV) Sterilizing, Medical	X–Rays Medical, Gauge Thickness, Inspection

$3 \times 10^{19} \rightarrow 5 \times 10^{20}$	$5 \times 10^{20} \rightarrow$
Gamma Rays Deeper Penetrating Than X–Rays, Detection of Radiation	Cosmic Rays Present in Outer Space

THE SOUND SPECTRUM

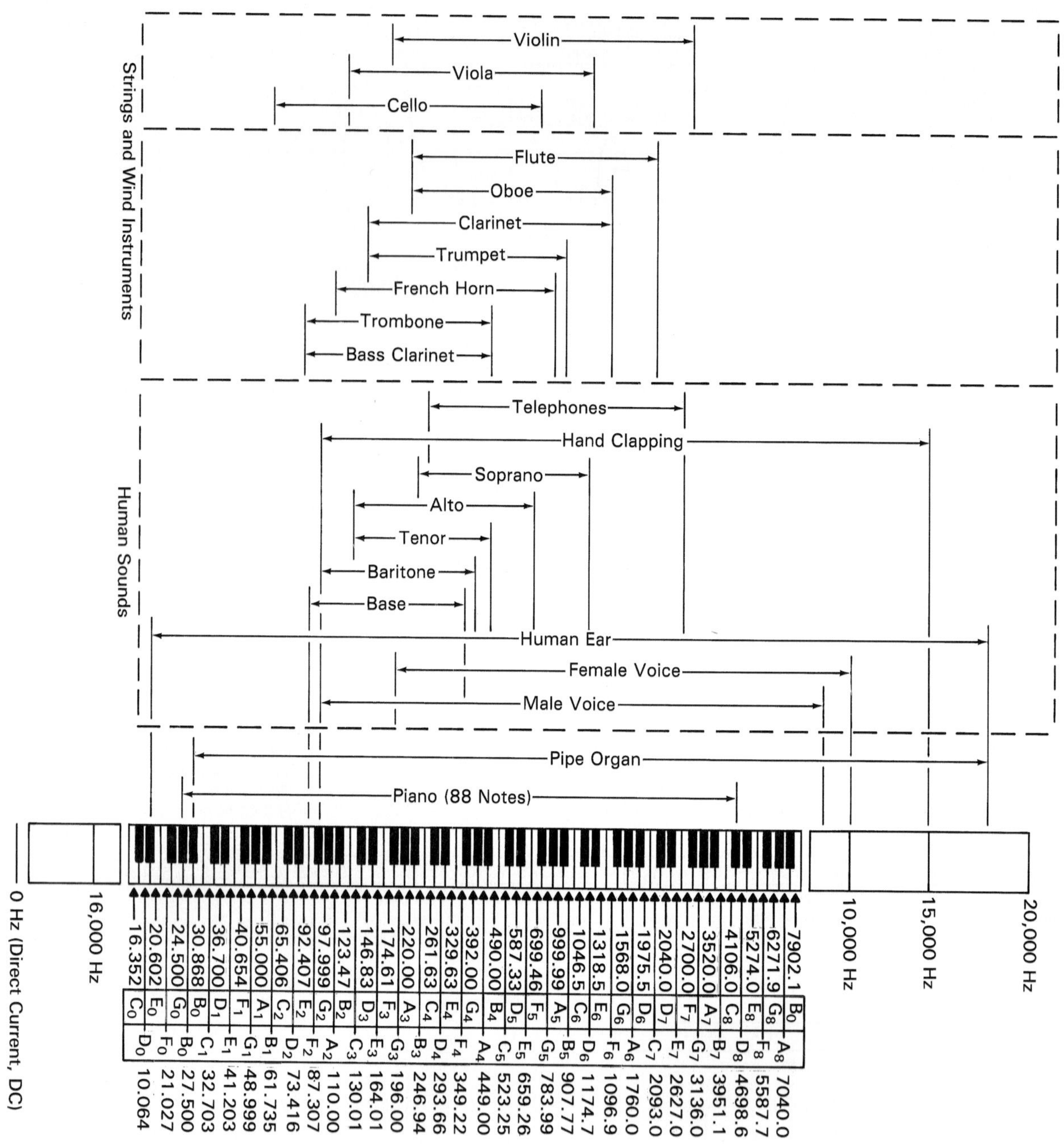

VLF (Very Low–Frequency Band)

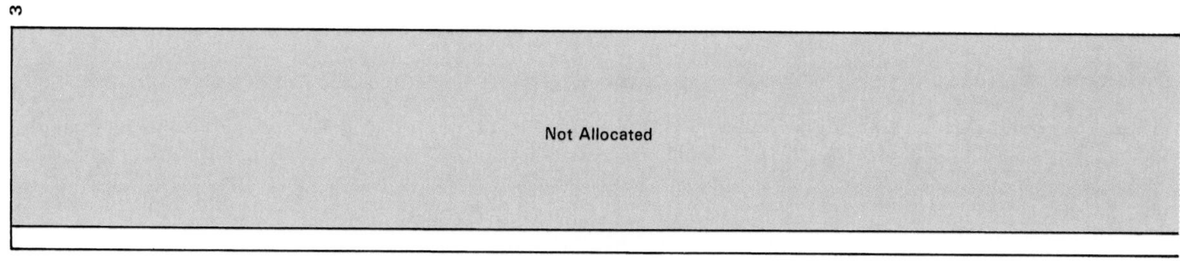

3 kHz

Not Allocated

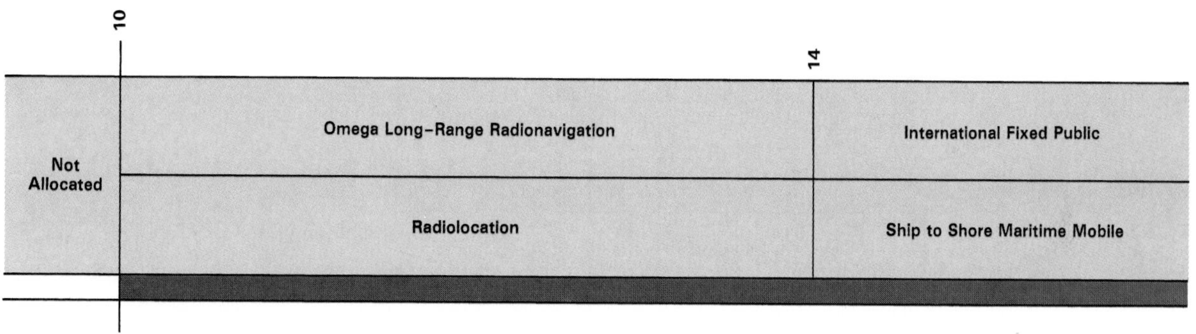

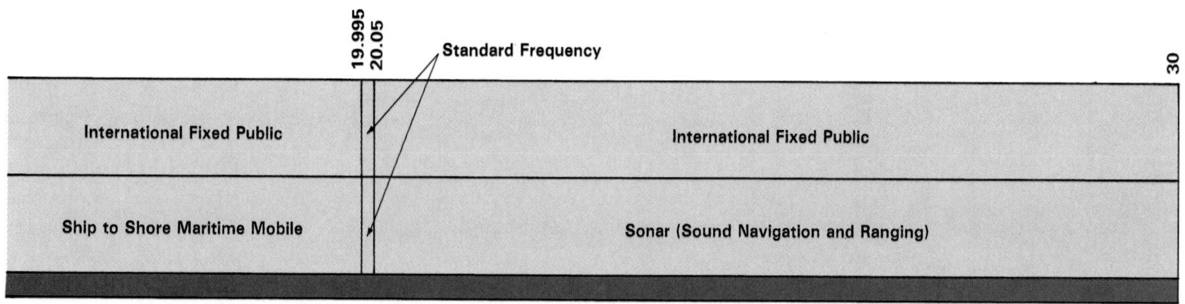

30 kHz

LF (Low–Frequency Band)

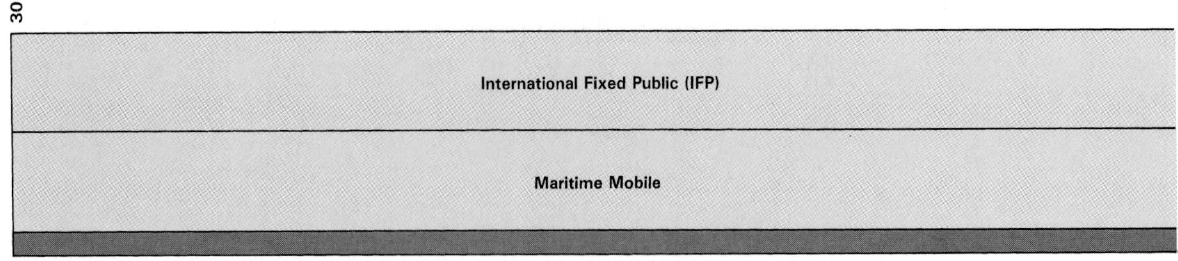

30 kHz

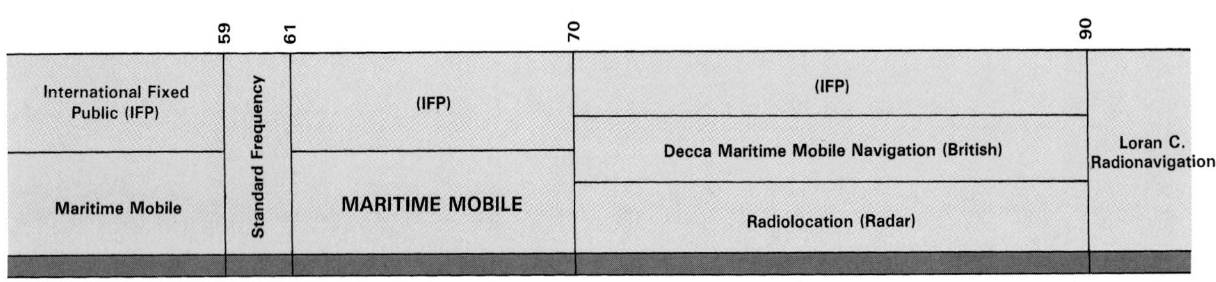

	110	130	160
Loran C. Radionavigation	(IFP)	(IFP)	(IFP)
	Maritime Mobile	Maritime Mobile (MM)	Maritime Mobile (MM)
	Radar (Radio Detection and Ranging)		

	200	285	300
(IFP)	Aeronautical Mobile Communications	Maritime Radio–navigation	
Maritime Mobile (MM)	Maritime Radionavigation	Aeronautical Radio–navigation	

300 kHz

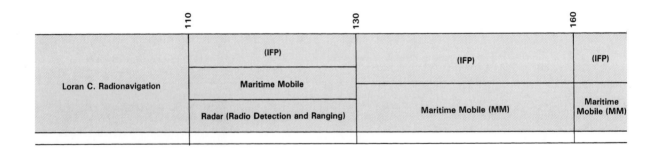

MF (Medium–Frequency Band)

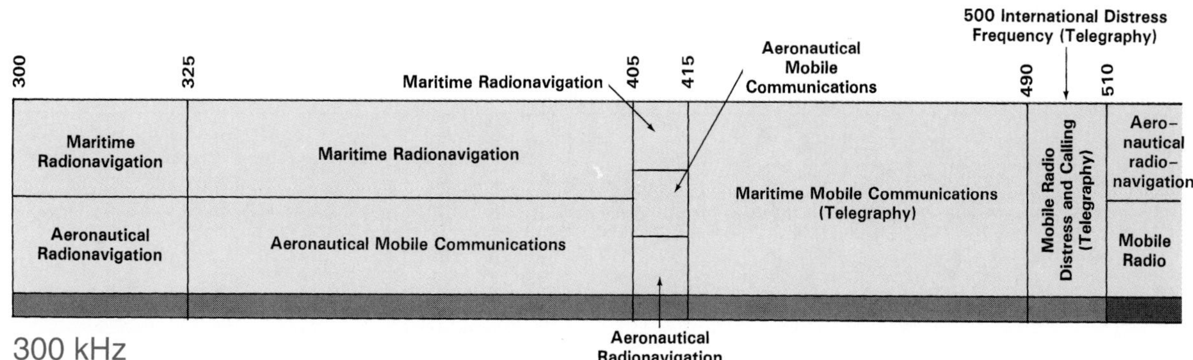

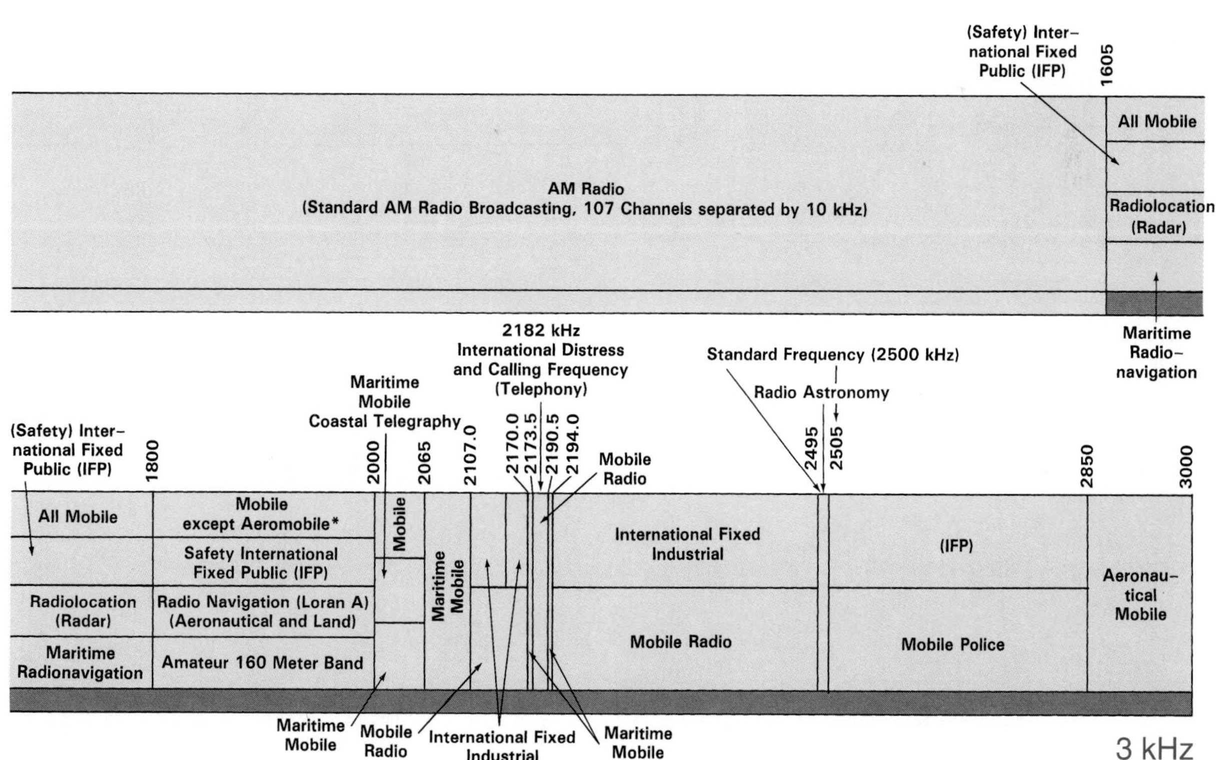

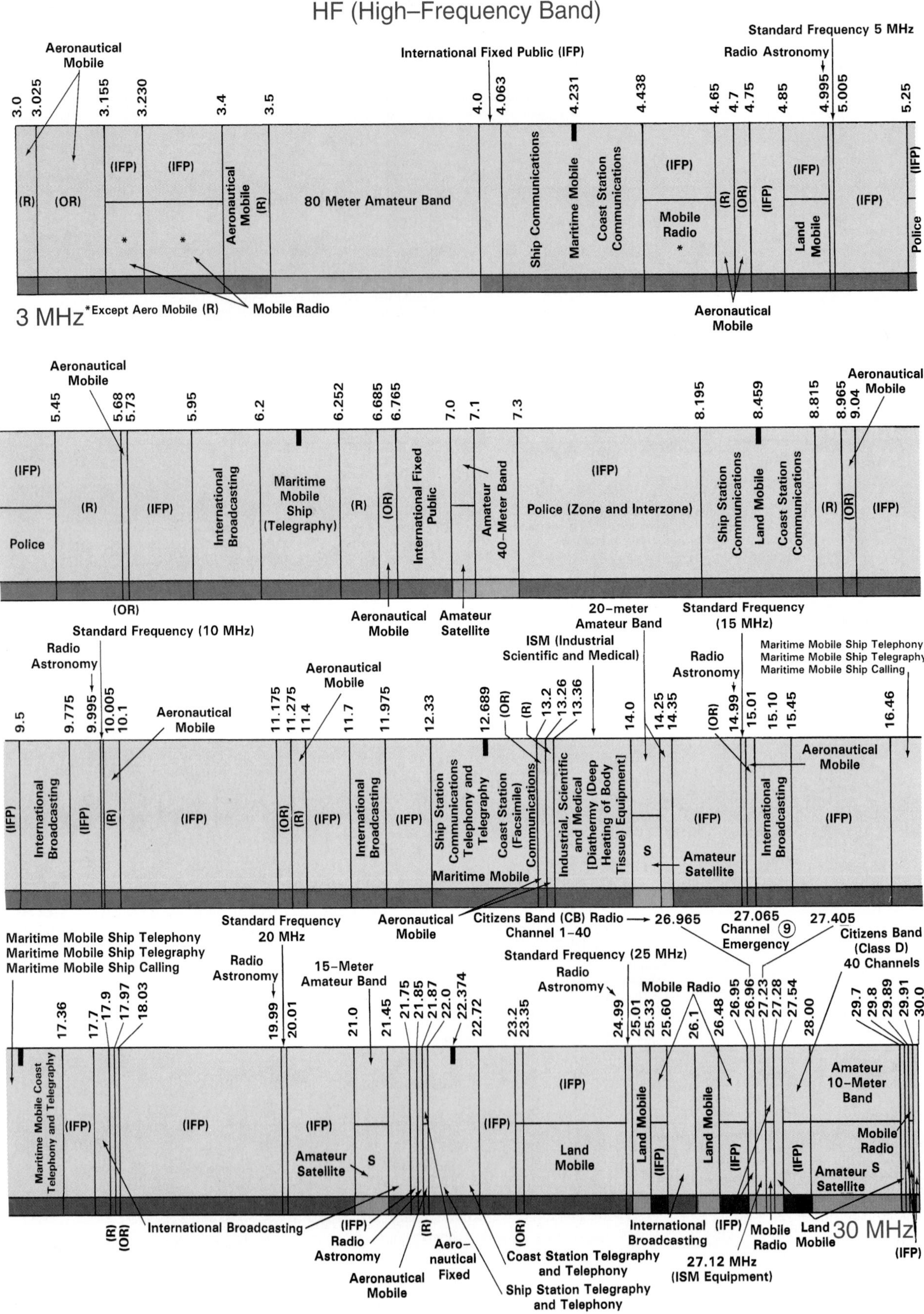
HF (High–Frequency Band)
3 MHz
*Except Aero Mobile (R)
Mobile Radio
Aeronautical Mobile
International Fixed Public (IFP)
Standard Frequency 5 MHz
Radio Astronomy
80 Meter Amateur Band
Ship Communications
Maritime Mobile
Coast Station Communications
Mobile Radio
Land Mobile
Police
Police (Zone and Interzone)
International Broadcasting
Maritime Mobile Ship (Telegraphy)
International Fixed Public
Amateur 40–Meter Band
Amateur Satellite
Ship Station Communications
Coast Station Communications
Standard Frequency (10 MHz)
20–meter Amateur Band
ISM (Industrial Scientific and Medical)
Standard Frequency (15 MHz)
Maritime Mobile Ship Telephony
Maritime Mobile Ship Telegraphy
Maritime Mobile Ship Calling
Ship Station Communications Telephony and Telegraphy
Coast Station (Facsimile) Communications
Industrial, Scientific and Medical [Diathermy (Deep Heating of Body Tissue) Equipment]
Standard Frequency 20 MHz
15–Meter Amateur Band
Citizens Band (CB) Radio Channel 1–40
27.065 Channel 9 Emergency
Citizens Band (Class D) 40 Channels
Standard Frequency (25 MHz)
Maritime Mobile Coast Telephony and Telegraphy
Amateur 10–Meter Band
International Broadcasting
Aeronautical Fixed
Coast Station Telegraphy and Telephony
Ship Station Telegraphy and Telephony
27.12 MHz (ISM Equipment)
30 MHz

VHF (Very High–Frequency Band)

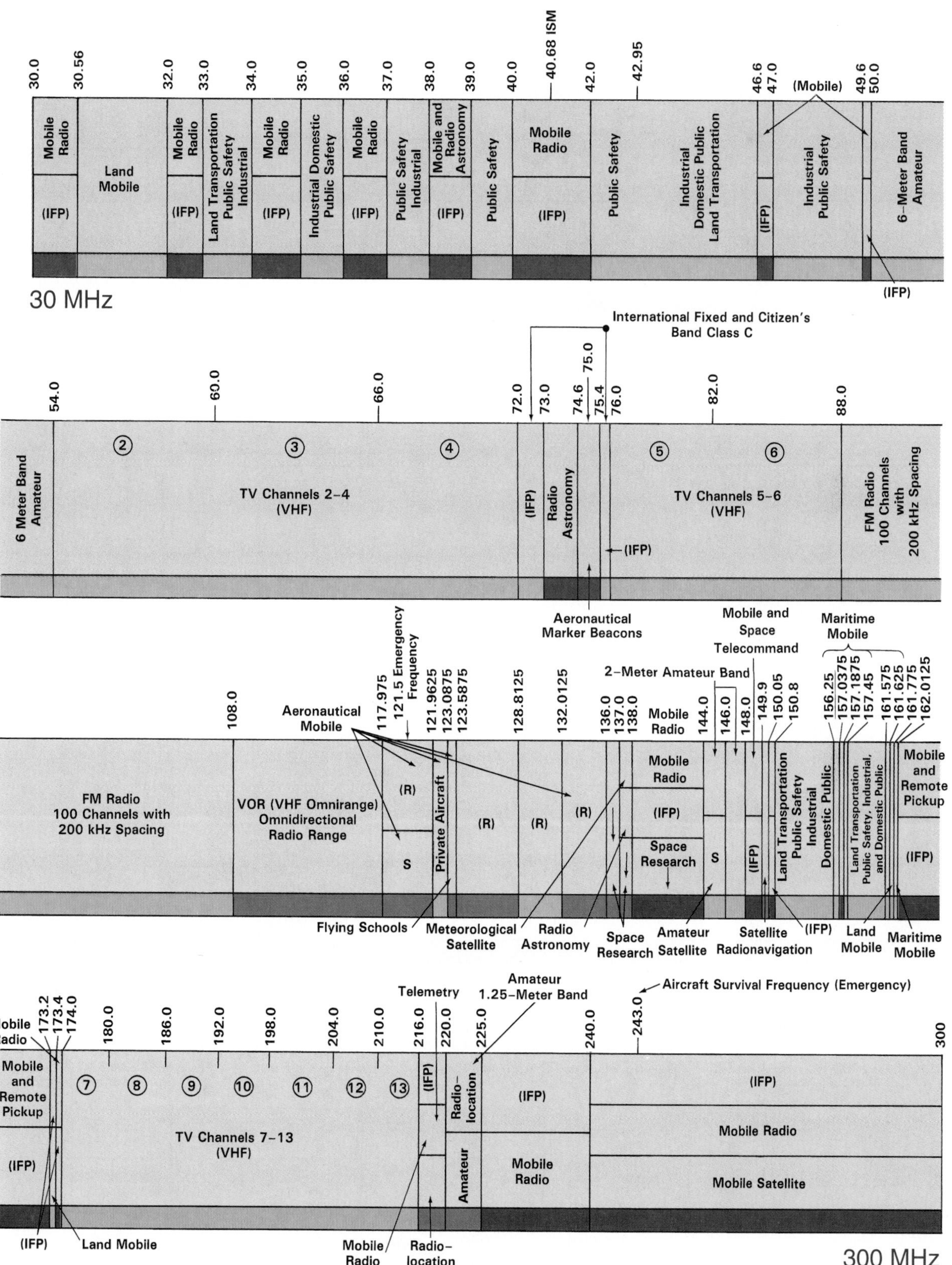

UHF (Ultra High–Frequency Band)

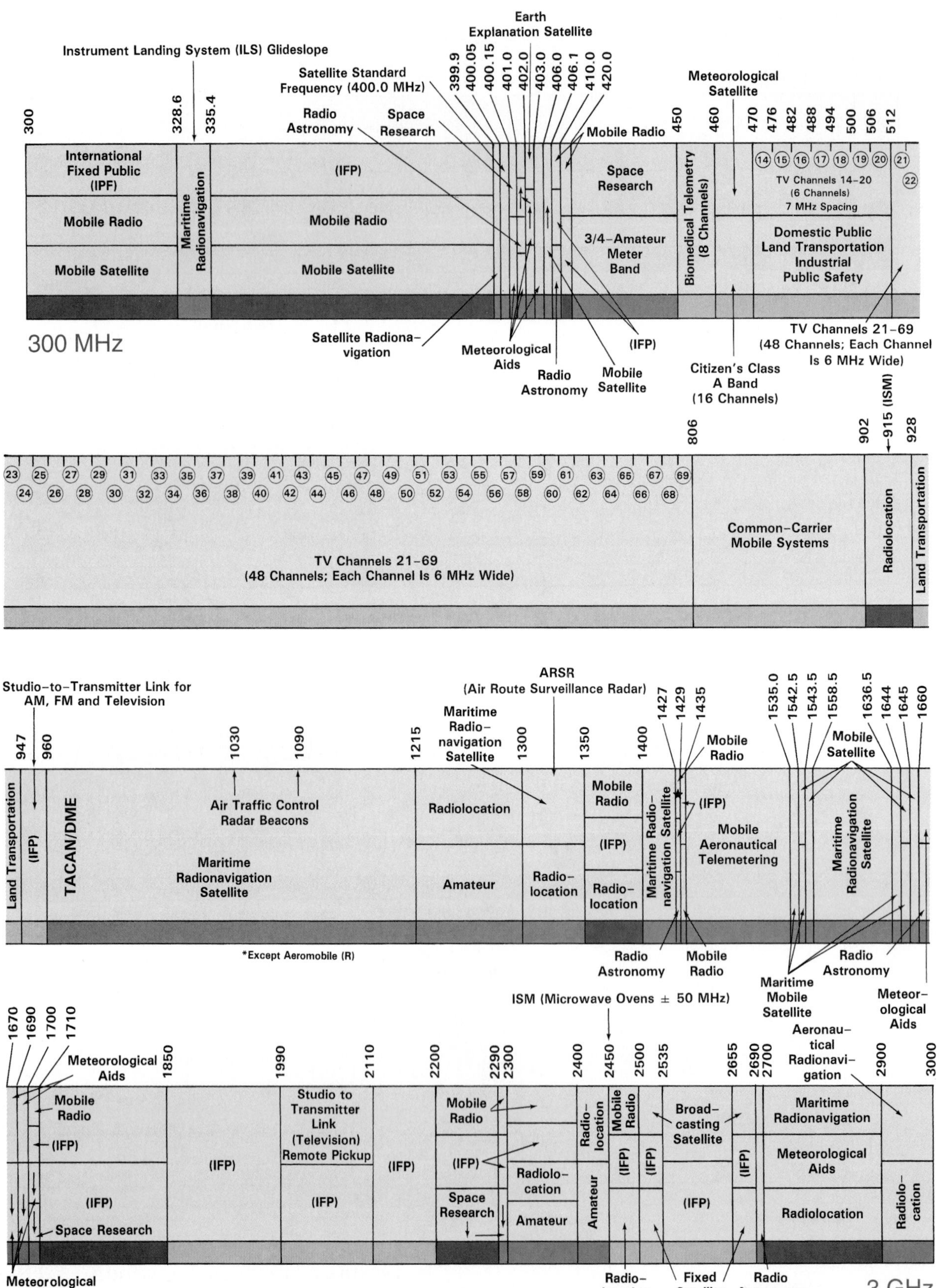

SHF (Super High–Frequency Band)

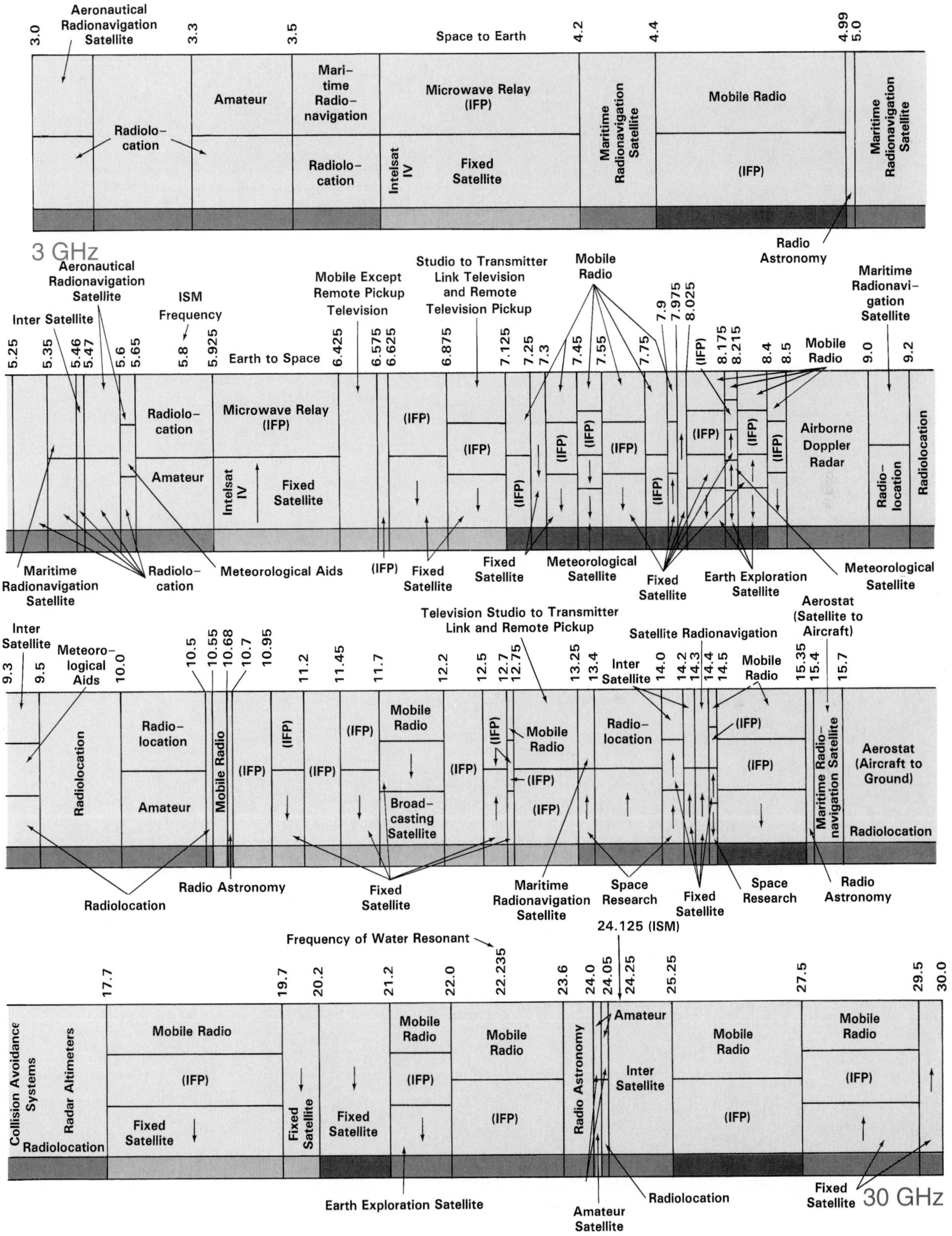

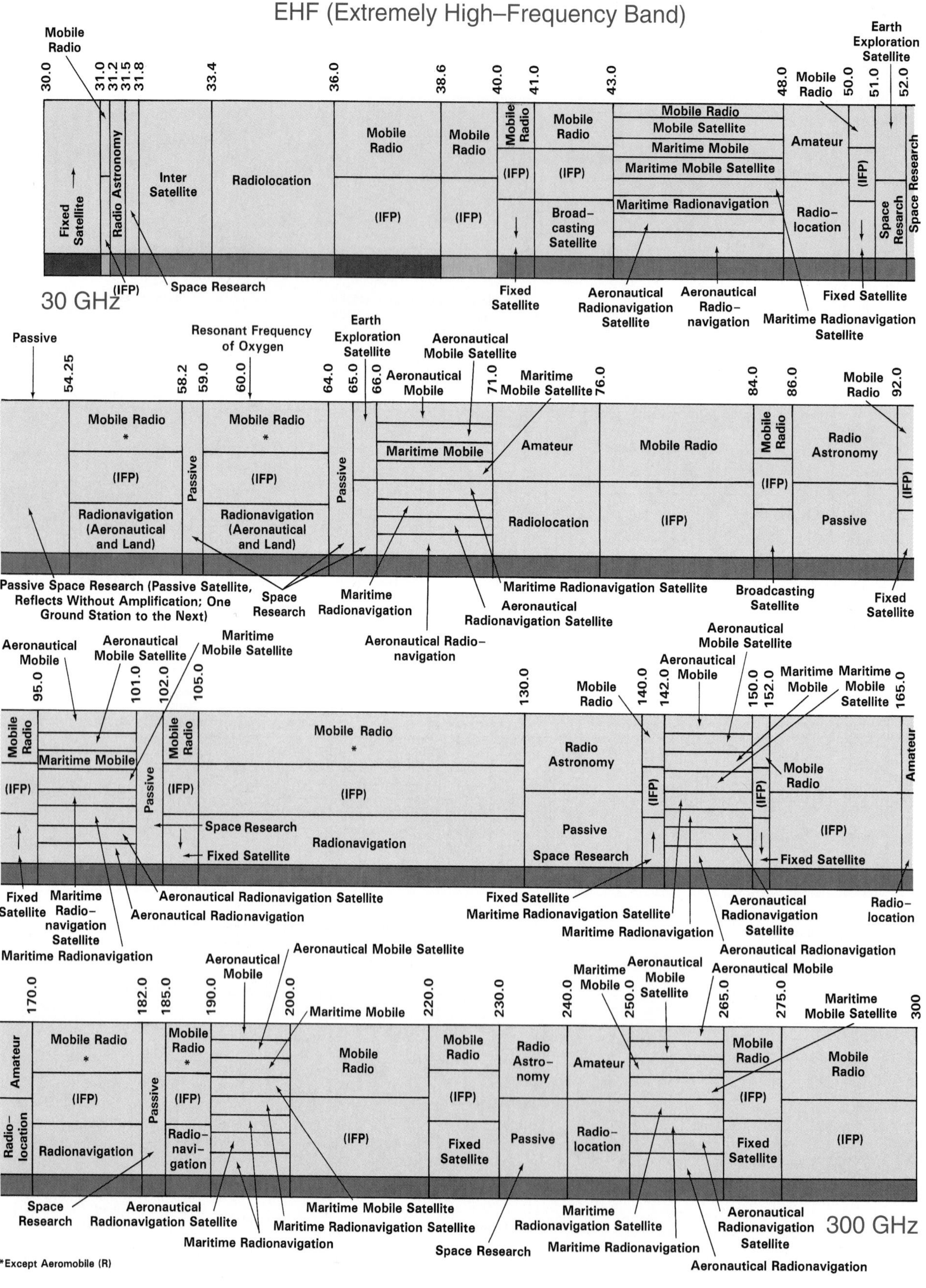
EHF (Extremely High–Frequency Band)
Mobile Radio
30.0
31.0
31.2
31.5
31.8
33.4
36.0
38.6
40.0
41.0
43.0
48.0
Mobile Radio
50.0
51.0
52.0
Earth Exploration Satellite
Fixed Satellite
Radio Astronomy
Inter Satellite
Radiolocation
Mobile Radio
(IFP)
Mobile Radio
(IFP)
Mobile Radio
(IFP)
Mobile Radio
(IFP)
Broad-casting Satellite
Mobile Radio
Mobile Satellite
Maritime Mobile
Maritime Mobile Satellite
Maritime Radionavigation
Amateur
Radio-location
(IFP)
Space Research
Space Research
(IFP)
Space Research
30 GHz
Fixed Satellite
Aeronautical Radionavigation Satellite
Aeronautical Radio-navigation
Fixed Satellite
Maritime Radionavigation Satellite
Passive
Resonant Frequency of Oxygen
Earth Exploration Satellite
Aeronautical Mobile Satellite
Aeronautical Mobile
Maritime Mobile Satellite
54.25
58.2
59.0
60.0
64.0
65.0
66.0
71.0
76.0
84.0
86.0
Mobile Radio
92.0
Mobile Radio
*
(IFP)
Radionavigation (Aeronautical and Land)
Passive
Mobile Radio
*
(IFP)
Radionavigation (Aeronautical and Land)
Passive
Maritime Mobile
Amateur
Radiolocation
Mobile Radio
(IFP)
Mobile Radio
(IFP)
Radio Astronomy
Passive
(IFP)
Passive Space Research (Passive Satellite, Reflects Without Amplification; One Ground Station to the Next)
Space Research
Maritime Radionavigation
Aeronautical Radio-navigation
Maritime Radionavigation Satellite
Aeronautical Radionavigation Satellite
Broadcasting Satellite
Fixed Satellite
Aeronautical Mobile
Aeronautical Mobile Satellite
Maritime Mobile Satellite
95.0
101.0
102.0
105.0
130.0
Mobile Radio
140.0
142.0
Aeronautical Mobile
Aeronautical Mobile Satellite
150.0
152.0
Maritime Mobile
Maritime Mobile Satellite
165.0
Mobile Radio
(IFP)
Maritime Mobile
Passive
Mobile Radio
(IFP)
Mobile Radio
*
(IFP)
Space Research
Radionavigation
Fixed Satellite
Radio Astronomy
Passive
Space Research
(IFP)
(IFP)
Mobile Radio
(IFP)
Fixed Satellite
Amateur
Fixed Satellite
Maritime Radio-navigation Satellite
Aeronautical Radionavigation Satellite
Aeronautical Radionavigation
Maritime Radionavigation
Fixed Satellite
Maritime Radionavigation Satellite
Maritime Radionavigation
Aeronautical Radionavigation Satellite
Radio-location
Aeronautical Radionavigation
Aeronautical Mobile
Aeronautical Mobile Satellite
Maritime Mobile
Aeronautical Mobile
Maritime Mobile
Aeronautical Mobile Satellite
170.0
182.0
185.0
190.0
200.0
220.0
230.0
240.0
250.0
265.0
275.0
Maritime Mobile Satellite
300
Amateur
Radio-location
Mobile Radio
*
(IFP)
Radionavigation
Passive
Mobile Radio
*
(IFP)
Radio-navi-gation
Mobile Radio
(IFP)
Mobile Radio
(IFP)
Fixed Satellite
Radio Astro-nomy
Passive
Amateur
Radio-location
Mobile Radio
(IFP)
Fixed Satellite
Mobile Radio
(IFP)
Space Research
Aeronautical Radionavigation Satellite
Maritime Radionavigation
Maritime Mobile Satellite
Maritime Radionavigation Satellite
Space Research
Maritime Radionavigation Satellite
Maritime Radionavigation
Aeronautical Radionavigation Satellite
Aeronautical Radionavigation
300 GHz
*Except Aeromobile (R)

Optical Spectrum

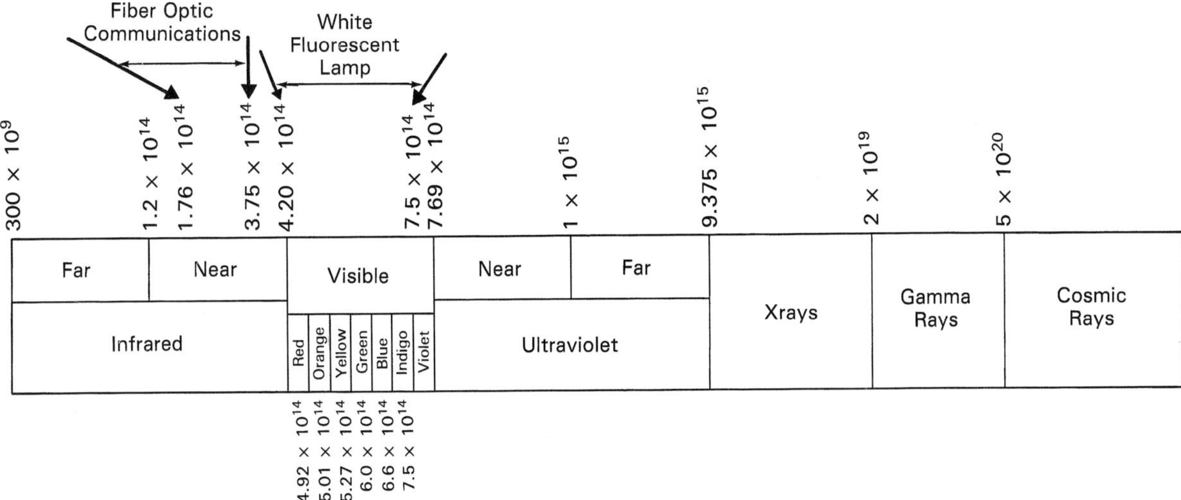

DEFINITIONS AND CODES

USAGE

Exclusively used by government, state, and federal agencies

Shared by government and nongovernment agencies

Nongovernment; publicly used frequencies

TYPES OF EMISSION

Amplitude (AM;Amplitude Modulation)
A0–Steady Unmodulated Pure Carrier
A1–Telegraphy on Pure Continuous Waves
A2–Amplitude Tone–modulated Telegraphy
A3–AM Telephony Including Single and Double Sideband with full, Reduced, or Suppressed Carrier
A4–Facsimile (Pictures)
A5–Television

Frequency FM;Frequency Modulation)
F0–Steady Unmodulated Pure Carrier
F1–Carrier–Shift Telegraphy
F2–Audio–frequency–shift Telegraphy
F3–Frequency or Phase–modulated Telephony
F4–Facsimile
F5–Television

Pulse (PM;Pulse Modulation)
P0–Pulsed Radar
P1–Telegraphy: On/Off Keying of Pulsed Carrier
P2–Telegraphy: Pulse Modulation of Pulse Carrier, Pulse Width, Phase or Position Tone Modulated
P3–Telephony: Amplitude (PAM), Width (PWM), Phase or Position (PPM) Modulated Pulses

ABBREVIATED TERMS

(R), International major air route (air traffic control)
(OR), Off route military (air traffic control)
S Assigned satellite frequency
↑ Earth to space communication (uplink)
↓ Space to earth communication (downlink)

ISM: industrial, scientific, medical
Passive communication equipment: generates no electromagnetic radiation (receive only)
Active communication equipment: generates electromagnetic radiation (transmits)

APPLICATIONS

- International Fixed: public radiocommunication service. Transmitters designed for one frequency.
- Aeronautical Fixed: a service intended for the transmission of air navigation and preparation for safety of flight.
- Fixed Satellite: satellites that maintain a geostationary orbit (same position above the earth)
- Radiolocation: radio waves used to detect an object's direction, position, or motion.
- Radio Astronomy: radio waves emitted by celestial bodies that are used to obtain data about them.
- Space Operations
- Space Research

APPLICATIONS (continued)

- Mobile: Radio service between a fixed location and one or more mobile stations or between mobile stations.
- Aeronautical Mobile: As above, except mobile stations are aircraft.
- Maritime Mobile: As above, except mobile stations are marine.
- Land Mobile: As above, except mobile stations are automobiles.
- Mobile Satellite
- Aeronautical Mobile Satellite
- Maritime Mobile Satellite
- Radionavigation (Aeronautical and Land): Navigational use of radiolocation equipment such as direction finders, radio compass, radio homing beacons, etc.
- Aeronautical Radionavigation
- Maritime Radionavigation
- Satellite Radionavigation
- Aeronautical Radionavigation Satellite
- Maritime Radionavigation Satellite
- Inter Satellite

APPLICATIONS (continued)

- Broadcasting: Transmission of speech, music, or visual programs for commercial or public service purposes.
- Broadcasting Satellite
- Amateur: a frequency used by persons licensed to operate radio transmitters as a hobby. Person is also called a radio ham.
- Amateur Satellite: Communication by radio hams via satellite.
- Citizen's: a radiocommunications service of fixed, land, and mobile stations intended for short-distance personal or business purposes.
- Standard Frequency: Highly accurate signal broadcasted by the national bureau of standards (NBS) radio station (WWV) to provide frequency, time, solar flare, and other standards.
- Standard Frequency Satellite: NBS broadcast via satellite.
- Meteorological Aids: a radio service in which the emission consists of signals used solely for meteorological use.
- Meteorological Satellite: Meteorological broadcast via satellite.
- Earth Exploration Satellite: Radio frequencies used for earth exploration.

APPENDIX

G

Conversions, Constants, Units, and Prefixes

Greek Alphabet

NAME	CAPITAL	LOWERCASE	DESIGNATES
Alpha	A	α	Angles, coefficients, attenuation constant, absorption factor, area.
Beta	B	β	Angles, coefficients, phase constant.
Gamma	Γ	γ	Specific quantity, angles, electrical conductivity, propagation constant, complex propagation constant (cap).
Delta	Δ	δ	Density, angles, increment or decrement (cap or small), determinant (cap), permittivity (cap).
Epsilon	E	ε	Dielectric constant, permittivity, base of natural (Napierian) logarithms, electric intensity.
Zeta	Z	ζ	Coordinate, coefficients.
Eta	H	η	Intrinsic impedance, efficiency, surface charge density, hysteresis, coordinates.
Theta	Θ	Θ	Reluctance, phase angle.
Iota	I	ι	Unit vector.
Kappa	K	κ	Susceptibility, coupling coefficient.
Lambda	Λ	λ	Wavelength, attenuation constant, permeance (cap).
Mu	M	μ	Prefix *micro-*, permeability, amplification factor.
Nu	N	ν	Reluctivity, frequency.
Xi	Ξ	ξ	Coordinates.
Omicron	O	o	—
Pi	Π	π	3.1416 (circumference divided by diameter).
Rho	P	ρ	Resistivity, volume charge density, coordinates.
Sigma	Σ	σ	Surface charge density, complex propagation constant, electrical conductivity, leakage coefficient, sign of summation (cap).
Tau	T	τ	Time constant, volume resistivity, time-phase displacement, transmission factor, density.
Upsilon	ϒ	υ	—
Phi	Φ	ϕ	Magnetic flux, angles, scalar potential (cap).
Chi	X	χ	Electric susceptibility, angles.
Psi	Ψ	ψ	Dielectric flux, phase difference, coordinates, angles.
Omega	Ω	ω	Angular velocity ($2\pi f$), resistance in ohms (cap), solid angles (cap).

Symbols and Units

QUANTITY	SYMBOL	UNIT	SYMBOL
Charge	Q	coulomb	C
Current	I	ampere	A
Voltage, potential difference	V	volt	V
Electromotive force	$\mathscr{E}$	volt	V
Resistance	R	ohm	Ω
Conductance	G	mho (siemens)	A/V, or mho (S)
Reactance	X	ohm	Ω
Susceptance	B	mho	A/V, or mho
Impedance	Z	ohm	Ω
Admittance	Y	mho	A/V, or mho
Capacitance	C	farad	F
Inductance	L	henry	H
Energy, work	W	joule	J
Power	P	watt	W
Resistivity	ρ	ohm-meter	Ωm
Conductivity	σ	mho per meter	mho/m
Electric displacement	D	coulomb per square meter	C/m^2
Electric field strength	E	volt per meter	V/m
Permittivity (absolute)	ε	farad per meter	F/m
Relative permittivity	ε_r	(numeric)	
Magnetic flux	Φ	weber	Wb
Magnetomotive force	$\mathfrak{F}$	ampere (ampere-turn)	A
Reluctance	$\mathscr{R}$	ampere per weber	A/Wb
Permeance	$\mathscr{P}$	weber per ampere	Wb/A
Magnetic flux density	B	tesla	T
Magnetic field strength	H	ampere per meter	A/m
Permeability (absolute)	μ	henry per meter	H/m
Relative permeability	μ_r	(numeric)	
Length	l	meter	m
Mass	m	kilogram	kg
Time	t	second	s
Frequency	f	hertz	Hz
Angular frequency	ω	radian per second	rad/s
Force	F	newton	N
Pressure	p	newton per square meter	N/m^2
Temperature (absolute)	T	Kelvin	K
Temperature (International)	t	degree Celsius	°C

SI Symbols and Units

UNIT	SYMBOL
ampere	A
ampere-hour	Ah
ampere per meter	A/m
angstrom	Å
attoampere	aA
bar	bar
barn	b
baud	Bd
bel	B
bit	b
bit per second	b/s
calorie	cal
candela	cd
candela per square meter	cd/m^2
centimeter	cm
circular mil	cmil
coulomb	C
cubic centimeter	cm^3
cubic foot per minute	ft^3/min
cubic meter	m^3
cubic meter per second	m^3/s
curie	Ci
decibel	dB
degree (plane angle)	°
degree Celsius	°C
degree Fahrenheit	°F
degree Rankine	°R
dyne	dyn
electronvolt	eV
erg	erg
farad	F
femtoampere	fA
femtovolt	fV
femtowatt	fW
foot	ft
foot per minute	ft/min
foot per second	ft/s
gauss	G
gigabit	Gb
gigaelectronvolt	GeV
gigahertz	GHz
gigaohm	GΩ
gigawatt	GW
gilbert	Gb
gram	g
henry	H
hertz	Hz
horsepower	hp
hour	h
inch	in.
inch per second	in./s
joule	J

UNIT	SYMBOL
kelvin	K
kiloampere	kA
kilobaud	kBd
kilobit	kb
kiloelectronvolt	keV
kilogauss	kG
kilogram	kg
kilohertz	kHz
kilohm	kΩ
kilometer	km
kilometer per hour	km/h
kilovolt	kV
kilovoltampere	kVA
kilowatt	kW
kilowatthour	kWh
lambert	L
liter	L
lumen	lm
lumen per square foot	lm/ft^2
lumen per square meter	lm/m^2
lumen per watt	lm/W
lumen second	lm•s
lux	lx
maxwell	Mx
megabar	Mbar
megabit	Mb
megaelectronvolt	MeV
megahertz	MHz
megampere	MA
megavolt	MV
megawatt	MW
megawatt-hour	MWh
megohm	MΩ
meter	m
microampere	μA
microbar	μbar
microfarad	μF
microhenry	μH
microhm	μΩ
micrometer	μm
microsecond	μs
microsiemens	μS
microvolt	μV
microwatt	μW
mil	mil
mile	mi
mile per hour	mi/h
milliampere	mA
millibar	mbar
milligauss	mG
milligram	mg
millihenry	mH

UNIT	SYMBOL
milliliter	mL
millimeter	mm
milliohm	mΩ
millirad	mrd
milliradian	mrad
milliroentgen	mR
millisecond	ms
millisiemens	mS
millivolt	mV
milliwatt	mW
minute (time)	min
nanoampere	nA
nanofarad	nF
nanohenry	nH
nanometer	nm
nanosecond	ns
nanovolt	nV
nanowatt	nW
nautical mile	nmi
neper	Np
oersted	Oe
ohm	Ω
ohm per volt	Ω/V
picoampere	pA
picofarad	pF
picosecond	ps
picowatt	pW
pound	lb
rad	rd
radian	rad
revolution per minute	r/min
revolution per second	r/s
roentgen	R
second (time)	s
siemens	S
square foot	ft^2
square inch	in.^{2}
square meter	m^2
square mile	mi^2
square yard	yd^2
steradian	sr
teraelectronvolt	TeV
terahertz	THz
teraohm	TΩ
tesla	T
var	var
volt	V
volt-ampere	VA
watt	W
watt-hour	Wh
watt per steradian	W/sr
weber	Wb
yard	yd

Metric Conversion Chart (Approximations)

WHEN YOU KNOW	MULTIPLY BY	TO FIND
	Length	
millimeters (mm)	0.04	inches (in.)
centimeters (cm)	0.4	inches (in.)
meters (m)	3.3	feet (ft)
meters (m)	1.1	yards (yd)
kilometers (km)	0.6	miles (mi)
	Area	
sq. centimeters (cm^2)	0.16	sq. inches ($in.^2$)
sq. meters (m^2)	1.2	sq. yards (yd^2)
sq. kilometers (km^2)	0.4	sq. miles (mi^2)
hectares (ha) (10 000 m^2)	2.5	acres
	Mass (weight)	
grams (g)	0.035	ounces (oz)
kilograms (kg)	2.2	pounds (lb)
tonnes (1 000 kg) (t)	1.1	short tons
	Volume	
milliliters (mL)	0.03	fluid ounces (fl oz)
liters (L)	2.1	pints (pt)
liters (L)	1.06	quarts (qt)
liters (L)	0.26	gallons (gal)
cubic meters (m^3)	35	cubic feet (ft^3)
cubic meters (m^3)	1.3	cubic yards (yd^3)
	Temperature (exact)	
Celsius (°C)	9/5(°C) + 32	Fahrenheit (°F)
	Temperature (exact) to Metric	
Fahrenheit (°F)	5/9(°F – 32)	Celsius (°C)
	Length	
inches (in.)	*2.5	centimeters (cm)
feet (ft)	30	centimeters (cm)
yards (yd)	0.9	meters (m)
miles (mi)	1.6	kilometers (km)
	Area	
sq. inches ($in.^2$)	6.5	sq. centimeters (cm^2)
sq. feet (ft^2)	0.09	sq. meters (m^2)
sq. yards (yd^2)	0.8	sq. meters (m^2)
sq. miles (mi^2)	2.6	sq. kilometers (km^2)
acres	0.4	hectares (ha)

*in. = 2.54 cm exactly

WHEN YOU KNOW	MULTIPLY BY	TO FIND
	Mass (weight)	
ounces (oz)	28	grams (g)
pounds (lb)	0.45	kilograms (kg)
short tons (2 000 lb)	0.9	tonnes (t)
	Volume	
teaspoons (tsp)	5	milliliters (mL)
tablespoons (tbsp)	15	milliliters (mL)
fluid ounces (fl oz)	30	milliliters (mL)
cups (c)	0.24	liters (L)
pints (pt)	0.47	liters (L)
quarts (qt)	0.95	liters (L)
gallons (gal)	3.8	liters (L)
cubic feet (ft^3)	0.03	cubic meters (m^3)
cubic yards (yd^3)	0.76	cubic meters (m^3)

MULTIPLICATION FACTOR	PREFIX	SYMBOL
$1\ 000\ 000\ 000\ 000\ 000\ 000 = 10^{18}$	exa	E
$1\ 000\ 000\ 000\ 000\ 000 = 10^{15}$	peta	P
$1\ 000\ 000\ 000\ 000 = 10^{12}$	tera	T
$1\ 000\ 000\ 000 = 10^{9}$	giga	G
$1\ 000\ 000 = 10^{6}$	mega	M
$1\ 000 = 10^{3}$	kilo	k
$100 = 10^{2}$	hecto	h
$10 = 10$	deka	da
$0.1 = 10^{-1}$	deci	d
$0.01 = 10^{-2}$	centi	c
$0.001 = 10^{-3}$	milli	m
$0.000\ 001 = 10^{-6}$	micro	μ
$0.000\ 000\ 001 = 10^{-9}$	nano	n
$0.000\ 000\ 000\ 001 = 10^{-12}$	pico	p
$0.000\ 000\ 000\ 000\ 001 = 10^{-15}$	femto	f
$0.000\ 000\ 000\ 000\ 000\ 001 = 10^{-18}$	atto	a

Standard Resistor Values

RESISTANCE TOLERANCE

0.1% 0.25% 0.5%	1%	2% 5%	10%	0.1% 0.25% 0.5%	1%	2% 5%	10%	0.1% 0.25% 0.5%	1%	2% 5%	10%
10.0	10.0	10	10	17.6	—	—	—	30.9	30.9	—	—
10.1	—	—	—	17.8	17.8	—	—	31.2	—	—	—
10.2	10.2	—	—	18.0	—	18	18	31.6	31.6	—	—
10.4	—	—	—	18.2	18.2	—	—	32.0	—	—	—
10.5	10.5	—	—	18.4	—	—	—	32.4	32.4	—	—
10.6	—	—	—	18.7	18.7	—	—	32.8	—	—	—
10.7	10.7	—	—	18.9	—	—	—	33.2	33.2	33	33
10.9	—	—	—	19.1	19.1	—	—	33.6	—	—	—
11.0	11.0	11	—	19.3	—	—	—	34.0	34.0	—	—
11.1	—	—	—	19.6	19.6	—	—	34.4	—	—	—
11.3	11.3	—	—	19.8	—	—	—	34.8	34.8	—	—
11.4	—	—	—	20.0	20.0	20	—	35.2	—	—	—
11.5	11.5	—	—	20.3	—	—	—	35.7	35.7	—	—
11.7	—	—	—	20.5	20.5	—	—	36.1	—	36	—
11.8	11.8	—	—	20.8	—	—	—	36.5	36.5	—	—
12	—	12	12	21.0	21.0	—	—	37.0	—	—	—
12.1	12.1	—	—	21.3	—	—	—	37.4	37.4	—	—
12.3	—	—	—	21.5	21.5	—	—	37.9	—	—	—
12.4	12.4	—	—	21.8	—	—	—	38.3	38.3	—	—
12.6	—	—	—	22.1	22.1	22	22	38.8	—	—	—
12.7	12.7	—	—	22.3	—	—	—	39.2	39.2	39	39
12.9	—	—	—	22.6	22.6	—	—	39.7	—	—	—
13.0	13.0	13	—	22.9	—	—	—	40.2	40.2	—	—
13.2	—	—	—	23.2	23.2	—	—	40.7	—	—	—
13.3	13.3	—	—	23.4	—	—	—	41.2	41.2	—	—
13.5	—	—	—	23.7	23.7	—	—	41.7	—	—	—
13.7	13.7	—	—	24.0	—	24	—	42.2	42.2	—	—
13.8	—	—	—	24.3	24.3	—	—	42.7	—	—	—
14.0	14.0	—	—	24.6	—	—	—	43.2	43.2	43	—
14.2	—	—	—	24.9	24.9	—	—	43.7	—	—	—
14.3	14.3	—	—	25.2	—	—	—	44.2	44.2	—	—
14.5	—	—	—	25.5	25.5	—	—	44.8	—	—	—
14.7	14.7	—	—	25.8	—	—	—	45.3	45.3	—	—
14.9	—	—	—	26.1	26.1	—	—	45.9	—	—	—
15.0	15.0	15	15	26.4	—	—	—	46.4	46.4	—	—
15.2	—	—	—	26.7	26.7	—	—	47.0	—	47	47
15.4	15.4	—	—	27.1	—	27	27	47.5	47.5	—	—
15.6	—	—	—	27.4	27.4	—	—	48.1	—	—	—
15.8	15.8	—	—	27.7	—	—	—	48.7	48.7	—	—
16.0	—	16	—	28.0	28.0	—	—	49.3	—	—	—
16.2	16.2	—	—	28.4	—	—	—	49.9	49.9	—	—
16.4	—	—	—	28.7	28.7	—	—	50.5	—	—	—
16.5	16.5	—	—	29.1	—	—	—	51.1	51.1	51	—
16.7	—	—	—	29.4	29.4	—	—	51.7	—	—	—
16.9	16.9	—	—	29.8	—	—	—	52.3	52.3	—	—
17.2	—	—	—	30.1	30.1	30	—	53.0	—	—	—
17.4	17.4	—	—	30.5	—	—	—	53.6	53.6	—	—

RESISTANCE TOLERANCE (continued)

0.1% 0.25% 0.5%	1%	2% 5%	10%	0.1% 0.25% 0.5%	1%	2% 5%	10%	0.1% 0.25% 0.5%	1%	2% 5%	10%
54.2	—	—	—	66.5	66.5	—	—	81.6	—	—	—
54.9	54.9	—	—	67.3	—	—	—	82.5	82.5	82	82
56.6	—	—	—	68.1	68.1	68	68	83.5	—	—	—
56.2	56.2	56	56	69.0	—	—	—	84.5	84.5	—	—
56.9	—	—	—	69.8	69.8	—	—	85.6	—	—	—
57.6	57.6	—	—	70.6	—	—	—	86.6	86.6	—	—
58.3	—	—	—	71.5	71.5	—	—	87.6	—	—	—
59.0	59.0	—	—	72.3	—	—	—	88.7	88.7	—	—
59.7	—	—	—	73.2	73.2	—	—	89.8	—	—	—
60.4	60.4	—	—	74.1	—	—	—	90.9	90.9	91	—
61.2	—	—	—	75.0	75.0	75	—	92.0	—	—	—
61.9	61.9	62	—	75.9	—	—	—	93.1	93.1	—	—
62.6	—	—	—	76.8	76.8	—	—	94.2	—	—	—
63.4	63.4	—	—	77.7	—	—	—	95.3	95.3	—	—
64.2	—	—	—	78.7	78.7	—	—	96.5	—	—	—
64.9	64.9	—	—	79.6	—	—	—	97.6	97.6	—	—
65.7	—	—	—	80.6	80.6	—	—	98.8	—	—	—

(Available in Multiples of 0-1, 10, 100, 1 k, and 1 m

APPENDIX

H

Answers to Self-Test Review Questions

STR 1-1 1. Electronics, electricity
2. (a) Analog electronic signal
3. (b) Digital electronic signal
4. Because the output of the circuit varies in direct proportion to the input (a linear response)

STR 1-2 1. Personal choice
2. Personal choice

STR 2-1 1. Diodes, transistors and integrated circuits (ICs)
2. a. Lower power consumption
b. Smaller and lighter
c. More rugged and reliable
d. Lower cost
3. Current or voltage
4. (c) Solid state devices

STR 2-2 1. Elements are made up of similar atoms—compounds are made up of similar molecules.
2. Protons, neutrons, electrons
3. Copper
4. Like charges repel—unlike charges attract.

STR 2-3 1. They all have 4 valence electrons.
2. 1, 4
3. Covalent bond
4. Negative, decreases
5. Increases
6. Intrinsic, positive, negative

STR 2-4 1. To increase their conductivity
2. Electrons, N
3. Holes, P
4. Electrons, holes

STR 2-5 1. Depletion region
2. (a) 700 mV
3. True
4. Open, closed

STR 3-1 1. Anode and cathode
2. Switch
3. Forward, reverse
4. OFF

STR 3-2 1. 0.7 V
2. 0.7 V
3. One
4. The peak inverse voltage or maximum reverse voltage for a silicon diode is normally about 50 V.
5. Decreases

STR 4-1 1. True
2. Reverse
3. Zener has a "Z" shaped bar instead of a straight bar.
4. No
5. Input voltage, load impedance

STR 4-2 1. Electrons and holes
2. The semiconductor compound used
3. (b) 2 V
4. Series current limiting resistor
5. Forward current

STR 4-3 1. Step 1: Diagnose
Step 2: Isolate
Step 3: Repair
2. Electronics, test equipment, troubleshooting techniques, and equipment repair
3. These charts are a graphical means to show the sequence of tests to be performed on a system or circuit that is malfunctioning. By completing these tests and following a path based on the results obtained, a technician can localize a problem.
4. An open control line will prevent power from reaching its respective LED and, therefore, that segment will never be able to turn on.

STR 5-1 1. Transformer, rectifier, filter, regulator
2. The rectifier
3. The transformer
4. Regulator

STR 5-2 1. Down
2. $V_S = \frac{N_S}{N_P} \times V_P$

$$V_S = \frac{1}{5} \times 120 \text{ V rms} = 24 \text{ V rms}$$

STR 5-3 1. ac, pulsating dc
2. Half
3. 50 Hz
4. Bridge

STR 5-4 1. (b) A low pass
2. Inversely proportional, C↑, % ripple ↓
3. R in series, C_{in} shunt.
4. Parallel

STR 5-5 1. To maintain a constant dc output voltage despite variations in the ac input voltage and the output load resistance
2. Small
3. Parallel
4. (a) +12 V (b) −5 V (c) +15 V (d) −12 V

STR 5-6 1. $V_{avg} = 0.636 \times 12\text{ V} = 7.632\text{ V}$, however, due to the capacitor at the regulator's input, the measured value of voltage applied to the regulator will be approximately equal to the peak value of 12 V.
2. Yes, since $V_{avg} = 0.318 \times 50\text{ V} = 15.9\text{ V}$; therefore, a 7815 would have an input voltage that is greater than its 15 V output. In this circuit example, the input regulator capacitor will ensure that the voltage applied to the regulator will be approximately equal to the peak value of 50 V.

STR 6-1 1. 325.2 V
2. The half-wave voltage doubler has two disadvantages. The first is that the ripple frequency of 60 Hz is difficult to filter, and the second is that an expensive output capacitor is required since it must have a voltage rating that is more than twice the peak of the ac input.
3. Voltage Tripler: $V_{dc} = 3 \times V_{Spk} = 3 \times 162.6\text{ V} = 487.8\text{ V}$
Voltage Quadrupler: $V_{dc} = 4 \times V_{Spk} = 4 \times 162.6\text{ V} = 650.4\text{ V}$
4. Half-wave voltage doubler circuit = 60 Hz
Full-wave voltage doubler circuit = 120 Hz
5. A high ripple frequency is easier to filter

STR 6-2 1. Series clippers and shunt clippers
2. Shunt
3. (d) Both the basic positive series and shunt clipper
4. Limiter

STR 6-3 1. DC restorer
2. Negative clamper and positive clamper circuits
3. Positive clamper

STR 7-1 1. An LED converts an electrical input into a light output, whereas a photodiode changes its electrical characteristics in response to a light input.
2. Photovoltaic action is a process by which a device generates a voltage as a result of exposure to light radiation.

Photoconduction is a process by which the conductance of a material is changed by electromagnetic radiation in the light spectrum.

3. Visible light band, orange/red
4. a. *p*-type, intrinsic layer, *n*-type (PIN) photodiode
   b. Integrated detector and preamplifier (IDP) receiver module
   c. Infrared emitting diode (IRED)
5. An increase in light will cause an increase in the conduction of a photodiode.

STR 7-2 1. False
2. In applications requiring a voltage-controlled variable capacitor
3. True
4. When forward biased, a varactor diode will operate like an ordinary junction diode. The depletion region, and therefore the anode to cathode capacitance, will be zero.

STR 7-3 1. True
2. True
3. Fast turn-on time
4. The metal oxide varistor (MOV)

STR 7-4 1. Voltage, current
2. Forward

STR 7-5 1. (b) Hot carrier or schottky diode
2. 0.4 V
3. Forward, negative resistance
4. True

STR 8-1 1. NPN and PNP
2. Emitter, base, and collector
3. As a switch, and as a variable-resistor
4. The two-state switching action is used in digital circuits, while the variable-resistor action is used in analog circuits.

STR 8-2 1. Current
2. Forward, reverse
3. (b)
4. Open switch
5. Closed switch
6. (a) Common-base
   (b) Common-collector
   (c) Common-emitter
7. Voltage-divider bias
8. Base-biasing

STR 9-1 1. Switching action and variable resistor action. The two-state switching action is used in digital electronics.
2. OR, AND, NOT, NAND, NOR
3. Transistor-transistor logic
4. 10
5. HIGH

6. HIGH, LOW, FLOATING (high impedance)
7. High speed
8. Small size

STR 9-2 1. Free-running or unstable
2. Monostable multivibrator
3. Bistable multivibrator
4. Because it will hold the circuit latched in its last state
5. The three 5 kΩ voltage divider circuit in the input
6. Square or pulse wave generator, one-shot timer

STR 9-3 1. Multimeter
2. Oscilloscope
3. Logic pulser
4. Current tracer, logic probe
5. They are not; IC has to be replaced.
6. Open, short

STR 10-1 1. $Gain(A) = \dfrac{Amplitude\ of\ Output\ Signal}{Amplitude\ of\ Input\ Signal}$
2. $V_{\text{OUT}} = A_V \times V_{\text{IN}} = 75 \times 0.5\text{ mV} = 37.5\text{ mV}$
3. Voltage, Power, Current
4. Exponent
5. $A_P = 10 \times \log \dfrac{P_{\text{OUT}}}{P_{\text{IN}}} = 10 \times \log \dfrac{15\text{ W}}{10\text{ mW}} = 31.76\text{ dB}$
6. 10,000
7. Class A
8. Frequency response curve

STR 10-2 1. True
2. Resistor R_E achieves circuit thermal stability through emitter feedback. With R_E included in circuit, V_{CC} is developed across R_C, Q_1's collector-emitter, and R_E. An increase in temperature will cause an increase in I_C, I_E, and therefore the voltage developed across R_E. An increase in V_{RE} will decrease the base-emitter bias voltage applied, causing Q_1's internal currents to decrease to their original values. A change in the output current (I_C) due to temperature will cause the emitter resistor to feed back a control voltage to the input.
3. False, transformers will block a dc signal.
4. The complementary dc amplifier arrangement enables us to cascade several stages to achieve high signal gain without distortion because the Q point, which is set by the previous stage, continually alternates between being more positive (slightly above the mid-point) or less positive (slightly below the mid-point).
5. Sum
6. Current
7. High, low

STR 10-3 1. 20 Hz to 20,000 Hz
2. Voltage and power
3. Voltage, power
4. Capacitor C_E is connected in parallel with R_E to prevent signal variations at the emitter decreasing the gain of the amplifier stage.
5. AC signal coupling and dc bias isolation
6. Class B
7. Fewer circuit components are needed to construct a power amplifier because a phase splitter is not needed due to the complementary nature of the NPN and PNP transistors.

STR 10-4 1. Power supply, load, or signal source
2. Disconnect the source from the load to see if the signal source returns to its normal level (load fault) or is still bad (source fault).
3. Because it indicates both the dc bias voltage and ac signal voltage at that test point
4. Because push-pull transistors are generally both damaged or their operating characteristics are altered when one of the transistors develops a fault

---

STR 11-1 1. (c)
2. (a) Radio frequency
   (b) Amplitude modulation
   (c) Federal Communications Commission
3. RF amplifier/multiplier circuit
4. $BW = 2 \times f_{AF(HI)} = 2 \times 3.5 \text{ kHz} = 7 \text{ kHz}$
5. Good
6. Increased, decreased

STR 11-2 1. (d)
2. RF amplifier/mixer
3. The IF amplifier is cheaper since it is fixed-tuned, and its narrow bandwidth gives it a higher gain.
4. Fixed-tuned

STR 11-3 1. Televisions, monitors, radar and sonar displays, and so on.
2. (b)
3. To frequency compensate for a circuit's inherent stray-wire and transistor junction capacitance.
4. True

STR 11-4 1. 1. First, connect a dc current meter between V_{CC} and the tuned circuit, as shown in Figure 11-11(c).
   2. Next, connect a frequency generator to the input of the amplifier, and set it to the desired frequency.
   3. Finally, adjust the variable inductor (L_1) until the meter shows a large dip or null in current.
2. When the resonant frequency of an amplifier's tuned circuit drifts off its designed setting

STR 12-1 1. *LC*, Crystal, *RC*
2. Positive
3. (a) Transformer
   (b) Grounded tapped inductor
   (c) Two capacitors grounded at common connection
4. 1
5. Parasitic oscillations are unwanted oscillations that occur within transistor circuits due to a circuit's inherent capacitance and inductance between input and output.
6. ON

STR 12-2 1. No frequency drift
2. High, low
3. Minimum, maximum
4. Series, parallel

STR 12-3 1. 10 kHz
2. 180
3. Wein-Bridge oscillator
4. 0

STR 12-4 1. An oscillator does not have a signal input.
2. Oscillator circuits are notorious as being difficult circuits to troubleshoot, due to their feedback system of generating an output and then feeding back part of that output to generate an output, and so on. Because every component plays such an important role, any break due to a component failure in the loop will shut down the whole circuit.

STR 13-1 1. (a) Bipolar junction transistor
   (b) Junction field effect transistor
   (c) Metal oxide semiconductor field effect transistor
2. *n*-channel and *p*-channel
3. Current, voltage
4. Reverse
5. True
6. False
7. Transconductance
8. Very high, reverse

STR 13-2 1. No need for $-V_{GG}$, temperature stability
2. R_S
3. Positive, 0 V, negative
4. Voltage divider

STR 13-3 1. Common-source
2. Common-drain
3. Common-source
4. Common-drain

STR 13-4 1. Its high input impedance
2. When Z_{IN} is high, circuit current is low ($I\downarrow$) and therefore power dissipation is low ($P_D\downarrow$). This condition is ideal for digital inte-

grated circuits (ICs), which contain thousands of transistors all formed onto one small piece of silicon. The low power dissipation of the JFET enables us to densely pack many more components into a very small area.

3. Because it responds well to the small signal voltages from the antenna, and because it is a low noise component

STR 13-5 1. (a) 0.3 V dc (min), 3.0 dc (max)
(b) 5 mA dc
2. Yes P+, N−, = LO Ω
P−, N+, = HI Ω

STR 14-1 1. Depletion, enhancement
2. True
3. ON
4. MOSFET
5. (d)
6. Dual-gate D-MOSFET

STR 14-2 1. False
2. OFF
3. (c) Drain feedback biased
4. True
5. (a)
6. MOS logic gates use less space due to their simpler construction, have a very high noise immunity, and consume less power.

STR 14-3 1. $I_{D(\text{ON})} = 3$ mA, and $V_{DS(\text{ON})} = 1.0$ V
2. (a) Turn ON delay = 45 ns
(b) Turn OFF delay = 60 ns
3. a. Ensure all MOS device pins are kept at same voltage level.
b. Use a wrist grounding strap.
c. All test equipment, soldering irons, and work benches should be properly grounded.
d. All power in equipment should be off before MOS devices are removed or inserted into printed circuit boards.
e. Any unused MOSFET terminals must be connected.
4. Infinite ohms

STR 15-1 1. Integrated circuit
2. Digital
3. Photolithography
4. Sand

STR 15-2 1. MOS
2. Bipolar
3. Hybrid
4. Hybrid IC

STR 15-3 1. a. Transistor outline can
b. Dual in-line package
c. Surface mount technology
2. SMT

STR 15-4 1. SSI
2. MOSFET, MOSFET

STR 16-1 1. Integrated circuit
2. –, +, $+V$, $-V$, Output
3. TO5 Can, DIP package
4. The prefix indicates the manufacturer, the following 3 digit code indicates the op-amp type, the letter after the part number indicates the operating temperature range, and the final suffix code indicates the package type.

STR 16-2 1. Yes
2. An open-loop op-amp circuit has no feedback, whereas a closed-loop op-amp circuit does have a feedback path.
3. High gain, very high input impedance, and very low output impedance
4. Differential amplifier, darlington-pair voltage amplifier, emitter-follower output amplifier
5. (a) Comparator
6. Negative or degenerative feedback will lower the op-amp's gain to
   a. prevent output waveform distortion,
   b. prevent the amplifier from going into oscillation, and
   c. reduce the gain of the op-amp to a consistent value.

STR 16-3 1. Voltage follower
2. Summing amplifier
3. The differentiator circuit has an input capacitor and a feedback resistor, while the integrator circuit has an input resistor and a feedback capacitor.
4. Differential or difference amplifier
5. See Figure 16-15(b)
6. An active filter circuit will filter and amplify the signal input, while a passive filter will only filter the signal input.

STR 17-1 1. Op-amps, voltage regulators, function generators, phase-locked loops
2. Input voltage, load resistance
3. Variable resistor
4. Switch

STR 17-2 1. Voltage controlled oscillator
2. Control voltage
3. Sine, square, triangular
4. Coarse frequency adjust is controlled by SW_1, fine frequency adjust is controlled by R_4.

STR 17-3 1. Phase comparator, amplifier and low-pass filter, and voltage controlled oscillator.
2. A phase-locked loop circuit consists of a phase comparator that compares the output frequency of a voltage controlled oscillator with an input frequency. The error voltage out of the phase com-

parator is then coupled via an amplifier and low-pass filter to the control input of the voltage controlled oscillator to keep it in phase, and therefore at exactly the same frequency as the input frequency.
3. A frequency divider circuit
4. Pin 7

STR 18-1 1. a. 10
b. 2
c. 16
2. 21_{10}, 15_{16}
3. $BF7A_{16}$
4. 100001_2, 21_{16}
5. $0111\ 0110\ 0010\ 1001_{BCD}$
6. Babbage

STR 18-2 1. Because it converts one code to another: BCD code to 7-segment code
2. True
3. MUX: one of several to one; DMUX: one to one of several
4. (b)
5. Latching
6. Frequency divider and binary counter

STR 18-3 1. The 74LS160As in the "count window generator circuit" function as decade frequency dividers, while the 74LS160As in the "counter, register, decoder, and display circuit" function as decade counters.
2. These outputs are what they should be.

STR 19-1 1. Decrease
2. False
3. Bidirectional
4. DIAC
5. True
6. One
7. False
8. Gate

STR 19-2 1. Resistor
2. Electrical, solar
3. Photoconductive and photovoltaic
4. Decrease
5. Electrical to light, electrical to light
6. True
7. False
8. PTC

APPENDIX

I

Answers to Odd-Numbered Problems

Chapter 1

1. b **3.** a **5.** c **7.** a **9.** b

(The answers to essay questions 11 through 20 can be found in the indicated sections that follow the questions.)

Chapter 2

1. a **3.** d **5.** d **7.** d **9.** b **11.** b **13.** b **15.** c

(The answers to essay questions 16 through 30 can be found in the indicated sections that follow the questions.)

31. **a.** Since the *p* region's terminal is positive (+1.3 V) with respect to the *n* region's terminal (0 V), the P-N junction is forward biased.
b. Since the *p* region's terminal is positive (+5 V) with respect to the *n* region's terminal (+3 V), the P-N junction is forward biased.
c. In this instance, the P-N barrier voltage (Si = 0.7 V) has not been overcome since the *p* region is only 0.5 V more positive than the *n* region's terminal (N = 0.2 V, P = 0.7 V, 0.7 – 0.2 = 0.5 V)
d. Since the *n* region's terminal is positive (+5 V) with respect to the *p* region's terminal (0 V), the P-N junction is reverse biased.

33. **a.** Forward biased Si, $V_F = 0.7$ V
b. Forward biased Ge, $V_F = 0.3$ V
c. Reverse biased P-N junction, $V_R = V_S = 12$ V

35. **a.** Forward biased, closed switch
b. Forward biased, closed switch
c. Reverse biased, open switch

Chapter 3

1. c **3.** d **5.** d **7.** a **9.** b **11.** a **13.** a **15.** a

(The answers to essay questions 16 through 30 can be found in the indicated sections that follow the questions.)

31. **a.** Reverse biased **c.** Reverse biased **e.** Forward biased
b. Reverse biased **d.** Forward biased **f.** Reverse biased

33. **a.** $V_{Diode} = 10$ V **b.** $V_{Diode} = 0.7$ V

35. SWITCH POSITION

SWITCH POSITION	A	B	C
1.	Logic 1	Logic 0	Logic 0
2.	Logic 0	Logic 1	Logic 0
3.	Logic 0	Logic 0	Logic 0

Logic 0 = 0 V, Logic 1 = +5 V.

a. With D_1 open, the code generated in position 1 will change since output line C cannot be pulled low. Therefore, the code generated when the switch is in position 1 with D_1 permanently open will be:

SWITCH POSITION	A	B	C
1.	Logic 1	Logic 0	Logic 1
2.	Logic 0	Logic 1	Logic 0
3.	Logic 0	Logic 0	Logic 0

Output line C will give a high in switch position 1.

b. Yes, since the maximum operation current will be 25.8 mA when the switch is in position 3 and D_5, D_6 and D_7 are on. Each parallel branch in this case will have the following value of current:

$$I = \frac{V_S - V_{Diode}}{R}$$

$$I = \frac{5\text{ V} - 0.7\text{ V}}{500\ \Omega}$$

$I = 8.6$ mA

Since there are three branches, the total current will be

3×8.6 mA = 25.8 mA.

A 200 mA fuse will therefore blow whenever the switch is put in position 3, because a current of 25.8 mA will be drawn.

Chapter 4

1. c **3.** d **5.** d **7.** c **9.** a **11.** c **13.** d **15.** d

(The answers to essay questions 16 through 40 can be found in the indicated sections that follow the questions.)

41. a. Polarity correct (+ → cathode, – → anode)
b. Polarity incorrect (+ → anode, – → cathode)
c. Polarity incorrect
d. Polarity correct
e. Polarity correct
f. Polarity for both D_1 and D_2 are correct

43. a. $$I_S = \frac{V_{in} - V_Z}{R_S} = \frac{10\text{ V} - 6.8\text{ V}}{200\ \Omega} = 16\text{ mA}$$

b. Since the zener diode is forward biased, we will assume a 0.7 V forward voltage drop. Therefore

$$I_S = \frac{V_{in} - V_Z}{R_S} = \frac{20\text{ V} - 0.7\text{ V}}{570\ \Omega} = 33.9\text{ mA.}$$

c. Since the zener diode is forward biased, we will assume a 0.7 V forward voltage drop. Therefore

$$I_S = \frac{V_{in} - V_Z}{R_S} = \frac{5\text{ V} - 0.7\text{ V}}{400\ \Omega} = 10.75\text{ mA.}$$

d. Since the input voltage is not large enough to send the zener into its reverse zener breakdown region, the circuit current will be equal to that of the reverse leakage current, which is almost zero.

e. $I_S = \dfrac{V_{in} - V_Z}{R_S} = \dfrac{6\text{ V} - 4.7\text{ V}}{200\ \Omega} = 6.5\text{ mA}$

f. For zener D_1, $I_S = \dfrac{V_{in} - V_Z}{R_S} = \dfrac{10\text{ V} - 6.8\text{ V}}{220\ \Omega} = 14.5\text{ mA}$

For zener D_2, since the input voltage is not large enough to send the zener into its reverse zener breakdown region, the circuit current will be equal to that of the reverse leakage current, which is almost zero.

45. $P_D = I_{ZM} \times V_Z = 6.5\text{ mA} \times 4.7\text{ V} = 30.55\text{ mW}$

A 50 mW zener diode would be adequate in this application.

47. By reversing the input voltage polarity, and the zener diode's orientation. For example, if the polarity of the 10 V supply voltage was reversed so that the negative terminal connects to R_S, and the zener's connection was also reversed so that its anode was connected to point *X*, the output voltage will be –5.6 V. Similarly, if the polarity of the 20 V supply voltage was reversed so that the positive terminal connects to R_S, and the zener's connection was also reversed so that its cathode was connected to point *Y*, the output voltage will be +12 V.

49. $V_{in} = 12\text{ V}, V_Z = 6.8\text{ V}, V_{RL} = 6.8\text{ V}, V_{RS} = V_{in} - V_Z = 12\text{ V} - 6.8\text{ V} = 5.2\text{ V}$

$I_{RS} = \dfrac{V_{RS}}{R_S} = \dfrac{5.2\text{ V}}{120\ \Omega} = 43.3\text{ mA}$

$R_L = 500\ \Omega$

$I_{RL} = \dfrac{V_{RL}}{R_L} = \dfrac{6.8\text{ V}}{500\ \Omega} = 13.6\text{ mA}$

$I_Z = I_{RS} - I_{RL} = 43.3\text{ mA} - 13.6\text{ mA} = 29.7\text{ mA}$

$V_{in} = 15\text{ V}, V_Z = 6.8\text{ V}, V_{RS} = V_{in} - V_Z = 15\text{ V} - 6.8\text{ V} = 8.2\text{ V}$

$I_{RS} = \dfrac{V_{RS}}{R_S} = \dfrac{8.2\text{ V}}{120\ \Omega} = 68.3\text{ mA}$

$R_L = 500\ \Omega$

$I_{RL} = \dfrac{V_{RL}}{R_L} = \dfrac{6.8\text{ V}}{500\ \Omega} = 13.6\text{ mA}$

$I_Z = I_S - I_{RL} = 68.3\text{ mA} - 13.6\text{ mA} = 54.7\text{ mA}$

51. **a.** Forward biased (+ → anode, – → cathode)

b. Forward biased

c. D_1 is reverse biased, D_2 is forward biased

53. Yes

55.

A	B	LED
0 V	0 V	OFF
0 V	+5 V	ON (D_2 turns ON and connects input power to D_3)
+5 V	0 V	ON (D_1 turns ON and connects input power to D_3)

57. Input = +10 V, Output = Green; Input = –10 V, Output = Red

59. *Input* = +10 V

$$I_{LED\text{–}GREEN} = \frac{V_S - V_{LED}}{R_1 + R_2} = \frac{10\text{ V} - 2.5\text{ V}}{300\ \Omega + 200\ \Omega} = 15\text{ mA}$$

Input = –10 V

$$I_{LED\text{–}RED} = \frac{V_S - V_{D1} - V_{LED}}{R_2} = \frac{10\text{ V} - 0.7\text{ V} - 2.5\text{ V}}{200\ \Omega} = 34\text{ mA}$$

61. Common-cathode. The cathodes of all the LEDs have a common connection.

63. Switch closed (CL) = LED ON, Switch open (OP) = LED OFF.

DIGIT	SW1	SW2	SW3	SW4	SW5	SW6	SW7
0	CL	CL	CL	CL	CL	CL	OP
1	OP	CL	CL	OP	OP	OP	OP
2	CL	CL	OP	CL	CL	OP	CL
3	CL	CL	CL	CL	OP	OP	CL
4	OP	CL	CL	OP	OP	CL	CL
5	CL	OP	CL	CL	OP	CL	CL
6	CL	OP	CL	CL	CL	CL	CL
7	CL	CL	CL	OP	OP	CL	OP
8	CL	CL	CL	CL	CL	CL	CL
9	CL	CL	CL	OP	OP	CL	CL

65. **a.** $I_{DIODE.} = \dfrac{V_S - 0.7\text{ V}}{R} = \dfrac{5.1\text{ V} - 0.7\text{ V}}{250\ \Omega} = 17.6\text{ mA}$

b. Position 1, since it brings into circuit five P-N junction encoder diodes and therefore current will equal 5 × 17.6 mA = 88 mA.

c. $I_{LED} = \dfrac{V_S - V_{LED}}{R} = \dfrac{5.1\text{ V} - 2\text{ V}}{250\ \Omega} = 12.4\text{ mA}.$

d. Maximum display current will occur when the display is showing 8, since this requires all seven segments ON. The value of current will be 7 × 12.4 mA = 86.8 mA.

e. An encoder diode, since it draws 17.6 mA compared to a display diode which draws 12.4 mA.

f. The load resistance changes as different digits are encoded and displayed since some circuit states require more current than others. Knowing load current and load voltage we can use Ohm's law to calculate the changes in load resistance.

$$R_L = \frac{V_L}{I_L} = \frac{5.1\text{ V}}{86.8\text{ mA}} = 58.8\ \Omega$$

$$R_L = \frac{V_L}{I_L} = \frac{5.1\text{ V}}{112.8\text{ mA}} = 45.2\ \Omega$$

67. Section 4-2-7

69. *Step 1:* DIAGNOSE. The first step is to determine whether a problem really exists. To carry out this step, a technician must collect information about the system, circuit, and components used, and then diagnose the problem.

Step 2: ISOLATE. The second step is to apply a logical and sequential reasoning process to isolate the problem. In this step, a technician will operate, observe, test, and apply troubleshooting techniques in order to isolate the malfunction.

Step 3: REPAIR. The third and final step is to make the actual repair, and then final test the circuit.

Chapter 5

1. b **3.** d **5.** b **7.** d **9.** a **11.** b **13.** d **15.** c

(The answers to essay questions 16 through 28 can be found in the indicated sections that follow the questions.)

29. Supply approximately a +5 V to +18 V dc output depending on the setting of R_1.

31. $V_P = V_{rms} \times 1.414 = 240 \text{ V} \times 1.414 = 339.4\ V_{peak}$

$$V_S = \frac{N_S}{N_P} \times V_P = \frac{1}{19} \times 339.4 \text{ V} = 17.86\ V_{peak}$$

33. DC ripple frequency out = AC ripple frequency in = 60 Hz

35. Full wave output ripple frequency = 2 × Input Freq. = 2 × 60 Hz = 120 Hz

37. Peak to Peak of ripple = 15 V to 21 V = 6 V pk-pk
Peak of ripple = 1/2 of pk-pk value = 1/2 of 6 V = 3 V
rms of ripple = 0.707 × 3 V = 2.12 V

$$\% \text{ of Ripple} = \frac{V_{rms} \text{ of ripple}}{V_{avg} \text{ of ripple}} \times 100 = \frac{2.12 \text{ V}}{18 \text{ V}} \times 100 = 11.78\%$$

39. Section 5-6

Chapter 6

1. c **3.** b **5.** b **7.** d **9.** c **11.** b **13.** d **15.** a

(The answers to essay questions 16 through 30 can be found in the indicated sections that follow the questions.)

31. **a.** Positive half-wave voltage doubler circuit

$$V_{primary\ pk} = 115 \text{ V} \times 1.414 = 162.6 \text{ V}$$

$$V_{S\ pk} = \frac{N_S}{N_P} \times V_P = \frac{1}{3} \times 162.6 \text{ V} = 54.2 \text{ V}$$

$$V_{out} = 2 \times V_{S\ pk} = 2 \times 54.2 \text{ V} = 108.4 \text{ V}$$

b. Negative half-wave voltage doubler circuit

$$V_{primary\ pk} = 115 \text{ V} \times 1.414 = 162.6 \text{ V}$$

$$V_{S\ pk} = \frac{N_S}{N_P} \times V_P = \frac{1}{2} \times 162.6 \text{ V} = 81.3 \text{ V}$$

$$V_{out} = 2 \times V_{S\ pk} = 2 \times 81.3 \text{ V} = 162.6 \text{ V}$$

33. A *dual output power supply circuit* is a circuit that provides both a positive and negative dc output voltage. In this example, the full-wave voltage doubler circuit will generate a positive voltage at TP1 (+162.6 V) with respect to ground, and a negative voltage at TP2 (–162.6 V) with respect to ground. The voltage between TP1 and TP2 will therefore be the difference between these two opposite voltages +162.6 V – (–162.6 V) = 325.2 V

35. $+V_{pk} = +V_{in\,pk} - 0.7\text{ V} = +12\text{ V} - 0.7\text{ V} = +11.3\text{ V}$
$-V_{pk} = 0\text{ V}$

37. A biased negative series clipper circuit
$+V_{pk} = +V_{in\,pk} - 0.7\text{ V} = +12\text{ V} - 0.7\text{ V} = +11.3\text{ V}$
$-V_{pk} = +V_{DC} = +5\text{ V}$

39. Basic negative series clipper

41. $+V_{pk} = +0.7\text{ V}$

$$-V_{pk} = \frac{R_L}{R_L + R_S} \times -V_{in} = \frac{56\text{ k}\Omega}{56\text{ k}\Omega + 23\text{ k}\Omega} \times -8 = -5.6\text{ V}$$

43. A diode symmetrical shunt clipper circuit
Due to the wide range of cassette player volume settings, the input signal may have a wide range of peak-to-peak values. The diode symmetrical shunt clipper circuit, made up of D_1 and D_2, is included to limit the HIGH and LOW input peaks to +0.7 (D_2) and –0.7 (D_1), producing an amplifier input that always has a peak-to-peak of 1.4 V.

45. A positive clamper circuit
$+V_{pk} = +V_{in\,pk\text{-}pk} = +30\text{ V}$
$-V_{pk} = -0.7\text{ V}$

47. Figure 6-40(a): There is no clipping action, possible D_1 short.
Figure 6-40(b): Input is not reaching output, possible D_1 open.

49. Figure 6-42(a): No clamping action, possible D_1 open.
Figure 6-42(b): Circuit is operating normally

Chapter 7

1. b **3.** a **5.** b **7.** d **9.** d **11.** a **13.** c **15.** c

(The answers to essay questions 16 through 35 can be found in the indicated sections that follow the questions.)

37. **a.** $f = \frac{c}{\lambda} = \frac{3 \times 10^{17}}{500} = 6 \times 10^{14}\text{ Hz}$

b. $f = \frac{c}{\lambda} = \frac{3 \times 10^{18}}{4550} = 6.5 \times 10^{14}\text{ Hz}$

c. $f = \frac{c}{\lambda} = \frac{3 \times 10^{14}}{850} = 3.5 \times 10^{11}\text{ Hz}$

39. **a.** Correctly reverse biased **c.** Correctly reverse biased
b. Correctly reverse biased **d.** Incorrectly forward biased

41. When dark, $I_D = 20\text{ nA}$, $V_R = 10\text{ mV}$ and $V_{D1} = 12\text{ V} - 10\text{ mV} = 11.99\text{ V}$:

$$R_{D1} = \frac{V_{D1}}{I_D} = \frac{11.99\text{ V}}{20\text{ nA}} = 599.5\text{ M}\Omega$$

When light, $I_D = 20\ \mu A$, $V_R = 10$ V and $V_{D1} = 12\ V - 10\ V = 2$ V:

$$R_{D1} = \frac{V_{D1}}{I_D} = \frac{2\ V}{20\ \mu A} = 100\ k\Omega$$

43. Diode is forward biased and therefore light will not affect operation.

45. Correctly (reverse) biased

47. Circuit 7-21(a) would use the bidirectional transient suppressor diode in Figure 7-21(e). Circuit 7-21(b) would use the unidirectional transient suppressor diode in Figure 7-21(c).

49. To begin with, the 6 V rms will have a peak of:
$V_{peak} = V_{rms} \times 1.414 = 6\ V \times 1.414 = 8.48$ V
Since two schottky diodes will be ON for each alternation of the input, the peak of the pulsating dc from the bridge rectifier will be
$V_{peak} - (2 \times 0.4\ V) = 8.48 - 0.8 = 7.68$ V

Chapter 8

1. d **3.** a **5.** b **7.** b **9.** a **11.** d **13.** d **15.** a **17.** c **19.** d

(The answers to essay questions 21 through 45 can be found in the indicated sections that follow the questions.)

47. a. NPN transistor is correctly biased. Base is positive relative to emitter (emitter diode ON), collector diode is positive relative to base (collector diode OFF).

b. NPN transistor is incorrectly biased. Base is positive relative to emitter (emitter diode ON), collector diode is negative relative to base (collector diode ON).

c. PNP transistor is correctly biased. Base is negative relative to emitter (emitter diode ON), collector diode is negative relative to base (collector diode OFF).

d. PNP transistor is incorrectly biased. Base is positive relative to emitter (emitter diode OFF), collector diode is positive relative to base (collector diode ON).

49. a. Any HIGH (+5 V) input will turn ON its associated diode, and therefore connect the HIGH (logic 1) to the output.

A	B	Y
0	0	0
0	1	1
1	0	1
1	1	1

(logic 0 = 0 V, logic 1 = +5 V)

b. A HIGH input will turn ON the transistor and connect the ground at the emitter through to the output. A LOW input will turn OFF the transistor and connect the $+V_{CC}$ voltage (HIGH) to the output.

A	Y
0	1
1	0

(logic 0 = 0 V, logic 1 = +5 V)

c. This circuit combines the circuit in (a) with the circuit in (b). Any HIGH input will turn ON its associated diode producing a HIGH output, which will be applied to the base of the transistor. This HIGH input will turn

ON the transistor and connect the ground at the emitter through to the output.

A	B	Y	(logic 0 = 0 V, logic 1 = +5 V)
0	0	1	
0	1	0	
1	0	0	
1	1	0	

51. **a.** $I_B = I_E - I_C = 1.98\text{ mA} - 1.8\text{ mA} = 0.18\text{ mA or }180\ \mu\text{A}$
b. $I_E = I_B + I_C = 120\ \mu\text{A} + 8\text{ mA} = 8.12\text{ mA}$
c. $I_C = I_E - I_B = 20\text{ mA} - 300\ \mu\text{A} = 19.7\text{ mA}$
d. $I_E = I_B + I_C = 500\ \mu\text{A} + 18.6\text{ mA} = 19.1\text{ mA}$

53. $A_V = \dfrac{\Delta V_{out}}{\Delta V_{in}} = \dfrac{11\text{ V} - 3\text{ V}}{3.25\text{ V} - 2.75\text{ V}} = \dfrac{8\text{ V}}{0.5\text{ V}} = 16$

The output voltage is 16 times greater than the input voltage.

55. **a.** NPN, base-biasing. **c.** NPN, voltage-divider biasing.
b. NPN, base-biasing. **d.** PNP, base-biasing.

57. Saturation Point, $I_{C(Sat)} = \dfrac{V_{CC}}{R_C} = \dfrac{20\text{ V}}{10\text{ k}\Omega} = 2\text{ mA}$

Q Point, $I_C = 1.16\text{ mA}$, $V_{CE} = 8.4\text{ V}$
Cutoff Point, $V_{CE(cutoff)} = V_{CC} = 20\text{ V}$

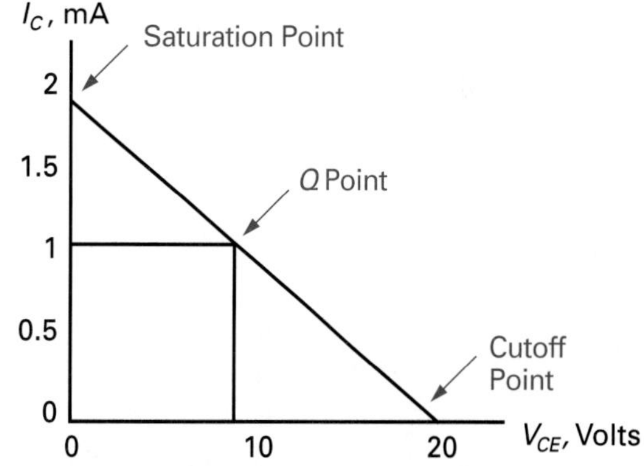
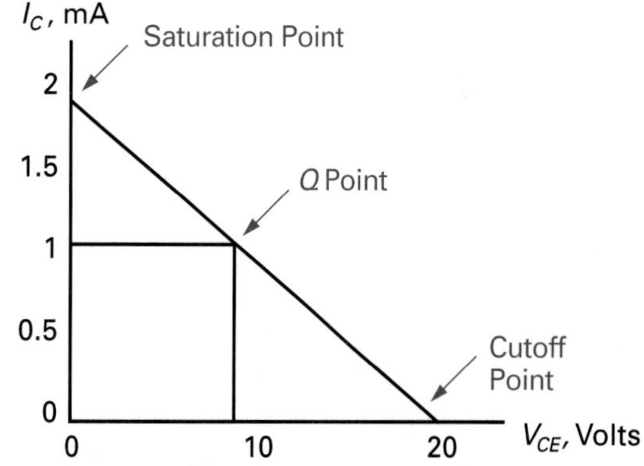

59. Saturation Point,

$$I_{C(Sat)} = \frac{V_{CC}}{R_C + R_E} = \frac{20\text{ V}}{4.7\text{ k}\Omega + 1.1\text{ k}\Omega} = 3.45\text{ mA}$$

Q Point,
$I_C = 1.7\text{ mA}$, $V_{CE} = 10.14\text{ V}$
Cutoff Point,
$V_{CE(cutoff)} = V_{CC} = 20\text{ V}$

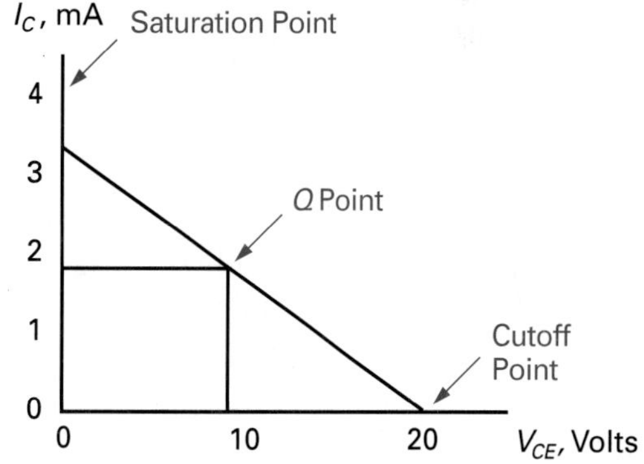

61. Section 8-2-4

63. **a.** Transistor's diodes both test OK.
b. Emitter diode is open, collector diode is OK—replace transistor.
c. Collector diode is shorted, emitter diode is OK—replace transistor.
d. Emitter diode is shorted, collector diode is OK—replace transistor.
e. Transistor's diodes both test OK.

65. a. Q_1 is obviously OFF because it is not receiving any base bias voltage. The lack of base bias voltage is probably due to an R_1 open, however the same symptoms could be caused by an R_2 short. Turning the circuit power OFF, and ohmmeter testing R_1 and then R_2 would isolate the problem. Probable root cause—R_1 open

b. The base voltage is much larger than normal (11.7 V instead of 6.7 V), and the difference between V_C and V_E is zero ($V_{CE} = 0$ V), indicating that the high base voltage has sent the transistor into saturation. The problem is once again the base bias. The probable cause is that R_2 is open, and therefore all of the V_{CC} voltage is being applied to the base. Since Q_1 is heavily ON, the base voltage is 0.7 V above the emitter voltage, and the emitter voltage is equal to $I_{C(sat)} \times R_E$. Turning the circuit power OFF and ohmmeter testing R_2 will isolate the problem. Probable root cause—R_2 open

c. The base voltage is at its correct voltage, however the collector voltage is equal to V_{CC} indicating that the transistor is cut off. Since V_E is equal to 0 V, it appears that the base-emitter junction is open. Turning the circuit power OFF, and ohmmeter testing Q_1 would isolate the problem. Probable root cause—$Q_{1_{B\text{-}E}}$ open

Chapter 9

1. c **3.** b **5.** a **7.** b **9.** d **11.** b **13.** b **15.** d **17.** d **19.** d **21.** d **23.** c **25.** a

(The answers to essay questions 26 through 55 can be found in the indicated sections that follow the questions.)

57. a. upper case "Z"
b. "?"
c. "="
d. decimal digit "0"
e. period "."

59. OR gate

61. When light is present, the photodiode will turn ON, and ground the input of the NOT gate resulting in a HIGH output at *Y*. On the other hand, when no light is present the photodiode will be OFF, the input to the NOT gate will be pulled HIGH, and so the output at *Y* will be LOW.

63. OR gate

65. Yes, the codes generated by the circuit in Figure 9-52 match the bits 4, 3, 2, and 1 of the same ASCII code.

67. $f = \dfrac{1}{1.4 \times (R \times C)} = \dfrac{1}{1.4 \times (5\text{ k}\Omega \times 2\ \mu\text{F})} = 71.4\text{ Hz}$

69. $P_W = 0.7 \times (RC) = 0.7 \times (20\text{ k}\Omega \times 0.1\ \mu\text{F}) = 1.4\text{ ms}$

71. $\tau(\textit{time constant}) = RC = 100\ \Omega \times 0.1\ \mu\text{F} = 10\ \mu\text{s}$
Since the capacitor will take 5 time constants to charge and 5 time constants to discharge, the negative spike will last for about 50 μs ($5 \times 10\ \mu$s).

73. **a.** $t_P = 0.7 \times C \times (R_A + R_B) = 0.7 \times 47\ \mu F \times (2\ k\Omega + 2\ k\Omega) = 131.6$ ms
b. $t_N = 0.7 \times C \times R_B = 0.7 \times 47\ \mu F \times 2\ k\Omega = 65.8$ ms
c. $t = t_P + t_N = 131.6$ ms $+ 65.8$ ms $= 197.4$ ms
d. $f = \dfrac{1}{t} = \dfrac{1}{197.4\ \mu s} = 5$ Hz

75. **a.** $P_W = 50\ \mu s$
b. $t = 500\ \mu s$
c. $f = \dfrac{1}{t} = \dfrac{1}{500\ \mu s} = 2$ kHz
d. Duty Cycle $= \dfrac{P_W}{t} \times 100\% = \dfrac{50\ \mu s}{500\ \mu s} \times 100\% = 10\%$

(The answers to troubleshooting questions 76 through 83 can be found in the indicated sections that follow the questions.)

85. The oscilloscope shows that the output of the NAND gate is permanently at a floating level despite the input changes. This indicates that the output of the AND gate is probably internally open, and therefore the 7408 should be replaced.

Chapter 10

1. b
3. d
5. c
7. b
9. d
11. a
13. d
15. a
17. d
19. d
21. b
23. b
25. d
27. b
29. a

(The answers to essay questions 31 through 76 can be found in the indicated sections that follow the questions.)

77. **a.** $A_V = \dfrac{V_{OUT}}{V_{IN}} = \dfrac{7.5\ mV}{150\ \mu V} = 50$
b. $A_V = \dfrac{\Delta V_{OUT}}{\Delta V_{IN}} = \dfrac{9\ V - 1\ V}{6\ mV - 2\ mV} = \dfrac{8\ V}{4\ mV} = 2000$
c. $V_{OUT} = A_V \times V_{IN} = 15 \times 2\ mV = 30\ mV$
d. $V_{OUT(pk)} = A_V \times V_{IN(pk)} = 50 \times 6\ mV = 300\ mV$ (pk)

79. **a.** $A_P = \dfrac{P_{OUT}}{P_{IN}} = \dfrac{15\ W}{25\ mW} = 600$
b. $P_{IN} = V_{IN} \times I_{IN} = 25\ mV \times 150\ \mu A = 3.7\ \mu W$
$P_{OUT} = A_P \times P_{IN} = 50 \times 3.7\ \mu W = 185\ \mu W$
c. $A_P = A_V \times A_I = 30 \times 15 = 450$
d. $P_{OUT(max)} = A_P \times P_{IN(max)} = 75 \times 10\ mW = 750\ mW$
$P_{OUT(min)} = A_P \times P_{IN(min)} = 75 \times 4\ mW = 300\ mW$

81. **a.** 1 **b.** 2 **c.** 1.4 **d.** 2.176 **e.** 4.02 **f.** 3.57 **g.** –0.456 **h.** 0.243

83. **a.** $A_{V(\text{dB})} = 20 \times \log \frac{V_{\text{OUT}}}{V_{\text{IN}}} = 20 \times \log \frac{7.5\text{ mV}}{150\ \mu\text{V}} = 33.98\text{ dB}$

b. $A_{V(\text{dB})} = 20 \times \log \frac{\Delta V_{\text{OUT}}}{\Delta V_{\text{IN}}} = 20 \times \log \frac{8\text{ V}}{4\text{ mV}} = 66.02\text{ dB}$

85. **a.** $A_{P(\text{dB})} = 10 \times \log \frac{P_{\text{OUT}}}{P_{\text{IN}}} = 10 \times \log \frac{15\text{ W}}{25\text{ mW}} = 27.78\text{ dB}$

b. $A_{P(\text{dB})} = 10 \times \log \frac{P_{\text{OUT}}}{P_{\text{IN}}} = 10 \times \log 50 = 16.99\text{ dB}$

87. **a.** $A_P = \text{antilog} \frac{A_{P(\text{dB})}}{10} = \text{antilog} \frac{20}{10} = 100$

b. $A_V = \text{antilog} \frac{A_{V(\text{dB})}}{20} = \text{antilog} \frac{-1.5}{20} = 0.84$

c. $A_I = \text{antilog} \frac{A_{I(\text{dB})}}{20} = \text{antilog} \frac{35.8}{20} = 61.66$

d. $A_P = \text{antilog} \frac{A_{P(\text{dB})}}{10} = \text{antilog} \frac{1.75}{10} = 1.5$

89. **a.** $R_C = \frac{R_3 \times R_L}{R_3 + R_L} = \frac{3.7\text{ k}\Omega \times 10\text{ k}\Omega}{3.7\text{ k}\Omega + 10\text{ k}\Omega} = \frac{\text{k}\Omega}{6.96\text{ k}\Omega} = 2.7\text{ k}\Omega$

$A_V = \frac{R_C}{R_E} = \frac{2.7\text{ k}\Omega}{1.2\text{ k}\Omega} = 2.25$

The output signal voltage will be 2.25 times greater than the input signal voltage.

b. $V_{\text{OUT}} = A_V \times V_{\text{IN}} = 2.25 \times 3.5\text{ V} = 7.9\text{ V}$

c. $V_{R2} = \frac{R_2}{R_1 + R_2 \times V_{CC}} = 1.8\text{ V}$

$I_{R2} = \frac{V_{R2}}{R_2} = \frac{1.8\text{ V}}{2.2\text{ k}\Omega} = 818\ \mu\text{A}$

$I_{R2} \cong I_{R1}$ (I_B can be ignored), therefore $I_{R1} = 818\ \mu\text{A}$

$V_E = V_B - 0.7\text{ V} = 1.8\text{ V} - 0.7\text{ V} = 1.1\text{ V}$

$I_E = \frac{V_E}{R_E} = \frac{1.1\text{ V}}{1.2\text{ k}\Omega} = 0.92\text{ mA}$

$I_E \cong I_C$, therefore $I_C = 0.92\text{ mA}$

$P_L = \frac{V_L^2}{R_L} = \frac{7.9\text{ V}^2}{10\text{ k}\Omega} = 6.24\text{ mW}$

$I_{CC} = I_{R1} + I_C = 818\ \mu\text{A} + 0.92\text{ mA} = 1.74\text{ mA}$

$P_S = V_{CC} \times I_{CC} = 10\text{ V} \times 1.74\text{ mA} = 17.4\text{ mW}$

$\eta = \frac{P_L}{P_S} \times 100\% = \frac{6.24\text{ mW}}{17.4\text{ mW}} \times 100\% = 35.86\%$

91. $\text{CMRR} = \frac{A_{VD}}{A_{CM}} = \frac{6000}{0.33} = 18{,}181.92$

93. $P_L = \frac{V_L^2}{R_L} = \frac{4.2\ V^2}{1\ k\Omega} = 17.64\ mW$

$P_S = V_{CC} \times I_{CC} = 5\ V \times 20\ mA = 100\ mW$

$\eta = \frac{P_L}{P_S} \times 100\% = \frac{17.64\ mW}{100\ mW} \times 100\% = 17.64\%$

(The answers to troubleshooting questions 95 through 98 can be found in the indicated sections following the questions.)

99. **a.** No input signal to Q_1
b. Q_3 will be OFF (0 V at base), no signal at base of Q_3
c. Signal at the base of Q_1, no signal at the collector
d. No signal at the base of Q_5, Q_5 is OFF

Chapter 11

1. c **3.** d **5.** c **7.** a **9.** b **11.** d **13.** a **15.** c

(The answers to troubleshooting questions 16 through 46 can be found in the indicated sections that follow the questions.)

47. **a.** $f_R = \frac{1}{2\pi\sqrt{LC}} = \frac{1}{2\pi\sqrt{(1.5\ mH) \times (100\ pF)}} = 410.9\ kHz$

b. $f_R = \frac{1}{2\pi\sqrt{LC}} = \frac{1}{2\pi\sqrt{(0.6\ mH) \times (60\ pF)}} = 838.8\ kHz$

49. **a.** $f_R = 410.9\ kHz$, $Q = 215$

$BW = \frac{f_R}{Q} = \frac{410.9\ kHz}{215} = 1.9\ kHz$

b. $f_R = 838.8\ kHz$, $Q = 126.4$

$BW = \frac{f_R}{Q} = \frac{838.8\ kHz}{126.4} = 6.6\ kHz$

51. Since the bandwidth is always twice the highest modulating frequency, $BW = 2 \times f_{AF(HI)} = 2 \times 20\ kHz = 40\ kHz$. Centered at 1360 kHz, the lower sideband will begin at 1340 kHz; the upper sideband will extend to 1380 kHz

(The answers to troubleshooting questions 52 through 56 can be found in the indicated sections following the questions).

Chapter 12

1. b **3.** c **5.** d **7.** a **9.** c **11.** a **13.** b **15.** a **17.** d **19.** d

(The answers to essay questions 21 through 45 can be found in the indicated sections that follow the questions.)

47. $C_T = \frac{C_1 \times C_2}{C_1 + C_2} = \frac{0.1\ \mu F \times 0.5\ \mu F}{0.1\ \mu F + 0.5\ \mu F} = 0.08\ \mu F$

$f_R = \frac{1}{2\pi\sqrt{L_1 \times C_T}} = \frac{1}{2\pi\sqrt{47\ mH \times 0.08\ \mu F}} = 2.6\ kHz$

Feedback factor $(B_F) = \frac{C_1}{C_2} = \frac{0.1\ \mu F}{0.5\ \mu F} = 0.2$ or 20%

$B_C = A_V \times B_F = 5 \times 0.2 = 1$

49. a. With the Hartley crystal oscillator shown in Figure 12-21(a), the crystal operates at its series-resonant frequency and therefore the output frequency will be 15 MHz.

b. With the Pierce crystal oscillator shown in Figure 12-21(b), the crystal operates at its parallel-resonant frequency and therefore the output frequency will be 33 MHz.

(The answers to troubleshooting questions 51 through 54 can be found in the indicated sections following the questions.)

55. Based on the voltmeter readings shown, the first probable fault would be that the radio frequency choke (RFC) has opened, disconnecting the collector supply voltage from Q_1. The emitter voltage is only present due to the small value of base current through Q_1 because the base-emitter junction is forward biased.

Chapter 13

1. d **3.** b **5.** b **7.** b **9.** d **11.** a **13.** b **15.** c

(The answers to troubleshooting questions 16 through 30 can be found in the indicated sections following the questions.)

31. $Z_{GS} = \frac{V_{GS}}{I_{GSS}} = \frac{20\ V}{1\ nA} = \frac{20\ V}{1 \times 10^9} = 20 \times 10^9\ \Omega$ or $20\ G\Omega$

33. a. $V_{GS} = V_{GG} = -1\ V$

b. $I_D = I_{DSS}\left(1 - \frac{V_{GS}}{V_{GS(OFF)}}\right)^2 = 6\ mA\left(1 - \frac{-1\ V}{-2\ V}\right)^2 = 1.5\ mA$

c. $V_{DS} = V_{DD} - (I_D \times R_D) = 10\ V - (1.5\ mA \times 3\ k\Omega) = 10\ V - 4.5\ V = 5.5\ V$

d. $I_{D(MAXIMUM)} = I_{DSS} = 6\ mA$

e. Q point; $V_{GS} = -1\ V$, $I_D = 1.5\ mA$, $V_{DS} = 5.5\ V$

35. a. $V_G = \frac{R_2}{R_1 + R_2} \times V_{DD} = \frac{1\ M\Omega}{3\ M\Omega + 1\ M\Omega} \times 12\ V = 3\ V$

b. $V_S = V_{RS} = I_S \times R_S = 2\ mA \times 3\ k\Omega = 6\ V$

c. $V_{GS} = V_G - V_S = 3\ V - 6\ V = -3\ V$

d. $V_{DS} = V_{DD} - (V_{RS} + V_{RD}) = 12\ V - [6\ V + (I_D \times R_D)]$
$= 12\ V - [6\ V + (2\ mA \times 1.5\ k\Omega)] = 12\ V - (6\ V + 3\ V) = 3\ V$

37. a. Common source

b. $V_{IN} = 0\ V$, JFET is fully ON, $V_{OUT} = 0\ V$

c. $V_{IN} = -6\ V$, JFET is cut OFF, $V_{OUT} = +6\ V$

d. Yes, negative going input is inverted to a positive going output.

39. Section 13-5-1

Chapter 14

1. b **3.** d **5.** c **7.** a **9.** c **11.** c **13.** c **15.** d

(The answers to troubleshooting questions 16 through 30 can be found in the indicated sections following the questions.)

31. **a.** $I_{D(\text{sat})} = \dfrac{V_{DD}}{R_D} = \dfrac{6\text{ V}}{1\text{ k}\Omega} = 6\text{ mA}$ **b.** $V_{DS(\text{OFF})} = V_{DD} = 6\text{ V}$

33. Figure 14-21 uses zero biasing
Figure 14-22 uses drain feedback biasing

35. **a.** Q_1 is always ON due to the +10 V drain feedback bias, and therefore it will always have a drain-to-source ON resistance $R_{DS(\text{ON})}$ of 1 kΩ.

b. When $V_{\text{IN}} = +5$ V, Q_2 is ON and has a drain-to-source ON resistance of 300 Ω. Therefore, V_{OUT} equals

$$V_{\text{OUT}} = \frac{R_{Q_2}}{R_{Q_1} + R_{Q_2}} \times V_{DD} = \frac{300\ \Omega}{1000\ \Omega + 300\ \Omega} \times 10\text{ V} = 0.23 \times 10\text{ V} = 2.3\text{ V}$$

c. When $V_{\text{IN}} = 0$ V, Q_2 is OFF and therefore $V_{\text{OUT}} = V_{DD} = 10$ V

37. The foam is conductive to ensure all pins are at the same potential, and therefore the resistance should be 0 Ω.

39. When $V_{GS} = 0$ V, Q_1 will be ON and I_D should be about half of its maximum since a V_{GS} of zero volts puts the Q point at a mid-point on the load line. Q_1 should therefore have some resistance, and therefore voltage drop, between drain and source. Since no voltage is reaching the drain and V_{DD} = 20 V, we would have to suspect that R_D is open.

Chapter 15

1. a **3.** b **5.** c **7.** a **9.** b **11.** c **13.** c **15.** c

(The answers to troubleshooting questions 16 through 25 can be found in the indicated sections that follow the questions.)

Chapter 16

1. b **3.** d **5.** a **7.** c **9.** e

(The answers to troubleshooting questions 11 through 32 can be found in the indicated sections following the questions.)

33. **a.** LM = National Semiconductor
318 = Op-Amp Type
C = Commercial 0 to 70°C
N = Plastic DIP with longer leads

b. NE = Signetics
101 = Op-Amp Type
C = Commercial 0 to 70°C
D = Plastic DIP

35. A single polarity supply (+5 V) inverting op-amp. The inverting amplifier would have to have a negative input because the output can only be between 0 V ($-V$ supply voltage) and +5 V ($+V$ supply voltage).

37. **a.** Differentiator (see waveforms in Figure 16-13)

b. Integrator (see waveforms in Figure 16-14)

39. **a.** Summing Amplifier

$V_{OUT} = -(V_{IN1} + V_{IN2}) = -(3\text{ V} + 6\text{ V}) = -9\text{ V}$

b. Difference Amplifier

$V_{OUT} = V_{IN2} - V_{IN1} = 10\text{ V} - 6.5\text{ V} = 3.5\text{ V}$

41. $f_C = \dfrac{1}{2\pi RC} = \dfrac{1}{2\pi \times 10\text{ k}\Omega \times 0.1\text{ }\mu\text{F}} = 159.2\text{ Hz}$

43. $A_{CL} = -\dfrac{R_F}{R_{IN}} = -\dfrac{12\text{ k}\Omega}{10\text{ k}\Omega} = -1.2$

(The answers to troubleshooting questions 45 through 47 can be found in 16-2-7)

Chapter 17

1. c **3.** b **5.** d **7.** c **9.** b

(The answers to troubleshooting questions 11 through 30 can be found in the indicated sections following the questions.)

31. **a.** $A = +15\text{ V}$ $B = -15\text{ V}$ $C = +5\text{ V}$ $D = -5\text{ V}$

b. 7815 and 7805 are fixed-positive output voltage regulators, while the 7915 and 7905 are fixed-negative output voltage regulators.

c. Positive and negative clipper circuit

d. The 555 schmitt trigger will sharpen the rise and fall times of the clipped sine wave input producing a square wave output that is at the same frequency as the input (60 Hz).

33. **a.** The 317 is a variable positive output voltage regulator.

b. Because the switch can be used to select one of three pre-set or programmed output voltages.

c. Switch SW_1 is used to connect a HIGH input to the base of one of three NPN switching transistors. When switched ON by a resistor in parallel with the 4.7 kΩ resistor, this parallel combination will determine the output voltage of the 317 voltage regulator, as listed in the table.

35. **a.** Phase-locked loop

b. Position 1 = 16 kHz, position 2 = 8 kHz, position 3 = 2 kHz.

c. Circuit is smaller and less costly than three crystal oscillator circuits.

Chapter 18

1. b **3.** b **5.** b **7.** d **9.** a **11.** b **13.** d **15.** c

(The answers to troubleshooting questions 16 through 40 can be found in the indicated sections following the questions.)

41. $1001_2 = 9_{10}$

$1001_{BCD} = 9_{10}$

43. $0011\ \ 1001_2 = 57_{10}$

$0011\ \ 1001_{BCD} = 39_{10}$

45. **a.** $0010\ \ 0011\ \ 0110\ \ 0101_{BCD}$

b. $0010\ \ 0100_{BCD}$

47. Because its 4-bit binary input is used to select one of 16 (hexadecimal) outputs.

49. Any LOW into a NAND gate will generate a HIGH output.

51. To generate the 4-bit BCD code that is equivalent to the decimal key pressed.

53. A multiplexer that will switch 1 of its 8 inputs through to the single output.

55. The 74151 could be described as a parallel-to-serial converter because it converts the parallel 8-bit word applied to its inputs into an 8-bit serial data stream.

57. (d) Toggle **59.** 4, 4. **61.** 16, 16.

63. 0100 (OFF, ON, OFF, OFF) or decimal 4.

65. **a.** The 100 kHz clock signal controls the operation of the entire circuit; when it is not present there will be no count window frequencies, and therefore no count, and therefore no display.
   b. A permanent LOW out of the 7476 Q output will mean that the AND gate is permanently disabled, and the unknown input frequency sample will be blocked from the counter. Display will be 0000.
   c. The frequency counter will not operate on the 1 Hz or 1 sec range.
   d. If the counter is not cleared after each frequency sample, it will continue to add each new frequency sample to the previous frequency sample count. As a result the counter will count higher and higher, and then reset and start again, giving an inaccurate reading.
   e. If the Q output from the 74121 is disconnected from the 74173 registers, the registers will not load the new value, and the display will show whatever value was previously stored in the register prior to this problem.

Chapter 19

1. e **3.** a **5.** e **7.** d **9.** c

(The answers to troubleshooting questions 11 through 30 can be found in the indicated sections following the questions.)

31. **a.** DIAC **e.** Photoresistor
   b. SCR **f.** Solar cell
   c. LED or ILD **g.** Phototransistor
   d. Photodiode **h.** LASCR

33. A LOW input will turn OFF Q_1, causing its collector output to go HIGH, triggering the SCR ON, connecting power to the relay's energizing coil.

35. Monostable

37. Because the TRIAC can be used to control the full ac cycle.

39. TheLASCR was used to connect dc power to a load, while the LATRIAC was used to connect ac power to a load.

41. Photoresistor

43. As the light level changes, the resistance of the photoresistor changes, changing the oscillator's RC frequency-determining value, and changing the frequency at the output of the 3909 being applied to the speaker.

45. To power a constant-frequency tone generator circuit using solar cells.

47. Phototransistor

49. When light is present, phototransistor is ON, NPN transistor is OFF, relay is de-energized, normally closed contacts are closed, normally open contacts are open. When light is not present, phototransistor is OFF, NPN transistor is ON, relay is energized, normally closed contacts are open, normally open contacts are closed.
51. To have a bargraph display represent the level of ambient light present.
53. To control the dc output offset of the op-amp.
55. When a voltage of 2 V is dropped across a resistor, its associated LED will turn ON. The larger value resistors in the bargraph voltage divider will develop a larger voltage drop. This means that the resistor at the bottom of the rung will turn ON first, when only a small voltage is applied from the 741. As the light level increases, the voltage out of the 741 increases, the voltage drop across all of the resistors increases proportionally, and the LEDs turn ON in order from the bottom up.

Index